3 1272 00338 6941
D1230509

The Encyclopedia of Ecology & Environmental Management

The Encyclopedia of Ecology & Environmental Management

Editor-in-Chief: PETER CALOW

Blackwell
Science

Editorial Offices:
Osney Mead, Oxford OX2 0EL
25 John Street, London WC1N 2BL
23 Ainslie Place, Edinburgh EH3 6AJ
350 Main Street, Malden
MA 02148 5018, USA
54 University Street, Carlton
Victoria 3053, Australia
10, rue Casimir Delavigne
75006 Paris, France

Other Editorial Offices:
Blackwell Wissenschafts-Verlag GmbH
Kurfürstendamm 57
10707 Berlin, Germany

Blackwell Science KK
MG Kodenmacho Building
7–10 Kodenmacho Nihombashi
Chuo-ku, Tokyo 104, Japan

First published 1998

Set by Excel Typesetters, Hong Kong
Printed and bound in Great Britain
by MPG Books Ltd, Bodmin, Cornwall

A catalogue record for this title
is available from the British Library

ISBN 0-86542-838-7

Library of Congress
Cataloging-in-publication Data

The encyclopedia of ecology
and environmental management/
editor-in-chief, Peter Calow;
editorial board, J. Grace . . . [*et al.*].
p. cm.
ISBN 0-86542-838-7
1. Environmental management–
Encyclopedias.
2. Ecology–Encyclopedias.
I. Calow, Peter.
GE300.E53 1998
363.7′003—dc21 97-43092
CIP

DISTRIBUTORS

Marston Book Services Ltd
PO Box 269
Abingdon, Oxon OX14 4YN
(*Orders*: Tel: 01235 465500
Fax: 01235 465555)

USA
Blackwell Science, Inc.
Commerce Place
350 Main Street
Malden, MA 02148 5018
(*Orders*: Tel: 800 759 6102
781 388 8250
Fax: 781 388 8255)

Canada
Login Brothers Book Company
324 Saulteaux Crescent
Winnipeg, Manitoba R3J 3T2
(*Orders*: Tel: 204 224-4068)

Australia
Blackwell Science Pty Ltd
54 University Street
Carlton, Victoria 3053
(*Orders*: Tel: 3 9347 0300
Fax: 3 9347 5001)

For further information on
Blackwell Science, visit our website:
www.blackwell-science.com

Contents

List of Contributors

EDITOR-IN-CHIEF

P. CALOW

EDITORIAL BOARD

P. CALOW [Ecology in Management]

D.A. FALK [Conservation and Management of Ecological Systems]

J. GRACE [Physical and Chemical Environment]

P.D. MOORE [Communities and Ecosystems]

B. SHORROCKS [Individuals and Populations]

S.C. STEARNS [Ecology and Evolution]

CONTRIBUTORS AND ADDRESSES

W.N.A. **W.N. ADGER** *CSERGE, School of Environmental Sciences, University of East Anglia, Norwich NR4 7TJ, UK*

J.K.A. **J.K. AGEE** *Division of Ecosystem Science and Conservation, University of Washington, Box 352100, Seattle, WA 98195, USA*

P.A. **P. ALBERCH** [Deceased] *Museo Nacional de Cienceas Naturales, J. Gutierrez Abascal 2, Madrid E-28006, Spain*

R.McN.A. **R. McNEILL ALEXANDER** *Biomechanics and Comparative Physiology Research Group, School of Biology, University of Leeds, Leeds LS2 9JT, UK*

P.M.S.A **P.M.S. ASHTON** *School of Forestry and Environmental Studies, Yale University, 360 Prospect Street, New Haven, CT 06511, USA*

J.A. **J. AUSTIN** *UK Meteorological Office, London Road, Bracknell RG12 2SZ, UK*

P.J.B. **P.J. BAILEY** *Department of Chemistry, University of Edinburgh, West Mains Road, Edinburgh EH9 3JJ, UK*

D.D.B. **D.D. BALDOCCHI** *Atmospheric Turbulence and Diffusion Division, US Department of Commerce, 456 South Illinois Avenue, PO Box 2456, Oak Ridge, TN 37831-2456, USA*

J.S.B. **J.S. BALE** *School of Biological Sciences, University of Birmingham, Edgbaston, Birmingham B15 2TT, UK*

C.J.B. **C.J. BARNARD** *School of Biological Sciences, University of Nottingham, University Park, Nottingham NG7 2RD, UK*

R.S.K.B. **R.S.K. BARNES** *Department of Zoology, University of Cambridge, Downing Street, Cambridge CB2 3EJ, UK*

R.W.B. **R.W. BATTARBEE** *Environmental Change Research Centre, University College London, University of London, 26 Bedford Way, London WC1H 0AP, UK*

B.B. **B. BAUR** *Department of Integrative Biology, Section of Conservation Biology (NLU), University of Basel, St Johanns-Vorstadt 10, CH-4056 Basel, Switzerland*

C.W.B. **C.W. BEAN** *Institute of Aquaculture, University of Stirling, Stirling FK9 4LA, UK*

J.B. **J. BEDDINGTON** *T.H. Huxley School of Environment, Earth Sciences and Engineering, Royal School of Mines, Prince Consort Road, London SW7 2BP, UK*

M.B. **M. BERAN** *OB Research Services, 1 The Croft, East Hagbourne, Didcot OX11 9LS, UK*

G.P.B. **G.P. BERLYN** *School of Forestry and Environmental Studies, Yale University, 370 Prospect Street, New Haven, CT 06511, USA*

D.A.B. **D.A. BERRIGAN** *Department of Zoology, Box 351800, University of Washington, Seattle, WA 98195, USA*

M.C.M.B. **M.C.M. BEVERIDGE** *Institute of Aquaculture, University of Stirling, Stirling FK9 4LA, UK*

I.J.B. **I.J. BEVERLAND** *Department of Public Health Sciences, University of Edinburgh, Medical School, Teviot Place, Edinburgh EH8 9AG, UK*

D.E.B. **D.E. BIGNELL** *Tropical Biology and Conservation Unit, Universiti Malaysia Sabah, 88999 Kota Kinabalu, Sabah, Malaysia*

S.J.B. **S.J. BISSELL** *Colorado Division of Wildlife, Denver, CO 80216, USA*

L.O.B. **L.O. BJÖRN** *Section of Plant Physiology, Lund University, Box 117, 221 00 Lund, Sweden*

J.F.B. **J.F. BLYTH** *Institute of Ecology and Resource Management, University of Edinburgh, Mayfield Road, Edinburgh EH9 3JU, UK*

C.B. **C. BOESCH** *Zoology Institute, University of Basel, Rheinsprung 9, CH-4051 Basel, Switzerland*

P.B. **P. BORRELL** *EUROTRAC ISS, Fraunhofer Institute, Kreuzeckbahnstr 19, D-82467 Garmisch-Partenkirchen, Germany*

A.D.B. **A.D. BRADSHAW** *58 Knowsley Road, Liverpool L19 OPG, UK*

P.M.B. **P.M. BRAKEFIELD** *Institute of Evolutionary and Ecological Sciences, University of Leiden, PO Box 9516, NL-2300 RA Leiden, The Netherlands*

D.R.B. **D.R. BROOKS** *Department of Zoology, University of Toronto, Toronto, Ontario M5S 3G5, Canada*

M.A.B. **M.A. BURGMAN** *School of Forestry, University of Melbourne, Creswick, Victoria 3363, Australia*

F.W.B. **F.W. BURLEY** *2400 NW 80th Street, Box 160, Seattle, WA 98117, USA*

G.L.B. **G.L. BUSH** *Department of Zoology, Michigan State University, East Lansing, MI 48824, USA*

J.C. **J. CAIRNS JR** *Department of Biology, Virginia Polytechnic Institute and State University, Blacksburg, VA 24061, USA*

T.V.C. **T.V. CALLAGHAN** *Sheffield Centre for Arctic Ecology, Department of Animal and Plant Sciences, University of Sheffield, Tapton Experimental Gardens, 26 Taptonville Road, Sheffield S10 5BR, UK*

P.C. **P. CALOW** *Department of Animal and Plant Sciences, University of Sheffield, Sheffield S10 2TN, UK*

J.N.C. **J.N. CAPE** *Institute of Terrestrial Ecology, Bush Estate, Penicuik EH26 0QB, UK*

T.C. **T. CARACO** *Biological Sciences, State University of New York at Albany, Albany, NY 12222, USA*

H.C. **H. CASWELL** *Biology Department, Woods Hole Oceanographic Institution, Woods Hole, MA 02543, USA*

R.C. **R. CATCHPOLE** *Ecology and Evolution Research Group, School of Biology, University of Leeds, Leeds LS2 9JT, UK*

W.G.C. **W.G. CHALONER** *Department of Geology, Royal Holloway, University of London, Egham Hill, Egham TW20 0EX, UK*

C.W.C. **C.W. CLARK** *Department of Mathematics, University of British Columbia, Vancouver, British Columbia V6T 1Z2, Canada*

J.A.C. **J.A. CLARK** *Department of Physiology and Environmental Science, University of Nottingham, Sutton Bonington Campus, Loughborough LE12 5RD, UK*

D.C. **D. CLODE** *Department of Zoology, University of Melbourne, Parkville, Victoria 3052, Australia*

S.G.C. **S.G. COMPTON** *Ecology and Evolution Research Group, School of Biology, University of Leeds, Leeds LS2 9JT, UK*

L.M.C. **L.M. COOK** *The Manchester Museum, University of Manchester, Oxford Road, Manchester M13 9PL, UK*

J.G.C. **J.G. COOKE** *Centre for Ecosystem Management Studies, Mooshof, 79297 Winden, Germany*

R.M.C. **R.M. COWLING** *Institute for Plant Conservation, University of Cape Town, Rondebosch 7701, South Africa*

R.M.M.C. **R.M.M. CRAWFORD** *Plant Sciences Laboratory, Sir Harold Mitchell Building, University of St Andrews, St Andrews KY16 9AL, UK*

M.J.C. **M.J. CRAWLEY** *Department of Biology, Imperial College of Science, Technology and Medicine, Silwood Park, Ascot SL5 7PY, UK*

J.Cr. **J. CRESSWELL** *Department of Biological Sciences, University of Exeter, Hatherby Laboratories, Exeter EX4 4PS, UK*

R.H.C. **R.H. CROZIER** *Department of Genetics and Human Variation, La Trobe University, Bundoora Campus, Bundoora, Victoria 3083, Australia*

P.J.C. **P.J. CURRAN** *Department of Geography, University of Southampton, Highfield, Southampton SO17 1BJ, UK*

K.D. **K. DAVIES** *SAC, Crops Division, Bush Estate, Penicuik EH26 0PH, UK*

R.W.D. **R.W. DAVIES** *Faculty of Science, Monash University, Wellington Road, Clayton 3168, Victoria, Australia*

A.J.D. **A.J. DAVIS** *Ecology and Evolution Research Group, School of Biology, University of Leeds, Leeds LS2 9JT, UK*

J.C.D. **J.C. DEUTSCH** *Crusade, 73 Collier Street, London N1 9BE, UK*

D.A.D. **D.A. DEWSBURY** *Department of Psychology, University of Florida, Gainesville, FL 32611, USA*

M.Di. **M. DICKE** *Department of Entomology, Wageningen Agricultural University, PO Box 8031, NL-6700 EH Wageningen, The Netherlands*

M.D. **M. DOEBELI** *Zoology Institute, University of Basel, Rheinsprung 9, CH-4051 Basel, Switzerland*

G.A.D. **G.A. DOVER** *Department of Genetics, University of Leicester, University Road, Leicester LE1 7RH, UK*

A.D. **A. DRIZO** *Institute of Ecology and Resource Management, University of Edinburgh, Mayfield Road, Edinburgh EH9 3JU, UK*

C.D. **C. DYTHAM** *Department of Biology, University of York, PO Box 373, Heslington, York YO1 5YW, UK*

D.E. **D. EBERT** *Zoology Institute, University of Basel, Rheinsprung 9, CH-4051 Basel, Switzerland*

P.E. **P. EGGLETON** *Department of Entomology, The Natural History Museum, Cromwell Road, London SW7 5BD, UK*

J.E. **J. EKMAN** *Department of Zoology, Uppsala University, S-752 36 Uppsala, Sweden*

M.A.E. **M.A. ELGAR** *Department of Zoology, University of Melbourne, Parkville, Victoria 3052, Australia*

J.R.E. **J.R. ETHERINGTON** *Parc-y-Bont, Llanhowell, Solva, Haverfordwest SA62 6XX, UK*

D.A.F. **D.A. FALK** *Society for Ecological Restoration, Department of Ecology and Evolutionary Biology, Biological Sciences West, University of Arizona, Tuscon, AZ 85721, USA*

J.F. **J.G. FARMER** *Department of Chemistry, University of Edinburgh, West Mains Road, Edinburgh EH9 3JJ, UK*

M.F. **M. FENNER** *Biodiversity and Ecology Division, School of Biological Sciences, University of Southampton, Bassett Crescent East, Southampton SO16 7PX, UK*

B.W.F. **B.W. FERRY** *School of Biological Sciences, Royal Holloway, University of London, Egham Hill, Egham TW20 0EX, UK*

E.A.F. **E.A. FITZPATRICK** *Department of Plant and Soil Science, University of Aberdeen, Cruickshank Building, St Machar Drive, Aberdeen AB24 3UU, UK*

T.L.F. **T.L. FLEISCHNER** *Environmental Studies Program, Prescott College, 220 Grove Avenue, Prescott, AZ 86301, USA*

V.F. **V. FORBES** *Department of Life Sciences and Chemistry, Roskilde University, PO Box 260, DK-4000 Roskilde, Denmark*

N.R.F. **N.R. FRANKS** *Department of Biology and Biochemistry, University of Bath, Claverton Down, Bath BA2 7AY, UK*

R.P.F. **R.P. FRECKLETON** *Schools of Environmental and Biological Sciences, University of East Anglia, Norwich NR4 7TJ, UK*

E.A.Fr. **E.A. FRIAR** *Rancho Santa Ana Botanic Garden, 1500 N. College Avenue, Claremont, CA 91711, USA*

W.G. **W. GABRIEL** *Zoologisches Institut der Ludwig-Maximilians-Universität, Postfach 20 21 36, D-80021 München, Germany*

J.G.F. **J. GARCIA FERNANDEZ** *Fundacion para la Conservacion de las Especies y el Medio Ambiente (FUCEMA), Pringles 10, Piso 3, 1183 Buenos Aires, Argentina*

B.A.G. **B.A. GARDINER** *Forest Research, NRS Roslin, Edinburgh EH25 9SY, UK*

J.H.C.G. **J.H.C. GASH** *Institute of Hydrology, Crowmarsh Gifford, Wallingford OX10 8BB, UK*

M.T.G. **M.T. GHISELIN** *California Academy of Sciences, Golden Gate Park, San Francisco, CA 94118-4599, USA*

O.L.G. **O.L. GILBERT** *Department of Landscape, University of Sheffield, Sheffield S10 2TN, UK*

P.S.G. **P.S. GILLER** *Department of Zoology and Animal Ecology, University College Cork, Lee Maltings, Prospect Row, Ireland*

H.C.J.G. **H.C.J. GODFRAY** *Department of Biology, Imperial College of Science, Technology and Medicine, Silwood Park, Ascot SL5 7PY, UK*

F.B.G. **F.B. GOLDSMITH** *Department of Biology, University College London, University of London, Gower Street, London WC1 6BT, UK*

J.G. **J. GRACE** *Institute of Ecology and Resource Management, University of Edinburgh, Mayfield Road, Edinburgh EH9 3JU, UK*

M.C.G. **M.C. GRAHAM** *Department of Chemistry, University of Edinburgh, West Mains Road, Edinburgh EH9 3JJ, UK*

A.N.G. **A.N. GRAY** *Forest Science Department, Oregon State University, Corvallis, OR 97331, USA*

P.J.G. **P.J. GRUBB** *Department of Plant Sciences, University of Cambridge, Downing Street, Cambridge CB2 3EA, UK*

E.O.G. **E.O. GUERRANT JR** *Berry Botanic Garden, 11505 SW Summerville Avenue, Portland, OR 97219, USA*

S.J.H. **S.J. HALL** *School of Biological Sciences, The Flinders University of South Australia, GPO Box 2100, Adelaide 5001, Australia*

A.H. **A. HALLAM** *School of Earth Sciences, University of Birmingham, Edgbaston, Birmingham B15 2TT, UK*

P.H.H. **P.H. HARVEY** *Department of Zoology, University of Oxford, South Parks Road, Oxford OX1 3PS, UK*

M.H. **M. HASSALL** *School of Environmental Sciences, University of East Anglia, Norwich NR4 7TJ, UK*

I.M.H. **I.M. HEAD** *Department of Fossil Fuels and Environmental Geochemistry, University of Newcastle, Newcastle-upon-Tyne NE1 7RU, UK*

M.R.H. **M.R. HEAL** *Department of Chemistry, University of Edinburgh, West Mains Road, Edinburgh EH9 3JJ, UK*

J.B.H. **J.B. HEALE** *Division of Life Sciences, King's College London, University of London, Campden Hill Road, London W8 7AH, UK*

G.A.F.H. **G.A.F. HENDRY** *Biological Sciences, University of Dundee, Dundee DD1 4HN, UK*

M.O.H. **M.O. HILL** *ITE, Monks Wood, Abbots Ripton, Huntingdon PE17 2LS, UK*

A.R.H. **A.R. HOELZEL** *Department of Biological Sciences, University of Durham, South Road, Durham DH1 3LE, UK*

J.J.H. **J.J. HOPKINS** *Joint Nature Conservation Committee, Monkstone House, Peterborough PE1 1JY, UK*

R.B.H. **R.B. HUEY** *Department of Zoology, University of Washington, Box 351800, Seattle, WA 98195-1800, USA*

R.N.H. **R.N. HUGHES** *School of Biological Sciences, University of Wales, Bangor, Deiniol Road, Bangor LL57 2UW, UK*

L.D.H **L.D. HURST** *Department of Biology and Biochemistry, University of Bath, Claverton Down, Bath BA2 7AY, UK*

H.A.P.I. **H.A.P. INGRAM** *Johnstonfield, Dunbog, Cupar, Fife KY14 6JG, UK*

M.I. **M. INGROUILLE** *Department of Biology, Birkbeck College, University of London, Malet Street, London WC1E 7HX, UK*

J.I. **J. IRVINE** *Institute of Ecology and Resource Management, University of Edinburgh, Mayfield Road, Edinburgh EH9 3JU, UK*

S.K.J. **S.K. JAIN** *Department of Agronomy and Range Science, University of California at Davis, Davis, CA 95616, USA*

A.J. A. JENKINS *British Antarctic Survey, High Cross, Madingley Road, Cambridge CB3 0ET, UK*

G.D.J. G. DE JONG *Department of Plant Ecology and Evolutionary Biology, Universiteit of Utrecht, Padualaan 8, 3584 CH Utrecht, The Netherlands*

W.R.J. W.R. JORDAN *University of Wisconsin Arboretum, 1207 Seminole Highway, Madison, WI 53711, USA*

T.J.K. T.J. KAWECKI *Department of Ecology and Systematics, University of Helsinki, PO Box 17, SF 00014 Helsinki, Finland*

J.K. J. KINDERLERER *Department of Molecular Biology and Biotechnology, University of Sheffield, Sheffield S10 2TN, UK*

T.B.L.K. T.B.L. KIRKWOOD *Department of Geriatric Medicine, University of Manchester, Oxford Road, Manchester M13 9PT, UK*

J.C.K. J.C. KOELLA *Department of Zoology, University of Aarhus, Universitetsparken B135, DK-8000 Aarhus C, Denmark*

W.D.K. W.D. KOENIG *Hastings Natural History Reservation, University of California at Berkeley, 38601 East Carmel Valley Road, Carmel Valley, CA 93924, USA*

J.Kz. J. KOZLOWSKI *Institute of Environmental Biology, Jagiellonian University, Oleandry 2a, 30-063 Krakow, Poland*

W.E.K. W.E. KUNIN *Ecology and Evolution Research Group, School of Biology, University of Leeds, Leeds LS2 9JT, UK*

R.L. R. LAL *School of Natural Resources, The Ohio State University, Columbus, OH 43210, USA*

J.L. J. LANCASTER *Institute of Ecology and Resource Management, University of Edinburgh, Mayfield Road, Edinburgh EH9 3JU, UK*

J.D.Lz. J.D. LAZELL *The Conservation Agency, 6 Swinburne Street, Jamestown, RI 02835, USA*

D.C.L. D.C. LEDGER *Institute of Ecology and Resource Management, University of Edinburgh, Mayfield Road, Edinburgh EH9 3JU, UK*

J.D.L. J.D. LEWINS *Director of Studies in Engineering and Management, Magdalene College, Magdalene Street, Cambridge CB3 0AG, UK*

R.I.L.S. R.I. LEWIS SMITH *British Antarctic Survey, High Cross, Madingley Road, Cambridge CB3 0ET, UK*

C.M.L. C.M. LIVELY *Department of Biology, Indiana University, Bloomington, IN 47405, USA*

J.Ly. J. LLOYD *Max-Planck-Institut für Biogeochemie, Sophienstrasse 10, D-07743 Jena, Germany*

J.W.L. J.W. LLOYD *School of Earth Sciences, University of Birmingham, Edgbaston, Birmingham B15 2TT, UK*

S.P.L. S.P. LONG *Department of Biological Sciences, John Tabor Laboratories, University of Essex, Wivenhoe Park, Colchester CO4 3SQ, UK*

L.L.L. L.L. LOOPE *US Geological Survey, Biological Resources Division, Pacific Islands Ecosystem Research Center, Haleakala National Park Field Station, PO Box 369, Makawao, HI 96768, USA*

N.L. N. LOPOUKHINE *Natural Resources Branch, Parks Canada, 25 Eddy Street, Hull, Quebec K1A 0M5, Canada*

W.L. W. LU *The Conservation Agency, 6 Swinburne Street, Jamestown, RI 02835, USA*

R.M.L. R.M. LUCAS *University of Swansea, Singleton Park, Swansea SA2 8PP, UK*

T.F.C.M. T.F.C. MACKAY *Department of Genetics, North Carolina State University, Box 7614, Raleigh, NC 27695, USA*

A.B.M. A.B. MACKENZIE *Scottish Universities Reactor Centre, East Kilbride, Glasgow G75 0QF, UK*

A.E.M. A.E. MAGURRAN *School of Environmental and Evolutionary Biology, Bute Building, University of St Andrews, St Andrews KY16 9TS, UK*

J.L.B.M. J.L.B. MALLET *Galton Laboratory, Department of Biology, University College London, University of London, 4 Stephenson Way, London NW1 2HE, UK*

A.M.M. A.M. MANNION *Department of Geography, University of Reading, Whiteknights, Reading RG6 6AB, UK*

N.M. N. MAWDSLEY *Ecology and Evolution Research Group, School of Biology, University of Leeds, Leeds LS2 9JT, UK*

A.M. A. McDONALD *Geography Department, University of Leeds, Woodhouse Lane, Leeds LS2 9JT, UK*

J.F.R.M. J.F.R. McILVEEN *Environmental Science Department, Institute of Environmental and Natural Sciences, Lancaster University, Lancaster LA1 4YQ, UK*

L.R.M. L.R. McMAHAN *The Berry Botanic Garden, 11505 SW Summerville Avenue, Portland, OR 97219, USA*

K.G.M. K.G. McNAUGHTON *Environment Group, HortResearch, PO Box 23, Kerikeri, New Zealand*

G.F.M. G.F. MEDLEY *Department of Biological Sciences, University of Warwick, Coventry CV4 7AL, UK*

J.M. J. MILBURN [Deceased] *Department of Botany, University of New England, Armidale, New South Wales 2351, Australia*

M.Mi. M. MILINSKI *Zoology Institute, University of Bern, Wohlenstrasse 50a, CH-3032 Hinterkappelen, Switzerland*

M.M. M. MOGIE *Department of Biology and Biochemistry, University of Bath, Claverton Down, Bath BA2 7AY, UK*

A.P.M. A.P. MØLLER *Laboratoire d'Ecologie, CNRS URA 258, Université Pierre et Marie Curie, Bât. A, 7ème étage, 7 quai St Bernard, Case 237, F-75252 Paris Cedex 05, France*

J.B.M. J.B. MONCRIEFF *Institute of Ecology and Resource Management, University of Edinburgh, Mayfield Road, Edinburgh EH9 3JU, UK*

P.D.M. **P.D. MOORE** *Division of Life Sciences, King's College London, University of London, Campden Hill Road, London W8 7AH, UK*

R.D.M. **R. DEL MORAL** *Department of Botany, University of Washington, Box 355325, Seattle, WA 98195-5325, USA*

J.I.L.M. **J.I.L. MORISON** *Department of Biological Sciences, University of Essex, Wivenhoe Park, Colchester CO4 3SQ, UK*

P.B.M. **P.B. MOYLE** *Department of Wildlife, Fish, and Conservation Biology, University of California at Davis, Davis, CA 95616, USA*

C.E.M. **C.E. MULLINS** *Department of Plant and Soil Science, University of Aberdeen, Cruickshank Building, Aberdeen AB24 3UU, UK*

C.J.N. **C.J. NAGELKERKE** *Institute for Systematics and Population Biology, University of Amsterdam, Kruislaan 320, 1098 SM Amsterdam, The Netherlands*

M.O'C. **M. O'CONNELL** *Palaeoenvironmental Research Unit, Department of Botany, National University of Ireland, Galway, Ireland*

P.O. **P. OLEJNICZAK** *Institute of Environmental Biology, Jagiellonian University, Oleandry 2a, 30-063 Krakow, Poland*

I.O. **I. OLIVIERI** *Institut des Sciences de l'Evolution, Universite Montpellier II, Place Eugene Bataillon, 34095 Montpellier, Cedex 05, France*

J.R.P. **J.R. PACKHAM** *School of Applied Sciences, University of Wolverhampton, Wulfruna Street, Wolverhampton WV1 1SB, UK*

P.P. **P. PAMILO** *Department of Genetics, Uppsala University, Box 7003, S-750 07 Uppsala, Sweden*

H.E.H.P. **H.E.H. PATERSON** *Department of Entomology, University of Queensland, St Lucia, Queensland 4072, Australia*

A.P. **A. PENTECOST** *Division of Life Sciences, King's College London, University of London, Campden Hill Road, London W8 7AH, UK*

G.F.P. **G.F. PETERKEN** *Beechwood House, St Briavels Common, Lydney GL15 6SL, UK*

D.L.P. **D.L. PETERSON** *US Geological Survey, Biological Resources Division, Field Station for Protected Area Research, University of Washington, Seattle, WA 98195, USA*

J.K.P. **J.K. PIPER** *Department of Biology, Bethel College, North Newton, KS 67117, USA*

T.D.P. **T.D. PRICE** *Department of Biology, University of California at San Diego, La Jolla, CA 92093, USA*

C.R.P. **C.R. PULHAM** *Department of Chemistry, University of Edinburgh, West Mains Road, Edinburgh EH9 3JJ, UK*

D.L.J.Q. **D.L.J. QUICKE** *Department of Biology, Imperial College of Science, Technology and Medicine, Silwood Park, Ascot SL5 7PY, UK*

G.B.R. **G.B. RABB** *Species Survival Commission, Chicago Zoological Society, Brookfield, IL 60513, USA*

P.B.R. **P.B. RAINEY** *Department of Plant Sciences, University of Oxford, South Parks Road, Oxford OX1 3RB, UK*

D.J.R. **D.J. RANDALL** *Department of Zoology, University of British Columbia, Vancouver, British Columbia V6T 1Z4, Canada*

J.D.R. **J.D. REYNOLDS** *School of Biological Sciences, University of East Anglia, Norwich NR4 7TJ, UK*

D.N.R. **D.N. REZNICK** *Department of Biology, University of California, Riverside, CA 92521, USA*

H.R. **H. RICHNER** *Zoology Institute, University of Bern, Wohlenstrasse, CH-3032 Hinterkappelen, Switzerland*

P.G.R. **P.G. RISSER** *Oregon State University, 646 Kerr Administration Building, Corvallis, OR 97331, USA*

R.H.R. **R.H. ROBICHAUX** *Department of Ecology and Evolutionary Biology, University of Arizona, Tucson, AZ 85721, USA*

D.A.R. **D.A. ROFF** *Department of Biology, McGill University, 1205 Dr Penfield Avenue, Montreal, Quebec H3A 1B1, Canada*

M.R.R. **M.R. ROSE** *Department of Ecology and Evolutionary Biology, University of California, Irvine, CA 92697, USA*

M.L.R. **M.L. ROSENZWEIG** *Department of Ecology and Evolutionary Biology, University of Arizona, Tucson, AZ 85721, USA*

G.R. **G. RUSSELL** *Institute of Ecology and Resource Management, University of Edinburgh, West Mains Road, Edinburgh EH9 3JG, UK*

M.W.S. **M.W. SABELIS** *Institute for Systematics and Population Biology, University of Amsterdam, Kruislaan 320, 1098 SM Amsterdam, The Netherlands*

P.S.H. **P. SCHMID-HEMPEL** *Experimental Ecology, ETH Zurich, ETH-Zentrum NW, CH-8092 Zurich, Switzerland*

R.S.S. **R.S. SCORER** *2 Stanton Road, London SW20 8RL, UK*

J.G.S. **J.G. SEVENSTER** *Institute of Evolutionary and Ecological Sciences, University of Leiden, PO Box 9516, NL-2300 RA Leiden, The Netherlands*

L.S. **L. SHEPPARD** *Institute of Terrestrial Ecology, Bush Estate, Penicuik EH26 0QB, UK*

B.S. **B. SHORROCKS** *Ecology and Evolution Research Group, School of Biology, University of Leeds, Leeds LS2 9JT, UK*

R.M.S. **R.M. SIBLY** *School of Animal and Microbial Sciences, University of Reading, Whiteknights, Reading RG6 6AJ, UK*

H.S. **H. SIEVERING** *Global Change and Environmental Quality Program, Geography and Environmental Sciences Department, University of Colorado at Denver, Box 172, Denver PO Box 173364, CO 80217, USA*

K.A.S. **K.A. SMITH** *Institute of Ecology and Resource*

Management, University of Edinburgh, West Mains Road, Edinburgh EH9 3JG, UK

R.H.S. R.H. SMITH *Department of Biology, University of Leicester, University Road, Leicester LE1 7RH, UK*

T.A.S. T.A. SPIES *USDA Forest Service, Pacific Northwest Forest Research Station, 3200 West Jefferson Way, Corvallis, OR 97331, USA*

J.M.S. J.M. STARCK *Institute of Zoology and Evolutionary Biology, University of Jena, Erberstrasse 1, D-07743 Jena, Germany*

J.S.S. J.S. STATES *2 Canyon Shadows Road, Lander, WY 82520, USA*

S.C.S. S.C. STEARNS *Zoology Institute, University of Basel, Rheinsprung 9, CH-4051 Basel, Switzerland*

P.A.S. P.A. STONE *Department of Biology, University of Central Oklahoma, Edmond, OK 73034, USA*

J.M.Sz. J.M. SZYMURA *Department of Comparative Anatomy, Jagiellonian University, ul Ingardena 6, 30-060 Krakow, Poland*

J.H.T. J.H. TALLIS *School of Biological Sciences, University of Manchester, Oxford Road, Manchester M13 9PT, UK*

K.C.T. K.C. TAYLOR *Desert Research Institute, Water Resources Center, University and Community College System of Nevada, PO Box 60220, Reno, NV 89506-0220, USA*

K.T. K. THOMPSON *NERC Unit of Comparative Plant Ecology, Department of Animal and Plant Sciences, University of Sheffield, Sheffield S10 2TN, UK*

J.R.G.T. J.R.G. TURNER *Ecology and Evolution Research Group, School of Biology, University of Leeds, Leeds LS2 9JT, UK*

M.V. M. VALERO *Laboratoire de Génétique et Evolution des Populations Végétale, Université de Lille 1, F-59655 Villeneuve d'Ascq cedex, France*

J.V.A. J. VAN ALPHEN *Institute of Evolutionary and Ecological Sciences, University of Leiden, PO Box 9516, NL-2300 RA Leiden, The Netherlands*

P.R.V.G. P.R. VAN GARDINGEN *Institute of Ecology and Resource Management, University of Edinburgh, West Mains Road, Edinburgh EH9 3JG, UK*

L.M.V.V. L.M. VAN VALEN *Department of Ecology and Evolution, University of Chicago, 1101 East 57th Street, Chicago, IL 60637, USA*

J.V.C. J. VANDE CASTLE *Department of Biology, University of New Mexico, Albuquerque, NM 87131, USA*

Y.V. Y. VASARI *Department of Ecology and Systematics, University of Helsinki, PO Box 7 (Unioninkatu 44), Fin-00014 University of Helsinki, Finland*

P.I.W. P.I. WARD *Zoologisches Museum, University of Zurich-Irchel, Winterthurerstrasse 190, CH-8057 Zurich, Switzerland*

P.H.W. P.H. WARREN *Department of Animal and Plant Sciences, University of Sheffield, Sheffield S10 2TN, UK*

W.W.W. W.W. WEISSER *Zoology Institute, University of Basel, Rheinsprung 9, CH-4051 Basel, Switzerland*

T.C.W. T.C. WHITMORE *Department of Geography, University of Cambridge, Downing Place, Cambridge CB2 3EN, UK*

H.W.W. H.W. WHITTINGTON *Department of Electrical Engineering, University of Edinburgh, Mayfield Road, Edinburgh EH9 3JL, UK*

A.J.W. A.J. WILLIS *Department of Animal and Plant Sciences, University of Sheffield, Sheffield S10 2TN, UK*

S.N.W. S.N. WOOD *Mathematical Institute, North Haugh, St Andrews KY16 9SS, UK*

S.R.J.W. S.R.J. WOODELL *Wolfson College, Linton Road, Oxford OX2 6UD, UK*

F.W. F. WORSFORD *The Transport Studies Group, University of Westminster, 35 Marylebone Road, London NW1 5LS, UK*

J.B.Z. J.B. ZEDLER *Botany Department, University of Wisconsin, Madison, WI 53706, USA*

Preface

Unlike Diderot in contemplating the great *Encyclopédie* (the first volume of which appeared in 1751), we can no longer aspire to write a compendium of all knowledge, nor to write a work of such intellectual and political propaganda as was coming to the boil in the mid-18th century. Indeed, even with a more focused view, on ecology and environmental management, it is likely that with the current detailed state of knowledge we shall have missed or under-represented the favourite topics of each and every one of our readers! Diderot *et al.* apart, most subsequent encyclopedists have been charged with some degree of idiosyncrasy.

Yet we have striven to achieve balance, for example between basic principles and their application, individuals, populations and communities, and animal and plant biology, by involving an experienced and broadly representative Editorial Board. They made selections on headwords and decisions on emphasis. As will be seen in the List of Contributors, we then engaged a large team of eminent specialists to write the several thousand entries.

Neither can we claim to be the first encyclopedia (cum dictionary) in ecology, the environmental sciences or environmental management. We hope, though, that we are distinctive in our breadth of coverage: from basic ecology to applying it in conservation and, very broadly, in environmental management. All headwords are defined in a succinct way, though for the sake of clarity and ease of use we have not been overzealous in editing out overlap and repetition between entries. Some headwords, the ones judged to be covering basic principles and/or that are especially influential, have been handled more extensively, usually critically and sometimes with illustrations and references. This work therefore seeks to adopt the twin attributes of dictionary and encyclopedia.

But who is the Encyclopedia intended for? Not quite, as was intended by the French encyclopedists, all the intellectual world; but certainly all those who have an interest in ecology and environmental management. It ought to be of use to students who, starting out on an ecological career, know little and want a broad coverage of the jargon and the principles. But equally, it ought also to be of interest to the professionals who might have become so specialist that they have lost contact with the words and principles that are being used in other branches of their discipline. So we hope that undergraduates, postgraduates, teachers, research professionals, conservationists, environmental managers in business and in government, and so on will all find something of use and of interest in what follows.

The work in putting together the many entries has been enormous and it is a pleasure to record thanks to the Editorial Board and contributors for hard effort and patience in facilitating this. Susan Sternberg, Commissioning Editor, encouraged us all in bringing the work to life. Special thanks, though, go to Julie Yeardley, my publications' secretary, whose hardwork, tenacity, and enthusiasm ensured that we got most of what was promised by contributors generally on time, and to Katrina McCallum, at Blackwell Science, who added considerable value to the final work by not only cross-referencing and picking up many solecisms but also in identifying inconsistencies in the science.

Unfortunately I have to bear the responsibility for all the deficiencies that might remain. But we are optimistic that the Encyclopedia will go into numerous editions and so I want to invite you, the reader, to play a part in ensuring the effective development of the work by letting me have any comments on any areas that will deserve more attention in future.

PETER CALOW
Sheffield, 1998

How to Use the Encyclopedia

General

The Encyclopedia contains almost 3000 entries of varying length arranged alphabetically by headword.

The core of the Encyclopedia consists of over 250 long entries, and these are supported by shorter entries ranging from one-line definitions to half-page commentaries. Any relevant headword that is mentioned within the text of any other appears in SMALL CAPITALS. In addition, obviously related headwords are linked by *See also* at the end of entries. This system makes the usual index superfluous. We hope that you will easily be able to navigate your way around the Encyclopedia.

Some entries are followed by references. These are not intended to provide a comprehensive bibliography, but indicate an entry into the literature.

Other features

Hyphenation When words such as life cycle, life history, etc. are used as a noun, no hyphen is necessary (e.g. the life cycle is semelparous), but when used adjectivally a hyphen is used to add clarity (e.g. the life-history pattern).

Alphabetical order of entries The entries are arranged word-by-word; thus hyphenated words are treated as separate words unless part of the term is a prefix or suffix which cannot stand alone. Numbers in headwords are ignored in the alphabetical ordering, and Greek letters are treated as if they were spelt out in full. Subscript and superscript characters are treated as normal characters. For example:

bet-hedging
beta diversity
β-particle

Plant nomenclature There are eight plant families for which there are two versions of their name. For these families, the 'new' family name is used in the Encyclopedia. These family names are as follows ('old' name in parentheses): Arecaceae (Palmae, the palms), Poaceae (Gramineae, the grasses), Brassicaceae (Cruciferae), Fabaceae (Leguminosae), Clusiaceae (Guttiferae), Apiaceae (Umbelliferae), Lamiaceae (Labiatae) and Asteraceae (Compositae).

Abbreviations/acronyms These are all defined at the first mention in each entry. Entries appear under the headword in full, rather than under the abbreviated form of the term. The following, common abbreviations are not defined: CAM, CFC, DDT, DNA, FAO, GIS, IUCN, NASA, OECD, RNA, UNCED, UNEP, USEPA, WHO.

The Encyclopedia of Ecology & Environmental Management

aapa fens *See* AAPA MIRE; MIRE.

aapa mire These are RHEOTROPHIC mires often occupying extensive areas of the BOREAL regions. They are typically found on gently sloping terrain and are distinguished by the pattern of ridges and pools across their surfaces. These are arranged along the contours of the MIRE at right angles to the slope. In North America the ridges are termed 'strings' and the mire system is often referred to as 'string bog'. This is not strictly accurate as the HYDROLOGY of the mire is of the FEN (rheotrophic) type, although individual strings can be elevated above the general groundwater table and hence develop ACID, mineral-poor, OMBROTROPHIC conditions. The main flow of water is through the linear pools, or 'flarks', and the vegetation of these areas is more distinctly hydrophilous and RHEOPHILOUS, although *Sphagnum* species and *Scheuchzeria* still tend to predominate.

There are many theories that attempt to account for the distinctive patterns on these and other boreal mires. Slope is clearly important and simple observation demonstrates that detritus carried down a gentle slope by water movement tends to be deposited in ridges at right angles to the slope. Once such an initial pattern is developed, the action of frost in the winter may accentuate the tendency, as will the biotic community, which is sensitive to small differences in the WATER TABLE and creates strips of vegetation that reflect the pattern and enhance it further. [P.D.M.]

abatement Reduction, lessening of POLLUTION, usually as a legislative requirement. [P.C.]

abiotic factors (physical factors) A descriptive collective term for components of the physical environment, for example TEMPERATURE, moisture and LIGHT. Distinguishes these features from BIOTIC FACTORS, which usually refers to other living organisms, for example competitors, natural enemies and host plants. Abiotic factors can be directly stressful to organisms (e.g. low temperature kills various animals and plants) or act as token stimuli (environmental cues) which trigger key behavioural or physiological responses, for example MIGRATION, DIAPAUSE, seasonal POLYMORPHISM. *See also* PHYSICAL FACTOR. [J.S.B.]

abortion, selective *See* SELECTIVE ABORTION.

Absolute population estimates

There are many techniques that can be employed to estimate the actual number of individuals in a unit of habitat or study area. These techniques differ from those for RELATIVE POPULATION ESTIMATES, where there is no attempt to calculate the actual DENSITY of individuals. The most appropriate techniques will differ depending on the types of organisms or habitat under investigation. Some of the more commonly encountered methods for the estimation of population densities are outlined below.

Nearest neighbour techniques

There are several nearest neighbour techniques available. The technique usually requires the investigator to select individuals at random and then find the distance from that individual to its nearest neighbour. If populations are at a high density, then the mean distance from a random set of individuals to their nearest neighbours will be smaller than in a low-density population. These methods are particularly appropriate for sessile organisms that are easy to locate, such as trees. The mean distance between individuals will produce an estimate for the absolute density of a randomly dispersed population:

$$D = \frac{1}{4x^2}$$

where D is the density per unit area and x is the mean distance between nearest neighbours.

Ideally each individual in an area will be identified so that each has an equal chance of being chosen at random. In practice, a point will be chosen and the nearest individual to that point selected as the random individual. This system will produce a biased set of individuals and therefore a biased estimate of density as those individuals on the edge of clumps, for example, are more likely to be selected.

More complex nearest neighbour techniques require the distance from an individual to the *n*th nearest neighbour. This will require more effort but will produce a more precise estimate. Another limitation is that the density estimate is affected by the pattern of the DISTRIBUTION. Nearest neighbour techniques are able to determine the absolute POPULATION SIZE only in a randomly (Poisson) distributed population (*see* POISSON DISTRIBUTION). However, if combined with another technique that estimates POPULATION DENSITY they can provide information about the pattern of dispersal.

Mark–release–recapture techniques

There are a variety of techniques for the estimation of the size of populations of mobile organisms that require at least two capture events. In most cases, individuals are marked at each capture except the last. Marking techniques will vary depending on the animals in question. If the number of individuals captured is expected to be small, then marks can be made so that individuals may be identified. This will be the case in studies of mammals and birds, where rings, ear tags or toe clips can be attached; or butterflies, where unique patterns of dots can be applied. If the number of captured individuals is large, then a mass-marking technique, such as radioactive markers or the application of fluorescent dust that can be later visualized using an ultraviolet (UV) light might be more appropriate. Whatever method is employed it is important that the process of marking or the presence of the mark has no effect on the survival, behaviour or mobility of the marked individuals. All mark–release–recapture techniques make other assumptions about the population being estimated; these include: equal catchability of individuals and random mixing of marked individuals within the whole population between captures.

The simplest method for the estimation of absolute population size using mark–release–recapture is the LINCOLN INDEX or PETERSEN ESTIMATE. This method requires only two capture events. It assumes that the proportion of marked individuals in the second capture is the same as the proportion of marked individuals in the total population. The size of the total population may be estimated:

$$\hat{E} = \frac{nm}{r}$$

where $\hat{E}$ is the estimated population size, n is the number of individuals in the second capture, m is the number of marked individuals in the population (i.e. number of individuals in the first capture that survived the marking process) and r is the number of marked individuals recaptured (marks in the second sample). Ideally n and m should be approximately equal.

Applying Bailey's correction produces a less biased estimate of population size when number of recaptures is low:

$$\hat{E} = \frac{(n+1)m}{r+1}$$

The Lincoln index not only assumes random mixing, equal catchability and no effect of marks on survival, but also assumes no IMMIGRATION or EMIGRATION between capture events and no MORTALITY. Because of these restrictions, estimates are only likely to be useful where the time between capture and recapture is very short and the population is relatively closed.

More sophisticated methods for estimating population size are available, although all require more than two capture events and therefore more effort. The FISHER–FORD METHOD requires a series of capture events and a marking system devised so that the capture events may be distinguished. For example, an individual captured during the fifth capture event may have been previously captured during the first and fourth captures and it must be possible to distinguish all previous captures for each recaptured individual. The method then constructs a series of population estimates from which not only an estimate of population size but also an estimate of the mean probability of survival from one capture event to the next is obtained. This method is appropriate if a series of capture events can be distinguished and SURVIVAL RATE is approximately constant.

The JOLLY–SEBER POPULATION ESTIMATION is particularly appropriate if a large proportion of the population is likely to be captured on each occasion. The method allows the calculation of a survival probability between each capture event thus allowing a picture of the variation in survival of the population to be constructed along with the usual population estimate.

There are many other mark–release–recapture methods available. All make slightly different assumptions about the population in question. Therefore the most appropriate method for the estimation of absolute population size will vary depending on the characteristics of the population under investigation.

Sampling fraction

The idea of sampling fraction is that all the individuals in one part of the study area are trapped and counted. The mean of several such samples is taken and then scaled up from the size of the sample to the size of the whole area to give an estimate of the total population. There are several problems with this approach. First, to get an accurate estimate of the population size it is important that all the individuals in the sampled areas are counted. When very small organisms are being considered, it is likely that the number counted will be considerably lower than the actual number in a sample. Second, individuals may evade the capture method in the sample area, again creating an underestimate. Third, in common with most sampling strategies, it is very difficult to design a strategy to cover a truly representative portion of the study area. However, in some habitats, where mark–release–recapture is impractical, this is certainly the most sensible approach to gain an absolute population estimate.

Removal trapping

This method is used quite rarely but is appropriate for particular circumstances. The idea is that individuals are trapped and removed from a population. Trapping is repeated through time and from the reduction in the number of individuals trapped using the same trapping effort the absolute number of individuals in the total population (before the first trapping) can be estimated. This method is particularly useful in circumstances where the population is relatively closed, such as a lake or an island. In an open population there is likely to be considerable dispersal into the study area as individuals exploit the opportunities provided by the artificially low population density created by the removal trapping. It is also possible to use this technique to extract population estimates from trapping, hunting or fishing records where captured individuals are removed from the population.

Population enrichment

There are various techniques based on the idea of adding individuals to a population. The study population is observed or trapped before and after addition. As a population has been augmented by a known number, the increase in captures or observations can be scaled to estimate the population size prior to the addition. *See also*

ABUNDANCE; CAPTURE–RECAPTURE TECHNIQUES; DISPERSION; SAMPLING METHODOLOGY/DEVICES. [C.D.]

REFERENCES
Caughley, G. (1977) *Analysis of Vertebrate Populations.* John Wiley & Sons, Chichester.
Southwood, T.R.E. (1978) *Ecological Methods.* Chapman & Hall, London.

---

absorption efficiency *See* ASSIMILATION EFFICIENCY.

absorption spectrum A graph or chart showing how the absorption of RADIATION by a medium depends on the wavelength of the incident energy. Absorption is defined as the fraction of the incident energy that is absorbed. A related quantity is the absorption coefficient (*see* BEER'S LAW). The energy not absorbed may be reflected or transmitted, and so it is often useful to present absorption, transmission and reflection spectra as one diagram.

Spectral properties may also be presented as length absorption coefficients (to apply to a unit thickness) or as mass absorption coefficients (applying to the thickness of the medium with unit mass per unit area). *See also* PHOTOSYNTHESIS. [J.G.]

REFERENCE
Gates, D.M. (1980) *Biophysical Ecology.* Springer-Verlag, Berlin.

abstraction Removal of WATER from natural supply: RIVERS, LAKES, ponds, groundwater. Usually controlled, often by licence, but there may also be some obligations in common law. [P.C.]

abundance The availability of a resource, or numbers in a population, often described in qualitative terms such as 'rare' or 'common'. Abundance may be used to describe the state of a species either in a restricted area of habitat or over its global distribution. A species that is found at high DENSITY where it occurs but does not occur in many places (i.e. is not very prevalent) would generally be thought of as less abundant than a more prevalent or widespread species that never achieved high densities in local populations. Conservation ecologists have extended the concept of abundance to include 'endangered'. The causes of mean levels of, and changes in, species abundance have been the focus of much ecological research and controversy; for example, disagreement about the relative importance of DENSITY DEPENDENCE and DENSITY INDEPENDENCE. *See also* ABSOLUTE POPULATION ESTIMATES. [R.H.S.]

REFERENCE
Begon, M., Harper, J.L. & Townsend, C.R. (1996) *Ecology: Individuals, Populations and Communities,* 3rd edn, Chapter 15. Blackwell Science, Oxford.

abyssal The bottom zone of the OCEAN at depths between about 4000 and 6000 m, constituting most of the DEEP SEA floor. Abyssal is also referred to as 'profundal'. The ocean floor consists of volcanic rock, covered by soft SEDIMENT. The sediment contains varying combinations of mineralized PLANKTON skeletons, CLAY and other minerals deposited from continental sources, volcanic rocks and precipitates, such as MANGANESE nodules. This part of the ocean is completely dark, physically stable, and (with the exception of areas surrounding HYDROTHERMAL VENTS) completely dependent upon the import of organic matter from the ocean surface.

Deep-sea species live at constant, low temperatures (i.e. 1–4°C) and at high pressure (e.g. the pressure is 400 atm at a depth of 4000 m), and both of these environmental variables have strong effects on organism function. The biomass of benthic organisms is much reduced in abyssal areas compared to the CONTINENTAL SHELF because of the lower input of organic matter to the deep sea. The reduced input of organic matter is also reflected in changes in the FEEDING TYPES of abyssal benthic animals. For example, carnivores and suspension-feeders become less abundant, whereas DEPOSIT-FEEDERS dominate in abyssal areas. Likewise the morphology and activity patterns of some deep-sea animals reflect the low food availability in this habitat (e.g. they are sluggish and have low metabolic rates). In addition, many abyssal species have slow or delayed sexual maturity, small gonads and slow embryological development. Although information on the life-history patterns of deep-sea animals is limited, SEMELPARITY and a lack of SEASONALITY in reproductive activity appear to be relatively common. In contrast to the monotonic decline in benthic biomass with increasing depth in the ocean, species diversity for a wide variety of animal groups increases with depth to a maximum at about 2000 m and then decreases with further increases in depth. Difficulties in sampling the deep sea have hampered studies of the biology and ecology of abyssal organisms. However, the relatively recent development of submersibles, remotely controlled vehicles, underwater photography and video has greatly increased our knowledge of the abyssal ocean. [V.F.]

abyssal cones Important topographic features of the plains of the deep OCEAN. Abyssal cones are depositional features, composed of SEDIMENT derived from RIVERS. They have their apices at river mouths and extend from shallow water across the CONTINENTAL SLOPE, rising to merge at their bases with abyssal plains. [V.F.]

acaricide A BIOCIDE intended to kill ticks and mites. *See also* INSECTICIDES; PESTICIDE. [P.C.]

acceleration As applied to EVOLUTIONARY RATES, a process of HETEROCHRONY where evolutionary transformation is achieved by speeding up the rate of development ('shape change') of a TRAIT, organ or whole organism, relative to the rate in the ancestral ONTOGENY. Acceleration results in the addition of new stages being terminally added to the ancestral sequence. Unlike HYPERMORPHOSIS, acceleration requires no change in the overall ontogenetic time. Forms generated by either acceleration or hypermorphosis are referred to as peramorphic. Phylogenetic transformations due to the action of acceleration conform to Haeckel's theory of RECAPITULATION. *See also* PERAMORPHOSIS. [P.A.]

acceptable risk In an environmental context, the perceived PROBABILITY that a process or substance may have an adverse EFFECT, but one that is deemed acceptable, usually because it is very low, by interested parties (e.g. politicians, policy-makers, regulators, public). *See also* RISK ASSESSMENT; RISK MANAGEMENT. [P.C.]

access to information (registers) Giving access to information on performance likely to influence the state of the environment. Is used as an instrument for environmental protection in both Europe and the USA. The rationale is that the information, once on public registers, is likely to influence the activities of pressure groups, customers and other interested parties such that the subjects are likely to try and anticipate this by altering their activities to avoid being placed on the registers or to present a more acceptable record. *See also* MARKET INSTRUMENTS; TOXIC RELEASE INVENTORY. [P.C.]

accident In an environmental context, an undesirable, unexpected event that causes CONTAMINATION and possibly POLLUTION. [P.C.]

acclimation The widespread phenomenon whereby living organisms adjust to the present environmental conditions, and in doing so, enhance their probability of survival. These adjustments are on time-scales of less than one generation, and may involve changes in physiological processes and structure. For example, when a growing plant is moved experimentally from a NITROGEN-rich rooting medium to a nitrogen-poor medium, an increase occurs in the fraction of assimilated carbon that is transported to root as opposed to shoot, thus tending to compensate for the scarcity of nitrogen in the medium. Similarly, when a human travels from the north temperate zone to the humid tropics, the rate of sweat production is increased over a period of several days of exposure to the new conditions, until a new rate is achieved, thus promoting cooling.

A distinction can be made between acclimatization, describing a longer-term seasonal ADAPTATION, and acclimation, where similar changes are induced more rapidly under laboratory conditions. Acclimatization in preparation for winter is cold HARDENING, and occurs in plants, animals and microbes. In many insects, exposure to temperatures in the range 0–10°C for 1–2 weeks (sometimes requiring a simultaneous decrease in PHOTOPERIOD) is sufficient to induce key biochemical changes (synthesis of ice nucleating agents, polyols, antifreeze proteins) that are central to the FREEZE TOLERANCE/FREEZE AVOIDANCE OVERWINTERING strategies.

Ability to acclimate to a changing environment is probably a characteristic of all organisms, though the extent to which acclimation is possible may vary from species to species, and undoubtedly depends on the rate at which the environment changes. *See also* PROXIMATE–ULTIMATE DISTINCTION. [J.G. & J.S.B.]

acclimatization *See* ACCLIMATION.

accreditation To make ENVIRONMENTAL MANAGEMENT SYSTEMS credible it is necessary to build in independent, third parties as verifiers of the audits. The verifiers have to be accredited. For this, they must be demonstrably competent in:

- environmental auditing methodologies;
- management information systems and processes;
- regulatory issues;
- environmental issues;
- relevant legislation and standards;
- relevant technical knowledge of the activities being verified.

See also ECOAUDIT. [P.C.]

accuracy/precision These terms commonly refer to how well a particular measurement has been made and are often used interchangeably, although strictly they do have slightly different meanings. Accuracy may be defined as 'the closeness of a measured or computed value to its true value', while precision may be defined as 'the closeness of repeated measurements of the same quantity' (Sokal & Rohlf 1995). Statistically, accuracy is determined by the size of a SAMPLE while precision is determined by the way in which the sample is taken. Samples are common as it is often impractical to measure the whole 'statistical population'. For instance, it might be possible to directly count the total number of oak trees in a given area but not their associated insect communities, which would need to be estimated indirectly. The accuracy of any sample is usually indicated by CONFIDENCE LIMITS, which are estimated from the sample's VARIABILITY and size. Generally the smaller the sample size, in relation to the total population, the less accurate it will be at describing its properties. The precision of a sample or measurement can be influenced by a wide range of factors, two of the most common being careless recording and the incorrect use of equipment. As the earlier definition suggests, the only way in which precision can be checked is through the repeated measurement of the same population or item. Statistically, this is usually achieved through replication, without which the precision of an estimate cannot be assessed. *See also* SAMPLING; STANDARD ERROR (OF THE MEAN); STATISTICAL POWER. [R.C.]

REFERENCE

Sokal, R.R. & Rohlf, F.J. (1995) *Biometry*, 3rd edn, pp. 13–16. W.H. Freeman, New York.

acid The Arrhenius definition of an acid is a substance that produces HYDROGEN ions in aqueous solution. Such protonic acids tend to be corrosive and are further classified as strong or weak. A strong acid (e.g. sulphuric, nitric and hydrochloric acids) is completely dissociated, whereas a weak acid (e.g. acetic, carbonic and citric acids) dissociates to form ions in aqueous solution. According to the Brønsted–Lowry concept (1923), an acid is a hydrogen ION donor. A more general definition provided by G.N. Lewis (1923) states that an acid is an electron pair acceptor, encompassing the Brønsted–Lowry definition and establishing a relationship between acid–base and OXIDATION–REDUCTION reactions. *See also* ACID RAIN AND MIST; BASES; pH; SOIL. [M.C.G. & J.G.]

acid precipitation, impact on ecosystems The effect of acid precipitation is to reduce pH in the media that surround organisms. The consequences are complex and multifarious. Reduction in pH can have a direct effect, for example by causing imbalances in acid–base relationships in tissues and at membranes, and/or can also have indirect effects, for example by LEACHING calcium from soils and leading to the mobilization of metals such as aluminium in watercourses. The effects will also depend upon the form of the deposition; thus acid rain washes rapidly from plant surfaces but can affect the soils in which the plants reside, whereas acid mist tends to coat leaves making it potentially more harmful than rain. Often acid precipitation will cause multiple stresses to an ECOSYSTEM at the same time. *See also* ACID RAIN AND MIST; ACIDIFICATION. [P.C.]

acid rain and mist RAIN is naturally acidic, containing carbonic acid formed by the solution of atmospheric CARBON DIOXIDE (CO_2), and has a pH of about 5.6 (Fig. A1). More acidic rain (lower pH) may be formed naturally from biogenic sources of SULPHUR-containing

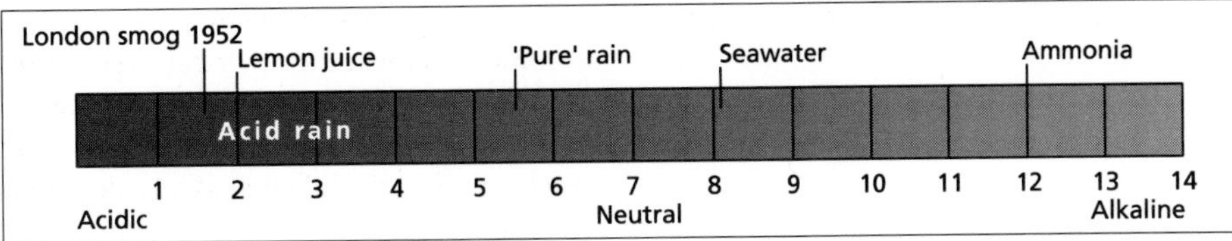

Fig. A1 The pH scale of alkalinity. Low pH represents acid conditions and high pH is alkaline, with pH 7.0 being neutral. It is a logarithmic scale, so one point drop in pH means 10 times more acid, two pH points drop means 100 times more acid, and so on. (After Moore, P.D., Chaloner, B. & Stott, P. (1996) *Global Environmental Change*. Blackwell Science, Oxford.)

gases, such as dimethyl sulphide ($(CH_3)_2S$), which is released by marine PHYTOPLANKTON. These gases are oxidized to sulphur dioxide (SO_2), and ultimately to sulphuric acid (H_2SO_4), which is incorporated into cloud and raindrops. Acidities 10 times those attributable to carbonic acid have been reported from remote marine sites. In polluted regions, anthropogenic EMISSIONS of sulphur and NITROGEN OXIDES from fossil fuel combustion are oxidized in the ATMOSPHERE to sulphuric and nitric acids, which cause widespread ACIDIFICATION of cloud and rainwater. These pollutants may travel for hundreds of kilometres (long-range transport) before falling to the ground as acid rain. Some of the acidity may be neutralized by AMMONIA gas (NH_3), released largely from agricultural processes. In parts of Europe and North America annual average rain pH may be as low as pH 4 (40 times 'natural' acidity), with individual rain events having acidities 10 times greater (pH 3). Cloud water is generally more acid than rain, by a factor of 2–5, as individual cloud droplets contain less water, and are therefore more concentrated, than raindrops. The wet deposition of pollutants in acid rain leads to the acidification of soils and freshwaters, and the ammonium and NITRATE ions may also lead to EUTROPHICATION. *See also* ACID PRECIPITATION, IMPACT ON ECOSYSTEMS; AEROSOLS. [J.N.C.]

acidification Reduction in the pH of the environment, mainly due to human actions; for example, ACID RAIN AND MIST and acid MINE DRAINAGE. *See also* ACID PRECIPITATION, IMPACT ON ECOSYSTEMS. [P.C.]

acidophilic Describing plants that are confined to, or more common on, ACID, but not necessarily CALCIUM-deficient, soil. For this reason the term acidophilic is not always synonymous with the term CALCIFUGE. Soils with acidophilic plants are commonly of soil pH 3–5. Acidophilic and ACIDOPHOBIC species are not rigidly exclusive but form a continuum of physiological variation. Acidophilic plants often absorb NITROGEN as the ammonium ion, rather than NITRATE, as do plants of neutral or alkaline soils. They may also tolerate TOXICITY due to IRON, MANGANESE and ALUMINIUM, which reach high concentrations in acid soils and waters. [J.R.E.]

acidophobic Describing plants that are excluded from ACID soil, usually by sensitivity to high concentrations of IRON, MANGANESE and ALUMINIUM, inadequacy of NITRATE supply, low concentration of PHOSPHORUS and general mineral nutrient deficiency. The term acidophobic should not be used as a synonym for CALCICOLE. *See also* ACIDOPHILIC. [J.R.E.]

acme zone *See* BIOSTRATIGRAPHIC UNIT.

acquired character A change in an organism due to environmental influence. It may be adaptive (due to selection for adaptative PHENOTYPIC PLASTICITY) or maladaptive. Thus loss of a tail by a lizard to escape a predator (autotomy) is adaptive, whereas accidental loss or impairment of a limb may be maladaptive. LAMARCKIAN INHERITANCE presumed that acquired characters were heritable. Plasticity can be inherited and is subject to normal Darwinian selection, but characters acquired by accident are not heritable. *See also* DARWINISM. [P.C.]

Acrisols *See* SOIL CLASSIFICATION.

acronyms Alphanumeric abbreviations for programmes, procedures and principles. They are common in environmental policy and legislation. Examples include: BATNEEC; BOD; BPEO; EPA; LCA; QA; QC; RCC; RIVPACS; SAR; TIE; and UNCED. [P.C.]

acrotelm The upper functional SOIL layer in a MIRE, in which plant litter is transformed into PEAT and beneath the base of which the WATER TABLE never sinks in an intact system. The term derives from the Greek *akros*, topmost, and *telma*, marsh (Ingram 1978). The *deyatel'nyy sloy* or *aktivnyy sloy* (active layer) of Russian telmatologists (Lopatin 1949). *See also* CATOTELM. [H.A.P.I.]

REFERENCES

Ingram, H.A.P. (1978) Soil layers in mires: function and terminology. *Journal of Soil Science* **29**, 224–227.

Lopatin, V.D. (1949) O gidrologicheskom znachenii verkhovykh bolot. *Vestnik Leningradskogo gosudarstvennogo universiteta* **2**, 37–49.

act of God *See* ACT OF NATURE.

act of nature An occurrence, without the involvement of any human agency, that cannot be foreseen or prevented and usually has disastrous consequences for humans and/or the environment. Also called an act of God. [P.C.]

actinium (Ac) Actinium (ATOMIC NUMBER = 89; relative ATOMIC MASS = 227.03; melting point = 817°C; boiling point = 2470°C) was discovered in 1899 by Andre Debierne. Actinium exists in nature only as the two radioisotopes ^{227}Ac ($t_{1/2}$ = 21.77 years) and ^{228}Ac ($t_{1/2}$ = 6.13 h), which are members of the natural radioactive decay series. It occurs in URANIUM- and THORIUM-bearing minerals and has similar chemical properties to SCANDIUM and YTTRIUM, existing in the 3+ oxidation state and having a very low solubility in natural waters. *See also* RADIONUCLIDE. [A.B.M.]

REFERENCE

Ivanovich, M. & Harmon, R.S. (eds) (1992) *Uranium-Series Disequilibrium.* Clarendon Press, Oxford.

action level The level or concentration of a substance that triggers active management (avoidance, ameliorative or remedial measures) to safeguard public health and/or the environment. *See also* THRESHOLD. [P.C.]

action spectrum In photobiology, a graph showing how the rate of a physiological process depends on the wavelength of the incident energy. Examples of action spectra are shown in Fig. A2. [J.G.]

activated charcoal Charcoal treated at high temperatures (*c.* 900°C) with an oxidizing gas (steam or carbon dioxide, CO_2) to greatly increase the surface area, up to 1500 $m^2 g^{-1}$. Coconut shell gives a hard microporous activated CARBON, used to remove gases from air (e.g. pollution control, solvent recovery, containment of radioactive gases in NUCLEAR POWER-STATIONS, and as the absorbent in gas masks). It can also be used to purify water, removing residual CHLORINE and organic compounds. Wood, peat and fabric-based activated carbons have much larger pores, and are used mainly in the liquid phase. [J.N.C.]

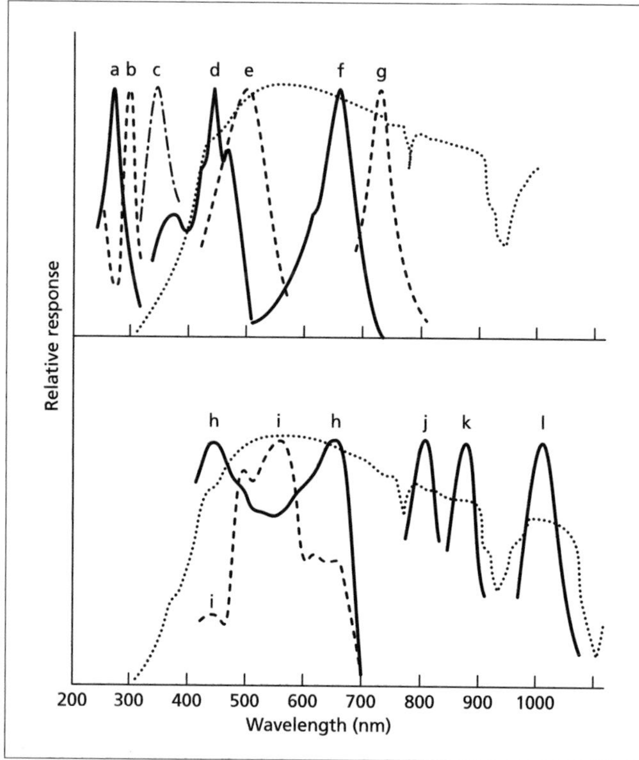

Fig. A2 Summary of action spectra: a, bactericidal (bacteria-killing) effect; b, sunburn; c, ultraviolet (UV)-sensitive visual cells in the honeybee; d, phototropism in plants; e, human twilight vision; f, phytochrome conversion ($P_R \rightarrow P_{FR}$); g, phytochrome conversion ($P_{FR} \rightarrow P_R$); h, photosynthesis in terrestrial plants; i, photosynthesis in red algae; j–l, infrared absorption maxima in different kinds of purple bacteria (probably coinciding with photosynthetic action maxima). The dotted curve shows the approximate spectrum of sunlight (photons per wavelength interval) at the surface of the Earth. The exact composition of the light depends on weather and the Sun's elevation. (After Björn, L.O. (1976) *Light and Life*. Hodder & Stoughton, Sevenoaks.)

activated sludge Sludge from the ACTIVATED-SLUDGE PROCESS—a method of treating SEWAGE by agitating and aerating a mixture of sewage and microorganisms (largely bacteria and protozoans). The biologically active sludge (hence the name) so formed is subsequently separated from the treated effluent by settlement. Most is recycled into the system. The excess is discharged for dumping or INCINERATION. *See also* SEWAGE SLUDGE. [P.C.]

activated-sludge process The biological oxidation of SEWAGE to reduce the ORGANIC LOADING that raw sewage would otherwise impose on receiving water. The process involves aeration in tanks by agitators. Aerobic bacteria grow on the raw sewage, converting some of this to their own biomass with the rest being transformed to WASTE gases. The products are a suspension with reduced organic loading, and a SEDIMENT which consists of bacteria and protozoans. Part of the latter is recycled to seed the raw input, with the rest being disposed of on land, at sea or even by INCINERATION. *See also* SEWAGE SLUDGE. [P.C.]

active dispersal The process by which organisms actively move away from one another as a means of increasing FITNESS. The function of active DISPERSAL is often to reduce COMPETITION with close relatives, for example dispersal of plant seeds away from the parent. Another function in species that exploit ephemeral resources is to find new resources for growth and reproduction. Active dispersal of individuals in sexual species also reduces the chances of mating with close relatives, thereby minimizing the fitness cost of INBREEDING DEPRESSION; in this case only one sex needs to disperse, and in animals it is most frequently the males. *See also* VAGILITY. [R.H.S.]

active drift-entry *See* DRIFT, FRESHWATER.

active ingredient In BIOCIDES, the component that kills the target organism. PESTICIDES are usually regulated primarily according to their active ingredients. [P.C.]

acute tests Testing, often referring to ECOTOXICOLOGY, for responses that are immediately debilitating; i.e. usually lethal over a short time-span (day(s)), and so involve relatively high concentrations of test substance. *Cf.* CHRONIC TESTS; LONG-TERM TESTS. *See also* PREDICTIVE ECOTOXICITY TESTS; SHORT-TERM TESTS. [P.C.]

adaptation A term generally used to mean within-species evolutionary response to a particular, often new environment. The implication is that NATURAL SELECTION has favoured alleles that gave their carriers an advantage in the specified environment. Thus adaptations to cold environments may include insulation, energy conservation and so on, provided these are genetically coded. In some cases the evolutionary process of adaptation has led to SPECIATION, producing different, but related species adapted to different ecological niches.

Confusingly 'adaptation' is also sometimes used by physiologists to describe reversible advantageous responses to new environments without any implication of genetic change. *Cf.* ACCLIMATION. *See also* ADAPTIVE LANDSCAPE; DARWINISM; EVOLUTION; EVOLUTIONARY OPTIMIZATION; PROXIMATE–ULTIMATE DISTINCTION. [R.M.S.]

adaptation to pollution POLLUTION represents anthropogenic changes to environmental factors that cause STRESS to organisms. Populations usually show genetic HETEROGENEITY and it is likely that different organisms show different responsiveness to the stress — as a result of physiological and/or morphological and/or behavioural differences. The POLLUTANT will therefore act as a SELECTION PRESSURE and possibly lead to the evolution of TOLERANCE. Thus plant strains sampled from metal spoil heaps are often more metal tolerant than strains of the same species from non-polluted soils. However, differing degrees of tolerance of organisms sampled from different populations, exposed to different levels of pollution, do not necessarily always signal genetic divergence. Exposure may induce the activation of genes that code for protection systems, such as METALLOTHIONEINES and HEAT-SHOCK PROTEINS, that are generally present irrespective of EXPOSURE. Such induction is often referred to as adaptability or ACCLIMATION. Transplantation and/or breeding experiments are needed to distinguish ADAPTATION from adaptability. Of course adaptability can itself be subject to selection and hence arise as an adaptation. [P.C.]

Adaptive landscape

1 The mean FITNESS of a population as a function of gene frequencies.
2 The mean fitness of a population as a function of the mean PHENOTYPE.

While these are formal definitions, the concept is also used heuristically to describe positions of alternative equilibria with respect to the GENOTYPE or phenotype frequencies. Empirical descriptions of adaptive surfaces are often based on estimates of individual fitnesses rather than mean population fitness.

Sewall Wright (*see* CHARACTERS IN ECOLOGY) introduced the concept of an adaptive landscape for gene frequencies in connection with his result:

$$\Delta p = \frac{p(1-p)}{\overline{W}} \frac{\partial \overline{W}}{\partial p},$$

where Δp is the change in allele frequency, p is the frequency of an allele, $\overline{W}$ is mean fitness, and $\partial \overline{W}/\partial p$ is a change in $\overline{W}$, holding fixed the relative frequencies of all other ALLELES at the locus. Mean fitness is the average fitness of all the individuals present in the population. Wright's formulation demonstrates that in the absence of FREQUENCY-DEPENDENT SELECTION mean fitness of the population always increases (the result does not hold under some extreme conditions, such as strong EPISTASIS between tightly linked loci). If the adaptive surface has multiple peaks and valleys, a population can become trapped at one peak, which may have lower mean fitness than alternative peaks. A simple adaptive surface of this sort is shown in Fig. A3. Wright argued that interactions among loci lead to complex multidimensional adaptive surfaces. He thought a central problem in EVOLUTION is the mechanism by which a population moves from a lower to a higher peak by crossing a valley of low mean fitness. In his 'shifting balance' theory GENETIC DRIFT results in occasional maladaptive evolution, and the crossing of a valley, setting the stage for a new round of adaptive evolution whereby a 'hill' of even higher fitness may be scaled. Much theoretical work has shown the shifting balance can work, although it is unclear how often conditions for it are met in nature. Alternative mechanisms for peak shifts are a change in the adaptive surface itself due to a (perhaps temporary) change in the environment or MACROMUTATION. If genotypic fitnesses are frequency dependent the rigorous concept of an adaptive surface breaks down, and populations may even evolve to positions of low mean fitness (i.e. valleys in the adaptive surface). The concept is still used heuristically, to envisage how populations may become stuck at one equilibrium, and shift between alternative equilibria.

By analogy with Wright's adaptive landscape for gene frequencies, G.G. Simpson developed the idea of an adaptive landscape for phenotypes. Simpson discussed several mechanisms for transition between alternative states, including genetic drift, a change in the features of the landscape, and the possibility that a peak shift in one TRAIT occurs as a correlated response to selection on other traits. A hypothetical adaptive surface for two phenotypic traits is illustrated in Fig. A3b.

Simpson's idea of an adaptive surface for phenotypic traits was formalized by Russell Lande. Mean fitness depends both on the

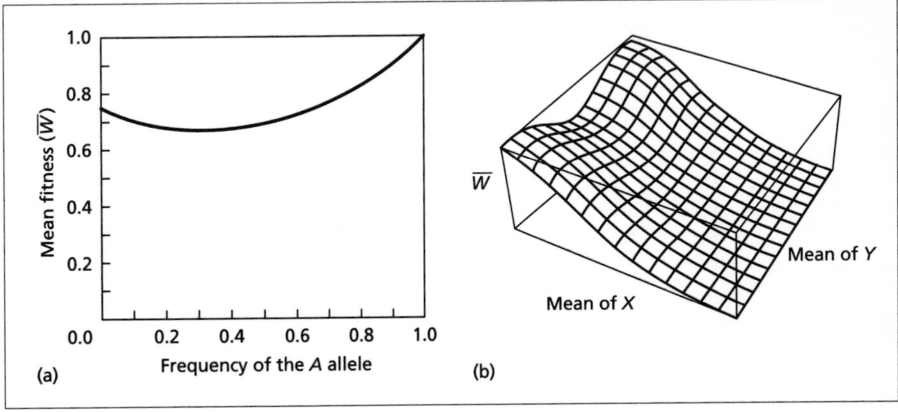

Fig. A3 (a) An example of a simple adaptive surface with multiple peaks. There are two alleles (*A*, *a*) at a single locus and fitnesses of the three genotypes are: *AA* = 1.0; *Aa* = 0.5; and *aa* = 0.75. The frequency of the *A* allele is shown on the *x*-axis. (b) An example of an adaptive surface for two continuously varying traits. *Y* is subject to disruptive selection, such that individuals with values *Y* = +1 or *Y* = −1 have the highest fitness and intermediates have lower fitness. *X* is subject to directional selection. (After Price, T., Turelli, M. & Slatkin, M. (1993) Peak shifts produced by correlated response to selection. *Evolution* **47**, 280–290.)

mean and variance of phenotypes, but the adaptive surface is usually described as a function of the mean phenotype, assuming a constant PHENOTYPIC VARIANCE, and constant phenotypic fitness (no frequency dependence). Assuming a NORMAL DISTRIBUTION of phenotypes Lande obtained the following result describing the evolution of a single trait:

$$\Delta\bar{z} = V_A \frac{1}{\overline{W}} \frac{d\overline{W}}{d\bar{z}},$$

where $\Delta\bar{z}$ is the change in the mean value of the trait from one generation to the next, $\overline{W}$ is the mean fitness, V_A is the ADDITIVE GENETIC VARIANCE of the trait, and $d\overline{W}/d\bar{z}$ is the tangent to the adaptive surface, measured at the population mean. If fitness is scaled so that mean fitness is 1.0, the direction and intensity of selection is measured by $d\overline{W}/d\bar{z}$. A similar result holds when many traits are simultaneously under selection; the intensity and direction of selection is described by a tangent to the mean fitness surface at the joint mean value for all traits, and pointing in the steepest uphill direction. As in Wright's formulation the mean fitness of the population always increases, and this implies that a population can become trapped at a peak of relatively low mean fitness.

Adaptive surfaces have rarely been estimated in nature. For genotypes the adaptive surface can be directly calculated, if all genotypic fitnesses have been measured (Fig. A3). Figure A4 shows the estimates of two adaptive surfaces for continuously varying phenotypic traits; such traits are often of more interest to ecologists. The first describes an estimate of expected population size as a function of mean beak size in a hypothetical species of Darwin's ground finches, derived from field observations on the finches. Finches with large beak and body size can crack large seeds, but among those finches able to crack a seed of a given size, those of small size maintain higher population densities for the same biomass of food. The predicted population size for a finch species of a given size thus varies with the abundance of seeds of different sizes on an island. For the adaptive surface shown for Isla Daphne Major there are several species of small seeds and two species of medium-sized seeds, and maxima in the adaptive surface occur for finches of small and intermediate body size. Construction of adaptive surfaces for 16 islands has been used to test hypotheses of COMPETITIVE EXCLUSION, CHARACTER DISPLACEMENT and ADAPTATION in Darwin's finches.

A two-dimensional illustration representing the ability of crossbills (*Loxia* spp.) of different morphologies to exploit four different cones is illustrated in Fig. A4b. Peaks in this surface are expected to approximate to peaks in a mean fitness surface if all four coniferous species are present in the habitat and food is limiting. *See also* DARWINISM; EVOLUTIONARY OPTIMIZATION; FITNESS LANDSCAPE. [T.D.P.]

REFERENCES

Benkman, C. (1993) Adaptation to single resources and the evolution of crossbill (*Loxia*) diversity. *Ecological Monographs* **63**, 305–325.

Lande, R. (1976) Natural selection and random genetic drift in phenotypic evolution. *Evolution* **30**, 314–334.

Price, T., Turelli, M. & Slatkin, M. (1993) Peak shifts produced by correlated response to selection. *Evolution* **47**, 280–290.

Schluter, D. & Grant, P.R. (1984) Determinants of morphological patterns in communities of Darwin's finches. *American Naturalist* **123**, 175–196.

adaptive radiation The evolutionary process by which an ancestral species gives rise to an array of descendant species exhibiting great ecological, morphological or behavioural diversity. Prominent examples are found in the biotas of oceanic archipelagos and continental rift lakes, where conditions appear to favour unusually high rates of SPECIATION and evolutionary innovation. Examples include the Hawaiian silversword alliance (plants), Hawaiian drosophilid flies, African cichlid fishes, Galapagos finches and Hawaiian honeycreepers (birds). *See also* ADAPTATION; ADAPTIVE LANDSCAPE; EVOLUTIONARY RADIATION AND POSSIBLE EXPLANATIONS. [R.H.R.]

additive experiments A design for experiments on interspecific plant COMPETITION, in which a constant density of individuals of species A is grown in MONOCULTURE and with various numbers of species B (e.g. yield is measured when a crop plant, sown at standard density, is grown with various densities of weeds). Thus the total number of individuals (species A and B together) differs from treatment to treatment, which means that any effects of INTERSPECIFIC

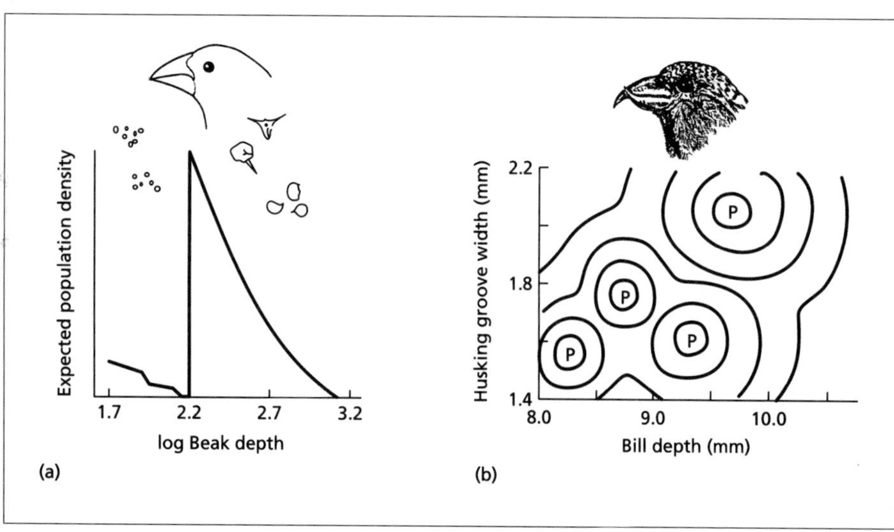

Fig. A4 (a) An adaptive surface (estimated as the expected population density) for beak depth of a finch species on Isla Daphne Major, Galapagos. Seeds contributing to each peak are shown above and to the right of the peaks. (After Schluter, D. & Grant, P.R. (1984) Determinants of morphological patterns in communities of Darwin's finches. *American Naturalist* **123**, 175–196, with permission from The University of Chicago Press.) (b) An approximation to the adaptive surface for crossbill beak and depth and groove width of the upper beak as estimated from studies of feeding efficiency. The four peaks (P) represent optimal morphologies; contour lines connect points of equivalent feeding efficiency. (After Benkman, C. (1993) Adaptation to single resources and the evolution of crossbill (*Loxia*) diversity. *Ecological Monographs* **63**, 305–325.)

COMPETITION are confounded with the effects of increased total POPULATION DENSITY. *See also* DE WIT REPLACEMENT SERIES; EXPERIMENTAL DESIGN. [M.J.C.]

additive genetic variance The additive GENETIC VARIANCE V_A is defined in two ways: in statistical empirical QUANTITATIVE GENETICS it is the variance of the BREEDING VALUES in the population, whereas in theoretical quantitative genetics it is that part of the genetic variance that is explained by linear regression of GENOTYPIC VALUES on number of A_1 ALLELES in the diploid GENOTYPE.

The first definition corresponds to the additive genetic variance as estimated by the regression of parent on offspring TRAITS (parent–offspring regression), and includes the additive-by-additive INTERACTION VARIANCES deriving from interactions between loci.

The second definition does not include components from interactions between loci. In theoretical quantitative genetics the additive genetic variance per bi-allelic locus *A*, $2pq\alpha^2$, equals the variance $2pq$ in number of A_1 alleles in the individuals of the population, times the square of the AVERAGE EFFECT OF A GENE SUBSTITUTION α representing the computed additive effect of an allele in this population. The total additive genetic variance over a number of (non-interacting) bi-allelic loci in LINKAGE DISEQUILIBRIUM equals:

$$V_A = \sum_{\text{loci } i} 2p_i q_i \alpha_i^2 + \sum_{\text{loci pairs } i,j} 4D_{ij}\alpha_i\alpha_j$$

where p_i is the allele frequency of allele A_1 at locus *i*, and D_{ij} is the coefficient of linkage disequilibrium between loci *i* and *j*. Linkage disequilibrium has a large effect on V_A. A population that has gone through a bottleneck and has high linkage disequilibrium as a result, might have much higher or much lower V_A than a population in linkage equilibrium.

The SELECTION RESPONSE of a quantitative character is predicted by the product of the additive genetic variance and the SELECTION GRADIENT. The selection gradient quantifies the strength of selection and the additive genetic variance quantifies what effect this selection will have on the QUANTITATIVE TRAIT values in the next generation. *See also* MEAN GENOTYPIC VALUE. [G.D.J.]

additivity The phenomenon whereby the combined ecotoxicological EFFECT of a mixture of chemicals is equal to the sum of the effects of the individual components in isolation. *Cf.* ANTAGONISM; SYNERGISM. *See also* TOXICITY OF MIXTURES. [P.C.]

adiabatic lapse rate The magnitude of the negative vertical gradient of atmospheric TEMPERATURE maintained by adiabatic vertical motion of air through surroundings in hydrostatic equilibrium. When the air is unsaturated it is known as the dry adiabatic lapse rate and has the value 9.8°C km^{-1} (g/C_p where g is the acceleration due to gravity and C_p is the specific heat of air at constant pressure), provided the temperatures of the moving and ambient air are nearly the same on the absolute scale, which is usually the case (McIlveen 1992).

If the air is kept saturated by condensation of vapour in rising air, and EVAPORATION of cloud in sinking air, a saturated adiabatic lapse rate results, whose numerical value varies considerably with pressure and temperature (as given in special thermodynamic tables or diagrams; Meteorological Office 1989). In the warm, low TROPOSPHERE saturated adiabatic lapse rates may be as little as half the dry adiabatic lapse rate, whereas in the cold, high troposphere they are only slightly less than the dry adiabatic value.

Near-adiabatic lapse rates are widely observed because CONVECTION triggered when they are exceeded holds actual lapse rates close to the appropriate adiabatic values. [J.F.R.M.]

REFERENCES

McIlveen, J.F.R. (1992) *Fundamentals of Weather and Climate*. Chapman & Hall, London.

Meteorological Office (1989) *Meteorological Office Tephigram*. Her Majesty's Stationery Office, London.

administrator In the USA, the head of the EPA, or any office or employee of the Agency to whom authority has been delegated. *See also* REGULATORY AGENCIES. [P.C.]

adoption The providing of all PARENTAL CARE to an immature individual by an individual that is not one of its biological parents. In biology, the adoptive parent can be unaware that it is caring for an unrelated individual, an example being the passerine bird working daily to feed the cuckoo chick that threw out of the nest all the nestlings of its HOST parents (*see* BROOD PARASITISM). Taken to the extreme, this applies to species with a low level of offspring recognition where nests have been parasitized by other birds, or where fathers feed offspring they did not sire, as has been observed in many monogamous bird species.

When the adoption follows the death of the parent providing the care, the adoptive individual is mostly a direct relative, for example an older brother or sister. Such adoptions are uncommon in mammals, in which there is generally a short period of dependency on the parent; consequently, older siblings are already gone when a youngster might need assistance. In primates, where the growth period lasts longer and sociality is more common, adoptions by older siblings are regularly observed. KIN SELECTION could explain the evolution of such adoptions, althougth some direct gain in terms of social partner and ally may also exist. Adoption by unrelated individuals is much rarer and is seen with regularity only in humans. *See also* PARENT–OFFSPRING CONFLICT; PARENTAL INVESTMENT. [C.B.]

advanced very-high-resolution radiometer (AVHRR) An imaging sensor used in environmental research. It is carried on the NATIONAL OCEANIC AND ATMOSPHERIC ADMINISTRATION (NOAA) series of satellites, has a swath width of 2400 km and a spatial resolution of 1.1 km. The current AVHRR records in five wavebands: red (0.58–0.68 μm); near infrared (0.725–1.1 μm); middle infrared (3.55–3.93 μm); and thermal infrared (10.5–11.5 μm, 11.5–12.5 μm). Images are available at least once daily in local area format (original spatial resolution) and approximately weekly in global area format (4 km degraded spatial resolution). Although designed for oceanographic and atmospheric applications, the imagery has also been used to estimate vegetation 'amount' at regional to global scales. *See also* REMOTE SENSING. [P.J.C.]

advection The transfer of energy or mass by horizontal movement. Advection is often contrasted to diffusion, which is random mixing and is expressed in units of (length)^2 per unit time. In the ATMOSPHERE local advection is a leading edge effect resulting from air movement across a surface discontinuity, for example a change in ROUGHNESS or EVAPORATION. During local advection the FLUXES of energy through the air above the surface vary in the horizontal and are not constant with height. An example is dry air blowing onto an irrigated field. Advection can be described by the equations of continuity for bulk air, water vapour, heat and momentum. However, this set of four equations has seven unknowns and can only be solved by making three assumptions. Regional advection occurs when there is large-scale transfer of energy from the atmosphere to the surface, as when warm oceanic air moves over a cold land mass.

In oceanography, advection describes the net movement of water along a major current, mathematically described by a velocity in units of length per unit time. [J.H.C.G. & V.F.]

REFERENCES

Brutsaert, W. (1982) *Evaporation into the Atmosphere*. D. Reidel, Dordrecht.

Itier, B., Brunet, Y., McAneney, K.J. & Lagouarde, J.P. (1994) Downwind evolution of scaler fluxes and surface resistance under conditions of local advection. Part I: A reappraisal of boundary conditions. *Agricultural and Forest Meteorology* **71**, 211–226.

adverse environmental effect The result of an action that is likely to be detrimental to the environment. [P.C.]

adversity selection *See* SELECTION PRESSURES, CLASSIFICATION.

advisory committees on environment Bodies set up by local, national or international authorities to provide expert (usually scientific) judgement on environmental issues. The composition, method of appointment and powers of such committees are often, though not always, defined by legislation. There is often a requirement that advice must be sought by the authorities, but not necessarily taken. In Britain, examples that advise the Government are:

- Advisory Committee on Business and the Environment (ACBE);
- Advisory Committee on Hazardous Substances (ACHS);
- Advisory Committee on Pesticides (ACP);
- Advisory Committee on Releases to the Environment (ACRE);
- Advisory Group on Eco-management and Audit (AGEMA).

[P.C.]

aerial (aircraft) spraying of pesticides *See* PESTICIDES AND SOIL CONTAMINATION.

aerial plankton Spores, bacteria and other microorganisms suspended and moving in the air. *See also* PLANKTON. [P.C.]

aerobic decomposition The gradual disintegration of dead organic matter by means of oxidation resulting from chemical processes or performed by organisms (detritivores and decomposers). Aerobic decomposition requires atmospheric OXYGEN and leads to the breakdown of complex organic molecules into CARBON DIOXIDE (CO_2), WATER and minerals. It is an important process in the flux of energy and matter within an ECOSYSTEM, and is crucial in nutrient cycling since the released NUTRIENTS can be assimilated by autotrophic organisms and repeatedly incorporated into BIOMASS. [P.O.]

aerobic respiration A metabolic process, occurring in the presence of OXYGEN, in which organic substances are broken down to yield energy with molecular oxygen acting as the final electron acceptor. During aerobic metabolism, 36 molecules of adenosine triphosphate (ATP) are produced for each molecule of glucose broken down (compared to generally only two molecules of ATP during ANAEROBIC METABOLISM). Aerobic respiration can be described by the idealized equation:

$$(CH_2O)_x(NH_3)_y(H_3PO_4)_z + (x+2y)O_2 \rightarrow$$
$$xCO_2 + (x+y)H_2O + yHNO_3 + zH_3PO_4.$$

See also RESPIRATION. [V.F.]

aerodynamic drag The DRAG on a body due to its motion relative to a gas, in particular air. The term is used extensively when describing the drag of the Earth's surface on the atmospheric BOUNDARY LAYER. A useful measure of the relative strength of the surface drag is provided by the aerodynamic ROUGHNESS LENGTH (z_0), which is defined in the LOGARITHMIC WIND PROFILE equation:

$$\bar{u}_z = \frac{u_*}{k}\ln\left(\frac{z}{z_0}\right)$$

where $\bar{u}_z$ is the mean wind speed at height z above the surface, u_* is the friction velocity and k is von Karman's constant (≈ 0.4). The roughness length is related to the height of surface features and vegetation, with values ranging from $0.1–10 \times 10^{-5}$ m over water to 1–5 m over forests. [B.A.G.]

REFERENCE

Kaimal, J.C. & Finnigan, J.J. (1994) *Atmospheric Boundary Layer Flows: Their Structure and Measurement*. Oxford University Press, New York.

Aerosols

The original definition of this term refers to suspended particles in the carrier medium air. However, it is now customary to apply the term aerosols, or aerosol particles, more broadly to include deposits of particulate matter since the atmospheric aerosol is collected on filters on particle size-separating impactor plates. The word 'aerosol' covers a wide range of materials but should be distinguished from DUST, which is larger pieces of solid material (≥20–30 μm in diameter).

Individual aerosol particles may be solid, liquid or mixed; they usually carry some moisture with them. Solid aerosol particles are primarily formed by soil erosion and enter the ATMOSPHERE by wind force. The major source of liquid aerosol particles is sea spray from the ocean surface, which upon EVAPORATION generally produces a concentrated aqueous solution of sea-salt crystals. The amount of WATER associated with the aerosol depends on the prevailing RELATIVE HUMIDITY. With increasing relative humidity more water condenses onto the particles; once the vapour pressure of water exceeds the SATURATION point some of the aerosol particles grow into fog or cloud droplets. These aerosol particles are called cloud condensation nuclei. Fogs and CLOUDS are not usually included in the term atmospheric aerosols. Due to the overlap in size in the two systems, any division is, however, somewhat arbitrary.

Two major aerosol types may be distinguished: primary and secondary. Primary aerosols are directly injected into the atmosphere from the Earth's surface, mainly from open water and soil areas, biological sources and anthropogenic processes. Secondary aerosols are formed after chemical conversion in the atmosphere, which generally involves gases, pre-existing aerosols and water vapour. Details about the chemistry of trace gas to aerosol conversion are somewhat limited. Recent interest has focused much attention on the conversion mechanisms associated with dimethyl sulphide (DMS, $(CH_3)_2S$) gas (emitted from the oceans) and sulphate (SO_4^{2-}) and methane sulphonic acid (MSA) aerosol products (Fig. A5). Secondary aerosols are almost always confined to the fine size range, defined as aerosol particles less than 1–2 μm in diameter. Aerosol particles larger than 1–2 μm in diameter are referred to as coarse.

The atmospheric aerosol, being a polydisperse system, cannot be described without taking into account the aerosol particle size spectrum. This spectrum typically includes macromolecules as small as 0.001 μm in diameter to coarse sea-salt and soil-derived particles up to 20–30 μm in diameter. Fine aerosol particles (< 1–2 μm in diameter) generally are observed to group into two distinct modes, usually referred to as the nucleation and accumulation modes. Whenever the aerosol accumulates material by condensation from the gas phase or by coagulation of smaller nucleation mode particles, the material is deposited in the size range of greatest aerosol particle surface area. This size range invariably is the 0.1–1.0 μm diameter range with a peak in aerosol volume near 0.3 μm, and is commonly referred to as the accumulation mode. The contributions of coagulation and condensation to aerosol volume or surface area in the accumulation mode must be distinguished. Condensation is the deposition of vapour-phase material onto pre-existing aerosols whereas coagulation is the formation of a new larger aerosol particle from the collision between two or more smaller ones. When the pre-existing particle concentration is very low, condensation may lead to the formation of many new particles. Provided the vapour pressure of the condensing substance is sufficiently high, condensation may still lead to the formation of some new particles despite the presence of pre-existing particles. Such new particles make up the nucleation mode, with diameters < 0.1 μm and with a peak in aerosol number at about 0.01 μm in diameter.

The removal of aerosol particles from the atmosphere is largely controlled by physical processes and varies greatly as a function of their size. The main mechanisms are removal in PRECIPITATION, and dry deposition to the Earth's surface. The mean RESIDENCE TIME in the atmosphere varies greatly with particle size. For aerosols < 0.1 or > 10 μm in diameter, tropospheric residence times are typically less than one day. For aerosols < 0.1 μm in diameter, coagulation is the main removal mechanism; for aerosols > 10 μm in diameter, SEDIMENTATION becomes increasingly important. Removal of aerosol particles in the 0.1–10 μm diameter range is primarily due to precipitation and dry deposition. Mean residence times are in the 1–10-day range, dependent mainly on whether precipitation has been active in a particular region of the TROPOSPHERE. For aerosols which are emitted into the STRATOSPHERE, such as by volcanic EMISSIONS, the mean residence time may be months to years since both precipitation and dry deposition are substantially reduced in this portion of the atmosphere.

Anions such as sulphate, NITRATE (NO_3^-) and CHLORIDE (Cl^-), and cations such as ammonium (NH_4^+), SODIUM (Na^+), CALCIUM

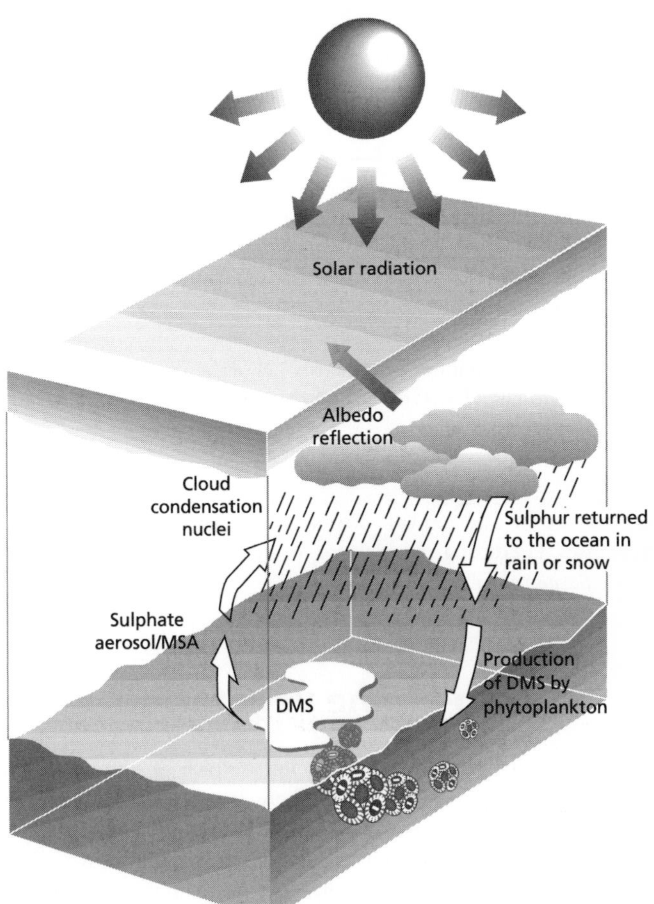

Fig. A5 Dimethyl sulphide (DMS)–cloud condensation nuclei–climate cycle. Sulphate/methane sulphonic acid (MSA, CH_3SO_3H) aerosol particles act as nuclei on which cloud droplets form. (After Fell, N. & Liss, P.S. (1993) *New Scientist* **139** (1887), 34–38.)

(Ca^{2+}) and MAGNESIUM (Mg^{2+}) tend to dominate the aqueous extracts of aerosol samples. Sulphate and ammonium are especially important in the accumulation mode whereas chloride, sodium and calcium are important in the coarse mode. Nitrate may be found in either the accumulation or coarse modes.

Sulphate is the most conspicuous constituent of aerosols, with ammonium being the principal cation associated with it over the continents. Over the oceans the major precursor of sulphate in the unperturbed marine environment is dimethyl sulphide, a biogenic compound which is emitted from the sea's surface. Oxidation pathways for dimethyl sulphide are not fully understood; a large fraction of it is converted to either sulphate in sea-salt aerosol particles or to sulphuric acid (H_2SO_4). These sulphuric acid aerosols may enhance the number of cloud condensation nuclei present over remote areas of the ocean (Fig. A5). Given that the emission rate of dimethyl sulphide varies with the ocean's surface water temperature, much effort has gone into understanding its emission and the production of cloud condensation nuclei as a part of global CLIMATE CHANGE research.

Aerosols cause haze in the atmosphere; this may originate from fine droplets or particles formed by the reaction of air POLLUTANT gases such as sulphur dioxide and NITROGEN OXIDES to give sulphates and nitrates. The 'blue haze' above large forests is caused by particles formed from the oxidation of organic gases (e.g. terpenes) emitted by trees. *See also* ACID RAIN AND MIST. [H.S. & J.N.C.]

---

aestivation A state of inactivity occurring in some animals, for example lungfishes, during prolonged periods of DROUGHT. Physiological processes slow down when an organism aestivates. *Cf.* DORMANCY; HIBERNATION. [P.O.]

affinity index An index of similarity (A) of the species composition of two samples. $A = C/\sqrt{(a+b)}$, where a and b are the numbers of species unique to each SAMPLE and C is the number of species common to both. *See also* SAMPLING. [P.C.]

afforestation The ESTABLISHMENT of FOREST either by natural processes, i.e. during SUCCESSION, or by planting. *Cf.* DEFORESTATION. *See also* FORESTATION. [P.C.]

Afro-alpine vegetation

The mountains of Africa have a flora which is very different from the surrounding lowlands. There is a strong ALTITUDINAL ZONATION, and above about 3000 m a montane and alpine vegetation. The Afro-alpine element is found on mountains in five major regions: Cameroon, Ethiopia, South Africa and on the western and eastern arc of the rift system of Central and East Africa. It is most distinct on the mountains of the Rift Valley system. Each region and mountain has a distinct flora related to its age and history. The best known East African volcanoes lie outside the main area of mountain endemism: Kilimanjaro is one of the newest mountains and has one of the least diversified floras. The main centre of endemism lies in the west of the rift system in the Ruwenzoris and Virungas.

Some of the most distinct life forms adapted to growing at high altitude are found on the drier mountains of the eastern rift. The Afro-alpine element is characterized by the presence of a tussock grassland within which there are scattered giant rosette plants of the genus *Lobelia* (woolly candle plant) and tree-like *Dendrosenecio* (giant groundsel). The latter are related to ragworts in the daisy family, but have thick-leafed rosettes on the ends of thick, weakly branching stems which are surrounded by a layer of dead leaves. They have a form adapted to the extreme diurnal temperature range, from −5°C at night to +15°C in the daytime. The vascular tissues of the stem of *Dendrosenecio* are protected by a thick overcoat of the dead leaves. The delicate growing buds are protected by the leaves closing around them at night and by the presence of a thick antifreeze slime. Similarly in the tussock grasses, species of *Festuca, Agrostis* and *Pentaschistis*, the sensitive heart of the plant is well insulated. The giant lobelias have a large rosette. Water collects at the base of the leaves, insulating the meristems. The huge thick inflorescence can grow up to 8 m in height, and the developing floral buds are protected by bracts and a thick white fur. A similar range of adaptations is found in the PÁRAMO of the Andes and to a lesser extent in the East Indies and Hawaii.

Widely isolated mountains share a similar range of life forms. Each mountain has its own endemic species of *Dendrosenecio: D. keniodendron* on Mt. Kenya; *D. erici-rosenii* on Ruwenzori; and several subspecies of *D. johnstoni* — ssp. *cottoni* on Mt. Kilimanjaro, ssp. *dalei* in the Elgon region of Kenya, and ssp. *adnivalis* on the Virunga volcanoes of eastern Zaire. *Lobelia wollastonii* is found in the west and *L. teleckii* in the eastern Rift. *Lobelia deckenii* has six different subspecies on different mountains. Distant mountains elsewhere have only a poorly represented Afro-alpine element.

Above 4000 m the tussock grassland is replaced by an alpine belt of low-growing shrubby species like *Helichrysum*. At the highest altitude this gives way to a moss-dominated community. Below the tussock grassland there is a zone of tree heathers, *Erica arborea* and *Phillipia* and heather-like forms such as *Stoebe kilmandscharica*. Bushes up to 10 m high may be coated with epiphytic lichens, *Usnea* spp.

At lower altitudes the flora has a greater and greater affinity to the local region. There may be bamboo gardens with species of *Arundo* or montane or CLOUD FOREST containing *Juniperus* (juniper), *Podocarpus* (a Southern hemisphere conifer), *Cyathea* (tree-fern), *Ocotea* and *Aningeria*. The ZONATION on the west and east of mountains differs, with SAVANNAH reaching up the mountains in the drier rain shadow. [M.I.]

REFERENCES

Crawford, R.M.M. (1989) *Studies in Plant Survival*. Blackwell Scientific Publications, Oxford.

Kingdon, J. (1990) *Island Africa*. Collins, London.

Takhatajan, A. (1986) *Floristic Regions of the World*. University of California Press, Berkeley.

Afro-Indian floral region A subunit of the Palaeotropical kingdom (Palaeotropical realm of some authors), extending from the Atlantic coast of North Africa across to the Indian subcontinent (see Fig. B6). This floral region approximates to the Saharan–Sindian region of Takhtajan (1969), which is nested within the African subkingdom of his Palaeotropical kingdom. It also coincides with the North African–Indian DESERT province of Neill (1969). This generally rather arid belt reaches from Morocco across North Africa, through the Sahara into Israel, Syria, southern Iraq and Iran, northern Arabia, Sind in western Pakistan and Rajasthan in India. These deserts or semi-deserts represent something of a barrier to fauna MIGRATION. As a result of the arid character of this region, it is floristically rather restricted, with about 50 endemic plant genera but no endemic families. Significantly, many of those endemics are in the Chenopodiaceae, a family characteristic of arid and saline habitats (e.g. *Nucularia, Fredolia, Agathophora*). Despite the fact that until the break-up of the great Palaeozoic southern continent of GONDWANA, Africa and India were juxtaposed at the core of that supercontinent, they do not have very close biogeographical affinity in their faunas and floras at the present day. As India broke away from Africa during the MESOZOIC era to drift northwards and eventually collide with the Asiatic plate, so it acquired elements of BIOTA from the north, west and east, and in consequence has relatively little biological distinctness of its own. [W.G.C.]

REFERENCES

Neill, W.T. (1969) *The Geography of Life*. Columbia University Press, New York.

Takhtajan, A. (1969) *Flowering Plants, Origin and Dispersal.* Oliver & Boyd, Edinburgh.

age and size at maturity From the point of view of NATURAL SELECTION for reproductive success, maturation is seen as dividing a life into two phases: prior preparation and subsequent fulfilment. Many organisms delay maturity, and variation in age at maturity can be detected among species, among populations within species, and among individuals within populations. There are strong phylogenetic effects on age and SIZE AT MATURITY. After the effects of size and phylogeny have been controlled, there remains a tendency in the birds and the mammals for organisms that delay maturity to have long lives and low fecundities. Bimaturism, where male and female ages and sizes at maturity differ, is also common. Without social structure, females mature later and larger than males because they continue to gain fecundity after males have grown into the region of diminishing returns. Where males compete with each other to control access to females, males mature later and larger than females. This observation ties LIFE-HISTORY EVOLUTION to SEXUAL SELECTION, which produces different life histories in the two sexes, as well as different body sizes and ornaments.

FITNESS is often more sensitive to changes in age and size at maturity than to changes in any other trait. The principal benefit to early maturation is the more rapid population growth implied by earlier maturity. With less time spent as juveniles, early-maturing organisms have a higher probability of surviving to maturity; they also have higher fitness because their offspring are born earlier and start reproducing sooner. This benefit is weaker in organisms that mature early and stronger in organisms that delay maturity.

To explain delayed maturity, the demographic pressure to mature early must be balanced by trade-offs with other fitness components. Three are generally important. First, if delaying maturity permits further growth and fecundity increases with size, then delayed maturity leads to higher initial fecundity. Maturity will be delayed for this reason until the fitness gained through increased fecundity is balanced by the fitness lost through longer GENERATION TIME and lower survival to maturity. Second, if delaying maturity means that the quality of the offspring produced or of the PARENTAL CARE provided will be higher, then delaying maturity reduces the instantaneous juvenile mortality rate of the offspring. Maturity will be delayed for this reason until the fitness gained through the increase in juvenile survival rate due to the mother's greater age at first birth balances the fitness lost through a longer generation time and that portion of survival to maturity that is attributed simply to the delay and not to any change in instantaneous juvenile mortality rate. Third, if delaying maturity means that the organism will live longer, grow larger, have more reproductive events, have higher fecundity later in life, or otherwise gain in LIFETIME REPRODUCTIVE SUCCESS, then maturity may be delayed beyond the point that one would predict when considering only the first two effects, which are the principal ones.

An optimal age and size at maturity are attained where the benefits and costs of maturation at different ages and sizes balance at a stable equilibrium point or along a REACTION NORM. One can successfully predict age and size at maturity by assuming that maturity is delayed to gain fecundity or to reduce instantaneous juvenile mortality. One can also predict reaction norms for the maturation event that are consistent with much of the evidence. The optimality approach does not, however, account for observed genetic variation in reaction norms, which is considerable. *See also* AGE AT MATURITY IN PLANTS; DEMOGRAPHY; LIFE-HISTORY TRAITS. [S.C.S.]

age at maturity in plants The period of time which has elapsed between the commencement of GROWTH and the attainment of a defined state of completed development. The age at physiological maturity is the time that has passed from the germination of the seed to the achievement of sexual reproduction (i.e. bearing of flowers and fruits). The mature phase of a tree also exists when the growth rate is reduced and energy is preferentially directed towards the maintenance and reproduction of the plant rather than to the increase in timber volume.

In natural forests, the age at maturity is the time period between the commencement of forest SUCCESSION and the development of a fully functional ECOSYSTEM consisting of a shifting mosaic of different sized patches cycling through all structural phases (e.g. gap, building, mature and senescent). A mature natural forest is also sufficiently large to contain all of these phases. Maturity of such forests is assumed when the annual increase in basal area or volume reaches an asymptote or such properties remain relatively invariant over time.

Commercial FORESTRY plantations, under traditional management, are deemed to be mature when the current annual increase in timber volume falls below the cumulative mean annual increment. However, this categorization is based upon financial gain and could be modified by economic criteria. The age of commercial forests at maturity is calculated from the year of planting and by accounting for the seedling age at the time of planting.

Due to variations in the definitions of maturity and the different time periods within which individual plants, species and forests attain maturity, the age at maturity can only be quantified for specific cases. *See also* AGE AND SIZE AT MATURITY. [R.M.L.]

age class Grouping of the organisms within a POPULATION by age instead of by life stage (e.g. in the construction of a LIFE TABLE). Such groupings allow more detailed analyses if the number of life stages is small or if life stages cannot be identified. They are inappropriate if age does not reflect biological stage (e.g. in insects exhibiting AESTIVATION or DIAPAUSE). *See also* DEMOGRAPHY. [A.J.D.]

age structure The relative numbers of individuals within different AGE CLASSES in a POPULATION. Age structure changes according to the birth and death rates acting at different ages currently and in the recent past of a population. Age structure is often summarized in a LIFE TABLE. *See also* STABLE AGE DISTRIBUTION. [R.H.S.]

Ageing

Ageing entails physiological deterioration with chronological age among the adults of most species of animals and plants, even when they are kept under good conditions. The explanation and analysis of ageing is now an active area of research within population biology. Here we consider further: (i) the definition of ageing; (ii) the occurrence of ageing in nature; (iii) alternative theories of ageing; and (iv) experimental research on ageing.

Definition of ageing

While ageing is often defined solely in terms of increasing age-specific MORTALITY rates, a parallel decline in age-specific fertility is often noted as well. Particularly in view of the likely evolutionary substratum of ageing (see below), a more appropriate definition for ageing should include declines in both types of character; one such definition is: a persistent decline in the age-specific components of FITNESS due to endogenous physiological deterioration. This definition excludes deterioration due to secular environmental change, even when that change is caused by the actions of the organism on its environment. It also excludes the chronic effects of persistent PATHOGEN infections, and the effects of changing mating opportunities on fertility.

Occurrence of ageing in nature

Ageing is not commonly observed in natural populations, although this may be partly because of the difficulty of distinguishing endogenous deterioration from that due to privation, injury or disease in the wild. There are two cases where ageing has been demonstrated to occur in nature with some certainty. Firstly, demographic analysis of the life tables of large wild vertebrates, such as African ungulates, indicates an excess of later mortality compared with that expected in the absence of SENESCENCE. Mortality in most human populations now exhibits this pattern. Secondly, and more definitively, a variety of semelparous (*see* SEMELPARITY) animals and plants exhibit spectacular deterioration after reproduction. Examples range from Pacific salmon to male marsupial mice to soybean. The endogenous nature of such deterioration has been revealed by castration experiments. When castrated these organisms live much longer, often years longer, than in their intact condition.

Alternative theories of ageing

Theories that attempt to explain ageing fall into two groups: non-evolutionary theories and evolutionary theories. Non-evolutionary, or 'physiological', theories of ageing attempt to explain ageing in terms of some endogenous problems of molecular, cellular or organismal machinery. There are many such theories. Among the physiological theories of ageing that have attained at least temporary credibility, are the SOMATIC MUTATION and error catastrophe theories of ageing. Both of these were based on accumulating errors in the synthesis of cell components with age, one by MUTATIONS in somatic DNA, the other by progressive loss of translation fidelity due to positive FEEDBACK of errors in the synthesis of the protein translation machinery. The problem faced by both of these theories is that there does not appear to be sufficient accumulation of consequential defects in PROTEINS with age to cause ageing, even among organisms that frequently synthesize protein as adults, like vertebrates. (Many organisms, like insects, do not appear to synthesize much protein as adults.) One possible exception to this general rule is the role of somatic mutation in carcinogenesis, given the age-associated incidence of cancer. However credible somatic mutation is as an explanation of cancer, these theories have not been successful at explaining the ubiquitous and diverse pathologies of ageing.

The second type of theory for ageing is that provided by evolutionary biology. Beginning with speculations from R.A. Fisher, J.B.S. Haldane and P.B. Medawar (*see* CHARACTERS IN ECOLOGY) in the middle part of the 20th century, evolutionary biologists have sought to explain ageing in terms of the force of NATURAL SELECTION, especially the idea that the force of natural selection should decline with age in organisms that do not reproduce by fissile reproduction. Mathematical analysis by W.D. Hamilton and B. Charlesworth in the 1960s and 1970s showed that the force of natural selection acting on age-specific survival probability does indeed decline with adult age, under reasonable evolutionary conditions. Less generally, the force of natural selection acting on age-specific fertility also tends to decline with adult age. This evolutionary theory provided a general explanation of ageing in a wide variety of organisms. Notably, strictly and symmetrically fissile organisms (*see* ASEXUAL REPRODUCTION), for which the evolutionary theory did not predict ageing, have been found to lack ageing.

As to physiological mechanism, the evolutionary theory of ageing does not provide any concrete hypotheses. Instead, the evolutionary theory allows that ageing may entail any of a variety of physiological mechanisms, whether simultaneously in one species, or among many species. In effect, the evolutionary theory is the only general theory that now seems credible, while at the same time it permits the partial validity of many particular physiological theories of ageing.

Experimental research on ageing

Most experimental research on ageing concerns the molecular or cellular biology of ageing in mammals, especially *in vitro* cell cultures of human tissue. At present, it is uncertain that any of this research has uncovered causal mechanisms controlling ageing.

The other major experimental approach has been the creation of organisms with postponed ageing, including both increased LIFESPAN and increased later fertility. Some of this work, particularly with the nematode *Caenorhabditis elegans* and the yeast *Saccharomyces cerevisiae*, has used mutagenesis followed by massive screening to find mutants with postponed ageing. Another approach to postponing ageing has been by MANIPULATION of diet. Reduction in caloric or other food intake is known to increase lifespan in various rodents and a wide variety of insects, including *Drosophila*. Many physiological changes occur with such dietary restriction, from molecular to organismal levels. It has been difficult to establish which of these changes play a causal role in the control of ageing.

The third approach to the experimental postponement of ageing has the greatest relevance to evaluating alternative theories of ageing. This approach uses natural selection in the laboratory to select for organisms that can reproduce at progressively later ages. This approach has been employed almost entirely in *Drosophila*, although there is also some work with mice. The experimental STRATEGY derives naturally from the evolutionary theory of ageing. The evolutionary theory predicts that the force of natural selection acting on survival probability will not start to fall until the beginning of the reproductive period. Thus, delaying the onset of a population's reproduction should lead to later and later onset of ageing, over evolutionary time. This protocol has now worked in the hands of multiple investigators working independently, leading to the production of longer-lived *Drosophila* and mice. This is another important corroboration of the evolutionary theory of ageing. *See also* ANTAGONISTIC PLEIOTROPY HYPOTHESIS; DEMOGRAPHY; LIFE-HISTORY EVOLUTION; MUTATION ACCUMULATION HYPOTHESIS. [M.R.R.]

REFERENCES

Finch, C.E. (1990) *Longevity, Senescence, and the Genome*. University of Chicago Press, Chicago, Ill.

Rose, M.R. (1991) *Evolutionary Biology of Ageing*. Oxford University Press, New York.

Agenda 21 Output of the UNCED, the so-called Earth Summit, held in Rio de Janeiro in June 1992. This reached a number of agreements, one of which was the Rio Declaration on Environment and Development which set out 27 principles for achieving SUSTAINABLE DEVELOPMENT. To support this general declaration the summit adopted Agenda 21, a comprehensive action plan for the pursuit of sustainable development into the 21st century. It contains 40 chapters of detailed recommendations aimed at all major players: international agencies; local and national governments; and non-governmental organizations (NGOs). A quinquennial review on progress was held in 1997. [P.C.]

Agent Orange *See* DIOXIN.

Agreement on Technical Barriers to Trade *See* GATT AND ENVIRONMENT.

aggregated distribution A term used to describe the spatial distribution of the individuals in a POPULATION. Other equivalent terms are contagious distribution, clumped distribution or OVERDISPERSION. When individuals in a population follow an aggregated distribution, the VARIANCE (s^2) of individuals per sample is greater than the MEAN number ($\bar{x}$) of individuals per sample ($s^2 > \bar{x}$). This is frequently summarized by saying that the variance to mean ratio is greater than 1 (i.e. $s^2/\bar{x} > 1$). *See also* AGGREGATION; NEGATIVE BINOMIAL DISTRIBUTION; RANDOM DISTRIBUTION; REGULAR DISTRIBUTION. [B.S.]

aggregation The term aggregation has several meanings in ecology. The first is a simple description of a group of individuals that are close together in space and is synonymous with other general terms such as clump or cluster, or more specific ones such as shoal or COLONY.

The second meaning is a description of the DISPERSION of individuals in space. If a POPULATION is distributed randomly in space then the MEAN and VARIANCE of samples from that population will be equal. If the variance/mean ratio is greater than 1 then this implies aggregation of individuals (i.e. they are more contagious, clustered or clumped than random). Other INDICES OF DISPERSION are available for the quantification of aggregation. The majority of organisms in their natural habitats have distributions that are more clumped than random: aggregation is the rule, not the exception. It is important to realize that a population may be simultaneously aggregated and ordered depending on the scale of the samples taken. For example, the number of ants per hectare may be very even in a wood but the number of ants per square metre will show a pattern of extreme aggregation.

The term aggregation is also used to describe the processes that lead to such clumped distributions. Such a process may be driven by a wide range of factors; these include the patchy DISTRIBUTION of essential resources, defence against predators, cooperative feeding and COMPETITION for mates. *See also* AGGREGATED DISTRIBUTION; SOCIAL ORGANIZATION. [C.D.]

REFERENCE

Southwood, T.R.E. (1978) *Ecological Methods*. Chapman & Hall, London.

Aggregation model of competition

Many insects exploit resources which are patchy, consisting of small, separate units, and which are ephemeral in the sense that they persist for only one or two generations. Such resources can include fruit, fungi, sap flows, decaying leaves, flowers, dung, carrion, seeds, dead wood and small bodies of water held in parts of terrestrial plants (i.e. PHYTOTELMATA). This general view of insect ecology inspired the aggregation model of competition (Shorrocks *et al.* 1979, 1984; Atkinson & Shorrocks 1981) that allows a competitively inferior species to survive in PROBABILITY REFUGES. These are patches of resource (a single fungus, fruit, etc.) with no or a few superior competitors, that arise because the competing stages (usually larvae) have an AGGREGATED DISTRIBUTION across the patches. An independent model proposed by Hanski (1981, 1983) uses different 'mechanics' to incorporate spatial variance into the environment and similarly promotes COEXISTENCE. These probability refuges are a permanent feature of such systems because patches, such as fungi, are ephemeral and aggregation increases MEAN CROWDING. Regional POPULATION DENSITY is limited by strong INTRASPECIFIC COMPETITION in patches with high local density while low-density patches still exist (e.g. POPULATION SIZE within a wood is limited by high density in some fungi, while other fungi still contain no, or a few, individuals). Coexistence is promoted because aggregation of the superior species increases its intraspecific competition and reduces INTERSPECIFIC COMPETITION.

In the aggregation model, the eggs of both insect species are independently distributed over the patches according to a NEGATIVE BINOMIAL DISTRIBUTION, which has an exponent, k, inversely related to the degree of INTRASPECIFIC AGGREGATION. The use of the negative binomial and the assumption of independence have been justified for drosophilid flies (Rosewell *et al.* 1990; Shorrocks *et al.* 1990). In the first version of the model (Atkinson & Shorrocks 1981) the parameter k (level of aggregation) was constant and independent of density. This is not valid for real populations, but relaxing this assumption does not prevent coexistence (Rosewell *et al.* 1990). Within each patch, competition is modelled by the Hassell and Comins model (Hassell & Comins 1976; *see* HASSELL MODEL AND HASSELL–COMINS MODEL):

$$N_i(t+1) = \frac{N_i(t)R_i}{\{1 + a_i[N_i(t) + \alpha_{ij}N_j(t)]\}^{b_i}} \qquad i \neq j$$

where $N_i(t)$ and $N_j(t)$ are the numbers of each species in a patch at time t, R_i is the net reproductive rate, α_{ij} is a COMPETITION COEFFICIENT and a_i and b_i are constants. The parameter a_i (equal to $(R_i-1)/K_i$, where K_i is the equilibrium density or CARRYING CAPACITY of a patch) is related to the population size at which DENSITY DEPENDENCE starts to act and b_i describes the type of competition. At $b=1$, competition is contest (*see* CONTEST COMPETITION) with density dependence exactly compensating. With increasing values of b, competition becomes increasingly scramble (*see* SCRAMBLE COMPETITION) with density dependence being overcompensating.

The predictions of the aggregation model are that with k of the negative binomial < 1 (strong aggregation), it is virtually impossible for the competitively 'superior' species to eliminate the competitively 'inferior' species. For drosophilid flies it has been demonstrated that k of the negative binomial is usually < 1 and that competition coefficients are not sufficiently large to prevent coexistence (Rosewell *et al.* 1990; Shorrocks *et al.* 1990). For many animals exploiting ephemeral and patchy resources, this model therefore provides a viable alternative to traditional RESOURCE PARTITIONING as an explanation for the coexistence of species. The two-species model has been extended to many-species (Shorrocks & Rosewell 1986, 1987) and predicts average group sizes of about seven species coexisting on identical resources. [B.S.]

REFERENCES

Atkinson, W.D. & Shorrocks, B. (1981) Competition on a divided and ephemeral resource: a simulation model. *Journal of Animal Ecology* **50**, 461–471.

Hanski, I. (1981) Coexistence of competitors in patchy environments with and without predation. *Oikos* **37**, 306–312.

Hanski, I. (1983) Coexistence of competitors in patchy environments. *Ecology* **64**, 493–500.

Hassell, M.P. & Comins, H.N. (1976) Discrete time models for two-species competition. *Theoretical Population Biology* **9**, 202–221.

Rosewell, J., Shorrocks, B. & Edwards, K. (1990). Competition on a divided and ephemeral resource: testing the assumptions. I. Aggregation. *Journal of Animal Ecology* **59**, 977–1001.

Shorrocks, B. & Rosewell, J. (1986) Guild size in drosophilids: a simulation model. *Journal of Animal Ecology* **55**, 527–541.

Shorrocks, B. & Rosewell, J. (1987) Spatial patchiness and community structure: coexistence and guild size of drosophilids on ephemeral resources. In: *Organisation of Communities: Past and Present* (eds J.H.R. Gee & P.S. Giller), pp. 29–51. Blackwell Scientific Publications, Oxford.

Shorrocks, B., Atkinson, W.D. & Charlesworth, P. (1979) Competition on a divided and ephemeral resource. *Journal of Animal Ecology* **48**, 899–908.

Shorrocks, B., Rosewell, J., Edwards, K. & Atkinson, W.D. (1984) Interspecific competition is not a major organising force in many insect communities. *Nature* **310**, 310–312.

Shorrocks, B., Rosewell, J. & Edwards, K. (1990) Competition on a divided and ephemeral resource: testing the assumptions. II. Association. *Journal of Animal Ecology* **59**, 1003–1017.

Agricultural meteorology

A branch of METEOROLOGY, also known as agrometeorology, that is concerned with the study of all aspects of the links between the ATMOSPHERE and AGRICULTURE, where agriculture is usually interpreted in its widest form to include most forms of managed land use (i.e. including horticulture, animal husbandry, pisciculture and FORESTRY). The links are considered in both directions to include both the obvious effects of WEATHER on agriculture, and the less obvious but increasingly important effects of agriculture on the atmosphere (e.g. emission of gases from cultivated soils, change of surface reflectivity or ROUGHNESS following clearance of forest for farming, EVAPORATION of irrigation water). The primary aim of the subject is to maximize the sustainable production of food and fibre by understanding atmospheric, hydrological and biological processes and the application of this understanding to prediction, management and forecasting of agricultural production. The field of interest of agricultural meteorology therefore extends from the SOIL at depth, through the air layer near the ground where plants and animals live and grow, up to the higher levels of the TROPOSPHERE. It is an interdisciplinary science as it merges with other subjects such as soil science, plant and animal physiology, HYDROLOGY, CLIMATOLOGY, MICROMETEOROLOGY and atmospheric science in general. Some workers distinguish agroclimatology as a sub-branch where activities rely on the analysis of CLIMATE records, for strategic planning, for example of erosion control and flood defences, prediction of pest outbreak frequencies, CROP suitability, etc.

Many different areas of agricultural meteorology exist on different time- and space-scales, from long-distance aerial transport of fungal spores, seeds, pollen and insects to plant MICROCLIMATE and leaf wetness prediction for fungal spore germination. The discipline can be divided into operational and research areas. Examples of operational work are: assessment of suitable crop types for a given climate; FROST frequency and timing predictions; hazardous weather prediction for livestock; forecasting suitable conditions for crop spraying; estimating evaporation rates and irrigation requirements; livestock fodder production predictions; prediction of odours dispersion from housed livestock; design of farm building ventilation; animal and crop disease forecasting; locust prediction and tracking; design of protected cropping systems; forest fire RISK ASSESSMENT; and prediction of environmental conditions during transport and storage of agricultural products.

Operational agricultural meteorology can be further divided into strategic or tactical activities. An example of strategic work is assessing how often irrigation might be required for a given crop at a given location characterized by a particular long-term climate and predicting what size of water supply is necessary, and what yield and financial benefits there might be. The tactical counterpart is the day-to-day forecasting of when irrigation is required, and how much, using weather forecasts, and estimates of crop water loss, soil water content and crop yield sensitivity to DROUGHT periods, etc.

Research in agricultural meteorology covers all the above activities, but primarily aims to understand and describe the mechanisms underlying the effects, and in developing practical methods for operational activities. In particular, much use is made of numerical models that can be used to predict the behaviour over time of processes (e.g. disease progress or crop water use) from the environmental conditions. In such models a simplified representation of the physiological and physical behaviour of the animal, plant and environmental system is formulated as a set of mathematical equations (generally time dependent) which take as input weather, soil and biological conditions (e.g. soil and air temperature, air humidity, crop type, density and stage in life cycle) and predict the time course of the system's behaviour. Agricultural meteorology, therefore, is making growing use of information technology, in particular in the measurement of weather conditions using automatic weather stations; in gathering and analysing remotely sensed data (especially from satellites to monitor weather, vegetation and ground conditions; *see* REMOTE SENSING); in the use of numerical models to integrate and process this information and in the dissemination of that information to the end users: meteorologists, agriculturalists and farmers. [J.I.M.L.]

REFERENCE

World Meteorological Organization (1981) *Guide to Agricultural Meteorological Practices*, 2nd edn (plus several recent supplements to this edition). World Meteorological Organization, Geneva.

---

agricultural pollution CONTAMINATION leading to adverse effects on ecological systems that arises from agricultural activities. Examples of these kinds of POLLUTION are wastes from farm animals, PESTICIDES, FERTILIZER treatment of CROPS and EMISSIONS from machinery. The term might also include the release of GENETICALLY MODIFIED ORGANISMS. Much of the release into surrounding ecosystems, both terrestrial and aquatic, will be diffuse and continuous (or semi-continuous) rather than from point sources, and so is difficult to control. Accidental releases, for example through the collapse of manure containment or spillage of pesticides, can also be a problem. End-of-pipe controls are rarely an option, and the best approach is an attempt to influence practice, through legislation and voluntary action. *See also* AGRICULTURAL WASTE; AGRICULTURE; DIFFUSE SOURCES. [P.C.]

agricultural waste Any mixture of faeces, urine, bedding material, spoiled feed, etc. derived from livestock farming. [P.C.]

Agriculture

Although the word originally meant the act of cultivating fields, i.e. growing CROPS, it has now been extended to include all aspects of growing crops and raising livestock. Indeed, the definition is now flexible enough to include such contrasting activities as intensive animal production in feedlots or barns, ranching, horticulture and the complex small-scale subsistence farming which characterizes parts of the humid tropics. The common element is conscious management of a biological resource. Although agriculture can be differentiated from FORESTRY on the basis of the time-scale of decision-making, these terms represent the ends of a continuum that includes agroforestry. Because of this diversity, it is important to recognize that the ENVIRONMENTAL IMPACT varies with the type of agriculture. Classifications can be based on the type and number of components (specialized arable, mixed livestock–arable, etc.), the degree to which the farm is integrated into the global market for agricultural produce (subsistence farming, cash cropping, etc.) or the level of inputs, such as fertilizer, that are applied (ORGANIC FARMING, high-input systems, etc.). If external inputs alone are considered, high-input farms are typical open systems.

Agriculture produces most of the world's foodstuffs together with raw materials for the fibre and vegetable oil industries. The three most important crops, on the basis of edible dry matter production, are wheat, rice and maize. Cattle are by far the most important class of livestock in terms of biomass. More than half the world's working population is still employed in agriculture and the associated industries. When permanent PASTURE is included, almost 40% of the world's land area is devoted to agriculture, of which about one-quarter is arable, i.e. cropped. These figures, however, disguise considerable variation from continent to continent and from country to country (see Table A1). The area of agricultural, and particularly arable, land per person is declining as the world population increases more rapidly than the agricultural land. Between 1977 and 1992, there was a net increase in the world's agricultural land area of just less than 5% (destruction of forests accounted for about 60% of this). This figure, however, represents a balance between gains from cultivating new land and losses from urbanization, erosion, DESERTIFICATION and SALINIZATION.

The origins of settled agriculture are complex and the subject of debate. What is clear is that the DOMESTICATION of plants and animals has taken place over a period of more than 12 000 years. During this time, agricultural systems have evolved which take advantage of the potential of the land while seeking to overcome the LIMITING FACTORS, which may be climatic, EDAPHIC or socio-economic in origin. The dawn of scientific agriculture and the start of major intensification are often considered to have coincided following the discoveries made in the 19th century by agricultural chemists in western Europe. However, many effective agricultural systems were developed long before then. For example, 2000 years ago, when the tribes of north-west Europe were fighting each other, sustainable irrigation-based agricultural systems had already been developed in south-east Asia and in Central America. By the close of the 12th century AD, the Arab agronomists of Andalucia in Spain had developed rules for solving many of the management problems associated with Mediterranean agriculture. In recent times many indigenous systems have been displaced by intensive agricultural operations tied closely to the global market economy. There is now a

Table A1 Proportional world agricultural areas by continent. Relative areas are given as a proportion of the total land area. Agricultural land includes arable land, land under permanent crops, such as olives, and permanent pasture. The Russian Federation is included with Europe. The data are for 1993 and have been calculated from FAO (1994) *1993 FAO Production Yearbook*, Vol. 47. Food and Agriculture Organization of the United Nations, Rome.

	Africa	North and Central America	South America	Asia	Europe	Oceania
Relative agricultural area	0.37	0.29	0.35	0.47	0.46	0.57
Relative arable area	0.06	0.12	0.06	0.16	0.26	0.06
Agricultural land per person (ha)	1.5	1.4	2.0	0.4	0.4	17.4
Arable land per person (ha)	0.23	0.60	0.31	0.13	0.24	1.86

growing realization that a considerable store of valuable indigenous knowledge is in danger of being lost.

A wide variety of agricultural techniques to grow crops and raise livestock have evolved over the centuries. These techniques include mechanization, PEST and disease control, plant and animal nutrition, and the breeding of better types of crops and livestock. Successful innovations have spread widely and rapidly around the world even though subsistence farmers cannot afford to take large risks and must do their utmost to protect the land resource on which the livelihood of their family depends.

Agriculture affects the environment in a number of ways. Direct AGRICULTURAL POLLUTION has caused serious local damage in some circumstances although its global importance has probably been overestimated. The risk of damage is greatest when high inputs of persistent or soluble AGROCHEMICALS are applied to a permeable SOIL overlying a WATER TABLE that is near the surface. In these cases, persistent PESTICIDES, nitrates and organic wastes can pollute watercourses and AQUIFERS. Where surface RUN-OFF is significant, eroded soil particles can also contaminate watercourses. Agronomists and soil scientists know how to minimize these problems but the introduction of control measures is difficult where there is either no political will to make changes or the farmers are poor and do not own their land. AGROECOSYSTEMS tend to show less BIODIVERSITY than the natural systems they replace, particularly since WEED species and many invertebrates and microorganisms are targeted for control. However, this reduction is not inevitable and many diverse and highly prized landscapes are the product of management for extensive agriculture. Although individual fields are often MONOCULTURES of single crop cultivars, farms usually contain fields with different crops. Continuous culture of the same crop year after year in the same field is largely restricted to those parts of the world where the climate inhibits the build-up of pests and diseases. Continuous culture may, for example, be ecologically appropriate in areas where there is a pronounced DRY SEASON after harvest.

Agriculture is important both as a source and a sink for GREENHOUSE GASES. Ignoring fossil fuel inputs, agriculture has a broadly neutral effect on the CARBON DIOXIDE (CO_2) flux at the Earth's surface. However, burning FOREST to create new agricultural land and cultivation of PRAIRIE and STEPPE soils, which are high in organic matter, has led to the release of large amounts of CO_2. Methane (CH_4) is emitted by RUMINANT animals—although these losses are partly offset by a reduction in the number of wild ruminants—and from rice paddies where the source is the anaerobic fermentation of organic matter. The emission of nitrous oxide (N_2O), which is formed from the natural processes of nitrification and DENITRIFICATION, is increased when land is cultivated for arable agriculture.

A more subtle effect of agriculture is on the regional energy balance. The replacement of natural vegetation by farmland is often marked by an increase in both the proportion of SOLAR RADIATION reflected (ALBEDO) and the latent heat flux (evapotranspiration) although it is difficult to generalize to the effect on regional climate.

Sustainability (*see* SUSTAINABLE DEVELOPMENT) is an increasingly important issue for agriculture. Soil erosion can be a major problem in parts of the world with a dry climate and soils of low organic matter. It is caused by the use of practices which are inappropriate for the soil type and climatic zone. Some areas have been permanently damaged. WATER has long been recognized as a major limiting factor to agricultural production. About one-third of the Asian crop area is irrigated, as are about 10% of the crops in the rest of the world. Barring major changes in climate, some of these systems are sustainable indefinitely. However, others such as those on the High Plains of the USA and in parts of Saudi Arabia depend on water from aquifers that are being depleted faster than they can be recharged. Elsewhere there is competition for water between agriculture and an increasingly urbanized population. Areas unsuitable for arable cropping may still support GRAZING by livestock, provided that the stocking density is balanced against the CARRYING CAPACITY of the land.

Table A2 Selected inputs for two countries with contrasting agricultural economies. These data have been presented per hectare of cropped land. The definition of agricultural workers excludes people employed in the ancillary industries. Note that the figure for nitrogen consumption in The Netherlands has been inflated by heavy applications to grassland. The support energy figures, which include the energy costs of fertilizer and machinery manufacture, are estimates.

	Agricultural workers per hectare	Tractors per hectare	Nitrogen fertilizer use (kg ha^{-1} year^{-1})	Support energy (MJ ha^{-1} year^{-1})
The Netherlands	0.54	0.20	412	12 000
Burkina Faso	3.11	Insignificant	2	100

Current intensive European and North American farming depends on a continual use of external inputs such as fertilizer and pesticides, although techniques such as INTEGRATED PEST MANAGEMENT are being developed to reduce the need for such inputs. Estimates of typical values of inputs are given in Table A2. Current intensive agricultural systems are characterized by a large fossil fuel input, mainly to support mechanization. However, the energy input to agriculture often represents only a small part of the total energy required to produce food, the major inputs occurring in the transport, processing and packaging sectors. Considerable research has been carried out into the feasibility of producing ETHANOL, rape methyl-ester and other BIOFUELS that could minimize the fossil fuel energy subsidy to agriculture. However, the land needed to produce energy crops will be unavailable for food production. One-quarter of the farm's arable area might be needed to produce all the farm's energy requirements, as was the case in the days of horse power. [G.R.]

Agriculture and soil conservation

AGRICULTURE involves management of SOIL, plants and animals to optimize PRODUCTIVITY and minimize risks of soil and environmental degradation. Land conversion from native vegetation cover or virgin land to agricultural use can lead to changes in WATER and energy balance and disruptions in cycles of carbon, nitrogen, phosphorus and other elements (*see* BIOGEOCHEMICAL CYCLE) Changes in water balance usually involve decrease in soil-water storage and deep drainage and increase in surface RUN-OFF. The latter is the principal cause of accelerated soil erosion (Fig. A6).

(a)

(b)

Fig. A6 Accelerated soil erosion on: (a) agricultural land; and (b) disturbed land and construction sites.

Soil erosion is the process of detachment, translocation and redeposition of soil particles over the landscape with the eventual transport to rivers and oceans. There are two types of soil erosion: natural, or geological; and accelerated. Geological erosion is a natural WEATHERING process and is responsible for soil formation and landscape evolution. The accelerated erosion is due to human-induced causes, or anthropogenic factors. Accelerated erosion is caused by the kinetic energy of run-off and rainfall. Agricultural activities usually lead to accelerated erosion, and the latter is responsible for soil and environmental degradation, decline in soil quality and productivity, EUTROPHICATION of natural waters, and possible EMISSIONS of radiatively active gases from soil and terrestrial ecosystems to the ATMOSPHERE. Therefore, a principal objective of sustainable agricultural systems is to conserve soil and water resources and minimize risks of accelerated soil erosion (*see* SUSTAINABLE DEVELOPMENT).

Mean annual global run-off is estimated at 47×10^3 km^3 or 92 mm out of the mean annual total PRECIPITATION of 119×10^3 km^3 or 233 mm received on the Earth's surface (UNESCO 1978). The surface run-off has a potential energy of 3×10^{20} J year^{-1} compared with total kinetic energy of about 0.7×10^{18} J year^{-1} for the RAIN received on the land surface. It is this combined kinetic energy of run-off and rainfall that is responsible for about 19×10^9 Mg of material transport to the oceans every year by the world's rivers. The material transport comprises 15×10^9 Mg year^{-1} of suspended and bed loads and 4×10^9 Mg year^{-1} of dissolved load (Walling 1987). Some estimates put gross erosion at 75×10^9 Mg year^{-1} (Pimentel *et al.* 1995). Assuming that the suspended and bed loads transported to the oceans represent only 10% of the total sediments that originated over the land surface due to the kinetic energy of run-off and rainfall, then gross erosion over the land surface is about 150×10^9 Mg year^{-1} from a total land area of 14.8×10^9 ha, giving global mean soil erosion rate of about 10 Mg ha^{-1} year^{-1}. Apparently, some ecologically fragile regions with harsh environments experience erosion rates an order of magnitude higher and others with mild climate and resilient soils lower than the global mean. Although the available statistics are vague and subjective, land area affected by erosion is estimated at 1.1×10^9 ha by water erosion and 0.55×10^9 ha by wind erosion (Oldeman 1994) The relative soil erosion in different continents is ranked thus: Asia > Africa > South America > Europe > North and Central America > Oceania and the Pacific (Table A3).

Expansion of agriculture and soil erosion

The expansion of agriculture, to meet the basic necessities of increase in human population, has accentuated risks of soil erosion and necessitated use of conservation-effective measures. World arable land area is estimated at 1.35×10^9 ha or about 9.1% of the total land area. Potentially cultivable land area is about 3.0×10^9 ha. Expansion of agricultural land usually occurs at the expense of FORESTS, WOODLANDS and GRASSLANDS. It involves removal of existing vegetation cover, burning of excess biomass, ploughing of the soil surface, and replacement of a complex and diverse biota with one or two species. This process exposes soil to climatic elements and accentuates the erosion hazard.

The area under forest and woodlands has been reduced during the postglacial era by 750×10^6 ha and 180×10^6 ha, respectively (Williams 1994). Estimated changes of areas of different types of natural vegetation cover to agricultural and urban uses from pre-agricultural times to the present era are shown in Table A4. This conversion from forest/woodlands to agricultural use is due to a

Table A3 Estimates of suspended sediment load and gross erosion from different continents. (After Lal, R. (1995) Global soil erosion by water and carbon dynamics. In: *Soils and Global Change* (eds R. Lal, J. Kimble, E. Levine & B.A. Stewart), pp. 131–142. CRC/Lewis Publishers, Boca Raton, Fla.)

Continent	Sediment transport to the ocean (×10^6 Mg year^{-1})	Suspended sediment yield rate (Mg ha^{-1} year^{-1})	Estimated erosion* (Mg ha^{-1} year^{-1})
Africa	530	0.4	4.0
Asia	6433	2.3	23.0
Europe	230	0.5	5.0
North and Central America	1462	0.8	8.0
Oceania/Pacific Islands	3062	5.9	59.0
South America	1788	1.0	10.0
World	13505	1.0	10.0

* Assuming delivery ratio of 10%.

Table A4 Estimates of change in land cover for different vegetation types. (Recalculated from Graetz, D. (1994) Grasslands. In: *Changes in Land Use and Land Cover: A Global Perspective* (eds W.B. Meyer & B.L. Turner), pp. 125–148. Cambridge University Press, Cambridge.)

Land cover	Pre-agricultural areas ($\times 10^9$ ha)	Present area ($\times 10^9$ ha)	Change (%)
Total forest	4.68	3.93	−16.0
Tropical forest	1.28	1.23	−3.9
Other forest	3.40	2.70	−20.6
Woodland	0.97	0.79	−18.6
Shrubland	1.62	1.48	−8.6
Grassland	3.40	2.74	−19.4
Tundra	0.74	0.74	0
Desert	1.59	1.56	−1.9
Agricultural land	0.00	1.35	?
Total	17.68	16.52	

gradual increase in population before modern times and a very rapid increase in population from about 1600 onwards. The world population has increased from 2.5 billion in 1950 to 5.6 billion in 1993. DEFORESTATION and conversion of land to agricultural use is very high in the tropics and subtropics where the rate of increase in population is also the highest and the problems of soil erosion are most severe.

Land use and management effects on erosion

Soil erosion is a symptom of land misuse and soil mismanagement. Therefore, soil conservation demands appropriate land use and judicious soil and crop management practices.

Methods of deforestation and land development

The methods of forest conversion and management systems affect the rate of soil erosion. In general, the more the soil is disturbed and the less the ground cover the greater is the soil erosion hazard. Therefore, erosion from agricultural land is generally more with mechanized than with manual methods of deforestation, more with plough-till than no-till methods of seedbed preparation, and more with clean-cultivated than mulched soil surface conditions. The data in Table A5 show significant differences in surface run-off and soil erosion between methods of deforestation and seedbed preparation. The least run-off and erosion were observed with manual deforestation and no-till system, and the maximum with tree-pusher/root-rake deforestation and plough-till system.

Ground cover and residue mulch

Ground cover is the most effective tool against the erosive forces of rainfall and run-off. In fact soil erosion decreases exponentially with increasing ground cover. Ground cover is most effective when it is close to the soil surface, i.e. low-canopy grass cover. A common strategy to achieve ground cover for erosion control is through the use of crop-residue mulch. Residue mulch from small-grain cereals (i.e. wheat, barley, rice, etc.) is more effective in providing a protective ground cover than that from coarse-grain cereals (i.e. maize, sorghum, millet, etc.). For the same mass, the residue of wheat provides more effective ground cover than that of maize.

The data in Table A6 show the impact of crop residue mulch on run-off and soil erosion from a tropical soil in western Nigeria. For the total annual rainfall of 741.5 mm, mean annual run-off was 393.1 mm (52.0% of rainfall) for unmulched control, 80.7 mm (10.9% of rainfall) for 2 Mg ha^{-1} of mulch, 30.1 mm (4.1% of rainfall) for 4 Mg ha^{-1} of mulch, and 12.9 mm (1.7% of rainfall) for 6 Mg ha^{-1} of mulch. The relative run-off was 100 : 20.5 : 7.7 : 3.3 for 0 : 2 : 4 : 6 Mg ha^{-1} mulch rates, respectively. A mulch rate of 6 Mg ha^{-1} was effective in reducing run-off for slope gradients of up to 15%.

Residue mulch was even more effective in erosion control than in run-off control. Mean annual soil erosion was 91.2 Mg ha^{-1} for unmulched control, 7.4 Mg ha^{-1} for a mulch rate of 2 Mg ha^{-1}, 1.4 Mg ha^{-1} for a mulch rate of 4 Mg ha^{-1}, and only 0.3 Mg ha^{-1} for a mulch rate of 6 Mg ha^{-1}. The relative soil erosion was 100 : 8.1 : 1.5 : 0.3 for 0 : 2 : 4 : 6 Mg ha^{-1} mulch rates, respectively. A mulch rate of 6 Mg ha^{-1} was extremely effective in erosion control on slope gradients of up to 15%.

Conservation tillage

Conservation tillage is a generic term that encompasses all tillage systems and methods of seedbed preparation that minimize the risks of soil erosion. A common form of conservation tillage is no-till

Deforestation method	Tillage method	Sediment density (g l^{-1})	Run-off (% of the rain)*	Erosion (Mg ha^{-1} per season)
Forest control	None	Traces	Traces	Traces
Partial clearing	Manual/ traditional	0.1	0.90	0.45
Manual	No-till	3.4	1.82	0.33
Manual	Plough-till	8.5	4.85	6.80
Shearblade	No-till	5.7	10.17	3.24
Tree-pusher/root-rake	No-till	5.6	18.27	15.00
Tree-pusher/root-rake	Plough-till	13.0	27.61	22.17

* Total seasonal rainfall received from April to August 1979 = 849.7 mm.

Table A5 The effects of methods of deforestation and seedbed preparation on run-off and soil erosion under maize grown in southern Nigeria in 1979. (After Lal, R. (1995) Deforestation and land-use effects on soil degradation and rehabilitation in western Nigeria. III. Runoff, soil erosion and nutrient loss. *Land Degradation and Development* **7**, 99–119.)

Table A6 Effect of mulch rate on run-off and soil erosion on an uncropped soil of different slopes in 1974 in western Nigeria. (Recalculated from Lal, R. (1976) *Soil Erosion Problems on Alfisols in Western Nigeria and their Control*. IITA Monograph 1. IITA, Ibadan, Nigeria.)

Slope (%)	Mulch rate (Mg ha^{-1} of straw) 0	2	4	6
Run-off (mm $year^{-1}$)				
1	411.7	36.2	6.7	0.0
5	483.0	126.1	28.3	10.7
10	302.9	73.8	34.7	21.1
15	374.7	86.8	50.6	19.9
Mean	393.1	80.7	30.1	12.9
Soil erosion (Mg ha^{-1} $year^{-1}$)				
1	9.4	0.9	0.3	0.0
5	123.0	6.3	1.5	0.2
10	137.0	5.6	1.0	0.2
15	95.5	16.8	2.7	0.8
Mean	91.2	7.4	1.4	0.3

Total annual rainfall = 741.5 mm.

(a)

(b)

Fig. A7 (a) Residue mulch from the previous maize crop decreases soil erosion in cowpeas. (b) Contour hedgerows of *Leucaena* are effective in erosion control on sloping lands in the sub-humid and humid regions.

Table A7 Run-off and erosion from ploughed and no-till watershed growing maize in Coshocton, Ohio. (From Harrold, L.L. & Edwards W.M. (1972) A severe rainstorm test of no-till corn. *Journal of Soil and Water Conservation* **27**, 30.)

Treatment	Slope (%)	Rainfall (cm)	Run-off (cm)	Erosion (Mg ha^{-1})
Ploughed, clean tilled	6.6	14.0	11.2	50.7
Ploughed, clean tilled	5.8	14.0	5.8	7.2
No-till residue mulch	20.7	12.9	6.4	0.1

(or plough-less) farming that leaves crop-residue mulch on the soil surface (Fig. A7). In addition to reducing cost, it may also reduce the adverse impact of agricultural practices on soil and the environment. The data in Table A7 from Ohio, USA, show the conservation-effectiveness of no-till farming. Despite a slope steepness of about 21%, a WATERSHED managed with no-till suffered soil loss of only 0.1 Mg ha^{-1}. Because of its conservation-effectiveness and other merits, the area under conservation tillage has been steadily increasing in the USA, from a mere 2.4 million ha (2.0% of the planted area) in 1968 to 40.0 million ha (35.5% of the planted area) in 1995. The area under conservation tillage in Ohio has increased from 0.7 million ha (19% of the planted area) in 1984 to 1.8 million ha (47% of the planted area) in 1994 (CTIC 1995).

In addition to erosion control, another merit of conservation tillage is CARBON sequestration in soil. In general, soil managed by conservation tillage contains more soil organic carbon than that ploughed with plough-based systems. Lal (1996) estimated that widespread adoption of conservation tillage in the world may lead to global carbon sequestration of about 1500 to 4900 Tg.

Crop management

Crop management, an important tool for soil conservation, involves a broad range of practices including timely sowing, appropriate seed rates, farm operations such as sowing on the contour, judicious and discriminate application of fertilizer and soil amendments, PEST management including WEED control, and following appropriate cropping systems or crop rotations and crop combinations. The adoption of improved management practices enhances crop growth and vigour, provides ground cover, and protects soil from the kinetic energy of raindrops and run-off.

An example of such improved practices is agroforestry, which involves growing seasonal/annual crops or raising animals along with woody shrubs and perennials. It is a useful technique of soil and water conservation, providing ground cover, and sustaining production. It includes establishing contour HEDGEROWS of woody shrubs or trees at 5–10 m spacing and growing crops or pastures between. The data in Table A8 show that contour hedgerows of *Leucaena*, a leguminous tree native to Hawaii, were extremely effective in soil and water conservation even with a plough-based system of seedbed preparation.

A widely adapted hedgerow plant for soil and water conservation on agricultural land is the vetiver grass (khas khas), which is native to southern India (Grimshaw & Helfer 1995). The bunch root

Table A8 The effects of agroforestry techniques on soil conservation on a tropical soil in western Nigeria in 1984. (After Lal, R. (1989) Agroforestry systems and soil surface management of a tropical Alfisol. II. Soil erosion and nutrient loss. *Agroforestry Systems* **8**, 97–111.)

Treatment	Run-off	Erosion
Plough-till	29.9	14.2
No-till	0.8	0.03
Leucaena—4 m	1.2	0.2
Leucaena—2 m	1.3	0.1
Gliricidia—4 m	4.9	1.6
Gliricidia—2 m	2.2	2.1
LSD (0.05)	3.0	1.5

Seasonal rainfall amount = 727 mm.
LSD (0.05) = least significant difference for 95% probability level.

system of the grass makes it ideal for establishing hedges. Well-maintained hedges of vetiver may be effective for a wide range of ECOREGIONS between 30° north and south of the Equator.

Reconciling agricultural sustainability with environmental quality

Agriculture, the most basic of all industries, has important impacts on human welfare and the environment. Inappropriate land use and mismanagement can lead to a decline in productivity, soil degradation and DESERTIFICATION, a decline in WATER QUALITY and the emission of radiatively active gases (i.e. CARBON DIOXIDE (CO_2), METHANE (CH_4), NITROGEN OXIDES (NO_x)) to the atmosphere. However, scientific agriculture can provide solutions to the problems of unsustainability and soil/environmental degradation. In addition to improving productivity, improved agriculture (based on mulch farming, conservation tillage, balanced and judicious use of AGROCHEMICALS, and improved cropping sequences and combinations) can lead to restoration of degraded soils, improvement in water quality, soil conservation, carbon sequestration in soil and mitigation of the greenhouse effect. *See also* CONSERVATION BIOLOGY; EROSION AND TOPSOIL LOSS. [P.G.R. & R.L.]

REFERENCES

CTIC (1995) *National Crop Residue Management Survey*. Conservation Tillage Information Center, West Lafayette, Ind.

Graetz, D. (1994) Grasslands. In: *Changes in Land Use and Land Cover: A Global Perspective* (eds W.B. Meyer & B.L. Turner), pp. 125–148. Cambridge University Press, Cambridge.

Grimshaw, R.G. & Helfer, L. (1995) *Vetiver Grass for Soil and Water Conservation, Land Rehabilitation, and Embankment Stabilisation*. World Bank Technical Paper No. 27. World Bank, Washington, D.C.

Harrold, L.L. & Edwards, W.M. (1972) A severe rainstorm test of no-till corn. *Journal of Soil and Water Conservation* **27**, 30.

Lal, R. (1976) *Soil Erosion Problems on Alfisols in Western Nigeria and their Control*. IITA Monograph 1. IITA, Ibadan, Nigeria.

Lal, R. (1989) Agroforestry systems and soil surface management of a tropical Alfisol. II. Soil erosion and nutrient loss. *Agroforestry Systems* **8**, 97–111.

Lal, R. (1995) Global soil erosion by water and carbon dynamics. In: *Soils and Global Change* (eds R. Lal, J. Kimble, E. Levine & B.A. Stewart), pp. 131–142. CRC/Lewis Publishers, Boca Raton, Fla.

Lal, R. (1995) Deforestation and land-use effects on soil degradation and rehabilitation in western Nigeria. III. Runoff, soil erosion and nutrient loss. *Land Degradation and Development* **7**, 99–119.

Lal, R. (1997) Conservation tillage for mitigating greenhouse effect and improving soil quality and resilience. *Soil Tillage Research* (in press).

Oldeman, L.R. (1994) The global extent of soil degradation. In: *Soil Resilience and Sustainable Landuse* (eds D.J. Greenland & I. Szabolcs), pp. 99–118. CAB International, Wallingford.

Pimentel, D.C., Harvey, P., Resosudarmo, K. *et al.* (1995) Environmental and economic costs of soil erosion and conservation benefits. *Science* **267**, 1117–1123.

UNESCO (1978) *World Water Balance and Water Resources of the Earth*. UNESCO Studies and Reports in Hydrology, No. 25.

Walling, D.E. (1987) Rainfall, runoff and erosion of the land: a global view. In: *Energetics of the Physical Environment* (ed. K.J. Gregory), pp. 89–117. John Wiley & Sons, Chichester.

Williams, M. (1994) Forests and tree cover. In: *Changes in Land Use and Land Cover: A Global Perspective* (eds W.B. Meyer & B.L Turner), pp. 97–123. Cambridge University Press, Cambridge.

agrochemicals Chemicals used in AGRICULTURE; for example to enhance YIELD or to protect against pests. *See also* FERTILIZERS; PESTICIDE. [P.C.]

agroecology *See* SUSTAINABLE AGRICULTURE.

agroecosystems Agricultural systems, particularly CROP plants and their weeds and associated human farming practices in the context of the wider semi-natural ECOSYSTEM. Agroecosystems develop within a larger ecosystem which determines the limits of form and variation of the particular agroecology. The concept of the agroecosystem has been used to explore the genetic origins and changes in genetic diversity of cultivated plants and their associated weeds. Examples include studies of the impact of changes to the plant GENE POOL following either the loss of traditional agroecosystems, as in western Europe, or following the introduction of new agroecosystems, as in 19th-century New Zealand. Changes to agroecosystems for the purposes of increasing PRODUCTIVITY may provide a stable environment for survival of the domesticated species but erode genetic diversity of the agroecosystem. *See also* AGRICULTURE. [G.A.F.H.]

REFERENCE

Rindos, D. (1984) *The Origins of Agriculture*. Academic Press, London.

agronomy *See* AGRICULTURE.

Air, chemical and physical properties

Air is largely NITROGEN (78% by volume) and OXYGEN (21%), with important but small concentrations of CARBON DIOXIDE (CO_2; 0.035%) and very low concentrations of several other gases (Table A9). The water vapour content is variable, with a fractional volume in the range 0.001–0.03 (*see* HUMIDITY). Common pollutant gases (*see* AIR POLLUTION) near ground level vary according to the presence of local sources and the meteorological conditions. Concentrations that become damaging to plants are: OZONE 300 ppbv;

Table A9 The composition of air, based on the US Standard Atmosphere 1976, specified here for sea-level dry air.

Gas species	Molecular weight	Fractional volume
N_2	28.0134	0.78084
O_2	31.9988	0.209476
Ar	39.948	0.00934
CO_2	44.01	0.00035
Ne	20.183	0.00001818
He	4.003	0.00000524
Kr	83.80	0.00000114
Xe	131.30	0.000000087
CH_4	16.04	0.000002
H_2	2.016	0.0000005

Table A10 Temperature-dependent properties of air at a pressure of 101 kPa.

	Temperature (°C)					
	−5	0	10	20	30	45
ρ_a (kg m^{-3})	1.31	1.29	1.25	1.20	1.16	1.11
ρ_{as} (kg m^{-3})	1.31	1.29	1.24	1.19	1.14	1.01
k (mW m^{-1} K^{-1})	240	243	250	260	264	275
κ (mm^2 s^{-1})	18.3	18.9	20.2	22.2	22.8	24.9
υ (mm^2 s^{-1})	12.9	13.3	14.2	15.5	16.0	17.4
D_V (mm^2 s^{-1})	20.5	21.2	22.7	24.9	25.7	28.0
D_C (mm^2 s^{-1})	12.4	12.9	13.8	15.1	15.6	17.0

Symbols: ρ_a, density of dry air; ρ_{as}, density of water vapour-saturated air; k, thermal conductivity; κ, thermal diffusivity; υ, kinematic viscosity; D_V, diffusion coefficient of water in air; D_C, diffusion coefficient of CO_2 in air.

Table A11 Solubility of air and its main constituents in water. Volumes of dissolved gases (cm^3) contained in a kilogram of water saturated with air at 101 kPa.

	Temperature (°C)			
	0	10	20	30
Oxygen	10.2	7.9	6.4	5.3
Nitrogen, argon, etc.	19.0	15.0	12.3	10.4
Sum of above (air)	29.2	22.9	18.7	15.7

nitrogen oxide ($NO+NO_2$) 30 μg m^{-3}; sulphur dioxide (SO_2) 30 μg m^{-3}. AEROSOLS, including DUST and pollens, are very variable.

Values for the physical properties of air are often needed, for example in calculations of HEAT TRANSFER and diffusion. Table A10 lists those that are temperature dependent. The specific heat of air at constant pressure is 1.01 J kg^{-1} K^{-1}. Air is sparingly soluble in cold water, and solubility declines with temperature (Table A11). The velocity of sound in air at sea level is 345 m s^{-1}. *See also* AIR QUALITY; ATMOSPHERIC CHEMISTRY; ATMOSPHERIC POLLUTION. [J.G.]

---

air mass A large body of air whose horizontal dimensions can be several hundred to several thousand kilometres in extent and in which at any vertical level the properties of TEMPERATURE and HUMIDITY are fairly uniform. Air masses are found over and near to the semi-permanent surface features of the atmospheric general circulation, for example subtropical high-pressure areas or snow-covered polar areas. Air in contact with these extensive uniform surface types can come to an equilibrium with the underlying surface through vertical mixing and radiative transfer, a process which can take place if the air mass is in place for periods of more than several days. Air masses are classified according to both their surface temperature, for example tropical (25°C), polar (5°C) or arctic (−10°C), and also their RELATIVE HUMIDITY, which reflects their source region, for example continental (40%) or maritime (80%). Air masses can be classified by combinations of these characteristics; for example Pm is a polar maritime air mass whereas Tc is a tropical continental air mass. Significant WEATHER occurs at the relatively narrow boundaries of air masses in regions known as FRONTS. *See also* METEOROLOGY. [J.B.M.]

air pollution The existence of CONTAMINANTS in the air at levels that interfere with human health and/or ecological systems. Major causes of POLLUTION are industrial and traffic EMISSIONS. Some important pollutants include SULPHUR OXIDES, CARBON MONOXIDE, NITROGEN OXIDES, organic substances of various kinds, and particulates. GREENHOUSE GASES and OZONE-depleting substances might also be described as causes of air pollution, but the term is usually restricted to substances likely to have 'substratospheric' effects. *See also* ATMOSPHERIC POLLUTION; CONTAMINATION. [P.C.]

air-pollution control POLLUTION control pertaining to EMISSIONS into, and quality of, the ATMOSPHERE. *See also* AIR POLLUTION; ATMOSPHERIC POLLUTION. [P.C.]

air quality Description of the extent to which the ATMOSPHERE in a defined locale is contaminated and polluted. It is usually expressed in terms of the concentration in air of one or more substances over a certain period of time and judged against standards. The latter, AIR-QUALITY STANDARDS, are concentrations, often legally prescribed, below which adverse effects on human health and/or the environment are unlikely. *See also* AIR POLLUTION; ATMOSPHERIC POLLUTION. [P.C.]

air-quality standards Prescribed concentrations of substances in ambient ATMOSPHERE that cannot legally be exceeded during a given period in a specified location. *See also* AIR QUALITY. [P.C.]

aircraft pollution Consists of chemicals and noise. The principal chemical products of aircraft engine combustion are WATER vapour and CARBON DIOXIDE (CO_2). In addition, smaller amounts of HYDROCARBONS, CARBON MONOXIDE (CO), oxides of NITROGEN (NO_x) and SMOKE particulates are produced. These potentially cause POLLUTION around airports but probably have more important impacts in flight. NO_x EMISSIONS in supersonic flight in the STRATOSPHERE are likely to cause OZONE destruction. On the other hand, NOISE POLLUTION is most important around airports. The nuisance depends upon perceived noise levels and on the number of aircraft heard in a given period; an index combining these two quantities (the NNI—noise and number index) is used as an indicator of this nuisance. *See also* AIR POLLUTION; ATMOSPHERIC POLLUTION. [P.C.]

ALAP *See* AS LOW AS PRACTICABLE.

ALARA *See* AS LOW AS REASONABLY ACHIEVABLE.

alarm calls Visual, oral or olfactory signals emitted by one individual to warn others in the group of danger. Examples include the movement of a white rump patch by some species of antelope; the call of a member of a flock of birds; and the chemicals emitted by some fishes and invertebrates (*see* INFOCHEMICALS; PHEROMONE). The signal can put the signaller at risk; however, this kind of altruism can usually be explained in terms of KIN SELECTION rather than GROUP SELECTION. [P.C.]

ALARP (as low as reasonably practicable) *See* AS LOW AS PRACTICABLE.

albedo The shortwave (300–1500 nm) reflectance of a natural surface, also known as the reflection coefficient, obtained by measuring the incident (S_i) and reflected (S_r) solar irradiance over the surface (land or water). Albedo (A) is defined as:

$$A = S_r / S_i$$

Generally, the albedo of vegetation varies with the height and ROUGHNESS of vegetation, and with solar angle. Often the albedo varies with season, because of changes in the amount or condition of foliage. Some average values are: coniferous forest, 0.15; tropical rain forest, 0.13; deciduous broadleaved forest, 0.18; tropical grassland, 0.25; Sahelian scrub, 0.2–0.3; heather, 0.14.

The albedo of the soil surface varies enormously, depending on the soil type. The published range is 0.1–0.3. Soil albedo falls when the soil is wetted. The albedo of a smooth water surface varies from 0.05 when the Sun is overhead, to nearly 1 when the solar beam is at grazing incidence.

The mean albedo of the Earth is about 0.39; for other planets the values are: Mercury, 0.06; Venus, 0.72; Mars, 0.16; Jupiter, 0.70; Saturn, 0.75; Uranus, 0.90; Neptune, 0.82; Pluto, 0.61. [J.G.]

REFERENCES

Allen, S.J., Wallace, J.S., Gash, J.H.C. & Sivakumar, M.V.K. (1994) Measurements of albedo variation over natural vegetation in the Sahel. *International Journal of Climatology* **14**, 625–636.

Monteith, J.L. & Unsworth, M.H. (1990) *Principles of Environmental Physics*. Edward Arnold, London.

albic horizons *See* SOIL CLASSIFICATION.

alcohols Alcohols are organic compounds containing one or more hydroxyl (OH) groups attached to a HYDROCARBON chain. Examples include:

Methanol	CH_3OH
Ethanol	C_2H_5OH
Propanol	C_3H_7OH
Butanol	C_4H_9OH
Pentanol	$C_5H_{11}OH$
Hexanol	$C_6H_{13}OH$
Heptanol	$C_7H_{15}OH$
Octanol	$C_8H_{17}OH$

As well as being widely used as solvents, alcohols are sometimes used as fuel for motor cars. In the 1980s the possibility of large-scale ETHANOL production from wood was opened up in New Zealand. The production system involved hot acid hydrolysis of wood chips, neutralization of the solution, followed by fermentation by yeast. Currently, the most successful scheme is that in Brazil, where ethanol obtained from sugar cane is used everywhere for motorized transport (cane production for this purpose occupies less than 10% of the agricultural land in Brazil). [J.G.]

aldehydes and ketones Aldehydes and ketones are compounds containing the carbonyl group (C=O). In aldehydes, carbonyl is at the end of a chain, as in ethanal (otherwise known as acetaldehyde), $CH_3(C{=}O)H$, whereas in ketones it occurs within chains, as in propanone (also known as acetone, or dimethyl ketone), $CH_3(C{=}O)CH_3$. [J.G.]

Alfisols *See* SOIL CLASSIFICATION.

algal bloom Dense populations of free-floating algae, often imparting a distinctive colour and odour to the water body. BLOOMS occur in marine and freshwaters and result from a period of intense, often monospecific growth in response to favourable nutrition and light. Most freshwater blooms are caused by planktonic cyanobacteria belonging to the genera *Anabaena*, *Aphanizomenon* and *Microcystis*. These photosynthetic organisms possess gas vacuoles allowing them to float in huge numbers at or near the water surface. Large blooms caused by another gas-vacuolate genus, *Trichodesmium*, occur in tropical seas. Among the eukaryotic algae, dinoflagellates of the genus *Gonyaulax* produce 'RED TIDES', often on a large scale. Huge blooms of oceanic coccolithophorid algae have been observed from space. Many blooms are toxic. They render the water unfit for drinking and unpleasant skin rashes may result from bathing. Some of the TOXINS are transported along the FOOD CHAIN. Poisoning from shellfish consumption is frequently associated with blooms of the algae *Gonyaulax* and *Prymnesium* in coastal waters. *See also* EUTROPHICATION; TOXIC BLOOM. [A.P.]

algorithm A series of instructions for carrying out a specific computational task; it might be in the form of a flow diagram or in lines of computer code. For example, an algorithm for determining the maximum likelihood estimate of k of the negative binomial (*see* NEGATIVE BINOMIAL DISTRIBUTION). [M.J.C.]

alien A non-indigenous organism; a species of plant or animal moved (typically by human agency) to a country outside its natural geographical RANGE. The status of alien organisms is described as 'naturalized' (when their ecological behaviour is indistinguishable from native species) or 'casual' (when they rely on continual reintroduction for their persistence). *See also* EXOTIC AND INVASIVE SPECIES. [M.J.C.]

ALIs (annual limits on intake) *See* MAXIMUM PERMISSIBLE LEVEL.

Alisols *See* SOIL CLASSIFICATION.

alkalinity The capacity of WATER to accept H^+ ions (protons). The species primarily responsible for alkalinity in water are bicarbonate ion (HCO_3^-), carbonate ion (CO_3^{2-}) and hydroxide ion (OH^-). It is important to distinguish between high basicity, i.e. an elevated pH, and high alkalinity, i.e. the capacity to accept H^+. Alkalinity measures the quantity of weak ACIDS and the cations balanced against these weak acids in a sample of water. Alkalinity is also described as the excess positive charge of seawater, measured in units of equivalents per kilogram, where an equivalent is equal numerically to a MOLE. The excess positive charge is balanced primarily by bicarbonate (HCO_3^-) and carbonate (CO_3^{2-}). Alkalinity has also been referred to as buffer capacity, titratable BASE, excess base or titration alkalinity. A general linear relationship between alkalinity and CHLORINITY has been described such that:

$$\left(\text{Alkalinity} \times 10^3\right) = \left(0.123 \times \text{Chlorinity } (‰)\right).$$

[V.F.]

REFERENCE

Manahan, S.E. (1994) *Environmental Chemistry*, 6th edn. Lewis Publishers, Boca Raton, Fla.

alkalis Alkalis are BASES that are readily soluble in WATER. The term is applied especially to hydroxides such as potassium hydroxide (KOH) and sodium hydroxide (NaOH). [J.G.]

alkanes Alkanes are HYDROCARBONS with no double bonds (i.e. saturated), otherwise known as PARAFFINS, with the general formula C_nH_{2n+2}. They are the chief constituents of PETROLEUM. Examples of the lower RELATIVE MOLECULAR MASS, unbranched alkanes are shown in Table A12. [J.G.]

alkenes Alkenes are HYDROCARBONS with at least one double bond (i.e. unsaturated), with the general formula C_nH_{2n}. They occur in crude PETROLEUM. Examples of the lower RELATIVE MOLECULAR MASS, unbranched alkenes are shown in Table A13. Ethene is the same as ethylene, a natural plant growth regulator, involved in ripening. [J.G.]

Table A12 Some common alkanes and their properties.

Name	Formula	Melting point (°C)	Boiling point (°C)
Methane	CH_4	−183	−164
Ethane	C_2H_6	−183	−89
Propane	C_3H_8	−190	−42
Butane	C_4H_{10}	−138	−0.5
Pentane	C_5H_{12}	−130	36
Hexane	C_6H_{14}	−95	69

Table A13 Some common alkenes and their properties.

Name	Formula	Melting point (°C)	Boiling point (°C)
Ethene	C_2H_4	−169	−104
Propene	C_3H_6	−185	−47
1-Butene	C_4H_8	−185	−6
1-Pentene	C_5H_{10}	−138	30
1-Hexene	C_6H_{12}	−140	63

alkyphenols *See* ENDOCRINE DISRUPTER.

Allee effect A special type of DENSITY DEPENDENCE which occurs in low-density populations. It was first identified by W.C. Allee (1931; *see* CHARACTERS IN ECOLOGY). Conventionally, density dependence is negative; as DENSITY decreases, survival and reproduction increase due to the lessening of COMPETITION between individuals. However, there may be a point at which reducing density actually decreases survival or reproduction — the Allee effect. For example, the effect may occur when density has become so low that mates are difficult to find or pollen is unlikely to be carried between plants. In Diptera the effect is often due to FACILITATION, where a critical number of larvae is required, perhaps within a carcass, to create appropriate conditions for survival. *See also* DILUTION EFFECT. [C.D.]

REFERENCE

Allee, W.C. (1931) *Animal Aggregations. A Study in General Sociology*. University of Chicago Press, Chicago, Ill.

Allee, W.C. *See* CHARACTERS IN ECOLOGY.

allele frequency Genetic variation in a population at a locus *A* implies the presence of more than one allele at that locus. For two ALLELES A_1 and A_2, the frequency (p) of allele A_1 is given by:

$$p = \left[\left(2 \times \text{number of } A_1A_1 \text{ homozygotes}\right) + \left(\text{number of } A_1A_2 \text{ heterozygotes}\right)\right] / \left(2 \times \text{total number of individuals}\right)$$

for diploid organisms. The genetical mechanics of EVOLUTION are given by allele frequency changes. *See also* POPULATION GENETICS. [G.D.J.]

alleles Two or more forms of a gene occupying the same locus on a chromosome. In somatic cells of diploid organisms each gene is represented by two alleles, except for the genes contained in sex chromosomes in the heterogametic sex. *See also* ALLELE FREQUENCY. [P.O.]

alleles, rare *See* ALLELES; MOLECULAR DRIVE THEORY OF EVOLUTION; NEUTRAL THEORY OF EVOLUTION.

allelochemical *See* INFOCHEMICALS.

allelopathy A form of INTERFERENCE COMPETITION by means of chemical signals; i.e. compounds produced by one species of plant which reduce the germination, establishment, growth, survival or fecundity of other species. It does not involve the depletion of a limiting resource (except in so far as habitable space is a limiting resource; *cf.* EXPLOITATION COMPETITION). Allelopathy represents a good example of extreme, asymmetric CONTEST COMPETITION, but in practice it is extremely hard to separate allelopathic effects from other processes that might produce a similar outcome (e.g. herbivory, resource depletion). *See also* INFOCHEMICALS. [M.J.C.]

REFERENCES

Harper, J.L. (1975) Review of *Allelopathy* by E.L. Rice. *Quarterly Review of Biology* **50**, 493–495.

Michelsen, A., Schmidt, I.K., Jonasson, S., Dighton, J., Jones, H.E. & Callaghan, T.V. (1995) Inhibition of growth, and effects on nutrient uptake or arctic graminoids by leaf extracts—allelopathy or resource competition between plants and microbes? *Oecologia* **103**, 407–418.

Rice, E.L. (1984) *Allelopathy*, 2nd edn. Academic Press, New York.

Allen curve A method for estimating secondary production in which N_t, the number of the survivors (usually in a COHORT) at age t is plotted against W_t, the mean mass of individuals at age t. The area under the curve between W_{t1} and W_{t2} gives the production in the interval between $t1$ and $t2$. *See also* PRODUCTIVITY; SECONDARY PRODUCTIVITY. [S.J.H.]

REFERENCE

Allen, K.R. (1951) The Horokiwi Stream. A study of a trout population. *New Zealand Marine Department of Fisheries Bulletin* **10**.

Allen's rule Allen's rule is based on the observation that anatomical extremities in endothermic animals become progressively shorter as latitude increases. This pattern has been attributed to the proportional relationship between heat loss and body surface area, particularly heat loss from the periphery of the vascular system. Another similar but more contentious rule (Bergmann's) also uses heat loss as an explanation for latitudinal increases in body mass of certain widely distributed mammals, such as deer. This has been attributed to the fact that larger animals experience a proportional decrease in heat loss as body mass increases because of the relationship between surface area and body mass. *See also* ALLOMETRY. [R.C.]

allochthonous Describing organic matter not generated within the COMMUNITY. Utilization of allochthonous material is one element characterizing the ENERGY FLOW through a community. It reflects how much a community depends on its own PRIMARY PRODUCTIVITY. For example, small streams depend almost exclusively on allochthonous material arriving as litter from surrounding vegetation, whereas in large rivers the productivity of the PHYTOPLANKTON is more important. *Cf.* AUTOCHTHONOUS. [P.O.]

allogenic This term was first applied to SUCCESSIONS which take place as a result of the action of forces originating outside the system, for example the sinking of land relative to the sea, or deposition of silt in a lake, and a contrast was made with AUTOGENIC successions that arise as a result of the effects of plants on the HABITAT (Tansley 1929).

Later, Tansley (1935, p. 287; *see* CHARACTERS IN ECOLOGY) applied the terms to factors rather than successions, and suggested that both allogenic and autogenic factors are present in all successions. Many successions are initiated by an allogenic event, for example landslide, deposition of a sandbank or retreat of a glacier, but are 'driven' by autogenic processes, for example NITROGEN FIXATION or change in soil pH. [P.J.G.]

REFERENCES

Tansley, A.G. (1929) Succession: the concept and its values. In: *Proceedings of the International Conference of Plant Sciences (Cornell 1926)*, pp. 677–686.

Tansley, A.G. (1935) The use and abuse of vegetational concepts and terms. *Ecology* **16**, 284–315.

allometric growth The kind of GROWTH in which the proportions of various body parts change. During allometric growth one measurement (y) increases non-linearly with the increase of another measurement (x). The relationship between x and y is most commonly described by the allometric equation $y=ax^b$. Variable x is usually body mass, and y can represent measurements of different body parts or organs. When x and y are plotted on logarithmic coordinates, the result is a straight line with slope b, because $\log y=\log a+b\log x$. Usually $b\neq 1$, which means that measurement y does not change linearly with body mass. There are two possible geometric reasons for this: (i) an organism's growth may be ANISOMETRIC, i.e. there is a linear change in proportions during ontogenesis because of the different growth rates of various organs; or (ii) y is a linear or surface measurement whereas x is a volume/mass measurement, which gives $b\neq 1$ even for isometric entities (e.g. the surface area of any isometric solid is proportional to its volume raised to the power $^2/_3$).

Changes in proportions during growth may result from mechanical or physiological CONSTRAINTS. For instance, because the metabolic rate increases more slowly than body mass, NATURAL SELECTION should adjust the size of such an energetically costly organ as a heart or a liver to meet but not to exceed metabolic needs. The heart of a human constitutes *c.* 0.55% of its body mass at birth but only 0.40% by age 20; the liver constitutes *c.* 3.8% of body mass at birth and 2.3% at age 20. Developmental constraints may also alter the proportions if the final size of an organ is the main target of selection (e.g. brain growth in NIDICOLOUS species: the human brain constitutes *c.* 10% of body mass at birth and only 2% at the age of 20). *See also* ALLOMETRY. [J.Kz.]

Allometry

Organisms that differ in size or mass also differ in the relative size of different organs, in the amount of time they spend performing particular behaviours, or when in the life cycle different events occur. Such size-related differences are termed 'allometric' and they can arise for a number of reasons. For example, neuronal division stops shortly after birth in humans, so that the newborn baby has a disproportionately large head which grows much more slowly than the rest of the body during postnatal development. Consequently, the brain constitutes a smaller proportion of body size as humans develop from birth to adulthood. Such developmental allometry often contrasts with interspecific adult allometry. For example, the brain size of adult mammals is relatively small in larger-bodied species so that, although in absolute size a whale has a much larger brain than a mouse, a whale's brain is about 0.05% of body mass whereas a mouse's brain is about 5% of adult body mass (Fig. A8). The reason for this difference is not understood. Other allometric scaling relationships are much better worked out. If we think we know what an organ has evolved to achieve in mechanistic or engineering terms, then it may be possible to estimate how we might expect the size of an organ to scale in relation to body mass. If the data do not fit expectations, we can proceed to test alternative hypotheses, or merely refine the one that we have.

As an example, consider the skeletons of land animals. Among geometrically similar animals, if m is body mass and l is a linear dimension (such as the length of the body or a limb bone), then $m \propto l^3$. But the cross-sectional area of any given bone, and hence its strength, scales as l^2, which means that beyond some critical size,

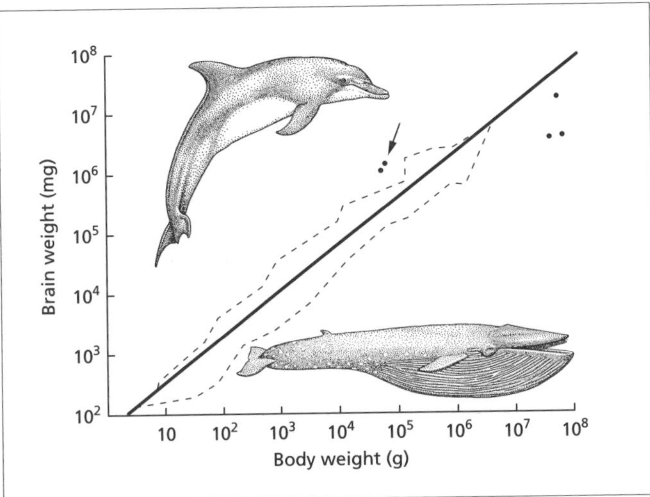

Fig. A8 Brain size in marine mammals. The regression line and ellipse are given for all mammals. Humans (arrowed) have the largest relative brain size, and the point nearest to humans is a dolphin. However, baleen whales have the smallest relative brain size. (After Cockburn, A. (1991) *An Introduction to Evolutionary Ecology.* Blackwell Scientific Publications, Oxford.)

such animals would collapse under their own mass. Bones face static (compressing) STRESS when the animal is stationary, and elastic (buckling) stress during locomotion. We can use mechanical principles to ask whether skeletons are adapted to prevent fracture by compression or by buckling. The leg bones of a quadruped with a vertical posture are like tall, thin cylinders. To withstand compression equally at all body masses (static stress similarity), their cross-sectional area must increase as m^1. Their length scales as l^1, or $m^{1/3}$, so bone mass should scale as $m^1 \times m^{1/3} = m^{4/3}$. But bone mass actually scales as $m^{1.08}$, so skeletons do not show static stress similarity. Perhaps bones are adapted to withstand buckling equally at all body sizes (elastic stress similarity)? Columns of diameter d buckle under loads proportional to d^4/l^2, so the ratio of buckling force to diameter is $(d^4/l^2)/d = d^3/l^2$. Elastic stress similarity therefore predicts $d^3 \propto l^2$, which is $l \propto d^{2/3}$. The limb bones of hoofed mammals fit the model but other mammals do not, perhaps because their limbs are less pillar-like and so face different stresses. Elastic stress similarity is also found further afield. Limb bones of bipedal theropod dinosaurs fit the theory. Tree trunks and limbs are also thin columns faced with buckling stresses: tree trunks retain elastic stress similarity as they grow, and branch thickness scales with length as predicted.

Behavioural patterns also change with body size and, as a consequence, it is possible to begin to understand how an organism exploits its environment by comparing expected and observed allometric relationships. For example, we know that metabolic turnover increases with body mass, such that basal metabolic rate (BMR) is proportional to $m^{0.75}$. If TERRITORY SIZE is determined by the food availability and food replaces itself at a constant rate, we should expect territory size also to increase proportional to $m^{0.75}$. In fact, territory size increases approximately with $m^{1.0}$ for territorial animals belonging to a wide range of taxonomic groups, including birds and mammals. This means that larger animals have disproportionately large territories. One possible explanation is that food exists in patches and that larger-bodied species need to include a higher proportion of non-productive land in their territories if they are to have enough patches of food in their territory to satisfy their metabolic demands. Interestingly, for non-territorial animals, maximum POPULATION DENSITY is proportional to $m^{-0.75}$. Since energy needs per individual increase proportionally with $m^{0.75}$, energy needs for a population of any species over a given area are proportional to $m^{0.75} \times m^{-0.75}$, which is m^0, i.e. a constant. This means that populations of different species in a community utilize roughly the same amount of energy. The reasons for this so-called 'energetic equivalence rule' remain a mystery. *See also* ALLOMETRIC GROWTH; ANISOMETRIC. [P.H.H.]

---

allomone *See* INFOCHEMICALS.

allopatric speciation The differentiation of geographically isolated populations into distinct species. GEOGRAPHICAL ISOLATION is generally believed to facilitate SPECIATION because separation decreases the likelihood of interbreeding, and separated populations can therefore evolve in different directions and achieve reproductive incompatibility. Geographically separated populations often differ, for example, in premating ISOLATING MECHANISMS, which may arise as pleiotropic effects (*see* PLEIOTROPY) of other genetic changes. Allopatric speciation is generally believed to be the most common mode of speciation, and spatially heterogeneous habitats often show higher species divergence compared to more uniform habitats. *Cf.* SYMPATRIC SPECIATION. *See also* ISOLATING MECHANISMS; SPECIES CONCEPTS. [V.F.]

allopolyploid An organism with more than two chromosome sets which are derived from different species (*cf.* AUTOPOLYPLOID). Allopolyploidy may be caused by multiplication of chromosome sets in a hybrid or by fusion of unreduced GAMETES of two different species. These processes are possible mechanisms of SPECIATION. Interbreeding between species often leads to sterile hybrids whose sterility results from the lack of homologous pairs of chromosomes, causing meiotic irregularities. After duplication of chromosome sets, meiosis proceeds without abnormalities and fertile hybrids are produced which can constitute a new species. [P.O.]

allotherm *See* ECTOTHERM.

allotropy The existence of a chemical element in two or more different physical forms (*see* SULPHUR as an example). [J.G.]

alloy *See* METAL.

alluvial Pertaining to SEDIMENT transported and deposited by a flowing RIVER (i.e. alluvium). [P.C.]

alluvium *See* ALLUVIAL.

Alm *See* ALPINE ZONE.

Alp *See* ALPINE ZONE.

alpha diversity *See* DIVERSITY, ALPHA, BETA AND GAMMA.

α-particle Particles of mass 4 amu (*see* ATOMIC MASS) and charge +2 (i.e. $^{4}_{2}$He nuclei), emitted during the radioactive decay of heavy, proton-rich nuclei. α-Particles typically have energies in the range 4–9 MeV. The α-particles emitted in the decay of a given RADIONUCLIDE generally all have a characteristic energy or group of energies. α-Particles cause intense ionization along straight-line paths in materials with which they interact, but have very limited penetrating power, with ranges of less than 0.1 mm in solids and only a few centimetres in gases. *See also* IONIZING RADIATION; SIEVERT. [A.B.M.]

Alpine zone

The alpine vegetation zone lies between the upper limit of tree growth and the snow line. Within the alpine zone vegetation communities vary depending on altitude, aspect, exposure, snow cover and water availability. Major discontinuities commonly recognized in temperate regions include:

1 a lower alpine zone with bush and tall-herb communities;

2 a middle alpine zone in which graminoid and heath species dominate;

3 an upper alpine zone marked by loss of dominance of the graminoid species, which are replaced by dwarf herbaceous and prostrate woody species together with lichens and mosses.

The lower boundary of the alpine zone is determined by the upper limit of the TREE LINE. At the interface between the tree limit and the lower alpine zone there can be a bush community often dominated by genetically dwarfed trees, frequently with a creeping or twisted growth form (KRUMMHOLZ—German *krumm* twisted, *holz* wood), which in the European Alps can be either the green alder (*Alnus viridis*) or the dwarf mountain pine (*Pinus mugo*) often associated with the hairy alpenrose (*Rhododendron hirsutum*). In Scandinavia krummholz is represented by *Betula tortuosa* (see Fig. K1), and in Scotland by rare remnant populations of a genetically dwarfed ecotype of *Pinus sylvestris*. In Scotland and Scandinavia common bush species in this zone are willows (*Salix glauca, S. phylicifolia, S. lanata, S. lapponum*). In some areas with ACID soils alpine heath develops in the lower alpine zone, dominated by ericaceous species (*Calluna vulgaris, Empetrum nigrum, Vaccinium* spp.).

Where there is no tree line, due to a long history of GRAZING, and particularly in oceanic areas of western Europe, alpine heath is contiguous and floristically similar to the MOORLAND that has developed in the subalpine zone. Within the alpine zone as a whole, late lying snow gives rise to azonal snow-patch communities where plants that can survive prolonged snow cover and short growing seasons profit from a habitat that is protected in winter from frost DESICCATION and in summer is less prone to DROUGHT. Species commonly found in European alpine snow-patch areas include *Sibbaldia procumbens, Veronica alpina, Gnaphalium supinum* and *Poa alpina*.

Wetland communities within the alpine zone can be considered under two headings.

1 Alpine flushes—where a continuous supply of moving groundwater, which seldom freezes, provides a suitable habitat for hygrophilous species such as the starry saxifrage (*Saxifraga stellaris*), blinks (*Montia fontana*), the opposite-leaved golden saxifrage (*Chrysosplenium oppositifolium*) and alpine willowherb (*Epilobium anagallidifolium*).

2 Alpine mires—which develop in hollows, often in basins left by retreating glaciers and where the flora depends on whether the area is a calcareous FEN or a RAISED BOG.

Throughout the alpine area the species composition of communities is strongly influenced by the pH of the bedrock (see below—alpine grasslands). This distinction is most clearly seen in rock outcrops, which have characteristic communities depending on whether the rock is siliceous or calcareous. In the Alps the calcareous rock outcrops are richer in species than those on acid rocks. Typical species on calcareous Alpine outcrops include *Primula auricula* and *Androsace helvetica*. Species typical of acid rock outcrops include *Sedum telephium* and *Asplenium adiantum-nigrum*.

Centuries of transhumance by pastoral peoples to the lower and middle alpine zones, to exploit the productivity of high-altitude summer meadows and pastures, has encouraged the development of herb-rich alpine meadows. Many of the populations of flowering plant species of these alpine meadows have evolved early synchronous flowering which allows cross-pollination and seed production before the alpine meadows are mown for hay or the pastures are severely grazed. Grazing and mowing together have prevented tree COLONIZATION and consequently the tree line is below that at which it would be found in the absence of grazing. Depopulation of rural mountainous areas is currently resulting in the re-establishment of trees on alpine pastures (Fig. A9) and meadows and the loss of many of the herb-rich communities that enriched the BIODIVERSITY of low to mid-alpine habitats. In German-speaking countries, *Alp* and *Alm* are both terms used to describe areas of transhumance movement to mountain summer pastures and are equivalent to *sæter* and *stol* in Norway, *shieling* in Scotland and *yaylak* in Turkey.

Ecologically, the species composition of alpine grasslands is strongly influenced by whether the soils are acid or calcareous. The first description of the floristic differences between calcareous and

Fig. A9 The mountain pine of central Europe (*Pinus uncinata*), colonizing alpine grazings in the Vercors (France) as a result of reduced summer grazing. (Courtesy of R.M.M. Crawford.)

siliceous soils was made by H.F. Link in 1789 in the Alps, where there are many examples of species pairs which occupy acid and calcareous soils respectively. Thus, *Rhododendron ferrugineum* occurs on acid soils while *R. hirsutum* is confined to calcareous soils. Similar vicarious pairing, with the first named species being the CALCICOLE species, is found in *Primula hirsuta* and *P. auricula; Sedum alpestre and S. atratum; Sempervivum arenarium* and *S. hirsutum; Ranunculus glacialis* and *R. alpestris*, etc.

In the Andes the equivalent high-altitude grassland vegetation type above the TIMBER LINE is dominated by bunch grasses (*Calamagrostis, Festuca* spp.), variously described as *PÁRAMO*. Typical *páramo* has high rainfall and humidity throughout the year and contrasts with a related plant formation, *puna*, which has a DRY SEASON that increases from north to south and from east to west (Troll 1959). These two types merge with each other and constitute the vast area of high-altitude Peru, Bolivia and northern Chile and northern Argentina referred to as the *altiplano*. In the Afro-alpine or tropical-alpine grasslands in the Old World the equivalent of the altiplano (*sensu lato*) is found in the Ethiopian tableland or high-altitude grassland STEPPE of Tibet (Vuilluemier & Monasterio 1988). In the Andes the expansion of these alpine grasslands below the climatic tree line due to grazing probably dates to some time after the rise of the Inca empire in the 11th century. At the time of the Spanish conquest (1528–1531) it was noted that the herds of llamas on the altiplano (Fig. A10) were so large that grazing could not be found for them (Garcilaso de la Vega 1608).

PACHYCAUL (thick-stemmed) species are a striking plant life form that has developed through CONVERGENT EVOLUTION in both South American and African alpine habitats. In the Afro-alpine region *Dendrosenecio* and *Lobelia* are typical examples as are *Puya* and *Espeletia* in South America (Fig. A11). The pachycaul construction is particularly suited to the occupation of sites with strong diurnal temperature fluctuations, as it retards both heating and cooling, which is further aided by the nyctinastic movements of leaves and bracts, the retention of leaf bases and the thickening and narrowing of leaves.

Alpine and Arctic-alpine are biogeographical qualifications which describe, respectively, whether species are limited to montane regions outside the ARCTIC or are widespread in both alpine and arctic habitats. Such is the difference in the nature of the alpine and arctic habitats that even within widespread and successful arctic-alpine species there are marked ecotypic differences (Billings *et al.* 1971). The term 'alpine tundra' is sometimes used when describing alpine habitats with similar plant communities to those in the true Arctic. This usage is misleading and the term 'alpine tundra' (*see* TUNDRA) should be restricted to denoting alpine regions where short cool summers and very low winter temperatures produce a PERMAFROST zone above the bedrock and below the surface layer subject to summer melt. *See also* AFRO-ALPINE VEGETATION; ALTITUDINAL ZONATION. [R.M.M.C.]

REFERENCES

Billings, W.D., Godfrey, P.J., Chabot, B.F. & Borque, D.P. (1971) Metabolic acclimation to temperature in alpine and arctic ecotypes of *Oxyria digyna. Arctic and Alpine Research* **3**, 277–290.

Garcilaso de la Vega (1608) *The Royal Commentaries of the Incas*. English edition (1869) by the Hakluyt Society, London.

Troll, C. (1959) Die tropischen Gebirge. *Bonner geographische Abhandlungen* **25**, 1–93.

Vuilluemier, F. & Monasterio, M. (1988) *High Altitude Tropical Biogeography*. Oxford University Press, Oxford.

alternation of generations The alternation of two different reproductive processes in the LIFE CYCLE of an organism. It is common and highly variable in its expression among animals, plants and fungi. In plants and many fungi, a sexual generation, which is usually diploid (DIPLOIDY) and spore producing, is followed by an asexual haploid generation (HAPLOIDY) during which GAMETES are produced. In mosses the haploid GAMETOPHYTE generation is the green plant and the SPOROPHYTE generation is reduced to a spore capsule. In ferns and higher plants the actual plant is the diploid sporophyte, while the gametophyte is often reduced. In contrast to

Fig. A10 Llamas grazing the Peruvian *altiplano* at an altitude of 3500 m. (Courtesy of R.M.M. Crawford.)

Fig. A11 *Puyo* sp. a pachycaul plant with the late Professor Cesar Vargas of the University of Cuzco above the Andean tree line at 4300 m. (Courtesy of R.M.M. Crawford.)

these heteromorphic examples, isomorphic alternations, in which both generations look similar, are found in many algae.

In contrast to plants, in most protozoans both generations are haploid, while in metazoans both generations are diploid. If the latter involves alternation between gamogamy and ASEXUAL REPRODUCTION through multicellular bodies (e.g. hydrozoans) it is called metagenesis, but if the asexual generation is parthenogenetic (*see* PARTHENOGENESIS), i.e. through unfertilized EGGS (e.g. rotifers, cladocerans), it is called heterogony. Heterogony is often FACULTATIVE, i.e. the alternation between generations depends on the environment. In some PARASITES, like the malaria parasite *Plasmodium* and many trematodes, the alternation of generations coincides with the alternation between HOSTS. *See also* COMPLEX LIFE CYCLES. [D.E.]

alternative energy/fuels Energy or fuels derived from non-traditional sources, such as solar, WIND, chemical feedstocks and wastes. [P.C.]

alternative hypothesis (H_1) The alternative to the NULL HYPOTHESIS, usually represented by H_1. Sometimes this is simply that the null hypothesis is not true, sometimes it is more precise. For example, when carrying out an *F* TEST, the null hypothesis is that the two sample variances (s_1^2 and s_2^2) are from populations with the same variances (H_0: $\sigma_1^2 = \sigma_2^2$), that is, if the sample variances are different it is only due to chance. The alternative hypothesis could simply be that the variances are not equal (H_1: $\sigma_1^2 \neq \sigma_2^2$), that is, the null hypothesis is not true. However, another alternative hypothesis is H_1: $\sigma_1^2 > \sigma_2^2$. *See also* ONE-TAILED TEST; STATISTICAL TABLES; TWO-TAILED TEST. [B.S.]

alternative technology Technology designed to stand in place of the complex, often polluting production processes of the developed world. It is often intended primarily for use in developing countries. Hence it should be easy to construct, use and maintain. An example is the use of WIND POWER, rather than a diesel-powered motor, to pump water. Such technology will generally be less polluting than the usual system. [P.C.]

altiplano *See* ALPINE ZONE.

Altitudinal zonation

Vegetation shows a marked ZONATION with altitude because of changing TEMPERATURE and rainfall (Fig. A12). In some ways the altitudinal zonation mirrors the latitudinal zonation (*see* LATITUDINAL GRADIENTS OF DIVERSITY) from the tropics to the polar regions, but the effects of both FROST and the LIGHT regime are quite different between the tropics and other latitudes. Temperature declines in tropical latitudes at the ADIABATIC LAPSE RATE of –0.55°C for every 100 m increase in altitude but below the snow line frost may form only nightly and does not deeply penetrate the SOIL. Light levels in tropical regions above the cloud layer are daily very high. In polar regions PERMAFROST is formed and light levels are lower and continuous in summer, dark in winter. These differences alter the nature of the altitudinal zonation and the PHYSIOGNOMY of plants in different zones at different latitudes. The AFRO-ALPINE VEGETATION has a different physiognomy to the alpine vegetation of the Alps or Rockies.

One of the most obvious boundaries between zones is the TIMBER LINE, above which it is impossible for WOODLAND or FOREST to grow. The subalpine zone lies below the timber line; in this area montane forest of various sorts (*see* MONTANE FORESTS, CONSERVATION), like CLOUD FOREST or ELFIN FOREST can be found. At the upper edge of

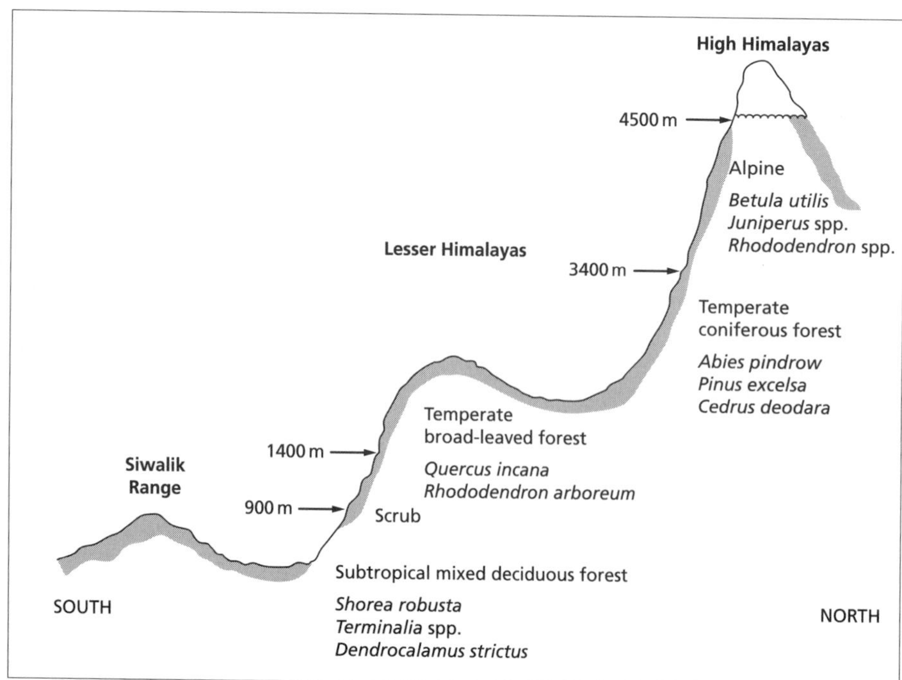

Fig. A12 Diagrammatic section of the western Himalayas in northern India, showing the approximate altitudinal limits of the major vegetation types. With increasing altitude one passes through vegetation belts similar to those found on passing into higher latitudes. The scrub zone (900–1400 m) is strongly modified by human deforestation. (After Cox, C.B. & Moore, P.D. (eds) (1993) *Biogeography: An Ecological and Evolutionary Approach*, 5th edn. Blackwell Scientific Publications, Oxford.)

the timber line there is a transition to shrubby plants. In some areas, especially in the BOREAL region, isolated trees may be low-growing, stunted or twisted, forming a KRUMMHOLZ zone (German for twisted wood) (Fig. A13). The nature of the transition zone depends on the species present. It may be sharp or gradual. Trees may peter out, without changing form, and there may be ericaceous shrubs—heathers or rhododendrons—or in the tropics, bamboos.

The level and distribution of rainfall are important factors in determining the altitudinal zonation. Hence, the zonation on different sides of mountains or mountain ranges whose slopes receive rain or are in the rain shadow have markedly different vegetation. Moreover, altitudinal zonations are depressed and compressed on small mountains compared to large ones. This is called the *Massenerhebung*, or mass elevation, effect. This effect may be associated with the lowering of the level at which cloud habitually forms.

The treeless ALPINE ZONES of the world form a discontinuous pattern. Many have been isolated for very long periods of time. This is reflected in floras with a high degree of endemism. The Afro-alpine flora has many unique characteristics but has some plants with a similar physiognomy to other tropical alpine regions in the Andes, East Indies and Hawaii. The alpine flora of New Zealand is a mixture of diversified immigrants from the north and locally derived relicts. Oscillations in the altitudinal limits of vegetation zones were marked in the Pleistocene with successive glacial and interglacial stages. In many places the TREE LINE was at its lowest 20 000–15 000 BP and at its highest most recently 10 000–8000 BP. In the Northern hemisphere at times of Pleistocene GLACIATION alpine areas became connected to the northern TUNDRA so that floristically the Northern hemisphere mountains show a marked relationship to the circumboreal ARCTIC tundra flora. They share some species and also have closely related vicariant species. The lower alpine zone is dominated by EVERGREEN shrubby species of various sorts. Many species are in the heather family. They are xerophilous, adapted to the dry cold winds of the uplands. Leaves are small and in-rolled with a thick waxy coat on the upper surface and hairs on the undersurface.

At higher altitudes the shrubs are replaced by sedges and grasses.

Fig. A13 Krummholz form of Scots pine (*Pinus sylvestris*) marking the upper boundary of the tree line and the beginning of the alpine zone in the Cairngorms, Scotland. (Courtesy of R.M.M. Crawford.)

They form small tussocks with the dead leaves providing protection from desiccating winds. Alpine flowers give the alpine meadows a beautiful show of colour when they flower in spring. Broad-leaved herbs are concentrated in areas where the underlying geology restricts the growth of grasses, because of wetness, a continuously eroding surface, or the presence of limestone. Some plants grow only in depressions, where snow cover protects them in winter. Perennial herbs and dwarf shrubs form compact cushions or dense, prostrate mats minimizing the exposure of their tender buds to cold and DROUGHT. They are protected by snow in winter and benefit from higher ground temperatures in summer. Growth is slow. Many are semi-evergreen or partly woody, preserving hard-won growth and carbohydrate reserves from one short growing season to the next. In spring they flower rapidly, maximizing the season for seeds to ripen. POLLINATION is chancy and many plants show adaptations for vegetative spread by the production of rhizomes, stolons or bulbils.

At the highest altitudes in the fellfield communities where conditions are very exposed, and snow cover limited, lichens and mosses dominate. Here the vegetation is similar to the high tundra. The mosses and lichens are poikilohydrous, having the ability to dry out and recover when moisture becomes available. The mosses have leaves with silvery bristle-like points which can collect moisture and protect from desiccating WIND. [M.I.]

REFERENCES

Archibold, O.W. (1995) *Ecology of World Vegetation*. Chapman & Hall, London.

Crawford, R.M.M. (1989) *Studies in Plant Survival*. Blackwell Scientific Publications, Oxford.

---

altricial The neonates of various species of birds and mammals differ markedly in the relative degree of maturation of many aspects of their behaviour, physiology and anatomy. The functional maturity of the chick at any given point in its postnatal development is closely tied to the care it receives from its parents, and probably also to many aspects of its environment. Variation among taxa in the developmental trajectory of this suite of attributes has led ornithologists to separate birds into altricial and PRECOCIAL developmental types. Designation of a species into one or the other category has been based largely on the condition of the hatchlings. The terms *Aves altrices* and *Aves precoces* were introduced into ornithology by Sundeval (1836). Altricial and precocial primarily refer to the developmental stage of the chick, whereas the alternatively used terms NIDICOLOUS and NIDIFUGOUS refer to nest attendance. Altriciality is characterized by a low functional maturity of tissues but high growth rates, closed eyes at birth and intense PARENTAL CARE for the young. The term altricial derives from the Latin *altrix*, wet-nurse. *See also* ALLOMETRIC GROWTH; DEVELOPMENTAL STATE OF NEWBORN; LIFE-HISTORY EVOLUTION. [J.M.S.]

REFERENCE

Sundeval, C.E. (1836) Ornithologiskt system. *Kongl. Vetenskops Acad, Handl.* For 1835, 43–130.

altruism *See* KIN SELECTION.

aluminium (Al) A light METAL (ATOMIC NUMBER = 13; relative ATOMIC MASS = 26.98; melting point = 660.37°C; boiling point =

2467°C) which, at 8.2%, is the most abundant metal and the third most abundant element in the Earth's CRUST. It is extracted from bauxite ore for use in packaging, construction, transport, pharmaceuticals and many other areas of daily life. Aluminium occurs widely in the silicate minerals of granites and CLAYS from which, in areas of low buffering capacity, it can be mobilized in ionic form by ACID rainfall. This has harmful consequences for vegetation, through interference with nutrient uptake, and for fish in lakes and streams, ultimately by suffocation through PRECIPITATION of gelatinous aluminium hydroxide ($Al(OH)_3$) on the gills. Deliberate addition, in the form of aluminium sulphate ($Al_2(SO_4)_3$), to remove colour and suspended COLLOIDS from reservoirs in upland catchments, can, in the absence of subsequent preventive measures, result in elevated dissolved aluminium concentrations in the water supply, causing dementia in kidney dialysis patients and perhaps contributing to Alzheimer's disease. [J.F.]

REFERENCE

Massey, R.C. & Taylor, D. (eds) (1989) *Aluminium in Food and the Environment.* Royal Society of Chemistry, Cambridge.

ambient Of, or relating to, surrounding environmental conditions. [P.C.]

amensalism An interaction, usually between two species, in which one has a detrimental effect upon the other, while the other has no effect upon the first species. Amensalism is an extreme form of ASYMMETRIC COMPETITION. *See also* SPECIES INTERACTIONS. [B.S.]

americium (Am) Americium (ATOMIC NUMBER=95; relative ATOMIC MASS = 243; melting point=994°C; boiling point=2607°C) is a man-made element which was discovered in 1944. The ISOTOPE ^{241}Am ($t_{1/2}$=423.7 years), produced by NEUTRON irradiation of nuclear reactor fuel, occurs in the environment mainly from DISCHARGES from nuclear fuel reprocessing plants. It is also produced from decay of fallout- or reactor-derived ^{241}Pu ($t_{1/2}$=14.4 years). Americium occurs dominantly in the +3 oxidation state and is highly insoluble, with a distribution coefficient (K_d) for sediment/water systems of about 10^6. *See also* RADIOACTIVE WASTE; RADIONUCLIDE. [A.B.M.]

Ames test A bacterial test for detecting point MUTATIONS. *See also* ECOTOXICITY; PREDICTIVE ECOTOXICITY TESTS. [P.C.]

ametabolous insects *See* INSECTS, DIVERSITY AND CONSERVATION.

amides Amides are organic compounds in which the HYDROGEN atoms of AMMONIA are replaced by organic acid groups. An example is acetamide, CH_3CONH_2. [J.G.]

amines Amines are organic compounds in which the HYDROGEN atoms of AMMONIA are replaced by hydrocarbon groups. Examples include primary amines (NH_2R), secondary amines (NHR_2) and tertiary amines (NR_3), where R is a hydrocarbon group. [J.G.]

amino acids A group of about 30 organic ACIDS that contain the amino (NH_2) group and which comprise the basic structural units of PROTEINS. They have the general formula:

$H_2N–CHR–COOH$

where R is a group of atoms. Over 20 amino acids occur naturally. The term 'essential amino acids' refers to those that an organism is unable to synthesize itself and so must obtain from food. In humans there are eight essential amino acids. [J.G. & V.F.]

REFERENCE

Hay, R.K.M. (1981) *Chemistry for Agriculture and Ecology.* Blackwell Scientific Publications, Oxford.

amixis ASEXUAL REPRODUCTION in which progeny are produced mitotically, receiving a copy of the maternal genome. Amictic processes include ameiotic APOMIXIS (including apospory in plants) and adventitious embryony. Vegetative proliferation is sometimes described as amictic although there is disagreement over whether it is a process of reproduction or of growth. [M.M.]

ammonia An important source of inorganic NITROGEN, ammonia, in the form of ammonium ion (NH_4^+), is a major nutrient element for many organisms. Although in aquatic systems, nitrogen is also found as nitrite (NO_2^-) and NITRATE (NO_3^-) ammonia is usually the preferred form of nitrogen for marine primary producers because it can be used directly for AMINO ACID synthesis without a change in oxidation state. Ammonia is also the principal form of organic nitrogen released from aquatic sediments. It is released during the process of microbial DECOMPOSITION of DETRITUS and is produced as a major WASTE product by many aquatic animals. The usual concentration range in the OCEAN is 0.1–5.0 μmol l^{-1}. Typical values of NH_4^+ in rivers are in the range 0.005–10 mg l^{-1} (0.28–5.6 μmol l^{-1}) and in streams 0.005–1.0 mg l^{-1} (0.28–0.56 μmol l^{-1}). [V.F.]

REFERENCE

Horne, A.J. & Goldman, C.R. (1994) *Limnology*, 2nd edn. McGraw-Hill, New York.

Amoco Cadiz *See* ECOLOGICAL/ENVIRONMENTAL DISASTERS.

Amphibians, diversity and conservation

The chordate class Amphibia goes back in the FOSSIL RECORD some 350 million years, to the mid-Devonian, and includes the first vertebrate organisms to occupy terrestrial habitats. The name, from Greek, refers to double life: characteristically, a LARVA or tadpole stage is followed by METAMORPHOSIS to a very different looking adult. Typically, the larval stage is aquatic and gill-breathing, and the adult more or less terrestrial. However, there are numerous deviations from this simplistic scheme. All living amphibians belong to the subclass Lissamphibia and are divided into three orders: Caudata (or Urodela), the salamanders; Gymnophiona, the caecilians; and Anura (or Salientia), the frogs and toads. Duellman and Trueb (1986) provide details and discussion of classification, anatomy, physiology and biogeography. There are more than 4000 known species and new ones are being discovered and described at a rate of 30–40 per year, at least. If there were more field biologists looking, that rate of discovery would certainly go up because many of the new species are being described from ostensibly well-known areas, such as eastern North America.

The salamanders, about 400 species, are divided into nine or ten families depending on the weight given to their defining characters. All are elongate and most have four legs. One family, the Sirenidae of the south-eastern USA and north-eastern Mexico, is eel-like, aquatic, retains feathery external gills throughout life, and has completely lost the pelvic girdles and hindlimbs. The greater siren is large, attaining over 97 cm in total length. The relationship of sirenids to other salamanders is unclear and controversial. Members of another eel-like family from the south-eastern USA, the Amphiumidae, have all four legs, but these are tiny and with reduced toes; one representative is very large, exceeding 116 cm. Other salamanders that never lose their gills or become terrestrial include the mudpuppies in North America and their cave-dwelling relative, the olm, in southern Europe (family Proteidae); and the giant salamanders of eastern North America, China and Japan (family Cryptobranchidae). One cryptobranchid, *Andrias davidianus*, of China, is certainly the largest living amphibian, exceeding 1.5 m and said to attain 2 m in the historic past. It is a popular food item, so human harvest probably precludes individuals surviving to such enormous size today.

A handful of smaller species of salamanders that metamorphose to gill-less, lung-breathing adults are spread over Eurasia; one reaches North Africa. The more terrestrial representatives are regarded as 'typical' salamanders. Those that dwell in water are often called newts. The members of a small Asian family, Hynobiidae (with one representative qualifying as European in Russia), metamorphose and typically have lungs, but retain external FERTILIZATION ('spawning') like most eel-like and giant aquatic forms.

North America is the capital of the salamander world, with about 300 known species. In addition to those noted above, there are typical salamandrids (called newts or efts even when they live on land) in both eastern and western USA. A radiation of mole, Pacific giant and Olympic salamanders includes fairly large forms (to over 35 cm), some of which are very colourful. All have internal fertilization, lay EGGS in water, and produce gilled aquatic larvae. Most metamorphose into more or less terrestrial, lung-breathing adults, but some (e.g. the olm, *Proteus anguinus*) are neotenic, remaining in a larva-like state throughout life. The greatest radiation of salamanders is the lungless Plethodontidae. There are three centres of diversity: Central America, with some species entering South America; the western USA and Canada largely west of the Rockies; and eastern North America, with scores of species, some of which are neotenic cave dwellers.

Members of the order Gymnophiona are pantropical, limbless and reminiscent of worms; some are colourful. The *c.* 180 known species have recently been divided into six families. The primitive forms have gilled larvae that metamorphose; some more advanced forms lay eggs on land and have young that skip the gilled stage. The most advanced forms bear live young.

The vast majority of modern amphibians (including all toads) are frogs. Dozens of families are recognized, but many or most of the anatomical characters used to diagnose them are modal or occasional features, not clear distinctions. Members of several nominal 'families' hybridize with members of other 'families', so their taxonomy has little to do with biological relationship. The vast majority of frogs begin life as aquatic tadpoles and all of those metamorphose into tailless, four-legged adults. A few species have direct development from eggs laid on land, and a very few produce live young, miniatures of their parents. Frogs occur on all continents except Antarctica and some have dispersed to oceanic islands. Most have lifestyles familiar to us all, but some are wholly aquatic, many are fossorial, and legions are arboreal.

Conservation

Frogs have become one of the most-studied groups of organisms on Earth, especially since *c.* 1985 when catastrophic population declines and amazingly rapid extinctions began to be reported from many different parts of the world. The die-offs are enigmatic because some species are affected in a given habitat while others are not. Many salamander species have undergone similar catastrophic declines. This is probably the 'hottest' topic in biological conservation research today and especially important when one considers the enormous ecological and economic importance of amphibians. In the forests of north-eastern North America alone, the biomass of salamanders often exceeds that of either birds or mammals, and salamanders are all carnivores or insectivores. The biomasses and ecological influences of frogs in TROPICAL RAIN FORESTS have not been quantified, but they are obviously prodigious; again, all species are carnivores or insectivores.

Catastrophic amphibian declines are all anthropogenic, directly or indirectly, and divisible into three aetiological categories: habitat destruction, POLLUTION and exploitation. Humans eat amphibians but do not farm them. All are harvested from the wild. Exploitation for food consumes something like 2 million frogs per year in European countries like Greece and Switzerland (one wonders how many in France). The USA imports between 6 and 7 million pounds of frog meat annually, and that is a small part of total consumption because most frogs eaten are domestic in origin. Frogs were so popular in California as food that 48 000 were harvested around San Francisco in 1888. Harvest peaked at 118 000 in 1895, and then plunged. To meet demand, eastern bullfrogs were imported in 1896. The current wholesale price of live bullfrogs is US$3–6 per pound, and consumption increased four-fold between 1990 and 1992.

Biological supply-houses sell millions of frogs and salamanders annually. Bait shops in the USA sell hundreds of thousands of salamanders to fishermen. Frog (and toad) hides are now in demand for leather; they are often stuffed and sold as toys. All of these uses take amphibians caught in the wild. So, until very recently, did the pet trade. At least 1 million frogs, toads and salamanders are collected in the USA for sale as pets or imported. All these, of course, are effectively dead from the viewpoint of wild populations, and most die quite quickly anyway for lack of proper care. However, some herpetoculturists have begun to breed captive stocks for the pet trade.

Habitat destruction boggles the mind (*see* HABITAT DESTRUCTION AND ALTERATION, GLOBAL). We estimate the clearing of tropical rain forest at something like 20 000 ha day^{-1}. As much as two-thirds of the planet's tropical forest area had been cleared or degraded to a fraction of bioproduction by 1990. Temperate forests are being destroyed too. Biologists estimate that US Forest Service clearcuts in western North Carolina alone destroy 14 million salamanders per year. When this is combined with Forest Service policies concerning recreational development and road building all over the USA, the effects on amphibian species appear horrific (Phillips 1994).

Impoundments flood entire valleys, wooded slopes, springs and ravine habitats. Drainage empties WETLANDS. Humans love to convert temporary ponds (usually fish-free and often therefore ideal for amphibian breeding) into permanent water holes; we yearn to convert every MARSH (left undrained) into a lake. Our edifices, like highways, hopelessly fragment remaining habitats and become death strips for all sorts of wildlife. Often, amphibian populations cannot persist in the remaining small habitat fragments.

Pollution effects are integrally tied to habitat destruction. PESTICIDES, metals and road salts—to list but a few—are chemicals inimical to amphibian life (and human life too, at higher doses). All are well documented to have taken terrible tolls. A form of pollution not often considered is the introduction of exotic species (*see* EXOTIC AND INVASIVE SPECIES). Exotics—animals not native to the area in which they are introduced—are infamous in the annals of conservation. Among those devastating to some amphibians have been cattle, pigs, cats, fire ants and even fellow amphibians. At least four frog species, the African pipid (*Xenopus laevis*), the tropical American cane or giant toad (*Bufo marinus*), the eastern US bullfrog (*Rana catesbiana*) and the Cuban treefrog (*Hyla septentrionalis*) have been widely introduced outside their natural ranges. All are voracious predators and all have been implicated in demises of other frog species.

The human consumption of FOSSIL FUELS produces an array of pollutants. Combustion of fuels produces acid molecules with SULPHUR and NITROGEN that are hygroscopic and precipitate as acid rain (*see* ACID RAIN AND MIST). Downwind of major temperate climate population centres, especially in Europe and North America, precipitation may be as acid as pH 2.5. Until the 1960s, when 'stack scrubbers' were mandated in many places, FLY ASH, which is highly alkaline, mitigated the acidity, at least locally. The fly ash, however, produced the caustic 'London fog', a terrible pollutant itself. By the mid-1970s tremendous mortalities were being reported in amphibians resulting from acidity. The principal difficulty seems to be the inability of embryos to properly form bone in acid waters. Major secondary effects result from soil ACIDIFICATION, which disrupts normal amphibian (especially salamander) behaviour, including foraging, mating and egg laying. Even if the animals are not killed outright, they may fail to reproduce.

Another combustion product is CARBON DIOXIDE (CO_2), a gas necessary for life. In large quantities, however, CO_2 becomes a GREENHOUSE GAS, insulating the planet and retarding heat loss. The result has been very rapid GLOBAL WARMING and disruption of normal weather patterns. From local DROUGHTS to the now global effects of EL NIÑO, global warming has been implicated in abrupt amphibian extirpations and even extinctions.

The CFCs are a group of organic chemicals synthesized for industrial uses. They deplete OZONE (O_3). Ozone forms an atmospheric mantle protecting the Earth from ultraviolet light, especially the highly mutagenic UV-B, the standard cause of human skin cancer. UV-B also devastates certain species of amphibians, especially some frogs, in three ways.

1 Behaviourally: some frogs apparently can 'see' UV-B and try to avoid it, thus disrupting their normal life patterns.

2 As a MUTAGEN: UV-B alters DNA, the genetic material; some frogs seem unable to produce sufficient DNA-repair enzymes, while some others in the same habitats are able to do so.

3 Finally, and worst, UV-B kills eggs. This is probably also a MUTATION effect, wrecking genes in the critical periods of tissue development and organ differentiation.

The above causes, all of them human, can act singly or in concert to cause stresses on amphibians, which are then exploited by microparasites, bacteria, fungi and other natural enemies. The results can be wholesale death with a proximate cause that belies the true aetiology. Catastrophic amphibian die-offs are not occurring in isolation. They are part of 'a global environmental disaster that merely includes frogs as early victims and could end with humans' (Phillips 1994). *See also* COMPLEX LIFE CYCLES; CONSERVATION BIOLOGY. [J.D.Lz.]

REFERENCES

Duellman, W.E. & Trueb, L. (1986) *Biology of Amphibians*. McGraw-Hill, New York.

Phillips, K. (1994) *Tracking the Vanishing Frogs*. St Martins Press, New York.

---

amphoteric Describing substances that react as ACIDS to strong BASES, and as bases to strong acids. An example is zinc oxide (ZnO), which gives rise to zinc salts of strong acids and zincates of the alkali metals. AMINO ACIDS and PROTEINS also have amphoteric properties. [J.G.]

anadromous Describing a specific annual migratory pattern in fishes when they move from marine to FRESHWATER environments. This movement is usually linked with the organism's reproductive cycle, as is the case for Atlantic salmon (*Salmo salar*), which move to well-oxygenated stretches of river where they spawn and subsequently die. When the migratory pattern is reversed and the fish move from freshwater to marine environments, as is the case for the European eel (*Anguilla anguilla*), then the behaviour is termed CATADROMOUS. *See also* ANADROMOUS FISH CONSERVATION; MIGRATION. [R.C.]

anadromous fish conservation Fish species that spend their juvenile stages in FRESHWATER, migrate to marine environments until maturity and then return to freshwater to spawn are termed 'ANADROMOUS'. Anadromy is associated with some of the more primitive fish taxa and is relatively rare, occurring in less than 1% of the world fish fauna. It is generally restricted to species which inhabit northern subpolar and cool temperate regions and rarely occurs in species inhabiting the warm temperate zone and tropics. Although the anadromous behaviour of salmonids (Salmonidae), sturgeons (Acipenseridae) and lampreys (Petromyzontidae) has been well documented, representatives of other groups (i.e. Osmeridae, Cottidae, Gasterosteidae and Percithyidae) have also utilized this pattern of habitat use (McDowall 1987). The importance of both aquatic and marine environments in the LIFE CYCLE of anadromous species means that conservation measures must embrace both types of habitat.

Although a number of commercially valuable anadromous species, such as sturgeon (*Acipenser sturio*), are also in decline, in global terms, it is the anadromous behaviour of the Salmonidae which has received most attention from conservationists. Salmon (both *Salmo* and *Oncorhynchus* spp.) are highly prized as both a table fish and for the sport that they provide to anglers. Quantifying marine exploitation rates of salmon populations has been a problem facing conservationists for several decades. For example, the observed decline in the return rates of Atlantic salmon (*Salmo salar*) from the SEA to RIVERS in recent years has led to the formation of the North Atlantic Salmon Conservation Organisation (NASCO). The primary aim of NASCO is to contribute through international consultation and cooperation to the conservation, restoration, enhancement and rational management of salmon stocks. This organization consists of representatives from all North Atlantic countries with Atlantic salmon interests.

Salmonids, and to a lesser extent sturgeon, are cultured commercially. Although the AQUACULTURE of both species has increased in importance over the last 20 years, this has had dramatic effects on the conservation of these fish. The most positive effect of anadromous fish aquaculture has been the reduction in the need to exploit wild stocks. However, wild stocks still appear to be declining and suggested reasons for this include: poor juvenile habitat and WATER QUALITY during the freshwater stage; the transmission of diseases and parasites from aquaculture sites; and the genetic dilution of indigenous stocks by aquaculture escapees. *See also* COMPLEX LIFE CYCLES; CONSERVATION BIOLOGY. [C.W.B.]

REFERENCE

McDowall, R.M. (1987) The occurrence and distribution of diadromy among fishes. *American Fisheries Society Symposium* **1**, 1–13.

anaerobic decomposition The partial disintegration of dead organic matter not involving atmospheric OXYGEN. The process yields less energy than its aerobic equivalent, and the products are only partially oxidized. *Cf.* AEROBIC DECOMPOSITION. [P.O.]

anaerobic metabolism Metabolism in the absence of OXYGEN. Anaerobic bacteria can oxidize organic matter, e.g. by utilizing the oxygen found in sulphate and NITRATE radicals, forming highly reduced compounds such as METHANE, hydrogen sulphide and AMMONIA. The most important anaerobic reactions are nitrate reduction (NO_3^- to N_2), MANGANESE reduction (MnO_2 to Mn^{2+}), IRON reduction ($Fe(OH_3)$ to Fe^{2+}), sulphate reduction (SO_4 to H_2S), methane production and fermentation. The reactions occur in sequence such that those reactions yielding the most energy occur first (or at shallowest depth in a SEDIMENT). *Cf.* AEROBIC RESPIRATION. [V.F.]

anagenesis Evolutionary change within a LINEAGE. It is often contrasted with CLADOGENESIS, evolutionary change that results in SPECIATION and the branching of phylogenetic trees. This dichotomy does not consider reticulate evolutionary change. The concept was popularized by B. Rensch to describe progressive EVOLUTION towards 'higher forms', a view of evolutionary change that is now discredited. [D.A.B.]

REFERENCE

Rensch, B. (1959) *Evolution Above the Species Level*. Columbia University Press, New York.

analogy The term analogy is used to classify characters that have evolved to perform a similar function in distantly related taxa. This functional change is accompanied by superficially similar structural changes in the characters associated with the performance of the function in question. Birds' wings and insects' wings are a classical example of analogy because the two types of wings serve the same function (flying), are superficially similar morphologically (elongated, elliptical, thin, movable, appendage-like extensions), and they are found in distantly related taxa. In modern usage, analogy implies functional HOMOPLASY but need not imply structural homoplasy. For example, birds' wings and insects' wings are analogous in the sense that they are independently evolved structures for flying; hence, they show functional convergence. They are, however, not examples of structural convergence because they are evolutionary modifications of vastly different character systems. Analogy is thus not the opposite of HOMOLOGY, but refers to convergent functional evolution (CONVERGENT EVOLUTION) from widely divergent structural backgrounds. Hence, the function is always homoplasious, but the structures are neither homologous nor homoplasious because they are derived from incommensurate characters. [D.R.B.]

analysis of covariance (ANCOVA) A procedure closely related to that of ANALYSIS OF VARIANCE in which a group of independent variables (y) are tested for homogeneity of their means, after they have been adjusted for the group's differences in the independent variable (x). The analysis therefore involves the comparison of several regression lines. Slopes are assumed equal and y-intercepts are compared (Sokal & Rohlf 1995). [B.S.]

REFERENCE

Sokal, R.R. & Rohlf, F. (1995) *Biometry*, 3rd edn. W.H. Freeman, New York.

Analysis of variance (ANOVA)

A technique for comparing the difference between several SAMPLE means by analysing the VARIANCE in the total data. It is a parametric technique (*see* PARAMETRIC STATISTICS) and assumes that observations are normally distributed (*see* NORMAL DISTRIBUTION), with approximately equal variances in all samples. If these two conditions are not met they can frequently be obtained by TRANSFORMATION OF THE DATA. Consider the example data below:

	Sample 1	Sample 2	Sample 3	Sample 4
	4	3	10	11
	4	3	10	11
	5	4	11	12
	8	7	14	15
	9	8	15	16
Mean ($\bar{x}$)	6	5	12	13
Variance (s^2)	5.5	5.5	5.5	5.5

Casual observation would suggest that there are significant differences between the four sample means, with samples 1 and 2 having lower values than samples 3 and 4. Each sample has the same variance and therefore the average variance within a sample is 5.5. However, the variance of the whole data set ($s^2 = 17.79$) is much higher, because of the difference between the average size of the observations in some samples compared to others. This is how

ANOVA compares several sample means. The statistical test used in ANOVA is the F TEST, used as a ONE-TAILED TEST. The test is analogous to a $z(d)$ TEST or t TEST used to compare the difference between just two means, and in fact an ANOVA carried out on two samples will indicate the same significance. With the two sample tests, a measure of the difference between the two means (simply $\bar{x}_1 - \bar{x}_2$) is divided by an appropriate measure of variation to obtain the test statistic (z or t). With ANOVA the appropriate measure of difference between several means is not a simple difference but the SUM OF SQUARES of the differences between individual sample means and the overall MEAN. This sum of squares ($SS_{treatment}$), is corrected for the number of means, by dividing by $n-1$, to give a mean-square or variance ($MS_{treatment}$). If there was no difference between the several sample means, this $MS_{treatment}$ would be zero. This measure of the overall difference between the several sample means is then divided by a pooled estimate of the variances within the samples ($MS_{residual}$) to finally give the test statistic F.

Of course it would be possible to compare several sample means using a z test or a t test. However, if many means are involved the number of comparisons would be great and the calculations rather tedious. For example, with 7 samples there would be 21 possible comparisons to be made. More importantly, if a significant difference between any pair of means was accepted at the 5% level, the PROBABILITY of not making a mistake (TYPE I ERROR) would be 0.95. With 7 means, and therefore 21 comparisons, the probability of not making any mistakes is $0.95^{21} = 0.34$. There would therefore be a 66% chance of making at least one type 1 error if such multiple comparisons were carried out between seven sample means. ANOVA is therefore the correct statistical procedure when more than two sample means are involved.

Analysis of variance is a large subject and readers should look at Sokal and Rohlf (1995) for a more detailed treatment. As an introduction, three examples are given below.

One-way ANOVA

This is used for any number of samples, subjected to different levels of one 'treatment' (the word treatment is frequently used in ANOVA because it was originally developed to analyse agricultural experiments in which different plots or fields were given different fertilizer treatments). Samples can contain different numbers of observations. The data below show the number of eggs laid by female flies (*Drosophila simulans*), in a 24-hour period, on fruit (a single grape) in the laboratory. The single 'treatment' is temperature and the levels of the 'treatment' are 15°C, 20°C and 25°C.

	15°C (Sample 1)	20°C (Sample 2)	25°C (Sample 3)	Total
	10	11	12	
	11	10	14	
	7	12	12	
	9	10	13	
	10	13	10	
n	5	5	5	15
$\bar{x}$	9.4	11.2	12.2	10.9
s^2	2.30	1.70	2.19	3.20
Σx	47	56	61	164
Σx^2	451	634	753	1838

Before carrying out the one-way ANOVA it is necessary to check that the variances are equal. To do this we need only confirm that the smallest and largest variance in the set are not significantly different. This is known as the F_{max} test. In the above data, the 15°C sample has the largest variance and the 20°C sample the smallest. Therefore $F_{max} = 2.30/1.70 = 1.35$. Looking at STATISTICAL TABLES of F_{max} we find that this calculated value is less than the critical value of 15.5 for number of samples = 3 and DEGREES OF FREEDOM = $n-1=4$, and therefore we can assume that the variances are homogeneous.

The next stage is to calculate various sums of squares. In the formulas below, the subscripts 1, 2, 3 and T indicate that the calculations are carried out using observations from samples 1, 2, 3 and the total data respectively. First we calculate the total sum of squares using the formula:

$$SS_{total} = \sum x_T^2 - \frac{\left(\sum x_T\right)^2}{n_T}$$

$$SS_{total} = 1838 - \frac{164^2}{15}$$

$$SS_{total} = 44.94$$

The next stage is to calculate the between-treatment sum of squares using the formula:

$$SS_{treatment} = \frac{\left(\sum x_1\right)^2}{n_1} + \frac{\left(\sum x_2\right)^2}{n_2} + \frac{\left(\sum x_3\right)^2}{n_3} - \frac{\left(\sum x_T\right)^2}{n_T}$$

$$SS_{treatment} = \frac{47^2}{5} + \frac{56^2}{5} + \frac{61^2}{5} - \frac{164^2}{15}$$

$$SS_{treatment} = 20.14$$

The final sum of squares to be calculated is the within-treatment or residual sum of squares. Sometimes this is referred to as the error sum of squares but this is misleading since it is not necessarily due to error. This sum of squares is the sum of the separate sum of squares for each sample. It is calculated as:

$$SS_1 = \sum x_1^2 - \frac{\left(\sum x_1\right)^2}{n_1} = 9.2$$

$$SS_2 = \sum x_2^2 - \frac{\left(\sum x_2\right)^2}{n_2} = 6.8$$

$$SS_3 = \sum x_3^2 - \frac{\left(\sum x_3\right)^2}{n_3} = 8.8$$

$$SS_{residual} = 9.2 + 6.8 + 8.8 = 24.8$$

It is usual to display these sums of squares in a table along with their respective degrees of freedom (v). Total $v = n_T - 1$, treatment v = number of treatments − 1, and residual $v = (n_1 - 1) + (n_2 - 1) + (n_3 - 1)$. Notice that treatment and residual SS, and treatment and residual v, can be added to obtain the total equivalents. Dividing $SS_{treatment}$ and $SS_{residual}$ by their respective degrees of freedom will give $MS_{treatment}$ and $MS_{residual}$, and these are also displayed in the table. Dividing $MS_{treatment}$ by $MS_{residual}$ produces the test statistic F.

The completed ANOVA table for the *Drosophila* data will therefore look like:

	SS	v	MS	F
Treatment	20.14	2	10.07	4.86
Residual	24.80	12	2.07	
Total	44.94	14		

Looking at tables of F (one-tailed test) we find that the calculated value of 4.86 exceeds the critical 5% value of 3.98 for $v=2$ and 12. It does not exceed the 1% value of 6.91. Therefore, there is at least one significant difference between the three means in the above data. Temperature is having a significant effect on the number of eggs that female *Drosophila simulans* lay.

If the test statistic, F, is not significant then the analysis stops at this point. Such a result would mean that there were no significant differences between any of the sample means. However, if the test statistic is significant (as in the above example) then it is desirable to make further comparisons to see which means are different. Such tests, carried out after the collection of the data and the analysis, are called a posteriori, or unplanned comparisons. The development of such methods has occupied statisticians since the 1950s and many different methods have been developed, but no complete consensus has been achieved on which method is the best. For a detailed account of these a posteriori tests see Sokal and Rohlf (1995). Most of these tests involve a quantity called the LEAST SIGNIFICANT DIFFERENCE (LSD), which is equal to 'a critical value' multiplied by 'a STANDARD ERROR'. The appropriate SE is derived from the $MS_{residual}$ value in the ANOVA table. One such test is the Tukey test, in which the 'critical value' is obtained from a statistical table of 'Tukey values' for the appropriate number of samples (3 in the above data) and degrees of freedom (residual $v=12$ in the above data). For the egg-laying *Drosophila* this critical value (5%), in the Tukey table, is 3.77. The appropriate $SE=\sqrt{(MS_{residual}/n)}=\sqrt{(2.07/5)}=0.64$. The LSD is therefore $3.77\times0.64=2.42$. Any two means that differ by at least this amount are significantly different at the 5% level. With the *Drosophila* egg-laying data at three temperatures this implies that only the 15°C mean and 25°C mean are significantly different. The result of such an a posteriori test is frequently summarized as follows:

15°C	20°C	25°C
9.4	11.2	12.2
______	______	
	______	______

with the means arranged in increasing size and the underlined groups indicating non-significance.

Two-way ANOVA

This is used for any number of samples, subjected to different levels of two 'treatments'. However, samples must contain equal numbers of observations; this is an important difference between one-way ANOVA and two-way ANOVA. The data below show the number of eggs laid by female flies (*Drosophila phalerata*), in a 24-hour period, on fungi in the laboratory. One 'treatment' is temperature, with three levels of the 'treatment' (10°C, 15°C and 20°C) and the other 'treatment' is species of fungus, with two levels of the 'treatment' (*Phallus impudicus*, the stinkhorn, and *Agaricus bisporus*, the domestic mushroom).

Phallus impudicus	10°C (Sample 1)	15°C (Sample 2)	20°C (Sample 3)	Total
	5	10	6	
	10	4	8	
	7	8	10	
	2	9	7	
	5	7	6	
n	5	5	5	15
$\bar{x}$	5.8	7.6	7.4	
s^2	8.70	5.29	2.79	
Σx	29	38	37	104
Σx^2	203	310	285	798
Agaricus bisporus	(Sample 4)	(Sample 5)	(Sample 6)	
	2	3	3	
	7	4	0	
	4	5	7	
	7	9	7	
	4	6	2	
n	5	5	5	15
$\bar{x}$	4.8	5.4	3.8	
s^2	4.7	5.3	9.7	
Σx	24	27	19	70
Σx^2	134	167	111	412
n	10	10	10	30
Σx	53	65	56	174
Σx^2	337	477	396	1210

Before carrying out the two-way ANOVA we must again check for homogeneity of variances using the F_{max} test. The largest variance is 9.70, and the smallest is 2.79. The ratio F_{max} is therefore 9.70/2.79 = 3.48. Looking at statistical tables of F_{max} we find that this calculated value is less than the critical value of 29.5 for number of samples = 6 and degrees of freedom = $n-1=4$, and therefore we can assume that the variances are homogeneous.

The next stage, as in the one-way ANOVA, is to calculate the various sums of squares. In the formulas below, the subscripts 1, 2, 3, . . . , 6 and T indicate that the calculations are carried out using observations from samples 1, 2, 3, . . . , 6 and the total data respectively. First we calculate the total sum of squares using the formula:

$$SS_{total} = \sum x_T^2 - \frac{\left(\sum x_T\right)^2}{n_T}$$

$$SS_{total} = 1210 - \frac{174^2}{30}$$

$$SS_{total} = 200.80$$

The next stage is to calculate the between-samples sum of squares using the formula:

$$SS_{samples} = \frac{\left(\sum x_1\right)^2}{n_1} + \frac{\left(\sum x_2\right)^2}{n_2} + \frac{\left(\sum x_3\right)^2}{n_3} + \frac{\left(\sum x_4\right)^2}{n_4} + \frac{\left(\sum x_5\right)^2}{n_5} + \frac{\left(\sum x_6\right)^2}{n_6} - \frac{\left(\sum x_T\right)^2}{n_T}$$

$$SS_{samples} = \frac{29^2}{5} + \frac{38^2}{5} + \frac{37^2}{5} + \frac{24^2}{5} + \frac{27^2}{5} + \frac{19^2}{5} - \frac{174^2}{30}$$

$$SS_{samples} = 54.80$$

The next stage is to calculate the between-temperature (columns) sum of squares using the formula:

$$SS_{temperature} = \frac{\left(\sum x_{column\,1}\right)^2}{n_{column\,1}} + \frac{\left(\sum x_{column\,2}\right)^2}{n_{column\,2}} + \frac{\left(\sum x_{column\,3}\right)^2}{n_{column\,3}} - \frac{\left(\sum x_T\right)^2}{n_T}$$

$$SS_{temperature} = \frac{53^2}{10} + \frac{65^2}{10} + \frac{56^2}{10} - \frac{174^2}{30}$$

$$SS_{temperature} = 7.80$$

The next stage is to calculate the between-fungi (rows) sum of squares using the formula:

$$SS_{fungi} = \frac{\left(\sum x_{row\,1}\right)^2}{n_{row\,1}} + \frac{\left(\sum x_{row\,2}\right)^2}{n_{row\,2}} - \frac{\left(\sum x_T\right)^2}{n_T}$$

$$SS_{fungi} = \frac{104^2}{15} + \frac{70^2}{15} - \frac{174^2}{30}$$

$$SS_{fungi} = 38.53$$

The last two sums of squares ($SS_{temperature}$ and SS_{fungi}) give an indication of the effect of different temperatures and different fungi on the egg-laying behaviour of these flies. If $SS_{temperature} = 0$ then there is no difference in the numbers of eggs laid at the three different temperatures, and if $SS_{fungi} = 0$ then there is no difference in the numbers of eggs laid on the two different fungi. However, in addition to these pure effects of either temperature or fungi there may well be interactive effects between temperature and fungi. That is, the effect of one will depend upon the level of the other. This interactive effect is indicated by the interaction sum of squares ($SS_{interaction}$). This is calculated as:

$$SS_{interaction} = SS_{samples} - SS_{temperature} - SS_{fungi}$$

$$SS_{interaction} = 54.80 - 7.80 - 38.53$$

$$SS_{interaction} = 8.47$$

The final sum of squares to be calculated is again the within-treatment, or residual sum of squares. It is calculated as:

$$SS_1 = \sum x_1^2 - \frac{\left(\sum x_1\right)^2}{n_1} = 203 - \frac{(29)^2}{5} = 34.8$$

$$SS_2 = \sum x_2^2 - \frac{\left(\sum x_2\right)^2}{n_2} = 310 - \frac{(38)^2}{5} = 21.2$$

$$SS_3 = \sum x_3^2 - \frac{\left(\sum x_3\right)^2}{n_3} = 285 - \frac{(37)^2}{5} = 11.2$$

$$SS_4 = \sum x_4^2 - \frac{\left(\sum x_4\right)^2}{n_4} = 134 - \frac{(24)^2}{5} = 18.8$$

$$SS_5 = \sum x_5^2 - \frac{\left(\sum x_5\right)^2}{n_5} = 167 - \frac{(27)^2}{5} = 21.2$$

$$SS_6 = \sum x_6^2 - \frac{\left(\sum x_6\right)^2}{n_6} = 111 - \frac{(19)^2}{5} = 38.8$$

$$SS_{residual} = 34.8 + 21.2 + 11.2 + 18.8 + 21.2 + 38.8 = 146$$

The degrees of freedom appropriate for all these sums of squares are:

$$\text{Total} = n - 1 = 29$$

$$\text{Samples} = \text{Number of samples} - 1 = 5$$

$$\text{Temperatures} = \text{Number of temperatures} - 1 = 2$$

$$\text{Fungi} = \text{Number of fungi} - 1 = 1$$

$$\text{Interaction} = (\text{Number of temperatures} - 1) \times (\text{Number of fungi} - 1) = 2$$

$$\text{Residual} = (n_1 - 1) + (n_2 - 1) + (n_3 - 1) + (n_4 - 1) + (n_5 - 1) + (n_6 - 1) = 24$$

The results of this analysis are again conveniently displayed in a table, which also contains the various mean-squares (variances) obtained by dividing a sum of squares by its corresponding degrees of freedom. The completed ANOVA table for the two-way analysis would therefore look like:

	SS	v	MS	F
Temperature	7.80	2	3.90	0.64
Fungi	38.53	1	38.53	6.33
Interaction	8.47	2	4.23	0.70
Samples	54.80	5		
Residual	146.00	24	6.08	
Total	200.80	29		

Looking at tables of F (one-tailed test) we find that the calculated value of 6.33 exceeds the critical 5% value of 4.26 for $v = 1$ and 24.

It does not exceed the 1% value of 7.82. Therefore there is a significant difference between the two means for the fungal treatment. The species of fungus available to a female is having a significant effect on the number of eggs that *Drosophila phalerata* lays. There is no effect of temperature on this species and there is no interactive effect between temperature and fungal species. That is, there is the same difference between fungal species at all three temperatures. As with one-way ANOVA, the initial analysis can be followed up by an a posteriori test that makes specific comparisons between means. In this particular example, since the only significant effect involved only two means such an a posteriori test is not necessary. Two-way ANOVA can be extended to three-way, four-way, etc., so that the effect of a large number of independent treatments can be analysed simultaneously (see Sokal & Rohlf 1995).

Nested ANOVA

This is used for any number of samples, subjected to different levels of several 'treatments'. However, unlike two-way ANOVA, the treatments are nested within each other so that within samples at one level of treatment, other treatments may be identified. The data below show the number of eggs laid by female flies (*Drosophila simulans*), in a 24-hour period, on fruit (a single grape) in the laboratory. The single 'treatment' is temperature and the levels of the 'treatment' are 15°C, 20°C and 25°C. However, unlike the data for the one-way ANOVA, each female (there are nine females in all) was observed for five consecutive days. Therefore, within each temperature, we have information on another factor that may influence number of eggs laid — the specific female involved. Of course, this effect is different in kind from temperature. Temperature is a planned treatment and can be repeated, 'difference between females' is not.

Before carrying out the nested ANOVA we must again check for homogeneity of variances using the F_{max} test. The largest variance is 0.8, and the smallest is 0.2. The ratio F_{max} therefore = 4. Looking at statistical tables of F_{max} we find that this calculated value is much less than the critical value of 41.1 for number of samples = 9 and degrees of freedom = $n - 1 = 4$, and therefore we can assume that the variances are homogeneous.

The next stage, as in the one-way ANOVA and two-way ANOVA, is to calculate the various sums of squares. In the formulas below, the subscripts 1, 2, 3, . . . , 9 and T indicate that the calculations are carried out using observations from samples (females) 1, 2, 3, . . . , 9 and the total data respectively. The subscripts 15, 20 and 25 indicate that the calculations are carried out using observations from the three temperatures. First we calculate the total sum of squares using the formula:

$$SS_{total} = \sum x_T^2 - \frac{\left(\sum x_T\right)^2}{n_T}$$

$$SS_{total} = 5404 - \frac{488^2}{45}$$

$$SS_{total} = 111.91$$

The next stage is to calculate the between-treatment (temperature) sum of squares using the formula:

$$SS_{treatment} = \frac{\left(\sum x_{10}\right)^2}{n_{10}} + \frac{\left(\sum x_{15}\right)^2}{n_{15}} + \frac{\left(\sum x_{20}\right)^2}{n_{20}} - \frac{\left(\sum x_T\right)^2}{n_T}$$

$$SS_{treatment} = \frac{139^2}{15} + \frac{164^2}{15} + \frac{184^2}{15} - \frac{488^2}{45}$$

$$SS_{treatment} = 46.12$$

As with one-way ANOVA, the total sum of squares (SS_{total}) can be partitioned into the between-treatment sum of squares ($SS_{treatment}$) and the residual sum of squares ($SS_{residual}$). Therefore $SS_{residual} = SS_{total} - SS_{treatment} = 111.91 - 46.12 = 65.79$. However, with nested

	15°C			20°C			25°C			
	1	2	3	4	5	6	7	8	9	Total
	10	8	10	10	13	11	14	12	12	
	10	9	11	9	12	10	13	12	11	
	9	8	10	10	12	11	12	13	11	
	9	8	10	10	12	11	14	13	11	
	9	9	10	9	12	12	13	12	11	
n	5	5	5	5	5	5	5	5	5	
$\bar{x}$	9.4	8.4	10.2	9.6	12.2	11.0	13.2	12.4	11.2	
s^2	0.3	0.3	0.2	0.3	0.2	0.5	0.8	0.3	0.2	
Σx	47	42	51	48	61	55	66	62	56	
Σx^2	443	354	521	462	745	607	874	770	628	
n		15			15			15		45
$\bar{x}$		9.3			10.9			12.3		10.8
s^2		0.9			1.4			1.1		2.5
Σx		139			164			184		488
Σx^2		1301			1814			2272		5404

ANOVA the residual sum of squares (perhaps better called the within-treatment sum of squares) can be further partitioned. We continue the analysis by calculating the sum of squares within each female.

$$SS_1 = \sum x_1^2 - \frac{\left(\sum x_1\right)^2}{n_1} = 443 - \frac{(47)^2}{5} = 1.2$$

$$SS_2 = \sum x_2^2 - \frac{\left(\sum x_2\right)^2}{n_2} = 354 - \frac{(42)^2}{5} = 1.2$$

$$SS_3 = \sum x_3^2 - \frac{\left(\sum x_3\right)^2}{n_3} = 521 - \frac{(51)^2}{5} = 0.8$$

$$SS_4 = \sum x_4^2 - \frac{\left(\sum x_4\right)^2}{n_4} = 462 - \frac{(48)^2}{5} = 1.2$$

$$SS_5 = \sum x_5^2 - \frac{\left(\sum x_5\right)^2}{n_5} = 745 - \frac{(61)^2}{5} = 0.8$$

$$SS_6 = \sum x_6^2 - \frac{\left(\sum x_6\right)^2}{n_6} = 607 - \frac{(55)^2}{5} = 2.0$$

$$SS_7 = \sum x_7^2 - \frac{\left(\sum x_7\right)^2}{n_7} = 874 - \frac{(66)^2}{5} = 2.8$$

$$SS_8 = \sum x_8^2 - \frac{\left(\sum x_8\right)^2}{n_8} = 770 - \frac{(62)^2}{5} = 1.2$$

$$SS_9 = \sum x_9^2 - \frac{\left(\sum x_9\right)^2}{n_9} = 628 - \frac{(56)^2}{5} = 0.8$$

The total sum of squares within all females ($SS_{\text{within female}}$) (an estimate of the variation in egg laying from day to day) is therefore 1.2 + 1.2 + 0.8 + 1.2 + 0.8 + 2.0 + 2.8 + 1.2 + 0.8 = 12. The sum of squares within treatments (SS_{residual}) is made up from two components, a sum of squares between females ($SS_{\text{between females}}$) (within temperatures) and a sum of squares within females ($SS_{\text{within females}}$) calculated above. Therefore $SS_{\text{between females}} = SS_{\text{residual}} - SS_{\text{within females}} = 65.79 - 12 = 53.79$.

The degrees of freedom appropriate for all these sums of squares are:

Total $= n - 1 = 44$

Temperatures = Number of temperatures $- 1 = 2$

Females (within temperatures) = Number of temperatures

$\times$ (Number of females within temperatures $- 1) = 3(3-1) = 6$

Within females = Number of females $\times$ (Number of days $- 1)$

$= 9(5-1) = 36$

The results of this analysis are again conveniently displayed in a table, which also contains the various mean-squares (variances) obtained by dividing a sum of squares by its corresponding degrees of freedom. The completed ANOVA table for the two-way analysis would therefore look like:

		SS	v	MS	F
	Between temperature	46.12	2	23.06	2.57
Within temperature	Between females	53.79	6	8.96	27.15
	Within females (days)	12.00	36	0.33	
	Total	111.91	44		

The first source of variation to be tested is that between females (within temperatures). It is important to remember the 'within temperature' since if we simply examined differences between the nine females this would include the effect of temperature also. We examine this difference between females (within temperature) by dividing the $MS_{\text{between females}}$ by the $MS_{\text{within females}}$. The F value of 27.15 is very significant (critical 1% value for $v = 6$ and $36 = 3.36$). The second source of variation to be tested is that between temperatures. We examine this difference by dividing the $MS_{\text{treatments}}$ by the $MS_{\text{between females}}$. The F value of 2.57 is not significant (critical 5% value for $v = 2$ and $6 = 5.14$). This indicates that there are significant differences between females (within temperature) in number of eggs laid, over and above that observed from day to day, and that temperature does not significantly add to this variation. Levels of nesting within nested ANOVA are not limited to one as in the example above (see Sokal & Rohlf 1995). *See also* ANALYSIS OF COVARIANCE; STATISTICS. [B.S.]

REFERENCE

Sokal, R.R. & Rohlf, F.J. (1995) *Biometry*, 3rd edn. W.H. Freeman, New York.

ancestral homology *See* PLESIOMORPHY.

ancient woodland/forest In Britain, WOODLAND which existed before 1600 (Peterken 1993) or before 1700 (Rackham 1980). Some ancient woodland has since been cleared and some survives in the modern landscape, having existed continuously since 1600. The sites and shapes of vanished ancient woods often survive in the modern field pattern with their original margin of mixed shrubs forming a hedge. All PRIMARY WOODLAND is ancient woodland, but ancient woodland also includes SECONDARY WOODLAND originating before 1600 (Fig. A14). The proportion of ancient woodland that is primary varies from region to region, but in some southern parts of Britain most ancient woods are secondary. The threshold date of 1600 is chosen to pre-date the main era of planting (so most ancient woods originated naturally) and to coincide with the general availability of maps (so there is a reasonable chance that individual woods can be proved to be ancient, or otherwise). *See also* OLD-GROWTH/ANCIENT FORESTS, CONSERVATION. [G.F.P.]

REFERENCES

Peterken, G.F. (1993) *Woodland Conservation and Management*, 2nd edn. Chapman & Hall, London.

Rackham, O. (1980) *Ancient Woodland*. Edward Arnold, London.

Andisols *See* SOIL CLASSIFICATION.

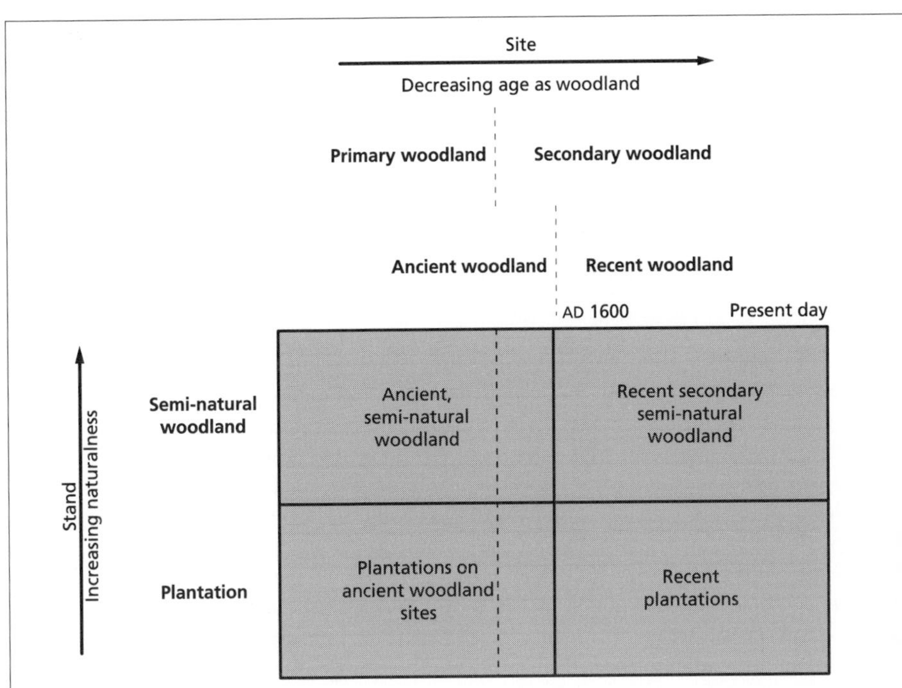

Fig. A14 Relationship between the terms primary, ancient, secondary and semi-natural as applied to woodland in Britain.

Andosols *See* SOIL CLASSIFICATION.

androdioecy A species is said to be androdioecious when some or all populations of the species contain both male (hence the prefix 'andro') and HERMAPHRODITE individuals (broad sense) (hence the suffix 'dioecious', referring to the existence of two sexes). In contrast to GYNODIOECY, in which genetic determinism of sex has been firmly established in a number of cases, there are very few studied examples of genetically based sex in androdioecious species. *See also* SEX ALLOCATION THEORY. [I.O.]

androgenic *See* ENDOCRINE DISRUPTER.

andromonoecy *See* HERMAPHRODITE.

anemometer An instrument for measuring the speed of the WIND. The traditional instrument is the cup anemometer, a set of cups on horizontal arms which are free to rotate on a vertical spindle. The rotation rate is recorded, either by a mechanical counter or by logging an electric signal as a voltage or as a train of pulses. Cup anemometers measure only the horizontal component of wind, and do not rotate at very low wind speed, but they are robust and still widely used for general survey.

Propeller, or vane anemometers were originally used for measuring airflow in ducts, but may also be used in the natural environment. Like cup anemometers, they stall at low wind speeds, but the most sensitive types have jewelled bearings and their stalling speed may be as low as $0.1\ \mathrm{m\,s^{-1}}$.

Sonic anemometers have no moving parts, but measure the air movement from the time of flight of pulses of ultrasound from a source to a receiver. They may be configured to measure the vertical, horizontal stream-wise, and horizontal cross-wind components of wind, and therefore to enable the wind vector to be found. Currently, they are mainly research instruments but are likely to become the standard instrument in meteorological stations in the future. They are especially useful in measurements of TURBULENCE (*see* EDDY COVARIANCE).

Hot-wire anemometers are made of a short length of platinum resistance wire stretched between two prongs to form a miniature probe. The wire is heated by a current, and when the wind flows the wire cools and undergoes a change in electric resistance. This may be detected electrically, and modern instruments produce a voltage proportional to the wind speed. These are fragile instruments, mainly used in research applications in WIND TUNNELS. However, variants for use in the natural environment are also available, for example: (i) where the platinum wire is replaced by a thermistor; and (ii) where the wire is replaced by a film deposited on a sphere, to make an omnidirectional anemometer.

Pressure tube anemometers are simply tubes that point into the wind, enabling the PRESSURE fluctuations caused by the wind to be registered on a chart. They are commonly used in meteorological stations. A similar device, the PITOT TUBE, is used to calibrate wind tunnels. *See also* METEOROLOGY. [J.G.]

REFERENCE

Grace, J. (1989) Measurement of wind speed near vegetation. In: *Physiological Plant Ecology: Field Methods and Instrumentation* (eds R.W. Pearcey, J. Ehleringer, H.A. Mooney & P. Rundel), pp. 57–74. Chapman & Hall, London.

aneuploidy The usual definition is possession of a chromosome number that is not an exact multiple of the usual haploid (HAPLOIDY) number. However, some individuals may have a reduced number of chromosomes through Robertsonian fusions, without having a reduced number of arms. Such individuals are not described as aneuploids. Thus a better definition might be possession of a number of chromosomal arms that is not an exact multiple of the usual haploid number of arms ('fundamental number'). [I.O.]

Angiosperms, global diversity and conservation

Flowering plants (angiosperms) are major components of the world's flora and its VEGETATION. Considered the culmination of evolution in the plant kingdom, they originated about 130 million years ago and evolved very rapidly, leading to an enormous progressive increase in plant DIVERSITY to the present day. The angiosperms are seed-producing vascular plants of extremely varied form; many are valuable as sources of food and other commodities. Dominant in virtually all plant communities—the exceptions being sphagnum BOGS and the BOREAL region — they characterize BIOMES and, although mostly terrestrial mesophytes, range from desert xerophytes to aquatics (both freshwater and seawater). Major vegetation types include: many kinds of FOREST (e.g. TROPICAL RAIN FOREST, DECIDUOUS FOREST, sclerophyll forest); SAVANNAH and parkland GRASSLANDS, STEPPES, PRAIRIES and llanos; WETLANDS and coastal vegetation including MANGROVES.

In lifespan, angiosperms vary from a few weeks (EPHEMERALS), 1 year (ANNUALS), 2 years (BIENNIAL PLANTS) to more than 2 years (PERENNIALS); some perennials live many hundreds of years (e.g. oak, *Quercus robur*). In size, angiosperms range from minute aquatics (e.g. duckweeds, *Wolffia* and *Lemna*), to trees more than 100 m tall and of enormous girth (e.g. the eucalypts). Among the great diversity of forms are thalloid representatives resembling algae or mosses (the tropical aquatics of the Podostemaceae), herbs, shrubs and trees. Some have underground perennating structures such as bulbs, corms and rhizomes, or have a climbing habit (e.g. tropical lianas). Flower structure ranges from simple wind-pollinated forms (many trees) to elaborate insect-pollinated types (orchids, labiates). POLLINATION may also be effected by, for example, water, birds, bats and molluscs. The COEVOLUTION of flowering plants and their pollinators (notably insects) is exemplified by many fascinating associations of organisms.

The number of species of extant angiosperms is usually estimated at around 250 000 (some 180 000 dicotyledons and 70 000 monocotyledons), with nearly 17 000 genera, classified by some authorities into as many as 600 families (many hundreds of species are extinct). Largest is the advanced monocotyledonous family Orchidaceae, of COSMOPOLITAN distribution, but with many tropical EPIPHYTES; it has about 30 000 species and nearly 1000 genera.

The dicotyledonous 'natural' unit Asteraceae (Compositae, daisy family), well represented in the temperate zone, has about 1500 genera and some 20 000 species. The Fabaceae (Leguminosae, pea family), abundant in the tropics and subtropics, has over 600 genera and over 14 000 species. The grasses of the cosmopolitan family Poaceae (Gramineae), characterizing extensive grasslands of many types, and including wheat, maize and rice as well as bamboos, are classified into some 680 genera and 10 000 species.

Some families are largely tropical, including major hardwood producers, for example Ebenaceae (ebony), Meliaceae (mahogany) and the Dipterocarpaceae, abundant in Malaysia; others yield important fruits, for example Musaceae (bananas), Moraceae (figs, breadfruit), Annonaceae (custard apples) and Arecaceae (Palmae, dates).

A number of families well represented in the flora of the British Isles have a widespread distribution, mainly in the Northern hemisphere, and each contains 2000 or more species worldwide. These families, in approximately decreasing order of their number of species, include (besides Orchidaceae, Asteraceae, Fabaceae and Poaceae): Euphorbiaceae (7500 species, spurge family, mainly SUBTROPICAL); Rubiaceae (many shrubs and trees, including coffee and *Cinchona*, yielding quinine); Scrophulariaceae (foxglove family); Cyperaceae (sedges); Ericaceae (heath family); Lamiaceae (Labiatae, mint family); Brassicaceae (Cruciferae, mustard family); Rosaceae (rose family); Apiaceae (Umbelliferae, carrot family); Solanaceae (potato family); Boraginaceae (borage family); Caryophyllaceae (carnation family) and Ranunculaceae (buttercup family, one of the most primitive). Many of these families are best represented in temperate parts of the Northern hemisphere, although the Solanaceae is important in tropical South America.

In contrast to these large families, more than 30 are MONOTYPIC (having only a single species). In the British Isles, three monotypic families are represented: Butomaceae (flowering-rush); Hippuridaceae (mare's-tail) and Scheuchzeriaceae (rannoch rush). The three plants so classified are, however, by no means confined to Britain; indeed, the aquatic mare's-tail (*Hippuris vulgaris*) occurs also in the Southern hemisphere.

The diversity of higher plants increases substantially from the poles to the Equator and from high to low altitudes. The continent of Antarctica supports only two species of angiosperms; in sharp contrast, approximately two-thirds of the world's flowering plants are in the tropics, even though less than one-third of the Earth's total land surface is in the tropics. Moist tropical forests are particularly rich in species, as also are areas with Mediterranean climate.

Eighteen so-called 'hot spots', most of which are moist tropical forests or have a Mediterranean climate, occupy only some 0.5% of the Earth's land surface but support over one-fifth of the world's angiosperms. Many plants of the 'hot spots' are endemic, i.e. entirely restricted to the particular location. South America, not well explored botanically, may contain as many as one-third of present-day angiosperms, Brazil being the richest country, floristically, in the world, with an estimated 55 000 species.

Oceanic islands in or near the tropics may also have very rich floras. Madagascar supports some 10 000 species, the majority endemic; eight families here are endemic, including the Didiereaceae, with its strange tubular cactus-like forms. Although small in total, the great majority of the angiosperms of St Helena are endemic; also over a third of the flowering plants of Mauritius are endemic. Small islands, distant from land, have limited floras but high endemism, reflecting their isolation and the difficulty of long-distance DISPERSAL.

In Europe, Mediterranean countries have the richest floras (Italy, with about 5500 species is the highest), with quite high endemism. The angiosperm flora of Britain, one of the very best known in the world, is limited to about 1550 species. Indeed the angiosperm flora of the whole of Europe, estimated at approximately 12 500, is small relative to that of some tropical countries.

The well-studied North American angiosperm flora is estimated at about 18 000 species. Mexico, however, has a greater number, including strong representation, in the drier parts, of the Cactaceae, a family believed native only, or very largely, to the New World.

The flowering plants of Australia number some 15 000 species, many of which are endemic. This flora is especially rich in eucalypts and acacias, and has strong representation of xerophytic Proteaceae and Epacridaceae of ericoid nature; the grass trees (*Xanthorrhoea* spp.) are distinctive components of Australian vegetation.

Tropical and subtropical African countries have smaller angiosperm floras than those of South American countries, probably reflecting differences in size, TOPOGRAPHY, climate and history. Dry tropical and subtropical regions, however, such as the Sahel, have a much more restricted flora, with few endemics.

The Southern hemisphere, with its much smaller land surface than the Northern hemisphere, supports many fewer angiosperms than in the north, but HEATHLAND communities of the Cape Peninsula of South Africa and of southern Australia are distinctly species-rich.

Conservation

Efforts to maintain the diversity of angiosperms may be based on the ECOSYSTEM or on the individual species. The most important form of conservation of BIODIVERSITY is widely considered to be the preservation of species as part of a functioning ecosystem: a habitat change often affects species diversity; also number and association of species often affect the habitat. Such an ecosystem-based policy is especially relevant to tropical rain forests, where diversity is very high and many species are undescribed. In the developed world, a network of protected area systems is a major mechanism of management to sustain biodiversity. Legislation may prohibit collection, and land-use controls and incentives help to maintain the natural environment. The Convention on International Trade in Endangered Species of Wild Fauna and Flora (CITES), to which the UK was a signatory in 1973, prohibits or controls commercial trade of many plants. World heritage sites and BIOSPHERE RESERVES promoted through the World Conservation Union (IUCN) give some protection to many important landscapes and floras. Efforts are made to restrict DEFORESTATION through the Tropical Forestry Action Plan (TFAP) and the International Tropical Timber Agreement (ITTA).

The species-based approach to conservation seeks to protect particular plants of high priority which are threatened (those in danger of extinction in the foreseeable future, in most urgent need of conservation). This may well be best achieved by protecting the habitat by controlling land use. In Britain, more than three-quarters of the threatened plants (over 300) are in national nature reserves (NNRs) or sites of special scientific interest (SSSIs), many established by the former Nature Conservancy and now managed by its succeeding country agencies. Red Data books provide information on threatened species and their conservation. More than half of the world's botanical gardens currently play an active role in the *EX SITU* CONSERVATION of plants, but few of these gardens are in the tropics. In 1989 an international Plant Conservation Programme was launched with botanic gardens taking the leading position in plant conservation. In Britain, rare native plants have been grown in botanical gardens and contributions to conservation of angiosperms are being made on an international scale by the Conservation Unit of the Royal Botanical Gardens, Kew; for example, culture of *Trochetiopsis melanoxylon* of St Helena, once thought extinct. SEED BANKS are an effective method of conservation of plants whose seeds are suitable for long-term storage; many botanic gardens have seed banks with wild-origin seed suitable for distribution.

Rare plants can be increased clonally from cuttings, from meristems by micropropagation (especially of orchids) or by seeds, and then reintroduced into their native habitats. For example, seed held for research permitted the reintroduction of *Stephanomeria malheurensis* (Asteraceae) to its only known site in the wild in Oregon, USA, from which it had become extinct.

The Convention on Biological Diversity, supported by many governments in 1992 at Rio de Janeiro, aims to slow the loss of biological diversity by the development of strategies for conservation and the sustainable use of biological resources. The Global Environmental Facility of the Convention seeks to promote the development of national programmes of biodiversity, most needed with respect to angiosperms in tropical countries with a rich flora where the rate of extinction of species is high. *See also* ARBORETUM; AUSTRALIAN FLORAL REALM; CONSERVATION BIOLOGY; MEDITERRANEAN CLIMATE REGIONS, CONSERVATION; MEDITERRANEAN SCRUB/FOREST. [A.J.W.]

REFERENCES

Airy Shaw, H.K. (ed.) (1973) *A Dictionary of the Flowering Plants and Ferns*, 8th edn. Cambridge University Press, Cambridge.

Anon. (1994) *Biodiversity: The UK Action Plan*. Her Majesty's Stationery Office, London.

Frankel, O.H., Brown, A.H.D. & Burdon, J.J. (1995) *The Conservation of Plant Biodiversity*. Cambridge University Press, Cambridge.

Groombridge, B. (ed.) (1992) *Global Biodiversity. Status of the Earth's Living Resources*. World Conservation Monitoring Centre; Chapman & Hall, London.

Myers, N. (1990) The biodiversity challenge: expanded Hot-Spots analysis. *The Environmentalist* **10** (4), 243–255.

angular transformation *See* TRANSFORMATION OF DATA.

animal ecology *See* ECOLOGY.

animal pollination *See* POLLINATION.

anion *See* ION.

anisogamy, evolution of The production of GAMETES of two different sizes. By definition, the smaller gametes are male, while the larger gametes are female. There are three main theories to account for the EVOLUTION of anisogamy (Bulmer 1994). The first, originally proposed by Parker *et al.* (1972), is that there is DISRUPTIVE SELECTION acting on gamete size. According to Maynard Smith (1978), the primitive condition is the production of microgametes of the smallest size, and anisogamy will evolve, through invasion by macrogamete producers, when adult size is such that it is difficult or impossible for a single motile cell to grow and differentiate into the adult form. An alternative starting point, followed by Bulmer (1994), is that anisogamy evolved from an isogamous population with two mating types. The models rely on the assumptions that gametes fuse randomly in pairs to form zygotes, and that the probability of survival of the zygote depends on total mass.

The second theory is that anisogamy is a by-product of selection for uniparental, vertical transmission of cytoplasmic parasites and ORGANELLES (Bulmer 1994). Law and Hutson (1992) showed that uniparental cytoplasmic inheritance prevents the spread of parasites that are harmful to the host, whereas biparental inheritance allows it. The production of small gametes is thus viewed as an ADAPTATION to ensure that cytoplasmic particles are inherited only through female gametes and are excluded from male gametes because they have very little cytoplasm.

According to Bulmer (1994), the two theories are not exclusive, and it is likely that both factors contributed to the evolution of anisogamy.

The third theory (Hoekstra 1984) is that anisogamy evolved as a side effect of selection for differences in gamete motility resulting in sessile eggs producing pheromones and motile sperm adept at chemotaxis. *See also* SEX ALLOCATION THEORY. [I.O.]

REFERENCES

Bulmer, M. (1994) *Theoretical Evolutionary Ecology*. Sinauer Associates, Sunderland, Mass.

Hoekstra, R.F. (1984) Evolution of gamete motility differences II. Interactions with the evolution of anisogamy. *Journal of Theoretical Biology* **107**, 71–83.

Law, R. & Hutson, V. (1992) Intracellular symbionts and the evolution of uniparental cytoplasmic inheritance. *Proceedings of the Royal Society of London, Series B* **248**, 69–77.

Maynard Smith, J. (1978) *The Evolution of Sex*. Cambridge University Press, Cambridge.

Parker, G.A., Baker, R.R. & Smith, V.G.F. (1972) The origin and evolution of gamete dimorphism and the male–female phenomenon. *Journal of Theoretical Biology* **36**, 529–533.

anisometric Relating to, or characterized by, dissimilar proportions. Any two entities are isometric if all their corresponding linear measurements have the same relative proportions; if not, they are anisometric. Living things of different sizes are usually anisometric, even when organized on a similar pattern, because of mechanical or physiological CONSTRAINTS. For instance, if animals were isometric, the strength of their bones (roughly proportional to bone cross-sectional area) would only increase with the square of their body length; but because the body mass that must be supported by the limbs increases approximately with the cube of body length, the limbs of large land animals are, in fact, disproportionately massive. *See also* ALLOMETRY. [J.Kz.]

annual LIFE CYCLE over 1 year, characteristic of many plant species. Annuals are often semelparous (*see* SEMELPARITY), i.e. breed once at the end of the year and die. However, semelparous species may be subannual.

A typical annual life cycle of a plant involves germination in autumn (in Mediterranean climate) or spring (in temperate zones), flowering and seeding in spring or summer, and dying in summer. The annual life cycle is usually considered as an adaptation to summer DROUGHT (Mediterranean species) or winter cold (temperate species). Annual species are particularly abundant in Mediterranean areas and desertic zones, in which the life cycle may be particularly short (EPHEMERAL species). Although annual species flower only once and all together, many of them may have OVERLAPPING GENERATIONS through the persisting SEED BANK. Moreover, some species are 'FACULTATIVE ANNUALS' (called 'biennials'; *see* BIENNIAL PLANTS): they may or may not flower during the first year. Such facultative annuals are, in fact, PERENNIAL semelparous species, which may wait for many years before flowering and then die. *Cf.* ITEROPARITY. *See also* LIFE-HISTORY EVOLUTION. [I.O.]

annual limits on intake (ALIs) *See* MAXIMUM PERMISSIBLE LEVEL.

anoxia (hypoxia) Environmental conditions with no (or low) OXYGEN. [P.C.]

antagonism The interaction of chemicals such that when mixed together they have less joint ecotoxicological EFFECT than expected from their individual effects in isolation. *Cf.* ADDITIVITY; SYNERGISM. *See also* TOXICITY OF MIXTURES. [P.C.]

antagonistic pleiotropy hypothesis Different ALLELES may have effects on multiple characters, including opposed effects on components of Darwinian FITNESS. This situation is called 'antagonistic pleiotropy', where PLEIOTROPY is used in its conventional genetic sense of one genetic difference having multiple phenotypic effects. There are two interesting consequences of antagonistic pleiotropy. The first is that it joins diverse characters together in evolutionary genetic 'TRADE-OFFS', such that not all components of FITNESS can be maximized. The second is that the algebra that defines fitness can give rise to overdominance (*see* BALANCE OF SELECTIVE FORCES), and its higher-order analogues, when there is antagonistic pleiotropy. Such overdominance can then lead to selective maintenance of GENETIC POLYMORPHISM.

A particularly important situation involving antagonistic pleiotropy arises when some alleles have beneficial effects early in life, but deleterious effects later in life. Since the force of NATURAL SELECTION is lower at later adult ages (*see* AGEING), the deleterious effects will not play much role in determining the outcome of selection. Hence, these alleles will tend to increase in frequency, enhancing early fitness-components, but aggravating later ageing.

Antagonistic pleiotropy is not common among new MUTATIONS. However, some of the alleles that segregate at high frequencies in outbred populations appear to involve antagonistic pleiotropy. For example, outbred populations of *Drosophila melanogaster* exhibit antagonistic pleiotropy between early FECUNDITY and LONGEVITY in a wide variety of experiments, though this result is not universal. [M.R.R.]

REFERENCE

Rose, M.R. (1991) *Evolutionary Biology of Ageing*. Oxford University Press, New York.

Antarctic ozone hole The seasonal depletion of OZONE in a large area of the ATMOSPHERE over Antarctica. *See also* OZONE HOLE; STRATOSPHERIC CHEMISTRY. [P.C.]

Antarctica, ecology and conservation

Politically, Antarctica is a stateless continent with offshore islands and surrounding ocean lying south of 60°S (the Antarctic Treaty Area). It has no permanent human population but is occupied by (in 1998) 36 year-round (and several additional summer-only) research stations belonging to 17 nations, although a further nine nations undertake intermittent research programmes. The transient summer population of scientists and support personnel of this 14.5 million km^2 land is barely 2000, reducing to less than 500 in winter. In fact, the summer population is boosted four-fold by a rapidly increasing industry bringing tourists into the biome for brief visits.

Ecology

In biological terms, Antarctica comprises all land and associated ICE cover (including ice shelves attached to the land) south of latitude 60°S, but including the South Sandwich Islands (56–59°S) and Bouvetøya (54°S). Throughout this southern polar biome mean summer (December–February) air temperatures remain low, reaching 1–2°C only in the northernmost coastal areas. This, together with daily freezing and thawing cycles and almost daily drying and wetting cycles, impose severe physiological STRESS on terrestrial organisms, inhibiting metabolism, growth and reproduction, as well as the processes of COLONIZATION and community development. Only those species which possess special physiological and biochemical adaptations can tolerate the exceptionally severe living conditions and survive. Besides climatic constraints on survival, species diversity is further restricted by isolation from all other Southern hemisphere landmasses which may serve as a source of biological immigrants in the form of spores, seeds, eggs or other viable propagules capable of establishing new populations. Less than 2% of the land is ice-free in summer, greatly limiting the range and extent of habitats available for colonization. Even for the very few organisms which do reach favourable habitats within Antarctica, they must be able to develop, reproduce, disperse and colonize new sites. An individual organism which becomes established but cannot increase in numbers or in range is not successful, although such species may become important if they can respond positively to periods of exceptional living conditions, such as the current trend in increasing temperatures in some regions of Antarctica.

Because of the climatic and geographical limitations, the terrestrial BIOTA of Antarctica is greatly restricted in number of life forms and species (Table A14). The greatest diversity occurs in the maritime Antarctic, i.e. along the western side of the Antarctic Peninsula and its associated archipelagoes (South Sandwich, South Orkney and South Shetland Islands) where the climate is somewhat milder and wetter, and the growing season longer than in the major mass of continental Antarctica. This region is also closest to a biologically rich landmass (southern South America) from where most of its flora and fauna has originated. However, the eastern side of the Peninsula is much colder and drier, ice-free habitats are relatively few and the biota is correspondingly sparser and with far fewer species. The decrease in species diversity and abundance continues along a latitudinal gradient, with the main landmass of continental Antarctica supporting very few species.

Table A14 Approximate numbers of species of the major macroscopic terrestrial flora and fauna of Antarctica.

Biological group	Maritime Antarctic	Continental Antarctic: Coastal	Continental Antarctic: Inland	Total*
Plants				
Flowering plants	2	0	0	2
Mosses	85	20	5	100
Liverworts	15	1	0	15
Lichens	250	50	20	300
Algae	20	10	5	20
Cyanobacteria (blue-green algae)	10	5	3	10
Fungi (toadstools)	20	0	0	20
Invertebrates				
Flies (midges)	2	0	0	2
Mites	30	27		54
Springtails	12	11	0	22
Worms (enchytraeids)	3	0	0	3

* The total number of species is usually less than the sum of the three regions as some species are common to more than one region.

Antarctica possesses only two native flowering plants, the hairgrass, *Deschampsia antarctica*, and pearlwort, *Colobanthus quitensis*, both restricted to the maritime Antarctic south to around 68°S. There are also only two 'higher' insects, a winged midge, *Parochlus steinenii*, found only in a few localities in the South Shetland Islands, and a wingless midge, *Belgica antarctica*, occurring rarely in these islands but more frequently on the west coast of the Peninsula; both are 2–3 mm in size.

By far the greatest diversity of species comprises cryptogams (mosses, liverworts, lichens, algae, cyanobacteria and fungi) and microarthropods (mites and springtails). Individual plants of some lichen species are exceedingly old. Disc-like thalli of *Umbilicaria antarctica* can reach 30 cm diameter and are estimated to be at least 500 years old; large bushy lichens and various encrusting species probably also reach this age or much more. At the micro-level there are numerous species of algae, fungi and bacteria, as well as protozoans, rotifers, tardigrades and nematode worms living in the soil and among plants. Freshwater ponds and lakes, despite their more stable but permanently cold environment, have an even more depauperate macroscopic flora and fauna—a few species of aquatic moss, numerous filamentous algae and a small number of crustacea, the largest of which is the fairy shrimp (*Branchinecta gaini*), which, at almost 2 cm in length, is the largest non-ocean-dependent animal in Antarctica.

The species-poor biota of Antarctica creates a network of interrelated and often interconnected communities, each dependent on certain combinations of environmental conditions. These create a biologically simple ECOSYSTEM with very few trophic levels (i.e. stages in the FOOD CHAIN). Of the *c.* 400 species of plants only about a dozen mosses, perhaps a dozen lichens, a single alga and two or three cyanobacteria achieve dominance on rock and soil over areas larger than about 25 m^2. The two flowering plants usually occur together, mainly in sheltered, moist, north-facing sites close to the shore, with the grass very occasionally forming swards of a few hundred square metres. Habitats kept moist by meltwater support various communities of mosses, especially in the wetter and milder maritime Antarctic, occasionally covering a hectare or more. Small, dense moss stands also occur in suitable places around coastal continental Antarctica. Wherever the terrain is exposed to strong winds, such as windswept plateaux, ridges and rock faces, availability of free water becomes the major factor controlling species distribution and abundance. Such fellfield communities of various lichens and compact cushion-forming mosses are relatively open with much rock and unstable soil remaining uncolonized, but quite commonly cover several hectares. It is these more extensive cryptogamic communities where most of the invertebrate fauna exists, but only two species (the mite, *Alaskozetes antarcticus*, and springtail, *Cryptopygus antarcticus*) occur in dense aggregations. Most are herbivores, subsisting mainly on algae and fungi, but a very few mites are predatory, exemplifying the simplicity of the Antarctic terrestrial FOOD WEB. Where cold and DESICCATION become more severe, as in inland continental mountain ranges and nunataks, the biota is restricted almost entirely to a few species of encrusting lichens and mites. In areas of coarse sandstone an association of bacteria, fungi, algae, cyanobacteria and lichens lives from 5 to 15 mm below the rock surface (ENDOLITHIC). This remarkable community obtains its moisture from the melting of very infrequent snow or FROST when water percolates into the crystalline structure, and gets its light through the quartz crystals of the rock.

Some plant communities are particularly prominent. Coastal rocks and cliffs used as breeding sites by penguins and various petrels are inundated with an AEROSOL of nitrogenous mist derived from the guano of the large aggregations of these seabirds. The rock is typically covered by numerous brightly coloured lichens (especially orange species of *Caloplaca* and *Xanthoria*), the existence and growth of which are dependent on high concentrations of ammonium-NITROGEN. Similar lichen communities occur on inland nunataks where there are colonies of snow and Antarctic petrels. In the maritime Antarctic, two species of moss (*Chorisodontium aciphyllum* and *Polytrichum alpestre*) are solely responsible for accumulating PEAT to depths of up to 2.5 m in the form of mounds raised above the ground surface, usually with eroding vertical edges. These are permanently frozen below the top 25 cm, and the bases of some of these have been radiocarbon-dated at up to 5500 years old. In lowland coastal areas adjacent to large aggregations of penguins and seals, trampling and toxic levels of nutrients in their excreta prevent all but the green alga *Prasiola crispa* from colonizing the disturbed terrain. In mid–late summer, many hectares of melting icefields in the maritime Antarctic often become strongly coloured by dense aggregations of unicellular red and green snow algae.

Mosses and lichens are exceptionally tolerant of extreme environmental conditions such as frequent freeze–thaw and drying and wetting cycles during the growing season. These plants, like the microarthropods, are well adapted to survive such physiologically stressful microenvironments. They contain within their cells many chemical substances (mainly CARBOHYDRATES and PROTEINS) which act as antifreeze protectants, preventing the cellular fluids from freezing and rupturing the cell walls, thereby killing the organism. While many of the arthropods can survive temperatures of −30°C or less in a dormant phase, many plants endure much lower temperatures in habitats unprotected by snow cover. Some lichens remain photosynthetically active down to −17°C, and mosses to around −10°C. While the fauna can retreat into safe niches when environmental conditions become life-threatening, plants must endure whatever they are exposed to. High levels of ultraviolet-B (UV-B) radiation can be potentially damaging to life processes, but most cryptogams possess dark photoprotective pigments which prevent physiological or cellular damage.

Conservation

The simple, virtually pristine, terrestrial ecosystem is highly adapted to the stresses (*see* STRESS) imposed by the Antarctic environment. Its component communities are very fragile and sensitive to change and perturbation, for example climatic warming especially in summer, increasing UV-B resulting from OZONE depletion in spring, and a rapidly increasing population of fur seals which devastate vegetation in the maritime Antarctic. For these and other reasons, protection of the Antarctic environment has been foremost in the policy of the Antarctic Treaty since it came into force in 1961; indeed, the entire Treaty Area was designated a 'special conservation area'. Although lacking any strict legal conditions or any form of policing, such protection has been remarkably successful under the Agreed Measures for the Conservation of Antarctic Fauna and Flora. In 1991 this was superseded by the Protocol on Environmental Protection. The Protocol, when adopted by all national governments involved in Antarctic activities, will play a crucial role in maintaining the unique ecological status of the biome. The five Annexes to the Protocol include: (i) Environmental Impact Assessment; (ii) Conservation of Antarctic Fauna and Flora; (iii) Waste Disposal and Waste Management; (iv) Prevention of Marine Pollution; and (v) Area Protection and Management.

Under the Agreed Measures numerous protected areas were designated: specially protected areas (SPAs) to protect areas of outstanding scientific interest and preserve their unique natural ecological system, and sites of special scientific interest (SSSIs) to protect the interests of scientific research within an area. Several other categories of protected area were introduced later. The Protocol has rationalized the Antarctic Protected Area System, and both SPAs and SSSIs were included in a new category, Antarctic specially protected areas (ASPAs). SPAs were originally designated only for biological reasons. They included representative examples of major land and freshwater ecosystems, areas with unique complexes of species, areas which are the type locality or only known habitat of any plant or invertebrate species, areas which contain exceptional breeding colonies of birds or mammals (seals), and areas which should remain inviolate so that in the future they may be used for purposes of comparison with localities that have been disturbed by human activity. In addition to the original criteria applying to any such site, ASPAs can now be designated for non-biological reasons, for example areas of outstanding geological, glaciological and geomorphological features, outstanding aesthetic and wilderness value, and sites or monuments of recognized historic value. Entry into both categories of ASPA is prohibited except in accordance with a permit, but the reason for entering an SPA must fulfil a scientific purpose which cannot be served elsewhere. For all sites, detailed management plans with maps must be prepared and the sites themselves provided with informative signs at the approved points of access.

There are currently (1998) 49 terrestrial ASPAs (19 SPAs and 30 SSSIs) very unevenly distributed around the continent, but reflecting the main regions of accessibility and scientific interest. Twenty-five are in the Antarctic Peninsula and offshore islands sector, and 15 in the Ross Sea sector; three other sectors have from two to four and three have none. Of the SPAs, only seven are designated specifically for their unique or diverse terrestrial biota, for example Lynch Island, South Orkney Islands (extensive grass community), Green, Litchfield and Lagotellerie Islands, western Antarctic Peninsula, and Moe Island, South Orkney Islands (extensive moss and lichen communities). Parts of Deception Island, South Shetland Islands and Cryptogam Ridge, Mt. Melbourne, have unique microbiological communities associated with geothermal activity.

A major concern as human activity increases in Antarctica is the possibility of NON-INDIGENOUS SPECIES being introducted. Many examples of ALIEN plants, both deliberately and accidentally introduced, and of alien invertebrates have been recorded near places of human occupation. Both the Agreed Measures and the Protocol stipulate that it is prohibited to bring live plants, animals and unsterilized soil into the Antarctic, to safeguard the pristineness of the environment. While it is impossible to prevent the introduction of microorganisms, every precaution is taken to minimize this risk. The other potential human impact on the fragile environment is the rapidly expanding tourist industry in Antarctica. Although over 8000 tourists visit a relatively small number of sites each year, their conduct has so far posed no serious threat to the terrestrial ecosystem. So long as the tour operators adhere responsibly to their largely self-imposed protocol of behaviour, and all visitors respect the uniqueness of the Antarctic environment, this situation could remain indefinitely. *Cf.* ARCTIC. *See also* CONSERVATION BIOLOGY. [R.I.L.S.]

REFERENCES

Anon. (1991) *Protocol on Environmental Protection to the Antarctic Treaty*. Her Majesty's Stationery Office, London.

Battaglia, B., Valencia, J. & Walton, D.W.H. (eds) (1997) *Antarctic Communities —Species, Structure and Survival*. Cambridge University Press, Cambridge.

Bonner, W.N. & Walton, D.W.H. (eds) (1985) *Key Environments. Antarctica*. Pergamon Press, Oxford.

Kerry, K.R. & Hempel, G. (eds) (1990) *Antarctic Ecosystems. Ecological Change and Conservation*. Springer-Verlag, Berlin.

Laws, R.M. (ed.) (1984) *Antarctic Ecology*, Vol. 1. Academic Press, London.

Lewis Smith, R.I., Walton, D.W.H. & Dingwall, P.R. (eds) (1994) *Developing the Antarctic Protected Area System. Conservation of the Southern Polar Region No. 1*. IUCN, Gland.

---

anther *See* POLLINATION.

anthropic horizons *See* SOIL CLASSIFICATION.

anthropocentric Either: (i) describing animals or plants that have become strongly associated with humans; or (ii) reasons or explanations that put humans in the centre of things.

For species associations, the term is synonymous with synanthropic. Examples are the house sparrow (*Passer domesticus*), town pigeon (*Columba livia*), Oxford ragwort (*Senecio squalidus*), weeds of arable land, and many species of invertebrate belonging to groups such as slugs, woodlice, cockroaches, centipedes and millipedes. Organisms often become anthropocentric when introduced beyond their normal RANGE (Fig. A15). The NATURAL habitats from which they have been recruited are often disturbed, such as strandlines or the side of volcanoes, or they may have been associated with nests, lairs or cadavers. If humans disappeared, most strongly anthropocentric species would become extinct. The St Kilda house mouse (*Mus muralis*) became extinct 2 or 3 years after the remote island where it lived was evacuated. Though the term is more usually applied to animals, a number of plants are decidedly anthropocentric. For example, London pride (*Saxifraga* × *urbium*), is a sterile hybrid not known in the wild but frequently seen semi-naturalized in towns.

An example of an anthropocentric reason is environmental protection that aims at preserving ECOSYSTEM SERVICES to humans. Similarly, an anthropocentric explanation would be one that seeks to understand ecological processes as services to humans. [O.L.G.]

REFERENCE

Gilbert, O.L. (1989) *The Ecology of Urban Habitats*. Chapman & Hall, London.

anthropomorphism Attributing human qualities, especially to animal behaviour and/or to the processes of nature. For example, NATURAL SELECTION has been presented, erroneously, as operating (as if) with creativity. [P.C.]

Anthrosols *See* SOIL CLASSIFICATION.

antibiosis The production by microorganisms or plants of antimetabolites (antibiotics) that, in small amounts, are lethal or inhibiting to other organisms (usually bacteria or fungi). [P.C.]

anticyclonic Moving in a clockwise direction in the Northern hemisphere and in an anticlockwise direction in the Southern

Fig. A15 Wild fig (*Ficus carica*) growing by the industrialized River Don, Sheffield, UK. At its northern limit this species becomes anthropocentric, only being found associated with heated water discharges, in this instance from the steel industry. (Courtesy of O.L. Gilbert.)

hemisphere. As a result of the combination of global WIND patterns and the Coriolis force (*see* CORIOLIS EFFECT), the major CIRCULATION patterns in the OCEAN basins are anticyclonic. *See also* ATMOSPHERIC CHEMISTRY; ATMOSPHERIC CIRCULATION. [V.F.]

antifouling Natural or artificial measures taken to prevent the growth of bacteria, plants and invertebrates on surfaces. Bottom-dwelling, SESSILE aquatic organisms are particularly susceptible to fouling and show a wide range of antifouling adaptations. Examples of natural antifouling measures include the pedicellariae of echinoderms, acid secretion and VANADIUM sequestering by tunicates, and saponin production by holothurians. Artificial approaches suggested or used with success include the application of metals (COPPER, tributyltin), adjustment of surface wettability (e.g. by applying Teflon), incorporation of adsorbed enzyme molecules that are active against fouling, BIOLOGICAL CONTROL with predators or parasites of fouling organisms, periodic heating, electrical discharge, ultraviolet illumination, ultrasonic cleaning and various mechanical means. *See also* TRIBUTYLTIN (TBT) POLLUTION. [V.F.]

antimony (Sb) A HEAVY METAL (ATOMIC NUMBER=51; relative ATOMIC MASS=121.75; melting point=630°C; boiling point=1750°C) which is a trace constituent (0.2 mg kg^{-1}) of the Earth's CRUST, although occurring at 1–2 mg kg^{-1} in soils derived from shales. It is non-essential and of moderate TOXICITY to plants, which can take up soluble oxyanions of antimony via their roots. The trivalent oxidation state is more toxic, interfering by bonding to SH groups. Often associated with ores of other metals, its main commercial source is stibnite (Sb_2S_3), from which it is extracted for use in alloys and in a variety of oxide, halide and organometallic compounds as a flame retardant, pigment, CATALYST and pharmaceutical. Atmospheric particulate concentrations of antimony are highest in areas of smelting and coal combustion. Stibine (SbH_3) is a highly toxic gas, analogous to AsH_3 for ARSENIC. [J.F.]

aphotic zone The region of the OCEAN and deep LAKES in which there is no LIGHT penetration and, hence, no PHOTOSYNTHESIS. It extends from the BASIN floor up to the PHOTIC ZONE. Animals that inhabit this zone permanently are usually carnivores or suspension- or detritus-feeders and depend entirely on a rain of organic matter from upper regions. *See also* DEEP SEA. [J.L.]

apomictic plants *See* POLLINATION.

apomictic thelytoky *See* PARTHENOGENESIS.

apomixis The production of females from unfertilized EGGS, without meiosis. The offspring thus produced are genetically identical to their mother. Some species can produce seeds through both apomixis and sex. Other species reproduce mostly asexually. Purely apomictic species, which reproduce only asexually (*see* ASEXUAL REPRODUCTION), are interesting from the point of view of the EVOLUTION OF SEX. Because such parthenogenetic species (*see* PARTHENOGENESIS) are evolutionary dead ends, it has been postulated that they should suffer extinction more often than species with sexual reproduction. This would provide a good example of a long-term advantage for sex, through the operation of SPECIES SELECTION. [I.O.]

apomorph A character that has evolved from an ancestral (plesiomorphic) form. It offers a relative description of the phylogenetic status of a character. For example, in the evolutionary sequence B1 → B2 → B3, B2 is apomorphic to B1 but plesiomorphic to B3. Similarly, the description of a character as apomorphic can vary as the taxonomic group being investigated is broadened or narrowed. For example, the possession of flowers is a unique angiosperm TRAIT but is found in all angiosperm families. Consequently, it is apomorphic to the angiosperms as a whole but it is plesiomorphic for each family.

In cladistic analysis (*see* CLADISTICS) the reconstruction of evolutionary relationships is based on the identification of apomorphic characters which are shared by taxa (synapomorphies), as sharing is most likely to be due to common ancestry (although it may be due instead to convergence). In ecological studies, a historical explanation (inheritance from an ancestor) can account for a taxon's plesiomorphic and synapomorphic traits, but the possession of apomorphic characters which are unique to the taxon (*see* AUTAPOMORPH) may require adaptationist explanations. However, because a character which is plesiomorphic at one taxonomic level will be (aut)apomorphic at another, all characters may require an adaptationist explanation at the level at which they are autapomorphic. Thus, the most parsimonious explanation for the possession of flowers by a rose is the historical explanation that it inherited the capacity to produce them, whereas an adaptationist explanation may be required to account for the origin of the flowering habit. *See also* HOMOLOGY; HOMOPLASY; PLESIOMORPHY; SYNAPOMORPHY. [M.M.]

REFERENCE

Wanntorp, H.-E., Brooks, D.R., Nilsson, T. *et al.* (1990) Phylogenetic approaches in ecology. *Oikos* **57**, 119–132.

aposematic (warning) coloration Distasteful, toxic or otherwise potentially dangerous prey items are frequently brightly coloured and contrast with their background (see Plate 20, facing p. 398). The reasons for the EVOLUTION of such aposematic, or warning, coloration are not immediately clear. In a population of cryptic distasteful prey, we might expect that the rare mutant which stood out against the background would be more likely to be preyed upon, thereby providing a selective premium against the evolution of warning coloration. There is some evidence that aposematic insect species frequently live in family groups. Copies of a rare mutant gene for aposematic coloration are more likely to be found in close KIN, so that if predators more easily learn to avoid prey that contrast with the background, the sacrifice of one member of a family group would effectively protect the others. Alternatively, it has been argued, aposematic coloration evolved in species that live solitarily and that, subsequently, mutual protec-tion was gained by group living. If the latter scenario is correct, the problem of the evolutionary origins of aposematic coloration remains. *See also* KIN SELECTION; MIMICRY. [P.H.H.]

apostatic selection A type of FREQUENCY-DEPENDENT SELECTION in which the common phenotypes are selected against and the rare phenotypes are favoured, for example a predator prefers the most frequent prey, which then suffers higher mortality. Apostatic selection may lead to a stable distribution of frequences of phenotypes. [P.O.]

apparent competition An ecological situation where two species appear to show the reciprocal negative effects associated with interspecific competition (*see* SPECIES INTERACTIONS), but this is, in fact, the result of predation by a third species. Consider the situation depicted in Fig. A16. A single species of PREDATOR attacks two species of prey. The predator–prey interactions are of a −+ type (competition) and therefore both species are adversely affected by the predator, and the predator is positively affected by both species of prey. This means that the positive effect that prey 1 has on the predator will, in turn, increase the negative effect upon prey 2, and vice versa. The overall consequence of this is that there will appear to be a −− interaction between prey 1 and prey 2, even if they are not competitors for any essential and limiting resource. That is, they will show apparent competition (Holt 1977, 1984, 1987).

Using a simple MATHEMATICAL MODEL, Holt (1977) showed that the condition allowing both prey species to coexist was:

$$r_1 > a_1 N_p$$

where r_1 is the intrinsic rate of natural increase of prey 1, N_p is the equilibrium density of the predator in the absence of prey 1, and a_1 is the attack rate of the predator on prey 1. Intriguingly, this implies that COEXISTENCE will be favoured by prey 1 reducing a_1 and this could be achieved by being 'less similar' to prey 2. Like true competition therefore, apparent competition could result in CHARACTER DISPLACEMENT or NICHE differentiation. However, in the case of apparent competition this is due to competition for 'ENEMY-FREE SPACE' (Jeffries & Lawton 1984) rather than competition for some essential and limiting resource. [B.S.]

REFERENCES

Holt, R.D. (1977) Predation, apparent competition, and the structure of prey communities. *Theoretical Population Biology* **12**, 197–229.

Holt, R.D. (1984) Spatial heterogeneity, indirect interactions, and the coexistence of prey species. *American Naturalist* **124**, 377–406.

Holt, R.D. (1987) Prey communities in patchy environments. *Oikos* **50**, 276–290.

Jeffries, M.J. & Lawton, J.H. (1984) Enemy-free space and the structure of ecological communities. *Biological Journal of the Linnean Society* **23**, 269–286.

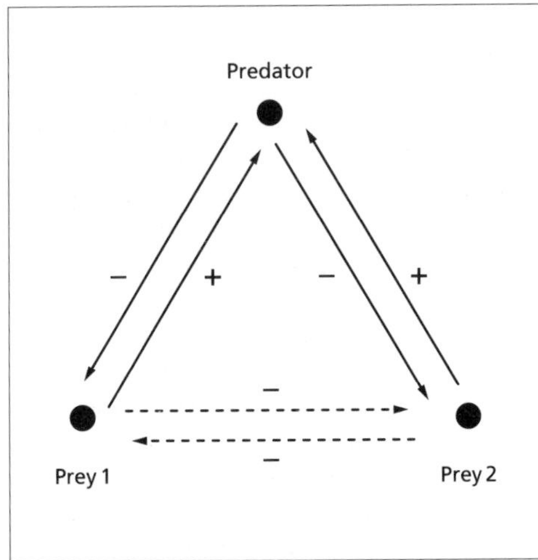

Fig. A16 Diagram showing apparent competition between two prey species. The solid lines indicate direct interactions between prey and predator, while the dashed lines show indirect (apparent) interactions between the two prey species.

appetite Appetite is commonly defined in terms of desire for, or inclination towards, certain commodities or other stimuli, most often in the context of food or sex. In the broader behavioural context, however, it is an anthropomorphic metaphor for the apparent goal-directedness of much activity. Thus a food-deprived animal moving around its environment in such a way that it eventually finds and consumes food is characterized as hungry and showing appetitive behaviour for food. While this may capture the functional significance of the behaviour, however, it cannot be used (as unfortunately it often is) to infer the existence in the animal's 'mind' of a goal to be achieved by the behaviour. To do this is to confuse function with cause and create a sterile, descriptive tautology. The notion of appetite is thus one aspect of the more general problems of ANTHROPOMORPHISM and TELEOLOGY inherent in motivational terminology. [C.J.B.]

applied ecology *See* ECOLOGY.

apposition beach A shingle structure formed by repeated development of stable shingle ridges which come to lie adjacent and parallel to each other. Processes involving longshore drift and both constructive and destructive wave action are involved in ridge generation. Where such shingle structures form under the influence of WAVES from two directions, a cuspate foreland may be developed. Although they may become subject to erosion at some later time, the ridges tend to become sufficiently permanent to allow long-term COLONIZATION by plants, leading to a PRIMARY SUCCESSION and associated ZONATION of plant communities across the ridge sequence. On an individual ridge, vegetation tends to be confined to where the habitat is most supportive of colonization and growth, usually on the ridge top and sides where the shingle texture is finest. The major gradients determining the SUCCESSION are, from the coast inland, increasing ridge age and decreasing SALINITY. Processes of SOIL build-up, changing nutrient status and pH, although subsidiary to these, are also important. These shingle structures may occasionally become inundated with sand to form DUNE systems. [B.W.F.]

aquaculture Culturing aquatic organisms for commercial purposes, either in artificial systems—for example tanks or channels—or in nature. The latter might include cultures in completely open systems or in enclosures (e.g. cages). FISH FARMING is one form of aquaculture. Often EFFLUENTS from unconsumed food, wastes and pharmaceuticals can cause POLLUTION. [P.C.]

aquatic ecotoxicology *See* ECOTOXICOLOGY.

aquatic systems, conservation and restoration All aquatic ecosystems (LENTIC and LOTIC) can be considered as RENEWABLE RESOURCES and thus available or potentially available for industrial and/or non-consumptive purposes such as recreation. With the use of even a renewable resource there is the danger that the aquatic ECOSYSTEM will be changed or even destroyed.

Conservation is the preservation of the ecosystem (or components of the ecosystem) either in its existing condition or in some determined ideal (optimum) condition (*see* CONSERVATION BIOLOGY). Conservation is frequently a matter of TRADE-OFFS between how much of the ecosystem (or component) can be used to achieve specific gains without loss of PRODUCTIVITY and/or DIVERSITY. Not all DISTURBANCES are deleterious as disturbance is often natural and essential for the development (succession, evolution) of the ecosystem.

Restoration is the bringing back to a former state or condition determined to be desirable for aesthetic or economic reasons to provide a SUSTAINABLE YIELD of resources (*see* ECOLOGICAL RESTORATION). Implicit in restoration is the ability to identify a desirable (optimum) condition for an aquatic ecosystem, and an ability to identify changes from that condition as either positive (beneficial) or negative (harmful). Also implicit in the concept of restoration is that aquatic ecosystems have homeostatic processes maintaining the system in a definable state. Many aquatic ecosystems may appear to have an equilibrium state determined by physical, chemical and biological variables but these temporary, dynamic equilibria change with the scale of temporal or spatial observation.

Conservation and restoration of aquatic ecosystems has frequently been linked to the concept of ECOSYSTEM INTEGRITY. However, integrity is not an objective, quantifiable property and has four major, sometimes incompatible, themes:

1 ecosystem structure and/or process staying at predefined baseline levels;
2 a system permitted to change without human influence;
3 preservation of an organizing or self-correcting ability with an end state defined as normal or optimal; or
4 maintenance of qualities desired by society or components of society.

[R.W.D.]

aquifer A geological formation containing sufficient saturated permeable material to yield WATER to wells and springs in significant quantities. Aquifers are defined as unconfined if their upper surface is open to the atmosphere and confined, or artesian, if they are overlaid by impermeable layers. In confined aquifers the water pressure is greater than atmospheric pressure and if the confining layer is pierced by a well the pressure may be enough to force water to the land surface. Most of the world's major aquifers are in rocks of sedimentary origin, notably chalk, sandstone, coarse sand and gravel. [D.C.L.]

arboretum A collection of plants, in particular trees, shrubs and woody climbers, cultivated for scientific, conservation, educational and/or aesthetic purposes. The term can also embrace the institution/organization responsible for administering and maintaining such a collection. 'Botanical garden' describes a collection that is not restricted to ligneous (woody) plants, although in practice the distinction is sometimes blurred. [P.C.]

arboriculture The cultivation of woody plants, particularly trees, primarily for ornamental purposes in human-dominated settings (as opposed to SILVICULTURE in forests). Trees may be used to provide shade, air purification, noise reduction, erosion control, educational opportunities, and fruit in addition to enhancing aesthetic value. Arboriculturists may be involved in designing plantings, landscaping and preparing planting sites, planting trees, pruning branches, mitigating construction damage, fertilizing, treating injuries or infestations, staking and bracing stems and branches, moving trees, or felling trees and removing stumps.

Modern-day arboriculture itself increasingly relies on specialist studies of genetics, plant breeding including cell culture techniques, seed technology including seed preservation, mycology particularly of MYCORRHIZAE, forest ecology, SOIL structure and nutrition, fungal, bacterial and insect-plant pathology, CLIMATOLOGY, landscaping as well as the technological advances in tree planting, HARVESTING and processing. The practice of arboriculture therefore requires intimate knowledge of tree biology and AUTECOLOGY. *See also* FORESTRY PRACTICES. [A.N.G. & G.A.F.H.]

arbuscules *See* ENDOTROPHIC MYCORRHIZAE.

arbutoid mycorrhizae *See* ENDOTROPHIC MYCORRHIZAE.

arcsine transformation *See* TRANSFORMATION OF DATA.

Arctic

Definition

Historically, the word Arctic refers to the region of the Earth below the stellar constellation of the Great Bear (Greek *Arktos*, bear). Climatically, the terrestrial Arctic can be considered to be that region of the Earth's land surface that is underlaid by PERMAFROST or permanently frozen ground. Permafrost (Fig. A17) underlies approximately 20% of the Earth's land surface and occurs not only at high latitudes but also in some non-Arctic locations at high elevations (Brown *et al.* 1995) and is the main LIMITING FACTOR to the northward extension of trees (*see* TAIGA; TUNDRA). Geographically, the Arctic can be defined simply as the portion of the Earth's surface that lies north of the Arctic Circle (66°33′N). This definition emphasizes the photoperiodic characteristics of the region: continuous periods of daylight in summer and continuous darkness in winter. Alternatively, the Arctic can be defined ecologically by the upper latitudinal limit to tree growth (Fig. A17), which approximates to areas where the mean temperature of the warmest month of the year does not exceed 10°C—Köppen's rule (Köppen 1931). Within this broad biological definition lies a wide range of terrestrial habitats over an extensive land area (5.5% of the Earth's land surface). Although the number of species present in Arctic communities is low, with only 892 species of flowering plants (Polunin 1959), there is a high degree of specialization in relation to microhabitats coupled with considerable BIODIVERSITY at the subspecies level (Crawford & Abbott 1994).

The Arctic can also be defined biogeographically on the basis of the species composition of the plant communities (Fig. A17) and as such is traditionally divided biogeographically into the low Arctic,

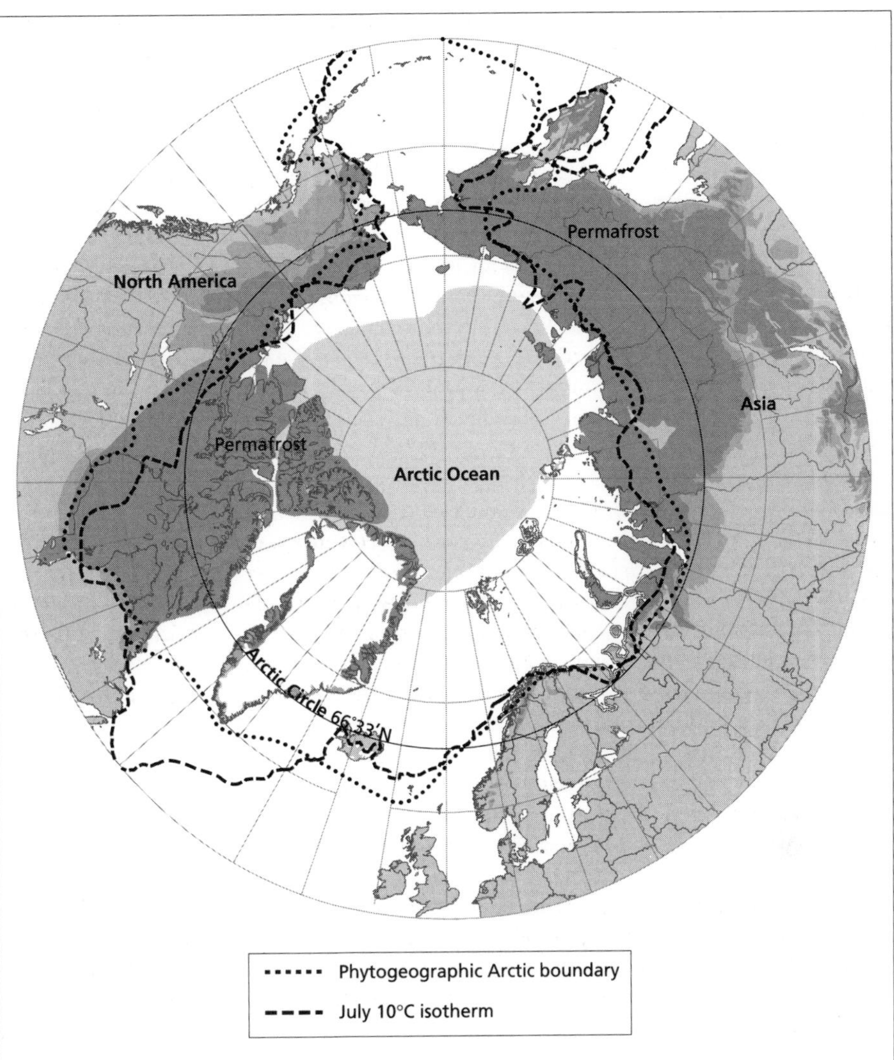

Fig. A17 The Arctic region as defined by (a) the Arctic Circle; (b) the July 10°C isotherm corresponding to the northern limit to tree growth; (c) the Arctic phytogeographic boundary derived from the distribution of arctic plant communities (see text); and (d) distribution of continuous permafrost. Map adapted from data compiled by the Norwegian Directorate for Nature Management.

with plant communities dominated by woody shrubs (e.g. *Alnus, Betula, Salix*), and the high Arctic (Fig. A18), which is characterized by much of the land surface and SEA being covered by permanent ICE, with terrestrial vegetation limited to a thin discontinuous cover of diminutive flowering plants, mosses and lichens. At present, polar desert (see below) is found mainly in the northern Canadian archipelagos, the Queen Elizabeth Islands, Ellesmere Island, northern Greenland, Svalbard in Nordaustlandet, Kong Karls Land, the northern tip of Novaya Zemlya, Franz Josef Land, Cape Chelyuskin and Severnaya Zemlya.

Arctic environment

In the high Arctic the growing season is extremely brief, but the lack of precipitation at these latitudes (polar desert) allows extensive areas of coastal plain to remain free of ice in summer long enough for a substantial flora of flowering plants to persist, with 57 species and subspecies of flowering plants above 80°N in Franz Josef Land (Aleksandrovna 1988). During the Weichselian (last) GLACIATION the increased aridity of this region due to the southern retreat of

Fig. A18 High Arctic scene at Magdalena Fjord, Spitsbergen (79°N). Despite the surrounding snow and ice-cover, the nunataks provide a polar desert refuge for many species of flowering plants. (Courtesy of R.M.M. Crawford.)

the warm ocean currents (the Atlantic conveyor) together with a reduction in sea level, variously estimated to have been between −55 and −120 m (Richards *et al.* 1994) would have caused the area of polar desert to expand. As the Pleistocene polar ICE SHEETS spread south, a zone of low precipitation developed on their northern edge which resulted in a shrinkage of the northern extensions of the ice sheets (Velichko *et al.* 1984; Fulton 1989). The existence of such extensive ice-free terrain at high latitudes for a considerable part of the Weichselian era would have given an opportunity for Pleistocene COLONIZATION (*see* CIRCUMPOLAR) by vegetation similar to that which exists today in polar deserts (Aleksandrovna 1988; Serebryanny *et al.* 1993).

Ecological zones

Recognition of various arctic and subarctic zones by ecologists (taiga, tundra, polar desert) varies between North America and Eurasia due to differences in climate, snow cover and geomorphology. The Canadian Arctic archipelago creates a climatic and biotic boundary between the mainland and the extensive land areas to the north (Bliss 1988), while the Russian Arctic has a large expanse of continental land with groups of islands in the Arctic Ocean, and consequently there is a more progressive transition from subarctic to high arctic conditions. Figure A19 provides a diagrammatic representation of the major zonations as recognized by North American and Eurasian ecologists.

The low Arctic of North America corresponds to the southern tundra and typical tundra of Eurasia and is a broad zone both ecologically and geographically within which there are many boundary limits to the northward extension of plant species. North of the TREE LINE in North America and Siberia there is a distinct northern distribution limit, sometimes called the 'tussock line', marked by the disappearance of the tussock-forming sedge *Eriophorum vaginatum*. The tussock line not only marks the disappearance of *Eriophorum vaginatum*, but is followed by the disappearance of dwarf birch (*Betula nana*) and Labrador tea (*Ledum groenlandicum*). This line, marking as it does the northern limit of shrub vegetation as well as the tussock sedge community, is taken as an indicator of the northern limit of the traditional low arctic (Young 1989). The intervening regions between this typical low arctic shrub and tussock vegetation and the polar desert of the high Arctic are occupied by a mosaic of communities; these communities are determined largely by TOPOGRAPHY and become progressively poorer in species the further north they occur. These zones are frequently described as various forms of tundra, which implies the existence of a permafrost layer at varying depths below the SOIL surface. There are many descriptions and definitions of various tundra communities. Two examples of contrasting types are described below.

Dwarf heath tundra, depending on geographical location, contains various species of stunted or creeping shrubs with the OVERWINTERING buds placed close to the ground or in the surface soil, which is usually sandy and poor in NUTRIENTS, and frequently has a layer of raw HUMUS. Typical species include: *Loiseleueria procumbens, Diapensia lapponica, Arctostaphylos alpinus, Rubus saxatilis, Salix glauca* and *S. lanata*.

Graminoid moss tundra, as seen in the Queen Elizabeth Islands, provides the major GRAZING habitat for musk oxen and is an important breeding area for waterfowl and shore birds. Dominant species include *Carex stans, Eriophorum scheuchzerii, E. triste, Dupontia fischeri* and *Alopecurus alpinus* (Bliss 1988).

High arctic (polar desert) habitats extend as far north as land is available and are characterized by a lack of a continuous vegetation cover. The main producers of organic matter in the high Arctic are not the higher plants, bryophytes or lichens but the unicellular algae, particularly cyanobacteria. Nevertheless, the higher plants

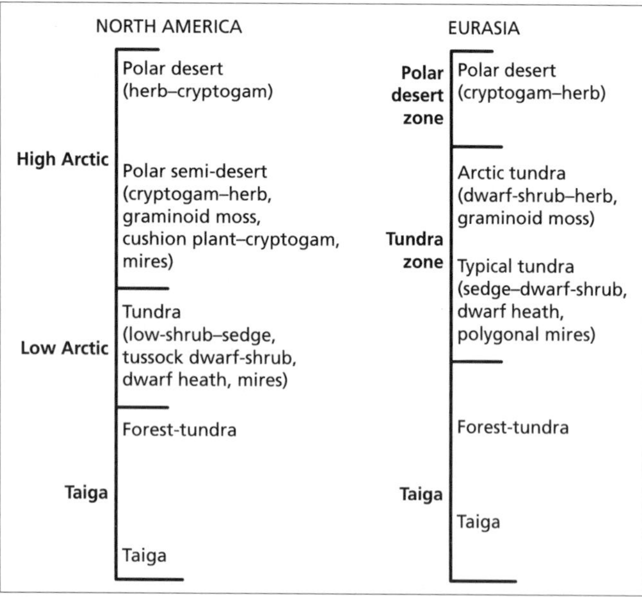

Fig. A19 Classification schemes for arctic vegetation in North America and Eurasia. (After Bliss, L.C. & Matveyeva, N.V. (1992) Circumpolar arctic vegetation. In: *Arctic Ecosystems in a Changing Climate* (eds F.S. Chapin III, R.L. Jefferies, J.F. Reynolds, G.R. Shaver, J. Svoboda & E.W. Chu), pp. 59–89. Academic Press, San Diego, Calif.)

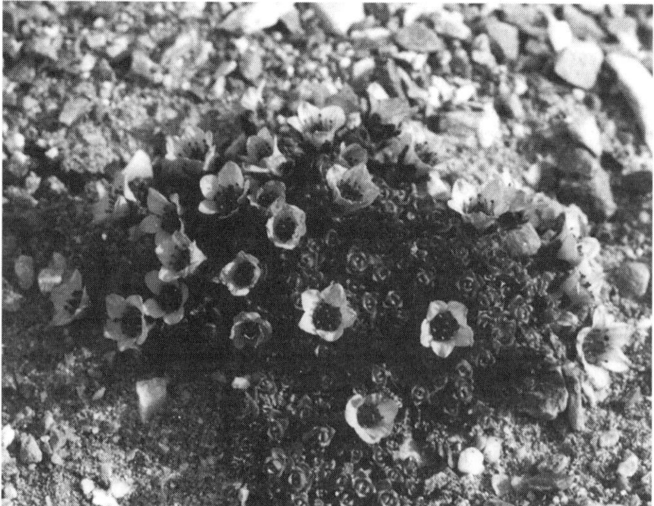

Fig. A20 Purple flowering saxifrage (*Saxifraga oppositifolia*) growing as a robust and healthy plant at 78°55′N in Spitsbergen. (Courtesy of R.M.M. Crawford.)

that do survive at these latitudes appear well adapted for survival in these ultra-short growing seasons. Purple flowering saxifrage (*Saxifraga oppositifolia*, Fig. A20) is to be found on Cape Morris Jesup in Peary Land (83°39′N; see Fig. A17), a mere 450 miles from the North Pole. The most characteristic life form is that adopted by the cushion plants (e.g. *Silene acaulis* and the tufted form of *Saxifraga oppositifolia*), which are ideally adapted to conserving heat and water and nutrients. Where water and nutrients are less limiting *S. oppositifolia* is found as a creeping form (Crawford & Abbott 1994). Even at this outer edge of plant survival *Saxifraga oppositifolia* shows no sign of loss in vigour, with two common robust ecotypes adapted respectively to late, wet snow beds and drier ridges, and a high level of biodiversity as shown by the existence of much variation in chloroplast DNA. Other diminutive species include *Draba* spp., *Braya purpurescens, Minuartia* spp. and *Cerastium arcticum*. *Cf.* ANTARCTICA, ECOLOGY AND CONSERVATION. [R.M.M.C.]

REFERENCES

Aleksandrovna, V.D. (1988) *Vegetation of Soviet Polar Deserts* (trans. D. Löve). Cambridge University Press, Cambridge.

Bliss, L.C. (1988) Arctic tundra and polar desert biome. In: *North American Terrestrial Vegetation* (eds M.G. Barbour & W.D. Billings), pp. 1–32. Cambridge University Press, New York.

Bliss, L.C. & Matveyeva, N.V. (1992) Circumpolar arctic vegetation. In: *Arctic Ecosystems in a Changing Climate* (eds F.S. Chapin III, R.L. Jefferies, J.F. Reynolds, G.R. Shaver, J. Svoboda & E.W. Chu), pp. 59–89. Academic Press, San Diego, Calif.

Brown, J., Ferrians, J.O.J., Heginbottom, J.A. & Melnikov, E.S. (1995) Circum-arctic map of permafrost and ground-ice conditions. In: *U.S. Geological Survey Circum-Pacific Map CP-45, 1:10,000,000*. Reston, Va.

Crawford, R.M.M. & Abbott, R.J. (1994) Pre-adaptation of Arctic plants to climate change. *Botanica Acta* **107**, 271–278.

Fulton, R.J. (ed.) (1989) *Quaternary Geology of Canada and Greenland. Geological Survey of Canada, No. 1*. Geological Society of America, v. K-1. Boulder, Colo.

Köppen, W. (1931) *Grundriss der Klimakunde*, 2nd edn. Walter Gruyter, Berlin.

Polunin, N. (1959) *Circumpolar Arctic Flora*. Clarendon Press, Oxford.

Richards, D.A., Smart, P.L. & Lawrence Edwards, R. (1994) Maximum sea level for the last glacial period from U-series ages of submerged spelotherms. *Nature* **367**, 357–360.

Serebryanny, L.R., Tishkov, A.A., Malyasova, E.S., Solomina, O.N. & Ilves, E.O. (1993) Holocene palaeoecology of the Arcto-Atlantic. *Izvestiya Akademii Nauk Rossii. Seriya Geographiya* **3**, 39–52. [In Russian.]

Velichko, A.A., Isayeva, L.L., Makeyev, V.M., Matishov, G.G. & Faustova, M.A. (1984) Late Pleistocene glaciation of the Arctic Shelf, and the reconstruction of the Eurasian ice sheets. In: *Late Quaternary Environments of the Soviet Union* (ed. A.A. Velichko), pp. 35–44. Longman, London.

Young, S.B. (1989) *To the Arctic*. John Wiley & Sons, New York.

Arctic soils *See* SOIL CLASSIFICATION.

area cladogram An area CLADOGRAM is a branching diagram that depicts the historical relationships among geographical areas based on (i) geological evidence or (ii) the phylogenetic relationships of the species inhabiting those areas (made by substituting areas for taxa on a PHYLOGENETIC TREE). Area cladograms are used to identify the origin of each species in a given ECOSYSTEM or COMMUNITY and to identify general biogeographic DISTRIBUTION patterns. When an area cladogram based on geological evidence is congruent with one based on the phylogenetic relationships of the species inhabiting the areas, we can hypothesize that there is a long history of association between the species and the areas in which they now occur. Species whose phylogenetic relationships are congruent with geological history form the resident core of communities. Points of incongruence between the two types of area cladograms indicate a more recent history of association between the species and the areas based on DISPERSAL by the species from their areas of origin to the areas in which they now occur. [D.R.B.]

area gradient *See* DIVERSITY GRADIENT.

arenes Arenes are aromatic HYDROCARBONS, the simplest being BENZENE. They are hydrophobic, strong-smelling, often toxic or carcinogenic, and good organic solvents. Their chemical reactivity comes from the groups attached to the benzene ring, which is itself unreactive. Examples include toluene (a benzene ring to which CH_3 is attached), naphthalene, anthracene and xylene. [J.G.]

Arenosols *See* SOIL CLASSIFICATION.

Argentina, conservation in

Major natural regions of Argentina

Throughout its vast territory of more than 2.7 million km^2, Argentina has several different kinds of ecosystems. Ten major life zones or bioregions can be identified (Fig. A21). These units and their main conservation problems are as follows.

1 *Patagonian steppe*. This arid and cold region occupies a large part of southern Argentina. Poor land management due to overgrazing during this century has given rise to an intense erosive process, which is increasing yearly.

2 *Sub-Antarctic forests*. These forests cover the southernmost section of the Andes mountain range. While a significant part of this region is protected by several National Parks, it is necessary to find sustainable productive methods for the rural population and especially for the indigenous *mapuche* communities settled there.

3 *Monte desert*. This is a wide strip, with arid climate and plenty of xerophilous vegetation, that cuts Argentina from the mountain range foothills in the north-west to the marine coasts in the south-east. As in other arid areas of the country, the main threat for its conservation derives from low-quality cattle raising and poor ranch management.

4 *Pampas grasslands*. Both the population and the industrial activity of the country are concentrated in the fertile grasslands of the PAMPAS, mainly in the axis of the Paraná–La Plata rivers. Therefore, the area has suffered significant modifications. This has become more serious due to the lack of protected areas in the region.

5 *Espinal and Mesopotamian savannah*. This bioregion surrounds the pampas grassland. Its main characteristic is the low and open forest of different *Prosopis* species. In the last 50 years these woodlands have been depleted by intense cutting practices, mostly as a result of land conversion for agriculture and timber exploitation.

6 *Chaco woodlands*. This is the main bioregional unit in the north

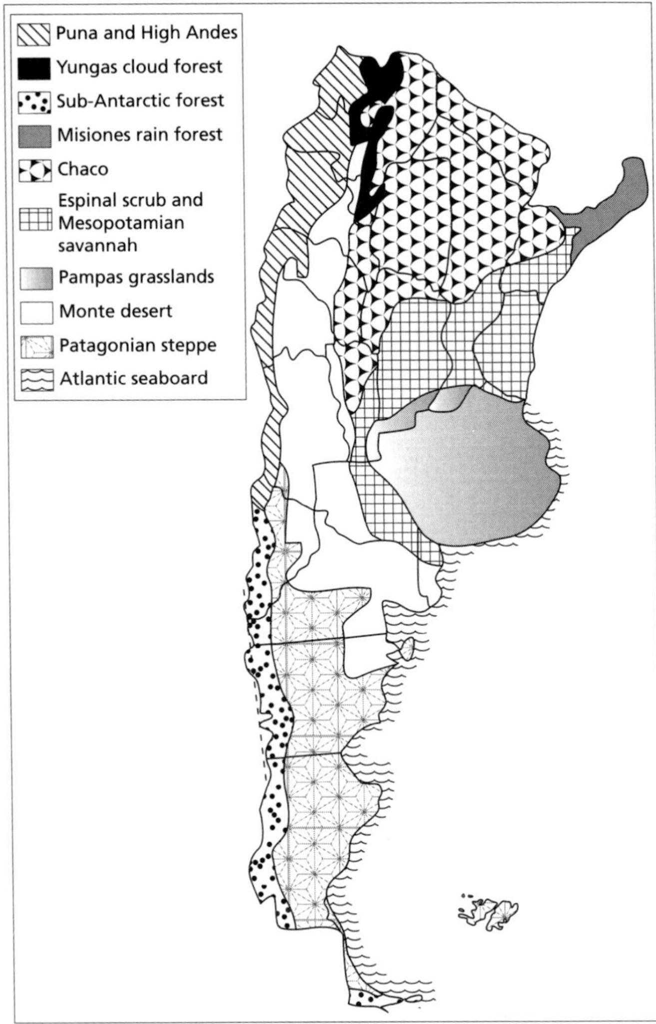

Fig. A21 Major natural regions of Argentina.

of the country. The climate in the eastern portion is more humid and it turns semi-arid westwards. Various factors, including timber extraction, clearing for agricultural purposes, hunting, overgrazing, the poor indigenous human population and the lack of protected areas, mean that this region of the country needs more urgent conservation measures.

7 *Misiones rain forest.* While it occupies only a small area of north-eastern Argentina, this region deserves special treatment because it is the main surviving remnant of the originally large humid forests that reached up to San Pablo, Brazil. It is necessary to enhance sustainable methods of development in order to conserve its high biodiversity.

8 *Puna and the High Andes.* This region occupies the northern part of the Andes mountain range in Argentina. In the higher grasslands of the Puna there are many communities of indigenous herders. The fragility of this environment and the poor conditions in which the peasants live require creative measures to achieve both conservation and development goals.

9 *Mountain cloud forests (Yungas).* In Argentina, this region forms a narrow strip that continues up to Bolivia and Peru throughout the eastern slopes of the Andes. Its lower parts have been heavily transformed due to urbanization, agriculture and related industries. In order to conserve its significant biodiversity, it is necessary to establish corridors to connect the existing protected areas.

10 *Marine coast.* Along the country's extensive coast there are many concentrations of birds and marine mammals which constitute a world-famous tourist attraction. There remain conflicts to be resolved between some productive activities, such as oil extraction and fisheries, and conservation efforts.

The national system of protected areas

Argentina was the first South American country to create (in 1904) a national park and to establish (in 1934) a National Parks Service under the federal government. The National Parks Service was originally responsible for six large areas: Nahuel Huapi; Lanín; Los Alerces; Los Glaciares and Perito Moreno (in the sub-Antarctic forests of the south-west); and Iguazú (in the Misiones rain forest of the north-east). The National Parks Service also established a training centre for park wardens, originally at Isla Victoria in Nahuel Huapi National Park, nowadays at Horco Molle Reserve in Tucumán Province.

Subsequently, many provincial states created new protected areas under their own jurisdiction. At present, there are 210 protected areas in Argentina, covering 4.6% of its surface. Most of them belong to the provincial state governments, as shown in Table A15. However, the lack of sound management threatens these laudable initiatives, especially at provincial level. About 75% of the provincial protected areas are poorly managed, or have no management at all.

Ecosystem representation within the national system of protected areas is varied. Figure A22 shows the surface area of each natural region and the percentage of protected areas. Two ecological units may be considered well represented (>15%) in the protected area system: sub-Antarctic forests and Puna/High Andes. Both are in the mountain regions on the border with Chile. Four other regions are poorly represented (3–6%). Two of them are high biodiversity spots and hold significant endemic biota: Yungas/mountain cloud forests and Misiones rain forests. The other two, monte desert and Patagonian steppe, are arid ecosystems unique to Argentina.

Chaco woodlands, espinal scrub/Mesopotamian savannah, pampas grasslands and the marine coast are the most unprotected regions (<0.5%). Pampas grasslands and Mesopotamian savannahs have suffered severe changes due to agriculture, especially livestock farming. There are almost no surviving remnants of the original vegetation communities of these plains.

Table A15 Federal and provincial protected areas of Argentina.

	Number of areas	Total surface area (km^2)	Percentage, of the protected area system	Percentage, of the country surface
Federal protected areas	26	28 000	21.5	1
Provincial states' protected areas	184	102 000	78.5	3.6
Total	210	130 000	100	4.6

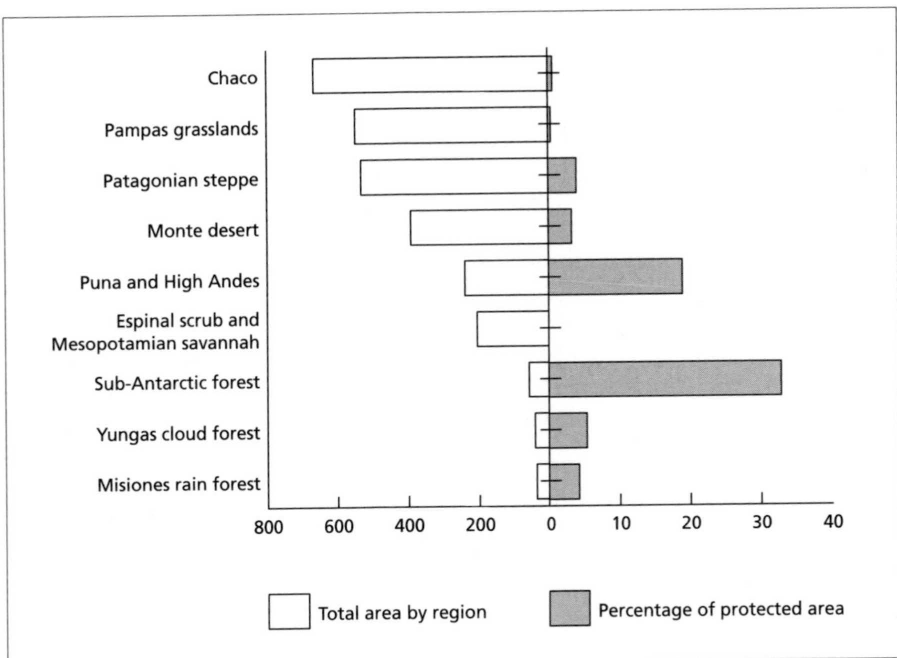

Fig. A22 Major natural regions of Argentina. Total area by region (×10³ km²) and percentages of protected areas.

Main threats to habitat conservation

DEFORESTATION, DESERTIFICATION, conversion of grasslands to agriculture and dam construction are the main threats to nature conservation in Argentina. Moreover, the country lacks a sound nationwide assessment of habitat loss by bioregion.

Deforestation rates may be deduced from scattered sources, with good local estimates. The best available data show that the rates of forest loss for Argentina during the first seven decades of the present century were higher than those measured for Brazilian Amazonia one decade ago. The woodlands of the southern Chaco region (in the provinces of Santiago del Estero and Santa Fe) were especially devastated.

Nowadays, remnant forests cover almost 400 000 km² of Argentina. annual deforestation amounts to 1600 km², giving a rate of 0.4% for the whole country. It must be pointed out that major forest loss is now centred on the ecotone between mountain cloud forests (Yungas) and the western Chaco woodlands. About 1.5 million ha of native forest land has been turned over to agriculture in this area during the past 20 years. This area alone accounts for half of the present annual forest loss of the country (Table A16). Nevertheless, the main forestal ecosystems of the country still have good remnants. This is the case for Misiones rain forests, mountain cloud forests/Yungas and northern Chaco woodlands.

Soil conservation needs to be addressed urgently. Almost 40% of the country's surface has some degree of soil erosion. Desertification is widespread throughout the western arid lands of the country. Patagonian steppe is the most affected area due to overgrazing by sheep in this century.

Another significant threat to habitat and species conservation is dam construction for energy production. ENVIRONMENTAL IMPACT ASSESSMENTS of dams are weak and mitigation actions scarcely implemented. A great number of dams have been built in the mountain range of the south-west, and more recently in the Paraná and Uruguay rivers basins, in the north-east.

Table A16 Present forest loss in Argentina (total) and in its north-west forests.

	Remnant forests (km²)	Annual deforestation (km²)	Annual loss rate (%)
Argentina (total)	400 000	1600	0.4
Umbral al Chaco (ecotone between Chaco woodlands and Yungas mountain forests)	20 000	750	3.75

Legal tools for nature conservation

Argentina is a federal country, so that each provincial state is responsible for the conservation and management of their natural resources within their boundaries. Laws and their enforcement vary greatly among states.

The main federal conservation laws are those related to national parks and wildlife conservation. Both were passed in 1980. The country lacks special federal laws on ENDANGERED SPECIES, while forestry laws are weak in their conservation goals.

In 1974, commercial exploitation of marine mammals and birds was banned. The main regional conservation treaty adopted in the 1970s was the Convention for the Conservation of the Vicuña (*Vicugna vicugna*). These two legal measures have been successful as marine mammals and vicuñas are the country's best examples of endangered species enjoying recovery.

Since 1989, Argentina has greatly improved the enforcement of the CITES Convention and has ratified other significant treaties, such as the Convention on Migratory Species, the Ramsar Convention, the Framework Convection on Climate Change, and the Biodiversity Convention. *See also* CONSERVATION BIOLOGY. [J.G.F.]

REFERENCES

Bárbaro, N. (1994) *Perfil Ambiental de la Argentina.* Comité Argentino de la UICN, Buenos Aires.

Burkart, R. (1994) *El Sistema Nacional de Areas Naturales Protegidas de la Argentina.* Administración de Parques Nacionales, Buenos Aires.

García Fernández, J. (1994) *Argentine Programme for Conservation & Sustainable Development.* Comité Argentino de la UICN, Buenos Aires.

Secretaría de Agricultura, Ganadería y Pesca de la Nacíon (1995) *El deterioro de las tierras en la República Argentina.* Buenos Aires.

---

argic horizons *See* SOIL CLASSIFICATION.

argillic horizons *See* SOIL CLASSIFICATION.

argon (Ar) A gaseous element of the noble gas group (ATOMIC NUMBER=18; relative ATOMIC MASS=39.948; melting point= −189.3°C; boiling point=−185.8°C). Argon is the third most abundant gas in the ATMOSPHERE, after NITROGEN and OXYGEN, and is present at 0.93% by volume. Although natural ISOTOPES are ^{36}Ar and ^{38}Ar, the isotope ^{40}Ar dominates by outgassing from decay of ^{40}K in the Earth's CRUST. Estimates suggest that all the ^{40}Ar released has been retained in the atmosphere. The ^{40}K to ^{40}Ar decay ($t_{1/2}$= 1.3 billion years) is the basis of an important geochemical DATING technique. The element is physically and chemically inert and has no other environmental significance. It is obtained from liquefaction of air and used as an inert atmosphere for welding and in filament bulbs. [M.R.H.]

Aridisols *See* SOIL CLASSIFICATION.

Aristotle *See* CHARACTERS IN ECOLOGY.

arithmetic mean A measure of central tendency or average. *See also* MEAN. [B.S.]

arms races A continuing evolutionary interaction between species in which changes in one species exert a SELECTION PRESSURE for changes in the other interacting species. Arms races refer to changes occurring on an evolutionary time scale rather than on the scale of individual lifetimes. These changes, analogous to a sequence in the development of weapons and defences by opposing military forces, characterize predator–prey, parasite–host and COMPETITION interspecific interactions in which defences evolved, for example, by the prey or host select for the EVOLUTION of new offensive strategies by the predator or parasite (*see* PREDATOR–PREY INTERACTIONS; PARASITE–HOST INTERACTIONS). Arms races driven by SEXUAL SELECTION have also been used to describe evolutionary changes in males and females of the same species. A symmetric arms race occurs when two species try to achieve the same goal (e.g. competitors); an asymmetric arms race occurs when the interacting species have different, opposing goals (e.g. predator–prey). Arms races may occur in any case of evolutionary conflict. *See also* COEVOLUTION; CONFLICTS; RED QUEEN'S HYPOTHESIS. [V.F.]

arrangement *See* ARTIFICAL CLASSIFICATION.

arrhenotoky A haplo-diploid genetical system (*see* HAPLODIPLOIDY) in which males arise from unfertilized EGGS, females from fertilized eggs. As a consequence males only transmit the genome of the mother. The females are diploid and the males generally haploid, but restoration of DIPLOIDY in males may occur (e.g. coccids in the genus *Lecanium*). Arrhenotoky can be easily identified because virgin females produce sons only. It may give the mother the potential to determine the sex of her offspring by controlling FERTILIZATION of eggs. Arrhenotoky has arisen (from diplodiploidy) about 12 times in arthropods, rotifers and nematodes. Arrhenotoky in arthropods occurs, for example, in ants, wasps and bees (Hymenoptera), thrips (Thysanoptera), scales and whiteflies (Homoptera) and various taxa of mites (Acari). *See also* PSEUDOARRHENOTOKY; SEX RATIOS. [M.W.S. & C.J.N.]

REFERENCES

Bull, J.J. (1983) *Evolution of Sex Determining Mechanisms.* Benjamin/Cummings, Menlo Park, Calif.

White, M.J.D. (1973) *Animal Cytology and Evolution,* 3rd edn. Cambridge University Press, Cambridge.

Wrensch, D.L. & Ebbert, M.A. (eds) (1993) *Evolution and Diversity of Sex Ratio in Insects and Mites.* Chapman & Hall, New York.

Arsenic (As)

A heavy metalloid element (ATOMIC NUMBER=33; relative ATOMIC MASS=74.92; melting point=817°C; boiling point=613°C) which is a trace constituent (1.5 mg kg^{-1}) of the Earth's CRUST. It is found in minerals such as arsenopyrite (FeAsS) and orpiment (As_2S_3), but mainly in the sulphide ores of other metals (especially COPPER). Biologically non-essential and (subject to chemical form) highly toxic, arsenic has had many uses throughout history. These have ranged, in the extreme, from homicidal to therapeutic for humans. For example, arsenic trioxide (As_2O_3, 'white arsenic') was used by poisoners from Roman times until the 19th century, during which time there were also many accidental poisonings from the use of inorganic arsenic compounds in pigments (copper arsenite, $CuHAsO_3$, 'Scheele's green'; copper acetoarsenite, $Cu(CH_3COO)_2 \cdot 3Cu(AsO_2)_2$, 'Paris green') and agricultural PESTICIDES (lead arsenate, $PbHAsO_4$). Medically, arsenic sulphide was used long ago by Hippocrates to treat ulcers and, by the 19th century, compounds of arsenic such as Fowler's solution (1% $KH(AsO_2)_2$) were used as 'tonics' to treat a variety of ailments. At the beginning of the 20th century, the synthesis of organoarsenic compounds such as atoxyl (for sleeping sickness) and the arsanilic acid derivative, arsphenamine (Salvarsan 606), by Ehrlich (for the treatment of syphilis), ushered in the chemotherapeutic age. Today, however, the pharmacological use of organoarsenicals is very limited, although some are used as animal feed additives to promote growth. The major (and declining) uses of arsenic compounds today are as HERBICIDES and cotton desiccants (sodium salts of monomethylarsonic acid (MMAA), $CH_3AsO(OH)_2$, and of dimethylarsinic acid

(DMAA), $(CH_3)_2O(OH)$; sodium arsenite, $NaAsO_2$), wood preservatives (copper chromium arsenate), GLASS manufacture (As_2O_3) and semiconductor production (arsine gas, AsH_3).

Released to the environment from anthropogenic activities such as the mining of fossil and mineral resources, smelting and the combustion of FOSSIL FUELS, the behaviour of inorganic arsenic in soils, waters and sediments is governed to a large extent by prevailing redox and pH conditions and by its oxidation state. Arsenic forms oxyanions (arsenate, As(V), $H_2AsO_4^-$, $HAsO_4^{2-}$, AsO_4^{3-}; arsenite, As(III), $H_2AsO_3^-$, $HAsO_3^{2-}$, AsO_3^{3-}), which can be adsorbed onto ferric oxyhydroxides. Under reducing conditions, such as may be found at depth in lake sediments, As(III) may be released into pore waters, taking part in a redox-controlled cycle of dissolution, diffusion and reoxidation to As(V) and readsorption on Fe(III) compounds in surface oxic layers and possible PRECIPITATION as As_2S_3 at depth.

Biological intervention, however, by moulds, fungi and bacteria, can lead to the methylation of inorganic As(V) and As(III) in soils, sediments, SEWAGE, etc., through intermediates MMAA and DMAA, to the highly toxic gas trimethylarsine ($(CH_3)_3As$). In the marine environment, however, where arsenate predominates in seawater, arsenic speciation in the biota is dominated by a number of structurally more complex organoarsenic compounds, of which by far the most important is the inert and non-toxic arsenobetaine ($(CH_3)_3As^+CH_2COO^-$), the principal form of arsenic found in fish and crustaceans. Essentially, within the marine FOOD CHAIN, this is the end-product of a series of steps to defend against the potentially toxic effects of arsenate, which, being chemically analogous to nutrient phosphate, can readily enter algal cells and uncouple oxidative phosphorylation.

When humans eat fish, arsenobetaine is readily absorbed but rapidly excreted, unchanged, in urine and can be considered non-toxic. Both arsenate and arsenite, on the other hand, are toxic, the former uncoupling oxidative phosphorylation while the latter, with its affinity for SH groups, interferes with enzyme activity, most notably the pyruvate oxidase system, which is essential to cellular metabolism. The human body, via the action of *S*-adenosyl methionine in the liver, methylates the inorganic arsenic to the less toxic MMAA and DMAA, the latter predominating in urine, the principal route of excretion. This is important as the acute TOXICITY of inorganic arsenic is about 1–4 mg kg^{-1} body mass, although SH-containing CHELATING AGENTS, such as dimercaprol, are usually administered. Symptoms are severe, including vomiting, shock, intense pain and peripheral vascular failure. Chronic effects from long-term EXPOSURE to small amounts of inorganic arsenic via inhalation, ingestion or absorption through the skin, include cancer of the lung, liver and skin. The gas arsine (AsH_3) is very toxic to humans (as are its methylated derivatives), liberating haemoglobin from red blood cells. An overall order of toxicity is: AsH_3 > As(III) > As(V) > MMAA, DMAA >> arsenobetaine. *See also* HEAVY METALS. [J.F.]

REFERENCES

Cullen, W.R. & Reimer, K.J. (1989) Arsenic speciation in the environment. *Chemical Reviews* **89**, 713–764.

Nriagu, J.O. (ed.) (1994) *Arsenic in the Environment*. John Wiley & Sons, New York.

artesian *See* AQUIFER.

artificial classification Any CLASSIFICATION that contains one or more PARAPHYLETIC or POLYPHYLETIC GROUPS, rendering it inconsistent with the PHYLOGENY of the organisms, is an artificial classification or special reference system. In such cases the PHYLOGENETIC TREE cannot be wholly reconstructed from the classification scheme. An arrangement is a classification of a group whose phylogenetic relationships have not yet been delineated so it can be either a natural or an artificial classification. The overwhelming majority of current classifications are arrangements, serving as necessary but interim vehicles for classifying organisms until their phylogenetic relationships have been determined. Neither artificial classifications nor arrangements have been constructed via a rigorous, phylogenetic methodology. It is therefore inappropriate to convert such classification schemes into phylogenetic trees because one cannot assume that taxonomic relationships are consistent with phylogenetic relationships. [D.R.B.]

artificial radionuclides *See* RADIONUCLIDE.

artificial selection The deliberate choice of a select group of individuals to be used for breeding (Hartl & Clark 1989). The response to artificial selection is usually quite large, especially during the first generations of selection. This suggests that the response is due to GENETIC VARIANCE already present, rather than to new MUTATIONS. Other characteristic features of response to artificial selection include the following (Maynard Smith 1989).

1 There is a response to selection in both directions (upward and downward), although the response is often asymmetric.

2 The population reaches a selection limit, or plateau, particularly in small populations, beyond which further progress is difficult or impossible, partly because of exhaustion of genetic variance, partly because NATURAL SELECTION may oppose artificial selection.

3 When selection is relaxed, however, populations usually tend to return towards the original state (this shows that genetic variance still existed in the selected population).

4 Correlated responses to selection, i.e. indirect selection of characters other than those selected, usually show up, as a result of pleiotropic effects (*see* PLEIOTROPY) of the selected ALLELES, or of LINKAGE DISEQUILIBRIUM between the selected alleles and loci affecting other TRAITS in the initial population. Such correlated responses are useful to study trade-offs among LIFE-HISTORY TRAITS, thus providing a method to test the ANTAGONISTIC PLEIOTROPY HYPOTHESIS for the evolution of SENESCENCE. *Cf.* NATURAL SELECTION. [I.O.]

REFERENCES

Hartl, D.L. & Clark, A.G. (1989) *Principles of Population Genetics*, 2nd edn. Sinauer Associates, Sunderland, Mass.

Maynard Smith, J. (1989) *Evolutionary Genetics*. Oxford University Press, Oxford.

artificial substrate A material placed in the environment by an experimenter so that patterns of COLONIZATION and COMMUNITY development can be observed. A commonly used method for studying marine hard-substrate communities is to deploy slates or similar materials to act as SETTLEMENT panels. Periodic inspection of these panels provides information on patterns of colonization and on the competitive interactions between occupiers of primary and secondary space. Use of artificial substrates has the advantage of allowing high levels of experimental replication and control. In studies of aquatic soft-sediment communities, artificial substrates are often prepared by defaunating natural sediments, often by drying, and then returning them to the environment in sediment trays. By placing substrates in suitable frames they can be easily recovered from environments where repeated observation would otherwise be very difficult (e.g the DEEP SEA). *See also* SAMPLING METHODOLOGY/DEVICES. [S.J.H.]

as low as practicable (ALAP) A regulatory principle that can be applied to control EXPOSURE to harmful substances. The exclusion of 'reasonably', as in 'as low as reasonably practicable' (ALARP) and AS LOW AS REASONABLY ACHIEVABLE, signals that the costs of the regulatory measure do not weigh too heavily. [P.C.]

as low as reasonably achievable (ALARA) A regulatory principle that can be applied to EXPOSURE to harmful substances. Inclusion of 'reasonably' (*cf.* AS LOW AS PRACTICABLE) signals that the costs of the proposed measure ought to be balanced against the benefits arising. out of regulations. *See also* COST (RISK)–BENEFIT APPROACH; MAXIMUM PERMISSIBLE LEVEL. [P.C.]

as low as reasonably practicable (ALARP) *See* AS LOW AS PRACTICABLE.

asbestos The name given to a group of fibrous silicate minerals which, on account of excellent resistance to fire, heat and chemical attack, have been used in structural materials, brake linings, insulation and pipes. The commonest forms are chrysotile (white asbestos, $3MgO.2SiO_2.2H_2O$), crocidolite (blue asbestos, $Na_2O.Fe_2O_3.3FeO.8SiO_2.H_2O$) and amosite (brown asbestos, $(FeMg)SiO_3$). Asbestos is a human CARCINOGEN, with inhaled small fibres of crocidolite in particular penetrating deep into the lungs, where their physical nature can give rise to mesothelioma and other diseases, often after lengthy latency periods of 20 years or more. Although uses are now severely curtailed and replacement programmes are common, major EMISSIONS of asbestos to the air can still occur from demolition works and LANDFILL sites. [J.F.]

asexual reproduction In a broad sense, any form of reproduction not causing rearrangement of maternal genes, so producing offspring that are genetically identical by descent. In this sense, asexual reproduction is synonymous with cloning (*see* CLONE) either by somatic division (agametic cloning) or by PARTHENOGENESIS (gametic cloning).

In a strict sense, however, 'asexual reproduction' is reproduction not involving sexually derived processes. Parthenogenesis is excluded because it is derived from sexual reproduction. In this sense, asexual reproduction is synonymous only with agametic cloning (*see also* VEGETATIVE REPRODUCTION). It involves successive phases of GROWTH and division of the body, the mechanism of division varying from fragmentation by exogenous environmental forces to endogenous FISSION or budding. The rare phenomenon of polyembryony is a form of fission or budding confined to the embryonic stage in the LIFE CYCLE.

Asexual reproduction is of great importance in the multiplication and SPREAD of plants. Spores produced asexually, mostly unicellular but in some species multicellular, are almost universal in lower plants, fungi (some entirely lacking a sexual stage) and bacteria. Spores are usually formed by the SPOROPHYTE generation following meiosis, giving the haploid GAMETOPHYTE. New individuals of plants are also often formed by fragmentation and vegetative reproduction.

Gametic and agametic cloning have different evolutionary and ecological implications. Gametic cloning originates as aberrant gametogenesis and evolutionarily is difficult to reverse. Apart from the highly successful, but taxonomically rare, cyclical parthenogens (e.g. aphids and cladocerans), gametic cloners are incapable of switching to normal sexual reproduction and consequently may have reduced long-term evolutionary potential. Agametic cloning is independent of gametogenesis and therefore does not preclude sexual reproduction. Gametic cloning incurs the time-consuming and energetically expensive process of embryogenesis, whereas agametic cloning is simply an extension of body growth, producing offspring at a relatively advanced state of development and often of a relatively large size.

Because they can reproduce as isolated individuals, clonal organisms often make good colonizers. Population growth is enhanced by the lack of males among the progeny of gametic cloners and by the avoidance of embryogenesis in agametic cloners. Good colonizing ability and high intrinsic rate of population growth enable many clones to exploit unstable habitats opportunistically. On the other hand, grouping and aggressive cooperation among clone-mates enable some species to dominate stable habitats competitively. *See also* EVOLUTION OF SEX. [R.N.H. & A.J.W.]

REFERENCE

Hughes, R.N. (1989) *A Functional Biology of Clonal Animals.* Chapman & Hall, London.

assemblage A general term for a collection of plants and/or animals, or the fossilized remains of these organisms in a geological SEDIMENT. Unlike the term 'COMMUNITY', it does not imply interrelationships between the organisms and, in the case of FOSSIL ASSEMBLAGES, many and various factors may have been involved in the assembling of the individuals. Often, however, the composition of an assemblage can provide information concerning the environment in which it is found and the past history of the site. [P.D.M.]

assemblage zone A BIOSTRATIGRAPHIC UNIT that is defined on the basis of the total fossil content, all or part of which constitutes 'a natural assemblage or association which distinguishes it in biostratigraphic character from adjacent strata' (Hedberg 1972). The upper and lower boundaries of assemblage zones have been traditionally placed on the basis of careful inspection of the fossil composition of the rock or sedimentary strata. In this, the experience gained by the investigator from study of similar strata at other sites may play a significant role. Various computer-based techniques for the analysis of multivariate data are often employed in defining assemblage zones, but mainly as a means of independent verification of the boundaries of assemblage zones already established by traditional quasi-subjective methods (Birks 1987).

Local assemblage zones are normally established on the basis of one or more local studies and, subsequently, regional assemblage zones are established which embrace one or more local assemblage zones. An example is the *Artemisia* regional pollen assemblage zone (PAZ) as defined by Watts (1977). This highly characteristic PAZ embraces the herb-dominated, *Artemisia*-rich pollen spectra which are characteristic of that part of pollen profiles from Ireland relating to the climatic deterioration known as the Younger Dryas and datable to *c.* 11 400–12 570 cal. BP (radiocarbon-calibrated (sidereal) years before present). In this instance there is close correspondence with the lithostratigraphy, which is invariably characterized by highly minerogenic sediments, the result of solifluction in the basin catchment. [M.O'C.]

REFERENCES

Birks, H.J.B. (1987) Multivariate analysis of stratigraphic data in geology: a review. *Chemometrics and Intelligent Laboratory Systems* **2**, 109–126.

Hedberg, H.D. (1972) Summary of an international guide to stratigraphic classification, terminology, and usage. *Boreas* **1**, 213–239.

Watts, W.A. (1977) The Late Devensian vegetation of Ireland. *Philosophical Transactions of the Royal Society of London, Series B* **280**, 273–293.

assembly rules A term originally used by Jared Diamond (1975) in his studies of the land-birds living on the islands of the Bismarck archipelago off the coast of New Guinea. Diamond suggested that compared with all the possible combinations of species, from the species pool of the archipelago, only certain 'permissible' combinations of species actually occurred on individual islands. He believed that these permissible combinations resist invasion by species that would transform them into 'forbidden' combinations. The forbidden combinations do not exist in nature because they would transgress one or more of three types of rule. These assembly rules he called:

1 compatibility rules;
2 incidence rules;
3 combination rules.

Diamond illustrates his ideas about assembly rules by reference to a GUILD of cuckoo-doves. These are a group of arboreal, long-tailed, fruit-eating pigeons that live in the middle storey of shaded forest from India to Melanesia. In the Bismarck archipelago there are four species/superspecies. These are the *Reinwardtoena* superspecies (R; 279–315 g), *Macropygia amboinensis* (A; 128–178 g), *M. mackinlayi* (M; 73–110 g) and *M. nigrirostris* (N; 73–97 g).

1 *Compatibility rules.* Distributional information combined with a knowledge of the ecology of species would suggest that certain pairs of species are incapable of coexistence on an island under any circumstances, That is, either alone as a pair, or as part of a larger combination. *Macropygia mackinlayi* and *M. nigrirostris* are such a pair. These species are of similar size, with similar ecologies, and they coexist on no island. Thus the combination MN as well as MNR, AMN, and AMNR are forbidden by compatibility rules.

2 *Incidence rules.* For each of the common species of land-bird in the Bismarck archipelago, Diamond (1975) produced what he called INCIDENCE FUNCTIONS. These are graphs in which the probability of occurrence on an island (J=(Number of islands with the species present)/(Total number of islands)) is plotted against the number of species (S) on that island. Because of the species–area effect, S is also a measure of island size. Figure A23 shows the incidence functions for the four cuckoo-doves. We can see from Fig. A23 that for all values of $J_N>0$, $J_A=1$. That is, all islands that have *nigrirostris* also have *amboinensis*. Therefore any combination containing N but not A (N, NR, MN and MNR) is forbidden.

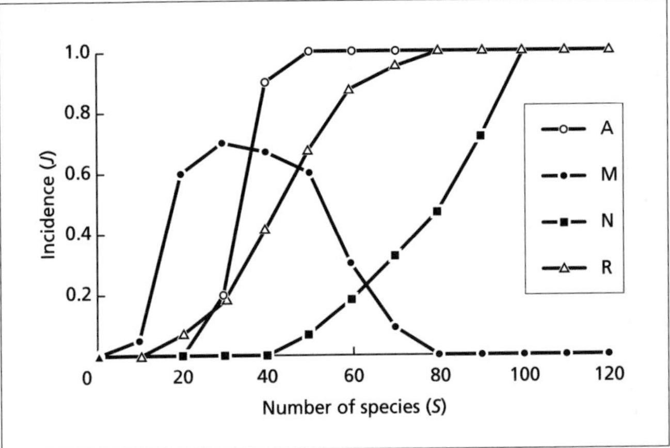

Fig. A23 Incidence functions for the four cuckoo-doves, *Reinwardtoena* superspecies (R), *Macropygia amboinensis* (A), *M. mackinlayi* (M) and *M. nigrirostris* (N).

3 *Combination rules.* Although not forbidden by compatibility or incidence rules, the combinations AMR, AN and R are not found on the Bismarck islands. In the latter two cases this may simply be a sampling problem, but in the case of AMR enough islands of the appropriate size have been sampled to expect its occurrence to be seen. For example, looking at the product $J_A J_M J_R$ in the region around S=40 (i.e. when J_N is also zero), would suggest that about one-third of all islands of this size should support this combination. Of course, if there were only two islands of this size, the combination could be missing by chance (as may be the case for the combinations AN and R), but since there are 16 sampled islands in this range the absence of the combination AMR may be of ecological significance. *See also* COMMUNITY ECOLOGY. [B.S.]

REFERENCE

Diamond, J.M. (1975) Assembly of species communities. In: *Ecology and Evolution of Communities* (eds M.L. Cody & J.M. Diamond), pp. 342–444. Belknap, Cambridge, Mass.

assessment end-point *See* END-POINT.

assimilation The absorption and use of simple NUTRIENTS and/or products of digestion in metabolism and building up the constituents of an organism. Assimilated energy is also called metabolizable energy (ME), i.e. that proportion of the digested energy that is not excreted as urinary waste but remains available to fuel resting metabolism, the animal's activities and the production of new somatic and reproductive tissues. It is measured as the difference between the amount or energy content of food consumed minus the sum of materials or energy defecated and excreted, or by adding the amount of energy lost during RESPIRATION to the amount incorporated due to production of new tissues. *See also* ENERGY BUDGET. [M.H.]

assimilation efficiency Assimilation EFFICIENCY is the amount of food assimilated expressed as a percentage of the amount of food consumed. However, assimilation is usually intended to include absorbed energy only (*cf.* ASSIMILATION), so this efficiency is more properly an absorption efficiency. *See also* ECOLOGICAL EFFICIENCY; ENERGY BUDGET. [M.H.]

assimilative capacity The extent to which a system can take up, and (by implication) process, XENOBIOTICS. [P.C.]

association Similar to CORRELATION, but a more general term. For example, it applies to the relationship between QUALITATIVE INFORMATION in a CONTINGENCY TABLE. [B.S.]

association hypothesis The argument that females are more likely to care for their young than males because they are more likely to be in a position to help the young at birth. It is supported by the observation that male PARENTAL CARE is more common in fishes with external rather than internal FERTILIZATION. However, this observation could also be explained by a reliability of PATERNITY argument. T.H. Clutton-Brock suggests that the hypothesis could be tested by studying species where females lay eggs shortly after fertilization. Male care should be as common in these species as it is in species with external fertilization. [D.A.B.]

REFERENCES

Clutton-Brock, T.H. (1991) *The Evolution of Parental Care*. Princeton University Press, Princeton, N.J.

Gross, M.R. & Shine, R. (1981) Parental care and mode of fertilization in ectothermic vertebrates. *Evolution* **35**, 775–793.

assortative mating A reproductive system in which matings are not at random, with respect to PHENOTYPE or GENOTYPE characteristics of mating pairs. Assortative mating (or homogamy) refers to cases in which mating occurs preferentially among similar phenotypes (e.g. there could be homogamy at the flower colour loci in flowering plants, if pollinators forage one colour at a time), whereas DISASSORTATIVE MATING refers to cases in which mating occurs preferentially among different phenotypes (e.g. there is disassortative mating at the incompatibility locus in the plant species *Trifolium repens*). Assortative mating only modifies HETEROZYGOSITY at the loci involved in the choice of sexual partners, and possibly statistically linked loci to these loci. It creates a FREQUENCY-DEPENDENT SELECTION at these loci. *See also* MATING SYSTEMS. [I.O.]

astatine (At) An element (ATOMIC NUMBER = 85; relative ATOMIC MASS = 210; melting point = 302°C; estimated boiling point = 337°C) that occurs in only minute amounts in nature as part of the radioactive decay series of ^{235}U; the half-life of ^{219}At is only 54 s. Even the most stable ISOTOPE, ^{210}At, has a half-life of 8.3 h, and so knowledge of the chemistry of astatine is extremely sparse. [C.R.P.]

Astronomical Almanacs *See* TIME.

asymmetric competition COMPETITION, usually between a pair of species, in which the adverse effect of one species on the other is much greater than the reciprocal effect. The extreme situation is called AMENSALISM. Lawton and Hassell (1981) suggest that for many pairs of insect species, strong asymmetric competition, indistinguishable from amensalism, may be the norm. They list 23 examples, from the literature, of such strong asymmetric competition. [B.S.]

REFERENCE

Lawton J.H. & Hassell, M.P. (1981) Asymmetrical competition in insects. *Nature* **289**, 793–795.

Atlantic North American floral region Part of the Boreal subkingdom of the Holarctic kingdom of Takhtajan (equivalent to the Atlantic North American PROVINCE of Neill (1969); see Fig. B6). It comprises a large part of North America, from the central part of Alaska, most of Canada south of 60°N, and the USA from east of the Rockies to the Atlantic seaboard. It has about 100 endemic vascular plant genera (*see* ENDEMIC SPECIES), including the insectivorous *Sarracenia* and *Dionaea*, the beach heath, *Hudsonia*, and the bellwort, *Uvularia*. The bulk of the flora consists of genera common to either the Asian or European parts of the Holarctic area. Within this region, three subregions are recognized.

1 The Canadian–Appalachian subregion, forming the northern and eastern part, extending from Alaska eastwards to Ontario and Quebec, and southwards as far as Georgia and Texas. The weakness of the MIGRATION barrier represented by the Bering Strait, and its impersistence within Quaternary time, is reflected in the floristic affinity of this subregion with the Euro-Siberian region.

2 The southern Atlantic–North American subregion is formed of the coastal plain and the states bordering on the Gulf of Mexico. This has, rather surprisingly, some common elements with the Sino-Japanese region.

3 The Central grasslands subregion, comprising the prairies and plains east of the Rocky Mountains.

[W.G.C.]

REFERENCES

Neill, W.T. (1969) *The Geography of Life*. Columbia University Press, New York.

Takhtajan, A. (1969) *Flowering Plants, Origin and Dispersal*. Oliver & Boyd, Edinburgh.

Atlantic Ocean The warmest, saltiest, and second largest of the world's oceans and the first to be explored. The Atlantic Basin has an S-shape and stretches between Europe and Africa in the east and America in the west; its northern and southern boundaries are less clearly defined (the broadest definition extends from the Antarctic continent in the south to the Bering Straits in the north). The area, volume and mean depth of the Atlantic are 8.4×10^7 km^2, 3.23×10^4 km^3, and 3844 m, respectively, or 10.6×10^7 km^2, 3.51×10^4 km^3, and 3293 m, if its marginal seas are included. The deepest part of the Atlantic occurs in the Milwaukee Depth in the Puerto Rico Trench at 9219 m. The basin of the Atlantic can be separated into eastern and western troughs, symmetrically distributed on either side of the Mid-Atlantic Ridge. In contrast to the South Atlantic, which has one marginal SEA (i.e. the Weddell Sea), the North Atlantic has many marginal seas (i.e. MEDITERRANEAN SEA, Gulf of Mexico, NORTH SEA, BALTIC SEA, English Channel, Irish and Celtic

Seas, NORWEGIAN SEA, Greenland Sea, Irminger Sea, Labrador Sea, Hudson Bay and Baffin Bay), some of which have a major influence on the CIRCULATION of the entire Atlantic.

The Atlantic has the most active vertical mixing of the world's oceans, and North Atlantic deep water is of critical global importance (the two major hydrodynamical sinks of the world OCEAN are both located in the Atlantic). In contrast to the nearly anaerobic deep Pacific, deep Atlantic water is highly oxygenated. The area draining into the Atlantic is approximately four times that draining into the Pacific or Indian Oceans. The bottom of the Atlantic Ocean is underlain by a layer of SEDIMENT twice as thick as found in the Pacific, and consists of TERRIGENOUS (or land-derived) material, biogenic material (e.g. remains of marine organisms), authigenic material (i.e. deposits formed in place such as MANGANESE nodules), and cosmic material (derived from extraterrestrial sources). Approximately 67.5% of the area of the Atlantic floor is covered by calcareous ooze, 6.7% by siliceous ooze and 25.8% by red CLAY. *See also* BARENTS SEA. [V.F.]

atmosphere The mainly gaseous envelope overlying the solid and liquid surface of a planet. The Earth's atmosphere has mass equivalent to a 10-m-deep envelope of WATER, 90% of which is concentrated in the first 16 km above sea level. It consists almost entirely of air—a mixture of gases, many of which (with the important exception of water vapour) are kept in highly uniform proportions by stirring. In order of diminishing proportion the gases are: dinitrogen (N_2), dioxygen (O_2), ARGON (Ar), water vapour (H_2O; concentrated in the lowest few kilometres by efficient cloud formation and PRECIPITATION), CARBON DIOXIDE (CO_2), and hundreds of TRACE GASES such as OZONE (O_3). The Earth's atmosphere also contains very small amounts of condensed matter. In order of diminishing total mass they are cloud (localized suspensions of *c.* 10 μm scale water droplets and ICE crystals widespread in the TROPOSPHERE), precipitation (RAIN, hail and snow falling through and from thicker CLOUDS), and AEROSOL (the diffuse suspension of submicrometre scale particles and solution droplets in the troposphere and STRATOSPHERE). *See also* ATMOSPHERIC CHEMISTRY; ATMOSPHERIC POLLUTION. [J.F.R.M.]

atmospheric boundary layer *See* PLANETARY BOUNDARY LAYER.

atmospheric chemistry In the TROPOSPHERE, the chemical reactions by which trace substances released into the ATMOSPHERE from natural and anthropogenic sources are oxidized. The reactions form stable compounds, such as CARBON DIOXIDE (CO_2) and WATER, or more soluble compounds, such as peroxides or ACIDS, which are rained out from the atmosphere. In the STRATOSPHERE, the term refers particularly to the chemical processes that determine the concentration of OZONE. *See also* STRATOSPHERIC CHEMISTRY; TROPOSPHERIC CHEMISTRY. [P.B.]

Atmospheric circulation

The patterns of air motion in the ATMOSPHERE, especially the TROPOSPHERE, after removing transient, localized distortions associated with individual WEATHER systems.

In low latitudes (roughly between 30°N and 30°S) the circulation and associated climatic zones are dominated by the Hadley circulation (Fig. A24), containing opposing Hadley cells of gentle but persistent meridional motion. The TRADE WINDS form the base of a shallow, relatively steady equatorward airflow from the flanks of the SUBTROPICAL high-pressure systems which picks up WATER vapour from the surface, and feeds warm, moist air into the intertropical convergence zone. There the air rises rapidly to the high troposphere in several thousand deep shower CLOUDS, scattered unevenly along the zone in various types of tropical weather system, and depositing heavy bursts of PRECIPITATION on areas experiencing their rainy season. The risen air, its POTENTIAL TEMPERATURE raised by release of latent heat during cloudy ascent, diverges polewards and sinks gently as it loses heat by net emission of terrestrial (infrared) RADIATION. Near latitudes 30° this subsidence is locally intensified, filling much of the troposphere from the top down with dry air. Cloud and precipitation are rare and sporadic in these subtropical high-pressure zones, which overland maintain the great hot DESERTS of North Africa, Australia, etc. Conservation of the Earth's westerly angular momentum in the upper arm of the Hadley circulation maintains strong westerlies in the upper troposphere at the poleward edge of each Hadley cell, forming the local subtropical jet stream. In the lower arm a similar tendency in reverse maintains the easterly component of the trade winds as they flow towards the Equator, though speeds are kept low by surface friction.

Poleward of the subtropical high-pressure zones, the strong poleward temperature lapse combines dynamically and thermodynamically with the local vertical component of the Earth's rotation to maintain the middle latitudes in a chronically disturbed state, with

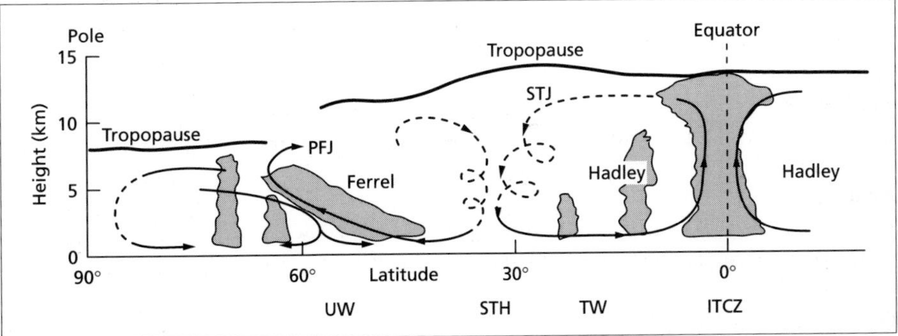

Fig. A24 A schematic vertical cross-section along a meridian showing the Hadley and Ferrel circulations and the intertropical convergence zone (ITCZ), subtropical jet stream (STJ), subtropical high pressures (STH), trade winds (TW), unstable westerlies (UW) and polar front jet stream (PFJ), together with cloudy weather systems (shaded), and the tropical and higher latitude tropopauses.

only a very feeble meridional Ferrel cell to compare with the strong opposing Hadley cell of lower latitudes. The disturbances of this zone of unstable westerlies appear as transient depressions and anticyclones, with lifetimes ≈ 10 days and horizontal dimensions ≈ 2000 km, in which tongues of warm air climb polewards from the poleward flanks of the subtropical high-pressure zones, and cool air sinks equatorwards from high latitudes, populated with showers as it flows over progressively warmer surfaces.

The rising moist airflows form extensive sheets of deep stratiform cloud along active polar FRONTS (zones of concentrated temperature gradient and weather), and veer eastwards by conservation of westerly angular momentum to maintain a vast CIRCUMPOLAR vortex of strong westerly flow throughout the mid-latitude troposphere. In this flow the individual depressions drift eastwards and slightly polewards as they develop, mature and die, with a strong westerly polar front jet stream in the high troposphere on the cold side of each front. Depressions lose their vigour when they move inland out of reach of the warm moist air which is their fuel. In continental interiors and eastern margins this decay leads to hot, sunny summers and bitterly cold winters, accentuating the typical mid-latitude annual cycle of temperature and contrasting with the moderate, changeable conditions over the oceans and in western CONTINENTAL MARGINS.

ANTICYCLONIC circulation in the highest latitudes is especially marked in the Southern hemisphere, where air subsides over the Antarctic plateau and spills over its edge to the poleward edges of the mid-latitude weather systems. *See also* AIR, CHEMICAL AND PHYSICAL PROPERTIES. [J.F.R.M.]

---

atmospheric pollution The addition of substances to the ATMOSPHERE that have a capacity to cause HARM to human health and/or ecological systems. It can involve the emission of chemicals from stacks, the emission of SMOKE and SOOT from chimneys, the emission of smoke, soot and chemicals from fire, explosions, leaks, etc. *See also* AIR POLLUTION; SMOG. [P.C.]

atoll *See* REEF.

atomic absorption spectrometer Analytical equipment used to quantify the concentrations of dissolved metals. Sample solutions for analysis, containing the element of interest, are rapidly heated in a flame or a graphite furnace, to convert the element into individual atoms, which emit and absorb LIGHT of a specific wavelength. When light of the specific wavelength is shone through the heated gases, the amount of light absorbed is proportional to the concentration of element in the sample. Variants include releasing ELEMENTS from solution as the gaseous hydride (e.g. ARSENIC) before heating to produce atoms, or using the cold elemental vapour (MERCURY). Commercial instruments may be fully automated, from the dissolution of the solid sample to the final quantification of the element of interest. [J.N.C.]

atomic mass The mass of an ISOTOPE of an element measured in atomic mass units (amu). In 1961 the amu was formally defined as $^1/_{12}$ of the mass of ^{12}C CARBON. The absolute mass of an amu is about 1.66×10^{-27} kg. Examples include: HYDROGEN, 1.008; LEAD, 207.19; THORIUM, 232.04. [J.G.]

atomic number The atomic number, Z, is the number of protons in the nucleus of an atom or the number of electrons associated with one atom. Examples include: for HYDROGEN, $Z=1$; for LEAD $Z=82$; for THORIUM $Z=90$. [J.G.]

attack rate *See* SEARCHING EFFICIENCY.

attenuation, light Diminution of light intensity with depth in the OCEAN due to absorption and scattering in the water column. Attenuation is expressed as the sum of the fraction of the radiant flux absorbed and the fraction scattered. Water is a poor absorber of visible RADIATION below about 560 nm wavelength, with minimal attenuance at about 475 nm. Thus, red light attenuates most rapidly in water whereas blue light penetrates deepest. Differences in colour between clean water (blue-green) and water containing DISSOLVED ORGANIC MATTER (yellow) are due to differential absorption of light by water molecules compared to organic substances. [V.F.]

attractor The solution to which a system tends after a period of time. An attractor can be described by a single point (a limit-point attractor) or limit-cycle (*see* STABLE LIMIT CYCLE). In chaotic systems the resulting attractors often have a much more intricate geometric structure in phase space than they do in non-chaotic systems and are called strange attractors. Strange attractors are characterized by having a fractal (*see* FRACTAL ANALYSIS). Dynamical systems can have multiple attractors, and which of these is approached depends on the initial conditions. The catchment area of each attractor is known as the basin of the attractor. [V.F.]

aufwuchs *See* BENTHIC HABITAT CLASSIFICATION.

Australian faunal region The principal land masses of this region are Australia and New Guinea (see Fig. B5). The most distinctive features are the occurrence of highly distinctive endemic lungfishes and mammals; there are no comparable endemic groups of amphibians, reptiles or birds (*see also* AMBHIBIANS, DIVERSITY AND CONSERVATION; REPTILES, DIVERSITY AND CONSERVATION).

Among the few freshwater fishes is a striking endemic RELICT ceratodontid lungfish, assigned to a separate family. The only Australian amphibians are frogs, some with South American affinities. Reptiles include many geckos and skinks, varanids and crocodiles (in the north) and a wide variety of snakes, including many elapids.

The birds of the Australian region include representatives of about 58 families, 12 of which are wholly or nearly endemic. These include cassowaries, emus, lyrebirds, honey-eaters, bell magpies, bowerbirds and birds of paradise. The region also has abundant and diverse pigeons and parrots, including important endemic groups. The proportion of Australian birds in endemic or nearly endemic families is larger than in any other continental FAUNAL REGION except the NEOTROPICAL. The native land mammals comprise a few monotremes, many marsupials belonging to six families, rather diverse rodents of the family Muridae, fruit bats and six families of insectivorous bats. *See also* AUSTRALIAN FLORAL REALM. [A.H.]

REFERENCE

Darlington, P.J. (1957) *Zoogeography: the Geographic Distribution of Animals.* John Wiley & Sons, New York.

Australian floral realm The floral realm corresponding to the Australian kingdom of Takhtajan (1969) (see Fig. B6). The biogeographic isolation of the Australian plate as it drifted away from the great southern continent of GONDWANA during the MESOZOIC era, prior to its contact with the Indonesian region during the CENOZOIC, has left a strong stamp on the plant life of that continent. There are about 400 endemic genera of flowering plants in Australia, and a number of families, or large parts of families, are endemic. These include most notably the Epacridaceae, Pittosporaceae, many members of the Proteaceae (including *Banksia*, which gets into New Guinea) and many of the Myrtaceae (including *Eucalyptus*, which extends into Malaysia). The Casuarinaceae (comprising the single genus *Casuarina*, which some authors would split into three separate genera) is also a characteristic element of this realm, but extends far outside it into Indonesia and Malaysia and the Mascarene Islands, and to New Guinea and Fiji. With a secure FOSSIL RECORD in Australia going back some 60 million years, it is likely that this family (like the Epacridaceae and the genera *Banksia* and *Eucalyptus*) radiated on the Australian plate in the course of its northward movement from Gondwana. Its members then spread to the north and east as it came into juxtaposition with south-east Asia.

This kingdom is subdivided by Takhtajan into three floristic regions. The North and East Australian region extends as a broad belt around the whole northern, north-eastern and eastern margins of Australia, down to and including Tasmania. The Central Australian region (with the central and southern DESERTS, and the SAVANNAHS of the north and east) occupies most of the remainder of the continent, while the extreme south-western corner constitutes a third, separate region. This includes chaparral (*see* CHAPARRAL SHRUBLANDS) and subtropical evergreen FOREST isolated by the dry central part of the continent from the comparable evergreen forests of its eastern side, and this isolation is reflected in the 75% level of endemism at the species level. In this respect (but not in the actual composition of the flora) it is comparable to the CAPE FLORAL REGION. *See also* ANGIOSPERMS, GLOBAL DIVERSITY AND CONSERVATION; AUSTRALIAN FAUNAL REGION. [W.G.C.]

REFERENCE

Takhtajan, A. (1969) *Flowering Plants, Origin and Dispersal*. Oliver & Boyd, Edinburgh.

autapomorph An apomorphic TRAIT (*see* APOMORPH) that has become fixed within a single species and is considered diagnostic for that species. Autapomorphs play a role in cladistic systematics (*see* CLADISTICS), which aims to reflect phylogenetic relationships between organisms in their CLASSIFICATION. Autapomorphy does not help resolve phylogenetic relationships, although it does give valuable information about different degrees of divergence among contemporaneous taxa. For example, spoken language is an autapomorphy of *Homo sapiens* among primates which, by itself, provides no evidence about who are our closest relatives. *Cf.* SYNAPOMORPHY. [D.R.B. & P.O.]

autecology The ECOLOGY of individual species as opposed to the ecology of whole communities. Autecology includes behavioural, physiological and population ecologies. [M.H.]

authorization Legal acceptance (usually in document form) of the release of EMISSIONS/EFFLUENTS of specified quality and quantity, usually with the aim of preventing POLLUTION. *See also* CONSENTS; POLLUTION PERMITS. [P.C.]

autochthonous Describing material originating in its present position; for example BIOMASS (leading to SEDIMENT) that is formed from PHOTOSYNTHESIS in a freshwater lake or river. *Cf.* ALLOCHTHONOUS. [P.C.]

autogenic The term was introduced by Tansley (1929, 1935; *see* ALLOGENIC; CHARACTERS IN ECOLOGY). He first applied it to successions that take place as a result of the effects of the plants on the HABITAT, and a contrast was made with allogenic SUCCESSIONS, which result from the action of forces originating outside the system. Later he applied the terms to factors rather than successions, and suggested that both allogenic and autogenic factors are present in all successions.

Autogenic effects 'drive' many successions, i.e. cause them to take a certain course, or cause them to take place much more rapidly. One clear example is fixation of NITROGEN by plants with root nodules containing symbiotic bacteria; this is especially important in primary successions and in post-fire SECONDARY SUCCESSIONS (fire commonly volatilizes much of the soil nitrogen while leaving phosphate and cationic NUTRIENTS in the ash). Other examples are rapid ACIDIFICATION of the soil, change of HUMUS-type from MULL to MOR, waterlogging of the system by bog mosses (*Sphagnum* spp.), stabilization of the substratum (as on young DUNES or the lower parts of SALT MARSHES and MANGROVE swamps), and raising of the level of the substratum relative to that of the water (as through accumulation of organic debris, and through the build-up of tussocks that emerge from the water in fens). [P.J.G.]

REFERENCES

Tansley, A.G. (1929) Succession: the concept and its values. In: *Proceedings of the International Conference of Plant Sciences (Cornell 1926)*, pp. 677–686.

Tansley, A.G. (1935) The use and abuse of vegetational concepts and terms. *Ecology* **16**, 284–315.

automictic thelytoky *See* PARTHENOGENESIS.

automixis The production of females from unfertilized EGGS, with a normal meiosis producing four haploid pronuclei. The diploid number is then restored by the fusion either of two of the haploid products of meiosis, or of two genetically identical nuclei produced by the mitotic division of the haploid cell nucleus. Offspring differ from their mother, in contrast to APOMIXIS and ENDOMITOSIS. The offspring are usually homozygous at some or all loci at which the parent was heterozygous. *See also* ASEXUAL REPRODUCTION; PARTHENOGENESIS. [I.O.]

automobile pollution *See* MOTOR VEHICLE POLLUTION.

autopolyploid An organism with more than two chromosome sets, all from the same ancestral species. Autopolyploidy can induce SYMPATRIC SPECIATION through postzygotic REPRODUCTIVE ISOLATION (e.g. a cross between an autotetraploid and its diploid ancestor will produce sterile triploids). Autopolyploids often have low fertility and may be rare in nature. Common causes of low fertility include meiotic irregularities, which can arise because each chromosome has more than one homologue with which it can pair (resulting in multivalent formation and irregular segregation patterns), and incomplete sexual differentiation, as POLYPLOIDY can disrupt the balance between sex chromosomes achieved by diploid ancestors. [M.M.]

autotroph Simply, an organism that can make its own food: the term is derived from the Greek *auto*, self, and *trophos*, feeder. Autotrophs can exploit simple inorganic compounds, using CARBON DIOXIDE (CO_2) or carbonates as the CARBON source for building their organic constituents. Two categories of autotroph are defined by the energy source used in the synthesis of their organic requirements from inorganic carbon species.

1 Photoautotrophs: these use LIGHT energy, and comprise all photosynthetic organisms, including most terrestrial plants, algae and photosynthetic bacteria (*see also* PHOTOSYNTHESIS).

2 Chemoautotrophs: these comprise bacteria that obtain energy from the oxidation of simple inorganic or 1-C organic compounds and can use the energy released to assimilate CO_2 and transfer the energy into organic compounds. For example, *Thiobacillus* species can obtain their energy by oxidizing hydrogen sulphide or elemental SULPHUR to sulphuric acid.

Cf. HETEROTROPH. *See also* CHEMOSYNTHETIC MICROBES. [S.P.L.]

average *See* MEASURE OF CENTRAL TENDENCY.

average effect of a gene substitution In QUANTITATIVE GENETICS, the average effect on a QUANTITATIVE TRAIT of gene substitution at a bi-allelic locus is defined as the average deviation from the population MEAN of those individuals that have received the allele from one parent, when the allele received from the other parent is a RANDOM SAMPLE of the population. Equivalently, it is the slope $\alpha = a + d(q-p)$ of the linear regression of GENOTYPIC VALUES on number of A_1 ALLELES in the diploid GENOTYPE. Alternatively, α is given by half the first derivative of the mean genotypic value at locus A towards the ALLELE FREQUENCY p of A_1. The average effect of a gene substitution plays the role of additive effect of an allele and appears in the ADDITIVE GENETIC VARIANCE of a quantitative TRAIT. *See also* MEAN GENOTYPIC VALUE. [G.D.J.]

AVHRR *See* ADVANCED VERY-HIGH-RESOLUTION RADIOMETER.

avian ecology *See* ECOLOGY.

aviation *See* AIRCRAFT POLLUTION.

avoidance *See* DEFENCE MECHANISMS.

a_x In a LIFE TABLE, the number of individuals in AGE CLASS x or at stage x (where x defines the order of the classes or stages). [A.J.D.]

Ayala's experiments

In the late 1960s and early 1970s, Francisco Ayala conducted a series of laboratory 'bottle' experiments (Kareiva 1989), each using a pair of *Drosophila* species (Ayala 1969, 1970, 1972). Although five species were used in different combinations, the basic design and outcome of all these experiments was very similar. One experiment, using *D. serrata* and *D. pseudoobscura*, will be sufficient to illustrate the apparent dilemma that emerged.

Drosophila serrata comes from Australia and *D. pseudoobscura* comes from North America. Notice, therefore, that since neither of these flies comes from the same continent, let alone the same community, Ayala was simply using the flies like analogue computers to find out what might happen between flies in real communities.

At 25°C, *D. serrata* eliminates *D. pseudoobscura* in a few generations, whereas at 19°C *D. pseudoobscura* eliminates *D. serrata*. Ayala performed his experiments at 23.5°C, at which temperature the two species coexisted. Populations were started with 300 adult flies of each species and maintained in a series of milk-bottles with standard *Drosophila* medium (a kind of yeasty porridge). Adult flies feed off the yeasty surface of the medium, female flies lay their eggs on to the surface and larvae feed within it. Third instar larvae pupate on the surface of the medium or on the side of the bottle. When an apparent equilibrium was reached in the mixed populations, single species populations were established and maintained under the same conditions. From the population data, and assuming that the LOTKA–VOLTERRA MODEL of COMPETITION describes what is happening between the two species, Ayala was able to calculate the COMPETITION COEFFICIENTS, α_{12} and α_{21}. By rearranging the Lotka–Volterra equations he obtained:

$$\alpha_{12} = \frac{K_1 - \dfrac{dN_1/dt \cdot K_1}{r_1 \cdot N_1} - N_1}{N_2}$$

$$\alpha_{21} = \frac{K_2 - \dfrac{dN_2/dt \cdot K_2}{r_2 \cdot N_2} - N_2}{N_1}$$

However, by assuming that the two species populations had reached an equilibrium ($dN_1/dt = dN_2/dt = 0$) these equations can be reduced to:

$$\alpha_{12} = \frac{K_1 - N_1}{N_2}$$

$$\alpha_{21} = \frac{K_2 - N_2}{N_1}$$

where K_1 and K_2 equal the equilibrium population densities of each species on their own and N_1 and N_2 equal the equilibrium population densities for each species in the mixed populations.

All the pairs of species that Ayala studied coexisted for many generations; however, in every case he found that:

$$\alpha_{12} > \frac{K_1}{K_2} \quad \text{and} \quad \alpha_{21} > \frac{K_2}{K_1}$$

This is not the 'case for coexistence' in the Lotka–Volterra model of competition. Intriguingly, Ayala (1969) did not conclude that the model was incorrect but rather that the principle of COMPETITIVE EXCLUSION was false.

Of course, the Lotka–Volterra model incorporates the LOGISTIC EQUATION, which assumes (as a first approximation) that the relationship between rate of increase and DENSITY is a straight line. However, Gilpin and Justice (1972) examined the relationship

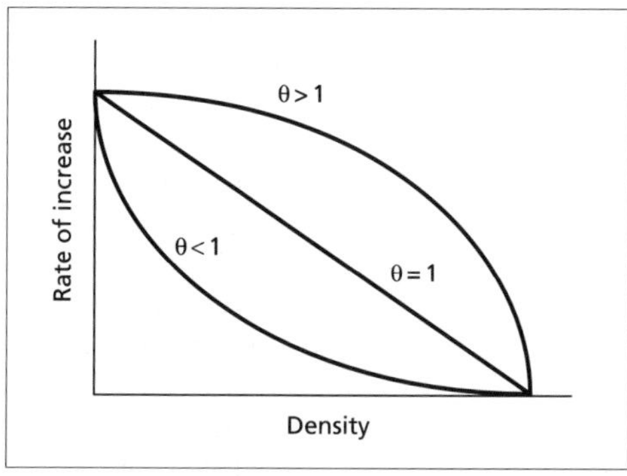

Fig. A25 The nature of density dependence for different values of θ in Gilpin and Ayala's modified Lotka–Volterra model.

between birth/death and density in Ayala's experiments and found that for these *Drosophila* populations it was frequently a curve. It is not surprising, therefore, that an apparent conflict had occurred between theory and practice. This led Gilpin and Ayala (1973) to develop a new Lotka–Volterra model that incorporated such 'non-linear' relationships. Their new model of competition has four parameters, one more than the Lotka–Volterra model, and explains 95% of the variance in Ayala's experiments. For two species, the model can be written:

$$\frac{dN_1}{dt} = r_1 N_1 \left[1 - \left(\frac{N_1}{K_1} \right)^{\theta_1} - \left(\frac{\alpha_{12} \cdot N_2}{K_1} \right) \right]$$

$$\frac{dN_2}{dt} = r_2 N_2 \left[1 - \left(\frac{N_2}{K_2} \right)^{\theta_2} - \left(\frac{\alpha_{21} \cdot N_1}{K_2} \right) \right]$$

The new parameter (θ) describes the 'non-linear' nature of DENSITY DEPENDENCE. When $\theta=1$, the new model is identical with the Lotka–Volterra model. Figure A25 shows the nature of density dependence when $\theta \neq 1$. [B.S.]

REFERENCES

Ayala, F.J. (1969) Experimental invalidation of the principle of competitive exclusion. *Nature* **224**, 1076–1079.

Ayala, F.J. (1970) Competition, coexistence and evolution. In: *Essays in Evolution and Genetics in Honor of Theodosius Dobzhansky* (eds M.K. Hecht & W.C. Steere). North Holland Publishing.

Ayala, F.J. (1972) Competition between species. *American Scientist* **60**, 348–537.

Ayala, F.J., Gilpin, M.E. & Ehrenfeld, J.G. (1973) Competition between species: theoretical models and experimental tests. *Theoretical Population Biology* **4**, 331–356.

Gilpin, M.E. & Ayala, F.J. (1973) Global models of growth and competition. *Proceedings of the National Academy of Sciences of the USA* **70**, 3590–3593.

Gilpin, M.E. & Justice, K.E. (1972) Reinterpretation of the invalidation of the principle of competitive exclusion. *Nature* **236**, 273–274.

Kareiva, P. (1989) Renewing the dialogue between theory and experiments in population ecology. In: *Perspectives in Ecological Theory* (eds J. Roughgarden, R.M. May & S.A. Levin), pp. 68–88. Princeton University Press, Princeton, N.J.

Richmond, R.C., Gilpin, M.E., Salas, S.P. & Ayala, F.J. (1975) A search for emergent competitive phenomena: The dynamics of multispecies *Drosophila* systems. *Ecology* **56**, 709–714.

Azonal soils *See* SOIL CLASSIFICATION.

B

BACI (before–after/control–impact) A SAMPLING design used in RETROSPECTIVE ANALYSIS/MONITORING, involving replicate samples before and after DISTURBANCE in each of an undisturbed control and putatively impacted location. An impact due to the disturbance (e.g. effluent to river from a POINT SOURCE) will cause the difference in mean ABUNDANCE from before to after, in the impacted location, to differ from any natural change in the control.

However, there might still be natural temporal variations in the control and putatively impacted site that confound the interpretation. For example, following the release of an effluent from a pipe to a river, adult insects may arrive and lay eggs in the upstream, control location but not in the downstream location, purely by chance. As a result, BACI would suggest an apparent impact. Hence, the interpretation is complicated and any effect of the disturbance cannot be distinguished from the natural variation. Therefore, if there is an ENVIRONMENTAL IMPACT it is *necessary* to demonstrate differences between control and putatively impacted sites. But, in itself, this is not *sufficient* evidence of impact (*see also* PSEUDOREPLICATION).

A possible solution to these problems is to replicate the sampling, for example between control and impacted sites in separate RIVERS, but this is rarely possible or even desirable for impacted sites. So only the control sites might be replicated, and this leads to asymmetrical sampling designs for which some statistical procedures are available. *See also* BEFORE–AFTER STUDIES; STATISTICS. [P.C.]

REFERENCES

Green, R.H. (1979) *Sampling Design and Statistical Methods for Environmental Biologists*. John Wiley & Sons, Chichester.

Underwood, A.J. (1996) Spatial and temporal problems with monitoring. In: *River Restoration* (eds G.P. Petts & P. Calow), pp. 182–204. Blackwell Science, Oxford.

background extinction rates *See* EXTINCTION MODELLING.

background level The concentration of a substance in the environment that is not attributable to a particular human activity. The term often refers to the concentration produced by NATURAL phenomena; i.e. of natural substances such as metals that can be added to by human activities. [P.C.]

backshore That part of the shore normally beyond the reach of the highest TIDES and subject only to very occasional inundation by saltwater. Nevertheless, the saline influence usually remains sufficient to encourage halophytic plants (*see* HALOPHYTE). [B.W.F.]

bactericide A BIOCIDE used to control or destroy bacteria. [P.C.]

Bailey's triple-catch method A method of estimating POPULATION SIZE from mark–recapture data, using the shortest sequence possible for an open population. *See also* CAPTURE–RECAPTURE TECHNIQUES. [L.M.C.]

balance of nature The idea that there are tendencies in the natural world towards equilibria in populations, communities, ecosystems, etc. An extreme form of this is the notion that there is some kind of natural design or predetermined goal states—a notion rejected by most ecologists. A less extreme concept is that there are dynamic steady states brought about by BIOLOGICAL CONTROL systems of various kinds. *See also* GAIA; TELEOLOGY. [P.C.]

Balance of selective forces

When two or more selective forces act on a population or a group of populations, a balanced GENETIC POLYMORPHISM may be maintained (*see* BALANCED POLYMORPHISM). Most theories rely on the concept of protection of an allele: a given allele at a locus is said to be protected if its frequency increases when rare (*see* ALLELE FREQUENCY; ALLELES). This concept is very close to that of the EVOLUTIONARILY STABLE STRATEGY (ESS) in evolutionary GAME THEORY (Maynard Smith 1989). A polymorphism at a locus is protected if at least two alleles are protected. Examples of protected polymorphisms at a bi-allelic locus can arise through various mechanisms, depending on the situation.

Strict overdominance

In the simplest case of strict overdominance the FITNESS of heterozygous individuals is greater than the fitness of either homozygote. When one allele is rare, it is mainly present in a heterozygous state, and thus increases in frequency. When frequent, it is mainly present in a homozygous state, and thus the alternative allele increases in frequency (*see* DOMINANCE).

Selection in opposing directions in variable environments

When environmental conditions are variable, several mechanisms for maintaining genetic variation have been described.

- Overdominance for GEOMETRIC MEAN fitness in a temporally

variable environment (Haldane & Jayakar 1963; Gillespie 1974). When one homozygote is sometimes fitter, sometimes less fit than the other homozygote, the polymorphism may be maintained because in each generation the heterozygote is intermediate between the two homozygotes, thus its VARIANCE across generations is lower and its geometric mean fitness is greater.

• Overdominance for HARMONIC MEAN fitness in a coarse-grained spatially variable environment. If one homozygote is favoured in one environment, and the other homozygote is selected for in another environment, a polymorphism could be maintained under some conditions (Levene 1953).

• In a temporally variable environment, when the dominant PHENOTYPE has a geometric mean fitness greater than the recessive GENOTYPE, and the recessive genotype has a relative ARITHMETIC MEAN fitness greater than the dominant phenotype (Haldane & Jayakar 1963). The dominant allele is protected if, although the fitness of the dominant phenotype is on (arithmetic) average lower, it varies less through time than the recessive genotype. Conversely, the recessive genotype is protected if, although its fitness might sometimes be very low (even zero), it is also sometimes very much better than the dominant phenotype. The reason why the recessive allele may be protected even when the fitness of the recessive phenotype is sometimes very low, is that it can 'hide' in the dominant phenotype when rare. This is not true of the dominant allele, which is always expressed. One could view the recessive allele as a 'gambler' allele, and the dominant one as a 'security' allele.

• In a coarse-grained spatially variable environment, a polymorphism between any two phenotypes (haploid or diploid) may be maintained if there is an inverse ranking of harmonic and arithmetic mean relative fitnesses (Levene 1953) (*see* BET-HEDGING).

• FREQUENCY-DEPENDENT SELECTION.

Other mechanisms for maintaining genetic variation

Apart from the nature of alleles *per se*, several other mechanisms may contribute to maintaining genetic variation in a population. These include:

• opposing viability selection in haploid and diploid phases;

• opposing MEIOTIC DRIVE and viability selection;

• opposing selection at more than one level. For instance, at the level of a METAPOPULATION, dispersal behaviour is selected for at the between-group level, at recolonization following a local extinction, whereas it is selected against at the within-group level (Olivieri *et al.* 1995). A POLYMORPHISM between genotypes with high and low dispersal rates may thus be maintained. *See also* SELECTION DIFFERENTIAL; SELECTION PRESSURE. [I.O.]

REFERENCES

Gillespie, J.H. (1974) Natural selection with varying selection coefficients—a haploid model. *Genetical Research (Cambridge)* **21**, 115–120.

Haldane, J.B.S. & Jayakar, S.D. (1963) Polymorphism due to selection of varying direction. *Journal of Genetics* **58**, 237–242.

Levene, H. (1953) Genetic equilibrium when more than one ecological niche is available. *American Naturalist* **87**, 331–333.

Maynard Smith, J. (1989) *Evolutionary Genetics*. Oxford University Press, Oxford.

Olivieri, I., Michalakis, Y. & Gouyon, P.-H. (1995) Metapopulation genetics and the evolution of dispersal. *American Naturalist* **146**, 202–228.

balanced diet Some evidence suggests that animals may somtimes achieve balanced diets by what looks like 'nutritional wisdom', i.e. decision rules that result in an animal acquiring its range of necessary dietary components. Early evidence from rats, for instance, suggested that, when presented with a variety of mineral sources, animals selected a balanced mineral diet. Moreover, rats deprived of a particular nutrient appeared to compensate by incorporating an appropriate new food source in their diet (such as faeces to replace yeast as a source of vitamins of the B group). Initially, this apparent ability was explained in terms of 'specific hungers' for essential dietary components, but now the weight of evidence suggests more complex underlying mechanisms which are difficult to generalize across trophic groups. In rats, only a restricted range of dietary components (e.g. salt and water) turn out to be selected directly in relation to need. Compensation for other deficits appears to occur on a trial-and-error basis, often by learning to avoid foods that are themselves deficient in the required commodity. Some other groups, for example vertebrate grazing herbivores, seem to be more limited in their ability to fine-tune their nutritional intake, possibly because of the nature of their food and the physical difficulty of harvesting sources of particular nutrients selectively. *See also* FEEDING; FOOD SELECTION. [C.J.B.]

balanced polymorphism POLYMORPHISM is a form of genetic variation in which different morphs, suites of behaviour or life histories within a population or species are observed at frequencies too common to be maintained by MUTATION. Balanced polymorphisms are polymorphisms that persist through time, maintained by differential mortality of GENOTYPES as a result of changing SELECTION PRESSURES. They act as a buffer against environmental UNCERTAINTY without loss of FITNESS arising from conflicting gene recombinations.

Balanced polymorphisms reportedly occur in a wide range of plants and animals, although they are more apparent among some taxa than others. The concept has been invoked frequently in studies of fish to account for the COEXISTENCE of sympatric morphs which differ in colour or body shape, or of different feeding or breeding behaviours within populations (Thresher 1978; Skúlason & Smith 1995). Sympatric morphs of charr (*Salvelinus* spp.), for example, differ markedly in body shape according to feeding habit (benthivory, planktivory and piscivory) and whether or not they are migratory (Skúlason & Smith 1995).

Increasingly, research suggests that not only are balanced polymorphisms more common than has been hitherto recognized, but also that there are underlying mechanisms which are similar. The issues surrounding balanced polymorphism may be relevant in determining the effectiveness of BIOLOGICAL CONTROL methods (Cilliers 1989) and in the selection of material for cultivation (e.g. Sugita & Fujio 1982) and may pose problems for management of wild populations (Skúlason & Smith 1995). *See also* BALANCE OF SELECTIVE FORCES. [M.C.M.B.]

REFERENCES

Cilliers, B. (1989) A model of balanced polymorphism in a host–parasite relationship. *Phytophylactica* **21**, 279–280.

Skúlason, S. & Smith, T.B. (1995) Resource polymorphisms in vertebrates. *Trends in Ecology and Evolution* **10**, 366–370.

Sugita, M. & Fujio, Y. (1982) Effects of genotypes at the AAT-1 locus on the survival and growth rate in the cultured oyster. *Tokyo Journal of Agricultural Research* **33**, 42–49.

Thresher, R.E. (1978) Polymorphism, mimicry, and the evolution of the hamlets (*Hypoplectrus*, Serranidae). *Bulletin of Marine Science* **28**, 345–353.

Baltic Sea The largest brackish-water SEA in the world, lying between north and central Europe and including the Gulf of Bothnia, the Gulf of Finland, the Gulf of Riga, the Baltic proper, the Sound, the Belt Sea and the Kattegat (up to 57°N). Considered a young sea, it was formed during the last ice age, approximately 10 000–15 000 BP. It has a mean depth of only 55 m, and is characterized by a number of deep BASINS separated by shallow sills. The deepest basin is the Landsort Deep (459 m). The total area of the Baltic Sea is about 390 000 km^2 and its volume is 20 000 km^3. The drainage area of the Baltic measures approximately 1 650 000 km^2. The surrounding countries engage in heavy industry and intensive agriculture, and sea traffic is considerable, including tankers, cargo vessels, large container ships and ferries. The Baltic is connected to the NORTH SEA through the Danish Straits (Great Belt, Little Belt) and the Sound.

Seventy per cent of the saline water entering the Baltic does so through the Great Belt; inflows of oxygenated saline water are sporadic, and the hydrographic features of the Baltic lead to a permanent water STRATIFICATION and periods of low OXYGEN concentration in the deeper basins. The general CIRCULATION pattern is counterclockwise; mixing and water exchange are primarily wind driven and the TIDES are minimal. SALINITY varies from 3–4‰ in the bottom water of the northernmost part of the Bothnian Bay, to 6–8‰ in the Baltic proper, to 8–20‰ in the Straits and Sound, to just under 35‰ in the northern Kattegat. The low biological diversity characteristic of the Baltic is a consequence of its youth as well as its intermediate and fluctuating salinities. *See also* ATLANTIC OCEAN. [V.F.]

REFERENCE

Aniansson, B. (1989) *Northern Europe's Seas—Northern Europe's Environment.* Report to the Nordic Council's International Conference on the Pollution of the Seas, 16–18 October 1989. Nordic Council, Copenhagen.

bar A shingle or sand structure joined to solid coastline at both ends. Shingle bars may comprise several parallel ridges. A bar forms across an ESTUARY or a coastline indentation, and typically originates as a spit attached at one end which extends, eventually to link with solid coastline at the other end. More rarely, two spits may develop from opposite directions and meet to form a bar. On its landward side a bar may enclose a lagoon, which may be brackish (*see* BRACKISH WATERS) or even FRESHWATER in nature. [B.W.F.]

Barents Sea A marginal SEA of the ATLANTIC OCEAN and part of the polar BASIN around the ARCTIC. This high-latitude sea (i.e. above 70°N) has 3 months of continuous light in the summer and 3 months of continuous darkness in the winter. It is open to the NORWEGIAN SEA (and thereby the Atlantic) to the west. The Barents Sea has an unstable climate and is one of the most stormy seas in the world. Its location between the Atlantic and Arctic Oceans results in a number of interesting hydrographic features. One of the more prominent features is the polar FRONT which occurs at the boundary of cold Arctic and warm Atlantic waters and which is an area of unusually high productivity. In contrast to temperate seas, peak diatom productivity occurs during the month of May and tapers off through July; ZOOPLANKTON abundance reaches its peak in July and remains high until October. Intensive vertical temperature mixing maintains highly oxygenated conditions throughout the Barents Sea. It has a relatively deep SHELF-SLOPE BREAK at about 450 m. It is partially covered by polar ICE for some of the year. Sedimentation rates are very low (i.e. ≤1–3 cm per 1000 years), with no accumulation in some areas. The Barents Sea is a breeding area for large numbers of fish and contains some of the world's largest populations of seabirds. [V.F.]

barium (Ba) An element (ATOMIC NUMBER = 56; relative ATOMIC MASS = 137.327; melting point = 725°C; boiling point = 1140°C) with an abundance of 0.039%, found as the mineral barite, $BaSO_4$. Barite is used as a heavy mud slurry in well-drilling. Barium METAL is prepared by reduction of its oxide with ALUMINIUM and has limited use alloyed with NICKEL for spark-plug wire. Barium salts are toxic, although the insolubility of barium sulphate allows it to be used in X-ray imaging of the gastrointestinal tract. [C.R.P.]

barometer An instrument for measuring atmospheric PRESSURE. The Fortin barometer consists of a glass tube, closed at one end, filled with MERCURY and inverted with its lower, open end in a bath of mercury. The vertical height of the mercury column which the atmospheric pressure can support is read with a vernier scale. This instrument is not portable, and in normal use is attached to a wall. The aneroid barometer is a portable instrument. It is an evacuated chamber covered by a corrugated metal diaphragm which distorts under the influence of atmospheric pressure. The small distortions are detected by a system of amplifying levers, or by an appropriate transducer such as a strain gauge. The Fortin barometer is an absolute instrument against which the aneroid instrument may be calibrated. *See also* METEOROLOGY. [J.G.]

barrier island A shingle structure formed where a large mass of shingle accumulates offshore and is not connected to the mainland, often being separated from it by shallow LAGOONS. Frequently, such structures are associated with SAND DUNES, which may overlie the shingle, and with adjacent SALT MARSHES formed on the sheltered landward side. [B.W.F. & V.F.]

barrier reef A REEF adjacent to a landmass and separated from it by a lagoon or channel of variable extent. Barrier reefs are not readily distinguishable from fringing reefs, though barrier reefs tend to be separated from the landmass by a greater distance and by a deeper water channel than fringing reefs. Barrier reefs are common throughout the CORAL REEF zones in all oceans. The largest barrier reef, extending nearly 200 km in length, is the Great Barrier Reef in Australia. [V.F.]

Bartlett's three-group method *See* REGRESSION ANALYSIS, LINEAR.

basal species A term used to denote those species in a FOOD WEB which are preyed upon, but which do not themselves have prey. Strictly speaking, all basal species should be AUTOTROPHS, but the incompleteness of food web data sets has sometimes led authors to mistakenly classify non-autotrophs as basal species. *See also* FEEDING TYPES, CLASSIFICATION. [S.J.H.]

baseline data Data describing some original, or 'normal', state of a system—for example POPULATION DENSITY, population AGE STRUCTURE, species composition, ENERGY FLOW — that can be used as standard/control against which changes are judged. This can often suffer from the problem of PSEUDOREPLICATION. [P.C.]

bases Substances that dissociate in WATER to yield hydroxyl ions, OH^-. An example of a strong base is potassium hydroxide, KOH:

$$KOH \rightarrow K^+ + OH^-$$

A weak base is one which does not completely dissociate, for example calcium hydroxide, $Ca(OH)_2$:

$$Ca(OH)_2 \Leftrightarrow Ca^{2+} + 2OH^-$$

Cf. ACID. *See also* ACID RAIN AND MIST; pH. [J.G.]

basic reproductive rate (R_0) The expected lifetime production of EGGS by a newborn female. Also called net reproductive rate or net FECUNDITY rate, it can be calculated from COHORT data as:

$$R_0 = \sum L_x M_x$$

where L_x is age-specific survivorship and M_x is age-specific fecundity. It is frequently used to describe the production of female offspring by females, and as a FITNESS measure suitable for modelling EVOLUTION in a stable population where R_0 for female offspring must equal 1. *Cf.* INTRINSIC RATE OF INCREASE. *See also* DEMOGRAPHY; LIFE-HISTORY EVOLUTION; LIFE TABLE. [D.A.B.]

REFERENCE

Carey, J.R. (1993) *Applied Demography for Biologists*. Oxford University Press, Oxford.

basin A geographical area that is effectively enclosed, referring either to terrestrial or oceanic locations. In terrestrial situations it is typically a region that collects the precipitation falling upon it (*see* CATCHMENT). [P.D.M.]

Batesian mimicry A form of MIMICRY in which one (or more) palatable species of animal exhibit the same colour pattern as an unpalatable species. The latter species acts as a model for the former, which is the mimic. Although the mimic is perfectly palatable to predators, it derives some protection because potential predators sometimes mistake it for the unprofitable prey which they have previously learned to avoid. *See also* MÜLLERIAN MIMICRY. [P.M.B.]

bathyal zone The bottom of the OCEAN between 200 m and about 4000 m depth, corresponding to the depth of the CONTINENTAL SLOPE and CONTINENTAL RISE. The bathyal zone occupies approximately 16% of the submerged ocean floor. The exact depth of the bathyal zone depends on the local depth of the shelf break, latitudinally controlled light penetration, CURRENTS, SALINITY, transparency, etc. The bathyal zone is mostly aphotic, although some weak light penetration may occur, especially in the tropics. This zone is further characterized by relatively constant salinity and temperature, low current activity (except in areas of UPWELLING) and often low OXYGEN. [V.F.]

bathypelagic The oceanic water column between about 1000 and 4000 m depth, seaward of the SHELF-SLOPE BREAK. [V.F.]

BATNEEC *See* BEST AVAILABLE TECHNIQUES NOT ENTAILING EXCESSIVE COSTS.

bats, diversity and conservation *See* CHIROPTERANS, DIVERSITY AND CONSERVATION.

battery A portable power supply, made from a set of electrolytic cells. A cell consists of two electrodes immersed in an electrolyte, one of which forms the positive pole (anode) and the other the negative pole (cathode). When the two poles are connected to an external circuit, current flows as a result of chemical reactions on the surface of the electrodes. Each cell produces a rather small voltage, but by connecting several cells together in series (i.e. the anode of one is connected to the cathode of another), a useful voltage is obtained. Batteries are specified by their voltage, their capacity (in ampere-hours (Ah)) and the type of storage cell from which they are composed.

The most common type of non-rechargeable battery until recently was the dry form of the Leclanché cell. The original Leclanché consisted of a positive electrode of CARBON surrounded by manganese dioxide and carbon in a porous pot. The pot stood in the electrolyte, which was an ammonium chloride solution. The negative electrode was ZINC. The dry form of this cell is portable. It has the ammonium chloride as a paste, and the negative zinc electrode is the case itself. The voltage produced is 1.5 V per cell.

The alkaline cell supersedes the 'flashlight battery' based on the dry form of the Leclanché cell. The latter has a low storage capacity and tends to leak ammonium chloride. Alkaline cells contain potassium hydroxide sealed in a steel case, and are 1.5 V for each cell; capacity is 0.5–8 Ah, depending on physical size.

Rechargeable cells are widely available, known as secondary cells. Batteries made of them are sometimes termed accumulators. Lead–acid batteries are the sort of rechargeable batteries used in motor cars. The negative and positive electrodes are LEAD and lead oxide, and the electrolyte is 10% sulphuric acid (specific gravity 1.20–1.28). The positive plate is lead oxide (PbO_2) and the negative is lead (Pb). The individual cells are 2.05 V, and the battery as used in a car is usually nominally 12 V with a capacity of about 40 Ah. Such batteries are charged using a constant voltage source, and charging takes several hours. They deteriorate when they have been deeply discharged, owing to the tendency for both plates to become lead sulphate ($PbSO_4$), which is not decomposed on charging. Modern, cleaner versions of lead–acid batteries have the electrolyte as a gel to avoid spillage.

Nickel–cadmium batteries have a nickel hydroxide cathode, CADMIUM anode and potassium hydroxide electrolyte, in steel containers. They develop 1.24–1.27 V, with a capacity of 0.15–10 Ah, have a long service life, and are lightweight. They do not suffer from deep discharge, but need a special charger as they are charged at constant current not constant voltage.

Specialist batteries include those based on SILVER or mercuric oxide, the familiar 'button' cells of calculators, cameras and watches; and LITHIUM cells, which have a high terminal voltage and a low self-discharge rate, used in computers and to power clocks. Discarded batteries can be a significant source of POLLUTION. [J.G.]

BCF *See* BIOCONCENTRATION FACTOR.

beach A rather imprecise term used in a narrow sense to describe the layer of unconsolidated materials overlying harder geology occurring between the tidal limits (the shore). More correctly it should be used to refer to the INTERTIDAL ZONE plus that region below extreme low tide where unconsolidated materials are subject to DISTURBANCE by WAVES. The beach, therefore, is usually a region of considerable instability where coarse-grained materials, from sand to pebbles or cobbles, are regularly churned up by waves and where COLONIZATION by attached or rooted plants is impossible. In more sheltered situations, where wave action is minimal and the limits of the beach coincide closely with tidal limits, mud or SILT can accumulate. Only in this latter situation can plants colonize, for example SALT MARSH communities in temperate regions and mangroves in the tropics. The term 'beach' tends not to be applied to ROCKY SHORES. [B.W.F.]

beat sampling *See* SAMPLING METHODOLOGY/DEVICES.

becquerel (Bq) A measure of the RADIOACTIVITY present in a body; 1 unit represents 1 nuclear disintegration per second. Thus, when applied to a sheep which has been grazing on land on which radioactivity has fallen in rain, it is measured as the number of nuclear transformations per second per kilogram of flesh. Physically, it is a measure of the emission rate, or energy deposited in a recipient of the emission, and not of the biological effect on the sheep or on a person who ingests its flesh. This is an SI UNIT and supersedes the former unit, rem. [R.S.S.]

bedload The particles being moved along the bed of a RIVER or stream. Once a particle resting on the bed is entrained it might go into suspension. However, if it is too heavy it can travel along the bed surface. Particles may roll or slide along the bed or hop as they rebound on impact with the bed. [P.C.]

Beer's law A law describing the attenuation of a parallel beam of RADIATION as it passes through a homogeneous medium (e.g. air or WATER). Suppose the flux density at a distance x inside the medium is I_x, then the absorption of the beam by a thin layer Δx is given by:

$$\Delta I = -kI_x\Delta x$$

where k is a wavelength-dependent property of the medium known as the absorption coefficient. On integration we obtain the flux density of the beam at any distance into the medium, given an incident beam of I_o:

$$I_x = I_o \exp(-kx)$$

The law is often used for non-ideal cases where it applies only approximately. For example, in turbid media such as the ATMOSPHERE and natural waters, the light beam has directional attributes, and scattering is important. In leaf canopies, the same applies but the system is markedly inhomogeneous and the flux density needs to be spatially averaged to overcome the point-to-point variability. [J.G.]

Beetles, diversity and conservation

'Beetle' is the common name for insects of the order Coleoptera. One feature that distinguishes beetles from other insects is the front pair of wings, which are hardened as elytra (singular elytron). The name Coleoptera combines the Greek words *keleo*, 'sheath', and *ptera*, 'wings'. These modified wings are lightweight, waterproof, and not used for flight. They overlie the membranous hindwings used for flight and usually conceal the abdomen. The abdominal spiracles do not open directly to the exterior but into the space between the elytra and the abdomen. The elytra provide protection through the hard cover, and—perhaps more significantly—water conservation for the spiracles. This may be the main reason why beetles are so successful, with more species than any other order in all the animal kingdom.

There are about 300 000 described species of beetles in the world, approximately one-quarter of all described animal species, and as many species as in the entire plant kingdom, including algae and fungi. However, the actual number in existence may be two to ten times more. Because we do not know how many of them exist, or where, or how so many of them live, there is great difficulty contemplating conservation of such DIVERSITY.

Life cycles

Beetles are holometabolous and undergo complete METAMORPHOSIS from EGG, via LARVA and pupa, to adult. A few cave-dwelling beetles skip the larval stage. They hatch from single large eggs and immediately pupate. Some beetles produce live young. While most beetles reproduce sexually, some lack males and exhibit parthenogenetic reproduction. Most beetles complete development annually, but some have two or three generations a year, usually correlated with their host's cycle. Some require a few years to complete a LIFE CYCLE (*see also* COMPLEX LIFE CYCLES).

Taxonomic diversity

There are more than 150 families of beetles worldwide, belonging to four suborders. Suborder Archostemata includes the most primitive living beetles in only two families, the reticulated beetles in Cupedidae, and a single rare species in Micromalthidae. The latter has a remarkable life cycle with larvae that are able to reproduce (both oviparously and viviparously) parthenogenetically (*see* OVOVIVIPARITY; PARTHENOGENESIS; VIVIPARITY). Members of this suborder are mostly found under decaying logs.

Suborder Myxophaga includes some minute (< 1.5 mm) aquatic or semi-aquatic beetles in four families, each containing one genus. They live among filamentous algae in streams or shallows or under rocks, roots or moss in boggy areas. One species of Hydroscaphidae (skiff beetles) is extremely tolerant of temperature variations. It lives in both icy rivers and hot springs.

Suborder Adephaga includes eight families of beetles, nearly all predacious except Rhysodidae (wrinkled bark beetles), which live on decaying wood. Cicindelidae (tiger beetles) and Carabidae (ground beetles) are mostly terrestrial; most actively pursue their prey both as adults and larvae, but a few ground-beetle larvae are parasites of other insects. Amphizoidae (trout-stream beetles), Haliplidae (crawling water beetles), Dytiscidae (predacious diving beetles), Noteridae (burrowing water beetles) and Gyrinidae (whirligig beetles) are aquatic or semi-aquatic. Some of them feed on other aquatic insects, others feed on shrimps, worms, leeches, snails, tadpoles and even small fish.

Suborder Polyphaga constitutes all the remaining families, and is recently evolved compared to the other suborders. Eleven of the most common and species-rich families are in this suborder: Staphylinidae (rove beetles), Scarabaeidae (scarab beetles), Buprestidae (metallic wood borers), Elateridae (click beetles), Cantharidae (soldier beetles), Cleridae (checkered beetles), Coccinellidae (ladybird beetles), Tenebrionidae (darkling beetles), Cerambycidae (longhorn beetles), Chrysomelidae (leaf beetles), and Curculionidae (snout beetles or true weevils). Together with the Carabidae of Adephaga, these 12 beetle families account for 70% of all beetle species in the world.

Most of the families in Polyphaga are terrestrial, but some are aquatic, like Hydrophilidae (water scavenger beetles) and Elmidae (riffle beetles). Lampyridae (fireflies) and Phengodidae (glow-worms) are not commonly thought of as beetles. Both are luminescent as larvae and adults, except the glow-worm adult male. Fireflies feed on snails and earthworms. Most beetles have strong mandibles to chew and bite, but weevils have their heads elongated to form a snout with their mouthparts located at the tip. This not only enables these beetles to nibble plants but also to bore into various fruits and seeds, which are sometimes quite hard. Some longhorn beetles have huge mandibles, but often their most difficult task is to gnaw their way out of the pupal cocoon.

Many beetles of Polyphaga feed on roots, stems, leaves, blossoms or fruits of plants. Some of them are serious agricultural pests. The Colorado potato beetle, *Leptinotarsa decemlineata* (Chrysomelidae), feeds voraciously and no pesticides have effectively controlled it. This beetle has ultimately caused the cessation of commercial potato farming in many places worldwide, and is a subject of quarantine in some countries. However, some of the plant eaters are used as effective biocontrol agents for weeds, as in the successful control in the Far East and Africa of an invasive Brazilian water weed, *Salvinia molesta*, mainly by the weevil *Cyrtobagous salviniae*. Many PHYTOPHAGOUS beetles are beneficial as pollinators. The weevil *Elaeidobius kamerunicus* has replaced hand POLLINATION of oil palm, saving US$13 million per year.

Some beetles of Polyphaga feed on wood itself, or on living phloem and cambium. Many feed on decaying plants, or animal faeces, or carcasses. These beetles are extremely useful in the ecological processes of DECOMPOSITION and energy recycling, especially those that feed on CELLULOSE and carrion. Many of them live together with other insects. Some species live in ant or termite nests, and are called myrmecophiles or termitophiles, respectively. They may be welcome guests, tolerated visitors or sometimes even unwelcome lodgers. Some beetles, such as the metallic wood borers, the longhorn beetles and the darkling beetles, harbour intestinal symbiotic yeasts or bacteria to digest wood, hair, wool, feathers, HUMUS or dry cereals. Subsocial behaviour like PARENTAL CARE appears among these beetles. The ambrosia beetles cultivate a fungus on which the larvae feed. Dung beetles of both sexes work together to roll a faecal ball and bury it in a suitable place, providing shelter and food for their larvae.

Predacious beetles of Polyphaga, especially ladybird and checkered beetles, play an important role in keeping populations of other insects in check. Many ladybird beetles are beneficial and used as biocontrol agents. *Rodolia cardinalis*, a ladybird beetle from Australia, has saved the citrus industry in the USA from the ravages of the cotton cushion scale, *Icerya purchasi* (Homoptera).

Several dozen parasitic beetles are known to science. This is a rare and exceptional lifestyle, when one considers that there are about 300 000 species of Coleoptera known. Some of these are ectoparasitic on other insects, like the cicada parasite beetles (Rhipiceridae); others parasitize beavers (Leptinidae, Platypsyllidae), rats (Staphylinidae) and reptiles (Chrysomelidae).

Larvae and female adults in Stylopidae are ENDOPARASITES of bees and leafhoppers. The adult stylopid female is larva-like and continues to be parasitic after the larval stage. The first-instar larvae hatch from eggs laid on flowers. They attach themselves to a suitable HOST when it visits the flower, and become parasitic. The stylopid male lives long enough to seek a mate when she protrudes her abdomen through the host's cuticle. Because of their bizarre morphology and life history, many treat this family as a separate order, Strepsiptera.

The blister beetles (Meloidae) can cause blisters on human skin by secreting a substance called cantharidin, which has been used variously as a counterirritant, diuretic and aphrodisiac. These beetles require up to three years to complete their life cycle. Adult blister beetles are phytophagous and can cause damage to host plants. Their larvae, however, eat the egg pods of grasshoppers, which makes them sometimes beneficial, or parasitize bees. The female adults lay their eggs in batches near the breeding ground of grasshoppers or nests of bees, or parts of a plant visited by bees. The young larvae actively search for and attach to the host as soon as they hatch—a relationship called phoresy (*see* PHORETIC ASSOCIATION). These larvae then undergo complex development to several distinctive forms from less active to inactive — a process termed heteromorphosis, which also occurs in Micromalthidae, Drilidae and Rhipiphoridae. Such complex behaviours and life cycles demand special conservation attention. *See also* BIODIVERSITY.

Distribution and habits

Beetles are almost everywhere, occupying all terrestrial and freshwater habitats, but not the open sea. Some beetles occur in the hot springs of the Himalayas at 5400 m above sea level, others survive in the searing heat of the Namib Desert. The greatest number of species occur in the tropics. A few species are even found in ARCTIC regions.

Beetles include some of the largest and smallest insects. The largest are the tropical American *Megasoma elephas* and *Dynastes hercules* (Scarabaeidae), which reach a maximum length of 16 cm. The smallest beetle is probably the Mexican feather winged beetle,

Nanosella fungi (0.04 mm, Ptiliidae). A scarab beetle, *Scarabaeus sacer,* was considered a symbol of resurrection by ancient Egyptians. Adult beetles of Dytiscidae and Hydrophilidae are considered a delicacy and of medicinal value in southern China.

Adaptation of beetles to the aquatic environment is most interesting. Bodies are streamlined or boat-shaped, and legs are fringed or flattened like oars. Adults of aquatic Adephaga take up air at the surface and store it beneath their elytra as they swim under water. Many aquatic Polyphaga keep a film of air around their bodies, trapped by dense microscopic hairs, for underwater breathing. Most aquatic larvae extract OXYGEN from water by means of abdominal or tracheal gills or through the thinner parts of the body cuticle. They usually leave the water to pupate.

The coloration of beetles ranges from dull and black to brightly metallic and luminous as a rainbow, often closely related to habitat. Beetles living in darkness are typically black or dark brown. Beetles living on plants glow with the spectrum of colours, sometimes mimicking leaves and flowers, and other times exhibiting flashy aposematic colours to warn off their enemies (*see* APOSEMATIC (WARNING) COLORATION). Many species of longhorn beetles living on branches or trunks of trees have greyish spots resembling the lichen scattered over the bark.

SEXUAL SELECTION also plays a role in beetles' diversity. The male stag beetles (Lucanidae) have spectacular, large antler-like mandibles. They are used to wage battles, driving weaker suitors away from the female. This enables the most successful male to transmit his FITNESS to future generations. Beetles that are big and colourful, or with bizarre appearance, often become trophies for collectors.

Conservation

Most human interactions with beetles have been detrimental for the beetles. As agricultural pests, beetles have been fed up to explosive population densities in a few cases, only to become targets of pesticides, often with catastrophic results for NON-TARGET SPECIES. Many large and colourful beetles provide human collectors with great pleasure and even monetary gain, but only in death. Landscape modification and rat predation have seriously depleted populations of a weevil, *Lyperobius luttoni,* in New Zealand. The American burying beetle, *Nicrophorus americanus* (Silphidae), is a large, orange-red and black carrion feeder that buries rodent-sized animals. Formally widespread over eastern and central North America, it has been reduced—presumably by pesticides—to small colonies in Rhode Island and Oklahoma. Introduction of exotic species has led to elimination of three large weevils in New Zealand. Biocontrol has sometimes backfired when the supposed controlling species preyed on non-target species.

Beetle conservation is a new concept emerging beside the larger field of beetle extirpation. The IUCN's RED LISTS of threatened animals and the WWF's RED (DATA) BOOK include authoritative lists of a few threatened and endangered beetles worldwide. As with most invertebrate conservation movements, beetle conservation happens mostly in developed countries, but the developing countries of the tropics are very much in need. Habitat conservation and SUSTAINABLE AGRICULTURE may provide the most inexpensive protection of beetle diversity at present. *See also* CONSERVATION BIOLOGY; INSECTS, DIVERSITY AND CONSERVATION. [W.L.]

REFERENCES

Borror, D.J., Triplehorn, C.A. & Johnson, N.F. (1989) *An Introduction to the Study of Insects,* 6th edn. Harcourt Brace, New York.

Collins, N.M. & Thomas, J.A. (eds) (1991) *The Conservation of Insects and Their Habitats.* Academic Press, London.

Evans, A.V. & Bellamy, C.L. (1996) *An Inordinate Fondness for Beetles.* Nevraumont Publishing Co., New York.

United States Congress, Office of Technology Assessment (1993) *Harmful Non-indigenous Species in the United States.* US Government Printing Office, Washington, D.C.

before–after studies A method of impact assessment that uses the state of the ENVIRONMENT and ecological systems within it as a STANDARD against which to judge the condition following development and/or release of EFFLUENTS/EMISSIONS. However, because ECOSYSTEMS are dynamic it cannot always be presumed that the properties of the system(s) before would have been like those after, even if there had been no impact. *See also* BACI. [P.C.]

behaviour *See* BEHAVIOUR, IMPLICATIONS FOR CONSERVATION; BEHAVIOURAL ECOLOGY.

Behaviour, implications for conservation

Unlike most subdisciplines of biology, CONSERVATION BIOLOGY does not hinge on the need to describe a discrete aspect of an organism's biology. Instead, conservation biologists must integrate all aspects of the biology of a target species, and consider the social, political and economic impacts on humans. Thus, successful conservation biologists must be broadly trained biologists and effective sociologists and politicians.

No aspect of biology is more important for a conservation biologist to understand than behaviour. Vast amounts of biological DIVERSITY have been lost or are in jeopardy because organisms often cannot alter their natural behaviour in environments that have been altered by humans. In addition, otherwise well-conceived conservation plans have been destroyed by unexpected behavioural responses of study organisms.

Organisms are well adapted to their natural environments. Behaviours are tuned to reliable environmental cues, and disrupting these cues can cause organisms to behave in ways that decrease their fitness. For example, hatchling sea turtles of many species find the ocean after emerging from their nests by moving towards the brightest horizon. Because water reflects light from the Moon, stars and clouds, the brightest horizon on natural beaches is where the ocean meets the sky. Thus, the ocean-finding behaviour of hatchling sea turtles makes use of a reliable environmental cue. However, human development of many coastal areas where sea turtles nest has altered this reliable environmental cue; in developed coastal areas, the brightest horizon is often directly away from the ocean and directly towards artificial lights. As a result, whenever sea turtles nest near human development, hatchling sea turtles are found wandering away from the relative safety of the ocean. Once conservation biologists unravelled the cause of these misdirected migrations, conservation programmes aimed at turning off beach lights were launched and the problem was lessened (Raymond 1984).

Another behaviour that became maladaptive following human-induced environmental change is the well-known tameness of animals that live on islands. Unfortunately, the conservation problem in this case has not been as easy to solve. Wherever humans have settled on islands they have brought with them a variety of organisms that eventually became established and proliferated in their new island habitats. These exotic organisms have ravaged native species on islands throughout the world. Exotic predators in particular can decimate native species in very short order because native species typically lack the appropriate behavioural responses to predators. Cats, dogs, rats and pigs are the most common exotic predators, and their introduction has radically altered many island faunas. Another exotic PREDATOR, the brown tree snake (*Boiga irregularis*), has in less than 40 years eaten nearly every native bird on the island of Guam, driving several species to extinction (Fritts 1988).

The preceding examples detail how behaviour can lead to conservation problems. Behaviour can also interfere with conservation plans. A familiar example is the frequent failure of CAPTIVE BREEDING programmes due to the failure of organisms to breed or to thrive in captivity. Giant panda (*Ailuropoda melanoleuca*) conservation is thus hindered; despite overwhelming public empathy and seemingly well-laid plans, this animal does not breed well in captivity (MacKinnon & MacKinnon 1991). Behaviour is implicated in two ways in this example. First, the failure to breed is a behavioural response by the giant panda. Conservation biologists must overcome this behaviour if the captive breeding programme is to succeed. Second, it seems likely that the failure of pandas to breed in captivity is due to the absence of some environmental cue that would trigger a key behaviour that initiates courtship or mating. This possibility underscores the need for detailed knowledge of the biology of any species we wish to conserve.

Behaviour is a key element of the biology of all animals. In order to create and carry out effective conservation policies, we must understand not only the behaviour but also the morphology, physiology, ecology and evolutionary biology of target species. In addition, we must consider the social and political ramifications of conservation policy on humans. Most failed conservation plans are the result of our failure to understand one or more of these factors. *See also* BEHAVIOURAL ECOLOGY. [P.A.S.]

REFERENCES

Fritts, T.H. (1988) The brown tree snake, *Boiga irregularis*, a threat to Pacific islands. *US Fish and Wildlife Service, Biological Report* **88** (31).

MacKinnon, K. & MacKinnon, J. (1991) Habitat protection and re-introduction programmes. *Zoological Society of London Symposia* **62**, 173–198.

Raymond, P.W. (1984) *Sea Turtle Hatchling Disorientation and Artificial Beachfront Lighting: a Review of the Problem and Potential Solutions*. A report for the Center for Environmental Education, Sea Turtle Rescue Fund. Washington, D.C.

---

behavioural drift *See* DRIFT, FRESHWATER.

Behavioural ecology

Behavioural ecology can be defined as the study of the EVOLUTION of behaviour in relation to ECOLOGY. Its principal tenet is that behaviour patterns and decision-making have evolved mainly by NATURAL SELECTION and are thus adaptive in relation to problems of survival and reproduction imposed by the organism's ENVIRONMENT. Behavioural ecology emerged as an identifiable discipline in Europe and the USA in the early 1970s. Its roots are diverse, but in broad terms it represented a fusion of three traditions developing through the 1960s: (i) a revisionist critique of evolutionary explanation exposing the fallacy of arguments based on GROUP SELECTION (ADAPTATIONS for the benefit of the species/population/social group, etc.) and asserting the individual or allele as the level at which NATURAL SELECTION acts; (ii) a growing awareness, through comparative studies, of the impact of ecology on SOCIAL ORGANIZATION and MATING SYSTEMS in different species; and (iii) a predictive, experimental approach to the study of behavioural adaptation pioneered initially by the Nobel Prize-winning ethologist Niko Tinbergen. These were complemented to some extent by the emergence of sophisticated mathematical modelling approaches in population and COMMUNITY ECOLOGY.

The upshot was an emphasis on rigorous, quantitative hypothesis testing based on an individual or gene selection view of evolution. Mathematical modelling became an integral part of formulating hypotheses and, as in ecology itself, some areas of the subject have remained entirely in that realm. Undoubtedly, however, it has been one of the strengths of behavioural ecology, and a major reason for its continued success 20 years on, that MATHEMATICAL MODELS have generally developed hand in hand with empirical tests of their predictions. An irony of this success, however, is that two of Tinbergen's four fundamental questions about behaviour — those of mechanism and development — were virtually eclipsed by a mass preoccupation with the other two, function and evolution.

Three conceptual frameworks form the cornerstones of behavioural ecology. W.D. Hamilton's now celebrated notion of inclusive FITNESS and its corollary, KIN SELECTION, revolutionized explanations of cooperation and conflict in social relationships by pointing to the implications of a simple and well-understood fact: that alleles are shared by given individuals as a function of their degree of relatedness. In combination with comparative social ecology, the inclusive fitness/kin selection paradigm launched the SOCIOBIOLOGY bandwagon of the late 1970s and the vacuous misunderstandings of its implications for humanity by anthropologists, sociologists and, regrettably, some psychologists and evolutionary biologists.

Optimality theory (*see* OPTIMALITY MODELS) has provided behavioural ecology with a productive theoretical framework for testing quantitative hypotheses. Borrowing many of its principles from economics and engineering, the optimality approach is predicated on selection itself being a natural optimizing agent. Any 'decision' an organism makes, whether behavioural, morphological or physiological, is likely to have both positive (benefit) and negative (cost) implications for its survivorship and reproductive success. SELECTION will be sensitive to differences in the cost : benefit ratio of alternative 'decisions' and favour those yielding the greatest benefit per unit cost. The task becomes one of identifying the problem an organism has to solve, deciding on the available alternative solutions and their attendant costs, benefits and CONSTRAINTS (assuming appropriate genetic variants will have arisen), then seeing whether the organism decides on the solution that maximizes its net reproductive benefit (the optimal solution).

It is important to emphasize, of course, that the approach does not assume organisms are clever mathematical modellers; optimal outcomes can be achieved using astonishingly simple 'rules of thumb', such as 'go for the biggest' or 'regard any large moving object as a predator'. The optimality approach has been fruitful in virtually every sphere of behavioural decision-making, from diet choice and egg-laying to aggressive conflict and the timing of copulation. However, perhaps because potential reproductive costs and benefits are easier to identify than in other contexts, and parallels with economic decision-making theory are closer, the optimality approach has been developed most fully in the context of foraging for food.

Related to optimality theory is the concept of EVOLUTIONARILY STABLE STRATEGIES (ESSs). Developed in the early 1970s by John Maynard Smith, ESS theory, like optimality theory, assumes that the decisions organisms make reflect heritable genetic strategies whose alleles compete with alternatives at their locus for a place in the GENE POOL. Using GAME THEORY (a branch of mathematics developed in economics), Maynard Smith showed that the optimal STRATEGY for an individual in any given situation depends on the strategies adopted by other individuals in the population. Thus it may pay to fight aggressively for a resource if everyone else around you is likely to retreat, but not if they are likely to fight back. Fighting aggressively is thus an advantage when it is rare in the population, but a disadvantage when it is common. The same is true for adopting a non-aggressive stance since it can be exploited by an aggressor when common but avoids the costly consequences of mutual escalation when rare. Neither the aggressive nor the non-aggressive strategy is therefore evolutionarily stable in its own right. Instead, the ESS is a mixture of aggressive and non-aggressive strategies in the population where the net reproductive benefit (the payoff) to both is the same. This can be achieved as a stable ratio of two GENOTYPES, one aggressive, the other non-aggressive, or by each individual being aggressive and non-aggressive for stable proportions of the time. Whether the stable outcome is a mixture of strategies as in this case, or a single strategy as in many other cases, depends critically on the field of alternatives (the strategy set) and the costs and benefits accruing to each competing strategy. The essence of an ESS is that it is robust to invasion by feasible alternatives so that stability rather than local optimality determines evolutionary success. ESS theory therefore provides a powerful means of generating testable predictions about the long-term outcome of natural selection for behavioural decisions.

Although behavioural ecology developed as an approach to studying the function of behaviour in the organism's natural environment, its philosophy has more recently crossed the interfaces with causal and developmental interests. Neuroanatomy, learning theory and immunology, for example, are three areas in which insights have sprung from asking functional questions about why things work the way they do. At the same time, dramatic advances in the techniques of molecular genetics have yielded powerful tools for testing functional hypotheses in the field. Paradoxically, therefore, after 20 years of dominating the study of animal behaviour, behavioural ecology has begun to unite Tinbergen's famous four questions into an integrated and mutually informative framework for the study of behaviour. *See also* EVOLUTIONARY OPTIMIZATION; OPTIMALITY MODELS; SOCIOBIOLOGY. [C.J.B.]

REFERENCE

Krebs, J.R. & Davies, N.B. (1993) *An Introduction to Behavioural Ecology*, 3rd edn. Blackwell Scientific Publications, Oxford.

---

benthic boundary layer The layer of water immediately above the SEA floor and extending upwards for a distance of 10 to several hundred metres from the bottom. It is the region above the bottom where flow is measurably slowed in comparison to the more vertically uniform mean velocity in overlying water. A laminar BOUNDARY LAYER (sometimes called a viscous sublayer) is a very thin layer adjacent to a fixed boundary in which the flow occurs in streamlines or sheets with no lateral mixing. At the sediment–water interface the boundary layer acts as a stagnant film and presents a barrier to molecular transport. The thickness of this film is estimated to vary between 30 and 3000 μm. A similar boundary layer occurs between the sea surface and overlying atmosphere.

Most lakes are not large enough to have the necessary bottom-water movement to make a benthic boundary layer a useful concept. Streams and rivers may have a boundary layer, but it will typically be of the order of millimetres to centimetres in thickness. [V.F.]

benthic community *See* BENTHOS.

benthic ecology *See* BENTHIC HABITAT CLASSIFICATION; ECOLOGY.

benthic habitat classification Organisms associated with the substrate–water interface of LAKES, RIVERS, ESTUARIES and SEAS form the BENTHOS. The benthos can be either phytobenthos (plant) or zoobenthos (animal), both of which are subdivided into macro- (visible with the naked eye) or micro- (visible with a microscope).

Historically, botanists developed a CLASSIFICATION of the phytobenthos, which is equally applicable to the zoobenthos. The rhizobenthos is anchored (or rooted) into or on the substrate, the haptobenthos is adnate to the surface and is synonymous with the more commonly used terms PERIPHYTON or *aufwuchs*. Haptobenthos can also be classified with respect to the substrate on which it grows, being EPILITHIC on rocks, epiphytic on plants, epipelic on sediments, epizoic on animals and episammic on sand.

In addition to the true benthos, organisms found deeper in the sediments are hyporheic. The hyporheos consists of the interstices between the particles composing the substrate.

Some organisms are associated with the substrate–water interface (benthos) on some occasions and at other times are found only in the water column, and are termed nektobenthic, or merobenthic. [R.W.D.]

benthic–pelagic coupling The functional linkage between benthic and PELAGIC subsystems. The supply of organic matter from the PLANKTON to the BENTHOS is balanced by MINERALIZATION of ORGANIC DETRITUS by the benthic microbial community and release of dissolved metabolites to the overlying water. Interactions between the two subsystems are largely determined by the amount of TURBULENCE in the water column and can have important controlling influences on both pelagic and benthic community structure. Most interest has been in nutrient dynamics, but recent work suggests that biotic interactions may be equally important. In experimental enclosure studies it has been demonstrated, for example, that changes in the pelagic community are correlated with different sediment communities and with the presence or absence of benthos. Results suggest that the actions of the benthic community shorten the pelagic FOOD WEB (reduce the numbers of pelagic ZOOPLANKTON) and inhibit the response of the pelagic fauna to nutrient enrichment in well-mixed COASTAL waters. The nutrient flux from sediment to overlying water can have a substantial influence on PRODUCTIVITY in shallow waters, but this influence diminishes with increasing water depth and is probably minimal below the CONTINENTAL SLOPE. *See also* MICROBIAL ECOLOGY; MICROBIAL LOOP. [S.J.H. & V.F.]

benthos Benthos refers to all the attached, creeping or burrowing organisms that inhabit the bottom of RIVERS, LAKES and the SEA. The term is derived from Greek, meaning 'depth of the sea'. It includes plants and animals living on the SEDIMENT surface, on the surface of HARD SUBSTRATA or protruding from the bottom (EPIFAUNA and epiflora), as well as those living buried in the sediment column (infauna). The term DEMERSAL has also been used for organisms (especially fish) living on or near the bottom. Many organisms that are benthic as adults have planktonic larvae (*see* LARVA) that may spend from a few hours to a few months in the PLANKTON. Benthic organisms have an important influence on the chemical and physical properties of sediments, such as porosity, grain size, oxygen content, microbial rate processes, etc.

Macrobenthic fauna include organisms > 1 mm in size, and in the oceans are represented by the polychaetes, molluscs, anthozoans, echinoderms, sponges, ascidians and crustacea. Meiobenthic fauna include organisms in the range 0.1–1 mm in size, and oceanic representatives of these are small polychaetes, bivalves, harpacticoid copepods, ostracods, nematodes, turbellarians and FORAMINIFERANS. Organisms < 0.1 mm in size are referred to as microbenthos, and include bacteria, benthic diatoms, ciliates, amoebae and flagellates.

The benthic macroflora are best represented by three major plant communities, namely SALT MARSHES, seagrasses and MANGROVES. Benthic microalgae can supply a substantial amount of primary production in shallow habitats and stabilize sediment via the secretion of mucous films.

Benthic animals may derive their food from PHYTOPLANKTON (suspension-feeders; *see* SUSPENSION FEEDING), or sediment-associated organic matter (DEPOSIT-FEEDERS). Others may be HERBIVORES (e.g. intertidal gastropods), predators (e.g. nereid polychaetes), scavengers (e.g. crabs) or parasites.

The VERTICAL ZONATION of benthic species depends primarily on light, moisture and pressure. The species diversity of benthic fauna increases and biomass decreases from the CONTINENTAL SHELF to the CONTINENTAL SLOPE. From the continental slope to the ABYSSAL plain, diversity again decreases. An exception to this pattern occurs in the HYDROTHERMAL VENT communities, which have biomass and diversity characteristics comparable to those of shallow-water communities. [V.F.]

benzene A colourless, flammable liquid (boiling point = 80.1°C) and the simplest aromatic HYDROCARBON (C_6H_6). Benzene (and derivatives, Fig. B1) are components of petrol, and car exhausts account for about 80% of total emission to the environment. It is an influential species within a large class of VOLATILE ORGANIC COMPOUNDS that are precursors to photochemical AIR POLLUTION and identified in water supplies. Urban atmospheric concentrations of benzene are typically between 1 and 50 ppbv, and significantly higher near major sources such as petrol stations. Cigarette smoke is an additional source for smokers, particularly indoors. Benzene is both toxic and a carcinogen and prolonged exposure has been linked to an increased risk of leukaemia. The aromatic ring forms the basic structure for another important class of organic compounds, the POLYCYCLIC AROMATIC HYDROCARBONS (PAH). [M.R.H.]

BEO *See* BEST ENVIRONMENTAL OPTION.

Berger–Parker dominance index *See* DIVERSITY INDICES.

Bergmann's rule A rule proposed by the German zoologist, C. Bergmann, in the mid-19th century, that HOMEOTHERMS within a single closely related evolutionary line increase in SIZE along a GRADIENT from warm to cooler CLIMATES. This is usually explained in terms of surface area to volume ratios; which reduce as body size increases in geometrically similar organisms. Less surface to volume means less heat loss — an advantage in cold conditions but not in warm conditions. *See also* ALLOMETRY. [P.C.]

Bering Sea An extension of the North Pacific, the Bering Sea lies between the Aleutian Island Arc to the south (51°N) and the Bering Strait to the north (66°N) and between 157°W and 163°E. It is the second largest of the marginal seas with an area of 2.3×10^6 km^2, a volume of 3.4×10^6 km^3, an average depth of 1491 m and a maximum depth of 4096 m. It is unusual in having an extremely large CONTINENTAL SHELF. The southern Bering Sea maintains a series of FRONTS off the SHELF-SLOPE BREAK which divides the PELAGIC realm into several distinctive trophic webs. Due to a lack of extensive lateral ADVECTION, the trophic webs remain distinct. *See also* PACIFIC OCEAN. [V.F.]

berkelium (Bk) A man-made element (ATOMIC NUMBER = 97; relative ATOMIC MASS = 247; melting point = 986°C; boiling point

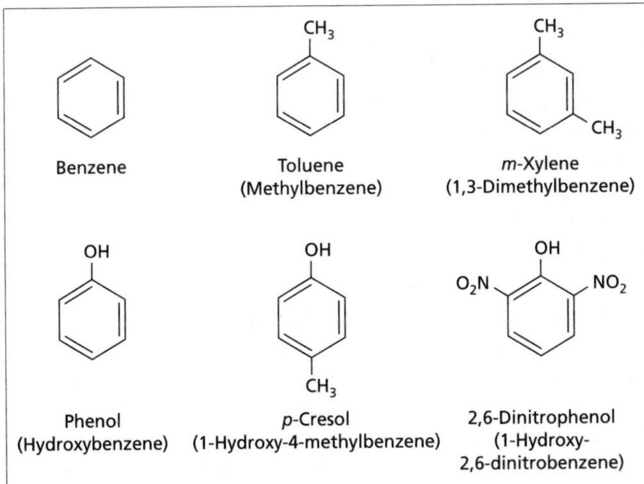

Fig. B1 Benzene and its derivatives.

unknown) which was discovered in 1949. Isotopes with mass numbers ranging from ^{240}Bk to ^{251}Bk have been produced by particle bombardment of other actinide elements, but only very small quantities have been generated. Consequently, berkelium does not occur in detectable quantities in the environment. Investigations of the properties of berkelium have mainly used ^{249}Bk ($t_{1/2}$ = 320 days). The most common oxidation state for Bk is +3 but it can also exist in the +4 form. [A.B.M.]

beryllium (Be) A very light, hard, corrosion-resistant metallic element (ATOMIC NUMBER = 4; relative ATOMIC MASS = 9.01; melting point = 1278°C; boiling point = 2970°C) which is a trace constituent (2 mg kg^{-1}) of the Earth's CRUST. Found in beryl (a beryllium aluminium silicate) and beryllium oxide (BeO), it can be released to the environment during refining for its use in alloys and components for the space and NUCLEAR POWER industries, but its main atmospheric input is via the combustion of coal. Beryllium is one of the most hazardous toxic ELEMENTS. Inhalation of the metal or its compounds can lead to berylliosis, a chronic form of pneumoconiosis of the lung. [J.F.]

best available control technology (BACT) *See* BEST AVAILABLE TECHNIQUES NOT ENTAILING EXCESSIVE COSTS.

best available controls (BAC) *See* BEST AVAILABLE TECHNIQUES NOT ENTAILING EXCESSIVE COSTS.

best available techniques not entailing excessive costs (BATNEEC) Legislation controlling possible POLLUTION from EMISSIONS from industrial plants sometimes not only specifies standards to be achieved but also that they have to be done by the best available techniques—a precautionary device. BATNEEC is an important element of the UK ENVIRONMENTAL PROTECTION ACT 1990. 'Best' is interpreted as meaning 'most effective'; 'available' means procurable (not just in development) from anywhere in the world; 'technique' embraces both plant (technology) and how it is used, recognizing that a poorly skilled workforce may not realize the potential of the best technology, whereas a well-trained workforce may compensate for technology that is not the best available; 'excessive cost' can theoretically be interpreted in a number of ways, for example as environmental return, since the returns in protection may reduce increasingly as more and more is invested so there is a point where further investment is not worthwhile, or from the point of view of the financial position of the business. *See also* COST (RISK)–BENEFIT APPROACH. [P.C.]

best available techniques reference document (BREF) Guidance from the European Commission on best available techniques to be used in the context of INTEGRATED POLLUTION PREVENTION AND CONTROL. *See also* INTEGRATED POLLUTION CONTROL. [P.C.]

best available technology (techniques) economically achievable (BATEA) *See* BEST AVAILABLE TECHNIQUES NOT ENTAILING EXCESSIVE COSTS.

best environmental option (BEO) An environmental management STRATEGY, usually used in connection with INTEGRATED POLLUTION CONTROL, designed to regulate releases to give most benefit or least damage, irrespective of cost. It thus differs from BEST PRACTICABLE ENVIRONMENTAL OPTION, which provides most benefit or least damage at acceptable cost. The concept was first developed in the 12th Report of the ROYAL COMMISSION ON ENVIRONMENTAL POLLUTION [P.C.]

best practicable environmental option (BPEO) Often in planning environmental protection measures more than one option is available. The BEST ENVIRONMENTAL OPTION (BEO) is the one that brings maximum gains. The best practicable environmental option is the one that brings the most gains in the most cost effective manner. Thus under INTEGRATED POLLUTION CONTROL legislation in Britain, for certain substances and processes it is necessary to put technology and techniques in place that produce best effects across all media not entailing excessive costs (*see* BEST AVAILABLE TECHNIQUES NOT ENTAILING EXCESSIVE COSTS). Very often more than one option will be available. Each option can be judged in terms of the environmental effects it allows and its costs. BEO is the option that causes minimum impact. BPEO is the option that causes minimum impact without entailing excessive cost (Fig. B2). *See also* BEST PRACTICABLE MEANS; COST (RISK)–BENEFIT APPROACH. [P.C.]

best practicable means (BPM) A strategy for the TECHNOLOGY-BASED CONTROL of release of CONTAMINANTS that emphasizes a pragmatic approach. It implies that while better EMISSION STANDARDS may be achievable, industry should not necessarily be required to implement them if this would not be practicable especially with respect to costs. *Cf.* BEST AVAILABLE TECHNIQUES NOT ENTAILING EXCESSIVE COSTS. *See also* BEST PRACTICABLE ENVIRONMENTAL OPTION; COST (RISK)–BENEFIT APPROACH. [P.C.]

Bet-hedging

A reproductive STRATEGY in temporally varying environments; a bet-hedging genotype's reduced among-generation VARIANCE in

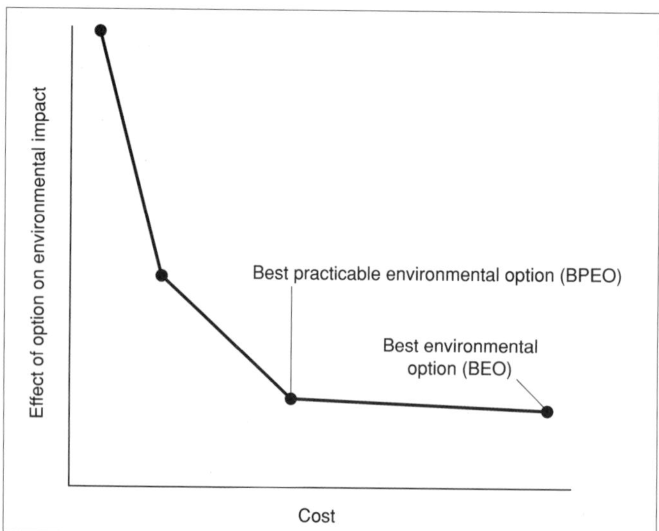

Fig. B2 Diagram showing the best practicable environmental option and the best environmental option.

FITNESS compensates for a reduction in its ARITHMETIC MEAN fitness.

Temporal variation in the environment can imply that the fitness a given GENOTYPE achieves will vary between generations. Since population growth is a multiplicative process, a genotype's GEOMETRIC MEAN fitness gives that genotype's expected long-term growth rate. Hence geometric mean fitness allows comparisons of different genotypes' demographic ADAPTATION to temporal variation. To define geometric mean fitness, suppose that an individual of genotype i achieves fitness λ_{ij} when j ($j=1, 2, \ldots, J$) is the state of the environment. Each generation independently experiences a single environmental state; that state is j with probability θ_j. Then the geometric mean fitness of genotype i is:

$$G_i = \prod_{j=1}^{J} \left(\lambda_{ij}\right)^{\theta_j}; \quad \lambda_{ij} > 0,\ 0 < \theta_j < 1$$

As long as there is some variation among the λ_{ij}, G_i will be less than the arithmetic average of the λ_{ij}. Geometric mean fitness (i) increases as the arithmetic mean of the λ_{ij} increases (fitness variance held constant); and (ii) decreases as the variance of the λ_{ij} increases (arithmetic mean held constant). Within this context, strategies that increase geometric mean fitness qualify as risk-averse adaptations to environmental variation.

To demonstrate bet-hedging, suppose a population of annual organisms inhabits an environment where wet and dry years occur randomly, independently and with equal probability (0.5). An individual of genotype 1 produces four surviving offspring in a wet year, but can expect only 0.5 offspring in a dry year. An individual of genotype 2 produces three surviving offspring in a wet year, but leaves only one offspring in a dry year. Genotype 1 has the greater arithmetic average fitness (2.25 > 2), but genotype 2 has the greater geometric mean fitness ($3^{1/2} > 2^{1/2}$). Genotype 2 has a sufficiently smaller variance in fitness between wet and dry years that its geometric mean fitness, hence its expected long-term growth, exceeds that of genotype 1. The lower variance more than compensates for genotype 2's lower arithmetic mean fitness. Genotype 2's less extreme (i.e. bet-hedged) response to temporal environmental variation indicates that its frequency in the population should increase relative to that of genotype 1.

Under certain circumstances a genotype that increases its within-generation fitness variance can, as a consequence, reduce its among-generation variance, and therefore increase its expected long-term growth rate. Suppose the above example's population now includes genotype 3. Half the individuals of genotype 3 express the PHENOTYPE of genotype 1; half express the phenotype of genotype 2. In the preceding example both genotypes 1 and 2 produce more offspring in wet years than in dry years. Retaining this assumption, genotype 3 has an intermediate arithmetic mean fitness (2.125), and an intermediate geometric mean fitness ($2.63^{1/2}$). Genotype 2 has the lowest fitness variance of the three strategies and the greatest geometric mean fitness. But a simple change in assumptions leads to a quite different result. Let genotype 1 remain the same, but suppose that the reproductive performance of genotype 2 is reversed with respect to wet vs. dry years. That is, an individual of genotype 2 now produces one surviving offspring in wet years, and produces three offspring during dry years. The arithmetic and geometric mean fitnesses of genotypes 1 and 2 remain exactly the same as in the preceding example. Genotype 3 has the same arithmetic mean fitness as in the first case (2.125). But genotype 3 now has the greatest geometric mean fitness ($4.375^{1/2}$). Genotype 3 mixes wet and dry 'specialist' phenotypes; as a LINEAGE, it never experiences a year when all its bearers do poorly. In this latter case, genotype 3 hedges its bets by generating fitness variance within each generation, since this strategy now reduces among-generation fitness variance.

The concept of bet-hedging in temporally varying environments extends to age-structured populations where individuals may reproduce more than once. Variances of age-specific vital rates, and the COVARIANCES between vital rates, influence the expected long-term growth rate of these populations. A number of empirical studies of plants, invertebrates and vertebrates suggest that certain life history characteristics may have evolved as bet-hedging responses to temporal environmental variation. *See also* LIFE-HISTORY EVOLUTION; RISK AVERSION. [T.C.]

REFERENCES

Philippi, T. & Seger, J. (1989) Hedging one's evolutionary bets, revisited. *Trends in Ecology and Evolution* **4**, 41–44.

Tuljapurkar, S. (1989) An uncertain life: demography in random environments. *Theoretical Population Biology* **35**, 227–294.

---

beta diversity *See* DIVERSITY, ALPHA, BETA AND GAMMA.

β-particle Either a negatively charged electron (β^--particle) or a positively charged positron (β^+-particle). β^--Particles are electrons emitted during radioactive decay of neutron-rich nuclei and typically have energies in the range 10 keV to 4 MeV. For a given RADIONUCLIDE, β^--particles have a continuous distribution of energies up to a defined maximum, since the decay energy is shared with a neutrino which is also emitted in the process. β^--Particles cause ionization in absorbers, but the specific ionization (amount of ionization per unit path length) is about 1000 times lower than that of α-PARTICLES of corresponding energy. β^+-Particles are positrons emitted in the decay of proton-rich nuclides. *See also* RADIOACTIVITY. [A.B.M.]

Bhopal *See* ECOLOGICAL/ENVIRONMENTAL DISASTERS.

bidirectional reflectance Reflectance defined in terms of the direction of the irradiating flux and the direction along which the reflected flux is detected. The specification of these directional components of reflectance is important as they determine the magnitude of flux recorded from the Earth's surface. [P.J.C.]

biennial plants PERENNIAL semelparous species (*see* SEMELPARITY) that may flower the first year or wait for many years before flowering, and then die. In some biennial species, the flowering date is genetically determined. For instance, in *Beta maritima*, one dominant gene determines bolting the first year (Boudry *et al.* 1994). Many ANNUAL plants can behave as biennials under adverse conditions by delaying flowering to the second season, or biennials may complete their LIFE CYCLE in a single favourable season. Biennials are widespread in the Apiaceae, Scrophularaceae (e.g. the genera *Verbascum* and *Digitalis*), Asteraceae (e.g. the thistle genera *Carduus* and *Cirsium*) and not uncommon in other unrelated families. *See also* LIFE-HISTORY EVOLUTION. [I.O. & G.A.F.H.]

REFERENCE

Boudry, P., Wieber, R., Saumitou-Laprade, P. *et al.* (1994) Identification of RFLP markers closely linked to the bolting gene B and their significance for the study of the annual habit in beets (*Beta vulgaris* L.). *Theoretical and Applied Genetics* **88**, 852–858.

bimodal distribution A DISTRIBUTION having two MODES. [B.S.]

binary fission The division of a body into two approximately equal parts. It occurs in PROKARYOTES and in ciliates. In prokaryotes the DNA molecule (which is attached to the plasma membrane) replicates and the replicates are separated by growth of the membrane between their attachment sites. When the cell has approximately doubled in size the plasma membrane pinches inwards between the replicate chromosomes and a cell wall forms, dividing the cell into two daughter cells. Binary FISSION in ciliates involves an amitotic splitting of the macronucleus (micronuclei divide mitotically) prior to cell division.

In multicellular organisms it involves splitting into two and replacement of the lost part by mitosis. *See also* ASEXUAL REPRODUCTION. [M.M.]

binomial classification The practice, due to Linnaeus (*see* CHARACTERS IN ECOLOGY), of describing taxa under a generic and a specific name (*see* LINNAEAN CLASSIFICATION). Thus, *Musca domestica* is the name of a fly, species *domestica*, belonging to the genus *Musca*. By convention, both names are italicized and the generic, but not the specific name has an initial capital. This arrangement was criticized for some time after it was proposed. If species are real entities, divinely created or evolved to form non-interbreeding lineages, what need is there to state the genus? The utility of the system for cataloguing biological organisms and as a basis for evolutionary investigation has, however, stood the test of time, even when considering fossils and, to a lesser extent, microorganisms. Generic and specific names were originally part of the description of the organism, indicating what it was like, as in the example of the domestic, or house fly, given above. As the number of described species has increased to some 1.5 million the names have become arbitrary, their importance lying in the fact that they provide a unique pair of words for each species.

CLASSIFICATION is bound by an international code and rules of nomenclature, based on the binomial system, which must be adhered to when publishing descriptions of new species. To be complete, the binomial should be followed by the name of the taxonomist who described it, and the date; hence, in our example, *Musca domestica* Linnaeus, 1758. *See also* SPECIES CONCEPTS. [L.M.C.]

binomial (positive binomial) distribution A discrete PROBABILITY DISTRIBUTION that describes situations where observations can fall into one of two categories (e.g. heads or tails when spinning a coin, and male or female offspring in a clutch or family). It is sometimes referred to as the positive binomial distribution (*cf.* NEGATIVE BINOMIAL DISTRIBUTION). If the probability of getting one of the categories (e.g. heads or male) is represented by p and the probability of getting the other category (e.g. tails or female) is represented by q, then in a binomial situation $p+q=1$ (there are only two categories possible). The probability distribution of the two categories of observation (according to the binomial distribution) are given by the expansion of $(p+q)^k$, where k is the number of events (e.g. number of coins spun or number of offspring). For example:

$$\text{for } k=1: (p+q)^1 = p+q$$

$$\text{for } k=2: (p+q)^2 = p^2+2pq+q^2$$

$$\text{for } k=3: (p+q)^3 = p^3+3p^2q+3pq^2+q^3$$

Therefore if we spin a single coin ($k=1$), the probability of getting a head $=p$ and the probability of getting a tail $=q$. If we spin two coins ($k=2$), the probability of getting two heads $=p^2$, the probability of getting a head and a tail $=2pq$, and the probability of getting two tails $=q^2$. If we spin three coins ($k=3$), the probability of getting three heads $=p^3$, the probability of getting two heads and one tail $=3p^2q$, the probability of getting one head and two tails $=3pq^2$, and the probability of getting three tails $=q^3$. Once p and q have specific values, the above probabilities can be shown graphically in the form of a discrete probability distribution (Fig. B3). Figure B3(a) shows the predicted outcome when $n=4$ and $p=q=0.5$. Figure B3(b) shows the predicted outcome when $n=4$ and $p=0.7$ and $q=0.3$. The individual terms of the binomial expansion $(p+q)^k$, can also be calculated by:

$$P_{(x)} = \frac{k!}{x!(k-x)!} q^{k-x} p^x$$

where $P_{(x)}$ is the probability of x events (i.e. male offspring in Fig. B3), and $k!$ means k factorial (e.g. $4! = 4\times3\times2\times1 = 24$). Therefore the probability of observing three females in a family of four (one male and three females) when $p=0.7$ and $q=0.3$ is:

$$P_{(3)} = \frac{4!}{3!(4-3)!} \times 0.3^{4-3} \times 0.7^3$$

$$P_{(3)} = \frac{24}{6} \times 0.3 \times 0.7^3$$

$$P_{(3)} = 0.41$$

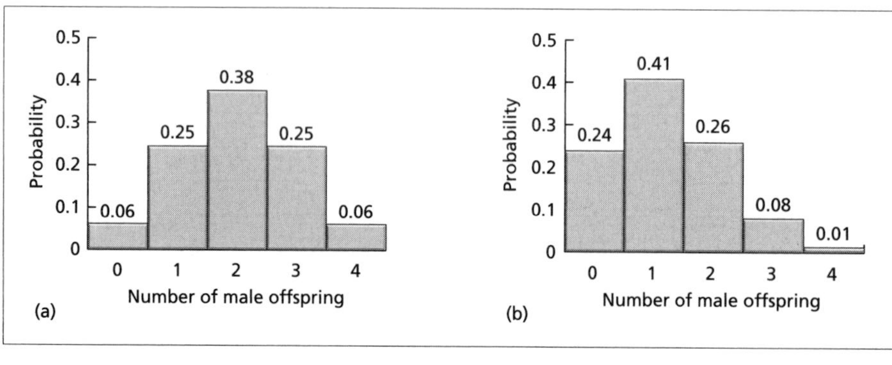

Fig. B3 The expected discrete probability distribution of male offspring in families of four ($k=4$). (a) $p=q=0.5$; (b) $p=0.7$, $q=0.3$.

which is equal to $4p^3q$, the second term in the expansion of $(p+q)^4$ (see Fig. B3(b)). *See also* DISTRIBUTION; STATISTICS. [B.S.]

binomial (sign) test A non-parametric test that may be used in lieu of the parametric paired *t* TEST to test for differences between MATCHED OBSERVATIONS, or to test for GOODNESS-OF-FIT. The principle of the test is very simple: the numbers of positive and negative signs among the differences between pairs of values are recorded. The NULL HYPOTHESIS that the numbers of positive and negative differences are equal is then tested. If the null hypothesis is true, then the sampling distribution of the signs of these distributions should follow a (positive) binomial distribution (*see* BINOMIAL (POSITIVE BINOMIAL) DISTRIBUTION). For example, if we have 10 pairs of values we would, on average, expect to observe 5 negative differences and 5 positive differences (i.e. mean probability=0.5 for either positive or negative differences). If, however, we observe 9 positive values, and 1 negative value then the binomial distribution (i.e. for $n=10$ trials, and mean probabilities of $\hat{p}=\hat{q}=0.5$) predicts that the probability of observing one or fewer negative differences is 0.00098. For a TWO-TAILED TEST this yields a probability of 0.00196, and we would therefore reject the null hypothesis, and conclude that a statistically significant difference exists. *See also* NON-PARAMETRIC STATISTICS. [R.P.F.]

bioaccumulation The progressive increase in a substance—usually in an organism or part of an organism—because the rate of intake via the body surface (*see* BIOMAGNIFICATION) or in food (*see* BIOCONCENTRATION FACTOR) is greater than the output from active or passive removal processes. [P.C.]

bioactivation The metabolic conversion of a XENOBIOTIC to a more toxic derivative. [P.C.]

bioassays Methods of analysis that use living tissues or whole organisms or collections of organisms to make quantitative and/or qualitative measurements of the amounts or activity of substances. A bioassay is not synonymous with an ECOTOXICITY TEST. [P.C.]

bioavailability In ECOTOXICOLOGY, the extent to which a substance can be taken up into an organism from that in food or the surrounding environment. In TOXICOLOGY, the term is more precisely defined as the extent to which a substance to which the body is exposed (ingestion, inhalation, infection or skin contact) reaches the systemic circulation and the rate at which this occurs. [P.C.]

biochemical oxygen demand (BOD) A measure of OXYGEN absorption from a sealed sample of water over a fixed period, usually five days (BOD_5). It is therefore an index of ORGANIC LOADING and microbial activity and hence WATER QUALITY. A suitable dilution of sample with a mineral salt medium is prepared. A suitable inoculation can be added as well. The sample is then incubated at 20°C for a defined period in the dark, after which time the DISSOLVED OXYGEN is determined. The initial dilution of the sample should be judged to achieve a depletion of between 30 and 70% SATURATION level. A number of reservations can be made about the relevance of BOD as a measure of water quality — mismatched microbes, irrelevant incubation period, DILUTION EFFECTS, interference by algae — but it continues to be used widely for this purpose. *Cf.* CHEMICAL OXYGEN DEMAND. *See also* SEWAGE POLLUTION. [P.C.]

REFERENCE

Adams, N. & Bealing, D. (1994) Organic pollution: biochemical oxygen demand and ammonia. In: *Handbook of Ecotoxicology*, Vol. II (ed. P. Calow), pp. 264–285. Blackwell Science, Oxford.

biocide A generic term for any substance that kills (or inhibits) organisms. Examples of biocides include: BACTERICIDES (bacteria); FUNGICIDES (fungi and moulds); HERBICIDES (weeds); INSECTICIDES (insects); rodenticides (rodents); and PESTICIDES (pests in general). [P.C.]

biocoenosis A term used, most frequently in the East European literature, to denote biotic communities of populations living together in a physically defined space at the same time. *See also* COMMUNITY; HISTORICAL EFFECTS (COMMUNITY ASSEMBLY); ZOOCOENOSIS. [M.H.]

bioconcentration factor (BCF) The ratio of the concentration of a substance in an organism or tissue to the concentration in the surrounding environment at apparent equilibrium. Hence it expresses the extent to which the substance concentrates in the organism or tissue. However, it is sensitive to conditions both outside and inside the organism. *See also* BIOACCUMULATION; OCTANOL–WATER PARTITION COEFFICIENT. [P.C.]

biocultural restoration *See* ECOLOGICAL RESTORATION.

biodegradation The processes of natural DECOMPOSITION/decay whereby synthetic and natural compounds are broken down to produce substances that can be used in biological renewal cycles. Thus in sewerage systems many synthetic chemicals are de composed by microbial metabolism into CARBON DIOXIDE (CO_2), WATER, dinitrogen (N_2), PHOSPHORUS (P), etc. However, some synthetic com- pounds are not amenable to biodegradation (e.g. PLASTICS) and so can accumulate within ECOSYSTEMS. *See also* SEWAGE TREATMENT WORKS. [P.C.]

biodestructible Describing some artificial materials that are designed to be broken down by biological action. For example, some PLASTICS have starch included in the polymers. These are subject to microbial metabolism, so the large POLYMER is broken into smaller fragments. [P.C.]

biodeterioration Damage to materials (e.g. foodstuffs) caused by biological activity. [P.C.]

biodiversity The number and variety of taxa in ecological systems ranging from parts of communities to ECOSYSTEMS, regions and the BIOSPHERE. There is a deep concern that human activities are leading to species' losses, i.e. reduction in biodiversity. There is considerable activity in biodiversity research and the following cross-references draw attention to this. The benefits of biodiversity to humans are dealt with in BIODIVERSITY, ECONOMIC VALUE. *See also* AFRO-ALPINE VEGETATION; AFRO-INDIAN FLORAL REGION; AGENDA 21; AGROECOSYSTEMS; AMPHIBIANS, DIVERSITY AND CONSERVATION; ANADROMOUS FISH CONSERVATION; ANCIENT WOODLAND/FOREST; ANGIOSPERMS, GLOBAL DIVERSITY AND CONSERVATION; ANTARC-

TICA, ECOLOGY AND CONSERVATION; AQUATIC SYSTEMS, CONSERVATION AND RESTORATION; ARBORETUM; ARBORICULTURE; ARGENTINA, CONSERVATION IN; ATLANTIC NORTH AMERICAN FLORAL REGION; AUSTRALIAN FAUNAL REGION; AUSTRALIAN FLORAL REALM; BEETLES, DIVERSITY AND CONSERVATION; BENTHOS; BIODIVERSITY GRADIENTS; BIOSPHERE RESERVES; BIOTA; BOREAL FOREST; CANADA, CONSERVATION IN; CAPE FLORAL REGION; CHIROPTERANS, DIVERSITY AND CONSERVATION; CLIMATE CHANGE, POTENTIAL IMPACTS ON BIOLOGICAL SYSTEMS; CORAL REEFS, DIVERSITY AND CONSERVATION; DEEP SEA; DIVERSITY; DIVERSITY GRADIENT; EAST AFRICAN FLORAL REGION; ENDANGERED SPECIES ACT; ETHIOPIAN FAUNAL REGION; EURO-SIBERIAN FLORAL REGION; EXOTIC AND INVASIVE SPECIES; EXTINCTION CRISIS, CURRENT GLOBAL; FISH, DIVERSITY AND CONSERVATION STRATEGIES; FLORAS, MAJOR PUBLISHED; FUNGI, DIVERSITY AND CONSERVATION; GEOGRAPHIC INFORMATION SYSTEMS (GIS) IN CONSERVATION; GLOBAL ENVIRONMENTAL FACILITY; GRADIENTS OF SPECIES RICHNESS; GYMNOSPERMS, GLOBAL DIVERSITY AND CONSERVATION; INSECTS, DIVERSITY AND CONSERVATION; INTERNATIONAL CONSERVATION CONVENTIONS; INTERNATIONAL SPECIES INFORMATION SYSTEM; INTERNATIONAL WHALING COMMISSION; ISLAND BIOGEOGRAPHIC THEORY IN CONSERVATION; ISLAND BIOGEOGRAPHY; IUCN; MACARTHUR–WILSON MODEL OF ISLAND BIOGEOGRAPHY; MANGROVE COMMUNITIES, ECOLOGY AND CONSERVATION; MEDITERRANEAN CLIMATE REGIONS, CONSERVATION; MEDITERRANEAN SCRUB/FOREST; MONOCULTURE; NEOTROPICAL FAUNAL REGION; NEOTROPICAL KINGDOM; PALAEARCTIC FAUNAL REGION; PANBIOGEOGRAPHY; RIVER INVERTEBRATE PREDICTION AND CLASSIFICATION SYSTEM; SPECIATION; SPECIES-ABUNDANCE MODELS; SPECIES–AREA RELATIONSHIP; SPECIES-LEVEL DIVERSITY, GLOBAL; SPECIES SURVIVAL COMMISSION; TROPHOSPECIES; TROPICAL MONTANE FOREST; TROPICAL RAIN FOREST; TROPICAL SEASONAL FOREST; UNITED NATIONS PROGRAMMES AND BODIES; WETLANDS (TIDAL AND COASTAL), CONSERVATION. [P.C.]

biodiversity, economic value The protection of species brings constraints on economic development. For example, restrictions on the use of AGROCHEMICALS have implications for agricultural yields and also knock-on effects to employment in both the agricultural and chemicals sectors. At the same time there are possible economic returns from WILDLIFE in terms of food yield (e.g. from fish) and other resources such as pharmaceuticals and natural pest-control chemicals. More fundamentally, species support ecological processes that in turn provide ECOSYSTEM SERVICES to the human economy, such as clean air, clean water, good soil and so on. There are various ways of valuing these contributions to the economy (*see* ENVIRONMENTAL (ECOLOGICAL) ECONOMICS), and of weighing these benefits derived from BIODIVERSITY with the costs associated with protecting it. *See also* COST (RISK)–BENEFIT APPROACH. [P.C.]

REFERENCE

Pearce, D. & Moran, D. (1994) *The Economic Value of Biodiversity*. Earthscan Publications, London.

biodiversity gradients Life is more 'abundant' at the tropics than at the poles. This applies not only to the BIOMASS and number of individuals, but in many taxonomic groups to the number of species as well. The cause of this planetary 'biodiversity gradient' in SPECIES RICHNESS remains in doubt. Explanations can be divided into dynamic theories and structural theories.

Dynamic theories explain BIODIVERSITY as an equilibrium between the birth and death rates of species.

1 Wallace's dynamic theory invokes global birth (speciation) and global death (extinction). The area occupied by a species is probably a major factor in its probability of extinction and also (in the ALLOPATRIC SPECIATION model) governs the probability that it will divide into two species. As the planet has a greater area at low latitudes, it follows that the tropics are a zone of high species birth, and the higher latitudes are a zone of high species death. This sets up the global biodiversity gradient from the Equator to the poles. The slope may further change over geological time: high latitudes may now have lost much of their diversity as a result of the Quaternary GLACIATIONS.

2 Hutchinson's dynamic theory (also known as the species-energy hypothesis) maintains that biodiversity levels are maintained by the cycle of local birth and death, that is local species extinction and local recolonization. The global biodiversity gradient is caused by the latitudinal gradient in thermal energy, through a mechanism (Fig. B4a) which can be summed up as: (i) the total level of thermal energy, chiefly in the form of air temperature or water temperature (as well as water availability and nutrient availability) controls the total biomass and therefore the population sizes of the species present, by influencing rates of growth and activity; (ii) POPULATION SIZE controls the rate of local extinction of any one species; and (iii) at higher population sizes fewer species, particularly the rarer ones, become locally and temporarily extinct. Therefore species richness is higher at low latitudes, where there is more energy. In species-rich communities evolution then causes greater specialization and the narrowing of ecological niches.

Hutchinson's structural theory (Fig. B4b) stands this idea on its head and proposes that there are more species at low latitudes because ecological niches are narrower. It is thus the structure of ecological communities that determines their species richness. The most widely believed variant maintains that the stable environment of the tropics (both seasonally and/or over geological time) encourages the evolution of more specialized species with narrower ecological niches. In turn, this permits more species to 'pack in' to the ecosystem.

These three theories have many features in common. All may be true: Wallace's theory seeks to explain how the diversity gradient evolves, and Hutchinson's theories to explain how it is maintained and modified. *See also* DIVERSITY GRADIENT. [J.R.G.T.]

REFERENCES

Hutchinson, G.E. (1959) Homage to Santa Rosalia *or* Why are there so many kinds of animals? *American Naturalist* **93**, 145–159.

Ricklefs, R.E. & Schluter, D. (eds) (1993) *Species Diversity in Ecological Communities: Historical and Geographical Perspectives*. University of Chicago Press, Chicago, Ill.

Rosenzweig, M.L. (1995) *Species Diversity in Space and Time*. Cambridge University Press, Cambridge.

Turner, J.R.G., Lennon, J.J. & Greenwood, J.J.D. (1996) Does climate cause the global biodiversity gradient? In: *Aspects of the Genesis and Maintenance of Biological Diversity* (eds. M. Hochberg, J. Claubert & R. Barbault), pp. 199–220. Oxford University Press, Oxford.

Wallace, A.R. (1878) *Tropical Nature, and other Essays*. Macmillan, London.

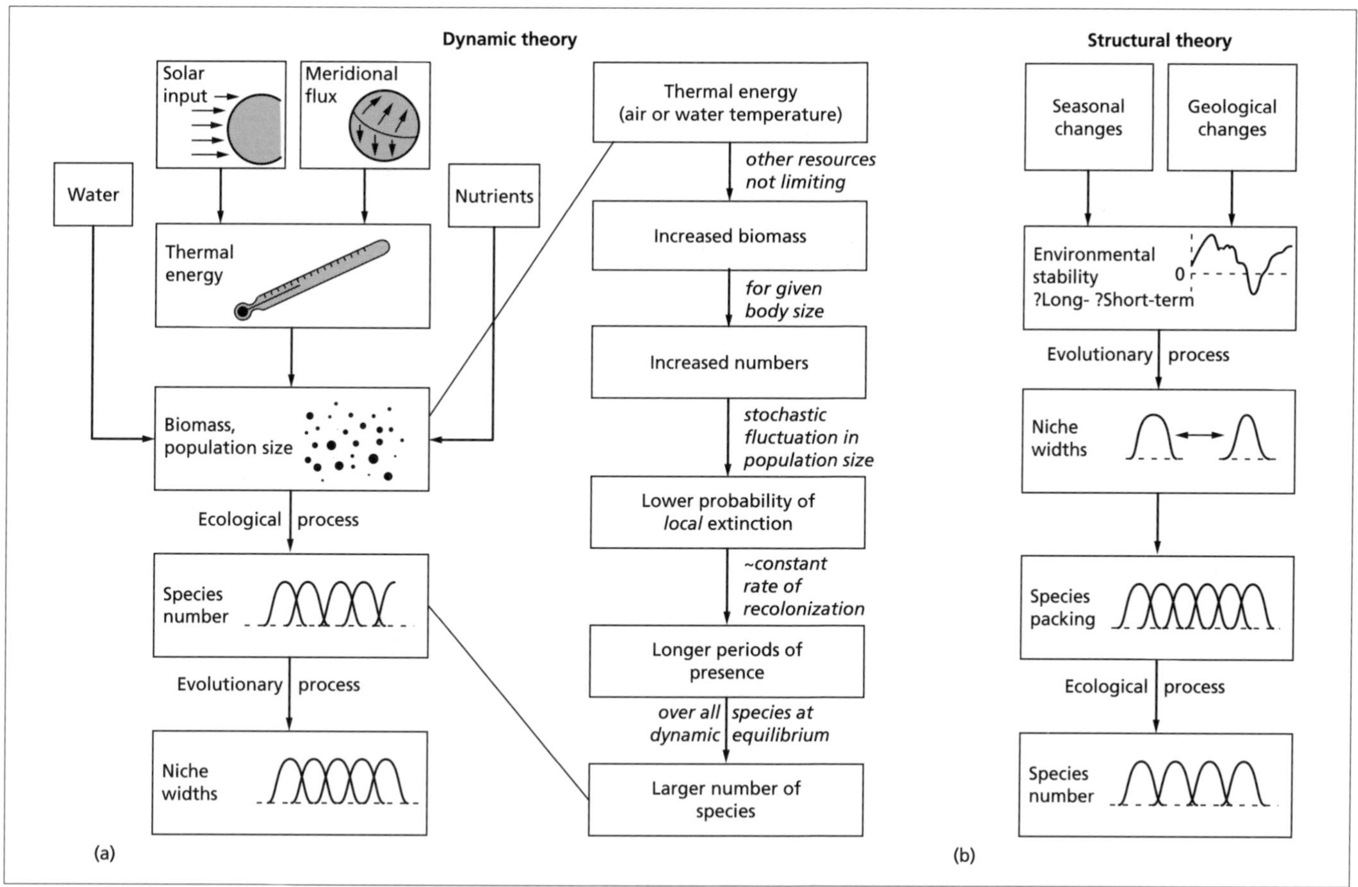

Fig. B4 Flow diagrams of Hutchinson's two theories of the biodiversity gradient. (a) The dynamic, or energy theory. The stochastic process of population turnover (local extinction), according to the dynamic theory, links the resource supply to species richness. (b) The structural theory. (After Turner, J.R.G. (1992) Stochastic processes in populations: the horse behind the cart? In: *Genes in Ecology* (eds R.J. Berry, T.J. Crawford & G.M. Hewitt), pp. 29–53. Blackwell Scientific Publications, Oxford.)

bioenergetics The study of ENERGY FLOW through biological systems. It includes both the physiological studies of energy transfer during metabolic processes and ecological studies of the rates and efficiencies of energy transfer through organisms in different trophic levels. *See also* ECOLOGICAL ENERGETICS. [M.H.]

biofilm, in aquatic systems The organic layer that coats all underwater surfaces. It is a heterogeneous collection of largely heterotrophic bacteria, microfungi, protozoans and micrometazoans such as rotifers. It can also contain photosynthetic algae. There is usually a polysaccharide matrix. It is considered by some as an initial successional stage in the COLONIZATION of FRESHWATER surfaces—one that never progresses beyond this stage in certain circumstances, for example the bottoms of disturbed boulders where no PHOTOSYNTHESIS is possible. *See also* EPILITHIC. [P.C.]

REFERENCE
Lock, M.A., Wallace, R.R., Costerton, J.W., Ventullo, R.M. & Charton, S.E. (1984) River epilithon: toward a structural-functional model. *Oikos* **49**, 10–22.

biofouling Unwanted accumulation of BIOTA on water-covered surfaces or appliances (e.g. hulls of boats, pipes) that impedes their effectiveness. Algae/barnacle growths on ships are classical examples. The zebra mussel, *Dreissena polymorpha*, has been introduced into North American freshwaters from Europe. As an exotic species, the POPULATION GROWTH is prolific and it has caused problems by fouling pipes. [P.C.]

biofuels Fuel/energy derived from extant BIOMASS (i.e. not fossil fuel). Examples are wood, bioethanol from sugar cane, and biogases from organic wastes. In principle, all biofuels represent RENEWABLE ENERGY sources. Historically, they have been the major source of energy for humans, but now supply less than 5% of the primary energy consumed by industrialized countries. [P.C.]

biogenetic law *See* NON-TERMINAL ADDITION.

biogenic sedimentation Deposition of material in the SEA consisting mainly of two substances produced as hard parts of organisms, namely calcium carbonate ($CaCO_3$) and silica (SiO_2), precipitated by PLANKTON living in the waters of the surface OCEAN. Calcium carbonate is formed mainly by coccoliths and foraminiferans, whereas silica is formed largely by diatoms and radiolarians. Sediment is considered to be biogenous if it contains at least 30% by volume of the skeletal remains of organisms. Biogenic SEDIMENTATION can also include the deposition of FAECAL PELLETS by benthic suspension-feeding or PELAGIC organisms. [V.F.]

biogenic sorting A regular change in SEDIMENT median grain size caused by the activities of benthic organisms. Many deposit-feeding organisms preferentially ingest small particles and deposit them on the sediment surface, so forming a biogenically sorted sediment with fine particles on the sediment surface and coarse particles at depth. *See also* BIOGENICALLY REWORKED ZONE. [V.F.]

biogenically reworked zone The depth zone within a SEDIMENT that is actively burrowed by benthic organisms. This zone can extend for several centimetres below the sediment–water interface. The burrowing activities influence the microtopography of soft bottoms, the rate of exchange of dissolved or absorbed substances across the sediment–water interface, the form of vertical gradients in oxidation potential (Eh), pH and Po_2, DECOMPOSITION rates of organic matter and pollutants, and the cycling of NUTRIENTS. *See also* BIOGENIC SORTING. [V.F.]

biogeochemical cycle Any of various natural cycles of ELEMENTS, involving biological and geological compartments. *See also* CARBON CYCLE; NITROGEN CYCLE; PHOSPHORUS CYCLE; SULPHUR CYCLE. [P.C.]

Biogeographical units

The world may be divided up into areas of similar character in terms of the animals and plants (the BIOTA) present in them. These biogeographical units are based on similarity of composition in terms of the systematics (and hence evolutionary history) of the biota. The extent and boundaries of these units have been determined by changes in CLIMATE and the movement of continents, and the accompanying changes in the physical and climatic barriers to MIGRATION.

A distinction must be made between such biogeographical units and the concept of BIOMES, which merely reflect the ecological character (and particularly the spectrum of life forms) in any given situation. Thus the RAIN FOREST biome is represented in humid tropical West Africa and in the Amazon Basin, but the actual composition of evergreen tropical forest in those two areas is very different at the level of species and genera. They are accordingly placed in two different biogeographical units (the Palaeotropical kingdom and the NEOTROPICAL KINGDOM respectively).

Zoologists, in delimiting zoogeographical units, have tended to put emphasis on the mammals and birds as the larger and more obvious features of the fauna, in recognizing faunal subdivisions of the world's biota (Fig. B5). Plant biologists, in drawing up comparable phytogeographical units, base them largely on the distribution of flowering plants (angiosperms) (Fig. B6). Because the centres of origin of major plant and animal groups have, of course, been different, and their capacity to migrate across different types of barrier (oceans, mountains, deserts) are equally diverse, the zoogeographical and phytogeographical units do not coincide, although they have obvious similarities (cf. Figs B5 and B6).

Several different hierarchies of biogeographical units are widely used, and there is no international agreement about their nomenclature (as there is for the organisms themselves!) A widely used nomenclature for phytogeographical units is that of Takhtajan (1969; see also Collinson 1988), which is closely similar to that of Good (1964), and both these authors recognize four levels in their hierarchy:

1 kingdom (of which they recognize only six);
2 subkingdom;
3 region;
4 subregion (see Fig. B6).

It should be noted that the terms kingdom and subkingdom have also been used for designating the highest level of taxa of organisms (e.g. animal kingdom). Also, only some of Takhtajan's kingdoms are actually subdivided into subkingdoms, and subkingdoms to regions. The Cape kingdom of South Africa, for example, has simply one region (*see* CAPE FLORAL REGION). This usage of four nested units is not universal, and Neill (1969), for example, uses the term PROVINCE instead of the more widely used region, as a subordinate unit of subkingdom. The terminology of Good and Takhtajan is preferred here.

The recognition of the corresponding zoogeographical units goes back to Wallace (and a number of other authors); these constitute a rather simpler hierarchy in which three levels are widely recognized:

1 realm;
2 region;
3 subregion (see Fig. B5).

Neill (1969) recognizes three zoogeographical realms and two transition zones, one between Australia and Asia, and one between North and South America, in Central America and the Caribbean area. Note that the use of region here is at a higher level than that in the phytogeographical hierarchy.

Some authors favour quantifying the different levels of the hierarchy in terms of percent endemicity. On this basis a region has been defined as having a 50–75% endemic flora or fauna, while a province, if treated as a subunit of a region, is then defined as having a 25–50% endemic flora or fauna (see e.g. Lincoln *et al.* 1982). While this approach seems to enhance the precision of the terminology, it is difficult if not impossible in practice. There is no region of the world for which the precise BIODIVERSITY, even in terms of the status of the larger organisms, can be accurately quantified, making it impossible to be precise about percentages of endemics in the flora or fauna. *Cf.* BIOMES, MAJOR WORLD. [W.G.C.]

REFERENCES

Collinson, A.S. (1988) *Introduction to World Vegetation*, 2nd edn. Unwin Hyman, London.
Good, R. (1964) *The Geography of the Flowering Plants*. Longman, Green & Co., London.
Lincoln, R.J., Boxshall, G.A. & Clark, P.F. (1982) *A Dictionary of Ecology, Evolution and Systematics*. Cambridge University Press, Cambridge.
Neill, W.T. (1969) *The Geography of Life*. Columbia University Press, New York.
Takhtajan, A. (1969) *Flowering Plants, Origin and Dispersal*. Oliver & Boyd, Edinburgh.

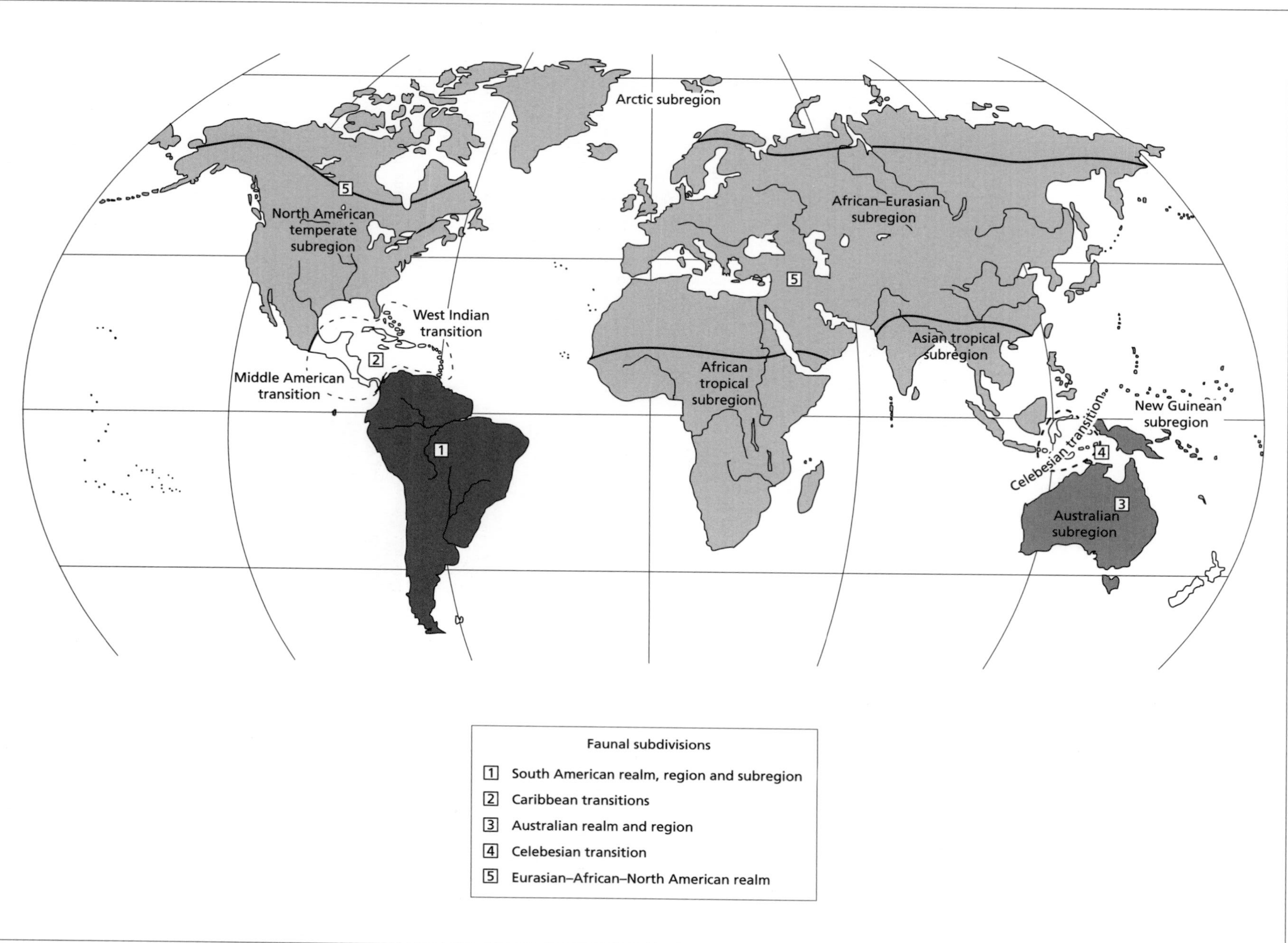

Fig. B5 Faunal subdivisions. (From *The Geography of Life*, by Wilfred T. Neill. Copyright © 1969, Columbia University Press. Reprinted with permission of the publisher.)

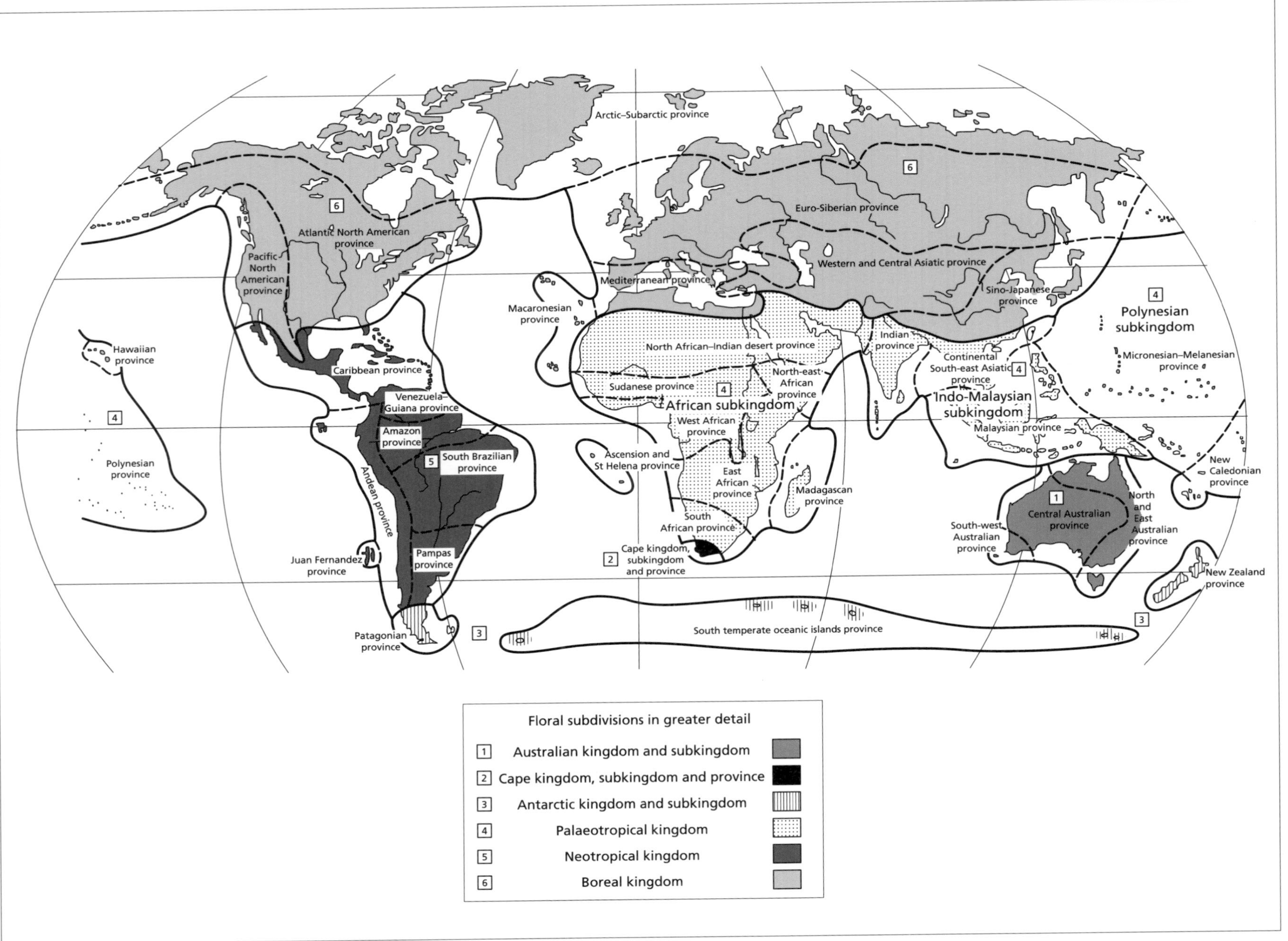

Fig. B6 Floral subdivisions. (From *The Geography of Life*, by Wilfred T. Neill. Copyright © 1969, Columbia University Press. Reprinted with permission of the publisher.)

Biogeography

Biogeography is concerned with the patterns of DISTRIBUTION of species over the face of the globe and understanding the origins and mechanisms which determine the distribution. Biogeography is also concerned with exploring the underlying changes (territorial expansions and contractions) that have taken place in the past or, increasingly, which may occur under CLIMATES of the future.

The origin of biogeography as a scientific discipline has been attributed to the great French naturalist of the Enlightenment, Georges Leclerc, Comte du Buffon. In 1761 he pointed out that the Old and New Worlds have no mammalian species in common. This led to a generalization sometimes known as Buffon's law, that different regions of the globe, though experiencing the same environmental conditions, were inhabited by different species of animal and plant. Such a 'law' implies that the key factor controlling organic distribution is history rather than ecology. It was actually the Swiss botanist A.P. de Candolle who was the first to make this generalization, early in the 19th century, and his 1820 publication can be taken as marking the true birth of biogeography. In this work, de Candolle listed 20 botanical regions across the world, defined by ENDEMIC SPECIES.

Charles Darwin was a pioneer of dispersalist biogeography. Chapters 12 and 13 of *The Origin of Species* are devoted to GEOGRAPHIC DISTRIBUTION. He concluded that:

> . . . all the grand leading facts of geographic distribution are explicable on the theory of migration, together with subsequent modification and the multiplication of near forms. We can thus understand the high importance of barriers, whether of land or water, in not only separating, but in apparently forming the several zoological and botanical provinces.

The dispersalist interpretation, implying MIGRATIONS of species from 'centres of origin' remained the dominant one until the 1970s. However, directly as a consequence of the general acceptance by geologists of CONTINENTAL DRIFT, the alternative interpretation was put forward that formerly widespread BIOTA have been split up (vicariated) by the establishment of barriers, such as newly opened oceans. In consequence, the degree of geographic proximity of organisms could be more a function of laterally migrating continents than migrating biota.

Much biogeography, however, is concerned with the Quaternary, involving only the last 2 million years, which is too short a time for the effects of continental drift to be important. Instead, the overwhelmingly dominant environmental leitmotif was global temperature change associated with recurrent GLACIATIONS and deglaciations, with concomitant sea-level fluctuations as water was alternately abstracted from or returned to the ocean. Virtually all the Quaternary fossils known belong to extant species, and thus their ENVIRONMENTAL TOLERANCES are often known very precisely. It has become quite clear that most organisms did not change their tolerances but tracked the environment. Thus cold-intolerant organisms have migrated towards lower latitudes during colder climatic cycles, and migrated towards higher latitudes during times of climatic amelioration.

The distribution of extant organisms or communities can often be explained by their coincidence with patterns of climate, particularly precipitation and temperature, of SOIL types, of TOPOGRAPHY, or by proximity to rivers and sea coasts or by BIOTIC FACTORS including human agricultural practices. Distribution patterns in most cases are a consequence of more than one of these. For example the stemless thistle, *Cirsium acaulon*, is relatively common throughout temperate Europe in closely grazed pastures (created as a result of agriculture) but in Britain its northern boundaries also closely match the 16°C minimum mean July ISOTHERM.

Contemporary distribution patterns often reveal apparent anomalies such as isolated (or disjunct) populations or absences from otherwise suitable territories. In many cases these anomalies reflect climatic conditions of the Pleistocene where the advance and subsequent retreat of icefields has isolated once-contiguous communities. An example is the alpine communities of mountain tops in mid-latitude Europe isolated by the postglacial SPREAD of forests. Isolation or disjunction can also follow as a result of incursion of the sea, as has occurred in the Pliocene to once-united coastal floras to the north and south of the Mediterranean sea. Biogeography can be extended even further back to reconstruct floras of once-united landmasses which have subsequently moved apart, over long periods of time. By reconstructing the GONDWANA flora of 70 million years ago it is possible to account for many of the similarities in the contemporary floras (and to some extent the faunas) of the eastern coast of South America with the west coast of Africa, or India with Madagascar, or the dissimilarities of the Australasian flora and fauna from that of south-east Asia. Comparisons of the biogeographical data of single species such as the giant redwood (*Sequoia* spp.), say 50 million years ago, with its contemporary Californian distribution can be used to predict with considerable precision the palaeoclimatic conditions of the now-arctic Spitsbergen (Svalbard). The presence on Spitsbergen of extensive fossil remains of conifers such as golden larch (*Pseudolarix*), chinese cypress (*Glyptostrobus*) and dawn redwood (*Metasequoia*) and the distribution of those same genera today, strongly suggests that the temperature of Spitsbergen 50 million years ago was frost-free with a mean annual temperature of 15–18°C, the climatic conditions found today in mid-China (*see* PALAEOCLIMATIC RECONSTRUCTION).

Biogeographical analyses have also been extended to the reconstruction of the distribution of species in past centuries and millennia. The construction of pollen distribution maps from postglacial deposits has provided evidence of the expansion and contraction of a number of plant species. From this it is also possible to pinpoint with considerable accuracy the advent of human interference in natural distribution patterns. Information from past distribution of plants and animals, combined with increasingly accurate data for climates of the past is being used to predict the distribution of species under climates of the future. [A.H. & G.A.F.H.]

REFERENCES

Candolle, A.P. de (1820) *Geographie Botanique*. In: *Dictionnaire des Sciences Naturelles*, Vol. 18, 359–422.

Cox C.B. & Moore, P.D. (1993) *Biogeography: An Ecological and Evolutionary Approach*, 5th edn. Blackwell Scientific Publications, Oxford.

Tallis, J.H. (1991) *Plant Community History*. Chapman & Hall, London.

---

bioindicator An organism or its metabolic system that is used to signal the presence of a CONTAMINANT and its effects. A bioindicator may be a microbe or metazoan; bacterium, fungus, plant or animal; a single-species or multi-species system. *See also* BIOMARKERS. [P.C.]

biological control The limitation of the abundance of living organisms by other living organisms. Biological control can refer to the naturally occurring REGULATION of plant or animal populations by HERBIVORES, predators or diseases (natural biological control), or (more commonly) to the manipulation of these natural enemies by humans. Applied biological control is used as a means of reducing populations of PEST species, either on its own, or in combination with other control methods as part of an INTEGRATED PEST MANAGEMENT (INTEGRATED CONTROL) system.

Classical biological PEST CONTROL involves the deliberate introduction and establishment of natural enemies as control agents, and has most commonly been directed against 'ALIEN' species which have become pests after being introduced to areas outside their natural RANGE. Alien species typically lack the full range of natural enemies that would be reducing their populations in their region of origin, and biological control seeks to eliminate this advantage. Typically this involves the importation of potential control agents that attack the pest within its natural range, but it may also include agents that would, under natural circumstances, be attacking closely related taxa, rather than the pest itself. Potential control agents are usually laboratory-reared in large numbers, prior to inoculation at several sites within the range of the alien pest. If these populations succeed in becoming established, and they have a significant impact on the abundance of the pest, then effectively permanent control can be obtained.

The concept of biological control has been broadened to encompass all manipulations of pest natural enemies. These include the release of control agents that cannot permanently establish their populations, and so must be released repeatedly (inundative release), and the use of biological PESTICIDES. *See also* BIOCIDE. [S.G.C.]

biological invasions Expansion in the distribution of certain species of plants, animals and microorganisms which are transported by humans and often competitively favoured by the DISTURBANCE around human settlements. Human activity is breaking down the distinctiveness of ecosystems through the movement of organisms to new locations beyond the limits of their native GEOGRAPHIC RANGES. The effect worldwide is one of homogenization of the until recently highly distinctive fauna and flora of individual landmasses.

Invasions pose serious problems worldwide to BIODIVERSITY, agricultural potential and human health. Ecosystem disruptions of the following types have been recorded in the literature in otherwise (except for biological invaders) near-natural areas:

1 acceleration of soil erosion rates (FERAL ungulates);
2 alteration of BIOGEOCHEMICAL CYCLES (feral pigs, invasive nitrogen fixers, salt accumulators);
3 alteration of geomorphological processes (*Casuarina* trees on beaches, dune and marsh grasses);
4 alteration of hydrological cycles (phreatophytes);
5 alteration of fire regimes (invasive grasses and shrubs);
6 prevention of RECRUITMENT of native species (invasive plants, mammals and ants).

As invasive species expand their populations, native species may be gradually crowded out, with some eliminated. Oceanic islands, such as the Hawaiian and Galapagos archipelagos, are among the most drastically affected areas: much of the local BIOTA is in the process of being reduced and displaced by aggressive invading species such as feral ungulates (goats, pigs, cattle, deer), rats, foreign ants and escaped plants (e.g. guava, lantana, pasture grasses). *See also* EXOTIC AND INVASIVE SPECIES. [L.L.L.]

Biological monitoring

Monitoring of biological organisms, species or ECOSYSTEMS is carried out with the aim of assessing the degree to which observations meet our expectations. This is usually conducted in the context of concern about loss of species, quality of HABITAT, increases in the levels of pollutants or some other problem. However, it is also important to monitor the effects of management carried out on reserves and other protected areas and to ensure that what we do is what we intended and is cost-effective.

There are several words in current usage with different shades of meaning in the context of MONITORING. These do not always have precise definitions and their meanings overlap to a certain extent. Biological recording is usually without motive, important and widely practised, whereas monitoring, strictly speaking, is conducted with clear objectives, a standardized procedure and clear rules for stopping (called termination by some people). Surveys are a one-off recording exercise whilst, strictly speaking, SURVEILLANCE refers to repeated SURVEYS. CENSUS involves recording all individuals and usually information about RECRUITMENT (births) and MORTALITY (deaths). It not only applies to animals but also to plants with discrete individuals such as orchids and trees.

There is a very real need for more and better monitoring of nature reserves and other protected areas such as national parks to ensure that the resources that most of us value are safeguarded. In order to achieve this, monitoring needs to be written into every management plan, which should also indicate the precise procedure, who is responsible, how and where the data will be stored, and what is to be considered an unacceptable change. This is often more difficult than it appears as organisms naturally fluctuate between fairly wide limits.

Other situations where there is currently an increasing interest in monitoring include aquatic environments such as streams and rivers. Concern about WATER QUALITY has led to the use of biological indices based on the presence or absence of a series of INDICATOR SPECIES. Indicator species have the advantage that they can detect a POLLUTION effect even if the observer was not present at the time of SAMPLING whereas sampling for a chemical substance, usually being intermittent, can miss pollution events.

Any of the criteria for conservation evaluation are suitable for monitoring. Two that are particularly widely used are rarity (*see* RARITY, BIOLOGY OF; RARITY, CLASSIFICATION) and BIODIVERSITY. Biodiversity, or SPECIES RICHNESS, is usually recorded in grid squares but the size of these can vary from 1 × 1 km for an English county to quarter degree squares ($^1/_4$° lat. × $^1/_4$° long.) in rural Africa.

There are a series of schemes for biological recording which are often cited as being monitoring. They are undoubtedly useful but were not necessarily initiated with clear objectives. These include the British Trust for Ornithology's Common Bird Census, which focuses on the DISTRIBUTION of breeding birds throughout Britain, and the Plant Atlases of the Botanical Society of the British Isles. All such biological recording is based on event recording, each event being the occurrence of a species at a place (its address), at a particular time, with a note of who made the record, and sometimes the details of the person who verified it. All this information is placed in a database to be subsequently processed to produce STATISTICS and maps.

One of the problems with these schemes is that different amounts of effort often go into different SAMPLE areas. For this reason it is better to use 'constant effort' schemes wherever possible.

Some schemes produce index values, usually annually, rather like the cost of living index or the Financial Times Stock Exchange (FT-SE, or 'Footsie') index. Discussion currently focuses on whether values should be logged or not and whether a run of three or five years should be used to give the baseline at the start of the scheme. Very different interpretations can ensue depending on whether the initial values happen to be at a high or low point of the natural FLUCTUATIONS. *See also* RETROSPECTIVE ANALYSIS (MONITORING). [F.B.G.]

REFERENCES

Goldsmith, F.B. (1991) *Monitoring for Conservation and Ecology*. Chapman & Hall, London.

Goldsmith, F.B. (1993) Monitoring for conservation. In: *Conservation in Progress* (eds F.B. Goldsmith & A. Warren). John Wiley & Sons, Chichester.

Spellerberg, I.F. (1991) *Monitoring Ecological Change*. Cambridge University Press, Cambridge.

---

biological recording *See* BIOLOGICAL MONITORING.

biological species concept The species is a category in the CLASSIFICATION of plants and animals, ranking below genus and family and containing SUBSPECIES. Ernst Mayr (1963) defined biological species as 'groups of actually or potentially interbreeding natural populations which are reproductively isolated from other such groups.' Categories above the species recognize degrees of similarity of organisms; they constitute a formal system which allows us to picture the hierarchical patterns of organization of living things. Within the species, there may be geographical variation, so that the species consists of subspecies, or races, occupying different parts of the RANGE, which can nevertheless interbreed (i.e. a polytypic species).

In practice, there are many difficulties in the application of the biological species concept. Capacity to interbreed cannot always be tested, nor can the potential for interbreeding. A single polytypic species may resemble a group of similar species occupying neighbouring areas. The museum taxonomist normally deals with dead material, with no information on breeding behaviour. The definition cannot readily be used for organisms, such as bacteria, which reproduce asexually for long periods, or plants which reproduce vegetatively. Nor can it be applied in palaeontology; the fossil sequence often lacks evidence of discontinuity within lineages, although different sections may differ as much as modern species. As a result, practical systematics often continues to be carried out using other definitions. *See also* SPECIES CONCEPTS. [L.M.C.]

REFERENCES

Mayr, E. (1963) *Animal Species and Evolution*. Harvard University Press, Cambridge, Mass.

O'Hara, R.J. (1993) Systematic generalization, historical fate, and the species problem. *Systematic Biology* **42**, 231–246.

bioluminescence LIGHT emission, often as flashes and without sensible heat production, produced by bacteria, fungi, plants and animals at a wide range of frequencies from ultraviolet to the red end of the spectrum. The light emission results from chemical reactions within certain cells or organs or in some form of secretion. The function of bioluminescence can be protection from predators. For example, some PELAGIC marine organisms produce light from organs on their ventral surfaces that obscures their appearance from below, while some dinoflagellates emit light that elicits escape responses in their copepod predators. In contrast, some other aquatic animals excrete a luminescent cloud that startles and confuses predators. Light emission may also be used by predators to attract prey, or individuals of the same species may communicate via light emission. Bioluminescence is of particular importance in the PHOTIC ZONE and is generally less common in the DEEP SEA. However, about 100 species of deep-dwelling angler fish and a number of deep-sea benthic invertebrates are luminescent. [V.F.]

biomagnification The process whereby the concentration of a chemical in the tissues of organisms increases as the chemical moves up the FOOD CHAIN. It is the result of BIOACCUMULATION and bioconcentration. Hence, only small amounts of a toxic chemical may accumulate per head of plant-eating animals; but many of these are eaten by predators, which therefore accumulate more of the chemical per head. [P.C.]

biomanipulation The practice of manipulating the densities of species (usually top predators) to push a COMMUNITY towards a desired state. It is of particular interest for management of eutrophic LAKES where reductions in PHYTOPLANKTON abundance can improve water clarity and prevent the development of anoxic conditions. Lake biomanipulation theory predicts that increasing the abundance of piscivorous species (usually fish) will decrease plantivorous fish abundance, thereby increasing ZOOPLANKTON which in turn leads to a reduction in phytoplankton abundance. This chain of events is an example of a trophic cascade (*see* TROPHIC CASCADE HYPOTHESIS), if the reduction in phytoplankton density occurs through increased grazing pressure by zooplankton (*see also* TOP-DOWN CONTROLS). Although a top-down effect through grazing is the most commonly cited explanation for changes observed after lake biomanipulations, other explanations are also possible. For example, phytoplankton reductions might also be caused when changes in the fauna lead to a shift in the nutrient dynamics, which in turn affects the phytoplankton. Not all attempts at biomanipulation are successful. An often considered alternative to biomanipulation in lakes is the direct reduction of nutrient inputs, although this may not always be possible. [S.J.H.]

REFERENCES

Carpenter, S.R. & Kitchell, J.F. (1992) Trophic cascade and biomanipulation: Interface of research and management — a reply to the comment by DeMelo *et al. Limnology and Oceanography* **37**, 208–213.

DeMelo, R., France, R. & McQueen, D.J. (1992) Biomanipulation: hit or myth? *Limnology and Oceanography* **37**, 192–207.

biomarkers Effects in biological systems used to indicate the presence of CONTAMINATION. The effects are usually at molecular and cellular levels. Some would argue that effects can be physiological, behavioural or even at the level of population and community—but then biomarker studies become synonymous with ECOTOXICOLOGY. The effects should be relatively long-lasting and specific. They may involve induction of protective systems (e.g. METALLOTHIONEINS) or an indication of damage (e.g. enzyme inhibition). Clearly they signal EXPOSURE to a substance, but do not necessarily indicate an adverse ecological EFFECT. This is particularly true of the induction of protection mechanisms since these are part of homeostatic systems that by definition will prevent expression of adverse effects. The advantages of biomarkers over chemical assays is that they can potentially integrate conditions over time thus providing increased opportunity for picking up low-level contamination and/or episodic effects of contamination. *See also* BIOINDICATOR; EXPOSURE. [P.C.]

biomass The total mass of living material within a specified area at a given time. It may be difficult to separate living tissues from those that are strictly dead (such as the wood tissues of a tree), so the term normally includes such non-living materials. An alternative expression, especially used of plant biomass, is 'STANDING CROP', which again may include the standing dead material often found in vegetation such as GRASSLANDS. Ideally, a value for the biomass of an ecosystem should include root material and soil organisms, but difficulties in the estimation of these has resulted in the use of 'above ground' biomass or standing crop data being extensively used. Biomass can be split into its various components, such as plant biomass, or species biomass, or into layers such as canopy, trunk and root biomass. It may also be split into various trophic levels.

Although they are closely related, it is important to distinguish between biomass and PRODUCTIVITY. Essentially, biomass is a quantity per unit area whereas productivity is a rate per unit area. The biomass at any particular time is an outcome of the productivity of the ecosystem, the time over which biomass has been accumulating, and the various sources of material loss (respiration, death, grazing, predation, emigration, etc.). Periodic assessment of the biomass of an ecosystem therefore provides a means of determining productivity, if the other sources of loss can be estimated.

Determination of biomass involves cropping SAMPLE areas of an ecosystem, separating into required components, drying and weighing. This provides an estimate of dry-mass biomass, but more valuable data may be obtained by determining the energy content of the different components so that biomass can be expressed in energetic terms. This is effected by the analysis of representative samples of tissues in a BOMB CALORIMETER and subsequent extrapolation. Alternatively, the levels of certain elements, such as carbon, nitrogen or phosphorus, can be determined so that their DISTRIBUTION within the system can be evaluated.

The possibility of using biomass as an energy resource has received increasing attention in recent times. Essentially, this involves the tapping of solar energy through the photosynthetic process and HARVESTING the outcome as organic material. Plantations of fast-growing vegetation, such as willow and poplar, provide an effective means of accumulating energy-rich organic materials that can be exploited for commercial energy production. *Cf.* NECROMASS. [P.D.M.]

biome Global VEGETATION can conveniently be divided into a series of major ECOSYSTEM types that relate to world CLIMATES (particularly temperature and water supply) and can be defined in terms of life forms and PHYSIOGNOMY of vegetation rather than taxonomic groups of organisms. These are called biomes. The biomes usually recognized include: TUNDRA; BOREAL FOREST (TAIGA); temperate forest; temperate GRASSLANDS (PRAIRIE, STEPPE, VELD, etc.); Mediterranean-type vegetation (CHAPARRAL SHRUBLANDS, FYNBOS, etc.); desert scrub; DESERT; tropical grassland (SAVANNAH); TROPICAL SEASONAL FOREST; and TROPICAL RAIN FOREST.

The aquatic ecosystems, marine and FRESHWATER, may also be regarded as separate biomes, although the operation of climatic and other factors and the ease of dispersal of organisms, means that they must be considered in a distinctive manner.

The impact of human beings on the land surface of the Earth is so extensive that many natural biomes have been considerably altered or even replaced by man-made ecosystems; some ecologists consider these replacements as biomes in their own right. Hence the use of terms such as 'urban biome' and 'agricultural biome' by some scientists. *Cf.* BIOGEOGRAPHICAL UNITS. *See also* BIOMES, MAJOR WORLD. [P.D.M.]

Biomechanical constraints

The same laws of physics apply to living things as to machines, vehicles and bridges. Nothing can evolve that is physically impossible. The structure of animals and plants, and the things they can do, are limited by the same design principles as apply in engineering.

For example, water skaters and some other insects can stand and walk on water, using the SURFACE TENSION of the water to support their mass. No large animal can do that because the force available is strictly limited by the properties of water, and is proportional to the length of the foot. A normal-sized water skater is supported on six feet, each 1 mm long, but a human-sized skater, one million times as heavy, would need feet a million times as long, or about 1 km. With feet that big, walking would not be feasible.

Few animals walk on water, but many fly, and their flying ability is limited by the principles of aerodynamics. One of the key design features is the wing loading: i.e. body weight divided by wing area. Birds with high wing loading cannot fly slowly, but must travel fast for their wings to generate enough lift force to support them. But relatively high wing loadings give the best performance if a bird wants to fly fast. Albatrosses have to fly fast over the sea to make headway against strong winds, and vultures circle slowly to gain height in thermals. Accordingly, albatrosses have wings of only about half the area of vultures of equal body mass. Each is well designed for its own style of flight.

Vultures and albatrosses travel mainly by soaring, using natural air movements to keep them airborne. Like man-made gliders, they sink in still air. It would be an advantage to sink as slowly as possible, so that they could be supported by the slightest air movements. From experience with man-made aircraft, it seems impossible to make a glider that will lose height at a rate less than about $0.5\ \mathrm{m\,s^{-1}}$, in still air. Observations on vultures show that they can lose as little as $1\ \mathrm{m\,s^{-1}}$, quite close to the engineering limit. Their failure to reach the limit is probably due to the need for compromise between gliding and other activities (including flapping flight). Man-made gliders are designed purely for gliding performance.

Shore-living animals such as limpets and sea anemones are buffeted by waves and must stick firmly to rocks to avoid being washed away. The force with which they can adhere to the rock is proportional to the area of the foot, but part of the force tending to dislodge them (the part due to accelerations of the water in breaking waves) is proportional to the animal's volume. The bigger an animal is, the smaller the ratio of its area to its volume (assuming the shape stays the same). That makes it harder for large animals than for small ones to hold on to rocks on wave-swept shores, which is probably why the molluscs and other animals that live in such places are fairly small. Limpets are unlikely to evolve to the size of the large corals that live in sheltered lagoons.

The final example concerns swimming. Salmon swim by beating their tails, and squid swim by jet propulsion, drawing water into their mantle cavities and squirting it out. Both are beautifully streamlined and look well adapted for swimming, but a comparison of salmon and squids of equal mass showed that the salmon could swim faster, yet used less energy. The explanation seems to be a limitation of the jet-propulsion system that squid have inherited from their ancestors. Propelling yourself by squirting out relatively small volumes of water is wasteful of energy for a reason grounded in simple physics. More kinetic energy has to be given to the water, to obtain the same thrust, than if propulsion depends on pushing on large volumes of water, as fishes' tails do. Their ancestry and the laws of physics combine to limit the squids' performance.

These few examples illustrate how EVOLUTION is constrained by biomechanics. A giant water skater would need unmanageably large feet; a small-winged vulture could not circle in thermals; birds cannot avoid losing height when they glide in still air; and inferior designs such as the squid jet-propulsion mechanism cannot evolve superior performance. In each case, performance limits are set by the laws of physical mechanics. *See also* CONSTRAINTS. [R.MCN.A.]

REFERENCE

Alexander, R.McN. (1985) The ideal and feasible: physical constraints on evolution. *Biological Journal of the Linnean Society* **26**, 345–358.

Biomes, major world

Communities of plants and animals of characteristic composition, distributed over extensive areas or regions of the world constitute global BIOMES. Although over 250 000 flowering plant species have been described, no single species occurs worldwide. Each species has its limits of DISTRIBUTION. The great majority of plant species, indeed, have a very localized distribution. Despite this, it is possible to discern quite distinct patterns of similarity and dissimilarity in the composition of plant and animal communities (Fig. B7). One example is that of the floral regions of the world.

Almost all biomes owe their origins to the distinctive CLIMATE of the region, particularly rainfall and temperature (Fig. B8). Contemporary examples of FOREST biomes include the TROPICAL RAIN FORESTS, SUBTROPICAL dry forests, TEMPERATE DECIDUOUS FORESTS, subpolar or BOREAL conifer forests. Non-forest dominated biomes include open SAVANNAH, semi-arid SCRUB, dry DESERT and TUNDRA. Many biomes are of relatively recent origins reflecting both the evolution and expansion of particular plant families in the latter half of the Tertiary, for example GRASSLANDS, Asteraceae-rich drylands and the retreat of the forests with increasing global DROUGHT in the late Miocene.

The classification of global communities can be used to highlight types of communities which are under threat from exploitation, though the threat may be more correctly used in the context of the ECOSYSTEM. Although there is no agreement on how widespread a COMMUNITY or ecosystem has to be before being raised to the status of biome, it is generally accepted that the major terrestrial global biomes include the following.

High-latitude to cold-temperate communities—tundra and taiga

Tundra (from the Lapp) is a treeless biome with a permanently frozen subsoil with sparsely distributed vegetation composed largely of lichens, mosses and dwarfed vascular herbaceous plants. Most tundra is found around the Arctic Circle, smaller areas being present in the sub-Antarctic islands. A distinct form of tundra is also present on the higher mountains of the temperate to tropical zones. The low PRODUCTIVITY of the tundra makes the communities particularly sensitive to DISTURBANCE (e.g. from oil exploration) and POLLUTION.

The TAIGA (from the Russian) is the extensive northern coniferous forest forming a belt of vegetation across Scandinavia, Eurasia (particularly Siberia) and North America. Taiga is also found on high mountains at lower latitudes. The greater part of the taiga vegetation comprises EVERGREEN conifers that are tolerant of the low temperatures of the extended winter and can withstand the physiological drought (with water frozen in the SOIL for many months). The taiga fauna includes more species of deer than any other biome and a characteristic assembly of large carnivores (wolves, lynx, bears). Blood-sucking mosquitoes and midges are abundant. The vastness of the taiga forests has led to profligate exploitation for timber.

Warm-temperate communities—the temperate forests and temperate grasslands

Of all contemporary global biomes, the temperate forests have undergone greatest exploitation and destruction by humans, largely within the past 5000 years. Of the original mixed conifer and deciduous broad-leaved forests of north-east America, central Europe and east Asia, little remains today beyond fragments. Of the conifer and evergreen broad-leaved forests of the Mediterranean region, most have been severely eroded if not destroyed. The deciduous broad-leaved forests of western Europe, northern Asia and eastern North America have long been exploited for hardwood timber. The evergreen broad-leaved forests of high rainfall areas (Florida, Japan and elsewhere) have been particularly vulnerable to human interference in the past century. The temperate forest biome is particularly rich in birds and smaller mammals. It is also probably the best studied of all global biomes. The much less extensive temperate forest biome of the Southern hemisphere has become particularly vulnerable to exploitation in the past 20 years.

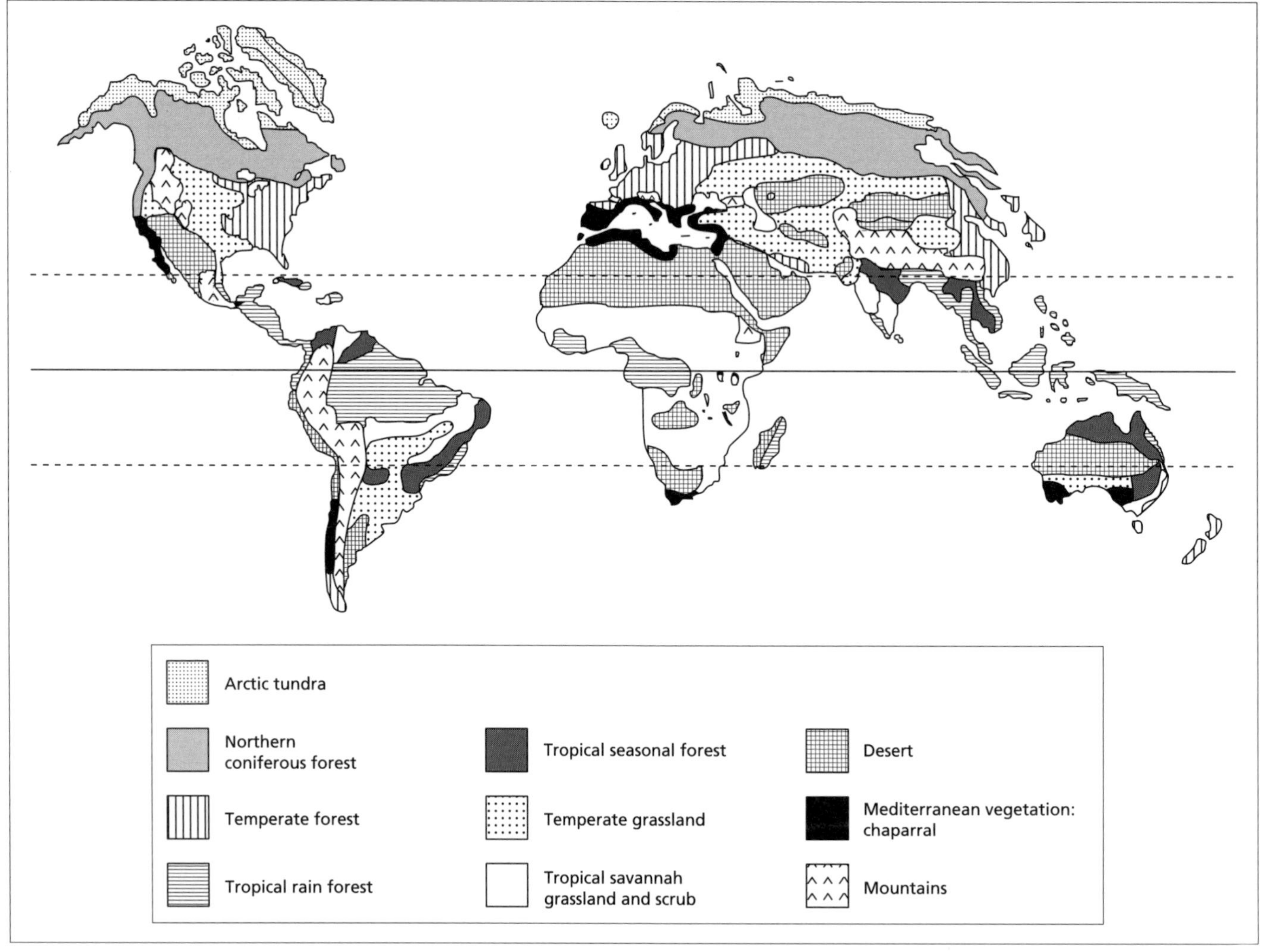

Fig. B7 Distribution of the major terrestrial biomes of the world. (After Cox, C.B. & Moore, P.D. (1993) *Biogeography: An Ecological and Evolutionary Approach*, 5th edn. Blackwell Scientific Publications, Oxford.)

Temperate grasslands arose relatively late in the Tertiary and reflect a climatic change to more seasonal rainfall with prolonged dry periods. The extensive prairies of North America, the steppes of Eurasia, PAMPAS of South America and VELD of southern Africa are dominated by grasses. However, in the drier areas domination by grass gives way to other drought-resistant herbaceous species, most notably of the Asteraceae family, and to scrub. The TEMPERATE GRASSLAND biome has been subjected to particularly intensive forms of AGRICULTURE in the past century with conversion of extensive areas to cereal CROPS. Overexploitation through intensive GRAZING and the use of fire has devalued the biome economically. Most of the original large mammal communities—bison, wild horse, antelope—have been destroyed.

Subtropical and tropical biomes—savannah, rain forests, tropical dry forests and deserts

Savannah is a term describing tropical vegetation ranging from near pure grassland, through grass with isolated trees, open dry forest with grass, grass with low shrubs, to grass with dry scrub vegetation. The characteristic climate is warm with an annual cycle of dry and wet seasons. Most of the vegetation has characteristic drought-resistant features. This is the biome of the large carnivores, such as lions and hyenas, and the once great herds of grazing mammals. Fire, both natural and latterly from human activity, is an important feature of this biome, one to which much of the vegetation is well adapted.

Extreme dryness or unpredictability of rainfall gives way to semi-arid to arid deserts (Fig. B8). Deserts may be hot (as in the Sahara) or cold (as in the Gobi). A characteristic of all deserts is the sparseness of vegetation with extensive areas of bare ground. Desert plants are adapted to severe water STRESS, often in unusual ways, including the ASSIMILATION of CARBON DIOXIDE (CO_2) at night, extreme reduction in leaf area and development of extensive tissues for the STORAGE of water. The desert biome is expanding in area as a result of changing patterns of precipitation and increasingly severe OVERGRAZING by domesticated animals.

Tropical dry forests and associated understoreys are particularly influenced by the distinctly seasonal but predictable rainfall. Such

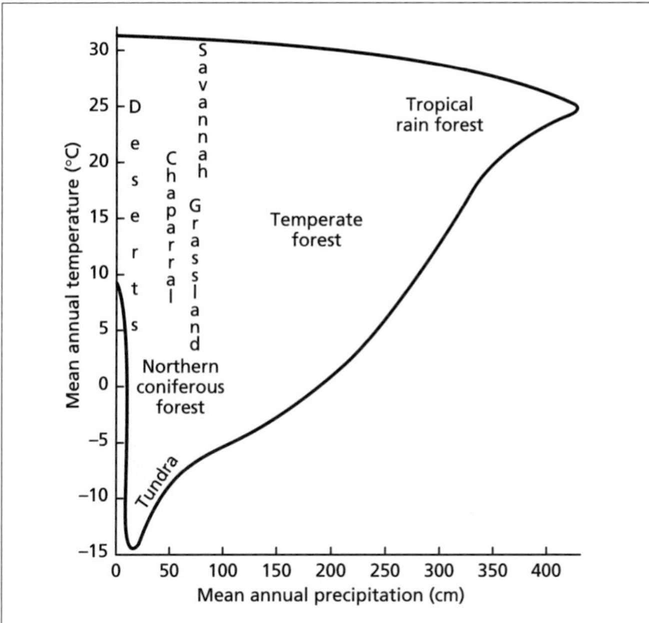

Fig. B8 The distribution of eight major terrestrial biomes with respect to mean annual temperature and mean annual precipitation. It is not possible to draw sharp boundaries between biome types—local climatic effects, soil types and fire can each shift the balance. (After Whittaker, R.H. (1975) *Communities and Ecosystems*, 2nd edn. Macmillan, London.)

forests are subjected to often frequent fires. In recent centuries this biome has been heavily exploited for timber.

Tropical rain forest develops naturally in areas of high temperatures and high rainfall and is characterized by the greatest number of species, animal, plant and fungal. Productivity is high as is the turnover of NUTRIENTS, though in many areas the soils are poor in nutrients. Many of the mammals are adapted to an arboreal life. The recent exploitation of this biome throughout the tropics is widespread and expanding, particularly in south-east Asia and central South America. Much of this has resulted in the conversion of extensive areas of this biome to a degraded class of forest or, increasingly, to often temporary grassland.

Aquatic biomes

The concept of biomes is also extended to aquatic systems, including WETLANDS, FRESHWATER bodies and marine systems. In the tropics many of the wetlands and shallow fresh to brackish water biomes are among the most productive on the globe, but at the same time are the most threatened. The catalogue of human activities in the destruction of wetlands includes: drainage for agriculture, mosquito control and flood protection; infilling for WASTE DISPOSAL; mining for gravel; ABSTRACTION of groundwater as well as the more insidious effects of EUTROPHICATION and pollution from industrial, agricultural and domestic activities. Added to these are the effects of natural erosion, of changes in sea level and longer-term effects of CLIMATE CHANGE. Recognition of the global threat to wetlands and associated shallow water bodies led to the 1971 Convention on Wetlands of International Importance — better known as the Ramsar Convention. The 75 countries ratifying the convention are obliged to protect designated wetlands to an internationally agreed standard.

The oceans cover 70% of the Earth's surface. It is often not practical to distinguish regional oceanic biomes because of the free movement of aquatic organisms and water bodies. Nevertheless, certain geographical zones, notably the northern Atlantic, polar and tropical seas, are treated as distinct biomes for specialist purposes, such as cold-water communities or CORAL REEF systems. Instead, marine biomes are often more usefully considered as the shore biome (subdivided into rocky, sandy, muddy sub-biomes) and the oceanic biome. The latter may be divided into sub-biomes covering the free-floating and subsurface planktonic communities, the nektonic (zone of actively swimming fish, reptiles and aquatic mammals) and the sea-floor or benthic communities. In the larger freshwater LAKES, similar biomes, each with their characteristic animal and plant life, correspond to the planktonic, nektonic and benthic communities of the oceans. *Cf.* BIOGEOGRAPHICAL UNITS. [G.A.F.H.]

biometeorology The study of the interactions between WEATHER and life. This is a multidisciplinary subject, drawing on CLIMATOLOGY, METEOROLOGY, biology, HYDROLOGY, physiology and medicine. Biometeorologists seek to record and understand the impact of weather on plants and animals, not only the more obvious and more extensively perceived effects (e.g. the influence of weather on the germination and establishment of CROPS, and the populations of insect pests) but also the more subtle and unexpected. These include the effect of weather on the incidence of heart attacks and suicides; the effect of sunspots on plant growth; human performance in extreme environments; the possible influence of NEGATIVE IONS on human well-being, and the design of cattle houses to provide shelter from heat stress. Biometeorology is all-pervading, though the elements of biometeorology are often taught as part of other disciplines such as ECOLOGY and ENVIRONMENTAL SCIENCE. The International Society of Biometeorology publishes a journal in four issues a year, the *International Journal of Biometeorology*. [J.G.]

biometrics The application of mathematical (usually statistical) techniques to biological systems. *See also* STATISTICS. [P.C.]

biomonitor Any system incorporating living organisms (ranging from microbes to vertebrates) that can be used to assess (often continuously) quality of the ENVIRONMENT. *See also* BIOPROBE; BIOSENSOR. [P.C.]

REFERENCE

Rawson, D.M. (1993) Bioprobes and biosensors. In: *Handbook of Ecotoxicology*, Vol. 1 (ed. P. Calow), pp. 428–437. Blackwell Science, Oxford.

biophilia A term coined in 1979 by the biologist E.O. Wilson of Harvard University, who defined it in 1994 as 'the inborn affinity human beings have for other forms of life, an affiliation evoked, according to circumstance, by pleasure, or a sense of security, or awe, or even fascination blended with revulsion'. Wilson derives this idea from the fact that 'humanity is ultimately the product of biological evolution', and 'the diversity of life is the cradle and greatest natural heritage of the human species'.

The reality of biophilia is incontestable. It is manifest in human DOMESTICATION of animals and cultivation of plants, in zoos and gardens, in our art, and in much of our mythology and philosophy—such as the *fungshui* traditions of China, traceable back thousands of years.

Whether biophilia is universal or even broadly persistent in our species is more problematical. Biophobia — fear and loathing of animals and wilderness—is also strong and seemingly gaining precedence. It seems unlikely that positive biophilia, or any of the arts and sciences we collectively refer to as the humanities, can survive the oppressive crush of overpopulation. Biophilia does not appear to be a characteristic of the urban poor, and this group is increasingly becoming the dominant mode of our species. It is difficult to share even a little of Wilson's optimism. (J.D.LZ.)

REFERENCES

Kellert, S.R. & Wilson, E.O. (eds) (1993) *The Biophilia Hypothesis*. Island Press, Washington, D.C.

Wilson, E.O. (1984) *Biophilia*. Harvard University Press, Cambridge, Mass.

Wilson, E.O. (1994) *Naturalist*. Island Press, Washington, D.C.

biophobia *See* BIOPHILIA.

bioprobe A form of BIOSENSOR located on a rod and used for taking measurements at specific points—like a pH or oxygen probe. *See also* BIOMONITOR. [P.C.]

bioremediation Remedying chemical POLLUTION using biological agents, usually microbes, with appropriate catabolic systems; for example using selected microbes to decompose oil after accidental spillage at sea. [P.C.]

biosensor A device that combines biological material and transducers, usually in intimate association, to signal the presence of a particular substance or group of substances. The general system is illustrated in Fig. B9; some examples of biosensors available for environmental MONITORING are given in Table B1. *See also* BIOMONITOR. [P.C.]

REFERENCE

Rawson, D.M. (1993) Bioprobes and biosensors. In: *Handbook of Ecotoxicology*, Vol. 1 (ed. P. Calow), pp. 428–437. Blackwell Scientific Publications, Oxford.

Biosphere

All the organisms on the planet, and their environment, viewed as a system of interacting components. The biosphere may be regarded as a thin film on the planet's surface, consisting of all the organisms and the WATER, SOIL and air surrounding them. The biosphere thus includes parts of the ATMOSPHERE, the LITHOSPHERE and most of the HYDROSPHERE.

Most organisms occur within 100 m of the surface of the hydrosphere and lithosphere, and are confined to the first 5000 m of the atmosphere, but of course there are exceptions. Some anaerobic bacteria are located at great depth, in caves, oil wells and in the deep OCEAN; spores and pollens are carried to great heights in the atmosphere. The upper and lower boundaries of the biosphere are thus indistinct.

The age of the Earth determined from the oldest rocks on Earth is slightly less than 4×10^9 years. The early atmosphere is thought to have been produced by outgassing from the Earth's crust, and was probably like that of Mars and Venus, largely CARBON DIOXIDE (CO_2) and devoid of OXYGEN. Early life forms are believed to have

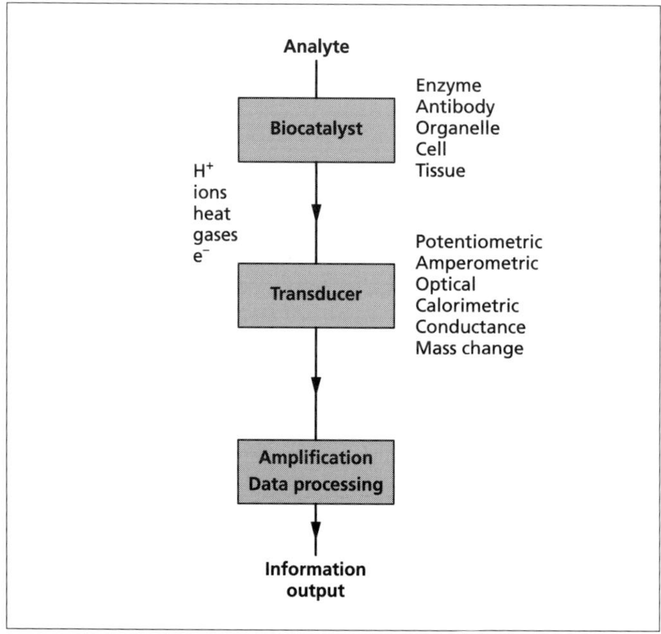

Fig. B9 Biosensor configuration. (After Rawson, D.M. (1993) Bioprobes and biosensors. In: *Handbook of Ecotoxicology*, Vol. 1 (ed. P. Calow), pp. 428–437. Blackwell Scientific Publications, Oxford.)

Table B1 Biosensors for environmental monitoring. (From Rawson, D.M. (1993) Bioprobes and biosensors. In: *Handbook of Ecotoxicology*, Vol. 1 (ed. P. Calow), pp. 428–437. Blackwell Scientific Publications, Oxford.)

Substance detected	Biological component	Sensor transducer
2,4-Dinitrophenol	Monoclonal antibody	Potentiometric electrode
Phenols	Polyphenol oxidase	Amperometric electrode
Nitrite	Nitrite reductase	NH_3 gas sensor
Naphthalene	*Pseudomonas* + *lux* coded plasmid	Photomultiplier
Triazine herbicides	Enzyme immunoassay	UV spectrophotometer
Formaldehyde	Formaldehyde dehydrogenase	Piezoelectric crystal
Mercury(II)	Urease	CO_2 electrode
Organophosphates	Acetylcholine esterase	pH electrode
Heavy metals	Urease	Microcalorimeter
Carbamate insecticides	Acetylcholine esterase	pH sensitive fibre-optics
Herbicides	*Synechococcus*	Mediated amperometry
Chlorophenols	*Escherichia coli*	Mediated amperometry

been anaerobic, resembling modern-day PROKARYOTES. The earliest fossils found are the stromatolites, as old as 3.5×10^9 years. The fossils appear as attached, laminated deposits, believed to be similar to modern intertidal cyanobacteria first found in the Bahamas, which form dense colonies by trapping particulate SEDIMENT. Cyanobacteria are photosynthetic, and the stromatolites may have been the first photosynthetic organisms. PHOTOSYNTHESIS fixes CO_2 by splitting water into HYDROGEN (reducing power) and oxygen, and the latter escapes to the atmosphere. However, during death and DECOMPOSITION, the BIOMASS formed by photosynthesis is oxidized and the fixed CARBON is returned to the atmosphere. To form an atmosphere containing oxygen in significant amounts, it is necessary to postulate that a proportion of biomass is protected from oxidation, by being buried in anaerobic deposits, and thus becoming coal, oil and gas.

The release of oxygen into the atmosphere changed the surface conditions from reducing to oxidizing—IRON and SULPHUR appear in the geological strata in their oxidized as opposed to reduced state – and much carbon was deposited in the ocean as calcium carbonate.

EUKARYOTES appeared 1.6×10^9 years ago, setting the scene for the evolution of multicellular organisms, including the land plants. Evidence for land plants, including fragments of cuticle and spores, first appears in the Silurian ($0.41–0.44 \times 10^9$ years ago), and the earliest evidence for a vascular plant (*Cooksonia*) comes from the Silurian. The ancestral forms of modern vascular plants, the Rhynopsida, are found in the Silurian and Early Devonian deposits. By the end of the Devonian (0.35×10^9 years ago) FOREST ecosystems were widespread, containing all types of vascular plant (except for the angiosperms), with true leaves and seeds. The angiosperms, with their insect-pollinated flowers, did not appear until the Early Cretaceous (0.14×10^9 years ago).

This COLONIZATION of the Earth by plants influenced the energy balance of the planetary surface, both by the changes it induced in the composition of the Earth's atmosphere, and also by the change in ALBEDO. It is presumed by many scientists that the conditions for life on the planet have been substantially influenced by the vegetational cover, and continue to be so influenced. In 1979 Lovelock proposed the GAIA hypothesis, whereby organisms, by their activities, maintain a more or less constant physical and chemical environment on the planetary surface, at least one sufficiently constant to ensure that life has prevailed over three billion years. The activities of the biosphere include BIOGEOCHEMICAL CYCLES, such as the carbon, suphur and NITROGEN CYCLES, and the hydrological cycle (*see* WATER (HYDROLOGICAL) CYCLE).

Despite Gaia, it is considered that the biosphere has been, and continues to be, under threat; from cosmic impacts and from various human influences. The latter include: NUCLEAR WINTER; OZONE HOLE caused by CFCS; CLIMATE CHANGE brought about by EMISSIONS of GREENHOUSE GASES; enhanced deposition of nitrogen; and release of synthetic organic chemicals into the environment.

'Biosphere 2' is the name given to an experiment in which humans and plants were sealed into a giant glass and steel bubble occupying 1.2 ha of desert north of Tucson, USA, to determine whether it is possible to establish life-support systems on Mars (see Odum 1993). [J.G.]

REFERENCES

Bradbury, I.K. (1991) *The Biosphere.* Belhaven Press, London.

Dansgaard, W., Johnsen, S.J., Clausen, H.B. *et al.* (1993) Evidence for general instability of past climate from a 250-kyr ice-core record. *Nature* **364**, 218–220.

Edwards. D. & Selden, P.A. (1992) The development of early terrestrial ecosystems. *Botanical Journal of Scotland* **46**, 337–366.

Lovelock, J.E. (1988) *The Ages of Gaia: a Biography of our Living Earth.* Norton, New York.

Margulis, L. & Schwartz, K.V. (1988) *Five Kingdoms: An Illustrated Guide to the Phyla of Life on Earth,* 2nd edn. W.H. Freeman, New York.

Odum, E.P. (1993) Biosphere 2: a new kind of science. *Science* **260**, 878–879.

Biosphere reserves

Biosphere reserves are a worldwide system of designated protected areas and surrounding landscapes that attempt to integrate conservation, research, education and SUSTAINABLE DEVELOPMENT.

Biosphere reserves are designated by local or national governments, following guidelines of the MAN AND THE BIOSPHERE PROGRAMME (MAB) of the United Nations Economic, Scientific and Cultural Organization (UNESCO). The designation was established in 1971 in recognition of the need to ascertain global priorities for protecting biological DIVERSITY in an increasingly human-dominated landscape. As of 1995, 324 such reserves have been designated in more that 75 countries around the world. MAB's goal is to protect at least one representative area from each of the world's 193 biogeographic provinces, in a global system.

Spatial organization is central to the biosphere reserve concept. The land base included in each biosphere reserve is organized (at least in theory) into three zones, which ideally are concentric. At the centre of the reserve is a core conservation zone, accorded the highest degree of protection from commercial resource exploitation and other forms of destructive interference. Core conservation zones provide maximum protection for species, communities and ecological processes in their natural state. In these areas, the only human activities normally permitted are non-disruptive research and MONITORING. A secondary 'shell' around the core is designated as the BUFFER ZONE, in which limited and regulated human activities are permitted, including ecologically-oriented tourism, training, low-impact experimental research and traditional low-intensity AGRICULTURE and resource-gathering activities. The guiding principle for these activities is that they must be conducted in a manner that is consistent with protection of the core. The tertiary landscape surrounding the buffer is the transition zone which, as its name implies, serves as a boundary between the biosphere reserve and the surrounding landscape. Human settlements and more intensive economic activities are permitted in the transition zone, again provided that they are fundamentally consistent with protection of the core.

The three zones of a biosphere reserve reflect the central objectives of the programme, and highlight its unique characteristics. Unlike other systems of protection (such as the US national parks) where essentially no resource extraction or human use other than visitation and low-impact research is normally allowed, the biosphere reserve concept is based on a recognition that humans are a natural part of the global landscape, and that this presence, if properly managed, can be consistent with the goals of conserving BIODIVERSITY. Indeed, most modern biosphere reserves are located in areas where humans have been living and using natural resources for tens of thousands of years.

The biosphere reserve concept was developed partly in response to serious problems with managing national parks and protected areas, especially in developing nations, that attempted to exclude human presence entirely. First, such efforts were rarely successful; indigenous peoples often continued to use the area to extract resources they needed for survival, despite attempts to discourage them. Second, the effort to create sharp boundaries between 'protected' and 'unprotected' areas led to economic impoverishment of many people who had lived in that area, which in turn contributed (and continues to contribute) to severe problems with poaching of animals from the parks. Third, excluding large numbers of indigenous peoples from their own lands led to their widespread alienation and antagonism toward park management authorities, and the concomitant loss of a potentially valuable constituency for protection of the land itself. Thus, the exclusionary approach to park protection appeared, in many cases, to fail its own objectives for both effective conservation and political acceptability.

Biosphere reserves attempt to overcome these polarities by integrating the objectives of conservation, education and economic development. The most highly protected area, the core conservation zone, is the most productive in terms of ecological research, and often (although by no means always) contains the highest measure of biodiversity. Buffer zones allow a wide range of activities beneficial to the inhabitants of the region. In addition to farming and resource collection, local residents often guide and provide services to research teams and ecotourism groups. Residents are also given high priority to receive training in land protection and management, and are frequently employed as park rangers, naturalists, enforcement officers, mangers and scientists. In this way, biosphere reserves attempt to defuse the sharp and unnecessary polarization between conservation and economic development.

In practice, the MAB biosphere reserve programme has not been without problems. First, many designated biosphere reserves were already protected national parks or protected areas, so there may have been little additional protection actually afforded by the United Nations designation. Second, in many cases only the core has been actually delineated, while buffer and transition zones remain unidentified. This leaves the core highly vulnerable to exploitation and violates the system's fundamental spatial concept. Third, a number of biosphere reserves are essentially 'paper reserves', designated and protected on paper only, but with little actual enforcement to prevent potentially harmful or illegal activities. Fourth, many (if not most) have inadequate budgets and staff to carry out the research, management and educational activities that are central to the concept. Fifth, economic development activities in the surrounding lands are not necessarily consistent with protection of the core, and may indeed undermine its long-term ecological sustainability. Some biosphere reserves have large mines, converted agricultural lands, forestry plantations, road systems, tourist developments, sources of pollution, and other elements nearby that threaten the viability of the core. Sixth, the global distribution of biosphere reserves is highly uneven, and does not reflect well the distribution of biogeographic provinces that it attempts to cover. Analysis of the present system reveals a heavy concentration in North America, western and central Europe, central Africa and western South America. While these areas are all unquestionably important for global ecological diversity, the absence of reserves in eastern South America (including almost the entire Amazon Basin), northern and southern Africa, China, the Indian subcontinent, northern Russia and Canada, south-east Asia (including New Guinea) and northern Australia leaves a great deal of work to be done. Seventh, at present only a small fraction of global ECOSYSTEM types—barely 2%—is represented in the MAB system overall. Lastly, many national governments have provided little or no tangible cooperation, political encouragement, or economic support for the system.

Despite these problems, a global system of protected areas, perhaps using some variant on the biosphere reserve system, is essential for the long-term maintenance of global biodiversity, and the system remains a landmark effort toward this vital goal. *See also* CONSERVATION BIOLOGY; INTERNATIONAL CONSERVATION CONVENTIONS. [D.A.F.]

REFERENCES

Batisse, M. (1986) Developing and focusing the biosphere reserve concept. *Nature Res.* **22**, 1–27.

Tangley, L. (1988) A new era for biosphere reserves. *BioScience* **38**, 148–155.

UNESCO (1974) *Task Force on Criteria and Guidelines for the Choice and Establishment of Biosphere Reserves.* Final Report, MAB Report Series No. 22. United Nations Education, Scientific and Cultural Organization, Paris.

UNESCO (1985) Action plan for biosphere reserves. *Environmental Conservation* **12**, 17–27.

UNESCO (1997) Home page: http://www.unesco.org/mab/activity/brfaq

---

biostratigraphic unit A biostratigraphical unit is defined on the basis of the fossil content or palaeontological character of a geological stratum (Hedberg 1972, 1976). Several types of biostratigraphic unit or zone are recognized based on the occurrence and/or ABUNDANCE of one, or more frequently several, selected taxa. A taxon-range zone is based on the occurrence of a particular taxon; concurrent-range zone, or overlap-range zone, is defined by that part of the range of two or more selected taxa that overlap; acme zone, or peak zone, is where maximum development of a taxon is recorded (this may be much narrower than the taxon's total range); and interval zone refers to a stratum between two distinctive biostratigraphical horizons but does not represent a distinct biostratigraphic ASSEMBLAGE. In Quaternary studies, where the relative abundance of several taxa is often as important as presence/absence of a particular taxon, the most frequently employed biostratigraphic unit is the ASSEMBLAGE ZONE. *See also* BIOSTRATIGRAPHY. [M.O'C.]

REFERENCES

Hedberg, H.D. (1972) Summary of an international guide to stratigraphic classification, terminology, and usage. *Boreas* **1**, 213–239.

Hedberg, H.D. (1976) *International Stratigraphic Guide.* John Wiley & Sons, New York.

biostratigraphy Biostratigraphy refers to the definition of geological strata on the basis of their distinctive fossil content. The units so defined are independent of the lithostratigraphy and constitute the traditional basis for correlation of rock and SEDIMENT types from within and between regions. There are several types of BIOSTRATIGRAPHICAL UNITS or zones. The most useful and widely used in palaeoecological studies is the pollen assemblage zone, which reflects the former VEGETATION cover and hence is indicative of ecological conditions at/near the sampling site. *See also* STRATIGRAPHY. [M.O'C.]

biota The total flora and fauna of a region — the POPULATION of living organisms in general. This will include microorganisms, fungi and invertebrates in and above the soil, plants rooted in the soil, and epiphytes (especially in humid regions), and all animals occupying every level in the system from the soil to the tree canopy. All the components of the biota, together with the inorganic ENVIRONMENT, comprise the ECOSYSTEM. [S.R.J.W.]

biotechnology and genetically engineered organisms MANIPULATION of GENETIC SYSTEMS by means other than selection and breeding and often with a view to commercial gain. *See also* GENETIC ENGINEERING; MOLECULAR ECOLOGY; TRANSGENIC ORGANISMS. [P.C.]

biotic factors Factors limiting the DISTRIBUTION of a species in space or time due to the effects of the animals, microorganisms and plants present. They are contrasted with the ABIOTIC FACTORS of climate, soil, wind, flood and fire. The separation is arbitrary because the MICROCLIMATE and soil properties affecting a given individual animal, microorganism or plant are modulated by the surrounding organisms. Historical factors, determining whether or not certain species are present, should also be considered. [P.J.G.]

biotic indices Biological criteria that are used to give information on ECOSYSTEM condition. Such indices have been used especially as indicators of riverwater quality. Use two observed characters of the effects of STRESS on communities: as POLLUTION stress increases the total number of species declines; moreover, as the extent of pollution stress increases, species tend to be selectively removed, with sensitive species (e.g. ephemeropterans) disappearing first and tolerant species (e.g. chironomids) last. Several indices are in general use in the UK: the Trent biotic index; the Chandler biotic score; and the Biological Monitoring Working Party (BMWP) scheme. A major weakness of all these indices is that the relative sensitivity of species is pollution specific. Sensitivity in the indices listed is largely by reference to ORGANIC POLLUTION and may not be appropriate for chemical pollution. *See also* DIVERSITY INDICES; RIVER INVERTEBRATE PREDICTION AND CLASSIFICATION SYSTEM. [P.C.]

REFERENCE

Abel, P.P. (1989) *Water Pollution Biology*. Ellis Horwood, Chichester.

biotic integrity *See* INDEX OF BIOTIC INTEGRITY.

biotic potential *See* CARRYING CAPACITY.

biotope A region that is distinguished by particular environmental conditions and therefore a characteristic ASSEMBLAGE of organisms. *See also* HABITAT. [P.D.M.]

bioturbation Physical effect on SEDIMENT, by the activities of organisms in it. [P.C.]

biotype A specific type of plant or animal defined within a species; a group of individuals with similar GENOTYPES. *Cf.* BIOTOPE. [P.O.]

bird banding *See* MARKING METHODS.

bismuth (Bi) A metallic element (ATOMIC NUMBER = 83; relative ATOMIC MASS = 208.98; melting point = 271.4°C; boiling point = 1564°C), with an average crustal abundance of 0.008 mg kg^{-1}, bismuth occurs as the native metal and in minerals, including bismite (Bi_2O_3), bismuthinite (Bi_2S_3) and bismutite [$(BiO)_2CO_3$]. ^{209}Bi is the only stable bismuth ISOTOPE. ^{210}Bi ($t_{1/2}$ = 5.013 days), ^{211}Bi ($t_{1/2}$ = 2.14 min), ^{212}Bi ($t_{1/2}$ = 1.01 h) and ^{214}Bi ($t_{1/2}$ = 19.9 min) occur as members of the natural radioactive decay series in URANIUM- and THORIUM-bearing minerals. Bismuth exhibits +3 and +5 oxidation states. [A.B.M.]

black-box system Any complex system (cell, organism, POPULATION, ECOSYSTEM) about whose organization nothing is known except what can be deduced from inputs and outputs and relationships between them. However, fundamental principles of systems theory indicate that whereas specific organizations lead to predictable input/output behaviour, specific input/output behaviour can arise from various organizations. Hence, what can be learnt about black-box systems by observation of their input/output relationships is limited. *See also* SYSTEMS ECOLOGY. [P.C.]

Black List Associated with environmental protection legislation, especially in Europe, are numerous LISTS of substances that are given some kind of priority in the way they are treated; for example, because of their potential to cause HARM, to bioaccumulate or to resist degradation. The European Community (EC) Existing Substances Regulation, for example, generates lists of priority substances on a regular basis for further more detailed RISK CHARACTERIZATION by the authorities.

Another good example of such lists is associated with the EC directive on POLLUTION caused by certain dangerous substances discharged into the aquatic environment. This established two lists for action: List I — sometimes referred to as the 'Black List' — contains a group of dangerous substances, pollution from which has to be eliminated; List II — sometimes referred to as the 'GREY LIST' — contains a group of substances for which it is necessary to reduce EMISSIONS to avoid pollution.

In compiling List I the European Commission selected 500 substances produced in quantities of > 100 tonnes per annum across Europe, and then on the basis of advice from consultants on HAZARDS to the aquatic environment, selected the initial list of 129 chemicals, to which a few others have now been added (Table B2). This is known as the candidate List I. Substances on this are treated as being on List II until their candidacy is confirmed.

For confirmation, the Commission selects substances for further consideration on the basis of criteria that are not altogether clear but which take account of particular circumstances and problems within the European Union (EU). The UK has been somewhat critical of this and has developed its own RED LIST (Table B2) of the most

Table B2 The European Commission priority candidate list; known as the Black List. (From Calow, P. (1997) *Controlling Environmental Risks from Chemicals*. John Wiley & Sons, Chichester. Copyright John Wiley & Sons Ltd. Reproduced with permission.)

2 2-Amino-4-chlorophenol
3 Anthracene
5 Azinphos-ethyl
6 Azinphos-methyl **R**
8 Benzidine
9 Benzyl chloride
10 Benzylidene chloride
11 Biphenyl
14 Chloral hydrate
16 Chloroacetic acid
17 2-Chloroaniline
18 3-Chloroaniline
19 4-Chloroaniline
21 1-Chloro-2, 4-dinitrobenzene
22 2-Chloroethanol
24 4-Chloro-3-methylphenol
25 1-Chloronaphthalene
26 Chloronaphthalenes (technical mixture)
27 4-Chloro-2-nitroaniline
28 1-Chloro-2-nitrobenzene
29 1-Chloro-3-nitrobenzene
30 1-Chloro-4-nitrobenzene
31 4-Chloro-2-nitrotoluene
32 Chloronitrotoluenes (other than 4-chloro-2-nitrotoluene)
33 2-Chlorophenol
34 3-Chlorophenol
35 4-Chlorophenol
36 Chloroprene
37 3-Chloropropene
38 2-Chlorotoluene
39 3-Chlorotoluene
40 4-Chlorotoluene
41 2-Chloro-*p*-toluidine
42 Chlorotoluidines (other than 2-Chloro-*p*-toluidine)
43 Coumaphos
44 Cyanuric chloride
45 2,4-D (including salts and esters)
47 Demeton (including demeton-0; -S; -S-methyl; -S-methyl sulphone)
48 1,2-Dibromoethane
49 Dibutyltin dichloride
50 Dibutyltin oxide
51 Dibutyltin salts (other than dibutyltin chloride and dibutyltin oxide)
52 Dichloroanilines
53 1,2-Dichlorobenzene
54 1,3-Dichlorobenzene
55 1,4-Dichlorobenzene
56 Dichlorobenzidines
57 Dichlorodiisopropyl ether
58 1,1-Dichloroethane
60 1,1-Dichoroethylene
61 1,2-Dichloroethylene
62 Dichloromethane
63 Dichloronitrobenzenes
64 2,4-Dichlorophenol
65 1,2-Dichloropropane
66 1,3-Dichloropropan-2-ol
67 1,3-Dichloropropene
68 2,3-Dichloropropene
69 Dichlorprop
70 Dichlorvos **R**
72 Diethylamine
73 Dimethoate
74 Dimethylamine
75 Disulfoton
76 Endosulfan **R**
78 Epichlorohydrin
79 Ethylbenzene
80 Fenitrothion **R**
81 Fenthion
86 Hexachloroethane
87 Isopropylbenzene
88 Linuron
89 Malathion **R**
90 2-Methyl-4-chlorophenoxyacetic acid
91 2-Methyl-4-chlorophenoxypropanoic acid
93 Methamidophos
94 Mevinphos
95 Monolinuron
96 Naphthalene
97 Omethoate
98 Oxydemeton-methyl
99 PAH (with special reference to 3,4-benzopyrene and 3,4-benzofluoranthene)
100 Parathion (including parathion-methyl)
101 PCBs **R**
103 Phoxim
104 Propanil
105 Pyrazon
106 Simazine **R**
107 2,4,5-T (including salts and esters)
108 Tetrabutyltin
109 1,2,4,5-Tetrachlorobenzene
110 1,1,2,2-Tetrachloroethane
112 Toluene
113 Triazophos
114 Tributyl phosphate
115 Tributyltin oxide **R**
116 Trichlorfon
119 1,1,1-Trichloroethane
120 1,1,2-Trichloroethane
122 Trichlorophenols
123 1,1,2-Trichlorotrifluoroethane
124 Trifluralin **R**
125 Triphenyltin acetate
126 Triphenyltin chloride
127 Triphenyltin hydroxide
128 Vinyl chloride
129 Xylenes (technical mixture of isomers)
131 Atrazine **R**
132 Bentazone

R, on UK Red List, which also contains other substances including pesticides.
PAH, polycyclic aromatic hydrocarbons; PCB, polychlorinated biphenyls.

dangerous chemicals from the list of 129 by explicit prioritization procedures.

In any event the Commission refers selected chemicals to an advisory committee of experts that, on the basis of information largely concerned with hazard, decides if substances should be retained on List I, and if they should, on the basis of existing information, proposes ENVIRONMENTAL QUALITY STANDARDS (EQSs). On this basis the Commission drafts daughter directives that specify controls including the EQSs.

The procedure has been somewhat slow — largely for political reasons. Though the legislation was adopted in 1976 still only a handful of the candidates on List I have been dealt with. In consequence, the legislation is under review. [P.C.]

Black Sea A relatively ancient SEA, located between south-eastern Europe and Asia. Age estimates for the Black Sea vary from Precambrian to early Quaternary, with a Jurassic–Cretaceous age being favoured. It has an area of 461 000 km^2, a volume of 537 300 km^3 and a mean depth of 1197 m (including the Sea of Azov). During the Quaternary, the Black Sea experienced repeated changes in level and associated fluctuations in SALINITY as its connection with the Mediterranean was cut off and restored. It is an oval BASIN with a relatively thick crust (18–24 km) and a SEDIMENT cover 812 km thick. Sedimentation rates are of the order of 30 cm per thousand years. An important feature of the Black Sea is that, due to limited water exchange with the adjacent MEDITERRANEAN SEA, it is anoxic below about 200 m depth. Precipitation greatly exceeds evaporation so that the salinity is considerably less than that of the adjacent Mediterranean, as is species diversity of both planktonic and benthic organisms. A layer of relatively fresh water traps a body of normal-salinity water out of contact with the atmosphere or the open OCEAN. Black Sea waters are supersaturated with respect to calcium carbonate. The tidal range is only about 8–9 cm and much of the mixing is thus wind-driven. [V.F.]

blank A SAMPLE designed to detect the introduction of artefacts in measurement procedures. [P.C.]

blanket bogs *See* BLANKET MIRE; MIRE.

blanket mire As its name implies, blanket mire is characterized by the widespread nature of the PEAT cover, often enveloping hill-tops, plateaux, slopes and valleys. It is found only in regions of high PRECIPITATION and low evapotranspiration, most typically in north-western Europe, the north-eastern and north-western extremities of North America, the southern tip of South America, and a number of mountainous districts and oceanic islands elsewhere. The abundant water supply permits the development of peat even on slopes of 35° in the wettest locations.

The VEGETATION, HYDROLOGY and STRATIGRAPHY of blanket mires are as varied as the TOPOGRAPHY they cover. Much blanket mire is truly OMBROTROPHIC and can therefore be termed 'BOG', but sloping sites and valleys may bear a more RHEOPHILOUS vegetation. The stratigraphic PROFILE (*see* STRATIGRAPHY) of slopes and plateaux often reveals fossil soils, frequently with the remains of woody vegetation, while the mire covering valley locations and depressions has often developed from former LAKES or VALLEY MIRES.

Although the primary limitation of the development and the distribution of blanket mire is climatic, there is much evidence from the British Isles and Norway to suggest that early human activity, in the form of burning and grazing, altered the hydrological regime of marginal areas and encouraged the spread of blanket mire. Its development consists of PALUDIFICATION rather than the more common terrestrialization (becoming increasingly elevated and therefore drier) in wetland SUCCESSION, and could be regarded as retrogressive (*see* RETROGRESSIVE SUCCESSION). *See also* MIRE. [P.D.M.]

bloom A sudden increase in density of algae in a water body, usually due to an excess of NUTRIENTS. *See also* ALGAL BLOOM. [P.C.]

blue-green algae toxins *See* CYANOBACTERIAL TOXINS.

BNFL *See* BRITISH NUCLEAR FUELS LIMITED.

BOD *See* BIOCHEMICAL OXYGEN DEMAND.

body burden The total amount of XENOBIOTIC in an organism at a given time. [P.C.]

body size *See* ALLOMETRY; GROWTH; SIZE AT MATURITY.

bog A major subdivision in the classification of PEAT-forming ecosystems, or MIRES. It comprises those mires in which the supply of rainfall forms the sole input of water to the mire surface and which consequently receive their input of chemical elements only from dissolved materials or suspended dust (OMBROTROPHIC). A consequence of this definition is that bog can develop only in situations where the PRECIPITATION is sufficiently high and the EVAPORATION/TRANSPIRATION (dependent partly on temperature) sufficiently low to permit waterlogging and hence the inhibition of microbial respiration that is required for peat formation. Bogs are therefore more frequently associated with high latitudes where either oceanicity provides heavy precipitation or low winter temperature restricts evapotranspiration. There are exceptions, however, particularly in high-altitude sites even in low latitudes. Some tropical lowland forested mires can also be termed bogs, such as the bog-forests of Sarawak, since they have elevated surfaces fed entirely by rainfall.

The development of a bog ecosystem is dependent on the accumulation of undecomposed litter to such a degree that the mire surface is elevated above the influence of groundwater. Water movement through such compacted peats is very poor (*see* CATOTELM), hence there is little or no capillary upward movement of water in such situations, but rather an impedance of gravitational water down the PROFILE, thus perpetuating waterlogged conditions at the surface.

The term 'bog' is often used loosely even by ecologists (e.g. 'valley bog'), and a stricter use of wetland hydrological terminology is to be encouraged. The term 'mire' should be used as a general expression for peat-forming ecosystems. [P.D.M.]

Bog-Podzolic soils *See* SOIL CLASSIFICATION.

Boltzmann's constant *See* RADIATION LAWS.

bomb calorimeter A device for measuring the heat of combustion of fuels or BIOMASS. One version consists of a steel vessel immersed in a well-lagged, stirred water-bath. A weighed sample of test material is placed inside the vessel, which is pressurized with OXYGEN. The sample is then ignited electrically, by passing a small current through a heating filament inserted into the sample. On combustion the energy liberated is transferred by CONDUCTION and CONVECTION to the water-bath, and the rise in TEMPERATURE is measured with a precision thermometer. The device is calibrated using a weighed quantity of a material whose heat of combustion is precisely known. *See also* ECOLOGICAL ENERGETICS. [J.G.]

bootstrap estimates of diversity *See* DIVERSITY INDICES.

Bordeaux mixture *See* PESTICIDES AND SOIL CONTAMINATION.

boreal A term meaning 'northern', 'of the North' (from *Boreas*, Greek and Latin, (God of) the north wind; *boreus* and *borealis* adj., Latin). It is used in BIOGEOGRAPHY to denote conditions intermediate between ARCTIC and temperate (*see* BOREAL FOREST). In Quaternary studies, Boreal is the name of a Holocene period in north-western Europe (9000–8000 BP), originally supposed to have had a dry (continental) climate warmer than subarctic. [Y.V.]

Boreal forest

Boreal forest, or TAIGA, is the world's largest vegetation formation, comprising coniferous forests that stretch around the Northern hemisphere interrupted only by the North Atlantic Ocean and the Bering Strait. It forms a 1000–2000 km broad zone between treeless TUNDRA in the north and either broad-leaved DECIDUOUS FORESTS (in oceanic areas) or dry GRASSLANDS and semi-deserts (in continental areas) in the south. Along the oceanic southern border of the boreal forests there is an intermediate hemiboreal zone with varying mixtures of coniferous and deciduous trees (Fig. B10). The boreal forests proper begin in areas where the summers become too short (less than 120 days with temperatures higher than +10°C) and the winters too long (over 6 months) and severe for the broad-leaved deciduous trees. The northern boundary of the boreal forest follows approximately the +10°C ISOTHERM of the warmest month where also the winter becomes 8 months or longer. Because of orographic factors, boreal forests can occur in mountainous areas outside the zone of boreal coniferous forests.

Within the large CIRCUMPOLAR zone of boreal forests the climate is by no means similar everywhere but varies from cool oceanic, with a rather small temperature amplitude, to cold continental, with temperature extremes from +30°C to −70°C. Consequently, there are also major differences in the forest vegetation. Also

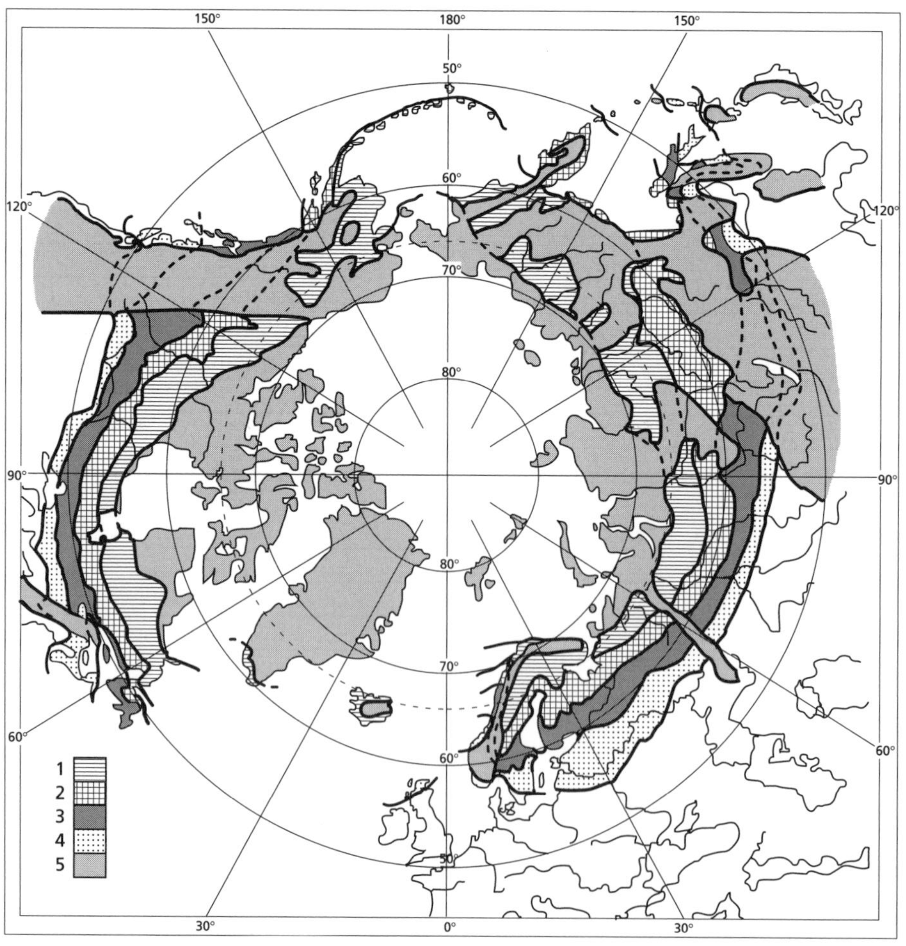

Fig. B10 The circumboreal zone and its transcontinental subzones: (1) northern boreal; (2) middle boreal; (3) southern boreal; (4) hemiboreal; (5) arctic and complex mountain oroarctic areas where the subzones are not indicated though they occur there as an outlier (e.g. the valley of the Yukon River in Alaska belongs widely to the middle boreal subzone (cf. Viereck, L.A. (1975) Ecology of the Alaska taiga. *Proceedings, Circumpolar Conference on Northern Ecology 1975*, 1. Toronto.); in the valleys of the Mackenzie River and its tributaries there are even southern boreal outliers, and Seymchan on the Kolyma River is middle boreal). (After Hämet-Ahti, L. (1981) The boreal zone and its biotic subdivision. *Fennia* **159**, 69–75.)

historical factors contribute to differences in the floristic composition.

The boreal forests are floristically poor compared to most other great vegetation formations of the Earth. The forests are dominated by coniferous trees belonging to a few genera only: pine (*Pinus*; 80–90 species); spruce (*Picea*; *c.* 40 species); fir (*Abies*; *c.* 40 species); Douglas fir (*Pseudotsuga*; 6 species); hemlock (*Tsuga*; 10 species); larch (*Larix*; 10 species) and juniper (*Juniperus*; *c.* 60 species). The main genera of the deciduous trees are: poplar (*Populus*; *c.* 30 species); willow (*Salix*; *c.* 200 species); birch (*Betula*; *c.* 50 species) and alder (*Alnus*; *c.* 20 species). On the basis of the arboreal flora the vast zone of boreal forests is divisible into three regions: the Euro-Siberian region, which has the smallest number of tree species, and the North American and East Asiatic regions, which have much greater richness of arboreal species.

The non-arboreal flora of the boreal forests is also quite poor compared to most of the other major vegetational zones. This is largely due to the EDAPHIC conditions, and the general prevalence of podzolic SOILS. The litter, composed mainly of conifer needles and twigs, decays only slowly resulting in an accumulation of raw, acid plant debris forming a layer on the soil surface. The humid climate (relatively high precipitation and low evaporation) favours LEACHING of the soluble chemical compounds (e.g. aluminium and iron oxides) in the upper layers of the soil. As a result, an ash-coloured whitish layer develops. Somewhat deeper down with decreasing acidity the iron and humic compounds are enriched and attached to soil particles forming a brown to blackish layer above the mineral material unaffected by soil-forming processes.

The field layer in the forests is mostly dominated by dwarf shrubs. In the bottom layer bryophytes dominate in moister conditions while lichens (*Cladonia* and *Cetraria* spp.) cover the soil in dry forests and in the light northern forest. The field layer, and especially the bryophyte and lichen floras, show great uniformity within the boreal forests, more so than the tree species.

Fire has always played a considerable role in the ecology of the boreal forests (*see* FIRE, ECOLOGY AND MANAGEMENT). It has been estimated that even under completely natural conditions the boreal forests experienced a fire every 80–90 years. A forest fire initiates a vegetational SUCCESSION: the first PIONEER trees are deciduous trees, the next immigrants are light-demanding trees like pine, then shade-tolerant trees such as spruce form the CLIMAX stage.

The zone of boreal forests is divisible into three climatically based subzones, called southern boreal, middle boreal and northern boreal. The hemiboreal zone is intermediate between boreal forests and temperate broad-leaved forests. The forests tend to become thinner and slower growing from south to north. Along the humidity–aridity gradient the boreal forests proper can be divided into several sectors. In the humid sectors, as in the western and eastern fringes of Eurasia and in southern Alaska, birch and/or alder replace the coniferous trees. The moderately humid sectors are generally dominated by coniferous forests. In the most continental sectors bordering the grasslands, aspen and larch are important trees. [Y.V.]

---

boring animals Animals that penetrate solid substrata by mechanical abrasion or chemical dissolution via the secretion of organic acids. Isopod crustaceans (e.g. gribbles) and bivalves (e.g. shipworms) bore into wood. Some shipworms have symbiotic bacteria that can degrade CELLULOSE; others bore into the wood but feed on material in suspension. Some species of sponges, sea urchins, bivalves and worms bore into coral, shell material or even concrete. [V.F.]

boron (B) A non-metallic element (ATOMIC NUMBER $=5$; relative ATOMIC MASS $=10.81$; melting point $=2300°C$; boiling point $=2550°C$) which is a trace constituent ($10\,mg\,kg^{-1}$) of the Earth's CRUST. Boron, which occurs primarily in hydrated sodium borate minerals, such as borax, is used as a NEUTRON absorber in nuclear reactors, as a hardening agent in steels and, through a variety of compounds, in GLASS manufacture and the preparation of washing powders and disinfectants. In the form of boric acid or borates, it is an essential TRACE ELEMENT for the growth of plants, deficiency giving rise first to discoloration and distortion of leaves and eventually to rot. Legumes are richest ($25–50\,mg\,kg^{-1}$) in the element. Elevated soil boron concentrations, as found in geothermal areas, can be harmful to plants. [J.F.]

botanical garden *See* ARBORETUM.

bottle effects Differences in growth or behaviour of aquatic organisms that are the result of being constrained in a relatively small volume where contact with the internal surface becomes important. For example, bacteria reach greater abundances when grown in small, relative to large, bottles of seawater. Such bottle effects can result in significant over- or underestimates of biological rate processes when measured in laboratory systems. [V.F.]

'bottle' experiments *See* AYALA'S EXPERIMENTS.

bottom-up controls Controls on the structure of a FOOD WEB operating through limitation of available RESOURCES, often NUTRIENTS and/or LIGHT in the case of plants or PRIMARY PRODUCTIVITY in the case of animals. It is contrasted with TOP-DOWN CONTROL, where predators at the top of the trophic hierarchy control web structure. The relative importance of top-down and bottom-up controls has been a source of considerable controversy, often with conflicting evidence for the same system. It has been suggested that some of this controversy may be resolved if one considers processes in terms of HIERARCHY THEORY. For example, in some aquatic systems there is evidence for both bottom-up control of web productivity, through limitation of PHYTOPLANKTON production, and top-down control of production through predation by fish. These data may not, however, be in conflict with one another because phytoplankton POPULATION DYNAMICS operate over short time-scales, which may explain intra-seasonal variability in production, whereas fish operate on longer time-scales, which may explain inter-seasonal variation. [S.J.H.]

REFERENCE

Carpenter, S.R. (1989) Transmission of variance through lake food webs. In: *Complex Interactions in Lake Communities* (ed. S.R. Carpenter). Springer-Verlag, New York.

boundary layer WATER movements in RIVERS and streams are driven by gravitational gradients. The structure of this flow is mediated by friction induced by the channel boundary. The region where

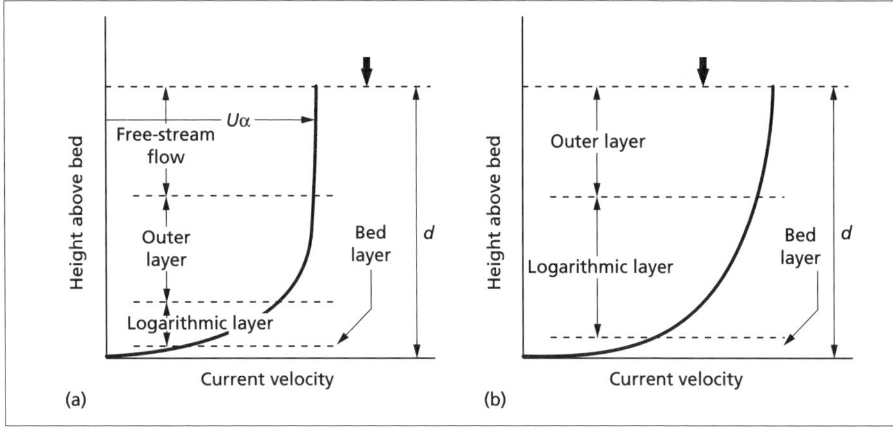

Fig. B11 (a) Structure of the boundary layer in deep flow. (b) Structure of the boundary layer in shallow water. *d*, Water depth; *Uα*, free-stream velocity. (After Carling, P.A. (1992) In-stream hydraulics and sediment transport. In: *The Rivers Handbook*, Vol. 1 (eds P. Calow & G.E. Petts), pp. 101–125. Blackwell Scientific Publications, Oxford.)

these frictional effects are felt is the boundary layer. In deep rivers it may occupy only a small proportion of total depth whereas in shallow rivers it may extend to the water surface. In turn it can be divided into component layers (Fig. B11) including a bed layer (closest to the bed, often a few millimetres thick, where there is often, but not always, LAMINAR FLOW), a logarithmic layer (not affected by the local ROUGHNESS of the bed nor by the free streamflow structure) and another layer that is influenced by the free-stream velocity. *See also* DEAD ZONES; PLANETARY BOUNDARY LAYER. [P.C.]

REFERENCE

Carling, P.A. (1992) In-stream hydraulics and sediment transport. In: *The Rivers Handbook*, Vol. 1 (eds P. Calow & G.E. Petts), pp. 101–125. Blackwell Scientific Publications, Oxford.

boundary layer meteorology and climatology *See* METEOROLOGY.

Boyle's law *See* GAS LAWS.

BPEO *See* BEST PRACTICABLE ENVIRONMENTAL OPTION.

BPM *See* BEST PRACTICABLE MEANS.

brackish habitats BRACKISH WATERS and soils contain ionic solutes at higher concentrations than those of FRESHWATER or normal soil solutions but at considerably lower concentration than seawater. They originate by dilution of seawater with rain or groundwater, or by evaporation of water and consequent enrichment with solutes in arid environments. Compared with freshwater, maritime brackish water contains more Na^+, Cl^-, Mg^{2+} and SO_4^{2-} (sulphate) while, in arid habitats, additional solutes such as Ca^{2+} and HCO_3^- (bicarbonate) may also be present. The concentration of solutes in seawater is usually about 35 $g\,kg^{-1}$. Brackish waters are much more dilute but considerably exceed the few $mg\,kg^{-1}$ of freshwaters.

The upper levels of SALT MARSHES and MANGROVE swamps experience brackish conditions as do coastal LAKES and ponds and tidal parts of RIVERS where seawater and freshwater mix. Salt desert habitats may be brackish where inflow of groundwater dilutes the salt-rich soil solution. In all of these cases a specific flora and fauna occurs, having physiological characteristics which allow them to cope with the osmotic effects of saline water and the potential toxicity of high concentrations of ionic solutes, particularly Na^+ and Cl^-. [J.R.E.]

brackish lakes *See* LAKES, TYPES OF.

Brackish waters

Brackish waters occur wherever seawater is diluted by FRESHWATER, or conversely where freshwater is contaminated by seawater. It is thus WATER of SALINITY intermediate between the salinities of these two other major aquatic environments: conventionally it is often taken to lie between 0.5‰ and some 30‰ by mass of dissolved salts. In most usages 'brackish' is also applied to seawater that has been concentrated by EVAPORATION to values of >40‰, although this is sometimes separately referred to as 'hypersaline water'. The term has even been applied as a catch-all for all non-marine and non-freshwater habitats, including those inland bodies of saltwater, such as the Dead Sea (Middle East), Lake Nakuru (Africa), Lake Eyre (Australia) and the Great Salt Lake of Utah (North America), that occupy depressions in the continental surface and have derived their salts from the land by evaporative concentration of the inflowing RIVERS. Since such landlocked salt LAKES possess what is essentially an impoverished freshwater BIOTA, however, they are normally excluded from the usual maritime concept of brackish and termed 'athalassic (i.e. non-marine) saltwaters'. Here, brackish is restricted to those waters in which the salts have been contributed by seawater. Major brackish-water environments are therefore estuaries, most coastal LAGOONS, the inland Eurasian seas, and regions of high-latitude coastal SEA diluted by ICE meltwater. Most coastal areas contain numerous other, more minor BRACKISH HABITATS, including rock pools, pools within SALT MARSHES and MANGROVE swamps, drainage ditch systems, the maritime WATER TABLE, etc.

Estuaries, the regions through which rivers DISCHARGE to the coastal sea, are by definition brackish. Many species are characteristically estuarine in their distributions although most, if not all, of them are nevertheless not restricted to brackish water. Estuaries are generally inhabited by a subset of coastal marine species that are typical of sheltered and often shallow-water muddy sediments.

They thus also occur in suitable habitats in areas of full-strength seawater, wherever such may be found, although in some regions there is evidence that they may be restricted to low-salinity REFUGES by local superior but more stenohaline marine competitors unable to tolerate salinities much below 30‰. By virtue of their size and of past periods of GEOGRAPHICAL ISOLATION, the Eurasian brackish seas such as the Baltic and the BLACK SEA do possess a number of ENDEMIC SPECIES and, again by definition, since the seas are brackish, their endemics are *de facto* restricted to brackish water.

The same occurs to an even greater extent with respect to the Caspian and Aral Seas. These once formed part of the large complex of brackish coastal seas that comprised the remains of the eastern Tethys Sea in the Eocene and Miocene after the northward movement of the African and Indian plates. Part of the current fauna of the Caspian and Aral, from which the endemics derive, dates back to this brackish period in the history of these seas, although most of their present salts have originated athalassically from land drainage. The Caspian/Aral basin has therefore had an extremely complicated history involving marine, brackish, freshwater and athalassic phases.

Brackish (and athalassic) waters are often extremely productive, notwithstanding the hostile salinity regime (Fig. B12). This is because they receive the discharge of freshwater, and freshwater contains more NUTRIENTS, volume for volume, than does seawater. Their shallowness also results in: (i) a lack of separation of surface and bottom waters (unless a HALOCLINE is present, see below), facilitating nutrient cycling; and (ii) the occurrence of benthic PHOTOSYNTHESIS not only from microalgae but from fringing salt marsh or mangrove vegetation and from intertidal or submerged seagrasses, together with their various epiphytes. Microalgal PRODUCTIVITY and the export of DETRITUS from the stands of angiosperms supports numerous individuals of a limited range of deposit-feeding annelid, crustacean and molluscan species that in turn achieve high productivity although BIOMASS is often low in comparison with other coastal habitats. This invertebrate production then supports a wide range of migratory fishes, crustaceans and birds that use brackish-water environments as feeding grounds as well as, in respect of the aquatic consumers, nursery grounds for their young. Brackish waters therefore often support fisheries for juvenile fish (e.g. 'whitebait') and for adult prawns and crabs, and form favoured areas for bird-watching.

Brackish waters are often subdivided on the basis of their salinity into oligohaline (*c.* 0.5–5‰), mesohaline (*c.* 5–18‰) and polyhaline (*c.* 18–30‰) systems or stretches, although the relevance of any such subdivision to biotic distributions has been much criticized. It is apparent, however, that within estuaries and to a somewhat lesser extent in other brackish waters, there is a marked decline in the species of freshwater ancestry as salinities increase through the oligohaline reach, and an equivalent decrease in essentially marine species as salinity falls through the polyhaline zone. Granted that few species are specifically brackish, mesohaline waters are then the region of minimum SPECIES RICHNESS. This situation has often been considered a direct affect of the environmental salinity regime, although attempts to correlate distributions with specific salt concentrations in the medium have been almost universally unsuccessful. The dominant invertebrates in estuaries are extremely euryhaline and appear able to survive and reproduce over a range of *c.* 4–50‰. Whilst it is self-evidently true that few marine species can also inhabit freshwater and vice versa, the low species richness of estuaries may be more a consequence of lack of spatial HETEROGENEITY in the monotonous muddy habitat, coupled with the frequency of environmental catastrophes, such as prolonged freshwater flooding, and the long periods of time required to recolonize range lost after such events. DIVERSITY is certainly relatively high in stable brackish waters and in the vicinity of bryozoan reefs, mangrove forests and other elements providing surface relief.

Since freshwater is less dense than seawater, the one tends to float on the other; the density interface between the two is the halocline. Haloclines are present in some estuaries, lagoons and other poorly mixed brackish systems, and they have the effects that (i) regions below the halocline may become anoxic (in exact correspondence to the position with respect to a THERMOCLINE in freshwater lakes), and that (ii) marine species, by keeping to deep waters, may penetrate further into estuaries than might be apparent from the prevailing surface salinities.

By virtue of the characteristic shelter from wave action, estuaries and lagoons form safe harbours, and throughout the world their shorelines have been developed into port facilities. Associated with these are the industry and housing associated with materials arriving by ship, and with processes requiring an abundant supply of cooling water. Inputs, both deliberate and accidental, of SEWAGE, oil and a wide range of other pollutants are therefore typically very high. All lagoons and the Eurasian inland seas are almost completely landlocked and hence flushing times are extremely long, and although the flushing times of estuaries are shorter, the loads discharged are in many areas so high that brackish waters in the developed and developing worlds contain extensive anoxic or micro-oxic stretches and are grossly polluted. Furthermore, RECLAMATION for industry, housing or AGRICULTURE has decreased the area of a number of systems virtually down to dredged shipping channels.

The soft muds that are anoxic immediately below the surface render the shallow waters of brackish habitats treacherous and smelly, and make access on to exposed MUDFLATS difficult. The smaller systems are hence often considered as unattractive 'waste-

Fig. B12 The estuarine Lake Butrinti, Albania, with racks for shellfish culture in the foreground. (Courtesy of R.S.K. Barnes.)

land': appropriate places to put to 'better' uses and into which to dispose of wastes. Up to 95% of the area of individual estuaries has been lost to concrete; all the area of several individual lagoons has been lost to agriculture; and the Caspian and Aral Seas have been reducing in area through water ABSTRACTION. Only their attractive birdlife has prevented many brackish habitats in the temperate zone from being totally destroyed, and their fisheries potential likewise in the tropics. [R.S.K.B.]

Braer *See* ECOLOGICAL/ENVIRONMENTAL DISASTERS.

breeding aggregations *See* GROUP SOCIAL STRUCTURE.

breeding mechanisms in plants *See* POLLINATION.

breeding systems in plants *See* POLLINATION.

breeding value The breeding value, or additive genetic value, of an individual with respect to a QUANTITATIVE TRAIT measures the heritable effects of its GENOTYPE on its PHENOTYPE. The breeding value of an individual can be estimated by mating it randomly with a number of individuals and measuring the trait in the offspring. Under random mating and in the absence of non-genetic correlations between parents and offspring (e.g. MATERNAL EFFECTS), the expected deviation of the phenotype of an individual's progeny from the population mean equals half the breeding value of the individual. Because of non-additivity in gene expression (*cf.* ADDITIVE GENETIC VARIANCE), breeding value constitutes only a part of genetic effects on the phenotype. Non-additive components (due to DOMINANCE and EPISTASIS) are not transmitted to progeny because of segregation and recombination accompanying sexual reproduction. The variance of breeding values (additive genetic variance) determines the trait's potential to respond to NATURAL SELECTION. *See also* QUANTITATIVE GENETICS. [T.J.K.]

REFERENCE

Falconer, D.S. (1989) *Introduction to Quantitative Genetics*, 3rd edn. Longman, London.

BREF *See* BEST AVAILABLE TECHNIQUES REFERENCE DOCUMENT.

Brillouin index *See* DIVERSITY INDICES.

British Nuclear Fuels Limited (BNFL) British Nuclear Fuels Limited operate a number of nuclear sites in the UK including four Magnox nuclear reactors and separate facilities for URANIUM enrichment, nuclear fuel production, nuclear fuel reprocessing and low-level solid RADIOACTIVE WASTE disposal. The Sellafield site, which is the location of the nuclear fuel reprocessing plant, also contained the now-defunct Calderhall NUCLEAR POWER-STATION and Windscale experimental piles. The authorized discharge of low-level liquid radioactive waste into the Irish Sea from Sellafield has been the focus of considerable attention in the context of marine environmental pollution. Radionuclides were also released to the environment from a fire in one of the Windscale experimental piles in 1957. *See also* WASTE DISPOSAL. [A.B.M.]

British scrub *See* SCRUB.

broad-sense heritability In QUANTITATIVE GENETICS, the broad-sense HERITABILITY is defined as the ratio of the GENETIC VARIANCE to the phenotypic (total) variance (*see* PHENOTYPIC VARIANCE) of a QUANTITATIVE trait: $h^2 = V_G / V_P$. The broad-sense heritability indicates what proportion of differences in the values of the quantitative trait in the population are caused by genetic differences. In the broad-sense heritability, differences between individuals due to DOMINANCE between ALLELES at a locus and due to interactions between loci are included in the genetic variance. Not all these genetic differences are heritable in the sense that they are found in the offspring. The broad-sense heritability is larger than the NARROW-SENSE HERITABILITY, which includes only the fraction of the variance between individuals that is due to additive genetic effects contributed by the parents to the offspring.

In clonally reproducing organisms broad-sense heritability determines the degree of similarity among individuals belonging to the same CLONE. *See also* ENVIRONMENTAL VARIANCE. [G.D.J. & T.J.K.]

REFERENCE

Falconer, D.S. (1989) *Introduction to Quantitative Genetics*, 3rd edn. Longman, London.

Broadbalk experiment An experiment involving the continuous cultivation of winter wheat, begun at ROTHAMSTED in 1844. It has provided important insights into NITROGEN FIXATION and the nutrient budgets of CROPS growing under different nutrient inputs. The experiment is of great historical interest because it was used to refute Liebig (*see* LIEBIG'S LAW OF THE MINIMUM; *see also* CHARACTERS IN ECOLOGY) who believed that plants could obtain all the nitrogen they needed directly from the air, and only needed to be fertilized with inorganic NUTRIENTS (P, K, Mg, etc.). The Broadbalk experiment shows the central importance of nitrogen fertilization (in the form of NITRATE or ammonium ions) in producing peak agricultural YIELDS. The experiment is also of current ecological interest, because one of its sections has never been treated with HERBICIDES and still supports an ancient arable WEED flora, including herbicide-susceptible species that have become national rarities as a result of universal chemical weed control. [M.J.C.]

REFERENCE

Johnston, A.E. (1994) The Rothamsted Classical Experiments. In *Long-term Experiments in Agricultural and Ecological Sciences* (eds R.A. Leigh & A.E. Johnston), pp. 9–37. CAB International, Wallingford.

broken stick model The broken stick model was proposed by MacArthur (1957; *see* CHARACTERS IN ECOLOGY) to predict the small-scale pattern of relative ABUNDANCE of species in an ASSEMBLAGE, assuming that species in a GUILD divide available RESOURCES among themselves and, in particular, that relative abundance is determined by one critical resource. MacArthur drew an analogy with a stick along which $n-1$ points were picked at random where n is the number of species in the assemblage. When the stick is broken at each point the length of each segment corresponds to the abundance of one species.

This procedure leads to the statement that the expected abundance (A_r) of the r the rarest species among n species and m individuals is given by:

$$A_r = \frac{m}{n}\sum_{i=1}^{r}\left(\frac{1}{n-i+1}\right)$$

When the segments are arranged on a logarithmic scale of decreasing rank, the expected DISTRIBUTION of stick lengths decreases with a characteristic, almost linear curve.

Broken stick distributions of COMMONNESS and rarity can be demonstrated for many guilds of closely related organisms living together in small areas with closely synchronized LIFE CYCLES, such as birds, fish and predatory gastropods. However, many small-bodied, short-lived organisms such as soil arthropods, nematodes and phytoplankton have a less equitable distribution (*see* EQUITABILITY) of abundances with fewer common and more rare species. Moreover, many similar distributions can be produced by models with different properties (Cohen 1968). So, while remaining an accurate predictor of patterns in some species assemblages, the broken stick model does not provide a fundamental explanation of these patterns. *See also* RANK–ABUNDANCE MODELS; SPECIES-ABUNDANCE MODELS. [M.H.]

REFERENCES

Cohen, J. (1968) Alternative derivations of a species–abundance relation. *American Naturalist* **102**, 165–172.

MacArthur, R.H. (1957) On the relative abundance of bird species. *Proceedings of the National Academy of Sciences of the USA* **43**, 293–295.

bromine (Br) A halogen element (ATOMIC NUMBER = 35; relative ATOMIC MASS = 79.904; melting point = –7.2°C; boiling point = 58.78°C), found as Br^- in an abundance of 0.0065% in seawater and in higher concentrations in brines, from which it is extracted by oxidation with CHLORINE. The main industrial use of bromine until recently has been the production of dibromoethane—an additive in gasoline which scavenges LEAD. Bromine compounds are also used as pesticides, flame retardants, drilling fluids, and in photography. Elemental bromine is a corrosive liquid whose FUMES are very toxic. [C.R.P.]

Brood parasitism

The laying of EGGS in the nests of other individuals and the rearing of offspring by these CONSPECIFICS (intraspecific brood parasitism) or HETEROSPECIFICS (interspecific brood parasitism).

Brood parasites lay their eggs in the nests of HOSTS, which subsequently care for and rear these eggs. Brood parasitism has evolved independently in insects, fishes and birds, with most cases being described for birds. Well-known examples include parasitic cuckoos, cowbirds and viduine finches (Viduinae). Intraspecific brood parasitism, which is considered an early step in the evolution of interspecific parasitism, has been reported for a range of insect, fish and bird species. Well-known examples are many geese and ducks, swallows (Hirundinidae), starlings (Sturnidae) and weavers (Ploceidae). The major advantage of brood parasitism for the PARASITE is a reduction in the COST OF REPRODUCTION, while the hosts of brood parasites pay a cost in terms of reduced reproductive success.

PARENTAL CARE is a costly activity in terms of energy and time use, and any individual that is able to persuade other individuals to provide parental care and hence avoid the costs would leave more descendants than a parental individual. Brood parasites obtain this benefit by laying their eggs in the nests of conspecifics or heterospecifics. In accordance with the hypothesis that parasites avoid the cost of parental care, interspecific brood parasitism has evolved independently more frequently among bird families with extensive parental care than among families with little or no care. PARASITISM will allow intraspecific parasitic individuals to rear more offspring by laying eggs in other nests, while simultaneously caring for their own offspring. Interspecific parasites only lay eggs in the nests of other species, and the entire REPRODUCTIVE EFFORT can be spent on egg production. Interspecific parasites therefore generally have larger clutches than their close relatives. The hosts of brood parasites pay a cost in terms of reduced reproductive success because parasitism frequently results in a considerable or complete loss of reproductive success for the host. The cost of reproduction for the host of interspecific brood parasites may also become elevated considerably because the duration of parental care for parasites often is considerably longer than that for conspecific offspring.

Hosts of intraspecific brood parasites generally breed in dense colonies with short distances to the nearest neighbour. Individuals that start laying later during the breeding season are more susceptible to parasitism attempts. Host individuals guard their nests less intensely than non-hosts, and the absence of a guarder considerably increases the risk of parasitism. Slightly asynchronous laying by a nearby neighbour reduces the risk of intraspecific parasitism, because of exploitation of the territorial and nest-guarding activity of neighbours and a reduced risk of parasitism.

Hosts of interspecific brood parasites are mainly young birds that are breeding for the first time. Since birds have to learn the appearance of their own eggs, and since the eggs of a clutch vary more in appearance among young than old birds, first-time breeders will have greater difficulty distinguishing their own eggs from those of a parasite than experienced breeders. When given a choice, parasites choose host individuals that have better parental abilities compared to individuals not chosen to rear a parasite.

Successful intraspecific brood parasites are often found in dense and large colonies with many potential hosts nearby. The close presence of naive reproducers may also facilitate reproduction, and recruitment of first-time breeders to the neighbourhood of an experienced breeder may facilitate later intraspecific brood parasitism. This behaviour is also beneficial for experienced male breeders because it increases the probability of successful EXTRA-PAIR COPULATIONS. Successful brood parasites therefore raise more offspring than nearby conspecifics.

Adaptations of brood parasites

Brood parasites have evolved a range of ADAPTATIONS that increase their probability of successful parasitism. Several interspecific brood parasites have flight behaviour that resembles that of accipitrid hawks, which are common predators of many passerine birds. This flight behaviour may keep hosts at a distance during laying. Egg laying by interspecific parasites is incredibly fast, and this reduces the likelihood of a parasite being seen in the vicinity of a host nest. At least one intraspecific brood parasite is able to transport its eggs to other nests in the beak, and this behaviour allows parasitism during a period that extends beyond the normal egg-laying period. Interspecific parasites frequently remove a host egg during parasitism (Fig. B13 & Plate 1, facing p. 398), and this reduces the level of COMPETITION for food encountered by the parasitic offspring, provides food for the parasitic female, and prevents the host from obtaining information on parasitism from the size of the clutch. Eggs of interspecific parasites are usually laid outside the usual diurnal time of laying by the host, and this prevents encounters with the host.

The egg of an interspecific parasite is typically relatively small for the size of bird, which reduces the risk of detection of parasitism by the host. Parasites' eggs have shells that are very thick for their size, and this thickness prevents the host from puncturing the parasite's egg and the egg from being damaged when dropped from the rim of the nest of the host. Eggs of interspecific brood parasites are often mimetic (*see* EGG MIMICRY), and this prevents or reduces the risk of egg ejection by the host.

Nestlings of interspecific brood parasites that are raised together with those of their host often mimic their hosts to an incredible extent. For example, the nestlings of viduine finches can only be distinguished from those of their hosts by minute differences in the colour patterns of the inside of the beak. Nestling MIMICRY extends to the vocalizations of nestlings, which resemble those of the host. Interspecific brood parasites develop very quickly with embryonic development being fast and hence incubation periods very short for a bird of the size of a brood parasite. Early hatching provides parasitic offspring with a head start, and when the nestlings of the host hatch, the parasite is usually much larger than the host nestlings, which therefore receive a disproportionately small amount of food. Their prospects for survival are therefore considerably reduced. The nestlings of several species of parasitic birds remove or kill all the offspring of the host shortly after hatching. For example, the nestling of the European cuckoo (*Cuculus canorus*) pushes host eggs and nestlings over the rim of the nest (Fig. B13), while nestling parasitic honeyguides (Indicatoridae) kill their nestmates with a very sharp hook on their beak. Parasitic nestlings develop very quickly, and they are able to leave the nest at or shortly after the normal time of fledging of the host's offspring. Fledged parasites are often cared for by the hosts for several weeks, and the duration of post-fledging care thus extends considerably beyond normal parental care by the host.

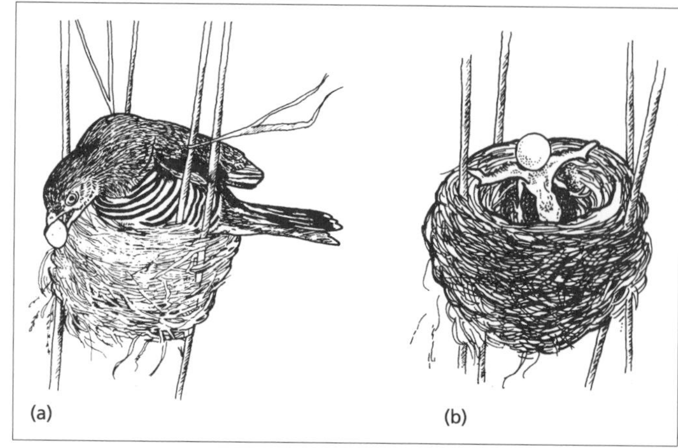

Fig. B13 (a) A female cuckoo (*Cuculus canorus*) laying in a reed warbler (*Acrocephalus scirpaceus*) nest. She has removed one of the host eggs, which she is holding in her bill while she lays her own egg in its place. (b) The newly hatched young cuckoo ejects the host eggs from the nest. (After Krebs, J. & Davies, N.B. (1993) *An Introduction to Behavioural Ecology*, 3rd edn. Blackwell Scientific Publications, Oxford.)

Host acceptance of parasites

Despite many cases of egg and nestling mimicry, the offspring of interspecific parasites frequently have an appearance that is very different from that of the host. Given that parasite eggs and particularly parasite nestlings can be readily distinguished, why are parasites accepted by their hosts?

Two basically different types of hypotheses may explain host acceptance of parasites. The evolutionary lag hypothesis suggests that ejection behaviour has not yet evolved among hosts because the interaction with the parasite is of relatively recent origin. Some host species are not currently parasitized, but still recognize parasite eggs, and these species may have been parasitized in the past. Extensive ejection behaviour by hosts during the past may have forced parasites to switch to other host species.

The evolutionary equilibrium hypotheses suggest that hosts accept parasite offspring because the benefits from doing so exceed the costs. For example, ejection of eggs with a dissimilar appearance sometimes result in a host species ejecting its own egg. Alternatively, if the offspring of the host are raised together with parasitic offspring, hosts may have higher reproductive success by accepting the parasite than from rejection. This is the case if the parasite destroys the clutch of the host, in case the egg of the parasite is missing after successful ejection.

Host antiparasite adaptations

Hosts do not passively accept becoming parasitized, and a number of antiparasite adaptations have evolved to prevent or reduce the risk of parasitism. Parasites often search for potential hosts by watching their nest-building behaviour, and nests placed away from vantage points run low risks of parasitism.

Guarding the contents of a nest against potential parasites is an efficient means of reducing the risk of parasitism, unless two parasite individuals cooperate during the parasitism event. Hosts often mob parasites (*see* MOBBING) by directly attacking them and giving ALARM CALLS, and this behaviour distracts parasites. When an individual of a host species has found a parasite near its nest, this elicits inspection of nest contents and may result in nest DESERTION. *See also* LIFE-HISTORY EVOLUTION; NEST PARASITISM. [A.P.M.]

REFERENCES

Rothstein, S. (1990) Avian brood parasitism. A model system of coevolution. *Annual Review of Ecology and Systematics* **21**, 481–508.

Rothstein, S. & Robinson, S.K. (eds) (1995) *Brood Parasitism*. Oxford University Press, New York.

Brown forest soils *See* SOIL CLASSIFICATION.

Brown semi-desertic soils *See* SOIL CLASSIFICATION.

brown tree snake (*Boiga irregularis*) *See* BEHAVIOUR, IMPLICATIONS FOR CONSERVATION; EXOTIC AND INVASIVE SPECIES.

browsing *See* FEEDING TYPES, CLASSIFICATION.

Brundtland Report The Report (published 1987) of the UN-sponsored World Commission on Environment and Development, chaired by Mrs Gro Harlem Brundtland (Norwegian Prime Minister at the time of the meeting) and more formally entitled *Our Common Future*. It has provided the most commonly used working definition of SUSTAINABLE DEVELOPMENT. *See also* UNCED. [P.C.]

Bryophytes, global diversity and conservation

A distantly co-related group of non-vascular terrestrial plants comprising the mosses and liverworts. The bryophytes are of ancient LINEAGE, probably evolving directly from an algal ancestor at an early stage in land-plant EVOLUTION and possibly independent of the line which gave rise to vascular plants. Unlike vascular plants, the bryophytes have remained dependent on a close association with water for sexual reproduction. Unfortunately, because of their small size and relatively fine structure, bryophytes have not been readily recognized in the earliest FOSSIL RECORDS. The majority of contemporary species are still found most commonly in continuously moist habitats (though there are a number of exceptions with extraordinarily high capacities for tolerance of DESICCATION). The bryophytes are classified into three taxa: the mosses (Musci), with about 9000 described species, the liverworts (Hepaticae), with about 8000 species, and a small and questionably related group of hornworts (Anthocerotae), of about 250 species.

Unlike the vascular plants, in bryophytes it is the haploid, or GAMETOPHYTE phase which is dominant and photoautotrophic. The minor SPOROPHYTE phase is usually EPHEMERAL and largely if not wholly dependent for nutrition and structural support on the gametophyte. Bryophytes possess chlorophylls, cell-wall components and storage CARBOHYDRATES similar to those of the green algae. However, while the majority of orders of bryophytes and green algae (Chlorophyta) store starch there are two exceptions — the largest classes of liverworts, the Jungermanniales, and of mosses, the Sphagnales, store a POLYMER of fructose or fructan. Fructan is found in the green algae only in the Dasycladales and Cladophorales; the former has fossil representatives in the Cambrian with contemporary forms showing an intriguing habit of withstanding desiccation on exposure in shallow tidal tropical LAGOONS, echoing perhaps the ecology of the first land plants some 450 million years ago.

Considered by one of the botanical founding fathers (F.O. Bower) to be an evolutionary dead end, the bryophytes failed to evolve roots and little beyond the simplest of internal water-conducting systems. Dead end or not, the bryophytes have actually persisted longer than any other terrestrial plant group, and today occupy almost all major global habitats. In tropical moist forests, bryophytes make up a significant element of the epiphytic floras (e.g. the moss family Orthotrichaceae and liverworts of the Plagiochilaceae), while in the extensive BOG formations of the northern temperate and polar regions the sphagna mosses may constitute virtually all of the actively growing BIOMASS.

The majority of bryophytes are small, often inconspicuous, compared with other terrestrial plants (though species of *Dawsonia* from the swamps of south-east Asia may exceed 1 m in length). Their conservation status is only now, probably belatedly, being assessed. The regions of highest bryophyte DIVERSITY include the CLOUD FORESTS of New Guinea, Borneo and Sulawesi and the oceanic Andes of South America, all regions with limited field conservation. The concept of rare mosses or moss communities has been evoked, however, as an important conservation consideration in attempts to block the development of one of Europe's largest proposed 'super-quarries' in the Hebridean island of Harris. In The Netherlands, Germany and the UK conservation concerns over the exploitation and drainage of moss-rich bogs has focused attention on this otherwise largely overlooked backwater of plant evolution. *See also* CONSERVATION BIOLOGY. [G.A.F.H.]

BS 7750 A STANDARD produced by the British Standards Institution (BSI) on the elements of an ENVIRONMENTAL MANAGEMENT SYSTEM. COMPLIANCE enabled companies to carry the BSI mark and standard number on all documentation. It has now been replaced by standards produced by the International Organization for Standardization (ISO), under the designation ISO 14 000. [P.C.]

BSI *See* BS 7750.

bubble concept A concept used in POLLUTION control that involves drawing an imaginary dome-shaped boundary (the bubble) around a source of pollution (which can be a stack, plant, area or even geographical region) and putting an upper limit on the total amount of CONTAMINATION allowed into it. The idea, which has its origins in the USA, is that although the total level of CONTAMINANT is kept constant, increased emission from one plant can be balanced against reductions from another. A market can thus be created in options to release (*see* MARKET INSTRUMENTS). [P.C.]

buffer solution A solution that can resist changes in pH brought about by adding ACID or ALKALI. Blood is an example of a natural buffer solution (pH is maintained at about 7.4). For the control of pH in the chemical laboratory, or for CALIBRATION of pH meters, buffers are made using proprietary 'buffer tablets' or by making up solutions according to well-established recipes (e.g. see Weast 1985). [J.G.]

REFERENCE

Weast, R.C. (1985) *CRC Handbook of Chemistry and Physics*. CRC Press, Boca Raton, Fla.

buffer zone A buffer zone is a boundary area surrounding or adjacent to a core conservation area, such as a park or nature reserve (*see* BIOSPHERE RESERVES). Buffer zones are generally intermediate between core conservation areas and the surrounding landscape in several respects, including their ecological function, extent of human impact, and degree of protection. Buffer zones can be composed of successional habitat, such as second-growth or managed forests, abandoned agricultural lands, grazing or range land, or other areas that have been modified to some extent in composition, structure and ecological function.

Buffers serve a variety of related purposes. They provide additional habitat and foraging area for many species that inhabit the core area. For example, many forest birds nest and roost in interior, closed-canopy forest but forage in edges, openings and other portions of a heterogeneous landscape. If buffer zones are composed of habitat types or SERES different from those found in the core area, they can provide important areas for seasonal MIGRATION. Many montane animals descend to lower elevations during winter, in search of water, warmer temperatures and food. These lower-elevation environments are often outside of protected reserves, so buffers in these habitats can serve important functions in maintaining habitat DIVERSITY and WILDLIFE resources on the landscape.

Buffer zones also allow populations of species resident in the core to expand and move across the landscape over time, in response to changing environmental conditions. The range of most species is affected over many generations by forces such as CLIMATE CHANGE and SUCCESSION; buffer zones provide areas for COLONIZATION and potential ESTABLISHMENT of new populations as conditions change.

Vegetation buffers can mitigate or moderate microclimatic effects of edges (*see* EDGE EFFECT) around forest fragments. If they are intermediate in structural characteristics (e.g. stem density, stem height, canopy lift, foliage density), buffer zones can significantly ameliorate many effects of forest fragmentation associated with edges.

Buffer zones play important roles in ecological DISTURBANCE regimes in managed landscapes. Under natural conditions, fires propagate across prairies, savannahs and forested landscapes in response to topography, weather and fuels. Where such habitats are adjacent to human settlements or structures, buffer zones provide a protective zone separating the core conservation area—where disturbances such as fires should be allowed to proceed uncontrolled—from the surrounding human landscape. Buffers can be designed to permit the continuation of natural disturbance regimes within core areas, thus contributing to the maintenance of important ecological processes.

Linear buffers are frequently established along RIPARIAN CORRIDORS. There they serve important ecological functions (such as organic matter recycling and moderating stream temperatures), provide sources of coarse woody debris and nesting sites for riparian species, and play an important role in erosion control. Where buffers are absent or removed (e.g. by clear-cut FORESTRY PRACTICES), severe damage to the riparian system can occur. In forests managed for sustainable timber production, buffers of various widths (generally 10–100 m) are prescribed to prevent damage both to the riparian area and the adjacent uplands.

Unlike core conservation areas, buffer zones are often used by people for resource extraction (timber, minerals, forage). However, such uses are generally more restricted in designated buffers than in the surrounding landscape. In many designated biosphere reserves, for example, subsistence and limited commercial hunting, fishing, timber extraction, collecting of special forest and grassland products (such as medicinal plants), and even human habitation are permitted in buffer areas, especially where there is a history of traditional cultural uses. [D.A.F.]

Buffon's law *See* BIOGEOGRAPHY.

buoyancy Loss of mass due to immersion in a liquid. In order to maintain their position in the water column, PELAGIC organisms have a number of adaptations to increase their buoyancy. Small size, the presence of spines, air or gas bubbles, and the accumulation of oils and fats can all act to increase buoyancy and thus decrease the rate of sinking. Many deep-sea species possess body tissues with a relatively low protein content, high water content and high lipid content as a means for maintaining buoyancy. [V.F.]

C

C_3/C_4 plants These two abbreviations define plants that assimilate CARBON via two different CARBON DIOXIDE (CO_2)-trapping mechanisms. C_3 plants bind CO_2 to a five-carbon acceptor to form two molecules of a three-carbon first-product. C_4 plants trap CO_2 by combining it with a three-carbon acceptor to form a four-carbon first-product. In both cases the trapping mechanisms are part of a cyclic process that regenerates the acceptor while converting surplus carbon into carbohydrate. Because of biochemical/biophysical differences, C_4 plants are capable of utilizing higher LIGHT intensities than C_3 plants, have greater maximum photosynthetic rates and use less water per unit of photosynthate. C_3 plants are the most widespread while C_4 species are only common in mid-latitude, warm, dry environments (Fig. C1). *See also* CAM; PHOTOSYNTHESIS. [J.R.E.]

C : N ratio In soils and BIOMASS, a close relationship exists between the ratios of the elements CARBON and NITROGEN. The C : N ratio in arable soils is often from 8 : 1 to 15 : 1, with a tendency for the ratio to be lower in arid climates and lower in subsoils. In biomass, the C : N ratio reflects the content of CELLULOSE and LIGNIN. For microorganisms the ratio is as low as 4 : 1 to 9 : 1, for leguminous plants it is 20 : 1 to 30 : 1, for straw residues it is 100 : 1 and for sawdust it is 400 : 1. Thus, the plant residues that become incorporated into the SOIL usually contain relatively little nitrogen in relation to the microbial population that they support.

Thus, when residues are being decomposed, it is necessary for the microorganisms to take up large amounts of nitrogen (ammonium or NITRATE), thus reducing the supply for higher plants. Only later, when carbon has been lost as CARBON DIOXIDE (CO_2) in microbial RESPIRATION and the residue material has been largely decomposed, is there a flush of nitrogen from the breakdown of the bodies of the microbes. *See also* DECOMPOSITION. [J.G.]

REFERENCE

Brady, N.C. (1990) *The Nature and Properties of Soils*, 10th edn. Macmillan, New York.

C-value paradox The amount of DNA in the nucleus varies enormously between different species. The total amount of DNA in the haploid genome is known as the C-value. C-values range from less than 10^7 base pairs for Gram-positive bacteria, 10^8 for annelid worms, 10^9 for crustaceans and up to 10^{11} for angiosperms, with a trend towards increased amounts of DNA with increased complexity of the organism. The amount of DNA varies 5000-fold among algae, 1200-fold among pteridophytes and 2500-fold among angiosperms. The term 'C-value paradox' refers to the extraordinary variation in DNA amount and our inability to explain the function of the additional DNA, particularly in closely related species. Much of the variation in DNA can be explained as non-protein-coding DNA but this does not explain the functional significance of the variation. Several attempts have been made to find links between variation in DNA in plants and DISTRIBUTION of the species (the same correlations are not found in animals). As a broad generalization, in angiosperms C-values increase with increasing latitude; the highest values of DNA are found at latitudes 50–70°, at least among cereal crops. At the two extremes, rice *Oryza sativa* has less than 1 pg of haploid DNA, while temperate-region wheat *Triticum aestivum* has approximately 6 pg. At the cellular level, C-values are positively correlated with cell size, with the volume of the nucleus and with the length of the mitotic and meiotic cycles. The amount of DNA in the nucleus also appears to influence the phenotypic development of the cell: cells with large amounts of DNA are contained in nuclei with large volumes and, in turn, in cells with large volumes. Species with large C-values have a long minimum GENERATION TIME (time from seed germination to production of new seed). The relationship between C-values and type of LIFE CYCLE reveals that, at least in the temperate floras, species with large values undertake growth early in the season by relying substantially on expansion of preformed cells (preformed in the previous autumn) rather than temperature-dependent mitosis. Species with low values undertake growth later, in the warmer months, largely by cell division. The mechanism of growth by cell expansion, at least in the grasses (Poaceae), the Liliaceae (and relatives) and among the Asteraceae, has been further correlated with the concentration of a soluble vacuolar carbohydrate, fructan, which largely replaces starch in these species. Thus, two woodland species with late winter–early spring growth (bluebell, *Hyacinthoides non-scriptus* (2C DNA 42.4 pg) and tufted hair-grass, *Deschampsia caespitosa* (2C DNA 18.0 pg)) have high C-values and store high concentrations of fructan in the vacuoles. In late winter the fructan reserve is rapidly depolymerized so increasing the number of molecules (regardless of size) of fructan in the vacuole, raising the osmotic potential and drawing water into the cell, which then inflates causing the plant to grow. Starch-storing species tend to have lower C-values and to delay growth (by mitosis) until the warmer months. The paradox associated with C-values is that none of this really explains why such species have such high amounts of DNA. Also, there is the lack of correlation between C-values and the presumed amount of genetic information contained within genomes: interspecific variation in genome sizes among EUKARYOTES seems to bear no relationship to organismic complexity. [G.A.F.H. & I.O.]

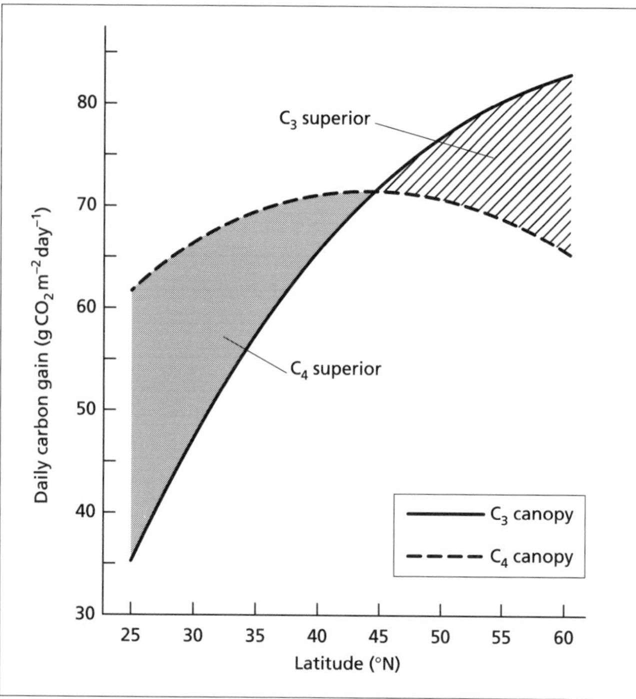

Fig. C1 Predicted levels of photosynthesis for C_3 and C_4 species over a range of latitudes in the Great Plains during July. The C_4 advantage is lost in latitudes greater than 45°N. (After Ehleringer, J.R. (1978) Implications of quantum yield differences on the distribution of C_3 and C_4 grasses. *Oecologia* **31**, 255–267.)

REFERENCES

Bennett, M.D. (1987) Variation in genomic form in plants and its ecological implications. *New Phytologist* **106** (Suppl.), 177–200.

Hendry, G.A.F. (1987) The ecological significance of fructan in a contemporary flora. *New Phytologist* **106** (Suppl.), 201–206.

CA model *See* CELLULAR AUTOMATA.

cadmium (Cd) A soft, HEAVY METAL (ATOMIC NUMBER = 48; relative ATOMIC MASS = 112.40; melting point = 320.9°C; boiling point = 765°C) that is a trace constituent (0.1 mg kg^{-1}) of the Earth's CRUST. Found as an impurity in ZINC minerals such as sphalerite (ZnS) and calamine ($ZnCO_3$) (with cadmium contents of up to 0.5% and 5% respectively), cadmium is used in corrosion prevention, POLYMER stabilization and electronics. A biologically non-essential element, cadmium is rather labile and bioavailable in soils, to which major contributors of the element are phosphatic FERTILIZERS, industrial EMISSIONS and SEWAGE SLUDGE.

Cadmium can cause kidney damage and, in extreme cases, skeletal deformities, such as occurred in the 1950s in Japanese people feeding on rice grown in paddy fields contaminated by cadmium from lead–zinc mining. This has become known as *itai-itai* (ouch-ouch) disease because of the excruciating pain. As cadmium exhibits similar properties to the ESSENTIAL ELEMENT zinc, it can displace the latter from many of the zinc-containing enzymes. Cadmium is often found associated with metallothionein, a soluble low-RELATIVE MOLECULAR MASS protein with a large number of sulphydryl (SH) groups. Metallothionein can help to reduce the TOXICITY of cadmium, as can a high zinc intake. Unfortunately, a high cadmium intake can aggravate zinc deficiency. Cadmium is also implicated as a contributory factor in hypertension and cigarette smoking is believed to be a major source of human exposure to the element. [J.F.]

REFERENCE

Nriagu, J.O. (ed.) (1980) *Cadmium in the Environment.* John Wiley & Sons, New York.

caenogenesis *See* NON-TERMINAL ADDITION.

caesium (Cs) Caesium is a soft, silver-white METAL (ATOMIC NUMBER = 55; relative ATOMIC MASS = 132.91; melting point = 28.5°C; boiling point = 705°C), with an average crustal abundance of 2.6 mg kg^{-1}. Caesium occurs in the +1 oxidation state and is soluble in natural waters. CLAYS, especially illites, adsorb caesium by ION exchange. ^{133}Cs is the only stable caesium ISOTOPE. The radioisotopes ^{134}Cs ($t_{1/2}$ = 2.06 years), ^{135}Cs ($t_{1/2} = 3 \times 10^6$ years) and ^{137}Cs ($t_{1/2}$ = 30 years) occur as contaminants from the nuclear industry. ^{137}Cs was also a major constituent of weapons-testing fallout. *See also* RADIOACTIVE WASTE; RADIONUCLIDE. [A.B.M.]

calcic horizons *See* SOIL CLASSIFICATION.

calcicole A plant or organism limited to, or more abundant on, calcareous soils or in water of high CALCIUM status (*cf.* CALCIFUGE). Such soils are usually alkaline, exceeding pH 7. Calcicolous plants may have a high demand for calcium, have uptake mechanisms allowing IRON absorption at high pH and may be excluded from ACID soils by ALUMINIUM toxicity and possibly MANGANESE and iron toxicity. They may also suffer NITROGEN deficiency if their seeds germinate in acid soils where nitrification is impeded. Calcicole and calcifuge species are not rigidly exclusive categories but form a continuum of physiological variation. [J.R.E.]

calcification The process whereby CALCIUM and carbonate ions are combined to form calcareous skeletal materials. The calcium carbonate COMPENSATION DEPTH occurs at the depth in the sea at which calcium carbonate dissolves because of high concentrations of CARBON DIOXIDE (CO_2) in the water (between about 3500 and 6000 m). Calcification also refers to the deposition of calcium and other soluble salts in soils where EVAPORATION greatly exceeds PRECIPITATION. [V.F.]

calcifuge A plant or organism limited to, or more abundant on, soils or in water of low CALCIUM status and usually of pH 5 or less (*cf.* CALCICOLE). Calcifuges do not survive on high pH soils as they fail to absorb IRON and sometimes MANGANESE. Excessive uptake of calcium in such soils may also interfere with the metabolism of iron and a common symptom of this failure, combined with iron deficiency, is LIME-induced chlorosis. Calcifuges are exceptionally tolerant of high concentrations of ALUMINIUM, iron and manganese and they are often able to assimilate NITROGEN as the NH_4^+ ion. [J.R.E.]

Calcisols *See* SOIL CLASSIFICATION.

calcium (Ca) Calcium is a soft, silver-white METAL (ATOMIC NUMBER = 20; relative ATOMIC MASS = 40.078; melting point = 839°C; boiling point = 1484°C). It has an abundance of 4.66% and is found as vast deposits of sedimentary calcium carbonate ($CaCO_3$). Calcium metal is produced by electrolysis of fused calcium chloride ($CaCl_2$). Important compounds include calcium oxide and hydroxide (CaO and $Ca(OH)_2$, used as a flux to remove PHOSPHORUS, SULPHUR and SILICON in the manufacture of steel, and as scrubbers of FLUE GASES from power-stations) and calcium carbonate ($CaCO_3$, used as an antacid and as an abrasive in toothpastes). Calcium salts play a crucial role in biological systems in the formation of structural materials such as bone and shell, and also as ions which trigger nerve signals. [C.R.P.]

calibration Translation of units of measurement recorded by instrument or system into units of interest and/or operation for determining values or errors of instrument or system. Thus ATTENUATION of LIGHT at a particular wavelength is often used to measure concentration of a substrate with appropriate optical properties in metabolic reactions. Calibrations are carried out with standards of known concentration. Calibration can also result in a correction factor or factors that can subsequently be used to correct readings from the system. [P.C.]

californium (Cf) Californium is a synthetic element (ATOMIC NUMBER = 98; relative ATOMIC MASS = 251; melting and boiling points unknown) that was discovered in 1950. Isotopes with mass numbers ranging from ^{239}Cf to ^{256}Cf have been generated by particle bombardment of other actinides, but only very small quantities have been produced. Consequently, californium does not occur in detectable quantities in the environment. Investigations of the properties of californium have mainly used ^{252}Cf ($t_{1/2}$ = 2.645 years), produced by multiple NEUTRON capture. ^{252}Cf undergoes decay by α-PARTICLE emission (96.91%) and spontaneous FISSION (3.08%) and can be used as a neutron source. The dominant oxidation state of californium is +3. [A.B.M.]

calorific value The calorific value of a material is a measure of its energy content expressed in JOULES per unit mass. Hence, it is probably more properly termed a 'joule equivalent'. It is most frequently determined by combustion in a BOMB CALORIMETER, which monitors the rise in TEMPERATURE when a known mass of dried material is ignited in an oxygen-enriched atmosphere. The instrument is calibrated with benzoic acid of which the calorific content is accurately known. Estimating the calorific values of materials enables us to express different components of FEEDING and PRODUCTIVITY equations in a common currency of energy units. *See also* ECOLOGICAL ENERGETICS. [M.H.]

CAM

An abbreviation for crassulacean acid metabolism, a CARBON-assimilation pathway now known to occur in a wide and unrelated range of families, including the eponymous Crassulaceae and extending to the Orchidaceae, Bromeliaceae, Cactaceae and many others. DROUGHT-tolerant CAM plants have a reversed stomatal cycle in which CARBON DIOXIDE (CO_2) is trapped at night via the C_4 pathway, producing the four-carbon compound malate, which is stored in the mesophyll-cell vacuoles until daytime.

At dawn, the stomata close and the malate is decarboxylated to release CO_2, which is assimilated by the C_3 photosynthetic cycle (Calvin cycle) of the chloroplasts. Water loss per unit of photosynthate is consequently much less than that of C_3 and C_4 plants (*see* C_3/C_4 PLANTS) because these photosynthesize with open stomata during the daytime when the WATER POTENTIAL gradient between leaf interior and atmosphere is much greater than at night. Drought-tolerant CAM plants usually have succulent tissues in which the mesophyll-cell vacuoles are much larger than normal, providing STORAGE volume for the nocturnally synthesized malate (Fig. C2).

Though CAM is most commonly associated with drought tolerance, it may also be interpreted as a mechanism that increases the efficiency of carbon trapping in the face of any environmental limitations on CO_2 availability, including the stomatal closure necessary to limit water loss. For example, some CAM plants are astomatal submerged aquatics, in which the high affinity of CAM for CO_2 permits its scavenging from the very low concentrations in water of low CALCIUM status. Aquatic CAM was first described in the pteridophyte *Isoetes* but is now known in a variety of submerged angiosperms. A very few cases of astomatal CAM are found in terrestrial plants, for example in the valamen tissue of the aerial roots of some epiphytic orchids where undue CO_2 loss is controlled by a much slower than normal decarboxylation of malate, which limits the CO_2 diffusion gradient between the interior of the valamen cells and the external atmosphere.

CAM plants are less photosynthetically productive than C_3 and C_4 plants and, predictably, some plants that usually photosynthesize in the C_3 mode are able to switch into CAM when water-stressed by drought or exposure to salt (Fig. C2).

The shared ecological characteristic of CAM plants is the highly carbon-conservative nature of CAM, which permits NICHE extension into habitats where CO_2 availability is limited by sporadic

Fig. C2 The hottentot fig (*Carpobrotus edulis*, Mesembryanthemaceae), a facultative CAM plant growing on a foreshore in northern Portugal. (Courtesy of R.M.M. Crawford.)

drought or physical–chemical factors in aquatic systems. CAM also permits recycling of respiratory CO_2 produced at night or during permanent stomatal closure under severe drought STRESS, further enhancing carbon conservation. The typical habitat of CAM plants is associated with recurrent climatic drought of relatively short duration or, in the special cases of EPIPHYTES and rock-crevice plants, drought caused by limited water-storage capacity of the rooting medium (Fig. C3). *See also* PHOTOSYNTHESIS. [J.R.E.]

---

cambic horizons *See* SOIL CLASSIFICATION.

Cambisols *See* SOIL CLASSIFICATION.

camouflage Concealing coloration (e.g. MELANISM) and morphology (e.g. stick insects) as defence against predation. [P.C.]

Canada, conservation in With the second largest landmass in the world, Canada contains significant portions of the world's BOREAL, WETLAND and ARCTIC ecosystems. Conservation in Canada has repercussions for tempering global CLIMATE CHANGE and other processes. As a ratifier of the Biodiversity Convention (*see* RIO SUMMIT; UNCED), Canada prepared a strategy for meeting its global and national obligations for the conservation of BIODIVERSITY. Completing a system of protected areas, encompassing 12% of Canada, is an integral element of the national strategy.

Public demands continue to make industry and governments face the challenges of conservation. Industries lead in devising and adopting codes of practices and statement of principles. Some of these industrial changes, AIR-POLLUTION abatement, soil and water conservation practices have brought economic benefits. Governments in Canada, at the federal, territorial, provincial and municipal levels, directly manage over 90% of Canada's land, air and water. The remaining is managed indirectly through legislation, policy formulation or financial contributions. In all jurisdictions, ecological-based management is increasingly recognized as the means of achieving higher levels of conservation.

However, contravening programmes or policies, flowing from overlapping government jurisdictions, poses the greatest challenge to conservation. Biodiversity loss is often the outcome of policies to encourage development or even to maintain unrealistic production quotas. The recent collapse of the Atlantic cod fishery provides a dramatic example.

In 1994, Canada proclaimed the Canadian Environmental Assessment Act. Under this law, projects are subject to environmental reviews if they affect federal jurisdictions, funding and or include federal land transfers (including the offshore). All provincial and territorial governments have similar environmental review policies. *See also* CONSERVATION BIOLOGY. [N.L.]

Fig. C3 The Joshua tree (*Yucca brevifolia*), one of the larger species to show CAM metabolism, growing in the Mohave desert. (Courtesy of R.M.M. Crawford.)

canalization Coined by C.H. Waddington, in recognition of the fact that developmental processes resist DISTURBANCES from the external environment and from within (e.g. MUTATIONS). Development is conceptualized as a ball rolling down canals or chreods. The depths of the canals (the extent of canalization) determines the extent to which the ball can be deflected (Fig. C4). An alternative term is developmental HOMEOSTASIS. Environmental canalization reduces the phenotypic effects of environmental disturbance and random events during development (developmental noise). GENETIC CANALIZATION buffers the PHENOTYPE against genetic disturbances caused by mutations, including those caused by insertion of viruses and P-elements. Canalization is expected to evolve for TRAITS that are under strong STABILIZING SELECTION. [P.C. & T.J.K.]

REFERENCES

Rendel, J.M. (1967) *Canalisation and Gene Control.* Logos Press, London.
Waddington, C.H. (1975) *The Evolution of an Evolutionist.* Edinburgh University Press, Edinburgh.

canalized reaction norm REACTION NORM denotes the systematic change of the GENOTYPIC VALUE of a GENOTYPE under a systematic change in environment. The sensitivity to minor environmental DISTURBANCE might be negligible or absent, leading to a very strict relation between environment and PHENOTYPIC VALUE of a genotype: the reaction norm for the genotype is buffered or canalized. Moreover, the differences in expression between the genotypes might have been minimized by optimizing selection, leading to phenotypic expression over environments that is similar for all genotypes. *See also* CANALIZATION; PHENOTYPIC PLASTICITY; QUANTITATIVE GENETICS. [G.D.J.]

Cannibalism

The killing and consumption of another member of the same species is widespread and often common. It has been reported in species as

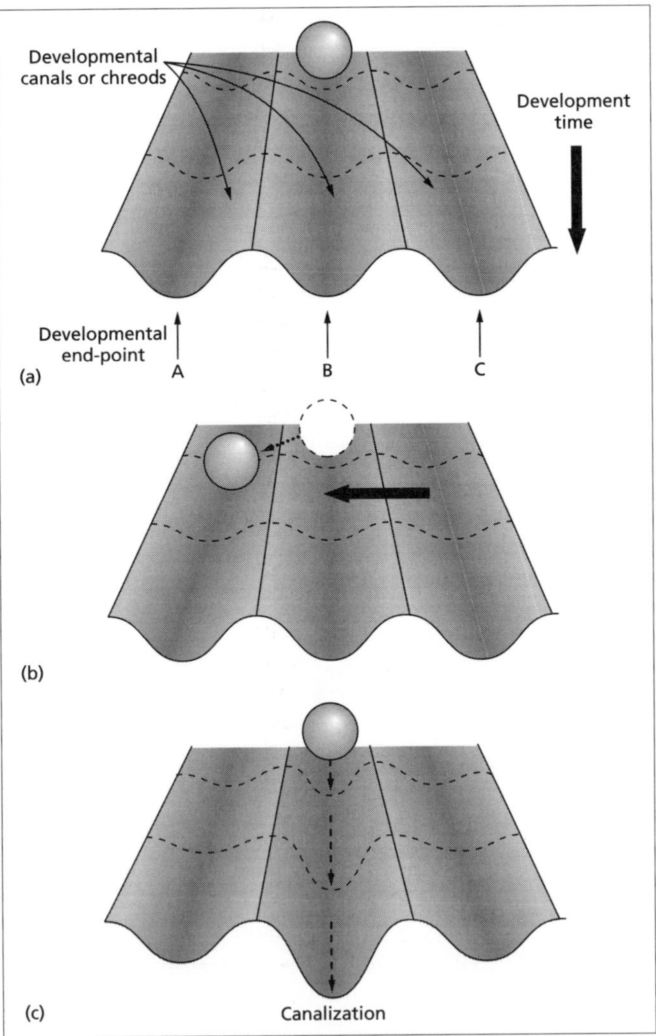

Fig. C4 (a) Waddington's model of development, conceptualized as a ball moving down a canal or chreod to an end-point. (b) Environmental disturbances (horizontal black arrow) can deflect development to an alternative end-point. (c) Selection can deepen canals and ensure that one end-point is favoured. (After Calow, P. (1983) *Evolutionary Principles*. Blackie & Son, Glasgow.)

diverse as protozoans and mammals, and occurs for many different reasons. Adults may cannibalize their offspring to reduce COMPETITION between their young. Juveniles may eat competing siblings or unrelated eggs and juveniles. More rarely, adult females may attempt to kill and eat courting males, perhaps as a means of assessing male quality.

Cannibalism is a potentially risky behaviour, since the intended victim can be similar to the prospective assailant. Consequently, there is usually an age, size or life-history asymmetry between cannibal and victim; most cannibalism occurs when there are large differences in vulnerability. It is more common for adults to eat juveniles or eggs than other adults; when adults do cannibalize other adults, they are more likely to eat smaller victims. Cannibalism is also costly, as it may increase the risk of PARASITISM, reduce inclusive fitness (*see* KIN SELECTION) and counteract the benefits of sociality. Parasites are more likely to be acquired by eating members of the same species than eating members of another species. Eating relatives, such as during brood reduction or siblicide, may reduce inclusive fitness and cannibalistic behaviour is frequently directed away from close KIN. The benefits of sociality may also be reduced by cannibalism. For example, the advantages of group-living for gulls or larval willow leaf beetles (*Plagiodera versicolora*) decline with intra-group cannibalism. Finally, sexually cannibalistic females that consume males before copulation risk losing their reproductive potential.

Cannibalism may be beneficial simply as a means of surviving periods of food shortage, particularly when cannibalism is directed at non-kin. Such behaviour has the additional advantage of reducing competition for the limited resource. Cannibalism of offspring enables a parent to redirect its REPRODUCTIVE EFFORT during food shortages. Reabsorbing offspring with low survival prospects enables the parent to increase its future reproductive potential when circumstances improve. Cannibalism also allows the parent to direct its reproductive effort selectively to offspring with the highest value, and to reduce intra-sibling competition. Similarly, siblicide in sharks and birds may be tolerated by parents as a means of provisioning cannibalistic offspring with more RESOURCES. Post-mating sexual cannibalism in spiders and other invertebrates is also a possible means by which the male can contribute to the FECUNDITY of the female and hence increase his reproductive output.

More direct reproductive competition favours cannibalism in a number of taxonomic groups. INFANTICIDE in mammals enables a new male to sire offspring more rapidly and redirect female resources towards his offspring rather than those of another male. Similarly, female fish consume the eggs of other females that are laid in male-tended nests in order to receive the maximum paternal care for their offspring. This competition for care-giving may favour cannibalism in many cooperative breeders. Female acorn woodpeckers (*Melanerpes formicivorus*) and both carpenter and euglossine bees cannibalize the eggs or offspring of co-breeding females in order to establish reproductive dominance, or to prevent another female establishing reproductive dominance. Similar behaviour occurs in cooperatively breeding mammals, such as African hunting dogs (*Lycaon pictus*) and prairie dogs (*Cynomys*), although such infanticide does not always involve cannibalism.

Cannibalism clearly has significant behavioural consequences. Cannibalism of young by non-parents, such as in seabirds, sticklebacks and rodents, is an additional source of predation and increases the necessity for PARENTAL CARE. Some of the elaborate COURTSHIP strategies employed by male spiders may be related to the risk of cannibalism by the female. Cannibalism of offspring, either by other adults or siblings, may have initiated the evolution of trophic eggs or nurse eggs or the production of excess young. In an extreme example, the Australian social thomisid spider (*Diaea socialis*) retains trophic oocytes within her body as a food reserve for her offspring when they cannibalize her during the winter food shortage. Reabsorption of embryos by rodents in response to unfamiliar males may reflect the anticipation of cannibalism and allow female rodent mothers, rather than the infanticidal male, to obtain the nutritive benefits. Cannibalism is not only responsible for the deployment of many avoidance strategies, but may be an important mechanism favouring the adoption of COOPERATIVE BREEDING strategies. The frequency with which cannibalism occurs in subsocial insects and cooperatively breeding vertebrates suggests that the reproductive suppression observed in eusocial species and many obligate co-operative breeders may be the result of the avoidance of cannibalism by subordinate individuals.

Cannibalism may play an important role in the dynamics of natural populations. While POPULATION DENSITY and food availability clearly affect the frequency of cannibalism in many laboratory and some natural populations, there is less information on the impact cannibalism has on population structure and FLUCTUATIONS. Cannibalism by flour beetle (*Tribolium*) larvae, for example, causes large OSCILLATIONS in POPULATION SIZE and structure, depending upon the age of the larvae and their ability to cannibalize. In other species, cannibalism of juveniles by adults tends to have a stabilizing effect on the population. Clearly these effects on the dynamics of natural populations depend upon the frequency of cannibalism, for which few data are available. [D.C. & M.A.E.]

REFERENCE

Elgar, M.A. & Crespi, B.J. (eds) (1992) *Cannibalism: Ecology and Evolution among Diverse Taxa.* Oxford University Press, Oxford.

---

canonical correlation A multivariate statistical method. Most of the statistical methods described in detail in this encyclopedia are univariate because they are concerned with single variables or observations. Even the complex methods of ANALYSIS OF VARIANCE are only concerned with explaining the variation in one variable. Multivariate methods analyse several related variables or observations simultaneously. For further details see Manly (1986). *See also* STATISTICS. [B.S.]

REFERENCE

Manly, B.F.J. (1986) *Multivariate Statistical Methods: A Primer.* Chapman & Hall, London.

canonical ordination *See* ORDINATION.

canopy The upper stratum of a WOODLAND or FOREST, comprising trees receiving full daylight over all or part of their crowns. The canopy is uniform if it comprises trees of approximately the same heights, or irregular if it is a mixture of trees of different heights, including 'emergents', i.e. trees markedly taller than general canopy level. [G.F.P.]

Cape floral region This is the smallest of the six kingdoms recognized by phytogeographers and is sometimes referred to as the South African kingdom, although Cape kingdom is to be preferred (see Fig. B6). It extends from Capetown northwards for only about 100 km and eastwards for 700 km. The cool dry CLIMATE of the region is bordered on the north by DESERT, and to the east and north-east by tropical GRASSLAND and FOREST. The rainfall is low and strongly seasonal, being largely confined to the Southern hemisphere winter. There are seven families of flowering plants endemic to the kingdom and many endemic genera and species. Of families familiar elsewhere, there are many endemic Aizoaceae, Ericaceae, Asclepiadaceae and Proteaceae. The latter family, peculiar to this region and Australasia, is one of the interesting links between those two regions, usually attributed to their earlier juxtaposition as part of GONDWANA. Despite the large number of endemics in the flora and the distinctive character of the vegetation, there is no recognizable zoological counterpart to the botanical Cape kingdom (compare Figs B5 and B6). [W.G.C.]

capital breeder An organism that uses stored energy for reproduction. The alternative is an INCOME BREEDER, that uses energy acquired during the reproductive episode. Examples of capital breeders include many species of temperate-zone fish and large mammals, which store energy in various body organs and then deplete these for the production of EGGS (fish) or to support lactation (mammals). *See also* COSTS OF REPRODUCTION; LIFE-HISTORY EVOLUTION. [D.A.R.]

captive breeding Breeding of animals, usually ENDANGERED SPECIES, in captivity, normally with a view to release back to nature and hence to the conservation of the species. This highlights the difference in approach in conservation versus preservation. The latter would breed in captivity without attempts to reintroduce species back to their natural habitats. Zoos have often emphasized preservation, but increasingly have become involved with programmes of conservation. *See also* CONSERVATION BIOLOGY; ZOO. [P.C.]

Capture–recapture techniques

When animals are marked, released and recaptured or otherwise re-examined, information may be obtained about their behaviour. For instance, distance moved may be estimated by recording displacement from release point to recapture point. A widely used application is for estimation of POPULATION SIZE. If a SAMPLE is taken, marked and released, marked and unmarked animals behave in the same way and the population is closed (i.e. with no input or loss), then the ratio of marked individuals to total population size is equal to the ratio of marked recaptures to total size of a second sample. For a population of unknown size P, if a sample n_1 is marked and released and there are m marked recaptures in a second sample n_2, then the population estimate is $P=n_1n_2/m$. It is not essential for the same effort to be put into capturing the two samples. This principle has been known at least since the time of Laplace who, in 1783, used it to estimate the population of France from the register of births and the number of births recorded in a sample of parishes of known size. Later, C.G.J. Petersen (1889 and later) used the mark and release technique in studies of fish populations and F.C. Lincoln (1930) in studies of wildfowl. The expression given above is commonly known as the LINCOLN INDEX or the PETERSEN ESTIMATE.

Since the population should be closed for application of the Lincoln index, the estimation refers both to the time when the first release occurred and the time of the second capture. In most populations there is likely to be loss (death, EMIGRATION) or gain (birth, IMMIGRATION) or both. The ratio in the second sample is not influenced by loss alone so long as marked and unmarked individuals are affected in the same way, so that the population at the first capture is correctly estimated, but afterwards the population declines by an unknown amount. With gain alone, the ratio is affected by dilution of the second sample by new unmarked individuals. The population size at the second SAMPLING is estimated and that at the first sampling is unknown. When there is both loss and gain, as is usually the case, an unreal estimate is made that refers to neither the first nor the second sampling time and can be quite unrepresentative. To give a numerical example, suppose 100 individuals are marked out of a population of 500, and a second sample examines 10% of the total. If sampling were exactly proportional, the second sample would contain 10 marked and 40 unmarked individuals. The estimate would be $P=100\times50/10=500$, the number on the first and second occasions. Now, imagine that only 60% survive (whether marked or unmarked), so that 300 are left. The estimate is $P=100\times(50\times0.6)/(10\times0.6)=500$, the number at the start. Alternatively, no individuals die but 150 enter. In that case, $P=100\times(50+15)/10=650$, the final number. If both events occurred together the second population would be 450, but $P=100\times(50\times0.6+15)/(10\times0.6)=750$, a number referring to neither occasion. The way round this problem is to estimate either the rate of loss or the rate of gain. If one or other can be done, then a suitable adjustment can be made to provide a correct estimate of population size. If SURVIVAL RATE were known for the numerical example, we could calculate $750\times0.6=450$, which is the number present on the second occasion. In order to estimate change, at least one more sampling must be made. On the third capture, individuals marked on the second occasion must be distinguishable from those marked on the first by means of distinct marks (it is not necessary for individuals to be identifiable individually). For this reason population estimations are usually based on at least three, and commonly many, samplings. N.T.J. Bailey (1951) was one of the first to state this explicitly and he provided a method for three sampling periods, known as the triple-catch method (it is not, however, the most efficient way of using the available data).

Multiple sampling techniques were used by C.H.N. Jackson in the late 1930s to estimate numbers of tsetse flies in the field. For practical reasons he had some estimations where a single release was made followed by several recaptures (and loss was estimated), while in others several releases were made on different occasions with distinct marks but all recapture took place at a final session (and gain was estimated). R.A. Fisher and E.B. Ford (1947) developed Jackson's techniques and produced an analysis in which capture, marking and recapture takes place on a sequence of occasions, survival is assumed to be constant and is estimated, and population estimates are obtained on all occasions but the first. The method provides a robust estimation procedure, applicable so long as constancy of loss rate throughout the period may be assumed. G.M. Jolly and G.A.F. Seber (1965) pointed out independently that it was often unreasonable to assume constancy of loss and that more efficient statistical estimation methods may be used than had hitherto been applied. These methods use only the final record of previous captures of a marked animal, whereas earlier methods recorded total number of previous captures among animals caught at each sampling (i.e. marks rather than individuals were employed in the calculation). Using the Jolly–Seber method, the population estimate is made for all but the first and last sampling occasions and estimates of loss and gain between sampling occasions are obtained, all with STANDARD ERRORS.

This might be the last word in estimation methods, except that if a parameter can realistically be assumed to be constant there is a loss of efficiency when it is treated as a variable. The standard error associated with a Lincoln index estimate of population is smaller than that of a Jolly estimate, but it is useless unless the restrictive assumption of a closed population is applicable. In that case, but only then, it should be used in preference to the Jolly–Seber method. Apart from loss rate, which may sometimes be constant and sometimes variable, the PROBABILITY of capture may vary during the sampling period. Jolly (1982) provides efficient estimates for the four combinations of assumptions, called methods A to D. Method A, the original Jolly–Seber method, assumes both loss and probability of sampling to be variables, while method D assumes both to be constant; one or other is constant in methods B and C. These methods cover all the general estimation conditions. Other methods are available that have been developed for specific situations and much work has been done to test the ROBUSTNESS OF MODELS when assumptions are incompletely met, often using computer simulations. *See also* ABSOLUTE POPULATION ESTIMATES; ABUNDANCE; BAILEY'S TRIPLE-CATCH METHOD; SAMPLING METHODOLOGY/DEVICES. [L.M.C.]

REFERENCES

Bailey, N.T.J. (1951) On estimating the size of mobile populations from recapture data. *Biometrika* **38**, 293–306.

Brownie, C., Anderson, D.R., Burnham, K.P. & Robson, D.S. (1985) *Statistical Inference from Band Recovery Data—A Handbook.* US Department of the Interior, Washington, D.C.

Burnham, K.P., Anderson, D.R., White, G.C., Brownie, C. & Pollock, K.H. (1987) *Design and Analysis Methods for Fish Survival Experiments Based on Release–Recapture.* American Fisheries Society, Bethesda, Md.

Jolly, G.M. (1982) Mark–recapture models with parameters constant in time. *Biometrics* **38**, 301–321.

Seber, G.A.F (1973) *The Estimation of Animal Abundance and Related Parameters.* Griffin, London.

Seber, G.A.F. (1986) A review of estimating animal abundance. *Biometrics* **42**, 267–292.

Skalski, J.R. & Robson, D.S. (1992) *Techniques for Wildlife Investigations: Design and Analysis of Capture Data.* Academic Press, New York.

An introduction to the methods will be found in:

Begon, M. (1979) *Investigating Animal Abundance: Capture–Recapture for Biologists.* Arnold, London.

Blower, J.G., Cook, L.M. & Bishop, J.A. (1981) *Estimating the Size of Animal Populations.* Allen & Unwin, London.

---

carbohydrates Carbohydrates are common natural substances synthesized by plants and animals, with a general formula $(CH_2O)_n$ where *n* can be from three to many thousands. The smallest units are monosaccharides such as glucose and fructose ($n=6$), ribose ($n=5$) and glyceraldehyde ($n=3$). Glucose is the product of PHOTOSYNTHESIS, and other CARBON compounds in the plant are synthesized from it. The C_5 and C_6 units exist in a five- or six-sided ring structure. These basic units can be joined together to make oligosaccharides (such as sucrose, which is glucose and fructose; and maltose, which is two glucose units). In plants, oligosaccharides are the form in which carbohydrate is transported, and so they are found in the stems especially (ordinary sugar, sucrose, is usually obtained from the stems of sugar cane).

Large chains (known as polysaccharides) are formed from numerous C_6 units. The most well-known and widespread example is starch, which accumulates as a food store in the roots, tubers, stems and leaves of plants, especially those species such as rice, potato and manoic which form the staple diet of humankind. Starch is composed of amylose, unbranched chains of glucose units with RELATIVE MOLECULAR MASSES from 4000 to 50 000, and amylopectin, where the glucose chains are branched and joined. In two floral families, Asteraceae and Poaceae, starch is often replaced by inulin, made of fructose units. CELLULOSE is a structural carbohydrate forming cell walls and a constituent of wood. It is also composed of linked glucose chains and has a relative molecular mass of up to 500 000. Glycogen is a polysaccharide found in the muscles of animals, and resembles amylopectin. [J.G.]

carbon (C) A non-metallic element (ATOMIC NUMBER = 6; relative ATOMIC MASS = 12.011; melting point = 3500°C; boiling point = 4827°C) widespread in the universe, occurring in the Sun, stars, planetary atmospheres and in some meteorites. There are two stable ISOTOPES, ^{12}C and ^{13}C, and four radioactive isotopes, ^{10}C, ^{11}C, ^{14}C and ^{15}C. The most common isotope is ^{12}C (98.9% of carbon in CO_2 on Earth) and is used as the STANDARD for atomic mass. ^{14}C has a $t_{1/2}$ of 5715 years and is commonly used to age archaeological specimens and as a radiotracer in biology.

Allotropic forms of carbon include the crystalline materials graphite and diamond, and the amorphous forms such as charcoal, carbon black and coke. Recently a new crystalline form has been synthesized, fullerine, which is a C_{60} spherical cage with no dangling bonds; it occurs naturally in rocks in Russia.

Graphite is soft (specific gravity = 2.3) and used as a lubricant. Diamond is hard (specific gravity = 3.5) and used in cutting and polishing as well as for gems. Charcoal, carbon black and coke are prepared by the ignition of carbon-containing materials such as BIOMASS, oil, natural gas and coal in the absence of OXYGEN. They have a remarkably high surface area per mass, for example carbon black used for pigments has a value of 300–500 $m^2 g^{-1}$. They have a variable density and many uses, for example adsorption of materials in the gas and liquid phases (*see* ACTIVATED CHARCOAL) and for reinforcement of RUBBER.

Carbon is usually tetravalent and forms nearly 10 million compounds, including CARBON DIOXIDE (CO_2), CARBON MONOXIDE (CO), METHANE (CH_4), ethylene (C_2H_4), BENZENE (C_6H_6), ETHANOL (C_2H_5OH), acetic acid (CH_3COOH), chloroform ($CHCl_3$) and carbon tetrachloride (CCl_4). It is the basis of life, occurring in macromolecules such as CELLULOSE, LIGNIN, PROTEINS and nucleic acids. Plants are typically 45–50% carbon, and on Earth the above-ground VEGETATION represents a stock of about 610 Gt carbon. The SOIL has 1500 Gt carbon in the form of organic matter derived from plant DECOMPOSITION and the ATMOSPHERE contains 750 Gt carbon as CO_2. The oceanic stock is about 1000 Gt carbon near the surface with 38 000 Gt in deeper waters (see Fig. C5). Large amounts of carbon occur as deposits of gas, oil, coal and carbonate rocks. The stocks of carbon are connected to each other by processes such as PHOTOSYNTHESIS and decomposition to form the CARBON CYCLE. The atmospheric pool of CO_2 turns over in about 4 years. [J.G.]

Carbon cycle

The carbon cycle is one of the four key BIOGEOCHEMICAL CYCLES on Earth: CARBON, NITROGEN, SULPHUR and PHOSPHORUS. All these

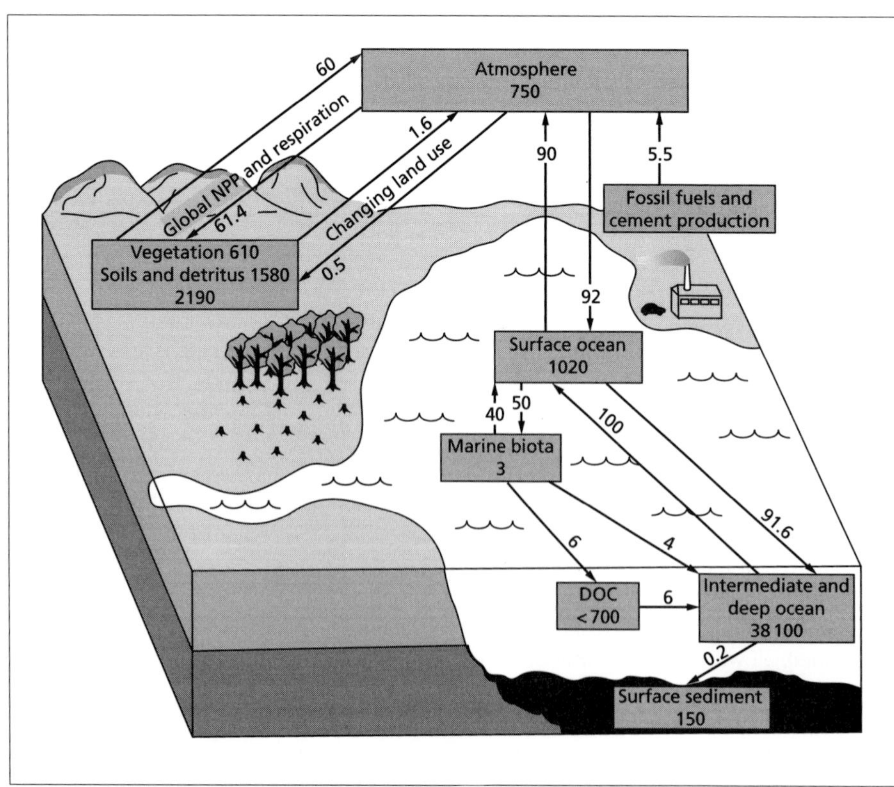

Fig. C5 The global carbon cycle, showing the reservoirs (Gt) and fluxes (Gt year^{-1}) as annual averages over the period 1980–1989. DOC, dissolved organic carbon; NPP, net primary production. By 1996 the fossil fuel emissions had increased to 6.6 Gt of carbon per year. (After Houghton, J.T., Meira Filho, L.G., Bruce, J. *et al.* (eds) (1995) *Climate Change 1994: Radiative Forcing of Climate Change and an Evaluation of the IPCC IS92 Emission Scenarios*. Cambridge University Press, Cambridge.)

ELEMENTS are cycled around the globe between different components (also known as reservoirs or compartments) by physical, chemical and biological processes. Carbon in particular is a key element in life on Earth, being a component of an enormous range of organic compounds, formed because of the stable, long-chain covalent bonding possibilities of the carbon atom. Elemental carbon is rare (diamonds, graphite) and the commonest carbon compounds are the oxidized forms of CARBON DIOXIDE (CO_2) and carbonate. The organic carbon compounds have been reduced by the action of organisms, for example in PHOTOSYNTHESIS. Carbon is exchanged between the major reservoirs of carbon: the ATMOSPHERE, the OCEANS and FRESHWATER (although the latter is only a very small component), the terrestrial BIOSPHERE, and the sediments and sedimentary rocks. On very long GEOLOGICAL TIME-SCALES, sedimentary rocks are exposed, allowing WEATHERING to return CO_2 to the atmosphere. However, the faster parts of the cycle are dominated by biological activity, removing carbon in the form of CO_2 from the atmosphere and from solution into living organisms (BIOMASS) during photosynthesis and releasing it during DECOMPOSITION (particularly the activity of microorganisms) as CO_2 and METHANE (CH_4) (Fig. C5).

The atmospheric carbon content is largely contained in CO_2 (currently at about 360 ppm by volume (ppmv), amounting to some 747 Gt carbon) with trace amounts of CH_4 (1.8 ppmv, representing 3 Gt carbon) and even smaller concentrations of CARBON MONOXIDE (CO, varying from about 0.05 to 0.20 ppmv). Atmospheric CO_2 is removed by photosynthetic uptake by vegetation both on land and in water, and by solution in the oceans and freshwater. CO_2 is returned to the atmosphere through oxidative processes such as plant and animal RESPIRATION, decomposition in SOILS and burning of vegetation. The oceans contain about 50 times as much carbon as the atmosphere in four forms: dissolved organic carbon (DOC), dissolved inorganic carbon (DIC), particulate organic carbon (POC) and marine organisms (plant and animal). About 95% of the ocean carbon is DIC, and at the pH of seawater (about 8) most of this is in the form of bicarbonate (HCO_3^-) and carbonate (CO_3^{2-}). At the temperature and salinities present, the ocean is well 'buffered', so that only a 1% change in total inorganic carbon would occur for a 10% change in atmospheric CO_2 concentration. Marine organisms (mainly PHYTOPLANKTON and ZOOPLANKTON) are very important, even though they only contain about 3 Gt carbon globally, because they are concentrated in the upper tens of metres, the PHOTIC ZONE, and distributed very variably over the oceans, with large amounts in the coastal upwelling regions where other NUTRIENTS are available to support growth. The PRODUCTIVITY of these coastal zones is some 30% of the whole, even though the area comprises only some 8% of the total. Organic material that falls into deeper water (e.g. below 100 m) tends to be composed of very long-lived (thousands of years) chemically resistant compounds. The deep-water organic and inorganic carbon content is relatively isolated from the surface, slow exchange taking place largely through vertical movements of the THERMOCLINE and large-scale mass mixing of water in currents.

The terrestrial biosphere contains large amounts of carbon in the live vegetation and in the soil organic matter, i.e. decomposing plant material, which contains over twice as much carbon as the living material. As the type, density and productivity of vegetation varies greatly over the world due largely to CLIMATE, and in particular to varying seasonal temperature and PRECIPITATION regimes, the carbon content of the terrestrial biosphere is highly variable and this causes some uncertainty in the exact amount, but it is about three times that in the atmosphere. Major BIOMES with the largest amounts of carbon are the broad-leaved humid forests, the TAIGA, the TUNDRA and the warm GRASSLANDS; the latter contain nearly 10 times as much carbon in the soil as in the live biomass. An additional important exchange of carbon from vegetation and soil to the atmosphere is through CH_4 release, particularly from anaerobic respiration of microorganisms, especially in WETLANDS.

Measurements of the atmospheric CO_2 concentration at remote, well-mixed sites (e.g. Mauna Loa, Hawaii) show that there is a pronounced annual cycle, of about 5 ppmv, due to the seasonal cycle of photosynthetic uptake and respiratory loss. This annual cycle is larger in the Northern hemisphere compared with the Southern hemisphere, due to the former's larger land area. It is thought that the global carbon cycle was in approximate balance before large-scale human intervention and that the seasonal uptake of carbon by vegetation matched the losses due to respiration and decomposition. Human combustion of FOSSIL FUELS (a product of past photosynthetic carbon uptake and SEDIMENTATION) and changes in land use, including DEFORESTATION, have changed this balance and altered exchanges between the different reservoirs, even though the annual amount of anthropogenic release is small in contrast to natural fluxes. The most obvious result has been an increase in atmospheric CO_2 since the Industrial Revolution (i.e. over the last 100 years) from a value of about 280 ppmv at a rate of 0.5–1.5 ppmv year^{-1}. This increase is the prime cause of the 'anthropogenic greenhouse effect', which threatens to modify the global climate and which may then cause additional FEEDBACKS affecting the vegetation and its exchange of CO_2 and CH_4. *See also* GREENHOUSE GASES; NITROGEN CYCLE; PHOSPHORUS CYCLE; SULPHUR CYCLE. [J.I.L.M.]

REFERENCES

Holmén, K. (1992) The global carbon cycle. In: *Global Biogeochemical Cycles* (eds S.S. Butcher, R.J. Charlson, G.H. Orians & G.V. Wolfe), pp. 239–262. Academic Press, New York.

Houghton, J.T., Meira Filho, L.G., Bruce, J. *et al.* (eds) (1995) *Climate Change 1994: Radiative Forcing of Climate Change and an Evaluation of the IPCC IS92 Emission Scenarios.* Cambridge University Press, Cambridge.

Carbon dioxide

Carbon dioxide (CO_2), a colourless gas 1.5 times heavier than air, is produced when CARBON is oxidized by ignition or during RESPIRATION. It is also produced from volcanoes and springs. Odourless, tasteless and non-toxic at the low concentrations generally found on Earth (0.03–0.05%), CO_2 is dangerous at the high concentrations sometimes found near volcanoes and springs (>20%) where it has a sharp taste and causes choking, asphyxiation and death. CO_2 is soluble in WATER (1 m³ gas dissolves in 1 m³ of water at 15°C) to form carbonic acid (H_2CO_3) with a pH of about 5.8.

CO_2 absorbs RADIATION in the infrared, with major absorption bands at 2.67, 2.77, 4.25 and 15 μm and lesser bands at 1.4, 1.6, 2.0, 4.8, 5.2, 5.4 and 10.4 μm. Consequently, it is an important 'GREENHOUSE GAS', and without it the temperatures at the surface of the Earth would be much lower than they are.

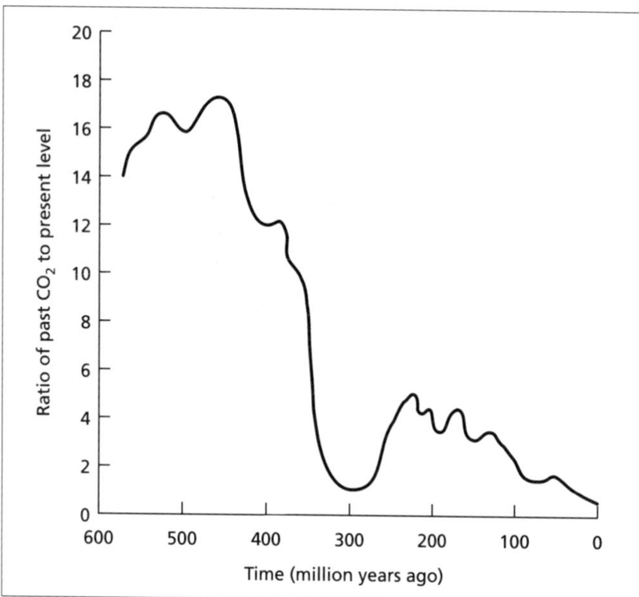

Fig. C6 Changes in atmospheric carbon dioxide (CO_2) through the last 600 million years of geological time; the vertical axis shows the CO_2 level expressed in terms of its preindustrial level, the horizontal axis is in millions of years before present. The main colonization of the land by plants took place around 400 million years ago; the great coal-forming swamps were widespread in the Northern hemisphere around 300 million years ago. (After Berner, R.A. (1994) 3Geocarb II: A revised model of atmospheric CO_2 over Phanerozoic time. *American Journal of Science* **291**, 56–91.)

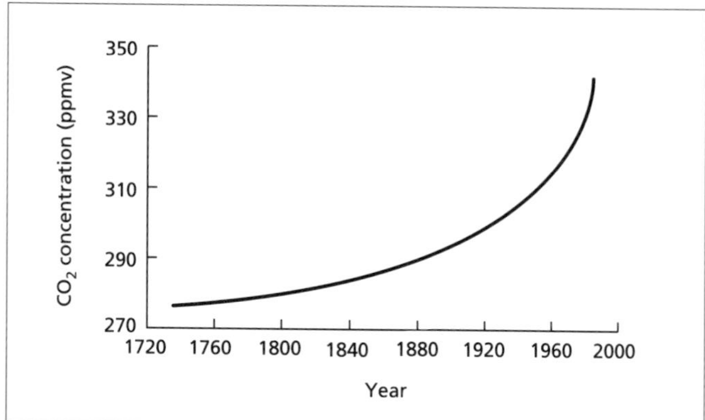

Fig. C7 The changing concentration of carbon dioxide in the atmosphere since the early 18th century, based upon analysis of air bubbles trapped in ice. Concentrations are expressed as ppm by volume. The most recent values are derived from atmospheric measurements at Mauna Loa. (After Moore, P.D., Chaloner, W.G. & Stott, P. (1996) *Global Environmental Change*. Blackwell Science, Oxford.)

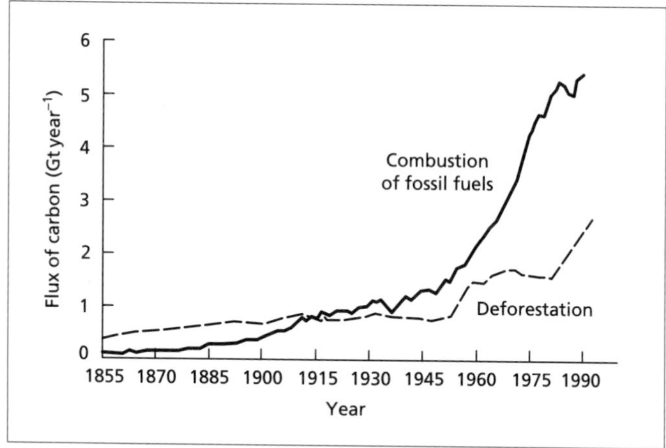

Fig. C8 Estimated annual global emissions of carbon in Gt year^{-1} derived from fossil burning and deforestation. (After Houghton, R.A. (1991) The role of forests in affecting the greenhouse gas composition of the atmosphere. In: *Global Climate Change and Life on Earth* (ed. R.L. Wyman), pp. 43–55. Routledge, Chapman & Hall, New York.)

The concentration of CO_2 on Mars and Venus is very high (>90%), and presumed to have been derived from volcanic activity. On Earth, before the advent of PHOTOSYNTHESIS the CO_2 concentration may have also been high, and has usually been higher than present-day values (Fig. C6). Concentrations of CO_2 have fluctuated considerably in more recent times, as can be seen by the analysis of gas bubbles in ice cores. Minimum concentrations of about 180 ppm by volume (ppmv) were reached 160 000 and 40 000 years ago. Over the last 200 years the concentration has increased from 280 to 355 ppmv as a result of combustion of fossil fuel and conversion of forested lands to AGRICULTURE, and is now rising by about 1.8 ppmv year^{-1} (Fig. C7). Detailed measurements at remote sites worldwide show pronounced annual fluctuation in the atmospheric concentration, which falls when it is summer in the Northern hemisphere, the time of maximum global photosynthesis, and rises in winter.

Currently (1997), about 2 Gt carbon is released as CO_2 to the ATMOSPHERE each year as a result of DEFORESTATION and 6.6 Gt from burning FOSSIL FUELS (Fig. C8). About 4 Gt is taken up by 'sinks' such as OCEAN and forests, and the remainder appears in the atmosphere. The relative importance of these two sinks is controversial and may vary from year to year. *See also* CARBON CYCLE; PRIMARY PRODUCTIVITY. [J.G.]

REFERENCES

Keeling, C.D., Whorf, J.P., Wahlen, M. & Van der Plicht, J. (1995) Interannual extremes in the rate of rise of atmospheric CO_2 since 1980. *Nature* **375**, 666–670.

Schimel, D.S. (1995) Terrestrial ecosystems and the global carbon cycle. *Global Change Biology* **1**, 77–91.

Carbon dioxide, response of plants to elevated

The concentration of CARBON DIOXIDE (CO_2) in the environment around plants has a major influence on their GROWTH. While there are a number of possible effects of CO_2 on plants, the main influence is because CO_2 is essential for all photosynthetic plants for the production of CARBOHYDRATES (sugars, starch, CELLULOSE, etc.) during photosynthetic CARBON metabolism.

The LIGHT energy absorbed by pigments in the 'light reactions' of PHOTOSYNTHESIS is used to reduce CO_2 in the photosynthetic cells and incorporate it into carbon compounds that form part of several linked metabolic cycles. The rate of the CO_2-assimilation reaction depends therefore on the light energy available, the availability of acceptor molecules for the CO_2, the concentration of CO_2 in the photosynthesizing cells, the activity and amount of primary photosynthetic enzyme present, and other conditions, especially temperature and the water status of the cells. However, there are two distinctly different primary photosynthetic enzyme systems distributed widely in different plant taxa, which can therefore be grouped biochemically into C_3 and C_4 species (*see* C_3/C_4 PLANTS). In C_4 species, the primary enzyme system is saturated with respect to CO_2 at or near present-day atmospheric CO_2 concentrations. In C_3 species (which comprise the majority of the world's biomass) in general the more CO_2 available, the higher the rate of CO_2 uptake, although this is an asymptotic relationship, increasing from zero at CO_2 concentrations between 50 and 100 ppm by volume (ppmv) with SATURATION at concentrations of 800–1200 ppmv (Fig. C9).

As plant growth (increase in DRY MASS) is the net result of the accumulation of carbohydrate through photosynthesis, less any losses due to RESPIRATION, increased CO_2 may lead to increased growth in C_3 plants. This growth response of plants to CO_2 is widely used in commercial horticulture, where glasshouse and polytunnel air is artificially enriched with CO_2 up to 800–1500 ppmv, to increase growth rates and give a shorter harvest interval for high-value crops such as tomatoes, lettuce, peppers, cucumbers and flowers. The CO_2 response also affects plant growth globally, due to the increased atmospheric CO_2 concentration since preindustrial times (before 1850) from 280 ppmv to the present 355 ppmv; future increases of between 560 and 800 ppmv are predicted over the next century. There is much experimental evidence indicating that increases to dry-mass accumulation of 10–50% occurs with a doubling of CO_2 concentration. However, the interaction of increased CO_2 with other environmental variables, such as light, temperature, nutrient supply, salinity, pollutants and water availability, is very important in determining the magnitude of the effect. In particular, in conditions of low NITROGEN supply the photosynthetic response to increased CO_2 may be reduced due to photosynthetic ACCLIMATION. In contrast, in conditions of water stress the response to CO_2 may be enhanced because of the action of stomata, which determine the exchange of CO_2 and water between leaf and air, and the concentration of CO_2 in the photosynthesizing cells. Generally, stomatal aperture reduces as the concentration of CO_2 rises inside the leaf, which reduces the EVAPORATION of water and increases the water use EFFICIENCY (ratio of CO_2 taken up to water lost). If water supply is limited, increased CO_2 therefore increases the dry-mass accumulation; this applies to both C_3 and C_4 plants. There may also be effects of increased CO_2 on leaf and root expansion due to the changed water relations caused by the reduced evaporation, and effects on vegetation water use. However, at these larger scales FEEDBACKS between local evaporation, surface temperature and local air humidity are complex and the net result of an increase in CO_2 on regional evaporation rates difficult to predict.

Other effects of increased CO_2 on plant growth have been noted: small effects on development rate (both positive and negative); effects on plant composition (especially an increase in the C:N RATIO) that are likely to have profound effects on herbivory; and effects on respiration (both positive and negative) that will also affect growth. Furthermore, differences in the effect on growth and seed production among individuals, populations and species are likely to result in substantial ecological impacts of elevated CO_2. [J.I.L.M.]

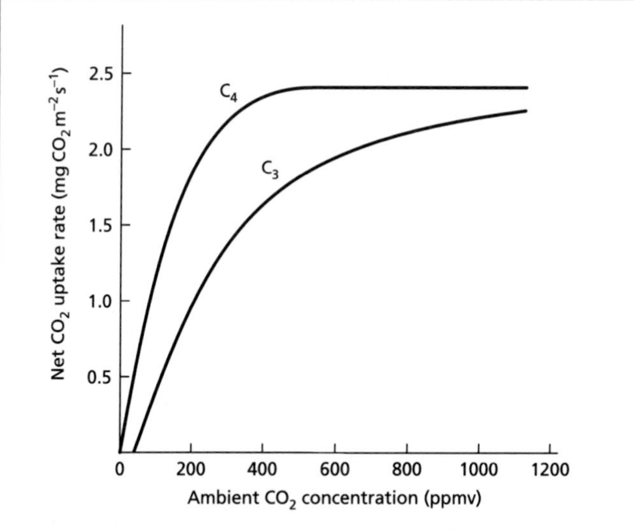

Fig. C9 Response of plant photosynthetic rate per unit leaf area to carbon dioxide.

REFERENCES

Lawlor, D.W. & Mitchell, R.A.C. (1991) The effects of increasing CO_2 on crop photosynthesis and productivity: a review of field studies. *Plant, Cell and Environment* **14**, 807–818.

Woodward, F.I., Thompson, G.F. & McKee, I.F. (1991) The effects of elevated concentrations of CO_2 on individual plants, populations, communities and ecosystems. *Annals of Botany* **67**, 23–38.

carbon dioxide springs CARBON DIOXIDE (CO_2) springs are globally widely distributed, often associated with areas of geothermal activity or in locations with extensive deposits of mineral carbonate. Major CO_2 springs have been commercially important for several centuries as sources of naturally carbonated mineral water, CO_2 or for their reputed beneficial effects on human health. Scientific interest in the effects of elevated atmospheric concentrations of CO_2 on ecosystems has stimulated research at sites with natural CO_2 emissions. CO_2 springs thus form important natural laboratories for studies of the long-term effects of elevated concentrations of CO_2 on plants, animals and other components of natural ecosystems. [P.R.V.G.]

carbon monoxide A highly flammable and toxic gas produced from incomplete combustion of CARBON compounds. Carbon monoxide (CO) has an average ambient concentration of ~0.1 ppm by volume (ppmv) through atmospheric oxidation of METHANE (CH_4) and other natural HYDROCARBONS. Anthropogenic sources (principally car exhausts) account for about half of total global CO

EMISSIONS. The major sink is complete oxidation to CO_2 by OH. CO is absorbed through the lungs and binds irreversibly to haemoglobin, causing drowsiness and response impairment. Urban levels of CO can exceed 10–20 ppmv, particularly inside vehicles, and the WHO has set limits of 9 ppmv and 27 ppmv for 8-hour and 1-hour means, respectively. *Cf.* CARBON DIOXIDE. [M.R.H.]

carbon tax Tax on FOSSIL FUELS, related to CARBON DIOXIDE-generating potential. A MARKET INSTRUMENT. [P.C.]

carboxylic acids Carboxylic acids are organic ACIDS containing the carboxylic group. Examples are methanoic acid, also called formic acid, H(C=O)OH, and ethanolic acid, also called acetic acid, CH_3(C=O)OH. [J.G.]

carcinogens, mutagens, teratogens (CMTs) Substances that can induce cancer, and/or MUTATIONS, and/or defects in development (sometimes also called reprotoxins). They are of considerable importance for individual organisms, but their ecological significance is open to question. Mutations in cells in the bodies of organisms, especially if they are somatic, and developmental abnormalities and cancers need not have significant effects on the population as a whole. [P.C.]

carnivory This is applied to an organism that only consumes the tissue of other heterotrophic organisms. This contrasts with herbivory, where only autotrophic organisms are consumed. A carnivore may be an animal or plant. *See also* FEEDING TYPES, CLASSIFICATION; PREDATOR. [R.C.]

carr A RHEOTROPHIC mire dominated by trees (often alders and willows). [P.D.M.]

carrion-feeders *See* FEEDING TYPES, CLASSIFICATION.

carrying capacity (*K*) This term describes a DYNAMIC EQUILIBRIUM around which a POPULATION fluctuates. Another analogous definition that is often used to describe this process is the equilibrium POPULATION DENSITY (N^*). This definition offers a more accurate reflection of natural populations as it does not imply that there is a maximum number of individuals that can be 'carried' in a particular habitat. The equilibrium will be such that when the population density is above it then the number of individuals will tend to decrease, and when the population density is below it then the number of individuals will tend to increase. In reality, the equilibrium density is determined by a range of physical and biological factors such as the availability of food resources and the presence of competitors. The exact nature of this relationship will vary between different organisms and habitats, although fundamentally it will be determined by basic demographic processes, such as NATALITY, as well as dynamic processes, such as COMPETITION and predation, which will interact with the surrounding physical environment to give the equilibrium density of a population. Carrying capacity is sometimes applied to a group of organisms rather than an individual population. These may consist of single-species groups or collections of similar species exploiting a common resource (e.g. insectivores). *Cf.* POPULATION GROWTH RATE. *See also* DENSITY DEPENDENCE; POPULATION DYNAMICS; POPULATION REGULATION. [R.C.]

carrying capacity, importance to conservation The maximum number of organisms of a given species that can be supported in a given habitat or geographic area, usually denoted as K (e.g. in the LOGISTIC EQUATION) or as the K-value. CARRYING CAPACITY has a direct impact on the amount and quality of habitat that must be preserved to sustain a MINIMUM VIABLE POPULATION. The problem for managers of wildlife parks and nature reserves is compounded by the fact that carrying capacity and habitat requirements differ for different species. Furthermore, the size of reserves is often restricted by other practical considerations. It is often necessary to strike a compromise between the needs of neighbouring human communities and the minimal requirements of the most ENDANGERED SPECIES in the park. *See also* CONSERVATION BIOLOGY. [A.R.H.]

Carson, R. *See* CHARACTERS IN ECOLOGY.

cascade model of food-web structure A model to explain patterns observed in food-web datasets, so called because its central assumption is that species in the web can be ordered a priori into a FEEDING hierarchy. The model assumes that a species can never feed on those above it and that feeding links with those below it are chosen at random with a probability c/n, where c is a positive constant less than n (the number of species in the web). The number of links per species must be specified from available data before the model can be used. *See also* FOOD CHAIN; FOOD WEB. [S.J.H.]

caste The term 'caste' is used in biology to draw attention to differences among members of eusocial colonies. The clearest such differences are between reproductive and worker forms. In TERMITES and naked mole rats (*Heterocephalus glaber*), workers may be female or male and are non-reproductive. In the eusocial Hymenoptera (certain bees and wasps and the ants) workers are females. (In a very small number of these species some males perform certain non-reproductive tasks but these cases are exceptional.) The term 'caste' is also used to discriminate between different worker forms or workers that exhibit different behaviour. Here the clearest differences are seen between different worker morphs. In termites, different larval instars may form different functional castes: they are mobile and can perform different sets of tasks for their society. In certain ants, adult workers within the same society can have recognizably different sizes and shapes. These may form a small number of so-called physical castes such as majors and minors.

The concept of worker castes is inextricably linked with the DIVISION OF LABOUR. (Polyethism is the term used to mean 'performing many tasks'.) Workers often change the sets of tasks they accomplish during their own lifetimes (e.g. moving from brood care to nest maintenance to foraging). The term 'temporal castes' has been applied to the 'stages' in this process. The term 'age polyethism' is often applied, in general, to this form of division of labour. In the majority of cases there is an association between the age of workers and the set of tasks they accomplish. However, such correlations are often weak, with workers being able to revert to earlier task sets or accelerate and skip into later task sets. Hence the term 'age polyethism' should not be used, as it often appears to be, to imply a causal relationship between age and task. For this reason it may be best to drop 'age polyethism' in favour of the less prejudicial term 'temporal polyethism'.

The problem with the term 'caste' is that it might be seen to imply that workers in eusocial colonies fall into naturally discrete categories. Much recent work has shown that in many cases variation within eusocial colonies is far more continuous than recognized by the early authors who brought the caste concept into social biology. For example, so-called temporal castes are arguably most often not discrete since different workers of the same age may be employed flexibly in a wide range of different tasks. Furthermore, in the case of hymenopteran worker and reproductive castes, even when these two castes differ markedly in external morphology certain workers may be able to reproduce. In many species of ants, bees and wasps, workers can produce their own sons parthenogenetically (*see* PARTHENOGENESIS) and some, remarkably, worker-like forms can mate and produce daughters. Hence, in many cases the term 'caste' has to be applied with caution. 'Caste' can be applied usefully to draw attention to phenotypic variability within the same extended PHENOTYPE, i.e. the eusocial COLONY. In other cases, however, its uncritical use can be a potentially misleading oversimplification. *See also* BEHAVIOURAL ECOLOGY; SOCIAL INSECTS; SOCIAL STATUS; SOCIOBIOLOGY. [N.R.F.]

caste systems *See* SOCIAL INSECTS.

castration, parasitic Infecting parasites cause HOST gonads not to function (e.g. by inhibiting development, causing loss once developed, causing malfunction). It occurs in a wide variety of PARASITE–host relationships, with hosts including plants, crustaceans, molluscs, echinoderms, insects and vertebrates. The host may live longer as an individual but, of course, does not contribute either at all, or as much, to RECRUITMENT. So host density may decline. The latter raises puzzles about the evolution of 'parasitic castration' since it may cause short-term benefits to the parasite in terms of more resources being available to it (i.e. not lost as GAMETES) for longer periods of time (host lives longer), but with the long-term consequence of less hosts to parasitize. This may not, therefore, be an EVOLUTIONARILY STABLE STRATEGY (ESS) but may nevertheless evolve due to short-term FITNESS gains. *See also* LIFE-HISTORY EVOLUTION. [P.C.]

casual species ALIEN species incapable of forming self-replacing populations and relying on continual reintroduction for persistence outside their native or NATURALIZED ranges. [M.J.C.]

CAT *See* CONTROL ACTION THRESHOLD.

catadromous A pattern of MIGRATION in which adults living in FRESHWATER migrate to breeding sites in saltwater. Offspring move in the opposite direction. Examples include the European eel (*Anguilla anguilla*) and the Hawaiian freshwater shrimp. *Cf.* ANADROMOUS. [P.O.]

catalyst A substance that increases the rate of a chemical reaction but is not itself altered chemically in the process. Finely divided metals and oxides at high temperature are often used as catalysts in chemistry. They work by providing a high surface area on which the reagents are adsorbed and therefore come together. In living cells, enzymes are the catalysts. They work by effecting the proper arrangement of the reacting molecules with respect to each other. [J.G.]

catalytic converter A term usually applied to the device for chemical transformation of POLLUTANTS in vehicle exhausts. Two-way oxidation CATALYSTS oxidize unburnt fuel HYDROCARBONS and CARBON MONOXIDE (CO) to WATER and CARBON DIOXIDE (CO_2) but do not affect NITROGEN OXIDE (NO_x) emissions. Three-way catalytic converters contain a ceramic filter coated with a mixture of PLATINUM, PALLADIUM and RHODIUM and also reduce NO_x to N_2. Operational mixture strength is regulated by an OXYGEN sensor in the exhaust manifold and electronic control of the engine. Unleaded petrol is required since LEAD poisons the catalyst. All new cars sold in the UK since 1993 have catalytic converters fitted. *See also* AIR POLLUTION; ATMOSPHERIC POLLUTION; MOTOR VEHICLE POLLUTION. [M.R.H.]

catastrophic agents *See* DENSITY INDEPENDENCE.

catastrophic drift *See* DRIFT, FRESHWATER.

catastrophism *See* UNIFORMITARIANISM.

catch level *See* FIXED-QUOTA HARVESTING.

catch per unit effort Return from fishing, hunting and SAMPLING. Usually follows law of diminishing returns; for example number of species sampled from a community increases but at a reducing rate as sampling effort is increased. Hence catch per unit effort varies with effort. [P.C.]

catchment A geographical region within which hydrological conditions are such that water becomes concentrated in a particular location, either a BASIN or a single river by which the catchment is drained. The American term 'watershed' may be used as a synonym, but is employed rather differently in European literature where it implies a ridge separating catchments. In ecology, the catchment has provided a valuable means of studying nutrient movements within and between ecosystems. If a site is based on impermeable rock and a catchment is drained by a single exit stream, then the measurement of chemical concentrations and stream flow rate provide an indication of the movement of water-borne elements out of the catchment ecosystem. Precipitation measurements and analyses supply corresponding input figures and differences between the two can provide information on the internal behaviour of elements as a result of such processes as soil WEATHERING, BIOMASS accumulation (or loss), animal movements, etc.

In well-documented catchment studies from the USA, at Coweeta and at Hubbard Brook, this approach to nutrient dynamic studies in ecosystems has been developed at an experimental level, certain catchments being subjected to specific management regimes and the impact on nutrient movements being recorded over long periods. [P.D.M.]

cation *See* ION.

catotelm Lower functional SOIL layer in a MIRE: the PEAT deposit proper, perennially waterlogged in an intact system. From the

Greek *kata*, down; *telma*, MARSH (Ingram 1978). The *inertnyy sloy* (inert layer) of Russian telmatologists. *See also* ACROTELM. [H.A.P.I.]

REFERENCE

Ingram, H.A.P. (1978) Soil layers in mires: function and terminology. *Journal of Soil Science* **29**, 224–227.

cavitation When a ship's propeller rotates fast in the water, the reduced hydrostatic PRESSURE behind the blades causes bubbles to form. This process is termed cavitation.

Cavitation is also important in plants. In the sporangia of ferns, the sudden release of tension by cavitation is the source of energy for spore DISPERSAL. Cavitation may be an everyday event in the stems of vascular plants. When leaves are actively transpiring, the water columns in the water-conducting tubes come under great tension (0.5–3 MPa), and they may either break or gas may be drawn in from outside. In either case, the result is a gas embolism that may persist; many of these may impair water transport, causing even more tension and more cavitation ('runaway' cavitation). When cavitation occurs, energy is released and the breakage of the water column can be detected with an appropriate acoustic sensor attached to the stem or petiole.

In vessel-bearing species, such as vine and maple, embolized vessels may be refilled in the spring by root pressure. In tracheid-bearing species, such as pine, the embolisms may persist for more than a year, and so DROUGHT in one year is likely to exacerbate the effect of drought in the next year. [J.G.]

REFERENCES

Borghetti, M., Grace, J. & Raschi, A. (1993) *Water Transport in Plants under Climatic Stress*. Cambridge University Press, Cambridge.

Grace, J. (1993) Consequences of xylem cavitation for plant water deficits. In: *Water Deficits, Plant Responses from Cell to Community* (eds J.A. Smith & H. Griffiths), pp. 109–128. Bios Scientific Publishers, Oxford.

cellular automata Also known as CA models, these are a class of models that are becoming widely used in many areas of science. These models divide the world into discrete patches or cells which usually have a small number of possible states. This will often be as few as two, representing 'occupied' and 'empty' for example. In a CA model, time is also discrete and the rules governing changes in state from one time to the next are exclusively local.

Conway's famous *Game of Life* is a simple CA model. Cells have only two states and the birth, death or survival of a cell is determined by the number of occupied neighbouring cells. The reason for the great interest in CA models is that such a simple model can show quite complex behaviour through time, often producing fascinating patterns. In many instances this complex behaviour, although deterministic, can only be predicted by direct simulation. It is possible to relax the deterministic rule and replace it with stochastic behaviour based on local rules and this approach is commonly used when modelling biological systems.

Cellular automata are becoming increasingly useful in the modelling of the effects of habitat fragmentation, where the spatial structure of a population, as well as its absolute population size, will be important in determining its continued survival. This has led to a linkage of cellular automata approaches to metapopulation models. *See also* COUPLED MAP LATTICE; PATCH DYNAMICS. [C.D.]

cellulase A complex enzyme system that hydrolyses CELLULOSE to sugars of lower RELATIVE MOLECULAR MASS, including cellobiose and glucose. At least two components are involved: one involves hydrolysis by preliminary activation or disaggregation of cellulose chains and the other components are responsible for depolymerization to soluble form. Obviously of great importance in DECOMPOSITION, the process is carried out by bacteria and fungi, but whether metazoans are capable of synthesizing cellulase, or maybe get microbes to do it for them, still remains problematic. [P.C.]

cellulose Cellulose is classified as a polysaccharide comprising linked monosaccharide units. It is a straight-chain natural POLYMER consisting of glucose units with average RELATIVE MOLECULAR MASS of more than 500 000. Cellulose forms the main structural units of all plants and is the most abundant polysaccharide occurring in nature. Cellulose is the main constituent of cell walls of all higher plants, many algae and some fungi.

The fibrous nature of cellulose extracted from the cotton plant has resulted in its use in the textile industry. Cotton fibres are composed almost entirely of cellulose. Unlike starch, cellulose is not a human food source. Cellulose contains only β-glucose units linked via OXYGEN atoms and cannot be hydrolysed by enzymes present in the human digestive system. The enzyme CELLULASE contained in certain bacteria is present in the digestive system of GRAZING animals, which utilize cellulose as a major food source.

Cellulose is often associated with non-polysaccharide compounds in living organisms, for example lignocelluloses are complex fibrous biopolymers that occur in the woody part of higher plants. Cellulose makes up about 50% of wood.

Other commercial uses of cellulose obtained from wood pulp or cotton linters include the production of artificial materials including viscose rayon or acetate rayon (from cellulose xanthate or cellulose acetate respectively). Additionally, cellulose acetate can be used in lacquers, varnishes and non-shatterable GLASS. Cellulose nitrate is highly flammable and has explosive properties but is also used for celluloid film. [M.C.G.]

Cenozoic Era of geological time occupying approximately the last 65 million years and including the Tertiary and Quaternary periods. [P.D.M.]

censuses Estimates of POPULATION SIZES. [P.C.]

central dogma *See* LAMARCKIAN INHERITANCE.

central limit theorem For large samples the sample ARITHMETIC MEAN, $\bar{x}$, of n independent observations, taken from a POPULATION with mean μ and VARIANCE s^2, has approximately a NORMAL DISTRIBUTION with mean μ and variance s^2/n. This theorem applies to any PROBABILITY DISTRIBUTION with a well-defined mean and variance (it would not apply, for example, to a Cauchy distribution which has an infinite mean and variance). This theorem is used to define the sample standard error ($\sqrt{s^2/n} = s/\sqrt{n}$), which is used in hypothesis testing, since it allows us to predict the probability of observing a sample estimate of $\bar{x}$ for a given population mean. Whilst the central limit theorem holds for large sample sizes, for smaller samples from normally distributed populations (approximately $n < 30$) it has been possible to predict the distribution of the sample mean relative to the population mean through Student's t distribution, which gives rise to the t TEST. *See also* STANDARD ERROR (OF THE MEAN). [R.P.F.]

centre of origin The area in which a taxonomic group originated and from which it dispersed. A centre of origin may be found by tracing back the GEOGRAPHIC DISTRIBUTION of a TAXON to the starting point, assuming that the radiation was accompanied by the EVOLUTION of particular TRAITS from primitive to more specialized. A centre of origin is usually characterized by the greatest differentiation of a taxon because it has been evolving in that location for the longest time. *See also* SPECIATION. [P.O.]

centrifugal speciation *See* SPECIATION; SPECIES CONCEPTS.

centrolecithal eggs *See* EGGS.

CEQ *See* COUNCIL ON ENVIRONMENTAL QUALITY.

CERCLA (Comprehensive Environmental Response Compensation and Liability Act 1980) *See* SUPERFUND LEGISLATION; UNITED STATES LEGISLATION.

CEREs (Coalition of Environmentally Responsible Economies) *See* VALDEZ PRINCIPLES.

cerium (Ce) Cerium is a lanthanide, the most abundant of the so-called rare earth metals (ATOMIC NUMBER = 58; relative ATOMIC MASS = 140.11; melting point = 798°C; boiling point = 3257°C), having an abundance of 68 mg kg^{-1} in the Earth's CRUST. It occurs in several minerals, of which monazite and bastnaesite are the most important. Pure cerium is a METAL: iron-grey, lustrous, malleable, and will ignite if scratched with a knife. It forms ceric salts, which are orange-red or yellowish, and cerous salts, which are usually white. It is used in pyrophoric alloys for cigarette lighters, as a constituent of gas mantles, arc-lighting and as a CATALYST in PETROLEUM refining and metallurgical applications. In alloys, it improves the malleability. Cerium salts are used in GLASS-making, and the oxide is used instead of rouge for polishing glass. [J.G.]

CFCs (chlorofluorocarbons) Synthetic substituted ALKANES with one or more HYDROGEN atoms replaced by CHLORINE or FLUORINE, existing as gases or low-boiling liquids at normal temperatures and pressures. Chemically inert and non-toxic, CFCs are used as refrigerants, solvents, AEROSOL propellants and in blowing plastic foam, but are gradually being replaced by more reactive hydrochlorofluorocarbons (HCFCs) and hydrofluorocarbons (HFCs). They are important 'GREENHOUSE GASES', contributing about 10% to radiative forcing of the CLIMATE. CFCs accumulate in the TROPOSPHERE with lifetimes of tens to hundreds of years, and are transported to the STRATOSPHERE where they are photolysed and perturb the chemistry of OZONE. The largest contributors are CFC-11 ($CFCl_3$), CFC-12 (CF_2Cl_2) and CFC-113 ($CF_2ClCFCl_2$). Production and use of CFCs and other HALOCARBONS is now regulated under the 'Montreal Protocol', which seeks to prevent further depletion of stratospheric ozone and consequent increases in ultraviolet (UV)-B radiation at the Earth's surface. *See also* CLIMATE CHANGE; OZONE HOLE; OZONE LAYER. [J.N.C.]

chalk and limestone grassland Limestones are rocks composed mainly of calcium carbonate ($CaCO_3$), laid down in shallow seas. Chalk is a particularly pure form of limestone. The soils formed on these rocks are shallow, often with fragments of parent rock at or near the surface, and well drained. They are high in free calcium carbonate and have a high pH. They are also low in essential NUTRIENTS, especially NITROGEN and PHOSPHORUS.

Chalk and limestone grassland is found in warm areas with low rainfall, often on steep slopes, and is maintained by GRAZING animals, mainly sheep and rabbits. Frequently no species becomes really dominant and there is often an abundance of broad-leaved herbs, many deep rooted. Disturbance caused by grazing animals creates bare sites that are often colonized by ANNUALS or short-lived PERENNIALS.

Since sheep were introduced very early into north-western Europe and Britain, some limestone GRASSLANDS, such as those on Salisbury Plain, are very ancient. They can be species-rich, with up to 40 grasses and flowering plant species in a square metre. They are very attractive, with many colourful and abundant flowers, such as rock-rose (*Helianthemum nummularium*), marjoram (*Origanum vulgare*), thyme (*Thymus polytrichus*) and stemless thistle (*Cirsium acaule*), together with many orchids. Many invertebrates, including several scarce butterflies, are dependent on this grassland for their existence.

Reduction in rabbit numbers by myxomatosis and conversion of most grassland to arable has resulted in drastic reductions in the area of this attractive habitat, and much conservation effort is devoted to maintaining what is left. [S.R.J.W.]

chamaephyte *See* PHYSIOGNOMY; RAUNKIAER'S LIFE FORMS.

channelization The process of converting a meandering natural stream to a rip-rapped ditch, in order to reduce flooding of surrounding land by accelerating RUN-OFF. Channelization results in drastic reductions in the abundance of fish and invertebrates because habitat diversity is eliminated and water velocities are greatly increased during high-flow periods. Wide-scale channelization of headwaters increases the size of flood events downstream by increasing both the amount and rapidity of run-off, as demonstrated by recent major flooding along the Mississippi River (USA) and Rhine River (Holland). [P.B.M.]

chaos Chaos is a term used to describe deterministic dynamic systems that are neither steady nor periodic, but which exhibit irregularity and complexity. Systems expressing chaotic dynamics display sensitive dependence on initial conditions. The sensitivity is exponential, meaning that as time goes on small errors in the solution (e.g. due to noise and computer round-off) grow exponentially. This means that short-term predictions may be made accurately, but it is not possible to make useful long-term predictions. Chaotic behaviour can be exhibited by deterministic equations and is called deterministic chaos. Chaotic behaviour exhibited by simple-looking equations was first described by the meteorologist Edward Lorenz in 1963. Since then, chaotic dynamics have been demonstrated in a wide variety of systems including fluids, plasmas, solid-state devices, circuits, lasers, mechanical devices, biology, chemistry, acoustics and celestial mechanics. An example of chaos in ecological systems is shown by the erratic, complex and unpredictable OSCILLATIONS in the size of populations having a high INTRINSIC RATE OF INCREASE (r). For example, if the net reproductive rate (R) in a discrete-time logistic POPULATION GROWTH MODEL is such that

3.57<R<4 the population exhibits chaotic FLUCTUATIONS in size. *See also* POPULATION GENETICS. [V.F.]

chaparral shrublands Specifically, the term is applied to the fire-prone SCRUB vegetation of those areas of the western USA and north-western Mexico where Mediterranean-type climate (hot dry summers, mild moist winters) prevails. Fire-adapted shrubs (*Adenanthos, Arctostaphylos, Ceanothus*) co-exist with other woody species to form dense, often impenetrable thickets. Many annuals and short-lived perennials appear in the early post-fire years.

Summer DROUGHT limits vegetation growth and plants cope with this either by means of leathery, EVERGREEN leaves (sclerophyllous) or by dropping their leaves when stressed (deciduous).

Similar vegetation occurs in the other Mediterranean-climate regions, namely kwongan and mallee in southern and south-western Australia; matorral in central Chile; matorral, maquis and garrigue in the Mediterranean Basin (see Plate 2, facing p. 398); and FYNBOS in south-western South Africa. Plant-species diversity in all these shrublands is very high and all have broadly similar ecological regimes of winter rain and summer drought, recurrent fire and low soil fertility. Their phylogenetically unrelated biotas show strong structural and functional similarities, or ecological convergence, at all levels of organization. The most striking convergence is between Australian kwongan and South African fynbos. Both of these heath-like shrublands are associated with ancient landscapes, extremely infertile soils and a relatively high frequency (4–40 years) of fires. Fynbos and kwongan are distinguished from the other shrublands by their relatively open shrub stratum; high shrub diversity; and the highest incidence in the world of woody plants that are killed by fire, have canopy seed storage (serotiny) and ant dispersal (MYRMECOCHORY).

The remaining shrublands occur in geologically younger and more fertile landscapes, have dense shrub canopies and burn at much longer intervals. Diversity in chaparral is mainly concentrated in fire-adapted annual and herbaceous perennial taxa that are conspicuous only in the years immediately following fire. Most Mediterranean Basin shrublands are derived, as a result of human disturbance, from ancient forests. Chilean matorral lacks a large fire-adapted flora since natural fires are extremely infrequent in this region. [R.M.C.]

REFERENCES

Davis, G.W. & Richardson, D.M. (eds) (1995) *Mediterranean-type Ecosystems. The Function of Biodiversity*. Springer-Verlag, Berlin.

Kruger, F.J., Mitchell, D.T. & Jarvis, J.U.M. (eds) (1983) *Mediterranean-type Ecosystems. The Role of Nutrients*. Springer-Verlag, Berlin.

character displacement A DISTRIBUTION pattern encountered in plant and animal populations where closely related sister species are recognizably different in zones of overlap (sympatric), but virtually indistinguishable where each occurs alone (allopatric). Two types are generally recognized. Ecological character displacement pertains to evolutionary change in the zone of overlap that results from INTERSPECIFIC COMPETITION for resources. Examples in nature and experimental confirmation are well established (Schluter 1994). Reproductive character displacement (REINFORCEMENT) involves changes in the prezygotic MATE RECOGNITION system that evolve as a response to selection against hybridization. The expected pattern of variation occurs in nature, but the process has yet to be demonstrated experimentally (Howard 1993). *See also* ALLOPATRIC SPECIATION; SYMPATRIC SPECIATION. [G.L.B.]

REFERENCES

Howard, D.J. (1993) Reinforcement: the origin, dynamics, and fate of an evolutionary hypothesis. In: *Hybrid Zones and the Evolutionary Process* (ed. R.J. Harrison), pp. 118–142. Oxford University Press, Oxford.

Schluter, D. (1994) Experimental evidence that competition promotes divergence in adaptive radiation. *Science* **266**, 798–801.

character polarity The plesiomorphic or apomorphic status of each TRAIT is called its polarity. The process of identifying the plesiomorphic and apomorphic traits is called character polarization. In PHYLOGENETIC SYSTEMATICS this is a logical process of assigning traits to plesiomorphic and apomorphic status based on a priori deductive arguments using OUTGROUP comparisons. *See also* PHYLOGENETIC INFERENCE. [D.R.B.]

characteristic return time A measure of the rate at which a population regains its equilibrium value after a perturbation. The characteristic return time is the inverse of the EXPONENTIAL GROWTH rate (r), i.e. the time taken for a population to increase in size by a factor of exp=2.718 growing exponentially with rate constant r. Populations with a high r can rapidly track changes in the CARRYING CAPACITY or can recover from density-independent effects on the POPULATION GROWTH RATE. When the characteristic return time is less than the period of an environmental fluctuation divided by 2π, the population will track changes in the environment closely. *See also* DENSITY DEPENDENCE; POPULATION DYNAMICS; POPULATION REGULATION. [V.F.]

Characters in ecology

A list of some of the major characters of the past who have played some part in moulding ECOLOGY. They are ordered alphabetically.

Allee, W.C. (1885–1955). Studied forces determining social groupings (AGGREGATIONS) in animals.

Aristotle (384–322 BC). Greek philosopher. As well as forming the foundations of philosophy, is regarded by many as the father of natural history. His *Historia Animalium* contains much information on many species gained through observation and dissection.

Carson, R. (1907–1964). Author of the influential best-seller *Silent Spring* (1967) that highlighted the blight of the environment by PESTICIDES. So captured the public imagination that it launched the popular environment movement.

Clements, F.E. (1874–1926). American ecologist. Developed a holistic philosophy of ecology that represented the plant COMMUNITY as a SUPERORGANISM with a predictable development, i.e. SUCCESSION. Initiated detailed quadrat studies of plant communities.

Darwin, Charles (1809–1882). Originator of the concept of EVOLUTION by NATURAL SELECTION that was expounded in his book *On the Origin of Species* (1859) and in which the interaction, 'struggle', between organisms and environment 'for survival' is all important.

Darwin, Erasmus (1731–1802). Grandfather of Charles. Early theory of evolution based on a struggle for existence, but with impetus for change coming from the acquisition of characters for improvement. Hence more similar to Lamarckism than DARWINISM.

Dobzhansky, Th. (1900–1975). Pioneering work on POPULATION GENETICS using laboratory cultures of *Drosophila.*

Elton, C. (1900–1992). As Director of Bureau of Animal Populations in Oxford had enormous influence on development of British ecology. Launched long-term ecological study on Wytham Wood, Oxfordshire. His *Animal Ecology* (1972) is a classic and was the major textbook in ecology in Britain for many years.

Fisher, R.A. (1890–1962). One of the founders of POPULATION GENETICS. Also very influential in the application of rigorous statistical procedures in ecology.

Gause, G.F. (1910–1986). Performed experiments with protozoans to test predictions from Lotka–Volterra equations. Made explicit COMPETITIVE EXCLUSION PRINCIPLE, which is often known as the Gause principle.

Haeckel, E. (1834–1919). Originator of the term 'ecology' (Oecologie) in his *Generelle Morphologie* (1866) to denote interactions between organisms and external world. Derived from the Greek *oikos,* referring to operations of the family household.

Haldane, J.B.S. (1892–1964). With R.A. Fisher was one of the founders of POPULATION GENETICS. Drew attention to MELANISM in the peppered moth, *Biston betularia,* as an example of rapid evolution in action.

Hooker, J.D. (1817–1911). Succeeded his father William as Director of Kew Gardens. Collected plants on expeditions to the southern polar regions and to the Himalayas.

Hooker, W. (1785–1865). First official Director of Kew Gardens. Succeeded by his son Joseph.

Humboldt, Alexander von (1769–1859). Founder of botanical geography. Showed how physical environment determined DISTRIBUTION of plants on a geographical scale. Observed an equivalence between belts of vegetation ascending a mountain and geographical zones that girdle the Earth.

Hutchinson, G.E. (1903–1991). Promoted view that ecological relationships should be seen as systems governed by causal relationships. Raised a number of provocative issues for causal analysis: the *n*-dimensional NICHE; why there are so many kinds of species; ecology as the theatre in the evolutionary play.

Lack, D. (1910–1973). Tested predictions from evolutionary theory on field populations of birds. Worked with Galapagos finches, where he showed that very similar species had significant differences in feeding habits.

Lamarck, J.B. (1744–1829). Did important work on the CLASSIFICATION of animals. But best known for his theory of evolution involving the INHERITANCE OF ACQUIRED CHARACTERS.

Leopold, A. (1886–1948). In an influential essay, 'The Land Ethic', expressed the need for humanity to treat nature as something to which we belong, not as a commodity to be exploited.

Lindeman, R. (1915–1942). In a sadly short career, pioneered the view of ECOSYSTEMS as dynamic systems that involve flows of energy and cycles of matter.

Linnaeus, C. (1707–1778). Father of modern systematics. Author of *Systema Naturae* (1735). Initiated the binomial system of nomenclature.

Lotka, A.J. (1880–1944). Developed mathematical theory of POPULATION SIZE. Associated with V. Volterra.

Lyell, C. (1797–1875). Father of modern geology. Author of *Principles of Geology* (1830–1833).

MacArthur, R. (1930–1972). Student of G.E. Hutchinson. Expanded the Hutchinsonian approach and made it more rigorous and mathematical.

Nicholson, A.J. (1895–1969). Explored the different kinds of factors that might affect POPULATION SIZE, emphasizing the distinction between density-dependent and density-independent factors.

Shelford, V.E. (1877–1968). Applied Clements' concepts of COMMUNITY and SUCCESSION to animals.

Tansley, A.G. (1871–1955). Developed studies on plant communities in Britain following Clements' techniques but not his philosophy. Deeply opposed to notion of COMMUNITY as SUPERORGANISM.

Volterra, V. (1860–1940). Eminent Italian mathematical physicist. Developed mathematical theory of POPULATION SIZE. Associated with A.J. Lotka.

Wegener, A. (1880–1930). First champion of theory of CONTINENTAL DRIFT.

White, Gilbert (1720–1793). English naturalist. Author of *The Natural History of Selbourne,* a collection of letters written to fellow naturalists documenting observations on the WILDLIFE in his parish in southern England.

Wright, Sewall (1889–1988). Became interested in selection through animal breeding. Saw the importance of structure within populations to form subpopulations. Introduced concept of ADAPTIVE LANDSCAPE.

[P.C.]

Charles's law *See* GAS LAWS.

Charter for Sustainable Development Following the BRUNDTLAND REPORT *Our Common Future* from the World Commission on Environment and Development (1987), the International Chamber of Commerce established a task force of business representatives to create this Business Charter for Sustainable Development. It comprises 16 principles for environmental management which, for business, is a vitally important aspect of SUSTAINABLE DEVELOPMENT. It was formally launched in April 1991 at the Second World Industry Conference on Environmental Management.

1 *Corporate priority.* To recognize environmental management as among the highest corporate priorities and as a key determinant to sustainable development; to establish policies, programmes and practices for conducting operations in an environmentally sound manner.

2 *Integrated management.* To integrate these policies, programmes and practices fully into each business as an essential element of management in all its functions.

3 *Process of improvement.* To continue to improve corporate policies, programmes and environmental performance, taking into account technical developments, scientific understanding, consumer needs and community expectations, with legal regulations as a starting point; and to apply the same environmental criteria internationally.

4 *Employee education.* To educate, train and motivate employees to conduct their activities in an environmentally responsible manner.

5 *Prior assessment.* To assess ENVIRONMENTAL IMPACTS before starting a new activity or project and before decommissioning a facility or leaving a site.

6 *Products and services.* To develop and provide products or services that have no undue environmental impact and are safe in their intended use, that are efficient in their consumption of energy and natural RESOURCES, and that can be recycled, reused or disposed of safely.

7 *Customer advice.* To advise, and where relevant educate, customers, distributors and the public in the safe use, transportation, storage and disposal of products provided; and to apply similar considerations to the provision of services.

8 *Facilities and operations.* To develop, design and operate facilities and conduct activities taking into consideration the efficient use of energy and materials, the sustainable use of renewable resources, the minimization of adverse environmental impact and WASTE generation, and the safe and responsible disposal of residual wastes.

9 *Research.* To conduct or support research on the environmental impacts of raw materials, products, processes, EMISSIONS and wastes associated with the enterprise and on the means of minimizing such adverse impacts.

10 *Precautionary approach.* To modify the manufacture, marketing or use of products or services or the conduct of activities, consistent with scientific and technical understanding, to prevent serious or irreversible environmental degradation.

11 *Contractors and suppliers.* To promote the adoption of these principles by contractors acting on behalf of the enterprise, encouraging and, where appropriate, requiring improvements in their practices to make them consistent with those of the enterprise; and to encourage the wider adoption of these principles by suppliers.

12 *Emergency preparedness.* To develop and maintain, where significant HAZARDS exist, emergency preparedness plans in conjunction with the emergency services, relevant authorities and the local community, recognizing potential transboundary impacts.

13 *Transfer of technology.* To contribute to the transfer of environmentally sound technology and management methods throughout the industrial and public sectors.

14 *Contributing to the common effort.* To contribute to the development of public policy and to business, governmental and intergovernmental programmes and educational initiatives that will enhance environmental awareness and protection.

15 *Openness to concerns.* To foster openness and dialogue with employees and the public, anticipating and responding to their concerns about the potential hazards and impact of operations, products, wastes or services, including those of transboundary or global significance.

16 *Compliance and reporting.* To measure environmental performance; to conduct regular environmental audits and assessment of COMPLIANCE with company requirements, legal requirements and these principles; and periodically to provide appropriate information to the board of directors, shareholders, employees, the authorities and the public.

See also ENVIRONMENTAL MANAGEMENT SYSTEMS. [P.C.]

chasmoendolithic *See* ENDOLITHIC.

chasmogamy *See* CLEISTOGAMY.

cheating In the context of MUTUALISM, means not providing reciprocal benefits. [P.C.]

checkerboard distribution A species distribution pattern in which different species, with similar HABITAT requirements exclusively occupy interdigitated patches of habitat. This has most often been observed among islands on an archipelago, where one, or a group of, species occupy each island in the group, but with no clear spatial trend in the pattern of occupancy. When the species involved are trophically similar, such distributions are strongly indicative of COMPETITION, because they rule out the possibility that the complementary species invaded the archipelago from different directions. *See also* ISLAND BIOGEOGRAPHY. [S.J.H.]

chelating agent The word 'chelate' is derived from the Greek *chela* meaning claw. A chelating agent or chelating ligand applies to a compound containing more than one atom with a lone pair of electrons that can be used to bind a metal ION, i.e. more than one OXYGEN, NITROGEN or SULPHUR atom with a lone pair. The formation of chelate complexes can be used to provide a metal ion in a protected form, for example chelation enables the transport of metal ions across a membrane in biological systems. Additionally, WATER softening can be achieved by the chelation of Ca^{2+} ions. *See also* DETERGENTS. [M.C.G.]

chemical control Attempt to regulate, restrict or remove species populations (of pests) by use of toxic substances. *See also* PESTICIDE. [P.C.]

chemical monitoring Judging the quality of a WASTE stream, effluent/emission, receiving body, particular habitat by analysis of chemical composition, which may be done for one or many chemicals. *Cf.* BIOLOGICAL MONITORING. [P.C.]

chemical oxygen demand (COD) Amount of OXYGEN consumed in the complete oxidation of carbonaceous material in a WATER sample as carried out in a STANDARD test, usually using potassium dichromate as oxidizing agent. *See also* BIOCHEMICAL OXYGEN DEMAND. [P.C.]

Chemical Release Inventory (CRI) List of pollution EMISSIONS from industrial plants in England and Wales controlled under INTEGRATED POLLUTION CONTROL (IPC). *See also* TOXIC RELEASE INVENTORY. [P.C.]

chemical time bomb The time-bomb analogy draws attention to conditions that are set now that might lead to a sudden and serious disaster in the future. Thus synthetic chemicals, especially organics, are known to accumulate in SEDIMENTS. Plausible scenarios can be imagined when changes in physicochemical conditions in the surrounding medium, for example lowering pH due to ACIDIFICATION, can lead to release of these substances and therefore cause sudden and dramatic problems. [P.C.]

chemolithotrophs These are anaerobic bacteria that use inorganic and organic compounds as sources of energy in the absence of light. Chemolithotrophic bacteria are important members of

HYDROTHERMAL VENT communities, where they sometimes live symbiotically in the tissues of filter-feeding bivalves and tube-worms, and oxidize reduced METAL and SULPHUR compounds that are released in the heated vent water. *See also* CHEMOSYNTHETIC MICROBES. [V.F.]

chemoreception The ability to perceive chemical signals. The response of living cells to chemical changes occurs via the plasma membrane and is associated with the opening and closing of ION channels and with changes in the transmembrane flow of ions. Chemoreception is involved in the monitoring of changes in the OXYGEN and CARBON DIOXIDE (CO_2) levels of the blood of vertebrates, and in the pH of their cerebrospinal fluid, such that reflexes are initiated to control the rates of blood flow and breathing. Chemoreception is also involved in the location of food and mates, in COMMUNICATION among individuals of the same species and in navigation through the environment. Smell is usually regarded as the detection of molecules from objects at a distance, and taste as the detection of molecules from objects in contact with some part of the body. However, the distinction between taste and smell is not always clear. *See also* INFOCHEMICALS. [V.F.]

chemosynthetic microbes The chemoautotrophic or chemolithotrophic bacteria obtain their energy by oxidizing reduced inorganic compounds. This energy is used to reduce CARBON DIOXIDE (CO_2) using the Benson–Calvin cycle. Most of these bacteria are true AUTOTROPHS, but some degree of heterotrophy is found among the SULPHUR-oxidizing forms. These bacteria occur primarily in aerobic environments and are very widely distributed, particularly in the SOIL. They are important in the geochemical cycling of NITROGEN and sulphur. Members of the genus *Nitrosomonas* obtain energy by oxidizing AMMONIA to NITRATE, while *Nitrobacter* oxidizes nitrite to nitrate. There are numerous colourless sulphur bacteria that oxidize simple compounds of sulphur, for example sulphide and thiosulphate. One species, *Thiobacillus denitrificans*, can also respire anaerobically. *Hydrogenobacter* is the only form known to oxidize HYDROGEN gas, which occurs as a product of DECOMPOSITION. *Gallionella*, together with some species of *Thiobacillus*, gain energy by oxidizing reduced forms of IRON and MANGANESE. *See also* CHEMOLITHOTROPHS. [A.P.]

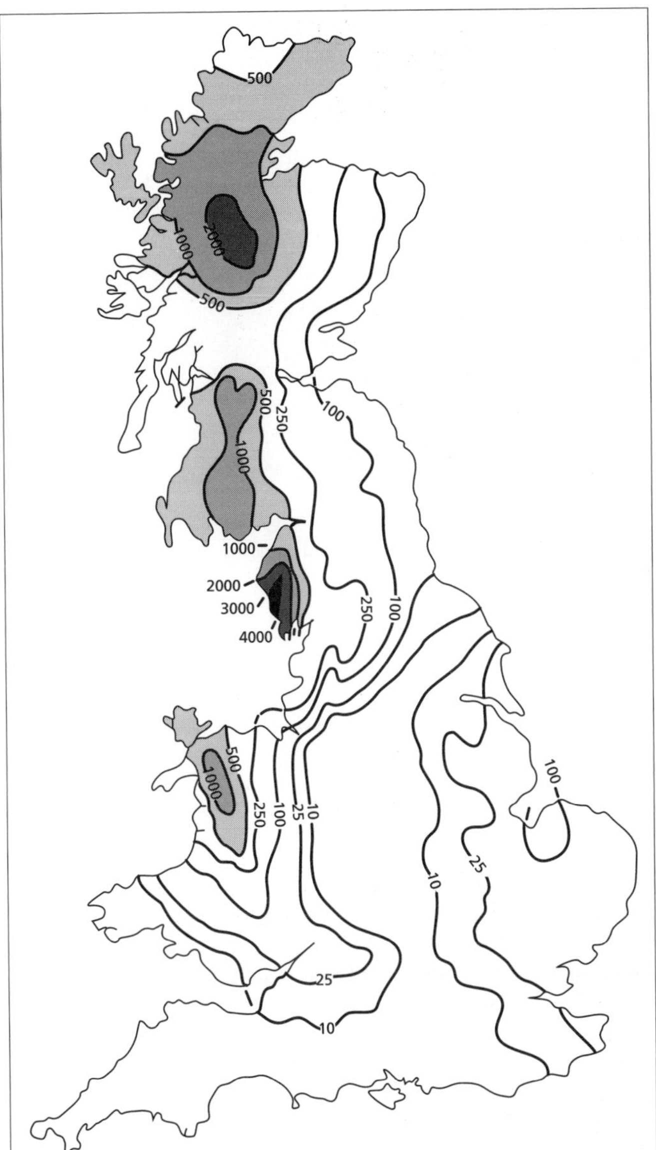

Fig. C10 The distribution in Great Britain of fallout of radioactive caesium (Bq m^{-2}) from the Chernobyl nuclear accident in the Soviet Union in 1986. The map shows the persistence of the pollutant on acid-upland soils where it is recycled through soils, plants and animals. Sheep in the upland areas contained more ^{137}Cs in 1987 and 1988 (after recycling) than in 1986. ^{137}Cs has a half-life of 30 years. On typical lowland soils it is more quickly immobilized and does not persist in the food chains. (After NERC (1990) *Our Changing Environment*. Natural Environment Research Council, London; courtesy of the Institute of Terrestrial Ecology.)

Chernobyl disaster Events at the water-cooled, graphite-moderated NUCLEAR POWER reactor at Chernobyl in the Ukraine in April 1986 proved the most serious ACCIDENT in the history of the peaceful applications of nuclear power. The effects of the explosion and subsequent graphite fire caused the immediate deaths of over 30 fire-fighters and the widespread distribution of radioactive FISSION products over many parts of Europe (Fig. C10). Although this led to restrictions on the use of agricultural land in many countries, these effects are generally minor compared to the local consequences in the Ukraine, Russia and Belorussia where there has been heavy CONTAMINATION, chiefly with CAESIUM and STRONTIUM, causing local authorities to evacuate most agricultural settlements within 30 km of the plant.

There is uncertainty about the direct effect of the RADIATION release on the local and wider populations of the former USSR. Ongoing international studies certainly record a substantial incidence of effects that might be attributable to IONIZING RADIATION but equally (and none the less real) to radiation phobia. Other consequences of the accident include the deterioration of the local diet due to the loss of food production and the disruption of distribution; and the shortage of electricity not only from the No. 4 reactor but from the other reactors on site, which have been closed for various periods since the accident. *See also* ECOLOGICAL/ENVIRONMENTAL DISASTERS; RADIATION POLLUTION. [J.D.L.]

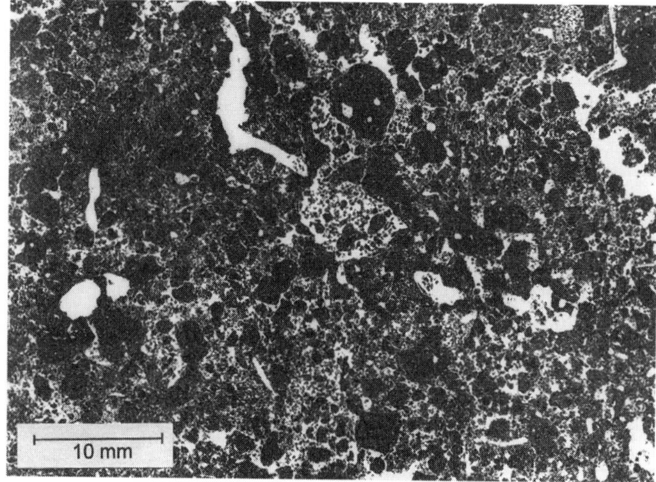

Fig. C11 Compound structure of fine and very fine granular in the topsoil of a Chernozem. The fine granular is produced by earthworms while the very fine granular is produced by enchytraeid worms.

REFERENCES

Borovoi, A. & Bogatov, S. (1997) *Consequences of Chernobyl*. Advances in Nuclear Science and Technology. Plenum Publishing Corporation, New York.

Worley, N., Lewins, J.D. & Lakey, J. (1991) *Chernobyl: Five Years On*. Watt Committee on Energy Paper, London.

Chernozem (Mollisols) This is a type of SOIL that has a thin (1–2 cm) litter at the surface, followed below by a dark-coloured horizon up to 100 cm or more in thickness. It is formed by vigorous incorporation of the litter and mixing by earthworms, enchytraeid worms and small mammals, such as the suslik in Russia (Fig. C11). Thus, this horizon is composed predominantly of faunal faecal material, which imparts a very good structure. This horizon grades down into a horizon of calcium carbonate ($CaCO_3$) accumulation. These soils are formed predominantly on calcareous loess and are very fertile. The term is derived from the Russian *chermi*, black; *zem*, soil. *See also* SOIL CLASSIFICATION. [E.A.F.]

Chestnut soils *See* SOIL CLASSIFICATION.

Chi-squared test (χ^2)

A non-parametric test that compares observed numbers, placed in categories, with those calculated (expected) on the basis of some hypothesis (*see* QUALITATIVE INFORMATION). It is used in two types of situation: GOODNESS-OF-FIT tests and CONTINGENCY TABLES.

Goodness-of-fit situations involve a single classification of categories. For example, imagine that we have spun a coin 100 times, with a result of 46 'heads' and 54 'tails'. Does this result indicate that the coin is biased in some way, i.e. that the occurrence of more than 50 'tails' cannot be attributed to chance? The NULL HYPOTHESIS here is that the coin is unbiased with a PROBABILITY of getting a 'tail' being 0.5. The data would therefore be:

	Heads	Tails
Observed	46	54
Expected	50	50

In this instance the expected values are easily calculated as 'probability × number of observations', i.e. 0.5 × 100 for both 'heads' and 'tails'. The formula for the test is:

$$\chi^2 = \sum \frac{(O-E)^2}{E}$$

where O is the observed number, E is the expected (calculated) number and Σ means sum all the terms $(O-E)^2/E$. Notice that chi-squared tests are carried out on numbers of observations, not frequencies or percentages. When $O=E$, for every category, $\chi^2=0$, indicating no difference between what is observed and what is expected. Increasingly large values of the test statistic indicate that observed and expected are different. At some point this difference could only have been obtained by chance 5% (1%) or less of the time, and it is then said to be significant at the 5% (1%) level. We have rejected the null hypothesis that $O=E$ and accepted the ALTERNATIVE HYPOTHESIS that $O \neq E$. For the chi-squared test, the appropriate relationship between the test statistic and probability depends upon DEGREES OF FREEDOM (v) which, for the above type of test, is $c-1$, where c is the number of categories. For the coin example, $v=2-1=1$ and the test statistic is:

$$\chi^2 = (46-50)^2/50 + (54-50)^2/50$$
$$\chi^2 = 16/50 + 16/50$$
$$\chi^2 = 0.64$$

Looking this value up in STATISTICAL TABLES of χ^2, we discover that for $v=1$ the 5% value of $\chi^2=3.84$. Since our calculated value does not exceed this, the result of the test was not significant. We cannot reject our null hypothesis and therefore we have no reason to believe that the coin is biased. However, a result of 40 'heads' and 60 'tails', which would have given $\chi^2=4.00$, would have been significant at the 5% level and we would have concluded that there had been more 'tails' than we could reasonably have expected by chance alone.

It should also be noted that with goodness-of-fit tests the calculated value of χ^2 can also be smaller than we would expect by chance. In other words, the agreement between observed and expected is too close than we would expect by chance. This is like spinning an unbiased coin 10 times and always getting exactly 5 'heads' and 5 'tails'. Under these circumstances we might begin to suspect that the experimental results have been contrived. Because of this, tables of χ^2 usually contain values of the test statistic that correspond to large probabilities such as $P=0.95$.

A more ecological example of data in categories is given under QUALITATIVE INFORMATION for reed warbler nests in different types of habitat. The calculation of expected values is more problematic with this example. The null hypothesis in this case would be that warblers are placing nests into habitats at random, i.e. they have no preference. However, we do not simply expect equal numbers of nests in all four habitats if the warblers are not choosing a particular one. A more reasonable calculation of expected numbers would be based upon the relative areas of the four habitats available to the birds. Let us assume that the percentage area for each type of habitat was: reeds 15%, wet scrub 25%, dry scrub 25% and woodland 35%. On this basis the table of observed and expected values would look like this:

	Reeds	Wet scrub	Dry scrub	Woodland
Observed	12	6	2	2
Expected	3.3	5.5	5.5	7.7

with $\chi^2=29.18$, with 3 degrees of freedom. This exceeds the 1% value of χ^2 in the table and therefore observed and expected are said to be very significantly different. This implies that reed warblers are selecting habitats for reeds and against woodland.

There are two complications with the chi-squared test that arise because the test statistic evaluated by the formula (a discrete quantity called by some authors X^2) is only approximately equal to the continuous PROBABILITY DISTRIBUTION given in the tables (χ^2). Usually this does not matter, but in two situations it does: when $v=1$ and when $E<5$. With 1 degree of freedom the solution is to use YATES' CORRECTION FOR CONTINUITY. This involves subtracting 0.5 from the numerator in the chi-squared formula, before squaring. The formula, including Yates' correction is therefore:

$$\chi^2 = \sum \frac{\left(\left|O-E\right|-0.5\right)^2}{E}$$

where $|O-E|$ indicates that it is the absolute value of the difference from which 0.5 should be subtracted. For the coin example above, to which Yates' correction should have been applied, the correct value of $\chi^2=0.49$, which is still not significant.

With expected values less than about 5 the only solution is to collect more data or amalgamate columns. For example, with the reed warbler data we have one expected less than 5 (3.3), which could be overcome by adding the data for the 'wet' habitats together. This would give three categories and a new $\chi^2=15.82$, which is still significant at the 1% level. It is worth noting that the traditional lower limit of $E=5$ was suggested by Fisher (1948) but that Cochran (1954) suggests that this is too conservative and recommends a lower limit of $E=1$.

Contingency table situations involve at least a double classification of categories. Imagine that we have placed 100 quadrats in a certain habitat and recorded the presence or absence of two species of plant, A and B. The results are recorded in the following contingency table:

		Species A +	Species A −
Species B	+	15	45
Species B	−	30	10

Thus 15 quadrats contained both species, 45 contained only species B, 30 contained only species A and 10 contained neither species. With contingency tables the expected values are calculated from the row and column totals, with a null hypothesis that the two classifications are not associated. For example, the probability of being in the top row is $60/100=0.6$ and the probability of being in the left column is $45/100=0.45$. Therefore the probability of being in the top-left cell of the table is $0.6\times0.45=0.27$, provided that the two classifications are independent (the null hypothesis). Therefore the expected number in this cell will be $0.27\times100=27$. In other words, the expected value for the top-left cell is:

$$E = 60/100 \times 45/100 \times 100$$

$$E = \left(60 \times 45\right)/100$$

$$E = 27$$

Each expected is therefore calculated by multiplying the row and column totals together and then dividing by the grand total. If we do this for each cell we finally have:

Observed

15	45
30	10

Expected

27.0	33.0
18.0	22.0

χ^2 is calculated using the formula already given, using Yates' correction if necessary. For a contingency table, degrees of freedom are (columns−1)(rows−1). For the above 2×2 table this equals 1 and therefore Yates' correction should be used. For each cell of the table we calculate $(|O-E|-0.5)^2/E$ and obtain the four values

4.89	4.01
7.35	6.01

which added together give $\chi^2=22.26$, which is significant at the 5% level. We therefore reject the null hypothesis of no association between the two species. As with goodness-of-fit tests, it is unwise to proceed if $E<5$ (but again note Cochran 1954). With a large contingency table it may be possible to amalgamate certain rows or columns, but with 2×2 tables, FISHER'S EXACT TEST should be employed.

For more advanced topics in chi-squared tests, such as tables with three classifications and more, the reader should refer to Conover (1971) and Sokal and Rohlf (1995). *See also* NON-PARAMETRIC STATISTICS. [B.S.]

REFERENCES

Cochran, W.G. (1954) Some methods for strengthening the common χ^2 tests. *Biometrics* **10**, 417–451.

Conover, W.J. (1971) *Practical Nonparametric Statistics*, 2nd edn. John Wiley & Sons, New York.

Fisher, R.A. (1948) *Statistical Methods for Research Workers*. Oliver & Boyd, Edinburgh.

Sokal, R.R. & Rohlf, F.J. (1995) *Biometry*, 3rd edn. W.H. Freeman, New York.

chill coma A state of cold-induced immobility accompanied by a lower body TEMPERATURE and reduced metabolism. Response is directly related to temperature and is reversible on a regular basis (in contrast to DIAPAUSE). Exothermic organisms such as insects may become inactive in a DIURNAL cycle with low night-time temperatures, or for a prolonged period of time through winter. Most insects are immobile below 0°C and take several hours to become fully active on return to favourable temperatures. [J.S.B.]

china clay China clay, or kaolin, is important for the manufacture of paper and china. It is a breakdown product of more complex CLAYS derived from feldspars and other minerals occurring in granites and related rocks. Major deposits occur in south-west England due to localized volcanic activity within ancient granitic masses. As a result the clay is mixed with silica and mica derived from the granite in the ratio 1 : 6 : 1.

The kaolin is extracted hydraulically, leaving large amounts of sand waste in conspicuous heaps, and smaller amounts of mica waste in lagoons. The resulting moonscape of pits and eroding heaps casts considerable environmental blight on the region.

In the 1970s a programme of RECLAMATION was begun, aimed at establishing a simple grass cover, in keeping with the surrounding farmland. This met with success once it was appreciated that the major problem to be overcome was extreme lack of NUTRIENTS, especially NITROGEN for which a soil capital of 1000 kg ha^{-1} is needed. This can only realistically be provided by using nitrogen-fixing legumes and equivalent species. Nitrogen-fixing trees are also possible.

In other areas of the world, such as the southern USA, kaolin occurs in secondary ALLUVIAL deposits. Extraction of these produces little waste. Nevertheless the worked-out pits have the same acute nutrient deficiencies. These are also treated by the use of legumes, especially *Lespedeza sericea,* combined with the local conifer, *Pinus taeda*. [A.D.B.]

REFERENCE

Wardell Armstrong (1993) *Landscaping and Revegetation of China Clay Wastes.* Department of the Environment, London.

Chiropterans, diversity and conservation

The order Chiroptera (from Greek, 'foot wing') comprises the only truly flying mammals: bats. In all bats the forelimbs are greatly enlarged and support wings of skin. In all forms there are five fingers. The first (the pollex or thumb) is very short and at least partially free of the flight membrane. The second digit supports the distal edge of the membrane and consists of the single metacarpal in advanced forms. The remaining three digits retain phalanges within the membrane. The wing membrane attaches to the ankle, but the foot bears at least vestiges of all five toes free of the membrane. In most forms there is an interfemoral membrane between the hindlimbs, supported by a sturdy strut, the calcar, extending from the heel and often including some or all of the tail, if one is present. All bats have prominent ear pinnae, and most have a leathery, antenna-like structure within each pinna called the tragus.

As many as 18 families of bats are recognized and the order is worldwide in distribution except for the polar regions and some remote oceanic islands. With nearly 1000 species, bats are second only to rodents in DIVERSITY among mammalian orders. The status of several nominal families is controversial. Koopman (1993) provides a detailed classification.

Arguably, the most primitive family is Pteropodidae, the Old World fruit bats or flying foxes, with more than 170 species. These are often put in their own suborder, Macrochiroptera, although some are smaller than the largest of other bats, the 'Microchiroptera'. It is said their brain structure resembles that of primates more than other bats, but biochemically they fit with the rest. They have big eyes and see very well, although a few species are capable of echolocation (the normal orientation system of most bats). They have a metacarpal and one phalange in the second digit on the forward edge of the wing, and usually a claw at this point. In addition to fruit, many eat insects and some are specialized feeders on pollen and nectar from flowers. The family occurs nearly throughout the Old World tropics, including Australia and many Indian Ocean and South Pacific islands. The largest flying foxes have wingspreads over 1.7 m and several species are regularly eaten by humans.

The Emballonuridae, sac-winged or sheath-tailed bats, include about 50 species in the warmer parts of all continents, most tropical islands including Indonesia, the Philippines and even the Seychelles and Solomon Islands, but are absent from most of the West Indies. These bats are named for the sac glands in the membrane anterior to the humerus near the shoulder. Their tails exit the interfemoral membrane dorsally and can slide in and out of a skin tube, allowing great expansion and contraction of the membrane.

The Rhinopomatidae, three species of rat-tailed bats, occur from North Africa to Thailand and Sumatra. They have very long tails, almost entirely free of the interfemoral membrane, which is not sheathed. They resemble emballonurids in many respects, but like pteropodids they have a phalange in the second digit of the wing. Another close relative is Kitti's bat (*Craseonycteris thonglongyai*), perhaps the smallest living mammal at 2 g, and in a family of its own, Craseonycteridae. Known only from Thailand, it lacks a tail and a calcar. Little bigger than a bumble-bee, it is insectivorous like emballonurids and rhinopomatids.

The Old World horseshoe and nose-leaf bats, Rhinolophidae, include more than 170 species, but the group is about equally split if one separates out all but the genus *Rhinolophus* as family Hipposideridae. All these bats are remarkable in lacking a tragus and all have fancy facework involving combinations of flesh folds, often forming horseshoe-shaped flanges, horn-like tridents or projections like leaves and spear blades. Perhaps the most bizarre is *Anthrops*, the flower-faced bat of the Solomon Islands.

Hipposiderids, as opposed to *Rhinolophus* (the Rhinolophidae in the strictest sense), have distinctive pectoral and pelvic girdles and only two, as opposed to three, phalanges in toes two to five. The whole group is insectivorous. Rhinolophids in the broad sense occupy Eurasia, the East Indies, the Philippines and some Pacific Islands.

The slit-faced or hollow-faced bats, Nycteridae, have a spectacular concavity in the front of the skull flanked externally by large skin folds. Their ears are united across their heads. Their most peculiar feature, unique among mammals, is their T-shaped terminal tail vertebral cartilage, which gives great strength to the interfemoral membrane. There are 13 species in Africa, Madagascar, north to the Mediterranean and east through southern Asia to Indonesia. All are insectivorous and some also catch and eat other prey, including small vertebrates.

The Megadermatidiae, or false vampires, also have very large, united ears, but no tail. They have enormous nose leaves and a unique, two-part, divided tragus. There are only five species, ranging from Africa through southern Asia to the Philippines, with the largest an Australian representative with a wingspread of 60 cm. These are vigorous carnivores, consuming various sorts of small animals and insects too, of course.

The New World leaf-nosed bats (Phyllostomidae) are diverse, with about 150 species ranging from the Florida Keys and southwestern USA to Argentina. They dominate the mammal faunas of the West Indies. Most, but not all, have prominent nose leaves or other fleshy projections, but the group is hard to characterize. The Mormoopidae, moustached or naked-backed bats, are eight species spread from Arizona and Texas through most of Brazil. They differ from phyllostomids in having chin leaves rather than nose leaves, or at least very modified lower lips. The mouth opens to form a funnel. In several species, the wing membranes, quite hairless, join across the back.

Another phyllostomid derivative is Noctilionidae, the fisherman bats, with but two species in the Caribbean, Central America and northern South America. They lack nose leaves but have enormous cleft lips, huge feet and rigidly ossified pelvic girdles. They eat insects but also catch fish. Their wingspreads may exceed 0.5 m.

The phyllostomoid radiation includes the greatest diversity of bat diets. Not only are they insectivores and fish-eaters, but some prey on small vertebrates like frogs. Within Phyllostomidae are included a radiation of fruit-eaters and another of nectar- and pollen-gathering flower-feeders; some are folivores (leaf-eaters). And it is to this family that the only true vampires, living wholly on fresh blood, genus *Desmodus*, belong.

A plurality of the world's bat species belong to the last radiation, the vespertilionoids, led by Vespertilionidae with over 350 species. Most are insectivores, but at least one catches fish. There are a number of widely recognized splinter families. Two in tropical America are the Furipteridae, or thumbless bats, and the Thyropteridae, or disc-winged bats. The former, with two species, do have thumbs, but they are vestigial and enclosed in the wing membrane except for the claw. Disc-wings, or American sucker-foot bats, have amazing, stalked, round, suction pads subtending each thumb and on the sole of each foot. There are also but two species. Disc-wings live inside rolled leaves of large size, like those of bananas. Both groups seem to eat mostly moths.

A single species on Madagascar is given its own family, Myzopodidae. It has disc suckers on wrists and ankles, but they are not stalked. Histologically, the suckers of *Myzopoda* are not homologous to those of thyropterids. This sucker-foot bat has very large ears and a long tail extending well beyond the interfemoral membrane. The toes are syndactylous: the phalanges are reduced to two and fused.

The Natalidae, funnel-eared bats, blend many features of furipterids and thyropterids with those of generalized vespertilionids. Even their ears are not really very distinctive. Their family rank is based on peculiar vertebral fusions. The four species, all insectivores, occur from Mexico to Brazil.

Free-tailed bats, Molossidae, are widespread and abundant throughout the warmer parts of the world; there are about 90 species. They have bristle-fringed hind feet and tails that extend well beyond the interfemoral membrane. These are the most advanced fliers of all mammals, with several structural modifications of the pectoral girdle to give great strength to the wing beats. Some of the commoner species may be among the world's most numerous mammals, with colonies in the millions. All insectivorous, molossids are KEYSTONE SPECIES in many ecosystems.

The New Zealand short-tailed bats are close to molossids, but have a secondary claw, called a talon, on the thumb (and toes too), and a broad hind foot with a grooved plantar pad. The tail perforates the interfemoral membrane dorsally. At the opposite pole from molossids, these bats called Mystacinidae may be the most specialized for climbing. The two very similar species are insectivores.

An overview of bat biology is provided by Kunz and Pierson in Nowak (1994), who provides species accounts. Ecology is treated in detail by Findley (1993).

Conservation

Kunz and Pierson state: 'Bat populations appear to be declining almost everywhere in the world'. Habitat destruction is probably the major threat to bats. DEFORESTATION not only eliminates foraging habitats but also removes roost sites in old, hollow trees. Mines and caves are often sealed as human safety measures, and many caves are commercialized or vandalized. Wherever numbers of bats come into close contact with humans, problems are apt to arise. Humans are usually offended by bat droppings (despite the economic importance of bat guano as fertilizer and a source of chemicals). Typically, steps are taken to extirpate, or at least drive out, bat colonies from edifices, including commercialized caves.

In recent years, at least in Europe and North America, there has been some reversal of the anti-bat trend: a few humans have put up bat houses. Although successful occupation of these is described as moderate, that can scarcely compensate for the tens of thousands of bats killed by pesticides, lumber-treating toxins, environmental pollutants, destruction of roost sites and outright human predation. No ENDANGERED SPECIES are known to use bat houses as yet. At least seven species of bats are believed to have been exterminated in the last few decades and eight species have been reduced to remnant populations.

In view of their enormous importance as insectivores, seed dispersers and pollinators in nearly every terrestrial ecosystem, bats deserve far more conservation effort. *See also* CONSERVATION BIOLOGY. [J.D.Lz.]

REFERENCES

Findley, J.S. (1993) *Bats*. Cambridge Studies in Ecology. Cambridge University Press, Cambridge.

Koopman, K.F. (1993) Chiroptera. In: *Mammalian Species of the World* (eds D.E. Wilson & D.M. Reeder), 2nd edn, pp. 163–175. Smithsonian Institution Press, Washington, D.C.

Nowak, R.M. (1994) *Walker's Bats of the World*. Johns Hopkins University Press, Baltimore, Md.

---

chloride A compound of CHLORINE with another element (e.g. calcium chloride, $CaCl_2$). Also, an atom of chlorine in solution, bearing a single negative charge. Chloride is, by mass, the most abundant constituent of seawater (followed by SODIUM) and is present in 35‰ seawater at a concentration of 19.353 g kg^{-1}. [V.F.]

chlorinated hydrocarbons *See* ORGANOCHLORINES.

chlorine (Cl) Chlorine, a greenish-yellow poisonous gas (ATOMIC NUMBER = 17; relative ATOMIC MASS = 35.453; melting point = −100.98°C; boiling point = −34.6°C) is found in abundance as sodium chloride (NaCl) in seawater (30 kg m^{-3}), from which it is extracted by electrolysis. It is used on a large scale to produce organochlorine compounds (e.g. polyvinylchloride, PVC), bleaches and inorganic chlorine compounds. Chlorine gas is highly toxic. Hydrochloric acid (HCl) is an important compound that is used in the pickling of steel, the neutralization of effluent and the desulphurizing of PETROLEUM. Chlorine atoms and chlorine oxides (formed from CFCs in the STRATOSPHERE) have been implicated in the depletion of the Earth's OZONE LAYER. [C.R.P.]

chlorinity Grams of CHLORIDE ions per 1000 g of seawater expressed as parts per thousand (‰). The constancy of the proportions of the major dissolved ELEMENTS in seawater permits the use of chlorinity as an index of SALINITY. For example, by titrating a seawater sample with silver nitrate ($AgNO_3$), the amount of chloride (plus a chloride equivalent of bromide and iodide) is obtained and relates to salinity according to the equation: S (‰) = 1.80655 × Cl (‰). Chlorinity measurements generally give a less accurate estimate of salinity than do electrical conductivity techniques. [V.F.]

chlorofluorocarbons *See* CFCs.

chlorophyll *See* PHOTOSYNTHESIS.

cholinesterase inhibitor Substance that inhibits the action of acetylcholinesterase (AChE) and causes hyperactivity in parasympathetic nerves. Includes carbamate and organophosphorous ESTERS that are contained in many INSECTICIDES. *See also* PESTICIDE. [P.C.]

choosiness Rejection of prospective mates. In most animal species females are more choosy than males. The magnitude and direction of this asymmetry arises from differences between the sexes in the effects of choosiness on reproductive output and survival. In many taxa, males provide less per brood in the form of GAMETES and PARENTAL CARE than females. Males therefore have a higher potential rate of reproduction; they are limited more strongly by access to mates than are females, which depend more on access to RESOURCES such as food and high-quality breeding sites. Thus, while both sexes may obtain better mates (including genes and resources offered), males that reject potential mating partners would suffer a greater loss in reproductive success. Males in many taxa have therefore evolved elaborate COURTSHIP DISPLAYS (*see* SECONDARY SEXUAL CHARACTERISTICS) and are less discriminating than females, which can afford to be choosy because receptive males are rarely in short supply. Of course, where the parental contributions of males and females are similar, potential rates of reproduction are also similar and the two sexes may be equally choosy. *See also* SEXUAL SELECTION. [J.D.R.]

REFERENCES

Clutton-Brock, T.H. & Parker, G.A. (1992) Potential reproductive rates and the operation of sexual selection. *Quarterly Review of Biology* **67**, 437–456.

Reynolds, J.D. (1996) Animal breeding systems. *Trends in Ecology and Evolution* **11**, 68–72.

chreod *See* CANALIZATION.

chromium (Cr) A HEAVY METAL (ATOMIC NUMBER = 24; relative ATOMIC MASS = 52.01; melting point = 1857°C; boiling point = 2672°C) that is a trace constituent (100 mg kg^{-1}) of the Earth's CRUST, although concentrations can be much higher in soils developed on serpentinites. Found in chromite ($FeO.Cr_2O_3$), the main use of chromium is in the IRON–chromium alloy of stainless steel. The dumping of chromium wastes from tanneries and electroplating results in the POLLUTION of soils, where the anionic hexavalent Cr(VI) form is more plant-available and phytotoxic than the relatively insoluble and less mobile trivalent Cr(III) form. While chromium can cause cancer and dermatitis in those exposed to chromium-containing dusts, it is also an important TRACE ELEMENT essential for carbohydrate metabolism in humans. [J.F.]

chronic tests Testing for responses that are not immediately debilitating, such as impairment of growth, reproduction, propensity for diseases including cancer. Therefore these tests are carried out at relatively low concentrations over long time-spans and hence at low concentrations *Cf.* ACUTE TESTS; SHORT-TERM TESTS. *See also* LONG-TERM TESTS; PREDICTIVE ECOTOXICITY TESTS. [P.C.]

chronospecies Also known as successional species, palaeospecies or evolutionary species, this is a palaeontological species concept used in the systematics of extinct organisms. When a LINEAGE reconstructed from fossils exhibits substantial evolutionary changes, it is convenient to divide its history into separate phases. The successive stages of the lineage, defined on the basis of morphological similarity, are called chronospecies. As opposed to SPECIATION, which refers to the splitting of lineages, the transition from one chronospecies to another is termed phyletic speciation. The division of the continuum into chronospecies is arbitrary. [P.O.]

chronostratigraphy Chronostratigraphy is based on measurement of the age of geological strata, normally by an independent method that frequently depends on quantifying the isotopic ratios in a decay series such as $^{14}C/^{12}C$, THORIUM–URANIUM and POTASSIUM–ARGON. In recent times, TEPHRA (volcanic DUST), which can be chemically fingerprinted to particular volcanic eruptions the age of which may be known, has been widely used to provide fixed time-points in sedimentary columns. Its usefulness as a chronostratigraphic marker depends largely on the dispersal of ash from large volcanic eruptions over wide geographical areas. Strata defined on strict chronological criteria may coincide, but not necessarily, with those established on the basis of BIOSTRATIGRAPHY or lithostratigraphy. Chronostratigraphical units overcome difficulties associated with those based on other criteria, which often result in units that are time-transgressive to varying degrees. It should be noted, however, that many of the widely used DATING techniques, such as ^{14}C dating, have a level of accuracy that sometimes falls short of what is required. Where possible, it is desirable that CLASSIFICATION be based on chronostratigraphical or well-defined biostratigraphical criteria rather than on imprecisely defined units such as the Blytt–Sernander Boreal, Atlantic, Subboreal and Subatlantic subdivisions of the Holocene. *See also* STRATIGRAPHY. [M.O'C.]

REFERENCES

Doyle, P., Bennett, M.R. & Baxter, A.N. (1994) *The Key to Earth History. An Introduction to Stratigraphy*. John Wiley & Sons, Chichester.

Mangerud, J., Andersen, S.T., Berglund, B.E. & Donner, J.J. (1974) Quaternary stratigraphy of Norden, a proposal for terminology and classification. *Boreas* **3**, 109–128.

chronozone A chronozone is a chronostratigraphical unit, the upper and lower boundaries of which are defined on the basis of age (*see* CHRONOSTRATIGRAPHY). Age is normally determined by a radiometric method, i.e. by an absolute or independent, rather than relative, method of DATING. The particular radiometric method employed depends largely on the age of the stratum being dated and its constituent components. The radiocarbon method, which is a measure of the ^{14}C content of the organic component, is most commonly used to date Late Quaternary organic-rich sediments (younger than *c.* 60 000 years) while other methods, such as URANIUM series dating (Th/U), thermoluminescence (TL) and AMINO ACID dating, are employed to date older Quaternary deposits (Smart & Frances 1991). Chronozones facilitate time correlation, and where the boundaries of chronozones coincide with those of local and especially regional biostratigraphic or lithostratigraphic units the possibilities for reconstruction of past environments over wide areas are greatly enhanced. [M.O'C.]

REFERENCE

Smart, P.L. & Frances, P.D. (eds) (1991) *Quaternary Dating Methods—A User's Guide*. Quaternary Research Association, London.

CI *See* CONDITION INDEX.

Cinnamonic soils *See* SOIL CLASSIFICATION.

circadian rhythms Endogenous self-sustained rhythm (OSCILLATION; sometimes called entrained rhythm) with a periodicity close to 24 h. Both the period and phase of the rhythm become entrained (synchronized) to 24 h by external cues (zeitgebers). Dominant entraining cue is the 24-hour light–dark cycle, which may be reinforced by the daily temperature cycle. Circadian rhythms of locomotory activity occur in individual organisms at different times in the daily cycle in a species-specific way: cockroaches are nocturnal, tsetse flies are DIURNAL. Circadian rhythms are temperature compensated; the Q_{10} is ~ 1. Once in a lifetime population events (egg hatch, insect emergence from pupae) may also show a circadian periodicity. Other biological rhythms have a synchronized periodicity of 12.4 h (tidal), 14.76 days (semi-lunar), 29.53 days (lunar) and 1 year (circannual). *See also* ENDOGENOUS RHYTHM. [J.S.B.]

REFERENCE

Saunders, D. (1977) *An Introduction to Biological Rhythms*. Blackie, London.

circulation, oceanic OCEAN circulation can be subdivided into two components: surface circulation and thermohaline deep circulation. The major surface CURRENTS of the oceans are primarily WIND driven and thus closely follow the major wind systems. However, due to the rotation of the Earth, currents are deflected to the right in the Northern hemisphere (resulting in a tendency towards clockwise circulation patterns) and to the left in the Southern hemisphere (resulting in counterclockwise circulation patterns). Currents tend to be narrower, deeper and faster along the western edges of the oceans compared with those along the eastern edges. Western boundary currents transport substantial amounts of heat toward the poles, whereas eastern boundary currents are often associated with UPWELLING. Circulation greatly influences ocean PRODUCTIVITY through the transport of NUTRIENTS and PELAGIC and benthic organisms. Circulation patterns of deep water are controlled primarily by differences in density, which occur as a result of differences in SALINITY and TEMPERATURE. Deep water originates primarily at the poles. Due to the high density of this very cold water, it sinks and flows toward the Equator. Originally high in OXYGEN, the deep water becomes gradually depleted in oxygen as it spreads across the ocean floor. The deep water is eventually returned to the surface by wind-driven mixing. [V.F.]

circumpolar Distributed around the regions of the North Pole (circumboreal) or South Pole (circumaustral or circumantiboreal). In the Northern hemisphere the circumboreal BIOTA represent a distinct flora and fauna, with many species exhibiting a complete circumpolar DISTRIBUTION. Several animal species, notably polar bear (*Ursus maritimus*) and arctic fox (*Alopex lagopus*), migrate for great distances around the entire polar region, while others are more regional and have developed local varieties as in the Spitsbergen (Svalbard) race of reindeer (*Rangifer tarandus*). Circumpolar plant communities are almost entirely composed of circumpolar genera with vicarious species occupying similar habitats in different geographical areas. Many arctic plant species are completely circumpolar but show distinct ecotypes depending on habitat and geographical region. [R.M.M.C.]

REFERENCE

Polunin, N. (1959) *Circumpolar Arctic Flora*. Clarendon Press, Oxford.

cirque lakes *See* LAKES; LAKES, TYPES OF.

CITES (Convention on the International Trade in Endangered Species) *See* INTERNATIONAL CONSERVATION CONVENTIONS.

civil liability for environmental damage *See* ENVIRONMENTAL LAW.

clade *See* MONOPHYLY.

cladism *See* PHYLOGENETIC SYSTEMATICS.

cladistic species concept The cladistic species concept is one of three major SPECIES CONCEPTS currently adopted by practitioners of phylogenetic systematics, the other two being the evolutionary species concept (*see* EVOLUTIONARY TAXONOMIC CLASSIFICATION) and the phylogenetic species concept (*see* PHYLOGENETIC SYSTEMATICS). The cladistic species concept has two major elements. First, each species to be analysed phylogenetically must be recognized a priori by non-phylogenetic criteria, such as reproductive cohesion within the putative species or lack of interbreeding with other putative species. Second, each branch of a PHYLOGENETIC TREE represents a distinct species. Terminal branches therefore represent species distinct from their non-terminal, or ancestral, antecedents, so no contemporaneous species are ever considered to be persistent ancestors. The evolutionary species concept as proposed by Wiley is a modification of the species concept of the same name orginally formulated by George Gaylord Simpson. This concept differs from the cladistic species concept by permitting some terminal taxa to be interpreted as persistent ancestors if they lack any apomorphic TRAITS. The phylogenetic species concept provides a purely phylogenetic criterion for recognizing species. Each population of organisms that exhibits a diagnosable trait (a putative autapomorphy) is a distinct species. *See also* CLASSIFICATION. [D.R.B.]

cladistics The process of determining the relationships between clades of organisms, i.e. their pattern of descent from closest common ancestors. The underlying theory was developed most explicitly in the 1950s by the German entomologist, Willi Hennig. His rationale was that only shared derived character states, called synapomorphies, were informative about the evolutionary relationships between organisms (Fig. C12). Nowadays the term 'cladistics' is often taken to refer to a suite of methods and principles (see below).

If it was always possible to distinguish derived, apomorphic, character states from primitive, plesiomorphic ones then reconstructing phylogenies would be a trivial matter. However, parallel and CONVERGENT EVOLUTION and character state reversals are extremely common, and it is often impossible to distinguish similar states resulting from these evolutionary events from truly homologous states that indicate PHYLOGENY. The noise in the data resulting from convergence, reversals, etc. is called HOMOPLASY. The most widely applied principle is that of PARSIMONY. The argument is that if character state changes occur rarely during evolution, then that relationship between taxa (tree) which requires the fewest evolutionary steps, i.e. is most parsimonious, constitutes the best hypothesis about relationships. The advent of fast computers has enabled parsimony to be applied to quite large datasets, but it is never likely that it will be possible to guarantee finding the shortest tree for more than about 25 taxa, and often considerably fewer. This is because of the very large number of possible ways that the members of a set of taxa can potentially be related to one another: 15 ways for five taxa but more than 10^{74} for 50 taxa.

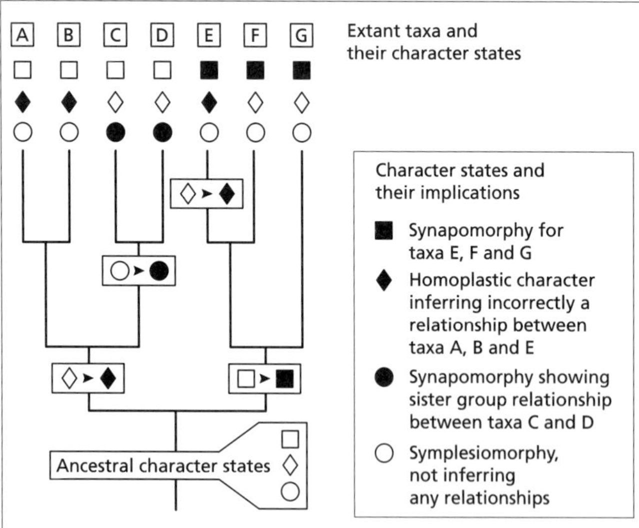

Fig. C12 Cladogram with superimposed character state changes. Some characters give a true picture of evolutionary relationships (square and circle) whilst others are homoplastic (diamond), i.e. they either display convergent evolution (as here) or reversals. The greater the proportion of homoplastic characters in the dataset, the more difficult phylogeny reconstruction becomes.

One alternative to parsimony is compatibility analysis. The underlying principle is that true characters, i.e. ones that do not display homoplasy, will always be compatible with one another. If two characters suggest different and conflicting evolutionary histories, then at least one of them has to be false. Compatibility analysis searches for the largest mutually compatible clique of characters in a dataset. However, in practice, the high levels of homoplasy in most datasets means that the largest clique will often comprise only a small proportion of the characters. As most homoplastic characters still provide some signal about relationships, compatibility analysis results in throwing away a considerable amount of information.

Another approach, which is largely restricted to various forms of molecular data, is maximum likelihood. One of the problems with parsimony is the so-called Felsenstein zone. This is the problem that when rates of evolution along different lineages differ widely, those lineages with the higher rates of change lead to long branches that tend to attract one another because homoplastic change falsely suggesting relationships will come to exceed the true signal which is due to HOMOLOGY. This is largely circumvented by MAXIMUM LIKELIHOOD METHODS, which seek to discover the tree most likely to have given rise to a given dataset. Further, maximum likelihood does not make the assumption that character state change is innately unlikely. Unfortunately, maximum likelihood requires a precise model of evolution, i.e. it requires the probabilities of all possible character state transitions. With DNA sequence data, and a small number of other molecular types of data, the probabilities of different character state transitions can be estimated empirically. *See also* CLASSIFICATION; NATURAL CLASSIFICATION. [D.L.J.Q.]

REFERENCES

Hennig, W. (1966) *Phylogenetic Systematics*. University of Illinois Press, Urbana, Ill.

Quicke, D.L.J. (1993) *Principles and Techniques of Contemporary Taxonomy*. Chapman & Hall, London.

cladogenesis Literally meaning branching divergent EVOLUTION, cladogenesis refers to the production of new species via the subdivision of an ancestral species. The term 'cladogenesis' encompasses an array of SPECIATION mechanisms, such as geographical subdivision of the ancestor or active movement of organisms followed by their subsequent isolation in a peripheral habitat. The end-result of such processes is the production of two or more species from one species. Although the term is usually applied to phylogenetic patterns exhibited by species, populations within a species may also show a branching pattern of interrelationships. In this case, cladogenesis is the result of geographic subdivision (vicariance) of the species producing different local populations. The populations are not thought to represent independent evolutionary lineages. Cladogenesis therefore increases the number of species of differentiated populations and differs from reductive speciation, which results in the origin of one new species from two previously existing species. *See also* ISOLATING MECHANISMS; SPECIES CONCEPTS. [D.R.B.]

cladogram In evolutionary biology, a DENDROGRAM is a branching diagram containing entities linked by some criterion. The term 'cladogram' was originally coined independently by Robert Sokal and Ernst Mayr as a perjorative term for the branching diagrams produced by phylogenetic systematic analysis. Phylogenetic systematists subsequently adopted the term. There are currently two different ideas about what a cladogram is within the systematic community. Traditional phylogenetic systematists, or phylogeneticists, consider a cladogram to be a branching diagram of entities in which the branching is based on the inferred historical connections among the entities. These historical connections in turn are indicated by synapomorphies. Phylogeneticists thus tend to use the terms 'phylogenetic tree' and 'cladogram' interchangeably. Advocates of the derivative transformed CLADISTICS movement, or cladists, have expanded the definition of cladogram to include not only the product of a phylogenetic systematic analysis (a PHYLOGENETIC TREE) but also the results of phylogenetic studies in COEVOLUTION and historical BIOGEOGRAPHY (e.g. AREA CLADOGRAMS or host cladograms). For some cladists, the term 'cladogram' is considered synonymous with any branching diagram of taxa derived in any manner (e.g. a phenogram). [D.R.B.]

class *See* TAXONOMIC CATEGORIES.

classification The process of arranging a set of objects in an order. The procedure is usually hierarchical. In the biological context there are two objectives. One is to produce a system allowing different types of organism to be referred to unambiguously. The other is to arrange them in an order that corresponds as nearly as possible to their evolutionary histories. The two processes sometimes conflict. A standard nomenclature is now used, starting from species and grouping progressively into more inclusive categories of genus, family, order, class, phylum and kingdom. There are rules governing the use of these names, so that reference to particular types is unambiguous (*see* LINNAEAN CLASSIFICATION). Phylogeny cannot be known for certain, and similarity and relatedness are not directly proportional to each other, so that the extent to which such a classification indicates the evolutionary relationship of the organisms is uncertain. *See also* BIOLOGICAL SPECIES CONCEPT; EVOLUTIONARY TAXONOMIC CLASSIFICATION; PHYLOGENETIC SYSTEMATICS; SYSTEMATIC METHODS; TYPOLOGY. [L.M.C.]

clay SOIL component consisting of mineral particles <2–4 μm, of low permeability and capable of being moulded when moist. [P.C.]

Clean Air Act *See* UNITED STATES LEGISLATION.

clean-up Action to move and remedy CONTAMINATION/POLLUTION. [P.C.]

Clean Water Act *See* UNITED STATES LEGISLATION.

clean(er) technology It is dubious if technology can ever be absolutely clean; invariably it uses up natural resources and generates WASTE. However, some options are likely to be better, i.e. cleaner, than others. This is the philosophy behind best available technology (BAT) and BEST PRACTICABLE ENVIRONMENTAL OPTION (BPEO). *See also* ENVIRONMENTAL INDUSTRIES SECTOR. [P.C.]

cleidoic *See* EGGS.

cleistogamy Cleistogamous flowers are those that never open, appearing not to develop beyond the bud stage. They may be produced in the absence of chasmogamous flowers, which open normally, for example *Salvia, Cleistogama*. However, a considerable number of plants produce both chasmogamous and cleistogamous flowers, sometimes together but more usually at different times (e.g. the first flowers of *Chenopodium chenopodisfolia* are subterranean and cleistogamous) or under different environmental conditions (e.g. poor nutrient supply, too little or too much water, deep shade). Cleistogamy is a process that promotes self-pollination and seed set in the absence of pollinators. Cleistogamous flowers tend to be small, often concealed among bracts or leaves. It occurs in at least 58 families of angiosperms. Cleistogamy is common in the genus *Viola*, particularly among flowers developed in the summer. Pseudocleistogamy is when the pollen matures early and is shed directly on to the receptive stigma while the flower is closed, although it later opens. *See also* FLORISTICS; POLLINATION. [I.O. & G.A.F.H.]

Clements, F.E. *See* CHARACTERS IN ECOLOGY.

cliff ecosystems Sea cliffs vary in geology, steepness, aspect and distance from the SEA. Where steep and exposed to the prevailing WIND they receive considerable quantities of salt spray and develop a distinctive flora. The most characteristic species around the coasts of western Europe are sea pink or sea thrift (*Armeria maritima*); rock samphire (*Crithmum maritimum*); sea campion (*Silene maritima*); rock sea spurrey (*Spergularia rupicola*); and in more southerly areas, golden samphire (*Inula crithmoides*). These can be very colourful and attractive, looking like natural rock gardens, which is exactly what they are.

Their fauna can also be distinctive, especially the concentration of nesting seabirds, such as gulls, auks (razorbill (*Alca torda*), guillemot (*Uria aalge*) and puffin (*Fratercula arctica*)), kittiwake (*Rissa tridactyla*) and gannet (*Sula bassana*), whose distribution is mainly controlled by inaccessibility to people and rats and the abundance of fish sources offshore. There are also some distinctive invertebrates, such as the large and distinctive amphipod, the sea slater (*Ligia oceanica*). Some cliffs also have large numbers of small mammals such as shrews.

Usually the maritime flora exhibits a marked gradient from the sea inland. The maritime species decrease in abundance (number of species and cover) and mesophytes increase. The number of species increases until the VEGETATION becomes dominated by grasses (especially cocksfoot), bracken, heather, blackthorn, gorse or windswept WOODLAND. Whilst the cliff vegetation itself is not usually affected by geology the inland vegetation is, so that heather is found on granite and other igneous rocks and certain types of GRASSLAND and SCRUB on chalk or limestone.

It has been shown that sea-cliff species do not have a requirement for salt spray but are negatively affected by COMPETITION from species usually found inland. It is for this reason that some cliff species, such as sea pink, are also to be found on the tops of mountains. Thus competition is the main factor controlling the distribution of these plants and not salinity or wind.

Sea cliffs are very attractive to people as a result of the spectacular views, good birdwatching, attractive flora and healthy exercise that cliff walking offers. Many areas have been made the subject of long-distance footpaths, for example Pembrokeshire, Devon and Cornwall. Shorter paths are also to be found in the North York Moors, North Wales and Jersey. The National Trust has protected extensive areas as a result of its Operation Neptune campaign, which commenced in 1971. Several areas of sea cliff around the coast of Britain are protected as national nature reserves, and by other planning designations such as heritage coast and areas of outstanding natural beauty.

In the north of England, but more especially in Northern Ireland and Scotland, arctic alpine plant species come down to sea level on sea cliffs. This is almost certainly because of the reduced competition from other plants. Their presence can add to the pleasure of botanizing in such locations. [F.B.G.]

REFERENCES

Goldsmith, F.B. (1977) Rocky cliffs. In: *The Coastline* (ed. R.S.K. Barnes), pp. 237–252. John Wiley & Sons, Chichester.

Hepburn, I. (1952) *Flowers of the Coast*. Collins, London.

Malloch, A. (1971) Vegetation of the maritime cliff-tops of the Lizard and Lands End Peninsulas, West Cornwall. *New Phytologist* **70**, 1155–1197.

Mitchley, J. & Malloch, A.J.C. (1991) *Sea Cliff Management Handbook for Great Britain*. University of Lancaster, Lancaster.

climate Climate is often regarded as 'average weather', i.e. a statistical description of the state of the ATMOSPHERE at any given location. The climate of an area can be described by its mean values of TEMPERATURE, rainfall, WIND speed or number of sunshine hours but it should also include some information on the extremes of the statistical distribution as it is often the extreme events that cause loss of life or economic hardship. Thus a climate description should mention, for example, the frequency of occurrence of hurricanes or DROUGHTS if relevant. To establish a proper statistical description of climate, WEATHER records for a reasonable length of time are required: a climatological average period is usually taken to be 30 years or so in length. Such long records of weather observations are by no means universal and there are many parts of the world for which only short-period or no weather records are available. The word climate is derived from the Greek word for slope, indicating that early ideas on how climates varied over Earth's surface sought the explanation in terms of the angle the Sun's rays made with the horizontal; thus the original climate classification was torrid for hot areas, temperate for mid-latitude climates and frigid for polar areas. Clearly there is more to climate than a simple classification essentially based on latitude; climate varies over Earth also according to TOPOGRAPHY, relief and proximity to large bodies of water such as lakes and oceans. Climate is non-stationary and has varied markedly in historical and prehistoric times. The reasons for this include external influences, such as the varying Earth–Sun distance, on time-scales of thousands to hundreds of thousands of years, coupled with variability in the orientation of Earth's axis of rotation. There is also variability within the Earth–atmosphere system as a result of PLATE TECTONICS at one end of the time-scale to volcanic eruptions and varying ocean currents at the other. Such variability through space and over a range of time-scales makes the prediction of climate very difficult and the detection of an unequivocal signal indicating CLIMATE CHANGE difficult to distinguish from the natural noise of the climate system. Some component parts of the climate system, such as the oceans, may operate on time-scales of thousands of years, whereas the atmosphere may react locally to changes in land-surface changes on the scale of days or months. The ICE SHEETS are another feature that introduce essentially long-term memory into the climate system. Humans also have the capacity to alter climate, albeit knowingly on a small scale but perhaps unknowingly on a global scale as a result of increased concentrations of radiatively active gases through inefficient combustion of FOSSIL FUELS. *See also* CLIMATE OBSERVING SYSTEMS. [J.B.M.]

Climate change

Major changes in the Earth's energy balance and CLIMATE have occurred over geological time, and smaller changes are evident over the last few hundred years of recorded history. Climate change is not well understood as it involves many interacting factors, some of which are influenced by human activities but many others are natural.

It is the balance between incoming and outgoing RADIATION that determines the surface TEMPERATURE and essentially drives the climate system. The incoming radiation from the Sun varies because of changes in the luminosity of the Sun and variation in the Earth–Sun geometry. The outgoing radiation flux depends on the reflectance of the surface cover, and on the emission of thermal radiation from the surface which depends on the fourth power of surface temperature. Both the incoming and outgoing radiation streams are influenced by atmospheric composition. WATER vapour and volcanic SMOKE cause a reduction in the SOLAR RADIATION reaching the ground. Gases absorb radiation at certain wavelengths; many diatomic gases absorb thermal radiation and so contribute to the greenhouse effect.

Human activity may influence climate change in several ways. The release of the gases CARBON DIOXIDE (CO_2), METHANE (CH_4), nitrous oxide (N_2O) and CFCs enhances the greenhouse effect and is expected to cause GLOBAL WARMING. Changes in reflectance of the surface, brought about by changes in land use such as the replacement of forests by farmland, are thought to be significant. Release of AEROSOLS in fossil-fuel combustion may have a small effect.

Evidence for climate change in the past comes from various sources. Ice cores from the Antarctic and elsewhere contain a time-series of gas concentrations, and associated oxygen isotope STRATIGRAPHY provides a record of temperatures. The geological record provides evidence for the ice ages when ICE covered a major part of the Earth. Over the last 10 000 years the record of pollens from PEAT bogs and lake sediments shows a warming until some 5000 years ago, followed by a cooling. History demonstrates that there have been recent changes, for example the depopulation of upland villages around AD 1700 as a result of a 'Little Ice Age'.

Evidence that climate change is presently occurring comes from the fact that glaciers everywhere in the world have been melting, and continue to melt. Another type of evidence comes from the data on temperature, collected from many sites over the world. This shows an increase of about 0.5°C since the year 1900. It also shows that most of the warmest years of the century have occurred since 1980.

The prediction that climate change will bring a warmer world is obtained by running GENERAL CIRCULATION MODELS (GCMs) with a prescribed assumption ('scenario') about the rate at which CO_2 from anthropogenic sources will increase in the future. Such models take days or weeks to run on the largest computers, yet are still considered to lack critical detail about the behaviour of CLOUDS and the representation of canopy surface conductance to water vapour. The predictions are that a doubling of CO_2 will cause an increase in temperature of 1.5–4.5°C. Warming will not be uniform over the world, but more pronounced at high latitudes. Patterns of rainfall and storms cannot be predicted with much confidence.

Future climate change, if it occurs as fast as has been predicted, will have profound impacts on the BIOTA, affecting AGRICULTURE, FORESTRY and nature conservation. Although species are able to migrate (and certainly have done in the past), the rate of change may be too great for them to keep pace. OUTBREAKS of pests may be more common, especially insect pests at high latitudes. Sea level has been rising by 1–2 mm $year^{-1}$ and will rise faster as the ice melts and as oceans undergo thermal expansion, posing problems for nations and people with low-lying ground.

The Intergovernmental Panel on Climate Change (IPCC) was established in 1988 with a mission to (i) assess the scientific information and (ii) formulate realistic response strategies for the management of climate change. *See also* CLIMATE CHANGE, POTENTIAL IMPACTS ON BIOLOGICAL SYSTEMS; GLOBAL WARMING; GREENHOUSE GASES. [J.G.]

REFERENCES

Dansgaard, W., Johnsen, S.J., Clausen, H.B. *et al.* (1993) Evidence for general instability of past climate from a 250-kyr ice-core record. *Nature* **364**, 218–220.

Houghton, J.T., Jenkins, G.J. & Ephraums, J.J. (1990) *Climate Change. The IPCC Scientific Assessment.* Cambridge University Press, Cambridge.

Climate change, potential impacts on biological systems

CLIMATE is, and always has been, a dominant influence on the distribution and abundance of plants and animals on the Earth. In the world after the ice age between 12 000 and 10 000 BP, a veritable blink of the eye in geological terms, 200 mammalian genera became extinct worldwide. Average Earth surface TEMPERATURE increased 5°C during this time as glacial ICE SHEETS and continental glaciers receded. TUNDRA was transformed into FOREST and many other changes in vegetation and animal habitat occurred at northern latitudes. Expansion of human populations may have contributed to the mammal extinctions, but even this expansion was induced by the warmer climate. Palaeoecological studies (of fossil pollen, plant parts and various invertebrate animals) from throughout the world have documented the dynamic relationship between climate and biological systems. Even submillenial variations such as the warm Medieval Optimum (*c.* AD 1000–1200) and cool Little Ice Age (*c.* AD 1600–1850) had significant impacts on vegetation distribution and fire frequency.

GENERAL CIRCULATION MODELS (GCMs) of large-scale atmospheric systems suggest that average Earth surface temperature will increase 1–3°C during the 21st century, although the exact magnitude and timing of climatic change are highly uncertain. This rate of warming would be up to 100 times greater than any previous climatic warming. If temperature increases, the largest increases will probably occur at northern latitudes. PRECIPITATION patterns may also change, but the direction of change is difficult to predict. In addition, the ATMOSPHERE of the 21st century will contain much higher concentrations of CARBON DIOXIDE (CO_2; a LIMITING FACTOR for plant growth) and may allow passage of greater amounts of ULTRAVIOLET RADIATION (a potential mutagenic factor) to the Earth's surface due to the depletion of stratospheric OZONE.

Plant species have different tolerance levels for temperature–precipitation combinations. A warmer and drier climate may produce increasing levels of physiological stress for species that are sensitive to low SOIL MOISTURE, while species that are tolerant of moisture stress may have a competitive advantage. Sensitive species may grow more slowly and have more difficulty regenerating at locations where they are now dominant. Displaced plant species may 'migrate' (e.g. through seed dispersal) to locations with a more favourable temperature–precipitation regime. Some ecologists have predicted that there will be a large-scale shift in plant species, with northern-latitude species generally moving northward and southern-latitude species moving southward. Species growing in mountainous regions may shift their distributions upward in elevation, particularly if snowpacks are reduced. In the future, there may be completely different combinations of plant species than are currently present, because species will shift their distributions independently of each other rather than as coherent associations. Palaeoecological data indicate that this phenomenon has happened many times in the past.

Changes in the physical environment may have minimal direct impacts on terrestrial animals. However, changes in vegetation distribution will have profound impacts on the abundance and distribution of many animal species. If habitat changes, animals will have to locate areas with the particular combinations of TOPOGRAPHY, vegetation and other organisms they need to survive. For example, moose (*Alces alces*) depend on forest vegetation and will tend to stay in areas that are heavily forested, while elk (*Cervus elaphus*) depend on meadow species and will tend to stay in areas with grass and herbaceous species. If the distribution of forest and meadows changes in the future, either latitudinally or altitudinally, the distribution of ungulates and other animals will also change.

Perhaps the greatest impacts on terrestrial fauna will occur in those species that have long-distance migrations. For example, caribou (*Rangifer tarandus*) migrate thousands of kilometres via well-established routes at northern latitudes. These migrations are a critical part of their life cycle and are closely tied to food availability. If ECOSYSTEMS change such that the availability and abundance of forage are significantly altered, MIGRATION routes must change or populations will decline. Migratory waterfowl, which spend their summers above the Arctic Circle and winters in temperate regions, require open water and food in the north and south as well as along their migratory route. If the availability of open water and vegetation change, then long-standing migrations could be disrupted.

The impacts of climatic change on aquatic systems are more difficult to predict. Many freshwater fish species, such as trout, are extremely sensitive to temperature and would suffer under increased summer temperatures. Other species that are more tolerant of higher temperatures would undoubtedly become more dominant. All fish species would be subject to the impacts of temperature changes on populations of prey organisms, which include other fish, invertebrates and PLANKTON. The very nature of aquatic FOOD WEBS could be altered dramatically. Marine mammals such as whales and seals are sensitive to the temperature of ocean currents and location of sea ice with respect to both physiological function and the availability of prey species. Altered ocean temperatures could affect migration routes and reproductive success.

A critical factor in determining the impacts of climatic change will be the role DISTURBANCE plays in ecosystem function under different climates. Infrequent but potentially severe wildfires produce dramatic changes in forest structure and processes over large land areas. Windstorms and insect OUTBREAKS also modify ecosystem characteristics, although normally over a smaller spatial scale. Disturbances essentially 'reset the clock' by returning systems to an earlier successional state in which COMPETITION for resources among regenerating organisms determines which organisms ultimately dominate a site. Increases in fire frequency and other disturbances would maintain more forest, shrubland and grassland areas in this early successional state. Increased amplitude and frequency of the EL NIÑO southern oscillation (ENSO) would have a profound effect on both terrestrial and aquatic systems, with specific impacts varying geographically. Variations in ENSO result in spatial variations in the temperature of ocean currents. This affects a wide range of fish and marine mammal species. More frequent or more prolonged DROUGHTS (or alternatively more frequent or more prolonged wet periods) in terrestrial systems could rapidly change the distribution and abundance of some plant species.

Agricultural systems would also be impacted by long-term changes in temperature and precipitation. Prolonged droughts would obviously have a negative impact on yields of many crop species. Increased moisture stress on crops may also make them more susceptible to insects and pathogens, thereby reducing yield and increasing the need for mechanical or chemical treatment. Some productive agricultural areas may become less productive, while areas that are currently too cold for most crops might become viable for AGRICULTURE in a warmer climate. An interesting possibility is that increased levels of tropospheric CO_2 will result in increased PHOTOSYNTHESIS and PRODUCTIVITY in some species.

The long-term impacts of climatic change on biological systems is highly speculative at the present time. The complexity of the Earth's atmospheric system cannot be accurately modelled by the current generation of climatic models and current 'predictions' are coarse estimates at best. Uncertainty about changes in the physical environment, coupled with the complexity of interactions between physical and biological systems, makes it difficult to confidently predict biological impacts. Careful MONITORING of different ecosystem components (e.g. air temperature, ocean temperature, tree growth, fish populations) will help determine if there are any trends that might be associated with changes in climate. *See also* CLIMATE CHANGE; GLOBAL WARMING; GREENHOUSE GASES. [D.L.P.]

---

climate observing systems CLIMATE is not observed directly but is deduced by averaging daily WEATHER observations over much longer periods, often 20 years. The climate may be characterized by one of several different measured elements: TEMPERATURE (at standard screen level), rainfall, duration of sunshine, spells of weather (e.g. extent of DROUGHT, or FROST), cloudiness, frequency of gales, etc. The recent increased interest in climatic change has generated new definitions of global temperature such as mean tropospheric temperature, for which it is necessary to generate routinely computed models, with the parameter routinely generated by being programmed into the model. Otherwise, comparisons with earlier years are made by calculating the same quantity, defined by a formula such as the mean between the daily maximum and minimum screen temperatures, for the same sites over the years. Such records reach back only in a very few, and unrepresentative, places before the year 1860.

The effects of climate on various indicators that remain as residues from previous eras may be used to assess past climates. Some are biological, such as pollen and the skeletal foraminifera deposited with other material on the OCEAN beds. These may be investigated by taking cores and working out a chronology that has to agree with other cores from the same general area, with the assumption that the species were sensitive enough to temperature (or whatever) in the same way as they are today. Cores have been taken particularly in the great ICE packs of Greenland and Antarctica, which are 2 or 3 km in thickness and go back close to 200 000 years ago. In these cases the indicator may be the electrical conductivity of the ice pack, which depends on the mineral content of the snow when it originally fell; this indicates something about the origin of the air and the nature of the surface (e.g. ice or desert) at that place of origin. Or it may be the ratio of two isotopes (e.g. HYDROGEN/DEUTERIUM, $^{12}C/^{14}C$ or isotopes of OXYGEN). In each case the lighter ISOTOPE is deposited preferentially and so appears in chemical compounds in which they have been captured from the ATMOSPHERE in greater relative abundance than in the atmosphere itself. This abundance depends on the temperature at the time it was deposited, and so can be used as a climatic indicator. Some analysts have claimed to be able to count the individual years by the texture of the snow, which melts partially in the height of summer. This does not mean that several thousand years have to be counted individually in order to fix the date because some dates can be fixed by other clues that are more accurately known from other cores. [R.S.S.]

Table C1 Scheme for classifying climate zones. (From Larcher, W. (1995) *Physiological Plant Ecology*, Springer-Verlag, Berlin; compiled from several sources.)

Climate type	Soils	Vegetation
Equatorial: warm and humid, little seasonality	Equatorial brown clays, ferralitic soils	Evergreen tropical rain forests
Tropical: summer rains with cooler dry season	Red clays or red earths, ferralitic savannah soils	Tropical deciduous forests or savannahs
Subtropical: arid desert climate	Desert soils, often saline	Succulents, desert ephemerals, reduced leaves
Mediterranean: winter rain and summer drought	Mediterranean brown earths, often fossil terra rossa	Sclerophyllous woody plants, sensitive to frost
Warm-temperate: often with summer rain maximum, or mild-maritime	Red or yellow forest soils, slightly podzolized	Temperate evergreen forests
Cool-temperate: short cold winters	Forest brown earths and grey forest soils	Broad-leaved deciduous forests
Continental: arid-temperature, cold winters	Steppe soils, Chernozems, Castanozems, Sierozems	Steppe to desert
Boreal: cold summer, long winter	Podzols	Coniferous forests
Polar: very short summer	Tundra, humus-rich soils with solifluction	Treeless vegetation on permafrost

climate zones The classification of the world into CLIMATE zones or regions, based partly on the mean annual rainfall and TEMPERATURE and also the soils and life-forms of the VEGETATION, was introduced by German scientists including Köppen (in the last century), Holdridge (1947) and Walter and Breckle (1985). Although the emphasis has turned away from classification towards empirical and partially mechanistic models that link climatic variation to vegetation PHYSIOGNOMY (Box 1981; Solomon & Shugart 1993), such classifications are still useful today in mapping and also in studies of CLIMATE CHANGE (Table C1). *See also* BIOGEOGRAPHY; BIOMES, MAJOR WORLD. [J.G.]

REFERENCES

Box, E.O. (1981) *Macroclimate and Plant Forms: An Introduction to Predictive Modelling in Phytogeography*. Junk, The Hague.

Holdridge, L.R. (1947) Determination of world plant formations from simple climatic data. *Science* **105**, 367–368.

Larcher, W. (1995) *Physiological Plant Ecology*. Springer-Verlag, Berlin.

Solomon, A.M. & Shugart, H.H. (1993) *Vegetation Dynamics and Global Change*. Chapman & Hall, London.

Walter, H. & Breckle, S.-W. (1985) *Ecological Systems of the Geobiosphere*. Springer-Verlag, Berlin.

climatology There are two meanings: (i) the climatology of a place is a description of the long-term CLIMATE there; and (ii) climatology is also the scientific study of the climate, by which is meant the long-term patterns in the WEATHER. Climatology is related to METEOROLOGY, but the latter gives more emphasis to physical processes in the short term, whereas climatology is concerned with the statistical patterns over periods of years and decades. [J.G.]

REFERENCE

Barry, R.G. & Chorley, R.J. (1992) *Atmosphere, Weather and Climate*, 6th edn. Routledge, London.

Climax

Climax was the name given by Clements (1916, p. 3; *see* CHARACTERS IN ECOLOGY) to the END-POINT of SUCCESSION. It has generally been defined as a kind of VEGETATION that is self-perpetuating in the absence of DISTURBANCE. However, it is doubtful whether regeneration of any 'climax' occurs without the intervention of gap-forming disturbances, for example treefall as a result of windblow in FOREST, or mound formation by vertebrates in grassland. For this reason the species that characteristically invade such gaps are sometimes regarded as part of the climax complex (*see* INTERNAL SUCCESSION).

Three problems lead to the term 'climax' being commonly rejected now. First, all soils tend to change with time—most often as a result of LEACHING, associated ACIDIFICATION and locking up of phosphate in unavailable forms (either organic or inorganic)—and the plant species best suited to a site will change with time, even if very slowly relative to the life of a human (Vitousek 1987) (*see also* RETROGRESSIVE SUCCESSION). Secondly, CLIMATE is not constant but changes on a time-scale that is short relative to the lives of many long-lived perennials such as large trees; species composition at any one time may reflect past climate rather than present (Davis 1986). Thirdly, some late-successional species have been slow to spread from the REFUGIA they occupied at the height of the glacial period (*c.* 18 000 BP) and it is doubtful whether they have yet reached the limits of the ranges they could occupy (Davis 1981). By analogy, recently evolved species may not have reached the limits of their potential ranges.

It is preferable to refer to individual species or collections of species as 'late successional' rather than climax.

Species are sometimes classified as 'climax' (late successional) or 'pioneer' (early successional), and there are many consistent differences in plant form and behaviour between these extremes (Wells 1976; Swaine & Whitmore 1988). However, it has long been known that one species may fill the two roles. For example, in very dry areas in western North America *Pinus ponderosa* may be the only tree species, while in less dry areas it may invade gaps in forest formed by Douglas fir (*Pseudotsuga menziesii*); in turn *Pseudotsuga* is early successional in yet moister areas, and so on (Daubenmire 1966). Similarly, the birch *Betula pubescens* is the only tree in some oceanic northern parts of Europe, but is early successional further south.

According to the original concept of climax (Clements 1916), the same one species or set of species would eventually come to occupy all sites in a given region, regardless of the substratum (*see* CONVERGENCE OF SUCCESSIONS). In other words, the climax was deemed to be determined solely by climate. A complex nomenclature of DISCLIMAX, POSTCLIMAX, PRECLIMAX and SUBCLIMAX was invented to cope with the fact that different site types in a given region are not, in fact, occupied by one group of species. It was once quipped that the climax of the Clementsian system was a terminological jungle! Before the term 'climax' fell from favour, most ecologists accepted the idea of POLYCLIMAX. *See also* REGENERATION AND CYCLIC CHANGE. [P.J.G.]

REFERENCES

Clements, F.E. (1916) *Plant Succession*. Carnegie Institution of Washington, Publication 242.

Daubenmire, R. (1966) Vegetation: identification of typal communities. *Science* **151**, 291–298.

Davis, M.B. (1981) Quaternary history and the stability of forest communities. In: *Forest Succession: Concepts and Application* (eds D.C. West, H.H. Shugart & D.B. Botkin), pp. 132–153. Springer-Verlag, New York.

Davis, M.B. (1986) Climatic instability, time lags, and community disequilibrium. In: *Community Ecology* (eds J. Diamond & T.J. Case), pp. 269–284. Harper & Row, New York.

Swaine, M.D. & Whitmore, T.C. (1988) On the definition of ecological groups in tropical rain forests. *Vegetatio* **75**, 81–86.

Vitousek, P.M. (1987) Colonization, succession and resource availability: ecosystem-level interactions. In: *Colonization, Succession and Stability* (eds A.J. Gray, M.J. Crawley & P.J. Edwards), pp. 207–223. Blackwell Scientific Publications, Oxford.

Wells, P.V. (1976) A climax index for broadleaf forest: an *n*-dimensional ecomorphological model of succession. In: *Central Hardwood Conference* (eds J.S. Fralish, G.J. Weaver & R.C. Schlesinger), pp. 131–176. Department of Forestry, Southern Illinois University, Carbondale, Ill.

cline A GRADIENT in the genetic make-up of a species, typically along a spatial gradient in an environmental factor like altitude, latitude, salinity, photoperiod, etc. The existence of a cline is usually established by common garden and reciprocal TRANSPLANT EXPERIMENTS: GENOTYPES typically grow, survive and reproduce best in the conditions on the gradient from which they were obtained and compete less well under different conditions (both higher up or lower down the gradient) (Fig. C13). *Cf.* ELLENBERG'S RULE. [M.J.C.]

REFERENCE

Clausen, J., Keck, D.D. & Hiesey, W.M. (1958) *Experimental Studies on the Nature of Species, III*. Carnegie Institution of Washington, Publication 581.

clock model *See* HETEROCHRONY.

clonal *See* MODULE.

clonal growth This type of GROWTH typically occurs in both plants and some lower invertebrates such as hydroids. It relies on basic cell division (by mitosis) to produce new genetically identical 'CLONES' of the original organism. Such 'modular' growth allows the organism to exploit a relatively large area without producing GAMETES and thus avoiding any parent–offspring COMPETITION for limited local RESOURCES. Vegetative growth in plants, particularly grasses, follows this pattern where a single genetic individual may occupy a relatively large area. This may consist of a single interdependent organism or a number of discrete subunits. [R.C.]

clone Genetically identical individuals, produced by mitotic cell division or physical division of viable plant parts (cuttings, rhizomes, grafted buds), which vary from one another only as a result of SOMATIC MUTATIONS. The term is also used to describe genetically uniform cell lines or DNA molecules; the products of VEGETATIVE REPRODUCTION by seed (APOMIXIS) or from fragmentation of body parts (e.g. break-up of plant rhizomes); and colonial animals and animals capable of reproduction by asexual means (e.g. a COLONY of aphids). *See also* ASEXUAL REPRODUCTION. [M.J.C.]

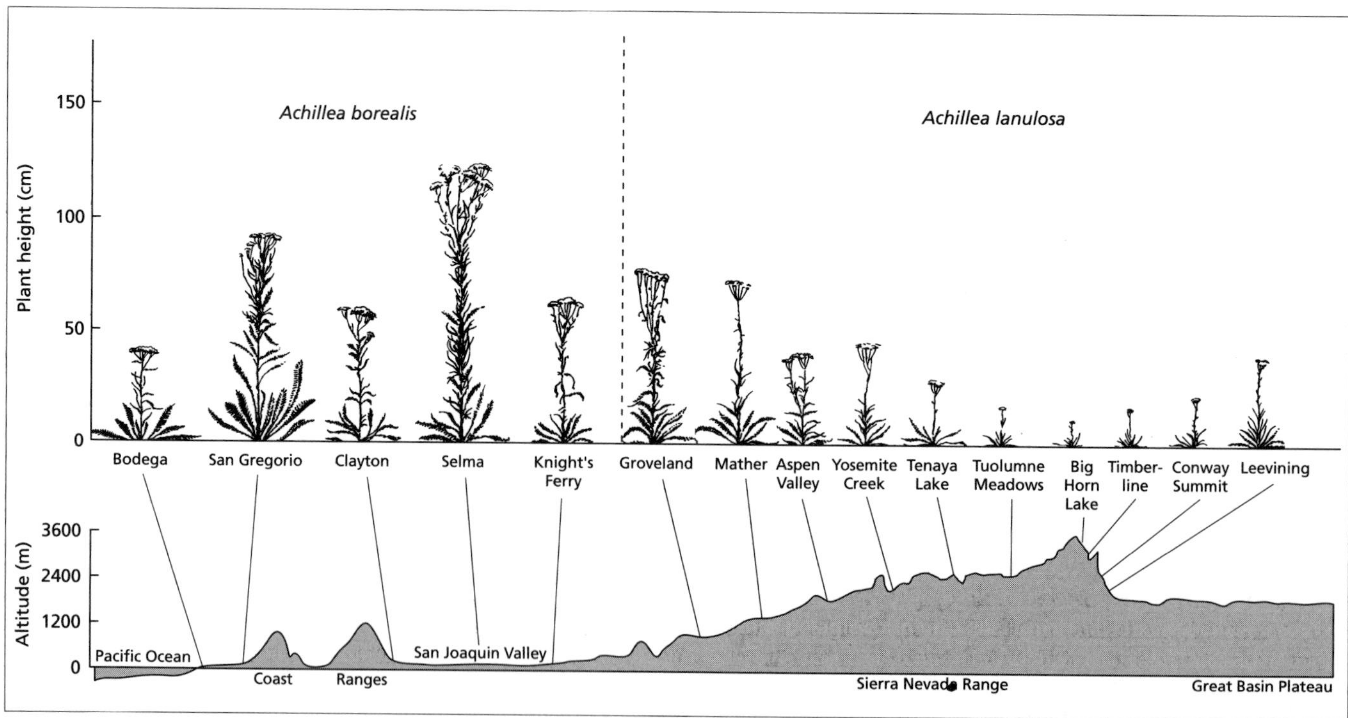

Fig. C13 A cline in genotypes of yarrow (*Achillea*, Asteraceae) as revealed from studies of phenotypes in common gardens at a range of altitudes through the Sierra Nevada mountains in California. Plants grew largest and flowered most profusely only at the altitude from which the genotypes were collected. (After Clausen, J., Keck, D.D. & Hiesey, W.M. (1958) *Experimental Studies on the Nature of Species, III*. Carnegie Institution of Washington, Publication 581.)

closed community An ASSEMBLAGE of organisms covering the complete ground area of a habitat (*cf.* OPEN COMMUNITY), effectively precluding COLONIZATION by other species, as all ecological niches are already occupied. In closed vegetation, plants make continuous lateral contact. Closed communities are characteristic of late stages of SUCCESSION. [A.J.W.]

closed forest Stands of trees whose CANOPY cover is complete or nearly so. Individual tree crowns may not actually touch or overlap, but there are no gaps in the canopy into which a new tree may grow. Usually applied at a small scale, i.e. to stands of trees, but applicable to even-aged forests on a larger scale if there has been no recent thinning and no natural disturbances. [G.F.P.]

closed system A system is closed if there is no exchange of matter or energy with the outside world. The converse is an OPEN SYSTEM. ECOSYSTEMS are almost invariably open, because they depend on an external energy supply for their persistence. However, systems that are closed to external inputs of matter and which are open only to light are frequently constructed for experimental purposes; examples are aquatic microcosms (Meador *et al.* 1993) and the Biosphere-2 experiment (Avise 1994). Materially closed systems are particularly relevant to space travel (which necessitates them) and can serve as a model for the functioning of the BIOSPHERE.

Systems that are physically open may, on occasion, be represented by closed or autonomous models, such as the Lotka–Volterra predator–prey equations. In this case, the behaviour of the system does not depend on external inputs or driving variables but is determined entirely by the modelled mathematical relationships. A closed population model lacks IMMIGRATION; EMIGRATION is represented simply as a sink comparable to death. Some authors extend this terminology so that a set of weakly connected subpopulations is called an open system and contrasted with the closed system of a single population without immigration. This use of language is not recommended: a model of a set of weakly connected subpopulations can perfectly well be a closed model. [M.O.H.]

REFERENCES

Avise, J.C. (1994) The real message from Biosphere 2. *Conservation Biology* **8**, 327–329.

Meador, J.P., Taub, F.B. & Sibley, T.H. (1993) Copper dynamics and the mechanism of ecosystem level recovery in a standardized aquatic microcosm. *Ecological Applications* **3**, 139–155.

cloud forest A kind of TROPICAL MONTANE FOREST. Hot moisture-laden lowland air rises up mountain slopes. As it rises it cools and at a certain altitude a flat-bottomed cloud layer is formed where the DEW POINT is reached. CLOUDS form and remain from mid-morning to evening, drastically reducing INSOLATION. The regular presence of the cloud layer produces a strong ALTITUDINAL ZONATION, a sharp discontinuity in the forest vegetation.

The trees are closely spaced and there is a dense shrubby UNDERSTOREY in which tree-ferns are common. There are hebaceous vines, but woody climbers are rare. EPIPHYTES are abundant, including diverse vascular epiphytes: orchids, aroids, ferns and, in the neotropics, bromeliads, but these are less abundant than in lower montane forest. In contrast, non-vascular epiphytes (mosses, liverworts and lichens) festoon the trees, even growing over the surface of leaves. Many species of trees have leaves with drip tips to keep their surface dry so that epiphyllous mosses, liverworts and fungi do not find a damp surface for growth. Light and nutrient levels are low and growth slow. In some places at higher altitudes the cloud forest gives way to ELFIN FOREST. [M.I.]

REFERENCE

Mabberley, D.J. (1992) *Tropical Rain Forest Ecology*, 2nd edn. Blackie, Glasgow.

Clouds

Most commonly formed as a result of cooling of the air to below its dew-point TEMPERATURE by adiabatic expansion on rising to a greater height above the Earth's surface. This may be by CONVECTION (cumulus cloud) or at a frontal surface in a larger WEATHER system.

Theoretically, and in practice, cloud is formed when two masses of saturated or nearly saturated air, initially at different temperatures, are mixed to acquire an intermediate temperature, which must be below the DEW POINT of the mixture. However, this is uncommon except in aircraft condensation trails (contrails), some large industrial chimney plumes, fog thickening as a result of TURBULENCE in the WIND over rough ground and, occasionally, 'arctic smoke' (steaming fog) in very cold air over a warm WATER surface.

The edges of clouds are usually evaporating as a result of mixing with surrounding drier air. Therefore, the mixture acquires a changed temperature and a density which is different from that of the cloud or the clear air and as a consequence generates more eddies, which promote further mixing. The outer layer of a cumulus cloud can therefore usually be seen to be descending.

Larger areas of cloud may be evaporated as a result of adiabatic warming when the air subsides, as in an anticyclone.

The shapes produced by the condensation of cloud and EVAPORATION are greatly influenced by fallout (RAIN, hail, snow or other PRECIPITATION), which forms streaks and descending air CURRENTS that evaporate the smaller droplets.

Condensation of cloud usually occurs if the air is cooled below its dew point; however, this requires the presence of condensation nuclei, which are almost always present in the air so that a significant degree of supersaturation is scarcely ever measured. The number of cloud droplets may be between 10 and 20 000 cm^{-3} of air, depending on the number of nuclei present.

Freezing to ICE crystals does not occur as soon as the air is cooled to below 0°C, and supersaturation with respect to ice often occurs because ice nuclei are uncommon. Supercooled water droplets in clouds are likely to remain unfrozen in the temperature range 0 to −39°C for long periods if no ice nuclei are present; however, if a few drops freeze, all the remaining drops may very soon freeze as a result of the multiplication of icy fragments emitted by droplets freezing on the outside first and shattering or squirting out fragments when the interior expands on freezing. Freezing becomes more likely as −40°C is approached and is usually spontaneous at or below that temperature. Before all the droplets become frozen, the freezing of supercooled droplets causes growth by condensation of the crystals and evaporation of the liquid droplets because the vapour PRESSURE over water exceeds that over ice (Table C2). However, ACID salts on the droplets may reduce the freezing temperature to below the actual temperature, so that even at −90°C the cloud will have the optical properties of a cloud of spherical particles

T_d	0	−4.4	−9	−14	−19.4	−24.5	−32.2	−40.4
$T_f - T_d$	0	0.5	1	1.5	2	2.5	3	3.5

Table C2 Difference between the frost point (T_f) and dew point (T_d) as a function of T_d (values are °C).

and not those of a cloud of ice particles, a special example being polar stratospheric clouds.

When cloud condensation nuclei are very plentiful and the initial number of nuclei is very large with the droplets being correspondingly small, unless the nuclei are hygroscopic salts SURFACE TENSION increases the interior pressure, which results in evaporation of the smaller droplets and growth by condensation of the largest. This begins a re-formation of the drop-size spectrum and a reduction in the ALBEDO of the cloud. In regions where condensation nuclei are rare the number of droplets is smaller when the cloud first forms, as may be the case over some parts of the oceans. The exhaust from a ship, which is usually a profuse source of nuclei, causes a dramatic rise in the number, and reduction in size, of droplets and the ship's trail consists of cloud with a greatly increased albedo, i.e. a trail of much brighter white cloud. [R.S.S.]

Club of Rome Study Body established in 1965 by an international group of economists, scientists, technologists, politicians and others. The aim is to study the interaction of economic, scientific, biological and social aspects of humanity, with a view to predicting impacts of proposed and alternative policies on environment and the survival of humanity. *See also* ENVIRONMENTAL (ECOLOGICAL) ECONOMICS. [P.C.]

clubmosses *See* PTERIDOPHYTES, GLOBAL DIVERSITY AND CONSERVATION.

cluster analysis A multivariate statistical method. Most of the statistical methods described in detail in this encyclopedia are univariate because they are concerned with single variables or observations. Even the complex methods of ANALYSIS OF VARIANCE are only concerned with explaining the variation in one variable. Multivariate methods analyse several related variables or observations simultaneously. For further details, see Manly (1986). *See also* STATISTICS. [B.S.]

REFERENCE

Manly, B.F.J. (1986) *Multivariate Statistical Methods: A Primer*. Chapman & Hall, London.

clutch size The NUMBER OF OFFSPRING produced during a single breeding episode. There is no problem in determining what is meant by clutch size in organisms such as birds and mammals, but the term becomes more difficult to use for organisms that produce EGGS continually. For example, female fruit flies, *Drosophila melanogaster*, like many invertebrates, lay their eggs singly, laying several hundred in a single day. However, many other insects may produce several discrete batches of eggs during the day, laying these in clumps; each discrete batch is most properly called the clutch. In general, it is preferable to reserve the term 'clutch' for a discrete grouping of offspring, between which there is interaction. Thus we speak of the daily FECUNDITY of *D. melanogaster* as being its clutch size if it lays these eggs in a single location. However, in the case of the female butterfly that lays, say, 100 eggs in a day, each egg being laid singly upon a separate leaf, it is more appropriate to speak of 100 clutches of size one egg.

Studies of the evolution of clutch size have centred principally on ENDOTHERMS and insects, and have developed somewhat differently. In the case of endotherms, considerable attention has been given to the cost of increased clutch size on offspring survival and future reproduction of the parents. D. Lack advanced the hypothesis that clutch size in birds is determined by the number of offspring that the parents can raise, this not affecting the future success of the parents. Experimental manipulation of clutch size has demonstrated clear effects on nestling survival and that, at least in some cases, the future success of the parents is reduced. The effect of the latter is to reduce the OPTIMAL CLUTCH SIZE below that predicted by Lack's hypothesis. In insects, much attention has been concentrated on the influence of search time and patch quality on the optimal clutch size, parasitoid insects being more commonly investigated. When search time is finite the optimal clutch size is also below the LACK CLUTCH size, approaching this value as search time increases. *See also* COSTS OF REPRODUCTION; LIFE-HISTORY EVOLUTION; SOME WORLD RECORDS FOR LIFE-HISTORY TRAITS. [D.A.R.]

CMTs *See* CARCINOGENS, MUTAGENS, TERATOGENS.

coadapted gene complex Several tightly linked loci with interrelated and integrated function forms a coadapted gene complex. For example, in the land snail *Cepaea nemoralis* the tightly linked loci for shell colour and shell banding are part of a gene complex. The allele for 'yellow colour' and the allele for 'banded shell' usually occur together in one GAMETE, while another gametic composition is the allele for 'brown colour' together with the allele for 'unbanded shell'. Brown–unbanded and yellow–banded seem to be adaptively coloured shell patterns. *See also* PLEIOTROPY. [G.D.J.]

Coalition of Environmentally Responsible Economies (CEREs) *See* VALDEZ PRINCIPLES.

coarse-grain exploitation Exploiting a series of RESOURCES disproportionately to their relative ABUNDANCE in the environment. In other words, specialization on a particular resource occurs. *See also* GRAIN. [V.F.]

coarse-grained habitats Refers to the way that a particular organism uses RESOURCES in its HABITAT. In a patchy or heterogeneous environment, an organism that spends time in patches of resource disproportionately to the occurrence of those patches is said to be exploiting the habitat in a coarse-grained manner. Other things being equal, smaller animals encounter habitats as coarse-grained more than do larger animals. Use of the concept of habitat GRAIN has largely fallen out of favour. *See also* FINE-GRAINED HABITATS. [R.H.S.]

REFERENCE

Pianka, E.R. (1988) *Evolutionary Ecology*, 4th edn, pp. 257–261. Harper & Row, New York.

coarse-particulate organic matter (CPOM) Plant DETRITUS in aquatic systems >1 mm in diameter. In freshwater will be of the form of decomposing leaves, needles and woody material. *Cf.* FINE-PARTICULATE ORGANIC MATTER (FPOM) (<1 mm but >0.45 µm) and DISSOLVED ORGANIC MATTER (DOM) (<0.45 µm). [P.C.]

coastal Referring to processes or features of the shallow portion of the OCEAN, generally overlying the CONTINENTAL SHELF, where CIRCULATION and other features are strongly influenced by the bordering land. Two features of particular importance in coastal areas are UPWELLING of deep nutrient-rich water and the formation of FRONTS (i.e. boundaries between horizontally juxtaposed water masses of dissimilar properties). [V.F.]

coastal and tidal wetlands, conservation *See* WETLANDS (TIDAL AND COASTAL), CONSERVATION.

coastal lagoons *See* LAGOONS.

coastal wetlands The mixing of fresh and saline waters in low-lying, flat coastal areas leads to the development of a series of wetland vegetation types, depending upon climate, water-flow patterns, SEDIMENTATION, etc. Fine-particle deposition in temperate regions leads to the development of SALT MARSHES, consisting largely of herbaceous and dwarf shrub vegetation. In tropical and SUBTROPICAL regions, arboreal vegetation dominates the SUCCESSION, forming mangroves.

BEACH and REEF features can lead to the formation of LAGOONS, and where evaporation losses are considerable (as in the Carmargue of southern France) these can develop into highly saline lakes. [P.D.M.]

coastline lakes *See* LAKES, TYPES OF.

cobalt (Co) A HEAVY METAL (ATOMIC NUMBER=27; relative ATOMIC MASS=58.93; melting point=1495°C; boiling point=2870°C) that is a trace constituent (20 mg kg^{-1}) of the Earth's CRUST. Although found in specific minerals such as linnaeite (Co_3S_4) and used in alloy manufacture, cobalt is more commonly associated with the ores of other metals: the ferromagnesian minerals of basic and ultrabasic rocks and, notably, at greatly elevated concentrations, with oceanic MANGANESE nodules. An essential TRACE ELEMENT, cobalt is required in the form of vitamin B_{12} by humans to avoid the deficiency condition of pernicious anaemia. Ruminants can convert dietary cobalt directly to vitamin B_{12}. Nitrogen-fixing bacteria and some plants need cobalt from soils, where cobalt availability may be limited by adsorption on manganese oxides. [J.F.]

cobweb graph This is a graphical method of analysis of POPULATION DYNAMICS that works well for organisms, such as ANNUAL plants, which have DISCRETE GENERATIONS. Using the NET RECRUITMENT CURVE it is possible to predict how many individuals there will be in the next generation. The result can be reflected about a line representing a stable population ($N_t=N_{t+1}$) and then the process is repeated allowing prediction of the POPULATION SIZE a further generation into the future. The technique gets its name because the resulting population trajectory may eventually resemble a web. If the net recruitment curve crosses the $N_t=N_{t+1}$ line with a shallow angle the population will stabilize at that point (a STABLE EQUILIBRIUM point). If the line is humped and crosses at a steeper angle, then the population may oscillate about the crossing point and perhaps diverge from it (an UNSTABLE EQUILIBRIUM point). Thus the technique will predict a range of different dynamics determined by the shape of the net recruitment curve. [C.D.]

COD *See* CHEMICAL OXYGEN DEMAND.

coefficient of community The relation of the number of species common to two COMMUNITIES to the total number of species in these communities. The Jaccard coefficient or index of similarity is often used, expressed as a percentage:

$$IS_J = \frac{a}{a+b+c} \times 100$$

where a is number of species common to both communities, b is number of species unique to one community and c is number of species unique to the other. Also frequently used is Sørensen's coefficient, which gives greater weighting to the species common to both samples than to those present in only one SAMPLE:

$$IS_S = \frac{2a}{a+b+c} \times 100$$

with notation as above. Both coefficients are used in calculations of similarity matrices for numerical CLASSIFICATION procedures. *See also* SIMILARITY COEFFICIENT. [A.J.W.]

coefficient of inbreeding Also known as the coefficient of kinship. This is the probability that two homologous genes, drawn randomly from the two individuals (one gene per individual), are identical by descent (i.e. are copies of the same gene in an earlier generation). The coefficient of inbreeding of an individual is the probability that the two genes at a locus in an individual are identical by descent. The coefficient of inbreeding of an individual is equal to the coefficient of kinship of its parents. The coefficient of relatedness of an individual A to an individual B is the probability that a gene drawn at random from B is identical by descent to a gene present in A. In a diploid, random mating population, the coefficient of relatedness of A to B is equal to that of B to A, and is equal to twice their coefficient of kinship. Hamilton's rule relates the coefficient of relatedness to the costs and benefits of altruistic behaviour. *F*-STATISTICS are another way of expressing inbreeding coefficients, estimated at various spatial scales. *See also* QUANTITATIVE GENETICS. [I.O.]

coefficient of kinship *See* COEFFICIENT OF INBREEDING.

coefficient of variation The sample STANDARD DEVIATION (s) is an estimate of the variation in a POPULATION (σ). If we wanted to compare the variation in the mass of mice from several populations we could use the standard deviation. However, we could not compare the value of s for mass in mice and elephants, since the scale of the measurements are so different. One solution is to use the coefficient of variation, $CV=s/\bar{x}$, usually expressed as a percentage by multiplying by 100. *See also* STATISTICS. [B.S.]

coenocline *See* ZONATION.

coevolution Coevolution is EVOLUTION in which the FITNESS of each GENOTYPE depends not only on the population densities and genetic composition of the species itself but also on the species with which it interacts. It is a process by which members of two (or more) species contribute reciprocally to the forces of NATURAL SELECTION that they exert on each other such that the evolution of each TAXON becomes partially dependent upon the evolution of the other. The concept of coevolution was originally proposed to explain the evolution of defences against herbivores by plants, and the responses of herbivores allowing them to circumvent those defences. Coevolution has also been used to describe patterns of ADAPTATION exhibited in predator–prey, host–pathogen, competitive and mutualistic relationships. When the two species generate selective forces that direct the evolution of the other, the interaction between the species (whether it be antagonistic or cooperative) may continually escalate. The role of coevolution as a widespread and important evolutionary force has not been fully established and there is disagreement among ecologists as to what exactly coevolution is. Coevolution is frequently used to describe apparently matching adaptations between pairs or small groups of species without evidence of the evolutionary history of the relationship. In practice, coevolution may be difficult to distinguish from the occurrence of matching adaptations between pairs or groups of organisms that arise from adaptation to shared features of the physical environment. The best evidence for coevolution comes from studies of mutualistic pairs of species. A more recent and expanded view of coevolution (referred to as diffuse coevolution) considers that the general overlapping of species relationships into webs of interaction results in corresponding webs of adaptation among species in a COMMUNITY. *See also* COEXISTENCE; CONFLICTS. [V.F.]

coexistence The permanent co-occurrence of two or more species in the same area. Mostly concerns potentially competing species. *See also* COEVOLUTION; COMPETITION; COMPETITIVE EXCLUSION. [J.G.S.]

cohort A group of CONSPECIFIC individuals belonging to the same generation and therefore usually of approximately the same age. Individuals of the same age descended from parents of different generations (e.g. from a female and from that female's mother) are not members of the same cohort. Often incorrectly used to mean group or POPULATION. *See also* DEMOGRAPHY; LIFE TABLE. [A.J.D.]

cohort life table A LIFE TABLE constructed using data derived from following a single COHORT from the time at which the constituent individuals were 'born' to that at which the last representative dies. Also known as horizontal, dynamic or age-specific life tables. *See also* DEMOGRAPHY. [A.J.D.]

coleopterans, diversity and conservation *See* BEETLES, DIVERSITY AND CONSERVATION.

Cole's paradox The INTRINSIC RATE OF INCREASE of an ANNUAL species with a CLUTCH SIZE of c is equivalent to the intrinsic rate of increase of a PERENNIAL with a clutch size of $c+1$. As stated, the result is paradoxical because it suggests that the perennial habit should be very rare, which it is not. The result arises because L. Cole failed to consider differences in juvenile and adult survival: Cole's result only applies to the case in which adult and juvenile survival are equal and less than or equal to 1. The perennial habit is favoured by high adult survival and a low rate of increase. *See also* LIFE-HISTORY EVOLUTION; REPRODUCTIVE EFFORT. [D.A.R.]

coliform count Test for purity of drinking waters, i.e. the number of coliform (i.e. gut organism) bacteria per 100 ml. *Escherichia coli* is used as an indicator. Though this species is not very pathogenic to humans, its presence indicates CONTAMINATION of water by faecal matter and hence the possibility of enteric disease. *See also* WATER QUALITY OBJECTIVE/STANDARD. [P.C.]

collectors *See* FEEDING TYPES, CLASSIFICATION.

colloids Colloids are suspensions of very small particles in a fluid. Examples of colloids include starch in water, albumin and paint. Colloids are important in soil chemistry, where very small particles of HUMUS and CLAY are suspended in the soil solution; they provide an enormous surface area to which ions are adsorbed. [J.G.]

Coloniality

The close spatial AGGREGATION of reproducing individuals of one or more species.

Coloniality is widespread among invertebrates, fish, amphibians, reptiles, birds and mammals and has evolved independently a large number of times. Colonies may reach considerable sizes and maxima of more than 1 million have been reported for seabirds. Mixed colonies of more than a single species such as occur in birds and pinnipeds are common and suggest that more than a single common resource may be responsible for the formation of colonies. Colonial reproduction may occur simply as a cause of a limited number of suitable breeding sites being available. There is relatively little evidence for this explanation, although breeding sites certainly are limited for some animals such as seabirds. The size of seabird colonies is reduced if there are neighbouring colonies within the foraging range of a focal COLONY, and the magnitude of this reduction is directly proportional to the combined size of all neighbouring colonies. Hence, there is INTRASPECIFIC COMPETITION for food among individuals breeding in different colonies and the combined food resource available within a specific area is divided among colonies relative to their size. However, the absence of breeding individuals from sites that have been used recently and variation in colony size among years, suggest that factors other than site limitation are responsible for the dense aggregation of reproducing individuals in colonies.

Coloniality is assumed to be associated mainly with an unpredictable temporal and spatial distribution of food because such a resource cannot be economically defendable. A number of individuals can therefore exploit such resources and breed together in a site of limited extension. Examples of such temporally and spatially clumped food sources include fruit, seeds, swarming insects and fish schools. There are a number of obvious costs of coloniality, including increased intensity of competition for resources, including mates, and increased risks of PARASITISM. The prevalence or intensity of nest parasites (*see* NEST PARASITISM) have been shown to increase with colony size in a number of studies and larger colonies therefore tend to produce a smaller NUMBER OF OFFSPRING per nest than smaller colonies. Competition for nest sites, food or mates also appears to be more intense in larger colonies. The frequency of intraspecific BROOD PARASITISM and EXTRA-PAIR COPULATIONS increase with increasing colony size. In the purple martin (*Progne subis*), experienced breeders attempt to attract first-time breeding males to their colonies by giving specific RECRUITMENT vocalizations, and the females of such inexperienced males frequently copulate with the more experienced males. MANIPULATION OF CONSPECIFICS by experienced reproducers thus may contribute to the formation of colonies, but this should simultaneously select against colonial breeding among first-time breeders. Colonies are also more conspicuous and this may render colonially breeding individuals more vulnerable as prey. Contrary to their solitarily breeding close relatives, colonially breeding bird species often have less concealed nests and their eggs are often non-cryptic in appearance. The number of predators has been shown to increase with increasing colony size, but there is no evidence that the risk of predation per nest increases with increasing colony size. The presence of many conspecifics increases the level of VIGILANCE and hence reduces the latency time to detection of a potential PREDATOR. The costs of coloniality must be balanced by benefits of a similar or larger magnitude in order to make colonial reproduction evolutionarily stable.

If coloniality mainly arises as a consequence of spatially and temporally unpredictable concentrations of food, it is reasonable to suggest that foraging and food may be involved in the evolution and the maintenance of colonial breeding. Two mechanisms of improved foraging success under colonial breeding have been suggested: SOCIAL FACILITATION and the INFORMATION CENTRE HYPOTHESIS. The first explanation suggests that a large number of individuals searching independently for an unpredictably distributed food source will be able to encounter food at a higher rate than a single or few individuals. This may be particularly important during adverse weather conditions, when food availability is low and unpredictable and the needs of offspring high. Since an unpredictable food source cannot be defended because of its unpredictability, individuals from the same colony are able to obtain information on the foraging success of conspecifics directly from observations of successful foragers. Successful foraging is readily discernible from particular behavioural patterns that cannot easily be concealed for conspecifics. Furthermore, foraging success in a number of species is facilitated by the presence of many foragers. The importance of social facilitation for variation in the reproductive success of individuals breeding in colonies of different sizes remains unknown.

The second foraging mechanism that may give rise to coloniality is the information centre hypothesis, which suggests that coloniality may provide conspecifics with indirect information on the location of profitable food sources. The foraging success of neighbours can be assessed directly from their feeding rates of nestlings, and unsuccessful foragers may follow successful conspecifics to the most profitable feeding grounds by simply waiting in the colony and assessing the success of neighbours. Information on the location of the most profitable food sources can be exploited by professional followers that do not search actively for food. The evolution of such social cheaters can be controlled if individuals benefit from being followed (e.g. due to prey being more easily captured by groups of predators) or if individuals regularly change their roles from follower to searcher. There is some direct evidence for the information centre hypothesis, suggesting that colonially breeding individuals actually do follow successful foragers. However, the relative importance of the information centre mechanism relative to the social facilitation mechanism remains unknown.

If previous use of a site is indicated by structures such as old nests, prospecting individuals that have not yet chosen a site for reproduction are able to indirectly assess the size of breeding aggregations from such signs of breeding activity before choosing a site. There is little direct evidence suggesting that the number of old nests or similar signs of previous reproductive activity are a reliable predictor of future colony size.

Direct assessment of reproductive success of conspecifics is a better predictor of future reproductive success than indirect assessment. Reproductive success integrates information on weather conditions, habitat quality, food availability, rates of parasitism and predation, and levels of intraspecific competition, information that would be very difficult or impossible to acquire by inexperienced breeders. The possibility of such prediction of future success is present because particular sites are consistent in their quality as breeding sites during different years. This is particularly the case if the time interval between assessment and reproduction is of relatively short duration. Many colonially breeding individuals visit several potential sites before choosing a particular site for reproduction; this is done mainly late during the breeding season, when the success of the present breeding season can readily be assessed. Extended periods of sexual immaturity provide individuals with extended opportunities to visit potential sites of reproduction and prospect for signs of reproductive success. The number of visits by non-breeders to colonies is related to their reproductive success and such visits are reliable predictors of future settlement in birds. Recruitment of first-time breeders to kittiwake (*Rissa tridactyla*) colonies and hence future colony size is reliably predicted from the relative success of breeders during the previous season relative to that of neighbouring sites. However, the number of breeding individuals in the previous breeding season is a less reliable predictor of future colony size.

The colony size that gives rise to maximum average reproductive success may rarely be reached because single individuals often will experience higher success by joining a colony rather than being on their own. Realized colony size may therefore often be larger than the size maximizing average reproductive success. Furthermore, the optimal colony size may differ for different categories of individuals. Individuals of poor phenotypic quality may benefit from being in a small colony, while high-quality individuals may acquire more mates or be able to lay more eggs in the nests of neighbours if the size of the colony is large. Realized colony sizes may be a compromise between the optimal size for participating individuals. *See also* SOCIOBIOLOGY. [A.P.M.]

REFERENCES

Danchin, E. & Wagner, R. (1997) The evolution of coloniality: the emergence of new perspectives. *Trends in Ecology and Evolution* **12**, 342–347.

Wittenberger, J.F. & Hunt Jr, G.L. (1985) The adaptive significance of coloniality in birds. In: *Avian Biology* (eds D.S. Farner, J.R. King & K.C. Parkes), Vol. 8, pp. 1–78. Academic Press, Orlando, Fla.

colonization The process of arrival of a species (as adults, juveniles or propagules) in: (i) a geographic region where it did not previously occur (*see* BIOGEOGRAPHY); (ii) an unoccupied patch of HABITAT (*see* METAPOPULATIONS); or (iii) a particular successional stage (*see* SERE). Establishment does not necessarily follow colonization; most colonists die without establishing a self-replacing population. *See also* ESTABLISHMENT. [M.J.C.]

colonizing species All species are capable of some form of DISPERSAL, but the phrase 'colonizing species' is used to describe species (often called RUDERAL or TRAMP SPECIES) that exhibit unusually effective long-distance dispersal (by wind, water or attached to, or in the guts of, animals). Colonizing species also exhibit the ability to establish with higher than average probability on arrival at their settling point, and are often associated with habitats created directly or indirectly by human activities (waste ground, arable fields, gardens, railway embankments, sea walls, etc.). Many *r*-strategists are colonizing species (*cf.* DORMANCY). *See also* COLONIZATION; ESTABLISHMENT; *r*- AND *K*-SELECTION. [M.J.C.]

REFERENCES

Hamilton, W.D. & May, R.M. (1977) Dispersal in stable environments. *Nature* **269**, 578–581.

Williamson, M.H. (1981) *Island Populations*. Oxford University Press, Oxford.

colony A POPULATION of organisms living in the same area. *See also* COLONIALITY; SOCIAL ORGANIZATION. [P.D.M.]

combined heat and power Productive use of heat generated from power production (e.g. in electricity generation) and, more generally, from INCINERATION. Some would count this as contributing to RECYCLING. Combined heat- and power-plants pipe hot water or steam to nearby industries, office blocks, institutions and housing. [P.C.]

command and control Use of law, in the form of specified prohibitions and CONSENTS that are monitored and infringement of which invites penalties, to protect the environment. *See also* REGULATORY INSTRUMENTS. [P.C.]

Commensalism

Commensalism forms part of a suite of associations between organisms of different species that includes symbiosis and PARASITISM, between which it approximately lies. The term was first coined in 1876 to describe the sharing by one species of food caught by another and in this sense would embrace, for example, all the scavenging mammals, birds, arthropods and others that might take a share of an antelope originally killed by a lion. It is currently used more restrictively to denote a situation in which two differing organisms characteristically and actively occur together, the one utilizing a food source provided mainly by the activities of the other, in which the utilizer therefore benefits from the association whilst the food provider is unaffected either positively or negatively.

This differential response is sometimes largely a consequence of markedly different individual sizes. Several small polychaetes and bivalve molluscs, for example, live commensally within the burrow systems or actually on the body surface of much larger polychaetes, echinoderms and members of some other groups. The leptonaceans (<2 cm long) are a characteristically commensal group of bivalves. The association is obligatory (or almost so) for the small species but not so for the larger. The ecology of these associations is poorly understood and the precise diet of the smaller partner is often only surmised. Indeed, the labelling of such associations as commensal can be merely an admission of ignorance; it simply appears unlikely that the large polychaete or echinoderm could derive any benefit or harm from its minute commensal, whilst the advantage gained by the commensal could be protection rather than food supply.

In other cases (e.g sharks and remoras, various SOCIAL INSECTS and the several arthropod inhabitants of their nests), the HOST species does appear to tolerate the presence of those commensals that scavenge within the protection of, and on the food resources afforded by, the potentially more dangerous partner. A similar situation applies to damsel fish and sea anemones, and to small spiders living within the pitchers of the insect-catching *Nepenthes*, although here the host does not so much tolerate the commensal's presence as find itself, for one reason or another, unable to prevent the association notwithstanding that some of its potential food is being taken. Since the amount of food being lost is relatively small, the disadvantage to the host is considered to be insignificant.

The dividing line between different types of ecological interaction, however, is almost always blurred. Thus associations between hermit crabs and sea anemones and between shrimps and fish that involve utilization of the same food sources are usually termed 'symbiotic' because some advantage to both partners can be envisaged. However, similar associations between bivalve molluscs and pinnotherid crabs and between the ant *Myrmica* and late instar larvae of the butterfly *Maculinea* are labelled parasitism since the crab may consume gill tissue whilst stealing the food collected by the bivalve's filter system and the caterpillar consumes some of the larval ants. The latter example illustrates the difficulty of forcing the diversity of nature into a series of discrete categories in that the ants acquire both gains and losses from the interaction whilst the butterfly only gains: if there is a net gain to the ants the interaction would become symbiotic; if there is a net loss it would be parasitic; and if gains equalled losses then it could be commensalism. *See also* MUTUALISM; SPECIES INTERACTIONS. [R.S.K.B.]

---

commercial fishing The commercial exploitation of fish accounts for roughly 5–10% of the protein consumed by the world's population. Approximately 86% of the total world fishery consists of marine species. Geographically, the major fisheries are concentrated in the waters overlying the continental shelves. The most important commercially exploited marine animals belong to four groups: bony and cartilaginous fishes, marine mammals, molluscs and crustaceans. Fish are by far the most important group by mass, and commercial landings are dominated by herring, sardine and anchovy. As of 1987, these three groups accounted for 28% of the total world catch of fish and shellfish. The second largest group of commercial fishes are the bottom-dwelling gadoids (i.e. cod, haddock, pollock and hake); the third largest group are the mackerels. Tuna are among the largest of the commercial fish species and the basis for the only major open ocean fishery. The introduction of modern fishing gear, including factory ships, larger nets and more accurate methods of locating fishes, has led to overexploitation of many FISH STOCKS. The signs of overexploitation are normally a change in AGE STRUCTURE, a decline in average size of fish and an increase in effort needed to land the same amount of fish. Attempts to manage the fisheries by limiting the allowable catch through the enforcement of quotas have generally proved disappointing. Problems in setting effective quotas include the difficulty in selecting a reasonable catch itself, the problem of dealing with 'by catches' (i.e. fish recovered unintentionally with the species being fished) and the problem of ensuring accurate reporting of catches. Classic examples of overexploited fisheries include the Peruvian anchoveta, the catch of which peaked in the 1960s at about 10 million tonnes and which, exacerbated by a severe EL NIÑO, crashed in 1972 following at least a decade of OVERFISHING. The dramatic decline of the whale fishery was the direct result of overexploitation and of unsuccessful attempts at regulating this fishery. It has been estimated that the total SUSTAINABLE YIELD of the world fishery is approximately 100 million tonnes. Since current catches are in the range of 84 million tonnes, it is unlikely that the ocean can supply much more of the world's food. Potential areas for new fisheries development include Antarctic KRILL and possibly some continental-slope fish species. [V.F.]

Commission for Sustainable Development Set up by the United Nations after the UNCED to monitor progress in implementing agreements made at the conference. *See also* RIO SUMMIT. [P.C.]

Commission on National Parks and Protected Areas *See* GAP ANALYSIS.

common property resources Natural RESOURCES that are the property of no one. Open to abuse and overexploitation. *See also* THE COMMONS; TRAGEDY OF THE COMMONS. [P.C.]

commonness Can apply to a species that is locally abundant and/or has a wide geographical SPREAD. *Cf.* RARITY, BIOLOGY OF. *See also* ABUNDANCE. [P.C.]

the Commons Major RESOURCES of the planet (ATMOSPHERE, WATER, SOIL) to which all have right of access and use, and which no one has a right to spoil. The concept was used more widely following a publication by Garrett Hardin in 1968. *See also* TRAGEDY OF THE COMMONS. [P.C.]

REFERENCE

Hardin, G. (1968) The integrity of the commons. *Science* **162**, 1243–1248.

communal species *See* SOCIALITY, TYPES OF.

communication The transfer of information between animals using visual, audible or chemical means. The term has been used of plants, which may release volatile chemicals into the atmosphere as a result of damage that can cause a response in neighbouring individuals. *See also* INFOCHEMICALS. [P.D.M.]

communication as manipulation *See* MANIPULATION.

community The total living biotic component of an ECOSYSTEM, including plants, animals and microbes. The term (unlike 'assembly' or 'ASSEMBLAGE') implies interaction between the individuals and species in the form of COMPETITION, predation, MUTUALISM, COMMENSALISM, etc. Plant ecologists often use the term specifically of the botanical component of the ecosystem and refer to 'plant communities'. The study of such plant communities has led to considerable debate concerning the interactions, integrity and CLASSIFICATION of VEGETATION. The American ecologist F.E. Clements (*see* CHARACTERS IN ECOLOGY) originally conceived the idea of plant communities as organic entities that were so rich in internal dependencies that they could be regarded as 'organisms'. This attractive idea led to the development of complex schemes of vegetation classification (*see* PHYTOSOCIOLOGY). An alternative concept, supplied by H.A. Gleason, regarded plants as essentially independent in their distribution, each species determined by its own requirements, i.e. an 'individualistic' concept. There are strong arguments on both sides, but studies of historic changes in the distribution of species indicate that each species has a distinctive biogeographical history, so communities as we now see them are a product of the current overlapping of ranges of species with similar ecological requirements. 'Communities' are therefore constantly changing their composition in the course of history.

The botanical argument has become somewhat sterile and is in need of a broader attitude involving the animal components of the ecosystem to form a more integrated approach to the study of communities, their internal structure, stability and management. *See also* BIOCOENOSIS; COMMUNITY ECOLOGY; CONTINUUM; DIVERSITY; SPECIES INTERACTIONS. [P.D.M.]

community assembly *See* COMMUNITY; HISTORICAL EFFECTS (COMMUNITY ASSEMBLY).

community continuum *See* CONTINUUM.

community ecology Community ecology is the study of the interactions between populations of organisms, and between populations and the physical ENVIRONMENT, in a particular community, and the effects that those interactions have on the behaviour and structure of that COMMUNITY. A community cannot simply be seen as the sum of its constituent species or populations, but has EMERGENT PROPERTIES that are not features of the component populations. The term, 'community' here encompasses a wide range of scales, and indeed one community can be described within another—thus a rotting tree stump may house a community of microorganisms, higher plants, and insects, whilst that stump might be part of a much larger community—a forest. *See also* ECOLOGY. [P.C.]

community matrix The matrix defining the strengths of all pairwise species interactions in a COMMUNITY. In the case of a competitive interaction, for example, the strength of interaction would be given by the COMPETITION COEFFICIENT. Used in models of community dynamics. *See also* SPECIES INTERACTIONS. [J.G.S.]

comparative method This is, in fact, a whole battery of techniques designed to help make sense of cross-taxonomic DIVERSITY in character states and ecology. Two overriding patterns emerge from comparisons of TRAITS or characters across species. The first is that closely related species tend to be similar: birds have beaks and feathers and lay eggs, while placental mammals have teeth and fur and give birth to live young. The second pattern is that species with similar lifestyles also tend to be similar in ways that adapt them to their specific lifestyles: plants from different taxonomic families that live in deserts have fewer and smaller stomata. Because closely related species tend to live in similar environments and to have similar lifestyles, it is very easy to find a relationship between variation in two traits, neither of which is responsible for the other. For example, if we took no account of evolutionary affinity and we had a collection of birds and placental mammals, on finding that some species had beaks and others had teeth we might attempt to identify the selective forces in the environment responsible for the difference by examing differences in diet between beaked and toothed species in our sample. Modern comparative methods deal with this problem by using the principle of evolutionary convergence, whereby similar evolutionary responses occur in independently evolving lineages that have moved into similar environments. Each time the same character evolves in response to the same environmental change, we have an independent evolutionary event. For example, whenever female primates evolve PROMISCUITY, males respond to this by producing more sperm per ejaculate because this increases their chances of fertilizing receptive eggs (as with having more tickets in a lottery). As a consequence, relatively large testes size accords with female promiscuity in the family tree of primates. This means that comparative biologists need a good PHYLOGENETIC TREE of the group being examined and need to locate with precision the evolutionary transitions on the branches of the tree. They can then examine whether evolutionary transitions between two characters or between a character and the environment that are thought to bear a functional relationship to each other tend to occur together more frequently than expected. It should also be borne in mind that biologists dealing with even the most sophisticated comparative methods are using arguments of inference based on correlational evidence; it is usually very difficult to separate cause from effect under such circumstances. [P.H.H.]

compartmentalization in communities Two extreme views of COMMUNITY organization are that: (i) every species interacts with every other; or (ii) species interact only with their food RESOURCES, specialist natural enemies and immediate competitors. In the second case, it is possible, in principle, to compartmentalize communities into a large number of relatively species-poor compartments. Interactions (FEEDING, COMPETITION, predation and PARASITISM) are predicted to be strong between members of the same compartment, but weak between members of different compartments. *See also* COMMUNITY ECOLOGY; COMPARTMENTS IN FOOD WEBS; SPECIES INTERACTIONS. [M.J.C.]

REFERENCES

Lawton, J.H. (1989) Food webs. In: *Ecological Concepts* (ed. J.M. Cherrett), pp. 43–78. Blackwell Scientific Publications, Oxford.

Pimm, S.L. & Lawton, J.H. (1980) Are food webs divided into compartments? *Journal of Animal Ecology* **49**, 879–898.

compartments in food webs Compartments exist in a FOOD WEB if the interactions within a web are grouped so that species interact strongly only with species within their own compartment and interact little, if at all, with species outside it. There are good biological reasons why food webs should be compartmentalized, as species within a habitat are more likely to interact with species within the same habitat than with those outside it. Theoretical predictions also suggest that model food webs have a better chance of being stable when interactions are arranged into compartments. If compartments arose as a result of dynamic CONSTRAINTS, it might be predicted that they would exist within habitats as well as between them. Empirical studies show that there is little evidence for compartmentalization within habitats suggesting that food-web compartments do not arise from dynamical constraints but more as a response to fundamental biological constraints. *See also* COMPARTMENTALIZATION IN COMMUNITIES. [M.H.]

REFERENCE

Pimm, S.C. (1982) *Food Webs*. Chapman & Hall, London.

compensating density dependence Compensating DENSITY DEPENDENCE is where demographic parameters vary with POPULATION DENSITY in such a way that there is REGULATION of POPULATION SIZE via direct or negative density dependence. Compensating density dependence does not necessarily mean that regulation is perfect; in other words, COMPENSATION may not lead to immediate restoration of an equilibrium population size. *See also* EXACT COMPENSATION; OVERCOMPENSATION; UNDERCOMPENSATION. [R.H.S.]

compensation

1 Plant GROWTH: regrowth following tissue loss (e.g. from herbivory or physical damage), such that the final loss in reproductive performance is less than would be expected on the basis of the amount of tissue originally destroyed. It may involve redistribution of reserves (e.g. from root to shoot), production of epicormic shoots, increase in the duration of the growing period or increased net photosynthetic rates of surviving green tissues.

2 POPULATION DYNAMICS: a MORTALITY factor is said to be compensated when an increase in the death rate does not lead to reduced breeding POPULATION DENSITY (e.g. mortality caused by predation of juveniles may be compensated if adult population density is limited by the availability of nest sites). Generally, the later in life that a mortality factor operates and the less intense the DENSITY DEPENDENCE of subsequent mortality factors, the less likely it is that compensation will occur.

[M.J.C.]

compensation depth The depth, in an aquatic ecosystem, at which ATTENUATION of LIGHT limits gross photosynthetic production so that it is equal to respiratory CARBON consumption; or, alternatively, the depth at which the amount of OXYGEN produced in PHOTOSYNTHESIS equals the oxygen consumed in RESPIRATION. It may be defined for a single photosynthetic organism, for all photosynthetic organisms or for all producer and consumer organisms in the ecosystem. The basis of definition must be stated. This depth represents the lower boundary of the EUPHOTIC ZONE and is controlled by season, latitude and transparency of the water column. Also sometimes called the oxygen compensation depth. *Cf.* COMPENSATION POINT. *See also* APHOTIC ZONE; CRITICAL DEPTH; PHOTIC ZONE. [J.R.E. & V.F.]

compensation light intensity The light intensity at which OXYGEN evolved from a photosynthesizing organism equals that consumed in its RESPIRATION. The compensation light intensity is the amount of light present at the COMPENSATION DEPTH. [V.F.]

compensation point The compensation point for LIGHT is defined as the photosynthetically active radiant flux density at which gross PHOTOSYNTHESIS of a CANOPY (or a whole plant or plant organ) is equal to its gross RESPIRATION. At the compensation point the canopy, plant or organ is thus neither accumulating nor losing CARBON, i.e. net photosynthetic carbon-fixation rate is zero. Light compensation point is widely variable, plants of open habitats having high values (5–10% full sunlight) and shade plants much lower values, down to 0.1% full sun or even less.

The compensation point for CARBON DIOXIDE (CO_2), at a given photosynthetically active radiant flux density, is the CO_2 concentration that reduces net photosynthetic carbon fixation to zero. Empirically, this concentration is the minimum to which a leaf or plant can reduce the CO_2 concentration in a sealed container. Plants that fix carbon by the C_3 pathway have a high CO_2 compensation point (*c*. 100 ppm by volume, ppmv), while the C_4 pathway permits reduction of CO_2 to 0 ppmv before net photosynthesis reaches zero. This is possible because CO_2 released by respiration in C_4 plants is internally recycled.

Plants with the C_4 pathway are ecologically differentiated from C_3 plants: they are particularly well represented in subarid grassland communities. CAM plants, which collect CO_2 at night, utilize the C_4 pathway and predominantly inhabit seasonally arid habitats. *Cf.* COMPENSATION DEPTH. *See also* C_3/C_4 PLANTS. [J.R.E.]

competent authorities In a European Union context, set up by member states to carry out certain functions with respect to European Commission legislation, for example in the context of CHEMICAL CONTROL legislation to receive, evaluate and make recommendations on premanufacture notifications and also to carry out RISK ASSESSMENTS of new and existing substances. Also referred to as rapporteurs. Usually will be a government department or agency. *See also* PREMANUFACTURE NOTICES. [P.C.]

Competition

Negative, i.e. detrimental, interaction between organisms caused by their need for a common resource. The effect of the interaction may be measured, for instance, by the change in equilibrium POPULATION DENSITY or by its consequences for the (absolute) FITNESS of the individuals involved. Competition may occur between individuals of the same species (INTRASPECIFIC COMPETITION) or of different species (INTERSPECIFIC COMPETITION).

Competition has also been defined as occurring when 'organisms utilize common RESOURCES that are in short supply'. However, resource limitation does not imply competition if it is not caused by the organisms themselves. (Thus, a water shortage caused by drought does not necessarily imply competition.) Some definitions of competition require that there is a reciprocal negative effect, others do not, and some authors classify every reciprocally negative interaction as competition (Arthur & Mitchell 1989) (*see also* APPARENT COMPETITION).

Forms of competition can be classified in various ways. The mechanism of competition separates INTERFERENCE COMPETITION, where competition takes place through direct interaction (e.g. fighting), and EXPLOITATION COMPETITION, where competition is mediated indirectly through the depletion of the shared resource. The effect of competition separates CONTEST COMPETITION, where a limited number of individuals obtains a sufficient share of the resource while the excess individuals get nothing at all, and SCRAMBLE COMPETITION, where all individuals obtain an equal share of the resource, which may or may not be sufficient. Obviously, intermediates between these categories exist. Some authors consider interference as a synonym of contest, and exploitation as a synonym of scramble. Indeed, these pairs often go hand in hand. Nevertheless, exploitation–contest (e.g. sequential exploitation of a limited resource) and interference–scramble (e.g. reduced feeding efficiency at high density) do occur.

'Competition theory' considers the effect of competition on niches and COMMUNITY structure. The theory can be unravelled into four major steps.

1 Competition is an important force in natural communities.
2 Interspecific competition should lead to extinction if species are too similar to their competitors (*see* COMPETITIVE EXCLUSION; LIMITING SIMILARITY).
3 The selective pressure exerted by interspecific competition should lead to divergence of the competing populations (*see* CHARACTER DISPLACEMENT). Both **2** and **3** should result in reduced NICHE OVERLAPS between actually coexisting species.
4 Given enough ecological and evolutionary time, competition should lead to communities that fully occupy the available niche space, i.e. there should be no VACANT NICHES.

While everyone would agree that competition is likely to be an important determinant of community structure in at least some cases but not in all, there is considerable debate as to the general importance of competition in this respect. Competition in natural populations is notoriously difficult to demonstrate (Connell 1980). Nevertheless, Schoener (1983) and Connell (1983), in their extensive searches of the literature, found evidence for competition reported in a large number of cases. The occurrence of competition seemed rather evenly spread across taxonomic groups and ECOSYSTEMS, but PHYTOPHAGOUS insects showed a lower proportion of positive results. It is not clear to what extent the literature is biased towards positive reports by the researchers' choice of objects and by the selection imposed by journal editors.

Another difficulty is that the strength of present-day competition and the historical impact of competition on community structure need not be coupled. Severe competition does not necessarily affect species composition (e.g. in spatially heterogeneous systems; *see* AGGREGATION MODEL OF COMPETITION) and the changes in community structure that are thought to be brought about by competition lead to the elimination of competition itself (the 'GHOST OF COMPETITION PAST'). Thus, if we want to investigate whether communities are structured by competition, we could look for traces that competition may have left in the structure of the community, such as reduced overlaps between coexisting species. However, this approach has led to considerable controversy because, on the one hand, patterns that are consistent with competition theory often have alternative explanations while, on the other hand, a lack of consistency can be explained as a failure to look at the relevant niche axes. *See also* AYALA'S EXPERIMENTS; NULL MODEL; SPECIES INTERACTIONS. [J.G.S.]

REFERENCES

Arthur, W. & Mitchell, P. (1989) A revised scheme for the classification of population interactions. *Oikos* **56**, 141–143.

Connell, J.H. (1980) Diversity and the coevolution of competitors, or the ghost of competition past. *Oikos* **35**, 131–138.

Connell, J.H. (1983) On the prevalence and relative importance of interspecific competition: evidence from field experiments. *American Naturalist* **122**, 661–696.

Schoener, T.W. (1983) Field experiments on interspecific competition. *American Naturalist* **122**, 240–285.

---

competition coefficient The relative per capita effect (on population growth) exerted by a competing species; in other words, the number of CONSPECIFICS needed to cause the same competitive effect as one individual of the competing species. The competition coefficient is used in models of competition to convert the competitive burden incurred by a competing species to the equivalent number of conspecifics. Sometimes, the competition coefficient is considered to be equivalent with NICHE OVERLAP, but see Hurlbert (1978) and Holt (1987). *See also* LOTKA–VOLTERRA MODEL. [J.G.S.]

REFERENCES

Holt, R.D. (1987) On the relation between niche overlap and competition: the effect of incommensurable niche dimensions. *Oikos* **48**, 110–114.

Hurlbert, S.H. (1978) The measurement of niche overlap and some relatives. *Ecology* **59**, 67–77.

Schoener, T.W. (1974) Some methods for calculating competition coefficient from resource–utilization spectra. *American Naturalist* **108**, 332–340.

competitive exclusion The (local) extinction of a species due to INTERSPECIFIC COMPETITION. The 'principle of competitive exclusion' (also called Gause's principle; *see* CHARACTERS IN ECOLOGY) states that species cannot coexist (*see* COEXISTENCE) as long as they occupy the same NICHE or, in more precise terms, as long as they are limited by the same RESOURCE. The principle is usually attributed to Hardin (1960), although it was formulated by other authors much earlier (see Vandermeer 1972). The principle of competitive exclusion is often taken to include the notion that divergence of the niches of the competing population (*see* CHARACTER DISPLACEMENT) is an alternative for competitive extinction. The theory that led to the principle of competitive exclusion assumes a homogeneous environment, i.e. an environment without temporal changes or spatial structure. The ubiquitous variation in natural systems therefore severely limits the applicability of the principle. *See also* LIMITING SIMILARITY. [J.G.S.]

REFERENCES

Hardin, G. (1960) The competitive exclusion principle. *Science* **131**, 1292–1297.

Vandermeer, J.H. (1972) Niche theory. *Annual Review of Ecology and Systematics* **3**, 107–132.

competitive release The expansion of the NICHE following the removal of a competing species. The definition may include immediate, phenotypic expansions, for example in response to changes in the exploitation level of RESOURCES. It may also include evolutionary, i.e. genotypic, expansions due to the change in SELECTION PRESSURES associated with the removal of the competing species. In the latter sense, competitive release can be considered as the reversal of CHARACTER DISPLACEMENT *sensu lato* (i.e. including non-morphological divergences). [J.G.S.]

competitors, good and bad

1 The superior competitor (*sensu* Tilman) is the species that can reduce the rate of resource supply to the lowest level (R^*, pronounced 'R star').

2 The superior competitor (*sensu* Grime) is the species that can sequester RESOURCES at the greatest rate.

Using Grime's terminology, Tilman's good competitor is a STRESS tolerator, the stress in this case being the low supply rate of the limiting nutrient. In Tilman's terminology, Grime's good competitor is actually a poor competitor; it is unable to persist under low supply rates of either light or soil nutrients. Although Grime's good competitors grow quickly and cast a dense shade at ground level when light and nutrients are abundant, they are intolerant of shade and/or low rates of nutrient supply.

Tilman's is a long-term, equilibrium view of COMPETITION, while Grime's is a short-term, transient view. Tilman's use has the advantage of making predictions about the identity of the species that will eventually come to DOMINANCE (it will be the species with the lowest R^*), while it is less clear how one might experimentally refute a prediction from Grime's model. [M.J.C.]

Complex life cycles

A LIFE CYCLE of an organism is considered to be complex if at least two of the phenotypes produced in the cycle differ dinstinctly in morphology. Complex life cycles (CLCs) are very prevalent among animals. The most common form of CLC is that of a larval stage transforming into a sexually mature adult form via METAMORPHOSIS. Well-known examples are holometabolous insects (insects that go through a pupal stage during development), marine invertebrates and amphibians. Apart from this single-generation CLC, there are multiple-generation CLCs. In this case, the life cycle involves a lineal succession of morphologically distinct individuals before a new cycle is initiated. For example, some tropical butterflies produce two different morphs in the wet and dry season. CLCs may also involve the alternation of one to many asexually produced generations with a sexually produced generation. In parasites with CLCs, the different generations often use distinct HOST taxa. Each generation is specialized for attacking its particular host. One example of such a CLC are the parasitic digeneans (Platyhelminthes) that include the liver fluke, a PARASITE of sheep and cattle. In the digeneans, sexual reproduction takes place in a vertebrate host and ASEXUAL REPRODUCTION occurs in a mollusc. In the vertebate host, the parasite produces eggs that leave the host in faeces, urine or sputum. Infection of the next host, a gastropod mollusc, occurs via a larvae called a miracidium. The miracidium either exists as a free-swimming larvae that actively enters the snail or it hatches after the egg has been accidentally ingested by the snail. Within the molluscan host the miracidium transforms into a sporoplast, which is a gutless sac containing germinal cells. These develop into either a second generation of sporoplasts or a new larval type with a gut called redia. There can be further generations of sporoplasts or redia within the snail but eventually a new form called cercariae is produced that leaves the molluscan host. The cercariae are free-swimming larvae whose head develops into the final worm in a vertebrate host after the tail has been shed. However, most digenean life cycles involve further intermediate hosts in which the tail-less cercariae (now called metacercariae) encyst themselves and wait until the final host ingests the intermediate host. The final host is therefore often predatory (e.g. fish) and the intermediate non-molluscan hosts can be vertebrate or invertebrate. Another example of a multiple-generation CLC are aphids (Homoptera: Aphidoidea). In temperate climates, aphids produce a number of asexual generations followed by a generation that sexually produces a diapausing egg in autumn. During the asexual phase, a number of morphs are produced. These can be winged or unwinged and differ in their morphology, physiology and behaviour. In the aphid family Aphididae, a morph called fundatrix hatches from the diapausing egg in spring. This fundatrix asexually produces a generation called fundatriginae. The following asexually produced generations (virginoparae) are winged (alatae) or unwinged (apterae). At the end of the season, before males and and egg-laying females (oviparae) are produced, there might be a special generation of morphs that produces either only females (gynoparae), only males (androparae) or both of the sexual morphs (sexuparae). In about 10% of all aphids, the life cycle involves the alternation between a primary host plant (a woody plant) where eggs are deposited and a SECONDARY HOST plant (mostly herbaceous plants). While alatae only switch between different individuals of the secondary host plants colonized in summer, there might be further morphs that are specialized in switching between the primary and secondary host plants. One example of a host-alternating aphid is the black bean aphid (*Aphis fabae*), whose primary host plant is spindle (*Euonymus europaeus*) and whose secondary hosts are a number of herbs including beans. CLCs also occur in plants (e.g. in rust fungi).

The different phases produced in a CLC often occupy different ecological niches. For example, in frogs and toads, the larvae are typically aquatic herbivores whereas the adults are terrestrial carnivores. This NICHE shift has led to the hypothesis that CLCs are an ADAPTATION for the exploitation of RESOURCES. According to this hypothesis, a CLC allows a population to acquire resources more efficiently. A more general hypothesis states that the different phases of a CLC are selected to perform different FITNESS tasks most efficiently. Because of metamorphosis, adaptations for conflicting tasks can evolve more independently. For example, in anurans, the larval phase is adapted to GROWTH whereas the adult phase is specialized for reproduction, DISPERSAL and the survival of cold or dry seasons. The third hypothesis is that CLCs result from developmental CONSTRAINTS. In this view, certain TRAITS, mostly those of early stages, have to be maintained because they are linked to traits of later stages, even if they are no longer adaptive in the current environment. For example, in this hypothesis the larval form of anurans would be maintained because developmental mechanisms make it impossible to produce an adult without it. *See also* LIFE-HISTORY EVOLUTION; POLYMORPHISM. [W.W.W.]

REFERENCES

Ebenman, B. (1992) Evolution in organisms that change their niches during the life-cycle. *American Naturalist* **139**, 990–1021.

Hille Ris Lampers, D. (1966) Polymorphism in Aphididae. *Annual Review of Entomology* **11**, 47–78.

Istock, C.A. (1967) The evolution of complex life-cycle phenomena: an ecological perspective. *Evolution* **21**, 592–605.

Moran, N.A. (1994) Adaptation and constraint in the complex life-cycles of animals. *Annual Review of Ecology and Systematics* **25**, 573–600.

Werner, E.E. (1988) Size, scaling, and the evolution of complex life-cycles. In: *Size-structured Populations* (eds B. Ebenman & L. Persson), pp. 60–81. Springer-Verlag, Berlin.

Wilbur, H.M. (1980) Complex life-cycles. *Annual Review of Ecology and Systematics* **11**, 67–93.

---

complexity of ecosystems Strictly speaking, the term 'ECOSYSTEM complexity' should denote some integrated measure of the complexity of the biological COMMUNITY, together with its physical ENVIRONMENT. In practice, however, it is often taken to refer only to the biological community, the complexity of which often reflects that of the physical environment (*see* HABITAT STRUCTURE). The mean number of species in the system, the number of interactions between species or some combination of these and related measures can be taken as indices of complexity. As ecosystems become more mature they generally increase in structural and biological complexity (*see* SUCCESSION). Understanding the determinants of complexity and the relationship between this property and others such as the ecosystem stability remain a central goal for ecology (*see* DIVERSITY–STABILITY HYPOTHESIS). *See also* COMPLEXITY–STABILITY. [S.J.H.]

complexity–stability The complexity of a COMMUNITY is measured as the number of species, the number of interactions between species, the average strength of interaction, or some combination of these. Stability is a measure of a community's sensitivity to DISTURBANCE. Stability may be measured in terms of the number and ABUNDANCE of its component species or by other community properties or processes. The concept of stability includes the property of RESILIENCE (the speed with which a community returns to its former state after it has been displaced from that state by a perturbation) and RESISTANCE (the ability of a community to avoid displacement in the first place). Communities may be described as locally stable (able to recover from small perturbations) or globally stable (able to recover from large perturbations). If a community is stable only within a narrow range of environmental or biological conditions it is said to be DYNAMICALLY FRAGILE; if it is stable over a broad range of conditions it is described as DYNAMICALLY ROBUST. During the 1950s and 1960s, ecologists believed that increased complexity within a community would lead to increased stability. More recent models of communities suggest that stability tends to decrease as complexity increases. However, evidence from real communities indicates that the relationship between community complexity and stability varies with the precise nature of the community, with the way in which the community is perturbed and with the way in which stability is assessed. Ecologists have observed a general tendency for the existence of complex, dynamically fragile communities in stable environments (i.e. tropics) and the existence of simpler, dynamically robust communities in relatively variable environments (e.g. temperate regions). *See also* COMPLEXITY OF ECOSYSTEMS; DIVERSITY–STABILITY HYPOTHESIS. [V.F.]

compliance Operating within the requirements of the law. In environmental terms this often means conforming with legally defined standards. [P.C.]

compost *See* COMPOSTING.

composting Biological breakdown of biological solids so as to stabilize them as humic substances (compost). [P.C.]

Comprehensive Environmental Response Compensation and Liability Act 1980 (CERCLA) *See* SUPERFUND LEGISLATION.

concentration–response The quantitative relationship between the concentrations of a TOXICANT to which a subject is exposed in the AMBIENT medium and the incidence or extent of an adverse EFFECT. *Cf.* DOSE–RESPONSE. [P.C.]

concerted evolution The EVOLUTION of all members of a GENE FAMILY as a block. Concerted evolution occurs because a MUTATION spreads across the gene family, for example by UNEQUAL CROSSING-OVER or by unequal GENE CONVERSION. The result is that all members of a gene family (including both expressed and interstitial sequences) may be identical in one species and different from those of a closely related species. [V.F.]

Concorde fallacy The error of throwing good money after bad. Named after the Concorde supersonic transport, an Anglo-French airliner notorious for its high cost, the term was coined by R. Dawkins and T.R. Carlisle in their critique of Trivers' analysis of the domestic bliss strategy. Trivers argued that females could ensure that males did not desert them following copulation by requiring substantial investment prior to copulation. Dawkins pointed out that this would only be true if most other females in the population used the same strategy, otherwise males could simply avoid females that required pre-copulatory investment. [D.A.B.]

REFERENCE
Dawkins, R. & Carlisle, T.R. (1976) Parental investment, mate desertion and a fallacy. *Nature* **262**, 131–132.

concurrent-range zone *See* BIOSTRATIGRAPHIC UNIT.

condition index (CI) Index of metabolic condition of an individual, usually in terms of mass per unit length. [P.C.]

conductance In physics, conductance is the current flowing through the conductor divided by the voltage between its ends. Conductance is the reciprocal of resistance. The units are reciprocal ohms or siemens (S) and the symbol used is usually Ω^{-1}. In ENVIRONMENTAL SCIENCE, the conductance/resistance analogy is used to analyse transfer processes in which an entity is transferred along a potential gradient. For example, vegetational surfaces are characterized by surface conductances for the transfer of WATER vapour and CARBON DIOXIDE (CO_2), with units of $m\,s^{-1}$ or $mol\,m^{-2}\,s^{-1}$. [J.G.]

REFERENCE
Jones, H.G. (1992) *Plants and Microclimates*. Cambridge University Press, Cambridge.

conduction, of electricity The movement of electricity through a conductor. The flow of current I (amperes) is proportional to the potential gradient V (volts), and the constant of proportionality is the reciprocal ohm or mho (Ω^{-1}):

$$I = \Omega^{-1} V$$

or, more usually stated as Ohm's law, the current is proportional to the voltage and inversely proportional to the resistance R:

$$I = V/R$$

The electrical resistivity is the resistance of a metre cube and is measured in ohm metre units (Ω m); examples are given in Table C3. *Cf.* CONDUCTANCE. [J.G.]

conduction, of heat The movement of heat through a solid, liquid or gaseous medium, without visible motion of the medium, in a continuous process whereby the energy is passed from molecule to molecule. The flow of energy J ($W\,m^{-2}$) is proportional to the TEMPERATURE gradient $\partial T/\partial z$ ($°C\,m^{-1}$), and the constant of proportionality K is known as the thermal conductivity ($W\,m^{-1}\,°C^{-1}$):

$$J = K(\partial T/\partial z)$$

Thermal conductivity declines slightly with temperature. The examples in Table C4 are quoted at 273 K. *Cf.* CONDUCTANCE. [J.G.]

confidence limit Confidence limits describe where we would expect a population parameter to lie in relation to a statistic estimated from a SAMPLE. For example, since the SAMPLING distribution of means (from samples of size $n \geq 30$) (*see* STANDARD ERROR) is a NORMAL DISTRIBUTION, we know that 95% of observed means will be within the interval $\mu \pm 1.96s/\sqrt{n}$, where $s/\sqrt{n}$ is the estimated standard error of the MEAN. Or, stated the other way round, we can be 95% confident that the interval $\bar{x} \pm 1.96s/\sqrt{n}$ will contain the population mean (μ). This interval is known as a confidence interval or, more precisely, as the 95% confidence interval. The interval $\bar{x} \pm 2.58s/\sqrt{n}$ is known as the 99% confidence interval. Clearly, if the confidence interval is large we can place less reliability on the sample mean as an estimate of the population mean.

Table C3 Examples of electrical resistivities.

Material	Resistivity (Ω m) at 273 K
Silver	1.47×10^{-8}
Copper	1.55×10^{-8}
Aluminium	2.50×10^{-8}
Mercury	94×10^{-8}
Carbon (amorphous)	6×10^{-5}

Table C4 Examples of thermal conductivities.

Material	K (W m^{-1} °C^{-1}) at 273 K
Silver	428
Copper	403
Aluminium	236
Mercury	8
Diamond	10–26
Carbon (amorphous)	1.5
Glass	0.8–1.1
Brick-wall	0.8–1.2
Water	0.6
Sulphur (amorphous)	0.2
Concrete	0.1–1.5
Cardboard	0.2
Air, atmospheric pressure	0.2
Cork	0.04
Expanded polystyrene	0.035

When $n<30$ the sampling distribution of means is no longer normal. Standardized normal deviates or z-scores are replaced by t, which reflect a different relationship between STANDARD DEVIATION and PROBABILITY. For example, the 95% confidence interval is now $\bar{x} \pm t.s/\sqrt{n}$, where t is the value, at the intersection of $P=0.05$ and $n-1$, from t tables (*see* STATISTICAL TABLES).

Note that the standard error of other sample statistics can be used to place confidence limits around them. [B.S.]

Conflicts

A conflict may be said to exist between two parties if an action by the first party increases the FITNESS of that party (and hence the gene for the TRAIT tends to spread in a population) and decreases the absolute fitness of a second party (which is not in direct COMPETITION with the first) such that spread of trait 1 creates the context for the spread of a suppressor of the action performed by the second party.

But when is an interaction competition rather than conflict? An economic argument may clarify the distinction. Consider, for instance, a shoe manufacturer that made new shoes that were much more comfortable than ordinary shoes and so started selling well (party 1). Alternative shoe makers must be losing market share if party 1 is gaining. Let us also consider, however, that the new shoes reduce wear on socks. As a consequence all sock manufacturers end up selling fewer socks. None loses their percentage of the market but all lose custom. When we ask about the spread of trait 1 (i.e. the increasing proportion of the market held by the new shoe), the alternatives to trait 1 (i.e. the production of old-type shoes) must by definition be having their market share reduced. It is these manufacturers that are in competition. In contrast, the second party, the sock makers, have their fitness reduced, but it is not necessarily the case that any particular sock manufacturer does worse than any other. These are not in competition with the shoe maker (they are not selling shoes), but are in conflict with the manufacturer of the new shoes.

The potential for conflict can be expressed in terms of the interests of parties. Those actions that a party might perform that were to the benefit of that which codes for the action one can refer to as the interests of that which codes for the action.

To say that a cytoplasmic genome (which is transmitted only through females) has 'no interest' in a male is not to imply that the cytoplasmic genome has thought the issue through and decided that it would be bored. Rather it is a condensed way to express the notion that cytoplasmic genes which could, for instance, turn males into females, could spread. Similarly, when two parties do not share precisely the same set of interests we might say that there is potentially a conflict of interests. Conflicts can exist between any two parties that are not perfectly related (i.e. not clonal derivatives or identical twins). Potential conflicts of interest may exist at any level of biological organization. Intragenomic conflict occurs when a SELFISH GENE (alias SELFISH DNA) manipulates the host GENETIC SYSTEM to its own ends. The hypothesized cytoplasmic factor described above (actually found in certain crustaceans) would be one such example. Alternatively, a MEIOTIC DRIVE gene (party 1) may act in heterozygous males to kill off all sperm that do not contain it. By so doing, the drive allele may achieve fertilization of more than 50% of the eggs of any given female. As non-driving ALLELES can at best gain only 50% fertilization on average, the drive allele can spread through the population even if host fitness is somewhat reduced. However, in so doing the context is created for alleles in unlinked components of the genome (party 2) that can suppress this drive activity to spread in the population. If these suppressors are cost-free, then the meiotic drive allele will be removed from the population. A meiotic drive gene is in competition with its allelic counterparts but in conflict with the rest of the genome.

Intragenomic conflict of this variety may be considered a subclass of host–parasite conflict, selfish genes being unusual PARASITES in that they receive no (or negligible) horizontal transmission. The logic of conflict is clearly illustrated, however, by reference to more conventional parasites. In general, parasites are deleterious to the host (party 2) and so the spread of a novel parasite strain will create the context for the spread of antiparasite alleles in the host population. Just such a process is thought to explain the rapid evolution of both parasite antigens and host antigen-recognition systems. In this instance, the novel parasite that successfully manipulates its host is in conflict with the host, but in competition with alternative parasite strains whose total frequency must go down as much as that of the given parasite strain goes up.

Host and parasite are not usually considered as part of the same individual (in contrast to intragenomic conflict) and hence is an instance of interindividual conflict. The most familiar examples of conflict exist between individuals. In mammals with promiscuous females, for example, mothers and fathers may have different relatedness to given individuals in a brood. Mothers have guaranteed relatedness to all the progeny, whereas a given father may be related to some but not all. A conflict then exists between father and mother. A father would prefer that only its progeny would receive the mother's milk, whereas the mother might prefer a more equitable resource allocation.

A similar conflict may also exist between mother and offspring. A given offspring's best interests might be to have more of mother's milk than its full sibs and half-sibs (paternally unrelated). A mother's best interest is served by, again, a more equitable distribution as all offspring are equally highly related to her. This conflict may be expressed by transplacental fetal MANIPULATION of the mother. Many curious details of placental physiology and gene expression can be interpreted in this light (Haig 1993). Conflicts between close relatives may also exist over the optimal sex-ratio allocation. In haplo-diploid insects (e.g. wasp, bees, ants), males are haploid and produced by the parthenogenetic development of unfertilized eggs. Females are diploid and the result of fertilization of eggs. A male is then related to his mother by $r=1$ (its genes must be from the mother), but the mother is related to him by $r=0.5$ (a given allele has a 0.5 chance of being in the son). Daughters are likewise related to their mother by 0.5. It also follows that females are related to full-sib sisters by $r=0.75$ (take a random allele in a female and the probability that her full sister will have a clonal copy of it is 0.75). In colonies of SOCIAL INSECTS, in which worker females are sterile, this collection of unbalanced relatedness creates the conditions for a large array of possible conflicts, not least of which is a queen–worker conflict over the sex ratio. Worker females are highly related to their full sisters and only poorly related to their brothers ($r=0.25$). Hence a mutant gene in workers that could skew the sex ratio towards sisters could spread if the sisters are usually full sisters. If this trait was successful, then its spread would bias the sex ratio in the population towards females and a gene acting in a queen that could produce more males than the average would spread (males would be in great demand if the sex ratio was biased). There is hence a conflict of interest between queen and worker. It transpires that much of the time the workers are in control and the sex ratio is about 3 : 1 in favour of females (Trivers & Hare 1976).

A conflict might also exist between worker and queen over the number of fathers that a brood might have. A given worker is only highly related to its full sisters and not to those from different fathers. If a female is to give up reproduction (as workers do), her best interests are typically served by minimizing the number of fathers that the queen utilizes. However, multiple mates may have fitness advantages to the queen. If multiple fathers have contributed to a brood there is a potential for inter-worker conflict over, for example, which eggs are to mature as queens.

Conflict is a term that, as employed above, is often of great interest in arenas where average relatedness might be high but the variance in relatedness between parties is also high. However, as in the host–parasite example, conflicts must often exist in arenas where the parties are unrelated.

Within tightly knit organizations the potential for conflict to disrupt the functioning of the COMMUNITY is intense. It is hence a paradox why both components of the genome and of social insect communities are not more selfish. In general, resolutions of differences in interests may not be predictable in an a priori manner and depend on intimate details of mechanisms. The concept of power has been introduced to attempt to provide a language with which to consider these issues. *See also* BEHAVIOURAL ECOLOGY; SOCIOBIOLOGY. [L.D.H.]

REFERENCES

Haig, D. (1993) Maternal–fetal conflict in human pregnancy. *Quarterly Review of Biology* **68**, 495–532.

Hurst, L.D., Atlan, A. & Bengtsson, B.O. (1996) Genetic conflicts. *Quarterly Review of Biology* **71**, 317–364.

Trivers, R.L. (1985) *Social Evolution*. Benjamin Cummings, Menlo Park, Calif.

Trivers, R.L. & Hare, H. (1976) Haplodiploidy and the evolution of the social insects. *Science* **191**, 249–263.

conformer An organism whose physiological state (e.g. body temperature (thermoconformer) or body fluid composition (osmoconformer)) is identical to, and varies identically with, that of the external environment. Some species may show conformance with respect to some environmental variables but REGULATION with respect to others. [V.F.]

connectance Fraction of all possible pairs of species within a COMMUNITY that interact directly as feeder and food. In other words, the number of actual connections in a FOOD WEB divided by the total number of possible connections. *See also* TROPHIC CLASSIFICATION. [P.C.]

consents Same as AUTHORIZATION, but usually refer to effluent DISCHARGES to public sewer or receiving waters. *See also* POLLUTION PERMITS. [P.C.]

Conservation biology

Conservation biology is an emerging, interdisciplinary field that seeks to establish a scientific basis for the conservation and management of populations, communities and ECOSYSTEMS. At the same time, conservation biology draws on the empirical observations and results of land management practices as a primary source of information and insight. Consequently, conservation biology may be thought of as the interface between ECOLOGY and allied disciplines on one hand, and the practice of conservation management on the other.

Global context

The impetus for the creation and growth of the science of conservation biology has been the continuing and accelerating loss of global biological DIVERSITY and the massive alteration of ecosystems by human activities in recent decades. Conservation biologists recognize a series of related factors that contribute to this ecological crisis. Continued EXPONENTIAL GROWTH of the global human population—currently more than 5.8 billion people and increasing at 1.6% annually—remains a dominant concern for conservation biologists. Per capita consumption of many natural RESOURCES continues to exceed rates of replenishment on a global scale, even for those resources that are supposedly 'sustainable'. The global inventory of natural forests continues to decline at alarmingly rapid rates in tropical, temperate and BOREAL regions. Other ecosystem types—GRASSLANDS, SAVANNAHS, DESERTS and the fragile coastal, lacustrine and marine communities—are also experiencing severe rates of loss and alteration.

At the same time, increasing inequity in the distribution of wealth has left billions of people living perpetually on the edge of starvation, while a small affluent minority enjoys a standard of living (and rate of consumption) unequalled in human history. Conservation biologists recognize severe poverty in less affluent nations as a major contributing factor to the destruction and degradation of ecosystems, including massive soil erosion, DEFORESTATION and poaching of endangered WILDLIFE from reserves. Indigenous peoples have suffered large losses of land and population in recent decades; conservation biologists view the global loss of cultural diversity as a phenomenon parallel to that destroying ecosystems and BIODIVERSITY.

Along with increased consumption has come increased POLLUTION of air and water, again on a global scale; ecologists working even in the areas most remote from any human population, such as Antarctic and ARCTIC regions and deep ocean trenches, find pollutants and toxic substances that are the result of industrial processes. Some forms of industrial technology have set in motion the most dramatic, and until recently, almost unthinkably large-scale, change in Earth's ecology, the actual modification of planetary CLIMATE as a result of combustion of FOSSIL FUELS and huge emissions of CARBON DIOXIDE (CO_2). Other by-products of the industrial age, such as atmospheric release of halogenated HYDROCARBONS, are contributing to the breakdown of upper atmospheric ozone layers that shield life on Earth from ULTRAVIOLET RADIATION.

Although the net effect of contemporary human civilization could be measured in many ways, to the conservation biologist perhaps none is so profoundly alarming as the current and projected global loss of species. Many scientists have estimated that present rates of species extinction are three to five orders of magnitude greater (1000 to 10 000 times as rapid) than the background extinction rate for most periods of life on Earth, including our own. Elevated rates of extinction in terrestrial ecosystems are estimated to be particularly high in the most species-rich regions on Earth, the great tropical forests of South America, Africa and south-east Asia. These areas are estimated to contain over half of all species on Earth, although they cover less than 7% of the land area. Areas of high coastal and marine diversity, such as CORAL REEFS and ESTUARIES, are similarly the sites of unprecedented loss of species diversity. International studies estimate that at least 15% of all species on Earth are at risk of immediate extinction if dramatic steps are not taken for their protection.

All of these problems require various forms of conservation action, both to mitigate their short-term effects and to provide strategies for long-term restoration. The goal of conservation biology is to provide a reasoned scientific basis for such actions.

Related disciplines

A number of fields provide the conceptual foundation for conservation biology. Of most immediate relevance are terrestrial and aquatic ecology, POPULATION ECOLOGY and genetics, HYDROLOGY, soil science, BIOGEOGRAPHY, paleontology, evolutionary biology and systematics. However, because the conservation biologist deals not in an academic vacuum but with real-world issues of pressing importance, a variety of other fields also figure prominently in conservation biology thinking, including law, economics, philosophy, sociology, and anthropology, as well as other natural and physical sciences. At the same time, conservation biology provides a direct interface with disciplines of natural resource management, such as FORESTRY, fisheries, WILDLIFE MANAGEMENT and agronomy.

Although conservation biology has roots that extend back literally thousands of years, the contemporary discipline was articulated less than 20 years ago. In the early 1980s a series of books and organizational steps led to the rapid articulation of conservation biology; groundbreaking publications included works by Dasmann (1959), Ehrenfeld (1970), Soulé & Wilcox (1980), Frankel & Soulé (1981), Schonewald-Cox *et al.* (1983), Wilson (1986) and Soulé (1986). In 1985 the Society for Conservation Biology was formed, and in 1987 the journal *Conservation Biology* was initiated.

Current priorities

Because of the dominant concern over species extinctions and the concomitant loss of evolutionary and ecological potential, many problems in conservation biology are related to the biology of rare organisms and small populations. Hence, conservation biology has been a central forum for discussion of consequences at the population level of inbreeding, GENETIC BOTTLENECKS and demographic and environmental stochasticity. These frequently have been evaluated in terms of their potential effect on extinction probabilities. Conservation biology drew heavily in its early years on elements of population biology, such as MINIMUM VIABLE POPULATION (MVP) estimation and population viability analysis (PVA), and on analysis by various techniques of the magnitude and distribution of genetic variation within and among populations. The field has also been influenced profoundly by island biogeographic theory (MacArthur & Wilson 1967), which relates species–area functions to the diversity of insular patches of HABITAT. This theory provided the central conceptual basis for analysis of the potential SPECIES RICHNESS of habitat 'islands' such as parks and remnant fragments, and has served as the driving force for emphasis on the necessity of preserving large contiguous blocks of habitat and creating BUFFER ZONES and CORRIDORS.

COMMUNITY ECOLOGY, vegetation analysis and LANDSCAPE ECOLOGY have also figured prominently in the field since its inception, providing a basis for understanding the impacts of ecosystem fragmentation, EDGE EFFECTS and the impacts of invasive introduced species. More recently, emphasis has been given to species interactions (such as POLLINATION, seed dispersal and COMPETITION) and to the fundamental ecological role of DISTURBANCE processes and SUCCESSION in shaping ecosystems. As a field, conservation biology has been attentive from the outset to the enormous fund of knowledge of the natural world possessed by indigenous peoples, and has helped both to document this knowledge and to highlight the crisis of cultural survival in many parts of the world, especially the global tropics.

Conservation biologists have also contributed substantially to the development of strategies for conserving populations, species, communities and ecosystems. Among the topics that have been addressed are the conservation use of buffer zones and corridors, techniques for managing small populations, sustainable utilization of animal and plant resources, new techniques for FOREST management, monitoring protocols for critical INDICATOR SPECIES and processes, the use of *ex situ* resources in INTEGRATED CONSERVATION STRATEGIES, legislative strategies for protecting species and ecosystems, design and management of ecologically sustainable reserves and the use of reintroduction as a conservation tool for animals and plants.

The eminent plant population biologist, Professor John Harper of University College of North Wales, has written (1992) that conservation biology aims ultimately 'to develop a predictive science of ecological management'. Whether the field ever attains that goal remains to be seen. However, as both a discipline of applied ecology and the scientific underpinning for ECOSYSTEM MANAGEMENT, conservation biology has already proven to be a source of profound insight into our threatened natural world. *See also* AGRICULTURE AND SOIL CONSERVATION; AMPHIBIANS, DIVERSITY AND CONSERVATION; ANADROMOUS FISH CONSERVATION; ANGIOSPERMS, GLOBAL DIVERSITY AND CONSERVATION; ANTARCTICA, ECOLOGY AND CONSERVATION; AQUATIC SYSTEMS, CONSERVATION AND RESTORATION; ARGENTINA, CONSERVATION IN; BEETLES, DIVERSITY AND CONSERVATION; BEHAVIOUR, IMPLICATIONS FOR CONSERVATION; BRYOPHYTES, GLOBAL DIVERSITY AND CONSERVATION; CANADA, CONSERVATION IN; CARRYING CAPACITY, IMPORTANCE TO CONSERVATION; CHIROPTERANS, DIVERSITY AND CONSERVATION; CORAL REEFS, DIVERSITY AND CONSERVATION; DISPERSAL AND COLONIZATION OF AQUATIC ORGANISMS, CONSERVATION IMPLICATIONS; DISPERSAL AND COLONIZATION OF TERRESTRIAL ORGANISMS, CONSERVATION IMPLICATIONS; ECOLOGICAL RESTORATION; ENVIRONMENTAL ETHICS AND CONSERVATION; EVOLUTION AND CONSERVATION; *EX SITU* CONSERVATION; FARMLAND AND FARMING COMMUNITIES, CONSERVATION; FISH, DIVERSITY AND CONSERVATION STRATEGIES; FISHERIES, CONSERVATION AND MANAGEMENT; FRAGMENTATION OF HABITATS AND CONSERVATION; FUNGI, DIVERSITY AND CONSERVATION; GEOGRAPHIC INFORMATION SYSTEMS (GIS) IN CONSERVATION; GYMNOSPERMS, GLOBAL DIVERSITY AND CONSERVATION; HETEROZYGOSITY, INTERPRETATION FOR CONSERVATION; INSECTS, DIVERSITY AND CONSERVATION; INTEGRATED CONSERVATION STRATEGIES; INTERNATIONAL CONSERVATION CONVENTIONS; ISLAND BIOGEOGRAPHIC THEORY IN CONSERVATION; IUCN; MANGROVE COMMUNITIES, ECOLOGY AND CONSERVATION; MEDITERRANEAN CLIMATE REGIONS, CONSERVATION; MICROCLIMATES AND CONSERVATION BIOLOGY; MONTANE FORESTS, CONSERVATION; OLD-GROWTH/ANCIENT FORESTS, CONSERVATION; PLANTS AND FUNGI, GLOBAL DIVERSITY AND CONSERVATION; POPULATION DYNAMICS; PTERIDOPHYTES, GLOBAL DIVERSITY AND CONSERVATION; RAINFORESTS AND CLOUD FORESTS, CONSERVATION; REPRODUCTION (ANIMAL) AND CONSERVATION; REPTILES, DIVERSITY AND CONSERVATION; RUSSIA AND THE FORMER SOVIET UNION, CONSERVATION IN; WETLANDS (TIDAL AND COASTAL), CONSERVATION. [D.A.F.]

REFERENCES

Dasmann, R.F. (1959) *Environmental Conservation*. John Wiley & Sons, New York.

Ehrenfeld, D.W. (1970) *Biological Conservation*. Holt, Rinehart & Winston, New York.

Frankel, O.H. & Soulé, M.E. (1981) *Conservation and Evolution*. Cambridge University Press, Cambridge.

Harper, J. (1992) Foreword. In: *Conservation Biology: The Theory and Practice of Nature Conservation, Preservation and Management* (eds P.L. Fiedler & S.K. Jain), pp. xi–xviii. Routledge, Chapman & Hall, New York.

Hunter, M.L., Jr. (1996) *Fundamentals of Conservation Biology*. Blackwell Science, Cambridge, Mass.

MacArthur, R.H. & Wilson, E.O. (1967) *The Theory of Island Biogeography*. Princeton University Press, Princeton, N.J.

Meffe, G.K. & Carroll, C.R. (1997) *Principles of Conservation Biology*, 2nd edn. Sinauer Associates, Sunderland, Mass.

Primack, R.B. (1993) *Essentials of Conservation Biology*. Sinauer Associates, Sunderland, Mass.

Schonewald-Cox, C.M., Chambers, S.M. MacBryde, B. & Thomas, L. (eds) (1983) *Genetics and Conservation: A Reference for Managing Wild Animal and Plant Populations*. Benjamin/Cummings, Menlo Park, Calif.

Soulé, M.E. (ed.) (1986) *Conservation Biology: The Science of Scarcity and Diversity*. Sinauer Associates, Sunderland, Mass.

Soulé, M.E. & Wilcox, B.A. (eds) (1980) *Conservation Biology: An Evolutionary–Ecological Perspective*. Sinauer Associates, Sunderland, Mass.

Wilson, E.O. (ed.) (1986) *Biodiversity*. National Academy Press, Washington, D.C.

---

conservation evaluation *See* RARITY, BIOLOGY OF.

conspecific An individual of the same species. [B.S.]

constant-effort harvesting This type of HARVESTING is theoretically attractive as it involves natural FEEDBACK between the level of catch and the population of the harvested resource. The idea is that the amount of MORTALITY inflicted remains effectively constant. This can be done by setting, for example, a restriction on the amount of hunters or, in the case of fishing, the number of fishing vessels of a certain power. The operation of this constant amount of effort means that the catch will vary in proportion to the POPULATION SIZE. This makes an implicit assumption that catch is related to effort and population size by a simple linear relationship. The constant of proportionality is usually termed in fisheries as the catchability coefficient. The basic equation is:

$$\text{Catch} = \text{Catchability coefficient} \times \text{Effort} \times \text{Population size}$$

In theory this is an ideal mechanism, as when the population is small the catches are small. However, there are problems in its practical application as technological improvements occur and the effective effort can increase undetected. Furthermore, in certain cases, particularly involving RESOURCES that aggregate spatially, the relationship between catch per unit of effort and population size is non-linear, which can mask overexploitation. *See also* HARVESTING; SUSTAINABLE YIELD. [J.B.]

constant final yield This involves the idea of a population equilibrium in which the removals by HARVESTING (the YIELD) exactly equal the surplus that comes from the balance of births, deaths and growth in the population. This constant final yield can occur at all levels up to the MAXIMUM SUSTAINABLE YIELD, but not beyond it. In practice, this theoretical concept is useful, but as all RESOURCES are subject to environmental variation and, indeed, demographic variation such yields will tend to fluctuate around some characteristic average. *See also* SUSTAINABLE YIELD.[J.B.]

Constraints

In its formal mathematical context, constraint refers to any internal factor that interferes with the optimizing function of NATURAL SELECTION by preventing the required variation to be expressed in the population under consideration. The term has gained popularity as an evolutionary concept very recently. Its origin can be traced to an influential article by Gould and Lewontin (1979). These authors introduced the concept of constraint as an alternative approach to extreme adaptationist programmes that predominated at the time, associated with the advent of the paradigm of SOCIOBIOLOGY. However, the concept was not given a concise definition until the issue of 'developmental constraint' was introduced (see Oster & Alberch 1982, as representative of a series of papers on the issue, with the most current retrospective in Alberch 1989), within a general conceptual framework of a new 'biology of constraint', inspired and popularized by S.J. Gould. Ever since, the concept has attained a fair amount of popularity, leading to a variety of quite different meanings for the term, which has led to the current stage of confusion and misunderstanding surrounding the word. This situation is further complicated by attempts to revise its original definition in the concept of generative rules and their effects on the expression of variation in populations. Thus, the definition put forward here will probably be more restrictive than others. It is a counterpoint to the abuse of the concept in the recent literature, where it is often utilized, in a colloquial manner, to define any state of non-optimality in adaptiveness. I divide the definition into several key concepts surrounding the issue of constraints on evolutionary processes.

1 *Constraints versus natural selection.* At its most fundamental level, constraint is only meaningful in the context of selection. The causal explanation of 'order' in nature must be the result of the combined effects of these two deterministic agents. It is important to emphasize that constraint is a concept that can be integrated within the conceptual scheme of NEO-DARWINISM. It dictates a distinct methodological approach to the study of EVOLUTION as well as forcing a reassessment of the nature of variation upon which natural selection operates.

2 *Global versus local.* Constraint, as any other evolutionary mechanism, operates at the level of the population. Hence, any effect is local by definition, i.e. any constraint has an 'actual' effect on future evolution, as well as being the result of a historical contingency. History is actually hauled along in the form of a system of constraints characteristic of the LINEAGE of an organism. There is a hierarchy of constraints correlated with the hierarchical structure of life as well as the hierarchical CLASSIFICATION of species. There are a few global constraints equated with the essence of life itself. Many of them have been referred to as 'material constraints'. For example, the biomechanical properties of an organism are constrained by carbon-centred biochemistry. Thus, the functional design of many organisms could be improved if other materials could be 'manufactured' by organisms, such as aluminium or plastic, but the ORIGIN OF LIFE on the planet already defined some generic constraints. There are other global constraints, not as general as the previous ones, which are related to features that developed at a slightly higher branching point in the hierarchy of the tree of life, for example the molecules used to encode the information to pass on to the next generation. The use of DNA as the molecule of inheritance is almost general to all living organisms. As organisms evolved, historical contingency has locked it within a set of invariant fundamentals. Nevertheless, there is a slightly higher probability that alternative molecules for the transmission of information, such as existing RNA-centred systems, could have evolved. There is nothing absolute about the DNA molecule that makes it functionally unique. The classification of organisms reflects the existence of historically dependent basic plans of structural organization. The ability to identify vertebrates, arthropods, annelids, etc., reveals that life has undergone ADAPTIVE RADIATIONS within a body plan that is bounded by how much remodelling can be done with it. The stabilizing influence of internal constraints, in the form of a specific set of genetic and developmental systems, is a historical constraint that is reflected by 600 million years of variation within the boundaries evidenced by the resilient body plans. Constraints, as the outcome of the dynamics of internal processes of an organism, must reflect the history of the lineage.

Empirical proof of constraint is scarce, given the novelty of the research programme as well as the mechanistic knowledge required. (See Alberch 1989, for a comprehensive review of examples of constrained variation in patterns, three-headed forms as physically forbidden due to some generic properties of dynamical systems, etc.) In general, constraints, as differential expressions of VARIABILITY, would be distinguished by phylogenetic parallelism and intrapopulational patterns of variability and are amenable to experimental testing.

3 *Generative versus functional (biomechanical) constraints.* A constraint affects the probability of expression of a given PHENOTYPE, as a 'mutant' in a population within a species. (Mutant means a genetically based morphology, but not necessarily the outcome of a particular single MUTATION; the focus is on the genesis of phenotypes and their morphogenetic rules above the level of genetic instructions; thus a given phenotype can be the result of a wide variety of underlying genetic perturbations.) It does not include any functional component in its definition; a constraint derives from the properties of a generative system. Lethal teratologies, for example, reflect a 'maladaptive' form but not a constrained one. If we accepted functional arguments in the definition of constraint, then natural selection acting on less functionally adapted forms would also be a constraint, rendering the concept trivial given its universality. This is an aspect that goes against some popular belief. It is often taken for granted that a structuralist approach to the issue of functional performance should be included in the definition of a constraint, given its organismal and integrative focus. For example, an analysis of biomechanical functions that generate a 'constraint' on optimality, given the existence of contravening functional demands, is often equated with the concept defined here. It is important to keep it as a separate concept, independently of its acknowledged quality and validity. It is centred around 'relative performance', which maps on to a relative FITNESS value. It is logically consistent to separate this treatment of morphological evolution through the analysis of functional performance from the issue of generative constraints dependent on internal structures of systems.

In conclusion, there are only two broad classes of interrelated, constraints: phylogenetic and systems. The first has already been discussed. 'Systems constraints' are EMERGENT PROPERTIES from the internal dynamics of a given generative system, which must be endowed with the following properties. It is dynamical, which means that its 'state' changes through time, as the result of the 'iteration' of a given set of 'rules'. Thus, the state of the dynamical system will depend on its initial conditions, plus the iterations of its specific rules. The effects of iteration will, in turn, be context dependent. Recent years have been characterized by major advances in the mathematics of complex dynamical systems and its properties. This body of theory demonstrates clearly that any biological system, with the very general characteristics mentioned above, will exhibit generative constraints. Examples of such processes are many, a selection from various levels in the hierarchy of biological organization including: a regulatory genetic network; a morphogenetic system, like Turing systems, of well-known complex patterns emerging from systems of biochemical interactions; a neural network controlling some cognitive function; or even the changes of an ecosystem under perturbation, as the result of the process fuelled by the interactions of the species that form characteristic trophic networks. Thus, the appearance of variability upon perturbation, at the most diverse biological levels, will be ordered, in the sense that certain outcomes will be more likely than others. Generative (systems) constraints postulate a different approach to evolutionary biology, one that focuses on the internal dynamics of the system in order to make predictions on the generative potential of a particular system (a property that has been defined as 'evolvability'). The internalist emphasis contrasts with the classical adaptationist approach, which focuses on the interactions between a phenotype and its environment. It is important to emphasize that ADAPTATION and internal constraint are not mutually exclusive but complementary in their focus on a most peculiar feature of biological systems: 'variation within themes'. *See also* LIFE-HISTORY EVOLUTION. [P.A.]

REFERENCES

Alberch, P. (1989) The logic of monsters: evidence of constraint in development and evolution. *Geobios* **12**, 21–57.

Gould, S.J. & Lewontin, R.C. (1979) The spandrels of San Marco and the Panglossian paradigm: a critique to the adaptationist programme. *Proceedings of the Royal Society of London, Series B* **205**, 581–598.

Oster, G.F. & Alberch, P. (1982) Evolution and bifurcation of developmental processes. *Evolution* **36**, 444–459.

Stearns, S.C. (1986) Natural selection and fitness, adaptation and constraint. In: *Patterns and Processes in the History of Life* (eds D.M. Raup & D. Jablonski), pp. 23–44. Springer-Verlag, Berlin.

---

constraints on foraging Optimality (*see* BEHAVIOURAL ECOLOGY) approaches to foraging behaviour have seen the advent of rigorous, quantitative models that generate predictive hypotheses about foraging decisions (e.g. where to start foraging, how long to stay there, what sort of prey to take, etc.). Like all OPTIMALITY MODELS, optimal foraging models make a number of basic assumptions, for instance about what selection has shaped a forager to do (e.g. maximize its net rate of energy intake, minimize the risk of doing badly) and about the reproductive costs and benefits accruing from alternative foraging strategies. One important set of assumptions concerns the CONSTRAINTS that limit the predator's options at any given decision point. Some constraints are imposed by the environment. For example, the rate at which a bumble-bee can extract nectar from a flower may be limited by the morphology of the flower and the viscosity of the nectar. The risk of predation may be another LIMITING FACTOR if, say, good food sources tend to occur in exposed areas where the forager is vulnerable to attack. Other constraints may be inherent to the forager itself. Thus perceptual limitations may constrain the ability to distinguish between certain kinds of prey, or wing-loading characteristics may set a limit to the amount of food that can be harvested before flying back to the nest. A troublesome possibility is that the absence of appropriate genetic variation in the forager's evolutionary past may act as a constraint on its present options. The judicious incorporation of constraints into foraging models is clearly a crucial prerequisite to testing the models fairly. However, it is important that convenient 'constraints' are not used uncritically to improve the match between prediction and empirical outcome. *See also* FORAGING, ECONOMICS OF; TROPHIC CLASSIFICATION. [C.J.B.]

consumer organisms Consumer organisms are all heterotrophic organisms that ingest (consume) other live plants or animals or parts of other live plants or animals and so include herbivores, carnivores and omnivores. Consumers are one of the four basic components of most ECOSYSTEMS, the others being the abiotic environment, the producers and the decomposers. The terms define a functional classification based on the energy source and type of nutrition of the organism. This is an ecological classification based on function, not on species. Some species can occupy positions in more than one compartment of this classification, for example the insectivorous plant, the sundew, *Drosera* (see Plate 3, facing p. 398), is both a primary producer, as it obtains most of its energy by PHOTOSYNTHESIS, but also a consumer when it catches and digests insects to increase its rate of NITROGEN uptake. Many more species eat both living and dead plant material, for example the woodlouse, such as *Porcellio scaber*, normally feeds on dead leaves and decaying material but will, on moist warm summer evenings, readily climb tree trunks to graze *Pleurococcus* algae growing on them; it thus occupies positions as both a decomposer and a primary consumer. Many species of slug are regarded by farmers as serious pests due to their very heavy predation of crop seedlings, leading them into competition with humans as primary consumers, but they feed for the majority of the time on dead plant material and so function mostly as decomposers.

Subdivisions within the consumer compartment are classified according to how far removed they are from the primary producers, the plants that trap energy in the system by photosynthesis. Those FEEDING directly on living plant tissues are primary consumers, i.e. the herbivores and granivorous species (e.g. elephants, water fleas (*Daphnia*), harvester ants); those that feed on these are secondary consumers or carnivores (e.g. lions, water boatmen (*Notonecta*), ladybirds); while tertiary consumers are the secondary carnivores (e.g. the pike *Esox*).

Parasitic FOOD CHAINS are also considered to be within the consumer compartment of ecosystems. Parasites do not kill the HOST that they consume and usually get progressively smaller the further the position they occupy is from the primary producers, for example from a flea to a protozoan PARASITE within the flea.

Organisms that feed on other live organisms which have derived their nutrition from dead organic matter, for example fungi or bacteria, are not normally considered as consumers but as microbivores or fungivores and form a part of parallel food chains that exist within the decomposer FOOD WEBS. *See also* FEEDING TYPES, CLASSIFICATION; HETEROTROPH; TROPHIC CLASSIFICATION. [M.H.]

consumer rank *See* CONSUMER ORGANISMS.

consumption Consumption is that part of the material removed by any trophic unit: individual, population, group of populations or trophic group, or even whole TROPHIC LEVEL, from a lower trophic level that passes into the body of the organism(s). For most organisms this refers to all potential food materials that are ingested, i.e. enter into the alimentary system, but for some parasites it also includes those NUTRIENTS absorbed directly through the exterior body wall. *See also* ECOLOGICAL ENERGETICS; ENERGY BUDGET; ENERGY FLOW. [M.H.]

consumption vector *See* TILMAN'S MODEL OF RESOURCE UTILIZATION.

contagious distribution *See* AGGREGATED DISTRIBUTION.

contaminant A non-natural substance in the natural environment, not necessarily causing HARM. *Cf.* POLLUTANT. [P.C.]

contaminated land Land which due to current or previous use has under or in it substances that are causing or might cause HARM to human health or the environment. There is a divide between a precautionary approach, which would define contaminated land as that containing CONTAMINANTS, and a more pragmatic approach, which defines contaminated land as containing substances of a kind and/or quantity likely to cause harm. Treatment of CONTAMINATION can be costly and so this distinction may be important. It may also have implications for the extent of CLEAN-UP: to the limits of detection of contaminants, or to levels unlikely to cause problems in the context of the way land may be used (suitable for use). Problems with this latter approach are that requirements for use may change (e.g. especially as pressure is put on development of brown field sites), and not all implications of contamination for harm may be understood.

An expression that has been used in British legislation is 'contaminative use', i.e. any use of land (past or present) that may cause it to be contaminated. This obviated the need to demonstrate actual contamination, which depends upon available techniques. Using this as a legal basis for action proved controversial and, although included in British legislation, was never implemented.

In the UK these issues are the subject of the 19th Report of the ROYAL COMMISSION ON ENVIRONMENTAL POLLUTION. *See also* LAND RECLAMATION. [P.C.]

contamination Release of a by-product of human activity, chemical or physical. Contamination sometimes causes HARM to human health and/or the natural environment, but need not. *Cf.* POLLUTION. [P.C.]

contaminative processes *See* LAND REGISTERS.

contaminative use *See* CONTAMINATED LAND.

contest competition A term originally coined by A.J. Nicholson in 1954 (*see* CHARACTERS IN ECOLOGY) to describe resource use, though frequently used now to describe both the behavioural process and the ecological outcome of INTRASPECIFIC COMPETITION. In contest competition 'the individuals may be said to compete for prizes . . . the individuals are either fully successsful, or unsuccessful; and the whole amount of the requisite obtained collectively by the animals is used effectively and without wastage in maintaining the population' (Nicholson 1954). Contest competition often involves some form of direct INTERFERENCE, for example aggressive interactions between male birds competing for territories. In pure contest competition, the number of 'winners' (e.g. territory holders or survivors) is constant. The ecological outcome of pure contest competition is that there is EXACT COMPENSATION and POPULATION DYNAMICS tend to exhibit dynamic stability, in contrast with SCRAMBLE COMPETITION. *See also* ASYMMETRIC COMPETITION. [R.H.S.]

REFERENCE

Nicholson, A.J. (1954) An outline of the dynamics of animal populations. *Australian Journal of Zoology* **2**, 9–65.

continental climate The type of CLIMATE characteristic of the interior of a continent with hot summers, cold winters, low rainfall and where the prevailing winds have travelled across large land areas. [V.F.]

Continental drift

From the birth of geology as an independent science in the late 18th century until early in this century the notion that the continents were geographically fixed with respect to each other, with the oceans being permanent features of the Earth, had not undergone serious challenge. This state of affairs changed when the German scientist Alfred Wegener (*see* CHARACTERS IN ECOLOGY) put forward his revolutionary hypothesis of continental drift. He proposed that several hundred million years ago the continents were united into one supercontinent that he called Pangaea, meaning all land. Late in the MESOZOIC era it began to split up, with the opening of the Atlantic and Indian oceans, although the final severance of North America and Eurasia did not take place until as late as the Quaternary (Fig. C14).

Wegener based his hypothesis on both theoretical arguments and a wide array of data. He noted, for example, the so-called jigsaw fit of the continents bordering the South Atlantic and pointed out that the close resemblances of certain geological structures and rock types on the two sides of the Atlantic made more sense if the intervening oceanic gap were closed up. Particularly striking was the DISTRIBUTION of certain Carboniferous and Permian deposits known as tillites, which signify the former existence of ICE SHEETS. Their present distribution, including tropical India, was inexplicable; however, if the continents were fitted together as Wegener suggested, a southern polar ice sheet comparable to the northern polar ice sheet of the Quaternary would very satisfactorily account for their distribution.

One particular argument of Wegener's warrants special attention because it involves BIOGEOGRAPHY. Wegener had recognized that there was consensus among palaeontologists that land connections must have existed between the southern continents in Mesozoic times. This was because of close similarities among the faunas known from fossils. If South America, Africa, Madagascar, India and Australia had been separated by ocean as they are today, these resemblances would have been inexplicable. However, palaeontologists explained this by invoking the existence of land bridges

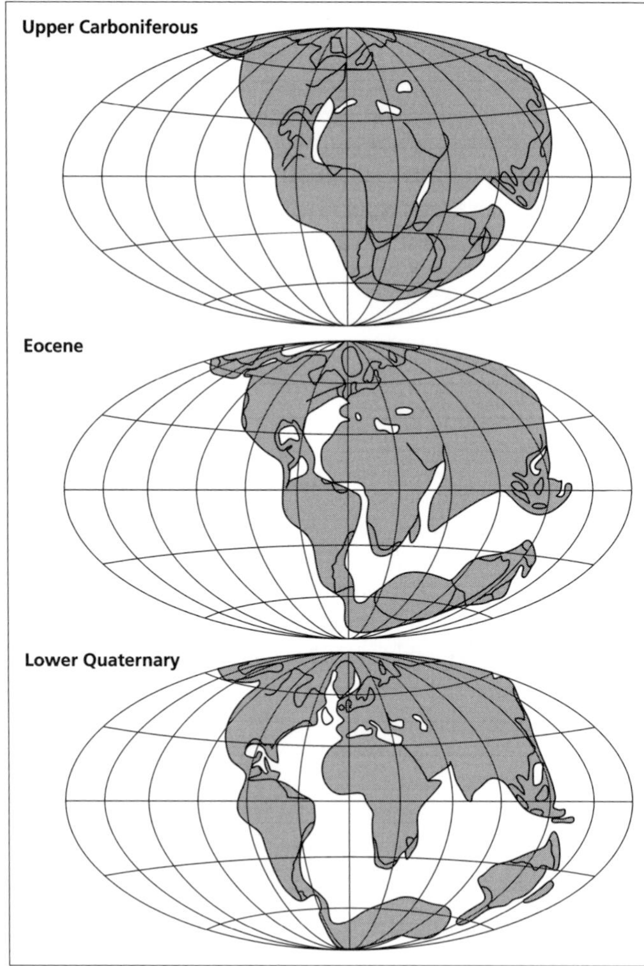

Fig. C14 Wegener's reconstruction of the various positions of the continents from the Carboniferous to the Quaternary. (After Wegener, A. (1929) *The Origin of Continents and Oceans*, 4th edn. Methuen, London.)

across the Atlantic and Indian oceans, which subsequently sank beneath the ocean. Wegener pointed out that, even given the limited geophysical and geological knowledge that we had of the ocean floor at that time, the concept of sunken land bridges was scientifically untenable, on grounds of gravity measurements, isostasy and the petrology of rocks on oceanic islands. The only plausible alternative to account for the fossil distributions was continental drift.

After initial sympathy from some German compatriots, a groundswell of opinion adverse to Wegener's hypothesis grew, and the general attitude to continental drift in the period between the two world wars was decidedly hostile, especially in North America. In particular, Wegener's failure to provide a mechanism plausible enough to satisfy the geophysicists was generally regarded as a telling point against his ideas. There was a minority of enthusiastic supporters, however, including a number of biogeographers, who were able to improve on Wegener's work and provide a more plausible mechanism based on subcrustal convection currents. It was not until the latter part of this century, however, following new geophysical research on the ocean floor, that the consensus among earth scientists was shifted in favour of the lateral mobility of continents. The relevant techniques involved the study of rock magnetism and borehole coring into the ocean floor. Within a few years in the 1960s there was a dramatic conversion of opinion to the new geological paradigm of PLATE TECTONICS, implying what was formerly called continental drift. The advent of plate tectonics marked a great advance in the earth sciences and forms the basis for present understanding of our planet. [A.H.]

REFERENCES

Hallam, A. (1973) *A Revolution in the Earth Sciences: from Continental Drift to Plate Tectonics*. Clarendon Press, Oxford.

Hallam, A. (1975) Alfred Wegener and the hypothesis of continental drift. *Scientific American* **232**, 88–97.

---

continental front Continental fronts occur along the edges of continents as a result of wind-driven oceanic CIRCULATION. Divergent continental fronts form in association with CURRENTS that flow toward the Equator along the western coasts of continents. Due to the Earth's rotation, the currents are driven away from shore and are thus associated with UPWELLING (e.g. Benguela Current, California Current, Peru Current). Convergent continental fronts are associated with currents flowing away from the Equator along the eastern coasts of continents. They are characterized by an accumulation of warm, nutrient-poor water (e.g. as occurs along the Great Barrier Reef). [V.F.]

continental margins The zones separating the emergent continents from the deep-sea bottom. Continental margins generally consist of a CONTINENTAL SHELF, slope and rise. Active margins occur where two lithospheric plates converge. They are characterized by crustal SUBDUCTION, active volcanoes, earthquakes and recently formed mountains. Passive margins occur in the interior of lithospheric plates and are formed when ocean BASINS are created by splitting continents. They typically have thick SEDIMENT deposits and show an absence of volcanic activity. [V.F.]

continental rise The part of the sea floor extending below the CONTINENTAL SLOPE and above the ABYSSAL plain, ranging to a depth of between 4000 and 5000 m. It is formed by the accumulation of SEDIMENT eroded from the continents and deposited at the base of the slope. [V.F.]

continental shelf A broad expanse of ocean bottom, representing the submerged edge of a continent. The continental shelf divides the open OCEAN from the inshore, or NERITIC, zone and accounts for about 7% of the sea floor. It extends from the line of permanent immersion to the depth (usually 200 m, but may be as deep as 700 m, e.g. around Antarctica) at which there is a marked increase in the slope. The width of the shelf averages 50 km, but it may be entirely absent (e.g. off Chile and south-western Alaska) or as wide as 1350 km (e.g. off the Arctic coast of Siberia). [V.F.]

continental slope The relatively steep downward slope extending seaward from the outer edge of the CONTINENTAL SHELF to the flat ocean floor. The continental slope usually extends to a depth of 2000–3000 m and varies in width between 20 and 100 km. It is usually covered by sediments of fine SILT and mud. [V.F.]

contingency table Most frequently a two-way table (at least two rows and two columns) in which QUALITATIVE INFORMATION is displayed prior to analysis for ASSOCIATION. *See also* CHI-SQUARED TEST; STATISTICS. [B.S.]

continuum Mathematically, a continuum is an infinite set of objects between any two of which a third can always be interposed. The real numbers form a continuum, whereas the integers are a discrete set. The mathematical concept has had a powerful influence on ecology, providing a paradigm according to which communities are seen as nature's response to continuous underlying environmental variation (Goodall 1963; Whittaker 1973; Austin 1985). The response, though not the underlying variation, may show discrete boundaries, for example a sharp boundary between a closed-canopy forest and a fire-maintained heathland. The continuum concept underlies the theory of ORDINATION and GRADIENT ANALYSIS (Ter Braak & Prentice 1988), and is opposed to the view that communities are discrete units with coevolved species populations. *See also* COMMUNITY. [M.O.H.]

REFERENCES

Austin, M.P. (1985) Continuum concept, ordination methods, and niche theory. *Annual Review of Ecology and Systematics* **16**, 39–61.

Goodall, D.W. (1963) The continuum and the individualistic association. *Vegetatio* **11**, 297–316.

Ter Braak, C.J.F. & Prentice, I.C. (1988) A theory of gradient analysis. *Advances in Ecological Research* **18**, 271–317.

Whittaker, R.H. (1973) Direct gradient analysis: results. In: *Ordination and Classification of Communities* (ed. R.H. Whittaker), pp. 33–51. Junk, The Hague.

contract (testing) laboratories Under contract to industry or regulators to make certain environmental measurements, for example contract ecotoxicological testing laboratories test NEW CHEMICALS according to regulatory GUIDELINES and generally are required to follow GOOD LABORATORY PRACTICE (GLP). [P.C.]

control action threshold (CAT) A term used in PEST CONTROL and defined as the pest density at which one should take action to prevent a pest outbreak. The CAT varies with the season, with the cost of the crop and of the preventive measures to be taken, and with the population densities of the natural enemies of the pest. [S.C.S.]

controlled waste *See* WASTE.

controlled waters In British law, those subject to protection under the ENVIRONMENTAL LAWS and including territorial waters, coastal and estuarine waters, surface freshwater and groundwater. [P.C.]

convection Transport of heat or mass by movement of a fluid. If a hot object is placed in a WIND TUNNEL, it will cool at a rate dependent on the wind speed. This is 'forced convection'. If the wind is turned off, CONDUCTION OF HEAT from the surface to the air will occur, followed by vertical movement of the warmed (i.e. less dense) air. As the warmed air rises, colder air from below will be brought into contact with the hot object and the process will continue. This is 'natural convection'. Many situations in nature, such as heat loss from leaves, involve both forced and natural convection, and are called 'hybrid convection'. [J.G.]

REFERENCE

Monteith, J.L. & Unsworth, M.H. (1990) *Principles of Environmental Physics*. Arnold, London.

Convention on the International Trade in Endangered Species (CITES) *See* INTERNATIONAL CONSERVATION CONVENTIONS.

Convention on Wetlands of International Importance *See* INTERNATIONAL CONSERVATION CONVENTIONS.

convergence of successions According to F.E. Clements's (*see* CHARACTERS IN ECOLOGY) original theory of CLIMAX, the VEGETATION on all sites under a given climate should converge with time on a single species or single group of species, even where the early successional species are quite different, for example in lakes, on rocks or on SOILS with different chemical properties, etc. Some studies have shown evidence for convergence but, in general, critical studies have led to the rejection of this viewpoint and replacement with the idea of POLYCLIMAX. *See also* SUCCESSION. [P.J.G.]

convergent evolution Appearance of similar PHENOTYPES or GENOTYPES as a result of similar SELECTION PRESSURES rather than as a result of common ancestry (HOMOLOGY). One of several kinds of HOMOPLASY along with parallelism and reversal. An older, more restrictive use of convergent evolution is for similar TRAITS arising from different ancestral backgrounds. Classic examples include wings in diverse lineages and sucking mouthparts in Diptera versus Hemiptera. Convergence can be used as evidence for ADAPTATION and can complicate PHYLOGENY reconstruction. *See also* PARALLEL EVOLUTION. [D.A.B. & D.R.B.]

conveyor-belt species Species of deposit-feeding animals that live infaunally (i.e. buried within the SEDIMENT) and that orient with their head down and their tail end at the sediment surface. Because they feed on sediment at depth and defecate at the sediment surface, they act as geochemical conveyor belts, redistributing sediment and altering its physical and geochemical properties in a predictable fashion. *See also* ECOSYSTEM ENGINEERS. [V.F.]

Cooperative breeding, evolution of

Cooperative breeding (or 'helping at the nest') occurs when individuals exhibit parent-like behaviour toward young that are not their own. Most commonly, cooperative breeding involves non-breeding individuals called 'helpers', 'auxiliaries' or 'workers' that forgo reproduction of their own while aiding in the reproductive efforts of others. More rarely, it involves two or more pairs of individuals collaborating to raise young or 'cooperative polygamy' in which more than a single male or female share breeding status within the same social unit. Cooperative breeding in some form is found in a small proportion of birds, mammals and other vertebrates. Cooperative breeding is also closely allied to eusociality in insects, differing primarily in that workers in eusocial societies may differ morphologically from others CASTES and are generally thought to be permanently sterile, while helping in cooperative breeders is a temporary condition engaged in by younger individuals that will later become breeders.

Most non-breeding helpers are offspring that delay DISPERSAL and independent breeding while remaining in their natal social unit and helping their parents raise younger siblings. Delayed dispersal is most commonly thought to be 'making the best of a bad job' by individuals that are restricted in their ability to disperse and breed on their own (*see* ECOLOGICAL CONSTRAINT HYPOTHESIS). Elegant experimental evidence for the importance of POPULATION DENSITY, often known as the 'habitat saturation' hypothesis, comes from cooperatively breeding Seychelles warblers (*Acrocephalus sechellensis*) transplanted to an unoccupied island. At first, all offspring dispersed and bred independently. Only after population density increased to the point that all high-quality territories were occupied did offspring again begin to delay dispersal and act as helpers. Other factors promoting delayed dispersal include high cost of dispersal, high ratio of optimal to marginal habitat, low environmental variability and low ability of individuals within a social unit to depress available RESOURCES.

Once having delayed dispersal, offspring living in their natal groups may gain indirect, or kin-selected, FITNESS benefits by increasing the production of non-descendant KIN. Indirect fitness benefits appear to be important in species in which helpers are more likely to assist relatives compared with non-relatives and in which reproductive success or survival of breeders is enhanced by the assistance of helpers. Such species offer some of the best examples of altruism to be found anywhere, especially if workers in eusocial species are included.

Helpers may also gain DIRECT FITNESS benefits by enhanced survivorship, higher future probability of breeding and increased future reproductive success. Such benefits are the only ones to be gained when helpers are unrelated, as in 'secondary' male helpers in pied kingfishers (*Ceryle rudis*) that form social relationships with the breeding female and thereby increase their chances of mating with her in subsequent years. However, direct fitness benefits may also be important in species where helpers are usually offspring, such as in Florida scrub-jays (*Aphelocoma coerulescens*), where TERRITORY SIZE is correlated with group size and helpers, by increasing reproductive success, increase their own chances of 'budding off' and inheriting part of their natal territory.

More problematical is the evolution of cooperative breeding systems involving cooperative POLYGAMY or mate-sharing. One important factor in many such systems is the advantage in acquiring territories or other critical resources gained by coalition formation, such as the control over prides of females by coalitions of male lions (*Panthera leo*) and the successful filling of reproductive vacancies by SIBLING groups of acorn woodpeckers (*Melanerpes formicivorus*; see Plate 4, facing p. 398). Other factors that appear to be important in some mate-sharing systems include CONSTRAINTS on nesting sites and CONFLICTS of interest between males and females. The latter has been explored particularly well in dunnocks (*Prunella modularis*), in which unrelated males frequently share one or more independently nesting females.

Cooperative breeding is relatively more common in the tropics than in temperate areas and appears to be particularly characteristic of the Australian avifauna. Cooperative breeders are usually non-migratory, which may in part be due to the relatively milder climate in these areas. In general, however, there is often no satisfactory explanation for differences in the frequency of cooperative breeding among different geographic areas or the reasons why some species are cooperative breeders and others with apparently similar ecological requirements are not. *See also* BEHAVIOURAL ECOLOGY. [W.D.K.]

REFERENCES

Davies, N.B. (1992) *Dunnock Behaviour and Social Evolution*. Oxford University Press, Oxford.

Emlen, S.T. & Wrege, P.H. (1989) A test of alternate hypotheses for helping behavior in white-fronted bee-eaters of Kenya. *Behavioral Ecology and Sociobiology* **25**, 303–319.

Koenig, W.D. & Mumme, R.L. (1987) *Population Ecology of the Cooperatively Breeding Acorn Woodpecker*. Princeton University Press, Princeton, N.J.

Komdeur, J. (1992) Importance of habitat saturation and territory quality for evolution of cooperative breeding in the Seychelles warbler. *Nature* **358**, 493–495.

Reyer, H.-U. (1980) Flexible helper structure as an ecological adaptation in the pied kingfisher (*Ceryle rudis rudis* L.). *Behavioral Ecology and Sociobiology* **6**, 219–227.

Stacey, P.B. & Koenig, W.D. (eds) (1990) *Cooperative Breeding in Birds: Long-term Studies of Ecology and Behavior.* Cambridge University Press, Cambridge.

Woolfenden, G.E. & Fitzpatrick, J.W. (1984) *The Florida Scrub Jay: Demography of a Cooperative-breeding Bird.* Princeton University Press, Princeton, N.J.

cooperative polygamy *See* COOPERATIVE BREEDING, EVOLUTION OF.

Cope's rule The observation that mammals and other lineages tend to increase in SIZE over evolutionary time. The explanation for this pattern is thought to be that smaller species survive MASS EXTINCTIONS better than large species. The ADAPTIVE RADIATIONS following an extinction event tend to involve evolutionary increases in size to fill newly vacated ecological niches. [D.A.B.]

REFERENCE

Stanley, S.M. (1979) *Macroevolution: Pattern and Process.* W.H. Freeman, San Francisco, Calif.

copper (Cu) A HEAVY METAL (ATOMIC NUMBER = 29; relative ATOMIC MASS = 63.54; melting point = 1083°C; boiling point = 2567°C) that is a trace constituent (70 mg kg^{-1}) of the Earth's CRUST. Found in ores such as chalcopyrite ($CuFeS_2$) and malachite ($CuCO_3.Cu(OH)_2$), copper is widely used in alloys, the transmission of electricity and pipework. Copper is an essential TRACE ELEMENT for humans, as part of several important enzymes including ferroxidases, cytochrome oxidase, superoxide dismutase and amine oxidases, and for animals and plants. Soil deficiency can lead to poor crop yields. Problems of TOXICITY can occur in crops and livestock on soils polluted by copper from mining, smelting and AGROCHEMI-

CALS. Copper sulphate ($CuSO_4$) is used to kill algae in ponds and reservoirs. [J.F.]

coppice A WOODLAND managed by the practice of coppicing. This involves cutting the tree at or near the ground and allowing it to sprout from the stump to generate a new crop of poles. The regrowth was traditionally cut again usually after 5–30 years, but both shorter and longer rotations have been recorded. Coppices usually included both coppiced trees (underwood) and timber trees (standards). Coppiced trees develop stools, which develop over centuries into rings of contorted, living stumps bearing the current crop of stems. Margins of coppices were defined by a bank and hedge. Farm animals were usually kept out, but in the medieval period deer and farm animals were often allowed to graze the herbage for 7 years after the last coppicing. [G.F.P.]

coprophagy A specialization in FEEDING on faeces. *See also* FEEDING TYPES, CLASSIFICATION. [P.O.]

coral reefs Coral reefs are named for that group of cnidarians (belonging to the class Anthozoa) that secrete an external skeleton of calcium carbonate. They are found in shallow waters surrounding tropical landmasses. They are restricted to high SALINITY, silt-free waters warmer than 18°C, generally in a band that lies between the Tropic of Cancer and the Tropic of Capricorn. There are three types of coral reefs: atolls, BARRIER REEFS and fringing reefs. An atoll is a horseshoe or circular type of coral reef that grows around a subsided island (typically an oceanic volcanic SEAMOUNT) and encloses a shallow lagoon. In contrast to a barrier reef, an atoll has no central island extending above the sea surface. Fringing reefs do not have a lagoon separating the reef from its associated landmass. Atolls may be formed either by subsidence of an island surrounded by a fringing reef or by an increase in sea level following GLACIATION. The idea that coral atolls were the tops of volcanoes that had subsided beneath the water's surface was first proposed by Charles Darwin. The world's largest coral reef is the Great Barrier Reef off eastern Australia, which measures almost 2000 km in length and 100 km in width. Coral reefs are characterized by a number of morphological features. The surface of the reef is called the terrace or the reef flat. The reef flat is that part of a coral reef that is partly or entirely dry at low tide. This dead portion of the reef may contain shallow pools, potholes, gullies and patches of coral debris and sand. The seaward edge of the reef is the fastest growing part of the reef and is composed of an algal ridge, a buttress zone and a reef face. The algal ridge extends slightly higher than the reef flat and takes the force of breaking WAVES. The buttress zone lies just below the algal ridge and acts to deflect and channel waves upward over the algal ridge. The reef face extends from just below the buttress zone to the depth at which living coral and coralline algae become rare (16–30 m). Coral-reef development is generally greater in areas subject to strong wave action but where suspended SEDIMENT concentrations are low. Corals are carnivorous, feeding on ZOOPLANKTON, but harbour endosymbiotic dinoflagellates called ZOOXANTHELLAE. The zooxanthellae, which may comprise up to 75% of the coral tissue, provide OXYGEN and may release up to 60% of the organic material produced during PHOTOSYNTHESIS for the coral's use. The coral, in turn, provides protection for the zooxanthellae and inorganic metabolic WASTE products that can be used by the zooxanthellae for photosynthesis. In addition to corals, several groups of red and green algae also constitute an important structural component of the reef. Coral reefs provide a habitat for a diverse array of vertebrates and invertebrates. *See also* CORAL REEFS, DIVERSITY AND CONSERVATION; SUBLITTORAL ZONE. [V.F.]

coral reefs, diversity and conservation CORAL REEFS are sites of extremely high biological DIVERSITY and may, in fact, be the most complex systems in the marine environment. There are many more species on a coral reef than in the surrounding water or sediments. Photosynthetic rates are an order of magnitude greater on the reef than in the surrounding waters due to rapid RECYCLING of NUTRIENTS, the fixation of NITROGEN gas to NITRATE (an inorganic source of nitrogen needed by photosynthetic organisms) by cyanobacteria associated with the reef, and an UPWELLING effect caused by the physical presence of the reef structure. The high PRODUCTIVITY and habitat complexity of coral reefs support a rich variety of benthic algae, epifaunal and burrowing invertebrates (sometimes referred to as cryptofauna), a variety of crustaceans and fish. Bioerosion by the grazing and boring activities of cryptofaunal species may be the most important of the processes that act to break down coral reefs. Coral reefs are notable for the widespread COEXISTENCE of species with strongly overlapping resource requirements. Explanations for such coexistence remain controversial. Pacific reefs are more diverse than Atlantic reefs, i.e. there are 85% more coral species in the western Pacific, where corals are believed to have originated. Despite the high productivity, reef-building corals are highly adapted to low-nutrient environments and become stressed and eventually outcompeted if nutrient levels are in excess. Since corals are killed by minor deviations from optimal conditions, the climatic stability of coral-reef environments is evidenced by the persistence of reefs in the tropics for over 50 million years. Occasional environmental DISTURBANCE (e.g. hurricanes), as well as biological disturbances (e.g. predation and grazing), are believed to be important in maintaining diversity. However, recent population explosions of the coral-eating sea star, *Acanthaster planci*, have resulted in catastrophic mortality of reefs throughout the western Pacific. The cause of such population explosions is not completely understood, but may be partly human-induced. Bleaching of coral reefs occurs from the loss of the symbiotic ZOOXANTHELLAE that live within the coral. Bleaching may occur as a result of increased water temperature or light-related effects. A particularly extensive bleaching episode (causing the death of at least 70% of the corals along the Pacific Central American coast) has been attributed to the EL NIÑO event of 1982–1983. Increased SEDIMENTATION from land RUN-OFF, dredging and in association with DEFORESTATION can damage corals by decreasing the amount of light available for PHOTOSYNTHESIS, by clogging the coral polyps and by preventing the settlement of coral larvae. Additional human-caused damage to coral reefs occurs from explosives used to kill fish and open channels, fishing with bleach and CYANIDE, shell collectors and tourism. *See also* CONSERVATION BIOLOGY. [V.F.]

core species These are species likely to be regionally common, locally abundant and relatively well spaced in NICHE space. Such species are very likely to remain in a COMMUNITY despite a general

TURNOVER of species in the system. They are also likely to dominate the interactions within a system. *See also* SATELLITE SPECIES. [C.D.]

REFERENCE

Hanski, I. (1982) Dynamics of regional distribution: the core and satellite species hypothesis. *Oikos* **38**, 210–221.

corers *See* SAMPLING METHODOLOGY/DEVICES.

Coriolis effect The component of the centripetal acceleration arising from the motion of a body relative to the rotating Earth. Its component in the local horizontal is given by fV (see McIlveen 1992), where V is the horizontal speed of the body relative to the Earth's surface and f is the Coriolis parameter ($2\Omega\sin\Phi$, where Ω is the Earth's angular velocity and Φ is the angle of latitude). The acceleration is perpendicular to V on the side favoured by the Earth's rotation (i.e. to the left in the Northern hemisphere) and may be alternatively described by a centrifugal Coriolis force (per unit mass) acting in the opposite direction.

Large-scale horizontal flows of air (and water) are dominated by geostrophic balance between the Coriolis and PRESSURE gradient forces, in which air flows along isobars at speeds inversely proportional to isobar spacing and low pressure lies to the left when facing downwind in the Northern hemisphere. [J.F.R.M.]

REFERENCE

McIlveen, J.F.R. (1992) *Fundamentals of Weather and Climate*. Chapman & Hall, London.

correlation When two sets of measurements (often called x and y) are related or show ASSOCIATION they are said to be correlated. If one (y) increases when the other (x) increases they are said to be positively correlated. If one decreases when the other increases they are said to be negatively correlated. Degrees of correlation are shown in Fig. C15 and can be summarized by STATISTICS called correlation coefficients. Both parametric (PRODUCT–MOMENT CORRELATION COEFFICIENT) and non-parametric (KENDALL'S RANK CORRELATION COEFFICIENT and SPEARMAN'S RANK CORRELATION COEFFICIENT) statistics are available. Remember that a correlation between two variables does not necessarily mean that one causes the other. It may be that the two variables are both effects of a common cause. *See also* COVARIANCE; NON-PARAMETRIC STATISTICS; PARAMETRIC STATISTICS. [B.S.]

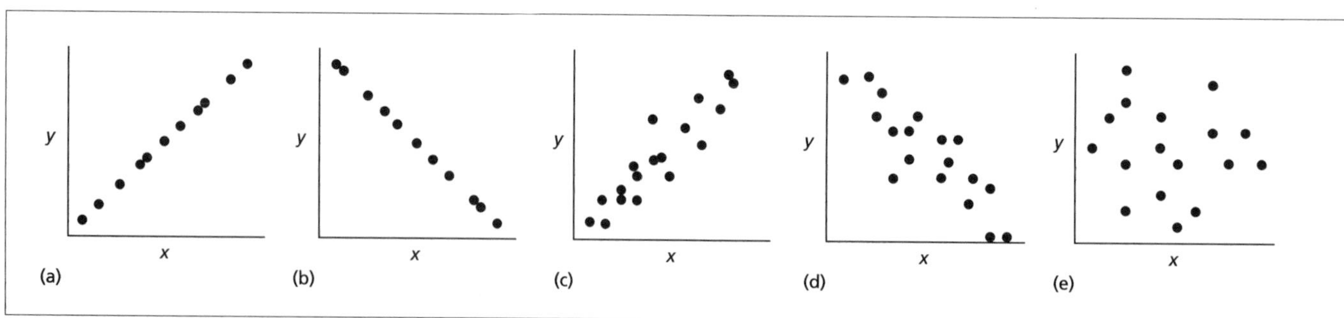

Fig. C15 Scatter plots of two variables, x and y, showing: (a) perfect positive correlation; (b) perfect negative correlation; (c) positive correlation; (d) negative correlation; and (e) no correlation.

correlogram A graphical depiction of the autocorrelation coefficients for a time series, calculated at increasing time delays, with the coefficients plotted on the y-axis and the corresponding time delays on the x-axis. A time series with a significant underlying periodicity will have a correlogram that displays a significant peak in the graph of autocorrelation coefficients at the time delay that corresponds to the periodicity. *See also* CORRELATION. [S.C.S.]

corridor A corridor is a linear patch of HABITAT, usually established or maintained to connect two or more adjacent habitat areas. Examples of corridors include HEDGEROWS, railroad and highway rights-of-way, forested shelter-belts, and GALLERY FORESTS along RIPARIAN zones.

The most widespread use of corridors is to facilitate the movement of organisms between areas of suitable habitat. Small mammals, insects and birds use forested corridors to move through agricultural landscapes with some degree of protection from predators. The connectivity functions of riparian corridors are often enhanced by the addition of linear BUFFER ZONES which expand the effective width of the corridor.

Because they are relatively narrow landscape features, a large proportion of most corridors is influenced strongly by EDGE EFFECTS. Consequently, corridors may be less likely to serve as primary habitat for species that reside in interior forests or at the centre of other large contiguous ecosystems.

Movement corridors play an important role in conservation planning for species that require the ability to move across the landscape. In North America, movement corridors have been proposed for the Florida panther (*Felis concolor coryi*), grizzly bear (*Ursos arctos*) and other species with large natural ranges. Many migratory species utilize different RESOURCES during MIGRATION than in their breeding or wintering grounds. For instance, North American species of crane (*Grus* spp.) utilize a complex ecological mosaic of seasonal WETLANDS, agricultural stubble and natural GRASSLANDS during migration. The protection of these species must thus include management of the proper habitat types in migratory corridors along the route. *See also* CONSERVATION BIOLOGY. [D.A.F.]

corridor dispersal When formerly isolated landmasses become linked by the emergence of an isthmus, terrestrial organisms can intermigrate across the newly created CORRIDOR. A classic example is provided by the so-called 'great American biotic interchange' in the late CENOZOIC, a few million years ago. This is exhibited most conspicuously by the mammal faunas of North and South America before and after the emergence of the Panama isthmus. Prior to the emergence, and indeed throughout the Cenozoic era, the mammal fauna of South America had been isolated by sea and consequently was highly endemic (Fig. C16). After the emergence there was considerable cross-migration between North and South America, with mammals of the former continent invading more successfully and many of the South American endemics going extinct (Fig. C17).

Another good example is the Bering LAND BRIDGE that intermittently linked Alaska and north-east Asia at times of low sea-level stand associated with Pleistocene GLACIATIONS within the last 2 million years. This allowed cross-migration of a wide variety of terrestrial populations including our own species, which invaded the New World a few tens of thousands of years ago to establish what European emigrants a few hundred years ago came to regard as 'natives'. [A.H.]

REFERENCE

Hallam, A. (1994) *An Outline of Phanerozoic Biogeography*. Oxford University Press, Oxford.

cosmic-ray-produced radionuclides *See* RADIONUCLIDE.

cosmic rays The Earth is continuously subject to bombardment by cosmic RADIATION, or cosmic rays, which originate in space and, to a lesser extent, in the Sun. Primary cosmic rays, consisting mainly of high-energy (up to 10^{18} MeV) protons plus smaller numbers of heavier particles, interact with nuclei in the upper atmosphere, gen-

Fig. C16 Divergence of mammals in the New World occurred during a period of millions of years when North and South America were unconnected. Mammals then unknown in the south included mastodons, tapirs, primitive camels and various carnivores. The mammals of the south included many that are now extinct. Illustrated here are (a) *Mylodon*, a giant sloth; (b) *Paedotherium*, a notoungulate; (c) *Prodolichotis*, a rodent; (d) *Macrauchenia*, an ungulate; and (e) *Plaina*, an early relative of the more successful armadillo. (After Hallam, A. (1972) Continental drift and the fossil record. *Scientific American* **227** (5), 57–66; with permission from Eric Mose, Jr.)

Fig. C17 Convergence of mammals began about the end of the Pliocene epoch, after a land bridge was established between North and South America. One mammal native to South America, the armadillo, migrated northward. However, so many mammals formerly unknown south of Panama moved to South America that the two continents soon came to have 22 families of mammals in common. At the same time many South American mammals were unable to compete with the immigrants from the north and became extinct. (After Hallam, A. (1972) Continental drift and the fossil record. *Scientific American* **227** (5), 57–66; with permission from Eric Mose, Jr.)

erating cosmic ray-produced nuclides (e.g. ^{3}H, ^{7}Be, ^{10}Be and ^{14}C) plus secondary cosmic rays (mainly mesons with smaller numbers of photons, electrons and positrons). Cosmic radiation contributes a natural background dose rate of about 0.28 mGy year^{-1} at sea level and higher dose rates with increasing altitude as atmospheric screening decreases. [A.B.M.]

cosmopolitan distribution There are few taxonomic groups of plants or animals that can be described as truly cosmopolitan, i.e. that occur throughout the world in suitable habitats. However, there are some plants and animals that are almost cosmopolitan since they are widely distributed geographically. Examples of plants in this category are the Asteraceae, Poaceae, Fabaceae and Cyperaceae. These represent four of the 300–400 families of the order of flowering plants known as the Angiospermae. Members of these families have achieved a near cosmopolitan distribution because they evolved before the large continental landmasses had completely separated. In the animal kingdom, the majority of taxonomic groups have local (endemic) ranges or discontinuous (disjunct) ranges. The exceptions, which can be considered as cosmopolitan, are the Muridae (rats and mice), Leporidae (hares and rabbits) and Canidae (dogs and wolves). These are found on all the continents except Antarctica and represent just three of the 89 families of animals. Bats (Chiroptera) and marine animals are also considered to have a cosmopolitan distribution.

Modern humans (*Homo sapiens sapiens*) are also cosmopolitan, even having reached Antarctica albeit in the last 100 years. The mobility, culture and technology of modern humans have also resulted in a number of other plants and animals achieving cosmopolitan status. For example, knotgrass (*Polygonum aviculare*) and charlock (*Sinapis arvensis*) can be so described because they are now widespread, having been dispersed in crop seed, silage and animal feedstuff. Similarly, the cat (*Felis cattus*) has become cosmopolitan as have many species of insects through their dissemination by humans. *See also* GEOGRAPHIC DISTRIBUTIONS. [A.M.M.]

cosms Artificial ECOSYSTEMS of various levels of complexity and SCALE. *See also* MACROCOSM; MESOCOSM; MICROCOSM. [P.C.]

cospeciation Studies of cospeciation attempt to uncover the patterns of geographical or ecological associations among clades that share a close and evident ecological association (such as host–parasite, predator–prey, herbivore–crop or host plant–pollinator). There are two components to these patterns. First, two or more species may be associated today because their ancestors were associated with each other in the past (association by descent). Second, two or more species may be associated today because at least one of them originated in some other context and subsequently became involved in the interaction by COLONIZATION of a host or DISPERSAL into a geographical area (association by colonization; dispersal or host-switching). This term has also been used more recently in a more restrictive sense to indicate clades that are so tightly linked biologically that they cause one another to speciate. Such dramatic coevolutionary interactions are hypothesized to produce congruent patterns of SPECIATION accompanied by a similar degree of genetic differentiation in the descendant species of both clades. *See also* COEVOLUTION; RED QUEEN'S HYPOTHESIS; SPECIES INTERACTIONS. [D.R.B.]

cost recovery Legal process when potentially responsible parties (PRPs) that have contributed in some way to CONTAMINATION can be required to pay for CLEAN-UP, for example under SUPERFUND LEGISLATION in the USA. [P.C.]

cost recovery charging Part of the polluter-pays philosophy is that the cost of implementing environmental controls should be recovered from potential polluters. These include the costs for site inspection, the bureaucracy associated with CONSENTS/AUTHORIZATION, substance MONITORING and audit, etc. Increasingly, regulations try to recover these costs by charging those seeking consents/authorizations. *See also* COST RECOVERY. [P.C.]

cost (risk)–benefit approach Industry/technology bring benefits (i) in terms of products that improve health, food supply, welfare and create wealth. At the same time they carry risks (ii) for human health and the environment. This can be turned around: controlling industrial activities and outputs brings economic costs (iii) but can bring benefits (iv) of increased environmental protection and reduced risks to human health. Trying to balance these elements in coming to a view about the need for controls is generally described as cost (risk)–benefit analysis. However, note that this can refer to definition (i) versus definition (ii) or definition (iii) versus definition (iv) and this can sometimes mislead. *See also* ECOLOGICAL VALUATION; ENVIRONMENTAL (ECOLOGICAL) ECONOMICS. [P.C.]

costs of reproduction Deleterious effects on survival or FECUNDITY that occur as a consequence of current REPRODUCTIVE EFFORT. Life-history theory depends on estimates of these costs and there is some controversy over the best method to estimate them. Four main approaches have been considered: phenotypic CORRELATIONS; genetic correlations estimated from breeding designs; genetic correlations estimated from correlated responses to selection; and experimental manipulations such as additions of EGGS to a nest. It is generally agreed that phenotypic correlations can be misleading. *See also* EVOLUTIONARY OPTIMIZATION; LIFE-HISTORY EVOLUTION; TRADE-OFF. [D.A.B.]

REFERENCE

Reznick, D. (1985) Costs of reproduction: an evaluation of the empirical evidence. *Oikos* **44**, 257–267.

costs of resistance Resisting adverse effects from outside agents is often metabolically costly: escaping a predator requires increased activity; resisting parasitism or disease involves the deployment of active defence mechanisms; resisting chemical poisoning may require avoidance behaviour, active transport and excretion processes, detoxification and repair of damaged tissues. These can all be considered as costs of RESISTANCE. It is presumed that some costs trade off with other elements of FITNESS in the evolution of TOLERANCE. *See also* LIFE-HISTORY EVOLUTION; TRADE-OFF. [P.C.]

costs of sex Although nearly all organisms have some form of sexuality, the widespread existence of sex has been considered paradoxical because of the measurable costs involved with this mode of reproduction. Although evolutionary advantages for the population and species may result from sexual recombination (and the resulting creation of genetic DIVERSITY), advantages to the individual are less clear. The costs of sex include the time and energy used to find a mate, to engage in COURTSHIP and to make and maintain secondary sexual organs (e.g. organs for sexual display). In addition, more energy is generally invested in the production of EGGS compared with sperm, and frequently the female invests more in her offspring than does the male. There is therefore a cost to the female of investing in offspring that only carry half of her genes. This is sometimes referred to as the cost of producing males. The cost of sex in dioecious populations depends on the sex ratio, the advantage of an asexual CLONE becoming two-fold when the proportion of males in the sexual population is 50%. The cost of male allocation (less efficient production of offspring) can be distinguished from the cost of meiosis (reducing the genetic contribution to one's offspring) and which of these costs is more important may vary depending on specific circumstances. A reduction in genetic FITNESS due to reassortment includes losses due to both segregation and recombination, and may put sexuals at a disadvantage compared to asexuals in which superior epistatic gene combinations are preserved. However, the increased FECUNDITY inherent in asexuality may be offset by the wider range of RESOURCES available to the more genetically heterogeneous sexuals. Although the maintenance of sex has long been one of the most hotly debated issues in evolutionary biology, the explanation for the relative preponderance of sexual reproduction has not been fully resolved. *See also* EVOLUTIONARY OPTIMIZATION; LIFE-HISTORY EVOLUTION. [V.F.]

REFERENCES

Lively, C.M. & Lloyd, D.G. (1990) The cost of biparental sex under individual selection. *American Naturalist* **135**, 489–500.

Maynard Smith, J. (1989) *Evolutionary Genetics*. Oxford University Press, Oxford.

Coulter Counter™ An electronic device used to estimate the BIOMASS of PLANKTON belonging to different size categories. Biomass is thus estimated as the diameter of a sphere equivalent in volume to the original particle multiplied by the number of particles. [V.F.]

Council on Environmental Quality (CEQ) Set up by the US National Environmental Policy Act (NEPA) of 1969 to advise the President on environmental enforcement measures and objectives. [P.C.]

coupled map lattice A form of spatially explicit POPULATION model which, in ecology, is used to model population growth and interactions within a series of patches linked by DISPERSAL. Whereas CELLULAR AUTOMATA models assume that individual patches exist in one of a finite number of discrete states (e.g. occupied or unoccupied), the state of a patch in a coupled map lattice model is not limited in this way. Instead patches may assume a range of states, such as a continuous range of densities. These models may generate aesthetically attractive spatial patterns, but are ecologically important in a number of respects. Using such models it has been possible to explore the consequences for population and COMMUNITY dynamics of habitat heterogeneity, varying dispersal ability and to determine the limits of simpler models that ignore spatial processes. [R.P.F.]

Courtship

Behavioural patterns preceding copulation or the formation of social bonds between individuals of the opposite sex.

Courtship is presumed to facilitate successful copulation or the formation of a pair bond. Courtship behaviour usually consists of a number of different sexual DISPLAYS that convey information on skills or endurance of performance. For example, males often perform displays that exaggerate their size, the frequency of their calls relative to the same calls used in other contexts or the rate of calls relative to the same calls used in other contexts, which may reflect the general tendency for females to prefer males with more exaggerated displays. Repeated behavioural sequences and sexual displays may allow females to assess particular skills or endurance of performance, and only males in prime condition may be able to perform the ritualized displays in exactly the same way a large number of times. For example, the intensity of sexual displays in salamanders has been shown to reflect their ability to produce a spermatophore. Courtship displays may also provide reliable information on general health or parasite infection status (as is the case for the intensity of male coloration), the size of male displays or the intensity or other features of vocalization. Females usually respond to courtship displays by showing coy behaviour, and courtship events generally reflect male salesmanship and female sales resistance. Reversed sex roles during courtship can be found in sex role-reversed species such as seahorses and pipefishes (Syngnathidae), phalaropes (Phalaropinae), jacanas (Jacanidae) and some crickets, in which males are the main providers of PARENTAL CARE or investment.

Courtship frequently involves transfer of food or other substances that contain important amounts of energy, such as spermatophores and ejaculate constituents. The transfer of courtship food may provide males with a DIRECT FITNESS benefit if their FERTILIZATION success increases with the size of nuptial gifts, while females benefit because the food may be used for egg production. For example, male scorpionflies (Panorpidae) and other insects transfer sperm to the female as long as the female is feeding on the nuptial gift, and the amount of sperm transferred is proportional to the number of EGGS fertilized by the current male copulating with the female. Alternatively, courtship may provide females with important reliable information on the ability of males to provide parental care for the offspring during subsequent reproduction. For example, courtship feeding rates of common terns (*Sterna hirundo*) reliably predict the feeding rate of males during the nestling period, and singing rates of male stonechats (*Saxicola torquata*) predict their later feeding rates.

Courtship has an optimal duration, so responses to courtship cannot be too fast because such behaviour may signal potential unreliability of a partner for reproduction. For example, female ring-doves (*Streptopelia risoria*) that have previously engaged in copulations with a male in the absence of the present male are courted for a longer period than control females that have just been isolated without the presence of a male. This extended courtship is adaptive, since females are able to store sperm for an extended period and a delay of reproduction will therefore increase the certainty of PATERNITY of the current male. Males that engage in EXTRA-PAIR COPULATIONS also perform courtship, but do so at a more intense rate than males engaged in ordinary pair formation. The time interval available for an extra-pair copulation is usually short, because of MATE-GUARDING by the partner of the female, and rapid courtship may facilitate the success of an extra-pair copulation attempt. *See also* SEXUAL SELECTION. [A.P.M.]

---

covariance An analogous term to VARIANCE. While the variance is a measure of how single observations (x) in a SAMPLE vary, the covariance is a measure of how pairs of observations (x and y) in a sample covary (i.e. the degree of ASSOCIATION between them). It contributes to the PRODUCT–MOMENT CORRELATION COEFFICIENT. *See also* STATISTICS. [B.S.]

cover The proportion of ground occupied by the perpendicular projection on to it of the aerial parts of individuals of each species present. Estimations of cover may be made visually (subjectively) or objectively by SAMPLING points from an area and recording which species (if any) cover the ground at these points, usually using a POINT QUADRAT. Percentage cover including also ABUNDANCE (from 100 to insignificant) may be expressed on the 10 (or 11)-point Domin scale or the 5 (or 6)-point Braun-Blanquet scale (see Kershaw & Looney 1985). [A.J.W.]

REFERENCE

Kershaw, K.A. & Looney, J.H.H. (1985) *Quantitative and Dynamic Plant Ecology*, 3rd edn. Arnold, London.

CPOM *See* COARSE-PARTICULATE ORGANIC MATTER.

cradle-to-grave analysis Assessment of ENVIRONMENTAL IMPACT of product from raw materials to disposal. *See also* LIFE-CYCLE ASSESSMENT. [P.C.]

***Crataegus monogyna–Hedera helix* scrub community** *See* SCRUB.

Crater Lake *See* LAKES.

crepuscular Animals that are active at dusk. [P.C.]

CRI *See* CHEMICAL RELEASE INVENTORY.

critical depth Depth at which the total photosynthetic PRODUCTION taking place in the water column is just balanced by the total respiratory losses of photosynthesizers in that same depth layer. *See also* COMPENSATION DEPTH. [V.F.]

critical habitats *See* HABITAT.

critical level *See* CRITICAL LOADS.

critical loads Quantitative estimate of EXPOSURE to one or more pollutants below which significant harmful effects on specified sensitive elements of the environment do not occur according to present knowledge. Usually refers to ATMOSPHERIC POLLUTION (*cf.* ENVIRONMENTAL QUALITY STANDARD). As usual there can be much debate about specifying appropriate sensitive elements of the environment and about the meaning of significant harmful effects. Also of relevance are the terms 'critical level' and 'target load'. A critical level is analogous to critical load but refers to a THRESHOLD of damage for gaseous pollutants acting upon species (usually vegetation). Target load is understood as the load determined by political agreement; it may be greater than the critical load and thus accepts a degree of damage or it may incorporate a safety factor and be less than the critical load. [P.C.]

REFERENCE

Horning, M. & Skeffington, R.A. (1993) *Critical Loads: Concepts and Applications*. ITE Symposium No. 28. Her Majesty's Stationery Office, London.

critical shear stress Shear stress at the sediment–water interface is the key parameter determining whether a sediment grain will be transported. When the critical shear stress for erosion of a sand grain is just exceeded, it rolls or saltates, entering a state called BEDLOAD transport. [V.F.]

crocodilians *See* REPTILES, DIVERSITY AND CONSERVATION.

crop Species used in AGRICULTURE and their YIELD; usually refers to plants (crop of maize, of rice; good crop from a harvest) but occasionally to animals (e.g. output/yield of fish farm in AQUACULTURE). [P.C.]

cross-environment genetic correlations In QUANTITATIVE GENETICS, the CORRELATION between the phenotypic values of given GENOTYPES in environment 1 and the phenotypic values of the same genotypes in environment 2 provides a measure of the genetic variation in PHENOTYPIC PLASTICITY. Genotypes that are identical in their sensitivity to the environments give rise to a cross-environment genetic correlation of 1. In that case no genotype*environment interaction exists and the genotype*environment interaction variance equals zero. Note that genotype*environment interaction might exist even if the cross-environment genetic correlation equals 1. A cross-environment genetic correlation of –1 indicates that the performance of the genotypes in the two environments is exactly opposite.

The response to selection in two environments depends upon the ADDITIVE GENETIC VARIANCE in each environment and the additive genetic COVARIANCE between environments. Selection on phenotypic plasticity over two environments is possible if the cross-environment genetic correlation is $-1 < r_{A12} < 1$ (Via & Lande 1985).

The cross-environment additive genetic correlation $r_{A12} = \mathrm{cov}_A(x_1, x_2)/\sqrt{(V_{A,1}, V_{A,2})}$, and the genotype*environment interaction variance both measure genetic variation in phenotypic plasticity. If the two environments occur with equal probability, the cross-environment additive genetic covariance $\mathrm{cov}_A(x_1, x_2) = \Sigma_{loci} 2p_i q_i \alpha_{i,1} \alpha_{i,2}$ at linkage equilibrium, and as the right-hand side equals $\Sigma_{loci} 2p_i q_i (\bar{\alpha}_i^2 - \sigma^2\{\alpha\}) = \Sigma_{loci} 2p_i q_i \bar{\alpha}_i^2 - \Sigma_{loci} 2p_i q_i \sigma^2\{\alpha\} = V_A - V_{AE}$ the cross-environment additive genetic covariance $\text{cov}_A(x_1,x_2)$ equals the difference between the main effect additive genetic variance V_A and the additive genetic genotype by environment interaction variance V_{AE}: $\text{cov}_A(x_1,x_2) = V_A - V_{AE}$. *See also* REACTION NORM. [G.D.J.]

REFERENCE

Via, S. & Lande, R. (1985) Genotype–environment interaction and the evolution of phenotypic plasticity. *Evolution* **39**, 505–522.

cross-resistance *See* MULTIPLE RESISTANCE TO PESTICIDES.

cruel bind hypothesis *See* GAMETE-ORDER HYPOTHESIS.

crust, Earth The solid outer portion of the Earth. Composed primarily of OXYGEN (46.60% by mass), SILICON (27.72%), ALUMINIUM (8.13%), IRON (5.00%), CALCIUM (3.63%), SODIUM (2.83%), POTASSIUM (2.59%), MAGNESIUM (2.09%) and TITANIUM (0.44%), which make up various silicate minerals. Crustal thickness and elastic properties vary widely and abruptly. Continental crust is lighter (average density = 2.7 g cm^{-3}) and thicker (average thickness = 50 km) than oceanic crust (average density = 3.0 g cm^{-3}; average thickness = 7 km). Beneath the oceans, the outermost layer of crust is composed of SEDIMENT deposits, weathered lavas and basaltic rocks. Continental crust has a relatively thin sedimentary top layer underlain by metamorphic rocks, thus forming two layers: the upper layer is rich in silicon and aluminium (called sial) and the lower layer is rich in silicon and magnesium (called sima). The crust beneath the oceans differs from that beneath the continents in that the former is generally thinner and lacks an upper layer rich in silicon and aluminium. New crust is generated at mid-ocean ridges and resorbed along SUBDUCTION zones (*see* PLATE TECTONICS). The oldest continental crust is nearly 4×10^9 years old, whereas the oldest oceanic crust is about 190 million years old. [V.F.]

Cryogenic taiga soils *See* SOIL CLASSIFICATION.

cryopreservation The keeping of organisms, or their propagules, at low temperatures and hence extending their normal lifespans, usually for purposes of selective breeding and conservation. [P.C.]

Cryosols *See* SOIL CLASSIFICATION.

crypsis A STRATEGY adopted by prey to reduce the probability of detection by a PREDATOR by increasing their resemblance to a random sample of the environment (Figs C18 & C19). Crypsis is one of several mechanisms selection may favour to reduce vulnerability to predators. Cryptic organisms may be brightly coloured and may have complex colour patterns. What matters is how easy it is for a predator to pick out the prey from the backgrounds the prey inhabits. Crypsis can only be defined with respect to an environment, a predator and a victim. Environments can be characterized with respect to the distribution of patches of varying size, shape, brightness and colour; predators by their behavioural or neural response to signals; and prey with respect to their behaviour in different backgrounds. Crypsis may be involved in TRADE-OFFS, with conspicuous coloration favoured by selection for attractiveness to members of the opposite sex. INDUSTRIAL MELANISM is an example of crypsis (Fig. C18). *See also* CAMOUFLAGE. [D.A.B.]

(a)

(b)

Fig. C18 Peppered moths (*Biston betularia*) naturally settle on the undersides of twigs in higher branches of trees, not on tree trunks. (a) Melanic forms are better camouflaged in polluted areas, (b) peppered forms are better camouflaged in unpolluted areas. (a, From Brakefield, P.M. (1987) Industrial melanism: do we have the answers? *Trends in Ecology and Evolution* **2**, 117–122; b, courtesy of T.G. Liebert.)

REFERENCES

Edmunds, M. (1974) *Defense in Animals: A Survey of Anti-predator Defenses.* Longman, London.

Endler, J.A. (1986) Defense against predators. In: *Predator–Prey Relationships: Perspectives and Approaches from the Study of Lower Vertebrates* (eds M.E. Feder & G.V. Lauder), pp. 109–134. University of Chicago Press, Chicago, Ill.

cryptobiont Literally, a hidden organism. Usually applied to dormant spores (often of microbes) within an ecosystem that can remain viable but difficult to detect for considerable periods. However, their presence can affect the long-term development of an ecosystem. [P.D.M.]

cryptoendolithic *See* ENDOLITHIC.

Fig. C19 A cryptic lizard, which blends in beautifully with its background. (From Krebs, J. & Davies, N.B. (1993) *An Introduction to Behavioural Ecology*, 3rd edn. Blackwell Scientific Publications, Oxford.)

cryptogamic soil crust Also called microbiotic SOIL crusts, these are delicate symbioses of cyanobacteria, lichens and mosses from a variety of taxa that inhabit arid and semi-arid ecosystems. Crusts perform several essential ecological functions: increase organic matter and available PHOSPHORUS, soil stability and water infiltration; provide favourable sites for the germination of vascular plants; and, most crucially, perform the major share of NITROGEN FIXATION in DESERT ecosystems. Crusts were once much more extensive, but are extremely fragile: nitrogenase activity can be reduced 80–100% in the crust under a single human footprint. Thus they have been largely eliminated by anthropogenic activities such as livestock GRAZING and recreation. [T.L.F.]

cryptophyte A plant life form in which the organ of perennation is hidden in SOIL or mud during a period unfavourable for GROWTH. It includes geophytes, where survival depends upon below-ground protection (e.g. plants with bulbs, corms, tubers, etc.), and HYDROPHYTES, where the perennating organs are protected by submersion. *See also* PHYSIOGNOMY; RAUNKIAER'S LIFE FORMS. [P.D.M.]

culling This term tends to be applied to the HARVESTING of large mammals, including marine mammals. In practice, it is usually applied to a situation when there is some strong selection in the harvesting towards particular groups characterized by sex, age or other attribute. For example, the harvesting of deer may involve the selective removal of certain males that have unattractive trophy heads or the removal of a proportion of the population, for example young males or senescent females that do not contribute to the reproductive process. The term is also used in situations where there is some form of conflict between the population that is being culled and some other activity. Culling in this sense is usually aimed at ensuring that the population lies below some THRESHOLD level. *See also* POPULATION REGULATION. [J.B.]

curie (Ci) A UNIT (now superseded) equal to the emission from 1 g of RADIUM, which is equivalent to 3.7×10^{10} Bq. [R.S.S.]

curium (Cm) Curium is a synthetic element (ATOMIC NUMBER = 96; relative ATOMIC MASS = 247; melting point = 1340°C; boiling point unknown) that was discovered in 1944. Isotopes from ^{238}Cm to ^{251}Cm have been produced by nuclear reactions. ^{242}Cm ($t_{1/2}$ = 162.94 days), ^{243}Cm ($t_{1/2}$ = 28.5 years) and ^{244}Cm ($t_{1/2}$ = 18.1 years) are produced by successive NEUTRON capture events in nuclear reactor fuel and occur in the environment mainly as a result of releases from nuclear facilities. Curium occurs dominantly in the +3 oxidation state and is highly insoluble in natural waters. [A.B.M.]

currents Important agents of erosion on land and in the SEA. Ocean currents are driven by two forces: WIND drives horizontal surface WAVES, whereas density differences among water masses drive deep-water CIRCULATION and vertical exchange. [V.F.]

$CV^2 > 1$ rule In many discrete-time MODELS OF PARASITOID–HOST INTERACTIONS derived from the Nicholson–Bailey model, stability requires that HOSTS vary in their risk of parasitoid attack and that the square of the coefficient of VARIANCE of risk among hosts is greater than unity. [H.C.J.G.]

REFERENCE

Hassell, M.P & Pacala, S. (1990) Heterogeneity and the dynamics of host–parasitoid interactions. *Philosophical Transactions of the Royal Society of London, Series B* **330**, 203–220.

cyanide Inorganic salts containing the cyanide ION, CN^-, are extremely poisonous. Cyanide ions form extremely strong complexes and inhibit the uptake of OXYGEN by haemoglobin molecules in blood. Cyanide ions form coordination complexes with many METAL ions, for example Fe^{3+} interacts with cyanide ions to form the ferricyanide ion $[Fe(CN)_6]^{3-}$. Cyanide forms a soluble, stable complex ion with GOLD and SILVER. Ores containing silver or gold are sprayed with dilute solutions of sodium cyanide (NaCN) or potassium cyanide (KCN). In this way, the solution containing the precious metal ions can be separated from the ore and recovered by reduction using ZINC metal. [M.C.G.]

cyanobacterial toxins In nutrient enriched FRESHWATER or saltwater many blue-green algae (cyanobacteria) reproduce rapidly by vegetative cell division. The consequent high-density populations are described as BLOOMS and usually occur following EUTROPHICATION with PHOSPHORUS and NITROGEN from agricultural FERTILIZERS or SEWAGE. When the cells die they may release TOXINS which poison fish, and birds or mammals drinking the water. The toxins include low RELATIVE MOLECULAR MASS polypeptides, tertiary amines and alkaloids. Most are neurotoxins which cause respiratory distress and convulsions in animals. Some are suspected of causing liver failure in livestock and others are dermatogenic, causing skin inflammation in humans. In potable water supplies the toxins have been blamed for outbreaks of gastroenteritis. The cyanobacterium *Microcystis* occurs widely as blooms in rice paddy, and its toxin, a ring of seven AMINO ACIDS, is believed to be the cause of a high incidence of primary liver cancer in rice-growing areas, where growth of cyanobacteria is encouraged to take advantage of their nitrogen-fixing abilities. [J.R.E.]

cyclic succession Cyclic SUCCESSION occurs where a given species or group of species is locally destroyed and there is succession back to that or those species. It is typical of species-poor vegetation. One example is the subalpine fir forest regenerated in wave-like bands resulting from the destruction of trees by winter winds, described by Sprugel (1976). In some cases cyclic succession has been called 'autosucession', as in Californian chaparral periodically disturbed by fire (Hanes 1971). [P.J.G.]

REFERENCES

Hanes, T.L. (1971) Succession after fire in the chaparral of southern California. *Ecological Monographs* **41**, 27–52.

Sprugel, D.G. (1976) Dynamic structure of wave-regenerated *Abies balsamea* forests in the north-eastern United States. *Journal of Ecology* **64**, 889–911.

cyclomorphosis Cyclic change in PHENOTYPE, such as seasonal changes in morphology, particularly conspicuous among cladoceran Crustacea and rotifers (phylum Rotifera), also described in dinoflagellates and protozoa. Its expression is based on environmentally induced (e.g. daylength, temperature, presence of predators) PHENOTYPIC PLASTICITY. Under natural conditions, seasonal clonal SUCCESSION may contribute to this phenomenon. Typical changes involve the formation of helmets and elongated tail spines in spring and summer in the cladoceran *Daphnia*, and the formation of spines in rotifers. It is believed that it reduces the susceptibility to predation. *See also* COMPLEX LIFE CYCLES. [D.E.]

cyclone A cyclone is a mass of air, ranging from a few metres to a few thousand kilometres in size, and often but not always extending from the Earth's surface up to the TROPOSPHERE, which is caused to rotate cyclonically by the convergence of air towards its centre and the upward divergence that must accompany it. The smallest are dust devils, waterspouts and tornadoes, with funnel CLOUDS ranging in width from a few metres to tens of metres at the surface. Small polar-air rotating storms range in size from a very few kilometres to tens or even hundreds of kilometres. Tropical cyclones, called hurricanes in the North Atlantic, the extreme western Pacific and some parts of the INDIAN OCEAN but known as typhoons in the PACIFIC OCEAN, range in size from a few hundred kilometres to well over 1000 km. Large cyclones of temperate latitudes that travel into polar regions may occupy a width of up to 2000 or 3000 km and often have well-defined FRONTS, which are the boundaries between discrete AIR MASSES of different geographical origin and along which the main cloud and RAIN systems are aligned. These latter systems can be well represented in computer-based mathematical models, while the smaller systems are on too small a scale in both space and time to be adequately represented by the grid points of the model, which are of the order of 150 km spacing and 3 h time output (although the computed time interval may be as small as 20 min).

Tropical cyclones do not persist over land because the CONVECTION upwards depends on the latent heat generated by the condensation of cloud and its fallout as rain. A land surface does not generate a persistent enough supply of heat to evaporate the WATER from the surface. Also the low PRESSURE at the centre produced by the rotation lowers the TEMPERATURE of the inflowing air, which increases convection over the warm SEA. The centrifugal forces also cause a downflow of air at the centre and occasionally stratospheric air reaches the surface in the eye of the storm. The sea temperature usually needs to be above 28°C, but occasionally cold continental air reaches the central Mediterranean and similar-looking storms appear there, usually significantly smaller, although still possessing an eye and very strong CYCLONIC winds.

Cyclones in temperate latitudes obtain their energy in the first place from the temperature contrast of the air masses, although they persist for several days or even weeks through the condensation of rain along the fronts, where the air passes in at the base and out at a higher level depositing its rain.

Frontal cyclones are common in temperate latitudes throughout the year. Typhoons have been observed in the western Pacific in every month of the year. They most commonly occur in the autumn when the sea has achieved its maximum temperature. Hurricanes do not normally occur in the South Atlantic, but are seen in the Arabian Sea and typhoons may be very severe in the Bay of Bengal and close to Madagascar. Central Mediterranean storms have been seen mainly in midwinter.

All cyclones have the same direction of rotation as the Earth beneath: counterclockwise in the Northern hemisphere but clockwise south of the Equator. An ANTICYCLONIC weather system is a region of sinking, and therefore diverging, air and rotates in the opposite direction relative to the Earth, with convectively stable air and cloud systems (if any); however, it still rotates in the same absolute direction in space but more slowly than the Earth beneath. [R.S.S.]

cyclonic Moving in a counterclockwise direction in the Northern hemisphere and in a clockwise direction in the Southern hemisphere. [V.F.]

D

D soils *See* SOIL CLASSIFICATION.

damped oscillations *See* OSCILLATIONS.

dams, effects of The effects of a dam on a stream and its biota depends on the size of the dam, its location in the WATERSHED and how the impounded WATER is used. Most dams alter the hydrological regime of the stream below them, reducing both total stream flow and variability in flow, as well as altering water temperatures and clarity. Typically, seasonal peak flows are captured, reducing downstream processes that alter channels and flood RIPARIAN areas. This results in major changes in the riverine biota, often reducing biotic diversity because habitat diversity has been reduced. Dams block migrations of salmon, eels, lampreys and other DIADROMOUS fishes, lowering their abundance by denying them access to upstream spawning and rearing areas. Construction of dams on large RIVERS usually results in major losses of fisheries, such as salmon fisheries on the Columbia River (USA), the Mediterranean sardine fishery at the mouth of the Nile River (Egypt) and the sturgeon fishery in the Volga River (Russia). [P.B.M.]

Darwin, Charles *See* CHARACTERS IN ECOLOGY.

Darwin, Erasmus *See* CHARACTERS IN ECOLOGY.

Darwinism The theory of EVOLUTION by NATURAL SELECTION as elaborated by Charles Darwin (1809–1882). Darwin became convinced that, although distinct, species were not separately created by a divine hand but that transition occurred from one to another. This idea was by no means unique at the time, but it was strongly contested. Darwin's contribution was to provide a mechanism by which evolution could take place, namely natural selection. A.R. Wallace also recognized that natural selection could drive evolution. Papers by both were presented at a meeting of the Linnean Society in London in 1858, to be followed in 1859 by publication of Darwin's book *The Origin of Species*. This immediately stimulated great public interest and dispute, at first about evolution itself rather than the mechanism, some scientists and thinkers supporting Darwin while other scientists and many clerics criticized the ideas. The term 'Darwinism' was first used in 1860 by T.H. Huxley, a combative and vociferous supporter, and in 1889 Wallace published his own work under the unselfish title *Darwinism, An Exposition of the Theory of Natural Selection*. Since then the term has become something of a semantic minefield, being used less by biologists than by philosophers, sociologists and those who oppose evolution on religious grounds as a label for what they construe to be a particular attitude to evolution. Social Darwinism was a movement of the beginning of the 20th century which argued that since there was COMPETITION in nature leading to elimination of the unfit, society should operate in the same way; people with socially desirable characteristics should be encouraged to have children and those with undesirable TRAITS discouraged. In the 1930s this thinking became associated with the obscenities of Nazism. NEO-DARWINISM is a term used to indicate a mid-20th century synthesis of Darwinian ideas and genetics, supporting the idea that evolution proceeds by the accumulation of MUTATIONS with small effects (micromutations) giving rise to ADAPTATION, SPECIATION and eventually to the major evolutionary innovations. An alternative view is that micromutations result in adaptation but speciation requires some major reorganization such as a chromosomal change to come into being. *See also* FITNESS. [L.M.C.]

data logger A programmable device for capturing data. Data loggers record signals from sensors over a period of time from seconds to weeks or even months whilst unattended, for subsequent inspection or downloading to a computer for further analysis. Data loggers are usually rugged, compact and weatherproof microprocessors, powered by batteries, to which several or many sensors can be connected. The system interogates the sensors at preprogrammed intervals, utilizes CALIBRATION tables to convert a measured voltage to appropriate engineering units (e.g. temperature or humidity) and stores the result as a series of spot readings or means of a series (e.g. the average temperature over an hour). [J.G.]

Dating, radiocarbon

The study of past environmental conditions (*see* PALAEOECOLOGY) demands the existence of a time-scale in which such studies can be set. It is necessary, therefore, to develop techniques of assigning dates to fossils or their matrices, usually sediments. Dating methods can be relative or absolute.

Relative dating techniques are ultimately dependent on absolute methods, but can provide a first indication of the age of a material. The most commonly used relative method is the FOSSIL ASSEMBLAGE itself and is based on the concept that certain fossils are found only with particular time periods and can therefore be used for the definition of such periods. Such fossils can be termed 'indicator fossils'. The method becomes more secure when whole assemblages of such fossils can be used; in other words, when fossil biotic communities are the basis of time definition. Microfossils (small fossils that require microscopy for their study) have proved particularly valuable in this respect, partly because they are often found in large numbers in sediments and can therefore be subjected to statistical techniques. Such microfossils include pollen and spores (*see* PALYNOLOGY), diatoms, foraminiferans, cladocerans, etc.

However, assemblages of indicator fossils themselves need to be assigned an absolute time-scale if they are to be used effectively in relative dating. Some of the most widely used absolute dating methods depend upon the predictability of radioactive decay in unstable isotopes. A number of such opportunities exist in rocks and sediments but their success requires the measurement of either the product of the decay or the residual ISOTOPE that remains undecayed. If the half-life of the isotope is known, the degree to which decay has proceeded can be calculated and hence the age inferred.

POTASSIUM–ARGON dating uses the decay of the isotope ^{40}K into argon, ^{40}Ar. It is complicated by the fact that ^{40}K can decay into ^{40}Ca, which can also arise in other ways, but in situations where argon is retained in the rocks (as is often the case in rocks of volcanic origin) the method is a useful one. It has been extensively used in the study of human origins in the Pliocene/Pleistocene rocks of East Africa.

URANIUM has many unstable isotopes with varying half-lives. As in the case of potassiun–argon dating, it is the product of decay that is measured; in corals, for example, uranium is accumulated and subsequently decays to THORIUM and PROTACTINIUM, which are retained in the coral skeleton and can be used to determine its age. Methods that measure end-products have the advantage that they are not limited in the time-scale over which they operate; the older the material, the greater the quantity of the decay product. Methods that measure the remnant of the isotope, on the other hand, are limited by the efficiency of detection of the declining resource and (depending upon the half-life of the isotope) can only be used over a defined time-scale. Such is the case with radiocarbon.

Radiocarbon dating is the most extensively used method for dating organic materials from the latter parts of the Pleistocene and the Holocene (in total approximately the last 40 000 years). It depends upon the decay of the radioactive carbon isotope ^{14}C into nitrogen atoms (from which they were originally generated in the atmosphere by cosmic bombardment with neutrons). The half-life of ^{14}C (the time taken for 50% of any given population of ^{14}C atoms to decay) is 5569 ($\pm$30) years. When an autotrophic organism dies, it no longer derives ^{14}C from the air so that the ratio of ^{14}C to the stable isotopes of CARBON (^{12}C and ^{13}C) begins to decline. A comparison of the ratio of ^{14}C to the stable isotopes of carbon in a dead organism with that ratio in the atmosphere is a measure of the time elapsed since the organism died. Dates are conventionally given in years before present (BP), where 'present' is taken to be 1950. This datum was selected because the advent of above-ground nuclear testing at that time confused all subsequent radiocarbon production. The determination of age requires the estimation of the residual ^{14}C in a sample and this can be achieved either by observing the rate at which decay is taking place (β-PARTICLES are emitted) or, preferably, by direct measurement of the ^{14}C atoms using a cyclotron. Simply observing decay is less effective, particularly in older materials, since large samples are needed for analysis (because of low ^{14}C concentration) and decay may need to be logged over long periods if ^{14}C is scarce and emissions infrequent. The use of a cyclotron permits the use of smaller samples and is more time-efficient.

The use of radiocarbon dating depends upon the assumption that the rate of production of ^{14}C in the atmosphere is uniform over time; this is not the case. The CALIBRATION of the technique against items of known absolute age, such as the growth rings of trees, has shown that there is a deviation away from the true age that increases with time. A wood sample with a known age of 7000 years, for example, gives a radiocarbon date of approximately 6000 years. In addition, the calibration curve is not a simple one; there are 'wiggles' and 'plateaux' that can lead to considerable difficulties in the use of this dating method at particular points in history. At about 10 000 years ago, for example, there is a plateau in the calibration curve, which means that samples differing in age by several centuries at this time may give the same radiocarbon age. This is a particularly unfortunate period for such a plateau to occur since it was a time of rapid CLIMATE CHANGE at the opening of the Holocene, when palaeoecologists concerned with environmental reconstruction would much appreciate a fine-resolution dating method.

It has now become conventional to refer to dates derived from radiocarbon techniques as 'corrected' if subjected to adjustment by reference to the calibration curve, or 'uncorrected radiocarbon years' if not. Sometimes uncorrected dates are given as 'bp' rather than 'BP', but this convention is not universally accepted. Although we can regard radiocarbon dating as 'absolute', it can be seen that it still requires calibration. The truly absolute dating methods are those that require no such adjustments, such as the counting of annual growth rings in trees (DENDROCHRONOLOGY), or the counting of annual laminations in lake sediments (varves) or in ICE cores. These are the most satisfactory dating techniques, but are available only for certain types of materials.

One further dating technique, which is becoming increasingly important for the study of events in the later part of the Pleistocene (the last 200 000 years), is AMINO ACID racemization. This depends upon the L-isomer of amino acids (the form found in living PROTEINS) reverting after the death of an organism to the D-isomer, with a decompositional half-life of approximately 15 000 years. Residual protein in bones and mollusc shells is often adequate for the provision of a date on this basis. [P.D.M. & V.F.]

REFERENCE

Lowe, J.J. & Walker, M.J.C. (1984) *Reconstructing Quaternary Environments*. Longman, London.

day-degrees Product of the number of days and number of degrees (°C) by which the TEMPERATURE exceeds an arbitrary or predetermined THRESHOLD. Day-degrees (DD) is a cumulative index that quantifies the thermal budget required for an organism to develop from one growth stage to another. If the temperature threshold for development of an insect is 5°C and the species takes 20 days at 20°C to develop from egg to adult, the day-degree requirement per generation is $(20-5)\times 20=300$ DD. At a mean temperature of 15°C the same species would require 30 days to develop from egg to adult. The day-degree concept can be used, for example, to forecast the date of emergence each year of pest insects and estimate the number of generations that are possible in particular climatic zones. [J.S.B.]

day number Day number is a chronology describing the number of days since the first day of January within a year: 1 January is defined as having a day number of 1. Day number is frequently confused with JULIAN DAY, which counts the number of days since noon on 1 January 4713 BC. *See also* TIME. [P.R.V.G.]

daylength The period of daylight between sunrise and sunset, which are defined as the times at which the true position of the centre of the solar disc passes over the horizon. Variation in daylength results from the Earth's rotation about its own axis, relative to the orbital motion of the Earth around the Sun. Daylength (N_d) at any location on Earth is a function of latitude (ϕ) and SOLAR DECLINATION (δ), which is itself a function of DAY NUMBER (all angles are given in degrees ($90° = \pi/2$ radians)):

$$N_d = \frac{2}{15}\cos^{-1}\left(-\tan\phi\ \tan\delta\right)$$

Several important observations follow from this equation. At the Equator, latitude (ϕ) is zero and daylength is always 12 h, independent of day number or solar declination. At the equinoxes, when solar declination (δ) is zero, daylength is independent of latitude and equal to 12 h. The precise dates of the equinoxes vary between years depending on the Sun–Earth geometry but are typically 21 March and 23 September. The longest and shortest days occur halfway between the equinoxes at the summer and winter solstices on about 21 June and 22 December. By convention, seasons and daylengths refer to the Northern hemisphere and should be reversed for the Southern hemisphere. The period of civil twilight is important for many biological processes and is defined as the interval between sunrise or sunset and when the centre of the solar disc passes 6° below the horizon. [P.R.V.G.]

REFERENCE

Iqbal, M. (1983) *An Introduction to Solar Radiation.* Academic Press, London.

DDT (dichlorodiphenyl-trichloroethane) DDT is a potent stomach and contact organochlorine INSECTICIDE, discovered in 1939, and considered a revolutionary development in PEST CONTROL. It was apparently much safer to humans than earlier insecticides and was widely used to control human lice and disease-transmitting insects in the Second World War. DDT was also used on a wide range of crops and on storage pests. However, it has been widely banned because of environmental PERSISTENCE and BIOACCUMULATION in animal body fats and the FOOD CHAIN, with concerns regarding reproduction in higher animals, notably thinning of egg-shells of raptorial birds. *See also* ORGANOCHLORINES; PESTICIDE. [K.D.]

REFERENCE

Mellanby, K. (1992) *The DDT Story.* British Crop Protection Council, Farnham.

de Wit replacement series An experimental protocol for studying interspecific plant COMPETITION, named after the Dutch ecologist and agronomist C.T. de Wit; also known as substitutive experiments. The technique involves holding total plant DENSITY constant and varying the proportion of two competing species. For example, a five-level de Wit series might have 100/0, 75/25, 50/50, 25/75 and 0/100 as percentages of individual plants belonging to species A and B respectively (Fig. D1). The technique is used for the computation of indices of competition. Replacement series have been criticized for their assumption that total plant density (A plus B) is constant and that the choice of this constant density is arbitrary. An ideal design would alter both density and frequency in a factorial combination. *See also* ADDITIVE EXPERIMENTS. [M.J.C.]

REFERENCES

Harper, J.L. (1977) *The Population Biology of Plants.* Academic Press, New York.

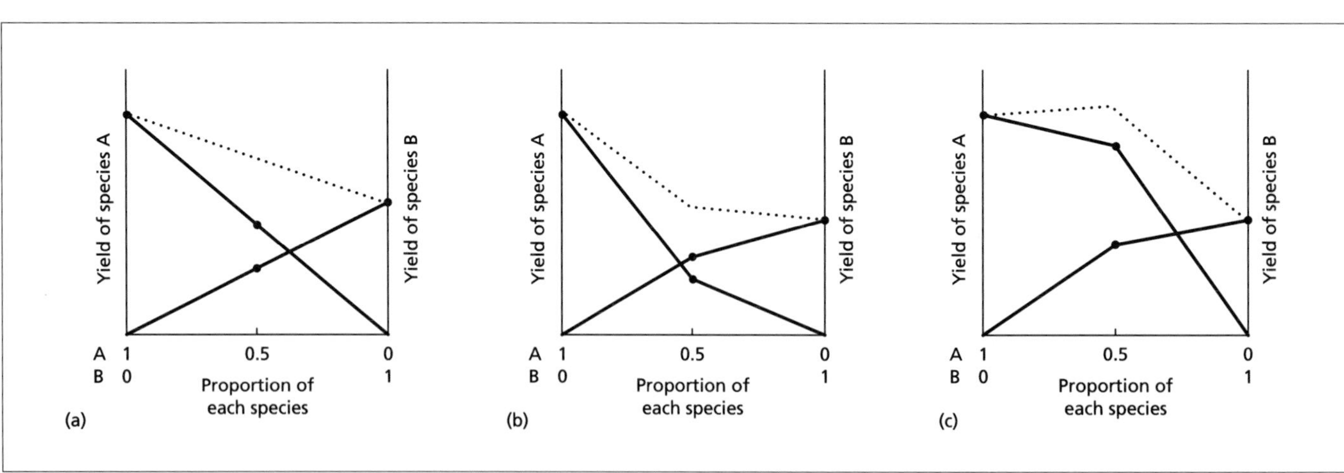

Fig. D1 De Wit replacement series. Two plant species, A and B, are grown in monocultures and in various proportional mixtures at the same total constant density. (a) The growth of individuals of the two species is unaffected by the presence of the other species, so that the yield of the mixture (dotted line) can be predicted by linear extrapolation from the yields of the two monocultures. (b) Underyielding, in which the total yield of the mixture is less than predicted on the basis of the two monocultures. (c) Overyielding, in which the yield of the mixture is greater than the monoculture yield of the highest yielding of the two species. In practice, overyielding is uncommon but has been observed in some special cases (e.g. some grass–legume mixtures).

Law, R. & Watkinson, A.R. (1989) In: *Ecological Concepts* (ed. J.M. Cherrett), pp. 243–284. Blackwell Scientific Publications, Oxford.

dead zones Zones of reduced flow that retain water in a flowing-water body. If the bottom is 'rough', there will be many such zones, chiefly on the downstream side of large boulders where flow is reduced. The retained water in these zones is nevertheless progressively renewed by fluid exchange with the main flow. These areas can support an extensive PLANKTON community. They also act as REFUGES for organisms and as temporary sinks for sediments under high flows. *See also* BOUNDARY LAYER. [P.C.]

REFERENCE

Reynolds, C.R. (1992) Algae. In: *The Rivers Handbook*, Vol. 1 (eds P. Calow & G. Petts), pp. 195–215. Blackwell Scientific Publications, Oxford.

debt for nature programmes These involve the purchase of a country's debt notes, which are discounted on the secondary market. These are presented to the debtor country in exchange for local currency in the amount of the face value of the debt, with the local currency being invested in conservation. [P.C.]

deciduous forest Forest whose trees reduce water loss by dropping their leaves during the unfavourable season of the year, which is winter in the case of summer-green deciduous forests (*see* TEMPERATE DECIDUOUS FOREST). TROPICAL SEASONAL FORESTS occur in humid tropical climates with a pronounced DRY SEASON, during which some, many or all the trees drop their leaves. The deciduous habit is, in this instance, related to a decrease in rainfall rather than of temperature. If the dry period extends beyond 4 or 5 months, fully dry-deciduous WOODLAND may develop. [J.R.P.]

decomposers *See* DECOMPOSITION.

decomposition Breakdown of chemicals to simpler products, i.e. nearer thermodynamic equilibrium. Often applied to breakdown of BIOMASS under physical and biological action. Complex organic systems are simplified, leading to a change in form (often from more to less organized) and ultimately the conversion of biomolecules to their inorganic constituents. Those organisms bringing about decomposition are known as decomposer organisms. They can involve bacteria, fungi and animals. Animal decomposers can effect decomposition by mechanical action (burrowing through SOIL and SEDIMENT) or by feeding processes. Those involved in decomposition by feeding are also known as DETRITUS-feeders, detritivores or detritophages. *See also* FEEDING TYPES, CLASSIFICATION. [P.C.]

decreaser A plant that declines in ABUNDANCE on heavily grazed PASTURE or rangeland. Decreasers tend to be palatable species, often more competitive than increasers in the absence of herbivory (this implies the existence of a TRADE-OFF between competitive ability and palatability, in which there are competing demands within the plant for investment in GROWTH versus defence). Decreasers tend to be the desirable (more nutritious) rangeland plants. *Cf.* INCREASER. [M.J.C.]

deep scattering layer Well-defined horizons in all parts of the world's oceans, usually between 100 and 500 m depth, that reflect sonar and that usually consist of fish, squid or other larger ZOOPLANKTON. These layers have been observed to rise to the surface of the OCEAN at dusk (100 m) and to descend to their daytime depths at sunrise (up to 800 m), reflecting the VERTICAL MIGRATION activities of zooplankton. In the ARCTIC, the scattering layers appear to show an annual rather than a DIURNAL pattern. [V.F.]

deep sea The part of the marine environment that lies below the level of effective LIGHT penetration for PHYTOPLANKTON photosynthesis in the open OCEAN and below the depth of the continental shelves (greater than *c.* 200 m). It is sometimes referred to as the APHOTIC ZONE. Although it is by far the most extensive habitat on Earth inhabited by living organisms, relatively little is known of the biology and ecology of the deep SEA. Environmental characteristics of the deep sea are remarkably constant. SALINITY, TEMPERATURE and OXYGEN levels vary little in space and time. Pressure is the most variable factor and increases by 1 atmosphere (101 kPa) for each 10 m in depth. Except for HYDROTHERMAL VENT communities inhabited by chemosynthetic bacteria, the deep sea is unique in lacking a source of primary PRODUCTION and most deep-sea organisms derive their food from material transported from the surface ocean. The absence of light and lack of food in the deep sea are associated with a number of ADAPTATIONS of deep-sea fauna including large eyes or reduced eyes, light-producing organs, large mouths, small body size in most fauna but gigantism in certain invertebrate groups and changes in biochemical composition. [V.F.]

defence mechanisms Organisms are subject to a variety of threats from MORTALITY agents: some are associated with external (extrinsic/ecological) agents such as accidents, predators, diseases and toxic chemicals; and some are due to internal factors associated with system deterioration or AGEING. Against each, organisms can deploy a variety of responses, collectively referred to as defence mechanisms. For threats from predators, they can involve avoidance and disuasions or repulsion. For diseases, they can involve exclusion of the disease-causing organisms and/or the deployment of immunological systems. For pollutants, they can involve avoidance, exclusion, excretion or neutralization (e.g. by induction of PROTECTIVE PROTEINS). These toxicants, and the processes that lead to ageing, can cause damage to tissues and macromolecules, so another form of defence is by repair processes which, within organisms, invariably involves cellular and molecular turnover.

A common feature of these defence systems, whether they be against predators, parasites, microbes or wear and tear, is that they are all expensive in material and energy and may therefore trade off against somatic growth and reproduction processes. This leads to thoughts about the optimum allocation of RESOURCES between defence (which enhances survival chances) and other aspects of FITNESS (e.g. associated with FECUNDITY) that, of course, will vary with species and ecological circumstances. *See also* COSTS OF RESISTANCE; EVOLUTIONARY OPTIMIZATION; PLANT DEFENCE THEORY; TRADE-OFF. [P.C.]

REFERENCE

Barnes, R.S.K., Calow, P. & Olive, P.J.W. (1993) *The Invertebrates. A New Synthesis*, 2nd edn, Chapter 13. Blackwell Scientific Publications, Oxford.

deflected succession Deflected SUCCESSION occurs where human intervention, for example grazing, mowing or burning, induces a new succession in place of the one that would have otherwise occurred (Godwin 1929), producing a distinctive COMMUNITY often called a PLAGIOCLIMAX. Widespread examples in the northern temperate zone are mown fens, and grazed or mown grasslands where the CLIMATE and SOIL could support forest. In the tropics, the equivalent communities are grasslands induced by burning for hunting in forest areas, and savannahs modified by burning that occurs more frequently than would occur without human impact. In general, grasses are favoured over woody plants in deflected successions, but there are also more subtle sorting effects among grasses and forbs. [P.J.G.]

REFERENCE

Godwin, H. (1929) The subclimax and deflected succession. *Journal of Ecology* **17**, 144–147.

defoliant Substance causing loss of leaves from plants. *See also* BIOCIDE; HERBICIDE. [P.C.]

deforestation Deforestation, the removal of CLOSED FOREST or open WOODLAND, has been taking place all over the world for millenia. Commonly, deforestation is to create land for AGRICULTURE; if this is abandoned, shrublands and tree COVER eventually re-establish unless the site has become very seriously eroded. The Mediterranean Basin was already substantially deforested in classical times and now has only 5% forest cover. Western and central Europe have lost most of their former, almost continuous, TEMPERATE DECIDUOUS FOREST. For example, England's forest cover had been reduced from 90% to 20% by Roman times and then declined to 5% by 1350, the time of the Black Death. In much of Europe the forest area has increased, largely by planting, during the past century or so: Britain now has about 10% cover; France, which was down to 17% forest in the mid-19th century, is now 25% forested. A pattern of rapid deforestation after European settlement, followed by slight recovery this century, is shown by eastern North America. European colonization of Australasia led to similar massive deforestation. Global forest and woodland loss 1700–1980 is estimated to have been 9%, and what remains is increasingly fragmented. The greatest tracts of remaining intact forest are the BOREAL conifer forests at high northern latitudes and the tropical moist forests that girdle the Equator (*see* TROPICAL RAIN FOREST). These last are currently undergoing massive and highly publicized deforestation, at a rate of about 0.9% per year during the decade 1980–1990. Peninsular Malaysia went from 74% to 40% forest cover in the 34 years from 1958 to 1990. The assault on tropical forests is thus the latest episode in mankind's continuing war on the world's natural vegetation cover, more widely publicized due to modern communications and more rapid because of powerful modern machinery. What is unique to tropical moist forests is their phenomenal SPECIES RICHNESS in all groups of plants and animals. There is consequently a much higher probability of species extinction as the forest shrinks and becomes fragmented compared with what occurred in other parts of the world, for example in Britain where the substantial forest loss has led to negligible species loss. As elsewhere, an ultimate pattern of forest patches in an agricultural landscape seems the probable fate of much of the lowland humid tropics, but with greater consequences for BIODIVERSITY than elsewhere. *Cf.* AFFORESTATION. [T.C.W.]

REFERENCES

FAO (1993) *Forest Resources Assessment 1990. Tropical Countries.* FAO Forestry Paper 112. Food and Agriculture Organization, Rome.

Mather, A.S. (1990) *Global Forest Resources.* Belhaven, London.

Whitmore, T.C. & Sayer, J.A. (eds) (1992) *Tropical Deforestation and Species Extinction.* Chapman & Hall, London.

degrees of freedom (ν) From a purely practical point of view this is a number that we need to calculate, in most statistical tests, before we can look up the significance of the test statistic in the appropriate STATISTICAL TABLE. For example, in the *t* TEST it is n_1+n_2-2 and in the chi-squared GOODNESS-OF-FIT test it is frequently $n-1$ (where n is the number of categories).

However, this term also refers to the number of independent items of information in an experiment or statistical test. For example, when calculating the VARIANCE we divide the SUM OF SQUARES by its degrees of freedom ($n-1$). It may appear that the correct value for the denominator should be n (as for the MEAN) rather than $n-1$. This is not so. Suppose that we had only one observation ($x=9$) in our SAMPLE. How much information would we have about the population mean and variance? Our best estimate of the mean would be 9 (from one piece of information), but we would have no information about variation. With two observations ($x=9$ and $x=11$) we have two (n) pieces of information about the likely value of the mean (our best estimate would be $(9+11)/2=10$) and one ($n-1$) piece of information about variation (9–11). With three observations ($x=9$, $x=11$ and $x=6$) we have three pieces of information about variation (9–11, 9–6 and 11–6), but only two ($n-1$) are independent. *See also* CHI-SQUARED TEST. [B.S.]

Delaney Clause Clause of the US Food, Drug and Cosmetic Act. States that food additives that cause cancer in humans or animals at any level shall be prohibited from use. This is important from an environmental point of view since it suggests the principle that substances with carcinogenic properties should not be allowed into the environment at large, i.e. there are no safe THRESHOLD levels. [P.C.]

delayed density dependence A change in POPULATION GROWTH RATE associated with a change in POPULATION DENSITY some time previously. The term was introduced by G.C. Varley in 1947 in the context of describing the interaction of a predator and a prey population: when prey and predator numbers are closely linked, an increase in prey ABUNDANCE may lead to an increase in predator numbers, which eventually causes prey numbers to decline. Delayed DENSITY DEPENDENCE tends to destabilize POPULATION DYNAMICS; thus closely linked prey and predator populations sometimes show coupled OSCILLATIONS. *See also* POPULATION REGULATION. [R.H.S.]

delayed semelparity Life-history pattern of long-lived organisms that reproduce only once. Examples include some salmon, yucca, lobelias and cicadas. Simple theory suggests that semelparity can evolve when the TRADE-OFF between adult survival and LONGEVITY is concave or when adult MORTALITY rates are high and juvenile mortality rates low. *See also* COLE'S PARADOX; LIFE-HISTORY EVOLUTION. [D.A.B.]

REFERENCE

Young, T.P. (1990) Evolution of semelparity in Mount Kenya lobelias. *Evolutionary Ecology* **4**, 157–172.

deleterious mutations *See* MUTATION.

delta carbon The proportional deviation of the CARBON molar abundance ratio of a compound from some standard. Usually it is the $^{13}C:^{12}C$ ratio that is of interest:

$$\delta^{13}C = \frac{R_A}{R_S} - 1$$

where R_A is the $^{13}C:^{12}C$ ratio of the compound being examined and R_S is the $^{13}C:^{12}C$ ratio of the standard. Traditionally, R_S for $^{13}C:^{12}C$ determinations has been CARBON DIOXIDE (CO_2) prepared from a fossil belemnite taken from the Pee Dee formation in South Carolina (denoted as PDB or PDB-CO_2). The R_S of PDB is traditionally taken as 0.01124, but some workers now claim a value of 0.01118. The supply of PDB has long been exhausted, but numerous secondary standards have been defined and compared to other well-known materials. All carbon ISOTOPE ratios are still referred to PDB. [J.Ly.]

REFERENCE

Craig, H. (1957) Isotopic standards for carbon and oxygen and correction factors for mass spectrometric analysis of CO_2. *Geochemica Cosmochemica Acta* **12**, 133–149.

delta oxygen The proportional deviation of the OXYGEN molar abundance ratio of a compound from some standard. Usually it is the $^{18}O:^{16}O$ ratio that is of interest:

$$\delta^{18}O = \frac{R_A}{R_S} - 1$$

where R_A is the $^{18}O:^{16}O$ ratio of the compound being examined and R_S is the $^{18}O:^{16}O$ ratio of the standard. Most chemists use standard mean ocean water (SMOW) as their standard for $^{18}O:^{16}O$ ratio determinations, for which R_S=0.002005. An exception is some atmospheric scientists who express $^{18}O:^{16}O$ ratios of atmospheric gases (oxygen and the oxides of CARBON and NITROGEN) with respect to the $^{18}O:^{16}O$ ratio of CARBON DIOXIDE (CO_2) prepared from a fossil belemnite from the Pee Dee formation in South Carolina, denoted PDB-CO_2 (this is also the standard for $^{13}C:^{12}C$ determinations; *see* DELTA CARBON). The $^{18}O:^{16}O$ ratio of PDB-CO_2 is traditionally taken as 0.002058. It should be noted that there is a fractionation during the preparation of CO_2 from calcite ($CaCO_3$). Thus, when considering $\delta^{18}O$ values expressed with respect to PDB, it should always be ascertained whether it is the $^{18}O:^{16}O$ ratio of the $CaCO_3$ or that of the CO_2 released from the $CaCO_3$ that is being used as the standard.

To convert from SMOW to PDB ($CaCO_3$):

$$\delta^{18}O\left(\text{SMOW}\right) = 1.03086\left[\delta^{18}O\left(\text{PDB}\right)\right] + 30.86$$

and to convert from SMOW to PDB-CO_2 scale:

$$\delta^{18}O\left(\text{SMOW}\right) = 1.04143\left[\delta^{18}O\left(\text{PDB-CO}_2\right)\right] + 41.43$$

There are, however, other conversion equations that differ slightly from those given above. [J.Ly.]

REFERENCE

Craig, H. (1957) Isotopic standards for carbon and oxygen and correction factors for mass spectrometric analysis of CO_2. *Geochemica Cosmochemica Acta* **12**, 133–149.

deme A local, randomly mating POPULATION, partially isolated from other such local populations; a basic unit of genetic population structure. *See also* LINEAGE; QUANTITATIVE GENETICS. [T.J.K.]

demersal Living on or near the bottom of a SEA or LAKE. *See also* BENTHIC HABITAT CLASSIFICATION. [P.O.]

demographic stochasticity The chance variation in the number of integer births and deaths in a population with time-invariant vital rates. Although individuals do not give birth to fractions of offspring (or die as fractions of individuals), it is often mathematically convenient to treat them as if they did using DIFFERENTIAL EQUATIONS to model populations. Such deterministic models describe continuous change in the population over time so that, in effect, each member of the population gives birth to a fraction of an individual (and dies by a fraction) at each infinitesimally small time interval. When only integer births and deaths can occur, as in the real world, stochastic variation arises because the production of 1, 2, . . . , *n* offspring in any given time interval is probabilistic. Models that describe population growth in this probabilistic way are termed 'stochastic'. For modelling populations with large numbers of individuals there is little difference in the behaviour of the two types of model and deterministic models are usually preferred owing to their simpler mathematics. However, for small populations STOCHASTIC MODELS may often be more appropriate. An important distinction is between demographic and environmental stochasticity, which arises through time-dependent variation in vital rates caused by environmental FLUCTUATIONS. In contrast to demographic stochasticity, population variation generated by environmental fluctuations is independent of the absolute POPULATION SIZE. *See also* POPULATION DYNAMICS. [S.J.H.]

REFERENCE

May, R.M. (1973) *Stability and Complexity in Model Ecosystems*. Princeton University Press, Princeton, N.J.

demography The processes of birth, death, IMMIGRATION and EMIGRATION that determine the size, FLUCTUATIONS and AGE STRUCTURE of populations. Also the study of these processes and their effects. *See also* LIFE TABLE; POPULATION DYNAMICS. [A.J.D.]

Dempster model In the single-locus multiple-niche POLYMORPHISM model of E.R. Dempster (1955), a total of *n* NICHES is present for larval development. The PROBABILITY DISTRIBUTION of zygotes over the niches is the same in each generation. Genotypic FITNESS depends upon the niche: $w_{11,k}$, $w_{12,k}$ and $w_{22,k}$ for the GENOTYPES A_1A_1, A_1A_2 and A_2A_2, respectively. Genotypic fitness within each niche determines the probability to reach adulthood from zygote. The contribution of each niche to the population random mating pool depends upon its mean fitness. Genotypic fitness of any genotype A_iA_j over the total ENVIRONMENT is given by the weighted ARITHMETIC MEAN genotype fitness over the n niches:

$$\overline{w_{ij}} = \sum_{k=1}^{k=n} f_k \, w_{ij,k}$$

where f_k is the probability that a zygote finds itself in niche k (k= 1, . . . , n). Allele frequency change between generations is computed from the weighted average genotypic fitnesses. Mean fitness per environment is given by:

$$\overline{w_k} = \sum\nolimits_{ij} p_i p_j w_{ij,k}$$

while overall mean fitness is given by the arithmetic mean of the mean fitnesses in each niche:

$$\overline{\overline{w}} = \sum_{k=1}^{k=n} f_k \overline{w_k} = \sum\nolimits_{ij} p_i p_j \overline{w_{ij}} = \sum_{k=1}^{k=n} \sum\nolimits_{ij} f_k p_i p_j w_{ij,k}$$

The necessary and sufficient conditions for stable polymorphism are:

$$\sum_k f_k w_{11,k} < \sum_k f_k w_{12,k} > \sum_k f_k w_{22,k}$$

This multiple-niche polymorphism model became known as the 'hard' selection model, referring to the 'hard selection' of Wallace (1970, 1975). This suggests that the model is restricted to a particular type of selection within each niche. However, the model is valid for any type of selection within the niches, including FREQUENCY- and DENSITY-DEPENDENT SELECTION. Using the arithmetic mean genotypic fitness in a model of spatial variation is not related to the type of selection within any niche, but follows from modelling the LIFE CYCLE starting with zygotes. *See also* HARD SELECTION. [G.D.J.]

REFERENCES

Dempster, E.R. (1955) Maintenance of genetic heterogeneity. *Cold Spring Harbor Symposia on Quantitative Biology* **20**, 25–32.

Wallace, B. (1970) *Genetic Load: its Biological and Conceptual Aspects*. Prentice-Hall, Englewood Cliffs, N.J.

Wallace, B. (1975) Hard and soft selection revisited. *Evolution* **29**, 465–473.

dendrochronology The use of data on the widths of annual tree rings for the purposes of dating past events. Variations in ring width are assumed to be caused mainly by variations in CLIMATE (wide in years that are good for tree GROWTH and narrow in poor years). Other ecological processes, however, can affect annual ring width and, in practice, these may be difficult to factor out from climate effects (periodic defoliation by insect herbivores, fires, production of unusually large seed crops by the tree, etc.). Characteristic runs of good and bad years for tree growth can be used to overlap timbers of different ages and hence to extend the dendrochronological record back through several generations of trees. [M.J.C.]

dendrogram A tree-like diagram designed to show postulated phylogenetic relationships between taxa. *See also* CLADISTICS; PHYLOGENETIC TREE. [P.O.]

denitrification The microbial conversion of NITRATE (NO_3^-) ions into NITROGEN gas occurs mainly in anaerobic soils, especially at high pH, freshwater and seawater. In conditions where DISSOLVED OXYGEN is unavailable, certain bacteria (e.g. some species of *Bacillus* and *Pseudomonas*) are able to use nitrate ions in RESPIRATION during the DECOMPOSITION of organic materials within the SOIL. In the process, nitrate is reduced to nitrogen gas or nitrous oxide. Approximately 180 million tonnes of atmospheric nitrogen are generated globally per year in this way, which is roughly equivalent to the total biological fixation of nitrogen. *See also* NITROGEN CYCLE. [P.D.M.]

density The number of individuals per unit area (terrestrial species) or per unit volume (aquatic or aerial species). Density is often used to quantify the quality or attractiveness of a HABITAT or a patch of habitat, for example prey density is a measure of food availability, while tree density might help to predict the attractiveness of an area for a tree-dwelling species. The factors that determine density have been a source of controversy. *See also* ABUNDANCE; DENSITY DEPENDENCE; DENSITY INDEPENDENCE; POPULATION DENSITY; POPULATION REGULATION; SAMPLING METHODOLOGY/DEVICES. [R.H.S.]

density dependence Density dependence is where one or more demographic parameters (birth, death, IMMIGRATION or EMIGRATION rates) is a function of POPULATION DENSITY, i.e. a change in population density allows prediction of a change in a demographic parameter. A.J. Nicholson argued strongly in the 1950s that density-dependent interactions were the main determinants of POPULATION SIZE, generally operating as stabilizing or regulating mechanisms. Direct or negative density dependence acts to reduce the POPULATION GROWTH RATE as population density increases, for example animals may produce less young or plants less seeds at high population density because of increased COMPETITION for food or space. Inverse or positive density dependence is where population growth rate increases as population density increases and is sometimes found over a range of low population densities, perhaps because it is hard to find mates (e.g. overexploited whale species) or because efficient use of resources requires sufficient organisms to be present (e.g. cooperative feeding in blowfly larvae). *See also* DELAYED DENSITY DEPENDENCE; EXACT COMPENSATION; OVERCOMPENSATION; POPULATION DYNAMICS; POPULATION REGULATION; UNDERCOMPENSATION. [R.H.S.]

density-dependent selection Density-dependent selection implies that the FITNESS differences underlying selection in the population depend upon the total number of individuals of a specified AGE CLASS or life stage. *See also* NATURAL SELECTION. [G.D.J.]

density independence Density independence is the absence of a relationship between any demographic parameter and POPULATION DENSITY. Andrewartha and Birch (1954) argued strongly in favour of density-independent factors determining POPULATION SIZE and against the importance of DENSITY DEPENDENCE. Density-independent factors were earlier termed 'catastrophic agents' by Howard and Fiske (1911). *See also* POPULATION DYNAMICS; POPULATION REGULATION. [R.H.S.]

REFERENCES

Andrewartha, H.G. & Birch, L.C. (1954) *The Distribution and Abundance of Animals*. University of Chicago Press, Chicago, Ill.

Howard, L.O. & Fiske, W.F. (1911) The importation into the United States of the parasites of the gipsy-moth and the brown-tail moth. *U.S. Department of Agriculture, Bureau of Entomology, Bulletin 91*.

Table D1 Approximate densities of some common materials and elements.

Material	Density (g cm^{-3})	Element	Density (g cm^{-3})
Acetone	0.8	Aluminium	2.70
Alcohol	0.8	Copper	8.93
Balsa wood	0.2	Gold	19.28
Beech wood	0.75	Iodine	4.95
Bone	1.9	Iron	7.87
Brass	8.4	Lead	11.34
Coal	1.4	Mercury	13.55
Cork	0.25	Silicon	2.33
Diamond	3.5	Sulphur	2.09
Ebony	1.2	Tin	7.28
Glass	2.5–4.0	Titanium	4.51
Ice	0.92	Zinc	7.13
Ivory	1.8		
Linseed oil	0.95		
Marble	2.7		
Mild steel	7.9		
Milk	1.03		
Oak	0.7		
Olive oil	0.9		
Paraffin oil	0.8		
Quartz	2.6		
Sand	2.6		
Teak	0.85		
Water	0.96–0.99		

density of materials Density is defined as mass per volume (kg m^{-3}). Densities of some common materials and a few of the ELEMENTS are given in Table D1. [J.G.]

REFERENCE

Weast, R.C. (1985) *CRC Handbook of Chemistry and Physics*. CRC Press, Boca Raton, Fla.

density of seawater The density of any substance is the mass per unit volume of that substance. The density of seawater is between 1 and 1.07 g ml^{-1} of fluid. Density may be expressed by a σ value, which is (density − 1) × 1000. The density of seawater is inversely related to TEMPERATURE, and positively related to SALINITY and, to a lesser extent, hydrostatic PRESSURE (which itself increases 1 atmosphere (101 kPa) for every 10-m increase in depth). In contrast to FRESHWATER, which is densest at 4°C, seawater continues to increase in density with decreasing temperature all the way to the freezing point (about −1.9°C). Density differences, arising from differences in salinity and temperature, drive the CIRCULATION patterns of the world's oceans. [V.F.]

density vagueness A concept pertaining to REGULATION of populations whereby density-dependent regulation is only manifest at very high (and possibly very low) densities, with populations fluctuating at random or in response to variation in the environment at intermediate densities. *See also* POPULATION DYNAMICS; POPULATION REGULATION. [S.J.H.]

deposit-feeder An aquatic animal that feeds on the surface layer of SEDIMENT, for example certain polychaete worms. *See also* FEEDING TYPES, CLASSIFICATION. [D.A.B.]

depth gradient The rate of change of a quantity with respect to depth, for example the rate of decrease of TEMPERATURE (or increase of PRESSURE) with depth in the OCEAN. Important depth gradients in the ocean are (i) the decrease in organic matter input to the sea floor, (ii) the decrease in microbial activity within the seabed, and (iii) the decrease in benthic BIOMASS with increasing depth and distance from shore. [V.F.]

depuration Metabolic process that results in elimination of XENOBIOTIC material from organisms. *See also* DETOXIFICATION. [P.C.]

derelict land Land so damaged by industrial activity or other development that it is incapable of beneficial use without treatment. [P.C.]

derived homology *See* APOMORPH.

derived units *See* SI UNITS.

desert Approximately 45% of the Earth's land surface is occupied by desert (which may be hot or cold), characterized by dry conditions and low BIOMASS and plant PRODUCTIVITY. Despite having a PRIMARY PRODUCTIVITY rate of only about 0.3 kg m^{-2} year^{-1}, the deserts support about 13% of the world's human population.

Defining deserts in terms of PRECIPITATION is difficult because the TEMPERATURE conditions will determine EVAPORATION rates and hence WATER availability in the SOIL. The United Nations Environmental Programme (UNEP) has devised an aridity index, calculated by dividing the annual precipitation by annual potential evaporation. On this basis, 10% of the Earth's surface can be regarded as dry (aridity index 0.50–0.65), 18% is semi-arid (0.2–0.5), 12% is arid (0.05–0.20) and 8% is hyper-arid (<0.05).

Deserts are found mainly concentrated around the latitudes 30° north and south of the Equator, where the AIR MASSES that have been forced up by convection currents over the equatorial regions descend once more, having lost much of their water content in equatorial rains. A high-pressure, dry climatic zone results, which shifts seasonally. In Asia, particularly north of the Himalaya mountain chain, deserts are found on some elevated plateaux, particularly where monsoonal rains are blocked by the mountain ridges.

Deserts are characterized by sparse vegetation (see Plate 5, facing p. 398), biomass diminishing with aridity index. In arid zones, plant life may become virtually restricted to drainage channels (wadis) that fill during rare RAIN events and hence periodically have the most humid soils (termed 'contracted vegetation'). Both plant and animal life is faced with problems of high INSOLATION, high temperatures in daytime and low water availability. Survival may entail living below ground and physiological techniques for water conservation (e.g. C_4 and CAM in plants, low excretion rates in animals). DROUGHT-resistant lichens on the soil surface and ENDOLITHIC algae living within the structure of porous stones and rocks are important in terms of both productivity and the stabilization of some desert soils.

Although the BIOME has a generally low BIODIVERSITY and is structurally simple, rendering it liable to DISTURBANCE, it is also resilient. Desert ecosystems have developed under severe conditions where soil instability and climatic irregularity are frequently experienced. Recovery from perturbation is therefore often rapid and complete. [P.D.M.]

desertification The general reduction in the BIOMASS and PRODUCTIVITY of the world's drylands that has become increasingly apparent over the past few decades has been termed 'desertification'. Many areas that were formerly capable of sustained AGRICULTURE and pastoralism have, as a consequence, been lost and widespread famines have periodically resulted. The role of human beings in the process of desertification has been widely stressed and there are clear situations where human impact via intensive cultivation, SALINIZATION through IRRIGATION, and OVERGRAZING have contributed to a general reduction in the productivity of a region. Goats are often blamed for the destruction of vegetation in semi-arid regions, although they are often simply the final stage in a process of degradation and are still able to survive when other animals would fail.

The United Nations Conference on Desertification (UNCOD) in 1977 defined desertification as 'the intensification or extension of desert conditions . . . leading to reduced biological productivity, with consequent reduction in plant biomass, in the land's carrying capacity for livestock, in crop yields and human welfare'. The process concerns the international community, therefore, because of its direct impact on our species, and much effort is now expended on methods for the control of DESERT spread. Before it is possible to combat desertification, however, one must be assured that it is human induced and not simply a product of fluctuating CLIMATE. In fact, both processes appear to be operative.

In the Sahel zone of Africa, south of the Sahara Desert, precipitation diminished substantially during the late 1960s and has continued low through the subsequent decades. This has induced such calamities as crop failure, human migrations, concentrations of starving livestock and consequent overgrazing, social and political tensions and famine. If current predictions of CLIMATE CHANGE prove even reasonably accurate, then many of the dry areas of the world will be brought under yet further STRESS and continued desertification appears inevitable (Fig. D2).

The innate RESILIENCE of the desert ECOSYSTEM may well provide it with the capacity to recover from the resulting damage. Meanwhile, the carrying capacity of the system is reduced and such recovery can only occur if the demands exerted upon it are diminished. The social, political and economic problems involved, however, are probably more difficult to solve than the ecological ones. [P.D.M.]

desertion 'Desertion' is usually used to describe abandonment of mating and parent–offspring associations, though the distinction between the two is not a clear one since desertion of a mate during offspring development amounts to desertion of the offspring as well. Despite its negative connotations, desertion may reflect adaptive life-history decisions on the part of the deserter (*see also* INFANTICIDE). In a mating context, abandonment of a pair bond, at least in long-term pairs, may be triggered by a poor reproductive season, for

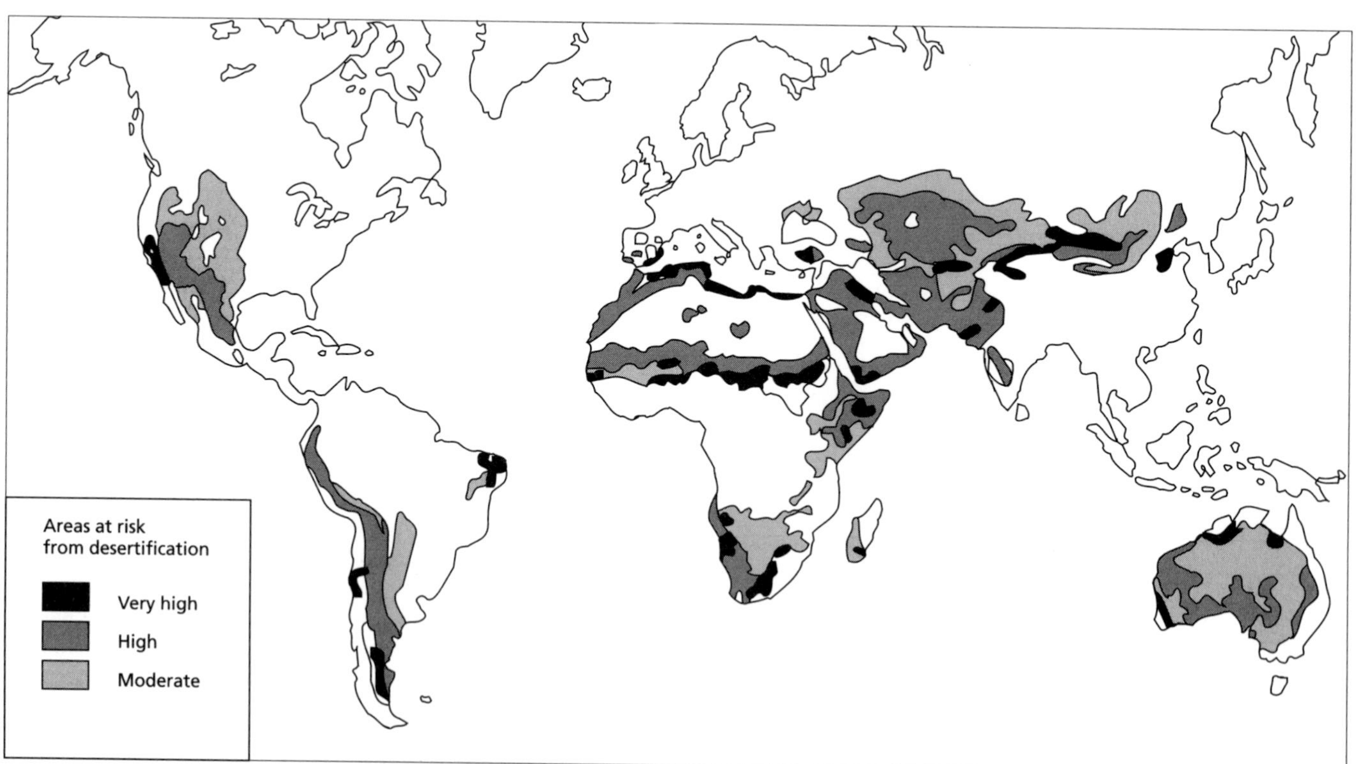

Fig. D2 The areas of the Earth most at risk from desertification, as determined by the 1977 United Nations Conference on Desertification (UNCOD). (After Moore, P.D., Chaloner, B. & Stott, P. (1996) *Global Environmental Change*. Blackwell Science, Oxford.)

instance in kittiwakes (*Rissa tridactyla*). Within breeding periods, however, desertion is predictable from the different degree of PARENTAL INVESTMENT on the part of the two sexes. As a broad generalization (many SELECTION PRESSURES cause departures from this general picture), the lower investment by males at the GAMETE stage (*see* SEXUAL SELECTION) means that they have less to lose by desertion if the offspring dies, for it costs them less than it does females in the future to get back to the same stage of offspring development. In addition, their cheap gametes and relatively low involvement time per FERTILIZATION mean that they can spend the time gained from desertion copulating with other females. While a male risks losing his current offpring if he deserts, there is still a chance that the female can rear them on her own. Indeed, the female is left in a 'cruel bind' since if she deserted the offspring too they would be sure to die. Selection has therefore favoured a number of strategies among females to offset the impact of desertion (e.g. choosing healthy males or males with well-provisioned territories to sire their offspring). Of course, the 'cruel bind' can cut both ways. If there comes a point where either parent could complete the rearing of offspring alone, then females may be tempted to desert first. One reason why males are often the caring parent in fish, for example, is that fish are usually external fertilizers and it is the sex that sheds its gametes last (often the male) that risks being left 'holding the baby'. *See also* BEHAVIOURAL ECOLOGY; LIFE-HISTORY EVOLUTION. [C.J.B.]

desiccation The drying of living tissues, generally caused by EXPOSURE to air, and exacerbated if the air is dry and hot. Living organisms are composed of approximately 90% WATER and body function can be impaired if body fluids become too concentrated as a result of water loss. Mechanisms for losing heat to compensate for temperature stress may involve water loss. Terrestrial and intertidal marine organisms show a wide variety of adaptations to avoid desiccation. The presence of a thick shell, the secretion of protective mucus around sensitive body tissues, the production of concentrated excretory products, burrowing and other behaviours that increase contact with cool or moist surfaces are all examples of adaptations that decrease desiccation. [V.F.]

detection limit *See* ZERO EMISSIONS.

detergents Cleaning agents that are composed of: SURFACTANTS, wetting agents that provide a link between 'dirt' and WATER molecules; builders, sequestering agents that sequester 'hard' water ions so water becomes alkaline, which is necessary for removal of dirt; and a variety of brighteners, enzymes, perfumes, etc. Detergents are used as 'spreaders' in crop-spraying formulations, so that the droplets of spray do not run off the leaves. They tend to persist in the environment and have been linked with EUTROPHICATION of LAKES and RIVERS and implicated in endocrine disruption (*see* ENDOCRINE DISRUPTER). [P.C. & J.G.]

REFERENCE

Hennes-Morgan, E.C. & DeOude, N.T. (1994) Detergents. In: *Handbook of Ecotoxicology*, Vol. 2 (ed. P. Calow), pp. 130–154. Blackwell Scientific Publications, Oxford.

determinate growth Increase in body SIZE that ceases with maturation, such that there is no correlation between body size and age in adult individuals. Found, for example, in some molluscs, holometabolic and hemimetabolic insects, mammals and most ANNUAL plants. Whether it is adaptive or not to cease GROWTH at maturity depends on the relationship of FECUNDITY, survival and resource acquisition with the body size of an organism. *Cf.* INDETERMINATE GROWTH. *See also* VON BERTALANFFY EQUATION. [D.E.]

deterministic models, communities A deterministic model is a mathematical model of a process in which changes over time are determined by the initial state of the system and by the supplied values of driving variables, and do not depend on the outcome of chance events. The converse is a stochastic model, in which some changes are attributed to chance. Even in a deterministic model chance may play a part, because there may be a driving variable such as the weather, which behaves in an essentially stochastic way. Only if chance is represented explicitly does the model become stochastic.

Deterministic models use DIFFERENCE EQUATIONS to represent changes in discrete time or DIFFERENTIAL EQUATIONS to describe continuous change. Examples are models of large populations, which may be represented by deterministic transitions between age states, for example Leslie matrices (Caswell 1989) and the Volterra equations for COMPETITION between *n* species (Hutchinson 1978).

Stochastic models can often be approximated by deterministic models in which random variables are replaced by their MEAN values. However, deterministic approximations may be impossible where processes depending on the configuration of individuals are important. A notable class of stochastic individual-based community models is the gap models of SUCCESSION typified by JABOWA (Botkin *et al.* 1972) and its descendants (Shugart 1984). These cannot readily be reduced to analogous deterministic models.

Deterministic models may be very sensitive to initial conditions and may show chaotic behaviour that appears to be random (May 1981). Chaotic behaviour is the norm in models of fluid dynamics and weather, but is rarely found in realistic ecological models. *See also* DETERMINISTIC MODELS, POPULATIONS. [M.O.H.]

REFERENCES

Botkin, D.B., Janak, J.F. & Wallis, J.R. (1972) Some ecological consequences of a computer model of forest growth. *Journal of Ecology* **60**, 849–872.

Caswell, H. (1989) *Matrix Population Models*. Sinauer, Sunderland, Mass.

Hutchinson, G.E. (1978) *An Introduction to Population Ecology*. Yale University Press, New Haven, Conn.

May, R.M. (ed.) (1981) *Theoretical Ecology*, 2nd edn. Blackwell Scientific Publications, Oxford.

Shugart, H.H. (1984) *A Theory of Forest Dynamics*. Springer-Verlag, New York.

deterministic models, populations Models in which a given input to the model always produces the same output. Models that predict the outcome of a process using PARAMETERS not subject to stochastic (random) variation. Deterministic models are generally mathematically more convenient than STOCHASTIC MODELS and their outcome is certain. Deterministic population models assume infinite POPULATION SIZE and ignore random FLUCTUATIONS in the environment with time. *See also* DETERMINISTIC MODELS, COMMUNITIES; POPULATION DYNAMICS; POPULATION REGULATION. [V.F.]

detoxification Metabolic reaction(s) that reduce(s) the possibility of XENOBIOTICS causing HARM within organisms. Often involves sequestering to reduce TOXICITY or increasing WATER solubility to facilitate excretion and/or reaction with other molecules leading to reduced toxicity. In medical TOXICOLOGY this term is used more strictly to describe treatment whereby toxicants are removed from intoxicated patients. *See also* DEPURATION. [P.C.]

detritivore *See* FEEDING TYPES, CLASSIFICATION.

detritophage *See* FEEDING TYPES, CLASSIFICATION.

detritus Non-living organic matter. Usually refers to particulate matter and, because it persists longer, to that of plant rather than animal origin, for example leaf litter. However, the term can also be applied to animal materials such as faeces and can sometimes be used to describe soluble organic materials. [P.C.]

detritus food chain DETRITUS is the primary source of energy and matter for some ecosystems, for example soils and sediments. Hence, detritivores are primary consumers in these systems. Often these detritivores rely on microbial communities, outside or within themselves, to render NUTRIENTS available from the hard-to-digest detritus. The detritivores may themselves be eaten by carnivores, that in turn could be eaten by secondary carnivores. So a detritus FOOD CHAIN might involve:

$$\text{detritus} \rightarrow [\text{microbe}] \rightarrow \text{detritivore} \rightarrow \text{primary carnivore} \rightarrow \text{secondary carnivore.}$$

See also FEEDING TYPES, CLASSIFICATION. [P.C.]

deuterium A stable ISOTOPE of HYDROGEN, ^2H or D (with a nucleus consisting of one proton and one NEUTRON), which has a relative ATOMIC MASS of 2.014 and is found in natural hydrogen compounds, especially WATER, to the extent of 0.015%. The mass difference between ^1H and D can lead to significant and useful isotopic fractionation in nature. For example, the vertical profiles of the D : H ratio in ICE at the poles contain a record of past TEMPERATURE variations on Earth over the past few hundred thousand years, largely because of the temperature-dependent differences in the EVAPORATION (and subsequent condensation) of H_2O and HDO from the oceans before eventual PRECIPITATION as snow at the poles. Deuterium oxide (D_2O, heavy water) is used as a moderator in nuclear reactors. [J.F.]

developmental constraints *See* CONSTRAINTS.

Developmental state of newborn

Neonates of mammals and birds exhibit a varying degree of developmental maturity reaching from relatively independent and functionally mature young (PRECOCIAL) to embryo-like newborns that are highly dependent on PARENTAL CARE. In birds, 'super-precocial' mound-builders (Megapodiidae) occupy the precocial extreme of the spectrum. The young are entirely independent of their parents and, in some species, the chicks can fly from the first day of postnatal life. Megapode super-precociaty continuously grades into the precociaty typical of galliform species, whose hatchlings seek their own food and depend on their parents only for protection and brooding. Nice (1962) distinguished megapodes as precocial grade 1 and galliforms as precocial 2. Ostriches, ducks, geese, jacanas and many shorebirds exhibit a similar independence and usually are placed in the precocial 2 category of Nice (1962). In some of the Galliformes, for example most turkeys (Meleagrididae) and grouse (Tetraonidae), parents show food to their chicks, and these taxa are placed in a lower category of precociality (precocial 3 of Nice 1962). Chicks of cranes, rails, grebes, loons, bustards and buttonquails display a precocial development of mobility and sensory organs, but their parents offer them food for some time (often 1–2 weeks) after hatching (precocial 4 of Nice 1962; see Fig. D3 for classification by different authors). The young of many auks (Alcidae), gulls and terns (Laridae) are described as semi-precocial because of their relatively less developed locomotor activity, stronger nest attendance and complete dependence on the parents for food. Semi-precocial chicks do undertake excursions around the nest within a few days after hatching to hide in vegetation, but they return to the nest to be fed by their parents. Species whose chicks remain in the nest for much or all of their development are referred to as ALTRICIAL, NIDICOLOUS or '*Nesthocker*'. Among them, raptors, storks, herons and

Nice 1962	Most others	1	2	3	4	5	6	7	8	9	a	b	
Precocial 1	Precocial	■	■	■		■					■	■	Mound-builders
Precocial 2		■	■	■	■	■						■	Pheasants, ducks, many shorebirds
Precocial 3		■	■	■	■	■							Rhea, guinea fowl, turkeys, grouse
Precocial 4	Semi-precocial	■	■	■	■		■						Buttonquail, rails, cranes
Semi-precocial		■	■	■			■						Many auks, gulls, terns
Semi-altricial 1	Semi-altricial	■	■				■	■					Raptors, storks and ibises
Semi-altricial 2	Altricial	■	■				■	■	■				Pigeons and doves, penguins
Altricial			■				■	■	■	■			Songbirds, parrots

1 Downy hatchling plumage
2 Motor activity
3 Locomotor activity
4 Follow parents
5 Search food and feed alone
6 Young fed by parents
7 Stay in nest
8 Eyes closed at hatching
9 Without external feathers at hatching
a No parent–chick interaction
b Contour feathers at hatching

Fig. D3 The altricial–precocial categories of avian hatchlings.

many other families are designated as semi-altricial 1 because neonates are densely covered by down and have open eyes; semi-altricial chicks are relatively active soon after hatching. Fully altricial hatchlings hatch with closed eyes and exhibit little motor activity other than begging. Some altricial chicks hatch with natal down (semi-altricial 2 after Nice 1962), others hatch without externally visible feathers. Presence (ptilopaedic, from the Greek *ptilon*, feather; *pais*, child) or absence (gymnopaedic, from the Greek *gumnos*, naked; *pais*, child) of feathers has been used to further distinguish between semi-altricial 2 and altricial in Nice's (1962) scheme (see Fig. D3). There are few apparent exceptions to the order of characters in traditional classifications of developmental state of the neonate. The South American hoatzin (*Opisthocomus hoazin*) is difficult to place because it hatches sparsely covered with down, but its eyes are wide open 24 h after hatching. When 2 weeks old and before they are feathered, chicks start to clamber about on branches of their nesting trees. Under normal conditions the chicks stay close to the nest, which is always built in trees over open water of rivers. However, when predators approach they jump out of the nest into the water, hiding among emergent vegetation. Later, the chicks climb back into the nest. Storm-petrels (Hydrobatidae) and many of the petrels (Procellariidae) provide another contrary example. These hatch after long incubation periods densely covered with down, with their eyes soon open, and stay in the nest/burrow for long periods unattended by their parents. Their thermogenic capacity and aspects of their internal anatomy resemble those of many precocial species. Among altricial birds, we find species that are similarly difficult to place in the traditional altricial–precocial classification. For example, many pigeons and doves (Columbidae), although highly altricial when hatching, leave the nest when only half grown (*Geopelia cuneata, Oena capensis*) and as little as one-quarter grown in the case of the purple-crowned fruit pigeon (*Ptilinopus superbus*).

The developmental pattern of birds appears to be evolutionarily conservative and is generally uniform within large taxonomic groups (i.e. orders). The altricial–precocial spectrum is unevenly occupied, with most bird hatchlings residing on the altricial side. However, a continuous gradient of functional maturity of bird hatchlings reaches from super-precocial to semi-altricial. Altricial neonates are clearly separated from the others (Starck & Ricklefs 1997). The evolutionary origin of mode of development is not finally resolved but there is some evidence supporting the view that precocial mode of development is ancestral for birds, whereas altricial neonates were in the stemgroup of mammals. Rare findings of embryonic skeletons of fossil birds and their sister group, the dinosaurs, allowed for some palaeontological speculation about the ancestral mode of avian development. Especially, accumulations of juvenile dinosaur bones of different size and AGE CLASSES were interpreted as representing nest with altricial young of different age (Weishampel & Horner 1994), thus suggesting that altriciality might already have evolved in the stemgroup of birds.

As in birds, mammalian neonates exhibit a wide spectrum of development ranging from extremely altricial marsupials and monotremes to the relatively independent young of ungulates. However, compared to birds, the mammalian embryo develops under fundamentally different physiological and structural conditions that place it into a different framework of selection and internal constraint. The most striking differences in reproductive strategies of birds and mammals are as follows.

1 Birds are oviparous while mammals are viviparous (except for the oviparous monotremes).

2 Nutrient supply to the mammalian embryo through the placenta is continuous and almost unlimited, as compared to the energy-limited egg compartment in birds.

3 The intrauterine development of the mammalian embryo allows relative independence of movement of the mother during almost all phases of her reproductive status as compared to birds, which incubate their EGGS through the whole embryonic period.

4 Lactation allows for nourishment of neonates and growing young mammals relatively independently of the actual environmental conditions, i.e. food abundance, weather conditions, etc., which have important effects on avian postnatal growth.

5 The COSTS OF REPRODUCTION are almost exclusively borne by the female in mammals, whereas they are more equally partitioned among sexes in many birds.

The evolution of altricial and precocial neonates under such different reproductive strategies makes an interesting contrast between birds and mammals.

Many insectivores (Insectivora), rabbits (Lagomorpha), many rodents (Rodentia, Muridae) and carnivores (Carnivora) give birth to small neonates with eyes and ears closed, without hair, generally poorly developed and dependent on maternal care for a long lactation period. Newborn marsupials and monotremes are extremes, resembling early embryos, even compared with other altricial mammals. Precocial neonates are found among the ungulates (Cetacea, Sirena, Proboscidea, Hyracoidea, Perissodactyla, Artiodactyla), the Pinnipedia and several rodent taxa (Rodentia). Precocial mammals, especially ungulates, may run swiftly and for long distances shortly after birth and some begin to feed on solid food within a few days. The neonates of some taxa are intermediate between altricial and precocial development, for example primates are born with well-developed fur and open sensory organs but depend to a much higher degree on parental care than precocial neonates. Different intermediate stages may be found within orders. For example, within the rodents, mice and rats (Muridae) produce highly altricial young, whereas spiny mice (*Acomys minous*) and guinea-pigs (*Cavia* spp.) have highly precocial young. Bats (Chiroptera) show a similar diversity of neonates ranging from altricial young (Megachiroptera), which are born naked, to more precocial taxa (Phyllostomatoidea), which have eyes open when born and are covered by fur. Derrickson (1992) developed a classification scheme for mammalian neonates employing four developmental categories representing neonatal independence in four areas: thermoregulation, sensory organs, locomotion and nutrition. Mammalian orders clearly exhibit greater flexibility of developmental mode than do avian orders, with Carnivora and Rodentia exhibiting the full range from altricial to precocial. *See also* LIFE-HISTORY EVOLUTION; NIDIFUGOUS. [J.M.S.]

REFERENCES

Derrickson, E.M. (1992) Comparative reproductive strategies of altricial and precocial mammals. *Functional Ecology* **6**, 57–65.

Nice, M.M. (1962) Development of behavior in precocial birds. *Transactions of the Linnean Society of New York* **8**, 1–211.

Ricklefs, R.E. (1983) Avian postnatal development. In: *Avian Biology* (eds D.S. Farner, J.R. King & K.C. Parkes), Vol. VII. Academic Press, New York.

Starck, J.M. (1993) Evolution of avian ontogenies. *Current Ornithology* **10**, 275–366.

Starck, J.M. & Ricklefs, R.E. (1997) Patterns of development. The altricial–precocial spectrum. In: *Avian Growth and Development. Evolution within the Altricial–Precocial Spectrum* (eds J.M. Starck & R.E. Ricklefs), pp. 3–30. Oxford University Press, New York.

Weishampel, D.B. & Horner, J.R. (1994) Life history syndromes, heterochrony, and the evolution of Dinosauria. In: *Dinosaur Eggs and Babies* (eds K. Carpenter, K.F. Hirsch & J.R. Horner), pp. 229–243. Cambridge University Press, Cambridge.

---

deviance The measure of residual variation in statistical analyses that use maximum likelihood rather than least squares methods. Deviance is analogous to the residual SUM OF SQUARES (SSE) in least squares regression or ANALYSIS OF VARIANCE. In general, the deviance is twice the log likelihood, and it differs from one ERROR STRUCTURE to another. For Poisson errors the deviance is $2\Sigma y \ln(y/fv)$ where y is an observed count and fv is the fitted value from the current model (the NULL MODEL would have $fv=\bar{y}$, the average value of y, and this would have maximum deviance; the full model, with a parameter for every data point, would have deviance$=0$ because $y=fv$ for all cases). For binomial errors the deviance is:

$$2\sum\left[x\ln\left(\frac{x}{fv}\right)+(n-x)\ln\left(\frac{n-x}{n-fv}\right)\right]$$

where there are x successes and $n-x$ failures out of n binomial trials ($p=x/n$, MEAN $=np$, variance$=npq$), and where fv is the fitted value for the number of successes from the current model. For the deviance of other error structures, see Crawley (1993). *See also* GENERALIZED LINEAR MODELS; LINK FUNCTION; STATISTICS. [M.J.C.]

REFERENCE

Crawley, M.J. (1993) *GLIM for Ecologists*. Blackwell Scientific Publications, Oxford.

dew point The TEMPERATURE of a chilled solid surface at which dew begins to condense from the adjacent saturated WATER vapour. Since the SATURATION vapour pressure (or density) is a known function of temperature only (as given in thermodynamic tables), dew point is a measure of the absolute HUMIDITY content of the immediately adjacent and more remote air. If the dew point is below 0°C it is called a frost point, as vapour condenses directly to the ICE phase to produce hoar frost on the cold surface. If a substantial volume of typical air is chilled to its dew point, further chilling produces rapid condensation on numerous AEROSOL droplets and particles, to produce a visible cloud or fog of droplets. If the dew point is below 0°C the result is a cloud of supercooled droplets rather than ice crystals. [J.F.R.M.]

diadromous Migrating between saltwater and FRESHWATER. *See also* ANADROMOUS; CATADROMOUS. [P.O.]

Diapause

Diapause is a condition of arrested GROWTH or reproductive development common in many organisms, particularly insects, that live in seasonally varying environments. Diapausing stages may also show reduced metabolism and enhanced resistance to adverse climatic factors, such as cold, heat or DROUGHT. In many species, a 'resting stage' may be induced by the same conditions that it is adapted to survive; for instance, insects enter CHILL COMA at low TEMPERATURE and become active again at higher temperatures. This 'quiescence' is therefore a temporary or intermittent interruption of growth or activity that is 'directly' controlled by environmental factors. In contrast, diapause is an actively induced state that usually involves a reduction or cessation of neuroendocrine functions; diapause is induced by environmental factors (token stimuli) that signal the approach of unfavourable conditions, but which are themselves not necessarily harmful.

Some species show an obligatory diapause that occurs in a particular stage in the LIFE CYCLE in each individual of every generation, i.e. the individual is predestined to enter a diapause irrespective of the environmental conditions experienced; most obligatory diapausing species are univoltine. FACULTATIVE diapause can be induced or averted by appropriate changes in environmental conditions and is usually found in multivoltine species that have two or more generations a year. Diapause manifests in the EGG, LARVA or pupa of insects as a halt in development, and in the adults as a failure to enlarge the reproductive organs, particularly the ovaries of females. In facultative diapause, the most common and reliable inducing factor is PHOTOPERIOD, followed by temperature, with some involvement in some species of moisture and food quality. In temperate species, a winter diapause is usually triggered by decreasing DAYLENGTH (increasing night length) in late summer–early autumn, coupled with decreasing temperature. Diapause is induced by the experience of the 'sensitive stage(s)' (e.g. larvae), though the diapause usually occurs in a later stage of development (e.g. pupa or adult).

At the hormonal level, diapause is induced or inhibited by the interaction between the brain, other neuroendocrine centres (suboesophageal glands, prothoracic glands) and specific hormones that vary between egg, larvae/pupa and adult diapauses.

The diapause stage is often characterized by a range of behavioural and physiological ADAPTATIONS including accumulation of fat reserves prior to diapause, economical utilization of RESOURCES while in diapause, reduced sensitivity to tactile stimuli, decline in metabolism (OXYGEN consumption) and qualitative adjustments in respiratory pathways.

In the early literature it was widely assumed that most species had both induction and termination stimuli, leading to the view that diapause was important as a 'synchronizing' mechanism, i.e. the winter diapause ended in spring when environmental conditions became favourable for growth and reproduction. Now it is known that in many species this 'winter diapause' is effectively completed by midwinter and it is the post-diapause development that is then prevented until environmental conditions improve. Diapause is therefore primarily a series of physiological and biochemical adjustments that either occur inherently or are triggered by trends of change in inducing stimuli that enable organisms to survive through periods of environmental STRESS when conditions are unsuitable for continuous growth and reproduction. *Cf.* DORMANCY. [J.S.B.]

---

diazo compounds Aliphatic diazo compounds are characterized by the presence of the group >C=N+=N−, for example diazomethane CH_2N_2, which is a useful methylating agent for hydroxyl groups. Aromatic diazo compounds include those in the form Ar–N=N–X where X is not a CARBON atom, the azo compounds Ar–N=N–Ar, and diazonium salts Ar—N+≡NX−. Diazonium salts, formed when a primary aromatic AMINE is treated with nitrous acid at low temperature, are important synthetic reagents in the preparation of other aromatic compounds, dyes and drugs, either involving displacement or retention of the diazo group. [M.C.G.]

dichlorodiphenyl-trichloroethane *See* DDT.

dichogamy *See* POLLINATION.

diel Referring to events or actions that occur with a 24-hour periodicity, for example migrations of planktonic animals, changes in oceanic photosynthetic potential and changes in near-shore PLANKTON communities in response to the tidal cycle. Often used synonymously with the term 'DIURNAL' (i.e. diurnal TIDES have one high tide and one low tide each day). [V.F.]

dieldrin A non-systemic and persistent ORGANOCHLORINE insecticide, with high contact and stomach activity in most insects. First described in 1949, dieldrin was widely used until residue problems were confirmed in the 1970s. Dieldrin tends to partition into animal fats, concentrating in the FOOD CHAIN and resulting in poisoning levels in predators, which has hastened its withdrawal from most uses. It is still used to control locusts and tropical-disease vectors, and in timber preservation and similar uses in some countries. RESISTANCE to dieldrin has also appeared in some insects. *See also* CHEMICAL CONTROL; INSECTICIDES; PESTICIDE. [K.D.]

diet switching Several aspects of potential prey influence their inclusion in a predator's diet (*see* CONSTRAINTS ON FORAGING; DIET WIDTH). One obvious factor is their abundance. The change in the number of prey attacked by a PREDATOR within a given period as the density of the prey increases is known as the FUNCTIONAL RESPONSE. Functional response curves can take various forms depending on the constraints on a predator's intake rate. Where the constraints are SATIATION and HANDLING TIME (*see* DIET WIDTH), functional responses describe either a sharply truncated (type 1 response) or smoothly decelerating (type 2 response) curve with increasing prey density. In some cases, however, the contribution of a given prey type to the predator's diet also depends on the relative abundance of other types of prey. Many predators, particularly vertebrates, can learn to concentrate their SEARCHING effort on the most abundant prey. The result can be a sigmoidal (type 3) functional response, in which the proportion of a prey type in the diet is lower than expected from random sampling when its relative abundance is low but greater than expected when its abundance is high. The flip from underrepresentation to overrepresentation in the diet has been termed 'SWITCHING'. One explanation of switching is based on the concept of search images. According to this hypothesis prey tend to be overlooked by the predator when they are scarce. As their numbers increase, however, they are encountered more often and the predator learns 'to see' them. Once the predator has a search image, the prey are detected more easily and the predator's attack rate increases.

Although it is possible to account for type 3 functional responses and switching in terms of search-image formation, alternative explanations can account for the relationship in various contexts, such as the idea that it reflects a shift in searching subtly different habitat types corresponding to different types of prey, or that the apparent formation of a search image is due to changes in searching intensity or economic thresholds of encounter rate for dietary specialization. *See also* FEEDING TYPES, CLASSIFICATION; OPTIMAL FORAGING THEORY; TROPHIC CLASSIFICATION. [C.J.B.]

diet width Food for most organisms does not come in neat, standard packages. Items vary in their energy value, nutrient content, ease of capture, digestibility and numerous other ways that affect the costs and benefits of SEARCHING for and consuming them. The range of items included in an organism's diet can be modelled using optimality theory (*see* BEHAVIOURAL ECOLOGY). This demands at least some knowledge of the organism's nutritional requirements (currencies to be (usually) maximized), the time, energy and other costs associated with alternative strategies for satisfying its requirements, and the CONSTRAINTS imposed on these alternatives by the environment and the organism's design and PHYLOGENY (*see* CONSTRAINTS ON FORAGING).

The basic prediction of optimal diet models is that organisms should prefer the most profitable food items, where profitability is defined as the ratio of nutritional value (in the appropriate currency and net of the costs of ASSIMILATION) and handling costs (the costs of catching, subduing and ingesting items). However, the wisdom of taking only the most profitable items depends on their abundance in the environment. If they are common, the searching or waiting costs between successive profitable items are small so it is unlikely to pay to take any less profitable items that might be encountered. As items become scarcer, however, the returns for including less and less profitable items in the diet are likely to increase. On this basis, therefore, it becomes a matter of seeking an optimal breadth of diet rather than an optimal preference for a particular class of food item.

Field studies and carefully designed laboratory experiments have generally borne out the predictions of this kind of OPTIMAL DIET MODEL, though usually on the assumption that food energy is the currency being maximized and often supporting predictions qualitatively rather than quantitatively. In many cases the problem is also simplified by considering differences in profitability within a single type of food item (e.g. different-sized individuals of a particular prey species). Nevertheless, it is possible to use the optimality approach to predict broader mixtures of foodstuff in diets. For instance, linear programming techniques, making use of physical and physiological constraints to identify optimal combinations of food types, have had some success in predicting the diets of various herbivorous mammals. In moose (*Alces alces*), for example, the combination of terrestrial and aquatic plant material in the diet appears to be determined by the animal's daily energy requirement (met more easily from terrestrial plants), its daily requirement for sodium (met only from aquatic plants) and the limited capacity of its rumen (which prevents the moose from satisfying its total food demand from bulky aquatic plant material). However, the match between prediction and outcome in these cases appears to be very sensitive to the model's assumptions about digestive constraints. *See also* OPTIMAL FORAGING THEORY. [C.J.B.]

REFERENCE

Stephens, D.W. & Krebs, J.R. (1986) *Foraging Theory*. Princeton University Press, Princeton, N.J.

difference equations Difference equations are equations that involve finite differences. In ecology these are often used as population models for organisms having relatively DISCRETE GENERATIONS. In this application, the population at one time is determined by the population some fixed time previously. An example is the logistic map, which relates population at time $t+1$ (N_{t+1}) to population at time t (N_t):

$$N_{t+1} - N_t = rN_t\left(1 - N_t/K\right)$$

where r is the INTRINSIC RATE OF INCREASE of the population and K is its CARRYING CAPACITY. In common with many simple difference equation models, the logistic map can display very complex apparently random FLUCTUATIONS from one time-step to the next, a phenomenon known as CHAOS (May 1974). Difference equations are often used for STOCHASTIC MODELS; models in which the state depends only on the immediately preceding state are known as Markov chains. *See also* HASSELL MODEL AND HASSELL–COMINS MODEL; LOGISTIC MODEL; POPULATION DYNAMICS. [S.N.W.]

REFERENCE

May, R.M. (1974) Biological populations with nonoverlapping generations: stable points, stable cycles, and chaos. *Science* **186**, 645–647.

differential equations Differential equations are equations involving not only variables but also some of the derivatives of those variables with respect to each other. A typical ecological example might be the ordinary differential equation:

$$\frac{dN}{dt} = f(N)$$

to describe the way in which the rate of change of population abundance with time, dN/dt, is related to abundance, N, through the function $f()$. Another example is the partial differential equation:

$$\frac{\partial N}{\partial t} = rN + D\left(\frac{\partial^2 N}{\partial x^2} + \frac{\partial^2 N}{\partial y^2}\right)$$

for a population growing exponentially, subject to diffusive movement. Stochastic differential equations involve additional stochastic terms, but their use is complicated by theoretical difficulties associated with taking limits of differences of stochastic processes. *See also* POPULATION DYNAMICS. [S.N.W.]

differential resource utilization The exploitation of different RESOURCES by the coexisting species. Differential resource utilization is a mechanism of NICHE differentiation that reduces INTERSPECIFIC COMPETITION. [P.O.]

differentiated disjunction *See* DISJUNCT POPULATIONS.

diffuse competition Community-wide INTERSPECIFIC COMPETITION among a group of species, subjecting each species to a range of competitive pressures exerted by other species. The term is applied to trees in tropical forests, for example, where the species composition of neighbouring trees usually differs from one individual of a species to another. While the DIRECTIONAL SELECTION that could be associated with pairwise competition may lead to CHARACTER DISPLACEMENT, diffuse competition should be much less effective in causing NICHE shifts. Some authors include the indirect effects that occur when groups of species compete in the concept of diffuse competition. *See also* POPULATION REGULATION. [J.G.S.]

diffuse sources CONTAMINATION of environment from a distributed area, for example from crop spraying, agricultural RUN-OFF in general. *Cf.* POINT SOURCE. [P.C.]

diffusion *See* ADVECTION.

diffusion equations Diffusion is a form of movement, brought about by the continual random motion of 'particles'. This can range from the random motion of molecules in a liquid (Brownian motion) to the random DISPERSAL of many animals. This diffusion or random walk usually results in a rate of spread from one region to another that is proportional to the concentration difference between the two regions. Equations that describe this process are called diffusion equations. Therefore, diffusion equations are closely related to random walk models and are important in modelling stochastic POPULATION DYNAMICS and movement. For movement models they arise by assuming that all individuals in a population move randomly and independently, so that on average there will be a net flow of individuals from areas of high DENSITY to areas of low density. By considering movement between adjacent small areas in a small time interval and allowing the areas and interval to shrink to zero a diffusion equation is obtained. For example, suppose that $N(x,y,t)$ is POPULATION DENSITY at time t at location (x,y): in the absence of births and deaths and assuming that individuals are moving at random the density is controlled by:

$$\frac{\partial N}{\partial t} = D\left(\frac{\partial^2 N}{\partial x^2} + \frac{\partial^2 N}{\partial y^2}\right)$$

where D is the 'diffusion coefficient'. Diffusion equations also appear as models of POPULATION SIZE in which the size of a population is considered to be varying at random or with a random component. [S.N.W. & B.S.]

dilution effect An advantage of living in a group, whereby a member of a group dilutes the impact of an attack by a PREDATOR that can kill only one group member per successful attack. *See also* ALLEE EFFECT. [P.O.]

dimension analysis The detailed measurement of plant or plant organ dimensions and DRY MASS in PRODUCTIVITY studies. Relationships of simple non-destructive measurements, such as length, height or diameter, to dry mass are established from a subsample of individuals or quadrats so that dry masses of other individuals, or the same individuals at future SAMPLING occasions, can be made without the need for further destructive harvests. The technique depends on the principle of ALLOMETRIC GROWTH, i.e. as the organisms grow, linear dimensions such as height and diameter increase in proportion and in defined relationships with each other, and with mass and volume.

In FORESTS, routine harvests of replicated plots at different times of year would rarely be practical and, here, dimension analysis is of particular value. For forests, the non-destructive measure commonly measured is the diameter at breast height (DBH), i.e. at about 1.2 m from the ground. For the major commercial forest trees the relationship between DBH and wood mass has been established in YIELD tables. However, the approach has also been applied to a wide range of woody plants, allowing estimates of dry matter in mixed woody communities. Whittaker and Woodwell (1968) showed a linear relationship, with very little scatter, between the logarithms of stem wood productivity ($g\,year^{-1}$) and DBH, constructed from a range of different trees and shrubs within a single forest. They showed similar relationships for predicting leaf and twig production. Because dimension analysis allows the non-destructive estimation of dry mass, it can be used, through repeated measures in time, to estimate the productivity of individuals. This is particularly useful in estimating productivity in highly heterogeneous communities, where variation between destructively harvested samples may greatly exceed the dry mass increments between harvest dates. Dimension analysis in productivity studies has most commonly concerned aerial parts of terrestrial plants. The increased use of minirhizotrons (apparatus for viewing the spatial and temporal variation in root numbers and dimensions *in situ*) is allowing the extension of dimension analysis to the estimation of below-ground net primary productivity. [S.P.L.]

REFERENCE

Whittaker, R.H. & Woodwell, G.M. (1968) Dimension and production relations of trees and shrubs in the Brookhaven forest, New York. *Ecology* **56**, 1–25.

dimethyl sulphide *See* AEROSOLS.

dimictic lakes *See* LAKE STRATIFICATION.

dimixis *See* LAKE STRATIFICATION.

dioecy A species is dioecious if its population contains both male and female individuals and no HERMAPHRODITES (in the broad sense). The genetic determinism of sex is often found linked to the existence of sexual chromosomes (e.g. in many animals and some plant species), but this is not necessarily the case. In humans and some insects, the males are heterogametic. Some species show biased SEX RATIOS, usually in favour of females, especially in animals. Many theoretical models have considered the evolution of biased sex ratios in dioecious species. [I.O.]

dioxin Generic term for a family of chlorinated HYDROCARBONS including polychlorinated dibenzo-*p*-dioxins and furans. They are formed primarily during combustion of chlorinated organic materials (e.g. chlorinated solvents, PLASTICS), although vehicle EMISSIONS and coal burning are also sources. The different congeners have very different toxicities, with some being among the most toxic substances known, so that concentrations as low as $10^{-12}\,g\,m^{-3}$ in air, or less, need to be measured for effective monitoring of human health risks. [J.N.C.]

diphasic life cycle *See* SEXUAL REPRODUCTIVE CYCLES.

diploid life cycle *See* SEXUAL REPRODUCTIVE CYCLES.

Diploidy

A descriptor of the diploid state (two copies of homologous genome) that primarily refers to the cell nucleus. The cell nucleus is haploid when it includes a single copy of the genome. Haploid and diploid nuclear phases alternate during the LIFE CYCLE of all sexualized EUKARYOTES. Haploidy and diploidy do not refer only to the ploidy level of the cell nucleus but also to that of a given organism or a group of individuals. For example, mammals are diploid because almost all their cells (except the GAMETES) are diploid.

The distribution of ploidy levels among eukaryotic kingdoms does not appear to be random. 'Complex' forms, like higher animals and higher plants, are almost exclusively diploid while less differentiated organisms, like unicellular protists, show the two levels of ploidy. Thus diploidy may be a necessary condition for cellular differentiation, but it is not sufficient. However, because of this correlation between increased size/structural complexity and diploidy, there is quite a consensus on the EVOLUTIONARY TREND from haploidy towards diploidy.

The problem of evolution towards diploidy is one of the important questions of evolutionary biology. Ideas concerning the costs and benefits of diploidy versus haploidy can be grouped in two main categories: genetic and ecological hypotheses.

Genetic hypotheses are derived from the fact that diploid cells contain twice as much genetic information as haploids. Among the genetic hypotheses, the most commonly used argument about the advantage of diploidy is that DOMINANCE can protect against the deleterious effects of a recessive MUTATION. In other words, if one copy of a gene is altered in a diploid individual, the homologous copy can take over by restoring the gene function. This argument for an obvious advantage of diploidy is in fact paradoxical. Because of the covering up of deleterious mutations in diploid individuals, these mutations will increase in frequency within their genome. In contrast, recessive mutations cannot accumulate so easily within the genome of a given haploid individual because they are fully expressed and then selected against. Therefore, if the two groups evolve separately, the diploid POPULATION will suffer a larger FITNESS reduction than the haploid. Consequently, in competition with haploid individuals, diploid forms cannot persist in long-term selection. This paradox was resolved recently by two kinds of approaches showing that diploidy can benefit from a long-term advantage and thus become fixed within a species. The first approach considers different levels of selection (in particular INDI-

VIDUAL SELECTION). The second approach considers that the effect of a deleterious mutation is not independent from the number of other deleterious mutations existing within the genome. The idea is to consider that many deleterious mutations are more detrimental for the individual than few mutations.

A second argument for the genetic advantage of diploidy is the repair hypothesis. In diploid individuals (in contrast with haploid ones), double-strand damage can be repaired because a second DNA molecule is present. A less obvious genetic argument suggests that diploidy shows the long-term advantage of retaining two copies of the same gene so that one copy is always available for the acquisition of a new function. This hypothesis for the evolution of diploidy proposes an explanation for the correlation between structural complexity and diploidy.

Ecological hypotheses are derived from the fact that diploid cells contain twice as much DNA material as haploids. The first argument relies on the strong positive correlation among DNA content, cell and nuclear volume, and cell-cycle length. According to this hypothesis, cellular DNA content depends on a compromise between two conflicting selective forces: haploidy should have evolved when selection favours high reproductive rates; and diploidy should be favoured when selection increases lifespan, in general positively correlated with cell size.

Another ecological argument states that haploidy could be a powerful mechanism for limiting necessary nutrient uptake and thus explain why most unicellular AUTOTROPHS are haploid while HETEROTROPHS are diploid. Ecological hypotheses propose a mechanism explaining the maintenance of haploidy as well as evolution towards diploidy. In contrast, genetic hypotheses fail to account for the maintenance of haploidy. It may be that diploidy does benefit from a general genetic advantage and that consequently haploidy should be maintained only for particular ecological reasons. *See also* SEXUAL REPRODUCTIVE CYCLES. [M.V.]

---

diplontic life cycle *See* SEXUAL REPRODUCTIVE CYCLES.

direct fitness In KIN SELECTION theory, that part of an individual's inclusive FITNESS gained through its direct investment in its own offspring (as opposed to the offspring of relatives it may have helped to raise). The direct fitness of an individual does not include the portion of its own offspring that it gained through the actions of others. [T.J.K.]

REFERENCE

Grafen, A. (1982) How not to measure inclusive fitness. *Nature* **298**, 425–426.

direct ordination *See* ORDINATION.

direct transmission, parasites The process by which a PARASITE is passed from one (infectious) HOST to another (susceptible) host without any intervening mechanism. Direct transmission can be via close contact (e.g. influenza), sex (e.g. gonorrhoea), environmental CONTAMINATION (e.g. cholera) and transfer of body fluids in a medical (e.g. blood transfusion) or non-medical (e.g. sharing needles amongst intravenous drug users) context, all of which are known as horizontal transmission (*see also* INDIRECT TRANSMISSION, PARASITES). These classifications are not mutually exclusive and there is special difficulty in separating close contact from environmental contamination. Vertical (or maternal) transmission occurs from mother to offspring (e.g. human immunodeficiency virus). Many parasites are able to utilize more than one route of transmission. *See also* PARASITE–HOST INTERACTIONS. [G.F.M.]

directed selection *See* SYMPATRIC SPECIATION.

directional selection NATURAL or ARTIFICIAL SELECTION favouring phenotypes at one extreme of the distribution. As a result of directional selection, the population MEAN of the selected TRAIT is expected to shift in the favoured direction (Fig. D4). The most efficient form of directional selection is truncation selection, which eliminates all individuals with phenotypes below (or above) a THRESHOLD value. Although often used by breeders and in experiments, it is rare in nature. *Cf.* DISRUPTIVE SELECTION; STABILIZING SELECTION. [T.J.K.]

disassortative mating Mating with phenotypically dissimilar individuals. *See also* MATING SYSTEMS. [P.O.]

discharge Release of CONTAMINANT(S); usually EFFLUENTS to aquatic environment; usually from point sources. [P.C.]

disclimax In the scheme of Clements (1936; *see* PROCLIMAX and CHARACTERS IN ECOLOGY) a kind of VEGETATION differing from the climatically determined CLIMAX as a result of the effects of humans and/or their domestic animals. His type example was the self-perpetuating ANNUAL grassland of central California, dominated by species of *Avena* and/or *Bromus* introduced from the Mediterranean Basin of Eurasia and Africa, which has virtually replaced the native perennial bunch-grassland dominated by *Stipa* spp. The annual grassland is of lower stature than the native perennial grassland, but the reverse characteristic can be found in a disclimax where DOMINANCE by shrubs is induced, for example where OVERGRAZING of grasses in PRAIRIE of the Great Plains results in a takeover by species of *Artemisia*. *See also* SUCCESSION. [P.J.G.]

REFERENCE

Clements, F.E. (1936) Nature and structure of the climax. *Journal of Ecology* **24**, 252–284.

discrete generations POPULATIONS of a single species in which generations do not overlap and are therefore distinguishable from each other. Typically occurring when all the births within a single-species population occur at approximately the same time and parents cease breeding before their offspring begin. Populations of a species with discrete generations therefore contain individuals of approximately the same age. COHORT LIFE TABLES and several simple population models assume discrete generations. *See also* HASSELL MODEL AND HASSELL–COMINS MODEL; LIFE-HISTORY EVOLUTION; LOGISTIC MODEL; POPULATION DYNAMICS. [A.J.D.]

discrete size-classified population model *See* SIZE DISTRIBUTIONS.

discriminant function analysis A multivariate statistical method. Most of the statistical methods described in detail in this encyclope-

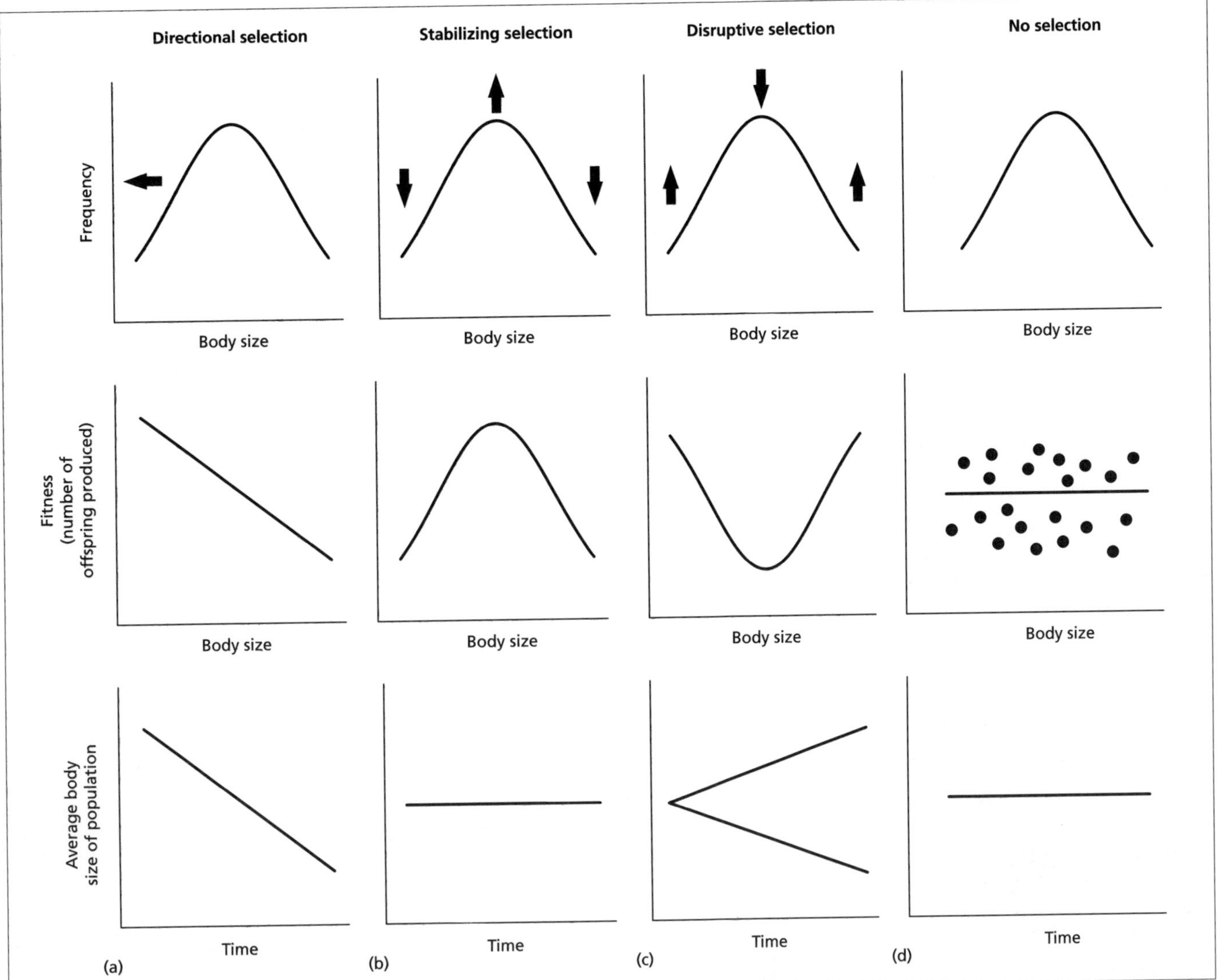

Fig. D4 Three kinds of selection. The top row shows the frequency distribution of the character (body size). For many characters in nature, this distribution has a peak in the middle, near the average, and is lower at the extremes. (The normal distribution, or 'bell curve', is a particular example of this kind of distribution.) The second row shows the relation between body size and fitness, within one generation, and the third row shows the expected change in the average for the character over many generations (if body size is inherited). (a) Directional selection. Smaller individuals have higher fitness, and the species will decrease in average body size through time. (b) Stabilizing selection. Intermediate-size individuals have higher fitness. (c) Disruptive selection. Both extremes are favoured. If selection is strong enough, the population splits into two. (d) No selection. If no relation exists between the character and the fitness, natural selection is not operating on it. (After Ridley, M. (1996) *Evolution*, 2nd edn. Blackwell Science, Cambridge, Mass.)

dia are univariate because they are concerned with single variables or observations. Even the complex methods of ANALYSIS OF VARIANCE are only concerned with explaining the variation in one variable. Multivariate methods analyse several related variables or observations simultaneously. For further details see Manly (1986). [B.S.]

REFERENCE

Manly, B.F.J. (1986) *Multivariate Statistical Methods: A Primer*. Chapman & Hall, London.

disharmony A characteristic of the fauna and flora of isolated islands, which often completely lack elements that are conspicuously present on continents, for example the only mammals endemic to the Hawaiian islands were one species of bat and one species of seal. *See also* ISLAND BIOGEOGRAPHY. [P.O.]

Disjunct distribution

A given TAXON (species, genus, family, etc.) is said to possess a disjunct distribution when its geographical RANGE comprises two or more populations that are separated by distributional gaps greater than the normal DISPERSAL capacity of the taxon. These may be

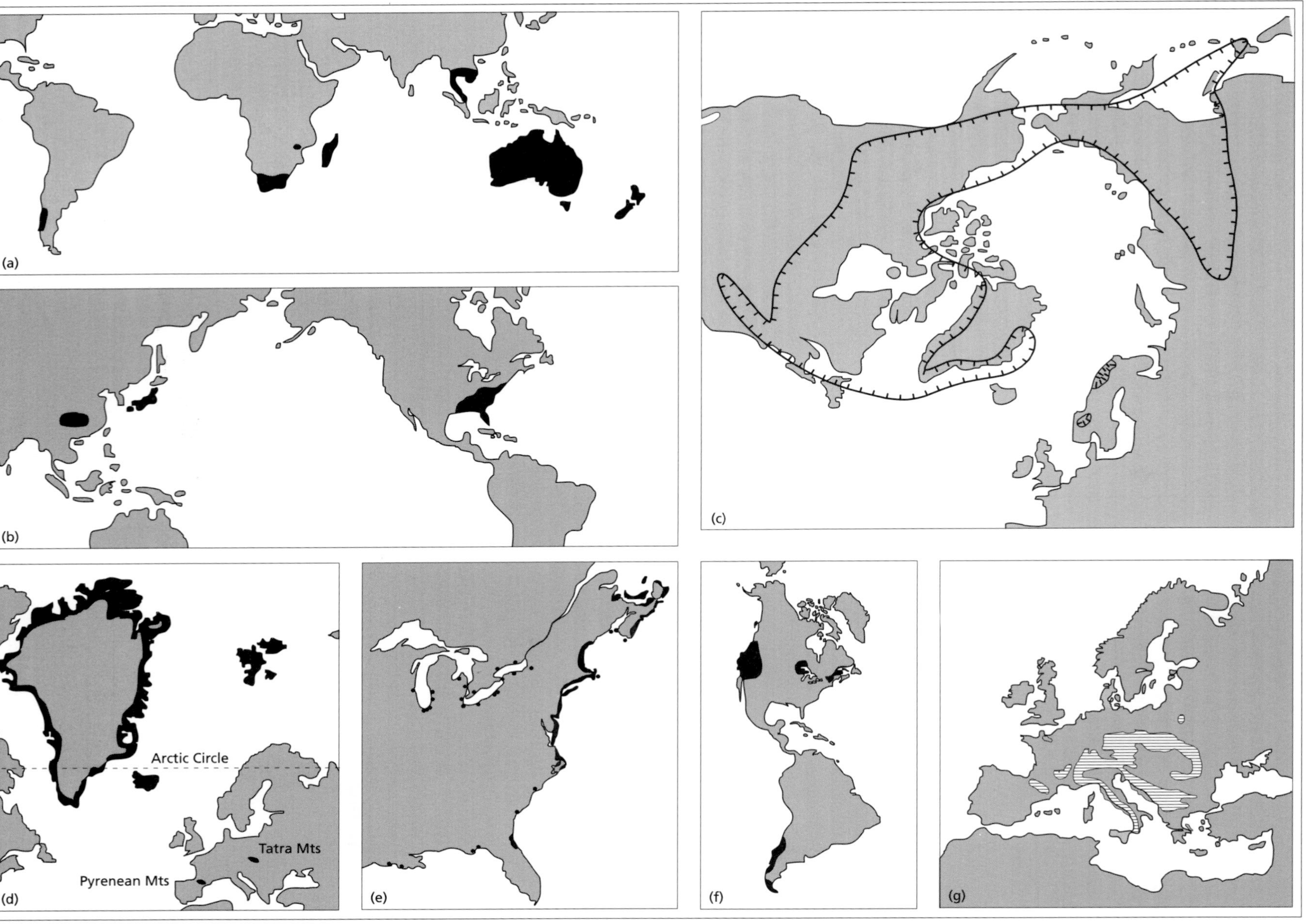

Fig. D5 Examples of different types of range disjunction, at various scales. (a) The angiosperm family Restionaceae: Southern hemisphere disjunct. (b) The evergreen shrub *Stewartia*: eastern Asia and North America. (c) Lapland rosebay, *Rhododendron lapponicum:* circum-Arctic. (d) The springtail, *Tetracanthella arctica*: centred in arctic regions, but with isolated populations in central and southern Europe. (e) The sea spurge, *Euphorbia polygonifolia*: disjunct inland populations along the shores of the Great Lakes, North America. (f) The woodland plant *Osmorhiza chilensis*: temperate North and South America. (g) The silver fir, *Abies alba*: disjunct European populations at middle altitudes in the Massif Central and Pyrenees. (a, b, c, e & f, After Pielou, E.C. (1979) *Biogeography*. John Wiley & Sons, New York, copyright © 1979, reprinted by permission of John Wiley & Sons, Inc.; d, after Cox, C.B. & Moore, P.D. (1993) *Biogeography: An Ecological and Evolutionary Approach*, 5th edn. Blackwell Scientific Publications, Oxford; g, after Tallis, J.H. (1991) *Plant Community History*. Chapman & Hall, London.)

large-scale, intercontinental gaps (often with an intervening ocean barrier) or smaller-scale, intra-regional gaps (with intervening ecological barriers). A broader definition of a disjunction (Pielou 1979) is a range discontinuity that is biogeographically interesting, and raises one or both of the questions, 'Why is the taxon absent from the areas between its subranges?' and 'How did the taxon come to have spatially separated subranges?'. Disjunct distributions at the intercontinental scale are commoner among higher-rank TAXONOMIC CATEGORIES. Thus Thorne (1972) estimates that 254 of the 324 families of angiosperms worldwide show major range disjunctions, but only about 3000 of the 12 500 genera and perhaps 2000 of the 225 000 species.

Disjunct distributions can be characterized and classified in several ways: geographically, by the overall range patterns; historically, by the time of origin of the disjunction; taxonomically, by the rank of the taxon involved; and causally, by the mechanisms inducing disjunction. An example of a geographical CLASSIFICATION is that by Thorne (1972) of the major disjunction patterns shown by angiosperm genera at the intercontinental scale. Classification by taxonomic rank broadly parallels that by age, as the longer a disjunction has existed the higher will be the rank of the taxa in the separate subranges. Although rates of EVOLUTION vary markedly between different plant and animal groups, as a rough guide one can expect range disjunctions at the generic and family level to have been produced by events in the time ranges 10^6–10^7 and 10^7–10^8 years ago, respectively, and at the species level by events in the time range 10^3–10^6 years ago, approximately. Very different causal mechanisms may thus be involved. Three quite separate processes can result in range disjunction: break-up of a formerly continuous range by separation, contraction or MIGRATION; long-distance dispersal; and separate origins of the populations in the subranges. The last process is inherently improbable, but is suggested to have occurred for *Macaranga quadricornis* in the Malay Peninsula (Stott 1981). Mechanisms of range fragmentation that have been identified include: changes in the configurations of land and sea resulting from PLATE TECTONICS; long-term climatic change through the Tertiary period; and the oscillatory climatic and SEA-LEVEL CHANGES of the Pleistocene period. Long-distance dispersal can be natural or human-aided.

Examples of different types of range disjunction, at various scales, are illustrated in Fig. D5 and considered briefly below.

- The angiosperm family Restionaceae: Southern hemisphere disjunct. Similar patterns are shown, for example, among plants by the family Proteaceae and the southern beeches of the genus *Nothofagus*, and among animals by the flightless birds of the superorder Ratitae and the earthworm genus *Microscolex*. The range disjunctions probably resulted from the break-up of formerly continuous ranges across the southern landmass of GONDWANA, beginning *c.* 140 million years ago.
- The evergreen shrub *Stewartia*: eastern Asia and North America. Similar patterns are shown by a large number of temperate tree genera and by the beetle genus *Calloides*. A formerly continuous belt of forest across Eurasia and North America through Beringia was fragmented by long-term climatic cooling during the Tertiary period.
- Lapland rosebay, *Rhododendron lapponicum:* circum-Arctic. Similar patterns are shown by many cold-tolerant plants and animals. The range occupies areas previously covered by Pleistocene ICE SHEETS. Whether subsequent spread of *Rhododendron* was from plants surviving on nunataks, from offshore REFUGIA or to the south of the ice sheets is still a matter for debate, as is the role played by long-distance dispersal across open frozen landscapes.
- The springtail, *Tetracanthella arctica*: centred in arctic regions, but with isolated populations in central and southern Europe. This is a distribution pattern shown by many so-called Arctic–Alpine plant and animal species and probably resulted from the contraction polewards in the Holocene of a more extensive glacial and late-glacial range. A comparable North American example is the white spruce, *Picea glauca*.
- The sea spurge, *Euphorbia polygonifolia*: disjunct inland populations along the shores of the Great Lakes, North America, are survivors from a period of higher sea level *c.* 10 000 years ago as the Laurentide Ice Sheet melted.
- The woodland plant *Osmorhiza chilensis*: temperate North and South America. This pattern is shown by many plant taxa, but perhaps few animal taxa, and is mostly easily explained by long-distance dispersal (probably by birds). However, for woodland taxa such as *Osmorhiza* the disjunction could have arisen by slow diffusion through the montane forests of the Rockies and Andes followed by extinction of the middle part of the range at lower latitudes.
- The silver fir, *Abies alba*: disjunct European populations at middle altitudes in the Massif Central and Pyrenees. The mid-Holocene spread of *Abies* was into already established montane forest communities and must have involved jump-dispersal across intervening lowland areas.

See also GEOGRAPHIC DISTRIBUTIONS. [J.H.T.]

REFERENCES

Cox, C.B. & Moore, P.D. (1993) *Biogeography: An Ecological and Evolutionary Approach*, 5th edn. Blackwell Scientific Publications, Oxford.

Pielou, E.C. (1979) *Biogeography*. John Wiley & Sons, New York.

Stott, P. (1981) *Historical Plant Geography*. Allen & Unwin, London.

Tallis, J.H. (1991) *Plant Community History*. Chapman & Hall, London.

Thorne, R.F. (1972) Major disjunctions in the geographic ranges of seed plants. *Quarterly Review of Biology* **47**, 365–411.

disjunct populations Populations that are geographically isolated from the rest of the TAXON. There are two types of disjunction: (i) undifferentiated disjunction refers to disjunct populations that are similar; and (ii) differentiated disjunction occurs when one or more of the disjunct populations diverge from the others. Disjunction is of considerable evolutionary interest because it is believed to facilitate SPECIATION. Disjunct populations may arise through DISPERSAL or through CONTINENTAL DRIFT. The classical example of speciation in disjunct populations is that of Darwin's finches. *See also* DISJUNCT DISTRIBUTION. [V.F.]

dispersal Dispersal is the movement of individuals away from where they were produced and may be active or passive (see Plate 6, facing p. 398). Dispersal, or lack of dispersal, is often the reason why organisms are absent from places or habitats that are wholly suitable for growth and reproduction. Dispersal by humans, either deliberate or accidental, has led to the ESTABLISHMENT of organisms in parts of the world from which they had been absent, often leading to them becoming pests or displacing native species. Some organisms have spectacular powers of dispersal, for example the directed MIGRATION of some birds from the tropics to the poles in pursuit of seasonally abundant food supplies. *Cf.* DISPERSION. *See also* ACTIVE DISPERSAL; PASSIVE DISPERSAL. [R.H.S.]

dispersal and colonization of aquatic organisms, conservation implications Both LENTIC and LOTIC ecosystems are characterized by discontinuities in space and time, with water bodies geographically isolated and usually short term and transitory on a geological scale. The insular nature of FRESHWATER and saline ecosystems divides species into isolated populations allowing a degree of independent evolution that results in different varieties or races. As a result of their transitory nature, evolution rarely leads to the development of new species so that ENDEMIC SPECIES found only in a single ecosystem are rare. Thus, the DISPERSAL and COLONIZATION abilities of aquatic organisms are of great importance, otherwise when an ecosystem disappears the species or variety (race) will become extinct and newly formed ecosystems will not be colonized.

Dispersal of hololimnic species (all life stages aquatic) occurs passively by other agencies such as wind, rain or animals, while merolimnic species disperse by independent locomotion by the non-aquatic life stages and/or passively. Successful dispersal and colonization is severely restricted by the distance between suitable ecosystems. Thus, any reduction in the number of aquatic ecosystems and their DISTRIBUTION will potentially affect dispersal and colonization as well as the elimination of varieties of species populations not found anywhere else. With PASSIVE DISPERSAL by animals, conservation of these vectors is also necessary for dispersal of aquatic organisms. *See also* CONSERVATION BIOLOGY. [R.W.D.]

Dispersal and colonization of terrestrial organisms, conservation implications

'Give me space and motion and I will give you a world' (Descartes, 1596–1650). DISPERSAL refers to the movement of plants, animals and their propagules (seeds, spores, larvae, etc.) to find new places to settle, food, shelter or simply independence from parents. Transoceanic seed dispersals to remote islands and seasonal MIGRATION of birds are dramatic popular examples but the spatial patterns of each species' DISTRIBUTION and its persistence and EVOLUTION also represent many research interests. For instance, island biota differ in the role of anemochorous (wind-dispersed) and zoochorous (animal-dispersed) species. Many predator–prey systems coevolve in relation to both local and global dispersal patterns. Accordingly, population structure and dynamics theory has increasingly emphasized several scales of variation in spatio-temporal environments. In POPULATION GENETICS, patterns (i.e. rates, distances) of dispersal are treated in terms of GENE FLOW within and among populations of a species or even across species barriers. Most of the features of variation, local or global ADAPTATION and foundation of new colonies depend on gene-flow characteristics as well.

COLONIZATION involves at least two different processes: (i) a series of populations regularly undergoing some local extinction–recolonization dynamics (also called metapopulation dynamics) with dispersal or gene flow to the nearby old or new vacant patches of habitat; and (ii) certain colonizing and weedy biota living through successional cycles, DISTURBANCE or geographical range expansion. Dispersability, HABITAT SELECTION, large PROPAGULE source, opportunistic early ESTABLISHMENT and highly plastic reproductive tactics are some of the common requirements for success.

The relative rates (distance-related) of dispersal or gene flow are estimated using a variety of methods. Release and recapture of marked animals are legendary and indeed help in many behavioural and other ecological studies. Marking is done using tags, radiometric devices, dyes, etc.; pollinator movement among flowers is scored by direct record-keeping; some rare earth elements have been used for seed-dispersal data. Indirect methods include the use of genetic markers, for example analyses of rare allele distribution, PATERNITY analysis, estimation of outcrossing patterns using parent–progeny GENOTYPE data, or simply inferences from spatial variation patterns. Unfortunately, most of these methods assay localized gene flow rather than long-distance dispersal events, which are relatively less common or subject to large estimation errors. This produced a biased view nearly 30 years ago; however, leptokurtosis, very localized dispersal and undue neglect of long-distance dispersal events in population biology as well as BIOGEOGRAPHY are now being de-emphasized. Is dispersal often bimodal along the distance axis? Are there morphological and behavioural dimorphisms underlying any such bimodality? And how do we compare relative dispersability of different related or coexisting species? These and many such questions are now áttracting serious attention. Clearly, all of these methods and various underlying theoretical models are needed to evaluate how dispersal affects genetic structure, evolution of certain traits that directly affect mobility or post-dispersal success, PARENT–OFFSPRING CONFLICTS, habitat cueing or community dynamics.

Many species living in highly subdivided habitats, with populations subject to local extinctions and recolonization, are treated using metapopulation models in which dispersal features play a critical role. For example, models for a desert plant suggest DORMANCY and dispersal to be alternative evolutionary outcomes depending on the patterns of yearly variation in survival. Some species might persist through spreading the risk as different populations face extinction asynchronously. A COLONIZING SPECIES might be relying more on frequent, repetitive dispersal as much as on the wide-ranging 'lottery' dispersal to many areas. Many plant–animal interactions of mutual benefit are cited as examples favouring specific plant-dispersal tactics. In population genetics theory, Sewall Wright's pioneering work relied heavily on multiple DEMES, local selection and interdeme gene flow. Likewise, theories of FOUNDER EFFECTS and random drift require dispersal and colonization events to be analysed in terms of both POPULATION DYNAMICS and genetics. Several recent studies of gynodioecious plant species, for example, have been advanced using patterns of relative dispersal of exnuclear male sterility factors and the nuclear restorer genes for explaining their variation among populations. Several studies are also exploring experimentally the current evolutionary shifts in dispersability of island species, newly derived autogamous taxa and of certain rare species of specialized habitats.

This entire topic has important implications in CONSERVATION BIOLOGY. A well-known debate centres on the issue of designing nature reserves for protecting a species or an ecosystem, i.e. one large or many small reserves (SLOSS); the choice depends on numerous practical issues, but equally so on the dispersal and connectivity of small reserves, either natural or managed, and with or without some newly created CORRIDORS allowing natural movements. Examples to support each side of this debate are widely cited but aside from a lack of generality, not enough is known about the dispersal rates. Population geneticists also point to the models for maintaining VARIABILITY in one large or many small, connected populations, a basic need for continuing evolution in protected taxa. *See also* METAPOPULATIONS. [S.K.J.]

REFERENCES

Chepko-Sade, B.D. & Halpin, Z.T. (eds) (1987) *Mammalian Dispersal Patterns*. University of Chicago Press, Chicago, Ill.

Crawley, M.J. (ed.) (1997) *Plant Ecology*, 2nd edn. Blackwell Science, Oxford.

Hanski, I.A. & Gilpin, M.E. (eds) (1997) *Metapopulation Biology*. Academic Press, San Diego, Calif.

Rice, K. & Jain, S. (1985) Plant population genetics and evolution in disturbed environments. In: *The Ecology of Natural Disturbance and Patch Dynamics* (eds S.T.A. Pickett & P.S. White), pp. 287–304. Academic Press, New York.

dispersal polymorphism DISPERSAL may result from the day-to-day activities of an organism or be a specific activity that leads the organism to move outside of its typical foraging range. In the latter case, it has been observed that the tendency to disperse is not randomly distributed among individuals within the population but is inherited to a greater or lesser degree. In the simplest case dispersal might be controlled by a single locus but, more generally, several to many genes are likely to be involved. As shown in Fig. D6, dispersal polymorphism is common amongst insects and flowering plants. [D.A.R.]

dispersion Dispersion is a statistical term meaning variation or SPREAD and is not to be confused with DISPERSAL, which is a biological process. In ecology, dispersion usually refers to the spatial DISTRIBUTION of organisms. Unfortunately, use of words based on dispersion has led to some confusion in the literature. The term 'overdispersed', for example, may imply one thing to a statistician (a distribution with a high VARIANCE, as when organisms occur in clusters) and the opposite to a behavioural ecologist (dispersal to such an extent that organisms are very spaced out); 'underdispersed' causes similar problems. For this reason, use of the terms 'overdispersed' and 'underdispersed' is best avoided.

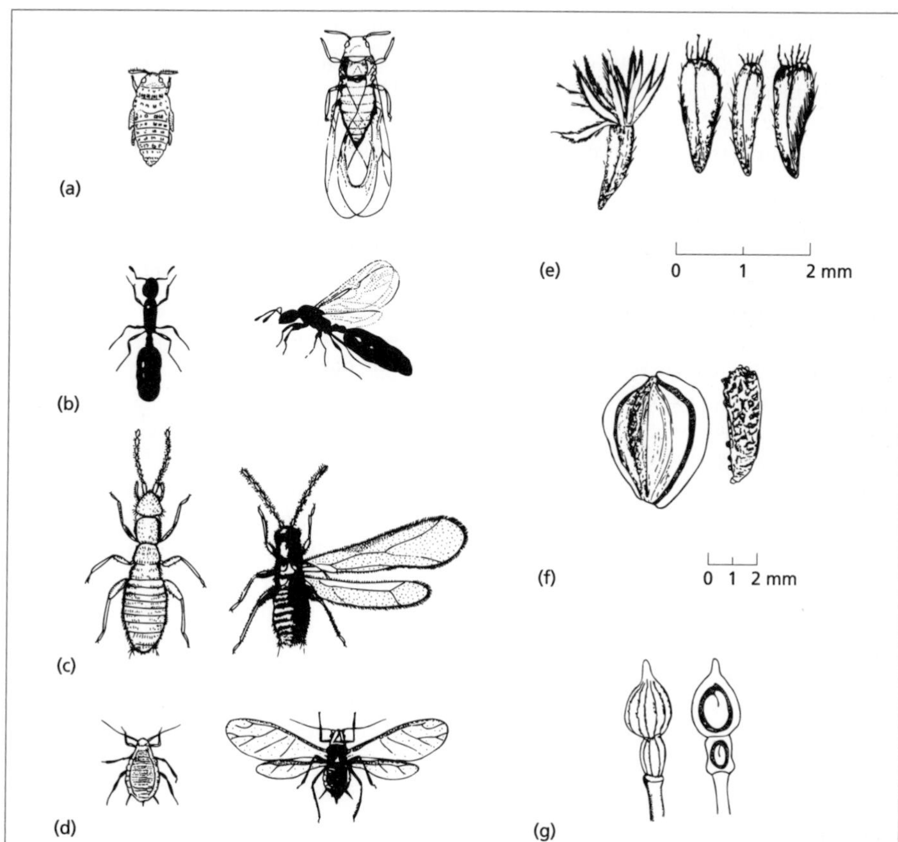

Fig. D6 Dimorphism with respect to dispersal in animals and plants. Winged and wingless females of: (a) the grape phylloxera (*Viteus vitifoliae*); (b) the little black ant (*Monomorium minimum*); (c) the zorapteran (*Zorotypu hubbardi*); and (d) the apple aphid (*Aphis pomi*). Contrasting sizes and dispersal features of seeds and fruits: (e) *Galinsoga parviflora*, fruits from the centre (on left) and margin (on right) of the capitulum; (f) *Dimorphotheca pluvialis*, again showing central and marginal fruits; (g) *Rapistrum rugosum*, a two-seeded pod with marked difference in seed size. (After Begon, M., Harper, J.L. & Townsend, C.R. (1996) *Ecology: Individuals, Populations and Communities*, 3rd edn. Blackwell Science, Oxford.)

Three main types of dispersion pattern are recognized. When individuals are very evenly spread, perhaps because any individual actively avoids being close to other individuals (e.g. red grouse defending territories), their distribution is said to be uniform or regular (*see* REGULAR DISTRIBUTION). When individuals occur in clumps or clusters, perhaps because they are attracted to a patch of food (e.g. carrion flies on a carcass), their distribution is said to be contagious or aggregated (*see* AGGREGATED DISTRIBUTION). A so-called RANDOM DISTRIBUTION occurs if there is an equal chance of any individual occupying any point in space and individuals are distributed independently of one another. Analysis of dispersion patterns is conventionally based on either distances between individuals, counts in randomly placed quadrats or 'units of habitat', such as fungal bodies, fruit or carcasses. *See also* STATISTICS. [R.H.S.]

disphotic zone The depth zone in the OCEAN between the EUPHOTIC ZONE and the APHOTIC ZONE, where there are small but measurable quantities of LIGHT. The amount of light is insufficient for PHOTOSYNTHESIS but is sufficient to elicit animal responses. The disphotic zone extends to a depth of about 1000 m. [V.F.]

Displays

Many communicatory acts take the form of elaborate, often bizarre, displays. Male fiddler crabs (*Uca* spp.) wave a grossly enlarged claw in the air to attract females, male ducks court by performing sham drinking and preening actions, and lizards of the genus *Sceloporus* bob their heads vigorously to threaten rivals. Such displays can be puzzling in their complexity and in the seeming irrelevance of some of their elements to the task in hand. However, as the ethologists Konrad Lorenz and Niko Tinbergen pointed out many years ago, much can be understood about displays by considering their motivational ancestry.

Lorenz and Tinbergen suggested that many displays have evolved from ancestral movements that incidentally gave away the future intentions of the performer. Thus a dog might bare its teeth immediately before biting an opponent. If this was usually the case, the opponent could predict that it was likely to get bitten as soon as it saw the glint of enamel. Selection would then favour dogs that bared their teeth to engineer a quick retreat by rivals, and teeth-baring would begin to evolve into a more exaggerated threat display. Studies suggest that many displays are consistent with an origin in INTENTION MOVEMENTS: the crouching and stretching postures adopted during COURTSHIP in many birds, for instance, are easy to derive from intention movements during take-off and lunging. Other displays can be derived from ambivalent postures or movements during moments of indecision: for instance when deciding whether to attack or flee. The zig-zag dance of the male three-spined stickleback (*Gasterosteus aculeatus*) may be an example, reflecting conflict between approaching and avoiding a female. Yet others may initially have reflected conflict between two mutually exclusive options such as attacking and mating. Apparently irrelevant actions, like the drinking and preening movements that punctuate courtship in some duck species, can be explained in these terms and have been referred to as displacement activities. However, the original thinking behind displacement activities stemmed from the long-abandoned concept of motivational energy (two conflicting motivational systems caused pent-up energy to 'spark over' or 'displace' into a third system irrelevant to the goals of the other two) and more recent evidence suggests that such activities may not be as irrelevant to the context in which they occur as at first appears. Autonomic nervous activity as a result of stress may be the origin of some components of displays associated with aggressive conflict. Piloerection during contests and urine-marking territory boundaries are two common aggression-related displays among mammals that are likely to have evolved from autonomic fear responses.

The evolutionary process by which behaviours originally associated with particular motivational states became elaborated into stylized displays has been dubbed 'ritualization'. Classical ethologists saw ritualization as a means of increasing the clarity of signals. The stereotyped behavioural and postural elements of threat displays, for example, are often the complete antithesis of those of appeasement displays. The stereotyped consequences of ritualization have resulted in what Desmond Morris referred to as 'typical intensity' in displays, an almost uniform pattern of performance across individuals performing them. Morris saw this as the inevitable end-point of reducing ambiguity in signal transmission. More recent thinking, however, sees it as a vehicle for MANIPULATION. In the manipulation scenario, elaborate stylized displays are the upshot of an evolutionary ARMS RACE between exploitative signallers trying to coerce another individual into responding in the signaller's reproductive interests and discerning receivers who resist being duped (*see* MANIPULATION). Far from reflecting mutualistic COMMUNICATION, therefore, displays become manipulative bluster. The extent to which a signaller's bluff is called is likely to depend on the cost of discrimination to the receiver, one of several features of receivers that are now regarded as important in constraining and potentiating the evolution of different kinds of signal and display. *See also* BEHAVIOURAL ECOLOGY; INTERSEXUAL SELECTION. [C.J.B.]

REFERENCE

Guilford, T. & Dawkins, M.S. (1991) Receiver psychology and the evolution of animal signals. *Animal Behaviour* **39**, 706–716.

Disposable soma theory

A theory to explain AGEING (SENESCENCE) that simultaneously addresses the questions 'Why does ageing occur?', a problem in evolutionary biology, and 'How is ageing caused?', a problem in physiology (Kirkwood 1977, 1981; Kirkwood & Holliday 1979). The disposable soma theory explains ageing through asking how best an organism should allocate its metabolic RESOURCES (primarily energy) between, on the one hand, keeping itself going from one day to the next (maintenance) and, on the other hand, producing progeny to secure the continuance of its genes when it itself has died (reproduction).

No species is immune to hazards of the environment, such as predation, starvation and disease. These hazards limit the average survival time, even if senescence does not occur. It follows that maintenance is only needed to an extent which ensures that the body (soma) remains in sound condition until an age when most individuals will have died from accidental causes. In fact, a greater investment in maintenance is a disadvantage because it eats into resources that in terms of NATURAL SELECTION are better used for reproduction. The theory concludes that the optimum course is to invest fewer resources in the maintenance of somatic tissues than would be necessary for indefinite somatic survival.

The disposable soma theory is named for its analogy with disposable goods, which are manufactured with limited investment in durability, on the principle that they have a short expected duration of use. The theory, as outlined above, applies to any species that exhibits an iteroparous (repeatedly reproducing) life-history plan and that has a clear distinction between soma and GERM LINE (*sensu* Weismann; see Kirkwood & Cremer 1982). The principle of optimizing the TRADE-OFF between maintenance and reproduction can be extended, however, to other life-history patterns (see Kirkwood 1981).

Central to the disposable soma theory is the idea that maintenance involves energy costs. Although it is not yet possible to quantify all of the costs of maintenance in detail, there is evidence that the overall maintenance cost is substantial. Basal metabolism accounts for the major part of the ENERGY BUDGET of a typical organism and basal metabolism is largely concerned with maintenance of one kind or another. In particular, there is an extensive network of intracellular processes (DNA repair, antioxidant defences, proofreading macromolecule synthesis, etc.) whose operation is essential for maintaining cellular HOMEOSTASIS. Each of these processes requires energy.

In addition to explaining why ageing occurs, the disposable soma theory also suggests how the genetic control of LIFESPAN is arranged and how different species have evolved different lifespans.

Natural selection operates on the genes that regulate key aspects of somatic cell maintenance in such a way as to secure the optimum balance between surviving long enough and spending too much on survival. The 'set point' of a maintenance function determines the average period of LONGEVITY assured. As a specific example, consider DNA repair. If it is supposed that the exposure of DNA to damaging agents is constant but the level of DNA repair is increased, then the rate at which DNA damage accumulates is reduced while the energy cost associated with DNA repair rises. The rate of DNA damage accumulation determines the average length of time taken before DNA defects reach harmful levels and, hence, longevity.

Now if we recall that it is the presence of EXTRINSIC MORTALITY that makes it not worthwhile to invest in better maintenance than is needed to preserve somatic functions through the normal expectation of life in the wild, we can see that it is the level of environmental risk that provides the basis of selection for a longer or shorter intrinsic lifespan. A species subject to high environmental risk will do better to invest in rapid and profilic reproduction at the expense of somatic maintenance, whereas a species that has a low level of environmental risk will tend to do the reverse.

Predictions of the disposable soma theory include the following: (i) longevity should be positively correlated with levels of somatic maintenance; (ii) multiple maintenance mechanisms contribute to ageing and longevity (this is because the theory applies with equal force to each of the different processes that forms a part of the somatic maintenance network); and (iii) the accumulation of somatic defects that leads eventually to ageing begins early in life, possibly soon after the differentiation of somatic cell lineages from the germ line.

The relationship of the disposable soma theory to other views on the evolution of ageing has been reviewed by Kirkwood and Rose (1991). *See also* DEFENCE MECHANISMS; EVOLUTIONARY OPTIMIZATION; LIFE-HISTORY EVOLUTION. [T.B.L.K.]

REFERENCES

Kirkwood, T.B.L. (1977) Evolution of ageing. *Nature* **270**, 301–304.

Kirkwood, T.B.L. (1981) Repair and its evolution: survival vs. reproduction. In: *Physiological Ecology: An Evolutionary Approach to Resource Use* (eds C.R. Townsend & P. Calow), pp. 165–189. Blackwell Scientific Publications, Oxford.

Kirkwood, T.B.L. & Cremer, T. (1982) Cytogerontology since 1881: a reappraisal of August Weismann and a review of modern progress. *Human Genetics* **60**, 101–121.

Kirkwood, T.B.L. & Holliday R. (1979) The evolution of ageing and longevity. *Proceedings of the Royal Society of London, Series B* **205**, 531–546.

Kirkwood, T.B.L. & Rose, M.R. (1991) Evolution of senescence: late survival sacrificed for reproduction. *Philosophical Transactions of the Royal Society of London, Series B* **332**, 15–24.

disruptive selection NATURAL or ARTIFICIAL SELECTION favouring phenotypes that deviate in either direction from the population MEAN (see Fig. D4). Disruptive selection leads to an increase in genetic variation and may produce a BIMODAL DISTRIBUTION or a POLYMORPHISM. Strong disruptive selection favours ASSORTATIVE MATING and may lead to SYMPATRIC SPECIATION. *Cf.* DIRECTIONAL SELECTION; STABILIZING SELECTION. [T.J.K.]

dissolved organic matter (DOM) Dissolved molecules derived from degradation of dead organisms or excretion of molecules synthesized by organisms. In practice, DOM is defined arbitrarily to include all organic matter passing through a 0.45-μm filter. In the ocean, seasonal maximum values of DOM occur about 1 month after the phytoplankton BLOOM in temperate waters. DOM includes readily metabolizable compounds (such as glucose, AMINO ACIDS, acetate) and refractory compounds (such as lignins, HUMIC ACIDS and PROTEINS). The CARBON fraction of DOM (DOC) may vary seasonally by at least three-fold. Dissolved organic carbon concentrations vary among rivers (10–20 $mg\,l^{-1}$), estuaries (1–5 $mg\,l^{-1}$), coastal waters (1–5 $mg\,l^{-1}$) and the open ocean (0.5–1.5 $mg\,l^{-1}$). Dissolved solids are total substances (both organic and inorganic) remaining after the water has been evaporated from a filtered sample. [V.F. & P.C.]

dissolved oxygen (DO) OXYGEN in aquatic environments that is freely available to support RESPIRATION. [P.C.]

dissolved solids *See* DISSOLVED ORGANIC MATTER.

distribution Within ecology this term in used in two slightly different contexts. It can be used to describe the spatial distribution of the individuals in a POPULATION (*see* AGGREGATED DISTRIBUTION;

RANDOM DISTRIBUTION; REGULAR DISTRIBUTION). It can also be used to describe the distribution of any set of measurements or observations (*see* PROBABILITY DISTRIBUTIONS). In some situations, a particular probability distribution can be used to describe the spatial distribution of individuals (*see* NEGATIVE BINOMIAL DISTRIBUTION; POISSON DISTRIBUTION). In this case the probability distribution attempts to describe the spatial distribution. *See also* DISPERSION; STATISTICS. [B.S.]

disturbance

1 Any process that destroys plant or animal BIOMASS (*sensu* Grime); effects of animals (GRAZING, trampling and burrowing), fire and bulldozer are all agencies of disturbance. One of the two important axes of GRIME'S TRIANGLE, the other being environmental FAVOURABILITY.

2 The creation of a seedbed by physical removal of vegetative cover, possibly accompanied by tillage of the soil (e.g. break-up of the soil crust in arid ecosystems). Natural agencies of disturbance include landslide, hurricane, fire, flood, silt deposition, GLACIATION, etc. Digging animals (rabbits), tunnelling species (moles, earthworms) and colony-dwelling species (gophers) are important biotic agents of disturbance, creating open conditions for seedling ESTABLISHMENT. In many ecosystems, the principal agents of disturbance are humans and their machines (agricultural cultivation, forest clearance, construction works, etc.).

3 The creation of microsites (safe sites) for RECRUITMENT of plant species, by whatever means (reduction of perennial plant cover, provision of bare soil or canopy gaps suitable for seedling establishment).

4 The creation of gaps in a forest canopy by removal or death of the dominant trees where sapling recruitment can occur (GAP PHASE REGENERATION).

Disturbance creates space in previously CLOSED COMMUNITIES and provides opportunities for establishment; thus disturbed communities are particularly susceptible to invasion by new species. [M.J.C. & K.T.]

disuasions *See* DEFENCE MECHANISMS.

diurnal Most terrestrial organisms are exposed to daily cycles of LIGHT and dark to which they become entrained, i.e. adapted. Many physiological (e.g. flowering) and behavioural (e.g. drosophilid eclosion) patterns are determined by specific periodic responses to such external cues. When a particular activity or response occurs during the light phase then it is described as diurnal (or circadian). Alternatively it may occur at dusk (CREPUSCULAR) or during the dark phase (nocturnal). [R.C.]

divergence The horizontal flow of WATER away from a coast or away from a common centre, usually resulting in UPWELLING. Divergence also refers to the moving apart of lithospheric (i.e. crust and upper part of mantle) plates. Divergence is the opposite of convergence. [V.F.]

divergence principle Darwin's insight that the continued evolutionary modification of independent (non-interbreeding) lineages will cause them to diverge morphologically. *See also* DARWINISM; EVOLUTION; SPECIATION. [P.O.]

Diversity

The terms 'diversity', 'ecological diversity', 'species diversity', 'biological diversity' and 'BIODIVERSITY' all refer to the variety and ABUNDANCE of species at a specified place and time. Diversity is a concept that is intuitively easy to understand but remarkably difficult to quantify (*see* DIVERSITY INDICES). The reason for this is that diversity consists of not one but two components: SPECIES RICHNESS, i.e. number of species, and EQUITABILITY (sometimes termed 'evenness'), which is a measure of how equally abundant those species are. Communities with a large number of species are obviously more diverse than species-poor ones. However, high equitability, which occurs when species are equal or virtually equal in abundance, is also equated with high diversity. For example, a sample of 100 moths with 10 individuals in each of 10 species is considered more diverse than another which also has 100 moths and 10 species, but in which one of the species is represented by 91 individuals. These two components of diversity may be evaluated either separately (*see* EQUITABILITY; SPECIES RICHNESS) or jointly by means of a composite measure such as the Shannon or Simpson index.

Some of the most interesting questions in ecology ask why certain spatial and temporal patterns of diversity exist in the natural world. Diversity in this context usually, but not invariably, means species richness. For instance, there are well-documented gradients, such as the latitudinal gradient of diversity, in which the number of species declines with distance from the tropics. Diversity also varies systematically with area (*see* DIVERSITY GRADIENT) and altitude. Temporal patterns of diversity can be detected on a variety of timescales. The FOSSIL RECORD reveals that diversity has increased over evolutionary time. This has occurred despite five MASS EXTINCTIONS. Diversity also varies markedly over ecological time, for example during SUCCESSION, following climatic change or after a major DISTURBANCE such as a fire. DIAPAUSE, MIGRATION or other short-term effects such as CIRCADIAN RHYTHMS may lead to shifts in diversity over behavioural time. In this latter case, samples taken at different times of the day or seasons can give quite different estimates of diversity, even though the underlying composition of the community may not change.

Considerable effort has been devoted to explaining widespread and predictable patterns of diversity. MacArthur and Wilson's *Theory of Island Biogeography* (1967) showed that larger areas can support more species than smaller ones through the interplay of IMMIGRATION and extinction rates. Robert MacArthur also found that habitat diversity can explain the level of diversity in associated organisms. There is no doubt that the impressively high diversity of insects in TROPICAL RAIN FORESTS is linked to the species and architectural diversity of trees that occur there. Other BIOTIC and ABIOTIC FACTORS have been invoked to account for patterns of diversity. For example, the number of species that are able to coexist will often be greatest at moderate levels of disturbance or PRODUC-

TIVITY. COMPETITION and predation are also thought to influence levels of diversity. See Rosenzweig (1995) for a full discussion of these issues.

Diversity, as the Rio Earth Summit in 1992 demonstrated, is an issue of global concern and governments worldwide have realized that it is vital to minimize extinctions. Current estimates reveal that diversity is being lost at an alarming rate. For example, it has been estimated that the average LIFESPAN of an animal species in the fossil record (from SPECIATION to extinction) is in the order of 10^6–10^7 years. Extinction rates at present are approximately four orders of magnitude faster than those that pertained over evolutionary time. This means that, if current trends continue, up to half of all terrestrial vertebrate species might be lost within the next millennium. This loss of diversity is almost entirely a consequence of human activity. See Lawton and May (1995) for more details.

Measures of diversity, such as species richness or a diversity index, play an important role in environmental assessment and can, for example, be used to assess the impact of pollution. *See also* SPECIES-LEVEL DIVERSITY, GLOBAL. [A.E.M.]

REFERENCES

Lawton, J.H. & May, R.M. (eds) (1995) *Extinction Rates*. Oxford University Press, Oxford.

MacArthur, R.H. & Wilson, E.O. (1967) *The Theory of Island Biogeography*. Princeton University Press, Princeton, N.J.

Rosenzweig, M.L. (1995) *Species Diversity in Space and Time*. Cambridge University Press, Cambridge.

---

diversity, alpha, beta and gamma Three measures of species DIVERSITY in space.

All species have a distinctive ecological specialty that allows them to co-occur with others. Some of these specializations are founded on temporal differences in the environment, some on spatial differences, some on having different victims or enemies or mutualists, and some on differences in the ability to tolerate poorer conditions or dominate better ones. MacArthur (1964) suggested that we could separate the contributions of some of these components to diversity by measuring how diversity accumulates in time or space. Whittaker (1970) named the spatial measures alpha, beta and gamma diversity.

The number of species at a single point in space and time is alpha diversity. It estimates how many species co-occur because they specialize on different interacting species. For example, differences in body size between species often lead to differences in which species consume them or which species they eat; so species of various body sizes often co-occur at a point and contribute to alpha diversity. Temporal differences can also contribute to alpha diversity because species may decline very slowly after their special season or year has passed.

As we sample areas larger than a point, additional habitats are included and diversity grows. The rate of growth is beta diversity. We have no standard formula to calculate beta diversity, but the most common uses the coefficients of the SPECIES–AREA RELATIONSHIP: $S=cA^z$, where S is the number of species and A is the area (*see* DIVERSITY GRADIENT).

The gamma diversity is the value of S in an area. It combines point diversity with the effect of having many habitats in space. [M.L.R.]

REFERENCES

MacArthur, R.H. (1964) Environmental factors affecting bird species diversity. *American Naturalist* **98**, 387–397.

Whittaker, R.H. (1970) *Communities and Ecosystems*. Macmillan Inc., New York.

Diversity gradient

A CORRELATION of DIVERSITY with another spatial or temporal variable.

1 *Area gradient*. Larger areas have more species. This, the most pervasive diversity gradient, is called the SPECIES–AREA RELATIONSHIP. It characterizes the diversity of all taxa. It is known from all BIOMES, cool and warm, wet and dry, saline and sweet, all over the world. It occurs on mainlands, and among islands and continents (Fig. D7). It is also the oldest known pattern in ECOLOGY, having first been noticed in 1805 by Alexander von Humboldt and soon after by H.C. Watson and by de Candolle. In most cases, this relationship fits an equation of the form: $S=cA^z$, where S is the number of species and A is the area sampled. This equation is most often transformed to logarithmic space for comparative study: $\log S=\log c+z\log A$. One of its two coefficients, the z-value, has been the focus of much work (see below).

2 *Temporal gradient*. As the time interval over which samples are collected increases, so does diversity. This happens in both ecological time, as habitats vary with the seasons and as environments change, and evolutionary time, as species vanish and new ones replace them.

3 *Habitat complexity gradient*. The more habitats in a fixed area, the more species live there. This gradient also pervades life in every form and place. It motivates much CONSERVATION BIOLOGY, whose goal is often to reserve as many habitats as possible and protect them from degradation.

4 *Latitudinal gradient*. The diversity of most taxa declines as samples are taken at higher latitudes. The extremely high diversity of tropical systems reflects this correlation. So does the low diversity of tundras. Latitudinal gradients are ancient. They have existed among fossil marine foraminifera for at least the last 40 million years and among flowering plants for 100 million years or more.

5 *Unimodal gradients*. In three diversity gradients, diversity peaks at intermediate values of the correlate instead of rising or falling monotonically with it. These are the ecological DISTURBANCE gradient, the ecological PRODUCTIVITY gradient (Fig. D8) and the elevational gradient along mountain slopes.

Understanding diversity gradients requires confronting their interdependence and their operation at several scales of time and space. The area gradient operates at four. Tiny areas produce so few individuals that they suffer a small-sample-size bias (*see* DIVERSITY INDICES). As area increases, more individuals are counted, the bias diminishes and diversity seems to rise. The same bias and pattern causes the increase of diversity over very small time intervals. But, in somewhat larger areas, diversity increases because more habitats occur in the sample. How large is the area required to escape the sample-size bias and be affected by increasing habitat diversity? That depends on the typical POPULATION DENSITY, the mobility and even the lifespan of the taxa being studied. Stout, sessile life forms like trees may yield only one individual in an area that harbours dozens of birds and tens of thousands of insects. Time again parallels the area pattern because habitats can be restricted temporally as well as spatially. Longer time intervals mean more seasons and more chance for habitats to change. The value of the exponent, the z-value, for relationships governed by habitat increase varies between 0.1 and 0.2 independently of the taxon or biome studied.

Data for larger-scale studies typically come from islands or separate, whole biogeographical provinces. These data are usually logged separately for each such unit instead of being accumulated. Thus, in a study of island land snails, Hawaii and Oahu are treated as separate areas, each with its separate species list. But in smaller scale studies, 1 km^2 of forest would contain both the areas and the lists of all smaller areas in the same study. Accumulating island data in the manner of small-scale studies imparts a convex-upward curvature to the species–area relationship plotted on logarithmic axes. However, it seems to have little influence on average z-value.

Island diversity depends on the balance between two rates: extinction and new COLONIZATION. Larger islands have more species because they contain more habitats and, were they to have the same diversity as a small island, their larger populations would suffer lower extinction rates. Island species–area curves typically have z-values of 0.25–0.35, indicating they have less diversity than similar areas of mainland. Islands are depauperate because they have no sink species, i.e. species requiring regular IMMIGRATION to maintain a presence in an area. The influence of immigration rate on island diversity also explains why z-values of islands rise to about 0.6 as an island becomes more and more remote. Its remoteness further limits colonization rates, thus depressing diversity still more.

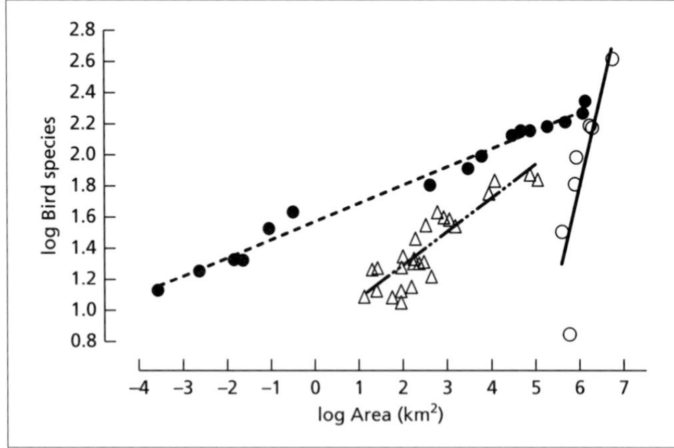

Fig. D7 Three scales of species–area curve for birds. The shallow curve (filled circles) comes from increasingly larger areas within the Mediterranean scrub of Chile; $z = 0.12$. (Data from Cody 1975.) The intermediate curve (open triangles) comes from Caribbean islands; $z = 0.22$. (Data from Wright 1981.) The steep curve (open circles) comes from the tropical rain forests of separate biogeographical provinces ranging in size from the Australian to the Neotropical; $z = 1.23$. It includes only the frugivorous species. (Data from Fleming *et al.* (1987) and Rosenzweig (1995) which see for complete citations.)

Diversities of continents, i.e. separate biogeographic provinces, depend on rates of SPECIATION and extinction. A larger province, were it to have the same diversity as a smaller one, would support a higher speciation rate and a lower extinction rate. Because such rates are usually small compared to immigration and local extinction, the latter usually dominate and increase diversity beyond the level it would reach from internal speciation alone. So, a subarea of a large province has more species than a similar-sized island, which in turn has more species than a similar-sized whole, small province. Consequently, among provinces, z-values exceed 0.6. For example, the z-value for tropical vertebrate frugivores in different provinces approximates to unity.

We do not understand why diversity falls over higher productivities, but it probably rises over lower productivities because low productivities support low POPULATION SIZES, low enough to have a dominant influence on extinction rates. For this reason also, the diversities of higher animal trophic levels are less than those of lower ones. Diversity seems to rise in each TROPHIC LEVEL only until each average species gets about the same energy as one in a lower level.

The provincial area pattern together with the productivity pattern appear to explain latitudinal gradients. Tropical biomes support large diversities because tropical biomes are larger. They cover more area than others for three reasons: (i) longitudes converge at the poles, so that the size of the circle described by each latitude diminishes from the Equator to the poles; (ii) northern and southern tropical latitudes abut and so combine to double the area of either alone; and (iii) climate changes little across all the latitudes of the tropics, but rapidly across other latitudes. Meanwhile, the low productivities of high northern latitudes diminish the effectiveness of the holarctic TUNDRA, the Earth's second most extensive terrestrial biome, to support diversity (otherwise we would expect it to have the second highest diversity of any terrestrial province). Other exceptions to the general relationship between provincial area and latitude (the northern and southern Pacific gyres) do show greater diversity than the smaller tropical Pacific province.

The elevational gradient may also depend on a combination of area and productivity, but its mechanics remain ill defined. In con-

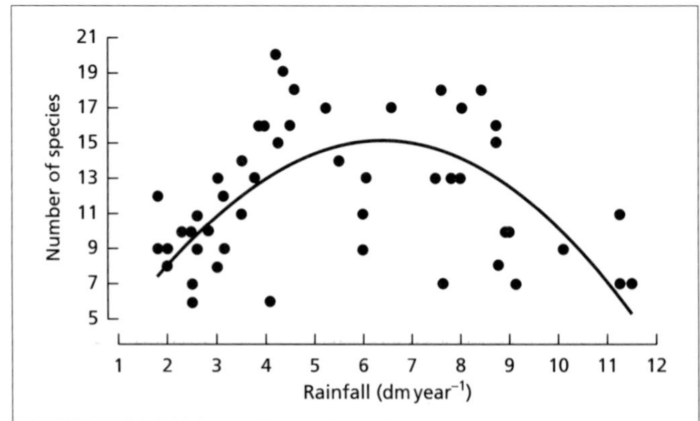

Fig. D8 Productivity–diversity curve for African herbivorous mammals larger than 10 kg. (After Rosenzweig, M.L. (1995) *Species Diversity in Space and Time*. Cambridge University Press, Cambridge.)

trast, the causes of the disturbance pattern are known. A disturbance is anything that destroys most or all life in a patch. Frequently disturbed patches rarely receive a full complement of species because they are usually interrupted by a disturbance before immigration has had the time to deliver them. Rarely disturbed patches usually contain only the species that have prevailed in COMPETITION and survived predation after a full complement of species has arrived. Intermediate patches usually receive most species but do not last long enough to see them winnowed away by population interactions. The disturbance gradient is a small-scale pattern that assumes no changes in the number or identity of species in a region.

The habitat gradient arises from the habitat specializations that evolve among species (*see* DIVERSITY, ALPHA, BETA AND GAMMA). However, the fineness of those specializations increases with the size of a province because large provinces contain more species at their steady-state diversity. Thus, species in a tract of TROPICAL RAIN FOREST within a large province recognize more habitats than those in a similar-sized tract of tropical rain forest within a smaller province. *See also* BIODIVERSITY GRADIENTS. [M.L.R.]

REFERENCE

Rosenzweig, M.L. (1995) *Species Diversity in Space and Time.* Cambridge University Press, Cambridge.

Diversity indices

Measures of species DIVERSITY play a central role in ECOLOGY and CONSERVATION BIOLOGY. Following the collection, sorting and species identification of samples from habitats, the ecologist is often left with large datasets including long lists of species and numbers of individuals of each species. A diversity index is technically a numerical expression or descriptive statistic that summarizes certain properties of such datasets and that can be used to allow comparisons to be made between the diversity of different groups of organisms within a COMMUNITY or between different communities themselves. Comparisons of this nature are useful in that they can potentially contribute to an understanding of processes that structure those communities. Diversity itself has two fundamental properties: (i) the number of different types of species in the sample or community; and (ii) the relative number of individuals in each species. The number of species is generally referred to as SPECIES RICHNESS (S). The relative abundance (p_i) or degree of DOMINANCE of individuals amongst species is usually referred to as evenness or EQUITABILITY. These two properties can be used to evaluate any component of biological diversity from the diversity of alleles for a particular genetic locus, to diversity of diet of predators, to the diversity of habitats in a landscape or of landscape types in a geographical region.

There are, in fact, three major groups of diversity index: (i) species-richness measures; (ii) SPECIES-ABUNDANCE MODELS with an associated diversity index; and (iii) indices based on proportional abundance of species that incorporate both properties of diversity (richness and relative abundance). The choice of appropriate index from the bewildering variety available depends on such factors as difficulty in appraisal of species abundance and the success in sampling and identifying all species present.

Species-richness indices

As a measure of diversity, species richness has great intuitive appeal, providing an instantly comprehensible expression of diversity and avoiding many of the pitfalls of the more complex indices described later.

Species DENSITY is the most commonly used measure and is simply given as the number of species per specified collection area. However, given that SPECIES–AREA RELATIONSHIPS occur and the species–area curves are not linear, it is important that comparisons amongst habitats or communities are based on standardized sampling areas. By sampling progressively larger areas of a habitat, species accumulation curves can be derived which, for well-known floras and faunas, rise to an approximate asymptote where the total number of species in the area has been recorded.

Numerical richness is an alternative, easily understood, measure of diversity based on species richness per fixed number of individuals, normally 1000.

Complete counts of species are possible for temperate and polar higher plants (thus the index is widely used by botanists) and is often possible for birds, mammals, reptiles and fish. For insects, soil invertebrates and tropical plants, only samples of community richness are normally possible. A rare species is likely to be present in a sample if the total number of individuals sampled is great enough so that at least one individual of the rare species is captured. Thus, the observed number of species will vary with sample size and in fact may greatly underestimate the actual number of species in the habitat or community. Where complete counts are not possible, methods are available to estimate the actual number of species by extrapolation from samples of various sizes (reviewed by Colwell & Coddington 1994). These include the following.

(a) Jackknife estimates: for example, $S^* = s + (n - 1/n)^k$, where s is the observed number of species in n quadrats/samples and k is the number of unique species (to a single sample/quadrat). This method tends to overestimate species richness.

(b) The alternative bootstrap procedures require simulation on a computer: $B(S^*) = s + \Sigma(1 - p_i)^n$. A RANDOM SAMPLE of n individuals is drawn from q quadrats/samples within the computer, s is the observed number of species in the original data, and p_i is the proportion of bootstrap quadrats that have species i present. The procedure is repeated usually 50–200 times to increase precision of the estimate of total species richness.

A number of easy-to-calculate indices have been derived using a combination of the number of species (S) and total number of individuals (N) sampled. Margalef's diversity index is the best known:

$$D_{MG} = \left(S - 1/\ln N\right)$$

and is similar to Gleason's index:

$$D_G = S/\ln N$$

These vary from 0 to >30 depending on sample size. Menhinick's index is a variant on the above:

$$D_{MN} = S/\sqrt{N}$$

These indices are all arbitrary measures, strongly influenced by sample size and thus little used today.

Species-abundance models with associated diversity index

Two of the major species-abundance distributions have been adopted as providing some measure of community or sample diversity.

Log series index alpha or Fisher's alpha

$$\alpha = N(1-x)/x$$

where N is the total number of individuals in the sample and x is estimated from the iterative solution of:

$$S/N = (1-x)/x[-\ln(1-x)]$$

(In practice, x lies between 0.9 and 1.0.) The disadvantage of alpha is that it is based solely on S (number of species) and N (total number of individuals), and thus cannot discriminate between communities where S and N remain constant. Because Fisher's alpha requires minimal information, it may be applied to older reports in the literature. Such reports often omit the number of individuals in each separate species, preventing one from using any other index to correct for sample-size bias.

Log normal index

$$\lambda = \ln 2 \Big/ \left[2a(\ln s_0)^{1/2}\right]$$

where a approximates to 0.2 and s_0 is the number of species in the modal abundance class of the LOG NORMAL DISTRIBUTION. This index is quite complex and not widely used.

Q statistic

This presents an interesting approach to the measurement of diversity, based on the inter-quartile slope of the cumulative species-abundance curve (Fig. D9).

$$Q = \frac{\frac{1}{2}n_{R1} + \sum n_r + \frac{1}{2}n_{R2}}{\log\left(\frac{R2}{R1}\right)}$$

where $^1/_2 n_{R1}$ is half the number of individuals in the class where the lower (25%) quartile (R1) falls, $^1/_2 n_{R2}$ is half the number of individuals in the class where the upper (75%) quartile (R2) falls and Σn_r is the total number of species between the quartiles. Q may be biased in small samples and is little used.

Indices based on proportional abundance of species

These diversity indices are a function of species richness and evenness. Diversity can thus increase as evenness increases while species richness actually falls slightly.

Shannon index (sometimes referred to as Shannon–Wiener index)

This is perhaps the best-known index but has rather a weak link to ecological theory and may have a substantial bias in species-rich communities. It is now less preferred to other indices such as Simpson's diversity (see below).

$$H' = -\sum p_i \log p_i$$

where p_i is the proportion of the total number of individuals or total biomass composed of species i. Values usually fall between 1.5 and 3.5 (rarely >4.5).

Shannon evenness measure is given by:

$$E_s = H'/\log s$$

where s is the number of species and $\log s$ is maximum diversity. E varies from 0 to 1. Any log base can be used, but consistency is needed when comparing between samples or estimating evenness.

Brillouin index

This is closely related to the Shannon index and is only appropriate where the community is completely censused with every individual accounted for (e.g. in the light trap sample):

$$\text{HB} = \ln N! - \sum \ln n_i / N$$

where N is the total number of individuals in the sample and n_i is the number of individuals in species i. Values rarely exceed 4.5. The Brillouin evenness value is also complicated:

$$E_{\text{HB}} = \text{HB}/\text{HB}_{\text{max}}$$

$$\text{HB}_{\text{max}} = \frac{1}{N}\ln\frac{N!}{\left\{\left(\frac{N}{S}\right)!\right\}^{s-r}\cdot\left\{\left[\left(\frac{N}{S}\right)+1\right]!\right\}^{r}}$$

where $r = N - S(N/S)$.

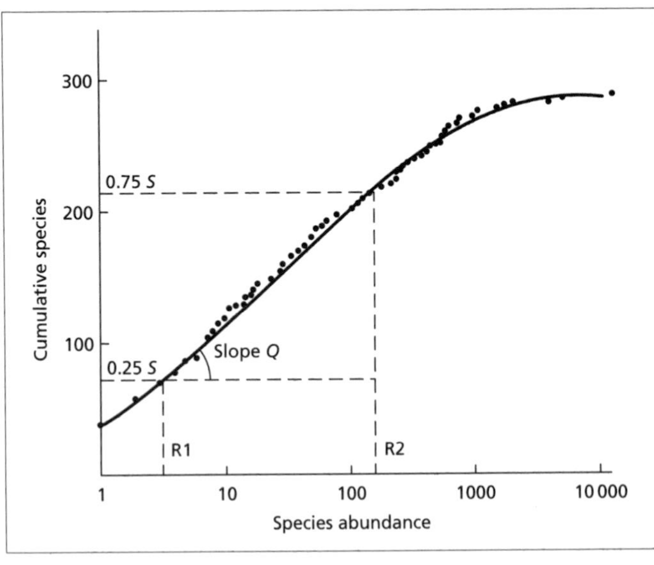

Fig. D9 An illustration of how to calculate the Q statistic of diversity. The plot shows a species-accumulation curve (species abundance on log scale). R1 and R2 are the lower and upper quartiles respectively, and represent the points where the cumulative number of species reaches 25 and 75% of the total. The Q index of diversity is given by the slope of the curve between the two quartiles. (After Magurran, A. (1988) *Ecological Diversity and its Measurement*. Princeton University Press, Princeton, N.J.)

Simpson's index

This is basically a dominance measure weighted towards abundances of the commonest species and less sensitive towards species richness. It is simpler to calculate than the Shannon index and is widely used:

Simpson's C or $\lambda = \Sigma p_i^2$

This measure is actually an index of dominance but can be expressed as a diversity index in the form of $D = 1 - \lambda$ (ranges from 1 to $1 - (1/S)$, where S is the number of species) or more usually as the reciprocal of Simpson's index (also known as Hill's N_2) where $D = 1/\lambda$ (ranges from 1 to S) such that the value increases as diversity increases and dominance reduces. Evenness is given by D/D_{max}, where $D_{max} = 1 - (1/S)$ for $D = 1 - \lambda$ and $D_{max} = S$ for $D = 1/\lambda$.

McIntosh index

This is a little-known index that was derived as a measure of dissimilarity between two communities, but an index of diversity can be calculated as:

$$D = N - U/N - \sqrt{N}$$

where $U = \sqrt{\Sigma n_i^2}$, n_i is the number of individuals in the ith species and N is the total number of individuals. This ranges from 0 to 1. U is MacIntosh's original index and represents the euclidean distance of the ASSEMBLAGE from the origin of an S-dimensional hypervolume.

Berger–Parker index

This is an intuitively simple dominance measure that is easy to calculate:

$$d = N_{max}/N$$

where N_{max} is the number of individuals in the most abundant species. However, diversity is usually expressed as $1/d$, such that the value increases as diversity increases and dominance reduces. This index is independent of S but influenced by sample size, yet it is considered by many authors to be one of the most satisfactory of the indices.

Sequential comparison index (SCI)

One of the more recent approaches to measuring diversity, this index has a strong intuitive appeal and reflects the essence of the species diversity concept. It is also related to Simpson's index:

$$SCI = \text{Number of runs}/X$$

X individuals are randomly drawn from the sample. A run consists of a sequence of individuals of the same species (i.e. each individual is identical to the previous individual selected; Fig. D10). The greater the number of individuals drawn from the sample, the more accurate the value of the index, which ranges from near 0 (where all individuals selected are identical) to 1 (where each individual selected is different from the previous one).

The various indices differ primarily in the degree to which they emphasize species richness or evenness. A single index cannot express all the information needed to describe/characterize something as complex as the diversity of an ecological community, for example which species are present, which ones are dominant and which rare. Many indices also give virtually the same value for very different patterns of species abundance and hypothetical species-abundance patterns can be ranked in reverse order by different indices. However, the indices are usually highly correlated when applied to real biological data. Some indices have more desirable properties than others (Table D2), but depending on the questions being asked and the type of data obtained, one index may be more appropriate than another. *See also* DIVERSITY, METHODS OF MEASUREMENT AND ANALYSIS; RANK–ABUNDANCE MODELS. [P.S.G.]

REFERENCES

Colwell, R. & Coddington, J. (1994) Estimating terrestrial biodiversity through extrapolation. *Philosophical Transactions of the Royal Society of London, Series B* **345**, 101–118.

Lande, R. (1996) Statistics and partitioning of species diversity, and similarity among multiple communities. *Oikos* **76**, 5–13.

Magurran, A. (1988) *Ecological Diversity and its Measurement.* Princeton University Press, Princeton, N.J.

Smith, B. & Wilson, B. (1996) A consumer's guide to evenness indices. *Oikos* **76**, 70–82.

Washington, H.G. (1984) Diversity, biotic and similarity indices. A review with special reference to aquatic ecosystems. *Water Research* **18**, 653–694.

Table D2 A summary of the characteristics and performance of the major diversity indices. The assessments are partly subjective and partly based on comments by various authors in the literature. (After Magurran, A. (1988) *Ecological Diversity and its Measurement.* Princeton University Press, Princeton, N.J.)

Index	Ability to discriminate communities	Sensitivity to sample size	Bias towards richness or evenness	Ease of calculation	Misleading results?	Widely used?
α (log series)	Good	Low	Richness	Simple	No	Yes
λ (log normal)	Good	Moderate	Richness	Complex	No	No
Q statistic	Good	Good	Richness	Complex	No	No
S (species richness)	Good	High	Richness	Simple	No	Yes
Margalef's index	Good	High	Richness	Simple	No	No
Shannon index	Moderate	Moderate	Evenness	Intermediate	Yes	Yes
Brillouin index	Moderate	Moderate	Evenness	Complex	Yes	No
McIntosh *D* index	Poor	Moderate	Dominance	Intermediate	Yes	No
Simpson's index	Moderate	Low	Dominance	Simple/intermediate	No	Yes
Berger–Parker index	Poor	Low	Dominance	Simple	No	No
Sequential comparison index	Moderate	Moderate	Evenness	Simple	No	No

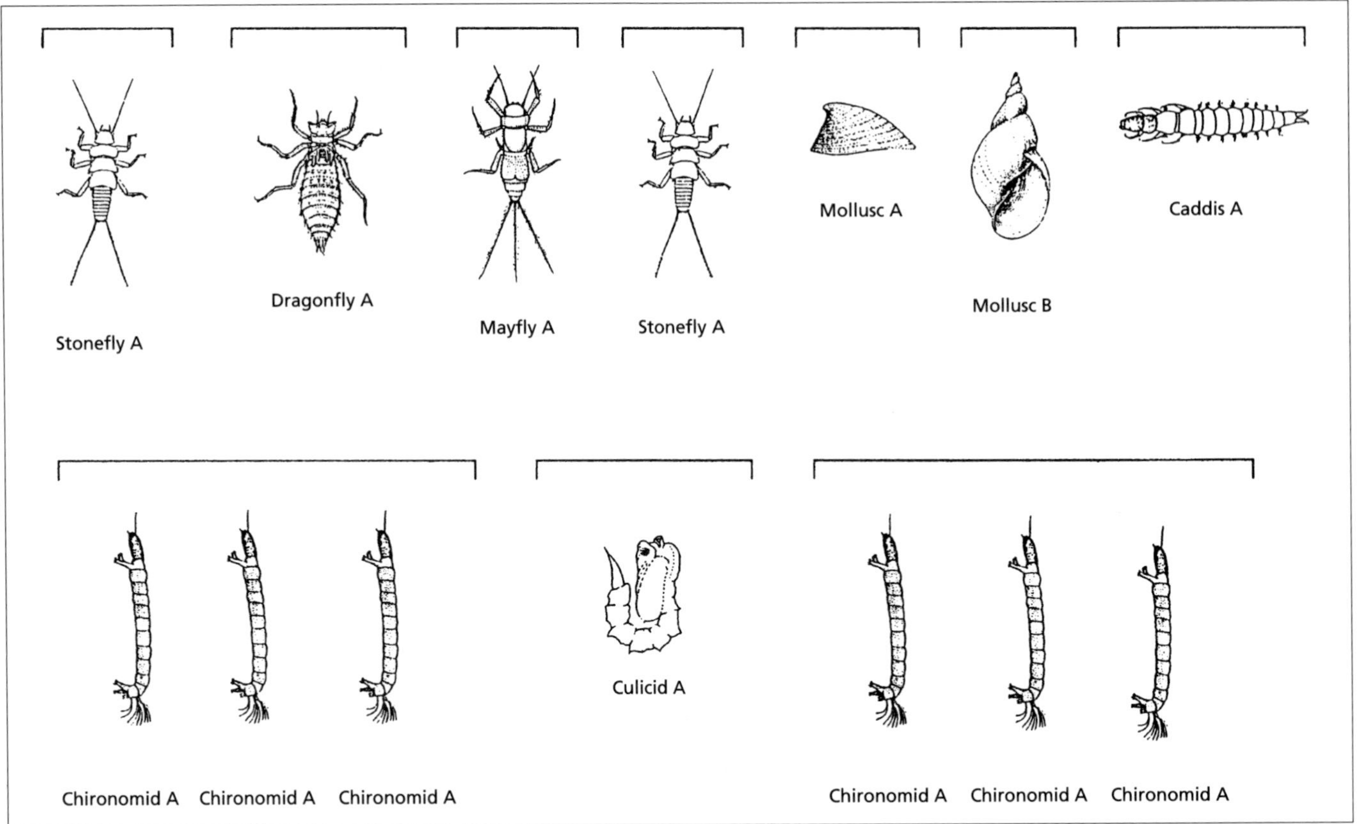

Fig. D10 The sequential comparison index of diversity. Randomly selected organisms from a stream sample are compared in sequence only with the previous specimen. A run consists of a sequence of identical individuals as shown by the brackets in the diagram. (Redrawn from Washington, H.G. (1984) Diversity, biotic and similarity indices. A review with special reference to aquatic ecosystems. *Water Research* **18**, 653–694, with kind permission from Elsevier Science Ltd, The Boulevard, Langford Lane, Kidlington OX5 1GB, UK.)

diversity, methods of measurement and analysis DIVERSITY estimates rely on techniques for censusing populations. Whatever SAMPLING devices or schemes work well on the target TAXA will also be useful for measuring their diversity.

Once raw data exist, many indices (*see* DIVERSITY INDICES) provide numbers that substitute for the number of species. However, Kemp's index may be read directly as 'the number of equally common species', and the rarefaction estimate is read as 'the number of species per sample of N individuals'.

Mark–recapture techniques help estimate the number of species when sampling is incomplete. A species (rather than an individual) observed in one sample is considered 'marked.' Open model analysis of its recaptures assumes that species may enter or leave a system but that all are observed with the same PROBABILITY. Closed model analysis makes the opposite assumptions. Pollock's robust method combines the two, permitting the estimation of diversity in dynamic communities of dissimilar species.

Chao provides two other methods to prevent overlooking rare species in estimating total diversity. The more promising, Chao 2, requires a set of samples:

$$S^* = S + \left(U^2/2D\right)$$

where S is the observed number of species, U the number occurring in only one of the samples and D the number occurring in exactly two samples. [M.L.R.]

REFERENCES

Colwell, R.K. & Coddington, J.A. (1994) Estimating terrestrial biodiversity through extrapolation. *Philosophical Transactions of the Royal Society of London, Series B* **345**, 101–118.

Nichols, J.D. (1992) Capture–recapture models. *Bioscience* **42**, 94–102.

diversity–stability hypothesis Up to the 1970s, our understanding of the relationship between DIVERSITY and stability in ecological systems was essentially based on arguments that used a mixture of observations of nature and reasonable logic. Both suggested that more diverse, or complex, systems were more stable than simple ones. For example, it was observed as early as 1927 that larger FLUCTUATIONS in POPULATION SIZES and more frequent pest outbreaks occurred in high-latitude (species-poor) compared with low-latitude (species-rich) systems. On theoretical grounds it was reasoned that the greater connectedness and SPECIES RICHNESS there is in a system, the more alternative trophic pathways there will be and the less likely it is that the loss of any particular species will result in a system collapse. In the 1970s this conventional wisdom was challenged, primarily by theoretical modelling studies which suggested that in fact the opposite may be true: more complex systems may be less stable than simple ones. Although it is now apparent that the relationship between stability and complexity varies among community types, the nature of the DISTURBANCE and the way stability is assessed, there does appear to be an overall tendency for simpler communities to be more stable. Associated with this conclusion is the observation that stable environments seem to support more complex but fragile communities (*see* COMPLEXITY OF ECOSYSTEMS) and that simpler more robust communities tend to occur in more variable environments. *See also* DIVERSITY INDICES; STABILITY, COMMUNITIES; STABILITY, POPULATION; STABILITY–TIME HYPOTHESIS. [S.J.H.]

division *See* TAXONOMIC CATEGORIES.

division of labour Different units/organisms (modules) in a colonial, modular or SOCIAL ORGANIZATION have different biological/ecological roles (e.g. in terms of FEEDING, DEFENCE and reproduction). *See also* MODULAR ORGANISM. [P.C.]

DNA banking Long-term storage of extracted genomic DNA, typically at −20 or −70°C in a buffered solution. Alternatively, cell lines derived from fibroblasts (from various tissues) are maintained in CRYOPRESERVATION (under liquid nitrogen) from which cells can be cultured and DNA extracted. The latter is the more efficient method as it is a renewable source of DNA; however, tissue culture involves greater expense and initial effort than cold storage of extracted DNA. DNA banking can be used as a strategy towards the conservation and analysis of BIODIVERSITY, and to facilitate population genetic, phylogenetic and biomedical research. [A.R.H.]

DO *See* DISSOLVED OXYGEN.

Dobzhansky, Th. *See* CHARACTERS IN ECOLOGY.

Dollo's law The proposition that EVOLUTION is irreversible or that past states of an evolutionary LINEAGE can only be at most partially re-created. Named after L. Dollo (1893). A variant on this proposition is that an organ, once lost, will never again be present in its original form. Evolution is a historical process and the totality of past evolutionary events can never be exactly repeated. To that extent, Dollo's law is almost a truism. There is, however, no reason why particular characters should not reappear. The simple teeth of toothed whales provide an example. These whales are derived from ancestors that had more complex teeth of mammalian type, but have redeveloped a tooth form seen in reptilian ancestors of the mammals. [L.M.C.]

DOM *See* DISSOLVED ORGANIC MATTER.

domestic bliss strategy *See* CONCORDE FALLACY.

domestication Historical and evolutionary changes in plants and animals when brought under human household (Latin *domus*, home) care and uses. Nearly 10 000 years ago, the so-called Neolithic Revolution took place when hunter-gatherer societies developed AGRICULTURE based on wheat, barley, pea, lentil, flax, sheep, cattle, pig and dog in the Middle and Near East. Evidence from archaeology, ethnobiology and biosystematics suggests three or more such centres of early domestication, with the evolution of maize and *Phaseolus* beans, for example, in the New World. Charles Darwin documented at length a vast array of breeds and varieties in many domesticated species; today much of this genetic VARIABILITY is recognized and conserved for use in breeding. Besides agricultural species, there are numerous exploited captives (cats, dogs, camels, llamas, elephants), game ranching and ZOO animals. Domestication provides an interesting chapter in evolutionary biology for its economic, genetic and ecological aspects of the balance between the exploitation and the conservation of RESOURCES. *See also* ARTIFICIAL SELECTION. [S.K.J.]

REFERENCES

Clutton-Brock, J. (1981) *Domesticated Animals from Early Times*. British Museum, London.

Harlan, J.R. (1993) *Crops and Man*. American Society of Agronomy, Madison.

dominance

1 In BEHAVIOURAL ECOLOGY the term is used to refer to the SOCIAL STATUS of an animal within a group. An individual that maintains high social status by aggressive behaviour towards others is exhibiting dominance. *See also* DOMINANCE, SOCIAL.

2 In VEGETATION studies, dominance refers to the capacity for one species to exert overriding influence upon others within the COMMUNITY. Usually the dominant plant species is the one that has the highest BIOMASS in the ECOSYSTEM and which therefore controls the microclimatic conditions for all others. Occasionally, however, dominance may be exerted by a plant of lower stature, which may control potential invaders or the regeneration of more robust species by competitive means. Therefore, it may be difficult to determine dominance by visual inspection and the use of the term then becomes speculative.

3 In genetics, refers to ALLELES whose effects are expressed in the PHENOTYPE. *Cf.* RECESSIVE GENE.

[P.D.M.]

dominance, behavioural *See* DOMINANCE, SOCIAL.

dominance-controlled communities Communities where competitive interactions between species form a strong hierarchy in which inferior competitors are displaced by the arrival of superior competitors resulting, in the absence of DISTURBANCE, in the development of the COMMUNITY to a predictable end-state dominated by the competitively superior species. The alternative model (FOUNDER-CONTROLLED COMMUNITY) is one in which species are of similar competitive ability and the initial occupier of a patch resists invasion by others. Community change proceeds in a more stochastic manner, depending on chance COLONIZATION by species of vacant patches (usually generated by disturbance). *See also* LOTTERY MODEL; STABILITY, COMMUNITIES. [P.H.W.]

dominance hierarchy Dominance hierarchies are group-level constructs that reflect competitive relationships between individuals within social AGGREGATIONS (where aggregations may be temporary gatherings or cohesive social groups). They arise where competitive interactions between a given pair of individuals are resolved, sometimes quickly and with little aggression, with a consistent bias in favour of one individual. Individuals can thus be ranked into a hierarchy according to their competitive priority (DOMINANCE) over other members of the aggregation. While terms like 'dominant', 'subdominant' and 'subordinate' can be used to characterize different RANK positions, their usefulness depends on the number and transitivity (linearity) of rank positions within the aggregation and the consistency of rank orders across different competitive contexts. In themselves, dominance hierarchies have no adaptive significance. Neither do they necessarily reflect adaptive decision-making by their component individuals: hierarchies can be generated simply because individuals differ in body size and bigger individuals beat smaller ones in random aggressive encounters. In many cases, however, hierarchies appear to reflect assessment of the chances of winning a dispute and decisions to persist or withdraw accordingly. Assessment may be based on indicators of competitive ability, such as SIZE, weaponry or 'badges of status' (e.g. androgen-related plumage markings), or simply on accumulated 'confidence' as a result of the balance of past wins and losses. Such assessment conventions may determine priority of access to important limited RESOURCES like food and mates with little if any aggression, thus conferring a SOCIAL STATUS on each individual. The rewards of high status are likely to set in motion evolutionary ARMS RACES between strategies of CHEATING (sporting appropriate cues of status but not investing resources in competitiveness) and assessment (distinguishing cheats from genuinely high status individuals) with the upshot that status conventions become honest reflections of competitive ability and thus costly to challenge. [C.J.B.]

Dominance, social

Social DOMINANCE may best be defined as a stable asymmetry or predictability in the outcome of aggressive interactions between two or more individuals. However, other definitions have been proposed. For some authors, dominance entails a privileged access to RESOURCES, such as food, nesting sites, or mating partners. The problems with such definitions are that: (i) a stable asymmetry may lead to priority of access to some, but not all, resources thus causing conceptual muddle; and (ii) often the adaptive significance of dominance is explained in relation to priority of access to resources, thus leading to circularity. In practice, one individual in a relationship usually wins any fights or, more often, is reliably displaced from its location by the other; the subordinate may give appeasement or submissive DISPLAYS during such encounters. The direction of the relationship is usually obvious to the observer.

The term dominance is applied to relationships between two or more animals. Thus, whereas an animal can be dominant, it cannot possess 'dominance', as implied by some authors; dominance as a relationship transcends the individual (Bernstein 1981). Although some authors (e.g. Rowell 1974) have suggested that one should speak of subordinancy relationships rather than of dominance, in fact, the relationship is the product of both the dominant and the subordinate individuals.

To qualify as dominance, the relationship must persist over time rather than exist only in brief interactions. The dominance relationship is the result of past interactions which have generated it. Often there is much aggression when a new group is formed; a DOMINANCE HIERARCHY may lead to a decrease in aggression as animals display dominance and subordinate behavioural patterns appropriate for their RANK. This can be to the benefit of all individuals concerned.

The term can be used for either the relationship between two individuals or, when referring to a dominance hierarchy, in a group of animals. Each individual in a group can have a rank in the hierarchy and maintains relationships with a number of individuals. The group hierarchy is sometimes called a peck order, because of the right that dominant chickens express in pecking at subordinates, whereas the subordinates cannot reciprocate. When the dominance relationships are transitive, as for example when A is dominant to B, B is dominant to C, and A is also dominant to C, there is said to be a linear hierarchy. In some cases, there is a non-linear hierarchy, with triangular relationships, as when A is dominant to B and B to C, but C is dominant to A.

Even dominance relationships are not always limited to two individuals. In some species, such as olive baboons (*Papio anubis*), two individuals may form a coalition, reliably backing each other up in dominance CONFLICTS whenever both are present. Thus, two individuals both lower in the hierarchy can, when together, dominate a higher ranking individual. Each animal, then, has not only its primary rank, but its derived rank or leverage, that when the ally is present.

Dominance hierarchies can be formed among the males and females as a group, but more often develop independently in one or both sexes, as in rhesus macaques (*Macaca mulatta*) and red jungle fowl (*Gallus gallus*). The hierarchies of males often are more obvious than those of females and have received more study. In many species, males are generally dominant to females, but in others, such as Syrian golden hamsters (*Mesocricetus auratus*), the females generally dominate.

Dominance and aggression should not be confused. The dominant individual need not be the most aggressive. Sometimes the individuals who are slightly lower in the hierarchy fight more in contesting position than does the alpha individual. What is critical is not that the dominant fights often but that he or she wins when engaging in a dispute.

Dominance interactions can be dramatic or subtle. Two male northern elephant seals (*Mirounga angustirostris*) battling during the rut produce a spectacular battle. By contrast, dominance relations in male bison (*Bison bison*) can sometimes be expressed with gestures that would be unnoticed by the casual observer. When female laboratory rats (*Rattus norvegicus*) pass each other in a narrow area, such as a burrow, either head-to-head or head-to-tail, the passed animal freezes while the passing animal continues walking, thus expressing a dominance relationship.

Young animals must develop the skills that can make them dominant adults over time. Young rhesus macaques often work out a linear hierarchy by the time they are just 1 year old. Body size, age, sex, duration of tenure in a group, subtle social skills, and subtle environmental variables can affect the ability of in individual to become dominant. Genetic factors, as well as non-genetic influences, are important in providing an animal with attributes that permit it to achieve a dominant position in a group. In some species of mammals, such as rhesus macaques and spotted hyenas (*Crocuta crocuta*) maternal rank is 'inherited' via non-genetic processes. That is, the ranks of daughters in a female hierarchy reflect those of their mothers (see Holekamp & Smale 1991). The young animal gets social support from its mother and siblings and thus has an advantage in acquiring high rank.

Among the factors affecting dominance in adult mammals are hormones. Various studies have shown dominant males to have higher testosterone titers than subordinates. The testosterone may be a factor in making a male dominant, but also may be increased as a result of high rank. Subordinate females in a group of primates may show an inhibition of ovarian function. In addition, adrenal function is affected by dominance interactions, with adrenal masses generally increasing in animals exposed to repeated aggression and adrenal function differing between dominant and subordinate individuals. There is thus a complex reciprocal relationship between the hormones and the behaviour.

As noted, the primary adaptive significance of achieving a dominant position is related to privileged access to resources. These in turn, can lead to increased levels of inclusive FITNESS. The dominant individual may have priority in obtaining food, protection, and mating partners. The literature on dominance rank, copulatory behaviour, and differential reproduction is a complex one (Dewsbury 1982). There are numerous studies showing that dominant individuals copulate more and sire more offspring than subordinates. For example, a dominant male deer mouse (*Peromyscus maniculatus*) in a group of two males and two females in a seminatural enclosure will generally copulate more than, and outreproduce, the subordinate. However, such relationships have not been found in numerous other species studied. Factors such as the species, sex, nature of the genetic marker used to indicate PATERNITY, duration of the study, and sampling procedures used in behavioural observations can affect results. Females may prefer mating with dominant versus subordinate males, thus complicating the relationship. Even if the dominant individuals do not produce more offspring, they may be better able to aid their young to survive and prepare them to reproduce than are subordinates.

The concept of dominance has been and remains an important one in understanding animal SOCIAL ORGANIZATION. It should be remembered, however, that there is much affiliative and reconciliation behaviour as well. Studies of animal social organization may have overemphasized the importance of dominance to some degree. Animal social relationships are complex and multidetermined. [D.A.D.]

REFERENCES

Bernstein, I.S. (1981) Dominance: The baby and the bathwater. *Behavioral and Brain Sciences* **4**, 419–429.

Dewsbury, D.A. (1982) Dominance rank, copulatory behavior, and differential reproduction. *Quarterly Review of Biology* **57**, 135–159.

Holekamp, K.E. & Smale, L. (1991) Dominance acquisition during mammalian social development: The 'inheritance' of maternal rank. *American Zoologist* **31**, 306–317.

Rowell, T.E. (1974). The concept of social dominance. *Behavioral Biology* **11**, 131–154.

---

dominance variance Dominance variance V_D and ADDITIVE GENETIC VARIANCE V_A are the two genetic variances deriving from within-locus variation in the QUANTITATIVE GENETICS of a metric trait. The dominance variance is defined in two ways: in statistical empirical quantitative genetics as the variance arising from the deviation of the individuals from their BREEDING VALUES; and in theoretical multilocus quantitative genetics as that part of the GENETIC VARIANCE that is not explained by within-locus linear regression of GENOTYPIC VALUES on the number of A_1 ALLELES per locus in the diploid GENOTYPE.

In theoretical quantitative genetics the dominance variance per bi-allelic locus *A* equals $(2pqd)^2$.

The dominance variance can be estimated from the variance in the mean values of full-sib families in a random mating population, as the expected value of the COVARIANCE of full sibs equals $^1/_2V_A + {}^1/_4V_D$. *See also* MEAN GENOTYPIC VALUE. [G.D.J.]

donor-controlled systems These are 'ideal' systems where the POPULATION SIZES of species at a particular TROPHIC LEVEL are entirely controlled by the ABUNDANCE of species in the trophic level below it. In a simple example of a donor-controlled system, the density of grasses controls the density of herbivores which, in turn, controls the density of predators. However, herbivore density is independent of predator density and grass density independent of herbivores. While such systems are not entirely realistic there are some that appear to follow this pattern quite well and they do make population modelling considerably easier. *See also* POPULATION DYNAMICS; POPULATION REGULATION. [C.D.]

dormancy Resting condition with relatively reduced metabolism that might involve the whole organism, as in higher plants and animals, or be confined to propagules, such as resting spores in fungi and bacteria, resting EGGS in some animals, non-germinating seeds and non-growing buds in plants. Also referred to as hypobiosis. *See also* AESTIVATION; DIAPAUSE; HIBERNATION. [P.C.]

dose–response The quantitative relationship between the amount of a TOXICANT administered or taken (usually by feeding or injection) or absorbed by a subject and the incidence or extent of adverse effects. *Cf.* CONCENTRATION–RESPONSE. [P.C.]

downwelling *See* UPWELLING.

drag The force exerted on a body by a fluid due to the fluid's motion relative to the body. Drag (*D*) is related to the relative fluid velocity (*U*) by:

$$D = \frac{1}{2}\rho C_D A U^2$$

where ρ is the fluid density, *A* is a representative area for the body and C_D is the drag coefficient. C_D is a non-dimensional measure of the relative resistance of bodies of the same area and is a function of REYNOLDS NUMBER.

Drag can be divided into PRESSURE or form drag, skin-friction drag and lift drag. Form drag results from pressure variations around the body due to the flow separating from the body surface, skin-friction drag is a result of the no-slip condition at the body surface and lift drag results from flow spillage from high- to low-pressure areas on lifting surfaces.

Reducing drag improves the efficiency of swimming and flying animals and reduces the risk of wind damage to plants. Form drag is usually the largest contributor to drag and animals use ingenious methods to stop flow separation, even at the expense of increased skin-friction drag. These include triggering TURBULENCE at the surface, generating vortices at leading edges and blowing fast-moving fluid into the surface flow. Plants reduce their form drag by streamlining of branches and the rolling up of leaves. Skin-friction drag can be reduced by the use of SURFACTANTS and bubbles, whilst behavioural adaptations such as 'porpoising' and coordination of the flapping of a fish's tail with the vortices it produces can have substantial drag benefits. [B.A.G.]

REFERENCES

Bushnell, D.M. & Moore, K.J. (1991) Drag reduction in nature. *Annual Review of Fluid Mechanics* **23**, 65–79.

Tritton, D.J. (1988) *Physical Fluid Dynamics*. Oxford University Press, Oxford.

drainage Drainage is a network of underground pipes or surface ditches that moves surface and near-surface waters efficiently to adjacent streams. They are employed either where the WATER TABLE is near the surface or where permeability is low causing surface water ponding. Waterlogging or ponding restricts AGRICULTURE, industry and human occupancy. Cities have highly impermeable surfaces and complex drainage systems (sewerage) occur below city streets. Sewerage may be combined with sewers (WASTE flows) and are a significant determinant of WATER QUALITY in receiving streams.

In agricultural areas, tile-drains were used first but today drains are made from corrugated plastic pipes. To improve efficiency and to reduce silting the pipe circumference is expanded by an envelope of coarse material. The envelope may be natural (gravel, shells), organic (peat, flax, cocoa matting) or synthetic (fibreglass, rock-wool, polypropylene/styrene grains). Drainage has significant environmental consequences because altered water conditions promote changes to vegetation communities. In some areas the drained water cannot flow by gravity and has to be pumped up to a higher level. Here, drainage forms a part of LAND RECLAMATION. [A.M.]

drainage lakes *See* LAKES, TYPES OF.

dredge sampling *See* SAMPLING METHODOLOGY/DEVICES.

dredgings Muds removed from the bottom of water bodies by machine. Both the process of removal and subsequent dumping can disturb environments. [P.C.]

drift, evolutionary Random FLUCTUATIONS of gene frequencies in populations caused by random sampling of GAMETES and/or GENOTYPES in reproduction. Drift is a key process in the 'shifting balance' theory and in the neutral theory of molecular EVOLUTION.

Because gamete production according to Mendel's law of segregation is a stochastic process, genotype frequencies in subsequent generations fluctuate even in perfectly constant environments, as well as in the absence of any selection. Besides in gametogenesis, random sampling can occur between genotypes, for example under clonal reproduction the genotype distribution fluctuates due to random birth and death processes. If selection operates, drift causes deviations from the expected shift in genotype frequencies.

Drift is most important in small populations, but the fate of rare ALLELES is determined by random sampling also in very large populations. Drift induces fixation and loss of alleles and, thereby, can cause a decay of genetic VARIABILITY. In the absence of new MUTATIONS, the HETEROZYGOSITY, h, in populations of diploid organisms decreases due to drift as $h(t)=h(0)(1-(2N_e)^{-1})^t$ with time t measured in generations and N_e denoting the (variance) EFFECTIVE POPULATION SIZE, which can be viewed most easily as a pool of $2N_e$ genes with equal probability of being transmitted. With an input of μ mutations per locus per individual per generation, a balance between drift and new mutations will be reached at an equilibrium heterozygosity of $h=4\mu N_e/(1+4\mu N_e)$.

Given that a neutral mutation becomes fixed, its expected time until fixation is about $4N_e \pm 2.15N_e$. The fixation rate of neutral mutations is independent of population size and equals the mutation rate μ. The probability of ultimate fixation of a neutral allele is equal to its initial frequency ($=1/(2N_e)$ for new occurring mutations). Non-neutral mutations with a SELECTION COEFFICIENT, s, become fixed with the probability of about $[1-\exp(-2sN_e/N)]/[1-\exp(-4sN_e)]$ in a population of actual size N. *See also* NATURAL SELECTION; QUANTITATIVE GENETICS. [W.G.]

REFERENCES

Crow, J.F. & Kimura, M. (1970) *An Introduction to Population Genetics Theory*. Harper & Row, New York.

Ewens, W.J. (1979) *Mathematical Population Genetics*. Springer-Verlag, Heidelberg.

Kimura, M. (1983) *The Neutral Theory of Molecular Evolution*. Cambridge University Press, Cambridge.

Nagylaki, T. (1992) *Introduction to Theoretical Population Genetics*. Springer-Verlag, Berlin.

Wright, S. (1968–1978) *Evolution and Genetics of Populations*, Vols 1–4. University of Chicago Press, Chicago, Ill.

drift, freshwater When benthic organisms of streams or RIVERS enter the water column and are moved downstream, the phenomenon is known as drift (reviewed in Brittain & Eikeland 1988).

Many benthic organisms in streams possess morphological and/or behavioural adaptations to remain adhered to, or closely associated with, the substrate. However, many of these organisms display distinct DIEL periodicities in drift rate, with the most commonly observed pattern being nocturnal maxima just after sunset and just prior to sunrise. Several different hypotheses have been suggested for the adaptive significance of drift, ranging from it being an antipredator escape response to a mode of ACTIVE DISPERSAL to escape overcrowded (i.e. food limited) or undesirable conditions.

The hypothesis that risk of predation constrains drift activity has gained support from the results of empirical studies which demonstrated that prey size classes with the greatest risk to size-selective predators also exhibit greater predisposition for night-time drift. Diel periodicity in drift activity has also been observed in streams that have had visual predators experimentally removed, suggesting that nocturnal activity is a fixed behavioural response to predation and is not simply a direct consequence of DIURNAL feeding by predators. However, chemical signals released by fish appear to reduce total drift activity of some species (e.g. *Gammarus*) rather than causing peak drift activity to shift to night-time hours.

Two major classes of drift have been recognized: catastrophic and behavioural (Waters 1972). Catastrophic drift most often results from direct physical DISTURBANCE of the substrate. It is indiscriminate with respect to the taxa involved and is particularly evident during periods of high DISCHARGE during the spring RUN-OFF period in temperate regions and during periods of flooding from excessive precipitation. Other ABIOTIC FACTORS, such as increasing temperature, anchor ice, POLLUTION and toxicants, have also been implicated in inducing catastrophic drift.

Two hypotheses have been proposed regarding the mechanisms involved in behavioural drift: passive and active drift-entry. Passive drift-entry involves the dislodgement of an organism into the water column and subsequent downstream transport that results from direct contact with the water current as an indirect consequence of an individual's behaviour. Increased movements associated with foraging, escape from predators and increasing water flow over respiratory surfaces have all been implicated in producing passive drift. Active drift-entry involves individuals actively releasing themselves from a substrate to enter the water column and is the direct result of a behavioural adaptation that increases an individual's fitness by having it disperse to a new area downstream. Thus, the major distinction between active and passive drift is whether fitness benefits are directly incurred by the individual.

Müller (1974) suggested that for members of the merolimnic fauna (only part of the LIFE CYCLE in the aquatic medium) with the ability to fly or walk during the non-aquatic phase, there should be compensatory upstream movements. He proposed the 'colonization cycle' for downstream drift and the compensatory upstream flight by adults. There are few data to support this hypothesis, although some species have been shown to have limited capabilities to move upstream in the BOUNDARY LAYER or by moving from one sheltered location to another, although comparisons of upstream and downstream within-substrate movement have generally shown that downstream movement is higher than upstream movement. *See also* DISPERSAL; MIGRATION. [R.W.D.]

REFERENCES

Brittain, J.E. & Eikeland, T.J. (1988) Invertebrate drift. *Hydrobiologia* **166**, 77–93.

Müller, K. (1974) Stream drift as a chronobiological phenomenon in running water ecosystems. *Annual Review of Ecology and Systematics* **5**, 309–323.

Waters, T.F. (1972) The drift of stream insects. *Annual Review of Entomology* **17**, 253–272.

drift nets *See* SAMPLING METHODOLOGY/DEVICES.

drillings Materials removed from drilling activities on land or in water. Both the removal of materials and their dumping can disturb environments. [P.C.]

drinking-water quality The quality of WATER delivered at the tap for the consumer. It should therefore be judged in terms of human consumer requirements for appearance, taste and safety. However, there are complications. Thus substances used in sterilization are, by definition, potentially toxic. Nevertheless, WHO (1993) guidelines state that 'The potential consequences of microbial contamination are such that its control must always be of paramount importance and must never be compromised'. [P.C.]

REFERENCE

WHO (1993) *Guidelines for Drinking Water Quality*, 2nd edn. World Health Organization, Geneva.

Drought

Definition of drought

Drought has no universally accepted definition, although it is normally recognized as a sustained and regionally extensive state of WATER deficiency. Frequently associated with famine and aridity, it should not be confused with either: famine can be caused by other natural and human agencies than drought; aridity is a permanent state whereas drought is ephemeral. Some regions of the Earth's surface are particularly prone to drought: sub-Saharan Africa (Sahel), Brazil's Nordest, the Midwest of the USA, southern Iberia and much of Australia. What characterizes such regions is the tendency for anomalous dry conditions to persist. However, with demand for water resources often finely tuned to its normal availability, few locations can be regarded as drought-proof.

Drought is a needs-related notion; it has to impact upon someone or something to be recognized as such. Figure D11 encapsulates three broad bases for defining drought. Meteorological drought is related to large negative excursions in P–PE, a measure of the inability of PRECIPITATION to meet the atmosphere's demand for water. Hydrological drought is measured by a shortfall in P–AE, which equates to the availability of water for RUN-OFF or recharge of soil and groundwater. Agricultural (equivalent to ecological) drought is related to large values of PE–AE, reflecting vegetation's inability to meet its intrinsic demand for water.

Mechanisms

Conditions for drought arise from a combination of local and remote events in the ATMOSPHERE. Locally there must be warm dry air aloft with slowly descending AIR MASSES. The condition is sustained by successive high-pressure cells, which can often be linked to remote factors such as long waves, sea-surface temperature anomalies and jet-stream displacement causing a blocking of anticyclones. Tropical-zone drought is most serious in areas with long DRY SEASONS where impairment of the monsoon mechanism brings drought. Persistence can be enhanced locally via the so-called Charney mechanism in which soil drying and vegetation die-back increase surface ALBEDO. The locality then loses more radiative

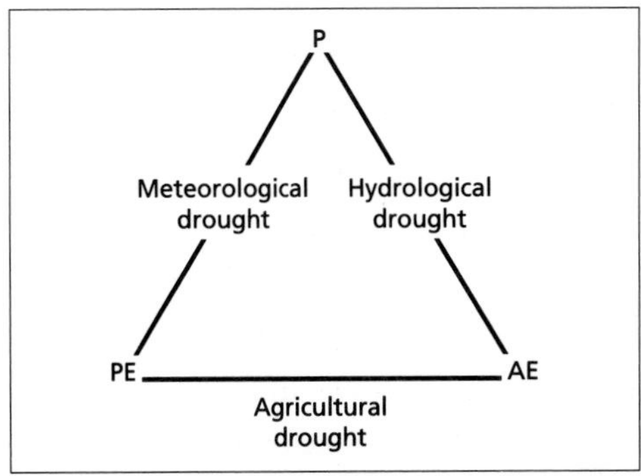

Fig. D11 Schema for water availability as used in broad-scale drought definition. P, precipitation; PE, potential evaporation; AE, actual evaporation.

energy than it gains from SOLAR RADIATION to create a local circulation that imports heat aloft to maintain thermal equilibrium against the descending warm air.

Numerical indexing of drought severity

P, PE and AE, evaluated for a specified area and duration, singly and in combination as in Fig. D11, provide the most frequently used indices of drought severity. Drought can be defined as a quantile, for example 80% exceedance frequency, or a departure from the average, for example less than 40% of the long-term average rainfall. The Palmer drought index is in common use in the USA and is based on a locally standardized monthly balance of P and PE. Because of its standardization a single value, for example –4, can be set as a drought threshold to initiate protection measures. In the UK, the Meteorological Office's MORECS (Meteorological Office Rainfall and Evaporation Calculation System) system provides similar data for SOIL MOISTURE deficit, by maintaining a daily soil moisture balance whose main features are a diminution of AE–PE as the deficit increases, and run-off occurring when P exceeds AE.

For general descriptive purposes it may suffice to index drought as the fraction of normal precipitation in the affected region and SEASON, or even, as the UK Meteorological Office does, by runs of rainless days. From the standpoint of a specific requirement such a 'portmanteau' index is not adequate. However, it can be justified on the grounds that the observations can be cheaply obtained, rapidly processed and readily understood. Also precipitation may be thought of as the primary driver for drought, even ecological drought, as AE is constrained by available moisture.

The numerical index can be considered as a random variable to be analysed statistically. This allows droughts to be ranked objectively and return periods to be calculated. Annual precipitation anomaly over the study region is likely to be approximately normally distributed. The Weibull distribution is designed for use with extreme values and is theoretically the best suited to fit bounded variables such as minimum lake level or river flow:

$$p(x) = (\gamma/\theta)x^{(\gamma-1)} \exp-(x^{\gamma}/\theta)$$

where $p(x)$ is the density of x, the variable of interest, and θ and γ are scale and shape parameters. The variate value x can be replaced by $x-\varepsilon$ to yield a three-parameter distribution with $\varepsilon<x<\infty$.

Drought forecasting and monitoring

Drought's precursors should permit a forecast of impending drought to be attempted. The simplest is based on 'teleconnections' between the precursor phenomenon and the event of interest. Most attention has been given to forecasting CROP yield from EL NIÑO/SOUTHERN OSCILLATION INDEX, sea-surface temperature and atmospheric pressure fields. There is recent interest in using the GULF STREAM position of Newfoundland to predict ecological phenomena in the UK. The forecast ability of this correlative approach is low.

Nowadays, pressure fields are increasingly used as starting conditions for running a GENERAL CIRCULATION MODEL to project conditions for the season ahead. This is best developed for the Sahel region although, up to now, seasonal precipitation can be forecast with only a moderate level of precision, with no more than 50% of the interannual variance explained.

Drought MONITORING is important for emergency management, for example provision of food, agistment of stock. Most make use of the network of hydrological and CLIMATE stations in the drought-afflicted region. A regional drought management centre can produce maps in near real-time of key variables such as soil moisture content or the necessary precipitation to make up the deficit. Increasingly, remote sensing is being used to augment the ground-based data. For example cloud-top temperature is used routinely in drought-prone regions to monitor rainfall. Fire and dust can be monitored, vegetation indices such as NDVI (normalized differential vegetation index; *see* REMOTE SENSING) and thermal imagery assist in delineating droughted areas. Prospects exist for the future application of visible and near-infrared radiometry to detect greenness and day–night differences, which in turn relate to chlorophyll amount and surface wetness.

Ecological significance of drought

The energy for CARBON fixation by vegetation derives from the photolytic splitting of water into HYDROGEN and OXYGEN during PHOTOSYNTHESIS. Much still remains to be learned of the mix of mechanisms used by plants to recognize and signal water shortage. It is not surprising, given the fundamental biotic role of water, that drought incidence has been instrumental in determining evolutionary directions and ECOSYSTEM properties. The C_4 and CAM photosynthetic pathways are believed to be in part an evolutionary response to conditions of repeated water shortage.

These are reflected in large-scale patterns of plant geography with drought-adapted vegetation structure reappearing at points of the globe even if taxonomic affinities are very different. Predictions from mathematical simulation models of global BIOME distribution and regional carbon balance are strongly conditioned on future moisture supply and its year-on-year variation. Gas exchange is moisture dependent, with biogenic gases such as METHANE and nitrous oxide being mediated by regional hydrological conditions acting through redox and oxic status.

At global scale, strong positive correlations have been established between moisture availability and BIODIVERSITY. Within an area, centres of DIVERSITY are associated with HABITAT transition, i.e. spatial. Though clearly a negative influence during the period of drought, it is speculated that episodic stresses such as drought also maintain diversity in the long term by providing opportunities for successional changes.

Drought and desertification

The land-surface mechanisms responsible for local persistence referred to previously can be induced or aggravated by human activities and lead to DESERTIFICATION. For example, OVERGRAZING, defoliation or destruction of soil structure through concentration of herds create artificial droughts. Localized run-on zones can erode topsoil, concentrate NUTRIENTS restricting its availability to the few remaining shrubs, and adversely and irreversibly affect radiation balance and momentum exchange. The product of this process is a man-made DESERT.

Drought and climate change

There have been speculations as to whether GLOBAL WARMING and the greenhouse effect will increase drought frequency. Global warming implies increased EVAPORATION within a strengthened hydrological cycle. The balance between the two opposing forces, and their seasonal phasing, suggests a complex spatial pattern of increased and diminished drought risk. Simulations by general circulation models reveal a tendency towards aggravated drought risk. This is due variously to the effect on evaporation demand outstripping the increase in moisture supply, altered SEASONALITY of soil moisture from a shortened snow season and a transient effect in which continental interiors warm up more rapidly than oceans. A further complicating factor is the influence of increased atmospheric CARBON DIOXIDE (CO_2) concentration. Elevated CO_2 increases water use efficiency in many species so may assist by reducing the impact of drought. [M.B.]

---

dry mass The mass of a part of an organism, a whole organism or all the organisms from a given area after removal of its moisture by EVAPORATION. It is used in preference to wet mass (i) because it rather than wet mass is the result of PRODUCTION, and (ii) because it is a more conserved quantity. Plant WATER contents can vary substantially over the course of a day, in some cases by as much as 20%. When rapidly transpiring plants are harvested, a rapid loss of wet mass can occur before weighing is complete leading to further substantial errors. Because an organism may include organic constituents with a boiling point below that of water, drying is normally conducted at 80°C in a forced-draught oven and continued until a constant mass is obtained; see Chiariello *et al.* (1989) for a detailed account of alternative methods. In production studies, plant dry mass is often used as a surrogate of energy content or mass of organic matter. In this case, the ash-free dry mass is more appropriate, i.e. the dry mass less the mass of residue of inorganic matter remaining after complete combustion of the plant material in a muffle furnace at *c.* 500°C. [S.P.L.]

REFERENCE

Chiariello, N.R., Mooney, H.A. & Williams, K. (1989) Growth, carbon allocation and cost of plant tissues. In: *Plant Physiological Ecology. Field Methods and Instrumentation* (eds R.W. Pearcy, J. Ehleringer, H.A. Mooney & P.W. Rundel), pp. 327–366. Chapman & Hall, London.

dry season A period of the year when PRECIPITATION is greatly reduced or absent. The term is particularly used of tropical and SUB-TROPICAL regions in which precipitation (often monsoonal) is confined to particular times of the year and the remainder of the year is dry. Such a dry season can induce leaf-fall in tropical DECIDUOUS FORESTS and general die-back in tropical GRASSLANDS. [P.D.M.]

dump Site used to dispose of solid wastes. Lack of controls are usually implied. [P.C.]

dumping at sea Transporting and disposal of domestic and/or industrial wastes in the open SEA, usually beyond COASTAL or tidal areas. Coming under increasing national and international controls. [P.C.]

dune An elevated landform produced by the activity of WIND upon sand. *See also* SAND DUNE. [P.D.M.]

Duplex soils *See* SOIL CLASSIFICATION.

durational stability *See* SELECTION PRESSURES, CLASSIFICATION.

duripan horizons *See* SOIL CLASSIFICATION.

dust Minute solids, light enough to be suspended in air (usually <25 μm); wind-blown soils; mechanically produced particles. [P.C.]

duty of care Legal requirement (either by statute or civil law liability) that person takes all reasonable steps to avoid problems to human health and environment that might arise from an operation, and/or steps to ensure certain requirements are met with regards to how the operation is managed. Thus, the UK ENVIRONMENTAL PROTECTION ACT 1990 specifies a legally binding duty of care with respect to solid WASTE DISPOSAL. This requires the appropriate documentation of transfer notes to licensed operators, so that the fate of waste can be effectively tracked from factory gates to ultimate destination. [P.C.]

d_x In a LIFE TABLE the proportion of individuals that die in each AGE CLASS or in each stage. Calculated as $l_x - l_{x+1}$. The values of d_x are additive so that the total MORTALITY at ages 3, 4 and 5 is $d_3 + d_4 + d_5$ but, unlike q_x and k_x, they give little indication of the intensity of mortality at each age since, even with the same MORTALITY RATE, d will be large if there is a large number of organisms present and at risk of dying. [A.J.D.]

dynamic equilibrium A dynamic equilibrium occurs when some characteristic of a system is in an unchanging state, despite the fact that its constituent parts may be changing. For example, the size of a population in which IMMIGRATION and birth exactly balances EMIGRATION and death will be in dynamic equilibrium. The individuals making up the population are changing constantly and the birth, death and MIGRATION rates may all be varying, but the size of the population is constant. By contrast, the spacing between songbird territories may be adjusted in spring, but as territories become fixed the spacing reaches a static equilibrium. *See also* POPULATION REGULATION. [S.N.W.]

dynamic meteorology *See* METEOROLOGY.

dynamic pool models These models were developed in the 1950s by two British scientists, Ray Beverton and Sidney Holt, developing earlier work by Russian and British authors. The basic idea of the models involves a full examination and documentation of the underlying demographic processes of birth, death and growth. In the simplest form, as developed by Beverton and Holt, the models follow a COHORT of young fish born at an assumed fixed time that suffer natural MORTALITY until they reach an age when they become subject to fishing. From this point the mortality inflicted is the combination of natural mortality and that due to fishing. During the entire period of the LIFESPAN of this cohort, growth is occurring according to some fixed model: in the original formulation by Beverton and Holt this was the VON BERTALANFFY EQUATION.

The utility of these models is that, for a fixed level of RECRUITMENT, the behaviour of the cohort mimics the behaviour of the whole population. At equilibrium it is then possible to examine two key parameters, which can determine the level of YIELD that can be taken from the population. These key parameters are the age at which fishing starts (determined, for example, by the size of net) and the mortality due to fishing (determined by the level of fishing). In the simplest formulation by Beverton and Holt, contours of yield can be generated by examining the way in which yield varies with the age at which fishing starts and the level of fishing mortality. In this way it is possible to assess what is the optimum combination of fishing effort and age at first capture, and examine individual fisheries to see the extent to which they are suboptimal.

Elaboration of this simple idea involves the relationship between recruitment and the size of the spawning stock. A central problem of the earlier approach was that recruitment was considered to be constant, while in fact there is a clearly defined relationship between recruitment and stock size. There is likely to be some maximum recruitment, and at low stock sizes, certainly at zero, recruitment has to be at zero.

Furthermore, recruitment in natural populations of fish is highly variable; hence elaboration of this type of analysis involves the consideration of stochastic variations around some basic stock and recruitment relationship. These models are extremely data hungry and require a good knowledge of the age composition of a fish stock.

The British scientist John Gulland developed an extension of these models to an estimation technique known as virtual population analysis (VPA). Sophisticated versions of this technique are still the basis of most stock assessments in developed countries. *See also* DEMOGRAPHY; FISHERIES, CONSERVATION AND MANAGEMENT; POPULATION DYNAMICS. [J.B.]

dynamic programming A method of modelling dynamic decision-making problems in which current choices are likely to influence options in the future. The approach relies on knowing the required end-state at the outset, then working back to the starting state seeking appropriate choices at each decision point. While the method is useful in principle, it can be difficult to use in non-trivial cases and becomes impractical when states have many dimensions. Despite this, it has been used with some success in modelling dynamic optimization problems in animal behaviour, good examples being risk-sensitive time and energy budgeting in high-energy-demand predators and optimizing the reduction of multiple physiological deficits through consummatory behaviours. *See also* EVOLUTIONARY OPTIMIZATION; OPTIMAL FORAGING THEORY. [C.J.B.]

dynamic state variable *See* OPTIMAL CONTROL THEORY.

dynamically fragile If the dynamics that a system displays are very sensitive to parameter changes or small perturbations then the system is dynamically fragile. *Cf.* DYNAMICALLY ROBUST. [S.N.W.]

dynamically robust Dynamics that are fairly insensitive to parameter changes or perturbations are robust: the same patterns will occur for a relatively wide range of circumstances. *Cf.* DYNAMICALLY FRAGILE. [S.N.W.]

dysprosium (Dy) Dysprosium is a lanthanide, one of the so-called rare earth metals (ATOMIC NUMBER = 66; relative ATOMIC MASS = 162.50; melting point = 1409°C; boiling point = 2562°C), having an average abundance of 3 mg kg^{-1} in the Earth's CRUST. Together with other rare earths it occurs in several minerals, especially monazite and bastnaesite. It has a bright metallic lustre, soft enough to be cut with a knife, and it is stable at room temperatures. It absorbs neutrons well and has potential applications in the nuclear industry. With VANADIUM and other rare earths, it is used in making laser materials. [J.G.]

dystrophic A term applied to a shallow FRESHWATER lake in which the presence of organic materials lends a dark brown colour to the water. Such LAKES are of low biological PRODUCTIVITY and have poor light penetration. Cf. eutrophic and oligotrophic (*see* EUTROPHICATION). [P.D.M.]

E

EA (Environmental Agency) *See* ENVIRONMENTAL (PROTECTION) AGENCIES.

early-life-stage tests ECOTOXICITY test carried out on GAMETES, embryos or juveniles. These stages are often more sensitive to chemical stress than other stages in the LIFE CYCLE. [P.C.]

Earth Summit *See* RIO SUMMIT; UNCED.

Earth–Sun geometry *See* RADIATION.

East African floral region

The East African floral region can be divided into two main areas: the Sudano-Zambezian flora and the flora of the Indian Ocean coastal belt. The flora of Madagascar and smaller Indian Ocean islands is so distinct, with over 80% of flowering plants not found on the mainland, that it is better treated separately.

The Sudano-Zambezian flora covers most of Africa south of the Sahara and outside the tropical forests. In East Africa it can be divided into three main provinces: the Somalo-Ethiopian PROVINCE of the Ethiopian plateau; a Sudanian province in the area of southern Sudan and northern Uganda; and the rest to the south, part of the huge Zambezian region. A separate flora of the AFRO-ALPINE VEGETATION has sometimes been recognized, scattered on the mountains and uplands throughout East Africa like an archipelago of islands, but the flora of each 'island' has its closest affinity with the surrounding area. The Somalo-Ethiopian province is found on the highly dissected Ethiopian plateau. It is an area of high endemism. Isolated from moister FOREST areas to the south by DESERTS and the grassy flood plains of the White Nile, woody species would have suffered most in the colder and drier conditions at the height of Pleistocene cold phases. Tenuous links to the north along the Red Sea Hills provided a route for COLONIZATION of Mediterranean elements. Contrasting assemblages can be recognized in areas at different altitude and rainfall: berha, *Acacia–Commiphora* bushland lies below 500 m; kolla, *Acacia*-dominated bush and WOODLAND lies between 500 and 1500 m; weyna-dega SAVANNAH, dominated by *Acacia* in drier areas and enriched with *Ficus* and *Cordia* in moist areas. Most of this province's natural woodland and forest lies between 2300 and 3200 m, from which *Juniperus, Hagenia, Podocarpus* and bamboo has been lost. At greater altitudes lie the subalpine heaths and GRASSLANDS of the Afro-alpine vegetation.

In the Sudanian province, important genera are *Anogeissus, Terminalia*, the tall grasses *Hyparrhenia, Andropogon* and trees *Isoberlina*. The African mahogany *Khaya senegalensis* is a characteristic species. The main Zambezian province to the south has woodlands, savannahs and grasslands. The very varied TOPOGRAPHY supports a wide variety of plant assemblages. Plant assemblages are transitional, ranging from typically tropical areas to more temperate areas of the highlands and mountains. Species of *Brachystegia, Julbernardia* and *Isoberlina*, tall flat-topped trees, are conspicuous or dominant. Light reaches below the CANOPY to support a rich grassland and herbaceous flora. On the moist lower slopes of the mountains an EVERGREEN forest gives way to the unique Afro-montane flora. At drier lower altitudes there are extensive savannahs of DROUGHT-tolerant grasses with thorn trees *Acacia* or thickets of *Commiphora*.

In the relatively narrow belt near the coast there is a different flora. The topography is varied and the vegetation is a mosaic of forest, bush forest or bush. The flora is particularly interesting because several species have a scattered wide distribution in Africa. This has encouraged some to suggest that the area has served as a refuge for many taxa throughout the Pleistocene changes. There are vicariant species of the walnut *Octoknema* and the aromatic *Polyceratocarpus*, which show an ancient link with the tropical forests of the Congolese Basin. The area has also provided an important source of colonists for the young volcanic mountains of the East African interior. Floristic diversity here is due in large part to the greater coastal rainfall. The relatively small East Usambara Mountains in Tanzania act as a rain-trap and are by far the most diverse part of the coastal belt in the north. A large proportion of the larger woody species are endemic with 92 forest trees endemic. In the southern coastal belt from Tongoland to Pondoland more than 200 woody species are endemic.

In the south it is a mixture of SUBTROPICAL forest and grassland. Although there are some links to the Sudano-Zambezian and the Madagascan floras, about 40% of species are endemic. They include *Stangeria*, a distinct kind of cycad. This is the centre of diversity of another genus of cycad, *Encephalartos*. [M.I.]

REFERENCES

Archibold, O.W. (1995) *Ecology of World Vegetation*. Chapman & Hall, London.
Takhatajan, A. (1986) *Floristic Regions of the World*. University of California Press, Berkeley, Calif.

EC_{50} Statistically derived concentration of XENOBIOTIC that has a defined adverse EFFECT (often behavioural, e.g. 'dancing' movement in *Daphnia*) in 50% of an observed POPULATION over a prescribed time in defined conditions. Also referred to as median effective concentration. Can also refer to exposures in terms of dose (ED_{50}). *See also* MEDIAN EFFECTIVE CONCENTRATION/DOSE. [P.C.]

ecdysone *See* METAMORPHOSIS.

ecoaccident zones *See* RECLAMATION.

ecoarchitecture Design of buildings with an intention to make them more in keeping with their ecological surroundings and/or to minimize their impact, and/or that are based on ecological principles. An example of the latter is the ecohouse or autonomous house, which is supposed to simulate ECOSYSTEMS by creating autonomous cycles of materials, for example wastes are converted to fuel by anaerobic digestion for methane production; solar and wind energy is used for creating power, etc. [P.C.]

ecoaudit Business management tool involving a systematic, documented, regular and objective evaluation of the performance of the business, management systems within it and measures designed to protect the environment. It may involve an independent and expert body but can also be internal. It requires information upon which auditors can express opinion against prescribed standards and/or targets. The opinion of the auditors may or may not be made public. Ecoaudits are a key part of ENVIRONMENTAL MANAGEMENT SYSTEMS. [P.C.]

ecobalance Same as LIFE-CYCLE ASSESSMENT. [P.C.]

ecocline *See* CLINE; ZONATION.

eco-controlling Way of making strategic choices about products and/or industrial production processes or about business options in a way that integrates economic and environmental considerations using a so-called 'ECO-RATIONAL PATH method'. [P.C.]

REFERENCE

Schaltegger, S. & Sturn, A. (1992) Eco-controlling. In: *Green Business Opportunities* (eds D. Koechlin & K. Müller), pp. 229–240. Pitman, London.

eco-efficiency More efficient use of materials and energy in order to reduce economic costs and ENVIRONMENTAL IMPACTS, i.e. 'more from less'. [P.C.]

ecofeminism *See* ENVIRONMENTAL ETHICS AND CONSERVATION.

ecofunds Finances made available on the basis of speculative investment from the private sector for industrial projects that will make a positive contribution to environmental protection, for example the development of CLEAN(ER) TECHNOLOGY. Stakes in these funds are often referred to as ecoshares. *See also* ETHICAL INVESTMENT; GREEN FUNDS. [P.C.]

ecolabel Label awarded to product deemed to have less ENVIRONMENTAL IMPACT per unit function, judged on the basis of a LIFE-

Fig. E1 Environmental labels in use in Germany, Canada, Japan, the Nordic countries and the European Union.

CYCLE ASSESSMENT, than others designed to carry out the same or similar function. Some labels are illustrated in Fig. E1. [P.C.]

ecological agents *See* DEFENCE MECHANISMS.

ecological constraint hypothesis The hypothesis that certain individuals aid in the reproductive efforts of others, either as 'helpers at the nest' in cooperative breeders or as 'workers' in eusocial species, because they are restricted in their ability to disperse or breed on their own and are thus 'making the best of a bad job'. The alternative is that helpers or workers experience higher FITNESS by aiding the breeding efforts of others compared to breeding independently because of intrinsic advantages derived via helping.

Ecological constraints may involve space, mates or the RESOURCES necessary to successfully raise young. The first proposes helping at the nest as the next best option for individuals failing to obtain territories suitable for breeding and includes the 'habitat saturation' hypothesis for delayed DISPERSAL in cooperative breeders (*see* COOPERATIVE BREEDING, EVOLUTION OF). Similarly, individuals failing to obtain mates may benefit by opting to delay dispersal and act as helpers. Reproduction may also be limited by the inability of some individuals to acquire the resources needed for successful breeding. This is most likely in species where conditions are highly variable and rearing young is especially costly (see Plate 4, facing p. 398). *See also* BEHAVIOURAL ECOLOGY. [W.D.K.]

REFERENCES
Emlen, S.T. (1982) The evolution of helping. I. An ecological constraints model. *American Naturalist* **119**, 29–39.
Koenig, W.D., Pitelka, F.A., Carmen, W.J., Mumme, R.L. & Stanback, M.T. (1992) The evolution of delayed dispersal in cooperative breeders. *Quarterly Review of Biology* **67**, 111–150.

ecological efficiency Ecological EFFICIENCY (also called LINDEMAN EFFICIENCY or transfer efficiency) is a measure of the proportion of the energy assimilated by one TROPHIC LEVEL that is then assimilated by the trophic level above it. This is thus a measure of the efficiency with which energy is transferred between levels of a trophic pyramid and is defined as:

$$\text{Ecological efficiency} = \frac{\text{Assimilation at trophic level}}{\text{Assimilation at trophic level}_{n-1}}$$

It was originally thought that this might be a predictable constant for different trophic levels in different ECOSYSTEMS and it was suggested that approximately 10% was a widespread value, based on laboratory studies of water fleas (*Daphnia* spp.) feeding on phytoplankton. Further research has shown that it is neither constant nor clearly predictable; values can vary from *c.* 1%, for example for wolves feeding on moose on Isle Royale in Lake Superior (Colinveaux & Barnett 1979), up to 70% in some marine FOOD CHAINS (Pepita *et al.* 1970).

One consequence of low ecological efficiencies is that organisms at the base of the FOOD WEB are much more abundant than those at higher trophic levels, which explains the shape of pyramids of numbers and biomass. These illustrate graphically the consequences of the rapid loss of energy between plants and herbivores and between herbivores and carnivores, which is a biological manifestation of the second law of thermodynamics.

The ecological 'inefficiency' is partly due to energy being lost from the system as heat, usually monitored by measuring the RESPIRATION of members of a trophic level, and partly because no trophic level is completely assimilated by the level above it: some energy is always diverted into the decomposer pathways either as dead bodies and/or as undigested faecal remains. The realization that a majority of the energy incorporated as net PRODUCTIVITY by members of one trophic level usually flows into these saprophytic pathways has promoted much research into the ecology of decomposers. *See also* ASSIMILATION EFFICIENCY; ENERGY FLOW; ECOLOGICAL PYRAMID; PRODUCTION EFFICIENCY. [M.H.]

REFERENCES
Colinveaux, P.A. & Barnett, D.B. (1979) Lindeman and the ecological efficiency of wolves. *American Naturalist* **114**, 707–718.
Pepita, T.S., Pavlova, E.V & Mironov, G.N. (1970) The food web structure, utilisation and transport of energy by trophic levels in the planktonic communities. In: *Marine Food Chains* (ed. J.H. Steale), pp. 142–167. University of California Press, Berkeley, Calif.

Ecological energetics

Ecological energetics concerns the transfer of energy that occurs within ECOSYSTEMS measured at the level of the individual organism, populations, group of populations or TROPHIC LEVEL. Detailed studies of the transformations of energy that occur within the cells of an organism are usually considered to be the realm of physiological energetics. The ecologist interested in energetics is primarily concerned with the quantity of incident energy per unit area of the ecosystem and the efficiencies with which this energy is converted by organisms into other forms. The term BIOENERGETICS has been taken by some to subsume ecological energetics, but by others it has been used more restrictively to refer to the energetics of domestic (especially agricultural) animals, and by others to refer to cell energetics.

The initial question in a study of ecological energetics might be to ask how efficiently the AUTOTROPHS in the system convert solar energy into the potential chemical energy of plant protoplasm during the processes of primary PRODUCTION. It may be of interest to know what proportion of gross primary productivity is then lost again as heat due to the metabolic processes of plants obeying the basic laws of thermodynamics and, thus, what proportion remains as net primary productivity to be potentially available for uptake by consumers, for example humans and their stock.

The partitioning of food materials and energy for an individual animal can be seen as a series of dichotomies (Fig. E2). This scheme can be used to derive a series of efficiencies of energy transfer through the animal.

The proportion of the net productivity from the lower trophic level ingested by an animal or group of animals is the exploitation efficiency:

$$\text{Exploitation efficiency} = \left(\text{Consumption}\left(C\right)/\text{Prey production}\right) \times 100$$

Of the energy consumed (C), some will be lost in the faeces (F) and some excreted with the urine (U) (often in practice inseparable, so

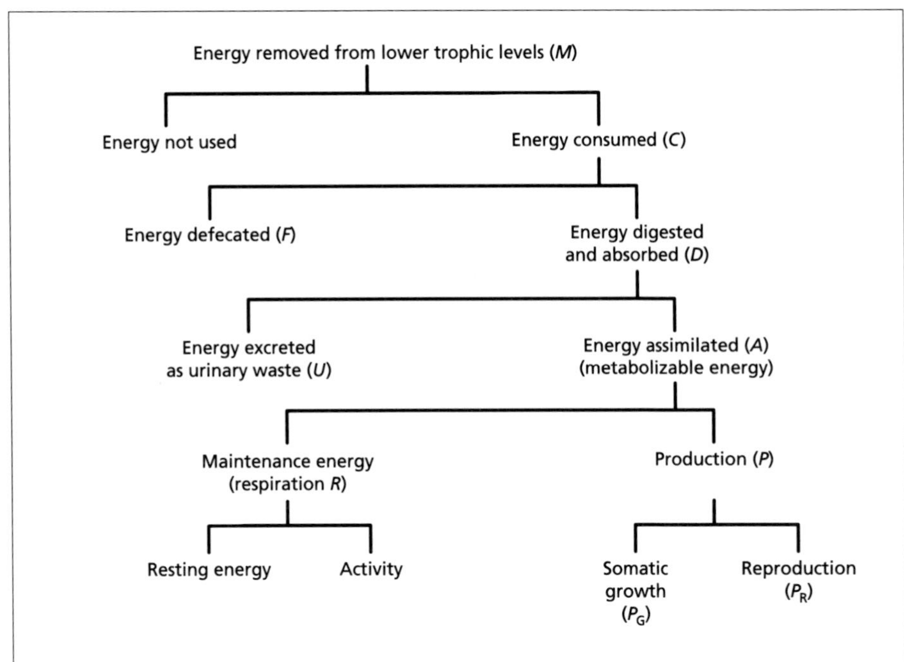

Fig. E2 Partitioning of energy within an animal. (After Krebs, C.J. (1972) *Ecology. The Experimental Analysis of Distribution and Abundance*. Harper & Row, New York.)

called rejecta and abbreviated FU). The remainder of the digested and absorbed energy is assimilated (A), so:

$$\text{Assimilation efficiency} = A/C \times 100$$

where A is usually measured as $A = C - (F + U)$. Some of the assimilated energy is used to fuel the body's basic metabolic and maintenance needs and some is converted to kinetic energy of movement to enable the animal to find food, avoid predators and undertake social interactions. The remainder is available for developing new tissues or STORAGE materials for both GROWTH (P_G) and reproduction (P_R), which together contribute to the net productivity (P) of the animal or trophic level. The efficiency of this process is the net PRODUCTION EFFICIENCY:

$$\text{Production efficiency} = (P_G + P_R)/A \times 100$$

Combining ASSIMILATION EFFICIENCY and net production efficiency gives the gross production efficiency:

$$\text{Gross production efficiency} = P/C \times 100$$

The values for these efficiencies vary enormously between and within ecosystems so that few broad generalizations are possible. One is that most predators have higher assimilation efficiencies than an average HERBIVORE feeding on plant tissues with a large proportion of structural components. The cell walls of plants are made from very recalcitrant molecules such as CELLULOSE and LIGNIN, which animals find much more difficult to digest than other animal tissues. Decomposer animals usually have even lower assimilation efficiencies as they consume that which has been rejected by the herbivores and discarded by the plants. Thus they have to 'scrape the very bottom of the nutritional barrel' for the most indigestible dregs. *See also* BOMB CALORIMETER; ECOLOGICAL EFFICIENCY; ENERGY FLOW. [M.H.]

---

ecological/environmental disasters The word 'disaster' connotes large and sudden adverse effects; when applied to the environment it usually has consequences for both human health and ecological systems. The following is a list of some of the much publicized large-scale disasters:

- the release of DIOXINS after a major chemical explosion at Givandan's plant at Seveso in 1976, mainly affecting human health;
- *Amoco Cadiz*, a supertanker that ran aground off Brittany in March 1978 discharging >200 000 tonnes of crude oil, causing the deaths of thousands of birds and extensive damage to marine life;
- leak of poisonous FUMES from Union Carbide's plant at Bhopal in 1984, mainly affecting human health;
- the explosion at the Soviet nuclear plant at Chernobyl in 1986, affecting both human health and ecological systems;
- a major industrial ACCIDENT at the Sandoz chemical plant in Basle, Switzerland, 1986 when the RUN-OFF from the water used in firefighting carried >13 tonnes of chemical into the River Rhine causing serious damage to flora and fauna and polluting the underlying AQUIFER;
- OIL SPILLS into Prince William Sound off Alaska from the *Exxon Valdez* in 1989, with immediate effect on the marine ecosystem;
- Saddam Hussein's deliberate oil releases in Kuwait during the Gulf War, with immediate effect on the marine ecosystem;
- oil spill from the *Braer* along the Shetland Island's southern coastline in 1993, with immediate effect on the marine ecosystem.

Cf. NATURAL DISASTERS. [P.C.]

Ecological experiments

Ecological, or field, experiments bridge the gap between laboratory experiments and observations, and simple observations and measurements of the natural world. The attraction of ecological experiments is realism: the full complexity of the environment in which animals and plants live remains, yet variables of interest are manipulated in a controlled and replicated manner so that hypotheses about their role can be tested. One of the most common types of ecological experiment is where the densities of a particular species is manipulated in some way and the responses of the COMMUNITY are observed. DENSITY manipulations are achieved in a variety of ways including, for example, fishing, removal or addition by hand or spraying with insecticides. An important objective for the experimenter is to minimize and control for the effects of the method itself. For plants and sessile or slow-moving animal species, or when the HABITAT itself has a natural boundary (e.g. islands or lakes), there is little difficulty in maintaining desired densities. In other cases, however, it is often necessary to use physical or chemical barriers, such as fences, cages or toxic paint, to surround experimental plots and keep species at higher and/or lower densities than those in the unmanipulated environment. Designing adequate controls for the effects of such structures is a key part of field experimentation. Other types of treatments in ecological experiments include the MANIPULATION of nutrient or light levels, or the degree of DISTURBANCE.

Carefully controlled field experiments have an important place in the ecologist's armoury because investigators are often more confident that the results observed are relevant for the real world than those which could be obtained in laboratory experiments where potentially important aspects of the system cannot be simulated. Despite these advantages, field experiments have undoubted drawbacks. Although experiments are rarely repeated because of the high cost, when it does occur the results can often be different, at least in their details. It has been argued that this apparent lack of REPEATABILITY results from the complex web of indirect interactions between species (*see* INDIRECT INTERACTION). Even when prey species densities increase with the removal of a predator as expected, it is not always the case that all prey species increase and it is often impossible to predict which ones will respond. Where repeatable results are obtained, it may be because the experiments are not run for long enough for the effects of indirect interactions to be seen. The apparent limits to the predictability of ecological interactions between species in a community have important implications for how ecological experiments are interpreted. Another important issue for experimental studies concerns the spatial and temporal scales of the experimental manipulations and the levels of treatment replication that can be achieved. These are often limited by both high costs and technical difficulty (*see* SCALE). Thus, many important ecological questions are simply not amenable to the experimental approach and must be tackled in other ways. *See also* EXPERIMENTAL DESIGN; EXPERIMENTAL ECOLOGY. [S.J.H.]

REFERENCES

Bender, E.A., Case, T.J. & Gilpin, M.E. (1984) Perturbation experiments in community ecology: theory and practice. *Ecology* **65**, 1–13.

Yodzis, P. (1988) The indeterminacy of ecological interactions as perceived through perturbation experiments. *Ecology* **69**, 508–515.

---

ecological footprint The area of land functionally required to support a human COMMUNITY, which lies beyond the land occupied by that community, i.e. 'appropriate CARRYING CAPACITY'. [P.C.]

Ecological genetics

Ecological Genetics was the title of an influential book written by Professor E.B. Ford of Oxford University. Ford had employed the term for many years in lectures and one of his colleagues, P.M. Sheppard, had previously used it as a chapter heading in his book *Natural Selection and Heredity* (1958). Ford and his book on ecological genetics were influential primarily because they provided the impetus for a 'school' of like-minded researchers employing the same general approach. In North America, Th. Dobzhansky and his colleagues provided a similar impetus both in terms of approach and direction.

Ecological geneticists combine field and laboratory studies to investigate the adjustments and ADAPTATIONS of wild populations to their environment. Their work has concentrated on the EVOLUTION of a number of adaptive TRAITS, for example INDUSTRIAL MELANISM in moths, shell colour and banding variation in *Cepaea* snails, insecticide RESISTANCE in insects, heavy-metal TOLERANCE in plants, BATESIAN MIMICRY in butterflies, wing spotting in meadow brown butterflies and chromosome inversions in fruit flies and mice.

Industrial melanism remains a prominent textbook example of the evolution by NATURAL SELECTION of an adaptive trait in response to a changing environment. The peppered moth *Biston betularia* was uniformly a pale-winged species before the Industrial Revolution and the effects of AIR POLLUTION began to change the resting background of the moths in industrial regions. Sometime around the middle of the 19th century in north-west England, a mutant *carbonaria* individual occurred with wings darkened by increased deposition of black melanin pigments. As in the majority of other industrial melanics the *carbonaria* allele is dominant to the wild type, implying that at least some offspring of melanic moths are always themselves black in colour. The consequence of DOMINANCE of a favourable mutant is that selection leads to a much more predictable and rapid increase in frequency than for a favoured, recessive mutant. In industrial regions melanic peppered moths tend to be more closely matched to the tree bark on which they rest than the pale 'wild-type' moths. During the Industrial Revolution, this bark surface had become darkened by particulate air pollutants and devoid of pale, epiphytic lichens and algae due to gases such as sulphur dioxide (SO_2). Strong natural selection then favoured both a rapid increase and a spread of the *carbonaria* allele (male moths can disperse over long distances) through the more effective background matching, or CRYPSIS, of the melanic moths reducing their mortality from insectivorous birds relative to the pale moths. Declines in smoke and SO_2 in the industrial regions of north-west Europe since the 1950s have led to new changes in the relative FITNESS of melanic and non-melanic phenotypes and rapid declines in melanism. This has been termed 'evolution in reverse', although it does not represent precisely a return to the situation that existed before air pollution arose.

Molecular techniques were not available in the formative years of ecological genetics. Consequently, there was an emphasis on working with species like the peppered moth that exhibit conspicuous GENETIC POLYMORPHISM. They were used to study geographical VARIABILITY among and between populations and the process of adaptation. POPULATION and QUANTITATIVE GENETICS became increasingly important in providing the theoretical framework within which to explore the consequences for evolutionary change of such variables as fitness, GENE FLOW (MIGRATION), GENETIC DRIFT and MUTATION. Several successful examples of the application of this framework to ecological genetics have involved surveys of CLINES in gene frequency in which a particular allele progressively changes in frequency along a TRANSECT. A particularly impressive example also involved the peppered moth. J.A. Bishop studied a cline from industrial Liverpool to rural North Wales in the frequency of the *carbonaria* allele; *carbonaria* declines in frequency with distance away from the industrial conurbation. A mathematical model based on field and laboratory data can provide a very close fit to the observed cline. The model incorporates details of population biology including rates of adult migration (gene flow), estimates of changes in relative fitness resulting from spatial variation in visual selection and CAMOUFLAGE (crypsis) along the gradient in air pollution, and differences between the phenotypes in other, 'non-visual', aspects of fitness.

Polymorphic variation

Many investigations in ecological genetics began with surveys of polymorphic variation among populations of a particular organism combined with an attempt to relate any spatial patterns to changes in some environmental gradient. This initial approach also required a good working knowledge of the natural history of the organism. The observation of some correlation between the polymorphic or phenotypic variation and an environmental gradient suggested hypotheses about natural selection (in the case of the peppered moth, that the agents of selection were bird predators hunting by sight in combination with historical or spatial variation in the resting background of moths producing differences in the effectiveness of their background matching and crypsis). The hypothesis can also be described in the context of a FITNESS LANDSCAPE: how does the variation in reproductive success correlate with spatial variation in an environmental factor? More detailed tests of this hypothesis usually necessitated additional work on the population biology of the organism, laboratory experiments to determine patterns of inheritance (the genetical basis of the polymorphic variation) and some combination of laboratory and field experiments to investigate the relationship between environment and fitness. Frequently, further work on a particular organism and phenotypic trait would be extended to studies of populations in other regions. Examples of parallel clines or patterns of variation in different regions (or species) can provide powerful evidence in support of adaptive explanations. Industrial melanism, for example, has evolved in an essentially similar manner in over 100 species of moth in Britain.

Some of the most successful studies in ecological genetics initiated 30 or more years ago are continuing to the present day. The early studies of industrial melanism in moths have been supplemented by work on other examples of melanism, notably in the two-spot ladybird, *Adalia bipunctata* (see Fig. M15). This beetle exhibits forms that are black with red spots (melanics) and others which are red with black spots (non-melanics). In these conspicuously coloured insects, there is little or no correlation between melanism and air pollution and no evidence for the involvement of visual selection and crypsis comparable to the moth examples (black two-spot ladybirds occurred in the 17th century). Rather, climatic selection occurs with melanic beetles, for example, being favoured in conditions of limited sunshine and low air temperatures because of the absorption properties of a predominantly black, as against red, cuticle. The black colour leads to more incident radiation being absorbed and thus to higher body temperatures and more activity in cool conditions. Genetic polymorphisms involving colour patterns have been intensively studied in other invertebrates, especially butterflies and snails. Selection by visual predators has frequently been implicated as an important factor resulting in observed patterns of spatial variation among populations.

In plants, numerous ecological geneticists have studied the evolution of heavy-metal tolerance. Particularly successful studies have been made of the process by which local populations of certain plant species have adapted to heavy-metal contamination of spoil deposits from mines. The boundary of contaminated soil, which represents an ECOTONE, can be associated with extremely steep changes (or stepped clines) in the frequency of genes conferring tolerance. This work has led to valuable contributions in landscape and restoration ecology. Other ecological studies that have made important contributions in an applied context include the evolution of resistance to insecticides in many agricultural pests and to the poison warfarin in rats, which is unsurprising because the activities of humans can have a rapid and dramatic effect on the environment of other organisms. Human activities frequently generate intense natural selection; if populations are to persist, rapid changes in adaptive traits are necessary. Such circumstances are precisely those in which the approach of ecological genetics can be especially successful.

Continuous variation and molecular data

Studies have also been made of continuous variation in the PHENOTYPE, which is not under the control of a few 'major' genes but is influenced by many genes and the environment during development. One early example involved the pattern of wing spotting in the meadow brown butterfly, *Maniola jurtina*. This study rose to prominence in the 1950s and 1960s because of the controversy over whether variation in natural populations was primarily influenced by natural selection or was effectively neutral in character and subject to stochastic processes such as random genetic drift. Ford and his colleagues argued that the patterns of variation in wing spotting they observed among populations in the Isles of Scilly archipelago were consistent with selection. Others argued that the data showing divergence among islands could be explained by founder events and genetic drift (Fig. E3).

This type of controversy was continued later when many ecological geneticists began to use gel ELECTROPHORESIS to survey patterns of variation associated with genetic polymorphisms for the enzymes involved in regulating biochemical processes. A few intensive and integrated studies of particular enzyme polymorphisms, for example of the enzyme phosphoglucoisomerase (PGI) in *Colias* butterflies, provide unequivocal evidence for functional differences among GENOTYPES *in vitro* and for variation in fitness and selection in the wild. Many other enzyme polymorphisms are probably effectively neutral in natural populations; in other words any differences in fitness are extremely small and variation in gene frequency is

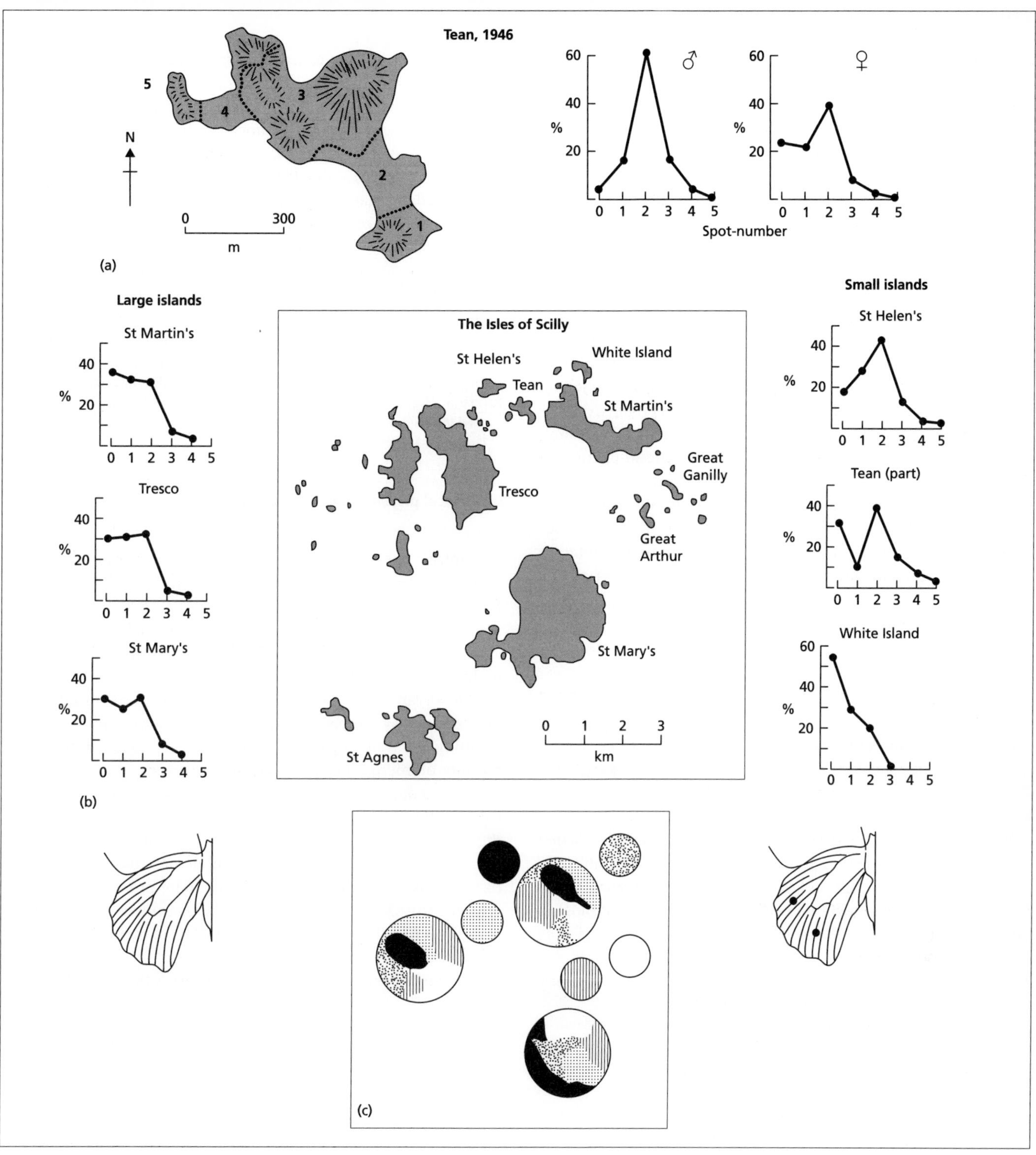

Fig. E3 Variation in the number of spots on the ventral surface of the hindwings of the meadow brown *Maniola jurtina* in the Isles of Scilly; two specimens with zero or two spots are illustrated. (a) A summary of the frequency of males and females with different spot numbers collected by Ford and his co-workers in their original study on Tean. Note the difference between the sexes. Their mark–release–recapture experiments showed that the population inhabiting areas 1, 3 and 5 of the island were largely isolated from one another by unfavourable habitat in areas 2 and 4. (b) Map of the Isles of Scilly showing the spot frequency distributions for females found on three large islands and on three small islands. (c) Diagram of the hypothesis put forward by Ford to account for the dissimilarity among small island populations and the uniformity on large islands. The proposed occurrence of specialized environments on small islands and of mixed habitats on large ones is illustrated by different shadings. (After Brakefield, P.M. & Shreeve, T.G. (1992) Case studies in evolution. In: *The Ecology of Butterflies in Britain* (ed. R.L.H. Dennis), Chapter 9. Oxford University Press, Oxford; by permission of Oxford University Press. Based on material in Berry, R.J. (1977) *Inheritance and Natural History*. Collins, London; Dowdeswell, W.H. (1981) *The Life of the Meadow Brown*. Heinemann Educational Books, London; Ford, E.B. (1975) *Ecological Genetics*, 4th edn. Chapman & Hall, London; Shorrocks, B. (1978) *The Genesis of Diversity*. Hodder & Stoughton, London.)

driven primarily by stochastic processes such as random genetic drift. Attention has now shifted to considering variation in the DNA nucleotide sequence itself.

There are two major legacies of the approach and development of ecological genetics in the fields of modern ecology and evolutionary biology. One is the much more pervasive attempt by ecologists to take genetic variation among individuals within a population into account when interpreting their results; all important ecological textbooks will include chapters on genetical variation and its evolutionary and ecological consequences. This field may be referred to as GENETICAL ECOLOGY. The other legacy is the integrated approach to studying the evolution of adaptive traits, which has its origin in the combination of laboratory and field studies central to early ecological genetics. It is not sufficent to demonstrate and measure natural selection; detailed studies are necessary to understand in biological terms why observed differences in fitness and reproductive success occur among phenotypes and genotypes. The latter question requires the integration of investigations of the gene-to-phenotype transformation with experiments on natural selection and fitness and a knowledge of population biology. Only such combined approaches can produce a complete understanding of the processes of natural selection and of any CONSTRAINTS involved in adaptation of organisms to variation in their environment. *Cf.* QUANTITATIVE GENETICS. [P.M.B.]

REFERENCE
Ford E.B. (1964) *Ecological Genetics*. Chapman & Hall, London.

ecological indicator Organism(s) whose presence indicates occurrence of particular set of conditions. *See also* BIOTIC INDICES. [P.C.]

ecological isolation

1 Two broadly coexisting species are said to be ecologically isolated when their normal environmental requirements are significantly different. In ecology, this usage often occurs in discussions of INTERSPECIFIC COMPETITION for resources. In such cases, ecological isolation is often invoked to account for the observed COEXISTENCE of closely related species.

2 Sympatric organisms of two species are said to be ecologically isolated when they are constrained from meeting, and hence conceivably mating, by their different environmental requirements. Ecological isolation is one of the pre-mating ISOLATING MECHANISMS recognized by the authors of the BIOLOGICAL SPECIES CONCEPT. Darwin was aware of ecological isolation in terms of REPRODUCTIVE ISOLATION (Darwin 1859, p. 103). Paterson (1985) has pointed out that ecological isolation in this sense is an incidental consequence of geographically DISJUNCT POPULATIONS adapting to distinct habitats to which they have been contingently restricted. Thus, ecological isolation does not evolve as an ADAPTATION with the function of preventing hybridization as the term 'isolating mechanism' might suggest. [H.E.H.P.]

REFERENCES
Darwin, C. (1859) *On the Origin of Species by Means of Natural Selection, or the Preservation of Favoured Races in the Struggle for Life*. John Murray, London.
Paterson, H.E.H. (1985) The recognition concept of species. In: *Species and Speciation* (ed. E.S. Vrba), pp. 21–29. Transvaal Museum Monograph No. 4. Transvaal Museum, Pretoria.

ecological physiology *See* PHYSIOLOGICAL ECOLOGY.

ecological pyramid Ecological pyramids are ways of conceptualizing the structure of natural communities of interacting species. They are all based on the fundamental premise that members of a COMMUNITY can be compartmentalized into trophic levels by classifying them according to how they derive their energy, in the case of animals by what they eat.

If the organisms in each TROPHIC LEVEL are scored in some way, for example by simply counting them (PYRAMID OF NUMBERS), counting and weighing them (PYRAMID OF BIOMASS) or measuring the rate of energy transfer through them (PYRAMID OF ENERGY), with the scores for each trophic level displayed along a horizontal axis and the levels stacked vertically with the primary producers as the base, it is usually found that the resulting diagram represents a pyramid (Fig. E4). Several generalizations stem from this observation. One is that prey items are usually more numerous or have a greater BIOMASS per unit area than the consumers that eat them. Another is that this illustrates the second law of thermodynamics by showing that the EFFICIENCY of energy transfer through trophic levels is never perfect but that each transformation is accompanied

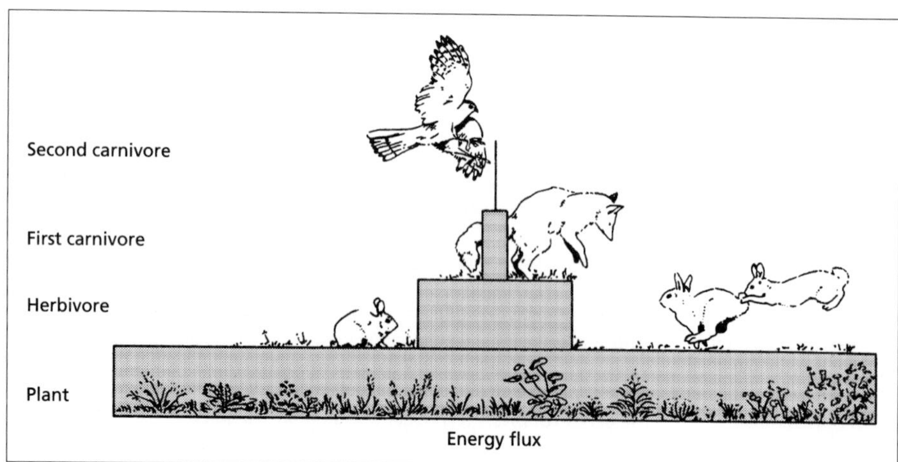

Fig. E4 An ecological pyramid representing the net productivity of each trophic level in an ecosystem. (After Ricklefs, R.E. (1979) *Ecology*, 2nd edn. Nelson, Middlesex.)

by a significant loss of energy as heat. *See also* ECOLOGICAL EFFICIENCY; ECOLOGICAL ENERGETICS. [M.H.]

REFERENCE

Ricklefs, R.E. (1979) *Ecology*, 2nd edn. Nelson, Middlesex.

Ecological restoration

Ecological restoration is the return of an ECOSYSTEM to an approximation of its structural and functional condition before damage occurred. Lost species cannot be merely replaced without ensuring that the system is functioning in a manner similar to its condition before DISTURBANCE. Similarly, restoring lost or damaged ecological functions while imposing a totally different structure (e.g. with exotic species) is not generally desirable either. Since each ecosystem is the result of a sequence of climatic and biological events that are unlikely to be repeated in precisely the same way, exact replication of the condition before disturbance will be rare. On the other hand, improvement through ecological restoration is highly probable. This amelioration is probable if carried out by persons with a knowledge not just of ecology but of the restoration process itself. The goal of ecological restoration is to produce a natural system that is integrated into the ecological landscape in which it occurs. Natural systems are self-maintaining despite the fact that they are dynamic and change over time. Another way to view this is that ecosystems have integrity, which transcends such characteristics as successional processes where some species are replaced by others. Ecosystems with integrity are resilient, and they return to a close approximation of their original form after such events as floods, natural fires, hurricanes, droughts, and temperatures beyond, but not too far from, the range of normal variability. Although the attributes of self-maintenance and resiliency are extremely important, they are not easy to measure with present methodology. An investigator must often wait for an episodic event, such as a drought, flood, heat wave, or long-lasting snow cover, to determine if the system will rebound. This scenario poses considerable difficulty for regulatory agencies that may require restoration following surface mining, clear-cutting, or accidental spill of oil or other hazardous materials.

In some cases, such as surface mining, restoration bonding is required before a permit for mining is approved. Specific requirements may vary from place to place but, generally in the USA, requirements include some time limit, such as 5 years, after which, if the restoration has been satisfactory, the mining organization is released from the bond. This requirement places a considerable burden on those who must judge the efficacy of restoration efforts because, even if an episodic event such as those just described has occurred within the 5-year period, natural systems may not recover for 1, 2 or more years, which may often go beyond the limits of the bonding period. Although some difficulties exist with restoration bonding, it is basically a good concept to use with those transporting hazardous materials or engaging in surface mining and other ecologically disruptive activities because of the guarantee that at least some level of ecological restoration will occur following the disturbance. Even if regulatory personnel do not have all the evidence needed to determine whether there has been complete restoration, they can still determine whether the basic structure and function of the system have been repaired. Some functions that may be important, at least in some ecosystems, are the ability to process fallen leaves and other DETRITUS, to remove CARBON from the ATMOSPHERE and to store it in wood or other forms of BIOMASS, or to hold the topsoil for preventing erosion.

The question of whether a restored ecosystem is integrated with the larger landscape in which it occurs also requires much judgement. If, for example, surface mining occurs in a heavily wooded area, the intuitive judgement is that, once the mining has ceased, trees of the same species as the surrounding woodlands should be growing on the former mining site. This revegetation is quite possible on relatively flat lands; however, in areas with precipitous slopes, re-establishment in an area of moderate to heavy rainfall is virtually impossible. The VEGETATION needed to hold the slopes together simply cannot be established rapidly enough. Additionally, re-creating the SOIL PROFILES requires much care. In surface mining, removing the 'overburden' under which a coal seam is located inevitably means altering the degree of compaction, which is not easily re-established, especially if the soil profile is also dramatically altered. Under these circumstances, regrading to produce more physically stable slopes so that a longer time is available for revegetation may be the best option. The regraded ecosystem, even though it is initially colonized by species dissimilar from those in the surrounding area, could be considered as superior to an unstable system with a steep slope because the regraded ecosystem is not exporting ecologically harmful materials, such as HEAVY METALS and SUSPENDED SOLIDS, to adjacent ecosystems at the rate that the unstable system would be. This system could be viewed as ecologically integrated because its relationship with adjacent ecosystems is harmonious, even though it is dissimilar to them. Since most landscapes are mosaics of habitat types, the question of beneficial interactions is probably more important than strict adherence to a particular habitat type for meeting this ecological criterion.

Regrettably, the word 'restoration' has been used in so many contexts that the resulting confusion is not surprising. In the USA, even federal agencies of the government use the word in a variety of ways, which are often quite dissimilar. In 1992, the National Research Council (the operating arm of the National Academies of Science and Engineering) produced the book *Restoration of Aquatic Ecosystems: Science, Technology and Public Policy* that urged a uniform definition for the word by all federal agencies; thus far this has not happened. In situations involving OIL SPILLS, the word 'restoration' has been used to mean that CLEAN-UP of spilled oil will be accomplished to the degree possible with present methodologies, but it does not mean ecological restoration. In other cases, the term 'restoration' has been used to describe the return of a particular species that is of interest to society, but it does not mean ecological restoration. Therefore, when the word is used in an ecological context, it should be accompanied by a specific description of exact goals and specific accomplishments thus far.

Another reason for requiring an exact description of what is meant by restoration is the virtual certainty that no landscape-level ecological restoration will endure, however successful the initial stages may be, if the project is neither understood nor supported by human society. This description means communicating in some detail precisely what damage has occurred, which parts of the damage can be repaired immediately and which parts will require many years for repair and healing. In this context, the terms 'biocultural restoration' and 'eco-societal restoration' have been used. Both terms have appeal: most ecosystems are damaged by inappropriate human behaviour and restoration will require behavioural change on the part of human society; a greater scientific effort is necessary for re-establishing conditions that existed before ecological damage occurred. Restoration may require reconstruction of antecedent physical, hydrological and morphological conditions and, in the case of spills or introduction of hazardous materials, chemical clean-up or adjustment of the environment. Last, but not least, once the physical/chemical conditions are suitable for biological recolonization, the reintroduction of missing species can begin, both by natural processes and, in some cases, by human-assisted recolonization. All ecological restoration projects are attempts to re-establish conditions known to exist before the damage occurred or which exist in other undamaged habitats that correspond closely with the predamaged habitat. For both scientific and economic reasons, human society must rest satisfied with a naturalistic assemblage of plants and animals rather than a precise replication of the species that once inhabited the area.

Some common failures of restoration projects are listed here.

1 Failure to set explicit goals. Vague generalities, such as 'restore wildlife habitat', are substituted for testable objectives. Even if explicit objectives are listed but fail to include both structural and functional components, only half of the explicitly stated goals are present.

2 Creation of a new habitat in a relatively natural system to replace a specific habitat destroyed by an airport taxiway, shopping mall or highway. Even if a habitat that is in plentiful supply is being destroyed to replace a lost habitat not in plentiful supply, some serious ecological and ethical questions still must be addressed about this procedure.

3 Restoration of a damaged site to replace 'in kind' habitat being destroyed by a development process. Sometimes the site is not actually restored. The project may be part of a mitigation agreement, but the destruction proceeds and the restoration never occurs.

4 Occurrence of unanticipated problems. Problems may make it impossible to follow original plans. Course corrections are made on site and without the same level of consultation that occurred when the project was originally endorsed. This on-site change may mean that biologists, hydrologists or others with necessary skills are absent and construction crews make decisions. For example, in one case, a contracting firm charged with constructing a wetland in mitigation for a lost wetland did so without consulting a hydrologist. The consulting firm was required to redo the project, but ecosystem functions were lost during the corrective period that occurred over a 2-year span.

5 Lack of MONITORING. No substantive monitoring is carried out of the progress and no monitoring of the site continues after the restoration has supposedly been completed. Thus, even if goals are explicitly stated, failure to determine whether they have been reached reduces the benefits of the goal-setting exercise.

6 Absence of reports. No reports are produced that can be reviewed by either regulatory agencies or restoration ecologists not on the project. Scientific review is an important quality-control or quality-assurance element that definitely should be present at all times.

7 Minimal staff. Regulatory agencies simply may not have the necessary staff to carry out SURVEILLANCE, and other groups in the area are not authorized to do so.

8 Denial of access. Regulatory agencies with appropriate staff may be denied access, or may have access delayed, when they wish to make on-site inspections. Competent professionals from academic organizations may also be denied access to the site. While the regulatory agencies do gain access, much valuable time has been lost through delays to legal tactics.

9 Uninformed public relations groups. Even if slides, photographs or videotapes are available for public relations, supporting data and/or measurable criteria to document any assertions about revegetation or wildlife restoration may not be a part of the presentation of a public relations group. Often public relations groups are not knowledgeable about ecological restoration and, therefore, are incapable of answering questions involving detailed evidence.

An exemplary ecological restoration project would be one in which all of these issues have been addressed adequately and substantial public involvement has been evident in the entire process from design through completion and follow-up.

Many exotic species are particularly adept at taking advantage of opportunities presented by ecological damage because the number of competitive species and predators have been reduced. This disruption has simultaneously made space and other RESOURCES, such as NUTRIENTS, more available than they otherwise would have been before the native species were stressed. Most exotics that are regarded as 'pests' are considered so because they grow without much attention and under conditions unsuitable for many native species. Often, exotics are deliberately introduced in the 'restoration' of closed landfills, previously surface-mined areas and the like. A number of good reasons are given for the introduction of exotics. When restoring a closed LANDFILL, obtaining seeds, cuttings or growing plants that are indigenous to the area may be difficult. Even when such seeds are available, they commonly cost an order of magnitude more than some of the commonly used species for lawns, AGRICULTURE, roadside embankments, golf courses and the like. In this era, when reducing costs and budgets is the norm, the cheapest way of revegetating is often the most attractive choice. The outcome may be more certain with an exotic, regularly used species than the less frequently used native species. Even if the decision is made to use indigenous BIOTA, exotics inevitably invade. Then the question must be faced of how much energy and resources should be devoted to controlling or eliminating these exotics. If the source of COLONIZING SPECIES is close to the area being restored, the probability is high that exotics will be a perpetual problem, although the COLONIZATION rate may diminish as the native vegetation becomes established. This is by no means certain. David Wingate's ecological restoration of Nonsuch Island in Bermuda (to re-establish species native to the Bermuda Island complex) is seriously endangered by birds flying out to roost at night from neighbouring islands rich in exotics and defecating the seeds of these exotics all over the island. Eliminating these exotics on an island as small as Nonsuch is labour intensive and, so far, is a never-ending activity. A good case can be made for doing so on Nonsuch, since one would hope that Bermuda and all other countries would have at least one area where native species dominate. If it is a country with a wide variety of habitats, such as China, the USA or Russia, one hopes that each of the major ECOREGIONS would have one area where the native biota is protected from exotic invaders as much as is reasonably possible. Island habitats, such as Bermuda and even those as large as Australia, are particularly vulnerable to displacement of native species by exotic species. In practical terms, exclusion of all exotics is probably not a viable option for most restoration projects. Exclusion of certain exotics, such as kudzu in the USA, may be necessary for the re-establishment of the native vegetation, particularly in the south-eastern area of the USA.

If an ecosystem has been disrupted by oil spills, intense industrial activity or partial mining of an area, should restoration be to the condition before disturbance when more disturbance is highly probable? For example, in areas where traffic of oil-bearing tankers is very heavy and spills have occurred in the past, is it not likely that more will occur in the future? Therefore, the area should be restored with a community more resistant to oil spills than the pre-oil spill community. There are some philosophical dangers in this design: precautions against spills are likely to be less rigorous if it is known that oil-resistant species are present; only those species tolerant of the worst aspects of a technological society will be selected and many species will be lost that are valuable in various ways.

Ecological restoration is referred to as eco-societal restoration because of the high probability that no restoration project will be approved unless it is endorsed by human society. Similarly, no successful restoration project will endure unless human society is environmentally literate and willing to change the behaviour that resulted in ecological damage so that it does not occur again. The types of ecosystems coexisting with human society will depend both on the tolerance of various species to societal activities and the understanding of human society of the behavioural changes necessary to protect the less-resistant species. *See also* CONSERVATION BIOLOGY; ECOSYSTEM MANAGEMENT; HISTORY OF ECOLOGICAL RESTORATION; RECLAMATION; REMEDIATION. [J.C.]

ecological rucksack The total mass of material flow 'carried by' an item of economic consumption (product) in the course of its life cycle. *See also* LIFE-CYCLE ASSESSMENT. [P.C.]

ecological species concept Van Valen (1976) described the ecological species as 'A lineage (or closely related set of lineages) which occupies an adaptive zone minimally different from that of any other lineage in its range and which evolves separately from all lineages outside its range'. *Cf.* BIOLOGICAL SPECIES CONCEPT. [L.M.C.]

REFERENCE

Van Valen, L. (1976) Ecological species, multispecies, and oaks. *Taxon* **25**, 223–239.

ecological trusts *See* ETHICAL INVESTMENT.

ecological valuation Putting a money value on ecological entities. This is not intended to represent intrinsic value, but to be a quantitative indicator of social preferences, i.e. it is the value that people put on environmental entities. The methods fall into two categories: (i) revealed preferences, which use information from real expenditure (e.g. where households spend money on insulation to avoid ENVIRONMENTAL IMPACT) and/or market-places (e.g. fish or lumber); and (ii) contingent valuation, where market-research techniques are applied that try to discover what value people put on ecological systems for use (pastimes, sport) or simply to know that they exist. Somewhere between these extremes is the hedonic pricing method, which values non-marketed environmental resources from observed variations in the prices of marketed goods, such as property prices in different areas. Ecological values are used in cost–benefit analyses of environmental protection measures to compare benefits derived from controls with costs arising from what is controlled and/or the costs of controls themselves. *See also* ENVIRONMENTAL (ECOLOGICAL) ECONOMICS. [P.C.]

ecologism Use of ecological terms and concepts (often superficially and/or naively) in political/moral debate. Also refers to the terms that are used in that way. [P.C.]

ecology Heinrich Haeckel is usually attributed with originating the term from the Greek *oikos*, house or dwelling place (*see* CHARACTERS IN ECOLOGY).

It began as a subject concerned predominantly with the ENVIRONMENT (Haeckel's 'house'), i.e. the study of the way that organisms are influenced by physicochemical conditions. However, as it has developed, it has become increasingly concerned with interactions between individuals, so that it is now probably best defined as that area of biology concerned with the study of collective groups of organisms. As such, it stands at the opposite end of the biological scheme of things from the study of cells and molecules.

The spheres of interest of ecology and ecologists therefore range from individual organisms, through populations, to communities and ECOSYSTEMS. Subdisciplines emphasize interests in different elements of these interactions: PHYSIOLOGICAL ECOLOGY looks at the way physiology influences DISTRIBUTION and ABUNDANCE and is influenced by environment; BEHAVIOURAL ECOLOGY is concerned with the way behaviour responds to ecological challenges in both the short term and the longer (evolutionary) term; POPULATION ECOLOGY describes and explains distribution and abundance; COMMUNITY ECOLOGY describes and explains species associations; ecosystem ecology describes and explains FLUXES of energy and matter as with communities. The study of individuals to populations is sometimes referred to as AUTECOLOGY, whereas the study of communities and ecosystems is sometimes referred to as synecology. There are numerous other descriptions: functional ecology is physiological ecology but emphasizes an experimental and hypothesis-testing approach; trophic ecology studies feeding ecology; MOLECULAR ECOLOGY uses techniques from molecular biology to address ecological questions; ECOLOGICAL GENETICS tries to understand gene frequencies in terms of ecological processes, etc. And, of course, it can also be subdivided with respect to HABITAT (marine, freshwater, soil, littoral, benthic, etc.) and organisms (microbial, animal, plant, avian, fish, etc.).

Applied ecology involves using understanding from all these various aspects of ecology in the protection, conservation and management of ecological systems. Lawton (1996) has recently made the point that theoretical ecology usually involves the development of broad understanding with emphasis on generality, whereas applied ecology is often concerned with the protection or exploitation of particular systems and with attempts to make precise quantitative predictions about them (Fig. E5); linking these two endeavours is not always easy. [P.C.]

REFERENCE

Lawton, J. (1996) Corncrake pie and prediction in ecology. *Oikos* **76**, 3–4.

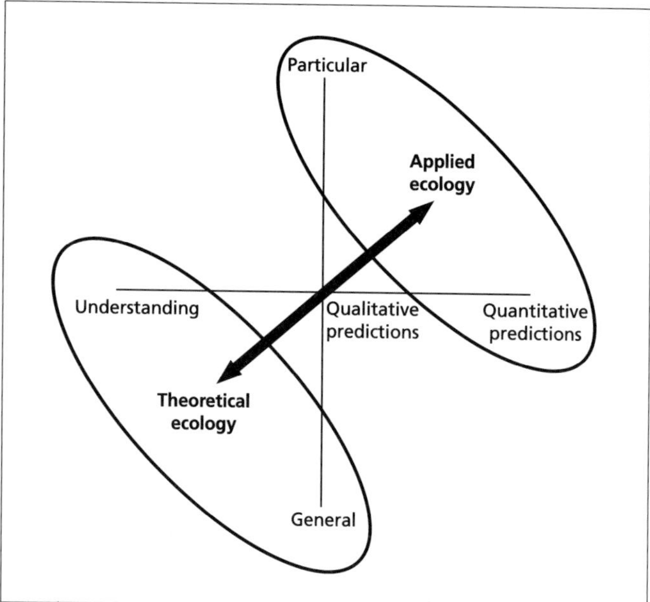

Fig. E5 A simple conceptual scheme linking theory and practice in ecology. (After Lawton, J. (1996) Corncrake pie and prediction in ecology. *Oikos* **76**, 3–4.)

Eco-management and Audit Scheme (EMAS) European Union voluntary environmental management programme: applied on an industrial-site basis. *See also* ENVIRONMENTAL MANAGEMENT SYSTEMS. [P.C.]

ecomargin Ratio between an organization's actual ENVIRONMENTAL IMPACTS and those compatible with SUSTAINABLE DEVELOPMENT. The latter is almost impossible to measure so proxy measures are used, for example ratio of actual impacts and those minimal impacts that are technically feasible and/or required by law. Just as a firm cannot survive without a reasonable profit margin, it cannot survive in the long term without a low or minimum ecomargin. [P.C.]

REFERENCE

James, P. (1992) The corporate response. In: *Greener Marketing* (ed. M. Charter), pp. 111–137. Greenleaf Publications, Sheffield.

ecomorphology The ecomorphological hypothesis states that morphometric data serve as an index of ecological characteristics such as NICHE WIDTH. In other words, differences in species morphology reflect differences in their ecology. This is a long-standing and pervasive idea in ecology and, perhaps, the most obvious support for the hypothesis that an individual's physical features are unresponsive to fluctuating environmental conditions compared with behavioural and physiological features, which respond quite quickly. Particularly good illustrations of the close relationship between physical form and ecological function come, for example, from work on fishes, where mouth structures often reflect the type of diet and body form often agrees with MICROHABITAT use. Similar relationships have also been found for other vertebrate groups, but it should be recognized that not all studies show close associations and doubts have been cast on the utility of the concept. *See also* ECOLOGICAL SPECIES CONCEPT; NICHE. [S.J.H.]

economic injury level (EIL) A level of PEST abundance. Above this level it costs less to control the pest than is saved by PEST CONTROL. Below this level, pest control costs more than is saved. [P.O.]

economic instruments *See* REGULATORY INSTRUMENTS.

economic thresholds *See* INTEGRATED PEST MANAGEMENT.

eco-organ Characteristic morphological features that reflect ADAPTATION to specific environmental conditions. Used especially in the context of classifying VEGETATION types ecologically rather than taxonomically, for example refers to leaf form, stomatal structures. [P.C.]

ecophysiology Use of physiological observations on organisms to help explain their DISTRIBUTION and ABUNDANCE. *See also* PHYSIOLOGICAL ECOLOGY. [P.C.]

ecopolitics Broadly, politics are about the implementation of policies that influence, to a greater or lesser degree, the creation of capital and its distribution. Ecopolitics attempts to base this on ecological principles and/or taking protection of the ENVIRONMENT as of paramount importance in developing the policies. Political parties have been formed to develop and implement these policies. The world's first Green Party was founded in New Zealand in 1972 and called the Values Party. The first green politicians to take seats in a legislature were in Switzerland; Swiss Greens won two seats in the Swiss Parliament in 1979. West German greens first won seats in the Bundestag in 1983. The British Green party, founded in 1973 as the People Party, was the first in Europe. In 1975 it changed its name to Ecology Party and in 1985 to the Green Party. *See also* ENVIRONMENTAL (ECOLOGICAL) ECONOMICS; GREEN(S); SUSTAINABLE DEVELOPMENT. [P.C.]

REFERENCE

McCormick, J. (1991) *British Politics and the Environment.* Earthscan Publications, London.

ecoprofile Another term for LIFE-CYCLE ASSESSMENT of manufactured products. *See also* ECOBALANCE. [P.C.]

eco-rational path Outcome of ECO-CONTROLLING. [P.C.]

ecoregion A relatively homogeneous area in terms of geography, HYDROLOGY and land use. *See also* ECOSYSTEM INTEGRITY. [P.C.]

ecosabotage Disruption of activity/process by legal or illegal means to draw attention to its putative ADVERSE ENVIRONMENTAL EFFECTS. [P.C.]

ecoshares *See* ECOFUNDS.

eco-societal restoration *See* ECOLOGICAL RESTORATION.

ecosphere Same as BIOSPHERE. [P.C.]

Ecosystem

A functional ecological unit in which the biological, physical and chemical components of the ENVIRONMENT interact. This term focuses attention on the complex interplay between plants and animals and ABIOTIC FACTORS of their HABITAT. It is a much wider concept than the COMMUNITY with which it is sometimes confused.

Although interactions between organisms and their environment have been recognized for centuries, the term 'ecosystem' was first used in print by Tansley in 1935 in a paper concerning vegetational concepts and terms (*see* CHARACTERS IN ECOLOGY). Tansley's basic concept of the ecosystem involved both the organism-complex and the whole complex of factors of the environment. He wrote 'Though the organisms may claim our primary interest, when we are trying to think fundamentally we cannot separate them from their special environment, with which they form one physical system. . . . These ecosystems, as we may call them, are of the most various kinds and sizes'.

Lindeman (1942) regarded the concept of the ecosystem to be of fundamental importance in interpreting data of dynamic ecology, drawing particular attention to the transfer of energy from one part of an ecosystem to another. He considered the ecosystem to be 'composed of physical–chemical–biological processes active within a space-time unit of any magnitude'. The systems approach has been applied, however, not only to energy relationships and trophic levels but also to the cycling of CARBON, MINERAL NUTRIENTS and WATER. Odum (1953) fostered the ecosystem concept, emphasizing interdependence and causal relationships. Functionally, two components of the ecosystem can be recognized (AUTOTROPHS and HETEROTROPHS) and usually four constituents (abiotic, producers, consumers and decomposers). The ecosystem approach has proved valuable in the study of mineral cycling in, for example, watersheds (Bormann & Likens 1967) and tropical forest (Golley *et al.* 1975).

In the 1960s the development of 'SYSTEMS ECOLOGY' (Odum 1964) provided a framework for the elucidation of the 'structure and function of levels of organization beyond that of the individual and species', involving aspects of the development, dynamics and disruption of ecosystems. Mathematical procedures applied to quantitative studies have facilitated the development of modelling of flow diagrams and of complex interactions between organisms and environmental factors. Dynamic models using DIFFERENTIAL EQUATIONS are considered particularly appropriate for the analysis of ecosystems (Mauersberger & Straskraba 1987; Jørgensen 1992).

The ecosystem approach provided a valuable framework for the INTERNATIONAL BIOLOGICAL PROGRAMME in the early 1960s and also the long-term multidisciplinary Hubbard Brook investigations in the USA (Likens *et al.* 1977; McIntosh 1985). The whole-ecosystem experiments run at Hubbard Brook, conducted against a control (reference) ecosystem, proved very informative and are considered a landmark to the development of 'big biology'. Indeed, the experimental MANIPULATION of ecosystems has been regarded as 'a very powerful analytical approach' (Likens 1985). A fairly recent development is that ecosystem studies have changed in emphasis from the descriptive mode to the predictive mode.

The scale of the ecosystem may vary enormously from, for example, an anthill or tiny pool to a huge forest or even the entire BIOSPHERE of the globe. It is desirable to specify the size of any ecosystem under study because some considerations are scale-dependent (Willis 1997). As many animals are very mobile and some may spend part of their life in water and part on land, boundaries being diffuse, the ecosystem concept may be of greater relevance to botanists than zoologists. However, another approach defines ecosystems spatially using the area of the food web of the largest predator at any location. Referred to as the ecosystem trophic module, it makes ecosystems into countable objects (Cousins 1990). With defined scales of space and time, the ecosystem is a valuable and highly used unit by all ecologists. [A.J.W.]

REFERENCES

Bormann, F.H. & Likens, G.E. (1967) Nutrient cycling. *Science* **155**, 424–429.

Cousins, S.H. (1990) Countable ecosystems deriving from a new food web entity. *Oikos* **57**, 270–275.

Golley, F.B., McGinnis, J.T., Clements, R.G., Child, G.I. & Duever, M.J. (1975) *Mineral Cycling in a Tropical Moist Forest Ecosystem.* University of Georgia Press, Athens.

Jørgensen, S.E. (1992) *Integration of Ecosystem Theories: A Pattern.* Kluwer Academic, Dordrecht.

Likens, G.E. (1985) An experimental approach for the study of ecosystems. *Journal of Ecology* **73**, 381–396.

Likens, G.E., Bormann, F.H., Pierce, R.S., Eaton, J.S. & Johnson, N.M. (1977) *Biogeochemistry of a Forested Ecosystem.* Springer-Verlag, New York.

Lindeman, R.L. (1942) The trophic–dynamic aspect of ecology. *Ecology* **23**, 399–418.

Mauersberger, P. & Straskraba, M. (1987) Two approaches to generalized ecosystem modelling: thermodynamic and cybernetic. *Ecological Modelling* **39**, 161–176.

McIntosh, R.P. (1985) *The Background of Ecology: Concept and Theory.* Cambridge University Press, Cambridge.

Odum, E.P. (1953) *Fundamentals of Ecology.* W.B. Saunders, Philadelphia, Pa.

Odum, E.P. (1964) The new ecology. *BioScience* **14**, 14–16.

Tansley, A.G. (1935) The use and abuse of vegetational concepts and terms. *Ecology* **16**, 284–307.

Willis, A.J. (1997) The ecosystem: an evolving concept viewed historically. *Functional Ecology* **11**, 268–271.

---

ecosystem engineers Organisms that directly or indirectly modulate the availability of RESOURCES (not themselves) to other species, by causing physical changes in the biotic or abiotic materials of the environment. In so doing they create or modify HABITATS. ALLOGENIC engineers change the environment by transforming living or non-living materials from one physical condition to another via mechanical or other means, for example beavers and earthworms. AUTOGENIC engineers alter environments by their own physical structure, such as the growth of a forest or coral reef.

The term 'engineer' needs care. Clearly it is intended as an analogy; however, 'engineers' work to 'design' to produce artefacts for the benefit of others. The dictates of NATURAL SELECTION are such that the ECOSYSTEM 'engineering' has to bring FITNESS benefits to the 'engineers' and not for the good of the group (*cf.* GROUP SELECTION). The analogy needs to be explicit that engineers are usually compensated by fees for their work. But even this is somewhat forced, since the fees will usually be related to the quality and quantity of the work, whereas the fitness returns to the ecosystem 'engineers' will not necessarily be proportional to the benefits gained by the 'beneficiary' species. [P.C.]

REFERENCE

Lawton, J.H. (1994) What do species do in ecosystems? *Oikos* **71**, 367–374.

ecosystem health It is relatively easy to know when people are ill; they develop disease-specific symptoms, feel off-colour and function abnormally. We can gauge the seriousness of this in terms of various general indicators such as body temperature and pulse rate, the extent to which normal function is impaired and also in terms of the likelihood of recovery. Increasingly, some scientists and policy-makers are describing the condition of ECOSYSTEMS in terms of states of health. Yet there are some important differences between human bodies and ecosystems. The latter are not as obviously and tightly organized as organisms. Where, for example, are the coordinating nerve centres of ecosystems? Individuals and species can be lost from ecosystems without apparently impairing normal function to an extent that cells, tissues and organs could not be lost from organisms. So ecological norms are not so easily defined as body norms; it is more difficult to identify the 'pulses' of ecosystems. Also it seems that their patterns of recovery after DISTURBANCE may not always be very predictable and the recovered systems may never be quite the same as the originals.

Ecosystem health is, of course, meant as an analogy between the human condition and that of ecosystems. Analogies are meant to use understanding in one area to help understanding in another. However, if the analogy is taken too literally, that ecosystems have similar properties to organisms, the analysis is not only unhelpful but misleading (*see* GAIA). If, on the other hand, the analogy is intended to draw attention to the similarities between the way we study human health and the condition of ecosystems, i.e. synthesizing observations on a variety of systems and using this in the light of previous experience to recognize symptoms of deterioration and diagnose possible causes, then it might be helpful. By using imagery that involves the human experience it is also helpful in promoting environmental protection needs amongst non-scientists. *See also* ECOLOGICAL RESTORATION; ECOSYSTEM MANAGEMENT. [P.C.]

ecosystem integrity The view that a number of key states (properties and processes) must be intact for an ECOSYSTEM to persist in a stable state. Under the Clean Water Act (*see* UNITED STATES LEGISLATION) the USEPA is required to 'evaluate, restore and maintain the physical and biological integrity of the Nation's waters'. The latter is judged in relation to the state of a suite of criteria of resident fish and bottom-dwelling invertebrates using baseline data obtained from reference sites within a defined ECOREGION. It is not based on first principles. As a concept it suffers similar problems to ECOSYSTEM HEALTH; and it suffers from the same shortcomings associated with RETROSPECTIVE ANALYSIS (MONITORING). [P.C.]

Ecosystem management

The science and art of directing human activities to sustain or restore the desired DIVERSITY and PRODUCTIVITY of terrestrial and aquatic ecosystems in an area. The essence of ecosystem management (EM) lies in its systematic approach and its multiscale spatial and temporal views, not in the use of specific management practices. Implementation of EM requires many areas of expertise, including ECOLOGY, economics, sociology and politics. Although the term has been used at least since 1970 and came into widespread use in the late 1980s, the general concept has been around at least since Leopold (1949; *see* CHARACTERS IN ECOLOGY) advocated treatment of the land as a 'whole organism'.

Statements of EM goals vary in emphasis. One viewpoint emphasizes maintenance of viable populations of all native species, maintenance of all native ecosystem types within their range of natural variation, maintenance of long-term ecological and evolutionary processes, and accommodation of human use within these constraints (Grumbine 1994). Alternative goals emphasize sustainable outputs of goods and services from ecosystems as life-support systems to humans. The former viewpoint may also be motivated by concerns for human welfare, but takes a more cautious, longer-term view of sustainability and maintenance of future management options. Many discussions of EM address only the technical aspects of developing and implementing new management plans, while others consider the need for organizational change or even changes in basic popular values. Many proponents argue that we must direct human resource-consumption behaviours to become more appropriate to the living systems of which they are a part and on which they depend.

The need for EM is motivated by the accelerated extinction of species in many parts of the world, the deterioration of important ecosystem 'services' to humans and by changing scientific knowledge and societal demands. Humans derive many goods and services from ecosystems, but the capacity of any ecosystem to meet human needs in perpetuity has limits. DESERTIFICATION in arid countries, soil loss from agricultural lands, tropical DEFORESTATION and the collapse of ocean fisheries around the world are a few indications that current human populations are not using ecosystems sustainably; continued human population growth makes long-term ecosystem sustainability unlikely.

Traditional management usually considers management effects over small areas on a few species of economic or cultural importance. In contrast, EM considers the entire range of organisms and abiotic components associated with ecosystems, as well as the maintenance of ecosystems of little immediate, direct value to human consumption. Managing on a species-by-species basis is logistically impossible due to the sheer number of organisms and lack of detailed autecological information on species. Managing for functioning ecosystems that many species use as HABITAT is thought to be more efficient and realistic. By adopting a broader view than traditional management, EM should be better able to sustain whole ecosystems, be more resilient to unexpected changes and keep open more future options.

EM usually focuses on landscapes (e.g. large watersheds) but takes a multiscale view, because ecosystems are influenced by smaller systems that comprise them and the larger systems of which they are a part. For example, a log on the forest floor is an ecosystem with many interacting microbes, fungi and insects. Forest patches containing such logs across a landscape represent an ecosystem for a population of woodpeckers dependent on insect prey in logs. The global abundance of forests with logs on the ground in turn represents an important storage compartment for CARBON in the global ecosystem. Because of interactions among ecosystems, maintenance of one ecosystem type may depend on, or influence, the maintenance of other ecosystem types. For example, management practices in upland forest ecosystems influence the condition of streams and river ecosystems.

Ecosystems are dynamic, and their composition, structure and function change in response to DISTURBANCE, MIGRATION, internal development processes (e.g. vegetation growth, nutrient accumulation) and CLIMATE CHANGE. EM often draws on knowledge of past dynamics of 'natural' ecosystems (i.e. those with few human impacts) to provide prototypes for managing human actions because ENDEMIC SPECIES and processes that maintain productivity are presumably adapted to disturbance regimes and environmental conditions of recent millennia. Thus EM often attempts to maintain, restore or mimic natural ecosystem processes. For terrestrial ecosystems, understanding the natural disturbance regime and/or the range of ecosystem conditions (e.g. the extent and spatial pattern of different vegetation seres and COMMUNITY types) is particularly useful. For example, restoring low-intensity ground fire to dry pine forests can maintain species composition (fire-resistant, shade-intolerant trees), HABITAT STRUCTURE (open canopy forests and dead wood) and ecosystem processes (nutrient cycling and hydrological regulation). Because the organisms and structures left after disturbance are often critical determinants of how ecosystems develop, EM usually emphasizes what is left behind following resource extraction as much as what is removed. Species populations may rise, fall or even go locally extinct as environments change on a given hectare, but regional stability is maintained by connections among habitats and the ability of species to migrate and re-establish in a landscape.

Most regional-scale EM plans advocate creation or maintenance of core reserve areas within which natural processes are emphasized, with low-impact management of the matrix around reserves and/or maintenance of habitat CORRIDORS between reserves (to allow for migration to maintain the distribution of species and because most existing reserves (e.g. national parks) are not large enough to sustain the full complement of species) (Clark 1977; FEMAT 1993). Watershed-scale EM plans usually specify different types and intensities of management on different ecosystems and topographic units to mimic natural disturbance and maximize both resource extraction and ECOSYSTEM INTEGRITY (Fig. E6). Because MONITORING thousands of species is logistically impossible, identification of 'indicator' or 'umbrella' species is often used with the premise that the viability of a given species (or group of species) can indicate the status of an entire ecosystem (e.g. tigers for intact tropical forest). Minimum thresholds for POPULATION SIZE are usually established to judge the likelihood of extinction. Complete reliance on one or a few INDICATOR SPECIES is not warranted however, because a species may be an indicator of only a narrow range of conditions within an ecosystem.

A particular problem for EM is dealing with the common mismatch of ecological and political boundaries. Managing ecosystems often requires working with multiple landowners, cultures and political jurisdictions. For example, the Columbia–Snake Rivers ecosystem in the north-western USA includes over 500 different jurisdictions and even more numerous land ownerships. Some wide-ranging organisms (e.g. migratory birds and whales) cross many international boundaries. Involvement and cooperation of landowners, policy-makers and other interested people is essential to the success of EM. Development of social processes to support planning and implementation is one of EM's greatest challenges.

Due to uncertainties from imperfect knowledge and unpredictable change, implementation of EM requires adaptive management (Lee 1993): design a plan based on the best available knowledge, implement it, monitor it, learn from it and adjust the original plan. Monitoring of ecosystems to assess whether explicit goals are being met is a critical component in this loop but has rarely

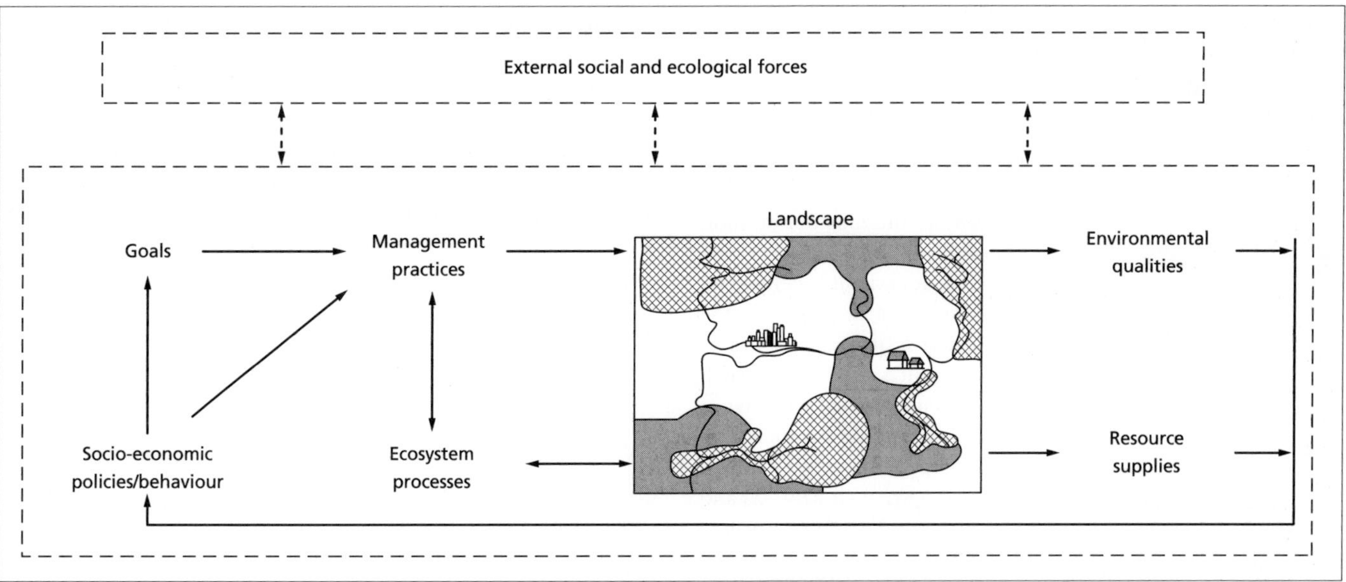

Fig. E6 Ecosystem management process for a landscape, including internal and external factors that influence environmental quality and resource supply.

been satisfactorily implemented. EM is more complicated than traditional resource management, requiring more sophisticated planning, cooperation, use of spatial data and implementation techniques. However, the ever-increasing demands of human populations on ecosystems requires more sophisticated approaches, such as EM, to sustain the diversity of ecosystem goods and services. *See also* CONSERVATION BIOLOGY; ECOLOGICAL RESTORATION; REMEDIATION. [A.N.G. & T.A.S.]

REFERENCES

Clark, J.R. (1977) *Coastal Ecosystem Management*. John Wiley & Sons, New York.

FEMAT (Forest Ecosystem Management Assessment Team) (1993) *Forest Ecosystem Management: An Ecological, Economic, and Social Assessment*. USDA Forest Service Pacific Northwest Region, Portland, Oreg.

Grumbine, R.E. (1994) What is ecosystem management? *Conservation Biology* **8**, 27–38.

Lee, K.N. (1993) *Compass and Gyroscope*. Island Press, Washington, D.C.

Leopold, A. (1949) *A Sand County Almanac*. Oxford University Press, New York.

ecosystem redundancy Removal of some species does not necessarily lead to the breakdown of function in an ECOSYSTEM. Hence it is presumed that their role is redundant (i.e. unneeded functional capacity) in 'normal' ecosystems. [P.C.]

REFERENCE

Schindler, D.W. (1987) Detecting ecosystem responses to anthropogenic stress. *Canadian Journal of Aquatic Sciences* **44** (Suppl.), 6–25.

ecosystem services Benefits that society obtains from properly functioning ECOSYSTEMS in terms of the supply of raw materials, food, clean water, clean atmosphere, etc. This is not to suggest that ecosystems are designed to provide these services, rather that humans have evolved in non-disturbed ecosystems and hence depend for their well-being on them. SUSTAINABLE DEVELOPMENT depends upon maintaining these services and environmental management should be directed towards this end. [P.C.]

ecoterrorism Threat or use of contrived ENVIRONMENTAL IMPACT/disaster to extort money and/or terms, or make political point. *See also* ECOSABOTAGE. [P.C.]

ecotone Defined by diCastri *et al.* (1988) as 'the zone of transition between (different) adjacent ecological systems (ECOSYSTEMS, COMMUNITIES, HABITATS) with characteristics uniquely defined by time and space and by the strength of the interaction between the adjacent ecological systems'. The ecological system as used in this context is approximately equivalent to patch in modern patch theory and ecotone can be defined as the boundary that separates or defines patches. Hence, the term 'ecotone' should be applied to more abrupt transitional gradients, where species TURNOVER is high (high beta diversity; *see* DIVERSITY, ALPHA, BETA AND GAMMA) and boundaries reasonably clearly defined, and the term 'ecocline' reserved for more gradual transitions. However, the term 'ecotone' has been applied to the whole range of scale from BIOME boundaries to very local boundaries of ecosystems or communities, for example woodland margin. Ecotones typically exhibit greater species DIVERSITY and often greater ABUNDANCE of species than their adjacent ecological systems, a phenomenon sometimes referred to as the EDGE EFFECT. Their species comprise both members from the adjacent ecological systems, which may derive only some of their resources from the ecotone, plus species unique to the ecotone. Fragmentation of habitat, typically by humans, often creates and expands associated ecotones; some ecotones, for example HEDGEROWS, are a direct product of human activity.

Less often the term is used for a boundary between two species, for example two species of conifer along an altitudinal TRANSECT, or any other type of environmental boundary, for example saltwater–freshwater, clay–sand substrate. A CLINE is a more-or-less gradual variation in any organism character along an environmental GRADIENT. The character may be phenotypic or genetic, for example height, colour, chromosome number, ALLELE FREQUENCY. [B.W.F. & K.T.]

REFERENCE

diCastri, F., Hansen, A.J. & Holland, M.M. (eds) (1988) *A New Look at Ecotones: Emerging International Projects on Landscape Boundaries*. Biology International, Special Issue 17. IUBS, Paris.

ecotope Term used in an evolutionary context to describe the full range of NICHE and HABITAT factors that affect a species and determine its survival. To survive, a species must have a unique ecotope, either with respect to niche or habitat factors or both. [B.W.F.]

ecotoxicity Capacity of chemicals to cause injury to ecological systems, i.e. populations and/or communities and/or ECOSYSTEMS. Expressed in terms of various indicators that can represent short-term/long-term, acute/chronic effects (e.g. LC_{50}, EC_{50}, IC_{50}, NOEC (NO OBSERVED EFFECT CONCENTRATION), LOEC (LOWEST OBSERVED EFFECT CONCENTRATION), MATC (MAXIMUM ACCEPTABLE TOXIC CONCENTRATION)). [P.C.]

ecotoxicity tests (predictive tests) *See* PREDICTIVE ECOTOXICITY TESTS.

ecotoxicology A subject area concerned with understanding where anthropogenic chemicals go in the environment (their FATE) and hence the extent to which ecological systems are exposed to them (EXPOSURE) and, in consequence, the ecological effects that they have (effects). It can be carried out retrospectively (are releases having effects?) and prospectively (are releases likely to have effects?). Retrospective analyses involve MONITORING programmes; prospective studies involve predictive tests. Ecotoxicology is therefore a multidisciplinary/interdisciplinary subject involving a combination of environmental chemistry, TOXICOLOGY and ecology. It is being used increasingly as a basis for environmental RISK ASSESSMENT in the development of environmental protection legislation. *See also* ECOTOXICITY; PREDICTIVE ECOTOXICITY TESTS. [P.C.]

REFERENCES

Calow, P. (ed.) (1993) *Handbook of Ecotoxicology*, Vol. 1. Blackwell Scientific Publications, Oxford.

Calow, P. (ed.) (1994) *Handbook of Ecotoxicology*, Vol. 2. Blackwell Scientific Publications, Oxford.

ectendotrophic mycorrhizae *See* ENDOTROPHIC MYCORRHIZAE; MYCORRHIZA.

ectomycorrhizae *See* ECTOTROPHIC MYCORRHIZAE.

ectoparasite A PARASITE that lives on the outside of the HOST (used only in reference to MACROPARASITES; *cf.* ENDOPARASITE). Ectoparasites include examples such as ticks, mites, lice, leeches and blood-feeding insects. Ectoparasites have varying degrees of host specificity, have variable degrees of dependence on their hosts, spend variable proportions of their LIFE CYCLE on the host and have more or less specificity to particular areas of the host. *See also* SPECIES INTERACTIONS. [G.F.M.]

ectoparasitoids *See* PARASITOID–HOST INTERACTIONS.

ectotherm An ectotherm (also called POIKILOTHERM or allotherm) is an organism that relies on the TEMPERATURE of its surrounding environment to regulate and maintain its own internal body temperature. The relationship may be either passive or active. Active REGULATION usually consists of behavioural modifications such as basking that enables a limited degree of temperature control. [R.C.]

Ectotrophic mycorrhizae (ectomycorrhizae)

In ectotrophic mycorrhizae, which are very common amongst temperate forest trees, there is a thick (20–40 μm) outer sheath or fungal mantle, consisting of closely interweaving and branching septate hyphae (forming a pseudotissue) around the outside of the root, with hyphal strands leaving its outer surface and traversing the RHIZOSPHERE zone (usually defined as 'the narrow, 1–2 mm zone of soil, immediately surrounding the actively growing root, in which the plant has a direct influence on the soil microflora') to colonize the adjacent soil particles and organic substrates. Characteristic ectomycorrhizal fungi include particular examples within the Basidiomycotina from 25 different families, for example the Agaricales (*Tricholoma*, *Boletus*, *Amanita*) and Russales (*Cortinarius*, *Russula*, *Lactarius*). Some Ascomycotina fungi are also involved, for example *Tuber*, the truffle fungus. Most of these fungi develop ectotrophic mycorrhizae with different plant species, but some are more specific with respect to their host, for example *Boletus elegans*, which is restricted to species of larch. Only about 3% of all plant species act as the autotrophic partner in ectomycorrhizae and these are almost all evergreen or deciduous trees and shrubs in 130 different genera in 43 families. Typical tree species are members of the families Betulaceae, Myrtaceae, Fagaceae and Pinaceae; many of these will grow in culture but only in a limited way because of complex nutritional requirements only satisfied in the dual relationship. The thick fungal sheath (mantle) surrounding the terminal, lateral root branches is connected to an intercellular network of hyphae (Hartig net) in the outer cortex; the Hartig net develops as a result of the hyphae penetrating between the epidermal cells and then the subepidermal cortical cells, which are pushed apart. Normally, the host cells in ectomycorrhizae are not penetrated by fungal hyphae, an important distinction when considering endomycorrhizae, where intracellular penetration commonly occurs. Whereas root hairs are produced on uninfected, lateral roots, the ectomycorrhizal lateral roots become thicker, sometimes pigmented, and root hairs are absent. The main roots continue to elongate without much obvious sign of the mycorrhizal partner, but their life is considerably extended. The fungus causes the extensive racemose or dichotomous root branching system to develop. The outer surface of the ectomycorrhiza is smooth, with only relatively few hyphae leaving the root for the soil. However, in *Pinus* (pine), there is an interconnecting mycelium and mycelial strands that traverse the soil between established ectomycorrhizal roots. New roots become 'infected' by hyphae extending from the sheath of existing mycorrhizal roots (containing dikaryotic, secondary mycelium) of the same tree or adjacent tree of the same species. Seedling roots become colonized either by secondary (dikaryotic) mycelium derived from hyphal fusions involving compatible, paired, primary (monokaryotic) mycelia produced by germinating basiodiospores of the particular fungal partner in the immediate vicinity of the root as it grows through the soil, or by extending dikaryotic hyphae from an appropriate established mycorrhiza in the vicinity (*see* MYCELIUM). In total, the fungal component can amount to 25% or more of the root dry mass. Ectomycorrhizal roots are found mainly in the litter layers of the forest floor, where they are important in the RECYCLING of mainly soluble nutrients arising from the extracellular breakdown/DECOMPOSITION (by free-living saprophytic microorganisms, e.g. fungi and bacteria) of fallen leaves, debris, etc. Here, substrate availability is discontinuous and the ability of the mycorrhizal fungus to take up and store phosphates especially is likely to be a vital benefit to the plant. Root EXUDATES and their diffusion gradients play a role in attracting hyphal growth in the direction of the plant's root tip. Fruiting bodies of the fungi concerned are eventually produced (often in autumn) above ground (except for truffle fungi) in various positions around the colonized tree and are connected via hyphal strands to nutrient sources within the mycorrhizal roots. Active uptake of phosphate (as the soluble orthophosphate) by the hyphae in the outer layers of the ectomycorrhizal fungal sheath, and its transport to root cells, is a major function in phosphorus-deficient soils; AMMONIA, sulphate, POTASSIUM and CALCIUM are also rapidly taken up. In general, ectomycorrhizae are most beneficial to the plant partner in soil types, climatic zones and time-periods where growth is otherwise restricted, when factors such as phosphate (and other nutrient) concentrations become limiting. For successful ESTABLISHMENT of young tree seedlings in forestry, development of ectomycorrhizae is often essential; for this, nursery soils can be selected that are known to contain the appropriate beneficial fungus or, in some cases, inoculation can be useful. The ectomycorrhizal system can interconnect many plants in the same community, facilitating nutrient movement within the network; this results in some plants being able to establish themselves in otherwise limiting conditions, such as shade. A further advantage is that ectomycorrhizae can improve RESISTANCE to infection by pathogenic root-infecting fungi, both in the nursery as well as in forest plantations. There are relatively few studies on the importance of ectomycorrhizae in tropical countries, but their widespread distribution in forests and woodlands within Africa has been demonstrated. *See also* ENDOTROPHIC MYCORRHIZAE; MYCORRHIZA. [J.B.H.]

ED_{50} *See* EC_{50}.

edaphic Edaphic factors are the physical, chemical and biotic characteristics of the SOIL that influence plant growth and distribution.

The term is now rarely used but appears frequently in the earlier ecological literature. [J.R.E.]

eddy A current of air, water or any fluid, often on the side of a main current, especially one moving in a circle. Eddy diffusion is the process by which a solute is transported by turbulent mixing from regions of high concentration to regions of low concentration. It is analogous to molecular diffusion but several orders of magnitude greater. Common features of the oceans, mesoscale eddies are characterized as time-dependent current and temperature patterns with horizontal dimensions of 100–500 km and periods of one to several months. The role that such eddies play in the dynamics of the oceans is not well understood. [V.F.]

eddy covariance Eddy covariance is a micrometeorological technique for calculating the net exchange of atmospheric properties such as heat, water vapour or carbon dioxide (CO_2) between the land's surface and the ATMOSPHERE. The atmospheric BOUNDARY LAYER is usually turbulent as a result of surface friction causing air to be tripped into eddying motion or by solar heating causing thermal expansion and ascent of air. Turbulence can be viewed as a series of eddies that move with the horizontal WIND over the surface and which transport heat, moisture, CO_2 or other parameters between the surface and the atmosphere. The eddies range in size from a fraction of a metre to about the size of the boundary layer itself, i.e. several hundred metres or more. It is important to measure the rate of vertical movement of all these eddies plus the concentration of heat, moisture, CO_2, etc., that they contain. Making measurements by eddy covariance involves sampling the vertical wind speed at rates up to 10 Hz, and when these FLUCTUATIONS are multiplied at the same instant by measurements of fluctuations in gas or heat concentration then the instantaneous FLUXES of that property are obtained. If these instantaneous fluxes are now averaged over, say, 30 min, the net flux above the ecosystem is obtained. One application of such a technique is to determine which types of land and OCEAN surface are net sources or sinks of CARBON, an understanding of which will be vital to model the enhanced greenhouse effect. *See also* MICROMETEOROLOGY. [J.B.M.]

edge effect Edges are structural or compositional discontinuities between two adjacent patch types. Among the most familiar form of edges in temperate zones are boundaries between forests and clearcut areas or agricultural fields. However, edges can also be of natural origin. Edge environments occur naturally at many ECOSYSTEM boundaries: along the perimeter of bodies of water, cliffs, outcrops of exposed rock, and sharp discontinuities in soil type or HYDROLOGY. Less permanent edges can also occur as a legacy of ecosystem disturbances such as fires, storms and insect or fungal infestation. These natural edges are recognizable on the basis of gross structural features such as tree height, VEGETATION density or UNDERSTOREY development. Edges can also be delineated by compositional characteristics such as species composition or COMMUNITY type.

Edge effects include both biotic and abiotic dimensions. Near FOREST edges, microclimatic and other ABIOTIC FACTORS, such as LIGHT intensity and vertical distribution, air speed and TEMPERATURE, RELATIVE HUMIDITY, soil temperature and moisture, may differ from both forest interior and the adjacent open area. Many of these parameters are also more variable near the edge than in the forest interior. The intensity of edge effect generally decays with distance into the interior, reaching parity with interior conditions at varying distances from the edge itself. In temperate forests, the microclimatic depth of edge effect may extend from 0 to 200 m or more depending on the parameter of measurement, SEASON, time of day, edge orientation, degree of edge closure, TOPOGRAPHY and other variables.

The biotic component of edge effects includes the response of organisms to edge proximity. Many species characteristic of early successional habitats are especially abundant and diverse in and around edges, taking advantage of protective cover from predators, increased abundance of some food sources and other attributes of a heterogeneous HABITAT. Conversely, many species restricted to (or modally associated with) forest interiors are less abundant and diverse near edges, in part due to increased COMPETITION from species that benefit from the edge. Interior species such as mosses may also be less abundant near edges as a result of altered microclimatic conditions (such as increased temperature and decreased SOIL MOISTURE) that may preclude growth or even survival. Consequently, the biotic depth of edge effect may vary significantly among organisms, and can only be defined for a specific group of species in a given situation.

A number of problems in CONSERVATION BIOLOGY and land management are associated with the ecology of edges. Since small protected areas have relatively high edge–interior ratios, the proportion of truly interior forest in a fragmented landscape may be exceedingly small. Declines in forest bird populations in small forest fragments have been linked to increased nest predation by edge-associated competitors. Populations of forest understorey herbs associated with old-growth or other late-successional conditions may decline or disappear altogether in fragmented landscapes. Edges with high structural contrast, such as the boundary between an old-growth forest and adjacent clear-cut area, are especially prone to severe wind-throw of trees and extensive penetration of edge effects into the forest remnant. *See also* ECOTONE; OLD-GROWTH/ANCIENT FORESTS, CONSERVATION. [D.A.F.]

Education, environmental

The branch of education that deals with teaching and learning about the relationships between living organisms and their non-living, natural and human-built ENVIRONMENT. This area of education primarily deals with the human role in ecological systems. In the broadest sense, environmental education allows human culture to flourish in diverse environmental conditions. It will include:

- raising general awareness of environmental issues and the need for a responsible approach to these issues;
- raising awareness on how the activities of individuals and other institutions/organizations operate in respect to the environment;
- imparting skills/knowledge to measure the environmental effects;
- imparting skills/knowledge to avoid or remediate our impacts on the environment;

• in all this, giving an awareness of the holistic context of environmental problems.

However, the term is usually meant to imply educational programmes that encourage wise decisions through an understanding of the evolutionary and ecological relationships of humans to their local environment. This might be done through courses, training programmes, media, magazines, books, etc.

Because environmental issues are complex and involve multiple areas of scientific, cultural and political expertise, environmental education is ideally interdisciplinary. The earth sciences, biology, METEOROLOGY, economics and sociology are all important, but not exhaustive, transdisciplinary elements of environmental education programmes. The most successful environmental education programmes are not based on individual curricula, but rather cross the confines of specific and defined areas of expertise and focus on the interactions of learning and action.

The element of action on the part of learners in environmental education carries practical as well as theoretical implications. Environmental education is often directed at critical thinking and problem-solving. It is also often stressed that environmental education requires special attention to process and context. Thus, diverse teaching methods and the use of numerous physical teaching conditions is often an element of environmental education. Environmental education has its origins in nature study, conservation education and outdoor studies. These primarily biological and agricultural based approaches that teach about certain aspects of the natural environment have progressed into programmes which emphasize the role of human culture as it affects the natural world. As such, environmental education focuses on the impacts of humans on the environment and includes an element of how to make decisions about the use of natural RESOURCES.

Contemporary environmental education often focuses on public health and the environment, sustainable human activities in diverse environments, WILDLIFE and ENDANGERED SPECIES, natural heritage issues and quality education. However, issues such as tourism based on nature studies (ecotourism), international environmental affairs, POLLUTION control, hazardous WASTE management, AIR QUALITY and urban development are all elements of some environmental education programmes. These programmes have traditionally stressed terrestrial environments, but there are currently efforts in environmental education to broaden the scope of programmes to include aquatic ecosystems. This is especially seen in the numerous national and international programmes based on the ecology of river systems.

This emphasis on the setting of education is particularly important to environmental programmes. Special attention to giving teachers and learners direct experience with natural systems in a variety of physical environments is considered vital to the educational experience. These settings do not necessarily have to be in particular environments or include specific types of biological communities, but rather the emphasis should be on exposure of the teachers and students to a diversity of environmental conditions. One issue here is that environmental education treats the local environmental conditions as an inherent component of the education experience. Environmental education also pays attention to the issues of environmental protection and policy within the educational institutions in which it is taught.

As an overall concept, environmental education addresses the process of education as a factor in the way learners will live their lives after the formal educational experience is completed. This emphasis on 'experience in' and 'empathy for' the natural world is considered, within environmental education, as the development of an ecological consciousness. Thus environmental education seeks to raise the overall competence of the learner in regard to the natural world rather than to simply increase the level of factual information.

In recent years there have been a large number of texts and curricula directed at environmental education. However, curriculum supplements and interdisciplinary aids have had the most success. These approaches emphasize activities and allow teachers to incorporate environmental education material into a comprehensive education programme. While there are approaches that treat environmental education as a discrete subject area, these are less widely used and require special programme design. The majority of environmental education programmes are interdisciplinary, action based and emphasize personal obligations to environmental learning. [S.J.B. & P.C.]

REFERENCES

Hanselman, D., Raghunathan, M. & Kartikeya, S. (eds) (1990) *Essential Learnings in Environmental Education*. The North American Association for Environmental Education, Troy, Ohio.

Orr, D. (1992) *Ecological Literacy: Education and Transition to a Postmodern World*. State University of New York Press, Albany, N.Y.

Toyne, P. (1993) *Environmental Responsibility. An Agenda for Further and Higher Education*. Her Majesty's Stationery Office, London.

Wilke, R.J. (ed.) (1993) *Environmental Education Teacher Resource Handbook*. Kraus International Publications, Millwood, N.Y.

EEA *See* EUROPEAN ENVIRONMENT AGENCY.

effect

1 In an ecotoxicological context, adverse consequences of CONTAMINANT. May be expressed at one or more levels: molecular structure and processes; cellular structure and function; organismic survival, development, GROWTH and reproduction; POPULATION DYNAMICS; COMMUNITY structure; ECOSYSTEM function.

2 More generally, change ascribed to a cause. There may be one or more factors bringing about an effect. It is usually presumed that specific causes produce specific effects. However, effects in complex systems may be brought about by more than one kind of causal factor.

[P.C.]

effective ecosystem size The 'effective size' of an ECOSYSTEM depends on its absolute size (extent) and on the degree of fine-grained physical structure within it. The term refers to the characteristics of an ecosystem that cause it to be functionally larger (for the organisms in it) than it actually is. Ecosystems of greater 'effective size' reduce the risk of predator or prey extinctions by limiting the amplitude of density OSCILLATIONS. [S.J.H.]

REFERENCE

Crowley, P.H. (1978) Effective size and persistence of ecosystems. *Oecologia* **35**, 185–195.

Effective population size

The effective POPULATION SIZE (Wright 1931) of an actual population is the number of individuals in a theoretically ideal population having the same magnitude of random GENETIC DRIFT as the actual population. An ideal population is a random mating population in which the sole evolutionary pressure is genetic drift, i.e. one in which variation of gene frequencies are due to SAMPLING errors from one generation to the next, because of small population size. The effective population size of an actual population can be seen as a measure of how many genetically distinct individuals actually participate in the formation of the next generation and as a predictor of the rate at which, because of drift, genetic VARIABILITY will be lost from a population or inbreeding will increase. It has thus applications in CONSERVATION BIOLOGY, as it might be used to produce the minimum population size under which a given species might lose its evolutionary potential. There are several kinds of effective population size, not all amounting to the same value (Hartl & Clark 1989). The 'inbreeding effective size' of a non-ideal population is the size of an ideal population that would show the same increase of the average inbreeding coefficient. The 'eigenvalue effective size' is the size of an ideal population that would show the same rate of loss of expected HETEROZYGOSITY (which measures genetic variability). Finally, the 'variance effective size' is the size of an ideal population that would show the same variance of the change in ALLELE FREQUENCY. It is also the size of an ideal population that, if replicated an infinite number of times, would show the same rate of increase of variance of gene frequencies among replicate populations as the actual population if it were replicated the same number of times. Usually, the variance effective size and the inbreeding effective size are the same, but there are exceptions.

There are many reasons why the effective size of a population might differ from the actual size of that population. One reason might be a biased sex ratio: if, for instance, in a population there are more females than males and if every female produces on average two offspring, then the effective size will be much smaller that the actual size. It can be shown that in a population made of M males and F females, the effective size is equal to $4MF/(M+F)$. The reason is that few males contribute many genes to the next generation, thus increasing the relatedness among offspring and thus the inbreeding coefficient. In the same way, if the population size fluctuates through time, it can be shown that the effective size is the GEOMETRIC MEAN of the effective sizes at each generation. Consequently, if a population goes through phases of bottlenecks, with very small population sizes from time to time, the average effective size is led by the smallest sizes. Thus in order for populations to maintain large amounts of variability, constantly large size is necessary as well as a small variance in the contribution of each individual to the next generation. In particular, if selection is strong, the effective size for selectively neutral loci might be extremely small, because actually very few individuals might contribute to the next generation. In a diploid population of constant size N, in which the mean contribution per individual is two gametes and its variance is σ^2, Wright (1938) showed that the effective population size was $(4N-2)/(\sigma^2+2)$. It is approximately equal to the actual population size N if the mean contribution per individual follows a POISSON DISTRIBUTION, with $\sigma^2=2$. If all individuals contribute exactly two gametes to the following generation, the effective size is almost twice the actual population size. If selection is strong, σ^2 is large and effective size may be much reduced, so that the effects of genetic drift increase. *See also* QUANTITATIVE GENETICS. [I.O.]

REFERENCES

Hartl, D.L. & Clark, A.G. (1989) *Principles of Population Genetics*, 2nd edn. Sinauer Associates, Sunderland, Mass.

Wright, S. (1931) Evolution in Mendelian populations. *Genetics* **16**, 97–159.

Wright, S. (1938) Size of population and breeding structure in relation to evolution. *Science* **87**, 430–431.

effects register Literally implies listing of possible ecological effects (usually adverse) of emanations from an industrial process. However, most often involves lists of chemicals in those emanations that are known to be hazardous. Usually an integral part of an ENVIRONMENTAL MANAGEMENT SYSTEM. [P.C.]

efficiency Ratio of a useful output to an input. Used widely to describe ecological systems at various levels. Thus for individuals, digestive (ASSIMILATION) efficiency is the ratio of fuel absorbed to food ingested and growth efficiency is the ratio of body increase to food used to achieve this. Population efficiencies are the sum of individual efficiencies, allowing for age and size structure. Trophic efficiences (sometimes known as ecological or LINDEMAN EFFICIENCIES) are the ratios of total intake for organisms at one TROPHIC LEVEL to the total intake of organisms exploiting them. These efficiencies are usually expressed as percentages. When expressed in terms of matter they can be 100% (laws on conservation of matter) but when expressed in terms of energy only a fraction of the input can be transformed into useful output (second law of thermodynamics). [P.C.]

effluents Waste release from point sources, usually as liquid into external environment. *Cf.* EMISSIONS. [P.C.]

effort-based sampling *See* SAMPLING METHODOLOGY/DEVICES.

egg mimicry The close resemblance of the appearance of the EGGS of some brood-parasitic birds to that of their HOSTS.

The function of egg mimicry is the prevention of egg ejection by a parasitized host. The degree of matching in terms of coloration and the pattern and position of spots and streaks is often remarkable. Brood parasites that produce eggs deviating the most from those of their hosts are presumed to suffer from an increased frequency of reproductive failure due to the recognition and ejection of their eggs by hosts. Hosts cannot evolve perfect discriminatory ability between their own and parasitic eggs because of learning of the appearance of their own eggs and errors in discrimination. The discriminatory ability of hosts depends on learning since there is no innate ability to recognize their own eggs. Host eggs are not completely similar and strict discrimination against a slightly odd egg may result in the host ejecting one of its own eggs. Experimental PARASITISM of the clutches of host species with eggs varying in resemblance to those of the host has shown that the use of non-mimetic eggs gives rise to a higher frequency of egg ejection than do mimetic eggs. Differential host responses dependent on the degree of mimicry of parasitic eggs are not found among bird species that are unsuitable as hosts, because there has been no selection on such species to discriminate between their own eggs and those of other species. *See also* BROOD PARASITISM; NEST PARASITISM. [A.P.M.]

egg-shell effects Refers to effects of pollutants on birds' egg-shell structure, usually thickness. ORGANOCHLORINES such as DDT and DDE (1,1-dichloro-2,2-bis-(*p*-chlorophenyl) ethylene), for example, have been implicated. Egg-shell thinning was an important consideration leading to the banning of DDT in the USA. *See also* PESTICIDE. [P.C.]

eggs Unicellular products of female reproductive system. Usually formed by meiosis and often requiring conjugation with male contribution to induce development. However, the latter need not be the case in PARTHENOGENESIS. The cytoplasm often contains some material (yoke) that supports early development. In cleidoic eggs, where there is a shell that provides isolation from the outside world, these stores are extensive. In viviparous species, the stores are less extensive. Birds and many fishes have a telolecithal egg in which yoke is so extensive that it compresses the egg cell against one pole. Many amphibians have eggs with a moderate amount of yoke and are called mesolecithal; eggs of viviparous animals may contain little or no yoke and are said to be alecithal. Some insects have a peculiar egg in which yoke is concentrated in the centre of the egg cell; this is said to be centrolecithal. These various conditions are illustrated in Fig. E7. *See also* LIFE-HISTORY EVOLUTION; PROPAGULE; SPERMATOZOA; VIVIPARITY. [P.C.]

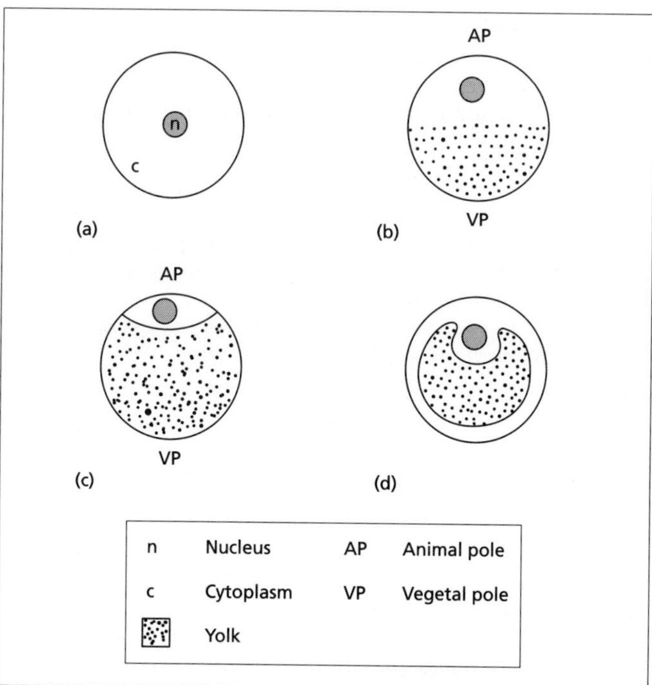

Fig. E7 Types of eggs: (a) alecithal; (b) mesolecithal; (c) telolecithal; and (d) centrolecithal.

EIA *See* ENVIRONMENTAL IMPACT ASSESSMENT.

eigenvalue A term from matrix algebra. In age-structured MATRIX MODELS, the dominant eigenvalue is the net multiplication rate of the population, once (and if) a STABLE AGE DISTRIBUTION has been reached. Thus, a population with three AGE CLASSES, with FECUNDITY of 0, 9 and 12 females per female per year and survivorship of $^1/_3$ from the first to the second age class and $^1/_2$ from the second to the third, has the following Leslie matrix:

$$\begin{vmatrix} 0 & 9 & 12 \\ \frac{1}{3} & 0 & 0 \\ 0 & \frac{1}{2} & 0 \end{vmatrix}$$

Once it has achieved a stable age distribution (EIGENVECTOR), this population doubles each year; its dominant eigenvalue is 2.0. The other (subdominant) eigenvalues ($-1+0.000577i$ and $-1-0.000577i$) are complex numbers that describe the nature of the transient dynamics through which the system passes before the stable age distribution is obtained. If the population consisted entirely of adults, it could increase faster than this; if it consisted entirely of juveniles, the population would decline for a year (until the first individuals reached reproductive maturity). The eigenvalue is obtained by finding the roots of the characteristic equation. A function to compute eigenvalues for particular parameterized square matrices is available in most mathematical and statistical computer packages. *See also* EQUATIONS. [M.J.C.]

REFERENCE

Caswell, H. (1989) *Matrix Population Models*. Sinauer, Sunderland, Mass.

eigenvector The age (or stage) structure of a model population once initial transient dynamics have damped away. For the Leslie matrix presented in the previous entry (*see* EIGENVALUE) the dominant eigenvector is {0.986, 0.164, 0.041}. This means that after transient dynamics have damped away, the proportion of individuals in the first AGE CLASS will be 0.828 (this is 0.986/1.191), 0.138 in the second and the remainder, 0.034, in the third and oldest age class. [M.J.C.]

EIL *See* ECONOMIC INJURY LEVEL.

EINECS *See* EUROPEAN INVENTORY OF EXISTING COMMERCIAL CHEMICAL SUBSTANCES.

einsteinium (Es) Einsteinium is a synthetic element (ATOMIC NUMBER = 99; relative ATOMIC MASS = 252; melting and boiling points unknown) that was discovered, along with FERMIUM, in debris from the first thermonuclear explosion at Eniwetok Atoll in 1952. Isotopes with mass numbers ranging from ^{243}Es to ^{256}Es have been produced by particle bombardment of other actinides or multiple NEUTRON capture events in nuclear reactors, but only in very small quantities (2 mg or less). Consequently, einsteinium does not occur in detectable quantities in the environment. Investigations of the properties of einsteinium have mainly used ^{253}Es ($t_{1/2}$ = 20.4 days). The dominant oxidation state of einsteinium is +3. [A.B.M.]

Einstein's law *See* RADIATION LAWS.

EIONET (Environmental Information and Observation Network) *See* EUROPEAN ENVIRONMENT AGENCY.

EIS *See* ENVIRONMENTAL INDUSTRIES SECTOR.

Ekman circulation Movement of surface water at an angle from the WIND as the result of the CORIOLIS EFFECT. An Ekman spiral is a theoretical description according to which the surface water layer moves 45° to the right of the wind direction in the Northern hemisphere (and 45° to the left of the wind direction in the Southern hemisphere). Water at successive depths moves in directions more to the right and at decreasing speeds until, at some depth, the water moves in a direction opposite to the wind. Ekman transport is the net transport of surface water set in motion by wind, due to the Ekman spiral phenomenon. Net transport of water is 90° to the right of the wind direction in the Northern hemisphere and 90° to the left of the wind direction in the Southern hemisphere. [V.F.]

El Niño

Episodic climatic changes that include warming of the equatorial PACIFIC OCEAN and suppression of UPWELLING into the EUPHOTIC ZONE off the coast of Peru by intrusions of warm, nutrient-poor, surface water. The southward-flowing tongue of the equatorial countercurrent is a regular phenomenon in northern Peru and occurs every year in February or March. Periodically the current extends further south to >12°S and displaces the cold northward-flowing Peru current. El Niño is associated with a weakening of the TRADE WINDS, which occurs typically in cycles of 7 years and may be caused by global circulation anomalies that are in turn influenced by SOLAR RADIATION. The El Niño has been responsible for mass mortality of PLANKTON and fish, has led to starvation of seabirds and has contributed to the collapse of the Peruvian anchovy fishery.

El Niño (derivation: Christ boy-child, as it typically occurs around Christmas) was named by Spanish-speaking fishermen off Ecuador and Peru who correlated its occurrence (warm surface-water temperatures) with disastrous fishing conditions. The Australian Bureau of Meteorology (1988) definition is: 'the occasional warming of the usually cool surface waters of the eastern equatorial Pacific'. Over the past 40 years, nine El Niños have affected the South American continent.

The concept has gradually been extended to cover the whole of the Southern Ocean (*see* ENSO; SOUTHERN OSCILLATION INDEX), and now beyond. El Niño now affects the entire Indo-Pacific oceanic region. However, in October 1997 there was an enormous incursion of El Niño driven air from the tropical Pacific into the Caribbean and North Atlantic, resulting in a 6-day reversal of the trade winds–to the south-west. In recent years there has been considerable interest in, and support for, the hypothesis that El Niño effects can explain the occurrence of DROUGHTS and also rainfall patterns in continents such as Australia. Australian rainfall patterns are strongly dependent on the fickle southerly incursions of tropical CYCLONES. Their genesis and frequency seem to be reduced by El Niño patterns, now generally monitored as a negative southern oscillation index (SOI).

Surface OCEAN temperatures are strongly correlated with atmospheric PRECIPITATION and are also strongly dependent on global marine CURRENTS generally. These are driven by relatively predictable forces such as lunar influences, water density (SALINITY), CONVECTION and Coriolis (Earth's rotational) forces but, in addition, by several more variable driving forces that are difficult to quantify, including sea levels, surface and stratospheric winds, solar exposure (cloud formation) and sunspot activity. Apparently this complex flow is further modified by its interaction with terrestrial TOPOGRAPHY to give rise to hydraulic resonance phenomena with periodicities of several years.

The notion is held that SOI periodicity shifts (i.e. oscillates) the cloud formation and rainfall patterns, and especially tropical cyclones, over neighbouring landmasses such as those of Australia. Causes and effects are difficult to discern. Also there have been at least two clear anomalies since 1880, in 1905 and 1987, when negative SOIs failed, i.e. falsely indicated, the occurrence of major droughts in Australia. There is no doubt that these patterns will continue to be monitored ever more closely and quantitatively. It remains to be seen if the anomalies can be explained or if, indeed, it will be possible to substitute an even more reliable predictive indicator than SOI.

Populations are very concerned to know forthcoming WEATHER patterns, but most conventional forecasts are of only relatively short durations (a few days). However, the benefits of longer-term predictions are potentially enormous because they impact on better planting regimes and avoidance of disasters, for example droughts and FLOODS in agriculture and horticulture, fires, cyclone damage, poor fishing and even mosquito plagues, which may spread diseases. Hence the great interest in El Niño as a long-term climatological predictive system. [J.M. & V.F.]

REFERENCES

Dudley, N.J. & Hearn, A.B. (1993) El Niño effects hurt Namoi irrigated cotton growers, but they can do little to ease the pain. *Agricultural Systems* **42**, 103–126.

Philander, S.G.H. (1983) El Niño Southern Oscillation phenomena. *Nature* **302**, 295–301.

Wallace, J.M., Vogel, S. *et al.* (1994) *El Niño and Climate Prediction. Reports to the Nation on our Changing Planet*, pp. 1–24. UCAR Office for Interdisciplinary Earth Studies, Boulder, Colo.

Wright, P.B. (1985) The Southern Oscillation: an ocean–atmospheric feedback system? *American Meteorological Society Bulletin* **66**, 398–412.

ELA *See* EXPERIMENTAL LAKE AREA.

elasticity *See* SENSITIVITY ANALYSIS.

elasticity analysis *See* SENSITIVITY ANALYSIS.

electromagnetic pollution Electromagnetic RADIATION from electronic equipment, for example office equipment, is thought by some to lead to lethargy and to SICK BUILDING SYNDROME. A possibly more dangerous form is the extremely low frequency radiation produced by high-voltage power lines, alleged by some to cause epilepsy, through heart palpitations, and eye cancers. [P.C.]

electromagnetic spectrum Electromagnetic RADIATION is an inductive coupling between electric and magnetic vibrations and is the means by which energy is transferred millions of miles through space at the speed of LIGHT (3×10^8 m s^{-1}). Examples of electromagnetic radiation include radio waves, microwaves, infrared, visible light, ultraviolet, X-RAYS and γ-RAYS (Fig. E8). The electromagnetic spectrum is a representation of these radiations with respect to wavelength or frequency. [J.G.]

electromorph *See* ELECTROPHORESIS.

electrophoresis A technique used to separate a mixture of charged molecules by their differential migration through a stationary phase in an electrical field. This method is typically used for distinguishing different genetic forms of the same enzyme or to separate nucleic acid molecules of different sizes and shapes. It may be used as an analytical or preparative technique. An electrical potential is applied across a gel (made typically of starch, agarose or polyacrylamide) on to one end of which samples have been placed. The mobility of the nucleic acid molecules or of different genetic forms of an enzyme (allozymes) varies as a function of net charge, conformation and size. After allowing time for migration the location of the molecules is detected by treating the gel with a substance to permit visualization of the molecules. For example, a fluorescent dye such as ethidium bromide, which binds strongly to nucleic acid, can be used to detect the location of different nucleic acids in a gel upon examination under ultraviolet light. For enzymes, a substrate upon which the enzyme can act and a dye that reacts with the product of the enzyme–substrate reaction are employed to detect the location of the different allozymes. It has been estimated that about 25–30% of all AMINO ACID substitutions can be detected by the usual techniques of enzyme electrophoresis.

Genotypes differing in electrophoretic properties as a result of amino acid (ultimately gene) substitutions are sometimes referred to as electromorphs. [V.F.]

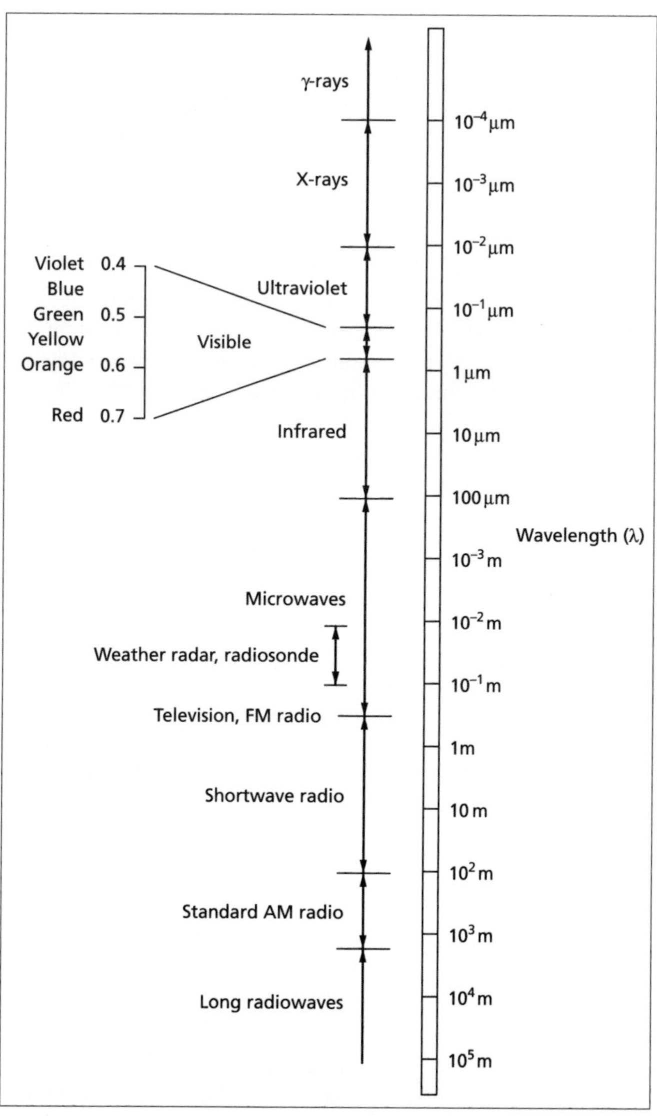

Fig. E8 The electromagnetic spectrum. (After Oke, T.R. (1978) *Boundary Layer Climates*. Methuen, London.)

elements The Aristotelian concept of the elements (earth, water, fire and air), rejected in the 17th century, was replaced by the present definition of an element as a substance that cannot be decomposed into simpler substances.

Atoms of an element have the same number of protons and electrons but may have differing numbers of neutrons (*see* ISOTOPE). Elements have been classified according to their various properties. The alchemists differentiated between noble and base metals whilst Jöns Jacob Berzelius made the distinction between metals and nonmetals. In the 19th century, Dmitri Mendeleev recognized that the properties of elements were a periodic function of their ATOMIC MASS. *See also* PERIODIC TABLE. [M.C.G.]

elfin forest Elfin or mossy FOREST is a kind of TROPICAL MONTANE FOREST or CLOUD FOREST in which the trees are short and gnarled, with a CANOPY in some cases only 2 m high. Growth is by equal branching, either sympodial or dichotomous, and long–short shoot patterns are common. This gives rise to a very dense subcanopy of slender branches. Many plants produce aerial and adventitious roots. Small leaves are crowded on twigs. The trees are festooned with EPIPHYTES, bryophytes, especially leafy liverworts, and lichens. Vascular epiphytes are less common than in montane forest.

Trees are stunted so that generally the same species of trees can be found lower down growing quite normally. Growth form is mainly a plastic response to environmental conditions. Conditions are cool and humid and there is seasonal cold. Light levels are low in the misty conditions. The rocky soils may be poor in NUTRIENTS, which are anyway difficult to obtain because of low TRANSPIRATION rates. Elfin forests are an expression of ALTITUDINAL ZONATION. They are generally found at above 3000 m in the tropics but are sometimes found much lower where low nutrient conditions or misty conditions prevail. Woodlands with a similar character and PHYSIOGNOMY can be found in temperate regions. [M.I.]

REFERENCE
Howard, R.A. (1968) The ecology of an elfin forest in Puerto Rico. *Journal of the Arnold Arboretum* **49**, 381–418.

ELINCS *See* EUROPEAN LIST OF NOTIFIED CHEMICAL SUBSTANCES. *See also* EUROPEAN INVENTORY OF EXISTING COMMERCIAL CHEMICAL SUBSTANCES.

Ellenberg's rule Ecological studies on plant GROWTH along experimentally created gradients demonstrate that most species perform best when they are grown in low-density MONOCULTURE under roughly similar conditions along most kinds of GRADIENT. In COMPETITION with other species, however, the inferior competitors tend to be forced out from the optimal environmental conditions and to persist in REFUGIA in suboptimal conditions. It is a mistake, therefore, to conclude that plant species are found in the field under conditions that are optimal for their growth. Optimal growth conditions appear to be remarkably similar for most plants and it is competition with other species that excludes the majority of plant species from optimal conditions in the wild. Exceptions to this rule include extreme CALCICOLE and CALCIFUGE species, which cannot be grown on ACID or calcareous substrates respectively; calcicoles and calcifuges are found in the field under conditions of soil pH close to their optima. *See also* CLINE. [M.J.C.]

REFERENCE
Ellenberg, H. (1954) Uber einige Fortschritte der kausalen Vegetationskunde. *Vegetatio* **5/6**, 199–211.

Elton, C. *See* CHARACTERS IN ECOLOGY.

Eltonian pyramid The idea of expressing the numbers in, and BIOMASS or mass of, different trophic levels as horizontal bars stacked in a pyramid is usually attributed to Charles Elton (1927), who first drew such diagrams following his formative expedition to Spitsbergen.

One aspect of trophic interactions that particularly interested him was the relative size of members of different trophic layers. He noted that there was often a quantum difference in size between members of different trophic levels, for example phytoplankton and water fleas (*Daphnia*) or midge larvae and fish, which he attributed to the advantages of larger predators being able to overcome smaller prey more easily. Thus if the numbers of organisms are plotted as bars against their size, the commonest are often found to be the smallest while the bigger species are relatively scarce (Fig. E9). *See also* ECOLOGICAL EFFICIENCY; ECOLOGICAL PYRAMID; PYRAMID OF BIOMASS; PYRAMID OF NUMBERS; TROPHIC LEVEL. [M.H.]

REFERENCES
Colinveaux, P. (1986) *Ecology.* John Wiley & Sons, New York.
Elton, C.S. (1927) *Animal Ecology.* Macmillan, New York.

elutriate test ECOTOXICITY test of aquatic SEDIMENT using a solution derived from it by elutriation. A slurry of sediment and WATER are mixed vigorously, then allowed to settle or filtered. The elutriate is then exposed to the test system. [P.C.]

REFERENCE
Luoma, S.N. & Ho, K.T. (1993) Appropriate uses of marine and estuarine sediment bioassays. In: *Handbook of Ecotoxicology,* Vol. 1 (ed. P. Calow), pp. 193–226. Blackwell Scientific Publications, Oxford.

EMAS *See* ECO-MANAGEMENT AND AUDIT SCHEME.

emergence marsh The upper part of a SALT MARSH, which is subject to only short periods of submersion (immersion) during spring TIDES and where the key environmental conditions are associated with long periods of emersion. The plants and animals of this level may be considered as essentially species of terrestrial origin that are tolerant of limited submersion (immersion). The physical boundary between emergence and SUBMERGENCE MARSH is, according to Chapman (1960), that level subjected to about 360 flooding tides per year, but this does not coincide consistently with

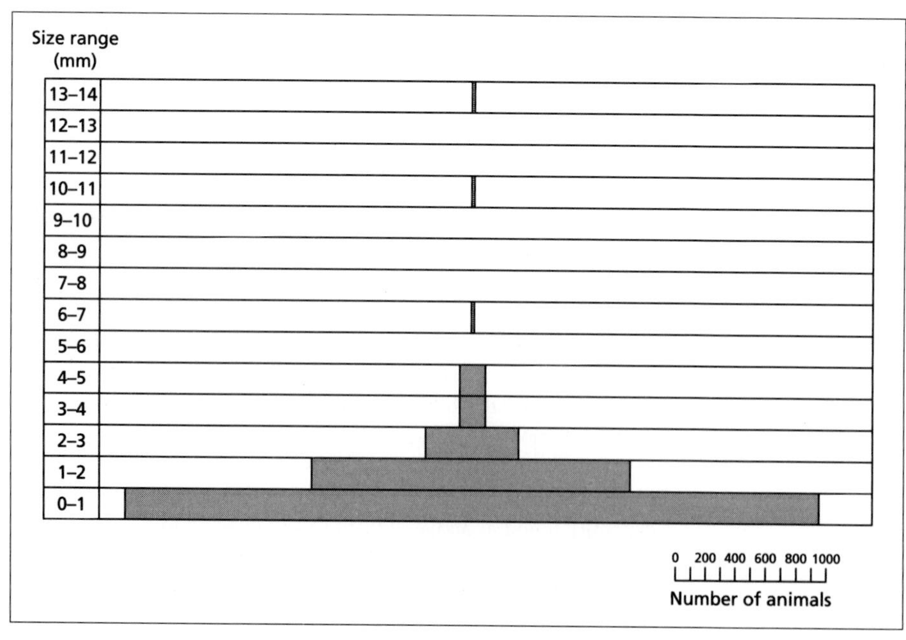

Fig. E9 Eltonian pyramid of numbers on the floor of a forest in Panama. Animals in small samples of the litter on the floor of a forest were counted and sorted into size fractions. The smallest and most numerous animals were Collembola (springtails) and mites, both of which are herbivores or scavengers feeding in the litter. The larger, rarer animals, such as ground beetles and spiders, are carnivores. (After Colinveaux, P. (1986) *Ecology.* John Wiley & Sons, New York; copyright © 1986, reprinted by permission of John Wiley & Sons, Inc.)

species boundaries, which seem to vary regionally, almost certainly reflecting regional climatic differences and perhaps also substrate variations. *See also* SUCCESSION. [B.W.F.]

REFERENCE

Chapman, V.J. (1960) *Salt Marshes and Salt Deserts of the World.* Interscience Publishers, New York.

emergent plant Aquatic plant species, usually growing in the shallower marginal regions of a freshwater body, which are rooted below the water surface but produce shoots that extend above the surface. Their structural characteristics make emergent plants particularly valuable for some aquatic insects that have an aerial phase in their life cycle, such as dragonflies. [P.D.M.]

emergent properties Features not fully (or at all) explicable in terms of the properties of the component parts. Thus the dynamics of collective groups of organisms involve interactions between them that are not fully explicable on the basis of understanding of individuals and their parts (cells and molecules). [P.C.]

emigration Exit of CONSPECIFICS from a POPULATION to elsewhere. Emigration leads to GENE FLOW out of the population. *See also* DISPERSAL; IMMIGRATION; MIGRATION. [P.C.]

emission spectrum The electromagnetic RADIATION emitted by atoms or molecules after they have been excited, shown as a spectrum projected on to a screen or plotted as a graph of energy (y-axis) against wavelength (x-axis) (see Fig. R2). [J.G.]

emission standards Amount of CONTAMINANT not to be exceeded in the DISCHARGE from a POINT SOURCE. Usually set to avoid POLLUTION, at least outside the MIXING ZONE. *See also* ENVIRONMENTAL QUALITY STANDARD. [P.C.]

emissions Waste release from point sources; usually refers to release into ATMOSPHERE through chimney stacks. *Cf.* EFFLUENTS. [P.C.]

EMS *See* ENVIRONMENTAL MANAGEMENT SYSTEMS.

end-of-pipe solution Removal of CONTAMINATION by treatment of EMISSIONS (*see* FLUE-GAS DESULPHURIZATION) and EFFLUENTS (e.g. by use of filters). [P.C.]

end-point Often used to refer to what is measured in ecotoxicological tests. These measurement end-points are presumed to relate to those aspects of ecological systems that are of interest, i.e. that we want to protect. These latter are sometimes referred to as assessment end-points. *See also* PREDICTIVE ECOTOXICITY TESTS. [P.C.]

endangered species Official designation is species that have 20% probability of becoming extinct in 20 years or 10 generations (US ENDANGERED SPECIES ACT). *See also* MINIMUM VIABLE POPULATION. [P.C.]

Endangered Species Act UNITED STATES LEGISLATION passed in 1973 to protect species in danger of becoming extinct in the wild. First of its kind, but many other nations now have similar legislation. *See also* ENDANGERED SPECIES. [P.C.]

endemic *See* EPIDEMIC.

endemic species An animal or plant that is native to a particular location and is restricted to that location in its DISTRIBUTION. An endemic species that is of ancient origin and has remained restricted because of problems of DISPERSAL is said to be a palaeoendemic. One that has recently evolved and is restricted simply because it has had no time to disperse is a neoendemic. [P.D.M.]

endemism *See* ENDEMIC SPECIES.

endobiont An organism such as a bacterium or alga that lives inside the cells of other organisms, often potentially an endosymbiont. *See also* MUTUALISM. [S.G.C.]

endocrine disrupter TOXICANTS (POLLUTANTS) that impair reproduction and development by interfering with normal endocrine control. Tributyltin appears to have its effect by interfering with hormones that inhibit maleness in female molluscs. Many higher plants synthesize chemicals that interfere with the endocrine controls of animals (especially insects) that feed upon them. Many PESTICIDES are designed to have this specific effect.

Those endocrine disrupters that interfere with male systems by simulating vertebrate female hormones are called oestrogenic and those causing female disruption by simulating male hormones are called androgenic. Substances implicated have been pesticides, alkyphenols (used in cleaning products), POLYCHLORINATED BIPHENYLS, DIOXINS and by-products of the contraceptive pill. They can have adverse effects at very low concentrations. [P.C.]

endogenous rhythm Internally generated rhythm (oscillation) that is self-sustained. A CIRCADIAN RHYTHM is an endogenous rhythm with a daily periodicity. Circannual rhythms have an annual periodicity. Endogenous rhythms have a natural free-running periodicity that becomes entrained by environmental cues (zeitgebers), such as the daily light–dark cycle. Heart beat is also a self-sustained internal rhythm. *See also* ENTRAINMENT; OSCILLATIONS. [J.S.B.]

endolithic Organisms living with rocks. Golubic *et al.* (1981) distinguish three endolithic niches for microorganisms: cryptoendolithic, chasmoendolithic and euendolithic (Fig. E10). Cryptoendoliths grow in spaces between the mineral components and are particularly prevalent in porous sandstones. Chasmoendoliths occupy minute fractures while euendoliths actively etch the mineral components, producing long tubes or hollows into which they grow. Most euendoliths belong to the cyanobacteria and are known mainly from coastal limestones, where they tolerate a wide range of salinities and grow to a depth of several millimetres. Several algae and fungi, including some lichens, are endolithic and occur in some of Earth's most inhospitable regions. Several molluscs are also endolithic and excavate holes several centimetres deep into soft rocks. [A.P.]

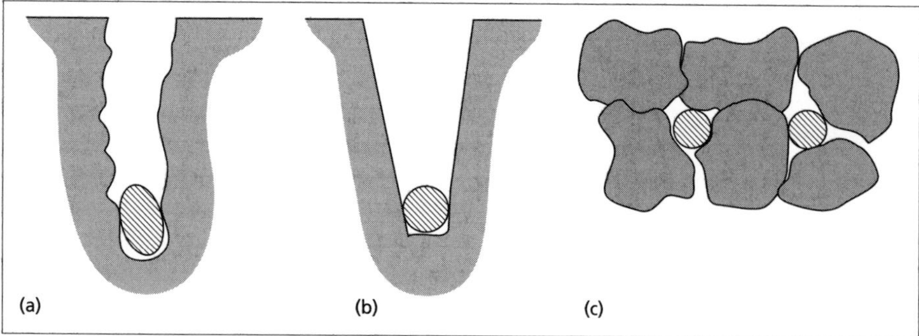

Fig. E10 Three endolithic niches for microorganisms: (a) euendolithic; (b) chasmoendolithic; and (c) cryptoendolithic. (After Golubic, S., Friedman, I. & Schneider, J. (1981) The lithobiontic niche, with special reference to microorganisms. *Journal of Sedimentary Petrology* **51**, 475–478.)

REFERENCE

Golubic, S., Friedman, I. & Schneider, J. (1981) The lithobiontic niche, with special reference to microorganisms. *Journal of Sedimentary Petrology* **51**, 475–478.

endomitosis Refers to a particular way of producing females from unfertilized EGGS, with meiosis. Meiosis is preceded by a round of chromosome replication, giving rise to a tetraploid cell. Pairing then occurs between sister chromosomes, i.e. chromosomes that have just arisen by the replication of the same parent chromosome. Offspring genetically identical to their mother are produced. *See also* PARTHENOGENESIS. [I.O.]

REFERENCE

Maynard Smith, J. (1989) *Evolutionary Genetics*. Oxford University Press, Oxford.

endomycorrhizae *See* ENDOTROPHIC MYCORRHIZAE.

endoparasite A PARASITE that lives within the HOST (used only in reference to MACROPARASITES; *cf.* ECTOPARASITE). There are many parasites of the gut (e.g. *Ascaris lumbricoides*), bloodstream (e.g. *Schistosoma mansoni*) and lymphatic system (e.g. *Wuchereria bancrofti*), but all body tissues may be parasitized. Endoparasites have varying degrees of host specificity, variable degrees of dependence on their hosts and spend variable proportions of their LIFE CYCLE in the host. [G.F.M.]

endoparasitoids *See* PARASITOID–HOST INTERACTIONS.

endotherm An endotherm is an organism that is able to maintain and regulate its internal body TEMPERATURE across a wide range of conditions present in the surrounding external environment. Hypothalamic REGULATION in such organisms has enabled colonization and exploitation of more variable environments. *Cf.* ECTOTHERM; POIKILOTHERM. *See also* HOMEOTHERM. [R.C.]

endothermic/exothermic reaction Chemical reactions may proceed with a net absorption of energy, in which case they are said to be endogonic, or with a net release of energy (exogonic). The energy transaction may include heat, LIGHT or electricity. The terms 'endothermic' and 'exothermic' refer to the heat exchange only: endothermic reactions are those that absorb heat, exothermic reactions are those that result in a net release of heat. [J.G.]

Endotrophic mycorrhizae (endomycorrhizae)

Widespread in fertile soils and more diverse than ECTOTROPHIC MYCORRHIZAE, endotrophic mycorrhizae in every case are typified by the fungal hyphae penetrating some of the plant HOST cells, i.e. intracellular penetration occurs. This is followed usually by a balanced and intimate relationship between the fungal structures and the colonized host cell of the plant.

Vesicular–arbuscular mycorrhizae

Vesicular–arbuscular (VA) endomycorrhizae are the most widespread and significant mycorrhizal types, occurring commonly in 90% of all vascular plant species, including legumes, grasses, shrubs and hardwood trees, as well as many lower plants. In the Southern hemisphere and tropics, they are the typical MYCORRHIZAE of forest trees, rather than ectomycorrhizae. They are so called because of two characteristic types of fungal intracellular haustoria produced by the intercellular hyphae colonizing the root cortical tissue: the smooth, balloon-like, 'vesicles' (also formed outside the cortical host cells) and the fine, tree-like, branching pattern of the 'arbuscules'; some 'coiling' of hyphae also occurs. A further distinguishing feature is the aseptate nature of the hyphae, since all the fungi involved are members of the Zygomycotina in the order Mucorales (family Endogonaceae); there are four genera: *Glomus*, *Gigaspora*, *Acaulospora* and *Sclerocystis*. Very large chlamydospores, and in some cases zygospores, are produced in the external SOIL and are important in survival. These can help to identify the fungus concerned, which is always difficult since none has yet been grown in axenic culture, i.e. they are OBLIGATE symbionts. There is an outer, very indistinct hyphal 'weft' traversing the RHIZOSPHERE (rather than an external sheath) and normal root hairs are still produced; the relatively few external hyphae connecting with those within the root cortex are nevertheless responsible for significant increases in the uptake of phosphorous compounds and other NUTRIENTS from the soil. The fungus never invades the stele or the above-ground organs of the plant. The arbuscules are the major site of nutrient and mineral exchange, having a high surface area and raised adenosine triphosphatase (ATPase) activity on the plasma membrane of both partners at their point of near contact, with a special matrix between the hyphal wall and the plant membrane. The arbuscules live for only 4–15 days, but new ones continue to form. Although spores and vesicles are produced on external hyphae (connected with the colonized root) in soil surrounding the root, there is no free-living saprophytic stage. A wide variety of different plant species are colonized by a single type of 'VA fungus' and thus the level of host specificity is very low. Seedlings of crop plants inoculated with spores, fragments of colonized root tissue or soil previously in contact with the fungus, show typical increases in growth rate and nutrient uptake as the VA mycorrhizal partnership develops.

Ericaceous/ericalean mycorrhizae

The ericaceous/ericalean types of endomycorrhizae include the well-studied and typical ericoid form found widely within the family Ericaceae, which comprises flowering plants and shrubs including heathers, bilberry and rhododendron/azaleas, often growing on relatively nutrient-poor, acidic soils. They play a dominant and vital role worldwide, particularly in the Northern hemisphere, in ecological plant communities such as peaty moors, heaths, BOGS and SWAMPS. In the case of heathers such as *Calluna* and *Erica* growing in cold regions and in upland zones, their very existence in such harsh conditions is attributed directly to their ericoid, endomycorrhizal condition. Their roots, which lack root hairs, end in very fine hair-like extensions in which the very narrow cortical layer (one to three cells in width) surrounding the small stele is extensively colonized by intracellular, septate, mycorrhizal hyphae, which occupy the majority of the cortical cell volume in the form of coils surrounded by the invaginated host plasmalemma. A matrix of pectic material is present between the hyphal wall and host plasmalemma, and there is a 'collar' at the site of penetration. There is a loose weft of hyphae on the narrow rootlet surface, connected to the penetrating hyphae growing into the host plant cells below. Up to 70% of the root cortical cells can be colonized, but meristematic cells or those within the stele are never penetrated. The host cell remains alive for 3–4 weeks following penetration (during which time nutrient exchange can occur) but then the plasma membrane breaks down, the cell accumulates PHENOLS and the fungus degenerates. The majority of fungi involved here are not host-specific. They are thought to be members of the Ascomycotina (e.g. *Pezizella*) or Deuteromycotina. They also grow as long-lived, free-living organisms in soil, being responsible for the initial infection of new rootlets; however, in culture they grow only very slowly as dark, septate hyphae, rarely fruiting, making identification very difficult. Ericoid mycorrhizal roots can utilize both AMINO ACIDS and PROTEINS as NITROGEN sources, giving a clear advantage to plants growing in acidic HEATHLAND soils for example, where MINERALIZATION (conversion of organic nitrogen to NITRATE and ammonium) is delayed. Furthermore, they can tolerate up to 100 ppm of HEAVY METALS, enabling such plants to colonize soils containing high levels of CADMIUM, ZINC and COPPER; the hyphae retain the metals, thus reducing levels in the whole plant.

Ectendotrophic mycorrhizae show several features typical of both endotrophic or ectotrophic forms of mycorrhizae. There is a Hartig net with intercellular septate hyphae in the arbutoid form (e.g. *Arbutus*, the strawberry tree) of ericalian mycorrhizae, as well as intracellular coils in the outer cortex; the fungi involved include Basidiomycotina species, for example *Cortinarius zakii*. The same type of fungi can also form ectomycorrhizae with conifers and other plants.

The final type of ericalean mycorrhizae is the ectendotrophic monotropoid form, involving the chlorophyll-lacking (achlorophyllous), saprotrophic, higher plant genus *Monotropa* and a Basidiomycotina fungus that colonizes the root. Both a fungal sheath composed of septate hyphae and a Hartig net are present, as well as intracellular fungal 'pegs' enclosed by the host cell wall; eventually, the nutrient contents of the 'peg' are dispersed into a sac formed by the host plasma membrane. Unlike other mycorrhizae, but similar to orchidaceous endomycorrhizae, the fungal partner supplies necessary organic CARBON requirements to the host plant.

Orchidaceous mycorrhizae

Endomycorrhizae are developed by all members of the orchid family (Orchidaceae), which are distributed worldwide, with the majority being tropical. Septate hyphae invade the very young embryo from the soil, after the seed coat has been ruptured, penetrating intracellularly into epidermal cells adjacent to the embryo's suspensor region. Later, outer cortical cells are colonized by inter- and intracellular hyphae. Within the host cell, the penetrating hypha invaginates the plasma membrane to form a 'coil' surrounded by host cytoplasm. The coils remain healthy for a period of 1–4 days, then later show signs of digestion by the host cell. As root growth occurs, the fungus colonizes the cells behind the growing tip; thus, exploitation and digestion continue. Infection of adult roots can also occur from hyphae growing in the adjacent soil, or from colonized roots in the same vicinity. There is a sensitive balance between symbiosis and the parasitic mode; in certain conditions, extensive digestion of the fungus by the plant occurs.

Unlike the great majority of mycorrhizae (with the exception of the achlorophyllous, monotropoid forms), the fungus supplies organic carbon compounds to the plant from the breakdown of insoluble materials in the surrounding soil (or from neighbouring plants which it has infected). Because orchid seeds are so small and carry only trace amounts of nutrients which are insufficient for development, they are incapable of germination in the absence of the appropriate mycorrhizal partner. Thus, the orchid host (some of which are achlorophyllous) is an obligate SYMBIONT, and will only germinate in culture if suitable sugars and other nutrients (e.g. VITAMINS) are present.

The fungi involved can be cultured axenically; some are sterile, but can be recognized as belonging to the Basiomycotina (often from the genus *Rhizoctonia*) because of their typical clamp connections. Many of these fungi are capable of degrading complex materials found in soil (e.g. starch, LIGNIN, CELLULOSE, etc.), and in other situations can attach to different plant hosts as pathogens, causing severe symptoms — for example, *R. solani* and *Armillaria mellea*. Others degrade wood and are found on rotting tree stumps — for example, *Coriolus versicolor*. [J.B.H.]

enemy-free space This term relates to those regions of NICHE space in a COMMUNITY that are free from PREDATORS. The dimensions of a niche are determined by a range of biological and PHYSICAL FACTORS in which an organism can survive, grow, reproduce and maintain a viable population. COMPETITION for limited RESOURCES, such as predator-free space, is thought to be a major factor in niche diversification (Jeffries & Lawton 1985). This enables a habitat to be divided in such a way that the widest possible range of organisms can exploit the available resources without coming into direct conflict with each other. The driving force behind such diversification in this instance appears to be through indirect competition between species that share a common predator. This occurs because increases in the abundance of one prey species causes the size of the predator population to increase, which then leads to an associated increase in the rate of predation of the other prey species that are present. The relationship between enemy-free space and this 'APPARENT COMPETITION' was first suggested in an elegant simulation model (Holt 1977) that showed that COEXISTENCE was favoured by reducing the rate of predation on only one of the species present in the model. *See also* PREDATOR–PREY INTERACTIONS. [R.C.]

REFERENCES

Holt, R.D. (1977) Predation, apparent competition and the structure of prey communities. *Theoretical Population Biology* **12**, 197–229.

Jeffries, M.J. & Lawton, J.H. (1985) Enemy free space and the structure of ecological communities. *Biological Journal of the Linnean Society* **23**, 269–286.

Energy budget

An energy budget is a statement of the balance between energy input and energy output at any point of energy transfer through an ECOSYSTEM, TROPHIC LEVEL, POPULATION or individual organism. It is usually formulated in the form of an equation, with the units usually used to form a common currency for different elements of the equation being joules or kilojoules.

The purpose is to calculate and cross-check calculations of the component quantities of an energy budget so that efficiencies of energy transfer through different species, perhaps of different systematic groups (e.g. vertebrates versus invertebrates), different physiological groups (e.g. homiotherms versus poikilotherms) or different trophic groups (e.g. carnivores versus herbivores), can be compared.

Some basic energy-budget equations are as follows.

$$P_n = \Delta B + L + G$$

where P_n is net primary productivity (which represents the energy fixed in PHOTOSYNTHESIS, i.e. gross primary productivity minus energy lost as a result of RESPIRATION by the plant), ΔB is the BIOMASS change or tissue accumulation in the plants over a fixed period of time, L is the biomass lost due to death of plants and G is loss due to CONSUMER ORGANISMS (GRAZING).

For animals or animal populations:

$$C = P + R + F + U$$

where C is consumption, the total intake of food by a heterotrophic organism during a defined time period; P is production due to growth and reproduction, i.e. the net accumulation of tissue during a defined time period; R is respiration, energy converted to heat and dissipated in life processes (metabolism) in a defined period of time, the cost of maintenance and activity; F is egesta, the part of consumption egested as faeces; and U is excreta, material derived from assimilated food and exreted as urine or through the skin.

In practice for many animals, for example insects and birds, it is very difficult to separate egested material and excreted urinary wastes, so these two categories are considered together as rejecta (J) where:

$$J = F + U$$

Further energy-budget equations for animals are:

$$A = C - J$$

where A is ASSIMILATION, that part of the food intake which is digested and absorbed minus that excreted as urinary waste. Assimilation is also, therefore, the sum of production and respiration:

$$A = P + R$$

and

$$D = P + R + U$$

where D is digested energy or material that is also absorbed into the animal's body including material later excreted as urine, i.e. assimilation plus excreta.

Where faeces and excreta can be separated, D is often calculated from consumption minus defecation or egestion rates:

$$D = C - F$$

These equations lead to the formulation of a number of efficiency ratios central to the comparative study of ENERGY FLOW between populations and ecosystems:

Assimilation efficiency = $A/C \times 100$
Production efficiency = $P/A \times 100$

See also ASSIMILATION EFFICIENCY; CONSUMPTION; ECOLOGICAL EFFICIENCY; ECOLOGICAL ENERGETICS; ENERGY, LAWS OF THERMODYNAMICS; PRODUCTION EFFICIENCY; SECONDARY PRODUCTIVITY. [M.H.]

REFERENCE

Petrusewicz, K. & Macfadyen, A. (1970) *Productivity of Terrestrial Animals: Principles and Methods*. Blackwell Scientific Publications, Oxford.

energy–diversity relationship Correlations have been found on a global or regional scale between SPECIES RICHNESS and indices of availability of energy to organisms such as PRIMARY PRODUCTIVITY, evapotranspiration and availability of food or limiting NUTRIENTS (e.g. for birds). This led D. Wright to propose the species–energy theory to explain geographical DIVERSITY patterns. The theory assumes that the likelihood of occurrence of a species increases and the probability of extinction decreases as mean ABUNDANCE increases. Productivity of a habitat determines the number of individuals per unit area; thus species diversity should increase with increasing productivity.

However, latitudinal diversity patterns in terrestrial plants cannot be explained by the theory. The relationship is stronger when based on higher taxonomic levels (order, class) than lower ones (genus, family) and better for animals than plants. On local scales, for a wide range of taxa in a wide range of habitats, the relationship between richness and productivity seems to be humped, peaking at some intermediate level of productivity. Similar unimodal relationships are found for terrestrial plants and PHYTOPLANKTON in enrichment experiments. Also, some highly productive habitats like SALT MARSHES have few species and when productivity is accompanied by a decrease in resource variety, as in polluted rivers, species richness actually declines. Productivity and the rate of flow of energy through the ECOSYSTEM therefore does not seem to have a simple relationship with species diversity. [P.S.G.]

REFERENCES

Rosenzweig, M. & Abramsky, Z. (1993) How are diversity and productivity related? In: *Species Diversity in Ecological Communities: Historical and Geographical Perspectives* (eds R. Ricklefs & D. Schluter), pp. 52–65. University of Chicago Press, Chicago, Ill.

Wright, D., Currie, D. & Mauer, B. (1993) Energy supply and patterns of species richness on local and regional scales. In: *Species Diversity in Ecological Communities: Historical and Geographical Perspectives* (eds R. Ricklefs & D. Schluter), pp. 66–74. University of Chicago Press, Chicago, Ill.

Energy flow

The idea of using energy flow to explain the pyramids of numbers and biomass (*see* PYRAMID OF BIOMASS; PYRAMID OF NUMBERS) was hailed as a major breakthrough in the understanding of ecological systems when it was first introduced by Lindeman (1942; *see* CHARACTERS IN ECOLOGY) in his paper 'The trophic dynamic aspects of ecology'. For the first time the structure of communities, as described by the number and mass of organisms in each TROPHIC LEVEL, was given a dynamic perspective by describing the rates at which energy flows through each layer of the TROPHIC STRUCTURE of a COMMUNITY. This thus encouraged formation of much wider ranging generalizations about the ways in which components of ECOSYSTEMS are organized; in particular it stimulated much research into the efficiency with which energy flows through the different components of an ecosystem. The concept of an energy flux being degraded as it passed through an ecosystem offered a much more unifying principle for ecosystem studies than simply the physical impacts of habitats on community life. The application of such fundamental laws as those of thermodynamics to biological systems enabled ecologists to create visual models, such as those depicting a hydraulic analogy of ecosystem functioning, as pioneered by Odum (1957) (Fig. E11). By comparing such diagrams as these, differences between ecosystems became apparent at a glance. They also drew attention very vividly to the importance of the decomposer component of ecosystems, even those that were intensively managed by humans, for example an agricultural grazing pasture in South Wales where Macfadyen (1961) showed that only 14% of NET PRIMARY PRODUCTION was eaten by the livestock for which the system was being managed but 89% entered the soil litter system as dead organic matter (Fig. E12).

This is further illustrated by comparing different communities (Fig. E13) all of which lose most of the energy trapped in net primary productivity through the decomposer rather than grazing FOOD CHAINS. The grazer system is most prominent in terms of energy flow in planktonic communities, where a large proportion of net primary productivity is consumed alive and assimilated at quite a high efficiency. In contrast, streams have a very small grazing system with most of the energy entering the community in the form of dead organic matter from terrestrial ecosystems blown or washed into the water.

Thus by comparing overall energy flows throughout populations it is possible to compare the relative importance of competing species within a trophic level of successive levels in an ecosystem and of one system with another. Energy flow can also be used as a powerful tool in assessing quantitatively the effects of more intangible interactions between species, for example in the promotion of microbial activity in decomposer systems through the stimulating effects of different components of the fauna. *See also* ECOLOGICAL EFFICIENCY; ECOLOGICAL ENERGETICS; ENERGY BUDGET; ENERGY-FLOW HYPOTHESIS. [M.H.]

REFERENCES

Begon, M., Harper, J.L. & Townsend, C.R. (1996) *Ecology: Individuals, Populations and Communities*, 3rd edn. Blackwell Science, Oxford.

Lindeman, R.L. (1942) The trophic dynamic aspects of ecology. *Ecology* **36**, 137–139.

Macfadyen, A. (1961) Metabolism of soil invertebrates in relation to soil fertility. *Annals of Applied Biology* **49**, 215–218.

Odum, E.P. (1959) *Fundamentals of Ecology*, 2nd edn. W.B. Saunders, Philadelphia, Pa.

Odum, H.T. (1957) Trophic structure and productivity of Silver Springs, Florida. *Ecological Monographs* **27**, 55–112.

energy-flow hypothesis

1 The energy-flow hypothesis was proposed to explain why in ELTONIAN PYRAMIDS of COMMUNITY structure there is a progressive reduction in PRODUCTIVITY in successively higher trophic levels. The hypothesis asserts that this is a consequence of the second law of thermodynamics, which states that changes in energy state are always accompanied by a loss of energy from the system as heat. Thus wherever assimilated energy is built into new BIOMASS there will always be energy expended in maintaining the basic metabolism and fuelling the activities of the animal, which will be lost as respiratory energy.

The hypothesis is often formulated as a visual model of ENERGY FLOW through the community, graphically depicting respiratory heat losses from the system (see Figs E11 & E12) as energy is transferred from one TROPHIC LEVEL to the next.

2 Another, related version of the energy-flow hypothesis is that the progressive loss of energy at each trophic level along a FOOD CHAIN will put a limit on the number of links in a food chain. However, this is now not without dispute.

[M.H.]

energy flux *See* ENERGY FLOW.

energy, laws of thermodynamics When energy is converted or transformed from one form to another the processes follow rigorous physical laws, called the laws of thermodynamics.

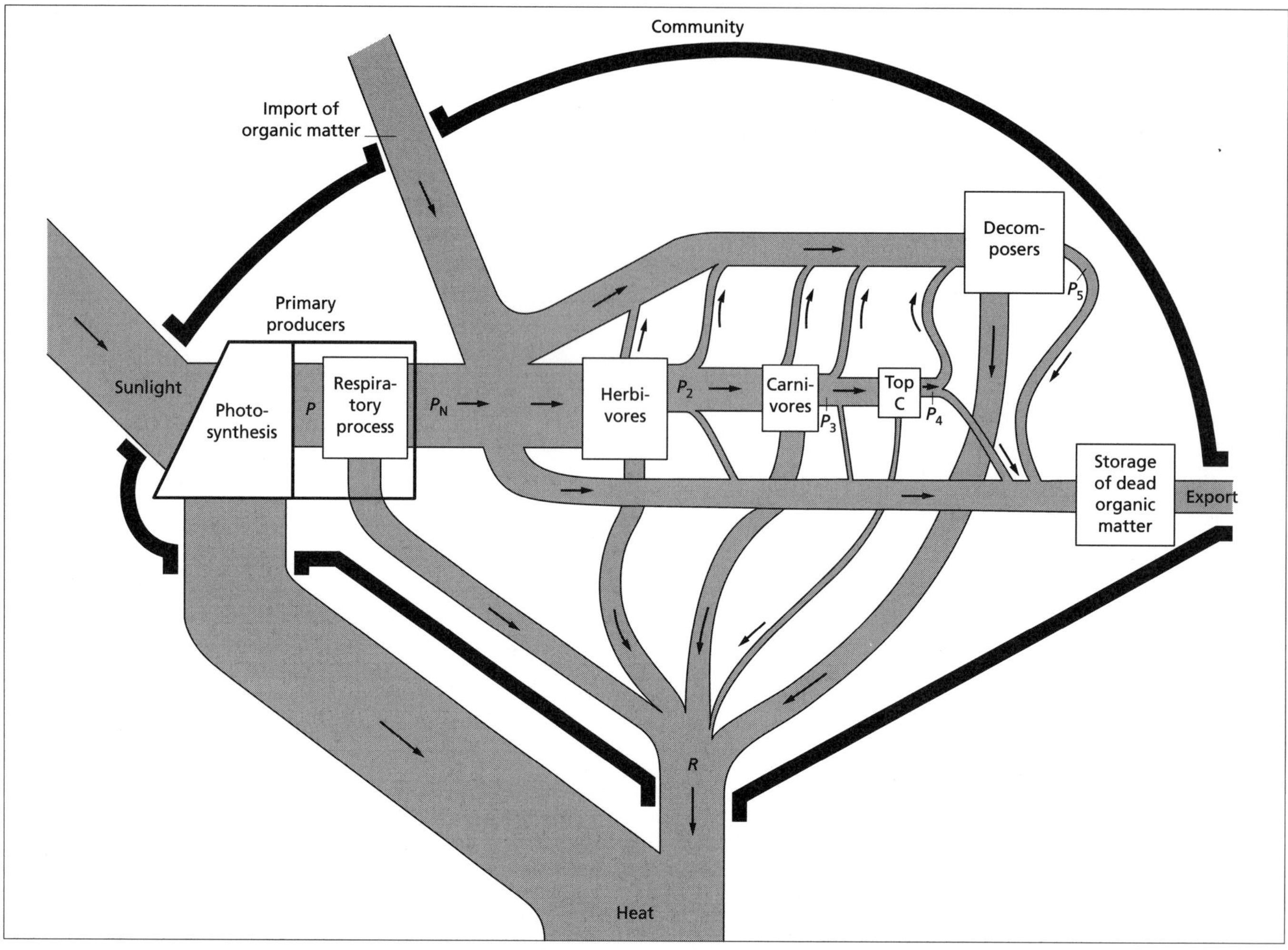

Fig. E11 Energy flow diagram of a community showing successive fixation and transfer by components and the large respiratory losses at each transfer. *P*, gross primary productivity; P_N, net primary productivity; P_2, P_3, P_4 and P_5, secondary productivity at the indicated levels; *R*, community respiration. (After Odum, H.T. (1957) Trophic structure and productivity of Silver Springs, Florida. *Ecological Monographs* **27**, 55–112.)

The first law of thermodynamics concerns the conservation of energy and states that 'energy may be transformed from one form into another but is neither created nor destroyed'.

This law underlines all the energy-budget equations that relate energy input to any ecological unit to energy accumulation and loss from that unit. Thus all the potential chemical energy consumed by an animal must equal the sum of all chemical energy lost as WASTE material, plus energy accumulated in reproductive and somatic tissues plus energy lost as heat (unless it is an electric eel when some energy may be emitted as electricity). Because of the universal applicability of the first law of thermodynamics, if all the elements but one of an ENERGY BUDGET have been carefully measured it is then possible to estimate the final one from the energy-budget equation, although independent measures of all elements for careful cross-checking is always desirable.

The first law of thermodynamics recognizes the interconvertibility of all forms of energy but does not predict how complete the conversions will be. For all energy transformations some energy will be lost as heat, which is a property of random molecular movement, whereas other forms of energy are more ordered and have nonrandom arrangements. The second law of thermodynamics is concerned with the conversion of other forms of energy to heat and states 'processes involving energy transformations will not occur spontaneously unless there is a degradation of energy from a nonrandom to a random form'. The conversion of energy from more ordered to more disordered form is often referred to as the production of entropy.

This law underpins the universal observation that ecological efficiencies (the ratio of PRODUCTION by one trophic unit to the PRODUCTIVITY of the trophic unit below it) are never 100%. In ecological systems the heat lost during energy transformation is usually measured as RESPIRATION. By measuring the amount of OXYGEN taken up and/or the amount of CARBON DIOXIDE (CO_2) given off, the heat lost in metabolic oxidation processes can be estimated. This is a vital component of many energy budgets and the ratio of energy incorporated in new tissues to energy lost during respiration is

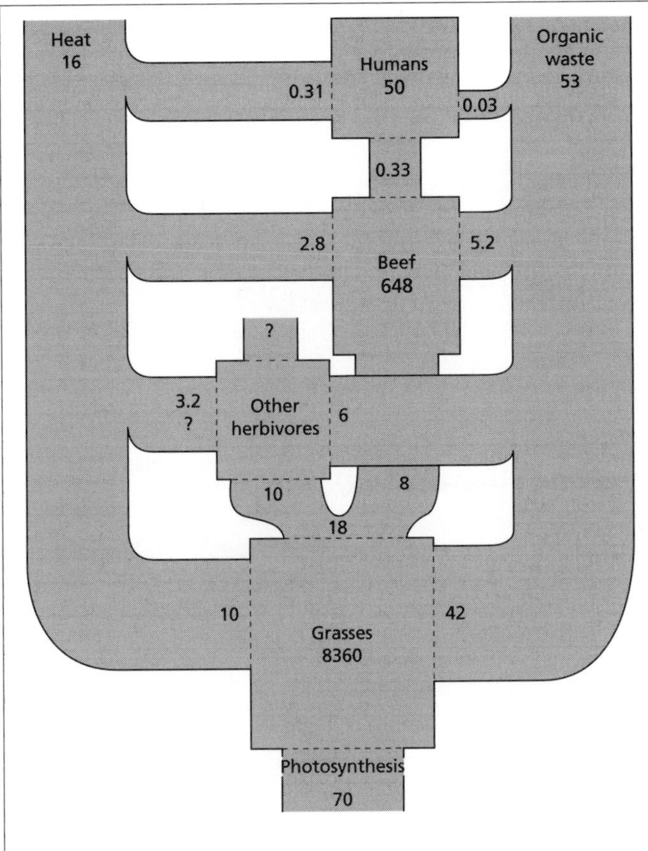

Fig. E12 Transfer of energy of photosynthesis through a meadow containing 0.5 humans per hectare and 1.1 cattle per hectare. Units are kJ for biomass and kJ m^{-2} day^{-1} for flows. (After Macfayden, A. (1961) Metabolism of soil invertebrates in relation to soil fertility. *Annals of Applied Biology* **49**, 215–218.)

an important statistic summarizing the EFFICIENCY with which that ecological unit converts energy from food into new tissues. *See also* ECOLOGICAL EFFICIENCY; ENERGY BUDGET; LINDEMAN EFFICIENCY. [M.H.]

REFERENCE

Phillipson, J. (1966) *Ecological Energetics*. Edward Arnold, London.

ENSO ENSO means EL NIÑO southern oscillation. *See also* SOUTHERN OSCILLATION INDEX. [J.M.]

Entisols *See* SOIL CLASSIFICATION.

entrained rhythms *See* CIRCADIAN RHYTHMS.

entrainment Process by which solid materials are caught into and carried by flowing waters in a river or stream, or by which air outside a growing cloud is caught into the rising CONVECTION movements within the cloud. [P.C.]

entropy *See* ENERGY, LAWS OF THERMODYNAMICS.

environment This word from the French *environer*, to encircle or surround, describes the supporting matrices of life: WATER, earth, ATMOSPHERE and CLIMATE. However, it projects two false impressions: one is that these matrices are separate from living things and the other is that they are designed to support life. On the first count, the chemical and physical compositions of the matrices that surround life are importantly influenced by the metabolism and behaviour of organisms. The composition of the atmosphere is crucially dependent on the gases that living things remove from it and add to it; the soils are largely dead plant materials that are importantly moulded by the microbes that decompose it and the animals that move through it; the quality of waters in the SEAS, LAKES and RIVERS depend to some extent on aeration by plants and a myriad of other transfers. So the environment of organisms can properly be said to consist not only of a dead but also a living matrix, i.e. other organisms as well.

On the second count, the ADAPTATION of the environment to living things, the same criticisms apply as are levelled at the GAIA hypothesis. [P.C.]

Environment Act 1995 *See* ENVIRONMENTAL (PROTECTION) AGENCIES.

Environmental Action Programmes Series of strategic programmes adopted by the Commission (bureaucracy) of the European Union (EU) and playing an important part in defining policy used as a basis for legislation proposals put before the Council (law maker) of the EU. There have been five programmes. The first and second, adopted respectively in 1973 and 1977, included reference to prevention being better than cure and that polluters should pay. The third, adopted in 1982, shifted the emphasis from control to prevention. The fourth, adopted in 1987, emphasized effective implementation of legislation, increased public access to and dissemination of environmental information and the need for integration with other policies. The latest and fifth action programme, adopted in 1992 and reviewed in 1995, was entitled *Towards Sustainability* in line with UNCED. It emphasized more shared responsibility, less COMMAND AND CONTROL and the need for an integrated approach within economic sectors. *See also* ENVIRONMENTAL LAW; ENVIRONMENTAL (PROTECTION) POLICIES. [P.C.]

Environmental Agency (EA) *See* ENVIRONMENTAL (PROTECTION) AGENCIES.

environmental asset management Taking account of environmental commodities in general or particular environmental holdings in business practice. *See also* ENVIRONMENTAL ASSETS; ENVIRONMENTAL CAPITAL. [P.C.]

environmental assets Assets can mean general commodity from the environment, such as raw materials or sites for dumping waters, or particular holdings by companies, such as estates, forests, gardens, water bodies, etc. [P.C.]

environmental assurance bonding Proposed levy of an amount of money equal to the best estimate of the largest potential future environmental damage that might arise from a particular industrial

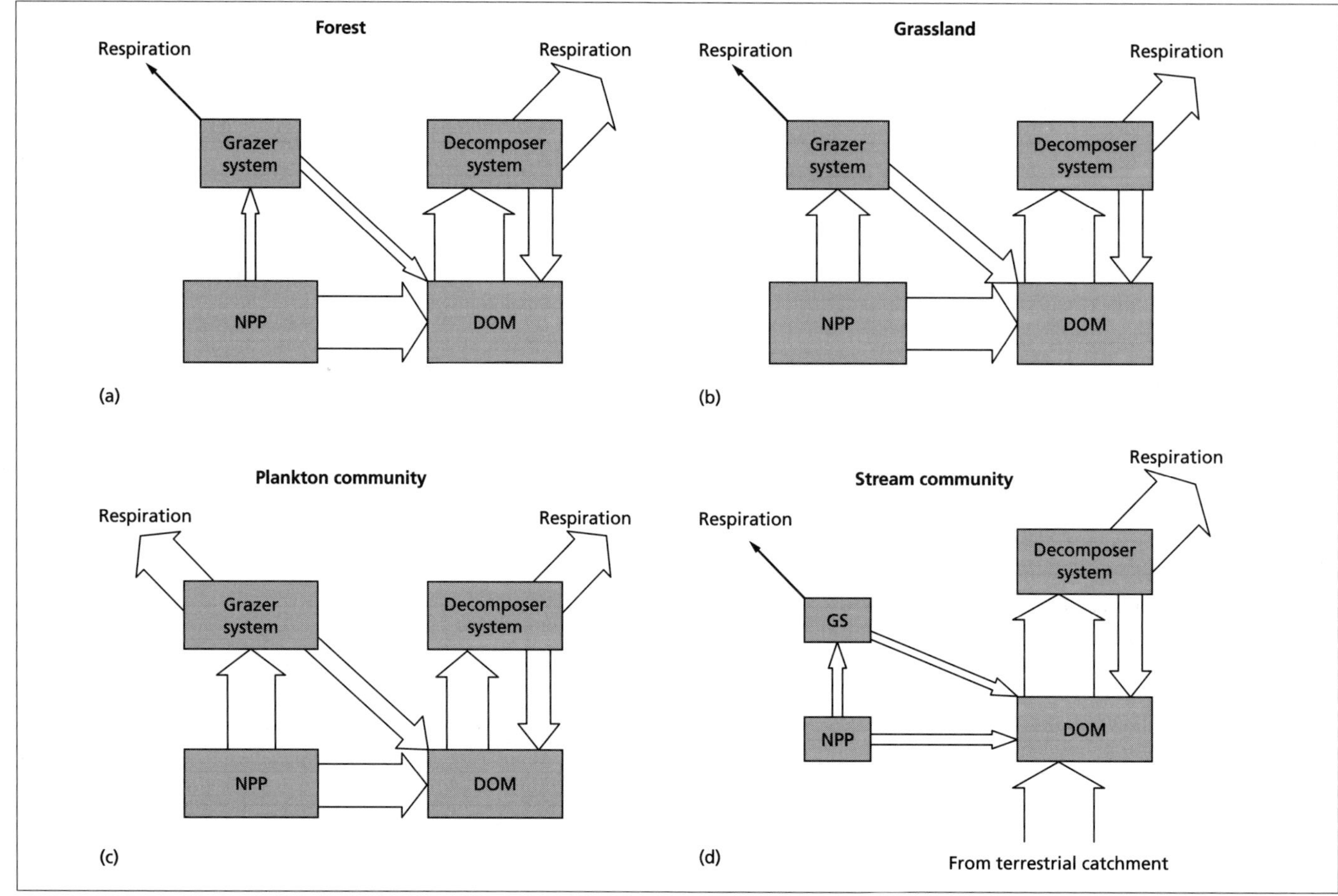

Fig. E13 General patterns of energy flow for (a) a forest, (b) a grassland, (c) a plankton community in a large lake or the sea and (d) the community of a stream or small pond. Relative sizes of boxes and arrows are proportional to relative magnitudes of compartments and flows. NPP, net primary production; DOM, dead organic matter. (After Begon, M., Harper, J.L. & Townsend, C.R. (1996) *Ecology: Individuals, Populations and Communities*, 3rd edn. Blackwell Science, Oxford.)

process or product. This would be kept in an interest-bearing escrow account for a predetermined period. Portions of the bond (plus interest) would be returned if and when it could be demonstrated that the suspected worst-case damages had not occurred or would be less than anticipated. In the event of damages the bond would be used in REMEDIATION. The cost of the bond would define the price of the right to use the environmental resource in some specified way and would therefore be tradeable. *See also* REGULATORY INSTRUMENTS. [P.C.]

REFERENCE

Costanza, R. & Perrings, C. (1990) A flexible assurance bonding system for improved environmental management. *Ecological Economics* **2**, 57–76.

environmental burdens Concentration of XENOBIOTIC in an environmental compartment. *See also* CRITICAL LOADS. [P.C.]

environmental business charter Voluntary charters that by signing and implementing business organizations can demonstrate a commitment to environmental protection. Some of these are: Chemical Industries Responsible Care Programme; Valdez Principles; International Chamber of Commerce (ICC) Business Charter for Sustainable Development (BCSD); and the Confederation of British Industry (CBI) Agenda for Voluntary Action. [P.C.]

environmental capital Natural resources potentially available for use by society (usually industry) and valued as such. Capital is usually thought of in terms of man-made measures, but total capital also includes natural, ENVIRONMENTAL ASSETS. It is argued by environmental (ecological) economists that these can be valued in money terms in the same way as man-made capital. *See also* ENVIRONMENTAL (ECOLOGICAL) ECONOMICS; SUSTAINABLE DEVELOPMENT. [P.C.]

REFERENCE

Pearce D., Markandya, A. & Barbier, E.B. (1989) *Blueprint for a Green Economy*. Earthscan Publications, London.

environmental deviation In QUANTITATIVE GENETICS, the deviation of the PHENOTYPIC VALUE of one individual from the average phenotypic value of all individuals of the same GENOTYPE, i.e. the individual's deviation from its GENOTYPIC VALUE due to environmental influences. The average environmental deviation over all individuals of the same genotype is, by definition, zero. The VARI-

ANCE in the population over the environmental deviation of all individuals equals the ENVIRONMENTAL VARIANCE V_E. [G.D.J.]

environmental disclosures Disclosure is information made available to the public through the formal accounting procedures of a company. Environmental disclosure is information on the environmental performance of a company through the same route. *See also* ENVIRONMENTAL MANAGEMENT SYSTEMS. [P.C.]

environmental due diligence Exercise conducted by a purchaser, underwriter or lender in a wide range of commercial transactions, including public offering of shares, refinancing, asset purchase, share purchases or merger. The objectives are to ascertain what actual and/or potential environmental liabilities and risks they might be assuming as a consequence of the transaction and to quantify any liabilities found. It is therefore an external examination of the environmental implications of the operation of another company. [P.C.]

environmental (ecological) accountancy Applying accounting techniques and principles to assess the ENVIRONMENTAL IMPACT of businesses. It involves:

- identifying environmentally related costs and revenues within the accounting system;
- devising new forms of financial and non-financial accounting systems, information systems and control systems to encourage environmentally sensitive management;
- developing new forms of financial performance indicators that take environment into account;
- experimenting with ways sustainability can be assessed and incorporated into usual procedures.

See also ENVIRONMENTAL (ECOLOGICAL) ECONOMICS. [P.C.]

REFERENCE

Gray, R. (1993) *Accounting for the Environment.* ACCA, Paul Chapman Publications, London.

environmental (ecological) economics Economics is the study of how people use limited RESOURCES to supply their unlimited needs. It is therefore concerned with trade-offs and social preferences. It can be applied internationally to trade and its consequences (macroeconomics) or to small units such as businesses and households (microeconomics). It is concerned with judging value (social preferences) in market-places by reference to willingness to pay. Until recently, resources from the environment were considered external to these processes (as externalities); this included raw materials and space available for dumping WASTE. For a long time, however, economists have recognized that environmental resources are finite and hence should be taken into account in terms of competing human needs (e.g. Thomas Malthus in 1797 published *Essay on the Principles of Population* that made this point). Environmental entities should therefore be valued so that they can be taken into account in the overall scheme of human economics. This internalization is an important requirement in the defining of SUSTAINABLE DEVELOPMENT. Valuing environmental entities and taking these into account in the development of policies is referred to as environmental economics. Valuing ecological entities, such as HABITATS, ECOSYSTEMS and species, can be viewed as a subset of this and referred to as ecological economics.

Environmental valuation techniques involve the following.

1 Use of actual market-places where commodities are traded, such as might occur with fish and lumber.

2 Establishment of hypothetical market-places, for example when the values of entities traded are likely to be influenced by environment, such as house prices.

3 Use of sociological techniques, such as questionnaires, to gauge value that people might put on environmental entities. These are usually divided into assessments that involve values which people put on use of environmental entities and values which people put on existence of entities that they do not use. The former might involve use of local parks; the latter might involve entities such as the Antarctic that may never be visited.

4 Values might also be estimated from the costs of preventing HARM or the costs of REMEDIATION.

There are those who argue that valuation of ecological/environmental entities, no matter how it is done, is unethical. The economist's rejoinder is usually that the values given are not intrinsic values but simply measures of social preferences that can be taken into account in weighing the costs of environmental protection against potential benefits of what is being protected. [P.C.]

environmental ethics and conservation The field of moral philosophy that deals with human obligations and duties to nonhuman organisms and to the natural world. At the elementary level, environmental ethics is a simple extension of human ethical systems to other living beings. This extension may be based on utilitarian concepts of pleasure and pain, which can be used to extend moral regard to sentient animals. A further extension of human moral duties may be based on the thought that all life is worthy of moral regard. This biocentric extension of human morality affords limited rights to plants and other non-sentient organisms as well as sentient animals. Further still, ethical regard may involve abstract concepts of natural process and biological relationships. This evolutionary ecological, or ecocentric, concept of ethics assigns moral conditions to the actions of humans as they affect the natural environment.

Areas of ethical concern commonly treated as part of the emerging field of environmental ethics include animal rights, feminine critiques of ENVIRONMENTAL HARM (ecofeminism), radical ENVIRONMENTALISM and several schools of philosophical thought often called 'land ethics' after the work of Aldo Leopold. However, the inclusion of animal rights within the body of environmental ethics is not universally accepted because the emphasis on the moral treatment of individual animals seems to preclude ethical concern in the treatment of relationships such as species or process such as ECOSYSTEMS. The ecofeminist agenda ascribes the domination of nature to the same psychological and historical root as the domination of women. The radical environmentalist movement finds connections between the traditions of civil disobedience, abolitionist politics, anarchy and other political forms that seek rapid, radical and occasionally violent changes in society because of the perceived inadequacy of environmental protection and/or harmful environmental policies. Lastly, there are several philosophical approaches that attribute ethical concerns to the relationship of humans to biotic communities and evolutionary process. This last concept may be more than even a radical extension of human morality; it may represent an original level of moral discourse. *See also* CONSERVATION BIOLOGY. [S.J.B.]

REFERENCES

Leopold, A. (1949) *A Sand County Almanac and Sketches Here and There.* Oxford University Press, New York.

Rolston III, H. (1988) *Environmental Ethics: Duties to and Values in the Natural World.* Temple University Press, Philadelphia, Pa.

environmental externality adders Surcharges to the prices of commodities, for example energy, in order to reflect the damage done to the environment in their production and use. Based on 1990 prices, the adders for various sources of electricity generation in the USA would be (US cents/kWh): coal (conventional), 0.058; oil (low sulphur), 0.0027; natural gas, 0.010; nuclear (including accidents and decommissioning), 2.910. *See also* ENVIRONMENTAL (ECOLOGICAL) ECONOMICS. [P.C.]

REFERENCE

Pearce, D. (1993) *Economic Values and the Natural World.* Earthscan Publications, London.

environmental factor *See* ABIOTIC FACTORS; BIOTIC FACTORS; PHYSICAL FACTOR.

environmental harm Adverse change in one or more components of the physical, chemical and biological compartments of the environment, i.e. the EFFECT of POLLUTION. [P.C.]

environmental impact Having an EFFECT, usually harmful (*see* ENVIRONMENTAL HARM), on one or more of the physical, chemical and biological components of the environment. [P.C.]

environmental impact assessment (EIA) This is generally used to refer to the evaluation of effects likely to arise from a major project, such as the construction of a dam or a power-station. However, it would also apply to smaller projects, such as the construction of a new factory or of new plant within an existing industrial operation.

The EIA should supply decision-makers with an indication of the likely outcomes of their decisions. It should help selection between alternatives (methods and sites of construction) and lead to avoidance of unacceptable options. It is thus an anticipatory and participatory approach to environmental management.

Every EIA system is unique with respect to the project being proposed, its location and the environmental/ecological systems at risk. In general, though, they involve detailed SURVEYS of flora and fauna in the areas concerned with an assessment of how these might be impacted by the proposals. More often than not, this is done subjectively but, at least in principle, might be the subject of more rigorous modelling.

Occasionally, but not often, the EIA and its predictions might be subject to FEEDBACK from MONITORING either during or after the project. *See also* RISK ASSESSMENT. [P.C.]

environmental impairment insurance Contract transferring liability for environmental impairment to another party, the insurer, in consideration of a premium. The contract will usually have exclusions: payment of fines (implicitly), often (explicitly) the consequence of POLLUTION due to gradual release, cost of cleaning up the insurer's own site. [P.C.]

environmental impairment liability Legal responsibility for environmental impairment in the form of penalties, damages and REMEDIATION. Very often is strict, i.e. independent of wrongful intent or negligence, as compared with fault based. However, foreseeability is often a relevant criterion (those that are held liable should have reasonably been able to have anticipated the outcome) and it may be context dependent. Thus what is foreseeable now as a result of the state of science will be different from what was foreseeable 20 years ago, which has implications for retrospective action. *See also* ENVIRONMENTAL LAW. [P.C.]

environmental index Numerical indicator of ENVIRONMENTAL IMPACT from industrial process or product. For example, Rhône Poulenc calculate indices on a monthly basis for their major plants for impact on each of the major environmental compartments. The WASTE index calculates total output of toxic materials, SUSPENDED SOLIDS, total NITROGEN, PHOSPHORUS and dissolved salts, together with an account of CHEMICAL OXYGEN DEMAND, each weighted by judgements of relative importance of impacts. These indices are used as FEEDBACKS on performance and as targets for future improvements. *See also* ENVIRONMENTAL MANAGEMENT SYSTEMS. [P.C.]

environmental industries sector (EIS) Those industries that produce cleaner techniques and technologies for use in other production processes, for example filters that remove toxic substances from emission outlets, CATALYTIC CONVERTERS for transforming toxic products to less- or non-toxic outputs, probes that can measure pollutants and hence contribute to controls on polluting processes, the production of less hazardous raw materials such as water-based rather than organic solvent-based paints and lubricants. Sometimes, consultants needed in the provision of environmental services such as ECOAUDIT are described as being members of the EIS, but usually the term is restricted to the manufacturing sector.

Because of the pressure from environmental legislation, customers and other stakeholders for the use of cleaner techniques and technologies, it is thought that environmental industry is a rapidly growing business sector. [P.C.]

Environmental Information and Observation Network (EIONET) *See* EUROPEAN ENVIRONMENT AGENCY.

Environmental law

Laws and legal principles governing the behaviour of persons, businesses, government agencies and other public utilities to protect the ENVIRONMENT and the health of humans from possible adverse effects arising from the environment (not usually including the domestic environment and work-place). Environmental law can therefore encompass public health as well as conservation, POLLUTION control and land-use control.

There are two sources of environmental legislation: one is from statutes (hereafter referred to as statutory control) and the other from common law, essentially by case law in the courts. Statutory controls are involved in protecting the general interests of the State and granting powers through Parliament to the Government or its agencies whereby this can be achieved. A subset of common law is concerned with disputes between individuals and this, which will be referred to as civil law, is of environmental importance. It should also be said, though, that certain civil rights can be defined in statutes. Failure to respond to statutory controls can bring criminal proceedings, so some criminal law is defined in statute. Though of no consequence for environmental protection, some criminal law is established by common law, for example relating to murder. A further complication is that some transgression of civil law can also bring criminal prosecution (see below). Finally, there is also law that governs the action of the Government and its agencies, and this is referred to as public or administrative law.

Some important features of statutory controls are as follows.

1 They are usually defined in terms in framework statutes that are given detail by statutory instruments, issuances from REGULATORY AGENCIES and interpretations in the courts.

2 They can prescribe standards and these can be of the following kinds:

(a) ENVIRONMENTAL QUALITY STANDARDS (EQSs), sometimes also referred to as target standards, because they relate to the quality of a target environmental system;

(b) EMISSIONS standards, sometimes referred to as environmental limit standards (ELSs), are measured with regard to what is emitted rather than the EFFECT on the receiving environment *per se*;

(c) process standards;

(d) product standards.

The standards may be precise (e.g. referring to a level of emission of a particular substance) or imprecise (e.g. referring to more or less ill-defined concepts such as BATNEEC (BEST AVAILABLE TECHNIQUES NOT ENTAILING EXCESSIVE COSTS), BPEO (BEST PRACTICABLE ENVIRONMENTAL OPTION), etc.).

3 They can also provide powers under which information can be required on various aspects of environmental performance.

4 Enforcement of statute law is through agencies and regulatory bodies; failure to comply can be handled by the variation or revocation of rights or criminal proceedings.

5 In statutory environmental law the vast majority of offences are of strict liability, i.e. there does not need to be proof of negligence or fault. However, there does need to be proof, beyond (reasonable) doubt, that an offence has occurred; in the UK the burden of finding that proof is with the 'prosecutor'.

6 Appeals against the application of statutory law are possible in terms of both challenging provisions of specific controls (e.g. AUTHORIZATIONS and CONSENTS) and challenging the decision of the courts. In the UK, the conduct of government ministers or regulatory bodies can also be challenged as being outside of the powers granted through the statutes (*ultra vires*) by a procedure of judicial review administered by the High Court. This is the realm of public law.

Some important features of civil law are as follows.

1 It seeks only to protect private interests. It has its roots in settling disputes concerning property and land, and therefore has obvious environmental implications. Being based largely on case law, it is even more devious and complex than statutory law.

2 It is also often referred to as TORT law, a tort being a crooked conduct (a wrong) that causes HARM to an individual.

3 In contrast to statutory controls that seek to prevent ENVIRONMENTAL HARM, civil law seeks to remedy harm. The threat of remedy can, of course, act as a deterrent in itself and there are powers of injunction that are preventative.

It is also important to note that international treaties and conventions make up the field of international environmental law; these can be binding on the signatories. [P.C.]

REFERENCES

Ball, S. & Bell, S. (1994) *Environmental Law*. Blackstone Press Ltd, London.

McGregor, G.I. (1994) *Environmental Law and Enforcement*. Lewis Publications, Boca Raton, Fla.

---

environmental limit standards (ELSs) *See* ENVIRONMENTAL LAW.

environmental management systems (EMS) Systems applied to business operations to control environmental effects. They usually entail: a systematic review of risks in an initial audit; the formulation of general policy aimed at controlling (usually reducing) risks; the development of procedures (techniques, technology and training) to effect this; methods of MONITORING and keeping records of achievements; and a system of auditing to ensure aims are being achieved. There may also be regular public statements that are generated by the EMS designed to keep the public informed on achievements. This system is formalized into various standards on environmental management systems (a British version, BS 7750, has been superseded by an international standard ISO 14001). An EC regulation (EEC 1863/93) allows voluntary participation by companies in the industrial sector in a community eco-management and audit scheme (EMAS). *See also* ECOAUDIT; ECO-MANAGEMENT AND AUDIT SCHEME; ISO 14000. [P.C.]

environmental manual Document used by businesses with ENVIRONMENTAL MANAGEMENT SYSTEMS that collates the environmental policy, objectives and targets, and programme; specifies key roles and responsibilities of staff; describes the management system and how its elements fit together and interact; and provides direction to related documentation. In addition to covering normal procedures, it should also cover procedures under abnormal conditions and in the event of emergencies. [P.C.]

environmental (policy) statement Environmental policy statement is a statement (usually public) by a business of its approach to environmental protection in its operation. Frequently addresses very broad aspects of the organization's activities, but should be tailored to the nature of the organization. Often it will state commitments on reduction of WASTE and POLLUTION, efficient use of energy and RESOURCES, control of environmental effects of obtaining raw materials, on HABITATS and BIODIVERISTY and minimizing the environmental effects of new developments. An environmental statement, on the other hand, is a report (usually public and may be subject to formal VALIDATION) of achievements by business with respect to predetermined targets and of future interests. *See also* ENVIRONMENTAL MANAGEMENT SYSTEMS. [P.C.]

environmental politics Broad term for involvement of environmental issues in political processes. At one extreme it can involve the principles of environmental protection and respect as a basis for all political action (*see* ENVIRONMENTALISM; GREEN(S)). At the other extreme it simply means reference to environmental issues as part of a general political agenda. [P.C.]

environmental pressure (lobby) groups Groups that put pressure on policy-makers and industry, especially by influencing public opinion. Early examples in the UK were the Fog and Smoke Committee of 1880 and, a derivative, the Coal Smoke Abatement Society of 1899 formed to put pressure on the British Government to take action to prevent smogs by pressing for stricter implementation of existing law against the nuisance of black smoke EMISSIONS.

Now there are a number of national and international pressure groups with substantial following. Well known amongst these are Friends of the Earth (founded in the USA in 1969) and Greenpeace (founded in Canada in 1972). The World Wide Fund for Nature, initially founded as the World Wildlife Fund in 1961, though less confrontational, also plays a major role in lobbying for environmental protection. Also noteworthy in the UK are the Ramblers' Association, National Trust, Council for the Protection of Rural England, Royal Society for Nature Conservation and Royal Society for the Protection of Birds. *See also* GREEN(S). [P.C.]

environmental profile Analysis of businesses' actual and potential interaction with the ENVIRONMENT. [P.C.]

Environmental Protection Act (EPA) UK legislation to make provision for the improved control of POLLUTION from certain industrial and other processes, enacted 1990. Covers INTEGRATED POLLUTION CONTROL (IPC) and local authority AIR-POLLUTION CONTROL (LAAPC), WASTE, litter, radioactive substances, genetically modified organisms, nature conservation and a miscellany of other things. IPC was an innovation. [P.C.]

environmental (protection) agencies Regulatory/enforcement authorities given powers under governments to implement environmental protection legislation. Those for England and Wales (Environmental Agency, EA) and Scotland (Scottish Environmental Protection Agency, SEPA) were formed recently under the Environment Act 1995. The EA represents an amalgam of the National Rivers Authority (NRA), Her Majesty's Inspectorate of Pollution (HMIP) and waste regulation authorities. Similarly, SEPA was an amalgam of Her Majesty's Industrial Pollution Inspectorate (HMIPI), river purification authorities and waste regulation authorities. The USEPA was formed earlier, in 1970, by an Executive Order signed by President Nixon and bringing together the Federal Water Quality Administration from the Interior Department, the Pesticides Regulation Division from the Agriculture Department and the Office of Pesticide Research from the Department of Health Education and Welfare.

The underlying philosophy in the formation and operation of all these bodies is that INTEGRATED POLLUTION CONTROL requires integrated agencies to effect the control. The British and US agencies work in significantly different ways. The British agencies are only able to interpret legislation under guidance from government departments and secretaries of state. The USEPA can elaborate and extend legislation, within limits defined by legislation, by rule-making that is subject to scrutiny and challenge in the courts. *See also* REGULATORY AGENCIES. [P.C.]

environmental (protection) policies Set of principles, premises and presumptions that guide actions of government and other international bodies in developing measures and legislation on environmental protection. They involve a complex interaction between political philosophy, perception of public interest, lobbying and involvements and commitments at an international level. They are usually articulated in manifestos, consultation papers, white papers and action programmes. *See also* ENVIRONMENTAL ACTION PROGRAMMES. [P.C.]

environmental quality guideline Non-mandatory form of standard/objective. *See also* ENVIRONMENTAL QUALITY OBJECTIVE; ENVIRONMENTAL QUALITY STANDARD. [P.C.]

environmental quality objective (EQO) In the UK, used to define the state of environments in terms of ecological/human health goals, for example whether water is suitable for fish stocks, suitable for drinking, suitable for bathing. Elsewhere, often synonymous with ENVIRONMENTAL QUALITY STANDARD. Usually mandatory. *Cf.* ENVIRONMENTAL QUALITY GUIDELINE. [P.C.]

environmental quality standard (EQS) Concentration of a substance that should not be exceeded in an environment if HARM (to humans and/or ecosystems) is to be avoided. It is therefore based on the presumption that CONTAMINATION does not necessarily cause POLLUTION. EQSs are usually obtained by applying a safety margin to no-effect or low-effect concentrations identified in DOSE–RESPONSE toxicological and ecotoxicological studies. Sometimes used synonymously with ENVIRONMENTAL QUALITY OBJECTIVE (EQO). However, in the UK a distinction has been made. An EQO is the overall state to be aimed for in an aspect of the natural environment, for example water in a river such that fish can survive and breed. Hence an EQO is often expressed in qualitative not quantitative terms. An EQS is also different from a LIMIT VALUE (LV). The latter is the limit that must not be exceeded in an emission/effluent. It is designed to achieve EQSs and EQOs, but often also takes account of what can be achieved by a reliable technique. [P.C.]

environmental quality target State of the environment towards which management and control strategies are directed, for example a targeted ENVIRONMENTAL QUALITY STANDARD/OBJECTIVE. [P.C.]

environmental reporting Production and release of environmental statements by businesses, typically to shareholders, but also to employees and public. *See also* ENVIRONMENTAL (POLICY) STATEMENT. [P.C.]

environmental science Literally, scientific endeavour applied in describing and understanding the ENVIRONMENT. It has been equated with 'earth science', namely the study of the ATMOSPHERE, land, oceans and freshwaters and the biogeochemical FLUXES within them. The emphasis has been on the physics and chemistry of these processes, but increasingly the importance of biological/ecological systems is being recognized and the interface with human society is also clearly important, leading to interaction with social sciences and economics. This, then, is a broad multidisciplinary/interdisciplinary venture that plays an important part in understanding and hence mitigating our impact on natural systems and processes. *See also* ECOLOGY. [P.C.]

REFERENCE

O'Riordan, T. (1995) *Environmental Science for Environmental Management.* Longman Scientific & Technical, London.

environmental tax Tax on commodities/products related to their actual or potential ENVIRONMENTAL IMPACTS. *See also* ENVIRONMENTAL (ECOLOGICAL) ECONOMICS; MARKET INSTRUMENTS. [P.C.]

environmental tolerance Response of a genotype's mean FITNESS across an environmental gradient. A tolerance curve depicts the environment to which a genotype is best adapted and quantifies the breadth of ADAPTATION or the degree of specialization. If one considers fitness as a TRAIT, a tolerance curve can be viewed as a special kind of norm of reaction: it relates the phenotypic expression of the trait 'fitness' of a genotype to its environment. The optimal shape and the position of the maximum of a tolerance curve are determined by the temporal and spatial variation of the environment and by the developmental noise of traits influencing the tolerance curve. The optimal breadth of adaptation usually increases with increasing HETEROGENEITY of the environment. The within-generation temporal variance has more influence than the between-generation temporal component. However, if the between-generation temporal variance is larger than the within-generation component, increasing spatial heterogeneity selects for specialists. The smaller the temporal within-generation variation compared with spatial and/or between-generation temporal variation, the greater the selective advantage for shifting the mode of the tolerance curve during development in the presence of reliable environmental cues. Further modifications can occur by reversible physiological ACCLIMATION and fitness-enhancing activities, which are coupled to a narrow range of the environmental gradient under consideration. *See also* ADAPTIVE LANDSCAPE; REACTION NORM. [W.G.]

REFERENCES

Gabriel, W. & Lynch, M. (1992) The selective advantage of reaction norms for environmental tolerance. *Journal of Evolutionary Biology* **5**, 41–59.

Gilchrist, G.W. (1995) Specialists and generalists in changing environments. I. Fitness landscapes of thermal sensitivity. *American Naturalist* **146**, 252–270.

Lynch, M. & Gabriel, W. (1987) Environmental tolerance. *American Naturalist* **129**, 283–303.

environmental toxicology Study of how human health is impacted by toxic substances through environmental EXPOSURE (usually excluding home and work-place). Sometimes used more broadly and synonymously with ECOTOXICOLOGY. [P.C.]

environmental trustees In the UK, body controlling funds from a rebate out of a LANDFILL levy on WASTE. Industry can opt to pay levy into these funds instead of as TAX, and funds can be used by the trust to enhance local environmental amenities. [P.C.]

environmental utilization space or ecospace The capacity of the BIOSPHERE's environmental functions to support human economic activities, sometimes defined at a national or PER CAPITA level according to a 'global fair shares' principle. [P.C.]

environmental valuation *See* ENVIRONMENTAL (ECOLOGICAL) ECONOMICS; SUSTAINABLE DEVELOPMENT. Not to be confused with ENVIRONMENTAL VALUE. [P.C.]

environmental value In QUANTITATIVE GENETICS, the overall influence of an environment is quantified as the mean PHENOTYPIC VALUE of all (relevant or investigated) GENOTYPES in that environment. This mean value over all genotypes is called the environmental value of the particular environment. The environmental sensitivity of any genotype over a range of environments is then expressed as the deviation of that genotype from the mean over all genotypes. For all environments, the phenotypic value of that particular genotype and the mean phenotypic value are tabulated. The regression line of the phenotypic values of the investigated genotype on the mean phenotypic values over all genotypes is calculated. Any genotype that is identical in expression to the mean over all genotypes will show a regression line with slope 1. Any genotype that is more sensitive to the environment will show a regression line with a slope higher than 1. Any genotype that is less sensitive to the environment than the genotypes on average will show a regression line with a slope lower than 1.

The method is occasionally used in an attempt to summarize the effect of various environmental variables (light, water, nutrients, etc.) into one variable, environmental value. This may only be justified if these environmental variables work on an additive scale. Not to be confused with ENVIRONMENTAL VALUATION. *See also* GENOTYPE*ENVIRONMENT INTERACTION; PHENOTYPIC PLASTICITY. [G.D.J.]

environmental variance In QUANTITATIVE GENETICS, that part of VARIANCE in a population that is attributed to random influences of the environment on the phenotypic values of individuals. As a consequence of random differences in the environment between individuals, their phenotypic values deviate from the GENOTYPIC VALUE by an ENVIRONMENTAL DEVIATION E. The environmental variance V_E is the variance of the environmental deviation E in the population. *See also* ENVIRONMENTAL VALUE; PHENOTYPIC VALUE. [G.D.J.]

environmental verifier Person/organization accredited to scrutinize and confirm company environmental statements. *See also* ENVIRONMENTAL MANAGEMENT SYSTEMS; ENVIRONMENTAL (POLICY) STATEMENT. [P.C.]

environmentalism Ideology that protection of, and respect for, environment should influence all that we do and that this is (usually) in contrast to the industrially driven, capitalist western societies. *See also* ENVIRONMENTAL ETHICS AND CONSERVATION. [P.C.]

environmentally friendly Shorthand for any action, industrial process or product that have, or are intended to have, reduced impact on the environment. Some would say that no industrial process or product can be without some impact, so it is better to refer to one as friendlier (more friendly) than others. [P.C.]

environmentally sensitive area Area designated under law as being particularly desirable to conserve, protect or enhance, for example by the adoption of particular agricultural methods. *See also* SITE OF SPECIAL SCIENTIFIC INTEREST. [P.C.]

EPA *See* ENVIRONMENTAL (PROTECTION) AGENCIES.

EPC *See* EXTRA-PAIR COPULATION.

ephemeral An organism capable of completing one or more generations in a calendar year. Ephemeral species are those that depend on rare and unpredictable environmental conditions for growth and reproduction, they tend to exhibit either fractional DORMANCY (i.e. some propagules enter long-term dormancy while others are capable of immediate germination under suitable conditions of moisture, temperature, daylength and light quality) or they possess adaptations for long-distance dispersal (often by wind), but they seldom exhibit both traits. Examples of ephemeral species include plants living in DESERTS that germinate from a buried SEED BANK only following rain, and tardigrades, which are invertebrates that can survive both extremes of temperature and DESICCATION for 10 years or more. Ephemeral insects (e.g. mayflies) are those with short-lived adult stages. Animals with multiple generations per year are referred to as multivoltine. [M.J.C. & R.H.S.]

ephemeral habitats Habitats that are transitory or short-lived. Strictly speaking, EPHEMERAL means lasting only a day (from the Greek, *ephemeros*). In ecology, habitats are described as ephemeral for species whose GENERATION TIME is longer than the typical duration of that habitat, for example fungi or fruits are ephemeral habitats for fruit flies, though they are long-lived habitats for many microorganisms. [R.H.S.]

ephemeris time *See* TIME.

epibenthic Living on the surface of (usually) the SEA bottom (although strictly it could also refer to freshwater systems). The term 'epibenthic' may include PELAGIC species that live in association with the sea floor. Sometimes used synonymously with EPIFAUNA, epibenthic organisms may be attached (SESSILE) or mobile. *See also* BENTHOS. [V.F.]

epibiont An organism of restricted RANGE whose DISTRIBUTION has contracted from a formerly more extensive range. [P.D.M.]

epidemic An increase in the number of cases of a disease per unit time (incidence) above that expected in a defined population over a period less than the life expectancy of the HOST. Disease may be continuously present in a host population when it is referred to as endemic. An endemic infection may show a pattern of recurrent epidemics. A global epidemic is referred to as a pandemic.

Epidemics of infectious disease are generally caused by a change in the ecology of the host population (e.g. increased STRESS or increase in the density of a vector species), a genetic change in the PARASITE population or the introduction of a new parasite to a host population (by movement of parasites or hosts). Generally, an epidemic occurs when host immunity to a parasite population is suddenly reduced below that found in the endemic equilibrium and the TRANSMISSION THRESHOLD is exceeded.

The characteristic pattern of an epidemic is caused by the density-dependent processes controlling the dynamics of transmission. At the start, the rates of infection and disease grow exponentially (*see* PREVALENCE OF INFECTION; TRANSMISSION RATES OF PARASITES), occurring at their highest rate determined by the basic reproduction number. When numbers of individuals infected is small, stochastic effects are important in determining the pattern of an epidemic and there is a probability that the epidemic will not occur. As immunity increases in the host population, transmission slows until the epidemic reaches a peak and then declines due to build-up of immunity in the host population (and death of hosts). For the infection to become endemic, more susceptible hosts are required either through birth or loss of immunity.

Not all individuals are infected during an epidemic. OUTBREAKS and epidemics may occur over long as well as short time-periods. Outbreaks and epidemics can be described mathematically using EPIDEMIC MODELS. There are many working definitions of epidemic, which is largely synonymous with outbreak. *See also* VECTOR, PARASITES. [G.F.M.]

epidemic models Mathematical descriptions of the dynamic processes of transmission (infection) of PARASITES within a community of HOSTS. Prevalence models are used for MICROPARASITES (*see* PREVALENCE OF INFECTION) and describe the rates of change of the proportion of hosts susceptible, infected, immune, etc. Intensity models (*see* INTENSITY OF INFECTION), used for MACROPARASITES, describe the rates of change of the parasite population and its distribution between hosts. These models form the basis for interpretation of EPIDEMIC patterns and the design of control strategies against infectious disease. [G.F.M.]

REFERENCES

Anderson, R.M. & May, R.M. (1991) *Infectious Diseases of Humans: Dynamics and Control*. Oxford University Press, Oxford.

Scott, M.E. & Smith, G. (eds) (1992) *Parasitic and Infectious Diseases*. Academic Press, New York.

epidemiology The study of patterns of disease (infectious and non-infectious) and health states in populations. Can be applied to both human and wildlife populations. Non-infectious disease epidemiology is primarily concerned with relating disease to hypothesized risk factors. Risk factors may be behavioural or environmental. This is achieved by statistical analysis of studies relating individual EXPOSURE to risk of disease development. These studies may be classified as (i) COHORT (a group of individuals is monitored prospectively or retrospectively), (ii) case–control (each case of disease is matched with an individual without disease) or (iii) cross-sectional (a population is sampled at one point in time). Such studies are also used to demonstrate the relationship between risks and specific diseases and can be closely related to trials of putative interventions designed to prevent or cure disease (clinical epidemiology).

Infectious disease epidemiology is concerned with the ecological interaction between PARASITES and their HOSTS and includes the study of EPIDEMIC MODELS. In infectious disease, the transmission rate of infection (and disease) is related to the PREVALENCE OF INFECTION and INTENSITY OF INFECTION, whereas no such direct relationship exists in non-infectious disease epidemiology (*see* TRANSMISSION RATES OF PARASITES). [G.F.M.]

REFERENCES

Hennekens, C.H. & Buring, J.E. (1987) *Epidemiology in Medicine*. Little, Brown & Co., Boston, Mass.

Scott, M.E. & Smith, G. (eds) (1992) *Parasitic and Infectious Diseases*. Academic Press, New York.

epifauna Animals that live on the bottom of aquatic environments (benthic fauna) and on the substrate surface, as opposed to living within the substrate (infauna) or resident in burrows. Mobile, free-ranging examples include crabs and starfish, whereas mussels, barnacles and anemones are more SEDENTARY. *See also* BENTHOS; EPIBENTHIC. [J.L.]

epiflora *See* BENTHOS.

epigean Above ground. Usually refers to seed germination, for example in which cotyledons are carried above the soil. Cf. hypogean, below ground. [P.C.]

epigenetics The study of the interactions of gene products, both among themselves and with their surrounding environment, during development. Often misused to refer to non-genetic aspects of development. Historically, the term derives from 'epigenesis', an idea where embryogenesis is postulated to be a process of gradual increase in morphological complexity, from a formless egg to a complex multicellular organism. Charles Bonnet (1720–1793) was one of the best-known advocates of this concept. At the time, epigenesis was antithetical to 'preformation', which claimed the complexity of the adult to be already fully present in the egg, albeit in miniature form. Thus, preformation claimed development to be a process of unfolding of an initially complex mould. This historical debate parallels the more modern confrontation of epigenesis versus 'mosaic' development, still found in most textbooks of embryology. Such a dichotomous approach is not real, since the two concepts describe the ends of a continuum of developmental processes. A contemporary 'epigenetic approach' emphasizes the regulatory, and feedback, interactions in processes such as morphogenesis, induction, pattern formation, PHENOTYPIC PLASTICITY, etc. Such a perspective contrasts with a view of embryogenesis as a process of unfolding the complex patterning already present in the early embryo, in the form of maternal factors heterogeneously distributed in the egg cytoplasm. As development proceeds by cell division, causing an uneven distribution of cytoplasmic determinants, a spatial pattern of visually similar, but genetically distinct, populations of cells (mosaics) will form. Such invisible 'spatial mosaics' will result in the experimentally observed autonomous expression of different developmental pathways (fates) depending on specific cell lineages. Reminiscent of preformism, heterogeneity is already present in the egg in mosaic development. It does not require an 'epigenetic' process of pattern formation as an emergent property of physicochemical interactions. *See also* CANALIZATION. [P.A.]

epilimnion *See* LAKE STRATIFICATION; LAKES.

epilithic Growing upon the surface of a rock or stone. [P.D.M.]

epilittoral zone *See* WITHIN-LAKE HABITAT CLASSIFICATION.

epimicropleuston *See* WITHIN-LAKE HABITAT CLASSIFICATION.

epipelagic The part of the OCEAN water column from 0 to about 200 m depth, seaward of the SHELF-SLOPE BREAK. [V.F.]

epipelic *See* BENTHIC HABITAT CLASSIFICATION.

epiphyte An organism that grows upon the surface of a plant, using its HOST only for support (Fig. E14). Usually used of plants growing upon trees and shrubs, but it can also be used of MICROFAUNA in aquatic situations. [P.D.M.]

Fig. E14 Spanish moss (*Tillandsia usneoides*, Bromeliaceae), an epiphyte which can absorb liquid water by specialized glands and also adapt from C_3 to CAM photosynthesis depending on the availability of water. (Courtesy of R.M.M. Crawford.)

epiphytic *See* BENTHIC HABITAT CLASSIFICATION.

episammic *See* BENTHIC HABITAT CLASSIFICATION.

episodic pollution EMISSIONS and EFFLUENTS that come in bursts, for example from STORMWATER DISCHARGES, RUN-OFF after storms, accidental spills and emissions. They can be characterized by their duration, amplitude (peak concentration) and frequency. [P.C.]

epistasis Refers to the interaction among non-homologous genes. When two loci interact, the PHENOTYPIC VALUE of a GENOTYPE cannot be predicted from the sum of the contributions of ALLELES at each of the two loci: an interaction term must be added. Epistatic variance refers to the part of the total variance in a quantitative character that is not explained by either additive or DOMINANCE effects. It is thus a 'residual' variance, a statistical parameter that is population dependent. *See also* PLEIOTROPY. [I.O.]

epistatic fitness Refers to the interaction among loci in determining the FITNESS of a multilocus GENOTYPE. In a single population, the Fisher fundamental theorem predicts an increase in the population mean fitness, proportional to the additive variance in fitness. When there is only epistatic variance in fitness left, the population might be stuck on an adaptive peak, with no possibility to move to another peak (with a different genotype with a higher fitness) because intermediate combinations are selected against. Sewall Wright proposed, in his shifting balance theory of EVOLUTION, that such epistatic variance in fitness might be released as ADDITIVE GENETIC VARIANCE through FOUNDER EFFECTS whereby, by chance, new combinations of genes would be in different frequencies and thus allow the population to reach a new peak. This provided a mechanism for GROUP SELECTION to occur through differential productivity of those populations that had reached a new peak. Recent experiments on *Drosophila* have shown that indeed founder effects might transform epistatic variance in a quantitative character into additive, thus heritable, variation. *See also* ADAPTIVE LANDSCAPE. [I.O.]

epizoic *See* BENTHIC HABITAT CLASSIFICATION.

epizootic An EPIDEMIC in a non-human animal population. [G.F.M.]

epoxy resins Epoxy resins are synthetic thermosetting polymers produced either by copolymerizing epoxide compounds with PHENOLS or by reaction of glycols and dihalides. They contain oxygen linkages and epoxide groups. Their properties can be varied by the addition of cross-linking agents, for example diamines or fillers. The resins can be further polymerized using CATALYSTS. They are extensively used as surface coatings, adhesives and also casings for electrical equipment due to their combined properties of toughness, chemical resistance, flexibility and insulation. [M.C.G.]

EQO *See* ENVIRONMENTAL QUALITY OBJECTIVE.

EQS *See* ENVIRONMENTAL QUALITY STANDARD.

equation of time *See* TIME.

equations Consider the relationship between two variables (x and y) summarized by the numbers in the table below.

y	5	8	11	14	17	20
x	1	2	3	4	5	6

The relationship can also be summarized by plotting values of y against values of x on a graph.

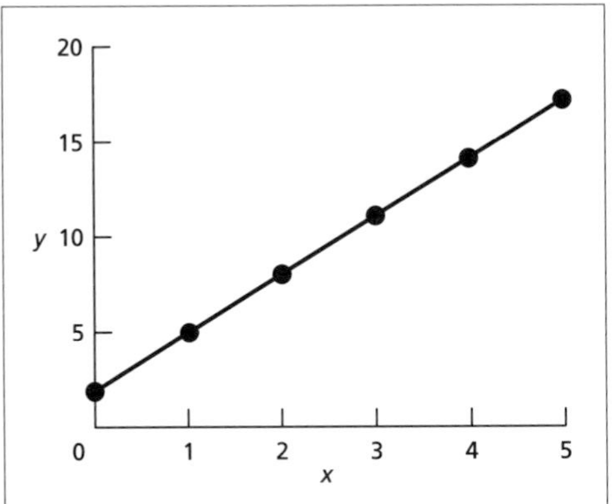

This relationship between the two variables (x and y) can also be summarized by writing an equation, $y=2+3x$. Specifically, this type of equation is called an integral equation and it allows any value of y to be calculated from any value of x. Since the slope of the line is equal to 3 in this example, another way of expressing this relation is to say that the difference between any two values of y will be equal to the corresponding difference between x multiplied by 3, or $y_2=y_1+3(x_2-x_1)$. This type of equation is known as a difference equation. Since the slope of the line is equal to 3, we can also say that the rate of change of y with respect to x is 3. In symbols these words are written $dy/dx=3$. This type of equation is known as a differential equation. The symbol, dy/dx, is called 'dy by dx' and is a rate of change. Therefore, a single relationship between two variables can be summarized by a variety of equations, depending upon what aspect of the relationship you wish to stress. [B.S.]

equilibrium, communities Equilibrium of communities is considered in two rather different contexts: indefinite COEXISTENCE of species in mixture, and ability of communities to return to a certain condition after DISTURBANCE. In the first case, equilibrium is said to occur where the species present can coexist indefinitely as a result of NICHE differentiation. Interpreted in the narrowest sense, this means indefinite coexistence without any impact of disturbance or year-to-year variation in weather (or other variables) and hence in RECRUITMENT of various species. The idea is believed to be applicable to many communities of birds and mammals. It is doubtful if it is ever applicable to communities of plants. Interpreted more broadly, to encompass the prevalent range of disturbances and year-to-year variation in weather and other factors, many plant communities may be of an equilibrium type. Particular interest has attached to the most species-rich communities, notably those in TROPICAL RAIN

FORESTS, and the commonest viewpoint now is that they are not at equilibrium.

A community's ability to return to a given state after disturbance is termed 'resilience'. Resilience may occur *in situ* (from root sprouts and the SEED BANK) or by MIGRATION (movement back into an area where the whole plant cover has been destroyed). *See also* RESILIENCE. [P.J.G.]

equilibrium population density *See* CARRYING CAPACITY.

equilibrium, populations Refers to a POPULATION, the size of which is not changing through time, or the characteristics of which are not changing through time. A population's size is at equilibrium when births and IMMIGRATION balance deaths and EMIGRATION. Characteristics other than size may also be at equilibrium, for example age-structured populations can reach an equilibrium AGE STRUCTURE (proportion of the population in each AGE CLASS), although the total population is changing. *See also* POPULATION DYNAMICS; POPULATION REGULATION. [S.N.W.]

equitability In no ecological community are all species equally common. Instead, some species will be very abundant, others moderately common, with the remainder infrequent or rare. Equitability (sometimes also termed 'evenness') is a measure of the extent to which species are equally represented in a community. For example, if one species is dominant and the others rare (as might occur, for instance, when only one species of moss thrives in the shaded conditions of a conifer plantation) then equitability is low. On the other hand, when the ABUNDANCES of species are almost equal (e.g. in certain bird communities) equitability is high.

Equitability can be described in a number of ways. One approach is to fit a SPECIES-ABUNDANCE MODEL to the data. The shape of the DISTRIBUTION is a measure of equitability. Equitability can also be calculated using the Shannon diversity index or Simpson's diversity index. However, the most satisfactory and intuitively simple measure of equitability is provided by the Berger–Parker index (which is simply the proportional abundance of the most abundant species):

$$d = N_{max}/N$$

where N_{max} is the number of individuals in the most abundant species and N is the total number of individuals in the sample.

The reciprocal form of the measure is conventionally adopted since in this form an increase in the value of the index reflects an increase in equitability. *See also* DIVERSITY INDICES. [A.E.M.]

REFERENCES

Magurran, A.E. (1988) *Ecological Diversity and its Measurement.* Chapman & Hall, London.

May, R.M. (1975) Patterns of species abundance and diversity. In: *Ecology and Evolution of Communities* (eds M.L. Cody & J.M. Diamond), pp. 81–120. Harvard University Press, Cambridge, Mass.

erbium (Er) Erbium is a lanthanide, one of the so-called rare earth metals (ATOMIC NUMBER = 68; relative ATOMIC MASS = 167.26; melting point = 1529°C; boiling point = 2863°C), having an average abundance of 2.8 mg kg^{-1} in the Earth's CRUST. It is found in the same minerals as DYSPROSIUM. Like dysprosium it has a bright metallic lustre, is soft and malleable and is stable in air at room temperatures. When combined with VANADIUM, it lowers the hardness and improves the workability. Natural erbium is a combination of six isotopes; nine radioactive isotopes are known. Erbium oxide (Er_2O_3) has been used to colour glasses and glazes. [J.G.]

ericaceous/ericalean mycorrhizae *See* ENDOTROPHIC MYCORRHIZAE.

erosion and topsoil loss Loss of topsoil as a result of physical removal by WIND and WATER, and physical, chemical and biological changes. It is a natural process but its intensity has been increased by human activities. For example, removal of natural VEGETATION for agricultural purposes exposes soil to increased action of wind and water. Worldwide, it has been estimated that >25 million tonnes of material are removed each year from topsoil by excessive erosion. *See also* AGRICULTURE AND SOIL CONSERVATION. [P.C.]

error structure In GENERALIZED LINEAR MODELS, the user can specify various sorts of non-normal sampling errors. The most commonly used error structures are:

1 Poisson errors for the analysis of count data and CONTINGENCY TABLES;

2 binomial errors for the analysis of proportion data and binary response variables;

3 gamma errors for the analysis of data where the VARIANCE increases with the MEAN but where the COEFFICIENT OF VARIATION is constant;

4 various kinds of survival analysis, where age at death is analysed using exponential, Weibull or other error structures.

See also DISTRIBUTION; STATISTICS. [M.J.C.]

REFERENCE

Crawley, M.J. (1993) *GLIM for Ecologists.* Blackwell Scientific Publications, Oxford.

ERS *See* EUROPEAN REMOTE SENSING (ERS) SATELLITE.

ESS *See* EVOLUTIONARILY STABLE STRATEGY.

essential elements There are 17 ELEMENTS essential for plant GROWTH. From the air and water, CARBON, HYDROGEN and OXYGEN are derived. From the SOIL the following are obtained: NITROGEN, PHOSPHORUS, POTASSIUM, CALCIUM, MAGNESIUM, SULPHUR and, in trace quantities, IRON, MANGANESE, BORON, MOLYBDENUM, ZINC, CHLORINE and COBALT. The last seven are termed 'MICRONUTRIENTS' or 'TRACE ELEMENTS'. A few other elements are apparently needed as micronutrients by some plant or animal species: SODIUM, FLUORINE, IODINE, SILICON, STRONTIUM and BARIUM. *See also* LIMITING FACTOR. [J.G.]

REFERENCE

Marschner, H. (1995) *Mineral Nutrition of Higher Plants.* Academic Press, London.

essential resource *See* RESOURCES.

establishment A technical term used in different ways in different subject areas.

1 METAPOPULATION dynamics. Following COLONIZATION of a patch, there are several possibilities: all the propagules might die; the propagules might produce individuals that live for a long time, but these individuals never reproduce; or the colonists may reproduce, giving rise to an expanding new population and producing migrants for the colonization of new patches. Only this latter case constitutes establishment. Species where the colonists survive but do not reproduce are called CASUAL SPECIES.
2 ISLAND BIOGEOGRAPHY. Establishment is the formation of a breeding population of a species new to an island. Species arrive on islands at a rate inversely proportional to the number of species already established and directly proportional to the size of the species pool in the source region from which the colonists originate.
3 POPULATION DYNAMICS. The formation of a local breeding population (*cf.* RECRUITMENT). An established population must exhibit the ability to increase when rare; technically, this property requires that $dN/dt>0$ when N is small (the INVASION CRITERION; *see also* ALLEE EFFECT). [M.J.C.]

esters Esters are obtained by replacing the HYDROGEN of an ACID by an organic group. For example, ethyl acetate ($CH_3COOC_2H_5$) is the ethyl ester of acetic acid (CH_3COOH). Esters are used for flavouring, solvents and textiles (polyesters). [J.G.]

estimated (predicted) environmental concentration Predicted concentration of XENIOBIOTIC in an environmental compartment based on information on patterns and amounts of production, use and disposal, and physicochemical properties of the chemicals. *See also* PREDICTED ENVIRONMENTAL CONCENTRATION. [P.C.]

estuarine lagoons *See* LAGOONS.

estuary A semi-enclosed body of water that has a free connection with the open SEA and within which seawater is diluted measurably with FRESHWATER derived from land drainage. Many estuaries were formed by the rise in sea level following the last glacial maximum, and because rather small changes in sea level can form or eliminate estuaries they are unlikely to persist for more than about 10 000 years. Water movement in estuaries depends on the amount of river discharge, tidal action and BASIN shape. The general pattern involves the downstream flow of low-density riverwater in the upper water layer and the upstream flow of higher-density seawater in the lower layer. Wind and tidal mixing cause vertical exchange in the water column and the SALINITY at any point in the estuary is usually dependent on seasonal factors (e.g. freshwater flow and input of land-derived particulate material). Estuaries are among the most biologically productive habitats in the world, provide important nurseries for the development of larval and juvenile fish and invertebrates, and support major fisheries. That fewer species tend to be found in estuaries compared with freshwater or marine habitats is due to the geological youth of most estuaries combined with the dramatic changes in salinity occurring in them. A minimum in species number occurs between 3 and 8‰ salinity, as both freshwater and marine species have difficulty surviving such conditions. [V.F.]

ET_{50} *See* MEDIAN TIMES TO EFFECT.

ethanol Ethanol is a colourless, inflammable ALCOHOL with chemical formula C_2H_5OH and boiling point 78.1°C. Its boiling point is lower than would be predicted from molecular mass due to hydrogen bonding between molecules. It is miscible with WATER due to formation of hydrogen bonds with the solvent water molecules and is also miscible with most organic solvents. It can be prepared by the hydration of ethene, oxidation of natural gas followed by fractional distillation, reaction of CO and H_2 gas under pressure and in the presence of a CATALYST followed by fractional distillation, or from the fermentation of sugar using enzymes contained in yeast, maltase and zymase. Maltase converts the maltose into glucose and zymase converts the glucose into ethanol.

Industrial alcohol is simply rectified spirit and contains 93–95% ethanol (5–7% BENZENE and water). Methylated spirit is either mineralized methylated spirit, which contains approximately 90% rectified spirit, 9% METHANOL, 1% PETROLEUM and a purple dye, or industrial methylated spirit, which contains approximately 95% rectified alcohol and 5% methanol. Absolute alcohol, which is about 99.9% ethanol, is obtained from rectified spirit by distillation over calcium oxide (CaO) and then over CALCIUM.

Ethanol is a useful solvent for gums, resins, paints, varnishes and in the preparation of ESTERS, ether, chloroform, etc. In the chemical industry, ethanol has been replaced by ethene as a raw material or feedstock compound.

Ethanol mixed with diesel is called biodiesel. With little alteration, cars may run on pure alcohol or a mixture of alcohol and traditional fuels, for example 10% ethanol : 90% diesel mix. Ethanol can be produced from sugars from almost any source, for example molasses, grain and fruits, although the highest grade fuel is obtained from grain. [M.C.G.]

ethers Ethers are organic compounds with the general formula R–O–R′, where R and R′ are organic groups. Common ether, used as a solvent and formerly as an anaesthetic, is diethyl ether $C_2H_5–O–H_5C_2$. [J.G.]

ethical investment Provides investors with investment opportunities that do not compromise their ethics, for example with respect to avoiding oppressive regimes, businesses connected with tobacco, alcohol, weapons, etc. Green or ecological trusts selectively invest in businesses that minimize their ENVIRONMENTAL IMPACTS. [P.C.]

Ethiopian faunal region The boundaries of the Ethiopian FAUNAL REGION (see Fig. F1) are the ocean, except in the north. Most Ethiopian vertebrates occur only south of the Sahara or extend north only along the Nile. The northern parts of Morocco and Algeria, north of the Atlas Mountains, has a depauperate fauna that is European in most of its relationships. In between, the great DESERTS of northern Africa have a very limited, specialized fauna with mixed relationships. Under these circumstances it is not important whether the northern boundary of the Ethiopian region is set at the Atlas Mountains, where P.L. Sclater put it, or across the middle of the Sahara at the northern boundary of the tropics, where A.R. Wallace and most later zoogeographers put it. South-western

Arabia has a vertebrate fauna that is in some ways Ethiopian, but depauperate. This part of Arabia can be included in the Ethiopian region or it can be considered part of a complex transition area.

The freshwater fishes of the region are a very distinctive assemblage, with some clear but distant relationships towards South America, some closer ones towards the Orient, and important exclusive groups such as the archaic bichirs. They include lepidosirenid lungfishes, many catfishes and characins, some cyprinids and cichlids and salt-tolerant cypriodonts. Amphibians and reptiles are less distinctive, with again some South American relationships (including pipid frogs and pelomedusid turtles) and more oriental ones. Characteristic amphibians are caecilians, pipid, bufonid and ranid frogs. The reptiles include pelomedusid turtles, testudinine land tortoises, many geckos and skinks, chameleons and lacertids. Among the snakes are pythons, boas, elapids, viperids and many colubrids.

Ethiopian birds belong to numerous families, most of which are widely distributed in other parts of the world and therefore less distinctive than the fish, amphibians and reptiles. Their main relationships are with the Oriental region, but six families are exclusively Ethiopian or nearly so; they include the ostrich, secretary bird, hammerhead, touracos, mousebirds and helmet shrikes, guinea fowls, tree hoopoes, bush shrikes, buffalo weavers and widow birds. The Ethiopian mammals are rather strongly oriental in their main affinities, but with a number of endemic groups. The larger animals are extremely familiar, but some of the less conspicuous mammals are more important zoogeographically. Endemic or almost endemic mammals include the otter shrew, elephant shrew, golden moles, hyraxes, hippopotamuses and giraffes. *See also* EAST AFRICAN FLORAL REGION; ORIENTAL FAUNAL REGION. [A.H.]

REFERENCE

Darlington, P.J. (1957) *Zoogeography: the Geographic Distribution of Animals.* John Wiley & Sons, New York.

EU/EC environmental protection

The Treaty of Rome (EEC Treaty) established the European Economic Community (EEC) in 1957 and was about a common trading market. However, harmonization of national laws to prevent trade barriers (Article 100 EEC) was used as a basis for justifying laws relating to environmental protection. Article 235 (EEC) relating to general and residual powers was also used as a justification for environmental protection (e.g. a directive on wild birds was legitimized on this basis in 1979). The Single European Act (SEA; came into force 1986) amended the EEC Treaty by introducing explicit environmental law-making powers in Articles 130r, 130s and 130t. It also introduced Article 100a, which requires that the issuing of legislation for the approximation of measures within member states with respect to the functioning of the internal market will take as a base a high level of protection concerning the environment. The Treaty on European Union (TU) extended environmental responsibility further. It was signed in Maastricht in February 1992, but not formally ratified by all member states and adopted until November 1993, and consists of two segments. The first, on common provisions, stands alone. The second amended the EEC Treaty in detail and renamed it the Treaty Establishing the European Community (EC Treaty). The second segment introduced a modification to Article 2 (EC Treaty) that stated that the community will promote 'a harmonious and balanced development of economic activities, sustainable and non-inflationary growth respecting the environment'. This is taken further by the Treaty of Amsterdam, adopted in 1997, but at the time of writing, not formally ratified by all member states. It also introduced the PRECAUTIONARY PRINCIPLE as a major consideration into Article 130.

The TU created the European Union (EU). This rests on three pillars: common foreign and security policy; home affairs and justice policy; all policies previously carried out under the terms of the EEC Treaty. The first two pillars were new. The last one is as redefined in the amended EC Treaty. It remains correct, therefore, to refer to the latter as EC policy. At the same time, since the EC is part of the European Union, it also remains correct to refer to the EU. The Treaty of Rome, on the other hand, initially created the European Economic Community and this was explicitly replaced in the TU by European Community, so EEC is no longer appropriate. One last complication: the European Commission, an institution of the EC/EU, is also sometimes abbreviated to EC, but more often to CEC.

The legal instruments that are formed under the treaties are directives, which have to be implemented by member states through their own legislation, and regulations that apply directly. These are formulated by the European Commission, the bureaucracy of the EU, but can only be adopted by the Council (composing heads of state but with an Environment Council comprising environment ministers). The European Parliament also has a voice and can block certain kinds of legislation.

There are now hundreds of pieces of legislation that have an influence on environmental protection covering water, WASTE, air, harmful substances, RADIOACTIVITY, WILDLIFE and countryside, noise, impact assessment and provision of environmental information. It includes not only COMMAND AND CONTROL instruments but also financial and economic instruments and makes some provision for voluntary activities. *See also* ENVIRONMENTAL (PROTECTION) POLICIES. [P.C.]

REFERENCE

Haigh, N. (1992) *Manual of Environmental Policy: the EC and Britain.* Longman, London.

---

euendolithic *See* ENDOLITHIC.

eukaryote Organism whose genome consists of chromosomes, in which DNA is complexed with histones, surrounded by a membrane to form a nucleus. Consists of protistans, algae, fungi, plants and animals. *Cf.* PROKARYOTE. [P.C.]

Euler–Lotka equation Also known as the Euler equation, the Lotka equation or the characteristic equation, this allows calculation of the intrinsic rate of population growth, r, from knowledge of the life histories of the individuals in the population. Suppose that individuals breed at ages $t_1, t_2, t_3, \ldots$ producing $n_1, n_2, n_3 \ldots$ female offspring and that survivorship from birth to ages $t_1, t_2, t_3, \ldots$ is $l_1, l_2, l_3, \ldots$, then the equation is:

$$1 = n_1 l_1 e^{-rt_1} + n_2 l_2 e^{-rt_2} + n_3 l_3 e^{-rt_3} + \cdots$$

If the individuals breed at ages 1, 2, 3, . . . then the equation can be written:

$$1 = \sum_{x=1}^{\infty} n_x l_x e^{-rx}$$

The value of *r* calculated from the equation can be used to forecast future POPULATION SIZE provided that the life histories and the relative numbers of individuals in each AGE CLASS do not change, i.e. the population has a STABLE AGE DISTRIBUTION. *See also* LIFE-HISTORY EVOLUTION; POPULATION DYNAMICS; POPULATION REGULATION. [R.M.S. & D.A.R.]

eulittoral zone Generally also called the LITTORAL ZONE (although sometimes classified as a subdivision of this zone) or INTERTIDAL ZONE. *See also* ROCKY SHORES; SANDY SHORES; WITHIN-LAKE HABITAT CLASSIFICATION. [V.F.]

euphotic zone Also called the photic or EPIPELAGIC zone. Refers to the lighted part of the OCEAN in which PRIMARY PRODUCTIVITY occurs. It extends from the surface to the depth at which PHOTOSYNTHESIS is no longer possible due to a lack of LIGHT. The lower boundary of the euphotic zone is generally between 100 and 200 m, but the exact depth varies with water clarity. [V.F.]

eupleuston *See* WITHIN-LAKE HABITAT CLASSIFICATION.

European Environment Agency (EEA) Created under European Community (EC) legislation to collect, collate and disseminate information relating to the quality of the environment, the pressures on the environment and its sensitivities to these pressures. It is required to establish an Environmental Information and Observation Network (EIONET). It currently has no regulatory or enforcement functions, *cf.* ENVIRONMENTAL (PROTECTION) AGENCIES. It is located in Copenhagen, Denmark. [P.C.]

European Environmental Bureau Federation of about 150 environmental non-governmental organizations. It is located in Brussels. [P.C.]

European Inventory of Existing Commercial Chemical Substances (EINECS) Listing of all commercial chemicals on the market in Europe to 18 September 1981 amounting to *c.* 100 000 chemical substances. It is required under European Community (EC) law. Chemicals on the list are designated as existing substances. Chemicals not on the list are new substances. Manufacturers of new substances must submit a NOTIFICATION of properties influencing environmental HAZARDS associated with the substance, on which basis it can be packaged, classified and labelled in a standard way. Substances on EINECS, under separate legislation, are prioritized according to likely environmental effects and then subject to RISK ASSESSMENT as a possible product for RISK MANAGEMENT. [P.C.]

European List of Notified Chemical Substances (ELINCS) A list of chemical substances notified under European Community (EC) new substances legislation. *See also* EUROPEAN INVENTORY OF EXISTING COMMERCIAL CHEMICAL SUBSTANCES. [P.C.]

European Remote Sensing (ERS) satellite A REMOTE SENSING satellite. The first European Remote Sensing satellite (ERS-1) has a Sun-synchronous polar orbit with an altitude of 780 km and a 3-day repeat cycle. It was launched in July 1991 and carries three sensors: a C band SYNTHETIC APERTURE RADAR (SAR) that produces images of the Earth's surface independently of solar illumination/atmospheric conditions with a spatial resolution of 30 m; a radar altimeter that measures the height of the satellite above the Earth to an accuracy of ±2 cm; and the Along-Track Scanning Radiometer that measures and maps sea-surface temperature to an accuracy of ±0.5°C. ERS-2 was launched in April 1995. [P.J.C.]

europium (Eu) Europium is the most reactive of the lanthanides, the so-called rare earth metals, and one of the rarest (ATOMIC NUMBER = 63; relative ATOMIC MASS = 151.96; melting point = 822°C; boiling point = 1597°C); having an average abundance of 1.2 mg kg^{-1} in the Earth's CRUST. It is as soft as LEAD, silvery-white and quickly oxidizes in air. As with other rare earth metals, it is found in the minerals monazite and bastnaesite. Europium has 17 isotopes. Europium-activated YTTRIUM vanadate is used as the red phosphor in television tubes; europium-doped plastic has been used as a laser material. [J.G.]

Euro-Siberian floral region A phytogeographic unit representing the major Old World components of the Boreal kingdom (the Boreal subkingdom of Takhtajan, 1969), shown as Neill's (1969) Euro-Siberian PROVINCE in Fig. B6. It lies to the south of the ARCTIC and Subarctic region and so extends across Europe from Britain and northern Spain in the west, Scandinavia and most of Siberia south of the Arctic Circle to the Pacific coast. Its southern boundary takes in northern Spain, the northern half of Italy and passes north of Greece but takes in the remainder of the Balkans; a finger then extends to include the southern margin of the BLACK SEA across to the Caspian Sea. It is then bounded to the south by the Western and Central Asiatic region and the Sino-Japanese region. The borders shown in Fig. B6 approximate to those of Good (1964) and others, but Takhtajan (1969) places the northern margin so as to include the whole of Scandinavia. He also includes the Aleutian Islands, so that technically the region on his interpretation reaches into the New World. On any interpretation, this region has an enormous east–west extent, reaching half-way around the world. Although this means that the actual floristic composition of the western and eastern extremes are significantly different, there is no very sharp change across the spread of the region, offering a basis for dividing it into two halves. As a result the Ural Mountains are taken, rather arbitrarily, as a boundary between the European and Asiatic subregions. Although this is a political rather than a floristic boundary, it is a convenient one since it represents the eastern margin of the coverage of the *Flora Europaea*.

Hardly any families are endemic to the Euro-Siberian region. A number of genera endemic to the region are familiar as garden plants (both inside the region and in the New World), such as lilac

(*Syringa*) and *Doronicum*. *Hedera*, ivy, is near-endemic to the region but just reaches into the western Mediterranean. *Lamium*, the dead nettle, is another endemic genus but, like many weedy plants, it has been spread far outside the area by human migration; for example, five species are widely NATURALIZED in North America. [W.G.C.]

REFERENCES

Good, R. (1964) *The Geography of the Flowering Plants*. Longmans, Green & Co., London.

Neill, W.T. (1969) *The Geography of Life*. Columbia University Press, New York.

Takhtajan, A. (1969) *Flowering Plants, Origin and Dispersal*. Oliver & Boyd, Edinburgh.

eury- From the Greek *eurus* meaning wide. Applied as prefix meaning wide range of tolerance for some environmental factors, for example eurybathic for hydrostatic PRESSURE, eurythermal for TEMPERATURE, euryhaline for SALINITY. *Cf.* STENO-. [P.C. & V.F.]

eurybathic *See* EURY-.

euryhaline *See* BRACKISH WATERS; EURY-.

eurythermal *See* EURY-.

eurytopic Eurytopic organisms are those able to tolerate a wide range of conditions and that consequently have very widespread distributions. An example from amongst the butterflies that occur in the British Isles is the meadow brown, *Maniola jurtina*, which occurs throughout England, Wales, Scotland and Ireland and is also very widespread in Europe. It is tolerant of a very wide range of temperature conditions, being found at altitudes from 0 to 2000 m, and flies on both sunny and cloudy days from June until September. It eats a wide range of larval food plants as it feeds on many different species of grass, although *Poa* species are preferred. *See also* STENOTOPIC. [M.H.]

eusociality *See* SOCIAL INSECTS; SOCIALITY, TYPES OF.

eutrophic *See* EUTROPHICATION.

eutrophication Biological effects of an increase in plant NUTRIENTS (usually NITROGEN and PHOSPHORUS, but sometimes SILICON, POTASSIUM, CALCIUM, IRON or MANGANESE) on aquatic systems. The term 'eutrophic' was first used by Weber, a German botanist, at the turn of the century to describe the conditions that determine the plant community in the initial stages of raised peat BOGS. He subsequently described the nutrient stages that control vegetation changes as leading from 'eutrophe' to 'mesotrophe' and then 'oligotrophe'. These three terms were used in LIMNOLOGY shortly after by the Swedish botanist Naumann to describe freshwater-lake types containing low (oligotrophic), moderate (mesotrophic) and high (eutrophic) concentrations of phosphorus, nitrogen and calcium, though precise levels were never specified. Indeed, there is still a lack of precision in the definition of these terms, possibly because the trophic nature of water bodies varies considerably from system to system.

The sequence of changes in LAKES subject to eutrophication involve increases in plant biomas and changes in the OXYGEN regime, largely resulting from the accumulation and decay of plant DETRITUS and oxygen consumption of microbial decomposers. Both suspended and settled particulate matter increase, leading to the attenuation of light and ultimately reduced PHOTOSYNTHESIS and plant BIOMASS. All these changes have knock-on effects to the species composition of aquatic flora and fauna at all levels.

Eutrophication may be due to the natural effects of in-filling and SUCCESSION slowly affecting lake morphometry and to the effects of accelerated nutrient enrichment from human sources, largely from SEWAGE inputs. *See also* ORGANIC POLLUTION. [P.C.]

REFERENCE

Harper, D. (1992) *Eutrophication of Freshwaters*. Chapman & Hall, London.

evaporation Evaporation in the general sense is the conversion of any liquid into a gas but, in ENVIRONMENTAL SCIENCE, is usually taken to mean the loss of WATER from a wet surface. The rate of evaporation depends on the amount and TEMPERATURE of water present, on the energy available, on the vapour pressure of the air and on the diffusion of water molecules away from the surface (which is therefore dependent on WIND speed), i.e. washing dries much quicker on a hot, sunny, dry and breezy day than on a still night. Energy is required to supply the latent heat of vaporization, which is the energy required for the molecules of water to leave the liquid surface. Evaporation therefore cools the evaporating surface (and the surrounding air in contact) and condensation warms it. Evaporation of water is a critical process at both the large scale of the global CLIMATE system and at the very small scale of the energy balance and physiological state of individual organisms. For the whole globe, approximately 76% of the annual solar energy reaching the surface is used in evaporation and about 86% of this is from the oceans. Evaporation takes place from the OCEAN, snow and ICE, the SOIL surface, the surface of plants and from inside plants (often referred to as TRANSPIRATION). On land, the transpiration component is very important as vegetation continues to take up water through roots distributed in the soil and evaporate it, even when the soil and vegetation surfaces have dried out. On the short time-scale of minutes, the energy used in transpiration from dense vegetation is often over half the energy received from the Sun. For terrestrial organisms, evaporation rates can be higher than reserves or uptake rates can supply and dehydration and water stress can result, causing physiological problems due to change in osmotic balance. [J.I.L.M.]

evenness *See* EQUITABILITY.

evergreen Evergreen leaves and other plant organs are not seasonally deciduous and have a photosynthetic life of a few to many years, thus allowing PHOTOSYNTHESIS at any time of year when temperature or water is not limiting. Evergreenness has been interpreted as a mechanism of mineral nutrient conservation (particularly NITROGEN and PHOSPHORUS), which allows more photosynthetic CARBON fixation relative to a given investment in NUTRIENTS that does not have to be replenished each year. HERBIVORE deterrence by well-developed chemical and physical defences is often linked to evergreenness for similar reasons of resource conservation. Sclerophylls have particularly well-developed mechanical tissue, which may play a part in both water conservation and herbivore resistance. Again, these plants are usually evergreen.

Equatorial plants are often evergreen as the photosynthetic period is almost year-long unless limited by flooding or DROUGHT. The deciduous habit appears to be more widespread in mid-latitudes and it is here that the disadvantages of winter damage to persistent leaves may be greater than the energy costs of gathering nutrients afresh each year. At higher latitudes, evergreen conifers become prominent: here it may be that prolonged cold limits nutrient uptake sufficiently for conservation by evergreenness to become cost-effective again. *See also* MINERAL NUTRIENTS; SCLEROPHYLLOUS VEGETATION. [J.R.E.]

evolution Evolution is a comprehensive framework that combines with physics and chemistry to explain all biological phenomena. It is the only part of biology containing principles not implicit in physics and chemistry. It has three major principles: NATURAL SELECTION, inheritance and history; and one fundamental property: selection acts on organisms but the response to selection occurs in stored information. The first principle, natural selection, is a great law of science, the only mechanism known that maintains and increases the complexity of organisms. Variation among organisms in reproductive success produces natural selection. Populations of organisms respond to selection when some of that variation is genetically based; this happens in stored information. The result is a genetically based change in the phenotypic design of offspring from the more reproductively successful parents.

Many, but not all, organisms appear to us as sets of related individuals that we call species. How two or more daughter species arise from a single ancestral species is an important problem, for SPECIATION has produced an evolutionary tree recording the history and describing the relationships of all life on Earth. To reconstruct that history and those relationships is a central question whose answers combine the logic of phylogenetic analysis with the data of molecular evolution. Hypotheses about the relationships of taxonomic groups provide the framework within which the evolutionary history of a group and its ADAPTATIONS can be interpreted.

Evolutionary biologists want to understand how the variation in reproductive success that causes selection arises, how the genetic variation that enables a response to selection originates and is maintained, and how that response is constrained by geography, time, inheritance, development and history. *See also* DARWINISM; FITNESS. [S.C.S.]

evolution and conservation The linkage between the natural processes that have given rise to the diversity of life on Earth and the human efforts to preserve and restore that diversity. Among the many processes that underlie life's diversity are MUTATION, recombination, NATURAL SELECTION, ADAPTATION and SPECIATION. Paradoxically, extinction is also an important evolutionary process. The evolutionary response to environmental change is influenced by the amount and pattern of genetic variation in populations of organisms. Thus, a key element of many conservation strategies is preservation of genetic variation in populations of endangered organisms. *See also* CONSERVATION BIOLOGY; EVOLUTION. [R.H.R. & E.A.Fr.]

evolution of sex Sex ought to be at a disadvantage because on average each female will produce equal numbers of sons and daughters. Parthenogenetic females produce all daughters and hence are twice as productive. Genes for PARTHENOGENESIS should therefore overwhelm genes for sex. Although most evolutionary biologists agree that the advantage of sex for populations arises from the creation and perpetuation of genetic diversity, there has been less consensus about those specific conditions under which the benefits of sex are likely to outweigh the obvious costs, and to identify benefits for the individual as well as the population. Consideration of the relative advantages and disadvantages of sex inevitably leads to questions about the FITNESS costs of generality versus specialism and the interplay between these costs and patterns of environmental variability. Several models have been proposed to predict the types of biotic and abiotic conditions that should favour one reproductive mode over another. Some of the more realistic models are:

1 the high fecundity (or 'Sisyphean') model of Williams (1975) in which strong selection due to large, unpredictable environmental differences favours sexually-produced, variable offspring;

2 the tangled bank model (Bell 1982), in which spatially varying abiotic conditions confer an advantage to sex in finite populations with incomplete RESOURCE COMPETITION among GENOTYPES; and

3 the Red Queen model (*see* RED QUEEN'S HYPOTHESIS), according to which temporally unpredictable biotic conditions (e.g. evolutionary changes in predators or parasites) favour sex due to the frequency- and density-dependent advantages to rare genotypes. [P.C.]

REFERENCES

Bell, G. (1982) *The Masterpiece of Nature: The Evolution and Genesis of Sexuality*. Croom Helm, London.

Williams, G.C. (1975) *Sex and Evolution*. Princeton University Press, Princeton, N.J.

evolutionarily stable strategy (ESS) Often referred to as 'evolutionary' but strictly should be 'evolutionarily', the point being that EVOLUTION is not 'goal-directed' and so should not be thought of as having strategies. Some complex ADAPTATIONS are more likely to succeed and persist in populations than others and these are evolutionarily stable. Maynard Smith (1972) defined them by reference to game theory: those moves in the evolutionary game that cannot be replaced by an invading mutant. ESS theory has therefore proved extremely useful in the analysis of animal behaviour. For an example of the approach, see GAME THEORY. *See also* BEHAVIOURAL ECOLOGY; EVOLUTIONARY OPTIMIZATION. [P.C.]

REFERENCE

Maynard Smith, J. (1972) *On Evolution*. Edinburgh University Press, Edinburgh.

evolutionary conflict *See* CONFLICTS.

evolutionary optimization A research programme that aims at interpretation of structures, behaviours and other properties of organisms in terms of the contribution they make to Darwinian FITNESS. It is based on the assumption that NATURAL SELECTION has led to the EVOLUTION of PHENOTYPES that confer the highest fitness possible under given circumstances. An optimization study makes assumptions about the functional CONSTRAINTS and TRADE-OFFS that delimit the set of possible phenotypes, about the optimization criterion (fitness measure) and about the pattern of inheritance.

Mathematical methods are then used to make predictions about the phenotype that should result in highest fitness if these assumptions were true. Empirical falsification of the predictions leads to changes in the assumptions. The optimization approach has been criticized for its Panglossian assumption that organisms are optimally adapted to their current environments, which itself is not under test. A population may not reach an optimum because of lack of appropriate genetic variation, GENE FLOW, other genetic constraints, or GENETIC DRIFT and MUTATION coupled with weak selection. The current state of the population may also represent ADAPTATION to past rather than present conditions (evolutionary lag). Despite these shortcomings, evolutionary optimization has proved successful in explaining patterns of differences among organisms, in particular in behaviour and life history. *See also* BEHAVIOURAL ECOLOGY; GAME THEORY; LIFE-HISTORY EVOLUTION. [T.J.K.]

REFERENCE

Maynard Smith, J. (1978) Optimization theory in evolution. *Annual Review of Ecology and Systematics* **9**, 31–56.

evolutionary radiation and possible explanations A natural question to ask is why a particular ancestral group has been so successful in terms of number of descendant species in a LINEAGE and/or their morphological or behavioural diversity? Had it evolved some key feature which led to its evolutionary success? Or did it just turn up at the right place at the right time? There seems to be reasonably well-documented examples of both. Cichlid fishes are an extraordinarily speciose group in African lakes. During the evolutionary history of the group a particular change in jaw musculature seems to have opened up a whole new range of feeding niches into which the group could radiate, despite COMPETITION from species belonging to several other fish families. The change in jaw musculature has been called a key-innovation. In contrast to cichlid fishes, there probably was nothing special about the species of finch that originally invaded the GALAPAGOS ISLANDS and led to the radiation known as Darwin's finches. That was just a case of turning up at the right time in an area with several unoccupied niches. *See also* ADAPTIVE RADIATION; SPECIATION. [P.H.H.]

evolutionary rate The speed of evolutionary change. For characters whose measurement has changed from x_1 to x_2 during a time-interval of length Δt (in years or million of years), the evolutionary rate, r, can be calculated as $r = \ln(x_2) - \ln(x_1)/\Delta t$ (in units of darwin). Logarithms are taken to make the evolutionary rate independent of scale (i.e. the size of an organism). Evolutionary rates have been estimated from the FOSSIL RECORD, from recent COLONIZATION events and from selection experiments. They vary between taxa and over geological time. In Gingerich's classic study (1983), rates varied over several orders of magnitude and seemed to be higher the shorter the interval over which change was measured. Problems of rate estimates are that short-term selection experiments apply unnaturally high SELECTION PRESSURES, that characters may change qualitatively rather than quantitatively during the large time-scales considered in the fossil record (e.g. a leg changes into a wing) or that the direction of EVOLUTION is sometimes reversed. For characters that change qualitatively, evolutionary rates can be calculated by dividing the character into states and measuring the rate of change in states.

Proponents of the theory of PUNCTUATED EQUILIBRIUM maintain that evolutionary rates have a specific pattern: they are high during SPECIATION events and low ('stasis') between LINEAGE splits. The incompleteness of the fossil record makes it difficult to test this theory.

The taxonomic rate of evolution is expressed as the number of species a lineage produces in 1 million years. It is often used to compare evolutionary rates among taxa and it depends on the rates of both speciation and extinction of lineages.

Rates of evolution can also be calculated for genomes and molecules. Rates of protein evolution are expressed as the number of AMINO ACID substitutions per site per year (in the order of 0.0–10.0 $\times 10^{-9}$) and vary among PROTEINS and among different sites within a protein. DNA sequence data of species, along with the time back to their common ancestor, are used to estimate the rate of nucleotide substitutions. To obtain rate estimates, models of molecular evolution are applied to correct for multiple substitutions.

The patterns of evolutionary rates, their causes and their consequences are an ongoing focus of research. *See also* EVOLUTIONARY TRENDS; MACROEVOLUTION. [W.W.W.]

REFERENCES

Gingerich, P.D. (1983) Rates of evolution: effects of time and temporal scaling. *Science* **222**, 159–161.

Levinton, J. (1988) *Genetics, Paleontology and Macroevolution*. Cambridge University Press, Cambridge.

Simpson, G.G. (1953) *The Major Features of Evolution*. Columbia University Press, New York.

evolutionary species concept *See* SPECIES CONCEPTS.

evolutionary stable strategy *See* EVOLUTIONARILY STABLE STRATEGY.

evolutionary taxonomic classification Evolutionary TAXONOMY is an approach to systematic biology in which taxa are classified in ways that highlight a combination of phylogenetic relationships and particular biological mechanisms or criteria. Evolutionary taxonomists feel that important aspects of PHYLOGENY cannot be reflected in CLASSIFICATIONS based solely on ancestor–descendant relationships. This is because they believe that evolutionary diversification has involved sweeping 'ADAPTIVE RADIATIONS' of different taxa at different times in history, and their subsequent replacement by other groups. If evolutionary history has proceeded in this way, they argue, biological diversity at any point in time will be a combination of species representing current adaptive radiations as well as those representing remnants of past adaptive radiations. The former will be recognizable as groups called 'clades', which include all the descendants of a single ancestral species, whereas the latter will be recognizable as groups called 'grades', which include the surviving descendants of more than one ancestral species. Evolutionary taxonomists refer to both clades and grades as monophyletic groups, and refer to clades as holophyletic groups. The more common practice is to call clades monophyletic groups and to call grades paraphyletic groups. Evolutionary taxonomists feel justified in naming and ranking paraphyletic groups, such as 'fish' or 'reptiles' or 'gymnosperms', as significant evolutionary entities. They consider classifications comprising only clades to be incomplete representations of evolutionary history. Studies of the empirical properties of classifications by advocates of classification based only on clades, however, have shown that classifications containing one or more paraphyletic or polyphyletic groups are logically inconsistent with the sequence of ancestor and descendant relationships among the species representing the group being classified. In such cases the PHYLOGENETIC TREE of the species involved cannot be wholly reconstructed from the classification scheme. *Cf.* CLADISTIC SPECIES CONCEPT; PHYLOGENETIC SYSTEMATICS. [D.R.B.]

evolutionary trends An evolutionary change in a given direction that is maintained for a long time. Evolutionary trends such as a general increase in SIZE (COPE'S RULE) are commonly observed in the FOSSIL RECORD. They can be seen either as a persistent character change within a single LINEAGE or as a change in the average over several related lineages. For example, the mean body size of North American horses has increased over the last 55 million years (from *Hyracotherium* (eohippus) to *Equus*), even though some of the lineages show a decrease in size over time. Evolutionary trends are seen in many characters but their causes are often not clear. COEVOLUTION, SEXUAL SELECTION or INTRASPECIFIC COMPETITION can result in evolutionary trends. Evolutionary trends may also result if a persistent directional change of the environment causes DIRECTIONAL SELECTION on characters. It is difficult to prove whether an evolutionary trend is adaptive. Trends can also be maladaptive, such as the exaggerated increase in antler size of the now extinct Irish elks caused by sexual selection (females prefer males with longer antlers).

Proponents of the theory of PUNCTUATED EQUILIBRIUM suggest that SPECIES SELECTION is another cause of evolutionary trends. For example, if in some lineages body size increases through NATURAL SELECTION whereas in other related lineages smaller individuals are selected for, and if big species are for some reason more likely to go extinct, there would be a trend among all of these lineages that average body size declines over time.

The existence of evolutionary trends does not imply that EVOLUTION is directed. There are no profound trends in evolution that would support vitalistic or teleological explanations of evolutionary change. [W.W.W.]

REFERENCE

McNamara, K.J. (1990) *Evolutionary Trends*. University of Arizona Press, Tucson, Ariz.

e_x In a LIFE TABLE the expectancy of further life of the individuals remaining alive at age or stage *x*. Calculated as T_x/l_x where:

$$T_x = \sum_{x}^{\infty}\left(\frac{l_x + l_{x+1}}{2}\right)$$

[A.J.D.]

***ex situ* conservation** *Ex situ*, or off-site, conservation is the practice of maintaining individuals or genetic material of species at risk at a site removed from their natural habitats. For plants, methods include storing seed (SEED BANKS), storing pollen, growing plants in cultivation (field gene banks), maintaining plant material as undifferentiated tissue (tissue culture or micropropagation), grafting on to stock plants, or even storing genetic material directly such as in gene banks of DNA. *Ex situ* methods for animals include maintaining breeding populations in captivity, which may be distributed among many geographically dispersed institutions, or storing ova and sperm.

Ex situ conservation is intended to complement *in situ*, or on-site, conservation. Maintaining material off-site can provide individuals for reintroduction, population enhancement or habitat restoration. *Ex situ* collections of plants can also provide plant material for CROP breeding. *Ex situ* populations are generally available for research such as behavioural or taxonomic studies. In the event that a species or population becomes extinct in the wild, individuals maintained *ex situ* can become the only genetic resource for that entity.

Any well-designed *ex situ* method will seek to adequately represent genetic diversity present in the wild populations. Users of the *ex situ* resource must be able to generate healthy, reproductively mature individuals from the off-site collection.

To be effective, *ex situ* methods must contribute to the survival of species in natural environments. *Ex situ* conservation is not yet an exact science. Uncritical reliance on *ex situ* methods can lead to a false sense of security. Obtaining and perpetuating over time an adequate degree of genetic diversity off-site, without harming natural populations in the process, are key challenges for the practice of *ex situ* conservation. *See also* CONSERVATION BIOLOGY. [E.O.G. & L.R.M.]

***ex situ* genetic reserves** Conservation of species (and often SUBSPECIES) is increasingly focused on botanic gardens, germplasm centres, microbial culture and gene library collections. Together with zoos, they constitute *ex situ* as opposed to *in situ* conservation (*see* *IN SITU* GENETIC RESERVES), the latter more the provenance of nature reserves or natural habitats protected by conservation legislation. The concept of *EX SITU* CONSERVATION is particularly well established in botanic gardens worldwide, with about 1000 institutions (as a conservative estimate) currently active in *ex situ* plant conservation. In some, the focus is on species of a defined area (e.g. the island flora maintained by the Conservatoire et Jardin Botanique, Réunion) or nation (Center for Plant Conservation, USA; Australian Network for Plant Conservation, Australia). More frequently the focus is on a particular taxa (gymnosperms by the Royal Botanic Gardens, Edinburgh; orchids by the Smithsonian Collection in Washington), specialist use (medicinal plants, Chinese Academy of Medicinal Plants, Beijing), in the form of seeds (dry tropical seed collection of the Royal Botanic Gardens, Kew) or as part of a systematic research collection (Gothenburg Botanical Gardens). Focused field work in recent years has resulted in extensive *ex situ* collections of, for example, the rare and threatened floras of New Zealand and South Africa or of families such as the cacti, cycads and palms. In AGRICULTURE, the International Board for Plant Genetic Resources has coordinated national programmes for the collection and conservation of many CROPS of traditional farming. Reintroduction of plants into the wild from *ex situ* collections has been particularly successful, though often limited by the reduced area of the existing wild habitat. In zoology, *ex situ* conservation of animal species, both domesticated and wild, has been more controversial and in terms of number of species far less effective than with plants. The difficulties of maintaining the adult animals as breeding populations has to an extent been relieved by advances in CRYOPRESERVATION of sperm. Government agency policies of funding the conservation of ENDANGERED SPECIES only in zoos has attracted considerable criticism, particularly in the USA. [G.A.F.H.]

exact compensation Density-dependent MORTALITY, birth or growth rate that leads to a return to population equilibrium in one generation, following DISTURBANCE from equilibrium. Exact COMPENSATION is sometimes said to be the ecological consequence of pure CONTEST COMPETITION. Density-dependent mortality is detected by plotting $k=\log_{10}$(initial density/final density) against $\log_{10}$(initial density); if the slope $b=1$, compensation is exact. *See also* OVERCOMPENSATION; POPULATION REGULATION; UNDERCOMPENSATION. [R.H.S.]

exaptation A word constructed, like 'adaptation', from an adjective plus 'apt', meaning fitted. An exaptation is a character currently selected that arose in some way unrelated to its present function. *See also* PREADAPTATION. [L.M.C.]

exceedance In an environmental context, the violation of environmental protection standards by exceeding authorized (consented) levels. [P.C.]

exclosure An exclosure is any area from which animals are excluded. Exclosures are typically used to prevent the entry of GRAZING mammals (wild or domesticated) into an area that is protected for scientific or ecological management purposes. Experimental exclosures are used in GRASSLAND, SAVANNAH and FOREST research to study effects of grazing or browsing on VEGETATION development and SUCCESSION. Small exclosures may be used to study (or prevent) seed collection by rodents and other small animals. [D.A.F.]

exclusion principle (Gause's principle) *See* COMPETITIVE EXCLUSION.

exclusive economic zone An area of territorial waters over which a coastal country assumes authority for controlling RESOURCES and POLLUTION and in which exploitation by other countries is officially banned. Such zones were included in the United Nations Law of the Sea Convention, completed in 1982, and may extend to distances of 200 miles (322 km) from shore. The first such zone was established by Canada in 1977 (followed shortly thereafter by the USA), largely in response to inadequate enforcement of fishing quotas by the International Commission for the Northwest Atlantic Fisheries. If exclusive economic zones for all coastal nations are considered, 35% of the world's oceans are accounted for. [V.F.]

existing chemicals Inventories of chemicals known to be in existence for commercial purposes prior to arbitrarily specified dates for legal purposes. Under US Toxic Substances Control Act (TSCA) this was December 1979 and amounts to *c.* 70 000 chemicals. For the European Community (EC), the date was September 1981 and amounts to *c.* 100 000 chemicals. This latter list is called the EUROPEAN INVENTORY OF EXISTING COMMERCIAL CHEMICAL SUBSTANCES (EINECS). Chemicals on these LISTS are subject to prioritization followed by RISK ASSESSMENT and possibly RISK MANAGEMENT. Any chemicals not on the lists are described as NEW CHEMICALS. [P.C.]

existing substances *See* EUROPEAN INVENTORY OF EXISTING COMMERCIAL CHEMICAL SUBSTANCES.

Exotic and invasive species

Species of plants and animals that have been transported, with human aid, beyond the limits of their native GEOGRAPHIC RANGES and which continue to expand their distributions by displacing species indigenous to the invaded areas. Other terms (e.g. introduced, immigrant, non-indigenous, ALIEN) are commonly used to describe such species, which may have been transported intentionally or inadvertently. BIOLOGICAL INVASION is not an entirely new phenomenon; it has occurred through the ages, but in recent decades has been accelerated by orders of magnitude. Invasive species are often competitively favoured during RANGE expansion by the DISTURBANCE caused by human activities but are frequently able to persist and thrive in their newly occupied habitats, often in natural disturbance niches of these habitats or in artificial ECOSYSTEMS. The immensity of the ecological and economic damage caused by invasive species can probably best be grasped through examples.

Golden apple snail in Asian rice systems

Native to a small wetland area on the Brazil–Paraguay border, the golden apple snail (*Pomacea canaliculata*) gradually spread and became noted as a PEST throughout South America with the large-scale cultivation of rice. Nevertheless, it was purposely introduced into Taiwan about 1980 with the intent of agricultural production for local food supply and export (*escargot*). Promoted as the 'golden miracle snail' because of its rapid reproduction and growth, it was quickly spread to all countries of south-east Asia from the Philippines, Indonesia and Thailand north to China and Japan. After this pest was widely established, it was discovered to be worthless as a human food source (partly because it is a carrier of human disease) and to be a major consumer of rice. Annual losses to south-east Asian rice growers are now measured in billions of dollars, with no remedy in sight.

Invasive grasses worldwide

Grasses have been moved actively by humans for agricultural purposes and often compete effectively with native BIOTA. Large areas of former tropical FOREST have been displaced by non-native grasses through their rapid ESTABLISHMENT after DEFORESTATION. Once established, invasive grasses fuel fire and respond to fire with rapid growth, leading to a grass–fire cycle, which continues to erode tropical forests and prevents forest re-establishment. D'Antonio and

Vitousek (1992) have recognized that grasses are a set of invasive species that may have already become sufficiently widespread and locally pervasive to alter regional and even global aspects of ecosystem function, through fire effects, nutrient loss, altered local MICROCLIMATE and prevention of SUCCESSION.

Brown tree snake

The brown tree snake (*Boiga irregularis*) is native to the Solomon Islands, Papua New Guinea and northern Australia, where it is not particularly abundant. It became established on the island of Guam in the eastern Pacific about 1950 and eventually has attained population densities of 4000–12 000 km^{-2}, feeding on birds, rats, shrews and lizards. Nine of the 11 native bird species present on Guam in 1945 have been eliminated by the snake. They also cause a host of economic and quality-of-life problems on Guam, such as damage to electrical utilities and frequent snake bites. The high densities of snakes on Guam and the snake's nocturnal habits make the probability of stowaways in air and ship cargo very high. The brown tree snake has been noted as a serious and imminent potential invader of Hawaiian and other Pacific islands. Seven brown tree snakes were found near airport runways in Hawaii between 1981 and 1994.

Miconia calvescens

Miconia calvescens, in the melastome family (Melastomataceae), is a tree up to 12 m tall, native to the neotropics, that deserves its reputation as the most invasive and damaging of exotic plant species to wet forests of Pacific islands. In its native habitat in Central and South America, it is a successional tree in patches of disturbed middle elevation RAIN FOREST. A tree with very large (to over 1 m in length) and strikingly attractive purple and green leaves, *M. calvescens* was introduced into two botanical gardens in Tahiti in 1937 and has now become widespread and overwhelmingly dominant on that island; it is in the process of crowding out native plant species to the point that many are on the brink of extinction. It has also spread to at least three other nearby islands of French Polynesia. Meanwhile, it was inadvertently introduced to the Hawaiian islands as an ornamental in the 1960s. In Hawaii, it ultimately poses a threat to all habitats receiving 2000 mm or more annual precipitation. As of 1995, *M. calvescens* had been found on four Hawaiian islands: Hawaii, Maui, Oahu and Kauai.

Zebra mussel, a mollusc

The zebra mussel (*Dreissena polymorpha*) was apparently restricted to the drainage basins of the Aral, Black and Caspian seas until the early 1800s, when construction of canals and increased commerce allowed it to invade the river systems of Europe. In Europe, it attained densities of 10 000 m^{-2} on surfaces to which it attached and quickly became recognized as a destructive agent, fouling water-intake pipes, hydroelectric installations and navigation locks. The zebra mussel was first recorded in North America in 1988 in Lake St Clair, a lake connecting Lake Huron and Lake Erie. It is believed to have reached there through release of ballast water by a ship from Europe. Within 3 years of its discovery in Lake St Clair, the zebra mussel had entered most of the major drainage basins in eastern North America, where COMPETITION is causing native benthic (attached) fauna of stream bottoms to be overwhelmed. Ecological damage aside, economic damage in North America is immense. In the Great Lakes and other bodies of water, zebra mussels adhere to water pipes, clogging flow and increasing SEDIMENTATION and erosion. Estimated annual costs to the electrical power industry are in the neighbourhood of US$300 million (1991 dollars).

These and other examples have an important factor in common: invasive species are normally not particularly abundant or aggressive in their native habitats. They exhibit explosions of populations, pest status and ecological and economic damage when exposed to new ecological conditions, which may at least partially involve release from predation or parasitism. As a result, it is not easy to predict which species may rise out of their native habitat to become pests in other parts of the world. However, once a species has successfully invaded one part of the world, its invasion of other climatically similar areas is likely.

With ever-increasing global trade, global CLIMATE CHANGE and changing land-use patterns, society is faced with the fact that invasive species can be expected to proliferate in the near future more than ever before.

How can invasions of exotic species be prevented? Most countries have quarantine regulations to restrict disease and agricultural pests. National quarantines need to be expanded and coordinated internationally, based on an improved information base. Better education of citizens about the nature and magnitude of the invasive-species problem is a prerequisite for effective action at local, national and international levels. Although fully effective prevention of further invasions is undoubtedly impossible, much can be done to reduce the magnitude of the problem. [L.L.L.]

REFERENCES

D'Antonio, C.M. & Vitousek, P.M. (1992) Biological invasions by exotic grasses, the grass–fire cycle, and global change. *Annual Review of Ecology and Systematics* **23**, 63–87.

Meyer, J.-Y. & Florence, J. (1996) Tahiti's native flora endangered by the invasion of *Miconia calvescens* DC. (Melastomataceae). *Journal of Biogeography* **23**, 775–781.

Rodda, G.H., Fritts, T.H. & Chiszar, D. (1997) The disappearance of Guam's wildlife. *BioScience* **47**, 565–573. [Includes references on invasion of the brown tree snake and other snakes and their impacts.]

Vitousek, P.M., D'Antonio, C.M., Loope, L.L., Rejmanek, M. & Westbrooks, R. (1997) Introduced species: a significant component of human-caused global change. *New Zealand Journal of Ecology* **21**, 1–16. [Includes many examples of, and references for, invasions, including golden apple snail, zebra mussel, grasses, etc.]

Experimental design

When an ecologist carries out an experiment, either in the laboratory or the field, five activities are involved:

1 formulating the hypothesis to be tested;
2 designing the experiment;
3 carrying out the experiment;
4 analysing the results, using graphical and statistical methods;
5 interpreting the analysed results.

Only **2** will be discussed in this entry, although the others are also important. In one sense **4** and **5** are less important, since if **1–3** are 'good' then **4** and **5** can always be redone. If **1–3** are flawed, then no amount of sophisticated statistical analysis or clever interpretation will save the experiment; **3** will frequently involve technical problems of measurement and the practical competence of the experimenter.

Experiments are susceptible to a number of confounding factors or sources of confusion (Hurlbert 1984). These confounding factors and the features of good experimental design that can help to eliminate or reduce them are outlined below.

Imagine that we have a woodland (shown as a rectangle in the diagram below) in which an ecologist wishes to test three different types of pitfall trap for collecting ground-dwelling insects. Pitfall traps are simply small containers sunk into the ground into which beetles will fall. We will call the different types of traps A, B and C. Below are shown two different experimental designs.

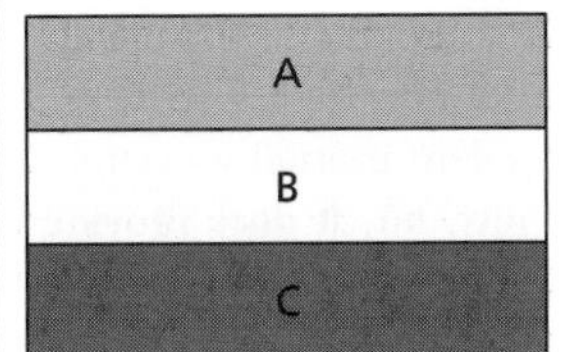

B	C	A
A	A	B
C	C	B
B	A	C
A	B	A
B	C	C

The design on the left is simple and traps of all three designs are present in the wood. However, it is a bad design and will not allow the ecologist to reach any conclusion about the characteristics of each trap. The problem is that in addition to the design (A, B or C) affecting the numbers and types of insects collected, the area of the wood may affect numbers and types of insect also. That is, trap type A, for example, may collect a greater diversity of insects not because of its design but because of its location. The design on the right minimizes this potential source of confusion using two elements of good design, replication and randomization. We now have six replicates of each trap design, randomly allocated to different areas of the wood. Location effects may still be present, but they will be randomly spread around the three designs, making comparison between them possible. A related element of good design is interspersion: the replicates of each experimental treatment (design of traps in the above example) must be mixed up. With a reasonable number of replicates (e.g. six in the above wood) it is usually sufficient to assign treatments at random in order to achieve adequate interspersion. (If we divide the above wood into three areas vertically, all three trap designs are present in each area of the wood. If we divide the wood into three areas horizontally, all three trap designs are present in each area of the wood.) However, in some ecological situations there are limits to the numbers of replicates available. For example, an ecologist wishes to introduce fish into ponds in order to examine the effect of predation on the BIODIVERSITY of freshwater invertebrates. Only four ponds are available for the experiment, located along a valley bottom. If the ecologist simply chooses the ponds for fish introduction (shaded in the diagram below) on a random basis there is a very high PROBABILITY (50%) of obtaining an arrangement like the one below.

Interspersion is poor in this design and therefore the traditional method of assigning treatments at random has failed. Because all the replicates of a treatment are not adequately interspersed, they may share confounding factors. In effect, we have returned to the left-hand wood in the pitfall trap example. We think we have replication but in fact we don't. This has been called PSEUDOREPLICATION (Hurlbert 1984). Adequate interspersion is the objective of the exercise. Randomization is just the method of obtaining this objective. Ecologists must always be vigilant to ensure, when planning experiments, that adequate interspersion has been achieved. Highly segregated layouts should be rejected and treatments reassigned at random until a satisfactory degree of interspersion is achieved. With the pond example above it would probably be better to assign the treatment to the extreme left-hand pond at random and then alternate treatments along the series of ponds. *See also* STATISTICS. [B.S.]

REFERENCE

Hurlbert, S.H. (1984) Pseudoreplication and the design of ecological field experiments. *Ecological Monographs* **54**, 187–211.

experimental ecology Ecology carried out according to the experimental ideal. Possible explanations of the structure of ecological systems and the processes within them are expressed as clearly defined hypotheses that are then tested in controlled circumstances. The hypotheses may be complex and expressed as models that can be explored using analytical and/or simulation techniques to expose specific predictions that can be tested. The testing can be carried out in the laboratory or the field. Tests cannot prove hypotheses to be true beyond doubt because other circumstances might turn up that lead to contrary conclusions. Once disproved (refuted), though, hypotheses can be rejected decisively. Hence, supporting results should be interpreted as not refuting the hypothesis under consideration and the more experiments that do this, the more confidence that can be put in the hypothesis. Sometimes 'experimental ecology' is simply used to describe studies that involve complex instrumentation. Very often these will be used in hypothesis testing, but sometimes they will be used to make observations without necessarily any reference to hypotheses. *See also* ECOLOGICAL EXPERIMENTS; EXPERIMENTAL DESIGN. [P.C.]

REFERENCE

Rigler, F.H. & Peters, R.H. (1995) *Science and Limnology*. Ecology Institute Nordbünte, Germany.

experimental lake area (ELA) Area containing a number of lake basins in north-western Ontario, Canada that have been subjected to comprehensive MANIPULATION to explore ECOSYSTEM responses to EUTROPHICATION, ACIDIFICATION and POLLUTION. For example, one interesting study involved the gradual acidification of Lake 223 from pH 6.8 to 5.0 between 1976 and 1983. The manipulation was carried out after a 2-year period of baseline studies. Primary production and phytoplankton diversity remained unchanged, but several species replacements occurred. Some fish populations, including young lake trout, increased slightly in intermediate stages of the experiment. However, recruitment failed at pH levels near 5.6 due to appreciable changes in benthic production. [P.C.]

REFERENCE

Schindler, D.W., Mills, K.H., Malley, D.F. *et al.* (1985) Long-term ecosystem stress: the effects of years of experimental acidification on a small lake. *Science* **228**, 1395–1401.

experimental manipulation and life-history evolution *See* TRADE-OFF.

expert judgement Basing decisions about environmental protection and/or conservation measures on the views of persons whose qualifications and/or experience enable them to assess and interpret complex evidence. An expert witness is a person testifying in legal proceedings who, on the basis of qualification and/or experience, is allowed to use hearsay, assumptions and hypothetical questions; cf. a lay witness who is allowed to testify only on the basis of facts within their personal knowledge. [P.C.]

expert system Software involving methods of solving problems and carrying out tasks by means of an information base containing rules and data based on human experience and previously encountered problems. [P.C.]

expert witness *See* EXPERT JUDGEMENT.

explanatory variable The *x*-axis in a regression analysis or ANALYSIS OF COVARIANCE, or the factors in an ANALYSIS OF VARIANCE. Combinations of continuous explanatory variables and categorical explanatory variables (factors) are used in GENERALIZED LINEAR MODELS to partition variation in the RESPONSE VARIABLE, and hence to form a view on the relative importance of the different explanatory variables in causing variation in the response variable. Formerly known as independent variables. *See also* EQUATIONS; STATISTICS. [M.J.C.]

exploitation competition COMPETITION that does not involve direct interaction between individuals, i.e. competition is passive because there is no INTERFERENCE. Under exploitation competition, individual FITNESS is determined by the level of different RESOURCES that remain in the zone where competition takes place after resources have been exploited by others. Examples of exploitation competition include competition by plants for mineral or water resources in the soil or blowfly larvae competing for flesh. If more of a limiting resource is supplied to a zone of competition, all individuals in that zone benefit and increase fitness. *See also* POPULATION REGULATION; SCRAMBLE COMPETITION. [R.H.S.]

REFERENCE

Birch, L.C. (1957) The meanings of competition. *American Naturalist* **91**, 5–18.

explosive breeding A phenomenon most commonly found in anurans (frogs and toads) and some insects in which all mating occurs during a very short breeding season, of the order of 1 or 2 days. Research on explosively breeding bullfrogs has demonstrated that variance in male mating success can be strongly constrained by breeding synchrony. Synchrony may be promoted because it reduces the risk of predation to EGGS and young by briefly swamping a resident population of predators with a glut of food. [J.C.D.]

exponential distribution A continuous PROBABILITY DISTRIBUTION that is commonly used to describe a variety of ecological processes, such as MORTALITY RATES and DISPERSAL distributions. In statistical analyses, GENERALIZED LINEAR MODELS rely on the exponential distribution or the family of distributions related to it. The exponential distribution arises when the probability of some discrete event occurring at any moment in time is constant. For example if, at any moment in time, the probability of an individual in a POPULATION dying is p (i.e. the HAZARD FUNCTION is constant), then over a small time interval, a proportion p of the population will be lost; in the next small time interval a further proportion p will be lost, and so on. The proportion of the population remaining, or the probability of observing a single mortality at a given time, therefore declines exponentially over time, hence the name of the distribution. It should be noted that this distribution does not describe processes where, for example age-dependent mortality occurs, or where individuals vary in dispersal ability, but it does provide a simple NULL MODEL. [R.P.F.]

exponential growth Unlimited population growth or population growth with no COMPETITION for RESOURCES. Exponential growth is only possible in the short term while conditions for population increase are favourable and effectively constant, for example during the log phase of bacterial population growth. Characterized by a constant average rate of increase per individual or POPULATION GROWTH RATE r (also known as the INTRINSIC RATE OF INCREASE). Exponential growth is used as the NULL HYPOTHESIS when testing statistically for DENSITY DEPENDENCE. In modular organisms, exponential growth is characterized by a straight-line relationship when the logarithm of number of individuals is plotted against time. In clonal organisms growing exponentially, the logarithm of biomass increases linearly with time. [R.H.S.]

export of pollution Provocative phrase sometimes used to describe transport of chemicals and radioactive wastes from one nation to another. [P.C.]

exported production The fraction of primary PRODUCTION that is transported from where it was produced to somewhere else; very often refers to transport from the EUPHOTIC ZONE to the DEEP SEA. Quantifying exported production is relevant to understanding the controls on BIOMASS, production and fate of PHYTOPLANKTON in the OCEAN. For example, it is assumed that the mass of particulate NITROGEN (i.e. phytoplankton) that can be exported from the surface layer cannot, on average, exceed the mass produced as new production (i.e. production based on newly supplied NUTRIENTS rather than recycled ones) without a decrease in average annual nitrogen concentrations. [V.F.]

exposure The contact concentration and/or intensity of exposure between a chemical or physical agent and a biological system. For chemicals, depends upon FATE. [P.C.]

extended phenotypic effects *See* MANIPULATION.

external fertilization *See* FERTILIZATION.

extinction coefficient The ratio between the intensity of LIGHT at a given water depth and the intensity at the surface of a water body. [V.F.]

Extinction crisis, current global

A total of about 490 animal extinctions and 580 plant extinctions have been recorded globally since 1600. The number of extinctions has increased dramatically in the last 100 years. There has been a preponderance of extinctions on islands compared to continents. For example, 75% of recorded animal extinctions since 1600 have been of species inhabiting islands, even though islands support a small fraction of the number of animal species found on continents. The apparent decline in extinctions in the last 30 years to 1990 is at least partly due to the time-lag between extinction events and their detection and recording. A number of species are likely to have become extinct recently without yet having been recorded as such. It is also possible that conservation efforts over the last 30 years have slowed the rate of extinctions.

Data such as those in Fig. E15 probably underestimate the true rate of extinction. The numbers ignore the possibility of the loss of BIODIVERSITY among other taxa including fungi, bacteria and viruses. Only named species are recorded as extinct; species are presumed to be extinct when a specific search has not located them, when they have not been recorded for several decades or when expert opinion suggests they have been eliminated. Most taxa have not been described or named and there is no reason to expect that undescribed groups are more immune to extinction than groups that are better described. The vast majority of even the described species are not monitored and species may be locally or globally eliminated without our knowledge of the event. The process of certification of extinction in most countries ensures that there is a considerable time-lag between the extinction of a species and its certification. These sources of uncertainty all contribute to the expectation that there have been many more extinctions in the recent past than have been documented by science.

Table E1 provides current estimates of the proportion of species in various taxonomic groups that are known to have become extinct. It suggests that extinction rates have been highest among birds and mammals, intermediate in reptiles, fish, amphibians, molluscs and plants, and low elsewhere. Unfortunately, it more accurately reflects the interest and taxonomic effort expended on different groups over the last several decades. The observed rates in Table E1 are minima and the true rates are likely to be substantially higher. There are likely to have been many more extinctions than have been recorded, especially amongst the most numerous groups such as the insects.

Even if observed extinction rates are assumed to be unbiased, they suggest that extinction rates are substantially higher than those observed globally over the last 50 million years. For example, there are about 4300 species of mammals on Earth. If the extinction rate of mammals was approximately the same as the background rate of extinction in geological time (between one and five species globally per year), if extinctions occur with more or less equal frequency among different taxa and if we assume, conservatively, that there are a total of 10 million extant species, we would expect there to have been no more than about one mammal extinction within the last 400 years, yet 69 have been documented. Comparing observed and expected current rates in three groups, it is evident that all three are substantially above background (Table E2).

There have been several attempts to estimate future extinction rates. The lines of evidence include extrapolation of past trends,

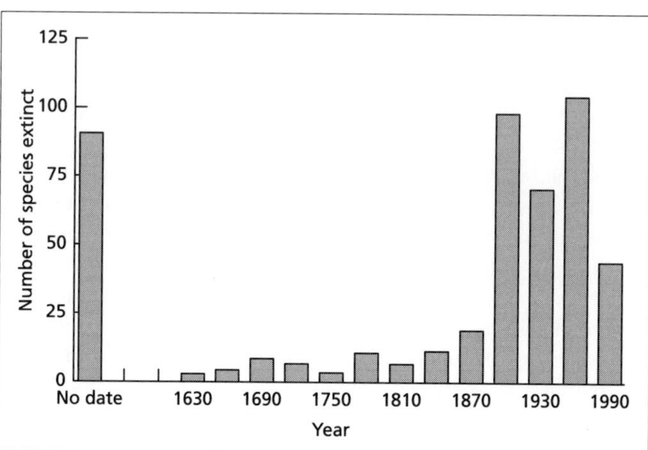

Fig. E15 The number of recorded animal extinctions in recent history. The values represent the number of extinctions recorded within the 30-year period up to the date labelling the class. (After Jenkins, M. (1992) Species extinction. In: *Global Diversity: Status of the Earth's Living Resources* (eds World Conservation Monitoring Centre), pp. 192–233. Chapman & Hall, London.)

Table E1 Species in major plant and animal taxa that are known to have become extinct since 1600 or are currently threatened with extinction. (After May, R.M., Lawton, J.H. & Stork, N.E. (1995) Assessing extinction rates. In: *Extinction Rates* (eds J.H. Lawton & R.M. May), pp. 1–24. Oxford University Press, Oxford; by permission of Oxford University Press.)

Taxon	No. of extinct species	No. of threatened species	Total no. of recorded species (×1000)	Per cent extinct
All animals	485	3565	1400	0.04
Molluscs	191	354	100	0.2
Crustaceans	4	126	40	0.01
Insects	61	873	1000	0.006
Vertebrates	229	2212	47	0.5
Fish	29	452	24	0.1
Amphibians	2	59	3	0.1
Reptiles	23	167	6	0.4
Birds	116	1029	9.5	1
Mammals	59	505	4.5	1
All plants	584	22137	240	0.2
Gymnosperms	2	242	0.8	0.3
Dicotyledons	120	17474	190	0.06
Monocotyledons	462	4421	52	0.9

Table E2 Observed number of extinctions globally in the last 400 years and predicted number of extinctions in the same period assuming current rates are equal to background rates in geological time.

Taxon	No. of species	Expected no. of extinctions if background rate is 1 per year	Expected no. of extinctions if background rate is 5 per year	Observed no. of extinctions
Mammals	4300	<1	<1	69
Birds	9600	<1	2	115
Molluscs	200000	8	40	191

assuming exponential increases in human impacts, extrapolations for land clearance and FOREST loss combined with species–area curves, and estimates of future losses of habitat in species-rich TROPICAL RAIN FORESTS and other areas with large numbers of ENDEMIC SPECIES. The results of these calculations provide estimates of global loss of species per decade of between 1 and 11%, with most estimates around 6%.

Estimates are uncertain for numerous reasons: we have only approximate indications of the total number of species, even within relatively well-known taxa; the POPULATION SIZE of most species is unknown; we have very little information on the geographical distribution of species, even for described species; information on TAXONOMY, distribution and population size is scarcer for species in RAIN FOREST than in most other regions; extrapolations from extinction rates in mammals and birds (for example) to other taxa are tenuous; extrapolations of rates of extinction from a forest in one part of the world to another are tenuous, even within the same forest type; we know very little about the capability of PRIMARY FOREST species to adapt to SECONDARY FOREST; we have no understanding of the impact that conservation measures have had on extinction rates; we are usually unable to confirm that extinction has taken place until long after the event; and for predictive models that assume equilibrium numbers of species (such as species–area curves) we do not know how long it takes for species assemblages to reach a new equilibrium (if at all) after an impact such as land clearance.

While it is difficult to estimate the absolute values of current and future extinction rates, there is little doubt that we are in the midst of a MASS EXTINCTION event of a magnitude matched only by five other such events in geological time. The current mass extinction event is unique because it is the product of the impact of a single species rather than the result of environmental change. *See also* EXTINCTION IN THE FOSSIL RECORD. [M.A.B.]

REFERENCES

Hammond, P. (1992) Species inventory. In: *Global Diversity: Status of the Earth's Living Resources* (eds World Conservation Monitoring Centre), pp. 17–39. Chapman & Hall, London.

Jenkins, M. (1992) Species extinction. In: *Global Diversity: Status of the Earth's Living Resources* (eds World Conservation Monitoring Centre), pp. 192–233. Chapman & Hall, London.

May, R.M., Lawton, J.H. & Stork, N.E. (1995) Assessing extinction rates. In: *Extinction Rates* (eds J.H. Lawton & R.M. May), pp. 1–24. Oxford University Press, Oxford.

Pimm, S.L., Russell, G.J., Gittleman, J.L. & Brooks, T.M. (1995) The future of biodiversity. *Science* **269**, 347–350.

Reid, W.V. (1992) How many species will there be? In: *Tropical Deforestation and Species Extinction* (eds T.C. Whitmore & J.A. Sayer), pp. 55–73. Chapman & Hall, London.

extinction curve A graphical representation of the relationship between extinction rate (species loss per unit time interval) and either time or the number of species on an island. R.H. MacArthur and E.O. Wilson's theory of ISLAND BIOGEOGRAPHY depicts extinction rates that increase monotonically as the result of a decrease in average population size in more speciose communities and therefore there is a greater probability of extinction. Extinction curves on small isolated islands rise faster than on large islands near a mainland because of their smaller population sizes and the 'rescue effect' through IMMIGRATION. The intersection of the extinction curve and the IMMIGRATION CURVE predicts the equilibrium SPECIES RICHNESS of an island. [N.M. & S.G.C.]

Extinction in the fossil record

The period since the Early Cambrian 600 million years ago marks the beginning of the diversification of multicellular organisms. Terrestrial diversification did not commence until about 450 million years ago. Most analyses of extinction in the fossil record rely on changes in the geological ranges of genera and families. The fossil record at the species level is very incomplete. Time-series comprised of additions and losses of genera and families are most useful for identifying periods of unusual extinction intensity and for estimating overall losses of BIOTA. The fossil record for marine invertebrates with shells is the most complete and reliable for examining extinction patterns, because marine environments provide continuous and extensive SEDIMENTATION and shelled invertebrates are very abundant and well studied. Relatively comprehensive data exist for some terrestrial vertebrates, insects and plant groups. Fossils come predominantly from areas where sedimentary material accumulates (swamps, river deltas and lake beds). Animals and plants from other environments and animals lacking relatively large, heavy bones are not well represented.

Most species that have ever existed are now extinct. Today's biotic diversity constitutes perhaps between 1 and 10% of the total that has ever been. Many species were not eliminated in the sense that we think of extinction today. Rather, NATURAL SELECTION and MUTATION transformed many species, giving the appearance of losses and gains in the fossil record. Others underwent change following geographic subdivision of an ancestral population or following hybridization between different species after DISTURBANCE or DISPERSAL.

The average lifetime of species in most taxa is a few million years. For example, invertebrates average 11 million years, marine invertebrates average 5–10 million years and mammals average 1 million years. Most species within a major taxonomic group tend to be short-lived relative to the average lifetime for species in the TAXON, while a few species persist for much longer.

Extinctions in the fossil record are generally typified as falling within one of two regimes: background periods and MASS EXTINCTION events. Mass extinctions are substantial BIODIVERSITY losses that are global in extent, taxonomically broad and rapid relative to the average duration of the taxa involved. The background rate of extinction is the average rate of loss of species globally in geological time, outside the periods of mass extinction. If there are 10 million species currently in existence and the average lifespan of a species is 5–10 million years, then very roughly the background extinction rate is of the order of one or two species per year globally. The background rate over the last 600 million years has been punctuated by five mass extinction events, each associated with relatively severe environmental change. Despite the severity and extent of mass extinction events, it is likely that more than 90% of all extinctions occurred in background times.

Even during background times, the rate of extinction has not been uniform. The durations of species and genera within higher taxa vary by more than an order of magnitude. The rate is highly variable among continents and at different times in the geological past. Because of this variation, global averages are not particularly informative. Variation in the background rate of extinction is related to GEOGRAPHIC RANGE and POPULATION SIZE. Many groups have undergone periods of unusually high extinction rates, after which ADAPTIVE RADIATION has increased global diversity. [M.A.B.]

REFERENCES

Jablonski, D. (1995) Extinctions in the fossil record. In: *Extinction Rates* (eds J.H. Lawton & R.M. May), pp. 25–44. Oxford University Press, Oxford.

Jenkins, M. (1992) Species extinction. In: *Global Diversity: Status of the Earth's Living Resources* (eds World Conservation Monitoring Centre), pp. 192–233. Chapman & Hall, London.

Sepkowski, J.J. (1993) Ten years in the library: new data confirm paleontological patterns. *Paleobiology* **19**, 43–51.

extinction modelling Extinction modelling is the process of building conceptual, statistical and MATHEMATICAL MODELS to summarize our understanding of the dynamics of extinction and to predict patterns of extinction.

At the level of communities within regions, the theory of ISLAND BIOGEOGRAPHY is sometimes used to predict numbers of extinctions. At equilibrium between IMMIGRATION and extinction, the number of species in a HABITAT patch is estimated by $S = cA^z$ where A is area. The proportional change in the number of species, ΔS, is given by:

$$\Delta S = 1 - (S_1/S_0) = 1 - (A_1/A_0)^z$$

where ΔS is a function of the remaining habitat. Normally, the value of z is taken to be about 0.25 with a range of $0.15 < z < 0.35$.

The likelihood of extinction of species has been estimated by developing statistical models of local extinction (extirpation) based on species characteristics. Some factors important to mammals and birds include POPULATION SIZE and range, body size, social structure and the ability to use disturbed habitat.

Extinction risks for individual species may be estimated by developing dynamic mathematical models that include stochastic elements representing sources of UNCERTAINTY, a process known as population viability analysis. It may be used to explore how management activities influence the systematic decline or PROBABILITY of extinction of a species. The probability of extinction is usually estimated by MONTE CARLO SIMULATION of the factors that govern the chances of persistence of the species including demographic parameters (survivorships and fecundities), density-dependent REGULATION, environmental variation, spatial correlation of environmental variation, DISPERSAL, changes in the status of habitat and the impact of infrequent, catastrophic events such as fire. Appropriate model structure depends on consideration of the ecology of the species, data availability and requirements, and an understanding of management needs. *See also* ENDANGERED SPECIES; EXTINCTION CRISIS, CURRENT GLOBAL; EXTINCTION IN THE FOSSIL RECORD; MASS EXTINCTION. [M.A.B.]

REFERENCES

Boyce, M.S. (1992) Population viability analysis. *Annual Review of Ecology and Systematics* **23**, 481–506.

Burgman, M.A., Ferson, S. & Akcakaya, H.R. (1993) *Risk Assessment in Conservation Biology*. Chapman & Hall, London.

Possingham, H.P., Lindenmayer, D.B. & Norton T.W. (1993) A framework for the improved management of threatened species based on population viability analysis. *Pacific Conservation Biology* **1**, 39–45.

Simberloff, D. (1988) The contribution of population and community biology to conservation science. *Annual Review of Ecology and Systematics* **19**, 473–511.

extinction rate *See* EXTINCTION CURVE.

extractive reserves Extractive reserves (ERs) are designated conservation areas in which natural resource extraction is carried out complementary to the objective of conserving biological diversity and the natural resource base. Today, many parks, reserves and other established protected areas worldwide include some form of resource extraction and consumption.

ERs are of many types and generally are not designated simply by that name. In the international system of protected areas monitored by the World Conservation Union's World Commission on Protected Areas, ERs are included in category VI of the revised international classification: the 'Managed Resource Protected Area'. Two important objectives for category VI designation are to protect and maintain the BIODIVERSITY and other natural values of the area in the long term and to promote sound management practices for sustainable production purposes.

Extraction of products can include anything from the natural resource base of the reserve: wildlife for meat or other products; plants for food and medicines; wood used as lumber for building, pulp for paper or as raw material for other products; many non-wood forest products such as flowers, fruits, nuts and leaves; or water for domestic and industrial use.

There are many examples of ERs in nearly all countries worldwide. Brazil has ERs designated primarily for RUBBER production and to maintain employment for rubber tappers and their families. In Ecuador, indigenous people have developed a small jewellery industry by HARVESTING tagua palm nuts, and in Guatemala the xate palm is being harvested for small-scale commercial development. In Papua New Guinea, the Tonda Wildlife Management Area is managed to protect biodiversity and to provide local people with a sustainable harvest of animals. The Rio Bravo Conservation Area in

Belize is managed for mahogany lumber production and other forest products such as chicle, in addition to the primary management objective of preserving its wildlife, extensive tropical forest and wetlands.

However, the long-term usefulness and effectiveness of ERs are not yet proven. There are few examples today where we can say that the natural resource base and biodiversity within the reserve are being maintained and managed on a demonstrated sustainable basis. Most ERs, as designated protected areas, are too new to ensure that they are working as planned.

Despite this caution, conservationists agree that ERs have great potential to effectively conserve biodiversity while generating revenues from useful goods and services. However, it is not easy to properly manage the exploitation of RESOURCES in ERs. It is particularly difficult to avoid overharvesting resources, especially wood for lumber or pulp. Except for timber, oil or minerals in ERs, it is likely that there will be few natural resources which will have high commercial value and be significant in national or international markets. Relying on small-scale harvesting of wood and other products by local people, although an attractive prospect, in most cases is not the answer to creating a permanent better standard of living for those people, especially if many of them wish to move into the cities. There is ample opportunity, however, for innovative resource management systems to be developed within ERs, including a wide variety of agroforestry techniques. [F.W.B.]

REFERENCES

Redford, K.H. & Padoch, C. (eds) (1992) *Conservation of Neotropical Forests: Working from Traditional Resource Use.* Columbia University Press, New York.

Smith, N.J.H., Williams, J.T., Plucknett, D.L. & Talbot, J. (1992) *Tropical Forests and Their Crops.* Cornell University Press, Comstock Publishing Associates, Ithaca, N.Y.

extra-pair copulation (EPC) A copulation between a mated individual and an individual of the opposite sex to which it is not socially bonded.

Copulations outside a social bond, as in socially monogamous or polygynous animals, sometimes involve the majority of all mated individuals. Extra-pair copulations are forced when one or more males behave aggressively and sometimes directly harmfully towards female non-mates. Unforced extra-pair copulations occur when females solicit copulations from males other than their social mate. Extra-pair copulations often peak in frequency during the time of the reproductive cycle, when FERTILIZATIONS take place and male non-mates often have considerable knowledge of the fertility status of neighbouring females. Proximate cues used by males for assessment of the fertility status of neighbouring females include elevated body mass of fertile females, intense MATE-GUARDING by the male mate and intense territorial display by mated males having fertile females. Mated males of asynchronously reproducing animals mainly pursue extra-pair copulations when their own mates are no longer fertile and males therefore are not constrained in their behaviour by intense guarding of their mate. Extra-pair copulations frequently result in sperm transfer and extra-pair PATERNITY. Males engaging in extra-pair copulations mainly benefit from siring an increased NUMBER OF OFFSPRING compared with sexually monogamous males. Females may benefit by receiving viable sperm if their social mate is infertile, by having their offspring fathered by males of superior genetic quality or by avoidance of INFANTICIDE, if extra-pair copulations prevent males from becoming infanticidal. *See also* BEHAVIOURAL ECOLOGY; MATING SYSTEMS; SPERM COMPETITION. [A.P.M.]

extra-pair mating *See* PATERNITY.

extra-pair paternity *See* PATERNITY.

extrinsic agents *See* DEFENCE MECHANISMS.

extrinsic mortality Mortality caused by external factors such as predation, accident (e.g. fire, drowning), famine or infection. The term is used to distinguish these kinds of death from those associated with INTRINSIC MORTALITY. [T.B.L.K.]

exudates Dissolved metabolites (i.e. AMINO ACIDS, short-chain ACIDS, glycerol, CARBOHYDRATES and polysaccharides) released by PHYTOPLANKTON. Amounts lost through exudation range from 15 to 50% of the CARBON DIOXIDE (CO_2) photosynthetically fixed. The highest loss rates occur at high light intensities and at the end of phytoplankton BLOOMS. Such exudates may stimulate bacterial growth. [V.F.]

Exxon Valdez *See* ECOLOGICAL/ENVIRONMENTAL DISASTERS.

F

F-statistics POPULATION GENETICS parameters and statistical tools used to study the partitioning of VARIANCE of a given TRAIT (frequency of an allele at a locus or value of a quantitative character) into hierarchical effects. For instance, F-statistics are often used to describe the variance of allele frequencies among populations, among subpopulations within populations, among individuals within subpopulations, and among ALLELES within individuals. The F parameters, introduced by Sewall Wright as 'correlation coefficients between uniting gametes', are usually calculated as deficits of heterozygotes relative to the expected frequency of heterozygotes at a given scale. For instance, the parameter F_{is} (where 'is' refers to 'individuals within subdivision') describes the mating system at the lowest spatial level of the POPULATION. In an infinitely large, unsubdivided population, with a mixed mating system (panmixia with a probability t and selfing with probability $1-t$), it can be shown that, at equilibrium, the deficit of heterozygotes F_{is} is equal to $(1-t)/(1+t)$. This parameter thus varies from 0 (PANMICTIC population) to 1 (selfing species).

The parameter F_{st} ('st' for subdivision within total) is equal to the deficit of heterozygotes due to the subdivision of the total population into subpopulations (Wahlund effect). It is also equal to the variance of ALLELE FREQUENCY among subpopulations, divided by the maximal variance, that would be observed if each subpopulation were fixed for one allele. It is a function of many other parameters, such as the GENE FLOW among subpopulations, the local effective size of subpopulations, the MUTATION rate, the time since the populations were founded by a common ancestor population, etc. Very often, the parameter F_{st} is used to calculate the average number of individuals exchanged among subpopulations, using the relation $F_{st}=1/(1+4Nm)$, where m is the MIGRATION rate and N the local EFFECTIVE POPULATION SIZE. This relationship is valid only under highly specific conditions, and should be applied with caution.

The parameter F_{it} ('it' for individuals in total) is the overall deficit of heterozygotes, due to both population subdivision and local non-panmictic mating system. It is related to F_{is} and F_{st} by the following equation: $1-F_{it}=(1-F_{is})(1-F_{st})$. The F coefficients can also be described as inbreeding coefficients, relative to the spatial level immediately above. They are then designated by f_{is}, f_{st} and f_{it}. The parameter f_{is} is then equal to $(f_0-f_1)/(1-f_1)$, where f_0 is the probability of identity of two genes drawn from the same individual, and f_1 is the probability of identity of two genes drawn at random from the gametic pool of the population. The parameter f_{st} is equal to $(f_1-f_2)/(1-f_2)$, where f_2 is the probability that two genes drawn at random from the whole population gametic pool are identical. Finally, the parameter f_{it} is equal to $(f_0-f_2)/(1-f_2)$, and still verifies the relation $(1-f_{it})=(1-f_{is})(1-f_{st})$.

Recently, many statistical tools have been developed to study F-statistics (Weir 1990), and their theoretical values under various ranges of demographic conditions (Whitlock 1992). Since evolutionary properties of populations and METAPOPULATIONS very much depend on the partitioning of variance, these studies are important in contemporary population genetics and CONSERVATION BIOLOGY. *See also* ANALYSIS OF VARIANCE; QUANTITATIVE GENETICS; STATISTICS. [I.O.]

REFERENCES

Weir, B. (1990) *Genetic Data Analysis*. Sinauer Associates, Sunderland, Mass.

Whitlock, M.C. (1992) Temporal fluctuations in demographic parameters and the genetic variance among populations. *Evolution* **46**, 608–615.

F test A statistical procedure that tests the NULL HYPOTHESIS (H_0) that two SAMPLE variances are from POPULATIONS with equal variances (H_0: $\sigma_1^2=\sigma_2^2$). One VARIANCE is divided by the other to produce a ratio that will equal one if the two variances are equal. This ratio (the variance ratio) has been called F, in honour of R.A. Fisher, an English statistical biologist, and its formula is simply:

$$F = \frac{s_1^2}{s_2^2}$$

If the ALTERNATIVE HYPOTHESIS (H_1) is: $\sigma_1^2 \neq \sigma_2^2$ it is a TWO-TAILED TEST. This is the case when performing this test before carrying out a z test or t TEST to see if the two variances are homogeneous. The test is two-tailed because we have no a priori reason to expect one particular variance to be greater. If the alternative hypothesis is H_1: $\sigma_1^2 > \sigma_2^2$ it is a ONE-TAILED TEST. This is the most frequent use of the F test, in ANALYSIS OF VARIANCE.

STATISTICAL TABLES of F show DEGREES OF FREEDOM for both numerator (ν_1) (usually columns) and denominator (ν_2) (usually rows). Therefore, unlike other statistical tables (t, r, χ^2), levels of PROBABILITY are shown in separate tables. Some statistical books (Fowler & Cohen 1992) show separate 5% tables for both one-tailed and two-tailed tests, while others (Rohlf & Sokal 1995) show only one-tailed tables, but have both 5% and $2^1/_2$% tables. The latter should be used as 5% tables in a two-tailed test. *See also* STATISTICS. [B.S.]

REFERENCES

Fowler, J. & Cohen, L. (1992) *Practical Statistics for Field Biologists*. John Wiley & Sons, Chichester.

Rohlf, F.J. & Sokal, R.R. (1995) *Statistical Tables*, 3rd edn. W.H. Freeman, New York.

facilitation Facilitation occurs during SUCCESSION when the chance (or speed) of ESTABLISHMENT of a given species is increased by the prior occupation of the site by another species (Connell & Slatyer 1977). Most simply it occurs through the provision of shelter for the germinating seed and seedling—the reduction in extremes of irradiance, humidity and temperature. Very often it occurs through the attraction of a dispersing agent (typically a bird or squirrel) to the facilitating plant, and less often by slowing of WIND so that seeds drop below the plant in question. Sometimes protection against HERBIVORES appears to be most important, for example by spiny bushes of *Juniperus*. In other cases the facilitating plant has some key effect on the substratum, for example stabilizing it or raising it, or changing the pH and/or availability of nitrogen and phosphorus. The most extreme cases of facilitation are sometimes distinguished as enablement, for example where a given species absolutely cannot invade unless a previous colonizer has changed the pH of the substratum (Grubb & Hopkins 1985). [P.J.G.]

REFERENCES

Connell, J.H. & Slatyer, R.O. (1977) Mechanisms of succession in natural communities and their role in community stability and organization. *American Naturalist* **111**, 1119–1144.

Grubb, P.J. & Hopkins, A.J.M. (1985) Resilience at the level of the plant community. In: *Resilience in Mediterranean-type Ecosystems* (eds B. Dell, A.J.M. Hopkins & B.B. Lamont), pp. 21–38. Junk, Dordrecht.

factor *See* ABIOTIC FACTORS; BIOTIC FACTORS; PHYSICAL FACTOR.

factor analysis A multivariate statistical method. Most of the statistical methods described in detail in this encyclopedia are univariate because they are concerned with single variables or observations. Even the complex methods of ANALYSIS OF VARIANCE are only concerned with explaining the variation in one variable. Multivariate methods analyse several related variables or observations simultaneously. *See also* EXPERIMENTAL DESIGN; STATISTICS. [B.S.]

REFERENCE

Manly, B.F.J. (1986) *Multivariate Statistical Methods: A Primer*. Chapman & Hall, London.

factorial experiment *See* EXPERIMENTAL DESIGN.

facultative A facultative factor is one which does not have to be present in order for an organism to survive in a particular environment. Every species has a range of requirements which may either be OBLIGATE or facultative. ADAPTATION to habitats without such features is usually transient and may carry an associated physiological cost that can reduce FITNESS if prolonged exposure occurs. The definition most commonly refers to the physical rather than the biological features of an environment. *See also* MUTUALISM. [R.C.]

facultative annual An ANNUAL plant species in which not all individuals complete their LIFE CYCLES within the year of birth or germination. A part of the population of facultative annuals postpones reproduction to their second or later year of life, representing a biennial or PERENNIAL strategy. *See also* LIFE-HISTORY EVOLUTION. [P.O.]

faecal pellets Defecated material (usually used in reference to aquatic invertebrates) that forms compacted aggregations of particles released by egestion. Due to their relatively large size (e.g. 100–300 μm), they sink quickly from the PELAGIC zone and, together with the bacteria that recolonize them, can provide an important source of organic matter to the DEEP SEA. The packaging of egested SEDIMENT into faecal pellets by benthic DEPOSIT-FEEDERS alters ambient sediment particle size, which in turn influences the availability of the particles for reingestion and the likelihood with which the sediment will become resuspended. [V.F.]

fallout *See* RADIOACTIVE FALLOUT.

family *See* TAXONOMIC CATEGORIES.

family groups *See* GROUP SOCIAL STRUCTURE.

FAO (Food and Agriculture Organization) *See* UNITED NATIONS PROGRAMMES AND BODIES.

farming practices, sustainable Agricultural activities that:

1 maintain and enhance the quality and PRODUCTIVITY of the SOIL;

2 conserve soil, WATER, energy, natural RESOURCES, species diversity, genetic diversity and WILDLIFE habitat;

3 maintain and enhance the quality of surface water and groundwater;

4 protect the health and safety of persons involved directly in AGRICULTURE as well as consumers;

5 promote the well-being of animals;

6 maintain the economic viability of farming.

This involves reducing, to the extent feasible, soil loss and dependence on non-renewable energy sources, and reducing the application of FERTILIZERS and BIOCIDES that contaminate air, soil, water and wildlife habitat, or harm people and other non-target organisms. Examples of sustainable farming practices include BIOLOGICAL CONTROL of weeds and insect pests, intercropping, CROP rotation, minimal tillage, and NITROGEN fertility supplied by legumes and/or animal manure. *See also* AGRICULTURAL POLLUTION; AGRICULTURE AND SOIL CONSERVATION; ORGANIC FARMING; SUSTAINABLE AGRICULTURE. [J.K.P.]

farmland and farming communities, conservation Activities that protect agricultural soils and adjacent natural areas from loss via erosion, CONTAMINATION or development. Protection of human communities supported by AGRICULTURE involves maintaining the economic viability of farming while preserving the integrity and self-sufficiency of agriculturally based communities. A variety of life and culture is maintained by preserving the agricultural practices and CROP varieties specific to local areas. *See also* CONSERVATION BIOLOGY; EROSION AND TOPSOIL LOSS. [J.K.P.]

fate The destiny of a XENOBIOTIC in the environment. It is influenced by transport, transformation and degradation. *See also* PREDICTED ENVIRONMENTAL CONCENTRATION. [P.C.]

Faunal region

The term faunal region is used for land and freshwater vertebrates. Six main faunal regions are now generally accepted. They were first proposed in 1858 by P.L. Sclater and were confirmed by

A.R. Wallace in 1876 for vertebrates in general and for some invertebrates also. The accepted system of continental faunal regions is as follows (Fig. F1):

1 Ethiopian region: Africa south of the Sahara, with part of southern Arabia.

2 Oriental region: tropical Asia, with associated continental islands.

3 Palaearctic region: Eurasia north of the tropics, with the northern limit of Africa.

4 Nearctic region: North America excepting tropical Mexico.

5 Neotropical region: South and Central America including tropical Mexico.

6 Australian region: Australia, with New Guinea, etc.

A number of general patterns of distribution may be discerned. Vertebrates of all classes are limited both northward and southward in cold latitudes, and outward from the continents on to islands. Freshwater fishes extend far northwards into the ARCTIC, but strictly freshwater ones scarcely extend outwards beyond the limits of the great continents and closely associated continental islands. Amphibians extend northwards to parts of the Arctic but not to the northern limit of fishes, but they extend further on to islands than strictly freshwater fishes do. Reptiles fall far short of the northern limit of amphibians in most places in the north, but they extend still further into islands. Birds extend both far northwards and far on to islands; they reach no final limits in either direction. Even birds, however, like other vertebrates, are limited in diversity in high latitudes and on remote islands. Mammals also extend far into high latitudes, but (excepting bats) they are about as limited on islands as amphibians.

There is a further limitation of different, dominant orders or families to different parts of the world. Among strictly freshwater fishes there is a profound complementarity, involving large parts of faunas, between the dominant cypriniforms of Eurasia and the dominant non-cypriniforms of South America, the North American and African faunas being mixed (and Australia being outside this pattern). Amphibians exhibit a complementarity which is less profound and more complex, but which includes dominance of ranids in most of the world and of leptodactylids in South America and Australia, and of rhacopterid tree frogs in Africa and the Oriental region, and of hylids elsewhere. Reptiles show a still less marked and still more complex complementarity in which the dominance of emydid turtles in Eurasia and North America and of side-necked turtles elsewhere, and of colubrid snakes in most parts of the world and of elapids in Australia are important features. Birds probably show the least profound, most complex complementarity of distribution of any vertebrates, although the Oscines (songbirds), dominant over most of the world, and Suboscines, dominant in South America, are incompletely complementary. Terrestrial mammals show a striking difference between the almost wholly placental fauna of the main part of the world and the chiefly marsupial fauna of Australia, the South American fauna being mixed.

It was formerly believed that the distribution of continental vertebrates could be explained by DISPERSAL from centres of origin, which were usually placed in central Eurasia. Within the last few decades, however, the general acceptance of CONTINENTAL DRIFT has led to the development of VICARIANCE BIOGEOGRAPHY, with the movement of continents being granted more importance than the MIGRATION of organic populations. Taking into account also the FOSSIL RECORD, a much more satisfactory account is now available about such otherwise peculiar phenomena as the distinctness of the Australian fauna and the close relationships between the faunas of Africa and South America.

Fig. F1 Wallace's (1876) Mercator projection of the world. Simplified outline with continental faunal regions as named and bounded by Wallace.

Another important biogeographic pattern is latitudinal ZONATION, which evidently relates to CLIMATE. This is strongest among freshwater fishes and amphibians. Among the latter, for instance, most families are mainly tropical; some of the tropical groups extend into the north temperate zone, but salamanders and a few small groups of frogs are mainly north temperate. Among reptiles, most families are tropical; there is no well-defined, zonal, north temperate group. This concentration in the tropics and extension northwards of some tropical families is also the main pattern of zonation of land-birds and mammals, although these animals extend northwards further and in greater numbers than reptiles do. *See also* AUSTRALIAN FAUNAL REGION; ETHIOPIAN FAUNAL REGION; NEARCTIC FAUNAL REGION; NEOTROPICAL FAUNAL REGION; ORIENTAL FAUNAL REGION; PALAEARCTIC FAUNAL REGION. [A.H.]

REFERENCES

Darlington, P.J. (1957) *Zoogeography: the Geographic Distribution of Animals.* John Wiley & Sons, New York.

Hallam, A. (1994) *An Outline of Phanerozoic Biogeography.* Oxford University Press, Oxford.

Faunas, major published As for Floras, but refer to animals. *See also* FLORAS, MAJOR PUBLISHED. [P.C.]

favourability The contrast is sometimes made between favourable (benign) and unfavourable (hostile or stressful) environments. The notion of environmental favourability (*see* GRIME'S TRIANGLE) is problematical, since all environments are hostile in one way or another. As Darwin recognized, the struggle for existence goes on everywhere. If abiotic conditions (temperature, water and nutrient availability) are non-limiting, then biotic processes (like COMPETITION or predation) are likely to be intense. If abiotic conditions are harsh (e.g. hot, dry deserts, exposed mountain tops) then population densities are likely to be low because of limited access to essential RESOURCES. It is a common mistake to assume that competition is less important or less intense in abiotically harsh environments, but there is no necessary relationship between absolute POPULATION DENSITY and the importance of competition in POPULATION REGULATION. The important parameter is the density of competitors relative to the availability of limiting resources. Thus, competition between desert shrubs can be extremely intense even though the plants are widely spaced and occur at low absolute population densities. *See also* FAVOURABLENESS. [M.J.C.]

favourableness A term sometimes used in the CLASSIFICATION of SELECTION PRESSURES. [P.C.]

fecundity A measure of the NUMBER OF OFFSPRING produced by an individual during a period of time. Fecundity is sometimes used interchangeably with fertility, but fecundity is the preferred term used to describe reproductive output in ecology—fertility has other meanings; for example, the potential of an animal to produce offspring, or the ability of soil to support plant growth. The concept of fecundity is relevant to modular organisms, but in clonal organisms reproduction is not measured so simply. Fecundity usually varies with the age of an individual and is often tabulated as the m_x column in a LIFE TABLE or the top row of a Leslie matrix (*see* POPULATION PROJECTION MATRIX). [R.H.S.]

fecundity schedule A tabulation of the pattern of offspring production (NATALITY) of individuals with respect to their age. Such a schedule, also called a FECUNDITY table, is frequently included in a LIFE TABLE, and contains F_x (total NUMBER OF OFFSPRING produced by the population at age x), $m_x=F_x/a_x$ (age-specific fecundity or mean number of offspring produced by each individual at age x), and $V_x=l_x m_x$ (the number of offspring produced at age x per individual in the initial COHORT). (In fecundity schedules, natalities are expressed in relation to females alone by halving a_x and l_x.) The BASIC REPRODUCTIVE RATE, R_0, is computed by summing the values of V_x. [A.J.D.]

feedback Information regarding a control mechanism or controlled system that enables actual performance to be compared with target performance. Performance can then be adjusted either positively or negatively, corresponding to, respectively, either positive or NEGATIVE FEEDBACK. Feedback can involve physical feedback loops, as with thermostats, reflex responses and management systems. Alternatively, feedback can be apparent, for example when liquid escapes from a leaking tank the rate of outflow reduces as volume reduces, giving the impression that there is feedback control on rate of loss, but there is no feedback hardware here. Similarly, as POPULATION SIZE reduces under predation, there may be more food per individual, which causes increased FECUNDITY and RECRUITMENT, so reducing population decline. Again this might appear to be feedback, but there is no active control here that helps keep population to any goal DENSITY in the same sense as a thermostat actively controls temperature to a goal temperature. *See also* SYSTEMS ECOLOGY. [P.C.]

Feeding

Feeding includes all behaviour directed towards acquiring, manipulating and consuming food. It is an essential activity for heterotrophic organisms (including non-photosynthetic parasitic plants), providing them with the energy and NUTRIENTS required for their GROWTH, maintenance and reproduction. It is also found in facultatively autotrophic organisms (mostly unicellular organisms, which are autotrophic in the presence of light but which switch to heterotrophy in its absence), in autotrophic hemiparasitic plants (which obtain soil nutrients and water by parasitizing the roots of their plant HOSTS), and in insectivorous plants (which obtain AMINO ACIDS from their prey).

Types of feeding

Feeding may take the form of CARNIVORY, DECOMPOSITION, detritivory, herbivory (also called phytophagy), omnivory (also called diversivory) and PARASITISM (including pathogenicity). HERBIVORES include FRUGIVORES (fruit-eaters), folivores (leaf-eaters), and pollen- and nectar-feeders (which a plant may utilize as pollinators). Feeders (consumers) may show different degrees of generality in their diet, consuming items from a single species or genus (monophagy), from a single family (oligophagy), or from different families (polyphagy). *See also* FEEDING TYPES, CLASSIFICATION.

Needs and preferences

Through feeding, an organism attempts to bring its current nutritional state to one which is ideal. The quality as well as the quantity of the food is important, requiring selectivity (preferences) in feeding which may be achieved by regulating the amount of a particular food type consumed or through selecting among different food types. For example, green leaves are more energy and nutrient rich than stems or dry or old material, and through selective GRAZING the proportion of green leaf in a ruminant's diet can be much greater than that in the sward. In general, preferences may be for one particular type of food item (e.g. one species, or one size class within a species), or for a range of food items which together produce a BALANCED DIET.

Often, nutritional needs will vary with age, requiring a switch in preferences. For example, early-instar stages of PHYTOPHAGOUS insects require a high-protein diet whereas later stages often require a diet high in CARBOHYDRATES, necessitating a change in feeding habits as an individual develops. Growth *per se* (rather than changing nutritional requirements) may also force a change in feeding habits. For example, the small early stages of individuals of many Lepidoptera feed by mining their host plants but later stages are too large to mine and must switch to external feeding.

A preference for one particular type of food item indicates that the available diet varies little in composition (e.g. in the ratio of carbohydrate to protein) but that the cost-benefit ratio of obtaining, manipulating and consuming different types of item exhibits considerable variation, for example in the amount of energy gain (energy gained : energy expended per unit HANDLING TIME). Within a single food item type, energy gain may be maximized by selectively feeding from a subset of size classes. For example, crabs (*Carcinus maenas*) reject both energy-rich but thick-shelled large mussels, and thin-shelled but energy-poor small mussels in favour of mussels of intermediate size, as these offer the best energy gain, comprising a good energetic reward while having shells that can be broken relatively easily.

Specialization may evolve if a single type of food item can provide all dietary requirements. Factors which favour specialization include the food resource having a suitable distribution in space and time, COMPETITION among feeders favouring RESOURCE PARTITIONING, a shortage of suitable alternative food choices, and a negative association between foraging efficiency and the number of choices. In addition, according to the sexual rendezvous hypothesis, specialist feeding behaviour may become established if mates are more likely to be found on the selected food item than elsewhere. For example, MONOPHAGOUS hummingbird flower mites, which live and mate in the corollas of flowers, are transported from flower to flower by hummingbirds, alighting in corollas of the host species; nectar odour provides the cue that the corolla is produced by the correct host species. Clearly, alighting on the wrong host would reduce FITNESS, favouring selection for a narrow host range and for host fidelity.

Among phytophagous insects, specialist feeders may be more successful than GENERALIST feeders at evading generalist predators, suggesting that generalist predators may select for specialist feeding among this group. The link between mating success and specialist feeding identified by the sexual rendezvous hypothesis, and the maintenance of the ability to evade generalist predators, may also function as mechanisms for reinforcing specialist feeding behaviour once it has evolved. Other mechanisms include the evolution of morphological and physiological adaptations which make the feeder more able to exploit the selected food item but less able to exploit other types of food item.

Preferences for a range of food items indicate that different items vary in composition and that no single type provides a balanced diet. Here, the optimal feeding STRATEGY (or nutritional target) is one which may maximize no single component of the diet (e.g. energy gain) but balances the different components. This will be the case whenever no single food type provides all of the nutritional requirements of the feeder, as in the case of caribou (*Rangifer tarandus*), which must overcome mineral deficiency resulting from a lichen-based diet by gnawing shed antlers, ingesting urine-contaminated snow or drinking seawater.

A feeder may also exhibit preferences for certain parts of a food item, and may need to consume different parts in different proportions in order to obtain a balanced diet. For example, the protein : carbohydrate ratio can vary among young and old leaves, a situation which may require a herbivore to balance its consumption of young and old leaves in order to achieve a balanced diet.

It will not be possible to approach a balanced diet by exercising preferences for one or a range of food items if no available food item type or combination can provide the different components of the feeder's nutritional requirements in a suitable ratio. Here, the feeder may have to consume large quantities of the food and differentially eliminate components of the diet in order to achieve a balance. This strategy is found in aphids, which must take in much more CARBON than they need (eliminating the surplus as honeydew) in order to obtain sufficient NITROGEN. Similarly, a lack of a suitable choice of food items will mean that a feeder has to select among what is available rather than for what is optimal. If low-quality food items are much more commonly encountered than high-quality items it may not be in the interests of the feeder to ignore them. *See also* FOOD SELECTION; OPTIMAL FORAGING THEORY.

Costs of feeding

There are costs to feeding. These include a greater risk of predation if a feeder is not able to maintain a high level of VIGILANCE, or if its food occurs in areas in which the feeder is more apparent to predators. Some predators may mimic food items, entrapping the feeder (e.g. the wormlike lure of angler fish, e.g. *Cryptopsaras couesi*). Feeding also carries with it the risk of PARASITE infection through CONTAMINATION of the food item with transmission stages of the parasite. Competition for food may be a cost of group feeding. It will occur whenever one individual deprives another of a food item, either by consuming the food item itself (competitive exploitation) or by acting in a way which warns the prey item that it should hide or flee (competitive INTERFERENCE). Here, competition may be reduced by increasing the distance between feeders. Finally, a feeder may accumulate TOXINS. Strategies for ensuring that toxin levels remain acceptable include: consuming a mixed diet of several species (each of which may contain a different toxin); or adopting a feeding pattern which involves consuming only parts of a food item which lack toxins or which possess them in small quantities. The feeder may also act to increase the palatability of a food item. For example, resin can make a leaf unpalatable; phytophagous insects may cut the resin canals, reducing the flow of resin to the leaf and thereby increasing its palatability.

Feeding and social behaviour

There can be close correspondence between feeding behaviour and SOCIAL ORGANIZATION. For example, forest-dwelling antelopes tend to feed on berries and shoots, both of which have a scattered distribution. This favours solitary feeding and the sharing of a territory by a monogamously mating pair. In contrast, plains-dwelling antelopes tend to bulk feed on low-quality, patchily distributed food and to graze in nomadic herds, a social system which provides the opportunity for polygamous mating. However, the causes of the close correspondence between feeding behaviour and social organization may not be obvious. For example, either a patchy food supply or predation pressure could favour the evolution of group feeding, because a group may be more efficient than an individual at locating patches but may also be more efficient at detecting and deflecting a PREDATOR, or because a group-living individual is able to put another individual between itself and a predator thereby reducing its risk of being predated (the selfish herd). *See also* BEHAVIOURAL ECOLOGY. [M.M.]

REFERENCES

Begon M., Harper, J.L. & Townsend, C.R. (1996) *Ecology: Individuals, Populations and Communities*, 3rd edn. Blackwell Science, Oxford.

Bernays, E.A. & Chapman, R.F. (1994) *Host-Plant Selection by Phytophagous Insects*. Chapman & Hall, London.

Krebs, J.R. & Davies, N.B. (eds) (1997) *Behavioural Ecology: An Evolutionary Approach*, 4th edn. Blackwell Science, Oxford.

Simpson, S.J. & Raubenheimer, D. (1993) A multi-level analysis of feeding behaviour: the geometry of nutritional decisions. *Philosophical Transactions of the Royal Society of London, Series B* **342**, 381–402.

---

feeding interference Reduction of food intake caused by the presence of other individuals (normally CONSPECIFIC). Decreased access to food can be due to aggressive displacement from food or prey disturbance. Aggressive displacement is normally linked to the development of a RANK order (within species as well as between species) reflected in a heavily biased and predictable outcome of aggressive contests between pairs of individuals. Overt aggression is rare and rank orders are largely upheld by ritualized behaviours. In flocks where the members meet repeatedly, rank orders develop into DOMINANCE HIERARCHIES where social rank determines the priority of access to food.

Interference at foraging sites forces flock members of low social rank to wait longer before FEEDING and to feed at higher risk of predation. Flock members of low rank feed in sites more exposed to predators and they return faster to food after a PREDATOR attack. Despite the priority to food for dominants there is no consistent influence of foraging INTERFERENCE on the amount of food ingested. Flock members of low rank do not always eat less, and they do not consistently carry smaller body reserves than high-rank flock members. Among birds, cases are known where low-ranked flock members carry the largest reserves in body fat.

Local depletion of food and prey disturbance are other mechanisms of foraging interference without aggression. Individuals in flocks are more likely to encounter areas already searched and depleted (Ranta *et al.* 1993). The local availability of prey can be reduced through predator evasion elicited by exposure to flock mates. *See also* COMPETITION. [J.E.]

REFERENCE

Ranta, E., Hannu, R. & Lindström, K. (1993) Competition vs. cooperation: success of individuals foraging alone or in groups. *American Naturalist* **142**, 42–58.

Feeding types, classification

Food occurs in many forms. The supply is nevertheless limited, so that NATURAL SELECTION will have tended to promote diversification in FEEDING habits; this should have resulted in the complete exploitation of all potential foods on this planet. A number of schemes of CLASSIFICATION have been used to give some order to the large quantity of information available on feeding. The main ones are listed below. *See also* FOOD SELECTION; OPTIMAL FORAGING THEORY.

Particle size

C.M. Yonge (1928) divided animals into three major groups on the basis of the particle sizes of the foods they are adapted to exploit. Examples are illustrated in Fig. F2.

Microphagous feeders

These are animals that ingest small particles (<1 mm diameter). They may feed on particles suspended in solution (suspension-feeders) or deposited on the substratum (DEPOSIT-FEEDERS).

Suspension-feeders

Suspension-feeders (Fig. F2a) are highly diversified in taxonomic and ecological status. Their main requirement is a filtering device. This may be formed from flagella, cilia, setae (which occur as fringes on various appendages of crustaceans and some insect larvae) or tentacles, which are used for filtration, for example in sea cucumbers. Mucous secretions play an important part in ciliary and tentacular feeding mechanisms as a final means of trapping food particles and in conveying them to the mouth. However, a few suspension feeders use mucus alone to trap food. For example, the sessile gastropod, *Vermetus gigas*, secretes a mucous veil from its pedal gland. This veil, used exclusively for the capture of food particles, acts as a kind of net; from time to time it is hauled into the mouth of the snail and a new one is secreted.

Deposit-feeders

Deposit-feeders (Fig. F2b) are also very varied. Some use specialized mechanisms for collecting sediments; for example the mobile ciliated tentacles of tubiculous polychaetes like *Terebella*; the mucoidal tentacles of sea cucumbers; and the specialized oral tube-feet of burrowing sea urchins and brittle-stars. Other deposit-feeders use non-specialized feeding mechanisms. Many burrowing polychaetes and oligochaetes, for example, use a simple mouth to eat their way through large quantities of sand, mud, silt and earth. The gastropod molluscs, intermediate in many respects between microphagous and macrophagous feeders, use a toothed ribbon, the radula, to scrape up small algae, DETRITUS and bacteria (Fig. F2c).

Macrophagous feeders

These can be divided into those animals that seize and swallow only and those that use some kind of preingestion processing mechanism.

Seizers and swallowers

Seizers and swallowers (Fig. F2d) occur in almost every phylum. Cnidarians use tentacles, bearing a battery of stinging cells (the cnidoblasts) to seize prey and to deliver it to the mouth. Nemerteans capture prey by means of a long, slender proboscis, which is sometimes barbed. Many polychaetes, like *Nereis* and *Glycera*, are carnivores and possess a muscular pharynx armed with jaws.

Seizers, masticators and swallowers

Members of this group (Fig. F2e) are more restricted in their taxonomic distribution. Cephalopods use a beak and a modifed radula for shredding prey. Pincer-like chelae are found in some crustaceans (e.g. crabs) for seizing and shattering shellfish. Mandibles are used for the mastication of both animal and vegetable material by insects, and modified parts of chelicerae and pedipalps are used for this purpose by those arachnids which masticate food. Vertebrates have evolved jaws and teeth.

Fluid-feeders

These use NUTRIENTS that are dissolved in aqueous solutions and take them by soaking, or sucking, or piercing and sucking.

Soaking

This mechanism (Fig. F2f) is used widely by ENDOPARASITES, and tapeworms are good examples. These lack any form of mouth or digestive system and absorb food from the gut of their HOST through a specially modified cuticle, which has microvilli to increase the absorptive surface area and absorb host enzymes. The cuticle is also rich in mitochondria, which supply the energy needed for the active transport of nutrients across the body wall.

Free-living organisms in aquatic habitats are also surrounded by a medium rich in DISSOLVED ORGANIC MATTER (DOM)—as much as 5 mg l^{-1} in seawater—and some can, at least in principle, take up nutrients through their outside body surfaces. However, it is still unclear how important DOM is as a food source for aquatic invertebrates.

Sucking

This mechanism (Fig. F2g) is developed in moths and butterflies where the insect mouthparts have been modified into a long proboscis for the exploitation of nectar. Muscid flies also have mouth-

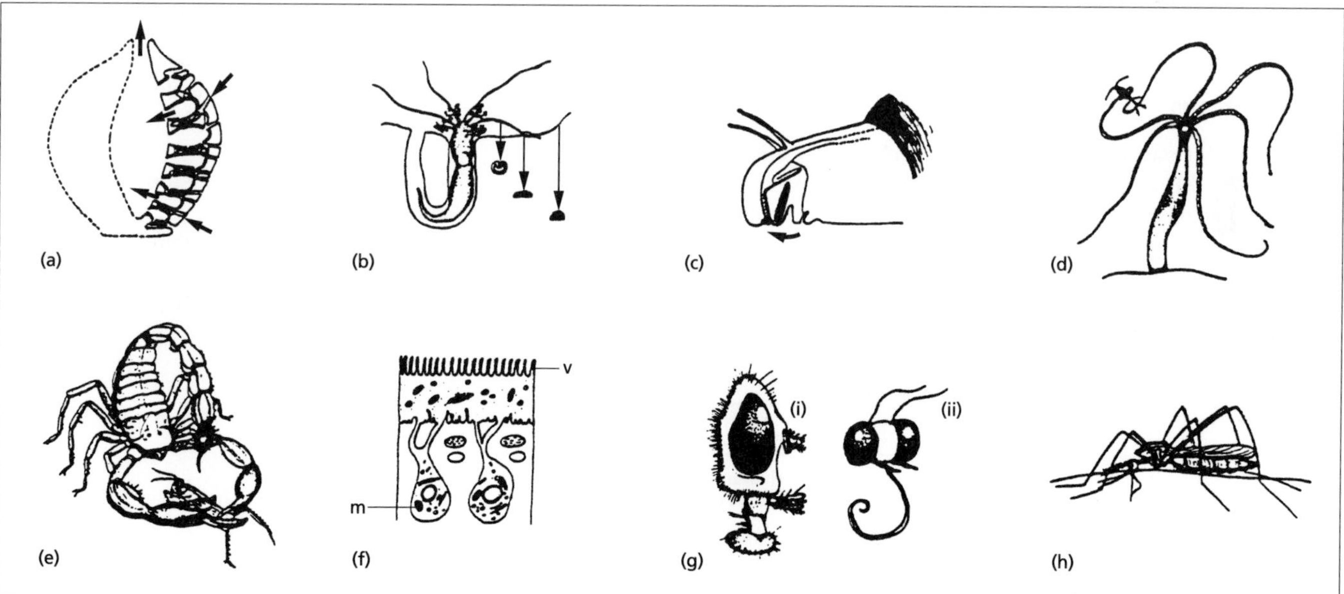

Fig. F2 Feeding mechanisms classified on the basis of the particle size of the food processed. (a) Filtering chambers of a sponge, lined with ciliated choanocytes—arrows show water movement. (b) *Terebella* using tentacles to take up deposits—arrows point to cross-sections of the tentacle. (c) Radula of a snail. (d) *Hydra*. (e) Scorpion. (f) Section of absorptive surface of a tapeworm (v, microvilli; m, mitochondria). (g) Sponging labellum of a fly (i), and (ii) suctorial proboscis of a butterfly. (h) Mosquito, piercing skin of host with sharpened mouthparts. (After Calow, P. (1981) *Invertebrate Biology. A Functional Approach*. Croom Helm, London. Illustration by L.J. Calow. (e) Redrawn from Jennings, J.B. (1972) *Feeding, Digestion and Assimilation in Animals*. Macmillan, London.)

parts modifed for sucking. A spongelike mechanism carried at the tip of the mouthparts of these flies soaks up organic fluids from the surfaces of plants and animals.

Piercing and sucking

Combined piercing and sucking (Fig. F2h) occurs in a variety of animal- and plant-eating invertebrates. Turbellarian platyhelminths use enzyme secretions to push a pharynx into prey, whereas lice, fleas and aphids depend more on physical pressure and sharpened mouthparts for insertion.

Functional feeding groups

A related classification has been developed for benthic aquatic invertebrates where the feeding types are referred to as functional feeding groups. These are:

- shredders — feed on COARSE-PARTICULATE ORGANIC MATTER (CPOM) or live macrophytes;
- collectors — feed on FINE-PARTICULATE ORGANIC MATTER (FPOM) either by filtering or gathering;
- scrapers—feed on attached PERIPHYTON;
- piercers—feed on macroalgae by piercing individual cells;
- predators—feed on other animals.

In flowing-water systems it is predicted that the frequency of functional feeding groups changes in a predictable way from source to sink, and this is part of the RIVER CONTINUUM CONCEPT (RCC). Shredders are supposed to be dominant in upper reaches, where ALLOCHTHONOUS detritus is a major source of food. Collectors and scrapers dominate in mid-reaches, where both allochthonous and AUTOCHTHONOUS impacts are important. Collectors dominate in the sediment-rich lower reaches. *See also* FUNCTIONAL GROUP.

Trophic status

Another way of classifying foods is in terms of their own trophic status (i.e. into AUTOTROPHS, which can manufacture organic foods from inorganic precursors, and HETEROTROPHS, which cannot) and their condition before being eaten (living or dead). Thus it is possible to construct the scheme of classification of feeders shown in Fig F3 (*see also* FUNCTIONAL GROUP).

Availability and mobility

Yet another way that food materials can be classified is according to their availability and mobility. This influences the mobility of the feeders themselves and therefore provides another basis for classifying them.

Autotrophic foods are usually immobile, since locomotory powers are not required for trapping light (exceptions being some PHYTOPLANKTON) whereas heterotrophic foods are more often mobile because this aids them in the capture of their own food and in the avoidance of being eaten. As a result of their distribution and abundance, some foods are readily available whereas others are scarce. This may alter seasonally.

The mobility and availability of the food will influence the effort that must be expended by the feeder in getting it. In general, natural selection should have acted to reduce the energy spent in food-getting because this will maximize the returns from each meal. The most economical strategy to deploy in the absence of food is a 'sit-and-wait' strategy, whereby the organism waits for food to be replenished. However, if in practice this is unlikely because the food is either scarce or immobile, then time and energy must be expended in adopting a 'seek-it-out' strategy. Sit-and-wait feeders approximate to what Schoener (1971) has described as a type 1 PREDATOR, and the seek-out feeders approximate to what he describes as a type 2 predator. This latter classification is widely used in the ecological literature.

Table F1 summarizes what type of feeder is expected to exploit different kinds of food. Microphagous invertebrates, particularly filter-feeders, which exploit foods moved by water currents, adopt the ultimate in sit-and-wait strategies, i.e. they are permanently attached. Alternatively most HERBIVORES, exploiting a food source which for most of the time is readily available, have to move to find food when it is in short supply because the food itself is immobile. Top carnivores, which exploit a precarious but mobile food, often adopt sit-and-wait strategies; spiders, which sit and wait for prey beside webs, are good examples. Alternatively carrion-feeders, or carnivores that exploit sessile prey, adopt seek-out strategies.

When food is in very short supply, extreme forms of strategies may be adopted. Animals may either enter a resting stage (undergo DIAPAUSE and AESTIVATION) or embark on extensive MIGRATION. Both strategies are usually adopted in anticipation of food shortage rather than as a direct result of poor feeding, since a food supply is

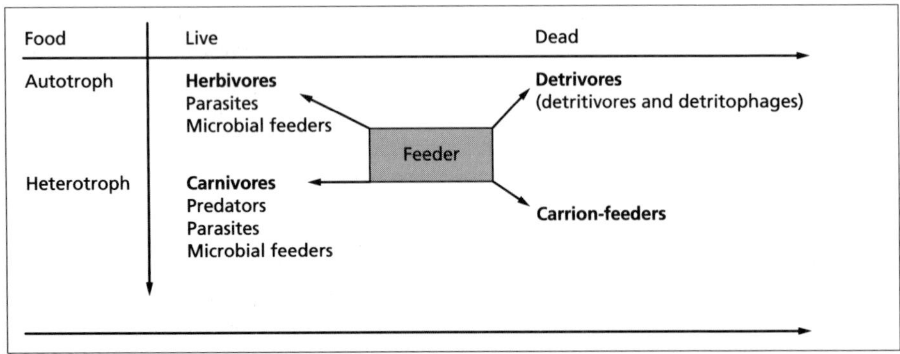

Fig. F3 Classification scheme of feeders in terms of trophic status and condition of food.

Table F1 Distribution of trophic types relative to the availability and mobility of food. (After Calow, P. (1981) *Invertebrate Biology. A Functional Approach*. Croom Helm, London.)

	Food availability		
Food mobility	High	Low	Very low
High	*Sit-and-wait feeders* Microphagous	Carnivores	*Migratory stages* *Resting stages*
Low	*Seek-out feeders* Herbivore Detrivore	Carrion-feeders	

necessary to lay down stores in preparation for the extended periods of inactivity or activity.

Finally, it is important to note that the adoption of a particular feeding strategy is also dependent on the physiology of the feeder, particularly in terms of its LIFESPAN and level of STORAGE reserves. Thus in the Insecta, short-lived dipterans with low-level reserves cannot 'afford' to sit and wait for food and invariably become more active as they are starved, whereas long-lived cockroaches with extensive fat reserves, though exploiting dead and immobile foods, nevertheless become quiescent upon starvation. *See also* TRAPPING FOOD; TROPHIC CLASSIFICATION. [P.C.]

REFERENCES

Calow, P. (1981) *Invertebrate Biology. A Functional Approach*. Croom Helm, London.

Cummins, K.W. & Wilzbach, M.A. (1985) *Field Procedure for the Analysis of Functional Feeding Groups in Stream Ecosystem*. Pymatuning Laboratory of Ecology, University of Pittsburgh, Pa.

Schoener, T.W. (1971) Theory of feeding strategies. *Annual Reviews of Ecology and Systematics* **2**, 369–404.

Yonge, C.M. (1928) Feeding mechanisms in the invertebrates. *Biological Review* **3**, 21–76.

Female preference

The phenomenon whereby a characteristic of a female makes her prefer to mate with one male rather than another on the basis of differences in a male TRAIT. If there is genetic variation in the population for the extent to which the male trait is developed, either natural or (inter) SEXUAL SELECTION will result in the trait developing in the direction of the female preference, resulting in SEXUAL DIMORPHISM if the trait is a physical characteristic. NATURAL SELECTION operates if the trait affects the viability or FECUNDITY of offspring, sexual selection operates if it affects male mating probability. As a preferred male trait may affect both sets of characteristics (e.g. disease-free males may be more attractive to females and may also increase offspring viability by donating to them genes for disease resistance), it may be subject to both kinds of selection. However, under sexual selection a male trait can become exaggerated past the point favoured by natural selection, the trait becoming costly, for example, in terms of male survivorship (e.g. long tail feathers used by males in COURTSHIP may make them more obvious to predators).

Fisher's theory

The selective advantage of female preference may be clear, as when a female prefers to mate with males who have large territories which can support many offspring. It is less clear how female preference can result in the production of costly forms of male traits, or why it should evolve in species in which males only provide sperm. R.A. Fisher argued that female preference for exaggerated male traits could evolve under a combination of natural and sexual selection (see, for example, Fisher 1915). In the presence of sufficient additive genetic variation for the trait (which influences its HERITABILITY), and of a positive association between male survivorship and the trait, natural selection would favour exaggerated forms of the trait. It would also favour a gene which caused females to prefer mates with the most exaggerated form of the trait, as they would produce sons who also exhibited the trait in exaggerated form and who would therefore have greater than average survivorship. Consequently, the gene for female preference would increase in frequency. As the proportion of females exhibiting the preference increases, preferred males would not only survive better than males with less exaggerated forms of the trait but would have a greater than average chance of mating, with the result that genes for female preference and for exaggerated forms of the male trait become positively associated. Eventually, the male trait will attain the size which is optimum for survivorship, and natural selection would operate to maintain it at this size. But female preference could drive the male character beyond this optimum (in runaway sexual selection) towards an equilibrium at which the decrease in survivorship of males with the most exaggerated (and now costly) form of the trait is exactly compensated by their increased reproductive success, at which point natural selection would operate to halt the process of runaway sexual selection.

The handicap theory

In Fisher's theory, once the preferred form of the trait becomes fixed in the male component of the population it ceases to serve any function for the female. That is, it was advantageous prior to fixation only because males exhibiting it were more successful at attracting mates than males with less exaggerated forms of the trait. It conferred no other benefit. This is not the case in the handicap theory. This theory, which was introduced by A. Zahavi, argues that sexual selection for exaggerated forms of a male trait will result in female preference for these forms if there is a positive association (up to and including the point of equilibrium) between the form of the trait and the benefit to the female. The theory assumes that males have good genes which impart high FITNESS and bad genes which do not, that they signal this to females, and that females can recognize this signal and preferentially mate with males with good genes. Females benefit by having the good male genes inherited by their offspring. There must be a strong positive association between the strength of the male signal and male genetic quality, to prevent CHEATING by males with bad genes. This association is realized by females preferring a male trait which handicaps the male when it is exaggerated (by imposing a considerable energetic or survival cost on the male predators) so that only males with good genes can overcome the handicap sufficiently to fully develop the trait or to survive to reproductive maturity.

The association between male genetic quality and the handicap may be developed in various ways. For example, investment in the handicap may be the same for all males but only high-quality males will survive the investment; or a male may only invest in the handicap to the extent that the investment optimizes his fitness (condition-dependent handicap), so that males who invest the most in the handicap are signalling their higher quality by surviving a level of handicap which lower-quality males would be unable to survive. W.D. Hamilton and M. Zuc have argued that genetically determined differences between males for health genes may favour the development of handicaps, as only a healthy male will be able to produce a form of the handicap which females prefer. For example, bright plumage in male birds may have been selected as a handicap because it is a sign of health, brightness being negatively associated with parasite load.

The models compared

The extent to which female preference for exaggerated male traits may be due to runaway sexual selection or selection for handicaps is not clear, although it is clear that the male traits selected under the two processes are likely to differ. This is because, initially, traits selected under runaway sexual selection are not costly to the male, but they must be initially costly if they are to be selected as a handicap. Similarly, it is becoming clear that when female preference imposes a cost on the female (e.g. a choosy female may take longer than a randomly mating female to find a mate, and therefore be exposed longer to the threat of predation) in taxa in which the male only provides GAMETES, female preference is only likely to become established if it results in the female acquiring good genes for her offspring. That is, female preference is only likely to arise for traits which handicap the male.

A general problem with these models is that genetic variation for the male trait is required to enable selection to operate but selection reduces the genetic variation (by getting rid of 'bad' genes). This problem is reduced when the preferred trait is a condition-dependent handicap, as this provides a general indication of the health of the male which will be a reflection of the quality of his genes at many different and independent loci. Here, genetic variation will be maintained even if there is intense selection.

Non-genetic traits

Heritability of, and genetic variation for, the preferred male trait are not essential requirements for the evolution of female preference. Preference could benefit the female as long as there was environmental variation for the male trait and different variants had positive or negative effects on the female or her offspring. This situation could arise, for example, if there was environmental variation among males for parasite loads, as a heavily parasitized male is more likely to heavily infect a female or her offspring, and is less likely to be able to provide adequate PARENTAL CARE, than a male with a low parasite load.

Female preference in plants

Female preference is expected in plants as well as in animals. However, its presence in plants can be difficult to demonstrate and the process by which it is realized can be difficult to discern. For example, the non-random abortion of seeds may reflect female preference for mates operating postzygotically, but it could also reflect female preference among genetically viable offspring GENOTYPES; however, it is also possible that it is not a consequence of female preference but of sib COMPETITION or zygote inviability.

Although female preference among mates is difficult to demonstrate in plants, conditions which would favour its evolution are widely distributed. For example, a female will often have access to several potential mates, and differences between males have been shown to influence seed size and seedling growth rates, indicating that there is variation among males in characters which affect female fitness. Indeed, an increase in the number of mates can be achieved by increasing the number of flowers or the period over which a flower remains receptive to pollen (both of which will increase the number of mating opportunities and consequently the opportunity for females to exercise mate choice). However, it is not clear whether selection for increased female choice has played a major part in the evolution of these traits. *See also* BEHAVIOURAL ECOLOGY; MATING SYSTEMS. [M.M.]

REFERENCES

Charlesworth, D., Schemske, D.W. & Sork, V.L. (1987) The evolution of plant reproductive characters; sexual vs. natural selection. In: *The Evolution of Sex and its Consequences* (ed. S.C. Stearns), pp. 317–335. Birkhauser Verlag, Basel.

Fisher, R.A. (1915) The evolution of sexual preference. *Eugenics Review* **7**, 184–192.

Hamilton, W.D. & Zuc, M. (1982) Heritable true fitness and bright birds: a role for parasites. *Science* **218**, 384–387.

Hamilton, W.D., Axelrod, R. & Tanese, R. (1990) Sexual reproduction as an adaptation to resist parasites (a review). *Proceedings of the National Academy of Sciences of the USA* **87**, 3566–3573.

Harvey, P.H. & Bradbury, J.W. (1991) Sexual selection. In: *Behavioural Ecology: An Evolutionary Approach*, 3rd edn (eds J.R. Krebs & N.B. Davies), pp. 203–233. Blackwell Scientific Publications, Oxford.

Marshall, D.L. & Folsom, M.W. (1991) Mate choice in plants: an anatomical to population perspective. *Annual Review of Ecology and Systematics* **22**, 37–63.

Zahavi, A. (1975) Mate selection — a selection for a handicap. *Journal of Theoretical Biology* **53**, 205–221.

femtoplankton Planktonic organisms of size in the range 0.02–0.2 μm, i.e. viruses. [V.F.]

fen A RHEOTROPHIC wetland ecosystem, dominated by herbaceous plants, in which the WATER TABLE approximately coincides with the soil surface during drier times of year. It represents a successional stage in MIRE development, between the wetter reed SWAMP vegetation and the drier CARR vegetation that is dominated by woody species.

The nutrient status of a fen depends upon the chemical quality of water feeding the mire and also the rate at which water flows through it. A rich fen normally has a rich supply of calcareous water, while a poor fen may have a slow flow-rate of nutrient-poor water.

The fact that water flows (i.e. rheotrophic) means that it is still regarded as a fen even if the low nutrient status allows the invasion of some plant species more typical of a BOG (i.e. OMBROTROPHIC).

Fens (particularly rich fens) are biologically diverse. They produce sediments that are well supplied with plant nutrients and which can result in agriculturally productive soils when drained. Consequently, many of the fenlands of highly developed countries have been reclaimed for agriculture and globally the habitat is under threat. [P.D.M.]

feral Living in a wild or NATURAL state. Used especially for domesticated animals and plants that become wild. [P.C.]

fermium (Fm) A man-made element (ATOMIC NUMBER = 100; relative ATOMIC MASS = 257; melting and boiling points unknown) which was discovered, along with EINSTEINIUM, in debris from the first thermonuclear weapon test at Einwetok Atoll in 1952. Isotopes with mass numbers ranging from ^{242}Fm to ^{259}Fm have been produced by particle bombardment of other actinides or multiple NEUTRON capture events in nuclear reactors but only in very small quantities. Consequently, fermium does not occur in detectable quantities in the environment. Investigations of the properties of fermium have used the ISOTOPE with the longest half-life, ^{257}Fm ($t_{1/2}$ = 100.5 days). The dominant oxidation state of fermium is +3. [A.B.M.]

ferns *See* PTERIDOPHYTES, GLOBAL DIVERSITY AND CONSERVATION.

Ferralsols *See* SOIL CLASSIFICATION.

fertility *See* FECUNDITY.

fertility schedule *See* FECUNDITY SCHEDULE.

fertilization The process whereby the nucleus of a male GAMETE (e.g. a SPERMATOZOA) fuses with that of a female gamete (e.g. an ovum) to form a cell (zygote) with a single nucleus. In animals, fertilization of EGGS can occur outside or within the maternal body. Both modes are often found within a phylum. External fertilization is ancestral, occurring in many aquatic invertebrates, most fishes and many amphibians.

Internal fertilization occurs in many aquatic invertebrates, all terrestrial invertebrates, some fishes, certain amphibians and in all other vertebrates. Sperm are transferred in one of several ways.

1 Broadcast into the water and taken up by females via inhalant currents, as for example in many colonial benthic invertebrates and certain bivalves.

2 Placed in packets (spermatophores) that are retrieved by the female or delivered to her by the male, as in many arthropods and certain amphibians.

3 Delivered by copulation, usually directly into the female's reproductive tract, as in cephalopods, crustacea, elasmobranchs, certain teleosts and most terrestrial animals.

Internal fertilization enables females to store sperm, which is important when receipt of sperm is unpredictable, for example at low population density, especially among animals restricted in mobility. Internal fertilization is a prerequisite for brooding and for terrestrial reproduction. SPERM COMPETITION probably promotes the evolution of internal fertilization. *See also* ANISOGAMY, EVOLUTION OF; POLLINATION. [R.N.H.]

Fertilizers

Materials added to the SOIL to increase the PRODUCTIVITY of farmland or forest by supplying chemical ELEMENTS that are at suboptimal levels for plant growth. Fertilizers are needed to replace the net loss of NUTRIENTS from the soil when the nutrient income from rain or the breakdown of the soil minerals is insufficient to balance the losses from LEACHING, erosion and the sale of CROP products and livestock. Fertilizers can be classed as inorganic (salts such as ammonium nitrate, NH_4NO_3) or organic (based on CARBON compounds). Manures are complex organic fertilizers derived from plant residues or animal excreta.

Fertilizers normally contain NITROGEN, PHOSPHORUS and POTASSIUM compounds but may also contain SULPHUR, CALCIUM and occasionally MICRONUTRIENTS. Lime, which is used to adjust the soil pH to increase the availability of nutrients and increase the effectiveness of rooting, is sometimes classed as a fertilizer.

Fertilizer usage is quoted either as the mass of fertilizer or of nutrients per unit area. In 1993, fertilizer consumption ranged from 21 kg ha^{-1} of arable or permanently cropped land in Africa to 145 kg ha^{-1} in Europe. However, these figures hide variation from less than 1 kg ha^{-1} in some African countries to more than 400 kg ha^{-1} in the intensively farmed countries of western Europe. Phosphorus (P) and potassium (K) in fertilizer can be quoted either as the elemental form or as the P_2O_5 or K_2O equivalent. Elemental P and K can be calculated from the P_2O_5 and K_2O equivalents by multiplying by 0.83 and 0.44 respectively.

Manures and composts have been used for many centuries. Roman farmers seem to have understood soil fertility well enough to use manures to improve soil fertility. In the late 18th and early 19th centuries, agricultural chemists such as Justus von Liebig in Germany investigated the link between plant nutrition and growth. At that time, bones were used as a source of phosphorus, although poor solubility led to responses varying markedly with soil type. In 1842 near London, England, John Bennett Lawes opened the world's first fertilizer factory to manufacture superphosphate, which is soluble, by treating bones with sulphuric acid. In western Europe, this innovation marked the start of a transition between the older, predominantly CLOSED SYSTEMS of AGRICULTURE to ones that were more open and market dominated. Until the 20th century, superphosphate was the only manufactured fertilizer, although sulphate of AMMONIA, a by-product of the gas industry, and nitrate of soda from Chile, were also used.

Major types and sources of fertilizers

Phosphorus and potassium fertilizers are largely derived from natural mineral rock deposits localized in a relatively few countries. Almost 90% of the rock phosphate comes from the USA, the countries of the former USSR, Morocco and China, and more than 90%

of the potash from salt deposits in France, Germany, the USA and Canada. The world reserves have been estimated at between 100 and 200 years' supply. Production of phosphate rock from the Pacific islands has already fallen to 20% of its 1972 peak because deposits have been worked out. Nitrogenous fertilizers are mainly derived from ammonia synthesized from atmospheric nitrogen using the Haber–Bosch process. The energy costs of extracting raw materials and manufacturing fertilizers are typically 5 MJ kg^{-1} (K), 15 MJ kg^{-1} (P) and 40 MJ kg^{-1} (N). The most advanced nitrogen fertilizer plants operate at close to the theoretical maximum chemical efficiency although there are still many old and inefficient factories. Fertilizers represent about 40% of the energy inputs to a typical arable farm.

Some fertilizers (straights) consist of single chemical compounds such as ammonium nitrate, urea or ammonium superphosphate. Compound fertilizer is designed to maximize the amount of nutrient per unit mass of fertilizer. This has the unfortunate side effect that the sulphur or micronutrients which formerly occurred as contaminants or as associated compounds are no longer present. The composition of fertilizers has to be declared on the bag as the percentage by mass of N, P_2O_5 and K_2O (or N, P and K) and farmers can buy a range of fertilizers (e.g. 13 : 13 : 20, 17 : 17 : 17, 32.5 : 0 : 0). Nitrogen and potassium fertilizers are generally soluble in the soil solution but phosphate fertilizers vary considerably in solubility, depending on soil type and pH. Most fertilizers are sold in the form of spherical granules which can be accurately distributed across the field. In some countries, concentrated liquid fertilizers are popular.

Environmental effects

Fertilizers are often cited, probably unfairly, as a potent source of AGRICULTURAL POLLUTION. Although agricultural land is the main source of nitrate in water supplies from rural CATCHMENTS, by no means all originates from fertilizer. For example, the application of excessive amounts of organic manures, particularly at times of year when the soil is wet and crops are not growing, can be a major cause of NITRATE POLLUTION. Techniques such as timing nitrogenous fertilizer applications by crop development stage and splitting doses into several smaller dressings can minimize losses of nitrate. Danish experiments have shown that losses of nitrate expressed per unit of grain produced decrease until the application rate approaches the economic optimum. Losses of phosphorus to watercourses are generally small. However, losses can be large enough to cause EUTROPHICATION where very large applications of manure are applied, as in The Netherlands, or where soil erosion is occurring, particularly where the soil is sandy or organic. Potassium losses are generally low.

Plant responses to fertilizers

When the elements needed for plant growth are present at levels that are too low, the plants show characteristic deficiency symptoms and YIELDS are depressed considerably below the potential set by the weather. Once deficiency symptoms have been removed by appropriate fertilization, the response of yield to further applications is generally small. Nitrogenous fertilizer is an important exception with responses being obtained at levels well in excess of those needed to remove deficiency symptoms. In the case of cereals, the response can reach 50 kg grain per kg nitrogen applied. However, there comes a point where yield starts to decline because excessive uptake of nitrogen leads to reduced partitioning of photosynthetic products to the harvested part of the crop. Phosphorus deficiency was formerly widespread in western Europe but years of application of phosphatic fertilizers has resulted in large areas now requiring only maintenance dressings. In parts of Africa and in other regions of the world, phosphate deficiency is still common and severely limits crop production.

The quality of many crops depends on their nutrient status and this depends on the natural fertility of the field and on fertilizer input. Although many people believe that the nutritional value of crops produced using fertilizers is less than that produced organically, there is no clear scientific evidence for this opinion. The level of protein and of some of the A and B VITAMINS tends to rise with nitrogenous fertilizer application and the levels of vitamin C and sugars tend to fall. Excessive applications of nitrogenous fertilizer can result in plants which are vulnerable to foliar pests and diseases due to the cell walls being thinner. It is sometimes asserted that inorganic fertilizers cause damage to soil structure and a reduced diversity of soil microorganisms. However, the former is more likely to be caused by the reduction of soil organic matter seen when agricultural production is intensified. Long-term (150-year) fertilizer trials carried out at the ROTHAMSTED Experimental Station in England have failed to show a difference in wheat yield between fields given farmyard manure and those given inorganic fertilizer.

Recommendations for fertilizer usage

Fertilizer trials have been carried out all over the world in order to determine the optimum rate, form and timing of fertilizer applications. Recommendations are made to farmers by governmental agencies and commercial companies. There is no single recommendation because the optimum depends on the soil, the weather and the crop. Nitrogen recommendations are particularly difficult because residual mineral nitrogen is easily lost from the soil by leaching and DENITRIFICATION when the soils are wet and the crop is not growing actively and because the rate of MINERALIZATION of the soil organic matter is a complex function of a range of environmental factors. In practice, empirical corrections are usually based on the cropping history, the rainfall regime and other factors. Even in intensively farmed systems, nitrogen derived from soil organic matter is extremely important for crop growth, and fields given no nitrogen fertilizer often give yields of about a third of those given the optimum rate. Techniques are now being developed to use soil or plant analysis to optimize the nitrogen application rate. Available P and K can be estimated using soil tests. Alternatively, applications can be based on a NUTRIENT BUDGET or balance sheet approach. It has recently been recognized that the optimum fertilizer rate can change across fields in response to natural soil differences. Modern technology now allows application rates to be varied across the field according to previously mapped requirements. This technique, which is called site-specific or precision farming, holds out the promise of greater efficiency of fertilizer use and reduced ENVIRON-

MENTAL IMPACT in the near future. *See also* FARMING PRACTICES, SUSTAINABLE; MACRONUTRIENTS; MINERAL NUTRIENTS. [G.R.]

---

fetal provisioning *See* SEXUAL CONFLICT.

FGD *See* FLUE-GAS DESULPHURIZATION.

FGN (flue-gas denitrification) *See* FLUE GAS.

field A term often used to describe NATURAL sites or systems, for example field experiments are carried out in the open as opposed to in laboratories. [P.C.]

field capacity The amount of WATER held by capillary forces of soil pores against the force of gravity. The field capacity is an indicator of the physical properties of a SOIL. The difference between the field capacity and the permanent wilting point of a given area yields the quantity of water available to plants after excess water has drained away. [P.O.]

filter *See* SWEEPSTAKE DISPERSAL ROUTE.

filterers, feeding type *See* FEEDING TYPES, CLASSIFICATION.

fine-grain exploitation The exploitation of a series of RESOURCES at rates proportional to their relative ABUNDANCES in the environment. In fine-grain exploitation, the relevant environmental resources occur in small patches with respect to the activity patterns of an organism, among which the organism cannot usefully distinguish. Examples of fine-grain exploitation include suspension-feeders feeding indiscriminately on several phytoplankton species, or deposit-feeders ingesting sediment without sorting for grain size. [V.F.]

fine-grained habitats Refers to the way an organism uses RESOURCES in a patchy or heterogeneous environment. A fine-grained HABITAT is one where an organism encounters and uses patches of resource in proportion to their occurrence in that habitat; for example, herbivores that graze pasture unselectively are using pasture as a fine-grained habitat. Larger animals will generally tend to use an environment in a relatively fine-grained way compared with smaller animals. This terminology is no longer used much. *Cf.* COARSE-GRAINED HABITATS. *See also* GRAIN. [R.H.S.]

REFERENCE

Pianka, E.R. (1988) *Evolutionary Ecology*, 4th edn, pp. 257–261. Harper & Row, New York.

fine-particulate organic matter (FPOM) Plant detritus in aquatic systems between *c.* 0.45 μm and 1 mm. *Cf.* COARSE-PARTICULATE ORGANIC MATTER. [P.C.]

fire, ecology and management The use of ecological responses of plants and animals to guide the management of fire on a landscape. Fire has been present during the evolution of the Earth's current biota, and many species have morphological or behavioural adaptations to its presence. Plant adaptations include thick bark, the ability to sprout, and the presence of seed banks. Seed banks allow plants to store seed for decades and release it when fire burns the FOREST. Some trees of the genus *Pinus* have serotinous, or late-opening, cones, which are sealed by resin until intense fires burn through the forest. Some shrubs, such as *Ceanothus* spp., have refractory seed that must be scarified, or cracked, in order to germinate; these germinate in profusion after fire has burned across the landscape. Beetles that lay their eggs in the bark of damaged or dying trees have the ability to sense infrared radiation from fires up to 80 km away. These adaptations may be to particular kinds of fire frequencies and intensities, rather than to fire presence alone, and are adaptations to a particular fire regime. Fire management plans may include the use of prescribed fire in order to perpetuate fire-adapted species on the landscape (*see* PRESCRIBED BURNS AND FIRE MANAGEMENT). [J.K.A.]

Fish, diversity and conservation strategies

Fishes are the most numerous of the vertebrates, a reflection of their antiquity and the extent and variety of habitats they occupy. Although estimates vary, there are probably around 24 000 species, the majority of which (14 500; 62%) are marine (Nelson 1994). Most are torpedo-shaped; however, round, flat and even angular body forms exist. Some attain sexual maturity when only 2–3 cm in length (e.g. the world's smallest commercially exploited fish, the Philippine goby, *Mistichthys luzonensis*), while others, such as the whale shark (*Rhincodon typus*), can attain lengths in excess of 20 m.

Loss of fish diversity

Fish DIVERSITY is now recognized to be diminishing at an unprecedented rate, losses being greatest in fresh waters where fishes are the most threatened group of vertebrates after the amphibians (*see* AMPHIBIANS, DIVERSITY AND CONSERVATION). It is estimated that 20% of the world's freshwater fish fauna is extinct or in danger of extinction, and that in Mediterranean climates this figure may be as high as 65% (see Moyle & Moyle 1995). Although only 20% or so of bony fishes show any form of PARENTAL CARE, one-third of recent extinctions is believed to be from this group, suggesting that members of more specialized GUILDS, perhaps because of their more sedentary behaviour, may be more prone to extinction (Bruton 1995). Among marine species, coastal and estuarine species are more at risk than those that inhabit the open oceans or deep-sea trenches.

Although the causes differ from one biogeographical area to another, losses of fish diversity have largely resulted from habitat degradation, species introductions and overexploitation. Habitat degradation, widely regarded as the single greatest threat to fish diversity, arises not only from changes in land use (e.g. DEFORESTATION) but also as a result of POLLUTION and EUTROPHICATION, changes in hydrological regimes and global CLIMATE CHANGE. Habitat destruction causes reductions in species ranges, increasing the vulnerability of populations to extinction by other causes, while pollutants can affect populations by reducing food intake, growth and reproductive success.

Fish introductions, accidental or for fisheries or AQUACULTURE development, have been widespread, resulting in habitat destruction and extinction of local species or populations through COMPETITION or predation (Welcomme 1988; Beveridge *et al.* 1994). Problems are particularly apparent in fresh waters. At the species level, the introduction of the piscivorous Nile perch (*Lates niloticus*) to Lake Victoria in the late 1950s is thought to have resulted in the elimination of some 200 ENDEMIC SPECIES of small cichlid (Moyle & Moyle 1995). This not only has had profound impacts on subsistence fisheries and those who depend on them, but also represents one of the greatest vertebrate extinction episodes of modern times. It is perhaps ironic that studies of the ADAPTIVE RADIATION of these fishes have made a significant contribution to our understanding of evolutionary biology (Bruton 1995). At the intraspecific level, introductions of exotic genetic material are feared to have had effects on wild populations of salmonids in Europe and North America.

Aquatic organisms are the only important food resource largely derived from hunting. There is a growing consensus that current levels of exploitation of many, if not most, commercially fished stocks are unsustainable. The problems are being compounded by modern fishing methods, through which as much as a third of nominal catches is caught and discarded or landed unrecorded (Pauly & Christensen 1995). However, the effect of significant reductions in POPULATION SIZE on genetic VARIABILITY, through OVERFISHING, has been poorly studied (Voigtlander 1994). Aquaculture, once believed to be a means by which pressures on wild stocks would be reduced, is now recognized as having some adverse impacts on fish diversity, not only through introductions of exotic species and strains and habitat destruction (see above), but also because fishmeal and fish oil are important ingredients in commercial aquaculture feeds (Beveridge *et al.* 1994).

Conservation efforts

The conservation of aquatic BIODIVERSITY remains comparatively neglected and there has been less effort on the conservation of fishes than any other vertebrate group. Traditionally, fish have been regarded from a largely utilitarian point of view, as a resource to exploit to the maximum for their food potential, a situation that has few parallels among terrestrial vertebrates. Fish are further disadvantaged by being largely invisible, and by being generally perceived as cold, wet and slimy: hence, the emotionally loaded publicity campaigns employed to save other threatened vertebrates are unlikely to be effective (Voigtlander 1994).

The causes of loss of fish diversity may be understood (in general terms, at least), but how is fish diversity best conserved? Tools are readily available: for example, there are guidelines to minimize risks associated with translocating fish species, and the technology for producing sterile fish for farming is well developed. In practice, however, the conservation of fishes is complicated both by political issues, such as national sovereignty and the fulfilment of societal needs, and by biological considerations, such as fishes' unusual potential for hybridization. While it may be accepted by fisheries scientists that the burden of proof should be switched from project opponents to proponents, so that developers would have to prove that stocking, shoreline development and hydrological regulation, etc., will have no effect on fish diversity, the change is unlikely to come about in the short term because decision-makers tend to be ignorant of fundamental ecological and evolutionary tenets (Voigtlander 1994). In the long term, however, such changes may be achieved through increased public awareness of the importance of biodiversity. *See also* CONSERVATION BIOLOGY. [M.C.M.B.]

REFERENCES

Beveridge, M.C.M., Ross, L.G. & Kelly, L.A. (1994) Aquaculture and biodiversity. *Ambio* **23**, 497–502.

Bruton, M.N. (1995) Have fishes had their chips? The dilemma of threatened fishes. *Environmental Biology of Fishes* **43**, 1–27.

Moyle, P.B. & Moyle, P.R. (1995) Endangered fishes and economics: intergenerational obligations. *Environmental Biology of Fishes* **43**, 29–37.

Nelson, J.S. (1994) *Fishes of the World*, 3rd edn. John Wiley & Sons, New York.

Pauly, D. & Christensen, V. (1995) Primary production required to sustain global fisheries. *Nature* **374**, 255–257.

Voigtlander, C.W. (ed.) (1994) *The State of the World's Fisheries Resources*. Oxford and IBH Publishing Co., New Delhi.

Welcomme, R.L. (1988) *International Introductions of Inland Aquatic Species*. FAO Fisheries Technical Paper 294. Food and Agriculture Organization, Rome.

fish ecology *See* ECOLOGY.

fish farming The culture of fish (often used broadly to include both fish and molluscs!), in indoor/outdoor tanks or enclosures in natural water bodies, for human consumption. *See also* AQUACULTURE. [P.C.]

fish kill Observed MORTALITY, usually extensive, in natural fish populations signalling POLLUTION from known, suspected or unknown causes. [P.C.]

fish stocks The standing stock is the total mass of fish present at a given time in a fishery. Understanding the dynamics of fish stocks is a primary goal of fisheries biology and includes study of the rates at which fish grow, die and reproduce. Fish stocks may be estimated by analysis of commercial fish catches, by scientific sampling of fish populations, or by extrapolating from estimates of PRIMARY PRODUCTIVITY. Fisheries managers attempt to estimate the MAXIMUM SUSTAINABLE YIELD, i.e. the largest number of fish that can be harvested year after year without diminishing the stocks. Stock RECRUITMENT models are fisheries models that predict the amount of juvenile recruitment as a function of the parent stock. In recent years, there have been numerous examples of declining fish stocks of all types in all areas of the world, in part due to OVERFISHING and in part due to poorly understood environmental factors. *See also* COMMERCIAL FISHING; FISHERIES, CONSERVATION AND MANAGEMENT. [V.F.]

fish tags *See* MARKING METHODS.

Fisher–Ford method A robust method for estimating numbers from a series of mark–release–recapture samplings of an animal POPULATION. Survival rate is assumed to be constant through the SAMPLING period and estimates are provided for all sample dates except the first. It was developed in the 1940s for studies of populations of Lepidoptera. Statistically, the method has now been superseded, but it is referred to widely in the literature. *See also* CAPTURE–RECAPTURE TECHNIQUES; SAMPLING METHODOLOGY/DEVICES. [L.M.C.]

REFERENCE

Fisher, R.A. & Ford, E.B. (1947) The spread of a gene in natural conditions in a colony of the moth *Panaxia dominula* L. *Heredity* **1**, 143–174.

Fisher, R.A. *See* CHARACTERS IN ECOLOGY.

Fisheries, conservation and management

Commercially exploited FISH STOCKS have become an increasingly important component of the world's food supply. This is best demonstrated by examining the 15 years between 1955 and 1970, when the total world catch more than doubled from 30 million tonnes to more than 70 million tonnes per year (Pitcher & Hart 1982). World catches have continued to increase and in 1992 this figure rose to 98 million metric tonnes (FAO 1994a). This need for protein has also given rise to an increase in world AQUACULTURE production, and has added a further 14 million metric tonnes to the wild fish exploitation figure (FAO 1994b). In addition to the capture of fish for human consumption, the demand for fishmeal as a high-protein diet for intensive AGRICULTURE and aquaculture has also increased.

Management of marine fisheries

Fisheries biologists have long realized the need to manage the commercial exploitation of aquatic and marine RESOURCES. This need for the conservation of marine fish stocks has been recognized by the international community, and this has led to the formation of advisory and legislative bodies, for example the International Council for the Exploration of the Sea (ICES). Such organizations have implemented a number of control measures, for example EXCLUSIVE ECONOMIC ZONES (EEZ) and total allowable catches (TAC).

Fisheries management encompasses four basic objectives: (i) to maximize annual YIELD from the resource; (ii) to prevent stock collapse; (iii) to maximize economic functions (i.e. profit); and (iv) to achieve a socio-economic balance (i.e. employment). The three principal means of achieving these objectives are:

1 Control of fishing effort, either indirectly by imposing quotas or closed seasons, or directly by effort control.

2 Controlling the age or size at which fish are first caught by having tighter control on the mesh sizes used or by excluding fishing vessels from fish nursery areas or known juvenile habitats.

3 Measures to enhance either stock or habitat.

Stock assessment models

Early stock assessment models designed to estimate stock sizes based on age profiles were published in the 1930s (Graham 1935). Many of these early single-unit stock, or 'surplus yield', models (Shaefer 1954) treated the stock as a single entity with a 'natural rate of increase'. Such models made several assumptions about the underlying laws of population growth and were concerned only with inputs and outputs to the BIOMASS of the population. However, more mathematically elaborate multi-unit stock models were developed which considered the fate of individual COHORTS or year classes from first exploitation until extinction (Beverton & Holt 1957). To operate effectively, these models required fishing and natural MORTALITY data as well as information on growth and FECUNDITY. These are termed 'dynamic pool' or 'yield per recruit' models.

The reliance of fisheries biologists on stock assessment models to manage and conserve marine fish populations has been criticized in recent years after they failed to predict declines in the Peruvian anchovy (*Engraulis ringens*) fishery (Pitcher & Hart 1982) and cod (*Gadus morhua*) stocks of the Canadian Grand Banks (MacKenzie 1995). The inaccuracy of these models appears to be due to a poor understanding of the factors which influence RECRUITMENT within fish populations, an inability to integrate sporadic climatological/oceanographic events (e.g. EL NIÑO) into the model, and the unavailability of unbiased fish population data.

Advances in quantitative echo-sounding technology and in the development of efficient fish-sampling gear have meant that the availability of reliable fish population data has improved in recent years. This may lead to a significant improvement in the efficacy of these predictive models. The provision of accurate predictive data may also aid the development of fishery simulation models which can then be used to forecast the effects of new management and conservation strategies.

Freshwater fisheries

FRESHWATER fisheries are less reliant on stock assessment models than their marine counterparts. Management strategies are based almost entirely on actual population parameters and CATCH PER UNIT EFFORT (CPUE) data. Historically, methods such as restocking, predator removal, habitat restoration, the implementation of closed seasons and fish size/catch limits have been used to augment freshwater fisheries. Current theory suggests that freshwaters should be managed on a whole-catchment scale. This integrated approach acknowledges the role that other users of the aquatic and terrestrial environment may have on the well-being of both riverine and lacustrine fish stocks. *See also* CONSERVATION BIOLOGY; HARVESTING; OPTIMAL HARVEST. [C.W.B.]

REFERENCES

Beverton, R.J.H. & Holt, S.J. (1957) On the dynamics of exploited fish populations. *Ministry of Agriculture, Fisheries and Food, Fisheries Investigations, UK, Series 2*, **19**, 1–533.

FAO (1994a) *Yearbook of Fishery Statistics*, Vol. 74. Food and Agriculture Organization, Rome.

FAO (1994b) *Aquaculture Production 1986–1992*. FAO Fisheries Circular No. 815. Food and Agriculture Organization, Rome.

Graham, M. (1935) Modern theory of exploiting a fishery and application to North Sea trawling. *Journal du Conseil International pour L'Exploration de la Mer* **10**, 264–274.

MacKenzie, D. (1995) The cod that disappeared. *New Scientist* **1995**, 24–29.

Pitcher, T.J. & Hart, P.J.B. (1982) *Fisheries Ecology*. Chapman & Hall, London.

Shaefer, M.B. (1954) Some aspects of the dynamics of populations important to the management of commercial marine fisheries. *Bulletin of the Inter-American Tropical Tuna Commission* **1**, 27–56.

Fisher's alpha *See* DIVERSITY INDICES.

Fisher's exact test This is a test for ASSOCIATION in a 2×2 CONTINGENCY TABLE. It is called an exact test because the exact probabilities of every similar 2×2 table, with the same marginal totals, are calculated. It is used when expected values are less than 5 and the CHI-SQUARED TEST (χ^2) becomes approximate. Suppose we have collected 9 fungi from a wood, 5 from shaded locations, 4 from open locations. We record the emergence of a certain fly species from some of these fungi. Is there any reason to suspect that the location of the fungus influences the pattern of emergence?

	No emergence	Emergence	Total
Open	4	1	5
Shade	0	4	4
Total	4	5	9

We need to know if this is an extreme, and unlikely, arrangement of the four observations within the central cells of the table, given the observed marginal totals. Five other combinations of the data would have been possible, and these are:

4	1
0	4

0.0397

3	2
1	3

0.3175

2	3
2	2

0.4762

1	4
3	1

0.1587

0	5
4	0

0.0079

Poor emergence from 'open' → Poor emergence from 'shade'

and their associated probabilities are given under each table. If the numbers in a 2×2 table are represented by:

a	b	a + b
c	d	c + d
a + c	b + d	n

then the formula for the appropriate PROBABILITY is:

$$P = \frac{(a+b)!(c+d)!(a+c)!(b+d)!}{a!b!c!d!n!}$$

where $x!$ means multiply together all the numbers from 1 to x (but $0! = 1! = 1$). Now we must decide which probability is appropriate. Is it a ONE-TAILED TEST or a TWO-TAILED TEST? If we could not imagine that 'open' sites would produce more emergence than 'shade' sites then it is a one-tailed test and the observed table is the most extreme result possible. Since the probability of getting this by chance is only $P = 0.0397$ (which is less than 0.05), this is significant. If we have no prior reason to expect 'open' or 'shade' to produce fewer emergences then it is a two-tailed test, and the probability of getting our extreme result, or anything worse, is $P = 0.0397 + 0.0079 = 0.0476$, which is still significant. *See also* STATISTICS. [B.S.]

Fisher's fundamental theorem The theorem, as stated by Fisher (1958, p. 37), that 'the rate of increase in FITNESS of any organism at any time is equal to its GENETIC VARIANCE in fitness at that time.' Despite a full interpretation of this theorem by Price (1972) it is still misunderstood to mean that selection will lead continually to an increase in average fitness. Numerous authors have shown that this interpretation of the theorem is incorrect since there are many circumstances (e.g. presence of EPISTASIS, LINKAGE DISEQUILIBRIUM) in which average fitness does not increase (Frank & Slatkin 1992). The change in fitness can be partitioned into two components, one due to NATURAL SELECTION and one due to environmental change. Fisher's theorem only refers to the first component and means that the change in this component of fitness is equal to the additive variance in fitness. Changes in environmental conditions may, however, cause total fitness to decline. [D.A.R.]

REFERENCES

Fisher, R.A. (1958) *The Genetical Theory of Natural Selection*, 2nd edn. Dover Publications, New York.

Frank, S.A. & Slatkin, M. (1992) Fisher's fundamental theorem of natural selection. *Trends in Ecology and Evolution* **7**, 92–95.

Price, G.R. (1972) Fisher's fundamental theorem made clear. *Annals of Human Genetics* **36**, 129–140.

Fisher's sex-ratio theory A theory proposed by R.A. Fisher in 1930 to explain the EVOLUTION of stable SEX RATIOS. The theory predicts that a stable ratio of males and females in a population will be that at which the parental expenditure on males and females is equal. If the costs of a son and a daughter are equal, the theory predicts a stable sex ratio of 1 : 1. Whenever the sex ratio deviates from 1 : 1, the rarer sex has a selective advantage (i.e. will pass relatively more of its genes to the next generation) and so the population is predicted to return to an equilibrium value of 1 : 1. In fact, most dioecious species have sex ratios close to one. Deviations from this can occur when, for example, a gene affects its own likelihood of being transmitted (i.e. MEIOTIC DRIVE), when not all males or females have an equal chance of mating (i.e. local mate or resource COMPETITION), in cases of self-fertilization, or when interference between functions selects for unequal allocations. *See also* SEX ALLOCATION THEORY. [V.F.]

REFERENCE

Michod, R.E. & Levin, B.R. (eds) (1988) *The Evolution of Sex—An Examination of Current Ideas.* Sinauer Associates, Sunderland, Mass.

fission

1 A mode of ASEXUAL REPRODUCTION.

2 NUCLEAR FISSION is a radioactive decay process in which a heavy, NEUTRON-rich nucleus splits to produce two lighter nuclei, known as fission fragments, or fission products, neutrons and energy. Spontaneous nuclear fission affects only nuclides with ATOMIC NUMBER

(*Z*) ≥90 and MASS NUMBER (*A*) ≥230 and occurs as a minor decay mode in competition with alpha decay for some nuclides (e.g. ^{238}U has a spontaneous fission half-life of about 10^{16} years) but, for a limited number of nuclides (e.g. ^{250}Cm, ^{256}Fm and ^{254}Cf), it is the main mode of decay. Induced fission by thermal neutron irradiation of ^{235}U or ^{239}Pu is used to produce a self-sustaining chain reaction in nuclear reactors and nuclear bombs. Fission of a single nucleus of ^{235}U or ^{239}Pu releases about 200 MeV of energy, which is translated mainly into kinetic energy of the fission fragments. Fission products from ^{235}U range from zinc to terbium, in a biomodal distribution with maxima at mass numbers about 95 and 138. Radioactive fallout from nuclear explosions and releases from nuclear facilities have resulted in low-level global contamination of the environment with fission products. Spontaneous nuclear fission of ^{238}U in minerals produces fission tracks along the paths of the fission fragments and quantification of the number of tracks can be used to estimate the age of minerals or the time since they were last subject to thermal annealing. *See also* RADIOACTIVE FALLOUT; RADIOACTIVE WASTE; URANIUM.

[A.B.M.]

REFERENCES

Frieleander, G., Kennedy, J.W., Macias, E.S. & Miller, J.M. (1981) *Nuclear and Radiochemistry*. John Wiley & Sons, New York.

Keller, C. (1988) *Radiochemistry*. Ellis Horwood, Chichester.

Fitness

NATURAL SELECTION on a PHENOTYPE or GENOTYPE involves a quantity known by the technical term 'fitness'. 'Fitness' indicates a short-term measure of expected reproductive success. Fitness is not a property of an individual: fitness summarizes a model for genotypic DEMOGRAPHY that makes certain assumptions about the population and the genotypic and individual interactions within it. In order to arrive at genotypic fitness in a population, the genotypic differences between fitness components such as age-dependent viability, age-dependent FECUNDITY and age at maturity should be known; one should also know whether individuals can be regarded as independent in survival and reproduction, or whether they interact.

In demographic recurrence equations, the number of independent units in the population is specified for two successive time units. If individuals are independent in reproduction and survival and mating is at random between GAMETES, the demographic recurrence equations for natural selection specify the number of each genotype born at time *t* given the number of each genotype present at time $t-1$. The demographic recurrence equations for genotypic numbers are accounted in the earliest life-history stage that lives independently of other individuals; the count is made at reproduction, and includes newborns, but before survival to the next year. The contribution of individuals of age *x* and genotype A_iA_j that are alive at time *t* to the gamete pool at time *t* is given by the summation over their number, $n_{ij,x,t}$, times their average number of gametes, $m_{ij,x}$:

$$\sum\nolimits_{x\geq 1} n_{ij,x,t} m_{ij,x}$$

The definition of demographic genotypic fitness, *W*, is (de Jong 1994): the total contribution by a genotype to the gamete pool at time *t*, divided by the number of newborns of the same genotype at time $t-1$:

$$W_{ij} = \sum\nolimits_{x\geq 1} n_{ij,x,t} m_{ij,x} / n_{ij,0,t-1}$$

Demographic genotypic fitness (with random mating of gametes and equal fitness of males and females) leads to:

$$n_{0,t} = n_{0,t-1}\overline{W}_{t-1}$$
$$\Delta p = \tfrac{1}{2} p_{t-1} q_{t-1} \mathrm{d}\overline{W}_{t-1} / \mathrm{d}p_{t-1}$$

The number of individuals of age *x* and genotype A_iA_j at time *t*, $n_{ij,x,t}$, can be written in the total number of their COHORT at birth, $n_{0,t-x}$, the genotype frequency in their cohort at birth, $x_{ij,t-x}$, and the survival from birth to age *x*, $l_{ij,x}$:

$$W_{ij} = \sum\nolimits_{x\geq 1} n_{ij,x,t} m_{ij,x} / n_{ij,0,t-1}$$
$$= \sum\nolimits_{x\geq 1} \left(n_{0,t-x} x_{ij,t-x} l_{ij,x} m_{ij,x} \right) / \left(n_{0,t-1} x_{ij,t-1} \right)$$

Demographic genotypic fitness is, in general, both density- and frequency-dependent and depends upon time *t*. At polymorphic equilibrium, and approximately with weak selection, $x_{ij,t-x} = x_{ij,t-1}$. Assuming weak selection, demographic genotypic fitness therefore becomes independent of the genotype frequency:

$$W_{ij} = \sum\nolimits_{x\geq 1} \left(n_{0,t-x} l_{ij,x} m_{ij,x} \right) / n_{0,t-1}$$

With a STABLE AGE DISTRIBUTION and a constant POPULATION GROWTH RATE, λ, demographic genotypic fitness becomes density independent and independent of time, *t*:

$$W_{ij} = \sum\nolimits_{x\geq 1} \left(n_{0,t} \lambda^{-x} l_{ij,x} m_{ij,x} \right) / \left(n_{0,t} \lambda^{-1} \right)$$
$$W_{ij} = \sum\nolimits_{x\geq 1} \left(\lambda^{-x+1} l_{ij,x} m_{ij,x} \right)$$
$$= \sum\nolimits_{x\geq 1} \left(e^{-r(x+1)} l_{ij,x} m_{ij,x} \right)$$

The expression for the growth rate of a population of one genotype is similar, and *r*, the rate of increase of that population, is widely felt to be appropriate as a summarizing measure of the fitness of the genotype (Partridge & Harvey 1988).

Numerically, the demographic genotypic fitness, *W*, therefore equals λ times the REPRODUCTIVE VALUE at birth, $\mathrm{RV}_{0,ij} = \Sigma_{x\geq 1} (\lambda^{-x} l_{ij,x} m_{ij,x})$. Reproductive value is, however, defined over a cohort, and involves summation of survival and reproduction over all ages the cohort reaches, while demographic genotypic fitness, *W*, is defined from the genotypic contribution to the newborn independent individuals in one time unit, and involves summation over reproducing individuals of all ages at the same moment of time. This difference in definition joins fitness to population growth rate, in contrast to reproductive value, RV_x, net reproductive rate, R_0, and LIFETIME REPRODUCTIVE SUCCESS, LRS; these three quantities refer to a cohort. However, at equilibrium population number $\lambda = 1$ and RV_0, R_0, and average LRS become identical in value to *W*. Population rate of increase, *r*, RV_0, R_0, and average LRS are widely used as measures of fitness (Kozlowski 1993).

When mating between individuals (rather than between gametes) is at random, and the combination of genotypes of male and female determines fecundity, the demographic fitnesses differ. The individual is no longer an independent unit in reproduction;

the pair is the independent unit and fitness in a sexual population has therefore to refer to the genotypic combination in the pair. Fecundity of the pair depends upon the genotypic combination in the pair. It is possible to define a demographic pair fitness and to define a male fitness up to reproducing, in a manner analogous to the definition of demographic genotypic fitness. ALLELE FREQUENCY change depends only upon demographic pair fitness. Demographic pair fitness and the growth rate of a population made up of that genotype are not equivalent. The growth rate of a population of one genotype becomes $\lambda = W_{pair}/W_{♂}$, including male fitness as well as demographic pair fitness; the division indicates that the growth rate of a population of one genotype mainly depends upon female fitness. The growth rate of a population of one genotype cannot be used to predict demographic pair fitness and the outcome of selection. Even if fecundity is only dependent upon the female's genotype, the pair fitness that has to be used in making life-history predictions is not equal to the growth rate of a population of one genotype (Abugov 1986; de Jong 1994).

The appearance of genotypic number in demographic genotypic fitness shows that fitness depends not only on individual fitness components like viability and fecundity, but acquires a component that derives from properties at the population level. For density-dependent regulation, this implies that density variation in space and in time influences overall fitness. Density variation in time has potentially a very large effect. Interest is in what would happen when POPULATION SIZE is not constant, but on average neither increases nor decreases. In a population showing some stochastic variation in number, assuming R_0 and fitness identical seems good enough for a qualitative prediction. Neither arithmetic nor geometric mean demographic genotypic fitness is predictive of the actual genotypic dynamics when density variation in time has non-linear dynamics. The full model of genotypic dynamics must then be developed.

To find fitness from fitness components means constructing a model for genotypic demography, in accordance with life-history CONSTRAINTS such as pair fitness or parent–offspring interactions in fitness, and in accordance with population characteristics such as inhomogeneous interactions between individuals. It is the detail of the model that is informative, as much as the summarizing quantity called fitness.

The fundamental statement about natural selection on a TRAIT is the existence of a genetic covariance between trait values and fitness values. Trait values can be measured occasionally for a genotype, demographic genotypic fitness values summarize a genotypic life history and its population background. The change in mean GENOTYPIC VALUES of a trait depends upon this genetic covariance; this is Robertson's secondary theorem of natural selection. *See also* AGE AT MATURITY IN PLANTS; FITNESS PROFILE; INTRINSIC RATE OF INCREASE. [G.D.J.]

REFERENCES

Abugov, R. (1986) Genetics of Darwinian fitness. III. A generalized approach to age structured selection and life history. *Journal of Theoretical Biology* **122**, 311–323.

de Jong, G. (1994) The fitness of fitness concepts and the description of natural selection. *Quarterly Review of Biology* **69**, 3–29.

Kozlowski, J. (1993) Measuring fitness in life-history studies. *Trends in Ecology and Evolution* **8**, 84–85.

Partridge, L. & Harvey, P.H. (1988) The ecological context of life history evolution. *Science* **241**, 1449–1455.

fitness components *See* LIFE-HISTORY TRAITS.

fitness landscape A FITNESS landscape can be a useful way of representing the variation in fitness or reproductive success of a PHENOTYPE or GENOTYPE across a surface representing environmental variation, time or space. Peaks in the landscape can be referred to as adaptive peaks, with intervening valleys, or troughs, of low fitness. *See also* ADAPTIVE LANDSCAPE; ECOLOGICAL GENETICS. [P.M.B.]

fitness profile The mathematical function, also called fitness function, describing the relationship between a QUANTITATIVE TRAIT and FITNESS (Fig. F4). Fitness profiles reflect the direction and strength of NATURAL SELECTION. Monotonic fitness profiles correspond to DIRECTIONAL SELECTION; those with an intermediate optimum within the population range of variability describe STABILIZING SELECTION. Fitness profiles with an intermediate optimum are often asymmetric, which results in higher fitness costs of deviating in one direction from the optimum than in the other. In these circumstances selection may shift the population mean off the optimum in the direction of slower decline of fitness. Flat fitness profiles facilitate maintenance of genetic variation. The fitness profile for a TRAIT often changes with the values of other traits and

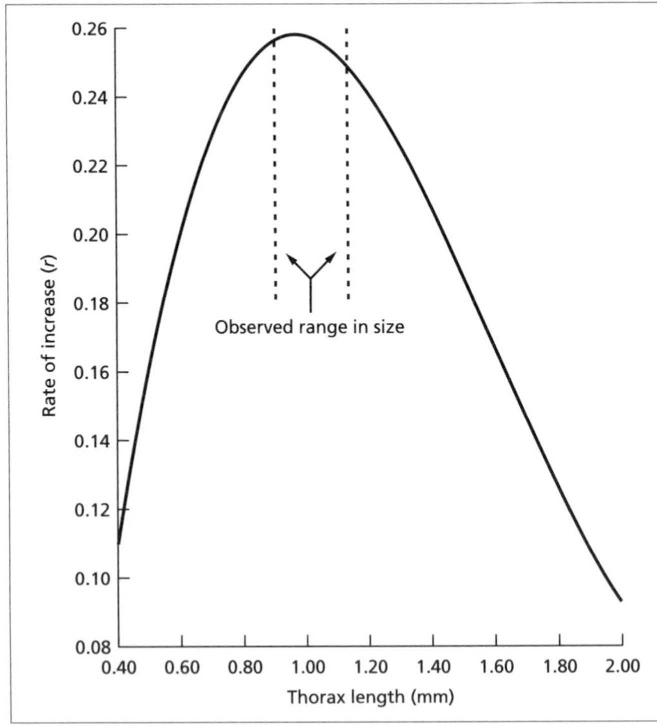

Fig. F4 A fitness profile. Estimated relationship between fitness, as measured by the logarithmic rate of increase (*r*), and thorax length in *Drosophila melanogaster*. (After Roff, D.A. (1981) On being the right size. *American Naturalist* **118**, 405–422, University of Chicago Press.)

the environment. It may also depend on the distribution of phenotypes in the population (FREQUENCY-DEPENDENT SELECTION) or on POPULATION DENSITY (DENSITY-DEPENDENT SELECTION). [T.J.K.]

fixed-quota harvesting A HARVESTING strategy in which harvesting activity is regulated by a fixed catch level. This catch level is usually set below the maximum that can be sustained and thus, in theory, should mean that the resource is protected from severe exploitation. The main problem with this type of harvesting is that it does not involve any natural FEEDBACK between the POPULATION SIZE and the level of harvesting. Thus, if for any reason, the population falls below the level at which its natural surplus is less than the fixed quota, then the decline will continue.

Although in theory this is a rather unattractive mechanism, it is by and large the main basis of managing FISH STOCKS in the developed world. Such stocks are managed under the regulations which involve a total allowable catch (TAC). In the major fisheries management bodies, regular scientific MONITORING of the stock size means that the TAC can vary from year to year in the light of information on the size of the stock. This avoids the problem of continued depletion of the stock by a constant quota. *Cf.* CONSTANT-EFFORT HARVESTING. *See also* FISH, DIVERSITY AND CONSERVATION STATEGIES. [J.B.]

fjord A narrow, deep, steep-walled inlet, formed either by the submergence of a glaciated mountainous coast or by entrance of the OCEAN into a deeply excavated glacial trough after the glacier melts. Fjords are often bounded on their seaward end by a bedrock sill which marks the approximate maximum advance of the ICE. The deep waters of fjords are almost completely isolated from the surface CIRCULATION; exchange occurs only when water at the sill is denser than the bottom water and can flow into the fjord to displace it. Due to this limited exchange, the deep waters of fjords tend to become depleted of OXYGEN. [V.F.]

flagship species Popular, and usually charismatic, species that provide a symbol and focus for conservation awareness and action, for example the panda and dolphins. *See also* CONSERVATION BIOLOGY. [P.C.]

flocculation The formation of clots or aggregates of CLAY minerals. Clay particles, which have a high surface area per unit volume, in FRESHWATER attract a shell of negatively charged IONS (e.g. OH^-) around their positive crystal surface which then results in the clay particles repelling each other. Upon contact with seawater, the negatively charged clay particles bind with the abundant positive ions present in seawater (e.g. Na^+, Mg^{2+}). The particles thus become neutralized and stick together into aggregates. [V.F.]

floods Flows of WATER on areas of land not normally covered with water and which cause or threaten to cause damage. Floods occur when river DISCHARGE exceeds bank capacity. These discharges arise from short-duration intense rainfall; prolonged heavy rainfall; rapid snowmelt; and catastrophic events such as dam failures or landslides into a water body. Worldwide on average there are *c.* 30 major flood disasters per year, more than any other form of NATURAL DISASTER (Table F2). However, loss of life from flooding is, in most years, lower than that caused by DROUGHT, CYCLONES and earthquakes. Loss of life due to flooding occurs predominantly in the third world, whereas monetary loss dominates such events in the First World. Floods can be controlled by CATCHMENT management to absorb water and retard flows; by engineering to enlarge flow and storage capacity; and by economic planning to minimize value at risk and to reinstate damage, through insurance. *See also* CLIMATE CHANGE; CLIMATE CHANGE, POTENTIAL IMPACTS ON BIOLOGICAL SYSTEMS. [A.M.]

REFERENCE

Swedish Red Cross (1984) *Prevention is Better Than Cure*.

floral province A BIOGEOGRAPHICAL UNIT based on the component plants; one that is subordinate to a subkingdom, in the usage of Takhtajan (1969). It is the exact equivalent of a floral PROVINCE of Neill (1969), but Takhtajan's usage is preferred here. [W.G.C.]

REFERENCES

Neill, W.T. (1969) *The Geography of Life*. Columbia University Press, New York.

Takhtajan, A. (1969) *Flowering Plants, Origin and Dispersal*. Oliver & Boyd, Edinburgh.

Floras, major published Floras give a taxonomic (systematic) treatment of plants of a particular geographical area, which may be a country, region or smaller area. Floras may cover one or several major plant groups; most include vascular plants only. Treatment may be general but more detailed for small areas, with full descriptions of taxa and information, for example on habitat, distribution and ecology. Many Floras have keys for identification; not all are illustrated. Some major Floras are complete (but may be subject to revision). Among these are the early *Illustrierte Flora von Mittel-Europa* (completed 1931), *Vascular Plants of the Pacific Northwest, Flora S.S.S.R., Flora Republicii Socialiste România* and *Flora of Turkey*. Five volumes of *Flora Europaea* were issued between 1964 and 1980; a much revised second edition of Volume 1 (Psilotaceae to Platanaceae) was published in 1993. Major Floras still incomplete include *Flora Malesiana, Flore de l'Afrique du Nord* and *Flora Neotropica*. Frodin (1984) gives a guide to the numerous Floras of the world. Coverage is uneven; the floras of some tropical and southern-hemisphere countries (e.g. Uruguay) are poorly known, but that of Europe is well documented. The treatment of non-vascular plants is more limited, although some major works exist, for example *Moss Flora of Britain and Ireland* and *Hepaticae and Anthocerotae of North America*. Vascular plants in Britain are covered in the *Flora of the British Isles* by Clapham, Tutin & Warburg (1952, 1962); 3rd edn, Clapham, Tutin & Moore (1987). Stace's *New Flora of the British Isles* was issued in 1991 and a revised second edition in 1997. [A.J.W.]

Table F2 Average annual number of major disasters in the world. Simplified from Swedish Red Cross (1984) *Prevention is Better Than Cure*. 1980s' data is incomplete.

	1960s	1970s	1980s
Floods	15	24	27
Storm	13	15	17
Earthquake/tsunami	7	8	12
Drought/famine	5	10	11

REFERENCE

Frodin, D.G. (1984) *Guide to Standard Floras of the World*. Cambridge University Press, Cambridge.

floristics The totality of plant species found in a given site. [P.D.M.]

flower structure *See* POLLINATION.

fluctuations Variations over time, for example in the size of a POPULATION or in environmental variables such as temperature, salinity, pH, etc. [V.F.]

flue gas The gases vented by a chimney to the ATMOSPHERE in static fossil-fuel combustion. The major POLLUTANT flue gases are sulphur dioxide (SO_2) (*see* SULPHUR OXIDES) and NITROGEN OXIDES (NO and NO_2), which are precursors to acid deposition (*see* ACID RAIN AND MIST). FLUE-GAS DESULPHURIZATION (FGD) systems can achieve 90% reductions in SO_2 by spraying a slurry of LIME into the flue gas to absorb and oxidize SO_2 to calcium sulphate. Flue-gas DENITRIFICATION (FGN) is less efficient and relies on burner redesign or injection of urea or AMMONIA into the combustion zone to effect chemical conversion of NO to N_2. [M.R.H.]

flue-gas denitrification (FGN) *See* FLUE GAS.

flue-gas desulphurization (FGD) A system for removing sulphur dioxide (SO_2) from gas emitted from a chimney after combustion in the burner that it is venting. SO_2 is absorbed by a limestone slurry spray to form calcium sulphite, which is then oxidized to calcium sulphate, which, together with WATER, forms gypsum ($CaSO_4.2H_2O$). Removal of SO_2 from the EMISSIONS of coal-fired power-stations makes a positive contribution to reducing acid rain (*see* ACID RAIN AND MIST). The main disadvantages are: (i) that considerable quantities of limestone are needed; and (ii) that not all the gypsum produced may be used efficiently, so creating difficulties with storage and problems from possible leachates. [P.C.]

fluid-feeders *See* FEEDING TYPES, CLASSIFICATION.

fluorine (F) A poisonous gaseous halogen element (ATOMIC NUMBER=9; relative ATOMIC MASS=18.998; melting point=−219.62°C; boiling point=−188.14°C) with an abundance of 0.0544% in the Earth's CRUST; it is found as the mineral fluorite, CaF_2. Fluorine gas is produced by electrolysis of potassium fluoride dissolved in anhydrous hydrogen fluoride. It is extremely reactive and combines (often explosively) with most ELEMENTS. It is used to manufacture uranium hexafluoride for enrichment of uranium in NUCLEAR POWER-STATIONS. Although the free element is extremely toxic, low concentrations of fluoride ions in water supplies provide protection for teeth. Higher concentrations of fluoride are toxic. Important fluorine compounds include CFCs, and hydrogen fluoride, which is used to increase the octane number of gasoline. [C.R.P.]

flushing rate *See* LAKES.

fluviatile lakes *See* LAKES, TYPES OF.

Fluvisols *See* SOIL CLASSIFICATION.

fluxes Rates of transfer; for example, of CARBON or energy in FOOD CHAINS, or of elements in BIOGEOCHEMICAL CYCLES. [P.C.]

fly ash The residue produced after burning pulverized fuel, for example coal in power-stations, or rubbish in incinerators. Although very efficiently removed from exhaust gases by electrostatic precipitators, pulverized fuel ash (PFA) is found ubiquitously as an AEROSOL in the ATMOSPHERE, and in RAIN, and may be transported for hundreds of kilometres from its source. PFA particles are characterized microscopically as spherical siliceous particles with diameters in the range 0.1–10 μm. Fly ash removed from exhaust gases by precipitators is often used as LANDFILL, or for LAND RECLAMATION at coastal sites. [J.N.C.]

fly tipping The illegal disposal of solid WASTE in an unauthorized manner and location. [P.C.]

fodder Dried, cured plant material (e.g. hay and straw) used as feed for animals. [P.C.]

fog *See* PHOTOCHEMICAL SMOG.

Food and Agriculture Organization (FAO) *See* UNITED NATIONS PROGRAMMES AND BODIES.

food chain A series of organisms linked by their FEEDING relationships. Each link in the chain feeds on, and obtains energy from, the preceding link and provides food and energy for the following link. The number of links in the chain is limited: for example primary producer → herbivore → primary carnivore → secondary carnivore. This could be related to energy constraints (*see* ENERGY-FLOW HYPOTHESIS), but in general there is no correlation between chain length and PRIMARY PRODUCTIVITY. Another possibility is that the interacting system becomes inherently less stable as links in the food chain increase. In ecological energy transformations between trophic units a large proportion of the potential chemical energy in the food is usually used for basic maintenance and to fuel activity so is lost via RESPIRATION. This leaves only a small proportion incorporated as BIOMASS of that link to support the following link. Thus food chains are rarely more than three to five links long.

Food chains are of three types.

1 The PREDATOR chain, which starts from a plant base then to herbivores and then to carnivores, and usually goes from smaller to larger animals.

2 The PARASITE chain, which goes from larger to smaller animals.

3 The saprophytic chain, which goes from dead organic matter through decomposer organisms.

Cf. TROPHIC CONTINUUM. *See also* FOOD WEB. [M.H.]

food selection Using the net rate of energy gain as currency, models predicting optimal diets have been applied to all major trophic categories except endoparasites. An early model developed for consumers of discrete items such as prey, fruit or seeds, has provided the foundation for subsequent development. This OPTIMAL DIET MODEL makes a number of restrictive assumptions, including the sequential, rather than simultaneous, evaluation of food items, the instantaneous recognition by the forager of different food 'types' and the lack of any influence of ingestive conditioning or learning on preferences and food-handling skills. Food items are assignable to types according to profitability, defined as the ratio of net energy yield to HANDLING TIME (time required for apprehension and to complete ingestion in readiness for renewed search). Available types can be ranked from highest to lowest in profitability. Foragers are predicted to always accept the most profitable items, but to accept lower-ranking items only when encounter rates with better items fall below a THRESHOLD. Expansion of the diet is predicted to be a step function, items either being always accepted or always rejected. This last prediction has been violated in most tests of the model, but such observed 'partial preferences' can be accommodated by relaxing some of the model's assumptions. The many ways of doing this have generated new insights into the CONSTRAINTS operating on foragers.

The optimal diet model does not apply to bulk processors, such as large, herbivorous mammals and deposit-feeding invertebrates, which require a different modelling approach. Food items are consumed simultaneously in large quantities, whence processing rate may be more important than SELECTIVE FEEDING in determining the rate of net energy gain. *See also* BEHAVIOURAL ECOLOGY; FORAGING, ECONOMICS OF; OPTIMAL FORAGING THEORY. [R.N.H.]

REFERENCES

Hughes, R.N. (ed.) (1993) *Diet Selection: an Interdisciplinary Approach to Foraging Behaviour*. Blackwell Scientific Publications, Oxford.

Stephens, D.W. & Krebs, J.R. (1986) *Foraging Theory*. Princeton University Press, Princeton, N.J.

Food web

Food webs are ways of describing the structure of ecological communities, and the trophic interactions which occur between the component units of a COMMUNITY. They illustrate that FOOD CHAINS are not isolated sequences but are interconnected with each other. A simple example is shown in Fig. F5, which not only illustrates the interlocking nature of food chains and the different trophic levels within them, but also that some organisms occupy intermediate positions between the three trophic levels; for example, the net-spinning caddis-fly larvae feed on both plant and animal material and are therefore intermediate between primary and secondary consumer levels. Other animals may change their FEEDING habits as they get older; for example, tadpoles of the common frog (*Rana temporaria*) change from HERBIVORES to carnivores as they grow. Thus, it is apparent that comprehensive food webs of real communities can become very complex. Even so, patterns can be discerned, even in very complex food webs, due to the common CONSTRAINTS of ENERGY FLOW, POPULATION DYNAMICS and the structural design of animals. The explanation and prediction of food-web patterns is fundamental to understanding the processes at work in an ECOSYSTEM.

One generalized hypothesis concerning food webs, which has stimulated a lot of work, was the suggestion that the more complex a community the greater its stability (MacArthur 1955). MacArthur argued that when predators have alternative prey their own numbers depend less on fluctuations in the number of a particular species, and that where energy can take many routes through a

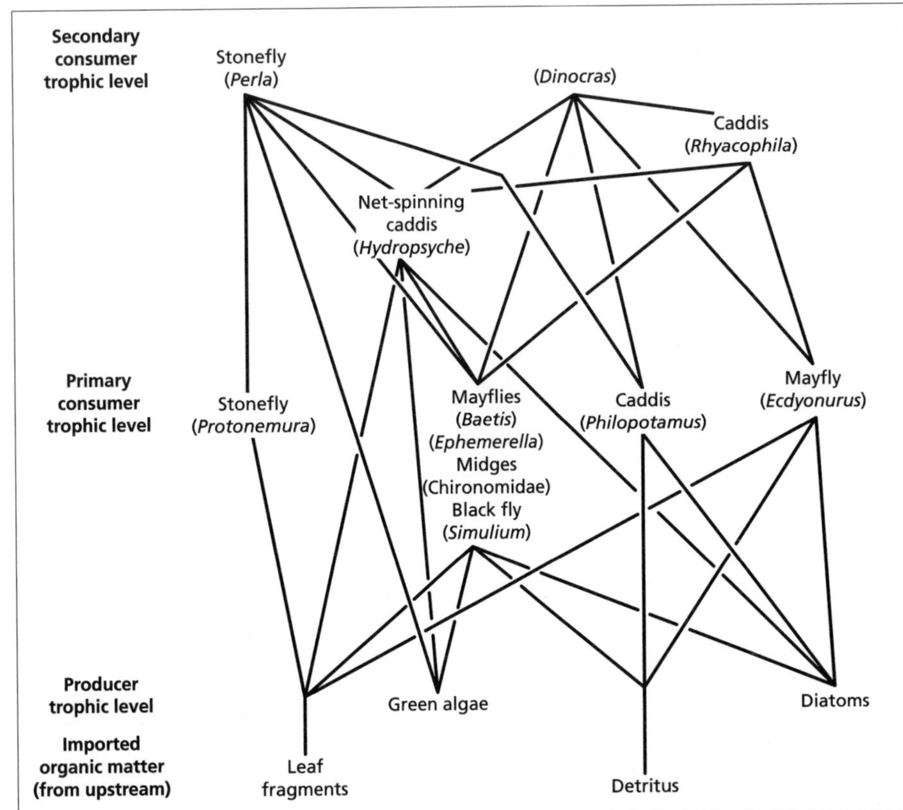

Fig. F5 A portion of a food web in a small stream community in South Wales. The diagram illustrates: (i) the interlocking of food chains to form the food web; (ii) three trophic levels; (iii) the fact that organisms, such as *Hydropsyche*, may occupy an intermediate position between major trophic levels; and (iv) an 'open' system in which part of the basic food is 'imported' from outside the stream. (After Jones, J.R.E. (1949) A further ecological study of calcareous streams in the 'Black Mountain' district of South Wales. *Journal of Animal Ecology* **18**, 142–159.)

system a disruption of one pathway results in it being diverted down another. MacArthur thus linked community stability with food-web complexity and species diversity.

Further analysis of community stability, i.e. its RESISTANCE to perturbation and ability to return to equilibrium, has shown that it is influenced by at least five attributes of food webs:

1 diversity of species within trophic levels;
2 complexity or CONNECTANCE between different food chains;
3 the length of the food chains;
4 the extent of omnivory between trophic levels; and
5 COMPARTMENTALIZATION, i.e. formation of discrete subdivisions of the food web (Pimm 1982).

The limits to pursuit of food-web analyses are usually set by the difficulties of making the empirical descriptions of what are often enormously complex systems. At best, our diagrammatic models of food webs are but caricatures of nature. Like caricatures they are distortions of reality, but they can nevertheless give insight into some fundamental truths about their subject: the structuring of natural communities. *Cf.* TROPHIC CONTINUUM. *See also* COMPARTMENTS IN FOOD WEBS. [M.H.]

RFEFERENCES

Jones, J.R.E. (1949) A further ecological study of calcareous streams in the 'Black Mountain' district of South Wales. *Journal of Animal Ecology* **18**, 142–159.

MacArthur, R.H. (1955) Fluctuations in animal populations and a measure of community stability. *Ecology* **36**, 533–536.

Odum, E.P. (1959) *Fundamentals of Ecology*, 2nd edn. Saunders & Co., Philadelphia, Pa.

Pimm, S.L. (1982) *Food Webs*. Chapman & Hall, London.

---

foraging, economics of The profitability of behaviours for acquiring energy and NUTRIENTS. Examples of decisions are selection of types of food (diet choice; *see* FOOD SELECTION) and the time before moving on to a new area (patch choice). Increasing (generalizing) or decreasing (specializing) the number of prey types eaten represents a behavioural decision influencing the economics of diet choice. Foraging economics is based on the 'lost opportunity' principle (Stephens & Krebs 1986), which states that one decision precludes opportunities of choosing potentially more profitable alternatives. For instance, handling a prey excludes SEARCHING for more profitable prey types. (*See also* TRADE-OFF.)

The economics of foraging is evaluated in a currency such as the net energy intake rate, or to what extent foraging provides an insurance against the risk of death through starvation. Protection against starvation is achieved through a reduced probability for dangerously low energy reserves. The size of energy reserves is then crucial to survival. Foraging policies to avoid starvation are state-dependent (Mangel & Clark 1988) and risk-sensitive. Foragers with low reserves should use a risk-prone behaviour, searching out risky options with the potential of a high reward; but when reserves are larger, foragers should be risk-averse and avoid uncertainty.

The lost opportunity principle assumes prey encounter to be sequential. The consequences of simultaneous encounter are discussed by Stephens and Krebs (1986). Economic foraging requires decision rules and information acquired through sampling. Decision rules can be based on environmental cues to foraging profitability such as a time limit since a prey was last encountered—the giving up time (GUT) — or a critical DENSITY of prey — the giving up density (GUD). *See also* BEHAVIOURAL ECOLOGY; FEEDING; OPTIMAL FORAGING THEORY. [J.E.]

REFERENCES

Mangel, M. & Clark, C.W. (1988) *Dynamic Modeling in Behavioral Ecology*. Princeton University Press, Princeton, N.J.

Stephens, D.W. & Krebs, J.R. (1986) *Foraging Theory*. Princeton University Press, Princeton, N.J.

foraminiferan Any member of an order (Foraminfera) of planktonic or benthic protozoans that (usually) secrete a calcareous test and possess pseudopodia. They range in size from about 30 μm to a few millimetres and feed on bacteria, PHYTOPLANKTON or small ZOOPLANKTON. There are approximately 40 known PELAGIC species and 4000 benthic species.

After they die, their skeletons accumulate on the sea floor. Sediments containing more than 30% foraminiferan skeletons are called foraminiferan ooze. Foraminiferans have been reported from rocks as old as the Cambrian, but are limited in the FOSSIL RECORD until early in the Carboniferous. The distribution and characteristics of fossil foraminiferans can provide important insights into past climates. For example, some foraminiferans have tests that coil either sinistrally (left) or dextrally (right). Variation in coiling direction with time has been shown to coincide with major CLIMATE CHANGES. [V.F.]

foreshore The portion of the shore lying between the normal high- and low-water marks; the INTERTIDAL ZONE. [V.F.]

forest Nowadays, this term usually refers to a large area of tree-covered land, but including other habitats in a matrix of trees. However, in Britain, it originally referred to unclaimed land beyond the bounds of farmland and habitation (cf. 'foreign', possibly from the same root). This land was commonly wooded, but not necessarily so. By the medieval period the term had been appropriated to describe large tracts of open land, including WOODLAND, subject to a particular form of law (i.e. Forest Law), which constituted the Crown hunting grounds and a source of fresh venison. Each forest was administered by a hierarchy of officials. Some medieval forests survive as forests (e.g. New Forest, Forest of Dean), whereas others survive in name only. Other ancient well-wooded tracts were also 'forests', though they were never Crown land (e.g. Forest of Arden, Leicester Forest). In the 20th century a forest has become the basic unit of administration of land managed by the Forestry Commission, much of which has been planted with Sitka spruce (*Picea sitchensis*) and other introduced conifers. Accordingly, 'forest' (and 'FORESTRY') has come to connote extensive plantations managed primarily for timber, whereas woodland has come to denote native tree COVER, managed traditionally or not managed at all. The original usage to describe open land as well as tree-covered land (e.g. Dartmoor and Exmoor Forests) survives in the modern 'deer forests' of the Scottish Highlands, most of which are virtually devoid of trees. [G.F.P.]

forest fragmentation *See* FRAGMENTATION OF HABITATS AND CONSERVATION.

forest soil Any SOIL that has developed under a FOREST cover, with characteristic properties resulting from the deep rooting of the trees, the activities of organisms characteristically associated with forest vegetation, and the influence of the litter layer (Pritchett 1979). In general, forest soils possess thinner and paler-coloured topsoils and are less fertile than those developed under grassland, which have higher contents of organic matter and bases.

The deep roots recycle nutrient ions by returning them to the surface in litter fall. They also alter the physical environment, contributing to soil WEATHERING by penetrating fissures in bedrock, and improving drainage and aeration by penetrating compacted layers. Unlike the situation in grassland or arable soils, microbial activity is dominated by symbiotic mycorrhizal associations between tree roots and fungi, which play a major role in tree nutrition and stabilize soil structure. Deciduous litter decomposes rapidly to form a base-rich HUMUS, whereas that from conifers is generally slower to decay and very ACID, encouraging the process of podzolization, by which IRON, ALUMINIUM and organic matter are translocated down the SOIL PROFILE.

Not all forest soils are still under forest today; up to one-third of the total area covered by these soils is now in agricultural use. However, many of the characteristic features of forest soils are still discernible, even though the change of use commonly results in loss of organic matter, deterioration of structure and increased erodibility. [K.A.S.]

REFERENCE

Pritchett, W.L. (1979) *Properties and Management of Forest Soils.* John Wiley & Sons, New York.

forestation The creation of FOREST conditions on an area of land which has not been forested for a long time, normally decades or centuries. The reasons for lack of forest COVER generally involve human activity, such as clearance of natural forest for AGRICULTURE, but the unforested state may have been maintained by natural agents or processes (e.g. GRAZING pressure). The reasons for forestation are normally economic, to establish a forest for production or protection purposes, HABITAT creation or the improvement of visual amenity. FORESTRY objectives will determine the type of forest and its management, the main points to consider being species choice and silvicultural system.

Since forest cover has an ameliorating influence on site conditions its ESTABLISHMENT involves the investment of effort (and therefore capital) to recreate forest conditions. Necessary operations may include: ground preparation by cultivation and drainage; tree planting; protection of trees from pests, diseases and vegetative competition by control of HERBIVORES and use of HERBICIDES; fertilizer application to rectify any growth-limiting soil nutrient deficiencies; and ongoing care of the young trees.

Most forestation has taken place in recent decades to create industrial plantations for timber production, often using selected GENOTYPES of single exotic species to maximize growth rates and financial return on investment. Recent initiatives in some industrialized countries have developed a much broader management remit, seeking to satisfy a range of non-production objectives in addition to timber YIELD. Such forests can be designed to mimic natural forest, using a mixture of native species including shrubs. In both cases forestation will result in a radical change of VEGETATION type with consequent effects on plant and animal communities, and recreational use. The forestation phase will in time succeed in re-establishing forest conditions and permit a wider range of management options. *Cf.* DEFORESTATION. *See also* AFFORESTATION; FORESTRY PRACTICES. [J.F.B.]

Forestry

Forestry concerns the interaction of trees, woodlands or forests with people. It comprises both practice and theory—the management of forest RESOURCES for a specific purpose and the scientific study of forest ECOSYSTEMS, including all aspects of their management. The terms 'woodland' and 'forest' refer primarily to smaller and larger areas of tree-dominated land respectively, although they can also refer to specific types of forest COVER.

Objectives of forestry

Since forestry implies management, the nature of forestry depends on management objectives. These may range widely, relating to the conservation of WILDLIFE habitat or genetic diversity, environmental protection, the production of domestic or industrial raw materials, the pursuit of scientific research or the meeting of social needs—for example, goods and services, both physical and spiritual, that trees and forests can provide. Thus forestry has the potential to satisfy multiple objectives and can be seen as both a science and an art.

The most common objective of forestry is timber production. On a worldwide basis, approximately half the timber is used for domestic fuel and buildings and half for industrial purposes, the latter proportion varying from 10 to 90% in less-developed and industrialized countries respectively. Non-timber forest products are numerous and varied, ranging from tree foliage used for FODDER or herbal medicines to the commercial production of maple syrup or gum arabic, for example.

Phases of forestry activity

Historically, for any given country, successive phases of forestry activity can be observed corresponding to man's perception of forest resources as exploitable, scarce or renewable. These phases are not necessarily determined by the level of technological development at the time, although this does affect the rate of change from one phase to another. For example, changes that took place over centuries in Europe have recently occurred over mere decades in West Africa. However, both the historical phase and technology influence the aspects of forestry activity considered important at any given place and time. Forestry first develops management practices when forest resources become locally scarce, as in the control of felling and regeneration in medieval Europe, which represents an early example of sustainability as a fundamental principle in the management of natural resources. Economic management tools were developed during the industrial revolution, and efficient HARVESTING machinery has been developed in the present 'renewable phase' context of financial and energy constraints.

It is interesting to note that the spiritual element of forestry—its aesthetic and recreational value, for example—is being rediscovered by the industrialized world at the same time that sophisticated technology has been developed capable of destroying significant areas of whole forest ecosystems. The resolution of this paradoxical dilemma is essentially a political matter and will constitute a major challenge to forestry in the 21st century. *See also* AGRICULTURE; FORESTATION. [J.F.B.]

forestry practices Activities used to create or maintain desired FOREST conditions and products. The types of tools used depend on landowners' silvicultural objectives and willingness to incur expenses. Because VEGETATION tends to aggressively occupy any given site and compete for available RESOURCES, forest development is usually influenced by removing some vegetation and promoting the survival and growth of desired plants.

Tree harvest systems cut trees to obtain wood products and create space for ESTABLISHMENT of new trees. Clear-cutting involves felling all mature trees within an area, while seed-tree systems leave a few trees on site to provide seed for the next generation (Fig. F6). Shelterwood systems leave several trees on site to moderate climatic extremes for tree regeneration; overstorey trees are removed once tree seedlings are well established. Selection systems remove trees singly or in groups. Greater levels of tree removal tend to promote rapid growth and DOMINANCE by light-demanding species, while selection cuts tend to favour shade-tolerant species.

After harvest, site preparation can enhance conditions for tree seedling survival. Methods include piling branches, burning vegetation and debris, mechanical ploughing or chopping, application of HERBICIDES, and GRAZING. Tree seedling regeneration may be accomplished by natural DISPERSAL of seed from local trees, by sowing seed of desired species, or by planting nursery-grown seedlings. Tending of young and intermediate age forests can include thinning (removing some trees to allow greater growth for remaining trees), pruning, fertilizer application, cutting diseased trees, prescribed burning, or killing of tree-damaging animals (e.g. with hunting or INSECTICIDES). *See also* ARBORICULTURE; NEW FORESTRY; SILVICULTURE. [A.N.G.]

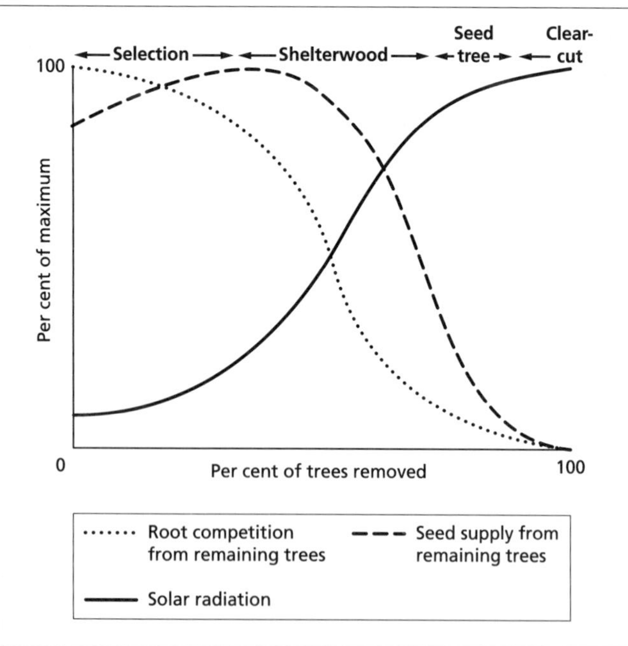

Fig. F6 Simplified representation of effects of different types of harvest cuttings on factors commonly important for vegetation response. (After Smith, D.M. (1986) *The Practice of Silviculture*, 8th edn. John Wiley & Sons, New York; copyright © 1986, reprinted by permission of John Wiley & Sons, Inc.)

REFERENCE

Smith, D.M. (1986) *The Practice of Silviculture*, 8th edn. John Wiley & Sons, New York.

fossil assemblage The fossilized remains of former plant and animal communities that are available for investigation by the palaeontologist or palaeoecologist. The fossil ASSEMBLAGE, while being the best available evidence of former animal and plant communities, is normally only a partial, and sometimes a biased, reflection of the original life assemblage or biocoenose for a variety of reasons. Transport, diagenesis and redeposition may have a major influence on the composition of the fossil assemblage. AUTOCHTHONOUS fossils, such as the remains of BOG-inhabiting animals and plants in a PEAT sample, normally bear close resemblance to the original biocoenose because of the close spatial relationship between the life and fossil assemblages. This contrasts with ALLOCHTHONOUS fossils in lake sediments, for several reasons. In lake sediments, the relative proportions of the various species are normally altered during transport from the site of production, i.e. mineral ground, to that of fossilization, i.e. lake SEDIMENT. During diagenesis, i.e. the various processes that take place immediately prior to fossilization, some differential destruction of fossils invariably takes place, which again affects the death assemblage. In addition, redeposition due to a variety of factors including BIOTURBATION at the sediment/water interface and resuspension, especially of littoral sediments due to wave action or changing lake levels, can further complicate the relationship between the biocoenose and the fossil assemblage. Reconstruction of former animal and plant communities, on the basis of the fossil assemblage, requires a thorough understanding of the processes and factors that operate at various stages between biocoenose, death assemblage and finally the fossil assemblage. [M.O'C.]

fossil fuels HYDROCARBON energy sources derived from fossil organic materials: coal, oil and natural gas. *See also* ACID PRECIPITATION, IMPACT ON ECOSYSTEMS; ACID RAIN AND MIST. [P.C.]

fossil record Fossils are the remains or imprints of animals and plants of the geological past, naturally preserved by burial in sediments. Fossils provide a record of the EVOLUTION of life on Earth, furnish clues to past environments and provide basic data for the history of the Earth. The fossil record is used to correlate geographically separated rock strata, in part to estimate their age. Fossils are first observed in abundance in rocks from the Cambrian period. Although sedimentary rocks from the Precambrian may contain primitive fossils, these are either sparse or lack characteristic features useful for correlation. [V.F.]

founder-controlled communities 'Possession is nine-tenths of the law.' The ecological processes involved in founder control are site pre-emption and INHIBITION OF SUCCESSION. In a COMMUNITY where the species possess roughly equal DISPERSAL abilities and exhibit similar competitive abilities, then COEXISTENCE is possible and COMPETITIVE EXCLUSION need not occur if there is founder control. The dynamics of this kind of INTERSPECIFIC COMPETITION, and the necessary and sufficient conditions for coexistence, have been worked out by the use of LOTTERY MODELS. Note, however, that coexistence through founder control does not imply that species with identical niches can coexist; there must be differences (e.g. in the kind of years in which peak RECRUITMENT occurs) between the coexisting species. The essence of coexistence in lottery models is the same as the Lotka–Volterra criterion (*see* LOTKA–VOLTERRA MODEL); INTRASPECIFIC COMPETITION must be more influential than interspecific competition, otherwise competitive exclusion will occur. *Cf.* DOMINANCE-CONTROLLED COMMUNITIES. [M.J.C.]

founder effect A term introduced by Ernst Mayr (1954) for the supposed effect of extreme reduction in numbers in an isolated population (a POPULATION BOTTLENECK). Mayr argued that the genome is a complex of loci with ALLELES selected not only for their direct effects but also to work successfully with alleles at other loci. Such a system is said to show strong epistatic interaction or to constitute a coadapted genome. A small population (at its smallest, one or two individuals) may be accidentally isolated; the descendants then increase in numbers to form a successful new population. In doing so, an opportunity is presented for a new coadapted GENE POOL to be formed from the subset of genes carried by the founders, augmented by those which subsequently enter the population by MUTATION. Evolution has taken a new course and the new complex may be at least as successful as, and become reproductively isolated from, the parental type. Mayr wished to draw attention to the importance his theory attached to the initial reduction in numbers. The founder effect/coadaptation theory has stimulated research, but observation and experiment do not support it in its original form. Coadaptation may be demonstrated, but not the highly organized interacting complexes required by the theory. *See also* SPECIATION. [L.M.C.]

REFERENCE

Mayr, E. (1954) Change of genetic environment and evolution. In: *Evolution as a Process* (eds J. Huxley, A.C. Hardy & E.B. Ford). Allen & Unwin, London.

Fourier analysis A form of analysis whereby the cyclic components of a process are determined. In classical Fourier analysis, one assigns to a periodic function *f(t)* a sum of simple trigonometric functions that represent the basic oscillatory components of *f(t)*. A function *f(t)*, defined for all real *t*, has period Φ if:

$$f(t) = f(t + \Phi)$$

for all *t*. Its Fourier series has the form:

$$\alpha_0 + \alpha_1 \cos\frac{2\pi}{\Phi}t + \beta_1 \sin\frac{2\pi}{\Phi}t + \ldots + \alpha_k \cos\frac{2\pi k}{\Phi}t + \beta_k \sin\frac{2\pi k}{\Phi}t + \ldots \qquad (1)$$

where $k=0, 1, 2, \ldots$. The *k*th summand:

$$\alpha_k \cos\frac{2\pi k}{\Phi}t + \beta_k \sin\frac{2\pi k}{\Phi}t$$

of the series is a simple wave function with period Φ and frequency k/Φ. The coefficients α_k and β_k depend on the function *f(t)* and determine the amplitude and the phase of the cyclic component with frequency k/Φ, hence its importance for the periodic pattern defined by *f(t)*. For example, *f(t)* might describe the FLUCTUATIONS of the air pressure in the ear that are caused by a musical instrument. Then the various summands in (1) represent the decomposition of the sound into pure tones, and the strength of these components determines its pitch. Expression (1) can be generalized to non-periodic functions. Fourier analysis is a very well-developed mathematical theory with many ramifications and applications. In particular, it is widely used in applied mathematics. It is an important tool for analysis of time series obtained from measurements, for which one wants to determine the oscillatory components of the signal. Other applications include its use in analysing PROBABILITY DISTRIBUTIONS and for solving differential, partial differential and integral equations. *See also* OSCILLATIONS. [M.D.]

FPOM See FINE-PARTICULATE ORGANIC MATTER.

fractal analysis Benoit Mandelbrot in his *Fractal Geometry of Nature* has demonstrated that the length of a complex structure such as a coastline depends upon the SCALE at which it is measured. At 1 km scale using an atlas we get one answer, at the scale of a human pace (1 m), walking around it, we get another (larger) answer. So geometrical objects such as this do not have one fixed dimension but a FRACTAL DIMENSION, *D*, such that the perceived length $L(\lambda)$ depends upon 'step length' of the measurement, λ, such that:

$$L(\lambda) = C\lambda^{1-D}$$

where C=constant. When $D=1$, we have the familiar Euclidean situation in which *L* is independent of measuring scale.

This concept has a number of implications and applications for ecology. For example, organisms at different sizes can perceive landscape and habitat in different ways. Alternatively, abrupt changes in the fractal dimension *D* can signal different ecological processes operating at different scales. [P.C. & V.F.]

fractal dimension Illustrating the concept of a fractal dimension is often done with reference to the problem of measuring the length of a coastline. The essence of the problem is that if one makes more detailed measurements by tracing around smaller coastal features, the coastline becomes longer. At the limit, with an infinitely small length-scale of measurement, the coastline becomes infinitely long despite bounding a finite area. In mathematical terms such curves can be described as having a dimension between one and two, which can be estimated by measuring the slope of a plot of estimated length versus length-scale of measurement. Fractal surfaces are also described as self-similar, because they have the same structure at all scales of observation.

Ecologists have noted that many of the features of natural systems appear to be fractal. For example, environmental HETEROGENEITY appears to be describable in fractal terms, and it has been suggested that examining variation in fractal dimension of the environment may lead to a deeper understanding of the causes of

VARIABILITY in species-area relationships. While fractal geometry appears to show promise as a tool for explaining ecological phenomena such as the DISTRIBUTION of animal body sizes, a complete picture of its utility has yet to emerge. *See also* FRACTAL ANALYSIS; SCALE. [S.J.H.]

REFERENCES

Kaye, B.H. (1989) *A Random Walk Through Fractal Dimensions*. Wiley VCH, New York.

Morse, D.R., Lawton, J.H., Dodson, M.M. & Williamson, M.H. (1985) Fractal dimension of vegetation and the distribution of arthropod body lengths. *Nature* **314**, 731–732.

fragile dynamics Dynamics which change dramatically with small parameter changes, or slight changes in conditions or model structure, or when subjected to small perturbations. For example, chaotic POPULATION DYNAMICS (*see* CHAOS) which could be stabilized with a tiny amount of IMMIGRATION could be described as fragile. Similarly, a population model which shows distinct generation cycles when the time from birth to maturity is a fixed constant, might show rapid approach to a fixed equilibrium point when the time to maturity is allowed to vary little between individuals within the population: the generation cycles are then an example of fragile dynamics, as they disappear with a small change in the model structure. [S.N.W.]

fragipan horizons *See* SOIL CLASSIFICATION.

Fragmentation of habitats and conservation

Fragmentation of natural systems is a pervasive ecological phenomenon and a central problem in conservation and restoration ecology (Harris 1984; Wilcove *et al.* 1986; Saunders *et al.* 1991). As forested landscapes become fragmented, changes occur in the physical environment of forests (Skole & Tucker 1993), the ability of organisms to migrate and disperse across the landscape (Harris 1984), and the spatial and temporal dynamics of DISTURBANCE. With increased spatial DISPERSION of suitable or accessible HABITAT, populations of some species may experience reduced success in dispersing to, or colonizing, new sites. Populations may also decline as habitat area is reduced and may become increasingly vulnerable to local extinction due to genetic, demographic or environmental stochasticity. Finally, fragmentation may favour some groups of species over others, resulting in changes in biotic composition of the forested landscape.

Consequently, fragmentation is a central issue in the conservation and management of parks and protected areas (Schelhas & Greenberg 1996; Laurance & Bierregard 1997). By virtue of their fragmentation, many protected ECOSYSTEMS become modified remnants of once large and more continuous systems (Skole & Tucker 1993). There is growing concern that over time, fragmentation will cause degradation of the ecological characteristics of many protected areas (Janzen 1983; Harris 1984; Skole & Tucker 1993; Shafer 1995).

Most studies of the ecological effects of fragmentation have focused on forest-interior dwelling birds, large carnivores, or other animals with large HOME RANGES, emphasizing reductions in interior forest habitat, increased INTERSPECIFIC COMPETITION, decreased genetic exchange among populations, and impediments to foraging or migratory movements across the landscape (Wilcove *et al.* 1986; Saunders *et al.* 1991). For example, a series of now-classic studies has documented increased nest (BROOD) PARASITISM of forest-dwelling NEOTROPICAL migratory birds by introduced European cowbirds (*Molothrus* spp.), as well as predation by mammals, in small FOREST fragments. Studies of the effects of habitat fragmentation on other organisms—such as arthropods and forest UNDERSTOREY plants—are comparatively rare. As a result, many of the effects of forest fragmentation are probably not yet realized by ecologists or land managers.

Small, disjunct patches of forest can originate through a variety of natural processes, including isolation by EDAPHIC, hydrologic or topographic features, or by disturbance. Wildfires or volcanic eruptions may leave isolated forest remnants surrounded by a burned, scorched or otherwise devastated landscape. Some fragments, such as those associated with edaphic or topographic features, can persist on the landscape for centuries or millennia. Others, such as those resulting from disturbance or natural successional processes, may be more transient, persisting for only a few decades.

Increasingly, forests are being fragmented anthropogenically, by expanding urbanization, transportation CORRIDORS, AGRICULTURE and activities associated with natural resource extraction (Skole & Tucker 1993). In many regions, a dominant cause of habitat fragmentation is commercial timber harvest. On the Olympic Peninsula of Washington State, USA, fragmentation of forests has been extensive; as much as 80% of the original old-growth forest outside of Olympic National Park has been cut since the 1940s. The loss and fragmentation of these ecosystems has also been accompanied by changes in the distribution of forest-patch sizes. Before European settlement of North America, large-scale disturbances in the windward Olympics occurred every 200–700 years. Consequently, at any given time, large areas of forest were comprised of older stands, with smaller portions of the landscape in earlier successional stages. With increasing frequency of harvest, a much larger proportion of the landscape consists of younger forests in patch sizes that are generally smaller. Most remaining old growth on the Peninsula exists as fragments of <400 ha; one-third is in patches of <75 ha. Even greater rates of fragmentation are occurring locally in tropical forests of Brazil, Ecuador, Zaire, Malaysia and Indonesia, associated with the spread of roads and large-scale agriculture into areas that have previously been continuously forested with primarily small, localized disturbance (Skole & Tucker 1993).

Various conceptual models have been proposed to account for changes in biological DIVERSITY of forest fragments. Island biogeographic theory (IBT) has been applied extensively to the ecology and conservation of habitat patches such as nature reserves (MacArthur & Wilson 1967; Harris 1984). As IBT is applied to terrestrial habitats, the diversity of fragments of once larger ecosystems is predicted to change over time in a fashion similar to islands of geologic origin. Patch ('island') size, distance from source of propagules and time since isolation, influence rates of IMMIGRATION, local extinction and hence affect ultimate species CARRYING CAPACITY. Habitat islands are hypothesized to behave more specifically like continental islands, inasmuch as they are remnants of once continuous systems that have become isolated by changes in the surrounding landscape (e.g. a forest remnant surrounded by clear-cuts). IBT predicts that SPECIES RICHNESS will decline following isolation because the insular area has a lower capacity to support species than does the original 'mainland'. IBT also predicts that smaller or more isolated fragments will contain fewer species than large fragments or those closer to a 'mainland', and that fragmented habitats will experience ongoing attrition of species, until an equilibrium is reached with extinction balanced by immigration.

In a fragmented landscape, species DISPERSAL and COLONIZATION may be problematic as interpatch distances exceed dispersal abilities and patches become separated by expanses of unsuitable habitat. Forest species with active POPULATION DYNAMICS involving frequent replacement of subpopulations would be especially sensitive to isolation of existing populations.

Although the habitat : island metaphor may apply to some biotic components of terrestrial ecosystems, its applicability to forest fragments remains a matter of debate for several reasons. First, relative to geologic islands, contemporary forest islands are generally smaller, closer together, and isolated for much shorter periods. Since the time necessary to observe population declines or local extinctions may be orders of magnitude longer than the lifespans of habitat islands that originate through disturbance events, there may be insufficient time for the consequences of insularization to be expressed. Second, unlike the relatively constant oceanic matrix, the terrestrial matrix surrounding forest fragments is compositionally heterogeneous and constantly changing (undergoing SUCCESSION). Finally, the habitat : island metaphor may violate an underlying premise of ECOLOGICAL ISOLATION from and by the matrix. In many instances the matrix of clear-cuts, plantations, or other disturbed areas, can support individuals that serve as sources of propagules for adjacent forest remnants (e.g. Janzen 1983). For example, many forest understorey species colonize old-regrowth forests from adjacent younger stands, provided that a dispersal vector is present. Although IBT remains widely cited as a conceptual model in CONSERVATION BIOLOGY, the validity of extending its spatial and temporal domain to relatively small forest fragments within a dynamic, terrestrial landscape requires further study.

An alternative approach for understanding the response of certain organisms (such as understorey or CANOPY plant species) to fragmentation emphasizes changes in the physical environment of forest fragments induced by abrupt edges. Edges are typically associated with reductions in interior forest area and (frequently) invasion of RUDERAL or edge species into forest interiors. Even if they retain their former overstorey composition and physical structure, small fragments experience greater variability in air and soil temperature, higher light levels, greater windspeeds, and lower humidity than large, continuous stands. Moreover, these microclimatic parameters may reach interior equilibrium at different distances from the edge, resulting in variation in the depth and strength of EDGE EFFECTS with edge orientation and closure, weather, time of day and other factors. The extent of such edge effects may be substantial. For example, in low-elevation conifer forests of the US Pacific North-West, these altered microclimates can extend 200 m or more from the edge into intact forest. Thus, depending on their size and geometry, small (e.g. <25 ha) forest fragments may have such extensive edge effects as to provide little or no interior habitat as defined by physical parameters. The extent to which gradients in forest understorey composition parallel those of microenvironment remains unclear.

Changes in forest MICROCLIMATE or local disturbance regime may also promote the invasion of exotic or ruderal taxa (Laurance 1991; Brothers & Spingarn 1992) or the regeneration/release of forest understorey trees or shrubs that can suppress the cover and diversity of forest herbs. Populations of plants near forest edges may also experience greater herbivory. Thus, changes in the abundance and diversity of forest-interior species may be mediated indirectly through effects on other taxa.

Environmental effects associated with forest fragmentation are likely to be strongest in regions in which the difference in physical conditions between large continuously-forested areas and fragments are most distinctive. In the US Pacific North-West, such differences are probably greatest in coastal forests of the Olympic Peninsula that are characterized by large, continuous expanses of closed canopy forest and infrequent disturbances. Under natural disturbance regimes, the physical and biotic environments associated with large edges occur infrequently in these forests. Species adapted to interior forest conditions may be affected more by fragmentation than would species found in more open or frequently disturbed forest types. Tropical forest types with dense and complex canopy structures and infrequent natural disturbance may similarly reflect effects of fragmentation most dramatically.

The depth and intensity of edge effects also vary with the structure of the forest edge and its contrast with the surrounding landscape. Fragment–matrix edge contrast is largely a function of the difference in height of vegetation and the degree of edge closure, although these differences can be amplified or diminished by topographic variation (slope and aspect). Release of advanced tree regeneration or tall shrubs can also mitigate edge effects by reducing light penetration or air movement into the forest interior.

In some forest types, the primary effects of fragmentation may concern disturbance regime. Temperate and tropical forest types with more frequent disturbance regimes may be affected differently by fragmentation, by altering the patterns of propagation of disturbance (e.g. forest-fire spread). For instance, an isolated fragment of *Pinus ponderosa* (Ponderosa pine) forest may have little opportunity to experience low-intensity ground fires on a normal interval (7–20 years). As a result, the normal thinning of seedlings may be inhibited, resulting in a forest STAND with substantially increased stem density and reduced stem size and mean age.

Since much forest fragmentation results from FORESTRY methods, recognition of its pervasive effects may result in improved silvicultural practices such as larger patch sizes, creation of structurally less abrupt edges, cutting patterns that reflect TOPOGRAPHY and forest stature, and longer rotation intervals. For ECOLOGICAL RESTORATION, forest fragmentation represents an important opportunity to re-establish the connectivity of forested landscapes after human disturbance. [D.A.F.]

REFERENCES

Brothers, T.S. & Spingarn, A. (1992) Forest fragmentation and alien plant invasion of central Indiana old-growth forests. *Conservation Biology* **6**, 91–100.

Harris, L.D. (1984) *The Fragmented Forest*. The University of Chicago Press, Chicago, Ill.

Janzen, D.H. (1983) No park is an island: increase in interference from outside as park size decreases. *Oikos* **41**, 402–410.
Laurance, W.F. (1991) Edge effects in tropical forest fragments: Application of a model for design of nature reserves. *Biological Conservation* **57**, 205–219.
Laurance, W.F. & Bierregard, R.O. (eds) (1997) *Tropical Forest Remnants: Ecology, Management, and Conservation of Fragmented Communities*. University of Chicago Press, Chicago, Ill.
MacArthur, R.H. & Wilson, E.O. (1967) *The Theory of Island Biogeography*. Monographs in Population Biology, No. 1. Princeton University Press, Princeton, N.J.
Saunders, D.A., Hobbs, R.J. & Margules, C.R. (1991) Biological consequences of ecosystem fragmentation: A review. *Conservation Biology* **5**, 18–32.
Schelhas, J. & Greenberg, R. (eds) (1996) *Forest Patches in Tropical Landscapes*. Island Press, Covelo, Calif.
Shafer, C.L. (1995) Values and shortcomings of small reserves. *BioScience* **42**, 80–88.
Skole, D. & Tucker, C. (1993) Tropical deforestation and habitat fragmentation in the Amazon satellite data from 1978 to 1988. *Science* **260**, 1905–1910.
Wilcove, D.S., McClellan, C.H. & Dobson, A.P. (1986) Habitat fragmentation in the temperate zone. In: *Conservation Biology: The Science of Scarcity and Diversity* (ed. M.E. Soule), pp. 237–256. Sinauer Associates, Sunderland, Mass.

francium (Fr) A radioactive element (ATOMIC NUMBER = 87; relative ATOMIC MASS = 223; melting point = 27°C; boiling point = 677°C) has 20 known isotopes, but occurs naturally only as ^{223}Fr (β^-; $t_{1/2}$ = 21.8 min), which is produced by branching in the decay of ^{227}Ac, a member of the ^{235}U decay series:

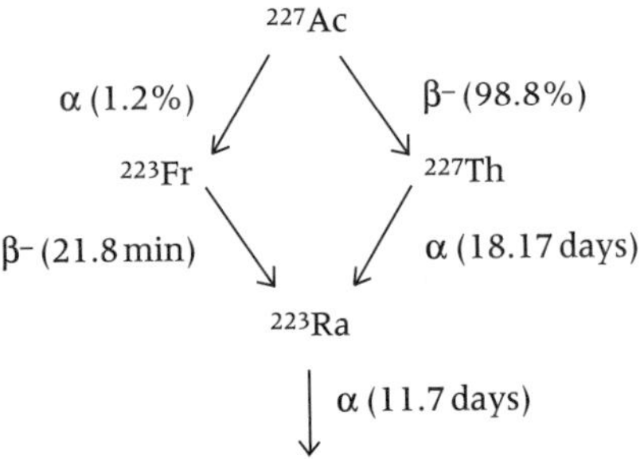

Francium, discovered by Perey in 1939, has an estimated crustal abundance of about 2×10^{-18} mg kg^{-1}, corresponding to approximately 15 g in the top kilometre of the Earth's crust. It has an oxidation state of +1. [A.B.M.]

freeze avoidance The STRATEGY by which freeze-intolerant (or freeze-susceptible, frost-susceptible, freeze-sensitive) organisms, notably insects, survive at low TEMPERATURE due to their ability to supercool (i.e. to depress their body's freezing temperature) in winter. Freeze avoidance is the dominant OVERWINTERING strategy of cold-hardy insects. The body freezing temperature, or SUPERCOOLING point (SCP), is progressively lowered from summer through autumn to winter by the process of cold hardening, or acclimatization, in which the insect: (i) evacuates or masks potential nucleators from the digestive system and other body compartments; and (ii) synthesizes polyhydroxy ALCOHOLS (polyols), to depress the SCP, and thermal hysteresis (antifreeze) PROTEINS, to stabilize the supercooled state, i.e. inhibit ICE formation. Extensive supercooling is not, though, a guarantee of winter survival. Many insects die before they freeze (pre-freeze MORTALITY) in either short (acute) or long (chronic) exposures. *See also* FREEZE TOLERANCE. [J.S.B.]

REFERENCE

Leather, S.R., Walters, K.F.A. & Bale, J.S. (1993) *Ecology of Insect Overwintering*. Cambridge University Press, Cambridge.

freeze tolerance The ability of animals and plants to tolerate the formation of ICE in their tissues and fluids. In insects, freeze tolerance is a minority, low-risk OVERWINTERING strategy most commonly found in species in the northern polar regions of Canada, Scandinavia, Russia and China. Freezing occurs extracellularly to prevent intracellular freezing, which is lethal. In freeze-tolerant insects, freezing is usually initiated at elevated subzero temperatures (−5 to −10°C) in extracellular areas (e.g. haemocoel) by ice-nucleating agents (PROTEINS), creating an osmotic gradient across the cell membrane. As TEMPERATURE further decreases, there is a gradual movement of WATER from the cells to the extracellular ice masses reducing the risk of spontaneous freezing in any cell. Freeze-tolerant insects also synthesize polyhydroxy ALCOHOLS (polyols), such as glycerol, which cryoprotect the partially frozen tissues, and antifreeze proteins, which inhibit secondary recrystallization during warming. Freeze-tolerant insects can survive to −40°C and some to −80°C. There are also freeze-tolerant amphibians (frogs) and reptiles (snakes and turtles) in which the lower lethal temperature is above −10°C. *See also* FREEZE AVOIDANCE; SUPERCOOLING. [J.S.B.]

REFERENCE

Leather, S.R., Walters, K.F.A. & Bale, J.S. (1993) *Ecology of Insect Overwintering*. Cambridge University Press, Cambridge.

Freon A generic tradename patented by the US chemical company Du Pont for the CFC class of compounds containing only CARBON, CHLORINE and FLUORINE. Equivalent compounds manufactured by ICI in the UK have been marketed under the name 'Arcton'. Freons are low boiling point liquids developed in the 1930s as non-toxic, non-flammable and chemically inert refrigerant replacements for sulphur dioxide and AMMONIA. Other major applications are AEROSOL propellents, degreasing solvents and blowing agents for plastic foams. The most widespread Freon compounds are dichlorodifluoromethane (Freon-12, CF_2Cl_2), trichlorofluoromethane (Freon-11, $CFCl_3$) and Freon-113 ($CF_2ClCFCl_2$), although Freon-114 (CF_2ClCF_2Cl) and Freon-115 (CF_2ClCF_3) have also been manufactured. Global production of the compounds accelerated rapidly to a peak of 470 000 and 380 000 tonnes for Freon-12 and Freon-11, respectively, in 1974. Measurements during the 1970s and 1980s demonstrated that Freons were accumulating in the ATMOSPHERE at the rate of total production. Destruction occurs by eventual transport to the STRATOSPHERE and ultraviolet (UV) photolysis to release chlorine atoms, which destroy stratospheric OZONE in a catalytic cycle, most emphatically demonstrated through the discovery of the ANTARCTIC OZONE HOLE. The GLOBAL WARMING potential of Freons is 12 000–16 000 times greater than CARBON DIOXIDE (CO_2) and calculations indicate they may contribute 30% to radiative forcing by 2015. International concern over the global impact of Freons culminated in the 1987 Montreal Protocol (and its subsequent amendments), where signatory countries agreed to complete cessation of Freon production by 1996. However, the extremely long atmospheric lifetimes (60–150 years) will ensure that Freon concentrations do not decrease until well into the 21st century, particularly if a global ban is not maintained. [M.R.H.]

frequency-dependent competition A situation in which the outcome of INTERSPECIFIC COMPETITION is dependent upon the frequency of the competing species. It is a stabilizing force and, although not explicitly stated, is present in all models/descriptions of competition that allow for a stable COEXISTENCE of species. *See also* POPULATION REGULATION. [B.S.]

REFERENCE

Begon, M. (1992) Density and frequency dependence in ecology: messages for genetics? In: *Genes in Ecology* (eds R.J. Berry, T.J. Crawford & G.M. Hewitt), pp. 335–352. Symposium of the British Ecological Society No. 33. Blackwell Scientific Publications, Oxford.

frequency-dependent selection Frequency-dependent selection implies that the FITNESS differences underlying selection depend upon the frequencies of the GENOTYPES in the population. *See also* NATURAL SELECTION. [G.D.J.]

freshwater WATER with no significant amount of dissolved salts or minerals. It generally contains <1000 mg l^{-1} of dissolved solids. *Cf.* SEA. [P.C.]

freshwater ecology *See* ECOLOGY.

front

1 Relating to WEATHER; *see* FRONTAL SYSTEM.

2 A boundary between horizontally juxtaposed WATER masses of dissimilar properties where CIRCULATION and mixing change abruptly. Fronts can be classified into six general types: (i) river plume fronts; (ii) UPWELLING fronts; (iii) shallow sea fronts; (iv) shelf break fronts; (v) fronts at the edge of major western boundary CURRENTS; and (vi) fronts of planetary scale, removed from major OCEAN boundaries. The first four types are COASTAL phenomena, whereas the last two types are open-ocean phenomena. Fronts act to concentrate NUTRIENTS and PLANKTON, and are thus sites of high PRODUCTIVITY. [V.F.]

frontal system A front is a transition zone between two AIR MASSES of different densities and usually temperatures. It may also be distinguished by a PRESSURE trough, major discontinuity, and change in the WIND direction. The frontal system is the pattern of fronts as illustrated on summary WEATHER charts. [P.C.]

frost/freezing damage Frost is a white dew coating and freezing damage to living organisms occurs when temperatures fall below 0°C and pure WATER freezes. Intracellular ICE formation is lethal to living cells unless the rate of freezing is extremely rapid, such as is caused by immersion in liquid NITROGEN (–196°C), a technique used to preserve many tissues and GERM LINES. Many organisms have evolved mechanisms of FREEZE AVOIDANCE (e.g. the antifreeze PROTEINS found in polar fishes and many insects). Many insects and some amphibians can tolerate freezing of extracellular water. These and many plants and unicells utilize free sugars and/or polyhydric ALCOHOLS to limit low-temperature damage to proteins and membranes (*see* FREEZE TOLERANCE).

Membranes and their associated transport systems are frequently the primary site of low-temperature damage, and measurement of ION leakage from cells has provided a simple and effective means of detecting such damage. Previously frozen tissue is immersed in pure water and the solution conductivity measured after a set time, autoclaved to release all intracellular ions and thus normalize for differences in size and composition, cooled, and left to stand before the conductivity is measured again. The extent of damage is related to the proportion of cellular ion content released, and in plants this may vary with tissue, species and time of year as well as experimental treatment. Tissue damage may also be indicated by secondary tissue browning, changes in vital staining, or a reduction in chlorophyll fluorescence or electrical impedance. The various methods examine/measure different cellular properties and reveal different aspects of low-temperature damage. No single method can be recommended for all species or conditions but a combination of two or more tests may improve the overall assessment of freezing damage. [L.S.]

frugivore An animal specializing in FEEDING on fruits. *See also* FEEDING TYPES, CLASSIFICATION. [P.O.]

fuelwood Wood was the first energy resource for humankind, and is still used as the prime energy supply for cooking and heating by the majority of the world's population. Probably half of all the wood harvested is used as fuelwood.

The energy obtainable from wood on combustion depends on the water content, and on the method of combustion. The following values relate to the heat of combustion obtained in a BOMB CALORIMETER, which does not vary much between species. Stoves are likely to be only 5–25% efficient in transferring this heat of combustion to a pot of water for cooking. The units are megajoules per kilogram (MJ kg^{-1}):

Green wood (50% moisture)	10
Air-dried wood (20% moisture)	16
Oven-dried wood (8% moisture)	20

These figures may be compared to those of some other fuels: coal (30), charcoal (34), fuel oil, PETROLEUM or gasolene (47), METHANE (56), HYDROGEN (121).

Forests vary enormously in their growth rates, mostly producing between 1 and 20 tonnes DRY MASS of BIOMASS per hectare per year. The 'fuelwood crisis' refers to the shortage of wood in many poor countries, especially those afflicted by DROUGHT. Villagers may have to walk great distances to gather fuelwood, or buy expensive bundles of it at the local market. Without a supply of energy, human health and performance is undermined, as cooking tenderizes food and renders it digestible, and also kills many pathogens. One of the strategies developed in response to this crisis is the use of 'multipurpose trees' in agroforestry. These trees are often fruit-bearing, or nitrogen-fixing. Another strategy is the use of 'improved' stoves, rather than the traditional three stones used to support the cooking pot. Fuelwood became popular again in developed countries in the 1970s, partly in response to the oil crisis, and as part of a 'back to nature' movement. In densely populated countries like the UK the population density and the per capita consumption of energy is much too great in relation to the capacity of any forest production system, but in Sweden much research was done on the possible use of fast-growing, short-rotation CROPS of willow (*Salix* spp.) as a major contributor to the energy supply. *Cf.* FOSSIL FUELS. [J.G.]

REFERENCE

White, L.P. & Plaskett, L.G. (1981) *Biomass as Fuel*. Academic Press, London.

fugacity The tendency for a substance to transfer from one environmental medium to another. It is analogous to chemical potential —the tendency of a chemical to escape from a phase. The concept is used in models that predict the environmental FATE of CONTAMINANTS. [P.C.]

fugacity models Predictions of EXPOSURE to CONTAMINANTS that model the partition of substances between environmental compartments on the basis of the FUGACITY of chemicals — essentially a chemical's partial pressure that can be viewed as an escaping tendency or pressure. It has units of pressure and can usually be related linearly to concentration. *See also* MACKAY FATE MODELS. [P.C.]

fugitive emissions Uncontrolled and unwanted releases of CONTAMINANTS to the environment through leaks. [P.C.]

fugitive species A competitively inferior species that coexists with a competitively superior species by having better powers of DISPERSAL. The fugitive species finds suitable patches of habitat first (on average), reproduces and leaves before the competitively superior species arrives. The idea of a species existing as a fugitive was first used by Elton (1929) in connection with the copepod *Eurytemora velox*. Hutchinson (1951), also inspired by copepods, used the term 'fugitive' to describe those species that are 'able to exist primarily by having good dispersal mechanisms, even though they inevitably succumb to competition'.

A nice example of a fugitive species is provided by the laboratory experiments of Armstrong (1976). The model laboratory system consisted of two species of fungi, *Aspergillus nidulans* and an unidentified species of *Penicillium*, competing for space on agar-filled Petri dishes. The fugitive species (*Penicillium*) matured quickly and produced daughter colonies in abundance, while the competitive dominant (*Aspergillus*) matured slowly and produced colonists at a lower rate. In the experiments, a fraction of the Petri dishes were replaced at intervals, and at an average patch (dish) age of 2.5 weeks COEXISTENCE was readily obtained.

A possible field example is provided by Hanski and Ranta (1983), who studied three species of *Daphnia* in small rock pools off the coast of southern Finland. There is some evidence to suggest that colonizing ability decreases in the order:

D. magna > *D. pulex* > *D. longispina*

with competitive ability decreasing in the reverse order. Another possible field example is provided by the parasitoids studied by Zwolfer (1979). He studied two hymenopterous parasitoids, *Eurytoma serratulae* and *E. robusta*, competing for a common host, the knapweed gall fly, *Urophora cardui*. *Eurytoma serratulae* is the more efficient searcher, finding potential hosts when they are younger. *Eurytoma robusta* is the dominant species within a patch (flower head), often consuming any *E. serratulae* larvae that it encounters.

A more detailed account of fugitive species, including theoretical models, is found in Shorrocks (1990). [B.S.]

REFERENCES

Armstrong, R.A. (1976) Fugitive species: experiments with fungi and some theoretical considerations. *Ecology* **57**, 953–963.

Elton, C.S. (1929) The ecological relationships of certain freshwater copepods. *Journal of Animal Ecology* **17**, 383–391.

Hanski, I. & Ranta, E. (1983) Coexistence in a patchy environment: three species of *Daphnia* in rock pools. *Journal of Animal Ecology* **52**, 263–279.

Hutchinson, G.E. (1951) Copepodology for the ornithologist. *Ecology* **32**, 571–577.

Shorrocks, B. (1990) Coexistence in a patchy environment. In: *Living in a Patchy Environment* (eds B. Shorrocks & I. Swingland), pp. 91–106. Oxford Science Publications, Oxford.

Zwolfer, H. (1979) Strategies and counterstrategies in insect population systems competing for space and food in flower heads and plant galls. *Fortschritte der Zoologie* **25**, 331–353.

fumes Airborne, suffocating POLLUTANTS; they may be gaseous and/or suspensions of solids and/or vapours. [P.C.]

functional constraints *See* CONSTRAINTS.

functional ecology *See* ECOLOGY.

functional explanation An explanation for a characteristic of an animal or plant based on FITNESS-related performance. The alternative would be phylogenetic, a carry-over from a previous state. [L.M.C.]

functional feeding group *See* FEEDING TYPES, CLASSIFICATION.

functional group A functional group is a set of biological entities that possess common functional or structural attributes. Such groups often comprise biological species, but groupings can be made at any level of biological organization. A related concept is that of the GUILD, which groups species on the basis of shared RESOURCES. By using functional groups, the general principles which govern the functioning ecological systems may often be more easily understood. When considering predation, for example, classifying animals into true predators, grazers, parasites and parasitoids offers ways of exploring and clarifying ecological relationships that would be obscured if these functional distinctions were not made.

The choice of criteria on which to group species depends on the objectives of the researcher, and the results of defining functional groups will often be quite different depending on the aim. If one is interested in ENERGY FLOW in a marine benthic system, for example, one might group species on the basis of feeding, which would place mussels and some polychaete species together under the general heading of 'filter-feeder'. In contrast, an interest in nutrient regeneration through the processes of BIOTURBATION would separate these same species into different groups.

Despite the attractiveness of functional groupings, there are difficulties with their use. For example, the feeding mode classification system for marine benthic invertebrates has considerable advantages, but many species are rather plastic in their behaviour and can adopt one of a number of feeding strategies. Thus, the same species might be placed in one of three or more groups, depending on food availability and other conditions. Similar difficulties exist with other functional classifications, often making it impossible to place entities into single categories. *See also* FEEDING TYPES, CLASSIFICATION. [S.J.H.]

functional parity (functional equivalency) Efforts to restore or construct HABITATS that mimic natural ecosystems will have functional parity when both COMMUNITY structure and ECOSYSTEM processes (functions) are equivalent to (indistinguishable from) one another. Traditionally, vegetation cover has been used to assess project 'success', but this is now recognized as inadequate. Attributes such as BIODIVERSITY support (ability to support native species), PRIMARY and SECONDARY PRODUCTIVITY, food-chain support (supplying food for higher trophic levels), nutrient processing and storage, and related functions are among the measures needed to assess functional parity. A human-modified or created ecosystem may be judged to have functional parity with a natural ecosystem when a scientific comparison (conforming to accepted norms of research design and statistical VALIDATION) of a range of ecosystem processes fails to show significant differences. [J.B.Z.]

REFERENCE

Zedler, J.B. (1996) Does mitigation provide wetland gardens or functionally equivalent, self-sustaining wetland ecosystems? In: *Restoring Diversity: Ecological Restoration and Endangered Species* (eds D. Falk, C. Millar & P. Olwell), pp. 331–342. Island Press, Washington, D.C.

functional response The number of prey (HOSTS) that are consumed (attacked) by a PREDATOR (PARASITOID) as a function of host DENSITY. Three types are normally considered.

• Type 1: the number consumed increases linearly with prey density until a THRESHOLD density, beyond which a constant number is consumed. It probably applies best to animals such as filter-feeders.

• Type 2: the number consumed increases with prey density at a decelerating rate until an asymptote is reached. Models such as HOLLING'S DISC EQUATION which constant encounter rates and a finite HANDLING TIME give rise to type 2 responses.

• Type 3: the relationship between numbers consumed and prey density is sigmoid (S-shaped). A type 3 functional response occurs if predators are inefficient at detecting rare prey (perhaps because they do not develop a search image) or if they switch to attacking more common items of prey.

Functional responses lead to inverse density-dependent predation except in the case of type 3 functional responses where MORTALITY is directly density dependent at densities below the point of inflexion. *See also* PREDATOR–PREY INTERACTIONS. [H.C.J.G.]

fundamental niche (potential niche) The NICHE in the absence of competitors and predators. *See also* REALIZED NICHE. [J.G.S.]

Fungi, diversity and conservation

Membership in the kingdom Fungi is generally assigned to organisms which are heterotrophic (i.e. those which cannot independently synthesize a nutritional energy source), nucleated, usually filamentous, and typically reproduce by sexual and asexual spores. As thus defined, fungi constitute an extremely diverse, polyphyletic assemblage of over 100 000 species classified in the phyla Zygomycota, Ascomycota and Basidiomycota. This definition is often extended to include fungal-like protists such as the aquatic moulds (Chytridiomycota and Oomycota) and the slime moulds (Myxomycota and Acrasiomycota).

Ecological roles

Despite their broad morphological DIVERSITY, fungi share similar functional roles in natural ECOSYSTEMS. They are the principal decomposers (saprotrophs) of terrestrial plants and plant products. They also form important symbiotic associations with plants as either parasites (pathogens) or as mutualists (mycorrhizae and lichens). The general association of fungi with plants, rather than with animals, is largely based on a strong metabolic preference for carbohydrate-based organic compounds like sugar, CELLULOSE and LIGNIN, which are the basic constituents of plant biomass. This is in contrast to the decomposer bacteria, which prefer the NITROGEN-rich protein molecules found in animals and animal products. Collectively, fungi and bacteria represent the decomposer component in terrestrial ecosystem models of nutrient and energy cycling. The fungi, in all their complexity and diversity, account for most of the DECOMPOSITION and CARBON flow in natural terrestrial ecosystems. Although they cause the greatest number of plant diseases, they also form mutually beneficial, in some cases indispensable, associations with plant roots as ENDOTROPHIC MYCORRHIZAE and ECTOTROPHIC MYCORRHIZAE. Nearly all vascular plants form mycorrhizae, and the species of mycorrhizal fungi number in the thousands (*see* MYCORRHIZA).

A wide variety of saprobic, parasitic and mutualistic fungi can be found within any given ecosystem. The presence of different species and strains has been observed to be related to the species of plant, its nutrient status and the environmental conditions that characterize different habitats. Fungal species composition in any given habitat is not constant since dynamic COMMUNITY interactions serve to both create and destroy niches fostering specific fungal associations and processes. In addition, the degree of specificity in fungus–plant associations ranges from narrow to broad (one to numerous HOST associates). Because environmental conditions strongly influence specificity, fungus–plant interactions in nature may be greatly different from those produced and observed experimentally.

Fungal systematics

Most fungal genomes are relatively small and conserved (low percentages of REPETITIVE DNA sequences) thus indicating that the genetic basis of species diversity would be relatively simple to determine. However, the wide array of sexual and asexual reproductive mechanisms, some peculiar to fungi alone, have resulted in considerable genomic diversity and an extraordinarily heterogeneous array of species and strains. Although little is yet known about evolutionary processes and relationships in the fungi, recent application of methods in molecular biology have resulted in some useful contributions to the understanding of fungal systematics, analysis of natural populations, and estimates of fungal BIODIVERSITY.

Conservation

Until recently the importance and value of fungi in ecosystem studies and environmental management has been frequently overlooked and/or underestimated. Without an awareness of their essential ecological role, there has been little impetus from the social conscience to generate and promote conservation measures. Efforts to incorporate fungi into contemporary conservation concerns have been stimulated by destructive and wasteful lumbering practices and industrial POLLUTION. Mycorrhizae associated with FOREST trees characteristically produce large, often edible, fruitbodies and are formed by some 5000–6000 species of fungi, including gilled (Agaricaceae) and pored (Boletaceae) mushrooms, the puffballs (Gasteromycetes) and their subterranean relatives, the false truffles. Studies of fruitbody production in Europe and western North America have revealed a significant decline in numbers and diversity of mycorrhizal species. This decline was associated with logging DISTURBANCES in old growth forests and in areas impacted by AIR POLLUTION. Therefore fungi have been included in ecosystem studies, where the primary objective is to gather information on their diversity, ecological importance and the effects of disturbance on their survival. Excessive commercial HARVESTING of edible mycorrhizal mushrooms is another potential cause of decline. Following the example of many European countries, states in the Pacific North-West are developing management guidelines for commercial mushroom harvesting and monitoring this resource for sustained productivity. *See also* CONSERVATION BIOLOGY. [J.S.S.]

REFERENCES

Arnolds, E. (1991) Decline of ectomycorrhizal fungi in Europe. *Agriculture, Ecosystems and Environment* **35**, 209–244.

Bruns, T.D., White, T.J. & Taylor, J.W. (1991) Fungal molecular systematics. *Annual Review of Ecology and Systematics* **22**, 525–564.

Harley, J.L. & Smith, S.E. (1983) *Mycorrhizal Symbiosis*. Academic Press, New York.

Molina, R., Massicotte, H. & Trappe, J.M. (1992) Specificity phenomena in mycorrhizal symbioses: Community-ecological consequences and practical implications. In: *Mycorrhizal Functioning, an Integrative Plant-Fungal Process* (ed. M.F. Allen), pp 357–423. Chapman & Hall, New York.

---

fungicides Pesticides designed to eradicate or control fungal pathogens on plants, animals and materials derived from organic sources. Fungicides have to cope with a wide range of pathogenic fungi. The latter can be classified according to the mode of infection.

1 Airborne infections, involving:

(a) motile spores that are water-dependent at infection stage, e.g. downy mildews;

(b) spores that move passively through water and mycelia that live under the cuticle (e.g. botrytis);

(c) spores that are not directly dependent on water and mycelia that live under the cuticle or deeper (e.g. powdery mildews and rusts).

2 Internal and external seedborne infections of many kinds.

3 Soilborne infections (e.g. club root, damping off).

The first agricultural fungicide to be formulated may have been the mixture of copper sulphate and LIME (Bordeaux mixture) used for control of downy mildew of vines from 1885. By the turn of the century, lime sulphur was being used on fruit trees and copper sulphate for control of potato blight. Organomercury seed dressings were introduced in 1913 to control a range of seedborne diseases, but is was not until 1934 that the first of the modern organic chemical fungicides, thiram (a non-systemic dithiocarbamate fungicide), was introduced. These novel compounds had a predominantly contact action; i.e. if plant surfaces are treated, the fungicide protects against the germination of fungal spores arriving after treatment.

Systemic fungicides appeared in the 1960s. These compounds act against internal fungal mycelia and haustoria. They attack already established diseases or offer protection from attack. They are not as subject to degradation as external protectants. The older protective compounds have tended to be generally non-specific, but most modern fungicides are rather more target- and site-specific (such as attacking single receptors or enzymes), and many of these are systemic in action. Systemic fungicides applied to roots or seed may be translocated to protect other parts of the plant from airborne infections.

In practice, most CROPS are subjected to simultaneous attacks by pathogens, and mixtures of two or more compounds, including systemic and contact materials, are often employed. Such mixtures of types of fungicide help to reduce the risk of RESISTANCE to fungicides, which is otherwise a major problem in the maintenance of established products, and in the development of new products. Plant breeding for disease resistance is an equally important technique in managing fungal diseases.

Many fungicides can be used very safely for the control of fungal diseases on animals because of their specificity of action, and in the control of fungal attacks on many materials derived from organic sources.

The earlier multiple-site protectants such as organomercurials and copper sulphate are more generally toxic to animals, and the organomercurials have been banned in a number of countries. As many modern fungicides tend to be very target specific, they tend to have low TOXICITY to other organisms. *See also* BIOCIDE. [K.D.]

fungicole A fungus living on another fungus, as either a SAPROTROPH or as a PARASITE (i.e. hyperparasite/mycoparasite). This is a very common phenomenon in all natural environments and ecological niches, i.e. in the general SOIL of all types, as well as within the RHIZOSPHERE, rhizoplane, mycorrhizae and phylloplane. Mycoparasites can be divided into either biotrophic types, existing in balance with the living hyphae of the HOST fungus, or necrotrophs, causing death of the host hyphae. The biotrophic mycoparasites absorb NUTRIENTS from their hosts by means of intrahyphal hyphae, achieving only relatively slow growth in the absence of the host. *Trichoderma* is a very important NECROTROPHIC mycoparasite living in soil which has a wide host range, being widely tested as a BIOLOGICAL CONTROL agent against destructive root-infecting plant-pathogenic fungi such as *Pythium* spp., which cause 'damping off' of seedlings in wet, poorly aerated soils. [J.B.H.]

fusion The fusion of two light nuclei to produce a heavier nucleus results in the release of binding energy. Natural nuclear fusion reactions provide the power of stars and, in the process of nucleosynthesis, have been responsible for the formation of the ELEMENTS that compose the Earth and, indeed, the universe. Present theories of nucleosynthesis suggest that element formation has occurred mainly in stars, with processes in young stars involving fusion of isotopes of HYDROGEN and HELIUM, while in older stars fusion of heavier elements, notably CARBON, NITROGEN and OXYGEN, takes place.

Controlled nuclear fusion reactions give the potential for the use of fusion reactors as a power source, and reactions that have so far been investigated in this context have involved fusion of the hydrogen isotopes DEUTERIUM (D) and TRITIUM (T) to produce:

$$D+D \rightarrow T+p$$

$$D+D \rightarrow {}^{3}He+n$$

$$D+T \rightarrow {}^{3}He+n$$

Fusion reactors have the potential for energy production with little waste and an almost unlimited supply of deuterium (natural abundance = 0.015 atom %) as fuel. The supply of tritium is more restricted since it is radioactive, with a half-life of 12.33 years, and must be produced by reactions such as:

$$^{6}_{3}Li+n \rightarrow T+\alpha$$

Conceptual designs so far investigated for fusion reactors have involved production of deuterium and tritium ions at temperatures in excess of 5×10^{7} K in a plasma enclosed in a magnetic field. [A.B.M.]

REFERENCES

Friedlander, G., Kennedy, J.W., Macias, E.S. & Miller, J.M. (1981) *Nuclear and Radiochemistry*. John Wiley & Sons, New York.

Henderson, P. (1982) *Inorganic Geochemistry*. Pergamon Press, Oxford.

Keller, C. (1988) *Radiochemistry*. Ellis Horwood, Chichester.

fynbos A flora unique to the southern and south-western coastal region of South Africa, consisting of over 8500 species, of which almost one-fifth are endemic. The almost treeless VEGETATION (*fynbos* is Afrikaans for fine bush) characteristically consists of tough, fine-leaved bushy plants and is dominated by the Proteaceae (on limestone soils and at lower elevations), Ericaceae (on sandy soils and at mid-elevations) and the monocotyledonous rush-like Restionaceae (at higher elevations). The contemporary CLIMATE of the area is essentially Mediterranean with warm, dry summers but with considerable east–west variation in PRECIPITATION and in SOIL type, elevation and aspect. The fynbos is subject to recurrent fires of variable frequency, and seed germination and seedling ESTABLISHMENT of the proteid and ericoid flora appear to be largely dependent on fires. The EVERGREEN restonoids undergo rapid vegetative regeneration following burning.

Elements of the fynbos vegetation show close affinities with the mountain vegetation of tropical Africa, and these observations suggest that one part of the fynbos flora may have originated in more temperate areas, further north, and that the fynbos communities represent a RELICT flora pushed southwards. However, the 300 species of Restionaceae show stronger affinities with the GONDWANA floras of Australia, South America and Madagascar, suggesting that this important element of the fynbos community has been in place for a considerable time. The fynbos supports a characteristic fauna, including insects with a possible Gondwana origin. *See also* CHAPARRAL SHRUBLANDS. [G.A.F.H.]

REFERENCE

Cowling, R. (1992) *The Ecology of Fynbos: Nutrients, Fire and Diversity*. Oxford University Press, Oxford.

G

G soils *See* SOIL CLASSIFICATION.

G test (log likelihood ratio test) A NON-PARAMETRIC TEST that is an alternative to the CHI-SQUARED TEST for analysing QUALITATIVE INFORMATION. It can be used in both GOODNESS-OF-FIT (one-way classification) and CONTINGENCY TABLE (two-way classification) situations. With one-way classifications the formula for the test is:

$$G = 2\sum O \ln \frac{O}{E}$$

where O and E are, respectively, observed and expected numbers, ln means natural logarithms and Σ means 'add up' all $O\ln(O/E)$ for all categories. For the reed warbler nests, in different types of habitat, analysed in the chi-squared test entry, the table of observed and expected values looked like this:

	Reeds	Wet scrub	Dry scrub	Woodland
Observed	12	6	2	2
Expected	3.3	5.5	5.5	7.7

Therefore, using the above formula we get:

$$G = 2\left[12 \times \ln(12/3.3) + 6 \times \ln(6/5.5) + 2 \times \ln(2/5.5) + 2 \times \ln(2/7.7)\right]$$
$$= 22.58$$

Notice that the last two terms are subtracted because the logarithm is negative. The test statistic, G, is examined for significance using the conventional chi-squared tables, entered at $c-1$ DEGREES OF FREEDOM (v) with c = the number of categories. For the above example, the 5% value of χ^2, for $v = 3$, is 7.81 and therefore the result is significant. Since the 1% value in the χ^2 table is 11.34, it is also very significant. Because the TYPE 1 ERROR tends to be higher than the intended level, Williams (1976) suggested that G should be divided by a correction factor. This correction factor (q) is equal to $1+(c^2-1)/6nv$, where c is the number of categories, n is the total number of observations and v is the degrees of freedom ($=c-1$). For the above table, $q=1+8/396=1.02$. The adjusted value of the test statistic is now:

$$G_{\text{adj}} = 22.58/1.02 = 22.13$$

which is still very significant at the 1% level.

In a contingency table, when assessing the degree of ASSOCIATION between two classifications the formula for computing G is even more straightforward. Unlike the chi-squared test there is no need to calculate expected values. As an example, consider a 2×2 table (calculations for a larger table follow exactly the same procedure). If the observed numbers are represented as below:

a	b	a + b
c	d	c + d
a + c	b + d	n

The formula is:

$$G = a\ln(a) + b\ln(b) + c\ln(c) + d\ln(d) \text{ [using the observed numbers]}$$
$$+ n\ln(n) \text{ [using the total number of observations]}$$
$$-(a+b)\ln(a+b) - (c+d)\ln(c+d) - (a+c)\ln(a+c)$$
$$-(b+d)\ln(b+d)] \text{ [using row and column totals]}$$

For the two plant species in quadrats analysed in the chi-squared test entry, the table of observed values looked like this:

		Species A +	Species A −
Species B	+	15	45
	−	30	10

Using the above formula we get:

$$G = 15\times\ln(15) + 45\times\ln(45) + 30\times\ln(30) + 10\times\ln(10)$$
$$+100\times\ln(100) - 60\times\ln(60) - 40\times\ln(40) - 45\times\ln(45)$$
$$-55\times\ln(55)$$
$$= 12.59$$

The Williams correction factor, usually only applied to 2×2 tables, is:

$$q = 1 + \frac{\left[\left(\frac{n}{a+b}\right) + \left(\frac{n}{c+d}\right) - 1\right]\left[\left(\frac{n}{a+c}\right) + \left(\frac{n}{b+d}\right) - 1\right]}{6n}$$

For the above 2×2 table this correction factor $=1.02$, and therefore:

$$G_{\text{adj}} = 12.59/1.02 = 12.34$$

which is significant at the 1% level, indicating that the two species of plant are associated (negatively) at the scale of the quadrat used.

Although chi-squared is the older, more traditional statistic for this type of analysis of qualitative information, many statisticians believe G to be a more reliable test statistic, being less susceptible to problems of small expected values. For further information on the G test see Sokal and Rohlf (1995). *See also* STATISTICS. [B.S.]

REFERENCES
Sokal, R.R. & Rohlf, F.J. (1995) *Biometry*, 3rd edn. W.H. Freeman, New York.
Williams, D.A. (1976) Improved likelihood ratio tests for complete contingency tables. *Biometrika* **63**, 33–37.

gadolinium (Gd) A lanthanide metallic element (ATOMIC NUMBER = 64; relative ATOMIC MASS = 157.5; melting point = 1313°C; boiling point = 3266°C) with an average abundance of 7.7 mg kg^{-1} in the Earth's CRUST. Gadolinium is used mainly in ferrites and in selenide in electronic equipment. It is also used as a CATALYST for olefin polymerization. [J.G.]

Gaia In Greek mythology, Gaia, or Gaea, is the goddess of the Earth. The name has been used by James Lovelock for his theory that the planet operates as a self-regulatory system — a SUPERORGANISM — to ensure that life is maintained. This does not necessarily mean that human life will be maintained.

A major problem with the Gaia theory is that it presumes that the activities of species are somehow coordinated for the common good, whereas Darwinian selection favours species that behave selfishly! Moreover, life on the planet does not appear to be as tightly coordinated and integrated as the cells, tissues and organs are within an organism. [P.C.]

REFERENCES
Lovelock, J.E. (1979) *Gaia*. Oxford University Press, Oxford.
Lovelock, J.E. (1989) *The Ages of Gaia*. Oxford University Press, Oxford.

Galapagos Islands A group of islands in the PACIFIC OCEAN lying on the Equator at about 90°W, some 1000 km from the coast of South America. The islands are volcanic in origin, with a maximum age of 5 million years. They belong to Ecuador and have both Spanish and English names, the latter used in relation to their significance in the history of evolutionary theory. The largest, Isabela (Albemarle), is 100 km long and rises to a maximum elevation of 1700 m, while the smallest are less than 1 ha in area. These islands, like other oceanic archipelagos, contain ENDEMIC SPECIES. The endemic fauna of the Galapagos belong to a number of animal groups, including iguanid lizards, giant tortoises (*galápagos* in Spanish) and birds. Not only are the species different from those of the nearest continent, but in some cases they differ from island to island.

The endemic fauna and flora of the Galapagos Islands served in part as inspiration for Darwin's theory of EVOLUTION by NATURAL SELECTION after Darwin had visited the Islands on his expedition with HMS *Beagle* in 1835. For example, the pattern of carapace of the tortoises varies between islands, suggesting that they have diverged from one another. Perhaps the most famous example is that of Darwin's finches, which consist of 13 endemic species on the Galapagos Islands and one on Cocos Island, lying between the Galapagos and Panama. They display a pattern of ADAPTATION on the islands, occupying niches that, on the neighbouring continent, are filled by species in other families. They are the classic example of insular ADAPTIVE RADIATION. An international biological research station has been established to support the study and conservation of the Islands' fauna and flora. *See also* DARWINISM; ISLAND BIOGEOGRAPHY. [L.M.C.]

REFERENCE
Grant, P.R. (1986) *Ecology and Evolution of Darwin's Finches*. Princeton University Press, Princeton, N.J.

gallery forest A type of FOREST vegetation found along the stream and river banks within dry SAVANNAH grasslands. GRAZING and an annual rainfall of less than 1500 mm excludes forest from the grassy plains of the savannah, but beside the watercourses the moist soils support forest. Unlike the savannah grasslands the gallery forest is relatively resistant to fire. The steep TOPOGRAPHY which is sometimes present limits the effectiveness of GRAZERS. On the flatter river terraces the soils are ALLUVIAL and rich in nutrients. The forest is multilayered and closed and an impenetrable undergrowth may be present. On occasionally flooded terraces a floristically rich forest can be found, which includes very tall buttressed trees that emerge above the CANOPY. At the lowest level beside the river or stream, in a fringe most exposed to seasonal flooding, the trees are highly branched and the undergrowth is reduced. Early explorers in Africa, like Stanley, who travelled by river, gained an impression that the whole Congo basin was dense tropical forest, because they did not see the grasslands beyond. [M.I.]

REFERENCES
Mabberley, D.J. (1992) *Tropical Rain Forest Ecology*, 2nd edn. Blackie, Glasgow.
Whitmore, T.C. (1990) *An Introduction to Tropical Rain Forests*. Clarendon Press, Oxford.

gallium (Ga) A silvery metallic element (ATOMIC NUMBER = 31; relative ATOMIC MASS = 69.723; melting point = 29.78°C; boiling point = 2403°C), with an average abundance of 18 mg kg^{-1} in the Earth's CRUST, it is extracted and concentrated from bauxite. Two of its compounds, gallium arsenide (GaAs) and gallium phosphide (GaP), find extensive use in the electronics industry as semiconductors. Manufacture of these materials involves the use of arsine and phosphine, both of which are highly toxic. Although gallium has no definite biological role, some of its compounds are used as antitumour agents. The radioisotope ^{67}Ga is used as an imaging agent for the location of certain types of tumours. [C.R.P.]

Game theory

Organisms, or at least the TRAITS that they carry, can be thought of as playing games for existence with each other, and game theory can be used to define the optimum moves in the game. These behavioural games are most vividly illustrated by aggressive interactions between players; when the outcome, in terms of winners (with highest FITNESS), depends upon what other individuals do.

Consider one such game that consists of two strategies (gene-determined behavioural traits): hawks always fight to injure and kill even though they risk injuring themselves; doves display and never engage in fights. Which is the optimum move? Assume that we can represent the effect of a particular move on fitness as a score and, for the sake of argument, let winners score +50 and losers 0; let the time wasted in display cost −10 and the cost of injury be −100. When a hawk encounters a hawk, half the time it is likely to win and half the time lose and sustain injury. So the overall payoff is 0.5(50)+ 0.5(−100)=−25. When a hawk encounters a dove the hawk always wins and gets +50; and dove gets 0. When a dove encounters a dove, DISPLAYS always occur and it wins half the time and loses half the time, so the payoff is 0.5(50−10)+0.5(−10)=+15. These results can be summarized in a payoff matrix:

	Opponent	
Attacker	Hawk	Dove
Hawk	−25	+50
Dove	0	+15

So in a population of all doves (average payoff +15) any invading hawk would do very well and get +50 at every encounter; so dove is not an EVOLUTIONARILY STABLE STRATEGY (ESS). Hawks would spread to take over the population, but then the average payoff would be −25, and now an invading mutant dove would do better. Hence hawk is not an ESS.

Intuitively it would seem that a mixture of strategies ought to be stable. Let H be the proportion of hawks in a population and $1-H$ the proportion of doves. The average payoff ($\bar{P}_H$) for hawk is the payoff for each encounter times the probability of meeting the contestant:

$$\bar{P}_H = -25H + 50(1-H)$$

and for dove:

$$\bar{P}_D = 0H + 15(1-H)$$

These can be solved by making $\bar{P}_D = \bar{P}_H$ (the ESS), and this gives $H=7/12$ with $1-H=5/12$. This stable condition can be achieved by: (i) each individual in a population being hawk or dove (playing pure strategies) and the population consisting of 7/12 hawks and 5/12 doves; or (ii) each individual playing hawk or dove (mixed strategy) at, respectively, an average of 7/12 and 5/12 of encounters.

A final caveat: it is not intended to portray EVOLUTION as a thoughtful player of games. Rather, because of the way selection works, strategies follow the principles of game theory. *See also* BEHAVIOURAL ECOLOGY; EVOLUTIONARY OPTIMIZATION. [P.C.]

REFERENCE

Maynard Smith, J. (1976) Evolution and the theory of games. *American Scientist* **64**, 41–45.

gamete Sexual reproductive PROPAGULE that is haploid (*see* HAPLOIDY) and that generally requires to fuse with another in FERTILIZATION to initiate development (*cf.* PARTHENOGENESIS). *See also* ANISOGAMY, EVOLUTION OF; EGGS; OVA; SPERMATAZOA. [P.C.]

gamete-order hypothesis A hypothesis, also called order of GAMETE release hypothesis, or cruel bind hypothesis, to explain which sex provides PARENTAL CARE in species in which only one parent attends the young. It was proposed in the 1970s that in species with internal FERTILIZATION, such as mammals, since females could not desert their young immediately after mating, males could abandon the female first. This leaves her in a 'cruel bind'—abandon the young and let them die, or remain with the young at a cost to future reproduction. This, it was argued, would select for female parental care. Conversely, with external fertilization, typical of most teleost fishes, females were thought to be able to produce gametes first, because sperm would tend to diffuse faster than EGGS, and therefore could not be produced earlier. This gave females the first opportunity to abandon males. Hence the predominance of male parental care in those fishes where any care is exhibited.

This idea was largely discredited in the 1980s because contrary to the hypothesis, externally fertilizing fishes typically produce gametes simultaneously or in rapid alternations as the batch of eggs is produced. Thus, the proposed order of gamete release (female first) often fails to match cases of male care. There are also numerous exceptions to the hypothesis even in species where gametes are produced successively by each sex in a clear order. *See also* BEHAVIOURAL ECOLOGY; MATING SYSTEMS. [J.D.R.]

gametic phase equilibrium The situation in which the frequency of a given HAPLOTYPE in the GAMETE phase can be calculated from the knowledge of single-locus GENOTYPE frequencies. For instance, loci A and B, with ALLELES A_1 and A_2 in frequencies p_1 and p_2 at locus A, and B_1 and B_2 with frequencies q_1 and q_2 at locus B, are said to be in gametic phase equilibrium, if $p(A_1B_1)$, the frequency of gametes with alleles A_1 and B_1, is equal to the product p_1q_1. There is thus statistical independence of the genotype at locus A and the genotype at locus B. The difference between $p(A_1B_1)$ and p_1q_1 is equal to a parameter, D, called LINKAGE DISEQUILIBRIUM (or gametic phase disequilibrium). At equilibrium of an infinitely large PANMICTIC population, with no selection, MUTATION and MIGRATION, D equals zero, whatever the physical linkage between the two loci. *See also* QUANTITATIVE GENETICS. [I.O.]

gametophyte The GAMETE-forming, and usually haploid, phase in the LIFE CYCLE of a plant—gametes are formed by mitosis. In bryophytes this is the dominant, conspicuous and independent phase. Fusion of two haploid gametes leads to formation of the sporophyte generation consisting of, in mosses, a thin stalk and capsule (or sporangium), dependent for support and nutrition on the gametophyte. In most vascular plants the gametophyte is the minor phase of the life cycle. In ferns the gametophyte is a relatively inconspicuous but free-living prothallus which may be capable of extended survival under continuously moist conditions. An extreme example occurs in the rare Killarney fern (*Trichomanes speciosum*) in which colonies of the gametophyte generation persist with no apparent capacity to form the sporophyte. In angiosperms and gymnosperms the gametophyte is reduced to the pollen, the pollen tube and the ovule, with no capacity for independent existence. The evolution of plants from the independent gametophyte, with its close association with water, to the dominance of the sporophyte, capable of withstanding drought, probably occurred shortly after the invasion of the land. [G.A.F.H. & I.O.]

gamma diversity *See* DIVERSITY, ALPHA, BETA AND GAMMA.

γ-rays Photons of electromagnetic RADIATION which are emitted during de-excitation of product nuclides in radioactive decay. The energies of γ-rays (typically in the range 10 keV to 5 MeV) are characteristic of the nuclide undergoing decay. They are highly penetrating, requiring 10 cm of LEAD to provide effective shielding, but have low linear energy transfer (LET) rates. γ-Rays interact with matter by: (i) the photoelectric effect, in which the PHOTON imparts all of its energy to an electron; (ii) Compton scattering, in which the photon imparts some of its energy to an electron and is scattered with a reduced energy; or (iii) pair production, in which the photon is converted into a positron-electron pair. *See also* RADIOACTIVITY; RADIONUCLIDE; SIEVERT. [A.B.M.]

gap analysis The process used by conservation biologists to determine priorities in conserving biological diversity. The first step in conservation planning is to identify what already is adequately protected. Gap analysis is a systematic method of analysing the current status of the elements of BIODIVERSITY (species, ECOSYSTEMS, HABITAT types, VEGETATION, etc.). Priorities for action, then, can be set based on identified gaps in protection. The process can be carried out at several levels and degrees of thoroughness. It can yield useful information at all levels of analysis.

Gap analysis as a conservation methodology has been used for many years, although the term itself is more recent. In Africa, for example, a 1974 analysis of protected area coverage and known conservation needs for many species of WILDLIFE resulted in identification of major conservation 'gaps', and priorities for protection were defined or adjusted accordingly. Similarly, in North America as early as 1926, a comprehensive gap analysis of major ecosystem types set the stage for conservation action for the next half century. In the UK, the long-standing system of SITES OF SPECIAL SCIENTIFIC INTEREST (SSSIs) has included the analysis of gaps in conservation coverage, and priorities were adjusted every few years as additional information accumulated. Since the 1980s the IUCN, through its World Commission on Protected Areas, has reviewed the protected areas system of all regions of the world, and priorities for future conservation now are reviewed and revised systematically.

In the late 1990s, gap analysis in many countries has become more rigorous with the use of GIS and vastly more detailed information on the distribution, abundance and threats to the security of species and ecosystems worldwide. *See also* CONSERVATION BIOLOGY. [F.W.B.]

REFERENCES

Burley, F.W. (1988) Monitoring biological diversity for setting conservation priorities. In: *Biodiversity* (ed. E.O. Wilson), pp. 227–230. National Academy Press, Washington, D.C.

Scott, J.M., Davis, F., Csuti, B. *et al.* (1993) Gap analysis: A geographic approach to protection of biological diversity. Wildlife Monographs No. 123 of The Wildlife Society. *Journal of Wildlife Management* **57** (1) (Suppl.).

gap phase regeneration A pattern of natural regeneration in FOREST dynamics in which trees recruit to the CANOPY layer only after the creation of a canopy gap. Gaps may be formed by tree fall, death of a standing tree or by hurricane, landslide or local tree felling by humans or larger animals like elephants. Different tree species may require gaps of different sizes (e.g. extreme light-demanding species may be restricted to the centre of the largest gaps), but there is little evidence of widespread NICHE specialization (e.g. gap edge vs. gap centre species) in most tropical forests studied so far. The plants which recruit following gap formation may be individuals that have germinated from seed (either from the SOIL SEED BANK or following recent seed DISPERSAL), or they may grow from established, but shade-suppressed saplings living on the forest floor (often for many years) prior to gap formation (this is known as advanced regeneration). The spatial pattern of mature canopy trees reflects the size distribution and spatial distribution of canopy gaps of different sizes, and hence captures the history of the frequency and intensity of DISTURBANCES of different kinds. [M.J.C.]

garrigue *See* CHAPARRAL SHRUBLANDS.

gas chromatography A technique used for the determination of chemical components in a wide range of environmental samples. An inert carrier gas transports the sample through a column containing an immobilized medium (referred to as the stationary phase). Separation and identification of components is achieved through differences in migration rates and known retention times within the column. The stationary phase may be a solid with a large surface area (gas–solid chromatography). However many compounds remain permanently adsorbed on solid media, requiring the use of a liquid stationary phase (gas–liquid chromatography). Samples are introduced into the carrier gas stream as a discrete 'plug' of vapour by means of a gas sampling valve or injection system. Non-gaseous samples are vaporized. In gas–liquid chromatography the stationary phase may be held either (i) on a finely ground inert solid (packed column), or (ii) as a sub-micrometre thick coating within a narrow capillary column.

Methods of detecting the elution of compounds from the column include:

1 pyrolysis in a hydrogen-air flame (flame ionization detector);

2 changes in the thermal conductivity of the gas stream (thermal conductivity detector);

3 ionization of the gas stream using β-radiation (electron capture detector).

An ecological application of gas chromatography is in the measurement of METHANE fluxes from wetlands. The efflux can be determined from the rate of increase in methane concentration in a sealed chamber placed on the wetland by taking samples of air for gas chromatographic analysis at regular intervals. [I.J.B.]

Gas emissions

Gas EMISSIONS associated with ECOSYSTEM processes lead to large-scale cycling of the ELEMENTS between the ATMOSPHERE and terrestrial and oceanic biota. These phenomena are components of the BIOGEOCHEMICAL CYCLES, via which substances in the natural environment are transported and transformed on a global scale.

The major global cycles are of CARBON, SULPHUR, NITROGEN and WATER. Human interventions have led to perturbations in these global cycles through industrial processes, energy use and in the terrestrial sphere through land-use change. The impacts of gas emissions on ecosystems are presently great, and potentially even greater in future.

CARBON DIOXIDE (CO_2) is currently increasing at 0.5% per annum in the atmosphere, and now constitutes approximately 360 ppm by volume (ppmv) compared to 280 ppmv in preindustrial times. Estimates of the global BIOMASS pool range from 550 to 830×10^9 tonnes (gigatonnes; Gt) of carbon (Gt C), with the other major pools being the atmosphere (720 Gt C), oceans (38 Gt C) and fossil-fuel reserves (6 Gt C). Estimates suggest global emissions from DEFORESTATION at 1.6 Gt C year^{-1}, the central estimate of fossil-fuel emission rates being 5.4 Gt C year^{-1} (Table G1). The estimates of emissions from natural sources are, however, subject to great uncertainty caused mainly by ranges in estimates of tropical deforestation. Moreover, emissions vary from year to year. In the period from 1990 to 1996 the CO_2 emissions rose from 6.1 to 6.7 Gt C year^{-1}.

The atmospheric concentration of METHANE (CH_4) has been rising steadily in the last 300 years. The current concentration of 1720 ppbv corresponds to an atmospheric reservoir of around 4900 Mt CH_4, which is increasing by around 30 Mt CH_4 year^{-1}. Indirect evidence suggests that there has been a marked acceleration in the atmospheric concentration in the last 100 years. However, atmospheric concentrations have only been measured systematically for the last 30 years, the results suggesting that concentrations have increased roughly 1% per year over that period, with a small decrease in the rate in the late 1980s. Around two-thirds of these emissions are related to human activities, and one-third to natural sources. Of the CH_4 sources, AGRICULTURE-related emissions form almost half the anthropogenic emissions, even excluding the land-use change-related emissions from biomass burning. Emissions from ruminants are estimated to provide an atmospheric source of 65–100 Mt CH_4 year^{-1}.

Atmospheric nitrous oxide (dinitrogen oxide; N_2O) emissions are currently rising at a rate of approximately 0.8 ppbv year^{-1}, and are presently $\approx$310–320 ppbv. The atmospheric N_2O cycle and in particular the capacities of the various sources and sinks are poorly understood. The principal sink for N_2O is destruction by ultraviolet (UV) light in the STRATOSPHERE, which accounts for its long atmospheric RESIDENCE TIME of approximately 150 years. Of the estimated 14 Mt of N_2O emitted to the atmosphere each year, only a small fraction is now thought to come from fossil-fuel use. The remainder arises from gas exchanges with the oceans, from biomass burning and deforestation, and emissions from soils through the processes of DENITRIFICATION and nitrification, both of which are associated with the application of agricultural FERTILIZERS.

Table G1 1980s global carbon fluxes (tonnes carbon $\times 10^9$). (After Houghton, J.T., Jenkins, G.J. & Ephraums, J.J. (eds) (1990) *Climate Change: the IPCC Scientific Assessment*. Cambridge University Press, Cambridge.)

Emissions from fossil fuels	5.4 ± 0.5
Emissions from deforestation and land use	1.6 ± 1.0
Accumulation in the atmosphere	3.4 ± 0.2
Uptake by the ocean	2.0 ± 0.8
Net imbalance	1.6 ± 1.4

Nitrification is microbial transformation of AMMONIA (NH_4^+) to NITRATE (NO_3^-), while denitrification is the transformation of NO_3^- to molecular nitrogen (N_2). Nitrification is dominant in aerobic soils, whereas denitrification occurs in anaerobic conditions. Factors influencing the speed of turnover of nitrogen are soil temperature, moisture, fertility, availability of organic matter, and drainage. Hence, N_2O fluxes vary considerably both spatially and temporally.

The NITROGEN CYCLE involves many important biological and abiotic processes. Human intervention and use of compounds of nitrogen also lead to a range of impacts, from the perturbation of stratospheric OZONE to the CONTAMINATION of groundwater through acid rain and other processes.

The global SULPHUR CYCLE is also important in biogeochemical terms, particularly its impacts on natural ecosystems. Biological sources of sulphur in the terrestrial BIOSPHERE, however, form only a small fraction of the total cycle.

Many significant perturbations in gas emissions from biological sources are connected to land-use changes. The single most important trend in global land use is the reduction in the world's forests and woodlands, which now may amount to only 81% of the total in 1700. The expansion of the cropped land area approximately offsets the loss in forest area over the period. The rate of conversion has increased dramatically this century, with the absolute changes in world cropland being greater in the period 1950–1980 than in the previous 250 years. Although tropical deforestation receives the greatest attention at present, particularly for its greenhouse gas implications, evidence suggests that over the time-frame of a few centuries, temperate and boreal deforestation is by far the most significant in terms of absolute area.

The implications of global environmental changes due to gas emissions are numerous. For example, the natural greenhouse effect keeps the Earth warmer than it would otherwise be, but gas emissions enhance this greenhouse effect, resulting in additional warming of the Earth's surface. The reasons for this concern are the potential for global CLIMATE CHANGE, sea-level rise and impacts on human and natural systems. Other impacts on ecosystems have been observed due to stratospheric ozone depletion and resulting enhanced UV irradiation. Forest-quality changes arise due to ACIDIFICATION of PRECIPITATION and impacts due to other gaseous pollutants. *See also* CARBON CYCLE. [W.N.A.]

REFERENCES

Adger, W.N. & Brown, K. (1994) *Land Use and the Causes of Global Warming*. John Wiley & Sons, Chichester.

Dixon, R.K., Brown, S., Houghton, R.A. *et al.* (1994) Carbon pools and flux of global forest ecosystems. *Science* **263**, 185–190.

Houghton, J.T., Jenkins, G.J. & Ephraums, J.J. (eds) (1990) *Climate Change: the IPCC Scientific Assessment*. Cambridge University Press, Cambridge.

Schlesinger, W.H. (1997) *Biogeochemistry: an Analysis of Global Change*, 2nd edn. Academic Press, London.

Turner, B.L., Clark, W.C., Kates, R.W. *et al.* (eds) (1990) *The Earth as Transformed by Human Action*. Cambridge University Press, Cambridge.

gas laws Laws describing the macroscopic behaviour of an ideal gas. Many gases in the atmosphere are sufficiently close to this ideal behaviour to enable the laws to be widely used in calculations that deal with conditions prevailing in the PLANETARY BOUNDARY LAYER. Suppose we take a fixed quantity of a gas in a tank, arranged so that the temperature and volume may be varied. The relationship between volume, temperature and pressure is then described by the gas laws.

1 Boyle's law, named after Robert Boyle (1627–1691), states that if the temperature is kept constant, the volume will be inversely proportional to pressure, so pressure (P) and volume (V) are related thus:

$$PV = k_1$$

This relationship is called Mariottes' law in some European countries, although Boyle's discovery was in 1662 and Mariottes' was later, in 1676.

2 Charles's law, named after Jacques Charles (1746–1823) — also called Gay-Lussac's law, after Joseph-Louis Gay-Lussac (1778–1850)—states that if the pressure is fixed, the volume of the gas will be proportional to temperature (T):

$$V/T = k_2$$

When V and T are plotted against each other, a straight line thus results, which can be extrapolated to demonstrate that the gas would have a nil volume at −273.15°C, or 0 K.

Combining the two laws we have PV/T = constant. In fact:

$$PV = nRT$$

where n is the number of moles of the gas, and R is the universal gas constant, with the value 8.314 J mol^{-1} K^{-1}. [J.G.]

GATT and environment Of the three international economic institutions proposed following the Second World War, only two — the International Bank for Reconstruction and Development (World Bank) and International Monetary Fund — were formally established. The third, the International Trade Organization (ITO), was only realized as a General Agreement on Tariffs and Trade (GATT). GATT came into force in 1948 with the objective of reducing tariffs and other barriers to, and eliminating discrimination in, international trade. By 1991, 103 states were formal Contracting Parties (CPs) to GATT, including the 24 industrialized countries that comprise the OECD and account for 75% of world trade. Another 29 states were applying GATT on a *de facto* basis.

GATT's rules are principally aimed at obtaining non-discriminatory treatment:

- of products, whether produced domestically or imported, or consumed domestically or exported; and
- of countries, specifically GATT CPs.

In general, domestic laws and regulations that treat like products equally, irrespective of their country of origin and destination, will not contravene GATT rules. Unequal treatment is likely to do so. GATT rules are amended, or their scope enlarged, by periodic 'Rounds' of negotiations.

The GATT rules do not mention the ENVIRONMENT, although they do allow exceptions to be made where 'necessary to protect human, animal or plant life or health', or 'relating to the conservation of exhaustible resources'. The environment is specifically acknowledged in the Agreement on Technical Barriers to Trade, concluded in 1979 as part of the Tokyo Round and accepted by 38 countries by 1991, as one of the considerations that can justify the setting of product standards higher than international norms.

Permitting general unilateral action for environmental reasons would, however, increase protectionism, encroach on national sovereignty and deny the exploitation of legitimate competitive advantage. Moreover, not only would such action almost certainly not be the most efficient way to address an environmental problem, it could even be counterproductive.

Environmentalists contend that the GATT rules amount to a serious obstacle to the attainment of SUSTAINABLE DEVELOPMENT. They also point out that the rules may also run counter to the trade measures envisaged for the enforcement of various international environmental agreements, such as the Montreal Protocol to protect the OZONE LAYER. They argue that GATT should be reformed specifically to allow environmental protection to take precedence over the promotion of international trade. [P.C.]

REFERENCES

Arden-Clarke, C. (1991) *The General Agreement on Tariffs and Trade, Environmental Protection and Sustainable Development.* World Wide Fund for Nature, Geneva.

Pearce, D. (1992) *Should the GATT be Reformed for Environmental Reasons?* CSERGE Working Paper, GEC 92–06. CSERGE, University of East Anglia/University College London.

Gause, G.F. *See* CHARACTERS IN ECOLOGY.

Gause's experiments

G.F. Gause was an ecologist who worked at the Laboratory of Ecology, in the Zoological Institute at the University of Moscow, in the early 20th century (*see* CHARACTERS IN ECOLOGY). He carried out a series of simple experiments with yeasts and protozoans that are instructive, are of historic interest and set the experimental protocol for many of the so-called 'bottle' experiments (Kareiva 1989) carried out in the 1940s, 1950s and 1960s. His early experiments are detailed in a classic book, *The Struggle for Existence* (Gause 1934). He was mainly interested in testing the LOTKA–VOLTERRA MODEL, which had been put forward a few years earlier as a description of the interspecific competitive process.

Gause's experimental design was simple. He grew two species, each alone and then together in mixed culture, to determine the course of COMPETITION under laboratory conditions.

In his first experiments he used two species of yeast, which he called *Saccharomyces* and *Schizosaccharomyces* (Gause 1932, 1934). However, we are not now sure of the exact taxonomic identity of the two species. He grew these species on an extract of brewer's yeast, water and sugar. The growth medium was not renewed, but growth ceased before the sugar resource was exhausted, apparently because of the accumulation of alcohol. These experiments with yeast are therefore an example of INTERFERENCE COMPETITION, rather than EXPLOITATION COMPETITION. The main interest of these experiments lies in the calculation of the COMPETITION COEFFICIENTS (α), which Gause did by estimating all the variables in the Lotka–Volterra competition equations except α. The INTRINSIC RATE OF INCREASE (r_1 and r_2) and the CARRYING CAPACITY (K_1 and K_2) were estimated by fitting a LOGISTIC EQUATION to the single-species growth curves. From the mixed cultures, he estimated both the numbers of each species (N_1 and N_2) and the growth rates (dN_1/dt and dN_2/dt). By rearranging the Lotka–Volterra equations he obtained:

$$\alpha_{12} = \frac{K_1 - \left[\dfrac{(\mathrm{d}N_1/\mathrm{d}t) \times K_1}{r_1 N_1}\right] - N_1}{N_2}$$

$$\alpha_{21} = \frac{K_2 - \left[\dfrac{(\mathrm{d}N_2/\mathrm{d}t) \times K_2}{r_2 N_2}\right] - N_2}{N_1}$$

Since all quantities on the right-hand side are known he could estimate the competition coefficients (α_{12} and α_{21}). Since N_1, N_2, $\mathrm{d}N_1/\mathrm{d}t$ and $\mathrm{d}N_2/\mathrm{d}t$ all vary with time, Gause selected three occasions during his experiments to estimate α. He then took the mean of these three calculations. His experiments were also repeated under two environmental conditions, which he called anaerobic (yeasts grown in test tubes) and aerobic (yeasts grown in flasks). Since α is a measure of the effect that one individual of species 1 has upon the population growth of species 2 (and vice versa), and with these species competition is thought to be mainly the result of alcohol poisoning, it should be possible to calculate α directly from the relative production of alcohol by the two species. The values of α calculated from population experiments and relative alcohol production, under both anaerobic and aerobic conditions, are shown below.

		Experiments	Alcohol
Anaerobic	α_{12}	3.15	2.19
	α_{21}	0.44	0.46
Aerobic	α_{12}	1.25	1.25
	α_{21}	0.85	0.80

Two important points come out of these yeast experiments. First, the competition coefficients are not constant — they vary (quite markedly) with different environmental conditions. Second, relative alcohol production and its effect on growth is sufficient to explain the interference competition between these two species (particularly under aerobic conditions). Therefore, in this experiment, the competition coefficients are relatively simple, pure quantities and $\alpha_{12} \approx 1/\alpha_{21}$.

In his later, and perhaps more famous experiments, Gause used two species of *Paramecium*. Initially he used what he called *P. caudatum* (1) and *P. aurelia* (2), although once again we are not sure about the precise taxonomy because these 'species' are now known to be closely related groups of species. The *Paramecium* were grown in 5 cm³ of Osterhaut's medium (a mixture of salts), in test tubes, at 26°C. Fresh food was added to the test tubes daily. The food was a bacterium grown on agar plates and added to the cultures using a platinum loop (to standardize the amount given). The bacterium used did not grow in Osterhaut's medium. Each experiment was started with 20 paramecia on day 0, and Gause estimated the number of individuals each day by sampling 10% of the medium. Since *P. aurelia* is only 39% of the volume of *P. caudatum*, he estimated the volume of a typical cell for each species and converted the estimated numbers to biomass (volume). Following the same protocol as the yeast experiments, the two species were grown alone (four cultures of *P. caudatum* and three cultures of *P. aurelia*) and in mixed culture (three cultures), and Fig. G1 shows the mean volumes for these experiments. In single culture, both species appear to grow to an equilibrium at around $K=200$. In mixed culture, *P. caudatum* appears to suffer more from INTERSPECIFIC COMPETITION than does *P. aurelia*, although *P. aurelia* also grows less well. Using the same procedure as for the yeast experiments, Gause was able to estimate the competition coefficients (α_{12} and α_{21}), and these are shown below.

	α_{12}	α_{21}
'First days'	−1.00	+0.50
Greater than fifth day	+1.64	+0.61

Notice that in the first few days the competition coefficient for the effect of *P. aurelia* upon *P. caudatum* is negative. This implies that the effect of *P. aurelia* is actually positive at this stage. Given that $K_1 = K_2 = 200$, $\alpha_{12} = 1.64$ and $\alpha_{21} = 0.61$ then we can calculate $K_1/\alpha_{12} = 122$ and $K_2/\alpha_{21} = 328$ and therefore construct the Lotka–Volterra phase-

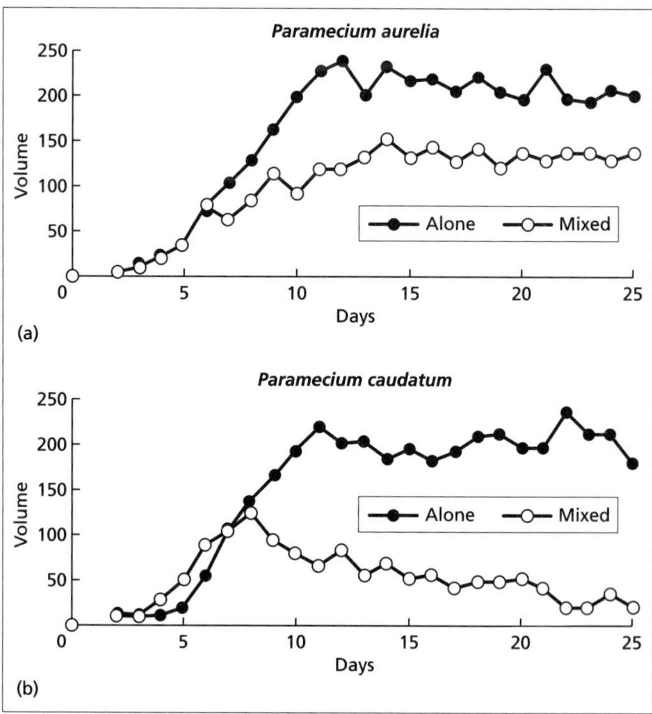

Fig G1 Growth in 'volume' of (a) *Paramecium aurelia* and (b) *P. caudatum* in both single and mixed culture. Volumes are converted from number of individuals per 0.5 cm³ in the culture. Since the volume of *P. caudatum* was set at 1.00 and that of *P. aurelia* at 0.39, the volume for *P. caudatum* = the number of individuals per 0.5 cm³. (After Gause, G.F. (1934) *The Struggle for Existence*. Williams & Wilkins, Baltimore, Md.)

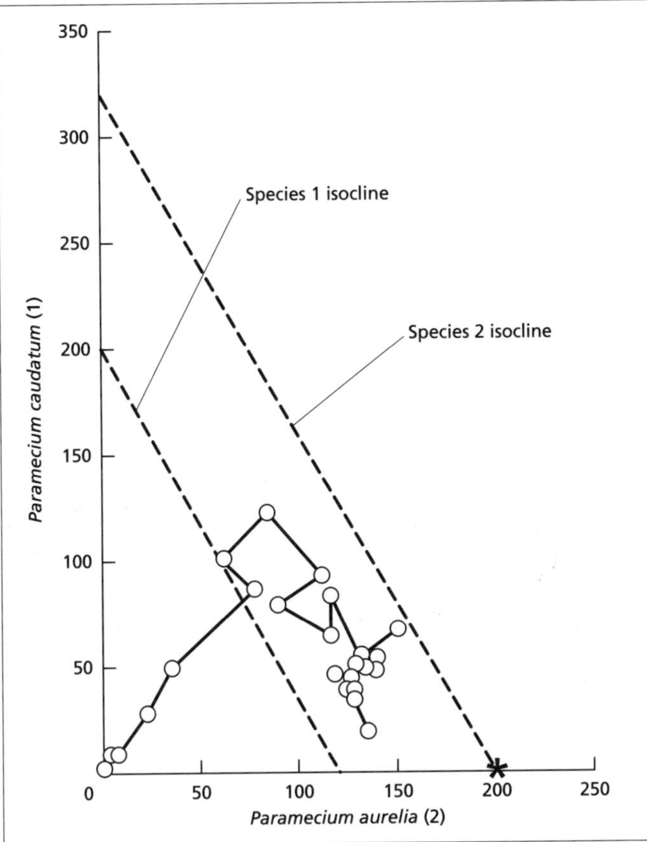

Fig. G2 Phase-plane graph showing the experimental trajectory from Fig. G1. Numbers are actually 'volume' of each species. Intercepts (and therefore zero isoclines) are worked out from Gause's experimental data (see text). The predicted equilibrium is indicated by an asterisk.

plane for this experiment (*see* LOTKA–VOLTERRA MODEL). This is shown in Fig. G2, along with the two-species population TRAJECTORY for the results in Fig. G1. These experiments with *Paramecium* appear to broadly confirm the predictions of the Lotka–Volterra model of interspecific competition, although the mortality imposed by the sampling procedure (10%) and the premature termination of the experiments result in the predicted equilibrium not being reached in 25 days. However, later experiments in which Osterhaut's medium was buffered to a pH 8 and the daily food concentration reduced, resulted in *P. caudatum* being excluded from the cultures in just 16 days. [B.S.]

REFERENCES

Gause, G.F. (1932) Experimental studies on the struggle for existence. I. Mixed populations of two species of yeast. *Journal of Experimental Biology* **9**, 389.

Gause, G.F. (1934) *The Struggle for Existence*. Williams & Wilkins, Baltimore, Md.

Kareiva, P. (1989) Renewing the dialogue between theory and experiments in population ecology. In: *Perspectives in Ecological Theory* (eds J. Roughgarden, R.M. May & S.A. Levin), pp. 68–88. Princeton University Press, Princeton, N.J.

Gause's principle exclusion principle *See* COMPETITIVE EXCLUSION.

GCM *See* GENERAL CIRCULATION MODEL.

GEF *See* GLOBAL ENVIRONMENTAL FACILITY.

Geiger counter An IONIZING RADIATION detector, also called Geiger–Müller counter, which consists of a glass tube filled with an inert gas, and fitted with a central anode and cathode concentric with the walls. A potential difference, lower than the discharge voltage of the gas, is applied across the electrodes and when a RADIATION particle or PHOTON enters the gas, the ionization produced gives rise to a small transient electrical pulse or 'count'. The count rate can be used to measure the activity of radioactive sources. In a Geiger counter, the pulse size is independent of the radiation energy. In a proportional counter, which is of similar construction, the pulse size is proportional to the radiation energy. Geiger counters are used as basic radiation monitors and for analysis of β^--emitting RADIONUCLIDES. [A.B.M.]

gelbstoff A German word meaning 'yellow substance', used as a name for poorly identified, dissolved organic compounds found in water and usually considered to consist at least in part of humic and fulvic compounds. [V.F.]

Gelisols *See* SOIL CLASSIFICATION.

gene bank *See* SEED BANK.

gene cluster A group of functionally and structurally related genes that occur together in certain regions of chromosomes. Examples include the human globin genes, histone genes, the ribosomal RNA genes, the immunoglobulin genes and the major histocompatibility genes. Eukaryotic genomes contain both simple gene clusters, composed of a single set of related genes, as well as compound gene clusters, composed of simple gene clusters repeated in tandem (e.g. histone genes in humans). The grouping together or clustering of genes with related function and structure occurs presumably in order to facilitate their coordinate expression. Gene clusters may arise by gene duplication followed by divergence during EVOLUTION. *See also* EPISTASIS; PLEIOTROPY. [V.F.]

gene conversion A process that occurs during crossing-over and by which the DNA sequence of one ALLELE is converted to that of the other allele. Gene conversion was first found in yeast by Carl Lindegren and in *Neurospora* by Mary Mitchell and is characterized by a genetic exchange ratio involving two closely linked genes that is non-reciprocal. In gene conversion an allele of one parent is lost while an additional allele is gained from the other parent, leading to a meiotic segregation ratio of 3 : 1 rather than 2 : 2. The process was so-named because it appeared that one allele had somehow been converted to another during an event where genetic exchange also occurred. Gene conversion relies on DNA repair enzymes correcting mismatches of bases during the process of recombination and is an important mechanism for generating allelic novelties. In fungi, this process involves meiosis and a recombinational step, but in other organisms meiosis may not be required. [V.F.]

gene, definitions of The functional unit of heredity; a hereditary determinant that behaves as a unit of inheritance occupying one (usually) contiguous region of a chromosome and which can mutate to various allelic forms. A gene is used in evolutionary theory to describe any hereditary information for which there is a favourable or unfavourable selection bias equal to several or many times its rate of endogenous change. A gene can refer specifically to the region of DNA that codes for a polypeptide. Used in this way it is synonymous with cistron or structural gene. Genes not only encode the amino-acid sequences of polypeptides, but can also specify the nucleotide sequences of structural RNA molecules, such as ribosomal RNA (rRNA). In EUKARYOTES, the coding regions of a gene may be interrupted by intervening sequences that are internal and non-coding. The word 'gene' was first coined in 1911 by the Danish geneticist, Wilhelm Johannsen, as a convenient term for the elementary unit of Mendelian genetics. In 1913, Alfred Sturtevant demonstrated that genes occupy limited but specific regions of chromosomes and published the first genetic map. *See also* MENDELIAN INHERITANCE; QUANTITATIVE GENETICS. [V.F.]

REFERENCES

King, R.C. (1974) *A Dictionary of Genetics*, 2nd edn. Oxford University Press, New York.

King, R.C. & Stansfield, W.D. (1990) *Encyclopedic Dictionary of Genetics*. VCH Publishers, New York.

gene family Genes of related function that have arisen by duplication and have been modified during the course of EVOLUTION; also termed multigene family. Such duplication plays a key role in the development of new functions, since one of the copies of the duplicated gene maintains the old function while others are free to evolve in new directions. The members of gene families are often arranged in clusters on chromosomes (*see* GENE CLUSTER). [V.F.]

gene flow The MIGRATION of genetic information occurring between individuals, populations or other spatial units of information (e.g. between two species) (*see* F-STATISTICS). Gene flow within a species is the main evolutionary force maintaining the cohesiveness of the species (the other main force being NATURAL SELECTION when it acts homogeneously throughout the range of the species). *See also* GENE, DEFINITION OF; GENE POOL; SPECIATION. [I.O.]

gene pool Two populations are said to share a common gene pool if they are actually or potentially capable of exchanging genes. The term is mostly used in describing the genetic VARIABILITY of CROP plants and their relatives. Harlan and de Wet (1971) divided these plants into:

1 the primary gene pool—cultivated species and those relatives that are easily crossable and between which gene transfer is simple;

2 the secondary gene pool—species that will cross with the crop but where gene transfer is difficult;

3 the tertiary gene pool—species that can cross with the crop only with difficulty and where gene transfer is virtually impossible.

According to Marshall (1989), new molecular genetic technologies may make this system obsolete. *See also* HARDY–WEINBERG LAW; POPULATION GENETICS. [I.O.]

REFERENCES

Harlan, J.R. & de Wet, J.M.J. (1971) Toward a rational classification of cultivated plants. *Taxon* **20**, 509–517.

Marshall, D.R. (1989) Crop genetic resources: Current and emerging issues. In: *Plant Population Genetics, Breeding, and Genetic Resources* (eds A.H.D. Brown *et al.*), pp. 367–388. Sinauer Associates, Sunderland, Mass.

general adaptation syndrome (GAS) A supposed recognizable suite of physiological responses to environmental STRESS that could, therefore, provide indicators of stress experience in organisms. It was proposed by the Austrian physician, Hans Selye, in the 1950s (see, for example, Selye 1973). However, such a generality of stress responses now seems questionable. [P.C.]

REFERENCE

Selye, H. (1973) The evolution of the stress concept. *American Scientist* **61**, 629–699.

general circulation model (GCM) Any very large computer model of the three-dimensional global CLIMATE system that describes the redistribution of momentum, heat and water vapour due to atmospheric motion; also called global circulation model, or climate model. Such models are based upon the fundamental EQUATIONS that describe the movement of air or water on the rotating Earth, as influenced by the pole to equator temperature gradient that is driven by the input of energy from the Sun. These non-linear partial DIFFERENTIAL EQUATIONS are solved numerically, by dividing the ATMOSPHERE into discrete layers and calculating solutions at discrete grid points. Even on the largest supercomputers the resolution is limited by computing power: typical horizontal resolutions are from 300 to 1000 km, with 2–20 vertical levels, and using approximately half-hour time steps. They are much used in studies of past and future CLIMATE CHANGE. The current emphasis is on linking ('coupling') atmosphere and ocean models—atmosphere and ocean general circulation models (AOGCMs)—so as to include FEEDBACKS between the ocean and the atmosphere. Two types of experiments, equilibrium and transient, are carried out with AOGCMs. In equilibrium experiments new climate conditions are imposed in a single step on the model and the experiment continued until a new equilibrium state is reached which can then be compared with the original prediction. This is very different from the more realistic transient experiments, where the model is continuously perturbed with changing input conditions to determine the time-dependent behaviour. [J.I.L.M.]

REFERENCE

Cubasch, U. & Cess, R.D. (1990) Processes and modelling. In: *Climate Change: the IPCC Scientific Assessment* (eds J.T. Houghton, G.J. Jenkins & J.J. Ephraums), pp. 69–91. Cambridge University Press, Cambridge.

general life-history problem Life histories show great variation. Such variation can be encapsulated in the Euler equation, which specifies how the age schedules of survival and reproduction interact to determine a population's or genotype's rate of increase. The general problem is to understand the CONSTRAINTS that determine particular combinations of the two schedules. *See also* LIFE-HISTORY EVOLUTION; REPRODUCTIVE EFFORT MODEL. [D.A.R.]

general-purpose genotype A GENOTYPE, or genome, that confers an ability to exploit a wide range of ecological conditions. The wide ecological and physiological tolerances of general-purpose genotypes may be associated with relatively inferior performance in favourable environments (i.e. the jack of all trades is the master of none). The observation that many asexually reproducing organisms have general-purpose genotypes, compared to their sexual relatives, has been used to explain the broader geographical DISTRIBUTION of asexuals and the proliferation of asexuals in environmentally extreme conditions. *See also* ASEXUAL REPRODUCTION; EVOLUTION OF SEX; OPTIMAL REACTION NORMS; PHENOTYPIC PLASTICITY; REACTION NORM. [V.F.]

REFERENCE

Hughes, R.N. (1989) *A Functional Biology of Clonal Animals*. Chapman & Hall, London.

generalist A species, an individual or (in BEHAVIOURAL ECOLOGY) a STRATEGY that uses a relatively large proportion, or in extreme cases all, of the available resource types. *Cf.* SPECIALIST. [J.G.S.]

generalized linear models (GLIM or GLM) Models that extend conventional parametric statistical analyses by allowing the user to specify non-normal errors (*see* ERROR STRUCTURE) and non-identity LINK FUNCTIONS. They are especially useful in environmental work for dealing with count data (e.g. log-linear models for CONTINGENCY TABLES), proportion data (e.g. logistic analysis of mortality data), binary response variables (e.g. dead or alive, infected or not, male or female, etc.) and the analysis of survival data (e.g. age at death, using exponential or Weibull errors). The general term used to describe residual variation in GLM is DEVIANCE; this is analogous to the familiar error SUM OF SQUARES used in regression and ANALYSIS OF VARIANCE. The models are called linear models because they are linear in their parameters and in their error term. It does *not* mean that they are restricted to describing linear relationships between the RESPONSE VARIABLE and the EXPLANATORY VARIABLES. The following curvilinear relationships can be represented as linear models, either by variable substitution (put $x^2=z$ or $\exp(x)=z$):

$$y=a+bx+cx^2$$

$$y=ae^x$$

or by choosing an appropriate transformation (link function):

$$y=\exp(a+bx)$$

$$y=\frac{1}{(a+bx)}$$

which are linearized by the log link and reciprocal link respectively. *See also* EQUATIONS; STATISTICS. [M.J.C.]

REFERENCES

Crawley, M.J. (1993) *GLIM for Ecologists*. Blackwell Scientific Publications, Oxford.

Dobson, A.J. (1990) *An Introduction to Generalized Linear Models*. Chapman & Hall, London.

generation time The time elapsed from one generation to the next. There exist several ways of calculating the 'length' of a generation from basic demographic information. Consider an age-classified population described by its SURVIVORSHIP CURVE $l(x)$ and maternity function $m(x)$, with a net reproductive rate:

$$R_0=\int_0^\infty l(x)m(x)\mathrm{d}x$$

and an INTRINSIC RATE OF INCREASE r. Then generation time can be measured by:

1 the time T required for the population to increase by a factor R_0, given by:

$$T=\ln R_0/r$$

2 the mean age of the parents of the offspring produced by a COHORT during its lifetime, given by:

$$\mu_1=\frac{\int_0^\infty xl(x)m(x)\mathrm{d}x}{R_0}$$

3 the mean age of the parents of the offspring produced by a population at the STABLE AGE DISTRIBUTION, given by:

$$\bar{A}=\int_0^\infty xe^{-rx}l(x)m(x)\mathrm{d}x$$

For populations described by stage- or size-classified population projection matrices (*see* POPULATION PROJECTION MATRIX), measures of generation time can be computed using methods based on Markov chains (see Cochran & Ellner 1992). *See also* DEMOGRAPHY; LIFE-HISTORY EVOLUTION; LIFE TABLE. [H.C.]

REFERENCES

Coale, A.J. (1972) *The Growth and Structure of Human Populations: A Mathematical Approach*. Princeton University Press, Princeton, N.J.

Cochran, M.E. & Ellner, S. (1992) Simple methods for calculating age-based life history parameters for stage-structured populations. *Ecological Monographs* **62**, 345–364.

generative constraints *See* CONSTRAINTS.

genet

1 A catlike viverrine woodland mammal of the genus *Genetta*.

2 A genetic individual plant or animal (in contrast to a RAMET); the product of a single zygote (contrast with a MODULE). Plant genets may consist of very large numbers of modules (e.g. tillers of a rhizomatous perennial grass species) and can extend to cover many hectares. The modules may be more or less physiologically independent, depending upon the extent of hormonal and resource transfer between them.

3 For animals like aphids, with multivoltine asexual generations, a single genet may consist of millions of individual, but genetically identical, insects living in many distinct habitats.

See also ASEXUAL REPRODUCTION; VEGETATIVE REPRODUCTION. [M.J.C.]

genetic bottleneck A substantial decrease in genetic DIVERSITY within a population that is associated with a drastic reduction in POPULATION SIZE, for example as often occurs during the initial stages of COLONIZATION. Because only a fraction of the original GENOTYPES survive a bottleneck, genetic divergence among populations is facilitated. Since the isolated populations will represent an impoverished and atypical sample of the species GENE POOL, the rate of genetic differentiation leading to the EVOLUTION of new species is enhanced (i.e. the FOUNDER EFFECT). If a population goes through a bottleneck, the VARIABILITY of mitochondrial DNA (mtDNA) is reduced to a much greater degree (one genome per breeding pair) than that of nuclear DNA (four genomes), and with subsequent expansion and division into daughter populations variation in mtDNA within populations will be high compared to that between populations. The patterns of mtDNA variability are particularly useful for studying the evolution of populations at or below the species level. *See also* POPULATION BOTTLENECK. [V.F.]

genetic canalization An evolutionary response to selection whereby TRAITS are buffered against genetic perturbations during development. In general, traits with canalized development are relatively insensitive to the disruptive effects of outside influences on development. Genetic CANALIZATION is used specifically to mean the selection process by which development becomes less sensitive to disruptive genetic events such as recombination or MUTATION, whereas environmental canalization is used to denote the stabilization of development to the effects of non-genetic, environmental perturbations (e.g. heat shock). [C.M.L.]

REFERENCES

Stearns, S.C. (1989) The evolutionary significance of phenotypic plasticity. *BioScience* **39**, 436–445.

Waddington, C.H. (1957) *The Strategy of the Genes*. George Allen & Unwin, London.

genetic correlations and life-history evolution *See* TRADE-OFF.

genetic distance A measure of the difference in frequency of ALLELES among different GENOTYPES, populations or species. It is often used as a measure of the time since the populations being compared diverged from a single ancestral population (assuming that the rate of gene substitutions is constant). Several formulae have been proposed for estimating genetic distance (or the reverse, genetic similarity), which vary in terms of their underlying genetic model. The most widely used measure is Nei's genetic distance, which is a measure of the accumulated number of codon differences per locus and which is defined as $D = -\ln I$, where I is the normalized identity of genes between two populations. The degree of genetic distance may or may not be tightly coupled to phenotypic differences (e.g. in morphology, developmental pattern, behaviour or ecological role), thus there is no fixed relationship between the degree of genetic distance *per se*, and the degree of morphological difference between species. *See also* QUANTITATIVE GENETICS. [V.F.]

genetic diversity in rare organisms Molecular genetic DIVERSITY in small populations diminishes as a consequence of GENETIC DRIFT and inbreeding (which could lead to INBREEDING DEPRESSION). Genetic drift is a random change of gene frequencies due to the gain and loss of individuals and GAMETES in a population. In small populations, some genetic loci are likely to become fixed (homozygous in all individuals) as a consequence of these stochastic FLUCTUATIONS. Lost genetic variation requires a long time to recover based on neutral MUTATIONS alone (approximately the reciprocal of the mutation rate). The consequences of lost variation include a reduction in the adaptability of a population to changes in the environment. At the same time, there can be an inverse relationship between molecular genetic diversity and the diversity of quantitative genetic characters (such as bone morphology). Such increases in the VARIABILITY of quantitative characters (or sometimes in the asymmetry of bilateral characters) are random, and therefore potentially maladaptive. *See also* EFFECTIVE POPULATION SIZE. [A.R.H.]

genetic drift Random FLUCTUATIONS of gene frequencies due to small POPULATION SIZES. The Wright–Fisher model (Fisher 1930; Wright 1931) derives the distribution of allele frequencies among populations undergoing random genetic drift. In a finite population, if drift is the sole evolutionary force, polymorphisms will eventually disappear from each population. The probability of ultimate fixation of a given allele is simply the initial frequency of the allele in the population. The relationship between random genetic drift and inbreeding is best viewed by considering an infinite population subdivided into an infinite number of isolated subpopulations: because the POLYMORPHISM is lost from each subpopulation, the average HETEROZYGOSITY among the subpopulations decreases through time (Fig. G3). The inbreeding coefficient f_{st} measures the amount of local inbreeding due to drift relative to the whole population (*see* F-STATISTICS). At equilibrium, the overall gene frequencies are unchanged, but each subpopulation is fixed for alternative ALLELES, and $f_{st} = 1$. *See also* EFFECTIVE POPULATION SIZE. [I.O.]

REFERENCES

Fisher, R.A. (1930) *The Genetical Theory of Natural Selection*. Oxford University Press, Oxford.

Wright, S. (1931) Evolution in Mendelian populations. *Genetics* **16**, 97–159.

genetic engineering MANIPULATION of the genome, usually DNA, with various techniques. It allows artificial manipulation of genetic VARIABILITY within species and even the combination of genetic materials across species. *See also* TRANSGENIC ORGANISMS. [P.C.]

REFERENCE

Shorrocks, B. & Coates, D. (1993) *The Release of Genetically-Engineered Organisms*. British Ecological Society, Ecological Issues, No. 4. Field Studies Council, Shrewsbury.

genetic fingerprinting A method used to quantify genetic differences among individuals, populations or species. A pattern, or so-called 'fingerprint', is obtained by enzymatically cleaving a protein or nucleic acid and subjecting the digest to two-dimensional chromatography or ELECTROPHORESIS. *See also* ECOLOGICAL GENETICS; MOLECULAR ECOLOGY; QUANTITATIVE GENETICS. [V.F.]

REFERENCE

Klug, W.S. & Cummings, M.R. (1994) *Concepts of Genetics*, 4th edn. Prentice–Hall, Englewood Cliffs, N.J.

genetic homeostasis As defined by Lerner in 1954, the property of a population to equilibrate its genetic composition and to resist sudden changes. Genetic HOMEOSTASIS may be brought about when: (i) the mean values of some characters in a population are preserved by NATURAL SELECTION; (ii) HETEROZYGOSITY is favoured; and (iii) selection for one TRAIT results in changes in other traits that lead to decreased FITNESS. Genetic homeostasis may explain the gradual loss of a selected trait once the selection is relaxed as is sometimes observed during artificial breeding. [V.F.]

REFERENCES

Dobzhansky, T. (1970) *Genetics of the Evolutionary Process*. Columbia University Press, New York.

Lerner, I.M. (1954) *Genetic Homeostasis*. Oliver & Boyd, Edinburgh.

genetic load A measure of the amount of NATURAL SELECTION associated with a certain amount of genetic VARIABILITY. The word load was first used by Muller in 1950 in assessing the impact of

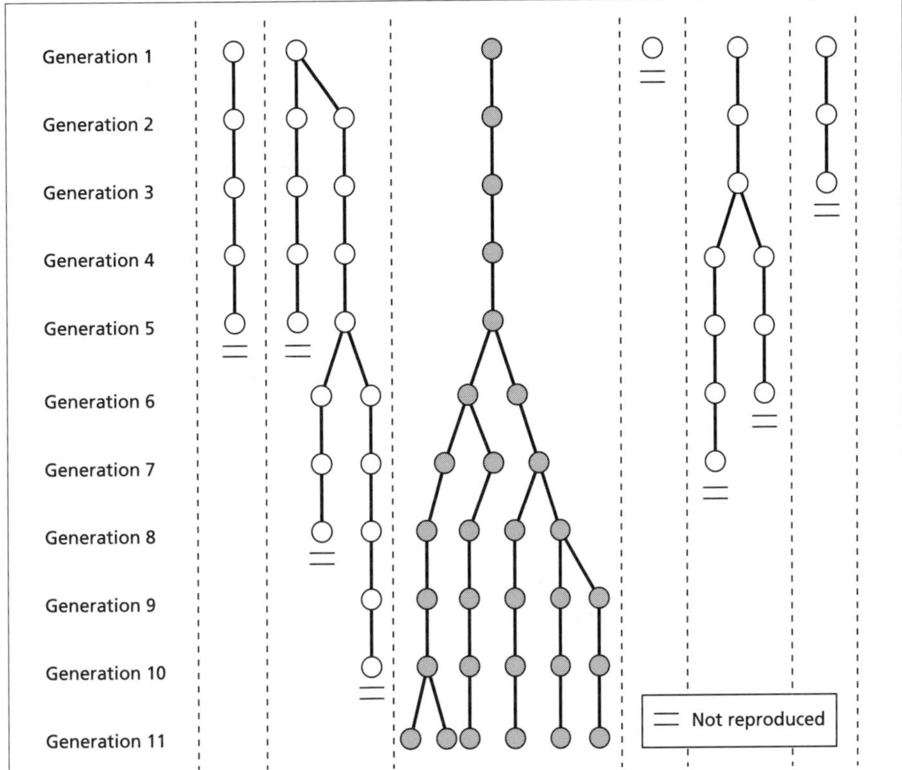

Fig. G3 The drift to homozygosity. The figure traces the evolutionary fate of six genes; in a diploid species these genes would be combined each generation in three individuals. Every generation, some genes may by chance fail to reproduce and others by chance may leave more than one copy. Because a gene that fails to reproduce its line is lost forever, over time the population must drift to become made up of descendants of only one gene from an ancestral population. In this example, the population after 11 generations consists of descendants of gene number 3 (shaded) in generation 1. (After Ridley, M. (1996) *Evolution*, 2nd edn. Blackwell Science, Cambridge, Mass.)

MUTATION on the human population. It can be seen as a measure of the cost a population must pay in order to let natural selection sort out the fittest GENOTYPES every generation. It is equal to the difference between the mean population FITNESS and the fitness of the fittest genotype, divided by the mean population fitness. For instance, in a population with mutation rate μ from a dominant, favourable ALLELE, A_1, towards a recessive, deleterious allele, A_2, the genetic load (L) is equal to the mutation rate μ, and is then called the mutation load. It can be shown that L is independent of the deleterious effect of A_2 and of the mating system. It depends, however, on the DOMINANCE relationship between alleles. If the deleterious allele is dominant, L is equal to twice the mutation rate. The genetic load concept was used by J.B.S. Haldane to study the paradox of natural selection on many loci: if selection was to occur on many loci at the same time, then the genetic load would be far too large for a population to be able to maintain itself. The paradox became even greater with the discovery of extensive protein POLYMORPHISM in most natural populations, and was more or less solved with the assumption that most of this polymorphism was actually neutral and transitory, at equilibrium between mutation and drift. Genetic loads may arise from allele substitution, segregation, and from temporal changes in genetic optima (i.e. LAG LOAD). *See also* SEGREGATIONAL LOAD; SUBSTITUTIONAL LOAD. [I.O. & V.F.]

genetic polymorphism The presence, within a population of a single species, of discrete genetically determined forms or phenotypes such that the occurrence of the rarest of them cannot be due to recurrent MUTATION (a frequency of 1% is usually taken as an arbitrary threshold). The genetics of this type of POLYMORPHISM is usually straightforward, depending on allelic variation at one or a small number of gene loci. Gel ELECTROPHORESIS provides a direct way of scoring this genetic polymorphism for genes specifying enzymes. *See also* ECOLOGICAL GENETICS. [P.M.B.]

genetic renegade *See* SELFISH DNA.

genetic revolution One of several terms used in evolutionary studies in connection with the relation of ADAPTATION to species formation. Microevolution by selection of small genetic differences between individuals is a continuous process adapting individuals to local conditions or changing environments. Similar but distinct species differ at many loci, including those which ensure REPRODUCTIVE ISOLATION. Arguably, the transition from one species to another requires some kind of major genomic reorganization—a genetic revolution—in addition to MICROEVOLUTION.

Some fossil lineages remain morphologically invariant for long periods, after which they change at a relatively very rapid rate. Such change, called punctuated EVOLUTION by Niles Eldredge and Stephen J. Gould, could indicate a genetic revolution. It is implicit in this explanation that loci interact (show EPISTASIS), so that response to selection at one locus may lower FITNESS conferred by interaction between that locus and another. Epistatic interaction would slow response to direct selection, and a genetic revolution could decouple the effects of the two loci. For example, suppose selection for larger size in a mouse allows survival at lower temperatures, but the metabolic changes involved also result in slower escape response. If the coupling is due to epistatic interaction it is possible to envisage a genetic change removing the correlation between body size and response speed so that evolution can take a new path. *See also* COADAPTED GENE COMPLEX; FOUNDER EFFECT; PUNCTUATED EQUILIBRIUM. [L.M.C.]

genetic systems For a system to evolve by NATURAL SELECTION there needs to be information transmission from one unit of selection to another, and hence 'memory' of that which is favourable. Most commonly this takes the form of a genetic system, which fundamentally involves information transmission in the form of nucleic acids, but may differ in a number of details. A genetic system refers to the organization and method of transmission of the genetic material for a given species. An alternative is the cultural information that is transmitted from one generation of humans to another. Some see this as being divisible into gene analogues — memes — which raises the possibility of having memetic systems. *See also* DARWINISM; GENE, DEFINITIONS OF; MENDELIAN INHERITANCE. [V.F.]

genetic variance In QUANTITATIVE GENETICS, the genetic variance in a population is that part of the total PHENOTYPIC VARIANCE of a character that is attributed to genetic causes. The genetic variance, V_G, is the variance of the GENOTYPIC VALUES of individuals in the population. The genotypic value, G, of an individual is considered to be the sum of statistically independent parts: $G=A+D+I$, where A equals the individual's BREEDING VALUE, D its dominance deviation, and I the contribution to the genotypic value of interactions between loci. Since A, D and I are independent, V_G equals the sum of the ADDITIVE GENETIC VARIANCE (V_A), the DOMINANCE VARIANCE (V_D) and the INTERACTION VARIANCE (V_I). *See also* BROAD-SENSE HERITABILITY. [G.D.J.]

genetical ecology The integration of genetics into ECOLOGY. Early ecological research treated all individuals in a POPULATION as identical. In genetical ecology, genetic variation among individuals which can result in variation in PHENOTYPE and in FITNESS or reproductive success is taken into account. It differs from ECOLOGICAL GENETICS, which is primarily the genetics of Mendelian TRAITS in natual populations, in also emphasizing the QUANTITATIVE GENETICS of polygenic traits. Most traits of ecological interest are polygenically determined. [P.M.B.]

genetically modified animals *See* TRANSGENIC ORGANISMS.

genetically modified organisms (GMO) *See* TRANSGENIC ORGANISMS.

genetically modified plants *See* TRANSGENIC ORGANISMS.

genome *See* GENOTYPE–PHENOTYPE RELATIONSHIP.

genomic imprinting The phenomenon in which genes have differential expression depending on the sex of the parent from which they are inherited: the ALLELES inherited from fathers behave differently from those inherited from mothers, even though they may be genetically identical. Thus, genes or chromosomes retain some memory, or imprint, of their GAMETE origin. Genomic IMPRINTING is seen in many taxa, but it is most dramatic in mammals. Haig and Westoby (1989) proposed that genomic imprinting has evolved in mammals (and flowering plants) because their offspring are nourished directly from maternal tissues. Genes that are expressed in an embryo (or seed) can influence the quantity of resources that the offspring receives from its mother. However, the greater the nutrient demand of the pregnancy, the greater the cost to the mother's potential future reproduction. This creates a CONFLICT between the interests of the maternal and paternal genes within an embryo: for example, a new mutant allele for a placental growth factor that causes increased nutrient demands on the mother, but only when the allele is paternally derived, has a selective advantage over an established non-imprinted allele. If a locus has preferential paternal expression it will function to increase the nutrient demands on the mother, whereas preferential maternal expression reduces those demands. For example, human triploid fetuses develop a large placenta if the extra genome is paternal, but have little placental tissue if the extra genome is maternal.

Haig and Westoby predict that most imprinted genes will affect how much an offspring receives from its mother, at the expense of sibs. Thus, imprinting is expected at loci that influence placental growth, suckling, neonatal behaviour, appetite, nutrient metabolism and postnatal growth. As a result of an evolutionary ARMS RACE, with each genome counteracting the anticipated actions of the other one, an embryo cannot develop if its genome comes exclusively from either a female or a male. By experimental MANIPULATION of mouse development, phenotypes of parthenogenetic and androgenetic embryos can be produced, which have whole diploid genomes derived from either a female or a male parent. Parthenogenetic embryos typically have poorly developed trophoblast but a relatively normal embryo proper, while androgenetic embryos have well-developed trophoblast but a poorly developed embryo proper. Both types of embryo abort. As a side effect of the evolution of genomic imprinting, parthenogenetic reproduction is prevented in mammals. *See also* PARENT–OFFSPRING CONFLICT; SEXUAL CONFLICT. [M.Mi.]

REFERENCES

Haig, D. & Westoby, M. (1989) Parent-specific gene expression and the triploid endosperm. *American Naturalist* **134**, 147–155.

Moore, T. & Haig, D. (1991) Genomic imprinting in mammalian development: a parental tug-of-war. *Trends in Genetics* **7**, 45–49.

Solter, D. (1988) Differential imprinting and expression of maternal and paternal genomes. *Annual Reviews in Genetics* **22**, 127–146.

genomic parasites *See* SELFISH DNA.

genotoxic A broad term that usually describes a CONTAMINANT that has a measurable/observable EFFECT on DNA and/or chromosome structures. Genotoxicity can be measured by a variety of techniques based on determinations of MUTATIONS, DNA damage/repair or chromosomal aberrations. [P.C.]

genotype

1 The genetic constitution of organisms.

2 A class of equivalence, employed to classify individuals according to their genetic information. Two individuals are said to have the same genotype if they possess genetic information that is more similar between them than compared to individuals with a different genotype. In classical POPULATION GENETICS, monolocus genotypes are considered (e.g. genotypes A_1A_1, A_1A_2 and A_2A_2 at locus A with ALLELES A_1 or A_2), but multilocus genotypes can also be described. In QUANTITATIVE GENETICS, a 'genotype' is often assigned to a group of individuals that are genetically related more to each other than to other groups.

See also GENE, DEFINITIONS OF; MENDELIAN INHERITANCE. [I.O.]

genotype*environment covariance In QUANTITATIVE GENETICS, genotype*environment covariance is a major experimental problem, especially in field work. The quantitative genetic model of a trait is written as $P=G+E$, i.e. the PHENOTYPIC VALUE (P) of an individual derives from genetic (G) and environmental (E) influences. The observed variance over all individuals, the PHENOTYPIC VARIANCE, V_P, is therefore given by $V_P=V_G+V_E+2\text{cov}(G,E)$, where V_G is the GENETIC VARIANCE (the variation between individuals deriving from their genetic difference), V_E is the ENVIRONMENTAL VARIANCE (the variation between individuals deriving from differences in their environments) and $\text{cov}(G,E)$ is the genotype*environment covariance. The term $\text{cov}(G,E)$ equals zero when the probability that a GENOTYPE finds itself in any environment is equal over genotypes; i.e. when the genotypes are randomly and independently distributed over the environment. Independent distribution of genotypes over the environment, $\text{cov}(G,E)=0$, is a fundamental assumption in applications of quantitative genetics. However, genotype and environment might covary, leading to $\text{cov}(G,E)\neq 0$. If 'genetically strong' animals would monopolize 'good territories', the distribution of genotypes over environments would not be random: HERITABILITY estimates and predictions of selection responses would be highly but unpredictably compromised. *See also* ENVIRONMENTAL DEVIATION; GENOTYPIC VALUE. [G.D.J.]

genotype*environment interaction In QUANTITATIVE GENETICS, GENOTYPE by ENVIRONMENT interaction denotes the existence of a difference in the sensitivity of different genotypes to different environments. Differences in the expression of the genotype in different environments leads to different phenotypes over environments, i.e. to PHENOTYPIC PLASTICITY. Genotypes that are phenotypically plastic might differ only in average expression, in which case the REACTION NORMS given by those genotypes are parallel: no genotype*environment interaction exists and the genetic variation is identical in each environment. If genotypes that are phenotypically plastic differ in sensitivity to the environments, the reaction norms given by those genotypes are not parallel and genotype*environment interaction does exist (Fig. G4).

The formal quantitative genetic model of a phenotypically plastic trait can be written as $P=G+E+I_{GE}$ where the standard quantitative genetics for a QUANTITATIVE TRAIT, $P=G+E$ is augmented by the interaction term I_{GE}. The interaction term I_{GE} leads to the genotype*environment interaction variance in a two-way ANALYSIS OF VARIANCE with factors genotype and environment. The genotype*environment interaction variance expresses the GENETIC VARIANCE in phenotypic plasticity, and is related to the genetic variance in the slopes of the reaction norms and to the cross-environment genetic covariances (Fry 1992; de Jong 1995).

The probability of occurrence of any particular (sub)environment x equals f_x. The additive genotypic variance over the total environment, $\Sigma f_x V_{A,x}$, can be divided into a main-effect ADDITIVE GENETIC VARIANCE, V_A, and the additive genotype*environment interaction variance, V_{AE}. In a multilocus model of a quantitative genetic trait that varies over the environments x, the total additive genetic variance will be:

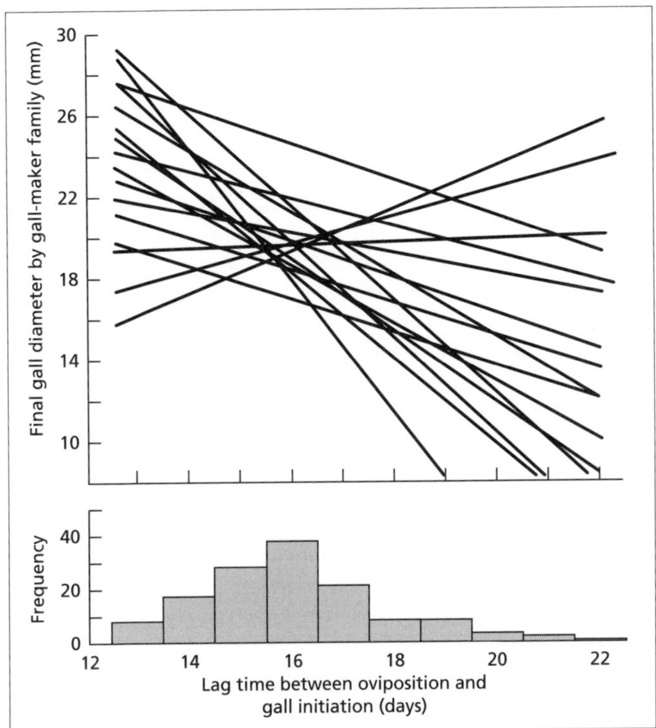

Fig. G4 Gall size as a function of plant reaction time. Shown are means of full-sib families divided over environments. The mean values of a full-sib family define a reaction norm. Reaction norms show differences in slope with the environment. Genetic variation in phenotypic plasticity, and genotype*environment interaction, exists. (After Weis, A.E. & Gorman, W.L. (1990) Measuring selection on reaction norms: an exploration of the *Eurosta–Solidago* system. *Evolution* **44**, 820–831.)

$$\sum f_x V_{A,x} = \sum_{\text{loci}} 2p_i q_i \left(\sum f_x \alpha^2_{x,i}\right)$$

From this expression:

$$\sum_{\text{loci}} 2p_i q_i \left(\bar{\alpha}^2_i + \sigma^2\{\alpha\}\right) = \sum_{\text{loci}} 2p_i q_i \bar{\alpha}^2_i + \sum_{\text{loci}} 2p_i q_i \sigma^2\{\alpha\}$$
$$= V_A + V_{AE}$$

V_{AE} therefore equals the product of the variance in number of A_1 alleles at each locus $2p_iq_i$ and the variance over environments in the AVERAGE EFFECT OF A GENE SUBSTITUTION $\sigma^2\{\alpha\}$. *See also* CROSS-ENVIRONMENT GENETIC CORRELATIONS; PHENOTYPIC VARIANCE. [G.D.J.]

REFERENCES

de Jong, G. (1995) Phenotypic plasticity as a consequence of selection in a heterogeneous environment. *American Naturalist* **145**, 493–505.

Fry, J.D. (1992) The mixed-model analysis of variance applied to quantitative genetics: biological meaning of the parameters. *Evolution* **46**, 540–550.

Via, S. & Lande, R. (1985) Genotype–environment interaction and the evolution of phenotypic plasticity. *Evolution* **39**, 505–522.

Weis, A.E. & Gorman, W.L. (1990) Measuring selection on reaction norms: an exploration of the *Eurosta–Solidago* system. *Evolution* **44**, 820–831.

genotype–phenotype relationship A PHENOTYPE refers to the structural and functional characteristics of an organism determined by its genome, subject to modification by the ENVIRONMENT. Strictly speaking, any part of the organism that is not DNA can be referred to as phenotypic. The selection of genes is mediated by the phenotype, and both genotypic and phenotypic VARIABILITY are necessary for NATURAL SELECTION to act. In order to be favourably selected, a TRAIT must be heritable and the gene(s) responsible for it must augment phenotypic reproductive success as the arithmetic mean effect of its activity in the population in which it is selected. Genes are favourably selected because their phenotypic expression favours the reproduction of their carriers (or close relatives of their carriers). GENOTYPE is commonly used to refer to some part under study of the total genetic complement of an individual, and phenotype as its physical manifestation. In contrast, genome refers to the total genetic complement of an individual, represented by all the chromosomes with their included genes in a somatic cell. *See also* GENOTYPE*ENVIRONMENT INTERACTION. [V.F.]

genotypic cluster species *See* SPECIES CONCEPTS.

genotypic value In QUANTITATIVE GENETICS, the genotypic value is defined as the average value of the character of interest over all individuals of the GENOTYPE considered. The genotypic value, G, of an individual is considered to be the sum of statistically independent parts: $G=A+D+I$, where A equals the individual's BREEDING VALUE, D is its dominance deviation, and I is the contribution to the genotypic value made by interactions between loci. In theoretical quantitative genetics, the genotypic values at a bi-allelic locus A are often denoted by: $+a$ for genotype A_1A_1, d for genotype A_1A_2, and $-a$ for genotype A_2A_2. The VARIANCE in the genotypic value is the GENETIC VARIANCE, V_G. [G.D.J.]

genus *See* TAXONOMIC CATEGORIES.

Geographic distributions

The DISTRIBUTION of organisms on a small spatio-temporal scale depends essentially on local environments, such as vary up the slopes of a mountain from a plain, or with depth of sea, when ecological factors exert a dominant control. The study of BIOGEOGRAPHY, however, is normally concerned with distributions on a regional or global scale, which also requires taking historical factors into consideration in an attempt to understand what has controlled the distributions.

Over extensive regions and through geologically considerable periods of time the overwhelmingly most important factors are PLATE TECTONICS, CLIMATE and sea level.

Since the 1970s the universal acceptance by geologists of plate tectonics, implying CONTINENTAL DRIFT, has proved a great stimulus to the study of biogeography. In evaluating the organic response to shifting continental positions and associated suturing and rifting apart of landmasses a fundamental principle from evolutionary biology must be adopted, that genetic isolation leads to phenetic divergence, and vice versa. SPECIATION is taken to be allopatric in all or at least the overwhelming majority of cases. It is evident that terrestrial organisms will respond to isolation of continents by evolving independently, but the same should also be true of marine NERITIC organisms, to varying extents depending on MIGRATION potential and the degree of separation of the continents.

Convergence and divergence

A number of simple models can be proposed that indicate the changing patterns that should develop through time. Figure G5 illustrates the phenomenon of convergence (not to be confused with the term used in evolutionary biology). As two continents approach one another, their associated marine shelf faunas progressively merge, with a time sequence depending on migration potential. The converse phenomenon, when continents move apart, is known as divergence, and is easily illustrated by reversing the time sequence of Fig. G5. A good example of convergence is the progressive merging of neritic invertebrate faunas during the closure of the Iapetus Ocean between 'North America' and 'Europe' in the Ordovician, between 500 and 400 million years ago. A good example of divergence relates to the opening of the ATLANTIC OCEAN in the MESOZOIC and CENOZOIC eras. The faunal similarity of the two sides of both the North and South Atlantic, as measured by Simpson coefficients, is inversely related to the geologically established width of the ocean.

Complementarity

Another type of pattern expresses the contrasting response to continental suturing and separation by terrestrial and marine organisms and is known as complementarity. Figure G6 illustrates how suturing of two continents leads to convergence of the terrestrial but divergence of the marine faunas. Good examples are the establishment of land CORRIDORS between North and South America in the Pliocene, about 3 million years ago, and between Africa–Arabia and Eurasia in the Miocene, about 20 million years ago.

Disjunct endemism

A fourth pattern is termed disjunct endemism. This refers to a type of regionally restricted distribution of a fossil taxon in which two or more component parts are separated by a major physical barrier and hence are not readily explicable in terms of present-day geography. The classic example is *Mesosaurus*, a small reptile of Permian age that has been found only in South Africa and southern Brazil. Although possibly an aquatic form, there is no doubt that it was incapable of crossing the South Atlantic. Indeed, if it were so mobile it should have spread far more widely around the world. A. Wegener considered this to be one of the strongest palaeobiogeographic arguments in favour of continental drift.

Climatic changes

Climate is also a major factor controlling organic distributions. One of the most striking biogeographic features of the present day is the strong DIVERSITY GRADIENT in most terrestrial and marine groups of organisms between the tropics and the poles. There is generally a systematic decline of familial, generic and species diversity with increasing latitude, the minimum value being at or close to the

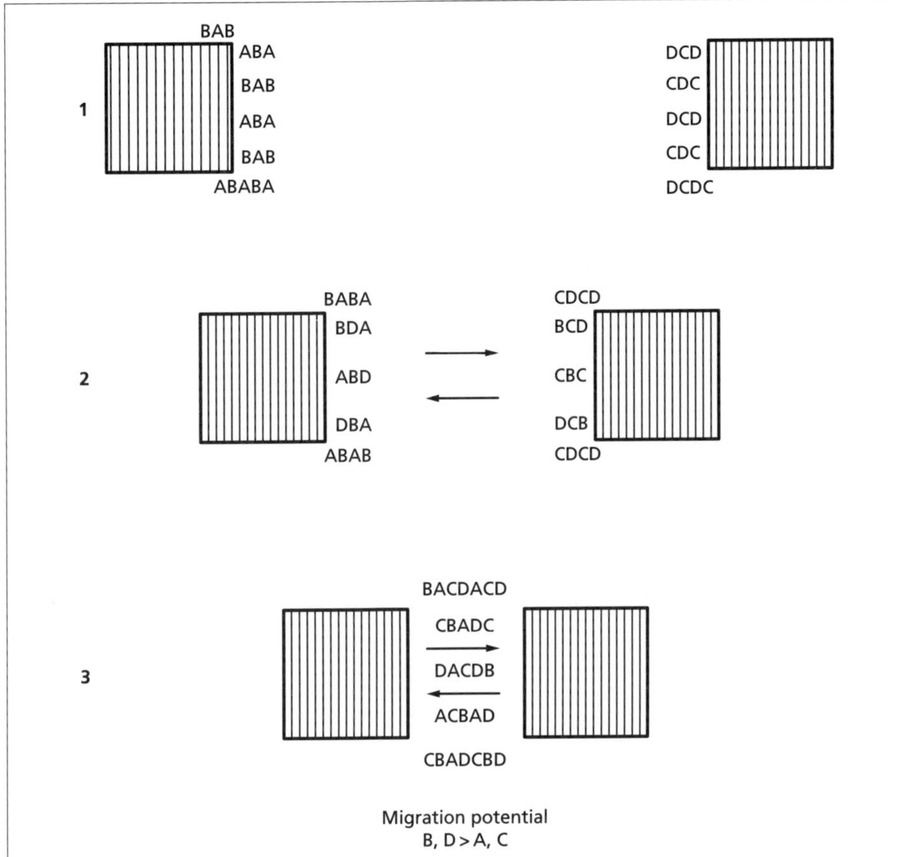

Fig. G5 Diagrammatic explanation of convergence of faunas of different continents, signified by vertical lines. As the continents approach each other as a result of sea-floor spreading, the faunas (symbolized by A, B, C and D) progressively merge. (After Hallam, A. (1974) Changing patterns of provinciality and diversity of fossil animals in relation to plate tectonics. *Journal of Biogeography* **1**, 213–225.)

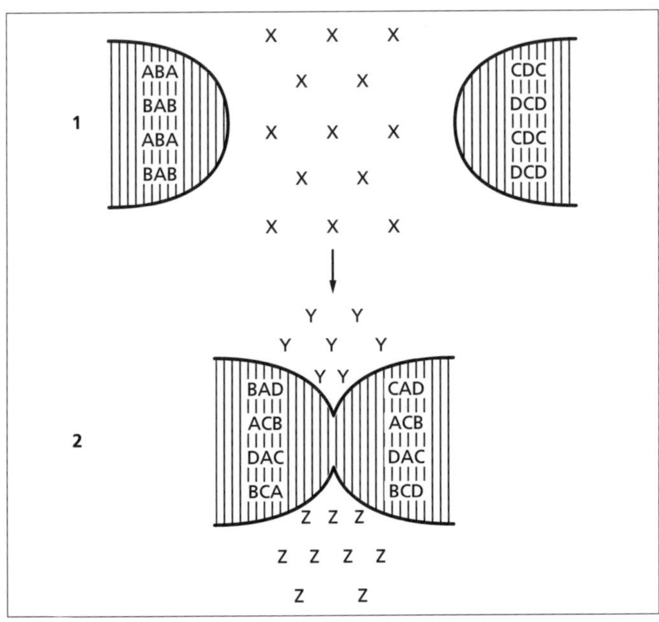

Fig. G6 Complementarity in the distribution of terrestrial and marine faunas as two continents approach each other and eventually become sutured together. The terrestrial faunas exhibit convergence because of the creation of a corridor, while the marine organisms diverge from common ancestors X to faunas Y and Z. (After Hallam, A. (1974) Changing patterns of provinciality and diversity of fossil animals in relation to plate tectonics. *Journal of Biogeography* **1**, 213–225.)

poles. It is natural to relate these latitudinal diversity gradients to the fall in temperature with increasing latitude, and indeed many plants are confined to the tropics because they are intolerant of frost. Many of the changes in flora that take place with latitude are mirrored in tropical land areas of strongly varying relief, with increase in altitude corresponding to increase in latitude. Within the marine realm certain types of organisms, such as reef-building corals, giant molluscs, mangroves and many green algae are also confined to the tropics. It must, of course, be borne in mind that the pattern of diversity change is a general rule to which there are exceptions. Thus penguins and sandpipers are most diverse at very high latitudes, and conifers, ichneumons, salamanders and voles are most diverse at intermediate latitudes. Strong climatic fluctuations of the recent geological past, producing the alternating glacial and interglacial episodes of the Quaternary, have provoked migrations of animal and plant life towards the tropics or the poles.

Sea-level changes

The stratigraphic records give abundant evidence of rises and falls of sea level with respect to the continents throughout the PHANEROZOIC, i.e. the time since metazoan life became established between 500 and 600 million years ago. Some of these were global in extent, others merely regional, but in either case the biogeographic consequences for both marine and terrestrial organisms must have been significant. At present, climate is a major control on biogeographic distributions because of the marked temperature gradient from the tropics to the poles, but in more equable times in Phanerozoic history the role of climate is likely to have assumed lesser importance. Expansions and contractions of epicontinental seas would have been correspondingly more important in controlling the degree of endemism of marine faunas. At times of low sea level and restriction of seas, faunal migrations between continental shelves would be rendered more difficult, and with less GENE FLOW more local speciation would take place among the less dispersible organisms that occupied habitats of shallower waters. Conversely, at times of high sea level there should be a reduction in marine endemism and greater spread of cosmopolitan species.

Changes of sea level can also have an important effect on the distribution of terrestrial organisms, even without lateral movements of continents. It is a matter of indifference to them whether the sea separating them from their relatives to another landmass is underlain by oceanic or continental crust. Thus at times of high sea level, emergent regions of continents may be isolated by epicontinental sea, leading to the evolution of independent terrestrial BIOTA. A good example is the North American mid-continent sea in the Late Cretaceous, which isolated western North American–Mongolian dinosaurs from eastern North American–European dinosaurs. [A.H.]

REFERENCE

Hallam, A. (1994) *An Outline of Phanerozoic Biogeography*. Oxford University Press, Oxford.

---

geographic information systems (GIS) in conservation The use of computer hardware and software tools for the acquisition, management, analysis and display of resource data for conservation. These data, for example, can include elevation, soil or vegetation types, road or river networks, as well as image data acquired from aircraft- and satellite-based cameras or sensors. The actual input data will depend on what conservation information is required. These diverse data are integrated within a common framework of the GIS by relating and referencing the data by their spatial location. The integration of these data permits evaluation, modelling, statistical analysis and visual display. GIS analysis can display the results of past, current and future outcomes of different conservation practices, providing information to assist in resource management planning and decisions. *See also* CONSERVATION BIOLOGY. [J.V.C.]

geographic range The extent of the global DISTRIBUTION of a species. *See also* GEOGRAPHIC DISTRIBUTIONS. [P.C.]

geographical barrier A physical barrier to free movement of individuals between populations and hence to GENE FLOW through interbreeding. *See also* SPECIATION. [P.C.]

geographical floral element A set of taxa (usually species) within the flora of a given region that show similar overall DISTRIBUTION (RANGE) patterns, either within the territory under consideration, or with regard to the total range, both within and outside the territory. A floral (or floristic) element is named by reference to the location of the main parts of the ranges of the component taxa. Although subdivision of a flora into component geographical floral elements is normally a subjective procedure, attempts have also been made to define groupings by objective quantitative techniques such as CLUSTER ANALYSIS (e.g. Proctor 1967).

For the British flora (Matthews 1937), seven floral elements have been recognized, based on distribution of species within the British Isles (British, English, Scottish, Highland, Germanic, Atlantic, local), and 16 based on the total ranges of the species (Northern hemisphere, Eurasian, General European, Continental, Continental Northern, Continental Southern, Mediterranean, Oceanic Southern, Oceanic West European, Oceanic Northern, Northern-Montane, Arctic-Subarctic, Arctic-Alpine, Alpine, American, Endemic). With larger, more complex floras, generic analysis may be more appropriate; thus for New Guinea, with 1465 genera of flowering plants, 15 geographical floral elements have been recognized (Stott 1981).

One interpretation of the coincident range patterns shown by the component taxa of geographical floral elements is that all the taxa within a particular floral element share the same evolutionary and migrational history, and are governed by similar ecological CONSTRAINTS. Proof of that supposition requires a detailed knowledge of the ecology and history of the component taxa, which is rarely available. Examples of studies utilizing palaeoecological data are those for Manitoba (Love 1959), and for eastern Washington and northern Idaho (Daubenmire 1975). However, Stott (1981) distinguishes ecological, historical, migrational and genetic elements of a given flora as components separate from the geographical elements. *See also* FLORAL PROVINCE. [J.H.T.]

REFERENCES

Daubenmire, R. (1975) Floristic plant geography of eastern Washington and northern Idaho. *Journal of Biogeography* **2**, 1–18.

Love, D. (1959) The postglacial development of the flora of Manitoba: a discussion. *Canadian Journal of Botany* **37**, 547–585.

Matthews, J.R. (1937) Geographical relationships of the British flora. *Journal of Ecology* **25**, 1–90.

Proctor, M.C.F. (1967) The distribution of British liverworts: a statistical analysis. *Journal of Ecology* **55**, 119–135.

Stott, P. (1981) *Historical Plant Geography*. Allen & Unwin, London.

geographical isolation The separation of a group of individuals or a population from other populations of the same species such that GENE FLOW through interbreeding is impaired. It is caused by GEOGRAPHICAL BARRIERS and can lead to SPECIATION. *See also* ISOLATING MECHANISMS. [P.C.]

geological time-scales The age of the solar system is believed to be 4.7×10^9 years, and the oldest terrestrial rocks are dated at 4.0×10^9 years. Geologists use a hierarchical chronology for the time-span from the origin of fossils to the present day, in which the largest unit is the era. There are only four eras covering the FOSSIL RECORD of life: CENOZOIC, $0–65 \times 10^6$ years ago; MESOZOIC, $65–245 \times 10^6$ years ago; PALAEOZOIC, $245–540 \times 10^6$ years ago; and Protozoic, $0.540–2.7 \times 10^9$ years ago. The eras are divided into periods, which are themselves divided into epochs. Table G2 depicts the time-scale spanning the Cenozoic, Mesozoic and Palaeozoic eras, with some of the most important events. *See also* MASS EXTINCTION; TIME. [J.G.]

REFERENCES

Edwards, D. & Selden, P.A. (1992) The development of early terrestrial ecosystems. *Botanical Journal of Scotland* **46**, 337–366.

Emiliani, C. (1992) *Planet Earth*. Cambridge University Press, Cambridge.

Table G2 The geological time-scale covering the Cenozoic, Mesozoic and Palaeozoic eras.

Era	Period	Epoch	Age (years × 10^6)	Events and noteworthy fossils (dates × 10^6 BP)
CENOZOIC	Quaternary	Holocene	0	*Homo sapiens* 0.4, Iron Age 0.003, Atomic Age 0.00005
		Pleistocene	1.6	*Homo erectus* 0.5–1.6, *Homo habilis* 2
	Neogene*	Pliocene	5.2	*Australopithecus* 2.3–4
		Miocene	23	
	Palaeogene*	Oligocene	34	Tree-dwelling primate
		Eocene	53	
		Palaeocene	65	**Mass extinction 65**
MESOZOIC	Cretaceous	Late	95	North America, Eurasia, India, Africa, Antarctic/Australia, South America, North America separate
		Early	135	Flowering plants, including herbaceous habit
	Jurassic	Late	152	*Archaeopteryx*
		Middle	180	Coccoliths
		Early	205	**Mass extinction 208**
	Triassic	Late	230	Dinosaurs
		Middle	240	Cycadales
		Early	250	**Mass extinction 245**
PALAEOZOIC	Permian	Late	260	Reptiles
		Early	300	
	Carboniferous	Silesian	325	Coal formation, winged insects
		Dinantian	355	Amphibia
	Devonian	Late	375	**Mass extinction 367**
		Middle	390	Secondary xylem; tree ferns, 'age of fishes'
		Early	410	Tall vascular plants
	Silurian			Small vascular plants on land
			438	**Mass extinction 439**
	Ordovician			Land plants with spores and cuticles, probable bryophytes
				Corals, first vertebrates
			510	**Mass extinction**
	Cambrian		570	Arthropoda, Echinodermata, Hemichordata, Chordata

* These two units are commonly combined as a single period, the Tertiary.

geometric mean The back-transformed MEAN of logarithmically transformed observations. The formula for calculating the geometric mean (GM) is:

$$\mathrm{GM} = \mathrm{antilog}\left(\frac{\sum \log x}{n}\right)$$

where x is each observation in the SAMPLE, n is the number of observations in the sample, and the Greek symbol Σ (capital sigma) is the mathematical notation for 'sum of'. *See also* STATISTICS. [B.S.]

geometric series

1 Also called a geometric progression, a series of terms in which each term is a constant multiple of the preceding term. The general form of a geometric series is $a, ar, ar^2, ar^3, \ldots, ar^n$, where a is a constant and the sequence has n terms.

2 One of the four commonly used SPECIES-ABUNDANCE MODELS (distributions) (*see* RANK–ABUNDANCE MODELS). See also Magurran (1988) and May (1975). [B.S.]

REFERENCES

Magurran, A.E. (1988) *Ecological Diversity and its Measurement.* Croom Helm, London.

May, R.M. (1975) Patterns of species abundance and diversity. In: *Ecology and Evolution of Communities* (eds M.L. Cody & J.M. Diamond), pp. 81–120. Harvard University Press, Cambridge, Mass.

geometric series model *See* RANK–ABUNDANCE MODELS.

GEOSAT *See* GEOSTATIONARY SATELLITE.

geostationary satellite (GEOSAT) A satellite that remains stationary relative to the Equator. The orbit of a geostationary satellite is synchronous with that of the Equator at an altitude of around 36 000 km. A scanning RADIOMETER on board provides meteorological images of the entire Earth disc every half hour in 3–4 wavebands. Examples of these satellites are the Geostationary Operational Environmental Satellites (GOES) of the NATIONAL OCEANIC AND ATMOSPHERIC ADMINISTRATION (NOAA), and Meteosat of the European Space Agency (ESA). [P.J.C.]

geostrophic current A current resulting from the balance between gravitational forces and the CORIOLIS EFFECT. As a result of the convergence and DIVERGENCE of water masses at the OCEAN surface, a subtle TOPOGRAPHY is formed, with hills in the areas of convergence and valleys in the areas of divergence. As water runs downhill in response to the pull of gravity it is deflected to the right (in the Northern hemisphere or to the left in the Southern hemisphere) by the Coriolis force such that most of its motion is parallel to the side of the hill. The balance between the gravitational force, which pulls the water downhill, and the Coriolis force, which deflects it, gives rise to geostrophic currents. [V.F.]

germ line Some time (usually early) in the ONTOGENY of multicellular organisms, cell lines giving rise to GAMETE-producing cells, tissues or organs become specialized. August Weismann and colleagues first noted this early sequestration of the genetic complement that can be transmitted to the next generation from the rest of the body. Their studies demonstrated that the germ line is buffered from environmental influences affecting the rest of the organism (the SOMA) during its life. This relative independence of the germ line from direct environmental influences, dubbed Weismann's barrier, renders EVOLUTION non-Lamarckian. Modern molecular studies have demonstrated that Weismann's barrier is not as impermeable as originally thought; some viruses, for example, may introduce novel genetic material into the germ line. Moreover, some multicellular organisms reproduce 'vegetatively', by budding and by FISSION. In this circumstance soma and germ line may be one and the same entities. *Cf.* SOMA. [D.R.B.]

REFERENCE

Buss, L.W. (1987) *The Evolution of Individuality*. Princeton University Press, Princeton, N.J.

germanium (Ge) A metalloid element (ATOMIC NUMBER = 32; relative ATOMIC MASS = 72.61; melting point = 937.4°C; boiling point = 2830°C) with an average abundance of 1.8 mg kg^{-1} in the Earth's CRUST, it is extracted from zinc sulphide ores. Germanium compounds find uses as semi-conductors, CATALYSTS and in scintillation materials for γ-radiation cameras. Some organogermanium compounds have been used as antitumour agents. Apart from the volatile germanium hydrides, germanium compounds are not especially toxic to animals and humans, but high concentrations of germanium in aqueous solution are detrimental to organisms which utilize SILICON in biosilicification processes, for example freshwater sponges. [C.R.P.]

Getz model *See* MODELS OF PREDATOR–PREY INTERACTION.

ghost of competition past This phrase was coined by Connell (1980) to indicate that the asserted effect of COMPETITION is to eliminate itself by causing NICHE divergence in competing species, and that, as a consequence, it is difficult to prove or disprove that competition is responsible for the divergence. In more general terms, the 'ghost of competition past' could be defined as characteristics of species' niches or distributions, or of COMMUNITY structure that (are believed to) have been caused by competition in the past, but are no longer subjected to competition at present. Next to niche divergence, the traces left by competition may include morphological divergence (i.e. CHARACTER DISPLACEMENT *sensu stricto*) and non-random distribution patterns. The ghost becomes even more elusive if we take into account that the failure to find such traces can be explained as a failure to look for the relevant aspects of the community. [J.G.S.]

REFERENCE

Connell, J.H. (1980) Diversity and the coevolution of competitors, or the ghost of competition past. *Oikos* **35**, 131–138.

GIS *See* GEOGRAPHIC INFORMATION SYSTEMS (GIS) IN CONSERVATION.

glacial lakes *See* LAKES, TYPES OF.

glaciation

1 The processes and the results of erosion and deposition arising from the presence of an ICE mass on a landscape. The term is applied to these phenomena whether on a continental scale or within the confines of a valley. Glaciers move at rates varying between a few centimetres and some tens of metres a day, the speed being affected by the nature of the ice and the retarding effects of the ice–rock interface at the sides and base of the glacier. There are three main types of glacier: (i) mountain or valley glaciers, which form from snow lying in that part of a mountain range above the snow line; (ii) piedmont glaciers, which develop when valley glaciers emerge from their confining channels and spread out over a lowland area; and (iii) ice sheets and ice caps, which are masses of radially spreading ice presently found only at high latitudes (e.g. Greenland, Antarctica).

2 The term glaciation may also refer to geological periods during which a large part of the Northern hemisphere was covered with an ice sheet (also called ice ages). Glaciation in this sense refers to ice cover of extensive parts of the landscape associated with climatic oscillation. The last 700 000 years have been dominated by major ice ages with an approximate 100 000-year cycle interrupted by relatively short interglacials such as we are currently enjoying. A widely accepted thesis proposes that the factor controlling the timing of glacial periods is the orbital eccentricity of the Earth around the Sun, causing major changes in INSOLATION.

These CLIMATE CHANGES have been expressed differently in different parts of the world, depending upon distance from the Equator, ocean position and currents, mountain ranges, etc. In the last ice age an ice sheet covered much of North America (as far south as 40°N) while the Scandinavian ice sheet only reached 52°N, covering parts of Britain and northern Europe. In the Southern hemisphere Australia and South America were not subject to extensive glaciation, but their CLIMATES were modified.

The primary effects of such climatic changes on a species are major changes in range (distribution) and abundance (both global and local) as the species tracks a suitable environment, with the prospect of some species being driven to extinction.

Glaciers can have profound effects on the morphological structure of the surfaces over which they travel (e.g. lead to the formation of glacial valleys and FJORDS) and leave distinctive deposits (e.g. moraines). [V.F. & P.C.]

REFERENCE

Hewitt, G.M. (1996) Some genetic consequences of ice ages and their role in divergence and speciation. *Biological Journal of the Linnean Society* **58**, 247–276.

glass Any disordered, non-crystalline solid; i.e. one whose constituent atoms have no long-range ordered pattern. A glass is formed when a solid compound such as silica is heated above its melting point and then rapidly cooled. A glass does not have a sharp melting point but tends to soften over a range of temperatures. Common glass, or soda glass, is used for windows and bottles and is produced when a mixture of sodium carbonate (Na_2CO_3) and calcium oxide (CaO) is added to the silica melt. The properties of a glass can be altered by varying the additive compounds. For example, borosilicate glass (B_2O_3 added to silica) is tougher than soda glass and expands and contracts little over a wide temperature range, making it useful for laboratory ware and cooking utensils. The addition of potassium monoxide (K_2O) and lead monoxide (PbO) to the silica melt results in an especially tough glass with high melting point, which can be ground with precision to produce optical glasses and contact lenses. The addition of the PbO produces a denser glass with a higher refractive index.

A further type of glass, termed bioglass, in which the additives are sodium monoxide (Na_2O), calcium oxide (CaO) and phosphorus pentoxide (P_2O_5), is compatible with bone and is used as a coating on surgical implants.

Melted glass can also be drawn into fibres which may be spun into threads and woven into fabrics. The main use of these fibres, in conjunction with resins, is in the production of strong, corrosion-resistant structures, for example boat superstructures, car bodies. [M.C.G.]

Gleysols *See* SOIL CLASSIFICATION.

GLIM *See* GENERALIZED LINEAR MODELS.

GLM *See* GENERALIZED LINEAR MODELS.

Global Change Master Directory *See* GLOBAL DATABASES.

global circulation model (GCM) *See* GENERAL CIRCULATION MODEL.

global constraints *See* CONSTRAINTS.

global databases Global meteorological, oceanographic and geophysical data are widely available, either over the Internet or on archived media such as tapes and CD-ROMs. Some data are free and directly available, but in many cases orders must be placed and fees must be paid to the provider agencies. An extensive summary of what is available, both over the Internet and on other media, is maintained at the National Center for Atmospheric Research (NCAR) in Colorado. This document, updated monthly, can be obtained by anonymous file transfer protocol (FTP) to rtfm.mit.edu, from the files weather/data/part1 and weather/data/part2 in the directory /pub/usenet/news.answers.

Also available by telnet or World Wide Web is the *Global Change Master Directory*, a multidisciplinary on-line information system, containing descriptions of earth and space science holdings available to the science community. These include data from NASA, the National Oceanic and Atmospheric Administration (NOAA), NCAR, the US Geological Survey (USGS), the US Department of Energy, Carbon Dioxide Information and Analysis Center (DOE (CDIAC)), the US Environmental Protection Agency (USEPA), the National Science Foundation (NSF) and other agencies, universities and research centres.

Surface atmospheric data over land include the PRESSURE, TEMPERATURE, SPECIFIC HUMIDITY, cloud cover and rate of PRECIPITATION as measured in the world meteorological network, which currently consists of about 7000 weather stations reporting on a daily basis. There are also over 800 rawinosonde stations, which report daily to the world communications network.

Over the oceans the observations include sea surface temperature, SALINITY, atmospheric temperature, pressure, HUMIDITY, WIND direction and wind speed. Some subsurface OCEAN data (e.g. bathythermograph data) are also available on a global scale.

Satellite data are becoming increasingly important, with remotely sensed estimates of RADIATION, WATER vapour and surface temperature now becoming available. Satellite data are now also aiding global estimates of the rates of processes in the terrestrial and marine biopshere, for example by providing estimates of PHYTOPLANKTON abundance and chlorophyll concentration, and via the use of the normalized density vegetation index (NDVI) as a surrogate for LIGHT interception by terrestrial VEGETATION.

Many available datasets consist of measurements at individual measuring stations or locations, but in many cases gridded datasets of the same PARAMETERS with global coverage are also available. Also available in gridded form are global maps of vegetation, SOIL properties, cultivation intensity and human population density, to name a few. [J.LY.]

REFERENCES

Gurney, R.J., Foster, J.L. & Parkinson, C.L. (eds) (1993) *Atlas of Satellite Observations Related to Global Change.* Cambridge University Press, Cambridge.

Peixoto, J.P. & Oort, A.H. (1992) *Physics of Climate.* American Institute of Physics, New York.

Global Environmental Facility (GEF) A fund, established in 1991 and administered by the World Bank and United Nations Development Programme, to promote environmental and conservation projects in developing countries. *See also* UNITED NATIONS PROGRAMMES AND BODIES. [P.C.]

global stability If a system has an equilibrium to which it will return from any state (other than zero state) then the system is globally stable. Global stability is hard to prove. However, for a system of DIFFERENTIAL EQUATIONS whose state is described by the vector $\boldsymbol{N}$ and which has an equilibrium $\boldsymbol{N}^*$ then global stability is guaranteed if there exists a function $L(\boldsymbol{N})$ such that: (i) $L(\boldsymbol{N}^*)=0$; (ii) $L(\boldsymbol{N})>0$ for $\boldsymbol{N}\neq\boldsymbol{N}^*$; and (iii) $\mathrm{d}L/\mathrm{d}t\leq 0$ with the equality only at equilibrium. L is a Lyapunov function. The most reliable means for finding one, is to wait for an angel to bring it in a dream. Equivalent definitions of Lyapunov functions also apply to DIFFERENCE EQUATIONS. *See also* STABILITY ANALYSIS. [S.N.W.]

global warming CLIMATE warming on a global scale, usually attributed to an enhanced greenhouse effect. There is no doubt that the TEMPERATURE of the Earth has increased since the end of the 19th century, by 0.3–0.6°C. A trend can be seen in the record from meteorological stations in both the Northern and Southern hemispheres (Fig. G7). It is known from historical records that sea levels have risen and that glaciers are receding. There is also no doubt that the atmospheric concentration of CARBON DIOXIDE (CO_2) has increased over the last century (it was 280 ppm and is now 360 ppm), and that the concentrations of other GREENHOUSE GASES have been increasing over the past few decades. This observed rise in the concentration of greenhouse gases is believed to be the underlying cause of recent global warming. The link between these two phenomena is explored by MATHEMATICAL MODELS of the energy balance of the Earth's surface in relation to the optical properties of the ATMOSPHERE. Generally, in such models, any overall

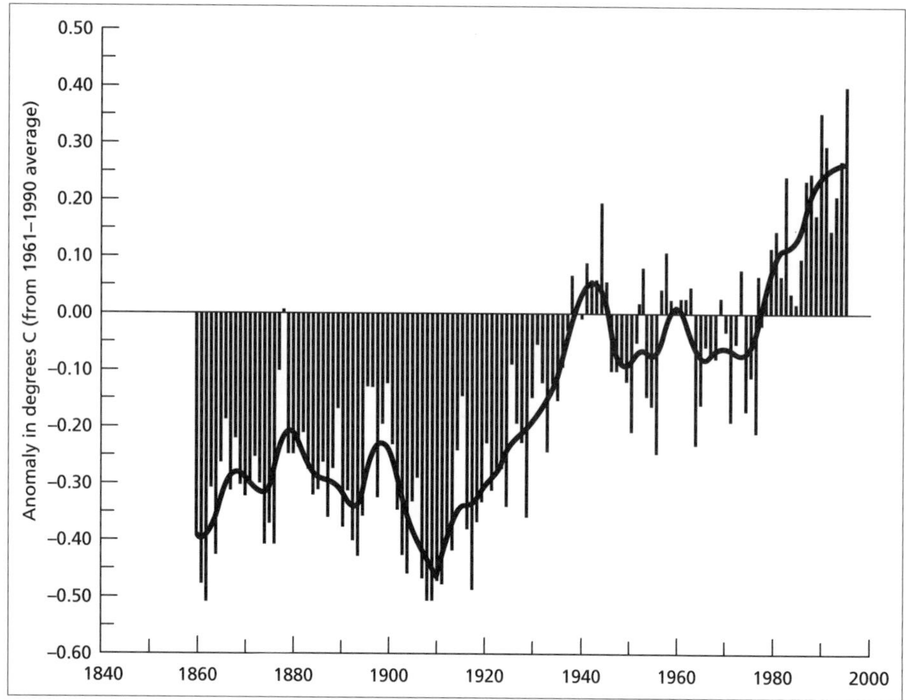

Fig. G7 Combined land, air and sea surface temperatures relative to 1961–1990 averages. (Data from IPCC (1996) *Climate Change 1995: The Science of Climate Change*. Cambridge University Press, Cambridge.)

increase in the concentration of greenhouse gases causes an increase in the global temperature. Quantitative predictions of the extent of this effect need very elaborate global circulation models (GCMs) that run on supercomputers. Long-term change in the surface air temperatures following a doubling in the CO_2 concentration is predicted to be between 1.5 and 4.5°C, with the 'best' estimate being 2.5°C. There is evidence that vegetation in the Northern hemisphere is already responding to global warming. The spring 'drawdown' of CO_2 caused by photosynthesis is occurring earlier, and so is the spring 'greening' as seen from satellite-borne sensors. *See also* CLIMATE CHANGE. [J.G.]

REFERENCES

Houghton, J.T., Jenkins, G.J. & Ephraums, J.J. (eds) (1990) *Climate Change: the IPCC Scientific Assessment*. Cambridge University Press, Cambridge.

IPCC (1996) *Climate Change 1995: The Science of Climate Change*. Cambridge University Press, Cambridge.

Keeling, C.D., Chin, J.F.S. & Whorf, T.P. (1996) Increased activity of northern vegetation inferred from atmospheric CO_2 measurements. *Nature* **382**, 146–149.

Maunder, W.J. (1994) *Dictionary of Global Climate Change*, 2nd edn. UCL Press, London.

glossic horizons *See* SOIL CLASSIFICATION.

Glossisols *See* SOIL CLASSIFICATION.

GLP *See* GOOD LABORATORY PRACTICE.

glycophyte A plant of non-saline soil or water. The term is used to differentiate the majority of salt-intolerant plants from the salt-tolerant HALOPHYTES. Most glycophytes are inhibited in growth by even moderate concentrations of sodium chloride (NaCl) and other ionic solutes which may be toxic and/or osmotically interfere with water uptake. The boundary between glycophytes and halophytes is indistinct as different plant species show a continuum of sensitivity to dissolved solutes. Within species there is a similar genotypic continuum of sensitivity which has been exploited by plant breeders in selection of salt-tolerant crop varieties. *See also* OSMOREGULATION. [J.R.E.]

glyphosate An organophosphorus HERBICIDE, introduced in 1971, that exhibits high activity on most vascular plants, and is now the most widely used herbicide in the world. It is inactivated in the SOIL, so has to be applied to actively growing foliage. Adsorption by leaves is followed by translocation in suitable conditions to the plant's storage regions. This allows good control of PERENNIAL plant species.

$$(HO)_2P(=O)\text{–}CH_2.NH.CH_2.COOH$$

Animal TOXICITY is very low, allowing use near and in aquatic environments, and in other areas of environmental sensitivity. Selective WEED control can be achieved by targeted 'wiping-on' of glyphosate on to the foliage of particular weeds. [K.D.]

GMT (Greenwich mean time) *See* TIME.

gNNP *See* GREEN NET NATIONAL PRODUCT.

gnotobiotic Describing a culture, COSM or laboratory animal(s) (e.g. used in TOXICITY/ECOTOXICITY tests) in which the exact composition of organisms is known, including, in principle, all associated microbiota. The word derives from the Greek *gnosis*, knowledge, and *bios*, life. [P.C.]

gold (Au) A soft, yellow metallic element (ATOMIC NUMBER = 79; relative ATOMIC MASS = 196.967; melting point = 1064.43°C; boiling point = 2807°C). Gold is widely, if sparsely, distributed in the Earth's CRUST (average crustal abundance = 0.0011 mg kg^{-1}), being found either in its native form or as sulphide or telluride ores, often associated with similar ores of IRON, COPPER, NICKEL, etc. Gold is also present in seawater to the extent of 10^{-3} ppm. The refining of gold ores via LEACHING with dilute CYANIDE solutions, followed by PRECIPITATION of metallic gold by the action of ZINC, presents a potential environmental HAZARD.

$$4Au + 8KCN + O_2 + 2H_2O = 4K[Au(CN)_2] + 4KOH$$

$$2K[Au(CN)_2] + Zn = K_2[Zn(CN)_4] + 2Au$$

Gold has long been used in chrysotherapy for its anti-arthritic activity. [P.J.B.]

golden apple snail (*Pomacia canaliculata*) *See* EXOTIC AND INVASIVE SPECIES.

Gondwana A supercontinent consisting of the landmasses now known as Australasia, Antarctica, India, Africa and South America, which split up during Cretaceous times. Their former connection explains several features of the shared fossil fauna and flora of these current continents and also some features of their modern biota. [P.D.M.]

gonochoristic Animals with separate male and female individuals. Plants with this characteristic are dioecious. Thus for animals gonochoristic is the antonym of hermaphroditic. [P.C. & S.C.S.]

good laboratory practice (GLP) A formalized process under which laboratory studies on ECOTOXICITY are carried out to ensure reliability and REPEATABILITY of results. This is essential in tests that are used in a legal context, where results might be open to challenge. Standardization of procedures should therefore avoid duplication, ensure mutual recognition and reduce the numbers of animals involved. The procedures for GLP involve following standard protocols and keeping thorough and accurate records of observation. They also involve appropriate training and MONITORING by independent observers (*see* QUALITY ASSURANCE AND QUALITY CONTROL). The principles of GLP were originally laid down by the OECD. [P.C.]

REFERENCE

OECD (1981) *OECD Guidelines for Testing Chemicals*. Organization for Economic Cooperation and Development, Paris.

goodness-of-fit The extent to which an observed and expected frequency DISTRIBUTION coincide. It is frequently assessed using non-parametric tests (*see* NON-PARAMETRIC STATISTICS) such as a CHI-SQUARED TEST or a *G* TEST. *See also* STATISTICS. [B.S.]

gradient The rate of change of a characteristic over distance. Topographic (slope) and climatic gradients (e.g. temperature, rainfall) may have pronounced effects on biota. Also soil gradients of pH, major nutrients and salinity may strongly affect the distribution of organisms, especially abrupt changes at ECOTONES. Ecological GRADIENT ANALYSIS orders vegetation samples by reference to environmental data; indirect gradient analysis examines variation within vegetation independently of environmental data, deriving ORDINATION axes from vegetation data. [A.J.W.]

gradient analysis The analysis of trends in COMMUNITY composition with respect to environmental gradients in PHYSICAL FACTORS such as pH or moisture. Gradient analyses can often be criticized on the grounds that the choice of environmental factor is almost always subjective, and trends in species composition which correlate with a particular factor do not necessarily imply that the chosen factor is the most important.

In terrestrial plant ecology, gradient analysis is sometimes given a wider definition as a synonym of ORDINATION. [S.J.H.]

gradients of species richness It is evident that some habitats are richer in species than others, and it has also been observed that low latitudes and low altitudes tend to support richer floras and faunas than high latitudes and high altitudes. In the case of bird species, for example, there are 603 species breeding in Costa Rica, 472 in Guatemala, 286 in California and 222 in Alaska. There is thus a latitudinal GRADIENT in RICHNESS from the tropics to the Arctic. Similar gradients can be observed in different types of organism. Among mammals, for example, Costa Rica has 140 species and Alaska has 40. Florida has approximately 180 species of tree, while the boreal regions of Canada have only 20–30 species.

There are several explanations of why these gradients exist. Since similar gradients have been observed with altitudinal change, it is apparent that climate plays an important role. Closer study of plant richness gradients shows that they are not simply related to latitude, but do relate strongly to climate. Where conditions are suitable for the highest levels of PRIMARY PRODUCTIVITY the richness tends also to be highest.

In the case of animals, habitats and food requirements may play an important part. There are many more species of bats in the tropics, reflecting the availability of many food sources (insects, fruit, vertebrates, blood, etc.). Also the much greater structural diversity of a tropical forest when compared with a scrub-tundra means that far more microhabitats are available for small organisms to survive.

Studies of fossils show that the tropical regions have always been diverse. The flowering plants, for example, appear to have evolved and diversified first in the tropics, and the low latitudes have continued to be rich in species ever since. Some have argued that the climatic stability of the tropics has aided this process, but evidence suggests that most tropical regions have suffered considerable climatic changes (both in temperature and precipitation) in the past, so stability as a cause of diversity is generally unproved. On the contrary, it may well be that DISTURBANCE, both long-term climatic changes and short-term changes due to local catastrophes—floods, fires, wind, etc.—may actually have diversified habitats and led to greater species richness.

Although the precise causes of richness gradients are still under debate, one interesting observation has emerged that helps to clarify the factors in operation. Species of low latitude and low altitude tend to have narrower ecological distributions than those at the other end of the spectrum. Thus low-altitude species may have a narrower altitudinal RANGE, for example, than species of higher altitude, enabling more species to be packed within a given area. This is known as 'Rapoport's rule' after the biogeographer E.H. Rapoport. *See also* BIOGEOGRAPHY; GEOGRAPHIC DISTRIBUTIONS; LATITUDINAL GRADIENTS OF DIVERSITY. [P.D.M.]

grain The sizes of environmental patches relative to the ecological perception of organisms. Fine-grained ENVIRONMENTS are those in which a single individual encounters all the HETEROGENEITY in a lifetime. A coarse-grained environment is one in which patches are large or long-lasting enough for a single individual to spend its entire life in one patch, followed each generation by DISPERSAL of offspring to the same or other patches. [P.C.]

REFERENCE

Levins, R. (1968) *Evolution in Changing Environments.* Princeton University Press, Princeton, N.J.

graph theory *See* INTERVALITY.

grasslands BIOMES dominated by herbaceous species in which grasses (Poaceae) or sedges (Cyperaceae) are abundant, accompanied by forbs, which account for the highest proportion of plant species found in grasslands, with sometimes scattered shrubs and trees. Grasslands can be classified as natural, semi-natural and artificial.

Natural grasslands occur where environmental factors prevent or suppress tree and shrub growth. Most often this is due to CLIMATE, particularly aridity; on coasts and mountains exposure to salt-laden or freezing winds is responsible. Some grasslands occur due to pedological factors, including shallow IRON pans, SALINITY or ALKALINITY, and HEAVY METALS or other phytotoxins. In areas of active deposition or erosion, as along natural river courses, geomorphological processes create natural seral grasslands. Wild HERBIVORES and natural fires are thought to have played a supplementary role in maintaining and extending the area of natural grasslands.

Semi-natural grasslands are found where DEFORESTATION and GRAZING has created open landscapes for COLONIZATION by species of naturally open habitats such as natural grassland and cliffs. These occur in the wet and dry FOREST zones of the world, and in many forest zones such grasslands now exceed the area of original forest.

Artificial grasslands are manipulated through FERTILIZER and HERBICIDE application, as well as reseeding, to favour a few commercially desirable forage species, particularly ryegrass (*Lolium* spp.) and white clover (*Trifolium repens*).

Natural grasslands have been extensively destroyed by cultivation, and most of those remaining are modified by grazing of domesticated animals and the introduction of ALIEN plant species. Most temperate grasslands of all types are modified by AGROCHEMICALS. [J.J.H.]

REFERENCE

Coupland, R.T. (ed.) (1979) *Grassland Ecosystems of the World.* Cambridge University Press, Cambridge.

gray (Gy) The SI UNIT of absorbed dose, used to quantify the interaction of IONIZING RADIATION with biological tissue. One gray is equal to the deposition of 1 JOULE of energy in 1 kg of matter. The gray has now replaced the former unit of dose, the rad; 1 rad = 10^{-2} Gy. *See also* SIEVERT. [A.B.M.]

grazer Broadly, an animal which eats plants (*see also* PHYTOPHAGOUS). However, the term has more specialist meanings and limitations. It may be used of large mammals and reptiles to distinguish FEEDING on herbaceous plants at ground level (GRAZING) from feeding at tree or bush level (browsing). In this sense, grazing is not confined to the consumption of grasses but to all herbaceous plants. The form of alimentary canal and dentition of vertebrate grazers differs from browsers in many ways. The evolution and radiation of the larger herbivorous mammals of the past 50 million years has seen a shift away from browsing towards grazing, seen most notably in the evolution of horses and the expansion of the bovids. The change from browsing to grazing, particularly grazing of GRASSLANDS, will also have been accompanied by a significant change in diet from the predominantly starch-storing leaves of trees to the fructan-storing tissues of grasses.

The term grazing is also extended to invertebrates, including molluscs and leaf-eating insects. The term is less commonly used of animals which eat the underground parts of plants, and is rarely used of seed-eating animals.

In the marine literature, grazing is widely used of a large number of animals, vertebrate and invertebrate, including ZOOPLANKTON, which use various mechanisms to consume PHYTOPLANKTON (single-celled algae through to seaweeds) and the DETRITUS from previously consumed phytoplankton.

The theory of plant defences rests on the observation that the widest range of plants are subjected to grazing (*see* PLANT DEFENCE THEORY). *See also* FEEDING TYPES, CLASSIFICATION. [G.A.F.H.]

grazing The consumption of the above-ground parts of plants by animals. When related to the larger mammals the term may be used to distinguish consumption of herbaceous plants from consumption of leaves and woody tissue of bushes and trees (i.e. browsing). In specialized usage, grazing may also be applied to the FEEDING activities of small animals on defined substrates. Examples include the feeding activities of molluscs, particularly gastropods, on leaves or in the laboratory on algal-impregnated plates.

Grazing, together with fire, is one of a limited number of manipulative devices derived from the natural ecosystems which have been adapted, via AGRICULTURE, for the management of plant and animal communities.

The ecological significance of grazing is often less concerned with the act of consumption (with its nutritional implications for the GRAZER) and more with the effects of removal of the aerial BIOMASS of the plant. In this sense, ecologists and environmental managers may extend the term grazing to cover the clipping or mowing or even hand-pulling of the aerial parts and their removal from the ecosystem. This form of simulated grazing is used, for example, to prevent COLONIZATION of grassland by non-grass species, including seedlings and saplings of otherwise invasive shrubs and trees; or simulated grazing may be practised in an endeavour to maintain BIODIVERSITY where the natural SUCCESSION may be towards a decline in the number of species. Apart from simulated mechanical grazing, controlled grazing (Fig. G8) for environmental management may also be achieved by the introduction of a limited number and type of grazers, usually domesticated mammals, often for a limited part of the year. A specialized development of this type of grazing has been adapted for the conservation of particular rare

Fig. G8 Grasses are particularly tolerant of grazing. The effects of grazing are the most important factors in the vegetation of temperate grasslands. Here, the difference between a field which has undergone predation from grazers, and one which has not, can be clearly seen. (Courtesy of M. Crawley.)

species or unusual communities which depend on the manipulation of VEGETATION often representative of former agricultural practices (e.g. hay-making on annually flooded wet meadows). In both traditional and modern agriculture and FORESTRY elaborate methods of excluding or controlling grazing have evolved and these are reflected in the structure of the landscape of western Europe. Examples include barriers such as stone walls, hedges, and banks with ditches (particularly in woodlands), many of which today form habitats of considerable antiquity and conservation value. [G.A.F.H.]

Grazing practices

Managing the GRAZING of wild plant communities by domesticated animals is one of the oldest human professions. Land chosen for this usage is typically among the least productive biologically; arid and semi-arid rangelands are often too marginal for AGRICULTURE, FORESTRY or other commodity extraction uses. Grazing by livestock (primarily cattle, sheep and goats) is the most widespread ecological influence in many dryland ecosystems; for example, approximately 70% of western North America, including DESERTS, shrub-STEPPE, FORESTS and alpine TUNDRA, is grazed by livestock.

In many parts of the world, grazing has occurred for millennia, and so it is difficult to discern its effects. Ecological ramifications of livestock grazing are more easily seen in sites such as the Intermountain West of North America, that evolved in the absence of heavy-bodied HERBIVORES (including bison), and to which millions of livestock were introduced relatively recently.

The ecological costs of livestock grazing can be summarized as follows.

1 Alteration of species composition of communities, including decreases in DENSITY and BIOMASS of individual species, reduction of SPECIES RICHNESS, and changing COMMUNITY organization.

2 Disruption of ECOSYSTEM functioning, including interference in nutrient cycling and ecological SUCCESSION.

3 Alteration of ecosystem structure, including changing vegetation STRATIFICATION, contributing to SOIL erosion and decreasing the availability of WATER to biotic communities.

Changes in species composition

Decreases in the density of native plant species and in the DIVERSITY of native plant communities as a result of livestock grazing have been observed in a wide variety of ecosystems. Grazing affects the species composition of plant communities in two ways: (i) active selection by herbivores for or against a specific plant taxon; and (ii) differential vulnerability of plant taxa to grazing. Recent estimates on federal grazing lands in the western USA concluded that less than half the vegetation was 50% similar to the presumed CLIMAX community. Grazing also can exert great impact on animal populations, usually due to indirect effects on HABITAT STRUCTURE and prey availability. Deleterious effects of grazing have been observed in all vertebrate classes. The relationship of grazing to insect populations remains unclear.

Grazing-induced changes in particular species can translate into major conversions of community organization, for example transforming grassland to desert. Grazing can delay plant PHENOLOGY, which in turn can have dramatic effects on communities of pollinators and dispersers. Grazing has also been observed to alter animal foraging GUILDS.

Grazing destabilizes plant communities by aiding the SPREAD and ESTABLISHMENT of ALIEN species in three ways:

1 dispersing seeds in fur and dung;

2 opening up habitat for weedy species;

3 reducing COMPETITION from native species by eating them.

Disruption of ecosystem function

Grazing also disrupts the fundamental ecosystem functions of nutrient cycling and succession. CRYPTOGAMIC SOIL CRUSTS perform the major share of NITROGEN FIXATION in desert ecosystems. However, crusts are most prominent where large-bodied herbivores are absent from recent evolutionary history; in these sites, they are easily damaged by livestock, and nitrogen fixation is decreased. Ecosystems also lose NUTRIENTS because they are tied up in livestock faeces, which in some cases cannot be recycled due to lack of appropriate decomposers (e.g. dung beetles in the Great Basin, USA). When livestock are exported to markets, nutrients are lost from ecosystems on a massive scale.

Grazing also can disrupt ecological succession. Long-term, continuing DISTURBANCE by livestock maintains seral vegetation in many areas. Livestock often prevent regeneration of RIPARIAN tree species by eating and trampling seedlings. Thus, riparian plant communities in the presence of grazing often become even-aged, non-reproducing systems.

Physical structural changes

The physical structure of ecosystems, on large and small scales, can also be changed by grazing. Shrub layers have been added to grazed forests, and extensive willow stands have been removed from streamcourses. Grazing causes structural changes to the habitat for ground-dwelling vertebrates, such as snakes and lizards, through the loss of low-height vegetation. Soil litter is removed, which can have both physical and biological effects. Grazing also contributes to the deterioration of soil stability and porosity. The removal of plant COVER causes more surface RUN-OFF, more soil erosion and increased flooding. Grazing has also repeatedly been shown to increase soil compaction and thus decrease water infiltration. The net effect of such interactions is a loss of available water to biotic communities, and DESERTIFICATION.

Livestock spend a disproportionate amount of their time in riparian zones, which are among the most productive habitats in arid lands, and provide essential wildlife habitat. In many areas, grazing is considered the greatest threat to these important habitats. Livestock affect four general components of riparian systems: (i) streamside vegetation; (ii) stream channel morphology; (iii) shape and quality of the water column; and (iv) structure of streambank soil.

Streamside vegetation is altered by livestock in several ways:

1 compaction of soil, which increases run-off and decreases water availability to plants;

2 herbage removal, which allows soil temperatures to rise, thereby increasing EVAPORATION;

3 physical damage to vegetation by rubbing, trampling and browsing;

4 altering the growth form of plants by removing terminal buds and stimulating lateral branching.

Livestock activities especially deleterious to native fishes are the removal of vegetative cover and the trampling of overhanging streambanks. Moreover, livestock decrease the WATER QUALITY of streams by changing water chemistry and temperature.

Indirect ecological costs

In addition to grazing *per se*, the industry of livestock production entails a number of indirect ecological costs. Livestock compete with native herbivores for forage ('usurpation'), and often consume the most nutritive species ('highgrading'). Fencing, which is a fundamental livestock management tool, creates obstacles for many native wildlife species, such as the pronghorn (*Antilocapra americana*). The livestock industry has played a large role in the elimination of native predators; some of the most vehement opposition to predator reintroduction proposals continues to come from livestock interests. Alien species, such as crested wheatgrass (*Agropyron cristatum*), are often planted as 'range improvements'. Additionally livestock-borne diseases can be transmitted to native wildlife, such as bighorn sheep (*Ovis canadensis*).

There has been considerable debate, especially in the North American southwest, concerning the respective roles of CLIMATE CHANGE and human land use (including livestock grazing) in changing native vegetation. The most recent historic evidence seems to support the idea that livestock grazing, interacting with fluctuations in climatic cycles, has been a primary factor in altering these (and presumably other) ecosystems. [T.L.F.]

REFERENCE

Fleischner, T.L. (1994) Ecological costs of livestock grazing in western North America. *Conservation Biology* **8** (3), 629–644.

Greek alphabet

Upper case	Lower case	Name
Α	α	Alpha
Β	β	Beta
Γ	γ	Gamma
Δ	δ	Delta
Ε	ε	Epsilon
Ζ	ζ	Zeta
Η	η	Eta
Θ	θ	Theta
Ι	ι	Iota
Κ	κ	Kappa
Λ	λ	Lambda
Μ	μ	Mu
Ν	ν	Nu
Ξ	ξ	Xi
Ο	ο	Omicron
Π	π	Pi
Ρ	ρ	Rho
Σ	σ	Sigma
Τ	τ	Tau
Υ	υ	Upsilon
Φ	φ	Phi
Χ	χ	Chi
Ψ	ψ	Psi
Ω	ω	Omega

[J.G.]

Green Alliance An independent, non-profit-making organization in the UK which aims to raise the prominence of the environment on the agendas of all key policy-making institutions. *See also* ENVIRONMENTAL PRESSURE (LOBBY) GROUPS. [P.C.]

green consumerism The use of criteria relating to ENVIRONMENTAL IMPACTS in making choices about the purchase of goods. It emerged as an important instrument in the 1980s. *See also* ECOLABEL. [P.C.]

REFERENCES

Elkington, J. & Hailes, J. (1988) *The Green Consumer Guide. High Street Shopping for a Better Environment.* Victor Gollancz, London.

Elkington, J. & Hailes, J. (1989) *The Green Consumer's Supermarket Shopping Guide.* Victor Gollancz, London.

green funds Money invested in the financial markets not only on the basis of potential financial returns but also on the basis of the environmental (more generally ethical) credentials of companies. These investment products can cover unit trusts, pension funds, personal equity plans (PEPs), investment trusts and venture capital funds. Like other shares, green shares are best bought during bear markets, but there is evidence that they tend to outperform in bull markets and underperform in bear markets (possibly because markets tend to sell off high-grade stock in times of uncertainty), and seem to be attractive for investors interested in long-term capital gains. *See also* ECOFUNDS; ETHICAL INVESTMENT. [P.C.]

REFERENCE

Joly, C. (1992) Green funds, or just greedy? In: *Green Business Opportunities* (eds D. Koechlin & K. Muller), pp. 131–153. Pitman, London.

green logistics Logistics is the spatial or geographical aspect of the flows of materials and products from manufacturers to markets and vice versa. Green logistics refers to devising distribution systems that minimize ENVIRONMENTAL IMPACT. A major criterion is to maximize distribution for minimum fuel use, using integrated strategies. [P.C.]

green marketing The use of good environmental features of a product to promote its sales. *See also* MARKET INSTRUMENTS. [P.C.]

REFERENCE

Charter, M. (ed.) (1992) *Greener Marketing*. Greenleaf Publishing, Sheffield.

green net national product (gNNP) Gross domestic product (GDP) is a measure of economic activity in a country. It is arrived at by adding the total value of a country's annual output of goods and services = private consumption + change in building-stock + investment + government expenditure + (exports – imports). The gross national product (GNP) is GNP plus residents' incomes from economic activity abroad and property held abroad minus the corresponding incomes of non-residents in the country. Some argue that these indices of economic well-being are at odds with environmental well-being because what GNP measures is a result of the consumption of natural RESOURCES, natural capital, and this may bring short-term gains but only at the expense of long-term prospects; it is not sustainable. GDP and GNP are not good indicators of a sustainable economy. They require also to take into account losses of natural capital to give a green net national product (gNNP). *See also* ENVIRONMENTAL (ECOLOGICAL) ECONOMICS; SUSTAINABLE DEVELOPMENT. [P.C.]

REFERENCE

Pearce, D. (1993) *Measuring Sustainable Development*. Blueprint 3. Earthscan, London.

green tourism Tourism is travelling for pleasure. It is one of the world's largest businesses; for example, in Europe it accounts for *c.* 5% of gross domestic product (GDP) overall, *c.* 5% of export earnings and *c.* 6% of total jobs, with obvious centres such as the Mediterranean Basin. The impact on the environment thus results from travel to and from countries, and from activities that cause physical damage to sites and create wastes; this impact can be enormous. Green tourism, also referred to as 'sustainable tourism' (and, sometimes, 'ecotourism', although this can sometimes mean travel to view ecologically interesting sites), is a recognition of the problem and strives to ameliorate or reduce the impact.

Such action could include:

- development of international and national management plans for centres;
- staggering of holidays;
- practical guides for the industry and for the tourists themselves;
- development of codes of conduct.

This is not just altruism on the part of the industry or the tourist centres: degradation of the environment can itself destroy the business and lead to the search for undisturbed centres, so spreading the problem. [P.C.]

green trusts *See* ETHICAL INVESTMENT.

greenhouse effect *See* GREENHOUSE GASES.

greenhouse gases Gases that contribute to GLOBAL WARMING by absorbing long-wave energy emitted from the Earth's surface—the so-called 'greenhouse effect'. The most widely discussed greenhouse gases are CARBON DIOXIDE (CO_2), METHANE (CH_4), nitrous oxide (N_2O) and the various CFCs, all of which have increased in the last hundred years or so as a result of human activities. WATER vapour is also a greenhouse gas, though often overlooked, and so is OZONE. These gases differ enormously in their effectiveness as greenhouse gases, and various methods have been devised to express this effectiveness, based upon spectroscopic data. The relative molecular forcing is the effectiveness per molecule of the gas, relative to the effectiveness of CO_2. A more useful index is the global warming potential (GWP), which takes into account the fact that some TRACE GASES persist for longer periods in the ATMOSPHERE than others. GWP is the warming that would result from the instantaneous release of 1 kg of the gas relative to that from the simultaneous release of 1 kg CO_2, integrated over a stated period of time (Table G3).

Although CO_2 is a relatively weak greenhouse gas per molecule or per unit mass, the total EMISSIONS of CO_2 from fossil-fuel burning and DEFORESTATION far exceed those of the other gases, and so the calculated warming resulting from one year's emissions, integrated over 100 years, shows CO_2 to be dominant (Table G4). [J.G.]

REFERENCE

Houghton, J.T., Jenkins, G.J. & Ephraums, J.J. (eds) (1990) *Climate Change: the IPCC Scientific Assessment*. Cambridge University Press, Cambridge.

Green(s) A term usually used to describe those associated with green (environmental/ecological) political parties or groups (*see* ECOPOLITICS), but it may be used more broadly to describe anyone who embraces ENVIRONMENTALISM. *See also* ENVIRONMENTAL PRESSURE (LOBBY) GROUPS; GREEN ALLIANCE. [P.C.]

Greenwich mean time (GMT) *See* TIME.

Grey-brown desertic soils *See* SOIL CLASSIFICATION.

Table G3 The global warming potentials (GWPs) of greenhouse gases, integrated over 20, 100 and 500 years, and the approximate lifetimes.

Gas	Lifetime (years)	GWP_{20}	GWP_{100}	GWP_{500}
CO_2	10	1	1	1
CH_4	10	63	21	9
N_2O	150	270	290	190
CFC-11	60	4500	3500	1500
CFC-12	130	7100	7300	4500

Table G4 Estimated contributions made to global warming by the main greenhouse gases, given the current rate of emissions, calculated from the global warming potentials with a 100-year integration time. (Tg = 10^{12} g.)

Gas	Current man-made emissions (Tg year^{-1})	Contribution to warming as a percentage of total warming
CO_2	26 000	61
CH_4	300	15
N_2O	6	4
CFC-11	0.3	2
CFC-12	0.4	7

Grey-cinnaminic soils *See* SOIL CLASSIFICATION.

Grey forest soils *See* SOIL CLASSIFICATION.

Grey List A list of potentially hazardous chemicals drawn up under EC Framework Directive 76/464/EEC. Formally known as List II. *See also* BLACK LIST; LISTS. [P.C.]

grey literature Literature not generally available through normal bookselling channels, and so difficult to identify and obtain. It includes reports, technical notes and specifications, conference proceedings, proposals, supplementary publications, data compilations, etc. A considerable amount of environmental information is conveyed in this way. Publications issued by pressure groups are often of this kind. Often grey literature does not have an international standard book number (ISBN) or an international standard serial number (ISSN). [P.C.]

Greyzems *See* SOIL CLASSIFICATION.

Grime's triangle An extension of the *r–K* continuum (*see* *r*- AND *K*-SELECTION) with particular reference to plant species. Grime's triangle is a two-dimensional ORDINATION whose axes are environmental FAVOURABILITY and DISTURBANCE (Fig. G9). The model is triangular because one corner of the square ordination, which represents high disturbance and low favourability, is essentially uninhabitable. Grime has recognized three primary plant strategies, which he calls competitors, RUDERALS and STRESS tolerators, and each is characteristic of one of the three apices of the triangle. In favourable conditions where disturbance is low, competitors are favoured. In unfavourable conditions where disturbance is low, stress tolerators are favoured. In favourable conditions with high disturbance, ruderals are favoured. SECONDARY SUCCESSIONS after periodic disturbance exhibit a characteristic trajectory through the triangle; there is a progression from ruderals (short-lived, often EPHEMERAL plants) which are competitively displaced by competitors (tall, rapidly growing plants) and these, in turn, are eventually displaced by stress tolerators (shade-tolerant plants).

Several aspects of Grime's triangle are controversial.

1 His definition of a good competitor is quite different from Tilman's (*see* TILMAN'S MODEL OF RESOURCE UTILIZATION).

2 The distinction between favourable and unfavourable environments is contentious (*see* FAVOURABILITY).

3 It is impossible to provide an objective general measure of stress.

It is an improvement over the *r–K* classification to the extent that it provides three pigeonholes into which species can be placed rather than just two (*see also* SELECTION PRESSURES, CLASSIFICATION). The triangle is perhaps best seen as a shorthand version of a multidimensional ordination of species in a statistical space defined by disturbance, nutrient availability and (potentially) many other environmental factors. In a multivariate analysis of floristic data from the Sheffield area (Grime *et al.* 1988), it is clear that the nutrient availability/stress tolerance axis is the most important dimension involved in ordering the species from high nutrient-demanding species like *Urtica dioica* to tolerators of low nutrient conditions like *Festuca ovina*. *See also* ADAPTATION; LIFE-HISTORY EVOLUTION. [M.J.C.]

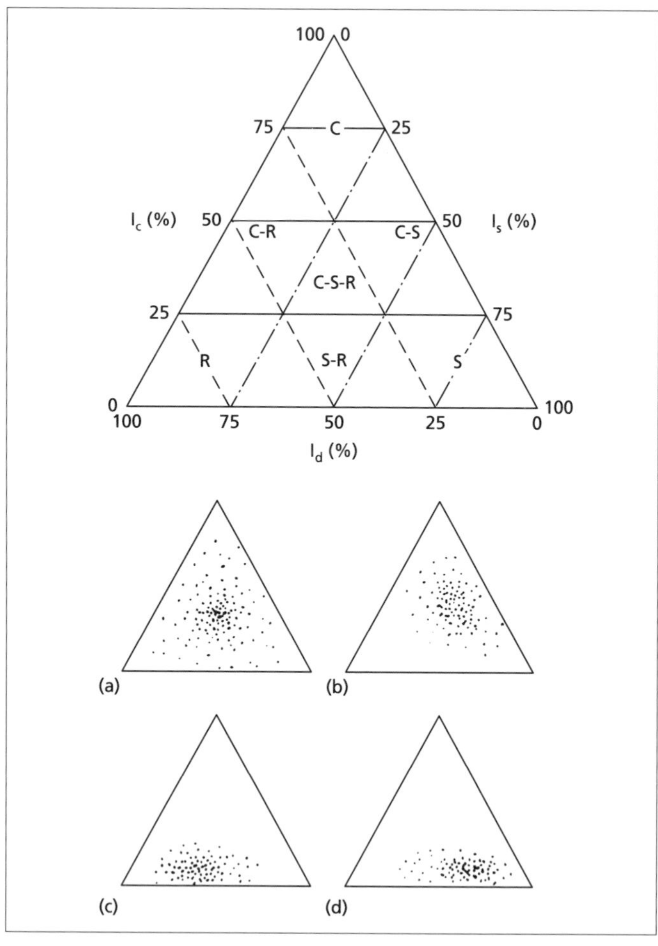

Fig. G9 Grime's triangle. A two-dimensional ordination of species on the basis of productivity and disturbance, showing the location of Grime's three fundamental plant strategies: C, competitors; R, ruderals; and S, stress-tolerators, with various intermediates (e.g. C-R means competitive ruderals). I stands for 'importance', so I_c means importance of competition. The strategic range of different plant life forms: (a) herbs; (b) trees and shrubs; (c) bryophytes; and (d) lichens. (After Grime, J.P., Hodgson, J.G. & Hunt, R. (1988) *Comparative Plant Ecology: A Functional Approach to Common British Species.* Unwin Hyman, London.)

REFERENCES
Grime, J.P., Hodgson, J.G. & Hunt, R. (1988) *Comparative Plant Ecology: A Functional Approach to Common British Species.* Unwin Hyman, London.
Tilman, D. (1988) *Plant Strategies and the Dynamics and Structure of Plant Communities.* Princeton University Press, Princeton, N.J.

grooming (preening) Animals always invest a certain proportion of their time in carefully searching and cleaning their skin for ECTOPARASITES and dirt. They either do it directly for themselves (auto-grooming) or do it to someone else (allo-grooming). Ectoparasites can be very abundant and this has been shown to have large negative consequences on their HOSTS. Therefore, it is not surprising that preening of the skin is an important activity in reducing the PARASITE load and it has been observed in most animal species.

Social grooming is regularly observed between mating and social partners. However, it often seems to last longer than would be needed for a purely hygienic function. Grooming has been shown to have a strong effect in reducing heart rates both in horses and in some primate species. It seems that grooming acquired a secondary social bonding function in primates, as time spent grooming increases with group size, grooming is more frequently observed after agonistic or sexual interactions, and animals invest more time in grooming when relationships are being established or are under threat by rivals. *See also* BEHAVIOURAL ECOLOGY. [C.B.]

gross primary production (GPP) The process of energy, matter or CARBON accumulation by the AUTOTROPHS in a community. Its rate, gross primary productivity (P_g or G), equals the sum of net primary productivity and RESPIRATION by the autotrophs. Essentially, gross primary productivity by plants, or other photoautotrophs, represents the rate of total transformation of light energy into chemical energy. In practice, it is estimated for terrestrial plants by correcting net primary productivity with estimates of respiratory losses, most commonly determined from the rate of CARBON DIOXIDE (CO_2) emission in darkness and the assumption that the rate during daylight hours will equal that during the dark. In PHYTOPLANKTON, gross primary productivity may be estimated by feeding $^{14}CO_2$ or $H^{14}CO_3^-$ over a short time period so that the radioactive label is assimilated in PHOTOSYNTHESIS, but does not have sufficient time to enter pools of respiratory substrates.

However, developments in plant physiology now question whether gross primary productivity can be measured or whether the concept has any real biological meaning. First, photorespiration is a consumption of light energy which results in the oxidation of CARBOHYDRATES synthesized through photosynthesis. It is an integral part, and probably an inevitable consequence, of C_3 photosynthesis. It cannot be determined simply by placing the plant in darkness. The rapid loss of a significant and variable proportion of the recently photo-assimilated energy limits accurate determination of gross primary productivity. Secondly, reducing (NADPH.H) and phosphorylating (ATP) power generated on the photosynthetic membrane are used not only to drive photosynthetic CO_2 reduction and photorespiratory oxidation, but can also replace some of the functions of dark respiration within the photosynthetic cell. For example, the reduction of NITRATE and nitrite to AMMONIA and its incorporation into AMINO ACIDS, can and does use the reducing and phosphorylating power of the chloroplast. These considerations all suggest that gross primary productivity, as measured traditionally, will seriously underestimate the true gain of energy, but do not point to any simple alternative method of measurement. *See also* ECOLOGICAL ENERGETICS. [S.P.L.]

groundwater *See* HYDROGEOLOGY.

groundwater pollution CONTAMINATION of groundwater to an extent that is likely to cause HARM to human health (if the groundwater is used as source of drinking water) and the environment. *See also* POLLUTION. [P.C.]

Groundwork Trust An organization that assists people from all sectors in the UK to improve their local environment through partnerships that involve business, local government and voluntary organizations. [P.C.]

group selection The NATURAL SELECTION of some groups of individuals at the expense of other groups. The concept was originally proposed in trying to understand the EVOLUTION of altruistic behaviour. Until quite recently, many social behaviours were interpreted as having evolved 'for the good of the species', with no proper appreciation of the importance of within-species CONFLICTS of interest. In particular, ALLELES producing non-altruistic behaviour may be favoured if their carriers do not pay the costs of altruism, but reap the benefits of the altruistic behaviour of other individuals. Clearly, non-altruists do not act for the good of the species. In an attempt to resolve this paradox, the process of group selection was proposed, according to which groups containing non-altruists are disadvantaged in comparison with groups of altruists. The process was clearly described by V.C. Wynne-Edwards in 1962. Necessary to the evolution of altruism by this process were that MIGRATION rates between groups be low, and that groups be short-lived. Most ecologists do not, however, find it plausible that these conditions obtain in the field. This, together with observational support for the rival theory of KIN SELECTION enunciated by W.D. Hamilton in 1964, has led to the general abandonment of group selection as an evolutionary explanation for altruistic behaviour.

Group selection continues, however, to play an important, if limited, role in our understanding of some evolutionary processes, for instance PARASITE evolution. Parasite strains which kill their HOSTS too quickly may not infect as many further hosts as less virulent strains. Here, important selection may occur between the groups parasitizing different hosts. This kind of process is believed to be responsible for the evolution of reduced VIRULENCE of the myxoma virus to rabbits in Australia and Britain. *Cf.* KIN SELECTION; NATURAL SELECTION. [R.M.S.]

REFERENCES
Harvey, P.H. (1985) Intrademic group selection and the sex ratio. In: *Behavioural Ecology: Ecological Consequences of Adaptive Behaviour* (eds R.M. Sibly & R.H. Smith). Blackwell Scientific Publications, Oxford.
Maynard Smith, J. (1989) *Evolutionary Genetics.* Oxford University Press, Oxford.
Wynne-Edwards, V.C. (1962) *Animal Dispersion in Relation to Social Behaviour.* Oliver & Boyd, Edinburgh.

group size *See* OPTIMAL GROUP SIZE.

Group social structure

The non-random social composition of groups of CONSPECIFICS or HETEROSPECIFICS. Groups of animals often have a non-random composition with respect to sex, age, relatedness or PHENOTYPE. These compositions are presumed to reflect the net advantages to the participant individuals in question.

Reproductive groups

The most common type of social group is that of unrelated reproducing individuals. The AGGREGATION of such individuals is obviously beneficial in terms of reproduction. Males of many animal species attend females both before and after copulation, and this attendance is termed pre- and post-copulatory MATE-GUARDING. Males enhance their certainty of PATERNITY by such attendance. The sex that gains the largest benefits in terms of increased reproductive success from mating with additional mates should be the sex that is the first to desert. In species with the most common sex roles with male reproductive success being limited by access to females, males are the main deserters.

Family groups

Male presence near the time of birth of offspring may predispose males for PARENTAL CARE, provided males do not gain larger net benefits from DESERTION. The benefits of parental care for reproductive success may result in males or females staying with the offspring, and groups of parents and their offspring are the second most common kind of social group. Groups of parents and their offspring frequently dissolve when offspring reach maturity, and offspring disperse when the benefits from doing so exceed the costs of staying. Some species of birds and mammals with cooperative breeding systems retain one or more offspring as helpers, which provide care for sibs. Such social groups are maintained because the offspring of the non-dispersing sex benefits in terms of acquisition of breeding experience, inheritance of the breeding territory, or its own reproduction, or benefits from the raising of close KIN by staying. Since siblings share half of their genes, helpers may benefit from raising sibs if the chances of reproducing successfully elsewhere are less than the reproductive success when helping relatives, devalued by the coefficient of relatedness (0.5 in the case of full siblings). There are many examples of matrilinear groups of mammals, with females and their daughters being spatially aggregated. Social groups of close relatives are common in many animals, apparently because close kin may benefit from the behaviour of close relatives. Altruism is uncommon among animals, because costly benefits should not be given freely to unrelated individuals. Apparent exceptions to this rule can be explained in terms of reciprocal altruism and similar kinds of repeated interactions between individuals.

Breeding aggregations

Breeding aggregations of animals are very common, and a number of explanations for such social groups has been proposed. COLONIALITY may arise as a consequence of the close aggregation of individuals in safe breeding sites, and the unpredictable distribution of RESOURCES in time and space may render the defence of resources unprofitable for the individual animal. Breeding aggregations, such as those of many insects and pinnipeds, may arise because females, by aggregating, increase the level of male–male COMPETITION and in that way allow females more easily to assess the phenotypic and genotypic quality of potential mates. Leks are close aggregations of displaying individuals of one sex in which the displaying males provide females with no resources other than sperm. Aggregations of displaying males may arise because less attractive males aggregate around the most attractive males (the 'hot-shot' hypothesis). Alternatively, males may aggregate in sites that are frequented especially often by females, and the close aggregation of a number of males will in such circumstances attract females at a higher rate (the 'host-spot' hypothesis).

Group size and fitness

The group size that gives rise to the maximum average FITNESS may never be reached because single individuals often will experience higher success by joining a group rather than being on their own. Realized group size may therefore often be larger than the size maximizing average fitness. Furthermore, the OPTIMAL GROUP SIZE may differ for different categories of individuals. Individuals of poor phenotypic quality may benefit from being in a small group, while high-quality individuals may acquire fitness if the size of the group is large. Realized group sizes may be a compromise between the optimal sizes for participating individuals.

Kleptoparasitism

Not all social groups are necessarily beneficial to all participants. Kleptoparasitism is a frequent mode of resource acquisition by conspecifics or heterospecifics in a wide array of animals. Groups can be considered to consist of producers, which detect and retrieve food, and scroungers, which parasitize the efforts of the producers. The relative frequency of scroungers is obviously limited by the relative frequency of producers, since very high frequencies of scroungers cannot exist on few producers. The spatial and temporal distribution of producers and scroungers is therefore presumed to follow an ideal-free distribution that allows net fitness benefits of scroungers to be similar in different groups. *See also* BEHAVIOURAL ECOLOGY; SOCIAL ORGANIZATION; SOCIALITY, TYPES OF; SOCIOBIOLOGY. [A.P.M.]

---

growth The net increase in size or mass of an organism. Growth can be further subdivided into production of somatic (i.e. non-reproductive) tissue and reproduction. Growth rate is typically expressed in units of mass, length, energy units or CARBON per unit time and is a product of the rate at which RESOURCES are made available and their efficiency of conversion to tissues. Growth efficiency is the amount of growth attained per unit of food ingested (gross growth efficiency) or assimilated (net growth efficiency). Energetic models predict that net growth can occur only if energy intake from food is greater than energy expenditure. Such models are used to relate somatic and reproductive growth to FEEDING rates, absorption efficiencies and metabolic costs (*see* SCOPE FOR GROWTH).

For many invertebrates FECUNDITY is positively related to body size, which would tend to favour investment in somatic growth prior to the onset of reproduction. Also, since organisms generally need to reach some threshold size before reproduction can occur, somatic growth can be defined as the developmental means of achieving a reproductive state. Below the threshold size, the benefits of rapid growth are that reproductive size is reached more quickly. However, there are expected to be certain costs associated with rapid growth rates, and such costs must be balanced with the benefits gained by rapid growth. Above the threshold size, TRADE-OFFS can be expected between the amount of resources devoted to further somatic growth versus the amount devoted to reproductive output. Similar energetic trade-offs provide the basis for the development of life-history models and are relevant to understanding the various reproductive modes observed in different groups of organisms (e.g. SEMELPARITY vs. ITEROPARITY; lecithotrophy vs. planktotrophy).

Somatic growth rates generally vary with body size, and most metazoan animals show a sigmoid relationship between body size and age. Some organisms exhibit above-normal growth rates (catch-up growth) following periods of growth retardation or inhibition, which in some cases may be advantageous in terms of survival or future reproductive output. *See also* DETERMINATE GROWTH; ECOLOGICAL ENERGETICS; INDETERMINATE GROWTH; LARVA; LIFE-HISTORY EVOLUTION; VON BERTALANFFY EQUATION. [V.F.]

guidelines Usually, documents that supplement legislation. Thus the European Commission produces technical guidance documents that facilitate interpretation of RISK ASSESSMENT and management procedures for existing and new substances; the OECD produces guidelines on how to carry out ecotoxicological tests; and the USEPA produces guidelines on risk assessment. Guidelines do not normally have the same legal force as legislation. [P.C.]

guild A group of species exploiting the same class of RESOURCES in a similar way (Root 1967). The species assigned to a guild need not be taxonomically related, although they often are. Species can be members of more than one guild; for example, a species can belong both to a guild with particular foraging habits and to a guild with particular nest-site requirements. The delimitation of a guild is nearly always somewhat arbitrary: while niches of some member species are in the centre of the guild, the niches of other species are more peripheral and/or broader, linking one guild to the next. *See also* FUNCTIONAL GROUP. [J.G.S.]

REFERENCES

Root, R.B. (1967) The niche exploitation pattern of the blue-gray gnatcatcher. *Ecological Monographs* **37**, 317–350.

Simberloff, D. & Dayan, T. (1991) The guild concept and the structure of ecological communities. *Annual Review of Ecology and Systematics* **22**, 115–143.

Gulf Stream The clockwise OCEAN current carrying warm water from the Gulf of Mexico across the Atlantic to the British Isles and Scandinavia. The Gulf Stream separates the relatively warm, saline water of the SARGASSO SEA from the colder, fresher waters near the coast of North America. It is the dominant GYRE in the North Atlantic and transports between 75 and 115 million $m^3 s^{-1}$ of seawater northwards, extending down to the sea floor in most regions. Maximum velocities in the main axis reach 4–5 knots. The Gulf Stream consists of three segments.

1 The Florida Current, which flows between Cuba and Florida. It has an average speed of 5 km h^{-1} and carries 20–40 million $m^3 s^{-1}$.

2 The Gulf Stream proper, which reaches its greatest volume off Chesapeake Bay, equal to 70 million $m^3 s^{-1}$, varies in width between 50 and 150 km (averaging 80 km) and travels at a speed of 5–12 km h^{-1} with an average depth of 500 m.

3 The North Atlantic Current, which represents the north-eastern Gulf Stream and is slower and less definite. It brings warm water as far north as the NORWEGIAN SEA and ARCTIC, and has a significant influence on the CLIMATE of the British Isles and Scandinavia. [V.F.]

guyot A flat-topped submerged mountain that is an extinct volcano. In the past, guyots stood above sea level, but were truncated by waves and have sunk to various depths. [V.F.]

Gymnosperms, global diversity and conservation

The gymnosperms are an ancient group of seed-producing trees whose origins go back at least to the Late Devonian some 400 million years ago. (In contrast the flowering plants, or angiosperms, only appear in the FOSSIL RECORD around 130 million years ago; *see* ANGIOSPERMS, GLOBAL DIVERSITY AND CONSERVATION.) The distinction between the gymnosperms and angiosperms lies in the reproductive organs and FERTILIZATION processes. In the gymnosperms the seeds are naked (covered only by an often paper-thin integument) and borne on a megasporophyll (the cone) which does not entirely enclose the developing seeds (gymno- derives from the Greek γυμνος, naked). In angiosperms (angio-, αγγειον, vessel) the megasporophyll (the carpel) entirely surrounds the seed. Unlike angiosperms, double fertilization does not occur in gymnosperms thus no true endosperm is formed. The structure of gymnosperm wood is also distinctly different from angiosperms, the gymnosperm water-conducting xylem is composed only of tracheids (though vessels are present in the Gnetales) and the phloem lacks companion cells.

Evolution and diversity

Gymnosperms were particularly abundant in the Carboniferous period and are a significant component of coal deposits. By the Jurassic period, around 200 million years ago, they had reached their greatest DIVERSITY and their most widespread, almost global, distribution. Their history thereafter has been one of geographical retreat and MASS EXTINCTIONS. Among those extinctions are the Cordaitales of the late PALAEOZOIC era; these were massive trees often exceeding 30 m in height. Further extinctions at the end of the Jurassic included the Caytonales; these showed reproductive features with tantalizing similarities to the angiosperms, such as a spherical cupule surrounding much of the seed. The Cretaceous extinctions included the once widespread cycad-like Bennettitales. The final epoch of the Cretaceous, the Maastrichtian, was a disaster for the gymnosperms (and much else) when up to 80% of all gymnosperm species disappear from the fossil record. The survivors today are the conifers (Coniferales and Taxales), cycads (Cycadales) and Gnetales. Of these, the conifers are the most widespread, forming the CLIMAX vegetation at high latitudes, while in warmer regions they are characteristic of high altitudes.

Some 600 species are all that remains of a once globally dominant group; these present-day species include the pines, cypresses, cedars, junipers, spruces, firs, larches and yews, many of which are immensely valuable for their timber and other products. A few conifers attain remarkable sizes and ages; specimens of *Sequoia sempervirens* may exceed 100 m in height and have over 2000 growth rings, while 5000-year-old examples of *Pinus longaeva* are known. Most conifers are trees, but a small number are bushy and a few (Australasian) species form heather-like shrubs.

The cycads, now reduced to just 120 species worldwide, are today confined to the tropics and include the sago palms (*Cycas* spp.). The third order, the Gnetales, is a fascinating group in its own right, consisting of three families of highly unusual plants with weak affinities to the other gymnosperms and with each other. The Gnetales probably go back no further than the Early Cretaceous period and more closely resemble the angiosperms (also of Early Cretaceous origins) than any other seed plant. The final order of gymnosperms is the Ginkgoales. Once widespread in the Permian, Jurassic and Cretaceous periods, the order is represented today by just a single species, *Ginkgo biloba*, the maidenhair tree. Fossil remains of leaves almost indistinguishable from modern *G. biloba* are common in Jurassic deposits.

There is probably no single reason why the gymnosperms are today of declining evolutionary significance. In cultivation gymnosperms are highly successful; for example, *G. biloba* has spread from a handful of Japanese and Chinese temple sites to the polluted air of many Western cities. The many possible explanations for the decline of the gymnosperms include failure to develop the herbaceous habit of the angiosperms, competition from angiosperms, the cost of dependence on mycorrhizal associations for nutrition, and failure to develop long-lived seeds. Competition from angiosperms, which may explain the territorial restriction of many species of gymnosperms, does not explain the demise of many of the pre-Cretaceous families. The particular susceptibility of the gymnosperms to the mass extinction event at the end of the Cretaceous cannot be explained by their tree habit (angiosperm trees survived the event). The dependence of probably most gymnosperms on establishing a mycorrhizal association is shared by many angiosperm trees. Gymnosperm seeds do not, in general, persist in the soil for many years but the same is true of the majority of tropical angiosperm trees.

What we see today is a once widespread, even ecologically dominant ancient order undergoing a prolonged evolutionary decline, if not into extinction then at least into evolutionary obscurity. Species such as *Sciadopitys verticillata*, widespread in the Tertiary and a significant element in European brown coal deposits, is now native to a few small areas of Japan. *Sequoia sempervirens*, the Californian redwood, was once widespread throughout the the Northern hemisphere in the Early Tertiary but is now confined in the wild to a few protected areas on the Pacific fringe of North America. *Cupressus macrocarpa* is reduced to two small groves in California—ironically this species is the male parent of the now widespread hybrid Leyland cypress (*Cupressus macrocarpa* × *Chamaecyparis nootkatensis*). *Ginkgo biloba*, *Glyptostrobus lineatus* and *Cephalotaxus harringtonia* are today only known from cultivation though there are reports of extant wild colonies in remote areas of China. The penultimate ginkgo, *G. adiantoides*, survived in Europe to the last ice age. And in the cycads, the future for *Encephalartos woodii* is today represented only by male trees.

Conservation

It is with these as examples that a considerable effort has been made, particularly in the last two decades, to undertake a comprehensive programme of conservation of gymnosperms. In part this has been stimulated by the pharmaceutical value of gymnosperm products, such as taxol (used in cancer treatment) and ephedrine (used in treating respiratory disorders), but also because of the all too visible effects of logging, particularly in North America. *See also* CONSERVATION BIOLOGY. [G.A.F.H.]

---

gynodioecy The condition when some or all populations of a species contain both female (hence the prefix 'gyno-') and HERMAPHRODITE individuals (broad sense) (hence the suffix '-dioecy', referring to the existence of two sexes). The genetic determinism of sex has been firmly established in a number of cases. It has usually been found that both cytoplasmic and nuclear genes are involved. In most species, sex has in fact been found to be a continuously varying character (with strictly female individuals and strictly hermaphrodite individuals, but also hermaphrodites that produce non-viable male GAMETES, or low amounts of viable male gametes). *See also* SEX ALLOCATION THEORY. [I.O.]

gynomonoecy *See* HERMAPHRODITE.

gypsic horizons *See* SOIL CLASSIFICATION.

Gypsisols *See* SOIL CLASSIFICATION.

gyre A circular or spiral form, usually applied to a very large semi-closed current system, in an open ocean BASIN. The major ocean gyres are centred in SUBTROPICAL high-pressure regions (i.e. 30°N and 30°S). The gyres circulate clockwise in the North Atlantic and North Pacific Oceans and counterclockwise in the South Atlantic, South Pacific and Indian Oceans. [V.F.]

H

H_0 *See* NULL HYPOTHESIS.

H_1 *See* ALTERNATIVE HYPOTHESIS.

habit

1 The regular performance of a particular activity acquired by an animal as a result of learning. In contrast to a conditioned reflex a habit does not require an external stimulus.

2 The characteristic form or mode of growth of an animal or plant. [P.O.]

Habitat

The ecological term 'habitat' refers to a place where a species normally lives, often described in terms of physical features such as TOPOGRAPHY and SOIL MOISTURE and by associated dominant forms (e.g. intertidal rock pools or mesquite woodland). Definitions in ecological literature vary widely but there seems to be a consensus for the following:

- 'habitat' entails the description of some key environmental features related to a species or COMMUNITY distribution;
- habitat and VEGETATION classifications may be concordant, but are not always so;
- habitats may have subdivisions as MICROHABITATS, varying in degrees of FAVOURABILITY related to or judged by a species' abundance;
- NICHES within a habitat refer to certain requirements for functional, behavioural and competitive well-being.

Clearly then, ecology being a science for studying the distribution and relative abundance of various taxa, individually and in ASSEMBLAGES, detailed study of habitat is essential. In the popular natural history writings also, habitats and their special adaptive challenges are fascinating topics. Likewise, biological aspects of current conservation efforts are also largely directed at the classification, evaluation and preservation of representative areas including threatened species or their habitats.

Newell (1979) defined various intertidal habitat types such as rock pools and subsaline pools; each one is recognized for specific physicochemical conditions (dissolved oxygen, pH, salinity, etc.) and faunal assemblages (*see* INTERTIDAL ZONE). ZONATION along the rocky shore, on the other hand, is described in terms of supralittoral, littoral and sublittoral zones determined by the flooding pattern due to spring tides; moreover, ROCKY SHORES and SANDY SHORES are two larger classes of habitats.

In dealing with the North American plant communities, alpine refers to a region as also does DECIDUOUS FOREST region within which several vegetation types and their habitats are described. The term strand ecosystem includes BEACH and DUNE plant communities which many authors list as habitat types, whereas COASTAL WETLANDS have numerous habitats defined by various combinations of temperature, salinity and tidal inundation!

The serpentine habitat is widely recognized for its soil characteristics (montmorillonite clay, low calcium, low levels of nitrogen and phosphorus, high magnesium and heavy-metal toxicity) and its specialized stress-tolerant vegetation (*see* SERPENTINE SOIL). The most common vegetation type is chaparral but the flora in California is divided by physiographic provinces along the coastal–inland, north–south or dry–mesic gradients. Various endemic and INDICATOR SPECIES show a diversity of physiological strategies for survival in the serpentine habitat.

Thus, habitat terminology varies according to an author's choice, and with the context and level of studies (ecotype, species or community/ecosystem). Contrast this with an attempt to define WETLANDS as:

> areas of marsh, fen, peatland or water whether natural or artificial, permanent or temporary, with water that is static or flowing, fresh, brackish, or salt, including areas of marine water the depth of which at low tide does not exceed six meters (World Conservation Monitoring Centre 1992).

This legalese becomes somewhat chaotic.

Increasingly, niche terminology is helping to clarify the use of habitat names or descriptors. The term BIOTYPE, often used as a synonym for niche, emphasizes functional environment in terms of food, shelter and breeding, thus including both abiotic and biotic variables. Niche partitioning within a habitat has been well illustrated for a set of warbler species by using tree height diversity used by each species, and for lizards showing horizontal niche partitioning. Evolution in Darwin's finches has been neatly demonstrated using vegetational, climatic and dietary niche variables. Both HABITAT SELECTION and niche selection in plants and animals are important in the studies of COEXISTENCE versus COMPETITIVE EXCLUSION.

Many sources of natural and human-induced disturbances (and catastrophes) require research on habitat changes, including fragmentation and area loss. Successional and open habitats tend to be colonized by weeds and other invasive organisms; exotics are found succeeding in level grassy swards, drier rocky beaches, roadside and urban vacant lots, and margins of agricultural fields.

Hence, habitats can be characterized in terms of their properties over time and through space. Through time they can be: constant, seasonal, unpredictable or EPHEMERAL. Through space they can be: continuous, patchy or isolated. Classifications based on these properties have been compiled by a number of ecologists. *See also* HABITAT STRUCTURE. [S.K.J. & P.C.]

REFERENCES

Begon, M. (1985) A general theory of life-history variation. In: *Behavioural Ecology* (eds R.M. Sibly & R.H. Smith), pp. 91–97. Blackwell Scientific Publications, Oxford.

Newell, R.C. (1979) Faunas of special habitats. In: *Biology of Intertidal Animals*. Marine Ecology Surveys, Faversham.

Southwood, T.R.E. (1976) Bionomic strategies and population parameters. In: *Theoretical Ecology: Principles and Applications* (ed. R. May). Saunders, Philadelphia, Pa.

Southwood, T.R.E. (1977) Habitat, the templet for ecological strategies. *Journal of Animal Ecology*, **46**, 337–365.

World Conservation Monitoring Centre (1992) *Global Biodiversity*. Chapman & Hall, London.

habitat complexity gradient *See* DIVERSITY GRADIENT.

Habitat destruction and alteration, global

HABITAT destruction is probably the most important cause of BIODIVERSITY loss. The world's habitats have been so significantly modified by humans that very few can be considered not to bear the mark of human action. Some estimates suggest that more than 70% of the Earth's land surface, other than rock, ice and barren land, is either dominated by human action or at least partially disturbed. Even that which is considered 'undisturbed' is probably subject to some impact.

These anthropogenic impacts take two main forms:

1 conversion of one habitat type to another (e.g. for AGRICULTURE or for dwellings);

2 modification of conditions within a habitat (e.g. managing natural FORESTS).

Forest loss

Forest clearance is widespread, and since forests are the most species-rich terrestrial habitats, this has important consequences for biodiversity. DEFORESTATION effects species loss through habitat destruction, isolation of formerly contiguous habitats, and EDGE EFFECTS in the boundary between forest and unforested areas. Brazil has the largest remaining area of tropical forest but it is under human pressure and subject to these kinds of changes.

Grasslands

Unlike forests, which are in decline with respect to areal coverage, the world's GRASSLANDS have maintained a rough balance over the past 300 years. It is the distribution that has changed, with losses through conversion to cropland being balanced by gains through deforestation. Experience within Europe suggests that the WILDLIFE value of grasslands reduces as PRODUCTIVITY is increased. DESERTIFICATION has also been a factor.

Wetlands and coral reefs

Other vulnerable habitats are the WETLANDS. These are complex ecosystems comprising RIVERS and their RIPARIAN zones, which can be extensive. But they also include ESTUARIES, MANGROVES, flood plains, PEATLANDS and SWAMPS. These are all subject to change by drainage, damming and 'engineering' of various kinds.

CORAL REEFS are another highly diverse ecosystem whose HABITAT STRUCTURE of inorganic matrices is subject to change and destruction by mining, fishing, drilling and aquatourism.

Habitat loss is, therefore, a major cause of biodiversity loss, though surprisingly little information is available on the extent of habitat loss worldwide. *See also* CONSERVATION BIOLOGY. [P.C.]

REFERENCE

McCloskey, J.M. & Spalding, H. (1989) A reconnaissance level inventory of the amount of wilderness remaining in the World. *Ambio* **18**, 221–227.

habitat selection Organisms tend to occur within a restricted range of HABITATS. Although the notion of habitat is open-ended and difficult to define precisely, most organisms are associated with particular biotic and abiotic environments which collectively can be referred to as a habitat. While such associations are apparent, however, it is not always obvious how they arise in practice.

A mundane possibility is that such associations simply reflect differential MORTALITY: individuals that end up in suitable habitats survive and reproduce, whereas those that fail to find such habitats die (passive habitat selection). A more interesting alternative is that organisms exercise an active preference for particular habitats (active habitat preference). Choice experiments with a variety of species have shown active preferences for habitats in which the species occur naturally. In some cases the preferences are present in naive, captive-bred individuals and are maintained despite enforced experience with different habitat types, thereby implying a genetic basis to the preference.

A smaller number of studies have related habitat preferences to particular behavioural skills. In tit species (Aves: Paridae), for example, preference for deciduous versus coniferous woodland habitats appears to reflect the foraging skills needed by different species in the two habitats. In many cases, habitat preferences vary adaptively within and between populations of the same species. Larvae of the polychaete *Spirorbis borealis*, for instance, settle on different algal species depending on water turbulence and the anchorage afforded by the algae. The difference in this case appears to reflect changes in the DOMINANCE of genes affecting preference across habitats. *See also* BEHAVIOURAL ECOLOGY. [C.J.B.]

habitat structure The term HABITAT structure, or 'architecture', refers to the physical structure of the ENVIRONMENT that individuals inhabit. When considering the effects of this structure, a distinction can be made between changes in the number of RESOURCES (i.e. microhabitats or NICHE space) associated with differences in the physical complexity of the environment and the effects of differences in architecture *per se*. The effects of architecture can be dramatic, particularly with regard to the interactions between predators and prey. The most often recorded effect is where increased structural complexity reduces the levels of predation by providing prey REFUGES or by reducing the search efficiency of predators. However, changes in the opposite direction can also occur. It has been shown, for example, that normal pea plants prove to be susceptible to infestation by aphids because ladybirds, which prey on aphids, tend to fall off the plant's slippery leaves. In contrast, genetic strains which possess no leaves enjoy lower levels of infestation in the presence of ladybirds. The distribution of body sizes within a COMMUNITY can also be affected by the structural complexity of the habitat (*see* FRACTAL DIMENSION). [S.J.H.]

REFERENCE

Bell, S.S., McCoy, E.D. & Mushinsky, H.R. (1991) *Habitat Structure: the Physical Arrangement of Objects in Space*. Chapman & Hall, London.

habitat templet *See* SELECTION PRESSURES, CLASSIFICATION.

hadal Of, or pertaining to the biogeographic region of the deep OCEAN. The HADAL ZONE lies directly below the ABYSSAL zone and encompasses all marine habitat below about 6000 m. Problems of studying hadal species and communities are truly formidable owing to the technical difficulties of obtaining specimens and keeping them alive. [S.J.H.]

hadal zone The benthic zone from a depth of about 6000 m to the deepest parts of the OCEAN (*c.* 10 000 m), including trenches and canyons. The total area of the HADAL zone is equal to 1.2% of the total ocean area (compared to 76% of the ocean area between 3000 and 6000 m). Four groups of organisms dominate the hadal zone: pogonophorans, echiurid worms, holothurians and isopods. [V.F.]

Haeckel, E. *See* CHARACTERS IN ECOLOGY.

hafnium (Hf) A silvery metallic element (ATOMIC NUMBER = 72; relative ATOMIC MASS = 178.49; melting point = 2222°C; boiling point = 4450°C), having a crustal abundance of 2.8 mg kg^{-1} and six naturally occurring isotopes: ^{174}Hf (0.18%), ^{176}Hf (5.20%), ^{177}Hf (18.50%), ^{178}Hf (27.14%), ^{179}Hf (13.75%) and ^{180}Hf (35.24%). It occurs in nature closely associated with ZIRCONIUM and the two metals are very difficult to separate. Hafnium has a high NEUTRON capture cross-section and can be used to produce control rods for nuclear reactors. The dominant oxidation state of hafnium is +4 and it is highly insoluble in natural waters. [A.B.M.]

Haldane, J.B.S. *See* CHARACTERS IN ECOLOGY.

halocarbons Chemical compounds of CARBON and one or more halogens (FLUORINE, CHLORINE, BROMINE, IODINE). The term usually refers to small molecules which are gases or liquids at normal temperatures and pressures. They are widely used as solvents, dry-cleaning and degreasing agents. Although CFCs are non-toxic, other halocarbons are cumulative TOXINS (e.g. carbon tetrachloride, chloroform). Methyl bromide is used agriculturally as a seed fumigant. Concern over the effects of chlorine and bromine on OZONE concentrations in the STRATOSPHERE, and over the role of halocarbons as GREENHOUSE GASES, has led to international controls on their production and use, defined by the MONTREAL PROTOCOL. [J.N.C.]

halocline A depth zone within which the rate of change of SALINITY with depth reaches a maximum. Haloclines may be permanent features of the water column or may occur only seasonally. [V.F.]

halophyte A plant of salt-rich habitats such as seawater, SALT MARSHES, MANGROVE swamps and salt DESERTS. The tidally flooded maritime habitats pose the additional problem of intermittent waterlogging with SOIL anaerobiosis. Most halophytes are FACULTATIVE, not actually needing high concentrations of solutes, though a few grow better on salt-enriched media and it is possible that some, for example *Suaeda aegyptica*, may flower only under conditions of high SALINITY.

Halophytes have been defined as those plants found growing where salinity exceeds 100 mol m^{-3}. In seawater and COASTAL habitats this derives from about equal molarities of Na^+ and Cl^- and *inter alia* lesser amounts of Mg^{2+}, Ca^{2+}, K^+, SO_4^{2-} and Br^-. In salt deserts where solute-enrichment is caused by evaporative concentration the composition may be more diverse and also includes carbonate (CO_3^{2-}) and bicarbonate (HCO_3^-).

Plants may respond to salinity in various ways, characteristic of:

1 very sensitive non-halophytes (i.e. GLYCOPHYTES);

2 halophytes (and non-halophytes) that tolerate salt and gradually decrease in growth to *c.* 200–300 mol m^{-3};

3 halophytes that show little or no reduction of growth to *c.* 300–500 mol m^{-3};

4 halophytes that are stimulated in growth up to *c.* 200 mol m^{-3}.

High concentrations of solutes may harm glycophytes by direct TOXICITY, ionic imbalance such as increase in Na : Ca ratio, and osmotic effects. Salt-tolerant plants may cope with these problems by:

1 tolerating the absorbed solutes;

2 preventing solute uptake, the roots differentially excluding some solutes;

3 control of internal salt content by excretion via salt glands, tissue loss or by sequestration of solutes into cell vacuoles.

This last mechanism is often combined with volume adjustment by formation of succulent tissues which provide additional vacuolar STORAGE volume.

Salt-tolerant metabolism is commonly found in halobacteria, but EUKARYOTES usually rely on the control of internal salt concentration. This may be achieved by salt glands, which resemble modified hydathodes (water-secreting glands) and occur in many unrelated halophyte species. They excrete salt by energy-dependent active pumping. A few plants lose salt sacrificially by the abcission of salt-laden organs or by exporting salt into bladder-hairs which are then shed or leached by rain (notably in some Chenopodiaceae).

Sequestration of solutes into vacuoles may reduce their concentration in the sensitive ORGANELLES of the cytoplasm. Many dicotyledonous halophytes are succulent, or develop succulence on exposure to salt, and have salt-rich vacuolar sap. The process consumes energy in active transport and may also establish harmfully large internal osmotic gradients.

It has been suggested that osmotic WATER POTENTIAL gradients between plant and soil solution caused by exclusion of solutes, or internally by sequestration of solutes to vacuoles, may be balanced by synthesis of non-toxic organic solutes. A wide range of solutes have been implicated, including proline, quaternary ammonium compounds, glycerol, mannitol and many others (Fig. H1). More hard evidence for the role of these compounds is needed. Halophytic modifications are costly in energy as some 20–40% of the plant dry mass may be needed as osmo-adjusting solutes when the external salt concentration is 100 mol m^{-3}.

The water relations of halophytes are complicated by the low water potential of the solute-rich external soil solution. To maintain a positive water balance and upward TRANSPIRATION stream the plant must generate an additional negative water potential sufficient to counterbalance the external solute potential. This may be achieved by uptake of inorganic solutes or internal synthesis of organic solutes as described above, or the plant may have to tolerate some level of tissue dehydration. *See also* BRACKISH HABITATS; BRACKISH WATERS; SALINIZATION; SALT FLAT. [J.R.E.]

halosere *See* ZONATION.

'handicap' hypothesis *See* SEXUAL SELECTION.

Fig. H1 Sea arrow grass (*Triglochin maritima* (Juncaginaceae)), an example of a halophytic species which is common in salt marshes and where proline is possibly the major compound synthesized in response to salt exposure. (Courtesy of R.M.M. Crawford.)

handling time The time between perceiving a food item and returning to search for further food — time typically occupied by catching, subduing, handling and sometimes digesting the food item. It is an important parameter, typically symbolized by T_h, in foraging and predator–prey theory (see, for example, HOLLING'S DISC EQUATION). *See also* FEEDING; FOOD SELECTION; OPTIMAL FORAGING THEORY. [H.C.J.G.]

haplo-dikaryotic life cycle *See* SEXUAL REPRODUCTIVE CYCLES.

haplo-diploid life cycle *See* SEXUAL REPRODUCTIVE CYCLES.

haplo-diploidy A term that generally refers to the GENETIC SYSTEM in which males are normally haploid and females are diploid. More rarely it is used for the alternation of haploid and diploid generations as seen in higher plants. The less ambiguous meaning of male-haploidy is preferable.

In the insect orders Hymenoptera and Thysanoptera all species, except for those secondarily possessing females-only PARTHENOGENESIS, combine male-haploidy with ARRHENOTOKY, in which males arise from unfertilized EGGS and females from fertilized ones. The same is true for some beetles, some homopterans, some nematodes, many mites and many rotifers. This combination is thus characteristic of about 20% of all animal species. Similar genetic systems occur in PSEUDO-ARRHENOTOKY (found in some homopterans and mites, in which both sexes arise from fertilized eggs but the paternal chromosome set is lost; the parasitic wasp, *Encarsia pergandiella*, sporadically does the same) and in scale insects such as *Lacansium cerasifex* (arrhenotokous, but with adult males diploid through a doubling of chromosome number).

Sex determination in arrhenotokous male-haploids differs from the general situation in which sex is determined by dosage effects (either an element's presence or absence, as in eutherian mammals and the *Sry* gene, or in the dosage of specific X-linked elements relative to autosomal ones). In many hymenopterans sex is determined by a single locus: heterozygotes become female and hemi- or homozygotes become male. This complementary sex determination imposes a load on the population, which becomes severe upon inbreeding, of essentially infertile diploid males. Inbreeding PARASITOID hymenopterans do not produce diploid males upon inbreeding and therefore must employ a different mechanism.

All genes in male-haploids are inherited similarly to sex-linked genes in organisms with both sexes diploid, but may occur on many chromosomes whereas sex-linked genes in other organisms generally occur on a single chromosome pair. Because males have only a single set of chromosomes to transmit, full sisters in male-haploids are more closely related (3/4) than in male-diploids (1/2), an important factor in the evolution of sociality (*see* SOCIALITY, TYPES OF). [R.H.C.]

REFERENCES

Bull, J.J. (1983) *Evolution of Sex Determining Mechanisms*. Benjamin Cummings, Menlo Park, Calif.

Cook, J.M. (1993) Sex determination in the Hymenoptera: a review of models and evidence. *Heredity* **71**, 421–435.

Wrensch, D.L. & Ebbert, M.A. (eds) (1993) *Evolution and Diversity of Sex Ratio in Insects and Mites*. Chapman & Hall, New York.

haplo-diplontic life cycle *See* SEXUAL REPRODUCTIVE CYCLES.

haploid life cycle *See* SEXUAL REPRODUCTIVE CYCLES.

haploidy The condition of having a single set of unpaired chromosomes in each nucleus, as in GAMETES, GAMETOPHYTES and many species of algae, fungi and vascular plants. Can also be a condition of parthenogenetic organisms. *Cf.* DIPLOIDY. [P.C.]

haplontic life cycle *See* SEXUAL REPRODUCTIVE CYCLES.

haplotype The multilocus GENOTYPE of a haploid genome. Such a genome could be that of a bacterium, or a haploid fungus, or a haploid GAMETE, or that of cytoplasmic ORGANELLES such as mitochondria and chloroplasts. Haplotypic information about non-recombinant mitochondrial DNA in animals has been used in phylogeography studies in humans, for instance. [I.O.]

haptobenthos *See* BENTHIC HABITAT CLASSIFICATION; WITHIN-STREAM/RIVER HABITAT CLASSIFICATION.

Hard selection

The term 'hard selection' has changed considerably in meaning since it was coined by Wallace (1968). Four stages in this evolution of meaning can be distinguished.

1 The term 'hard selection' originated at the end of the 1960s in the context of a discussion about whether all newly discovered enzyme polymorphisms could possibly be maintained by selection due to heterozygote advantage, as this would seem to impose a GENETIC LOAD no population could ever sustain. Wallace (1970) realized FITNESS had to be placed into an ecological context. He distinguished therefore two forms of selection, 'hard' selection and 'soft' selection. Hard selection requires that an individual of a given GENOTYPE meet certain requirements in order to function; if it fails to meet these to any degree, its fitness and its genotype's fitness is correspondingly decreased. Whether the individual meets the requirements depends only upon its own properties, and in no way upon the population the individual finds itself in. Hard selection thus involves an absolute criterion for selection. Hard selection has a direct influence on numbers, as it involves death from genetic causes, and represents a genetic load.

2 Wallace (1975) reconsidered his definition of 'hard' and 'soft' selection in the light of the models of DENSITY-DEPENDENT SELECTION and FREQUENCY-DEPENDENT SELECTION that had appeared between 1970 and 1975. His definition of 'hard' selection remained the same, but he added that 'hard' selection would refer to ways of determining fitness that would be independent of both density (total number) and frequency (of the various genotypes) in the population. This follows from the definition under **1**.

3 To Wallace, 'hard' and 'soft' selection occur within one environment. Christiansen (1975) connected the 'hard' selection of Wallace (1970) with the model of selection in a subdivided environment as described by Dempster (1955; *see also* DEMPSTER MODEL). The PROBABILITY DISTRIBUTION of zygotes over the subenvironments is the same in each generation. Genotypic fitness within each subenvironment determines the probability to reach adulthood from zygote. The contribution of each subenvironment ('NICHE') to the total population depends upon the mean fitness of its individuals, weighted by the probability of that subenvironment. Genotypic fitness of any genotype A_iA_j over the total environment is given by the weighted average genotypic fitness over the n subenvironments:

$$\overline{w}_{ij} = \sum_{k=1}^{k=n} f_k w_{ij,k}$$

where f_k is the probability that a zygote finds itself in environment k ($k=1, \ldots, n$) and $w_{ij,k}$ is the fitness of genotype A_iA_j in environment k. ALLELE FREQUENCY change between generations is computed from the weighted average genotypic fitnesses. Mean fitness per environment is given by:

$$\overline{w}_k = \sum_{ij} p_i p_j w_{ij,k}$$

where p_i is the frequency of allele A_i, while overall mean fitness is given by the ARITHMETIC MEAN of the mean fitnesses in each subenvironment:

$$\overline{\overline{w}} = \sum_{k=1}^{k=n} f_k \overline{w}_k = \sum_{ij} p_i p_j \overline{w}_{ij} = \sum_{k=1}^{k=n} \sum_{ij} f_k p_i p_j w_{ij,k}$$

Following Christiansen (1975), this multiple niche polymorphism model, starting with Dempster (1955), became known as the 'hard' selection model.

4 Via and Lande (1985) interpreted Christiansen's (1975) 'hard' and 'soft' selection and the multiple niche polymorphism models of Dempster (1955) and Levene (1953; *see also* LEVENE MODEL) as differences in number regulation. In hard selection, they supposed the population to be regulated globally. However, the difference in number regulation between 'hard' and 'soft' selection is not further worked out and is secondary to the difference in mean fitness and adaptive topography in Via and Lande's development of a model for the evolution of PHENOTYPIC PLASTICITY. Mean fitness in Via and Lande (1985) depends upon PHENOTYPIC VALUE rather than upon genotypic identity, but the adaptive topography of their hard selection model is the same as the adaptive topography in the one-locus multiple-niche polymorphism Dempster model. With hard selection, mean fitness over all environments is given by the arithmetic mean over the per environment mean fitnesses:

$$\overline{\overline{w}} = \sum_{k=1}^{k=n} f_k \overline{w}_k$$

Most usage of 'hard' and 'soft' selection refers at the moment (1995) to multiple-niche models, both one-locus models and QUANTITATIVE GENETICS models. The difference between 'hard' and 'soft' selection has become a difference in formal modelling. In 'hard' selection, genotypic fitness is averaged over environments, and from this overall genotypic fitness allele frequency change for the total environment is computed. Mean fitness over all environments is given by the arithmetic mean over the per environment mean fitnesses, even if within each environment selection would be 'soft' in the meaning of Wallace (1970, 1975). Ecological considerations show this way to proceed to compute allele frequency change over the total environment and the arithmetic mean fitness to be appropriate for a subdivided population. *See also* SOFT SELECTION. [G.D.J.]

REFERENCES
Christiansen, F.B. (1975) Hard and soft selection in subdivided populations. *American Naturalist* **109**, 11–16.
Dempster, E.R. (1955) Maintenance of genetic heterogeneity. *Cold Spring Harbor Symposia on Quantitative Biology* **20**, 25–32.
Levene, H. (1953) Genetic equilibrium when more than one ecological niche is available. *American Naturalist* **87**, 331–333.
Via, S. & Lande, R. (1985) Genotype–environment interaction and the evolution of phenotypic plasticity. *Evolution* **39**, 505–522.
Wallace, B. (1968) *Topics in Population Genetics*. Norton.
Wallace, B. (1970) *Genetic Load: its Biological and Conceptual Aspects*. Prentice-Hall, Englewood Cliffs, N.J.
Wallace, B. (1975) Hard and soft selection revisited. *Evolution* **29**, 465–473.

---

hard substrata Those substrata that present a surface to which an organism must attach or into which it must bore. Such substrata are composed of a single material, such as rock, coral or wood. Substratum type is the major controlling factor in the distribution of benthic species in that adaptations to hard surfaces usually preclude optimal functioning on SOFT SUBSTRATA (i.e. sand or mud). [V.F.]

hardening, in plants The process whereby plants become resistant to STRESS by gradual EXPOSURE to increasing levels of stress. Hardening is an example of ACCLIMATION. The most well-studied type of hardening is in relation to low temperatures, and many examples are given in Larcher (1995). For example, plants that are exposed to temperatures far below the freezing point of water develop, at the onset of winter, cold-stable phospholipids in their cell membranes and they accumulate materials that act as antifreeze agents. This enables them to withstand severe freezing. [J.G.]

REFERENCE
Larcher, W. (1995) *Physiological Plant Ecology*. Springer-Verlag, Berlin.

Hardy–Weinberg law What is known as the 'Hardy–Weinberg law' is made of two parts, which were discovered independently in 1908 by G.H. Hardy and W. Weinberg. The first part states that if there is no evolutionary force (MUTATION, SELECTION, drift or MIGRATION) acting on allele frequencies ('ideal population'), they do not change. It applies to diploid as well as haploid or n-ploid species. The second part states that if, moreover, there is random mating, the GENOTYPE frequencies attain equilibrium frequencies in a single generation in a diploid species.

In mathematical terms, let p and q $(p+q=1)$ be the frequencies of ALLELES A and a at a given locus. The first part of the law states that, in an infinitely large (no drift), isolated POPULATION (no migration), in the absence of mutation and selection at that locus, p and q are constant. The second part of the law states that, under random mating, if p and q are the gametic allele frequencies, then, in the following generation, there will be p^2 *AA*, $2pq$ *Aa*, and q^2 *aa* (Fig. H2). The Hardy–Weinberg law is currently used to test for random mating. These genotypic frequencies will be conserved over the following generations, provided the first set of hypotheses are valid. In that sense meiosis is 'fair'— all alleles have equal chances. Notice that although the observation of p^2 *AA*, $2pq$ *Aa*, and q^2 *aa* among zygotes can be taken as evidence that mating is at random, it says nothing about the first set of hypotheses, unless one knows the gene frequencies in the previous generations of zygotes. Conversely, the observation of different genotypic proportions does not automatically exclude random mating.

A generalized form of the second part of the Hardy–Weinberg law is p^2-2pqF *AA*, $2pq(1-F)$ *Aa*, and q^2+2pqF *aa*, where F is the inbreeding coefficient (*see* F-STATISTICS). This says that the mating system influences genotypic, not gene, frequencies.

Historically, the Hardy–Weinberg law has proved very important: it showed for the first time the fact that DOMINANCE of an allele (the fact that the phenotype of the heterozygote is the same as that of the homozygote for the dominant allele) has no effect on gene and genotypic frequencies in an 'ideal population'. It is the basic law of POPULATION GENETICS, and can be extended to multiple alleles and to n-ploid species $(n>2)$. In the last case, however, it takes more than a single generation to attain equilibrium genotype frequencies. It also conserves gene frequencies at the level of the population, just as DNA replication conserves it at the level of the genome, and this provides a stable structure on which selections can act. *See also* GENE POOL; MENDELIAN INHERITANCE; QUANTITATIVE GENETICS. [I.O.]

harem A group of two or more females reproductively paired simultaneously with the same male. This term is fading from the literature on animal breeding systems because many researchers are concerned with inappropriate connotations of a lack of an active role by females in decisions about mating. *See also* FEMALE PREFERENCE; MATING SYSTEMS. [J.D.R.]

harm An adverse EFFECT on individuals or an ecological system (populations, communities, ecosystems). For individuals, the adverse effects are judged in terms of probability of dying, health, development, growth, reproduction, etc. For populations, adverse effects are judged in terms of reduction in density over time. For

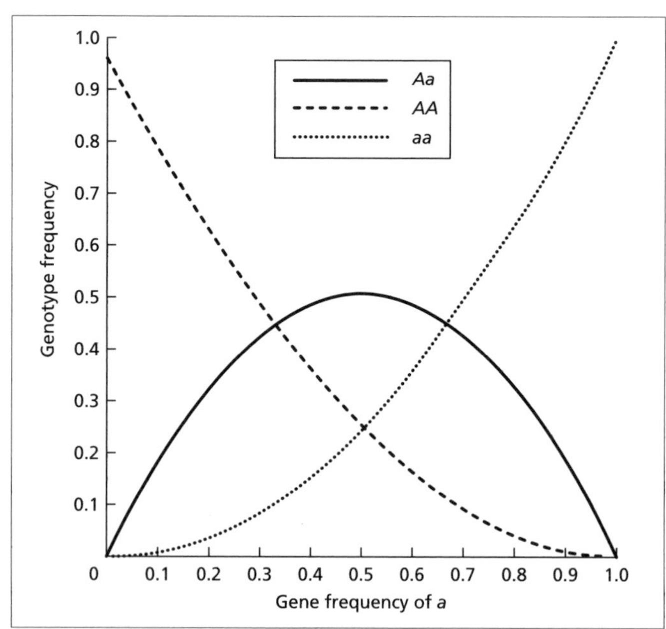

Fig. H2 Hardy–Weinberg frequencies of genotypes *AA*, *Aa* and *aa* in relation to the frequency of the gene *a* (q). (After Ridley, M. (1996) *Evolution*, 2nd edn. Blackwell Science, Cambridge, Mass.)

communities and ecosystems, adverse effects are judged in terms of diversity, energy and matter processing. Often specified in the aim in ENVIRONMENTAL LAW, but rarely defined very precisely. [P.C.]

harmonic mean The reciprocal of the MEAN of reciprocals. The formula for calculating the harmonic mean (H) is:

$$\frac{1}{H} = \frac{1}{n}\sum\frac{1}{x}$$

where x is each observation in the SAMPLE, n is the number of observations in the sample, and the Greek symbol Σ (capital sigma) is the mathematical notation for 'sum of'. *Cf.* GEOMETRIC MEAN. *See also* STATISTICS. [B.S.]

harmonized electronic dataset (HEDSET) A system for collecting information on chemicals, under the EXISTING CHEMICALS legislation of the European Union. *See also* EUROPEAN INVENTORY OF EXISTING COMMERCIAL CHEMICAL SUBSTANCES. [P.C.]

harvest method A technique for measuring PRODUCTIVITY, usually the net primary productivity of herbaceous terrestrial vegetation, for example, grasslands. Sample areas are selected, normally by a randomized block sampling design, and harvested at intervals through the growing season. Harvests are made by clipping the vegetation to ground level. BIOMASS (B) in the harvested material is separated from the dead material (D), and DRY MASS determined. The technique is most suited to annual crops and wild annuals. During the growing season, net primary productivity (P_n) will be given by the increment in biomass (ΔB) for the time interval (Δt):

$$P_n = \Delta B / \Delta t$$

The technique may also be used for perennial herbaceous vegetation, providing that either grazing can be excluded, is insignificant or can be determined with precision, and providing that losses of biomass through death are also taken into account. This can be achieved by adding the increment in the amount of dead vegetation or plant NECROMASS (ΔD) and in turn correcting this for any losses through DECOMPOSITION. [S.P.L.]

REFERENCE

Roberts, M.J., Long, S.P., Tieszen, L.L. & Beadle, C.L. (1993) Measurement of plant biomass and net primary production of herbaceous vegetation. In: *Photosynthesis and Production in a Changing Environment: A Field and Laboratory Manual* (eds. D.O. Hall, J.M.O. Scurlock, H.R. Bolhàr-Nordenkampf, R.C. Leegood & S.P. Long), pp. 1–21. Chapman & Hall, London.

harvesting The exploitation of a renewable natural resource, including FORESTRY, fisheries, game ranching and AGRICULTURE, although in agriculture a large number of other factors enter into the assessment of the process.

The basic activity in the scientific assessment of harvesting is to assess how the removal of individuals from the population affects the dynamics of that population. Sustainable harvesting involves harvesting at levels that do not drive a population to extinction.

A central idea of any harvesting is that of SUSTAINABLE YIELD. If harvesting exceeds the MAXIMUM SUSTAINABLE YIELD (MSY), then the population will decline, and if that harvesting continues, extinction will occur.

Species vary dramatically in their ability to sustain harvesting and there are very general relationships which determine the level of harvests that can be sustained. In general, the longer the LIFESPAN of the resource, the lower the level of harvest that can be taken. Hence, whales and tropical hardwoods have far lower levels of sustainable yield than, say, shrimps and fast-growing conifers.

Selective harvesting can greatly increase the levels of YIELD that can be harvested sustainably from the population. For example, in agricultural systems the age and size structure of the harvested population is substantially distorted so that the population has the highest possible reproductive potential. A typically managed harvest of a natural population would probably aim to ensure that there were fewer males than females and certain animals above reproductive age would be culled.

In complicated ecosystems the interaction between species can be utilized to generate additional benefits from harvesting. For example, the harvesting of a predator can permit greater levels of harvest of its prey. A classic example occurs in the Southern Ocean, where the depletion of baleen whale stocks through the operation of the long-distance WHALING fleets, has arguably led to a very significant potential for the harvesting of KRILL, the major food of the whales. *See also* CONSTANT-EFFORT HARVESTING; CONSTANT FINAL YIELD; FISHERIES, CONSERVATION AND MANAGEMENT; OPTIMAL HARVEST. [J.B.]

Hassell model and Hassell–Comins model

The Hassell model (Maynard Smith & Slatkin 1973; Hassell 1975) is a discrete model of single-species population growth (INTRASPECIFIC COMPETITION). The Hassell–Comins model is an extension of the single-species growth model to two-species competition (Hassell & Comins 1976).

The single-species model is intended to describe the growth of a species with discrete breeding seasons. Such discrete models express numbers at one time (N_{t+1}) in terms of numbers at a previous time (N_t). The notation, t and $t+1$, represents time (t) moving in discrete steps. The simplest situation is that in which there is a constant net reproductive rate (R), which simply represents how many individuals are produced by each individual in the population. If $R=2$ (the population doubles each generation) and we start with 10 individuals, we would progressively produce 20, 40, 80, 160, etc. individuals. In mathematical notation this would be:

$$N_{t+1} = N_t R$$

and would produce the familiar exponential or unlimited population growth. Intraspecific competition can be introduced into this equation by making the size of R dependent upon POPULATION DENSITY. This is shown in Fig. H3, in which the inverse of population change is plotted against density. Consider the two points, A and B. At point A (a very very small number, $N_t \approx 0$) there is no intraspecific competition and therefore $N_{t+1} = N_t R$ (exponential, unlimited growth) or $N_t / N_{t+1} = 1/R$. At point B (the equilibrium point (K)) population growth will have stopped ($N_{t+1} = N_t$) and therefore $N_t / N_{t+1} = 1$. Joining these two known points by a straight

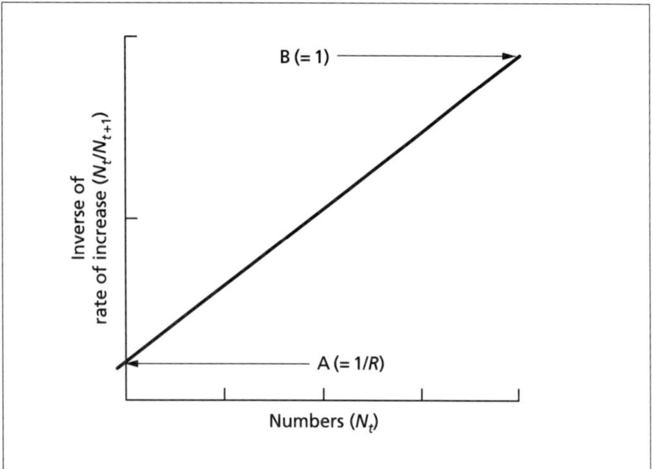

Fig. H3 Simple density-dependent relationship between the inverse of rate of increase (N_t/N_{t+1}) and numbers (N_t) in a population.

line assumes the simplest relationship possible and is not based on any empirical evidence. (Scientists, including ecologists, always develop their descriptions of the natural world by initially assuming the simplest relationships. Only when this proves inadequate do they make the explanation more complex (*see* AYALA'S EXPERIMENTS).)

We now have a relationship of the form: $y=a+bx$, where a is the y-axis intercept, and b is the slope of the line. In this particular case, $a=1/R$, $b=(1-(1/R))/K$, $y=N_t/N_{t+1}$ and $x=N_t$. Therefore the equation describing this relationship between the inverse of rate of increase (N_t/N_{t+1}) and density (N_t) is:

$$\frac{N_t}{N_{t+1}}=\frac{1}{R}+\left[\frac{1-(1/R)}{K}\right]N_t$$

Rearranging, this becomes:

$$N_{t+1}=\frac{N_tR}{1+[(R-1)/K]N_t}$$

and in most texts $(R-1)/K$ is replaced by the symbol a, giving the equation:

$$N_{t+1}=\frac{N_tR}{1+aN_t}$$

The reproductive rate is now $R/(1+aN_t)$ and is not a constant value but a variable value dependent upon density. This equation produces sigmoid population growth that reaches an equilibrium at $N=K$. It is in fact the difference form of the differential LOGISTIC EQUATION. Unfortunately, this equation only models CONTEST COMPETITION, but Hassell (1975) incorporated another parameter, b, that allows the full range of intraspecific competition, including SCRAMBLE COMPETITION, to be described. This final equation has the form:

$$N_{t+1}=\frac{N_tR}{(1+aN_t)^b}$$

where N_{t+1} and N_t are numbers in successive generations (or time intervals), R is the PER CAPITA net reproductive rate and a and b are constants. The parameter a ($=(R-1)/K$, where K is the equilibrium density or CARRYING CAPACITY) is related to the POPULATION SIZE at which DENSITY DEPENDENCE starts to act, and b describes the type of competition. At $b=1$, competition is contest with density dependence exactly compensating. With increasing values of b, competition becomes increasingly scramble with density dependence being overcompensating (Nicholson 1954; Hassell 1975). This type of difference equation has been shown to model insect competition very well (Gilpin & Ayala 1973; Stubbs 1977; Bellows 1981).

In the two-species model, there are two equations, one for species 1 and one for species 2. In the equation for species 1, N_t (now written $N_1(t)$) is replaced by $N_1(t)+\alpha_{12}N_2(t)$, and in the equation for species 2, N_t (now written $N_2(t)$) is replaced by $N_2(t)+\alpha_{21}N_1(t)$. The quantities α_{12} and α_{21} are COMPETITION COEFFICIENTS, and convert the numbers of one species into the equivalents of the other species (*see* LOTKA–VOLTERRA MODEL for more information on competition coefficients). The two equations describing competition between two species are therefore written:

$$N_1(t+1)=\frac{N_1(t)R_1}{\{1+a_1[N_1(t)+\alpha_{12}N_2(t)]\}^{b_1}}$$

$$N_2(t+1)=\frac{N_2(t)R_2}{\{1+a_2[N_2(t)+\alpha_{21}N_1(t)]\}^{b_2}}$$

These equations have been used to model the dynamics of insect populations in both the laboratory and field (*see* AGGREGATION MODEL OF COMPETITION). *See also* POPULATION DYNAMICS; POPULATION REGULATION. [B.S.]

REFERENCES

Bellows, T.S. (1981) The descriptive properties of some models for density dependence. *Journal of Animal Ecology* **50**, 139–156.

Gilpin, M.E. & Ayala, F.J. (1973) Global models of growth and competition. *Proceedings of the National Academy of Sciences of the USA* **70**, 3590–3593.

Hassell, M.P. (1975) Density dependence in single species populations. *Journal of Animal Ecology* **42**, 693–726.

Hassell, M.P. & Comins, H.N. (1976) Discrete time models for two-species competition. *Theoretical Population Biology* **9**, 202–221.

Maynard Smith, J. & Slatkin, M. (1973) The stability of predator–prey systems. *Ecology* **54**, 384–391.

Nicholson, A.J. (1954) An outline of the population dynamics of natural populations. *Australian Journal of Zoology* **2**, 9–65.

Stubbs, M. (1977) Density dependence in the life-cycles of animals and its importance in *K*- and *r*-strategies. *Journal of Animal Ecology* **46**, 677–688.

hazard A measure of the potential of a substance or process to cause HARM to human health and the environment. As far as chemicals are concerned, hazards are assessed from results of TOXICITY and ECOTOXICITY tests. This is sometimes referred to as HAZARD IDENTIFICATION. Cf. risk (*see* RISK ASSESSMENT). *See also* ECOTOXICOLOGY. [P.C.]

hazard function A statistical term for the age-specific, instantaneous rate of death in life-table studies. For the Deevey type II SURVIVORSHIP CURVE, where the log of the number of individuals in a COHORT born at time 0, still alive at time x, declines linearly with time, the age-specific risk of death is constant. In this case, the hazard is constant: $h(x)=c$, and the mean age at death (u) is related to the hazard by $u=1/c$, so that if the mean age at death is 10 years, then the hazard (per annum) is 0.1. Where the hazard changes with age (as in Deevey type I and III curves) then different statistical models are employed (e.g. Weibull distribution). *See also* LIFE TABLE. [M.J.C.]

hazard identification Assessment of potential to cause HARM from results of TOXICITY and/or ECOTOXICITY tests. *See also* HAZARD. [P.C.]

hazardous substance A substance that, due to its intrinsic properties, has the potential to cause HARM to human health and ecological systems given appropriate circumstances. *See also* HAZARD IDENTIFICATION. [P.C.]

hazardous waste *See* WASTE.

heat balance The balance of heat energy inputs and outputs that determines body TEMPERATURE. Inputs may be from irradiation and from exothermic metabolic reactions within the organism. Outputs are due to RADIATION, CONVECTION, conduction and EVAPORATION. *See also* ENERGY BUDGET. [P.C.]

heat-shock protein (hsp) Any of a subset of stress proteins, induced in what was originally referred to as a heat-shock response —because it occurred in response to exposure to elevated temperatures — but is now known to be induced by a variety of stressors; these include: HEAVY METALS; XENOBIOTICS; ANOXIA; SALINITY; and teratogens. Most are referred to by their apparent RELATIVE MOLECULAR MASS on polyacrylamide gels, for example hsp 90, hsp 70, hsp 58. They are used as BIOMARKERS of STRESS. *See also* METALLOTHIONEINS; MIXED-FUNCTION OXIDASE. [P.C.]

heat transfer Processes resulting in the transport of thermal energy (heat) within a system. In natural environments solar energy is the ultimate source of the great majority of heat, although geothermal energy is important locally. The laws of thermodynamics are fundamental to understanding heat transfer:

- the law of equilibrium (zeroth law) means that flows of heat energy require gradients of thermodynamic potential;
- the first law states that the total quantity of energy (enthalpy) must be conserved;
- the second law tells us that energy must always be degraded (entropy increased). (*See* ENERGY, LAWS OF THERMODYNAMICS.)

Within this framework, environmental heat transfer can be understood as the main determinant of the CLIMATES and MICROCLIMATES experienced by living organisms.

Heat transfer is often divided into two categories, though both are components of enthalpy transport.

1 Sensible, or dry heat transfer, which involves changes in TEMPERATURE. Sensible heat transfer takes place by three mechanisms:

(a) CONDUCTION — transport of heat through solids or fluids by molecular diffusion;

(b) CONVECTION — the transport of heat by moving WATER, air or bodily fluids;

(c) radiative heat transfer — due to the exchange of thermal RADIATION between bodies.

Only the last can take place without an intervening medium.

2 Latent heat transfer, which accompanies the absorption or release of heat energy when water changes state, between ICE and liquid water or between liquid water and water vapour. Latent heat transfer always involves convection, the bulk transport of water, water vapour or ice.

See also ENERGY BUDGET. [J.A.C.]

heathland This term is generally used to denote areas of land dominated by dwarf shrubs, usually of the family Ericaceae. Traditionally, it has been applied to the lowland regions of western Europe in which the heather, *Calluna vulgaris*, is the dominant plant, but its use has extended to related habitats of acidic soil and low vegetation; thus terms such as grass-heath and lichen-heath have evolved. The similar PHYSIOGNOMY, and sometimes shared species, of some Mediterranean vegetation types has led to the application of the term heath to these regions also.

Heathland is an anthropogenic landscape, produced as a result of forest clearance in north-west Europe, often in prehistoric times. It has been maintained by various management practices, including burning, GRAZING and cutting for thatch or bedding, or for turf. Upland heaths are termed 'moors'. [P.D.M.]

heavy-metal pollution CONTAMINATION by substances described as HEAVY METALS to an extent that HARM is caused to human health and/or ecological systems. The term 'heavy metal' has been used loosely in this context. *See also* TRIBUTYLTIN (TBT) POLLUTION. [P.C.]

REFERENCE

Depledge, M.H., Weeks J.M. & Bjerregard, P. (1994) Heavy metals. In: *Handbook of Ecotoxicology*, Vol. 2 (ed. P. Calow), pp. 79–104. Blackwell Scientific Publications, Oxford.

heavy metals A widely used but rather poorly defined term commonly adopted for the metals associated with environmental POLLUTION. It is often incorrectly taken to be synonymous with 'toxic metals' (such as the biologically non-essential metals LEAD (Pb), CADMIUM (Cd) and MERCURY (Hg)) and, even less satisfactorily, with 'trace metals' (i.e. metals present at concentrations of less than 100 mg kg^{-1} in the Earth's CRUST, which include those, such as ZINC and COPPER, that are biologically essential).

One definition of the term 'heavy metals' is that it covers the block of ELEMENTS in the PERIODIC TABLE bounded at its corners by TITANIUM (Ti, period 4, group 4), HAFNIUM (Hf, period 6, group 4), ARSENIC (As, period 4, group 15) and BISMUTH (Bi, period 6, group 15), with relative densities (the loose origin of the term) ranging from 4.5 g cm^{-3} for titanium to 22.5 g cm^{-3} for OSMIUM. This enables the inclusion of environmentally and toxicologically important metalloids like arsenic and SELENIUM, and excludes metals which are certainly not 'heavy', on the basis of relative density, like ALUMINIUM and BERYLLIUM, but are associated with toxic effects. The major drawback of the term 'heavy metals', however, is that it encompasses a heterogeneous array of elements with diverse chemical and biological properties.

It is particularly important to distinguish those elements which are unequivocally essential, and can give rise to deficiency problems, in animals (including, most probably, humans) — for example COBALT (Co), CHROMIUM (Cr), copper (Cu), IRON (Fe), MANGANESE (Mn), selenium (Se) and zinc (Zn) — and plants — for example Cu, Fe, Mn, MOLYBDENUM (Mo) and Zn — from those which are non-essential—for example cadmium, mercury and lead. All elements are potentially toxic in high concentrations, although additional major controlling factors of TOXICITY (and benefit) are duration of EXPOSURE, chemical form, route of absorption, SYNERGISM or ANTAGONISM of other agents, and biological individuality.

The most important anthropogenic releases of heavy metals to air, water and soils are from mining, agriculture, fossil-fuel combustion, metallurgical, electronic and chemical industries, and WASTE DISPOSAL. [J.F.]

REFERENCES

Alloway, B.J. (ed.) (1995) *Heavy Metals in Soils*, 2nd edn. Blackie, Glasgow.

Fergusson, J.E. (1990) *The Heavy Elements. Chemistry, Environmental Impact and Health Effects*. Pergamon Press, Oxford.

Nriagu, J.O. (1990) Global metal pollution: poisoning the biosphere? *Environment* **32**, 7–33.

hedgerow A row of shrubs and saplings forming a boundary or barrier; a type of managed natural fencing. Hedgerows act as harbours for natural vegetation and animal life. They have been subject to widespread destruction as a consequence of the mechanization of AGRICULTURE—leading to ecological impoverishment of the English countryside. [P.C.]

HEDSET *See* HARMONIZED ELECTRONIC DATASET.

helium (He) A gaseous element (ATOMIC NUMBER = 2; relative ATOMIC MASS = 4.0026; melting point = −272.2°C; boiling point = −268.9°C) of the noble gas group. Helium is present in the ATMOSPHERE at a mixing ratio of 5.24×10^{-6} (5.24 ppm by volume (ppmv)). The ISOTOPE ^{4}He is almost a million times more abundant than ^{3}He and is produced in the radioactive decay of heavier ELEMENTS. Although abundant in some natural gas reservoirs, most helium released to the atmosphere has escaped to space because of the low mass. The element is physically and chemically inert and is used for inert atmospheres, as a cryogenic coolant and as a non-flammable gas for balloons. [M.R.H.]

helping at the nest *See* COOPERATIVE BREEDING, EVOLUTION OF.

hemicryptophyte *See* PHYSIOGNOMY; RAUNKIAER'S LIFE FORMS.

hemimetabolous insects *See* INSECTS, DIVERSITY AND CONSERVATION.

hemiparasite

1 In mycology, a fungal hemiparasite (more usually described as a FACULTATIVE parasite) is a fungus which can grow in or on a living HOST from which it obtains its essential energy/NUTRIENTS, whilst having the optional ability to obtain these from non-living sources (and thus it can be cultured on laboratory media). This contrasts with OBLIGATE parasitic fungi, which are biotrophic and must obtain their necessary energy/nutrients from living host cells. Necrotrophic parasitic fungi derive their food energy from damaged or dead cells/tissues of the host plant, whilst secreting extracellular TOXINS/enzymes that kill adjacent living host cells/tissues. Hence, these also are facultative parasites.

2 In botany, angiosperm hemiparasites are photosynthetic plants that supplement their nutrition by being parasitic upon another plant; for example the Loranthaceae (mistletoes) and some Scrophulariaceae (e.g. *Euphrasia*).

See also PARASITISM; SPECIES INTERACTIONS. [J.B.H.]

hemispherical photography Photography using hemispherical, or 'fish-eye' lenses, which are wide-angle lenses giving a 180° field of view, originally developed for studies of CLOUDS. Hemispherical photography is used in biology to characterize the structure of plant canopies and the subcanopy light environment. Photographic images are recorded beneath plant canopies viewing upwards, or from above viewing downwards. The circular image obtained from a hemispherical lens is essentially a projection of the hemisphere on to a plane. The most common projection is equi-angular with distance from the centre of the image being proportional to SOLAR ALTITUDE. Digital image analysis of hemispherical photographs has now largely replaced previous manual methods of interpretation. CANOPY structure is described in terms of gap fraction, LEAF AREA INDEX and leaf angle distribution. The subcanopy light environment can be described in terms of estimates of direct and diffuse RADIATION. Estimates of the direct and indirect site factors derived from hemispherical photography can be correlated with parameters describing plant growth. Hemispherical photographs also provide a permanent non-destructive record of plant canopy structure. [P.R.V.G.]

REFERENCE

Rich, P.M. (1990) Characterising plant canopies using hemispherical photographs. In: *Instrumentation for Studying Vegetation Canopies for Remote Sensing in Optical and Thermal Infrared Regions. Remote Sensing Reviews*, Vol. 5 (eds N.S. Goel & J.M. Norman), pp. 13–29. Harwood Academic Publishers, London.

Hennigian systematist *See* HOMOLOGY.

Hennig's inclusion/exclusion rule *See* PHYLOGENETIC INFERENCE.

Her Majesty's Industrial Pollution Inspectorate (HMIPI) *See* ENVIRONMENTAL (PROTECTION) AGENCIES.

Her Majesty's Inspectorate of Pollution (HMIP) *See* ENVIRONMENTAL (PROTECTION) AGENCIES.

herbaria Collections of dried, pressed plants, mounted on sheets, labelled and stored for reference. All plant groups may be represented, most commonly higher plants and pteridophytes, but also bryophytes and lichens. Spirit-preserved specimens of fleshy structures may also be included, and boxes of bulky items (e.g. cones, fruits). Labels ideally give the plant name, collection locality, habitat, date, name of collector and collection number, and sometimes other details (but old collections are often deficient). Storage, in alphabetical or systematic order, sometimes combined with geography, is preferably in metal cabinets; curation involves protection against insect and fungal attack. Methods of plant drying have been developed to give the best possible quality of specimens (Bridson & Forman 1992) and, recently, to preserve colour as well. Herbaria, which may be worldwide or local, serve mainly for identification of specimens; preservation of voucher (representative) material (notably type specimens); for research, including preparation of

Floras (dry specimens yield, e.g. morphological, anatomical and chemical data); and for education. Over 2600 herbaria are listed in *Index Herbariorum* (Holmgren *et al.* 1990); 5 million or more specimens are held in herbaria at the Royal Botanic Gardens, Kew; the Natural History Museum, London; and at Paris, St Petersburg (Leningrad), Stockholm, New York and Geneva. Kent and Allen (1984) list locations of herbaria in the British Isles. *See also* ARBORETUM. [A.J.W.]

REFERENCES

Bridson, D. & Forman, N. (1992) *The Herbarium Handbook.* Royal Botanic Gardens, Kew.

Holmgren, P.K., Holmgren, N.H. & Barnett, L.C. (eds) (1990) *Index Herbariorum. Part I: The Herbaria of the World,* 8th edn. (*Regnum Vegetabile,* Vol. 120). New York Botanical Garden, New York.

Kent, D.H. & Allen, D.E. (1984) *British and Irish Herbaria,* 2nd edn. Botanical Society of the British Isles, London.

herbicide A BIOCIDE that kills weeds, i.e. unwanted plants. Herbicides can operate selectively or non-selectively. They are sometimes known as DEFOLIANTS. [P.C.]

herbivore Any consumer of living plant materials. *See also* GRAZER; GRAZING; PHYTOPHAGOUS. [P.C.]

heritability A population-dependent parameter, that measures how much of the observed phenotypic variation in a quantitative character is influenced by genetic variation. Heritability, however, says nothing about the mode of inheritance of the QUANTITATIVE TRAIT.

The BROAD-SENSE HERITABILITY measures what proportion of the total variation is genetic, whether or not this variation is heritable, i.e. whether or not the variations are transmitted to offspring. It is equal to V_G/V_P, where $V_G=V_A+V_D+V_I$ is the total GENETIC VARIANCE, V_D and V_I being the dominance and interaction variances respectively.

The NARROW-SENSE HERITABILITY provides an estimate of the similarity of progeny to their parents. It thus allows us to determine the response to selection. It is equal to V_A/V_P, where V_A is the ADDITIVE GENETIC VARIANCE, and V_P is the total PHENOTYPIC VARIANCE.

To estimate heritabilities, it is necessary to design experiments in which the environment is controlled, so that all individuals are measured in the same environmental conditions. Narrow-sense heritability is estimated by studying parent–offspring or half-sib COVARIANCES. Broad-sense heritability is the only parameter that can be estimated by rearing replicates of CLONES, for instance.

One challenge of LIFE-HISTORY EVOLUTION is to explain the fact that FITNESS components usually show quite large heritabilities in natural populations (Stearns 1992). Usually, however, fitness-related traits show lower heritabilities than fitness-independent traits (Falconer 1981), either because they have a lower genetic variance, or because, being complex characters determined by many genes of small effects with several interactions, their expression is more sensitive to small environmental variations. *See also* QUANTITATIVE GENETICS. [I.O.]

REFERENCES

Falconer, D.S. (1981) *Introduction to Quantitative Genetics,* 2nd edn. Longman, London.

Stearns, S.C. (1992) *The Evolution of Life Histories.* Oxford University Press, Oxford.

herkogamy *See* POLLINATION.

hermaphrodite A hermaphrodite animal is one that produces both EGGS and sperm. A hermaphrodite plant is one whose flowers produce both ovules and pollen (in contrast to a monoecious plant, which produces separate male and female flowers on the same individual; *see* MONOECY). However, following Charnov *et al.* (1976), we may call hermaphrodite (broad sense) any individual in which both sexes are combined. Thus, gynomonoecious plants (which produce separate female and hermaphrodite flowers on the same plant), as well as andromonoecious plants (which produce separate male and hermaphrodite flowers on the same plant) are both hermaphrodites. [I.O.]

REFERENCE

Charnov, E.L., Maynard-Smith, J. & Bull, J.J. (1976) Why be an hermaphrodite? *Nature* **263**, 125–126.

hermaphroditism A condition in which a single organism is both male and female. It is the opposite of gonochorism, in which there are separate males and females. In SIMULTANEOUS HERMAPHRODITISM the organism is both male and female at the same time. In SEQUENTIAL HERMAPHRODITISM there is change of sex. In protandrous hermaphroditism (*see* PROTANDRY) the change is from male to female, whereas in protogynous hermaphroditism (*see* PROTOGYNY) the change is from female to male. In alternating hermaphroditism there are repeated changes of sex. The term 'bisexual' should be avoided, for it can mean either that the organisms are HERMAPHRODITES or that they occur as sexual populations with males as well as females.

The botanical terminology is best restricted to plants and to colonial animals. In hermaphroditic plants all of the flowers on every organism are both male and female. In monoecious plants each organism has both male flowers and female ones (*see* MONOECY). In dioecious plants each organism has exclusively male or exclusively female flowers, not both (*see* DIOECY).

The adaptive significance of hermaphroditism can be explained in evolutionary terms by means of 'sex allocation theory', which treats the organisms as producing an economic 'optimum' with respect to investment in male and female functions. Older notions that hermaphroditism is somehow 'primitive' or 'advantageous to the species' have been rejected. Gonochorism is advantageous because it is more expensive for an organism to maintain both male and female organs. Simultaneous hermaphroditism is, however, advantageous when mates are hard to find (low-density model), and it is common among SESSILE or slow-moving animals as well as plants. Hermaphrodites may fertilize themselves, but this generally happens only if mates are not available. Sequential hermaphroditism is favoured where there is size-dependent or age-dependent reproductive success (size-advantage model). Protogyny, for example, has evolved in some animals in which large males have an advantage in combat that monopolizes the females. Protandry has evolved for similar reasons, for example where a large female enjoys an advantage in caring for offspring. *See also* EVOLUTION OF SEX; LIFE-HISTORY EVOLUTION; SEX ALLOCATION THEORY. [M.T.G.]

REFERENCES

Charnov, E.L. (1982) *The Theory of Sex-Allocation.* Princeton University Press, Princeton, N.J.

Ghiselin, M.T. (1969) The evolution of hermaphroditism among animals. *Quarterly Review of Biology* **44**, 189–208.
Ghiselin, M.T. (1987) Evolutionary aspects of marine invertebrate reproduction. In: *Reproduction of Marine Invertebrates*, Vol. 9 (eds A.C. Giese, J.S. Pearse & V.B. Pearse), pp. 609–665. Blackwell Scientific Publications, Palo Alto, Calif.

Hess samplers *See* SAMPLING METHODOLOGY/DEVICES.

heterochromatization *See* PSEUDO-ARRHENOTOKY.

heterochrony The 'process' of evolutionary change through changes in the timing and rates of development within a conserved ontogenetic sequence (also the scientific discipline that studies such phenomena). The concept can be extended to various levels of biological complexity. In consequence, beside the traditional comparative studies of morphological change through ONTOGENY and PHYLOGENY, heterochrony is currently also invoked to explain evolution in such diverse systems as behaviour and SOCIAL ORGANIZATION, the developmental role of genes that control timing of developmental events (e.g. *Lin-14*; Ambros 1988), or, among many others, the changes in shape due to differential growth rates — a subject of distinguished tradition known as ALLOMETRY (therefore, allometry should be considered a special case of heterochrony).

Historically, the term has been used in several different ways, since being coined by the German scholar Ernst Haeckel in 1866, in the context of his biogenetic law (Gould 1977). The original definition, quite distinct from the modern one, referred to changes in the timing of the development of an organ relative to others within the same organism (e.g. early formation of the heart in vertebrates). The noted British embryologist, Sir Gavin de Beer altered the definition of heterochrony in 1930 to describe any change in timing of a feature relative to the same feature in the ancestral ontogeny. Therefore, the term acquired an evolutionary connotation rather than the original intra-organismal one. Gould (1977) modified de Beer's cumbersome scheme of heterochronic processes, proposing the current, more mechanistic (process-oriented) definition of heterochrony. Gould's so-called 'clock model' was further modified into a more dynamic formalism by Alberch *et al.* (1979).

Besides a methodology that has become established as the basis of most current studies of heterochrony, the Alberch *et al.* scheme allowed for the derivation of a logical nomenclature of heterochronic processes and patterns. This current methodology divides heterochronic processes into the following major categories: PROGENESIS and NEOTENY generate a paedomorphic morphology (*see* PAEDOMORPHOSIS) while HYPERMORPHOSIS and ACCELERATION cause the appearance of a morphology characterized by PERAMORPHOSIS.

Heterochronic analysis assumes that ontogenetic sequences are conserved through evolutionary time. Evolution must proceed mostly by terminal addition and/or deletion. The ubiquity of NON-TERMINAL ADDITION, the result of events known as 'dissociation'— in the case of changes in the sequence of ontogenetic events—and 'caenogenesis' — when new embryonic stages are inserted into an ontogeny — precludes the comparison among ontogenies. *See also* LIFE-HISTORY EVOLUTION; RECAPITULATION. [P.A.]

REFERENCES

Alberch, P., Gould, S.J., Oster, G.F. & Wake, D.B. (1979) Size and shape in ontogeny and phylogeny. *Paleobiology* **5**, 296–317.
Ambros, V. (1988) Genetic basis of heterochronic variation. In: *Heterochrony in Evolution* (ed. M.L. McKinney). Plenum Press, New York.
Gould, S.J. (1977) *Ontogeny and Phylogeny.* Harvard University Press. Cambridge, Mass.
McKinney, M.L. & McNamara, K.J. (1991) *Heterochrony: The Evolution of Ontogeny.* Plenum Press, New York.

heterogeneity

1 The spatial DISTRIBUTION of RESOURCES or some other feature of the ENVIRONMENT of an organism. Thus if an environment is a patchwork of different kinds of resource, this is often described as heterogeneity. Similarly, the physical structure of the environment is said to show heterogeneity if it is diverse or unpredictable from one location to another. This sort of heterogeneity is relative and will differ between organisms. *See also* COARSE-GRAINED HABITATS; FINE-GRAINED HABITATS.

2 Another name for species DIVERSITY. This has two components: the number of species, or SPECIES RICHNESS; and the relative abundances of different species. Heterogeneity is higher in communities with more species and with similar abundances of different species. Several measures of heterogeneity are used in ecology (see Magurran 1988).

[R.H.S.]

REFERENCE

Magurran, A.E. (1988) *Ecological Diversity and its Measurement.* Croom Helm, London.

heterogony *See* ALTERNATION OF GENERATIONS.

heteromorphic *See* ALTERNATION OF GENERATIONS.

heterosis HYBRID VIGOUR such that an F_1 hybrid falls outside the range of the parents with respect to some character or characters (Allard 1960). It is usually applied to size or growth rate. The opposite of heterosis is INBREEDING DEPRESSION, which may be defined as the loss of vigour of inbred offspring compared to the parents. Heterosis may arise due to DOMINANCE of favourable ALLELES and/or overdominance. Overdominance is the situation in which heterozygotes at a given locus have a PHENOTYPE outside the range of the parents. Overdominance is the only explanation for heterosis involving a single locus. Dominance is the situation in which heterozygotes at a given locus have the same phenotype as one of the two homozygotes. At least two loci must be involved for dominance to explain such heterosis. For instance, assume that loci *A* and *B* contribute to vigour, with parent 1 of GENOTYPE *AAbb* and parent 2 of genotype *aaBB*. Further assume that each locus contributes 0 or 1 to the vigour, depending on whether it does or does not have the dominant allele, *A* or *B*. Thus, both P1 and P2 have a value of 1. The hybrid, with genotype *AaBb*, has a value of 2. In this example, it would be very difficult to distinguish between dominance and overdominance. [I.O.]

REFERENCE

Allard, R.W. (1960) *Principles of Plant Breeding.* John Wiley & Sons, New York.

heterospecific An individual of a different species. [B.S.]

heterostylous *See* POLLINATION.

heterotroph An organism requiring a supply of organic material from its ENVIRONMENT as food. Heterotrophic organisms cannot obtain their necessary energy input solely from sunlight as autotrophic organisms do (*see* AUTOTROPH), or from inorganic chemical reactions as chemotrophic organisms do, but rely on an uptake of the potential chemical energy in organic materials synthesized by other organisms. HERBIVORES, carnivores, PARASITES and decomposers (saprophytic organisms) are all heterotrophs. All animals and fungi, most bacteria and a few flowering plants are strongly heterotrophic, requiring organic food substances to make most of their own organic constituents. *See also* FEEDING TYPES, CLASSIFICATION. [M.H.]

heterozygosity The heterozygosity, *H*, of an average individual in a population estimates the probability that any individual is heterozygous at any of a specified set of loci. It is calculated as the average frequency of heterozygotes over loci. Heterozygosity differs between populations, species and taxonomic groups. In a number of populations and species, highly heterozygous individuals grow faster. This points to a positive relation between heterozygosity and FITNESS, but not necessarily to heterozygote advantage at each locus *per se*. Heterozygosity can be used to quantify population subdivision. Population subdivision leads to lower heterozygosity: the increase in HOMOZYGOSITY can be compared to inbreeding, and estimated by computing Wright's *F*-STATISTICS. *See also* CONSERVATION BIOLOGY; POPULATION GENETICS. [G.D.J.]

heterozygosity, interpretation for conservation HETEROZYGOSITY is the state of having one or more pairs of non-identical ALLELES, and a heterozygote is an organism having two different alleles at a locus. Together with POLYMORPHISM, which is the percentage of loci in a given species for which more than one allele is observed, heterozygosity is used to quantify the amount of genetic variation in a species. Heterozygous refers to the state of a GENOTYPE at a polymorphic locus, when two alleles occur together, one on each homologous chromosome. Among vertebrates the most polymorphic loci yet known are those comprising the major histocompatibility complex, for which heterozygosity may approach 100% in outbred populations.

The average degree of heterozygosity of a population is known to affect its average FECUNDITY and survivorship. There is evidence that individuals within a population that has recently lost some of its heterozygosity are less fit on average than individuals of the same species within populations that have not suffered a recent loss of heterozygosity (possibly due to selection against individuals with recessive semi-lethals exposed by the increased HOMOZYGOSITY). In small populations, the heterozygosity declines due to inbreeding and GENETIC DRIFT and, in the absence of NATURAL SELECTION for rare genes, only slowly builds back up by MUTATION and drift. The reduced FITNESS of inbred populations (INBREEDING DEPRESSION) may lead to a positive feedback loop between the size of the population and the average fitness of its members that can accelerate the population's trend towards extinction. Successful conservation of a population depends in part on maintaining a minimum POPULATION SIZE at which genetic VARIABILITY (i.e. heterozygosity) is maintained above some critical level. *See also* CONSERVATION BIOLOGY. [V.F.]

hibernation A state of greatly reduced metabolic activity employed by certain mammals as a way of surviving adverse winter conditions. The animal enters a deep sleep, surviving on its body resources. *Cf.* AESTIVATION; DORMANCY. *See also* OVERWINTERING. [P.C.]

hierarchy theory Hierarchy theory provides a framework for thinking about and understanding the interactions between processes which operate on different scales. In an ecological context the theory views the spatial and temporal processes in nature as continuous, but with small-scale fast processes at the bottom and large-scale slow processes at the top. It is argued that the relative intensities of interactions between levels can be used to decompose the system into reasonably discrete semi-autonomous parts, or 'holons', which can be viewed within an organizational hierarchy. This hierarchical view places emphasis on the organizational properties which emerge at higher levels in the hierarchy as a consequence of processes operating at other levels. It is argued that, in the same way that the properties of a working clock could not simply be thought of as the sum of its constituent parts, so the EMERGENT PROPERTIES at various levels of ecological organization cannot be derived from studies on the lower-order systems in isolation. *See also* SYSTEMS ECOLOGY. [S.J.H.]

REFERENCE

Allen, T.F.H. & Starr, T.B. (1982) *Hierarchy: Perspectives for Ecological Complexity.* University of Chicago Press, Chicago, Ill.

histic horizons *See* SOIL CLASSIFICATION.

historical effects (community assembly) Communities are not created instantaneously as complete entities but develop, or assemble, from much humbler beginnings. This history is important because the principles which govern the progressive ESTABLISHMENT of species in a system may explain some of the properties we observe. Increasing attention is now being given to understanding these historical effects and determining the rules which govern the processes of COMMUNITY assembly.

The importance of the order of arrival of species in a system is now well recognized. For example, predators are more likely to be absent from islands, because their prey would have to arrive first. Moreover, the more specialist the predator the lower the likelihood that it could establish — when a new plant colonizes an area, it is more likely to be attacked by POLYPHAGOUS insect species which feed externally on the plant than by more specialist leaf miners and gall-formers. The presence of a GENERALIST predator can also have effects by reducing the likelihood of establishment by a new prey species. A good example of this latter effect occurred in experimental studies on small Bahamian islands where efforts to introduce spiders were shown to be less likely to succeed when predatory lizards were present. *See also* COMMUNITY ECOLOGY; ISLAND BIOGEOGRAPHY. [S.J.H.]

REFERENCE

Drake, J.A. (1990) Communities as assembled structures: do rules govern pattern? *Trends in Ecology and Evolution* **5**, 159–163.

history of ecological restoration ECOLOGICAL RESTORATION has deep roots in more generally conceived forms of environmental management such as FORESTRY, game management and certain forms of AGRICULTURE and landscape design. Restoration in the fully modern sense, however, has been a comparatively recent development, which has reflected the application of the principles of the relatively young science of ECOLOGY to traditional forms of rehabilitative environmental management. In some cases this resulted in a form of rehabilitative management in which objectives were defined in ecological and historical rather than in utilitarian or aesthetic terms, and the emphasis was on creating a system that resembles the original, 'model' system as closely as possible. It is this form of management that is properly referred to as ecological restoration. Some of the earliest projects to reflect this concern for ecological and historical authenticity were attempts beginning in the 1920s and 1930s to re-create examples of native ecosystems at Vassar College in Poughkeepsie, New York, and at the University of Wisconsin Aboretum in Madison. The early work at Wisconsin was followed by increasingly ambitious PRAIRIE restoration efforts in Illinois and elsewhere in the American Midwest, with the result that today the tallgrass prairies are widely regarded as the birthplace of modern restoration, and the best work of prairie restorationists is widely regarded as a model for the practice of restoration in its purest and most environmentally sensitive form.

Restoration as an active attempt actually to reinstate the species and processes characteristic of a historical (or 'model') ecological system is a distinctively American, or New World conception of the more general task of environmental healing or rehabilitation, which clearly reflects certain aspects of American experience and character, including the recent, highly self-conscious experience of a landscape conceived as wild or primitive; the rapid, purposeful transformation or DOMESTICATION of that landscape, in many cases within living memory; and also a tendency to idealize the pre-contact landscape as an unspoiled Eden and a repository of innocence and value.

In other parts of the world, in Europe, Asia and Africa, there has also been growing concern about the condition of the environment during the past few decades, and the development of practices to reverse or otherwise ameliorate various forms of environmental damage. In these areas, however, the emphasis has generally been on rehabilitation or ecological healing in a more general sense, and goals are commonly defined abstractly, in terms of the health, integrity or functioning of the ECOSYSTEM, rather than concretely and historically in terms of how closely it resembles a particular historical landscape with respect not only to integrity and functioning, but also to species composition, structure and dynamics. Despite this it is becoming clear that restoration in the strict sense does have a significant role to play in the conservation of natural ecosystems in virtually all parts of the world. *See also* REMEDIATION. [W.R.J.]

Histosols *See* SOIL CLASSIFICATION.

HMIP (Her Majesty's Inspectorate of Pollution) *See* ENVIRONMENTAL (PROTECTION) AGENCIES.

HMIPI (Her Majesty's Industrial Pollution Inspectorate) *See* ENVIRONMENTAL (PROTECTION) AGENCIES.

Holarctic realm A biogeographic subdivision comprising the Palaearctic and Nearctic regions, i.e. mid- and high-latitude Eurasia and North America north of tropical Mexico. *See also* NEARCTIC FAUNAL REGION; PALAEARCTIC FAUNAL REGION. [A.H.]

holism The doctrine that whole systems cannot fully be understood in terms of the properties of their parts and the laws governing them: for example, populations cannot be understood solely in terms of their constituent individuals, nor communities in terms of their populations. There are important EMERGENT PROPERTIES that arise out of interaction of parts that have to be taken into account in explaining the properties and behaviour of whole systems and the laws governing them. Thus POPULATION DYNAMICS have to take account of birth rates and death rates, which may depend upon social encounters and interactions with predators, parasitoids and competitors, while COMMUNITY/ECOSYSTEM processes (e.g. energy fluxes) depend upon trophic interactions between species, none of which may be deducible from the study of the parts in isolation.

These whole coordinated and integrated systems are sometimes likened to organisms or superorganisms; this is referred to as organismic philosophy. The alternative view is that the whole can be understood completely in terms of its parts; this is referred to as REDUCTIONISM. Sometimes the latter is taken further to liken biological systems to machines; a position known as mechanism. Emergent properties have sometimes been portrayed as mysterious vital forces, and this philosophy is known as vitalism. *See also* SYSTEMS ECOLOGY. [P.C.]

Holling's disc equation A behavioural model of foraging that assumes the consumer encounters prey at a constant rate and requires a fixed HANDLING TIME to process each prey item. It predicts a type 2 FUNCTIONAL RESPONSE. The equation was originally parameterized for blindfolded human predators SEARCHING for discs of sandpaper that required a fixed handling time; hence the name. *See also* FEEDING; OPTIMAL FORAGING THEORY; PREDATOR–PREY INTERACTIONS. [H.C.J.G.]

holmium (Ho) A lanthanide metallic element (ATOMIC NUMBER = 67; relative ATOMIC MASS = 164.9; melting point = 1474°C; boiling point = 2695°C) with an average crustal abundance of 1.2 mg kg^{-1}. [J.G.]

holometabolous insects *See* INSECTS, DIVERSITY AND CONSERVATION.

holomictic lakes *See* LAKE STRATIFICATION.

holoplankton Organisms that spend all their life in the PLANKTON. *See also* PELAGIC. [V.F.]

home range Individuals of most animal species spend their lives within a circumscribed area which can be thought of as their home range. Empirical determination of home ranges usually involves following an animal about and plotting its movements or points at which it is sighted on a map, then joining the outermost points to form the minimum convex polygon. Of course, such estimates are heavily dependent on how long or often the animal is observed. Nevertheless, between-species comparative studies have been reasonably successful in predicting variation in home range size from life-history parameters such as body size, metabolic requirements and foraging strategy. Home range has a temporal as well as a spatial dimension, often changing on a seasonal basis (some migratory species inhabit different geographical regions at different times of year) or with shifts in resource distribution.

Home ranges may have considerable internal structure. Activity is often concentrated in core areas which contain high-quality feeding sites, resting places, water holes and so on, and much of the remaining home range may be used simply to travel from one core area to another. Parts of the home range may even be defended as a territory (*see* TERRITORIALITY) by particular individuals or groups.

Although home ranges may simply be a product of resource distribution, animals may benefit from the increased familiarity with an area that results from a restricted range of activity, particularly in terms of foraging efficiency and escaping from predators. *See also* BEHAVIOURAL ECOLOGY. [C.J.B.]

homeorhesis The maintenance of a pattern of change — as in development—despite environmental disturbances. It is equivalent to HOMEOSTASIS but is concerned with a dynamic rather than a static condition. *See also* CANALIZATION. [P.C.]

homeostasis The tendency of a system (organism, POPULATION, COMMUNITY or ECOSYSTEM) to change its properties in a way that minimizes the impact of outside factors. Homeostasis is shown at the level of the community or ecosystem where early-successional plants quickly restore the canopy COVER in gaps caused by DISTURBANCE ('scab-formers') and minimize potentially degenerative effects such as soil erosion. *See also* GAIA. [P.J.G.]

homeotherm An organism that maintains an approximately constant body TEMPERATURE as environmental temperature varies by the REGULATION of physiological processes, for example shivering and sweating (in ENDOTHERMS), or by appropriate behaviour, for example exposure to solar radiation (mostly in ECTOTHERMS). *Cf.* POIKILOTHERM. [P.O.]

Homeotic mutations

Historically, the name of this class of genes comes from William Bateson, in his classical 1894 study 'Materials for the Study of Variation', where he identified a particular class of variation that he defined as 'homoeosis': a phenomenon where the variant structure is not merely altered, but rather is transformed 'into the likeness of something else'. These peculiar transformations, already known as a special class of MUTATIONS in the classical Mendelian genetics of the early 20th century, were thoroughly studied in *Drosophila*, where the most detailed work has been done.

Initially, numerous homeotic mutations were described mainly on the basis of their morphological effect; for example the mutation causing the expression of a structure, identifiable as a leg, in the place where an antenna should have developed. Usually, but not invariably, these mutants are associated with segmentation or serial repetition in body parts. Unlike most mutations that tend to generate malformation or non-development of the affected PHENOTYPE, homeotic mutations induce the development of a fairly complete feature in the wrong region of the body of an organism. Thus, homeotic mutants appear to affect global pattern determination rather than morphogenesis. Such non-random transformation of one body part, or segment, into another, suggests that these genes play a regulatory role at the level of controlling the expression of other genes.

More recently, molecular studies have shown that all homeotic genes share a short, 180 base-pair motif of highly conserved DNA sequence (the homeobox region). This motif was initially found in both homeotic, as well as segmentation, genes in *Drosophila*. Given the pattern of spatial specification, and the spectacular effects of these genes during development, it was postulated that the homeo domain was essential for the presumed regulatory function of these genes. Further studies of the molecular structure of this region have shown that the homeobox encodes a domain of about 60 AMINO ACIDS, shown to be specific of protein-binding PROTEINS. Therefore, homeobox-containing genes (*Hox* genes) probably act as transcription regulators during development.

Comparative analysis of sequences have unequivocally shown that homeotic genes are not just specific to arthropods, since homologues have been found in most phyla, such as echinoderms, annelids and molluscs as well as in vertebrates. In these other taxa their mutations are not necessarily as dramatic in terms of phenotypic transformation as in insects.

Historically, homeotic mutants were invoked by some authors, like R. Goldschmidt, as evidence in favour of the discontinuous nature of evolution (*see* MACROMUTATION). Lewis (1978) executed a series of now classical genetic analyses to map the so-called bithorax complex, a group of genes involved in the determination of the posterior segments of the body of the fruit fly. Garcia-Bellido (1993) further extended the study to propose that segment determination in insects is the result of the interactions within a hierarchical regulatory network where homeotic genes act as top 'switches', activating a cascade of other genes involved in the execution of specific developmental pathways. The evolutionary diversification of *Hox* genes, as well as the regulatory nature of their interactions, indicate that gene duplication has played a major role in the evolution of homeotic gene complexes. *See also* HOPEFUL MONSTERS. [P.A.L.]

REFERENCES

Akam, M., Dawson, I. & Tear, G. (1988) Homeotic genes and the control of segment diversity. *Development* **104**, 123–133.

Carroll, S.B. (1994) Developmental regulatory mechanisms in the evolution of insect diversity. *Development* (Suppl.), 217–223.

Garcia-Bellido, A. (1993) A coming of age. *Trends in Genetics* **100**, 102–103.

Lewis, E.B.A. (1978) Gene complex controlling segmentation in *Drosophila*. *Nature* **276**, 565–570.

Ruddle, F.M., Bortels, J.L., Bentley, K.L. *et al.* (1994) Evolution of *Hox* genes. *Annual Review of Genetics* **28**, 423–442.

homing The ability of some animals to navigate over long distances to find their way back to a home area (e.g. a nest or breeding site). There are two elements of homing behaviour: (i) orientation towards the general direction of the home area while travelling over unfamiliar ground; and (ii) recognition of familiar features of the home area. Long-distance navigation is known in several animal taxa (e.g. insects, fishes, reptiles and birds). Birds use the Earth's magnetic field and the positions of the Sun and the stars for orientation. Insects, fishes and some other groups navigate by olfaction, and many other cues (e.g. sound, light polarization, carbon dioxide concentration) have been shown in different species to be used in navigation. *See also* MIGRATION. [R.H.S.]

REFERENCE
Baker, R.R. (1982) *Migration: Paths Through Time and Space*. Hodder & Stoughton, London.

homogamy *See* ASSORTATIVE MATING.

Homology

It has been recognized for a long time that most of the TRAITS exhibited by each species also occur in at least one other species. Since the advent of evolutionary theory, two general types of such interspecific similarity have been recognized. Either the species share the traits because they have inherited them from a common ancestor that possessed the trait, or all the species in question acquired the trait independently. When the explanation is common ancestry, we say that the traits are homologous. Thus, a homology is any character that is transmitted genetically from an ancestral species to its descendant species.

Homology and transformation

A homologous character may be transmitted relatively unaltered from ancestor to descendant. For example, the webbed feet of ducks and geese are highly similar. A homologous character may also change during the course of transmission from ancestors to descendants. For example, the forelimbs of shrews, bats, birds and lizards are considered evolutionary modifications of a common ancestral tetrapod forelimb, thus forming a homologous series, or TRANSFORMATION SERIES. This second aspect emphasizes that homology does not imply character STASIS, but is compatible with evolutionary change.

Homologous or not?

Although the definition of homology is not particularly problematical for modern biologists, the means of identifying homology have been hotly contested for decades. Because the definition of homology involves common ancestry, homologous traits shared by two or more species, by definition, indicate phylogenetic relationships among those species. Non-homologous traits shared by two or more species thus, similarly, suggest relationships that are not indicative of PHYLOGENY. If, as Darwin suggested, the goal of TAXONOMY should be the production of CLASSIFICATIONS that reflect phylogeny, those classifications should be based on homology. For nearly a century following the publication of *Origin of Species*, systematic biologists developed empirical criteria for comparing similar traits among species and determining homology. Their initial expectation was that such homology assessments would lead all systematists to the same set of relationships among species. When they did not, systematists fell to disputing each others' homologies on the basis of preconceived notions of phylogeny, exposing an inherent circularity in the research programme.

The German dipterist Willi Hennig provided a solution to this problem based upon two points.

1 His recognition that all homologies covary because they are products of a single evolutionary history.

2 His belief that phylogenetic influences in EVOLUTION are so pronounced that homologous characters will outnumber any covarying homoplasious characters (*see* HOMOPLASY) within any given group.

Hennig suggested that researchers should begin their analysis with the assumption that all characters which conform to traditional homology criteria are, in fact, evolutionarily homologous. In some cases the use of such criteria will lead to the incorrect, and initially undetectable, identification of homoplasious traits as homologues. Hennig reasoned that when a phylogeny was reconstructed by grouping taxa according to many shared homologies these misidentifications would be revealed because the homoplasious characters would not covary with the majority of the other characters. These traits could then be recognized, using the phylogenetic hypothesis, as homoplasies. This approach to determining homology is called the principle of taxonomic congruence.

The distinction between the non-Hennigian and Hennigian approaches is a subtle, but vital one. Consider the following example. Suppose that all of the females in five species of salmonids dig nests with their tails. A non-Hennigian systematist would say 'Since this behaviour looks the same in these five taxa and since this behaviour is performed by different members of a closely related group ('salmonids'), it is a homologous trait. We can use this trait to assess the phylogenetic relationships among these organisms.' A Hennigian systematist would say 'Since this behaviour looks the same in these five taxa, it is the same (is homologous). We can use this trait to assess the phylogenetic relationships among these organisms.' In the first case, homology is assumed because of similarity among characters, coupled with presumed relatedness among the taxa bearing the characters. Hence, there is an underlying assumption of prior knowledge of phylogeny. In the second case homology is assumed solely on the basis of similarity among characters. Hence, the approach advocated by Hennig is not circular because homologies, which indicate phylogenetic relationships, are determined without a priori reference to a phylogeny while homoplasies, which are inconsistent with phylogeny, are determined as such by reference to the phylogeny. [D.R.B.]

---

homoplasy Any character that appears to be the same in two or more taxa, but has not arisen from a single common ancestor. Homoplasious similarity may arise from a variety of processes, including common natural and sexual SELECTION PRESSURES or developmental phenomena. Homoplasious TRAITS are identified in phylogenetic analyses as those which support groupings that are inconsistent with phylogenetic relationships based on the overall weight of evidence. They appear on phylogenetic trees in two forms. The first is the independent occurrence of the same, or a highly similar, apomorphic trait (*see* APOMORPH). The second is the reappearance of the plesiomorphic condition in a species descended from a species having the apomorphic condition. This is called an evolutionary reversal. *See also* PLESIOMORPHY. [D.R.B.]

homozygosity In a random mating population, the probability that an individual is homozygous at any of n ALLELES at a locus is given by Σp_i^2 ($i=1, \ldots, n$). In a finite population of size N, the homozygosity equals the equilibrium value of the inbreeding coefficient, F, $=1/(4N\mu+1)$, where μ is the MUTATION rate. The

HETEROZYGOSITY, H, and homozygosity, F, are related by $H = 1 - F$. *See also* POPULATION GENETICS. [G.D.J.]

honest signalling *See* STOTTING.

Hooker, J.D. *See* CHARACTERS IN ECOLOGY.

Hooker, W. *See* CHARACTERS IN ECOLOGY.

hopeful monsters A metaphor used initially in 1933 by the geneticist R. Goldschmidt, in the context of his controversial, and mostly discredited, theory of evolution by sudden large discontinuous morphological changes (*see* MACROMUTATION). In his best-known treatise, *The Material Basis of Evolution* (p. 390), Goldschmidt justifies the term 'to express the idea that mutants producing monstrosities may have played a considerable role in MACROEVOLUTION. A monstrosity appearing in a single genetic step might permit the occupation of a new environmental NICHE and thus produce a new type in one step'. *See also* HOMEOTIC MUTATIONS; PUNCTUATED EQUILIBRIUM. [P.A.L.]

REFERENCE
Goldschmidt, R. (1940) *The Material Basis of Evolution*. Yale University Press, New Haven, Conn.

horizontal resistance *See* MULTIPLE RESISTANCE TO PESTICIDES.

hormesis Phenomenon in which benefits (e.g. to survival, growth, reproduction) appear to be derived from EXPOSURE to small doses/concentrations of substances that are toxic at larger dose/concentrations. [P.C.]

horsetails *See* PTERIDOPHYTES, GLOBAL DIVERSITY AND CONSERVATION.

host A living organism that acts as a 'medium' for the growth and development of another organism. The relationship may be mutually beneficial (e.g. vesicular arbuscular root associations; *see* MYCORRHIZA); neutral (e.g. phoretic arthropods on mammals; *see* PHORETIC ASSOCIATIONS); or detrimental (e.g. hymenopteran PARASITOIDS). Detrimental effects may range from a slight decrease in FITNESS to the death of the host organism. The PARASITE may spend the whole or part of its LIFE CYCLE in a particular host. When it utilizes more than one host during its life cycle (e.g. trypanosomes) then the hosts are classified as primary, secondary, etc. *See also* MUTUALISM; SPECIES INTERACTIONS; SYMBIONT. [R.C.]

hotshot model An explanation for the evolution of aggregations of males at sites visited by females solely for the purpose of mating (*see* LEK). The model has two main elements: an initial ability of particular males ('hotshots') to attract females, and a tendency for subordinate males to cluster around the hotshot males to mate with females that have been attracted. This hypothesis differs from the 'hotspot' model in its emphasis on an initial element of mate choice by females, and it differs from most other ideas on lek formation by downplaying the subsequent role of female choice in favour of male–male competition. Thus, female mating preferences are hypothesized to be sufficiently weak (*see* CHOOSINESS), that females are willing to mate with subordinate males that have clustered their display territories around those of dominant males. There is empirical support for this hypothesis for some species. For example, female natterjack toads (*Bufo calamita*) are attracted to males with the loudest calls, which tend to be the largest individuals. Smaller males position themselves near large males and attempt to intercept visiting females. This hypothesis does not explain the initial FEMALE PREFERENCES for particular males in LEKKING species, nor the locations and traditionality of lekking sites. *See also* MATING SYSTEMS. [J.D.R.]

REFERENCES
Beehler, B.M. & Foster, M.S. (1988) Hotshots, hotspots, and female preference in the organization of lek mating systems. *American Naturalist* **131**, 203–219.
Höglund, J. & Alatalo, R.V. (1995) *Leks*. Princeton University Press, Princeton, N.J.

Hubbard Brook Experimental Forest A site in New Hampshire, USA, for research on FOREST ecosystems. Here the ECOSYSTEM is defined as the WATERSHED where, among other things, the effects of cutting trees or burning UNDERSTOREY shrubs on the HYDROLOGY and water chemistry of the system have been determined. [S.J.H.]

REFERENCE
Bormann, F.H. & Likens, G.E. (1979) *Patterns and Processes in a Forested Ecosystem: Disturbance, Development, and the Steady State Based on the Hubbard Brook Ecosystem*. Springer-Verlag, New York.

Huffaker's experiment An experiment that demonstrated the importance of spatial HETEROGENEITY in stabilizing PREDATOR–PREY INTERACTIONS. C.B. Huffaker studied herbivorous mites feeding on oranges; these mites were themselves attacked by predatory mites. With few oranges present, the predatory mites normally drove their prey to extinction. Increasing the number of oranges and making the system more spatially complex (with Vaseline (petroleum jelly) barriers, etc.) led to a persistent interaction. The herbivore was able to escape extinction by colonizing empty patches and reproducing before discovery by the predator. This was an important experiment relevant to METAPOPULATION theory. *See also* POPULATION REGULATION. [H.C.J.G.]

REFERENCE
Huffaker, C.B. (1958) Experimental studies on predation: dispersion factors and predator–prey oscillations. *Hillgardia* **27**, 343–383.

Humboldt, Alexander von *See* CHARACTERS IN ECOLOGY.

humic acid A general term for a wide variety of organic compounds resulting from DECOMPOSITION of vegetable material. Such compounds impart acidity and a dark coloration to waters containing them. They can be extracted from SOIL with dilute alkali and precipitated by ACIDIFICATION to pH < 2. [P.C.]

humidity The WATER content of the air expressed as a vapour pressure, or mass of water. The maximum vapour pressure of water in air, e_s, depends on TEMPERATURE, T (kelvin). There are several empirical expressions to describe this dependency, the most common being the Tetens equation:

$$e_{s(T)} = 0.611\ \exp[17.27(T - 273)/(T - 36)]$$

where $e_{s(T)}$ is in kilopascals and T is in kelvin.

The dryness of an air sample with a water-vapour pressure of e_a

is conveniently expressed as the saturation deficit, $e_{s(T)} - e_a$, also referred to as the vapour-pressure deficit (VPD—but note that the latter abbreviation is used by plant physiologists to refer to a related quantity, the leaf–air vapour-pressure difference).

Relative humidity is the percentage saturation, $100 \times e_a/e_{s(T)}$.

Absolute humidity, χ, is the mass of water per volume of air ($g\,m^{-3}$):

$$\chi = \frac{M_W e}{RT} = \frac{2165e}{T}$$

where e is the vapour pressure of water (kPa), M_W is the molar mass of water ($18\,g\,mol^{-1}$) and R is the molar gas constant ($8.31\,J\,mol^{-1}\,K^{-1}$).

The SPECIFIC HUMIDITY, q, is the mass of water per mass of moist air, and may be found from the vapour pressure e, and the air pressure P by the approximation:

$$q = 0.622e/P$$

When air is gradually cooled, the temperature at which water in the gas phase condenses to make droplets is the DEW POINT, the point at which the air is saturated with water.

Humidity is measured by several methods:

- PSYCHROMETER—using wet and dry bulb thermometers;
- hair hygrometer—based on the expansion of hair with humidity;
- dew-point meter—a surface is cooled until dew is formed, then the temperature of the surface is recorded;
- INFRARED GAS ANALYSER—measures the absorption of infrared radiation by water molecules in the gas phase;
- 'humidity probes' — incorporating resistive or capacitative elements.

[J.G.]

REFERENCE

Monteith, J.L. & Unsworth, M.H. (1990) *Principles of Environmental Physics*, 2nd edn. Arnold, London.

humification The microbial DECOMPOSITION of organic matter in soil involves the progressive breakdown of complex molecules into simpler, more soluble ones. HUMIC ACID is a component of these breakdown products and its production is termed humification. The process is most clearly observed in PEAT deposits, where organic materials predominate in the soil. The water squeezed from surface-growing *Sphagnum* mosses on a BOG is practically clear and colourless, but samples of decomposing moss from a depth of a few centimetres provide water that is discoloured, brown and more opaque. Such samples are said to be more humified. Scales of humification have been devised that permit objective field estimation of this decay process.

Humification in such wetland situations is inhibited by waterlogging. Wet surface conditions, which may be related to moister climates, therefore lead to lower rates of humification than dry ones. An examination of the degree of humification of organic matter at different levels in a peat PROFILE may thus provide indications of past changes in surface wetness and hence, possibly, past climates. [P.D.M.]

'hump-backed' model of diversity A model which describes an observed relationship between PRIMARY PRODUCTIVITY and species DIVERSITY, where highest levels of diversity occur at intermediate productivity levels. The rising limb of this curve may be explained because increases in productivity can be expected to lead to an increase in the range of RESOURCES available, thereby supporting more species. A number of possible explanations for the falling limb of the curve have been proposed. One of these is that high levels of productivity may lead to high rates of population growth, which may in turn lead to rapid COMPETITIVE EXCLUSION—in lower productivity systems the environment is more likely to have changed before competitive exclusion can take place. An alternative explanation is based on the argument that it is the range of resource ratios which promotes the COEXISTENCE of species; at high productivity, the average concentration of available resources in the environment is higher, but the range of concentrations may remain the same, such that the range of resource ratios falls. This latter explanation seems to be most appropriate for plants. It has been noted that other factors, such as levels of DISTURBANCE, can be correlated with productivity. In some cases, it may be one of these correlates, not productivity itself, which leads to the observed relationship. [S.J.H.]

humus The natural organic product of DECOMPOSITION of plant material in the SOIL. It is colloidal, composed largely of humic groups — aromatic structures, including polyphenols and polyquinones, which are the products of decomposition, synthesis and polymerization. The compounds vary in RELATIVE MOLECULAR MASS from a few hundred to a few thousand, and are dark in colour. They are very resistant to attack ('recalcitrant') and they form a substantial stock of the world's soil CARBON. Examples are fulvic acid, HUMIC ACID and humin. Humus also contains non-humic groups, such as polysaccharides from higher plants and polyuronides from microorganisms, that tend to stick soil particles together and give soil some of its structure.

Humus COLLOIDS are negatively charged (especially so at high pH) and usually constitute a large part of the soil's cation exchange capacity. [J.G.]

REFERENCE

Lynch, J.M. (1983) *Soil Biotechnology: Microbiological Factors in Crop Productivity*. Blackwell Scientific Publications, Oxford.

hunger An abstract motivational response to a physiological need that is commonly restricted to higher animals. It is directly regulated by the animal's requirement for a specific level of nutrition. The intensity of hunger fluctuates temporally and is directly proportional to the activities that are undertaken. *See also* FEEDING; OPTIMAL FORAGING THEORY. [R.C.]

Hutchinson, G.E. *See* CHARACTERS IN ECOLOGY.

Hutchinson theories *See* BIODIVERSITY GRADIENTS.

Hutchinsonian niche *See* HYPERSPACE; NICHE.

Hutchinson's size-ratio rule According to this rule, coexisting species that have qualitatively similar ecological roles (have to) differ in size by at least a factor of about 1.3 (i.e. if a linear measurement is taken; whereas for masses, the factor should be $1.3^3 \approx 2$). The rule fits in the framework of COMPETITION theory and may be seen as a specification of the principle of COMPETITIVE EXCLUSION, providing a value of the LIMITING SIMILARITY in size between coexisting competitors. It was inspired by observations on corixid water bugs on a pool below a sanctuary of Santa Rosalia, Sicily (Hutchinson 1959). A wide range of examples have been put forward in support of the rule, including the size differences between musical instruments of similar general 'morphology'. Nevertheless, the rule has been the subject of considerable discussion, often due to disagreement on the appropriateness of the methods used to test it (e.g. Simberloff & Boecklen 1981). *See also* ASSEMBLY RULES. [J.G.S.]

REFERENCES

Hutchinson, G.E. (1959) Homage to Santa Rosalia *or* Why are there so many kinds of animals? *American Naturalist* **93**, 145–159.

Simberloff, D. & Boecklen, W. (1981) Santa Rosalia reconsidered: size ratios and competition. *Evolution* **35**, 1206–1228.

hybrid vigour Crosses between inbred lines often lead to hybrid individuals that are larger in size, produce more offspring or yield more seeds, are hardier and stronger. The phenomenon is called hybrid vigour (*see also* HETEROSIS). Both parental lines might suffer from INBREEDING DEPRESSION due to HOMOZYGOSITY in deleterious recessive ALLELES at different genes; in such a case, the hybrid individuals would not be homozygous for any of the deleterious alleles, and therefore would have greater vigour and FITNESS. A second explanation is that heterozygotes have intrinsically greater fitness, due to interaction between alleles at one locus. However, between natural populations OUTBREEDING DEPRESSION (reduced fitness of hybrids) might also occur, due to homozygosity of genes responsible for local ADAPTATION of the populations. [G.D.J.]

Hybrid zones

Narrow zones where genetically differentiated populations meet, mate and produce hybrids. Hybrid zones are common; over 200 cases have been investigated in a variety of organisms, with variations in the taxonomic rank and degree of genetic divergence of the hybridizing populations. Apparent hybrid zones separate either species or SUBSPECIES that differ in many characteristics (morphology, anatomy, behaviour, mating calls, allozymes, nuclear or mitochondrial DNA sequences). In contrast, cryptic zones can easily escape detection; these include, for example, contacts between PARAPATRIC chromosomal races within a morphologically uniform species. The zones are narrow relative to the DISPERSAL of individuals, though their widths differ in absolute terms.

Well-known European examples include: the house mouse (*Mus musculus/domesticus*), crows (*Corvus coronae/cornix*), fire-bellied toads (*Bombina bombina/variegata*), newts (*Triturus cristatus/marmoratus*) and grasshoppers (*Chorthippus parallelus parallelus/p. erythropus*) as well as chromosomal races of *Mus domesticus*, shrew (*Sorex araneus*) and grasshopper (*Podisma pedestris*).

Taxonomists, who drew attention to hybrid zones, tend to view them as borders between two unlike types, whereas population geneticists regard hybrid zones as transitions from one genetic equilibrium to another.

Main features

Two main types of hybrid zones are distinguished: smooth CLINES and mosaic zones. Smooth clines involve gradual change in either a single or multiple TRAITS (loci), whereas mosaic zones entail a patchy distribution of hybridizing forms in the contact area. Many intermediate situations are possible; hybrid zone structure can also change along the contact.

Often, the clines tend to be coincident and concordant. They occur in the same location, and have the same shape and width; i.e. the clines are highly correlated. In other instances, individual clines follow their own pattern and form a series of scattered clines with varying widths. The geographical scale of the hybrid zone matters, since superficially uniform clines may appear more mosaic on a finer scale, and, vice versa, mosaic zones can, in fact, be composed of low-scale smooth clines.

The complexity of situations found in nature makes identification of taxa within hybrid zones difficult, if not impossible. Hybrid indexes based on many independent traits are used to describe the composition of local populations and the extent of variation, while geneticists strive to gain insight into the genetic structure of the hybrid zone and the forces which operate within it by studying the distribution of multilocus genotypes.

In central populations parental types may be absent and populations consist exclusively of hybrids of various generations. In such populations, GENOTYPES at single loci usually conform to Hardy–Weinberg proportions but show non-random associations of ALLELES at multiple loci. There is an excess of parental combinations of alleles in GAMETES, with recombinant combinations being less common. This condition, known as linkage, or gametic, disequilibrium, results from matings between individuals which migrate into the zone from genetically different populations.

With respect to the origin, hybrid zones may be either primary or secondary. In theory, primary zones form within a continuous population under selection favouring distinct types at two ends of an environmental GRADIENT, in contrast to secondary zones, which result from contacts between genetically dissimilar populations that diverged in allopatry. Distinguishing between these two possibilities is difficult, but most well-studied temperate hybrid zones can be related to postglacial RANGE expansion, i.e. they are secondary.

Selection and stability

Despite former suggestions, there is now good evidence from theoretical and experimental work that hybrid zones can be stable and persist through time without populations undergoing fusion or evolving mating barriers. Stability of the zone results from a balance between selection against hybrids and GENE FLOW into the zone caused by dispersal.

Selection may act in two, not mutually exclusive ways. Hybrids often have lowered viability (*Bombina, Podisma*) and/or increased sterility (*Triturus, Chorthippus*) apparently due to malfunction of divergent genomes, and are therefore at a disadvantage relative to the parental forms. As many zones are positioned at environmental gradients with each type being superior at the opposite end of the gradient, selection may also favour alternative ADAPTATIONS in different habitats. Patchy distribution of habitats in environmentally complex areas would therefore produce a mosaic hybrid zone.

An alternative explanation for hybrid zone persistence relies on hybrid superiority in a narrow ECOTONE or hybrid habitat formed by human activity. Though applicable to some cases, this interpretation has less generality since most hybrid zones are thousands of years old and have formed independently of human DISTURBANCE.

In *Heliconius* butterflies of South America, narrowness of hybrid zones at boundaries between geographic races of different wing pattern results from STABILIZING SELECTION exerted by avian predators, which learn to recognize parental forms as distasteful Mülerian mimics and therefore remove atypically patterned hybrids since they do not associate them with unpalatability (see Plate 20, facing p. 398).

Hybrid zones are barriers to gene flow. Selection against hybrids is strong as it typically acts on a large number of genes spread throughout the genome. Recombination of alleles into a foreign genetic background is difficult and thus introgression is severely limited. Two continuously hybridizing taxa can therefore maintain their identity outside a narrow zone. The dynamic nature of hybrid zones also prevents evolution of mating barriers.

Scientific importance

Interest in hybrid zones comes from many disciplines. They are a taxonomist's nightmare and an evolutionist's delight, since they highlight the difficulties inherent in the BIOLOGICAL SPECIES CONCEPT. They are likely areas of reinforcement of ISOLATING MECHANISMS. Hybrid zones are centres of phenotypic and genotypic DIVERSITY, 'natural laboratories' in which one can study interactions between divergent genomes under natural conditions and the long-term effects of hybridization to gain insight into SPECIATION as well as the origin and maintenance of diversity. Hybrid zones, unique phenomena as they are, deserve both the attention of conservationists and legal protection. [J.M.SZ.]

---

hydraulic residence time *See* LAKES.

hydric Means wet. *Cf.* XERIC. [P.C.]

Hydrocarbons

An extremely large class of organic compounds composed of the elements CARBON and HYDROGEN only. Hydrocarbons divide into three categories.

1 Aliphatic: open chains of carbon atoms, e.g. ethane (C_2H_6) and ethene (C_2H_4).

2 Cyclic: at least one closed ring of carbon atoms, e.g. cyclohexane (c-C_6H_{12});

3 Aromatic: a special class of cyclic hydrocarbons specifically containing six-membered rings in which carbon bonds are intermediate between single and double bonds, e.g. BENZENE (C_6H_6).

A carbon with the maximum four separate bonds to other atoms is saturated (ALKANE) while a carbon that is bonded to another by double bonds (ALKENE), triple bonds (alkyne) or aromatic bonds is unsaturated. The type of carbon centre confers different chemical reactivities on the compound.

Sources

The primary source of hydrocarbons for industrial use is liquid PETROLEUM (crude oil) and its associated natural gases. Petroleum originates from hydrocarbons synthesized by living organisms (e.g. marine organisms with high lipid content) but few of the hydrocarbons are species specific. The relative proportions of hydrocarbons in petroleum differ greatly from those of biogenic hydrocarbons indicating that petroleum must have derived predominantly from abiogenic processes following deep burial of the original organic material in sediments. However, a small proportion of hydrocarbon material in sediments can be linked unambiguously to an organism or group of organisms and thereby act as useful BIOMARKERS. These become increasingly chemically altered (e.g. deoxygenated) following incorporation into sediments, and by the end of diagenesis are aliphatic or monoaromatic hydrocarbons, albeit ones whose skeletons can still be related to the precursor molecules, for example steranes, hopanes and phytanes from bacteria and algae. Petroleum is subsequently formed during catagenesis under extremes of temperature and pressure, on a time-scale of millions of years. The wide variety of low-molecular-mass hydrocarbons arises by loss of side chains from larger biomolecules. Formation under suitable rock structures results in deposits of oil and natural gas. Natural gas is composed mostly of METHANE, with about 5–10% ethane and propane, and is used almost exclusively for fuel. Petroleum is refined by distillation into fractions separated according to boiling-point range. Liquid fractions with boiling point (b.p.) <90°C contain C_5–C_7 hydrocarbons; the petrol (gasoline) fraction (b.p. 50–200°C) contains C_6–C_{10} compounds; and higher fractions are kerosene (b.p. 175–325°C, C_8–C_{14}), gas oil (b.p. >275°C, C_{12}–C_{18}), lubricating oils (>C_{18}) and asphalt. In addition to use as fuels, individual hydrocarbons form essential feedstock for a worldwide petrochemical and PLASTICS industry.

Volatile organic compounds

Hydrocarbons up to C_4 are gases at ambient temperatures but liquid hydrocarbons from C_5 to C_{12} also have significant vapour pressures. The most widespread natural and anthropogenic hydrocarbons fall within these categories and, together with their many functional group derivations, are generically classified as VOLATILE ORGANIC COMPOUNDS (VOCs). The bulk of hydrocarbon release to the environment is through VOCs to the ATMOSPHERE.

Methane

Methane, the simplest hydrocarbon (CH_4), dominates the abundance of all others, which are classified together as non-methane hydrocarbons (NMHC). Natural sources of methane include bacterial decay of organic matter in swamps, marshes and paddy fields, and enteric fermentation in RUMINANT animals. Geogenic methane is released by seepage from natural gas reservoirs. Atmospheric concentrations have more than doubled to 1.7 ppm by volume (ppmv) since industrialization, and this may reflect both increases in sources and decreases in chemical destruction rates. Methane impacts the environment directly as a GREENHOUSE GAS and indirectly through many chemical processes that lead to variation in concentrations of other atmospheric species such as OZONE (O_3), CARBON MONOXIDE (CO) and the hydroxyl radical (OH).

Non-methane hydrocarbons

Natural NMHC emission is dominated by vegetation. Over 400 different compounds (including oxygenated hydrocarbon derivatives such as aldehydes, ketones, organic ACIDS and ALCOHOLS) have been identified. The most abundant species are ISOPRENE (2-methyl-1,3-butadiene, C_5H_8) and a variety of monoterpene isomers ($C_{10}H_{16}$) emitted from grasses and trees. These are often partially responsible for characteristic aromas associated with particular vegetation, for example α- and β-pinenes from conifers. Emission rates are highly dependent on ambient temperature, light intensity or 'stress' caused, for example, by water deficit. Concentrations in rural air are generally a few ppb by volume (ppbv). The unsaturated double bonds confer high reactivity of biogenic hydrocarbons towards OH radicals and ozone, and in strong sunlight can lead to rapid formation of 'natural' PHOTOCHEMICAL SMOG and bluish AEROSOL hazes directly above a forest canopy.

In contrast, anthropogenic release of NMHC to the atmosphere comprises predominantly C_2–C_8 isomers of alkanes and alkenes and alkyl-substituted benzenes. The major sources are unburnt fuel in petrol and diesel combustion (827 kt, UK estimate for 1991), EVAPORATION of solvents (780 kt), and losses during petrochemical refinement and distribution (277 kt). Average concentrations of individual hydrocarbons in the atmosphere are typically in the range sub-ppb to a few ppb, but may reach a few hundred ppb during POLLUTION episodes.

Global biogenic EMISSIONS of NMHC exceed global anthropogenic emissions, but the former are distributed over all continental surfaces. In particular, hydrocarbons derived from fossil fuel predominate in areas of widespread industrialization. Estimates of 50–100 kt year^{-1} for biogenic emissions in the UK are insignificant compared with total anthropogenic emissions of over 1900 kt year^{-1}.

The ENVIRONMENTAL IMPACT of NMHC in the atmosphere varies considerably from compound to compound. Some are CARCINOGENS (e.g. benzene and 1,3-butadiene), others are chemically toxic, but all are precursors to tropospheric OZONE formation through photochemical oxidation. Degradation is initiated principally by the OH radical during the day and by the nitrate radical (NO_3) at night. In the presence of NITROGEN OXIDES, daytime hydrocarbon oxidation leads to net production of ozone and other photochemical smog pollutants such as PAN (peroxyacetylnitrate) and organic aerosol. The potential of individual hydrocarbons to generate ozone varies widely, correlating approximately with rate of oxidation. To formulate legislative controls for effective cutbacks on emissions, relative photochemical ozone creation potentials (POCPs) can be defined for each species.

Polycyclic aromatic hydrocarbons

An important class of hydrocarbons is POLYCYCLIC AROMATIC HYDROCARBONS (PAHs) consisting of fused benzene rings, for example naphthalene (two rings), anthracene and phenanthrene (three rings), pyrene and benz[a]anthracene (four rings) and benzo[a]pyrene (five rings). *See also* PAH POLLUTION.

Distribution and transport processes

The hydrophobicity of hydrocarbons is an important factor in transport through environmental compartments. Partition coefficients measure the distribution between gas or aqueous phases and organic phases, as for example, in the BIOACCUMULATION in LIPIDS of living material. Atmospheric hydrocarbons are preferentially scavenged by particles or droplets containing cores of organic material, which leads to enrichment of droplets and transport of hydrocarbon to soil and hydrological compartments. Additional inputs include dry deposition from the atmosphere (e.g. PAHs) and direct CONTAMINATION (e.g. oil and solvent spillage, waste DISCHARGE). Hydrophobic hydrocarbons interact strongly with components of soils and sediments by sorption or exclusion through solubilization into lipid-like organic surface layers on particulate materials. The hydrophobic bonding forces increase in strength with decreasing water solubility of the hydrocarbon. Migration of organic CONTAMINANTS down a PROFILE is facilitated by highly permeable sandy soils with low stationary organic content. The distribution between phases is described empirically in terms of interactions with organic matter by simple relationship to organic carbon content, i.e.

$$K_{oc} = K_d / \text{fraction of organic content}$$

where K_{oc} is the PARTITION COEFFICIENT normalized to organic carbon content and K_d is the specific partition coefficient. The normalized coefficient is constant for a wide range of substrates. Degradation of hydrocarbons in soils and sediments is accelerated by microorganisms, often specially adapted for the purpose in cases of direct treatment of contamination.

Marine hydrocarbon pollution

Petroleum hydrocarbons constitute a major POLLUTANT to coastal marine systems and it is estimated that more than 3 Mt are discharged annually. The main sources are oil transportation, municipal and industrial wastes, and RUN-OFF. Natural sources (marine seeps and sediment erosion) comprise only a small component (0.25 Mt annually). Atmospheric deposition to the marine environment is similarly small (0.3 Mt annually), although evidence suggests this is increasing. The relative rates of hydrocarbon evaporation to atmosphere, dissolution and COLLOID formation, partitioning to suspended particulates, and microbial and photochemical degradation, all affect the FATE of aqueous hydrocarbon discharge. Degradation processes generally act more rapidly on lighter molecular mass components, while PAHs are more persistent.

Analytical methods

The most flexible and discriminative technique for analysis of hydrocarbons is GAS CHROMATOGRAPHY coupled with flame ionization and/or mass spectrometric detection. Gaseous samples are concentrated on to a capillary column by trapping to an adsorbent and thermal desorption and liquid samples by solvent extraction or headspace analysis. Detection limits of ppt or lower are achieved. [M.C.G. & M.R.H.]

REFERENCES

Duce, R.A., Mohnen, V.A., Zimmerman, P.R. *et al.* (1983) Organic material in the global troposphere. *Reviews of Geophysics and Space Physics* **21**, 921–952.

Killops, S.D. & Killops, V.J. (1993) *An Introduction to Organic Geochemistry*. Longman/Wiley, New York.

Quality of Urban Air Review Group (1993) Organic compounds. In: *Urban Air Quality in the United Kingdom*, Chapter 9, pp. 139–157. Department of the Environment, London.

hydrodynamics A branch of fluid mechanics dealing with WATER in motion. The mathematical theory of the motion of ideal, inviscid fluids. [V.F.]

hydrogen (H) The lightest chemical element (ATOMIC NUMBER = 1; relative ATOMIC MASS = 1.00794; melting point = −259.14°C; boiling point = −252.87°C), occurring as three isotopes: protium (atomic mass = 1.007825, 99.983%), DEUTERIUM (atomic mass = 2.014102, 0.017%) and TRITIUM (atomic mass = 3.016049, 0.0003%). In the ATMOSPHERE hydrogen exists as a diatomic gas (H_2), present in concentrations of a few tenths of a part per million (by volume). It is present in the Earth's CRUST (including surface waters and the atmosphere) mostly as WATER, at 0.9% (by mass), although it is the commonest element in the universe, thought to account for over 90% of all atoms. The hydrogen ION in water (H^+) confers acidity, often measured in terms of solution pH, defined as $pH = -\log_{10}[H^+]$, where the square brackets denote the activity of the hydrogen ion in solution (*see* PH). [J.N.C.]

Hydrogeology

Hydrogeology is the study of WATER in the ground. It relates to both the saturated zone in the ground, in which all of the void space (porosity) is filled with water, and the unsaturated zone, in which the voids are only partially filled with water. The water is referred to as groundwater. Most groundwater is derived from PRECIPITATION directly as infiltration, or indirectly via a RUN-OFF mechanism, and therefore forms part of the hydrological cycle (*see* WATER (HYDROLOGICAL) CYCLE). When water infiltrates into the ground some will be taken up by plant use (evapotranspiration) and the remainder will flow to depth (recharge), through the unsaturated ground. Where significant recharge occurs the ground becomes saturated, a hydraulic head gradient is created, and under suitable topographical and geological conditions groundwater flow will occur. The groundwater will eventually DISCHARGE to a spring, a river, a lake, or the SEA, having moved through the ground via either a primary porosity (intergranular), or secondary porosity (fracture and/or fissure) route, or probably a combination of both. The movement or flow transmission through the ground is controlled by the resistance to flow in the ground, which reflects the porosity and is termed the hydraulic conductivity, and the hydraulic head that exists between recharge and discharge points. The flow is three-dimensional within area-limiting boundaries such as a river, a recharge mound divide, or a marked geological change.

Water resources

Qualitative descriptions are usually applied to the ability of rocks to transmit groundwater, and as it can provide very important water RESOURCES, the ability to yield water through wells is included. The term AQUIFER is applied to the ground that can provide adequate groundwater for a particular purpose. The term is therefore relative depending upon purpose.

In many geological situations strata that form aquifers are interbedded with non-aquifer materials, frequently referred to as aquicludes, that can form upper and lower hydraulic boundaries to an aquifer. For groundwater resources, the relationship between the hydraulic head developed in the aquifer and the upper hydraulic boundary is important. Where no boundary is present (i.e. below aquifer outcrop), or the boundary is at a higher elevation than the groundwater head, the aquifer condition is considered to be unconfined. Where the head is above the boundary, the aquifer condition is confined. The aquifer condition reflects the character of the groundwater storage, most easily described through head decline. In the unconfined area, if groundwater heads decline, the porosity will drain, or dewater under gravity, but will not fully drain because of grain surface tensional effects; the operable drainage or storage is therefore less than the porosity and is referred to as the specific yield. In a completely confined condition with head decline above the boundary no dewatering will occur in the aquifer. Aquifer storage will decrease, however, because of stress reduction in the water column and the resultant compaction of the aquifer material. The storage term used to describe this feature is the storage coefficient; this is orders of magnitude smaller in size than specific yield (e.g. 10^{-6} vs. 10^{-2}). In groundwater resources management the manipulation of specific yield is an important factor so that unconfined aquifers can normally have more operable potential than confined aquifers.

The use of groundwater implies that it is of suitable quality. However, through natural chemical processes it can acquire a chemistry in the form of solutes and particulate matter, and become unsuitable for use. It can also acquire undesirable chemical pollutants through human-induced activities. The movement of naturally occurring or POLLUTANT solutes in groundwater is referred to as solute transfer. The dominant control in transport is usually ADVECTION, whereby the solute moves concordantly with the groundwater velocity. Mixing of solute concentrations can occur, however, because of porosity heterogeneities and resultant velocity variations. Such mixing is therefore a function of the geology and is referred to as DISPERSION. Where significant concentration differences occur between juxtaposed groundwaters, as for example at the coast, chemical diffusion can be an additional transport mechanism.

Hydrogeology applications

Although most emphasis in hydrogeology is directed towards resources, there is also extensive study of groundwater CONTAMINATION, in construction and foundation engineering, in LANDFILL operations, oilfield hydraulics, and for RADIOACTIVE WASTE repositories, etc. In order to study the hydrogeology of an area, the initial requirement is to define the geometry of the groundwater flow system. This usually requires the application of geological interpretation, surface geophysical interpretation and borehole drilling with downhole geophysical investigations. The hydraulic and hydrochemical parameters that are required are usually determined by borehole testing and water sampling. Transient data, such as long-term groundwater heads or solute distributions, are important. Once a basic understanding is obtained, the system will usually be represented by numerical modelling procedures that can simulate both the hydraulic and hydrochemical distributions with time. Such procedures are iterative and refine the parametric understanding. Once an acceptable numerical representation of the historical performance of a system is obtained, the model developed is used to study the management of the system for future induced changes. *See also* GROUNDWATER POLLUTION. [J.W.L.]

---

hydrograph A graph of DISCHARGE (y-axis) against time (x-axis) in rivers and streams. Also FLUCTUATIONS of water levels with time in water (Fig. H4). [P.C.]

hydrographics The arrangement and movement of bodies of WATER, such as CURRENTS and water masses. Hydrography is the science and art of studying, surveying and mapping seas, LAKES, RIVERS and other waters, including study of their physical features, TIDES, currents, etc. [V.F.]

hydrography *See* HYDROGRAPHICS; HYDROLOGY.

hydrological cycle *See* WATER (HYDROLOGICAL) CYCLE.

hydrology The study of WATER, both the quantity and the chemicals and SEDIMENT which it carries, as it moves through the land phase of the hydrological cycle (*see* WATER (HYDROLOGICAL) CYCLE). *See also* HYDROGRAPHICS. [J.H.C.G.]

hydrophobic chemicals *See* OCTANOL–WATER PARTITION COEFFICIENT.

hydrophyte Any plant which is adapted to live in WATER either continuously or for certain seasons of the year. Plants that alternate between wet and dry habitats are, however, more precisely described as amphibious. Plants adapted to living in water have characteristic morphological ADAPTATIONS such as finely divided submerged leaves and large intact floating leaves. Most hydrophytes have well-developed aeration tissue (aerenchyma) in stems and roots to facilitate the downward diffusion of OXYGEN. Many hydrophytes show a remarkable tolerance of ANOXIA and can even extend and continue new shoot extension over a period of weeks in an entirely anaerobic environment such as occurs in the muds and sediments in pond bottoms (Crawford & Braendle 1996).

Hydrophyte *sensu stricto* usually refers to the life-form CLASSIFICATION (Raunkiaer 1937) in which higher plants are classified by the position of the dormant buds in the adverse season. In the case of hydrophytes the dormant bud passes the adverse season under water. The term macrophyte is sometimes used to distinguish aquatic flowering plants from microscopic aquatic vegetation. [R.M.M.C.]

REFERENCES

Crawford, R.M.M. & Braendle, R. (1996) Oxygen deprivation stress in a changing climate. *Journal of Experimental Botany* **47**, 145–159.

Raunkiaer, C. (1937) *Plant Life Forms* (trans. H. Gilbert-Carter). Clarendon Press, Oxford.

hydroponics The practice of growing plants without soil in which the roots are suspended in WATER or supported by inert material such as lava rock and covered in water. The water may or may not be aerated and GROWTH is sustained by adding MINERAL NUTRIENTS to the water. The method has been extensively used for growing certain vegetable CROPS such as tomatoes and cucumbers and also for the cultivation of crops for pharmaceutical products. [R.M.M.C.]

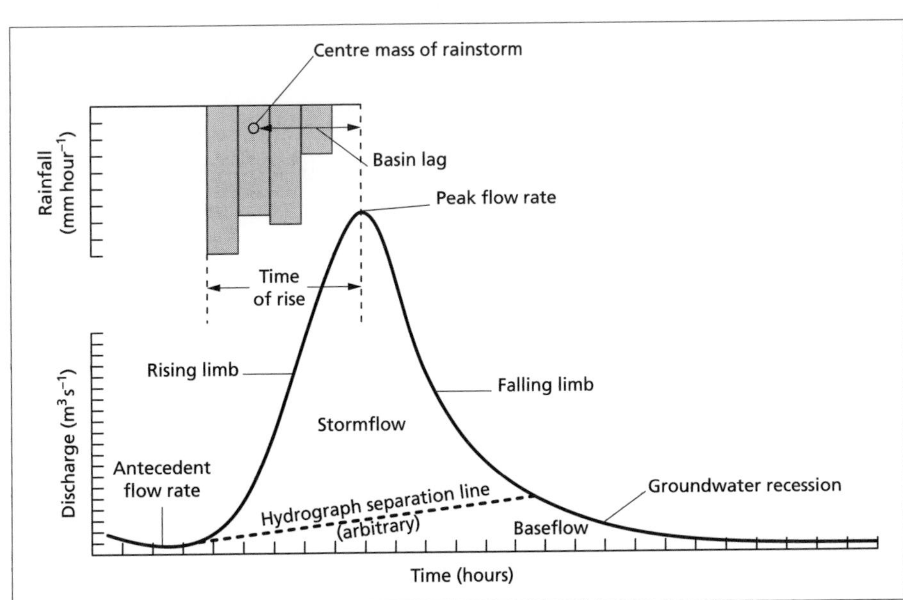

Fig. H4 Terms commonly used to describe a storm hydrograph. (After Burt, T.P. (1992) The hydrology of headwater catchments. In: *The Rivers Handbook*, Vol. 1 (eds P. Calow & G.E. Petts), pp. 3–28. Blackwell Scientific Publications, Oxford.)

hydrosere The SUCCESSION from open water via FEN to FOREST or BOG. It is important as a kind of succession which leaves a permanent record below it, and so shows evidence for RETROGRESSIVE SUCCESSION and VARIABLE PATHWAYS. *See also* ZONATION. [P.J.G.]

hydrosphere The part of the Earth covered by liquid or frozen WATER. *See also* ATMOSPHERE; BIOSPHERE; LITHOSPHERE. [J.H.C.G.]

hydrothermal vent A place on the OCEAN bed, on or near to a mid-ocean ridge, from which water heated by molten rock issues. This is often rich in dissolved sulphides, which are oxidized by chemosynthetic bacteria in the fixation of CARBON DIOXIDE (CO_2) and synthesis of organic compounds. Near to the vents there are animal communities that utilize these compounds and may even live symbiotically with the bacteria. [P.C.]

hypermorphosis A form of HETEROCHRONY consisting of the addition of new morphological stages at the end of the ancestral ONTOGENY. The result is a new morphology generated by terminal addition. Therefore, given the mechanism, such new morphology is 'peramorphic' (*see* PERAMORPHOSIS). Hypermorphosis causes RECAPITULATION of the species' ancestral PHYLOGENY during the individual's ontogeny. In violation of Haeckel's 'biogenetic law', however, it does not abbreviate the total time needed for the unfolding of the descendant hypermorphic ontogeny, which is now longer. [P.A.]

hyperparasitoid A PARASITOID that parasitizes another species of parasitoid. If the parasitoid can only develop in the parasitoid species it is termed an OBLIGATE hyperparasitoid; if it can develop in both the parasitoid and the parasitoid's HOST it is called a FACULTATIVE hyperparasitoid. Hyperparasitoids are sometimes called secondary parasitoids, and the rare cases of parasitoids that attack hyperparasitoids are called tertiary parasitoids. *See also* PARASITISM; SPECIES INTERACTIONS. [H.C.J.G.]

hyperspace A multidimensional space which has more than three dimensions so that its geometry cannot be represented by a simple physical model. The hyperspace concept has been employed by ecologists as a means of envisaging data with many attributes (which may be either environmental variables or species). According to this view, a sample can be represented as a point in a hyperspace whose axes correspond to the attributes.

Where the attributes are environmental variables, the samples in which a given species occurs define a region (hypervolume) of occupancy in hyperspace; this region is called the (Hutchinsonian) NICHE of the species (Giller 1984). In practice, a niche may often be represented in less than four dimensions, so that a graphical representation is possible.

With species as attributes, the representation of samples or species lists as points in a hyperspace provides a means of measuring sample dissimilarity. In this context, it is natural to use Euclidean distance as a measure of sample dissimilarity and to use PRINCIPAL COMPONENT ANALYSIS as a means of ORDINATION or dimensionality-reduction. There may, however, be difficulties of interpretation because of the non-linear response of species to environmental gradients (Austin 1976).

Hyperspaces can also be used to represent system dynamics, especially the dynamics of SUCCESSION or species change. In this case the hyperspace is sometimes called a phase space, and temporal change in species composition is represented by the motion of a point along a TRAJECTORY in the space. The position of the point at any time defines the species composition of the system at that moment. [M.O.H.]

REFERENCES

Austin, M.P. (1976) On non-linear species response models in ordination. *Vegetatio* **33**, 33–41.

Giller, P.S. (1984) *Community Structure and the Niche.* Chapman & Hall, London.

hypobiosis *See* DORMANCY.

hypolimnion The lower layer of deep, cold and relatively undisturbed water in LAKES with THERMAL STRATIFICATION. It often has reduced OXYGEN, poor LIGHT penetration and, therefore, a limited flora and fauna. Temperature differences between the epilimnion (surface layer) and hypolimnion may be as large as 20°C to 4°C in temperate continental lakes, or only a few degrees from 29°C to 24°C in the tropics. *See also* LAKE STRATIFICATION. [J.L.]

hypomicropleuston *See* WITHIN-LAKE HABITAT CLASSIFICATION.

hyporheic *See* BENTHIC HABITAT CLASSIFICATION.

hyporheic fauna *See* WITHIN-STREAM/RIVER HABITAT CLASSIFICATION.

hyporheos *See* BENTHIC HABITAT CLASSIFICATION; WITHIN-STREAM/RIVER HABITAT CLASSIFICATION.

I

IBI *See* INDEX OF BIOTIC INTEGRITY.

IBP *See* INTERNATIONAL BIOLOGICAL PROGRAMME.

IC_{50} Statistically derived concentration of a XENOBIOTIC that inhibits some metabolic property (e.g. growth) in 50% of an observed population over a prescribed time in defined conditions. Also expressed in terms of dose, ID_{50}. *See also* EC_{50}; LC_{50}; MEDIAN TIMES TO EFFECT; PREDICTIVE ECOTOXICITY TESTS. [P.C.]

ice The solid phase of WATER, stable at temperatures below 273.15 K at atmospheric pressure. It has the unusual property of being less dense than the liquid phase, a result of the open crystal structure caused by hydrogen bonding. The strength of the hydrogen bonds accounts for the abnormally high melting temperature, compared with that for the hydrides of heavier elements in the same group as oxygen.

Only 2.3% of the Earth's water is in the solid form, but this accounts for 99% of the FRESHWATER. Most ice forms in the ATMOSPHERE as snow, accumulating on land surfaces as glaciers, ice caps and ICE SHEETS. The snow is transformed to solid ice by a process of either melting and refreezing, which operates in temperate regions, or compaction and recrystallization, which operates in polar regions where no melting occurs. Ice also forms by freezing at the surface of RIVERS, LAKES and the OCEAN. Impurities in the water are expelled from the solid phase, but sea ice is slightly saline because brine is trapped in intercrystalline voids.

Ice deforms under an applied stress, much like a very viscous fluid, leading to the macroscopic phenomenon of glacier flow. Velocities in glaciers and ice sheets range from a few centimetres per year to a few kilometres per year. Where ice deformation is very rapid, fracture can occur that yields crevasses. Glacier ice appears blue because pure ice has a greater absorption of LIGHT at the red end of the visible spectrum. [A.J.]

ice ages *See* GLACIATION.

ice-core projects In some locations, the amount of snow that falls in an average year exceeds the amount of snow that melts in an average year. The snow that does not melt is compressed by subsequent snow fall and forms a layered sedimentary sequence of ICE, which contains PRECIPITATION and associated atmospheric particulates. Atmospheric gases are entrapped in the ice at a depth of 50–90 m when the pore spaces are sufficiently compressed that they are no longer interconnected, thus ceasing the exchange of gases. Specialized drilling equipment is used to recover individual ice cores, which are typically 5–20 cm in diameter and up to 6 m long. By recovering adjoining cores in a single hole, a continuous sampling of the ice is obtained. The chemical, physical and isotopic characteristics of the ice, particulates and gases are used to determine the environmental conditions when the material was entrapped in the ice. The length of time covered by the record depends on the characteristics of the drilling site. Subannual resolution has been obtained at many sites; however, the temporal resolution decreases for older ice due to ice flow. In favourable situations, annual layers can be identified by seasonal variations in the chemistry, isotopic concentration, particulates and bubble content of the ice. This allows layer-counting methods to be used to determine the age of the ice. Ice-flow modelling and comparison to other dated records is used to date ice cores when annual layers cannot be identified. Ice-core records are used to determine many different environmental properties for the same location and time, which allows many different aspects of the palaeoclimate to be reconstructed. Ice-core records can have higher temporal resolution than other long palaeoclimate records, which permits investigation of rapid or short duration changes in CLIMATE. Table I1 lists some of the environmental parameters that can be reconstructed as a function of time.

Recent polar projects include the Vostok ice core from Antarctica, which has provided a 225 000-year record of atmospheric gases and temperature (Jouzel *et al.* 1993), and the companion GRIP (Greenland Ice Core Project; Johnson *et al.* 1992) and GISP (Greenland Icesheet Project 2; Mayewski *et al.* 1994b) projects in Greenland, which have provided a high temporal resolution record of the last 80 000 years. Ice cores have also been used to develop records for low-latitude high-elevation sites (Thompson *et al.* 1986). At least six deep ice-core projects are underway at this time. Environmental records from ice cores have played a key role in enhancing our understanding of what the Earth's climate has been and what it might become. *See also* CLIMATE CHANGE. [K.C.T.]

REFERENCES

Alley, R.B., Meese, D.A., Shuman, C.A. *et al.* (1993) Abrupt increase in Greenland snow accumulation at the end of the Younger Dryas event. *Nature* **362**, 527–529.

Grootes, P.M., Stulver, M., White, J.W.C., Johnsen, S. & Jouzel, J. (1993) Comparison of oxygen isotope records from the GISP2 and GRIP Greenland ice cores. *Nature* **366**, 552–554.

Johnson, S.J., Clausen, H.B., Dansgaard, W. *et al.* (1992) Irregular glacial interstadials recorded in a new Greenland ice core. *Nature* **359**, 311–313.

Jouzel, J., Barkov, N.I., Barnola, J.M. *et al.* (1993) Extending the Vostok ice-core record of palaeoclimate to the penultimate glacial period. *Nature* **364**, 407–412.

LeGrand, M.R., Lorius, C., Barkov, N.I. & Petrov, V.N. (1988) Vostok (Antarc-

Table I1 Environmental parameters that can be reconstructed as a function of time.

Inferred environmental property	Measured parameter	Reference
Surface temperature	$^{18}O/^{16}O$ of ice	Jouzel *et al.* (1993)
Wind speed	Particle concentration, size and composition	Mosley-Thompson & Thompson (1982)
Atmospheric CO_2, NH_4	Entrapped gases	Lorius *et al.* (1993)
Annual snow accumulation	Thickness of annual layers	Alley *et al.* (1993)
Atmospheric circulation	Major ions and cations	LeGrand *et al.* (1988), Mayewski *et al.* (1994a)
Sea-ice extent	Major ions and cations	Mayewski *et al.* (1994a)
Volcanic activity	SO_4, H^+	Zielinski *et al.* (1994)
Rapid climate changes	H^+, major ions and cations	Taylor *et al.* (1993), Mayewski *et al.* (1994a)

tica) ice core: Atmospheric chemistry changes over the last climatic cycle (160,000 years). *Atmospheric Environment* **22**, 317–331.
Lorius, C., Jouzel, J. & Raynaud, D. (1993) Glacials-interglacials in Vostok: Climate and greenhouse gases. *Global and Planetary Change* **7**, 131–143.
Mayewski, P.A., Meeker, L.D., Whitlow, S. *et al.* (1994a) Changes in atmospheric circulation and ocean ice cover over the North Atlantic during the last 41,000 years. *Science* **263**, 1747–1751.
Mayewski, P.A., Wumkes, M., Klinck, J. *et al.* (1994b) Record drilling depth struck in Greenland. *EOS* **75**, 113–117.
Mosley-Thompson, E. & Thompson, L.G. (1982) Nine centuries of microparticle disposition at the South Pole. *Quaternary Research* **17**, 1–13.
Taylor, K.C., Lamorey, G.W., Doyle, G.A. *et al.* (1993) The 'flickering switch' of late Pleistocene climate change. *Nature* **361**, 432–436.
Thompson, L.G., Mosley-Thompson, E., Dansgaard, W. & Grootes, P.M. (1986) The little ice age as recorded in the stratigraphy of the tropical Quelccaya ice cap. *Science* **234**, 361–364.
Zielinski, G.A., Mayewski, P.A., Meeker, L.D. *et al.* (1994) Record of volcanism since 7000 B.C. from the GISP2 Greenland ice core and implications for the volcano–climate system. *Science* **264**, 948–952.

ice sheets The largest bodies of ICE on the planet, sometimes covering entire continents. The term is generally applied to ice masses greater than 50 000 km² in extent. Topographic features in the bedrock are often completely buried so that the ice surface has a simple dome shape with the thickest ice at the centre. Ice accumulates here and flows outwards to be lost at the margins by melting or iceberg calving. Flow is organized into separate drainage BASINS, separated by ice divides. Discharge is often focused into narrow fast-moving sections of the ice sheet, called ice streams.

At present there are two ice sheets on the Earth, one in Greenland and the other in Antarctica. The latter is by far the largest, containing 30 million km³ of ice, compared with a more modest 2.6 million km³ in Greenland. If all this ice were to melt, global sea level would rise by 70 m. The thickest ice (4.8 km) is found in Antarctica, where the mass of the ice sheet depresses the underlying bedrock by up to 1 km. Parts of the Antarctic ice sheet rest on bedrock that would be up to 2 km below sea level even after removal of the ice load. These areas are regarded as being the most sensitive to environmental change.

At the time of the last glacial maximum (about 20 000 years ago), ice sheets covered much of the northern parts of Eurasia and North America. Global sea level was 120 m lower as a result. *See also* ICEBERGS. [A.J.]

icebergs Floating ICE calved from a glacier or ICE SHEET that terminates in the SEA or less commonly a lake. The term is generally applied to pieces with a freeboard greater than 5 m. Between one-tenth and one-sixth of the total volume shows above the waterline, depending on the mean density of the iceberg, so the maximum freeboard depends on the shape. The largest icebergs, called tabular icebergs because their horizontal dimensions greatly exceed their thickness, are those calved from ice shelves, the floating extensions of ice sheets. A cycle of ice-shelf advance, followed by calving of a giant iceberg, may take more than a century. The resulting iceberg may be more than 100 km long, have a thickness of up to 500 m and may survive more than a decade following calving. In September 1986 a single calving event in Antarctica produced three icebergs totalling 13 000 km² in area. One of these began drifting north in 1990 and finally broke up off the coast of Argentina in 1992, leaving the other two still grounded in shallow water near the site of the calving.

The quasi-periodic nature of iceberg calving makes it difficult to determine the mean annual flux of icebergs from the ice sheets of Greenland and Antarctica. Recent estimates suggest that about 200 km³ year⁻¹ calve from Greenland (about 40% of the total mass loss) and about 2000 km³ year⁻¹ calve from Antarctica (about 80% of the total mass loss). [A.J.]

ICM *See* INTEGRATED CATCHMENT MANAGEMENT.

ICZN *See* INTERNATIONAL CODE OF ZOOLOGICAL NOMENCLATURE.

Ideal free distribution

'Ideal free' theory is concerned with the DISTRIBUTION of competitors across RESOURCES. In its simplest form, it considers a range of HABITATS that vary in the quality of resource (say food or mates) available to a population of competitors. Each competing individual is free to go where it chooses (i.e. is not excluded from particular habitats by dominant or territorial individuals) and 'ideal' in the sense of having perfect information about the quality of each of the habitats (thus the phrase 'ideal free'). The assumption is that competitors will exploit the habitats so as to maximize their reproductive potential. Since the effects of their decisions on reproductive potential will not usually be measurable directly (but see below), it is assumed that the rate of resource acquisition provides an indirect index of reproductive potential. The first competitors to arrive will therefore go to the richest habitat where the resource-acquisition rate will be greatest. As the number of competitors in the best habitat increases, however, the resource begins to deplete and acquisition rate drops. There will therefore come a point at which individuals would do just as well in what was originally the second-richest habitat. The same process occurs in this habitat until it becomes profitable to exploit the third-richest habitat. The upshot is that habitats will eventually be filled in such a way that the rate of resource acquisition by competitors is the same in each. This stable ideal free distribution could be achieved in one of two ways, either as the number of competitors that becomes distributed across habitats in the ratio of initial resource value or through individual competitors visiting all habitats but spending time in each in the ratio of initial resource value.

Experimental evidence from a variety of species has supported the predictions of ideal free theory, though usually under relatively simple conditions of continuous resource input (i.e. maintaining constant resource availability at alternative sites) and using food intake as an indirect measure of reproductive potential. In dung flies (*Scatophaga stercoraria*) and poplar aphids (*Pemphigus betae*), however, where individuals have been studied competing for copulations (male dung flies) or reproductive galling sites on leaves (female aphids), ideal free distributions have emerged for more direct measures of reproductive potential.

While competitors may distribute themselves in an ideal free fashion and on average secure equal rewards, the reward rate experienced by individuals may vary within habitats because of local HETEROGENEITY in resource availability and COMPETITION for the best local sites or because some individuals generally command greater access to resources regardless of resource heterogeneity. This highlights a major problem with ideal free theory in its simple form: its assumption that all individuals in a population have equal competitive ability is unlikely ever to be met in the real world. More sophisticated models have therefore based predictions on the distribution of 'competitive units' rather than individuals. Thus an individual that can acquire three times as much resource as another is worth three times the number of 'competitive units'. The reason the distribution of competitors often conforms to the predictions of simple ideal free theory while violating its basic assumption of uniform competitive ability is that, even with unequal competitors, the number of distributions of 'competitive units' that by chance result in an apparent ideal free distribution of numbers is much greater than the number that do not. This means that apparent ideal free distributions are statistically more likely to be observed and emphasizes the importance of testing conformity with assumptions as well as predicted outcomes. [C.J.B.]

REFERENCES

Fretwell, S.D. (1972) *Populations in a Seasonal Environment*. Princeton University Press, Princeton, N.J.

Parker, G.A. & Sutherland, W.J. (1986) Ideal free distributions when individuals differ in competitive ability: phenotype-limited ideal free models. *Animal Behaviour* **34**, 1222–1242.

idiobiont parasitoids *See* PARASITOID–HOST INTERACTIONS.

IGBP (International Geosphere–Biosphere Programme) An international programme that coordinates research into BIOSPHERE dynamics with the stated goal 'to describe and understand the interactive physical, chemical and biological processes that regulate the

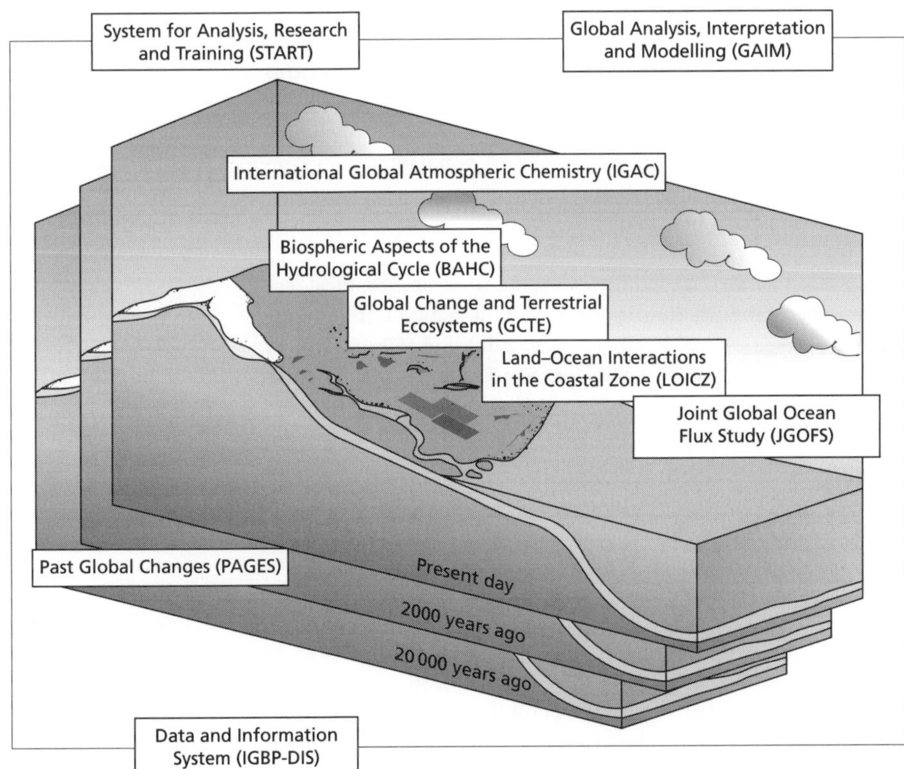

Fig. I1 The International Geosphere–Biosphere Programme has developed a set of linked projects to investigate the life-driven interactions within and between the land, atmosphere and ocean components of the Earth system. (After IGBP (1992) *Global Change: Reducing Uncertainties*. The International Geosphere–Biosphere Programme, The Royal Swedish Academy of Sciences, Stockholm.)

total Earth system, the unique environment that it provides for life, the changes that are occurring in the system, and the manner in which they are influenced by human actions'. The IGBP was instituted by ICSU (International Council of Scientific Unions) in 1986 as an open-ended programme and the science plan of the programme as a whole was published in 1988. The IGBP secretariat is located at the Royal Swedish Academy of Sciences.

The programme comprises a set of linked 'core projects' as illustrated in Fig. I1. Each core project has a broadly similar structure with a staffed project office and with a science plan that divides the research into broad foci and specific activities. Ecology permeates the entire IGBP and is especially prominent in the Global Change and Terrestrial Ecosystems (GCTE) core project. GCTE's aim is to be able to predict effects of changes in CLIMATE and land use on terrestrial ecosystems and their feedbacks into the climate system. GCTE has four foci on ecosystem physiology, ecosystem structure, agriculture, forestry and soils, and ecological complexity.

In common with the other core projects GCTE does not fund research directly, but achieves its aims through facilitating meetings and identifying gaps. It engages the support of individual scientists who are nationally funded and who elect to contribute their research to the project.

In addition to the thematic core projects, IGBP has three cross-cutting 'framework' activities for data and information (DIS), global analysis (GAIM) and training and developing country involvement (START). Close ties are maintained with the World Climate Research Programme and the Human Dimensions of Global Environment Change Programme. Starting in 1995 a system of transects has been developed, initially five in number and around 1000 km long. Study sites are to be located along the gradients, which will serve the needs of several core and cross-cutting projects. *See also* UNITED NATIONS PROGRAMMES AND BODIES. [M.B.]

ILO (International Labour Organization) *See* UNITED NATIONS PROGRAMMES AND BODIES.

immigration Entry of CONSPECIFICS into a population from elsewhere. Leads to GENE FLOW into the population. *Cf.* DISPERSAL; EMIGRATION; MIGRATION. [P.C.]

immigration curve Like extinction curves, these are graphical representations of the relationship between immigration rate and time or SPECIES RICHNESS of an island. MacArthur and Wilson's theory of ISLAND BIOGEOGRAPHY depicts immigration decreasing monotonically from an initial peak when the island is barren to zero when all species from the source pool have colonized. The immigration rate of species is predicted to be less on to small, isolated islands than for large, less-isolated islands. The intersection of the EXTINCTION CURVE and the immigration curve predicts the equilibrium species richness of a particular island and predicts a SPECIES–AREA RELATIONSHIP. [N.M. & S.G.C.]

immunity, to parasites The specific RESISTANCE of a HOST to reduce the number or survival of parasites. All living organisms are capable of a kind of immunity against parasites, but the precise mechanisms vary considerably. Most of the knowledge comes from vertebrates, some insects, crustaceans and plants. Research in immunity emphasizes physiological processes; however, behavioural strategies that reduce the risk of infection or alternative life-history patterns that avoid or minimize the effects of parasitic infection are also part of the host defence system.

A central problem for immunity is to distinguish between self and non-self. In the simplest case, immunity may be based on unspecific components of the host environment (e.g. PROTEINS) that are not favourable for the infection, establishment, growth, multiplication or reproduction of parasites. The more complicated immunity in insects is based on the interaction of humoral and cellular processes and a recognition of non-self. The cellular response involves the recruitment of haemocytes that attach to the foreign object (the antigen), which is thus eventually encapsulated and isolated. Vertebrates have an exceedingly complex immune system with an array of specialized cells. It is anticipatory, since antibodies to virtually all possible foreign objects are in stock. The existence of autoimmune diseases demonstrates that immunity is not without a cost and risk to the host itself. *See also* DEFENCE MECHANISMS; PARASITISM; SPECIES INTERACTIONS. [P.S.H.]

REFERENCES

Klein, J. & Horesji, V. (1997) *Immunology*, 2nd edn. Blackwell Science, Oxford.

Wakelin, D. (1984) *Immunity to Parasites*. Edward Arnold, London.

IMO (International Maritime Organization) *See* UNITED NATIONS PROGRAMMES AND BODIES.

imposex Imposition of male sexual organs on the female. Can occur in marine gastropods as a result of POLLUTION from tributyltin contained in ANTIFOULING paints. *See also* ENDOCRINE DISRUPTER; TRIBUTYLTIN (TBT) POLLUTION. [P.C.]

REFERENCE

Gibbs, P.E., Pascoe, P.L. & Burt, G.R. (1988) Sex change in the female dog-whelk *Nucella lapillus*, induced by tributyltin from antifouling paints. *Journal of the Marine Biological Association* **68**, 715–731.

imprimitive life history A population with an imprimitive life history does not converge to the STABLE AGE DISTRIBUTION even in a constant environment. This category includes all species with non-overlapping generations, as well as species with OVERLAPPING GENERATIONS that are semelparous and have a fixed juvenile period. Examples of the latter include BIENNIAL PLANTS without seed banks and insects with a fixed developmental period of two or more years. In these species, coexisting COHORTS are virtually demographically isolated. Periodical species are species with imprimitive life histories and overlapping generations characterized by large abundance differences among the cohorts or complete elimination of some of them (e.g. periodical cicadas, *Magicicada* sp.). *See also* DEMOGRAPHY; LIFE TABLE. [T.J.K.]

imprinting Ethological imprinting is a special learning process during a sensitive phase of an individual's ONTOGENY.

1 Imprinting upon an object. Konrad Lorenz maintained that a young animal becomes irreversibly imprinted upon whatever moving object is encountered during the sensitive phase and that it subsequently directs its filial and sexual behaviour toward that object. For example, newly hatched ducklings separated from their

mother will follow a person, a crude model duck or even a cardboard box. The function of this filial imprinting is obviously the build-up of an attachment between offspring and parent. The sensitive phase for sexual imprinting occurs later than that for filial imprinting. In cross-fostering experiments carried out, for example, with ducks and geese or with zebra finches (*Taeniopygia guttata*) and Bengalese finches (*Lonchura striata*) the offspring showed a strong sexual preference for the foster-parent species when they became sexually mature themselves. The function of sexual imprinting is not only species recognition as Lorenz had suggested but also to achieve an optimal compromise between inbreeding and outbreeding. The offspring learn the detailed characteristics of their mother and siblings as a basis for their choice of sexual partner when they are mature. Bateson showed experimentally with quail that the birds choose a mate that appears slightly different, but not too different, from its parents and siblings: a first cousin is preferred to both a SIBLING and a second cousin. Both filial and sexual imprinting can be regarded as a special form of associative learning.

2 Imprinting as perceptual learning of a behaviour pattern such as bird song. In zebra finches, as in many other birds, song learning takes place during a sensitive phase long before the birds sing themselves. Although song from a foster species is learned if there is no choice, young birds show some bias toward learning the song of their own species and, at a more detailed level, they also prefer to learn a song similar to that of their father. This suggests that they have an inherited template that guides learning. Song learning may have some relation to sexual imprinting because females might use the similarity between the songs of related males as a cue in mate choice so as to achieve optimal outbreeding. *See also* GENOMIC IMPRINTING. [M.Mi.]

REFERENCES

Bateson, P. (1990) Is imprinting a special case? *Philosophical Transactions of the Royal Society of London, Series B* **329**, 125–131.

McFarland, D. (1985) *Animal Behaviour.* Pitman, London.

Slater, P.J.B., Eales, L.A. & Clayton, N.S. (1988) Song learning in zebra finches: progress and prospects. *Advances in the Study of Behavior* **18**, 1–34.

imprinting, genomic *See* GENOMIC IMPRINTING.

***in situ* bioassays** Biological monitors that are either planted out in the environment to be monitored or in close proximity such that continuous or semi-continuous samples can be taken from it. Thus fish fitted with appropriate sensors can be caged and immersed in the habitat to be monitored or they can be housed in tanks on the bank through which water from the habitat to be monitored is run. *See also* BIOASSAYS. [P.C.]

***in situ* genetic reserves** Contemporary (and much favoured) practices in the conservation of genetic resources include the protection and management of animals and plants in their natural habitats (*in situ* reserves) as opposed to protection in zoos and botanic gardens (*see* *EX SITU* GENETIC RESERVES), often far removed from the natural habitats. In those parts of the world where *in situ* genetic reserves have been set up (often under various guises such as safari parks), the great majority of larger terrestrial animal species are likely to be present. For example, in a survey of 12 African countries over 80% of native birds were present in protected areas. In South Africa, 97% of bird species (along with 93% of mammals and 92% of amphibians and reptiles) native to the country were represented by breeding populations in the established *in situ* reserves.

However, the presence of a species in an *in situ* reserve is no guarantee of long-term survival, particularly with the highly varied standards of protection offered. In many instances the standards are directly related to the economic vagaries of income from tourism and hunting.

In situ conservation of plants through control of land use, with or without the status of reserve, is an important aspect of plant conservation practices. However, the effectiveness of these controlled areas varies considerably even between neighbouring countries. In Poland, over 90% of threatened plant species are present in reserves. In the UK, the figure is about 75% but in southern Europe declines to less than 20%. In Norway, until recently, the figure was even lower. In the UK, as elsewhere, *in situ* conservation is often most effective in areas of little value to agriculture and forestry (e.g. mountain tops, tidal marshes).

A more specialized development of *in situ* conservation has been the conservation of CROPS and wild relatives of crops, designed to slow down the genetic erosion of the global crop GENE POOL (*see* AGROECOSYSTEMS). Some limited success has been achieved through funding of farms to maintain traditional practices, for example in West Africa where the region holds extensive crop species and varieties both of traditional and more modern agriculture. One benefit of *in situ* genetic reserves has been through the Plant Genetic Resource Centre in Addis Ababa, which runs a programme to reintroduce traditional varieties and landraces to areas where recent decades of drought and war have destroyed much of the genetic stock. Elsewhere in the tropics the focus has turned recently to *in situ* conservation of tropical tree species, many of whose seeds cannot be stored at low temperatures. *See also* CONSERVATION BIOLOGY. [G.A.F.H.]

REFERENCE

Groombridge, B. (ed.) (1992) *Global Biodiversity: A Report Compiled by the World Conservation Monitoring Centre*, pp. 543–575. Chapman & Hall, London.

inbreeding depression A reduction of FITNESS in a normally outbreeding population as a consequence of increased HOMOZYGOSITY due to inbreeding. Increased homozygosity decreases fitness through the expression of deleterious homozygous recessive TRAITS. Close relatives are more likely to share these ALLELES in the heterozygous condition by chance and therefore more likely to produce homozygous recessive offspring (Fig. I2). [A.R.H.]

Inceptisols *See* SOIL CLASSIFICATION.

incidence functions Incidence functions describe how the PROBABILITY of occurrence of a species varies with selected characteristics of islands, usually their area or SPECIES RICHNESS. They are constructed by plotting the proportion of islands of a given area or species richness that are occupied against their area or species richness. Species incidence generally increases with area, but great variation exists in the shape of incidence functions between species, leading to the (arbitrary) distinction of 'high-S', 'tramp' and 'supertramp' species. Incidence functions have recently been used to parameterize equilibrium incidence-function metapopulation models

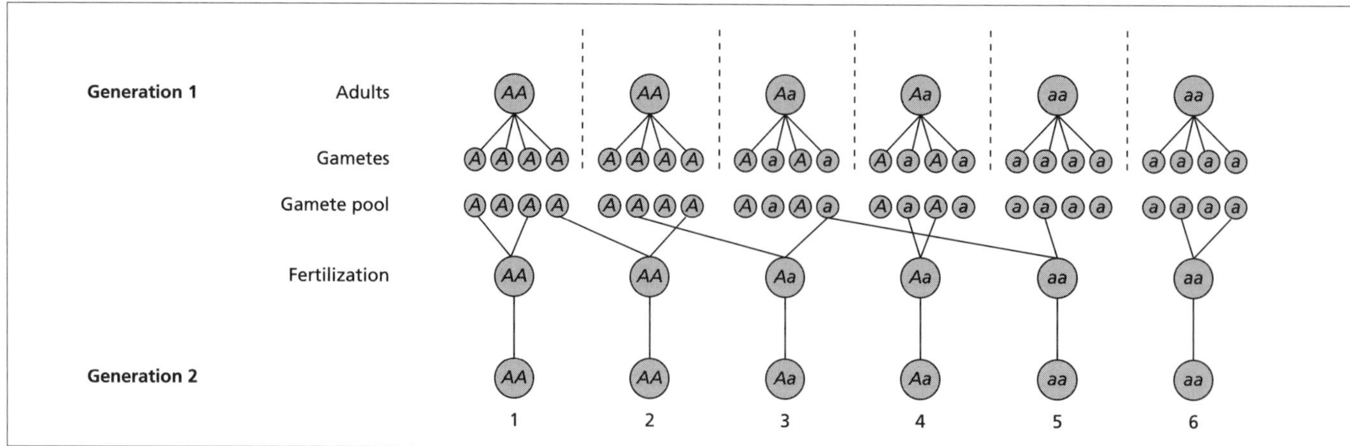

Fig. I2 Inbreeding in a small population produces homozygosity. A homozygote can be produced either by combining copies of the same allele from different individuals, or by combining two copies of the same allele at a given locus from a single hermaphrodite. Here we imagine that the population contains six adults, which are potentially self-fertilizing hermaphrodites, and each adult produces four gametes. Homozygotes can then be produced by the kind of cross-mating assumed in the Hardy–Weinberg theorem (e.g. offspring number 2) or by self-fertilization (offspring number 1). Self-fertilization necessarily produces a homozygote only if its parent is homozygous (compare offspring 1 and 4). For sexual reproduction, relatives will be more likely to share alleles, and therefore more likely to produce homozygous offspring (see offspring 2 and 5). (After Ridley, M. (1996) *Evolution*, 2nd edn. Blackwell Science, Cambridge, Mass.)

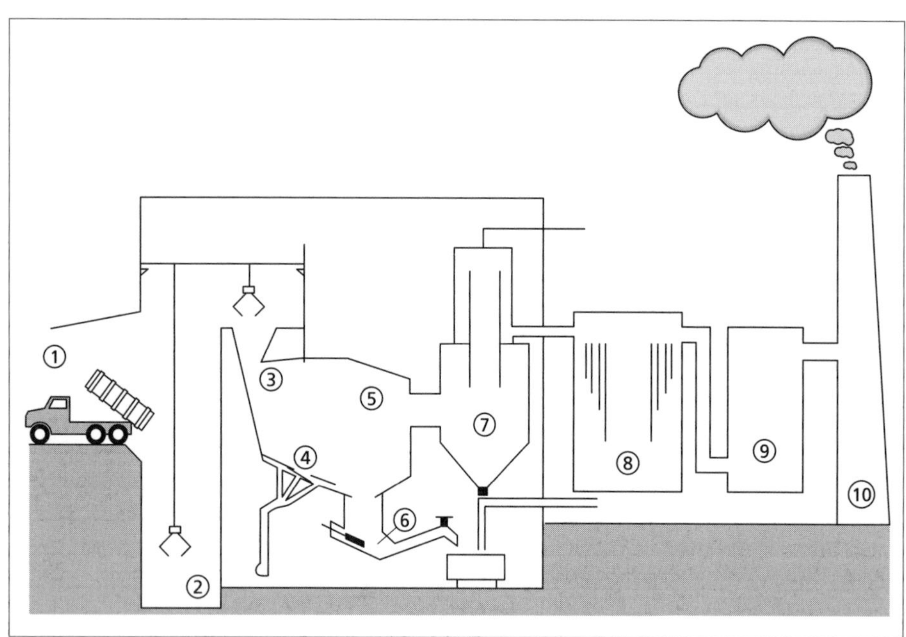

Fig. I3 How does incineration work? 1, Reception hall; 2, waste pit; 3, incinerator feed hopper; 4, combustion grate; 5, combustion chamber; 6, quench tank for bottom ash; 7, heat recovery boiler; 8, electrostatic precipitator; 9, acid gas scrubbing equipment; 10, incinerator stack. (After White, P. & Franke, M. (1993) *Integrated Solid Waste Management: A Lifecycle Inventory*. Chapman & Hall, London.)

to investigate the persistence of populations in fragmented habitats. *See also* ASSEMBLY RULES; ISLAND BIOGEOGRAPHY. [N.M. & S.G.C.]

REFERENCE

Diamond, J.M. (1975) Assembly of species communities. In: *Ecology and Evolution of Communities* (eds M.L. Cody & J.M. Diamond), pp. 342–444. Harvard University Press, Cambridge, Mass.

incineration The controlled burning of usually solid, but sometimes liquid or gaseous, combustive wastes to produce gases and solid residues containing little or no combustible materials. Burning solids has four distinct objectives:

1 volume reduction (depending upon composition, volume reductions can be >90% and mass reductions >70%);
2 stabilization, for example incinerated output (ash) is considerably more inert than the input of mixed municipal solid WASTE;
3 recovery of energy; and
4 sanitation.

The key factors for high levels of combustion and destruction of organics are TEMPERATURE (high), RESIDENCE TIME (long) and TURBULENCE (high). The majority of municipal solid waste incinerators have a furnace with a moving grate design (Fig. I3). The moving grate keeps waste moving through the furnace. Primary air for combustion is pumped through the grate; secondary air is introduced over the fire to ensure good combustion in the gas phase (e.g. of DIOXINS). Alternative designs are fluidized bed incinerators. The combustion chamber contains a fluidized bed, created by air forced through a bed of inert material such as sand. This hot fluidized bed ensures a high level of combustion and benefits from having no moving parts. Emissions need to be cleaned (dust filtered; acid gases removed; NO_x gases removed; dioxins removed by maintaining high temperatures). The bottom ash is cooled, usually exposed to a magnet to recover ferrous metal and then disposed of together with the FLY ASH and residues from gas cleaning. Bottom ash is usually dumped in LANDFILLS. *See also* WASTE DISPOSAL. [P.C.]

inclusive fitness *See* KIN SELECTION.

income breeder An organism that does not use stored body reserves to support reproduction. The alternate mode is a CAPITAL BREEDER. Many small mammals such as shrews are income breeders; because of their high metabolic rate such animals cannot accumulate excess food for STORAGE in their tissues and thus must support lactation by continuous foraging. *See also* LIFE-HISTORY EVOLUTION; TRADE-OFF. [D.A.R.]

increaser A plant that increases in ABUNDANCE in heavily grazed grassland or rangeland, generally as a result of competitor release of unpalatable or GRAZING-tolerant plant species. This competitor release is typically due to the reduction in COVER (or height) of previously superior but more palatable competitors as a result of selective herbivory. Increasers are usually regarded as noxious plants by graziers (they tend to be toxic to livestock, spiny, indigestible or woody). *See also* COMPETITIVE RELEASE. [M.J.C.]

indeterminate growth GROWTH that continues throughout the LIFESPAN of an individual, such that body size and age are correlated. It is usually characterized by a reduction in growth rate past maturation and is typical for many molluscs, many crustaceans, fish, amphibians, turtles and most PERENNIAL plants. The VON BERTALANFFY EQUATION is frequently used to model this type of asymptotic growth curve. It has been shown that often a TRADE-OFF exists between resource allocation into growth and reproduction. Whether continued growth ultimately enhances the organism's reproductive success depends on the effects of increased body size on survivorship and the capacity to acquire RESOURCES. *Cf.* DETERMINATE GROWTH. [D.E.]

index of biotic integrity (IBI) Attempt to quantify divergences from 'normality' in ECOSYSTEMS (ecosystem harm/disturbances) by measuring a number of attributes of populations and species, in particular: fish and/or macroinvertebrate SPECIES RICHNESS; ratios of native versus non-native species; trophic composition; overall ABUNDANCE and condition. Relatively disturbed sites are compared with IBIs from undisturbed sites in the region. Though multifactorial, this index suffers from similar problems to other BIOTIC INDICES. [P.C.]

REFERENCE

Karr, J.R. (1991) Biological integrity: a long-neglected aspect of water resource management. *Ecological Applications* **1**, 66–84.

Indian Ocean The youngest (*c.* 125 million years old) and third largest of the world's oceans with an area of 73 million km², a volume of 2840 km³ and an average depth of 3872 m (or 74 million km², 2840 km³ and 3840 m if its marginal seas are included). The deepest part of the Indian Ocean occurs in the Sunda Trench at a depth of 7455 m. It has a narrower CONTINENTAL SHELF compared with the ATLANTIC OCEAN and is marked by a large number of microcontinents (e.g. Madagascar, Chagos, etc.). The Mid-Indian Ocean Ridge is the most conspicuous feature of the bottom TOPOGRAPHY and is cut by several prominent fracture zones. CORAL REEFS are prominent features, particularly in the western parts of the Indian Ocean. Current systems in the north are strongly influenced by the monsoons and change seasonally. In the south, the South Equatorial Current and the Westwind Drift are the dominant ANTICYCLONIC currents. The TIDES are semi-diurnal along the East African coast, but are of a mixed type on the coasts of Arabia, Iran and India. Pure diurnal tides occur along the south-west coast of Australia, the southern part of the South China Sea and in the Java Sea. Approximately 54.3% of the Indian Ocean floor is covered by calcareous ooze, 20.4% by siliceous ooze and 25.3% by red CLAY. [V.F.]

indicator species This can be used as a general term to indicate species that may be indicative of a COMMUNITY or set of environmental conditions or, very precisely, to refer to species objectively defined as indicators of a phytosociological class, for example sea pink (*Armeria maritima*) for sea cliffs, salad burnet (*Sanguisorba minor*) for chalk grassland and bluebell (*Hyacinthoides non-scriptus*) for oceanic oak woods. The former approach is widely encountered in many areas of biology and is then based on the experience of the person talking or writing or their collective field experience. In the more objective case, the indicator species are derived by some statistical or multivariate technique to specially derive such groupings. There are many techniques ranging from association analysis, which was frequently used in the 1970s and early 1980s, to indicator species analysis and its variants, for example two-way indicator species analysis, which is in use today. It is the basis of TWINSPAN, the computer package used to produce the National Vegetation Classification. This is a comprehensive nationwide scheme for classifying British VEGETATION and is the basis for the production of a series of books (e.g. Rodwell 1991). The use of indicator species enables anyone to locate his/her own vegetation samples in the national scheme and to see how they compare with the regional and national variation in that community. On the continent of Europe, similar subjective methods permit the identification of community types by means of indicator species. These have been associated with the names of phytosociologists, such as Braun-Blanquet in France, and were widely applied in the 1960s, 1970s and 1980s, but today there has been a convergence of the two methods. They both end up with similar results and the widespread availability of electronic computing has facilitated the use of the more quantitative approaches. Often bryophytes, and other small or fairly obscure species, seem to be ideal indicators, which can lead to identification problems even for professional botanists. Neither approach has been found to be particularly easy to apply in the tropics because of the difficulty of identifying all the species.

Presence of tolerant species or absence of sensitive species can signal environmental STRESS from natural or human causes. There are two potential problems with this approach.

1 Specificity: organisms that are sensitive/tolerant to one stressor may not respond similarly to others so their presence/absence may not provide an indication of stress.

2 Generality: organisms showing general sensitivity/TOLERANCE do not give sufficiently precise information about environmental conditions.

See also BIOTIC INDICES; INDEX OF BIOTIC INTEGRITY; SENTINEL SPECIES. [F.B.G. & P.C.]

REFERENCES

Hill, M.O., Bunce, R.G.H. & Shaw, M.W. (1975) Indicator Species Analysis, a divisive polythetic method of classification and its application to a survey of native pinewoods in Scotland. *Journal of Ecology* **63**, 597–613.

Rodwell, J.S. (1991) *British Plant Communities. 1. Woodlands and Scrub.* Cambridge University Press, Cambridge.

Indices of dispersion

There are many ways of encapsulating the DISTRIBUTION of individuals in space into a single number (*see* DISPERSION). Most make some assumptions about the distribution and therefore different indices will be appropriate on different occasions. Some of the more commonly encountered indices are described below.

Variance: mean ratio

This is the simplest index of dispersion. If a set of equivalent SAMPLE units of a POPULATION are available (e.g. number of plants in a QUADRAT of fixed size) then calculation is straightforward. The index is easy to interpret as this ratio will be equal to 1 if the population is randomly distributed (i.e. conforms to a POISSON DISTRIBUTION). If the index is greater than 1, then the population is aggregated (i.e. more clumped than random) and if less than 1, the distribution tends towards uniformity (i.e. less clumped than random). Departures from 1 may be tested for significance.

A problem with the use of this index is that it is influenced by the size of the MEAN in aggregated populations. A simple modification of this index of dispersion is to express it as:

$$\text{Index} = \left(\frac{s^2}{\text{Mean}}\right) - 1$$

where s^2 is the sample VARIANCE. This is often called the index of clumping. The advantage of this index is that random patterns will give a value of 0. The disadvantage is that the value of the index is dependent on the number of individuals in the total sample.

A further modification to avoid the problem of dependence on sample size is called Green's index:

$$\text{Green's index} = \frac{\left(s^2/\text{Mean}\right) - 1}{n - 1}$$

where n is the total number of individuals in the sample. The value of this index will also be 0 for RANDOM DISTRIBUTIONS and will be 1 for maximum clumping (e.g. all individuals are found in one quadrat).

Taylor's power law

This index of dispersion makes use of the fact that in natural populations the variance is not independent of the mean. In most biological populations variance increases with increasing mean. TAYLOR'S POWER LAW states that there is a linear relationship between log(mean) and log(variance) and that the slope of this relationship can be used as an index of dispersion. If populations follow a random (Poisson) distribution at all densities then the slope of the line will be 1. An aggregated population will have a slope greater than 1.

k of the negative binomial

Distributions that are more clumped than random may often be approximated by a NEGATIVE BINOMIAL DISTRIBUTION. In nature, distributions fitting the negative binomial are very common and this index is consequently very useful, although the biological significance of the distribution is still a subject of some debate. The negative binomial is a discrete PROBABILITY DISTRIBUTION described by two parameters: the mean and k. The value of k can be used as an index of the clumping of a distribution. If the value of k is very high (in practice anything above about 10), then the distribution approaches a Poisson. If k is low, then it indicates AGGREGATION. A k of 1 or below implies strong aggregation. As this index is only applicable to AGGREGATED DISTRIBUTIONS (the negative binomial cannot describe a distribution more ordered than random) its reciprocal is often expressed instead, giving higher values for more aggregated distributions and a value approaching 0 for a Poisson distribution.

Accurate calculation of the parameter k of the negative binomial requires an exhaustive iterative technique. A rough estimate of k can be gained using only the sample mean and variance:

$$k \approx \frac{(\text{Mean})^2}{s^2 - \text{Mean}}$$

Positive binomial

As for the negative binomial but used in the rare situations where the variance of a set of samples is less than the mean (i.e. individuals are more even than random).

Lloyd's index of mean crowding

This is an index of dispersion that uses data from samples to calculate not the mean number of individuals in each sample but the mean number of individuals encountered by each individual. In other words, it measures the amount of crowding experienced by individuals. If a population follows a Poisson distribution, then MEAN CROWDING will equal the sample mean. If a population is aggregated, then mean crowding will exceed the sample mean. Other indices are based on the concept of mean crowding.

Nearest neighbour analysis

There are several NEAREST NEIGHBOUR TECHNIQUES available. Accurate positional data for every individual in a sample area may be required. The technique usually requires the observer to select individuals at random and then find the distance from that individual to its nearest neighbour. If populations are clumped, then the mean distance from a random set of individuals to their nearest neighbours will be smaller than if the population is randomly distributed. This sort of method is particularly appropriate for SESSILE organisms or analysis of photographic or satellite data. *See also* INTRASPECIFIC AGGREGATION. [C.D.]

REFERENCES

Pielou, E.C. (1969) *An Introduction to Mathematical Ecology*. John Wiley & Sons, New York.

Southwood, T.R.E. (1978) *Ecological Methods*. Chapman & Hall, London.

indirect fitness *See* KIN SELECTION.

indirect interaction When the DENSITY of a species alters in a COMMUNITY, the consequent changes in other species result from both direct and indirect interactions. Indirect effects occur when changes in one species affects another through a chain of intermediate species. Changes occurring through COMPETITION between two species for the same resource (EXPLOITATION COMPETITION) is an indirect effect, as is a change in the density of a plant species that results from an increase in a carnivorous PREDATOR which reduces HERBIVORE numbers. With the exception of exploitation competition, ecologists have for the most part ignored indirect effects, although it is increasingly recognized that these can be very important and can lead to large effects. Moreover, theoretical studies suggest that when indirect interactions occur they make predicting the outcome of density changes an almost impossible task. It appears, for example, that it is not even possible to predict whether, in the long term, the density of a prey species will go up or down when the density of its predator is reduced. These findings have important implications for the interpretation of field experiments where species densities are manipulated and the effects on the remaining community are observed (*see* ECOLOGICAL EXPERIMENTS). The uncertainties regarding the outcome of species manipulations also have important implications when environmental management options are being considered. *See also* SPECIES INTERACTIONS. [S.J.H.]

REFERENCE

Yodzis, P. (1988) The indeterminacy of ecological interactions as perceived through perturbation experiments. *Ecology* **69**, 508–515.

indirect ordination *See* ORDINATION.

indirect transmission, parasites The process by which a PARASITE is passed from one (infectious) HOST to another (susceptible) host through at least one, second species (*see* DIRECT TRANSMISSION, PARASITES). The major categories of intermediate species are vectors (*see* VECTOR, PARASITES) and SECONDARY HOSTS (*see also* RESERVOIR HOSTS, PARASITES). Many parasites are able to utilize more than one route of transmission. *See also* SPECIES INTERACTIONS. [G.F.M.]

indium (In) A rare, silver-white METAL (ATOMIC NUMBER = 49; relative ATOMIC MASS = 114.82; melting point = 156.61°C; boiling point = 2000°C), indium has an estimated abundance in the Earth's CRUST of 0.05 mg kg^{-1}. Commonly associated with ZINC ores, it is extracted as a by-product of zinc refining. The metal is used in low-melting alloys and as gaskets in high-vacuum apparatus. Indium phosphide (InP) and indium antimonide (InSb) are semiconductors and are used as infrared detectors. Indium tin oxide is an optically transparent, electrical conductor and so is used in solar cells and liquid crystal displays. Very little is known about the toxicities of indium compounds other than that they appear to be low. [C.R.P.]

individual selection A form of NATURAL SELECTION consisting of non-random differences among GENOTYPES within a population, which leads to changes in the proportions of different genotypes (genes) in the population. *Cf.* GROUP SELECTION. [V.F.]

individualistic hypothesis This proposes that species of plants are distributed individually with respect to biotic and ABIOTIC FACTORS rather than as groups or communities. Gleason (1926, p. 23) wrote: 'the vegetation of an area is merely the resultant of two factors, the fluctuating and fortuitous immigration of plants and an equally fluctuating and variable environment'. The hypothesis was proposed to replace Clements' view of the plant COMMUNITY embodied in the ORGANISMIC HYPOTHESIS. The individualistic hypothesis has gained much support from studies on the ORDINATION of records of species occurrence or abundance in surveys of vegetation, from unbiased studies of species occurrences in SUCCESSION (*see* VARIABLE PATHWAYS) and from evidence on the past abundances and distributions of plant species gained from studies of POLLEN ANALYSIS. Nevertheless, many instances are known of plants having beneficial or unfavourable effects on others (*see* FACILITATION; INHIBITION OF SUCCESSION) and of particular species having to be complementary to others in their behaviour if they are to coexist. It is also generally accepted now that, while no two areas of vegetation are exactly the same, there are sufficient similarities between stands for CLASSIFICATION of vegetation to be both practicable and worthwhile. [P.J.G.]

REFERENCE

Gleason, H.A. (1926) The individualistic concept of the plant association. *Bulletin of the Torrey Botanical Club* **53**, 7–26.

industrial ecosystem Designing industrial operations on the basis of natural ecosystems; RESOURCES are shared and WASTE recycled, for example spare heat from one factory is sold or bartered with another. Important features of industrial ecosystems are:

- compatibility with natural systems;
- maximum internal REUSE of materials and energy;
- selection of processes with reusable waste;
- extensive interconnection among companies and industries;
- sustainable rates of natural resource use;
- waste intensity matched to natural process cycle capacity.

See also SUSTAINABLE DEVELOPMENT. [P.C.]

REFERENCE

Stilwell, E.T., Cantry, R.C., Kopf, P.W. & Montron, A.M. (1991) *Packaging for the Environment*. Arthur D. Little, New York.

industrial melanism Many animals, especially insects, have melanic or black colour forms. Industrial MELANISM is a term used to describe the situation in which there is a correlation between the frequency of these forms within populations of a species and some measure of the level of AIR POLLUTION or industrial development (Fig. I4). *See also* ECOLOGICAL GENETICS. [P.M.B.]

inertial forces *See* REYNOLDS NUMBER.

Infanticide

Infanticide usually refers to the killing of young CONSPECIFICS, including in many cases the infanticidal individual's own offspring. It has been observed in a wide range of species from mammals (including humans), birds and fish to insects, arachnids and crustaceans. In some cases, it is clearly accidental, as in the case of elephant seal (*Mirounga angustirostris*) pups being crushed to death by fighting bulls. In others it appears to be pathological, arising from distortions of parental bonding and care behaviour. However, somewhat counter-intuitively from a functional/evolutionary perspective, infanticide may be an adaptive STRATEGY reaping reproductive benefits for the perpetrator. Following Huntingford and Turner (1987), there are at least three ways in which infanticide may be adaptive.

(a)

(b)

Fig. I4 Peppered moths (*Biston betularia*). (a) Melanic forms are better camouflaged in polluted areas, (b) peppered forms are better camouflaged in unpolluted areas. (a, From Brakefield, P.M. (1987) Industrial melanism: do we have the answers? *Trends in Ecology and Evolution* **2**, 117–122; b, courtesy of T.G. Liebert.)

Sexual selection

In some species, males may gain a reproductive advantage by killing young sired by other males because it accelerates the onset of female receptivity. Infanticide by male lions (*Panthera leo*) when they take over a pride of females is a well-known example. In other cases, the same ends may be achieved less directly (and bloodily). In some rodents, the presence of a strange male can prevent implantation of fertilized ova or cause abortion (the Bruce effect), again potentially increasing the receptivity of females. However, the extent to which the Bruce effect operates in natural populations is not clear.

Parental investment

Paradoxically, infanticide may be an adaptive PARENTAL INVESTMENT strategy. Following Trivers' (1972) parental investment argument, the wisdom of investing in current offspring depends on the impact of the investment on the ability to invest in future offspring. At times when food is short, say, LIFETIME REPRODUCTIVE SUCCESS may be enhanced by eating current young (whose chances of survival are likely to be slim anyway). Energy and nutrients are thus recouped, potentially high costs of PARENTAL CARE are saved and the parent is likely to survive to breed again when times get better. In some cases, however, CANNIBALISM of young appears to occur as part of a GENERALIST predatory repertoire, as when herring gulls (*Larus argentatus*) cannibalize the chicks of other breeding adults.

Competition

While the example of cannibalism in herring gulls may reflect catholic tastes in prey, it can also be regarded as a means of reducing COMPETITION. Under some conditions, competition for resources can lead to cannibalism among young themselves. Larval cannibalism among some beetle and anuran species, for example, may enhance the growth rate and later fecundity of the cannibals. Sometimes siblings compete for resources within their own brood. In a number of bird species, older, larger nestlings kill their smaller siblings. In large raptors, it is common for females to lay two EGGS but for the first hatchling eventually to kill the second (the so-called Cain and Abel effect). Death may result from injury or from the smaller chick being prevented from feeding. The parents do not intervene in this SIBLING competition, which appears to reflect an adaptive brood reduction strategy ensuring the survival of the stronger chick in the face of limited resources.

Many cases of infanticide and associated aggressive abuse of infants in humans can be interpreted in functional terms, often relating to certainty of parentage, age- and sex-dependent reproductive potential and environmental (e.g. economic) STRESS. *See also* SELECTIVE ABORTION. [C.J.B.]

REFERENCES

Huntingford, F.A. & Turner, A.K. (1987) *Animal Conflict.* Chapman & Hall, London.

Lenington, S. (1981) Child abuse: the limits of sociobiology. *Ethology and Sociobiology* **2,** 17–29.

Trivers, R.L. (1972) Parental investment and sexual selection. In: *Sexual Selection and the Descent of Man* (ed. B. Campbell), pp. 139–179. Aldine, Chicago, Ill.

infauna *See* BENTHOS.

Infochemicals

The involvement of chemicals in conveying information in intraspecific and interspecific interactions between organisms is widespread. Examples can be found in interactions among and between vertebrates, invertebrates, plants and microorganisms in the same or different trophic levels. They involve many behavioural and physiological responses, such as SEARCHING for food or oviposition sites, locating a mate, marking a resource area, etc. The chemicals involved are termed 'infochemicals', being defined as chemicals that, in the natural context, convey information in an interaction between two individuals, evoking in the receiver a behavioural or physiological response that is adaptive to either one of the interactants or to both (Dicke & Sabelis 1988). It is important to note that the term 'infochemical' refers to an aspect of chemicals, namely conveying information, that by itself does not confer benefit or HARM to a receiver. It is the response that is beneficial or detrimental. In addition, it should be stressed that the term 'infochemical' is not chemical specific but context specific. This is to say that a certain chemical that mediates interactions between individual organisms can be classified differently according to the aspect of the chemical taken into consideration. For instance, interactions between organisms can also be mediated by other classes of chemicals, such as NUTRIENTS and TOXINS. These classes refer to aspects of chemicals that by themselves are to the benefit or detriment of a receiver. Yet a nutrient or toxin may in some cases, but certainly not all, also confer information to an organism. If information conveyance is considered, the chemical is called an infochemical; if the value of the chemical itself is referred to, terms such as 'nutrient' and 'toxin' should be used.

Costs and benefits

That organisms invest in producing chemicals that transmit information is an interesting phenomenon. Why they do this requires a clear understanding of the benefits to the producer. The evolution of their production is less obvious than that of nutrients and toxins, because of the indirect way infochemicals confer benefit and detriment. After the information is disseminated, the producer is dependent on the translation of the information by the receiver (Dicke & Sabelis 1992). For instance, if a plant responds to being damaged by a HERBIVORE by producing a volatile chemical that attracts carnivorous enemies of the herbivore, the plant is dependent on the enemy of the herbivore decoding the message and responding by moving towards its source. Suppose that the carnivore does respond to the information, then some plant mutant may emit the information while not being damaged so as to ensure early RECRUITMENT of the carnivore as a preparation to a potential attack by herbivores. Such actions by plant mutants will reduce the information value of the infochemical. Thus, as soon as COMMUNICATION is established, the value of the infochemical is dependent on strategies of receivers and other emitters. To a large extent, communication through infochemicals depends on the costs involved in their production and the benefits that they yield through a response in a receiver. This holds for the producer side of the communicative interaction. At the receiver end, costs and benefits are also important factors in determining the fate of the communicative interaction. Among the costs for the producer are not only biosynthetic costs but also ecological costs, in terms of use by 'illegal' receivers and mimicked production by 'illegal' producers. It is exactly this aspect of infochemicals that makes them so special compared with toxins and nutrients that themselves are beneficial or detrimental to the receiver.

Pheromones, allomones, kairomones and synomones

Infochemicals can be further classified according to the interaction they mediate. Two criteria are important: (i) whether the interaction mediated by the infochemical is between two individuals from the same species (pheromones) or (ii) between individuals from different species (allelochemicals). In both cases, three subclasses can be distinguished, depending on whether the emitter, the receiver or both benefit from the infochemical (Table I2). Again, it should be realized that these terms are not chemical specific but context specific. One chemical can function as PHEROMONE, kairomone and synomone, depending on the context and the interactants consid-

Table I2 Definitions of infochemical classes.

Infochemical
A chemical that, in the natural context, conveys information in an interaction between two individuals, evoking in the receiver a behavioural or physiological response

Pheromone
An infochemical that is pertinent to the biology of an organism (1) and that evokes in a receiving organism of the *same* species (organism 2) a behavioural response that is adaptively favourable to:
- organism 1 but not to organism 2: (+,–) pheromone
- organism 2 but not to organism 1: (–,+) pheromone
- organism 1 and 2: (+,+) pheromone

Allelochemical
An infochemical that is pertinent to the biology of an organism (1) and that evokes in a receiving organism of a *different* species (organism 2) a behavioural response that is adaptively favourable to:
- organism 1 but not to organism 2: allomone
- organism 2 but not to organism 1: kairomone
- organism 1 and 2: synomone

ered. For instance, aphids may produce a volatile infochemical in response to being attacked by a PREDATOR. This infochemical initiates DISPERSAL in nearby aphids that are closely related individuals of the same CLONE and thus functions as a (+,+) pheromone. In addition, the infochemical may attract other predators to an aphid COLONY and thus functions as a kairomone. Furthermore, it may attract myrmecophilous ants that use the aphids as source of honeydew and protect them against predators; then the infochemical functions as a synomone.

Origin of infochemical

Infochemical terminology is based on 'semiochemical' terminology. Semiochemicals basically comprise infochemicals and toxins, but exclude nutrients (Nordlund & Lewis 1976). Moreover, infochemical terminology clearly differs from semiochemical terminology in that the origin of the infochemical does not play a role (Dicke & Sabelis 1988). In semiochemical terminology, one of the interactants has to be the producer of the chemical, while this is not necessarily so in infochemical terminology. There are two reasons for excluding the origin as a criterion in infochemical terminology. First, the origin is not always clear. The observation that an organism emits the infochemical does not guarantee that it also produced it. The infochemical may be produced by a symbiotic microorganism or the compound may have been sequestered from a food source. Yet if the compound mediates an interaction between the emitter and another organism this mere fact is important in an ecological sense. Thus, the second reason for excluding the origin as a criterion in infochemical terminology is that if one of the interactants must be the producer, as is the case in semiochemical terminology, ecologically important interactions will be excluded. In infochemical terminology, reference to the origin of the chemical has been replaced by reference to the chemical being 'pertinent to the biology of an organism'.

An overwhelming mix of infochemicals from essentially all organisms in the environment is available in nature. For any organism to employ this information, sensitive chemoreceptors tuned to decode the information are a prerequisite. Evidence for all major groups of organisms shows that such chemoreceptors are present. Yet the availability of such chemoreceptors may also incite an organism to produce infochemicals that are meant to manipulate another organism to do things that it would otherwise never do. Infochemicals mediate a wide range of interactions between individuals of the same or different species. Consequently, infochemicals play an essential role in the functioning of ecosystems and are subject to strong selection forces. Understanding these intriguing phenomena will give us a better understanding of the environment. [M.Di.]

REFERENCES

Dicke, M. & Sabelis M.W. (1988) Infochemical terminology: based on cost benefit analysis rather than origin of compounds? *Functional Ecology* **2**, 131–139.

Dicke, M. & Sabelis, M.W. (1992) Costs and benefits of chemical information conveyance: proximate and ultimate factors. In: *Insect Chemical Ecology: An Evolutionary Approach* (eds B.D. Roitberg & M.B. Isman), pp. 122–155. Chapman & Hall, New York.

Nordlund, D.A. & Lewis, W.J. (1976) Terminology of chemical releasing stimuli in intraspecific and interspecific interactions. *Journal of Chemical Ecology* **2**, 211–220.

information centre hypothesis The hypothesis holds that the exchange of information between individuals at a communal site about the location of rich and EPHEMERAL food patches has been the driving force for the evolution of colonial roosting or breeding behaviour, particularly in birds but also in other taxonomic groups. A foraging individual that is unsuccessful at one occasion could thereby follow a successful individual from the COLONY, and when it is successful it will be followed in turn by unsuccessful ones when leaving the colony. The hypothesis rests, therefore, on the assumption that the individuals participating in an information centre benefit and lose in turn, and that the altruistic act of an individual is reciprocated by others in the future. The key problem of the functioning of colonies as information centres concerns the question of why a successful forager should return to the colony, and thereby pay the time and energy cost of the trip from the food patch to the colony and back, and additionally suffer from the costs of being followed by parasitic, unsuccessful foragers. The stringent conditions posed by cooperation models and the high mobility of animals in colonies will most likely preclude stable cooperation between individuals in colonies. Without reciprocated cooperation, the originally proposed information centre mechanism rests on GROUP SELECTION. Although very popular, the information centre hypothesis therefore cannot provide a viable explanation for the evolution of colonial behaviour. More than 100 studies have attempted to test the hypothesis. None of them have provided convincing support and many of the results could be explained by simpler mechanisms such as local enhancement. Furthermore, other hypotheses have been largely neglected. A recent review can be found in Richner and Heeb (1995).

The key to understanding the evolution of COLONIALITY in organisms that exploit rich but highly patchy and ephemeral food sources must be found in either the benefits that arise to successful foragers at the colony or benefits that arise at the food patch by foraging in larger groups. Benefits at the colony may include safety from predators, thermal advantages, opportunities for mate finding and extra-pair copulations, or inclusive FITNESS benefits arising from informing genetically related individuals, as is the case for bees. Benefits at the food patch include reduced risk of predation by the simple DILUTION EFFECT or by earlier detection of a PREDATOR, reduced risk of starvation, reduced time spent in predator VIGILANCE, and increased FEEDING time and feeding rate by various mechanisms such as prey confusion, social learning or use of public information. The individuals that successfully localized a rich food patch could then use colonies to recruit other individuals and benefit from these group foraging advantages. Thus, the evolution of colonial behaviour may be better explained by the RECRUITMENT centre function of colonies. So far there are no empirical studies testing this recruitment centre hypothesis. The recruitment centre hypothesis could also account for the spectacular calling and flight DISPLAYS that are often seen at colonies of birds, which may be honest displays of food patch quality performed by the birds seeking recruits (Richner & Heeb 1996). *See also* SOCIOBIOLOGY. [H.R.]

REFERENCES
Richner, H. & Heeb, P. (1995) Is the information center hypothesis a flop? *Advances in the Study of Behaviour* **24**, 1–46.
Richner, H. & Heeb, P. (1996) Communal life: honest signalling and the recruitment center hypothesis. *Behavioral Ecology* **7**, 115–118.

infracommunity (parasite ecology) The COMMUNITY of PARASITE species that coexist within an individual HOST. The infracommunity comprises all the infrapopulations in the individual, where an infrapopulation is defined as all members of a given species of parasite within a single community. [S.J.H.]

REFERENCE
Esch, G.W., Bush, A.O. & Aho, J.M. (1990) *Parasite Communities: Patterns and Processes*. Chapman & Hall, London.

infrared gas analyser (IRGA) Infrared gas analysers measure the density (moles per unit volume) of gas components such as CARBON DIOXIDE (CO_2) and WATER vapour. Heteroatomic gases absorb RADIATION in characteristic wavebands, particularly in the infrared region of the spectrum. Infrared gas analysers for CO_2 are routinely used in biological applications to measure rates of PHOTOSYNTHESIS and RESPIRATION; infrared analysers for water vapour are used to estimate TRANSPIRATION rates of plants. Rapid-response CO_2 and water vapour infrared gas analysers are used as part of EDDY COVARIANCE systems used to estimate FLUXES of CO_2 and water vapour above communities of vegetation. Industrial applications of infrared gas analysers for process control and environmental monitoring cover a wide range of gases including HYDROCARBONS. Analysers are designed to measure the density of gas either on an absolute scale or as a differential relative to a reference gas of known composition.

Infrared gas analysers are designed as either dispersive (DIR) or non-dispersive (NDIR) instruments. DIR analysers are based around a gas cell in the optical path of a scanning spectrometer, allowing determination of the composition of complex gas mixtures. NDIR analysers measure the transmission of broad-band infrared radiation through an optical cell. NDIR measurements are made specific for one or two selected gases using a combination of optical filters and/or gas-filled detectors, although all NDIR analysers retain some degree of cross-sensitivity between interfering gases such as CO_2 and water vapour. NDIR analysers are used in most ecological applications since they tend to be more robust and can be made field portable. DIR analysers are reserved for analytical applications where their cost, larger size and complexity are considered acceptable. [P.R.V.G.]

REFERENCE
Jarvis, P.G. & Sandford, A.P. (1985) The measurement of carbon dioxide in air. In: *Instrumentation for Environmental Physiology* (eds B. Marshall & F.I. Woodward), pp. 28–57. Cambridge University Press, Cambridge.

inheritance of acquired characters This is the proposition that characters acquired through new use, in the lifetime of an individual, are transmissible from one generation to another. Similarly, characters lost through disuse in one generation do not reappear in future generations. This is known as LAMARCKIAN INHERITANCE, after the French biologist Jean Baptiste Lamarck (1744–1829), who used it as a central principle in his theory of EVOLUTION. Following the rediscovery of Mendel's work, it was increasingly abandoned. A central tenet of evolutionary biology now is that germ lines (genes) are isolated from SOMA (PHENOTYPE) so that the inheritance of ACQUIRED CHARACTERS is not possible; this is known as Weismann's principle (or barrier), after the German biologist August Weismann (1834–1914) who first made it explicit. In fact, towards the end of his life, Charles Darwin also espoused the inheritance of acquired characters because, not having the benefit of Mendel's insights, he presumed that inheritance worked by blending male and female characters, although this would tend to dilute ADAPTATIONS and work against NATURAL SELECTION. Thus he thought the only way that evolution could work was if advantageous characters, acquired during the lifetime of an individual, could be passed on.

However, there were two other features of Lamarckian theory that Charles Darwin would not accept:

1 all organisms were supposed to have innate powers to progress towards a more complex and perfect form;

2 all organisms had an inner disposition that caused the performance of actions sufficient to meet the needs created by a changing environment.

Using these principles, Lamarck explained the evolution of tentacles in the head of a snail as follows: as it crawls along it finds the need to touch objects in front of it and makes an effort to do so; in so doing it sends subtle fluids into extensions of its head; ultimately nerves grow into these and the tentacles become permanent; the acquired tentacles are transmitted from parent to offspring. *Cf.* DARWINISM. *See also* GERM LINE. [P.C.]

inhibition of succession Occurs where an earlier-invading species reduces the chance (or mean rate) of ESTABLISHMENT of another; this was contrasted by Connell and Slatyer (1977) with FACILITATION. Examples are found where dense invasion by shrubs effectively prevents invasion by trees (at least for decades, as shown by Niering & Egler 1955), and where LIGHT-demanding trees such as alders or birches suppress the growth of more shade-tolerant trees such as spruces (Chapin *et al.* 1994). The extreme cases of inhibition are sometimes distinguished as exclusion, for example where one species changes the pH in such a way that another cannot become established (Grubb & Hopkins 1985). [P.J.G.]

REFERENCES
Chapin, F.S., Walker, L.R., Fastie, C.L. & Sharman, L.C. (1994) Mechanisms of primary succession following deglaciation at Glacier Bay, Alaska. *Ecological Monographs* **64**, 149–175.
Connell, J.H. & Slatyer, R.O. (1977) Mechanisms of succession in natural communities and their role in community stability and organization. *American Naturalist* **111**, 1119–1144.
Grubb, P.J. & Hopkins, A.J.M. (1985) Resilience at the level of the plant community. In: *Resilience in Mediterranean-type Ecosystems* (eds B. Deil, A.J.M. Hopkins & B.B. Lamont), pp. 21–38. Junk, Dordrecht.
Niering, W.A. & Egler, F.E. (1955) A shrub community of *Viburnum lentago*, stable for twenty-five years. *Ecology* **36**, 356–360.

inhibitory resource *See* RESOURCES.

initial phase *See* LIFE-CYCLE ASSESSMENT.

innate capacity for increase *See* INTRINSIC RATE OF INCREASE.

insecticides The problem of insect pests has an ancient history, with classical Greek, Roman and biblical references. Until 1939 the insecticides available were often very poisonous, such as ARSENIC and nicotine; safer natural products such as derris and pyrethrum, which were difficult to produce in quantity, lacked persistency and potency and were not stable in light. The discovery of the insecticidal properties of DDT in 1939 was revolutionary, changing the view of researchers that such useful and apparently safer materials could be synthesized on a large scale. DDT was welcomed as a major breakthrough that saved millions of lives by preserving food CROPS and stocks and protecting humans and livestock from parasites and disease-carrying insects. Since that discovery, a very large range of insecticides have been found, mostly within a few main classes of compounds: organophosphates (*see* ORGANOPHOSPHORUS PESTICIDES), ORGANOCHLORINES (including DDT), carbamates and pyrethroids; other groups include organotins, acylureas, formamides and growth regulators (including PHEROMONE attractants).

The TOXICITY of insecticides to other animal life has always been of concern, but particularly attracted attention in the 1960s with the publication of *Silent Spring* by Rachel Carson. It became evident that although a number of insecticides were immediately highly toxic, others were less directly toxic but accumulated in the environment, while yet others, notably some organophosphates and organochlorines, accumulated in animal fats and thus were concentrated in higher levels of organism in the FOOD CHAIN.

Such concerns have led to the withdrawal or banning of a number of compounds in all or some sectors of the market, and there is increased emphasis on discovering analogues and new chemical groups of greater selectivity. For example, the synthetic pyrethroids, first introduced in 1976, have rapidly made an impact as more selective insecticides with reduced environmental PERSISTENCE, although they are not as effective soil-acting materials as members of the other chemical groups; however, more recent analogues are showing greater soil activity.

The non-agricultural uses of insecticides may be as important as the agricultural uses. They are important not only in human health, helping to control diseases such as malaria, sleeping sickness, typhus and yellow fever that are carried by insect vectors, but also in the preservation of organic materials, such as wood and clothing, from insect attack. *See also* BIOCIDE; INTEGRATED PEST MANAGEMENT. [K.D.]

Insects, diversity and conservation

Insects eclipse all other forms of life by sheer numbers and DIVERSITY, constituting about three-quarters of all animal species. They belong to phylum Arthropoda and share the features of an exoskeleton composed of chitin and sclerotin and appendages divided into segments separated by joints. Insects constitute class Insecta. They differ from all other arthropods by having three specialized body segments (head, thorax and abdomen), three pairs of appendages (six legs) on the thorax and a single pair of antennae on the head. Insects often have one or two pairs of wings on the thorax in addition to their legs.

Insects owe their great success to their highly effective external skeletal structure, which is lightweight, tough and waterproof. Although this exoskeleton has limited their size, small size has helped them exploit every terrestrial habitat on Earth and has allowed many of them to fly. Flight has permitted efficient SEARCHING and MIGRATION for food, mates and habitats for reproduction. The exoskeleton also contributes to the myriad forms and colours of insects. The survival advantage of MIMICRY of plants and other insects, both in form and colour, has driven diversification in many insect groups. The constraining skeleton requires METAMORPHOSIS for insects to grow, punctuated by moults. Some insects lack males and exhibit parthenogenetic reproduction. Others may reproduce sexually (biparental) at times and parthenogenetically at other times during their COMPLEX LIFE CYCLES. Developmental POLYMORPHISM is diverse in insects and has contributed to their dominance.

Classification and characteristics

There are at least 28 orders in class Insecta worldwide. They are distinguished mainly by wings, mouthparts and the kind of metamorphosis. Collembola (springtails), Protura, Diplura and Thysanura (silverfish and bristletails; Fig. I5) are the four most primitive insect orders. Most of them are small SOIL dwellers with simple mouthparts and with abdominal appendages in addition to the three pairs of thoracic legs, a primitive feature throughout the class. They are primitively wingless and 'ametabolous' (i.e. without external metamorphosis) and have three basic life stages of egg, nymph and adult. The nymph and adult look the same except for size and, in the case of Protura, the number of abdominal segments. These insects are not usually thought economically important, although some silverfish can be troublesome because they feed on starchy materials such as books. In 1 ha of soil and leaf litter in RAIN FOREST more than half the arthropods are springtails. Like many other insects, their conservation value in releasing and recycling NUTRIENTS locked up in decaying VEGETATION is undoubtedly enormous.

As insects evolved, wings and external metamorphosis developed. Gradual metamorphosis is a feature for 16 orders. Nymphs may have external wing pads that eventually develop into wings of adults. Among these orders, Ephemeroptera (mayflies), Odonata (dragonflies and damselflies) and Plecoptera (stoneflies) are 'hemimetabolous', with aquatic nymphs called naiads. Naiads use gills to breathe and are either detritus-feeders or predaceous. They differ from adults not only by appearance but also in habitat and often food preference.

The other 13 orders with gradual metamorphosis are 'paurometabolous': their nymphs and adults are similar in habitat and diet. Orthoptera (grasshoppers, crickets, katydids) and Phasma-

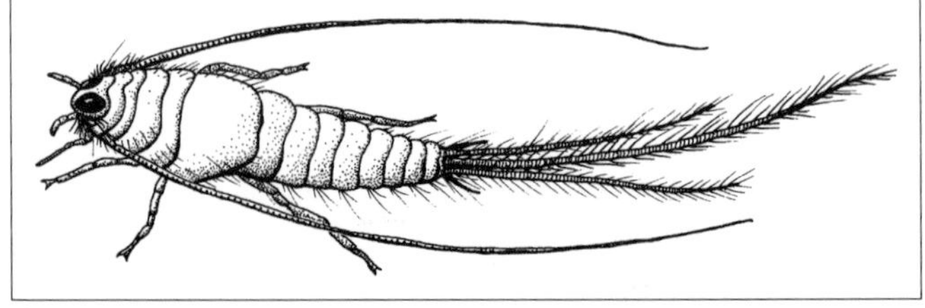

Fig. I5 A bristletail, order Thysanura, a primitive ametabolous insect.

todea (walkingsticks) are mostly herbivorous. Some grasshoppers, the true 'locusts', are harmless in their solitary phase; however, they can cause severe damage to agricultural CROPS in their gregarious phase. Mantodea (mantids) are an order mostly beneficial because they prey on other insects.

Embiidina (web-spinners) are chiefly tropical in distribution. They live in silken galleries spun by glands on their front legs and only males are winged; both are unusual characters in Insecta. Zoraptera and Isoptera (TERMITES) are gregarious. These orders, with Blattaria (cockroaches), Grylloblattodea (rock crawlers), Dermaptera (earwigs) and Psocoptera (booklice and barklice), all live in crevices under wood, rocks, bark, detritus, in caves or in shelters of their own construction in the case of some termites. They are mostly scavengers and play an important role in breaking down wood and other CELLULOSE materials otherwise indigestible by most animals. Termites, for example, have symbiotic protozoans and bacteria in their alimentary tract that convert cellulose to sugars essential for metabolism. This remarkable symbiosis, combined with their advanced social life, contributes to their great biological success in tropical and subtropical regions.

Mallophaga (chewing lice) and Anoplura (sucking lice) are secondarily wingless like some insects in the aforementioned orders. As a result of COEVOLUTION, most chewing lice and all sucking lice have specialized as ectoparasites of birds and mammals, respectively. Sucking lice, such as the body louse, may also act as vectors of disease, including relapsing fever, trench fever and particularly typhus.

Unlike most insects, Hemiptera (true bugs), Homoptera (cicadas, leafhoppers, aphids, scale insects) and Thysanoptera (thrips) do not have chewing mandibles. Instead, their mouthparts have specialized into a beak or proboscis. All homopterans, many thrips and true bugs use their mouthparts to pierce plant tissues to suck juices, a much more efficient way to utilize plants than chewing. A considerable number of true bugs, however, are predatory on other insects and invertebrates. A few of these have developed blood-sucking habits (like the infamous bed bug) and some of them are important vectors of human diseases. The majority of true bugs are terrestrial, but some are aquatic. There are even marine water striders in Hemiptera, the only insects living on the open ocean far out past the INTERTIDAL ZONE. These remain one of the mysteries of evolution.

The remaining, more advanced eight orders of insects undergo complete metamorphosis. They are 'holometabolous' and their LIFE CYCLES include egg, larval and pupal stages before becoming adults. The LARVA is worm-like, actively feeds and grows through moults. The different larval stages between moults, called instars, are similar in form but differ in size. The wings develop internally during the larval stage and are not everted until the end of the last instar. Then, the insect transforms into a pupa. The pupa does not feed and is usually inactive. Pupae are often covered by some protective structure like a cocoon, chrysalis or puparium. This has allowed holometabolous insects to evolve high diversity not only in form but also in life history. Most species of Neuroptera (Dobsonflies, lacewings, antlions; Fig. I6) are carnivorous in the larval stage and some as adults. Some of these, and most larvae of Trichoptera (caddisflies), are aquatic. Many caddisfly larvae build the most delicate and intricate cases or nets of various shapes under water, with bits of leaves, twigs, sand grains and pebbles. They are plant or detritus feeders. The free-living caddisfly larvae are generally predaceous, while flying adults are liquid-feeders. All insect orders with aquatic larvae are important in food chains of fish and other aquatic animals.

Mecoptera (scorpionflies, hangingflies) are either scavengers or predators, with larvae being soil-dwellers and with flying adults. Siphonaptera (fleas), a distinct order, may be secondarily wingless descendants of Mecoptera. Most flea larvae live and feed on debris and the bloody faeces of adults. Adult fleas feed on blood and may act as disease vectors.

Coleoptera (beetles, weevils) have the front pair of wings hardened as elytra to provide protective covers. This may be the major reason for their success as the largest order in all the animal kingdom. They occupy all terrestrial and freshwater habitats. Some are ectoparasitic, while others are endoparasites of other insects and higher animals as either larvae or adults. Many feed on algae, fungi, dead wood, plants and grains. Some have become serious pests of AGRICULTURE; some are predators, like the ladybird beetles, on other insects and used as natural enemies in BIOLOGICAL CONTROL of pests. Stylopidae, a group of insects with highly reduced elytra and large hindwings, have larva-like females parasitic on grasshoppers and bees throughout life. Many treat this family as a separate order, Strepsiptera.

Lepidoptera (butterflies, moths, skippers) rank second numerically in the class. Many of them are PHYTOPHAGOUS as larvae and cause damage to a variety of crops. However, this is the most popular and attractive group of insects. The brilliant colours of many species have caught the eye of collectors and the attention of conservationists. Silk produced by the silkworm (*Bombyx mori*) has been a source of wealth and beauty for thousands of years.

Hymenoptera (sawflies, wasps, ants, bees) range from solitary species to species with highly elaborated social systems such as honeybees, truly remarkable biological phenomena. They constitute the third largest group of insects and are chiefly associated with plants, as pests in a few cases but primarily as pollinators, where they exceed all other organisms in importance. Many females can sting with their needle-like ovipositors, which function as an effective organ of offence and defence. This makes many of them extremely beneficial as parasites or predators of other arthropods. There is bizarre diversity of PARASITISM in this order, for example adelphoparasitism, where the male parasitizes the female of the same species.

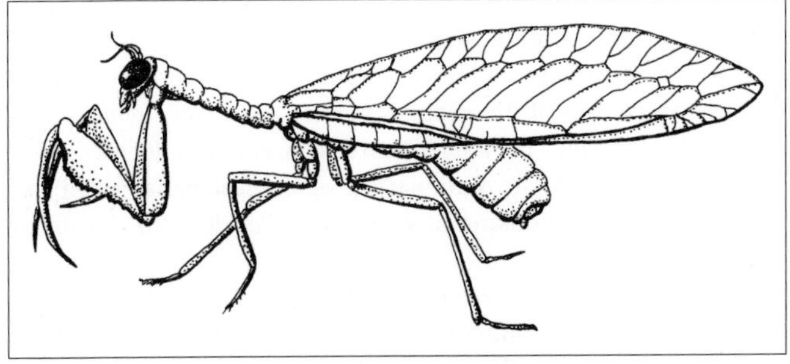

Fig. I6 A mantispid, order Neuroptera, an advanced holometabolous insect.

No animals other than people themselves are responsible for greater loss of human life and economic destruction than Diptera (flies, mosquitoes, midges, gnats). More than half of the world's population suffers from fly-borne diseases. Many fly and mosquito larvae are aquatic. Many flying adults feed on blood with their piercing and sucking mouthparts. These may inject saliva into their hosts and thus transmit diseases. On the other hand, many are extremely beneficial parasites and predators, as valuable as parasitic wasps. The role of fly larvae in the detritus cycle, processing dead plants and animals, is exceeded only by that of bacteria and fungi. Some species are found in freshwater, brine or even oil. Many are stem- and fruit-borers, others leaf-miners and gall-makers, causing crop losses.

Conservation and importance

The number of insect species described exceeds 750 000; a reasonable estimate of the total number is about 10 million. Being small has not only conferred success but also difficulty in being recognized. The huge gap between described and existing insect species imposes difficulties for insect conservation. Emphasis on discovery and description is the avenue of the future. The conservation of insects' incredible genetic diversity must overcome the traditional bias towards charismatic megavertebrates and against pests (only a few species of insects). It is essential to increase public recognition of insects' ecological services to our society in terms of food supply, silk and wax production, medicine, POLLINATION, biocontrol, energy recycling and DECOMPOSITION, in addition to the awareness of nature's worth.

Insects are virtually everywhere, excluded only by the extremes of climate at the poles and on the peaks of the highest mountains. The global human population explosion has fragmented and disturbed landscapes for housing development and farming. Only habitat protection can slow the loss of genetic diversity. In the process of adapting to living in such a vast variety of habitats, insects have also been able to consume a wide range of food resources. In addition to carnivores and scavengers that feed on dead animal bodies, bone, faeces, hair, feathers and ashes, 15% of insects are parasitic on other insects and animals. Extinction of hosts for ectoparasites, such as birds and mammals, may cause extinction of these insects. Probably more than half of insects are herbivores, feeding on plant material like seeds, leaves, flowers and stems; multitudes feed on algae, bore into wood or live on leaf litter. This implies that plant conservation underpins many aspects of insect conservation. As pollinators in both agricultural and natural ecosystems, insects are also critical to the survival of many plants. Conservation programmes must preserve such MUTUALISM. Compounded by the diversity inherent in metamorphosis, conservation of insects requires awareness of the changes in food resources, habitats and reproductive cycles. Therefore, insect conservation focuses on a vast array of variables: evolutionary, geographical, behavioural and in life history. Overlooking any of these may lead to disastrous consequences.

Species COMPETITION, weather, pathogens, parasitism and predation are factors that influence insect abundance and distribution. Above all, human modification of the landscape inevitably shifts the competitive advantage of one species towards another through changes in plant composition. Land use changes habitats and climates and is generally adverse for most species, though sometimes benefits a few. Species diversity is generally highest in warm and humid tropical regions, associated with plant diversity under the same conditions. However, both tropical and temperate species frequently show great variations in seasonal abundance. This means that preservation of population variation is just as important in the tropics as in the temperate zones. An obvious difficulty is that there is insufficient personnel to catalogue even a fraction of endangered tropical species, although some long-term strategies have been developed in affluent temperate areas. Landscape conservation and preservation, and MONITORING of indicator groups such as butterflies and dragonflies, are the most important and urgent umbrella options at the present time.

Concern arises from the potential impacts of GLOBAL WARMING on insect SPECIES RICHNESS. In particular, temperate lands in the Southern hemisphere are fragmented, avoided much of the perturbation of Pleistocene GLACIATION and, as a result, harbour very high diversity and endemicity. Ozone depletion and increasing aridity in this hemisphere could disproportionately imperil a large fraction of global BIODIVERSITY. The same logic may apply to tropical species when compared with temperate counterparts. Less climatic adversity in the tropics may have exposed them to less tolerance of climatic changes. Great strides have been made in the conservation of commercially valuable, large, colourful species such as butterflies (especially tropical swallowtails) and some beetles. A 1995 conference on conservation of Europe's bees highlighted the effects of declining distribution and abundance on their vital pollination services to natural plant communities and agriculture. These are only beginnings, however, and the broad issues of insect conservation have yet to be addressed. *See also* BEETLES, DIVERSITY AND CONSERVATION; BIODIVERSITY; CONSERVATION BIOLOGY. [W.L.]

REFERENCES

Nabhan, G. & Buchmann, S. (1995) *Forgotten Pollinators*. Island Press, Washington, D.C.

Samways, M.J. (1994) *Insect Conservation Biology*. Chapman & Hall, London.

Williams, C.S. (1995) Conserving Europe's bees: why all the buzz? *Trends in Ecology and Evolution* **10**, 309–310.

---

insolation Exposure to SOLAR RADIATION. [J.G.]

integral equations *See* EQUATIONS.

integrated catchment management (ICM) River management based on the understanding that a river system cannot be divided into stretches—there is a continuum of interaction throughout the system—and cannot be isolated from surrounding land. The quantity of water in a river is influenced by the surrounds (by RUN-OFF) and influences the surrounds (by flooding). Moreover, there is a continuum between riverwaters and lateral groundwaters. Thus management of WATER QUALITY has to be carried out in an integrated way, with all these factors in mind. Integrated CATCHMENT models try to capture this complexity and are used, for example, to simulate the input of run-off and its effects on water quality. [P.C.]

integrated conservation strategies Integrated conservation strategies are plans created to conserve species, particularly plants, using all available resources to meet a goal, such as species recovery. For a particular species, integrated strategies might include *in situ* conservation activities (land acquisition, management, habitat restoration), *EX SITU* CONSERVATION (seed or pollen banks, field gene banks or other germplasm storage), research (e.g. reproductive, germination and/or demographic studies) or enhanced legal protection such as through land-use decisions, regulatory listings as protected, or control of plants in commerce. *See also* CONSERVATION BIOLOGY; SEED BANK. [E.O.G. & L.R.M.]

REFERENCES

Falk, D.A. (1987) Integrated conservation strategies for endangered plants. *Natural Areas Journal* **7**, 118–123.

Falk, D.A. (1990) Integrated strategies for conserving plant genetic diversity. *Annals of the Missouri Botanical Garden* **77**, 38–47.

integrated environmental management Control of business operations to minimize ENVIRONMENTAL IMPACT by reference to the whole operation, including personnel, and the interactions within it. An alternative is to address isolated issues, such as 'bolting on' CLEAN-UP systems to the 'end of pipe'. An integrated approach, on the other hand, would involve managing the whole process, including the actions of personnel (e.g. through training) and even the design of the product to minimize WASTE and EMISSIONS. Integrated environmental management ought, therefore, to be more cost-effective in the long term, especially since it should also take account of the fact that business operations are complex integrated systems. [P.C.]

integrated pest management (integrated control) Use of different methods in combination to suppress PEST populations below their economic THRESHOLD. Techniques to be used are selected rationally with an emphasis on their economic and ENVIRONMENTAL IMPACT. Most successful integrated pest management (IPM) schemes combine chemical and BIOLOGICAL CONTROL with additional methods.

Overuse of pesticides, particularly INSECTICIDES, through routine insurance spraying creates a series of related problems: pest RESISTANCE, resurgence, secondary pest OUTBREAKS, killing of natural enemies, pollinating species and non-target organisms, and accumulation of toxic residues in the soil, waterways and food chains. Increasing selection for resistance exhausts available insecticides, leading to uncontrollable losses on major world CROPS (cotton, sugar-cane). IPM aims to integrate suitable techniques, minimize economic damage and limit disturbance to the environment.

The most critical element in IPM is the restricted use of selective pesticides, which are applied only when pest densities exceed their economic threshold. Economic thresholds are expressed as a DENSITY of pests per sampling unit (number of plants, flower heads, volume of soil) and relate expected crop loss at harvest to an estimate of pest density at an earlier stage in crop growth. The economic threshold is thus the density at which control measures should be implemented to prevent a population rising to the ECONOMIC INJURY LEVEL, the lowest density that causes economic damage.

If a pest population remains below the economic threshold, the cost of control (chemical, machinery, labour) will be greater than the expected value of the crop loss and PESTICIDE application would be uneconomic and vice versa. Decision-making in IPM is thus based on cost–benefit analysis. Economic thresholds for individual crops/pests vary from year to year depending on weather conditions, market value of the crop and other economic impact factors. IPM schemes have been developed for the sugar-cane borer (*Diatraea saccharalis*) in the southern USA, spotted alfalfa aphid (*Therioaphis trifolii*) on lucerne in California and cotton pests (*Heliothis virescens*) in Peru.

Common features of the most successful IPM schemes include comprehensive ecological information and identification of the pest generations/life-cycle stages that cause damage; information on migratory behaviour and accurate sampling techniques; replacement of broad-spectrum with selective chemicals; establishment of economic thresholds; conservation of natural enemies and introduction of biological control agents; use of resistant cultivars; and rationalization of cultural techniques. *See also* BIOCIDE. [J.S.B.]

REFERENCES

Apple, J.L. & Smith, R.F. (1976) *Integrated Pest Management*. Plenum Press, New York.

van Emden, H.F. (1989) *Pest Control*, 2nd edn. Edward Arnold, London.

Woods, A. (1974) *Pest Control: A Survey*. McGraw-Hill, London.

integrated pollution control (IPC) It follows from the laws of thermodynamics and conservation of matter that substances removed from one WASTE stream are likely to turn up in others, albeit in transformed states. A consequence of this is that POLLUTION from one environmental compartment may be shifted to another; removing pollutants from chimney EMISSIONS may create solids that need disposal in LANDFILL or that leach into natural waters. An integrated approach to pollution control recognizes this and, while trying to remove emissions in the first place, attempts to find the optimum solution for dealing with unavoidable emissions. This is generally done with respect to principles of best (practicable) environmental options (B(P)EO) and best available technology (BAT). IPC is now a central feature of UK environmental pollution legislation since the ENVIRONMENTAL PROTECTION ACT 1990, but in Europe has its roots in an EC Council Directive (87/217/EEC) on the prevention and reduction of pollution by ASBESTOS, which requires emission for all environmental compartments to be taken into account. Integrated pollution prevention and control (IPPC) has now been adopted as a European Union measure (a Directive adopted in 1995). *See also* BEST AVAILABLE TECHNIQUES NOT ENTAILING EXCESSIVE COSTS; BEST AVAILABLE TECHNIQUES REFERENCE DOCUMENT; BEST PRACTICABLE ENVIRONMENTAL OPTION. [P.C.]

integrated pollution prevention and control (IPPC) *See* INTEGRATED POLLUTION CONTROL.

integrated waste management There is a variety of management options in terms of dealing with solid WASTE: source reduction, REUSE, RECYCLING, COMPOSTING, INCINERATION (possibly with energy recovery), LANDFILL. For complex waste, such as municipal solid waste, no one of these options may be suitable for all components. Thus integrated waste managment refers to the complementary use of a variety of waste management options to optimize the overall result. [P.C.]

REFERENCE

White, P.R., Franke, M. & Hindle, P. (1995) *Integrated Solid Waste Management. A Lifecycle Inventory*. Chapman & Hall, London.

intensity of infection The level (number) of MACROPARASITES within an individual HOST (*see also* MEAN INTENSITY OF INFECTION). It is often not possible to measure the intensity of ENDOPARASITES directly without destruction of the PARASITE (and possibly host), so indirect methods (such as counting eggs released into the environment) are used. The intensity is usually positively correlated to the morbid consequences for the host and the strength of density-dependent CONSTRAINTS on the parasite. [G.F.M.]

intensity of selection *See* SELECTION INTENSITY.

intention movements Behaviour providing information on the intentions of an individual.

Intention movements are common signals of intention used in COMMUNICATION among animals. Examples include initiated flight behaviour when an animal is approached by a potential PREDATOR. The intention movement may communicate the ability of flight of the sender to the predator, which may then decide not to further pursue the potential prey. Another example concerns aggressive intention movements when an animal in possession of a limited food source is approached by a potential competitor. The intention movements may provide information on the willingness of an individual to escalate a fight, while the receiver may obtain information on whether to proceed with the acquisition of the food source or leave. Animals should only provide information on their intentions when there is a coincidence of interests between the receiver and the sender of a signal. Intention movements are therefore likely to be reliable signals and only likely to evolve in contexts where senders and receivers have congruent interests.

Intention movements can be considered to reflect the internal conflict of an individual with respect to whether to continue with a particular behaviour or switch to a novel behaviour. The explicit expression of an internal conflict by an animal to a receiver is likely to result in the acquisition of additional information on the future behaviour of the receiver (and the sender). The sender will therefore gain additional information on which to base its future decisions. [A.P.M.]

interaction variance The interaction variances constitute that part of the GENETIC VARIANCE that arises from interactions between loci in the determination of a QUANTITATIVE TRAIT. Additive-by-additive interaction variances contribute to the parent–offspring COVARIANCE, the two-locus V_{AA} by $^1/_4$, the three-locus V_{AAA} by $^1/_8$, etc., leading to cov(offspring,parents) $= {^1/_2}V_A + {^1/_4}V_{AA} + {^1/_8}V_{AAA} + \ldots$. However, interaction can also be non-additive, for example leading to DOMINANCE effects and EPISTASIS, when the interaction of ALLELES at two or more loci lead to a combined effect that differs from the sum of the individual loci. *See also* ADDITIVE GENETIC VARIANCE; DOMINANCE VARIANCE; QUANTITATIVE GENETICS. [G.D.J.]

interception Precipitation caught by the leaves and stems of a plant CANOPY and evaporated directly back into the ATMOSPHERE without reaching the soil. The high aerodynamic ROUGHNESS of forests leads to large interception losses. Interception can be separated into EVAPORATION during storms, which depends primarily on rainfall duration, and evaporation after storms, which depends on the amount of WATER left on the canopy after RAIN ceases. Interception ranges from 10% of incident rainfall in the tropics, where there are short-duration high-intensity storms, to more than 30% of the rainfall in temperate, maritime climates, with long-duration low-intensity rainfall. [J.H.C.G.]

interference A density-dependent reduction in the per capita efficiency of a PREDATOR or PARASITOID due to direct behavioural interactions. For example, as predator density increases the probability of individual predators encountering each other will go up; if each encounter leads to time-wasting behaviour such as fighting or dispersal, per capita efficiency will decline with increasing density. This is an important component of some MODELS OF PARASITOID–HOST INTERACTION. *See also* PSEUDOINTERFERENCE. [H.C.J.G.]

interference competition A process of direct or aggressive COMPETITION where individuals interact directly with others in a way that reduces FITNESS, even when RESOURCES necessary for growth and reproduction may not be limited. A major function of interference competition may be to keep POPULATION SIZE below a level at which EXPLOITATION COMPETITION becomes important, for example the production by a plant of root exudates that inhibit the growth of another plant. In mobile animals that establish and defend territories, the territory itself is the resource that is competed for. However, the territory may give access to other resources associated with fitness such as mates or food for offspring. *See also* CONTEST COMPETITION. [R.H.S.]

REFERENCE

Birch, L.C. (1957) The meanings of competition. *American Naturalist* **91**, 5–18.

intergenerational equity *See* SUSTAINABLE DEVELOPMENT.

Intergovernmental Panel on Climate Change (IPCC) A group of scientists convened in 1988 to review and summarize the literature on CLIMATE CHANGE and present policy options. [J.G.]

intermediate disturbance hypothesis Hypothesis that species DIVERSITY in a COMMUNITY is maximized at intermediate levels of DISTURBANCE (where level may refer to frequency, extent or intensity). If a disturbance removes organisms from a patch then diversity will initially be low but, in the absence of further disturbance, will increase through time as species recolonize. Eventually, diversity will begin to decline as competitive displacement of early colonists by later-arriving species occurs until, finally, the patch is just occupied by a small number of competitively dominant species. In a habitat comprising multiple patches, if the frequency of disturbance is very high most patches will be in the early successional stage (low diversity within a patch) and will all contain very similar sets of species (the early colonists). Hence overall diversity across the community will be low. If disturbance is of low frequency, then most patches will reach the CLIMAX state (low diversity within a patch) and will contain just the few competitively dominant species, again giving low diversity across the community as a whole. If disturbance is of intermediate frequency, there will be many patches that have had sufficient time since they were last disturbed to be colonized by substantial numbers of species (high diversity within many patches) but there will also be patches at all the different stages of SUCCESSION, resulting in high diversity across the community as a whole. The hypothesis has been applied to a wide variety of plant, and some animal (e.g. coral), communities and to various sorts of disturbance, including predation/grazing under some circumstances at least. It is related to other non-equilibrium explanations of species diversity. *See also* NON-EQUILIBRIUM COMMUNITIES. [P.H.W.]

internal fertilization *See* FERTILIZATION.

internal succession A term coined by Curtis (1959, p. 292) for a succession that occurs within a COMMUNITY after the creation of gaps by DISTURBANCE, for example in GRASSLANDS on mounds made by gophers or moles, in forests after treefalls, or in Mediterranean-climate scrubland after fire. [P.J.G.]

REFERENCE
Curtis, J.T. (1959) *The Vegetation of Wisconsin*. University of Wisconsin Press, Madison, Wis.

International Biological Programme (IBP) One of the first attempts to study ecosystems on a world scale through international cooperation. It was initiated by the International Council of Scientific Unions and extended from 1964 to 1974, involving some 40 countries. The sections were: PRODUCTIVITY of terrestrial communities, production processes on land and water, conservation of terrestrial communities, productivity of freshwater communities, productivity of marine communities, human adaptability and the use and management of biological resources. [J.G.]

International Code of Botanical Nomenclature The code is an agreed set of rules and recommendations that govern the scientific naming of plants and fungi. There are also codes for animals and bacteria. The code of nomenclature ensures that the name used by scientists for each organism, or group of organisms, is the same one, so that they can communicate effectively with each other. Organisms are placed in a CLASSIFICATION, a hierarchy of named taxonomic groups (taxa; singular, TAXON). The classification represents relationship. The most important taxon is the species. Every organism belongs to a species with only one correct name, the binomial. The binomial is underlined or italicized and consists of the name of the genus to which the species belongs, written with a capital initial letter, and a specific name, written with a lower case initial letter. Correct names are those that have been effectively and validly published. Valid publication must include a description of the organism and the citation of a nomenclatural type. The 'type' of a species is usually a pressed and dried herbarium specimen. The correct name of a species is the one with priority, the first effectively and validly published name after the publication of Linnaeus' *Species Plantarum* in 1753. Name changes occur as the result of taxonomic research, which discovers better ways to represent relationships in the classification by splitting or lumping together species or genera. [M.I.]

REFERENCE
Jeffrey, C. (1977) *Biological Nomenclature*. Edward Arnold, London.

International Code of Zoological Nomenclature (ICZN) These are the regulations governing the scientific naming of animals. They are drawn up and supervised by the International Commission on Zoological Nomenclature (ICZN). [P.C.]

international conservation conventions These are agreements that are concerned with the protection of species or natural areas involving two or more nations.

- Convention on International Trade in Endangered Species of Wild Fauna and Flora (CITES). Multilateral, signed in 1973. Designed to control commercial trade in species in danger of extinction.
- Convention on Wetlands of International Importance. Often referred to as Ramsar after the Iranian town in which it was formulated in 1971. It was the first multinational convention. It remains the only one that is concerned with protection of one general type of ecosystem.
- UN Convention on Biological Diversity. Adopted in 1992 at the Earth Summit (*see* RIO SUMMIT). Deals comprehensively with the protection of natural areas, species and genetic material for the purposes of global SUSTAINABLE DEVELOPMENT.
- Man and the Biosphere Programme (MAB). International, not legally binding, agreement with UNESCO (*see* UNITED NATIONS PROGRAMMES AND BODIES) with the aim of establishing an international network of reserves (BIOSPHERE RESERVES) for the purpose of studying human effects on natural systems.

See also CONSERVATION BIOLOGY; UNITED STATES LEGISLATION. [P.C.]

international environmental law *See* ENVIRONMENTAL LAW.

International Geosphere–Biosphere Programme *See* IGBP.

International Labour Organization (ILO) *See* UNITED NATIONS PROGRAMMES AND BODIES.

International Maritime Organization (IMO) *See* UNITED NATIONS PROGRAMMES AND BODIES.

International Monetary Fund *See* GATT AND ENVIRONMENT.

International Organization for Standardization (ISO) *See* BS 7750; ISO 14000.

international organizations and programmes associated with environmental protection *See* UNITED NATIONS PROGRAMMES AND BODIES.

International Species Information System (ISIS) Organizes and facilitates sharing of basic information on animals in zoos and other facilities worldwide. Involves more than 400 institutions in more than 50 countries and has a total inventory of >720 000 organisms. Developed microcomputer-based programs (VORTEX and SPARKS) for population management. [P.C.]

International Union for the Conservation of Nature and Nature Resources *See* IUCN.

International Whaling Commission (IWC) This body was set up shortly after the Second World War to regulate the WHALING industry. In its early days, its major function was to regulate the market and ensure that excessive HARVESTING of whales did not lead to marketing problems in whale products. In the 1960s, the major problem of the overexploitation of whales was realized and the IWC became more concerned with the sustainable harvesting of whale stocks. A special committee of three leading fisheries scientists, later expanded to four, addressed the question of the level of harvest that was sustainable. Their report led to an agreed ban on harvesting of the blue whale (*Balaenoptera musculus*).

Since that time, the appreciation of the relatively low YIELDS that can be taken sustainably from whale populations has grown, and under various management procedures introduced by the IWC the level of allowable whaling has been reduced dramatically. In 1987, the IWC agreed a moratorium on all commercial whaling, and whaling only continues now in some small aboriginal activities in the Northern hemisphere and under scientific permits. *See also* FISHERIES, CONSERVATION AND MANAGEMENT. [J.B.]

intersexual selection Refers to an individual's ability to attract members of the opposite sex. If the variation in mate quality is similar in both sexes, the sex with the lower potential reproductive rate, usually the female, is the choosy sex. Darwin thought that extravagant male adornments that obviously impair survival, such as the peacock's tail, had evolved because females prefer males with these characters. FEMALE PREFERENCE for luxury male TRAITS has been experimentally demonstrated in many species. The problem, however, is to understand the EVOLUTION of a preference for traits that would impair the survival of a choosy female's offspring. Three different hypotheses exist that are not mutually exclusive.

1 Fisher's runaway process: females have a sensory bias for male adornments because male traits would disappear in the absence of the female preference, but the female preference could SPREAD by drift in the absence of the male trait. Mutant males with the preferred trait are immediately preferred. Females with that preference have an advantage because they have attractive sons. Through a genetic correlation between male trait and female preference (which is produced by ASSORTATIVE MATING) both traits can 'runaway' quickly or reach an equilibrium at almost any expression of the male trait. At equilibrium, a son's lower viability is exactly compensated for by his higher attractivity.

2 Good-gene models: since the adornment is costly to produce and to maintain, males with good genes can have a large handicapping ornament. For example, a male that has a genetic RESISTANCE against the current PARASITE is in good physical condition and can therefore develop a large ornament. Choosy females receive the male's resistance genes for their offspring. Low-cost sexual ornaments such as odours can be reliable signals of good genes if females (or males) seek resistance genes that are complementary to their own genes. For example, in humans (*Homo sapiens*) and mice (*Mus musculus*), females prefer males that reveal complementary MHC (major histocompatibility complex) genes in their body odour.

3 Direct benefits: a male that is in good physical condition can develop a relatively larger ornament and still provide a better territory and more paternal care than a male in poorer condition with a smaller ornament. Choosy females would have stronger but not necessarily genetically better offspring.

Cf. INTRASEXUAL SELECTION. *See also* DISPLAYS; SEXUAL SELECTION. [M.MI.]

REFERENCES

Andersson, M. (1994) *Sexual Selection*. Princeton University Press, Princeton, N.J.

Wedekind, C. (1994) Mate choice and maternal selection for specific parasite resistances before, during and after fertilization. *Philosophical Transactions of the Royal Society of London, Series B* **346**, 303–311.

interspecific aggregation (C) To measure interspecific AGGREGATION, Ives (1988, 1991) derived an index, *C*, which measures the proportionate increase in the number of HETEROSPECIFIC competitors relative to a random ASSOCIATION. This measure is the interspecific equivalent of the intraspecific aggregation measure (*J*). The quantity *C* is calculated using the formula:

$$C_{12} = \frac{\sum_{i=1}^{P} \frac{n_{1i} n_{2i}}{m_1 P} - m_2}{m_2} = \frac{\mathrm{cov}_{12}}{m_1 m_2}$$

where cov is the COVARIANCE in numbers between a pair of species over habitat patches, the subscripts (1 or 2) indicate the species involved, n_i is the numbers in habitat patch *i*, *P* is the number of patches and *m* is MEAN numbers. When $C=0$ the two species are randomly associated with each other across patches. When $C>0$ they are positively associated; when $C<0$ they are negatively associated. Note that for each pair of species there are two different values of intraspecific aggregation (J_1 and J_2) but only one value of interspecific aggregation ($C_{12}=C_{21}$). The simple hypothetical example below illustrates the computation of *C*.

Species	Numbers per habitat patch						*m*	*V*	*J*	Cov	*C*
1	2	0	1	9	10	14	6	27.7	0.60	+0.83	+0.03
2	7	3	5	8	1	6	5	5.6	0.02		

Across the six patches, species 1 shows an AGGREGATED DISTRIBUTION while species 2 shows a RANDOM DISTRIBUTION. This is reflected in the measure of intraspecific aggregation (*J*) which, relative to random, shows a 60% increase in CONSPECIFIC 'contact' for species 1 and only a 2% increase for species 2. The measure of interspecific aggregation (*C*) shows only a slight increase (3%) over that of random association. Both Ives (1991) and Shorrocks and Sevenster (1995) illustrate use of the index on field-collected data. *Cf.* INTRASPECIFIC AGGREGATION. [B.S.]

REFERENCES

Ives, A.R. (1988) Aggregation and the coexistence of competitors. *Annales Zoologici Fennici* **25**,75–88.

Ives, A.R. (1991) Aggregation and coexistence in a carrion fly community. *Ecological Monographs* **61**, 75–94.

Lloyd, M. (1967) Mean crowding. *Journal of Animal Ecology* **36**, 1–30.

Shorrocks, B. & Sevenster, J.G. (1995) Explaining local species diversity. *Proceedings of the Royal Society of London, Series B* **260**, 305–309.

interspecific competition COMPETITION between individuals of different species (HETEROSPECIFICS). [B.S.]

interspecific hybridization *See* SYMPATRIC SPECIATION.

interspersion *See* EXPERIMENTAL DESIGN.

interstitial The water-filled spaces between the individual particles that comprise aquatic sediments or terrestrial soils form the interstitial habitat, and interstitial organisms are those that can move between the particles without displacing them. Accordingly, interstitial organisms are small and often elongate. Bacteria, protists and animals are all represented, the animals by some 22 phyla including the gastrotrichs, gnathostomulans, kinorhynchs and loriciferans, which are almost exclusively interstitial in habitat. Amongst the animals, however, nematodes and, in aquatic systems, copepods dominate the fauna. Average abundances are measured in millions, with average biomasses up to $2\,\mathrm{g\,m^{-2}}$. [R.S.K.B.]

interstitial water *See* POREWATER.

intertidal zone The shoreward fringe of the seabed between the highest and lowest extent of the TIDES. The intertidal zone is characterized by widely varying physical conditions and, particularly on ROCKY SHORES, by VERTICAL ZONATION of the dominant species. *See also* SANDY SHORES. [V.F.]

interval graphs *See* INTERVALITY.

intervality If, instead of drawing a line to link predators to prey as in a conventional food-web graph, one links elements that share common prey, the graph obtained is termed a predator, or NICHE, overlap graph. Such graphs can often be collapsed into one-dimensional representations known as interval graphs. FOOD WEBS with predator overlap graphs that can be collapsed in this way are described as interval and this property seems to occur more often in real webs than would be expected by chance. [S.J.H.]

REFERENCE
Cohen, J.E. (1978) *Food Webs and Niche Space*. Princeton University Press, Princeton, N.J.

intragenerational equity *See* SUSTAINABLE DEVELOPMENT.

intransitive competition This term is applied to multispecies interactions when a network of INTERSPECIFIC COMPETITION is present. It is used to classify the pattern of interaction present between all directly competing species. For example, species A may displace species B which may in turn displace species C, while species C may also be able to displace species A. Indeterminate unpredictable networks are produced that may promote more diverse and resilient assemblages than if a strict hierarchical structure were present. Tropical REEF communities epitomize this type of system, where DISTURBANCE creates an unpredictable spatial environment that prevents the development of fixed competitive hierarchies. Such networks are commonly associated with unpredictable highly heterogeneous environments. *See also* TRANSITIVE COMPETITION. [R.C.]

intrasexual selection COMPETITION among members of the same sex (usually the males) for access to members of the other sex. The EVOLUTION of TRAITS that increase the probability of winning a fight, such as physical strength, body size and shape, special weapons such as antlers or spurs and defensive organs, is favoured by intrasexual selection. Intrasexual ARMS RACES can lead to exaggerated male armaments at the expense of lowered male viability. Traits that are reliable signals of an individual's fighting ability (such as roaring in red deer stags) threaten opponents and therefore help to win a contest at a low risk of injury. *Cf.* INTERSEXUAL SELECTION. [M.Mi.]

intraspecific aggregation (*J*) Following on from the ideas of Lloyd (1967) about MEAN CROWDING, Ives (1988, 1991) has suggested a measure of intraspecific AGGREGATION, J, that is the proportionate increase in the number of CONSPECIFIC competitors 'experienced' by a random individual, relative to that 'experienced' in a RANDOM DISTRIBUTION. The quantity J is calculated using the formula:

$$J = \frac{\sum_{i=1}^{P} \frac{n_i(n_i - 1)}{mP} - m}{m} = \frac{\frac{V}{m} - 1}{m}$$

where n_i, P, m and V are the numbers in habitat patch i, number of patches, mean numbers and the VARIANCE in numbers of a species, respectively. In this particular case, the variance is obtained by dividing the sums of squares by n, since we are interested in the variance of the SAMPLE, not the POPULATION. Note that the index J is equal to $1/k$ of the NEGATIVE BINOMIAL DISTRIBUTION. A value of $J=0$ ($k=\infty$) indicates that individuals are randomly distributed, while a value of $J=0.5$ ($k=2$) indicates a 50% increase in the number of conspecifics expected in a patch, i.e. an increase in mean crowding of 50%. The simple example below illustrates the computation of J.

Numbers per habitat patch						m	V	J
1	1	1	9	9	9	5	16.0	0.44
1	3	5	6	7	8	5	5.6	0.02

For the numbers in the top row, which show an AGGREGATED DISTRIBUTION, the value of J indicates an increase of mean crowding of 44% relative to that 'experienced' in a random distribution. In the bottom row, the numbers follow an approximately random distribution and J indicates an increase in mean crowding of only 2%. Both Ives (1991) and Shorrocks and Sevenster (1995) illustrate use of the index on field-collected data. *Cf.* INTERSPECIFIC AGGREGATION. *See also* INDICES OF DISPERSION. [B.S.]

REFERENCES
Ives, A.R. (1988) Aggregation and the coexistence of competitors. *Annales Zoologici Fennici* **25**,75–88.
Ives, A.R. (1991) Aggregation and coexistence in a carrion fly community. *Ecological Monographs* **61**, 75–94.

Lloyd, M. (1967) Mean crowding. *Journal of Animal Ecology* **36**, 1–30.
Shorrocks, B. & Sevenster, J.G. (1995) Explaining local species diversity. *Proceedings of the Royal Society of London, Series B* **260**, 305–309.

intraspecific competition Refers to COMPETITION between individuals of the same species. Competition is recognized at the population level by a reduction in the PER CAPITA rate of increase as POPULATION DENSITY increases. Two extreme forms of intraspecific competition were recognized by Nicholson (1954): contest or INTERFERENCE COMPETITION and scramble or EXPLOITATION COMPETITION (see also Birch 1957). All forms of intraspecific competition contribute to POPULATION REGULATION, though CONTEST COMPETITION is associated with COMPENSATING DENSITY DEPENDENCE and SCRAMBLE COMPETITION with OVERCOMPENSATION. [R.H.S.]

REFERENCES

Birch, L.C. (1957) The meanings of competition. *American Naturalist* **91**, 5–18.
Nicholson, A.J. (1954) An outline of the dynamics of animal populations. *Australian Journal of Zoology* **2**, 9–65.

intraspecific victim mutualism *See* PREDATOR SATIATION.

Intrazonal soils *See* SOIL CLASSIFICATION.

intrinsic bioremediation Stated simply: self-purification. This involves encouraging a system's own capacity to clean-up CONTAMINATION and POLLUTION. For example, soils often have the capacity to degrade organic CONTAMINANTS through the action of indigenous microbes, but rates might be enhanced by appropriate management. Thus toluene contamination of soils in industrial sites is subject to microbial degradation, but this can be enhanced by good aeration of the soil since the process appears to be aerobic. *See also* REMEDIATION. [P.C.]

intrinsic mortality Mortality due primarily to internal physiological or metabolic causes such as cancer, heart disease, reproductive stress and degenerative conditions associated with AGEING (SENESCENCE). *Cf.* EXTRINSIC MORTALITY. [T.B.L.K.]

intrinsic rate of increase The POPULATION GROWTH RATE, r, first defined by the demographer A.J. Lotka in 1925. The intrinsic rate of increase is an instantaneous rate and represents the average rate of increase per individual (female) in a population, normally in the absence of crowding effects and shortage of resources; r is also known as the innate capacity for increase and may be calculated from age-specific birth and death rates using the EULER–LOTKA EQUATION or a Leslie matrix. Use of the term 'intrinsic' may be misleading because r depends on both the genetic constitution of a population and the environment in which birth and death rates are expressed, and the intrinsic rate is only achieved when a population has reached its STABLE AGE DISTRIBUTION. Another related measure is population multiplication rate (often specified as λ), which is the factor by which a population multiplies or recedes after a given time-period. (So if $\lambda=2$, it doubles; if $\lambda=0.5$, it halves.) *See also* FITNESS; LIFE-HISTORY EVOLUTION; NET RATE OF INCREASE; POPULATION PROJECTION MATRIX. [R.H.S.]

REFERENCE

Krebs, C.J. (1994) *Ecology: The Experimental Analysis of Distribution and Abundance*, 4th edn. Harper Collins, New York.

Introduction and reintroduction

Reintroduction and introduction are emerging as important elements in conservation strategies for rare and ENDANGERED SPECIES. Although traditional conservation prescriptions rely primarily on preservation and management of naturally occurring populations in their native habitats, populations of many species have become so seriously affected by human activities as to require more direct intervention. However, in responsible hands introduction and reintroduction are always practised as elements of overall strategies based ultimately on the recovery and return to natural POPULATION DYNAMICS *in situ*.

The primary circumstance driving the need for introduction and reintroduction is the continued high rate of destruction of natural habitats worldwide. On every continent and in every major BIOME, the remaining inventory of unaltered native HABITAT dwindles with every passing year. Many species are declining in the number, size and distribution of populations as these major changes take place on the global landscape. Parks and other protected areas are becoming surrounded by anthropogenic landscapes and hence ecologically isolated from other components of their larger ECOSYSTEM. As a result, it can no longer be assumed that protected and other natural areas will be large or numerous enough, or distributed in the correct locations, to ensure the survival of all, or even a majority, of species on Earth. Introduction and reintroduction are among the measures that can be used when populations are reduced or extirpated from their native habitat under these circumstances.

Historical considerations

Introduction and reintroduction are distinguished primarily by the variables used to determine the native HOME RANGE of a species. In traditional parlance, 'reintroduction' means the return of a species to its native range, while 'introduction' indicates the placement of individuals of a species not known to have occurred in that location previously. The ends of the introduction–reintroduction continuum are relatively clear, as in the case of species newly introduced to a continent where they have never occurred or the reintroduction of a recently extirpated population. Increasingly, however, these terms are recognized as problematic and poorly distinguished. For example, many species are naturally patchy in their distribution, even within the larger envelope of their overall range. Subpopulations of species with active metapopulation dynamics may occur ephemerally in various locations from decade to decade. Moreover, the ranges of most species changes over time in response to biotic and abiotic forces in the environment.

The ambiguity can be resolved by defining spatial and temporal scales explicitly. The spatial scale used to delineate a reintroduction may be defined broadly (e.g. as an occurrence anywhere within the overall known range of a species) or narrowly (e.g. only at specific sites from which the species has been previously recorded). Any placement of living individuals outside the defined spatial frame would then by definition be considered an introduction. Likewise, temporal scale can be defined broadly (e.g. using location data over thousands of years to define the species' range) or narrowly (using only location data at the time of extirpation). As these examples show, discriminating introduction from reintroduction involves a degree of judgement and is hence in some respects arbitrary. Explicit definition of spatial and temporal parameters at least makes these assumptions available for critique.

Biological considerations

A variety of biological considerations apply to designing successful plant and animal introductions and reintroductions. Plants may be introduced or reintroduced by seed, seedlings, rooted cuttings, mature individuals or other life stages. The receptor site must be analysed for appropriate SOIL, TOPOGRAPHY, HYDROLOGY and CLIMATE, as well as a suitable plant COMMUNITY type. Beyond the ESTABLISHMENT phase, successful introduction or reintroduction requires presence of pollinators, DISPERSAL agents, mycorrhizal symbionts, nodulating bacteria and other organisms with which the population may interact. Outplanted populations must also satisfy demographic criteria for sustainability and population-level RESILIENCE, including age and size structure, SEX RATIOS and a sufficient number of individuals to ensure a self-sustaining population. Several episodes of outplanting may be required to satisfy these criteria.

Additional considerations apply to animal introductions and reintroductions, including availability of wintering and summering habitat, availability of appropriate food sources, suitable habitat with appropriate physical structures for nesting or denning, locations for escape from predation and the presence of potential competitors. Animal introductions and reintroductions are further frequently complicated by the absence of social structures found in flocks, herds or packs, which provide essential mechanisms by which juveniles learn vital survival and social skills. CAPTIVE BREEDING programmes in zoos and aquaria are frequently the best source of healthy individuals for reintroduction, although in some cases wild-caught individuals are relocated to new sites.

Genetic considerations play an important role in both plant and animal reintroduction efforts. Little quantitative research has been done to guide genetic planning of plant reintroductions. Some species appear sensitive to the genetic bottlenecks that may occur in propagation and reintroduction programmes, while other species appear to be affected more strongly by demographic limitations. Research and practice in genetic considerations for animal reintroduction are presently more advanced; most captive breeding programmes record detailed information about the LINEAGE of each individual and extensive efforts are made to avoid inbreeding among closely related individuals. Monitoring of genetic changes in reintroduced populations of either plants or animals over multiple generations has been carried out for only a handful of species in the wild.

Potential reintroduction sites may be evaluated for the degree to which they meet criteria for ecological suitability, protection and historic locality. Preferred sites will meet all three criteria to a high degree. Secondary sites may meet only two criteria, one of which must be ecological suitability. Tertiary sites pass only the test of habitat suitability but are unprotected and possibly outside a defined GEOGRAPHIC RANGE. All other sites are presumed to be non-viable for both ecological and conservation purposes. *See also* CONSERVATION BIOLOGY; *EX SITU* CONSERVATION. [D.A.F.]

REFERENCE

Falk, D.A., Millar, C.I. & Olwell, M. (eds) (1996) *Restoring Diversity: Strategies for the Reintroduction of Endangered Plants*. Island Press, Covelo, Calif.

---

invasion criterion The ability to increase when rare. All persistent populations must exhibit $dN/dt>0$ when N is small. If they did not, then a series of environmental calamities would drive their numbers inexorably downwards to extinction. A key concept in the analysis of theoretical models of species COEXISTENCE. [M.J.C.]

invasive grasses *See* EXOTIC AND INVASIVE SPECIES.

invasive species *See* BIOLOGICAL INVASIONS.

inverse stratification *See* LAKE STRATIFICATION; LAKES.

iodine (I) A member of the halogen family (ATOMIC NUMBER = 53; relative ATOMIC MASS = 126.9; melting point = 113.5°C; boiling point = 184°C), iodine has an abundance of 0.46 mg kg^{-1} in the Earth's CRUST, although it is found in brines at a considerably higher concentration and it is from these that it is extracted. Iodine is used to make organoiodine compounds, CATALYSTS, dyestuffs and photographic chemicals. Iodine is essential for the thyroid gland and deficiencies result in an enlarged thyroid gland. ^{131}I is a FISSION product of URANIUM or PLUTONIUM and so can be released into the environment in the event of a nuclear explosion or a leak from a nuclear reactor. [C.R.P.]

ion An ion is formed from an atom by loss of one or more electrons making it positively charged (cation), or by gain of one or more electrons making it negatively charged (anion). The number of protons and neutrons in the nucleus of an atom is unchanged. Many atoms gain or lose electrons in order to achieve the electron configuration of the closest noble gas. More generally, completely filled or half-filled electron shells are associated with greater stability. Groups of atoms that are chemically joined together and have a net positive or negative charge are termed polyatomic ions. [M.C.G.]

Ionizing radiation

Ionizing RADIATION is the name that is collectively applied to high-energy electromagnetic and particulate radiations that deposit their energy in matter mainly by interactions with electrons, resulting in ionization and breaking of chemical bonds. The energy of ionizing radiation is normally expressed in units of electron volts (eV), where 1 eV = 1.6×10^{-19} J. A number of types of ionizing radiation arise from radioactive decay of unstable nuclei, including α-radiation (particles of mass 4 amu (atomic mass unit) and charge +2, with energies normally in the range 4–9 MeV), β-radiation (β^--particles are electrons and β^+-particles are positrons, with energies normally in the range 10 keV to 4 MeV) and γ-radiation (photons of electromagnetic radiation with energies normally in the range 5 keV to 7 MeV). X-RAYS also constitute a form of ionizing radiation and can be produced by two distinct processes. 'Characteristic X-rays' are photons of electromagnetic radiation, with energies in the approximate range 1–150 keV, which originate from rearrangements of electrons in atoms following the creation of a vacancy in an inner shell. Such inner-shell vacancies are produced when accelerated electrons, or other types of radiation (e.g. γ-photons), interact with matter. The X-rays produced in this way have discrete energies that are characteristic of the atom from which they were emitted. Alternatively, X-rays known as bremsstrahlung can be produced by deceleration of an electron in the electromagnetic field of a nucleus. Bremsstrahlung energy spectra have a continuous distribution extending from the energy of the electron, which can be as a high as 100 MeV, down to zero. Neutrons are not inherently ionizing but give rise to ionization as a secondary effect via NEUTRON capture or other reactions with nuclei, resulting in the emission of ionizing radiation. Additional forms of ionizing radiation, such as protons, deuterons and other high-energy charged particles generally occur only in special experimental facilities and are not encountered in the environment.

Ionizing radiation is a natural feature of the environment, arising from:

1 primordial radionuclides such as ^{238}U, ^{235}U, ^{232}Th and their descendants;

2 cosmic radiation, comprising primary COSMIC RAYS (predominantly high-energy protons), which are mainly absorbed in the upper atmosphere, and secondary cosmic rays (such as mesons), which penetrate to the surface of the Earth;

3 cosmic-ray-produced radionuclides, such as ^{3}H and ^{14}C, which are generated by interactions between cosmic rays and the gases of the upper atmosphere.

The total natural background radiation exposure from all these sources to an individual at sea level is estimated to be about 2.2 mSv year^{-1}.

Man-made radionuclides also make a significant contribution to the level of ionizing radiation in the environment. The largest source of man-made radionuclides to the environment was from the testing of nuclear weapons in the 1950s and 1960s. Releases of radionuclides to the environment also occur at all stages of the nuclear fuel cycle, from mining through to RADIOACTIVE WASTE disposal. Authorized discharges of radionuclides generally result in slightly enhanced levels of ionizing radiation in the environment around nuclear facilities. However, accidental releases of radionuclides from nuclear facilities have occasionally had a major effect on environmental radiation levels over a wide area (e.g. the CHERNOBYL DISASTER in 1986). Ionizing radiation is also an essential feature of some fields of employment, such as radiography and NUCLEAR POWER production, and can result in occupational exposure of workers to radiation.

Ionizing radiation presents a potential health hazard to living organisms as a result of bond breaking in biological tissue. Such effects are classed as: (i) somatic effects, which involve health detriment to the exposed individual; or (ii) genetic effects, which involve damage to the offspring of the exposed individual. Somatic effects are divided into acute effects, which occur directly as the result of exposure of the whole body or individual organs, and delayed effects, which occur some time, often years, after exposure. Acute effects have a recognized THRESHOLD exposure and the severity of the effect increases with increased dose above the threshold. Acute effects include nausea, vomiting, diarrhoea, loss of hair, sterility, opacity of the lenses of the eyes, depression of blood-forming processes and, ultimately, death. For whole-body irradiation, the LD_{50} (the dose at which 50% of the exposed population will die) is about 4 Gy. The most common delayed effects involve the development of cancers, the occurrence of such effects being stochastic, with no recognized threshold, and the probability, as opposed to the severity, of the effect increasing with dose. The efficiency of interaction with electrons varies considerably between different types of ionizing radiation, resulting in large differences in the density of ionization per unit path length. This is characterized as the rate of linear energy transfer (LET) or specific ionization. α-PARTICLES, with a double charge and relatively low velocity (compared to lighter particles of the same energy), have a highly efficient interaction with electrons, resulting in high LET values. Therefore α-radiation causes intense ionization but has a maximum range of about 0.1 mm in solids or a few centimetres in air, so α-emitting radionuclides represent a hazard only if they are inhaled or ingested (i.e. an internal hazard). In contrast, β-PARTICLES, being singly charged and for a given energy having greater velocity than α-particles, have a less-efficient interaction with electrons and consequently have low LET values. Thus β-radiation has a correspondingly greater range than α-radiation and can constitute both an internal and external hazard. γ-Photons have a low efficiency for interaction with electrons, so γ-radiation has a low LET value but a large range and consequently represents an internal and external hazard. The LET of radiation determines the distribution of energy within absorbing tissue and this must be taken into account when estimating the effect of RADIATION DOSES. Dose limits for exposure to radiation from industrial sources are recommended by the International Commission for Radiological Protection (ICRP) and the present limit for members of the general public is 1 mSv year^{-1}, while the limit for occupational exposure of workers is 5 mSv year^{-1}. *See also* RADIOACTIVE FALLOUT; RADIOACTIVITY; RADIONUCLIDE; RADON; SIEVERT; WASTE DISPOSAL. [A.B.M.]

REFERENCE

Kiefer, J. (1990) *Biological Radiation Effects*. Springer-Verlag, Berlin.

IPC *See* INTEGRATED POLLUTION CONTROL.

IPCC *See* INTERGOVERNMENTAL PANEL ON CLIMATE CHANGE.

IPPC (integrated pollution prevention and control) *See* INTEGRATED POLLUTION CONTROL.

IRGA *See* INFRARED GAS ANALYSER.

iridium (Ir) A white, heavy, brittle METAL (ATOMIC NUMBER = 77; relative ATOMIC MASS = 192.2; melting point = 2446°C; boiling point = 4428°C), iridium is a member of the PLATINUM family and is an ultratrace constituent (0.001 mg kg^{-1}) of the Earth's CRUST. Its presence at elevated concentrations at many locations around the world, in a sedimentary layer marking the end of the Cretaceous period some 65 million years ago, has been taken as indirect evidence of the impact of an asteroid hitting Earth, which may have led to the demise of the dinosaurs. *See also* HEAVY METALS. [J.F.]

iron (Fe) A malleable HEAVY METAL (ATOMIC NUMBER = 26; relative ATOMIC MASS = 55.85; melting point = 1535°C; boiling point = 2750°C), iron (at 6.2%) is the second most abundant metal and the fourth most abundant element in the Earth's CRUST. It is extracted from iron ores such as haematite (Fe_2O_3) and magnetite (Fe_3O_4) for use mainly in steel production but is also found in substantial quantities as pyrite (FeS_2). Essential to all organisms, iron is a constituent of haemoglobin, the oxygen-carrying molecule in blood. Iron is also essential to the function of chloroplasts in PHOTOSYNTHESIS, trace amounts being captured from SOIL by siderophores in plants. The environmental chemistry of iron is dominated by the Fe(II)/Fe(III) redox couple, for example soluble Fe^{2+} is oxidized to orange-coloured insoluble $Fe(OH)_3$ on exposure of acid MINE DRAINAGE waters to air. [J.F.]

Irrigation

On a world scale plant growth is frequently limited by the supply of WATER, a fact known to early civilizations, who introduced irrigation to enhance agricultural production. Arid lands are distributed in broad latitudinal zones between 30°N and 30°S, where the potential EVAPORATION far exceeds the annual PRECIPITATION, and it is in these zones that irrigation has been developed historically. Irrigation was used 5000 years ago in the Nile Valley, utilizing the annual flood and the associated rich ALLUVIAL soil. The Bible records the consequent reliability of Egypt's irrigated AGRICULTURE in contrast to Israel's rain-fed agriculture (Deuteronomy 11, vv. 10–17). Today, irrigation is important in over 271 million ha of the Earth's surface, distributed as follows: Asia, 184; North America, 34; Europe, 29; Africa, 13; South America, 9; Australasia, 2.

Irrigation methods vary from place to place. Water can be stored using large dams to provide an essentially constant flow of water that is conveyed by canals, conduits and furrows to perennial CROPS, as in the case of the Aswan High Dam in Lake Nasser, Egypt. Irrigation in Israel utilizes water from Lake Kinneret (Sea of Galilee) and the coastal AQUIFER that feeds into a nationwide pressurized supply to sprinklers and drip systems covering much of the cultivated land area. In Saudia Arabia, 'fossil' water is pumped from great depths and sprayed on the land using single-centre pivot systems that provide highly productive circular fields on which is grown crops such as cereals and alfalfa. High-technology scheduling of irrigation, such as that used for high-value crops in California USA, depends on the sensing of the water requirement, based on data from a network of weather stations, and the computer control of pumps and valves to deliver the required quantity of water each day. Irrigation is not only used in dry climates: it is commonly used in the humid tropics to grow rice and in north-west Europe in dry summer periods and in DROUGHT years.

Water requirement is related to the evapotranspiration rate, which depends on the CLIMATE and the vegetational cover. Evaporation from an open-water surface at meteorological stations has often been measured using an open pan, for example in the Negev DESERT the daily evaporation was 9.1 mm and the COEFFICIENT OF VARIATION 6.5%, while in Wellsbourne in the English Midlands the corresponding figures were 3.1 mm and 21%. Water use by a crop depends on the stomatal and aerodynamic resistance as well as on the weather and may be estimated using the PENMAN–MONTEITH EQUATION. Often it is not very different from that of an open-water surface.

Irrigation benefits crops directly, by relieving water stress and thus permitting the plant to maintain a positive internal hydrostatic pressure (turgor), and indirectly by facilitating uptake of NUTRIENTS. Once crops in arid lands are irrigated, the primary limitation to plant growth is removed and then it 'pays' to use FERTILIZERS and BIOCIDES to control pests. Even crop plants in a cool moist climate benefit from irrigation and it is especially useful in the establishment phases. Trees benefit from irrigation too, but it is not usually economic except for high-value tree crops such as apples.

There is a general relation between the rate of water use and the growth rate of plants, resulting from the fact that both CARBON DIOXIDE (CO_2) and water diffuse through the stomatal pores of the leaf surface. Thus, uptake of CO_2 inevitably causes loss of water vapour and so there is usually a good linear relationship between the BIOMASS accumulated at a site and the water used (another reason for this relationship may be that biomass is dependent on the accumulated number of photons absorbed by the leaves, which is closely related to absorbed energy, which in turn is a determinant of evapotranspiration). Penman was one of the first to suggest the relationship:

$$Y = kE_t$$

where E_t is the evapotranspiration from the crop (mm year^{-1}), Y is the YIELD of biomass by the crop (t ha^{-1} year^{-1}) and k measures the response of the crop to irrigation. For central England, there is variation according to the type of crop but the average k is about 0.3 t ha^{-1} cm^{-1}. In the USA, similar set of data for crops at Logan give corresponding k values of 0.15 and 0.6 t ha^{-1} cm^{-1} for lucerne and maize, respectively. It can be expected that the value of k will depend on the species and photosynthetic type (whether C_3 or C_4), on the HUMIDITY of the atmosphere and on the fertility of the site (at some sites, fast growth may be prevented by a lack of nutrients, for example).

When plants are deprived of water, the stomata may shut (preventing the photosynthetic uptake of CO_2), leaf temperature may rise, metabolic processes may be impaired, assimilates may be partitioned towards the root as opposed to the shoot and CAVITATION of water in the xylem may cause embolisms with consequent reduced capacity to transport water.

Irrigation schemes cannot always be sustained and are often hard to manage. Over decades of evaporation from the surface layers of irrigated soil, the salt content gradually increases and may become harmful to plants (*see* SALINITY; SALINIZATION). There may be waterlogging as a result of poor drainage and irrigation schemes may have a very heavy maintenance cost as a result. Several human diseases are associated with irrigation schemes in the tropics, including Japanese encephalitis and malaria transmitted by mosquitoes, and *Bilharzia* transmitted by snails. [J.G.]

REFERENCES
Penman, H.L., Pereira, C., Hamlin, M.J. & Mansell-Moullin, M. (eds) (1986) *Scientific Aspects of Irrigation Schemes*. Royal Society, London.
Stanhill, G. (1986) Irrigation in arid lands. *Philosophical Transactions of the Royal Society of London, Series A* **316**, 261–273.

ISIS *See* INTERNATIONAL SPECIES INFORMATION SYSTEM.

island biogeographic theory in conservation Island biogeographic theory (*see* ISLAND BIOGEOGRAPHY) has been adapted to the management of nature reserves, which are in effect islands located in oceans of ALIEN habitat. In theory, a single large nature reserve would conserve more species than a number of small ones of the same total area. Large nature reserves may also be needed to maintain viable populations of key species, such as predators. However, a number of small islands, spanning a range of habitat types, may in fact accommodate more species than a single large one and small islands may help contain the spread of EPIDEMICS. These considerations have led to the SLOSS (single large or several small) controversy. If the source area is lost then the DIVERSITY of nature reserves will dwindle, since IMMIGRATION will not be able to counterbalance the inevitable extinctions. *See also* CONSERVATION BIOLOGY. [A.E.M.]

REFERENCE
MacArthur, R.H. & Wilson, E.O. (1967) *The Theory of Island Biogeography*. Princeton University Press, Princeton, N.J.

Island biogeography

The theory of island biogeography was first proposed by Robert MacArthur and Edward Wilson in 1967. This is an equilibrium theory based on the observation that there is a constant relationship between the area of the world's islands and the number of species contained therein. The SPECIES RICHNESS of an island is the result of a balance between the rate of IMMIGRATION of species to the island and the rate of extinction on the island (Fig. I7). The resulting equilibrium is dynamic in so far as species are continually arriving and continually becoming extinct, though as time progresses the rates of arrival and extinction diminish. This theory involves a stochastic DYNAMIC EQUILIBRIUM that operates over long time-scales. It also considers island size and distance from the mainland. Wilson (1992) states that in relation to land-birds there is an average of 50 species on islands of 1000 km² and *c.* 100 species on islands of 10 000 km².

This relationship is expressed by the equation $S=cA^z$ where S is the number of species, c is a constant, A is the area and z is a biological constant relating to the group of organisms under consideration and which depends on whether or not the islands/archipelagos are close to, or distant from, the mainland. The inclusion of area in this equation also means that it can be used to predict what the impact of decreasing the size of an island will be on species extinction rates. The increase of BIODIVERSITY as island size increases is known as the 'area effect', whilst the further an island is from the mainland and/or other islands the less biodiversity it will contain, which is called the 'distance effect'.

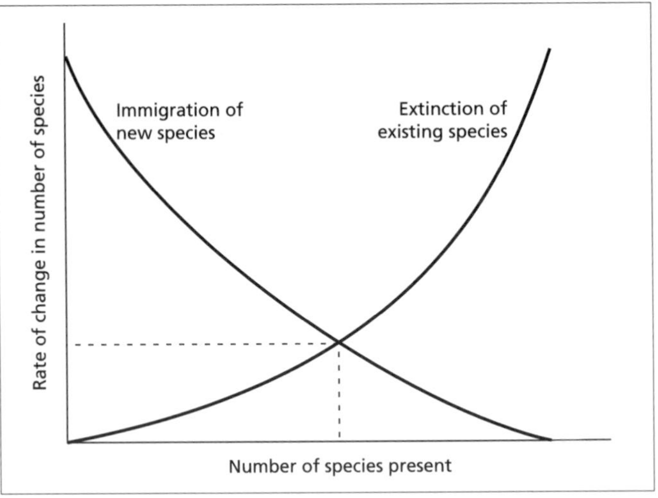

Fig. I7 Equilibrium model of the biota of an island. The curve of the rate of immigration of new species and the curve of the rate of extinction of species already on the islands intersect at an equilibrium point. The dashed line drawn vertically from this point indicates the number of species that will then be present on the island, while that drawn horizontally indicates the rate of change (or turnover rate) of species in the biota when it is at equilibrium. (After MacArthur, R.H. & Wilson, E.O. (1963) An equilibrium theory of island biogeography. *Evolution* **17**, 373–387.)

In order to test this theory, Wilson (with Daniel Simberloff) undertook an experiment involving four MANGROVE islands off the coast of Florida, USA. Inspired by Krakatau, which was formed by volcanic activity in 1883 in the Indonesian archipelago and which has subsequently been the source of much information on plant and animal invasions, Wilson and Simberloff selected four islands between 2 and 533 m distant from a large island. Their inventory of insect and other arthropods confirmed their supposition that the nearest island contained the largest number of species whilst the most distant island housed the fewest species. Then, in order to monitor immigration and extinction rates, a pest control company was engaged to fumigate the islands with methyl bromide (CH_3Br) and so eradicate the insect and arthropod populations. Over the next 2 years monitoring of recolonization was undertaken. This commenced within days of the fumigation and within a year the faunas had regained their original status. Over the 2-year period there was abundant evidence to confirm the operation of a dynamic equilibrium: total numbers remained consistent but species composition was continually changing.

Several other experiments have since been undertaken that confirm the basic relationships expressed by the theory of island biogeography. For example, the Minimum Critical Size of Ecosystems Project (MSC) which began in the late 1970s, set out to test the importance of the size of 'islands' in relation to biodiversity. The 'islands' in this experiment comprised patches of TROPICAL RAIN FOREST of between 1 and 1000 ha in size. In the first decade after the commencement of monitoring, the DIVERSITY of the smaller patches of rain forest had decreased most rapidly when compared with the larger patches. Other important findings concerned POPULATION SIZE and population FLUCTUATIONS, with implications for extinction rates. In particular, the smaller the average population of a species and the greater it fluctuates between generations, the more susceptible it is to extinction. This study illustrates the applicability of the theory of island biogeography to the design of nature reserves and its relevance to 'islands' that are not marine but which represent fragments of natural ecosystems within a human-dominated or cultural landscape. During the late 1970s and early 1980s island biogeography theory was considered to be not only an intellectual milestone in ecological studies, which include theories that are difficult if not impossible to validate experimentally, but also a means of defining management strategies for nature reserves. It was thus considered to be an important tool for conservationists.

However, it is now acknowledged that the theory is not as applicable to non-marine 'islands' as was once thought, nor is it as valuable as a conservation tool (*cf.* ISLAND BIOGEOGRAPHIC THEORY IN CONSERVATION). The drawbacks have been examined by Shafer (1990) who suggests that only the relationship between the number of species and the size of area is valid, i.e. the bigger the area the more species will be present. He believes that the other relationships suggested by the theory of island biogeography have not been tested sufficiently to establish that they are unequivocal. Even in relation to size there is room for debate; for example an island of small area can exhibit a large range of habitats, especially if there is a steep altitudinal gradient characterized by a series of climatic zones, which will support a wide range of species. This lends support for another equally contentious theory, that of environmental determinism. Moreover, the issue of size in relation to biological reserves has given rise to considerable debate. For example, two small reserves may be better for the conservation of species than one large reserve. This has been referred to as SLOSS (single large or several small) by Shafer (1990). Again, habitat diversity is a key issue in determining SLOSS. Two small reserves, when compared with one large reserve, are likely to contain more species because they will probably have an increased array of habitats. It is also important to consider the objective(s) of the reserve. Small species, i.e. fungi, invertebrates and bacteria, may best be preserved in small reserves. Other factors of importance include the shape and spacing of reserves and the existence of CORRIDORS. [A.M.M.]

REFERENCES

MacArthur, R.H. & Wilson, E.O. (1967) *The Theory of Island Biogeography*. Princeton University Press, Princeton, N.J.

Shafer, C.L. (1990) *Nature Reserves: Island Theory and Conservation Practice*. Smithsonian Institution, Washington, D.C.

Wilson, E.O. (1992) *The Diversity of Life*. Penguin Books, London.

ISO (International Standards Organization) *See* BS 7750; ISO 14000.

ISO 14000 The International Standards Organization (ISO) has developed a series of standards, under the designation ISO 14000, that establish a framework for businesses to manage the present and potential future effects of their operations and their products on the environment. The core is ISO 14001, which defines a model ENVIRONMENTAL MANAGEMENT SYSTEM (EMS). A specification of the elements of an EMS should be in place to manage environmental matters effectively. It provides a basis whereby a company's EMS can be reviewed by a third party and a determination made on COMPLIANCE. In the event of satisfactory compliance a registration is issued that states that the STANDARD has been met and this can be displayed by the company. However, the specification does not set a standard for what a company's performance should be. These only establish what management systems should be in place for a company to meet its specified performance standards and improve upon them. ISO standards are voluntary and applicable to companies of all sizes. ISO 14001 is compatible with EMAS (ECO-MANAGEMENT AND AUDIT SCHEME). *See also* BS 7750. [P.C.]

REFERENCE

von Zharen, W.M. (1995) *ISO 14000. Understanding Environmental Standards*. Government Institutes, Inc., Rockville, Md.

isogamy *See* ANISOGAMY, EVOLUTION OF.

Isolating mechanisms

The reproductive characteristics that prevent species from fusing. Isolating mechanisms are particularly important in the BIOLOGICAL SPECIES CONCEPT, in which species of sexual organisms are defined by REPRODUCTIVE ISOLATION, i.e. a lack of gene mixture. Two broad kinds of isolating mechanisms between species are typically distinguished, together with a number of subtypes (modified from Mayr 1970).

1 Pre-mating isolating mechanisms: factors that cause species to mate with their own kind (ASSORTATIVE MATING).

(a) Temporal isolation: individuals of different species do not mate because they are active at different times of day or in different seasons.

(b) Ecological isolation: individuals mate in their preferred habitat and therefore do not meet individuals of other species with different ecological preferences.

(c) Behavioural isolation: potential mates meet, but choose members of their own species.

(d) Mechanical isolation: copulation is attempted, but transfer of sperm does not take place.

2 Post-mating isolating mechanisms: genomic incompatibility, hybrid inviability or sterility.

(a) Gametic incompatibility: sperm transfer takes place, but egg is not fertilized.

(b) Zygotic mortality: egg is fertilized, but zygote does not develop.

(c) Hybrid inviability: hybrid embryo forms, but of reduced viability.

(d) Hybrid sterility: hybrid is viable, but resulting adult is sterile.

(e) Hybrid breakdown: first-generation (F_1) hybrids are viable and fertile, but further hybrid generations (F_2 and back-crosses) may be inviable or sterile.

An alternative classification of isolating mechanisms contrasts prezygotic isolation (see 1 and 2a above) with postzygotic isolation (see 2b–e above). As an example of the application of isolating mechanisms, the apple-feeding host race of the tephritid fruit fly (*Rhagoletis pomonella*) differs from the hawthorn-feeding race in pre-mating isolation: the apple race emerges earlier in the year (see 1a above) and each host race preferentially chooses to rest, lay eggs and mate on its own host plant (see 1b above). On the other hand, laboratory experiments show that there is little behavioural, mechanical or post-mating isolation (see 1c, d, 2a–e above).

The term 'isolating mechanisms' was introduced by Th. Dobzhansky in the 1930s and has been popularized in a number of books by E. Mayr. Both authors originally proposed that isolating mechanisms were group traits beneficial at the level of the species; today, this is generally disbelieved. Recent authors have pointed out that the word 'mechanism' is particularly misleading as pre-mating and post-mating isolation are likely to evolve as a by-product of NATURAL SELECTION or GENETIC DRIFT within species, rather than as a direct result of their utility as barriers to FERTILIZATION and gene mixing between species (a process known as REINFORCEMENT). A leading critic of the biological species concept and of the term 'isolating mechanisms' is H.E.H. Paterson, who argues that species are cohesive wholes as a result of prezygotic sexual signalling within species rather than due to isolating mechanisms between species. Paterson therefore introduced a competing idea of species, the RECOGNITION CONCEPT OF SPECIES, in which isolating mechanisms were replaced by specific mate-recognition systems as an alternative. Unfortunately, the word 'system' has as many group-benefit connotations as 'mechanism', and the recognition concept of species has not gained universal acceptance.

There is also the terminological problem that 'reproductive isolation' combines traits that reduce GENE FLOW, such as mate choice or fertilization barriers, with traits that select against genes that have flowed, such as hybrid incompatibility. Lumping these two antagonistic features is confusing, since they are unrelated and evolve in very different ways. For instance, whereas it is conceivable that reinforcement might evolve to reduce an individual's tendency to mate with another species and produce inviable offspring, it is almost impossible to imagine that hybrid inviability itself would evolve as an ADAPTATION. This reproductive isolation terminology leads also to a muddled use of the term 'gene flow' as the opposite of 'reproductive isolation'; in other words, gene flow comes to include not only the flow of genes but also the effects of any natural selection on the frequency of such genes within each population.

Perhaps the most fundamental problem with isolating mechanisms (and specific mate-recognition systems) is that species are implied to be qualitatively different from SUBSPECIES, races or forms by their possession of these traits. Races cannot, in theory, differ in either type of trait because only species are defined by their possession. Arguably, by making species seem qualitatively different from races, these terms have spawned a number of special models of SPECIATION, where geographic isolation (also known as allopatry) or sudden bursts of EVOLUTION in small founder populations (founder events or punctuated equilibria) play important roles. Only such unusual conditions were thought to be able to give rise to new species that differ in isolating mechanisms (or specific mate-recognition systems). In reality, there is little to distinguish mate choice and disruptive natural selection commonly observed within species from pre-mating and post-mating isolation between species; indeed, it is hard to distinguish species from races in many organisms (*see* SPECIES CONCEPTS).

Most serious research on speciation now avoids using the term 'isolating mechanisms' because of these unwanted connotations. Instead, researchers clearly distinguish between mate choice, hybrid incompatibility and other forms of reproductive isolation. It is probable that the term 'isolating mechanisms' is becoming extinct and even today is confined mostly to undergraduate textbooks. *See also* ALLOPATRIC SPECIATION; FOUNDER EFFECT; GEOGRAPHICAL ISOLATION; SPECIFIC-MATE RECOGNITION. [J.L.B.M.]

REFERENCE

Mayr, E. (1970) *Populations, Species, and Evolution*. Harvard University Press, Cambridge, Mass.

isometric *See* ANISOMETRIC.

isomorphic *See* ALTERNATION OF GENERATIONS.

isoprene Isoprene (2-methyl-1,3-butadiene), a volatile organic molecule emitted by plants, has an important role in ATMOSPHERIC CHEMISTRY. Emissions of CARBON as isoprene are believed to be quite significant, up to 2% of PHOTOSYNTHESIS in leaves of oak and aspen, amounting to 3×10^{14} g year^{-1} globally, about the same as METHANE. Isoprene is believed to protect plants from the effects of overheating. [J.G.]

REFERENCE

Sharkey, T.D. & Singaas, E.L. (1995) Why plants emit isoprene. *Nature* **374**, 769.

isotherm Line on a map joining points of equal TEMPERATURE. [J.G.]

isotope Atoms of the same ELEMENT that differ in their MASS NUMBER are termed 'isotopes'. Isotopes may be stable or unstable (*see* CARBON for examples). *See also* RADIOACTIVITY. [J.G.]

isotope discrimination (Δ) The deviation of the isotopic fractionation (α) from unity, i.e.

$$\Delta = \alpha - 1 = \frac{R_a}{R_p} - 1$$

where, for the CARBON isotopes and during PHOTOSYNTHESIS, R_p is the $^{13}C:^{12}C$ ratio of photosynthetically assimilated CARBON DIOXIDE (CO_2) and R_a is the $^{13}C:^{12}C$ ratio of the AMBIENT air. To a good approximation (when $R_a \sim R_p$):

$$\Delta = \delta_p - \delta_a$$

(*see* DELTA CARBON). Plants are generally depleted in ^{13}C compared with atmospheric CO_2 and hence show a positive discrimination against ^{13}C. Typically, C_3 plants have a discrimination of about 20×10^{-3}, which is normally expressed in the literature as 20‰ ('per mille'). This discrimination arises due to slower diffusion of $^{13}CO_2$ into and within leaves and because of discrimination against $^{13}CO_2$ by the primary carboxylating enzyme of photosynthesis, ribulose 1,5-bisphosphate carboxylase/oxygenase (Rubisco); the extent of ISOTOPE discrimination by C_3 plants gives an indication of the relative importance of stomatal CONDUCTANCE versus PHOTOSYNTHETIC CAPACITY in influencing photosynthetic rate. It can also be used as an indicator of plant water use EFFICIENCY.

C_4 plants show less discrimination than C_3 plants. This is because phosphoenolpyruvate carboxylase, the primary carboxylating enzyme in C_4 species, has a different intrinsic isotope effect and uses a different species of inorganic carbon that has an isotopic composition at equilibrium different from that of Rubisco. Isotopic screening provides a ready method for the separation of C_3 and C_4 plants. Plants exhibiting CAM often have values intermediate between C_3 and C_4 plants.

There is widespread variations in the isotopic fractionations of algae and other marine plants. At least part of this variation is due to the presence of mechanisms for raising the concentration of CO_2 and the sites of carboxylation above that of the environment.

It is also possible to express all other isotopic effects involving a reactant and product in terms of a discrimination, for example the effects of photosynthetic gas exchange on the $^{18}O:^{16}O$ ratio of the ambient CO_2. [J.Ly.]

REFERENCES

Farquhar, G.D. & Lloyd, J. (1993) Carbon and oxygen isotope effects in exchange of carbon dioxide between terrestrial plants and the atmosphere. In: *Stable Isotopes and Plant Carbon–Water Relations* (eds J.R. Ehleringer, A.E. Hall & G.D. Farquhar), pp. 47–70. Academic Press, New York.

Farquhar, G.D., Ehleringer, J.R. & Hubick, K.T. (1989) Carbon isotope discrimination and photosynthesis. *Annual Review of Plant Physiology and Molecular Biology* **40**, 503–537.

Lloyd, J. & Farquhar, G.D. (1994) ^{13}C discrimination by the terrestrial biosphere. *Oecologia* **99**, 201–215.

IT_{50} *See* MEDIAN TIMES TO EFFECT.

iteroparity Breeding more than once. Iteroparous organisms have a moderate REPRODUCTIVE EFFORT and continue to invest in the maintenance of somatic tissues during and after reproduction, which results in a substantial chance to survive and reproduce again.

There are several types of iteroparous life histories (see Fig. S12). Iteroparous ANNUALS include most univoltine invertebrates, as well as some annual plants that continue to grow and flower after first seeds have been set. Seasonal breeders live for many years and reproduce repeatedly during specific breeding seasons (e.g. PERENNIAL plants and most vertebrates of seasonal environments). Continuous breeders lack specific reproductive seasons and thus reproduce asynchronously. This category includes species living in aseasonal environments (e.g. TROPICAL RAIN FOREST and deep-sea species, internal parasites), species little affected by SEASONALITY (e.g. humans) and small organisms that complete many generations within one vegetative season (e.g. fruit flies, planktonic crustaceans). For the evolutionary advantages of iteroparity *see* SEMELPARITY. *See also* LIFE-HISTORY EVOLUTION. [T.J.K.]

REFERENCES

Roff, D.A. (1992) *The Evolution of Life Histories*. Chapman & Hall, London.

Stearns, S.C. (1992) *The Evolution of Life Histories*. Oxford University Press, Oxford.

IUCN (International Union for the Conservation of Nature and Nature Resources) Now renamed the World Conservation Union, although it still goes by the abbreviation IUCN. Created in 1948, it is a multinational body that compiles lists of rare species and protected natural areas. It also provides advice/expertise to national governments. *See also* CONSERVATION BIOLOGY. [P.C.]

IUCN Red (Data) Book *See* RED (DATA) BOOK.

IUCN Red Data categories *See* RARITY, CLASSIFICATION.

J

jackknife estimates of diversity *See* DIVERSITY INDICES.

Jackson's method Extension of LINCOLN INDEX for estimation of POPULATION SIZE from marked recaptures in sequences of samplings. Named after C.H.N. Jackson, a medical entomologist working in East Africa in the 1930s and 1940s. *See also* CAPTURE–RECAPTURE TECHNIQUES. [L.M.C.]

Jeltozem *See* SOIL CLASSIFICATION.

Jolly–Seber population estimation Method of estimating animal POPULATION SIZE from multiple mark–release–recapture data. A stochastic method providing successive estimates of population size, loss rate and gain and their standard errors. *See also* CAPTURE–RECAPTURE TECHNIQUES. [L.M.C.]

joule (J) SI UNIT of energy, work and heat, defined as the work done when a force of 1 N acts through 1 m. Relation to other units: 1 erg = 10^{-7} J; 1 eV = 1.602×10^{-19} J; 1 calorie = 4.18 J; 1 Btu = 1055 J. [J.G.]

joule equivalent *See* CALORIFIC VALUE.

Julian day Julian day is a chronology based on the Julian era with the epoch defined at noon on 1 January 4713 BC. The Julian day is frequently confused with that of DAY NUMBER, which counts the number of days from 1 January within a year. [P.R.V.G.]

juvenile hormone *See* METAMORPHOSIS.

K

K-dominance curve A curve that illustrates the DISTRIBUTION of individuals (or BIOMASS) among species in order to graphically represent the species DIVERSITY of a COMMUNITY. Obtained by plotting cumulative per cent ABUNDANCES (or cumulative per cent biomasses) versus species RANK or log species rank. Steepest curves have the lowest diversity. The advantage of K-dominance plots is that the distribution of species among individuals can be directly compared with the distribution of species' biomasses, because the units are the same for both plots. *See also* DIVERSITY INDICES. [S.J.H.]

REFERENCE

Lambshead, P.J.D., Platt, H.M. & Shaw, K.M. (1983) The detection of differences among assemblages of marine benthic species based on an assessment of dominance and diversity. *Journal of Natural History* **17**, 859–874.

k of the negative binomial *See* INDICES OF DISPERSION.

K-selection Selection that operates to maximize POPULATION SIZE. The terms 'r-selection' and 'K-selection' are derived from the LOGISTIC EQUATION

$$N(t) = \frac{K}{1 + e^{c - rt}}$$

where $N(t)$ is population size at time t, r is the rate of increase, c is a constant and K is the CARRYING CAPACITY of the environment. Selection that maximizes the rate of increase is termed r-selection. It has been hypothesized that there is a TRADE-OFF between selection for TRAITS that maximize r and those that maximize K. However, evidence does not support such a hypothesis. *See also* FITNESS; LIFE-HISTORY EVOLUTION; OPTIMALITY MODELS. [D.A.R.]

k-value Also known as 'killing k' or 'killing power', this is a measure of the loss of individuals from a population during a single AGE CLASS or life stage. It is usually followed by a subscripted number that indicates the relevant age or stage. The loss of individuals is usually an actual MORTALITY but may also include a loss due to EMIGRATION. Some authors use a pseudo k-value (k_0) to indicate failure to produce the maximum possible NUMBER OF OFFSPRING. A k-value is calculated as $\log_{10}$(no. entering a stage) $-$ $\log_{10}$(no. surviving a stage). All k-values may be combined to produce k_{total} (or K_{total}) as a measure of generation mortality.

The k-values are also used to determine the type of DENSITY DEPENDENCE exhibited by a population. This is achieved by plotting k-value against log density on entering a stage. If there is no relationship there is no density dependence, whereas a positive relationship indicates negative density dependence. If the positive relationship has a terminal slope of 1, then the density dependence is exactly compensating (CONTEST COMPETITION); if the slope is less than 1 it is undercompensating; and if greater than 1 it is overcompensating. In the unlikely event that the terminal slope of the line is infinity, this would indicate that pure SCRAMBLE COMPETITION was operating. *See also* KEY-FACTOR ANALYSIS; LIFE TABLE. [C.D.]

kairomone *See* INFOCHEMICALS.

kamikaze hypothesis *See* SPERM COMPETITION.

kamikaze sperm *See* SPERM COMPETITION.

kandic horizons *See* SOIL CLASSIFICATION.

Kastanozems *See* SOIL CLASSIFICATION.

Kendall's rank correlation coefficient (τ) A non-parametric statistic that summarizes the degree of CORRELATION between two sets of observations (x and y). The correlation coefficient (τ) varies between +1 (perfect positive correlation), through 0 (no correlation) to −1 (perfect negative correlation). The two variables need not follow a NORMAL DISTRIBUTION; it is only necessary that they can be ranked. The data below examines the possibility that the diversity of invertebrates (number of species) is related to pH in streams.

pH (x)	Diversity (y)	Rank x	Rank y
5.0	18	1.5	1
5.0	23	1.5	3
5.5	25	3	4
5.8	22	4	2
6.0	33	5	6
6.1	34	6	7
6.4	40	7.5	10
6.4	36	7.5	8
6.5	32	9	5
6.9	39	10	9

We now write down the ranks of one variable paired with the ranks of the other variable. If one variable has ties, the pairs are ordered by the variable without ties ('diversity' in the example above). If both variables have ties it does not matter which variable is ordered. Next obtain the sum of counts (ΣC_i) by obtaining a value of C_i for each rank in the column. Examine each value of R_2 in turn, and write down all the subsequent values of R_2 that are lower in rank. Ranks that are equal count $^1/_2$. This is set out in the table below for the example above.

Diversity (R_1)	pH (R_2)	Subsequent values of R_2 that are greater	C_i
1	1.5	4, (1.5), 3, 9, 5, 6, 7.5, 10, 7.5	8.5
2	4	9, 5, 6, 7.5, 10, 7.5	6
3	1.5	3, 9, 5, 6, 7.5, 10, 7.5	7
4	3	9, 5, 6, 7.5, 10, 7.5	6
5	9	10	1
6	5	6, 7.5, 10, 7.5	4
7	6	7.5, 10, 7.5	3
8	7.5	10, (7.5)	1.5
9	10		0
10	7.5		0

We now calculate the quantity $N=4\Sigma C_i-n(n-1)$, where n is the number of paired observations and 4 is a constant. For the data in the above table, $\Sigma C_i=37$ and $N=4(37)-10(9)=58$. The Kendall rank correlation coefficient is now calculated as:

$$\tau = \frac{N}{\sqrt{\left[n(n-1)-\sum T_1\right]\left[n(n-1)-\sum T_2\right]}}$$

where ΣT_1 and ΣT_2 are the sums of correction terms for ties in R_1 and R_2 respectively. A T value equal to $t(t-1)$ is computed for each group of t tied observations and summed over all groups. In the above data, there are no ties among R_1 and therefore $\Sigma T_1=0$ and two groups, each of two tied observations, among R_2. Therefore $\Sigma T_2=2(2-1)+2(2-1)=4$. Therefore the calculated test statistic for the above data is:

$$\tau = \frac{58}{\sqrt{\left[10(9)-0\right]\left[10(9)-4\right]}}$$

$$\tau = 0.66$$

This value of the test statistic is compared with the DISTRIBUTION of τ in the appropriate STATISTICAL TABLES for $n=10$ (DEGREES OF FREEDOM are not used in this test). However, the usual tables (e.g. Rohlf & Sokal 1995) are exact only if there are no ties. If there are ties, a special table can be consulted (see Burr 1960). For $n=10$ the 1% value for τ is 0.664 and therefore our computed value of τ indicates that there is a very significant positive correlation between pH and invertebrate diversity in these particular streams. *See also* STATISTICS. [B.S.]

REFERENCES

Burr, E.J. (1960) The distribution of Kendall's score S for a pair of tied rankings. *Biometrika* **47**, 151–171.

Rohlf, F.J. & Sokal, R.R. (1995) *Statistical Tables*, 3rd edn. W.H. Freeman, New York.

key-factor analysis Key-factor analysis uses the *k*-VALUES calculated from life-table data collected over a number of sampling events (often years) on the same populations to determine which MORTALITY factors have the greatest influence on the variation in the size of a particular population through time. This is achieved by calculating a coefficient of regression for each key factor on the total of all key factors, k_{total}. Unfortunately this is not a precise statistical test as k_{total} is not independent of its constituent key factors. A key factor is one that has a very high positive coefficient of regression. It is not necessarily the case that the key factor is the most significant agent of mortality. For example, a constant large mortality from year to year or a mortality factor that varies at random will not give a high value for a coefficient of regression of k_{total}. The key factor will be a mortality factor that fluctuates with k_{total}.

Identification of the key factor controlling the variation in k_{total} can be very important in the control of populations. If a population is a PEST, it will often be most cost-effective to attack it immediately after the effect of the key factor. *See also* DEMOGRAPHY; LIFE TABLE. [C.D.]

REFERENCE

Varley, G.C. & Gradwell, G.R. (1960) Key factors in population studies. *Journal of Animal Ecology* **29**, 399–401.

key organisms A term sometimes used by those involved in environmental management and conservation, which might refer to: importance in dynamics and structure of communities (e.g. as in KEYSTONE SPECIES); important representative of, for example, trophic groups in COMMUNITY; particularly sensitive species in terms of STRESS and/or DISTURBANCE and/or POLLUTION. None of these criteria is exclusive of the others, so that all might be intended. [P.C.]

keystone species A keystone species is one upon which many other species in an ecosystem depend and the loss of which could result in a cascade of local extinctions. The keystone species may supply a vital food resource, as for example in the case of a plant with energy-rich seeds on which many animals rely for food. Or it may be a predatory species that holds in check the populations of certain herbivores which would otherwise overgraze, reduce primary production potential and lead to the loss of other herbivorous animals, together perhaps with certain specific parasites and predators.

Conservation biologists should be particularly anxious to identify and give special attention to such keystone species. *See also* COMMUNITY ECOLOGY; FOOD WEB. [P.D.M.]

kick sampling *See* SAMPLING METHODOLOGY/DEVICES.

kilogram (kg) SI UNIT of mass. The international prototype kilogram is at the Bureau International des Poids et Mesures, Sèvres, near Paris. Relation to other units: $1\,g=10^{-3}\,kg$; $1\,lb=0.453\,kg$. [J.G.]

kin Relatives; i.e. sharing recent common ancestry. *See also* KIN SELECTION. [P.C.]

kin bias *See* KIN SELECTION.

Kin selection (altruism, indirect fitness)

Given ALLELES are usually shared by a number of individuals within a population. In most cases an allele is shared through recent common ancestry and the probability of two individuals being co-bearers is determined by their degree of relatedness. Thus a diploid sexually reproducing parent shares a particular allele with its offspring with a probability of 0.5. The same is true between siblings. However, half-siblings (sharing only one parent) and grandparents and grandoffspring share alleles with a probability of only 0.25.

As W.D. Hamilton was the first to acknowledge formally, the probability of sharing a particular allele has profound implications for the EVOLUTION of social, sexual and developmental relationships between organisms. Hamilton pointed out that an allele could enhance its transmission to the next generation either directly, by improving the reproductive success of its own bearer, or indirectly by improving the reproductive success of other individuals carrying a copy of it. The FITNESS of an allele (its rate of SPREAD relative to alternatives in the population GENE POOL) is thus the net result of both these direct and indirect components, thereby constituting what Hamilton termed 'inclusive fitness'. (Although Hamilton originally defined inclusive fitness at the level of the individual, it is less confusing, and indeed more accurate, to define it at the level of the allele.) The importance of inclusive fitness, and its corollary kin selection (the NATURAL SELECTION of alleles through their indirect fitness consequences), is that phenotypic TRAITS that appear to act against an individual's own short-term interests, such as cooperation and 'altruistic' caring and self-sacrificial behaviours, become explicable in terms of genetic selfishness and no longer encourage explanations in terms of GROUP SELECTION (*see* BEHAVIOURAL ECOLOGY) or other ill-founded evolutionary mechanisms.

The conditions for the spread of an allele for phenotypic altruism, set out in Hamilton's rule, are determined by the cost of the altruism and its benefits in terms of enhanced transmission of the allele (approximated as the improvement in reproductive success of the recipient devalued by its coefficient of relatedness to the altruist). Suppose an altruistic trait is specified by a single pair of alleles, A and a, such that A makes an individual more likely to be altruistic, for example when an adult risks itself for the survival of juveniles by diverting the attention of a PREDATOR. In performing the act, the adult reduces its survival chances by C and enhances those of juveniles by B. Hamilton's rule is that A will increase in frequency relative to a only if the relationship $rB - c > 0$ holds.

The calculation of inclusive fitness itself must be undertaken with care since only those individuals directly affected by the altruist's actions, and not all the altruist's existing relatives, should be taken into account. Moreover, the calculation should include only those additional offspring produced by recipients as a result of the altruist's efforts and discount the extra offspring produced by the altruist itself as a result of help from other altruists in the population. However, the nightmarish practical problems of calculating inclusive fitness in this way can be circumvented by the simple expedient of counting only the altruist's offspring (since this already takes account of the potential offspring lost through altruisitic effort and those gained as a result of help from others).

While kin selection provides a powerful framework for interpreting social relationships, associations between relatedness and behaviour (kin bias) can arise for other reasons that have nothing to do with kin selection. Mating preferences based on genetic similarity may have some inclusive fitness implications when local mate COMPETITION is intense, but are more likely to reflect trade-offs between inbreeding and outbreeding for genes affecting critical components of life history (e.g. immune function) or adaptedness to local environments. Kin bias may also arise because similar genotypes confer similar resource requirements or microclimatic preferences, or because SOCIAL ORGANIZATION results in a degree of inbreeding so that discrimination of own versus foreign group members becomes confounded with relatedness.

Kin selection and reciprocity are often discussed as alternative evolutionary routes to phenotypic altruism. However, at the level of alleles coding for the altruistic response they amount to the same thing, even though in reciprocity alleles may be shared by chance rather than recent common ancestry and, in principle, need not even be copies of the same DNA sequence. The difference is simply that in kin selection relatedness biases the probability that the beneficiary carries a copy of the altruism allele, while in reciprocity bias is effected by limiting responses to individual's demonstrating the altruistic trait and thus the possession of an appropriate allele. *Cf.* GROUP SELECTION. [C.J.B.]

REFERENCE

Grafen, A. (1984) Natural, kin and group selection. In: *Behavioural Ecology: An Evolutionary Approach* (eds J.R. Krebs & N.B. Davies), 2nd edn, pp. 62–84. Blackwell Scientific Publications, Oxford.

---

kingdom *See* TAXONOMIC CATEGORIES.

Kirchhoff's radiation law *See* RADIATION LAWS.

kleptoparasites *See* TRAPPING FOOD.

kleptoparasitism *See* GROUP SOCIAL STRUCTURE.

koinobiont parasitoids *See* PARASITOID–HOST INTERACTIONS.

Kolmogorov–Smirnov two-sample test A non-parametric test that examines differences between two distributions, rather than using ranks and comparing medians as in the MANN–WHITNEY *U* TEST. See Sokal and Rohlf (1995) for the details of calculation. *See also* NON-PARAMETRIC STATISTICS. [B.S.]

REFERENCE

Sokal, R.R. & Rohlf, F.J. (1995) *Biometry*, 3rd edn. W.H. Freeman, New York.

krill Derived from Norwegian *kril* (young fish) but now used as a common term for euphausids: PELAGIC, marine shrimps distributed globally. Key animal in Antarctic marine ecosystem because of its great abundance and its position in the FOOD WEB between the microalgae upon which it feeds and the large vertebrate predators, whales, seals and penguins, that feed on it. Now exploited by fisheries that lead to a range of end-products for human and animal consumption. *See also* FISHERIES, CONSERVATION AND MANAGEMENT. [P.C.]

krummholz Trees growing at (or just above) the TREE LINE on mountain slopes that exhibit a prostrate growth form, pruned by WIND and ice-blast (Fig. K1). Sometimes this growth form may be genetically determined, while in other cases it may be entirely due to PHENOTYPIC PLASTICITY. Some krummholz plants migrate slowly downwind across the mountainside, rooting at their sheltered side and dying back at their exposed windward side. A vital source of shelter for high alpine animals. [M.J.C.]

Fig. K1 Development of krummholz form in stems of mountain birch, *Betula pubescens* ssp. *czeraponovii* Hämet-Ahti (*B. tortuosa* Auct.) growing at the timber line in western Norway. (Courtesy of R.M.M. Crawford.)

Kruskal–Wallis test This is the non-parametric equivalent of a one-way ANALYSIS OF VARIANCE. However, it is used to compare several medians rather than several means. The procedure is very similar to the MANN–WHITNEY *U* TEST. The table below illustrates the basic calculations. An ecologist has placed baited traps within four woodlands in order to obtain relative estimates of POPULATION DENSITY in a species of fly. The numbers of flies caught at each trap, over a 24-h period, are shown below. Three woods had six traps, the other had seven traps.

	Wood 1	Wood 2	Wood 3	Wood 4
	10 (11)	18 (18)	0 (2)	16 (14)
	0 (2)	12 (12)	4 (6)	17 (16)
	6 (8)	17 (16)	2 (4)	28 (24)
	17 (16)	26 (23)	6 (8)	30 (25)
	14 (13)	20 (20)	0 (2)	19 (19)
	8 (10)	6 (8)	3 (5)	21 (21)
	22 (22)			
n	7	6	6	6
R	82	97	27	119
R^2/n	960.6	1568.2	121.5	2360.2

The number in parentheses is the rank for each observation within the table as a whole. Tied ranks (e.g. the three zeros) are given their average rank (e.g. $(1+2+3)/3=2$). For each SAMPLE (wood) we need to know the number of observations (n), the sum of the ranks (R) and R^2/n. These are also given in the table. We also require the two quantities $N=\Sigma n=25$ and $\Sigma R^2/n=5010.5$. The test statistic is then calculated using the formula below, with the numbers 12 and 3 as constants.

$$K=\left[\sum\left(R^2/n\right)\times\frac{12}{N\left(N+1\right)}\right]-3\left(N+1\right)$$

$$K=\left[5010.5\times\frac{12}{25\left(26\right)}\right]-3\left(26\right)$$

$$K=14.50$$

Since in this example there are ties (three 0s, three 6s and three 17s) this value of K should be corrected by dividing by:

$$D=1-\frac{\sum\left(t^3-t\right)}{\left(N-1\right)N\left(N+1\right)}$$

where t is the number of tied observations in each set of tied values. For the above example there are three groups of tied observations and within each group there are three tied values. Since in each group of ties $t=3$, $\Sigma(t^3-t)=(3^3-3)+(3^3-3)+(3^3-3)=72$, and $D=0.9954$. The corrected value of K is therefore 14.57. This test statistic is then compared with the tabulated values in a χ^2 table, with the DEGREES OF FREEDOM equal to the number of samples minus one ($4-1=3$ in the above example). For $v=3$ the 1% value of χ^2 is 11.34, which our calculated value exceeds. Therefore we reject the NULL HYPOTHESIS and conclude that there is a very significant difference between the numbers of flies caught in the four woods. Note that this test (like the statistic F in analysis of variance) applies to the group as a whole and tells us that there are some very significant differences between at least two of the four samples. We cannot make statements about particular pairs of samples, although we could safely conclude that the two woods with extreme sum of ranks (3 and 4) were significantly different. *See also* MEDIAN; NON-PARAMETRIC STATISTICS. [B.S.]

krypton (Kr) A gaseous element (ATOMIC NUMBER=36; relative ATOMIC MASS=83.80; melting point= −156.6°C; boiling point= −152.3°C) of the noble gas group, krypton is a minor constituent of the Earth's ATMOSPHERE with an average volume mixing ratio of

1.1×10^{-6} (1.1 ppm by volume). Krypton has a number of stable isotopes as well as short-lived radionuclides produced in radioactive processes. Krypton is physically and chemically inert and has no environmental significance. It is obtained from liquefaction of air and used in electric discharge signs. [M.R.H.]

kurtosis A statistic that measures one type of departure from a NORMAL DISTRIBUTION. Kurtosis is the 'peakedness' of the distribution. A leptokurtic distribution has more observations in the centre and tails, relative to a normal distribution. A platykurtic distribution has less in the centre and tails, relative to a normal distribution. An extreme platykurtic distribution would be a BIMODAL DISTRIBUTION. *See also* STATISTICS. [B.S.]

kwongan *See* CHAPARRAL SHRUBLANDS.

k_x Logarithm of the PROBABILITY of survival from one AGE CLASS to another. *See also* *k*-VALUE. [P.C.]

L

Lack clutch The CLUTCH SIZE that maximizes the number of young fledged assuming that MORTALITY is a function of clutch size, i.e. that the only important interactions are negative density-dependent interactions between siblings within a clutch. If adult survival, survival of offspring subsequent to fledging or the future FECUNDITY of the parents declines with clutch size, then selection will favour a clutch size below the Lack value. *See also* LIFE-HISTORY EVOLUTION; REPRODUCTIVE ALLOCATION; REPRODUCTIVE EFFORT. [D.A.R.]

Lack, D. *See* CHARACTERS IN ECOLOGY.

lag load A type of GENETIC LOAD defined as the difference between the optimal GENOTYPE and the actual mean FITNESS of a population due to continually changing fitness optima. Although in principle a species will evolve toward its optimum genotype, the actual fitness lags behind the optimum, which continually changes due to the COEVOLUTION of competing species, parasites or hosts. The lag load controls the chance that a species will become extinct (i.e. the more a species lags behind its competitors, the more likely it is to go extinct). As the lag load increases, the rate of EVOLUTION of the species will increase because it is subject to stronger SELECTION PRESSURE. *See also* QUANTITATIVE GENETICS. [V.F.]

lagg *See* MIRE.

Lagoons

A lagoon is a shallow body of coastal brackish or saltwater that is semi-isolated from the open SEA by some form of natural barrier (see Plate 7, facing p. 398). The term is used to describe two rather different types of environment: coastal lagoons and those impounded by CORAL REEFS.

Coastal lagoons, which occupy 13% of the world's coastline, are virtually tideless, pond- or lake-like bodies of coastal saline or brackish WATER that are partially isolated from the adjacent sea by a sedimentary barrier, but which nevertheless receive an influx of water from that sea. Only Canada, the western coast of South America, the eastern coast of Australia, the China Sea coast from Korea to south-east Asia and the Scandinavian peninsula lack significant lagoonal coastlines. Characteristically, coastal lagoons are floored by soft sediments, fringed by reed beds, mangroves or salt-marsh vegetation, and support dense beds of submerged macrophytes such as seagrasses, pondweeds or *Ruppia*.

Coastal lagoons are most characteristic of regions with a tidal range < 2 m, since large tidal ranges result in powerful water movements capable of breaching and destroying incipient sedimentary barriers. Further, the lake-like nature of lagoons requires the permanent presence of water and large tidal ranges are likely to result in loss of water from open systems during periods of low tide in the adjacent sea (although see below). Hence in Europe, for example, lagoons are particularly abundant around the shores of the microtidal Baltic, Mediterranean and Black seas but they are also present along some macrotidal coasts, such as the northern North Atlantic where offshore deposits of shingle are to be found as a result of past glacial action. Here, shingle can replace the more characteristic sand of microtidal seas as the barrier material because it is less easily redistributed by tidal water movements. In many cases, the SALINITY of the lagoonal water mass is exactly the same as that of the adjacent sea, although where FRESHWATER discharges into a lagoon its water is brackish and a (usually relatively stable) salinity gradient occurs between river mouth and lagoonal entrance channel. In regions where EVAPORATION exceeds PRECIPITATION for all or part of a year, lagoons are often hypersaline.

Five broad types of coastal lagoon can be recognized.

1 Estuarine lagoons, which as their name implies merge into estuaries, have formed where barriers have moved onshore and have partially blocked existing drowned river valleys. For this reason they usually have their long axis perpendicular to the coastline. Even if blockage of the river is complete, lagoonal status may remain if seawater can enter by overtopping of the barrier during high water of spring TIDES. Nevertheless, many former estuarine lagoons are now completely freshwater habitats with seawater entry being prevented by the barrier. With the gradual landward movement of sedimentary barriers that is currently taking place, this is probably the ultimate fate of many of the world's lagoons.

2 Bahira lagoons (from the Arabic for 'little sea') are pre-existing partially land-locked coastal embayments, drowned by the post-glacial rise in sea level, that have their mouths almost completely blocked by the later development of sedimentary barriers. Also included here are systems in which the sea has broken through a pre-existing sedimentary barrier to flood part of the hinterland, but in which the entrance/exit channel remains narrow. Bahira lagoons clearly merge into semi-isolated marine bays, such as the Dutch–German–Danish Waddenzee.

3 Typical coastal lagoons are regions that have become partially isolated from the parent sea by the development of a spit (as in the Fleet, Dorset, England and the Vistula and Kursk lagoons of the Polish–Lithuanian Baltic coast) or by one or more offshore/

longshore barrier BARS or islands (as in the lagoon of Venice and the majority of the world's lagoons). Seawater then enters and leaves these lagoons through the channels between the islands or at the end of the spit. More rarely, the lagoons are enclosed by multiple tombolos connecting an offshore island to the mainland. With the exception of tombolo lagoons, the long axis of typical lagoons—and many are extremely elongate—lies parallel to the coastline. Typical coastal lagoons are especially characteristic of microtidal seas. The abundant lagoons that occur within river deltas form a special case of this category; several deltas, for example those of the Danube and Nile, have developed within, and have now largely or completely obliterated, former lagoons enclosed by barrier-island chains.

4 Percolation lagoons occur where low-lying land is situated behind a longshore barrier through which seawater can percolate. They are mostly small pools owing their existence to human activities, although some of the natural limans of the BLACK SEA coast receive their salt input via seepage. Regions to landward of longshore shingle ridges that were once SALT MARSH but are now reclaimed, for example, usually receive an influx of water from the barrier WATER TABLE and this collects in depressions that were once part of the creeks draining the marshes. Equivalently, gravel pits or borrow pits in coastal shingle masses, from which building materials have been extracted, similarly receive an influx of water from out of the shingle. In both cases, the shingle water table is derived from seawater soaking into the barrier during high tide in the adjacent sea together with such rainfall as has also soaked in.

5 Sandy sedimentary barriers do occasionally develop in some relatively macrotidal areas (e.g. southern Iceland and Portugal), but the regions impounded to landward retain water only during high tide for the reasons outlined above; water drains away to leave exposed MUDFLAT over low tide. These environments that are true lagoons only during high tide are often referred to as 'tidal-flat lagoons'. They are therefore the relatively rare macrotidal coast equivalent of the typical lagoons of microtidal seas.

Coastal lagoons are heavily used by waders, wildfowl and marshland birds as feeding areas, and contain invertebrate and fish faunas that comprise mixtures of three different elements: (i) essentially freshwater and (ii) essentially marine or estuarine species capable of withstanding a degree of brackishness, as well as (iii) specialist lagoonal or 'paralic' species that are not in fact restricted to lagoons but also occur in the Eurasian inland seas as well as in (usually artificial) tideless brackish ponds and drainage ditches that are abundant in reclaimed coastal regions. The main feature of the lagoonal species category is that they are species of marine ancestry that seem to be restricted to shallow, non-tidal maritime habitats; over their ranges as a whole they are also clearly capable of inhabiting a wide range of salinity, including full-strength seawater, but not freshwater. Their relatedness to marine species is indicated by the fact that they often form species pairs with marine or estuarine species. Lagoons also form the nursery grounds and adult feeding areas for a large number of commercially important fish and crustaceans that migrate between lagoons and the sea. Throughout the world they therefore support (often artisanal) fisheries, as well as being the location of bivalve shellfish culture (mussels, oysters and clams). Within the Mediterranean, there is an average lagoonal fish yield of some 100 kg ha^{-1} $year^{-1}$, principally from eels, grey mullet, sea-bass and sea-bream. Many of the world's lagoons are threatened habitats, however, suffering deterioration as a result of POLLUTION and destruction through RECLAMATION (in addition to the natural loss resulting from SUCCESSION to freshwater and SWAMP habitats and from landward barrier migration). For this reason they are a 'priority habitat' under the European Habitats Directive.

In so far as is known, although the species may be different, the ecology of coastal lagoons shows the same general pattern as seen in estuaries and other regions of coastal soft-sediment. The action of predators as important forces in structuring the communities, for example, is reflected by the young stages of many species occurring within the dense beds of submerged macrophytes, as the hunting success of predatory species is lower there.

The second environment to which the word 'lagoon' is applied is associated with coral reefs; coral here replaces the unconsolidated sedimentary barrier of the coastal lagoon. Circular atoll reefs enclose the 'lagoon' within their perimeter, whilst BARRIER REEFS are separated from the mainland by an equivalent although less isolated body of water. Indeed, barrier-reef lagoons are virtually sheltered stretches of coastal sea. However, atoll lagoons are distinctive in being floored by coral sand that supports submerged beds of seagrasses and fringing MANGROVE swamps just as the coastal lagoons of similar latitudes. The similarity between the two types of lagoon is nevertheless purely physiographic, since the atoll lagoon fauna is that typical of coral reefs in general and not related in any way to those of coastal lagoons. [R.S.K.B.]

LAI *See* LEAF AREA INDEX.

Lake stratification

Two layers of WATER in a lake at different temperatures tend to remain static, with the denser cooler water lying beneath the less dense warmer water. Energy must be expended to mix the two layers and such a temperature-layered lake shows thermal resistance to mixing. The amount of energy needed to mix two layers of water at different temperatures increases as the temperature difference increases, particularly at high temperatures.

In climates cold enough for ICE formation the water column will be at a uniform temperature of approximately 4°C after ice melt in spring. Wind action on the lake surface will move surface water downwind, with a compensatory upwind movement of deeper water. As a result of these contrary water movements, mixing will occur since thermal resistance is low. The whole lake water will thus be circulated and mixed by the WIND and will gradually increase in temperature as a result of SOLAR RADIATION. As spring progresses, more heat enters the lake bringing the surface water temperature to a level substantially higher than that of the lower waters so that thermal resistance to mixing will be too great to be overcome by the wind. The lake is thus stratified into two principal regions, an upper circulating warmer region (epilimnion) and a lower cooler more-or-less stationary layer (HYPOLIMNION). Little heat is transmitted to the hypolimnion, which thus remains at the temperature it had at the time of STRATIFICATION. Between the epilimnion and hypolimnion there is a transitional layer (metalimnion or THERMO-

CLINE), in which there is a steep change of temperature with depth (as high as 3°C m−1). The metalimnion is not only a boundary between water masses of different temperatures but also a barrier to free movement between the contents of the epilimnion and the hypolimnion because of density differences.

As summer progresses, the rate of heat income to the lake decreases resulting in cooling of the epilimnion which thus sinks and erodes the metalimnion. With the epilimnion cooling while the hypolimnion remains at a constant temperature, the thermal resistance to mixing decreases gradually; eventually resistance to mixing is overcome by winds and the whole lake once again mixes and circulates. Thus until the water temperature of the whole lake drops to 4°C, the water temperature will be uniform from top to bottom.

If heat loss from the water surface is such that the surface freezes, the lake will remain insulated under a layer of ice for the rest of the winter. Such LAKES show inverse stratification, with the surface ice and water immediately below at 0°C. Beneath this layer there is a gradual increase of water temperature to the winter maximum somewhere between 0 and 4°C. The following spring, heat income increases, the ice melts and the lake again mixes and circulates. Such a lake showing two periods of total circulation or mixing (TURNOVER) per year and two periods of stratification or stagnation is termed 'dimictic'.

At the time of circulation, most lakes are mixed completely from top to bottom and are termed 'holomictic'; however, some lakes have a permanently stagnant water layer at the bottom and are termed 'meromictic'. Meromixis occurs because the energy available from the wind at the water surface is insufficient to overcome the resistance to mixing. In very deep lakes this occurs because the wind energy is insufficient to set the whole water mass in motion, but another cause of meromixis is the presence of a layer of saline water (monimolimnion) at the bottom of the lake. Under these circumstances, wind cannot produce sufficient energy to disturb the equilibrium because of the density difference between the saline layer and the overlying FRESHWATER (mixolimnion). Meromictic lakes are found under a wide range of climatic conditions, their existence being determined by the distribution of salts rather than by the climate. In some cases the salts in the saline layer are derived from underlying salt beds or from saline inflow, while in others they arise as a result of normal processes of DECOMPOSITION in the lake sediments.

For dimixis to occur, the climate must result in the water temperature ranging both above and below 4°C. In consequence, dimictic lakes occur principally at moderate to high latitudes (from around 40 to 75°), provided that the lake is deep enough. Shallow lakes may never acquire sufficient thermal resistance to mixing and are thus termed 'polymictic'.

At high latitudes and at high elevations in lower latitudes, the climate may never permit the water temperature to rise above 4°C. Thus, lakes will never attain summer stratification and are termed 'cold thereimictic'. In climates where the difference between summer and winter is relatively slight and the winter temperatures not much below 4°C (as in many oceanic temperate regions), lakes never develop ice cover and circulate throughout the winter; these are termed 'cheimomictic'.

In contrast, lakes where there is inverse stratification during the winter and the water warms appreciably above 4°C in the summer but does not stratify due to the lack of strong summer heating usually coupled with high winds are termed 'warm thereimictic'. Thus, these lakes fall near the boundary between dimictic and polymictic lakes. A little more summer heating would presumably convert warm thereimictic lakes to dimictic and a little less winter cooling would convert them to polymictic.

In warm water conditions the density difference is high, so that a temperature difference between the epilimnion and hypolimnion that would not confer stability under temperate climatic conditions make tropical lakes highly stable and resistant to mixing. Thus, many lakes of moderate or great depth in the tropics remain stratified most of the time. The stratification of such lakes is occasionally upset by freak weather conditions such as exceptionally cool weather coupled with high winds, causing the epilimnion to cool and reducing the thermal resistance to mixing. Such occasional overturns are catastrophic to the BIOTA because the hypolimnetic water, which has been isolated for a long period of time, is often anoxic.

Shallow tropical lakes, particularly if they are in windy climates and at high elevations (where nocturnal heat loss is greater than at low elevations), circulate freely most of the time, with stratification occurring only under unusually calm conditions. Polymixis similar to this is also frequently found in small ponds in temperate latitudes which may, under very warm and calm weather conditions, stratify daily and circulate every night.

The most commonly used classification of lakes according to their circulation patterns is that developed by Hutchinson and Loffler (1956) and modified by Bayly and Williams (1973) and Lewis (1983). *See also* VERTICAL ZONATION. [R.W.D.]

REFERENCES

Bayly, I.A.E. & Williams, W.D. (1973) *Inland Waters and their Ecology*. Longman, Camberwell.

Hutchinson, G.E. & Loffler, H. (1956) The thermal classification of lakes. *Proceedings of the National Academy of Sciences of the USA* **42**, 84–86.

Lewis, W.M. (1983) A revised classification of lakes based on mixing. *Canadian Journal of Fisheries and Aquatic Science* **40**, 1779–1787.

Lakes

A hole or BASIN containing a volume of WATER that is relatively large compared with its annual inflow and outflow. Lake basins can be small with EPHEMERAL water levels, such as temporary rain-fed rock pools, or extremely large and long-lived, such as Lake Baikal which contains 20% of the world's FRESHWATER and has had a continuous lacustrine history for at least 65 million years.

Lakes originate from a variety of geomorphological processes. Tectonic basins result from movements of the Earth's CRUST; most occur in the depressions between masses of a single fault displacement or in downfaulted troughs (e.g. Lake Baikal). Volcanic activity can generate lakes termed 'maars' and 'calderas' in volcano craters (e.g. Crater Lake, Oregon) or behind lava flows damming RIVERS. Landslides into river valleys can also cause dams and create lakes. The majority of the world's lake basins were formed during recent periods of GLACIATION: glacial ICE movements in mountainous areas created cirque lakes at the tips of valleys and paternoster lakes in lower valleys; massive ice scouring in non-mountainous piedmont areas created the Great Lakes of North America. Blocks of ice buried in glacial drift eventually melted to form kettle lakes. The erosive power of running water can produce plunge-pool lakes at the base of waterfalls, or oxbow lakes where river meanders become isolated.

Physical features

Lake water levels are a balance between inputs from PRECIPITATION on the lake surface, RUN-OFF from the surrounding CATCHMENT, groundwater seepage and springs, and water loss through EVAPORATION, drainage from outflows and seepage into groundwater through basin walls. Hydraulic residence time and its reciprocal, flushing rate, describe the relationship between lake volume and annual inflow.

Lake basins are divided into zones that reflect physicochemical and/or biological processes (Fig. L1).

SOLAR RADIATION is the major source of energy in lakes, as LIGHT for PHOTOSYNTHESIS and as heat. Light penetration and hence photosynthesis is restricted to the upper layers, the PHOTIC ZONE. Similarly, more heat is absorbed at the surface than at depth. Strong vertical temperature gradients in summer can result in THERMAL STRATIFICATION into a surface epilimnion overlying the HYPOLIMNION, with an intermediate metalimnion and THERMOCLINE where the rate of change is maximal. Autumn turnover occurs when stratification is disrupted and the entire water column is mixed through declining air temperatures and loss of heat from surface waters, resultant CONVECTION currents and WIND action. Inverse stratification can occur in winter with cold surface water under ice and warmer water at depth; spring turnover occurs when the ice breaks up and surface waters heat. Meromictic lakes have more complex stratifications and do not undergo complete circulation. Thermal stratification often has large effects on water chemistry.

Water movements within lake basins produced by wind result in a variety of rhythmic movements (OSCILLATIONS) on the water surface, such as surface WAVES, surface currents and LANGMUIR CIRCULATION or streaks, and internal oscillations within the basin. Displacement of the whole water mass gives rise to very long-standing wave patterns in which the water surface (surface seiche) or the thermocline (internal seiche) oscillate up and down. Incoming water from rivers will produce different inflow movements (overflow, underflow, interflow) depending on density differences between lake and river water.

Water chemistry

Dissolved OXYGEN is essential to the metabolism of all lake organisms capable of AEROBIC RESPIRATION. Oxygen solubility is strongly dependent on temperature; oxygen profiles often reflect thermal stratification and light penetration patterns. Diel cycles in oxygen concentrations are associated with high densities of macrophytes, attached algae and PHYTOPLANKTON: oxygen levels increase during the day from photosynthesis, but decrease at night from respiration. Extremely high oxygen demands during intense DECOMPOSITION can reduce lakes to almost complete ANOXIA and result in massive die-offs of many species of animals. Die-offs can also occur during turnover, when deoxygenated hypolimnetic water reaches the surface. Oxygen deficit indicates the relationship between metabolism in the trophogenic zone, the surface stratum where photosynthesis occurs, and the tropholytic zone, the aphotic stratum of heterotrophic decomposition.

The ionic composition of lake waters is dominated by four major cations (CALCIUM, MAGNESIUM, SODIUM and POTASSIUM) and by three anions (carbonate, sulphate and CHLORIDE). SALINITY is governed by WEATHERING of rocks and run-off from the catchment, atmospheric deposition, and the balance between evaporation and precipitation. Most lakes have relatively low salinities compared with marine systems, although there are inland saline lakes with specially adapted flora and fauna. Vertical stratification of fresh surface water and denser saline water at depth can produce a HALOCLINE.

Three elements comprise the major nutritional components of lake BIOTA: CARBON, NITROGEN and PHOSPHORUS; essential MICRONUTRIENTS include IRON, SULPHUR and silica. These exist dissolved in the water, bound in organic matter, or in the sediments. Complex mechanisms and processes govern their solubility, availability and biogeochemical cycling. Phosphorus and, to a lesser extent, nitrogen are often limiting NUTRIENTS to algal and higher plant growth in lakes. Overall nutrient levels indicate lake trophic status, ranging from eutrophic (very high nutrients) to oligotrophic (very low nutrients). Shallow waters of the LITTORAL ZONE are usually more productive than the limnos: deep, steep-sided lake basins are often unproductive, whereas shallow, low-relief basins are more productive. Two important human-induced processes can change lake PRODUCTIVITY: (i) artificial EUTROPHICATION through additions of phosphorus and nitrogen from SEWAGE and run-off from arable catchments increases productivity; and (ii) ACIDIFICATION through ATMOSPHERIC POLLUTION and run-off from afforested conifer plantations reduces productivity.

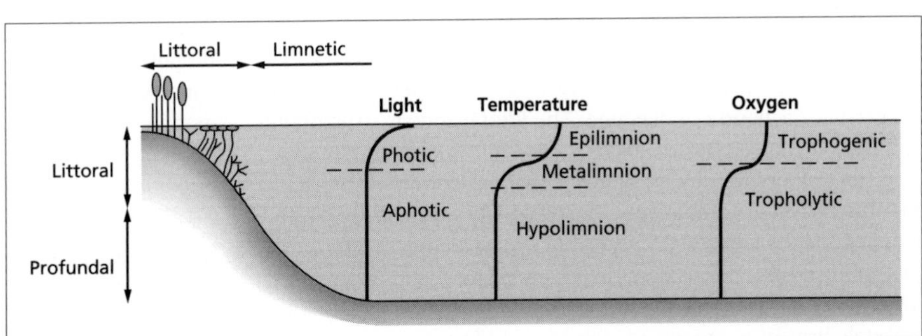

Fig. L1 A general scheme of lake zonation and idealized vertical distribution of light, temperature and oxygen concentrations in a summer stratified, eutrophic lake.

Lake biota

The biota of lakes can be divided into several major components depending on their metabolic status and what zone of the lake they occupy: phytoplankton, ZOOPLANKTON and fish of the limnetic zone, macrophytes and attached algae of the littoral zone, animals of the littoral zone and BENTHOS, and heterotrophic decomposers. Within any of these groups, many species may co-occur simultaneously in any one lake. The assemblage structure, dynamics and spatial distribution will depend on species-specific tolerances to physical and chemical parameters and on the outcome of biotic interactions such as competition, predation and parasitism.

Phytoplankton absorb nutrients directly from the water and are largely restricted to the photic zone. Populations display distinct seasonal patterns and periodicity, but the variability among species and lakes makes generalizations difficult. In polar and temperate regions, growth is reduced or negligible during winter; numbers and biomass may peak in spring, summer or autumn, or there may be two peaks in spring and autumn. In tropical lakes, maxima often occur in winter. In any one lake, successional seasonal periodicity of phytoplankton is reasonably consistent between years, barring outside perturbations. The major consumers of phytoplankton are herbivorous filter-feeding zooplankton, although large-bodied phytoplankton, such as some cyanobacteria, may not be available to grazers.

Zooplankton crops are set by phytoplankton productivity, but in complex ways. Zooplankton populations also display patterns of seasonal SUCCESSION that vary among species, and within species among lakes. CYCLOMORPHOSIS (seasonal POLYMORPHISM) is conspicuous among zooplankton, especially the Cladocera where a pronounced spike or crest can develop on the head in response to predation pressure from fish and, especially, the phantom midge *Chaoborus*. Some zooplankton undergo marked DIEL vertical migrations upwards from deep to more surficial waters at night, possibly to avoid predation from visual predators such as fish.

Virtually all lake fish feed on zooplankton at some stage in their life history. Some are OBLIGATE planktivores; others are benthic feeders or piscivores. Fish predation can alter the size structure of lake assemblages by selectively removing large-bodied zooplankton, with the knock-on effect of reducing grazing pressures and increasing phytoplankton biomass in a trophic cascade.

Macrophytes and macroalgae are largely restricted to the littoral zone. Their distribution within a lake is influenced by wind and wave action, and their ZONATION with depth by light penetration and water pressure. Depth limits for emergent forms are <2 m and for submerged forms <11 m; bryophytes and charophytes may go deeper. In temperate lakes, littoral vegetation can show marked seasonal variations, from maximum biomass in late summer to almost total dieback in winter; tropical lakes may show little or no SEASONALITY.

Animals of the lake littoral and benthos are extremely diverse and their dynamics poorly understood. The littoral zone is structurally complex: the presence of macrophytes, substrate and SEDIMENT type are important determinants of animal communities, both quantitatively and qualitatively. The profundal zone is more homogeneous and the fauna less diverse, but no less abundant. Profundal communities are often dominated by oligochaete worms and chironomid midge larvae that eat organic sediments.

Metabolism of DETRITUS (dissolved and PARTICULATE ORGANIC MATTER) from allochthonous, littoral and limnetic sources results in a complex CARBON CYCLE involving shredding invertebrates and heterotrophic decomposers. Much of the decomposition occurs in the sediment and in anoxic conditions. This MICROBIAL LOOP is poorly understood, but can dominate lake structure and function. If inputs of particulate matter exceed decomposition rates, lakes can become shallower. *See also* LAKE STRATIFICATION; LAKES, TYPES OF. [J.L.]

Lakes, types of

Of whatever origin, LAKES may be either 'open' with WATER losses from outlets or seepage, or 'closed' where the only water losses occur through EVAPORATION. Open lakes with outlets are termed 'drainage lakes', while those without outlets are termed 'seepage lakes'. The ionic composition, conductivity and SALINITY of closed, seepage lakes are invariably higher than in open lakes.

Hammer (1986) defined saline water and FRESHWATER as those having salinities ≥ 3 g l^{-1} and ≤ 0.5 g l^{-1} respectively. Saline lakes (also called athalassohaline, athalassic or poikilosaline lakes) occur on every continent and depend on: evaporation exceeding PRECIPITATION; endorheic (having no outflow) drainage BASINS; and the availability of soluble salts. The term 'brackish' is restricted to mixtures of seawater and freshwater.

Lakes grouped together in close geographic proximity often have similar origins. The classification presented takes this regional view into account following Hutchinson (1957), who distinguishes 76 types of lakes grouped into the following classes.

Tectonic lakes are produced by some form of earth movement other than volcanic as follows.

1 Upward movements of land bordering the SEA cutting off RELICT lakes of marine origin.

2 Lifting of the seabed above sea level forming usually shallow lakes within the irregularities. (Relict lakes of marine origin or lakes formed from the uplifting of the seabed frequently have higher salinities and their ionic composition reflects their saltwater origins.)

3 Subsidence of the CRUST to form graben lakes, which are often very large and geologically long-lasting. (Because of their large size and temporal persistence on a geological scale, graben lakes frequently contain ENDEMIC SPECIES.)

4 Earth movements blocking a valley.

Volcanic lakes are formed in the craters of volcanoes, sometimes from a single cone when they are usually relatively small and circular in shape, or from multiple craters in which case the lakes are large and irregularly shaped. Because the morphometry of these basins is very steep, the LITTORAL ZONE is narrow and forms a low proportion of the total surface area. Volcanic lakes often have low concentrations of NUTRIENTS and low PRODUCTIVITY due to the granitic nature of the lake basin and its CATCHMENT area. Some lakes are formed on collapsed lava flows, which are variable in size but tend to be shallow.

Landslide lakes are formed when debris from rockfalls, mudflows or any other landslide block the course of running water and a lake, often transitory, collects behind the landslide dam.

Glacial lakes are formed by the action of ICE. A lake can be formed at the foot of a glacier (proglacial), by the moraine, by ice scour or by retreat up a valley (cirque). Kettle lakes are formed by the deposition of large blocks of ice covered in mud and/or gravel when the main ice sheet retreats and tend to be shallow and EPHEMERAL. Alternate freezing and thawing of the ground can result in subsidence and many shallow arctic and antarctic ponds are formed in this way. If several of these ponds merge they form a thermokast lake. Temporary bodies of water are occasionally found occupying basins composed entirely of ice. In regions of perennially frozen ground, local thawing of the PERMAFROST will produce temporary thaw lakes. Where large glaciers occupied valleys at low elevations, FJORD lakes may be formed.

Solution lakes are formed by the percolation of water through soluble rocks either on the surface or underground; in the latter case, collapse of the chambers results in surficial depressions. Limestone is one of the commonest soluble rocks involved, but solution lakes also occur in sodium chloride (NaCl), calcium sulphate ($CaSO_4$), ferric hydroxide and aluminium hydroxide bearing rocks and consequently have high total dissolved solids and high conductivity.

Fluviatile lakes are formed by the erosion and/or deposition of sediments by running water. A waterfall has the erosive power to excavate a depression and if the river subsequently changes course the remaining basin forms a plunge or evorsion lake. Eddies in a river or stream can produce a pool, which can either be isolated seasonally or permanently.

RIVERS also deposit SEDIMENT that can form dams, either blocking the main watercourse to form a fluviatile lake or more commonly blocking the tributary valleys to form a lateral lake. In the lower reaches of rivers, particularly where there is a wide flood plain, meandering of rivers can produce oxbow lakes. In large rivers, sediments tend to be deposited in the marginal zones, forming levees where water can accumulate to form lakes.

Wind-formed lakes are formed either by the deposition of windborne material such as sand damming a valley or in the parallel valleys produced between SAND DUNES (slacks). Wind erosion can produce deflation basins that fill with water.

Coastline lakes are formed by the deposition of a sediment BAR (tombolo) across an indentation or irregularity in the coast of either the sea or a large lake or between the coast and an island.

Organic-accumulation lakes are formed by the accumulation of less readily decomposable parts of organisms (usually plant in origin but animal material, e.g. coral, can have the same effect), which act as a dam or form a ring around the basin.

Meteoric lakes occur when a meteor strikes the surface and forms a crater, partially from the impact itself but also frequently as a result of the explosion of the water and gases in the rocks due to the intense heat produced by the impact. Since the pressure of this explosion is equal in all directions, craters are usually circular and the floor of the crater consists of crushed or fused rocks.

Lakes produced by animals such as beavers (*Castor*) are well known. Humans have produced artificial lakes (reservoirs) by building dams since 2000 BC (in Egypt and Sri Lanka); lakes are also formed in disused quarries, clay pits and peat excavations. [R.W.D.]

REFERENCES

Hammer, U.T. (1986) *Saline Ecosystems of the World*. Junk, Dordrecht.

Hutchinson, G.E. (1957) *A Treatise on Limnology. Volume 1. Geography, Physics and Chemistry*. John Wiley & Sons, New York.

Lamarck, J.B. *See* CHARACTERS IN ECOLOGY.

Lamarckian inheritance Jean Baptiste de Monet, Chevalier de Lamarck (1744–1829) in his *Philosophie Zoologique* (1809) proposed that species evolve. He recognized that organisms are adapted to their environments and that they change in response to new needs as the environment changes. At the time almost nothing was known about inheritance and the idea that characters acquired during the lifetime of an individual are inherited by its offspring came to be associated with Lamarck's theory. Thus, Lamarckian inheritance is the INHERITANCE OF ACQUIRED CHARACTERS. However, it is virtually impossible for acquired TRAITS to become incorporated in the genetic material for two reasons. First, information flows from the genes to the PHENOTYPE rather than in the opposite direction (the 'central dogma' of Francis Crick; *see also* WEISMANNIAN ORGANISMS) and, more importantly, environmental effects on the phenotype cannot cause purposive or directed changes in the genes. Examples have been proposed that appear to show inheritance of acquired characters, but experiment usually fails to support the claim. For example, bacteria very rapidly become resistant in the presence of an antibiotic and this has sometimes been claimed to be a directed change. It can be shown, however, that the populations already contain mutants, which remain at a negligible frequency in the absence of the antibiotic. *Cf.* DARWINISM. [L.M.C.]

laminar flow Flow of any fluid in which the lines of flow (as revealed by injecting a tracer, such as smoke, dye or particles) are parallel. Laminar flow tends to occur at low flow rates. The tendency for laminar flow to break up into chaotic motion (TURBULENT FLOW) is measured by the REYNOLDS NUMBER:

$$Re = ud/\nu$$

where u is the speed of the fluid, d is the characteristic dimension of the surface and ν is the kinematic VISCOSITY of the fluid. When a fluid flows over a smooth plate or inside a smooth tube, the critical Reynolds number for the transition from laminar to turbulent flow is about 10^5 but it may be much lower if the original flow is disturbed or if the surface is rough (Fig. L2). The transition from laminar to turbulent flow may be observed by running a tap, slowly at first to achieve laminar flow (when the water appears clear), then rapidly. At the critical Reynolds number the water becomes turbulent and appears opaque. Most flows in biological systems (blood, water in plants) are laminar, whereas flows of water and air in the natural environment are turbulent or at least partially turbulent. [J.G.]

land bridge This is a land connection between continents and between areas within continents. Such connections and their changing temporal distribution have been important factors in the MIGRATION routes of terrestrial organisms. Land bridges have therefore influenced patterns of life on Earth today. Moreover, the existence of land bridges in the past was related to the distribution of continental land masses and the movements between them, i.e. CONTINENTAL DRIFT.

One example is the 'World Continent' embracing Africa, Europe, Asia and North America during the last 65 million years, i.e. from the opening of the Tertiary period after the Age of Mammals had begun. This contiguity facilitated an ADAPTIVE RADIATION of placental mammals, which reproduce by carrying their unborn progeny attached to a uterine placenta. The subsequent

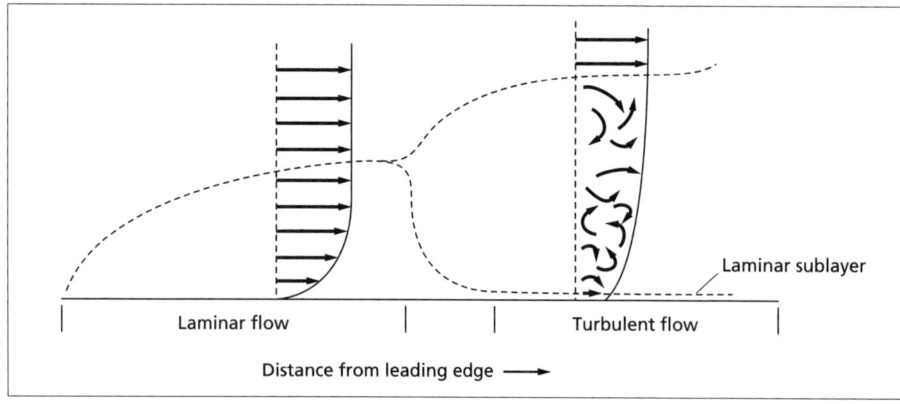

Fig. L2 Transition from laminar to turbulent boundary layer. Flow is from left to right. The arrows show the magnitude and direction of flow as the air passes over a surface from the leading edge (left-hand side). (After Grace, J. (1977) *Plant Response to Wind*. Academic Press, London.)

Table L1 The range of factors hostile to plant growth likely to occur in different types of degraded land: the most common and serious problems are soil structure and nitrogen content. (After Bradshaw, A.D. & Chadwick, M.J. (1980) *Restoration of Land: The Ecology and Reclamation of Derelict and Degraded Land.* University of California Press, Berkeley, Calif.)

	Physical problems				Chemical problems				
Materials	Texture and structure	Stability	Water supply	Surface temperature	Nitrogen	Other nutrients	pH	Toxic materials	Salinity
Colliery spoil	– – –	– – –/0	–/0	0/+ + +	– – –	– – –	– – –/0	0	0/++
Strip mining	– – –/0	– – –/0	– –/0	0/+ + +	– – –	– – –/0	– – –/0	0	0/+ +
Fly ash	– –/0	–/0	0	0	– – –	– – –	+/+ + +	+ +	0/+ +
Oil shale	– –	– – –/0	– –	0/+ +	– – –	– –/0	– –/0	0	0/+
Iron ore	– – –/0	– –/0	–/0	0	– – –	– –	0	0	0
Bauxite	0	0	0	0	–/0	–/0	0	0	0
Heavy metals	– – –	– – –/0	– –/0	0	– – –	– – –	– – –/+	+ + +/+	0/+ + +
Gold wastes	– – –	– – –	–	0	– – –	– – –	– – –/–	0/+ +	0
Kaolin wastes	– – –/–	– –	– –	0	– – –	– – –	–	0	0
Acid rocks	– – –	0	– –	0	– – –	– –	–	0	0
Calcareous rocks	– – –	0	– –	0	– – –	– – –	+	0	0
Sand/gravel	–/0	0	– –/+ +	0	– –/0	–/0	–/0	0	0
Coastal sand	– –/0	– – –/0	– –/0	–/0	– – –/0	– –/0	0	0	0/+
Urban land	– – –/0	0	0	0	– – –/0	0	0	0/+ +	0
Roadsides	– – –/0	– – –	– –/0	– –/0	– – –/0	– –/0	0	0	0
Ski runs	– – –	0	– –	0/+	– – –	–/0	0	0	0

– – –, severe deficiency; –, slight deficiency; 0, adequate; +, slight excess; + + +, severe excess. (For pH: – indicates low; + indicates high.) Actual score can vary between limits shown owing to site conditions and history of disturbance.

establishment of a land bridge between North and South America via the Panama isthmus 2.5 million years ago coalesced two landmasses that had been separated for 200 million years. This conjunction facilitated the SPREAD of South America's unique fauna northward and the southward movement of World Continent fauna, which came to dominate the Americas. More recently, and at more local scales, the presence or absence of land bridges has influenced organism distribution. For example, the repeated severance of Britain from Europe during the warm stages of the Quaternary period caused each stage to be characterized by a distinct arboreal assemblage. Similarly, the presence of a land bridge between Alaska and Siberia across the Bering Strait during the last cold stage allowed human groups to enter America from Asia. [A.M.M.]

land contamination *See* CONTAMINATED LAND; LAND RECLAMATION.

land decontamination *See* LAND RECLAMATION.

land drainage *See* DRAINAGE.

Land reclamation

In modern societies land is continuously being degraded by human activities, especially by industrial development and mining (Table L1). When these activities cease the land is often left derelict. This land is not only an eyesore but also leads to loss of PRODUCTIVITY and to environmental threats due to erosion and POLLUTION. 'Reclamation' is the term used for any activity by which this land is brought back into some sort of beneficial use. It can include full restoration, where the area is returned to the pre-existing land use, rehabilitation, in which this end-point is only partly achieved, or replacement, in which something new is created (Fig. L3). Remediation is often used as a synonym for RECLAMATION although the emphasis is more on the operations involved. Since it is ecosystems that are always damaged there is now considerable interest in what is called 'ecological restoration', in which the specific aim is to restore ecological structures and functions, normally of the ECOSYSTEM that existed immediately before degradation occurred but sometimes of the original indigenous ecosystem.

Restoration of this sort presents a considerable intellectual and practical challenge. Nature, given time, through the processes of PRIMARY and SECONDARY SUCCESSION, can usually restore what was originally present without assistance, but this usually takes 100 years or more. The aim of restoration, or reclamation in general, is to create the desired end-point within a much shorter period, usually in no more than a decade. Ecology as a science usually involves taking the natural world to pieces to understand how it works. Restoration, by contrast, involves putting it together again in such a way that it works satisfactorily and appropriately. As such, it is an acid test of our ecological understanding and has recently aroused considerable interest.

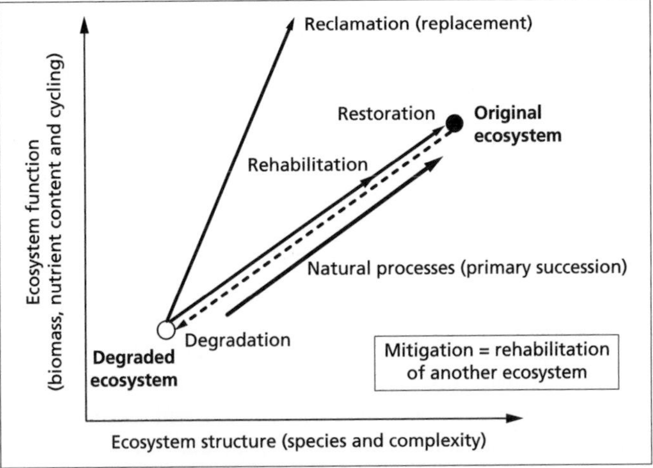

Fig. L3 The different options for the improvement of degraded land and ecosystems in terms of the two ecosystem characteristics of structure and function.

The important first step in treating a degraded site is to decide what should be the final end-point. It may seem appropriate to put back what was there before. However, it may be more sensible to take advantage of the DISTURBANCE and create something new. Sand and gravel pits are often left to fill with water to become WETLANDS, valuable in areas where there is a scarcity of such sites. Agricultural land disturbed by mining for coal may be best turned into WILDLIFE and amenity areas if the site is near where people live. In other regions FORESTRY may be more appropriate.

The next important step is to determine what factors might limit or retard ecosystem development. These will depend on the nature of the degradation and the nature of the situation; in the case of mining the nature of the wastes will be important. It will be easier to restore VEGETATION on a disturbed subsoil than on the wastes produced by mining for coal or HEAVY METALS because of severe toxicities in the latter. However, if each factor is treated in an appropriate manner its effects can be overcome. There is usually a short-term and a long-term solution. The long-term solutions are usually those nearest to what would happen naturally and are most likely to give self-sustaining results, in which ecosystem functions are both active and maintained. Although different plant species will be required for different climatic regions, the basic approaches given in Table L2 are widely applicable.

Usually the important plants are sown or planted using normal techniques and in ecologically suitable combinations. For wild species, SOIL or litter containing dormant seed, or shredded turf,

Table L2 The physical and chemical problems commonly found in degraded terrestrial ecosystems and the short-term and long-term treatments most likely to lead to self-sustaining restoration.

Category	Problem	Immediate treatment	Long-term treatment
Physical			
Texture	Coarse	Organic matter or fines	Vegetation
	Fine	Organic matter	Vegetation
Structure	Compact	Rip or scarify	Vegetation
	Loose	Compact	Vegetation
Stability	Unstable	Stabilizer or nurse	Regrade or vegetation
Moisture	Wet	Drain	Tolerant vegetation
	Dry	Irrigate or mulch	Tolerant vegetation
Nutritional			
Macronutrients	Nitrogen	Fertilizer	Nitrogen-fixing species
	Others	Fertilizer and lime	Fertilizer and lime
Micronutrients	Deficient	Fertilizer	—
Toxicity			
pH	Low	Lime	Lime or tolerant species
	High	Pyritic waste or organic matter	Weathering
Heavy metals	High	Organic matter or tolerant plants	Inert covering or tolerant plants
Salinity	High	Weathering or irrigate	Tolerant plants

may be used. Usually, animals and microorganisms can be left to colonize naturally, but this is not always the case. It may be necessary to reintroduce important but relatively immobile animals such as earthworms and some arthropods. There is increasing evidence for the need to introduce mycorrhizal species.

In new situations it is necessary to explore the efficacy and economy of such treatments in practice, in relation to the behaviour of desired species and their particular requirements. This usually entails a combination of both ecological experience and field experiments. In unfamiliar situations, the experiments may need to be continued for a number of years in order to confirm that the treatments are leading to satisfactory long-term results. There is always the possibility that the fundamental problems of the degraded land have not been properly understood.

With this information a final decision can be made on what should be the end-point. Unless full restoration is an absolute requirement, it may be sensible to choose a more ecologically satisfactory replacement ecosystem and land use that takes into account the needs of the local region and community. Even when the aim is to enhance wildlife and BIODIVERSITY, it may well be wiser to think of creative conservation and begin a new community and ecosystem likely to be in equilibrium with the new conditions

Once the main restoration operations have been completed the site cannot be left to its own devices. The ecosystem must be monitored to see if development is proceeding well or whether aftercare management is required. This may be anything from fencing to control unwanted grazing to additions of phosphorus to overcome phosphorus adsorption by the soil; forests may require thinning. It may also involve further species introduction if the ecosystem is complex and has ASSEMBLY RULES, with some of the required species only able to be established once the ecosystem has developed.

Whatever happens, it will not be possible to create a fully mature ecosystem immediately and no one should imply such a result. Any ecosystem requires time for growth and development; this is all too obvious where trees are involved. However, the essential qualities of many aquatic and simple terrestrial ecosystems based on herbaceous vegetation can be restored within 10 years. The biological aspects of soils, leading to the fertility supporting plant and ecosystem growth, need take no longer than this to develop, although the pedological features may take much longer. The aim of successful reclamation should be to achieve as quickly as possible the type of ecosystem functioning that allows the final desired ecosystem and landscape solution to develop unhindered. *See also* AGRICULTURE AND SOIL CONSERVATION; ECOLOGICAL RESTORATION; PRIMARY SUCCESSION; REMEDIATION; TOXICITY. [A.D.B.]

REFERENCES

Bradshaw, A.D. (1987) The reclamation of derelict land and the ecology of ecosystems. In: *Restoration Ecology* (eds W.R. Jordan, M.E. Gilpin & J.D. Aber), pp. 53–74. Cambridge University Press, Cambridge.

Bradshaw, A.D. & Chadwick, M.J. (1980) *Restoration of Land: The Ecology and Reclamation of Derelict and Degraded Land.* University of California Press, Berkeley, Calif.

Schaller, F.W. & Sutton, P. (1978) *Reclamation of Drastically Disturbed Lands.* American Society of Agronomy, Madison, Wis.

land registers Lists of actually or potentially CONTAMINATED LAND intended to guide the need for CLEAN-UP, the possibilities for development, the decisions of investors, purchasers, etc. Because it is never possible to prove zero CONTAMINATION on a site (evaluation in the wrong place or for the wrong substance; improving analytical techniques might expose contamination where it was previously undetectable), it is usually not wise to compile lists on the basis of actual contamination. Instead, potential contamination, for example identified as a result of contaminative proccesses that might have occurred historically on the site, is more reliably defined. Though helpful to some, these registers are frowned on by others because of the possible effects on land prices, i.e. so-called land blight. *See also* LAND RECLAMATION. [P.C.]

land rehabilitation *See* LAND RECLAMATION.

land remediation *See* LAND RECLAMATION.

land replacement *See* LAND RECLAMATION.

land restoration *See* LAND RECLAMATION.

landfill WASTE DISPOSAL method that involves dumping WASTE below or on the surface (and sometimes in the sea to create more land, as in Hong Kong). It essentially involves long-term storage of inert materials and relatively uncontrolled DECOMPOSITION of biodegradable waste. It stands alone as the only waste disposal method that can deal with all materials in the solid waste stream and so there will always be a place for it. The outputs from landfill involve gases and LEACHATE emissions, both of which can be polluting and dangerous and so some containment and control of releases is required. One extreme of this is complete containment with not only impervious lining to floor and sides but also capping. This dry containment has been described as long-term storage, since it slows down biological decomposition processes that depend crucially on the availability of water. This also means that it reduces the initial production of landfill gas. Decomposition will, of course, ultimately occur but possibly at a time in the future when the site is in a poorer state of repair and management. An alternative, therefore, is to admit water to accelerate decomposition and gas production at an early stage when the outputs can be more easily managed. This is tantamount to operating the site as a large bioreactor. A problem with this is that understanding of the process is rudimentary, which means that managment cannot be too precise. The bioreactions are likely to vary depending on the quality of the waste and also local climatic conditions. In both systems, leachates are contained until leaks ultimately occur or are removed in a controlled way. It is presumed that bioreaction will remove organic pollutants. However, the understanding of this process is again rudimentary and inorganic pollutants such as HEAVY METALS will remain a problem. So there is a need to monitor, and probably treat, controlled releases. [P.C.]

landfill sites *See* RECLAMATION.

landrace A CROP cultivar or animal breed that has been improved by traditional agriculturalists, but has not been influenced by modern breeding practices. [P.C.]

Landsat series *See* SATELLITE REMOTE SENSING.

landscape architecture *See* LANDSCAPE ECOLOGY.

landscape ecology The study of the interactions between the temporal and spatial aspects of a landscape and its flora, fauna and cultural components. This covers the academic side of the subject, which focuses on the large-scale pattern of landscape elements and their function, but not the applied work of the professional landscape ecologist, which is more practical and site based. At both scales it involves the study of habitats where the effect of people is a major influence, such as public open space, cemeteries, allotments, town parks, precincts, industrial areas, intensively cultivated regions, construction sites, roads and gardens. Knowledge of these habitats lags far behind that of more natural communities as the powerful anthropogenic influences in operation have rendered them unattractive to ecologists, who find the high proportion of exotics and mixtures of planted and spontaneous vegetation bewildering. In addition to studying these habitat patches much of the work of a landscape ecologist is involved with habitat creation and how to set about greening cities. They are expert at creating amenity GRASSLAND, wild-flower meadows, broad-leaved WOODLAND including the ground flora, MOORLAND and naturalistic river engineering. Recently they have started to produce landscape maps at a county or regional scale. These are concerned with identifying and reinforcing local distinctiveness.

Landscape ecologists need to have a thorough understanding of introduced taxa, the importance of DISTURBANCE, DISPERSAL mechanisms of plants and animals, artificial SOIL PROFILES, POLLUTION and the needs of people. Aesthetic conflicts and social conflicts have to be resolved. Sometimes these are too complex to be solved by ecologists; thus they need to team up with social scientists and environmental psychologists, who know how people use their environment and what they want from it. The output of professional landscape ecologists is more frequently reports, plans, schedules and landscape schemes on the ground rather than papers in scientific journals. They have to be able to communicate with and understand the jobs of architects, planners, engineers, contractors and quantity surveyors and be conversant with legislation, particularly, in the UK, the Wildlife and Countryside Act 1981. Advice often has to be simplified and, due to the pressures of time, important decisions have to be made in the light of incomplete information. This is particularly difficult for those coming into landscape ecology from more academic fields. The work of a professional landscape ecologist involves site survey, analysis of the results, master plan preparation, detailed design, preparation of planting schedules, costing, supervision of contractors and the preparation of management schedules and plans. Increasingly, on-going MONITORING is involved.

The academic side of landscape ecology has a geographical foundation. It focuses on (i) structure, the spatial pattern of landscape elements (patches, tessera, CORRIDORS, matrix) and ecological objects; (ii) function, the flow of objects between landscape elements; and (iii) change, alterations in the mosaic through time. It encompasses seven general principles: structure and function, biotic DIVERSITY, species flow, nutrient redistribution, ENERGY FLOW, landscape change and landscape stability. [O.L.G.]

REFERENCE

Forman, R.T.T. & Godron, M. (1986) *Landscape Ecology*. John Wiley & Sons, Toronto.

landslide lakes *See* LAKES, TYPES OF.

Langmuir circulation A pattern of water CIRCULATION that occurs when WIND blows steadily across the surface of relatively calm seas and which is characterized by vortices (several metres in diameter) that revolve around vertical axes, causing UPWELLING and downwelling of water. The vortices, or cells, tend to be regularly spaced and are often arranged in staggered parallel rows with their long axes parallel to the wind. The alternating areas of DIVERGENCE and convergence tend to cause aggregations of PLANKTON, which can enhance nutrient regeneration. Langmuir cells provide an important mechanism for transporting heat, momentum and matter from the water surface to layers a few centimetres deep. *See also* LAKES. [V.F.]

lanthanide metals The lanthanide (rare earth) elements. The lanthanides are a series of 14 consecutive elements of the PERIODIC TABLE, comprising LANTHANUM (ATOMIC NUMBER = 57) to lutetium (atomic number = 71). The name 'rare earths' is misleading since the least common of the naturally occurring lanthanides, THULIUM and lutetium, have a natural abundance of 0.5 mg kg^{-1} of the Earth's CRUST (comparable to IODINE), whilst the most abundant, CERIUM, is present to the extent of 60 mg kg^{-1} and is the 26th most abundant of all elements. One of the elements, PROMETHIUM, has no stable isotopes and therefore is not naturally occurring. The most important sources of the lanthanides are monazite sands, a mixture of lanthanide phosphates with 5–10% of the actinide THORIUM as its dioxide (ThO_2), and bastnaesite from which the heavier lanthanides and ThO_2 are virtually absent. Although thorium itself is only weakly radioactive (^{232}Th, α, $t_{1/2} = 1.4 \times 10^{10}$ years), it is contaminated with daughter nuclei such as ^{228}Ra, which is considerably more active, and careful handling is therefore required during the processing of monazite. Because of their chemical similarity, the lanthanides are difficult and expensive to separate and purify. However, technological applications of the elements in the metallurgical, electronics, nuclear and chemical industries are continuing to develop. Owing to the low TOXICITY of the lanthanides they are not considered to be a significant health risk, although the replacement of CALCIUM ions (Ca^{2+}) *in vivo* by these elements, in particular EUROPIUM which may form Eu^{2+} ions, may constitute a mode of action. [P.J.B.]

lanthanum (La) Lanthanum is a LANTHANIDE METAL (ATOMIC NUMBER = 57; relative ATOMIC MASS = 138.9; melting point = 921°C; boiling point = 3457°C) with an average abundance of 32 mg kg^{-1} in the Earth's CRUST. It is used in steel and MAGNESIUM and ALUMINIUM alloys. Its oxide is used in GLASS, as a glass polish, and in fluorescent phosphors. [J.G.]

larva A discrete stage in many species, beginning with zygote formation and ending with METAMORPHOSIS. Larvae may develop from planktonic GAMETES, emerge from egg cases or emerge from a parent. Planktonic larvae of benthic adults (i.e. MEROPLANKTON) must be able to locate a suitable site for SETTLEMENT at the appropriate time during development. Two types of meroplanktonic larvae can be distinguished. Lecithotrophic larvae hatch from large EGGS produced in small numbers and spend hours to about 2 weeks in the PLANKTON, during which time they do not feed but depend instead on yolk in the egg. Planktotrophic larvae hatch from small eggs produced in large numbers and spend weeks to months in the plankton, during which time they rely on planktonic food sources. Planktotrophic larvae tend to be more widely dispersed, with respect to their parents, relative to lecithotrophic larvae or non-planktonic larvae. Several distinctive larval types can dominate the plankton during certain times of the year. These include the trochophore larvae of polychaetes, which have several bands of cilia and are top-shaped, and the molluscan veliger larvae, which have a shell and a distinctive ciliated membrane called a velum. [V.F.]

Fig. L6 A concentration of concretions formed by differential erosion and then cemented together to form a type of laterite.

Fig. L4 A block of vesicular laterite: the outer surface is characteristically weathered.

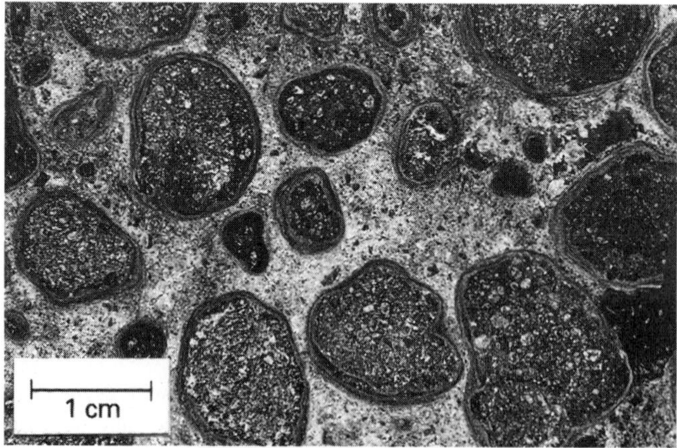

Fig. L5 Laterite containing mainly gibbsitic concretions.

laterite This term (derived from the Latin *later*, brick) was introduced to describe the vesicular, mottled, dark-red and cream, clayey material that was dug out of the ground in tropical countries, cut into bricks and dried in the sun. It hardened irreversibly when dry and was used to construct buildings and walls. Subsequently, the term has been applied very broadly and extended to include all highly weathered soils. As a result, attempts have been made to introduce other terms, such as plinthite, but this term is also very vague and many workers prefer to keep the term 'laterite' (see Figs L4–6). [E.A.F.]

latitudinal gradient *See* DIVERSITY GRADIENT.

Latitudinal gradients of diversity

Perhaps the most striking pattern in BIOGEOGRAPHY is the dramatic increase in the DIVERSITY of BIOTA between the poles and the Equator. Although there are conspicuous exceptions to the general rule, for example penguins are much more diverse at high latitudes and most species of salamanders and voles occur in temperate zones, the vast majority of taxa from all the major habitats show the same qualitative trend. This increase in diversity towards the tropics is apparent when either wide geographic regions are considered or when small communities are examined. For example, TROPICAL RAIN FORESTS contain 40–100 species of tree per hectare compared with deciduous forests in the USA (10–30 species) or coniferous forests in northern Canada (1–5 species). Also, many species groups are only found in the tropics, for example New World fruit bats and the Indo-Pacific giant clam.

Possible explanations

A variety of explanations for this biogeographic pattern have been put forward, none of which is entirely satisfactory when considered in isolation. It seems likely, however, that most if not all of them are relevant to some degree.

Plate 1 Brood parasitism
A female cuckoo parasitizing a reed warbler nest. First the female removes a host egg, holding it in her bill, then she lays her own. Her time at the nest is less than 10 seconds. (Courtesy of Ian Wyllie.)

Plate 2 Chaparral shrublands
Maquis vegetation in south-west Crete. Maquis, a chaparral-like shrubland, occurs in the Mediterranean Basin, and is characterized by drought-resistant, hard-leafed scrub of low-growing woody plants, which may take the form of a thicket of low-growing oaks or dense tangled brushwood. (Courtesy of P.B. Rainey.)

Plate 3 Consumer organisms
The sundew *Drosera rotundifolia*, an insectivorous plant that is both a primary producer and a carnivore. (Courtesy of M. Hassall.)

Plate 4 Cooperative breeding, evolution of
A female (upper) and male (lower) acorn woodpecker (*Melanerpes formicivorus*), a cooperatively breeding species in western North America that simultaneously exhibits helping at the nest and cooperative polygamy by both males and females. (Courtesy of R.L. Mumme.)

Plate 5 Desert
Namib Desert, Namibia, after good rains. (Courtesy of Michael Fogden, Oxford Scientific Films.)

(a)

(b)

Plate 6 Dispersal
For sessile organisms, it is external forces that control the distance and direction of dispersal—this is known as 'passive' dispersal. The seeds of this thistle plant (*Carduus nutans*) rely on wind for their dispersal. (b, Courtesy of D. Kelly.)

Plate 7 Lagoons
A coastal lagoon in southern Portugal. (Courtesy of R.S.K. Barnes.)

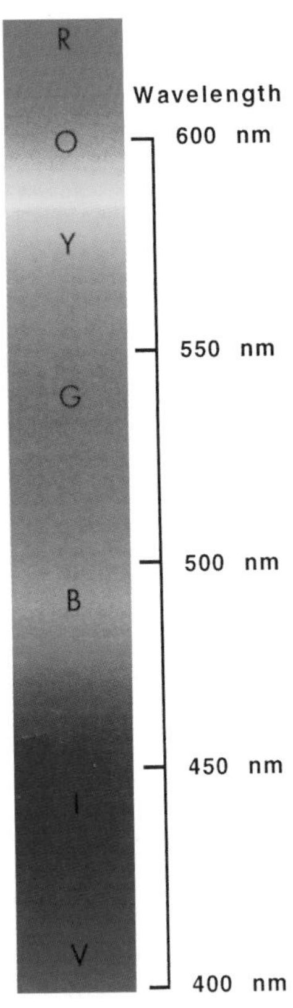

Plate 8 Light
Colour spectrum with approximate wavelength values indicated: R, red (620–700 nm); O, orange (600–620 nm); Y, yellow (580–600 nm); G, green (500–580 nm); B, blue (450–500 nm); I, indigo (430–450 nm, seldom designated as a separate colour nowadays); V, violet (400–430 nm). The region below 400 nm is called ultraviolet radiation and divided into UV-A (315–400 nm) and UV-B (280–315 nm). Shorter wavelength components do not occur in daylight at ground level. The region with wavelength longer than 700 nm is called infrared radiation, with 700–750 nm sometimes separated (especially by plant scientists) as far-red.

Plate 9 Mudflat
An estuarine mudflat—Aulne Estuary, France. (Courtesy of R.S.K. Barnes.)

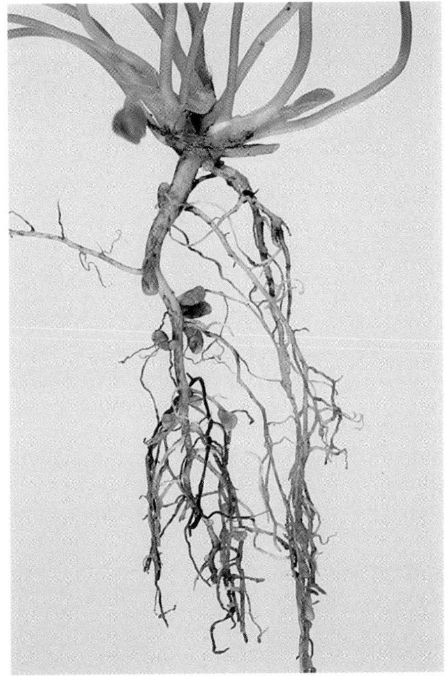

Plate 10 Nitrogen fixation
Nodules containing nitrogen-fixing rhizobia on clover (*Trifolium repens*) roots. (Courtesy of J. Baker.)

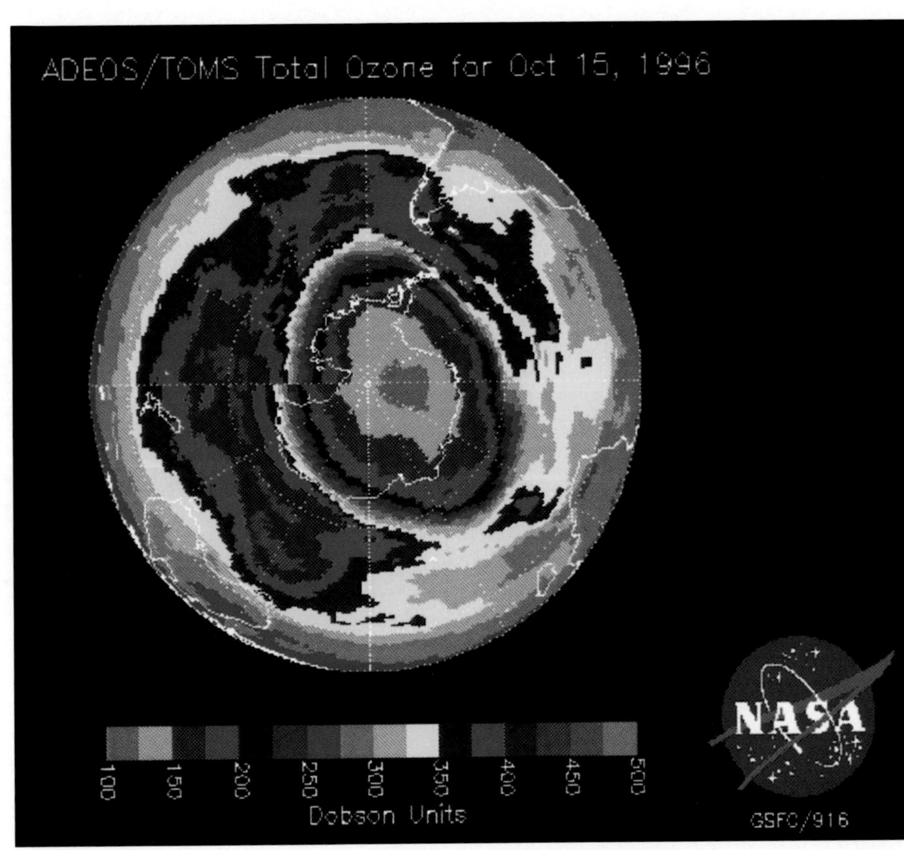

Plate 11 Ozone hole
Monthly mean total ozone for October 1996 from satellite measurements. (Courtesy of Ozone Processing Team, NASA/Goddard Space Flight Center, 1997.)

(a) 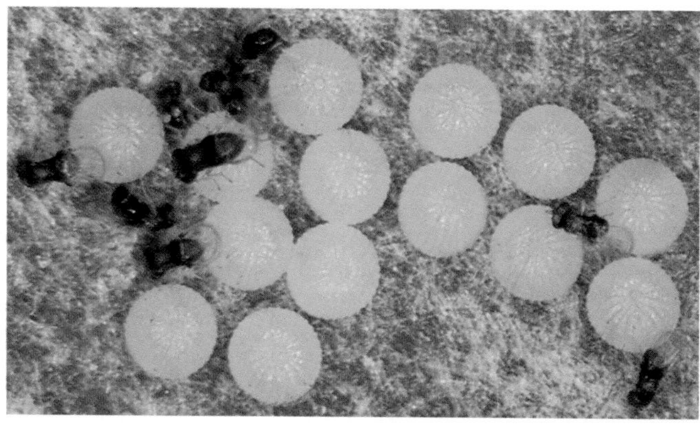

 (b)

Plate 12 Parasitoid–host interactions
(a) *Trichogramma* sp.: egg parasitoids of moth pests (here, the cabbage moth, *Mamestra brassicae*). The most widely used parasitoid in biological control today, it is mass produced and released to control various moth pests around the world. (b) *Exerastes cintipes*, an ichneumonid parasitoid of larvae of *M. brassicae*. Eggs are laid in young larvae (as here). (Courtesy of CAB International, Institute of Biological Control.)

(a) 

(b) 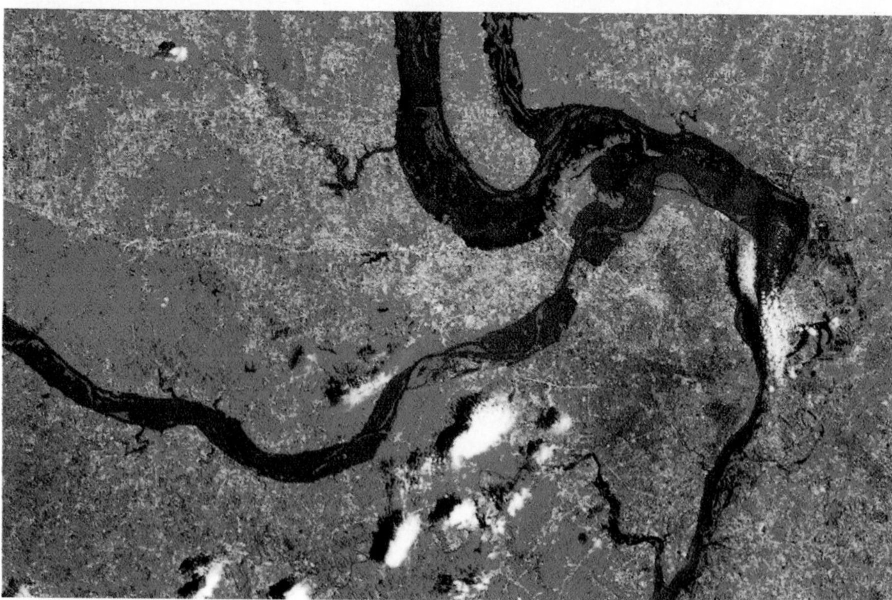

Plate 13 Rivers
Landsat images of the St Louis flood area: (a) 4 July 1988; and (b) 18 July 1993 (after extensive flooding). (Courtesy of the Earth Observation Satellite Company, Lanham, Md, USA.)

Plate 14 Rocky shores
Kelp forest on a rocky shore at low tide. (Courtesy of R.S.K. Barnes.)

Plate 15 Salt marsh
A salt marsh (low tide), Scolt Head barrier island, UK. (Courtesy of R.S.K. Barnes.)

Plate 16 Sand dune
A sand dune, Scolt Head barrier island, UK. (Courtesy of R.S.K. Barnes.)

Plate 17 Sandy shores
Sandy beach, Norfolk, UK. (Courtesy of R.S.K. Barnes.)

Plate 18 Savannah
Savannah grassland in Kenya with wildebeest and scattered trees. (Courtesy of A. Sternberg.)

Plate 19 Social insects
A subterranean nest of the common bumble-bee *Bombus terrestris.* (Courtesy of M. Pirker.)

Plate 20 Species concepts
Species and subspecies of mimetic butterflies. Warningly coloured *Heliconius* butterflies, showing local Müllerian mimicry and geographic differences between subspecies and species. *Heliconius erato* is shown in the left-hand column, *H. melpomene* on the right. Each row is an example of a local mimicry ring of these two species from a different area of Ecuador or northern Peru. The different geographic forms within each column are considered subspecies, in spite of the strong differences, because there are abundant intermediates where they meet in narrow hybrid zones. The only exception is *H. himera* (fourth row), a form closely related to *H. erato*, and which it replaces in dry Andean valleys in the border region of southern Ecuador and northern Peru where *H. melpomene* is generally absent. *Heliconius himera* is considered a separate species because hybrids between *himera* and *erato*, known from contact zones with three separate subspecies of *H. erato*, are much rarer than the pure forms.

Plate 21 Temperate deciduous forest
Beech (*Fagus sylvatica*) woodland in Autumn, Exmoor, UK. (Courtesy of P.B. Rainey.)

Plate 22 Temperate grassland
Artificial grasslands in Canterbury, New Zealand. (Courtesy of P.B. Rainey.)

(a)

(b)

Plate 23 Termites
(a) Soldiers (with the large, brown heads) and workers of the west African soil-feeding termite *Cubitermes severus*. The species produces methane actively at *c.* 0.3 μmol $g^{-1} h^{-1}$ *in vitro*. (b) Physogastrous queen and attendant workers of the African fungus-growing termite *Macrotermes bellicosus*. Fully grown queens are 10–15 cm in length.

Plate 24 Tidal flat
A tidal flat, Faro, Portugal. (Courtesy of R.S.K. Barnes.)

Plate 25 Tropical rain forest
(*Above*) Highland tropical rain forest landscape, showing dense foreground foliage framing distant forest-clad hilltops. Kuala Lumpur, West Malaysia. (Courtesy of Harold Taylor ABIPP, Oxford Scientific Films.)

Plate 26 Tundra
(*Left*) Arctic tundra, Greenland. Tundra is characterized by the presence of permafrost, water permanently frozen in the soil. Liquid water is present for only short periods of the year. (Courtesy of J.A. Vickery.)

(a)

(b)

Plate 27 Vernal pools
(a) A 'fairy ring', near Sacramento, California; with *Limnanthes douglasii* flowering. (b) *Limnanthes douglasii* var. *rosea*, Colusa County, California.

Early hypotheses

The earliest explanations centred on the idea that the physical environment in the humid tropics is less variable and has been subjected to less DISTURBANCE (e.g. GLACIATION) than higher latitudes. The argument goes that this benign physical environment has allowed greater specialization by species, with consequently greater species packing, and that the long-term stability has meant fewer species extinctions compared with temperate regions. There is little doubt that the past few million years (a reasonable time-frame for considering the evolution of existing biogeographic patterns) have seen large, recurrent and, in geological terms, rapid environmental change; this change has displaced and modified the biota throughout the world, including the tropics. However, teasing apart the likely effects of essentially the same basic disturbance for different regions now appears much more complicated than was hitherto imagined.

Although the argument presented above is quite plausible and probably has some merit, the ideas have been challenged in more recent years and in some cases even reversed completely. It has been proposed, for example, that the abundance of many tropical species could best be explained by repeated dissection and coalescence of the RAIN FOREST as a result of repeated environmental change over the past 2 million years. What now seems certain is that much work remains to be done in comparing the directions, magnitudes and frequencies of environmental changes with both the movement rates and evolutionary response rates of animals and plants. It seems unlikely that satisfactory explanations for the role of such environmental forces in determining observed gradients will be arrived at in the near future.

Seasonality

Seasonality in the physical environment provides another potential explanation for observed trends; it is argued that greater seasonal changes in CLIMATE require wider physical tolerances, which favours animals with broader niches. Conversely, more stable climates allow greater degrees of specialization and narrower niches. This hypothesis has received some empirical support. It has been observed, for example, that some insect prey types are more predictably abundant in the tropics, thereby supporting specialist avian predators. In contrast, however, studies on other groups have shown greater specialization in temperate latitudes.

Biotic interactions

Biotic interactions have also been put forward to explain latitudinal gradients and both predation and COMPETITION have been considered. One predation hypothesis assumes that predation is greater in the tropics than in temperate zones and that increased predation reduces the density of competitive dominants and allows more prey species to coexist. There is good evidence, however, that this increase in species is reversed when predation is very severe and there are few data available to test the assumptions of the hypothesis in relation to latitudinal gradients. A second predation hypothesis has been advanced to explain the diversity of tree species. The theory goes that disproportionately high MORTALITY of young trees will occur close to established adults, because these older trees harbour host-specific predators. This high density of specialist predators would increase the likelihood that different species would establish close to adults, thereby resulting in a greater diversity of species. While there is some evidence that this mechanism does indeed operate, neither this hypothesis nor the previous one address the question of why predation should be higher in the tropics in the first place. Related hypotheses advance the idea that biotic interactions may also have a part to play, with respect to the number of predator species that can be supported by a prey assemblage. The apparently greater number of prey species in the tropics should, in principle, represent a wider resource spectrum, which might allow a greater number of specialist predators to coexist. Again, however, the explanation for the higher diversity of prey remains unanswered.

Productivity

Theories that centre on differences in PRODUCTIVITY between the tropics and the poles have also been put forward. For heterotrophs, the argument seems quite plausible: there is a greater amount and wider range of RESOURCES at lower latitudes that could support richer assemblages compared with higher latitudes. Once again, however, we have a theory which begs the question: why is the resource base (plant community) richer? Productivity explanations for this question are somewhat more complicated but can be summarized as follows. The physical regime of the tropics (high light, temperature and humidity) leads to high plant biomass. This high biomass leads to low nutrient levels in the soil (because nutrients are locked in the vegetation) and a high diversity of light regimes; these two factors lead in turn to higher plant SPECIES RICHNESS.

Area

It has also been proposed that the greater area occupied by the tropics might account for the pattern. Among other things, increased area increases the chance that populations in one region might become ecologically isolated, thereby enhancing SPECIATION rates.

All the above hypotheses have some merit and it is quite reasonable to suppose that they have all operated to some degree. Indeed, one of the overwhelming features of the explanations of latitudinal gradients is the degree of interconnectedness between them and the hierarchical way in which they can be organized. In fact, viewing the explanations in a hierarchical way can simplify the picture considerably (*see* HIERARCHY THEORY). It has been proposed, for example, that despite the overall explanation being multifaceted, a single extrinsic factor, leading to even a slight DIVERSITY GRADIENT in plants, could lead to a cascade of effects which results in the strong latitudinal gradients we see today. For example, if plants were slightly more diverse in the tropics, the increase in plant structural diversity, food abundance and diversity would increase the richness of herbivores. This increased herbivore richness would, in turn, increase predation pressure on plants (thereby promoting plant species diversity) and would provide a richer resource for carnivores, which would in their turn increase predation pressure on the herbivores. It is easy to see how such positive gearing effects could leave their mark. *See also* GRADIENTS OF SPECIES RICHNESS. [S.J.H.]

REFERENCES

Pianka, E.R. (1978) *Evolutionary Ecology*, pp. 290–296. Harper & Row, New York.

Rosenzweig, M.L. (1995) *Species Diversity in Space and Time*. Cambridge University Press, Cambridge.

---

law of constant final yield Farmers have long known that sowing more seed does not lead to increased YIELDS at harvest. Above a relatively low THRESHOLD seed density, final yield is independent of seeding density; the higher the seed density, the lower the SIZE and FECUNDITY of the adult plants (Fig. L7). If density-dependent death of plants occurs when initial plant density is very high, then the yield COMPENSATION may be imperfect (*see* $-^3/_2$ POWER LAW). It is possible that extremely high seed densities cause a reduction in total final seed production (e.g. as a result of outbreaks of fungal pathogens like damping-off disease, *Pythium debaryanum*).

The law of constant final yield has important consequences for plant POPULATION DYNAMICS, because it means that total seed production is independent of the size structure of the breeding population; a few small plants will produce the same seed rain as a large number of small plants. Note that the effect of size distribution on plant MORTALITY is not simple; small plants are much more likely to die than are large plants.

The form of DENSITY-DEPENDENCE resulting from the relationship in Fig. L7 yields a relationship that exactly corresponds to the density-dependent function in Hassell's model (*see* HASSELL MODEL AND HASSELL–COMINS MODEL). This has facilitated the development of deterministic models for a variety of plant populations, especially arable weeds. [M.J.C.]

lawrencium (Lr) Lawrencium is a synthetic element (ATOMIC NUMBER = 103; relative ATOMIC MASS = 262; melting point and boiling point unknown) that was discovered in 1961. Particle bombardment of the other actinides has produced very small quantities of isotopes with mass numbers from ^{253}Lr to ^{260}Lr, the longest lived being ^{260}Lr ($t_{1/2}$ = 3.0 min). Consequently, lawrencium does not occur in detectable quantities in the environment. The symbol Lw was originally suggested but Lr was adopted by IUPAC (International Union of Pure and Applied Chemistry) in 1965. Investigations of the properties of lawrencium have mainly used ^{256}Lr ($t_{1/2}$ = 27 s), the ISOTOPE that can be produced in greatest abundance. The dominant oxidation state of lawrencium is +3. [A.B.M.]

laws of radiation *See* RADIATION.

layering The process by which a branch of a tree in contact with the ground produces roots at the point of contact and eventually forms a new individual. Natural layering is common in the underwood of long undisturbed WOODLAND and in extreme environments, such as on the coast or at the TREE LINE. Layering has been used to propagate trees in COPPICES by pegging slender stems to the ground. [G.F.P.]

LC_{50} Statistically derived concentration of XENOBIOTIC that is lethal to 50% of an observed population over a prescribed time in defined conditions. Also referred to as MEDIAN LETHAL CONCENTRATION. Also expressed in terms of dose (LD_{50}). Common indicator of ECOTOXICITY because when deaths are normally distributed with respect to concentration most of the population will be affected at this concentration and so it can be defined with greatest confidence. *See also* EC_{50}; IC_{50}; PREDICTIVE ECOTOXICITY TESTS. [P.C.]

LCA *See* LIFE-CYCLE ASSESSMENT.

LCI *See* LIFE-CYCLE INVENTORY.

leachate Liquid that seeps through WASTE and hence contains CONTAMINANTS; may therefore lead to hazardous substances entering surface water, groundwater or SOIL. *See also* FATE; LANDFILL; LEACHING; LYSIMETER. [P.C.]

leaching Removal by water of soluble components from SOIL or WASTE tip. The output is called a LEACHATE. [P.C.]

Lead (Pb)

Sources

A heavy, malleable, bluish-grey METAL (ATOMIC NUMBER = 82; relative ATOMIC MASS = 207.19; melting point = 327.5°C; boiling point =

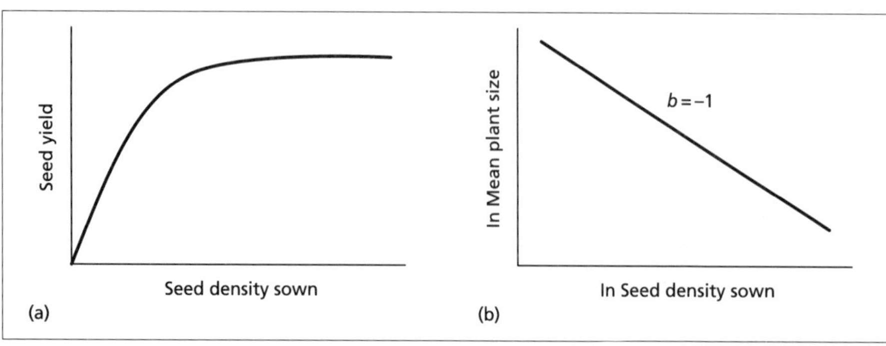

Fig. L7 Law of constant yield. (a) As the number of seeds sown is increased, final seed yield of an annual crop increases quickly to an asymptote. (b) Plotted on a log–log scale, the mean seed yield per plant (reflected in mean plant size) declines linearly with the density of seeds sown in a perfect rectangular hyperbola. The mechanism assumes that there is no limit to individual plant plasticity and that all increases in density can be accommodated by reductions in plant size without any increase in mortality (cf. Fig. P26).

1740°C) that is a trace constituent (13 mg kg^{-1}) of the Earth's CRUST. Found in galena ore (PbS), lead has had many uses, including lead–acid BATTERIES, solders, cable sheathing, piping, roofing, pigments (e.g. 'white lead', $2PbCO_3.Pb(OH)_2$) and chemicals (e.g. lead arsenate PESTICIDE, $PbHAsO_4$), anti-knock petrol additives (e.g. tetraethyl lead $(C_2H_5)_4Pb$), etc. As a result of these uses and major releases from the mining and smelting of lead and the combustion of FOSSIL FUELS, lead has been widely distributed throughout all phases of the environment. As a biologically non-essential and toxic metal, lead has therefore been one of the most intensively studied elements in terms of its environmental distribution, behaviour and impact upon health.

Historical records

Ice cores preserve historical records of the widespread dispersion and deposition of atmospheric lead. Data from Greenland show that lead concentrations increased from ~0.5 pg g^{-1} 3000–5000 years ago (natural sources) to ~2 pg g^{-1} some 1700–2500 years ago (reflecting use of lead in Roman times) before declining, after the fall of the Roman Empire, to ~0.7 pg g^{-1} around AD 500; during the mining renaissance in Europe levels increased again to ~4 pg g^{-1} by AD 1500 and to ~10 pg g^{-1} by AD 1750. Thereafter, the Industrial Revolution saw the lead concentration increase to ~50 pg g^{-1} by about 1930, whereupon the introduction of leaded petrol additives produced a dramatic rise to ~200 pg g^{-1} by the 1960s. Since then, with the increasingly restricted use of leaded petrol, especially in North America, lead concentrations fell to ~25–30 pg g^{-1} by the 1990s. Increases in Antarctica have been much lower, to only 2–3 pg g^{-1}, compared with Greenland, reflecting the latter's proximity to other continents (the sources of particulate material), the occurrence of 90% of industrial activity in the Northern hemisphere, interhemispheric and polar air circulation blocking systems, and tropospheric residence times of only a few weeks for particles.

Exposure pathways

Anthropogenic influences on environmental lead concentrations over the past few hundred years are also recorded in aquatic sediments, PEAT and soils, while in population centres there is a vast amount of direct information on lead levels in air, dust, SOIL, food and WATER. The principal pathways of human EXPOSURE to lead reflect, to some extent, the lead CONTAMINATION of these different materials, although there can be interconnections and differing routes of intake. For example, airborne lead, in addition to direct inhalation, can reach humans, after deposition, through ingestion of dust, water, aquatic biota, soil, plants, animal products, etc.

In the urban ATMOSPHERE, at least in those countries still using leaded petrol, most of the lead comes from car-exhaust EMISSIONS of PbBrCl particulates, which are usually soon converted to $PbSO_4.(NH_4)_2SO_4$. This compound is also found in dust and soil, where contributions from old paint and primers (e.g. 'white lead', 'red lead', Pb_3O_4, and yellow $PbCrO_4$) can occur as well. Lead deposited from the combustion of fossil fuels, especially coal, can also be significant.

While general exposure of humans to lead in soil includes intake of that absorbed by plants and animals, acute problems can arise from the unusual practice of direct ingestion of soil, known as pica. Similarly, in areas where plumbosolvent (i.e. slightly acidic, soft) water dissolves lead (Pb^{2+}) from lead pipes and tanks in the domestic plumbing of older houses, intake can be very high. Adventitious sources include lead-glazed and -soldered domestic vessels and utensils, eye cosmetics and medicinal preparations.

Health effects

The deleterious effects of lead upon human health have been known since Roman times, when it was used not only in plumbing and pottery but also, in the form of lead acetate, as a sweetener in wine. Today, effects are recognized as ranging from non-specific symptoms such as moodiness and irritability, through severe abdominal pains and anaemia, to peripheral nerve involvement, paralysis of fingers, hands and wrists, and brain damage. Rather more controversial, although fairly widely accepted now, has been the claim that lead exposure of young children in particular can result in behavioural problems and intelligence deficits. Children absorb ingested lead rather more readily than adults. Once absorbed via the gut or, if inhaled, via the lungs, lead is carried on the surface of red blood cells, which bring it into contact with soft tissues, including the brain, before ultimately being stored in bones and teeth. Certainly, subtle biochemical effects such as inhibition of the enzyme δ-aminolaevulinic acid dehydratase can be observed at blood lead concentrations of about 5 μg dl^{-1}, not untypical of the Western world today. Further effects, for example inhibition of the biosynthesis of haem and fetal TOXICITY (lead can cross the placenta, the blood–brain barrier is more permeable and there is less bone to act as a storage site), occur as concentrations increase to ~40 μg dl^{-1}, which is generally considered a benchmark for clinical signs of lead toxicity, eventually resulting in encephalopathy at ~100 μg dl^{-1}. The toxic mode of action of organolead compounds, which can be absorbed through the skin as well as the lungs, is rather different. For example, tetraethyl lead $(C_2H_5)_4Pb$ is converted in the liver to $(C_2H_5)_3Pb^+$, which attacks the central nervous system. In addition to anthropogenic emissions of organolead compounds during the manufacture and use of tetraalkyl lead petrol additives, there is evidence for some natural alkylation (methylation) of inorganic lead in the environment.

Regulatory control

Despite the ongoing debate about the existence and/or importance of subclinical effects of lead upon human health, regulatory bodies in many countries have taken strong measures to reduce exposure of the public to lead. These include reduction and/or elimination of lead additives from petrol; the removal, or the imposition of strict limits upon the use, of lead compounds in paints; the banning of lead-soldered cans and the introduction of more stringent limits upon the lead content of food; and measures to reduce the lead content of water in susceptible areas that have lead pipes and tanks, lead-soldered COPPER pipes, etc. The latter include liming of the domestic water supply to increase the pH to >9 and so reduce uptake of lead from pipes (by stabilizing compounds such as $Pb(CO_3)_2(OH)_2$ and $PbCO_3$ that may coat pipes) and the addition of orthophosphate to form insoluble lead compounds such as

$Pb_3(PO_4)_2$ and $Pb_5(PO_4)_3OH$ on the inner surfaces of pipes. Such steps have helped to reduce lead concentrations in both water and previously affected consumers, some of whose blood lead concentrations were worryingly high (>25 μg dl^{-1}). While these chemical approaches have enabled COMPLIANCE with, for example, the European Community (EC) limit of 50 μg l^{-1} for drinking water, it is unlikely that they will be sufficient to satisfy the latest WHO recommendation of 10 μg l^{-1}. Physical replacement of lead pipes will then be essential.

Although preventive and remedial measures are helping to reduce exposure to lead in air, water and food, much of the lead previously released remains in dusts and soils, from which it could eventually be mobilized and/or become bioavailable. Furthermore, present BODY BURDENS of lead can only be regarded as typical of today; they are neither normal nor natural. The latter were probably a factor of several hundred less than today's levels, which are within a factor of just four or five of levels associated with clinical lead poisoning. As lead has no known benefits to the body at any level, the history of its usage and effects is seen by many to serve as an exemplary warning to society about the possible long-term dangers of dispersing potentially harmful metal contaminants/pollutants over the surface of the planet. *See also* HEAVY METALS. [J.F.]

REFERENCES

Boutron, C.F., Gorlach, U., Candelone, J.-P., Bolshov, M.A. & Delmas, R.J. (1991) Decrease in anthropogenic lead, cadmium and zinc in Greenland snows since the late 1960s. *Nature* **353**, 153–156.

Nriagu, J.O. (ed.) (1978) *The Biogeochemistry of Lead in the Environment.* Elsevier/North Holland Press, Amsterdam.

Royal Commission on Environmental Pollution (1983) *Lead in the Environment.* Her Majesty's Stationery Office, London.

Rutter, M. & Jones, R.R. (eds) (1983) *Lead versus Health. Sources and Effects of Low Level Lead Exposure.* John Wiley & Sons, Chichester.

---

leaf area index (LAI) A dimensionless parameter measuring the ratio of leaf surface area (one side only) to ground area. For example, in a pasture there might be 2.5 m^2 of grass leaf area to every square metre of ground area; thus LAI = 2.5. At low LAI, net PRODUCTION is limited by the area of leaf available to intercept incoming SOLAR RADIATION (a lot of light is wasted by falling on bare ground). At high LAI, net production might fall because of self-shading of leaves (the shaded leaves respire more than they photosynthesize). The LAI at which net primary production is greatest is called the optimal LAI (LAI_{opt}). When LAI is greater than LAI_{opt}, defoliation (by herbivores or by mechanical cutting) can lead to an increase in net primary production by reducing the prevalence of self-shading. More usually, however, herbivory reduces LAI well below LAI_{opt}, thereby reducing net PRODUCTIVITY until the leaf area has been restored by regrowth. [M.J.C.]

least significant difference (LSD) The smallest difference between any two means that will allow them to be significantly different from each other. Calculated as part of an a posteriori test, after an ANALYSIS OF VARIANCE. [B.S.]

least squares method The method of least squares was discovered independently by two mathematicians, K.F. Gauss and A.-M. Legendre, about 200 years ago. It is used widely as a method for minimizing the residual variation when fitting a relationship or function to a set of observations. For example, when we fit a straight-line function to a set of paired observations (*x* and *y*) we are in fact splitting each 'observed *y*' into two components, a 'fitted *y*' reflecting the underlying straight line and a 'residual' reflecting the random departure from the straight line:

Observed y = Fitted y + Residual

where the residual is obtained by subtracting the observed value from the predicted value ($=a \pm bx$). In fitting the straight line we wish to make these residuals as small as possible on average. To do this, we must sum the residuals in some way and the most useful method is to sum the squares of the differences:

$$\sum_{i=1}^{n}\left(\text{Observed } y_i - \text{Fitted } y_i\right)^2$$

$$\sum_{i=1}^{n}\left(y_i - a \pm bx_i\right)^2$$

Fitting a straight line using the least squares method simply means making this quantity the smallest it can be. The method can also be used to fit curved relationships. With polynomials, the calculations are broadly similar to those for fitting a straight line. For non-linear relationships, the criterion of finding the fit with the least squares is the same but the estimation of PARAMETERS can be more problematic. *See also* REGRESSION ANALYSIS, CURVILINEAR. [B.S.]

lecithotrophic larvae *See* LARVA.

Lefkovitch matrix *See* POPULATION PROJECTION MATRIX.

lek A traditional mating ground where adult males defend very small territories devoid of obvious RESOURCES to which females move for the purpose of mating. About seven species of mammals, 30 species of birds and an unknown number of amphibians, fishes and insects mate on leks. Typically between five and several hundred males compete to hold territories on a lek, and females are thought to be free to choose between these prospective partners or their territories. *See also* LEKKING; MATING SYSTEMS. [J.C.D.]

lek-breeding birds *See* LEKKING.

lek-breeding mammals *See* LEKKING.

Lekking

The MATING SYSTEM or reproductive behaviour of species that mate on traditional mating grounds called leks. Leks are clusters of between five and several hundred small male territories devoid of RESOURCES. They have been documented in about seven species of mammals, 30 species of birds and a number of amphibian, fish and insect species. In lekking populations, sexually receptive females move to these arenas to mate and, once there, appear free to choose between mating partners or territories. Their choice often appears remarkably unanimous, with a few of the males monopolizing the matings. Leks have interested behavioural ecologists for two reasons.

1 Since females seem free to choose mates on leks, their choice is so unanimous and males offer them little but sperm, lekking species would seem to be the ideal systems in which to test whether females choose mating partners in order to obtain superior genes to pass on to their offspring, as envisaged by good gene models of SEXUAL SELECTION.

2 In most species with female-only PARENTAL CARE, the mating system can be understood as the outcome of males competing for access to sexually receptive females by defending these females directly (HAREM defence) or defending resources that attract them (RESOURCE-DEFENCE POLYGYNY). The behaviour of lekking males and females is not easily understood within this framework.

Research on lekking mammals has shed some light on each of these subjects. In a number of lek-breeding ungulate species (Uganda kob, Kafue lechwe, fallow deer), it appears that females pay more attention to the characteristics of territories than characteristics of males. They are attracted directly to other females and to territories that have been heavily used in the past, and may also move around on leks in order to reduce the risk of predation or harassment from juvenile males. The tendency of most females to mate with the same males seems to stem more from female AGGREGATION than from choice of mating partners, but it is not known whether females gain genetic benefits from their behaviour.

In lek-breeding mammals, males advertise for females from lek territories in part because the peculiar ecology of females makes the more common male reproductive strategies, i.e. harem or RESOURCE DEFENCE, impractical. Lekking species are characterized by very large female home ranges within which females move unpredictably in unstable herds. The unpredictability of female movements apparently rules out male defence of static resources and the unstable herds preclude defence of female groups. Moreover, females appear to resist mating in herds because of the risk of multiple copulation attempts, preferring instead small male territories where adult males can exclude disruptive juveniles. It is still unclear, however, why male territories become aggregated together on leks or why females move to aggregated territories to mate. Suggested explanations include improved female safety from predation or harassment, greater opportunities to compare between mating partners or the 'black hole' effect, a tendency for females moving frequently between adjacent territories to accumulate at territory clusters.

Research on lek-breeding birds has generated more evidence for female choice compared with work on mammals. Females in different species, populations and even years seem to vary in the cues they use to select mates or territories, in some cases preferring males with particular plumage characteristics, in others choosing previously popular territories, and often apparently preferring males with more vigorous DISPLAYS. As in lek-breeding mammals, females are commonly attracted to other females and some authors have suggested this may represent an adaptive form of mate-choice copying. There is some evidence, too, that by choosing brighter or more energetic males, females may gain access to superior genes.

However, the evolution of lek-breeding in birds is less well understood compared with mammals. Two popular models are the aggregation of males at 'hotspots', sites of maximal female density or RANGE overlap, and the accumulation of subordinate males around superior 'hotshot' ones. Neither of these mechanisms appears sufficient to generate the degree of male aggregation observed in lekking species. One alternative is that females move to the largest available cluster of male territories because males are distributed between leks in an IDEAL FREE DISTRIBUTION for unequal competitors, so that larger clusters contain the most competitive males. This has received some support from experiments on waders showing that females prefer larger male aggregations.

Many amphibians, fish and insects mate in aggregations and it is often unclear whether to call these 'leks'. While terminology is less important than biology, it is probably useful to restrict the term to those species or populations in which territorial males are more aggregated than either resources or non-receptive females. It is in these instances that the clustering of males requires an unusual explanation and that the possibility of sexual selection playing a role in the mating system arises. *See also* FEMALE PREFERENCE; POLYGYNY. [J.C.D.]

REFERENCES

Clutton-Brock, T.H., Deutsch, J.C. & Nefdt, R.J.C. (1993) The evolution of ungulate leks. *Animal Behaviour* **46**, 1121–1138.

Höglund, J. & Alatalo, R.V. (in press) *Leks*. Princeton University Press, Princeton, N.J.

Wiley, R.H. (1991) Lekking in birds and mammals: behavioral and evolutionary issues. *Advances in the Study of Behavior* **20**, 201–291.

lentic Still water system, i.e. ponds and LAKES. *Cf.* LOTIC. [P.C.]

lentic ecosystems *See* LIMNOLOGY.

Leopold, A. *See* CHARACTERS IN ECOLOGY.

leptokurtic distribution *See* KURTOSIS.

Leptosols *See* SOIL CLASSIFICATION.

Leslie matrix *See* MATRIX MODELS; POPULATION PROJECTION MATRIX.

Levene model In the single-locus multiple-niche POLYMORPHISM model of Levene (1953), a total of n niches is present for larval development. The fraction of adults contributed by each larval niche to the total adult breeding population is constant over generations, f_k for niche k ($k=1, \ldots, n$). Genotypic fitness depends upon the niche: $w_{11,k}$, $w_{12,k}$ and $w_{22,k}$ for the genotypes A_1A_1, A_1A_2 and A_2A_2, respectively. ALLELE FREQUENCY change in the total population is computed from the weighted average allele frequency change per niche:

$$\Delta p = \sum_k f_k \Delta p_k$$

$$= \sum_k f_k \frac{pq}{\overline{w}_k}\left[\left(pw_{11,k} + qw_{12,k}\right) - \left(pw_{12,k} + qw_{22,k}\right)\right]$$

Since:

$$\Delta p = \sum_k f_k \frac{pq}{2}\frac{\mathrm{d}\ln \overline{w}_k}{\mathrm{d}p} = \frac{pq}{2}\frac{\mathrm{d}\sum_k f_k \ln \overline{w}_k}{\mathrm{d}p}$$

it follows that the overall mean fitness in the Levene model is given by:

$$\bar{\bar{w}} = \prod_k \bar{w}_k^{\,f_k}$$

The condition for protected polymorphism is given by:

$$\frac{1}{\sum_k f_k\left(w_{12,k}/w_{22,k}\right)} < 1$$

$$\frac{1}{\sum_k f_k\left(w_{12,k}/w_{11,k}\right)} < 1$$

Polymorphism is protected in the Levene model if the HARMONIC MEAN fitness of each homozygote is less than the harmonic mean fitness of the heterozygote.

The Levene SOFT SELECTION model has played an important role in the investigation of the maintenance of genetic variation in POPULATION GENETICS. It has at least three drawbacks: (i) a constant contribution of adults by each niche to the total population together with constant fitness is not possible; (ii) it computes allele frequency change in adults rather than in zygotes; and (iii) it has no ecological basis. The Levene model would seem to involve 'soft selection', but this is frequency and density dependent (Wallace 1970, 1975). Since Prout (1965), it is agreed that models should be formulated on the basis of a life stage before selection not after selection as in the Levene model. Ecological models of selection in a subdivided environment start with the number of zygotes per niche and never yield the Levene model, even if soft selection is present within each niche. [G.D.J.]

REFERENCES

Levene, H. (1953) Genetic equilibrium when more than one ecological niche is available. *American Naturalist* **87**, 331–333.

Prout, T. (1965) The estimation of fitnesses from genotype frequencies. *Evolution* **19**, 546–551.

Wallace, B. (1970) *Genetic Load: its Biological and Conceptual Aspects*. Prentice-Hall, Englewood Cliffs, N.J.

Wallace, B. (1975) Hard and soft selection revisited. *Evolution* **29**, 465–473.

licences *See* POLLUTION PERMITS.

Liebig's law of the minimum The only important factor is the most LIMITING FACTOR. This is the archetypal single-factor explanation and predicts that the nutrient supplied at the lowest relative rate (i.e. the lowest ratio of supply to demand) is the only nutrient that is rate-limiting. This is generally tested by adding a series of potentially limiting NUTRIENTS, one at a time, and observing whether or not an increase in GROWTH rate occurs. Thus, one might speak of the growth of a given plant as being NITROGEN-limited, PHOSPHORUS-limited or LIGHT-limited in different habitats, but the implication is that a single factor (the 'minimum' factor) limits growth or proliferation in any given environment.

Justus von Liebig was in dispute with two eminent English agronomists, Sir John Bennett Lawes and Joseph Henry Gilbert, in the mid-19th century over the nitrogen nutrition of CROP plants. Liebig believed that plants could obtain all the nitrogen they required directly from the air and that full crop YIELDS required only the addition of mineral elements (phosphorus, POTASSIUM, MAGNESIUM, etc.). Lawes and Gilbert gathered masses of experimental evidence showing that this was not the case; they proved beyond all doubt that agricultural yields could only be maximized by the input of substantial quantities of nitrogen (of the order of 150 kg ha^{-1} year^{-1}, applied in the form of nitrates or ammonium ions as inorganic salts or organic farmyard manures). It was probably the intensity of the dispute between Liebig and Lawes which ensured that the classical experiments at Rothamsted Experimental Station were continued for as long as they were. *See also* BROADBALK EXPERIMENT; PARK GRASS EXPERIMENT. [M.J.C.]

life cycle The cycle of developmental events from PROPAGULE through to propagule, for example EGG, LARVA, juvenile, reproductive adult, more eggs, etc. This is sometimes distinguished from life history which is taken to be the sequence of developmental events from one stage in the life cycle to another, for example egg to adult. However, the distinction is often fuzzy, and the two terms are used interchangeably. Another problem encountered in usage is inclusion of a hyphen. The following is most appropriate. When used as a noun, no hyphen is necessary (e.g. the life cycle is semelparous); when used adjectively the hyphen aids clarity (e.g. the life-cycle pattern), indeed in these circumstances some advocate running the words together (e.g. lifecycle pattern). *See also* LIFE-HISTORY EVOLUTION; LIFE-HISTORY TRAITS. [P.C.]

life-cycle assessment (LCA) Commercial products have life cycles: from raw materials through production to use and ultimate disposal. It is myopic to consider environmental effects at just one point in this sequence because it might ignore more important effects at others. A more sound approach is to consider effects across all stages of a LIFE CYCLE 'from cradle to grave' (*see* CRADLE-TO-GRAVE ANALYSIS). The four main stages of an LCA are as follows.

1 Initial phase: defining the problem and establishing an inventory of important elements, i.e. LIFE-CYCLE INVENTORY (LCI).

2 Inventory phase: detailed description of raw material and energy inputs used at all points and the EMISSIONS/EFFLUENTS and solid WASTE outputs.

3 Impact assessment phase: relating inventoried inputs and outputs to real-world environmental problems.

4 Improvement phase: using information collected in the other phases to improve overall environmental importance.

Major challenges are to incorporate potential POLLUTION impacts and to find ways of representing and judging the relative merits of input and output effects of different components. Also of considerable importance is defining the boundaries of the study, which otherwise can extend indefinitely. Thus, a particular product might use wood as raw material: does the LCA stop at the forest or take into account the fuel used to run the lumber operation, the construction of the equipment such as saws, etc.? A rule of thumb is that if effects are likely to be less than 5% of the total they can be ignored, although knowing that this is the case is not straightforward, especially before the LCA is done.

LCA is used in establishing ECOLABEL criteria. [P.C.]

REFERENCE

Fava, J.A., Denison, R., Jones, B. *et al.* (1990) *A Technical Framework for Life-cycle Assessments*. SETAC Foundation, Washington, D.C.

life-cycle inventory (LCI) Inventory of use of RESOURCES (including energy) and creation of polluting WASTE along the life cycle of a product from raw materials, through manufacture and use, to disposal. *See also* LIFE-CYCLE ASSESSMENT. [P.C.]

life-cycle strategies *See* LIFE-HISTORY TRAITS.

life–dinner principle A metaphor used to explain why prey may always be one step ahead of predators in the evolutionary ARMS RACE of capture and escape. For example, rabbits are running faster than foxes because the rabbit runs for its 'life', while the fox is only after its 'dinner'. If the rabbit fails to escape it will no longer reproduce, while a fox may still achieve reproduction when it loses a race against a rabbit. Therefore, selection is stronger on rabbits than on foxes to improve running ability. The metaphor may apply to any other asymmetric situation. *See also* RED QUEEN'S HYPOTHESIS. [P.S.H.]

Life-history evolution

The EVOLUTION of those TRAITS that contribute to the major features of the LIFE CYCLE, namely the age schedules of birth and death. More specifically, life-history theory centres upon the evolution of the age and size at first reproduction, the age schedule of REPRODUCTIVE EFFORT and its attendent MORTALITY regime, and the number and size of offspring. Since virtually every aspect of an organism's biology ultimately affects its FITNESS and hence impinges upon the aforementioned traits, there is no precise boundary defining exactly what can be considered within the purvue of life-history theory. Thus, variation in behavioural traits that alter the likelihood of obtaining a mate or the amount of energy devoted to reproduction could be considered under the general umbrella of life-history evolution. A central feature of the theory of life-history evolution is the hypothesis that variation is circumscribed by the existence of trade-offs. For example, increasing the size of offspring can only be done at the expense of the NUMBER OF OFFSPRING. Given the existence of trade-offs, evolution will lead eventually to that combination which maximizes fitness. In order to avoid the problem of tautology, the measure of fitness must be defined a priori. The most frequently used measures are the INTRINSIC RATE OF INCREASE, r, or the net reproductive rate, R_0. The former defines the rate of population increase in the absence of density-dependent factors, while the latter refers to the specific situation of a stable population. The values of these parameters are obtained by solving the EULER–LOTKA EQUATION. In both cases density-dependent factors are assumed not to influence the trait under study. This does not necessarily mean that the population is not under density-dependent control. Suppose, for example, that in some organism POPULATION SIZE is determined by density-dependent processes that occur in the larval stage but the characters under selection at this stage are genetically uncorrelated with adult traits. In this circumstance analysis of the evolution of adult traits can proceed using the net reproductive rate, but analysis of larval traits must take into account the effects of density. Given a measure of fitness and a set of trade-offs we can predict which combination of traits will maximize fitness. The approach can be illustrated using the size at reproduction in the fruit fly, *Drosophila melanogaster*. Because this is a COLONIZING SPECIES, the appropriate measure of fitness is r. Three relationships are required: the relationship between adult size and the age schedule of reproduction; the relationship between development time and adult size; and the relationship between survival and adult body size (there are actually two survival functions, that for the larval period and that for the adult, the latter being assumed to be constant and independent of adult size). Putting these equations into the Euler–Lotka equation we obtain a curve that has a maximum at a thorax length of 0.95 mm, within the observed size range of 0.90–1.15 mm.

An alternative approach to the assumption that some specific fitness measure is maximized is to use a model incorporating the mode of transmission of the traits from generation to generation (e.g. single locus or quantitative genetic models) plus a criterion for the relative contribution of different genotypes to the next generation (i.e. a SELECTION COEFFICIENT). Both approaches have strengths and weaknesses but, in general, both lead to the same predicted evolutionary equilibrium. *See also* AGEING; EVOLUTIONARY OPTIMIZATION; OPTIMALITY MODELS; REPRODUCTIVE VALUE; SOME WORLD RECORDS FOR LIFE-HISTORY TRAITS. [D.A.R.]

REFERENCES

Roff, D.A. (1992) *The Evolution of Life Histories: Theory and Analysis*. Chapman & Hall, New York.

Stearns, S.C. (1992) *The Evolution of Life Histories*. Oxford University Press, Oxford.

life-history patterns *See* LIFE-HISTORY TRAITS.

life-history strategies *See* LIFE-HISTORY TRAITS.

life-history theory *See* LIFE-HISTORY EVOLUTION.

life-history traits These are major features of an organism's LIFE CYCLE, most directly related to the schedule of birth and death rates and thus to Darwinian FITNESS (often referred to as life-history or life-cycle strategies). This somewhat loosely defined set includes body size, patterns of GROWTH and STORAGE, age at maturity and METAMORPHOSIS, rate of SENESCENCE, LIFESPAN, REPRODUCTIVE EFFORT, mating success, number, size and sex ratio of offspring, PARENTAL INVESTMENT, etc. Because ADAPTATIONS at other levels (biochemistry, physiology, morphology, behaviour, etc.) must be reflected in survival, FECUNDITY or reproductive timing, life-history TRAITS play a key role in studying EVOLUTION by NATURAL SELECTION. At the same time, most life-history traits can be viewed as consequences of the way individuals allocate their limited RESOURCES to different functions (maintenance and repair, growth, defence, reproduction, storage, etc.). The resulting TRADE-OFFS between life-history characters are central to life-history theory. A part of the adaptationist research programme, life-history theory aims at identification of ecological circumstances and functional relationships underlying trade-offs that favour a particular life history. Despite problems inherent in the adaptationist programme (*see* EVOLUTIONARY OPTIMIZATION), life-history theory has proved quite successful in explaining the enormous diversity of life-history patterns. *See also* AGE AT MATURITY IN PLANTS. [T.J.K.]

REFERENCE

Stearns, S.C. (1992) *The Evolution of Life Histories*. Oxford University Press, Oxford.

life table A tabulation of the patterns of MORTALITY, and thus also of survivorship, of the individuals in a POPULATION, equivalent to an actuarial table used by life insurance companies to determine the life expectancy of an applicant. There are two distinct forms, the COHORT LIFE TABLE and the STATIC LIFE TABLE. Both contain measures of the POPULATION SIZE and of mortality with respect to stage, age or AGE CLASS (see Table L3 for definition of terms usually used in life tables) and frequently incorporate fertility schedules or FECUNDITY SCHEDULES. Life tables embody, together with RECRUITMENT curves and survivorship curves, the data necessary for demographic analysis (e.g. calculation of BASIC REPRODUCTIVE RATE R_0, *k*-factor analysis). *See also* a_x; DEMOGRAPHY; d_x; e_x; l_x; q_x. [A.J.D.]

lifespan Measure of the duration of life, used variously to describe either the maximum or average duration of life in a population or the duration of life of an individual organism. *See also* AGEING; DEMOGRAPHY; SOME WORLD RECORDS FOR LIFE-HISTORY TRAITS. [T.B.L.K.]

lifetime reproductive success The total NUMBER OF OFFSPRING produced by an individual over its lifetime, often taken as a measure of FITNESS. This is exact only in the case of average lifetime reproductive success of a GENOTYPE in a population that is constant in size. [G.D.J.]

Light

Electromagnetic RADIATION in the wavelength range 400–700 nm, where the human eye is most sensitive. Often the adjacent wavelength ranges, ULTRAVIOLET RADIATION on the shortwave side and infrared radiation on the longwave side, are included and the terms 'ultraviolet light' and 'infrared light' are often used but not recommended. To specify the range 400–700 nm the term 'visible light' can be used. It is also called photosynthetically active radiation (PAR), since it is the most important spectral band for plant PHOTOSYNTHESIS. A common misconception is that 'infrared radiation' is synonymous with 'heat radiation'. All daylight, including visible light, is heat radiation from the Sun, while infrared laser radiation is not heat radiation (although things absorbing infrared laser radiation will be heated by it).

Table L3 Definition of terms usually used in life tables.

x	Age (or stage) of individuals
a_x	Number of individuals of age (or stage) x
d_x	Proportion of individuals dying at age (or stage) x
e_x	Life expectancy of individuals at age (or stage) x
l_x	Proportion of the organisms at age 0 surviving to age x
q_x	Mortality rate at age (or stage) x

When light propagates it behaves as a wave motion with a certain wavelength, with oscillating electric and magnetic fields perpendicular to one another and to the direction of light propagation, hence the term 'electromagnetic radiation'. When light interacts with matter, however, it behaves as particles, photons. Early scientists (Isaac Newton, Christian Huygens) advanced either particle or wave theories for light and regarded such theories as mutually exclusive. The classical wave theory for light and other electromagnetic radiation was perfected by James Clerk Maxwell, who in 1864 summed it up in four famous DIFFERENTIAL EQUATIONS. The modern view, based on work by Max Planck and Albert Einstein, is 'dualistic' in the sense that neither a wave nor a particle theory alone is consistent with all aspects of the behaviour of light. Light is a form of energy and one PHOTON constitutes a quantum of energy. The energy of a photon is proportional to the ratio between the velocity of the light and the wavelength of the light (this ratio is the frequency, the rate of oscillation of the electric and magnetic fields). The proportionality constant is called Planck's constant (h). When light interacts with matter it can be reflected, scattered (see below), or absorbed and converted to another form of energy. The velocity of light in a vacuum is exactly 299 792 458 km s^{-1}, in line with a new definition of the METRE adopted in 1983. When light propagates in transparent matter, such as air, WATER or GLASS, the velocity is lower. The ratio between the velocity in a material medium and the velocity in a vacuum is called the refractive index of the medium.

Different wavelength bands have the following approximate colours to people with normal colour vision: 400–440 nm, violet; 440–500 nm, blue; 500–580 nm, green; 580–595 nm, yellow; 595–615 nm, orange; >615 nm, red. The sensitivity of the human eye does not end abruptly at 700 nm, although for practical reasons this is taken as the end of the visible wavelength band. Given sufficient intensity, even light of 1000 nm wavelength can be seen and appears more yellow than the radiation of 700 nm. The wavelength band 700–750 nm is designated as far-red, a term used mainly by plant scientists since it is important for the regulation of plant GROWTH (see Plate 8, facing p. 398). A mixture of all kinds of visible light is often said to give a white impression, but this depends very much on various conditions such as the previous light exposure. To a person adapted to green light a mixture may appear red, while to a person adapted to red light the same mixture appears green.

Natural light

Most natural light emanates from the Sun (*cf.* SOLAR RADIATION) and reaches Earth either directly (in about 8 min 18 s from the time it leaves the surface of the Sun) or via reflection from the Moon or other heavenly bodies. The terrestrial ATMOSPHERE only slightly modifies the spectrum of the total visible daylight ('global radiation', consisting of direct sunlight and scattered light or skylight) by slightly attenuating the spectral bands where OXYGEN (red light), water vapour (red light) and OZONE (green to red light) absorb. Other atmospheric gases, like CARBON DIOXIDE (CO_2), do not absorb visible light but absorb strongly in the ultraviolet and infrared spectral regions. The main effect of the terrestrial atmosphere on visible light is to scatter it, i.e. change its direction. Both

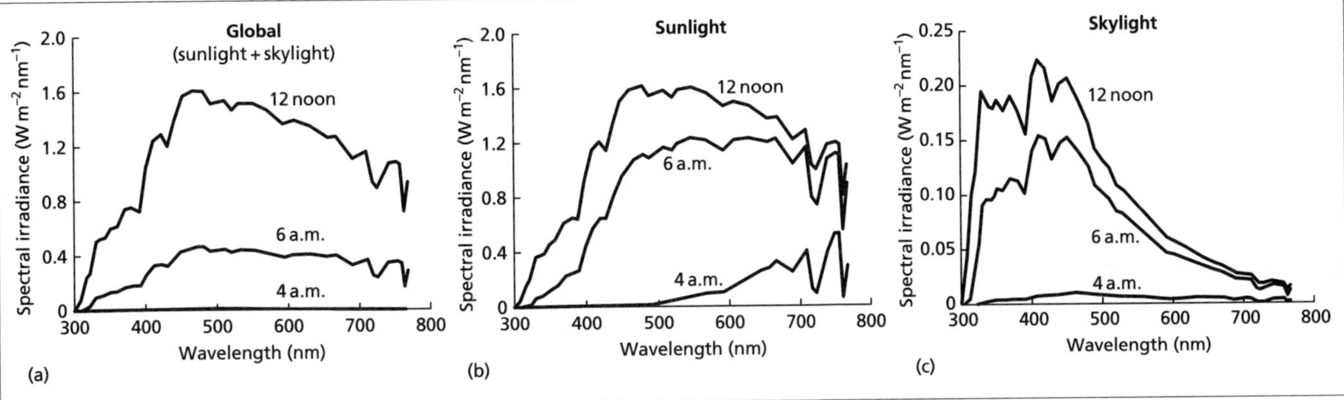

Fig. L8 The daylight spectrum at 60° latitude under a cloudless sky at 4 a.m. (with the Sun 0.6° above the horizon), 6 a.m. (27° above the horizon) and 12 noon (63° above the horizon), in all cases solar time. The graphs show (a) global radiation (sunlight plus skylight) measured on a horizontal surface; (b) sunlight only (measured in the direction of the Sun); and (c) skylight only (measured on a horizontal surface). These spectra were measured with such a bandwidth that the Fraunhofer lines are not visible.

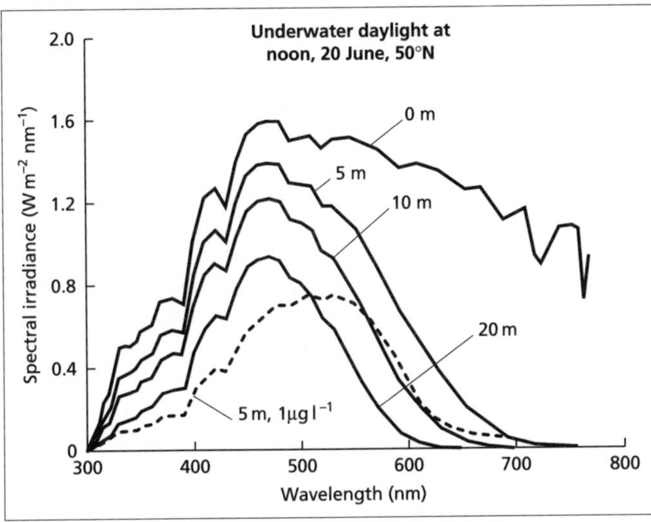

Fig. L9 The global daylight spectrum (horizontal irradiance) under a cloudless sky just below the water surface and at various depths. Most of the graphs are for clear ocean water, while one graph for water containing phytoplankton at a chlorophyll concentration of 1 µg l⁻¹ is also included.

the gases and particles, such as AEROSOL and CLOUDS, scatter light. Scattering by molecules and other small particles is stronger the shorter the wavelength. This is the reason that, under cloudless skies, blue light (and to an even larger extent ultraviolet radiation) reaches us not only along the line from the Sun but from all parts of the sky and thus the sky appears blue (Fig. L8). When the sky is low on the horizon, the light has to pass through more air before it can reach us than when the Sun is high, most of the blue light is scattered out of the direct beam and the Sun thus appears red (Fig. L8b). In clouds, the particles are so large that the scattering of light is rather independent of wavelength and therefore the clouds appear white or grey. Daylight under dense plant canopies, such as in forests, has a high proportion of far-red as this, in contrast to visible light, is not absorbed by chlorophyll. Water absorbs both shorter and, in particular, longer wavelength bands (Fig. L9).

Biological importance

Light is important for life on Earth in many respects. Plants, algae and some bacteria convert light to chemical energy by the process of photosynthesis. In the case of plants, algae and cyanobacteria (but no other bacteria), photosynthesis results in the production of molecular oxygen and in the same process CARBON atoms from CO_2 or bicarbonate ions are converted to organic carbon. Schematically the process can be written:

$$\text{Light} + H_2O + CO_2 \rightarrow \text{Organic matter} + O_2$$

Although the chemical energy to which light is converted in the process is sometimes regarded as residing in the organic matter, it is in fact the combined system of organic matter and oxygen that is energy rich. Non-photosynthetic organisms are dependent on the organic matter or on the oxygen produced in photosynthesis, or (as we are) on both. There are bacteria living at the bottom of the DEEP SEA where hydrogen sulphide emerges from hot vents. These bacteria obtain energy for their life processes from oxidizing hydrogen sulphide with oxygen, which has been produced by photosynthesis in the light above. Thus deep-sea bacteria, although not dependent on organic nutrients, are dependent on photosynthetic organisms.

Light is important to living creatures also because it makes vision possible, for many the most important of the senses. All organisms are not restricted to the light that is visible to us. Many insects, birds and fish also see ultraviolet radiation.

Another way in which light is important to most organisms is as a synchronizer of internal rhythms ('setting of the biological clock' and 'photoperiodic almanac'). It is important for organisms to keep track both of the time of day and of the time of year. All eukaryotic organisms (i.e. animals, plants, algae, fungi) and some bacteria possess an inner rhythm with a period of approximately, but not exactly, 24 h, the so-called biological clock or CIRCADIAN RHYTHM. This rhythm persists even under completely constant external conditions (in fact, humans and some other organisms possess two such rhythms, which may become uncoordinated under constant conditions). Under natural conditions, this rhythm is adjusted daily by the change between light and darkness. For example, it allows bees to forage specific flowers at the times of day when their nectaries are full, and migrating birds and fish to use the Sun as a compass. In humans, it regulates the alternation between activity and sleep, the excretion of urine, body temperature and many other functions. When we travel east or west, our internal rhythm becomes uncoordinated with respect to the external day–night rhythm ('jet lag'), but is soon reset by the changes in light. In humans and other mammals, the light for resetting the internal clock is sensed by the eyes (although the signal is not transmitted via the visual pathway); in many other animals (such as birds and insects) it is sensed by parts of the brain. In plants the internal circadian rhythm is also synchronized to the external day–night rhythm by light; in this case the light signal is sensed by the chromoprotein PHYTOCHROME. Detailed knowledge is now available for how the human circadian rhythm can be reset at once by suitable light exposure; both the timing and amount of light are important.

Both animals and plants keep track of the time of year by measuring the length of the days (or nights). This so-called photoperiodism was discovered around 1920 by scientists studying why plants flower at certain times. In plants, photoperiodism also regulates the alternation between active state and frost-resistant state, leaf fall and the deposition of reserve CARBOHYDRATE in underground storage organs. In animals it regulates many functions, such as gonad growth and reproductive behaviour, growth of antlers and MIGRATION instinct.

Some organisms produce their own visible light by BIOLUMINESCENCE. This is particularly important in the deep sea, where there is no other light. The light can be used for signalling to the opposite sex, for finding food, attracting prey or to confuse or deter predators. Luminescent animals occur in most of the phyla. Some fish house symbiotic luminescent bacteria rather than produce their own light. Examples of terrestrial luminescent organisms are some kinds of earthworm and toadstool and many insect species. Among the latter, such as glow-worms and fireflies, luminescence is used mostly for attracting members of the opposite sex, although some fireflies use it also for attracting prey (males of other firefly species).

Humans have profoundly changed their own light environment and to some extent that of other organisms. The discovery of fire in the Stone Age did not provide sufficient light to upset the biological clock; the big change has come with the introduction of electric light. Different kinds of electric lamp have very different spectra. Incandescent lamps, the first to be introduced, emit heat radiation as the Sun does. However, their filaments are not as hot as the Sun and therefore they emit less blue and other shortwave light and relatively more red. Most of their emission is in the infrared. Since the infrared is useless for most purposes, their efficiency is low and they are gradually being replaced by other lamp types. Of more powerful lamps available, low-pressure sodium lamps give the highest output of visible light per unit of electrical energy input, i.e. in a sense the highest efficiency. They are often used as street illumination but have the disadvantage that they produce only yellow light. In such light the colours of vehicles cannot be distinguished. High-pressure sodium lamps emit a wider range of wavelength components. Another common kind of street illumination is produced by medium-pressure mercury lamps, which emit more blue light. For indoors illumination, the most common lamp type after incandescent lamps is the fluorescent lamp. It is a kind of mercury lamp with a low pressure of mercury vapour and a fluorescent coating on the inside of the usually tubular envelope. The mercury gas emits mainly ultraviolet radiation, which is converted to visible light by the fluorescent coating. Depending on the coating, the light produced has a different spectral composition for different lamp types. The white-light types can be classified as either cool white (containing much blue and little red) or warm white (containing much red light). It has been found that in some people the cool-white lamp light has a very negative psychological effect (in some cases it has even been suggested as an important cause of suicide). The same can be said about the 100 or 120 Hz flicker that the early fluorescent lamps produced.

The transition from natural to partly artificial light has affected humans in ways that we are now only starting to understand. Some psychological disorders can be traced to disturbances of the internal rhythm and temporary remedy can be achieved by special light treatment (in 'light parlours').

Visible light is important for the human body in a number of other ways. Damage to cells caused by the shortwave ultraviolet component of daylight is partly repaired by enzymes called photolyases, which are active only under illumination by visible light (or long-wavelength ultraviolet radiation). Some forms of porphyria (disorders in which breakdown products of haemoglobin poison the body) make the body extremely light-sensitive, as do certain drugs such as tetracyclines (a group of antibiotics). Some other diseases are ameliorated by light. AGEING of the skin is accelerated by the ultraviolet component in daylight through phosphorylation of certain skin PROTEINS. [L.O.B.]

---

light traps *See* SAMPLING METHODOLOGY/DEVICES.

lightning When WATER vapour is forced high in the ATMOSPHERE by powerful updrafts, it forms ICE particles that collide with each other and with water droplets. Electrical charge separation by friction occurs and a thundercloud is formed. Typically, a thundercloud is several kilometres in diameter and stores tens to hundreds of coulombs, corresponding to a voltage of around 10^8 V: under such conditions lightning occurs. Although about 80% of lightning strokes are cloud to cloud (sheet lightning), most is known about cloud-to-ground lightning. Figure L10 illustrates the development of a cloud-to-ground discharge.

Generally, the lower surface of the cloud is negatively charged and partial electrical breakdown of the surrounding air (breakdown strength of approximately 3 MV m^{-1}) leads to a 'pilot streamer' (Fig. L10a). This develops towards Earth, creating an electrically conducting path along which an arc, the 'stepped leader', can be formed. Progress is by a series of steps about 50 m in length at a velocity of 1.5×10^5 m s^{-1}, with pauses of 50 μs or so: so far there is no discharge to ground.

When the pilot streamer reaches Earth, it forms a path for the 'return stroke' (Fig. L10b). A highly luminous electrical discharge

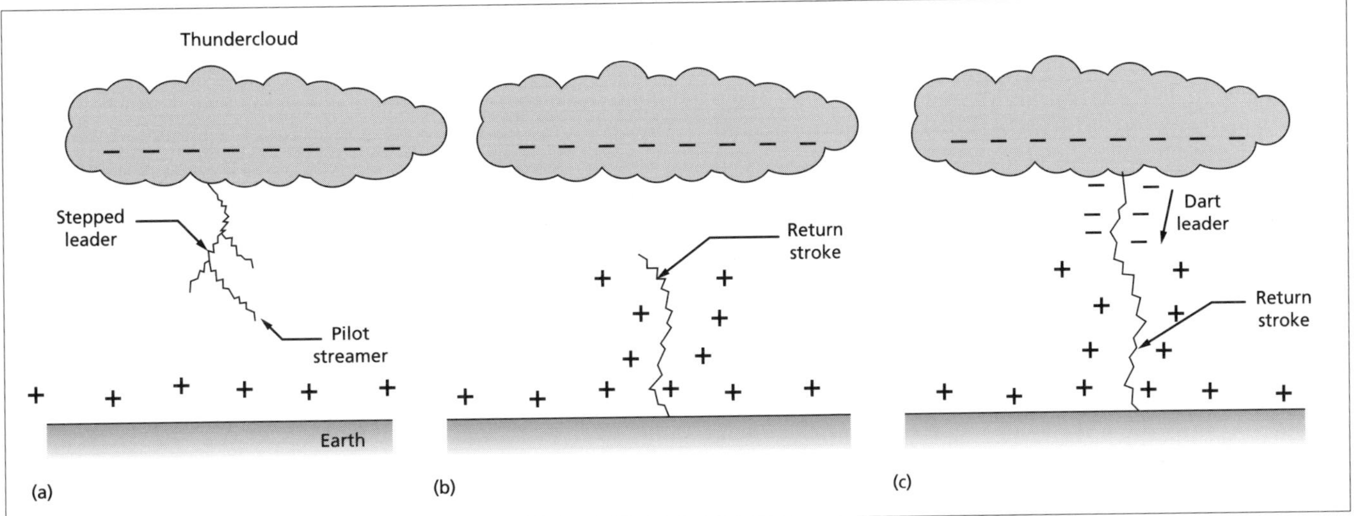

Fig. L10 Development of cloud-to-ground lightning stroke.

results, with current peaking up to 10 kA in less than 10 μs and releasing stored energy of typically 10^9 J. The discharge moves at about one-fifth the speed of LIGHT, connecting ground to cloud in typically 70 μs. The high associated temperature causes very rapid expansion of the air and a shock wave (thunder) results.

After the initial flash, a continuous luminous section about 50 m long and travelling at 2×10^6 m s^{-1}, a 'dart leader', usually occurs, leading to a second (or subsequent) stroke (Fig. L10c). In total a lightning flash lasts up to 0.5 s: there are usually three to four component strokes, each stroke lasting only a few hundred microseconds and separated by about 40 ms. [H.W.W.]

lignin Lignins are polymeric natural products arising from an enzyme-initiated polymerization of three primary precursors: *trans*-coniferyl 1, transinapyl 2 and *trans-p*-couramyl 3 alcohols. They are intimately associated with, and bonded to, hemicelluloses in plant cell walls and give wood its characteristic strength. Lignin is resistant to bacterial decay but is attacked by some fungi. [J.G.]

REFERENCE

Sarkanen, K.V. & Ludwig, C.H. (1971) *Lignins: Occurrence, Formation, Structure and Reactions*. John Wiley & Sons, New York.

lime Calcium oxide (CaO) is known as lime or quicklime and is formed by heating CALCIUM in the presence of OXYGEN or commercially by the thermal decomposition of calcium carbonate ($CaCO_3$). The latter reaction is reversible but can be driven to completion by the removal of the by-product, CARBON DIOXIDE (CO_2). CaO reacts with WATER with the evolution of much heat, in a process known as slaking, to form calcium hydroxide ($Ca(OH)_2$), which is a cheap ALKALI for treating ACID soils. CaO is also used as an additive in the production of GLASS and the manufacture of mortar for construction purposes. [M.C.G.]

limit cycles Oscillation of PREDATOR and PREY densities in which stabilization and destabilization of their interactions balance. [P.C.]

limit of detection (LOD) No CONTAMINATION detected within the limits of the analytical procedure. *See also* LIMIT OF QUANTIFICATION. [P.C.]

limit of quantification (LOQ) CONTAMINATION detected but at levels too small to quantify within the limits of the analytical procedure. *See also* LIMIT OF DETECTION. [P.C.]

limit value Limit concentrations or absolute amounts of CONTAMINANTS set to protect the environment and which must not be exceeded in EMISSIONS. Usually refer to end-of-pipe or end-of-stack levels. *Cf.* ENVIRONMENTAL QUALITY STANDARD. [P.C.]

limiting factor

1 Physiology: the (usually single) factor that currently limits the rate of a physiological process; the factor whose rate of supply is lowest relative to demand; the factor which, if experimentally manipulated, would lead to an increase in the rate of the process. For example, with photosynthetic rate, potentially important factors might include TEMPERATURE, LIGHT intensity, WATER availability, NITROGEN supply and CARBON DIOXIDE (CO_2) concentration. In any given circumstance, experimental increase in only one of these (the most limiting factor; *see* LIEBIG'S LAW OF THE MINIMUM) is likely to increase the photosynthetic rate. Thus if the plants were grown in full sun at their optimal temperature and provided with unlimited soil water and NUTRIENTS, then their photosynthetic rate might be CO_2-limited. Interaction effects occur when two or more factors simultaneously determine the rate of a physiological process.

2 POPULATION DYNAMICS: a MORTALITY factor (or a factor affecting the degree of realized FECUNDITY) for which changes in the level of the factor cause changes in breeding POPULATION DENSITY. Limiting factors may be density dependent or density independent; density-dependent limiting factors are generally referred to as REGULATING

FACTORS. Density-independent limiting factors cannot regulate population density. Contrast with compensated factors.

3 Ecosystems: although each of the species in an ecosystem might be limited in its performance by the availability of the single resource most limiting to the GROWTH of that particular species, the ecosystem as a whole is likely to be limited simultaneously by many different factors. In the case of soil nutrients, for example, the dominant plant species may be nitrogen-limited, so supplying extra POTASSIUM would have no impact on its performance. However, a currently minor component of the vegetation might be potassium-limited (e.g. a legume). In such a case, experimental addition of potassium might allow this species to increase in ABUNDANCE and perhaps eventually to replace the former dominant. Thus while the ecosystem might be said to be nitrogen-limited in its present form, altering the supply of a different, apparently non-limiting nutrient (like potassium) can change the identity of the dominant species and hence have a profound effect on long-term ecosystem structure and function.

4 BIOGEOGRAPHY: the factor that ultimately limits the location of the boundary of the GEOGRAPHIC RANGE of a species, separating areas within the range (where $dN/dt > 0$ when N is low) from areas outside the range (where $dN/dt < 0$); the factor limiting the growth and proliferation of the species in environments close to the edge of its geographic range.

See also COMPENSATION; DENSITY DEPENDENCE; DENSITY INDEPENDENCE. [M.J.C.]

limiting similarity The smallest difference between competitors that allows their COEXISTENCE. In simple, deterministic models of COMPETITION, such as the LOTKA–VOLTERRA MODEL, even the slightest deviation from complete overlap between competing species can lead to mathematically stable coexistence. However, the forces driving the system back to its equilibrium would then be too weak to maintain coexistence under the disturbances caused by natural environmental variability. Larger differences between competing species will lead to greater stability, but exactly how much difference is needed for lasting coexistence? Some models suggest that the probability of coexistence rapidly increases as the difference exceeds a certain value, which could be called the limiting similarity.

The idea of limiting similarity suffers from the same problems as other elements of competition theory: it depends largely on the often unrealistic assumption of a spatially and temporally homogeneous environment and the validity of its (empirical) evidence is hotly debated. *See also* COMPETITIVE EXCLUSION; HUTCHINSON'S SIZE-RATIO RULE. [J.G.S.]

REFERENCE

Abrams, P. (1983) The theory of limiting similarity. *Annual Review of Ecology and Systematics* **14**, 359–376.

limits of tolerance Maximum level (dose, concentration) of a STRESS that an organism can tolerate. Injurious effects are initially sublethal, affecting development and reproduction, then increasingly lethal. Measurable stresses include TOXICANTS (PESTICIDES), low temperatures and DESICCATION. Tolerance is usually normally distributed within a population, producing a characteristic sigmoidal DOSE–RESPONSE curve. An appropriate data transformation analysis (probit) is used to estimate doses required to kill a given proportion of a population; LD_{50} is the dose/concentration of an insecticide required to kill 50% of a target population. With abiotic stresses such as low temperature, MORTALITY is a function of the temperature and the duration of EXPOSURE. $LTemp_{50}$ is the temperature required to kill 50% of a population in an 'instantaneous' exposure; $LTime_{50}$ is the time required to kill 50% of a population at a constant temperature. For organisms that die only when they freeze, the body freezing temperature (SUPERCOOLING point) is an estimate of their lower lethal temperature. Freeze-tolerant insects are killed at much lower temperatures (–40 to –80°C) but mortality is again affected by the duration of exposure. [J.S.B.]

limnetic zone *See* LAKES; WITHIN-LAKE HABITAT CLASSIFICATION.

limnology Derived from the Greek *limnos* meaning SWAMP, lake or MARSH, the word 'limnology' was used originally to define the study of LAKES. Limnology was later expanded to include all inland waters (fresh and salt) contained within continental boundaries including standing (LENTIC) ecosystems such as lakes and ponds and running (LOTIC) waters (springs, streams, creeks, RIVERS). BRACKISH WATERS in river estuaries are also included in limnology as are EPHEMERAL (temporary) water bodies such as pools, tree holes, etc. Limnology is the study of the functional relationships and PRODUCTIVITY of inland waters and the effects of the physical, chemical and biotic environment on their communities. It includes disciplines such as geology, chemistry, physics, biology and ecology. In recent usage, limnology has been defined by some authors to include only the study of fresh waters excluding saline and brackish waters. [R.W.D.]

Lincoln index A basic equation used to estimate numbers in a mobile POPULATION in which marked animals are released and recaptured. Named after F.C. Lincoln who developed it in relation to the American bird banding programme established in the 1930s. *See also* CAPTURE–RECAPTURE TECHNIQUES; PETERSEN ESTIMATE; SAMPLING METHODOLOGY/DEVICES. [L.M.C.]

REFERENCE

Lincoln, F.C. (1930) *Calculating Waterfowl Abundance on the Basis of Banding Returns.* US Department of Agriculture Circular No. 118, pp. 1–4.

lindane An organochlorine INSECTICIDE first reported in 1935; the generic name of >99% pure γ-hexachlorocyclohexane (gamma-HCH). It is ingested, with some contact and soil fumigant activity, and is active on soil-dwelling and PHYTOPHAGOUS insects, public-health pests and animal ectoparasites. It was used, along with DDT, to kill human lice without apparent ill effects during the Second World War. However, it is very persistent in the environment, building up in the FOOD CHAIN and affecting a wide range of higher animals. It may be carcinogenic in some species. Although use is now restricted, the degradation characteristics of lindane mean that disappearance in the environment has been slow. *See also* BIOCIDE. [K.D.]

Lindeman efficiency Lindeman efficiencies are the efficiencies with which energy is transferred between trophic levels (also known as ecological or transfer efficiencies) and are the fundamental efficiencies that set the availability of energy to the next TROPHIC LEVEL.

For plants the Lindeman efficiency is:

$$\frac{\text{Gross primary productivity (rate of photosynthesis)}}{\text{Solar flux}} \times 100$$

For herbivores the Lindeman efficiency is:

$$\frac{\text{Rate of assimilation by herbivores}}{\text{Gross productivity of plants}} \times 100$$

For carnivores the Lindeman efficiency is:

$$\frac{\text{Rate of assimilation by primary carnivores}}{\text{Rate of assimilation of herbivores}} \times 100$$

See also CHARACTERS IN ECOLOGY; ECOLOGICAL EFFICIENCY. [M.H.]

Lindeman, R. *See* CHARACTERS IN ECOLOGY.

lineage Biological diversity is composed of lineages of one or a series of DEMES that share a common history of descent not shared by other demes. Lineages are thus cohesive groups of organisms, whose cohesion can be based on common history, developmental CONSTRAINTS, STABILIZING SELECTION and/or interbreeding. The most inclusive evolutionary lineages are species. *See also* CLASSIFICATION (*see* CLADISTICS); SPECIATION. [D.R.B.]

lineage-specific effect A group of species sharing a common ancestor that is shared by no other species can be referred to as a LINEAGE. Sometimes characters evolve just once in a PHYLOGENETIC TREE and are then shared by all descendant species; these are called lineage-specific effects. Such shared derived characters are termed 'synapomorphies' by cladists and form the basis of cladistic classifications (see CLADISTICS). [P.H.H.]

linear models Models in which the MEAN response is a linear combination of the PARAMETERS and which follow the general form:

$$\begin{aligned} y = \alpha &+ \beta_1(\text{something involving just } x \text{ values}) \\ &+ \beta_2(\text{second quantity involving just } x \text{ values}) \\ &+ \cdots + \beta_k(k\text{th quantity involving just } x \text{ values}) + \varepsilon \end{aligned}$$

Linear models do not have inherent scaling in that any solution for x and y can be scaled up or down and still be a solution. Linear differential models have the advantage that they can be solved analytically. *See also* EQUATIONS; REGRESSION ANALYSIS, CURVILINEAR. [V.F.]

link function The RESPONSE VARIABLE is related to the linear predictor in a GENERALIZED LINEAR MODEL by the inverse of the link function. In analysis of count data, the fitted values are constrained to be non-negative; this is handled most simply by making the fitted values the antilog of the linear predictor, $a+bx$:

$$y = \exp(a + bx)$$

Such an analysis is carried out using the log line function. For proportion data the logit link function is employed, where the probability of 'success' in a binomial trial, p, is related to the explanatory variables by the logistic:

$$p = \frac{\exp(a + bx)}{1 + \exp(a + bx)}$$

which has the necessary property of being bounded (p can never be greater than 1 or less than 0) and is linearized by the following simple transformation:

$$\text{logit}(p) = \ln(p/q)$$

where $q=(1-p)$. Other commonly used link functions are the reciprocal, square root and complementary log–log. [M.J.C.]

REFERENCE

Crawley, M.J. (1993) *GLIM for Ecologists*. Blackwell Scientific Publications, Oxford.

linkage disequilibrium Linkage disequilibrium means that ALLELES of different gene loci occur in GAMETES at frequencies that differ from the expected frequencies under random combination of alleles into gametes. Consider two loci, A and B, with alleles A_1 and A_2 and alleles B_1 and B_2, at ALLELE FREQUENCIES p and q, and r and s, respectively ($p+q=1$, $r+s=1$). Let the observed frequencies of the gametes A_1B_1, A_1B_2, A_2B_1 and A_2B_2 be x_{11}, x_{12}, x_{21} and x_{22}; the expected frequencies under random combination of alleles in gametes are pr, ps, qr and qs. The coefficient of linkage disequilibrium D indicates the deviation from random combination:

$$D = x_{11}x_{22} - x_{12}x_{21}$$

and $x_{11}=pr+D$, $x_{12}=ps-D$, $x_{21}=qr-D$, $x_{22}=qs+D$

The value of D is between –0.25 and +0.25. The linkage disequilibrium coefficient D can be regarded as the COVARIANCE between the allelic values in the gametes. Linkage disequilibrium has a strong influence on the ADDITIVE GENETIC VARIANCE and therefore on the response to selection. *See also* QUANTITATIVE GENETICS. [G.D.J.]

Linnaean classification The system of CLASSIFICATION of organisms devised by the Swedish polymath Carl von Linné or Carolus Linnaeus (1707–1778) and originally presented in his *Systema Naturae* (*see also* CHARACTERS IN ECOLOGY). The position in the system of a new species is determined using a set of characters chosen for their discriminatory value. In his *Species Plantarum* (1753) Linnaeus used characters drawn from the sexual parts of the flowers, without trying to elucidate general similarity from the ensemble of all types of character that could be scored. That classification is described as artificial, to distinguish it from a NATURAL CLASSIFICATION based on general similarity. For the purpose of naming a specimen the two approaches are equally valid, but a natural system is to be preferred if phylogenetic relationship is of interest. A feature of outstanding importance, which Linnaeus established in his *Species Plantarum*, is the binomial, consisting of a generic and a specific name, as a unique identifier for each species. *See also* BINOMIAL CLASSIFICATION. [L.M.C.]

Linnaeus, C. *See* CHARACTERS IN ECOLOGY.

lipids Esters of fatty ACIDS that are insoluble in WATER but soluble in organic solvents. They include fats, oils and waxes (simple lipids), phospholipids and glycolipids (complex lipids) and steroids (derived lipids). [J.G.]

lipophilic chemicals *See* OCTANOL–WATER PARTITION COEFFICIENT.

lists Within European Union (EU), US and UK legislation there are numerous lists of hazardous substances. These are usually compiled on the basis of HAZARD assessments. Examples are the EU Black and Grey lists of substances dangerous to the aquatic environment. The UK has a similar RED LIST. There are ENDANGERED SPECIES lists (in US federal legislation) and lists of violating facilities compiled by the USEPA. There are also many and various other environmental data lists. [P.C.]

lithium (Li) A soft, silver-white METAL (ATOMIC NUMBER = 3; relative ATOMIC MASS = 6.94; melting point = 180.54°C; boiling point = 1347°C), lithium is found in seawater at a concentration of about 180 ppb and as minerals in combination with ALUMINIUM and SILICON (e.g. spodumene, $LiAlSi_2O_6$). The metal is extracted electrolytically from a fused mixture of the CHLORIDE. The metal is used to make lightweight alloys with aluminium and MAGNESIUM, and in BATTERIES. It also plays a role in thermonuclear FUSION for civilian and military applications. Lithium salts are used as lubricants and in the treatment of manic depression. Lithium salts have only a low TOXICITY. [C.R.P.]

lithogenous (red clay) A SEDIMENT consisting primarily of mineral grains. A sediment is classified as a red CLAY if it contains < 30% of each of calcium carbonate ($CaCO_3$) and siliceous organic material. Thus, red clay predominates either where calcareous PLANKTON is sparse (in high latitudes) or in deep water high in CARBON DIOXIDE (CO_2) where $CaCO_3$ is mostly dissolved (i.e. most of the very deep ocean floor). It consists chiefly of quartz, mica, several kinds of clay minerals, chlorite and several of the hydrous aluminium silicate minerals of the zeolite group. Although some red clay is thought to form in place, from ions in solution in seawater, most of this material was derived from the continents. It varies in colour from brownish red or chocolate brown to blue, grey or green nearer to shore. [V.F.]

lithosphere This is simply defined as the Earth's CRUST, although this definition does not indicate the wide diversity of materials that are present. Specifically, it refers to the rocks and ICE that cover the surface of the Earth. Table L4 shows the main rock types and their chemical composition. [E.A.F.]

lithostratigraphy *See* STRATIGRAPHY.

littoral ecology *See* ECOLOGY.

littoral zone

1 For LAKES, the edge, sometimes defined imprecisely as the extent to which SAMPLING is possible before water overflows the boots, and sometimes more precisely as the area in which PHOTOSYNTHESIS is possible by benthic primary producers. *See also* WITHIN-LAKE HABITAT CLASSIFICATION.

2 For the seas, also called the eulittoral or INTERTIDAL ZONE. The limits of this zone are sometimes considered to lie between mean high-water and mean low-water levels. Alternatively, the zone may be considered to extend from the level of the highest spring high tide to the lowest spring low tide. Additionally, the supralittoral zone (splash or SPRAY ZONE) may be used to refer to those levels on the shore only submerged during spring high TIDES.

[V.F.]

littoriprofundal zone *See* WITHIN-LAKE HABITAT CLASSIFICATION.

living fossils Species of animals and plants that have changed little over long periods of evolutionary history to the present. Examples are the lungfishes, alligators, snapping turtles and the aardvark. They are thought by some to present a thorny problem for traditional gradualistic EVOLUTION: if NATURAL SELECTION is continuously reshaping species, why have some species apparently escaped? Darwin seems to have orginated the term 'living fossil' (first edition of *On the Origin of Species by Means of Natural Selection*, p. 107) suggesting that they were preserved because they have inhabited confined areas and thus avoided COMPETITION. However, many species of living fossil have broad distributions. Some believe they have persisted because of narrow specialized ADAPTATIONS; others because they have unusually broad unspecialized adaptations. Possibly no one answer applies. Equally, to suggest that they have remained unchanged because some aspect of their ecology has enabled them to avoid SELECTION PRESSURES is, in itself, somewhat vacuous if not circular. *See also* PUNCTUATED EQUILIBRIUM. [P.C.]

Lixisols *See* SOIL CLASSIFICATION.

lizards *See* REPTILES, DIVERSITY AND CONSERVATION.

Lloyd's index of mean crowding *See* INDICES OF DISPERSION.

Local Agenda 21 Initiative developed by local governments in the UK; Chapter 28 of AGENDA 21 encourages them to define their own SUSTAINABLE DEVELOPMENT strategies. [P.C.]

local constraints *See* CONSTRAINTS.

local stability A system is locally stable if it returns to an equilibrium when subjected to a very small perturbation away from that equilibrium. The concept is useful because it is much easier to demonstrate local stability than GLOBAL STABILITY and locally stable systems are often (but not always) globally stable. Local stability of models is usually proven by linearizing the model around the equilibrium state to obtain equations describing the fate of a small perturbation. *See also* POPULATION ECOLOGY. [S.N.W.]

local stability lists *See* STABILITY ANALYSIS.

LOD *See* LIMIT OF DETECTION.

LOEC *See* LOWEST OBSERVED EFFECT CONCENTRATION.

Table L4 Chemical composition of some common rocks. (From FitzPatrick, E.A. (1980) *Soils: Their Formation, Classification and Distribution*. Longman, Harlow. Reprinted by permission of Addison Wesley Longman Ltd.)

	Composition (%)											
Rock type	SiO_2	Al_2O_3	Fe_2O_3	FeO	MgO	CaO	Na_2O	K_2O	TiO_2	P_2O_5	MnO	H_2O
Basalt Labradorite, augite and pseudomorphs in chlorite and goethite after olivine; and opaque minerals (Guppy & Sabine 1956)	53	18	5	2	4	7	4	2	1	< 1	< 1	2
Diorite Hornblende, biotite, titanite, oligoclase, quartz and opaque minerals (Guppy & Sabine 1956)	58	17	3	4	3	5	4	3	2	1	< 1	1
Granite Fine-grained grey granite, composed of quartz, microcline, oligoclase, biotite, muscovite, accessory minerals, apatite and opaque minerals (Guppy & Sabine 1956)	76	13	< 1	1	< 1	< 1	3	5	< 1	< 1	< 1	1
Garnetiferous mica-schist Muscovite, biotite, chlorite, garnet, quartz and oligoclase, iron ore and apatite (Guppy & Sabine 1956)	58	19	1	7	2	2	2	4	1	< 1	< 1	2
Limestone Calcite (Clark 1924)	1	< 0.1	Nil	0.1	< 1	55*	< 0.1	< 1	Nil	< 0.1	Nil	< 0.1
Slate Quartz, chlorite, muscovite, apatite, tourmaline, albite and opaque minerals (Guppy & Sabine 1956)	57	20	10†		3	1	3	3	1	< 1	Nil	Nil

Clark, F.W. (1924) *The Data of Geochemistry*. United States Geological Survey Bulletin 770.
Guppy, E.M. & Sabine, P.A. (1956) *Chemical Analysis of Igneous Rocks, Metamorphic Rocks and Minerals*. Memoirs of the Geological Survey of Great Britain. Her Majesty's Stationery Office, London.
* CO_2 = 43%.
† Total Fe calculated as Fe_2O_3.

log normal distribution One of the four commonly used SPECIES-ABUNDANCE MODELS (distributions). For more details see Magurran (1988) and May (1975). *See also* DISTRIBUTION; DIVERSITY INDICES. [B.S.]

REFERENCES

Magurran, A.E. (1988) *Ecological Diversity and its Measurement*. Croom Helm, London.

May, R.M. (1975) Patterns of species abundance and diversity. In: *Ecology and Evolution of Communities* (eds M.L. Cody & J.M. Diamond), pp. 81–120. Harvard University Press, Cambridge, Mass.

log normal index *See* DIVERSITY INDICES; RANK–ABUNDANCE MODELS.

log normal model *See* RANK–ABUNDANCE MODELS.

log series distribution One of the four commonly used SPECIES-ABUNDANCE MODELS (distributions). For more details see Magurran (1988) and May (1975). *See also* DISTRIBUTION; DIVERSITY INDICES; RANK–ABUNDANCE MODELS. [B.S.]

REFERENCES

Magurran, A.E. (1988) *Ecological Diversity and its Measurement*. Croom Helm, London.

May, R.M. (1975) Patterns of species abundance and diversity. In: *Ecology and Evolution of Communities* (eds M.L. Cody & J.M. Diamond), pp. 81–120. Harvard University Press, Cambridge, Mass.

log series index *See* DIVERSITY INDICES; RANK–ABUNDANCE MODELS.

log series model *See* RANK–ABUNDANCE MODELS.

logarithmic transformation *See* TRANSFORMATION OF DATA.

logarithmic wind profile Measurements of the velocity of any viscous fluid close to a boundary reveals that the layers of fluid closest to the boundary move more slowly than layers further from the boundary because of the effects of surface friction. The fluid velocity at the boundary is at rest relative to the boundary; fluid velocity increases asymptotically with distance from the boundary. The same principle applies to measurements of horizontal wind speed made within a few metres of the Earth's surface. Wind speed increases exponentially with height and the relationship between wind speed and the logarithm of height can be described by a straight line. The slope of this line and its intercept are important measures of the degree of TURBULENCE in the air and the aerodynamic ROUGHNESS of the ground surface respectively. *See also* BOUNDARY LAYER; MICROMETEOROLOGY. [J.B.M.]

logistic equation Simplest equation relating POPULATION GROWTH RATE to POPULATION DENSITY. If N represents population density and t is time, then $1/N\, dN/dt$ represents population growth rate and this is assumed to be a straight-line function of N. This is the logistic equation:

$$\frac{1}{N}\frac{dN}{dt} = r - \frac{r}{K}N$$

or:

$$\frac{dN}{dt} = rN\left(1 - \frac{N}{K}\right)$$

where r is a constant representing population growth rate at low population density and K is the equilibrium population density, which is referred to as the CARRYING CAPACITY of the environment in which the population lives. The resulting sigmoidal curve of population density against time (Fig. L11) is referred to as the logistic curve. Without DENSITY DEPENDENCE the population would grow exponentially, $N=N_0e^{rt}$, where N_0 is population density at time zero (Fig. L11). The logistic equation is sometimes used in a discrete-time form, generally written as:

$$N_{t+1} = \frac{N_t R}{1 + aN_t}$$

where $R=e^r$ and $a=(R-1)/K$.

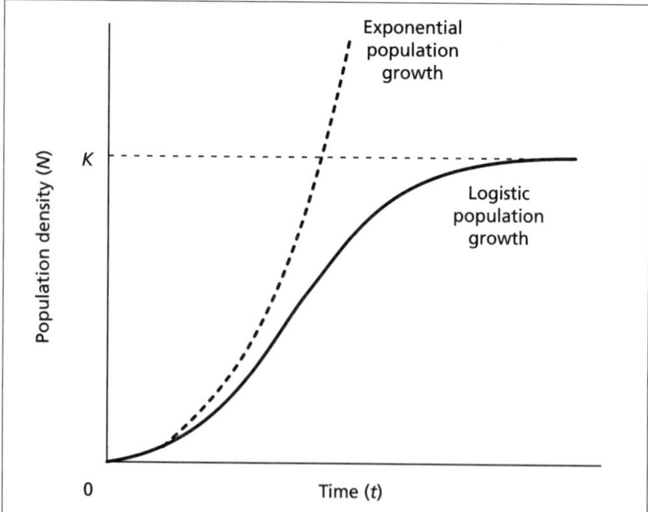

Fig. L11 the two simplest types of population growth, with (logistic) or without (exponential) density-dependence. K is the carrying capacity of the environment.

This simplistic treatment assumes the population has no AGE STRUCTURE, so that all individuals are equally capable of reproducing at any time. *See also* LOGISTIC MODEL; POPULATION DYNAMICS; POPULATION REGULATION. [R.M.S.]

logistic model The simplest model of density-dependent population growth, in which the POPULATION GROWTH RATE decreases in a straight-line relationship with increasing POPULATION SIZE. The two parameters of the logistic model are the INTRINSIC RATE OF INCREASE r and the equilibrium population level or CARRYING CAPACITY K. The continuous-time (differential equation) version of the logistic model shows SIGMOID GROWTH with a smooth approach to the carrying capacity. The discrete-time (difference equation) version in contrast shows a range of dynamic behaviour, from STABLE EQUILIBRIUM through regular cycles to CHAOS, as a consequence of DELAYED DENSITY DEPENDENCE. *See also* HASSELL MODEL AND HASSELL–COMINS MODEL; LOGISTIC EQUATION; LOTKA–VOLTERRA MODEL. [R.H.S.]

logistics *See* GREEN LOGISTICS.

Long-term Ecological Research Program (LTER) Programme in the USA that was initiated in the 1970s by the National Science Foundation. It focused on examining ecological processes and problems with temporal or spatial scales that could not be addressed by more traditional research projects with shorter time horizons. It now involves 18 sites that range from the Arctic/Antarctica to temperate and tropical. [P.C.]

long-term tests TOXICITY/ECOTOXICITY tests that are carried out over periods of time that are long relative to the LIFESPAN of the organism. Thus 1 day would be short term for a vertebrate but long term for a microbe. [P.C.]

longevity The length of life. [A.J.D.]

LOQ *See* LIMIT OF QUANTIFICATION.

lost opportunity principle *See* FORAGING, ECONOMICS OF.

lotic Running waters, such as RIVERS and streams. *See also* LIMNOLOGY. [J.L.]

Lotka, A.J. *See* CHARACTERS IN ECOLOGY.

Lotka–Volterra model

Vito Volterra was an Italian mathematician who, inspired by his son-in-law Umberto D'Ancona (an eminent hydrobiologist), published work on competitive interactions and PREDATOR–PREY INTERACTIONS in the 1920s and 1930s. A.J. Lotka was an American mathematician who published on similar topics, again in the 1920s and 1930s. Their names are associated with two basic models that have had a major impact on how ecologists think about INTERSPECIFIC COMPETITION and predation. (*See also* CHARACTERS IN ECOLOGY.)

Competition

The Lotka–Volterra model of interspecific competition is based upon two other models of population growth, the EXPONENTIAL GROWTH model and the LOGISTIC EQUATION. Let us start with exponential growth, i.e. no competition of any kind. Under such conditions the rate at which numbers change with time (dN/dt) can be represented by a per capita (per head) rate of increase multiplied by the number of individuals (N) in the population. The rate of increase used is the intrinsic rate of natural increase (r) and is equal to the difference between the per capita birth rate and the per capita death rate. In mathematical symbols this can be written:

$$\frac{dN}{dt} = rN$$

In this simple model, the rate of increase does not change with DENSITY, i.e. there is no DENSITY DEPENDENCE and growth is unlimited. However, in most real populations growth is a function of density ($f(N)$) and therefore the growth equation should be written:

$$\frac{dN}{dt} = rN\big(f(N)\big)$$

The simplest relationship that this function could have with density is shown in Fig. L12. We know that the end-points of this relationship are quite sensible. At very low POPULATION SIZE ($N \cong 0$), $f(N)$ must $\cong 1$ because this would give unlimited exponential growth. At high population size (more specifically the equilibrium population size or CARRYING CAPACITY, K), $N=K$, $f(N)=0$ and growth stops ($dN/dt=0$). Joining these two known points by a straight line assumes the simplest relationship possible and is not based on any empirical evidence. (Scientists, including ecologists, always develop their descriptions of the natural world by initially assuming the simplest relationships. Only when this proves inadequate do they make the explanation more complex. *See* AYALA'S EXPERIMENTS.) The slope of the line is equal to $1/K$ and the equation of this straight line is therefore:

$$f(N) = 1 - N/K$$

Substituting the value of $f(N)$ into the previous equation now produces

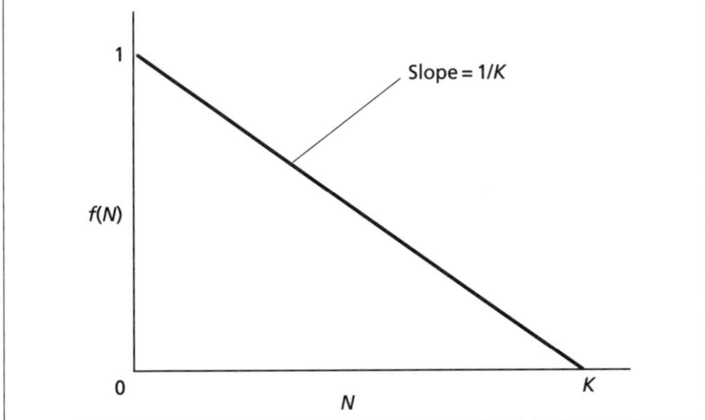

Fig. L12 The density-dependent function used to produce the logistic equation.

$$\frac{dN}{dt} = rN\left(1 - \frac{N}{K}\right)$$

which is the well-known logistic equation, producing sigmoidal limited population growth. It is important to remember that the term in parentheses represents intraspecific density-dependent regulation. The inhibitory effect of one individual on population growth is $1/K$ and the inhibitory effect of N individuals on population growth is N/K.

Now let us consider two species, each of which when living separately have their growth described by a logistic equation. However, when living together this logistic equation must be modified. In the following equations the two species are denoted by the subscripts 1 and 2. Let us consider first of all the population growth of species 1 when it competes with species 2. In terms of its own population growth, as described above by the logistic equation, the inhibitory effect of one individual of species 1 upon its own population growth is $1/K_1$, where K_1 is the carrying capacity of species 1. The inhibitory effect of every individual (N_1) of species 1 upon its own population growth is therefore N_1/K_1. If species 2 was identical to species 1 in its effect upon the population growth of species 1 then we could simply write the effect of every individual of species 2 upon the population growth of species 1 as N_2/K_1. However, this is unlikely. It is more realistic to imagine that each individual of species 2 would have a greater or lesser effect upon the population growth of species 1 and therefore we have to multiply N_2/K_1 by a constant that expresses this difference. This constant (usually called the COMPETITION COEFFICIENT and denoted by α_{12}) describes the relative effect of an individual of species 2 compared with an individual of species 1. If an individual of species 2 was identical (from an ecological point of view) then $\alpha_{12}=1$. Similarly, the effect of every individual of species 2 upon its own population growth will be N_2/K_2 and the effect of every individual of species 1 upon the population growth of species 2 will be N_1/K_2 multiplied by α_{21} (note the reversal of the subscripts).

To understand the competition coefficients (α_{12} and α_{21}) more clearly, let us take an imaginary example in which competition is solely for a single food resource. Let us imagine that *Paramecium* 1 consumes 10 bacteria per hour, while *Paramecium* 2 consumes 20 bacteria per hour. In terms of resource units (bacteria) an individual of species 1 equals half an individual of species 2, or two individuals of species 1 equals one individual of species 2. Therefore, $\alpha_{12}=2$ (the effect of adding one more individual of species 2 is like adding two more individuals of species 1) and $\alpha_{21}={}^1/_2$.

We can now rewrite the two logistic growth equations for the two species (which incorporate INTRASPECIFIC COMPETITION) to take into account interspecific competition:

$$\frac{dN_1}{dt} = r_1N_1\left[1 - \left(\frac{N_1}{K_1}\right) - \left(\frac{\alpha_{12}N_2}{K_1}\right)\right] = r_1N_1\left[\frac{K_1 - N_1 - \alpha_{12}N_2}{K_1}\right]$$

$$\frac{dN_2}{dt} = r_2N_2\left[1 - \left(\frac{N_2}{K_2}\right) - \left(\frac{\alpha_{21}N_1}{K_2}\right)\right] = r_2N_2\left[\frac{K_2 - N_2 - \alpha_{21}N_1}{K_2}\right]$$

This system of two species (the Lotka–Volterra model of two-species competition) will be at equilibrium when $dN_1/dt = dN_2/dt = 0$, i.e. when both species are at their carrying capacities and population growth has stopped. Just considering species 1, this means that:

$$r_1 N_1 \left[\frac{K_1 - N_1 - \alpha_{12} N_2}{K_1} \right] = 0$$

Since r_1, N_1 and K_1 cannot equal 0 (mathematically possible, but ecologically uninteresting if species 1 cannot reproduce, is not present or cannot exist in that habitat), this means that for species 1 $K_1 - N_1 - \alpha_{12}N_2 = 0$ at equilibrium. Rearranging this equation we get $N_1 = K_1 - \alpha_{12}N_2$, which is the equation of a straight line and is called the species 1 zero isocline. On a graph of the numbers of species 1 (N_1) plotted against the numbers of species 2 (N_2), it represents all those combinations of N_1 and N_2 for which the population growth of species 1 (dN_1/dt) = 0.

Following a similar procedure for species 2 will produce the species 2 zero isocline, which is $N_2 = K_2 - \alpha_{21}N_1$. On a graph of the numbers of species 2 (N_2) plotted against the numbers of species 1 (N_1), it represents all those combinations of N_1 and N_2 for which the population growth of species 2 (dN_2/dt) = 0.

There are four ways that these two straight lines can be placed on a graph of N_2 versus N_1 and these are illustrated in Fig. L13. These are known as phase plane graphs/diagrams. Each point on the graph represents a joint density of the two species. To each point in this graph we can attach an arrow, or vector, whose length and direction will indicate the dynamics of the system. Below its zero isocline, a species can increase in numbers. Above its zero isocline, a species will decline in numbers. Exactly on its zero isocline, a species will neither increase or decrease. This means that the four graphs predict different outcomes for interspecific competition. In (a) either species can win depending upon which area between the zero isoclines the population TRAJECTORY reaches. This depends upon starting numbers and/or relative growth rates (r_1 and r_2). Where the two lines cross represents an UNSTABLE EQUILIBRIUM point. In (b) the two species coexist at the values of N_1 and N_2 where the two lines cross. This represents a STABLE EQUILIBRIUM point. In (c) species 1 always wins and in (d) species 2 always wins.

Laboratory examples of one species winning (c and d) and CO-EXISTENCE (b) are provided by GAUSE'S EXPERIMENTS. Examples of one species winning (c and d) and either species winning (a) are provided by PARK'S EXPERIMENTS. It is important to make two comments about the Lotka–Volterra model. First, they contain hidden variables. The competition coefficients (α) are simply conversion factors to allow one species to be represented in numerical equivalents of the other species. Species may compete for food and space and poison each other, but all these mechanisms of competition are summarized and hidden within α. Second, because the real world (unlike most laboratory environments) is heterogeneous and patchy, even (a), (c) and (d) may result in coexistence. If within a habitat patch (a), (c) and (d) apply there may only be one species per patch, but looking over many patches both species will be seen to apparently 'coexist'.

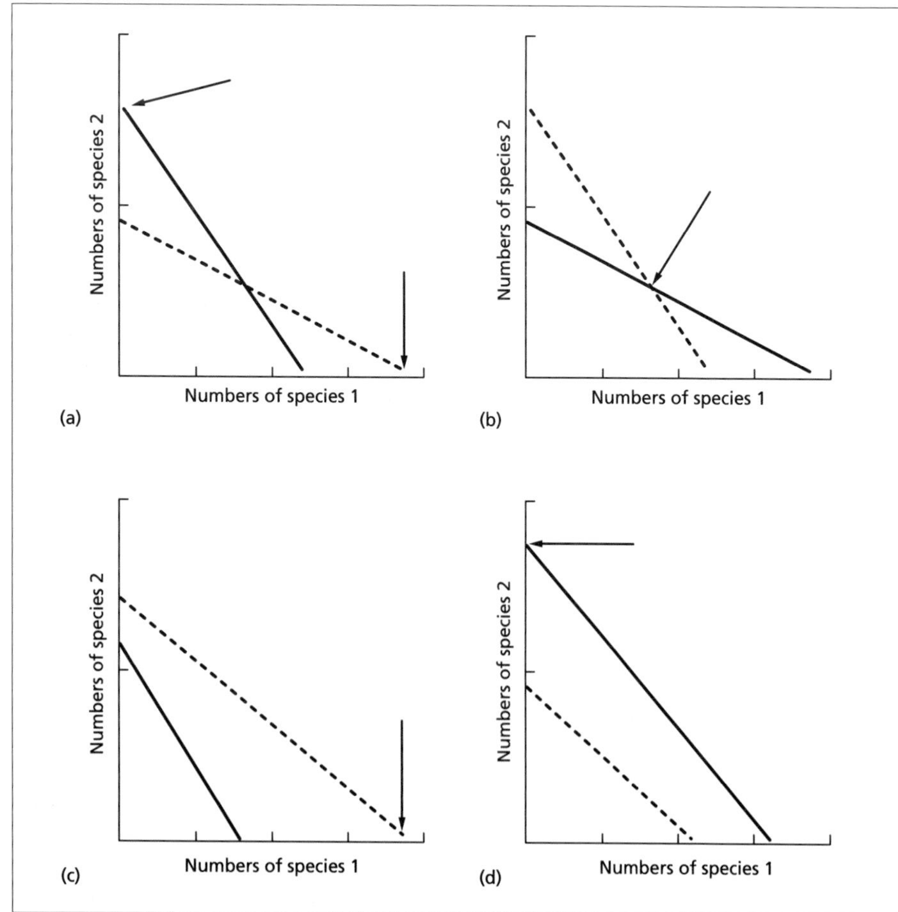

Fig. L13 The four predicted outcomes of the Lotka–Volterra model. Arrows indicate equilibrium points. Dashed line, species 1 zero isocline; solid line, species 2 zero isocline.

Predator–prey interaction

The Lotka–Volterra model of predator–prey interaction is based upon the exponential growth model. In addition, the model is based upon two very simple assumptions:

1 the birth rate of the predator will increase as the number of prey increases;

2 the death rate of the prey will increase as the number of predators increases.

In the following equations, numbers of prey will be denoted by N_1 and numbers of predators by N_2. This simple model assumes that prey get eaten by predators according to how often they bump into each other. Let us first consider the population growth of the predator. According to the simple exponential model, the population growth of the predator can be written:

$$\frac{dN_2}{dt} = r_2N_2$$

where r_2 is the intrinsic, per capita, growth rate of the predator. Since r_2 represents the difference between the per capita birth rate (b_2) and death rate (d_2) this could be rewritten

$$\frac{dN_2}{dt} = (b_2 - d_2)N_2$$

Assumption (1) above links the birth rate of the predator (b_2) with the numbers of the prey (N_1) and therefore we can rewrite this equation as:

$$\frac{dN_2}{dt} = (b_2N_1 - d_2)N_2$$

At equilibrium ($dN_2/dt=0$), therefore:

$$0 = (b_2N_1 - d_2)N_2$$

and rearranging successively gives:

$$0=b_2N_1N_2-d_2N_2$$

$$d_2N_2=b_2N_1N_2$$

$$N_1 = \frac{d_2N_2}{b_2N_2}$$

Since $N_2/N_2=1$ this finally becomes $N_1=d_2/b_2$. This equation says that population growth of the predator (N_2) will stop ($dN_2/dt=0$) when the population size of the prey (N_1) is equal to the ratio of predator death and birth rates. It is the zero isocline for the predator and can be drawn as a vertical straight line on a graph of predator numbers plotted against prey numbers (Fig. L14). To the left of this line, prey numbers (N_1) are so low that the predator's numbers (N_2) will decline. To the right of this line, prey numbers are high enough for the predator to increase its numbers.

Let us now consider the population growth of the prey. According to the simple exponential model, the population growth of the prey can be written:

$$\frac{dN_1}{dt} = r_1N_1$$

where r_1 is the intrinsic, per capita, growth rate of the prey. Since r_1 represents the difference between the per capita birth rate (b_1) and death rate (d_1) this can be rewritten:

$$\frac{dN_1}{dt} = (b_1 - d_1)N_1$$

Assumption (2) above links the death rate of the prey (d_1) with the numbers of the predator (N_2) and therefore we can rewrite this equation as:

$$\frac{dN_1}{dt} = (b_1 - d_1N_2)N_1$$

At equilibrium ($dN_1/dt=0$), therefore:

$$0 = (b_1 - d_1N_2)N_1$$

and rearranging successively gives:

$$0=b_1N_1-d_1N_1N_2$$

$$d_1N_1N_2=b_1N_1$$

$$N_2 = \frac{b_1N_1}{d_1N_1}$$

Since $N_1/N_1=1$ this finally becomes $N_2=b_1/d_1$. This equation says that population growth of the prey (N_1) will stop ($dN_1/dt=0$) when the population size of the predator (N_2) is equal to the ratio of prey birth and death rates. It is the zero isocline for the prey and can be drawn as a horizontal straight line on a graph of predator numbers plotted against prey numbers (Fig. L14). Above this line, predator numbers (N_2) are so high that the prey's numbers (N_1) will decline.

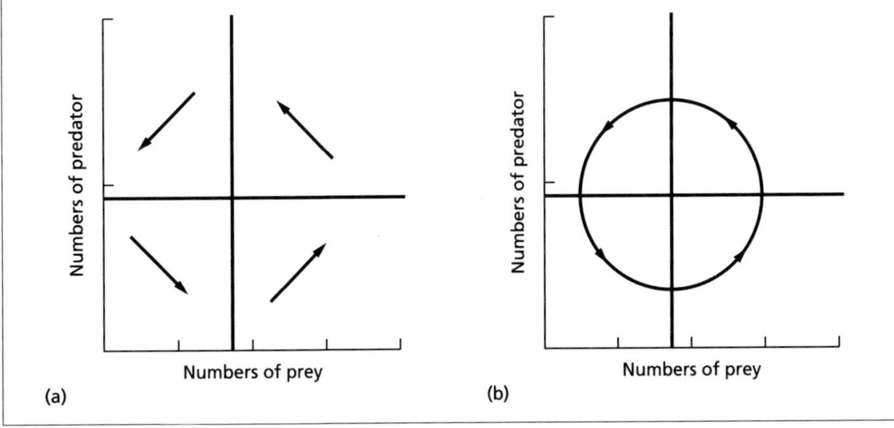

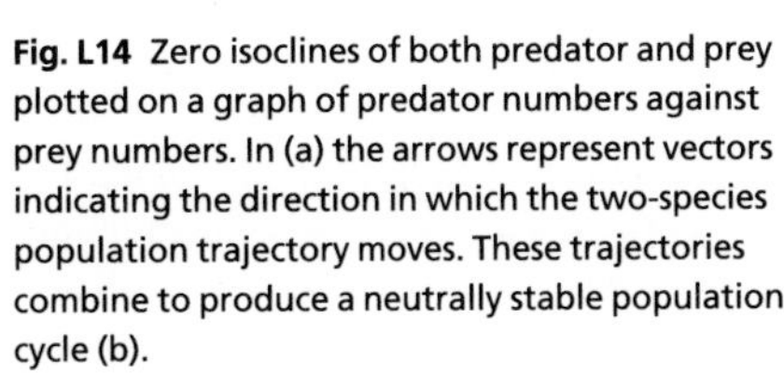
Fig. L14 Zero isoclines of both predator and prey plotted on a graph of predator numbers against prey numbers. In (a) the arrows represent vectors indicating the direction in which the two-species population trajectory moves. These trajectories combine to produce a neutrally stable population cycle (b).

Below this line, predator numbers are low enough for the prey to increase its numbers.

Since the two zero isoclines in Fig. L14 define when predator and prey will increase or decrease, we can draw arrows (vectors) that show the direction of movement of the two-species population in each of the four areas of the graph (Fig. L14a). In effect, the two-species population trajectory will describe a circular pattern and the resulting population numbers will show a cyclical pattern over time, with the OSCILLATIONS of the predator being just behind those of the prey. Laboratory systems of two mite species (prey, *Eotetranychus sexmaculatus*; predator, *Typhlodromus occidentalis*) living on oranges (Huffaker 1958) and field systems of lynx (*Lynx canadensis*) and snowshoe hare (*Lepus americanus*) living in Canada (Elton & Nicholson 1942) show this type of cyclical POPULATION DYNAMICS.

In addition to suggesting that predator–prey dynamics may involve oscillations of both species, the Lotka–Volterra model also predicts other interesting behaviour. For example, if the predator and prey are destroyed at the same rate by some outside factor then the prey might benefit and increase its numbers. Consider again the two equations:

$$\frac{dN_1}{dt} = b_1N_1 - d_1N_1N_2 \qquad \text{(prey)}$$

$$\frac{dN_2}{dt} = b_2N_1N_2 - d_2N_2 \qquad \text{(predator)}$$

If both N_1 and N_2 are reduced by the same proportionate amount, the effect will be greater in the product (N_1N_2) than in either N_1 or N_2 separately. Therefore the predator will have its birth rate more drastically reduced and the prey will have its death rate more drastically reduced. The advantage will be to the prey. If the prey species is an insect pest (e.g. an aphid) and we spray it with insecticide that also kills its insect predator (e.g. a ladybird), then we might actually see the prey species increase in abundance. *See also* INTRINSIC RATE OF INCREASE; ZERO GROWTH ISOCLINE. [B.S.]

REFERENCES

Elton, C.S. & Nicholson, A.J. (1942) The ten year cycle in numbers of the lynx in Canada. *Journal of Animal Ecology* **11**, 215–244.

Huffaker, C.B. (1958) Experimental studies on predation: dispersion factors and predator–prey oscillations. *Hilgardia* **27**, 343–383.

Lotka, A.J. (1925) *Elements of Physical Biology*. Williams & Wilkins, Baltimore, Md.

Lotka, A.J. (1932) The growth of mixed populations: two species competing for a common food supply. *Journal of the Washington Academy of Science* **22**, 461–469.

Volterra, V. (1926) Variazioni e fluttuazioni del numero d'individui in specie animali conviventi. *Mem. R. Accad. Naz. dei Lincei Ser. VI* **2**. Translation in: Chapman, R.N. (1931) *Animal Ecology*. McGraw-Hill, New York.

Volterra, V. (1931) *Lecons sur la Theorie Mathematique de la Lutte pour la Vie*. Paris.

---

lottery model A model of COMPETITION for primary space, originally proposed to explain the COEXISTENCE of the large numbers of territorial coral reef fish species. The model assumes that space is limiting and that there are many more larvae trying to settle and establish territories than there are spaces. Those larvae that do manage to find suitable territories are assumed to hold them for the rest of their lives, so a territory only becomes available when an individual dies. Larvae are highly dispersive, so that the number of a given species seeking territories in any one area is not thought to be related to the number of resident adults and newly available territories are assumed to be allocated to the first larva to arrive. Analyses show that the lottery model can account for coexistence on a local scale and, if it applies in nature, RESOURCE PARTITIONING between species ought to be weak or absent. Data indicate that there are certainly some habitats where this appears to be the case, but this cannot be used to argue that partitioning of RESOURCES does not occur at larger scales. *See also* RECRUITMENT. [S.J.H.]

REFERENCES

Abrams, P.A. (1984) Recruitment, lotteries, and coexistence in coral reef fish. *American Naturalist* **123**, 44–55.

Sale, P.F. (1982) Stock-recruitment relationships and regional coexistence in a lottery competitive system: a simulation study. *American Naturalist* **120**, 139–159.

lottery system *See* PRIORITY EFFECT.

lower infralittoral zone *See* WITHIN-LAKE HABITAT CLASSIFICATION.

lowest observed effect concentration (LOEC) The lowest concentration of chemical in an ECOTOXICITY test series to cause a statistically significant difference in an observed variable (EFFECT) from control. Clearly, this will depend upon the concentration intervals used in the experimental series, i.e. there could be lower unobserved effect concentrations between the LOEC and the NO OBSERVED EFFECT CONCENTRATION (NOEC). This is why the term 'observed' is used. LOECs should always be reported with the NOEC. [P.C.]

LSD *See* LEAST SIGNIFICANT DIFFERENCE.

LT_{50} *See* MEDIAN TIMES TO EFFECT.

LTER *See* LONG-TERM ECOLOGICAL RESEARCH PROGRAM.

lunar cycles Patterns of movement, FEEDING and reproduction of marine organisms of the seashore are often tightly linked to the tidal regime and hence to lunar cycles. Many shallow-water tropical reef fishes have lunar-periodic reproduction. Examples for the use of lunar cycles for the synchronization of GAMETE release, the SPAWNING of EGGS, or the hatching of larvae are found in brown algae (*Dictyoata dichotoma*), polychaetes (*Platynereis dumerilii*), intertidal spawning fishes, coral reef fishes, isopods (*Excirolana chiltoni*), crinoids, crabs (*Uca, Sesarma*) and marine insects (*Clunio*). Because the relative constellation of Sun, Earth and Moon modulates the amplitudes of the TIDES, marine organisms of the seashore often make use of the biweekly (14.77 days) cycle of the differences between the heights of low and high tide. This semi-lunar (or lunar semi-monthly) cycle between spring and neap tides can be used to synchronize reproductive events with a particular tidal situation. The California grunion, *Leurestes tenuis*, invades beaches on full and new Moon to spawn eggs at the high-tide line. Larvae hatch 2 weeks later on nocturnal maximum-amplitude high tides. In some European populations of the intertidal midge *Clunio*, adults eclose on local afternoon low water on days with spring tides, the only time when the habitat of the larvae, brown algae, falls dry for a few hours. The adults live for about 1 hour during which they mate and deposit their eggs.

The mechanisms by which organisms synchronize their activities to lunar cycles are still under investigation. Several physiological clocks (circadian and those with long-term rhythms) seem to be involved that are set by receptors for LIGHT or for water disturbances (in *Clunio*). [W.W.W.]

lutetium (Lu) *See* LANTHANIDE METALS.

Luvisols *See* SOIL CLASSIFICATION.

l_x In a LIFE TABLE, the proportion of the number of individuals in AGE CLASS or stage zero surviving to the start of age class or stage x. Calculated as a_x/a_0. [A.J.D.]

Lyell, C. *See* CHARACTERS IN ECOLOGY.

lysimeter Experimental system consisting of block of soil to which an aqueous liquid of known quality is added and from which the routes and rates of loss, by percolation, evapotranspiration and RUN-OFF, can be budgeted. The quality of percolate is often tested ecotoxicologically. Amongst other things, it may be used to test the potential of chemicals sprayed on land and LANDFILL materials to contaminate and pollute ground and surface waters. *See also* LEACHATE. [P.C.]

M

maars *See* LAKES.

MAB (Man and the Biosphere Programme) *See* INTERNATIONAL CONSERVATION CONVENTIONS.

MacArthur, R. *See* CHARACTERS IN ECOLOGY.

MacArthur–Wilson model of island biogeography A model proposed to explain the numbers of species observed on islands based on the idea that there is a dynamic balance between rates of IMMIGRATION and extinction (Fig. M1). For immigration, the theory states that the rate of species ESTABLISHMENT for any given island will depend upon the number of species already resident, the size of the regional species pool from which colonists can come and the proximity of the island to colonizing sources. For extinction, rates are thought to be low when only a few species are resident, but to rise disproportionately as the number of residents rises. This is thought to occur because COMPETITIVE EXCLUSION becomes more likely. Also, when the number of residents is high or the island is small, species POPULATION SIZES are likely to be smaller, thereby making chance extinctions more likely.

The theory predicts that (i) the number of species on an island should become roughly constant over time, (ii) at equilibrium there should be a continual TURNOVER of species, (iii) large islands should support more species than small islands and (iv) species numbers should decline with distance from sources of colonists. Unfortunately, only the second prediction can be considered as characteristic of this theory alone, although it has received some empirical support. *See also* ISLAND BIOGEOGRAPHY. [S.J.H.]

REFERENCE

MacArthur, R.H. & Wilson, E.O. (1967) *The Theory of Island Biogeography.* Princeton University Press, Princeton, N.J.

Mackay fate models A series of models that predict where chemicals go in the environment and to what extent, on the basis of EMISSIONS and FUGACITY of chemical. They vary in complexity from a simple equilibrium distribution of a conservative chemical to steady-state and time-varying descriptions of reactive components. *See also* EXPOSURE; FATE. [P.C.]

REFERENCE

Mackay, D. (1994) Fate models. In: *Handbook of Ecotoxicology,* Vol. 2 (ed. P. Calow), pp. 348–367. Blackwell Scientific Publications, Oxford.

macrobenthic fauna *See* BENTHOS.

macrocosm Large, isolated and controlled multispecies system, often outdoors, used for EXPERIMENTAL ECOLOGY and ECOTOXICITY testing. Some attempt has been made at being more precise in specifying size of system for which the term applies, but the separation of macrocosm from MESOCOSM is not sharp and probably not really helpful. *See also* COSMS; MICROCOSM. [P.C.]

macroevolution The EVOLUTION of great phenotypic changes, usually great enough to allocate the changed LINEAGE and its descendants to a distinct genus or higher TAXON. Much of the MODERN SYNTHESIS of the 1930s and 1940s was devoted to demonstrating that the mechanisms of MICROEVOLUTION (i.e. genetic changes within populations and species) can adequately account for macroevolutionary phenomena. More recent debate surrounding the PUNCTUATED EQUILIBRIUM model of evolution has renewed interest in the relationship between microevolutionary and macroevolutionary processes. Macroevolutionary processes include changes in diversity, changes in the frequency distribution of TRAITS, the extinction of major taxa, apparent trends in evolution and the development of phenotypic gaps between taxa. Evidence for macroevolutionary processes comes from the study of comparative anatomy, embryology, TAXONOMY, palaeontology and geographical distributions of species. The evolution of new species is an important macroevolutionary process. As members of a species are defined by their ability to interbreed, study of SPECIATION includes examination of reproductive ISOLATING MECHANISMS between species. Although it is likely that most speciation events occur in allopatric, rather than sympatric, populations, REPRODUCTIVE ISOLATION is critical for maintaining distinct species should they come into contact with each other. Common trends observed in evolutionary lineages include increases in body size for many groups and changes in skeleton, skull and teeth in related vertebrate lineages. Analysis of the FOSSIL RECORD indicates the occurrence of several massive extinctions, the two most recent of which occurred at the end of the Permian and the end of the Cretaceous. The causes for such extinctions remain poorly understood. Another important macroevolutionary process is the COEVOLUTION of interacting species, in which coupled evolutionary changes occur in species that are competitors, mutualists, predator and prey or parasite and host. *See also* ALLOPATRIC SPECIATION. [V.F.]

macrofauna

1 Generally, fauna in SAMPLES that are visible unaided by lens.

2 More specifically, the extraction of animals from marine sediments almost invariably requires the use of sieves and hence the mesh size of the sieve used determines the SIZE of animal retained—

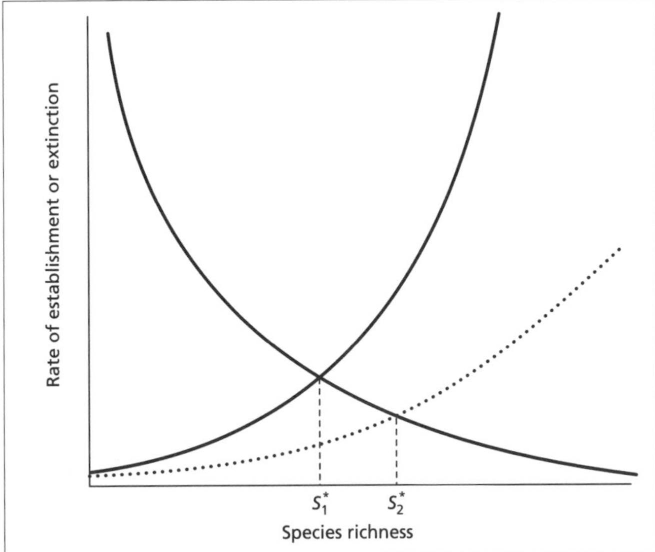

Fig. M1 MacArthur and Wilson's equilibrium theory of island biogeography. The rate of establishment of new species declines exponentially as species richness on the island increases, largely because the pool of potential immigrant species is depleted. The rate of extinction increases as species richness increases because of increased competition, reduced average population sizes and increased impact of natural enemies. The curve rises less steeply on larger islands (dotted line), so they have higher equilibrium species richness (S_2*) than is found on smaller (or more distant) islands (S_1*).

the macrofauna are defined as those benthic animals retained by a 0.5-mm mesh and, by extension, the same term is often used for animals of the same size from other benthic habitats (*cf.* MEIOFAUNA; MICROFAUNA). A different size-based classification system is used for PLANKTON. The SESSILE, and often modularly colonial, macrofauna that predominates on hard marine substrata is composed largely of sponges, cnidarians, bryozoans and tunicates, whilst polychaetes, gastropod and bivalve molluscs, crustaceans, echinoderms and fish are the major groups of the mobile component that is the dominant element on and in soft sediments. In freshwater, oligochaetes and insects occur in addition, whereas polychaetes and echinoderms do not. [R.S.K.B.]

macromutation Genetic alteration that causes a major phenotypic effect and initiates a process of SPECIATION. It was assumed that macromutations caused not only a large phenotypic alteration but also appeared to result in the expression of some degree of phenotypic reorganization, for example generating a homeotic, or atavistic, phenotype. Thus they were postulated to be mutations in regulatory loci. Currently, such a sharp dichotomy between structural genes, which code for proteins, and regulatory genes, which control developmental rates or switches among alternative developmental pathways, is being questioned. HOMEOTIC MUTATIONS would be examples of macromutation; they were often invoked to defend a discontinuous model for EVOLUTION by many authors, mainly in the early days of classical genetics between 1900 and 1940. A major advocate of macromutations was the unorthodox evolutionary biologist R. Goldschmidt, who postulated in 1934 that macromutations affected genes 'that controlled the rates of a chain of reaction during development'. They would generate dramatically distinct phenotypes (HOPEFUL MONSTERS) that constituted the material basis of speciation, in a process that he defined as MACROEVOLUTION, distinct from loci that had mutants with minor effects, micromutations, responsible for the observed patterns of geographic variability and MICROEVOLUTION. *See also* PUNCTUATED EQUILIBRIUM. [P.A.]

macronutrients These are the ELEMENTS (other than CARBON, HYDROGEN and OXYGEN) required by living organisms in substantial amounts: NITROGEN, PHOSPHORUS, POTASSIUM, CALCIUM, MAGNESIUM, SULPHUR and CHLORINE (*cf.* MICRONUTRIENTS). They mainly enter the FOOD CHAIN by plant uptake from the SOIL, as simple cations and anions. In addition, nitrogen enters by biological fixation of molecular nitrogen in the ATMOSPHERE (*see* NITROGEN CYCLE; NITROGEN FIXATION). The main reservoir of nitrogen in soils is in organic matter (>95%), from which it is released by MINERALIZATION as ammonium and microbially oxidized to NITRATE. Both forms are important for plant nutrition. Similarly, mineralization of organic phosphorus (as orthophosphate) and sulphur (as sulphate) releases these two elements in nutrient form. However, one-third to two-thirds of the phosphorus is present in the mineral fraction, together with the bulk of the potassium, calcium and magnesium. The natural supply of these macronutrients is often supplemented in agricultural soils by the addition of mineral FERTILIZERS and manures. The macronutrients are available as ions in the soil solution and as the much larger quantities held on ion-exchange sites on CLAY minerals, hydrous oxides and organic matter. When the equilibrium is disturbed by plant absorption from solution, it is restored by release from the exchange sites. Generally, macronutrient availability is at an optimum in soils with a pH around neutral. Acid conditions reduce mineralization and promote fixation of phosphorus as insoluble inorganic phosphates. Nitrogen and phosphorus are often the limiting factors in determining biological PRODUCTIVITY, in both terrestrial and aquatic environments. Phosphorus commonly controls the development of undesirable ALGAL BLOOMS in freshwater ecosystems; nitrogen is the usual LIMITING FACTOR for marine algae and, as nitrate, often constitutes a POLLUTION problem in waters used as sources of drinking water (*see* NITRATE). [K.A.S.]

REFERENCE

Wild, A. (ed.) (1988) *Russell's Soil Conditions and Plant Growth*, 11th edn. Longman, Harlow.

macroparasite The terms 'macroparasite' and 'MICROPARASITE' have no strict definition, but are used to divide PARASITES into two general categories of population dynamic behaviour requiring different EPIDEMIC model structures. Macroparasites are generally species in which the INTENSITY OF INFECTION is measurable and related to parasite-induced MORBIDITY and MORTALITY. They tend to be large parasites and have no direct reproduction within the primary or definitive HOST (*see* SECONDARY HOST). Macroparasites can be ECTOPARASITES or ENDOPARASITES. They tend to have COMPLEX LIFE CYCLES that include free-living or vector-borne stages and may include one or more secondary hosts. [G.F.M.]

REFERENCE

Anderson, R.M. & May, R.M. (1979) Population biology of infectious diseases: Part 1. *Nature* **280**, 361–367.

macrophagous feeders *See* FEEDING TYPES, CLASSIFICATION.

magnesium (Mg) A light, silver-white METAL (ATOMIC NUMBER = 12; relative ATOMIC MASS = 24.3; melting point = 650°C; boiling point = 1107°C), magnesium constitutes 2% of the Earth's CRUST and is widely distributed as minerals such as dolomite, $MgCa(CO_3)_2$. Magnesium is found in seawater as magnesium chloride ($MgCl_2$) at a concentration of about 1.7 kg m^{-3}, from which it is extracted by electrolysis. A significant proportion (20%) of magnesium is recycled. Magnesium is extensively used in lightweight alloys. Magnesium salts are non-toxic unless administered in very high doses. Magnesium plays a critical role in biological systems, for example the chlorophyll molecule contains a chelated magnesium atom. Magnesium oxide (MgO) FUMES can cause respiratory complaints. [C.R.P.]

major axis method *See* REGRESSION ANALYSIS, LINEAR.

malathion A non-systemic organophosphorous INSECTICIDE and ACARICIDE of diminished TOXICITY to mammals compared with earlier chemicals of the type. Malathion is the generic name for $C_{10}H_{19}O_6PS_2$. It has a wide range of agricultural and horticultural CROP uses and additionally is used to control animal ectoparasites, flies, household insects, human lice and mosquitoes. It is dangerous to the aquatic environment, so considerable care is needed with use near water. It has short to moderate PERSISTENCE in the environment. Excessive doses lead to poisoning and chromosome abnormalities in mammals, including humans. [K.D.]

mallee *See* CHAPARRAL SHRUBLANDS.

Malthusian parameter The rate of increase of a population growing at its STABLE AGE DISTRIBUTION with constant age-specific survivorship (l_x) and reproduction (m_x). The Malthusian parameter r satisfies:

$$1 = \int_0^\infty l(z)m(z)e^{-rz}dz$$

[H.C.]

Man and the Biosphere Programme (MAB) *See* INTERNATIONAL CONSERVATION CONVENTIONS.

Managed Resource Protected Area *See* EXTRACTIVE RESERVES.

mangal *See* MANGROVE.

manganese (Mn) A greyish-white, HEAVY METAL (ATOMIC NUMBER = 25; relative ATOMIC MASS = 54.94; melting point = 1245°C; boiling point = 1962°C), manganese is a trace constituent (950 mg kg^{-1}) of the Earth's CRUST but can occur at concentrations of 35% in manganese nodules on ocean floors. Found primarily in manganese dioxide (MnO_2) ore minerals, manganese is used in steels, alloys and dry cells. An ESSENTIAL ELEMENT for most animals and plants, where it is involved in PHOTOSYNTHESIS, manganese can be toxic to plants at higher levels, especially under conditions such as waterlogging, which can lead to the reduction of insoluble MnO_2 to soluble Mn^{2+}. The environmental chemistry of manganese is dominated by the Mn(II)/Mn(IV) redox couple and associated microbial reduction of manganese oxides (which have high sorptive capacities), thereby releasing adsorbed metal cations. [J.F.]

mangrove The word can be applied either to the plant belonging to a group of trees that dominate tropical intertidal forests or to the forest habitat itself. Some ecologists prefer to use the term 'mangal' specifically for the habitat in which mangrove plants grow rather than for the plants themselves.

Mangrove vegetation grows throughout the tropics in suitable intertidal zones. The mouth of the Ganges, the Mekong Delta, the seaward margins of the Florida Everglades and some estuaries in Papua New Guinea are particularly notable examples. The lack of oxygen in the muddy substrate has led to the evolution of looping aerial roots (pneumatophores, Fig. M2) by some of the dominant species of tree, such as *Rhizophora*. The number of plant species present is generally quite low, but those found are distinctive members restricted to this type of community. Several characteristic animals are also associated with the habitat, from mudskippers to proboscis monkeys.

The habitat is of considerable commercial importance, acting as a physically protected area for fish breeding and also as a coastal defence system in parts of the world often subjected to storms and flooding (e.g. Bangladesh). Despite this great value to human beings, mangal is a habitat that is often threatened by a wide range of coastal developments in the tropics. *See also* MANGROVE COMMUNITIES, ECOLOGY AND CONSERVATION. [P.D.M.]

mangrove communities, ecology and conservation MANGROVES are a group of taxonomically unrelated trees that can tolerate having their roots submerged in seawater for at least part of a tidal cycle. The fauna of mangroves is a unique mixture of terrestrial and marine organisms. Mangrove trees are distinguished by having shallow roots sometimes with aerial extensions, tough succulent leaves, adaptations to regulate salt and viviparous reproduction with water DISPERSAL of seedlings. Of the mangrove species, 90% occur in the Indo-Pacific and the remaining 10% in the Atlantic. The most widespread genus is red mangrove, *Rhizophora*. Mangroves occur in tropical and SUBTROPICAL latitudes, along sheltered

Fig. M2 *Avicennia* with pneumatophores, Natal, South Africa. (Courtesy of R.S.K. Barnes.)

shores, across a wide SALINITY range from estuarine to true marine. Approximately 60–75% of the coastline of the Earth's tropical regions is lined with mangroves. Mangroves show their greatest development in areas of high tidal range. Tidal exchange is important for importing NUTRIENTS, dispersing seedlings, inhibiting establishment of competing freshwater species and reducing salinity in areas of high EVAPORATION. Mangrove swamps tend to accumulate SEDIMENT and are therefore important in the development of new coastal land. Mangroves are also very productive (carbon fixed 370–450 $g\,m^{-2}\,year^{-1}$), although most of the plant material is not eaten directly but decays and enters the DETRITUS food chain. Mangroves show distinct floral and faunal ZONATION patterns, though species interactions in mangrove forests are poorly understood. Mangroves provide nursery areas for penaeid shrimp, spiny lobsters, crabs and various species of fish, many of which are of commercial importance. Mangroves are also important in providing habitat for a number of ENDANGERED SPECIES and for providing a BUFFER ZONE to protect coastal areas from storm-related damage.

The most important causes of large-scale mortality of mangroves are typhoons and hurricanes, which act to uproot trees, dump large amounts of sediment and alter salinity. Mortality may also be caused by the isopod *Sphaeroma terebrans*, which bores into the roots of the mangrove trees, although the importance of this source of mortality remains controversial. Massive mortality of mangroves has also been caused by human activities such as HERBICIDE spraying, filling, dredging, channel digging, woodcutting and clearing. Mangroves may be particularly susceptible to OIL SPILLS because their sheltered nature prevents oil from being dispersed by wave action and because the oil becomes adsorbed by the fine sediment particles that accumulate in mangrove swamps.

Successful conservation efforts include the establishment of 17 reserves in 14 states and Puerto Rico by the National Estuarine Reserve Research System of the USA. These reserves cover 1200 km^2 of estuaries, SALT MARSHES and mangrove forests. Efforts at restoring mangrove forests through the transplant of seedlings have also met with success. *See also* CONSERVATION BIOLOGY. [V.F.]

Manipulation

Manipulation in an evolutionary context is a metaphor for the effects of one entity on the behaviour (or other phenotypic attribute) of another that cause the second entity to respond in the apparent interests of the first. Manipulation can occur at a number of different levels, hence the hedging term 'entity'. Where it occurs at the level of the individual it is an example of what Richard Dawkins terms 'extended phenotypic' effects: adaptive phenotypic consequences of ALLELES outside the physical boundary of their immediate bearer. Manipulation in this sense occurs in many interspecific and intraspecific interactions, from host–parasite and predator–prey associations to relationships between parents and offspring and mated pairs.

Probably the best-known interspecific examples concern various parasites that appear to influence the behaviour or physiology of their hosts for their own purposes. Thus many helminth parasites that require more than one host species to complete their life cycle, such as some flukes, nematodes and acanthocephalans, cause changes in the behaviour of their intermediate hosts (often crustaceans, insects or gastropods) that make them more vulnerable to predation by the definitive host (usually fish, birds or mammals). Other parasites affect the host's metabolism and development, diverting its resources away from reproductive activity and into growth and sustenance of the parasite.

At both interspecific and intraspecific levels, the entire field of COMMUNICATION can be regarded as manipulation. The classical view that communicatory systems reflect mutual COEVOLUTION between the sender and receiver of signals, reducing ambiguity and enhancing the transfer of information, has largely been superseded by the view that they represent exploitation of the receiver in the interests of the sender. Signalling thus becomes a cost-effective means of usurping the muscle-power of other individuals. If the manipulation is not in the interests of the receiver, selection will favour the ability to 'see through it' and a manipulator–'mind-reader' ARMS RACE is likely to ensue. Such arms races could help explain, for example, the evolution of epigamic signals ('sales pressure' by males versus 'sales resistance' by females; *see also* SEXUAL SELECTION; SEXY SONS HYPOTHESIS) and cases of often astonishingly detailed MIMICRY between potential prey species ('false identity' deception by mimics versus discrimination by predators). Anecdotal, and some experimental, evidence suggests that sophisticated deceptive manipulation can pervade socio-sexual relationships, particularly among primates. The complex, context-dependent consequences of acceding to or resisting manipulation in socially structured groups (*see also* DOMINANCE HIERARCHY) are likely to have been among the SELECTION PRESSURES honing what Byrne and Whiten (1988) refer to as the Machiavellian intelligence characterizing our own species and quite possibly several others.

Of course, manipulation is not always to the detriment of the manipulated party. Many species of small bird that regularly feed in flocks emit characteristic ALARM CALLS when a predator is detected. One suggested function of alarm calling is that it manipulates companions to take flight so that the caller (who knows where the predator is) can gain protection while flying to cover. Although the startled companions have less information than the caller and may thus be at greater risk, they are nevertheless better off than they would have been had a warning not been given. However, there is evidence that alarm calls can be used dishonestly, an example being male barn swallows (*Hirundo rustica*) who apparently do it to scare off rivals while they leave their mates unguarded on the nest to forage. An obvious question in the swallows' case is: why don't rival males just ignore the false alarm? The reason is probably because the cost of checking the truthfulness of the call is too high. Predators can attack very swiftly, so a moment's hesitation could be fatal. Under these circumstances, therefore, selection for counterstrategies against manipulation is likely to be weak.

As well as these individual-level, extended phenotypic examples, manipulation can also occur at the cellular level. For example, segregation distorters are alleles that subvert the normal processes of cell division to achieve a disproportionate representation in subsequent cells. Since they act in their own interest and potentially against the interest of alleles elsewhere in the genome, segregation distorters are one form of so-called 'outlaw' genes. Just as manipulation at the individual level can result in selection for counterstrategies, so at the level of the genome it may have contributed to some of the various modifying interactions within and between loci, such as dominance/recessiveness and other forms of EPISTASIS. *See also* BEHAVIOURAL ECOLOGY; SELFISH GENES; SOCIOBIOLOGY. [C.J.B.]

REFERENCES

Byrne, R. & Whiten, A. (1988) Tactical deception in primates. *Behavioural Brain Sciences* **11**, 233–273.

Dawkins, R. (1982) *The Extended Phenotype*. W.H. Freeman, Oxford.

Dawkins, R. & Krebs, J.R. (1984) Animal signals: mind-reading and manipulation. In: *Behavioural Ecology: An Evolutionary Approach* (eds J.R. Krebs & N.B. Davies), 2nd edn, pp. 380–402. Blackwell Scientific Publications, Oxford.

Mann–Whitney *U* test (Wilcoxon rank sum test) This is the non-parametric equivalent of the *z*(*d*) TEST or *t* TEST. However, it is used to compare two medians, rather than two means, and is suitable for UNMATCHED OBSERVATIONS. The test procedure is outlined below. An ecologist has placed baited traps within two woodlands in order to obtain relative estimates of POPULATION DENSITY in a species of fly. The numbers of flies caught at each trap, over a 24-h period, are shown below; one wood had six traps, the other seven traps.

Wood 1	10	0	6	17	14	8	22
Wood 2	18	12	17	26	20	6	

First we must list all observations in ascending order and assign ranks to them. When there are tied ranks, the average rank is calculated. In the table below, the observations from wood 1 are underlined in order to distinguish them.

Observation	0	6	6	8	10	12	14	17	17	18	20	22	26
Rank	1	$2^1/_2$	$2^1/_2$	4	5	6	7	$8^1/_2$	$8^1/_2$	10	11	12	13

Now sum the ranks of each SAMPLE (wood) separately, with R_1 the sum for wood 1 and R_2 the sum for wood 2:

$$R_1 = 1 + 2^1/_2 + 4 + 5 + 7 + 8^1/_2 + 12 = 40$$

$$R_2 = 2^1/_2 + 6 + 8^1/_2 + 10 + 11 + 13 = 51$$

Now calculate the test statistic *U* for each sample, with n_1 and n_2 being the number of observations in each sample:

$$U_1 = n_1 n_2 + \frac{n_2(n_2+1)}{2} - R_2$$

$$U_1 = 12$$

$$U_2 = n_1 n_2 + \frac{n_1(n_1+1)}{2} - R_1$$

$$U_2 = 30$$

The smaller of the two values is taken as the test statistic and compared with the values in the statistical table of *U* for the appropriate values of n_1 and n_2. For $n_1 = 7$ and $n_2 = 6$ the critical value is 6. In contrast to many test statistics, the calculated value must be less than the critical value in the table to be significant. Since in this case the smaller value of *U* (12) is not less than the critical value (6), the two medians are not significantly different. *See also* MEDIAN (SECOND QUARTILE); NON-PARAMETRIC STATISTICS. [B.S.]

manometer An instrument for measuring the PRESSURE of a gas relative to that of the atmosphere or some other reference. The simplest manometer is a U-tube half-filled with a fluid of known density, often MERCURY or coloured water. The difference in pressure between two vessels can be determined by connecting the arms of the U-tube to ports in the vessels and noting the vertical displacement of the manometer fluid. Very sensitive manometers (micromanometers) use an inclined U-tube, in which the fluid level is measured with a microscope. Electronic manometers, sometimes called pressure cells, utilize a metal diaphragm soldered between two chambers; displacements of the diaphragm when the pressure changes are detected as a change in capacitance between the diaphragm and a metal plate. *See also* BAROMETER. [J.G.]

mantle, Earth's The largest layer of the Earth, the mantle has a volume of 898×10^9 km^3. It lies between the CRUST and the core, from 40 to 3500 km depth. The mantle consists of dense rock (4.5 g cm^{-3}). The upper part of the mantle, or athenosphere, is partially molten and flows plastically, whereas the lower mantle is rigid. The LITHOSPHERE includes the crust and the uppermost part of the mantle above the athenosphere. The mantle provides the source for new crust that forms along the MIDOCEANIC RIDGES. Convective movements occurring within the mantle, due to escaping heat from deep within the Earth, result in movement of the lithospheric plates and thus the movement of the continents and formation of ocean BASINS. *See also* PLATE TECTONICS. [V.F.]

maquis *See* CHAPARRAL SHRUBLANDS.

marginal value theorem The marginal value theorem (MVT) arose as a result of a question asked by John Krebs to Rick Charnov, late one evening in Oxford when both were 'under the influence of alcohol'. While the first experimental test was published in 1974, the theorem itself was not published by Charnov until 1976 because earlier attempts to publish it met with sceptical reviews. Now the paper by Charnov is possibly the most cited paper in BEHAVIOURAL ECOLOGY! The MVT predicts how long an animal should remain exploiting a specific source of a resource before leaving to exploit another. Applicable to any patchily distributed RESOURCES such as food, hosts or mates, the MVT is best known in the context of OPTIMAL FORAGING THEORY, where it is concerned with the exploitation of food. Resources are assumed to occur in patches separated by unproductive habitat. The patches are encountered randomly but are not revisited and are depletable within a single visit by the exploiter. Optimization is based on the ratio of gain (resource acquisition) to cost (time spent travelling to the patch and exploiting the patch itself). The optimal exploitation (patch-residence time) maximizes the instantaneous rate of resource acquisition (marginal value). This occurs when, because of depletion (or any other cause of diminishing returns), the instantaneous gain rate falls to the average for all patches within the habitat. Graphically, the solution is obtained by locating the tangent of a straight line, whose slope equals the average instantaneous gain rate among patches, to the cumulative exploitation curve. Assuming equal travel times, more productive patches should be exploited for longer than less productive ones (Fig. M3a). Assuming equal productivities, patches requiring longer travel times, because they are spaced further apart or are

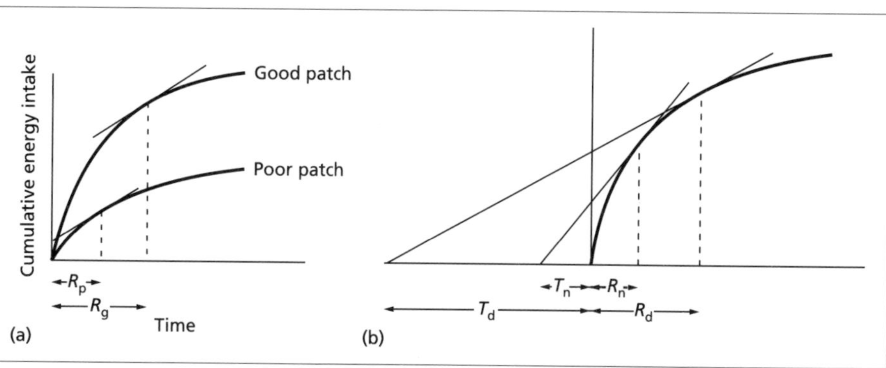

Fig. M3 (a) Patch quality. R_p, optimal residence time in poor patch; R_g, optimal residence time in good patch. The parallel tangential lines have a slope equalling the average rate of energy intake among patches. (b) Travel time. T_n, average travel time in habitat n, where patches are close together; T_d, average travel time in habitat d, where the distance between patches is larger. R_n, optimal residence time in habitat n; R_d, optimal residence time in habitat d. The slopes of the straight tangential lines equal the average rate of energy intake in the two habitats.

otherwise less readily accessible, should be exploited for longer than more readily accessible patches (Fig. M3b).

Much effort has been devoted to behavioural mechanisms, or 'rules of thumb', enabling foragers to identify the optimal exploitation time. One possible mechanism, for example, is the giving-up time rule, whereby the exploiter leaves the patch when the time elapsed since the last encounter with resource reaches a THRESHOLD (the giving-up time). Also of concern are mechanisms of information acquisition, storage and retrieval, necessary for apprising the forager of the status of patches in relation to accessibility and PRODUCTIVITY, so bringing the MVT into a psychological context. By adjustments to assumptions, the MVT may be extended to encompass the influence of COMPETITION and predators. Risk of predation has been accommodated by changing the currency used in the MVT from maximization of energy gained per unit time to maximization of energy gained per unit risk. Here, the optimal decision depends on patch productivity and risks of predation within and between patches. *See also* EVOLUTIONARY OPTIMIZATION; FOOD SELECTION; TRADE-OFF. [R.N.H. & J.V.A.]

REFERENCE

Charnov, E.L. (1976) Optimal foraging: the marginal value theorem. *Theoretical Population Biology* **9**, 129–136.

mariculture The artificial cultivation of marine species. Methods include transplantation of wild stocks to more favourable environments, trapping of wild populations in enclosures without artificial feeding, culturing of eggs followed by the release of young to their natural environment, and more intensive culturing in enclosures with attempts to control food, predators and disease. The most economical species to culture are those that occupy lower trophic levels, feed on naturally available food and produce high BIOMASS under crowded conditions. China and Japan currently produce the highest quantities of mariculture products in the world. *See also* AQUACULTURE. [V.F.]

marine ecology *See* ECOLOGY.

marine snow Amorphous particulate matter derived from living organisms. It is formed from mucus released by large planktonic filter feeders, which produces flakes. The flakes become colonized by bacteria and their flagellate and ciliate grazers. Sinking of this material, especially in dense concentration, looks like a snowfall. [V.F.]

maritime climate A CLIMATE typical of coastal areas and characterized by relatively little seasonal change, with warm moist winters and cool summers. It occurs as a result of prevailing winds flowing from OCEAN to land. [V.F.]

mark–release–recapture methods *See* CAPTURE–RECAPTURE TECHNIQUES; LINCOLN INDEX; SAMPLING METHODOLOGY/DEVICES.

market instruments A way of using consumer choice to effect environmental protection. Information on the relative environmental 'friendliness' of a product is fed into the market-place, either as information (labels) or by differential application of taxes. The latter are referred to as economic instruments. *See also* ECOLABEL; REGULATORY INSTRUMENTS. [P.C.]

marking methods Marked individuals have been used in the study of MIGRATION, of social structure and of POPULATION SIZE, birth and death rates. The essential feature is that the marking process does not change the chance of survival of the individual marked or alter its behaviour. Techniques of marking are limited only by the ingenuity of the investigator. In recent years two trends have assumed increasing importance. The amount of information obtained from marking programmes has increased, as a result of improved methods of statistical analysis and the use of developments such as radiotelemetry and satellite tracking. There has also been increasing concern that the marking process should be ethical and not damage or disfigure the animal. Toe clipping, which was at one time widely used in small mammal studies, is not now recommended.

The following examples give an indication of the range of approaches used. Lepidoptera were marked with paint spots in classical estimations of movement and population size; felt pen is now more likely to be employed. Individual numbers can be written on to hard parts of insects, such as wings; codes may be incised on the shell of snails and the carapace of chelonians. Sometimes the individuals themselves bear easily read unique patterns: the bills of Bewick's swans (*Cygnus bewickii*) have been recorded by drawing and the ventral patterns of crested newts (*Triturus cristata*) photocopied. Small insects have been marked with fluorescent dust and with radioactive isotopes. By far the largest systematic marking programme is bird banding, controlled by national authorities in most countries of the world, in which standard numbered and sometimes colour-coded bands are placed on the legs and a standard set of measurements recorded from the birds before release (Fig. M4). Radios

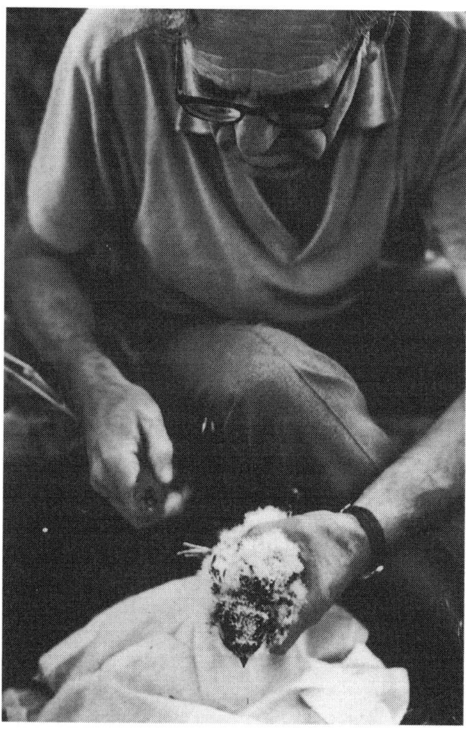

Fig. M4 Population size can be estimated using capture–recapture techniques. A random sample is caught, marked and released back into the population. Another random sample is taken and the population size is estimated from the proportion of this sample which bear a mark. Roughly speaking, the higher the proportion which bear a mark, the smaller the population size. In this photograph, a young bird is being marked by ringing its leg. (Courtesy of M. Campbell.)

have been used for tracking every type of animal from polar bears and geese to sea mammals and snails. Information is provided not only on location but also on temperature, heart rate and other physiological variables and, for sea mammals, depth. Fish tags, originally metal and now plastic, have long been used in studies of fish dynamics; tattooing is an option for mammals. Unique signatures may be obtained for territorial birds, courting amphibia and whales using sound recording. Fingerprinting, a classical method of recording individuals in human populations, has given its name to molecular methods that are increasingly important in ecological studies. *See also* CAPTURE–RECAPTURE TECHNIQUES; LINCOLN INDEX; SAMPLING METHODOLOGY/DEVICES. [L.M.C.]

REFERENCES

Southwood, T.R.E. (1978) *Ecological Methods with Particular Reference to the Study of Insect Populations*. Chapman & Hall, London.

Stonehouse, B. (ed.) (1978) *Animal Marking. Recognition Marking of Animals in Research*. Macmillan, London.

marking, territorial The marking of the boundaries of an exclusive area (a territory) by visual, vocal, olfactory or other signals.

TERRITORIALITY is common during both the reproductive and non-reproductive season among invertebrates and vertebrates. The boundaries of territories are usually delimited to potential competitors by territorial marking. Territorial marking may provide information on individual identity, sex, age and origin with respect to social group of territory owners. Territory owners generally intrude on the territories of neighbours at a low rate and respond less aggressively to the territorial markings of known neighbours compared with the markings of complete strangers. Responses to territorial marking thus tend to become reduced after continuous exposure. Examples of territorial marking include: (i) vocalizations, such as song in birds and calls in territorial primates like gibbons; (ii) olfactory cues, such as pheromones in many rodents and carnivores; and (iii) the presentation of visual DISPLAYS, like conspicuous colour patterns of the territory owner or faecal piles as in many large mammals.

The function of territorial marking is obviously to prevent intrusions by competitors. Experiments using playback of bird song have shown that vocalizations from a loudspeaker are sufficient to prevent or reduce the rate of intrusions. The intrusion rate is considerably lower in the presence of the owner and thus the territorial marker, in terms of the vocal signal, is insufficient to completely prevent intrusions. [A.P.M.]

Markov model/process A Markov process is a stochastic process where future development is determined only by the present state of the system and is independent of the modes of development. The use of Markov models has a long tradition in ecology and has played an important part in recent research using CELLULAR AUTOMATA and individual-based modelling techniques. *See also* TRANSITION MATRIX MODEL OF SUCCESSION. [S.J.H.]

marsh This term causes confusion in wetland nomenclature because it is used in very different senses in different parts of the world. In North America, the expression is equivalent to the European term 'SWAMP' and is discussed under that entry in this encyclopedia. In Europe, 'marsh' is used to refer to ecosystems with periodically waterlogged mineral soils but with little or no PEAT formation. The soils often take the form of groundwater gleys and the vegetation, which is usually herbaceous, is characteristically capable of tolerating waterlogging.

Marshes are particularly associated with habitats that are grazed by large herbivores and these not only serve to limit scrub invasion but also diversify the habitat in its microtopographic profile. Water erosion of the surface tends to amplify the development of hummocks and hollows. In many temperate regions, the intensity of GRAZING or the soil instability of marshes has kept them relatively free from forest cover and they are often associated with RELICT assemblages of plants and invertebrate animals from ice-age times. Their faunal and floristic diversity gives this habitat high conservation interest but its management requires the maintenance and control of grazing pressure. [P.D.M.]

mass balance Accumulation of total mass in a system is equal to the difference between input and output. May also apply to particular substance, and will apply to elements. Based on the principle of conservation of mass. [P.C.]

mass extinction Mass extinctions are substantial BIODIVERSITY losses that are global in extent, taxonomically broad and rapid relative to the average persistence times of species in the taxa involved. Five major mass extinction events have been identified for marine invertebrates, the group for which the FOSSIL RECORD is relatively extensive, reliable and well studied. Terrestrial vertebrates and plants exhibit perturbations that correlate approximately with the major mass extinctions detected in the oceans. The magnitude and timing of land-based and marine extinctions is still controversial. The five mass extinctions differed from one another and each one affected different taxonomic groups differently.

The most severe extinction event was 245 million years ago (during the late Permian period), at which time about 50% of the families of marine animals in the fossil record were lost (the number of genera declined by about 80% and the number of species by 95%). At the same time, the number of fish families declined by 44% and the number of tetrapod families by 58%. The late Permian event is generally accepted to have occurred rapidly (over 5–8 million years) and losses were experienced globally. The losses appear to have been associated with global physical changes including CLIMATE CHANGE and volcanic activity (Fig. M5 & Table M1).

The most recent mass extinction, which marks the boundary between the Cretaceous and Tertiary periods 65 million years ago, is the best documented. There is some evidence that it was associated with the impact of a meteorite on the Earth. The late Cretaceous event resulted in a decline of about 15% of marine families in the fossil record. A total of 40% of tetrapod families and perhaps 75% of plant species in the fossil record were lost at that time. The loss of plant species was unusually high compared to other mass-extinction events.

A number of other important extinction events have been recorded but they do not have the severity, geographic scale or taxonomic breadth of the five large events. Recovery of global DIVERSITY to the levels before mass extinction has varied and has taken between 10 and 100 million years. The rules that govern the likelihood of extinction of species during background times, such as POPULATION SIZE and RANGE, appear to be unimportant in determining likelihood of extinction during mass-extinction events.

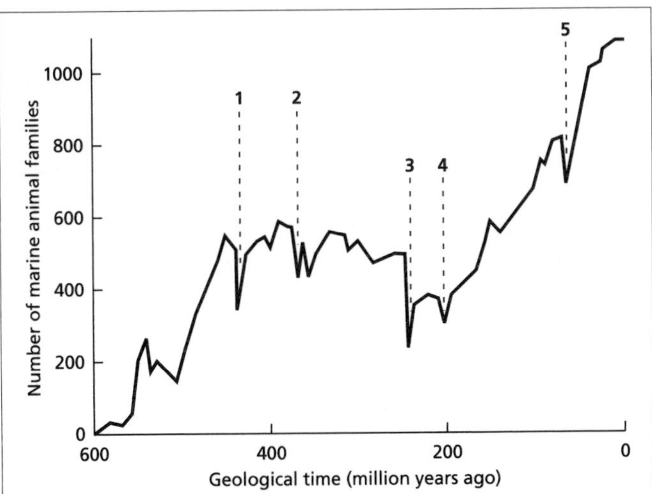

Fig. M5 The extinction rates in marine animal families over the last 600 million years including the five major mass extinction events. (After Sepkowski, J.J. (1993) Ten years in the library: new data confirm paleontological patterns. *Paleobiology* **19**, 43–51.)

Table M1 Extinction rates among marine invertebrates during the five major mass extinction events. The numbers for the mass extinction events refer to the events labelled in Fig. M5. Species-level estimates were inferred from extinctions of genera. (After Jablonski, D. (1995) Extinctions in the fossil record. In: *Extinction Rates* (eds J.H. Lawton & R.M. May), pp. 25–44. Oxford University Press, Oxford; by permission of Oxford University Press.)

Mass extinction event	Families lost (%)	Genera lost (%)	Species lost (%)
1 (439 mya)	26	60	85
2 (367 mya)	22	57	83
3 (245 mya)	51	82	95
4 (208 mya)	22	53	80
5 (65 mya)	16	47	76

mya, million years ago.

See also EXTINCTION CRISIS, CURRENT GLOBAL; EXTINCTION IN THE FOSSIL RECORD. [M.A.B.]

REFERENCES

Jablonski, D. (1995) Extinctions in the fossil record. In: *Extinction Rates* (eds J.H. Lawton & R.M. May), pp. 25–44. Oxford University Press, Oxford.

Jenkins, M. (1992) Species extinction. In: *Global Diversity: Status of the Earth's Living Resources* (ed. World Conservation Monitoring Centre), pp. 192–233. Chapman & Hall, London.

May, R.M., Lawton, J.H. & Stork, N.E. (1995) Assessing extinction rates. In: *Extinction Rates* (eds J.H. Lawton & R.M. May), pp. 1–24. Oxford University Press, Oxford.

Sepkowski, J.J. (1993) Ten years in the library: new data confirm paleontological patterns. *Paleobiology* **19**, 43–51.

mass number Mass number (A) is the integer nearest to the ATOMIC MASS of an element (equal to the number of protons plus neutrons). [J.G.]

mass spectrometer Mass spectrometry is used for the identification of chemical ELEMENTS and compounds on the basis of their charge to mass ($e:m$) ratio. Gaseous atoms or molecules of the analyte are bombarded by high-energy electrons to produce positively charged ions. These ions are accelerated and focused into a beam by an electromagnetic field. The ions pass through a further magnetic field that deflects their path according to their $e:m$ ratio. The beam impacts on a photographic detector plate allowing accurate and precise identification of ions with different $e:m$ ratios: ions with higher ratios are deflected further. A wide range of $e:m$ ratios can be measured by altering the strength of the deflecting magnetic field. Mass spectra are produced by varying the magnetic field so that ions of progressively higher masses strike the detector. The electron ionization step is a high-energy process that often leads to fragmentation of molecules, which means that compounds are usually identified by examination of their fragment ions. An example of the use of mass spectrometry in ecology is in the measurement of the isotopic abundance of soil gases. The measurement of ^{15}N in nitrous oxide (N_2O) and ^{13}C in METHANE (CH_4) and CARBON DIOXIDE (CO_2) can be used to determine the sources and metabolic pathways by which these gases are formed. [I.J.B.]

Massenerhebung *See* ALTITUDINAL ZONATION.

master horizons *See* SOIL PROFILE.

masticators *See* FEEDING TYPES, CLASSIFICATION.

MATC *See* MAXIMUM ACCEPTABLE TOXICANT CONCENTRATION.

matched observations Pairs of observations made on the same experimental or observed unit. For example, if we measured egg production (per female) in a group of fruit flies on one day (under one set of conditions) and then measured daily egg production in the same group of females on another day (under perhaps another set of conditions), the data on egg production would be paired because two measurements had been made on each female. *Cf.* UNMATCHED OBSERVATIONS. [B.S.]

mate-defence polygyny A MATING SYSTEM where individuals compete for members of the opposite sex without using fixed territories. Typically, this occurs where the male provides little or no PARENTAL CARE, and breeding resources are not sufficiently clumped in space or time to select for male TERRITORIALITY. Instead, males may move with groups of females in home ranges that overlap with those of other males. Examples include red deer (*Cervus elaphus*), gelada baboons (*Theropithecus gelada*) and wild turkeys (*Meleagris galapavo*). *Cf.* RESOURCE-DEFENCE POLYGYNY. [J.D.R.]

mate-guarding The close following of females by their mates during the female's period of fertility. The function of mate-guarding is to protect the PATERNITY of the guarding male. Mate-guarding occurs in a variety of organisms including molluscs, crustaceans, insects, fish, amphibians, reptiles, birds, mammals and humans.

Contact guarding is an extremely effective method of preventing females from copulating with another male. After copulation, males of dung flies (*Scatophaga stercoraria*) remain on the female's back, guarding her from other males, while she lays her eggs. This behaviour ensures that the guarding male's sperm are used to fertilize 80% of the eggs in the current batch, the remainder being fertilized using previously stored sperm carried by the female. In several species of damselfly the male retains his grip on the female's thorax after copulation until she has finished laying her eggs.

In several species of birds and mammals, males spend a large fraction of their time close to their mates during the period of fertility. In the Idaho ground squirrel (*Spermophilus brunneus*), CONSPECIFIC males that approach a consort pair are attacked by the guarding male. Males that can repel successfully the intruders sire all offspring in the litters of the female guarded.

The STRATEGY of male behaviour and the intensity of mate-guarding can vary between individuals of a population, between populations and between species depending on the quality of females and males, the distribution of mates in space and time and the effects of intrasexual and SPERM COMPETITION. Males of the soapberry bug (*Jadera haemotoloma*) guard females more frequently when they are exposed to a high risk of sperm competition (2.7 males per female) compared with a low risk of sperm competition (1.1 males per female). In the barn swallow (*Hirundo rustica*), mate-guarding intensity increases with increasing number of close neighbours in the colony.

Benefits of mate-guarding may be counterbalanced by costs for the male, which can be reduced food intake, reduced number of copulation opportunities with other females and increased predation risk. *See also* MATING SYSTEMS. [B.B.]

REFERENCES
Birkhead, T.R. & Møller, A.P. (1992) *Sperm Competition in Birds*. Academic Press, London.
Thornhill, R. & Alcock, J. (1983) *The Evolution of Insect Mating Systems*. Harvard University Press, Cambridge, Mass.

mate recognition Organisms that form a pair bond during the reproductive period or even for life, such as many birds, have the ability to recognize their partners as individuals. This is mate recognition and should be differentiated from SPECIFIC-MATE RECOGNITION, where an appropriate mate is recognized.

Because each individual is different to some degree, mate recognition involves learning. [H.E.H.P.]

Maternal effect

The PHENOTYPIC VALUE of an individual for a certain trait might be influenced by its mother's GENOTYPE or PHENOTYPE. If so, the QUANTITATIVE TRAIT shows a maternal effect, rather like a phenotypic time-lag.

The simplest maternal effect is when the mother constitutes part of the environment of the offspring. If so, the ENVIRONMENTAL VARIANCE V_E in the offspring population is divided into a part $V_{E,common}$ that the offspring of one mother have in common and a part $V_{E,within}$ that is specific to each offspring individual.

Maternal effects *sensu stricto* can be divided into environmental, genotypic and phenotypic effects. An environmental maternal effect occurs when the ENVIRONMENTAL DEVIATION in the mother's phenotypic value influences the phenotypic value of her offspring; even without genetic variation in the population, a maternal effect exists. A genotypic maternal effect occurs when the mother's genotype rather than the phenotype determines the trait value in the offspring.

In QUANTITATIVE GENETICS, phenotypic maternal effects are regularly observed. In a phenotypic maternal effect, the mother's phenotype partly determines the phenotype of the offspring, i.e. a large mother capable of feeding her offspring well might have large offspring capable of feeding their offspring well, just because the offspring themselves are in good condition. This is an example of a positive maternal effect. An empirical positive maternal effect might be exemplified by hind condition in the red deer (*Cervus elaphus*) on the island of Rhum (Clutton-Brock *et al.* 1984). Negative maternal effects are possible too: for instance, a large mother might have many offspring that heavily compete amongst themselves on a fixed food supply and grow into small adults; but a small daughter might have few offspring that scarcely need compete on a similar fixed food supply and consequently grow into large adults. This example refers to weaning mass in pigs (Falconer 1989). A negative maternal effect leads to a negative mother–offspring regression if it is large enough. An ecological example of a negative maternal effect is given by Janssen *et al.* (1988). Schluter and Gustafsson (1993) showed in a field study of the collared flycatcher (*Ficedula albicollis*) the simultaneous presence of large positive and negative maternal effects on adult condition and CLUTCH SIZE.

Formally, the phenotypic value P of an individual is written as the sum of the individual's own GENOTYPIC VALUE G, its environmental deviation E and the maternal effect M:

$$P = G + M + E$$

The maternal effect may depend upon the mother's genotype only, $M(G')$; on the mother's environmental deviation only, $M(E')$; or on the mother's phenotypic value, i.e. including the maternal effect in the mother's phenotype, $M(P') = M(G' + M' + E')$. If the latter, the maternal effect forms an infinite series of influences of all mother's mother's phenotypes.

The parent–offspring COVARIANCE estimates $^1/_2$ times the ADDITIVE GENETIC VARIANCE, $^1/_2V_A$, and the covariance half-sibs estimates $^1/_4$ times the additive genetic variance, $^1/_4V_A$, in the absence of a maternal effect. Assuming the maternal effect is a linear function, with slope m, of the maternal genotypic value, environmental deviation or phenotypic value, the PHENOTYPIC VARIANCE, parent–offspring covariances and the covariance half-sibs in the presence of a maternal effect are given in Table M2. Genotypic and environmental maternal effects do not add to the phenotypic maternal effect, due to the recurrent nature of the phenotypic maternal effect. If $-1 < {}^1/_2m < 1$, the infinite series of maternal effects in the case of the phenotypic maternal effect can be summed. The father–offspring regression shows a maternal effect originating with the father's mother. Both mother–offspring and father–offspring regression can be negative due to a negative maternal effect.

Selection may give unexpected results if a maternal effect, especially a negative maternal effect, is present. Suppose that FITNESS of an individual with phenotypic value P is given by $w = \beta P$. On the basis of the mother–offspring regression, the expected SELECTION RESPONSE will be $R = (1/\bar{w})\beta\ \mathrm{cov}(P, P'_{♀})$. With a negative maternal effect m, the mother–offspring covariance $\mathrm{cov}(P, P'_{♀})$ might well be negative, and selecting large mothers then implies obtaining small offspring. Kirkpatrick and Lande (1989) showed in a one-locus selection model that the maternal effects lead to persistent changes in the phenotypic mean of the population even after selection stops. *See also* FREQUENCY-DEPENDENT SELECTION. [G.D.J.]

REFERENCES

Clutton-Brock, T.H., Albon, S.D. & Guinness, F.E. (1984) Maternal dominance, breeding success and birth sex ratios in red deer. *Nature* **308**, 358–360.

Falconer, D.S. (1989) *Introduction to Quantitative Genetics*. Longman, Harlow.

Janssen, G.M., de Jong, G., Joosse, E.N.G. & Scharloo, W. (1988) A negative maternal effect in springtails. *Evolution* **42**, 828–834.

Kirkpatrick, M. & Lande, R. (1989) The evolution of maternal characters. *Evolution* **43**, 485–503.

Schluter, D. & Gustafsson, L. (1993) Maternal inheritance of condition and clutch size in the collared flycatcher. *Evolution* **47**, 658–667.

mathematical models Mathematical models have been used in ecology primarily to identify the essential elements of complex systems, to structure problems and to clarify key questions in an attempt to elucidate general rules governing ecological systems. Models may be used primarily for descriptive (predictive or empirical) purposes or for increasing insight (mechanistic). Mechanistic models attempt to describe observations in terms of fundamental postulates about the underlying biological processes. By contrast, empirical models provide quantitative descriptions of the patterns in the observations without attempting to describe the underlying processes or mechanisms involved. The building of mathematical models involves the idealization of a phenomenon by making simplifying assumptions concerning the nature of the phenomenon and by assuming that certain physical influences on the problem will have such a small effect that they can be neglected. Then certain mathematically defined laws are imposed on the idealized problem so that unknown quantities can be calculated or estimated. The solution to the model is obtained by applying mathematical reasoning and is used to draw conclusions for a real-world situation. The predictions of the model are compared with observation of real-world phenomena to determine the model's validity. For predictive models, an important step is VALIDATION, i.e. determining if and under what conditions the model is a useful predictor. For mechanistic models the goal is not validation (sometimes called corroboration) but rather refutation, i.e. attempting to reject the model (or parts of it) in favour of alternative hypotheses. Models that have been widely and successfully employed in ecology include: models

Table M2 Phenotypic variance and covariances between relatives including maternal effects. The maternal effect is written as a function of the maternal environmental deviation, the maternal genotypic value or the maternal phenotypic value.

	$P = G + M(E') + E$ $G = A + D$ $M(E') = mE'$	$P = G + M(G') + E$ $G = A + D$ $M(G') = mG'$	$P = G + M(P') + E$ $G = A + D$ $M(P') = mP'$
V_P	$V_A + V_D + (1 + m^2)V_E$	$(1 + {}^1/_2 m)^2 V_A + V_D + V_E$	$V_A + V_D + m^2V_{P'} + mV_A/(2 - m) + V_E$
$\mathrm{cov}(P, P'_{♀})$	${}^1/_2V_A + mV_E$	${}^1/_2V_A + {}^1/_4 mV_A + {}^1/_2 m^2V_A + mV_G$ $= {}^1/_2 V_A(1 + {}^1/_2 m)(1 + 2m) + mV_D$	${}^1/_2 V_A(1 + {}^1/_2 m + {}^1/_4 m^2 + \cdots) + mV_P$ $= V_A/(2 - m) + mV_P$
$\mathrm{cov}(P, P'_{♂})$	${}^1/_2V_A$	${}^1/_2V_A + {}^1/_4 mV_A$	${}^1/_2V_A + {}^1/_4 mV_A$
$\mathrm{cov}_{HS,♀}$	${}^1/_4V_A + m^2V_E$	${}^1/_4V_A + mV_A + m^2V_G$	${}^1/_4V_A + 2mV_A/(2 - m) + m^2V_P$
$\mathrm{cov}_{HS,♂}$	${}^1/_4V_A$	${}^1/_4V_A$	${}^1/_4V_A$

of community complexity–stability (e.g. May); life-history models (e.g. Stearns); demographic models (e.g. Caswell); models of population growth (e.g. Lotka–Volterra); foraging models (e.g. MacArthur and Pianka); and systems models (e.g. Odum). *See also* AGGREGATION MODEL OF COMPETITION; BROKEN STICK MODEL; CASCADE MODEL OF FOOD-WEB STRUCTURE; DEMPSTER MODEL; DETERMINISTIC MODELS, COMMUNITIES; DETERMINISTIC MODELS, POPULATIONS; DYNAMIC POOL MODELS; EPIDEMIC MODELS; FUGACITY MODELS; GENERAL CIRCULATION MODEL; GENERALIZED LINEAR MODELS; HASSELL MODEL AND HASSELL–COMINS MODEL; HOTSHOT MODEL; 'HUMP-BACKED' MODEL OF DIVERSITY; LEVENE MODEL; LINEAR MODELS; LOGISTIC MODEL; LOTKA–VOLTERRA MODEL; LOTTERY MODEL; MACARTHUR–WILSON MODEL OF ISLAND BIOGEOGRAPHY; MACKAY FATE MODELS; MARKOV MODEL/PROCESS; MATRIX MODELS; MENGE–SUTHERLAND MODEL; MODELS OF EXPONENTIAL GROWTH; MODELS OF INTERSPECIFIC COMPETITION; MODELS OF INTRASPECIFIC COMPETITION; MODELS OF MUTUALISM; MODELS OF PARASITOID–HOST INTERACTION; MODELS OF PREDATOR–PREY INTERACTION; NON-LINEAR MODELS; NULL MODEL; OPTIMAL DIET MODEL; OPTIMALITY MODELS; RANDOM WALK MODEL; RANK–ABUNDANCE MODELS; REPRODUCTIVE EFFORT MODEL; ROBUSTNESS OF MODELS; SPECIES-ABUNDANCE MODELS; STOCHASTIC MODELS; TILMAN'S MODEL OF RESOURCE UTILIZATION; TRANSITION MATRIX MODEL OF SUCCESSION (MARKOV MODEL). [V.F.]

REFERENCES

Caswell, H. (1989) *Matrix Population Models*. Sinauer Associates, Sunderland, Mass.

Lotka, A.J. (1932) The growth of mixed populations: two species competing for a common food supply. *Journal of the Washington Academy of Sciences* **22**, 461–469.

MacArthur, R.H. & Pianka, E.R. (1966) On optimal use of a patchy environment. *American Naturalist* **100**, 603–609.

May, R.M. (1973) *Stability and Complexity in Model Ecosystems*. Princeton University Press, Princeton, N.J.

Odum, E.P. (1969) The strategy of ecosystem development. *Science* **164**, 262–270.

Stearns, S.C. (1992) *The Evolution of Life Histories*. Oxford University Press, Oxford.

Volterra, V. (1926) Variations and fluctuations of the numbers of individuals in animal species living together. Reprinted in: Chapman, R.N. (1931) *Animal Ecology*. McGraw Hill, New York.

Mating systems

The numbers of mates obtained by males and females and the nature of reproductive relationships between them in species or populations of animals. Males may practise MONOGAMY (one mate, e.g. gibbons) or POLYGYNY (many mates, e.g. sage grouse); females too may be monogamous (one mate) or engage in POLYANDRY (many mates, e.g. marmosets and tamarins). Bonds between males and females may be lifelong (e.g. swans), seasonal (e.g. mallards) or non-existent (e.g. fallow deer). The majority of birds are monogamous, the majority of mammals polygynous and many fish and invertebrates promiscuous (many reproductive partners but no social bonds).

Where one of the two sexes expends much more energy in producing offspring than the other, as in most mammals, the mating system is quite readily understood as the outcome of members of the sex that expends less (usually males) competing for mating access to the sex that expends more (usually females). Where female home ranges are small enough to be defended by males or females are predictably drawn to RESOURCES, males frequently defend resource-based territories, generating RESOURCE-DEFENCE POLYGYNY if females are in groups (e.g. impala) or monogamy if females are solitary (e.g. dik dik). Otherwise males may defend cohesive groups of females (harems, e.g. red deer). Where resources or female groups are defensible but the female groups are very large, males may band together to share the defence of these resources or groups (e.g. baboons). If neither resources nor female groups are defensible, males may roam in search of females (e.g. polar bears) or advertise for them from display territories clustered on traditional mating grounds (leks, e.g. Uganda kob).

Females in species exhibiting resource-defence or HAREM polygyny may have little opportunity to choose mates or to mate with multiple males without leaving their home ranges or social groups, but those in LEK-breeding species or species with multi-male groups may have ample opportunity for mate choice or multiple mating. In many species of birds and some fish, females have been shown to choose mates based on display characters such as long tails but in mammals there is still little strong evidence for mate choice. In some species where they have the opportunity, females mate with several males during a single period of receptivity (Uganda kob, Malawi cichlid fish) but in others they mate only once (e.g. fallow deer); the causes of this variation are not known.

Where both sexes expend energy in the provisioning or care of offspring, as in most birds, the mating system may be more complicated. In the small sparrow-like dunnock, for example, both sexes care for the young and both males and females seek to acquire more than one partner. The outcome may be monogamy, polygyny, polyandry or polygynandry. Where males first establish resource-based territories and then females choose where to nest, the extent of polygyny may depend on whether a late-arriving female would be better off being the second partner of a male with a high-quality territory or the first partner of a male with a poorer territory. The formal model of this situation is termed the 'polygyny threshold model'. In a minority of bird species, including most raptors, no young can survive unless both sexes care for them full time; the resulting mating system has been termed 'OBLIGATE monogamy'.

Though 'mating system' has often been treated as synonymous with 'social organization', recent studies of birds making use of DNA fingerprinting techniques have shown that the actual patterns of FERTILIZATION or PATERNITY in a population may differ considerably from its SOCIAL ORGANIZATION. In many socially monogamous species, such as moorhens, fingerprinting studies have been used to determine the biological father of chicks, demonstrating that both males and females frequently sneak out of their territories to copulate with neighbours. As a result of these EXTRA-PAIR COPULATIONS (EPCs), the fertilization system of moorhens may be closer to polygyny or polygynandry than to monogamy.

The mating system of a species has profound implications for other aspects of that species' biology. For example, in polygynous species, SELECTION PRESSURES on males and females differ markedly, giving rise to differences in body size or weaponry (SEXUAL DIMORPHISM), behaviour and MORTALITY, whereas selection generally exerts similar pressures on the two sexes in monogamous species. In species where females have the opportunity to mate with several partners during one receptive period, males are commonly found to have relatively large testes, presumably because individuals able to produce larger quantities of sperm in the past have been more likely to fertilize females who have mated with several males. The mating system also influences the DISPERSAL of juvenile males and females, with implications for conservation and MACROEVOLUTION. *See also* LEKKING; PARENTAL INVESTMENT; POLYGAMY; PROMISCUITY; PSEUDOPREGNANCY. [J.C.D.]

REFERENCES

Clutton-Brock, T.H. (1989) Mammalian mating systems. *Proceedings of the Royal Society of London, Series B* **236**, 339–372.

Davies, N.B. (1991) Mating systems. In: *Behavioural Ecology: An Evolutionary Approach* (eds J.R. Krebs & N.B. Davies), 3rd edn, pp. 263–300. Blackwell Scientific Publications, Oxford.

Emlen, S.T. & Oring, L.W. (1977) Ecology, sexual selection, and the evolution of mating systems. *Science* **197**, 215–223.

---

matorral *See* CHAPARRAL SHRUBLANDS.

matrix models These are potentially a very powerful tool for the investigation of POPULATION DYNAMICS that take into account the age or stage structure of the population in question. The earliest uses of matrix algebra in the study of population dynamics come from the use of life tables, where the survival of different ages or stages in the life history of an organism are calculated separately. The simplest method is to divide age, which is a continuous variable (usually denoted x), into discrete AGE CLASSES all of the same duration, to determine survival from one age class to the next and to determine reproductive output of individuals of each age. This method is often known as the Leslie matrix. Matrix models can be used to model more than age-structured populations. Common extensions are for stage- or size-structured populations. It is also possible to incorporate space into the matrix approach to produce a spatially extended, age- or stage-structured matrix model (stage-structured models are also known as Lefkovitch models). *See also* AGE STRUCTURE; LIFE TABLE; POPULATION PROJECTION MATRIX. [C.D.]

REFERENCE

Caswell, H. (1989) *Matrix Population Models*. Sinauer Associates, Sunderland, Mass.

matrix population model *See* POPULATION PROJECTION MATRIX.

matter flux Matter flux is one of two fundamental processes that occur in the BIOSPHERE, the other being energy flux. Alternative descriptions for matter FLUXES include BIOGEOCHEMICAL CYCLES and nutrient transfers. Matter fluxes occur continually as elements and compounds are exchanged between the abiotic (non-living) and biotic (living or dead organic matter) components of the biosphere and the ATMOSPHERE. They can be classified as gaseous, sedimentary or hydrological depending on the dominant milieu in which the exchange occurs. For example, CARBON, OXYGEN and NITROGEN all have important atmospheric pools and many exchanges of these elements occur either in the atmosphere or at the boundary between the atmosphere and the biosphere. Consequently, they can be described as gaseous biogeochemical cycles. Metal and PHOSPHORUS fluxes occur predominantly in the biosphere with little if any involvement of the atmosphere and comprise exchanges between rocks, soils and biota. They are sedimentary biogeochemical cycles, whilst fluxes of HYDROGEN and oxygen are components of the hydrological cycle (*see* WATER (HYDROLOGICAL) CYCLE). The flux of matter between pools, i.e. between the atmosphere, the biosphere and the LITHOSPHERE (the Earth's CRUST), involves the use of energy. Thus, energy fluxes and matter fluxes are closely associated. The fundamental difference between them is that energy is degraded as entropy increases, i.e. the energy, usually as heat energy, is dispersed and cannot readily be reused (a law of thermodynamics) whilst matter retains its integrity. Many of the world's POLLUTION problems are caused by direct and inadvertent human modification of matter fluxes. *See also* ENERGY BUDGET. [A.M.M.]

REFERENCE

Butcher, S.S., Charlson, R.J., Orians, G.H. & Wolfe, G.V. (1992) *Global Biogeochemical Cycles*. Academic Press, London.

maximal food chain FOOD CHAINS, a series of organisms linked together by FEEDING relationships, are limited in length. [M.H.]

maximum acceptable toxicant concentration (MATC) Hypothetical THRESHOLD, somewhere between NO OBSERVED EFFECT CONCENTRATION and LOWEST OBSERVED EFFECT CONCENTRATION. [P.C.]

maximum likelihood methods Maximum likelihood is the most general method of making a statistical estimate; it is an alternative to the LEAST SQUARES METHOD. Given a model (M) and some data (D), the likelihood of a given value of an estimated parameter (V) is the PROBABILITY of the data given the model and that value of the parameter, $P(D\!: V,M)$. The probability of the data is thus considered as a function of the parameter. The probability of all possible datasets must add up to 1, but when the data are held constant and the parameter is varied the different values of $P(D\!: V,M)$ do not have to add to 1; they are called likelihoods to distinguish them from probabilities. The maximum likelihood method chooses that value V that maximizes the probability that the observed data would have occurred.

Maximum likelihood estimators have several desirable properties. Under general conditions they are consistent, which means that the probability that the estimate converges to the true value converges to 1 as the sample size goes to infinity. They are also asymptotically efficient, which means that the variance in the maximum likelihood estimate, as the sample size goes to infinity, is generally less than the variance in estimates made by other means, including least squares. Maximum likelihood estimates are often biased, which means that they differ systematically from the true value at small to intermediate sample sizes, although the bias can often be removed by a simple adjustment. Their main disadvantage lies in their calculation, which involves summing large numbers of logarithms. For moderately complex models, their calculation still takes a long time despite impressive advances in computer speed. *See also* MATHEMATICAL MODELS; STATISTICS. [S.C.S.]

maximum permissible level Maximum permissible concentrations of CONTAMINANTS in environmental materials are normally set by national regulatory bodies at levels that are estimated to give an acceptable degree of risk to public health. In addition to concentration limits, annual limits on intake (ALIs) may be defined for contaminants. Releases of RADIOACTIVE WASTE are also subject to the ALARA principle, which is that human RADIATION exposure or RADIONUCLIDE intake resulting from any waste DISCHARGE must not only be within limits but must also be AS LOW AS REASONABLY ACHIEVABLE (ALARA). The corresponding philosophy for non-radioactive pollutants is that waste releases must be within limits and their effects must be minimized by employing the BEST AVAILABLE TECHNIQUES NOT ENTAILING EXCESSIVE COSTS (BATNEEC). [A.B.M.]

maximum sustainable yield This key concept is largely theoretical but has a sensible underlying idea. A population's ability to sustain a harvest is limited: that limit is the maximum sustainable yield (MSY). The key to this process is DENSITY DEPENDENCE, i.e. as a population is reduced its demographic performance improves and the surplus accrued by that improved performance (more births, less deaths, more growth) can be used to generate a SUSTAINABLE YIELD. Clearly there are limits to the density-dependent response that can be achieved and such a response will vary with individual species. *See also* FISHERIES, CONSERVATION AND MANAGEMENT; HARVESTING. [J.B.]

McIntosh index *See* DIVERSITY INDICES.

McKendrick–von Foerster equation *See* SIZE DISTRIBUTIONS.

Meadow Chernozemic soils *See* SOIL CLASSIFICATION.

Meadow steppe chestnut *See* SOIL CLASSIFICATION.

mean This is the most common MEASURE OF CENTRAL TENDENCY or average, familiar to most people. It is calculated by summing all the individual observations in a SAMPLE and dividing this sum by the number of observations in the sample. The formula for calculating the mean ($\bar{x}$ or x-bar) is therefore:

$$\bar{x} = \frac{\sum x}{n}$$

where x is each observation in the sample, n is the number of observations in the sample and the Greek symbol Σ (capital sigma) is the mathematical notation for 'sum of'. In the following sample of seven observations or measurements:

2, 3, 3, 5, 9, 15, 16

the mean is therefore $(2+3+3+5+9+15+16)/7=53/7=7.57$. *Cf.* GEOMETRIC MEAN; HARMONIC MEAN; MEDIAN (SECOND QUARTILE); MODE. [B.S.]

mean crowding Mean crowding (m^*) is a term used by Lloyd (1967) and is the average DENSITY experienced by a randomly chosen individual in a population. It is equal to $m+V/m-1$, where m and V are the MEAN and VARIANCE of the distribution of numbers over samples or patches of habitat. With a RANDOM DISTRIBUTION (POISSON DISTRIBUTION) $V=m$ and therefore $m^*=m$, but with an AGGREGATED DISTRIBUTION (with variance greater than the mean) mean crowding is greater than the actual mean density. This has the important effect that individuals in a population showing an aggregated distribution are actually experiencing a greater degree of crowding than suggested by simply looking at the average POPULATION DENSITY. This can have important consequences for density-dependent processes such as INTRASPECIFIC and INTERSPECIFIC COMPETITION. The simple example below illustrates this effect.

Numbers per habitat patch						m	V	m^*
1	1	1	9	9	9	5	16.0	7.2
1	3	5	6	7	8	5	5.6	5.1

On the first line are numbers from six patches of habitat that clearly show an aggregated distribution. Some patches only have a single individual, while others have 9. The variance (16) is much greater than the mean (5) and the mean crowding experienced by an individual is 7.2 individuals. This is calculated using the above formula but can be seen to be correct by counting up the density experienced by every individual. Three individuals experience 0 CONSPECIFICS, while 27 individuals (9×3) experience 8 conspecifics. This means that 30 individuals experience 216 other individuals, or 7.2 (216/30) on average. Notice that in the bottom line the numbers from another six patches show a random distribution (approximately). Here m^* is approximately the same (5.1) as the ARITHMETIC MEAN (5). [B.S.]

REFERENCE
Lloyd, M. (1967) Mean crowding. *Journal of Animal Ecology* **36**, 1–30.

mean genotypic value The mean GENOTYPIC VALUE is identical to the mean PHENOTYPIC VALUE in the population, as the mean of the ENVIRONMENTAL DEVIATION over individuals equals zero. In theoretical QUANTITATIVE GENETICS, the genotypic values at a bi-allelic locus A are often denoted by: $+a$ for GENOTYPE A_1A_1, d for genotype A_1A_2, $-a$ for genotype A_2A_2. The mean genotypic value in a population at Hardy–Weinberg frequencies becomes $\bar{g}=ap^2+d.2pq-aq^2$. additivity of ALLELES implies that the DOMINANCE deviation $d=0$ and the AVERAGE EFFECT OF A GENE SUBSTITUTION $\alpha=a$. Linkage disequilibrium does not appear in the mean genotypic value over loci if the loci work additively $\bar{g}=\Sigma_{\text{loci } i}\alpha_i(q_i-p_i)$. If two additive loci A and B work multiplicatively, $\bar{g}=\alpha_A(q_A-p_A)\,\alpha_B(q_B-p_B)+\xi D_{AB}$, where ξ equals the recombination rate and D_{AB} the LINKAGE DISEQUILIBRIUM between the loci. [G.D.J.]

mean intensity of infection Average number of MACROPARASITES per HOST in a defined host population (*see* INTENSITY OF INFECTION; PREVALENCE OF INFECTION). Macroparasites typically have a skewed DISTRIBUTION between hosts, such that most PARASITES are found in a small proportion of the host population; thus the VARIANCE in intensity of infection is greater than the MEAN. The abundance of infection is the mean intensity in those hosts that are infected. [G.F.M.]

mean solar time *See* TIME.

measure of central tendency (average) A single number that is representative of all the numbers in a set of observations. Three types of average are commonly used: the MEAN, the MEDIAN and the MODE. When the observations follow a symmetrical DISTRIBUTION then the mean, median and mode have the same value. In a non-symmetrical (skewed) distribution the mean is moved towards the extended tail of the distribution (Fig. M6). [B.S.]

measurement end-point *See* END-POINT.

mechanism philosophy *See* HOLISM.

median effective concentration/dose Statistically derived concentration (dose) expected to produce an EFFECT (lethality, inhibition, other effect) in 50% of the population. *See also* EC_{50}; IC_{50}; LC_{50}; PREDICTIVE ECOTOXICITY TESTS. [P.C.]

median lethal concentration (MLC) Same as LC_{50}. *See also* PREDICTIVE ECOTOXICITY TESTS. [P.C.]

median (second quartile) This MEASURE OF CENTRAL TENDENCY or average is calculated as the middle observation in a SAMPLE of observations that have been placed in numerical order. Therefore the median divides a set of observations into two halves. In the following sample of seven observations or measurements:

2, 3, 3, 5, 9, 15, 16

the median value is 5. When the number in the sample is even, the median is conventionally taken as the MEAN of the middle two observations. Therefore, in the following sample of six observations:

2, 3, 3, 5, 9, 15

the median is $(3+5)/2=4$. *See also* PERCENTILE; QUARTILE. [B.S.]

median times to effect Times for 50% of test population to manifest the adverse EFFECT(S) under consideration at a specified concentration of TOXICANT. Might be in terms of lethality (LT_{50}), or inhibition (IT_{50}), or any other effect end-point (ET_{50}). *See also* PREDICTIVE ECOTOXICITY TESTS. [P.C.]

Mediterranean climate regions, conservation The special conservation importance of the world's five Mediterranean CLIMATE regions lies in their extraordinarily diverse and endemic-rich floras (Table M3). In contrast, vertebrate faunas are relatively depauperate and endemism is low. Very little is known about their invertebrate faunas. Mediterranean climate regions support some 46 000 species of vascular plants (some estimates range up to 70 000 species), 58% of which are endemic to these areas (Table M3). Thus these regions, comprising a mere 2.2% of the Earth's surface, are the exclusive home of 10% of the world's vascular plant flora. These endemic floras include both low-ranking (neoendemic) and high-ranking (palaeoendemic) taxa. The flora of the South African region is recognized as a floral kingdom (Cape).

The conservation status of Mediterranean climate regions is poor. AGRICULTURE and animal husbandry have been practised in the Mediterranean Basin for 10 000 years and most ecosystems are now human modified. Extensive transformations of the other regions began only after colonization by Europeans, about 500 years ago in Chile (earliest) and only 170 years ago in south-western Australia

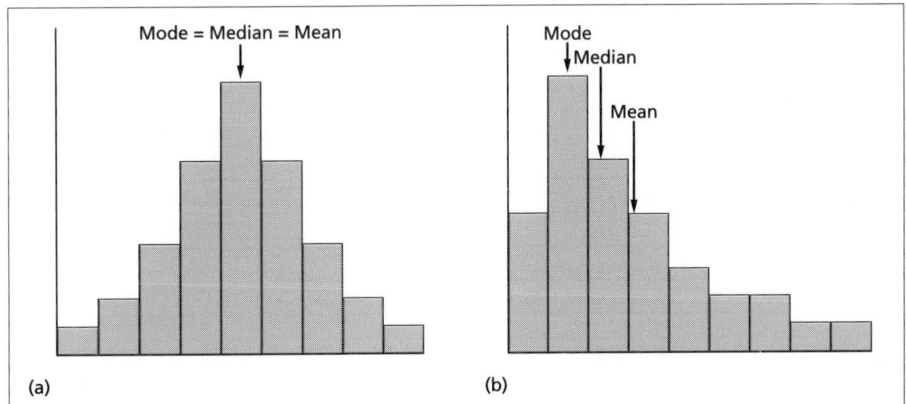

Fig. M6 (a) Symmetrical distribution. (b) Non-symmetrical (skewed) distribution.

Table M3 Plant biodiversity and conservation status of Mediterranean climate regions.

Region	Area (10^6 km^2)	Native flora* (% endemic)	Threatened taxa†	Conserved area (%)	Major threats‡
California	0.32	4300 (35)	718	1.6	Urbanization, agriculture
Chile	0.14	2100 (23)	?	2.6	Deforestation, grazing, agriculture
Mediterranean Basin	2.30	23 300 (50)	4251	3.1	Deforestation, agriculture, grazing, urbanization
South Africa	0.09	8550 (68)	1300	14.4	Invasive alien plants, agriculture, urbanization
South-west Australia	0.31	8000 (75)	1451	6.5	Agriculture, deforestation, introduced pathogens

* Vascular plant species.
† Include taxa at and below the species level.
‡ Agriculture includes afforestation using alien trees.

(latest). All the regions have very high concentrations of rare plant taxa whose survival is threatened mainly by clearing for agriculture and urbanization, but also by ALIEN plants and pathogens (Table M3). With the exception of the montane areas of South Africa, the reserve networks for these regions are inadequate. Currently, some 142 vascular plant taxa of Mediterranean climate regions are presumed extinct, 55 of these from south-western Australia. Although extinction rates may have peaked in some regions (e.g. California, Australia), they are probably accelerating in others (e.g. Chile, South Africa). *See also* CONSERVATION BIOLOGY. [R.M.C.]

REFERENCES

Davis, G.W. & Richardson, D.M. (eds) (1995) *Mediterranean-type Ecosystems. The Function of Biodiversity*. Springer-Verlag, Berlin.

Greuter, W. (1994) Extinctions in Mediterranean areas. *Philosophical Transactions of the Royal Society of London, Series B* **344**, 41–46.

World Conservation Monitoring Centre (1992) *Global Biodiversity: Status of the Earth's Living Resources*. Chapman & Hall, London.

Mediterranean scrub/forest Mediterranean-type climates are typified by hot dry summers and mild moist winters. This type of climate is found in five different parts of the world: the Mediterranean Basin, South Africa, southern Australia, California and Chile. The vegetation is characterized by dwarf shrubs, scrub or woodland. The Mediterranean region itself has been a centre of human occupation throughout the current interglacial (last 10 000 years) and has experienced prolonged and intensive human impact. Despite this, or perhaps in part because of it, the Mediterranean Basin shares with its climatic counterparts a remarkably high floristic DIVERSITY (*see* MEDITERRANEAN CLIMATE REGIONS, CONSERVATION).

Perhaps one of the main causes of such floristic diversity is the degree of patchiness in the vegetation, resulting from a range of disturbances including both natural and human-induced catastrophes. Fire is particularly important and most of these vegetation types are subjected to periodic burning, with a successional cycle of rebuilding intervening between fires. Climatic variability adds to the unpredictability of the environment and the nutrient poverty of soils prevents robust or productive species attaining physical DOMINANCE.

Many of the shrubs of this vegetation type retain their seeds in the CANOPY, often shedding them after fire and thus coinciding with a period of low competition and elevated nutrient availability. There is also a high rate of ant-mediated secondary DISPERSAL of seeds once they are shed.

The Mediterranean Basin, with its long human history, demonstrates that this vegetation type can maintain its diversity even under human pressure. Modern clearance, intensive agricultural methods and urban developments, however, present much more serious threats to the survival of the Mediterranean shrublands and their rich endemic floras. [P.D.M.]

Mediterranean Sea The SEA between Europe, Asia Minor and Africa. The area of the Mediterranean (including the BLACK SEA) is 3.02 million km^2, the average depth is 1450 m and the maximum depth is 5092 m. It is completely surrounded by land except for a narrow passage to the North Atlantic through the Strait of Gibraltar and to the Black Sea through the Bosporus. The Mediterranean is divided into a number of smaller seas (e.g. Adriatic, Aegean, etc.) and contains numerous small islands. The CONTINENTAL SHELVES are relatively narrow (less than 40 km on average) and the CONTINENTAL SLOPES are steep with many submarine canyons. Despite the fact that three major rivers empty into the Mediterranean, the Rhône, the Nile and the Po, EVAPORATION generally exceeds PRECIPITATION, causing vertical mixing as dense, saline surface water sinks. The SALINITY of deep water is between 38.4 and 39.0‰. Nutrient concentrations in the Mediterranean are low relative to the North Atlantic and TIDES are primarily semi-diurnal. [V.F.]

medium number systems, problems of prediction Biological communities are often termed 'medium number systems' because of the number of species they contain. The reason for the distinction lies primarily in the mathematical approaches that can be used to describe species dynamics. For large number systems describing, for example, the behaviour of molecules in a chemical reaction, average or probabilistic approaches can be used. In contrast, when there are very few interacting elements, models can be constructed using a set of EQUATIONS that specifies the behaviour of each individual element. For medium number systems, it has been argued that neither of these approaches can be used: there are too many elements (species) to construct and solve equations using the small number approach and too few for probabilistic behaviour to provide accurate predictions on the time-scales of interest. The absence of a satisfactory framework for modelling medium number systems can be a principal limitation for quantitatively predicting the responses of species populations in a COMMUNITY context for some classes of problem. *See also* MATHEMATICAL MODELS. [S.J.H.]

REFERENCE

Allen, T.F.H. & Starr, T.B. (1982) *Hierarchy: Perspectives for Ecological Complexity*. University of Chicago Press, Chicago, Ill.

megaplankton ZOOPLANKTON of between 20 and 200 cm in size. All are metazoans. [V.F.]

meiobenthic fauna *See* BENTHOS.

meiofauna The extraction of animals from marine sediments almost invariably requires the use of sieves and hence the mesh size of the sieve adopted determines the SIZE of animal retained. The meiofauna are defined as those multicellular benthic animals that pass through the 0.5-mm mesh used to collect the MACROFAUNA and, by extension, the same term is often used for animals of the same size from other benthic marine habitats, although it is not used in respect of the PLANKTON or terrestrially. The meiofauna are too small to construct burrow systems and therefore live in the INTERSTITIAL spaces between the sedimentary particles or in equivalent microhabitats. Broadly two types of animal comprise this fauna: the juvenile and/or larval stages of many members of the macrofauna; and wholly microscopic forms such as the gastrotrichs, gnathostomulans, turbellarians, nematodes, kinorhynchs, tardigrades, halacarid mites and copepods, as well as minute species of otherwise generally larger animals such as hydroids, priapulans, molluscs and bryozoans. Nematodes and harpacticoid copepods often numerically dominate the meiofauna. Besides each other, meiofaunal animals feed on interstitial DETRITUS, bacteria and algae; however, the extent to which they form a significant part of the diet of the benthic macrofauna is still contentious, although they are certainly consumed by the larger animals with which they share the habitat. [R.S.K.B.]

Meiotic drive

The process in which an ALLELE gains more than 50% representation in the progeny of a heterozygous individual through the ability of that allele to outcompete sister GAMETES bearing the wild-type allele. The process is usually due to the allele's activity during and/or after meiosis. The allele that gains overrepresentation is said to drive (rapidly spread) through the population. The process can be considered as the expression of intraindividual inter-gamete COMPETITION. The term should not be applied to interindividual inter-gamete competition (e.g. SPERM COMPETITION). The term is often restricted to those instances in which the driving/distorting allele inhibits wild-type gametes (as opposed to instances in which the allele gaining excess representation does so through increased gametic FITNESS, for example faster growing pollen, or through choice of gamete, exercised for instance by an egg on sperm or by a mother on sperm or on her own oocytes). When sister gamete inhibition is involved the drive allele is often deleterious. A deleterious drive allele can spread in a population just so long as (i) the inhibition of sister gametes containing the wild-type allele leads to a higher frequency of zygotes with the drive allele and (ii) the fitness of drive heterozygotes is not so low as to outweigh the advantage accrued through inhibition of competitor gametes. Genes that spread through overrepresentation are, in general, referred to as SELFISH GENES (alias selfish DNA) and meiotic drivers are hence one category of these.

Taxonomic distribution is quite wide (e.g. instances in ascomycete fungi, insects, mammals and strongly suspected in angiosperms) but incidences are usually considered to be relatively rare. The best described are *Segregation Distorter* (SD) on chromosome II of *Drosophila melanogaster* and *t-complex* on chromosome 17 of mice. If these alleles are represented as the *D* (distorter/drive) allele, then both have in common that nearly all the progeny of a *D*+ male (+ being wild type) inherit *D* because the *D* gene interferes with the activity of +-bearing sperm. In flies, half the sperm of a *SD*+ male do not mature and these all contain the + allele. In contrast, those sperm that do mature tend to contain the *D* allele. These and other well-characterized meiotic drive genes are usually not one gene but two that typically segregate as one linkage group. A general scheme for drive activity is as follows. One gene produces a toxin whilst the other codes for the antitoxin. At meiosis the toxin is produced and shared by all pro-gametes. After meiosis the antitoxin is produced but is restricted to the gametes bearing *D*. Hence + gametes are inhibited by the toxin whilst *D*-bearing gametes survive. Although two genes are involved, crossing-over between them is very rare and whenever a gamete contains the toxin allele it nearly always also has the antitoxin allele (they are in LINKAGE DISEQUILIBRIUM). Often this gene pair is held together by being in a chromosomal inversion and is often found near the centromeres on chromosomes, these being regions of low recombination. Drive is not restricted to intraindividual sperm–pollen competition. 'Spore killers' in ascomycete fungi kill spores not containing them and incidences of preferential incorporation into the oocyte rather than the polar body in females' meiosis may exist (e.g. an allele on chromosome I of mice).

Meiotic drive genes can be autosomal or on sex chromosomes. In the latter case their spread through a population creates a biased sex ratio. Y-linked distorters are known in mosquitoes but are otherwise unusual. X-linked distorters are relatively common, reported in numerous insects and a few mammals and strongly suspected in angiosperms. Both types of driver, but most particularly sex-linked ones, tend to create the conditions for the spread of unlinked suppressors of distortion. The spread of these in turn can return the population to a state at which all alleles segregate in a Mendelian fashion and/or create the context for enhancers of drive to spread. The latter will spread only if the enhancer is adequately tightly linked to the drive allele. This difference in 'interest' between suppressor and enhancer is indicative of a conflict of interest between drive allele and the rest of the genome.

The importance of meiotic drive to EVOLUTION is a contentious issue. It is implicated as an important force in the evolution of novel sex-determining systems in mammals (e.g. wood lemming) and scale insects (coccoids). The expression of meiotic drive genes in species hybrids is implicated in a few instances of hybrid sterility and hence may have a role in the process of SPECIATION. In instances of biased primary SEX RATIOS, sex chromosome-linked drive will be a candidate explanation. If found, an adaptive ('good for the organism') explanation is not necessarily required.

Meiotic drive should not be confused with molecular drive, a loosely defined set of hypotheses concerned with processes of genome evolution. *See also* CONFLICTS; QUANTITATIVE GENETICS; SEGREGATION DISTORTION; SELFISH DNA. [L.D.H.]

REFERENCE

Lyttle, T.W. (1991) Segregation distorters. *Annual Review of Genetics* **25**, 511–557.

---

melanic horizons *See* SOIL CLASSIFICATION.

melanism Many animals, especially species of moths and other insects, exhibit melanic forms that are black or dark in colour due to the deposition of melanin pigment in the cuticle or skin. *See also* ECOLOGICAL GENETICS; INDUSTRIAL MELANISM. [P.M.B.]

meme *See* GENETIC SYSTEMS.

memetic systems *See* GENETIC SYSTEMS.

mendelevium (Md) Mendelevium is a synthetic element (ATOMIC NUMBER = 101; relative ATOMIC MASS = 258; melting point and boiling points unknown) that was discovered in 1955. Particle bombardment of other actinides has been used to produce very small quantities of isotopes with mass numbers from ^{247}Md to ^{259}Md, with the longest lived being ^{258}Md ($t_{1/2}$ = 56 days). Consequently, mendelevium does not occur in detectable quantities in the environment. Investigations of the properties of mendelevium have mainly used the ISOTOPE that can be produced in greatest abundance, ^{256}Md ($t_{1/2}$ = 1.27 h). The dominant oxidation state of mendelevium is +3 but the +2 state can also be produced. [A.B.M.]

Mendelian inheritance The laws of Mendel concerning inheritance are the law of segregation (first law) and the law of independent assortment (second law).

The first law concerns the particulate nature of the genes. Genes inherited from the parents keep their separate identity when occurring together in one diploid individual. Alternate versions of the genes, called ALLELES, account for variation in inherited characters. The two genes segregate during GAMETE production. A cross between purple-flowered and white-flowered peas led to all purple-flowered F_1 offspring; crossing the F_1 amongst themselves led to purple-flowered and white-flowered pea plants in the ratio 3 : 1 in the F_2. The purple colour in the F_1 flowers indicates that the colour purple is dominant over the colour white: the allele *P* for purple is dominant, the allele *p* for white is recessive. Probability theory predicts that in a cross between F_1 heterozygotes the probability of a white *pp* individual is $^1/_4$, of a purple *Pp* individual is $^1/_2$ and of a purple *PP* individual is $^1/_4$, giving the 3 : 1 phenotypic ratio in colour.

The second law concerns crosses of two or more characters. If two characters, like seed colour and seed shape in peas, are inherited independently, the ratios of the combination of characters can be found by combining the ratios from the monohybrid crosses. If in a dihybrid cross between yellow round and green wrinkled peas the F_1 peas are yellow round, the expected ratio in the F_2 of yellow to green is 3 : 1 and of round to wrinkled 3 : 1. The expected proportions under independent assortment of the genes are 9 : 3 : 3 : 1 for yellow round, yellow wrinkled, green round and green wrinkled, respectively. Lack of independent assortment implies that the genes are linked, i.e. occur near to each other on the same chromosome (Fig. M7). *See also* MENDELIAN TRAIT; NEO-DARWINISM. [G.D.J.]

Mendelian trait Gregor Mendel was the first to show that TRAITS are passed from parent to offspring in discrete units, which we now call genes. He developed two laws: the law of segregation and the law of independent assortment. These laws apply to traits controlled by a single locus with two ALLELES. When there are more than two loci, several loci may occur on the same chromosome and hence may not assort independently. Even traits that show quantitative variation, such as body SIZE or FECUNDITY, can be considered Mendelian traits in which there are a large number of loci that interact. Most generally, however, the term is reserved for traits controlled by a few genes of large effect. *See also* MENDELIAN INHERITANCE. [D.A.R.]

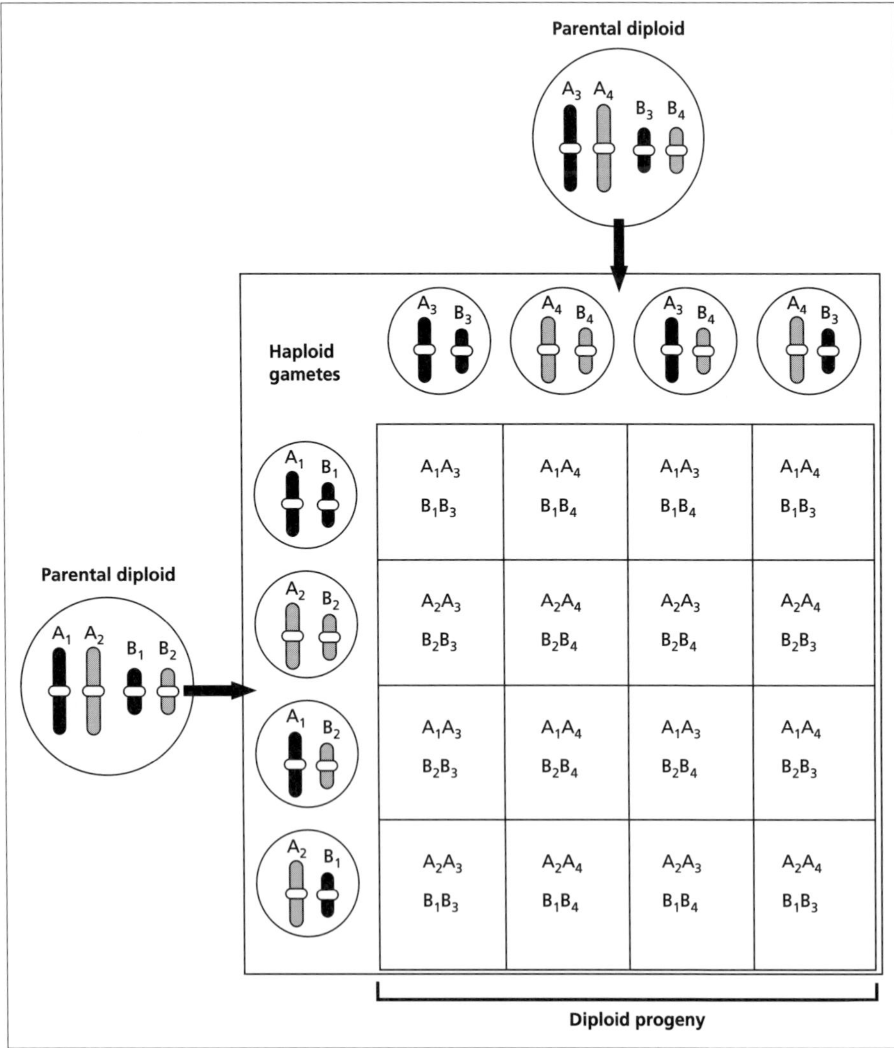

Fig. M7 Independent assortment of genes located on two chromosomes. This figure illustrates the variation that is generated by the independent assortment of chromosomes in a sexually reproducing diploid organism. Only two pairs of chromosomes (A and B) are represented. Maternal chromosomes (A_1, B_1, A_3 and B_3) and paternal chromosomes (A_2, B_2, A_4 and B_4) present in the diploids are shown in black and grey, respectively. Each diploid gives rise to four different haploid gametes and these fuse to produce 16 possible types of diploid progeny. As long as there are differences between the chromosomes A_1–B_4, the 16 progeny types will have different recombinant genomes. This reassortment is built into the reproductive system and does not require any physical exchange within chromosomes. (After Leach, D.R.F. (1996) *Genetic Recombination*. Blackwell Science, Oxford.)

Menge–Sutherland model A theoretical model of community REGULATION that attempts to describe how the relative importance of structuring process (i.e. COMPETITION, predation and DISTURBANCE) alter along gradients of environmental STRESS and for different trophic levels within the community. A later elaboration of the model includes the effects of differing levels of RECRUITMENT. *See also* COMMUNITY ECOLOGY. [S.J.H.]

REFERENCE

Menge, B.A. & Sutherland, J.P. (1987) Community regulation: variation in disturbance, competition, and predation in relation to environmental stress and recruitment. *American Naturalist* **130**, 730–757.

mercaptans A group of organosulphur compounds with a characteristically disagreeable odour found with other SULPHUR compounds in crude PETROLEUM. They are largely used as chemical intermediates, for example methylmercaptan CH_3SH (thiol), in the manufacture of dyes, pharmaceuticals, RUBBER chemicals, INSECTICIDES, molluscicides and FUNGICIDES. [K.D.]

mercury (Hg) A heavy, silver-white METAL (ATOMIC NUMBER = 80; relative ATOMIC MASS = 200.59; melting point = −39°C; boiling point = 357°C), mercury is an ultratrace constituent (0.08 mg kg^{-1}) of the Earth's CRUST. Found in cinnabar (HgS) and in association with lead–zinc–silver ores, it is the only metal that is liquid (quicksilver) at normal temperatures and is used in chlor-alkali production, electrical BATTERIES and CATALYSTS. The high volatility of this non-essential element renders it extremely dangerous, inhaled mercury vapour being capable of passing the blood–brain barrier and, with its great affinity for the SH groups of enzymes, severely affecting the central nervous system. The best-known case of industrial mercurialism was that of the 'mad hatters' in the 19th century, who used mercuric nitrate in the manufacture of felt. Inorganic mercury salts can also damage the kidneys. Released to the environment from volcanoes, mining, fossil-fuel combustion, the use of mercury compounds and the discharge of mercury-containing wastes, mercury is not greatly taken up by plants but can undergo methylation to highly toxic ORGANOMERCURY compounds, which are found in fish. *See also* HEAVY METALS. [J.F.]

REFERENCE

Nriagu, J.O. (ed.) (1979) *The Biogeochemistry of Mercury in the Environment.* Elsevier/North Holland Press, Amsterdam.

merolimnic fauna *See* DRIFT, FRESHWATER.

meromictic lakes *See* LAKE STRATIFICATION; LAKES.

meroplankton Organisms that spend part of their LIFE CYCLE in the water column, usually the EGGS and LARVAE of benthic or nektonic adults. *See also* PELAGIC. [V.F.]

mesocosm As MACROCOSM, but 'medium sized' and often indoors. [P.C.]

mesohaline *See* BRACKISH WATERS.

mesolecithal eggs *See* EGGS.

mesopelagic The part of the OCEAN water column extending from the bottom of the EPIPELAGIC zone (*c*.200–300 m) to about 1000 m depth. It corresponds approximately with the dysphotic (twilight) zone and is located seaward of the SHELF-SLOPE BREAK. [V.F.]

mesoplankton ZOOPLANKTON species that are between 0.2 and 20 mm in size. [V.F.]

mesotrophic *See* EUTROPHICATION.

Mesozoic Lying between the PALAEOZOIC and CENOZOIC eras, the Mesozoic era is the period of geological time between approximately 250 million years and 65 million years ago. It encompasses the Triassic, Jurassic and Cretaceous periods. [P.D.M.]

metagenesis *See* ALTERNATION OF GENERATIONS.

metal A dense substance having a metallic lustre, being a good conductor of heat and electricity, ductile and malleable. With the exception of mercury, metals are solid at room temperature. See the entries under the name of the metallic element for specific details of each metal.

Metal surfaces exposed to the air react with OXYGEN to form a film of oxide. Once formed, the film protects the surface from further corrosion in some cases (ALUMINIUM, COPPER, IRON, NICKEL, ZINC) but not in others (CALCIUM, POTASSIUM, MAGNESIUM, SODIUM). When two metals touch or are electrically connected in aqueous solution, electrons flow from one metal to another. The electrochemical series is a list of the metals organized with the strongest electron donor first (the most readily oxidized) and the strongest electron acceptor last:

Na Mg Al Ti Zn Fe Ni Sn Pb Cu Ag Pt Au

In the presence of WATER, this tendency of electrons to flow from one metal to another causes corrosion. Thus, iron in the presence of copper and water loses electrons to form Fe^{3+}, which reacts with OH^- from water to form $Fe(OH)_3$ and ultimately $Fe_2O_3.H_2O$, known as rust. As the copper is not used up, the metal object will be turned completely into rust.

Metals have been useful to humans in their elemental forms to make tools, water pipes, utensils, vehicles for transport by land, water and air, for scientific and musical instruments and decorations (especially iron, LEAD, zinc, copper, MERCURY, aluminium, GOLD, SILVER and PLATINUM). Often, they are used in mixtures or alloys. Examples of alloys are brass (copper and zinc), bronze (copper and TIN), pewter (tin and lead, with some ANTIMONY) and constantan (copper and nickel). Mercury combines loosely with other metals to form amalgams, used in dental fillings and for extracting gold. Advanced technologies have made use of other metals, such as CADMIUM, GERMANIUM, LITHIUM, TITANIUM and URANIUM.

Several metals are required in trace quantities by plants and animals as cofactors in enzyme reactions (*see* MICRONUTRIENTS), but at high concentrations metal salts and vapours are poisonous and are used as BIOCIDES and antiseptics. Metals often reach toxic levels in the environment (*see* CADMIUM; COPPER; LEAD; MERCURY; ZINC). Metalloids are elements with some of the characteristics of metals, such as ARSENIC and antimony. *See also* HEAVY METALS; PERIODIC TABLE. [J.G.]

metalimnion *See* LAKE STRATIFICATION; LAKES.

metallothioneins Low RELATIVE MOLECULAR MASS proteins rich in cysteinyl residues. They have affinity for certain metals that bind to sulphydryl groups. Synthesis is induced by the presence of CADMIUM, ZINC, COPPER, MERCURY or GOLD in the diet or ambient medium. They are thought to play a protective role in organisms as a sink for toxic metals and as part of 'normal physiology' in transport and storage of essential metals. Sometimes abbreviated as MT. *See also* DEFENCE MECHANISMS; HEAT-SHOCK PROTEIN; PROTECTIVE PROTEINS. [P.C.]

metamorphosis Transitional stage in organisms with a COMPLEX LIFE CYCLE, involving the remodelling of a larval form into an adult morphology. Some groups, like holometabolous insects, have more than one transitional stage during a complex ONTOGENY, involving larva, pupa and imago stages. Most endocrinological work on the regulation of metamorphosis has been done in insects and amphibians. In insects, this transition appears to be controlled by levels of juvenile hormone, secreted by the corpus allatum, which interacts with ecdysone to control the timing of onset of developmental events. In amphibians, thyroid hormones are involved in triggering the metamorphic process, although there is considerable controversy regarding the precise biochemical dynamics of the hormonal control as well as the mosaic nature of the responses of the various tissues to hormonal cues. From an evolutionary perspective, it is difficult to define metamorphosis as a specific and discrete developmental process. This is because of the extraordinary range of variability in terms of control mechanisms and degree of remodelling involved in the various groups of invertebrates as well as in amphibians and a few primitive vertebrate taxa, such as lampreys and hagfishes where 'metamorphosis' is postulated to occur. The definitional problem is compounded when the conceptual framework attempts to include the many taxa that have secondarily evolved direct development from an ancestral biphasic developmental pattern, such as described in the existing detailed work on echinoderms, ascidians and many amphibian species. In a comprehensive review of metamorphic remodelling in vertebrates, Rose and Reiss (1993) defend a quantitative, rather than qualitative, definition more consistent with an evolutionary perspective. Thus, vertebrate metamorphosis can be described as a pattern emerging from the temporal clustering of developmental events between two or more relatively stable periods in ontogeny. The accumulation of events within a short period results from selective pressures to minimize the time spent in the maladaptive transitional interval between the two functional morphologies (larva and adult). As indirect support for this definition of metamorphosis as a heterochronic pattern, a comparison with secondarily 'direct-developing' relatives is illustrative, since metamorphic processes inside the egg or in the reproductive tract of the mother are much less pronounced, given a broader distribution of the timing of remodelling events throughout ontogeny. Hence, it suggests that the new pattern is a consequence of a relaxation of selective pressures in direct-developing species. *See also* LIFE-HISTORY EVOLUTION; SEXUAL REPRODUCTIVE CYCLES. [P.A.]

REFERENCE

Rose, C.S. & Reiss, J.O. (1993) Metamorphosis and the vertebrate skull: ontogenetic patterns and developmental mechanisms. In: *The Skull. Vol 1. Development* (eds J. Hanken & B.K. Hall), pp. 289–346. University of Chicago Press, Chicago, Ill.

metapopulations The DISTRIBUTION of a species is often more or less fragmented, most obviously by being confined to isolated patches of suitable habitat. An assemblage of such separate populations can itself be treated as an entity with its own EMERGENT PROPERTIES: a metapopulation. The crucial feature of its dynamics is the balance between the rate of extinction of the local populations of which it is made and the rate at which empty habitat patches are recolonized by individuals or propagules migrating from the extant populations. This determines both the proportion of suitable habitats occupied and the time for which the whole metapopulation can exist; metapopulations have a THRESHOLD in their survival and are effectively immortal if the extinction–recolonization balance is above critical: below this value the whole metapopulation is itself highly vulnerable to extinction. This concept can be applied to the problem of species conservation under increasing anthropogenic habitat destruction or in networks of nature reserves. As always, real populations may differ in significant ways from the ideal model. *See also* POPULATION; POPULATION DYNAMICS. [J.R.G.T.]

REFERENCES

Hanski, I.A. & Gilpin, M.E. (1997) *Metapopulation Biology. Ecology, Genetics and Evolution.* Academic Press, San Diego, Calif.

Lennon, J.J., Turner, J.R.G., & Connell, D. (1997) A metapopulation model of species boundaries. *Oikos* **78**, 486–502.

Symposium on metapopulations (1991) *Biological Journal of the Linnean Society* **42** (1–2), 1–327.

meteoric lakes *See* LAKES, TYPES OF.

Meteorology

Originally the science of things aloft (from the Greek), it has come to mean the science of the physical nature of the lower ATMOSPHERE, especially the behaviour and forecasting of WEATHER systems.

Observational meteorology

Observational meteorology developed rapidly around 1850, when the invention of the electric telegraph encouraged the establishment of national networks of observers making synoptic (simultaneous) hourly observations and measurements at the surface. Since 1950 these have been coordinated by the WORLD METEOROLOGICAL ORGANIZATION in what is now called the surface network of World Weather Watch (WWW), an international network of many thousands of stations with separations ranging from a minimum of about 50 km overland in industrially developed nations to many times more than this in developing nations and to even larger values over oceans, where moving instrumented merchant ships and drifting buoys replace fixed stations. Large-scale weather systems, such as depressions and anticyclones, are resolved by this network and are therefore called synoptic-scale systems.

The WWW upper air network of radiosondes developed rapidly after the Second World War, making 6-hourly measurements of the synoptic-scale airflows, temperatures and humidities in the middle and upper TROPOSPHERE for weather forecasting (of surface and upper air conditions). Cost of manning and equipment enforces much greater station separation compared with the surface network and a completely inadequate oceanic coverage. Automatic sensors in large civil aircraft are being added to provide additional data.

Meteorological satellites have been in use since 1960 and the WWW satellite network combines low polar orbiting and high equatorial orbiting (geostationary) satellites to give nearly continuous global observation of cloud and some upper air temperatures using visible and infrared wavelengths. Networks of weather radars are being established in developed countries to give continuous semi-quantitative measurements of rainfall rate within about 100 km of each station. An international telecommunication network carries data to the national forecasting offices within an hour of the standard observation times.

As well as larger features, satellite and RADAR data reveal extensive mesoscale behaviour, i.e. activity on space scales between the 100-km minimum of the synoptic scale and the small-scale behaviour (individual showers, etc.) visible from individual surface stations. Although often very important (e.g. in revealing localized concentrations of PRECIPITATION), mesoscale behaviour is not yet well understood or integrated with the synoptic scale.

Dynamic meteorology

Dynamic meteorology seeks to explain and predict the behaviour of the lower atmosphere in terms of established physical laws. Pioneers at the turn of the century focused mainly on fluid dynamics and thermodynamics, but electromagnetic RADIATION and the complex microphysics of cloud and precipitation are now included (as well as being studied in their own right in cloud physics). Until the advent of computer modelling in the mid 1950s, forecasting was a subtle combination of observation and manually applied theory: now it is a complex exercise involving the largest available computers and continual interaction between theoretical advance and operational application. Since about 1970, computerized forecasting models have been increasingly used to model global CLIMATE and climatic consequences of volcanism, nuclear war and industrial POLLUTION in what are called general circulation models (GCMs). Severe problems remain in modelling behaviour on space scales less than that of the finite grid used in all computer models (currently about 50 km in the horizontal and 1 km in the vertical).

The behaviour of extremely simple dynamic models of CONVECTION, etc. has provided seminal examples of deterministic CHAOS; similar extreme sensitivity of forecasting models to initial conditions strongly suggests that useful forecasting of individual weather systems will always be limited to a few weeks (currently about 1 week).

Boundary layer meteorology and climatology

BOUNDARY LAYER meteorology and climatology examines the particular behaviour of the surface and first couple of hundred metres of the atmosphere, where direct interactions with the underlying surface are particularly strong and locally important. Major problems persist with the theory and observation of TURBULENCE, which is endemic in the layer. AGRICULTURAL METEOROLOGY studies the boundary layers above and in stands of vegetation and the conditions in the top metre of soil beneath.

As implied above, earlier distinctions between meteorology and climatology have blurred in recent years and the scientific study of present, past and future climate makes extensive use of direct and indirect observations (often very complex) of the atmosphere, together with comprehensive theoretical models, all of which overlap heavily with techniques established for weather forecasting.

Meteorology has developed exponentially since its speculative origins in the ancient world. Early optimism in the 17th century (when most pioneers of the scientific revolution had something to say on the subject) was frustrated by the complexity and scale of the subject but was kept alive by the inspired intuition of people like George Hadley and Benjamin Franklin. The practical need for weather forecasting triggered the necessary national (ultimately global) observational networks around 1850, and from 1880 to 1920 a series of pioneers like Lord Kelvin, Léon Teisserenc de Bort, Vilhelm Bjerknes, Sir Napier Shaw and Lewis Richardson laid a solid foundation for scientific study that has supported the huge expansion in the quantity and range of meteorological work since 1945. Although the dismal necessities of war have mothered invention and some terminology (e.g. the FRONT), the international scope of the subject is inescapable and currently well served by worldwide cooperation under the World Meteorological Organization in Geneva. *See also* MICROMETEOROLOGY. [J.F.R.M.]

---

methane The simplest hydrocarbon, methane (CH_4) exists as a gas at normal temperatures and pressures (boiling point −164°C); it is also known as coal gas, marsh gas and fire damp and is the major constituent of domestic 'natural gas'. It is found as a fossil fuel, often associated with oil and coal, where it may form an explosion hazard in coal mines and is actively vented to the atmosphere. It is also produced by microbes in waterlogged soils and LANDFILL sites (METHANOGENESIS) and by insects (e.g. TERMITES) and animals (especially ruminants). Natural sources account for about 160 Mt of the estimated annual EMISSIONS of 540 Mt. CH_4 is a 'greenhouse gas', contributing about 20% to the radiative forcing of the atmosphere. Concentrations in the atmosphere were 1.71 ppm by volume (ppmv) in 1992 and are currently rising by about 0.8% per year. Preindustrial concentrations (i.e. prior to 1800) were around 0.7 ppmv. The steady increase in CH_4 concentrations was interrupted in 1991 and 1992, but data from 1993 indicate a resumption of the upward trend. Over 90% of CH_4 in the lower atmosphere is removed by chemical reaction with the hydroxyl radical (OH) to form CARBON MONOXIDE (CO) and, eventually, CARBON DIOXIDE (CO_2) and WATER. The remainder is either absorbed by SOIL or is transported to the STRATOSPHERE, where it is broken down and acts as a major source of water vapour. [J.N.C.]

methanogenesis The production of METHANE (CH_4) from an organic substrate occurs via the following pathways.

1 By the action of microbes in anaerobic conditions in SOIL. Under these conditions, organic material derived from decaying VEGETATION is decomposed as a source of OXYGEN, with methane released as a by-product. Methane produced in this way is also known as marsh gas. The largest global methane sources are natural WETLANDS (*c.* 25%) and flood-irrigated AGRICULTURE, notably rice paddies (*c.* 12%). Large quantities of methane are also generated in LANDFILL sites, where buried organic waste is converted by microbes to methane, which may escape into the ATMOSPHERE or be tapped as a source of fuel.

2 By the action of microbes in the gut of animals, especially ruminants. This source, largely agricultural, accounts for *c.* 16% of global methane production. Insects (e.g. TERMITES) are also a small but significant global source of methane (*c.* 4%).

3 By the chemical transformation of organic matter at high temperatures and pressures in the earth, leading to the formation of natural gas, oil and coal. The methane (and other gases) so formed may migrate through rocks to accumulate in porous structures beneath impermeable strata, to form reservoirs that can be exploited for the extraction of natural gas. The extraction and use of FOSSIL FUELS contributes about 20% of global methane EMISSIONS. *See also* WASTE DISPOSAL. [J.N.C.]

methanol Methanol is a colourless inflammable liquid with boiling point 64.7°C. The boiling point is higher than predicted from its molecular mass due to the strong hydrogen bonding between molecules. It is miscible with WATER due to strong hydrogen bonds formed between the methanol and solvent water molecules. It is also miscible with most organic solvents. The chemical formula of methanol is CH_3OH. It is highly toxic, with ingestion often leading to blindness or even death.

Methanol or methyl alcohol was originally obtained by the destructive distillation of wood in the absence of air and was therefore named wood spirit. Catalytic oxidation of METHANE (CH_4) in the presence of air at high pressure and at 200°C also results in the formation of methanol. The industrial production of methanol involves its synthesis by the hydrogenation of CARBON MONOXIDE (CO) using a ZnO/Cr_2O_3 CATALYST at 400°C.

Its uses in industrial processes include the production of ethanoic acid (CH_3COOH), synthetic fibres, adhesives and PLASTICS. Methanol has a very high octane rating when mixed with automobile fuels and has sufficient volatility and low freezing point to be a viable fuel. It is often used in the high-performance engines of racing cars; 85% methanol/15% traditional fuel mixtures have been introduced in small-scale projects for use in specially adapted automobile engines. Methanol has been proposed as a long-term alternative energy (*see* ALTERNATIVE ENERGY/FUELS) supply that could help alleviate air-quality problems resulting from the combustion of petrol. [M.C.G.]

metre (m) SI UNIT of length, originally conceived as 10^{-7} of the distance on the Earth's surface from the North Pole to the Equator; now defined as the distance that LIGHT travels in 1/299 792 458 of a second. Relation to obsolete units: 1 m = 39.3701 inches; 1 yard = 0.914 m. [J.G.]

MFO *See* MIXED FUNCTION OXIDASE.

Miconia calvescens *See* EXOTIC AND INVASIVE SPECIES.

microbial ecology The science that specifically examines the relationships between microorganisms and their biotic and abiotic environments. Although the term 'microbial ecology' was first defined by Ernst Haeckel in 1866 (*see* CHARACTERS IN ECOLOGY), the field of microbial ecology did not begin to emerge until the early 1960s. The field became established during the late 1970s and has since continued to expand. Microbial ecologists study the role of microorganisms in the global cycling of materials and the interactions of microorganisms with each other and with plants and animals. The group 'microorganisms' includes a diversity of taxa such as viruses, prions, viroids, bacteria, fungi, algae and protozoa. Microorganisms (i.e. bacteria) are believed to have been the first living organisms of this planet and a key step in the EVOLUTION of higher organisms appears to have been the occurrence of symbiotic relationships between various microbes. Relationships of microorganisms with plants include interactions with plant root systems and the provision by microbes of NUTRIENTS to the plants, mutualistic relationships between nitrogen-fixing bacteria and leguminous plants (*see* NITROGEN FIXATION), epiphytic growth of microbes on plant surfaces (which may either be beneficial or harmful) and various pathogenic relationships. Microorganisms interact with animals by contributing an important nutritional source. For example, deposit- and detritus-feeding animals derive great nutritional benefit from microbes associated with the non-living organic matter on which they feed and most filter feeders depend on microalgae suspended in the water column as their primary food source. Microbes residing in animal guts are essential for the digestion of certain foods (e.g. CELLULOSE) and for synthesizing essential vitamins. Mutualistic relationships between animals and microbes are also common (e.g. between reef-forming coral and photosynthetic ZOOXANTHELLAE) as are pathogenic relationships; microbial disease can be an important selective force that influences the evolution of animal populations.

Features of microbial populations that are key in determining their ecology are their number, BIOMASS and metabolic activity. Abiotic factors that have been shown to have a major influence on microorganisms are electromagnetic radiation, temperature, atmospheric, hydrostatic and osmotic pressure, redox potential, pH, water availability and the concentrations and chemical forms of organic and inorganic compounds, which may be either essential nutrients or act as toxicants. Microorganisms dominate in the production and breakdown of oxygen-, hydrogen- and carbon-containing compounds in aquatic environments and in the catabolism of these compounds in terrestrial environments. In addition, they play key roles in the cycling of the major elements, nitrogen, phosphorus and sulphur, and of many metals.

Microbial ecology has important practical applications in WASTE management, in the maintenance of uncontaminated water and food supplies, in AGRICULTURE, in the recovery of metals from low-grade ores, in fuel production, in BIOREMEDIATION and, of course, in medicine. [V.F.]

microbial feeders *See* FEEDING TYPES, CLASSIFICATION.

microbial loop This refers to the regeneration of NUTRIENTS and their return to the FOOD CHAIN that is mediated by bacteria and protozoans. Bacteria can utilize both particulate DETRITUS (FAECAL PELLETS, dead organisms) and DISSOLVED ORGANIC MATTER (excretory products and microbial EXUDATES). The bacteria are in turn fed upon by protozoans, which are fed upon by higher trophic levels (e.g. copepods). Of the total PHYTOPLANKTON production converted to dissolved organic matter, 10–50% is consumed by bacterioplankton. Much of this material is cycled through the microbial loop. The microbial loop occurs, but is less well understood, in the BENTHOS where dominant bacterial grazers are probably flagellated protozoans and where ciliates feed on larger phytobenthos and flagellates. *See also* BENTHIC–PELAGIC COUPLING. [V.F.]

microclimate The CLIMATE in small places, usually considered over a spatial scale of a few centimetres or metres and over short periods of time (hours). Microclimates are studied with miniaturized sensors, such as thermocouples and hot-wire and miniature-cup ANEMOMETERS. The microclimates of plants and animals may be quite different from the climate measured at meteorological stations. For example, the air close to the ground in alpine vegetation may be poorly mixed with the air above, with the result that the surface properties of the vegetation exert a profound influence on the TEMPERATURE, HUMIDITY and CARBON DIOXIDE (CO_2) concentration of the air. In such cases, it is not uncommon to record surface and air temperatures that are 10°C warmer than the air aloft on sunny days and several degrees colder at night under clear skies. Environmental variables show a rapid rate of change with height above the ground, the vertical profile being determined to a large extent by the WIND speed.

Animals exploit microclimates and may migrate vertically within the vegetation on a seasonal or DIURNAL cycle to maintain body temperature. *See also* METEOROLOGY; MICROCLIMATES AND CONSERVATION BIOLOGY. [J.G.]

REFERENCES

Grace, J., Allen, S. & Wilson, C. (1989) Climate and meristem temperatures of plant communities near the tree-line. *Oecologia* **79**, 198–204.

Jones, H.G. (1992) *Plants and Microclimates*. Cambridge University Press, Cambridge.

Rosenberg, N.J., Blad, B.L. & Verma, S.B. (1983) *Microclimate: the Biological Environment*. John Wiley & Sons, New York.

microclimates and conservation biology Conservation biologists seek to preserve the BIOTA in many ways. Prime among the methods is to establish nature reserves. There are many non-biological constraints to this endeavour, including fiscal and cultural ones, but within most settings there is some leeway for decisions based on biological factors to be incorporated into the decision-making process. Much has been written about the 'design' of nature reserves, mostly from the biogeographic perspective. A single large reserve may benefit wide-ranging species and be more readily managed, while several smaller reserves may capture more of the landscape diversity. Larger reserves may be at greater risk from DISTURBANCES and disease; smaller ones may suffer higher extinction rates. Often overlooked in this debate is the importance of microsite HABITAT variation in maintaining SPECIES RICHNESS, permitting multiple successional paths, reducing extinction rates and offering multiple COLONIZATION sites.

The species diversity of any particular site is a function of many factors. COMPETITIVE EXCLUSION will reduce species richness under homogeneous stable conditions. Microtopographic variation provides the opportunity for similar species to establish in separate 'REGENERATION NICHES', then persist as adults in balanced COMPETITION. Such variation, for example tip and hollow TOPOGRAPHY that develops in old-growth forests, provides suitable conditions for quite different species. Variable or contrasting microclimates provide REFUGIA for species that would be excluded or perish in a more homogeneous habitat. For example, several Central American birds have become extinct on Barro Colorado Island, Panama Canal, because the island lacks wet sites that offer refuge during unusual droughts.

Two small reserves may be capable of supporting very different levels of species richness depending on microtopographic and microclimatic factors. If a choice is to be made, then it should favour the site with the greater habitat complexity. Managers should understand the role of small-scale HETEROGENEITY, as well as appropriate disturbance, to manage reserves for greater species richness and population stability. *See also* CONSERVATION BIOLOGY. [R.D.M.]

microcosm As MACROCOSM, but 'small' and usually indoors, in laboratories. [P.C.]

microevolution EVOLUTION of small changes within populations, given sufficient time, can lead to considerable divergence between them and ancestral groups and possibly once closely related but now isolated populations. Changes occur by NATURAL SELECTION, and microevolution is often considered to be synonymous with (neo) Darwinian evolution. *Cf.* MACROEVOLUTION. [P.C.]

microevolutionary trade-off Since before Darwin, it has been recognized that there is often a correlation among the different parts of an organism so that the EVOLUTION of one part is often associated with the evolution of some other part. For example, Simpson (1944) demonstrated that the evolution of larger body size in horses has necessarily been associated with the evolution of a change in the shape of the skull. The estimation of the genetic basis for this relationship among TRAITS has been formalized by QUANTITATIVE GENETICS via the estimation of genetic COVARIANCES among traits. One possible mechanism for such correlations is the pleiotropic effects of individual genes to influence multiple aspects of an organism's morphology. 'Microevolutionary TRADE-OFF' refers to a special case of correlated evolution for which there is a balance between the perceived costs and benefits derived from the evolution of a given trait. One class of examples comes from functional morphology. For instance, in the evolution of the limbs of tetrapod vertebrates there is a necessary trade-off between the potential output velocity of the limb, which might be adaptive for running, and the potential output force of the limb, which might be adaptive for jumping or digging. This trade-off is a consequence of the contributions of the relative lengths of the 'in' and 'out' arms of the limb, when the limb is modelled as a lever system. A relatively long 'in' arm can increase the output force of the limb, but it necessarily decreases the potential output velocity. The morphology of an organism thus represents a compromise associated with the importance of these different aspects of performance in its lifestyle. A 'specialist', such as a cheetah (specialized for running) or an armadillo (specialized for digging), is an organism for which one aspect of performance has had an overriding influence on FITNESS. This aspect of performance has been enhanced by NATURAL SELECTION at the expense of some other aspects of performance. Similar trade-offs are associated with many other aspects of morphology, such as the body shape of a fish and its influence on different aspects of swimming performance (e.g. sustained swimming versus acceleration) or the shape of a bird's wing and how it influences flight (e.g. gliding versus flapping flight). A second class of examples is associated with the evolution of life-history patterns. Natural selection can favour the evolution of earlier maturity and a higher rate of investment of available resources into reproduction; one such circumstance is high adult MORTALITY rates, attributable to some external factor in the environment such as disease or predation. It has been shown that this type of ADAPTATION may be accomplished at the cost of a reduced ability to produce young later in life or a reduction in LONGEVITY. Early-life reproductive performance therefore 'trades off' with late-life reproductive performance. Such a constraint may well be adaptive in the face of high adult mortality rates due to external causes, since the probability of surviving to successfully reproduce late in life is low anyway. This perspective of the life history has resulted in a fundamental shift in our understanding of the term 'fitness'. In Darwin's original treatise, fitness was primarily associated with longevity, giving the organism more opportunities to reproduce and hence increasing the frequency of its genes in subsequent generations. The theory of LIFE-HISTORY EVOLUTION, which is based on the premise that there are microevolutionary trade-offs among the different components of the life history, views fitness instead as the age-specific allocation of resources that results in the largest number of surviving offspring. Evolution under such CONSTRAINTS can be associated with the seemingly counterintuitive result that natural selection can favour a reduction in LIFESPAN, providing that there are sufficient COMPENSATIONS elsewhere in the life history. *See also* EVOLUTIONARY OPTIMIZATION. [D.N.R.]

REFERENCE

Simpson, G.G. (1944) *Tempo and Mode in Evolution*. Columbia University Press, New York.

microfauna The extraction of animals from marine sediments almost invariably requires the use of sieves and hence the mesh size of the sieve used determines the SIZE of organism retained. The microfauna are defined as those benthic 'protozoans' (more correctly the animal-like protists) that pass through a 0.5-mm mesh and, by extension, the same term is often used for heterotrophic protists from other benthic habitats, although it is not used in respect of the PLANKTON. The microfauna are therefore distinguished from the MEIOFAUNA not on the basis of size but by the kingdom (Protista) to which they belong; their ecology is essentially the same as that of the meiofauna. Ciliates and flagellates dominate this fauna, although amongst the non-photosynthetic protists only some foraminiferans and the Xenophyophora are too large to be considered microfaunal. The equivalently sized photosynthetic protists (and in some usages the bacteria) comprise the MICROFLORA. [R.S.K.B.]

microflora (symbionts) Usually refers to microscopic algae and possibly also bacteria and fungi. *Cf.* MICROFAUNA. [P.C.]

microgeographic variation Phenotypic or genetic variation among patches in a local HABITAT that may arise from MIGRATION, behavioural selection for appropriate substrates or selective MORTALITY among GENOTYPES. [V.F.]

microhabitat Precise physical location of a species. For example, EPILITHIC algae and benthic invertebrates live on submerged stones and pebbles, their HABITAT. However, the algae live on the upper surfaces, where they are exposed to light needed for PHOTOSYNTHESIS, and the invertebrates often occur on lower surfaces, where they obtain shelter from water movements and predators. Thus upper surfaces are the microhabitats of algae and lower surfaces the microhabitats of benthic invertebrates. [P.C.]

micrometeorology Micrometeorology is the branch of METEOROLOGY concerned with the lowest layers of the ATMOSPHERE at heights within a few tens of metres of the surface and the interaction of this surface BOUNDARY LAYER with Earth's surface. Essentially it is concerned with the layer of the atmosphere with which we are most familiar, which we breathe and live in, and in which our crops grow. It is the study of how the surface and the atmosphere interact through the exchange of heat, moisture and other properties such as CARBON DIOXIDE (CO_2) or air pollutants. The primary exchange mechanisms are RADIATION and CONVECTION in the atmosphere and conduction in the SOIL. The time-scales of most relevance to micrometeorology are on the order of seconds to hours; space scales of most relevance are on the order of a few square metres to a few square kilometres. Turbulence in the air can be generated by solar heating of the ground, for example thermals, or by mechanical means as the air is forced into an eddying motion via the effects of surface friction. The processes of micrometeorology bring about distinct microclimates that are the result of the ways in which different surface types (VEGETATION, TOPOGRAPHY and soil) influence the atmosphere on a small scale. *See also* CONDUCTION, OF HEAT; EVAPORATION. [J.B.M.]

micronutrients These are elements required by living organisms only in small amounts (*cf.* MACRONUTRIENTS). The great majority, for example BORON, MANGANESE, COPPER, ZINC and MOLYBDENUM in plants and FLUORINE, manganese, COBALT, copper, zinc, SELENIUM and IODINE in animals, are trace elements; IRON is also an essential micronutrient even though it is very abundant in the LITHOSPHERE. Below the micronutrient concentration range where normal growth occurs (Fig. M8), plants and animals exhibit deficiency symptoms and in severe cases may die. Conversely, above the optimum range, toxic effects may occur. Availability for uptake

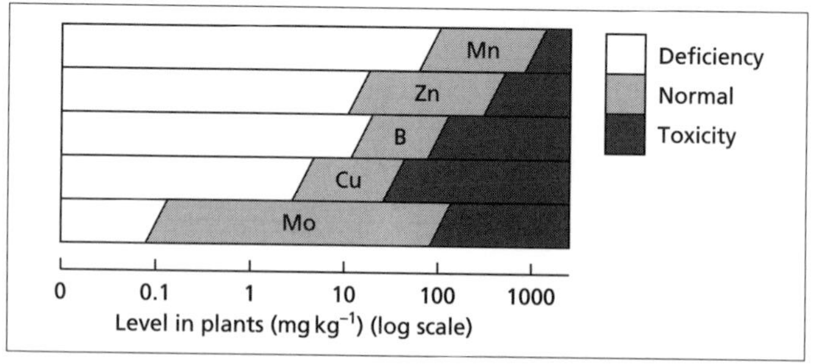

Fig. M8 Deficiency, normal and toxicity levels in plants for five micronutrients. Note that the range is shown on a logarithm base and that the upper limit for manganese is about 10 000 times the lower range for molybdenum. (After Brady, N.C. (1990) *The Nature and Properties of Soils*, 10th edn. Macmillan, New York.)

by plants is affected by soil pH, REDOX POTENTIAL and complex formation with organic matter.

Most micronutrients are present in the range 1–1000 mg kg^{-1}, one to three orders of magnitude less than macronutrients (1–50 g kg^{-1} plant dry matter). Because such small amounts are required, micronutrients may easily become physiologically limiting through deficiency or toxicity. Iron and manganese compounds are much less soluble at high soil pH and their uptake may thus be limited or excess CALCIUM may interfere with their activity within the plant. Copper and zinc are frequently deficient in the geological parent materials of some soils. At the other extreme, some micronutrients may be absorbed excessively, perhaps causing toxicity. Examples are iron and manganese in waterlogged or acid soils (redox or pH effects on solubility), CHLORINE in saline soils and copper and zinc in contaminated soils such as mining spoil. [K.A.S. & J.R.E.]

REFERENCE

Brady, N.C. (1990) *The Nature and Properties of Soils*, 10th edn. Macmillan, New York.

microparasite Microparasite is the term usually given to parasitic viruses, bacteria, protozoa and fungi, where the INTENSITY OF INFECTION is not usually measurable or meaningful in terms of MORBIDITY and MORTALITY associated with the infection (*see* MACROPARASITE). The PREVALENCE OF INFECTION is used as the principal measure of the parasite POPULATION SIZE. Microparasites tend to be small and reproduce directly in the HOST. [G.F.M.]

REFERENCE

Anderson, R.M. & May, R.M. (1979) Population biology of infectious diseases: Part 1. *Nature* **280**, 361–367.

microphagous feeders *See* FEEDING TYPES, CLASSIFICATION.

microtine cycles The enormous changes in population abundance characteristic of certain small rodents, particularly voles and lemmings, that lead to regular population explosions and crashes. The cycles occur with a fairly regular periodicity of between 2 and 5 years, but may vary widely in amplitude. Factors that have been proposed to explain the cycles include weather, food, predators, parasites, hormonal changes and behavioural changes. Despite intensive study, no clear explanation for the occurrence of these cycles has been found. *See also* POPULATION DYNAMICS; POPULATION REGULATION. [V.F.]

middle infralittoral zone *See* WITHIN-LAKE HABITAT CLASSIFICATION.

midoceanic ridge The longest mountain chain on Earth, extending through the ATLANTIC, INDIAN, Antarctic and South Pacific oceans, the NORWEGIAN SEA and the Arctic Basin to a distance of more than 56 000 km. The midocean ridges are areas in which new crustal material is forming as regions of the Earth's outer shell move apart. The midocean ridges extend to a height of 1–3 km above the sea floor and are generally more than 1500 km wide. They are roughly symmetrical, lie generally parallel to the CONTINENTAL MARGINS, are cut transversely by faults and exhibit high seismic activity and high heat flow, especially near the axial regions. The middle part of the ridges contain a rift valley, a deep notch or cleft in the ridge crest. [V.F.]

migration Usually refers to directional, synchronized movement of large numbers of a species towards a particular location or resource. Migration thus defined is different from DISPERSAL, which may involve only a few individuals and is often directed away from, rather than towards, a particular location. Migration may occur regularly over short distances on a very short time-scale (e.g. DIURNAL patterns of VERTICAL MIGRATION of marine and lake PLANKTON). The most spectacular examples of migration are animals that move seasonally over very long distances (e.g. migrating birds, ocean-going fish, whales, caribou and some butterfly species). Golden plovers, for example, fly over 13 000 km from their breeding grounds in the Arctic to winter-feeding ranges in South America. Studies of individually marked birds and fish have shown remarkable HOMING abilities. *See also* EMIGRATION; IMMIGRATION; OVERWINTERING. [R.H.S.]

REFERENCE

Baker, R.R. (1982) *Migration: Paths Through Time and Space*. Hodder & Stoughton, London.

Mimicry

Mimicry refers to striking similarities in the colour patterns and morphology of different species. Mimetic resemblances have been described in many groups of organisms. A dramatic example in plants involves the mimicry of the seeds of rice by a number of WEED species. The similarity in form and other characteristics increases the chance that the weed seeds contaminate the rice seed used to germinate a subsequent CROP; it therefore enhances their reproductive success and FITNESS. Several orchids have flowers that mimic females of specific species of insects, the males of which act as pollinators while attempting to copulate with the flowers. Certain species of *Passiflora* vines have specialized structures visually similar to the eggs of *Heliconius* butterflies. Experiments have demonstrated that the presence of these mimetic eggs reduces the chance that a female butterfly will lay her own eggs on the vine. The benefit for the vine is reduced herbivory due to fewer caterpillars hatching from the eggs.

It is in the animal kingdom that mimicry is most evident and has been best studied. In birds there is now much interest in the topic of EGG MIMICRY. Numerous bird species, including cuckoos and cowbirds, are brood parasites. They lay their eggs in the nests of other species, which then raise their young. Frequently the eggs of the parasitic species are closely similar in colour pattern to those of the host. This apparently reduces the chance that they will be rejected by the host parents. However, it is among the invertebrates, especially the insects, that mimicry can be especially dramatic and close. The description 'dramatic' is particularly apt in those examples that involve both intricate similarities in colour pattern and species known to have diverged from some common ancestor a long time ago and which thus have long independent evolutionary histories. The mimicry must then have involved processes of CONVERGENT EVOLUTION. We know now that these processes can involve very different genetic and developmental pathways to achieve the similarity in pattern in the two (or more) organisms.

Two fundamental forms of mimetic resemblance have long been

recognized. MÜLLERIAN MIMICRY occurs between two or more species (called a mimicry ring), each of which has a closely similar warning, or aposematic, colour pattern. Their colour patterns act as conspicuous signals to visually hunting predators of some form of TOXICITY, noxiousness or distastefulness. Insectivorous birds and lizards are especially important predators. Wasps provide an excellent example of a group of species with a warning colour pattern and chemical protection, in this case a toxic sting. Many laboratory experiments have confirmed that when so-called naive predators, which have been raised without any experience of a particular colour pattern, are presented with such unpalatable prey, they rapidly learn to avoid all individuals with the same pattern. The conspicuous nature of the aposematic colour pattern frequently involves a contrasting of patches or stripes of a bright red or yellow colour on a black background. A combination of red on black or of yellow on black may enhance the effectiveness of the processes of visual perception and avoidance learning in the predator. Bright or conspicuous colour patterns are sometimes hidden by an individual when at rest. They can then be abruptly exposed if the prey is disturbed by a potential predator, producing a STARTLE EFFECT that tends to fighten the predator away. A ritualized display is frequently associated with this STRATEGY. The display may include auditory responses and/or the production of offensive odours.

Müllerian mimics exhibiting the same warning coloration each derive some mutual benefit in terms of lower mortality because the efficiency of avoidance learning by predators increases with the rate of encounter of that particular colour pattern: the more individuals or species that express a particular pattern, the lower the probability of any one individual being killed during the learning process involving naive predators. This explains why selection favours the evolution of convergence among unpalatable species with some similarity in colour pattern to yield mimicry rings in which the member species show a close similarity to each other. Examples of this phenomenon can be found in bumble-bees, wasps and ladybird beetles in temperate regions and in numerous groups of butterflies in the tropics. Of the latter, *Heliconius* butterflies are an especially striking example (see Plate 20, facing p. 398). These brightly coloured butterflies are abundant in the tropical forests of Central and South America. The adults are unpalatable to birds principally because of TOXINS accumulated by their caterpillars from the *Passiflora* vines upon which they feed. The bright colour patterns, the gregarious nature of the butterflies and their slow, gliding flight ensure that these insects are conspicuous in the forest. Numerous species of insectivorous birds, for example jacamars, hunt in the forest. Genetic research on species of *Heliconius* has revealed how dramatic differences in colour pattern can result from comparatively small numbers of gene substitutions. The latter is said to occur when at a particular gene locus a MUTATION occurs, followed eventually by fixation of the new mutant allele (giving a new state of HOMOZYGOSITY and an ALLELE FREQUENCY of 100%).

The second major form of mimicry is BATESIAN MIMICRY. In contrast to the mutual benefit derived by the protected, unpalatable co-mimics within a Müllerian relationship, Batesian mimicry involves one (or more) palatable species exhibiting the same colour pattern as an unpalatable species. The latter species acts as a model for the former, which is the mimic. Although the mimic is perfectly palatable to birds or lizards, it derives some protection because potential predators sometimes mistake it for the unprofitable prey they have previously learnt to avoid. The phenomenon is of interest to evolutionary biologists partly because there is a conflict. This arises because while the mimic derives some protection in the presence of the model, the model species is more likely to be sampled by predators when a Batesian mimic is present. The degree of protection derived by the mimic is also frequency dependent: the more abundant the mimic relative to the model, the less well protected the mimic.

In temperate regions, wasps and bumble-bees represent the models for a number of palatable Batesian mimics, including species of moth, sawflies, beetles and hoverflies (Fig. M9). However, Batesian mimicry is most widely distributed in tropical ecosystems. Some Batesian mimics, such as the African mocker swallowtail butterfly, *Papilio dardanus*, are polymorphic, with colour forms mimicking different unpalatable models.

This GENETIC POLYMORPHISM is interesting for two reasons. Firstly, it appears to be a consequence of the FREQUENCY-DEPENDENT SELECTION associated with Batesian mimicry. In situations where unpalatable models with different colour patterns occur, it is beneficial to a Batesian mimic to evolve forms that mimic more than one of the models since each mimetic individual will then tend to be better protected. Secondly, the alleles of the genes determining the different polymorphic forms are tightly linked together on a chromosome so that they are effectively inherited as a single unit. This arrangement of the genes, known as a SUPERGENE, is necessary to avoid the production of mismatched phenotypes by recombination in meiosis and sexual reproduction. Recombinants of the alleles from two different polymorphic forms have sometimes been recovered both in the field and in laboratory crosses. Such individuals possess characteristics of more than one mimetic form and are consequently imperfect mimics with a lower probability of survival. It was initially considered that the tight linkage of the individual gene loci within such a supergene had evolved from a state of loose linkage by selection favouring any movement of the genes along the chromosomes to yield fewer recombinants. However, it is more likely that evolution has favoured those combinations of genes influencing the development of colour pattern that happen to be tightly linked together on a particular chromosome.

Fig. M9 The syrphid fly, *Metasyrphus americanus* (Diptera) is harmless to predators but gains protection by mimicry of the wasp's yellow and black pattern on the abdomen. (From Krebs, J. & Davies, N.B. (1993) *An Introduction to Behavioural Ecology*, 3rd edn. Blackwell Scientific Publications, Oxford.)

Batesian mimicry is thought to have evolved in a two-step process. The first step in the non-mimetic ancestor is that a mutation in a 'major' gene produces a discrete shift in colour pattern. This shift is sufficient to produce an imperfect mimicry of a model in which the novel mimic is sometimes mistaken by predators for the unpalatable species and not attacked. The second step then involves selection favouring 'modifier' genes of smaller effect that act to improve the resemblance, eventually producing close mimicry. Evidence for this scenario includes the finding that in crosses between geographical races of *P. dardanus* exhibiting different mimetic colour forms there is breakdown of mimicry: the major gene (or supergene) only produces perfect mimicry when present in the appropriate or coevolved genetic background. [P.M.B.]

---

mimicry ring *See* MIMICRY.

Minamata Classic case of environmental POLLUTION that affected human health. Minamata is a town on the west coast of Kyushu Island, Japan where methyl mercury contaminants of fish, the staple diet of the local population, caused severe disablement (numbness and seizures) and death. The source was a PVC plant that discharged an effluent containing both inorganic and organic mercury. The inorganic form was converted by fish into methyl mercury. *See also* ECOLOGICAL/ENVIRONMENTAL DISASTERS. [P.C.]

mine drainage These waters often emerge at low pH and rich in HEAVY METALS. Water ACIDIFICATION is due to the oxidation of pyrite (Fe_2S) and other sulphides exposed by the mining activities. H^+ is generated from sulphides in a series of complex reactions. The oxidation process is catalysed by bacteria that use pyrite as a source of energy. The extremely ACID conditions generated at the site lead to high dissolution of minerals, including heavy metals. Acid mine drainage is neutralized downstream of the source; as this happens FLOCCULATION and PRECIPITATION, especially of iron hydroxides, occur that are themselves damaging to the BENTHOS. Potentially toxic metals may be remobilized from such deposits if there is a subsequent drop in pH, for example during periods of high flow. *See also* POLLUTION. [P.C.]

REFERENCE

Hester, R.E. & Harrison, R.M. (1994) *Mining and its Environmental Impact.* Royal Society of Chemistry, London.

Mineral nutrients

Those NUTRIENTS that originate mainly by WEATHERING of minerals in the SOIL, including the MACRONUTRIENTS NITROGEN, PHOSPHORUS, POTASSIUM, CALCIUM, MAGNESIUM and SULPHUR, and the MICRONUTRIENTS IRON, MANGANESE, BORON, MOLYBDENUM, ZINC, CHLORINE and COBALT. Of these, nitrogen comes partly from the ATMOSPHERE but is still regarded as a mineral nutrient. The term is synonymous with 'inorganic nutrient'. All of these nutrients are found in the soil and the study of their behaviour in relation to the very complex chemical and physical environment of the soil is a large part of the discipline of soil science. In all cases, only a small fraction of the total nutrient content of the soil is available for uptake by higher plants. The nutrient may be present in organic form, its immediate origin being dead organic matter. In this case, it is generally unavailable until the organic molecules have been decomposed by the soil MICROFLORA. Alternatively, the nutrient may be in the form of a primary mineral and therefore unavailable, or attached to colloidal materials like HUMUS and CLAY particles and therefore only partially available. Nutrients in the soil solution are readily available and rapidly taken up by the roots.

Mineral nutrients are in the form of ions, either cations (positively charged) or anions (negatively charged); these ions are capable of 'sticking' to charged sites in the soil, of which there are usually many in the form of colloidal materials. In cation exchange, H^+ ions from the roots displace cations that are adsorbed on the soil COLLOIDS, including K^+, Ca^{2+} and Mg^{2+}, so that these nutrients get displaced and become part of the soil solution and freely available. Not all the colloids are cation exchangers; there are positive charges associated with some other colloids such as hydrous oxides of iron and ALUMINIUM and some clays, so anion exchange is also possible. Thus the colloidal substances in the soil (humus, clays, organic matter) are important reservoirs for mineral nutrients.

Mineral nutrients participate in nutrient cycles: the ELEMENTS taken up by the vegetation are finally returned to the soil as organic matter, then released by DECOMPOSITION only to be taken up again. The same elements are used over and over again in stable vegetation, such as an undisturbed forest or grassland, but where the cycles are broken by harvesting and cultivation then some of the nutrients are lost in LEACHING or as volatile forms (in the case of nitrogen). The most widely studied nutrient cycle is the NITROGEN CYCLE.

Historically, the manipulation of mineral nutrients, by the addition of plant and animal residues (manure), FERTILIZER from a bag or simply by the addition of LIME to raise the pH, has been one of the main methods of improving the yield of CROPS.

Nitrogen

Nitrogen is a chemical constituent of many organic molecules, especially AMINO ACIDS and PROTEINS. Leaf biomass contains 0.5–2% nitrogen on a dry mass basis. Much of the nitrogen in a leaf is in enzymes, with a very large proportion in the enzyme ribulose bisphosphate carboxylase oxygenase, involved in the ASSIMILATION of CARBON DIOXIDE (CO_2) by PHOTOSYNTHESIS. Nitrogen is a constituent of DNA, in which the genetic code is contained. When nitrogen, as NITRATE or ammonium, is added to plants they often grow faster, developing a higher proportion of leaf to root and exhibiting an enhanced rate of photosynthesis per area of leaf. Plants grown with excessive nitrogen may lack resistance to disease and may be easily damaged by stress. Nitrogen-deficient plants are stunted and their leaves turn yellow prematurely.

Phosphorus

Phosphorus is an important component of adenosine diphosphate (ADP) and adenosine triphosphate (ATP), used in energy transformations as the biochemical energy-storage molecule. Phosphorus also occurs in DNA. Cell membranes are composed of phospholipids. Leaf biomass contains 0.1–0.2% phosphorus. Phosphorus-deficiency symptoms include delayed flowering, spindly appearance and bronze-violet pigmentation of leaves and stalks.

Potassium

Potassium is an activator of many enzymes. K^+ is very mobile in the plant and participates in pumping ions across cell membranes, especially in the guard cells during stomatal opening. Leaf biomass typically contains 0.2–0.8% potassium, but it can be present in extremely high concentrations. Deficiency symptoms include drying of the leaf tips and curling of the edges of young leaves.

Calcium

Calcium forms calcium pectate in the middle lamella between plant cells and therefore cements cells together and influences the expansion of cells. It is also important as a chemical messenger, operating in association with calcium-pumping adenosine triphosphatases (Ca-ATPases). Leaves contain about 0.05–0.3% calcium. In calcium-deficient leaves, growth is disturbed and leaves and roots are deformed; leaf tips may dry out.

Magnesium

Magnesium is a constituent of chlorophyll and plays a part in the association of the two subunits of ribosomes. Foliage contains 0.03–0.1% magnesium. When deficient, old leaves turn yellow and growth is reduced. In conifers, magnesium deficiency is one of the signs of some forms of forest decline, in which leaves turn yellow or brown and lower leaves are shed prematurely.

Sulphur

Sulphur is a constituent of many amino acids and has an important role in protein structure. It is contained in the VITAMINS biotin, thiamine and B_1. It is typically present at a concentration of 0.02–0.2%. Sulphur-deficiency symptoms include chlorosis (yellowing) of young leaves.

Iron

Iron is present in peroxidase, catalase and cytochrome oxidase enzymes; it is found in ferredoxin, a molecule involved in OXIDATION–REDUCTION reactions. It is needed in chlorophyll formation and iron chlorosis in extreme cases can make young leaves look white (veins remain green). Leaves contain 0.01–0.4% iron.

Manganese

Manganese activates decarboxylase, dehydrogenase and oxidase enzymes; it is important in photosynthesis and nitrogen metabolism. Leaf tissue typically contains 0.01% manganese.

Boron

Boron activates some dehydrogenase enzymes, facilitates sugar translocation and synthesis of nucleic acids and is essential for cell division. Leaf tissue contains 0.001–0.01% boron.

Molybdenum

Molybdenum is present in nitrogenase and nitrate reductase enzymes; it is essential for NITROGEN FIXATION in leguminous species and for nitrogen assimilation. Leaf tissue contains less than 0.01% molybdenum.

Copper

Copper is present in oxidase enzymes and has some importance in photosynthesis and in protein and CARBOHYDRATE metabolism. Copper is usually less than 0.005% of biomass.

Zinc

Zinc is a constituent of several dehydrogenase, proteinase and peptidase enzymes; it promotes starch formation and seed production. Leaf tissue usually contains less than 0.03% zinc.

Chlorine

Chlorine influences photosynthesis and seems to be important in the growth of roots. It is necessary in minute quantities.

Cobalt

Cobalt is required for the symbiotic fixation of nitrogen. It is necessary in minute quantities.

Horticulturists, agronomists and foresters have often proposed optimal nutrient concentrations based on foliar analysis and crop-specific deficiency symptoms. For example, from a study of young pine trees (*Pinus sylvestris*) in nutrient solutions, Ingestad (1962) proposed the following foliar nutrient concentrations (per cent of dry mass) as optimal for growth: nitrogen, 2.4–3.0; phosphorus, 0.15–0.40; potassium, 0.9–1.6; calcium, 0.04–0.30; magnesium, 0.12–0.18; sulphur, 0.12–0.25. Growth is reduced and deficiency symptoms occur when foliar concentrations are below the following: nitrogen, 0.7–1.6; phosphorus, 0.06–0.10; potassium, 0.3–0.4; calcium, 0.05; magnesium, 0.06–0.09; sulphur, 0.07. Below these concentrations, deficiency symptoms may be as follows:

nitrogen: needles pale and small, yellow-green;
phosphorus: needles dull bronze green or darker green;
potassium: yellow needle tip merging gradually into a yellow-green or green base;
magnesium: yellow needle tip, with abrupt transition to green base;
calcium: young needles chlorotic and faded at tips, becoming brown-tipped.

[J.G.]

REFERENCES
Brady, N.C. (1990) *The Nature and Properties of Soils*, 10th edn. Macmillan, New York.
Hewitt, E.T. & Smith, T.A. (1975) *Plant Mineral Nutrition*. Unibooks, London.
Ingestad, T. (1962) Macro element nutrition of pine, spruce, and birch seedlings in nutrient solutions. *Meddelanden från Statens Skogsforskningsinstitut* **51**, 1–50.
Jordon, C. (1985) *Nutrient Cycling in Tropical Forest Ecosystems*. John Wiley & Sons, New York.
Marschner, H. (1995) *Mineral Nutrition of Higher Plants*. Academic Press, London.
Underwood, E.J. (1981) *The Mineral Nutrition of Livestock*, 2nd edn. Commonwealth Agricultural Bureaux, Farnham.

mineralization The natural process whereby MINERAL NUTRIENTS such as NITROGEN, SULPHUR and PHOSPHORUS are released from immobile organic forms in the SOIL and thus become available as inorganic ions for uptake by plants. [J.G.]

REFERENCE
Brady, N.C. (1990) *The Nature and Properties of Soils*, 10th edn. Macmillan, New York.

minimal area The smallest area of vegetation that can be examined and still supply an acceptably representative sample of the FLORISTICS of the site. It is determined by a sequential enlargement of a sample plot and determining the plant species present at each successive sample size. Graphical expression of these data usually displays a rapid initial rise in species number with plot size that gradually diminishes for larger samples. A failure to become level, or a further phase of increasing species, indicates heterogeneous vegetation and the inclusion of a new habitat type. The minimal area is defined as the plot size corresponding to the point on the graph where levelling-off commences. [P.D.M.]

minimum viable population (MVP) This is a popular concept in CONSERVATION BIOLOGY and is defined as the threshold POPULATION SIZE below which rapid extinction is virtually guaranteed. Clearly, in general, smaller populations will be more vulnerable than larger ones, although it has not been established, either theoretically or empirically, that there is some critical population size below which the vulnerability to extinction increases suddenly. It is perhaps more useful to estimate extinction probability as a function of time for different population sizes than to identify some specific MVP.

Shaffer (1981) defined an MVP size as that which would have at least a 99% chance of surviving 1000 years. Mace and Lande (1991) suggested that 'endangered' could be defined as a 20% chance of extinction in 20 years. *See also* ENDANGERED SPECIES; EXTINCTION MODELLING. [P.C.]

REFERENCES
Mace, G.M. & Lande, R. (1991) Assessing extinction threats: towards a re-evaluation of IUCN threatened species categories. *Conservation Biology* **5**, 148–157.
National Research Council (1995) *Science and the Endangered Species Act*. National Academy Press, Washington, D.C.
Shaffer, M.L. (1981) Minimum viable population sizes for species conservation. *Bio Science* **31**, 131–134.

Mire

This term has been used in a variety of ways, some specific and many broad and general. Over the past 30 years it has been used increasingly of PEAT-forming habitats and ecosystems and has proved a most valuable general term in this area.

Peat-forming ecosystems are those in which gross primary production is not equalled by community respiration, leading to an excess of organic material accumulating within the ecosystem as peat. In an equilibrium ecosystem (if such a thing exists) one would expect the respiratory activities of plants, animals and microbes (producers, consumers and decomposers) to account for all the energy entering the system via photosynthesis. In a successional ecosystem, this totality of ecosystem respiration does not equal the production rate and hence energy is added to the reservoirs of the system, either in biomass or in soil organic matter, or both. In mire ecosystems, detritivore and decomposer respiration is suppressed by waterlogging, leading to an accumulation of waterlogged soil organic material, the peat. The accumulation of this organic reservoir in the ecosystem is more closely related to the reduced respiration of decomposers rather than high productivity on the part of the vegetation. The proportion of net primary production that accumulates as peat varies considerably; in sheep-grazed upland BLANKET MIRE in England it was estimated at 15%.

DECOMPOSITION within the peat occurs largely, but not exclusively, within the upper, partially aerobic layer (ACROTELM). The fact that decomposition continues as a consequence of the activity of anaerobic bacteria in the deeper, permanently waterlogged layers (CATOTELM) implies that there is a limit to the depth of peat that can accumulate under any given set of environmental conditions. When the respiratory activity of decomposers within the total peat column added to the respiration of the autotrophs and heterotrophs in the peat-forming community on the surface equals the gross primary productivity, then peat formation will cease. Only a change in conditions (such as a shift in climate or a change in vegetation) can reinstate peat formation.

As a form of wetland, therefore, mire ecosystems are dependent on hydrological conditions for their formation and maintenance. An adequate water supply is a prerequisite for mire development in order to produce the waterlogging necessary for peat formation and this water may be derived from groundwater input and flow or directly from rainwater. Mires that derive their water from both sources are termed 'RHEOTROPHIC', i.e. they are literally fed by the flow of water through the system. Mires that have their surface elevated above the influence of groundwater (*see* RAISED BOG) are termed 'OMBROTROPHIC', fed by precipitation. Clearly, the latter type of mire can occur only under climatic conditions where the supply of precipitation (usually rainfall) is adequate, after allowing for EVAPORATION and TRANSPIRATION losses, to maintain a high WATER TABLE. Such mires are also dependent on the water-holding, impermeable nature of the underlying peats to ensure that drainage is slow (usually lateral through the surface layers of the growing peat). In tropical countries, ombrotrophic mires can exist only in highland areas, where evaporation is lower, or in lowlands where conditions for water retention are particularly good, as in the BOG forests of Sarawak.

Rheotrophic mires normally have a richer supply of MINERAL NUTRIENTS than ombrotrophic mires because of their groundwater supply, but this varies considerably with the chemical quality of the groundwater (depending upon the geology of the CATCHMENT) and the velocity of flow of water through the ecosystem. A fast water supply will provide a better nutrient supply to vegetation than a slow-moving one, even where the nutrient concentrations in the water are low. Rheotrophic mires are given the general term 'FEN' and they can be rich or poor according to the quality and flow rate of their water supply. Strictly, fen ecosystems need to have a sufficiently elevated peat surface for the water table to be at or below it during the dry (summer) season. In successional sequence fen replaces SWAMP (American terminology would use the expression 'marsh'), which is a productive wetland ecosystem often dominated by tall herbaceous plants, such as reeds (*Phragmites australis*), cattails (*Typha* spp.), sedge (*Carex* spp.) or papyrus (*Cyperus papyrus*), and which has a water table permanently above the SEDIMENT surface.

Sloping fens, especially when extensive as in Fennoscandia and North America, can develop complex structures of transverse ridges and elongate pools. In Finland these are known as AAPA MIRES and in America they have often been called string bogs, although the use of the term 'BOG' here is inadvisable since the expression should be reserved for ombrotrophic mires.

Rheotrophic mires may become invaded, and even dominated, by trees, such as willows (*Salix* spp.), alders (*Alnus* spp.) or swamp cypress (*Taxodium distichum*), when they can be termed 'CARR' (American terminology uses the expression 'swamp' for such forested WETLANDS). These are high biomass mires, but while the summer groundwater is relatively low they cannot accumulate peat rapidly because of high microbial (especially fungal) decomposition. In coastal locations, geological subsidence can operate to maintain a high water table and accelerate peat formation.

Ombrotrophic mires succeed rheotrophic mires in successional development and the resulting peat accumulation elevates the mire surface above the influence of groundwater, leading to their being entirely supplied by rainfall. Therefore, their nutrient supply is inevitably more limited and is closely related to the source of precipitation, those mires in coastal or oceanic regions receiving a more abundant supply of elements than more inland or continental sites. The term 'bog' should technically be reserved for ombrotrophic mires. It is occasionally used of rheotrophic mires that happen to have a poor nutrient supply (such as 'valley bog'), but this is a misuse of the term (VALLEY MIRE is preferable). True bogs include raised bogs and blanket bogs, the former being generally confined to specific basins or valley floors, whereas the latter are unconfined by topography.

Mires often occur in complex mixtures containing both fen and bog elements and these at a range of different scales. Raised bog systems have lateral areas of drainage and groundwater movement forming fen and carr vegetation in what is often termed 'lagg'. Aapa fens have elevated ridges that remain permanently out of reach of the moving waters and become dependent on rainfall; hence they develop a bog vegetation. Areas of blanket mire often contain valleys with saddle-shaped peat systems with raised surfaces and sloping areas of poor-fen vegetation. The fact that mire types often develop from one another in a successional sequence can add to their complexity. Thus the classification of mires on the basis of their morphology and HYDROLOGY can be difficult and often consists of drawing lines across a CONTINUUM of variation. [P.D.M.]

REFERENCES

Clymo, R.S. (1984) The limits to peat bog growth. *Philosophical Transactions of the Royal Society of London, Series B* **303**, 605–654.

Lindsay, R. (1994) *Bogs: the Ecology, Classification and Conservation of Ombrotrophic Mires*. Scottish Natural Heritage, Edinburgh.

Moore, P.D. & Bellamy, D.J. (1974) *Peatlands*. Elek Science, London.

---

mitigation, compensatory Compensatory mitigation is a political/legal concept that presumes the destruction of natural habitats or populations can adequately be offset by actions elsewhere. Examples include the creation of artificial wetlands to compensate for the destruction of natural wetlands and attempts to translocate affected populations to another site. The degree to which an artificially created habitat or population is ecologically equivalent to, and adequately performs the same ECOSYSTEM functions as, the naturally occurring ones destroyed in the process is a matter of active debate and may differ significantly from site to site and species to species. *See also* RECLAMATION. [E.O.G. & L.R.M.]

mixed-function oxidase (MFO) Mixed-function oxidases (or monooxygenases) are a group of enzymes that catalyse the conversion of lipophilic substrates (organic xenobiotics) to more polar, i.e. water-soluble, products. They occur in smooth endoplasmic reticulum. They require OXYGEN, NADPH and a cytochrome P450 isozyme; they are inducible. MFO activity is used as a BIOMARKER. *See also* DEFENCE MECHANISMS. [P.C.]

mixed strategy *See* EVOLUTIONARILY STABLE STRATEGY; GAME THEORY.

mixing zone Body of water (or air) immediately next to DISCHARGE in which dilution is taking place and hence in which the concentration of a CONTAMINANT may exceed the maximum allowable level. Usually defined by the regulatory authority. [P.C.]

mixolimnion *See* LAKE STRATIFICATION.

MLC *See* MEDIAN LETHAL CONCENTRATION.

mobbing The close attendance and alarm behaviour by certain vertebrates directed towards potential predators that serves to distract the attention of the PREDATOR and direct the attention of CONSPECIFICS and HETEROSPECIFICS to the predator.

Mobbing is a specific behavioural pattern performed by fish, reptiles, birds and mammals directed towards potential predators or mimics of predators. Mobbers approach predators closely, emit alarm vocalizations and perform other kinds of conspicuous behaviour. The close approach of a potentially dangerous predator entails a cost to the mobbing individual in terms of risks of predation, since mobbers sometimes are killed by the predator. Experiments with models of predators have shown that mobbing is elicited particularly by the beak and claws of predatory birds but less so, or not at all, by less-dangerous parts of the predator. Mobbing is a special kind of predator–prey COMMUNICATION, signalling that the predator has been detected and that further pursuit by the predator of the agile and aware mobber is unprofitable. Mobbing also functions to direct the attention of naive KIN to potentially dangerous predators and future dangers can thereby be avoided. The amount and intensity of mobbing can be considered a measure of the willingness of individuals to run risks and should therefore reflect PARENTAL INVESTMENT in reproduction. Consistent with this view, the sex that invests the most in reproduction also mobs predator models most intensely, the intensity of mobbing increases as offspring becomes large and parents mob predator models more when they have larger broods. *See also* BEHAVIOURAL ECOLOGY. [A.P.M.]

mode This MEASURE OF CENTRAL TENDENCY or average is simply the value of the most common observation. It is the 'peak' of the DISTRIBUTION. Most distributions in ecology have one mode (unimodal), although with SKEWNESS the mode may be asymmetrical (see Fig. M6). A distribution with two modes is called bimodal (*see* BIMODAL DISTRIBUTION) and one with many modes is called multimodal. *Cf.* MEAN; MEDIAN (SECOND QUARTILE). [B.S.]

model A simplified representation of the real world (e.g. a plastic crocodile heart) that aids understanding. In all models, some components of the real world are realistically represented and some relationships between these components are realistically represented. Notice, however, that only some components and relationships are realistically represented (e.g. the correct number and size of chambers in the model crocodile heart), while some aspects will not be realistically represented (e.g. the plastic of the crocodile heart). Notice that the plastic heart can be viewed as a perfectly good model of the real heart and provide useful insights into how a real heart might work, even though it is not real in all aspects. In ecology, models of the real world are usually constructed from EQUATIONS (*see* LOTKA–VOLTERRA MODEL) rather than from plastic. *See also* MATHEMATICAL MODELS. [B.S.]

REFERENCE

Kingsland, S.E. (1995) *Modelling Nature*, 2nd edn. University of Chicago Press, Chicago, Ill.

models of exponential growth EXPONENTIAL GROWTH is characterized by a constant INTRINSIC RATE OF INCREASE r and a constant NET RATE OF INCREASE R. The simplest models ignore AGE STRUCTURE. Continuous breeding may be represented by a simple differential equation, $dN/dt = rN$, where N is POPULATION SIZE at time t. Discrete generations are better represented by a difference equation, $N_{t+1} = RN_t$ where N_t and N_{t+1} represent population sizes in successive generations. Both of these models are characterized by an exponential increase in population size N with time t or a linear increase in the logarithm of population size with time. Age structure may be included in models of exponential growth, as in the Leslie matrix model where an exponential growth curve with time is only observed as the numbers in different AGE CLASSES in the population approach STABLE AGE DISTRIBUTION. *See also* MODEL. [R.H.S.]

models of interspecific competition Models, either mathematical or graphical, that attempt to describe (and therefore to help understand) the process and consequences of INTERSPECIFIC COMPETITION. *See also* AGGREGATION MODEL OF COMPETITION; HASSELL MODEL AND HASSELL–COMINS MODEL; LOTKA–VOLTERRA MODEL; TILMAN'S MODEL OF RESOURCE UTILIZATION. [B.S.]

models of intraspecific competition Models of INTRASPECIFIC COMPETITION incorporate a relationship between the PER CAPITA rate of increase and POPULATION DENSITY N. The simplest model is the LOGISTIC EQUATION, where the instantaneous rate of increase decreases linearly with increasing N, reaching the value 0 at the CARRYING CAPACITY or equilibrium population density when $N = K$. The differential equation form of the LOGISTIC MODEL shows SIGMOID GROWTH when N is plotted against time t, with a smooth approach to the carrying capacity. Logistic models for populations with DISCRETE GENERATIONS are expressed as DIFFERENCE EQUATIONS, for example $N_{t+1} = N_t\exp[r(N_t - K)/K]$, where the instantaneous POPULATION GROWTH RATE decreases linearly with N_t, or $N_{t+1} = N_tR/(1 + aN_t)$, where the finite population growth rate decreases linearly with N_t and $a = (R-1)/K$. Both of these models show a range of dynamic behaviour as the INTRINSIC RATE OF INCREASE r or the NET RATE OF INCREASE R increase, from STABLE EQUILIBRIUM K through LIMIT CYCLES to deterministic CHAOS.

Hassell (1975) proposed a model of intraspecific competition that incorporated an additional parameter b to describe a range of competition effects, from scramble to contest. Hassell's equation is $N_{t+1} = N_tR/(1 + aN_t)^b$. The new parameter b describes the slope of the relationship between the generation k-VALUE ($=\log_{10}(N_t/N_{t+1})$) and N_t. When $b = 1$, there is EXACT COMPENSATION (contest) and increasing b leads to OVERCOMPENSATION (scramble). *See also* CONTEST COMPETITION; HASSELL MODEL AND HASSELL–COMINS MODEL; SCRAMBLE COMPETITION. [R.H.S.]

REFERENCES

Hassell, M.P. (1975) Density dependence in single-species populations. *Journal of Animal Ecology* **40**, 473–486.

May, R.M. (1981) Models for single populations. In: *Theoretical Ecology: Principles and Applications* (ed. R.M. May), 2nd edn, pp. 5–29. Blackwell Scientific Publications, Oxford.

models of mutualism Models of MUTUALISM have not yet generated the same level of interest as those of other biotic interactions, such as COMPETITION or predation. Simple models of mutualism, such as modifications of Lotka–Volterra equations to incorporate reciprocal positive effects on growth rates between species, give unrealistic results as paired populations expand without constraint. Equilibrium conditions can none the less be generated when negatively density-dependent mutual rewards are included in the models, a modification that has gained empirical support with the realization that non-symbiotic mutualisms are typically 'conditional', varying in the strength of the benefits accrued depending on local conditions. [S.G.C.]

Models of parasitoid–host interaction

The first models describing the population-dynamic interactions between PARASITOIDS and their HOSTS were developed by Thompson and Nicholson in the 1920s and 1930s. While Thompson thought that parasitoids were primarily limited by their egg supply, Nicholson considered host limitation to be more important and it is his ideas that have proven more influential. The Nicholson–Bailey model applies to hosts and parasitoids with discrete, perfectly synchronized generations, as is often found in temperate PARASITOID–HOST INTERACTIONS. The model assumes that parasitoids search randomly and that the average number of parasitoid encounters per host over the season is the product of parasitoid density (P) and a scaling constant (a) normally called the SEARCHING EFFICIENCY or attack rate. The proportion of hosts escaping PARASITISM is given by the zero term of a POISSON DISTRIBUTION with mean aP (which is $\exp(-aP)$). So the proportion of hosts parasitized is simply $1-\exp(-aP)$. The numbers of hosts in the next generation is given by the number of hosts escaping parasitism multiplied by net host FECUNDITY. The number of parasitoids in the next generation is given by the number of parasitized hosts multiplied by a constant representing the number of parasitoids emerging per host. The Nicholson–Bailey model is a coupled difference equation whose equilibrium and stability properties are easy to analyse. The model always has an UNSTABLE EQUILIBRIUM and predicts diverging OSCILLATIONS. That the simplest host–parasitoid model does not predict a persistent interaction has led to much research on factors that may stabilize this type of interaction.

Functional response and interference

Early work focused on the parasitoid FUNCTIONAL RESPONSE and INTERFERENCE. Type 1 and type 2 functional responses tended to increase the tendency for diverging oscillations, as they weaken the ability of the parasitoid to control an expanding host population. Type 3 functional responses could contribute to stability, as over some ranges of host densities parasitoids respond to increasing host densities by an accelerating rate of host parasitism. However, for most realistic type 3 functional responses, their effect on stability is likely to be small. Interference is a reduction in parasitoid PER CAPITA efficiency with increasing parasitoid density. It exerts a stabilizing influence because it acts to prevent high densities of parasitoids from overexploiting their host population. Although interference can be demonstrated in laboratory experiments, most workers today believe that interference is unlikely to be strong enough in the field to be a major influence on host–parasitoid POPULATION DYNAMICS.

Heterogeneity of risk

The major factors contributing to persistent host–parasitoid interactions are currently considered to revolve around processes that tend to contribute to increased VARIANCE in the risk of parasitism across host populations. If some hosts, for whatever reason, are protected or partially protected from parasitoid attack, then the host population will suffer less catastrophic MORTALITY when parasitoid densities are high and this acts to damp the tendency for the system to display diverging oscillations. A number of factors have been shown to promote stability by increasing the variance in host risk.

1 Some hosts may exist in a physical refuge from parasitoid attack.
2 Some hosts may have the ability to defend themselves physically or physiologically from parasitism (*see* PARASITOID–HOST INTERACTIONS).
3 Hosts may be distributed heterogeneously across patches in the environment and parasitoids may tend not to visit certain patches.
4 The host and parasitoid populations may not be perfectly synchronized in time so that some hosts enjoy a reduced risk of parasitism.

Most of these processes can be studied by models derived from the Nicholson–Bailey model; for a wide class of phenomena, stability is guaranteed if the coefficient of variance (squared) in the risk of parasitoid attack is approximately greater than unity (the $CV^2 > 1$ RULE). It was once thought that if hosts were distributed unequally across patches and if parasitoids tended to aggregate to patches of high host density, then stability would be enhanced. This is true if parasitoids aggregate to high-density patches at the beginning of the season and then remain in that patch. However, if parasitoids move between patches during the season then the stabilizing effect of density-dependent AGGREGATION is weakened or destroyed. Mechanisms that increase the variance in host risk simultaneously introduce a density-dependent reduction in per capita parasitoid efficiency. Because interference has a similar effect, these mechanisms are sometimes described as examples of PSEUDOINTERFERENCE.

Spatial processes

Recent work has begun to consider spatial models of host–parasitoid interactions. A metapopulation of individually unstable subpopulations (for example described by the Nicholson–Bailey model) connected by MIGRATION can be persistent. As long as different subpopulations oscillate out of synchrony then a rescue effect is possible, whereby a particular subpopulation near extinction can be rescued by migration from out-of-phase subpopulations. Any factor that tends to synchronize subpopulations (external environmental forcing, very frequent migration) leads to instability, as does very low levels of migration which destroys the rescue effect. Spatially explicit models consisting of a locally connected array (lattice) of host–parasitoid interactions have also been studied. These models show waves of (parasitoid) pursuit and (host) evasion that may be spatially well organized into discrete spirals or that exhibit more complex, apparently chaotic, spatial structures.

Age-structure

A second recent trend has been the development of continuous-time models of host–parasitoid interactions, which assume overlapping as opposed to DISCRETE GENERATIONS. The continuous time equivalent of the Nicholson–Bailey model is the LOTKA–VOLTERRA MODEL, which predicts neutral as opposed to divergent oscillations. However, the Lotka–Volterra model implicitly assumes no host and parasitoid developmental lags: their incorporation into a modified Lotka–Volterra model (expressed as time-delayed differential or integral equations) leads to dynamic behaviour that has much in common with their discrete generation analogues. One new feature predicted by these models is the stabilizing influence of long-lived host stages that are immune from parasitoid attack. These 'storage stages' act as a refuge, allowing the host population to survive periods of heavy parasitoid attack. Where hosts and parasitoids have different developmental periods, the two TIME-LAGS can interact dynamically to give persistent cycles. In particular, if the parasitoid developmental lag is approximately half that of the host (and if adult hosts and parasitoids are relatively short-lived), cycles in both populations with a period of approximately one host generation are predicted. Such cycles have been observed in a number of tropical and semi-tropical host–parasitoid systems. Temperate systems with a number of semi-overlapping generations per year can be studied using a hybrid between the two approaches.

Classical host–parasitoid models consider a single species of host and parasitoid. These models have been extended to consider interactions between hosts, parasitoids and hyperparasitoids; between two hosts and a common parasitoid (*see also* APPARENT COMPETITION); and between two parasitoids attacking a common host (*see also* COMPETITION). Many host–parasitoid interactions in nature are embedded in a complex web of interactions, with most parasitoids attacking a number of hosts and most hosts subject to many parasitoid species. Attention is turning to models of host–parasitoid communities, although the subject is still in its infancy.

Much interest in host–parasitoid interactions has stemmed from the importance of parasitoids as BIOLOGICAL CONTROL agents in agriculture and forestry. Models of the type described above have been parameterized to describe specific host–parasitoid interactions and to act as decision tools in biological control. In addition, another school of modellers have used techniques from systems theory to build large simulation models designed to mimic the dynamic behaviour of specific systems. Such models frequently include the explicit representation of the dynamics of the host's food plant, as well as external processes such as the climate and even the economy. *See also* HYPERPARASITOID; METAPOPULATIONS; NICHOLSON–BAILEY EQUATIONS. [H.C.J.G.]

REFERENCES

Hassell, M.P. (1978) *The Dynamics of Arthropod Predator–Prey Systems*. Princeton University Press, Princeton, N.J.

Hassell, M.P. & Godfray, H.C.J. (1992) Insect parasitoids. In: *Natural Enemies* (ed. M.J. Crawley), pp. 265–292. Blackwell Scientific Publications, Oxford.

Models of predator–prey interaction

Symbolic abstractions of natural, exploitative relationships between species. The exploiter is the PREDATOR or enemy; its victim is the prey. Predator–prey models allow deductions about the POPULATION DYNAMICS and POPULATION SIZES of the interacting species.

Herbert Spencer produced the first model of PREDATOR–PREY INTERACTION in the mid-19th century. It was purely verbal, but from it he deduced that predatory interactions would likely lead to population OSCILLATIONS in predator and prey. Alfred Lotka and Vito Volterra independently produced the first mathematical model of predator–prey interaction in the 1920s. They stripped away all but four of the two species' natural features: b, PER CAPITA birth rate of prey; d, per capita death rate of predators; k, predation rate; and β, per capita birth rate of predators per predation event. Then they combined the birth and death rates of each species by assuming their additivity and wrote a pair of simultaneous DIFFERENTIAL EQUATIONS:

$$dV/dt = bV - kVP$$

$$dP/dt = \beta kVP - dP$$

where V and P are the population densities of predator and victim, respectively. In this model, notice that in the absence of the predator, the prey density grows infinitely, whereas in the absence of the prey, the predator becomes extinct.

Inadequate though it may be, Volterra used this model to produce a robust theory of BIOLOGICAL CONTROL. That theory has weakened only a tiny amount as the model has been improved. Also, the model already suggests many of the central questions that ecologists have asked of predator–prey dynamics. How do the densities behave? What is the equilibrium point of the system? What is the dynamic stability of the equilibrium point? The answers for this model are as follows.

1 Equilibrium point $(V^*,P^*) = (d/\beta k, b/k)$, where asterisks indicate equilibrium values.

2 Densities of both species oscillate (unless the system is at equilibrium).

3 The dynamics are neither stable nor unstable. The system tends to stay in whatever oscillatory trajectory it finds itself.

In the 1930s, G.F. Gause tested this model in his laboratory with mixed results. His systems did oscillate. However, the oscillations of *Paramecium* and yeast followed a LIMIT CYCLE: they had a stable trajectory rather than maintaining their initial trajectory; and the oscillations of *Paramecium* and *Didinium* were unstable unless he reduced the predation rate by adding habitat complexity (in the form of pebbles) to his aquaria.

Adding intraspecific victim COMPETITION greatly improved the basic model:

$$dV/dt = bV\left[(K - V)/K\right] - kVP$$

$$dP/dt = \beta kVP - dP$$

where K is victim CARRYING CAPACITY. This model may not even have an equilibrium point in the positive quadrant: predators may become extinct deterministically, leaving their victims at K. If there is an equilibrium, it will be oscillatory like the basic model. However, the oscillations will converge on equilibrium; the model is stable. Because it is relatively easy to solve, this model is still used in complex multispecies computer simulations of food webs.

A much better model results from changing the predator FUNCTIONAL RESPONSE (i.e. the rate at which an individual predator kills prey). A non-linear response models predatory HUNGER and SATIATION. There are two alternatives.

1 Half-saturation model: $kV/(\chi+V)$, where χ, the half-saturation constant, is the victim density at which a predator kills at $k/2$, i.e. half its maximum rate. This response model was conceived independently at least three times. Its most popular form, HOLLING'S DISC EQUATION, is known as the type 2 response, although the type 2 and the half-saturation forms are mathematically identical. They are also identical to the Michaelis–Menten equation of enzyme chemistry.

2 Ivlev model: $k(1 - e^{-cV})$, where c is a constant proportional to the predator's search speed.

Incorporating the half-saturation response into the whole model yields:

$$dV/dt = bV\left[\left(K - V\right)/K\right] - kVP/\left(\chi + V\right)$$

$$dP/dt = \beta kVP/\left(\chi + V\right) - dP$$

This model introduces the possibility of an UNSTABLE EQUILIBRIUM. The Ivlev model may be incorporated in the same way and also introduces this possibility. Instability in these cases would arise if predators keep the victim population so low at equilibrium that the primary effect of adding more victims would be to satiate predators rather than to add competing victims (*see* PREDATOR SATIATION).

Many other algebraic elaborations of predation models attempt to reconcile empirical work with theory. Some of these concentrate on particular interactions, like sheep and blowflies. Some are more general. The three most noteworthy general models deal with issues of wide application.

Andrei Kolmogorov introduced TIME-LAGS to predator–prey models. These make reproduction sensitive to preceding rather than current conditions. Time-lags can introduce instability to predator–prey dynamics. Kolmogorov also analysed a very general predator–prey model and showed that predator–prey oscillations can never touch the density axis of either species; oscillations are thus confined to the positive quadrant where neither species becomes extinct. This result depends on strict determinism in the model and its restriction to two species. Stochastic models do allow for extinction of either species. Deterministic multispecies models can also produce extinctions even without stochasticity.

Another class of predator–prey models allows direct predator interaction. Previous models tacitly assumed that predators interacted only indirectly by eating each other's food supply. To model direct interactions, a term θP is added to the half-saturation constant:

$$dV/dt = bV\left[\left(K - V\right)/K\right] - kVP/\left(\chi + V + \theta P\right)$$

$$dP/dt = \beta kVP/\left(\chi + V + \theta P\right) - dP$$

If predators compete directly (by fighting over food or defending territories), $\theta > 0$, but if $\theta < 0$ the interaction is mutualistic. Thus, the model also handles predators that exhibit features like cooperative hunting. Direct competition amongst predators adds stability to the interaction; MUTUALISM reduces stability.

Adding an exponent λ to V in the victim self-limiting term allows that term to fit a variety of curvatures:

$$dV/dt = bV\left[\left(K - V^{\lambda}\right)/K\right] - kVP/\left(\chi + V + \theta P\right)$$

Models to this point had limited the death rate of predators to a constant d. However, without food, some predators can all die very quickly. To allow this possibility, Wayne Getz revolutionized the predator's equation. He discarded its additivity and included the effect of the predation in an expanded version of a LOGISTIC EQUATION:

$$dP/dt = \rho P\left\{1 - \left[\zeta\left(\chi + V + \theta P\right)/kV\right]\right\}$$

where ζ is the value of V at which $dP/dt = 0$. In the laboratory and the field, the oscillations of predator and prey are asymmetrical—populations grow slowly and decline precipitously. But most simple models result in symmetrical waves—populations take just as long to decline as they did to grow. Without adding to the number of coefficients, the Getz model frees predator–prey interactions from such unrealistic wave forms. It also pays more attention to the biological functions that determine the dynamics (Fig. M10).

Predation models may result in an additional dynamical behaviour called CHAOS. Chaotic equilibria, belying their name, are fully deterministic, but trajectories around them are so dependent on initial conditions and so variable from oscillation to oscillation that they resemble those dominated by stochastic influences.

Isocline analysis constitutes a parallel approach to modelling predation with explicit equations. Instead of specific equations, it combines properties of the interaction and inequalities of the interaction into robust models that apply to whole classes of equations. The first such model used the definition of predation and the assumptions that (i) the underlying, unspecified equations are autonomous and continuous; (ii) victims compete directly with each other; (iii) predators do not; (iv) predators are limited only by their victim supply; (v) victims have a positive K value; and (vi) predators reproduce successfully if, and only if, V exceeds a threshold density specific to each system. These assumptions lead by deduction to a graph of the shapes and relationships of its zero isoclines. (The zero isocline of x in a dynamic system is the set of points (x,y) at which $dx/dt = 0$. For the y zero isocline, $dy/dt = 0$.)

Isocline analysis can combine with linearized STABILITY ANALYSIS, i.e. analysis based on properties of the isoclines in the neighbourhood of their equilibria. The combination predicts many of the properties of predatory systems. An equilibrium will be stable if victims compete with each other in its neighbourhood. It will be unstable if they are mutualistic. Thus, the better a predator at reproducing on a short supply of victims, the more likely the equilibrium will be unstable. Equilibria will often be oscillatory, but strong

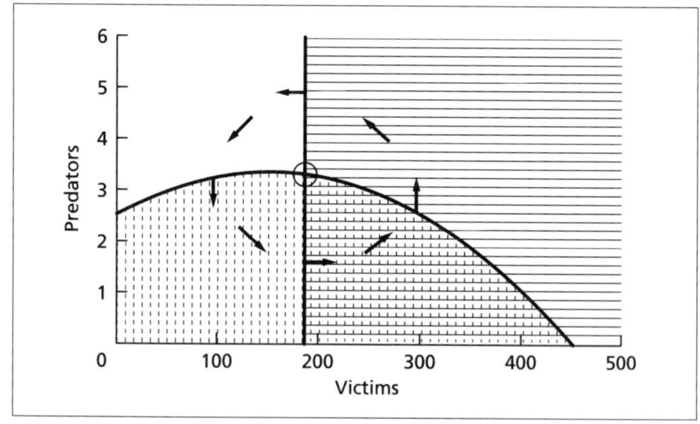

Fig. M10 An isocline diagram for a Getz model of predation with $\theta = 0$. The thick lines are the zero isoclines: $dP/dt = 0$ for the vertical one; $dV/dt = 0$ for the arched one. Fill patterns mark regions of population growth: the predator population grows to the right of its isocline; the victim population grows under its isocline. There is an equilibrium (V^*, P^*) at the circled intersection of the isoclines. Vectors reveal the general direction of population movement in this graph. Linearized stability analysis shows that the equilibrium is a stable focus: trajectories should converge on the equilibrium in a series of oscillations.

competition amongst victims eliminates the oscillations. The same interactions amongst predators have similar but weaker effects. Intraspecific interactions of the two species are not additive; instead, there are thresholds of mutualism that confer instability no matter how strong the competition in the other species. Finally, optimal use of habitats that differ in their predation coefficients can serve to stabilize the interaction (the prey-refuge model). If patches vary in victim density and predators aggregate where victims are dense, the effect on victims is like a PARTIAL REFUGE; such aggregative predator responses allow very low victim densities to be stable and contribute greatly to the success of biological control.

Isocline analysis predicts the period of a predator–prey interaction:

$$T = 4\pi \Big/ \left(4\alpha\gamma - \alpha^2 k_1^{\,2}\right)^{0.5}$$

where k_1 is the slope of the victim isocline in the neighbourhood of the equilibrium and:

$$\alpha = \partial\left(\mathrm{d}V/\mathrm{d}t\right)/\partial P$$

$$\gamma = \partial\left(\mathrm{d}P/\mathrm{d}t\right)/\partial V$$

This formula succeeds if, as in laboratory systems, enough data can be obtained to estimate the isoclines and the partial derivatives. *See also* LOTKA–VOLTERRA MODEL. [M.L.R.]

REFERENCES

Begon, M., Harper, J.L. & Townsend, C.R. (1996) *Ecology: Individuals, Populations and Communities*, 3rd edn. Blackwell Science, Oxford.

Berryman, A.A. (1992) The origins and evolution of prey–predator theory. *Ecology* **73**, 1530–1535.

Getz, W.M. (1993) Metaphysiological and evolutionary dynamics of populations exploiting constant and interactive resources: *r*–*K* selection revisited. *Evolutionary Ecology* **7**, 287–305.

Ginzburg, L.R. (1986) The theory of population dynamics: I. Back to first principles. *Journal of Theoretical Biology* **122**, 385–399.

Kolmogoroff, A.N. (1936) Sulla teoria di Volterra della lotta per l'esistenza. *Giorn. Instituto Ital. Attuari* **7**, 74–80. Translation in: Scudo, F.M. & Ziegler, J.R. (eds) (1978) *The Golden Age of Theoretical Ecology*, pp. 1923–1940. Springer-Verlag, Berlin.

May, R.M. (1981) Models of two interacting populations. In: *Theoretical Ecology* (ed. R.M. May), 2nd edn, pp. 78–104. Sinauer Associates, Sunderland, Mass.

Rosenzweig, M.L. (1977) Aspects of biological exploitation. *Quarterly Review of Biology* **52**, 371–380.

Yodzis, P. (1989) *Introduction to Theoretical Ecology*. Harper & Row, New York.

modern species concept *See* SPECIES CONCEPTS.

modern synthesis Refers to the synthesis of DARWINISM with the modern understanding of genetics, pioneered by R.A. Fisher, Sewall Wright and J.B.S. Haldane and carried out in large part by Th. Dobzhansky, E. Mayr, G. Simpson and G.L. Stebbins. Synonymous with NEO-DARWINISM. *See also* CHARACTERS IN ECOLOGY. [P.C.]

modular organism A modular organism has an indeterminate morphological form that changes throughout its LIFESPAN. It is constructed from a number of basic subunits (e.g. reproductive and vegetative plant GROWTH), which in turn produce further 'clonal' copies. The exact form (e.g. number of branches) is strongly influenced by the surrounding environment in which the organism develops. Physiologically, independent subunits are known as ramets, while the whole 'genetic individual' is known as a GENET. Such organisms are typically SESSILE in their adult stages. Although plants are the most obvious examples of modular organisms, there

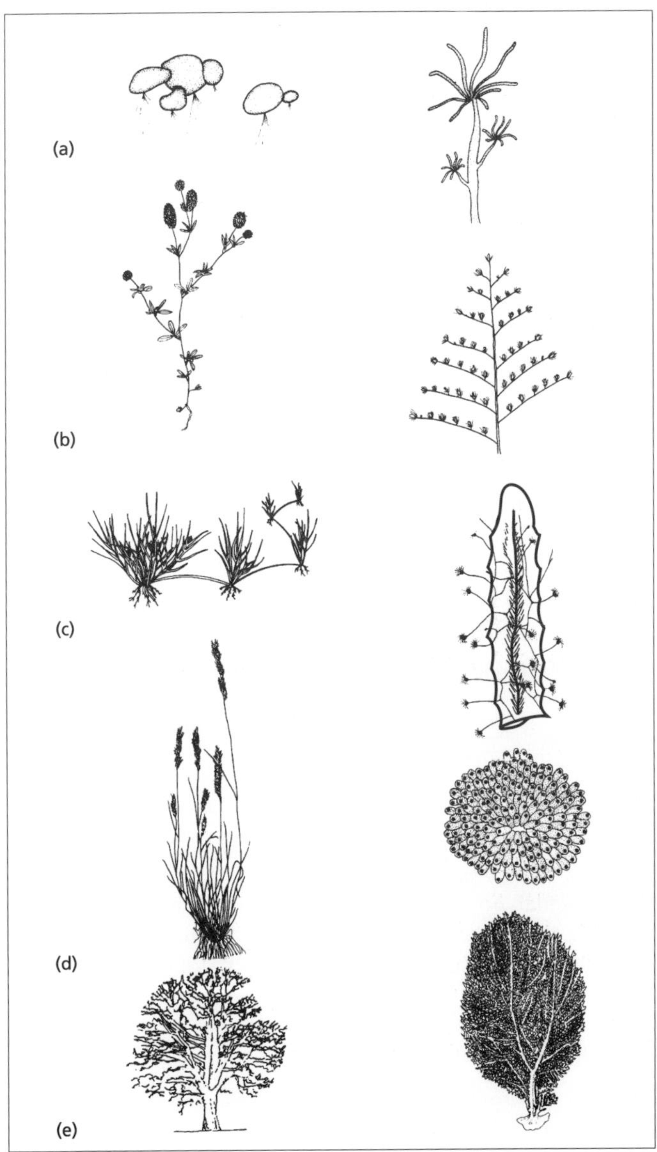

Fig. M11 A range of modular organisms: plants to the left and animals to the right. (a) Modular organisms that fall to pieces as they grow: duckweed (*Lemna* sp.) and *Hydra* sp. (a non-colonial cnidarian). (b) Freely branching organisms that are relatively short lasting: the annual hare's foot clover (*Trifolium arvense*) and *Pennaria* sp. (Cnidaria: Hydrozoa). (c) Rhizomatous and stoloniferous organisms in which clones spread laterally: buffalo grass (*Buchloe dactyloides*) and *Campanularia* sp. (Cnidaria: Hydrozoa). (d) Tussock-formers comprising tightly packed modules: six-weeks fescue grass (*Festuca octoflora*) and *Cryptosula* sp. (an encrusting bryozoan). (e) Persistent, multiple-branched organisms: an oak tree (*Quercus* sp.) and *Gorgonia* sp., the common sea fan (a coral, Cnidaria: Anthazoa). (After Begon, M., Harper, J.L. & Townsend, C.R. (1996) *Ecology: Individuals, Populations and Communities*, 3rd edn. Blackwell Science, Oxford.)

are also a number of modular animals such as corals and bryozoans (Fig. M11). *See also* ASEXUAL REPRODUCTION; MODULE; UNITARY ORGANISM; VEGETATIVE REPRODUCTION. [R.C.]

module Genetically identical parts of a modular (clonal) individual (the GENET) that might be used as a basis for counting when assessing the ABUNDANCE of an organism (e.g. the number of tillers of a grass plant or the number of workers in an ant COLONY). Modules may be capable of independent existence if they become physically separated from the genet (VEGETATIVE REPRODUCTION). Plant modules can be ramets, shoots or tillers. Animal modules can be individual asexual insects or parts of a clonal animal (e.g. polyps of a coral). MODULAR ORGANISMS grow by processes directly analogous to those operating in populations, i.e. by the dynamics of birth and death of modules. *See also* RAMET; SUPERORGANISM. [M.J.C.]

Moh scale of hardness The Moh scale of hardness is based on a ranked list of minerals, such that each one can be scratched by the one below: 1, talc; 2, gypsum; 3, calcite; 4, fluorite; 5, apatite; 6, orthoclase; 7, quartz; 8, topaz; 9, corundum; 10, diamond. [J.G.]

mole (mol) SI UNIT for the amount of a substance based on the Avagadro number (6.022×10^{23}) of particles it contains, whether atoms, molecules, ions or photons. For example, 1 mol of the pure ISOTOPE ^{12}C contains 6.022×10^{23} atoms and has a mass of 12 g, whereas 1 mol of $^{12}CO_2$ has a mass of 44 g. [J.G.]

Molecular drive theory of evolution

Molecular drive is a process that can change the average genetic composition of a population with the passing of the generations. It is operationally distinct from NATURAL SELECTION and neutral GENETIC DRIFT (*see* NEUTRAL THEORY OF EVOLUTION) in that it is consequential on a number of internal mechanisms of DNA turnover that place genomes in a state of continual flux on an evolutionary time-scale. There are a variety of distinctly different turnover mechanisms: GENE CONVERSION, UNEQUAL CROSSING-OVER, DNA slippage, transposition, retrotransposition and others. They all have in common the potential to increase or decrease the representation of variant DNA sequences in the lifetime of an individual. For example, gene conversion occurs when one ALLELE of a gene converts another allele to its own sequence. Hence, an original *Aa* heterozygote can become an *AA* or *aa* homozygote. Unbiased gene conversion results in parity between *AA* and *aa*. Biased gene conversion might preferentially convert, say, *A* to *a*. Gene conversion can occur between alleles of a gene, as in the above example, or between repetitive copies of a gene (multigene family) that might exist on the same or other chromosomes (Fig. M12). Stochastic fluctuations in the frequencies of variant gene sequences can lead, over time, to the concomitant homogenization of any given variant gene through a GENE FAMILY and the fixation of the variant gene through a sexual population. Biased gene conversion will drive a variant sequence to fixation that much quicker. The other mechanisms of turnover can achieve the same results as gene conversion. All mechanisms, whether biased or unbiased, underpin the molecular drive process.

The general rates of DNA turnover are about 10^{-4} per kilobase of DNA per generation. Hence, they are about two orders of magnitude faster than the rate of MUTATION of 10^{-6} per kilobase that generates the DNA differences in the first place. The rate of turnover is very much slower than the rate at which the sexual process shuffles chromosomes into new combinations at each generation. If, for example, an *Aa* heterozygote has been gene converted to *AA* then these two *A* alleles enter two different individuals at the next generation, in each of which a further round of gene conversion might occur, generating more *A* alleles from *a* alleles. Gradually *A* will spread through the population. The disparities in rates between mutation, turnover and sex ensures that, on average, all individuals at any given generation contain the same ratio of *A* genes to *a* genes in a multigene family during the dual process of homogenization and fixation. Eukaryotic genomes contain hundreds of REPETITIVE DNA families that are subject to turnover by one mechanism or another, often operating one on top of another. Both genic and non-genic DNA families are known to be homogeneous for mutations that are specific for individual species. Species-specific homogeneity patterns are called CONCERTED EVOLUTION (Fig. M13). Molecular drive is the process underlying such patterns.

A molecular drive of mutations is known to have occurred in many genes that have important cellular functions, leading to species-specific phenotypes. When such mutations are being homogenized and fixed in critical regions, at what level is selection

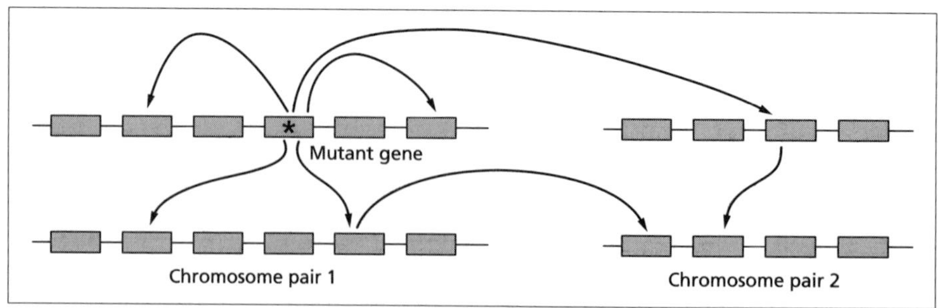

Fig. M12 Boxes represent copies of a gene spread over two pairs of non-homologous chromosomes. A variety of mechanisms of DNA turnover can spread a mutant copy amongst the other copies. Boxes can also represent internal repetitions within a gene or multiple controlling elements in front of a gene. After the sexual process, chromosomes enter different individuals at the next generation, in which further rounds of homogenization can take place, leading to the ultimate homogenization/fixation (molecular drive) of a variant repeat throughout a population.

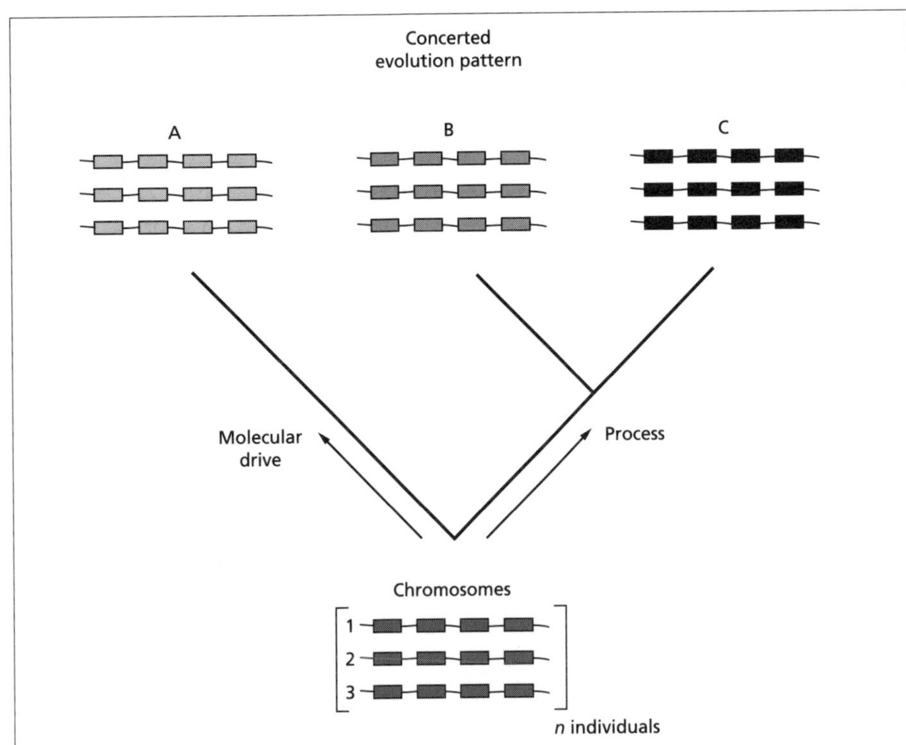

Fig. M13 Small boxes represent multiple copies of a gene spread over three chromosomes in each species A, B and C. The concerted evolution pattern is the observation that copies within all individuals of a species have been homogenized for diagnostic mutations. Molecular drive is the *process*, resulting from a variety of mechanisms of DNA turnover, that leads to the concerted evolution *pattern*. Small boxes can also represent regions of internal repetition within the body of a gene or multiple controlling elements in front of a gene.

expected to discourage or encourage the process? Genetic redundancy ensures that individual organisms can be buffered against the effects of one mutant gene in a background of many repetitive original genes. Homogenization might need to proceed to an appreciable extent (depending on the gene family in question) before individual FITNESS might be affected. By that time, however, all individuals have, on average, the same number of homogenized mutant copies. Selection would then have the opportunity to promote variant alleles of other genes whose products are known to functionally interact with the gene family undergoing homogenization. Such promotion would be specifically for variant alleles carrying mutations that can compensate for the homogenized mutations in the gene family. Hence, the interaction between molecular drive and natural selection ensures molecular COEVOLUTION between functionally involved genes. There is widespread evidence that molecular coevolution has occurred in such a fashion, leading to species specificity in the molecular mechanisms by which interactive genes bring about cellular functions. An incompatibility between two functionally interactive genetic components, each derived from a different species, has been observed in experimental 'mix and match' situations, providing evidence of molecular coevolution. Molecular coevolution ensures that key cellular processes are maintained despite the long-term genetic instability following on from mechanisms of genomic turnover. The incessant diversification of life forms can continue apace as a consequence of the facility for molecular coevolution in slowly homogenized and buffered redundant GENETIC SYSTEMS. The ever-increasing number of genes that are known either to be internally repetitive or to exist as multigene families testifies to the increasing role of the interaction between molecular drive and natural selection in the promotion of molecular coevolution and, ultimately, in the establishment of species-specific biological novelties. *Cf.* MEIOTIC DRIVE. [G.A.D.]

REFERENCES

Dover, G.A. (1982) Molecular drive: a cohesive mode of species evolution. *Nature* **299**, 111–117.

Dover, G.A. (1986) Molecular drive in multigene families: how biological novelties arise, spread and are assimilated. *Trends in Genetics* **2**, 159–165.

Dover, G.A. (1992) Observing development through evolutionary eyes: a practical approach to molecular coevolution. *Bioessays* (Special Issue) **14**, 281–287.

Dover, G.A. (1993) The evolution of genetic redundancy for advanced players. *Current Opinion in Genetics and Development* **3**, 902–910.

---

molecular ecology Molecular ecology is a generic term loosely applied to ecological studies that employ the tools of molecular biology. 'Molecular ecology' might suggest the existence of a branch of ecology devoted to the study of organisms and their ENVIRONMENT at a molecular level. Unfortunately, ecology is not yet so advanced and most studies in molecular ecology, with a few notable exceptions, use molecular biology primarily as a source of genetic markers.

Since the late 1960s ecologists and evolutionary biologists have used macromolecules such as protein, DNA and RNA as molecular markers. Studies employing these markers have had a significant impact in ecology and EVOLUTION. For example, the study of protein POLYMORPHISM in natural populations revealed abundant genetic diversity and triggered the many and important debates that continue today concerning the significance of molecular variation, its maintenance and distribution in populations. Molecular markers have enabled the tracing of organismal phylogenies which led to the discovery of a third kingdom of life (Archaebacteria) and provided support for the serial endosymbiosis theory. In addition, molecular markers have been used for sex identification in organisms where sexing is not possible on morphological grounds, to monitor and track genetically engineered organisms and pathogens, determine POPULATION and COMMUNITY structure, quantify MIGRATION and GENE FLOW, and assess the role of MUTATION and recombination in populations.

There is little doubt that molecular biology will continue to impact upon ecological studies, especially as molecular techniques become ever more tractable The challenge for ecologists is not to lose sight of the fact that molecules have important functions in their own right—that they are more than simply markers. In this respect an important future goal is to understand the ecological consequences of molecular variation in natural populations. *See also* GENETIC FINGERPRINTING; POPULATION GENETICS. [P.B.R.]

REFERENCES

Avise, J.C. (1994) *Molecular Markers, Natural History and Evolution*. Chapman & Hall, New York.

Lenski, R.E. (1995) Molecules are more than markers: New directions in molecular microbial ecology. *Molecular Ecology* **4**, 643–651.

Skibinski, D. (1993) Genetic markers. *Nature* **365**, 578.

molecular weight *See* RELATIVE MOLECULAR MASS.

mollic horizons *See* SOIL CLASSIFICATION.

Mollisols *See* SOIL CLASSIFICATION.

molybdenum (Mo) A very hard, silver-white, HEAVY METAL (ATOMIC NUMBER = 42; relative ATOMIC MASS = 95.94; melting point = 2617°C; boiling point = 4612°C), molybdenum is a trace constituent ($1.5\,mg\,kg^{-1}$) of the Earth's CRUST. Found in ores like molybdenite (MoS_2) and wulfenite ($PbMoO_4$), molybdenum is used in alloys, pigments and CATALYSTS. Existing as an anion (MoO_4^{2-}) in association with iron oxides in soil, molybdenum is essential for the activity of the enzyme nitrogenase in nitrogen-fixing microorganisms. Atmospheric deposition of molybdenum to soil can occur through smelting, fossil-fuel combustion, brick-making, etc., excess concentrations giving rise to molybdenosis, a molybdenum-induced COPPER deficiency, in livestock. [J.F.]

monimolimnion *See* LAKE STRATIFICATION.

monitoring Continuous or regular assessment of the quality of EMISSIONS or EFFLUENTS, or sources that lead to these, or specific places in the environment possibly subjected to them. For the latter, this can involve both chemical and biological assessment. *See also* CONTAMINATION; POLLUTION; RETROSPECTIVE ANALYSIS. [P.C.]

monoclimax The single type of CLIMAX vegetation expected, under Clementsian theory, in a given climatically determined area, irrespective of substratum. Now regarded as an inappropriate concept. *See also* POLYCLIMAX; SUCCESSION. [P.J.G.]

monoculture The agricultural, arboricultural and horticultural practice of cultivating one plant species, often a single cultivar, in one location. The commercial benefits from large-scale production and HARVESTING are intended to be sufficient to overcome the considerable risk, implicit in intensive agricultural monocultures, of changes in market value as well as the risk of fungal disease and insect devastation such as locust attacks. In relation to ecology, the term covers GRASSLAND planted with one species or a small number of species exhibiting successional monoculture, such as annual clovers with perennial ryegrass (*Lolium perenne*), or near monocultures of species that have exploited habitats created in abandoned farmlands. The definition does need to be modified in that a pure (single-species) monoculture is likely to be difficult to achieve in practice. Many weeds of agricultural importance exist as transient communities in monocultures or exploit the regular DISTURBANCE of the soil needed to maintain the monoculture. Monocultures may be deliberately modified by maintenance of wild relatives of the dominant CROP in order to extend the range of genus-related chemical defences. For example, the destructive activities of the Colorado beetle (*Leptinotarsa decemlineata*) in monocultures of potato (*Solanum tuberosum*) can be reduced by allowing the growth of other species of *Solanum* still attractive to the beetle but which contain alkaloids not detoxified by the beetle or its larvae. [G.A.F.H.]

monoecy The reproductive process where separate male and female flowers are developed on the same plant (*see also* HERMAPHRODITE). Monoecy decreases the likelihood of self-FERTILIZATION, particularly where the formation of male and female flowers is separated in time. The formation of monoecious floral structures is particularly common among wind-pollinated species, including many trees (e.g. the hazels *Corylus* spp.), the sedges *Carex* spp. and some grasses. A modified form of monoecy occurs in the Asteraceae, where the central florets may be entirely female and the outer florets male or hermaphrodite. Genetic control of monoecy is complex and involves combinations of genes that suppress development of one sex or the other. *See also* MATING SYSTEMS. [G.A.F.H.]

monogamy A mating system in which a single male and female mate only with each other during a breeding season, common in birds but rarer in mammals and other animals. The two individuals may or may not develop strong social bonds to each other and may remain together for life or for one breeding season. Monogamy seems to have evolved in two ways. In species exhibiting OBLIGATE monogamy, such as most raptors and some rodents and canids, no offspring will survive unless both parents work together to care for the young. In species with FACULTATIVE or economic monogamy, males attempt to gain mating access to as many females as possible by defending resource-rich territories or female home ranges, but because females are solitary or avoid males who are already paired, males are able to gain access to only one female. Recent DNA fingerprinting studies of socially monogamous birds have shown that males and females frequently sneak copulations with neighbours (so-called EXTRA-PAIR COPULATIONS, EPCs), so that the FERTILIZATION system more closely approximates POLYGAMY than monogamy. *See also* MATING SYSTEMS. [J.C.D.]

monophagous Monophagous organisms are only able to exploit a single type of food resource. This may consist of either a single species or a number of closely related species that provide very similar food types. Monophagy is common between species that have been subjected to prolonged periods of COEVOLUTION or where habitat fragmentation has produced associated 'host shifts'. *See also* FEEDING TYPES, CLASSIFICATION; POLYPHAGOUS. [R.C.]

monophasic life cycle *See* SEXUAL REPRODUCTIVE CYCLES.

monophyly Members of a monophyletic group are bound together by common ancestral relationships that they do not share with any other taxa. Each monophyletic group begins as a single species, the ancestor of all subsequent members of the clade. Because of the nature of SPECIATION, groups of species cannot give rise to other groups or to a single species. One species can give rise to one or more descendant species (or two species can produce a third species of hybrid origin), but no process can produce a genus from a genus (geniation) or a family from a family (familiation). This occurs because species are the largest units of taxic EVOLUTION; they are real, evolutionary entities. Higher-level categories, on the other hand, are human inventions. Monophyletic taxa include an ancestral species and all of its descendants and are diagnosed by synapomorphies. Monophyletic groups are also called holophyletic groups by evolutionary taxonomists, who include paraphyletic groups in their definition of monophyly. Examples of monophyletic groups are Mammalia (mammals), Aves (birds), Angiospermae (flowering plants) and Acanthocephala (thorny-headed worms). *See also* LINEAGE. [D.R.B.]

monotonic damping The complete lack of fluctuations in POPULATION SIZE as the population approaches its equilibrium. The lack of fluctuation results from perfect COMPENSATION of density-dependent responses, for example as population density increases, individual reproductive output or survival decreases at a rate that exactly compensates for the increased density. *See also* POPULATION DYNAMICS. [V.F.]

monotropoid mycorrhizae *See* ENDOTROPHIC MYCORRHIZAE.

monotypic Taxa (often families) having only a single species. *See also* ANGIOSPERMS, GLOBAL DIVERSITY AND CONSERVATION. [P.C.]

monsoon forest Tropical and SUBTROPICAL forests of certain areas, such as northern India and Burma, are supplied with rainfall during a wet season as a result of monsoon rains penetrating the continent as warm, moist winds blow from the ocean. During the wet season growth is lush, but in the following DRY SEASON the forest may suffer drought. During the drought many trees lose their leaves, thus avoiding excessive transpiration. The SEASONALITY of these forests distinguishes them from the consistently wet TROPICAL RAIN FORESTS, where there is no significant interruption to water supply.

Many of the trees of monsoon forests are of great value as sources of hardwood timber, such as teak (*Tectona grandis*). This has led to intensive exploitation of this vegetation type. [P.D.M.]

Montane forests, conservation

High-elevation forests are those forests below the TIMBER LINE (subalpine) but at least 1000 m (3280 feet) above sea level. These elevations apply generally, although higher elevations are required at lower latitudes before montane forests can occur.

At high elevations, there are many stresses from CLIMATE and SOIL that promote the specialized group of plants composing a montane FOREST. Montane forests have reduced CARBON DIOXIDE (CO_2) concentrations with a thinner ATMOSPHERE and are subjected to increased ULTRAVIOLET RADIATION. They usually have increased PRECIPITATION, but this varies with location on the surface of the Earth; some mountains become drier with altitude, especially above the cloud line. Biologically, they are characterized as having a relatively high RESPIRATION at a given temperature, lowered NET PRIMARY PRODUCTION, reduced tree height, reduced LEAF AREA INDEX (leaf area per unit of ground area) and reduced lignification and differentiation of wood and other tissues in proportion to the shortened growing season. As a consequence, montane forests are slow to regenerate. Research has shown that, once trees are of a certain size, prevailing winds in temperate montane forests promote lines of tree death that over time recede downslope from the windward edge of the mountain. Subsequently, young regenerates establish, grow to become a new forest until the whole DISTURBANCE process starts again. In other regions with steep physiography, landslides are important disturbance regimes that promote recolonization and growth of new forests.

Montane forests are of practical importance for WATERSHED functions, protection from avalanches and erosion, GRAZING by wild and domestic animals, lumber and FUELWOOD. In many forest areas, local residents rake out the litter to use as fertilizer for their field CROPS and gardens. Examples of this are the montane oak–pine forests of the Himalayas and Mexico. These practices can further impoverish soils that are frequently of relatively low fertility. Where lands are less populated, such as in the oak–pine forests of northern Thailand, swidden cultivation has been practised: fields lie fallow for periods of time allowing soil, organic matter and NUTRIENTS to build before recultivation. In many montane forests of tropical regions (Sri Lanka, Malaysia, Java, southern India, Kenya, Tanzania), lands have been permanently cleared for the cultivation of cool-loving vegetables in market gardens. In temperate climates, montane forests have been partially denuded and tree lines have been lowered for the provision of alpine pastures for sheep, goats and cattle (Pyrenees, Alps, Scandinavia, Himalayas).

Montane forests are valued for their recreational and aesthetic values. As society develops within a region, subsistence farming and PASTURE give way to commercial gardening and to tourist resorts for skiing and hiking, such as on many US Forest Service lands in the Appalachian and Rocky mountains. Recreation in montane forests is usually accompanied by roads, trails, housing, ski lifts and heavy equipment for maintenance.

In some regions, such as parts of the Andes, Rockies and the Cascade and Coast ranges of western North America, the mountains are rich in minerals. This has led to mining and in some cases enormous surface damage as well as toxic HEAVY-METAL pollution.

Montane forests are also collectors of pollution from downwind sources, sometimes hundreds of miles away, and thus their pollution load may be high despite the absence of industry and lower human populations in the vicinity. Montane forests therefore have a disproportionate burden of acid rain (*see* ACID RAIN AND MIST), fog mist, gaseous pollutants and dry deposition of heavy metals.

All these activities (grazing, cultivation, mining, industrial pollution) are real and are potential threats to the welfare of montane forests. They are less resilient compared with lower-elevation forests because of the inherent stresses that they already endure from the vagaries of climate. Human usage from all sources must be adjusted to a level where a montane forest is sustainable and capable of self-regeneration. Conservation must fit the unique physical and biological features of the forest when planning for the needs and culture of the people who live nearby as well as those who visit from elsewhere. *See also* CONSERVATION BIOLOGY. [P.M.S.A. & G.P.B.]

Monte Carlo simulation Method used to explore UNCERTAINTY in mathematical simulation models. It involves an iterative process in which model parameter values are selected at random from a specified frequency distribution. Output values, following simulation runs, are used to determine the PROBABILITY of occurrence of particular values, given the uncertainty in the parameter. *See also* RISK ASSESSMENT. [P.C.]

Montreal protocol UN instrument, developed under the auspices of the UNEP, on control of substances that deplete the OZONE LAYER. Adopted in 1987 and revised in 1990, it provides for the successive phasing out of CFCs and other specified substances likely to deplete the ozone layer. *See also* UNITED NATIONS PROGRAMMES AND BODIES. [P.C.]

moorland Upland areas of heather-dominated vegetation in north-western Europe are termed 'moorland'. The floristic distinction between HEATHLAND and moorland is imprecise. Generally, the wetter moorlands are richer in species of *Vaccinium*, *Trichophorum* and *Eriophorum*. As in the case of heathlands the soils are acid, but the higher levels of precipitation in the uplands lead to the development of thicker layers of peaty material at the surface of the soil. In the wettest types of moorland the vegetation merges into blanket bog.

Moorland, like heathland, is maintained by a management regime of burning and grazing, the latter being of economic interest as a result of both domestic animal husbandry (especially sheep) and the sporting value of some wild animals (such as red deer and red grouse). This has ensured the survival, maintenance and management of this anthropogenic vegetation type. [P.D.M.]

mor A surface HUMUS horizon that forms beneath conifers and open HEATHLAND and MOORLAND in temperate climates. It is acid, low in most microbial activities (not from fungi) and consists of several separated layers in various degrees of decomposition. [P.C.]

morbidity Any departure from a state of physiological (and/or psychological in people) well-being. Synonymous with sickness, illness or ill health. It is more difficult to measure than MORTALITY but can be done, for example in terms of proportion of population 'ill', frequency of periods of illness, duration of illness. [P.C.]

morphological species concept *See* SPECIES CONCEPTS.

mortality At its most simplistic level, mortality describes the number of individuals dying in a particular population. Most frequently it is expressed as a relative value or rate of change. A relative measure describes the proportion of individuals dying relative to the whole population, while a rate describes the pattern of mortality over time. These values are often combined to describe the proportional rate of mortality at key life stages (e.g. juvenile mortality). Mortality and its underlying causes are major factors in the REGULATION, DISTRIBUTION and ABUNDANCE of natural populations. *See also* LIFE TABLE; NATALITY; POPULATION DYNAMICS. [R.C.]

mortality rate The proportion of a population dying in each time interval, equivalent to the average PROBABILITY of an individual dying in that time interval. The value q_x in a LIFE TABLE. *See also* POPULATION DYNAMICS. [A.J.D.]

Motor vehicle pollution

No other mechanical product has made the same impact on the average person as the motor vehicle, especially the private car. It is not difficult to see why. For the individual, the motor vehicle provides much greater freedom of movement, choice and convenience. In a wider sense it has changed society's pattern of lifestyle in terms of where people work, live and carry out leisure activities.

For many, the motor vehicle has become a status symbol, a statement about how they express and see themselves and are seen by others. Certainly no other industrial product represents the same degree of individual freedom and mobility to the extent as the motor vehicle. However, attitudes are now changing and where the motor vehicle was once viewed as a sign of prosperity and happiness it is now being viewed in a new light as an environmental and health threat.

Growth in motor vehicles

For the present generation it is difficult to believe that less than 100 years ago there were virtually no motor vehicles in existence, nor the huge supporting infrastructure such as motorways, garages and oil industry. Neither was there the concern about environmental and health effects from POLLUTANT exhausts. That has all changed as a result of the phenomenal growth in the number of motor vehicles this century and increasing scientific and medical knowledge about the consequences.

Motor vehicle growth really became noticeable in the years following the Second World War and in many ways was a visible manifestation of the greater economic prosperity among wide sections of society. For example, according to the Washington-based World Watch Institute, in 1950 there were 53 million cars in the world. By the late 1980s this figure had increased to over 423 million. If commercial vehicles, such as trucks, buses and coaches, are added the figure rises to over 550 million. It is predicted, on current trends, that this figure will double by the year 2010: a frightening prospect.

Unfortunately, the picture has been mirrored in Britain. In 1950 there were just under 4 million motor vehicles on the roads. Each year since then the number of motor vehicles has steadily increased. In 1996 official Department of Transport statistics revealed that there were over 25.6 million vehicles on Britain's roads, a 540%

increase since 1950. It is predicted that the motor vehicle population will rise by 50% by the year 2015.

Pollution impact

In addition to the environmental problems caused by growing road traffic congestion and noise, especially in urban areas, and land used for road building, motor vehicles are responsible for producing a significant amount of pollutants at both a global and regional level. For example, the safe disposal of old motor vehicles, millions of used tyres, old BATTERIES and used oil will litter and contaminate the landscape if not properly controlled and recycled.

At the same time, the millions of road vehicles in daily use pollute the ATMOSPHERE with potential health and environmental risk. Motor vehicles only harness 10–20% of the potential energy in their fuel; the rest is converted to heat and exhaust pollutants. It is this aspect of motor vehicles where most official, media and public attention has focused. And with good reason.

It is now generally accepted that the EMISSIONS generated by fossil fuel (petrol and diesel)-powered motor vehicles have become a major source of ENVIRONMENTAL IMPACT, especially poor AIR QUALITY. Much AIR POLLUTION is produced by the combustion of FOSSIL FUELS. Motor vehicle exhaust emissions generate CARBON DIOXIDE (CO_2), CARBON MONOXIDE (CO), NITROGEN OXIDES (NO_x), sulphur dioxides (SO_x), HYDROCARBONS (HC) and particulate matter (PM10s). These pollutants are harmful and have a deleterious effect on society. Motor vehicles are responsible for about 85% of CO pollution, NO_x is a major contributor to acid rain (*see* ACID RAIN AND MIST) and CO_2 is one of the main gaseous pollutants from motor vehicles and the largest single contributor to GLOBAL WARMING.

In western Europe, motor vehicles are responsible for more than 80% of all CO emissions, 51% of NO_x, 45% of HC, 8% of particulates and 3% of SO_x. According to the European Union (EU) 1992 White Paper *The Future Development of the Common Market Transport Policy*, there has been a substantial increase over the last two decades in the number of transport-related pollutants, the most important of these being CO_2. The White Paper pointed out that between 1971 and 1989 CO_2 emissions from motor vehicles within the EU had increased by 76%. These findings are echoed in official British figures.

At national and international levels governments are committed to limiting CO_2 emissions in line with international agreements, notably the Climate Change Convention signed at the 1992 Rio de Janerio Summit. As a result of motor-vehicle usage, current concern has focused not only on CO_2 effects but also on a health-damaging PHOTOCHEMICAL SMOG known as ground-level OZONE. Sunny and heavily trafficked cities like Los Angles, Mexico City, Bangkok, Athens and Sydney are susceptible. The same applies to cities like London during hot days. Recently, there has been heightened concern about the particulate matter contained in motor-vehicle SMOKE (PM10s, ultrafine particles <2 μm in diameter that can easily penetrate the lung) in terms of it being carcinogenic and also contributing to rising asthma levels.

It is widely acknowledged throughout the world that constraints must be imposed on road vehicles: in their growth, where they travel and the pollution they produce. In the USA, the California Air Regulation Board (CARB) has taken the radical step of proposing that in the immediate future a certain percentage of cars must be ultra-low-emission vehicles; this will probably mean the use of electricity (batteries) or natural compressed gas. Other measures are currently being put into place in many countries, such as the mandatory fitting of CATALYTIC CONVERTERS, the use of unleaded petrol and low-sulphur diesel and Euro II exhaust standards in an attempt to curb the worst effects of motor-vehicle pollutant exhausts. [F.W.]

---

mountain environments In mountainous areas the vegetation shows pronounced ZONATION (see Fig. T5), brought about by the rate of change in environmental variables with respect to height above sea level. Forests at low altitude give way to CLOUD FORESTS, ELFIN FORESTS and KRUMMHOLZ vegetation as the altitude increases. Important variables in this environmental gradient are temperature, cloud cover and wind speed. The limit to tree growth, known as the TREE LINE, occurs naturally where the mean summer temperature is about 10°C. Above this limit, woody shrubs and other dwarf life forms (chamaephytes) thrive by exploiting the CLIMATE near the ground, including the small-scale topographic shelter afforded by boulders and crevices. Generally, such species become increasingly dwarf, often with shorter petioles and peduncles, and rosette forms and cushion-plants are frequent in the flora. In bright sunlight, surface temperatures of alpine vegetation often exceeds air temperature by as much as 10°C. Life cycles of alpine plants are often adapted to the harsh environments: PERENNIAL life forms often have a large OVERWINTERING rootstock, flowers are often profuse to attract pollinating insects. In other cases young plants are produced viviparously and alpine species have a well-developed capacity to harden to the low temperatures of winter. Snow cover is important for some species, allowing them to avoid the lowest air temperatures and protecting them from DESICCATION and abrasion by ICE and snow.

Air and soil temperature increases with altitude at about 6°C km^{-1}, actual gradients varying somewhat according to aspect and cloud formation. Pressure decreases, falling from 0.101 MPa at sea level to 0.898 MPa at 1000 m and 0.798 MPa at 2000 m; the partial pressure of CARBON DIOXIDE (CO_2) and other gases decreases proportionately. Breathing whilst climbing becomes noticeably difficult above 3000 m. Wind speed increases with elevation, but not as regularly as temperature and pressure. Even the windiest peaks have leeward spots, useful to climbers for shelter. Nevertheless, wind is an important factor in the safety of climbers, due to the cooling effect (*see* WIND-CHILL). Famous mountain winds include the föhn, in the valleys on the north side of the Alps; the mistral, occurring between the Alps and the Cévennes; the bise, at Lake Geneva between the Alps and the Jura Mountains; the chinook, on the eastern flank of the Rocky Mountains; and the zonda, in the Argentinian Andes. Up to 1000 m there is an increase in the frequency of cloud cover and rainfall, but at the highest elevations the peaks are usually above the cloud and solar INSOLATION can be very high, with a relatively high flux of ULTRAVIOLET RADIATION. This is because the effective pathlength for the Sun's rays through the atmosphere is diminished by the low pressure. Many plant species have pubescent reflective leaf surfaces and the ultraviolet flux may be sufficiently high to shape evolution of the plant surface. Humans do not escape the effects of elevated ultraviolet:

alpine skiers may become sunburnt. Total SOLAR RADIATION in the Alps sometimes exceeds the solar constant, as a result of the combined effect of direct sunlight and reflective CLOUDS.

EVAPORATION and TRANSPIRATION tend to fall with elevation. This is because the effect of a reduction in temperature on the saturated vapour pressure of the air is large and overrides other effects, such as the increase in wind speed and the increase in solar irradiance (above 1000 m).

Upland and mountain environments are not as clean as they seem to human observers. Cloud water deposited on high-altitude forests contains more dissolved solids than RAIN and may contain more acidity, NITROGEN and SULPHUR OXIDES and HEAVY METALS than rainwater; OZONE concentrations tend to increase with elevation. High-altitude forests are more inclined to suffer symptoms of forest decline than those at low elevations, partly as a result of these atmospheric inputs and partly as a result of the harsh climate (Fig. M14). [J.G.]

REFERENCES

Gerrard, J. (1990) *Mountain Environments*. Belhaven Press, London.

Langmuir, E. (1995) *Mountaincraft and Leadership*. Mountain Leader Training Board, Manchester.

Larcher, W. (1995) *Physiological Plant Ecology*, 3rd edn. Springer-Verlag, Berlin.

Pearsall, W.H. (1950) *Mountains and Moorlands*. Collins, London.

Tranquillini, W. (1979) *Physiological Ecology of the Alpine Timberline*. Springer-Verlag, Berlin.

μ/G selection *See* SELECTION PRESSURES, CLASSIFICATION.

mud *See* MUDFLAT.

Mudflat

'Mud' is not a precise concept: it generally denotes any ground that is soft and wet. Likewise 'mudflat' is applied to any TIDAL FLAT that does not have the usual clean appearance of sand (see Plate 9, facing p. 398). In practice, this means that appreciable quantities of SILT (particles 0.002–0.06 mm diameter) are present even though the major sedimentary component may still be sand (0.06–2.0 mm diameter) or even gravel (2–60 mm diameter). 'Sandy mud' and 'muddy sand' are terms often used to indicate muddy sediments dominated by silts or sands, respectively.

The necessary presence of significant quantities of deposited silt requires that local water movements are weak during high tide. It takes nearly 1 min for a silt particle of 0.005 mm diameter to fall 1 mm through still water and hence such particles will be maintained in suspension in turbulent river or sea water. However, riverborne silt particles flocculate when encountering seawater in estuaries and the larger effective size of the aggregate increases its sinking rate. Then, on reaching the bottom, the aggregates may encounter, and be bound by, mucilage released by benthic diatoms. Hence FLOCCULATION, shelter from turbulent wave action and mucilage all facilitate the accumulation of muds in estuaries and similar sheltered coastal regions; mudflats especially characterize such areas.

The shelter similarly leads to the deposition or stranding of waterborne organic particles and muds are generally rich in DETRITUS. Being small, silt particles pack tightly together and also clog the interstices of otherwise sandy sediments. Muds therefore have low

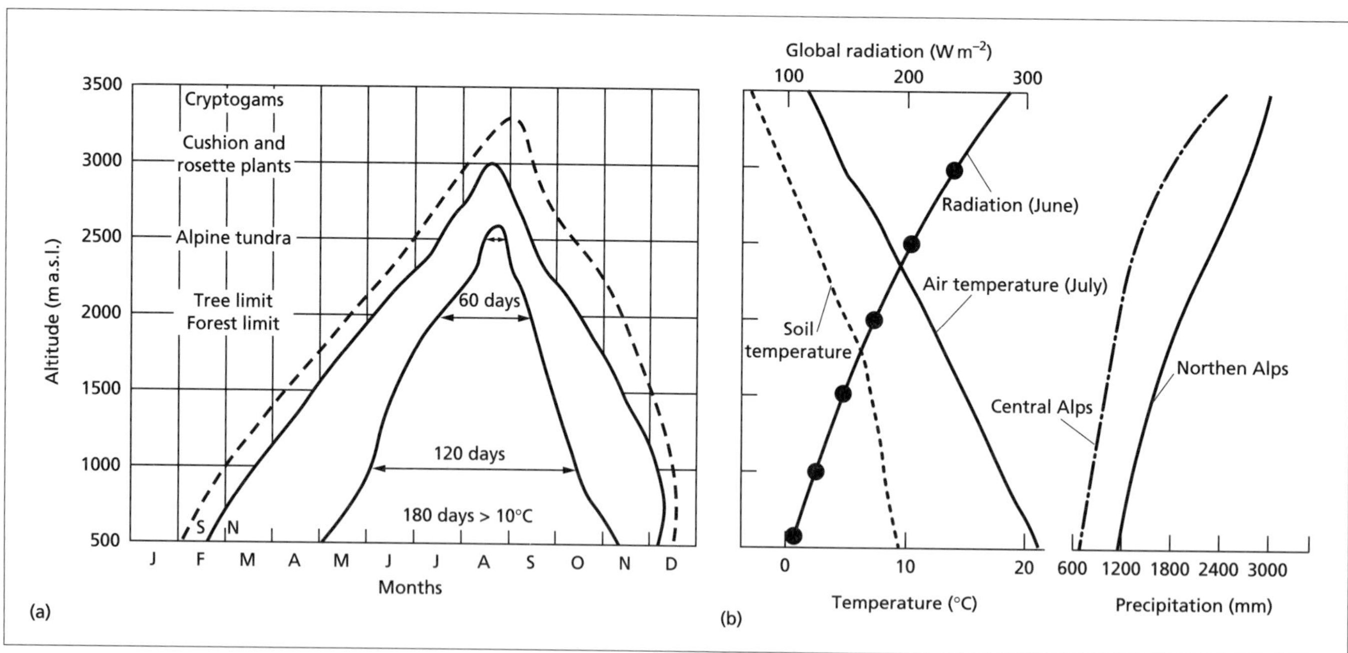

Fig. M14 Changes in climatic conditions with increasing altitude and altitudinal limits of vegetation in the Alps. (a) Altitudinal belts and plant distribution; duration of snow cover (gridded area) on shady north slopes (N) and slopes facing south (S); and length of the growing season (days with mean temperature above 10°C). (After Larcher, W. (1995) *Physiological Plant Ecology*, 3rd edn. Springer-Verlag, Berlin.)

porosities, water percolates through them very slowly and diffusion pathways are constricted. Under these circumstances, bacteria aerobically metabolizing the contained organics utilize interstitial OXYGEN as fast as it can diffuse through the SEDIMENT and thereby create anoxic conditions. However, some bacteria respire the oxygen present in the sulphate molecule (which is abundant in seawater) and in doing so they liberate sulphides. Hence, except for the surface layer exposed to the water/air, muds are not only anoxic but also black with poisonous sulphides. Interstitial organisms are therefore less abundant than on SANDY SHORES and animals dwelling within a mudflat have to occupy open burrow systems that they can irrigate with a stream drawn from the overlying water. Their burrows then stand out as brown (aerobic) tubes through the black (anaerobic) matrix. These burrow walls provide a specialist MICROHABITAT for interstitial species. Mangrove roots, faced with the same problems, are extremely shallow (and correspondingly extensive) and bear devices whereby atmospheric oxygen can be imported, for example pneumatophores. Not surprisingly, therefore, the most productive and diverse mudflats tend to be those with a large sandy component and relatively large surface aerobic layer.

Whereas sandy shores are perceived by the public as attractive and are therefore much frequented, mudflats are treacherous to walk across and are perceived as wasteland. Sandflats also typify exposed areas away from harbours and estuaries, whilst the characteristic mudflats of the latter are ideally situated to be 'put to good use' by being reclaimed for the industry and conurbations attendant on major ports. The majority of the mudflats of many estuaries and enclosed bays in industrialized countries have been so reclaimed and the pressure on the remaining ones is high.

Nevertheless, mudflats are productive systems, a fact not unappreciated by shorebirds, wildfowl and bird-watchers. The PRODUCTIVITY fuelling the system is in part an import in the form of detritus from elsewhere, including from fringing SALT MARSHES or MANGROVE swamps, but *in situ* production also occurs in the form of rooted seagrasses (especially *Zostera* spp.) and surface algal or cyanobacterial mats. In the sandier mudflats, interstitial algal protists may also be a significant source. The invertebrate consumers have usually been labelled as detritus-feeders and assumed to rely mainly on the bacterial decomposers (since the bulk of the detrital materials is refractory carbohydrates), but it is increasingly becoming apparent that, where available, they are subsisting mainly on surface diatoms. Suspension-feeders generally become rarer as the silt content of the sediment (and water) increases because of the high proportion of non-nutritive material that would be captured by their filter systems, although mussels and oysters may occur on suitable substrata.

Primary consumers within the sediment are, in the softer of muds, largely polychaetes, bivalve molluscs and amphipod and decapod crustaceans (with the fiddler crab, *Uca*, and other ocypodids dominating this niche in the tropics), although as the proportion of sand increases so do representatives of most other marine groups, including burrowing anemones, echinoderms, inarticulate brachiopods and various worm phyla. Densities of individual species can approach, and at times exceed, $100\,000\,m^{-2}$ with wet-mass biomasses of $100–200\,g\,m^{-2}$. In addition, gastropod molluscs graze over the surface in similar abundances. Control of these populations seems rarely to be effected by the larger and more obvious bird and fish predators. Because these prey only on the largest individuals of each prey species as well as foraging (optimally) over areas containing greater than average prey densities, they consume only the tip of the prey iceberg. Rather, control is effected by a combination of invertebrate predators living within the mud (particularly various worms) and small, indiscriminately feeding crustaceans and juvenile fish roaming over the surface during high tide.

Both groups prey on the abundant juveniles of the primary consumers and effect up to 99% mortality of the post-settlement young. Further large-scale mortality can be caused by adverse environmental conditions: discharge of abnormally large quantities of FRESHWATER down the RIVERS, ice-winters in the north temperate zones, etc. Several wildfowl species feed on seagrasses and in doing so disturb the surface layers to the detriment of the invertebrate inhabitants, as do rays and flatfish whilst feeding and/or burying themselves for concealment. The sea-grasses, however, appear to restrict the freedom of movement of some of the consumers of the juvenile 'detritus-feeders' and thereby ameliorate their effects.

Numerous interactions of the INTERFERENCE COMPETITION type have been recorded between different pairs of mudflat species. Burrowing species A can interfere with similar species B during its movements, causing it to leave the area to seek to burrow elsewhere; tube-dwellers and burrowers may interfere with each other by rendering the surface muds unstable or conversely by restricting freedom of movement; some species react aggressively to the presence of others; and so on. Species attempting to emigrate are then at the mercy of predators in the water column. Adult–larval interactions are also common: larvae endeavouring to settle are trapped by adult filtration systems and even if not consumed are in danger of being bound in mucus and cast out within a pseudofaecal pellet. Overt competition for limited food resources is known but appears to be not so evident as in several other habitat types, perhaps because of the heavy mortality on the juvenile invertebrates referred to above. All these interactions are a cause of the small-scale patchiness that here replaces the classic ZONATION of rocky intertidal regions.

The primary predators, as well as the larger 'detritus-feeders', are themselves the prey of the larger invertebrate and especially vertebrate secondary predators. Short-billed waders pick over the surface and their long-billed relatives probe into the mud for burrowing worms, crabs and molluscs during low tide; at the same time periophthalmid fish, crab-eating frogs, snakes, monkeys and others forage over the mudflat surface; whilst numerous fish consume surface individuals and the projecting parts (e.g. bivalve siphons and polychaete tentacles) of burrowing species during tidal emersion. The fish species are themselves the target of other fish and of seabirds. Many of these larger predators are migratory and use mudflats solely as seasonal feeding grounds; several fish use mudflats as nursery areas. Mudflat productivity thus provides a subsidy to other marine, and even terrestrial, systems. [R.S.K.B.]

---

mulch *See* AGRICULTURE AND SOIL CONSERVATION.

mull A surface HUMUS that forms in deciduous forests. It is neutral or alkaline. It provides generally favourable conditions for decomposition and in consequence it is well mixed. [P.C.]

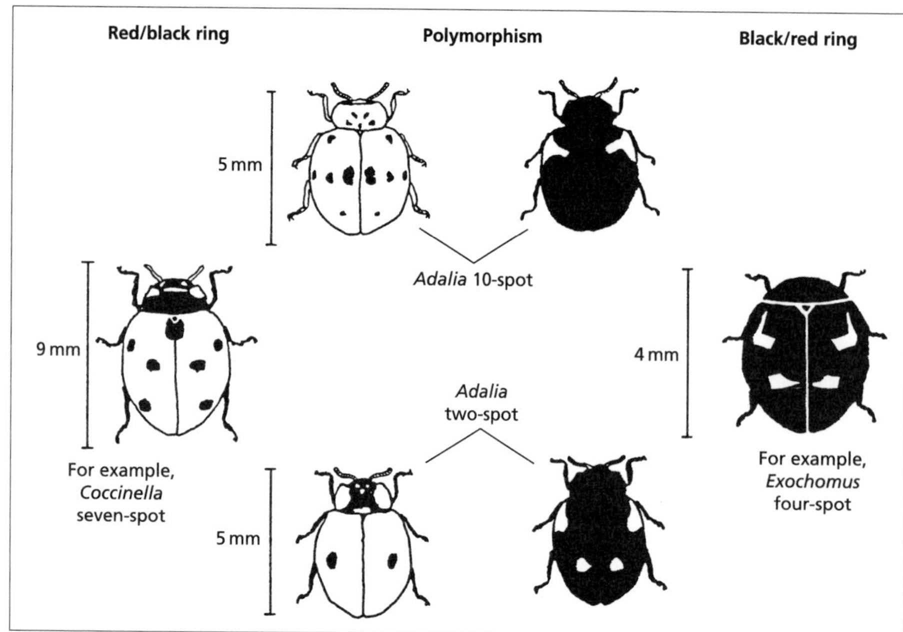

Fig. M15 The most abundant melanic and non-melanic forms of two *Adalia* ladybird beetles in western Europe, together with examples of monomorphic species of Müllerian mimicry rings. The polymorphic species appear to be Batesian mimics of the well-protected monomorphic species. Red areas of ladybirds are depicted in white. (After Brakefield, P.M. (1985) Polymorphic Müllerian mimicry and interactions with thermal melanism in ladybirds and a soldier beetle: a hypothesis. *Biological Journal of the Linnean Society* **26**, 243–267.)

Müllerian mimicry Müllerian MIMICRY occurs between two or more species (called a mimicry ring), each of which has a closely similar warning, or aposematic, colour pattern (Fig. M15 & Plate 20, facing p. 398). Each species derives some protection from the mimetic resemblance through a reduced rate of predation. The co-mimics are all protected by some form of noxious defence, for example by an unpalatable taste or smell or by a toxic sting. *See also* BATESIAN MIMICRY. [P.M.B.]

multidimensional scaling A multivariate statistical method. Most of the statistical methods described in detail in this encyclopedia are univariate because they are concerned with single variables or observations. Even the complex methods of ANALYSIS OF VARIANCE are only concerned with explaining the variation in one variable. Multivariate methods analyse several related variables or observations simultaneously. For further details see Manly (1986). [B.S.]

REFERENCE

Manly, B.F.J. (1986) *Multivariate Statistical Methods: a Primer.* Chapman & Hall, London.

multiparasitism The deposition of one or a clutch of EGGS by a PARASITOID on a HOST that has already been parasitized by a member of a different parasitoid species. *See also* PARASITE; SPECIES INTERACTIONS; SUPERPARASITISM. [H.C.J.G.]

multiple climax In the early 1900s the scientific world was reacting to Darwin's revelations on EVOLUTION. This spawned the formal philosophy of environmental determinism, which espouses the view that CLIMATE is the underlying control on plant and animal distributions (and on human activity). The manifestation of such ideas in ecology lies in the contemporary work of F.E. Clements, C.E. Moss and A.G. Tansley, who were eminent ecologists (*see also* CHARACTERS IN ECOLOGY). All advocated, in a holistic approach, the concept of the CLIMAX community. This was the culmination of changes in VEGETATION communities from their inception on newly formed or abandoned substrates and their progression through intermediate stages (the seral stages) in a unidirectional way. Clements believed that in any given region with a consistent macroclimate a uniform or climatic climax community would develop. Moss and Tansley advocated modifications of this idea based on the likelihood that variations in other environmentally significant factors, for example geology and soils, would also influence the character of vegetation community. Thus was born Tansley's POLYCLIMAX theory, which is also referred to as multiple climax theory. It did not undermine the significance of climate nor did it depart from the notion of the vegetation community as a functional organism. Inevitably, these theories had, and continue to have, many critics, for example H.A. Gleason's concept, formulated in 1926, of the 'individualistic concept of the PLANT ASSOCIATION'. Such criticisms are well founded but the idea of climax community development, by incorporating dynamism and being controversial, has made a major contribution to the science of ecology. *See also* HOLISM; SUCCESSION. [A.M.M.]

multiple equilibrium Alternative stable states in the ABUNDANCE of predators and their prey. The existence of multiple equilibria may provide an explanation for certain complicated patterns of organism abundance in nature. For example, prey populations may move from a low-density equilibrium to a high-density equilibrium in response to interactions with a predator. This may result in an outbreak of prey abundance that cannot be attributable to environmental factors. [V.F.]

multiple fission *See* ASEXUAL REPRODUCTION; FISSION; VEGETATIVE REPRODUCTION.

multiple mating *See* SPERM COMPETITION.

multiple paternity *See* SEXUAL CONFLICT.

multiple resistance to pesticides Also called multiresistance, this occurs when organisms evolve RESISTANCE to several pesticides. For example, *kdr*, a gene that helps confer resistance to DDT, also contributes to resistance to pyrethroids. Alternatively, it can occur as a consequence of selection for resistance to several different classes of pesticides simultaneously or in sequence. Where a single mechanism is responsible for resistance, the terms 'horizontal resistance' or 'cross-resistance' are sometimes used. There has been considerable theoretical and empirical work on techniques to prevent the evolution of resistance and to avoid multiple resistance. Work has focused on the sequential application of several different pesticides or on the application of combinations of pesticides. Another possibility is the use of substances with negative cross-resistance, chemicals that decrease resistance to a target chemical as organisms evolve resistance to them. It seems likely that the further evolution of multiple resistance can only be slowed by the implementation of well-designed programmes of INTEGRATED PEST MANAGEMENT (IPM). *See also* DEFENCE MECHANISMS. [D.A.B.]

REFERENCE

Priester, T.M. & Georghiou, G.P. (1978) Induction of high resistance to permethrin in *Culex pipiens quinquefasciatus*. *Journal of Economic Entomology* **71**, 197–200.

multiple stable states When the STATE SPACE of a system contains multiple stationary points. At any given time, a dynamic system is described by a point in its state space. In the course of time, this point moves on a trajectory through state space as the state of the system changes due to the forces that act upon it. A stable state is represented by a point in state space that is stationary. In such a state, the sum of the forces acting on the system is zero and the state of the system remains constant. A stable state is also called an equilibrium. The system has multiple stable states if there is more than one stationary point in its state space. In particular, there may be infinitely many stable states. For example, the state space may be the surface of a landscape describing the spatial location of a ball under gravitational force. Then all maxima and all minima, as well as all saddle points, represent stable states. A plateau in the landscape consists of infinitely many stable states. A stable state is called locally stable if all trajectories from nearby points end up in this state: after small perturbations, the system is forced back to the stable state. In the example above only the minima are locally stable. The term 'multiple stable states' is sometimes used to describe the situations in which there is more than one locally stable state. *See also* LOCAL STABILITY; POPULATION DYNAMICS. [M.D.]

multiresistance *See* MULTIPLE RESISTANCE TO PESTICIDES.

multispecies (eco)toxicity tests Systems for assessing and/or predicting the ecological effects of substances or EFFLUENTS involving more than one species. They can range in size or complexity from two-species interactions, to collections of several species in indoor/outdoor enclosures, to samples of multispecies systems either extracted from nature or isolated *in situ*. They can be small scale (*see* MICROCOSM), medium scale (*see* MESOCOSM) or large scale (*see* MACROCOSM). Claimed by some to be more ecologically relevant, since they incorporate between-species interactions that, by definition, are not a feature of single-species studies. However, the isolates may not be representative and isolation can destabilize the dynamics of the interacting populations. Moreover, the maintenance and deployment of multispecies sytems is usually more time-consuming and costly than single-species ones. *See also* PREDICTIVE ECOTOXICITY TESTS. [P.C.]

multivoltine *See* EPHEMERAL.

muskeg The PEATLANDS of Canada are often referred to as muskeg. They have been classified into the confined peatlands of the southerly areas and the unconfined cover of peatland in the higher latitudes. [P.D.M.]

Mussel Watch Globally coordinated programme that uses tissue analysis of mussels, in particular *Mytilus* spp., as an indicator of CONTAMINATION. These animals are thought to be particularly suitable sentinel (BIOINDICATOR) organisms because they are SESSILE filter-feeders. *See also* SENTINEL SPECIES. [P.C.]

REFERENCE

Cossa, D. (1988) Cadmium in *Mytilus* spp. World-wide survey and relationships between seawater and mussel content. *Marine Environment Research* **26**, 265–284.

mutagen A mutagen is any agent, such as IONIZING RADIATION, that can cause MUTATION of the genetic material of living organisms. *See also* GRAY; SIEVERT. [A.B.M.]

REFERENCE

Kiefer, J. (1990) *Biological Radiation Effects*. Springer-Verlag, Berlin.

mutation In his book *The Mutation Theory*, De Vries distinguished between continuous, individual variation and discrete saltational variations. He applied the term to the latter only. Now the term is used for any spontaneous, random change, large or small, in the genome. However, a distinction is sometimes made between a point mutation (impact on a single gene or even base substitution) and a MACROMUTATION (change with major, usually obvious visible, effect on the PHENOTYPE). Random mutation is intended to imply non-directional with respect to the course of EVOLUTION. Mutation generates variety upon which NATURAL SELECTION operates. Hence, it is selection and not mutation that gives evolution direction. Most mutations, especially those having large effects, are deleterious. Deleterious mutations decrease fertility or the probability of survival, i.e. FITNESS. [P.C.]

REFERENCE

De Vries, H. (1909) *The Mutation Theory*. The Open Court, Chicago, Ill.

mutation accumulation hypothesis A universal feature of EVOLUTION is the balance between MUTATION and SELECTION. Few spontaneously occurring mutations are beneficial, leaving neutral and deleterious mutations as the predominant classes of mutation. Selection ignores neutral mutations but acts to reduce the frequency of deleterious mutations, a process thereby called 'purifying'. ALLELES that have highly deleterious effects at early ages are kept at very low frequencies, such as those that cause fatal juvenile diseases in humans. One such disease is Hutchinson–Gilford's progeria, which is known to occur in only dozens of patients worldwide, despite dramatic clinical features.

A different situation arises with deleterious alleles that are expressed at later ages: the force of NATURAL SELECTION is much weaker, if not absent (*see* AGEING). Given that deleterious mutations will recur, they will not be pushed to such low frequencies by selection when they have effects confined to later ages, which in turn will lead to a greater frequency for any particular genetic disease. The type specimen of this pattern is the human genetic syndrome called Huntington's disease, which is caused by a single dominant mutation with deleterious effects primarily in middle age. This fatal disease afflicts some tens of thousands of individuals, several orders of magnitude more than are afflicted by Hutchinson–Gilford's disease. Thus Huntington's disease is a plausible example of the accumulation of a deleterious mutation. This is a process that may be the basis for many other human age-related disorders, such as Alzheimer's disease, as well as many of the mechanisms of ageing in other organisms. *See also* MUTATION–SELECTION BALANCE. [M.R.R.]

REFERENCE

Rose, M.R. (1991) *Evolutionary Biology of Ageing*. Oxford University Press, New York.

Mutation–selection balance

Recurrent MUTATION will prevent the elimination of deleterious ALLELES by SELECTION. The main factors affecting the equilibrium between mutation and selection are the intensity of selection, the mutation rate and presence or absence of selection against heterozygote carriers of the deleterious allele.

For alleles that lower the relative FITNESS of heterozygotes by a fraction hs, the equilibrium frequency for the deleterious allele ($\hat{q}$) is approximately $\hat{q}=\mu/hs$, where μ is the mutation rate from alternative selectively equivalent alleles (symbolized collectively as A) to the a allele. The relative fitness of the genotypes are $AA=1$, $Aa=1-hs$ and $aa=1-s$. This approximation, which is independent of both the fitness of the homozygote and the mutation rate from a to A, is accurate as long as $hs>>\mu$. When $hs=1$, the mutation is lethal and its frequency in the population is the mutation rate.

Recessive deleterious alleles are maintained at a higher frequency than dominants, because at low frequency they are largely hidden from selection in heterozygotes. Mutation–selection balance in this case leads to an equilibrium at $\hat{q}=\sqrt{(\mu/s)}$. Here the fitnesses are $AA=1$, $Aa=1$, $aa=1-s$. Thus, even if a recessive allele is lethal ($s=1$), it can be maintained at an appreciable frequency. For example, with a mutation rate of μ equal to 10^{-6} per generation, the frequency of the lethal allele will be 10^{-3} (which implies that about 1 in 500 of all individuals will carry it). About 1 in 25 people carry the gene for cystic fibrosis, a recessive human disease.

The formulas for mutation–selection balance can be used to estimate mutation rates, if selection coefficients are accurately known. In the first application, J.B.S. Haldane estimated the frequency of mutation to dominant dwarfism to be 4.3×10^{-5}, based on the observed frequency of dwarfism in Denmark, with an estimated SELECTION COEFFICIENT $s=0.8$, based on the relative NUMBER OF OFFSPRING produced when comparing individuals carrying the gene with their sibs who were not.

Much theoretical work has addressed the role of mutation in maintaining genetic variation in continuously varying TRAITS. Heritabilities for such traits are typically about 0.5 (the HERITABILITY is the ratio of the ADDITIVE GENETIC VARIANCE to the PHENOTYPIC VARIANCE). From studies on maize, *Drosophila*, *Daphnia* and mice, for quantitative-trait characters such as bones and bristles, the amount of new genetic variation from mutation each generation has been estimated to be about $V_M=10^{-3}V_E$, where V_E is the ENVIRONMENTAL VARIANCE. Mechanisms of selection in nature are not well known, although STABILIZING SELECTION, in which intermediates have the highest fitness and extremes are selected against, is thought to be common. Stabilizing selection depletes additive genetic variation. Theoreticians have therefore set out to address the question: can observed estimates of mutational variance maintain observed levels of genetic variation, in the face of stabilizing selection tending to decrease such variation? (Bulmer 1989).

One way to model the question is to extend the results from a single locus with two alleles to that of many loci each with two alleles. Using this approach, M.G. Bulmer obtained a result $\widehat{V_G}=4U(V_E+V_W)$, where $\widehat{V_G}$ is the equilibrium genetic variance, U the total mutation rate per GAMETE at all loci affecting the trait, V_E the environmental variance and V_W a measure of the intensity of selection (small V_W implies strong selection). Even if selection is not very strong, a heritability of 0.5 requires that 1% of all gametes are carrying a new mutant. This may be reasonable but does suggest a high per locus mutation rate or large number of loci (e.g. if 100 loci affect the trait, the mutation rate must be 10^{-4} per locus).

An alternative model popularized by R. Lande is to consider a continuous range of allelic effects at each locus (rather than just two alleles). The allelic effect is the average effect of the allele on the PHENOTYPE. In Lande's model allelic effects are assumed normally distributed, so that the regression of allelic effects on the phenotype is linear. The amount by which genetic variance at the locus is depleted by selection in a generation can be calculated in terms of the change in the phenotypic variance and the slope of the regression (which is directly related to the amount of expressed genetic variance due to that locus). The loss due to selection is then equated to the new mutational variance. The resulting equilibrium is:

$$\widehat{V_G}=\sqrt{\left[2nV_M\left(V_W+V_E+nV_M/2\right)\right]}+nV_M$$

where n is the number of loci. With 100 loci and mutational variance $V_M=10^{-3}V_E$ per generation, heritabilities of 0.5 can be maintained even under reasonably strong selection. The essential assumption of this model is that the allelic effects are normally distributed. This will never be exactly true, because the allelic effects are a mixture of two NORMAL DISTRIBUTIONS with different variances: the non-mutant alleles and the mutant alleles. M. Turelli has shown that the assumption will be valid if per locus mutation rates are high (about 10^{-3}). If per locus mutation rates are lower, the distribution of mutant alleles has a much larger variance than the distribution of the non-mutants. Turelli has constructed alternative continuous-allele models on the assumption that the per locus mutation rate is lower than 10^{-3} and finds results broadly in agreement with Bulmer.

The current state of opinion seems to be that the requirement of high per locus mutation rates, found in all the above models, means that selection–mutation balance alone cannot account for heritabilities as high as 0.5 and other mechanisms maintaining variation need to be invoked. The way in which selection works in nature is unclear and there are a number of ways it can act to maintain genetic variance (e.g. if SELECTION PRESSURES are geographically varying, if there are strong epistatic interactions among loci).

There is an alternative class of models that consider most of the variation to be maintained by mutation–selection balance, but selection is not acting on the trait itself. Instead, mutations that are always deleterious have pleiotropic effects on the trait in question. Such deleterious mutations are common and they may have pleiotropic effects on many traits of ecological interest, although the parameters involved are difficult to estimate. These models lead to maintenance of some genetic VARIABILITY in the trait. It remains unclear how much of the genetic variation typically observed in ecologically important traits is attributable to this mechanism.

Estimates of mutational variance are derived from the rate of divergence between inbred or clonal lines, selection responses in inbred lines and changes in genetic variance in lineages subject to inbreeding. Considerable empirical work is currently devoted towards obtaining better estimates for both the mutational and selection parameters. *See also* QUANTITATIVE GENETICS. [T.D.P.]

REFERENCE

Bulmer, M.G. (1989) Maintenance of genetic variability by mutation–selection balance: a child's guide through the jungle. *Genome* **31**, 761–767.

---

mutualism A widespread form of interaction between individuals of two or more species that results in mutual benefit (sometimes abbreviated to a ++ interaction). Where the participants are in close physical contact the mutualism can be described as a symbiosis. Symbiotic mutualisms are typically OBLIGATE, with one partner unable to persist in the absence of another, whereas non-symbiotic mutualisms are often FACULTATIVE, with only a proportion of the individuals of a species involved. Examples of non-symbiotic mutualisms between plants and animals include most forms of POLLINATION of flowers by animals (the plants benefit through the dispersal of their pollen, while the animals gain through a variety of 'rewards' provided by the plants, most commonly nectar); the dispersal of seeds by fruit-eating animals (dispersed seeds are more likely to establish new plants successfully, while the animals have a source of food); and the use of ant 'bodyguards' by plants (the plants provide food and/or accommodation for the ants, which in turn defend the plant from herbivory). Ants enter into similar mutualisms with some homopteran bugs and butterfly caterpillars (the ants provide protection from natural enemies, while feeding on the sugar-rich honeydew the other insects exude). A mutualism involving humans is still in existence in Africa, where honeyguides (birds related to woodpeckers) attract hunters to wild honeybee nests. After raiding the nests, the bounty is shared between them.

Some non-symbiotic mutualisms are characterized by extreme specificity (e.g. with one or two known exceptions, each of the 750 or so species of fig tree is pollinated exclusively by its own unique species of fig wasp), but diffuse relationships are more common, involving groups of interchangeable species. Many of these mutualisms appear to be 'conditional' (Bronstein 1994), involving only a proportion of the individuals within each species and varying in the quality of the rewards gained (if any) according to local circumstances. *See also* SPECIES INTERACTIONS. [S.G.C.]

REFERENCE

Bronstein, J.L. (1994) Our current understanding of mutualism. *Quarterly Review of Biology* **69**, 31–51.

MVP *See* MINIMUM VIABLE POPULATION.

mycelium The fungi are characterized by a LIFE CYCLE in which their dispersed propagules (spores) germinate on a favourable substratum by formation of a germ tube, which develops into a radiating system of filamentous, branching, tubular threads termed 'hyphae', producing a three-dimensional network collectively known as the mycelium (representing the thallus). Hyphae can be regularly septate, i.e. having cross-walls dividing adjacent cells or hyphal compartments and with either simple or complex septa, or they may lack septa and are aseptate/coenocytic. In the edible mushroom, *Agaricus bisporus*, the starter cultures (spawn) consist of the secondary, dikaryotic (two genetically distinct haploid nuclei per hyphal cell) mycelial phase, formed by the fusion (hyphal anastomosis) of two compatible, monokaryotic (one haploid nucleus per cell), primary mycelia. The secondary mycelium is capable of 'fruiting' only after an extended period of vegetative growth within the substrate, i.e. compost. In a wide variety of related mushrooms and toadstools, the secondary mycelia grow within SOIL, infected trees or in a symbiotic relationship (MYCORRHIZA) with roots of trees and other plants. Hyphae grow apically with branch initials being produced behind the tip; their walls usually comprise about 80% polysaccharide (with a ratio of 1 : 7 chitin : glucans, chitin being a polymer of *N*-acetylglucosamine), 10% lipid and 10% protein. Each young hypha on an agar medium extends at a constant linear rate; as a branching system develops, forming a small circular 'COLONY', the young mycelium as a whole grows exponentially for a relatively short period until growth-limiting factors begin to operate. This leads to the onset of maturation processes, spore production and, eventually, to cessation of growth and DORMANCY. Under natural conditions, the degree of ANTIBIOSIS from other microorganisms, particularly actinomycetes and bacteria, as well as availability of WATER, OXYGEN and NUTRIENTS, all affect the ultimate pattern of mycelial growth, which invariably departs from being uniform. Modifications of individual hyphae for particular functions include:

1 appressoria, for attachment of germ tubes and production of penetration/infection hyphae by plant pathogenic fungi;

2 haustoria, intracellular nutrient-absorbing structures within plant HOST cells; and

3 rapidly constricting loop-like structures or adhesive lateral branches/knobs to capture nematode worms (as formed by 'predatory fungi').

In many fungi, the mycelium takes on a specialized form, resulting in 'tissues' or multihyphal organs produced by extensive hyphal fusions and relatively complex developmental patterns, for example sclerotia, mycelial strands/rhizomorphs or, particularly in the higher fungi, sexual 'fruit bodies'. Extracellular enzymes, released from actively growing hyphae in soil and decaying organic matter and wood, digest complex substrates (e.g. CELLULOSE, starch, pectin, LIGNIN, chitin, keratin) followed by absorption of the soluble breakdown products. Mycelia of different fungal species usually develop over a wide range of pH and temperatures, but the presence of water, or at least a damp environment, is essential for fungal growth. [J.B.H.]

Mycorrhiza

Characteristics

The term 'mycorrhiza' (literally 'fungus root') was first used by A.B. Frank in 1885 to describe a long-lived association of a fungal MYCELIUM and plant roots that results in readily observable morphological complexes. The association is an example of a symbiotic relationship, i.e. COEXISTENCE of different species for at least part of their LIFE CYCLE. More importantly, it is also mutualistic, with both partners deriving direct benefit (e.g. an exchange of NUTRIENTS and/or metabolites), although in some cases a controlled, weak PARASITISM may also be involved. In general, the partners survive better together, i.e. they are more 'fit' as dual organisms than when apart. The main benefit to the fungus as a heterotrophic organism is a supply of organic CARBON, mainly in the form of hexose sugars that are obtained by selectively increased plant cell permeability (probably directed by the fungus), and which are then converted to sugar alcohols, for example arabitol, mannitol, etc. For the autotrophic plant partner, there is an increase in root branching and root surface area, resulting in more effective utilization of a given volume of SOIL, the mycorrhizal fungus taking up selected soil nutrients (especially PHOSPHORUS- and NITROGEN-containing compounds) and water more effectively than non-mycorrhizal roots. In other cases, such as orchids, the seed may only germinate if the appropriate fungal partner is present and the plant derives its carbon requirements from the fungus (as also for monotropoid mycorrhizae). Mycorrhizae have considerable ecological significance, as well as in agriculture and forestry. The large majority of the roots of plants (including trees) growing under natural conditions develop such intimate mutual partnerships with particular soil fungi and, once established, the mycorrhiza supports vigorous plant growth, increased yield (in the case of crop plants and trees) and greater longevity. They are the main organs of nutrient uptake in terrestrial ecosystems: 76% of plants sampled in a survey by Harley and Harley (1987) were found to be mycorrhizal. Nearly all angiosperms including all saprophytic higher plants, i.e. both monocotyledons and dicotyledons (but excluding crucifers and some chenopods like sugar beet), are known to be mycorrhizal, as are gymnosperms, some pteridophytes and certain bryophytes. There are fungi from all the major groups (i.e. Zygomycotina, Ascomycotina, Basidiomycotina and Deuteromycotina) involved in the dual relationship; in some forest trees, mycorrhizae are formed with more than one fungus, but in others the association is specific. The particular fungi concerned, when surviving as separate organisms, are weak competitors in the soil environment; others are incapable of an independent existence, i.e. they are OBLIGATE symbionts. In general, since the fungus lives 'in balance' with the HOST cells/tissues and does not usually kill them, it is described as being biotrophic.

Types of mycorrhizae

There are two main types of mycorrhizae with further categories as follows.

1 Ectotrophic/sheathing mycorrhizae (or ectomycorrhizae; *see* ECTOTROPHIC MYCORRHIZAE).

2 ENDOTROPHIC MYCORRHIZAE (or endomycorrhizae), a diverse group including:

(a) vesicular-arbuscular (VA) mycorrhizae;

(b) ericaceous/ericalean, including typical/ericoid and some intermediate/ectendotrophic mycorrhizae (arbutoid and monotropoid) types;

(c) orchidaceous mycorrhizae.

[J.B.H.]

REFERENCES

Brundrett, M. (1991) Mycorrhizas in natural ecosystems. *Advances in Ecological Research* **21**, 171–313.

Harley, J.L & Harley, E.L. (1987) A check list of mycorrhizae in the British flora. *New Phytologist* **105** (Suppl.), 1–102.

myrmecochory The DISPERSAL of seeds by ants. This phenomenon has been recorded in over 80 plant families and in a wide range of plant communities. It is particularly characteristic of dry habitats. For example, the Australian heathlands contain more than 1500 ant-dispersed plant species. Usually, the seeds are first dispersed on to the ground beneath the plant and are then found by the ants. The insects are attracted by the presence of a nutritious oil-body (elaiosome), which they use as a source of food. Experiments show that chemical constituents of the elaiosome trigger stereotypical carrying behaviour by the ants. The insects carry the intact seeds to the nest, where the elaiosomes are removed and the seed discarded undamaged, often below ground. Possible benefits to the plant include (i) dispersal from parent plant, though the distance involved is usually no more than a few metres; (ii) protection from granivorous rodents and from fire; and (iii) provision of a suitable MICROHABITAT for germination and ESTABLISHMENT. Energetically, myrmecochory represents a relatively inexpensive investment on the part of the plant (adding about 6–7% to the seed mass). *See also* MYRMECOPHILE. [M.F.]

REFERENCE

Beattie, A.J. (1985) *The Evolutionary Ecology of Ant–Plant Mutualisms.* Cambridge University Press, Cambridge.

myrmecophile A plant (or animal) that lives in close association with ants. Ant-plants have a symbiotic relationship with the insects. In many cases (e.g. some *Acacia* species), the ants are provided with food in the form of secretions from extrafloral nectaries, as well as shelter in the form of modified swollen shoots (domatia) in which the insects live. The ants effectively protect the plant from herbivory by other organisms by repelling any intruders. In some cases the ants also provide protection from COMPETITION and fire by removing other plants within a short radius of the base of the HOST plant. *See also* MYRMECOCHORY. [M.F.]

N

***n*-dimensional niche** *See* NICHE.

naiads *See* INSECTS, DIVERSITY AND CONSERVATION.

narrow-sense heritability In QUANTITATIVE GENETICS, the ratio of the ADDITIVE GENETIC VARIANCE to the phenotypic (total) variance (*see* PHENOTYPIC VARIANCE) of a QUANTITATIVE TRAIT: i.e. $h^2 = V_A/V_P$. The narrow-sense heritability indicates what proportion of differences in the values of the quantitative trait in a population are both caused by genetic differences and heritable in offspring, in contrast to the BROAD-SENSE HERITABILITY, which indicates what proportion of differences in the values of the quantitative trait in a population are caused by genetic differences. The narrow-sense heritability varies between 0 (differences between individuals are not found back in the offspring, there is no heredity at all) and 1 (all differences between individuals are due to genetics, and are inherited by the offspring). The narrow-sense heritability of a quantitative trait is most often estimated by the regression of the mean trait value in the offspring on the mean value of the trait in the two parents. The regression of the mean trait value in the offspring on the value of the trait in one of the parents estimates half the narrow-sense heritability, if the parents mated at random. [G.D.J.]

REFERENCE

Falconer, D.S. (1989) *An Introduction to Quantitative Genetics*. Longman, London.

NASA (National Aeronautics and Space Administration) The US organization responsible for conducting space research and exploration, and for supporting many associated research programmes of universities and other independent organizations. It has pursued exemplary cooperation with the WORLD METEOROLOGICAL ORGANIZATION both directly and through other national meteorological services to provide extensive routine observations by satellite. Used for research and data-gathering in the meteorological, oceanographic and terrestrial fields, these satellite observations have enormously extended human knowledge. [R.S.S.]

natality (fecundity) A measure of the production of offspring in a population. As with MORTALITY, this major demographic factor may also be expressed in a number of different ways. Three of the most commonly used measures are the BASIC REPRODUCTIVE RATE (R_0), the net reproductive rate (R) and the INTRINSIC RATE OF INCREASE (r). R_0 describes the average NUMBER OF OFFSPRING produced within a particular age-specific group; R describes the fundamental PER CAPITA rate of increase in a population (i.e. numbers born relative to the numbers surviving); and r describes the continuous rate of change in a population. Both the basic reproductive rate and the net reproductive rate are discrete measurements that are usually made over long time intervals, for example annually. In contrast, the intrinsic rate of increase is an instantaneous measurement which is usually made over shorter periods. There is a general relationship between these different quantities where:

Instantaneous rate = Natural log of finite rate

i.e.:

$r = \ln R$

When the interval of measurement is very small then the instantaneous and finite rates are nearly equal; this changes as the difference increases. Ultimately all populations are regulated by density-dependent balances between natality and mortality, which interact with dynamic processes such as COMPETITION and predation. *See also* DEMOGRAPHY; LIFE TABLE. [R.C.]

National Aeronautics and Space Administration *See* NASA.

National Oceanic and Atmospheric Administration (NOAA) Satellite The name of the satellite carrying the ADVANCED VERY-HIGH-RESOLUTION RADIOMETER (AVHRR) sensor. The US National Oceanic and Atmospheric Administration has operated several generations of meteorological satellites. These have a Sun-synchronous polar orbit and an altitude of 833 km. The even-numbered satellites (e.g. NOAA 12) cross the Equator at 7.30 a.m. (0730 hours) and 7.30 p.m. (1930 hours), and the odd-numbered satellites (e.g. NOAA 13) cross the Equator at 2.30 p.m. (1430 hours) and 2.30 a.m. (0230 hours). For environmental applications the most useful have been NOAA 6 (1979) onwards as these have carried the AVHRR. [P.J.C.]

National Priority List (NPL) *See* SUPERFUND LEGISLATION.

National Rivers Authority (NRA) *See* ENVIRONMENTAL (PROTECTION) AGENCIES.

native and naturalized species Native species are those that have evolved in place or have become established independently of human activity. Naturalized species were brought to a place by human activity, either intentionally or not, and have subsequently established self-sustaining populations. Weedy plant species that accompanied the spread of European peoples across the temperate regions of the globe, and rats that escaped from the ships of early explorers, are classic examples of NATURALIZED species. However, the distinction between native and naturalized is not always clear, and may rest on one's philosophical position about the relation of human activity to what is NATURAL. *See also* BIOLOGICAL INVASIONS. [E.O.G. & L.R.M.]

natric horizons *See* SOIL CLASSIFICATION.

Natura 2000 The collective name for sites in Europe that are subject to legislation on the conservation of natural habitats and wild fauna and flora. Provides for the establishment of a coherent network of designated sites on land and at sea, including special areas of conservation (SACs). The overall aim is the conservation of BIODIVERSITY by protecting habitats. [P.C.]

natural Literally, of nature. The term usually implies a situation in which there has been no modification by humans: for example, a natural landscape, a natural community. But there is another view, namely that humans are also 'of nature', being products of evolution, so 'natural' should not exclude humans and the consequences of their actions. Clearly, it is a difficult term to pin down unambiguously to the satisfaction of all. [P.C.]

natural classification Any CLASSIFICATION that contains only monophyletic groups and is thus logically consistent with the evolutionary relationships of the organisms. The genealogical relationships depicted on the PHYLOGENETIC TREE can be reconstructed from such a classification scheme. Other concepts of natural classification have been proposed. G. Gilmour suggested that a natural classification should contain taxa that would explain the distributions of as many characters as possible, even if that classification did not reflect accurately the phylogenetic relationships of the taxa involved. A natural TAXON in this view would be distinguished by tightly correlated characters largely or exclusively confined to it and no other taxon. P. Sneath and R. Sokal suggested that phenetic classifications best approximated Gilmour natural groupings, denoted by a measure called the cophenetic correlation coefficient. J. Farris showed that PHYLOGENETIC SYSTEMATICS maximizes the cophenetic correlation coefficient better than phenetic clustering methods. This empirical demonstration that one systematic approach maximized very different conceptions of 'natural' was a key event in establishing phylogenetic systematics as the method of choice for the general reference system in comparative biology. *See also* CLADISTICS. [D.R.B.]

natural disasters Upheavals leading to impacts on human society and/or ecological systems and/or the abiotic environment by phenomena that are not attributable to human causes. Examples include earthquakes, volcanoes, storms, tornadoes, hurricanes, FLOODS, forest fires, DROUGHT and meteorite impacts. *Cf.* ECOLOGICAL/ENVIRONMENTAL DISASTERS. [P.C.]

natural resource accounting and green gross domestic produce (GDP) Alternative systems of national accounting and performance measures, which incorporate ecological and human welfare considerations. *See also* ENVIRONMENTAL (ECOLOGICAL) ECONOMICS; GREEN NET NATIONAL PRODUCT. [P.C.]

Natural selection

Definition

A process involving differences between individuals in their rates of survival or reproduction, leading to some types of organism being represented more than others in the next generation. The process of natural selection is generally held to explain most if not all evolutionary change.

The precise meaning differs between authorities. To Darwin and his contemporaries the term, coined as an analogy for the 'ARTIFICIAL SELECTION' practised by animal breeders, covered the whole of the 'survival of the fittest' leading to evolutionary change. It thus included the origin of the variation which is the raw material of EVOLUTION, and the consequent change in the appearance of organisms. Modern evolutionists use the term 'natural selection' more strictly, to exclude both the origin of variation (MUTATION) and the final outcome.

History

The principle was first clearly stated around 55 BC by Lucretius, to demonstrate a philosophical point: the supposition that organisms had been created by the deities, because their ADAPTATION to their environments could be explained in no other way, was false. Organisms first occurred by spontaneous generation in all sorts of monstrous and misshapen forms; only those which functioned, and furthermore had adaptive qualities such as cunning or fleetness of foot, had survived. For two millennia, despite the extensive reading of the classics required in western education, this principle was not recognized, probably because nothing interesting seemed to follow. The point, recently much espoused by Dawkins, that this disposes of the necessity for a creator spirit, was generally unpopular.

Its scientific value emerged only during the early 19th century: a few writers used the idea to explain biological phenomena, such as the differences between human races (William Charles Wells) or the constancy of species (Edward Blyth). Darwin, A.R. Wallace and Patrick Matthew invoked it as the mechanism of unlimited organic change, or evolution. The concept is now a unifying principle in the whole of evolutionary biology. It provides the only scientifically satisfactory explanation for the adaptation of organisms (its only rivals in logical terms are wise creation and the inherited effects of use and disuse according to Lamarck), and has a startling power to generate objects of extremely high a priori thermodynamic improbability (such as advanced organisms) in a relatively short time.

Manifestations of natural selection

Formally, natural selection is COMPETITION between genetically different CLONES of nucleic acid occurring within a single population of a single species. Natural selection is a necessary consequence of the following two properties of nucleic acid: (i) that it replicates itself (thus inevitably causing Darwin's 'struggle for existence'); and (ii) that the replication involves errors (thus giving rise to competing clones). A gene or a type of nucleic acid can then be a successful competitor, via the organism that contains it, in several ways:

- the organism has a higher probability of survival up to the end of its reproductive life;
- the organism reaches reproductive maturity in a shorter time;
- the organism is more fertile;
- the organism is more successful at securing matings (or, in plants, attracting pollinators) — this is usually described as SEXUAL SELECTION.

Further, DNA sequences or genes may propagate themselves differentially as a result of molecular mechanisms while having no effect on the FITNESS of the individual organism, or indeed while having a detrimental effect. This selective molecular mechanism, for somewhat arbitrary and historical reasons, is not regarded as natural selection, but goes under such names as 'molecular drive' (*see* MOLECULAR DRIVE THEORY OF EVOLUTION), MEIOTIC DRIVE or SELFISH DNA.

Consequences of natural selection

The outcomes of natural selection are adaptation and evolution, but the latter can be either a long-term change in the organism or evolutionary STASIS, when natural selection actually prevents any change from occurring.

Evolutionary change is most simply modelled as the outcome of differing survival rates of two genetically different forms in a population of haploids. Commencing with the fitter of the two forms at low frequency (which would be the case if it were a new mutation), its frequency will increase in a sigmoid curve whose form in each half is logarithmic (Fig. N1). If the relative fitnesses or survival probabilities of the two forms are 1 and $1+s$, then this curve has the simple property that the number of generations required for the frequency to change by an order of magnitude (say from 0.001 to 0.01, or 0.9 to 0.99) is given by $t=2.3/s$ (eqn 1). For this purpose the intervals from frequencies of 0.1 to 0.5 and from 0.5 to 0.9 are each treated as an order of magnitude. Thus the speed of this kind of evolution is, in theory, highly predictable. In a diploid population, provided there is no genetic DOMINANCE between the two forms, the speed is half this ($t=4.6/s$) (eqn 2).

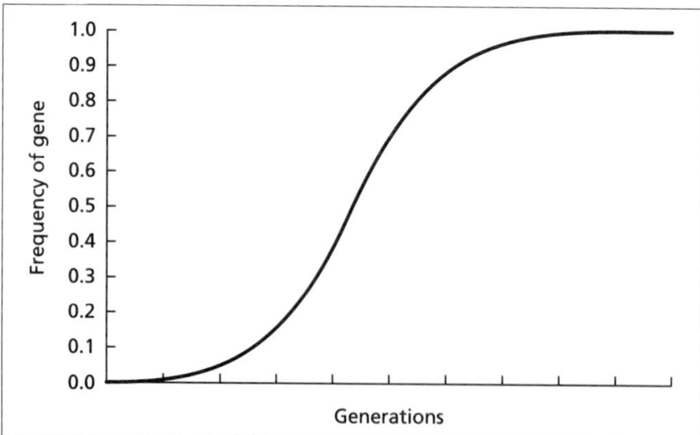

Fig. N1 The simplest example of the genetic consequence of natural selection: the change in frequency over time of a gene (allele) showing no dominance (or in a haploid population). The number of generations (x-axis) is given by equations 1 and 2, and is shorter the greater the strength of selection.

Full dominance of the genes radically alters this speed of change: an advantageous dominant gene progresses very quickly initially, but then fails to displace its recessive counterpart for an extremely long time at the end. Conversely, an advantageous RECESSIVE GENE takes such a long time to rise in frequency initially, that it is likely that such genes are only rarely successful in bringing about evolutionary change: for instance, all the genes producing INDUSTRIAL MELANISM in moths are dominant.

When the new gene is totally dominant or recessive, the number of generations required to spread the new PHENOTYPE from a frequency of 10^{-x} to $(1-10^{-x})$ is given closely by $t=[\sqrt{(10^x)}+3.5x]/s$ (eqn 3); this is about one-fifth the rate of equation 2.

Stable, static outcomes are also possible. The most widely recognized (although there are only three good examples) arises when the heterozygote for a pair of genes is fitter than either homozygote: instead of one gene replacing the other, there is then a STABLE EQUILIBRIUM in which both genes persist indefinitely at a constant frequency. The haemoglobin-*S* variant is maintained in human populations in the Old World because it confers enhanced resistance to malaria when heterozygous but causes a fatal (under Third World conditions) anaemia when homozygous. Similarly, a balanced equilibrium of two or more genes can occur if the fitness of the different genetic types is not constant but changes according to their frequency in the population.

When a character which gives rise to a fitness difference is controlled by several genes, the outcome requires a much more elaborate mathematical treatment, and a simple general description becomes impossible. It is in this case useful to distinguish DIRECTIONAL SELECTION, in which individuals with extreme values of a character (say greater size) are fitter than individuals at the other extreme, and STABILIZING SELECTION (or normalizing selection), in which individuals with an average value of the character are fitter than those at either extreme (see Fig. D4). This latter form of selection is potentially able to maintain evolutionary stasis for the character concerned, but in theory should lead to genetic uniformity in the population. That it does not is probably at least partly due to the steady occurrence of mutations in the genes involved. The converse form of selection, in which extreme individuals are fitter than those at the mean, is called DISRUPTIVE SELECTION. All three forms of selection have been recorded in natural populations.

If the variation which is selected is not heritable, then natural selection is without any further consequences (*see* HERITABILITY).

Problems with the concept

Whether natural selection finely moulds every feature of organisms is open to doubt. The 'opportunity for selection' is given by the VARIANCE in offspring number, at maturity, of every zygote in the population. The total 'intensity of selection' cannot exceed this quantity, and is divided between all the features of the organism that are adapting to the environment and all the genes that are being selected. The way in which this partition operates, and the outcome of possibly conflicting forms of selection on different genes in the same organism (as the varieties of nucleic acid do not compete directly but only through the organisms which they create), are major unsolved questions. While organisms are clearly adapted, there is a limit to the rate at which they can totally adapt, and therefore doubt about the further role of mutation coupled with GENETIC DRIFT in causing further evolutionary change, particularly in the sequences of PROTEINS and nucleic acids (*see* NEUTRAL THEORY OF EVOLUTION).

Classic examples of natural selection

Although the detection and measurement of selection in natural populations is not easy, a substantial number of instances have now been documented. The classic case of natural selection on a single-gene character is H.B.D. Kettlewell's work on the moth *Biston betularia*. Visual predation by birds placed the black form at a 50% advantage in industrial areas where the moths rested by day on polluted, lichen-free backgrounds, compared with an equal advantage to the original 'peppered' form in unpolluted areas where the moths rested on lichen. A complete evolutionary change in the expected direction (replacement of the 'peppered' by the black form) had occurred over 50 generations (roughly the time predicted from eqn 3) in industrial areas of Britain (*see* INDUSTRIAL MELANISM and Fig. I4).

A necessary property of natural selection, that there is a causal connection between the gene that is favoured and an increase in the fitness of the individual organism, is brought out even more clearly in later studies on the development of pesticide resistance in insects, or antibiotic resistance in bacteria, where the biochemical basis of the improved fitness is also understood.

One of the most intense episodes of directional selection on a continuously varying character was recorded in the Darwin's finch species *Geospiza fortis* on one of the Galapagos islands during a prolonged drought which reduced the amount of small soft seeds. Only relatively large individuals (hence with larger beaks) were able to eat the larger, harder seeds that remained, and the average size of the birds increased by one-third of a standard deviation in one generation. *Cf.* LAMARCKIAN INHERITANCE. *See also* DARWINISM; THEORY OF NATURAL SELECTION. [J.R.G.T.]

REFERENCES

Bowler, P.J. (1984) *Evolution: the History of an Idea.* University of California Press, Berkeley, Calif.

Dawkins, R. (1982) *The Extended Phenotype.* W.H. Freeman, Oxford.

Dawkins, R. (1986) *The Blind Watchmaker.* Longman, London.

Endler, J.A. (1986) *Natural Selection in the Wild.* Princeton University Press, Princeton, N.J.

Ford, E.B. (1975) *Ecological Genetics,* Chapter 14. Chapman & Hall, London.

Grant, P.R. (1986) *Ecology and Evolution of Darwin's Finches.* Princeton University Press, Princeton, N.J.

Kettlewell, H.B.D. (1973) *The Evolution of Melanism.* Clarendon Press, Oxford.

McKenzie, J.A. & Batterham, P. (1994) The genetic, molecular and phenotypic consequences of selection for insecticide resistance. *Trends in Ecology and Evolution* **9**, 166–169.

Ridley, M. (1996) *Evolution,* 2nd edn. Chapters 4, 5 & 13. Blackwell Science, Cambridge, Mass.

Taylor, C.E. (1986) Genetics and the evolution of resistance to insecticides. *Biological Journal of the Linnean Society* **27**, 103–112.

---

naturalized Describing species that have established themselves and are flourishing following their introduction to a non-native region. *See also* NATIVE AND NATURALIZED SPECIES. [P.O.]

nature versus nurture Shorthand for the relative involvement of, respectively, heredity (genetics) and ENVIRONMENT in the control and development of TRAITS. It is usually applied to the human condition, concerning which there have been heated debates on the relative involvement of these elements in determining such things as intelligence and criminality. From QUANTITATIVE GENETICS we know that the variation in all traits can be ascribed to GENOTYPE, environment and genotype × environment components. It is the relative importance of these that varies from trait to trait, and from species to species. [P.C.]

Nautical Almanacs *See* TIME.

Nearctic faunal region North America north of the tropics comprises the Nearctic region (see Fig. F1). The fauna in general is a mixture of families shared with the Palaearctic region and other parts of the Old World, and with the American tropics (NEOTROPICAL groups)—the former are more numerous northwards, the latter southwards. Among the few important groups characteristic of this region are the freshwater fishes. These include an endemic order represented by the bowfin, an endemic family of basses, many percids, one subfamily of which is exclusive to North America, and some mooneyes. Secondary-division fishes include garpikes, also found in Central America.

Dominant Nearctic amphibians belong to the same groups as Palaearctic ones (salamanders, *Rana, Bufo* and *Hyla*) but a few frogs have South American relationships. Nearctic reptiles are a mixture of Palaearctic and tropical American groups, the latter especially lizards, with few important endemic groups. Regionally notable are crocodiles and alligators in the south-east and geckos in the south-west. Among the snakes are many colubrids together with coral snakes and pit vipers.

Nearctic birds belong to 49 families, about 39 of which are more or less widely distributed, in the warmer parts of the region, at least in summer. Only one of these, however, the turkeys, is endemic or nearly so. The mammals include one marsupial, the opossum, with South American affinities, like the armadillo, both found in Texas. Many rodent genera are endemic, as is the pronghorn antelope. *See also* FAUNAL REGION; PALAEARCTIC FAUNAL REGION. [A.H.]

REFERENCE

Darlington, P.J. (1957) *Zoogeography: the Geographic Distribution of Animals.* John Wiley & Sons, New York.

nearest neighbour techniques There are several nearest neighbour techniques available and they may be used both to estimate POPULATION SIZE and to describe the type of DISTRIBUTION of a population. These techniques usually require selection of random individuals and then determination of the distance from that individual to its nearest neighbour. The most effective techniques require accurate positional data for every individual in a SAMPLE area, something that will require too much effort in most SURVEYS.

If populations are dense and clumped then the MEAN distance from a random set of individuals to their nearest neighbours will be smaller than if the population is sparse and randomly distributed. It is difficult to separate the effects of DENSITY and distribution using nearest neighbour techniques alone, so a second technique is often required.

Using the simplest nearest neighbour technique of measuring the distance from a randomly determined individual to its nearest neighbour and assuming a RANDOM DISTRIBUTION, the POPULATION DENSITY may be determined as follows:

$$D = \frac{1}{4x^2}$$

where D=density per unit area, and x=mean distance between nearest neighbours.

If density has been measured using another method, the mean distance between nearest neighbours can be used to describe the type of distribution:

$$\frac{D}{x} = q$$

where D=density per unit area estimated using a different technique (*see* ABSOLUTE POPULATION ESTIMATES); x=mean distance between nearest neighbours; and q is an index of DISPERSION. If the population is distributed at random then q will equal 0.25. This value will be greater in more ordered populations and smaller in more clumped ones.

More sophisticated techniques require use of the distance of a random individual to its second, third or more nearest neighbours. Clearly this will require much greater sampling effort but will allow much more accurate measures of population density to be made.

Other related techniques use a method where a point is selected at random and the distance from that point to the nth nearest individual is determined. As more individuals are sampled the estimates of population density become more accurate. This type of approach has some theoretical advantages over the simple distance between a random individual and its nearest neighbour but, again, it may require a greater sampling effort.

Yet more sophisticated and powerful ALGORITHMS based on nearest neighbour analysis are now available in computer packages. They provide INDICES OF DISPERSION and absolute population estimates using datasets derived from large-scale surveys.

Nearest neighbour techniques are particularly appropriate for SESSILE organisms, such as trees and barnacles, or analysis of photographic or satellite data. They are also widely used in situations where a population may be digitized so that a computer program may be used to find nearest n neighbours. *See also* SAMPLING METHODOLOGY/DEVICES. [C.D.]

REFERENCES

Ludwig, J.A. & Reynolds, J.F. (1988) *Statistical Ecology: a Primer on Methods and Computing*. John Wiley & Sons, Chichester.

Southwood, T.R.E. (1978) *Ecological Methods*. Chapman & Hall, London.

necromass Dead organic matter accumulated in a COMMUNITY. Unlike BIOMASS, necromass is incapable of transforming energy into new biomass. Necromass provides a source of energy to detritivores and decomposers. It consists of dead bodies, including dead wood that is part of living trees, and faeces. [P.O.]

necrotrophic Describing PARASITES that are able to continue to live on a dead HOST. Necrotrophic parasites—ecological pioneers in utilizing dead matter—are the first decomposers to get access to the body of the killed host. [P.O.]

NED (normal equivalent deviate) *See* PROBIT ANALYSIS.

Negative binomial distribution

A discrete PROBABILITY DISTRIBUTION that describes situations where observations (e.g. individuals in a SAMPLING unit) follow an AGGREGATED DISTRIBUTION in either space or time. A spatial example (the most frequent for ecologists) would be the number of barnacles in a QUADRAT. A temporal example would be the number of bats leaving a roost per minute. The sampling unit in the first case is a fixed amount of space (the quadrat), in the second a fixed amount of time (the minute).

Because the negative binomial distribution describes situations where observations are clumped, the VARIANCE is greater than the MEAN. The probability terms of the negative binomial distribution are given by the expansion of $(q-p)^{-k}$, where $p=\bar{x}/k$ and $q=1+p$. The PARAMETERS of this distribution are therefore the mean ($\bar{x}$) and the exponent k. Unlike the positive binomial (*see* BINOMIAL (POSITIVE BINOMIAL) DISTRIBUTION), k is not necessarily an integer (whole number) and can be used as a measure of the degree of AGGREGATION, or clumping, of the individuals in a population. As k approaches ∞ (infinity), the negative binomial distribution converges to the POISSON DISTRIBUTION. With smaller values of k, particularly below 1, the negative binomial distribution describes situations where observations are extremely aggregated or clumped.

The individual terms (probabilities) of the negative binomial are usually calculated by:

$$P_{(x)} = \left(1+\frac{\bar{x}}{k}\right)^{-k} \frac{(k+x-1)!}{x!(k-1)!} \left(\frac{\bar{x}}{\bar{x}+k}\right)^{x}$$

where $P_{(x)}$ is the probability of observing a certain number of individuals in a sampling unit, $\bar{x}$ is the mean number of individuals per sampling unit, k is the exponent, and $x!$ is x factorial ($4!=4\times3\times2\times1=24$, etc. and $0!=1$). The complete set of probabilities would therefore be:

$$P_{(0)} = \left(1+\frac{\bar{x}}{k}\right)^{-k}$$

$$P_{(1)} = P_{(0)} \times \left(\frac{k}{1}\right)\left(\frac{\bar{x}}{\bar{x}+k}\right)$$

$$P_{(2)} = P_{(1)} \times \left(\frac{k+1}{2}\right)\left(\frac{\bar{x}}{\bar{x}+k}\right)$$

$$P_{(3)} = P_{(2)} \times \left(\frac{k+2}{3}\right)\left(\frac{\bar{x}}{\bar{x}+k}\right)$$

to

$$P_{(x)} = P_{(x-1)} \times \left(\frac{k+x-1}{x}\right)\left(\frac{\bar{x}}{\bar{x}+k}\right)$$

Both $\bar{x}$ and k are calculated from the observed data. The mean is calculated by the usual method ($\Sigma x/n$), and k can be estimated using the formula:

$$k = \frac{s^2}{s^2 - \bar{x}}$$

where s^2 is the variance, again calculated in the usual way (*see* VARIANCE). However, this method for estimating the exponent k is only approximate and is not very reliable for values of k below 4, unless $\bar{x}$ is also below 4 (Anscombe 1950). A more accurate method involves substituting values of k into the equation:

$$n\ln\left(1+\frac{\bar{x}}{k}\right)=\sum\left(\frac{A_{(x)}}{k+x}\right)$$

until both sides are equal (i.e. the equation is solved by iteration or trial and error). In this equation, n is the total number of sampling units, ln is the natural logarithm (logarithms to the base e) and $A_{(x)}$ is the total number of samples exceeding x. Below are data on the number of eggs laid by fruit flies (*Drosophila subobscura*) on 85 fruit, sampled from the wild.

x	0	1	2	3	4	5	6	7	8	9	10	11
Freq.	44	14	8	6	2	4	2	1	1	0	2	1
$A_{(x)}$	41	27	19	13	11	7	5	4	3	3	1	0

Each fruit is a sampling unit, and 44 fruits had no eggs, 14 had one egg, 8 had two eggs and so on. For these data, $\bar{x}=1.55$ and $s^2=6.34$. Since the variance is much larger than the mean it would suggest that we are dealing with an aggregated or clumped distribution. An initial estimate of the exponent is:

$$k=\frac{6.34}{6.34-1.55}=1.32$$

However, substituting values of k into the more accurate iterative method (best performed on a computer) we obtain a value of $k=$ 0.46, illustrating that the first estimate was indeed rather unreliable. With this value of k, the left-hand side of the above equation is:

$$n\ln\left(1+\frac{\bar{x}}{k}\right)=125.35$$

and the right-hand side of the equation is:

$$\sum\left(\frac{A_{(x)}}{k+x}\right)=124.93$$

which is close to equality. Taking k to more places of decimals would increase the equality. Now that we have $\bar{x}=1.55$ and $k=0.46$ we can calculate the probabilities for the negative binomial distribution, and the corresponding expected number of fruit in a SAMPLE of 85:

$$P_{(0)}=\left(1+\frac{1.55}{0.46}\right)^{-0.46}=0.51 \quad =43.3\text{ fruit}$$

$$P_{(1)}=0.51\times\left(\frac{0.46}{1}\right)\left(\frac{1.55}{1.55+0.46}\right)=0.18 \quad =15.3\text{ fruit}$$

$$P_{(2)}=0.18\times\left(\frac{1.46}{2}\right)\left(\frac{1.55}{1.55+0.46}\right)=0.10 \quad =8.5\text{ fruit}$$

$$P_{(3)}=0.10\times\left(\frac{2.46}{3}\right)\left(\frac{1.55}{1.55+0.46}\right)=0.06 \quad =5.1\text{ fruit}$$

$$P_{(4)}=0.06\times\left(\frac{3.46}{4}\right)\left(\frac{1.55}{1.55+0.46}\right)=0.04 \quad =3.4\text{ fruit}$$

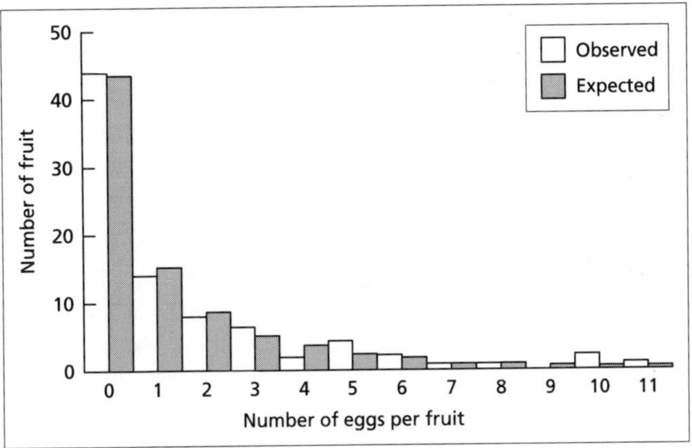

Fig. N2 Observed and expected numbers of eggs per fruit for *Drosophila subobscura*. The expected distribution is a negative binomial with $\bar{x}=1.55$ and $k=0.46$, both calculated from the observed data.

$$P_{(5)}=0.04\times\left(\frac{4.46}{5}\right)\left(\frac{1.55}{1.55+0.46}\right)=0.03 \quad =2.3\text{ fruit}$$

$$P_{(6)}=0.03\times\left(\frac{5.46}{6}\right)\left(\frac{1.55}{1.55+0.46}\right)=0.02 \quad =1.8\text{ fruit}$$

$$P_{(7)}=0.02\times\left(\frac{6.46}{7}\right)\left(\frac{1.55}{1.55+0.46}\right)=0.01 \quad =1.2\text{ fruit}$$

$$P_{(8)}=0.01\times\left(\frac{7.46}{8}\right)\left(\frac{1.55}{1.55+0.46}\right)=0.007 \quad =0.6\text{ fruit}$$

$$P_{(9)}=0.007\times\left(\frac{8.46}{9}\right)\left(\frac{1.55}{1.55+0.46}\right)=0.005 \quad =0.4\text{ fruit}$$

$$P_{(10)}=0.005\times\left(\frac{9.46}{10}\right)\left(\frac{1.55}{1.55+0.46}\right)=0.004 \quad =0.3\text{ fruit}$$

$$P_{(11)}=0.004\times\left(\frac{10.46}{11}\right)\left(\frac{1.55}{1.55+0.46}\right)=0.003 \quad =0.2\text{ fruit}$$

These expected numbers of fruit containing 0, 1, 2, 3, . . . , 11 eggs of this fly are shown in Fig. N2, along with the observed numbers for comparison. A negative binomial distribution certainly appears to be a good description of the real distribution of eggs for this species. *See also* STATISTICS. [B.S.]

REFERENCE

Anscombe, F.J. (1950) Sampling theory of the negative binomial and logarithmic series distributions. *Biometrika* **37**, 358–382.

negative feedback The coupling of an output process to an input process in such a way that the output process is inhibited. For example, high birth rate leads to high population, which leads to crowding and reduced birth rate, thus tending to reduce population growth (in this example birth rate is the input process and population increase the output process). Negative feedbacks tend to be stabilizing, although this is not always the case if the FEEDBACK is delayed. *See also* POPULATION REGULATION. [S.N.W.]

negative ions Molecules or atoms which have become negatively charged by gaining one or more electrons. Many people believe that an excess of negative ions in the atmosphere, as occurs near running water, is therapeutic. [J.G.]

nektobenthic *See* BENTHIC HABITAT CLASSIFICATION.

nekton Pelagic animals whose swimming abilities permit them to move actively through the water column and to move against CURRENTS. They include small crustaceans, cephalopods, fish, reptiles, birds and marine mammals. Nektonic species show adaptations for maintaining their BUOYANCY (e.g. gas or swim bladders in fish) and for effectively moving through water (e.g. a streamlined body form, paddle-shaped appendages, undulatory body movements). The warm-blooded marine mammals and birds are specially adapted to survive in OCEAN waters, and have special features to maintain their body temperature and osmotic balance, and adaptations for diving. Many nektonic species (especially mammals, birds and reptiles, but also fish) undertake substantial migrations for breeding or other purposes, and much study has been directed towards determining how these species navigate successfully over such great distances. Generally, nekton are carnivorous as adults and feed mainly on other nekton. [V.F.]

neo-Darwinian fitness *See* FITNESS.

neo-Darwinism Darwin proposed that EVOLUTION was driven by NATURAL SELECTION of heritable TRAITS. Yet his theory of inheritance proposed that traits from each parent blended, with more adaptive traits blending with less adaptive traits, so this would tend to dilute the effects of selection. The solution to this problem came after Darwin's death, with the rediscovery of Mendel's work on genetics, which showed how traits do not blend but are controlled by genes that remain distinct. The association of DARWINISM with Mendelian genetics is often referred to as neo-Darwinism. *See also* MENDELIAN INHERITANCE; MODERN SYNTHESIS; POPULATION GENETICS; QUANTITATIVE GENETICS. [P.C.]

neodymium (Nd) A LANTHANIDE METAL (ATOMIC NUMBER = 60; relative ATOMIC MASS = 144.2; melting point = 1021°C; boiling point = 3068°C) with an average abundance in the Earth's CRUST of 28 mg kg^{-1}. It is used in the manufacture of some forms of GLASS and also in capacitors. [J.G.]

neon (Ne) A gaseous element (ATOMIC NUMBER = 10; relative ATOMIC MASS = 20.179; melting point = −248.6°C; boiling point = −245.9°C) of the noble gas group. Neon is a minor constituent of the Earth's ATMOSPHERE with an average volume mixing ratio of 1.8×10^{-5} (18 ppm by volume, ppmv). There are three stable isotopes of relative atomic mass 20, 21 and 22. The element is physically and chemically inert and has no environmental significance. It is obtained from liquefaction of air and used in electric discharge signs. [M.R.H.]

neoteny A form of HETEROCHRONY that involves the slowing down of the rate of development generating an ONTOGENY that exhibits juvenile traits relative to the ancestral morphology. The term was originally coined in the 19th century to define the phenomenon of larval reproduction found in several taxa, the best known being salamanders such as the axolotl (*Ambystoma mexicanum*). This usage has remained in the specialized literature leading to confusion. In the absence of information about causal mechanisms, larval reproduction should be referred to as PAEDOMORPHOSIS, which is the current name used to define any morphology reminiscent of a juvenile or embryonic ancestral form. The reason is that larval reproduction could also be achieved by the heterochronic process of PROGENESIS. [P.A.]

neotropical The biogeographical region which comprises the tropics and subtropics of the New World (in contradistinction to palaeotropical, i.e. of the Old World tropics). On a stricter interpretation (as in the NEOTROPICAL KINGDOM as a phytogeographical unit of Takhtajan (1969); *see* BIOGEOGRAPHICAL UNITS, Fig. B6) it has been taken to include not only the Caribbean region, southern Mexico, southern Florida, Central America and the northern part of South America, but also extending as far south as Argentina and Chile, far outside the tropics.

Historically, the two components of the neotropics (broadly, Central America with the Caribbean, and South America) were separated through the greater part of the CENOZOIC era, only becoming connected at the Panama isthmus in the Pliocene epoch (about 5 million years ago). The mammals of South America in the early Cenozoic were principally marsupials together with two groups of primitive placental mammals. These were later joined by rodents and primates, arriving either from Africa or North America.

Continuous land connection between North and South America established in the Pliocene resulted in a southward MIGRATION of elements of the North American fauna (especially the placental mammals), and a northward spread of the South American fauna (especially marsupial mammals) into what is now Central and North America. Over the 2 million years since the end of the Pliocene, typical South American marsupials, adapted to a tropical environment, migrated as far as the southern margin of the Mexican Plateau. This is accordingly taken as the northern margin of the Neotropical region of zoogeographers, and coincides with that of the Neotropical kingdom of phytogeographers (see Figs B5 & B6).

In the early Tertiary period, some 60 million years ago, when a tropical climate extended further both northwards and southwards than at present, many plants typical of the present-day Neotropical flora extended northwards far beyond their present RANGE. They reached to about 50°N on the Pacific side of the continent and 37°N on the Atlantic side. This geographically more extended version of the Neotropical flora of the early Tertiary has been referred to as the Neotropical geoflora (see Crane 1987).

The flora of the present-day neotropics has a number of characteristic flowering plant families endemic to the New World, two of the better known being the Bromeliaceae and Cactaceae (although both of these families have isolated single genera occurring in Africa). The Cactaceae, with their adaptations to arid environments (e.g. photosynthetic and often water-storage stems, and leaves replaced by defensive spines), show striking convergence with similarly adapted members of the Euphorbiaceae of the Old World tropics. A number of other sizeable families such as the Cannaceae and Lecythidaceae are also endemic to the neotropics. [W.G.C.]

REFERENCES
Crane, P.R. (1987) Vegetational consequences of angiosperm diversification. In: *The Origins of Angiosperms and their Biological Consequences* (eds E.M Friis, W.G. Chaloner & P.R. Crane), pp. 107–144. Cambridge University Press, Cambridge.
Takhtajan, A. (1969) *Flowering Plants, Origin and Dispersal.* Oliver & Boyd, Edinburgh.

Neotropical faunal region A mostly tropical FAUNAL REGION extending from the southern temperate zone of South America through Central America to the tropical lowlands of Mexico (see Fig. F1). The West Indian fauna is limited and is in some ways transitional, forming a complex fringing pattern rather than part of a continental one.

NEOTROPICAL freshwater fishes are very numerous and highly endemic, with general African relationships. They are dominated by characins, gymnotid eels and catfishes. Some of the amphibians and reptiles, such as pipid frogs and pelomedusid turtles, also have African affinities. Others, such as the caecilians, *Testudo* and crocodiles, are primarily tropical, or North American in affinities (e.g. plethodontid salamanders, iguanid and teiid lizards).

The Neotropical bird fauna is very rich and distinct. Indeed, South America has the richest of all bird faunas, with many more species, especially of small birds, than any other equal area. Nearly half of some 67 families are endemic, or at least confined to the New World. These include rheas, New World vultures, trumpeters, hummingbirds, toucans, American flycatchers, manakins and cotingas. Endemic or near-endemic mammals include two groups of marsupials, the opossums and caenolestids, two families of monkeys, anteaters, sloths, armadillos, hystricomorph rodents and vampire bats. *See also* NEOTROPICAL KINGDOM. [A.H.]

REFERENCE
Darlington, P.J. (1957) *Zoogeography: the Geographic Distribution of Animals.* John Wiley & Sons, New York.

Neotropical kingdom A phytogeographical subdivision based on floral evidence and covering Central and most of South America (with the exception of the southernmost tip). The distinctiveness of the area has probably resulted from its isolation from other landmasses during the CENOZOIC era. *See also* NEOTROPICAL FAUNAL REGION. [P.D.M.]

REFERENCE
Takhtajan, A. (1986) *Floristic Regions of the World.* University of California Press, Berkeley, Calif.

neptunium (Np) An artificial radioactive element (ATOMIC NUMBER=93; relative ATOMIC MASS=237; melting point=640°C; boiling point=3902°C), discovered in 1940. Isotopes from ^{228}Np to ^{242}Np have been produced. ^{237}Np ($t_{1/2}=2.14\times10^6$ years) occurs in the environment as a result of release from nuclear facilities and from decay of ^{241}Am. As a consequence of its long half-life ^{237}Np is also of significance with respect to high-level RADIOACTIVE WASTE disposal. Neptunium occurs in the +5 oxidation state (although +3 to +7 states can be produced) and exhibits relatively soluble behaviour in natural waters. *See also* WASTE DISPOSAL. [A.B.M.]

neritic Referring to inshore waters between mean low water to 200 m in depth that overlie CONTINENTAL SHELVES (i.e. the zone landward of the SHELF-SLOPE BREAK). The neritic zone is characterized not by a fixed depth but by the fact that it is euphotic and thus an area in which PHOTOSYNTHESIS can occur. [V.F.]

Nest parasitism

The exploitation of dependent offspring in nests by invertebrate parasites that use blood or other tissue of the offspring as a source of nutrition (*cf.* BROOD PARASITISM). The nests of insects, birds and mammals often contain large numbers of invertebrates that either exploit the nest, its contents or its owners as a resource. Nest parasites include a large number of species of mites and ticks, fleas, lice, louseflies and blowflies.

Characteristics

The number of parasites per nest depends on the degree of nest reuse and the GENERATION TIME of the PARASITE relative to the duration of the nesting cycle of the HOST. While the number of mites may multiply many times during the breeding season of a bird host, the number of fleas remains relatively constant because of a long generation time. The number of mites in a nest may reach many thousands, while fleas usually are restricted to less than a few hundreds.

Parasites are transferred between nests by their hosts, or by their own movements when host nests are aggregated. Nest parasites are usually negatively phototactic during the reproductive stage of their hosts, but as soon as host offspring leave the nest their nest parasites become positively phototactic. Aggregation of parasites on the rim of the nest or around the hole of the nest of hole-nesting bird species facilitates DISPERSAL to new host nest sites, when a nest site is visited by a host. Nest parasites often remain in the nests of their hosts after the breeding season, and some parasitic species associated with birds' nests are known to be able to survive several years without food.

Damage to hosts

Nest parasites can inflict damage on their hosts by direct removal of resources such as blood or other tissue, damage of tissue through their bites, hypersensitivity reactions, costs of GROOMING and preening and other antiparasite defences, and costs of elevated immune responses. Parasitized bird nestlings have an elevated metabolic rate caused by PARASITISM, and unless the increased energy cost of parasitism is covered by an increased parental feeding rate, this will give rise to a reduction in nestling growth.

Parasites are detrimental to their hosts by exploiting resources that could otherwise be used by hosts for maintenance, growth or reproduction. A large number of experiments have shown that hosts suffer FITNESS costs from nest parasitism by experiencing delayed reproduction, reduced clutch and brood sizes, a smaller number of broods per season, and reduced nestling growth rate. Nest parasites are particularly detrimental to their hosts under adverse environmental conditions such as outside the main breeding season, in poor habitats, and during years with poor weather.

Nest parasites can extract resources from either offspring or parent hosts, but observations and experiments suggest that mainly host offspring are negatively affected by nest parasitism. This is shown by larger reductions in the survival of offspring in comparison to that of their parent in the presence of parasites. The age difference in MORTALITY costs of parasitism may be due to offspring having a less-developed and less-efficient immune system than their parents.

Parasite virulence

Parasite VIRULENCE reflects the amount of damage by parasite reproduction imposed on hosts in terms of reductions in host fitness, and hypothesized determinants of virulence are the relative frequency of horizontal transmission of parasites and the frequency of multiple infestations. Parasites that are horizontally transmitted can severely damage their hosts without restricting their own reproductive success, while that is not the case for vertically transmitted parasites. Similarly, parasites of different genetic origin will compete intensely with each other for access to host resources. Parasites of host species that frequently reuse their nest sites, such as hole and cavity nesters and colonially breeding birds, are easily transmitted horizontally, and single nests often contain parasites of multiple genetic origins, and are therefore more virulent than parasites of a single genetic origin in nests of host species that never or only rarely reuse nest sites. Hosts may be forced to coexist with their virulent nest parasites because there is a limited availability of safe nest sites such as those used by hole and cavity nesters and colonially breeding bird species.

Consequences for host reproduction

The reduction in the reproductive success of hosts subject to nest parasitism can be a result of either the cost of parasitism or altered PARENTAL INVESTMENT. Nest parasites obviously often reduce the reproductive success of hosts by simple removal of resources from the host. If the risk or the intensity of nest parasitism is unpredictable at the start of reproduction, parents may be unable to adjust their current reproductive decisions on parental investment. If a parent finds its nest heavily infested with parasites, and if the infestation rate differs among nests, it may reduce its investment in current reproduction partially by reducing its feeding rate, or completely by nest DESERTION. Parents may avoid investing in an already doomed brood and increase future reproductive success by such a reduced parental investment.

Antiparasite strategies

Hosts regularly suffering from nest parasitism have developed a range of antiparasite behaviours that reduce the negative impact of nest parasites on host fitness. Hosts are often able to assess the infection status of old nest sites, and they have been shown to change breeding sites before or during the start of reproduction in response to the presence of nest parasites.

Hosts of nest parasites can partly control the number of parasites by preening and grooming. Experimental restriction of the ability of hosts to preen or groom has shown that the number of parasites increases dramatically when mechanical antiparasite defence is prevented or restricted. Some host species such as European starlings (*Sturnus vulgaris*) directly eliminate nest parasites by removal of all nest material at the start or the end of the breeding season. The complete absence of nest material in the nests of some bird species such as fairy terns (*Gygis alba*), woodpeckers, nuthatches and others may facilitate the search for and direct removal of nest parasites.

Hosts of nest parasites can also reduce the negative effects of nest parasitism by the use of plants with high contents of secondary chemical compounds. Plants use chemicals in their defence against herbivores, and a number of compounds has been shown to have negative effects on the growth, survival and reproduction of nest parasites. A range of bird species use fresh plant material for their nests, and new material is incorporated in the nest throughout the breeding cycle. Bird species that use fresh plants in their nest are mainly reusing nest sites, and they include birds of prey, titmice, starlings and many others. European starlings choose specific plant species with high contents of secondary compounds as nest material, and *in vitro* experiments have shown that mites die after exposure to such plant material.

Hosts also appear to have evolved mechanisms of immune defence (*see* DEFENCE MECHANISMS) that alleviate the negative impact of nest parasites on reproductive success. The organs involved in immune defence such as the bursa of Fabricius and the spleen are relatively larger in bird species that frequently reuse their nest sites compared with closely related species that rarely or never reuse sites. Nest reuse selects for increased parasite virulence, and affected hosts have retaliated in the coevolutionary ARMS RACE between parasites and hosts by developing relatively larger immune defence organs. *See also* SPECIES INTERACTIONS. [A.P.M.]

REFERENCES

Clayton, D.H. & Moore, J. (eds) (1997) *Host–Parasite Evolution*. Oxford University Press, Oxford.

Loye, J.E. & Zuk, M. (eds) (1991) *Bird–Parasite Interactions*. Oxford University Press, Oxford.

Nestflüchter *See* NIDICOLOUS.

Nesthocker *See* NIDICOLOUS.

net growth isocline *See* ZERO GROWTH ISOCLINE.

net photosynthetic rate (NPR) The difference between gross PHOTOSYNTHESIS (gain) and RESPIRATION (loss). The plant grows when NPR > 0 and loses mass when NPR < 0 (e.g. in prolonged darkness). *See also* NET PRIMARY PRODUCTION. [M.J.C.]

Net primary production (NPP)

The process of energy, matter or CARBON accumulation net of respiratory losses by the AUTOTROPHS of a community. Its rate, net primary productivity (P_n or N), equals the difference between gross primary productivity and RESPIRATION by the autotrophs. Although the term applies to all autotrophs it is most commonly used in the context of photosynthetic organisms. The term net primary production has often been used synonymously with 'net primary productivity'. However, by precedence, net primary productivity should be used to describe the rate of net primary production, whereas PRODUCTION is the process itself (MacFadyen 1948).

Although often defined to represent ENERGY FLOW (Begon *et al.* 1996), net primary production is more commonly reported and more practically determined as matter accumulated. Although the flows of energy, matter and carbon in ecosystems follow fundamentally different pathways, at the primary producer level this has less importance since BIOMASS from one autotroph cannot, by definition, be used by another. Therefore at the primary producer level of an ECOSYSTEM, neither matter nor energy can be recycled within this layer of the ecosystem. Further, ratios of energy : biomass : carbon are relatively constant within similar primary producers, where for higher plants 1 kg plant dry matter ≈ 17 000 kJ ≈ 400 g C (Roberts *et al.* 1993). Because DRY MASS is a simpler and more direct measure than energy content, a more practical definition of net primary productivity is therefore the rate of new mass formed by a plant community per unit surface area of ground or ocean, reported as g dry matter m^{-2} $year^{-1}$, or for shorter-term studies g m^{-2} day^{-1}.

Estimation procedures

Net primary productivity can be estimated for terrestrial vegetation and stands of macrophytes by two basic groups of measurements.

Harvest methods

Net primary productivity is estimated from the net outputs, i.e. increments in plant biomass, corrected for any losses through GRAZING, LEACHING, exudation or death through causes other than grazing. For forests and other stands of woody plants, DIMENSION ANALYSIS may be used to estimate biomass without the need to destructively remove several trees.

Gas exchange

Net primary production is estimated from the net input, i.e. the net loss of carbon dioxide (CO_2) from the atmosphere. Enclosure methods seal a portion of the vegetation into a large chamber and measure the decrease in CO_2 concentration across the chamber. Chamberless methods, either aerodynamic or EDDY-COVARIANCE measurements, allow the calculation of the net loss of carbon from the atmosphere to the vegetation. However, these techniques cannot separate heterotrophic release of CO_2 by other trophic levels, for example the soil microorganisms, therefore a separate and simultaneous measurement of soil respiration losses is necessary. Roberts *et al.* (1993) provide an overview of the different approaches.

Phytoplankton

PHYTOPLANKTON require different techniques, which include measurement of uptake of radioactively labelled CO_2/HCO_3^-, or net rates of O_2 evolution.

Typical values

Annual net primary productivities have been tabulated for the BIOMES and for the globe as a whole (Table N1). Whilst the highest average is 2500 g m^{-2} $year^{-1}$ for coastal algal beds and reefs, the highest recorded value for a single community is 9970 g m^{-2} $year^{-1}$ for the monotypic stands of *Echinochloa polystachya*, an emergent C_4 macrophyte, on the *Varzea* flood plains of the Amazon region (Piedade *et al.* 1991). [S.P.L.]

REFERENCES

Begon, M., Harper, J.L. & Townsend, C.R. (1996) *Ecology: Individuals, Populations and Communities*, 3rd edn. Blackwell Science, Oxford.

MacFadyen, A. (1948) The meaning of productivity in biological systems. *Journal of Animal Ecology* **17**, 75–80.

Piedade, M.T., Junk, W.J. & Long, S.P. (1991) The productivity of the C4 grass *Echinochloa polystachya* in the Amazon floodplain. *Ecology* **72**, 1456–1463.

Roberts, M.J., Long, S.P., Tieszen, L.L. & Beadle, C.L. (1993) Measurement of plant biomass and net primary production of herbaceous vegetation. In: *Photosynthesis and Production in a Changing Environment: A Field and Laboratory Manual* (eds D.O. Hall, J.M.O. Scurlock, H.R. Bolhàr-Nordenkampf, R.C. Leegood & S.P. Long), pp. 1–21. Chapman & Hall, London.

Whittaker, R.H. (1975) *Communities and Ecosystems*, 2nd edn. Macmillan Inc, New York.

---

net radiometer A type of RADIOMETER that measures the difference between the energy transferred by RADIATION of all wavelengths to the Earth's surface and that which is reflected at or emitted by the surface. A net radiometer is usually a thermopile device, in which the imbalance in the radiation incident on an upward facing plate and a downward facing plate results in a TEMPERATURE difference between them. Incoming direct and diffuse shortwave radiation plus atmospheric longwave radiation will heat the top surface of the thermopile, whereas reflected shortwave and emitted longwave radiation will warm the lower surface of the thermopile. The difference in temperature across the thermopile will generate a potential difference via the Seebeck effect, and this can be measured with a voltmeter. The sensitive surfaces of the thermopile are usually protected from the elements by a thin polythene dome which permits both shortwave and longwave energy to pass through to the sensing surface. The net amount of radiative energy which this device measures essentially sets the limit for the total energy which can be used up in the processes of evaporating WATER or heating the air or soil, and as such is of fundamental importance in dictating microclimates. *See also* MICROMETEOROLOGY; RADIATION LAWS. [J.B.M.]

net rate of increase The multiplication rate per generation of a population (also known as the 'net reproductive rate'), defined as the ratio of total female births in successive generations and usually designated *R*, or R_0. *R* is calculated from a COHORT LIFE TABLE containing age-specific MORTALITY and FECUNDITY schedules. If l_x is the proportion of females surviving to age *x* and m_x is their fecundity (female offspring) at that age, then the total number of female offspring produced during a single generation is the sum of the l_xm_x products over the reproductive life of females:

$$R = \sum l_x m_x$$

Table N1 Annual net primary productivities in dry mass (P_n), dry mass of standing crop biomass (B), and productivity to biomass ratios ($P:B$) for the communities of the world. (After *Communities and Ecosystems*, 2nd edn, by R.H. Whittaker, © 1975. Reprinted by permission of Prentice-Hall, Inc., Upper Saddle River, N.J.)

Ecosystem type	Area ($\times 10^6$ km^2)	P_n (g m^{-2} year^{-1}) Normal range	Mean	P_n World ($\times 10^{12}$ kg)	B (kg m^{-2}) Normal range	Mean	B World ($\times 10^{12}$ kg)	$P:B$ (year^{-1}) Mean
Tropical rain forest	17.0	1000–3500	2200	37.4	6–80	45	765	0.049
Tropical deciduous forest	7.5	1000–2500	1600	12.0	6–60	35	263	0.046
Temperate evergreen forest	5.0	600–2500	1300	6.5	6–200	35	175	0.037
Temperate deciduous forest	7.0	600–2500	1200	8.4	6–60	30	210	0.040
Boreal forest	12.0	400–2000	800	9.6	6–40	20	240	0.040
Woodland and shrubland	8.5	250–1200	700	6.0	2–20	6	51	0.12
Savannah	15.0	200–2000	900	13.5	0.2–15	4	60	0.23
Prairie and steppe	9.0	200–1500	600	5.4	0.2–5	1.6	14	0.38
Tundra and alpine	8.0	10–400	140	1.1	0.1–3	0.6	5	0.23
Hot desert and semi-desert	18.0	10–250	90	1.6	0.1–4	0.7	13	0.13
Extreme desert: rock, sand and ice	24.0	0–10	3	0.07	0–0.2	0.02	0.48	0.15
Cultivated land	14.0	100–3500	650	9.1	0.4–12	1	14	0.65
Swamp and marsh	2.0	800–3500	2000	4.0	3–15	15	30	0.13
Lake and stream	2.0	100–1500	250	0.5	0.001–0.1	0.02	0.04	12.5
Continental subtotal	149.0		773	115.2		12.3	1841	0.063
Open ocean	332.0	2–400	125	41.5	0–0.005	0.003	1.00	41.7
Upwelling zones	0.4	400–1000	500	0.2	0.005–0.1	0.02	0.008	25.0
Continental shelf	26.6	200–600	360	9.6	0.001–0.04	0.01	0.27	36.0
Algal beds and reefs	0.6	500–4000	2500	1.5	0.04–4	2	1.20	1.25
Estuaries	1.4	200–3500	1500	2.1	0.01–6	1	1.40	1.50
Marine subtotal	361.0		152	54.9		0.01	3.88	15.20
Total	510.0		333	169.8		3.6	1845	0.09

The net reproductive rate used to be widely used in life-table studies because the INTRINSIC RATE OF INCREASE, r, is related to R as follows: $r = \ln(R)/T$ where T is the mean length of a generation. Calculation of r from R requires substitution of the mean GENERATION TIME, T, which can be estimated approximately for a population that is neither increasing nor decreasing too rapidly, using:

$$T \approx \left(\sum x l_x m_x\right) / \left(\sum l_x m_x\right)$$

Nowadays, the practical value of R as a means of estimating r has been reduced because of the possibility of estimating r directly from the EULER–LOTKA EQUATION or the Leslie matrix (*see* POPULATION PROJECTION MATRIX) by iterative computation. *See also* FITNESS; POPULATION DYNAMICS. [R.H.S.]

REFERENCE

Andrewartha, H.G. & Birch, L.C. (1954) *The Distribution and Abundance of Animals*. University of Chicago Press, Chicago, Ill.

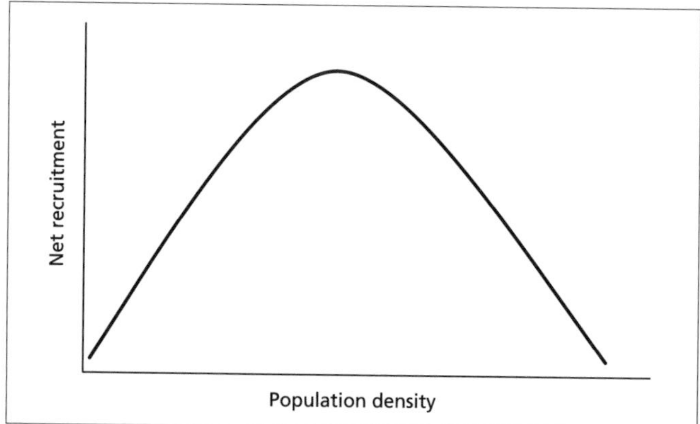

Fig. N3 Plot of net recruitment against population density showing the 'n' shape frequently found for net recruitment plots.

net recruitment curve A graph of RECRUITMENT in relation to another factor, in particular in relation to the POPULATION DENSITY (Fig. N3). In general, the curves are 'n'-shaped since when density is low NATALITY is usually high and MORTALITY is low, but when density is high natality is low and mortality is high; however, other forms are possible. Real, or theoretical, net recruitment curves are used to estimate the MAXIMUM SUSTAINABLE YIELD possible when HARVESTING organisms from a population. *See also* DEMOGRAPHY; LIFE TABLE. [A.J.D.]

neuston Planktonic organisms of the PELAGIC zone that live at or very near the surface. [V.F.]

neutral models *See* NULL MODEL.

Neutral theory of evolution

The neutral theory of EVOLUTION considers that the majority of the genetic differences between individuals or between species are due to the accidental accumulation of mutant DNA sequences that have little or no effect on individual reproductive success. Such sequences are said to be neutral with respect to the factors that underlie NATURAL SELECTION. In the absence of natural selection, a single MUTATION, occurring in a single gene on a single chromosome in a single individual is not expected to spread through a population with the passing of the generations. This is because it can be shown, mathematically, that an infinitely large, randomly mating population of individuals would be genetically stable (i.e. in Hardy–Weinberg equilibrium). Such assumed stability is based on the known features of random chromosome assortment during meiosis and random gametic fusion that govern Mendel's rules of genetic inheritance. Hence, the Mendelian rules, in themselves, cannot bring about a change in the average genetic composition of a population (Darwin's 'modification by descent').

Sampling error: fixation and loss of alleles

Real populations are not infinitely large but are mosaics of smaller populations within which local, non-random mating may take place. Importantly, continual FLUCTUATIONS in the sizes of natural populations at both the haploid (gametic) and diploid (zygotic) levels ensures that the relative frequencies of variant DNA sequences may alter from one generation to the next. This is due to SAMPLING error: if a small sample of, say, coloured marbles is drawn at random from an initial larger sample, then the frequencies of the different colours in the small sample might be unrepresentative of the original frequencies. If the small sample is allowed to re-expand to the original number and a small sample taken again, another accidental shift in frequencies might occur. Repeated sampling can lead both to the complete loss of some colours and to the fixation, or near-fixation, of other colours (Fig. N4).

Verification of the cumulative effects of sampling error at each generation in biological populations is derived from the comparative analysis of DNA sequence variation within and between species. Sequences that are known to have little or no effect on reproductive success (e.g. gene introns; third positions of triplet codons; and intergenic 'junk' DNA) show higher levels of within-species polymorphisms and between-species diversity relative to the first and second codon positions and to sequences involved in the regulation of gene expression or gene processing.

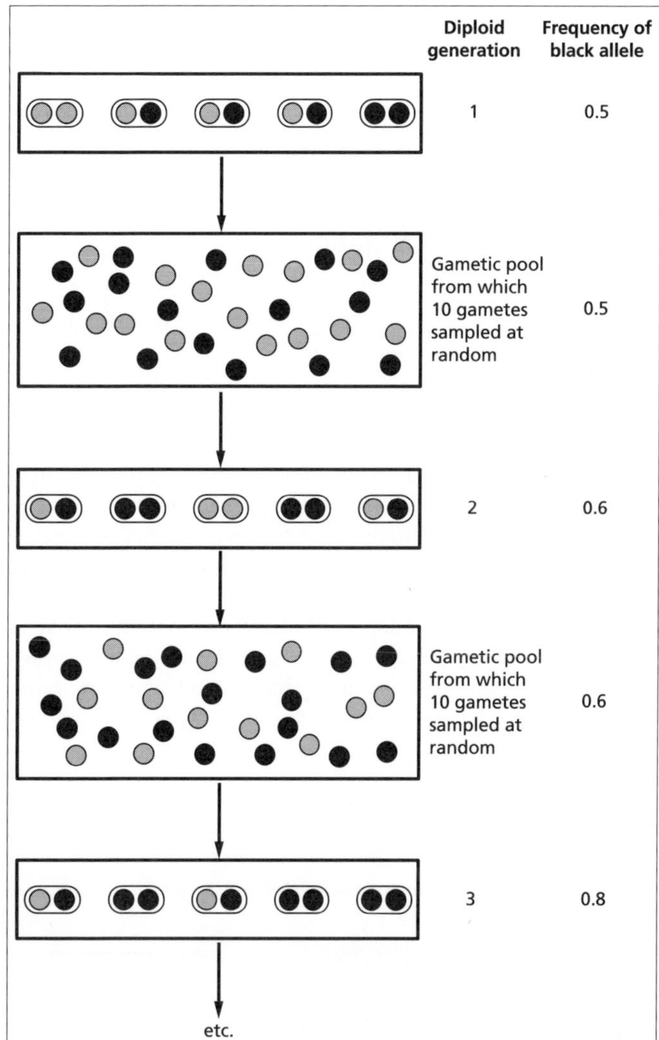

Fig. N4 Cumulative effects of sampling error. The meiotic process is expected to give rise to gametic pools with frequencies of alleles representative of those in the parental diploid generation. Random sampling of a few gametes to form the next generation of diploids can alter relative allelic frequencies.

Estimating genetic drift

The rate of accumulation (K) of mutations in neutral DNA is directly proportional to the rate of production (μ) of mutations. The probability (p) that a given mutation will drift to fixation via sampling error is $1/2N$, where N is the number of individuals in a sexual population and 2 accounts for the two ALLELES in each diploid individual. The number of neutral mutations in a population of size N is $2N\mu$. Hence $K=2N\mu p$ where $p=1/2N$. Therefore $K=\mu$.

Since not all individuals necessarily take part in reproduction, then the evolutionarily significant number of reproducing individuals might be smaller than the census population (N). The so-called 'EFFECTIVE POPULATION SIZE' (N_e) is given by:

$$(N_e) = 4N_m N_f/(N_m + N_f)$$

where N_f is the number of reproducing females and N_m is the number of reproducing males. For example, species in which a few dominant males fertilize many females have small N_e sizes; here, neutral drift can take place relatively quickly.

The 'molecular clock'

The simple formulation of $K=\mu$ gives rise to the concept of a 'molecular clock': neutral mutations accumulate at a constant rate depending only on the constancy of the mutation rate. When the mutation rate differs between different sections of DNA, or when the strength of selection for or against specific mutations varies depending on the functional biology of different sections of DNA, then the rate of accumulation of neutral mutations (the 'tick' of the molecular clock) will differ between DNA sections; although the rate could be roughly constant for any given section.

Patterns of neutral mutations

When neutral mutations occur near a sequence that is under frequent positive or negative selection then the expected distribution patterns of mutations within and between species can be distorted: such neutral mutations are either being eliminated by negative selection or hitching a ride in a positive selective sweep. Measurements of such distortions from neutral expectations are often considered to provide indirect evidence for the operation of selection in the evolutionary past of a given DNA sequence. However, such distortions can also occur through the activities of a variety of mechanisms of DNA turnover that underpin the molecular drive process of evolution (*see* MOLECULAR DRIVE THEORY OF EVOLUTION). The relative roles of natural selection, neutral drift and molecular drive in the evolution of any given biological TRAIT cannot be determined solely through the measurement of standing levels of genetic variation within and between species. *See also* DARWINISM; FITNESS; MENDELIAN INHERITANCE. [G.A.D.]

REFERENCES

Holzel, A.R. & Dover, G.A. (1991) *Molecular Genetic Ecology*. Oxford University Press, Oxford.

Kimura, M. (1983) *The Neutral Theory of Molecular Evolution*. Cambridge University Press, Cambridge.

---

neutral theory of molecular evolution *See* DRIFT, EVOLUTIONARY.

neutralism A situation in which two species, living in the same habitat, have no effect upon each other (0 0 interaction). *See also* SPECIES INTERACTIONS. [B.S.]

neutron An uncharged, subnuclear particle (nucleon) of rest mass 1.0087 amu. Free neutrons decay by β^- emission with a 10.6 min half-life. Neutrons emitted during nuclear FISSION can be used to propagate nuclear chain reactions by inducing further fission events. Nuclear reactors can be used as neutron sources, and reactor-produced neutrons are classed according to energy as:

- thermal (energies < 0.5 eV; average energy $= 0.025$ eV);
- epithermal (energies 0.5 eV to 1 MeV);
- fast (energies > 1 MeV).

Generators, utilizing the $^3H(d,n)^4He$ reaction, can be used to produce beams of neutrons of energy of 14 MeV. Neutron activation analysis entails neutron irradiation of samples to induce neutron capture by nuclei of ELEMENTS of interest and assay via characteristic RADIATION emitted during decay of the product radionuclides. [A.B.M.]

neutron probe An instrument for measuring soil water content. The NEUTRON probe provides non-destructive, accurate measurements of changes in soil water content with depth. A probe containing a radioactive source of high-energy fast neutrons is lowered down an access tube permanently installed in the soil. The fast neutrons collide with HYDROGEN atoms in the soil and in the water, and lose energy. These slow neutrons are sampled by a detector also housed in the probe. Because the strength of the signal is a measure of the number of hydrogen atoms in both the soil and the water, the soil must be calibrated. This is best done by gravimetric sampling close to the site under study. The water content is determined from the difference between wet and oven-dry mass, over a range of moisture contents. Any long-term drifts in the instrument CALIBRATION are removed by calculating SOIL MOISTURE content with respect to a water STANDARD. The sampling volume of a neutron probe is a sphere around the source, with 95% of the signal being the result of interactions within a radius of some 300 mm. [J.H.C.G.]

REFERENCE

Hodnett, M.G. (1986) The neutron probe for soil moisture measurement. In: *Advanced Agricultural Instrumentation. Design and Use* (ed. W.G. Gensler), pp. 148–192. Martinus Nijhoff, Dordrecht.

new chemicals Chemicals, produced commercially, that are not listed on the inventories of EXISTING CHEMICALS. These are subject to a NOTIFICATION procedure leading to RISK ASSESSMENT and possibly RISK MANAGEMENT. [P.C.]

new forestry Forest practices designed to retain more ECOSYSTEM complexity than is found in traditionally managed forests while also removing commodities (usually wood) (Swanson & Franklin 1992). The development of new forestry was motivated by the desire to sustain biological DIVERSITY and maintain soil PRODUCTIVITY, and by the perception that FOREST reserves (e.g. national parks) are not large enough, and are too isolated from one another, to conserve many forest species. Management practices vary according to forest type and objectives, and are derived from an understanding of natural DISTURBANCE regimes, ecosystem dynamics and species' habitat requirements.

Many new forestry prescriptions maintain mature-sized dead trees and logs, which are often important for WILDLIFE habitat, nutrient cycling and long-term soil productivity. Mature trees may also be left in harvested areas while a new cohort of trees develops in order to provide local seed sources, habitat complexity for wildlife, and future large (high-value) timber. A greater diversity of tree and shrub species might be promoted or allowed than would be common under intensive FORESTRY. In forest types that naturally experienced low-intensity disturbances (e.g. ground fires or treefall gaps), selection harvest may be used instead of clear-cutting. Landscape-level interactions are also considered in new forestry. Large trees may be provided along creeks for a future supply of logs to create in-stream structural complexity. Harvest areas might also be clumped together to minimize habitat fragmentation and provide large patches of interior forest habitat. *See also* ARBORICULTURE; ECOSYSTEM MANAGEMENT; FORESTRY PRACTICES; SILVICULTURE. [A.N.G.]

REFERENCE

Swanson, F.J. & Franklin, J.F. (1992) New forestry principles from ecosystem analysis of Pacific northwest forests. *Ecological Applications* **2**, 262–274.

new substances *See* EUROPEAN INVENTORY OF EXISTING COMMERCIAL CHEMICAL SUBSTANCES.

newton (N) The SI UNIT of force, defined as the force required to give 1 kg an acceleration of 1 m s^{-2}. Its relation to the obsolete British unit is: 1 N = 0.225 lb. [J.G.]

niche This term is often used loosely by both ecologists and laymen, but defining the niche concept formally is rather complicated. An old interpretation considers it to be the functional role or 'occupation' of a species in a COMMUNITY. Presently, the generally used concept is that of the 'Hutchinsonian niche' (Hutchinson 1957), which is defined as the requirements that the ENVIRONMENT has to meet to allow the persistence of (a POPULATION of) a species. The requirements consist of the tolerable ranges of conditions, i.e. those characteristics of the environment that are not influenced by the organism (such as salinity), and the tolerable ranges of RESOURCES that are depletable (such as food types and nesting sites). The distinction between conditions and resources is not always clear. Light and carbon dioxide (CO_2) may be conditions for a solitary plant, but may be resources in dense vegetation.

The tolerable ranges of conditions and resources are each envisaged as a section of an axis of a multidimensional 'niche space'. Together, the sections define the niche as a multidimensional hypervolume in niche space. Inside the hypervolume, the requirements for persistence of the species or population are met. The requirements along different axes may interact. For example, a high temperature may decrease the upper tolerance limit for salinity, or one type of food may be replaced by another. Such cases will produce rounded or flattened corners of the hypervolume.

For an older review of the niche concept see Vandermeer (1972). *See also* FUNDAMENTAL NICHE; REALIZED NICHE. [J.G.S.]

REFERENCES

Hutchinson, G.E. (1957) Concluding remarks. *Cold Spring Harbor Symposium on Quantitative Biology* **22**, 415–427.

Vandermeer, J.H. (1972) Niche theory. *Annual Review of Ecology and Systematics* **3**, 107–132.

niche overlap Quantification of the similarity of niches and, in particular, of the similarity in the use of resource types. The RESOURCES involved can be food, but also, for instance, may be aspects of the HABITAT. Many measures of NICHE overlap have been proposed (Hurlbert 1978; Holt 1987; Krebs 1989), some of which are similar to the measures used to express the similarity of COMMUNITIES. Some of the measures are symmetrical, i.e. O_{xy} (the overlap of species y on species x) = O_{yx}. For other measures this is not necessarily true. One of the simplest measures is the symmetrical 'percentage overlap':

$$O_{xy} = \sum_{i=1}^{n} \text{minimum}\left(p_{ix}, p_{iy}\right)$$

where p_{ix} is the proportion of total resources used by species x belonging to resource i, n is the number of resources, and O_{xy} is usually expressed as a percentage. The graphical interpretation of the percentage overlap is the area of overlap between the RESOURCE-UTILIZATION CURVES.

Overlap measures rarely take into account the differential abundances of resource types, although this would commonly be sensible. Some of the proposed overlap measures can be used as COMPETITION COEFFICIENTS, but generally measures of overlap and competition coefficients have incompatible properties (Hurlbert 1978). A measure of overlap should be chosen carefully for any particular problem. The ongoing quarrels in the literature show that this choice is far from straightforward. [J.G.S.]

REFERENCES

Holt, R.D. (1987) On the relation between niche overlap and competition: the effect of incommensurable niche dimensions. *Oikos* **48**, 110–114.

Hurlbert, S.H. (1978) The measurement of niche overlap and some relatives. *Ecology* **59**, 67–77.

Krebs, C.J. (1989) *Ecological Methodology*. Harper & Row, New York.

niche width (niche breadth) Quantification of the degree of specialization of a species (or a POPULATION), i.e of the degree to which different resource types are used. Many measures of NICHE width have been proposed, some of which are similar to measurements of DIVERSITY in communities. The measurement of niche width faces complications similar to those applying to NICHE OVERLAP, such as the differential availability of RESOURCES. [J.G.S.]

REFERENCES

Hurlbert, S.H. (1978) The measurement of niche overlap and some relatives. *Ecology* **59**, 67–77.

Krebs, C.J. (1989) *Ecological Methodology*. Harper & Row, New York.

Nicholson, A.J. *See* CHARACTERS IN ECOLOGY.

Nicholson–Bailey equations A discrete-time development of the Lotka–Volterra predator–prey model (*see* LOTKA–VOLTERRA MODEL), usually couched in terms of the numbers of hosts and parasitoids. The model was developed by the Australian entomologist A.J. Nicholson and physicist V.A. Bailey in 1935. Let H_t be the number of hosts at time t, and P_t the number of parasitoids. The Nicholson–Bailey equations are:

$$H_{t+1} = H_t \lambda \exp\left(-aP_t\right)$$

$$P_{t+1} = H_t\left[1 - \exp\left(-aP_t\right)\right]$$

where λ is the net per capita FECUNDITY of the hosts and a is a measure of the parasitoids' SEARCHING EFFICIENCY. The model assumes parasitoids and hosts have discrete OVERLAPPING GENERATIONS and that parasitoids search randomly for their hosts. The model predicts diverging OSCILLATIONS in population densities and is an important normative case with which more complicated and more realistic models are compared. *See also* POPULATION DYNAMICS; POPULATION REGULATION. [H.C.J.G. & R.M.S.]

nickel (Ni) A metallic element (ATOMIC NUMBER = 28; relative ATOMIC MASS = 58.71; melting point = 1453°C; boiling point = 2732°C), one of the HEAVY METALS, that is a trace constituent (100 mg kg^{-1}) of the Earth's CRUST. Found in ores associated with oxides and sulphides of IRON such as pentlandite ($(Fe,Ni)_9S_8$), nickel is used in the production of stainless steel and in electroplating. Although essentiality for humans remains unproven, nickel occurs as a cofactor in urease enzymes in animals and plants. Released to the environment from the burning of fuel and residual oils, nickel is rather toxic for most plants and fungi. Contact dermatitis is a common problem in nickel-related industries. [J.F.]

nidicolous (nest-squatters) The term nidicolous (from the Latin *nidus* nest, and *colere* to inhabit) is used to describe the nest attendance of neonates of birds and mammals. The contrasting terms nidicolous and nidifugous (from the Latin *fugere* to flee) were introduced by Oken (1837) and referred to *Nestflüchter* (nest-fleers, or nidifugous birds) and *Nesthocker* (nest-squatters, or nidicolous birds). Originally, nidicolous and ALTRICIAL were used as synonyms, as were nidifugous and PRECOCIAL. Today, these pairs of terms are used in slightly differing contexts although their general meanings overlap in some aspects. Altricial and precocial primarily refer to the developmental stage of the chick, whereas nidicolous and nidifugous refer to nest attendance. Additional terms have since been introduced to describe different syndromes of developmental maturity of hatchlings, parent–chick interactions and chick–environment interactions. *See also* DEVELOPMENTAL STATE OF NEWBORN; LIFE-HISTORY EVOLUTION; REPRODUCTIVE ALLOCATION. [J.M.S.]

REFERENCE

Oken, V.L. (1837) *Allgemeine Naturgeschichte für alle Stände*. Bd. 7 Abt. 1. Hoffmann'sche Verlagsbuchhandlung, Stuttgart.

nidifugous (nest-fleers) A term referring to the developmental state of neonates in mammals and birds. *Cf.* PRECOCIAL. *See also* NIDICOLOUS. [J.M.S.]

NIMBY *See* NOT IN MY BACK YARD.

NIMTO *See* NOT IN MY TERM OF OFFICE.

niobium (Nb) A metallic element (ATOMIC NUMBER = 41, relative ATOMIC MASS = 92.906; melting point = 2468°C; boiling point = 4927°C) occurring in the Earth's CRUST to the extent of 20 mg kg^{-1}, comparable to NITROGEN (19 mg kg^{-1}). The chemical properties of niobium are very similar to its congener TANTALUM (Ta) due to the effects of the lanthanide contraction which results in the two elements having very similar atomic radii, and consequently the two elements are invariably found together in ores such as (Fe,Mn)M_2O_6 (M = Nb,Ta), which is known as columbite or tantalite depending on which metal predominates. Niobium finds uses as an alloying metal for the production of high-temperature stainless steels; as an alloy with ZIRCONIUM (Zr) it is used in superconducting magnets. [P.J.B.]

Nitosols *See* SOIL CLASSIFICATION.

nitrate An anionic form of NITROGEN (NO_3^-) which has an important role in terrestrial and aquatic NITROGEN CYCLES. Nitrate is one of the two main forms (ammonium being the other) in which nitrogen is absorbed by plant roots. It is produced in soils and waters by microbial oxidation of ammonium (nitrification), and occurs in acid rain (*see* ACID RAIN AND MIST) due to the ionization of nitric acid, following the atmospheric oxidation of nitrogen monoxide (NO) and nitrogen dioxide (NO_2).

Nitrate concentrations are generally much higher in agricultural soils (5–250 mg kg^{-1} soil) than in those of natural terrestrial ecosystems (where they are usually less than 10 mg kg^{-1}), because of fertilizer and manure applications in the former, including the deposition of dung and urine by grazing livestock. Nitrate ions are readily leached from soils (except from some acid tropical soils), into surface and ground waters. Many areas of high population are dependent on pumped water from AQUIFERS for their potable water supplies, but in some aquifers which underlie intensively managed farmland the nitrate concentrations are rising towards, or even exceeding, the European Union's maximum acceptable level of 50 mg l^{-1}. The UK's Nitrate Sensitive Area scheme is an example of measures taken to reduce this problem. In this scheme, which commenced in 1990, 10 areas were designated in which a series of measures were introduced on a voluntary basis, including limits on fertilizer applications, sowing of cover crops in the autumn, and limits on ploughing of grassland, in return for financial compensation (Burt *et al.* 1993).

Denitrification can be exploited as a mechanism for removing unwanted nitrate from agricultural field drains. If the drain outflows are onto a river flood plain, where the soils are characteristically wet and partially anaerobic, as well as being CARBON-rich, this can result in a large reduction in nitrate content. The same concept is also being exploited for wastewater (e.g. SEWAGE effluent) treatment, using artificially constructed 'wetlands'.

Nitrate LEACHING may cause other environmental problems, such as the promotion of excessive algal growth, in both freshwater and marine ecosystems, at times when nitrate rather than phosphate is the nutrient limiting growth. Many ALGAL BLOOMS are toxic, and both they and excessive concentrations of nitrate itself have an adverse effect on WATER QUALITY (Fig. N5). *See also* NITRATE POLLUTION. [K.A.S.]

REFERENCE

Burt, T.P., Heathwaite, A.L. & Trudgill, S.T. (eds) (1993) *Nitrate: Processes, Patterns and Management*. John Wiley & Sons, Chichester.

nitrate pollution Run-off from agricultural land that has been worked with chemical fertilizer, or from AGRICULTURAL WASTE, which can be rich in nitrates. Such POLLUTION is a major cause of EUTROPHICATION in receiving waters. The CONTAMINATION of groundwater with nitrates is an important problem in Europe and North America. Groundwater is extensively used as a source of drinking water, and nitrates can cause problems for human health—not because they are toxic in themselves but in the mouth can be transformed by bacterial action to nitrites, which can induce methaemoglobinaemia (a reduction in oxygen-carrying capacity of the blood) especially in infants. According to WHO, water becomes unpotable when nitrates exceed 45 mg l^{-1}. The European Union has issued a directive requiring any area where nitrates exceed 50 mg l^{-1} in surface or groundwater to be declared a vulnerable area in which compulsory restrictions on farming must be applied. [P.C.]

REFERENCE

WHO (1993) *Guidelines for Drinking Water Quality*. World Health Organization, Geneva.

nitrogen (N) A gaseous element (ATOMIC NUMBER = 7, relative ATOMIC MASS = 14.0067). In the free elemental state nitrogen exists as an unreactive, diatomic gas, dinitrogen (N_2; melting point = −209.9°C; boiling point = −195.8°C), which is the most abundant component of the ATMOSPHERE (78.1% by volume). Global recy-

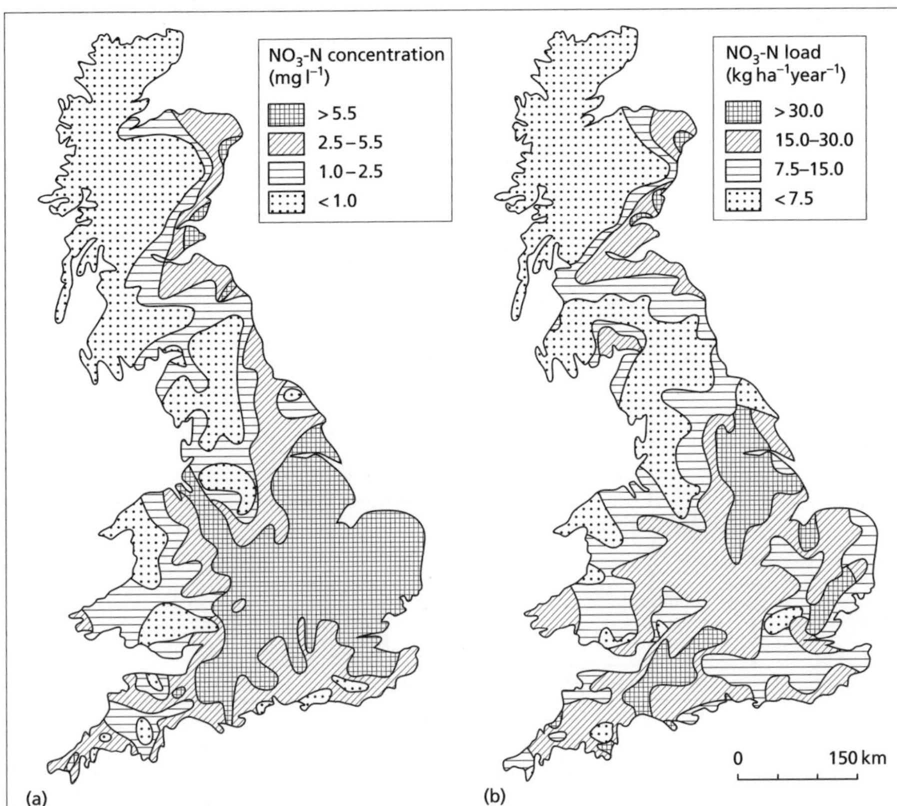

Fig. N5 (a) Mean nitrate concentrations and (b) mean nitrate loads in British rivers. (After Betton, C., Webb, B.W. & Walling, D.E. (1991) Recent trends in NO$_3$-N concentrations and loads in British rivers. *IAHS* **203**, 169–180.)

cling of the element is described by a NITROGEN CYCLE in which the major FLUXES are within terrestrial and marine BIOSPHERES and dead organic matter. Nitrogen is essential for plant GROWTH and is fixed from the atmosphere principally by bacteria in the soil (*see* NITROGEN FIXATION). The element takes oxidation states from –3 to +5 and combines to form a variety of important TRACE GASES and ions, for example: dinitrogen oxide (or nitrous oxide; N_2O); nitrogen monoxide (or nitric oxide; NO); nitrogen dioxide (NO_2); AMMONIA (NH_3); nitrite (NO_2^-); NITRATE (NO_3^-); and ammonium (NH_4^+). [M.R.H.]

Nitrogen cycle

The movement of NITROGEN between the ATMOSPHERE, the BIOSPHERE and the HYDROSPHERE, and the associated transformations between different chemical forms (Fig. N6). The Earth's atmosphere contains about 10^{15} t of nitrogen, as the relatively inert gas dinitrogen (N_2). This quantity is nearly 10 times greater than that contained in the oceans and in sediments. Biological nitrogen fixation, the conversion of atmospheric N_2 to AMMONIA by organisms, is the main natural mechanism for transfer of N to the biosphere (*c.* 150 Tg N year^{-1}) and hydrosphere (*c.* 40 Tg N year^{-1}). Industrial fixation is another major source (*see* NITROGEN FIXATION). The ammonia is either used directly, or converted to urea, ammonium nitrate or ammonium phosphates, for fertilizer use.

Soil processes

The microbial breakdown of plant residues in soils—MINERALIZATION—releases ammonium, which is available for uptake by plant roots or microorganisms (immobilization), or may be oxidized by nitrifying bacteria to nitrate (Haynes 1986). In temperate forest ecosystems there is normally a low concentration of mineral nitrogen present in the SOIL, and most of that is in the ammonium form. The tree root systems are an efficient sink for N, and the N cycle is said to be fairly closed. The exception is after clear-felling of an area of forest, when a large release of N from decomposing plant residues can result in substantial LEACHING losses (Vitousek *et al.* 1982). In contrast, in an intensive agricultural system, there is generally a relatively high concentration of nitrate present, particularly after fertilization and after the mineralization of crop residues in the autumn. Some of the nitrate is taken up by plants, but an appreciable amount may be lost from the soil in drainage water and accumulate in ground or surface waters (*see* NITRATE).

Sources of reactive gaseous nitrogen compounds

A natural source of nitrogen monoxide (NO) is microbial oxidation (nitrification) and chemical DECOMPOSITION in aerobic soils. In areas such as the tropical savannah regions of Africa and South America, this process, and biomass burning, are the major sources of NO. A small amount of atmospheric nitrogen, estimated at 5 Tg year^{-1}, is converted to NO and nitrogen dioxide (NO_2), by

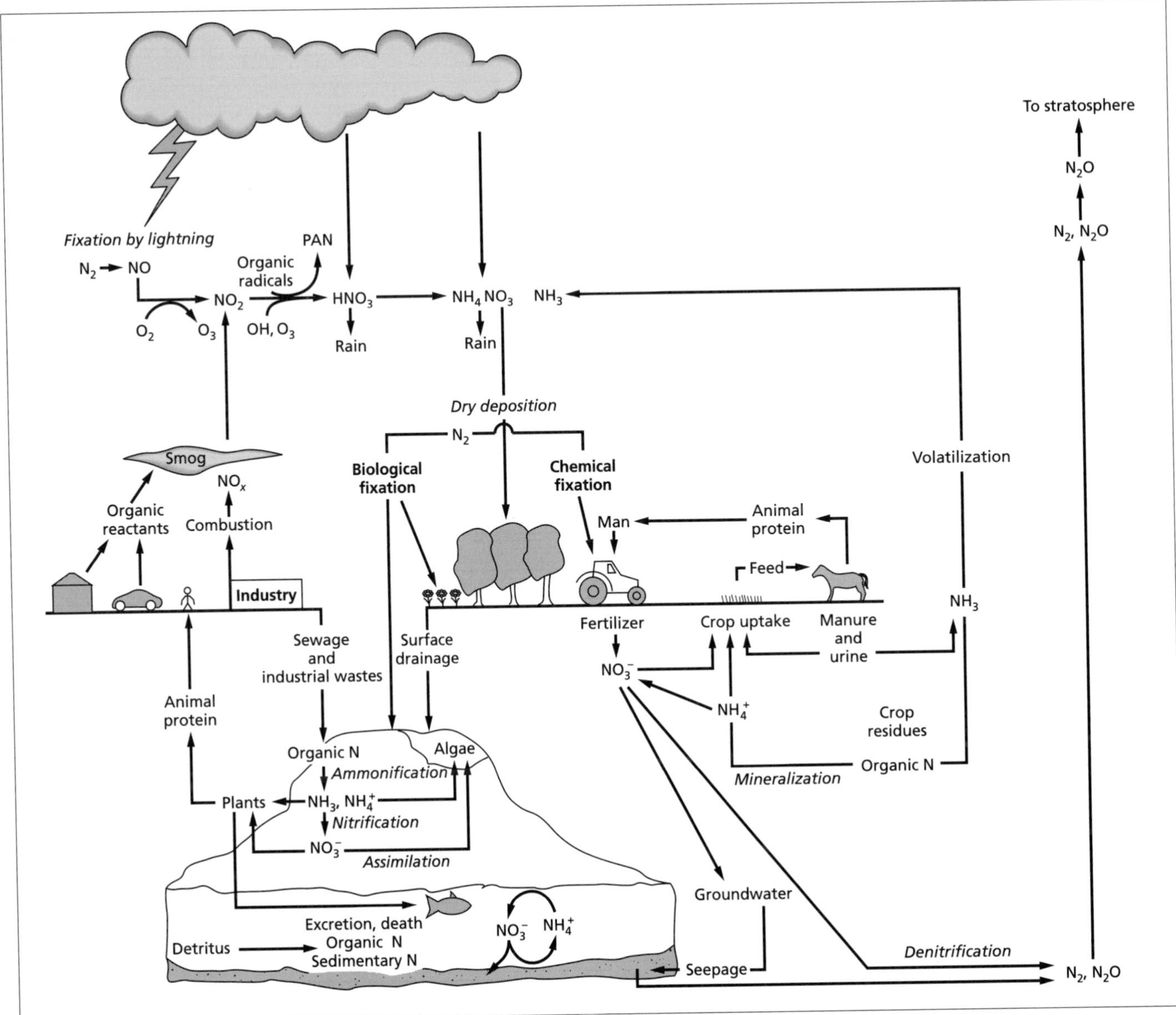

Fig. N6 The nitrogen cycle showing interactions in and between the atmosphere, aquatic ecosystems and terrestrial ecosystems. PAN, peroxyacetylnitrate. **Bold** type signifies major inputs; *italic* type signifies chemical or biochemical transformation processes. (Based on diagram in: *The Nitrogen Cycle of the United Kingdom: A Study Group Report* (1983). The Royal Society, London.)

LIGHTNING. The combustion of FOSSIL FUELS, in power-stations and internal combustion engines, is a larger source of these oxides (*c.* 20 Tg N year^{-1}) (Jaffe 1992). They contribute to PHOTOCHEMICAL SMOG and tropospheric OZONE formation, through a series of reactions involving VOLATILE ORGANIC COMPOUNDS and ultraviolet (UV) light. Oxidation in the atmosphere produces dinitrogen pentoxide (N_2O_5), which dissolves in cloud water to produce nitric acid. This nitric acid is a component of acid deposition, but usually a minor one compared with that derived from oxides of SULPHUR.

The other main source of reactive gaseous N compounds and contributor to acid deposition is the emission of ammonia (NH_3) from animal manures, fertilizer production and use, and from vegetation. Estimates of total worldwide EMISSIONS of NH_3 vary considerably, but are of the order of 100 Tg N year^{-1}. In Europe, emissions of N as NH_3 roughly equal those of nitrogen oxides collectively (NO_x). In some locations the deposition of N compounds is comparable with agricultural fertilizer inputs, and is damaging to forests and other natural ecosystems. For example, heathlands are becoming converted to grasslands because species such as *Calluna* are unable to compete with grasses at the enhanced level of N supply.

Denitrification and nitrous oxide

In anaerobic environments, for example waterlogged soils and sediments, microbial DENITRIFICATION reduces nitrate to nitrous oxide gas (N_2O) and N_2, thus returning fixed N to the atmosphere and completing the N cycle. Nitrous oxide is also released, along with NO (see above) by microbial nitrification in aerobic environments. Nitrous oxide is environmentally important: it is at present the largest contributor to ozone destruction in the STRATOSPHERE, through the production of nitrogen oxide free RADICAL catalysts by solar ULTRAVIOLET RADIATION (Prinn 1994). Nitrous oxide is also a significant GREENHOUSE GAS, contributing about 6% of the enhanced greenhouse effect. The major source of nitrous oxide worldwide has been thought to be wet tropical forests, where the rate of cycling of N in the soil-plant system is particularly rapid, and a proportion of the N being transformed by nitrification and denitrification (and possibly by fixation) is emitted as N_2O. However, it now appears likely that the release from intensive AGRICULTURE is of the same order, for similar reasons—i.e. the rapid cycling of N, and the relatively high amounts present in mineral form. *See also* BIOGEOCHEMICAL CYCLE. [K.A.S.]

REFERENCES

Haynes, R.J. (1986) *Mineral Nitrogen in the Soil-Plant System*. Academic Press, Orlando, Fla.

Jaffe, D.A. (1992) The nitrogen cycle. In: *Global Biogeochemical Cycles* (eds S.S. Butcher, R.J. Charleson, G.H. Orians & G.V. Wolfe), pp. 263–289. Academic Press, London.

Prinn, R.G. (ed.) (1994) *Global Atmospheric-Biospheric Chemistry*. Plenum Press, New York.

Vitousek, P.M., Gosz, J.R., Grier, C.C., Melillo, J.M. & Reiners, N.A. (1982) A comparative analysis of potential nitrification and nitrate mobility in forest ecosystems. *Ecological Monographs* **52**, 155–177.

Nitrogen fixation

Biological nitrogen fixation

The natural process of NITROGEN (N) fixation involves the conversion of atmospheric dinitrogen gas (N_2) to AMMONIA (NH_3) by organisms containing the enzyme complex nitrogenase. The dinitrogen molecule is very unreactive, and its reduction normally requires extreme conditions: elevated PRESSURE and TEMPERATURE and the presence of a CATALYST (see below). Yet nitrogenase achieves this reduction at ambient pressure and temperature. It is made up of two PROTEINS: a larger one with a relative molecular mass of over 200 000 and containing IRON, MOLYBDENUM and SULPHUR atoms, and a smaller one containing iron and sulphur. Neither of these proteins is active without the other (Postgate 1987). N fixation is an energy-demanding process. In order to function, nitrogenase needs the consumption of 16 adenosine triphosphate (ATP) molecules for the reduction of one N_2 molecule.

Soils contain free-living N-fixing bacteria such as *Azotobacter* spp. and in marine sediments fixation is principally by sulphate-reducing bacteria of the genus *Desulfovibrio*. However, most biological fixation of N occurs in symbiotic associations between bacteria and plants. The best known of these associations are those between *Rhizobium* and *Bradyrhizobium* bacteria and legumes, including major crops such as peas, beans, clover (see Plate 10, facing p. 398) and lucerne (alfalfa). Nearly all soils contain *Rhizobium* bacteria, which invade the legume roots via the root hairs and colonize certain cells, around which nodules develop. The cells become filled with the bacteria, which then become dormant as so-called bacteroids, capable of fixing N but not multiplying. Estimates of the N fixed by leguminous crops range from 70 to 100 kg ha^{-1} $year^{-1}$ for peas and beans to 300 kg ha^{-1} $year^{-1}$ for clover and lucerne (Postgate 1982). The legumes are exploited agriculturally for various purposes: (i) for human food, providing an important part of the diet particularly in developing countries; (ii) as animal FODDER of high nutritional value; and (iii) as green manure to improve the N supply to the following crop in the rotation.

In the natural environment, legumes are often PIONEER plants, colonizing N-deficient sites. Good examples are brooms (*Genista*) and gorse (*Ulex*). However, nodule-forming actinorhizal associations, such as that between alder (*Alnus*) and *Frankia* and cyanobacterial associations are more common in this role. Lichens — symbioses between fungi and green algae found in such locations as tree trunks, stones and walls — often have a nitrogen-fixing cyanobacterium as an additional partner. *Azolla*, a minute tropical water fern, is colonized by *Anabaena*, and the large amounts of N fixed by this system are very important in paddy rice production. In recent years it has been established that N-fixing biocoenoses (systems where there is some interdependence between the partners though both can grow satisfactorily on their own) exist in many grasses (Poaceae), including the grain crops maize, sorghum and millet, and important tropical forage species. These species form associations with bacteria such as *Azotobacter paspali* and *Azospirillum lipoferum*, and gain significant N inputs (up to 20 kg N ha^{-1} $year^{-1}$) from the fixation process. This is a small amount in terms of intensive western AGRICULTURE, but is very significant for the sustainability of low-input and subsistence farming systems.

Industrial nitrogen fixation

The industrial counterpart to biological fixation is the reaction of N_2 with HYDROGEN at high temperature and pressure in the presence of a catalyst to produce ammonia (Haber–Bosch process). This industrial fixation is mainly for fertilizer production — now *c.* 80 Tg N $year^{-1}$, and growing steadily in spite of reducing demand in western Europe, because of the need for increased food production in the developing world, especially in Asia. Other outlets are the production of nitric acid (HNO_3) and adipic acid (used in nylon production).

Managing nitrogen fertility

In intensive livestock-based agricultural systems, it has been shown that clover–grass, or lucerne–grass, PASTURE systems introduce sufficient nitrogen into the system to make large inputs of mineral N FERTILIZERS unnecessary. However, higher levels of management are required to maintain the legume-grass swards in a productive state than to maintain ryegrass-only pastures. In the developing world, much experimentation is going on with agroforestry and alley cropping systems, in which agricultural crops are planted between N-fixing trees or shrubs, and leaf litter and prunings from the legumes add nitrogen to the SOIL to the benefit of the non-leguminous crops. *See also* NITROGEN CYCLE. [K.A.S.]

REFERENCES

Postgate, J.R. (1982) *The Fundamentals of Nitrogen Fixation*. Cambridge University Press, Cambridge.

Postgate, J.R. (1987) *Nitrogen Fixation*, 2nd edn. Edward Arnold, London.

---

nitrogen oxides Gases formed from NITROGEN and OXYGEN, including nitrous oxide (N_2O), nitric oxide (NO) and nitrogen dioxide (NO_2). N_2O is inert in the TROPOSPHERE, but is decomposed in the STRATOSPHERE. NO and NO_2 are collectively known as NO_x, and are formed at high temperatures from the direct combination of atmospheric oxygen and nitrogen, as well as from the combustion of nitrogen in fuel. Worldwide, estimates of NO_x EMISSIONS vary between 80 and 230 Mt per year (expressed as NO_2), predominantly from anthropogenic sources. NO_x reacts in air to form nitric acid (HNO_3), a component of acid rain (*see* ACID RAIN AND MIST), and NITRATE aerosols. It is a major constituent of PHOTOCHEMICAL SMOG, with OZONE and PEROXYACETYL NITRATE (PAN). NO_x affects plant growth at concentrations above 15 ppb by volume (ppbv) and NO_2 may cause breathing problems to humans above 100 ppbv. Deposition of the gases from the ATMOSPHERE, and as nitrate in rain, contributes to EUTROPHICATION. *See also* MOTOR VEHICLE POLLUTION. [J.N.C.]

NMHCs (non-methane hydrocarbons) *See* HYDROCARBONS.

no observed effect concentration (NOEC) In an ECOTOXICITY test series, the highest concentration of a chemical to show no statistically significant difference from the control in an observed variable (the EFFECT). The NOEC clearly will depend upon the concentration intervals used; i.e. there could be a higher unobserved no effect concentration between the NOEC and the LOWEST OBSERVED EFFECT CONCENTRATION (LOEC). Moreover, proving a negative is always tenuous. It is for these reasons that the term 'observed' is used. NOECs should always be reported with LOECs. *See also* PREDICTIVE ECOTOXICITY TESTS. [P.C.]

NOAA *See* NATIONAL OCEANIC AND ATMOSPHERIC ADMINISTRATION (NOAA) SATELLITE.

nobelium (No) An artificial radioactive metallic element (ATOMIC NUMBER = 102) discovered in 1957. Particle bombardment of other actinides has been used to produce very small quantities of isotopes with mass numbers from ^{251}No to ^{259}No, with the longest lived being ^{259}No ($t_{1/2}$ = 58 min). Consequently, nobelium does not occur in detectable quantities in the environment. Investigations of the properties of nobelium have mainly used the ISOTOPE that can be produced in greatest abundance, ^{255}No ($t_{1/2}$ = 3.1 min). The most common oxidation state of nobelium is +2 but the +3 state can also be produced. [A.B.M.]

nocturnal *See* DIURNAL.

NOEC *See* NO OBSERVED EFFECT CONCENTRATION.

noise pollution Sound, including vibration, that is socially or medically undesirable. Excessive noise could also have an impact on ecological systems, for example by scaring wildlife, but this is not well documented. What constitutes noise POLLUTION is complex as not only the physical properties of the sound have to be taken into account, but also the psychological responses of those exposed to the noise. Environmental controls on noise and vibration most usually refer to conditions at the site boundary and beyond. Conditions within factories are usually subject to 'conditions at work' legislation. [P.C.]

nominal concentration In an ECOTOXICITY test, the concentration of chemical calculated from how solutions are made up; i.e. from the extent of dilution of an original concentrate. It contrasts with the actual concentration, which is measured at some time during the test. The two often differ, for example due to error or chemical 'sticking' to glassware. *See also* PREDICTIVE ECOTOXICITY TESTS. [P.C.]

non-equilibrium communities In essence the NON-EQUILIBRIUM THEORY of COMMUNITY organization is predicated on the idea that community structure is primarily determined by the vagaries of the environment such that density-independent controls operate and that COMPETITION between species is minimal. In contrast, the importance of competition is central to the equilibrium viewpoint where communities are considered to be saturated with species, which partition all of the available RESOURCES (*see* EQUILIBRIUM, COMMUNITIES; SATURATION). In non-equilibrium communities, DISTURBANCES must occur sufficiently frequently to prevent the process of COMPETITIVE EXCLUSION running its course. Here the focus is on time, variation and the transient behaviour of the system. This temporal component of the theory is absent from the equilibrium view, which emphasizes the properties of a community at an equilibrium point to which the system is attracted and at which it will remain if the environment remains constant over periods measured in multiples of the generation times of the characteristic species in the community. In reality, for most communities the distinction between equilibrium and non-equilibrium states is largely a matter of degree.

There is considerable evidence that species in many communities are not at equilibrium and that disturbances play an important part in controlling communities. Falling trees, bush fires, hurricanes and many other disturbances all serve to disrupt patches within communities, and the entire community can often be viewed as a mosaic of patches each with a different history (*see* HISTORICAL EFFECTS) and at a different stage of development. Theoretical studies show that, by migrating between patches, species which would otherwise be outcompeted in the absence of disturbance are able to persist within the community as a whole (*see* INTERMEDIATE DISTURBANCE HYPOTHESIS). *See also* COMMUNITY ECOLOGY. [S.J.H.]

REFERENCE

Giller, P.S. & Gee, J.H.R. (1987) The analysis of community organisation: the influence of equilibrium, scale and terminology. In: *Organization of Communities Past and Present* (eds J.H.R. Gee & P.S. Giller), pp. 519–542. Blackwell Scientific Publications, Oxford.

non-equilibrium theory The theory of an interaction where stable end-points, or equilibria, are not predicted but where the system changes constantly. An example is the genetic interaction between hosts and their parasites, where the theory predicts continuous cycling of the relative frequencies of different genotypes in both populations over evolutionary time. [P.I.W.]

non-indigenous species A species occurring outside its NATURAL geographical RANGE; also known as an ALIEN species, in contrast to a native species (*see* NATIVE AND NATURALIZED SPECIES). Geographical barriers (mountains, oceans, deserts, etc.) restrict the native range of most plant and animal species. The movement of species (usually by humans) across or around these barriers brings species into new environments where they may prosper when freed from competitors and natural enemies, or they may succumb for want of suitable resources (*see* BIOLOGICAL INVASIONS). Hybridization between native and non-indigenous species is commonplace and can lead to the origin of new sexual species (e.g. by chromosome doubling; sterile hybrids like the grass *Spartina* × *townsendii* can give rise to fertile new species like *Spartina anglica*). [M.J.C.]

non-interactive community A COMMUNITY in which interactions between species at the same TROPHIC LEVEL within a local habitat (*see* SCALE) are weak or absent. Although most emphasis in ecology is on interactive communities there is evidence that, for some groups, non-interactive communities may be more common. Phytophagous insects are an example of such a group. In a non-interactive community, the local-scale processes such as COMPETITION and predation are weak and are, in consequence, likely to be less important than the history of COLONIZATION from the surrounding area (*see* HISTORICAL EFFECTS). It is argued, therefore, that SPECIES RICHNESS and the taxonomic composition of a non-interactive community will be more dependent on the size of the regional pool of colonists than on processes of NICHE packing and competitive DOMINANCE. This in turn means that such communities are less likely to be saturated with species (*see* SATURATION). [S.J.H.]

REFERENCE

Cornell, H.V. & Lawton, J.H. (1992) Species interactions, local and regional processes, and limits to the richness of ecological communities: a theoretical perspective. *Journal of Animal Ecology* **61**, 1–12.

non-linear models Models in which the mean response is not a linear combination of the input PARAMETERS. For example, the model $Y=\alpha e^{\beta x}+\varepsilon$ is non-linear in its parameters since the mean response, *Y*, is an exponential, rather than a linear, function of the parameter β. Non-linear models may take many forms and may sometimes be converted to linear models by transformation of *Y* or *x* or both. Non-linear DIFFERENTIAL EQUATIONS cannot be solved analytically, but must be solved numerically. *See also* REGRESSION ANALYSIS, LINEAR. [V.F.]

non-Mendelian inheritance This most commonly refers to the transmission of continuously varying (also called quantitative, metric or polygenic) TRAITS from one generation to the next. This is in contrast to the transmission of traits that fall clearly into distinct phenotypic classes (i.e. MENDELIAN TRAITS). Whereas Mendelian traits are controlled by single genes having large phenotypic effects, non-Mendelian traits are controlled by many genes each having a small phenotypic effect. Non-Mendelian traits tend to follow a bell-shaped or normal frequency DISTRIBUTION which reflects the fact that many factors (genes and environmental influences) of small effect all combine to determine a character. Some traits that are controlled by many genes of small effect may nevertheless appear as discrete phenotypes and can be explained by a THRESHOLD model.

Early controversy over whether continuously varying traits were indeed heritable was resolved by R.A. Fisher (*see also* CHARACTERS IN ECOLOGY) who developed mathematical and statistical arguments to demonstrate that the segregation of many genes with small phenotypic effect could account for the pattern of similarity in quantitative characters observed among relatives. For quantitative traits, the GENOTYPE cannot be ascertained from the PHENOTYPE because different genetic combinations (in combination with environmental influences) may give rise to the same phenotype. Any given genotype determines a potential range of phenotypes (dependent on environment) called a norm of reaction.

Study of the inheritance of non-Mendelian traits is the subject of QUANTITATIVE GENETICS. Quantitative geneticists express the relation between phenotype (*P*), genotype (*G*), and environment (*E*) as follows: $P=G+E+I_{GE}$, where the last term represents non-linear interactions between genes and environment (GENOTYPE*ENVIRONMENT INTERACTION effects). Such a model is used to determine the extent to which phenotypic variation is caused by genetic variation. The GENETIC VARIANCE can be further partitioned into additive, DOMINANCE and interaction components, and only the additive variance is useful in predicting responses to selection.

As well as this, non-Mendelian inheritance can also be used to refer to cytoplasmic inheritance (chloroplasts, mitochondria, spindle apparatus); the cell wall acting as template in transmitting 'physiological information' between parent and offspring; and various maternal factors (including messenger RNA and transcription factors) transmitted in the egg cytoplasm. It may also mean LAMARCKIAN INHERITANCE. [V.F.]

non-methane hydrocarbons (NMHCs) *See* HYDROCARBONS.

non-parametric statistics Statistical tests are frequently classified into parametric and non-parametric tests. Non-parametric ones do not require QUANTITATIVE INFORMATION/observations to follow a NORMAL DISTRIBUTION and to have similar VARIABILITY (homogeneous variances). Non-parametric tests can be carried out on QUALITATIVE INFORMATION, i.e. ranked information or information in categories. Examples of non-parametric tests/techniques are Mann–Whitney tests (*see* MANN–WHITNEY *U* TEST), KENDALL'S RANK CORRELATION COEFFICIENT, SPEARMAN'S RANK CORRELATION COEFFICIENT, KRUSKAL–WALLIS TEST and the BINOMIAL (SIGN) TEST. [B.S.]

non-planktonic larvae *See* LARVA.

non-point source pollution *See* DIFFUSE SOURCES.

non-renewable resources Resources, such as minerals or FOSSIL FUELS, that are replaced only on geological time-scales, or species of organisms, which once extinct cannot be replaced. *Cf.* RENEWABLE RESOURCE. [P.C.]

non-target species Any species that is adversely affected by the application of a BIOCIDE that is not intended to affect that species. [P.C.]

non-terminal addition (caenogenesis) The insertion of a new stage into the intermediate stages of an ontogenetic sequence. Such alteration results in a violation of the 'biogenetic law' of RECAPITULATION, as postulated by E. Haeckel. Proposed examples of non-terminal addition have included larval traits and embryonic features, such as the mammalian placenta, which are not present in the adult nor in the ancestral ONTOGENY. Much controversy exists regarding the frequency of occurrence of such non-terminal modifications, since ubiquity of these phenomena would prevent the use of ontogeny as a criterion for distinguishing primitive and derived character states in systematic analysis. [P.A.]

REFERENCE

Mabee, P.M. (1993) Phylogenetic interpretation of ontogenetic change. *Zoological Journal of the Linnean Society* **107**, 175–291.

normal distribution A continuous PROBABILITY DISTRIBUTION that is symmetrical, unimodal and bell-shaped. The equation defining the shape and position of the curve is:

$$y = \frac{1}{\sigma\sqrt{2\pi}} \exp\left[-\frac{1}{2}\left(\frac{x-\mu}{\sigma}\right)^2\right]$$

where x is the measured variable, μ is the MEAN and σ is the STANDARD DEVIATION. The notation exp[ ] represents e[ ], with e the base of natural, or Napierian, logarithms. Both e (=2.71828) and π (=3.14159) are constants. The shape and position of any normal distribution is therefore defined by the mean and standard deviation (Fig. N7). For all normal distributions, the following percentages of observations lie within the indicated number of standard deviations, either side of the mean:

68.26% of observations within the $\mu \pm \sigma$

95.44% of observations within the $\mu \pm 2\sigma$

99.73% of observations within the $\mu \pm 3\sigma$

More conveniently:

95% of all observations lie within the $\mu \pm 1.960\sigma$

99% of all observations lie within the $\mu \pm 2.576\sigma$

Therefore, in a normal distribution, 5% of all observations fall outside the limits of $\mu \pm 1.960\sigma$, and 1% of all observations fall outside the limits of $\mu \pm 2.576\sigma$. These limits are important when carrying out statistical tests, such as the *F* TEST and *t* TEST. *See also* DISTRIBUTION; STANDARDIZED NORMAL DEVIATE; STATISTICS. [B.S.]

normal equivalent deviate (NED) *See* PROBIT ANALYSIS.

North Sea The largest marginal SEA of the Atlantic; a shallow marine region on the European CONTINENTAL SHELF, formed by the flooding of a landmass. This epicontinental sea extends northwards to 62°N and includes the Skagerrak, with a southern limit extending east from Cape Skagen, roughly at 57°N to the English Channel and its approach east of 5°W. It is surrounded by the British Isles, Norway, Denmark, Germany, Holland and Belgium. It has a mean depth of about 90 m and a maximum depth of 725 m in the Norwegian Trench.

The North Sea covers an area of 5.8×10^5 km^2 and has a volume of 50 000 km^3. The surface SALINITY increases towards the north and west; in the central North Sea salinity ranges between 34.8 and 35.1‰, whereas along the German shore the salinity ranges from 29‰ (in summer) to 32‰ (in winter). CIRCULATION is counterclockwise in the upper layer but irregular at the bottom. Approximately 90% of the water inflow enters between the Shetland and Orkney Islands. The TIDES are primarily semi-diurnal, and tidal height decreases towards the east. The North Sea is one of the world's most productive fishing areas and is rich in oil, gas and minerals. [V.F.]

North Sea Conference This began in 1984 to address marine pollution of the NORTH SEA on an international level. The third conference was held in 1990 and the fourth in 1997. Each produces a ministerial declaration applied to the protection of the coastal waters of the nine North Sea states (Belgium, Denmark, Germany, France, The Netherlands, Norway, Sweden, Switzerland and the UK). The European Union (EU) also participates and there are observers from the PARIS COMMISSION (PARCOM) and the International Maritime Organization (IMO). [P.C.]

Norwegian Sea A large intercontinental SEA, constituting the transition of the north-east Atlantic to the ARCTIC. It lies between 62°N and 70°N. In the north-east it passes into the BARENTS SEA, in the west into the Greenland Sea, in the south into the NORTH SEA, and in the south-west into the main ATLANTIC OCEAN. Oxygen SATURATION in the Norwegian Sea is high due to active vertical mixing. Sediments consist largely of foraminiferal and diatomaceous oozes

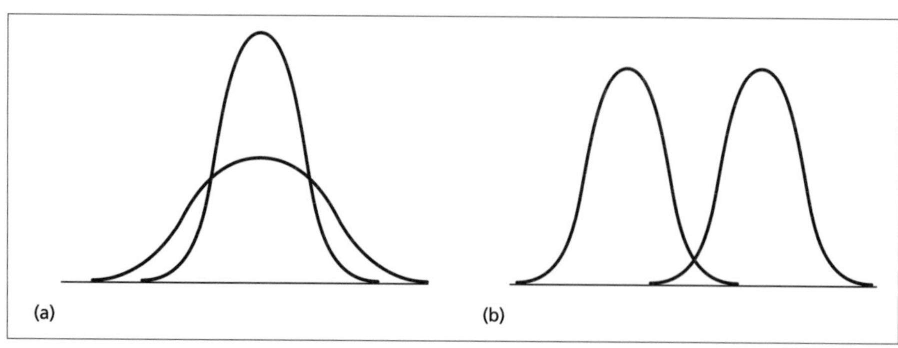

Fig. N7 Normal distributions. (a) Same μ different σ. (b) Different μ same σ.

in the deeper parts of the Norwegian Basin, and volcanic sediments in the west, around Iceland. [V.F.]

not in my back yard (NIMBY) A phrase characterizing the position of individuals or bodies who agree to developments that have ADVERSE ENVIRONMENTAL EFFECTS as long as such developments are not in close proximity. A variant is NIABY: not in anybody's back yard! *See also* NOT IN MY TURN OF OFFICE. [P.C.]

not in my term of office (NIMTO) A phrase describing politicians or other officials who only oppose a locally unpopular development —usually one with ADVERSE ENVIRONMENTAL EFFECTS—if it coincides with their term of office. *See also* NOT IN MY BACK YARD. [P.C.]

notification A dossier of physicochemical and ECOTOXICITY information on substances given by a producer/importer to a regulatory authority as required by legislation—or the act of submitting such information. [P.C.]

notification of new chemicals The supply of information relevant to human health and environmental quality, concerning synthetic chemicals designated as new, prior to AUTHORIZATION for the manufacture or marketing of such chemicals. The information covers the physicochemical properties of the substances and their TOXICITY and ECOTOXICITY. In the USA the latter is usually obtained from QUANTITATIVE STRUCTURE–ACTIVITY RELATIONSHIP (QSAR) analysis, whereas in the European Union actual testing is required according to defined procedures. The level of details required in the NOTIFICATION depends upon the intended volume of production. [P.C.]

NO_x emissions Emissions of various NITROGEN OXIDES, for example from fossil fuel-powered power-stations and automobiles. These compounds participate in PHOTOCHEMICAL SMOG. *See also* MOTOR VEHICLE POLLUTION. [P.C.]

NPL (National Priority List) *See* SUPERFUND LEGISLATION.

NPP *See* NET PRIMARY PRODUCTION.

NPR *See* NET PHOTOSYNTHETIC RATE.

NRA (National Rivers Authority) *See* ENVIRONMENTAL (PROTECTION) AGENCIES.

Nuclear fission

Nuclear FISSION refers to a process in which the nucleus of an atom is split into two or more major fragments accompanied by the emission of small charged particles and intense electromagnetic RADIATION known as γ-RAYS and, usually, some neutrons. The release of neutrons in induced NEUTRON fission implies the possibility of a self-sustaining chain reaction. The fragments are themselves radioactive and will decay with characteristic half-lives through a sequence of daughter products until ultimately becoming a stable ISOTOPE.

The fission process can release significant quantities of energy (*see* NUCLEAR POWER; NUCLEAR POWER-STATIONS), principally in the kinetic energy of the fission fragments, which manifests itself as heat as the fragments are slowed down in the fuel of a reactor, but also in the environmentally significant energies of the associated α-, β- and γ-radiation. The latter is the most penetrating and generally calls for the greatest care in shielding a nuclear reactor.

Hazards of fission products

The RADIOACTIVITY of the fission fragments is highly significant. Although this might represent only 10% of the energy release, the time-scale is different from the fission process itself. Fission can be 'turned off' by straightforward manipulation of the nuclear reactor, but the activity of the fission fragments is uncontrollable and makes them a HAZARD for hundreds or thousands of years. Assessment of this hazard, however, must be related to the natural activity of the parent fissile material (such as URANIUM), which is itself measured in thousands of years, so that after some 500 years or so the activity of the fission products has been reduced to the order of activity of the uranium. The fission fragments represent an immediate hazard in the reactor, and it is essential to provide coolant to deal with this 10% power level that cannot be turned off, especially in the event of an ACCIDENT denying the usual cooling arrangements to the reactor core.

A further important environmental consequence of an uncontrolled release of fission fragments is the occurrence of certain isotopes of IODINE, CAESIUM and STRONTIUM, which behave chemically as the natural and stable isotope or, in the case of strontium, mimic CALCIUM. They are thus liable to be taken up in biological systems. Iodine is concentrated in the thyroid where it may well cause cancer, while strontium accumulates in bones with similar consequences. PLUTONIUM is regarded by some as even more dangerous as a radiotoxic element. But the justification for this belief is limited: it is not readily ingested in the gut and is hazardous more on inhalation or via the (unlikely) intravenous route. While there is no doubt it can cause cancers, on a mass basis it is far less poisonous (and delayed in its effects) than, say, the botulinum toxin responsible for botulism. The reputation of plutonium as the most poisonous substance known is unwarranted.

In assessing the risks involved in IONIZING RADIATION, natural or otherwise, it should be remembered that its nuclear nature makes the measurement of radiation extremely sensitive so that the CONTAMINATION or the hazard it represents can be monitored more precisely than most other hazards. [J.D.L.]

REFERENCE

Harms, A.A. (1987) *Principles of Nuclear Science and Engineering*. Wiley — Research Studies Press, Taunton.

Nuclear power

A major application of NUCLEAR FISSION is as a source of power in the generation of electricity (*see* NUCLEAR POWER-STATIONS). Nuclear power provides, for example, some 10% of the land-based electricity production in the USA, 30% in the UK and around 80% in France. Whilst the programme of building further reactors in these countries is limited (partly as a consequence of the CHERNOBYL DISASTER), the developing countries of the Pacific rim have major programmes to expand their use of nuclear power.

Minor uses of nuclear power include the propulsion of submarines and some satellites or spacecraft. Side-products include the provision of radionuclides, which are valuable in medical diagnosis and treatment, in studying physiological systems; in equipment for testing welds, etc. in industry, and in the irradiation of foods to prolong shelf-life or eliminate pests.

Advantages and disadvantages

Advocates of nuclear power, while recognizing the substantial capital investment needed—greater than conventional fossil fuel-fired generators — emphasize the advantage of producing large amounts of energy from relatively small, and therefore cheap, amounts of fuel. Furthermore, URANIUM is widely dispersed throughout the world, in a pattern different to that of reserves of coal, oil and natural gas. It therefore has a strategic attraction, especially to countries that would otherwise be dependent on the vagaries of 'geo-oil politics'.

The WASTE products represent a substantial cost, perhaps 10% of the prime cost of fuel, and the reactors themselves become radioactive in use and hence impose a decommissioning cost. This is better appreciated in nuclear power, perhaps, than in competing sources of energy supply. Proponents of nuclear power will also point out that it produces vastly smaller quantities of CARBON DIOXIDE (CO_2) and other 'greenhouse' gases than FOSSIL FUELS for the same electricity.

Military links

No account of nuclear power could divorce the civilian use, mainly for the production of electricity, from the military use, mainly for nuclear weapons, with all their historical and practical links. Fundamental to this association is the nature of PLUTONIUM—used as both a fissile material for good and as a weapons material for ill.

It is undeniable that nuclear reactors can make plutonium, which can serve as weapons material. The plutonium formed over the extended periods of operation of power-generating reactors is less suited to this purpose, being mixed with higher isotopes than ^{239}Pu. There is, furthermore, little historical justification to suppose that nations who wished to develop nuclear weapons have done so through any means other than the direct approach of purpose-built reactors and uranium enrichment facilities. And in this respect, enrichment plants are more compact and more readily operated in secrecy than any complex of plutonium-producing reactors, with their fuel fabrication and reprocessing plants to separate the weapons material. [J.D.L.]

REFERENCES

Fremlin, J.H. (1989) *Power Production: What are the Risks?*, 2nd edn. Insitute of Physics, Bristol.

Lau, F.-S. (1987) *Radioactivity and Nuclear Waste Disposal*. Wiley — Research Studies Press, Taunton.

Lau, F.-S. (1987) *A Dictionary of Nuclear Power and Waste Management*. Wiley—Research Studies Press, Taunton.

Nuclear power-stations

Principal types

The chief type of nuclear power-station in use worldwide today is the pressurized water reactor (PWR). This was developed initially in the USA as an offshoot of the necessarily compact nuclear-powered submarine programme and has led to power-stations—each generating some 3000 MW thermal power, converted to, say, 1000 MW of electrical power—in many developed and developing countries. A variant, which has not achieved the dominance of the PWR, is the boiling water reactor (BWR). Obsolescent types include gas-cooled reactors.

The history of nuclear reactors, however, started with the use of graphite rather than water as the moderator, or slowing-down element, and some countries, notably the UK and France, pursued the development of graphite reactors for some decades before converting to the PWR. Graphite is less efficient as a moderator and consequently such power-stations are larger for the same capacity than either PWR or BWR stations, at least in the 'nuclear island' if not in the ancillary equipment or conventional islands of turbines, generators, etc. Therefore, one can say that the compactness is essential to the economy of producing more from less, but represents a larger risk from the greater power density.

Environmental considerations

A large amount of energy is released from the NUCLEAR FISSION process. The complete fission of 1 kg of URANIUM per day will serve to fuel a plant producing about 1 GW thermal power or some 300 MW electrical power. The vastly greater quantities of coal or oil needed to generate equivalent amounts of power impose ENVIRONMENTAL BURDENS in the extraction and transport of fuel to the power-station, and in the products of combustion. Like most other producers of electricity, the thermal stages of nuclear power-plants create the need to dispose of waste heat to the environment on substantial scales, as warmed water or air.

Fuel processing and storage

The processing of the fuel after use in a reactor involves dealing with intensely radioactive products, even allowing for the usual minimum delay of, say, 90 days to allow for natural decay. There is international controversy over whether reprocessing should take place immediately, whether only after centuries of decay, or perhaps never at all, with the disposal of used fuel elements directly in underground (or undersea) sites. Generally, the UK and France (more equivocally Germany) have undertaken reprocessing, whereby not only are the active fission products concentrated to make for easier storage but the unused uranium and, more tendentiously, the created PLUTONIUM, are available for reuse. This reuse with uranium as a mixed oxide fuel (MOX) has already been demonstrated commercially; it may be the most satisfactory way of disposing of the large quantities of 'spare' plutonium resulting from scaling down of the superpowers' weapons programmes.

The storage and/or disposal of the radioactive products, principally from the fuel, has engendered much controversy, a prime of example of the 'NIMBY' approach—NOT IN MY BACK YARD. Most countries have encountered political opposition when they have sought to establish a disposal site. Ideally, perhaps, there should be no measurable release of activity. But this is an ideal that must be matched against the risks associated with other treatments of the material, with other forms of electricity production or, indeed, with the consequences of not producing electricity. To search for disposal facilities 'guaranteeing' no release for, say, 10^5 years is unrealistic; no one could accept a 'proof' of such a claim. Perhaps some realism can be brought to bear if it is recognized that underground disposal would be nullified by the probable consequences of the next ice age, due within some 20 000 years, in which it is expected that the top kilometre of Europe will be removed and deposited in the Atlantic.

Decommissioning

The decommissioning of nuclear power-stations is technically straightforward but again not without political controversy. The major reactivity resides in the fuel elements, which are routinely removed during operation and as readily removed at the end of station life. The reactor vessel, however, will be radioactive from NEUTRON exposure. Unless there is some pressing necessity to reuse the site, this activity could be allowed to decay within the original containment for up to a hundred years. Disposal at that stage represents little HAZARD and can be undertaken largely by remote devices if desired. Power-stations in the USA have already been successfully decommissioned. There are immense economic advantages in funding the disposal now but spending the money one hundred years later.

Accidents

Proper concern surrounds the consequences of a severe ACCIDENT in a nuclear reactor. In the UK a non-electrical reactor caught fire at Windscale (now Sellafield), Cumbria, in 1957 when the procedure for dealing with the controlled release of Wigner energy (energy stored in displacements of the graphite lattice) was misapplied. Radioactive IODINE was released. Although the consequences were at that time not readily measurable, we would probably now find such a release unacceptable. The medical consequences have not been identified as rising above the normal fluctuating incidence of this class of diseases, for example cancer of the thyroid. In the USA, in 1978, a reactor at Three Mile Island suffered a failure which led to the partial melting of the core but essentially no immediate RADIATION release. The results are equivocal in that the accident certainly cost perhaps a billion dollars in loss of plant and electricity, several billion in CLEAN-UP and marked the abandonment, for the time being, of nuclear power in the USA. But it can also be said to have demonstrated that in a severe and unlooked-for failure, the containment successfully protected the environment in a dramatic demonstration. The CHERNOBYL DISASTER of 1986 had far more serious environmental consequences, and perhaps more serious political consequences as a paradigm of the failure of the Soviet system as well as the reactor containment. [J.D.L.]

REFERENCE

Murray, R.L. (1993) *Nuclear Energy*, 4th edn. Pergamon Press, Oxford.

nuclear winter A term coined to express the effect of the increased SMOKE and DUST burden in the ATMOSPHERE which would result from the explosion of thermonuclear devices at the surface and in the air as a result of a nuclear war. The effect of the smoke and dust has been modelled by computer simulation and it is expected that the ATTENUATION of sunlight would be such as to reduce the normal levels of daylight to levels more typical of present dusk or dawn, or midwinter. The reduction in SOLAR RADIATION would result in widespread cooling of the Earth's surface. *See also* CLIMATE CHANGE. [J.B.M.]

nuisance species Non-indigenous species, usually introduced as a result of human activity, that threaten the density and/or abundance of native species. [P.C.]

null hypothesis (H_0) The hypothesis that forms the basis of a statistical test. A neutral, or null, hypothesis is one proposing no difference, and is usually symbolized by H_0. For example, when carrying out a *t* TEST, the null hypothesis is that the two SAMPLE means ($\bar{x}_1$ and $\bar{x}_2$) are from populations with the same means (H_0: $\bar{\mu}_1 = \bar{\mu}_2$); i.e. if the sample means are different it is only due to chance. Similarly, when carrying out an *F* TEST, the null hypothesis is that the two sample variances (s_1^2 and s_2^2) are from populations with the same variances (H_0: $\sigma_1^2 = \sigma_2^2$); i.e. if the sample variances are different it is only due to chance. There is always an ALTERNATIVE HYPOTHESIS to H_0. *See also* EXPERIMENTAL DESIGN; STATISTICS. [B.S.]

null model A term introduced by Caswell (1976) in his investigation of COMMUNITY structure, and used mainly in COMMUNITY ECOLOGY and comparative analyses for a more precise, usually quantitative, NULL HYPOTHESIS. One of the main aims in these fields is to determine the ecological or evolutionary processes causing the patterns in nature which we can observe and measure. A good introductory discussion of null models can be found in Harvey *et al.* (1983). A MODEL predicts the magnitude of a particular effect and takes into account its VARIABILITY; i.e. it gives a MEAN and VARIANCE. This allows a more realistic comparison of predicted and observed patterns.

Community ecologists often wish to know which of the important processes, such as COMPETITION, predation or MUTUALISM, have shaped community structure or composition. Null models for the structure of a particular community can be constructed using different sets of assumptions; for example, that competition does or does not occur between a particular pair of species, or whether a particular species is an important predator of another. The null model approach to community composition is to examine why particular species are found together more frequently (or less frequently) than would be expected if communities were derived at random from a species pool of all species in the neighbourhood. Chains of islands have a readily identifiable species pool and so have often been used to examine whether the species in a particular community (e.g. on a single island) is a RANDOM SAMPLE from the pool.

The models can then be tested against actual community structure or composition. However, care should be taken when the empirical data are used to construct the relevant null model, as there may be hidden interactions between species which could markedly affect the results. The best studies compare real communities with a number of null models. This gives us a better chance of determining the actual processes forming the community.

In comparative studies of ecological relationships, such as between the body sizes and dispersal abilities of individuals of different species, the influence of common ancestry must be taken into account. Otherwise, conclusions could be biased by examining many close-related species which share characteristics for phylogenetic rather than ecological reasons (Nee *et al.* 1994). The goal here is to determine how many comparisons are phylogenetically independent and can be legitimately used for statistical examination. Null models can be constructed using different assumptions about the branching patterns of different clades, for example that a species produces two daughter species with a constant PROBABILITY over time or with increasing probability with increasing age. Actual patterns of species DIVERSITY or ecological relationships can then be compared to the null models. [P.I.W.]

REFERENCES

Caswell, H. (1976) Community structure: a neutral model analysis. *Ecological Monographs* **46**, 327–354.

Harvey, P.H., Colwell, R.K., Silvertown, J.W. & May, R.M. (1983) Null models in ecology. *Annual Review of Ecology and Systematics* **14**, 189–211.

Nee, S., May, R.M. & Harvey, P.H. (1994) The reconstructed evolutionary process. *Philosophical Transactions of the Royal Society of London, Series B* **344**, 305–311.

number of offspring Organisms may produce all their offspring in a single bout of reproduction (SEMELPARITY) or over several bouts (ITEROPARITY). The number of offspring is obviously a trait under selection. Continued selection for an increased production is balanced by the COSTS OF REPRODUCTION as, for example, when an increase in CLUTCH SIZE decreases the future reproduction of the parents. [D.A.R.]

numerical response The relationship between the number of predators or parasitoids in an area and the DENSITY of their prey or HOST. The numerical response may consist of a short-term behavioural and a longer-term reproductive component. The numerical response and the FUNCTIONAL RESPONSE act together to determine rates of predation. [H.C.J.G.]

Numerical taxonomy

Although this term might be applied to any numeric approach to TAXONOMY and CLASSIFICATION, it is now often taken to refer primarily or even exclusively to those numerical methods that group individuals or taxa on the basis of some function of their similarity to one another—often referred to as distance methods or phenetics. This contrasts with the cladistic approach (*see* CLADISTICS) that was developed by W. Hennig, in which only derived (apomorphic) character states are used to infer relationships — so-called character-based methods.

Phenetic algorithms

The phenetic side of numerical taxonomy was largely developed by two workers, Sneath and Sokal, during the 1960s and was further made possible by the increased accessibility of computers. The main problem with the phenetic side of numerical taxonomy is that clustering ALGORITHMS such as UPGMA (unweighted pair-group method using arithmetic averages) or the weighted equivalent (WPGMA), which do not take into account the direction of character evolution, can be shown to fail to reconstruct phylogenies under the majority of possible scenarios, and in fact they can be guaranteed to go wrong in many cases, particularly where EVOLUTIONARY RATES vary significantly along different lineages. Some more sophisticated techniques such as neighbour joining, are less likely to be misled by uneven evolutionary rates, and are commonly used for analysing DNA sequence data.

Most systematists now prefer to base classifications on cladistic methodology, but nevertheless, some still adhere to the idea that polythetic classifications—i.e. those based on the maximum number of characters—are inherently more stable and therefore desirable. It should be noted that some data types can only be analysed using distance methods, such as data from DNA–DNA hybridization or from certain immunological procedures, and in these cases there is no cladistic option.

Multivariate ordination methods

In addition to the various phenetic tree-building algorithms, a number of multivariate ORDINATION methods (e.g. PRINCIPAL COMPONENT ANALYSIS and DISCRIMINANT FUNCTION ANALYSIS) also fall under the umbrella of numerical taxonomy. Several of these have utility in taxonomy, not for determining relationships, but for identifying the presence of (usually) morphologically discrete entities such as geographic races, SUBSPECIES or SIBLING SPECIES.

These methods can be broadly separated into those that make no prior assumptions about groupings within the dataset, and which therefore simply search for discrete clusters among the taxa, and those that do and are used primarily for identifying which previously recognized cluster a particular specimen belongs to.

Principal components analysis falls into the first category, and is perhaps the most widely used of such techniques though it is by no means the only one—each technique has its advantages and disadvantages. Discriminant analysis, which does require some prior knowledge of groupings, can be helpful in situations where there is no single, simply observable character for recognizing a particular TAXON or strain. Further, rather than entering the data for an unknown entity into the whole analysis, which might entail making many measurements, discriminant functions produced by stepwise elimination programmes, make use of a reduced subset of characters that will still provide an acceptable degree of accuracy in identification.

The above two methods both involve finding sets of perpendicular axes in character space. The first axis is selected to explain as much of the VARIANCE as possible, the second explaining as much of the remainder as possible, and so on. Plotting the individuals against these principal component axes may reveal clusters of data points that are not immediately apparent by visual scrutiny of the data. *Cf.* PHYLOGENETIC SYSTEMATICS. *See also* PHENETIC SPECIES CONCEPT; SYSTEMATIC METHODS. [D.L.J.Q.]

REFERENCES

Quicke, D.L.J. (1993) *Principles and Techniques of Contemporary Taxonomy.* Chapman & Hall, London.

Sneath, P.H.A. & Sokal, R.R. (1973) *Numerical Taxonomy.* W.H. Freeman, San Francisco, Calif.

Sokal, R.R. & Rohlf, F.J. (1995) *Biometry,* 3rd edn. W.H. Freeman, New York.

Wiley, E.O. (1981) *Phylogenetics. The Theory and Practice of Phylogenetic Systematics.* John Wiley & Sons, New York.

nurse eggs *See* CANNIBALISM.

Nusselt number A non-dimensional quantity, commonly used in environmental engineering, that expresses the rate of HEAT TRANSFER to or from an object. [J.G.]

REFERENCE

Monteith, J.L. & Unsworth, M.H. (1990) *Principles of Environmental Physics.* Arnold, London.

nutrient budget The nutrient budget, or the nutrient balance, is an attempt to account quantitatively for the inputs ('income') of NUTRIENTS to an ECOSYSTEM, and the losses. In natural ecosystems, income is from RAIN, wet and dry deposition, NITROGEN FIXATION and WEATHERING of rocks. Losses are by LEACHING and RUN-OFF, and by volatilization during fire and DENITRIFICATION. In farmed systems, the natural FLUXES may be dwarfed by additions of fertilizer and losses at HARVESTING. Some examples of ecosystem gains and losses of nutrients are given in Jordan (1985). *Cf.* ENERGY BUDGET. *See also* BIOGEOCHEMICAL CYCLE. [J.G.]

REFERENCE

Jordan, C.F. (1985) *Nutrient Cycles in Tropical Forest Ecosystems.* John Wiley & Sons, New York.

nutrients Raw materials needed for life. Their exact nature varies according to whether the living system is heterotrophic or autotrophic — and even varies between species within these groupings. The term can include WATER, minerals, inorganic salts and various organic molecules (PROTEINS, CARBOHYDRATES, fats and VITAMINS). *See also* MACRONUTRIENTS; MICRONUTRIENTS. [P.C.]

nymphs *See* INSECTS, DIVERSITY AND CONSERVATION.

O

obligate Describing a factor that must be present in order for an organism to survive in a particular environment. Every species has a range of requirements which may either be obligate or FACULTATIVE. When the requirement is obligate then survival is not possible if the factor is not present; for example, some bacteria can only be cultured in conditions that exclude oxygen, and are consequently described as obligate anaerobes. The definition most commonly refers to the physical rather than the biological features of an environment. [R.C.]

observational meteorology *See* METEOROLOGY.

ocean The vast body of WATER covering 70.8% of the surface of the globe, or any one of its principal divisions: the Antarctic, ATLANTIC, Arctic, INDIAN and PACIFIC OCEANS. The average depth of the ocean is 3729 m and the deepest parts are 11 022 m from the ocean surface. The total area of the world ocean is 3.62×10^8 km^2, and its volume is 1.35×10^9 km^3. The ocean is believed to have formed between 4400 and 3000 million years ago. All known phyla originated in the ocean.

The deep water of the ocean is remarkably uniform in temperature (*c.* 3°C) and SALINITY (35‰) whereas the surface water is more variable and much warmer (average 18°C). CIRCULATION of the deep water occurs because surface water moves towards the poles, cools and sinks. The rate of circulation of the deep water is such that it takes about 1000 years for water to go through a complete cycle of sinking at the poles, circulation through the deep zone, and return to the surface. The present ocean BASINS began to form about 180 million years ago with the breaking up of Pangaea. [V.F.]

ochric horizons *See* SOIL CLASSIFICATION.

octanol–water partition coefficient (K_{ow}) The ratio of the concentrations of a chemical in *n*-octanol (C_o) and WATER (C_w) at equilibrium; i.e. K_{ow} (sometimes P or P_{ow}) $= C_o/C_w$, measured in a standard way (OECD 1981). The higher the value of K_{ow} the more hydrophobic and lipophilic is a chemical, and this correlates positively with its BIOACCUMULATION potential and non-specific TOXICITY in some animals. *See also* ECOTOXICOLOGY. [P.C.]

REFERENCES

Connell, D.W. (1994) The octanol–water partition coefficient. In: *Handbook of Ecotoxicology*, Vol. 2 (ed. P. Calow), pp. 311–320. Blackwell Scientific Publications, Oxford.

OECD (1981) *Guidelines for Testing of Chemicals. Subject No. 107, Partition Coefficient (*n*-octanol/water).* Organization for Economic Cooperation and Development, Paris.

OECD (Organization for Economic Cooperation and Development) A 'club' of some 20 member countries that gathers information and promotes cooperation with a view to promoting economic development. It collects and collates data on environmental degradation and spending on environmental protection as well as promoting a number of cooperative ventures aimed at environmental protection. Of particular importance are:

- Guidelines for Testing Chemicals, which seek standardization in ecotoxicological procedures;
- Principles of GOOD LABORATORY PRACTICE.

The OECD is also coordinating a High Production Volume (HPV) programme, in which member countries are now investigating in a coordinated way the risks associated with chemicals produced at high volumes. This may lead to recommendations on risk reduction.

Recommendations and decisions of the OECD are not mandatory, but they have had a major influence on the development of environmental protection measures in member countries and the European Union. *See also* UNITED NATIONS PROGRAMMES AND BODIES. [P.C.]

oestrogenic *See* ENDOCRINE DISRUPTER.

OFT *See* OPTIMAL FORAGING THEORY.

oil smoke plume Any system of continuous combustion produces a plume of its gaseous combustion products. These are primarily WATER vapour and CARBON DIOXIDE (CO_2) from the burning of hydrocarbons, with an admixture of sulphur dioxide (SO_2) according to the amount of sulphur present in the fuel, together with a few other components peculiar to the source including compounds of CHLORINE. Oxides of NITROGEN (NO_x) may be present, primarily nitric oxide (NO) which, within an hour or two, becomes oxidized in the ATMOSPHERE to nitrogen dioxide (NO_2) if there are normal amounts of OZONE (O_3) present.

Given a sufficiently high temperature and pressure of combustion the oxidation of atmospheric nitrogen is increased in the presence of excess OXYGEN (more than enough to oxidize the CARBON and HYDROGEN), but in the presence of an oxygen deficit some of the fuel appears as particulate carbon, a black SMOKE, and some CARBON MONOXIDE (CO) is then present in the plume. Furthermore, when the combustion is incomplete due to oxygen deficit the vaporized fuel condenses as very small smoke particles, which appear as a blue smoke if the particles are sufficiently small.

The presence of smoke and NO_x is determined by the mixing ratio of air and oil vapour at the time of combustion. This ratio is well controlled in the combustion chambers of aviation gas-turbine engines, but not in oil-well gas flares. Such plumes are seen to rise vertically with a conical outline in the absence of WIND, but are bent over according to the profile of the horizontal wind. They rise to an altitude determined by the ambient STRATIFICATION and the level at which their BUOYANCY is diluted by mixing with the ambient air to become zero. *See also* ACID RAIN AND MIST; AIR POLLUTION; ATMOSPHERIC POLLUTION. [R.S.S.]

oil spills Spills of crude oil from tanker accidents account for around 12% of the 3.2 Mt of PETROLEUM hydrocarbons released to the oceans every year. This is about one-third of the quantity entering the seas from sources such as atmospheric fallout and other inputs associated with the transportation of crude oil such as washings from tanks. Oil spills from tankers, in contrast to other sources of crude oil in the marine environment, produce locally very high levels of crude oil. It is this factor, coupled with the emotive images of oiled seabirds and mammals, that raises most public awareness of oil POLLUTION. Oil spills not only cause acute effects, such as oiling and poisoning of marine birds, mammals and invertebrates, but also have less tangible long-term effects such as changes in population structure of an ecosystem and reductions in reproductive success. Concomitant with this, damage to commercial fisheries and tourism may also occur.

As soon as oil has been spilled there are a number of natural mechanisms that come into play to dissipate an oil slick. These include spreading, EVAPORATION, dissolution, emulsification, degradation by microorganisms and photo-oxidation. Quantitatively the most important mechanisms are evaporation and BIODEGRADATION. The most effective means of removing oil pollution rapidly is by enclosing the slick with inflatable booms and removing the oil from the water surface. This is not often feasible due to remote location or prevailing weather conditions at a spill site. When oil is beached it is much more difficult to clean up and, until recently, REMEDIATION involved physical removal with mechanical diggers and high-pressure water. The relatively successful use of BIOREMEDIATION in the wake of the *Exxon Valdez* spill in March 1989, has demonstrated that a further weapon is now available for the effective removal of oil pollution from contaminated BEACHES. *See also* ECOLOGICAL/ENVIRONMENTAL DISASTERS. [I.M.H.]

Old-growth/ancient forests, conservation

Characteristics of old-growth forests

The term 'old-growth' describes forests that have developed over a long period of time without experiencing catastrophic DISTURBANCE. The age at which old-growth develops and the specific structural attributes that characterize old-growth vary with FOREST type, climate, site conditions and disturbance regime. For example, an alder forest may attain old-growth characteristics at 80 years, while a fir forest might not be considered old-growth until age 200 years or more (Fig. O1). While old-growth has been defined simply on the basis of tree age, size or growth rate, most ecologists (e.g. Franklin *et al.* 1981) use a combination of structural characteristics for distinguishing old-growth from younger forests (not all of which may be present in a given forest):

- large old trees (for a species and site);
- trees with broken or deformed tops, or with stem and root decay;
- wide variation in tree size and spacing;
- tree foliage well distributed vertically;
- large dead trees (both standing and on the ground) in a range of decay states;
- CANOPY openings from tree-fall gaps;
- patchiness of UNDERSTOREY vegetation.

Inventories of amounts of old-growth forest may differ widely even in well-surveyed regions due to differences in the characteristics or minimum standards emphasized in the definitions. The term 'old-growth' has been applied to BOREAL and temperate forest types; similarly undisturbed tropical forests, which do not contain many of the same structural features, are usually called 'primary' forest.

Development and changes

Forests are not static entities; forest development continues during the old-growth stage, although rates of change in composition and structure are relatively slow. Early in the old-growth stage, forests may be dominated by shade-intolerant species which established after catastrophic disturbances (e.g. fire, flood, or hurricane). If forests escape catastrophic disturbance long enough, they may become dominated by shade-tolerant trees which established under

Fig. O1 Old-growth Douglas-fir (*Pseudotsuga menziesii*) forest (Oregon, USA) illustrating large and small trees, multiple foliage layers, and standing and down dead wood. (Courtesy of A. Levno.)

existing forest, and forest composition may be relatively stable and self-perpetuating. Sporadic, low-severity disturbances (e.g. patchy mortality of canopy trees from fire, wind or disease) are an integral part of the dynamics of many old-growth forests. Some old forests may 'disintegrate', with most of the dominant trees dying in a short period of time due to SENESCENCE or disease, to be replaced by an even-aged forest of young trees. Old-growth forests are not necessarily wilderness areas, but may have developed following human disturbance (e.g. aboriginal-set fires, abandoned agricultural fields, partial cutting).

Habitat value

Relatively little is known about the compositional uniqueness of old-growth forests, but they do appear to represent important HABITATS for many life forms. Although relatively few species may depend exclusively on old-growth forests, abundance and diversity of many plant, animal and microbial species are often greater during the old-growth stage of forest development than during intermediate stages. Large standing dead and fallen trees and multiple layers of vegetation, which are commonly lacking from natural or managed young forests, provide nesting and roosting sites and hiding and thermal cover for many species. Large wood from old-growth forests can also become important habitat features for fish in streams, estuaries and even the open ocean (Maser *et al.* 1988).

Ecosystem function

Functionally, production of photosynthate is typically high in old-growth forests compared to younger forests, but tree BIOMASS is generally stable over time due to losses to RESPIRATION, decay and mortality. High biomass levels in old-growth forests can represent a considerable storage component in the global CARBON CYCLE. Old-growth forests tend to cycle NUTRIENTS with very low losses and typically have low SOIL erosion potential, yielding very high-quality WATER.

Economic and cultural value

Old-growth forests represent a valuable economic resource in many parts of the world. The dense, strong wood created by slow-growing trees is highly prized for structural and ornamental carpentry. Large volumes of wood fibre per hectare are also valuable for many other wood products. Economic use of old-growth forest is usually a one-time event; the greater value of old-growth trees rarely compensates for the long time-spans required to grow them.

Old forests have had a special place in human culture for millennia. In Neolithic, Greco-Roman and Buddhist mythologies, for example, groves of large trees have spiritual significance as sacred sites. Old-growth forests also have aesthetic appeal, with words like 'cathedral-like' often used to describe columnar tree stems and filtered light coming through tall canopies. Visitors to old-growth forests may express reverence and awe at the size (as great as 10 m in diameter and 100 m tall) and age (as much as 2500 years old) of the trees. Areas with exceptional groves of large trees often become popular attractions for tourists (e.g. national parks). Old-growth forests also have scientific value for studies of forest development and species' requirements and behaviour in natural habitats.

Destruction

The amount of old-growth forest on the planet has probably declined sharply over the last 2000 years, although no estimates exist of past or current amounts. The current area of 'primary' (or original) forest is estimated to be 24% of its pre-agricultural amount; percentages are much lower in many temperate regions (USA, 15%; China, 1%; Europe, 0%) (Postel & Ryan 1991). Most of the remaining old-growth forests exist either as fragments in greatly altered landscapes or on relatively unproductive or unpopulated lands (e.g. in boreal regions or high mountains). Felling to yield timber for construction (particularly for wooden sailing vessels) or for firewood accounted for the loss of most of the old-growth forests, and many forested areas (as much as a third of the total) were converted to non-forest land by GRAZING, farming or soil degradation. Many existing old-growth forests may not reflect natural ECOSYSTEM processes. Some old-growth forests may have been selectively logged or grazed over the centuries, and changes in populations of natural grazers in other forests due to game management and predator eradication may have led to marked shifts in the composition of understorey plants (i.e. future overstorey trees).

Conservation strategies

Strategies for conservation of old-growth forests will depend to some extent on species composition and natural disturbance regimes. For the short term, old-growth forests may be conserved by preservation. Eventually, severe disturbances may destroy remaining stands, or further development due to lack of disturbance may change forest structure (e.g. loss of shade-intolerant species typical of many old-growth stands). Thus conservation requires consideration of larger forested areas where disturbances and STAND development are continually generating old-growth forests. If disturbance frequency is too high, some types of old-growth fail to develop (e.g. fir, *Abies*) and if frequency is too low, some might be lost (e.g. aspen, *Populus*). If disturbances are too small, shade-intolerant types of old-growth (e.g. alder, *Alnus*, or pine, *Pinus*) will be lost. Thus, where disturbance regimes have been greatly modified by human control (e.g. fire or flood), managers may have to specifically introduce disturbances which mimic natural events.

Old-growth conservation strategies should consider several landscape-scale issues. Conservation of species which require large areas of old-growth habitat may entail maintenance of large areas of forest in the old-growth stage. Prevention of local extinctions or genetic isolation would require facilitation of organism travel between habitat areas. DISPERSAL among old-growth areas may be facilitated by: (i) short distances between areas; (ii) forests favourable to dispersal between areas; and/or (iii) CORRIDORS or 'stepping-stone' patches of old-growth between areas. The relative values of these approaches are generally unknown. It may also be useful for conservation purposes to maintain old-growth habitat elements in managed forests (*see*, e.g. NEW FORESTRY). Some attempts are being made to accelerate the development of old-growth characteristics by thinning young forests and retaining large live and dead trees (Spies *et al.* 1991). *See also* ANCIENT WOODLAND/FOREST; CONSERVATION BIOLOGY; ECOSYSTEM MANAGEMENT; LANDSCAPE ECOLOGY. [A.N.G.]

REFERENCES

Franklin, J.F., Cromack, K. Jr, Denison, W. *et al.* (1981) *Ecological Characteristics of Old-growth Douglas-fir Forests.* General Technical Report PNW-118. USDA Forest Service, Pacific Northwest Forest and Range Experiment Station, Portland, Oreg.

Maser, C., Tarrant, R.F., Trappe, J.M. & Franklin, J.F. (1988) *From the Forest to the Sea: A Story of Fallen Trees.* General Technical Report PNW-GTR-229. USDA Forest Service, Pacific Northwest Research Station, Portland, Oreg.

Postel, S. & Ryan, J.C. (1991) Reforming forestry. In: *State of the World* (ed. L.R. Brown), pp. 74–92. W.W. Norton, New York.

Spies, T.A., Tappeiner, J., Pojar, J. & Coates, D. (1991) Trends in ecosystem management at the stand level. In: *Transactions of the 56th North American Wildlife and Natural Resources Conference*, pp. 628–639.

oldfield ecosystems Extensive areas of abandoned farmland in the north-east USA that have been recolonized in the past half-century by now-maturing trees. These fields, originally forested then cleared and farmed before being abandoned, have been the subject of many short-term studies from the 1930s to provide an understanding of the processes of COLONIZATION and the SUCCESSION of VEGETATION following abandonment. Because the sequences—in broad terms—appear to be repeated in other productive habitats, data from the oldfield ecosystems have been used to compare and predict vegetation processes in other ecosystems following NATURAL DISASTERS (flood, fire or insect predator devastation). The limitations of short-term studies of oldfield ecosystems have, however, become apparent; succession is greatly influenced by the often unknown history of the original clearance and of the subsequent husbandry. Succession from annually fertilized arable fields, for example, differs from that of PASTURE land. However, because the oldfield ecosystems are so extensive they, together with analogous systems elsewhere, form an important element in the studies of natural succession and of human impact on natural ecosystems. [G.A.F.H.]

oligohaline *See* BRACKISH WATERS.

oligophagous Describing herbivorous organisms, typically insects, that utilize a range of food RESOURCES. These food items may consist of a number of different species, genera or orders. *Cf.* MONOPHAGOUS; POLYPHAGOUS. *See also* FEEDING TYPES, CLASSIFICATION. [R.C.]

oligotrophic *See* EUTROPHICATION.

ombrogenous Describing peatland that has developed under conditions where the sole input of water (and therefore nutrient elements) is derived from precipitation falling directly upon it. Literally, originating under the influence of rainfall. [P.D.M.]

ombrophilous Describing habitats, vegetation or individual species that require a water source (and therefore nutrient element source) that is solely derived from direct precipitation; literally, 'rain-loving'. [P.D.M.]

ombrotrophic Describing peatland that derives its nutrient input entirely from the atmosphere via direct precipitation; literally, 'rain-fed'. *See also* MIRE. [P.D.M.]

omnivore An animal that feeds on both animal and vegetable matter. [P.O.]

one-tailed test In the NORMAL DISTRIBUTION 95% of observations are found to one side of $\mu \pm 1.65\sigma$ (where μ is the MEAN and σ is the STANDARD DEVIATION) and the other 5% of observations are found in the remaining tail of the distribution (Fig. O2). When we carry out a $z(d)$ TEST to examine the difference between two means, and we have a priori reasons to expect a value of the test statistic above or below $z = 0$ (one mean should be larger), we use the 5% level of significance that corresponds to $z = 1.65$. This is a one-tailed test, and corresponds to a NULL HYPOTHESIS (H_0): $\bar{x}_1 = \bar{x}_2$ and an ALTERNATIVE HYPOTHESIS (H_1): $\bar{x}_1 > \bar{x}_2$. Any statistical test that uses the PROBABILITY in one tail of a distribution is therefore a one-tailed test. The alternative is to use the value of the test statistic that cuts off a total of 5% in both tails. For the normal distribution this would correspond to $z = 1.96$. Using this value as the 5% significance point would imply a TWO-TAILED TEST. *See also* POPULATION; STATISTICS. [B.S.]

ontogeny The LIFE CYCLE of an individual, both embryonic and postnatal. The term was coined by the German scholar Ernst Haeckel, in parallel with PHYLOGENY (the evolutionary lineage of a species), as part of his influential 'biogenetic law'. This claimed that 'Ontogeny is the short and rapid recapitulation of phylogeny. . . . During its own rapid development . . . an individual repeats the most important changes in form evolved by its ancestors during their long and slow paleontological development' (Haeckel in 1866; quote from Gould 1977, p. 77). Therefore, to Haeckel, both ontogeny and phylogeny were temporal sequences of forms describing the 'development' of either an individual (ontogeny), or of a specific lineage (phylogeny). These early theories about the relationships between ontogeny and phylogeny gave rise to the field of HETEROCHRONY. Although 'ontogeny' is synonymous with

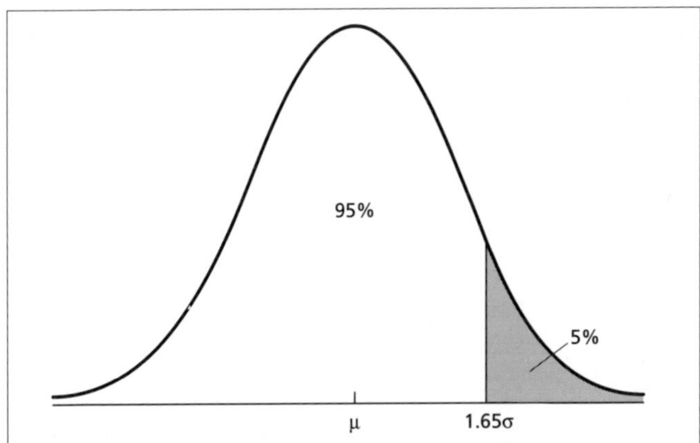

Fig. O2 Normal distribution showing the number of standard deviations (σ) that cut off 5% of the distribution in one tail.

'development', the term 'ontogeny' tends to be used in comparative descriptions of embryogenesis or life history, while 'development' is preferred in a more mechanistic context. *See also* RECAPITULATION. [P.A.]

REFERENCE

Gould, S.J. (1977) *Ontogeny and Phylogeny*. Harvard University Press, Cambridge, Mass.

ooze, calcareous and siliceous A SEDIMENT containing 30% or more FORAMINIFERAN or radiolarian skeletons. A sediment is defined as a calcareous ooze if it contains > 30% calcium carbonate ($CaCO_3$) and as a siliceous ooze if it contains < 30% $CaCO_3$ and > 30% siliceous organic material. Calcareous ooze is generally white or pinkish-white in colour and covers most of the floor of the western INDIAN OCEAN, the mid-ATLANTIC OCEAN and much of the equatorial and southern PACIFIC OCEAN. It consists primarily of shells of foraminiferans, coccolithophorids (calcareous algae) and pteropods with lesser amounts of CLAY, siliceous shells, minute fragments of meteorites, and minerals crystallized from ocean waters. Siliceous oozes accumulate where the solution of carbonate exceeds its production (in deep water high in CARBON DIOXIDE) and where the supply of land-derived material is relatively low. Siliceous oozes are restricted to relatively narrow strips in the North Pacific and equatorial Pacific and to a somewhat broader band in all the southern oceans. In the equatorial regions, it is composed of radiolarians (and is yellowish brown in colour) whereas in high latitudes it consists mainly of diatoms (and is yellowish, straw- or cream-coloured). *See also* LITHOGENOUS (RED CLAY). [V.F.]

open canopy Stands of trees in woodlands and forests in which the density of trees in the upper stratum is low enough to allow significant gaps to be present between the crowns of neighbouring trees. Collectively, a STAND has an open CANOPY if the crowns of trees in the tallest stratum cover less than about 50% of the crown by vertical projection. Open canopies form naturally after non-catastrophic fires, windthrow, drought and disease, and by silvicultural thinning. [G.F.P.]

open community An ASSEMBLAGE of organisms with incomplete ground COVER (*cf.* CLOSED COMMUNITY), readily invaded by immigrant organisms, some ecological niches being unoccupied. In open VEGETATION, gaps exist in the CANOPY which may separate the plants by up to as much as twice the diameter of the predominant individuals. If the vegetation is even more scanty—the ground space exceeding the plant cover—the vegetation may be termed 'sparse'. Open communities are characteristic of early stages of SUCCESSION. [A.J.W.]

open system An open system is commonly used to describe a physical system in which energy and matter can enter and leave without restriction. The definition can, however, also be applied to biological systems. In a physical sense the BIOSPHERE is an open system when radiant energy is considered but closed when geochemical cycles (e.g. carbon) are considered. In the same sense a biological system may be either open or closed. The key process that determines this is MIGRATION. When individuals move between suitable HABITAT patches then the system may be described as open. For each species this process will occur on different scales. Most functional natural communities are open and consist of a number of interacting subpopulations or compartments. *Cf.* CLOSED SYSTEM. *See also* SYSTEMS ECOLOGY. [R.C.]

opportunistic species Species that can successfully exploit new RESOURCES as and when they arise. Such species are also termed COLONIZING SPECIES, RUDERAL species or *r*-selected species (*see* *r*- AND *K*-SELECTION). Conditions suitable for the growth of opportunistic species occur unpredictably in time and space and last only for several generations. Opportunistic species include:

1 early successional (fugitive) species, which colonize disturbed habitats and are later replaced by superior competitors;

2 species using rich but unpredictable resources (e.g. fruit flies, some decomposers);

3 organisms that live in inherently unstable habitats (e.g. river banks, hydrothermal vents).

Opportunistic species tend to have a short GENERATION TIME and a high REPRODUCTIVE EFFORT, and to produce many small offspring. They are usually poor competitors and good dispersers, with propagules that can be dormant for a long time. *See also* BIOLOGICAL INVASIONS. [T.J.K.]

optimal age at maturity In evolutionary biology, age at maturity is typically defined as the age at first reproduction. Coupled with SIZE AT MATURITY, it is an important FITNESS component, usually under STABILIZING SELECTION with an optimum determined by the balance of costs and benefits of delayed maturity. Maturation is often accompanied by a slow-down or cessation of GROWTH because of RESOURCES diverted from growth to reproduction. Delayed maturity therefore usually results in larger adult SIZE, and is thus advantageous in terms of better adult survival, improved competitive ability, higher fertility or better offspring quality. There are two general sources of costs of delayed maturity. First, individuals maturing later spend more time as juveniles and have thus a higher probability of dying before reproduction. Therefore, a high MORTALITY RATE of individuals that are about to mature favours early maturity at a small size. Second, in populations with OVERLAPPING GENERATIONS, offspring born earlier start reproducing sooner and, if the population is growing, they contribute more to the parent's fitness than offspring born later. Therefore, conditions leading to rapid population growth favour early maturity if generations are overlapping. Other factors affecting the evolution of age at maturity include: SEASONALITY; intermittent availability of a resource essential for reproduction; and consequences of discontinuous growth. *See also* AGE AT MATURITY IN PLANTS; EVOLUTIONARY OPTIMIZATION; LIFE-HISTORY EVOLUTION; TRADE-OFF. [T.J.K.]

REFERENCES

Roff, D.A. (1992) *The Evolution of Life Histories*. Chapman & Hall, New York.
Stearns, S.C. (1992) *The Evolution of Life Histories*. Oxford University Press, Oxford.

optimal clutch size The NUMBER OF OFFSPRING produced per breeding attempt that maximizes FITNESS. The optimal CLUTCH SIZE is a function of TRADE-OFFS between FECUNDITY and survival in the parents, and SURVIVAL RATE of offspring resulting from changes in provisioning due to variation in clutch size and/or earlier breeding patterns of the parents. *See also* EVOLUTIONARY OPTIMIZATION; LACK CLUTCH; LIFE-HISTORY EVOLUTION. [D.A.R.]

Optimal control theory

Optimal control theory (or dynamic optimization theory) is a body of theory and computational techniques that is useful for solving dynamic optimization models. It is an essential ingredient in much of modern technology, particularly in the aerospace and communications industries. In evolutionary biology optimal control theory is being used increasingly in the study of behavioural and life-history ADAPTATIONS. A particular version, stochastic DYNAMIC PROGRAMMING, has proven especially popular, being relatively easy to use in a wide variety of applications (Mangel & Clark 1988).

A stochastic dynamic programming model of optimal behaviour has the following components.

1 A dynamic state variable $X(t)$, generally multidimensional, representing the state of the organism and its environment at time t.

2 A decision set $D(X,t)$, depending on the current time and state, representing the set of feasible decisions, or actions.

3 A dynamic equation of state, $X(t+1)=G[X(t), d(t), w(t)]$ describing how the state variable changes over time in response to the decision $d(t)$ and stochastic environmental influences $w(t)$.

4 A specification of FITNESS, usually defined in terms of expected LIFETIME REPRODUCTIVE SUCCESS.

The ALGORITHM used to calculate the optimal STRATEGY is known as dynamic programming (Bellman 1957).

Dynamic programming models have been widely used to analyse behaviour and LIFE-HISTORY TRAITS that involve TRADE-OFFS between foraging benefits and predation risks. The following abstract model serves to illustrate the logic and technique of dynamic programming in this setting.

Imagine a forager faced with the choice of two foraging habitats H_i. Each habitat is characterized by two parameters: g_i=net gain in body mass per unit time, and m_i=risk of predation per unit time. We suppose that habitat 2 is both more productive and more dangerous than habitat 1, i.e. $g_2>g_1$ and $m_2>m_1$ (if one of these inequalities is reversed the problem becomes trivial). The forager reproduces once, at the end of the foraging season, and reproductive output is a known function of terminal body mass. Thus choosing H_2 leads to greater reproduction, but incurs a greater risk of not surviving to reproduce, relative to H_1. The problem is to determine the optimal choice of habitat as a function of the forager's current age and body size. Though highly abstract, this model encapsulates behavioural (or developmental) choices faced by many kinds of animals.

To solve the model by dynamic programming we first divide time into discrete intervals, denoted by $t=1, 2, \ldots, T$, and assume that one of the two habitats will be selected in each time interval t. Let $X(t)$, the state variable, denote body mass at the beginning of period t. Reproductive success of a forager alive at the beginning of the terminal period T is denoted by $R[X(T)]$.

We next define a function $F(x,t)$ by: $F(x,t)$=maximum expected reproduction, given that $X(t)=x$. This function (closely related to the concept of REPRODUCTIVE VALUE) is initially unknown, but can be computed, together with the optimal strategy, from the dynamic programming algorithm:

$$F(x,T) = R(x)$$

$$F(x,t) = \max(1-m_i)F(x+g_i,t+1), \quad t<T, \; i=1,2$$

These equations are immediate logical consequences of the model assumptions and the definition of $F(x,t)$. The equations are easily programmed for computer solution. Figure O3 shows the optimal habitat choice for a specific set of parameter values. This is a typical 'feedback control diagram', specifying the forager's optimal decision at each point in time-state space (t,x). Typical trajectories through this space can be obtained by simulation; two examples are shown.

Empirical applications of dynamic programming models are now common. Examples of TRAITS that have been studied include: defence strategies of marine organisms; fledging behaviour of seabirds; MIGRATION strategies of birds and fish; winter hoarding strategies of birds; and oviposition strategies of parasitic insects. Such models often help in understanding the adaptive significance of complex behavioural traits. The models also often suggest that patterns of variation observed in natural systems and previously attributed to inexplicable randomness, may in fact reflect subtle evolutionary forces. *See also* BEHAVIOURAL ECOLOGY; EVOLUTIONARY OPTIMIZATION; OPTIMAL FORAGING THEORY. [C.W.C.]

REFERENCES

Bellman, R. (1957) *Dynamic Programming*. Princeton University Press, Princeton, N.J.

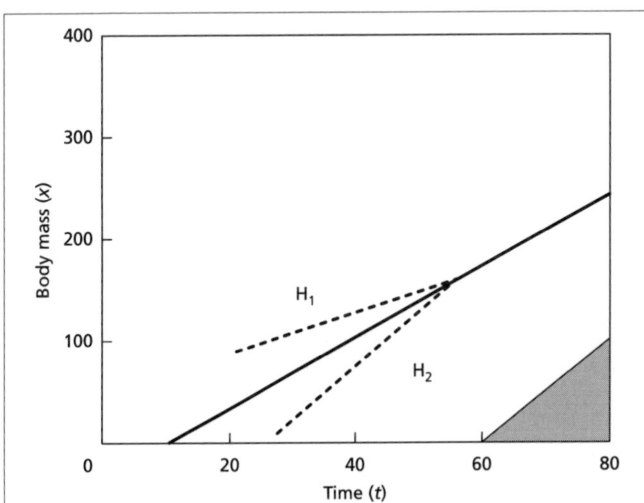

Fig. O3 Optimal habitat choice for a model of foraging under predation. Parameter values: $m_1=0.02$; $m_2=0.04$; $g_1=2$; $g_2=5$; $R(x)=x-100$. The shaded area is a region of zero fitness: foragers having body mass x at time t lying within this region cannot grow fast enough to reproduce. Otherwise foragers should use H_2 (risky habitat) or H_1 (safer habitat) if their body mass x at time t lies in the indicated region. Typical growth curves are shown as dashed lines. (After Clark, C.W. (1993) Dynamic models of behavior: An extension of life history theory. *Trends in Ecology and Evolution* **6**, 205; with permission from Elsevier Science.)

Mangel, M. & Clark, C.W. (1988) *Dynamic Modeling in Behavioral Ecology*. Princeton University Press, Princeton, N.J.

---

optimal diet model One of the principal models of OPTIMAL FORAGING THEORY (Stephens & Krebs 1986), the optimal diet model treats the decisions of a forager SEARCHING for food in a homogeneous environment. Several distinct types of food are encountered at random, each type being characterized by its relative abundance, its energy content and its HANDLING TIME (time required to consume the item). Upon encountering an item the forager may decide to eat it or to reject it: items having a low ratio of energy content to handling time may be rejected in the hope of encountering more profitable items. Based on the assumption that foragers will attempt to maximize the long-term average rate of energy gain, the standard optimal diet model predicts a THRESHOLD type of behaviour—items above the threshold are always accepted, those below rejected. Most published data, however, display partial preferences. Alternative diet models capable of explaining partial preferences have been considered by McNamara and Houston (1987). [C.W.C.]

REFERENCES

McNamara, J.M. & Houston, A.I. (1987) Partial preferences and foraging. *Animal Behaviour* **35**, 1084–1099.

Stephens, D.W. & Krebs, J.R. (1986) *Foraging Theory*. Princeton University Press, Princeton, N.J.

Optimal foraging theory (OFT)

Optimal foraging theory was born in 1966, when *American Naturalist* published two papers, one by MacArthur and Pianka and one by Emlen, both suggesting that NATURAL SELECTION would favour FOOD SELECTION behaviour that maximizes net energy gained per time spent foraging. MacArthur and Pianka (1966) formalized their ideas in a graphical model. Mathematical versions of this model were later derived more or less independently by others, including Schoener, Charnov, M.E. Timin, Maynard Smith, and E.E. Werner and D.J. Hall.

An important extension of optimal foraging theory was marked by the publication by Krebs, Ryan and Charnov in 1974 of the experimental test of a model. This model, known as the MARGINAL VALUE THEOREM, predicts for a heterogeneous environment — where RESOURCES are divided over patches of variable profitability —how long an animal must forage in each patch to maximize intake rate of food. Foraging is defined as all activities carried out to obtain resources; i.e. the search, the pursuit of prey, and handling (killing and eating). In both the prey choice and the patch time model, costs are measured in time and/or energy spent per unit time, while the benefits are measured as net energy gain per unit time spent foraging.

The seminal papers resulted in a boom during the 1970s and 1980s, and also stimulated research in other areas of BEHAVIOURAL ECOLOGY. Optimal foraging theory has been successful because it provides simple predictive models to explain why organisms prefer to eat certain items out of a set of available kinds of food, and because the predictions of the models are open to experimental tests.

The classical models considered only maximization of intake rate in an unchanging world with complete information of the forager. Later models have explored other strategies, for example minimizing energy expenditure, risk minimization, avoiding predation. Likewise, stochastic dynamic foraging models have explored the effects on optimal foraging of changes in the environment, in the information an animal has about its environment, and in the physiological condition of the foraging animal. John MacNamara, Alisdair Houston and Marc Mangel are the leading theorists in the development of this class of models.

Although net rate of energy gain is the 'currency' most commonly used as the basis for optimization, other currencies have been recognized as more appropriate to particular situations. The appropriateness of any currency may depend both on ecological circumstances and on the physiological state of the forager. For example, in a habitat containing predators that pose a risk to the forager, an individual with a secure ENERGY BUDGET may be expected to maximize energy gain per unit risk by choosing relatively safe places to feed, whereas the priority for a starving individual may be to maximize its net rate of energy gain even if this entails foraging in a relatively dangerous place. Here, optimal decisions based on the new currency, energy gain per unit risk, may be mutually exclusive of those based on the net rate of energy gain. Time itself may impose a severe constraint on foraging if other activities, such as mating or brooding, are of pressing importance. Instead of maximizing net energy gained during a foraging bout, such foragers may be expected to minimize the time spent gaining some required amount of energy.

Optimal foraging theory has been criticized on the ground that animals cannot be expected to be optimal because of genetic CONSTRAINTS preventing animals from adapting fast enough to changes in the environment. However, protagonists of optimal foraging theory have argued that its aim is not to prove that animals are optimal but, rather, to make testable predictions about animal behaviour and so advance our knowledge about the great diversity of behaviours seen in nature (e.g. Maynard Smith 1974). In directing and stimulating empirical and experimental research, the theory has been very successful indeed. *See also* EVOLUTIONARY OPTIMIZATION; OPTIMALITY MODELS; TRADE-OFF. [J.V.A. & R.N.H.]

REFERENCES

Emlen, J.M. (1966) The role of time and energy in food preference. *American Naturalist* **100**, 611–617.

Krebs, J.R., Ryan, J.C. & Charnov, E.L. (1974) Hunting by expectation or optimal foraging? A study of patch use by chickadees. *Animal Behaviour* **22**, 953–964.

MacArthur, R.H. & Pianka, E.R. (1966) On optimal use of a patchy environment. *American Naturalist* **100**, 603–609.

Maynard Smith, J. (1974) *Models in Ecology*. Cambridge University Press, Cambridge.

---

optimal group size Many species of animals live in groups, having either relatively permanent membership (as in family groups), or having a more ephemeral nature. The main advantages of grouping are thought to be: (i) increased protection from predators, resulting

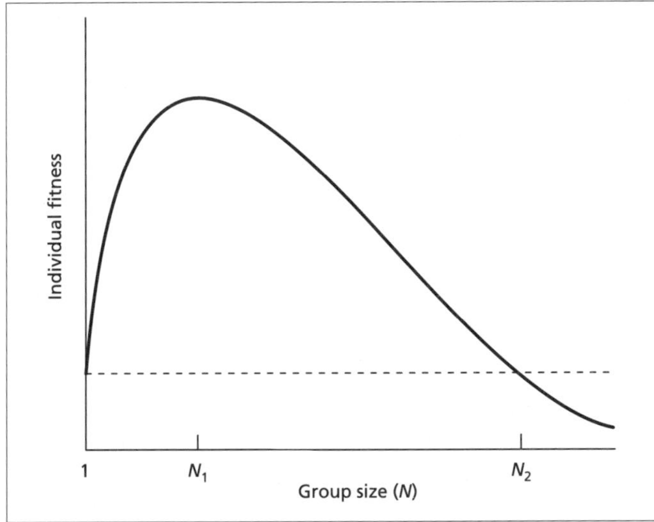

Fig. O4 Individual fitness as a function of group size, N. The optimal group size, N_1, maximizes individual fitness. The stable group size, $N_2 >> N_1$, equates fitness with that of groups of a single individual ($N = 1$).

from combined VIGILANCE, or merely from safety in numbers; and (ii) increased foraging EFFICIENCY.

The FITNESS of an individual group member can be envisaged as a function of the size of the group, N. The optimal group size is then the size (N_1) at which individual fitness is maximized (Fig. O4). In practice, however, it may happen that actual group sizes are much larger than optimal: the optimal group size may not be stable. Conference attenders often encounter this phenomenon, in attempting to establish a small group of people to go to a local restaurant. Once the optimal group has been formed, newly arriving acquaintances would prefer to join the group, rather than dine alone. According to the figure, fitness for a group slightly larger than optimal size exceeds fitness for a group of size 1. The stable group is only reached when fitness is reduced to the level of a group consisting of one individual. Complex behaviour, including DOMINANCE and aggression, may be required to keep groups close to the optimal size. *See also* AGGREGATION; ALLEE EFFECT. [C.W.C.]

optimal harvest This idea involves a consideration of economic factors as well as the purely biological. The simple idea of a MAXIMUM SUSTAINABLE YIELD (MSY) is a purely biological concept. It may well be that the level of harvest which achieves, for example, maximum profit, may be greater or less than the MSY. Indeed, in certain situations it may be optimal, in the sense that profits are maximized, for a population to be driven to extinction. This occurs when the SUSTAINABLE YIELD is relatively small compared to the size of the stock. In such a situation, the yield is lower than that which can be obtained in market interest rates and hence it becomes attractive to overexploit the resource and turn it into money which can earn interest. *See also* FISHERIES, CONSERVATION AND MANAGEMENT; HARVESTING. [J.B.]

optimal reaction norms The optimal REACTION NORM for a TRAIT describes the optimal value of this trait for each environment in the range of environments encountered by the population. PHENOTYPIC PLASTICITY can be a mechanistic by-product of environmental influence. However, for a trait under STABILIZING SELECTION, the optimum will often change with the environment, and plasticity may thus be an adaptive response to environmental cues. The optimal reaction norm for such a trait describes the ideal adaptive plasticity: it is the reaction norm of a hypothetical GENOTYPE that in each environment produces the optimal PHENOTYPE.

Studies of optimal reaction norms extend EVOLUTIONARY OPTIMIZATION to heterogeneous environments, in which the population encounters a range of environmental conditions, although each individual typically experiences only a small part of this range. They have concentrated in particular on LIFE-HISTORY TRAITS like age and size at METAMORPHOSIS and maturity, number and size of offspring, and investment in defence. The FITNESS PROFILES and optima for these characters are particularly sensitive to environmental parameters like resource abundance, COMPETITION intensity or MORTALITY RATE.

Qualitative predictions of these studies have been successfully tested in many organisms; water fleas (*Daphnia* spp.) are a good example. As predicted by the theory, they mature early and at a small size when exposed to chemical cues from planktivorous fish, but delay maturity and invest in defensive morphological structures in the presence of invertebrate predators that prey exclusively on the smallest individuals. They also respond to low food concentration by producing larger offspring, confirming another theoretical prediction.

A population may be prevented from evolving an optimal reaction norm by physiological or genetic CONSTRAINTS, costs of plasticity, or lack of reliable environmental cues. In addition, if some environments are encountered only rarely, selection on the corresponding parts of the reaction norm will be weak. Therefore the phenotypes expressed in a population should approach the optimal reaction norm only in the environments that are most frequently encountered. *See also* GENERAL-PURPOSE GENOTYPE. [T.J.K.]

REFERENCES

Kawecki, T.J. & Stearns, S.C. (1993) The evolution of life histories in spatially heterogeneous environments: optimal reaction norms revisited. *Evolutionary Ecology* **7**, 155–174.

Stearns, S.C. (1989) The evolutionary significance of phenotypic plasticity. *BioScience* **39**, 436–445.

Stibor, H. (1992) Predator-induced life-history shifts in a freshwater cladoceran. *Oecologia* **92**, 162–165.

optimal repair Allocation of metabolic resources to somatic repair (and maintenance) to maximize the FITNESS of an organism. This allocation is determined with regard to the TRADE-OFF that must be made between, on the one hand, protecting against age-associated increases in INTRINSIC MORTALITY by means of slowing the accumulation of somatic damage and, on the other hand, utilizing resources for growth and reproduction. The DISPOSABLE SOMA THEORY predicts that the optimal level of repair for iteroparous (i.e. repeatedly reproducing; *see* ITEROPARITY), sexually reproducing species is less than would be required for indefinite somatic survival, and that this results in AGEING (SENESCENCE). *See also* COSTS OF RESISTANCE; DEFENCE MECHANISMS. [T.B.L.K.]

optimal reproductive effort The amount of effort that should be devoted to reproduction to maximize FITNESS. The two primary components are the age schedule of reproduction and the allocation of energy per reproductive episode. At one extreme the optimal REPRODUCTIVE EFFORT may be so high that only one breeding episode is possible (SEMELPARITY): such a situation is favoured by low adult survival and high rate of increase. Iteroparity, or repeated reproduction, is the more common situation. The optimal reproductive effort will be determined in large measure by the trade-offs between survival and reproductive effort, and between GROWTH and reproductive effort. The importance of the former is readily apparent; the importance of the latter resides in the fact that energy channelled into reproduction is not available for growth, and growth may be more strongly favoured because of the increased FECUNDITY or mating success that frequently accompanies an increase in SIZE. *See also* LIFE-HISTORY EVOLUTION; REPRODUCTIVE EFFORT. [D.A.R.]

optimal size at maturity AGE AND SIZE AT MATURITY are among the traits most closely related to the evolutionary success of individuals. Body SIZE is closely associated with FECUNDITY and survival in many organisms, so that there is SELECTION PRESSURE for large size. On the other hand, to become large, many organisms must delay their maturity, in particular if their GROWTH slows after they have matured. But delaying maturity decreases reproductive success because it decreases the likelihood that an individual will reach maturity, and increases its GENERATION TIME. Thus there is selection pressure for early maturity and therefore a small body size. Assuming that EVOLUTION will lead to a balance of these opposing selection pressures, life-history theory can predict the age and size at maturity that maximize reproductive success. This approach helps us to understand several patterns of age and size at maturity, such as bimaturism, where females and males mature at different body sizes and different ages. The females of many species mature later and larger than the males, because their fecundity depends more on size than that of the males does. In those species where the males mature later and larger, it is often because they compete with one another for access to females and their competitive success depends on body size. *See also* LIFE-HISTORY EVOLUTION; TRADE-OFF. [J.C.K.]

REFERENCES

Roff, D.A. (1992) *The Evolution of Life Histories.* Chapman & Hall, New York.

Stearns, S.C. (1992) *The Evolution of Life Histories.* Oxford University Press, Oxford.

optimality models Optimality models are commonly used in evolutionary biology and in economics. They are based on a working assumption that the individual members of an ecological or economic COMMUNITY act as optimizers of some specified objective. In biological applications the optimality objective is usually taken to be Darwinian FITNESS, a measure related to survival and reproduction. In economics the optimization objective usually assumed for individuals is called utility, whereas for firms the objective is profits. Thus: biological organisms maximize fitness, individuals maximize utility, and firms maximize profits.

The use of such an optimality assumption should not be misconstrued as a claim that real-world organisms actually do maximize fitness, utility or profits. Optimality is merely a working hypothesis upon which a useful body of theory can be based (Maynard Smith 1978). Attempts to apply an optimality model outside its proper scope can lead to folly (Daly & Cobb 1989).

Optimality models are normally expressed in mathematical terms. Such a model requires the specification of:

1 state variables that describe the ENVIRONMENT in which decisions are made;

2 decision variables that describe the set of possible decisions or actions;

3 CONSTRAINTS that limit the scope of state and decision variables;

4 an optimality objective that describes the payoff, or reward, obtained from each decision under given circumstances.

Mathematical techniques (often coded as computer ALGORITHMS) used to solve optimality models include calculus, linear programming, OPTIMAL CONTROL THEORY and DYNAMIC PROGRAMMING.

Optimality models that include UNCERTAINTY or incomplete information are addressed in decision theory (Luce & Raiffa 1957). Models that involve CONFLICTS of interest between individuals are the subject of games theory; in evolutionary biology game-theoretic models have been widely used in the study of evolutionarily stable strategies (Maynard Smith 1982). *See also* EVOLUTIONARY OPTIMIZATION; GAME THEORY; TRADE-OFF. [C.W.C.]

REFERENCES

Daly, H.E. & Cobb, J.B., Jr. (1989) *For the Common Good: Redirecting the Economy Toward Community, the Environment, and a Sustainable Future.* Beacon, Boston, Mass.

Luce, R.D. & Raiffa, H. (1957) *Games and Decisions: Introduction and Critical Survey.* John Wiley & Sons, New York.

Maynard Smith, J. (1978) Optimization theory in evolution. *Annual Review of Ecology and Systematics* **9**, 31–56.

Maynard Smith, J. (1982) *Evolution and the Theory of Games.* Cambridge University Press, Cambridge.

optimum pollution This concept stems from the recognition that it is often necessary to compromise between the benefits of POLLUTION control—in terms of human health and the environment—and the costs of the restrictions—in terms of lost social benefits to health, food production and general welfare. Thus many pesticides pollute; but they bring benefits to food production that may be crucial in the developing world.

The ROYAL COMMISSION ON ENVIRONMENTAL POLLUTION, in a report published in 1972, describes the concept as 'unavoidably troublesome'. The Sixteenth Report (1992) comments:

> the concept of 'optimum' is 'coherent' (only) in the context of the analysis of a whole range of public decisions in which the harm caused by a certain amount of pollution needs to be compared with the harm which would be caused if *resources* were directed from other important purposes, not connected with pollution.

So optimality arguments are certainly difficult; objectively defining costs and benefits is intellectually challenging, and does not always lead to comfortable conclusions. But these trade-offs have to be kept in the public mind; it is important to realize that pollution control has to be paid for and that very often difficult decisions have to be made about the appropriateness of alternative products or courses of action. *Cf.* PRECAUTIONARY PRINCIPLE. *See also* COST

(RISK)–BENEFIT APPROACH; ENVIRONMENTAL (ECOLOGICAL) ECONOMICS; POLLUTER PAYS PRINCIPLE. [P.C.]

orchidaceous mycorrhizae *See* ENDOTROPHIC MYCORRHIZAE.

order *See* TAXONOMIC CATEGORIES.

order of gamete release hypothesis *See* GAMETE-ORDER HYPOTHESIS.

ordered transformation series *See* TRANSFORMATION SERIES.

Ordination

The arrangement of samples or sites along gradients on the basis of their species composition or environmental attributes. Ordination is the mathematical expression of the CONTINUUM concept in ecology. The basic ideas were developed in the 1920s simultaneously by the botanists L.G. Ramensky (Russia) and H.A. Gleason (USA). The German word *Ordnung* (Ramensky 1930) was translated into English by Goodall (1954) as ordination, a new technical term. Gradient analysis is often treated as a synonym by plant ecologists (Whittaker 1967; Ter Braak & Prentice 1988). Ordination and numerical CLASSIFICATION are the main methods for analysis of multivariate species data (Greig-Smith 1983; Legendre & Legendre 1983).

Data for ordination typically consist of a table of values specifying the ABUNDANCE or presence of species in samples. Environmental data are usually also available. A convenient way to envisage the structure of the data is as two matrices, **A** and **B**, stacked one beside the other:

$$\mathbf{C} = \left[\mathbf{A}\middle|\mathbf{B}\right] = \left[a_{ij}\middle|b_{ik}\right] \quad \left(i = 1, \ldots, m;\ j = 1, \ldots, n;\ k = 1, \ldots, p\right)$$

Rows of the matrix represent samples. The first block of n columns represents species; a_{ij} is the abundance of species j in sample i. The second block of p columns represents environmental variables; b_{ik} is the value of the kth environmental variable in sample i.

Ordination in two dimensions produces a diagram showing relations among samples and/or species (Fig. O5). Samples that are near to one another in the ordination diagram are inferred to resemble one another in species composition and environmental attributes. There is a tacit assumption that samples with similar species have similar environments.

Approaches to ordination

Given data on species and environment, there are two major approaches to ordination. Direct ordination is designed to reflect a known environmental variable. Indirect ordination is made on the basis of the species information alone; any environmental interpretation must come later, by regression of ordination axes on environmental variables or by plotting ENVIRONMENTAL VALUES on the ordination diagram.

Direct ordination

Direct ordination has two stages. As an example, consider direct ordination of altitude by weighted averaging. At stage 1, the altitudinal preference of each species is calculated as a weighted average:

$$y_{j1} = \frac{\sum_{i=1}^{m} a_{ij} b_{i1}}{\sum_{i=1}^{m} a_{ij}}$$

where, in the notation used above, a_{ij} is the abundance of species j in sample i and b_{i1} is the altitude (environmental variable 1) of sample i. At stage 2, an index of altitude for the sample is calculated, also as a weighted average:

$$x_{i1} = \frac{\sum_{j=1}^{n} a_{ij} y_{j1}}{\sum_{j=1}^{n} a_{ij}}$$

The weighted average x_{i1} can be calculated for additional samples to estimate the altitude where this is unknown. For this purpose a further CALIBRATION stage is required to estimate b_{i1} as a function of x_{i1}. The use of multivariate species data to estimate the values of environmental factors is reviewed by Ter Braak (1995).

Indirect ordination

Indirect ordination uses only species data, i.e. matrix **A** above. Many methods of indirect ordination are explicitly or implicitly based on a measure of the difference in species composition between the samples; examples are PRINCIPAL COMPONENT ANALYSIS, principal coordinate analysis and non-metric MULTIDIMENSIONAL SCALING. Correspondence analysis and its variant, detrended correspondence analysis (Hill & Gauch 1980), are based on a different principle, namely that of simultaneously ordinating both species and samples. The main purpose of ordination is to

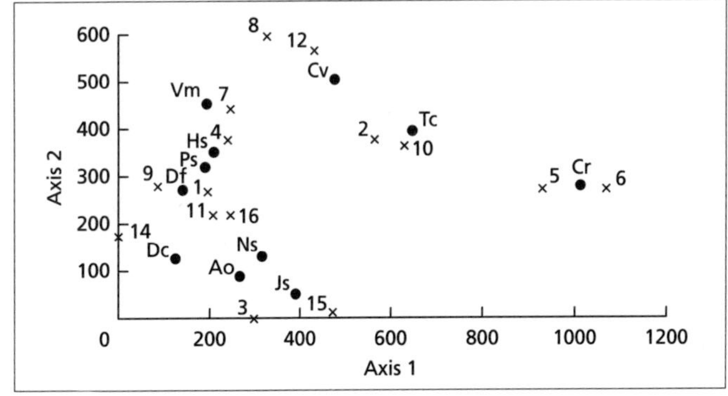

Fig. O5 Simultaneous indirect ordination of vegetation samples (crosses) and species (dots) by detrended correspondence analysis (unpublished data from the Cairngorm Mountains). The ordination shows two gradients in composition from dry (low values on axis 1) to wet (high values on axis 1) and from grassy (low values on axis 2) to shrubby (high values on axis 2). Sample 2, intermediate between the extremes of wet and shrubby, was described by the fieldworker as a sheltered boggy saddle with 12% shrub cover. The species ordination confirms the interpretation: Cv and Vm are the dwarf shrubs *Calluna vulgaris* and *Vaccinium myrtillus*; Cr is the semi-aquatic sedge *Carex rostrata*.

make inferences about the environmental characteristics of sites and the environmental preferences of species. When the controlling environmental factors have not been measured, indirect ordination can indicate how many factors are needed to explain the observed pattern of variation in species composition.

Canonical ordination

A powerful new technique, canonical correspondence analysis (Ter Braak 1986), has become popular in ecology. This has features of both direct and indirect ordination. It is an example of a group of mathematical methods known as canonical ordination (Ter Braak 1987) or constrained ordination (Ter Braak & Prentice 1988). Canonical correspondence analysis simultaneously ordinates the samples, the species and the environmental variables. By so doing, it removes much of the drudgery required to make environmental interpretations of ordinations.

Merits and drawbacks

Ordination has its limitations and defects. It can distract attention from individual species responses by focusing on the multivariate response. Being a correlative method, it can assist with the process of hypothesis generation, but it can rarely demonstrate causality. [M.O.H.]

REFERENCES

Goodall, D.W. (1954) Objective methods for the classification of vegetation. III. An essay in the use of factor analysis. *Australian Journal of Botany* **2**, 304–324.

Greig-Smith, P. (1983) *Quantitative Plant Ecology*, 3rd edn. Blackwell Scientific Publications, Oxford.

Hill, M.O. & Gauch, H.G. (1980) Detrended correspondence analysis: an improved ordination technique. *Vegetatio* **42**, 47–58.

Legendre, L. & Legendre, P. (1983) *Numerical Ecology*. Elsevier, Amsterdam.

Ramensky, L.G. (1930) Zur Methodik der vergleichenden Bearbeitung und Ordnung von Pflanzenlisten und anderen Objekten, die durch mehrere, verschiedenartig wirkende Faktoren bestimmt werden. *Beiträge zur Biologie der Pflanzen* **18**, 269–304.

Ter Braak, C.J.F. (1986) Canonical correspondence analysis: a new eigenvector technique for multivariate direct gradient analysis. *Ecology* **67**, 1167–1179.

Ter Braak, C.J.F. (1987) Ordination. In: *Data Analysis in Community and Landscape Ecology* (eds R.H.G. Jongman, C.J.F. Ter Braak & O.F.R. Tongeren), pp. 91–173. Pudoc, Wageningen.

Ter Braak, C.J.F. (1995) Non-linear methods for multivariate statistical calibration and their use in palaeoecology: a comparison of inverse (*k*-nearest neighbours, partial least squares and weighted averaging partial least squares) and classical approaches. *Chemometrics and Intelligent Laboratory Systems* **28**, 165–180.

Ter Braak, C.J.F. & Prentice, I.C. (1988) A theory of gradient analysis. *Advances in Ecological Research* **18**, 271–317.

Whittaker, R.H. (1967) Gradient analysis of vegetation. *Biological Reviews* **49**, 207–264.

---

organelles Structures within a cell that have a characteristic structure and function, such as the nucleus, mitochondria, chloroplasts and Golgi apparatus. [P.O.]

organic detritus Particulate material derived from dead organisms or from parts shed by living organisms. Detritus may be derived from the breakdown of land plants, benthic algae and grasses, PHYTOPLANKTON, and ZOOPLANKTON and zoobenthos. It provides a food source for certain organisms, referred to as detritivores. *See also* BIOMASS; NECROMASS. [V.F.]

organic farming AGRICULTURE that avoids, as far as possible, chemical support through FERTILIZERS and BIOCIDES. *See also* FARMING PRACTICES, SUSTAINABLE; SUSTAINABLE AGRICULTURE. [P.C.]

organic loading The addition of abnormal amounts of dead organic matter to the environment—usually to a watercourse. For example, the addition of SEWAGE to a river. *See also* BIOCHEMICAL OXYGEN DEMAND. [P.C.]

organic pollution Usually means CONTAMINATION from WASTES derived from living things that causes adverse effects on human health and/or ecological systems (*see also* SEWAGE POLLUTION). Can sometimes be intended as POLLUTION from organic chemicals. [P.C.]

organismic hypothesis According to Clements (1916; *see* CHARACTERS IN ECOLOGY and CLIMAX):

> The unit of vegetation, the climax formation, is an organic entity. As an organism, the formation arises, grows, matures and dies. Its response to the habitat is shown in processes or functions and in structures that are the record as well as the result of these functions.
>
> Furthermore, each climax formation is able to reproduce itself, repeating with essential fidelity the stages of its development. The life-history of a formation is a complex but definite process, comparable in its chief features with the life-history of an individual plant.

This viewpoint was based on three specific ideas: (i) that species occur with characteristic companions in distinct communities; (ii) that there is at any site-type a closely repeatable SUCCESSION of species ('single pathway'); and (iii) that FACILITATION, as it is now called, is a general phenomenon in succession These ideas have been largely displaced by the INDIVIDUALISTIC HYPOTHESIS, the concept of VARIABLE PATHWAYS, and the idea that many successions result from differences in life histories and tolerances without a major involvement of facilitation. The parallel between a plant 'formation' and an organism is generally rejected. [P.J.G.]

REFERENCE

Clements, F.E. (1916) *Plant Succession*. Carnegie Institution of Washington, Publication 242.

organismic philosophy *See* HOLISM.

Organization for Economic Cooperation and Development *See* OECD.

organochlorine pesticides *See* INSECTICIDES.

organochlorines Organic compounds containing one or more CHLORINE atoms. The first commercially produced organochlorine was carbon tetrachloride (1907), and other industrial solvents including 1,1,1-trichloroethane followed in the 1920s. These compounds display a range of toxic and carcinogenic properties; for example, carbon tetrachloride causes extensive damage to the liver whilst chloroform causes kidney damage and induces cancer in rats and mice. The toxic properties of polychlorinated compounds led to their widespread use as pesticides and herbicides; examples include DDT (dichlorodiphenyl-trichloroethane); and 2,4-D (2,4-dichlorophenoxyacetic acid). An awareness of the toxic effects of these chemicals on other organisms has led to imposed limitations on their use. *See also* PERSISTANT ORGANIC POLLUTANTS; VOLATILE ORGANIC COMPOUND. [M.C.G.]

organomercury A form of MERCURY in which it is combined with an organic moiety. Organomercury compounds may arise:

1 from deliberate chemical synthesis, as in alkylmercury halides (e.g. CH_3HgX, C_2H_5HgX) and phenylmercury halides (C_6H_5HgX), which are used in seed dressings, FUNGICIDES and slimicides;

2 from natural, methylcobalamin-mediated, biomethylation of inorganic mercury in aquatic sediments to produce highly toxic methylmercury (CH_3Hg^+), which can be bioaccumulated and biomagnified along the aquatic FOOD CHAIN, and the less soluble and more volatile dimethylmercury ($(CH_3)_2Hg$).

Organomercury is lipid-soluble and much more readily absorbed than inorganic mercury, allowing penetration of the blood–brain barrier and passage through the placenta. Many deaths and deformities resulted from the consumption of CH_3Hg^+-treated grain in Iraq in the 1960s and 1970s, and of CH_3Hg^+-contaminated fish by residents of the Japanese fishing village of MINAMATA in the 1950s when a chemical plant manufacturing acetaldehyde discharged mercury salts (being used as a CATALYST) and, possibly, methylmercury into the bay. *See also* HEAVY METALS. [J.F.]

organometal pollution A few metals and metalloids are capable of forming compounds with organic moieties via metal–carbon bonds to form organometal compounds. Of special relevance in the aquatic environment are compounds involving MERCURY (Hg), LEAD (Pb) and TIN (Sn). However, the production of synthetic organometals is at an early stage, and there is considerable scope for the formation of new molecules, so possible future environmental problems from these compounds cannot be ruled out.

A common form of organolead is alkyl lead, used as a petrol additive to increase octane rating. The most toxic forms of alkyl lead degrade rapidly in the environment. Methyl mercuric compounds are the most common forms of mercury in the aquatic environment, deriving from methylation of inorganic mercury in solution. They are highly toxic, being 4–30 times more toxic than inorganic forms, and highly bioaccumulative. A wide variety of organotin compounds are used in industrial processes (e.g. for stabilization of PVC, GLASS strengthening and various catalytic reactions, and in BIOCIDES). Tributyltin (TBT) has been used extensively in antibiofouling paints for boats and caused POLLUTION of the marine environment, which has been studied extensively. Many fish and crustaceans are sensitive to TBT, with lethal thresholds in the low microgram per litre range and chronic effects at submicrogram per litre levels. Molluscs are particularly sensitive to TBT, with sublethal effects at levels of low to subnanograms per litre. *See also* IMPOSEX; TRIBUTYLTIN (TBT) POLLUTION. [P.C.]

REFERENCE

Waldock, M.J. (1994) Organometallic components in the aquatic environment. In: *Handbook of Ecotoxicology*, Vol. 2 (ed. P. Calow), pp. 106–129. Blackwell Scientific Publications, Oxford.

organophosphorus pesticides Organophosphorus pesticides form a very large group of AGROCHEMICALS, mostly insecticidal compounds, stemming from research by Lange and von Krueger (in 1932) that showed that ESTERS of monofluorophosphoric acid were very toxic. G. Schrader developed the first useful insecticide, parathion, in 1944:

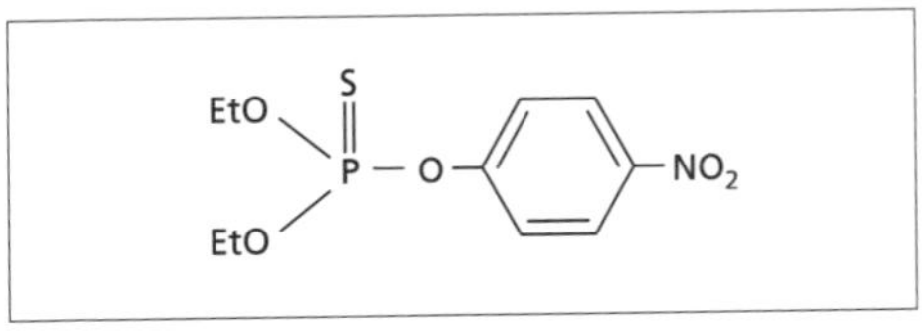

The success of this chemical stimulated agrochemical companies to look further at PHOSPHORUS-based pesticides. There are now about 250 such compounds in use in the world.

The compounds exhibit a very wide range of selectivity, stability and PERSISTENCE, and may be used as contact, soil-active and plant systemic INSECTICIDES (i.e. can move within plants). They are used in most of the major arable CROPS such as cotton, rice, maize, soybean and small-grain cereals, and horticultural crops such as soft fruit and vegetables. Products such as dichlorvos find important application as agents for control of ecto- and endoparasites in livestock.

Organophosphorus pesticides vary considerably in their TOXICITY to higher animals; the earliest analogues were extremely toxic, but now the LD_{50} (rat) ranges from 5 mg kg^{-1} body mass for parathion up to 5000 mg kg^{-1} for tetrachlorvinphos, and many of the earlier products have been withdrawn from use. The search for even more selective materials continues.

Organophosphorus insecticides are relatively unstable in biological systems compared with organochlorine insecticides, so are generally not persistent. However, varying sublethal effects have been noted in birds, fishes and some mammals, including humans. Teratogenic effects, with some indication of potential mutagenicity, has been noted for certain compounds, such as diazinon, and parathion, and possibly MALATHION.

Organophosphorus insecticides phosphorylate the hydroxyl group of acetylcholine hydrolase (acetylcholinesterase), which controls the functioning of acetylcholine, a chemical transmitter of nerve impulses in the insect nervous system, as well as in vertebrates. They also affect cholinergic neuromuscular functions in mammals. Death in mammals is mainly due to paralysis of the respiratory mechanism; it is not so certain what is the cause of insect death.

Organophosphorus insecticides can be classified according to their practical application.

1 Soluble in WATER, rapidly hydrolysed. Contact insecticides; for example mevinphos, tetrachlorvinphos.

2 Soluble in oils, moderate or high stability. Persistent contact

insecticides, moving into plant foliage but not systemic; for example diazinon, malathion, trichlorphon.

3 Able to enter plants and translocate; moderate to high stability. Systemic insecticides; for example dimethoate, phorate, disulfuron.

4 High vapour pressure. Fumigants; for example dichlorvos.

5 Effective in soils; for example bromophos, chlorfenvinphos. [K.D.]

Oriental faunal region The Oriental FAUNAL REGION essentially consists of tropical Asia and closely associated continental islands including Indonesia, Taiwan and the Philippines (see Fig. F1). However, by no means all of its boundaries are sharply defined, and the oriental fauna enters into complex transitions in several directions. The Himalayas form a natural northern boundary while in south-east Asia the so-called Wallace's Line, running between Borneo and Sulawasi, and Bali and Lombok, forms a relatively sharp boundary with the AUSTRALIAN FAUNAL REGION.

The freshwater fishes of the Oriental region form a rich and dominant assemblage, notable for a lack of archaic groups and for the dominance of cypriniforms. The amphibians and reptiles of the Oriental demonstrate both differences and similarities to their counterparts in the ETHIOPIAN FAUNAL REGION. For example, some notable differences are that the Orient lacks pipids and has pelobatid frogs, lacks pelomedusids and has emydine turtles, has gavials, and lacks or has only a few chameleons. In general, Africa has a greater diversity of lizards, the Orient of snakes, notably colubrids.

Oriental birds include almost all the widely distributed Old World tropical families that occur in the Ethiopian region, but in somewhat different proportions. The Orient has fewer representatives of some groups, for example shrikes, hoopoes, sunbirds and finches, and more of others, for example woodpeckers, babblers and corvids. The Orient also has many pheasants (cf. only one in Africa), many pigeons but, like Africa, only a moderate number of parrots. Of 66 families of birds, 53 are more or less widely distributed elsewhere, three are shared only or mainly with Africa, and five with the Australian region.

Oriental mammals have rather strong Ethiopian affinities, as shown for example by the apes and monkeys, lorisids, chevrotains, scaly anteaters, viverrids and one hyaena. Flying lemurs, tarsiers, hairy hedgehogs and tree shrews are endemic. [A.H.]

REFERENCE

Darlington, P.J. (1957) *Zoogeography: the Geographic Distribution of Animals*. John Wiley & Sons, New York.

Origin of life

The primeval Earth

The Earth is believed to be about 4.6 billion years old. The oldest fossil evidence of early life comes from sedimentary deposits in Australia and Africa that are 3.4–3.5 billion years old. These deposits are the first that were not subjected to extensive metamorphosis from the Earth's early geological activity (i.e. the first rocks that could possibly contain fossils) suggesting that life arose very soon after the Earth cooled.

The early ATMOSPHERE of the Earth was composed of gases such as AMMONIA, METHANE and HYDROGEN—gases that are toxic to most living things. In addition, there is believed to have been little free-standing WATER, no OXYGEN, intense solar and ULTRAVIOLET RADIATION, violent thunderstorms and volcanic activity on the early Earth.

Possible beginnings

The organic molecules necessary for life may have arisen by one of two processes: (i) extraterrestrial sources, or (ii) *in situ* formation. The composition of certain (chondritic) meteorites suggests that several of the precursor molecules necessary for the EVOLUTION of life may have formed early in the history of our solar system as a result of non-biological processes, and may have been introduced to Earth by falling meteorites. Despite such evidence, it is presently considered more likely that the organic molecules necessary for life formed on Earth.

Experimental evidence

The experimental synthesis of complex organic molecules from an inorganic 'soup' of molecules exposed to an oxygen-free atmosphere and energy has provided evidence for the *in situ* formation of life's building blocks. The famous experiments of Stanley Miller and Harold Urey in the mid-1950s attempted to re-create the conditions of the early Earth, and succeeded in abiotically synthesizing a number of AMINO ACIDS. Synthesis of other molecules, particularly purines, pyrimidines and sugars, has been somewhat less successful and requires that the inorganic soup be highly concentrated.

From non-living to living

The molecular and biochemical similarity of all living organisms indicates that all life developed from a single ancestral stock. Defining the point at which organic molecules become living entities can be based on three fundamental features that characterize all living things: organization, information and function. Thus the evolution of life requires the organization of informational molecules with functional molecules. In modern living organisms RNA acts primarily to transmit information between DNA (an information molecule) and PROTEINS (functional molecules). It is believed that ribosomes—which are composed of RNA and both contain information and can carry out function (i.e. catalyse certain reactions)—played a key role in the evolution of the first living organisms. [V.F.]

REFERENCES

Futuyma, D.J. (1979) *Evolutionary Biology*. Sinauer Associates, Sunderland, Mass.

Knox, B., Ladiges, P. & Evans, B. (eds) (1994) *Biology*. McGraw-Hill, Sydney.

---

orthogenesis From 1890 to 1940, a period sometimes called the 'eclipse' of DARWINISM, a number of theories of EVOLUTION competed to replace Darwinian ideas. One of these theories was formulated by biologists who were impressed by what they thought were recurring and parallel large-scale patterns of evolutionary change in many different lineages. Orthogeneticists believed that different lineages of species came into existence independently and evolved along predictable and parallel lines. The evolution of such lineages was marked by increasing ecological specialization and dependence on other species leading to the adoption of PARASITISM and, finally, self-imposed extinction. The 'new synthesis' restored Darwinian concepts to centre stage in evolutionary theory, but absorbed a number of concepts developed by orthogeneticists, including adaptive zones, ADAPTIVE RADIATIONS, taxonomic grades, the presumed secondarily simplified nature of parasitic organisms, and macroevolutionary (supraspecific) trends. *See also* MACROEVOLUTION. [D.R.B.]

oscillations Regular fluctuations through a fixed cycle above and below some mean value. *See also* POPULATION DYNAMICS. [V.F.]

osmium (Os) A rare metallic element (ATOMIC NUMBER = 76; relative ATOMIC MASS = 190.2; melting point = 3045°C; boiling point = 5027°C), occurring at a level of 0.005 $mg\,kg^{-1}$ in the Earth's CRUST. Its major sources, along with the other PLATINUM group metals, are the nickel-copper sulphide ores found in South Africa, Canada and the river sands of the Urals. The element is refined from the anode slimes produced during the electrolytic refinement of nickel. Industrial applications are limited although the extreme hardness of certain of its alloys is exploited in specialist applications. Osmium is the densest element known (22.57 $g\,cm^{-1}$) and possesses a characteristic smell due to its atmospheric oxidation to the volatile osmium tetroxide (OsO_4), which is a severe mucous-membrane irritant. [P.J.B.]

osmoregulation The mechanisms whereby an organism maintains relatively constant water content and salt concentrations inside its body. This process requires the exchange of salts and water between the extracellular fluid and the environment to compensate for obligatory losses and gains. These losses and gains depend on the osmotic GRADIENT across the body wall and its permeability, the surface-to-volume ratio of the organism, the intake of food and water, the evaporative losses associated with thermoregulation, and the production of urine and faeces.

The extracellular fluid in many animals is similar in composition to the primeval seas, which were probably the setting for the EVOLUTION of life. The ability of many animals to regulate the composition of their internal environment is closely correlated with their ability to occupy ecological environments that are at odds with the osmotic requirements of their tissues. Marine and terrestrial animals are faced with dehydration, whereas FRESHWATER animals must counteract hydration caused by uncontrolled uptake of water. Marine birds, reptiles and teleosts replace lost water by drinking seawater and actively removing salt through secretory epithelia in their nasal gland and gills, respectively. Freshwater fishes produce large volumes of urine and replace lost salts by active uptake. Birds and mammals conserve water by secreting hypertonic urine. Marine mammals cannot drink seawater and DESERT animals with no available drinking water must conserve limited supplies derived from metabolism and taken in with food. *See also* OSMOSIS. [D.J.R.]

osmosis Diffusion of a solvent through a semipermeable membrane into a more concentrated solution. The osmotic pressure is the pressure that must be applied to oppose the osmotic flow. The osmotic potential, Ψ_s, is the tendency of a solution or solid to attract solvent through a semi-permeable membrane, on a scale where pure water at sea level = 0, and with the sign convention that solutions are negative.

For solutions of a substance that does not dissociate $\Psi_s = -cRT$ where c is the concentration, R is the molar gas constant (*see* GAS LAWS) and T is the absolute TEMPERATURE. [J.G.]

osmotic pressure *See* OSMOSIS.

outbreaks

1 Outbreaks are (small) EPIDEMICS of infectious disease localized in time and space. The term is usually used to describe an increase in MORBIDITY and MORTALITY due to an endemic infection. Outbreaks may often occur by chance.

2 Can also refer to PEST species.

[G.F.M.]

outbreeding depression A decrease in FITNESS in offspring relative to their parents due to the crossing of CONSPECIFIC individuals that are too dissimilar. Two mechanisms have been proposed: local ADAPTATION and intrinsic coadaptation. Individuals from reproductively separated populations may become adapted to their local environments, and the disruption of those TRAITS through interbreeding could result in an individual that is poorly adapted to either environment. Intrinsic coadaptation includes such phenomena as chromosomal races, where different populations of the same species have different karyotypes. [A.R.H.]

outgroup A species or a higher monophyletic TAXON that is examined in the course of a phylogenetic study to determine which of two homologous characters may be inferred to be apomorphic; any group used to assess the evolutionary sequence or appearance of homologous characters in a phylogenetic analysis. One or several outgroups may be examined for each decision. The choice of the outgroup is constrained by the rule that it cannot contain any members which are part of the study group. The most important outgroup in any study is the sister group to the taxa being investigated. [D.R.B.]

outwelling The outflow of NUTRIENTS from an ESTUARY or SALT-MARSH system to shelf waters. Though it was earlier believed that the outwelling of dissolved nutrients and PARTICULATE ORGANIC MATTER from estuaries provided a principal nutrient source for shelf PRODUCTIVITY, recent estimates suggest that the importance of outwelled particulate organic matter to nearshore shelf PRIMARY PRODUCTIVITY is probably minimal. The outwelling of dissolved nutrients (especially NITROGEN) appears to be geographically variable, and thus its importance remains controversial. [V.F.]

ova Female GAMETES, adapted for supporting early development. *Cf.* SPERMATOZOA. *See also* ANISOGAMY, EVOLUTION OF; EGGS; FERTILIZATION. [P.C.]

overcompensation The phenomenon occurring when density-dependent MORTALITY leads to POPULATION SIZE overshooting an equilibrium following disturbance. In the longer term, overcompensation is associated with an oscillatory approach to equilibrium, sustained cycles around equilibrium or CHAOS. Overcompensation is often the ecological consequence of SCRAMBLE COMPETITION. Density-dependent mortality is detected by plotting the *k*-VALUE $k = \log_{10}$(initial density/final density) against $\log_{10}$(initial density); overcompensation occurs if the slope $b > 1$. *See also* POPULATION REGULATION. [R.H.S.]

overcrowding effect This term usually refers to DENSITY DEPENDENCE or INTRASPECIFIC COMPETITION, demonstrated by a reduction in the PER CAPITA rate of increase as POPULATION DENSITY increases. [R.H.S.]

overdispersed *See* DISPERSION.

overdispersion If a population is distributed randomly then samples from that population will follow a POISSON DISTRIBUTION and the MEAN and VARIANCE of the samples will be equal. If a population deviates from this distribution it is said to be either underdispersed—more ordered than random, giving a variance smaller than the mean of a SAMPLE — or overdispersed (or overdistributed) — more clumped than random, giving a variance greater than the mean. Therefore, overdispersion is, effectively, a synonym for AGGREGATION. The term is now used rather infrequently.

It is important to realize that a population that is overdispersed at one scale may be simultaneously underdispersed at another. An obvious example is a flock of birds: on a large scale individuals are clearly clumped but within a flock individuals may be almost equidistant. Sets of data using different sizes of sample unit may provide information about the pattern and scale of the clumping in an overdispersed population. Such techniques are called quadrat-variance methods.

One problem with the use of the term 'overdispersion' is that some authors have used it to describe the opposite type of distribution, where the variance is smaller than the mean (i.e. individuals are less clumped, or more ordered, than random). *See also* DISTRIBUTION. [C.D.]

overdominance *See* BALANCE OF SELECTIVE FORCES.

overfishing The practice of HARVESTING fish in quantities greater than can be replaced by RECRUITMENT to the population. *See also* FISHERIES, CONSERVATION AND MANAGEMENT. [P.C.]

overgrazing GRAZING by domestic or wild animals can provide a management tool both for habitat conservation and for economic gain. When the consumption rate of the GRAZER is matched by the primary production of the vegetation, or where the grazers feed only seasonally on a particular area, an equilibrium can be maintained. If, however, the population of herbivores exceeds the CARRYING CAPACITY of the system (i.e. if the food demand in the long term is greater than the primary productivity), then overgrazing results and the plant BIOMASS of the ecosystem goes into decline.

In addition to a general loss of biomass in such a situation, overgrazing will also result in compositional changes in the flora. Species of plants that are more attractive to the grazer will be removed first and other, less attractive species will increase in abundance as a result of reduced competition. The types of plant favoured in the early stages of overgrazing will be those that are unpalatable, either because of their chemical composition or as a result of structural features. Spiny plants may prove more difficult for some animals; dense, fibrous species may offer poorer rewards; rosette species with leaves flattened close to the soil surface may prove more difficult to graze. A grassland or range manager with experience of vegetation responses should be able to detect overgrazing in its early stages and reduce stock density to prevent undue damage.

The population structure of grazing-sensitive plants can provide information on past patterns of overgrazing. In the case of the Australian shrub, *Acacia burkitii*, both rabbits and sheep consume the juvenile individuals, so intense grazing results in reduced RECRUITMENT of young shrubs. Examination of the population structure of these shrubs in an area can provide evidence of recruitment gaps corresponding to past periods of overgrazing.

A fluctuating climate can exacerbate overgrazing problems since unexpected droughts and consequent reductions in primary productivity can lead to a pastoralist having larger herds than can be supported during a particular year. This has been the case in the Sahel region, south of the Sahara in Africa, where low rainfall in recent years has resulted in poor vegetation growth, overgrazing and the subsequent starvation of grazing animals (*see* DESERTIFICATION). [P.D.M.]

overgrowth competition COMPETITION for space on a two-dimensional surface shown, for example, by lichens and corals. One species (the competitive dominant) grows over another species and kills it. Overgrowth competition is an example of AMENSALISM. An excellent study, in which this type of competition was a major factor, is that of Buss and Jackson (1979) on CORAL REEFS off the north coast of Jamaica. [B.S.]

REFERENCE
Buss, L.W. & Jackson, J.B.C. (1979) Competitive networks: nontransitive competitive relationships in cryptic coral reef environments. *American Naturalist* **113**, 223–234.

overlapping generations The simultaneous occurrence in a single-species population of individuals of two or more different AGE CLASSES or life stages. Overlapping generations can arise in several ways.

1 By adults surviving after they reproduce, particularly if they reproduce again (ITEROPARITY; e.g. seasonal breeders like deer or trees) or reproduce continuously during the LIFESPAN of their offspring (e.g. murine rodents).

2 By variable length DORMANCY of one or more life stages (e.g. plants whose seeds may remain dormant for 1, 2 or more years, and some insects where, even among siblings, some individuals enter DIAPAUSE but others do not).

3 By asynchronous SEMELPARITY (e.g. BIENNIAL PLANTS where populations apparently contain two age classes: plants in their first year growing vegetatively, and plants in their second, reproductive, year).

COHORT LIFE TABLES can be constructed for populations with

overlapping generations only if the age of individuals is known or can be easily estimated. Otherwise STATIC LIFE TABLES are required.

Under most circumstances, an age-structured population in a constant environment converges to a STABLE AGE DISTRIBUTION. Some authors distinguish between overlapping and continuous generations, using these terms for seasonally and continuously breeding organisms, respectively. The distinction between overlapping and DISCRETE GENERATIONS is sometimes confused with the distinction between continuous and discrete time models. *See also* DEMOGRAPHY; POPULATION DYNAMICS. [A.J.D. & T.J.K.]

REFERENCE

Charlesworth, B. (1980) *Evolution in Age-Structured Populations*. Cambridge University Press, Cambridge.

overlapping iteroparity *See* OVERLAPPING GENERATIONS.

Overwintering

A generalized term describing the way in which an organism passes through winter. Low TEMPERATURE is injurious and potentially lethal to most forms of life. Survival in climatic regions with a winter season is achieved by one of three main options: MIGRATION, HIBERNATION, or FREEZE TOLERANCE/SUPERCOOLING.

Migration

Migration is movement to a warmer winter climate. For example, barnacle geese breed in the high Arctic in summer, taking advantage of the continuous daylight and limited numbers of predators, then fly south to overwinter in various sites in temperate Europe. Migration is more common in birds than mammals because of the relatively lower energy costs per unit distance required for flight compared to land travel for similar sized animals. A notable exception is the caribou, which migrates from the Arctic tundra in summer to overwinter in more southerly forests.

Hibernation

In the hibernation strategy, warm-blooded animals (endotherms), particularly small mammals, reduce their metabolic activity greatly and lower their body temperature to track AMBIENT (2–5°C), although selection of thermally buffered overwintering sites protects against subzero temperatures which would often be lethal. Hibernation is characterized by intermittent arousal, when the animal becomes active and the body temperature rises to 35–40°C. Periodic arousal may be an essential response to raise the body temperature away from the potentially lethal lower limit, or to replenish energy reserves. Hibernation also occurs in cold-blooded animals (ectotherms), which are unable to regulate their body temperature. Snakes (reptiles) and frogs (amphibians) make use of underground burrows and caves, although some frogs and toads show a limited freeze tolerance.

Freeze tolerance/supercooling

Many plant species and immobile or SEDENTARY ectotherms (insect eggs, larvae and pupae) have no alternative but to face up to the extremes of the winter climate. These organisms either tolerate the formation of ICE in extracellular areas of their bodies (freeze tolerance) or avoid freezing by supercooling, often overwintering under leaf litter and snow cover.

Overwintering state

Most organisms in the temperate, subpolar and polar regions overwinter in a distinct physiological state (DORMANCY, hibernation, DIAPAUSE), a specific stage of development or LIFE CYCLE (plant seed or corm, species-specific stage in insect METAMORPHOSIS), and often in a selected overwintering site. The vertical STRATIFICATION of temperature that occurs on calm, frosty nights through nocturnal radiant cooling has a marked influence on the temperatures experienced by insects in different overwintering sites (Fig. O6).

Environmental cues

Progressive changes in abiotic and BIOTIC FACTORS from summer through autumn to winter act as environmental triggers which induce specific physiological and biochemical adaptations associated with different overwintering strategies. Decreasing DAYLENGTH and water STRESS increases the concentration of abscisic acid in leaves, which inhibits growth and increases membrane permeability to water, an important feature in plant survival at subzero temperatures. Decreasing daylength induces the production of sexual morphs in aphids which lay the cold-hardy overwintering eggs. Short PHOTOPERIOD is also the dominant induction factor in insect diapause, leading to the accumulation of cryoprotectants such as glycerol, which is further enhanced by EXPOSURE to low temperatures.

Implications of overwintering survival

Overwintering survival is the major determinant of the year-to-year abundance and economic importance of some PEST insects. There is a strong negative correlation between winter temperatures and the timing of the spring migration of the aphid virus vector *Myzus persicae* (Fig. O7). Migration is earliest after the mildest winters, increasing the likelihood of a virus EPIDEMIC when aphids colonize young, susceptible plants (potato, sugarbeet) soon after germination. *See also* ACCLIMATION; CHILL COMA; FREEZE AVOIDANCE. [J.S.B.]

REFERENCES

Davenport, J. (1992) *Animal Life at Low Temperature*. Chapman & Hall, London.

Leather, S.R., Walters, K.F.A. & Bale, J.S. (1993) *Ecology of Insect Overwintering*. Cambridge University Press, Cambridge.

Marchand, P.J. (1987) *Life in the Cold: An Introduction to Winter Ecology*. University Press of New England, Hanover, N.H.

---

overwintering state *See* OVERWINTERING.

oviparity (egg-laying) The production of a haploid GAMETE (an egg) complete with sufficient NUTRIENTS to support the early GROWTH and development of the diploid offspring. DIPLOIDY is

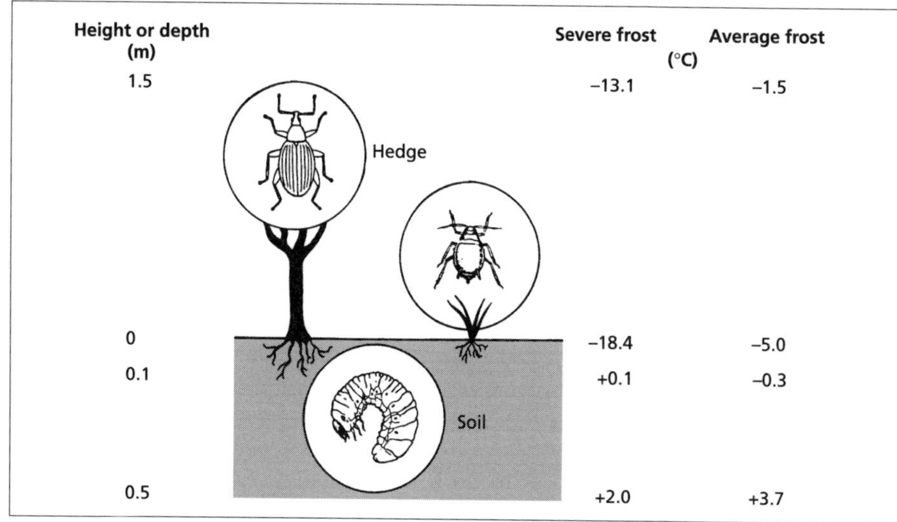

Fig. O6 Vertical stratification of temperature (°C) on nights with a severe and average frost. (After Bale, J.S. (1991) Insects at low temperature: a predictable relationship? *Functional Ecology* **5**, 291–298.)

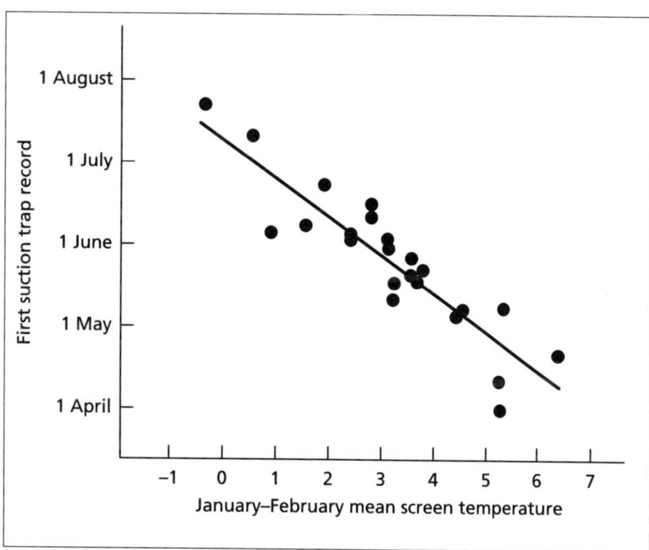

Fig. O7 Date of first record of *Myzus persicae* in the aphid suction trap at Rothamsted (1968–1991) in relation to the January–February mean temperature. (After Bale, J.S. (1991) Insects at low temperature: a predictable relationship? *Functional Ecology* **5**, 291–298.)

restored via fertilization by a sperm that is usually motile and contributes little or no nutrients to the embryo. One variation on this theme is for the meiosis of the egg to be triggered by fertilization. At least some development prior to hatching takes place after the egg has been laid, and hence external to the body of the female. Oviparity originally evolved as part of sexual reproduction, as an alternative to asexual forms of reproduction such as vegetative budding. It is present in virtually all phyla in the animal kingdom. It is not necessarily associated with sexual reproduction in all derived species; for example, asexual forms of reproduction such as PARTHENOGENESIS have evolved secondarily in many sexual lineages. [D.N.R.]

ovoviviparity (live-bearing) The production of a haploid GAMETE (an egg) complete with sufficient NUTRIENTS to support the early GROWTH and development of diploid offspring. EGGS are internally fertilized, retained within the female throughout development, then young are born live. There is no further transfer of nutrients from mother to young after FERTILIZATION, so the female functions as an egg retainer. Ovoviviparity is generally considered the primitive state in the evolution of live-bearing. It evolved independently in many invertebrate phyla, and has had multiple, independent origins in Chondrichthys, Osteichthys, Amphibia and Reptilia. A consequence of egg retention is that the female controls the environment of the young. This shift in the LIFE CYCLE appears to be associated with increased survival through the early stages of development and is often associated with the production of fewer, larger offspring. [D.N.R.]

ovule *See* POLLINATION.

oxic horizons *See* SOIL CLASSIFICATION.

oxidation–reduction Describing a reversible chemical reaction in which electrons are transferred from one molecule to another. Biologically these reactions are carried out by oxidoreductase enzymes, which include dehydrogenases and oxidases. Oxidases, however, transfer electrons or HYDROGEN to OXYGEN only, while dehydrogenase-mediated reactions are reversible. Dehydrogenases play a fundamental role in both aerobic and anaerobic RESPIRATION in removing hydrogen from activated sugars. In anaerobic respiration ALCOHOL and lactic dehydrogenase play an important role in transferring hydrogen from reduced nicotinamide adenine dinucleotide (NADH) to the anaerobic terminal hydrogen acceptors, lactic acid and ETHANOL, respectively. So essential is the role of dehydrogenases for maintaining living systems that assays of total dehydrogenase activity (e.g. using tetrazolium tetrachloride) can be used for estimating whether or not senescing or long-dormant tissues (e.g. seeds) are still alive. [R.M.M.C.]

Oxisols *See* SOIL CLASSIFICATION.

oxygen (O) A gaseous element (ATOMIC NUMBER=8; relative ATOMIC MASS=15.9994) and the most abundant element in the Earth's CRUST (49.2% by mass). In the free elemental state oxygen is a colourless, diatomic gas (dioxygen, O_2; melting point= −218.4°C; boiling point=−183.0°C.) and constitutes 20.9% by volume of the ATMOSPHERE. It is a powerful oxidant and is essential for combustion and in RESPIRATION of plants and animals to release energy. The non-equilibrium mixture of oxygen and oxidizable material on the Earth is maintained by release of oxygen through PHOTOSYNTHESIS. An important allotrope of oxygen is OZONE (O_3), which absorbs solar ULTRAVIOLET RADIATION and prevents it reaching the Earth's surface. [M.R.H.]

oxygen compensation depth *See* COMPENSATION DEPTH.

oxygen minimum layer A depth zone, usually below the THERMOCLINE, where DISSOLVED OXYGEN is minimal. This zone usually occurs between 400 m and 800 m depth and reflects the combination of minimal physical mixing of water at this depth and high RESPIRATION from biological activities (e.g. bacteria and ZOOPLANKTON tend to congregate here and feed on the accumulated PARTICULATE ORGANIC MATTER). [V.F.]

oxygen sag The fall and recovery in DISSOLVED OXYGEN downstream of a major source of organic effluent, such as a SEWAGE works, brewery or paper mill (Fig. O8). *See also* BIOCHEMICAL OXYGEN DEMAND. [P.C.]

ozone A gaseous allotrope of OXYGEN with formula O_3 and boiling point −112°C. It is a very powerful oxidizing agent, sparingly soluble in WATER, and is used industrially for water sterilization. In the STRATOSPHERE, ozone is formed by photolysis of molecular oxygen, and it absorbs ULTRAVIOLET RADIATION (UV-B) in the OZONE LAYER, thereby protecting life on Earth. Several synthetic chemicals, notably CFCS, perturb the chemistry of the stratosphere and can destroy ozone, leading to large increases in the transmission of UV-B radiation, particularly in polar regions in winter.

In the TROPOSPHERE, ozone occurs naturally, both from transport from the stratosphere, and from reactions of NITROGEN OXIDES (NO_x). Background concentrations in the Northern hemisphere have doubled in the past century as a result of pollutant EMISSIONS, rising from 15 ppb by volume (ppbv) to around 30 ppbv. Ozone is a minor GREENHOUSE GAS, but is important in controlling the rates of oxidation and removal of many other TRACE GASES in the ATMOSPHERE. Large concentrations, over 50 ppbv, are formed by reactions of sunlight with pollutant nitrogen oxides and HYDROCARBONS. Ozone is a major component of PHOTOCHEMICAL SMOG, and is toxic to plants. For long-term EXPOSURE, some plants show effects at concentrations as low as 40 ppbv, and visible injury to leaves of sensitive species can be observed after several hours' exposure to 60 ppbv. Ozone is also toxic to animals. The UK THRESHOLD LIMIT VALUE (TLV) for people at work is 100 ppbv over 7 h. *See also* OZONE HOLE. [J.N.C.]

ozone hole Thinning of stratospheric OZONE (O_3) over the Antarctic is a climatological feature, but since the late 1970s severe depletion has occurred each year in September and October, forming the ozone hole (Farman *et al.* 1985). Ozone depletion is most severe between altitudes of 16 and 22 km (Solomon 1990) where the Antarctic lower STRATOSPHERE is extremely cold. On the surfaces of frozen cloud particles, chemical reactions take place which transfer CHLORINE from inactive to active species. Ozone is then catalytically destroyed, predominantly by the following cycle (Molina & Molina 1987):

$$ClO + ClO + M \rightarrow Cl_2O_2 + M$$

$$Cl_2O_2 + h\nu + M \rightarrow 2Cl + O_2 + M$$

$$2Cl + 2O_3 \rightarrow 2ClO + 2O_2$$

where M is any molecule and $h\nu$ is a PHOTON. The chlorine is largely produced from anthropogenic CFCS.

Plate 11 (facing p. 398) shows the 1996 ozone hole from satellite data (method described by Stolarski 1993). In the pink and grey regions the October total ozone columns have been reduced by more than 50% since the mid-1970s and this feature is expected to occur annually for the next 50 years. Only limited ozone depletion has so far been measured in the ARCTIC (Stolarski *et al.* 1991) but conditions may change allowing an Arctic ozone hole to form in the future (Austin *et al.* 1992). *See also* STRATOSPHERIC CHEMISTRY; TROPOSPHERIC CHEMISTRY; ULTRAVIOLET RADIATION. [J.A.]

REFERENCES

Austin, J., Butchart, N. & Shine, K.P. (1992) Possibility of an Arctic ozone hole in a doubled-CO_2 climate. *Nature* **360**, 221–225.

Farman, J.C., Gardiner, B.G. & Shanklin, J.D. (1985) Large losses of total

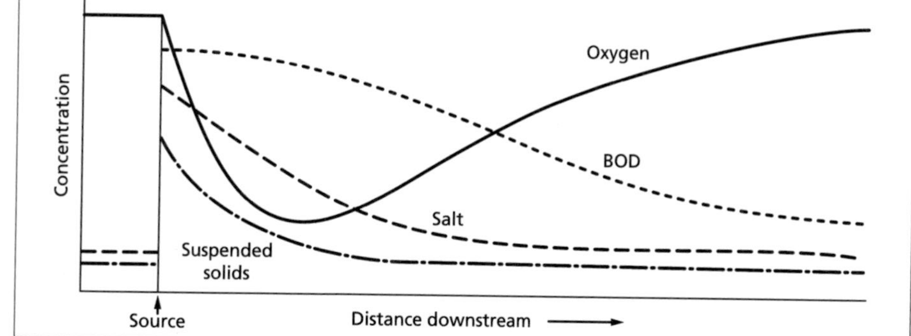

Fig. O8 Changes in water quality typically found in a river below the discharge of an organic effluent. BOD, biochemical oxygen demand. (After Hynes, H.B.N. (1960) *The Biology of Polluted Waters*. Liverpool University Press, Liverpool.)

ozone in Antarctica reveal seasonal ClO_x/NO_x interaction. *Nature* **315**, 207–210.

Molina, L.T. & Molina, M.J. (1987) Production of Cl_2O_2 from the self-reaction of the ClO radical. *Journal of Physical Chemistry* **91**, 433–436.

Solomon, S. (1990) Progress towards a quantitative understanding of Antarctic ozone depletion. *Nature* **347**, 347–354.

Stolarski, R.S. (1993) *Monitoring Stratospheric Ozone from Space. The Role of the Stratosphere in Global Change.* NATO ASO 18, pp. 319–346. Springer-Verlag, Berlin.

Stolarski, R.S., Bloomfield, P. & McPeters, R.D. (1991) Total ozone trends deduced from Nimbus 7 TOMS data. *Geophysical Research Letters* **18**, 1015–1018.

ozone layer Approximately 90% of atmospheric OZONE (O_3) occurs in the ozone layer, in the STRATOSPHERE 10–50 km above the Earth's surface. Ozone absorbs ULTRAVIOLET RADIATION, thereby protecting the BIOSPHERE from its harmful effects. It is formed naturally by the action of sunlight on OXYGEN molecules (Chapman 1930). The sunlight also heats the stratosphere giving rise to an increase in TEMPERATURE with height. The temperature structure of the stratosphere determines atmospheric motions, which transport ozone from the region of high production in the tropics to high latitudes. The highest ozone columns are found in high northern latitudes in late winter when ozone production is small (Dobson *et al.* 1929).

Ozone is chemically destroyed by a number of anthropogenic and natural constituents, by chemical reactions primarily of the form:

$$X + O_3 \rightarrow XO + O_2$$

$$XO + O \rightarrow X + O_2$$

The RADICAL X may be hydroxyl (OH) (Bates & Nicolet 1950), nitric oxide (NO) (Crutzen 1970), CHLORINE (Cl) (Molina & Rowland 1974) or BROMINE (Br) (Wofsy *et al.* 1975). The above process is catalytic, enabling several thousand ozone molecules to be destroyed for each molecule X. Cl and Br are largely anthropogenic from fluorocarbons and brominated HYDROCARBONS. *See also* CFCS; HALOCARBONS; OZONE HOLE; STRATOSPHERIC CHEMISTRY; TROPOSPHERIC CHEMISTRY. [J.A.]

REFERENCES

Bates, D.R. & Nicolet, M. (1950) The photochemistry of atmospheric water vapour. *Journal of Geophysical Research* **55**, 301–327.

Chapman, S. (1930) A theory of upper atmosphere ozone. *Memoirs of the Royal Meteorological Society* **3**, 103–125.

Crutzen, P.J. (1970) The influence of nitrogen oxide on the atmospheric ozone content. *Quarterly Journal of the Royal Meteorological Society* **96**, 320–325.

Dobson, G.M.B., Harrison, D.N. & Lawrence, J. (1929) Measurements of the amount of ozone in the Earth's atmosphere and its relation to other geophysical conditions: Part III. *Proceedings of the Royal Society of London, Series A* **122**, 456–489.

Molina, M.J. & Rowland, F.S. (1974) Stratospheric sink for chlorofluoromethanes: chlorine atom catalyzed destruction of ozone. *Nature* **249**, 810–812.

Wofsy, S.C., McElroy, M.B. & Yung, Y.L. (1975) The chemistry of atmospheric bromine. *Geophysical Research Letters* **2**, 215–218.

P

***P*:*B* ratio** The rate of energy or matter accumulation per unit BIOMASS per unit time; i.e. the ratio of PRODUCTIVITY to biomass. Because biomass may increase over the period of measurement of net primary productivity the mean or average biomass for the period is used to calculate $P:B$. In early successional stages plant biomass and net primary production may increase in parallel; however, a point will be reached where further increase in biomass will not result in proportionate increases in net production, and may eventually lead to a decrease in net production. The same changes are seen at other trophic levels and with population growth, when food supply is limited.

Within higher plants, increase in biomass per unit ground area will increase the amount of energy or matter lost in RESPIRATION, but once the vegetation has reached a size where virtually all the available light is intercepted, PHOTOSYNTHESIS cannot increase further. Additional growth will simply increase the proportion of the plant which consumes energy, but cannot increase the input. In a mature ECOSYSTEM, where B is approximately constant because production of new biomass balances losses through death, $P:B$ ratios indicate the rate of TURNOVER of the biomass. A COMMUNITY or population may have a low biomass because of low primary production; alternatively the low biomass may indicate a rapid turnover and high primary production, as in the case of some PHYTOPLANKTON communities. The average $P:B$ ratio of the phytoplankton communities of the open OCEAN is 1000 times greater than that of TROPICAL RAIN FOREST (see Table N1). Interpreting the causes of variation in $P:B$ therefore requires simultaneous measurement of respiration.

Odum (1969) suggested that more energy in a complex highly mature system is tied up in maintenance of order and less is used for production of new material, thus $P:B$ ratios decrease as ecosystems mature through the processes of SUCCESSION (see above). This implies that agricultural practices should be directed specifically to exploiting immature ecosystems where there is greatest potential for high CROP yields rather than more mature ones. Even multistorey tropical forests support a relatively low density of humans relying on hunting compared with the numbers that can be supported by AGRICULTURE in temperate GRASSLANDS, which have a far higher $P:B$ ratio. *See also* ECOLOGICAL EFFICIENCY; ECOLOGICAL ENERGETICS. [S.P.L. & M.H.]

REFERENCE

Odum, E.P. (1969) The strategy of ecosystem development. *Science* **164**, 262–270.

***P*:*R* ratio** The ratio of PRODUCTION rates (P), which represent the 'benefits' of energy transfer in terms of formation of new somatic and reproductive tissues, to RESPIRATION (R), which represents the energy 'costs' of metabolism and activity. This is an ecological cost/benefit statistic indicating how efficiently a particular trophic unit is able to accumulate BIOMASS from the energy it assimilates from the TROPHIC LEVEL below.

The costs of maintenance metabolism are influenced strongly by the environment and how an organism is adapted to it. Temperature and activity have a very strong influence on metabolic rate. Warm-blooded animals increase their metabolic rate to generate heat at low temperatures. For most mammals, peak activity requires expenditure of about eight times as much energy as resting. Rapid flight in birds requires 10 to 16 times as much energy as base metabolism.

$P:R$ ratios can be conveniently summarized in graphical form, as in Fig. P1, which shows that non-social insects are the most efficient in accumulating new biomass while insectivorous mammals, which are both endothermic and frenetically active, have the highest costs and so are least efficient. *See also* ECOLOGICAL EFFICIENCY; ECOLOGICAL ENERGETICS; PRODUCTION EFFICIENCY. [M.H.]

REFERENCES

Begon, M., Harper, J.L. & Townsend, C.R. (1996) *Ecology: Individuals, Populations and Communities*, 3rd edn. Blackwell Science, Oxford.

Humphreys, W.F. (1979) Production and respiration in animal populations. *Journal of Animal Ecology* **48**, 427–454.

pachycaul Thick-stemmed plant. *See also* ALPINE ZONE. [P.C.]

Pacific Ocean The oldest, deepest and largest of the major ocean BASINS. The Pacific, so named by Ferdinand Magellan (1480–1521), covers an area of 1.7×10^8 km² or 1.8×10^8 km² including its marginal seas (i.e. more than one-third of the Earth's surface). The volume and average depth of the Pacific are 7.0×10^8 km³ and 4188 m, respectively, or 7.1×10^8 km³ and 3940 m, if its marginal seas are included. The greatest depth of the Pacific is 11 022 m, in the Mariana Trench. The most striking features of the Pacific Ocean TOPOGRAPHY are the great island arcs that lie along its northern and western sides (e.g. Aleutian, Japanese and Philippine archipelagos), many of which are active volcanoes. The Pacific is surrounded by young mountain ranges, particularly along its eastern side, which inhibit the inflow of sediment and freshwater from land. Approximately 36.2% of the Pacific Ocean floor is covered by calcareous ooze, 14.7% by siliceous ooze, and 49.1% by red clay. [V.F.]

paedomorphosis The expression of a TRAIT as an adult with a morphology reminiscent of an embryonic, or juvenile stage, in the

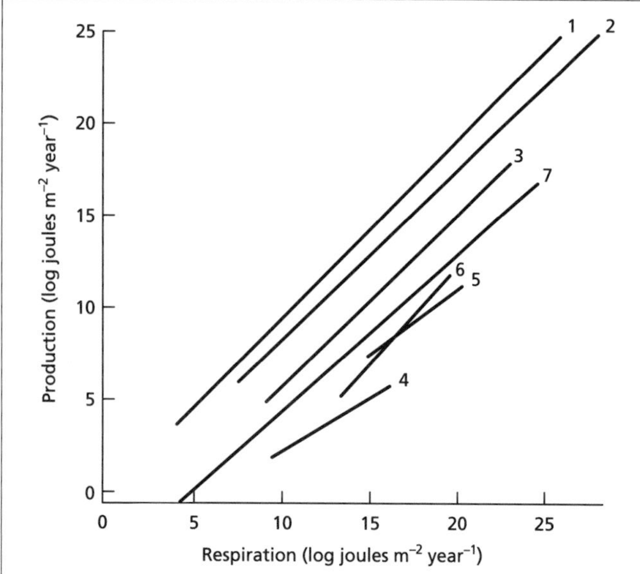

Fig. P1 Graphs of animal production against animal respiration in seven groups of animal: 1, non-social insects; 2, invertebrates other than insects; 3, social insects and fishes (not significantly different from each other); 4, insectivores; 5, birds; 6, small mammals; and 7, mammals other than insectivores and other small mammals. (After Humphreys, W.F. (1979) Production and respiration in animal populations. *Journal of Animal Ecology* **48**, 427–454.)

ancestral ONTOGENY. Such a phenomenon of HETEROCHRONY can result from a variety of underlying processes, particularly NEOTENY and PROGENESIS. Such a retention of ancestral juvenile characters in the descendant's ontogeny contradicts Haeckelian theories of RECAPITULATION; hence, the phenomenon was extensively used by W. Garstang in his influential critique of the biogenetic law. A much quoted example of such a phenomenon is the classical theory, dating from the early 20th century, claiming that many morphological traits in humans are characteristic of young apes. Hence humans would be paedomorphic apes, through a process of neoteny. The hypothesis would be supported by such paedomorphic features as lack of body hair, reduced chin outgrowth, etc. This extension of the juvenile stage in the human ontogeny would also be associated with a prolonged learning period and the possibility for greater intellectual development as a consequence. [P.A.]

REFERENCE

Garstang, W. (1922) The theory of recapitulation: a critical restatement of the biogenetic law. *Zoological Journal of the Linnean Society* **35**, 81–101.

PAH *See* POLYCYCLIC AROMATIC HYDROCARBONS.

PAH pollution The introduction by humans, directly or indirectly, of POLYCYCLIC AROMATIC HYDROCARBONS (PAH) to the environment resulting in deleterious effects such as: hazards to human health (e.g. some PAH are carcinogenic, some are very acutely toxic); hindrance of activities such as fishing (e.g. fish may develop lesions or other toxic effects); impairment of the quality for use of environmental resources (e.g. PAH-polluted water would not be suitable for drinking); and reduction in amenities such as recreation (e.g. tourists may be deterred from using beaches contaminated with PAH-containing oil and grease). PAH are produced naturally by microorganisms and plants, by natural burning of forests and prairies, and by volcanic activity. However, the main sources of PAH to the environment are from anthropogenic activities such as:

- combustion of wood and FOSSIL FUELS;
- sludge EFFLUENTS or OIL SPILLS;
- coke production in the IRON and steel industry;
- catalytic cracking in the PETROLEUM industry;
- carbon black;
- coal-tar pitch;
- asphalt hot road mix processes;
- heating and power generation;
- controlled refuse INCINERATION.

PAH do not generally occur as single substances but in combination, and the profile of PAH can provide information on the sources of CONTAMINATION. Photo-oxidation is the major DECOMPOSITION process by which PAH are degraded, BIODEGRADATION by microorganisms being a secondary reduction mechanism. Biodegradation is primarily an aerobic process, and PAH become increasingly refractory under reducing conditions (e.g. when buried in sediments). PAH released to the ATMOSPHERE become associated with AEROSOLS of various sizes and types. PAH enter the OCEAN mainly from marine transportation and river and urban RUN-OFF, and because of their hydrophobic nature, become rapidly associated with particulate material and concentrated in sediments. Decomposition of PAH in aquatic environments is influenced by oxygen concentration, temperature, salinity, hydrodynamics, PAH concentration, organic matter concentration, microbial composition and BIOTURBATION by infaunal invertebrates. *See also* BENZENE; MOTOR VEHICLE POLLUTION. [V.F.]

Palaearctic faunal region The Palaearctic vertebrate fauna occupies a region that essentially corresponds to Eurasia north of the tropics (see Fig. F1). It is much less rich than the more tropical Ethiopian and Oriental faunas. In terms of existing relationships, it is a transitional fauna, a complex mixture of Old World tropical and North American groups, the former most numerous southwards, the latter northwards, with a few important endemic groups of its own.

The freshwater fishes are predominantly cyprinids. Palaearctic amphibians comprise salamanders, frogs of three widely distributed dominant genera (*Rana, Bufo, Hyla*) and more or less localized representatives of four other families of frogs with Oriental and/or North American relationships. Palaearctic reptiles are few, with most being related at the generic or specific level to the faunas of the Old World tropics or North America. Those that are distributed widely include emydine turtles, skinks (Scincidae), lacertids, anguids, many colubrids, true vipers and, in Asia, pit vipers.

Palaearctic birds include representatives of 53 families, not many fewer than occur in Africa or the Oriental region, but some of them are only marginal or local in the Palaearctic. Only about 37 of the families are more or less widely distributed in the region, and some of them are mostly migratory there. Most of the widely distributed families are also widely distributed elsewhere, but loons and creepers (as well as auks, which are seabirds) are Holarctic; grouse and waxwings are additional Holarctic subfamilies, and hedge sparrows are exclusively Palaearctic.

Palaearctic mammals are only moderately diverse; endemics include the panda subfamily (south-west China), mole rat and the dormouse subfamily. Most of the genera are shared with the Old World (especially Oriental) tropics or with North America. The North American relationships are very strong in the north. *See also* ETHIOPIAN FAUNAL REGION; HOLARCTIC REALM; ORIENTAL FAUNAL REGION. [A.H.]

REFERENCE

Darlington, P.J. (1957) *Zoogeography: the Geographic Distribution of Animals*. John Wiley & Sons, New York.

Palaeobotany

The science of past plant life, as seen in the form of fossils preserved in sedimentary rocks. It includes the study of those aspects of past plant life which can be seen in living forms (their anatomy, their reproductive biology, their ecology, their geographical distribution, their chemical composition), when the necessary evidence is present in fossils.

Palaeopalynology

The study of fossil spores and pollen grains (or at least the resistant exine which forms their outer coat) is designated palaeopalynology, and is usually regarded as a separate field of study. Such microscopic fossils are referred to as microfossils in contradistinction to the macrofossils (or megafossils) of whole leaves and stems. The study of pollen and spores from PEAT and lake deposits of the Quaternary period, which has thrown much light on the response of VEGETATION to climatic change through that time, is sometimes referred to as 'pollen analysis'.

Fossilization processes

Plants which live on the land stand relatively little chance of fossilization, since the organic material of which they are formed is easily degraded by microorganisms. This contrasts with many marine animals with hard parts (shells and bones), which are more likely to survive as fossils in environments where they may become buried in SEDIMENT, and so get into the 'fossil record'.

Land plants need a series of rather unusual circumstances for them to become preserved in the FOSSIL RECORD. Only plants growing in a setting where they may become buried in sediment more or less where they grew (so-called AUTOCHTHONOUS preservation — e.g. plants growing on a delta surface) stand a fair chance of becoming preserved in that way. Such *in situ* preservation can also include the rather rare circumstances in which terrestrial plants have become buried in a volcanic ash fall, or as at Rhynie (a Lower Devonian silicified peat deposit in Aberdeenshire, Scotland), in other extruded volcanic material. Terrestrial plants growing well away from any wetland environment will need to be transported to a setting where they may become buried rapidly, before DECOMPOSITION occurs. This so-called allochthonous preservation may be quite remote from the original site in which they were growing.

The most informative type of plant fossil is one in which the internal anatomy is preserved by infiltration of mineral matter soon after its burial. This rather rare form of fossilization can preserve the size, shape and arrangement of the cells in the plant tissue, much as it was in life. An example of this type of preservation is seen in coal balls, formed of calcium and magnesium carbonate, which are present in some Carboniferous coal seams and which have given detailed knowledge of many of the plants of that period. The early land plants fossilized in the Lower Devonian Rhynie Chert are similarly preserved in silica derived from volcanic activity.

More commonly, plants are simply buried in CLAY or SILT, and collapse under the load of overlying sediment as they undergo partial BIODEGRADATION. Such a compression fossil (adpression fossil of some authors) is really a thin film of coal representing the remains of the plant's tissue, showing only the outline of its shape, and some features such as the positions of veins in a leaf. The cuticle sometimes survives remarkably unaltered in such a fossil, and may be prepared by oxidizing away the rest of the coaly material. A cuticle preparation may show the outlines of epidermal cells and stomata, which can be of value in establishing the identity of the plant.

Sometimes all remains of the original plant tissue have been entirely removed, and only an impression of the organ remains on the rock matrix. There are many records of angiosperm leaves preserved only in this way, on fine-grained lake muds or volcanic ash. Sometimes hollow stems or tough seed-coats may become infilled with sediment, leaving a 'natural cast' of the cavity when the plant structure degrades or is weathered away.

Fossil history of plants

Figure P2 shows a simplified picture of the fossil history of terrestrial plant life through time. Each of the shaded areas represents the changing fortunes of the major plant groups as we see them in the fossil record. The width of each shaded bank gives a somewhat subjective measure of both the ecological abundance and the diversity (number of taxa) of the major groups shown. Through the course of time, many groups (such as the progymnosperms, the seed-ferns and the bennettites) have become extinct, while others such as the living horsetails (*Equisetum* spp.) have a fossil history going back to the Triassic (over 200 million years ago). This time-span of *Equisetum*, together with that of the gymnosperm genus *Ginkgo*, which has a similarly long fossil record, illustrates a remarkable degree of STASIS in vascular plant EVOLUTION in those groups.

The vascular plants first appear in the fossil record in late Silurian times (over 400 million years ago). There is evidence from fossil spores that plants which were probably adapted to terrestrial life in some form may have pre-dated the earliest macrofossil remains by some tens of millions of years. Certainly the land surface must have had some cover of photosynthetic plants (algae, cyanobacteria) in

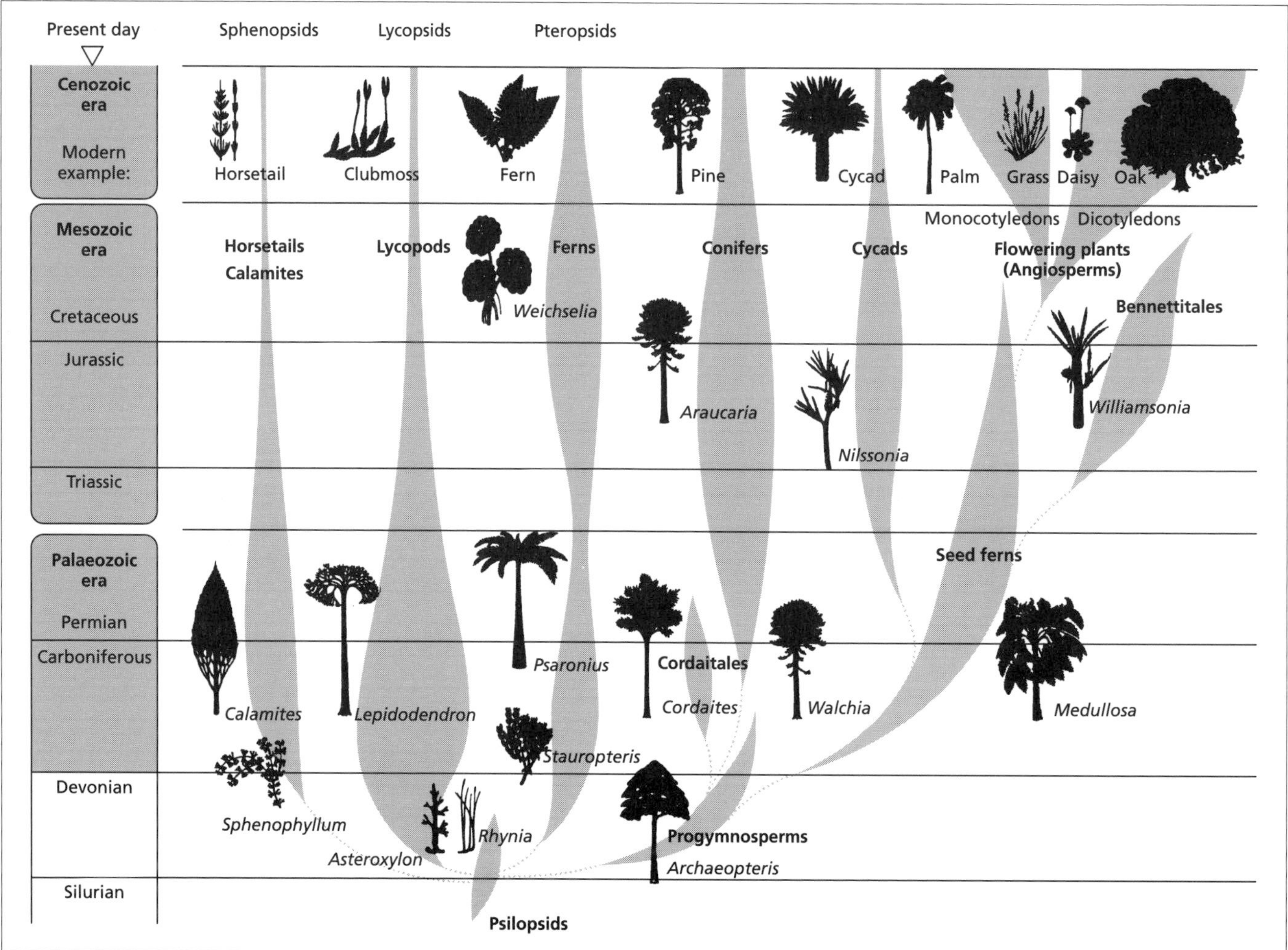

Fig. P2 Family tree of higher plants. (After Chaloner, W.G. & MacDonald, P. (1980) *Plants Invade the Land*, pp. 1–16. Published by Her Majesty's Stationery Office, for the Royal Scottish Museum, Edinburgh; copyright Trustees of the National Museums of Scotland.)

suitably moist habitats, long pre-dating the vascular plant 'COLONIZATION of the land' (see Chaloner & MacDonald 1980; Selden & Edwards 1989).

The Early Devonian saw a relatively rapid diversification of the homosporous vascular plants ('pteridophytes') into at least four classes (Psilopsida, Sphenopsida, Lycopsida and Pteropsida; Fig. P2). The heterosporous type of LIFE CYCLE, which is less dependent on a moist habitat than homospory, had evolved independently in three of these classes by early Carboniferous times, and the earliest seeds are known from about the end of the Devonian period. Several plant groups had developed secondary thickening of the conducting tissue and cortex before the end of the Devonian, and the Progymnospermopsida (notably the genus *Callixyon)* were forming large trees up to 1 m in diameter by that time. It is interesting to note that those FOREST-forming trees still had a heterosporous life cycle, and so were not seed plants. They were therefore exhibiting a spore-based reproductive biology unknown in any forest trees of the present day.

The Carboniferous coal-swamps of the European/American area were dominated by spore-bearing plants (especially lycopods, ferns and calamites) but the seed plants, in the form of seed-ferns and cordaites, were also important constituents of the vegetation. By the end of the Permian period, seed-bearing plants (gymnosperms) had expanded and diversified with a corresponding reduction and extinction of many of the spore-bearing plants. The conifers, cycads and the extinct cycad-like group, the bennettites, became the major components of the arborescent vegetation of the MESOZOIC era. By Cretaceous times floras of superficially similar aspect, composed of conifers, cycads, bennettites, ferns and horsetails, extended into high palaeolatitudes in both hemispheres. Floras ranging in age from Late Triassic to Early Cretaceous, of similar composition at least at the family level, are known from Greenland, across Eurasia, including India and Australia, and into the Antarctic peninsula. Although the seeming similarity of these floras may be in part a product of incomplete knowledge, it is inescapable that the Mesozoic terrestrial vegetation developed under a more uniform global CLIMATE, with a far more gradual climatic gradient between the poles and tropics than that seen at present.

The flowering plants (angiosperms) appear in the fossil record about 100 million years ago, in the mid-Cretaceous, and become the dominant feature of floras in all latitudes by the end of the Cretaceous. Their diversity of form and habitat quickly became immense, exceeding that of all pre-existing groups of vascular plants. All other land plant groups underwent a reduction in diversity and ecological status following the angiosperms' appearance, with only the ferns and conifers retaining major representation through to the present day. This swift rise of the angiosperms constitutes the most striking 'evolutionary explosion' in the history of plant life. A brief account of the origin and diversification of the angiosperms is given in Crane *et al.* (1995), and of the ecological impact of the rise of the angiosperms in Crane (1987). [W.G.C.]

REFERENCES

Chaloner, W.G. & MacDonald, P. (1980) *Plants Invade the Land*, pp. 1–16. Published by Her Majesty's Stationery Office, for the Royal Scottish Museum, Edinburgh.

Crane, P.R. (1987) Vegetational consequences of angiosperm diversification. In: *The Origins of Angiosperms and their Biological Consequences* (eds E.M. Friis, W.G. Chaloner & P.R. Crane), pp. 107–144. Cambridge University Press, Cambridge.

Crane, P.R., Friis, E.M. & Pedersen, K.R. (1995) The origin and early diversification of angiosperms. *Nature* **374**, 27–33.

Seldon, P.A. & Edwards, D. (1989) Colonisation of the land. In: *Evolution and the Fossil Record* (eds. K.C. Allen & D.E.G. Briggs), pp. 122–152. Belhaven Press, London.

Palaeoclimatic reconstruction

Palaeoclimatic reconstruction is normally based on proxy evidence for past CLIMATE CHANGE rather than on instrumental readings, largely because reliable instrumental measurements are only available from the earlier part of the 18th century onwards and then only at a very limited number of locations worldwide until the 20th century (Lamb 1995). Evidence for past climate change may be gleaned from historical sources but, generally, such sources are at best patchy with respect to both spatial and temporal distribution (Frenzel 1992). Most long-term palaeoclimatic reconstructions are, therefore, based on indirect or proxy evidence that may be biological, isotopic, physical or geomorphological (Lowe & Walker 1997).

Biological evidence

Macrofossil data

Biological evidence for past climate change is contained mainly in the fossil record. Both macro- and microfossils of animals and plants have been used to track and quantify palaeoclimate change through time. For example, severe climatic deterioration at the end of the Younger Dryas stadial in Ireland led to the extinction of the giant Irish deer (*Megaloceros giganteus*), the remains of which (usually male and with antlers) are frequently found at the transition from Allerød to Younger Dryas sediments (*c.* 10 600 cal. BC (radiocarbon-calibrated (sidereal) years BC)). Extinction can be attributed with confidence to a severe climatic downturn for which there is abundant supporting evidence. In this instance, it is possible to exclude, with a high degree of certainty, a human involvement since there is no evidence for the presence of Palaeolithic peoples in Ireland (Barnosky 1986). On the other hand, climatic amelioration, such as occurred in mid-Holocene times (*c.* 5500–4000 cal. BC), facilitated the well-documented northern extension of the RANGE of certain thermophilous plants and animals in Europe, for example hazel (*Corylus*) and the freshwater tortoise (*Emys orbicularis*).

Mires, and especially raised BOGS, contain a record in their peats of past MIRE vegetation, HYDROLOGY and rates of HUMIFICATION, which can be linked, in varying degrees, to past climate change. A 260-year periodicity in shifts from wet to dry cycles in the climate over the past *c.* 6000 years has been deduced on the basis of changing relative importance of hummocks and pools as preserved in the STRATIGRAPHY of raised bogs in Denmark (Aaby 1976), and, more recently, the potential of PEAT macrofossils (changes in the constituent species and isotopic composition, e.g. $^{2}D:^{1}H$ ratio) for reconstructing late Holocene climate has been explored with some success (Dubois & Ferguson 1985; Barber *et al.* 1994). Timbers preserved in peat provide another form of proxy evidence for postglacial climate change. It has been convincingly argued, for instance, that consistently narrow rings in dendrochronologically dated fossil oak (*Quercus*) timbers from Ireland, England and Germany (also bristlecone pine (*Pinus aristata*) from western USA) are attributable to wide-scale climatic deterioration caused by volcanic eruptions that injected large quantities of volcanic dust into the STRATOSPHERE (Baillie 1995).

Pollen data

Fossil pollen data, in the form of pollen diagrams, are an important source of evidence of past vegetation distribution patterns which, to varying degrees, are under the control of climate. Several pollen records from south-eastern France, extending back to the penultimate GLACIATION (*c.* 180 000 BP), provide a record of climate change in terrestrial habitats situated in mid-northern latitudes. More recently, a record extending back to the Holsteinian interglacial has been reported (Table P1; see also Reille & Beaulieu 1995).

Variations in selected climatic parameters, including seasonal changes in TEMPERATURE and PRECIPITATION, have been reconstructed on the basis of these fossil pollen data, to which climatic characteristics have been attached based on comparison with modern pollen samples from Eurasia and North Africa, the climate controls on which are known (Pons *et al.* 1992). At higher latitudes, long continuous terrestrial records do not appear to have survived the glaciations, but interglacial and interstadial deposits, normally intercalated in drift, not only facilitate the chronostratigraphical ordering of glacial deposits and hence glacial events but also provide snapshots of vegetation and hence climate conditions in particular intervals at these latitudes. Comparison with terrestrial records from lower latitudes enables important conclusions to be drawn relating to thermal and other gradients in earlier times (Behre 1989).

Pollen diagrams that provide a record of long-term change have also been constructed from ocean cores. Those from the north-western coast of Africa, for instance, have provided important evidence of shifts in the main vegetation belts in north-west Africa during interglacial/glacial cycles. This evidence complements that

Table P1 Marine oxygen isotope stages, ages in thousands of years before present, non-calibrated years (BP), and the terrestrial equivalent in various parts of the Northern hemisphere during the Late and part of the Middle Quaternary. Int. = interstadial; St. = stadial. (After Šibrava, V., Bowen, D.Q. & Richmond, G.M. (eds) (1986) *Quaternary Glaciations in the Northern Hemisphere*. Quaternary Science Reviews, Vol. 5. Pergamon Press, Oxford.)

Oxygen isotope stages (OIS)	Age (ka BP*)	Glacial/interglacial periods	Terrestrial equivalents of marine OIS		
			Great Britain	Central Europe†	North America and South America
1	0–12	Postglacial and part of Late-glacial	Holocene and part of Late-glacial	Holocene and part of Late-glacial	Holocene and part of Late-glacial
2 3 4	12–24 24–59 59–71	Late and middle pleniglacial including interstadial periods	Late and Middle Devensian	Late and Middle Weichselian (Upper and Middle Würm)	Late, Middle and Early Wisconsin
5a–5d	71–114	Early pleniglacial with well-developed stadials and interstadials in south-east France (see column, Central Europe)	Early Devensian. Includes the interstadials: Upton Warren, Chelford	Early Weichselian. Includes: St Germain II Int., Melisey II St., St Germain I Int., Melisey I St.	Eowisconsin
5e	114–133	Interglacial	Ipswichian	Eemian (Würm/Riss)	Sangamoan
6	133–186	Pleniglacial	Wolstonian‡	Saale‡ (Riss)	Illinoian
7	186–245	Interglacial	Hoxnian‡	Holsteinian‡ (Riss/Mindel)	Pre-Illinoian

* Almost 2000 years should be added to the conventional ^{14}C years BP dates quoted above (until *c.* 15 ka BP, beyond which information on calibration is lacking) to obtain age in sidereal years. Note that OIS 1 includes most of the Late-glacial.
† Classical system applicable to the Alpine region is given in parentheses.
‡ There is uncertainty regarding age and correlation of glacials/interglacials older than the last interglacial. Recent evidence suggests that the Holsteinian interglacial is equivalent to OIS 11 (Reille & Beaulieu 1995).

for climate reconstruction based on pollen and lake-level records from that continent (Dupont & Agwu 1992; Street-Perrott & Perrott 1993).

Pollen data, used in combination with other proxy evidence of climate change such as OSCILLATIONS in lake levels, have also permitted important climatic deductions to be made regarding climate change during the postglacial when, especially at middle and higher latitudes, oscillations were much smaller than those associated with glacial/interglacial or indeed most stadial/interstadial transitions, and hence are more difficult to detect and quantify. Syntheses of the evidence for past climate change from all the major land divisions on the globe since the maximum of the last glaciation have been recently collated and evaluated (Wright *et al.* 1993). These reconstructions have provided a basis for the first critical testing of a GENERAL CIRCULATION MODEL (GCM) of climate change, and especially the validity of the critical controls or 'boundary conditions' used in the construction of the model. (*See also* PALYNOLOGY; POLLEN ANALYSIS.)

Corroborative evidence

Because of the complexities in interpreting plant macro- and microfossil evidence due to the control exerted by factors other than climate, such as proximity of PROPAGULE source, successional processes and EDAPHIC developments, corroboration based on other proxy indicators of climate change is important. Fossil beetle (coleopteran) evidence, for instance, has enabled the quantification of rapid changes in the thermal environment on land, particularly during the Late-glacial period (*c.* 14 000–9500 cal. BC), to be quantified (Atkinson *et al.* 1987). It has been shown that variations in OXYGEN isotope values, i.e. $\delta^{18}O$ or the $^{18}O:^{16}O$ ratio, from calcareous terrestrial sediments provide a reliable measure of change in the local thermal environment (Lotter *et al.* 1992).

Records from ocean cores

Deep ocean cores, drilled from the 1960s onwards, have provided long and continuous records of palaeoclimate change, in some instances extending back to the early Quaternary, i.e. 2 million years. Changes in $\delta^{18}O$, as recorded in the testae of benthic foraminiferans, reflect primarily the degree of glaciation rather than variation in ocean temperature which, in deep waters, varies by less than 2°C between glacial and interglacial cycles. During interglacials, $\delta^{18}O$ values are typically depressed from *c.* –1‰ to less than –2‰ because of the release of large volumes of WATER from the ICE SHEETS that are 'light' with respect to ^{18}O ($\delta^{18}O$ values normally less than –30‰). During glacial periods, $\delta^{18}O$ values increase due to the cessation of 'light' meltwater input. Consistency in the pattern of movement of the $\delta^{18}O$ curve in ocean cores from various parts of the world has led to the recognition of oxygen isotope stages (OIS), with odd/even-numbered stages representing periods with low/high $\delta^{18}O$ values (warm/cold periods), respectively (Table P1). The more or less regular cyclicity of warm/cold periods, that these studies have demonstrated, has played an important part in the acceptance of the Croll-Milankovitch theory of astronomical forcing. This theory proposes that cyclical changes in the precession of the equinoxes, the Earth's tilt and the eccentricity of its orbit, underlies long-term climate change during the Quaternary.

In addition to isotopic studies, changes in species composition (foraminifera and diatoms) and physical and chemical composition of ocean cores—for example the presence of iceberg-rafted sand and calcium carbonate ($CaCO_3$) concentration – have also provided important indicators of palaeoclimate change. From these and other investigations, it has become clear that ocean CIRCULATION patterns *per se* and, in particular the North Atlantic deep water circulation system, play a major role in HEAT TRANSFER from lower to upper latitudes in the North Atlantic region so that when it is shut down, as happened during the Younger Dryas stadial, a major climatic deterioration ensued (Broecker & Denton 1990).

Records from ice cores

Ice cores, especially the Vostok core drilled in Antarctica during the early 1980s, and GRIP (Greenland Ice-core Project) and GISP2 (Greenland Ice Sheet Project) cores drilled at/near the summit of the Greenland ice sheet in the early 1990s, have provided detailed evidence of climate change dating back to at least the penultimate interglacial (*c.* 250 000 BP). Values of $\delta^{18}O$ from ice cores, which reflect the temperature prevailing as water vapour changed to ice, enable changes in temperature over time at the core site to be reconstructed. Analysis of air bubbles trapped in the ice cores provides evidence for sharp increases in the GREENHOUSE GASES, CARBON DIOXIDE (CO_2) and METHANE (CH_4), at the transition from glacial to interglacial conditions. Another important parameter is electrical conductivity in the ice core. This reflects the dust content of the ATMOSPHERE, which is normally high during cold periods due to increased storminess and/or increased levels of aeolian activity (see 'Loess' below) in continental areas beyond the limits of glaciation (Taylor *et al.* 1993).

Ice cores have the advantage over ocean cores in that a precise chronology can be attached, at least to the upper part where annual ice layers can be recognized. Beyond *c.* 14 000 cal. BP, however, due to compression annual layers are not easily distinguished and, near the base of the core, the possibility exists that deformation due to ice flow may distort the stratigraphy and hence the record of past change. Where a precise chronology is established, rates of change can be estimated. A striking feature, visible in most of the measured parameters, is the rapidity with which change from glacial to interglacial conditions takes place. The increase in $\delta^{18}O$ values as the temperature rose at the Late-glacial/postglacial transition (*c.* 9500 cal. BC), took less than 50 years, while doubling of snow accumulation at this time—an indicator of changing precipitation patterns—may have taken place in as little as 3 years, a more conservative estimate being 41 years (Alley *et al.* 1993).

Loess as an archive of past climate

In most continental regions, which were not subject to glaciation, thick loess deposits have accumulated which contain an important record of past climate change. In northern China, the classic loess area, exceptionally thick deposits are present, in places exceeding a depth of 300 m; these form a more or less continuous record of climate variation which can extend back for over 2 million years, i.e. to the beginning of the Quaternary (Derbyshire 1995). Loess deposits typically consist of wind-blown, silt-sized particles forming more or less unleached mineral horizons, which are interrupted by palaeosols or fossil soils in which, like most mature soils, the magnetic susceptibility is greatly enhanced. The pristine loess deposits are usually interpreted as forming during cold, arid periods, i.e. glacial periods, while the palaeosol horizons develop during warmer and wetter periods, i.e. interglacials, when the quantities of wind-blown dust were greatly reduced. In essence, the loessic deposits of central Asia represent a complex record of the interplay between the variations in the intensity of the monsoonal system and fluctuations in the distribution of STEPPE and DESERT in that extensive region.

It has been recognized for some time that the shape of the magnetic susceptibility curve from loess sections closely resembles the marine oxygen isotope curve, and the available evidence suggests that these two records match, i.e. each registers, in different ways, the same glacial/interglacial cycles. The strong linkage that has existed in the past between high-altitude Northern hemisphere and Asian climate history is thereby emphasized (Bloemendal *et al.* 1995) and another piece in the jigsaw of our understanding of the regulation of climate on a global scale has been put in place. [M.O'C.]

REFERENCES

Aaby, B. (1976) Cyclic climatic variations in climate over the past 5,500 yr reflected in raised bogs. *Nature* **263**, 281–284.

Alley, R.B., Meese, D.A., Shuman, C.A. *et al.* (1993) Abrupt increase in Greenland snow accumulation at the end of the Younger Dryas event. *Nature* **362**, 527–529.

Atkinson, T.C., Briffa, K.R. & Coope, G.R. (1987) Seasonal temperatures in Britain during the past 22,000 years, reconstructed using beetle remains. *Nature* **325**, 587–592.

Baillie, M.G.L. (1995) *A Slice Through Time. Dendrochronology and Precision Dating.* Batsford, London.

Barber, K.E., Chambers, F.M., Maddy, D., Stoneman, R. & Brew, J.S. (1994) A sensitive high-resolution record of late Holocene climatic change from a raised bog in northern England. *Holocene* **4**, 198–205.

Barnosky, A.D. (1986) 'Big Game' extinction caused by Late Pleistocene climatic change: Irish Elk (*Megaloceros giganteus*) in Ireland. *Quaternary Research* **25**, 128–135.

Behre, K.-E. (1989) Biostratigraphy of the last glacial period in Europe. *Quaternary Science Reviews* **8**, 25–44.

Bloemendal, J., Liu, X.M. & Rolph, T.C. (1995) Correlation of the magnetic-susceptibility stratigraphy of Chinese loess and the marine oxygen-isotope record – chronological and paleoclimatic implications. *Earth Planetary Science Letters* **131**, 371–380.

Broecker, W.S. & Denton, G.H. (1990) What drives glacial cycles? *Scientific American* **262**, 42–50.

Derbyshire, E. (ed.) (1995) *Aeolian Sediments in the Quaternary Record.* Quaternary Science Reviews, Vol. 14, pp. 639–809. Pergamon Press, Oxford.

Dubois, A.D. & Ferguson, D.K. (1985) The climatic history of pine in the Cairngorms based on radiocarbon dates and stable isotope analysis, with an account of the events leading up to its colonization. *Review of Palaeobotany and Palynology* **46**, 55–80.

Dupont, L.M. & Agwu, C.O.C. (1992) Latitudinal shifts of forest and savanna in N.W. Africa during the Bruhnes chron: further marine palynological results from site M 16415 (9°N 19°W). *Vegetation History and Archaeobotany* **1**, 163–175.

Frenzel, B. (ed.) (1992) *European Climate Reconstructed from Documentary Data: Methods and Results.* Gustav Fischer, Stuttgart.

Huntley, B. & Prentice, I.C. (1993) Holocene vegetation and climates of Europe. In: *Global Climates since the Last Glacial Maximum* (eds H.E. Wright Jr., J.E. Kutzbach, T. Webb III, W.F. Ruddiman, F.A. Street-Perrott & P.J. Bartlein), pp. 136–168. University of Minnesota Press, Minneapolis, Minn.

Lamb, H.H. (1995) *Climate, History and the Modern World,* 2nd edn. Routledge, London.

Lotter, A.F., Eicher, U., Siegenthaler, U. & Birks, H.J.B. (1992) Late-glacial climatic oscillations as recorded in Swiss lake sediments. *Journal of Quaternary Science* **7**, 187–204.

Lowe, J.J. & Walker, M.J.C. (1997) *Reconstructing Quaternary Environments,* 2nd edn. Addison Wesley Longman, Harlow.

Pons, A., Guiot, J., Beaulieu, J.L. & Reille, M. (1992) Recent contributions to the climatology of the last glacial–interglacial cycle based on French pollen sequences. *Quaternary Science Reviews* **11**, 439–448.

Reille, M. & Beaulieu, J.-L. de (1995) Long Pleistocene pollen records from the Praclaux Crater, south-central France. *Quaternary Research* **44**, 205–215.

Šibrava, V., Bowen, D.Q. & Richmond, G.M. (eds) (1986) *Quaternary Glaciations in the Northern Hemisphere.* Quaternary Science Reviews, Vol. 5. Pergamon Press, Oxford.

Street-Perrott, F.A. & Perrott, R.A. (1993) Holocene vegetation, lake levels, and climate of Africa. In: *Global Climates since the Last Glacial Maximum* (eds H.E. Wright Jr., J.E. Kutzbach, T. Webb III, W.F. Ruddiman, F.A. Street-Perrott & P.J. Bartlein), pp. 318–356. University of Minnesota Press, Minneapolis, Minn.

Taylor, K.C., Hammer, C.U., Alley, R.B. *et al.* (1993) Electrical conductivity measurements from the GISP2 and GRIP Greenland ice cores. *Nature* **366**, 549–552.

Wright, H.E., Jr., Kutzbach, J.E., Webb, T., III., Ruddiman, W.F., Street-Perrott, F.A., & Bartlein, P.J. (eds) (1993) *Global Climates since the Last Glacial Maximum.* University of Minnesota Press, Minneapolis, Minn.

Palaeoecology

Palaeoecology is the study of the ecology of past communities and past environments. It involves the reconstruction of the conditions and the flora and fauna of the past on the basis of a wide range of evidence, some biological, some geological. It is dependent upon the existence and survival of 'fossil' evidence of such past communities and conditions upon which the palaeoecologist can operate. The value of such studies is partly academic, i.e. the documentation and understanding of the past, but such work also has extensive implications in the study of modern ecology, much of which can only be appreciated as a single moment in a long history of evolutionary and ecological change.

Geological evidence

Stratified geological sediments (in the broadest sense, including everything from Precambrian rocks to infilled ditches in archaeological sites) are the most valuable source of palaeoecological information. Within such materials are preserved a range of different forms of evidence of past conditions. Some of the material is 'AUTOCHTHONOUS'; it has been derived from the immediate locality, such as the community of organisms that once occupied that site. Other material is likely to be 'ALLOCHTHONOUS', derived from other places and transported through the atmosphere (pollen, certain fruits, flying insects, volcanic DUST, etc.), or by water movements from surrounding catchments (trunks of trees, water-borne bones, sorted sands, human-derived detritus, etc.). Having extracted, identified and obtained quantitative data concerning the abundance of the different components of the sample, the sources of materials need to be considered prior to the reconstruction of conditions at the time of SEDIMENTATION.

Analysis may involve geological techniques, such as mineral analysis of particles, together with determination of particle size, physical arrangement and derivation. The chemical analysis of volcanic ash, for example, can provide information concerning not only the source, but often the precise eruption that gave rise to the fossil material. Particle size and bedding can provide information about the hydrological conditions at the time of sedimentation.

Biological evidence

Biological evidence, in the form of fossils, allows both local and regional reconstruction of the vegetation and fauna of the time. In the case of ancient ASSEMBLAGES of fossils, the sources of which are no longer extant, the process of interpreting their arrangement in communities (upland, lowland; wetland, dryland) may be difficult and may require the analysis of many sites. It may even prove difficult to demonstrate an association of two organs (such as pollen grains and fossil leaves) in a single species. For this reason, the palaeoecological reconstruction of the Carboniferous coal-forming swamps has been more difficult than charting the course of modern peatland development. In the case of relatively recent sediments, many of the species recorded as fossils may still survive, so interpretation is easier. Such work often rests on the assumption, however, that modern examples of a TAXON are similar in their physiology and ecology to those discovered in a fossil state. To a certain extent this can be justified. A fossil water lily, for example, is likely to have been associated with aquatic habitats in the past, but one cannot be fully assured that its chemical and temperature requirements were the same then as now.

Species assemblages

Since the use of single species as indicators of past conditions is suspect, the use of whole assemblages is an approach that seems more acceptable. Although the ecology of a single species may have changed over thousands of years, it is unlikely that this should be the case for collections of species. The use of this reasoning has proved valuable and has permitted the biological reconstruction of the pattern of the major environmental changes within the last, very changeable, 2 million years. Using multivariate statistics it has been possible to erect very plausible climatic records based on assemblages of pollen in lakes, and temperature records based on FORAMINIFERAN assemblages in ocean sediments. It is possible to relate the assemblage directly to the environmental variables instead of having to make deductions from known physiological requirements. The calculation of 'response surfaces' has provided direct means of translation from community to conditions through the past.

There are problems, however, even with the use of whole assemblages, because past collections of species are often unlike the current sets of species. Modern communities frequently have no precise analogues in the past, and vice versa. The reason for this state of affairs becomes apparent when the histories of given taxa are investigated, for species seem to have behaved very independently of one another and current coincidences of species' ranges may simply reflect biogeographical tracks that happen currently to overlap. Communities, in other words, do not exist as discrete entities that remain intact through time. Assemblages of species do not expand and contract their ranges in concert, but operate on an individualistic principle. New conditions in the future will undoubtedly produce new combinations of species and the communities with which we are currently familiar will disappear.

This discovery does not totally undermine palaeoecology, but it does mean that the study and the interpretation of relatively recent sediments must still be handled with care. Quite apart from the problems of deciding the origin of fossil material within a sample and the species assemblages that gave rise to them, we still have to be cautious about the environmental interpretations we place upon those assemblages. *See also* PALAEOBOTANY. [P.D.M.]

REFERENCES

Birks, H.J.B. & Birks, H.H. (1980) *Quaternary Palaeoecology*. Edward Arnold, London.

Moore, P.D., Webb, J.A. & Collinson, M.E. (1991) *Pollen Analysis*, 2nd edn. Blackwell Scientific Publications, Oxford.

Traverse, A. (1988) *Paleopalynology*. Unwin Hyman, Boston, Mass.

palaeoendemic *See* ENDEMIC SPECIES.

palaeolimnology The study of lake history based on the analysis of lake sediments. Organic sediments occur in most LAKES, usually accumulating in deep water, often at rates higher than 1 mm per year. Most of the SEDIMENT is derived from the lake and its CATCHMENT, but trace components come from the ATMOSPHERE. Sediment cores can be analysed using a wide variety of physical, chemical and biological (microfossil) techniques and the results used to reconstruct and explain changes in lakes and their environment through time. In recent years there has been a special interest in using lake sediment records to provide evidence for the timing and extent of lake ACIDIFICATION and EUTROPHICATION and as a means of detecting changes in CLIMATE.

Lake sediments can be sampled in a wide variety of ways using gravity, piston and freezer corers, depending on water depth, the nature of the sediment and the length of core required. Some sediments have seasonally and annually repeated bands (called varves) that can be used for direct dating (as for tree rings), but in most cases dating requires the use of radioisotopes such as ^{210}Pb and ^{14}C dating techniques.

Evidence for environmental change, caused by POLLUTION or natural events, can be derived in particular from an analysis of the microfossil ASSEMBLAGES in the sediment. Of these the diatom, cladoceran and chironomid fossil records are the most important. Using modern CALIBRATION datasets, which relate biological distributions to environmental gradients, transfer functions can be generated to enable quantitative reconstructions of lakewater pH, PHOSPHORUS, SALINITY and TEMPERATURE to be made from the FOSSIL RECORD. *See also* PALAEOBOTANY; PALAEOCLIMATIC RECONSTRUCTION. [R.W.B.]

palaeopalynology *See* PALAEOBOTANY.

Palaeotropical kingdom *See* AFRO-INDIAN FLORAL REGION; BIOGEOGRAPHICAL UNITS. *Cf.* NEOTROPICAL KINGDOM.

Palaeotropical realm *See* AFRO-INDIAN FLORAL REGION.

Palaeozoic The first part of the PHANEROZOIC, lasting from the end of the Precambrian period (about 590 million years ago) until the beginning of the Triassic (about 250 million years ago). It includes the Cambrian, Ordovician, Silurian, Devonian, Carboniferous and Permian periods. [P.D.M.]

palladium (Pd) A metallic element (ATOMIC NUMBER = 46; relative ATOMIC MASS = 106.43; boiling point = 2970°C) with six stable isotopes ranging in mass from 102 to 110 and an average abundance in the Earth's CRUST of 0.01 mg kg^{-1}. It is the least dense, and has the lowest melting point (1552°C) of the PLATINUM group of metals. It is used widely as an industrial CATALYST (e.g. in the hydrogenation of oils), and as a component of the three-way catalysts used on vehicle exhausts to reduce EMISSIONS of CARBON MONOXIDE (CO) and NITROGEN OXIDES (NO_x). [J.N.C.]

palsa mire A tundra mire type, whose name is derived from Finnish, found only within climates where the average air temperature remains at or below 0°C for more than 200 days in the year, roughly corresponding to the southerly limit of discontinuous PERMAFROST. Palsa mires are often associated with northerly AAPA MIRES, being found mainly on the flatter regions. The mire consists of elevated mounds, often 50 m or more in diameter and several metres in height, consisting of a core of frozen SILT and PEAT, with a peat and vegetation cover over their summits. These palsa hummocks are interspersed with pools of similar area, many of which become infilled by vegetation to form extensive wet lawns.

The developmental cycle of palsa mounds has been extensively studied. Their initiation from cotton-grass (*Eriophorum* spp.) lawns is associated with slight irregularities in the surface of the mire that result in winter snow being removed by wind and failing to accumulate on those areas elevated a few centimetres above the surroundings. The lack of snow leads to deeper FROST penetration and longer survival of these frozen areas in the spring. Eventually, a patch of permafrost develops within the peat. This permafrost core grows year by year to form the palsa mound with its capping of peat-forming vegetation. It is possible that the drying of the surface leads to increased colonization by lichens leading to increased ALBEDO and accelerated ice-mound growth. Eventually, however, the peat capping begins to erode, exposing bare peat, whose dark colour leads to the absorption of solar radiation in summer and finally a meltdown of the ice and the formation of a pool. The cycle then begins again. Palsa mires therefore consist of a mosaic of mounds and pools in different stages of development and regression. [P.D.M.]

paludification The process by which soils become waterlogged and PEAT formation commences and/or spreads. Many factors may contribute to this process, including the blocking of drainage by surrounding peat development, changing HYDROLOGY consequent upon modifications in catchment vegetation (including forest removal), charcoal input to soil leading to poor drainage, etc. *See also* PEATLANDS. [P.D.M.]

Palynology

Palynology is the study of pollen grains of flowering plants and gymnosperms and of spores from pteridophytes, bryophytes, algae and fungi. Such work can involve studies of structure, development, taxonomy, evolution, dispersal, preservation and STRATIGRAPHY. The applications of the studies are many, including allergies, forensic science and PALAEOECOLOGY.

Pollen

Seed plants (angiosperms and gymnosperms) produce pollen grains as a means of transferring male GAMETES from the anther of a flower to the stigma (and hence to the ovary) of the same or a different flower. Genetic advantages, of course, often accompany the transfer of gametes to a separate plant. The process of transfer is termed POLLINATION and is a passive process involving a physical (such as wind or water) or a biological vector (such as insects, birds, bats or mammals). The transfer is both hazardous and chancy (especially in the case of pollen carried by air movements). The major hazard is DESICCATION, which is counteracted by the presence of a protective outer coat, while the lack of directionality in such transfers is overcome by an abundance of pollen production. More directional transfer, such as by insects which, after visiting a flower, are likely to seek another similar flower, is more reliable and hence requires less duplication of effort and fewer pollen grains are produced.

The production of pollen grains involves meiotic cell division, so pollen constitutes the male GAMETOPHYTE generation of the seed plants. A pollen mother cell in the anther divides first meiotically and then mitotically to produce a tetrad of pollen which usually dissociates on maturation. As the grains mature their outer layers become impregnated with a polymer, sporopollenin, that will eventually form the major constituent of the outer part of the pollen wall (the exine). The inner wall (intine) is made of cellulose and is similar to normal plant cell walls. The exine needs to be desiccation resistant, yet flexible to cope with changes in cell volume as water is absorbed or lost. Its form may be smooth or highly sculptured with protrusions in the form of warts (verrucae), spines (echinae), club-shaped protrusions (clavae), etc. Generally, but certainly not invariably, the more elaborately sculptured pollen grains are associated with animal pollination; one proposal is that such projections are electrostatically charged and thereby attach more easily to the bodies of vectors.

Pollen grains also develop apertures — areas lacking the robust outer wall. These apertures may take the form of pori (circular pores) or colpi (elongate grooves), and they may be scattered over the surface or arranged in relation to the polar axis of the grain (which is determined by the orientation of the grain in the original tetrad). Some grains have just one aperture, many have three, and some have many. Alternatively, there may be none at all, while others may have arrangements of both pori and colpi. The number and arrangement of apertures, together with the nature of the exine sculpturing, provides the basis for the classification and hence the identification of pollen grains.

The function of the apertures is not entirely clear. Germination evidently requires water uptake, which may necessitate the presence of gaps in the exine, and the emergence of a pollen tube is clearly facilitated by such weak points, but this does not fully explain the duplication of apertures in many types. Recognition reactions between pollen and stigma involve the exudation of proteins and these probably leave the grain via the apertures. Germination then proceeds if the recognition process proves receptive. It is this aspect of pollen behaviour that creates health problems in humans, since the same proteins exuded by pollen on the mucous membranes of nose and throat can result in allergic reactions and the symptoms of hay fever.

Spores

Spores from pteridophytes and bryophytes differ from pollen grains in their function. They are also a part of the gametophyte generation, are produced by meiosis and are formed in tetrads, but their function is one of dispersal rather than genetic transfer. They are of similar size to pollen grains (*c.* 10–100 μm) and they are also frequently dispersed by air movements, but on arrival at an appropriate location they then germinate to produce the vegetative phase of the gametophyte generation.

Applications

Palynology has often proved a valuable tool for the plant taxonomist since it provides a criterion that may, like general flower structure, show a greater degree of evolutionary inertia than more vegetative features. Pollen grains may, however, show a high degree of uniformity within plant families (such as the Poaceae), or may reflect the diversity of the family (as in the Rosaceae). In some families pollen grains have proved useful in elucidating phylogenetic relationships (e.g. the Combretaceae).

One of the most frequently used applications of palynology is in the study of fossil plants and vegetation. The fact that pollen grains are produced in relatively large numbers and that they have a tough outer coat that is relatively resistant to microbial attack, means that they are preserved in some quantity in geological sediments. Sporopollenin is found even in Precambrian rocks. Spores provide valuable evidence of such processes as the invasion of the land by plants and, eventually, the evolution of the flowering plants. Their diversity and abundance has also been of great value in providing a means of relative dating of geological sediments, which is of importance in surveying, particularly in the oil industry.

The study of Quaternary palynology (covering approximately the last 2 million years of geological time) has received much attention for a variety of reasons. It is the geological period that leads up to and includes the present, so is of particular relevance to modern ecological studies. The taxa found in Quaternary sediments are largely still extant (or are represented by close relatives), so identification and interpretation of data in terms of vegetation and environment are greatly facilitated. The Quaternary has been characterized by frequent and rapid climatic changes, and palynology has provided important information on the responses of vegetation to such changes, which may prove useful in predicting future changes. The study of modern ASSEMBLAGES of pollen and spores and the application of multivariate statistical analyses has increasingly provided precision in the interpretation of FOSSIL ASSEMBLAGES, and this approach is available only for relatively recent geological materials.

One further application of palynological research has been the area of forensic science, where it can provide information concerning the geographical origin of soil or mud samples. Pollen analysis can also supply evidence concerning the origin of honey samples and has often been used for the detection of undeclared importation or blending of this foodstuff. [P.D.M.]

REFERENCES
Faegri, K. & Iversen, J. (1989) *Textbook of Pollen*, 4th edn (revised by K. Faegri, P.E. Kaland & K. Krzywinski). John Wiley & Sons, Chichester.
Moore, P.D., Webb, J.A. & Collinson, M.E. (1991) *Pollen Analysis*, 2nd edn. Blackwell Scientific Publications, Oxford.
Traverse, A. (1988) *Palaeopalynology*. Unwin Hyman, Boston, Mass.

---

pampas The extensive areas of temperate GRASSLANDS in South America. They are similar to the prairies of North America and the steppelands of Eurasia. Tall, medium and short grasses occur throughout the pampas, which dominate the Rio de la Plata lowlands in Argentina, Uruguay and Brazil. The most common grass genus is *Stipa* and clumps of trees occur in water-retaining depressions. The origin of the pampas is controversial as climatic characteristics alone are insufficient to explain the general absence of trees. Much modification has occurred due to irrigated farming and livestock ranching together with associated fire management. [A.M.M.]

PAN *See* PEROXYACETYL NITRATE.

panbiogeography A biogeographic method that focuses on the spatiotemporal analysis of DISTRIBUTION patterns of organisms. It is distinct from phenetic BIOGEOGRAPHY, which investigates similarities between BIOTAS in terms of the number of taxa they have in common. The method involves drawing lines on a map, known as tracks, which connect the known distribution of related taxa in different areas. When two or more tracks of unrelated taxa coincide, they are combined into generalized tracks, which are thought to indicate the distribution of an ancestral biota before it vicariated. [A.H.]

REFERENCE
Craw, R. (1988) Panbiogeography: method and synthesis in biogeography. In: *Analytical Biogeography* (eds A.A. Myers & P.S. Giller), pp. 405–435. Chapman & Hall, London.

Panglossian assumption *See* EVOLUTION OF SEX.

panmictic Describing a population where mating occurs at random with respect to the distribution of GENOTYPES in the population. Panmictic populations are probably very rare in nature because spatial divisions in the environment result in inbreeding and preferences for males or females with certain PHENOTYPES result in ASSORTATIVE MATING. Totally random mating is one of the assumptions of the Hardy–Weinberg theorem, and deviations from Hardy–Weinberg equilibria can sometimes be attributed to non-random mating. For example, inbreeding results in an excess of homozygotes. *See also* POPULATION GENETICS. [D.A.B.]

Paracelsus' principle The principle that any chemical can be toxic at a high enough concentration. It was recognized by the Swiss physician Philippus Paracelsus (1493–1541), who pioneered the medical use of chemicals. This principle of TOXICOLOGY is also applicable for ECOTOXICOLOGY. [P.C.]

paradox of enrichment The danger to a PREDATOR-limited population of victims that comes from increasing its supply of the very RESOURCES whose shortage is killing victim individuals or reducing their birth rates. The increase in resources can sometimes destabilize the interaction with predators and lead to the extinction of the entire victim population. [M.L.R.]

REFERENCE
Rosenzweig, M.L. (1977) Aspects of biological exploitation. *Quarterly Review of Biology* **52**, 371–380.

paraffins HYDROCARBONS of the series C_nH_{2n+2}, otherwise called ALKANES. Paraffin oil (kerosene in North America) is a mixture of some larger paraffins with boiling point 150–300°C, used as fuel for lamps and heaters. [J.G.]

parallel bottom community The similarity of species composition in temperate and arctic bottom communities of widely separated locations. In the mid-1950s, the Danish biologist, Gunnar Thorson, first noted that similar SEDIMENT types at the same depths around the world contained ecologically and taxonomically similar species. There are no similar parallel communities in tropical latitudes. [V.F.]

parallel evolution The occurrence of HOMOPLASY (independently evolved traits) in paraphyletic taxa. An older, more restrictive use of the term is for homoplasy derived from the same plesiomorphic background. *See also* CONVERGENT EVOLUTION; PARAPHYLETIC GROUP. [D.R.B.]

parameters The summary measurements or characteristics of a POPULATION. These are represented by Greek letters, for example, the MEAN (μ) and the VARIANCE (σ^2). *See also* STATISTICS. [B.S.]

parametric statistics Statistical tests are frequently classified into parametric and non-parametric tests. Parametric tests are older and require QUANTITATIVE INFORMATION/observations to follow a NORMAL DISTRIBUTION and to have similar VARIABILITY (homogeneous variances). Parametric tests cannot be carried out on QUALITATIVE INFORMATION, i.e. ranked information or information in categories. If the underlying conditions are met then parametric tests are usually more powerful (*see* STATISTICAL POWER). Examples of parametric tests/techniques are *t* TEST, PRODUCT–MOMENT CORRELATION COEFFICIENT and ANALYSIS OF VARIANCE. *Cf.* NON-PARAMETRIC STATISTICS. *See also* CENTRAL LIMIT THEOREM. [B.S.]

páramo Humid arctic/alpine meadows and SCRUB, occurring in the Andes and kept moist by mist. This type of vegetation is much altered by clearing and burning for PASTURE. The scrub is characterized by stunted tree-like species (e.g. *Espeletia*) with rosettes of hairy evergreen leaves. *See also* ALPINE ZONE. [P.C.]

REFERENCE

Crawford R.M.M., Wishart D. & Campbell R.M. (1970) A numerical analysis of high altitude scrub vegetation in relation to soil erosion in the eastern cordillera of Peru. *Journal of Ecology* **58**, 173–191.

parapatric Describing the pattern of DISTRIBUTION of two populations, species, etc. in which their ranges meet without overlapping along a narrow contact zone. The parapatric distribution of closely related species may result from their origination by the parapatric model of SPECIATION or by a secondary contact after a period of geographic isolation. *See also* ALLOPATRIC SPECIATION; PARAPATRIC SPECIATION; SYMPATRIC SPECIATION. [P.O.]

parapatric speciation Though there is potential GENE FLOW between populations which are PARAPATRIC, sufficient isolation can occur for SPECIATION if gene flow is low and the difference in SELECTION PRESSURES between semi-isolated groups is high. [P.C.]

paraphyletic group An incomplete evolutionary unit in which one or more descendants of a particular ancestor have been excluded from the group. For example, a group that placed all cartilaginous fishes, ray-finned fishes, coelacanths and lungfishes together in one taxon (called 'fish') would be paraphyletic because lungfishes and coelacanths share a common ancestor with tetrapod vertebrates that is not also shared with cartilaginous and ray-finned fishes. In order to make the taxon 'fish' monophyletic, we would have to include in it the tetrapods (amphibians and amniotes, including mammals, turtles, crocodilians, birds, the tuatara, lizards and snakes). Other paraphyletic groups are the Reptilia, the Protista, the Gymnospermae and the Bryophyta. Evolutionary taxonomists refer to some paraphyletic groups as evolutionary grades, and may include some paraphyletic groups in their CLASSIFICATIONS. Grades are considered important indicators of past ADAPTIVE RADIATIONS, which have been supplanted by more recent adaptive radiations indicated by monophyletic groups (clades) in classifications. [D.R.B.]

parasite An organism that interacts with another organism (the HOST) to reduce the FITNESS (survival and/or reproduction) of the host, and increase the fitness of the parasite, such that the parasite does not have to kill the host in order to derive the benefit (although the host may die as a result of the interaction). The term parasite is traditionally used for MACROPARASITES, although increasingly it is used to include MICROPARASITES. *See also* PARASITISM; SPECIES INTERACTIONS. [G.F.M.]

parasite–host interactions PARASITES reduce the FITNESS of their HOSTS. Ecologically they play the same role as PREDATORS and PARASITOIDS, and may act as density-dependent controls on their host population. *See also* PARASITISM; SPECIES INTERACTIONS. [G.F.M.]

Parasitism

An intimate and usually obligatory relationship between two organisms in which, essentially, one organism (the PARASITE) is exploiting RESOURCES of the other organism (the HOST) to the latter's disadvantage. In general, parasitism involves a long-lasting association between the two parties. However, it is difficult to give a universal definition, as parasitism is a widespread and extremely diverse phenomenon. Parasitism is different from predation (where the prey has no lasting association with the PREDATOR), from COMPETITION (where both parties may suffer), and from COMMENSALISM (where the negative effects are absent or negligible).

Parasitism is a highly complex and diverse phenomenon and occurs in all known groups of plants and animals. Host and parasite show many distinctive features of mutual ADAPTATION and COEVOLUTION. Parasitism can occur within the same species (e.g. BROOD PARASITISM in birds) or, more generally, between different species. Parasitism may be long-lasting or even permanent (e.g. tapeworms found in the intestines of mammals) or transient (e.g. blood-sucking leeches and ticks). More than one parasite, either from the same or from different species (multiple parasitism), can infect a host — a phenomenon known as SUPERPARASITISM. Hyperparasitism refers to parasitism where the host is a parasite itself. Parasites can live on (ECTOPARASITE) or in (ENDOPARASITE) their hosts.

Microparasites and macroparasites

A useful classification is into MICRO- and MACROPARASITES. Microparasites multiply within the host and are often intracellular. Examples are viruses (the causative agents of, for instance, measles, rabies and yellow fever), bacteria (e.g. tuberculosis, leprosy, Lyme disease) and protozoa (e.g. malaria, sleeping sickness, leishmaniasis) (Fig. P3).

Macroparasites (as endoparasites) grow within their host and multiply through production of EGGS or other infectious stages; they are often extracellular, for example, in the intestinal lumen. Examples include helminths (roundworms, e.g. spool worms; flatworms, i.e. tapeworms and flukes) and acanthocephalans, as well as groups such as mites, fleas, ticks and leeches. Somewhat separate are the PARASITOIDS, usually insects (parasitic wasps and flies), where the immature stages live in or on a host while the adult is a

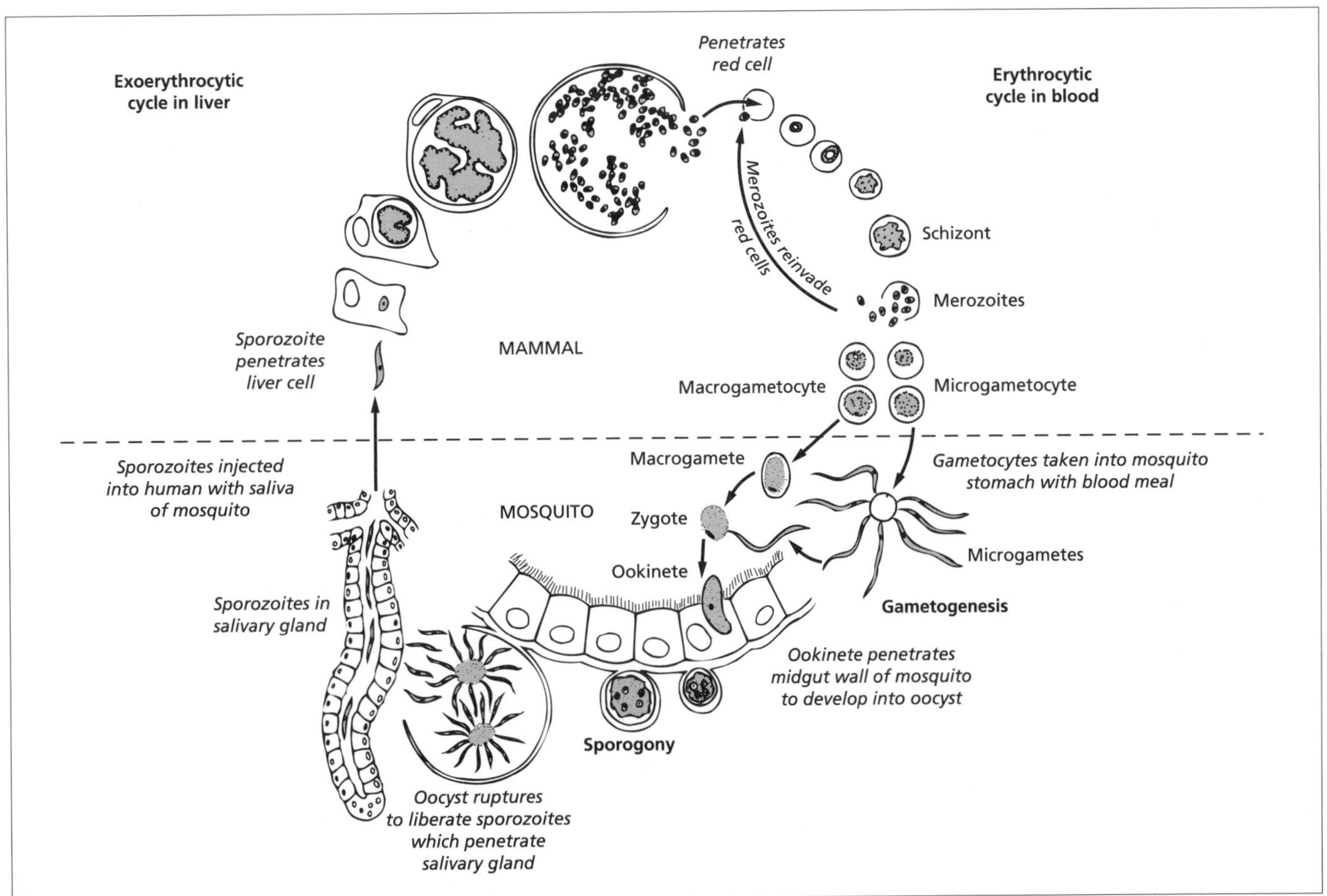

Fig. P3 The life cycle of *Plasmodium* spp. in mammals. The infection begins when sporozoites are injected directly into the bloodstream from the salivary glands of a mosquito. The sporozoites enter liver cells where they begin a phase of multiplication called 'exoerythrocytic schizogony', during which thousands of uninucleate merozoites are formed. These enter red blood cells in which they undergo a second phase of multiplication or erythrocytic schizogony, during which fewer than 24 merozoites are formed. These merozoites invade new red blood cells and the cycle may be repeated many times. Some of the merozoites are capable of developing into sexual stages or gametocytes. These are taken up by a mosquito. In the gut of the mosquito microgametes are produced and these fertilize the macrogametes and the resulting zygote or ookinete bores through the gut wall to come to lie on the outer surface where it forms an oocyst. Within the oocyst a third stage of multiplication occurs resulting in the formation of sporozoites that enter the salivary glands of the mosquito. (After Vickerman, K. & Cox, F.E.G. (1967) *The Protozoa*. John Murray, London.)

free-living stage. Well-known parasitoids are sphecid wasps, which parasitize larvae of butterflies and moths.

Parasite life cycles

The LIFE CYCLES of parasites are very diverse. In the simplest case, parasites are directly transmitted from host to host, via air, water or direct skin contact (e.g. the common cold virus is transmitted via coughing and expelling virus-loaded droplets). In many cases, long-lasting durable stages are deposited in the environment, from where they are eventually picked up by a passing host (e.g. the anthrax bacillus in the soil).

Still others rely on vectors (*see* VECTOR, PARASITES), often blood-sucking insects, to be transmitted to the next host. Examples include *Plasmodium* (the agent of malaria), which is transferred by mosquitoes during blood-sucking (Fig. P3), or *Trypanosoma* (sleeping sickness), transmitted by tsetse flies. Vectors are primarily functioning as carriers of the parasite, although occasionally the parasites may undergo development and reproductive processes inside the vector.

Some parasites have complicated life cycles whereby more than one host species has to be infected. In such cases, the parasite attains sexual maturity in the definitive (final) host, while intermediate hosts serve as temporary but essential environments for the completion of the life cycle. For example, the liver fluke *Dicrocoelium* lives in the gall-bladder and bile ducts of its final host (sheep, cattle, deer, etc.). Eggs pass out to be ingested by the first intermediate host, a land snail. The parasite develops inside the snail to form sporocysts that are shed in slime balls and ingested by the second intermediate host, a wood ant (*Formica* spp.). After further development of the parasites into metacercariae, the ant shows aberrant behaviour that will make it more likely to be eaten by the parasite's final host, thus completing the life cycle.

Effects on host

Parasitism has many effects on the host; the negative effects are often summarized as the parasite's VIRULENCE. Macroscopically, parasites can cause debilitation of the host, increased MORTALITY, castration (*see* CASTRATION, PARASITIC), gigantism or altered behaviour to mention just a few. Such damage can result from unavoidable side effects of the parasite's invasion, establishment, growth and reproduction, or arise when the parasite invades the 'wrong' organs and tissues. However, modern views consider virulence to be an adaptive trait that increases parasite FITNESS by facilitating its growth, multiplication and in particular its transmission. Consequently, parasites transmitted mainly to offspring of the current host (vertical transmission) are expected to evolve towards lower degrees of virulence than parasites being transmitted to other hosts (horizontal) or those possessing durable stages. Hosts in turn can mount defences against parasites at different stages of their life cycle.

Biological significance

Parasitism is not only important for human and veterinary medicine, but plays a very important, although as yet poorly known role in the ecology and EVOLUTION of natural populations in the wild. Parasitism has been implicated in the evolution of sexual reproduction, SPECIATION, the timing of life cycles and social behaviour. It may also be a factor in wider ecological aspects of POPULATION CYCLES and the geographical distribution or structure of local communities. The study of host–parasite interactions is based on principles of POPULATION DYNAMICS (EPIDEMIOLOGY). As a rule, host RESISTANCE against parasites and parasite virulence are based on the interaction of host and parasite genotypes. *See also* SPECIES INTERACTIONS. [P.S.H.]

REFERENCES

Anderson, R.M. & May, R.M. (1991) *Infectious Diseases of Humans.* Oxford University Press, Oxford.

Cheng, T.C. (1986) *General Parasitology.* Academic Press, Orlando, Fla.

Godfray, H.C.J. (1994) *Parasitoids.* Princeton University Press, Princeton, N.J.

Price, P.W. (1980) *Evolutionary Biology of Parasites.* Princeton University Press, Princeton, N.J.

parasitoid Any insect whose larvae develop by FEEDING on the bodies of other arthropods, usually insects. It is estimated that 0.8–2 million species of insects are parasitoids. The parasitoid LARVA or larvae eventually kill their HOST, and a single host provides all the food required for development into an adult. After attacking a host, the female parasitoid does not attempt to move the host to a protected site or nest (thus distinguishing parasitoids from certain solitary wasps which have an otherwise similar life history). Most parasitoids belong to the Hymenoptera (wasps), but this feeding habit also occurs commonly in the Diptera (true flies), sporadically in the Coleoptera (beetles) and rarely in other orders. Some authors adopt a wider definition of parasitoid, to include other carnivores and also herbivores which require a single resource individual upon which to complete development and which they invariably kill. *See also* HYPERPARASITOID; PARASITOID–HOST INTERACTIONS. [H.C.J.G.]

Parasitoid–host interactions

The interactions between PARASITOIDS and the arthropods that act as hosts for their developing young (see Plate 12, facing p. 398).

Host location and oviposition

In most species, the HOST is located by the adult female parasitoid. The most important cues used in long-range host location are chemical, and frequently parasitoids respond to a hierarchy of stimuli of progressively greater information value. Thus some parasitoids are known to respond to stimuli emanating from the MICROHABITAT of the host (independent of host presence); some to stimuli associated with the presence of the host (e.g. chemicals in host frass or emitted by host commensals); and some to stimuli from the host itself (e.g. host pheromones). Visual cues may also be important. At closer range, parasitoids frequently show an arrestant response to host-associated chemicals. After a host is discovered, tactile and contact chemical stimuli are used in deciding whether a host is suitable for oviposition. Parasitoids often learn to associate novel stimuli with the presence of hosts. In a small minority of parasitoids, EGGS are deposited away from the host and a modified first instar LARVA is responsible for host location. OPTIMAL FORAGING THEORY has been used extensively to understand host choice and PATCH USE in parasitoids.

Parasitoid life histories

At oviposition, the host is either killed or permanently paralysed (idiobiont parasitoids), or not or only temporarily paralysed (koinobiont parasitoids). The hosts of koinobiont parasitoids recover and continue feeding while the parasitoid remains quiescent as an egg or first-instar larva. Koinobionts can thus attack the host when it is still too small to support full larval development. Parasitoids may feed externally, with their mouthparts embedded in the host (ectoparasitoids), or they may feed internally (endoparasitoids). Parasitoids can also be classed as solitary (one parasitoid develops per host) or gregarious (two to several hundred parasitoids develop per host). CLUTCH SIZE theory has been widely used to study the oviposition behaviour of gregarious parasitoids. A special category of gregarious parasitoids comprises polyembryonic species, in which the female lays one or a few eggs that divide asexually to produce a large clutch. Parasitoids are often categorized by the life-cycle stage of the host that is attacked (egg, larval, pupal and adult parasitoids); some species oviposit into one stage but the developing parasitoid kills and emerges from another stage (egg-larval parasitoids, etc.). Parasitoids that attack other parasitoids are called HYPERPARASITOIDS, and PARASITISM of hosts previously attacked by other parasitoids is termed SUPERPARASITISM or MULTIPARASITISM.

Host defences and parasitoid countermeasures

Parasitoids that have to survive inside a still-living host (koinobiont endoparasitoids) must avoid their host's defences. The main defence is cellular encapsulation, in which certain cells circulating in the host haemocoele (blood-filled body cavity) recognize a body as foreign and cause other cells to form a capsule around it; the capsule subsequently melanizes (hardens and darkens), probably leading to the death of the parasitoid through suffocation. Humoral (non-cellular) defences, though less well understood, are also involved.

To defeat host defences, parasitoids have adopted both passive and active countermeasures. Passive defences include: the placement of eggs in tissue outside the haemocoele, where they are protected from the host immune system; a protective coating of the egg or larva that leads to the host's immune system mistaking it as self; and tolerance of host attack (many dipteran endoparasitoids anchor themselves in the host skin or major trachea and so, with access to air, can prevent themselves being suffocated by the host).

Active defences typically involve the disabling of the host immune system by TOXINS injected during oviposition or by substances released by the parasitoid larva and its membranes, by teratocytes, or by polydnaviruses. Teratocytes (found in certain hymenopteran parasitoids) are cells derived from embryonic membranes that circulate in the host haemocoele and are involved in immune suppression as well as possibly having a trophic function (they are eventually consumed by the parasitoid). Polydnaviruses (found in one superfamily of hymenopteran parasitoids) are multi-stranded DNA viruses (hence poly-DNA-viruses) that are stably incorporated in the parasitoid genome. In specialized cells of the female reproductive tract, the virus DNA replicates and is incorporated in protein capsules before injection into the host. In the host, the virus enters the cells of the immune system and other tissues where its genes are expressed (although it does not replicate). Viral products disable the immune system, for example by causing apoptosis of immune cells. Viruses with similar functions are being discovered in other parasitoid groups.

Sex ratio strategies

Hymenopteran parasitoids are haplodiploid, which means that females are able to control the sex of their offspring by choosing whether or not to fertilize an egg. This has allowed adaptive sex ratio strategies to evolve, and parasitoids have proved very important in testing many aspects of sex ratio theory. Parasitoids that emerge in the close proximity of siblings (either gregarious parasitoids, or solitary parasitoids of gregarious hosts) frequently show female biased SEX RATIOS due to local mate competition. The normal frequency-dependent advantage of producing an equal sex ratio is lost if sons compete strongly for mates. Solitary parasitoids often lay female eggs in large hosts and male eggs in small hosts. This has been explained as a case of conditional sex expression. Both sexes gain from being in a larger host as host size is related to parasitoid size and hence parasitoid FITNESS. However, females gain proportionally more from being large than males as female FECUNDITY is more strongly size-dependent than male mating success. Sex ratio in hymenopterous parasitoids is known to be influenced by a variety of non-Mendelian inherited elements including B chromosomes and bacteria. Infection by *Wolbachia* (bacteria) appears to be widespread and causes strain incompatibility and in some cases PARTHENOGENESIS.

Ecological significance

Parasitoids are known to be an important factor influencing and/or regulating the POPULATION DYNAMICS of their host (*see* MODELS OF PARASITOID–HOST INTERACTION). It has been suggested that parasitoids may be important in determining host COMMUNITY structure through APPARENT COMPETITION and competition for ENEMY-FREE SPACE. Parasitoids have been introduced to combat pests in many BIOLOGICAL CONTROL programmes, often with great success. *See also* POPULATION REGULATION; SPECIES INTERACTIONS. [H.C.J.G.]

REFERENCE
Godfray, H.C.J. (1994) *Parasitoids, Behavioral and Evolutionary Ecology*. Princeton University Press, Princeton, N.J.

PARCOM *See* PARIS COMMISSION.

Parent–offspring conflict

There is said to be parent–offspring conflict if a trait that increases the reproductive success of an individual offspring can only do so at the expense of the reproductive success of its parent(s) and vice versa. In short, parents and their offspring need not have precisely the same set of interests. When their interests are not identical there is said to be a potential conflict between the two (Trivers 1985).

Bases for conflict

Consider, for example, an offspring and its sibs. Sibs and the mother are related to this progeny. If there is no multiple PATERNITY then on the average they are related by $^1/_2$ (take a random allele in the progeny of interest and there is a 50% chance that this allele will also be in the mother and likewise for the other progeny). There is, however, a 100% chance that this allele is in the progeny in question (by definition). Hence, a gene acting in an offspring that forced a mother to give that particular infant more RESOURCES than its brood members could spread in a population, as those individuals that are receiving less resources are less related to the progeny than the progeny is to itself. In contrast, the mother is equally related to all progeny. As a consequence she may favour a more equitable distribution of resources. It is not only in the offspring's interest to demand more food than the mother would prefer to allocate, it is also in their interests for parental provisioning to be extended longer into development than the mother would prefer.

This conflict is all the more exaggerated as the relatedness between progeny diminishes. So consider what should happen if the progeny of a given mother all have different fathers. Again the mother is equally related to all progeny—she should hence provide an equal distribution of resources. However, any given progeny is related to its sibs only down the mother's line. So the average relatedness between sibs may be as low as 0.25. None the less, a given individual is still related to itself 100%. So a given progeny should be prepared to invest more in trying to obtain more from the mother under this circumstance than it should when the sibs are always full sibs.

Just this sort of theory is thought to account for the fact that the strength of call of chicks in a nest scales with the degree of multiple paternity (Briskie *et al.* 1994). In those species that are considered to be monogamous or nearly so, it is found that the calls for chicks demanding food from the mother are relatively weak when compared to those in which the progeny (and potential future progeny) are relatively unrelated. This sort of effect can be evoked to explain numerous interactions between parents and progeny.

Maternal–fetal interactions

Perhaps most interesting are hypotheses of the nature of maternal–fetal interactions across the mammalian placenta, for in this instance we can get a molecular handle on the nature of the interaction (Haig 1993). For example, if fetal demands went unopposed, it has been argued, the fetus would remove more glucose from maternal blood than was in maternal interests (the mother would prefer to save some to give to future progeny and to optimize her chances of having future progeny). It is thus suggested that the mother's best interests are served by reducing her blood sugar to limit fetal uptake. Further, it is argued, that a mother and her fetus will compete after every meal for control of blood sugar. Whereas the mother would prefer to take sugar out of the blood rapidly, so the fetus would prefer that the sugar remain in circulation so it can be taken across the placenta. This conflict between mother and fetus has resulted, it is proposed, in an evolutionary ARMS RACE in which a trait permitting a fetus to increase its output of anti-insulin hormone (keeping blood sugar high) will spread, and in response a trait which sees the mother increase her production of insulin (reducing blood sugar levels) can also spread. The outcome of this genetic arms race is a very high fetal production of an anti-insulin hormone and an equally high maternal production of insulin. The high production of these two hormones has practically no net effect on the flow of sugars into the fetus as the two effects cancel each other out.

In mechanistic terms human placental lactogen (hPL) and human placental growth hormone (hPGH) are proposed to be the fetal anti-insulin hormones. hPL is the most abundant peptide hormone produced by primates. Its concentration in maternal serum increases throughout pregnancy, reaching the remarkably high titre of 5–15 μg ml^{-1} near term. At this stage secretion from some tissues is as high as 1–3 g day^{-1}. For comparison, the plasma concentration of hGH, integrated over a day, is about 0.003–0.006 μg ml^{-1} in young non-pregnant adults. Levels of hPGH are much lower than those of hPL, but follow a similar temporal pattern, and at term exceed 0.015 μg ml^{-1}. As expected, concentrations of hPL and hPGH in fetal serum are much lower than concentrations in maternal serum.

If placental lactogens are the fetal means to manipulate sugar uptake, then we might expect rapid EVOLUTION of these sequences as they would be coevolving in an arms race with the maternally derived suppressor. This, indeed, appears to be the case. In, for example, a comparison of placental lactogens between mice and rats, of a sample of the 363 proteins only 15 (4%) had a lower percentage identity than that for PLI and 23 (6.3%) had a lower identity than PLII (note these rodents have two related placental lactogens).

Intensity of conflict

Classical parent–offspring conflict theory can also predict that the intensity of conflict should decline with parental age. The older the parent, so the lower probability of having future progeny, so the less her FITNESS can be reduced by giving offspring what they demand. It is this sort of theory which may explain why, in humans, there is a higher rate of Down's syndrome children in older parents. The difference between old and young mothers seems to be that young mothers are more fussy about which early embryos should be provisioned with resources (i.e. there is expected to be a high rate of natural abortion in young females). Old mothers have a low probability of future copulation and hence have an incentive to be less selective about which progeny to invest in.

Father–offspring conflict

All of the examples given so far are of conflict between a mother and its offspring. In instances in which fathers do the provisioning the same potential CONFLICTS remain. In cases of multiple paternity within a brood a particularly intense conflict can ensue, for in many species (e.g. lions) fathers may try to kill progeny that are not their own. Clearly this is not in the interests of the progeny which are to be killed or of the mother (to whom the offspring are related). This would hence be an example of SEXUAL CONFLICT (between mothers and fathers), but also a conflict between progeny and the father to whom they are not related. Note, however, that as the progeny that are killed are not related to the father, some might consider it inadvisable to apply the term 'parent-offspring conflict'. *See also* GAME THEORY; OPTIMALITY MODELS; TRADE-OFF. [L.D.H.]

REFERENCES

Briskie, J.V., Naugler, C.T. & Leech, S.M. (1994) Begging intensity of nestling birds varies with sibling relatedness. *Proceedings of the Royal Society of London, Series B* **258**, 73–78.

Haig, D. (1993) Genetic conflicts in human-pregnancy. *Quarterly Reviews in Biology* **68**, 495–532.

Trivers, R.L. (1985) *Social Evolution*. Benjamin/Cummings, Menlo Park, Calif.

---

parent–offspring regression *See* ADDITIVE GENETIC VARIANCE; MATERNAL EFFECT.

parental care Any form of behaviour by an adult which benefits its offspring. Forms of parental care vary greatly, and include: nutrient enrichment of eggs by females; nest-building and provisioning of food by birds and centipedes; lactation and preparation of a den site or nesting burrow in mammals; provision of nutrients in spermatophores by male insects; and guarding of eggs by fishes.

There are differences among higher taxa in the predominance of care by either or both sexes: over 90% of bird species have biparental care, in contrast with only 5–10% of mammal species (the rest of mammals have care provided by the female alone). Parental care beyond the egg stage is rare in insects, whereas in approximately 21% of families of bony fishes (teleosts) and 71% of families of amphibians some form of care occurs. Although patterns of parental care are phylogenetically conservative in some taxa (e.g. universal lactation by females in mammals), some species differ from close relatives in provision of care by males, females or both. Examples include fishes in the family Cichlidae and shorebirds in the genus *Calidris* (infraorder Charadriides) (Fig. P4).

Parental care is expected to evolve when the benefits from enhanced offspring survival exceed the costs to the parent, including reduced growth, survival or ability to breed again. Experimental and theoretical studies have shown that a parent's best option will depend in part on the response of the other parent to changes in its input. Thus, a joint evolutionary stability may be achieved, where both parents are selected to provide care. This is expected when the efforts of both adults are required to rear young successfully. Examples include geese, seabirds and some cichlid fishes, where the benefits of biparental cooperation during extended periods of brood-rearing are thought to exceed the benefits of deserting offspring and attempting to breed again with a new partner. Uniparental male care, exhibited by many teleost fishes, is thought to have evolved because defence of a nesting site does not greatly compromise mating success. Males can typically guard and ventilate the eggs of more than one female at a time. In contrast, uniparental female care may be the norm in mammals because internal FERTILIZATION and the evolution of lactation may constrain the females' opportunities to desert the brood. Male mammals may gain more from deserting their mates attempting to breed again than they gain from remaining with the female to rear the young.

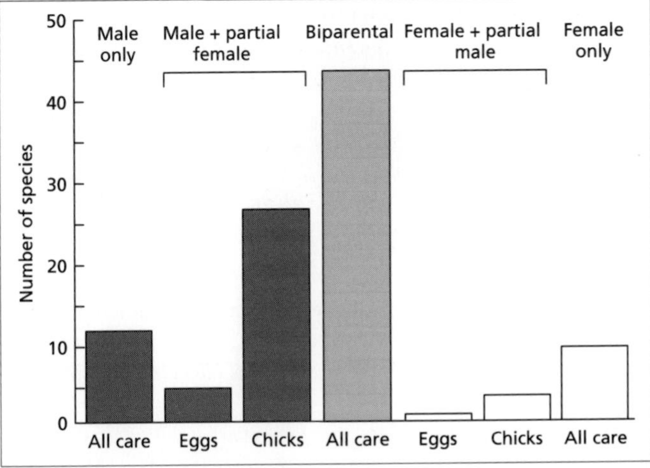

Fig. P4 Numbers of species of shorebirds (infraorder Charadriides) where each sex provides different amounts of parental care. (After Székely, T. & Reynolds, J.D. (1995) Evolutionary transitions in parental care in shorebirds. *Proceedings of the Royal Society of London, Series B* **262**, 57–62.)

Thus, differences among taxa in parental care have evolved as adjustments to the competing interests of each parent according to CONSTRAINTS and opportunities set by differences in life histories and ECOLOGY. *See also* ADOPTION; BEHAVIOURAL ECOLOGY; LIFE-HISTORY TRAITS; PARENTAL INVESTMENT; PARENT–OFFSPRING CONFLICT. [J.D.R.]

REFERENCES

Clutton-Brock, T.H. (1991) *Parental Care*. Princeton University Press, Princeton, N.J.

Gross, M.R. & Sargent, R.C. (1985) The evolution of male and female parental care in fishes. *American Zoologist* **25**, 807–822.

Székely, T. & Reynolds, J.D. (1995) Evolutionary transitions in parental care in shorebirds. *Proceedings of the Royal Society of London, Series B* **262**, 57–62.

parental investment A concept defined by Trivers (1972) as any behaviour by a parent which increases the FITNESS of that parent's offspring at a cost to the parent's own future reproductive success, including activities such as care of offspring or the provisioning of EGGS. Trivers' definition requires that parental investment be measured in fitness benefit to the offspring, but the term is often used loosely to mean energy invested in PARENTAL CARE. The level of parental investment necessary to maximize a parent's inclusive fitness will usually differ from the level required to maximize the fitness of the recipient offspring, leading to PARENT–OFFSPRING CONFLICT. Levels of investment in the two sexes strongly influence SEXUAL SELECTION and the MATING SYSTEM of the species: if one sex invests much more than the other, then the non-investing sex will usually compete over access to the investing one, and the investing sex will usually be choosier about mates. Species with substantial male parental investment thus often exhibit MONOGAMY, POLYANDRY or SEX ROLE REVERSAL, while those with little male investment are often polygynous (*see* POLYGYNY). *See also* ADOPTION; LIFE-HISTORY EVOLUTION; OPTIMALITY MODELS; REPRODUCTIVE EFFORT; TRADE-OFF. [J.C.D.]

REFERENCE

Trivers, R.L. (1972) Parental investment and sexual selection. In: *Sexual Selection and the Descent of Man* (ed. B. Campbell), pp. 136–179. Aldine Press, Chicago, Ill.

parental manipulation hypothesis The hypothesis that the EVOLUTION of eusociality in animals is based on the retention of mature offspring due to manipulation by their parents. Parental manipulation of offspring has been involved in the evolution of eusociality and in sociality in general because offspring forego reproduction by staying with their parents and providing help for their parents' reproduction rather than reproducing on their own. Other factors such as MUTUALISM and KIN SELECTION also appear to have played a role in the evolution of eusociality.

MANIPULATION by parents of their offspring increases parental inclusive FITNESS at the expense of that of their worker offspring. Manipulation involves one party being pushed by another party into performing certain activities that do not result in maximization of fitness of the manipulated party. Hence, unreliable signals are used by one party in COMMUNICATION to make other individuals perform activities to their own disadvantage. Queens of some species of SOCIAL INSECTS manipulate their daughters or other females into foregoing reproduction. This is done by behavioural DOMINANCE, for example by means of pheromonal communication that controls the reproduction and other activities of workers. The queen may also directly manipulate CONSPECIFICS by consumption of any eggs laid by workers. Similar cases of manipulation have been reported for bird species with helper breeding systems. Manipulation includes restriction of the amount of food provided to offspring of social insects, and this influences their size as adults. A small daughter is less likely to benefit by own reproduction, and her reproductive success may be maximized by providing help for raising near relatives rather than breeding herself. *See also* BEHAVIOURAL ECOLOGY; SOCIOBIOLOGY. [A.P.M.]

Paris Commission (PARCOM) The body responsible for the implementation of the Paris Convention for the Protection of the Marine Environment for the north-east Atlantic. All countries bordering on this area are signatory. It prepares controls on specific substances through binding 'Decisions' and non-binding 'Recommendations'. The Convention contains general provisions to reduce POLLUTION from land-based sources and offshore installations and to ban all types of DUMPING AT SEA. It recognizes the need for MONITORING and assessment as a basis for decisions. *See also* UNITED NATIONS PROGRAMMES AND BODIES. [P.C.]

Park Grass Experiment The longest-running field experiment in plant ecology, at ROTHAMSTED in south-east England. Begun in 1856 as a demonstration of the fertilizer requirements of a hay crop, it quickly became clear that different nutrient inputs have profound consequences on plant COMMUNITY structure and function ('NUTRIENTS matter'). The 93 plots (each more than 200 m^2) are treated with one of four levels of NITROGEN (applied in two forms: acidifying ammonium sulphate and non-acidifying sodium nitrate), and one of two levels of each of POTASSIUM, PHOSPHORUS and farmyard manure. There are four target levels of soil pH (4, 5, 6 and 7) adjusted by periodic application of LIME. The hay crop is cut twice per year; in early June and late October. SPECIES RICHNESS varies from more than 40 on the unfertilized, limed control plots, to just a single species on the most acid, high-nitrogen plots. The main determinant of annual net PRODUCTIVITY is rainfall just before and during the rapid growth phase (April and May). The main interest in the experiment is in the determination of species richness and relative ABUNDANCE, and in MONITORING the effects of long-term environmental change (GLOBAL WARMING, acid rain, etc.). The plots are as close to equilibrium plant communities (*see* EQUILIBRIUM, COMMUNITIES) as are known anywhere. *See also* EXPERIMENTAL ECOLOGY. [M.J.C.]

REFERENCE

Tilman, D., Dodd, M.E., Silvertown, J., Poulton, P.R., Johnston, A.E. & Crawley, M.J. (1994) The Park Grass Experiment: insights from the most long-term ecological study. In: *Long-term Experiments in Agricultural and Ecological Sciences* (eds. R.A. Leigh & A.E. Johnston), pp. 287–303. CAB International, Wallingford.

Park's experiments

In a series of studies, Thomas Park (1948, 1954, 1957, 1962) carried out 'bottle' experiments (Kareiva 1989) on two species of four beetle, *Tribolium confusum* (1) and *T. castaneum* (2). Both species are frequently found as stored-product pests and they can be conveniently kept under laboratory conditions. Populations were kept in glass tubes with 8 g of medium (95% flour + 5% yeast). All adults, pupae and larvae were counted every 30 days, and both single-species and mixed-species populations were kept. An interesting aspect of the EXPERIMENTAL DESIGN, which proved instructive, was that populations were kept at a number of temperatures (24°C, 29°C and 34°C) and relative humidities (30% and 70%), and each type of population (single or mixed), under all six environmental conditions, was replicated 30 times.

Single-species results

Apart from *T. castaneum* at 24°C and 30% humidity, all single-species populations persisted and the tables below show the average population densities during the course of the experiment. The value of 3, for *T. castaneum* at 24°C and 30% humidity, is the average population density until extinction.

T. confusum (1)	34°C	29°C	24°C
70%	41	33	29
30%	24	30	30

T. castaneum (2)	34°C	29°C	24°C
70%	30	50	45
30%	10	19	3

Using these average densities as an indication of success, Park compared the relative performance of these two species (alone) at each temperature/humidity combination. The results of this comparison are shown below, with an asterisk indicating a statistically significant difference between the two species.

	34°C	29°C	24°C
70%	1 > 2	2 > 1*	2 > 1*
30%	1 > 2*	1 > 2*	1 > 2*

Except for 34°C/70%, each of the physical environments significantly favours one species over another. These results are due to INTRASPECIFIC COMPETITION and the effect of the physical environment upon it.

Two-species results

Park found that while single-species populations persisted (with the exception of *T. castaneum* at 24°C/30% humidity), mixed-species populations did not. In each population only one species persisted, the other was eliminated. However, unlike most previous 'bottle' experiments (*see* GAUSE'S EXPERIMENTS) it was not always the same species that was eliminated, even under the same environmental conditions. This 'indeterminate' outcome of COMPETITION is, however, one of the predicted possible outcomes of the LOTKA–VOLTERRA MODEL. The table below shows the percentage of replicates that *T. castaneum* won.

	34°C	29°C	24°C
70%	100	86	29
30%	10	13	0

Significance of results

These experiments of Thomas Park illustrate two interesting and possibly universal features of competition. First, that the outcome of two species' competition can sometimes be predicted from the performance of each species on its own (i.e. 29°C/70% and 24°C/30%), but also sometimes it cannot (i.e. 34°C/70% and 24°C/70%).

Second, that the outcome of competition between two species can vary along an environmental gradient. Park's experiments go from hot-humid conditions (34°C/70%) to cool-dry conditions (24°C/30%). At the hot-humid end, *T. castaneum* eliminates *T. confusum* all the time. At the cool-dry end, *T. confusum* eliminates *T. castaneum* all the time. In the middle region either species can win, but *T. castaneum* wins more often towards the hot-humid end, and *T. confusum* wins more often towards the cool-dry end.

If we imagine that the isoclines in the Lotka–Volterra phase plane graphs will move their position depending upon environmental conditions, then these *Tribolium* results are still understandable in terms of the Lotka–Volterra model. Figure P5 shows a series of phase plane graphs, going from hot-humid (left) to cool-dry (right). The arrow(s) in each diagram indicate the equilibrium point. Notice that the species predicted to win moves from always *T. castaneum* (left), through either species, to always *T. confusum*. *See also* EXPERIMENTAL ECOLOGY. [B.S.]

REFERENCES

Kareiva, P. (1989) Renewing the dialogue between theory and experiments in population ecology. In: *Perspectives in Ecological Theory* (eds J. Roughgarden, R.M. May & S.A. Levin), pp. 68–88. Princeton University Press, Princeton, N.J.

Park, T. (1948) Experimental studies of interspecific competition: 1. Competition between populations of the flour beetles *Tribolium confusum* and *Tribolium castaneum*. *Ecological Monographs* **18**, 265–307.

Park, T. (1954) Experimental studies of interspecies competition. II. Temperature, humidity, and competition in two species of *Tribolium*. *Physiological Zoology* **27**, 177–238.

Park, T. (1957) Experimental studies of interspecies competition. III. Relation of initial species proportion to competitive outcome in populations of *Tribolium*. *Physiological Zoology* **30**, 22–40.

Park, T. (1962) Beetles, competition, and populations. *Science* **138**, 1369–1375.

parsimony There is an infinite number of hypotheses that might be suggested by a particular set of data. Scientists and philosophers rely on the principle of simplicity (parsimony) to pick the simplest hypothesis; the hypothesis that explains the data in the most economical manner. Simplicity, economy and parsimony all refer to the same general concept. However, the nature of this concept is not well understood and has been debated by philosophers without resolution for many years. Karl Popper viewed the simplest hypothesis as the most falsifiable and tied simplicity to information content. Elliot Sober suggested that simplicity and information content are in some ways antagonistic—that we strive for a balance between parsimony and information content. Both agree that no really informative hypothesis can be limited by the data then available.

In phylogenetic systematic analysis, for example, one begins with the assumption that all similar characteristics among the species being studied are homologous; i.e. that they have only a single evolutionary origin. If all hypotheses of HOMOLOGY are correct, the resulting PHYLOGENETIC TREE will list each characteristic only once; hence, the entire dataset will be described based on the initial input. In most actual studies, at least some of the characteristics conflict with each other by providing support for incompatible phylogenetic groupings. Under such circumstances the preferred hypothesis is the one requiring the fewest *ad hoc* statements of independent evolution necessary to explain the full array of available data. This is the most parsimonious, or simplest hypothesis. [D.R.B.]

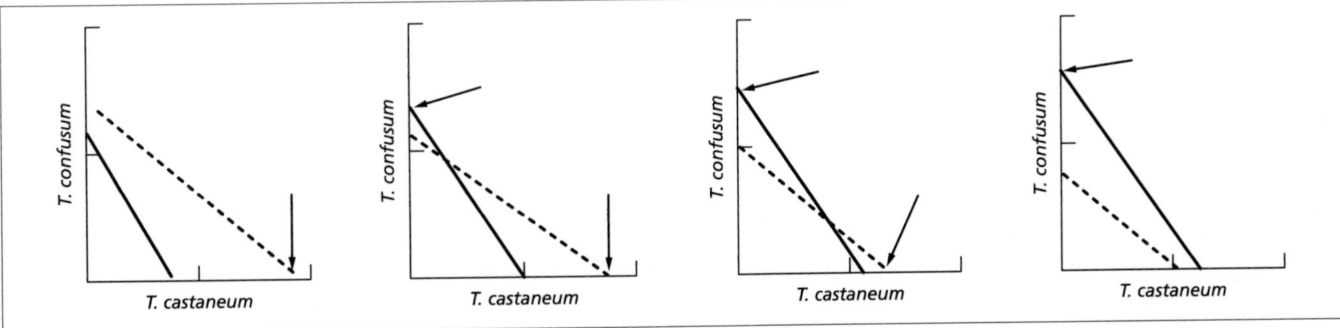

Fig. P5 A series of phase plane graphs along an environmental gradient (from hot-humid on the left to cool-dry on the right) in which two beetles compete.

parthenogenesis The production of a viable embryo from an unfertilized EGG. This may occur sporadically (ptychoparthenogenesis) or as a permanent feature of the LIFE CYCLE. Having many independent origins in higher plants and animals, parthenogenesis occurs through a wide variety of cytogenetic mechanisms. Parthenogenetic eggs may be haploid, as a result of normal oogenesis, or diploid owing to aberrant oogenesis.

Haploid eggs develop into males (i.e. ARRHENOTOKY). Arrhenotoky occurs in haplo-diploid hymenopterans, where it can be used by females to control SEX RATIO, and in some cyclical parthenogens, such as aphids and cladocerans, when they switch from agametic cloning to normal sexual reproduction.

Diploid eggs develop into females (i.e. THELYTOKY). Apomictic thelytoky, as occurs for example in dandelions and cyclical parthenogens, involves a single mitotic maturation division, so preserving the maternal genome (Fig. P6). In automictic thelytoky, meiosis is accompanied by some mechanism of restoring DIPLOIDY. Such 'restitution' may be pre-, intra- or postmeiotic. In premeiotic restitution, chromosomes undergo mitotic doubling but the nucleus fails to divide (ENDOMITOSIS). Meiosis then begins with twice the usual number of chromosomes and bivalents are formed by the pairing of sister chromosomes (Fig. P7), so nullifying the effects of recombination and segregation. Premeiotic restitution is found, for example, in the grasshopper *Warramaba virgo* and in the earthworm *Dendrobaena rubida*.

Intrameiotic restitution is achieved by suppression of either the first or second meiotic division. When first arising in a heterozygous mother, intrameiotic restitution generates OVA that differ genetically among themselves but also are more homozygous than the maternal nuclear genome. If repeated in successive generations, intrameiotic restitution would produce a series of completely homozygous CLONES.

In postmeiotic restitution, meiosis is followed either by pairing and fusion of the pronuclei or by endomitosis at the first cleavage division in embryogenesis (Fig. P8). Fusion of sister pronuclei generates HOMOZYGOSITY and occurs, for example, in the brine shrimp, *Artemia salina*, certain scale insects, sawflies and oribatid mites. Fusion of non-sister pronuclei occurs in the dipteran, *Lonchoptera dubia*, and in a race of the bagworm moth, *Solenobia triquetrella*. As chiasmata are suppressed in female lepidoterans, postmeiotic restitution in *S. triquetrella* preserves the maternal genome. Endomitosis at the first cleavage division destroys all HETEROZYGOSITY and is known to occur only in a few white flies, scale insects, coccids and mites. *See also* ASEXUAL REPRODUCTION. [R.N.H.]

REFERENCES
Bell, G. (1982) *The Masterpiece of Nature: the Evolution and Genetics of Sexuality*. Croom Helm, London.

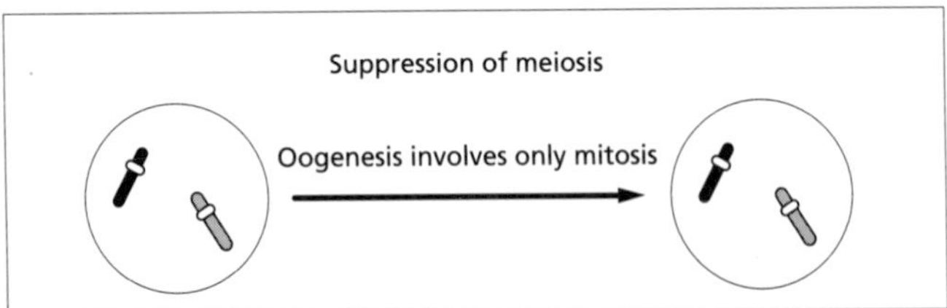

Fig. P6 Apomictic parthenogenesis. (Figures P6–P8 after Hughes, R.N. (1989) *A Functional Biology of Clonal Animals*. Chapman & Hall, London.)

White, M.J.D. (1973) *Animal Cytology and Evolution*, 3rd edn. Cambridge University Press, Cambridge.

partial refuge A REFUGE that is only available for part of the time or for a proportion of the population. [B.S.]

particle-size selectivity The ingestion of a size range of particles (usually of PHYTOPLANKTON, or SEDIMENT particles) in a proportion different from that available in the surrounding medium. SUSPENSION-FEEDING organisms that feed on phytoplankton may selectively ingest larger-sized species because food value is often proportional to phytoplankton cell size. Deposit-feeding organisms often selectively ingest smaller sediment particles because these have a larger surface area for the attachment of bacteria or nutritional organic molecules, and can therefore have a higher food value per unit mass than larger sediment particles. Particle-size selectivity can occur via behavioural mechanisms or as a result of fixed properties of the feeding appendages. *See also* FEEDING TYPES, CLASSIFICATION. [V.F.]

particulate organic matter (POM) Dead organic material (DETRITUS) either suspended in the water column or settled on the substratum of aquatic habitats. It may be divided into COARSE-PARTICULATE ORGANIC MATTER (CPOM), such as leaf fragments, and FINE-PARTICULATE ORGANIC MATTER (FPOM), such as sediments. *Cf.* DISSOLVED ORGANIC MATTER. [P.C.]

partition coefficient The ratio of the equilibrium concentrations of a substance distributed between two phases in contact with each other (often symbolized as K or P) (e.g. *see* OCTANOL–WATER PARTITION COEFFICIENT). *See also* FUGACITY; MACKAY FATE MODELS. [P.C.]

parts per billion *See* PPB.

parts per million *See* PPM.

pascal (Pa) The SI UNIT of PRESSURE (i.e. force per unit of area). 1 Pa = 1 N m^{-2}. Relationship with old units: 1 Pa = 0.01 millibar = 0.02089 pound force per square foot. Atmospheric pressure is about 10^5 Pa. [J.G.]

passive dispersal A process that leads to the movement of individuals away from others with which they had been associated (often parents or siblings) without any locomotory effort by the individuals concerned. In contrast with ACTIVE DISPERSAL, the individual is carried passively, for example by wind or water. Many plant seeds and larvae of marine animals are described as passive dispersers, though the distinction between active and passive dispersal is not clearcut because many organisms have evolved adaptations that enhance the chances of passive dispersal occurring. *Cf.* DISPERSION. *See also* MIGRATION; VAGILITY. [R.H.S.]

passive drift-entry *See* DRIFT, FRESHWATER.

passive suspension feeding The trapping of suspended particles by animals that do not propel water more than a few particle

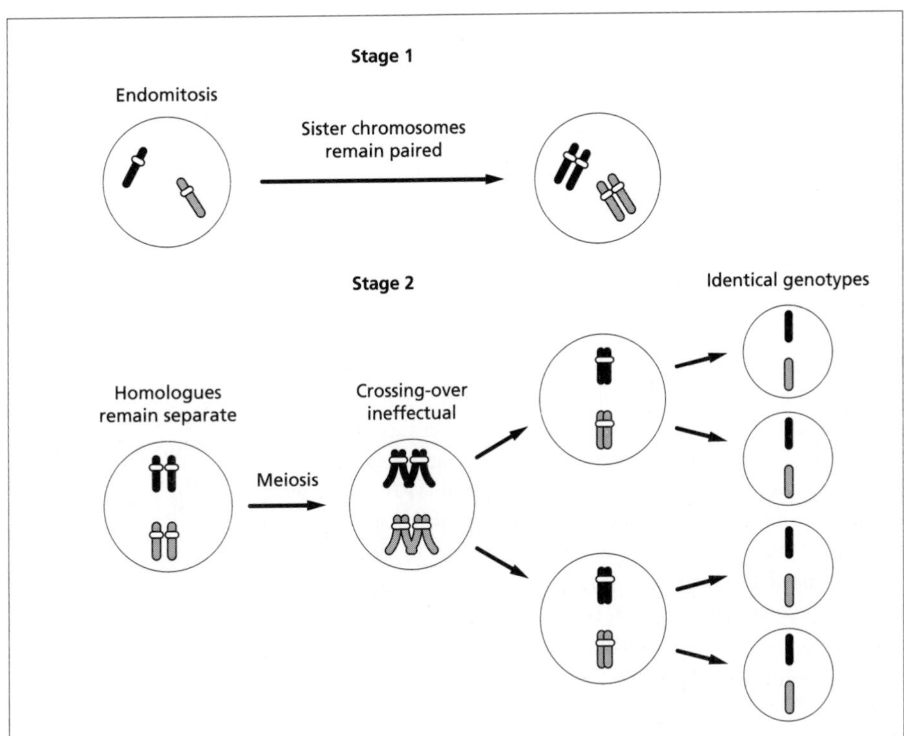

Fig. P7 Automictic parthenogenesis: premeiotic restitution.

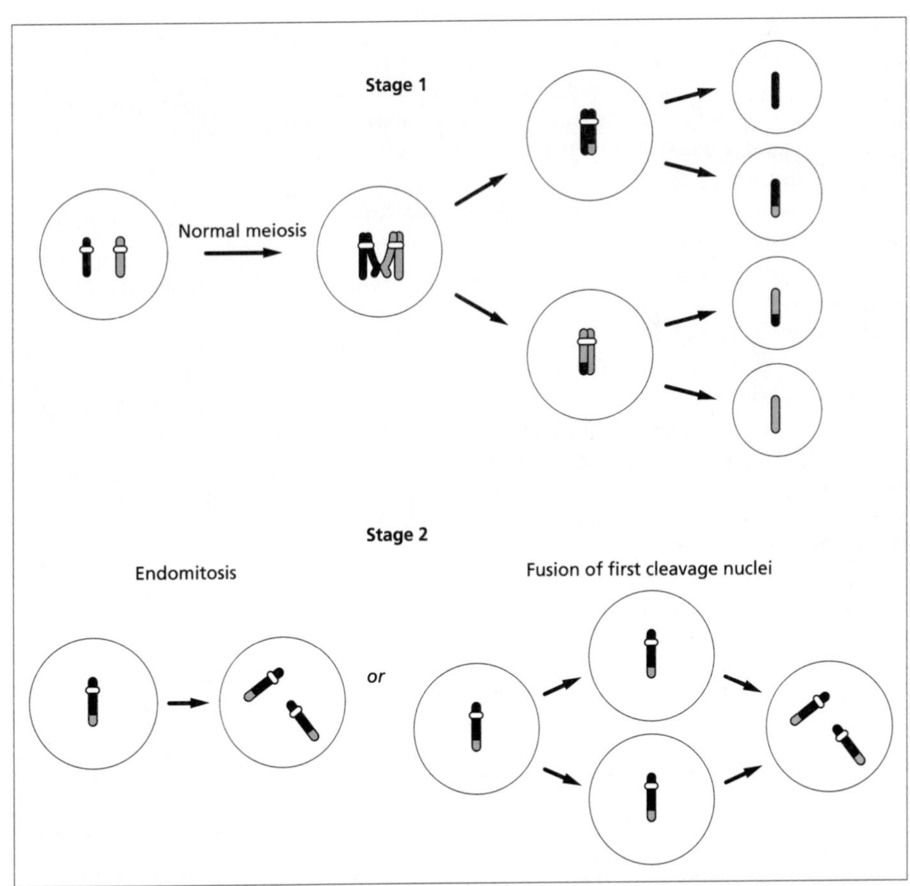

Fig. P8 Automictic parthenogenesis: postmeiotic restitution.

diameters from their collection appendages and that appear to trap particles by direct interception and by positioning themselves in BOUNDARY LAYER situations where particle fluxes are high. In contrast, active suspension-feeders move water to create their own feeding CURRENTS. *See also* FEEDING TYPES, CLASSIFICATION. [V.F.]

pasture GRAZING land that is managed to provide appropriate YIELD in terms of quantity and sometimes quality of herbs and grasses. *See also* AGRICULTURE. [P.C.]

patch dynamics An aspect of METAPOPULATION dynamics that describes the proportion $p(t)$ of HABITAT patches that is occupied at time t. Two parameters describe the key processes: the rate of COLONIZATION of empty patches (m); and the rate of extinction of local populations within patches (e). S. Levins, in 1970, proposed a model describing the dynamics of the set of local populations in which the occupied patches (fraction p) contribute to recolonization of empty patches (fraction $(1-p)$) at a rate $mp(1-p)$, and occupied patches become empty at a rate ep. This model leads to an equilibrium fraction of occupied patches, $p^*=1-e/m$. Levins' model assumes that local extinction occurs independently in different patches — synchronized local extinction may eliminate the metapopulation, which has obvious implications for pest management. Increasing the MIGRATION rate decreases the fraction of empty patches.

A simple message of Levins' model for conservation is that increasing links between patches will increase the fraction of occupied patches of habitat. Consideration of interacting species, however, forces us to modify this message because the within-patch dynamics of local populations sum to give the POPULATION DYNAMICS of species across the metapopulation. Using INTERSPECIFIC COMPETITION as an example, there are three main features of a metapopulation structure for pairs of competing species.

1 Patches are of four types:
  (a) empty;
  (b) one species present (INTRASPECIFIC COMPETITION);
  (c) another species present (intraspecific competition);
  (d) both species present (interspecific competition).

2 Pairs of competing species are likely to form independent AGGREGATIONS within patches, provided patches are sufficiently isolated (m not too large).

3 A consequence of **1** and **2** is reduced interspecific competition overall, because few patches have both species present; hence an unstable competitive interaction may be stabilized because there are effectively REFUGES from interspecific competition.

Thus a metapopulation structure may stabilize a competitive interaction, provided migration between patches is not too common. A more sophisticated message for conservation, therefore, is that COEXISTENCE and stability may be lost if there are too many links for migration — the adverse effects of patch extinction and COMPETITIVE EXCLUSION need to be balanced. *See also* AGGREGATION MODEL OF COMPETITION. [R.H.S.]

REFERENCES

Atkinson, W.D. & Shorrocks, B. (1981) Competition on a divided and ephemeral resource: a simulation model. *Journal of Animal Ecology* **50**, 461–471.

Shorrocks, B. & Swingland, I.R. (1990) *Living in a Patchy Environment*. Oxford University Press, Oxford.

patch use It is assumed that foragers perceive food as occurring in discrete patches separated by unproductive habitat. Two important behavioural decisions considered by OPTIMAL FORAGING THEORY (OFT) are which patches to exploit and how long to remain foraging in a patch before moving to another. The decision of which patches to exploit has been studied in relation to risk, of starving, of being outcompeted or of being killed by predators while foraging.

Risk of starvation may be a severe problem in animals, such as small homeotherms, with precarious energy budgets. 'Risk-sensitive foraging' models predict that starving foragers should be 'risk-prone' by choosing variable patches that, while posing the risk of low yield, also offer the chance of a high yield that could restore the ENERGY BUDGET into positive balance. Conversely, energetically secure foragers should be 'risk averse', choosing less-variable patches that maximize the long-term rate of net energy gain.

COMPETITION among foragers has been modelled by derivatives of the ideal-free distribution and ideal-despotic models, originally developed to predict the distribution of birds competing for nesting sites and employing aspects of GAME THEORY. Patch quality is devalued by the presence of competitors and this, together with competitive status, can be incorporated as a constraint on energy-maximizing decision-making.

Risk of predation has been accommodated by changing the currency used in the MARGINAL VALUE THEOREM (see below) from maximization of energy gained per unit time to maximization of energy gained per unit risk. Here, the optimal decision depends on patch productivity and on risks of predation within and between patches.

The decision of how long to stay in a patch has most successfully been addressed by a particular model, the marginal value theorem (MVT), which has emerged as one of the most enduring and influential models in BEHAVIOURAL ECOLOGY. The MVT uses the energy maximization premise (EMP) to predict optimal patch-residence time on the basis of relative productivity of the exploited patch and on distances between patches.

A different approach is required to model 'central-place foragers', such as birds provisioning a nest of young, limpets making foraging excursions from a refuge, or deposit-feeding invertebrates centred upon a burrow. For provisioners, behavioural decisions include how much food to carry in a single load back to the central place, how much time to spend foraging at successive distances from the central place, and how selectively to forage at these distances. It is predicted that food load, foraging time and selectivity should decrease further away from the central place, and that these trends should be reinforced by increasing productivity of the habitat and decreasing energy demands on the provisioner. Predictions for grazers and deposit-feeders foraging from a refuge or burrow include the optimal apportionment of time between foraging or refuging and the degree to which overlap should be avoided in successive foraging excursions. [R.N.H.]

patchy habitat An environment in which organisms tend to aggregate in patches where a resource (such as food or a breeding site) is at high density. The areas between patches may have no usable resource at all. A patchy HABITAT is defined in relation to a particular species. A particular stage in the life cycle of a species (usually juveniles) may be confined to a patch. Examples of patchy habitats include seeds (infested by weevils), fruits (infested by fruit flies), carcasses (infested by blowflies) and puddles (infested by mosquitoes). *See also* METAPOPULATIONS; PATCH DYNAMICS. [R.H.S.]

Paternity

The assessment of paternity of broods is made by genetic methods such as external polymorphisms, enzyme POLYMORPHISM or GENETIC FINGERPRINTING including microsatellite methods (*see also* MOLECULAR ECOLOGY).

Extra-pair paternity

The occurrence of extra-pair paternity (offspring fathered by males other than the male socially bonded with a female) has been studied in a large number of species during recent years. It varies from complete certainty of paternity in some species of birds, to species in which more than two-thirds of all offspring are fathered by males other than the resident male.

Recent methodological developments have allowed assessment of the frequency of extra-pair paternity and the unambiguous assignment of offspring to sires. Behavioural observations of copulation behaviour have suggested that extra-pair paternity should be less frequent than determined by biochemical paternity assessment methods. This apparent discrepancy between behavioural and genetic methods of assessment of paternity can be explained in at least two different ways. First, extra-pair males may transfer a larger number of sperm during EXTRA-PAIR COPULATIONS, and this may result in a larger proportion of offspring sired by extra-pair males. Such males may be able to transfer more sperm because they allocate more sperm to extra-pair females, or because they have copulated less frequently during the recent past and therefore are able to transfer more sperm. The probability of FERTILIZATION is directly proportional to the relative number of sperm contributed by a male, when the sperm are of approximately similar age.

Second, extra-pair males of a number of bird species appear to be able to time their copulations more optimally with respect to the timing of fertilization than pair males. This may be due to extra-pair males being better able to determine and respond to female signals of fertility, or to females explicitly soliciting copulations from extra-pair males relatively more often than from pair males when the probability of fertilization is high.

The frequency of extra-pair paternity has been related to a number of characteristics of extra-pair and pair males. The most attractive males in terms of physical or behavioural attributes appear often to father a disproportionate NUMBER OF OFFSPRING in neighbouring nests, and this is due to females soliciting extra-pair copulations from such males. Recent studies of primates have challenged the prevalent paradigm that dominant males should sire most offspring by monopolizing copulations during the peak of female fertility. A number of studies have shown that subordinate males fertilize a large proportion of offspring. Furthermore, older males apparently experience an advantage in terms of extra-pair paternity. Males that have managed to survive their first years thus appear to be preferred as sires for extra-pair offspring.

Control of paternity

The control of paternity appears generally to be determined by females because females rather than males determine the fate of sperm within the female reproductive tract. Females also often control which male is able to copulate shortly before the time of fertilization. For example, experimental studies of widowed female tree swallows (*Tachycineta bicolor*) have shown that females that were sexually faithful to their mates during the laying of their first EGGS, remained faithful during the laying of eggs after male removal, while females that had extra-pair offspring when laying their first eggs, had a higher proportion of extra-pair offspring when laying eggs after male removal. Female choice could therefore be considered a continuously reinforcing process with the effects of female choice of mates being enhanced during the fertilization process. Similar results for blue tits (*Parus caeruleus*) have shown that females mated to attractive males have few extra-pair offspring, and such females are faithful to their mates despite not being guarded intensely by their males. Extra-pair offspring in this species have higher probabilities of survival and local RECRUITMENT than pair offspring. This suggests that the main benefit of extra-pair paternity in this species is indirect FITNESS benefits in terms of general viability genes.

Male assessment of paternity

Pair males appear to be able to assess their certainty of paternity to some extent. For example, males of some species engage in copulations with their mates relative to the sexual interest that neighbouring males have shown for the female. Pair males also provide PARENTAL CARE for offspring relative to their certainty of paternity (*see* PATERNITY CERTAINTY HYPOTHESIS).

Implications of extra-pair mating

Traditionally, studies of reproductive success have simply recorded the number of offspring produced by particular individuals without considering certainty of paternity. Extra-pair paternity is prevalent in a large number of species in different groups, including invertebrates, fishes, reptiles, birds and mammals. The variation in realized reproductive success after accounting for certainty of paternity is frequently considerably higher than when considering only apparent reproductive success. The intensity of SEXUAL SELECTION as assessed by realized reproductive success is therefore higher than was previously thought. Comparative studies of extra-pair paternity in birds have shown that extra-pair paternity is particularly common among species with sexual differences in appearance. For example, differences in plumage coloration are associated with sexual selection, and extra-pair paternity through female choice of extra-pair mates thus appears to be a major component of sexual selection. *See also* MATING SYSTEMS; SPERM COMPETITION. [A.P.M.]

REFERENCE

Birkhead, T.R. & Møller, A.P. (1992) *Sperm Competition in Birds: Evolutionary Causes and Consequences*. Academic Press, London.

paternity certainty hypothesis The hypothesis that the evolution of male PARENTAL CARE is promoted by a high certainty of PATERNITY. Parental care is costly in terms of time and energy, and thus it is disadvantageous for parents to provide care for non-KIN. Therefore, a high certainty of paternity is assumed to promote the evolution of male parental care. Extra-pair paternity is frequently high, even in species with extensive male care, but it remains unknown whether males that care for non-kin have any better options available to avoid paying the cost of alloparental care. A number of studies of socially monogamous birds with extra-pair paternity have failed to demonstrate a reduction in the relative amount of male feeding of offspring, when a larger proportion of offspring are sired by extra-pair males. Comparison of the relative amount of male care provided for first and second broods that differ in extra-pair paternity has shown that males of the reed bunting (*Emberiza schoeniclus*) and the barn swallow (*Hirundo rustica*) provide more care when they are more certain of their paternity. Comparative studies of extra-pair paternity in birds have shown that males generally provide more parental care when they are more certain of their paternity. This may arise as a consequence of males of species with high uncertainty of paternity on average providing less care, or males intraspecifically adjusting their parental care to their certainty of paternity. Species with exclusive male parental care, such as sea horses and pipefishes, and phalaropes and other polyandrous birds, generally have a very high certainty of paternity. [A.P.M.]

paternoster lakes *See* LAKES.

Path analysis

Path analysis was introduced by Sewall Wright. It resembles multiple regression as a technique for disentangling CORRELATION and causation, but is more general, allowing tests of alternative models of causation. It is also used to predict correlation when causal pathways are postulated, and for purposes of graphical illustration. For a good introduction, see Sokal and Rohlf (1995, pp. 634–639).

Basic concepts for causation, correlation, multiple regression and path analysis are illustrated with a simple example. In humans there is a correlation between offspring and parent height of about $r=0.25$ (where r is the correlation coefficient). Consider a number of widows who remarry after their children are fully grown. There is a tendency for people to marry assortatively by height. The resulting correlation between mother and stepfathers also leads to a correlation between stepson and stepfather. This is clearly not a result of any direct influence of stepfather on stepson's height, but rather due to the (causal) influence of mother's height on son's height, and the correlation between mother and stepfather.

Multiple regression

Path analysis is an extension of multiple regression, and we first consider methods of multiple regression. In the example there are two variables, mother ($X1$) and stepfather ($X2$), which could affect a third variable, son (Y). Least squares regression fits an equation:

$$Y=\alpha+\beta_1 X1+\beta_2 X2$$

using as input the measured correlations among the three variables. α is the intercept and β_1 and β_2 are partial regression coefficients, describing the effect of each variable, $X1$ and $X2$, on Y when the other is held fixed; in the example the partial regression coefficient for stepson on stepfather is expected to equal zero. The main assumptions of multiple regression are that the independent variables, $X1$ and $X2$, act additively, and affect Y linearly. In this example, the path coefficients are standardized partial regression coefficients, obtained from multiple regression after values for the traits have been standardized by dividing by their STANDARD DEVIATION (for some applications the unstandardized coefficients are employed).

Path models

The path model for the example is shown in Fig. P9(a). The double-headed arrow reflects the correlation between $X1$ (mother) and $X2$ (stepfather), and single-headed arrows represent potential causal pathways. The correlation between any two variables in the path is the sum of values for all paths between the two variables, where the value for a path is the product of all coefficients. In the example, the correlation between $X1$ and Y is:

$$r_{y1}=p_{y1}+r_{12}p_{y2}$$

The correlation between $X2$ and Y is:

$$r_{y2}=p_{y2}+r_{12}p_{y1}$$

The basic method of path analysis (and multiple regression) is to solve such a set of simultaneous equations to calculate path coefficients from measured correlations.

Relationships between true parents and offspring under ASSORTATIVE MATING are shown in Fig P9(b). Since parental genotypes affect offspring, but offspring genotypes do not affect parents, the chain of causality is obvious. The correlation between mother and offspring is $^1/_2h^2+r_{mf}{}^1/_2h^2$, where h^2 is the HERITABILITY and r_{mf} is the correlation between mothers and fathers. The example illustrates how heritability can be estimated using measured correlations, and it is also easy to see how the alternative experimental method of disentangling causation and correlation to heritability could be employed. In an experiment the correlation between mothers and fathers could be removed by randomly mating, hence $r_{mf}=0$. Alternatively, all variation through one sex (e.g. the male) could be experimentally removed by mating females to a single male, or to males with the same PHENOTYPE.

Testing theories

Both the above examples are straightforward applications of multiple regression, and the main value of using path diagrams is graphi-

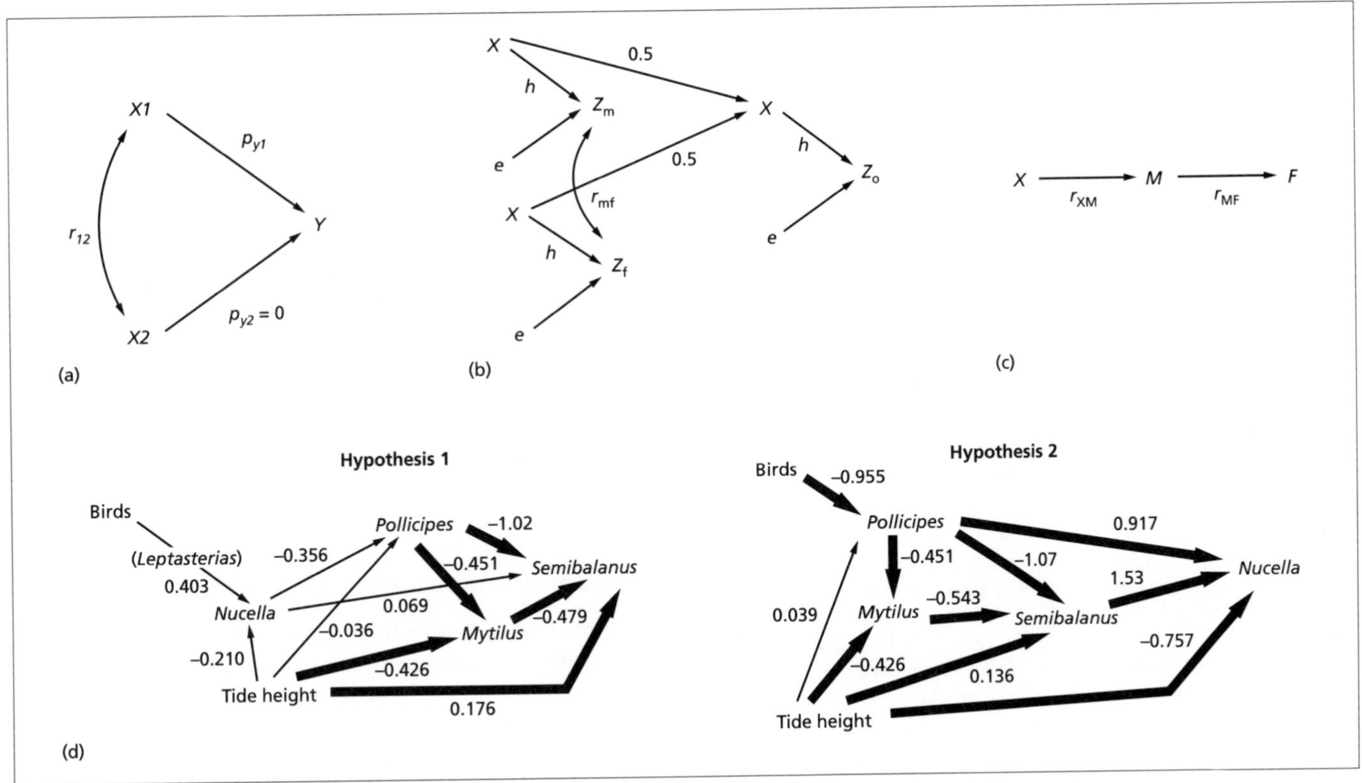

Fig. P9 (a) Path model for mothers (*X1*), their sons (*Y*) and the sons' stepfathers (*X2*) for height. (b) Path model for mothers (Z_m), fathers (Z_f) and their offspring (Z_o) when there is a correlation between mothers and fathers (r_{mf}). *X* represents the genotype and e represents the environmental value (subscripts omitted for clarity); *h* (the square root of the heritability) and 0.5 are values of path coefficients. (c) Path model for the chain of causality between genotype (*X*), morphology (*M*) and life-history trait (*F*). In the absence of other correlated variables, the path coefficients are identical to correlations. (d) Two alternative models to explain effects of birds on populations of intertidal invertebrates. Thick arrows represent statistically significant paths.

cal. To illustrate wider uses of path analysis we consider two more examples. The first concerns theory. It has long been known that life history traits—such as FECUNDITY—tend to have lower heritability than morphological traits. To explain this observation, Price and Schluter (1991) postulated a simple causal model (Fig. P9c): genetic variation affects morphology and morphological variation affects fecundity but genetic variation does not directly affect fecundity. If there are no other causal connections between variables, the correlation between GENOTYPE and fecundity ($r_{XF} = r_{XM}r_{FM}$) is necessarily lower than that between genotype and morphology (r_{XM}), and from this it follows that the heritability of fecundity will be lower than that of morphology.

Hypothesis testing

The second example illustrates methods of hypothesis testing in path analysis. Wootton (1994) describes interactions between birds, predatory snails (*Nucella*), acorn barnacles (*Semibalanus*), goose barnacles (*Pollicipes*) and mussels (*Mytilus*) in the rocky intertidal of Washington State, USA. He set up five cages to exclude birds, and compared population densities between them and five control plots. There were several results, and we present an abbreviated version for illustration. After 2 years there were more *Nucella* in the areas exposed to bird predation than in the exclosures, despite the fact that birds eat *Nucella*. This implies that effects operating through other species were affecting the bird–*Nucella* interaction.

Two alternative hypotheses are contrasted in the path diagrams of Fig. P9(d). In one, birds eat starfish (*Leptasterias*, not measured in the experiment) that prey on *Nucella*. In the other, birds eat *Pollicipes*. The consequent available space causes *Mytilus* and *Semibalanus* to increase. *Semibalanus* are an important food source for predatory *Nucella*, which therefore increase.

The two hypotheses were evaluated after including the physical variable tide height in both models, to control for any differences this may cause. Path coefficients were calculated using multiple regression on standardized variables (e.g. the path coefficients for *Nucella* are calculated by regressing *Nucella* abundance on presence/absence of birds for the first hypothesis, and on *Pollicipes*, *Semibalanus* and tide height in the second hypothesis). Hypotheses in path analysis are tested by comparing alternative path models in their ability to reproduce the observed correlation structure. In this study, observed and predicted correlations agree very well for hypothesis 2 ($R^2 = 0.997$), but poorly for hypothesis 1 ($R^2 = 0.555$).

The study represents a particularly good example of the application of path analysis, because the conclusions were corroborated by experimental manipulations of *Nucella*, *Semibalanus* and *Pollicipes*. In 11 tests the direction predicted by the paths of hypothesis 2 were corroborated by experimental MANIPULATION.

Evaluation

Path analysis is only as good as the data and the variables included. Incorporation or omission of variables may affect the whole interpretation. This can be seen in the example of Fig. P9(a), where omission of the variable *X1* would have led one (incorrectly) to infer a causal relationship between *X2* and *Y*. Computer software packages are available to compare alternative path models. One is LISREL (LInear Structural RELations), available in SPSS. Another is TETRAD II, which provides a more exhaustive search of alternative models. [T.D.P.]

REFERENCES

Price, T. & Schluter, D. (1991) On the low heritability of life history traits. *Evolution* **45**, 853–861.

Sokal, R.R. & Rohlf, F.J. (1995) *Biometry*, 3rd edn. W.H. Freeman, New York.

Wootton, J.T. (1994) Predicting direct and indirect effects: an integrated approach using experiments and path analysis. *Ecology* **75**, 151–165.

---

path models *See* PATH ANALYSIS.

pathogen A medical (human and veterinary) term for PARASITE, especially one associated with a high degree of MORBIDITY or MORTALITY. [G.F.M.]

Pattern analysis

A procedure used to detect the departure from randomness of the spatial arrangement of plants within an area (pattern). Distributions may be random, regular or, most commonly, clumped (*see* REGULARITY).

Detection of non-randomness

Pattern may be detected by comparing, for example, the observed number of individuals per QUADRAT with the expected number derived from the Poisson series (based on a POPULATION dispersed at random). In nearest-neighbour analysis, effectively distances between individual plants and their nearest neighbours and between random points and their nearest plants are determined: in a random population, plant-to-plant and point-to-plant distances are equal; in a regular population the plant-to-plant distance is greater; and in a clumped population it is smaller.

Grid or transect sampling

A more informative procedure of pattern analysis involves the measurement of the representation of a species in contiguous systematic samples, either in a grid or, more advantageously, along a TRANSECT. Values of COVER, frequency or DENSITY are taken at regular intervals (often by means of a POINT QUADRAT) on a transect and are grouped into 'basic units' frequently of five adjacent readings. Data from these 'basic units' are grouped into blocks of 2, 4, 8, 16, 32, 64, etc. Total VARIANCE between grid units is then apportioned by the usual procedure of ANALYSIS OF VARIANCE into amounts derived from the differences between units within blocks of two, between blocks of two within blocks of four, between blocks of four within blocks of eight, and so on. The results are plotted as graphs of mean square against block size. At the block size which corresponds to the SCALE of the pattern the variance is at a peak; in some cases more than one peak is shown. Strong peaks reflect intense patterns. The length of transect monitored will depend on the maximum scale of pattern it is desired to detect, and the size of the 'basic unit' will depend on the smallest scale of pattern to be detected (the 'basic unit' should be not more than half of the size of the smallest scale of pattern to be detected).

Causes and implications of patterns

Pattern may be caused by the morphology of the plant, HETEROGENEITY of the environment or by interaction of several causal factors. In plants of rosette or tussock form, the smallest pattern reflects the mean size of the individual. Pattern may also arise by spatial exclusion of one species by another. Pattern analysis may give information on morphology. The rhizome system of *Eriophorum angustifolium*, for example, gives rise to double peaks at small block sizes, reflecting the clustering around the parent shoot of shorter backwards-growing rhizomes and longer forwards-growing rhizomes, and also a pronounced peak at large block sizes showing grouping of shoots linked in a single rhizome system. The scale of pattern may reflect the vigour of a plant in different sites; scale is reduced under unfavourable conditions. Pattern may be environmentally determined as a result of general variation of a major factor (e.g. soil depth, oxygen diffusion rate) so elucidating ecological behaviour. Sociological pattern may result from the interaction of species on species and may also be affected by the microenvironment. *See also* AGGREGATED DISTRIBUTION; DISPERSION; DISTRIBUTION. [A.J.W.]

REFERENCES

Greig-Smith, P. (1983) *Quantitative Plant Ecology*, 3rd edn. Blackwell Scientific Publications, Oxford.

Kershaw, K.A. & Looney, J.H.H. (1985) *Quantitative and Dynamic Plant Ecology*, 3rd edn. Edward Arnold, London.

---

paurometabolous insects *See* INSECTS, DIVERSITY AND CONSERVATION.

PCB *See* POLYCHLORINATED BIPHENYLS.

peak zone *See* BIOSTRATIGRAPHIC UNIT.

peat An unconsolidated, stratified organic material derived largely from undecomposed or partially decomposed vegetation. It varies in its degree of DECOMPOSITION (HUMIFICATION) depending upon the nature of its content and the degree of waterlogging at the time of and since its formation. Peats formed under very wet conditions grow more quickly and are less decomposed than those formed under drier conditions. This offers some opportunity for the use of peat humification in long peat profiles as a proxy climate indicator (*see* RECURRENCE SURFACE).

Peat derived from *Sphagnum* mosses is usually poorly humified and consists of complete moss stems and leaves. In other types of peats, roots often predominate, forming a fibrous matrix. Peat may also contain wood fragments, fruits, seeds, leaves and stems, together with chitinous remains of fungi and the exoskeletons of insects and arachnids, such as mites. The analysis of peat contents provides a means of reconstructing the local vegetational and environmental history of a MIRE. The presence of smaller fossils, such as pollen, and fine dust particles of siliceous phytoliths or volcanic TEPHRA offer additional opportunities for the use of peat profiles in historical monitoring of the environment (*see* PALAEOCLIMATIC RECONSTRUCTION).

Peat has a high water-retaining capacity (often 95% of its mass consists of water) and it also has efficient cation-exchange properties. It is therefore much in demand as a soil conditioner in horticulture, and this has resulted in extensive draining of mires for peat extraction. The high organic content of peat also makes it valuable as an energy resource and it is widely used for fuelling power-stations in Ireland and in Russia. [P.D.M.]

Peatlands

Areas occupied by peat-forming vegetation (*see* MIRE; PEAT). Such areas are at least partially waterlogged and the rate of microbial DECOMPOSITION fails to keep up with that of plant litter formation, hence peat accumulates.

Exploitation

The energetic value of peat and its use as a soil conditioner has led to extensive drainage and exploitation of peatlands worldwide. Once stripped of a peat cover, the resulting soils can also be agriculturally productive, or valuable for forestry or energy farming (growth of poplars or willows as an energy resource). Hence peatland can be regarded as a particularly threatened habitat in conservation terms.

The management of virgin peatlands and the rehabilitation of damaged peatlands is dependent upon the control of HYDROLOGY. Often this entails the management of entire CATCHMENTS in the case of RHEOTROPHIC mires. For OMBROTROPHIC mires, the main problem is that damage and drainage in a portion of the mire (such as the sloping edge) can affect the hydrological balance of other parts of the mire.

Significance of conservation

The conservation of peatlands can be justified both on the basis of their distinctive wildlife and also because of their involvement in global processes. Since accumulating peat represents a reservoir for CARBON, peatlands play a role in the global cycling of carbon and must be regarded as a sink for atmospheric carbon. This is a valuable asset in a situation where we are supplying the ATMOSPHERE with more CARBON DIOXIDE (CO_2) than can be absorbed by physical and biological processes, leading ultimately to the greenhouse effect and GLOBAL WARMING. Although the quantity of carbon absorbed by peatlands is small (probably less than 10^{15} g each year globally) any reduction in the process can only make matters worse.

Perhaps even more important is the role of peatlands in METHANE (CH_4) production. Methane is a powerful greenhouse gas that is generated by incomplete oxidation of organic materials by microbes in anaerobic situations (e.g. in animal guts or in waterlogged soils). Methane is generated by peatlands and it is possible that a continued rise in global TEMPERATURE may stimulate yet greater methane production, leading to a positive FEEDBACK situation. Drainage could further stimulate this in the short term. The conservation and management of peatlands is thus of global concern. [P.D.M.]

PEC *See* PREDICTED ENVIRONMENTAL CONCENTRATION.

pedogenesis The myriad of processes leading to the formation of soils. Generally, these processes can be considered under the following headings.

- Chemical processes — solution, hydration, hydrolysis, oxidation and reduction.
- Physical processes — aggregation, translocation, freezing/thawing and expansion/contraction.
- Biological processes — HUMIFICATION, translocation and element cycling.

However, only a restricted number of these processes operate at any one time in any particular SOIL.

Hydrolysis is the main chemical process whereby the SILICATES such as feldspars, amphiboles and pyroxenes are decomposed and transformed into oxides, oxyhydroxides and CLAY minerals, as in Ferralsols (Oxisols).

Translocation by WATER is one of the main pedogenetic processes. The middle horizons of PODZOLS (Spodosols) are formed by the translocation and deposition of HUMUS and/or IRON and/or ALUMINIUM. Many theories have been suggested for this process, but the most likely possibility is that bacteria are responsible. The middle horizons of Luvisols (Alfisols) are formed by the translocation of clay (Fig. P10), to form clay coatings, while in some soils, such as CHERNOZEMS (Mollisols), calcium carbonate is translocated to the lower part of the soil. In arid and semi-arid areas, soluble salts are translocated, and may accumulate at any position in the soil.

Translocation by the soil fauna is also very significant. Earthworms, enchytraeid worms, arthropods including TERMITES and mites, and various insect larvae incorporate organic matter and translocate materials. This faunal translocation can be over a short distance, within the soil itself, as in the case of Chernozems, or, for example, material may be taken to the surface by termites to form termitaria. *See also* EROSION AND TOPSOIL LOSS. [E.A.F.]

REFERENCES

FitzPatrick, E.A. (1980) *Soils, Their Formation, Classification and Distribution*. Longman, Harlow.

Fig. P10 A clay coating showing the characteristic crescentic laminations.

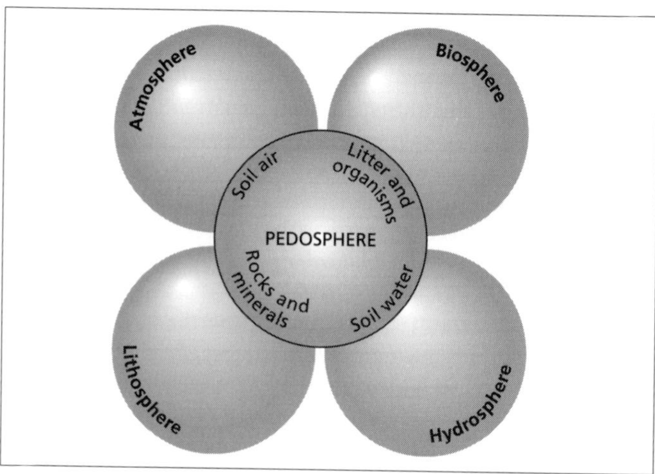

Fig. P11 The pedosphere is the amalgam of the atmosphere, the biosphere, the lithosphere and the hydrosphere.

Lundström, U.S. (1994) Significance of organic acids for weathering and the podzolization process. *Environment International* **20**, 21–30.

pedology The study of SOIL in all its aspects. Most mineral exchange between the BIOTA and the inorganic components of the ECOSYSTEM occurs in the soil. Soils vary in relation to the conditions under which they are formed. The factors that create soils and soil variation are: CLIMATE, parent material, TOPOGRAPHY (which affects drainage), and in later stages VEGETATION, microorganisms and animals. The age of a soil affects its nature, as it modifies the influence of the parent rock. Mature soils appear to be in a steady state, but are actually in a DYNAMIC EQUILIBRIUM in which removal of minerals by WATER or WIND is balanced by continuous input from erosion of the parent rock. Over ACID rocks, soils may remain shallow with little organic content. In the wet tropics, WEATHERING produces very deep soils. Temperate soils are often a few metres deep. They exhibit different layers, or horizons, which range from dead organic litter on the surface, through a HUMUS-rich layer (with many microorganisms) to a layer where rapid LEACHING of minerals takes place and where plant roots are concentrated because minerals are readily available. Below this is partly weathered parent rock. Soils vary enormously in their organic content, depending on the substrate and climate. *See also* SOIL CLASSIFICATION; SOIL PROFILE. [S.R.J.W.]

pedosphere The upper part of the Earth's CRUST that is occupied by soils. It is often easy to recognize the pedosphere, but sometimes it is extremely difficult because of the difficulties in defining the term or concept of SOIL. Some workers consider vegetated dryland surfaces as the pedosphere, while others include subaqueous situations. The relationships between the pedosphere, ATMOSPHERE, BIOSPHERE, LITHOSPHERE and HYDROSPHERE are shown in Fig. P11. [E.A.F.]

pelagic Living in the water column (for marine organisms), seaward of the SHELF-SLOPE BREAK. Pelagic organisms are adapted to maintain their position in the water column by having a shape (i.e. high surface area) or body composition (i.e. rich in lipid) that enhances BUOYANCY or by being able to swim actively. Some organisms are only pelagic for a part of their life cycle (MEROPLANKTON), whereas others are pelagic for their entire life cycle (HOLOPLANKTON). Many pelagic organisms migrate vertically in the water column, sometimes covering great distances, on a daily or seasonal basis in order to meet energy requirements or escape predators. Pelagic organisms that cannot swim actively, and whose movement in the water column depends on water CURRENTS, TURBULENCE and their own bulk density, under the control of viscous forces, are called PLANKTON. Pelagic organisms that can swim, thereby overcoming the action of water currents, but are influenced by inertial forces, are called NEKTON. [V.F.]

pelagic zone *See* WITHIN-LAKE HABITAT CLASSIFICATION.

Penman–Monteith equation The most widely accepted and generally applicable approach to the calculation of TRANSPIRATION or EVAPORATION rate is represented by the Penman–Monteith equation:

$$\lambda E = \frac{sA + \rho c_p D g_a}{s + \gamma(1 + g_a/g_c)}$$

where A is the available RADIATION, equal to the net radiation for a dense stand of vegetation ($W\,m^{-2}$); c_p is the specific heat capacity of air at constant PRESSURE ($J\,kg^{-1}\,K^{-1}$); D is the vapour pressure deficit (kPa); E is the evaporation rate ($kg\,m^{-2}\,s^{-1}$); g_a and g_c are, respectively, the canopy aerodynamic and stomatal CONDUCTANCES ($m\,s^{-1}$); s is the slope of the relationship between saturated vapour pressure and TEMPERATURE ($kPa\,K^{-1}$); λ is the latent heat of vaporization of WATER ($2.5\,MJ\,kg^{-1}$); γ is the psychrometric constant ($0.0666\,kPa\,K^{-1}$); and ρ is the density of air ($kg\,m^{-3}$).

The equation may be used to estimate the transpiration from a leaf or a stand of vegetation under a given climate, by using typical surface conductances (Table P2).

Table P2 Typical values for canopy stomatal conductance (g_c) and canopy aerodynamic conductance (g_a) for different vegetation types.

Vegetation	Maximal g_c (m s^{-1})	Range of g_a (m s^{-1})
Grassland/heathland	0.02	0.02–0.05
Agricultural crops	0.05	0.02–0.05
Plantation forest	0.02	0.1–0.3

When used in conjunction with an INTERCEPTION model, the Penman–Monteith equation may be used to estimate the daily water balance of a forested catchment, for example, in any weather conditions. Its use as a predictive tool is limited only by the tendency of the stomatal conductance to change over the day, and seasonally; and by the tendency for the evaporated water vapour to exert feedback on the stomata, which tend to open more widely when they experience a high HUMIDITY. Some of these problems are discussed in the references given. [J.G.]

REFERENCES

Jarvis, P.G. & McNaughton, K.G. (1986) Stomatal control of transpiration. *Advances in Ecological Research* **15**, 1–49.

McNaughton, K.G. & Jarvis, P.G. (1983) Predicting effects of vegetational changes on transpiration and evaporation. In: *Water Deficits and Plant Growth*, Vol. 7 (ed. T.T. Kozlowski), pp. 1–49. Academic Press, New York.

Monteith, J.L. (1965) Evaporation and environment. *Symposia of the Society for Experimental Biology* **19**, 205–234.

Rutter, A.J. (1975) The hydrological cycle in vegetation. In: *Vegetation and the Atmosphere*, Vol. 1, *Principles* (ed. J.L. Monteith), pp. 111–154. Academic Press, New York.

Thom, A. (1975) Momentum, mass and heat exchange of plant communities. In: *Vegetation and the Atmosphere*, Vol. 1, *Principles* (ed. J.L. Monteith) pp. 57–109. Academic Press, New York.

peptides Compounds made from two or more AMINO ACIDS by condensation. They constitute the building blocks of PROTEINS. Large numbers of amino acids linked together often form a helical structure. Peptides are also abundant in their own right, as hormones (oxytocin, vasopressin, glucagon) and antibiotics. [J.G.]

per capita Latin: literally, according to heads. This term is used to express a quantity as a function of an individual. It is essentially an average measurement which is commonly associated with dynamic processes or rates of change. For example, the per capita birth rate describes the average NUMBER OF OFFSPRING produced per individual in the population. This is often used as an alternative to finite measures when comparisons between populations are necessary as it is independent of POPULATION SIZE. [R.C.]

peramorphosis A neologism coined by Alberch *et al.* (1979) to describe the morphological outcome corresponding to the conceptual opposite of PAEDOMORPHOSIS. As part of the terminological review, and classification scheme, initiated by Gould (1977), there was a category of processes, such as NEOTENY, that generate a morphological 'outcome' called paedomorphosis. The corresponding opposite phenotypic outcome of heterochronic processes, such as HYPERMORPHOSIS and ACCELERATION, that results in a descendant form 'going beyond' (overdeveloping) relative to the ancestral sequence, is the term peramorphosis. Since then, it has become a 'standard' component in the terminology of HETEROCHRONY, where it is essential to discriminate between pattern and process. Thus, peramorphosis is a morphological outcome resulting from the terminal addition of stages to the ancestral ontogenetic series. Such a process of generating peramorphic forms by terminal addition results in the classical, Haeckelian phylogenetic pattern of RECAPITULATION. *See also* ONTOGENY. [P.A.]

REFERENCES

Alberch, P., Gould, S.J., Oster, G.F. & Wake, D.B. (1979) Size and shape in ontogeny and phylogeny. *Paleobiology* **5**, 296–317.

Gould, S.J. (1977) *Ontogeny and Phylogeny*. Harvard University Press, Cambridge, Mass.

percentile Any of the 99 actual or notional points that divide a DISTRIBUTION or set of observations into 100 equal parts. *See also* QUARTILE. [B.S.]

percolation lagoons *See* LAGOONS.

perennial A plant living for a number of years. Woody perennials have a permanent aerial form which continues to grow year after year. In herbaceous perennials the shoots die annually and are replaced by new ones from underground structures. Perennials can be either semelparous (*see* SEMELPARITY) or iteroparous (*see* ITEROPARITY). *See also* ANNUAL; BIENNIAL PLANTS. [P.O.]

periodic table A table of all known ELEMENTS arranged in order of increasing ATOMIC NUMBER (Fig. P12). The arrangement is such that elements in the same vertical columns (called groups or families) have similar chemical properties. This is because they have the same arrangement of outermost electrons. The horizontal rows of elements in the table are called periods. The letters in the boxes are the symbols for the elements and the number shown above each symbol is the atomic number (i.e. number of protons) for the element.

The periodic table was constructed to represent the patterns observed from the chemical properties of elements. J.W. Döbereiner first noted several groups of three elements (triads) with similar properties. The expansion of this theory to all known elements showed that this concept was severely limited. In 1864, J.A.R. Newlands proposed the law of octaves based on the observation that certain properties were repeated for every eighth element, in a way similar to a musical scale. In 1869 D.I. Mendeleev discovered that the properties of the elements were a periodic function of their ATOMIC MASSES, and the original published version of the table (1872) listed the elements according to relative atomic mass. The present form was conceived independently by J.L. Meyer and Mendeleev but was attributed to the latter due to his correct prediction of the properties of then undiscovered elements on the basis of this table. [M.C.G.]

peripatric speciation SPECIATION by evolution of ISOLATING MECHANISMS in populations located at the periphery of a species range. It is a special case of ALLOPATRIC SPECIATION, and is based on the observation that marginal, isolated populations often differ

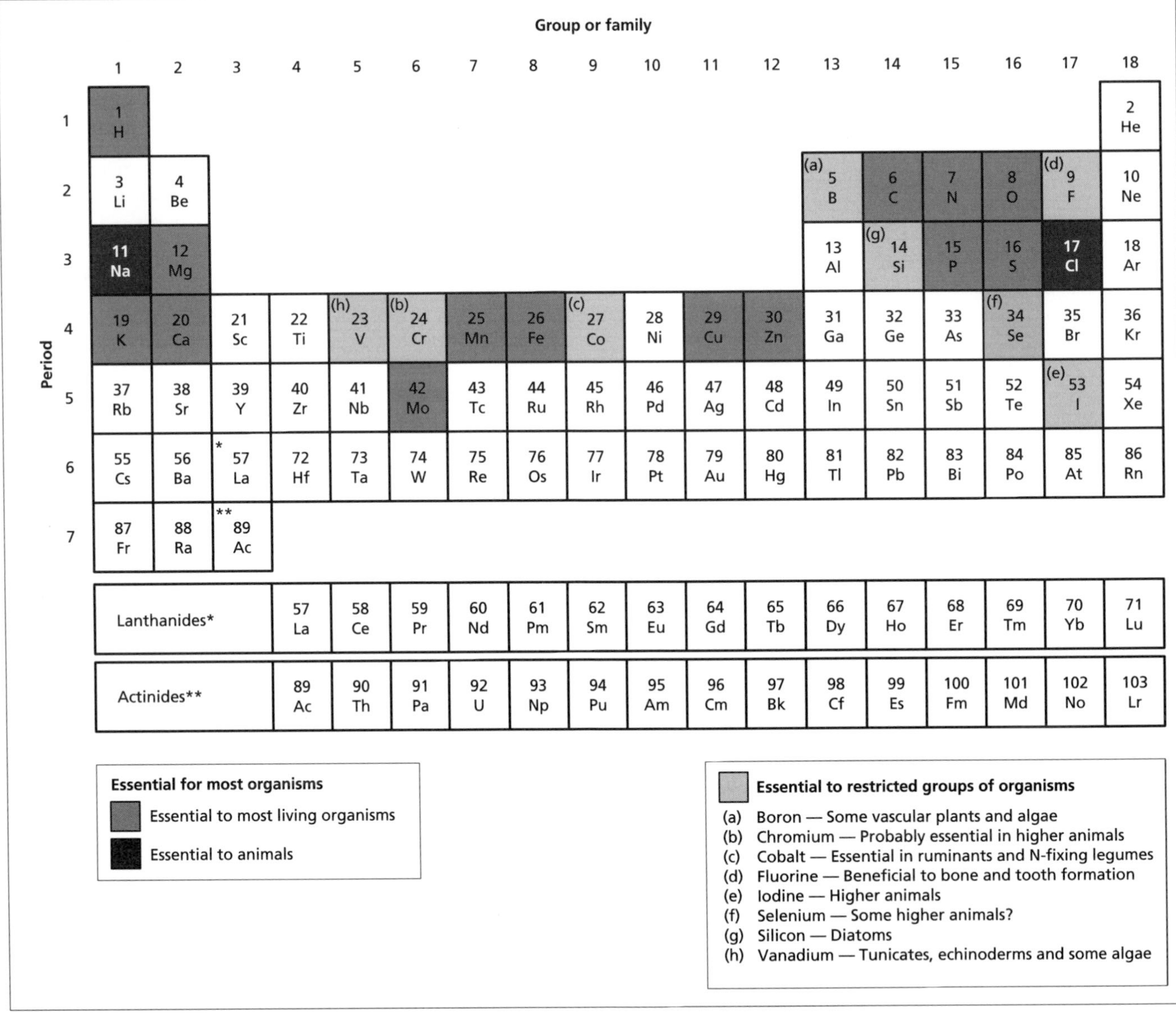

Fig. P12 The periodic table of the elements showing those that are essential resources in the life of various organisms. (After Begon, M., Harper, J.L. & Townsend, C.R. (1996) *Ecology: Individuals, Populations and Communities*, 3rd edn. Blackwell Science, Oxford.)

markedly from the main-range populations. It involves subdivision of a species range into two unequal parts by an extrinsic barrier (Fig. P13). The extreme version is Ernst Mayr's speciation by FOUNDER EFFECT, whereby a small group of individuals, or a single fertilized female, separates from the rest of the range to colonize a new area. Such a small population will have reduced variation (Fig. P14) and GENETIC DRIFT, which may set off reorganization of the genome through 'genetic revolution' — breaking down COADAPTED GENE COMPLEXES and creating new ones. Evolution in a small population is thought to be accelerated since establishment of new genotypes is facilitated because GENE FLOW opposing divergence is restricted due to spatial isolation from the rest of the range. Thus adaptive peak shift from one genetic equilibrium to another takes place. Founder-effect speciation, which forms a theoretical basis for PUNCTUATED EQUILIBRIUM, is a controversial idea, incompatible with mathematical models, which require very restrictive conditions unlikely to be met in nature. Moreover, experimental studies report high rather than low levels of variation in groups, like Hawaiian fruit flies, which were cited as examples of peripatric speciation. It is more likely that divergence of peripheral populations is due to selection for local ADAPTATION in a novel, often extreme environment, or SEXUAL SELECTION, which could operate differently in isolated populations. [J.M.SZ.]

periphyton A term sometimes applied to the community of bacteria and algae that occurs in the littoral or benthic zones of LAKES,

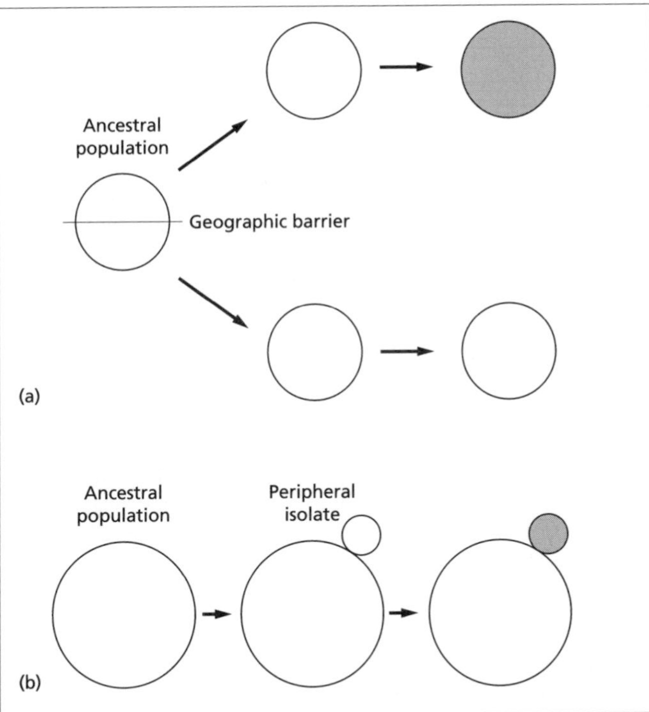

Fig. P13 Two models of allopatric speciation. Shading indicates evolution of a new species. (a) The 'dumb-bell' model, in which the ancestral species is divided into two roughly equal halves, each of which forms a new species. (b) Peripatric speciation, in which the new species forms from a population isolated at the edge of the ancestral species range. (After Ridley, M. (1996) *Evolution*, 2nd edn. Blackwell Science, Cambridge, Mass.)

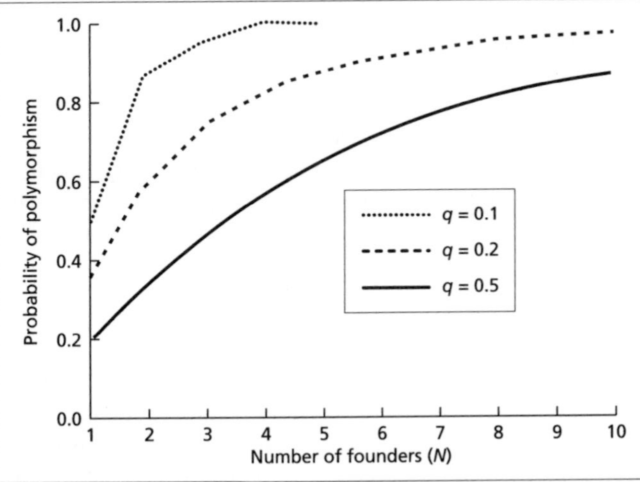

Fig. P14 The chance that a founder population will be homozygous depends on the number of founders and the gene frequencies. If less variation and fewer founders are present, the chance of homozygosity is higher. Here the chance of homozygosity is shown for three different gene frequencies at a two-allele locus. (After Ridley, M. (1996) *Evolution*, 2nd edn. Blackwell Science, Cambridge, Mass.)

and found attached to natural substrates such as larger plants (epiphyton), stones and rocks (epilithon), and sand grains (epipsammon). It also includes communities of motile algae that live on and beneath mud surfaces (the epipelon). [R.W.B.]

permafrost A naturally occurring earth material whose temperature has been below 0°C for several years regardless of the state of any moisture that might be present. The ground may be almost any kind of material including bedrock, CLAY, gravel, sand, SILT or organic SOIL. Dry permafrost has little excess ICE and is associated with inorganic substrates, whereas wet permafrost is found in fine-grain soils. Continuous permafrost covers about 7.64 million km^2 of the Northern hemisphere.

The depth of permafrost can reach 1000 m, with temperatures as low as −12°C. Above the permafrost the seasonally thawing soil is called the active layer. This is the biologically active area of the ground available for plant roots and soil flora and fauna. In many areas of the ARCTIC the active layer may be only 20–40 cm thick, while in northern Yakutia, only *c.* 11 cm of the SOIL PROFILE thaws each summer.

The presence of permafrost, in addition to contributing to the thermal dynamics of the soil, also controls HYDROLOGY in that it impedes drainage. Areas of discontinuous permafrost on the southern boundary of the Arctic are particularly sensitive to CLIMATE warming. Active layer detachment from the permafrost on slopes can lead to great DISTURBANCE at the landscape level. Permafrost contains and covers large amounts of the greenhouse gas METHANE. [T.V.C.]

permanence A concept of global community stability (*see* STABILITY, COMMUNITIES). Whereas LOCAL STABILITY measures examine whether a community tends to return to its original state following perturbation, permanence simply examines whether populations tend to coexist, irrespective of the stability of the equilibrium point. This distinction is important, since proving that an equilibrium is unstable does not necessarily imply that species cannot coexist; species may coexist, for example, on periodic or chaotic attractors. [R.P.F.]

REFERENCE

Hutson, V. & Schmitt, K. (1992) Permanence and the dynamics of biological systems. *Mathematical Biosciences* **111**, 1–71.

peroxyacetyl nitrate (PAN) A photochemical oxidant gas, formula $CH_3CO.O_2NO_2$, formed in polluted air by the action of sunlight on a mixture of HYDROCARBONS and NITROGEN OXIDES (NO_x), and associated with the formation of OZONE (O_3) and PHOTOCHEMICAL SMOG. The highest concentrations are found in urban areas with high traffic densities, in warm, sunny weather. PAN is thermally unstable, but at the low temperatures found high in the TROPOSPHERE and in polar regions, PAN may be transported for many hundreds of kilometres to act as a source of POLLUTANT NO_x. It is a powerful lacrimator, and causes characteristic bronzing of plant leaves at high concentrations (> 10 ppb by volume). [J.N.C.]

persistence The capacity of a commercial chemical to endure, without change, in environmental compartments. *Cf.* BIODEGRADATION. [P.C.]

persistent organic pollutants (POPs) Organic chemicals that, because of their resistance to natural destruction, can last for long periods in the environment and accumulate. They are toxic and/or ecotoxic. It is further often required as part of the definition that they are capable of regional or global transport in the ATMOSPHERE and/or HYDROSPHERE. They include DDT, aldrin, DIELDRIN, endrin, chlordane, heptachlor, hexachlorobenzene, mirex, toxaphene, POLYCHLORINATED BIPHENYLS (PCBs), DIOXINS and furans. They are under consideration for international regulation through the United Nations. [P.C.]

pest An animal (or, occasionally, a plant) that consumes and/or damages living or dead plant materials (less usually animal materials) intended for human consumption or use; for example, animals that consume CROPS, animal feeds or stored products. Such pests are effectively competitors with humans. [P.C.]

pest control Attempts to eradicate or reduce densities of pests to levels where they are no longer effective competitors with humans. Control can be by chemical means (PESTICIDE) or biological means (*see* BIOLOGICAL CONTROL), or a combination of these (*see* INTEGRATED PEST MANAGEMENT). [P.C.]

pesticide A substance that kills pests. Pesticides can be specific or non-specific. *See also* BIOCIDE. [P.C.]

pesticide formulation A particular PESTICIDE-containing product designed for sale. It generally consists of a complex mixture of ACTIVE INGREDIENT(S), carrier(s) and adjuvant(s). These are put together in such a way that the product is effective for the purpose claimed. [P.C.]

pesticide treadmill The seemingly endless search for new PESTICIDES that is required to counter the development of pest RESISTANCE to existing pesticides. Pests can, potentially, adapt (i.e. develop resistance) to any excessively used pesticide. This can necessitate the design of different pesticides against which there is no initial resistance. But pest resistance against these NEW CHEMICALS can potentially evolve—and so on. The treadmill is similar to the Red Queen dilemma (*see* RED QUEEN'S HYPOTHESIS). [P.C.]

pesticides and soil contamination PESTICIDES — HERBICIDES, FUNGICIDES, INSECTICIDES and molluscicides—are used extensively to increase plant productivity by regulating unwanted biotic interference, and in control of human and animal diseases. The application of pesticides almost inevitably involves contact with the soil surface, particularly where drifting of the pesticide occurs through aerial applications or where AEROSOLS and fine dusts remain airborne for extended periods. The development of SURFACTANTS, the control over aerial (aircraft) spraying, advice on suitable weather conditions for applications, and an awareness of the toxic effects of pesticides have combined to greatly reduce the unintentional targeting of pesticides in AGRICULTURE, horticulture and FORESTRY.

Once in the SOIL, pesticides are removed from the degradative processes of photochemistry. Persistence thereafter is a function of temperature, rainfall, soil pH, soil structure and aeration, activity of microbial and other biotic degradative agencies as well as the chemical composition, solubility and stability of the pesticide molecule. The widespread application of organochlorine insecticides between 1948 and 1968 has left a legacy of persistent and toxic compounds in the BIOSPHERE, including the soil. The average time for 95% disappearance of DDT, DIELDRIN and LINDANE from the soil is 10, 8 and 6.5 years, respectively. The relative insolubility of DDT in the soil and its slow release into water bodies meant that it took 7 years from 1977 to achieve a two-fold reduction in DDT in Lake Michigan — despite the ban on the agricultural use of DDT in the USA in 1973. The discovery of persistent POLYCHLORINATED BIPHENYLS, originally formulated as pesticides, in soils led to the evacuation of one town in Missouri 10 years after the original CONTAMINATION. Legislation, at least in western Europe and North America, today controls the use of persistent organic pesticides, such that most are degraded in the soil within weeks of application. However, inorganic pesticides based on toxic metals (e.g. COPPER, ARSENIC and MERCURY)—such as Bordeaux mixture—may leave a toxic residue which persists, under suitable soil conditions, for decades. [G.A.F.H.].

Petersen estimate A basis for estimating POPULATION SIZE from the proportion of marked individuals in a recaptured SAMPLE. It is named after C.G.J. Petersen, a Danish fishery scientist, who recognized its importance when studying plaice marked with brass tags. *See also* CAPTURE–RECAPTURE TECHNIQUES; LINCOLN INDEX. [L.M.C.]

REFERENCE

Petersen, C.G.J. (1894) On the biology of our flat-fish and on the decrease of our flat-fish fisheries. *Reports of the Danish Biological Station*, No. IV (1893–4), 1–146.

petrocalcic horizons *See* SOIL CLASSIFICATION.

petrogypsic horizons *See* SOIL CLASSIFICATION.

petroleum A mixture of aromatic and aliphatic HYDROCARBONS and non-hydrocarbon compounds such as carbazoles, thiophenes, furans and PHENOLS, that is produced in deep subsurface sediments and can be extracted in a liquid (crude oils) or gaseous form (natural gas). It is widely believed that petroleum is formed originally from biologically produced organic matter deposited in aquatic sediments. Petroleum can be released to the environment as a result of human activities (oil extraction and transport) and from natural sources such as naturally occurring hydrocarbon seeps where petroleum-bearing rock strata have been exposed near the surface of the Earth or the sea floor.

Since petroleum is produced by natural processes and petroleum hydrocarbons have been present in the environment for many millennia, low levels of petroleum hydrocarbons do not constitute a serious threat to the environment. It is only when uncontrolled release of large quantities of petroleum occurs that serious environmental problems may ensue.

The aliphatic components and many of the aromatic components of petroleum are readily dissipated in the environment by EVAPORATION and BIODEGRADATION — hydrocarbon-degrading microorganisms are all but ubiquitous. Some components of petroleum, generally those not readily removed by natural processes, are highly toxic. These include compounds like dibenzofurans and POLYCYCLIC AROMATIC HYDROCARBONS (PAHs). PAHs are lipophilic and can accumulate in animal tissues; they are subject to BIOMAGNIFICATION in food chains. In addition, oxygenase enzymes present in the liver can partially oxidize PAHs to form hydroxylated derivatives that are highly mutagenic. *See also* MOTOR VEHICLE POLLUTION. [I.M.H.]

pH A logarithmic scale indicating the activity (or, approximately, the concentration) of HYDROGEN ions in solution:

$$pH = -\log_{10}[H^+]$$

where $[H^+]$ is the concentration of H^+ in moles per litre.

The pH of RAIN in clean environments is about 5, due to the carbonic acid/CARBON DIOXIDE equilibrium:

$$CO_2 + H_2O \leftrightarrow H_2CO_3 \leftrightarrow H^+ + HCO_3^-$$

In polluted environments it may be much lower, about 4; this 'acid rain' contains sulphuric and nitric acids formed from the atmospheric oxidation of NITROGEN and SULPHUR OXIDES. The pH of surface and ground waters is very variable, ranging from <4 to >8, and that of soils even more so (2 to >10).

In climates where the PRECIPITATION exceeds EVAPORATION, allowing LEACHING to occur, the slight acidity of rain even in clean environments is sufficient in the long term to displace basic cations from exchange sites in soils and replace them with hydrogen ions. This reduces the pH in the SOIL solution and drainage WATER, in poorly buffered soils, leading to increased concentrations of ALUMINIUM in solution, which may be toxic to plants and especially to fish in LAKES and RIVERS. Atmospheric oxidation of acid sulphate soils, following drainage, can lower the pH to 2. The same process in mine spoil can result in leachates with high concentrations of toxic metals. Landscaping of long-undisturbed spoil heaps can regenerate the problem, and heavy and repeated applications of LIME may be needed before plants can be grown. Soils developed on chalk and limestone, and those in arid areas where salts accumulate, are alkaline in reaction, with pH values of 7–8 (calcareous soils) and up to 10.5 for sodic soils containing sodium carbonate. [K.A.S.]

Phaeozems *See* SOIL CLASSIFICATION.

phanerophyte A plant life form in which perennating organs (in this case buds) are held well above the ground. The exposure of these organs carries with it risks of FROST and/or DESICCATION damage, and this life form is most abundant in climates where frost and DROUGHT are infrequent or unknown (e.g. the equatorial rain belt). Trees and shrubs fall into this life-form category and may be deciduous or EVERGREEN. The buds of those phanerophytes that are exposed to a degree of drought or frost stress may be protected by bud scales, but the buds of most tropical phanerophytes have no such protection. Phanerophytes can be subdivided into nanophanerophytes (0.3–2 m), microphanerophytes (2–8 m), mesophanerophytes (8–30 m) and megaphanerophytes (>30 m). *See also* PHYSIOGNOMY; RAUNKIAER'S LIFE FORMS. [P.D.M.]

Phanerozoic The period of geological time consisting of the PALAEOZOIC, MESOZOIC and CENOZOIC. It begins at the end of the Precambrian period (about 590 million years ago) and continues up to the present time. The word means 'visible life', contrasting with the microscopic forms of life of the preceding Precambrian period. [P.D.M.]

phase plane analysis A graphical technique for investigating the dynamics of interaction between the POPULATION DENSITIES of two taxa (e.g. predator and prey; two competing species; host and pathogen; etc.). The idea is that conditions are favourable for species A at some densities of species B and unfavourable at others (Fig. P15). Population densities increase exponentially

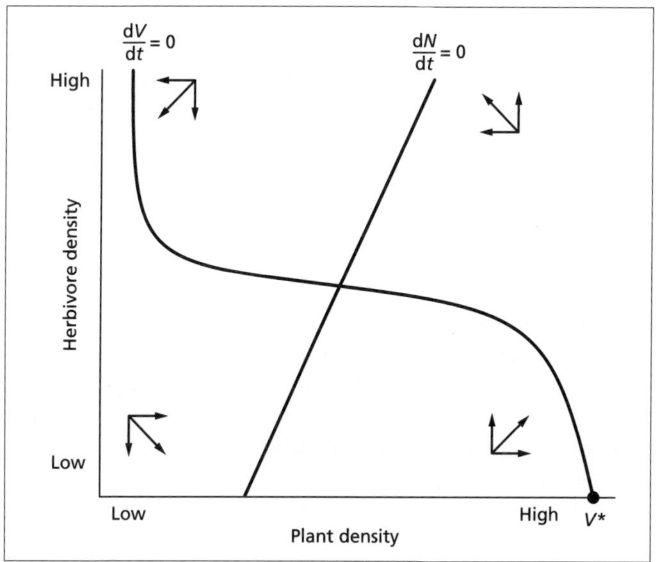

Fig. P15 Phase plane analysis. The axes show the densities of two interacting populations; these can be two competitors at the same trophic level or, as here, populations of exploiters and resources from two adjacent trophic levels. The lines in phase space are known as zero-growth isoclines (ZGI), and separate combinations of densities where one of the species increases from combinations where it decreases. The plant ZGI is where $dV/dt = 0$ (V = vegetation) and separates areas where herbivores are scarce (plants increase) from those where herbivores are abundant (plants decline). The curvature of the plant ZGI in this example incorporates a refuge for the plant (the herbivore cannot drive the plant to extinction), plus intraspecific competition among plants in the absence of the herbivore (giving a herbivore-free plant equilibrium population density of V^*). The herbivore ZGI is linear but rotated from the vertical in a clockwise direction, indicative of stabilizing density dependence acting on the herbivore population (e.g. competition for food). The four zones of phase space defined by the intersection of the ZGIs each have a characteristic trajectory for plant and herbivore numbers (e.g. in the top right corner both plants and herbivores are abundant, so herbivore numbers increase and plant numbers decline, with a resultant motion towards the top left). Moving through phase space (cobwebbing) leads to the prediction of a stable point equilibrium at the point where the two ZGIs cross. The equilibrium is not globally stable, however, since despite the existence of a refuge for the plant, the herbivore population can go extinct, if herbivore numbers are reduced to very low levels (i.e. the motion in the bottom left-hand corner is capable of intersecting the x-axis).

during favourable conditions and decline exponentially during unfavourable conditions. For example, in a plant–herbivore phase plane analysis, conditions would be favourable for plant growth when herbivore density was low, and unfavourable when herbivore density was high. Likewise, conditions would be favourable for herbivore growth when plant abundance was high, but unfavourable when plant abundance was low.

This simple graphical analysis provides answers to several important questions, including:

1 Will an equilibrium occur?
2 At what densities of the two taxa will the equilibrium occur?
3 By what kind of transient dynamics will the equilibrium be approached?
4 Will the system exhibit persistent cyclic behaviour?
5 Is the system globally stable?

See also LOTKA–VOLTERRA MODEL; ZERO GROWTH ISOCLINE. [M.J.C.]

REFERENCE
Crawley, M.J. (1992) Population dynamics of natural enemies and their prey. In: *Natural Enemies: The Population Biology of Predators, Parasites and Diseases* (ed. M.J. Crawley), pp. 40–89. Blackwell Scientific Publications, Oxford.

phase plane graphs/diagrams (two-dimensional state space graphs) *See* PHASE PLANE ANALYSIS.

phenetic algorithms *See* NUMERICAL TAXONOMY.

phenetic species concept A species concept based on phenetics, or NUMERICAL TAXONOMY. A set of organisms is classified as a species when its overall phenotypic similarity (and/or its phenetic distance to another set of organisms) exceeds a certain threshold. The phenetic species concept attributes no special significance to whether or not individuals of a species interbreed or share a common ecological NICHE. *See also* CLADISTICS; PHYLOGENETIC SYSTEMATICS. [W.W.W.]

phenology The study of periodic biological phenomena, especially in relation to seasonal environmental cycles. Phenomena typically studied in this regard include life-history features (GROWTH, reproduction, death), MIGRATION and DORMANCY (HIBERNATION, DIAPAUSE).

The term phenology also refers to the actual pattern or timing of recurrent phenomena. In the temperate and arctic zones, seasonal variations in energy availability and environmental TEMPERATURE are the primary factors influencing an organism's phenology. In the tropics, seasonal cycles of rainfall often drive organismal phenologies. In aquatic systems, tidal and LUNAR CYCLES, in combination with seasonal ones, control reproductive phenologies of certain fishes and invertebrates.

The timing of phenological events reflects complex interactions between innate and environmental influences. Seasonal changes in PHOTOPERIOD often trigger internal hormonal changes that in turn stimulate a phenological activity such as reproduction. Photoperiod here serves as an environmental cue (proximate factor), but temperature and food availability may be the ultimate (evolutionary) factors behind the particular pattern.

BIOTIC FACTORS also drive phenological patterns. For instance, the timing of reproduction in many herbivorous insects is closely linked to the phenology of their host plants. Furthermore, phenological timing can even be modulated by social factors within species: for example, the onset of reproduction in certain female birds and lizards is accelerated by displays of conspecific males.

By controlling the timing of key life-history events relative to key environmental cycles, phenology is strongly subject to NATURAL SELECTION. For example, the precise time of arrival at a breeding ground often determines the reproductive success or failure of a migratory bird. Not surprisingly, phenological patterns often differ between populations separated by latitude or altitude: the 'critical' photoperiod that triggers a given phenological event (e.g. flowering, diapause) may differ strikingly between northern and southern populations of a species. These geographic differences may reflect genetic or phenotypic (or both) responses to the ENVIRONMENT. *See also* LIFE CYCLE; LIFE-HISTORY TRAITS; ONTOGENY. [R.B.H.]

phenols Aromatic compounds with one or more hydroxyls attached to the BENZENE ring. In some respects they are like ALCOHOLS (in which the hydroxyls are attached to an aliphatic compound rather than an aromatic one), and are weak ACIDS. Phenol itself (containing a single hydroxyl group) is a strong antiseptic, in use since the 1890s. Other examples are *p*-cresol, resorcinol, 1-naphthol, pyrogallol and phloroglucinol. Phenols are widely used in the chemical industry, especially in the synthesis of drugs, dyes and PLASTICS. [J.G.]

phenotype Any structural or functional feature (i.e. TRAIT) of an organism other than genetic information. This definition encompasses physical and chemical characters of the organism (biochemistry, physiology, morphology), its behaviour and the life history. The 'extended phenotype' is a broader category, including an individual's artefacts, as well as its influence on other organisms and the environment. The phenotype is a product of the interaction between the GENOTYPE and the environment. *See also* OPTIMAL REACTION NORMS; PHENOTYPIC PLASTICITY; REACTION NORM. [T.J.K.]

phenotypic altruism *See* KIN SELECTION.

phenotypic correlations and life-history evolution *See* TRADE-OFF.

Phenotypic plasticity

The capacity of organisms for morphological, physiological or life-history modifications arising from either strictly environmental influences or GENOTYPE*ENVIRONMENT INTERACTIONS. Such modifications are usefully partitioned into two categories: phenotypic modulation and developmental conversion.

Phenotypic modulation

Phenotypic modulation results from exposure of developing organisms to different environmental conditions and is manifested as a

CONTINUUM of PHENOTYPES. The capacity for this kind of modulation may be heritable and thus subject to NATURAL SELECTION, or it may be an unselected consequence of developmental or physiological CONSTRAINTS in a continuously variable environment. The outcome of phenotypic modulation is known as the REACTION NORM, and has been found to vary among individuals in both plants and animals with different genotypes.

Developmental conversion

Developmental conversion (also termed conditional choice or developmental switch) results from an environmentally induced switch between alternative developmental programmes, resulting in discontinuous phenotypes (usually two). Such a switch is often associated with discrete differences in the environment and is of apparent adaptive value. Typically, the switch is made in the juvenile stage in response to a specific environmental cue, committing the adult to an irreversible morphological type. For example, the cue is sometimes associated with an increased risk of attack by predators, and induces individuals to develop a predator-resistant phenotype. Individuals that do not receive the cue develop according to a default developmental programme which confers no special resistance to predation, but may be superior in other ways. The occurrence of discrete environmentally induced morphs is often referred to as a POLYPHENISM or environmentally cued POLYMORPHISM. *See also* GENERAL-PURPOSE GENOTYPE; OPTIMAL REACTION NORMS. [C.M.L.]

REFERENCES
Lloyd, D.G. (1984) Variation strategies of plants in heterogeneous environments. *Biological Journal of the Linnean Society* **21**, 357–385.
Smith-Gill, S. (1983) Developmental plasticity: developmental conversion vs. phenotypic modulation. *American Zoologist* **23**, 47–55.

---

phenotypic value In QUANTITATIVE GENETICS, the measured value of the character of interest in an individual. The phenotypic value, P, of an individual is assumed to be under genetic and environmental influence: i.e. $P=G+E$, where G is the GENOTYPIC VALUE of the individual and E is its ENVIRONMENTAL DEVIATION. The mean of the environmental deviations in individuals of the same GENOTYPE equals zero. Therefore, the mean phenotypic value of a genotype equals the genotypic value, and the mean phenotypic value in a population equals the MEAN GENOTYPIC VALUE. [G.D.J.]

phenotypic variance In QUANTITATIVE GENETICS, the VARIANCE in a population of the PHENOTYPIC VALUE of the character of interest, symbolized as V_P. The phenotypic value, P, is written as $P=G+E$; therefore, the phenotypic variance, $V_P=V_G+V_E+2\mathrm{cov}(G,E)$, with V_G the GENETIC VARIANCE, V_E the ENVIRONMENTAL VARIANCE and $\mathrm{cov}(G,E)$ the GENOTYPE*ENVIRONMENT COVARIANCE. [G.D.J.]

pheromone A chemical compound emitted by one organism which alters the behaviour or physiology of another member of the same species. For example, a physiologically effective pheromone is 9-keto-2-decanoic acid, released by honeybee queens (*Apis mellifera*): worker honeybees refrain from egg-laying when it is present and lay EGGS when it is absent. Sex pheromones are powerful attractants famous for their ability to elicit SEARCHING behaviour at very low concentrations. A noted example is bombykol, released by female silkworm moths (*Bombyx mori*). Such pheromones have a high $Q:K$ ratio, where Q=quantity of pheromone emitted, and K= threshold concentration required for a response. Selection can also favour low $Q:K$ ratios, such as in ant-trail pheromones where response to low concentrations would inhibit the ability to follow a discrete trail.

Pheromone-like COMMUNICATION may occur between species via allelochemics, either to the benefit of the sender (allomone) or to its detriment (kairomone). CARBON DIOXIDE (CO_2) is a kairomone when emitted by mammals and attracting mosquitoes. (E)-β-farnesene is an aphid (*Myzus persicae*) alarm pheromone, but it also acts as an allomone when emitted by wild potato, repelling aphids.

The term pheromone is often restricted to an identified single compound, but behaviourally active odours often include several component pheromones. Such mixtures are variously termed odours or multicomponent pheromones.

Human pheromones are poorly known and their current adaptive significance is unknown, but there is evidence for them: young women living together show menstrual synchrony via unknown water-soluble substances.

Pheromones, especially sex attractants, are widespread, perhaps universal, in eukaryotes, and hence preceded hormones evolutionarily. *See also* INFOCHEMICALS. [R.H.C.]

REFERENCES
Masson, C. & Mustaparta, H. (1990) Chemical information processing in the olfactory system of insects. *Physiological Reviews* **70**, 199–245.
Wilson, E.O. (1975) *Sociobiology. The New Synthesis.* Harvard University Press, Cambridge, Mass.

philopatry Tendency of an animal to return to, or stay in, its home area. [P.C.]

phoresy *See* PHORETIC ASSOCIATION.

phoretic association Phoresy is a method of DISPERSAL whereby a species is transported by clinging to the body of another species and then drops off after a period of time. Phoretic association is a term which describes this relationship between two species. *See also* PASSIVE DISPERSAL; VAGILITY. [S.J.H.]

phosphorus (P) A non-metallic element (ATOMIC NUMBER=15; relative ATOMIC MASS=30.97; melting point=44.1°C; boiling point=280°C) that is the 12th most abundant element (0.11%) in the Earth's CRUST. Found in apatites ($Ca_{10}(PO_4)_6(F,OH)_2$), calcium phosphate ($Ca_3(PO_4)_2$) and other metal phosphates, phosphorus is used in FERTILIZERS, such as ammonium phosphate and soluble superphosphates ($CaH_4(PO_4)_2{\cdot}H_2O$), in polyphosphate DETERGENTS and in ORGANOPHOSPHOROUS PESTICIDES. It is an ESSENTIAL ELEMENT for all living organisms, in which it is a constituent of energy-transferring, genetic and information-carrying molecules. Phosphorus is often a biolimiting factor in SOIL and aquatic systems. The presence of residual soluble phosphates in LAKES, however, sometimes via internal loading from remobilization and release from bottom sediments, can stimulate the growth of unsightly photosynthetic ALGAL BLOOMS. *See also* EUTROPHICATION. [J.F.]

phosphorus cycle The movement of PHOSPHORUS (P) between the BIOSPHERE, LITHOSPHERE and HYDROSPHERE, and the associated transformations between its different chemical forms, from the release of P from its main reservoir in apatites (complex calcium phosphates) and other minerals to the ultimate formation of new minerals. Weathering converts some P into exchangeable phosphate anions, which are absorbed by plants and pass into the biological cycle. The uptake is often promoted by the association between roots and mycorrhizal fungi. P is present in all living organisms, for example in DNA and phospholipids. Organic residues returned to the SOIL (and aquatic sediments) are mineralized by microorganisms, releasing phosphate again for return to the biological cycle or PRECIPITATION as insoluble phosphates of IRON, ALUMINIUM or MANGANESE (ACID conditions) or of CALCIUM (alkaline conditions). P is often the limiting nutrient in aquatic ecosystems, and increased supply can result in ALGAL BLOOMS and other problems. *See also* EUTROPHICATION. [K.A.S.]

photic zone The uppermost zone of a water body in which the penetration of LIGHT is sufficient for PHOTOSYNTHESIS to exceed RESPIRATION. The photic zone can vary in depth from very shallow to up to 200 m, depending on TEMPERATURE, TURBIDITY and water chemistry. *See also* EUPHOTIC ZONE; LAKES. [J.L.]

photochemical smog Atmospheric haze. It appears to be initiated by nitrogen dioxide (NO_2), especially from motor vehicle EMISSIONS. Visible or ultraviolet (UV) energy of sunlight converts NO_2 to nitric oxides and free OXYGEN atoms (O). The latter combine with molecular oxygen (O_2) to form OZONE (O_3), which causes a variety of reactions to take place with HYDROCARBONS and certain other organic compounds. The low-volatility organic products contribute to the haze of small droplets that constitutes the SMOG. The phenomenon was first recognized in Los Angeles, but now occurs in urban conurbations worldwide. Clearly this kind of smog is likely to be most intense when the sunlight intensity is greatest, i.e. early afternoon. It contrasts with smog, which is most intensive in early morning. *See also* AIR POLLUTION; ATMOSPHERIC POLLUTION; MOTOR VEHICLE POLLUTION. [P.C.]

photon A quantum of LIGHT or other electromagnetic RADIATION. [L.O.B.]

photon flux The rate of flow of PHOTONS through a plane. In photobiology and ECOPHYSIOLOGY, the flux of photons is measured using a suitable photovoltaic or photoresistive cell ('quantum sensor'). The units employed are photons $m^{-2} s^{-1}$ or, more usually—by dividing the number of photons by Avogadro's constant ($6.02 \times 10^{23} mol^{-1}$) — mol photons $m^{-2} s^{-1}$. Full sunlight is about 2.0 mmol photons $m^{-2} s^{-1}$. The waveband over which the measurement is stated should also be given: for example, the photosynthetic photon flux, or the photosynthetically active radiation (PAR), refers to the waveband 0.4–0.7 μm, over which the energy can be utilized in PHOTOSYNTHESIS. *See also* LIGHT. [J.G.]

REFERENCE

Pontailler, J.-Y. (1990) A cheap quantum sensor using a gallium arsenide photodiode. *Functional Ecology* **4**, 591–597.

photoperiod The length of the day (*see* DAYLENGTH); or the length of the LIGHT period in experimental studies. Many organisms have the ability to detect changes in the photoperiod, and this enables them to adjust their LIFE CYCLES to the SEASONS. The phenomenon is called photoperiodism. Temperate trees, for example, form OVERWINTERING buds when shortening daylengths have been detected. Other plants use photoperiod as a signal for the production of flowers at the right time of year. Short-day plants such as tobacco, strawberry, coffee and *Chenopodium album* are those that require a long, uninterrupted night to stimulate flowering; long-day plants such as oats and ryegrass require the long days of the northern temperate summer to induce flowering. In plants, the detection of daylength involves the pigment PHYTOCHROME. In animals, photoperiodic responses include the control of the life cycle in insects, migratory response of birds, reproductive cycle of some mammals and the change of fur colour in the snowshoe hare (*Lepus americanus*). *See also* CIRCADIAN RHYTHMS. [J.G.]

REFERENCES

Saunders, D.S. (1982) *Insect Clocks*, 2nd edn. Pergamon Press, Oxford.

Wareing, P.F. & Phillips, I.D.J. (1981) *The Control of Growth and Differentiation in Plants*. Pergamon Press, Oxford.

Photosynthesis

Photosynthesis is the process in which LIGHT energy is used to reduce CARBON DIOXIDE (CO_2) to sugars and other CARBOHYDRATES and in which molecular OXYGEN (O_2) is evolved. It is central to all life, in that photosynthesis ultimately produces all sources of food. Past photosynthetic activity also produced the present oxygen-rich atmosphere, and the oil, gas and hydrocarbon-rich sediments.

The basic photosynthetic process takes place in all green higher plant and algal cells, within ORGANELLES termed chloroplasts, which have the necessary pigments (chlorophylls) to absorb and utilize light energy. A similar process also occurs in the cyanobacteria and in photosynthetic bacteria. The term photosynthesis is also used generally to indicate the uptake of CO_2 by plants, which involves the diffusion of CO_2 from the atmosphere into the leaves through the stomatal pores, and into the photosynthetic cells. Because of this diffusion, photosynthesis is intimately linked with TRANSPIRATION — the loss of WATER by EVAPORATION from cells in the leaf, and diffusion of the water vapour out through the pores into the air.

Biochemistry

Photosynthesis is actually the coupling of two processes.

1 Light capture, electron transfer and photophosphorylation on the thylakoid membranes within the chloroplasts to generate reduced assimilatory power in the form of NADPH and ATP.

2 Use of these reduced products to drive the 'dark reactions' that make carbohydrates in the stroma of the chloroplasts.

For the 'light reactions' the pigments are organized into two photosystems, termed PSI and PSII, and a light-harvesting complex (LHCP). Light absorption by chlorophyll molecules raises their excitation state, and electrons are passed to a reaction centre in the photosystem. In PSII, oxygen is evolved and ATP produced, while in PSI, NADPH is produced.

All green plants use the NADPH and ATP from the light reactions for the carbon fixation in the dark reactions, in the photosynthetic carbon reduction (PCR) or Calvin cycle. The key PCR enzyme is ribulose bisphosphate carboxylase-oxygenase (Rubisco), which catalyses the uptake of CO_2 through incorporation with a 5-carbon compound, ribulose bisphosphate, into a 3-carbon compound, phosphoglyceric acid. The enzyme can also oxygenate the ribulose bisphosphate, producing phosphoglycollate and releasing CO_2 in the process known as photorespiration. At typical temperatures and at the CO_2 and O_2 concentrations in most plant cells this oxygenase activity loses about 20–50% of the carbon fixed by carboxylation. This is the situation for the major part of the world's vegetation, which is of the C_3 photosynthetic type.

C_4 plants

Some species of plants, the C_4 group, have evolved a secondary carbon fixation pathway, using the enzyme phosphoenolpyruvate carboxylase (PEP carboxylase), which has no oxygenase activity, and a higher affinity for CO_2. In these species, CO_2 is first combined into a 4-carbon compound, in the mesophyll leaf cells, and then passed to the cells around the veins in the leaf (the bundle sheath cells) where the CO_2 is released at high concentrations for recarboxylation in the usual PCR pathway. These high CO_2 concentrations inside the bundle sheath suppress photorespiration.

These C_4 species (which are of diverse genera) are typically found in warmer and drier environments and, while not important in terms of the overall proportion of the world's vegetation, are locally and agriculturally very important (e.g. maize, sorghum, millet, sugar-cane and many subtropical and tropical forage grasses). As their photosynthetic rates are typically higher than those of the C_3 types, they also lose less water per molecule of CO_2 taken up, and therefore have a higher water use EFFICIENCY.

CAM plants

Another locally important group of plants with a different photosynthetic physiology are the CAM plants, such as many cacti and succulents, typical of arid zones. The problem in these environments is that CO_2 uptake involves the risk of dehydration. These species therefore harvest light energy during the day, while maintaining the stomata shut, but providing CO_2 for fixation in the PCR cycle from the decarboxylation of stored malate. At night they open the stomata, and accumulate CO_2 into malate; they have therefore separated the CO_2 fixation part of photosynthesis from the CO_2 uptake part, and thereby conserve water.

Efficiency

Photosynthesis is somewhat inefficient in energy terms. Firstly, only about 47% of the solar spectrum — the region between 400 and 700 nm (termed photosynthetically active radiation) — can be absorbed by the photosynthetic pigments. Secondly, between 10 and 30% of the light reaching the leaf is reflected. Thirdly, although the theoretical minimum quantum yield (the number of absorbed quanta required to fix one molecule of CO_2) is 8, reflection, absorption by non-photosynthetic pigments, fluorescence and decay of excited pigment molecules increase the value to about 12. In addition, substantial amounts of CO_2 are lost due to RESPIRATION and photorespiration, so that the real quantum requirement at low light levels is about 15–20 quanta. At higher light levels this requirement increases, because other factors in addition to light start to influence the photosynthetic rate. The overall energy efficiency of even highly productive agricultural CROPS is only around 2%, in terms of the amount of energy in the crop as a fraction of the solar energy incident. The rest of the solar energy incident on the crop is dissipated in heating the crop and the air in contact, and in evaporation. *See also* C_3/C_4 PLANTS; ECOLOGICAL EFFICIENCY; ECOLOGICAL ENERGETICS; PRIMARY PRODUCTIVITY. [J.I.L.M.]

REFERENCE

Hall, D.O. & Rao, K. (1994) *Photosynthesis*, 5th edn. Cambridge University Press, Cambridge.

---

photosynthetic capacity The maximum rate of CARBON DIOXIDE (CO_2) uptake by plants under optimal conditions in the natural environment. It is normally expressed as μmol CO_2 m^{-2} s^{-1}. Table P3 gives ranges of values for various types of plants, and further examples can be found in Larcher (1995). *See also* PHOTOSYNTHESIS; PRIMARY PRODUCTIVITY. [J.G.]

REFERENCE

Larcher, W. (1995) *Physiological Plant Ecology*. Springer-Verlag, Berlin.

Table P3 Typical ranges of values for photosynthetic capacity of leaves, expressed as the CO_2 uptake per projected area of leaf.

Plant type	Photosynthetic capacity (μmol CO_2 m^{-2} s^{-1})
C_4 grasses	30–70
C_3, various types	
Herbaceous crops	20–40
Tree crop	10–15
Arctic plants	8–20
Mountain	15–30
Mesophyte weeds	20–30
Deciduous trees, sun leaves	10–15
Deciduous trees, shade leaves	3–6
Evergreen conifers	3–15
Ferns	2–10
Mosses	2–3
Lichens	0.3–2
Freshwater macrophytes	5–10
Seaweeds in the tidal zone	2–6
CAM plants	
In light	5–12
In dark	6–10

phototaxis Orientation or movement in response to LIGHT. Organisms can either be positively phototactic and move towards light, or negatively phototactic and move away from light. [V.F.]

phototrophic Describing organisms that utilize LIGHT as an energy source. [V.F.]

phyletic gradualism A model of EVOLUTION in which species are presumed to diverge gradually after SPECIATION. *Cf.* PUNCTUATED EQUILIBRIUM. [V.F.]

REFERENCE
Futuyma, D.J. (1986) *Evolutionary Biology*, 2nd edn. Sinauer Associates, Sunderland, Mass.

phyletic speciation *See* CHRONOSPECIES.

phylogenetic autocorrelation If a phenotypic TRAIT, say body size, is measured across a group of species, then phylogenetic autocorrelation provides a means for partitioning the VARIANCE in that trait to different levels of the taxonomic hierarchy plus the COVARIANCE between the phylogenetic and specific terms. The model represents the trait y as a linear combination of phylogenetic and specific effects according to:

$$y = \rho W y + e$$

where y is the vector of length n containing the n species' data points, ρ is a scalar 'phylogenetic autocorrelation coefficient', W is an $n \times n$ 'phylogenetic connectivity matrix', ρWy is a vector of predicted y values representing the phylogenetic portion of y, and e represents the vector of residual values of the trait that cannot be predicted by the vector ρWy. The term e is used to analyse whether there is covariation between traits that is independent of phylogenetic effects. The masses in the phylogenetic connectivity matrix can be given arbitrarily (being large for closely related species and small for distantly related ones) or assessed according to the variance in the data. Potentially, the majority of the variation is assigned to phylogenetic components, the interpretation of which can be open to debate—it may be considered as non-adaptive or possibly adaptive, according to the whims of the investigator. *See also* COMPARATIVE METHOD. [P.H.H.]

phylogenetic constraint Evolutionary CONSTRAINTS are processes or phenomena that restrict the range of future phenotypes in a given population or species. Any ancestral legacy that constrains the range of phenotypes in descendant species is a phylogenetic constraint. Almost any evolutionary process can become a phylogenetic constraint: canalizing effects of developmental phenomena in which the requirements of functional integration constrain the range of viable evolutionary novelties or the stable states determined by principles of self-organization or by evolutionarily stable strategies. Phylogenetic constraints thus can be a source of long-term stability (*see* STASIS) for evolving lineages. They can also lead to extinction if they reduce the rate at which a LINEAGE can adapt to drastic environmental change. TRAITS appearing on phylogenetic trees that pre-date the species being studied are excellent indicators of phylogenetic constraints. These are traits that have not changed for a long time, and for which there is little evidence of change presently (in the case of contemporaneous species). At present, it appears that the majority of traits exhibited by any species are such ancestral legacies; hence, phylogenetic constraints seem to have played an important role in the production of biological DIVERSITY. *See also* CANALIZATION; EVOLUTIONARY OPTIMIZATION; OPTIMALITY MODELS. [D.R.B.]

phylogenetic fixation Some characteristics are variable and some, known as fixed characteristics, are relatively invariant within any species. Any fixed character in a species whose invariance is due to inheritance from an ancestor in which the TRAIT was also invariant can be called a phylogenetically fixed trait.

Fixed character states are easily recognized from a phylogenetic perspective; they are character states which are passed unchanged from an ancestor to its descendants. Although such patterns highlight instances in which a character has been fixed over long periods of time, the patterns by themselves do not tell us which of several processes were responsible for the character's evolutionary stability. Such processes include various forms of selection, as well as FOUNDER EFFECT and GENETIC DRIFT. Phylogenetic fixation is sometimes referred to as phylogenetic inertia, with an implication, usually undemonstrated, that there is something about the evolution of certain characters such that, once they become fixed, it is relatively difficult for modifications to occur. [D.R.B.]

Phylogenetic inference

The empirical process of using characteristics of taxa as evidence of their evolutionary relationships. Since the advent of Darwinian evolutionary theory, the primary criterion for recognizing evolutionary relationships has been HOMOLOGY. Until the advent of phylogenetic SYSTEMATIC METHODS, phylogenetic inference was plagued with problems of circular reasoning. That is, phylogenetic relationships were indicated by homologous TRAITS, but at the same time homologous traits were defined as such with respect to phylogenies. Phylogenetic systematic analysis attempts to remove this circularity from the process of phylogenetic inference while maintaining the fundamental evolutionary principle of homology through a five-step iterative process.

First, assume a priori that traits appearing the same or highly similar between two or more species are homologous (have a common origin). Assuming a priori that traits that look the same in different species are homologous is an important first step in phylogenetic inference because when you assume that all similar traits have evolved independently there is no basis for postulating any evolutionary relationships. In this first step, the identification of presumptive homologies is based upon biological criteria such as similarity of structure or position, and has nothing to do with the assumed evolutionary relationships among the taxa displaying the traits. Because the initial hypotheses of homology are erected independently of underlying PHYLOGENY, this is not an exercise in circular reasoning.

Second, determine the level of homology, apomorphic (derived) or plesiomorphic (ancestral), for each character by comparing the state of the character in the group of interest (the ingroup) to the state of the character in taxa outside the study group (outgroups). This comparison is known as Hennig's auxiliary principle, or character polarization.

Third, group taxa based on shared APOMORPHIES (synapomorphics) using Hennig's inclusion/exclusion rule. Steps 2 and 3 are important because homologous traits shared by two or more species that evolved prior to the common ancestor of those species (plesiomorphies) do not actually indicate particular phylogenetic relationships among those species; only apomorphic homologous traits provide direct evidence supporting particular phylogenetic relationships (called sister-group relationships).

Fourth, when there are conflicts in groupings supported by putative synapomorphies, choose the grouping that explains the data most efficiently (the principle of PARSIMONY). Finally, interpret a posteriori the characters that support the conflicting (less-parsimonious) groupings as homoplasies.

The identification of HOMOPLASY in phylogenetic systematic analysis falsifies our initial hypothesis that two characters in different taxa that met some biological criterion for being 'the same' are in fact homologous. The analysis, based upon the weight of evidence provided by many characters, indicates that these two characters have evolved independently in the taxa that display them. At this point, we need to look further for an explanation of why the characters are so similar. The final two steps are justified in two complementary ways. First, by determining homoplasies rather than homologies with respect to the phylogeny, systematics avoids being an exercise in circular reasoning. Second, by providing an empirical mechanism for assessing homoplasies, PHYLOGENETIC SYSTEMATICS provides a means for avoiding recognition of groups in a PHYLOGENETIC TREE that are not closely related to each other, but simply share one or more homoplasious traits.

There are forms of comparative data that are not easily, if at all, amenable to phylogenetic systematic methods. In some cases, this is because the data themselves cannot be interpreted in terms of homology and homoplasy. For example, a variety of statistical approaches exist that convert raw morphological or molecular data to some sort of 'distance' measure based on overall similarity or difference. These approaches attempt to reconstruct phylogenetic groupings on the basis of those distances, assuming that the more distantly related the taxa are, the greater will be the genetic distances between them. These distance estimates represent a combination of homoplasious, plesiomorphic homologous and apomorphic homologous similarity, and cannot be interpreted in terms of particular statements or hypotheses of homology. Because phylogenetic relationships can only be reconstructed on the basis of shared special homologies (synapomorphies), distance measurements can give us estimates of overall phenetic similarity among taxa, but can tell us nothing reliably about the phylogenetic relationships among taxa.

Other types of data are not amenable to phylogenetic systematics because there are difficulties in obtaining defensible estimates of PLESIOMORPHY derived from OUTGROUP comparisons. Some nucleotide sequence data, for example, suffer from the problem of rampant convergent evolutionary change within a group and in the outgroups. Because there are only three possible evolutionary changes for every base site, after three mutational events at any one site the probability that the next MUTATION will produce homoplasy is 100%. In this case, methods other than phylogenetic systematic inference must be used to analyse the data.

One such method is called maximum likelihood estimation. This method uses information about likely modes of molecular evolution to infer the most probable phylogenetic tree given a set of, for example, nucleotide sequences. It is important to remember that the resulting phylogenetic hypothesis cannot be used as an independent test of different models of molecular evolution because the phylogenetic tree is constructed according to a priori parameters specifying details about the mode of molecular evolution. Phylogenetic systematic inference, on the other hand, is not hampered by such built-in methodological problems because the inference process is not predicated upon any assumptions of evolutionary process.

Methodological differences aside, it is interesting to note that these different approaches to phylogenetic inference usually produce highly similar results. *See also* CLADISTICS; CLASSIFICATION; PHENETIC SPECIES CONCEPT. [D.R.B.]

---

phylogenetic species concept *See* SPECIES CONCEPTS.

phylogenetic subtraction Closely related species tend to be similar to each other in almost any characteristic we care to examine. Many historical and contemporary selective factors are likely to be responsible for that similarity. Phylogenetic subtraction extracts correlates of phylogenetic affinity from an analysis of character state variation to see if the residual variation could be related to variation in the states of other characters. For example, in an analysis of mammalian life-history variation, species within orders tended to be similar to each other. When species values were expressed as deviations from the mean value for the order, it was possible to compare species from different orders with that particular phylogenetic component removed. The analysis revealed that there was substantial covariation among LIFE-HISTORY TRAITS, suggesting which components were likely to be responding in similar ways to contemporary SELECTION PRESSURES. [P.H.H.]

Phylogenetic systematics

Phylogenetic systematics is the name used by Willi Hennig, a German specialist in dipteran (fly) systematics, for an approach to producing what he termed a 'general reference system for comparative biology'. Hennig contrasted special interest reference systems (CLASSIFICATIONS) with a general reference system. He reasoned that there are many possible criteria for constructing special interest classifications, any one of which could produce a classification that would be maximally efficient for a particular purpose, but would have limited applications. A general reference system, by contrast, would provide the maximum amount of utility to the maximum number of biologists interested in a wide variety of topics.

Hennig reasoned that if biological EVOLUTION was responsible for producing the biological DIVERSITY around us, then the logical criterion for the general reference system would be one that classified species according to their evolutionary relationships. He then proposed a logical method by which biologists could use similarities among species as evidence of relationships.

Types of similarity

Systematists have always used 'similarity' as a working criterion for recognizing relationships. Hennig recognized that 'similarity' comes in a variety of forms in biological systems. Two or more species can be similar because they share TRAITS that have evolved independently in each. Hennig called this type of similarity 'homoplasy', encompassing phenomena such as parallel and CONVERGENT EVOLUTION. Alternatively, two or more species may be similar because they share traits that they have inherited from a common ancestor. Following conventional terminology, Hennig called such similarity HOMOLOGY. Furthermore, Hennig recognized that two or more species may share two major types of homologous similarity: PLESIOMORPHY and apomorphy. Plesiomorphic traits do not provide evidence of phylogenetic relationships because they represent ancient, or persistent ancestral, homologous similarity that evolved before the taxa being studied. Apomorphic traits provide evidence of phylogenetic relationships because they represent traits that evolved in the unique common ancestor of a restricted group of species within those being studied.

Hennig chose the terms 'homoplasy', 'plesiomorphy' and 'apomorphy' purposely. He felt that the term 'ANALOGY' as an antonym of 'homology' carried with it connotations of functional equivalence, and Hennig supported a posteriori assessments of functional equivalence based on putative homoplasies indicated by a phylogenetic analysis. Hennig also felt that the terms 'ancestral' and 'derived' or 'advanced' carried with them connotations that certain traits could be considered absolutely ancestral or derived; he felt that all such assessments should be considered relative to the scope of a particular phylogenetic analysis. For example, having four limbs is apomorphic for analysis of all vertebrate groups, whereas it is plesiomorphic for analysis of all tetrapod groups. Hennig thus recognized that evolution itself changed apomorphic traits into plesiomorphic traits, and proposed terms intended to have relative rather than absolute connotations.

Phylogenetic systematic methodology is based on three critical steps:

1 construct initial hypotheses of homology for characters based upon biological criteria such as similarity of form or position;

2 determine the relative plesiomorphic and apomorphic status of the presumptive homologues via OUTGROUP comparisons;

3 reconstruct phylogenetic relationships among taxa based upon shared special homologies, or synapomorphies.

For a detailed discussion of this process *see* PHYLOGENETIC INFERENCE. Of these three steps, the last two, outgroup comparisons and grouping by synapomorphies, represent a major breakthrough in systematic biology.

Historical context

Hennig presented his approach in detail in 1950 in a German-language text. Phylogenetic systematics, however, did not attract a large following until the English-language publication of Hennig's book *Phylogenetic Systematics* in 1966. Since that time, this approach to the study of systematics has evolved considerably. In the past 30 years, phylogenetic systematics has passed through a period of vibrant theoretical development and expansion followed by the formation of a variety of groups unified only by their use of Hennig's method (for a detailed history see Hull's *Science as a Process*).

Applications

The logical nature of phylogenetic systematics has lent itself to computerization, and there now exist a number of user-friendly computer programs that permit the analysis of large numbers of taxa and characters. In addition to studies of a purely systematic nature, phylogenetic systematics has been used in comparative, statistical and experimental studies of topics including BIOGEOGRAPHY, SPECIATION, ADAPTATION, functional morphology, ECOLOGY, community evolution, COEVOLUTION and comparative ethology. This range of comparative biological studies based on the results of phylogenetic systematics seems to substantiate Hennig's notion that phylogenetic systematics could serve as the general reference system for comparative biology. *See also* CLADISTICS; CLASSIFICATION; PHENETIC SPECIES CONCEPT. [D.R.B.]

REFERENCES

Hennig, W. (1966) *Phylogenetic Systematics*. University of Illinois Press, Urbana, Ill.

Hull, D.L. (1988) *Science as a Process*. University of Chicago Press, Chicago, Ill.

phylogenetic tree A branching diagram depicting the sequence of SPECIATION events within a group (Fig. P16). As a graphical repre-

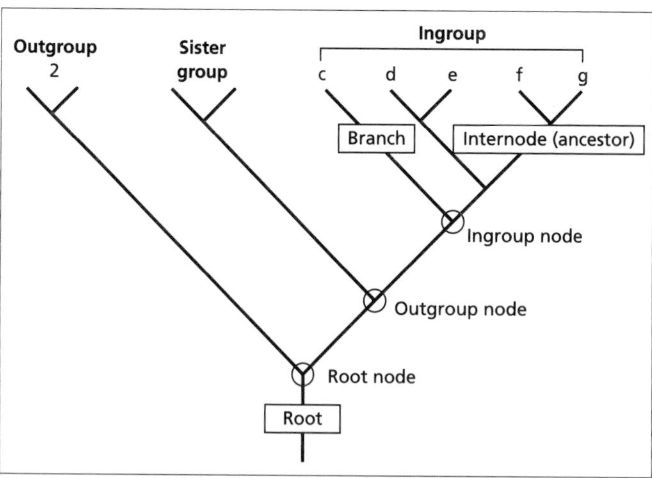

Fig. P16 The major elements of a phylogenetic tree.

sentation of genealogy, a phylogenetic tree is a hypothesis of the genealogical relationships among taxa. Since phylogenetic trees are hypotheses and not 'facts', they are dependent upon both the quality and quantity of data which support them.

A tree is composed of several parts:

- branch—the line connecting a branch point to a terminal TAXON, representing the terminal taxon;
- ingroup—any group of theoretically closely related organisms of interest to an investigator. Ingroups must contain more than two species because it is impossible to determine phylogenetic relationships for only two taxa;
- ingroup node—the ancestor of the ingroup;
- internode—the line connecting two speciation events, representing an ancestral species;
- node or branch point—an individual speciation event;
- OUTGROUP node—the node immediately below the ingroup node;
- root—the internode at the bottom of the tree;
- root node—the lowest node on the tree;
- sister group—the taxon that is most closely related genealogically to the ingroup. The ancestor of the ingroup cannot be its sister because it is a member of the ingroup.

Cf. CLADOGRAM. [D.R.B.]

phylogeny The history of life. *See also* PHYLOGENETIC AUTOCORRELATION; PHYLOGENETIC CONSTRAINT; PHYLOGENETIC FIXATION; PHYLOGENETIC INFERENCE; PHYLOGENETIC SUBTRACTION; PHYLOGENETIC SYSTEMATICS; PHYLOGENETIC TREE. [D.R.B.]

phylum *See* TAXONOMIC CATEGORIES.

physical constants See Table P4. For physical properties of water and air, *see* WATER and AIR, CHEMICAL AND PHYSICAL PROPERTIES. [J.G.]

REFERENCE

Kaye, G.W.C. & Laby, T.H. (1986) *Tables of Physical and Chemical Constants*, 15th edn. Longman & Wiley, New York.

physical constraints *See* BIOMECHANICAL CONSTRAINTS.

physical factor Any non-living, non-chemical factor that influences biological processes. Physical factors include: heat, LIGHT, electricity, magnetism, RADIATION, gravity and PRESSURE. *Cf.* BIOTIC FACTORS. *See also* ABIOTIC FACTORS. [P.C.]

Table P4 Physical constants.

Name	Symbol	Numerical value
Acceleration due to gravity on Earth	g	$9.8067\ m\ s^{-2}$
Avogadro constant	L or N_A	$6.022 \times 10^{23}\ mol^{-1}$
Molar gas constant	R	$8.3145\ J\ K^{-1}\ mol^{-1}$
Gravitational constant	G	$6.67 \times 10^{-11}\ N\ m^2\ kg^{-2}$
Planck constant	h	$6.6261 \times 10^{-34}\ J\ s^{-1}$
Speed of light	c	$2.998 \times 10^{8}\ m\ s^{-1}$
Stefan–Boltzmann constant	σ	$5.6705 \times 10^{-8}\ W\ m^{-2}\ K^{-4}$
Molar volume of ideal gas at 0°C 100 kPa		$2.27106 \times 10^{-2}\ m^3\ mol^{-1}$
Molecular mass of air	M_a	$28.96 \times 10^{-3}\ kg\ mol^{-1}$

physiognomy Form or appearance. The physiognomy of a plant can be used to measure indirectly the ENVIRONMENT because it represents a suite of ADAPTATIONS to that environment. The physiognomic method was commonly used to characterize different vegetations but has been largely replaced by the phytosociological method, which uses the associations of different species. Nevertheless, physiognomy is still useful in characterizing the environment of ancient habitats from the physiognomies of plant macrofossil ASSEMBLAGES. For example, well-defined growth rings are found in the wood tissue of trees which have experienced a seasonal climate.

Several different systems for characterizing physiognomy have been proposed. The Danish botanist Christen Raunkiaer characterized plants on the position and form of their resting buds (*see* RAUNKIAER'S LIFE FORMS):

- phanerophytes (tall trees and shrubs);
- chamaephytes (low shrubs);
- hemicryptophytes (herbaceous perennials);
- CRYPTOPHYTES (plants with bulbs or rhizomes);
- THEROPHYTES (ANNUAL plants).

The tropics have a high proportion of phanerophytes; regions with seasonal cold have a high proportion of hemicryptophytes; while regions with Mediterranean-type climates have a high proportion of therophytes.

Features like the shape, size and texture, and margin of leaves have also been used. For example, sclerophylls are EVERGREEN plants with hard, thick leaves; they are found in habitats with seasonal drought or cold or low nutrient levels.

The physiognomy of plants changes with the ALTITUDINAL ZONATION. [M.I.]

REFERENCES

Ingrouille, M.J. (1992) *Diversity and Evolution of Land Plants*. Chapman & Hall, London.

Upchurch, G.R. & Wolfe, J.A. (1987) Mid Cretaceous to Early Tertiary vegetation and climate: evidence from fossil leaves and woods. In: *The Origins of Angiosperms and their Biological Consequences* (eds E.M. Friis, W.G. Chaloner & P.R. Crane), pp. 75–106. Cambridge University Press, Cambridge.

physiological ecology Physiological ECOLOGY has traditionally been concerned with explaining distributions in terms of the interaction between physiological processes and physical and chemical conditions; for example, how TEMPERATURE influences RESPIRATION; how chemical quality of the environment influences the *internal milieu* of organisms; explaining the EVOLUTION of physiological ADAPTATIONS in terms of environmental conditions. Exploring these issues has relied upon a comparative approach that has analysed divergences between related organisms in different environments (e.g. contrasting physiologies of organisms in different zones on the seashore) or the convergence between unrelated organisms in similar environments (e.g. the evolution of ANAEROBIC METABOLISM in endoparasites).

Recently, physiological ecology has become more concerned with issues of POPULATION DYNAMICS: explaining how physiological processes influence POPULATION SIZE changes through time; i.e. by influencing growth rates and hence time to breeding, reproductive output and survival probability; explaining how different physiologies and the dynamics that are associated with them are favoured under different ecological circumstances so leading to physiological and life-cycle adaptations. Exploring these issues has relied on an interaction between the development of theoretical models and the design of experiments to test the hypotheses incorporated into the models and the predictions that emerge from them.

For examples of the more classical comparative approach, which is sometimes referred to as environmental physiology, see Schmidt-Nielson (1979). For examples of the newer, more predictive approach, which is sometimes also referred to as functional ecology, see Sibly and Calow (1986). [P.C.]

REFERENCES

Schmidt-Nielson, K. (1979) *Animal Physiology: Adaptation and Environment*. Cambridge University Press, Cambridge.

Sibly, R.M. & Calow, P. (1986) *Physiological Ecology of Animals*. Blackwell Scientific Publications, Oxford.

physiological time Physical TIME is invariant. However, whereas 1 minute is a long time with respect to the physiological processes and total LIFESPAN of a bacterium, it is a short time in the physiological processes and total lifespan of an elephant. Physiological time probably scales with body mass in the same way as metabolic rate. [P.C.]

physiological trade-off TRADE-OFFS are important in considering microevolutionary processes, especially in terms of life-cycle/life-history evolution (*see* LIFE-HISTORY EVOLUTION). If there were no trade-offs one would expect NATURAL SELECTION to maximize growth, reproduction and survival simultaneously; that it does not is powerful evidence for trade-offs even though their experimental demonstration has been elusive.

Physiological trade-offs are themselves based on the laws of thermodynamics, which specify that energy can neither be destroyed nor created, just transformed, and that these transformations lead to more entropy, which is usually manifest as heat loss. Thus if the organism is represented as a system that takes in energy (and other RESOURCES) as food and then allocates it between physiological demands such as maintenance, growth and reproduction, then the laws of thermodynamics require that if more energy is allocated to one process, less will be available for another. This is the basis of physiological trade-offs, sometimes known as the principle of allocation. It is used to explain trade-offs between life-cycle traits such as growth and survival (related to maintenance), growth and reproduction, and reproduction and survival. Detailed knowledge of the physiological processes involved may allow precise functional relationships to be established between physiological and life-cycle traits that are thought to be trade-offs, and hence to construct the trade-off curves that are so important in the rigorous development of life-cycle theory.

Various complications may blur these physiological trade-offs. For example: the input of resource (in FEEDING and PHOTOSYNTHESIS) may not be fixed but variable; body stores may be used to supplement the supply of resources to metabolism; some resource other than energy (e.g. phosphorus, trace elements) may be limiting; and the genetic basis of physiological trade-offs may be complicated by environmental effects. *See also* BIOMECHANICAL CONSTRAINTS. [P.C.]

REFERENCE

Sibly, R.M. & Calow, P. (1986) *Physiological Ecology of Animals*. Blackwell Scientific Publications, Oxford.

phytoalexins A series of chemical substances produced by plants, often in response to wounding, that have antifungal properties. *See also* DEFENCE MECHANISMS. [P.D.M.]

phytobenthos *See* BENTHIC HABITAT CLASSIFICATION.

phytochrome A protein pigment present in trace quantities in the leaves, stems and seeds of green plants, which changes its form depending on the ratio of red to far-red LIGHT. It is involved in the detection of: (i) shadelight; (ii) DAYLENGTH; and (iii) depth of the seed in the soil, and enables the plant to respond to the presence of competitors, and to adjust its LIFE CYCLE to the seasons. [J.G.]

phytogeography The study of the geographical DISTRIBUTION of plants on the Earth's surface. The word derives from the Greek words *phyton* plant, and *geographia*, to describe the distribution of Earth-surface features. Originally, phytogeography was concerned only with plant and plant-community distributions. Since the 1950s the scope of phytogeography has enlarged to encompass the factors and processes which contribute to producing and maintaining plant and plant-community distributions and characteristics. Phytogeography is, thus, an analytical as well as a descriptive subject with strong links to ECOLOGY. *See also* FLORAL PROVINCE. [A.M.M.]

phytophagous FEEDING on plants. In current usage the term is usually confined to smaller sap-sucking insects, particularly aphids, and to leaf-eating insects. It is not commonly extended to HERBIVORES in general. Phytophagous insects are increasingly used, because of their size and relative ease of handling, in studies of plant–animal interactions, in studies of plant defence chemistry and as a means of studying the dynamics of herbivory, usually in a CLOSED SYSTEM. *See also* FEEDING TYPES, CLASSIFICATION. [G.A.F.H.].

phytoplankton The collective name for groups of algae and bacteria that are adapted to living suspended in the open waters of marine and freshwater environments. Diatoms (Bacillariophyta) are the most important of the marine phytoplankton groups, contributing up to 25% of world net primary productivity, an amount equivalent to tropical rain forests. In fresh waters the most important groups are usually diatoms, green algae (Chlorophyta), blue-green algae (Cyanobacteria) and chrysophytes (Chrysophyta).

Phytoplankton can be classified according to their size (*see* PLANKTON), and according to the proportion of their LIFE CYCLE spent in the open water. Consequently the term euplankton is given to those organisms where all stages of their life cycle are completed in suspension. In contrast, many planktonic organisms (the MEROPLANKTON) spend a greater or lesser part of their life cycle associated with littoral or deep-water sediments, often in the form of resting spores. *Cf.* ZOOPLANKTON. [R.W.B.]

Phytosociology

The CLASSIFICATION of VEGETATION into units based upon floristic analysis. It is necessary to devise systems of vegetation classification for a number of reasons.

1 Communication between ecologists requires a shorthand form for referring to types of vegetation.

2 Mapping of vegetation demands that subdivisions should be delineated.

3 Conservationists require representatives of as many habitats as possible, and this must be based upon a system of vegetation classification.

Conceptual viewpoints

The methodological approach to the problem of classification depends upon the view of vegetation taken by those involved. If vegetation is actually a CONTINUUM of variation, with different 'types' of vegetation blending together as various component species overlap in their ranges and their ecological demands, then any subdivision must be artificial and essentially pragmatic, depending upon the task for which it is devised. If, on the other hand, vegetation consists of a series of plant communities, or associations, in which there is considerable interdependence of one species upon another, or a coincidence of ecological requirements between species resulting in certain combinations of species recurring frequently, then the task of the phytosociologist is that of detecting and describing associations rather than inventing them.

Both schools of thought (and many intermediates) exist. Phytosociology has largely grown as a result of the activities of the latter group since a belief in the reality of communities clearly provides a greater incentive for their description. Atheists are unlikely to make good theologians!

Bases for description

Several systems of phytosociology have evolved and have often hybridized. They are typically based purely upon floristic (rather than environmental) attributes of a vegetation SAMPLE.

Dominance

Some of the simplest systems are based upon the concept of dominance — the association being defined and named in terms of the species occupying most BIOMASS in the vegetation and hence regarded as the most influential. This has the advantage of simplicity in concept and use, but is somewhat crude in that the dominant species are often ones with wide ecological tolerances, thus providing very poor resolution if the use of floristics is to have any meaning in environmental and habitat terms.

Constancy and fidelity

The most successful approach to phytosociology has been achieved using the concepts of constancy and fidelity. A constant species is one which is almost always found within a particular association. Its presence is therefore demanded if a particular COMMUNITY type is to be identified. It may also, of course, occur in other communities, but it needs to be present if identification is to be assured. A faithful species, on the other hand, is confined to a particular association. These are often species of very narrow ecological amplitude and their presence immediately indicates a set of strict environmental requirements, and sometimes even a particular history of site development or management. In combination, the concepts of constancy and fidelity have led to the development of complex, but relatively robust, systems of vegetation classification in many parts of the world, but most especially in the continental mainland of Europe. Computerized systems of data sorting and matching have added further strength to phytosociology of this type and have removed much of the subjectivity from the process of matching a field dataset with extensive tabulations of vegetation descriptions.

Statistical treatments

Even those vegetation scientists who are unhappy with the idea of communities (*see* COMMUNITY) have long recognized the need for classification schemes and much effort has been spent very profitably on devising methodologies (often based on multivariate statistical analysis) for sorting floristic datasets. ORDINATION and clustering techniques have the theoretical advantage that they permit the recognition of intermediate types in addition to recurrent species combinations. There is still, however, a need for the vegetation mapper to draw lines, even if these may be somewhat arbitrary. *See also* FLORAL PROVINCE; FLORAS, MAJOR PUBLISHED. [P.D.M.]

---

phytotelmata Phytotelmata are small water bodies that occur in or on plants. Good examples are water-filled tree holes and the water contained in pitcher plants. The communities of animals that these water bodies support have been the subject of considerable interest, much of which stems from the relative ease of studying what are, in effect, small, bounded and—most importantly for experimental purposes — abundant communities. Phytotelmata provide an ideal model system and have been used to examine broader ecological issues such as how communities assemble or the role of energy input (in this case leaf litter) as a determinant of food-chain length. [S.J.H.]

PIC *See* PRIOR INFORMED CONSENT.

picoplankton PLANKTON, mostly bacteria, measuring 0.2–2.0 μm in size. [V.F.]

PICT *See* POLLUTION-INDUCED COMMUNITY TOLERANCE.

piercers *See* FEEDING TYPES, CLASSIFICATION.

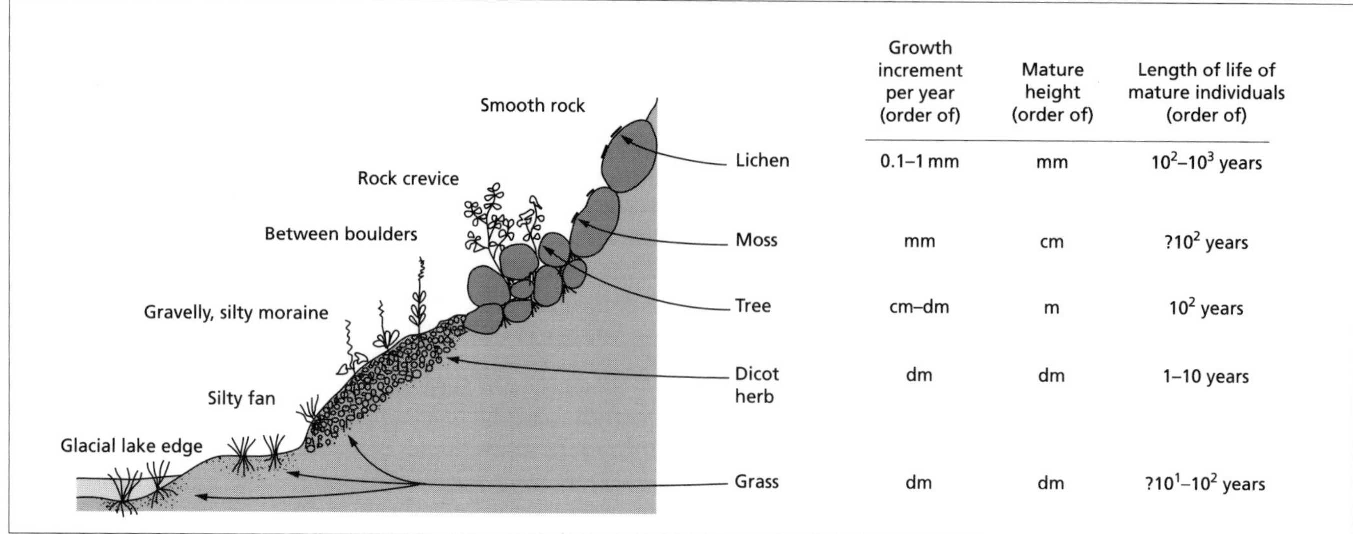

Fig. P17 The types of pioneer plant characteristic of different types of substratum in a recently deglaciated area within the range of present-day forest, and selected properties of the plants. (After Grubb, P.J. (1986) The ecology of establishment. In: *Ecology and Design in Landscape* (eds A.D. Bradshaw, D.A. Goode & E. Thorpe), pp. 83–97. Blackwell Scientific Publications, Oxford.)

piercing and sucking *See* FEEDING TYPES, CLASSIFICATION.

pioneer A term implicitly defined by Clements (1916; *see* CLIMAX and CHARACTERS IN ECOLOGY) as applying to one of the first species to become established in PRIMARY SUCCESSION. That usage persists with some present-day ecologists, but the term is now often applied to any early-successional plant, whether the SUCCESSION is primary or secondary (Wells 1976; Swaine & Whitmore 1988; *see* CLIMAX). The nature of pioneer plants, in the strict sense, depends on the substratum (Fig. P17): typical pioneers include lichens on the smoothest rocks and walls, mosses where there are slight indentations that catch water and NUTRIENTS, trees and shrubs on aggregations of boulders, dicotyledonous herbs on gravelly materials, and grasses on the finest-textured substrata (Grubb 1986). Rarely are NITROGEN-fixing plants pioneers—they usually arrive later, perhaps because it is hard to produce a tiny highly dispersible seed which contains sufficient starter nitrogen for root-nodule formation. [P.J.G.]

REFERENCES

Clements, F.E. (1916) *Plant Succession*. Carnegie Institution of Washington, Publication 242.

Grubb, P.J. (1986) The ecology of establishment. In: *Ecology and Design in Landscape* (eds A.D. Bradshaw, D.A. Goode & E. Thorpe), pp. 83–97. Blackwell Scientific Publications, Oxford.

Swaine, M.D. & Whitmore, T.C. (1988) On the definition of ecological groups in tropical rain forests. *Vegetatio* **75**, 81–86.

Wells, P.V. (1976) A climax index for broadleaf forest: an *n*-dimensional ecomorphological model of succession. In: *Central Hardwood Conference* (eds J.S. Fralish, G.J. Weaver & R.C. Schlesinger), pp. 131–176. Department of Forestry, Southern Illinois University, Carbondale, Ill.

pioneer species Early-successional FOREST-tree species. The PIONEER species germinate only in full sun or in the largest gaps; they often persist as dormant seeds in the SOIL SEED BANK (*cf.* PRIMARY SPECIES). They exhibit high photosynthetic and growth rates, have simple branching patterns, and require high temperature and/or high LIGHT intensity for germination. *See also* SUCCESSION. [M.J.C.]

pioneer stage *See* STAND CYCLE.

pistillate *See* POLLINATION.

pitfall traps *See* SAMPLING METHODOLOGY/DEVICES.

pitot tube A simple device to measure WIND speed. It consists of concentric tubes (Fig. P18). The opening of the inner tube points directly into the wind, while the outer tube is sealed at its

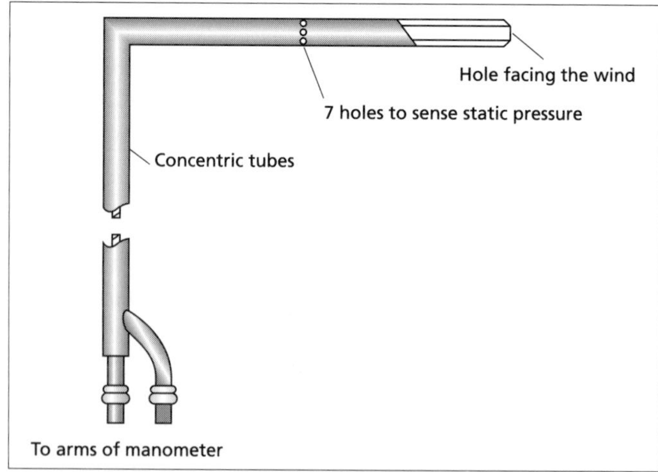

Fig. P18 Pitot-static tube, National Physical Laboratory (NPL) standard type.

front, but has small holes in the wall of the tube. At the base of the device, the tubes communicate with the two arms of a sensitive MANOMETER, so that the difference in pressure between the two tubes is sensed.

The total PRESSURE, P, exerted when the air at velocity u is arrested at the mouth of a tube is:

$$P=0.5\,\rho u^2+p_0$$

where ρ is the density of air and p_0 is the pressure exerted in all directions, called the static pressure. The manometer effectively subtracts p_0 from ρu^2, and the pressure recorded can therefore be used to measure the wind speed.

The pitot tube is widely used in WIND TUNNELS and ducts, but rarely in the natural wind because it works less well in turbulent flows and at low wind speeds. [J.G.]

PLA *See* PRODUKTLINIENALYSE.

placic horizons *See* SOIL CLASSIFICATION.

plaggen horizons *See* SOIL CLASSIFICATION.

plagioclimax A plant COMMUNITY produced in a DEFLECTED SUCCESSION that is the result of human intervention (Tansley 1935). [P.J.G.]

REFERENCE

Tansley, A.G. (1935) The use and abuse of vegetational concepts and terms. *Ecology* **16**, 284–307.

Planck constant *See* RADIATION LAWS.

Planetary boundary layer

The turbulent layer of the ATMOSPHERE closest to the ground (often called the atmospheric BOUNDARY LAYER) that is influenced by processes at the Earth's surface, such as friction, heating and cooling, on time-scales less than a day. The height of the layer can vary widely. Over land the layer is typically 1–2 km deep in the afternoon, while at night it can be less than 100 m deep. Over the oceans it displays little DIURNAL variation and is typically about 0.5 km deep. Above the planetary boundary layer lies the free atmosphere, where TURBULENCE is weak, intermittent or absent and which is little affected by processes at the ground.

Importance

The planetary boundary layer is important for several reasons. It is the atmospheric part of the BIOSPHERE, where life exists and must adapt to its extremes. Turbulent motions within the boundary layer are very effective in transferring momentum from the atmosphere to the Earth's surface — as much as half the kinetic energy of the atmosphere is dissipated in this way. Much of the heat and virtually all of the WATER vapour in the atmosphere enters it through the boundary layer. These have appreciable effects on formation of low-level CLOUDS and the evolution of WEATHER systems, so numerical weather models must take account of boundary-layer processes if the weather is to be forecast accurately more than a day or two ahead.

Pollutants, DUST and TRACE GASES released into it at the ground or from smokestacks are trapped within the boundary layer. This often makes the boundary layer visible as the layer of haze which aircraft must climb through after take-off before reaching clear air in the free atmosphere. The height of the planetary boundary layer affects the dilution of pollutants, so its height is an important determinant of AIR QUALITY.

Climatic changes

The CLIMATE within the planetary boundary layer responds to the daily cycle of surface heating and cooling. Over land, this is characterized by warm days and cool nights. These follow the regular cycle of solar heating during the day and net energy loss to the sky at night by thermal radiation. During the day, solar energy also goes to evaporate water from soils and from plants (TRANSPIRATION), so HUMIDITY is higher in the daytime boundary layer than in the free atmosphere. Higher humidity frequently causes clouds to form at the top of the layer, or for fog to form at night. The WIND blows less strongly in the boundary layer, slowed by friction with the ground, and its direction turns from the near-geostrophic wind, parallel to the isobars of the weather map in the free atmosphere, inwards towards centres of low pressure or outwards from centres of high pressure.

Neutral boundary layer

The height of the planetary boundary layer is one of its most important characteristics. It depends very much on the time of day and the prevailing weather and surface conditions. Under ideal conditions of high winds and overcast skies its height is determined only by wind speed and surface ROUGHNESS. The height of this so-called neutral boundary layer is given by the formula $h=0.25u_*/f$, where h is height of the boundary layer, u_* is the friction velocity, and $f=2\Omega\sin\phi$, is the Coriolis parameter where $\Omega=7.29\times10^{-5}\,s^{-1}$, the Earth's rate of rotation, and ϕ is latitude. The large-scale convective motions which develop in these conditions are in the form of elongated rolls, aligned with the mean wind. Cumulus clouds can form in these rolls, appearing as cloud streets spaced at about twice the height of the boundary layer.

Convective boundary layer

Neutral conditions are the exception. Usually BUOYANCY and large-scale atmospheric subsidence influence both the height and turbulent structure of the boundary layer. During the day over land, heating and, to a lesser extent, humidification of the air near the ground lowers the air's density and gives it a tendency to rise. This buoyancy contributes energy to the turbulence in the boundary layer. Near the ground these turbulent motions appear as random fluctuations, but higher up they become organized into walls and, higher still, plumes of buoyant air which rise until they impinge on the capping inversion, where they are stopped by opposing buoyancy forces. Thermal CONVECTION is very efficient at carrying substances from near the ground up into the boundary layer, so the main part of a convective boundary layer is often described as a 'mixed layer'. Pollutants or other substances released into a convective boundary layer typically become well distributed throughout its depth in less than an hour.

Height growth

A convective boundary layer grows during the day. Shortly after dawn, when convection begins, it may be only a hundred metres deep. As heating begins, convection actively erodes the base of the capping inversion, entraining air from the inversion into its bulk. This erosive action strengthens the capping inversion and would be self-limiting if surface heating did diminish the strength of the capping inversion by warming the boundary layer from below. Entrainment continues so long as heating is maintained at the ground.

Usually the convective boundary layer grows until mid-afternoon, when its upward advance is matched by general subsidence of the AIR MASS into which it grows. Thereafter height remains constant or decreases slowly. The maximum height reached varies widely, depending on the amount of surface heating, the stability of the atmosphere above and the rate of subsidence. It may reach only a few hundred metres in winter or under heavy overcast skies, or it may grow to perhaps 5 km over sunny deserts where, with no EVAPORATION, the available energy is all released as sensible heat. Though convective boundary-layers characteristically occur during the day, they can also occur at night over cities when heat stored in buildings and streets during the day is released at night along with heat from motor vehicles and air conditioning.

Cloud formation

Cumulus clouds often form in the humps where plumes strike and distort the capping inversions of convective boundary layers. The clouds themselves are not considered part of the boundary layer because they have their own dynamics, being powered by release of latent heat of vaporization of the condensing water droplets. These clouds modify the rate of growth of the boundary layer by affecting its radiation balance and therefore the energy available to sustain the surface heat and vapour FLUXES. If cumulus clouds generate sufficient buoyancy they may break right through the capping inversion and grow to considerable height, carrying moisture and other substances from within the boundary layer up to higher levels.

Stable boundary layer

At night a stable boundary layer forms when the land cools by loss of thermal radiation to the sky. This begins near sunset when a shallow layer of cooled air forms near the ground, while turbulence dies away in the residual mixed layer formed during the previous day's convection. With cooler air at the bottom, buoyancy is negative, resisting convection. This reduces turbulence and lessens momentum transfer to the ground. Winds slow near the ground but can accelerate higher up as the retarding DRAG is reduced. A velocity maximum, or nocturnal jet, frequently occurs at the top of the stable boundary layer. With little interaction between levels, wind can flow in different directions at different altitudes, so layers of moist or polluted air can intrude from various directions at different levels, creating complex night-time profiles. Fog or cloud may form within a stable boundary layer if the TEMPERATURE drops below the DEW POINT as the surface cools.

The height of the stable boundary layer is difficult to define exactly since there is no distinctive capping inversion to mark it. Rather, there are levels where turbulence dies out, downward momentum transfer becomes negligible, the lapse rate becomes adiabatic and where the low-level jet has its maximum speed. These need not be the same height. Whatever measure is chosen, the stable boundary layer is always shallower than the neutral boundary layer and it grows during the night as cooling proceeds. By dawn the stable boundary layer is typically a few hundred metres deep.

Oceanic boundary layer

Over oceans, the boundary layer does not exhibit the marked daily cycles that characterize boundary layers over land. Instead, the marine boundary layer adjusts to the sea-surface temperature, which remains steady. Higher humidity near the sea surface maintains buoyancy so the marine boundary layer is usually weakly convective with a capping inversion at its top. Continuing evaporation at the sea surface creates high humidity within marine boundary layers so clouds and fog are common. [K.G.M.]

REFERENCE

Garratt, J.R. (1992) *The Atmospheric Boundary Layer.* Cambridge University Press, Cambridge.

plankton The term plankton refers to very small, mainly microscopic organisms—bacteria (bacterioplankton), algae and cyanobacteria (PHYTOPLANKTON) or animals (ZOOPLANKTON) — that live suspended in the open water of freshwater and marine environments. Zooplankton and some phytoplankton are capable of swimming or can control movement and BUOYANCY but their vertical and horizontal distribution is substantially dependent on water movement generated by winds, CURRENTS and TIDES. In this way a distinction can be made with the NEKTON, which comprises larger organisms, mainly fish, that are capable of more active swimming through the water.

Although the existence of plankton had been recognized earlier, the term was first used in 1887. At this time only larger organisms that were retained by a silk plankton net of mesh size *c.* 65 μm were described. Smaller organisms that passed through the net were later discovered and given the term nanoplankton. Today this term refers to plankton in the 5–60 μm size range, whilst smaller organisms (≤5 μm) are referred to as ultraplankton; the term PICOPLANKTON has been introduced for the smallest algae (≤1 μm), which are now known to be important primary producers in the OCEANS and in oligotrophic LAKES.

Plankton nets of varying construction and mesh size are still used today for qualitative SAMPLING, with the net being towed either vertically or horizontally through the water. For quantitative analysis water-samplers are used, where a known volume from a prescribed water depth can be obtained. [R.W.B.]

planktonic larvae *See* LARVA.

planktotrophic larvae *See* LARVA.

Planosols *See* SOIL CLASSIFICATION.

plant association An ASSEMBLAGE of plants that frequently occurs in combination in nature—a plant COMMUNITY. It is used as a basic unit for the classification of vegetation types (*see* PHYTOSOCIOLOGY). [P.D.M.]

Plant defence theory

Plants as AUTOTROPHS provide an almost complete nutritional base for numerous animals, and for pathogenic and saprophytic fungi and bacteria. Plants synthesize and accumulate a range of readily digestible CARBOHYDRATES, fats and PROTEINS, and provide the majority of VITAMINS and essential minerals. The basis for plant defence theory is the assumption that, without defences, plants would be eaten out of existence by animals or would be eliminated by ubiquitous pathogenic microbes. Support for a defence theory comes from the observation that some plant species are subjected to heavy GRAZING (e.g. pasture grasses) while others consistently seem to escape predation (e.g. ragworts, *Senecio* spp.).

Defences may be structural or chemical. Structural defences include tough epidermis, thick bark, spines — in which the whole leaf is modified for defence (taken to extreme in the Cactaceae)—or prickles — which typically emerge from the leaf or stem epidermis, a feature of the Rosaceae. Several thousand compounds with some defensive function have been described, including alkaloids, PHENOLS and quinones, terpenoids and steroids (*see* PLANT TOXINS).

Function of defence mechanisms

The function of these defences, structural or chemical, is to limit or discourage grazing and other forms of FEEDING and to control microbial challenge. The defences act by inducing indigestion or pain, or repel through taste and smell, or influence the future generation through abortion or mutagenicity. Much less commonly chemical defences may bring about death of the HERBIVORE or microbial PATHOGEN.

All structural defences, and probably most chemical defences, are constitutive, formed in advance of any actual grazing or challenge. A small, but important number of chemical compounds are also formed only following microbial challenge, and these PHYTOALEXINS are an important element in fungal disease RESISTANCE in many species of agricultural importance. A further group of chemical defences, the coumarins and LIGNIN, are formed during the death throes of the plant following mechanical injury or fungal penetration.

Defence strategies

Plant defence theory recognizes that there are metabolic costs to synthesizing and accumulating and maintaining chemical and structural defences. In the 1970s Feeny and others proposed an apparency theory whereby some plant species are poorly defended because they rely on escaping discovery by herbivores (escape being achieved by smallness of size or by flowering and setting seed before or after the seasonal presence of the herbivore). This apparency theory predicts that only species which are easily found by herbivores need to invest in defence and that unapparent species will be poorly defended.

The major contribution of the apparency theory has been to provide experimenters with the intellectual opportunity to disprove it! Coley and colleagues subsequently proposed that plants differ in their defences because they differ in their growth rates. Slow-growing species can be defended at relatively little cost compared to fast-growing species where the cost of diverting photosynthate to defence would impede growth. The growth theory further predicts that the consequences of herbivory for a slow-growing species are great because the grazed parts take longer to replace, hence the need for defence, while the cost for fast-growing species is less as the grazed parts are quickly replaced, so the need for defence is less.

Seeds

When it comes to defence of the seed, however, the theories need to be modified. Many fast-growing, or unapparent species, form long-lived seeds which, when buried, may persist for many years to emerge as weeds following DISTURBANCE of the soil. Unlike the vegetative plant such seeds are often small, smooth, readily buried and heavily defended against subterranean herbivores and microbes. Examples include the docks (*Rumex* spp.) and willowherbs (*Epilobium* spp.). In contrast, long-lived apparent plants, such as many European trees and PERENNIAL grasses of rough grazing are heavily defended in the vegetative stage but form seeds with little capacity for persistence. Such seeds tend to be relatively poorly defended. *See also* COSTS OF RESISTANCE; DEFENCE MECHANISMS; EVOLUTIONARY OPTIMIZATION; TRADE-OFF. [G.A.F.H.]

REFERENCES

Coley, P.D. (1987) Interspecific variation in plant anti-herbivore properties: the role of habitat quality and rate of disturbance. *New Phytologist* **106** (Suppl.), 251–263.

Feeny, P. (1975) Plant apparency and chemical defense. *Recent Advances in Phytochemistry* **10**, 1–40.

---

plant ecology *See* ECOLOGY.

plant–herbivore interactions Like all biological, especially trophic interactions, the relationships between animal feeder and plant food can lead to responses in each or both on an ecological (short) time-scale and on an evolutionary (long) time-scale. Because the land is generally green, there has been a presumption that herbivores are not food limited, but instead are controlled by organisms that feed on them. This is more generally true for terrestrial than aquatic systems; in the former the tissues necessary for providing support in air (CELLULOSE/LIGNIN) are effective in preventing animal GRAZING and digestion. So more plant PRODUCTION goes to litter and soil on land than to sediment in waters. Nevertheless, terrestrial herbivores have evolved methods for ingesting plant materials (e.g. grinding teeth; or sharp suctorial mouthparts for removing sap) and can have local effects on survival, growth, reproduction and hence density of plant cover. In turn, plants (both terrestrial and aquatic) have evolved defence systems including physical spines and chemical TOXINS.

It is also of considerable importance that herbivory of reproductive tissues, or materials produced by them, can be of mutual benefit. Herbivores that consume pollen and nectar also transfer pollen from plant to plant; and fruit-feeders may disperse seeds.

An interesting question is why insects have not successfully colonized the sea. One answer is that the insects depend somewhat mutualistically on the angiosperms and these have not been able to colonize the sea because it provides adverse conditions for the dispersal of pollen. This illustrates how mutualisms between food and feeder might play an important part in the geographical distribution of taxa. *See also* SPECIES INTERACTIONS. [P.C.]

plant–microflora interactions *See* RHIZOSPHERE.

Plant toxins

Toxic compounds are widespread throughout most plant groups, including angiosperms, gymnosperms, pteridophytes and bryophytes. Of the 10 000 or so characterized terpenoids, 8000 phenolic compounds and several thousand alkaloids, probably the majority induce toxic responses, at least *in vitro*. Whether this toxic effect is the natural function has only been confirmed for a few hundred compounds. Even fewer TOXINS are known, with certainty, to confer a significant ecological benefit to the producer. However, intuitively and logically it is most likely that the majority of plant toxins function in the control of herbivory — ranging from the GRAZING of large mammals (Fig. P19) and larvae of moths to the FEEDING of aphids. Plant toxins also act as a potent defence against fungal, bacterial and viral challenge. The major plant toxins may be classified, for convenience, into three groups, based on structural affinities (Table P5).

Evolution of toxins

Following the invasion of the land by plant-algal ancestors around 450 million years ago, plant toxins evolved to control herbivores and fungal and microbial pathogens, so leading to the establishment of numerous and complex plant–animal, plant–fungal and plant–microbial interactions. These interactions would have included the evolution of non-toxic compounds involved in the initiation and regulation of mycorrhizal symbioses and NITROGEN FIXATION by bacteria, and later the attraction of animals, visually and by scent, for POLLINATION.

These non-toxic and toxic compounds share the same biosynthetic pathways with plant growth regulators, and the demarcation between toxin, pollinator attractant, symbiotic regulator and growth regulator is not always distinct. Phenolic stilbenes act as DORMANCY-inducing growth regulators and potent FUNGICIDES; monoterpenes function as odours for pollinator attraction and, in the form of pyrethroids, as potent INSECTICIDES.

From earliest times toxic compounds have evolved to keep pace with the evolution of DETOXIFICATION mechanisms present in insects, molluscs, vertebrates and fungi. This coevolutionary ARMS RACE may explain why today there are, at the very least, between 30 000 and 60 000 secondary compounds among 300 000 extant land plants, including lichens.

Production

The majority of toxins are constitutive — i.e. present in the plant regardless of the actual threat from PATHOGEN challenge or from herbivores. However, one important group of plant toxins are formed in response to a specific challenge, such as fungal infection or following mechanical injury. Analogous to the immune system of mammals, these toxic compounds are known as PHYTOALEXINS

Fig. P19 *Veratrum lobelianum* (Liliaceae), a conspicuous and ungrazed herb of alpine pasture which, due to its alkaloid content, is highly toxic to cattle. Extracts of rhizomes of the related *V. album* were used in past times as arrowhead poisons. (Courtesy of R.M.M. Crawford.)

Table P5 The major groups of plant toxins.

Group or generic name	Structure based on	Chemical example	Mode or type of action	Plant source
Phenolic	Phenylpropenoic acid	Coumarins	Haemolysis	*Melilotus alba*
	Gallic acid	Tannins	Indigestion	Nuts/bark of trees
	Naphthaquinone	Juglone	Antifungal	*Juglans nigra*
	Anthraquinone	Hypericin	Photosensitivity	*Hypericum* spp.
	Flavonol	Kaempferol	Allelopathic	Widespread
	Iso-flavone	Rotenone	Respiration inhibitor	Legumes
Terpenoid	Monoterpene	Pinene	On insect antennae	*Pinus* spp.
	Sesquiterpene	Gossypol	Taste repellent	*Gossypium barbadense*
	Diterpene	Rhodojaponin	Taste repellent	*Rhododendron* spp.
	Steroid	Progesterone	Mammal infertility	*Tulipa* bulbs
	Saponin	Digitonin	Haemolysis	*Digitalis* spp.
Nitrogen-based	Non-protein amino acid	Dihydroxyphenyl alanine	Insect cuticle assembly	*Mucuna* spp.
	Cyanoglycoside	Hydrogen cyanide	Respiration inhibitor	*Lotus corniculatus*
	Pyrrolidine alkaloid	Atropine	Vasodilation	*Atropa belladonna*
	Indole alkaloid	Hordenine	Taste repellent	*Phalaris* spp.
	Diterpene alkaloid	Aconitine	Respiratory inhibitor	*Aconitum* spp.
	Protein	Trypsin inhibitor	Inhibits digestion	*Glycine max*

Table P6 Plant toxins and their mode of action.

Response	Description	Chemical example	Plant source
Taste repellent	Compounds avoided by herbivores, at least in laboratory trials	Sesquiterpenes	Asteraceae
Indigestion	Interference with digestion resulting from, e.g. protein precipitation in gut	Low-molecular-mass tannins	Widespread
Intestinal injury	Intestinal bleeding in herbivores within hours of feeding	Coumarins	*Melilotus alba*
Death	Usually a rapid effect on primary metabolism such as inhibition of respiration	Cyanoglycosides	Legumes
Allelopathic	Inhibit growth of neighbouring plants	Naphthaquinones	Heartwood of trees
Fungicidal, bactericidal	Often ill-defined in terms of mechanism of antimicrobial response	Lignin	Widespread
Infertility	Disruption of oestrous cycle or abortion in mammals	Iso-flavones, steroids	Bulbs of tulips
Teratogenic	Producing abnormal (malformed) development in fetus of the herbivore	Alkaloids	Blighted potato tubers

(from the Greek words *phyton* plant, and *alexos* to ward off). An example is the phenolic stilbene resveratrol, the basis of disease RESISTANCE in the grape *Vitis vinifera* and also present in the unrelated clover family. With the advance of gene transfer technology, the genes for phytoalexin biosynthesis are being introduced into a number of disease-susceptible species.

Effects

Toxic responses are varied, not only in the severity of TOXICITY but in the mode of action (Table P6). Toxins do not necessarily kill the HERBIVORE; a potent toxic effect may be exerted as an unpleasant taste, impaired digestion, a dermatological response or temporary blindness, or a toxin may affect the offspring of the herbivore by impairing normal development of the embryo.

Although many toxins are effective against a wide range of organisms (e.g. cyanoglycosides), others appear to have a more selective effect. Several of the widespread flavonols, such as quercetin, appear to be toxic, at least at physiological concentrations, to insects but not to fungal pathogens. Horses and cattle avoid the ragworts (*Senecio* spp.) probably because of the presence of the toxic alkaloid retronecine. The same ragworts, however, are the food plants of the cinnabar moth (*Tyria jacobaeae*), where retronecine is accumulated in the adult, the larvae and the eggs. These moths have evolved, to their advantage, mechanisms to detoxify or sequester the alkaloid.

Storage and metabolism

Toxins, by definition, are likely to be toxic to the plant synthesizing and accumulating the toxin. However, the greater part of plant toxins are stored in the cell vacuole, isolated from primary processes of RESPIRATION and PHOTOSYNTHESIS. Only on disruption of the vacuolar membrane would the toxin come into direct contact with primary metabolism leading to self-toxicity. This suicide reaction probably occurs whenever plant tissues are mechanically damaged, for example during grazing by herbivores. Metabolism in the damaged tissue is inactivated and so isolated from the neighbouring undamaged cells. This may be important in preventing further incursion of fungal pathogens.

Most toxins are stored in a non-toxic form and require an activating step before exhibiting their toxic potential. A common form of inactivation is glycosylation, in which one or more sugar molecules are attached to the toxin. When the toxin-plus-sugar is, for example, ingested by a herbivore, the glycoside linkage is broken as part of the digestive process so releasing the now-activated toxin (Fig. P20).

Significance to humans

The significance of the terms plant toxins and plant defence chemistry implies that the natural function of the compound is to induce a toxic response in a herbivore, a microbial pathogen or a neighbouring plant. This assumption is based all too often not on studies of molecular responses in fungal pathogens or behaviour of herbivores but on the physiological effects on humans.

A number of toxins act as narcotics or hallucinogens in humans, including the alkaloids cocaine (from *Erythroxylum coca*) and morphine (from *Papaver somnambuliferum*) and the diterpene cannabinoids (from *Cannabis sativa*). The advantage to the plant of these narcotics is less clear. The tiger moth (Arctiidea) larva, for example, consumes the leaves and sequesters the cannabinoids without harm. The medical use of curare from *Strychnos* spp. as a muscle relaxant in anaesthesia can be traced to its use as a poison in hunting by South American Indians. From this, it may be logical to assume that the natural function of curare is in defence against vertebrate herbivores. However, because curare is present in the bark it may be that its real ecological significance is to protect the tree against wood-boring beetles. Or again, the alkaloid quinine, from *Cinchona* spp., which has an extremely bitter taste to humans, might by logical extension act to deter all herbivores. But quinine turns out to be significantly less of a taste repellent to herbivorous reptiles. The natural function of quinine is certainly not in the treatment of malaria, despite its long-established place in traditional medicine. Likewise, while the Athenians of 399 BC may have benefited by eliminating such political undesirables as Socrates by forced administration of the piperidine alkaloid coniine, it is not at all clear what benefit, ecologically, is conferred on the plant *Conium maculatum*. One certainty, however, is that among the several tens of thousands of plant toxins there are numerous compounds of considerable but unexplored pharmaceutical potential, and it is this which drives the growing interest in traditional medicines and also the concerns over the destruction of the Earth's BIODIVERSITY. *See also* COSTS OF RESISTANCE; DEFENCE MECHANISMS; PLANT DEFENCE THEORY. [G.A.F.H.]

REFERENCES

Chadwick, D.J. & Marsh, J. (eds) (1990) *Bioactive Compounds from Plants*. CIBA Foundation/John Wiley & Sons, Chichester.

Harborne, J.B. (1993) *An Introduction to Ecological Chemistry*, 4th edn. Academic Press, London.

Fig. P20 Alpine meadow with the poisonous globe flower (*Trollius europaeus*). In common with other members of the Ranunculaceae, this species contains the inocuous glycoside ranunculin which readily hydrolyses to form the vesicant (blister-inducing) protoanemonin. (Courtesy of R.M.M. Crawford.)

Palo, R.T. & Robbins, C.T. (eds) (1991) *Plant Defenses against Mammalian Herbivory*. CRC Press, Boca Raton, Fla.

plantation A term used in FORESTRY to distinguish areas of deliberately planted trees from FORESTS which are natural or have regenerated after clear-felling (or burning) without further human intervention. Plantations in the temperate regions are usually developed for timber production, though in the tropics they include other products such as RUBBER, coffee, tea or cotton. In forestry, plantations often develop in distinct patterns, with trees in straight lines to elevations dictated by commercial rather than ecological considerations; these features can persist for many decades or centuries beyond the original establishment of the plantation. Plantation trees, unlike natural populations, are usually of a closely related genetic stock. There is an implication, not always justified, that plantations have a reduced value for wildlife compared to natural communities. [G.A.F.H.]

Plants and fungi, global diversity and conservation

Consideration is given here to all plants except the angiosperms (*see* ANGIOSPERMS, GLOBAL DIVERSITY AND CONSERVATION).

Included are algae of different colours (not cyanobacteria, formerly known as Cyanophyta, blue-green algae), lichens, liverworts and mosses, ferns and fern allies, and gymnosperms. Also included are fungi (non-green), now widely regarded as a group distinct from plants.

Fungi

Some 70 000 species of fungi have been described, with about 12 000 recorded in the British Isles. A conservative estimate of the total number of fungal species worldwide is at least 1 500 000, although, unlike angiosperms, evidence of great tropical DIVERSITY is limited. However, fungi probably constitute the second largest group of organisms, after insects, in total number of species. An uncertain number of fungi are associated with insects, huge numbers of which are undescribed.

Fungi may be classified into four major divisions: Mastigomycotina, Zygomycotina, Ascomycotina and Basidiomycotina (a fifth group, the rather heterogeneous Deuteromycotina or Fungi Imperfecti, comprises mainly Ascomycotina but its members lack the perfect (sexual) stage or form this only rarely). The Mastigomycotina, regarded as the most primitive, and now often considered taxonomically closer to algae than other fungi, includes microscopic aquatic forms, the parasitic downy mildews and related fungi such as *Phytophthora infestans* (causing potato blight). The Zygomycotina is a small group containing the pin moulds (e.g. *Mucor*) and pea-truffles (e.g. *Glomus*).

Most members of the Ascomycotina have eight-spored asci; they include forms ranging from yeasts and parasitic powdery mildews to saprophytic larger fungi with fleshy fruit bodies (e.g. *Peziza*and the morels, *Morchella*). Members of the Basidiomycotina bear spores on basidia. They include the parasitic smut and rust fungi (e.g. *Puccinia graminis*, wheat rust), the agarics (mushrooms and toadstools), which are almost all saprophytic or mycorrhizal and include many edible species, and the polypores, some of which cause serious damage to trees (e.g. *Heterobasidion annosum*).

In form and habitat, fungi vary extremely widely, but only a few are marine. The LIFESPAN of some agarics is very long and a single individual may extend over a great area, whereas other fungi are minute and short-lived. Spores, of enormous diversity, are the chief propagules.

Algae

About 40 000 species of algae, normally considered the most primitive green plants, have been described; the global total is estimated to be at least 200 000, which may be a gross underestimate. The diversity of form in the algae is considerable, ranging from simple unicells to colonial, filamentous and thalloid structures, with major differences in pigments and ultrastructure. Algae are abundant in the sea and freshwater as PHYTOPLANKTON, while larger species are found on rocky seashores. Algae may also form a film on the soil surface, especially in the humid tropics, and some live on permanent snow and ice.

Most numerous but least known are the microalgae, including the diatoms (Bacillariophyta) and dinoflagellates (Pyrrophyta); these are very important in the marine FOOD WEB but are also present in freshwater. Many of these algae are abundant in tropical oceans, but some parts of the world are much under-collected and the degree of endemism is uncertain.

The most fully described are the Chlorophyta (green algae), Phaeophyta (brown algae), Rhodophyta (red algae) and Charophyta (stoneworts). These groups comprise some 1000 genera and more than 6000 species. The Chlorophyta vary considerably in structure and habitat; of global distribution are the Ulotrichales (mostly freshwater), Ulvales (marine and brackish), Cladophorales (marine) and some members of the Bryopsidales (includes tropical and subtropical frondlike forms). High endemism is shown in the green seaweed flora of southern Australia.

The Phaeophyta, with more than 1500 species, are mainly marine, members of the more primitive groups (Ectocarpales, Chordariales) having a global distribution. The kelps (Laminariales) are disjunctly distributed in temperate waters of both hemispheres. The most advanced order, Fucales, includes *Sargassum* of tropical waters and the Fucaceae of the temperate zone of the Northern hemisphere, plentiful and zoned on European ROCKY SHORES—different fucoids occupy corresponding habitats in the Southern hemisphere. Species-depauperate floras of this and other algal groups occur in coldwater UPWELLING areas and isolated islands.

The Rhodophyta, with over 555 genera, are mostly marine, especially in the intertidal and subtidal zones, with fewer than 5% in freshwater. Their form ranges from unicells to 'coralline' (resembling lumps of rock), branched filaments and even an organization resembling stem and leaves. Species-rich areas of red and other algae include the Japanese Pacific, California, Chile, the subtropical western Atlantic and the North Atlantic. In southern Australia 75% of the red algal flora is endemic. The marine algal flora of the British Isles (more than 700 species) is fairly rich.

The stoneworts, a small group of macroscopic algae with nodal structure, occur in freshwater or brackish conditions. About half of the 450 species are endemics.

Lichens

Lichens are composite organisms in which a fungal partner (mycobiont) is in symbiosis with a photosynthetic partner, a green alga or cyanobacterium (photobiont), forming an independent entity. They are successful PIONEER colonizers, of importance in, for example, cold (alpine and polar) regions, deserts and environments with toxic metals. On a world scale, lichens are one of the better known groups of organisms; some 15 000 species have been described, estimated as more than half of the world's total. However, tropical areas, where lichens are important in the CANOPY, are under-recorded. Temperate rain forests in the Southern hemisphere are rich in lichens, but the greatest number (about 480 genera and 3600 species) are known from the USA and Canada. About 280 genera and 1660 species are recorded in the UK.

Bryophytes

Bryophytes — mosses, liverworts and hornworts — number some 14 000 species, more than half of these being mosses; hornworts (Anthocerotales) comprise only five genera but have a widespread distribution, including epiphytes and epiphyllous species as well as terrestrial taxa. These non-vascular plants are most diverse in consistently moist cool and temperate climates. The greatest SPECIES RICHNESS is in the Indo-Australian archipelago and temperate and montane parts of South America. About 1000 species of bryophytes are known in the British Isles.

The CALCIFUGE moss *Sphagnum* is a major component of peat BOGS worldwide; this large genus is most diverse in cool oceanic parts of the Northern hemisphere. Also cosmopolitan are the moss families Dicranaceae (chiefly calcifuge), Pottiaceae (includes xerophytes), Bryaceae (notable in polar deserts) and Hypnaceae (well speciated regionally). Pterobryaceae are represented in tropical montane rain forests as frondose epiphytes with high endemism, whereas members of the Grimmiaceae are mainly alpine and subarctic.

Liverworts are not drought-tolerant, being most diverse in highly oceanic areas. Some thalloid forms are cosmopolitan; leafy liverworts (Jungermanniales) are highly diverse in tropical montane rain forests, many species being epiphyllous.

Pteridophytes

Spore-producing vascular plants with two free-living stages in their life-cycle. The ferns and fern allies (Pteridophytes), comprise some 300 extant genera and an estimated 12 000 species. Very many fossil forms are known, for example the simple psilophytes of the Silurian and Devonian periods, lycopods (*Lepidodendron* and allies), as well as the jointed *Calamites*, some of which were tall trees dominating in Carboniferous times and acting as major formers of coal.

Present-day members of these groups are RELICT genera, with species of wide distribution and small size. They include the club-mosses (e.g. *Lycopodium*) and their relatives *Selaginella* spp. — both are cosmopolitan genera with many species—the horsetails (*Equisetum* spp.), the extremely primitive *Psilotum* of the tropics and *Tmesipteris* of the Southern hemisphere.

The ferns are by far the largest living group of pteridophytes, with over 280 genera and some 9000 species. In form, pantropical ferns range from the filmy ferns (Hymenophyllaceae) to tall tree ferns (Cyatheaceae and Dicksoniaceae), as well as a few water ferns. Although worldwide, ferns are most plentiful in moist shady sites and are much less common in dry areas. Of about 35 families of ferns, only five (Polypodiaceae, Thelypteridaceae, Adiantaceae, Aspleniaceae and Woodsiaceae), both temperate (with British representation) and pantropical, account for nearly half of the world's fern species. Many have restricted distribution, but *Pteridium aquilinum* (bracken) is found worldwide except for very cold regions. Particularly rich in ferns are Papua New Guinea and India. In the tropics both bryophytes and pteridophytes tend to be most diverse at mid-elevations.

Gymnosperms

Gymnosperms are nearly all trees. They are characterized by bearing seeds that, unlike those of angiosperms, are not formed within a closed fruit. By far the largest group are the conifers, with about 600 extant species. They arose in the Carboniferous period and dominated vegetation until the rise of the angiosperms in the Cretaceous. The biggest genera are *Pinus* (pines), *Abies* (firs) and *Picea* (spruces), which form major forests in the BOREAL zone of Eurasia and North America, although conifers are taxonomically most diverse in Oceania. Conifers include the world's tallest tree (*Sequoia sempervirens* at 110 m) and the oldest tree (*Pinus aristata*, about 4900 years old). Important timber trees include the podocarps (Podocarpaceae), characteristic of tropical and subtropical forests of the Southern hemisphere, notably South America; kauri pines (*Agathis*) of New Zealand and Malaysia; and *Cunninghamia* of China. The ancient palmlike cycads, with strong fossil representation, are reduced to about only 100 present-day species, which occur in Central and South America, South Africa and south-east Asia to Australasia. Other gymnosperms include the Chinese sacred tree *Ginkgo*, a relict species, and the most angiosperm-like genera *Ephedra* (switch plants of dry areas), *Gnetum* (includes tropical lianas) and *Welwitschia* (long-lived plant of the Namib desert with only two leaves).

Conservation of plant biodiversity

Approximately 10–15% of world plant species are considered rare, threatened or endangered. Many plant species, formerly widespread, have now been greatly reduced in extent, number and mean population size, with largely unknown effects on population and species viability. Plant populations can also decline owing to less obvious threats, such as loss of pollinators, DISPERSAL vectors or other mutualists. Plants are also affected by habitat fragmentation, soil erosion, CLIMATE CHANGE, alteration of DISTURBANCE regimes and modification of successional trajectories.

Conservation of the BIODIVERSITY of the plants covered here is essentially similar to that of the angiosperms, but the emphasis for the more primitive forms is on conserving the habitat rather than the individual species. While the ECOSYSTEM approach is by far the most significant, conservation of particular species of some gymnosperms, for example the cycads, and certain members of primitive groups, such as the charophyte *Lamprothamnium* (protected in the British Isles by legislation), is also important.

Preservation of habitat, not always containing notable angiosperms, is essential for the continued existence of many species and ecosystems; the unique coastal fog lichen communities of northern Chile are an example. Not infrequently a particular habitat supports more than one rare non-flowering plant of conservation value. A general loss of aquatic habitats by drainage has adversely affected the populations of many lower plants, some of which are also sensitive to EUTROPHICATION. *See also* BRYOPHYTES, GLOBAL DIVERSITY AND CONSERVATION; CONSERVATION BIOLOGY; FUNGI, DIVERSITY AND CONSERVATION; GYMNOSPERMS, GLOBAL DIVERSITY AND CONSERVATION; PTERIDOPHYTES, GLOBAL DIVERSITY AND CONSERVATION. [A.J.W.]

REFERENCES

Falk, D.A. & Holsinger, K.E. (eds) (1991) *Genetics and Conservation of Rare Plants*. Oxford University Press, New York.

Given, D.R. (1994) *Principles and Practices of Plant Conservation*. Chapman & Hall, London.

Groombridge, B. (ed.) (1992) *Global Biodiversity. Status of the Earth's Living Resources.* World Conservation Monitoring Centre. Chapman & Hall, London.

---

plasticity *See* PHENOTYPIC PLASTICITY.

plastics Synthetic polymers that can be moulded into shape by using heat and/or pressure. They may be broadly classified into thermoplastics and thermosetting plastics. Thermoplastics retain their plasticity after manufacture, and include polyethylene, polypropylene, polystyrene, polyester and polyvinyl chloride (PVC). They are more or less inert and impermeable to WATER, and thus useful as thin films or sheets for packaging all kinds of goods including food. Thermosetting plastics retain their shape after manufacture and are good for making 'hard' components such as light-switches; they include bakelite and materials derived from epoxy resin (in epoxy molecules, OXYGEN is bound to two CARBON atoms to form a three-membered ring).

Plastics are derived from petrochemicals, are often cheap to make and form compared to metals, and have good thermal and electrical insulating properties. They are relatively inert, and are strong relative to their mass.

Plastics form a large fraction of domestic refuse, and they are non-degradable by microorganisms and light. Moreover, they are not easily recycled. Thus they pose a formidable environmental problem. Attempts to produce plastics that are biodegradable or photodegradable are being made, but the bulk of plastics used today do not have these properties and have to be buried in LANDFILL sites or incinerated.

Chemicals added to the plastic polymers to improve the properties of the finished plastic may enter the FOOD CHAIN. Plasticizers such as phthalates are not polymerized and may be volatile, leachable and biologically active, and so are a potential HAZARD when plastics containing such compounds are used for food wrapping or water pipes. [J.G.]

plate tectonics The process by which lithospheric plates are moved across the Earth's surface to collide (i.e. at SUBDUCTION zones), slide by one another (i.e. to form transform faults), or move away from each other (i.e. at DIVERGENCE zones) to produce the topographic configuration of the Earth. Plate movements are driven by heat (arising from radioactive decay) escaping from the Earth's core and mantle, which causes convective movements within the mantle. Plates move away from each other along midocean ridges, and new CRUST forms at the ridges as magma erupts through fissures in the floor. As the new crust cools, it becomes denser and sinks. Crust is destroyed at the ocean trenches as gravity draws one plate under another and the old, dense LITHOSPHERE is subducted.

The independent movements of the lithospheric plates are associated with the formation and destruction of ocean BASINS. Ocean basin formation begins when a continent remains in one position for about 100 million years. Thick continental crust impedes heat flow, causing the underlying mantle to heat up and to expand. Such expansion stretches the continental crust, causing long, narrow rift valleys and eventually narrow ocean basins to form. The ATLANTIC OCEAN is presently growing at a rate of about 5–10 cm $year^{-1}$. In contrast, the PACIFIC OCEAN is decreasing in size. New crust may also form at hot spots unassociated with mid-oceanic ridge systems. As the plates move across the hot spots, they form chains of volcanic islands and seamounts (e.g. Hawaiian Islands), that record the direction of plate movement. *See also* BIOGEOGRAPHY. [V.F.]

REFERENCES

Gilluly, J., Waters, A.C. & Woodford, A.O. (1968) *Principles of Geology*, 3rd edn. W.H. Freeman, San Francisco, Calif.

Pickering, K.T., Hiscott, R.N. & Hein, F.J. (1989) *Deep-Marine Environments: Clastic Sedimentation and Tectonics*. Unwin Hyman, London.

platinum (Pt) A metallic element (ATOMIC NUMBER = 78; relative ATOMIC MASS = 195.08; melting point = 1772°C; boiling point = 3827°C), found in deposits along with the other platinum group metals (average abundance in the Earth's CRUST = 0.001 mg kg^{-1}). Its use in CATALYSTS for automotive CATALYTIC CONVERTERS has led to an increase in demand for the metal over recent years. Platinum is also used as a catalyst in a variety of industrial applications. Platinum, in the form of cisplatin and its derivatives, is also extensively used in cancer chemotherapy. Approximately half of the population are susceptible to some allergic reaction to certain platinum compounds—so-called platinum allergy or platinosis—but usually only after prolonged EXPOSURE. In general, platinum must be considered to be carcinogenic and mutagenic. [P.J.B.]

platykurtic distribution *See* KURTOSIS.

playas Shallow, circular basins of ALLUVIAL fans in plains and DESERT landscapes, they are also known as bajadas. Fine-textured SOIL and accumulation of salts make playa lakes a poorly drained habitat; precipitation provides ephemeral water inflow, with each of the playas and its WATERSHED developing vegetation in a CLOSED SYSTEM. Over 20 000 playas are found on the southern High Plains in Texas and New Mexico, with an average size of 6.3 ha but high variance in size. As many as 14 physiognomic types of vegetation are described; for example, wet meadow type is dominated by *Echinochloa* and *Leptochloa* species, and the broad-leaved emergent type is dominated by *Polygonum* spp. For their ISLAND BIOGEOGRAPHY settings, high alpha and beta diversity (*see* DIVERSITY, ALPHA, BETA AND GAMMA), and rather high year-to-year variability in plant species composition, playas resemble VERNAL POOLS. However, playas allow run-offs to be used as IRRIGATION, and with their high biomass productivity in the narrow-leaved emergent type, agricultural and other land uses are common. Inundated large playa lakes serve wildlife interests in the winter; in other seasons they provide for CATCHMENT, sewage treatment and recreation. Such habitats serve as good models for ECOSYSTEM studies because of their well-known distribution, distinctive BIOTA, currently increased human impact and rather well-defined conservation agenda. [S.K.J.]

REFERENCES

Bolen, E.G., Smith, L.M. & Schram, H.L. (1989) Playa lakes: Prairie wetlands of the Southern High Plains. *BioScience* **39**, 615–624.

Louw, G. & Seely, M. (1982) *Ecology of Desert Organisms*. Longman, London.

pleiotropy The genetic phenomenon whereby a single allelic change has effects on multiple characters, rather than on one character only. To some extent, this term exists because of the unwarranted generalization of the 'one gene–one enzyme' theory to a spurious 'one gene–one effect' assumption. This way of thinking is also implicit in some early Mendelism. By contrast, many evolutionary geneticists, led by Sewall Wright, have argued that pleiotropy will be the rule, rather than exceptional.

Pleiotropy can generate a range of different phenomena in evolutionary ecology. Its role in shaping the evolution of AGEING is discussed in ANTAGONISTIC PLEIOTROPY HYPOTHESIS. More generally, pleiotropy can connect characters together so that they evolve in an allometric fashion (*see* ALLOMETRY), increasing in parallel following well-defined ratios. Conversely, pleiotropy can also connect characters so that they evolve according to a TRADE-OFF pattern. Moreover, pleiotropically interconnected suites of characters may evolve by combinations of allometry and trade-off, in ways that will often be unpredictable. Finally, pleiotropy may in some respects generate the cohesion and coordination that define the unitary nature of the organism itself. *See also* DISPOSABLE SOMA THEORY. [M.R.R.]

REFERENCE

Wright, S. (1968) *Evolution and the Genetics of Populations*. Vol. 1. *Genetic and Biometric Foundations*. University of Chicago Press, Chicago, Ill.

plesiomorphy The situation in which characteristics exhibited by members of a species or group of species have a common evolutionary origin (i.e. are homologous), but for which that origin occurred in a species ancestral to those presently exhibiting the trait. Such persistent ancestral traits may characterize species belonging to a variety of otherwise distantly related groups but are not evidence of CONVERGENT EVOLUTION. For example, humans, gorillas and elephants all have vertebrae, and vertebrae appear to have evolved only once; thus, the vertebrae in all three of those species are homologous. And yet, the presence of vertebrae in those species does not help us assess their phylogenetic relationships relative to each other because vertebrae evolved long before the origin of the ancestors of humans, gorillas and elephants.

In the methodology of PHYLOGENETIC SYSTEMATICS, the relatively ancestral trait, determined by OUTGROUP comparisons, of a pair or more of homologous traits linked evolutionarily in a TRANSFORMATION SERIES is called the plesiomorphic trait, and is not used to diagnose monophyletic groups within the study taxa. Evolutionary taxonomists may use some plesiomorphic traits to diagnose grades, or paraphyletic groups, in their classifications. [D.R.B.]

Plinthisols *See* SOIL CLASSIFICATION.

ploidy *See* DIPLOIDY.

Plume

A continuous stream of fluid emerging from a localized source and penetrating and mixing with an AMBIENT fluid under the influence of initial momentum and/or BUOYANCY. Mixing with ambient fluid causes the plume to diverge away from its source and dilutes the momentum and buoyancy of the plume, and any POLLUTANT it contains. In natural and artificial ATMOSPHERE plumes, the mixing is effected by TURBULENCE within and around the plume, and the ambient air is almost always turbulent as well. Usually the temperature and momentum of the plume become indistinguishable from those of its surroundings at a relatively short distance from the point of injection, whereas the presence of a pollutant may be visible or chemically distinguishable to a much greater distance. For example, the main radioactive plume from Chernobyl was detectable several thousand kilometres from its source. A plume which is not effectively continuous with its origin is known as a puff, and mixes even more rapidly with its surroundings on account of its larger surface to volume ratio.

Natural plumes

Large atmospheric plumes with substantial initial momentum and buoyancy are produced by erupting volcanoes, and may rise well into the STRATOSPHERE. The TROPOSPHERE is full of relatively weak CONVECTION plumes of air rising from relatively warm surfaces (Scorer 1978). Usually these are driven by near-surface buoyancy, though this may be enhanced by the liberation of heat as air rises above cloud base.

In the oceans, plumes of hot water issuing from hydrothermal vents can maintain anaerobic life by their heat and supply of vital chemicals.

Artificial plumes

Nuclear bombs produce highly energetic puffs which inject radioactive material through the troposphere and into the stratosphere if they are sufficiently powerful (i.e. equivalent energy yield >*c*. 1 Mt TNT).

However, most artificial atmospheric plumes are much smaller and less energetic than these, emerging from a wide range of industrial and domestic chimney stacks and cooling towers, and often rising initially only a few tens of metres above their injection level. Extensive studies to optimize their efficiency in dispersing locally unwelcome POLLUTION are discussed in Pasquill and Smith (1983) and Scorer (1978), and outlined below.

Though initial momentum and buoyancy are usually small, they may add usefully to the efficiency of the chimney stack, particularly by preventing downwash in the lee of the stack and adding to the effective level of injection (Fig. P21a). Prevention of downwash reduces local CONTAMINATION by the concentrated plume, while injection level is important because the pollutant concentrations in the plume are inversely proportional to the ambient wind speed at the injection level, which in turn increases significantly with height in the first 50 m or so. After initial mixing, typically in a horizontal distance an order of magnitude less than the stack height, the plume moves and disperses passively with the surrounding air.

In convectively stable air, such as develops in the first 100 m of height over land on cloudless nights, ambient turbulence is suppressed, and the plume is horizontal and shallow, fanning and meandering slowly in the horizontal with residual turbulence and wind shifts (Fig. P21b). Pollutant concentrations may be high in the

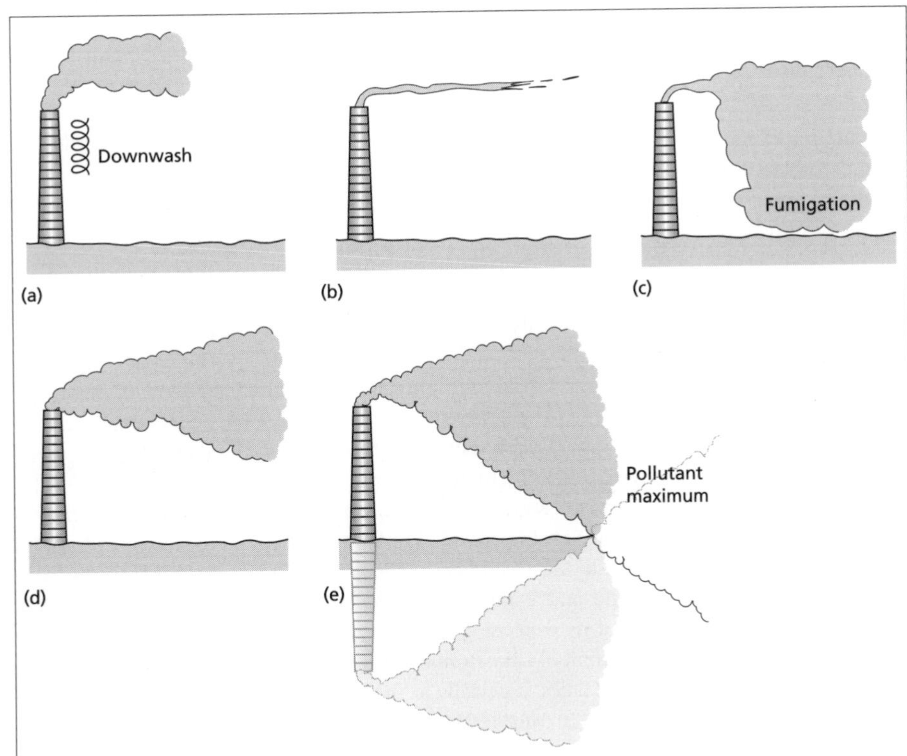

Fig. P21 Representations of a smoke or other plume leaving a chimney stack in a range of meteorological conditions. (a) Might have surface fumigation near the stack if buoyancy or inertia did not raise the plume sufficiently to keep pollutant clear of the downwash in the lee of the stack. (b) Shallow plume in quiet nocturnal conditions. (c) Surface fumigation by downward mixing after dawn. (d) Classic plume broadening by ambient turbulent mixing. (e) Surface fumigation and the mirror image construction.

plume, but are almost zero above and below it, including the surface. When sun-driven convection reaches up to the plume next morning, the surface beyond a minimum distance downwind of the stack will experience a burst of maximum fumigation (Fig. P21c) before the plume reaches a new DYNAMIC EQUILIBRIUM with the increased ambient turbulence. Something similar will happen if the ambient wind speed rises (as a windy weather system moves in) to the point where shear-driven turbulence erupts in the layer containing the plume.

When atmospheric turbulence is fully established, the plume adopts a classic shape, spreading vertically and horizontally in response to mixing by turbulent eddies smaller than the plume beneath, and meandering horizontally and vertically in response to larger eddies (Fig. P21d). Classical treatments (Scorer 1978; Pasquill & Smith 1983) concentrate on the plume in this condition, and usually assume Gaussian vertical and horizontal distributions of pollutant about the plume core at any given distance from the source.

Lack of understanding of natural atmospheric turbulence means that most mathematical descriptions of plumes are semi-empirical constructs resulting from measurements of pollution DISPERSION in a variety of conditions, rather than the first-principle models they would appear to be, but they usefully model the increase of dispersion with turbulent intensity, and the important characteristic maximum of surface pollution at a distance downwind of the stack, which increases with average wind speed and decreases with increasing turbulence (Fig. P21e). In the case of a pollutant which does not interact or attach to the surface, the concentration around and beyond the surface maximum is modelled using the concept of an inverted mirror-image stack and plume vertically below the actual stack. *See also* AIR POLLUTION; ATMOSPHERIC POLLUTION; FATE; OIL SMOKE PLUME; POINT SOURCE. [J.F.R.M.]

REFERENCES

Pasquill, F. & Smith, F.B. (1983) *Atmospheric Diffusion*. Ellis Horwood, Chichester.

Scorer, R.S. (1978) *Natural Aerodynamics*. Ellis Horwood, Chichester.

plutonium (Pu) An artificial element (ATOMIC NUMBER=94; melting point=640°C; boiling point=3230°C), first produced in 1940. Isotopes from ^{232}Pu to ^{246}Pu have been produced by nuclear reactions. ^{239}Pu ($t_{1/2}=2.41\times10^4$ years) is fissile and is used in nuclear weapons and reactors. Breeder reactors produce ^{239}Pu by fast NEUTRON irradiation of ^{238}U:

$$^{238}U + n \rightarrow {}^{239}U \xrightarrow[22.3\ min]{\beta^-} {}^{239}Np \xrightarrow[2.35\ days]{\beta^-} {}^{239}Pu$$

Plutonium occurs in the environment as a consequence of releases from nuclear facilities and fallout from weapons testing and satellite ACCIDENTS. Minute quantities of ^{239}Pu also occur naturally in URANIUM ores. Lower oxidation states of plutonium (+3, +4) are particle reactive and highly insoluble in natural waters whereas higher oxidation states (+5, +6) are more soluble. *See also* FISSION; RADIOACTIVE WASTE; RADIONUCLIDE. [A.B.M.]

REFERENCE
Katz, J.J., Seaborg, G.T. & Morss, L.R. (eds) (1986) *The Chemistry of the Actinide Elements*. Chapman & Hall, London.

pluvial Describing conditions in which PRECIPITATION exceeds, often considerably, evapotranspiration. In the tropics and subtropics especially, intervals of wet, pluvial climate have alternated with drier interpluvial conditions during the Quaternary period. These changes have been accompanied by shifts in the boundaries of tropical FOREST, SAVANNAH and more arid vegetation zones. Alternating pluvial and interpluvial conditions also produce rises and falls in lake levels. LAKES that have in the past undergone considerable expansion during wet periods are referred to as pluvial lakes. One of the best researched examples is the former Lake Bonneville, western Utah, USA, which today is represented by the shallow Great Salt Lake. This lake system has fluctuated considerably and rapidly during the last 500 000 years and, at one point, was almost 500 km in greatest length (Great Salt Lake is 120 km long). Other well-known pluvial lakes include the Dead Sea in the Near East, Lake Baikal in southern Siberia and Lake Chad in the southern Sahara, which, at the height of its expansion attained *c.* 300 000 km^2, or about three times the extent of Lake Bonneville. In general, it has been supposed that pluvial lakes, especially in the subtropics, relate to glacial periods but there is considerable inter- and intra-regional variation. [M.O'C.]

PM10s *See* MOTOR VEHICLE POLLUTION.

PMN *See* PREMANUFACTURE NOTICES.

PNEC *See* PREDICTED NO EFFECT CONCENTRATION.

Podzol The Podzol concept was developed in Russia, where the principal criterion is the presence of a leached horizon, as indicated by the origin of the term from the Russian words *pod* under, and *zola* ash; translated it means 'ashes under'—when these soils are cultivated, the plough turns up the leached horizon, which resembles ashes. Therefore all soils with leached horizons are Podzols, including Acrisols (Ultisols) and Luvisols (Alfisols).

In western Europe and North America, the general sequence of horizons in Podzols consists of litter and partially decomposed organic matter, overlying a dark, humose horizon (ochric horizon), followed by a leached horizon (albic horizon), which in turn overlies a middle horizon of IRON and/or HUMUS and/or ALUMINIUM accumulation (spodic horizon), which grades down into relatively unaltered material; the emphasis here is placed on the horizon of accumulation. Thus, there are two contrasting concepts, with the western European/North American concept being more specific.

Podzols are very common soils in the cool-temperate parts of the world, having developed mainly under coniferous FOREST or HEATHLAND. They also occur in other humid climates on highly quartzose sands. In these situations, they may be many metres thick, as in certain parts of Australia. On the other hand, in the ARCTIC, they are very shallow, and may be no more that 15 cm thick. Generally, Podzols are of low fertility, but with proper treatment, they can be very productive soils. *See also* SOIL CLASSIFICATION. [E.A.F.]

REFERENCE
FitzPatrick, E.A. (1980) *Soils, Their Formation, Classification and Distribution*. Longman, Harlow.

Podzolic soils *See* SOIL CLASSIFICATION.

poikilotherm An organism with a highly variable body TEMPERATURE. In contrast, a HOMEOTHERM has a relatively constant body temperature. These terms are extremes on a continuum describing the stability of body temperature. According to Bligh and Johnson (1973), homeotherms must maintain body temperatures that vary by less than ± 2°C. Parallel terms are eurytherm (an organism with a variable body temperature) or stenotherm (one with a constant body temperature).

The variability of body temperature depends strongly on an organism's physiology. Most poikilotherms (typically microorganisms, plants, invertebrates, fishes, reptiles and amphibians) have an ectothermal physiology: thus their mode of temperature REGULATION depends primarily on heat absorbed from the environment. Most homeotherms (birds and mammals, some insects and plants) are endotherms: their mode of temperature regulation depends primarily on heat produced by their own metabolism. Homeothermy results from the adjustment of metabolic heat production to balance heat lost to the environment.

Variability in body temperature is also influenced by the environment and by behaviour. An ECTOTHERM living in the deep sea may be homeothermic even by mammalian standards, merely because deep-sea temperatures are stable. Moreover, many ectotherms in fluctuating environments are able to maintain relatively constant body temperatures during active periods by behavioural thermoregulation (e.g. by shuttling between sun and shade).

Endotherms are sometimes poikilothermal. Humans show slight daily variation in core temperature, and regulate temperatures of their extremities much less carefully than that of the brain. Many endotherms (e.g. hummingbirds (*Eulampis*), pocket mice (*Perognathus*)) go into daily torpor during periods of STRESS, and may experience considerable daily and seasonal variation in temperature. [R.B.H.]

REFERENCE
Bligh, J. & Johnson, K.G. (1973) Glossary of terms for thermal physiology. *Journal of Applied Physiology* **35**, 941–961.

point mutation *See* MUTATION.

point quadrat A frame, often wooden, housing a row of pins (frequently 10), which are lowered vertically on to VEGETATION, the points of contact giving an objective measure of COVER. The pins resemble knitting needles, and are best guided through two laths to prevent parallax effects. The tip diameter is as small as possible (tip size correlates with exaggeration of cover values). A succession of measurements (e.g. 10 positions of the pin-frame along a line) are taken. In layered vegetation several contacts (hits) may be made as a pin is progressively lowered, giving values exceeding 100%. Such a technique provides a measure of the LEAF AREA INDEX. The method is well suited to grassland but is difficult to use in tall vegetation. *See also* SAMPLING METHODOLOGY/DEVICES. [A.J.W.]

point source An identifiable, discrete origin of CONTAMINATION or POLLUTION, such as the outlet of a pipe or chimney stack. *Cf.* DIFFUSE SOURCES. *See also* PLUME. [P.C.]

Poisson distribution

A discrete PROBABILITY DISTRIBUTION that describes situations where rare observations (e.g. individuals in a sampling unit) follow a RANDOM DISTRIBUTION in either space or time. A spatial example (the most frequent for ecologists) would be the number of barnacles in a QUADRAT. A temporal example would be the number of bats leaving a roost per minute. The sampling unit in the first case is a fixed amount of space (the quadrat), in the second a fixed amount of time (the minute). The term 'rare' is always used in statistics books, but may be confusing to ecologists. Observations should be rare compared to the number of times they could theoretically occur in that sampling unit. For most organisms this is usually true (the rocks on the seashore are not totally covered with barnacles or the evening sky completely full of bats!) so ecologists can safely employ the Poisson as a possible description of the distribution of individuals in space and time. In a random distribution the MEAN and VARIANCE are equal, and therefore, unlike the NORMAL DISTRIBUTION, for example, the Poisson only has one defining parameter, m, which is usually estimated by calculating the ARITHMETIC MEAN ($\bar{x}$), but which is also equal to the variance (s^2). By random we mean that observations occur independently of each other. That is, if we see one barnacle or bat within a sampling unit (quadrat or minute) it does not increase or decrease the probability of seeing another barnacle or bat within that same quadrat or minute. Increasing the probability would produce an AGGREGATED DISTRIBUTION, while decreasing the probability would produce a REGULAR DISTRIBUTION. The terms of the Poisson distribution, or Poisson series, can be calculated from:

$$P_{(x)} = e^{-m}\frac{m^x}{x!}$$

where $P_{(x)}$ is the probability of observing a certain number of individuals in a sampling unit, e is the base of natural logarithms (2.7183), m is the mean number of individuals per sampling unit and $x!$ is x factorial ($4! = 4\times3\times2\times1 = 24$, etc. and $0! = 1$). Therefore the probabilities of 0, 1, 2, 3, . . . , x individuals per sampling unit are:

$$e^{-m},\ e^{-m}\frac{m^1}{1!},\ e^{-m}\frac{m^2}{2!},\ e^{-m}\frac{m^3}{3!},\ \ldots\ e^{-m}\frac{m^x}{x!}$$

or more conveniently for calculation:

$$P_{(0)},\ P_{(0)}\frac{m}{1},\ P_{(1)}\frac{m}{2},\ P_{(2)}\frac{m}{3},\ \ldots\ P_{(x-1)}\frac{m}{x}$$

For example, consider the situation where an ecologist has been recording the number of aphids per leaf on a sycamore tree. Using the standard formula ($\bar{x} = \Sigma x/n$) the mean is calculated as 3.5 aphids per leaf. If the aphids are distributed over leaves at random, what would their distribution look like? Since $m = 3.5$, the probabilities of observing leaves with different numbers of aphids are:

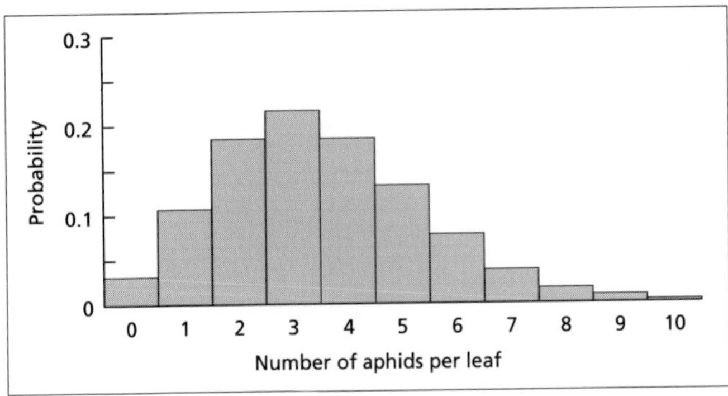

Fig. P22 Expected probability distribution of aphids per leaf if they followed a Poisson distribution with a mean = 3.5.

$P_{(0)} = e^{-3.5} = 0.030$

$P_{(1)} = 0.030\times3.5 = 0.105$

$P_{(2)} = 0.105\times1.75 = 0.184$

$P_{(3)} = 0.184\times1.17 = 0.215$

$P_{(4)} = 0.215\times0.87 = 0.187$

$P_{(5)} = 0.187\times0.70 = 0.131$

$P_{(6)} = 0.131\times0.58 = 0.076$

$P_{(7)} = 0.076\times0.50 = 0.038$

$P_{(8)} = 0.038\times0.44 = 0.017$

$P_{(9)} = 0.017\times0.39 = 0.007$

$P_{(10)} = 0.007\times0.35 = 0.002$

It would be possible to carry on computing the probabilities of even more aphids per leaf, but these would be increasingly small and therefore unlikely to be seen. It is usual, therefore, to stop calculating probabilities at some convenient point when the most common probabilities are known. The above probabilities sum to 0.99, and therefore the sum of all the remaining probabilities (equal to the probability of observing any leaf with more than 10 aphids) is only 0.01. The probability distribution for the aphid example is shown in Fig. P22. To obtain the number of leaves expected to have these numbers of aphids, we simply multiply these probabilities by the total number of observed leaves. If the ecologist had sampled 500 leaves, then the expected number having three aphids (if they followed a Poisson distribution with a mean of 3.5) would be $0.215\times500 = 107.5$ leaves. *See also* DISPERSION; DISTRIBUTION. [B.S.]

polarized transformation series *See* TRANSFORMATION SERIES.

pollard A tree whose top has been cut off well above the ground and which has sprouted again from the stump. Pollards are usually cut at 2–4 m height, high enough for new growth to be out of the reach of cattle, horses and sheep. The trunk, known as a bolling, becomes large, knarled and usually hollow with age. Pollarded trees often live longer than maiden (uncut) trees of the same species. Ancient woodlands (*see* ANCIENT WOODLAND/FOREST) and old hedges may include stub trees, which have been cut at heights intermediate between COPPICE and pollards. [G.F.P.]

pollen analysis The extraction, identification and quantitative assessment of fossil pollen grains in stratified lake sediments or PEAT deposits as a means of reconstructing the past history of vegetation in a region. *See also* PALYNOLOGY. [P.D.M.]

pollen tube *See* POLLINATION.

Pollination

Pollination is the process of transferring pollen from anthers to receptive stigma in the flowers of seed plants. It is, in essence, sexual intercourse by post. A pollen grain contains a sperm cell within a DESICCATION-resistant coat. In order to achieve FERTILIZATION, it must be deposited on the stigmatic surface, germinate, and send a long, narrow pollen tube through the tissue of the style to the recipient's ovary. The nucleus of the sperm cell must then pass down this tube to fertilize an ovule (containing an egg cell) in order to produce a fertile seed.

The process is complicated because most plants, being SESSILE, must depend on some intermediary (whether biotic or abiotic) to transfer their GAMETES to potential sexual partners. The variety of processes by which this is accomplished makes pollination biology an area rich in natural history, and creates an unusually fertile testing ground for theories in ECOLOGY and evolutionary biology. The paragraphs below provide an overview of the diversity of pollination systems, followed by a discussion of their application to ecological and evolutionary studies.

The diversity of pollination systems

Breeding systems

Perhaps the most fundamental way in which pollination systems differ is in the choice of potential sexual partners. At one extreme are apomictic plants that set seed asexually, without any pollination whatsoever. The vast majority of plants, however, produce seeds sexually, through the fusion of male and female gametes. As most plants are hermaphrodites (with each individual producing both male and female gametes) this sexual union does not necessarily require another partner. Some plants routinely engage in self-pollination; indeed some bear cleistogamous flowers that never open and pollinate themselves automatically. Such routine selfing results in very high levels of inbreeding, which may result in a loss of FITNESS for individuals and an inability for the population to adapt to changing conditions. The genetic costs of inbreeding are likely to be most severe, however, in species that seldom engage in the activity (Lande & Schemske 1985), and a number of mechanisms have evolved that reduce the likelihood of such unions.

In many flowers, male and female phases are separated in time (termed 'dichogamy'); in protandrous flowers the male ('staminate') phase comes first, whereas protogynous ones begin with the female ('pistillate') phase. Another mechanism reducing the probability of self-pollination is herkogamy, the spatial separation of male and female tissues. In its simplest form, this involves anthers positioned well away from a flower's stigma, but more extreme forms can involve separate male and female flowers on the same plant ('monoecy') or even separate male and female plants ('dioecy').

Many plants are self-incompatible, displaying any of a number of mechanisms that block self-fertilization even if selfed pollen is deposited on a flower's stigma. Heterostylous plants reduce the number of their potential mates much further, having separate mating types (typically long-styled 'pin' and short-styled 'thrum' morphs) which may only fertilize one another's ovules.

Flower structure and mode of pollination

Most of the breeding mechanisms described above require pollen transfer (usually between flowers of different plants) to effect fertilization. Amongst flowering plants, a wide variety of vectors are used to provide this service, resulting in an equally wide variety of apparent adaptations to facilitate pollination. A perusal of any flora or botanic garden will reveal a tremendous diversity of floral morphologies. These extraordinary structures perform an extraordinary function—channelling the impersonal forces of physics and the self-interested activities of other living things to the task of furthering a plant's reproductive success.

Wind pollination

At first glance, the simplest flowers are those pollinated by abiotic vectors—plants using the movements of the surrounding medium (air or, more rarely, water) to move their pollen about. Wind pollination may provide an effective mode of fertilization, so long as CONSPECIFIC individuals are not too widely separated; the density of pollen grains tends to fall with the square of distance downwind from a pollen source. The distance to which pollen will be carried depends on wind speed and on the terminal velocity (the settling rate) of the pollen, which in turn depends on its size and density (mass per unit volume). It is not surprising, then, that wind-pollinated plants typically live in exposed, windy environments (or flower before the surrounding vegetation leafs out) and have relatively small pollen grains. Some species (e.g. *Pinus* spp.) have air sacs or other adaptations to increase the BUOYANCY of their pollen in the air column. Tiny, light pollen grains may carry the furthest, but they are also the most difficult to intercept, as they are likely to ride air currents around any obstacle in their path—including a receptive stigma. Consequently, many wind-pollinated plants have finely branched, feathery stigmas, which have thin BOUNDARY LAYERS and thus trap airborne particles most effectively. Some wind-pollinated flowers have been shown to collect pollen non-randomly from passing air, channelling air currents and creating eddies in ways that send particles of appropriate size and density directly towards receptive surfaces (e.g. Niklas 1984). In general, wind-pollinated plants have relatively inconspicuous flowers, producing copious pollen and relatively few ovules. Water-pollinated plants also tend to have inconspicuous flowers, but the physics of pollen transport on and in water have selected for a very different set of adaptations in other respects.

Animal pollination

In contrast to abiotically pollinated plants, most animal-pollinated flowers produce a conspicuous visual and/or olfactory display, typically involving large and often brightly coloured petals and often including powerful odours as well. Pollen transfer may be carried out by any of a range of flower-visiting animals, including bees, flies, beetles, ants, thrips, birds, bats and even slugs and lemurs. Animal pollinators generally effect pollen transfer while foraging for floral rewards, which may consist of nectar, pollen or other inducements (oils, waxes, or even perfumes to be used as sexual pheromones). The original reward may have been the flower itself; most cases of animal pollination are thought to have evolved from floral herbivory, or florivory, which may have resulted in incidental pollen transfer to surviving flowers or flower parts. Floral rewards can thus be seen as a way of deflecting consumption away from reproductively vital flower parts, while maintaining incentives for floral visitation and the resulting pollen transfer. Some plants, however, rely on deception for pollination, with non-rewarding flowers that mimic a pollinator's mates or food sources (e.g. carrion, or rewarding floral species.).

It is also possible to attribute some of the variation among animal-pollinated flowers to the diversity of animal pollinators. Plants relying on different pollinators tend to display different suites of floral traits appropriate to the sensory abilities and FEEDING methods of the floral visitor. Thus bird-pollinated flowers are typically red (a colour beyond the visual spectrum of many insects) and tubular, whereas bat-pollinated plants tend to have large white or green flowers which dangle in ways that make them conspicuous to echo-locating foragers. Moth-pollinated flowers are typically white, night-scented, and possess a narrow corolla or spur in which a dilute nectar is stored, whereas bee flowers may be yellow, blue or purple (or indeed almost any colour except red), day-scented, more open in architecture and produce a nectar that is somewhat richer in sugars. Many plants do not, however, fit into a clearly defined floral syndrome and attract a wide range of pollinators to their flowers. On the other hand, some pollination systems (e.g. many figs and their associated fig wasps) display extreme levels of specificity, with single pollinator and plant species depending almost exclusively on one another.

The role of pollination in ecological research

Population biology

The two central parameters in population biology are survival and reproductive success. Pollination biology is vital to understanding the latter—the reproductive success of many plants is dependent on pollination. The fact that sexual activity in such plants is mediated by external forces makes pollination an ideal system for studying and experimentally manipulating reproductive performance in the wild. By blocking or augmenting pollination, for example, experimenters can effectively demonstrate trade-offs between reproductive success and survival, trade-offs that can only be inferred in many other systems. Experimental pollinations have also been used to assess whether reproductive performance in natural populations is limited by resources or by the availability of sexual partners, and in so doing have called into question the generality of an element of 'Bateman's principle' — that female reproductive success should always be limited by the availability of resources (Burd 1994). In addition, pollination has proved to be an ideal system for the study of 'ALLEE EFFECTS'—the inverse DENSITY DEPENDENCE in reproduction that may affect sparsely populated species. Reproductive failure at low density has been demonstrated both in wind-pollinated species (caused by the decline in pollen density with distance) and in those relying on animal pollinators (with density-related declines both in visitation rates and in the pollination effectiveness of visitors).

Community ecology

The study of interactions between species is a cornerstone of ecological science. The plant–pollinator interaction is typically a MUTUALISM, where the plant receives pollination services and the pollinator receives a reward. There are exceptions, however, either where the pollinator is duped and receives no reward or where the flowers are robbed of rewards by floral visitors that do not pollinate. In addition to the interactions between plants and their pollinators, however, community ecologists have devoted significant attention to the interactions between different species of pollinators (which may compete for floral resources) and to the indirect interactions between plant species that share pollinators. Plant species that bloom synchronously may compete for pollinator services. A highly rewarding plant species may monopolize the available pollinators at the expense of other plants, resulting in a form of EXPLOITATION COMPETITION. Moreover, where two species are visited by the same pollinators, interspecific transfer of pollen may occur, impeding each plant's stigmatic functions in an interaction analogous to INTERFERENCE COMPETITION. Thus, the usual range of competition-related ideas (e.g. RESOURCE PARTITIONING, CHARACTER DISPLACEMENT, ASSEMBLY RULES) can be considered in the context of pollination systems. Alternatively, coexisting plants may interact mutualistically, with the combined floral display attracting greater pollinator activity than either alone would have done. Floral mutualisms are also particularly likely in sequentially flowering species, where each helps maintain a pollinator population which is of value to the other.

Population genetics

The fact that most plants are rooted in place, combined with the diversity of their reproductive systems, makes them ideal subjects for many POPULATION GENETICS investigations. For example, the dynamics of pollination set the boundaries for GENE FLOW, determining to what extent individuals within a plant population are likely to share a common GENE POOL. As the distance between plants increases, the likelihood of pollen transfer declines whatever the mode of pollen transfer, so that restricted gene flow results and panmixis is prevented. Generally, it is difficult to quantify the spatial extent of either pollen DISPERSAL or the resulting gene flow because it is not easy to determine the origin of pollen or the PATER-

NITY of seeds. Pollen has been tracked by a variety of methods, including the addition of visual or even radioactive markers. It is also possible to estimate the extent of pollen dispersal from a source plant by reference to the movements of pollinating animals and to the amount of pollen deposited as the animal visits subsequent plants (pollen carryover). The determination of seed paternity has been achieved by assaying various genetic markers assessed either phenotypically, among expressed PROTEINS (isozymes), or in the DNA itself (e.g. microsatellites, RAPDs). A great deal remains to be done, however, before we will be able to predict with confidence the extent of gene flow in specific instances, especially in spatially heterogeneous plant populations.

Evolutionary ecology

Plant–pollinator systems are often held to be among the best examples of COEVOLUTION. For example, the long nectar spur of the orchid *Platanthera bifolia* and the matching elongate tongue of its moth pollinator (*Sphinx ligustri*) seem a clear example of coevolved traits. It is difficult, however, to demonstrate the coevolutionary history of even such a case unequivocally. It is possible, however, to experimentally test the selective pressures that control floral form. When the elongate nectar spur of the orchid is tied off to make it shallower, the pollinating efficacy of moth visits is reduced because moths can then remove nectar without pressing against the sexual parts of the flower (Nilsson 1988). Thus, we can envisage NATURAL SELECTION acting to favour plants with longer nectar spurs. Strictly, however, this does not demonstrate reciprocal coevolution; it is still only a conjecture that natural selection favours moths with longer tongues, presumably because they are more efficient at removing the nectar. None the less, the evidence of coevolution in this and similar cases is increasingly strong.

The last decade has seen a continuous increase in the number of floral traits that have been shown experimentally to be set by natural selection: flower size, anther–style relationships, nectar production, allocation to gender (male vs. female flower parts), daily flower number, etc. Yet more have been implicated in theory (e.g. floral longevity). When combined with studies on the HERITABILITY of the relevant traits, it is possible to compile a powerful case for the role of natural selection in floral EVOLUTION by extension from its role in the present. Indeed, there is a tendency to view flowers as structures that are finely tuned exclusively to the requirements of pollen transfer. However, one must also consider that plant CLASSIFICATION relies on the knowledge that certain aspects of floral architecture are highly conserved within plant families, for example. Thus, it is now possible to ask what proportion of a flower's form is the result of adaptation and what is the result of ancestral genetic inertia. As yet, there is no clear answer. *See also* LIFE-HISTORY EVOLUTION; MATING SYSTEMS; SIZE–NUMBER TRADE-OFF. [W.E.K. & J.Cr.]

REFERENCES

Burd, M. (1994) Bateman's principle and plant reproduction—the role of pollen limitation in fruit and seed set. *Botanical Review* **60**, 83–139.

Lande, R. & Schemske, D.W. (1985) The evolution of self-fertilization and inbreeding depression in plants. I. Genetic models. *Evolution* **39**, 24–40.

Niklas, K.J. (1984) The motion of windborne pollen grains around conifer ovulate cones: implications on wind pollination. *American Journal of Botany* **71**, 356–374.

Nilsson, L.A. (1988) The evolution of flowers with deep corolla tubes. *Nature* **334**, 147–149.

---

pollutant A substance that in the environment has an adverse EFFECT on human health and/or ecological systems. *Cf.* CONTAMINANT. *See also* POLLUTION. [P.C.]

Polluter pays principle (PPP)

The 'polluter pays principle' (PPP) is at the heart of environmental legislation in both the UK and the rest of the European Union (EU). It is one of the most widely advocated and, because of its appeal to equity, widely accepted principles in environmental policy and legislation. But there are some sources of ambiguity that, in practice, mean PPP has not been as widely and consistently applied as it might.

Pays for what?

This question can elicit a whole hierarchy of possible answers, which can include:

1 for the administration involved, for example in producing CONSENTS and AUTHORIZATIONS;
2 for costs of monitoring COMPLIANCE;
3 for costs of REMEDIATION following an illegal incident;
4 for costs of remediation of damage caused even when EMISSIONS are at a level deemed acceptable by public authorities.

PPP is taken as far as 1 and 2 in the UK issuance of authorizations, and 3 now has a statutory basis in UK and EU legislation. Moreover, the spectre of 4, as a very strict definition of liability for WASTE, has been raised in the EU proposals on civil liability for waste and general liability for environmental damage and is applied in the SUPERFUND LEGISLATION. The latter US instrument also raises the question of whether PPP should be applied retrospectively.

Who are the polluters?

The answer to this question is not always as obvious as it seems. For example, it has been argued that since it is the community at large, rather than specific dischargers, that benefits from enforcement activities, the community should pay the costs. The counterargument is, of course, that most of the costs can be passed back anyway via product prices to the community at large and in so doing provide more effective signals on environmental hazards. More serious questions can, nevertheless, be raised about the costs of establishing standards in the first place and of maintaining public registers. And there are, of course, the thorny questions about how jointly and severally the liabilities might be shared amongst managers, directors, lenders, insurers, etc. *See also* ENVIRONMENTAL ACTION

PROGRAMMES; ENVIRONMENTAL (PROTECTION) POLICIES; UNITED NATIONS PROGRAMMES AND BODIES. [P.C.]

REFERENCE
Parliamentary Office of Science and Technology (1992) *The Polluter Pays Principle and Cost Recovery Charges*. Post, London.

---

pollution The release of a by-product of human activity—chemical or physical—that causes HARM to human health and/or the natural environment; CONTAMINATION causing adverse effects. [P.C.]

pollution avoidance Measures taken to prevent POLLUTION either from ACCIDENTS or over the long term. It is based on the doctrine that prevention is better than cure. *See also* POLLUTION PREVENTION PRINCIPLE. [P.C.]

pollution-induced community tolerance (PICT) This is an ecotoxicological method based upon the premise that pollutants will act as SELECTION PRESSURES excluding sensitive and favouring more tolerant forms. This selection can operate between genotypes within species and between species in communities. Hence, communities that have been exposed to a POLLUTANT should show more TOLERANCE to it than communities that have not. The approach is applicable generally, but has largely been used on periphytic algae. Empirical evidence suggests that the community response is relatively pollutant specific. [P.C.]

REFERENCE
Blanck, H. & Wängberg S.Å. (1988) Validity of an ecotoxicological test system: short-term and long-term effects of arsenic on marine periphyton communities in laboratory systems. *Canadian Journal of Aquatic Science* **45**, 1807–1815.

pollution permits AUTHORIZATIONS, CONSENTS and licences that, normally, are issued by a regulatory agency to allow the release of potentially harmful substances into the environment. However, the permitted releases are usually set at levels predicted to cause no HARM to human health or the environment—hence 'pollution' is something of a misnomer in this context. *See also* POLLUTION QUOTAS. [P.C.]

pollution prevention principle (PPP) The principle that it is best to avoid POLLUTION in the first place rather than rely on REMEDIATION; i.e. what is repaired is rarely as good as the original, and the costs of repair may be considerable. This principle is now made explicit in the environmental policy of the European Union (EU) and is therefore used to guide the development of environmental legislation. It implies anticipating risks, and has connections with the German *VORSORGEPRINZIP*. An extreme interpretation is to require ZERO EMISSIONS, but this is practically impossible and often undesirable—CONTAMINATION is tolerable whereas pollution is not. Another related implication of this principle is a shift from attending to effects (cleaning up) to a consideration of causes; i.e. shifting emphasis from end-of-pipe/stack solutions to solutions that address what goes on within factories and, indeed, society at large. A valid slogan is that 'pollution prevention pays'—based upon the fact that preventing pollution can save energy and WASTE and hence potentially increase profitability of firms. *See also* ENVIRONMENTAL ACTION PROGRAMMES; ENVIRONMENTAL (PROTECTION) POLICIES. [P.C.]

pollution quotas Rights to release specified quantities of potentially polluting substances, issued usually by regulatory authorities, and generally set so that overall releases to the environment are below thresholds that cause harmful effects (so the term 'pollution' is something of a misnomer in this context). Such quotas may be tradeable. *See also* MARKET INSTRUMENTS. [P.C.]

polonium (Po) A radioactive metallic element (ATOMIC NUMBER = 84; melting point = 246–254°C; boiling point = 962°C), discovered in 1898 by Marie Curie. It has a crustal abundance of about 3×10^{-10} mg kg^{-1}, and naturally occurring isotopes are: ^{210}Po ($t_{1/2}$=138.4 days); ^{212}Po ($t_{1/2}$=2.0×10^{-7} s); ^{214}Po ($t_{1/2}$=1.6×10^{-6} s); ^{215}Po ($t_{1/2}$=1.8×10^{-3} s); ^{216}Po ($t_{1/2}$=0.15 s) and ^{218}Po ($t_{1/2}$=3.11 min), which are members of the natural radioactive decay series. Polonium can exist in oxidation states −2, +2, +4 and +6. It exhibits particle-reactive, insoluble behaviour in natural waters. *See also* RADIONUCLIDE. [A.B.M.]

polyandry An unusual MATING SYSTEM, found in birds and mammals, in which some females in a population mate with multiple males. Examples include African jacanas or lilytrotters, in which females defend resource-territories where several males build nests and rear the female's young; callitrichid primates, in which mature females produce twins at frequent intervals and are accompanied by several male partners who help carry the young; and African wild dogs, in which a single mature female is the sole reproductive female in a large den, with other individuals supplying meat to the many pups. Polyandry is strongly associated with male PARENTAL CARE and SEX ROLE REVERSAL, but its evolution is not well understood. In species without male care or strong bonds between males and females, females sometimes mate with multiple partners (e.g. Uganda kob, Malawi cichlid fish), but it may be more precise to term this behaviour 'multiple mating' rather than 'polyandry'. [J.C.D.]

polychlorinated biphenyls (PCB) Polychlorinated organic compounds containing two BENZENE rings linked by a single bond between ring CARBONS, also known as polychlorobiphenyl compounds. In total there are 209 PCB compounds. They were first produced on a commercial basis (in 1929) for use in printing inks and paints, as softeners in PLASTICS and insulators in transformers. PCBs were produced by the chlorination of the essential precursor compound, biphenyl, catalysed by iron (Fe(III)) chloride. Free RADICAL reactions lead to products containing different numbers of CHLORINE substituents (up to 10). The unseparated mixtures of products were marketed under a variety of tradenames. They are highly toxic to wildlife, and bioaccumulate and persist in the environment, especially in SEDIMENTS, for long periods. [M.C.G.]

polyclimax The situation where the persistent late-successional VEGETATION in a climatically determined area is composed of an array of communities occupying different site-types, dependent on position in the landscape (e.g. ridge, slope, gulley, broad valley floor), aspect and soil parent materials. The factors accepted as equal in rank in determining the elements in a polyclimax complex include not only microclimatic and EDAPHIC factors, which act continuously, but also proneness to disturbances, which are periodic—for example fire, which spreads all over rolling country on finetextured soils but stops at rocky scarps. In contrast to the MONOCLIMAX viewpoint, no one type of vegetation in an area is seen as 'the CLIMAX'; there is therefore no need for the profusion of PROCLIMAX types spawned by F.E. Clements (*see also* CHARACTERS IN ECOLOGY). The value of the polyclimax concept was argued by A.G. Tansley (*see also* CHARACTERS IN ECOLOGY), and by Whittaker (1953). The ideas inherent in it are now generally accepted, although the term climax itself has fallen from favour. *See also* SUCCESSION. [P.J.G.]

REFERENCE

Whittaker, R.H. (1953) A consideration of climax theory: the climax as a population and pattern. *Ecological Monographs* **23**, 41–78.

polyculture *See* MONOCULTURE.

polycyclic aromatic hydrocarbons (PAH) Non-polar, generally planar, organic compounds, composed of two or more fused BENZENE or other aromatic rings. They are of environmental concern because of their widespread distribution, PERSISTENCE, potential to bioaccumulate, acute TOXICITY, mutagenicity and carcinogenicity. Vapour pressure, water solubility, and susceptibility to oxidation and reduction all tend to decrease with increasing RELATIVE MOLECULAR MASS. As a general rule, the lower relative molecular mass PAH tend to be more acutely toxic but less environmentally persistent due to their relatively greater water solubility and high vapour pressure. The higher relative molecular mass PAH are less soluble but more persistent and show greater carcinogenic effects. Vertebrates, bacteria and some invertebrates can metabolize PAH to polar, water-soluble metabolites, and exposure to PAH in such organisms induces the activity of cytochrome-dependent, mixed function oxidase enzymes. In humans, exposure to PAH occurs from inhalation of tobacco smoke (by both active and passive smoking) and polluted air, and from ingestion of various vegetables (containing naturally produced PAH), fried and smoked foods, and contaminated seafood and water. *See also* MOTOR VEHICLE POLLUTION; PAH POLLUTION. [V.F.]

polyethism *See* CASTE.

polygamy A MATING SYSTEM in which (i) some of the males in a population mate with more than one female during a single breeding season (POLYGYNY); or (ii) some of the females mate with more than one male (POLYANDRY); or (iii) some members of each sex have multiple partners (polygynandry, or PROMISCUITY); the opposite of MONOGAMY. [J.C.D.]

polygenes *See* QUANTITATIVE TRAIT LOCI.

polygenic trait A TRAIT for which genetic variation between individuals is caused by the segregation of ALLELES at multiple loci (QUANTITATIVE TRAIT LOCI, or polygenes). [T.F.C.M.]

polygyny A MATING SYSTEM in which some of the males in a population mate with more than one female during a breeding season. The majority of mammalian species are polygynous, as are a sizeable minority of bird species and many other animals. Polygyny is often associated with male defence of resource-based territories which attract females (RESOURCE-DEFENCE POLYGYNY, e.g. waterbuck) or male defence of stable groups of females (HAREM polygyny, e.g. red deer). The degree of polygyny may depend on the extent to which a minority of males can monopolize RESOURCES or access to females. Extreme polygyny, with a small fraction of males monopolizing most matings, is characteristic of species which mate on traditional mating grounds, or LEKS (e.g. Uganda kob), and of other species, such as elephant seals, where females gather in dense groups to breed. The large variance in male reproductive success in these species increases the opportunity for SEXUAL SELECTION, and these species often exhibit pronounced differences between the sexes in body size or weaponry (SEXUAL DIMORPHISM) as well as behaviour and mortality. [J.C.D.]

polyhaline *See* BRACKISH WATERS.

polymer The product of a polymerization reaction, i.e. a chemical reaction in which two or more (usually many) identical molecules join together to form a material with new physical and chemical properties. For example, ethylene (more correctly called ethene), $CH_2{=}CH_2$, is a gas under normal conditions but can be polymerized to form the material polyethylene, otherwise known as polythene (*see* PLASTICS). [J.G.]

polymictic lakes *See* LAKE STRATIFICATION.

polymorphism The occurrence of discrete PHENOTYPES or forms within a population(s) of a species. It can refer to: a LIFE CYCLE (e.g. medusae/polyps of cnidarians); individuals with different GENOTYPES in a population (e.g. melanic and non-melanic moths); individuals with different functions within a COLONY/social group (e.g. feeding and reproductive individuals in colonial hydroids; workers and queens in social insects). Polymorphism is frequently applied to examples of variation in colour pattern or morphology but many studies have been made of biochemical polymorphisms including the alternative forms of enzymes known as isozymes.

Such variation is usually genetic in nature. However, the occurrence of discrete forms which are due to variation in some environmental factor in development is more properly referred to as POLYPHENISM. The latter phenomenon is an example of PHENOTYPIC PLASTICITY and not GENETIC POLYMORPHISM. *See also* ECOLOGICAL GENETICS; ELECTROPHORESIS; MIMICRY. [P.M.B. & P.C.]

polymorphism, genetic *See* GENETIC POLYMORPHISM; NEUTRAL THEORY OF EVOLUTION.

polyphagous Describing organisms that are able to exploit a wide range of food RESOURCES. Such GENERALIST behaviour allows selective foraging as the temporal abundance of food items changes. *Cf.* MONOPHAGOUS. *See also* FEEDING TYPES, CLASSIFICATION. [R.C.]

polyphenisms Multiple environmentally cued morphological types (*see* PHENOTYPIC PLASTICITY). Usually the morphs are discontinuous, and overlap in space; however, they may overlap in time if different generations are exposed to different seasons of the year, as in cyclomorphic *Daphnia* (also called seasonal polyphenism). *See also* POLYMORPHISM. [C.M.L.]

REFERENCE
Mayr, E. (1963) *Animal Species and Evolution.* Harvard University Press, Cambridge, Mass.

polyphyletic group An artificial evolutionary unit in which taxa that are separated from each other by more than two ancestors are placed together without including any other descendants of their ultimate common ancestor. The relationships among the taxa in a polyphyletic group are so distant that it misrepresents the evolutionary relationships which have arisen from SPECIATION events following the divergence of the shared common ancestor. This makes such groups invalid evolutionary units. Polyphyletic groups are diagnosed by convergent homoplasious traits. For example, placing birds and mammals in one TAXON because both groups are homeothermic would create a polyphyletic group. *See also* CLADISTICS; CLASSIFICATION; PHENETIC SPECIES CONCEPT; PHYLOGENETIC SYSTEMATICS. [D.R.B.]

polyploid hybrids Hybrids possessing more than two chromosome complements. Polyploid hybrids result from interspecific matings. Interspecific hybrids are usually completely or highly sterile due to chromosome mispairing in meiosis and production of aneuploid GAMETES, yet some hybrids may very rarely produce diploid gametes by failure of the first meiotic division. Union of two such diploid gametes results in a tetraploid progeny with normal meiosis because chromosomes form bivalents. Polyploidization in hybrids thus circumvents the problem of hybrid sterility. Such polyploid hybrids are called ALLOPOLYPLOIDS; an allotetraploid possesses two unlike diploid chromosome sets (from two parental species), in contrast to AUTOPOLYPLOID organisms, which possess the same chromosome set in multiple copies. Polyploid hybrids are often more robust than diploids, most likely due to doubled gene dosage and high HETEROZYGOSITY. They are also automatically reproductively isolated from parental species since fusion of a diploid and a haploid gamete yields a triploid, infertile offspring. However, such progeny may sometimes also undergo polyploidization or give rise to asexual lineages thereby forming polyploid complexes.

In nature, allopolyploid species are common in plants — some 40% of plant species have originated this way—but rare in animals, what seems to be related to disturbance of function of sex determining gene complexes. In plants, because fertility of a hybrid can be restored artificially by doubling chromosome complement by treatment with colchicine, polyploid hybrids provide an inexhaustible source of varieties for plant breeding. [J.M.SZ.]

polyploidy The condition of possessing more than two entire chromosome complements. Most eukaryotic organisms are diploid (i.e. have two chromosome sets) in the majority of their cells but produce haploid GAMETES (with one set of chromosomes). Polyploidy can arise by the addition of one or more extra sets of chromosomes, identical to the normal haploid complement of the same species (i.e. autopolyploidy) or by the fusion of genomes of two different species, for example by hybridization (i.e. allopolyploidy). Polyploidy can occur, for example, when a diploid ovum arises, as a result of suppression of meiosis, and is fertilized by a normal haploid sperm, or when a normal haploid ovum is fertilized by two sperm. The resulting offspring will have three of each kind of chromosome, and is called a triploid. An organism with four of each kind of chromosome is called a tetraploid. Polyploidy involving an even number of sets of chromosomes is more common than that resulting in an odd number of sets of chromosomes.

Polyploids are often reproductively isolated from their diploid ancestors, and particularly in plants (in which the polyploid descendants can also reproduce asexually) polyploidy has played an important role in the EVOLUTION of new species. Polyploidy is rare in most animal groups, presumably because it tends to interfere fatally with meiosis by generating chromosome imbalance, but is well known in lizards, amphibians and fishes. In addition to arising in germinal tissue, polyploidy can also occur in somatic tissue (e.g. in the mammalian liver), probably as a result of abnormal cell division cycles. *See also* POLYPLOID HYBRIDS. [V.F.]

polyterritoriality The simultaneous possession of two or more spatially separate breeding territories by FACULTATIVE polygynists (*see* POLYGYNY). Males of a proportion of polygynous temperate bird species defend two or more spatially separate breeding territories. When a female has been attracted to the primary territory, the male may establish a new territory elsewhere, and attempt to attract a secondary female. Since females that become the secondary mate of an already mated male achieve reduced reproductive success due to the absence of full male PARENTAL CARE, polyterritorial males may experience higher mating success because of concealment of their mating status for prospecting females. [A.P.M.]

polytypic species *See* BIOLOGICAL SPECIES CONCEPT.

POM *See* PARTICULATE ORGANIC MATTER.

pools, in streams *See* WITHIN-STREAM/RIVER HABITAT CLASSIFICATION.

poorly (sparingly) water soluble Describing toxicants or other materials, usually organic, that do not dissolve readily in WATER and hence create difficulties for presenting in TOXICITY/ECOTOXICITY tests. In these cases it is not uncommon to read reports where toxic concentrations exceed the solubility of the substance. Here test animals may have been exposed to much lower than expected concentrations with most of the chemical being on the bottom of the test vessel. This can be overcome, to some extent, by adding small amounts of relatively low-toxicity solvents such as acetone, which are miscible with water and which can maintain high concentrations of the chemical in solution. These are known as carriers. *See also* PREDICTIVE ECOTOXICITY TESTS. [P.C.]

POPs *See* PERSISTENT ORGANIC POLLUTANTS.

population
1 In ecology, a group of organisms of the same species, present in one place at one time. This definition is not always so simple to put into practice because the 'place' occupied by a population is not clear-cut. A population may extend over a large area, with limited interaction (e.g. occasional mating) between individuals in different parts of the population. POPULATION DYNAMICS nowadays stresses the METAPOPULATION concept, where local populations form part of a larger, regional population.
2 In STATISTICS, a term referring to all the observations that could be made on a particular item. The summary measurements of populations (PARAMETERS) are represented by Greek letters, for example, the MEAN (μ) and the VARIANCE (σ^2). *See also* SAMPLE.
[R.H.S. & B.S.]

population bottleneck A substantial decrease in POPULATION SIZE that may be associated with a loss of genetic HETEROZYGOSITY and with SPECIATION events. In very small populations, genes that are lost by chance (i.e. GENETIC DRIFT) are less likely to reappear by MUTATION compared to large populations because mutation rates are generally low. Because chance events play a larger role in small populations, this increases the likelihood that such small populations will come to diverge from each other. Population bottlenecks often occur during COLONIZATION of a new habitat by some fraction of a source population and result in a so-called 'FOUNDER EFFECT'. Population bottlenecks can be important in conservation in that certain minimum population sizes (termed MINIMUM VIABLE POPULATION sizes) are necessary to ensure long-term survival. The minimum viable size depends on the ecology (e.g. age structure, life history, population spatial pattern) and behaviour of a species. *See also* GENETIC BOTTLENECK; HETEROZYGOSITY, INTERPRETATION FOR CONSERVATION. [V.F.]

population cycles Changes in the numbers of individuals in a population involving repeated oscillations between periods of high and low density. Cycles may be driven by periodic fluctuations in the environment or may occur as the result of internal demographic processes which involve delayed or overcompensating density-dependent processes. *See also* OSCILLATIONS; POPULATION DYNAMICS; STABLE LIMIT CYCLE. [V.F.]

population density The local ABUNDANCE of a species expressed quantitatively as numbers per unit area or numbers per unit volume. Population density is more useful than POPULATION SIZE for many purposes because it provides a better indication of local abundance relative to the availability of resources. Population size, however, provides information important in predicting FLUCTUATIONS and extinction of ENDANGERED SPECIES where populations are small in absolute terms because such species are often not widespread. Small populations are more vulnerable to random or stochastic events, INBREEDING DEPRESSION and local habitat disturbance. [R.H.S.]

population dynamics The variations in time and space of the sizes and densities of populations. The dynamics of populations are controlled by rates of birth, death, IMMIGRATION and EMIGRATION, which reflect the LIFE CYCLE of the organisms comprising the POPULATION. POPULATION SIZE and DENSITY are usually estimated by counting individuals within plots of known area or by marking individuals and recapturing them. Changes in environmental factors, existence of a seasonal breeding period, and delayed responses by limiting factors can all influence population dynamics.

Population density is a key variable because it indicates the potential intensity of interactions between individuals and the extent to which necessary RESOURCES may become limiting. Time delays can also play a key role in controlling dynamics. In particular, if the time delay in response to a LIMITING FACTOR (T) times the INTRINSIC RATE OF INCREASE (r) is such that $0<rT<e^{-1}$ the population will approach equilibrium smoothly (i.e. exhibit MONOTONIC DAMPING); if $e^{-1}<rT<\pi/2$ the population will exhibit damped OSCILLATIONS to equilibrium; if $\pi/2<rT$ the population will sustain regular oscillations or LIMIT CYCLES with amplitude and frequency depending on rT.

Populations of predators and their prey may experience coupled oscillations as the density of each responds (following a time delay) to changes in density of the other. Damped oscillations are cycles having progressively smaller amplitude, as in some populations approaching their equilibria. Simple logistic models of population growth predict limit cycles to occur in populations for which the intrinsic rate of increase is greater than 2, such that close to the equilibrium population size the oscillations increase, and at high amplitude the oscillations are damped so that the population oscillates regularly between a low and high density. Oscillations of predator and prey populations occur when stabilizing and destabilizing tendencies of their interaction balance. *See also* POPULATION CYCLES; STABLE LIMIT CYCLE. [V.F.]

population ecology The study of the relationships of populations of organisms with the members of the POPULATION, with other populations and with external environmental factors. *See also* ECOLOGY; POPULATION DENSITY; POPULATION DYNAMICS; POPULATION GENETICS; POPULATION GROWTH RATE; POPULATION REGULATION; POPULATION SIZE. [R.H.S.]

population explosion *See* POPULATION (HUMAN), GLOBAL PROJECTIONS AND IMPACTS.

population genetics The branch of genetics concerned with the occurrence, patterns and changes of genetic variation in natural populations. Population genetics combines the study of the population consequences of Mendel's laws (for instance the HARDY–WEINBERG LAW) with the study of the ecological influences that impinge upon populations and lead to NATURAL SELECTION. Segregation and recombination of genes, together with the mating pattern between GENOTYPES and between groups, lead to the distribution of genotypes within the population. MUTATION and gene duplication provide new genetic variation, and population genetics studies the fate of this novel variation. Ecological and physiological influences on different genotypes lead to their different fitnesses, and to natural selection. The exact description of ALLELE FREQUENCY change under natural selection is a major concern of population genetics. Changes in frequency of selectively neutral ALLELES form an important aspect of the study of molecular evolution. Indeed, population genetics forms a core discipline for the study of evolution. *See also* FITNESS; MENDELIAN INHERITANCE; QUANTITATIVE GENETICS. [G.D.J.]

REFERENCE

Hartl, D.L. & Clark, A.G. (1989) *Principles of Population Genetics*, 2nd edn. Sinauer Associates, Sunderland, Mass.

population growth rate The number of individuals present at a particular time unit, divided by the number of individuals present one time unit earlier. When a POPULATION is in stable age distribution, each AGE CLASS grows in number at the same constant rate, and the growth rate equals the PER CAPITA birth rate minus the per capita death rate. *See also* DEMOGRAPHY; POPULATION DYNAMICS. [G.D.J.]

population (human), global projections and impacts The world's human population was *c.* 3.6 billion in 1970, passed 5 billion in the mid-1980s, and 6 billion in the mid-1990s. UN projections up to 2150 give the estimates shown in Fig. P23. On the medium projection (considered most probable), world population will increase to 10 billion in 2050, levelling off at around 11.6 billion in 2150. The low projection is based on the presumption that births will fall globally below replacement—this predicts that population will peak in 2050 and then fall. The high projection presumes that there will be a global average of 2.2 children per woman worldwide, so the world population will continue to rise indefinitely (12.5 billion by 2050 and 20.8 billion by 2150). This corresponds to a modest increase in PER CAPITA birth rate from 2.2 to 2.5.

More than 95% of the growth in population is taking place in developing countries (Fig. P24). The consequences, in terms of demand for RESOURCES—not least of which will be living space—and the generation of wastes, are likely to be enormous, putting particular emphasis on the search for sustainable methods of economic and social development.

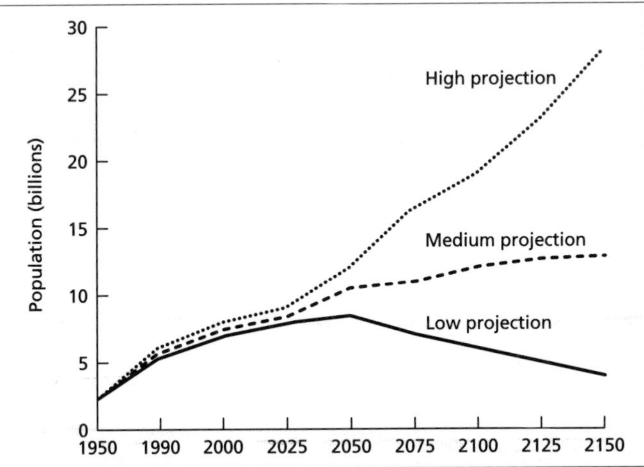

Fig. P23 World population growth to the year 2150 (high, medium and low scenarios). (After Tolba, M.K. & El-Kholy, O.A. (1992) *The World Environment 1972–1992*. Chapman & Hall, London, on behalf of the United Nations Environment Programme.)

Since T. Malthus (*see* CHARACTERS IN ECOLOGY), there has been much speculation about the CARRYING CAPACITY of the Earth in terms of human population, but this still remains a matter for speculation (Cohen 1995). [P.C.]

REFERENCES

Cohen, J.E. (1995) *How Many People Can the Earth Support?* Norton, New York.
Tolba, M.K. & El-Kholy, O.A. (1992) *The World Environment 1972–1992*. Chapman & Hall, London, on behalf of the United Nations Environment Programme.

population multiplication rate *See* INTRINSIC RATE OF INCREASE.

Population projection matrix

A matrix used to describe the dynamics of a POPULATION in which individuals are classified by age, size, developmental stage or other properties. Let $\boldsymbol{n}$ be a vector whose ith entry, n_i, is the number (or DENSITY) of individuals in category i. The dynamics of the population can be described by:

$$\boldsymbol{n}(t+1) = \mathbf{A}\boldsymbol{n}(t)$$

where the square matrix $\mathbf{A}$ projects the population from time t to $t+1$. The entry a_{ij} in the ith row and jth column of $\mathbf{A}$ gives the number of stage i individuals at $t+1$ per stage j individual at t. These coefficients may represent movement of individuals from one stage

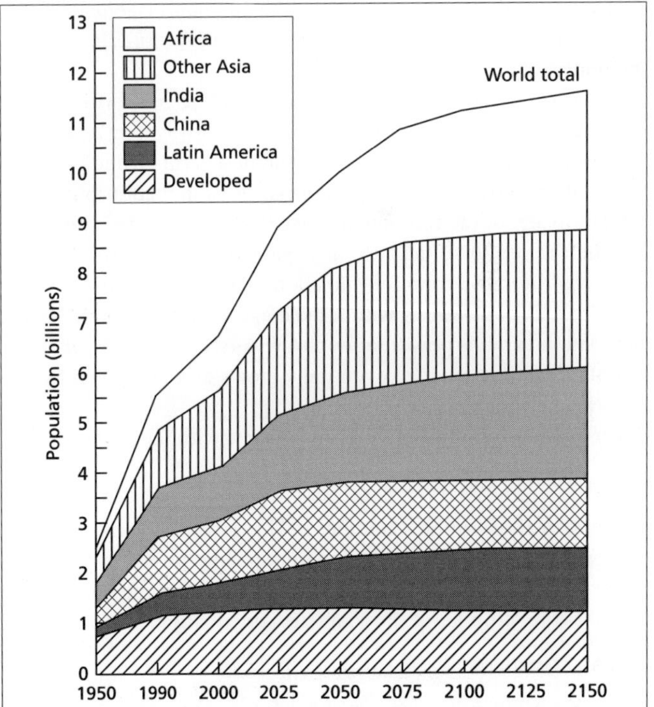

Fig. P24 Population projections to the year 2150 by region (medium variant). (After Tolba, M.K. & El-Kholy, O.A. (1992) *The World Environment 1972–1992*. Chapman & Hall, London, on behalf of the United Nations Environment Programme.)

to another (e.g. by growth) or production of new individuals (e.g. by reproduction).

Matrix models may be classified by the nature of **A**. In the simplest case the a_{ij} are constants, and the resulting model is linear and time invariant. It is also possible to write models in which the a_{ij} are functions of density or vary, either deterministically or stochastically, with time.

If individuals are classified by age into AGE CLASSES whose width is the same as the projection time interval, the resulting matrix is called a Leslie matrix, named after P.H. Leslie who developed its use in the 1940s (see below). Its only positive entries are age-specific fertilities (F_i) in the first row and age-specific survival probabilities (P_i) on the subdiagonal; for example:

$$\mathbf{A} = \begin{pmatrix} F_1 & F_2 & F_3 \\ P_1 & 0 & 0 \\ 0 & P_2 & 0 \end{pmatrix}$$

Classification by criteria other than age may produce more complex patterns of transition from stage to stage, and thus more complex arrangements of positive coefficients in the matrix. Such matrices are sometimes called Lefkovitch matrices, after L.P. Lefkovitch.

A linear, time-invariant matrix model leads to exponential population growth. The POPULATION GROWTH RATE is given by the maximum EIGENVALUE of **A**. If **A** satisfies two technical conditions (is irreducible and primitive), then the population converges to a stable STAGE DISTRIBUTION and eventually grows at the rate given by the dominant eigenvalue from any non-zero initial condition. The stable stage distribution is given by the right EIGENVECTOR of **A** corresponding to the dominant eigenvalue. The corresponding left eigenvector gives the REPRODUCTIVE VALUE distribution.

A matrix is irreducible if, in the corresponding LIFE CYCLE, every stage can contribute individuals, by some pathway, to every other. The most common violation of this condition occurs when there are post-reproductive age classes; these classes cannot contribute individuals to any of the reproductive age classes. A matrix is primitive if, when raised to sufficiently high powers, all its entries become positive. The most common violation of this condition occurs in semelparous populations with a fixed age at reproduction. Such a population exhibits persistent OSCILLATIONS in its age distribution, although its time average eventually grows at a rate given by the dominant eigenvalue.

An eigenvector of a matrix **A** is a vector $\boldsymbol{w}$ with the property that:

$$\mathbf{A}\boldsymbol{w}=\lambda\boldsymbol{w}$$

for some real or complex number λ. That is, multiplying an eigenvector by a matrix is equivalent to multiplying the eigenvector by a scalar. The scalar λ is called the eigenvalue. The eigenvalues are found as the solutions of the characteristic equation:

$$\det(\mathbf{A}-\lambda\mathbf{I})=0$$

where **I** is the identity matrix and det denotes the determinant.

Matrix population models were first studied in detail by Leslie at the Bureau of Animal Populations, Oxford, in the 1940s. Leslie, a self-taught mathematician of wide-ranging interests, also pioneered the use of life-table methods and was the first to calculate the INTRINSIC RATE OF INCREASE for a non-human species. Matrix models were almost completely ignored until the 1960s when they were rediscovered by both ecologists and human demographers. After Lefkovitch showed that they could be generalized to size- or stage-classified populations, they were seized upon by plant ecologists, who had discovered in the 1970s that size was often more important than age in plants.

Although it is obvious that rates of survival, growth and reproduction are not constant, this simple demographic model can still give valuable information about a population. The population growth rate, stable stage distribution and reproductive value must be treated as hypothetical projections of what the population would do if current environmental conditions were maintained, not of what it will do. As such, they provide information about the relation between the population and its current environment. This information can be useful in CONSERVATION BIOLOGY and ECOTOXICOLOGY, as well as in quantifying the effects of natural environmental factors. [H.C. & C.D.]

REFERENCES

Caswell, H. (1989) *Matrix Population Models: Construction, Analysis and Interpretation*. Sinauer Associates, Sunderland, Mass.

Lefkovitch, L.P. (1965) The study of population growth in organisms grouped by stages. *Biometrics* **21**, 1–18.

Leslie, P.H. (1945) On the use of matrices in certain population mathematics. *Biometrika* **33**, 183–212.

Tuljapurkar, S.D. (1990) *Population Dynamics in Variable Environments*. Springer-Verlag, New York.

Tuljapurkar, S.D. & Caswell, H. (eds) (1997) *Structured Population Models in Marine, Terrestrial, and Freshwater Systems*. Chapman & Hall, New York.

population pyramid POPULATION pyramids are a way of diagrammatically illustrating the age and sex structure of populations. This is done by plotting two histograms with the bars arranged vertically back to back, as in kite diagrams. The age categories of the population are arranged vertically in ascending order and the percentage of the population in each category of each sex is plotted on either side of the centre line.

For developing countries such as Brazil (Fig. P25a), the bars do take the shape of a pyramid showing how the young age categories do contain higher proportions of the population than older ones (more than half the population is under 20 in Brazil and less than 3% over 64). However for industrialized countries such as Great Britain the shape is quite different (Fig. P25b). Here the most numerous age categories are between 15 and 34 years. Only 20% of the population are under 15 and approximately 15% are over 64.

The structure of populations has many implications with respect to disease and death rates. Populations with the triangular pyramidal profile are characterized by high birth rates, high MORTALITY and substantial rates of increase in overall POPULATION SIZE. [M.H.]

REFERENCE

Open University (1985) *The Health of Nations*. Open University Press, Milton Keynes.

population regulation REGULATION implies a tendency for a POPULATION to decrease if numbers are above a certain level, or

Fig. P25 Population pyramids based on census data. (a) Brazil 1980, population = 123 million; (b) Great Britain 1981, population = 53.6 million. (After Open University (1985) *The Health of Nations*. Open University Press, Milton Keynes.)

conversely to increase if numbers are low. In other words, if population regulation occurs, POPULATION SIZE is controlled by density-dependent variation in birth, death, IMMIGRATION or EMIGRATION rates. Density-independent factors, such as climatic variation or random catastrophic MORTALITY, operate at a significant level in many natural populations and may obscure patterns of population regulation. Field evidence for population regulation has been hard to find in animal populations but is more common in plant populations. It is now recognized that population regulation may occur at different scales and that detection of population regulation requires study at the appropriate spatial scale. *See also* DENSITY DEPENDENCE; DENSITY INDEPENDENCE; POPULATION DYNAMICS; REGULATING FACTORS. [R.H.S.]

population size The total number of individuals in a POPULATION. *Cf.* POPULATION DENSITY. [P.C.]

population structure *See* AGE STRUCTURE.

porewater The WATER that surrounds the solid particles in SEDIMENT and SOIL, also referred to as INTERSTITIAL water. It is thought to be an important source of exposure to toxicants for sediment (soil)-dwelling organisms. [P.C.]

positive binomial *See* INDICES OF DISPERSION.

positive feedback *See* FEEDBACK.

postclimax In the scheme of Clements (1936; *see* PROCLIMAX and CHARACTERS IN ECOLOGY), a kind of VEGETATION characteristic of a CLIMATE moister or warmer than that obtaining in the general area where the vegetation occurs. The commonest types of vegetation deemed to be postclimax were the various kinds of forests found along rivers and in gorges in dry country dominated by grassland. Vegetation requiring high temperatures, and confined to steep south-facing slopes in the Northern hemisphere, was also called postclimax. The concept has been abandoned following the general acceptance of the ideas embodied in the POLYCLIMAX concept. *See also* CLIMAX; SUCCESSION. [P.J.G.]

REFERENCE

Clements, F.E. (1936) Nature and structure of the climax. *Journal of Ecology* **24**, 252–284.

post-displacement A change in the timing of onset of the ONTOGENY of a specific trait. If the timing of offset of the ontogenetic trajectory remains constant, post-displacement (delay in the timing of onset) would result in a shorter (in time) ontogenetic sequence. The outcome of this ontogenetic process is the generation of a paedomorphic morphology. *See also* HETEROCHRONY; PAEDOMORPHOSIS. [P.A.]

post-disturbance succession SUCCESSION is 'the directional change with time of the species composition and vegetation physiognomy of a single site' (Miles 1979). Such 'directional changes' are frequently initiated by a DISTURBANCE to an existing COMMUNITY. The course of post-disturbance succession will depend on the nature of the pre-existing community and the scale and severity of the disturbance. Small or low-intensity disturbances often lead rapidly to the re-establishment of the original community from vegetative fragments or juveniles. The same can also happen on a larger scale, for example when seed release is triggered by fire in fire-adapted conifer forests. Large or severe disturbances often lead to a more-or-less predictable succession. The initial stages are determined largely by the availability of propagules and are dominated by species with persistent SOIL SEED BANKS or effective wind DISPERSAL. In the absence of further disturbance, later stages reflect the outcome of COMPETITION and are dominated by large, long-lived woody species. Empirical studies show that most COLONIZATION usually takes place relatively soon after the disturbance, and that once a complete PERENNIAL cover is established further invasion occurs only very slowly. Post-disturbance succession is often referred to as SECONDARY SUCCESSION. Heterotrophic and regressive successions also occur but have received much less attention from ecologists. [K.T.]

REFERENCE
Miles, J. (1979) *Vegetation Dynamics*. Chapman & Hall, London.

postzygotic conflict *See* SEXUAL CONFLICT.

potamon *See* RIVERS AND STREAMS, TYPES OF.

potassium (K) A metallic element (ATOMIC NUMBER = 19; relative ATOMIC MASS = 39.0983; melting point = 63.65°C; boiling point = 774°C) with an abundance of 18.4 g kg^{-1} in the Earth's CRUST. The METAL is extracted by treatment of the CHLORIDE with SODIUM. Potassium superoxide, KO_2, is used as an auxiliary supply of OXYGEN in breathing apparatus, submarines and space vehicles; it yields oxygen according to the equation:

$$4KO_2 + 2CO_2 \rightarrow 2K_2CO_3 + 3O_2$$

Potassium phosphates are used in liquid DETERGENTS. Potassium manganate(VII), $KMnO_4$, is used as an oxidizing agent and in the manufacture of saccharin. Potassium salts play a key role in biological systems and are extensively used as FERTILIZERS. [C.R.P.]

potential temperature The TEMPERATURE (ϕ) of air taken dry adiabatically from its initial PRESSURE p and absolute temperature T to 1000 hPa (= 1000 mbar) and given by $\phi = T(1000/p)^{R/c_p}$, where R is the specific gas constant for dry air (287 J kg^{-1} K^{-1}) and c_p is the specific heat capacity for dry air at constant pressure (1004 J kg^{-1} K^{-1}). Because a dry adiabatic process is assumed in its definition, vertical profiles of ϕ immediately reveal the convective condition of unsaturated air: thus ϕ lapses with increasing height in a superadiabatic (unstable) layer, is uniform in a dry adiabatic (neutral) layer, and increases with height in a subadiabatic (stable) layer. [J.F.R.M.]

potentially responsible parties (PRPs) *See* COST RECOVERY.

−3/2 power law An empirical rule, stating that the slope of a graph of the logarithm of mean plant shoot DRY MASS (on the *y*-axis) against the logarithm of plant POPULATION DENSITY (*x*-axis) is typically about −1.5 (Fig. P26). The pattern is observed in experiments on even-aged, high-density plant monocultures undergoing density-dependent MORTALITY (*see* SELF-THINNING). It is also known as Yoda's rule, or the self-thinning law. The observed constancy of slope at $-^3/_2$ is probably nothing more than a reflection of dimensional CONSTRAINTS; since population density has dimensions L^{-2} (where L = length), and plant mass has dimensions L^3, it is evident that a graph of log mass against log density should have a slope of $-^3/_2$.

The $-^3/_2$ power law is not consistent with another strong empirical relationship, the LAW OF CONSTANT FINAL YIELD, which produces a graph of log plant mass against log plant density with a slope of −1 (direct proportionality) rather than −1.5. In self-thinning, the points on the graph represent the same plant populations recorded at different times over the course of the experiment, as density-dependent mortality reduces POPULATION SIZE. In contrast, each point in graphs of the law of constant final yield represents the final BIOMASS of different populations, initiated by sowing different densities of seeds.

In most cases, the $-^3/_2$ power law graph is drawn the wrong way round, in that it is presented with the RESPONSE VARIABLE (plant density) on the *x*-axis and the EXPLANATORY VARIABLE (mean plant size, or more correctly, total biomass) on the *y*-axis. In the $-^3/_2$ power law, population size declines because mean plant size has increased (i.e. because the large plants get even larger). This reversal of the conventional arrangement of the graph axes has been the cause of considerable misunderstanding, especially since it is plausible that changes in initial density should cause changes in mean plant size (as in the law of constant final yield).

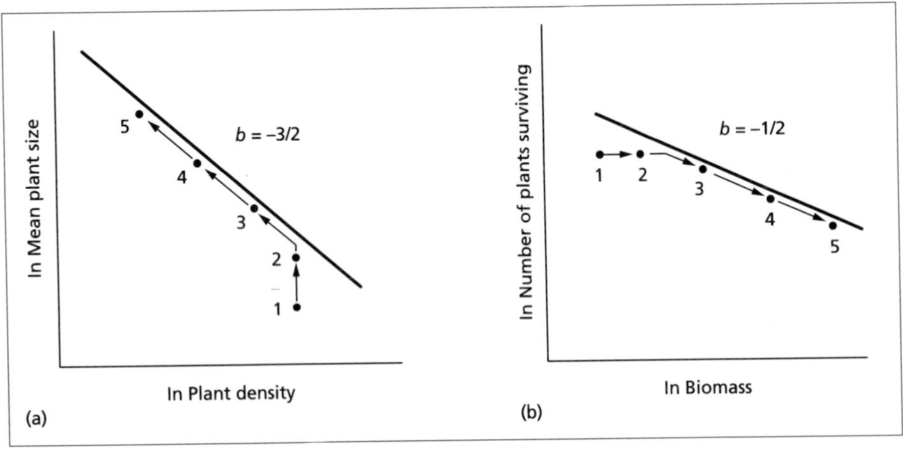

Fig. P26 The self-thinning rule ($-^3/_2$ rule). (a) A plot of mean plant size against mean surviving plant density typically has a slope of roughly $-^3/_2$. As time passes from 1 to 2 the plants increase in size but there is no mortality as a result of competition because total biomass is still small. Once mean plant size reaches the self-thinning line, any increases in plant size must be accompanied by reduced plant numbers. The self-thinning population (times 3 to 5) follows the $-^3/_2$ line. (b) The graph is better presented with the response variable, plant numbers, on the *y*-axis and with total plant biomass as the explanatory variable. Presented in this way, the predicted slope of the log–log plot is $-^1/_2$.

After maturity, most growth can be accommodated by size plasticity. Hence the change from a slope of $-3/2$ to -1 (*see* LAW OF CONSTANT FINAL YIELD). [M.J.C.]

REFERENCES

Westoby, M. (1984) The self-thinning rule. *Advances in Ecological Research* **14**, 167–225.

White, J. (1981) The allometric interpretation of the self-thinning rule. *Journal of Theoretical Biology* **89**, 475–500.

power spectrum The power spectrum (also called the VARIANCE spectrum or energy spectrum) quantifies the amount of variance (or energy) associated with particular frequency or wavelength scales. The power spectrum of a scalar or a wind velocity component is derived through a Fourier transformation (*see* FOURIER ANALYSIS) of a temporal (or spatial) series of a given variable. The Fourier transform converts the time (or space) series into a frequency (or wavelength) domain by representing it as an infinite sum of sine and cosine terms. Integration of spectral energy densities across the whole range of significant frequencies (or wavelengths) yields the total variance of the velocity component or scalar quantity under scrutiny. [D.D.B.]

REFERENCE

Stull, R.B. (1988) *An Introduction to Boundary Layer Meteorology*. Kluwer Academic, Dordrecht.

power transmission lines Electricity transmission is predominantly overhead because this is much cheaper (up to 20 times) than the equivalent underground cable option. Bare METAL conductors (lines) of ALUMINIUM, with a steel core for strength, are normally used, suspended from lattice towers (pylons), generally constructed from angled sections of galvanized steel, as illustrated in Fig. P27. The high operating voltage, used to reduce power losses in the resistance of the conductors, dictates the height of the conductors above ground, the length of the insulator strings and the width of the cross arms.

The high electric field strength at the surface of the conductors leads to partial electric breakdown of the surrounding air (corona), causing emission of broadband radio interference and acoustic noise. To reduce corona, conductor diameter must be increased but this would cause excessive mechanical loading on the towers. The solution is to space conductors using mechanical links into bundles to an 'effective' diameter which reduces both electric field and mechanical stress: for example, a two-conductor bundle is used at 275 kV and a four-conductor one at 400 kV.

The conductors are suspended from cross-arms on the towers, using strings of insulating GLASS or ceramic discs. Because right-of-way (wayleave) for transmission lines is often difficult to obtain, the power transmitted along a given route (power corridor) is maximized by installing two three-phase circuits (a double-circuit line). This results in six sets of conductors, suspended in a symmetrical fashion. An additional earth conductor runs between the tops of adjacent towers to provide earth continuity and to help protect the lines from LIGHTNING strikes. [H.W.W.]

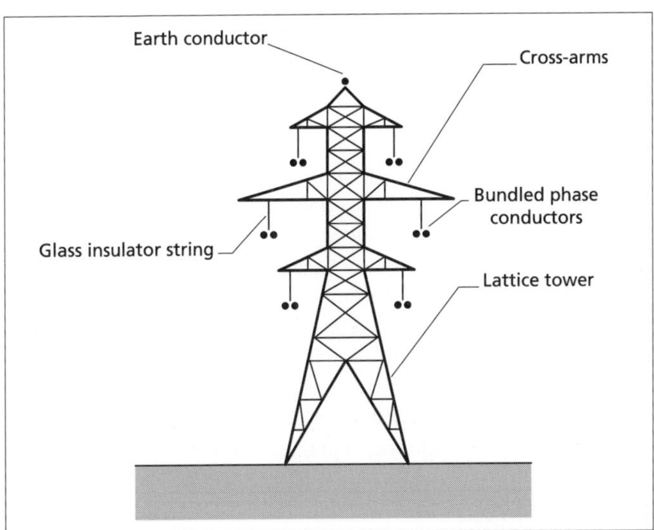

Fig. P27 A high-voltage transmission tower.

PP *See* PRECAUTIONARY PRINCIPLE; PRIMARY PRODUCTIVITY.

ppb (parts per billion) A UNIT of measurement of concentration, used especially in describing the concentrations of substances occurring in trace amounts, for example certain environmental pollutants. For example, 1 ppb = 1 part in $10^9 \equiv 1\ \mu g\ kg^{-1}$ (by mass) or $1\ \mu g\ l^{-1}$ (by volume; sometimes ppbv). Note that ppb was formerly used in the UK to mean parts per million million (i.e. 10^{12}), but US usage is now generally accepted. Where there is the possibility of doubt the unit should be defined as mass per unit mass, or mass per unit volume. [P.C.]

ppm (parts per million) A UNIT of measurement of concentration used especially in describing the concentrations of substances occurring in minute amounts. 1 ppm = 1 part in $10^6 \equiv 1\ mg\ kg^{-1}$ (by mass) or $1\ mg\ l^{-1}$ (by volume; sometimes ppmv). [P.C.]

PPP *See* POLLUTER PAYS PRINCIPLE; POLLUTION PREVENTION PRINCIPLE.

Prairie

The plains GRASSLANDS of North America, named from the French word for meadow. At the time of the first European settlement such grasslands are estimated to have covered 770 million ha, mainly in the USA, with smaller prairie areas in Canada and desert grasslands of the southern USA extending into Mexico. The prairies developed between the Middle Miocene and Early Holocene as increasing aridity led to the replacement of semi-open FOREST and WOODLAND with open grassland. Fires, both natural and deliberate (set by native Americans), are thought to have contributed to the expansion of grassland beyond climatic limits, particularly in the eastern part of the Great Plains where at the time of European settlement grassland–forest ecotones were abrupt, suggesting CLIMATE was not the only controlling influence.

Since the settlement of Europeans there has been widespread ploughing of grasslands; destruction of the natural GRAZING herds of buffalo, antelope, elk and deer; grazing by domesticated animals; introduction of invasive ALIEN flora; and protection from fire. All these factors have caused dramatic changes to the prairies, and historical documentation is required to interpret the prairies' original character.

Despite a limited physiognomic variation the prairies exhibit a wide range of floristic variation. There is a distinct north–south gradient in the composition of the grasses (Poaceae). In SUBTROPICAL southern prairies, warm-season C_3 grasses of the subfamilies Chloridioideae and Panicuoideae dominate. In grasslands found north of latitude 42°N there is a rapid replacement of cool-season C_4 grasses of subfamilies Arundinoideae and Pooideae.

The Rocky Mountains divide North America into two grassland provinces: (i) the Great Plains prairies to the east; and (ii) the Palouse prairies and Californian prairies to the west.

Eastern prairies

The Great Plains grasslands are divided into three main prairie types: tall-grass, mixed-grass and short-grass prairies.

Tall-grass prairies

Tall-grass prairies occupy the most humid eastern part of the Great Plains and are bounded to the east by deciduous forests. Tall rhizomatous and stoloniferous species, and tussock-forming bunch grasses characterize this grassland and give a 50 cm high canopy at the end of the growing season.

Particularly after a series of wet growing seasons, when PRODUCTIVITY is high and the leaf litter is abundant, such grassland is prone to fire. Many species are fire adapted, and early spring fires especially can have a beneficial effect upon maintaining the balance of grass species. In prairies protected from fire there is an increase of woody species.

The fertile Molisols of this grassland are highly suited to a range of AGRICULTURE and horticulture. Only a small proportion of the original tall prairie survives, often in churchyards, on road verges and similar REFUGIA. Where tall-grass prairie is managed as rangeland, grazing by livestock has resulted in profound changes. The more productive and palatable species of the original prairie are reduced and less-palatable tall-grass prairie species increase. With the most intensive grazing, invading species from short-grass prairies and alien species become dominant.

Mixed-grass prairies

Mixed-grass prairies occupy an intermediate ecological and geographical position between tall- and short-grass prairies, and contain a mixture of species typical of these types. DROUGHT cycles have a profound effect upon mixed-grass prairies not observed in other prairies. In drought years, short-grass species may reduce by up to 80% and tall grasses may be almost extinguished, while annuals and invasive perennials come to dominate the vegetation. Conversely, short-grass and bunch-grass species increase in more humid periods. These fluctuations are intensified by OVERGRAZING. The mixed-grass prairie BIOME is fire adapted, and fire has been used to control invasive species of grasses and forbs.

Short-grass prairies

Short-grass prairies are found in the driest areas, in which tall grasses are reduced to small pockets of wet soils. Low-growing grasses such as *Buchloe dactyloides* and *Agropyron smithii* account for all but a small part of the grass cover. Overgrazing of these low-productivity grasslands results in the increase of weedy PERENNIAL species (e.g. *Opuntia* and *Yucca*), and invasion of ANNUAL species (e.g. *Bromus* and *Hordeum*). Fire has damaging effects upon short-grass prairie and appears to have played a limited role in its evolution as a biome. In southern Texas, New Mexico and Mexico, on the margins of true deserts, short-grass prairies grade into DESERT grassland dominated by short bunch grasses, although many such areas have now reverted to mesquite (*Presopsis*)-dominated shrublands, possibly as a result of overgrazing.

Western prairies

West of the Rocky Mountains the prairies of California once covered 5 million ha and were dominated by cool-season perennial bunch grasses such as *Stipa pulchra*, *Elymus* spp. and *Koeleria cristata*. Today, after more than two centuries of ranching, such grasslands have been destroyed and replaced by vegetation of aliens, mainly annuals of the genera *Avena*, *Bromus*, *Lolium* and *Erodium*.

Similarly the Palouse prairies, found in the dry area between the Coastal Ranges and Rocky Mountains in south-west Canada, Washington and Oregon, are thought to have originally been dominated by *Agropyron*, *Festuca* and *Elymus* species. Overgrazing of a biome which developed under very low natural grazing pressures has resulted in massive change to the grassland, and *Bromus tectorum* and *Poa pratensis* are now the dominants. [J.J.H.]

REFERENCE

Sims, P.L. (1988) Grasslands. In: *North American Terrestrial Vegetation* (eds M.G. Barbour & W.D. Billings), pp. 265–286. Cambridge University Press, Cambridge.

---

praseodymium (Pr) A LANTHANIDE METAL (ATOMIC NUMBER = 59; relative ATOMIC MASS = 140.9; melting point = 931°C; boiling point = 3512°C) with an average abundance in the Earth's CRUST of 8.2 mg kg^{-1}. It is used in glasses and thermoelectric materials. [J.G.]

preadaptation A characteristic or feature of an organism that may or may not be beneficial in its present environment but may enhance its ability to function in a new environment or under different circumstances. It has been renamed 'EXAPTATION' by Gould and Vrba (1982). For example, certain primitive fishes are considered to have been preadapted for life on land because they possessed fins with skeletons that could be modified to form a limb for locomotion on land. The term does not imply the EVOLUTION of structures in anticipation of future need. Likewise, a preadaptive MUTATION is a mutational event which later becomes of adaptive significance. [V.F.]

REFERENCE

Gould, S.J. & Vrba, E.S. (1982) Exaptation—a missing term in the science of form. *Paleobiology* **8**, 4–15.

Precautionary principle (PP)

This principle is used to justify the imposition of controls to protect the environment where there is incomplete understanding of the relationship between human actions/products and their effects on the environment, yet there is a perception of likely major problems. It has its roots in the German VORSORGEPRINZIP.

Precaution and scientific RISK ASSESSMENT are different responses to uncertainties about human impacts on the environment. An extreme version of the precautionary principle argues that we should always presume the worst, i.e. that there will be a significant environmental EFFECT from a CONTAMINANT or development, and the onus to prove otherwise should be with those who stand to gain from what is being proposed. The problem is that this sets science an impossible task—we can never be sure that there will be no effect (this is both a matter of logic and practicality)—so taken to extreme this could be a recipe for no new developments. Risk assessment, on the other hand, uses scientific analysis of likely exposures and effects to calculate the probabilities that adverse effects will occur. The onus is then with society and its representatives in the form of government, COMPETENT AUTHORITIES, regulators, etc., to take decisions about the tolerability of risks; this is usually done, implicitly if not explicitly, by reference to the benefits that derive from what is being considered for control.

Of course, the distinction between the precautionary and scientific approaches need not be presented quite so sharply as this; there are other less extreme interpretations of precaution. For example, one interpretation is that precaution simply implies the need for anticipation and prevention, which could be completely compatible with using scientific risk assessment to anticipate and prevent. Precaution, it could be interpreted, is anticipation that goes beyond the strict requirements of demonstrating causation in developing preventative measures.

Another situation in which precaution is applied is in recognition of either the shakiness of science and/or the perceived seriousness of the potential problems that might arise if the science were wrong. So UNCERTAINTY margins are often built into risk assessments as application or safety factors. But even here it is worth noting that these factors are not rationally defined; if they were we would not need to apply the precautionary principle for we would then be applying margins on the basis of scientific understanding and hence demonstrating our confidence in it.

One reason why the precautionary principle is becoming more obvious in environmental protection legislation is that it is often presented as being a major pillar of SUSTAINABLE DEVELOPMENT, especially where sustainability is defined strictly as outlawing any development that might cause adverse change. On the other hand, sustainability is often defined less strictly as recognizing a need for a balance between the requirements for economic development and environmental protection and hence in defining optimal relationships in this context. And there are possible arguments here that precaution may not lead to optimally sustainable situations. If, for example, it is presumed that we have only a limited amount of resource to use on environmental protection (admittedly not accepted by all) overuse in one area may lead to underuse in another. And there is always the niggle that controls based on precaution might needlessly be denying benefits to society.

The Treaty of the European Union (EU) (Maastricht Treaty) for the first time made the precautionary principle explicit within EC legislation. Article 130r of the modified Treaty of Rome now requires that 'Community policy on the environment . . . shall be based on the precautionary principle and on the principles that preventative action should be taken.' *See also* ENVIRONMENTAL ACTION PROGRAMMES; ENVIRONMENTAL (PROTECTION) POLICIES. [P.C.]

REFERENCES

Milne, A. (1994) The perils of green pessimism. *New Scientist* 12 June, 31–37.

O'Riordan, T. (ed.) (1994) *Environmental Science for Environmental Management*. Longman Scientific and Technical, Harlow.

O'Riordan, T. & Cameron, J. (eds) (1994) *Interpreting the Precautionary Principle*. Cameron May, London.

Stonehouse, J.M. & Mumford, J.D. (1994) *Science, Risk Analysis and Environmental Policy Decisions*. Environment and Trade Series, United Nations Environment Programme, Geneva.

Wynne, B. & Meyer, S. (1993) How science fails the environment. *New Scientist* 5 June, 33–35.

---

precipitation

1 WATER in liquid or solid form, falling to earth under gravity as RAIN, snow or hail. The amount of precipitation is usually measured as a precipitation depth (mm) equivalent to the amount of liquid water in litres per square metre. Precipitation may be the vector for dissolved or suspended material from the ATMOSPHERE to the Earth's surface, as 'wet deposition'.

2 The coagulation and agglomeration of suspended particles (usually solids in water) to a size where the rate of gravitational settling becomes greater than the rate at which TURBULENCE maintains suspension in the fluid. The coagulation process may result from chemical changes, for example a change in pH or REDOX POTENTIAL.

[J.N.C.]

preclimax In the scheme of Clements (1936; *see* PROCLIMAX and CHARACTERS IN ECOLOGY), a kind of VEGETATION characteristic of a CLIMATE drier or cooler than that obtaining in the general area where the vegetation occurs. This type was exemplified most often in high-rainfall regions by vegetation on rocky or sandy soils which have very little water-retaining capacity, and in low-rainfall areas by vegetation on CLAYS into which there is little penetration after dry spells. Also classifiable as preclimax was vegetation found where the ALTITUDINAL ZONATION is reversed, as is found in the bottoms of deep frost-hollows, for example in many dolines in Croatia, where subalpine EVERGREEN coniferous FOREST occurs below montane DECIDUOUS FOREST.

The concept has been abandoned following the general acceptance of the ideas embodied in the POLYCLIMAX viewpoint. *See also* CLIMAX. [P.J.G.]

REFERENCE

Clements, F.E. (1936) Nature and structure of the climax. *Journal of Ecology* **24**, 252–284.

precocial Describing functional maturity and relative independence of the neonates at birth. Precocial (from the Latin *praecoquis* early maturing) development of young as a relative contrast to ALTRICIAL development has been recognized in mammals and birds. The terms 'aves altrices' and 'aves precoces' were introduced into ornithology by Sundeval (1836). Traditionally, the altricial–precocial spectrum has been characterized by a heterogeneous set of characters including behaviour (nest attendance, feeding behaviour), parent–chick relationships and purely anatomical traits (eyes open or closed, presence of natal down). Expression of some of the traits, particularly nest-leaving and self-feeding, depend on nest site and food supply and may therefore be independent of development *per se*. *Cf.* ALTRICIAL. [J.M.S.]

REFERENCE

Sundeval, C.E. (1836) Ornithologiskt System. *Kongl. Vetenskops Acad. Handl. For 1835*, pp. 43–130.

predator Any heterotrophic organism that consumes any other living organism. In its narrowest sense it is often used to describe primary consumers. More broadly it may refer to PARASITES, PARASITOIDS, HERBIVORES or carnivores. Predators have a key regulatory function in natural ASSEMBLAGES and profoundly influence the structure, distribution and abundance of prey items. Predators are, in turn, often regulated by the availability of prey and are thus linked in an interactive rather than deterministic way to their food source. *See also* FEEDING TYPES, CLASSIFICATION. [R.C.]

Predator–prey interactions

The relationships that obtain between certain species (predators) that take advantage of other species (prey). Mathematically, let V be the density of the prey (or victim) species and P be the density of the PREDATOR species. Then a predator–prey interaction exists if, and only if, $\partial \ln V / \partial P < 0$ and $\partial \ln P / \partial V > 0$. (*See also* MODELS OF PREDATOR–PREY INTERACTION.)

Originally, predator–prey interactions referred exclusively to species, one of whose individuals killed those of the other for food. Soon it was recognized that such species shared many dynamic properties with others such as parasitoids and their hosts, parasites and their hosts, and herbivores and plants. In fact, the taking of individuals for food is neither necessary nor sufficient to establish the existence of a predator–prey interaction. Some food species, such as predator-resistant gastropods, actually become extinct in the absence of their consumer species, such as starfish, thus receiving a net benefit from their consumer. On the other hand, Batesian model species are damaged by Batesian mimic species although the latter never consume a single molecule of the former.

Predators reduce the populations of their victims. That has led to the persecution of some predator species that compete with humans for resources. But it has also led to the practice of BIOLOGICAL CONTROL to contain pests and weeds. Often the latter are species that have been introduced to a foreign biological province and escaped their predators.

Population dynamics

Predator–prey interactions tend to promote OSCILLATIONS in the POPULATION SIZES of the interacting species. But these depend on the presence of a high degree of PREDATOR SATIATION, when predator and victim populations are in the neighbourhood of their equilibrium values. If instead, victims compete strongly for their own resources in this neighbourhood, predator–prey interactions result in populations that are stable.

Oscillatory dynamics threaten the existence of both predator and victim. Two forces can reduce or eliminate this threat. Predators may require other resources such as shelter. This requirement can limit the amplitude of oscillations. Prey may also have access to REFUGIA where predation rates are diminished and a small population of victims can persist at a steady state.

Coevolution

Predators and prey are engaged in a coevolutionary game (*see* GAME THEORY) of counter-adaptation. This leads to the development of features with antipredator effects such as poisons, thorns, tissues difficult to digest, CAMOUFLAGE, speed, hard shells, exquisite sensory organs, special life cycles and other life-history adjustments. Predators play the game the same way, often with some of the same features. Sometimes this game may lead to a revolution in ADAPTATION, as when durophagous predators evolved in the Palaeozoic. (A durophagous predator eats armoured prey.) Their victims soon evolved thick armour or highly ornamented shells that were difficult for a predator to crush. Otherwise they disappeared or found themselves, like brittle-stars, restricted to refugia.

Theories of the coevolutionary game suggest that in most instances neither predator nor victim outdistances the other. One possible outcome sees both at a coevolutionary steady state. At such a state, any further improvement of a species' abilities to succeed in the interaction would severely reduce its ability to fulfil some other function. NATURAL SELECTION would thus produce STASIS in both predator and prey because of the compromises that every species must reach when facing trade-offs.

Another outcome suggests that the COEVOLUTION never stops. Relative rates of predator and prey improvement are balanced but exceed zero. Since neither species ever makes a net improvement in its position relative to the other, this outcome has been called the RED QUEEN'S HYPOTHESIS (because in her races it took all the running you could do merely to stay in the same place). But the Red Queen's hypothesis depends on special characteristics of species, i.e. features independent of all others, and with an infinite supply of genetic VARIABILITY.

Some predators and victims seem to fulfil the Red Queen's assumptions. Their coevolutionary game is gene matching. Predators (such as Hessian flies) cannot succeed unless they have the right gene to counter the genetic defence of their prey (wheat). There is no evidence that such genes play any other role in either species. Moreover, once the prey's defence is countered, both species should abandon the ALLELES and move on to others. And once abandoned, the alleles become available for future reuse, thus supplying each species with effectively infinite variation. [M.L.R.]

REFERENCES

Crawley, M.J. (ed.) (1992) *Natural Enemies*. Blackwell Scientific Publications, Oxford.

Frank, S.A. (1995) Statistical properties of polymorphism in host–parasite genetics. *Evolutionary Ecology* **10**, 307–317.

Rosenzweig, M.L., Brown, J.S. & Vincent, T.L. (1987) Red Queens and ESS: the coevolution of evolutionary rates. *Evolutionary Ecology* **1**, 59–96.
Taylor, R.J. (1984) *Predation*. Chapman & Hall, London.

---

predator : prey ratio *See* TROPHIC RATIOS.

predator satiation (swamping) Reduction of the PER CAPITA loss rate to predators because of a large ratio of prey to PREDATOR individuals. Swamping occurs because each individual predator can pursue, kill and digest prey at a limited rate. Swamping sometimes causes an increase in the net per capita growth rate of victims (V) in the presence of larger victim populations: $\partial \ln V / \partial V > 0$. This is equivalent to intraspecific victim MUTUALISM, which, paradoxically, can destabilize predator–prey dynamics. Periodical cicadas have evolved life cycles lasting 13 or 17 years because these prime numbers allow large synchronized cohorts to swamp generalized vertebrate insectivores. [M.L.R.]

predicted environmental concentration (PEC) The concentration of any particular substance that is anticipated to occur in the environment. It is usually derived from FATE models, which take into account levels and patterns of production, use and disposal, and physicochemical properties of the substances that influence the ways they are likely to partition between environmental compartments, and how likely they are to be transformed or degraded. *See also* EXPOSURE; MACKAY FATE MODELS. [P.C.]

predicted no effect concentration (PNEC) The concentration of a substance at and below which no ecological EFFECT is expected. It is used to set limits (e.g. ENVIRONMENTAL QUALITY STANDARDS) by regulators. However, proving negatives and guaranteeing no effects is notoriously risky, so the PNEC is better understood as the level below which effects are unlikely. [P.C.]

Predictive ecotoxicity tests

The use of test systems to attempt to predict the likely effects of substances on ecological systems and to determine probable no EFFECT levels (*see* ENVIRONMENTAL QUALITY STANDARD). The test systems can involve single species or many species; within species they may involve organismic or physiological/biochemical effects. They can be carried out at high concentration over short periods (acute) or lower concentrations over longer periods (chronic). Some are standardized, routine tests, while others may be specific to a particular problem and/or site (Table P7). They can be carried out under laboratory conditions or in field or near-field situations. In any event it is important that all relevant variables are controllable so that the link between dose/concentration of substance and effect can be defined as precisely as possible.

Design criteria

There are two major, and to some extent conflicting, challenges in designing tests: they have to be standardized so that they give repeatable and reproducible results and yet they have to be ecologically relevant. Designing for REPEATABILITY (expectation of getting the same results in replicates of the same test in one laboratory) and REPRODUCIBILITY (expectation of getting the same results in the same tests carried out in different laboratories), requires controllability and hence usually simplification of systems. Yet ecological systems are often complex. The aim should therefore be to improve repeatability and reproducibility by controlling variables, while understanding how the responses observed relate to the VARIABILITY inherent in natural systems. Often this will only be addressed by applying large safety margins to results that try to take into account extrapolation from short- to long-term effects, from single-species to multispecies systems, and from laboratory to field, although such safety margins are often somewhat arbitrary. [P.C.]

Table P7 Types of tests and a crude indication of their frequency of use. (From Calow, P. (ed.) (1993) *Handbook of Ecotoxicology*, Vol. 1. Blackwell Scientific Publications, Oxford.)

Time (*re* generation time)	Short, i.e. acute		Long, i.e. chronic
Form	Quantal		Continuous
When measured (*re* exposure)	During	After	During
Level of biological organization			
Suborganismic (molecules and cells)	Common	None	?
Single-species	Very common*	Some	Common
Multispecies	Uncommon	?	Reasonably common

* Some short-term, subchronic (sublethal) tests attempt to assess chronic effects by concentrating on sensitive stages, e.g. sensitive (usually early) life-stage tests.

REFERENCES
Calow, P. (ed.) (1993) *Handbook of Ecotoxicology*, Vol. 1. Blackwell Scientific Publications, Oxford.
Forbes, V.E. & Forbes, T.L. (1994) *Ecotoxicology in Theory and Practice*. Chapman & Hall, London.

---

predisplacement A change in the timing of onset of a specific development trajectory. If all other ontogenetic parameters remain constant, this early onset of development may result in a peramorphic morphology, since the longer ONTOGENY could 'overdevelop', adding terminal stages in relation to the ancestral ontogeny. *See also* HETEROCHRONY. [P.A.]

premanufacture notices Under the US Toxic Substances Control Act all NEW CHEMICALS must undergo premanufacturing NOTIFICATION (PMN) review. Notices must contain specified information on the properties of the chemicals that enables the USEPA, among other bodies, to assess ECOTOXICITY. [P.C.]

prescribed burns and fire management Prescribed fire is the skilful application of fire under known conditions of fuel, weather and TOPOGRAPHY to achieve resource management objectives. It includes fire ignited by managers, and fires ignited by natural sources, such as LIGHTNING, that are managed after ignition by daily monitoring. The use of fire is based on principles of fire ecology (*see* FIRE, ECOLOGY AND MANAGEMENT), so that the application of fire will meet land management goals.

Fire management may include fire prevention, fire suppression and prescribed fire. Prescribed fires ignited by managers usually have precise fire behaviour parameters, such as flame length (m), rate of spread ($m\,s^{-1}$), or fireline intensity ($kW\,m^{-1}$). These can be predicted through computer programs, which require inputs of windspeed, fuel moisture levels, slope per cent, and the amounts and arrangement of several size classes of live and dead fuels, such as leaf litter and small dead branches near the ground. Naturally ignited prescribed fires, most typically allowed only in very large wilderness areas, usually have less precise fire behaviour parameters but may have prescription criteria that include maximum allowable size, the amount of SMOKE being produced, and the ability to suppress the fire if the prescription parameters are exceeded. [J.K.A.]

prescribed processes Industrial processes identified for special attention under environmental protection legislation. In the UK these include industrial operations, as defined under the ENVIRONMENTAL PROTECTION ACT 1990, that are subject to INTEGRATED POLLUTION CONTROL. The UK list is a successor to scheduled processes defined under the earlier pollution prevention legislation. It includes some of the most polluting industrial processes; for example fuel and power, WASTE DISPOSAL, and metal and chemical industries. It is extended by EC Integrated Pollution Prevention and Control legislation. [P.C.]

pressure Force per unit area. The SI UNIT of pressure is the PASCAL (symbol: Pa), equal to 1 NEWTON of force distributed over 1 square METRE; i.e. $1\,Pa = 1\,N\,m^{-2}$. Relation to old units: 1 bar = 0.1 MPa; 1 pound per square inch = 6.984 kPa; 1 atm = 0.101 MPa; 1 mmHg (torr) = 0.133 kPa. [J.G.]

pressure, atmospheric The force per unit area exerted on a body or surface by the ATMOSPHERE. The rate at which PRESSURE and density of the atmosphere change with altitude is seen in Table P8. [J.G.]

prevalence of infection The proportion (or percentage) of a HOST population infected with a PARASITE at any one time. The prevalence is determined by the incidence of infection (transmission rate of infection, *see* TRANSMISSION RATES OF PARASITES) and the rate of recovery (or death) of the infected hosts. For MACROPARASITES, the INTENSITY OF INFECTION is also required to define a parasite population. *See also* PARASITISM. [G.F.M.]

prezygotic conflict *See* SEXUAL CONFLICT.

primary consumer *See* CONSUMER ORGANISMS.

primary effects of pollution The direct metabolic effects of POLLUTION on an organism, or the consequential effects on that organism, such as impaired survival, growth, or reproduction. *Cf.* SECONDARY EFFECTS OF POLLUTION. [P.C.]

primary forest Virgin FOREST that has not been subjected to human clearance or interference. It may, however, have experienced natural catastrophes, such as fire, flood, hurricane, etc., but recovered from these and re-established a forest CANOPY. Within primary forest, there are likely to be small areas—gaps in the canopy—where minor damage has occurred or where an old tree has fallen and recovery is taking place. The primary forest should not, therefore, be regarded as a uniform cover of mature trees.

In temperate forest areas, where human impact has played a significant role in forest development and modification, the term primary forest (or WOODLAND) has also been used for situations where timber extraction and periodic clearance may have taken place, but where the history of the woodland is uninterrupted by periods of alternative land use, such as arable or pastoral agriculture. [P.D.M.]

primary producer *See* CONSUMER ORGANISMS.

primary productivity (PP) The rate at which BIOMASS is produced per unit surface area of land or ocean by plants or chemoautotrophs. Where biomass may be described by its chemical energy, DRY MASS or CARBON. 'Primary' indicates that this is the first step in the ENERGY FLOW through an ECOSYSTEM; 'productivity' indicates the rate of manufacture or output. Primary PRODUCTIVITY is often, but incorrectly, used as a synonym of net primary productivity. See Table N1; GROSS PRIMARY PRODUCTION; NET PRIMARY PRODUCTION and PHOTOSYNTHESIS for further information. [S.P.L.]

primary sex ratio *See* SEX RATIOS.

Table P8 Pressure and density of the Earth's atmosphere at various altitudes. (From NOAA/NASA (1976) *US Standard Atmosphere*. Government Printing Office, Washington D.C.)

Altitude (m)	Pressure (MPa)	Density ($kg\,m^{-3}$)
0	0.101	1.225
500	0.0954	1.167
1000	0.0898	1.112
1500	0.0846	1.058
2000	0.0795	1.006
2500	0.0746	0.957
3000	0.0701	0.909
3500	0.0657	0.863
4000	0.0617	0.819
5000	0.0540	0.736
6000	0.0472	0.660
7000	0.0411	0.590
8000	0.0356	0.525
9000	0.0308	0.467
10000	0.0265	0.413

primary species Late-successional FOREST tree species, capable of ESTABLISHMENT in shade and/or possessing long-persistent, shade-tolerant saplings *(cf.* PIONEER SPECIES). The CLIMAX dominant species are typically primary species. They exhibit low COMPENSATION POINTS, low growth rates and relatively complex branching structure. However, the forest CANOPY typically consists of a mix of primary and pioneer species at any given time. *See also* GAP PHASE REGENERATION; SUCCESSION. [M.J.C.]

primary succession Primary SUCCESSION is most often defined as succession occurring on substrata that either have had no previous plant cover or that contain no organic matter. There is a problem with either definition. Succession is generally regarded as 'primary' when it occurs on a landslide where rock and subsoil, virtually free of organic matter, are exposed, despite the fact that before the landslide was formed the substratum was just beneath a developed SOIL carrying VEGETATION. Succession is also generally regarded as 'primary' where it occurs on the fresh deposits of meandering RIVERS or accreting SALT MARSHES or MANGROVE swamps, despite the fact that some remains of animals, algae and microorganisms are present in the deposits. The essential point is that primary successions start with very little organic matter in the substratum or none at all, while SECONDARY SUCCESSIONS start with some kind of developed soil.

Primary succession merges with REGENERATION AND CYCLIC CHANGE when the commonest form of DISTURBANCE removing individual trees brings about primary succession, for example landslides in forests on steep slopes, as in parts of the Chilean Andes, or meandering of a large river across its flood plain, as in the case of the Peruvian Amazon. Primary succession may also be deemed to occur on a smaller scale within an established COMMUNITY, for example on the raw mineral subsoil on the underside of a single uprooted tree in a FOREST, or on the floor of the hole left by the uprooted tree.

Natural primary succession occurred over vast areas on glacial deposits at the end of the last glacial period (*c.* 10 000 BP), and occurs over much smaller areas today where glaciers have been retreating since the 'Little Ice Age' (since *c.* 1700 AD), on deposits at the edges of lakes and rivers, landslips and mudflows, silt on salt marshes and mangrove swamps, sand on dunes, lava flows (Fig. P28) and beds of volcanic ash, screes, cliffs and rock outcrops. Human-induced primary succession occurs in quarries and gravel pits, on abandoned roads and railways, on cuttings and embankments made for roads and railways, and on heaps of spoil from mines for coal, china clay or various minerals.

A recent summary of knowledge of mechanisms in primary succession is given by Miles and Walton (1993). [P.J.G.]

REFERENCE

Miles, J. & Walton, D.W.H. (eds) (1993) *Primary Succession on Land*. Blackwell Scientific Publications, Oxford.

primary woodland Woodland occupying a site which has remained wooded throughout the period since the original FORESTS of a region were fragmented. This definition refers to temperate regions. Continuity of WOODLAND is broken neither by natural catastrophes nor clear-felling. Many primary woods have been coppiced

Fig. P28 Natural disturbance by larva flows in Hawaii leave isolated refugia of moist forest which serve as sources of organisms which recolonize the larva during primary succession. (Courtesy of M. Crawley.)

(*see* COPPICE) for millennia. *See also* ANCIENT WOODLAND/FOREST. [G.F.P.]

primordial radionuclides *See* RADIONUCLIDE.

principal component analysis A multivariate mathematical technique, which was first used as a method of ORDINATION in ecology by Goodall (1954). At that time, computational difficulties prevented it from being widely used, but it is now a standard item in statistical software packages.

Principal component analysis arises in many mathematical contexts and is essentially the prototype of most multivariate ordination methods. It is most easily envisaged as a means of summarizing the variation in a number of variables. Consider the scattergram of occurrence of two species in a set of samples (Fig. P29). Each plotted point represents a SAMPLE, and the ABUNDANCE of the two species is specified by the coordinates of the point. The first principal component of the species may be defined as that direction in the species space along which the dispersion of the samples is maximized. The position of each sample on the principal component is derived by projecting the points on to the line defining the principal component.

Generalizing to a larger number of dimensions, a formal definition of the principal components can be given as follows. For a set of individuals (typically samples) a series of attributes are recorded. The principal components of the attributes $x_1, \ldots, x_p$ are a derived set of attributes $y_1, \ldots, y_p$ with the following properties.

1 Each y_i is a linear combination of the original attributes:

$$y_i = a_{i1}x_1 + \cdots + a_{ip}x_p$$

2 The sum of squared coefficients is 1:

$$a_{i1}^2 + \cdots + a_{ip}^2 = 1$$

3 The *i*th principal component is uncorrelated with all higher principal components, i.e. component i is uncorrelated with components $1, \ldots, i-1$, or is unconstrained if $i=1$.

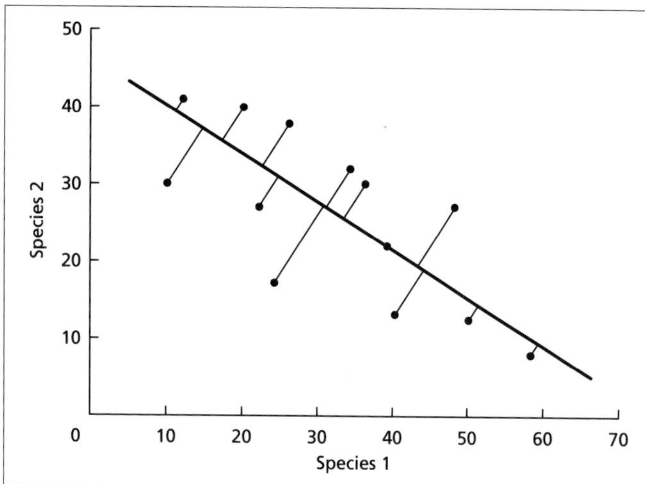

Fig. P29 The first principal component is the line on which the projections of data points have maximum variance and to which they have minimum mean-square distance. It passes through the centroid of the points. In the example, the points represent samples and their coordinates are specified by the abundance of two species.

4 Subject to the foregoing constraints, the VARIANCE of y_i is a maximum.

By convention, the principal components are centred to zero MEAN, i.e. for each i the sum of the values of the attribute y_i over all data points is zero. Mathematically, the derivation of principal components is equivalent to solving for the eigenvectors of the COVARIANCE matrix of the variables x_j. The sum of the eigenvalues is equal to the sum of the variances of x_j. It is possible to quantify the goodness of fit of the first i components by saying that they explain a proportion q of the variance. What this means is that the sum of their eigenvalues is q times the total sum of variances. The full set of components $y_1, \ldots, y_p$ necessarily explains the whole variance.

In ecological contexts, many variants of principal component analysis have been used. The commonest variant is when the original variables are themselves standardized to unit variance. In this case, the principal components are the eigenvectors of the CORRELATION matrix, and the sum of variances is p. A less-frequently encountered variant is non-centred principal component analysis (Noy-Meir 1973), sometimes accompanied by varimax rotation (Noy-Meir 1971).

A valuable aid to interpretation is the principal component biplot (Gabriel 1971; Ter Braak 1987). This allows the individuals (samples) and attributes to be displayed in the same scattergram.

Principal component analysis is closely related to, but not identical to the statistical technique of FACTOR ANALYSIS. A tacit assumption of both methods is that there should exist one or more underlying gradients which induce linear responses in the individual variables. Where this assumption is violated, difficulties result. The difficulties are particularly severe with the analysis of species data from widely differing environments. Correspondence analysis (reciprocal averaging) is then to be preferred (Ter Braak & Prentice 1988). *See also* STATISTICS. [M.O.H.]

REFERENCES

Gabriel, K.R. (1971) The biplot graphic display of matrices with application to principal component analysis. *Biometrika* **58**, 453–467.

Goodall, D.W. (1954) Objective methods for the classification of vegetation. III. An essay in the use of factor analysis. *Australian Journal of Botany* **2**, 304–324.

Noy-Meir, I. (1971) Multivariate analysis of the semi-arid vegetation in south-eastern Australia. I. Nodal ordination by component analysis. *Proceedings of the Ecological Society of Australia* **6**, 159–193.

Noy-Meir, I. (1973) Data transformations in ecological ordination. I. Some advantages of non-centering. *Journal of Ecology* **61**, 329–341.

Ter Braak, C.J.F. (1987) Ordination. In: *Data Analysis in Community and Landscape Ecology* (eds R.H.G. Jongman, C.J.F. Ter Braak & O.F.R. Van Tongeren), pp. 91–173. Pudoc, Wageningen.

Ter Braak, C.J.F. & Prentice, I.C. (1988) A theory of gradient analysis. *Advances in Ecological Research* **18**, 271–317.

prions *See* MICROBIAL ECOLOGY.

prior informed consent (PIC) A procedure established by UNEP and FAO where before allowing export of hazardous substances for the first time to a country, the exporting state is required to provide a suite of information, including identity, precautions necessary, summary of restrictions and reasons for them, to appropriate authorities in the country of destination. *See also* UNITED NATIONS PROGRAMMES AND BODIES. [P.C.]

REFERENCES

UNEP (1989) *London Guidelines for the Exchange of Information on Chemicals in International Trade* (as amended). United Nations Environment Programme, Geneva.

UNEP/FAO (1990) *The International Code of Conduct on the Distribution and Use of Pesticides* (amended version). United Nations Environment Programme, Geneva; Food and Agriculture Organization, Rome.

prioritization of chemicals Given the large number of existing commercial chemical substances it would not be possible to carry out an exhaustive assessment of the environmental risks associated with most of them. One approach is to find out what is known about them and to identify those in need of most urgent attention. This is referred to as prioritization. In fact many priority LISTS have been compiled on this basis for regulatory purposes. Criteria used include production levels, HAZARD IDENTIFICATION, indicators of PERSISTENCE and BIOACCUMULATION. These can be used to identify priority substances in at least two non-mutually exclusive ways: decision trees and scoring systems.

Decision trees classify on the basis of levels of TOXICITY, persistence, bioaccumulation and production. They therefore generate lists without ratings. Scoring, on the other hand, ascribes quantities to levels of response and production and can therefore rank priorities, but this can give a false impression that the process is more precise than it is. A very common difficulty in all these procedures is lack of adequate data. Some systems exclude substances in which this is the case (with the risk of excluding important substances) whereas others use worse-case default values (with the risk of including unimportant substances). Above all, it is important that the system favoured is simple to use and transparent. *See also* EXISTING CHEMICALS. [P.C.]

REFERENCE

Hedgecott, S. (1994) Prioritization and standards for hazardous chemicals. In: *Handbook of Ecotoxicology*, Vol. 2 (ed. P. Calow), pp. 368–393. Blackwell Scientific Publications, Oxford.

priority effect The increased competitive effect shown by one species on another species when it arrives first (has priority) at a habitat patch. In its most extreme form (Sale 1977) the first species prevents the second species from even establishing itself in the patch. Initial theoretical work (Levin 1974, 1976) suggested that such a system of priority effects, with two species, might promote COEXISTENCE. Later studies have suggested that extended persistence (delayed exclusion of one species by another), rather than permanent coexistence is more likely (Christiansen & Fenchel 1977; Shorrocks & Bingley 1994). Further discussion of priority effects is found in Shorrocks (1990).

The presence of a priority effect has been demonstrated in a number of field studies. Platt and Weis (1985) studied a group of five wind-dispersed PERENNIAL plants that colonize DISTURBANCE patches, produced by badgers, in tall-grass PRAIRIE in the USA. Immigrants on patches that are already colonized survive less and grow more slowly than those on unoccupied sites, with the number of leaves per plant at the end of the growing season being significantly lower. Kneidel (1983) investigated the priority effect upon dipteran species breeding in slug carrion. Twenty-one slug carcasses were exposed to dipteran oviposition in the field for 2 days. Seventy-five newly hatched larvae of the phorid *Megaselia scalaris* were transferred to each of 14 of these carcasses, seven before going into the field (high priority) and seven after the 2-day exposure (low priority). Both groups and the seven controls were then allowed to produce emergences in the laboratory. Both survival and body size for *M. scalaris* showed an increase in the high-priority treatment. There was no significant increase in the total number of flies produced in the high- and low-priority treatments compared to the controls, suggesting that the addition of *M. scalaris* precluded successful COLONIZATION by other species. A similar study using two species of drosophilid flies that breed in fungi has recently been carried out by Shorrocks and Bingley (1994). Flies arriving late at patches (fungi) had decreased survival, decreased body size and increased developmental time.

An extreme form of priority effect is the lottery system, proposed by Sale (1977, 1979, 1982) for communities of coral-reef fishes which have widely dispersing larvae and tenaciously territorial adults. When adults disperse or die, the site that they had originally occupied becomes available for colonization. Each vacated site is colonized by the first new recruit to arrive and once resident this fish, regardless of its species, will defend its patch against all subsequent colonists. *See also* COMPETITION. [B.S.]

REFERENCES

Christiansen, F.B. & Fenchel, T.M. (1977) *Theories of Populations in Biological Communities*. Springer-Verlag.

Kneidel, K.A. (1983) Fugitive species and priority during colonisation in carrion-breeding Diptera communities. *Ecological Entomology* **8**, 163–169.

Levin, S.A. (1974) Dispersion and population interactions. *American Naturalist* **108**, 207–228.

Levin, S.A. (1976) Population dynamic models in heterogeneous environment. *Annual Review of Ecological Systems* **7**, 287–310.

Platt, W.J. & Weis, I.M. (1985) An experimental study of competition among fugitive prairie plants. *Ecology* **66**, 708–720.

Sale, P.F. (1977) Maintenance of high diversity in coral reef fish communities. *American Naturalist* **111**, 337–359.

Sale, P.F. (1979) Recruitment, loss, and coexistence in a guild of territorial coral reef fish. *Oecologia* **42**, 159–177.

Sale, P.F. (1982) Stock-recruitment relationships and regional coexistence in a lottery competitive system: a simulation study. *American Naturalist* **120**, 139–159.

Shorrocks, B. (1990) Coexistence in a patchy environment. In: *Living in a Patchy Environment* (eds B. Shorrocks & I. Swingland), pp. 91–106. Oxford Science Publications, Oxford.

Shorrocks, B. & Bingley, M. (1994) Priority effects and species coexistence: experiments with fungal-breeding *Drosophila*. *Journal of Animal Ecology* **63**, 799–806.

Prisoner's dilemma *See* TRAGEDY OF THE COMMONS.

probability The chance of an event or measurement being observed. Probability is measured on a scale of 0 (impossible) to 1 (absolutely certain). For example, the probability, p, of getting a 'head' when you spin an unbiased coin is $p=0.5$ (or $p={}^1/_2$, or $p=$ 50%). Probabilities can be calculated in two ways: a priori (before the event) and a posteriori (after the event). Before spinning the coin (a priori) we can imagine that the probability will be ${}^1/_2$ since there are two sides to the coin and we are only interested in one of them (heads). We exclude the possibility of the coin landing on its edge! After spinning the coin several times (a posteriori) we could estimate the probability by dividing the number of 'heads' actually observed by the total number of spins. The accuracy of this estimate will depend upon the total number of spins (n). For example, if by chance we get one more 'head' than we expect when $n=10$ then $p=$ 60%, but if $n=100$, $p=51\%$. Some probabilities can only be estimated a posteriori because they encapsulate complex and unknown causes; for example, the probability that the egg of a particular species of bird will not hatch. *See also* PROBABILITY DISTRIBUTION; STATISTICS. [B.S.]

probability distribution A theoretical or observed distribution of probabilities or frequencies. Theoretical distributions use a priori probabilities, while observed distributions use a posteriori probabilities (*see* PROBABILITY). Some people restrict the term 'probability distribution' to the theoretical, while using the term 'frequency distribution' for the observed. However, both can be used and interpreted in the same way. The total area under a curve (for a continuous variable) or histogram (for discrete observations) is equal to one, and the area under part of the curve or histogram is equal to the probability or frequency of a particular event. For example, Fig. P30 shows the expected probability distribution of the number of 'heads' when four unbiased coins are spun; that is, with the probability of a 'head' (p) = 0.5 and the probability of a 'tail' (q) = 0.5 for an individual coin. We can see that the probability of getting no heads = 0.06 while the probability of getting 'at least three heads' (3 and 4) = 0.25 + 0.06 = 0.31 (the shaded area). This is an example of a binomial distribution, which is a discrete probability distribution (*see* BINOMIAL (POSITIVE BINOMIAL) DISTRIBUTION). Other discrete distributions used in ecology are the POISSON DISTRIBUTION and the NEGATIVE BINOMIAL DISTRIBUTION. An example of a continuous probability distribution is the NORMAL DISTRIBUTION, which is fundamental to PARAMETRIC STATISTICS. [B.S.]

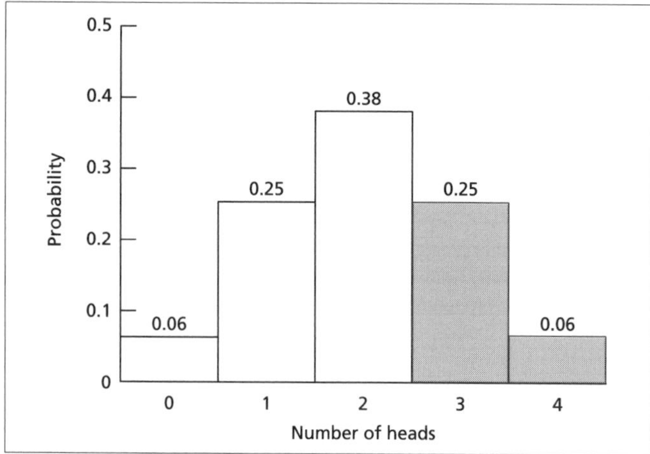

Fig. P30 The expected discrete probability distribution resulting from the spinning of four unbiased coins. The shaded area = probability of getting at least 3 heads.

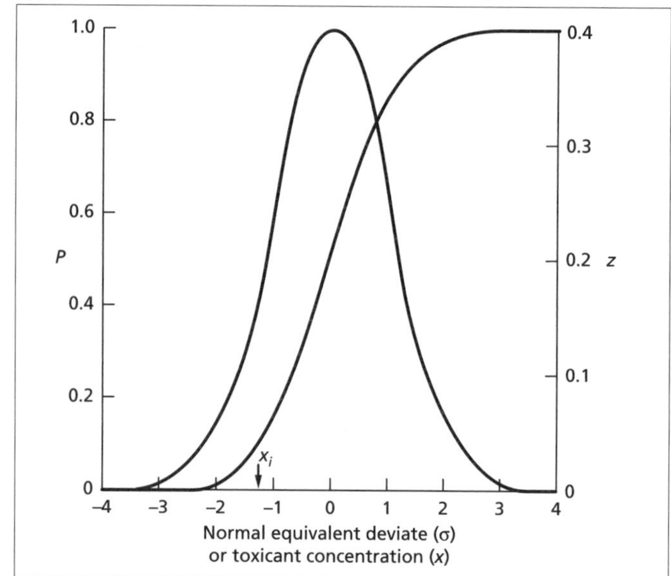

Fig. P31 Plot of the standardized normal frequency distribution (*P*) along with its integral, the cumulative normal frequency function (*z*); x_i is a specified value (see text). (After Forbes, T.L. (1994) The design and analysis of concentration–response experiments. In: *Handbook of Ecotoxicology*, Vol. 1 (ed. P. Calow), pp. 438–459. Blackwell Scientific Publications, Oxford.)

probability refuges Low-density REFUGES, available to an inferior interspecific competitor, that arise because of the AGGREGATED DISTRIBUTION of population numbers across habitat patches. Such refuges are an important part of the AGGREGATION MODEL OF COMPETITION. They are not identified with a particular patch type, but vary from generation to generation depending upon which patches happen, by chance, to hold the low densities. [B.S.]

probit analysis This is widely used in toxicological and ecotoxicological analyses. Assume that the DISTRIBUTION of a POPULATION of individual tolerances (in terms of mortality) or a population of continuous responses (say growth rates) is bell-shaped or normal (Fig. P31). Then any particular TOXICANT concentration (x_i in Fig. P31) must split the population into two groups: those affected (to the 'left' of the concentration) and those unaffected (to the 'right' of the concentration). So the area under the bell-shaped area to the left of an abscissa value corresponding to the particular concentration equals the fraction or proportion of individuals affected (say P_i) and the area to the right equals the proportion of individuals unaffected (say Q_i); where $P_i + Q_i = 1$. If the response indeed conforms to a NORMAL DISTRIBUTION (it is often necessary to transform dose/concentrations logarithmically to achieve this; i.e. a LOG NORMAL DISTRIBUTION) then the abscissa can be represented in units of STANDARD DEVIATIONS (δ_T). It follows that at a particular step in standard deviations away from the MEAN in a perfect normal distribution the fraction of population affected or unaffected is completely specified. To turn this around, a given fraction of population affected should be equivalent to a predictable step in standard deviations away from that concentration representing the mean (and median) response. This expected deviation can be specified as a normal equivalent deviate (NED) where NED = (concentration causing P_i – concentration causing the median effect)/δ_T. Thus if the actual concentrations at which particular fractions of a population are affected are plotted against NED values, and if the distribution is indeed normal, a straight-line relationship should be obtained. In fact NED can take negative values when median concentration is > P_i and this has traditionally been avoided by the addition of 5. NED + 5 is what is known as a probit value, and can be obtained from standard STATISTICAL TABLES. With the dose/concentration–effect relationship as a straight line it is more straightforward to estimate EC_x (e.g. EC_{50}) values. Calculating population statistics after probit transformation is known as probit analysis. [P.C.]

REFERENCES

Bliss, C.J. (1935) The calculation of the dosage—in mortality curves. *Annals of Applied Biology* **22**, 134–167.

Forbes, T.L. (1993) The design and analysis of concentration–response experiments. In: *Handbook of Ecotoxicology*, Vol. 1 (ed. P. Calow), pp. 438–459. Blackwell Scientific Publications, Oxford.

process standards *See* ENVIRONMENTAL LAW.

proclimax In the scheme of Clements (1916, revised 1936; *see* CHARACTERS IN ECOLOGY) this was defined as 'any more or less permanent COMMUNITY resembling the CLIMAX in one or more respects, but gradually replaceable by the latter when the control of CLIMATE is not inhibited by disturbance'.

Clements' concept of 'disturbance' differs from the concepts in use today; he used it to embrace the effects of all factors other than macroclimate, i.e. MICROCLIMATE, soil water and mineral relations, natural processes which destroy or modify VEGETATION such as fire, and human-induced processes such as grazing, mowing or changed fire regime. Four types of proclimax were recognized: DISCLIMAX, POSTCLIMAX, PRECLIMAX and SUBCLIMAX. This nomenclature has virtually disappeared following acceptance of the general ideas embodied in the POLYCLIMAX concept. [P.J.G.]

REFERENCE

Clements, F.E. (1936) Nature and structure of the climax. *Journal of Ecology* **24**, 252–284.

product–moment correlation coefficient (*r*) A PARAMETRIC STATISTIC that summarizes the degree of CORRELATION between two sets of observations (x and y). The correlation coefficient (r) varies between +1 (perfect positive correlation), through 0 (no correlation), to −1 (perfect negative correlation). Both variables should approximately follow a NORMAL DISTRIBUTION, and are assumed to be measured with error. This is in contrast to some types of regression analysis (*see* REGRESSION ANALYSIS, CURVILINEAR; REGRESSION ANALYSIS, LINEAR), in which only the y variable (dependent variable) is considered to follow a normal distribution and to be measured with error. Consider the following data, collected to examine the ASSOCIATION between body size and daily egg production in a small fly. Body size is actually measured as wing length in millimetres.

Wing length (mm) (x)	Eggs per day (y)
1.89	10
1.91	11
2.00	12
1.78	8
2.03	17
1.99	12
1.96	10
2.06	16
1.71	6
1.92	10

Several quantities are now calculated, as set out below.

n (number of paired observations) = 10

$$\sum x = 19.25 \qquad \sum y = 112$$

$$\sum x^2 = 37.16 \qquad \sum y^2 = 1354 \qquad \sum xy = 218.56$$

$$\sum x^2 - \left(\sum x\right)^2/n = 0.10 \qquad \sum y^2 - \left(\sum y\right)^2/n = 99.60 \qquad \sum xy - \left(\sum x \sum y\right)/n = +2.96$$

On the last line, the first quantity is the SUM OF SQUARES of x, the second quantity is the sum of squares of y, and the third quantity is the SUM OF PRODUCTS. The first two quantities are always positive, the third quantity can be positive or negative depending upon the relationship between x and y. Dividing each quantity by $n-1$ (DEGREES OF FREEDOM) will give the VARIANCE of x, the variance of y, and the COVARIANCE of x and y. The product–moment correlation coefficient is calculated by:

$$r = \frac{\sum xy - \left(\sum x \sum y\right)/n}{\sqrt{\left(\sum x^2 - \left(\sum x\right)^2/n\right)\left(\sum y^2 - \left(\sum y\right)^2/n\right)}}$$

that is, by dividing the sum of products by the square root of the product of the two sums of squares. In the example above this is:

$$r = \frac{+2.96}{\sqrt{0.10 \times 99.60}}$$

$$r = +0.94$$

Notice that the sign of the sum of products is retained in the correlation coefficient.

The significance of this calculated value of r is assessed by comparing it with values of r in STATISTICAL TABLES, for $n-2$ degrees of freedom (v). For $v=8$, the 1% value of $r=0.76$ and therefore the computed value of $r=+0.94$ indicates a very significant positive association between body size and egg production in our small fly. [B.S.]

product standards *See* ENVIRONMENTAL LAW.

product stewardship The responsible design, production and marketing of products, together with after-sales care, that takes into account possible HARM to humans and impact on the environment. It should obviously include LIFE-CYCLE ASSESSMENT as an integral part. *See also* ECOLABEL. [P.C.]

production Sometimes a distinction is made between the process generating new BIOMASS and the output of new biomass. Some refer to the process as production and the output as PRODUCTIVITY. Occasionally the reverse senses are intended. Usually, however, the two terms are used interchangeably. [P.C.]

production efficiency (PE) The proportion of assimilated energy (A) which is incorporated into new BIOMASS (PRODUCTION, P) of an organism due to somatic GROWTH and development of reproductive tissues:

$$\frac{P}{A} \times 100$$

The rest of the assimilated energy is lost as respiratory heat.

Production efficiency varies with taxonomic group and the homeostatic strategies of the animal. Invertebrates have higher efficiencies — usually of the order of 30–40% — than vertebrates, of which the ectotherms have values of approximately 10% and endotherms, which expend a lot of energy maintaining body temperature, have the lowest values, often converting only 1–2% of ASSIMILATION to production.

The smallest endotherms have the lowest values of all; for example, small insectivorous shrews, which spend a lot of their lives in frenetic activity, have production efficiencies of <1%. Representative values for different groups of animals are shown in Table P9.

Growth of tissues requires more than the energy within them as the processes involved in biosynthesis are not 100% efficient, as would be predicted from the second law of thermodynamics. The biosynthesis of tissues from assimilated material has an energetic efficiency of at most 40–85%, so setting an upper limit to the production efficiency of an organism. Even a nestling ring dove during its maximum rate of growth only has a production efficiency of approximately 20%, which declines rapidly as it matures (Ricklefs 1980). *See also* ECOLOGICAL EFFICIENCY; ENERGY BUDGET; ENERGY, LAWS OF THERMODYNAMICS. [M.H.]

Table P9 Production efficiency ($P/A \times 100$) of various animal groups ranked in order of increasing efficiency. (From Begon, M., Harper, J.L. & Townsend, C.R. (1996) *Ecology: Individuals, Populations and Communities*, 3rd edn. Blackwell Science, Oxford; after Humphreys, W.F. (1979) *Journal of Animal Ecology* **48**, 427–454.)

Group	Production efficiency (%)
1 Insectivores	0.86
2 Birds	1.29
3 Small mammal communities	1.51
4 Other mammals	3.14
5 Fish and social insects	9.77
6 Non-insect invertebrates	25.0
7 Non-social insects	40.7
Non-insect invertebrates	
8 Herbivores	20.8
9 Carnivores	27.6
10 Detritivores	36.2
Non-social insects	
11 Herbivores	38.8
12 Detritivores	47.0
13 Carnivores	55.6

REFERENCES

Begon, M., Harper, J.L. & Townsend, C.R. (1996) *Ecology: Individuals, Populations and Communities*, 3rd edn. Blackwell Science, Oxford.

Humphreys, W.F. (1979) Production and respiration in animal populations. *Journal of Animal Ecology* **48**, 427–454.

Ricklefs, R.E. (1980) *Ecology*, 2nd edn. Nelson, Sudbury-on-Thames.

productivity In ecological studies, it is common to refer to the growth and NET PHOTOSYNTHETIC RATE of an ecosystem as net primary productivity (P_n or N) and gross primary productivity (P_g or G) respectively. P_g is typically presented on a time-scale of months or years and is equivalent to what plant physiologists normally call CARBON DIOXIDE (CO_2) ASSIMILATION (photosynthetic) rate. That is, it is the net rate at which CO_2 is fixed into new photoassimilate by the ecosystem. Even during the day this is substantially less than the actual rate of gross CO_2 fixation, due to photorespiratory CO_2 release, and mitochondrial CO_2 release associated with RESPIRATION by the leaves. P_n is defined as PHOTOSYNTHESIS minus plant respiration, i.e.:

$$P_n = P_g - R_p$$

where R_p is the respiration by plants, including that by leaves at night. Although it is now possible to measure P_g and R_p directly using aerodynamic and chamber gas-exchange techniques most earlier studies concentrated on P_n and measured it as:

$$P_n = dM_p/dt + L$$

where dM_p/dt is the change in the mass of plant BIOMASS in the system studied and L represents the losses of material produced in the time studied. It has been most common to measure P_n as the change in above-ground biomass plus the loss of plant parts due to abscission and mortality (including herbivory) over a specified time interval. Thus, even though there may be no change in the net biomass of an ecosystem, P_n will be greater than zero due to litter fall and root turnover during the measurement period.

Ecologists also use the term 'net ecosystem productivity'. This is the difference between P_g and total ecosystem (i.e. plant + HETEROTROPH) respiration. When net ecosystem productivity is positive, the total amount of CARBON in the ecosystem (plants + soil) must be increasing.

In many ecosystems, P_n is notoriously difficult to measure, as it is necessary to measure the rate at which dead matter or 'litter' is shed from the plant and add it to the increment in the standing biomass. The data of Whittaker (1975) should be regarded as provisional, but they do provide an estimate of the net primary productivity and net secondary productivity of the major biomes (see Table N1). SECONDARY PRODUCTIVITY is defined as the rate of dry matter or carbon accumulation by heterotrophic organisms. The term 'PRODUCTION' refers to yield, an accumulation of mass per area of land, whereas 'productivity' is strictly a *rate* of accumulation. Typical units for P_n or P_g are g dry biomass m^{-2} year^{-1}, g carbon m^{-2} year^{-1}, or mol C m^{-2} year^{-1}. For conversions: 1 g dry biomass $\approx$ 0.5 g carbon; 1 mol C = 12 g carbon.

Agricultural systems often consist of annual crops, in which the rate of litter production is small and the biomass of heterotrophic organisms is minimal. Here, the biomass yield is an approximate measure of P_n, and secondary productivity is very small. [J.Ly. & J.G.]

REFERENCES

Olson, J.S. (1975) Productivity of forest ecosystems. In: *Productivity of World Ecosystems* (eds D.E. Reichle, J.F. Franklin & D.W. Goodall), pp. 33–43. National Academy of Sciences, Washington, D.C.

Whittaker, R.H. & Likens, G.E. (1975) The biosphere and man. In: *Primary Productivity of the Biosphere* (eds H. Lieth & R.H. Whittaker), pp. 305–328. Springer-Verlag, Berlin.

produktlinienalyse (PLA) A form of LIFE-CYCLE ASSESSMENT, pioneered in Germany, that takes a broad qualitative approach in describing adverse effects of products, including social and economic aspects, from cradle to grave. It completely avoids AGGREGATION of the inventory data, which makes it somewhat difficult to interpret. [P.C.]

profile A cross-sectional view of a HABITAT or of a sedimentary deposit. It can apply to the physical structure and TOPOGRAPHY of a habitat, as in the case of a sand-dune profile or a beach profile, or the structure and composition of the VEGETATION itself, as in the profile of a forest. In the case of sedimentary deposits (such as lake sediments and peat deposits), a profile provides a representation of the sequence of stratified materials and hence allows a reconstruction of the geological history of the site. [P.D.M.]

profundal *See* ABYSSAL; WITHIN-LAKE HABITAT CLASSIFICATION.

progenesis (paedogenesis) A process of HETEROCHRONY in which there is evolution of a new ontogenetic sequence by the truncation of the ancestral one. Such a process causes the elimination of terminal stages in the ancestral development series. It results in the generation of a paedomorphic form. Due to the predominantly global action of agents controlling the offset of development, such as hormones regulating sexual maturation, paedomorphic forms evolved through progenesis tend to be characterized by small size. [P.A.]

prohibition notice A legal device, served by a regulator, requiring the cessation of a practice that is deemed to be adversely affecting the environment and contravening legislation. *See also* ENVIRONMENTAL LAW; ENVIRONMENTAL (PROTECTION) AGENCIES. [P.C.]

prokaryote Lacking internal, membrane-bound organelles and a nuclear membrane. Bacteria are prokaryotes. *Cf.* EUKARYOTE. *See also* PROTOEUKARYOTE. [P.C.]

promethium (Pm) A LANTHANIDE METAL (ATOMIC NUMBER = 61; relative ATOMIC MASS = 145; melting point = 1080°C; boiling point = 2460°C). Promethium has never been isolated from naturally occurring minerals but ^{147}Pm is a FISSION product of URANIUM. All known ISOTOPES are radioactive. ^{147}Pm may have applications as a nuclear-powered battery in space, and as a phosphor. [J.G.]

promiscuity (promiscuous breeding) A MATING SYSTEM in which members of both sexes mate or reproduce with multiple members of the opposite sex and do not form stable pair-bonds. Examples include chimpanzees, in which both males and females frequently mate with many different members of their communities; broadcast-spawning fishes in which pairwise mating does not occur but the GAMETES of any one individual may join with those of many different members of the opposite sex; and many species of invertebrates. [J.C.D.]

propagule A reproductive body (including spore, seed, bulb, cyst, EGG, bud, LARVA, etc.) that separates from the parental organism and gives rise to a new individual. The term is traditionally applied to plants and fungi, sometimes to invertebrates. Propagules often constitute the main dispersing stage in the LIFE CYCLE. In life-history theory, propagule size usually means the size of the offspring at the end of the period of PARENTAL INVESTMENT. *See also* LIFE-HISTORY EVOLUTION. [T.J.K.]

proportional sensitivity *See* SENSITIVITY ANALYSIS.

protactinium (Pa) A metallic element (ATOMIC NUMBER = 91; relative ATOMIC MASS = 231.036; melting point = 1552°C; boiling point = 4227°C), having known isotopes from ^{216}Pa to ^{238}Pa, all of which are radioactive. ^{231}Pa ($t_{1/2} = 3.28 \times 10^4$ years) and ^{234}Pa ($t_{1/2} = 6.75$ h) are members of the natural radioactive decay series and occur in trace quantities in URANIUM- and THORIUM-bearing minerals. Protactinium exists in the environment in the +5 oxidation state (although +4 can be produced) and has a low solubility in natural waters. ^{231}Pa is a useful tracer of environmental processes via natural decay series disequilibrium (i.e. disruption of the state of radioactive equilibrium in the ^{235}U decay chain). *See also* RADIONUCLIDE. [A.B.M.]

REFERENCES

Ivanovich, M. & Harmon, R.S. (eds) (1992) *Uranium-Series Disequilibrium.* Clarendon Press, Oxford.

Katz, J.J., Seaborg, G.T. & Morss, L.R. (eds) (1986) *The Chemistry of the Actinide Elements.* Chapman & Hall, London.

protandry A kind of SEQUENTIAL HERMAPHRODITISM in which an organism begins its reproductive career as a male, then switches sex and functions as a female. According to SEX ALLOCATION THEORY, one would expect protandry to evolve where organisms of small body size function more effectively as males, those of larger body size function more effectively as females, and can profitably switch sex as they grow. A small male is able to fertilize a female with a modest amount of sperm and is relatively mobile. A female that cares for developing eggs and embryos may be able to accommodate more offspring as she grows larger. This kind of explanation seems to fit the facts for a substantial number of marine animals, such as the gastropod *Crepidula*. *See also* ASEXUAL REPRODUCTION; POLLINATION; PROTOGYNY. [M.T.G.]

protective proteins A generic term for PROTEINS that provide defence against abiotic stressors such as HEAVY METALS (*see also* METALLOTHIONEINS) or excessive heat (*see also* HEAT-SHOCK PROTEIN). Some have evolved as specific adaptations. Others may have evolved for other functions but, fortuitously, also provide protection when needed. They may operate to ameliorate the stressors (e.g. metallothioneins) or the damage done by them (e.g. heat-shock proteins). [P.C.]

proteins A class of macromolecules synthesized in plants and animals, and composed of PEPTIDES and polypeptides, which are themselves composed of AMINO ACIDS. Their relative molecular masses range from a few thousands to a few millions. They have a key role in biochemistry where, as enzymes, they control metabolism. In addition, they form muscle in animals, the OXYGEN-carrying pigment haemoglobin in blood, and certain structural materials such as keratin in hair and horn, and silk.

Proteins are characterized by their primary, secondary and tertiary structure. Primary structure refers to the sequence of amino acids, which for many proteins has now been determined. Secondary structure refers to the tendency of the sequence of amino acids to form a distorted helix, due to the shape of the constituent amino-acid units and the hydrogen bonding that occurs between these units. The tertiary structure is the overall shape of the molecule, which is influenced by chemical reactions between functional groups in the molecule. Thus the shape often resembles a highly convoluted sausage, seemingly tangled with itself. There may also be a quaternary structure in complex proteins where several such tertiary units are joined together by a functional group.

Proteins are denatured by extremes of temperature or pH, which change the shape of the molecule without affecting its primary structure (the familiar example is that of boiling egg white). [J.G.]

protoeukaryote Hypothetical cell which, together with certain endosymbiotic PROKARYOTES, formed the first EUKARYOTE. [P.C.]

protogyny A kind of SEQUENTIAL HERMAPHRODITISM in which an organism begins reproducing as a female, then switches sex to become a male. This kind of sex switch is quite common among coral-reef fishes, and is readily explained by what is called the 'size-advantage model' (*see* SIZE-ADVANTAGE HYPOTHESIS). This theory makes a few simple assumptions. First, it is better to reproduce at all sizes above a certain minimum than not to reproduce at all. Second, larger animals enjoy greater reproductive success than do smaller ones. Third, the benefit enjoyed through larger size is not the same for males and females. For protogyny this implies that small females will tend to have more offspring than small males, large males more offspring than large females.

These conditions are realized where there is a kind of SEXUAL SELECTION: the males fight as a means of monopolizing the opportunity to mate with the females, and the big males generally win out. Accordingly, the appropriate stratagem is to reproduce as a female until large enough to win the fights, and then switch. Often the males are polygynous, and when the male dies, the largest female in the group becomes a male. *See also* ASEXUAL REPRODUCTION; POLLINATION; PROTANDRY. [M.T.G.]

province A BIOGEOGRAPHICAL UNIT, generally used in PHYTOGEOGRAPHY rather than ZOOGEOGRAPHY. For many authors it is subordinate to the subkingdom (see Fig. B6). However, Takhtajan (1969) uses 'region' as the exact equivalent of 'province' of other authors. [W.G.C.]

REFERENCE

Takhtajan, A. (1969) *Flowering Plants, Origin and Dispersal.* Oliver & Boyd, Edinburgh.

proximate–ultimate distinction How an organism responds to environmental factors can be ascribed to at least two kinds of interrelated causes. These are the immediate changes in physiology and behaviour and the control exerted on them, for example by endocrine and neural systems. There are also long-term processes that have led to the evolution of certain physiological and/or behavioural responses as ADAPTATIONS. Explanations of how physiological and behavioural events are caused by environmental factors are described as proximate (mechanistic) explanations. Explanations of how NATURAL SELECTION is likely to have favoured the evolution of those reponses in the environments concerned are described as ultimate (functional) explanations. [P.C.]

REFERENCE

Waddington, C.H. (1962) Proximate and ultimate explanations in biology. *Science* **135**, 976–977.

PRPs (potentially responsible parties) *See* COST RECOVERY.

prudent predator concept The idea that PREDATORS moderate their individual predation rates to avoid causing the extinction of their prey. In its extreme form, the concept requires rates that maximize the production rate of victims. Given an additive model of predation, this rate is achieved when predators maintain their victims at that population generating the point of inflection of the victim's SIGMOID GROWTH curve.

However, the evolutionary interests of individual predators usually clash with those of the predator population. If NATURAL SELECTION forces predators to become increasingly proficient as individuals, EVOLUTION will reduce populations of victims until the predators (and perhaps their victims) become extinct. In such circumstances, the evolution of prudent predators would require GROUP SELECTION. That weak force—and the more potent evolutionary opposition of their victims—protects such predators from extinction (*see* PREDATOR–PREY INTERACTIONS).

Prudent predation or an approximation of it can evolve when the FITNESS of the individual predator benefits from moderate predation. For example, the myxoma virus evolved a profoundly reduced lethality to European rabbits (*Oryctolagus cuniculus*) in Australia; lethal strains of myxoma virus kill so quickly that they have relatively few successful offspring. Predators, like giant California limpets (*Lottia gigantea*), that control, as individuals, at least a subpopulation of their victims for a considerable length of time can also evolve moderate predation rates. *Lottia* establish territories from which they exclude other grazers and within which they allow algae to flourish. Smaller, wandering limpet species rapidly denude the rocks tended by *Lottia* after it is removed experimentally. *See also* POPULATION REGULATION. [M.L.R.]

***Prunus spinosa–Rubus fruticosus* scrub community** *See* SCRUB.

psammon A term used to describe the communities of microscopic flora, mainly unicellular algae, and associated fauna which inhabit the INTERSTITIAL spaces between sand grains on a seashore or lakeshore. The active flora tends to be confined to the top few millimetres where LIGHT is available, but storm conditions can lead to burial of cells to greater depths. [B.W.F.]

psammosere The PRIMARY SUCCESSION associated with sand-dune formation, typically in coastal areas. It is the process whereby wind-blown sand accumulates to form dunes which become gradually larger and more stabilized by colonizing vegetation. The classic stages (seres) in the psammosere are:

1 embryo dunes (formed on the STRANDLINE), which are very temporary, low-profile structures colonized by a very restricted range of ANNUAL plants;

2 foredunes (formed from previous embryo dunes), which are a little less transient and more elevated, and are generated as a result of colonizaton by such species as *Elytrigia juncea*, a PERENNIAL grass;

3 mobile dunes (formed from previous foredunes), which reach maximal height and are generated as a result of COLONIZATION by *Ammophila arenaria*, a perennial grass;

4 stable dunes (formed from previous mobile dunes), which are formed as *Ammophila* declines in vigour when mobile dunes mature and when a wider range of species, including woody forms, enter the succession.

The mobile dunes are mobile by virtue of loose sand being blown from the front and deposited behind, whilst stable dunes have been stabilized by the formation of a 'turf'.

The processes of SOIL formation and maturation (profile development) accompany the vegetation succession, and involve incorporation of organic material into the sand and changes in soil nutrient status, pH and water regime. The oldest parts of dune systems may become WOODLAND, HEATHLAND or grassland depending particularly on soil factors (such as the presence of shell fragments in the sand) and grazing pressures. All stages in the psammosere are likely to be present together in the form of a ZONATION pattern, extending from the coast inland, together with low-lying areas between the dunes (slacks), which may also show a succession from young to old. [B.W.F.]

pseudo-arrhenotoky A haplo-diploid genetical system (*see* HAPLO-DIPLOIDY) in which both males and females arise from fertilized EGGS but where males transmit only the genome of the mother. Males become effectively haploid after the inactivation (heterochromatization) and possible elimination of the paternal genome. Exactly when the inactivation and loss of paternal chromosomes occurs depends on the organism under study. In some taxa (e.g. diaspidid scale insects and plant-inhabiting predatory mites in the family Phytoseiidae), paternal genome loss occurs early in embryogenesis, but not until after several mitotic divisions have taken place. In other taxa (e.g. fungal gnats in the genus *Sciara* and scale insects in the genus *Comstockiella*), somatic tissues retain the paternally derived chromosome set in an active or inactive state, whereas it is eliminated in the GERM LINE just before or during spermatogenesis. *See also* ARRHENOTOKY. [M.W.S. & C.J.N.]

REFERENCES

Bull, J.J. (1983) *Evolution of Sex-Determining Mechanisms*. Benjamin/Cummings, Menlo Park, Calif.

Wrensch, D.L. & Ebbert, M.A. (eds) (1993) *Evolution and Diversity of Sex Ratio in Insects and Mites*. Chapman & Hall, New York.

pseudocleistogamy *See* CLEISTOGAMY.

pseudogamy A form of PARTHENOGENESIS in which male GAMETES are required. In animals, entrance of a sperm is necessary to activate development of the egg, but the sperm pronucleus does not fuse with the egg nucleus and the male genome is never incorporated into the egg. It may be viewed as a primitive form of parthenogenesis. In some species (e.g. certain spruce-breeding *Ips* species in North America) pseudogamous females coexist with normal sexual females of the same species, with whom they compete for sperm. Sometimes (e.g. the fish *Poeciliopsis*) the sperm of a closely related species triggers development, whereas in other instances (e.g. the hermaphroditic mollusc *Lasaea*) autosperm (sperm for self-fertilization) is used.

In plants the endosperm is fertilized and contains maternal and paternal genes, but the embryo is asexual and contains maternal genes only. *See also* ASEXUAL REPRODUCTION. [M.W.S. & C.J.N.]

REFERENCES

Asker, S.E. & Jerlin, L. (1992) *Apomixis in Plants*. CRC Press, Boca Raton, Fla.

Stenseth, N.C., Kirkendall, L.R. & Moran, N. (1985) On the evolution of pseudogamy. *Evolution* **39**, 294–307.

Suomalainen, E., Saura, A. & Lokki, J. (1987) *Cytology and Evolution in Parthenogenesis*. CRC Press, Boca Raton, Fla.

pseudogene A nucleotide sequence that is very similar to that of a functional gene but which is non-functional, frequently because it lacks a promoter for transcription. Pseudogenes may be located near related functional genes, but they have also been found at locations far removed from the related functional genes. Pseudogenes possess some MUTATION that affects their ability to produce normal RNA or protein. They are transmitted passively and are not affected by NATURAL SELECTION. As a result, pseudogenes show higher EVOLUTIONARY RATES than their functional counterparts. Pseudogenes differ from their functional relatives by insertions or deletions and by the presence of flanking direct repeat sequences of 10–20 nucleotides. Pseudogenes are designated by the prefix ψ (the Greek letter psi), followed by the symbol of the gene they most resemble. [V.F.]

pseudointerference A density-dependent reduction in the per capita efficiency of a PREDATOR or PARASITOID due to a mechanism that does not involve direct behavioural interactions (INTERFERENCE). The term is frequently used to describe the reduction in efficiency with density of predators and parasitoids that concentrate on certain patches of their prey or HOST. Pseudointerference is an important stabilizing factor in many models of host–parasitoid dynamics. *See also* PARASITOID–HOST INTERACTIONS; POPULATION REGULATION; PREDATOR–PREY INTERACTION. [H.C.J.G.]

pseudopleuston *See* WITHIN-LAKE HABITAT CLASSIFICATION.

pseudopregnancy A syndrome (also termed pseudocyesis), occurring in some mammals, in which uterine and hormonal changes following oestrus resemble certain characteristics of pregnancy but without implantation of an embryo. Pseudopregnancy frequently occurs in subordinate females of some cooperatively breeding mammals, and in these cases may have been selected as a mechanism enabling subordinate females to lactate and suckle the offspring of dominant females without producing offspring themselves, a form of helping or reproductive altruism. *See also* COOPERATIVE BREEDING, EVOLUTION OF. [J.C.D.]

pseudoreplication For statistical purposes in making comparisons between situations it is necessary to make more than one observation per situation to consider if the variation of observations within a situation is more or less than that between situations. So observations are replicated, and it is presumed that the SAMPLING is carried out such that the replicates are independent. When they are not truly independent, pseudoreplication is said to have been effected with the likelihood of underestimating VARIANCE within situations. There are three cases (Fig. P32), all common in survey work.

1 Spatial: 'upstream/downstream' comparisons around, for example, an effluent outlet pipe often involve comparisons of means from replicate samples taken either side of the POINT SOURCE. But this is not true replication because the replicates are nested within each site. Here, true replication would involve making observations on a number of similar 'upstream/downstream' sites.

2 Temporal: samples are often taken through time at the same place to test for 'before and after treatment'. But samples taken through time from the same place are likely to be correlated and therefore not independent. Again, true replication would involve comparison across a number of similar treatment situations.

3 Sacrificial: this occurs when researchers do have a number of the replicates per treatment but then inappropriately analyse units nested within experimental systems or ignore replicates by pooling data.

See also EXPERIMENTAL DESIGN; STATISTICS. [P.C.]

REFERENCES

Hurlbert, S.J. (1984) Pseudoreplication and the design of ecological field experiments. *Ecological Monographs* **42**, 269–299.

Rosenberg, D.M. & Resh, V.H. (1993) *Freshwater Biomonitoring and Benthic Macroinvertebrates*. Chapman & Hall, London.

psilophytes *See* PTERIDOPHYTES, GLOBAL DIVERSITY AND CONSERVATION.

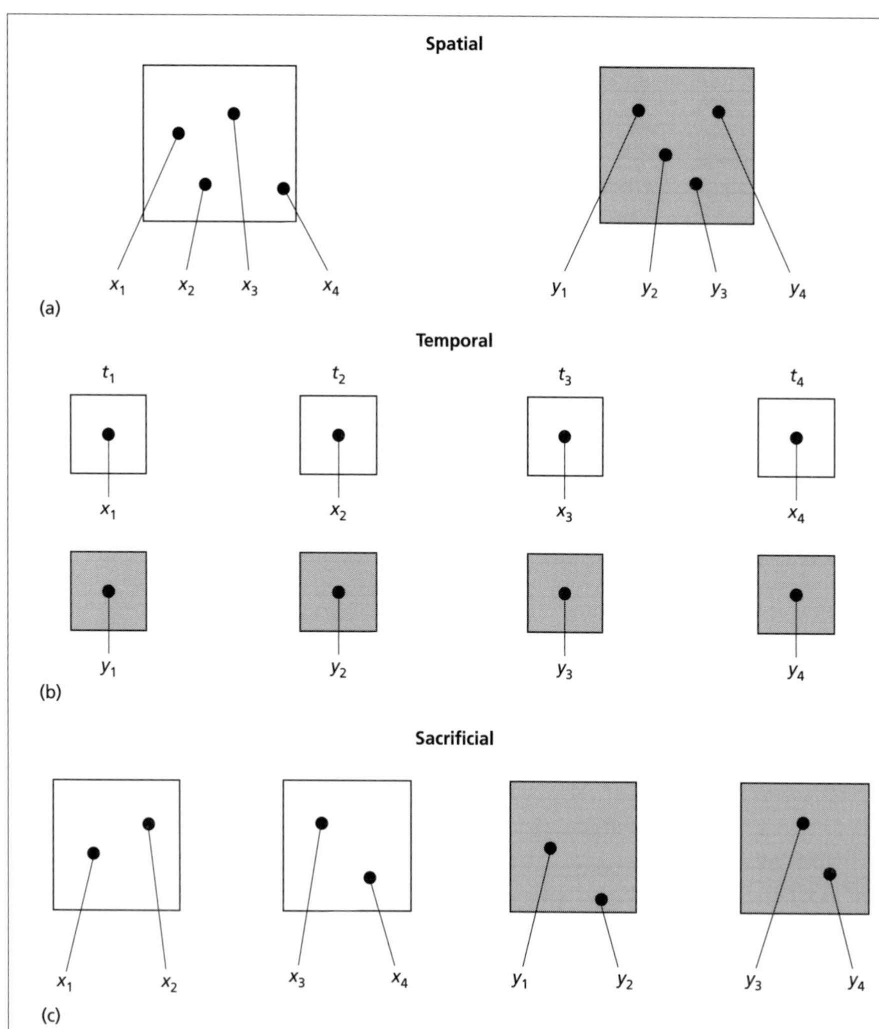

Fig. P32 Examples of three kinds of pseudoreplicated experimental designs. Pseudoreplication results from treating the sample units ($x_1, \ldots, x_4; y_1, \ldots, y_4$) as independent samples taken from each treatment (represented by shaded and unshaded boxes). In (a), the sample units are only nested within each experimental unit; in (b), the same experimental units are sampled repeatedly at consecutive times in such a way that successive observations are not independent of previous observations; in (c), the sample units either are regarded incorrectly as independent replicates or are pooled prior to analysis. (After Hurlbert, S.J. (1984) Pseudoreplication and the design of ecological field experiments. *Ecological Monographs* **42**, 269–299.)

psychrometer An instrument for measuring atmospheric water vapour pressure from the TEMPERATURE recorded on a dry and wet bulb thermometer. If the thermometer bulb is surrounded by wet fabric, EVAPORATION cools the bulb below the air temperature, T_a, to the 'wet bulb temperature', T_w. Then, the vapour pressure (e) may be calculated from the two readings as follows:

$$e = e_{s(T_w)} - \gamma(T_a - T_w)$$

and the RELATIVE HUMIDITY at air temperature is given by:

$$e_a / e_{s(T_a)}$$

where $e_{s(T_w)}$ is the saturated vapour pressure and γ is the psychrometric constant, equal to 66.1 Pa K^{-1} when the wet bulb is well ventilated and the atmospheric pressure is 100 kPa at 20°C. ($\gamma = Pc_p/0.622\lambda$ where P is pressure, c_p is the specific heat of air at constant pressure, and λ is the latent heat of vaporization; see Appendix 3 in Jones 1992.) Alternatively, the vapour pressure may be read from tables or a graph.

There are several practical instruments, the most widespread being the Assmann psychrometer. This comprises a pair of identical, mercury-in-glass thermometers, with bulbs well shielded from irradiation. One of the bulbs is kept wet by a wick. A clockwork or electric motor draws air over the bulbs at a rate of about 4 m s^{-1}. The rapid ventilation ensures the 'full' wet bulb depression ($T_a - T_w$). Wicks must be clean and free of grease or dirt, and the water must be distilled to avoid the build-up of salts on the wick.

Psychrometers may also be used to measure WATER POTENTIALS of materials, and the solute potential of solutions. They have a particular research use in ECOPHYSIOLOGY, where the thermocouple psychrometer has been used to measure leaf or stem water potential. In this case, the sensors are built into a very small chamber which is attached to the leaf or stem, or which receives leaf discs, and the air in the chamber is allowed to reach an equilibrium with the tissue. [J.G.]

REFERENCES

Allen, S.J., Brenner, A.J. & Grace, J. (1994) A low-cost psychrometer for field measurement of atmospheric humidity. *Plant, Cell and Environment* **17**, 219–226.

Jones, H.G. (1992) *Plants and Microclimates*. Cambridge University Press, Cambridge.

***Pteridium aquilinum–Rubus fruticosus* scrub community** *See* SCRUB.

Pteridophytes, global diversity and conservation

The ferns and their allies, or pteridophytes, as presently constituted, are not a natural grouping but represent, today, the end-points of several distinct evolutionary lines of non-seed-forming vascular plants. Five of these lines are considered here.

Rhyniophytes and psilophytes

The Rhyniophyta, now entirely extinct, include the earliest vascular land plants. Known only from fossils their importance today is as a key to understanding the origins of land plants and to the significance of conserving fossil plants and geological sites. The Psilophyta are the living link with the rhyniophyte flora of 450 million years ago, and once formed a dominant element of the Devonian flora. Today the group is reduced to two RELICT genera: *Psilotum* (tropics and Pacific) and *Tmesipterus* (Australasia).

Clubmosses and quillworts

The Lycopododophyta and allies (clubmosses, quillworts) are cosmopolitan in distribution though many are highly intolerant of POLLUTION. In the Carboniferous period the lycopod trees, growing up to 40 m high, would have formed a dominant component of the wet tropical forests, and thereafter became an important element of coal measures. Today the group is reduced to five relict genera, containing about 800 species; all are herbaceous and nowhere common. One genus, *Stylites*, of high Andean lakes, was only discovered in 1954. The largest genera are *Selaginella* (mainly tropical) and *Lycopodium*. With no respect for their ancient lineage, lycopods have been used to stuff upholstery, to dye wool and for dry-burning fireworks.

Horsetails

The Sphenophyta also once constituted one of the dominant tree groups of the Carboniferous, but are today reduced to the single, largely herbaceous genus *Equisetum* (horsetails). The 30 extant species are today cosmopolitan, and can become weeds of economic significance in wet soils. As with the lycopods, so the ancient horsetails have been employed in mundane domestic tasks, for example as a scouring aid to clean pewter. Several species are accumulators and BIOINDICATORS of GOLD.

True ferns

The Pteridophyta, or true ferns, today comprise four distantly related orders totalling some 10 000 species. Most present-day ferns are herbaceous although a small number form trees, notably the 20 m high, palmlike members of *Cyathea*. Although widespread, ferns are relatively scarce in arid zones and have their greatest concentration in the moist tropics; around 12% of the world's fern flora is thought to be located in Papua New Guinea.

The ferns probably reached their maximum distribution, abundance and species numbers in the Carboniferous, though they were still the dominant flora as recently as 70 million years ago, particularly in warm-temperate zones. Their demise—coincident with that of the dinosaurs—in favour of the angiosperms dates from the last epoch of the Cretaceous period, the Maastrichtian. However, one exceptional fern, *Pteridium aquilinum* (bracken, or hog brake), has become one of the world's most cosmopolitan species, having spread to all temperate and subtropical landmasses.

Conservation

The conservation status of ferns has attracted concern for some decades. The claim by one late-19th century horticulturist to have sent to market an entire goods-train load of rootstocks of the now uncommon *Osmonda regalis* (royal fern) illustrates the mass depredation of wild plants that supplied the Victorians' craze for parlour ferns. Many ferns (and their allies) have now been recognized as rare or threatened, at least within the well-studied western European flora. Five per cent of plant species enjoying statutory protection in Great Britain are ferns. *See also* CONSERVATION BIOLOGY. [G.A.F.H.]

---

ptychoparthenogenesis *See* PARTHENOGENESIS.

pulp-mill pollution Pollution caused by effluent from pulp mills. Pulp is the raw material for paper, usually obtained from trees by either physical means (grinding the wood to separate the fibres) and/or chemical means (use of sulphite, sulphate or soda processes to dissolve lignins and release the CELLULOSE fibres). Consequently the effluent can have a high BIOCHEMICAL OXYGEN DEMAND, and contain SUSPENDED SOLIDS with a mix of hazardous chemicals. [P.C.]

puna *See* ALPINE ZONE.

punctuated equilibrium A model of EVOLUTION in which a species is presumed to change rapidly as it comes into existence but quite slowly thereafter. When it subsequently undergoes SPECIATION it remains largely unchanged while producing a daughter species that diverges. Thus, the evolutionary pattern for a given species will be one of long periods of little or no change (STASIS) interrupted by periods of rapid change (punctuation) occurring during times at which speciation occurs.

Proponents of the punctuated equilibrium model of evolution (e.g. N. Eldredge, S.J. Gould, S.M. Stanley) argue that macroevolutionary phenomena (*see* MACROEVOLUTION) cannot be explained by the summed effects of changes of the kind that can be observed within populations (i.e. microevolutionary processes, *see* MICROEVOLUTION). In its extreme form, the punctuationist model views transitions between species as effectively instantaneous and non-adaptive, and that therefore SPECIES SELECTION is the major directing force in evolution. Others have explained the observed evolutionary patterns to arise from normalizing selection for a fixed optimum (appearing as stasis) punctuated by periods of DIRECTIONAL SELECTION — a pattern that is readily explainable by microevolutionary mechanisms.

Supporters of the punctuated equilibrium model of evolution emphasize the importance of developmental CONSTRAINTS that limit the evolution of taxa, whereas those favouring the alternative model (which is sometimes referred to as PHYLETIC GRADUALISM) emphasize the importance of selective forces, such as COEVOLUTION between competitors, between predators and prey, or between hosts and parasites. Much of the evidence for the punctuated equilibrium model derives from examination of morphological features of fossil taxa, and it is important to note that substantial morphological changes can sometimes arise from fairly simple genetic changes in critical control mechanisms of development. [V.F.]

REFERENCE

Futuyma, D.J. (1986) *Evolutionary Biology*, 2nd edn. Sinauer Associates, Sunderland, Mass.

pure strategies *See* EVOLUTIONARILY STABLE STRATEGIES; GAME THEORY.

pycnocline A zone in the water column within which the rate of change of density of seawater with increasing depth reaches a maximum. The pycnocline acts as a barrier to vertical water CIRCULATION and influences the distribution of certain chemicals in the SEA. It is often associated with OXYGEN MINIMUM LAYERS. [V.F.]

pyramid of biomass Pyramids of BIOMASS summarize in a graphical form the STANDING CROPS of organisms in different TROPHIC LEVELS of communities (Fig. P33). Pyramids of biomass give a rough picture of the overall effect of the food-chain relationships. The size of the organisms concerned, which can confuse the picture with pyramids of numbers, has less effect on pyramids of biomass. For example, one large oak tree may provide food for hundreds of thousands of herbivorous insects, but when expressed in terms of mass 30 t of tree may be supporting 50 g of insect being eaten by 5 g of predators.

Inverted pyramids of biomass can still occur where the rate of TURNOVER of a lower trophic level is much higher than that of the levels above. An example is in the English channel (Fig. P33b,c), where the biomass of PHYTOPLANKTON supports a higher biomass of ZOOPLANKTON (herbivores), which supports an even higher biomass of primary carnivores (pelagic fish). *See also* ECOLOGICAL ENERGETICS; ECOLOGICAL PYRAMID; ELTONIAN PYRAMID; PYRAMID OF ENERGY; PYRAMID OF NUMBERS. [M.H.]

REFERENCES

Odum, H.T. (1957) The ecosystem approach in the teaching of ecology illustrated with sample class data. *Ecology* **38**, 531–535.
Price, P.W. (1975) *Insect Ecology*. John Wiley & Sons, New York.
Stilling, P.D. (1992) *Introductory Ecology*. Prentice-Hall, Englewood Cliffs, N.J.

pyramid of energy The ECOLOGICAL PYRAMID which has the most consistently pyramidal shape is the pyramid of energy, which depicts the rates of energy transfer through each TROPHIC LEVEL of a COMMUNITY (Fig. P34). Unlike other forms of ecological pyramid this incorporates time into the diagram because the stacked bars

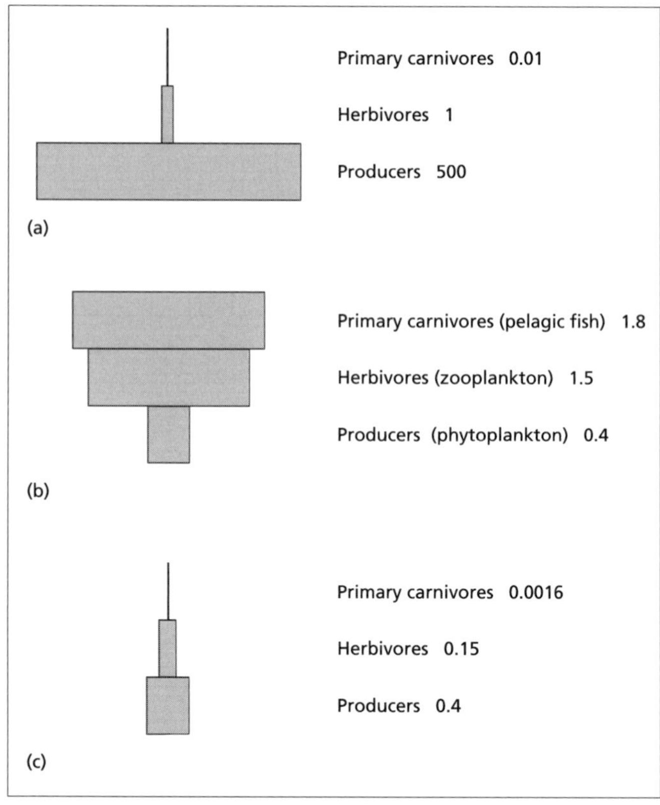

Fig. P33 Biomass pyramids. (a) A pyramid of biomass seen in an old field in Georgia (g dry wt m^{-2}) (Odum 1957). (b) An inverted pyramid of biomass in the English Channel (g dry wt m^{-2}). Phytoplankton has a lower standing crop than the zooplankton, and the zooplankton biomass is less than that for the fish. (c) Production rates of organisms in the English Channel (g dry wt m^{-2} day^{-1}). The inversion in (b) is made possible by the high rate of production by producers in that system. (After Price, P.W. (1975) *Insect Ecology*. John Wiley & Sons, New York; copyright © 1975, reprinted by permission of John Wiley & Sons, Inc.)

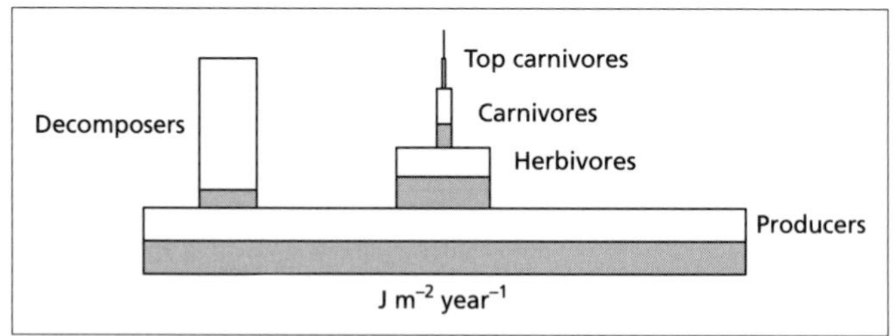

Fig. P34 Energy pyramid for Silver Springs, Florida, on an annual basis. The portion of the total energy flow which is actually fixed as organic biomass, and which is potentially available as food for other populations in the next trophic level, is indicated by the shaded portion of each tier. Note that about half of the total assimilation ends up in the bodies of organisms (the rest being lost as heat or export), except for the decomposer level where per cent loss in respiration is much greater. (After Odum, E.P. (1959) *Fundamentals of Ecology*, 2nd edn. Saunders, Philadelphia, Pa. Data from Odum (1957).)

represent rates of ENERGY FLOW or measures of PRODUCTIVITY per unit time. The reason why this pyramid of energy must retain its upright pyramidal shape is that ecosystems, like all other systems, obey the second law of thermodynamics, so there is always a loss of energy during the processes of energy transformation; thus each higher bar must be shorter than the bars below it.

This realization gave rise to great hopes that a whole new paradigm in ecology had been discovered by applying thermodynamic principles to communities. Unfortunately our understanding of the fundamental concepts of distribution and abundance was not revolutionized by this discovery as much as was originally hoped. *See also* ECOLOGICAL ENERGETICS; ENERGY, LAWS OF THERMODYNAMICS; PYRAMID OF BIOMASS; PYRAMID OF NUMBERS. [M.H.]

REFERENCES

Odum, E.P. (1959) *Fundamentals of Ecology*, 2nd edn. Saunders, Philadelphia, Pa.

Odum, H.T. (1957) The ecosystem approach in the teaching of ecology illustrated with sample class data. *Ecology* **38**, 531–535.

pyramid of numbers A generalization that emerged from the studies of food chains and food webs made by Charles Elton at the beginning of the 20th century was that top predators tend to be large and sparsely distributed, herbivores are often smaller and more common, and primary producers are more abundant still. Elton suggested that if each TROPHIC LEVEL in the COMMUNITY were represented as a block whose length corresponded to the numbers in the trophic level, and if all the blocks were stacked on top of each other with the primary producers at the bottom, the result would be a characteristic pyramid-shaped structure (Fig. P35).

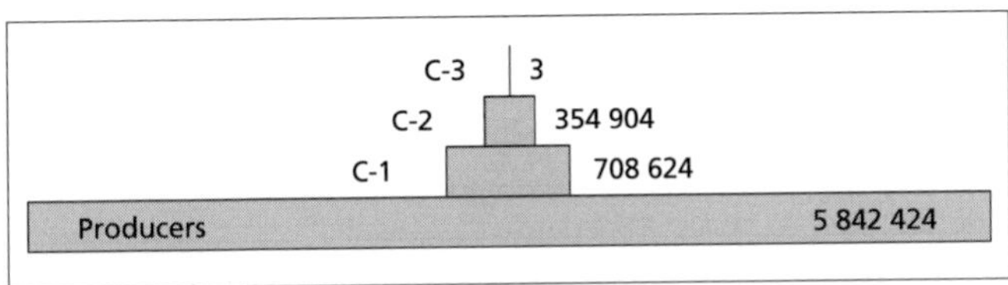

Fig. P35 An example of the pyramid of numbers. The numbers of organisms (exclusive of decomposers) in a bluegrass field are arranged according to trophic levels as follows: producers, green plants; C-1, herbivorous invertebrates; C-2, spiders, ants, predatory beetles, etc.; C-3, birds and moles. (After Odum, E.P. (1959) *Fundamentals of Ecology*, 2nd edn. Saunders, Philadelphia, Pa.)

The form of the pyramid varies widely between communities depending on the size of individuals. While many primary producers, such as PHYTOPLANKTON or grasses, are small relative to the size of herbivores that feed on them, so that it takes a larger number of the plants to support one animal, in other communities this is not the case: for instance, one sycamore tree provides food for hundreds of thousands of aphids. For this reason the pyramids of BIOMASS or energy are more widely used to compare the structures of different communities. *See also* ECOLOGICAL PYRAMID; ELTONIAN PYRAMID; PYRAMID OF BIOMASS; PYRAMID OF ENERGY. [M.H.]

REFERENCE

Odum, E.P. (1959) *Fundamentals of Ecology*, 2nd edn. Saunders, Philadelphia, Pa.

Q

Q_{10} A measure of the TEMPERATURE sensitivity of a chemical reaction or any other process. It was first proposed by J.H. van't Hoff in 1898, who noted that 'the ratio of the rate constant at one temperature to that at another differs little from reaction to reaction'. A Q_{10} can be calculated by measuring the ratio of the reaction rates at one temperature to that 10°C higher or lower. Using natural logarithms the rate of a reaction (R) with a Q_{10} temperature dependence, Q, should have a temperature dependence of:

$$R = A\exp\left(\frac{\ln Q}{10}T\right)$$

where A is a constant and T is the temperature. Despite its widespread use in biology, it is clear that the Q_{10} values of many processes (such as RESPIRATION) are not constant across physiological temperature ranges. [J.Ly.]

REFERENCES

van't Hoff, J.H. (1898) *Lectures on Theoretical and Physical Chemistry.* Part 1. *Chemical Dynamics* (trans. R.A. Lehfeldt), pp. 224–229. Edward Arnold, London.

Lloyd, J. & Taylor, J.A. (1994) On the temperature dependence of soil respiration. *Functional Ecology* **8**, 315–323.

***Q* statistic** *See* DIVERSITY INDICES.

QA *See* QUALITY ASSURANCE AND QUALITY CONTROL.

QC *See* QUALITY ASSURANCE AND QUALITY CONTROL.

QSAR *See* QUANTITATIVE STRUCTURE–ACTIVITY RELATIONSHIP.

QTLs *See* QUANTITATIVE TRAIT LOCI.

quadrat A small area, delimited by a frame, used to study the DISTRIBUTION and ABUNDANCE of species, commonly of VEGETATION but also fauna. A square frame of 1 m^2, which can be subdivided by strings or wires, is often employed in grassland. In WOODLAND or sparse vegetation larger areas are monitored, for example 25 m^2 or more. Quadrat size needs to be related to size and spacing of individuals; optimum size for particular communities may be estimated using the concept of MINIMAL AREA, by finding where the curve of species number plotted against area levels off. SAMPLING error in determining the quantitative composition of the vegetation is reduced by MONITORING more quadrats; normally at least 25 are desirable and often 50 recorded. For some vegetation, rectangular quadrats give more information (VARIANCE is less than that between squares of the same area) and make access easier. Occasionally, circular quadrats (metal rings) are used.

Records may be of the shoot frequency, including all foliage in the quadrat, or rooted frequency (not affected by plant size), including only species rooted in the quadrat. Quadrats may be placed randomly or at regular intervals, for example when used in conjunction with a pinframe (point frame). Above-ground PRODUCTION can be estimated by clipping all vegetation in quadrats at soil level, sorting into species, drying and weighing. For long-term studies of vegetation dynamics, SUCCESSION or the effects of management practices, permanent quadrats may be used. *See also* SAMPLING METHODOLOGY/DEVICES. [A.J.W.]

REFERENCE

Greig-Smith, P. (1983) *Quantitative Plant Ecology*, 3rd edn. Blackwell Scientific Publications, Oxford.

qualitative information (categorical information) Observations or information that can be placed into a number of different categories or classes. For example, the table below shows the number of reed warbler (*Acrocephalus scirpaceus*) nests in different types of habitat.

Reeds	Wet scrub	Dry scrub	Woodland
12	6	2	2

The information (number of nests) is placed into one of several categories (habitat types). Qualitative information can be analysed using non-parametric methods (*see* NON-PARAMETRIC STATISTICS). [B.S.]

quality assurance (QA) and quality control (QC) In ECOTOXICOLOGY/TOXICOLOGY, quality control may refer to procedures incorporated in protocols to reduce the possibility of error. Quality assurance can refer to the independent VALIDATION of the quality-control systems and their operation. Both QA and QC are required by GOOD LABORATORY PRACTICE. [P.C.]

Quantitative genetics

Continuous variation for quantitative TRAITS is caused partly by segregation of ALLELES at multiple loci (QUANTITATIVE TRAIT LOCI, or QTLs) affecting the trait, and partly from variation in the environment. Because the effects of the alleles at one QTL are not large relative to segregating variation at other loci and the environmental variation, classical Mendelian genetic methods for analysing genetic variation at single loci by observing segregation ratios in pedigrees cannot be used to study the genetic basis of quantitative traits. The only property of a quantitative trait that is directly observable on an individual is its measured value, or PHENOTYPIC VALUE (P).

At the level of the POPULATION we can observe the MEAN and VARIANCE of individual phenotypic values. Populations are composed of families, so in addition we can observe the COVARIANCE of phenotypic values between different sorts of relatives. If multiple quantitative traits are measured, the CORRELATION between phenotypic values of the traits can be estimated. The goal of quantitative genetics is to determine the relationship between the observable statistical properties of quantitative traits to the unobserved properties of the genes concerned (e.g. DOMINANCE, EPISTASIS and PLEIOTROPY), and non-genetic influences. The principles of quantitative genetics are often applied in practice to plant and animal breeding and to the study of EVOLUTION, because most traits of agricultural importance (e.g. milk production in dairy cattle; yield of crops), and traits that are acted on by NATURAL SELECTION (e.g. fertility) are quantitative.

Genotypic and breeding values

The basic assumption of quantitative genetics is that an individual's phenotypic value is caused in part by the aggregate effects of all alleles at all loci affecting the trait (G, the GENOTYPIC VALUE), and in part by the effects of the environment (E, the ENVIRONMENTAL DEVIATION). Algebraically, $P = G + E$. Since we do not know what the genotypic constitution of an individual is, we cannot tell by simply observing the PHENOTYPE to what extent a superior phenotypic value is due to a superior GENOTYPE and to what extent it is caused by an optimal environment. If, however, we have information on the mean phenotypic value of offspring of the individual obtained by mating it at random with a number of other individuals in the population, we can determine its BREEDING VALUE (A); i.e. the average value of genes passed on to the progeny.

The breeding value depends on the frequencies of alleles affecting the trait in the population to which the individual is mated as well as the individual's own genotype. The genotypic and breeding values will be the same if the genotypic values of heterozygotes at each locus affecting the trait are exactly intermediate between the homozygotes, and if the total genotypic value is the sum of the genotypic values at all loci affecting the trait; these two conditions constitute additive gene action. With non-additive gene action at a locus (dominance, D) or between loci (epistasis, I):

$$G = A + D + I$$

Components of variance

The variation in phenotypic values among individuals in a population (V_P) is thus partly attributable to variation in their genotypic values (V_G) and partly attributable to variation in environmental deviations (V_E):

$$V_P = V_G + V_E$$

The variation in genotypic values is further partitioned into variation in breeding values (V_A, also called additive genetic variation), DOMINANCE VARIANCE (V_D) and epistatic variance (V_I):

$$V_P = V_A + V_D + V_I + V_E$$

These are causal components of variance.

Heritability

The relative importance of heredity versus environment, or NATURE VERSUS NURTURE, in determining the phenotypic value is called the heritability. The ratio V_G/V_P is the BROAD-SENSE HERITABILITY (H^2), and refers to the extent to which individuals' phenotypes are determined by their genotype. The ratio V_A/V_P is the NARROW-SENSE HERITABILITY (h^2), and refers to the extent to which phenotypes are determined by breeding values. The two meanings are the same in the absence of dominance and epistasis. Heritability in either sense is a ratio, so its numerical value varies between 0 and 1. The interpretation of a heritability of 0 is that all variation in phenotypes is attributable to environmental variation. A heritability of 0 does not mean that genes do not affect the trait, but rather that there is no variation in genotype among individuals in that population, i.e. all individuals are genetically identical. Neither does a heritability of 1 imply that the trait is insensitive to environmental perturbations. Heritability depends on gene frequencies at the loci affecting the trait and the specific environment in which it is measured, and thus applies only to one particular population, place and time.

The heritability of a trait is not a fixed quantity; factors that change gene frequencies, such as selection, GENETIC DRIFT and inbreeding, will change the heritability; also increasing or reducing environmental sources of variation will affect the heritability. Despite these caveats, the narrow-sense heritability of a quantitative trait is in practice its most important property, as it determines the degree of resemblance between relatives and response to selection for the trait.

Resemblance between relatives

Related individuals have genes in common, by definition. The theoretical relationship between relatives (the genetic covariance, *cov*) is a function of the probability that two relatives have the same genes (r) or genotypes (u), which can be worked out easily from pedigrees, and the additive and dominance variance:

$$cov = rV_A + uV_D$$

For example, the genetic covariance between offspring and one parent is $^1/_2 V_A$, since parents pass on $^1/_2$ of their genes to each offspring. Similarly, the genetic covariance between half-sibs, with one parent in common, is $^1/_4 V_A$, and between full sibs, with both parents in common, is $^1/_2 V_A + ^1/_4 V_D$. The genetic covariance between identical (monozygotic) twins is $V_A + V_D$.

If relatives experience common environmental conditions, this can cause members of a family to be more similar than expected on the basis of the genetic relationship. This environmental source of covariance, V_{EC} (common environment), has to be added to the genetic covariance when necessary to give the phenotypic covariance between relatives. The ENVIRONMENTAL VARIANCE that occurs regardless of the relationship between individuals is V_{EW}; therefore:

$$V_E = V_{EC} + V_{EW}$$

In practice, the observed phenotypic relationship between relatives is estimated by the regression or intraclass correlation (from ANALYSIS OF VARIANCE) of phenotypic values. The regressions and correlations are the appropriate covariances divided by the phenotypic variances and are fractions of V_A/V_P. For example, the regression of offspring on one parent is $\frac{1}{2}V_A/V_P$; the correlation of half-sibs is $\frac{1}{4}V_A/V_P$; and the correlation of full sibs is $(\frac{1}{2}V_A + \frac{1}{4}V_D + V_{EC})/V_P$ (the V_{EC} term is included in the latter case because the common maternal environment is difficult to account for experimentally). Because the narrow-sense heritability determines the degree of phenotypic resemblance between relatives, the observed phenotypic regressions or correlations can be used to estimate the heritability. Given estimates of phenotypic covariances among different sorts of relatives with different expected causal components of variance, a full partitioning of variance can be made, separating $V_A : V_D : V_{EC} : V_{EW}$. Reliable estimates of heritability and the various components of variance require very large sample sizes and tend to have large STANDARD ERRORS. Nevertheless, there is a trend across many species for heritabilities of traits associated with viability and fertility (components of FITNESS) to have lower heritabilities than traits that are not obviously closely connected with fitness, because gene frequencies are closer to their limits for traits under strong natural selection.

Selection

The mean value of a quantitative trait can be changed by artificial and natural selection, or by inbreeding and crossbreeding. ARTIFICIAL SELECTION, put simply, means breeding the best to the best. The SELECTION DIFFERENTIAL, S, is the difference between the mean phenotype of the selected parents and of the entire population from which they were selected, and the response to selection, R, is the difference between the mean phenotype of the offspring of the selected parents and the mean of the unselected population. The relationship between R and S is thus:

$$R = h^2S$$

since the heritability predicts the extent to which the phenotypic values of the selected group are determined by their breeding values (i.e. heritability is the regression of breeding value on phenotypic value). The ratio R/S is the REALIZED HERITABILITY. The selection differential depends on both the proportion selected and the PHENOTYPIC VARIANCE of the unselected population. It is often standardized by the phenotypic STANDARD DEVIATION, σ_P, as $i = S/\sigma_P$, where i is the intensity of selection, which depends only on the proportion selected. Then $R = ih^2\sigma_P$. This equation is for response based on phenotypic values of individuals.

In applied animal and plant breeding, information on other relatives (full sibs, progeny) is usually available and can (or must, in the case of traits like milk yield that can only be measured in one sex) be used to predict breeding values and select individuals with the highest predicted breeding values. This is done by putting the phenotypic information from the individual (if available) and/or relatives together in a selection index. The relative weights placed on the information from relatives are the partial regressions of the individual's breeding value with the phenotypic value of the relative, which in turn depends on the heritability of the trait, and the coefficient of relationship (r) and the phenotypic correlation between the individual and the relative, and for the case of family means, the number of individuals in the family.

The equation for response to selection is strictly true for only the first generation of selection, since selection is expected to change the gene frequencies at loci affecting the selected trait and hence the heritability. In this sense it is not so much a prediction equation as it is a description of what has already occurred. In practice, however, the equation is a good predictor of short-term (i.e. 5–10 generations) response to artificial selection. In the longer term, SELECTION RESPONSE is expected to attenuate and reach a limit, as gene frequencies at loci affecting the trait become more extreme. The theoretical maximum response at the limit, R_{MAX}, is when a large number of loci, each with small effects, cause the genetic variation for the trait. It depends on the EFFECTIVE POPULATION SIZE (N_e) as well as the expected response in the first generation:

$$R_{MAX} = 2N_e ih^2\sigma_P$$

In practice, long-term selection response is not predictable. Often, limits are reached but genetic variation is not exhausted, as is the case when natural selection opposes artificial selection. Furthermore, in the very long term there may not be a limit to selection response, as MUTATIONS at the loci affecting the trait continually introduce new variation that can be acted on by selection.

Genetic correlation

Selection for one quantitative trait usually causes a response in other, correlated characters. The magnitude of the correlated response depends on the genetic correlation between the traits, which is a measure of the extent to which the characters are controlled by pleiotropic effects of the same genes. The correlated response to selection also depends on the heritabilities of the two traits. In applied breeding programmes, the goal is usually to improve multiple traits simultaneously. This is done by constructing a multiple trait selection index, where the weights for the different traits are determined by their relative economic values, their heritabilities and their genetic and phenotypic correlations with other traits. The theoretical consequence of continued multiple trait selection is that the genetic correlations between the index traits tend to become negative, as alleles affecting the traits in the same direction are fixed or eliminated by selection. The only genetic variation remaining at the limit should be for alleles with opposite pleiotropic effects on the selected traits (antagonistic pleiotropy), which cause a negative genetic correlation.

Inbreeding

The mean value of a quantitative trait can also be changed by inbreeding and by crossing inbred lines. If one starts with an outbred population and inbreeds many independent sublines from it, two consequences of inbreeding on the mean value of the trait may be observed. The first is that the means of the inbred lines will become differentiated over time as random changes in gene frequency occur, since the mean value of the trait depends on the frequencies of alleles at the QTLs affecting it. For the simple case of additive gene action (no dominance or epistasis), the mean of the entire population of inbred lines will be the same as that of the original outbred population, but the variance between and within inbred lines is $2FV_A$ and $(1-F)V_A$, respectively, where F is the inbreeding coefficient and V_A is the initial GENETIC VARIANCE in the base outbred population. As the inbreeding coefficient approaches 1, all individuals in each inbred line become homozygous at all loci, and the genetic variance within lines consequently approaches 0; all of the initial variance has been redistributed between the lines.

With dominance, the changes of variance between and within lines on inbreeding cannot be predicted as a simple function of the initial genetic variance. Furthermore, with dominance, the mean of the entire population of inbred lines does not equal the mean of the outbred population, but is reduced by an amount that depends on the average genotypic value of heterozygotes (directional dominance), the inbreeding coefficient and the average gene frequencies. The reduction in mean is towards the value of the more recessive alleles; if alleles that increase the value of the trait are dominant, then inbreeding reduces the population mean. Inbreeding depression is commonly observed for traits closely associated with fitness.

The converse of INBREEDING DEPRESSION is HETEROSIS: the restoration of fitness on crossing inbred lines. Although the mean value of trait exhibiting inbreeding depression is expected to be equal to the mean of the outbred population prior to inbreeding if all inbred lines are crossed at random, particular pairs of lines can show heterosis; i.e have means that exceed the mid-parental value. The amount of heterosis in any cross of inbred lines depends on directional dominance and the square of the difference in gene frequency at the loci affecting the trait between the two parental lines. In commercial plant breeding, inbred lines are often produced in order to utilize heterosis for crop improvement.

Fitness

The genetic properties of quantitative traits we observe in natural populations are the consequence of the action of natural selection, GENETIC DRIFT and mutation on them. Quantitative traits can be divided into three categories on the basis of their relationship to fitness. Major components of fitness, like viability and fertility, have high genetic correlations with fitness itself. Natural selection for the composite trait, fitness, will have pushed gene frequencies at loci affecting these traits close to their limits, so heritabilities for fitness components are relatively low. Genetic variance for fitness components is likely to come from alleles with antagonistic pleiotropic effects on viability and fertility, from loci exhibiting true overdominance, and from deleterious mutations maintained at low frequencies by mutation-selection balance. These traits typically show inbreeding depression when recessive deleterious alleles or overdominant alleles are forced to HOMOZYGOSITY. Because gene frequencies in equilibrium natural populations are those that produce the highest fitness, artificial selection lines usually exhibit a loss of fitness as a correlated response, as the ALLELE FREQUENCIES at loci affecting the selected trait are changed from their equilibrium frequencies.

At the opposite end of the spectrum are truly neutral or nearly neutral traits, with genetic correlations with fitness close to zero (i.e. variation in trait phenotypes is not associated with variation in fitness). These traits have high heritabilities and show little inbreeding depression. Genetic variance for neutral traits is maintained by a balance between mutation and genetic drift.

Most quantitative traits fall in the third category, and show an intermediate optimum with respect to fitness, whereby individuals with intermediate values of the trait are more fit than individuals that deviate from the optimum in either direction. In this case the genetic correlation between fitness and the absolute value of the trait is high. The consequence of this relationship to fitness on the genetic variance of the trait depends on the cause of the intermediate optimum. If selection operates on the trait value directly ('real' STABILIZING SELECTION), genetic variance will be reduced. Alternatively, loci affecting extreme values of the trait may have deleterious pleiotropic effects on fitness components, so selection is not directly on the trait ('apparent' stabilizing selection). In this case one cannot predict the amount and nature of the genetic variance maintained without knowing what components of fitness are being selected and how the trait is correlated with these components.

The extent to which variation for quantitative traits under real or apparent stabilizing selection can be maintained by a balance between mutation and selection is a subject of current debate. Unequivocal answers can only be obtained when we know mutation rates and allelic effects on the trait and fitness at individual QTLs. *Cf.* ECOLOGICAL GENETICS. *See also* MODERN SYNTHESIS; NEO-DARWINISM; POPULATION GENETICS. [T.F.C.M.]

REFERENCES

Bulmer, M.B. (1980) *The Mathematical Theory of Quantitative Genetics*. Oxford University Press, Oxford.

Falconer, D.S. & Mackay, T.F.C. (1996) *Introduction to Quantitative Genetics*, 4th edn. Longman Higher Education, Harlow.

quantitative information Observations or information that consists of measurements, counts, scores, etc., that take the form of a numerical assessment; for example, the length of a fly's wing, mass of a fledgling bird, POPULATION SIZE of trout in a stream, or IQ scores of undergraduates. Quantitative information can be analysed using parametric or non-parametric methods depending on how the data are distributed. *See also* NON-PARAMETRIC STATISTICS; PARAMETRIC STATISTICS. [B.S.]

quantitative structure–activity relationship (QSAR) An equation, often derived by regression analysis, that relates environmental EFFECT and FATE properties of chemicals to measures and indices of their physicochemical and structural properties. For example, with chemicals that cause non-specific narcosis there is often a good correlation between TOXICITY and OCTANOL–WATER PARTITION COEFFICIENT. Hence the latter can be used as a predictor of the former. This is therefore a short-cut technique that avoids the use of TOXICITY TESTS. More complex toxicological effects require more complex equations, which may not provide reliable predictions. *See* *also* ECOTOXICITY; ECOTOXICOLOGY; EXISTING CHEMICALS; NEW CHEMICALS; STRUCTURE–ACTIVITY RELATIONSHIP. [P.C.]

quantitative trait (quantitative character, metric trait) A TRAIT (character) for which variation between individuals in a population is continuous, and often normally distributed. Most morphological, physiological and behavioural characters that distinguish one individual from another are of this sort; for example, height, blood pressure, longevity. Threshold characters are a special class of quantitative trait for which variation appears discontinuous (e.g. affected or not affected with a disease) but that exhibit continuous variation on an underlying scale of liability. Continuous variation of quantitative traits is caused by segregation at many loci (QUANTITATIVE TRAIT LOCI) affecting the trait, as well as environmental effects. [T.F.C.M.]

quantitative trait loci (QTLs) The genes (also called polygenes), or loci, at which polymorphic ALLELES are responsible for genetic variation of quantitative TRAITS within or between populations and species. Occasionally, genes with large effects on quantitative traits have been discovered in natural populations or ARTIFICIAL SELECTION lines, but these are often recessive and have deleterious pleiotropic effects on FITNESS, and so contribute little to GENETIC VARIANCE for the trait in the natural population. By definition, most quantitative genetic variation is caused by alleles with effects that are too small to be followed by Mendelian segregation analysis. Therefore, QTLs are mapped by linkage to marker loci that can be followed by segregation analysis in crosses of lines, or populations, that differ for mean values of the QUANTITATIVE TRAIT and are also polymorphic for the marker loci.

Recently, the availability of abundant polymorphic molecular markers has facilitated QTL mapping in many species. These methods require large sample sizes, and identify genomic fragments associated with the trait that potentially contain multiple loci, not individual loci in the genetic sense.

In general, quantitative traits are determined by multiple QTLs. The number of QTLs found is a lower limit, since increasing the size of the experiment will always increase the power to detect more QTLs, with smaller effects. Furthermore, any one cross is limited to finding the QTLs that are polymorphic in the parental lines used, which will be only a subset of the QTLs that affect the trait.

Not all QTLs have equal effects, as is generally assumed when modelling quantitative traits theoretically. A few QTLs typically account for most of the difference in mean between the strains, with larger numbers of QTLs with smaller effects accounting for the remainder of the difference. ('Large' effects, however, are still too small to be detected in a segregating population.)

Most QTLs have additive effects on the trait, but the degree of DOMINANCE does vary throughout the range, from strictly dominant/recessive, to overdominant. Epistatic interactions between QTLs are found in experiments that are large enough to detect them, even for traits thought to be strictly additive.

Two approaches are possible to resolve QTLs to the level of genetic loci. The first is positional cloning, which requires fine-scale meiotic mapping. This method becomes increasingly difficult to apply as the effect of the QTL diminishes. The second is to identify candidate loci known to be functionally related to the trait, and search for ASSOCIATIONS of the trait PHENOTYPE with molecular polymorphisms at the candidate loci. This approach assumes that QTLs are not a special category of locus that are constrained to having only small effects. Rather, candidate loci are assumed to have a distribution of mutant effects, some large enough to qualify as a major MUTATION and hence identify the gene, whereas others have smaller effects on the trait and segregate in natural populations, contributing to quantitative genetic variation. This method has been successful in identifying loci affecting quantitative genetic variation for bristle number in *Drosophila*. *See also* QUANTITATIVE GENETICS. [T.F.C.M.]

REFERENCES

Falconer, D.S. & Mackay, T.F.C. (1996) *Introduction to Quantitative Genetics*, 4th edn. Longman Higher Education, Harlow.

Tanksley, S.D. (1993) Mapping polygenes. *Annual Review of Genetics* **27**, 205–233.

quantum requirement In photochemistry, the number of photons absorbed by a system can be usefully compared with the number of moles of the products of the reaction. For example, it has been found that under ideal conditions, met only in the laboratory, a minimum of 8 photons are needed for an algal cell to fix by PHOTOSYNTHESIS one molecule of CARBON DIOXIDE (CO_2) into glucose ($C_6H_{12}O_6$). This represents a quantum requirement of 8. Another way to express this result is to divide the energy fixed as glucose by the energy absorbed as photons (i.e. energy out/energy in). The average energy in a PHOTON of sunlight is about 220 kJ per mol photon; and the free energy content of glucose is about 480 kJ per mol C, so the maximal EFFICIENCY of the process (energy out/energy in) must be 27%. *See also* LIGHT. [J.G.]

quantum sensor A sensor (transducer) for LIGHT or other electromagnetic RADIATION (e.g. ULTRAVIOLET RADIATION) that is almost equally sensitive to all photons (quanta) within its spectral sensitivity range. Usually the term is applied to sensors for photosynthetically active radiation (PAR), i.e. radiation in the spectral band 400–700 nm. Quantum sensors are usually designed to either register light falling on a flat surface (so-called irradiance sensors, or cosine-corrected sensors), or to be equally sensitive in all directions and register fluence rate.

The distinction between irradiance (sometimes called vectorial irradiance) and fluence rate (sometimes called scalar irradiance) is important, since the two do not vary proportionally in natural daylight. Using a special technique it is possible to measure fluence rate using a quantum sensor for irradiance. [L.O.B.]

REFERENCE

Björn, L.O. (1995) Estimation of fluence rate from irradiance measurements with a cosine corrected sensor. *Journal of Photochemistry and Photobiology. B. Biology* **29**, 179–183.

quantum speciation The idea that species formation takes place by a single jump. The relation between MICROEVOLUTION, which improves adaptedness, and species formation, resulting in REPRODUCTIVE ISOLATION, has been debated since genetics was first studied. H. de Vries argued in 1901 that species were formed by single macromutations, distinct from micromutations, producing what was then called a 'HOPEFUL MONSTER'. This idea was extended in the 1930s by R. Goldschmidt. It is not true as originally stated, but chromosome reorganization has replaced MACROMUTATION as one mechanism proposed for reproductive isolation. Suitable conditions are most likely to be found in very small populations. Grant (1971) defines quantum speciation as 'the budding off of a new and very different daughter species from a semi-isolated peripheral population of the ancestral species in a cross fertilizing organism'. Lewis (1966) has used the term 'saltatory', and White (1978) the term 'stasipatric' to describe related ideas. *See also* GENETIC REVOLUTION. [L.M.C.]

REFERENCES

Grant, V. (1971) *Plant Speciation*. Columbia University Press, New York.

Lewis, H. (1966) Speciation in flowering plants. *Science* **152**, 167–172.

White, M.J.D. (1978) *Modes of Speciation*. Freeman, San Francisco, Calif.

quartile Any of the three points that divide a DISTRIBUTION or set of observations into quarters, or four equal parts. The first quartile (Q1, or lower quartile) has 25% of observations below it. The second quartile is the median (Q2, or middle quartile) and the third quartile (Q3, or upper quartile) has 75% of observations below it. Some rounding is necessary when the number of observations is not exactly divisible by 4. *See also* MEDIAN (SECOND QUARTILE); PERCENTILE. [B.S.]

quasi-social species *See* SOCIALITY, TYPES OF.

quillworts *See* PTERIDOPHYTES, GLOBAL DIVERSITY AND CONSERVATION.

q_x In a LIFE TABLE the age-specific or stage-specific MORTALITY RATE; the fraction of the proportion of individuals that survive to age x that die in age-class x. It is equivalent to the average probability of an individual dying at that age, and is calculated as d_x/l_x. High values of q_x indicate the age or stage with the most intense mortalities but, unlike d_x and k_x, q_x is not additive over stages. [A.J.D.]

R

r- and K-selection The terms *r*- and *K*-selection are derived from the LOGISTIC EQUATION:

$$N(t) = \frac{K}{1 + e^{c-rt}}$$

where *N(t)* is POPULATION SIZE at time *t*, *r* is the rate of increase, *c* is a constant, and *K* is the CARRYING CAPACITY of the environment. Selection that maximizes the rate of increase is termed *r*-selection. It has been hypothesized that there is a TRADE-OFF between selection for traits that maximize *r* and those that maximize *K*. However, evidence does not support such a hypothesis. *See also* CLASSIFICATION; EVOLUTIONARY OPTIMIZATION; LIFE-HISTORY EVOLUTION; OPTIMALITY MODELS; REPRODUCTIVE EFFORT; SELECTION PRESSURE. [D.A.R.]

r-selection *See* K-SELECTION; r- AND K-SELECTION

race formation *See* SYMPATRIC SPECIATION.

radar (remote sensing) A microwave remote sensor used to derive images of the Earth's surface. The radar senses the terrain to the side of the aircraft or satellite track. It does this by pulsing out electromagnetic RADIATION of long (up to radio) wavelengths and recording first the strength of the pulse return to the sensor, to detect objects, and second the time it takes for the pulse to return, to give the range of the objects from the sensor. (The name 'radar' is the ACRONYM of these functions: radio detection and ranging.) These pulses are emitted at right angles to the direction of travel and pulse lines can be built up to form an image. A SYNTHETIC APERTURE RADAR (SAR) is the commonly used radar for environmental applications. [P.J.C.]

radian *See* SI UNITS.

Radiation

Radiation is the transfer of energy by rapid oscillations of electromagnetic fields. Electromagnetic radiation travels in straight lines at very high speed (3.0×10^8 m s^{-1}), and is characterized by its wavelength, flux density and direction. In normal conditions, most of the radiant transfer of energy in the BIOSPHERE is in the waveband 0.2–100 μm, this being a small part of the ELECTROMAGNETIC SPECTRUM.

Laws of radiation

Radiation emitted by the Sun and the planetary surface follows a small number of well-understood RADIATION LAWS. Both the Sun and the land behave as nearly perfect radiators, emitting energy at a flux density (F) which is proportional to the fourth power of their absolute temperature (T):

$$F = \varepsilon\sigma T^4$$

where σ is the constant of proportionality, called the Stefan–Boltzmann constant ($= 5.67 \times 10^{-8}$ W m^{-2} K^{-4}) and ε is the emissivity, a property of the surface. For a perfect radiator $\varepsilon = 1.0$, and most natural surfaces are almost perfect radiators with an emissivity between 0.9 and 1.0.

The peak wavelength (λ_{max}) is found from Wien's displacement law:

$$\lambda_{max} = 2.88 \times 10^{-3}/T$$

where λ_{max} is in metres and T is the temperature in kelvin.

Solar radiation

Solar radiation varies geographically and seasonally, being influenced by cloud and by the Earth–Sun geometry (Fig. R1). Radiation from the Sun is distributed in the wavelength range 0.2–5.0 μm, and has a flux density above the Earth's ATMOSPHERE of 1350 W m^{-2}, a figure known as the solar constant. The spectral distribution outside the atmosphere is smooth and follows the distribution expected from a perfect radiator with a surface temperature of 5800°C (Fig. R2). Even with a clear sky and the Sun at the zenith, much of the radiation is absorbed on passage through the atmosphere by molecular absorption and scattering, reducing the flux to less than solar 1000 W m^{-2} at the Earth's surface.

The distribution of solar radiation between major bands is:

- ultraviolet (<0.380 μm): 5.779%;
- visible light (0.380–0.765 μm): 48.572%;
- infrared (>0.765 μm): 45.649%.

Colours perceived by the human eye are:

- violet: 0.380–0.455 μm;
- blue: 0.455–0.492 μm;
- green: 0.492–0.577 μm;
- yellow: 0.577–0.597 μm;
- orange: 0.597–0.622 μm;
- red: 0.622–0.765 μm.

The shape of the solar spectrum is modified by molecular absorp-

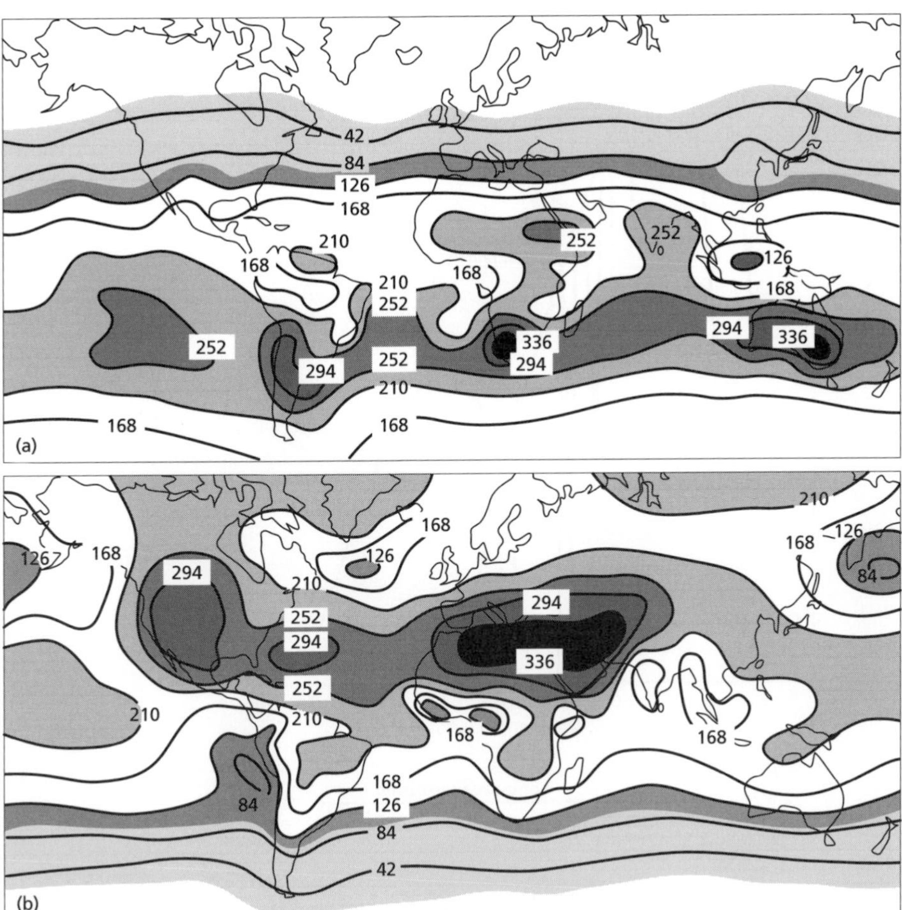

Fig. R1 Global patterns of monthly mean solar irradiance in (a) January and (b) July. The numbers show the irradiance in units of W m^{-2}. (After Jones, H.G. (1992) *Plants and Microclimate*. Cambridge University Press, Cambridge.)

tion by some of the gases in the atmosphere. This is especially significant at low solar angles when the Sun's rays have a longer path through the atmosphere. The ultraviolet component is absorbed by OZONE in the STRATOSPHERE, and ozone continues to show weak absorption throughout the solar spectrum. Nitrogen gas (N_2) absorbs weakly at 0.41 and 0.68 μm; oxygen gas (O_2) absorbs at 0.76 and 1.27 μm, and water vapour absorbs at 0.72, 0.81, 2.74, 4.25 and 6.3 μm. The entire waveband of solar radiation is often termed shortwave radiation (Fig. R2).

At the Earth's surface, scattering is a more important cause of loss of irradiance than molecular absorption. The highly polarized blue light from the clear sky is caused by the scattering of short wavelengths by molecules and small (<0.1 μm) particles, termed Rayleigh scattering. The white or grey appearance of CLOUDS is caused by the scattering of all wavelengths by water droplets, AEROSOLS and larger particles, and is termed Mie scattering.

Small variations in the solar constant are caused by changes in the Earth–Sun geometry and sunspots (*see* SUNSPOT CYCLES). Major volcanoes, such as Pinatubo in 1992, are capable of injecting sufficient DUST into the atmosphere to reduce the solar irradiation over large areas of the globe by up to 5%.

Solar radiation is measured routinely at major meteorological stations, sometimes using two solarimeters to separate the direct and diffuse components. When the Sun is shining, the diffuse component, measured as energy incident on a horizontal surface, is typically only one-tenth that of the direct beam of the Sun. Averaged over a year, the diffuse radiation is very important in cloudy climates. Solar radiation may also be estimated from data on 'hours of sunshine' and 'clear-sky' values may be obtained from tables compiled for various latitudes and atmospheric conditions.

Terrestrial radiation

Terrestrial radiation is the term given to the radiation emitted by the Earth. The spectral distribution follows the same law as the radiation emitted by the Sun, but as the Earth is much cooler, much longer wavelengths are involved and the flux density is much less. This radiation is distributed in the infrared region of the electromagnetic spectrum, 5–25 μm. It is strongly absorbed by various atmospheric gases, particularly water and CARBON DIOXIDE (CO_2), but also those present in the atmosphere as TRACE GASES such as ozone (O_3), CARBON MONOXIDE (CO), METHANE (CH_4) and nitrous oxide (N_2O). It is this absorption of terrestrial radiation by gases that is responsible for the greenhouse effect. Without the greenhouse effect, the Earth's surface would be considerably cooler; but if more CO_2 is discharged into the atmosphere the climate may be expected to warm. Terrestrial radiation is also called longwave, in contrast to shortwave (solar radiation). Terrestrial radiation may be sensed

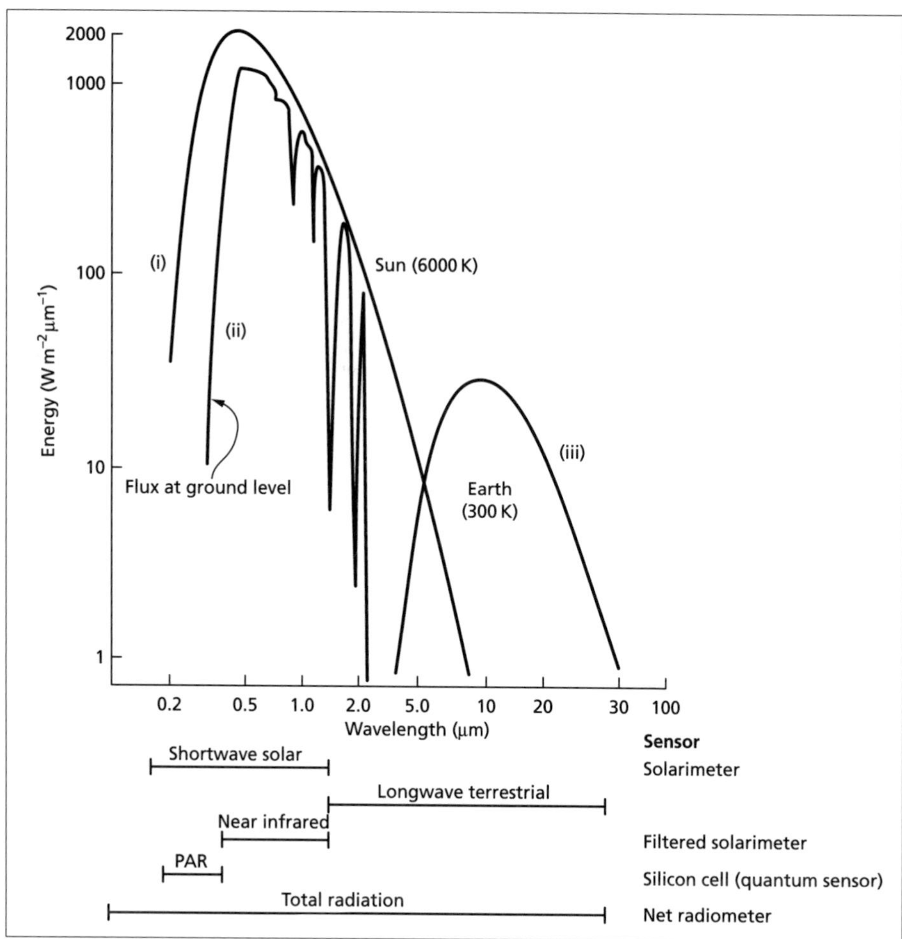

Fig. R2 The spectral energy distribution of: (i) the solar flux outside the Earth's atmosphere; (ii) the solar flux at ground level after attenuation by gases in the atmosphere; and (iii) the flux of radiation emitted by the Earth's surface (terrestrial flux). Listed at the bottom of the figure are the sensors used to measure components of the solar and terrestrial fluxes. PAR, photosynthetically active radiation. (After Grace, J.G. (1983) *Plant–Atmosphere Relationships*. Chapman & Hall, London.)

from satellites, and because of the relationship between surface temperature and radiation flux (Stefan's law), it is possible to measure ground temperature from space or from an aircraft.

Longwave radiation also emanates from the sky, which can be thought of as a cold surface. The apparent temperature of a clear sky is about 20 degrees lower than that of the ground. Thus, on cold nights, radiation flows from the ground to the sky, and the ground consequently cools.

Radiation balance

The net radiation to the land surface can be expressed in terms of the shortwave and longwave components:

$$R = S(1-\beta) + L_d - L_u$$

where R is the net radiation, β is the ALBEDO, S is the shortwave, L_d is the downward longwave and L_u is the upward longwave.

On sunny days S is large compared to the longwave component. In practice, a near-linear relationship between shortwave and net radiation is observed, the slope being strongly dependent on the albedo. Consider a typical case of a clear sky radiation over a forest in the tropics, with $S = 1000\ \mathrm{W\,m^{-2}}$, $\beta = 0.15$, surface temperature = 25°C and apparent sky temperature = 5°C. From the above equation we can expect R to be $742\ \mathrm{W\,m^{-2}}$. Now replace the forest by a farm where $\beta = 0.25$, and assuming the same surface and sky temperatures we have $R = 642\ \mathrm{W\,m^{-2}}$.

At night, the net radiation balance becomes negative, by tens of $\mathrm{W\,m^{-2}}$.

To determine the radiation balance of a surface such as vegetation, it is not necessary to measure separately the components S, L_u and L_d. Instead, net radiation may be measured directly with a NET RADIOMETER. *See also* IONIZING RADIATION; LIGHT. [J.G.]

REFERENCE

Coulson, K.P. (1975) *Solar and Terrestrial Radiation*. Academic Press, London.

radiation dose For ACCIDENTS, such as that at Chernobyl in 1986, the personal risk is expressed in terms of detrimental effects, and long-term effects which are regarded as stochastic, which have been accurately observed at Hiroshima and Nagasaki. The setting of dose limits for civil purposes involves assumptions about the equivalence of a dose received instantaneously and the same dose over a long period at a lower intensity.

The absorbed dose, measured in energy per unit of body mass ($\mathrm{J\,kg^{-1}}$), depends also on the type of RADIATION (α-, β-, γ-, X-), and the organ or tissue irradiated. This is allowed for by the use of weighting factors from which the PROBABILITY of detriment is computed, and may include hereditary and age effects which are particularly relevant in respect of length of loss of life in the case of fatal cancers. *See also* BECQUEREL; CURIE; ROENTGEN; SIEVERT. [R.S.S.]

radiation laws Physical laws for electromagnetic RADIATION. Several of them (Wien's radiation law, Rayleigh's radiation law, Stefan–Bolzmann's law and Wien's displacement law) can be regarded as special cases or derivations of the more general Planck's radiation law, which states that the heat radiation from a black body of absolute temperature T and unit surface is given by:

$$dE = 8\pi h\nu^3 / c^3 / \left[\exp(h\nu/kT) - 1\right] d\nu$$

where ν is the frequency (LIGHT velocity divided by wavelength), dE is the energy radiated in the frequency interval $d\nu$, h is the Planck constant (6.626176×10^{-34} J s), and k is Boltzmann's constant (1.380662×10^{-23} J K^{-1}). The law may be stated in several other mathematical forms.

Other important laws concerning electromagnetic radiation are Einstein's law, which states that in a photochemical reaction one PHOTON interacts with exactly one molecule, and Kirchhoff's radiation law, which states that the spectral emissivity of a certain surface at a certain frequency is equal to the spectral absorptivity at that frequency. Although not having a specific name, the relation (derived by Max Planck) that the energy of a photon is proportional to the frequency (the proportionality factor being Planck's constant) is another important radiation law. The frequency is the velocity of propagation divided by the wavelength. [L.O.B.]

radiation pollution IONIZING RADIATION consists of 'fast-flying' particles or WAVES that come naturally from the nuclei of unstable atoms or from nuclear facilities. Human beings and ecological systems are exposed daily to a certain amount of low-level radiation, or background radiation. Humans may also be exposed to radiation from medical uses. There may also be EXPOSURE due to accidental releases from the nuclear industry.

The major effects on biological systems are impacts on and consequent denaturing of macromolecules, leading in genetic material to MUTATIONS that may cause cancer. These are of considerable importance for human health, but their ecological importance is more problematic. Mutations that occur in the somatic cells of individuals are likely to be of little consequence unless they cause increased MORTALITY, but this mortality (e.g. due to tumours in a few individuals) may not have population consequences. Similarly, impairment of reproductive output would only have effects at a population level if it were appreciable. There is also the possibility of irradiation increasing the chances of more rapid evolution by enhancing the rate of mutation and hence the chances of beneficial mutation being produced. But POPULATION GENETICS would suggest that either mutation rates have to be very high or POPULATION SIZES small for this to be a likely prospect. *See also* RADIOACTIVE FALLOUT; RADIOACTIVE WASTE; RADON. [P.C.]

radical In chemistry, the term radical formerly referred to a group of atoms which maintained its identity in reactions involving the rest of the molecule. Now the term radical or free radical is applied to species which have one or more free valencies. Thus the atom at the centre of the radical does not have its full complement of electrons and so radicals are often reactive species. Examples of radicals include single atoms (e.g. •Cl, •Br) and those formed by the rupture of chemical bonds in organic compounds; for example, peroxides (ROOR) decompose on heating to give a pair of radicals (2R–O•). Radicals formed from chlorinated compounds in the ATMOSPHERE are important in the depletion of OZONE. Radicals are also industrial CATALYSTS in polymerization reactions. Radicals have a wide range of stabilities; examples of more transient free radicals include $\bullet C_2H_5$ and •OH. [M.C.G.]

radioactive fallout Radioactive fallout, mainly from atmospheric testing of nuclear weapons in the 1950s and 1960s, has resulted in worldwide low-level radioactive CONTAMINATION. Maximum atmospheric releases of radionuclides from weapons tests occurred in 1963, after which inputs were much smaller following the introduction of the atmospheric test-ban treaty. Weapons tests from 1954 to 1970, involving FISSION of some 10^{28} atoms of URANIUM or PLUTONIUM, released approximately 7.8×10^5 TBq of ^{137}Cs, 1.3×10^6 TBq of ^{90}Sr, 8.9×10^2 TBq of ^{239,240}Pu and 2.1×10^2 TBq of ^{238}Pu to the atmosphere. In 1964 the atmospheric burn-up of a US satellite, containing a system for nuclear auxiliary power (SNAP), released 6.3×10^2 TBq (1 kg) of ^{238}Pu, giving a composite ^{238}Pu/^{239,240}Pu activity ratio for fallout (weapons testing plus SNAP) of about 0.05 in the Northern hemisphere. Inventories of fallout radionuclides in soils depend upon rainfall, with values for the UK in 1977 being 3.7×10^3 Bq m^{-2} of ^{137}Cs, 1.3×10^3 Bq m^{-2} of ^{90}Sr and 68 Bq m^{-2} of ^{239,240}Pu per metre of rainfall.

Significant radioactive fallout also occurred as a result of the Chernobyl reactor disaster in April 1986 (*see* CHERNOBYL DISASTER), when about 2×10^{18} Bq of volatile fission and activation products, including about 4×10^{16} Bq of ^{137}Cs, were released to the atmosphere. Chernobyl fallout had a ^{134}Cs/^{137}Cs activity ratio of about 0.5, distinguishing it from weapons-testing fallout, which did not contain ^{134}Cs. The Chernobyl release of radionuclides from a POINT SOURCE resulted in significantly higher contamination levels throughout much of Europe and Scandinavia than those from weapons-testing fallout. Weapons-testing and Chernobyl-fallout radionuclides provide useful chronological markers in accumulating sediments. *See also* RADIATION POLLUTION; RADIONUCLIDE. [A.B.M.]

radioactive waste Radioactive WASTE is classified as high-, intermediate- or low-level depending upon the activity of RADIONUCLIDES present. Intermediate- and high-level radioactive wastes arise mainly from the nuclear industry in the form of spent fuel rods, highly irradiated reactor components and reprocessing wastes. The nuclear industry is also the dominant producer of low-level radioactive waste, but smaller quantities arise from other sources including hospitals, universities and industrial facilities. Selective concentration of natural decay series radionuclides generates radioactive waste in some non-nuclear industries (e.g. oil extraction, coal combustion and metal refining). RADIUM- or TRITIUM-containing wastes can also occur in the environment from the disposal of manufactured luminescent materials.

Gaseous and liquid low-level radioactive wastes are disposed of by authorized discharges to the environment, with the aim of dispersion in the atmosphere or aquatic environment to give dilution to levels that present a negligible radiological HAZARD. Solid low-level wastes are disposed of by burial in near-surface engineered repositories.

Designs for intermediate- and high-level waste repositories are based upon the concept of isolating the radionuclides for a sufficient length of time to allow the activity to decay to a non-hazardous level. In such designs, it is envisaged that the engineered section of the repository (known as the near field) will contain a series of barriers which will, for an extended period of time, prevent ingress of groundwater and subsequent releases of radionuclides. A typical design of this type would be to place the vitrified waste in a stainless-steel container which would be surrounded by bentonite clay and concrete. Upon the eventual failure of the near-field barriers, radionuclides will migrate through the host rock (known as the far field) and the rock type and repository depth would be selected to give long-term stability and a low rate of radionuclide migration. *See also* RADIATION POLLUTION. [A.B.M.]

REFERENCE

Chapman, N.A. & McKinley, I.G. (1987) *The Geological Disposal of Radioactive Waste*. John Wiley & Sons, Chichester.

radioactivity Radioactivity is a spontaneous process in which an unstable nucleus undergoes a transition to produce a new nucleus of lower internal energy plus radiation. One such transition is called a decay or disintegration, and the number of decay events per unit time is called the activity (A). The SI UNIT of activity is the BECQUEREL (Bq), which is a decay rate of one disintegration per second. The former unit of activity, the CURIE (Ci), corresponds to a decay rate of 3.7×10^{10} disintegrations per second (the activity of 1 g of ^{226}Ra). The main radioactive decay processes are as follows.

- α-Decay: emission of an α-PARTICLE (mass 4 amu, charge +2), giving a product nuclide two places to the left in the PERIODIC TABLE.
- β^--Decay: emission of a β^--particle (i.e. an electron) and a neutrino (ν), giving a product nuclide one place to the right in the periodic table.
- β^+-Decay: emission of a β^+-particle (i.e. a positron) and a neutrino, giving a product nuclide one place to the left in the periodic table.
- γ-Emission: de-excitation of a nucleus by emission of photons of electromagnetic radiation known as γ-radiation or γ-RAYS.
- Electron capture: capture of a 1s electron by the nucleus, giving a product one place to the left in the periodic table.
- Spontaneous NUCLEAR FISSION: splitting of the nucleus into two lighter nuclei.

For a pure, singly decaying RADIONUCLIDE, the activity is related to the number of atoms present (N) by:

$$A = \lambda N$$

where λ is a constant called the decay constant. The activity of a given radionuclide will halve with a regular periodicity, and the time required for the activity to halve is called the half-life ($t_{1/2}$). The half-life and the decay constant are related by:

$$t_{1/2} = \ln 2/\lambda$$

Radioactive decay follows first-order kinetics, so the activity A_t, at time t is related to the activity A_0 at time zero by:

$$A_t = A_0 e^{-\lambda t}$$

See also β-PARTICLE; GRAY; IONIZING RADIATION; SIEVERT. [A.B.M.]

REFERENCES

Friedlander, G., Kennedy, J.W., Macias, E.S. & Miller, J.M. (1981) *Nuclear and Radiochemistry*. John Wiley & Sons, New York.

Keller, C. (1988) *Radiochemistry*. Ellis Horwood, Chichester.

radiocarbon dating *See* DATING, RADIOCARBON.

radiometer A sensor for measuring the flux radiation in W m^{-2}. The most basic instrument is the SOLARIMETER, for measuring SOLAR RADIATION. It consists of a flat, blackened thermopile which warms up when radiation strikes it. To avoid convective loss of heat, the thermopile is covered with two concentric, optically ground hemispheres of quartz. The particular design used throughout Europe is the Moll–Gorczynski pyranometer, manufactered in The Netherlands by Kipp & Zonen, and called the Kipp solarimeter. The conspicuous reflective disc that surrounds the sensor itself serves to minimize solar heating of the underside of the instrument, which would otherwise adversely affect the temperature compensation. The hemispheres are concentric 'double glazing' to minimize the influence of the wind, which would otherwise transfer heat between the thermopile and the atmosphere. The signal produced by the solarimeter is a small voltage (millivolts) generated by a series of thermojunctions, and the response time of the instrument is about 10 s.

There are many other types of radiometer, well described in standard texts to which the reader is referred: Robinson (1966), Coulson (1975) and Iqbal (1983). *See also* RADIATION. [J.G.]

REFERENCES

Coulson, K.P. (1975) *Solar and Terrestrial Radiation*. Academic Press, London.

Iqbal, M. (1983) *An Introduction to Solar Radiation*. Academic Press, New York.

Robinson, N. (1966) *Solar Radiation*. Elsevier, Amsterdam.

Radionuclide

A radionuclide is a nuclide which has an unstable combination of neutrons and protons in its nucleus and spontaneously undergoes radioactive decay to produce a new nuclide of lower energy plus RADIATION. The main radioactive decay processes are α-decay, β^--decay, β^+-decay, γ-emission, electron capture and spontaneous NUCLEAR FISSION (*see* RADIOACTIVITY). The particular process by which a given radionuclide will decay depends upon its nuclear mass and the number of protons and neutrons in its nucleus.

Environmental occurrence

Radionuclides occur in the environment both as natural species and as pollutants arising from human activities.

Primordial radionuclides

Primordial radionuclides are naturally occurring species, formed by nucleosynthesis, and have been present in the Earth since its formation 4.5×10^9 years ago. Primordial radionuclides have very long half-lives ($> 10^8$ years) and are distributed throughout the Earth as a function of their geochemical behaviour. ^{238}U, ^{235}U and ^{232}Th are a particularly important group of primordial radionuclides, being the parent members of the three natural radioactive decay series (Fig. R3). The different radionuclides in these decay chains exhibit marked differences in geochemical behaviour, resulting in contrasting aqueous-phase/solid-phase partitioning. Uranium in the +6 oxidation state, RADIUM and RADON are relatively soluble in natural waters, whereas the other nuclides, plus URANIUM in the +4 oxidation state, are all highly insoluble. In the environment, separation of parent–daughter pairs in the decay chains as a consequence of differences in solubility can result in situations of radioactive disequilibrium. In such cases, observation of radioactive decay and ingrowth processes that occur after the initial separation of the radionuclides can often be used to study the mechanisms and rates of environmental processes, with examples being listed in Table R1.

Natural decay series radionuclides can present a potential health HAZARD either through high natural concentrations in the environment (e.g. ^{222}Rn) or through selective enrichment of the radionuclides during industrial processing of materials in which they occur as trace constituents (e.g. bauxite, rare earth element ores and TIN ores).

Primordial radionuclides which decay to give a stable daughter nuclide can be used in geochronological studies to determine the ages of minerals, with examples being ^{238}U–^{206}Pb, ^{87}Rb–^{87}Sr and ^{40}K–^{40}Ar dating techniques.

Cosmic-ray-produced radionuclides

Natural radionuclides are also continuously produced in the upper ATMOSPHERE by nuclear reactions induced by interaction of cosmic radiation with nuclei of atmospheric gases. The environmental distribution of cosmic-ray-produced radionuclides is controlled by their geochemical behaviour and half-lives. ^{14}C ($t_{1/2}$ = 5730 years) is a cosmic-ray-produced radionuclide which is generated at a relatively constant rate in the atmosphere by the reaction:

$$^{14}_{7}N + ^{1}_{0}n \rightarrow ^{14}_{6}C + ^{1}_{1}p$$

The ^{14}C is rapidly oxidized to $^{14}CO_2$ and subsequent uptake in plants by PHOTOSYNTHESIS and transfer through the FOOD CHAIN gives rise to a constant natural concentration of 0.226 Bq of ^{14}C per gram of CARBON in living organisms.

Observation of the decay of ^{14}C in organic matter following the death of the organism constitutes the basis of the well-known radiocarbon dating method. Complications in radiocarbon dating arise

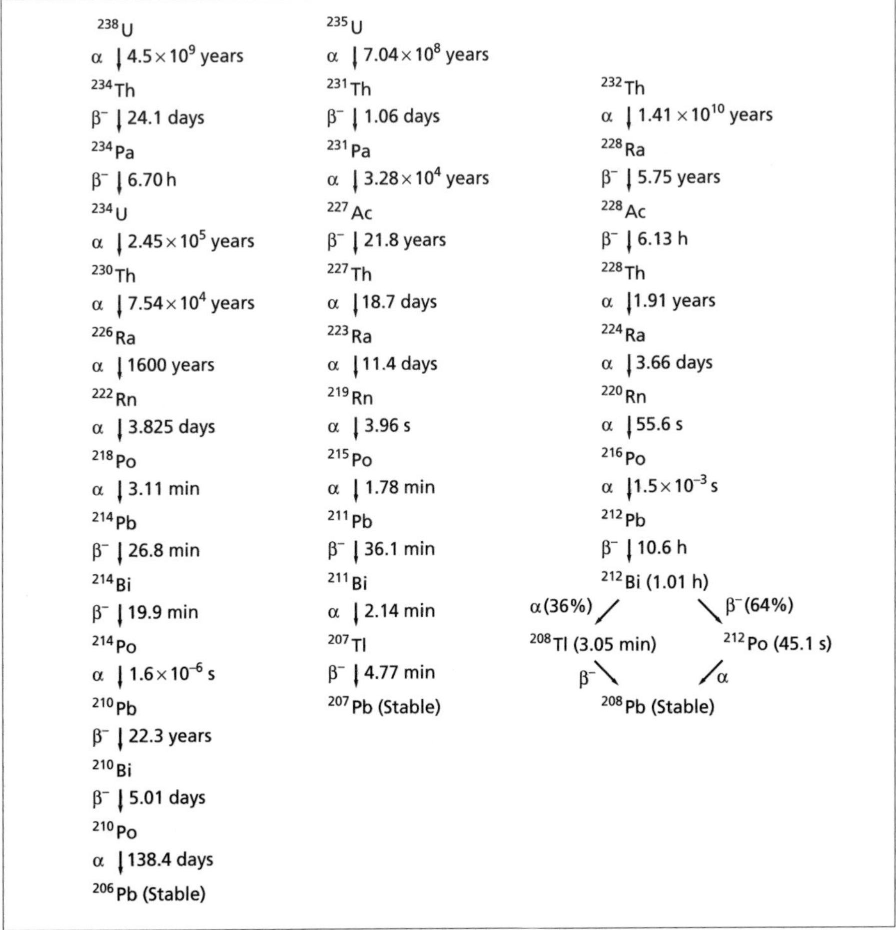

Fig. R3 The natural radioactive decay series (only major decay modes are shown).

Table R1 Examples of applications of natural decay series disequilibrium in the study of environmental processes.

Daughter nuclide	Parent nuclide	Process	Time-scale
^{234}Th	^{238}U	Euphotic zone production. Particle scavenging and settling in seawater	1–100 days
^{234}U	^{238}U	Carbonate growth rates	10^5–10^6 years
^{230}Th	^{234}U	Accumulation rates for carbonates and deep-sea sediments	10^3–10^5 years
^{226}Ra	^{230}Th	Ocean water movement	10^2–10^3 years
^{222}Rn	^{226}Ra	Ocean water movement	1–30 days
^{210}Pb	^{226}Ra	Accumulation and mixing in nearshore marine sediments, lake sediments and peats. Bioturbation and redistribution of deep-sea sediments	1–150 years
^{231}Pa	^{235}U	Accumulation rates for deep-sea sediments	10^4 years
^{228}Ra	^{232}Th	Seawater movement	1–10 years

from non-constancy of the COSMIC RAY flux with time, burning fossil-fuel carbon that does not contain ^{14}C (the Suess effect), nuclear explosions which gave an additional source of ^{14}C (the bomb effect), and changes in the accepted value of the ^{14}C half-life.

Artificial radionuclides

Artificial radionuclides are generated by fission of ^{235}U or ^{239}Pu nuclei, giving radioactive fission products, and NEUTRON capture processes, generating nuclei that are unstable with respect to β-decay. The dominant source of artificial radionuclides in the environment has been atmospheric fallout from nuclear weapons testing and releases from NUCLEAR POWER facilities. Smaller quantities of artificial radionuclides are produced for research, medical and other purposes by bombarding nuclei with particles in accelerators to induce nuclear reactions. The properties of selected radionuclides that occur in the environment are summarized in Table R2. *See also* COSMIC RAYS; IONIZING RADIATION; RADIATION POLLUTION; RADIOACTIVE FALLOUT; RADIOACTIVE WASTE. [A.B.M.]

REFERENCES

Brown, E. & Firestone, R.B. (1986) *Table of Radioactive Isotopes*. John Wiley & Sons, New York.

Friedlander, G., Kennedy, J.W., Macias, E.S. & Miller, J.M. (1981) *Nuclear and Radiochemistry*. John Wiley & Sons, New York.

Ivanovich, M. & Harmon, R.S. (eds) (1992) *Uranium-Series Disequilibrium*. Clarendon Press, Oxford.

radiosonde *See* SONDE.

radiotelemetry *See* TELEMETRY.

radium (Ra) A radioactive metallic element (ATOMIC NUMBER = 88; relative ATOMIC MASS = 226; melting point = 700°C; boiling point =

Table R2 Decay characteristics and origins of selected radionuclides that are present in the environment (only main decay modes are shown; maximum energies are shown for β⁻-decay). C, cosmic ray-produced; F, fallout; R, reactor-produced; N, natural decay series; P, primordial.

Nuclide	Half-life	Decay mode and energy (MeV)	Origin
^{3}H	12.33 years	β^- (0.0186)	C, F, R
^{7}Be	53.3 days	β^- (0.862), γ (0.478)	C
^{14}C	5730 years	β^- (0.156)	C, F, R
^{40}K	1.277×10^9 years	β^- (1.710), γ (1.461)	P
^{60}Co	5.27 years	β^- (1.491), γ (1.173, 1.333)	F, R
^{90}Sr	28.5 years	β^- (0.546)	F, R
^{90}Y	2.67 days	β^- (2.282)	Decay of ^{90}Sr
^{95}Zr	64.0 days	β^- (1.124), γ (0.724, 0.757)	R
^{95}Nb	3.61 days	β^- (1.161), γ (0.766)	Decay of ^{95}Zr
^{99}Tc	2.13×10^5 years	β^- (0.294)	R
^{106}Ru	1.02 years	β^- (0.039), γ (0.622)	R
^{129}I	1.57×10^7 years	β^- (0.154)	R
^{131}I	8.04 days	β^- (0.807), γ (0.364)	R
^{134}Cs	2.06 years	β^- (1.454), γ (0.605, 0.796)	R
^{137}Cs	30.0 years	β^- (1.175), γ (0.662)	F, R
^{208}Tl	3.05 min	β^- (2.377), γ (0.583, 2.614)	N
^{210}Pb	22.3 years	β^- (0.063), γ (0.046)	N
^{211}Pb	36.1 min	β^- (1.373), γ (0.350)	N
^{212}Pb	10.64 h	β^- (0.537), γ (0.239)	N
^{214}Pb	26.8 min	β^- (1.024), γ (0.296, 0.352)	N
^{210}Bi	5.013 days	β^- (3.270)	N
^{211}Bi	2.14 min	α (6.623, 6.279)	N
^{212}Bi	1.01 h	β^- (2.246), α (6.051, 6.090)	N
^{214}Bi	19.9 min	β^- (3.270), γ (0.609, 1.120)	N
^{210}Po	138.4 days	α (5.304)	N
^{212}Po	2.0×10^{-7} s	α (8.784)	N
^{214}Po	1.6×10^{-6} s	α (7.687)	N
^{215}Po	1.8×10^{-3} s	α (7.386)	N
^{216}Po	0.15 s	α (6.779)	N
^{218}Po	3.11 min	α (6.003)	N
^{219}Rn	3.96 s	α (6.819)	N
^{220}Rn	55.6 s	α (6.288)	N
^{222}Rn	3.825 days	α (5.490)	N
^{223}Ra	11.4 days	α (5.716, 5.607), γ (0.269)	N
^{224}Ra	3.66 days	α (5.686)	N
^{228}Ra	5.75 years	β^- (1.110)	N
^{226}Ra	1600 years	α (4.784), γ (0.186)	N
^{227}Ac	21.8 years	β^- (0.044)	N
^{228}Ac	6.13 h	β^- (2.079), γ (0.911)	N
^{227}Th	18.7 days	α (5.757, 5.987, 6.038), γ (0.236)	N
^{228}Th	1.91 years	α (5.423, 5.341)	N
^{230}Th	7.54×10^5 years	α (4.688, 4.621)	N
^{231}Th	1.06 days	β^- (0.390)	N
^{232}Th	1.41×10^{10} years	α (4.010, 3.952)	P
^{234}Th	1.17 min	β^- (2.281), γ (0.063)	N
^{231}Pa	3.276×10^4 years	α (5.013, 5.029, 4.951)	N
^{234}Pa	6.7 h	β^- (1.994), γ (0.131)	N
^{234}U	2.45×10^5 years	α (4.776, 4.724)	N
^{235}U	7.04×10^8 years	α (4.395, 4.364)	P
^{238}U	4.47×10^9 years	α (4.196, 4.147)	P
^{237}Np	2.14×10^6 years	α (4.788, 4.772)	R, F
^{238}Pu	87.7 years	α (5.768, 5.721)	R, F
^{239}Pu	2.41×10^4 years	α (5.156, 5.143)	R, F
^{240}Pu	6.56×10^3 years	α (5.168, 5.124)	R, F
^{241}Pu	14.4 years	β^- (0.021)	R, F
^{241}Am	432.7 years	α (5.486, 5.443), γ (0.060)	R, F
^{242}Cm	163 days	α (6.113, 6.070)	R, F
^{243}Cm	28.5 years	α (5.786, 5.742)	R, F
^{244}Cm	18.1 years	α (5.805, 5.763)	R, F

1737°C) having a crustal abundance of about 10^{-6} mg kg^{-1}. It was discovered by the Curies (Marie and Pierre) in 1898. Isotopes from ^{213}Ra to ^{230}Ra are known; ^{223}Ra ($t_{1/2}$ = 11.43 days), ^{224}Ra ($t_{1/2}$ = 3.66 days), ^{226}Ra ($t_{1/2}$ = 1600 years) and ^{228}Ra ($t_{1/2}$ = 5.75 years) occur in the natural radioactive decay series. Radium has an oxidation state of +2 and exhibits soluble behaviour in natural waters. It has similar chemical properties to CALCIUM and can present a radiological HAZARD via incorporation in bone. It is used in isotopic tracer studies of environmental processes, radiotherapy and, formerly, the production of luminescent materials. *See also* RADIONUCLIDE. [A.B.M.]

REFERENCE

Ivanovich, M. & Harmon, R.S. (eds) (1992) *Uranium-Series Disequilibrium.* Clarendon Press, Oxford.

radon (Rn) A radioactive gaseous element (ATOMIC NUMBER = 86; relative ATOMIC MASS = 222; melting point = −71°C; boiling point = −62°C) that is chemically inert and has known isotopes from ^{211}Rn to ^{224}Rn. It occurs naturally in the URANIUM and THORIUM decay series as ^{219}Rn ($t_{1/2}$ = 3.96 s), ^{220}Rn (sometimes called thoron) ($t_{1/2}$ = 55.6 s) and ^{222}Rn ($t_{1/2}$ = 3.825 days). Of the natural isotopes, only ^{222}Rn lives for a sufficient length of time to allow significant environmental transport. Movement of ^{222}Rn in groundwater and soil gas can result in its accumulation in mines and buildings, causing increased radiation EXPOSURE to humans and other organisms, and raised incidences of lung cancer. *See also* IONIZING RADIATION; MUTAGEN; RADIATION POLLUTION; RADIONUCLIDE; SIEVERT. [A.B.M.]

REFERENCES

Gates, A.E. & Gundersen, L.C.S. (eds) (1992) *Geological Controls on Radon.* GSA Special Paper 271. The Geological Society of America, Boulder, Colo.

Graves, B. (ed.) (1987) *Radon in Groundwater.* Lewis Publishers, Chelsea, Mich.

rain WATER droplets falling under gravity from CLOUDS to the Earth's surface, as the liquid phase of PRECIPITATION. Cloud droplets grow by collision, or by condensation of water vapour, until gravitational forces exceed the turbulent forces retaining droplets in suspension. There is no clear size boundary between droplets of cloud and those of rain; the distinction depends on WIND speed and TURBULENCE, but droplets bigger than 0.2 mm in diameter are usually regarded as rain.

Rain is not pure water, but contains dissolved salts (SODIUM (Na^+), CHLORIDE (Cl^-) and MAGNESIUM (Mg^{2+}) ions from sea spray; CALCIUM (Ca^{2+}) and magnesium ions from suspended soil) and gases (e.g. CARBON DIOXIDE (CO_2), as carbonic acid), so that unpolluted rain has a pH around 5.6. 'Acid rain' also contains sulphuric (H_2SO_4) and nitric (HNO_3) acids from the atmospheric oxidation of the sulphur dioxide (SO_2) and NITROGEN OXIDES ($NO+NO_2$) produced during fossil-fuel combustion, and ammonium (NH_4^+) ions, largely from agricultural EMISSIONS. Rain may also act as the vector for the deposition of organic materials from the ATMOSPHERE, scavenged from the air by cloud droplets or falling rain.

Amounts of rain are measured as a rainfall depth in millimetres, equivalent to litres of water per square metre of ground. Not all rain falling on the Earth's surface finds its way through soil into waterways; some is intercepted by vegetation, and evaporates to the atmosphere as 'INTERCEPTION loss'. Rainwater dripping through vegetation canopies is known as 'throughfall', which may have a very different chemical composition from incident rain. Material deposited from the atmosphere in rain is known as 'wet deposition'. *See also* ACID RAIN AND MIST. [J.N.C.]

rain forest Any FOREST in either tropical or temperate latitudes that receives particularly high levels of rainfall and whose structure and composition is determined by the high HUMIDITY of the environment. The TROPICAL RAIN FORESTS lie in the equatorial regions of the world and are provided with high PRECIPITATION as a result of the development of convective AIR MASSES rising and releasing their load of water as they cool with height. The TEMPERATE RAIN FORESTS are mainly located in oceanic temperate regions, such as the Pacific North-West of North America, and parts of New Zealand and Japan, where local conditions lead to exceptionally high rainfall through much of the year.

Both types of rain forest are characterized by high BIOMASS and PRODUCTIVITY. They have high tree canopies, with some exceeding 40 m in height. Their structure is complex, especially in the tropical ecosystems, and the high humidity of the region beneath the CANOPY leads to a rich growth of epiphytes, including higher plants (especially in the tropics), ferns, mosses and lichens.

The high economic value of the timber from these vegetation types has led to extensive clearance, and old-growth, PRIMARY FOREST is becoming increasingly scarce, especially in the temperate forests where exploitation has been more intensive and prolonged. *See also* RAIN FORESTS AND CLOUD FORESTS, CONSERVATION. [P.D.M.]

Rain forests and cloud forests, conservation

Importance

The conservation of the world's RAIN FORESTS and CLOUD FORESTS is considered by many to be the most compelling conservation issue of the late 20th century. Scientists estimate that more than half of the world's BIODIVERSITY, in terms of genetic material, species and ECOSYSTEM types, is found within these two broad FOREST types. Many studies have demonstrated the enormous DIVERSITY of species within small areas of TROPICAL RAIN FOREST, as for example, in Ecuador, where one 10 ha plot of lowland rain forest contains more than 300 tree species alone. The temperate rain forests, though less diverse in total numbers of species, also contain interesting and often poorly-known BIOTAS. They, too, are under severe threat of reduction or extermination, except in a few regions where parks or other reserves have long been established and are well protected.

Scientists at The World Conservation Monitoring Centre in Cambridge, UK, and at the UN FAO in Rome, have estimated that less than 10% of the remaining rain forests and cloud forests worldwide are found within some form of managed protected area, and that many of these forests are not adequately managed to ensure the long-term protection of their biodiversity.

Management approaches

It is very difficult to determine accurately the extent of the world's rain forests and cloud forests or the proportion that is under effective conservation management. In Colombia, for example, a country that is extensively forested, more than 25% of the country has been formally designated as territory of indigenous peoples. Much of this land contains tropical rain forest of unusually high BIODIVERSITY. A form of *de facto* conservation exists, but it is not yet clear whether these lands will be managed to conserve biodiversity in perpetuity, in addition to serving the needs of the indigenous communities. Also, changes in land management over time may result in less-effective conservation of these ecosystems unless governments and private individuals continue to take an interest in conserving this biodiversity.

Community involvement

Increasingly, conservation of all forests will require active participation and involvement of local people and the broader community as a whole. This is certainly true for rain forests and cloud forests in the tropical developing countries, where there is enormous demand for the natural RESOURCES within these forests. However, this also applies to developed nations in the temperate latitudes. No longer can we afford to lock away extensive forests simply for their amenity value or to preserve their biodiversity. EXTRACTIVE RESERVES increasingly will become an important method of conserving and managing these forests.

In the Pacific North-West of the USA and Canada, for example, management of the few remaining TEMPERATE RAIN FORESTS, by the 1990s, had become a painfully contentious issue that divided entire communities. Land-management decisions were taken out of the realm of science and into the realm of local and national politics. The same situation has developed for the temperate rain forests in Chile, New Zealand and Norway. Compromise between resource extraction and biodiversity conservation is nearly always necessary.

Private reserves

Although the conservation of rain forests and cloud forests today must take place primarily on publicly-owned lands, with management by public agencies, there remains ample opportunity for private individuals and organizations to influence conservation of these ecosystems worldwide. In Costa Rica, the Monte Verde Cloud Forest Reserve is on privately-owned forest land. Monte Verde has helped stimulate nature tourism in Costa Rica, an industry which brings in important revenues to the country. In the process, a small but significant area of natural cloud forest in the tropics is being preserved. The same type of conservation management on private land can be seen in several countries, including Peru, Brazil, Belize, Malaysia, Indonesia, South Africa, Australia, Norway and New Zealand. Private reserves can be effective as long as some safeguards are built into the ownership and management, in order to prevent drastic or sudden changes in land use.

Outlook

In any effort to determine gaps in ECOSYSTEM protection (*see* GAP ANALYSIS) and to establish world conservation priorities, the protection and management of the world's rain forests and cloud forests will continue to be recognized as a high priority. We can expect this to remain the case well into the future, as human demands on these ecosystems increase and as their extent on all continents continues to decrease. Global CLIMATE CHANGES — not yet predictable — also may affect conservation priorities by shifting the locations or shrinking the extent of rain forests and cloud forests. [F.W.B.]

REFERENCES

Myers, N. (1984) *The Primary Source: Tropical Forests and Our Future.* W.W. Norton, New York.

Redford, K.H. & Padoch, C. (1992) *Conservation of Neotropical Forests: Working from Traditional Resource Use.* Columbia University Press, New York.

Sandved, K.B. & Emsley, M. (1979) *Rain Forests and Cloud Forests.* Harry N. Abrams, New York.

---

raised bog A domed MIRE found in temperate regions. The central dome of PEAT is surrounded by lower-lying wetland habitats through which groundwater flows. The main cupola of peat can extend over several square kilometres and can be elevated in excess of 6 m above the groundwater table. Often they are developed over former lake basins which have passed through a successional sequence of emergent aquatic colonization (SWAMP), through forested swamp and FEN communities (CARR) and eventually into a *Sphagnum*- or *Eriophorum*-dominated vegetation that has contributed to the OMBROTROPHIC growth of the peat dome. Raised BOGS may also develop over estuarine sites in which the early sediments are of SALT MARSH preceding reed swamp.

The dome of OMBROGENOUS peat is surrounded by a sloping 'rand' and a region of RHEOTROPHIC conditions in which carr and fen vegetation persists (the 'lagg'). Groundwater from the surrounding CATCHMENT drains through the lagg and circumvents the raised peat cupola.

The maintenance of a perched WATER TABLE on the mire surface is dependent on the low hydraulic conductivity of the bulk of the peat mass (CATOTELM). Only the surface layers (ACROTELM) of unconsolidated, uncompressed material permit lateral movement of water derived from rainfall. So rainwater drains laterally from the peat dome.

The surface vegetation of raised bogs is typically treeless in more oceanic sites but a forest cover is frequent in more continental locations. Pools are a common feature of the mire surface and these are often arranged in a concentric pattern when viewed from the air. In the case of raised mires developed over sloping ground, the pool systems may be excentric, with their focus situated in the upper part of the slope. [P.D.M.]

ramet A potentially independent part of a modular plant — for example, a rooted shoot arising from a bud on a rhizome (below ground) or stolon (above ground) — that is capable of persisting when (and if) the connection to the parent plant (the rhizome or stolon) is severed. Parts of a clonal plant (or GENET), which are genetically identical except for SOMATIC MUTATION among them. *See also* MODULAR ORGANISM; MODULE. [M.J.C.]

RAMSAR *See* INTERNATIONAL CONSERVATION CONVENTIONS.

random distribution A term used to describe the spatial DISTRIBU-

TION of the individuals in a POPULATION. When individuals in a population follow a random distribution, the VARIANCE (s^2) of individuals per SAMPLE is equal to the MEAN number ($\bar{x}$) of individuals per sample ($s^2 = \bar{x}$). This is frequently summarized by saying that the variance to mean ratio is equal to 1 ($s^2/\bar{x} = 1$). *See also* AGGREGATED DISTRIBUTION; DISPERSION; POISSON DISTRIBUTION; REGULAR DISTRIBUTION. [B.S.]

random mutation *See* MUTATION.

random sample This is an unbiased subset of a POPULATION under investigation and is required for many statistical tests. A truly unbiased absolute estimate of POPULATION SIZE is only possible if the SAMPLE units are representative of the total population. The easiest way of achieving this is for each sample unit to contain a true random sample of the population. However, selecting a true random sample is not an easy task. If quadrats are being used, then the position of a QUADRAT within the area under investigation should be chosen at random using random numbers to generate two coordinates that are then used to position a corner of the quadrat. Although this method of choosing a position using random numbers often requires an area to be marked out, it is preferable to quasi-random techniques, such as throwing a quadrat, which are certain to introduce some involuntary observer bias.

Random walks are another way to produce random samples that do not require the area to be marked out. The observer walks a number of paces determined by a random number and then makes an observation or places a quadrat. Then another random walk is taken before the next observation, and so on. The advantages of this method are that SAMPLING can be very rapid and that it requires little preparation. The drawback is that sampling may be severely biased by the observer.

True random samples are ideal in a perfectly homogeneous HABITAT, but in a heterogeneous habitat they are likely to produce a biased sample with an estimated VARIANCE greater than that of the total population. A simple method used to minimize this problem is to take a stratified random sample: the total area is divided into equal plots and an even number of sample units are taken at random from each plot. It is possible to divide the total area into plots of different sizes if there are known to be different habitats in the total area. In this second case, the number of sample units from each plot should be proportional to its size. *See also* EXPERIMENTAL DESIGN; SAMPLING METHODOLOGY/DEVICES; STATISTICS. [C.D.].

random walk model A MODEL in which variables change by finite, possibly random amounts, in random directions as time progresses. Random walk models occur in two main areas of ECOLOGY: as the underlying model for diffusive movement and as stochastic POPULATION models. For example: insect foraging movement might be modelled as a series of short steps of random length, each one in a random direction (*see also* OPTIMAL FORAGING THEORY). The change in spatial DENSITY of a population of thousands of similar insects moving in the same way could then be described by a diffusion equation model. Another example of the use of a random walk model might be as the NULL MODEL against which to test hypotheses about DENSITY DEPENDENCE: changes in POPULATION SIZE being modelled as a series of random steps up or down. *See also* DIFFUSION EQUATIONS; POPULATION REGULATION. [S.N.W.]

randomization *See* EXPERIMENTAL DESIGN.

range

1 Spatial extent of species and/or POPULATION.

2 A measure of the spread or variation of a set of observations that is simply the difference between the smallest and the largest single observation. In the SAMPLE of eight observations:

3, 21, 22, 24, 28, 29, 29, 40

the range is $3 - 40 = 37$. This is clearly overdependent upon just two observations, and can therefore be misleading. A better measure of variation is the VARIANCE or the STANDARD DEVIATION. [B.S.]

range management The utilization of extensive areas of GRASSLANDS for pastoral purposes in both temperate and tropical areas demands careful management to avoid possible destructive effects (*see* OVERGRAZING). Range management has traditionally been applied to the control of stock density of domestic animals, usually cattle or sheep, to ensure the continued productivity of the ecosystem. Large, wild herbivores, however, are often the focus of attention for range managers since these have a number of advantages over introduced animals. They are often more resistant to the diseases and pests, particularly of tropical grassland areas, and they often have a wider range of GRAZING preferences, so are able to exploit the productive potential of grasslands more fully. This is particularly the case when a number of native herbivores can be cropped, since there is often a degree of NICHE differentiation between the HERBIVORE species as they draw upon different vegetation resources, or tap the resources at different times of the year. Theoretically, therefore, the harvesting of native grazing animals is more efficient in ecological terms and should, if carefully controlled, result in lower levels of damage to the grasslands.

Apart from the direct impact of grazing, range managers must also concern themselves with trampling impact on soils, especially in the vicinity of watering places where animals may congregate. Fire can also be a useful management tool. It is extensively used in the tropical SAVANNAH grasslands where it liberates nutrients locked up in litter and stimulates renewed plant growth. Periodic low-intensity fires prior to the build-up of high levels of BIOMASS and litter ensure that more damaging intense fires are forestalled, thus conserving the habitat.

Range management will need particular attention in many parts of the world if the expected rise in global temperature materializes. Even more important will be changes in precipitation and evaporation patterns, which will affect the soil water and hence the productivity of many tropical and temperate rangelands. Stock densities may well have to be modified to adjust the balance of the system to these new factors. [P.D.M.]

range size *See* RANGE; RARITY, BIOLOGY OF; RARITY, CLASSIFICATION.

rank Can refer to species in a COMMUNITY (*see* RANK–ABUNDANCE

MODELS) or individuals in a social grouping (*see* DOMINANCE, SOCIAL). [P.C.]

Rank–abundance models

Ecological communities comprise species that vary markedly in their ABUNDANCES. Some species are extremely common, some have moderate abundance and some are rare; this is their RANK. The DISTRIBUTION of the abundances for these species can be used to characterize the DIVERSITY of a COMMUNITY more effectively than is possible by composite diversity measures such as the Shannon or Simpson indices (*see* DIVERSITY INDICES).

Four models are traditionally used to describe patterns of species abundance. These are the GEOMETRIC SERIES, the log series, the log normal and the broken stick; they represent a CONTINUUM from low to high EQUITABILITY. In certain cases, such as the log series, the models have arisen as empirical fits to widespread patterns. Other models have been devised on the basis of specific assumptions about the allocation of RESOURCES or the degree of COMPETITION amongst community members. However, the biological basis of some of the models, most notably the broken stick, has now been discredited, and in any case there is no reason why a good fit by a particular MODEL vindicates the ecological assumptions that it is based upon.

Species-abundance data can be presented in a rank–abundance plot. In these graphs the abundance of each species is plotted on a logarithmic scale against the species' rank, in order, from the most abundant to the least abundant species. Species abundances may, in some instances, be expressed as percentages to provide a more direct comparison between communities with different numbers of species.

Characteristics

The four main models each have a characteristic shape on a rank–abundance plot (Fig. R4). The geometric series is denoted by a straight line with a steep GRADIENT. Similarly, the log series has a steep gradient, but here the curve is only approximately linear. Both of these models are ones in which equitability is low. By far the shallowest curve is produced by the BROKEN STICK MODEL. This is a consequence of the high degree of equitability that the broken stick model represents. In between the broken stick and the log series comes the log normal with its sigmoid curve.

Geometric series model

The geometric series represents a situation of extremely low evenness and is only likely to be found in species-poor communities occupying harsh habitats. Because the ratio of the abundance of each species to the abundance of its predecessor is constant through the ranked list of species, the series appears as a straight line when plotted on a rank–abundance graph (see Fig. R4).

Log series model

R.A. Fisher's log series was the first attempt to describe mathematically the relationship between the number of species and the number of individuals in those species. The log series is characterized by a large number of rare species and few common ones. A parameter of the model, α, makes an excellent diversity index. One of the virtues of α is that, unlike SPECIES RICHNESS or even the Shannon diversity index, it is not strongly dependent on SAMPLE size.

Log normal model

The majority of ecological communities follow a log normal pattern of species abundance. The LOG NORMAL DISTRIBUTION was initially applied to species-abundance data by F.W. Preston in 1948. Preston showed that many of log normal distributions found in natural communities follow a particular pattern, which he termed 'canonical'. A full discussion of what is meant by a canonical log normal, and reasons why it is so prevalent, are provided by May (1975) and Rosenzweig (1995). The index λ, which is calculated when the

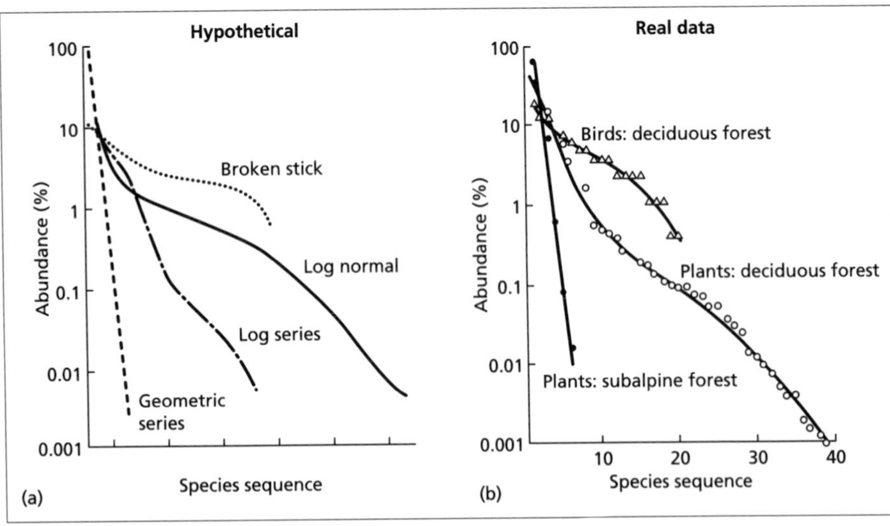

Fig. R4 Rank–abundance plots illustrating the typical shape of the four species-abundance models: geometric series, log series, log normal and broken stick. (a) Hypothetical curves for the four models. (b) Three examples of rank–abundance curves from real communities (after Whittaker, R.H. (1970) *Communities and Ecosystems*. Macmillan Inc., New York). The three communities are: nesting birds in a deciduous forest, West Virginia; vascular plants in a deciduous cove forest in the Great Smoky Mountains, Tennessee; and vascular plant species from subalpine fir forest, also in the great Smoky Mountains. As comparison with (a) suggests, the best descriptions of the three communities are, respectively, the broken stick, log normal and geometric series models. (After Magurran, A.E. (1988) *Ecological Diversity and Its Measurement*. Chapman & Hall, London.)

log normal distribution is being fitted, is a useful measure of diversity.

If samples, rather than complete CENSUSES of communities are obtained, rare species can be missed, and genuinely abundant species may seem to be uncommon. In such cases the underlying log normal distribution of the community is manifested as a truncated log normal distribution in the sample. In its truncated form the log normal can be virtually indistinguishable from the log series. Since many datasets will be described equally well by both the log series and the truncated log normal it may be difficult for the ecologist to decide which is more appropriate. In practice this does not matter greatly since the shape of the two distributions is similar and both α and λ are effective diversity indices.

Broken stick model

The fourth model, the broken stick, was devised by R.H. MacArthur. It is little used these days. Good fits for the broken stick model seem to be found primarily in narrowly defined communities of taxonomically related organisms. No diversity index has been derived for the distribution: as it represents a highly equitable state of affairs species richness is an appropriate measure of diversity.

Fitting a model

Although it is possible in many cases to deduce the underlying distribution from the shape of the rank–abundance plot, formal tests of fit are also available. Further details of the models, along with instructions for fitting them and calculating the diversity indices α and λ, are provided by Magurran (1988). [A.E.M.]

REFERENCES

Magurran, A.E. (1988) *Ecological Diversity and Its Measurement.* Chapman & Hall, London.

May, R.M. (1975) Patterns of species abundance and diversity. In: *Ecology and Evolution of Communities* (eds M.L. Cody & J.M. Diamond), pp. 81–120. Harvard University Press, Cambridge, Mass.

Rosenzweig, M.L. (1995) *Species Diversity in Space and Time.* Cambridge University Press, Cambridge.

Whittaker, R.H. (1970) *Communities and Ecosystems.* Macmillan Inc., New York.

Rapoport's rule *See* GRADIENTS OF SPECIES RICHNESS.

Rarity, biology of

Rare species have low ABUNDANCE, or small ranges, or frequently both. Rarity is highly SCALE-dependent, both temporally and spatially, and may frequently have one apparent cause at the large scale (e.g. climatic restriction) and another at a smaller scale (e.g. soil type). Naturalists, ecologists and conservationists pay particular attention to rare species because they are more likely to become extinct and because they are considered to be of higher value.

Theories of rarity

Positive correlations between local abundance and RANGE size are common, but their biological basis is still obscure. Two major competing theories are resource usage and metapopulation dynamics. Variants of the former assume either that rare species have narrower niches, or simply that common and widespread species utilize common and widespread RESOURCES. The latter assumes either that common species have higher carrying capacities, and hence higher IMMIGRATION and/or lower extinction rates, or that immigration rate increases with local patch occupancy and that high immigration reduces the probability of local extinction. There is support for the second variants of both hypotheses.

Recent patterns of land use in Great Britain have increased the area of fertile, disturbed land, and there is good evidence that plants which exploit these conditions are generally common and increasing. Metapopulation theory suggests that effectively dispersed taxa should have larger ranges than poorly dispersed taxa, and there is support for this from organisms as diverse as beetles, mice and higher plants.

In animals, body SIZE and rarity are frequently positively correlated, but establishing the causal basis of this relationship has proved to be extraordinarily difficult, if only because body size is itself correlated with numerous other potentially ecologically significant variables.

Causes of rarity

Today, rarity is frequently a direct consequence of human activities. In north-west Europe, the common and increasing plant species of densely populated England, Germany and The Netherlands are fast-growing and tolerant of an increasingly disturbed and eutrophic landscape. Rare and declining species in these countries are slow-growing and intolerant of modern land use. In sparsely populated Scotland and Ireland, COMMONNESS and rarity do not have such an obvious basis. Beyond an inability to get along with humans, however, the causes of rarity are generally idiosyncratic, and the search for illusory 'rules' may actually be a barrier to progress.

Rarity and conservation

Rarity is inextricably linked with conservation; much of the impetus to study and classify rare species comes from the need to allocate resources to conservation priorities. A consideration of such priorities reveals just how ignorant we are of rarity on a global scale; we have little or no idea of the composition of the 10% of species most threatened with extinction at the present time; the fragmentary evidence available suggests that some taxa characteristically have smaller geographical ranges than others. Allocation of conservation priorities to sites is inevitably based on one or a few well-known taxa, and it is far from clear whether this strategy will serve to conserve other taxa.

Rarity and 'value'

The interest in rare species can be explained as people valuing the chance of finding the unexpected. Many people derive pleasure from seeing a rare butterfly or orchid (although this presupposes an appreciation of rarity and attractiveness). So, for example, bird-watchers will travel hundreds of miles for the chance of seeing a rare bird. Such a bird is highly valued in much the same way as rare postage stamps or coins are highly valued. This topic is part of the subject of conservation evaluation. Ratcliffe (1977) describes the main criteria used by the UK government's Conservation Agency, and rarity is indeed one of the ten criteria. There has been considerable debate about these criteria. It continues in terms of how it can be turned into monetary value. Rarity, as a criterion, has the advantage that it can be measured more or less objectively; for example, it is possible to determine in how many grid squares or vice-counties a species occurs. The final irony of this kind of evaluation process is that a species increases in value until the day it becomes extinct.

Rarity continues to provide enormous opportunities for future study, but the urgent requirements of conservation of rare species mean that this work must be better focused. Urgent topics include, *inter alia*, the functional significance of BIODIVERSITY in general and of rare species in particular, and compilation of inventories of rare species and accumulation of comparative data (however simple) about them. *See also* CONSERVATION BIOLOGY; RARITY, CLASSIFICATION. [K.T. & F.B.G.]

REFERENCES
Gaston, K.J. (1994) *Rarity*. Chapman & Hall, London.
Goldsmith, F.B. (1983) Evaluating nature. In: *Conservation in Perspective* (eds A. Warren & F.B. Goldsmith), pp. 233–246. John Wiley & Sons, Chichester.
Goldsmith, F.B. (1994) The preparation of biodiversity strategies. *Science Progress* **77**, 221–232.
Ratcliffe, D.A. (1977) *A Nature Conservation Review*. Cambridge University Press, Cambridge.

rarity, classification A common TYPOLOGY of rarity is that of Rabinowitz (1981), who classified plant species according to GEOGRAPHIC RANGE (large or small), local POPULATION SIZE (large or small) and HABITAT specificity (wide or narrow). Seven of the eight possible combinations of these states (i.e. in which any attribute is small or narrow) are recognized as forms of rarity. Gaston (1994) considers that this approach requires arbitrary and unnecessary division of too many variables, and defines rarity simply in terms of ABUNDANCE and/or range size. Other workers have defined rarity in various ways, with varying objectives, and the outcome has been to classify anything from 1% to over 90% of the species in an ASSEMBLAGE as 'rare'. Gaston proposes that the least-abundant 25% of species in an assemblage should be defined as rare.

The distinction is sometimes made between species which are 'naturally' rare and those whose rarity is a consequence of human activities. The latter are increasingly frequent in the modern world. 'Rare' is frequently employed as a threat category; for example, the IUCN Red Data categories are: extinct, endangered, vulnerable and rare. However, rarity is a poor predictor of risk of extinction, and is better employed as a parameter in the assignment of species to categories, and not as a category itself. [K.T.]

REFERENCES
Gaston, K. (1994) *Rarity*. Chapman & Hall, London.
Rabinowitz, D. (1981) Seven forms of rarity. In: *The Biological Aspects of Rare Plant Conservation* (ed. H. Synge), pp. 205–217. John Wiley & Sons, New York.

Ratcliffe index An index of eggshell thickness, equal to:

Mass of eggshell/(Length + Breadth)

All measurements are determined after washing and drying overnight. The ratio is useful in gauging the effects of chemicals on eggshell thickness; for example, dichlorodiphenyldichloroethylene (ppDDE), a highly persistent metabolite of DDT, causes serious thinning of shells in predatory birds. *See also* ORGANOCHLORINES; PESTICIDE. [P.C.]

REFERENCE
Ratcliffe, D.A. (1967) Decrease in eggshell weight in certain birds of prey. *Nature* **215**, 208–210.

rates of evolution *See* EVOLUTIONARY RATE.

rates of population increase *See* INTRINSIC RATE OF INCREASE.

Raunkiaer's life forms A CLASSIFICATION of plants based on the height of the perennating buds, a character which reflects the ADAPTATION of a species to CLIMATE. There are five main life forms in the classification (Fig. R5 & Table R3), each class being divided further. The main life forms are:

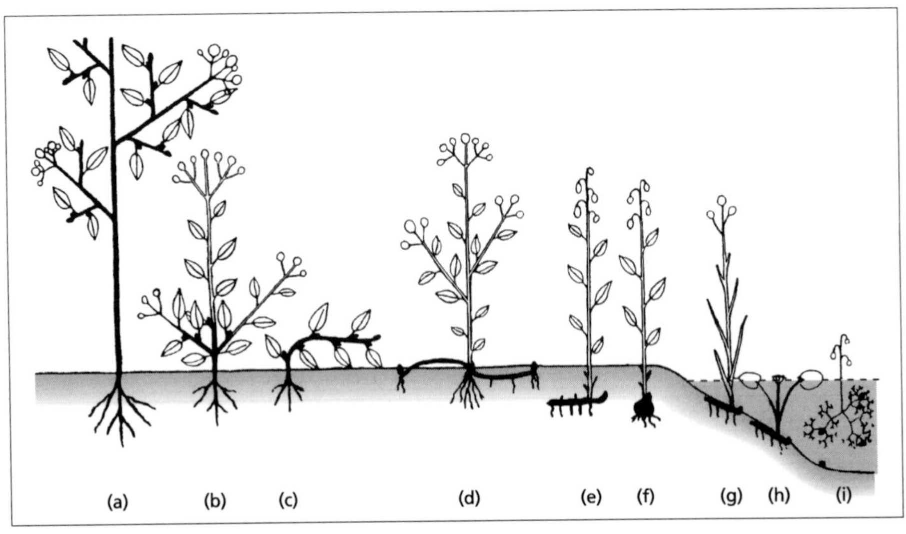

Fig. R5 The relative positions of the perennating parts of four life forms. (a) Phanerophytes; (b, c) chamaephytes; (d) hemicryptophytes; and (e–i) cryptophytes. The persistent axes and surviving buds are shown in black. (After Raunkiaer, C. (1934) *The Life Forms of Plants and Statistical Plant Geography*. Oxford University Press, Oxford; by permission of Oxford University Press.)

Table R3 The life-form classification of Raunkiaer (1934). (From Begon, M., Harper, J.L. & Townsend, C.R. (1996) *Ecology: Individuals, Populations and Communities*, 3rd edn. Blackwell Science, Oxford.)

Phanerophytes
The surviving buds or shoot apices are borne on shoots which project into the air
(a) Evergreens without bud covering } More than 2 m high
(b) Evergreens with bud covering } More than 2 m high
(c) Deciduous with bud covering } More than 2 m high
(d) Less than 2 m high

Chamaephytes
The surviving buds or shoot apices are borne on shoots very close to the ground
(a) Suffruticose chamaephytes, i.e. those bearing erect shoots which die back to the portion that bears the surviving buds
(b) Passive chamaephytes with persistent weak shoots that trail on or near the ground
(c) Active chamaephytes that trail on or near the ground because they are persistent and have horizontally directed growth
(d) Cushion-plants

Hemicryptophytes
The surviving buds or shoot apices are situated in the soil surface
(a) Protohemicryptophytes with aerial shoots that bear normal foliage leaves, but of which the lower ones are less perfectly developed
(b) Partial rosette plants bearing most of their leaves (and the largest) on short internodes near ground level
(c) Rosette plants bearing all their foliage leaves in a basal rosette

Cryptophytes
The surviving buds or shoot apices are buried in the ground (or under water)
(a) Geocryptophytes or geophytes which include forms with:
  (i) rhizomes
  (ii) bulbs
  (iii) stem tubers
  (iv) root tubers
(b) Marsh plants (helophytes)
(c) Aquatic plants

Therophytes
Plants that complete their life cycle from seed to seed and die within a season (this group also includes species that germinate in autumn and flower and die in the spring of the following year)

- phanerophytes: buds or shoot apices borne on aerial shoots, i.e. trees;
- chamaephytes: buds or shoot apices very close to the ground;
- hemicryptophytes: buds at ground level, all above-ground parts dying back at the onset of unfavourable conditions;
- CRYPTOPHYTES: buds below ground or submerged in water;
- therophytes: annual species completing their life cycle in a short time during a favourable period for growth, as with many DESERT plants.

In a tropical FOREST there are many life forms, but in climatically marginal zones like mountains and deserts, there are relatively few, i.e. the biological spectrum is attenuated (Fig. R6). In MOUNTAIN ENVIRONMENTS, plant growth is limited by TEMPERATURE, and the vegetation is dominated by chamaephytes, plants with their buds close to the ground, where air and tissue temperatures are elevated by several degrees during the day (though they may be very low at night).

Raunkiaer's book was published in 1909 and translated into English in 1934; there is a concise account in Kershaw (1973). [J.G.]

REFERENCES

Kershaw, K.A. (1973) *Quantitative and Dynamic Plant Ecology*, 2nd edn. Edward Arnold, London.

Raunkiaer, C. (1934) *The Life Forms of Plants and Statistical Plant Geography*. Oxford University Press, Oxford.

Rayleigh's radiation law *See* RADIATION LAWS.

RCC *See* RIVER CONTINUUM CONCEPT.

RCEP *See* ROYAL COMMISSION ON ENVIRONMENTAL POLLUTION.

reaction norm In QUANTITATIVE GENETICS, the phenotypic expression of a GENOTYPE over different environments; i.e. the systematic change of the GENOTYPIC VALUE of a genotype under a systematic change in environment. The reaction norm specifies the extent and type of the PHENOTYPIC PLASTICITY of a genotype. Imagine many individuals of identical genotype developing in different environments. The reaction norm is the function relating a particular environmental variable to the realized genotypic value of an individual of that genotype. Given the function that is the reaction norm, the value of the environmental variable can be regarded as the function argument, and the genotypic value of the individual the function value (Fig. R7). The mean PHENOTYPIC VALUE of a genotype over environments can be used to determine the reaction norm.

Reaction norms are parallel if genotypes are phenotypically plastic but differ only in average expression; then no GENOTYPE* ENVIRONMENT INTERACTION exists (Fig. R8a). If reaction norms are not parallel, genotypes that are phenotypically plastic differ in sensitivity to the environment, and genotype*environment interaction exists (Fig. R8b). Selection on the slope as well as on the height of the reaction norm is possible. *See also* OPTIMAL REACTION NORMS. [G.D.J.]

reactor response *See* COMMUNICATION.

realized heritability HERITABILITY (h^2, the regression of BREEDING VALUE on PHENOTYPIC VALUE) estimated from response to ARTIFICIAL SELECTION for a QUANTITATIVE trait. Starting with a base population with a mean value of the quantitative trait of $\bar{x}_P$, a proportion of individuals are selected with high (or low) values of the trait. The mean of the trait in the selected parents is $\bar{x}_S$. The SELECTION DIFFERENTIAL, *S*, is $\bar{x}_S - \bar{x}_P$. The mean of the trait in the offspring of the selected parents is $\bar{x}_O$, and the response to selection, *R*, is $\bar{x}_O - \bar{x}_P$. Realized heritability is estimated as *R*/*S*, where *R* and *S* are estimated from a single generation of selection or cumulated over several generations. *See also* QUANTITATIVE GENETICS. [T.F.C.M.]

realized niche The NICHE in the presence of competitors and predators. Note that the realized niche of a species may differ between populations, depending on the particular species composition of the COMMUNITY. *Cf.* FUNDAMENTAL NICHE. [J.G.S.]

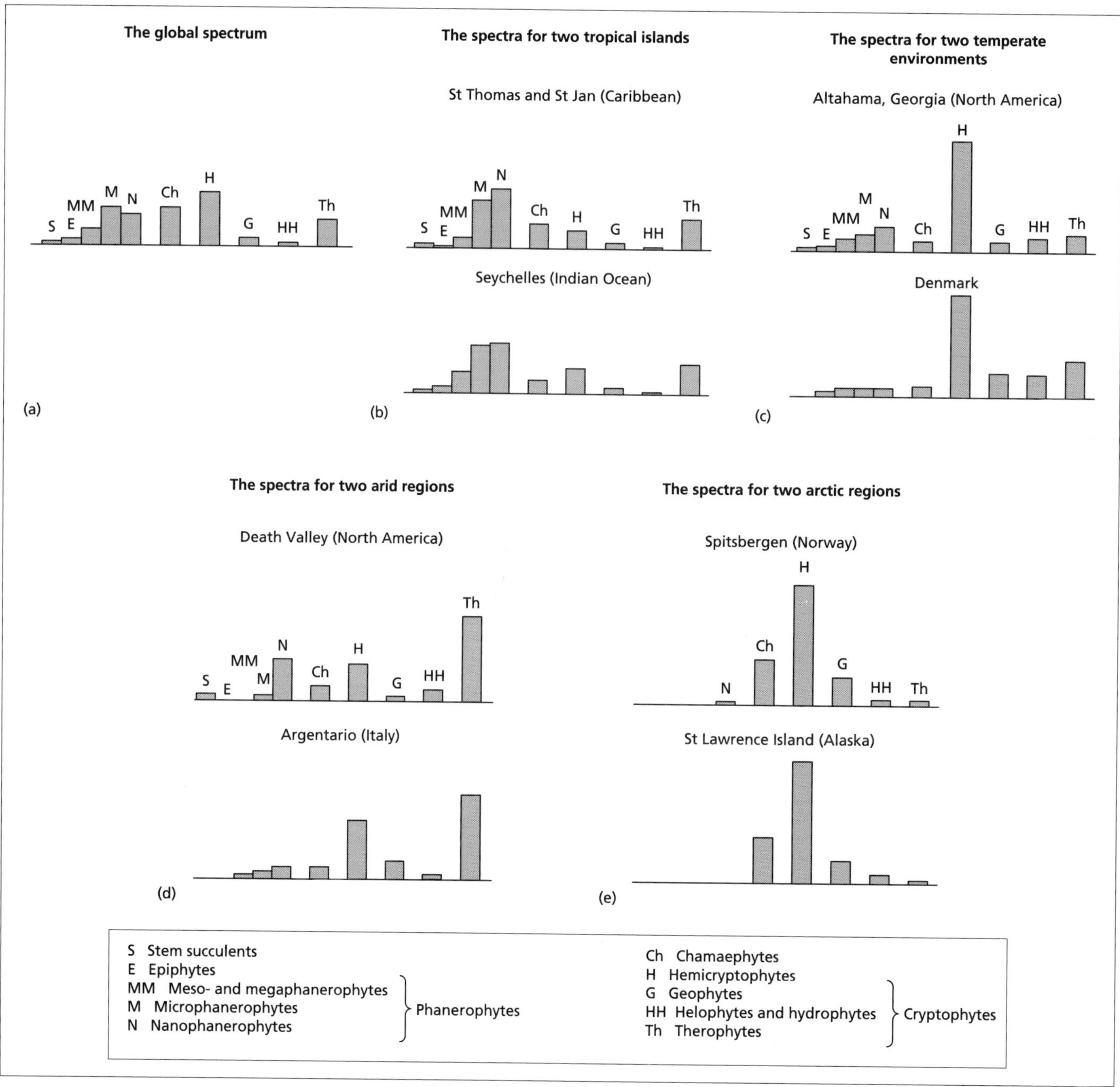

Fig. R6 Whole plant communities can be compared by the method of Raunkiaer (1934). (a) The global life-form spectrum obtained by sampling from *Index Kewensis*. (b) Life-form spectra for the vegetation of two tropical islands. (c) Life-form spectra for two temperate environments. (d) Life-form spectra for two arid environments. (e) Life-form spectra for two arctic regions. The various categories of 'life form' are described in Table R3. (After Begon, M., Harper, J.L. & Townsend, C.R. (1996) *Ecology: Individuals, Populations and Communities*, 3rd edn. Blackwell Science, Oxford.)

recapitulation The notion that an organism reiterates aspects of its PHYLOGENY during its embryological development. The term is derived from the motto 'ONTOGENY recapitulates phylogeny', a popular summary of the classic postulate of the 19th century German zoologist, Ernst Haeckel, who claimed that any organism, as it undergoes embryonic development, goes through (i.e. 'recapitulates') its phylogenetic history in a temporally abbreviated manner. For example, the transient formation of rudimentary gill arches in human embryos would be considered a recapitulation of the fishlike stage in our evolutionary LINEAGE. *See also* HETEROCHRONY; NON-TERMINAL ADDITION. [P.A.]

recessive gene A gene expressed phenotypically in diploid organisms only when it is present in both homologous chromosomes (a homozygous state). The recessive gene is masked in heterozygotes by the presence of its dominant ALLELE. The dominant gene often produces a functional product, while its recessive allele does not. *See also* DOMINANCE. [P.O.]

Reclamation

Terminology

Reclamation is defined by the National Research Council (1992) as a 'process designed to adapt a wild or NATURAL resource to serve a utilitarian human purpose, putting a natural resource to a new or altered use — often used to refer to processes that destroy native ecosystems and convert them to agricultural or urban uses.' Regrettably, the term 'reclamation' is often used interchangeably with the term 'ecological restoration', which aims to return a damaged ECOSYSTEM to a former natural condition. The terms 'reclamation' and 'creation' imply putting a landscape to a new or altered use to serve a particular human purpose. 'Preservation', of course, is the maintenance of the integrity of a natural system.

'Mitigation', another term commonly used in conjunction with ecosystem damage, is simply the intent to alleviate any or all detrimental effects arising from a given action. Reality may not always follow intention, however. For example, mitigation for filling in a wetland in order to build an airport taxiway may involve restoring a nearby, or even somewhat distant, wetland that had been filled in for some other reason, or it could involve creating a wetland on an adjacent area that was formerly upland vegetation. Before human society and REGULATORY AGENCIES appreciated the ecological importance of WETLANDS, they were often reclaimed for housing developments (e.g. Fort Lauderdale, Florida), filled in for agricultural purposes, or filled in for an airport taxiway or highway.

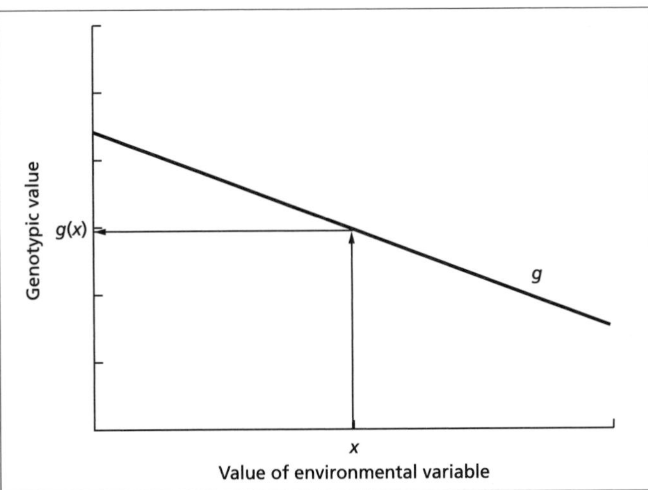

Fig. R7 The genotypic value in a particular environment is given by the genotypic reaction norm, *g*.

Reclamation in practice

Mining areas

One might legitimately ask, why not always restore an ecosystem to some semblance of its predisturbance condition? One reason may be that it is impossible to do so for physical reasons. For example, strip-mined areas on precipitous slopes in southwest Virginia, West Virginia, and parts of other US states, such as Kentucky, present formidable obstacles to re-establishing the predisturbance condition. On steep hillsides, the overburden (SOIL, rock and other materials covering the coal or other materials being mined) is simply pushed over the hillside because if it were moved uphill, it would immediately slide down into the mined area. Once the mining is completed, even if the material can be retrieved, it cannot be readily compacted to fit its original contours, resulting in a bulge in the hillside quite vulnerable to erosion by heavy rains, etc. Furthermore, vegetation is exceedingly difficult to re-establish under these conditions because the SOIL PROFILES have been disrupted and are virtually impossible to re-establish. Even more importantly, seeds and small plants are likely to be washed away during a moderate rainstorm on such steep slopes.

As a consequence, recontouring, a form of reclamation, is often the method of choice. Recontouring permits the establishment of slopes that are less vulnerable to erosion and, consequently, more amenable to the re-establishment of vegetation. Furthermore, reducing RUN-OFF means that HEAVY METALS and objectionable materials such as ACID water are less likely to be transported to adjacent, less-damaged ecosystems.

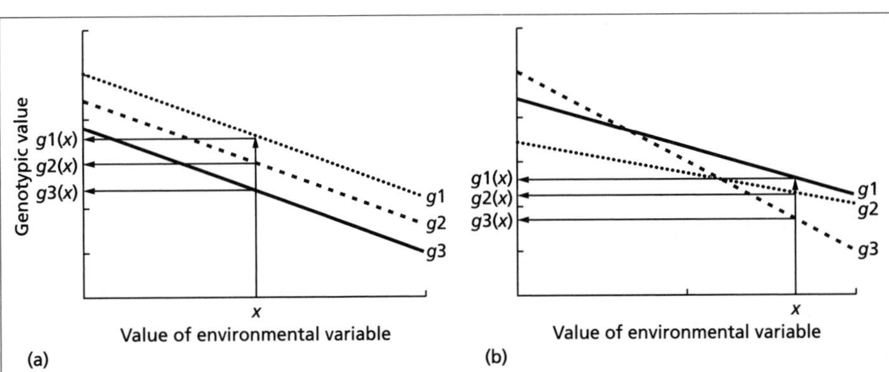

Fig. R8 (a) Reaction norms with identical phenotypic plasticity. (b) Reaction norms differing in phenotypic plasticity, leading to genotype*environment interaction.

Hazardous waste sites

A second situation where reclamation is often more appropriate than ECOLOGICAL RESTORATION is the decommissioning of hazardous waste sites. Many of these have long-term persistent chemicals or radioactive substances. CLEAN-UP of these to the extent that would meet general safety standards for the public or for sensitive WILDLIFE is technologically impossible or difficult, or financially prohibitive. Furthermore, technologies that are superior to present ones, or technologies of comparable effectiveness but that are considerably cheaper, may be developed in the near future. If developed, these technologies may be suitable for substantially reducing risk to human health and indigenous organisms. However, human society, at least for the foreseeable future, will require certain sites to which public access is restricted or prohibited, for carrying out risky procedures. Risks include both security ones and those likely to be encountered by people who are unknowledgeable about hazards of various kinds—in short, most of the general public.

The USA is dotted with Superfund sites (*see* SUPERFUND LEGISLATION) that could be reclaimed for other restricted uses once their present use has been completed. Since some of these sites already have an air, water and land MONITORING system, provisions for security, and professional personnel sufficiently prudent to work on the site, there are compelling reasons for reclamation of this type. Additionally, new areas would not, then, be sacrificed for other activities with hazardous waste risks. This may provide an added incentive for clean-up of the present Superfund sites. For example, government restrictions of use of uncontaminated land need not be applied to these areas, at least not as stringently as for totally new developments. Therefore, an organization might well save both time and money by helping with the reclamation. Battles with environmentalists should be reduced because putting these areas to such use means that uncontaminated systems are not put at risk.

Urban reclamation

Another category where reclamation seems appropriate is in deteriorating urban and suburban areas. A number of communities have initiated strategies to incorporate more naturalistic systems, such as RIPARIAN walkways with native flora, into the urban environment. As a result of such efforts, habitats for a limited number of species of both animals and plants have been developed. Unquestionably, such an effort has economic value because most people enjoy the combination of a naturalistic setting with the advantages of urban or suburban life. Commercial shops in such areas are willing to pay premium prices to be there, and the clientele are willing to pay premium prices to enjoy this restful setting.

HABITAT reclamation not only provides naturalistic settings in urban surroundings but also enlightens and educates people about the importance of ecosystems and the relationship of humans to ecosystems. Due to the mass migration of people into urban and suburban settings, much contact with nature has been lost, at least on a regular basis. While these reclaimed systems are far short of full ecological restoration, they nevertheless provide an increased opportunity for people to interact with other species, thereby increasing their regard for wild species. Since increased environmental literacy is essential to the sustainable use of the planet, and since much of humanity is isolated from natural systems, interacting with naturalistic systems on a regular basis close to home is almost certain to increase the level of environmental literacy and to promote greater respect for natural systems.

Ecoaccident zones

Reclamation may also be more appropriate than full ecological restoration in those areas with frequent ecoaccidents such as OIL SPILLS, releases of toxic chemicals, or intense pressure of one sort or another from human society. Attempting restoration to the predisturbance condition makes little sense if the time-frame is great and another ecoaccident is likely to occur before ecological restoration has been completed. This does not mean that reclamation should be treated more casually than ecological restoration but, rather, that species for the reclaimed area where ecoaccidents are likely, or human pressures are severe, might be more appropriately limited to those species more tolerant of human activities and thus less likely to be harmed by such ACCIDENTS. However carefully the reclaimed site is equipped with tolerant species, it will not be immune to the stress of ecoaccidents, but it should be more resistant to them. Furthermore, since the cost of reclaiming the area is known, the liability for further damaging the system can be determined fairly precisely.

Much consideration has been given to extending the bonding for ecological restoration to all organizations likely to create ecological damage. This is already in use for surface mining in most US states and could easily be extended to other types of reclamation or restoration. As in other areas for which bonds are issued, the integrity and past history of the applicant strongly influence the price of the bond.

Landfill sites

Last, but far from least, reclamation is applicable to LANDFILL sites, which exist wherever human society generates solid waste. Despite the well-intentioned and frequently highly successful efforts toward RECYCLING and diminishing the amount of solid waste generated, society has not yet reached the stage where landfills are no longer necessary. Since landfills store solid waste, they eventually fill up, even if, as in many cases, the mound of solid waste raises the elevation of the area substantially. Once the landfill is capped and closed, it can be reclaimed for other uses, such as a recreational area for human society. Re-establishing the predisturbance vegetation is often not considered a good option because the roots of some trees will penetrate the soil and CLAY caps, resulting in infiltration of the landfill by surface water from RAIN and a subsequent increase in LEACHATE (the result of running water over solid waste). Although most trees have roots that grow horizontally and are not likely to penetrate deeply, laws prohibiting the planting of trees on landfill sites exist in most US states. This makes true ecological restoration difficult and perhaps impossible. However, reclamation of the landfill for a picnic ground, miniature golf course, park, etc., is a desirable option where open space is at a premium. This provides residents of urban areas with limited naturalistic recreation closer to home than most truly natural areas and provides opportunities for interactions with birds and certain other animals, and plants.

Conclusions

Reclaiming surface mining areas, decommissioning hazardous waste sites, establishing natural systems in urban environments, maintaining ecoaccident areas and using landfill sites are but a few illustrative examples of reclamation. Where full ecological restoration is possible, it is often, but not inevitably, a good option to follow. When this is not possible, the resource might well be put to many other uses if all the opportunities for reclamation are examined. Any site that is damaged ecologically, or otherwise, is a candidate for one or more reclamation opportunities. If sustainable use of the planet is our goal, we cannot afford to let areas be permanently damaged so that they are deemed totally unattractive for all uses. Re-establishing their potential for use is what reclamation is about. *See also* REMEDIATION. [J.C.]

REFERENCE

National Research Council (1992) *Restoration of Aquatic Ecosystems: Science, Technology, and Public Policy*. National Academy Press, Washington, D.C.

---

recognition concept of species This concept is derived directly from the fundamental process of eukaryotic sex, which comprises meiosis, FERTILIZATION and syngamy. Only fertilization involves interaction between the sexes, and is thus centrally important in bringing about positive ASSORTATIVE MATING ('like fertilizing like').

Every sexual EUKARYOTE must have a fertilization system to achieve syngamy. The system comprises all structures and processes which together bring about fertilization. In motile organisms this includes a well-developed SPECIFIC-MATE RECOGNITION system (SMRS). The SMRS comprises a signal–response reaction chain which leads to organisms recognizing each other as appropriate mates. In sessile organisms the SMRS may be much reduced. In plants, fungi and primarily sessile animals it is restricted to the exchange of chemical signals between those cells of the organisms that are involved in 'mate recognition' (e.g. pollen/stigma). In motile animals the SMRS includes the aspects of 'COURTSHIP behaviour' concerned with mate recognition.

Thus, the positive assortative fertilization which characterizes species in undisturbed nature is fully accounted for by the positive action of the characters of the fertilization system. Species in the recognition concept can be defined as follows: 'a species is that most inclusive population of biparental organisms which share a common fertilization system' (Paterson 1993).

A critical point for ecology is that the fertilization systems determine species as by-products of the positive assortative mating they lead to. Thus, species are not adaptive entities, but epiphenomena. *See also* MATING SYSTEMS. [H.E.H.P.]

REFERENCE

Paterson, H.E.H. (1993) *Evolution and the Recognition Concept of Species*. Johns Hopkins University Press, Baltimore, Md.

recombinational load The reduction in FITNESS that is a consequence of recombination. Consider two loci A and B, each with two ALLELES per locus (A_1, A_2 and B_1, B_2, respectively). Suppose that the maximal fitness occurs for the GENOTYPE $A_1A_2B_1B_2$. Because of recombination in each generation other genotypes will be formed even though these have lower fitness. Selection would be expected to favour mechanisms that reduce recombination in this case; the inversions found in several taxa, notably *Drosophila*, represent such a mechanism. [D.A.R.]

recovery (waste option) *See* WASTE DISPOSAL.

recreation ecology The branch of ecology concerned with the impact of recreational and leisure activities on natural populations and the environment. The leisure activities of human beings are increasing and their impact upon the environment is being felt more and more. As a result, the new discipline of recreation ecology is emerging and will hopefully form a basis of knowledge that can be applied to the problems of leisure pressures.

Recreation can impact on the environment in many different ways. Sometimes whole areas may be taken over for leisure activities that are incompatible with WILDLIFE conservation. For example, the construction of playing fields, covered leisure complexes, or motor-cycle scramble areas provides little opportunity for the survival or encouragement of wildlife. On the other hand, golf courses, and lakes for angling or sailing, may provide habitats in which plant and animal life can maintain itself.

Wildlife itself is now a major leisure industry, and so-called 'ecotourism' results in pressures that may threaten the very habitats or wildlife supporting the industry. Access to wild and remote areas, for example, can create DISTURBANCE and even physical destruction of habitats as roads, hotels, campsites, etc., are constructed. Sensitive animal species, such as the gorillas (*Gorilla gorilla*) of western Uganda, need to be protected from excessive exposure to human contact.

HABITAT damage may be controlled by restricting access to certain locations or confining visitors to marked trails. Limitation of the number of visitors given access to a site within any particular day can also keep damage within acceptable bounds. The control of visitor numbers may be combined with economic returns by charging tourists for entry into parks. Such gains need to be fed back into local communities so that these can appreciate the economic value of conservation as opposed to other forms of land use or game poaching.

The confinement of visitors inevitably results in particularly heavy pressure on specific areas of a reserve. It may be necessary, therefore, to protect soils from potential erosion by paving paths or creating wooden walkways over sensitive vegetation (such as wetlands).

Aquatic (including marine) habitats present particular problems and are especially attractive for leisure activities, such as boating, swimming and angling. Measures need to be taken to protect water bodies from the side effects of leisure use, such as POLLUTION, EUTROPHICATION, disturbance of sediments or marginal vegetation and, in the case of tropical marine situations, the effects of trampling and boat damage to CORAL REEFS. [P.D.M.]

recruitment The addition of individuals to a POPULATION. The term is often treated as synonymous with NATALITY, but properly it also includes IMMIGRATION. Net recruitment is the difference between the gains and losses of individuals from a population; in a closed population this is equivalent to births (natality) minus deaths (MORTALITY) (Fig. R9). [A.J.D.]

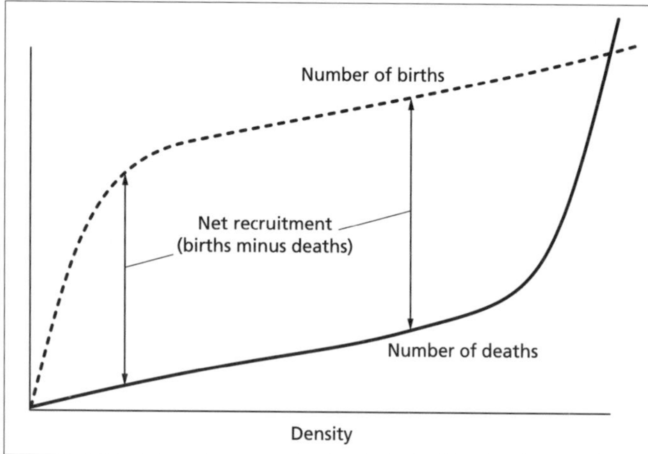

Fig. R9 Diagram indicating the derivation of recruitment in a closed population as the difference between the number of births (dashed line) and the number of deaths (solid line). Because of the relationships of births and deaths to density, indicated recruitment will be small at low densities because the difference between the numbers of births and deaths is small, large at intermediate densities where the difference is large, and small again at high densities where the difference between births and deaths is again small. Plotting recruitment against density will therefore, in this case, produce the typical n-shaped net recruitment curve.

recurrence surface The stratigraphic profiles of BOGS often reveal changes in the degree of DECOMPOSITION of the PEAT layers, reflected in the colour and texture of the peat. The replacement of a dark, well-humified peat by a light, poorly humified layer indicates an increase in the rate of peat formation in response to increased wetness. Such a transition is termed a recurrence surface and, if found extensively across a peat deposit, may be the product of general climatic change, although the possibility of local pool development or surface flooding also needs to be considered in the interpretation of these features. Recurrence surfaces in OMBROTROPHIC mires (raised and blanket bogs) are being investigated as proxy climate indicators. [P.D.M.]

recycling The processing of materials so they are available for REUSE by living organisms, manufacturing industries, etc. All matter cycles in nature (*see* BIOGEOCHEMICAL CYCLE) and so, by definition, is renewable. But such transformations are not always reversible; moreover they require energy and may lead to POLLUTION. So the recycling of human wastes requires care to optimize environmental protection. The most effective form of recycling from this point of view is usually reuse. Other forms of recycling involve reuse of materials but not necessarily in their original form. Direct recycling involves the reconversion of materials into a similar functional form; for example, pulp from WASTE paper being made into new paper. Indirect recycling involves conversion to a different functional form; for example, pulp from waste paper being made into packaging board. This term might even be applied to INCINERATION when heat is recovered and used productively (*see* COMBINED HEAT AND POWER). *See also* WASTE DISPOSAL. [P.C.]

Red (Data) Book A catalogue, published by the IUCN, listing species that are rare or in danger of becoming extinct locally or nationally. [P.C.]

Red List A list of 23 dangerous substances designated under the UK ENVIRONMENTAL PROTECTION ACT 1990 as chemicals whose DISCHARGE to water should be minimized by application of the BATNEEC (BEST AVAILABLE TECHNIQUES NOT ENTAILING EXCESSIVE COSTS) principle. Discharge CONSENTS are required for these substances. *See also* BLACK LIST; LISTS; PRIORITIZATION OF CHEMICALS. [P.C.]

Red Queen's hypothesis

The Red Queen's hypothesis, introduced by L.M. Van Valen in 1973, is that the sum of the momentary absolute fitnesses of interacting species in a BIOTA is constant. This zero-sum constraint for ecological or evolutionary change is a strong constraint; it is not meaningful in some approaches and it is counterintuitive in others. Many applications of what is supposed to be the hypothesis omit the constraint which lies at its heart. Such ambiguity does not promote clarity.

Another ambiguity applies to the name itself, which is often given without the possessive. However, the Queen herself stated her hypothesis, in *Through the Looking Glass*: 'Now here, you see, it takes all the running you can do, to keep in the same place.'

The Red Queen's hypothesis initially was an explanation for an unexpected pattern of extinction, namely that (except sometimes for MASS EXTINCTION) extinction is independent of the age of the TAXON (Fig. R10) but is selective in other ways. As has been demonstrated, the probability of extinction in an adaptively unified group is stationary over geological time, varying around a constant mean except usually for mass extinctions. Nevertheless, because some whole groups have a characteristically higher rate and tend to be eliminated, the probability per taxon per time for the world's

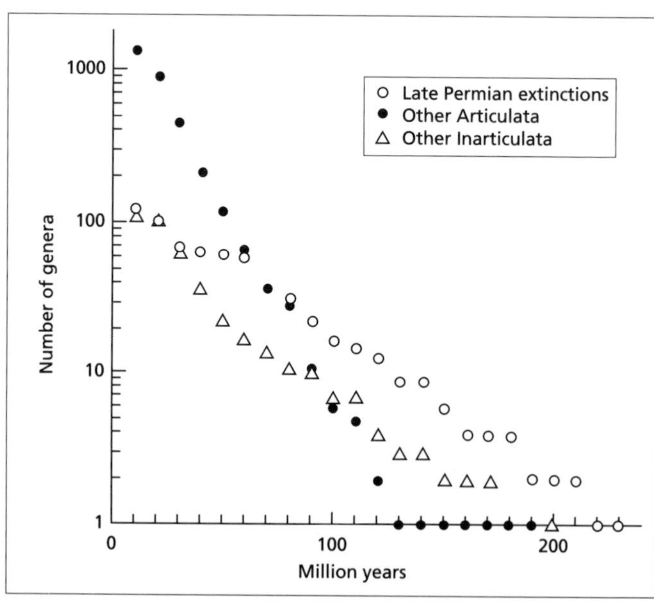

Fig. R10 The number of brachiopod genera surviving at different intervals (in million years) after their origination. (After Van Valen, L.M. (1973) A new evolutionary law. *Evolutionary Theory* **1**, 1–30.)

marine biota decreased exponentially through the PALAEOZOIC era and again after the Permian extinction, which reset the process to a higher value.

Even within an adaptive zone more susceptible taxa go extinct first. New taxa originate fast enough to prevent a decline in DIVERSITY, or else the adaptive zone itself shrinks proportionally. Because of the zero-sum constraint, some surviving taxa then become less fit than before, giving about the same overall distribution of susceptibility (itself determined otherwise, as by the causes of abundance distributions) and thus reconstituting the initial conditions. If successful origination is negligible and the adaptive zone is unchanged, the Red Queen's zero-sum constraint then requires that some surviving taxa become more, rather than less, fit, but again with the result of reconstituting the original distribution. The mean probability of extinction is thus predicted to be stochastically constant in both survivorship time and real time, as is observed.

The entire process can, in principle, be self-perpetuating, by biotic interactions. This has been misinterpreted to mean that the Red Queen uses only biotic interactions. Insults of the physical environment, however, can also have a stationary distribution over time, especially when the metric is merely severity of effect. (Again, mass extinctions can be exceptional.) These insults may, at the opposite extreme, entirely cause the individual extinctions, and the Red Queen proceeds as before.

It is impossible to estimate, even roughly, the relative importance of physical and biotic effects by such methods as correlating extinction rate with physical stresses. This is because the two kinds of effects interact strongly. For example, it is common for one species to outcompete another in one environment and lose elsewhere. Thus a single set of extinctions could be caused 90% physically and also 90% biotically. *See also* COMPETITION; PREDATOR–PREY INTERACTIONS. [L.M.V.V.]

REFERENCE

Van Valen, L.M. (1973) A new evolutionary law. *Evolutionary Theory* **1**, 1–30.

---

red tide Red, brown or yellowish discoloration of COASTAL waters during ALGAL BLOOMS of dinoflagellate PHYTOPLANKTON, due to high concentrations of accessory pigments other than chlorophyll *a*. Some Dinophyceae produce potent TOXINS that can cause illness or death of marine organisms and/or humans. TOXIC BLOOMS fall into three categories according to the organisms affected:

1 fish but few invertebrates killed (*Ptychodiscus brevis*);

2 primarily invertebrates killed (several species of *Gonyaulax*);

3 few marine organisms killed, but toxins concentrate within the siphons and digestive glands of shellfish, and can kill humans if eaten (*Gonyaulax catenella* and *G. tamarensis*).

The latter is called paralytic shellfish poisoning (PSP) and is caused by a powerful neurotoxin, saxitoxin. In extreme cases, paralysis of the diaphragm may result in respiratory failure and death. No specific antidote is known.

It is not clear what factors cause red tides, although they often occur in association with surfacing of NUTRIENTS by UPWELLING or tidal TURBULENCE, or with heavy surface run-offs that wash nutrients into the SEA. Toxic blooms may be initiated from benthic resting stages (cysts), which serve as seed populations when appropriate environmental factors cause resuspension and/or excystment. [J.L.]

Redfield ratio The ratio of the elemental composition of CARBON, NITROGEN and PHOSPHORUS in living matter and material of biogenic origin is commonly 106 : 16 : 1 (C : N : P). This ratio is named after A.C. Redfield, who brought it to the attention of the scientific community. [S.J.H.]

REFERENCE

Redfield, A.C. (1958) The biological control of chemical factors in the environment. *American Scientist* **46**, 205–222.

redox potential (Eh) Redox potential is a measure of the ability of an electron carrier to act as a reducing (addition of electrons) or oxidizing (removal of electrons) agent.

$$\mathrm{Eh} = E_0 - 59(a/n)\mathrm{pH}$$

Where E_0 is the standard potential at equal activities of reduced and oxidized species, a is the number of protons transferred, and n is the number of electrons involved in the reaction. The capacity for different carriers to transfer electrons is measured in volts or millivolts on a negative scale, as the yield of electrons is equivalent to that of liberated protons and can be measured electrochemically. The STANDARD reference is a HYDROGEN ION concentration of unity (pH = 0) and against this standard the hydrogen electrode at pH 7 has a potential difference of –0.42 V.

A redox system consists of two substances which differ from each other only in their state of oxidation; a definite electrical potential exists between such partners of a redox system, which is usually measured at a standard reference pH value of 7.0. The more negative the value the greater the capacity of the electron carrier to act as a reducing agent. The redox potential of the biological electron carrier system $NADH + H^+/NAD^+$ at equilibrium is –0.32 V. The actual redox potential measured in any system other than standard ones depends on the concentration of the oxidized and reduced species.

In the electron transfer system, biological electron carriers (e.g. ferredoxin, plastoquinone) are arranged in order of increasing potential (negative to positive). The approximate E_0 values at pH 7 are: hydrogen, –0.42; nicotinamide coenzymes, –0.32; flavoprotein, –0.06; cytochrome *b*, +0.04; cytochrome *c*, +0.026; cytochrome *a*, +0.29; OXYGEN, +0.82.

The redox potential of certain environments such as anaerobic muds can be measured against a PLATINUM electrode which responds to the electrochemical potentials of all redox couples present in the system and serves as an indication of the state of reduction of the SOIL. In oxidized soils, the redox potential is reported to range between +400 and +700 mV; in anaerobic soils, NITRATE reduction begins at +338 mV, followed by MANGANESE at +200 mV, IRON at around +120 mV and sulphate reduction between –75 and –150 mV (Gambrell & Patrick 1978). [R.M.M.C.]

REFERENCE

Gambrell, R.P. & Patrick, W.H. Jr. (1978) Chemical and microbiological properties of anaerobic soils and sediments. In: *Plant Life in Anaerobic Environments* (eds D.D. Hook & R.M.M. Crawford), pp. 375–423. Ann Arbor Science, Ann Arbor, Mich.

redox potential discontinuity (RPD) The depth below the SEDIMENT–WATER interface marking the transition from chemically oxidative to reducing processes. The depth of the RPD can be measured by a change in the Eh potential from positive to negative and can be detected visually as a change in sediment colour from brown (oxidized) to grey (RPD layer) to black (anoxic). [V.F.]

redox reactions Reactions involving both reduction of an oxidizing reagent and oxidation of a reducing agent. Oxidation always involves electron loss, and reduction involves electron gain. For example, in well-aerated soil, any IRON in the ferrous state is oxidized to the ferric state:

$Fe^{2+} \rightarrow Fe^{3+} + e^-$

whereas in waterlogged soil the lack of OXYGEN creates a reducing environment in which Fe^{2+} is formed, and the soil changes colour from reddish to greenish-grey.

The REDOX POTENTIAL measures the tendency of a cell, molecule or ION to donate electrons, and is expressed as a voltage relative to the voltage produced by a standard HYDROGEN/PLATINUM cell. [J.G.]

reduced major axis method *See* REGRESSION ANALYSIS, LINEAR.

reductionism The opposite to HOLISM. [P.C.]

reed-bed purifiers Artificially constructed WETLANDS created for the sole purpose of wastewater or stormwater treatment. During the last decade, they have been put into operation as a means of low-cost treatment applicable to urban and industrial areas across the world.

Constructed wetlands may be classified according to the life form of the dominating macrophyte into: free-floating, submerged and emergent macrophyte-based systems (Brix & Schierup 1989). The design of the systems also varies according to the flow regime and the medium used. They can be constructed with free surface flow (water surface is exposed to the atmosphere), subsurface horizontal flow (water level is maintained below the surface media placed in the beds), or vertical (non-saturated) flow (water is usually applied at time intervals, and allowed to percolate through the medium). The medium can be SOIL, gravel or, more recently, industrial substratum.

A typical arrangement of a reed bed consists of a basin, sealed to prevent percolation of effluent to the subsoil, and containing a bed matrix in which a stand of *Phragmites* is grown (Findlater *et al.* 1990). The objective of the engineering bed is to reduce greatly the concentration of pollutants during the passage of the effluent water through the bed.

The major system components involved are:

1 the media—physical filtration and chemical interactions;

2 the reeds (*Phragmites*)—plant uptake and transport of OXYGEN to the RHIZOSPHERE;

3 the rhizosphere organisms — aerobic and anaerobic bacteria influence NITROGEN transformations by nitrification followed by DENITRIFICATION to dinitrogen (N_2) gas as the end product.

See also BIOREMEDIATION; SEWAGE TREATMENT WORKS. [A.D.]

REFERENCES

Brix, H. & Schierup, H. (1989) The use of macrophytes in water pollution control. *Ambio* **18**, 100–107.

Findlater, B.C., Hobson, J.A. & Cooper, P.F. (1990) Reed bed treatment systems: performance evaluation. In: *Constructed Wetlands in Water Pollution Control* (eds P.F. Cooper & B.C. Findlater). Proceedings of the International Conference on the Use of Constructed Wetlands in Water Pollution Control, Cambridge, 24–28 September 1990, pp. 193–205. Pergamon Press, Oxford.

reef An off-shore consolidated rock, often presenting a hazard to navigation. *See also* CORAL REEFS. [V.F.]

reforestation The replacement of FOREST cover which has been removed either recently or in the past. Modern methods of forest management often involve planting seedlings, now frequently produced in enormous numbers in forest nurseries. Continuous-cover FORESTRY aims to avoid erosion and other problems of reforestation by felling trees in small groups, in many cases having already initiated a fresh crop beneath them. Reforestation of long-cleared areas may result when agricultural land is abandoned and woody VEGETATION develops as a result of successional processes, or by deliberate planting, or by management policies designed to extend existing areas of native WOODLAND into neighbouring GRASSLAND or HEATHLAND through natural seeding. *Cf.* DEFORESTATION. *See also* AFFORESTATION. [J.R.P.]

refuges Locations (in time or space) where one species can escape the competitive or predatory interactions of another species. [B.S.]

refugia

1 Populations of animals and plants that survive in HABITATS of limited area in regions which would not otherwise support such organisms or communities. A refugium typically constitutes the flora and fauna of an ENVIRONMENT or COMMUNITY that was once widespread but is now considerably diminished in area. For example, the BOREAL refugia found today at the higher elevations of mountains in temperate Europe are genetically isolated representatives of the early Holocene communities of lowland western Europe. Such refugia are valued in WILDLIFE conservation. Increasingly, refugia are being created not only as a result of long-term CLIMATE CHANGE but as a consequence of climatic trends over recent decades and by rapid developments in land exploitation, particularly into tropical, and hitherto remote areas.

2 Related, but somewhat different or distinctive habitat and/or NICHE space that ameliorates COMPETITION.

[G.A.F.H.]

regeneration and cyclic change Plant communities are subject to both small- and large-scale damage and degeneration; their regeneration involves a spectrum of responses varying from the production of vegetation very different from that originally present to one which is broadly the same. Cyclic change involving gap regeneration (*see* GAP PHASE REGENERATION) is the sum of the processes which lead to a community having approximately the same composition in the long term and over a relatively large area despite the death of individuals through old age and periodic destruction of the biomass by outside influences.

The exact processes of regeneration vary greatly between communities. In some cases regeneration occurs after large-scale DISTURBANCES, such as fire, wind-blow or (less often) flood over large areas. In others it occurs on patches that are small relative to the dominant plants, for example in forests in gaps made by the fall of single trees, or in grassland on mounds made by gophers or moles.

In relatively species-poor communities there may often be a high degree of REPEATABILITY in the SUCCESSION of species after the disturbance, leading back to the late-successional species (*see* CYCLIC SUCCESSION). In very species-rich communities the species occupying an area the size of an average mature individual of the tallest species may vary greatly through time, for example in a tropical lowland RAIN FOREST. Then only the composition of relatively large areas remains relatively constant through time.

Replacement of WOODLAND trees which are felled or die naturally is often from seed; native trees which arise in this way are known as maidens. Woodland regeneration by this means is often hindered by the destruction of seed by microorganisms and animals, notably birds and small mammals, and the effects of shade, disease and GRAZING on the seedlings and young trees. Aspens (*Populus tremula*) and most elms reproduce from suckers, while other species including Norway spruce (*Picea abies*) and western red cedar (*Thuja plicata*) sometimes form new individuals when roots grow from low-growing branches (LAYERING). Wood-producing trees can be managed by coppicing, pollarding or shredding; the tree regenerating after the crop has been taken. If only one shoot is allowed to develop from a COPPICE stool a new tree is formed.

Before 1950 many studies in vegetation dynamics concerned succession, often PRIMARY SUCCESSION. Since then much more attention has been given to studies of cyclic change, which were greatly stimulated by the presidential address of Watt (1947), and of the regeneration of badly damaged communities. [P.J.G. & J.R.P.]

REFERENCE

Watt, A.S. (1947) Pattern and process in the plant community. *Journal of Ecology* **35**, 1–22.

regeneration complex The surfaces of many RAISED BOGS bear mosaics of pools and hummocks and it was once thought that these replace one another through time in a regular sequence, pool–hummock–pool–hummock, and so on, forming a 'regeneration complex' over the MIRE surface. Many recent stratigraphic studies, however, have revealed that pools, once initiated, usually persist because DECOMPOSITION is relatively fast at the PEAT–WATER interface, so the concept of a regeneration complex is not currently in favour. [P.D.M.]

regeneration niche An expression of the requirements for a high chance of success in the replacement of one mature individual plant by a new mature individual of the same species (Grubb 1977). The idea requires a statement of the requirements at every stage of the LIFE CYCLE (production of viable seed, dispersal of seed, germination, establishment and onward growth). It thus embodies consideration of not only different kinds of gaps that may be filled, but also variation from year to year in the amount of seed reaching a site, germinating there and surviving both predation and physical hazards ('variable recruitment'). [P.J.G.]

REFERENCE

Grubb, P.J. (1977) The maintenance of species-richness in plant communities: the importance of the regeneration niche. *Biological Reviews* **52**, 107–145.

region of compliance Part of an ecosystem in which COMPLIANCE with an environmental STANDARD is required and/or exists; for example, outside the MIXING ZONE. [P.C.]

register of effects LISTS of potential ENVIRONMENTAL IMPACTS compiled as part of an ENVIRONMENTAL MANAGEMENT SYSTEM and taking into account all aspects of an operation from 'cradle to grave', including past activities that might have led to, for example, land CONTAMINATION, and considering normal, abnormal and accidental situations. In practice, the register usually consists of lists of potentially polluting EMISSIONS (potential causes of POLLUTION) rather than actual effects on ecological systems. This is not surprising given the complexity of making precise linkages between complex business operations and complex ecological processes, especially at a global level. [P.C.]

registers, environmental

1 Under environmental (protection) legislation, any of various LISTS of environmental information that are open to public scrutiny. Such registers are usually compiled by authorities as a statutory requirement.

2 In the context of ENVIRONMENTAL MANAGEMENT SYSTEMS, any of various lists of applicable legislation and possible adverse environmental effects.

[P.C.]

Regosols *See* SOIL CLASSIFICATION.

regression analysis, curvilinear Many relationships in ecology can be more appropriately represented by a curved line of some kind, rather than a straight line (*see* REGRESSION ANALYSIS, LINEAR). One family of curves (of which the straight line is just one example) are known as polynomials. Three are shown below.

$y = a + bx$	straight line (linear)
$y = a + b_1x + b_2x^2$	quadratic
$y = a + b_1x + b_2x^2 + b_3x^3$	cubic

Although the first equation is traditionally called linear, it should strictly speaking be simply called the equation of a straight line. This is because mathematicians refer to all the above as linear relationships, because they are made up from a linear combination of quantities involving x. Alternatively, EQUATIONS such as that for exponential increase:

$$y = ae^{bx}$$

are said to be non-linear or, more accurately, non-linear in their PARAMETERS. For more details about such curvilinear relationships see Brown and Rothery (1993) and Sokal and Rohlf (1995). [B.S.]

REFERENCES

Brown, D. & Rothery, P. (1993) *Models in Biology: Mathematics, Statistics and Computing*. John Wiley & Sons, Chichester.

Sokal, R.R. & Rohlf, F.J. (1995) *Biometry*, 3rd edn. W.H. Freeman, New York.

Regression analysis, linear

Linear regression analysis is a technique that fits a straight-line relationship (a regression line) to a set of paired observations, using the simple straight-line equation $y = a \pm bx$. The quantity a denotes the point that the regression line crosses the y-axis (the y intercept) and the quantity b is the slope (steepness) of the regression line. If b is positive then increasing x means increasing y. If b is negative then increasing x means decreasing y. Figure R11 illustrates different values of a and b.

The purpose of all regression is to enable the ecologist (or other user) to predict values of y from values of x. Ecologists may also use regression to help them explore the possible causation of changes in y by changes in x, or explain some of the variation in y by x. Note, however, that the fitting of a regression line to a set of paired observations, and calling one the dependent variable, does not mean that y is caused by x. From the point of view of causation there are two fundamentally different ways of obtaining a set of paired observations. First, we could manipulate or specifically choose values of an independent variable (x), and subsequently measure the response of a dependent variable (y). This is the more usual situation for regression and is more correctly termed model I regression. Second, we could simply collect pairs of covarying observations and call one of them y and one of them x. This second situation is often more correctly analysed using CORRELATION, unless one specifically wants to predict one variable from another, when you should use a model II regression.

Model I regression

An ecologist investigating the ABUNDANCE of a limpet (*Ancylus fluviatile*) in a freshwater tarn obtained varying numbers of snails from 1 m² quadrats placed at different depths of water, along two transects (I and II).

	Depth (cm) (x)				
	5	10	15	20	25
No. of snails (y) I	76	34	12	23	4
II	40	50	55	2	0

Several quantities are now calculated, as set out below (note that these are the same quantities as used in the PRODUCT–MOMENT CORRELATION COEFFICIENT). Because there are two transects, each with the same chosen values of x, there are in fact 10 pairs of observations (5 76, 5 40, 10 34, 10 50, etc.).

n (number of paired observations) = 10

$$\sum x = 150 \qquad \sum y = 296$$

$$\sum x^2 = 2750 \qquad \sum y^2 = 14750 \qquad \sum xy = 3025$$

$$\sum x^2 - \left(\sum x\right)^2/n = 500 \qquad \sum y^2 - \left(\sum y\right)^2/n = 5988.4 \qquad \sum xy - \left(\sum x \sum y\right)/n = -1415$$

On the last line, the first quantity is the SUM OF SQUARES of x, the second quantity is the sum of squares of y, and the third quantity is the SUM OF PRODUCTS. The first two quantities are always positive, the third quantity can be positive or negative depending upon the relationship between x and y. Dividing each quantity by $n-1$ (DEGREES OF FREEDOM) will give the VARIANCE of x, the variance of y, and the COVARIANCE of x and y. The slope of the regression line is:

$$b = \frac{\sum xy - \left(\sum x \sum y\right)/n}{\sum x^2 - \left(\sum x\right)^2/n}$$

$$b = \frac{-1415}{500} = -2.83$$

The y intercept (a) is now calculated by substituting the value of b in the equation:

$$a = \bar{y} - b\bar{x}$$

$$a = 29.6 - (-2.83)15$$

$$a = 29.6 + 42.4$$

$$a = 72$$

The regression line is therefore $y = 72 - 2.83x$, so that, for example, at a depth of 12 cm (x) we would predict, on average, 38 snails. The data and the fitted regression line are shown in Fig. R11(b). The regression line will pass through the points $(0,a)$ and $(\bar{x}\bar{y})$ and this is the simplest way to put the line on to the graph. This regression line 'fits' the data in a specific way. It is the line which minimizes the sum of the squared differences between each data point and the line — it is the line with the least squares (*see* LEAST SQUARES METHOD).

The next stage in the analysis is to see if the fitted regression line is statistically significant. This can be done in two ways: by ANALYSIS OF VARIANCE (ANOVA), or by computing the STANDARD ERROR of the slope to determine if b is significantly different from zero.

Using ANOVA, we partition the total sum of squares of y (SS_{total})

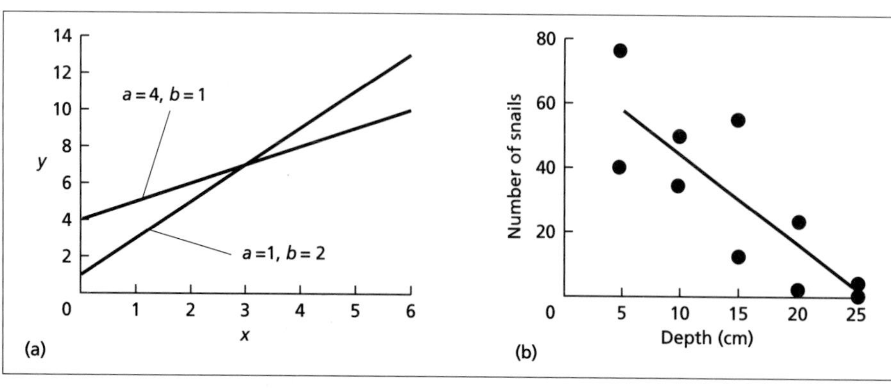

Fig. R11 (a) Illustrative regression lines for $y = 1 + 2x$ and $y = 4 + x$. (b) Data and regression line for number of snails against depth of water in a freshwater tarn. Notice that the regression line is not extended beyond the data.

into two components: that explained by the linear regression ($SS_{regression}$) and that left unexplained ($SS_{residual}$). Since no other source of variation is considered we have:

$$SS_{total} = SS_{regression} + SS_{residual}$$

The regression sum of squares is calculated by dividing the square of the sum of products by the sum of squares of x. The unexplained or residual sum of squares is then simply obtained by subtraction. The equations for the three quantities, and the values for the snail example are:

$$SS_{total} = \sum y^2 - \left(\sum y\right)^2 / n = 5988.4$$

$$SS_{regression} = \frac{\left[\sum xy - \left(\sum x \sum y\right)/n\right]^2}{\sum x^2 - \left(\sum x\right)^2 / n} = \frac{\left(-1415\right)^2}{500} = 4004.4$$

$$SS_{residual} = \left[\sum y^2 - \left(\sum y\right)^2 / n\right] - \frac{\left[\sum xy - \left(\sum x \sum y\right)/n\right]^2}{\sum x^2 - \left(\sum x\right)^2 / n}$$

$$= 5988.4 - 4004.4 = 1984$$

The complete ANOVA can now be set out in a table, as below. Notice that degrees of freedom are also additive and for the regression component is always equal to 1. Total degrees of freedom = $n-1$. The quantities called 'mean squares' are obtained by dividing a sum of squares by its appropriate degrees of freedom (although traditionally in ANOVA they are called 'mean squares' they are actually variances).

Source of variation	Sum of squares	Degrees of freedom	Mean square	*F*
Regression	4004.4	1	4004.4	16
Residual	1984.0	8	248.0	
Total	5988.4	9		

The final column shows the result of an F TEST. This ratio is obtained by dividing the regression mean square by the residual mean square. Because the ALTERNATIVE HYPOTHESIS is H_1: $\sigma_1^2 > \sigma_2^2$ it is a ONE-TAILED TEST. If we look at STATISTICAL TABLES of F, the 1% value, for 1 and 8 degrees of freedom, is 11.26 and therefore the above value of 16 is very significant. Hence, there is a very significant negative, linear relationship between numbers of snails and depth of water.

Computing the standard error of the slope, to determine if b is significantly different from zero, involves the same quantities as used in the ANOVA. We first calculate the residual variance, as in the table above, using the equation

$$s_R^2 = \frac{1}{n-2}\left\{\left[\sum y^2 - \left(\sum y\right)^2 / n\right] - \frac{\left(\sum xy - \left(\sum x \sum y\right)/n\right)^2}{\sum x^2 - \left(\sum x\right)^2 / n}\right\} = 248$$

The standard error of b (SE_b) is then computed as:

$$SE_b = \frac{s_R}{\sum x^2 - \left(\sum x\right)^2 / n} = 0.0315$$

To see if the calculated slope is significantly different from zero, we divide b by its standard error to obtain the test statistic t. For the snail data this is $t = 2.83/0.0315 = 90$, which far exceeds even the 1% value of t for $n-2=8$ degrees of freedom.

Model II regression

This is a controversial subject. There are several methods (e.g. reduced major axis, major axis and Bartlett's three-group method) which can be used depending upon the intentions of the ecologist and the nature of the observations. Of course, it is possible mathematically to apply the equations for a model I or least squares regression to a model II situation, but the regression coefficient a and b will be biased. [B.S.]

REFERENCES

Mandel, J. (1964) *The Statistical Analysis of Experimental Data.* John Wiley & Sons, New York.

Sokal, R.R. & Rohlf, F.J. (1995) *Biometry,* 3rd edn. W.H. Freeman, New York.

---

regular distribution A term used to describe the spatial DISTRIBUTION of the individuals in a POPULATION. Other equivalent terms are uniform distribution, even distribution or underdispersion. When individuals in a population follow a regular distribution, the VARIANCE (s^2) of individuals per SAMPLE is less than the MEAN number ($\bar{x}$) of individuals per sample ($s^2 < \bar{x}$). This is frequently summarized by saying that the variance : mean ratio is less than 1 ($s^2/\bar{x} < 1$). *See also* AGGREGATED DISTRIBUTION; DISPERSION; RANDOM DISTRIBUTION. [B.S.]

regularity The type of DISPERSION pattern shown by organisms, which may be random, regular or clumped (aggregated, contagious). Records from QUADRAT or pinframe SAMPLING can be used to determine the type of DISTRIBUTION, if MEAN values ($\bar{x}$) and VARIANCE (s^2) are calculated. If the ratio $s^2/\bar{x} \approx 1$, the individuals are randomly distributed; if this value is significantly greater than 1, distribution is clumped (overdispersed); and if the value is significantly less than 1, individuals are in a regular pattern (underdispersed).

In RANDOM DISTRIBUTIONS, individuals occur strictly according to chance and independently of all other individuals present, with no single factor having a significant effect on dispersion. Randomly distributed populations are occasionally found where a species invades a homogeneous bare area, but they are not common. In REGULAR DISTRIBUTIONS individuals occur in a uniform order, such as that in which trees are often planted. In natural situations, regular distribution is rare, often because of HETEROGENEITY in the environment or nearness to parents, but it may be found where COMPETITION for a certain factor is intense (e.g. for water in deserts). In clumped distributions individuals are present in groups, being clustered or aggregated together. The great majority of plant populations are clumped, exhibiting what is termed pattern.

Clumping often results from the mode of growth and type of vegetative propagation (e.g. spread by rhizomes), as well as proximity to parent plants and environmental differences. [A.J.W.]

regulated-escapement harvesting A form of HARVESTING that aims to ensure that the number of adults escaping harvest is sufficient to provide (potentially) satisfactory RECRUITMENT of young to the population. It thus depends entirely on the relationship between the size of the adult stock and the recruitment of young. Below some natural THRESHOLD the recruitment will tend to decline, reaching zero at some crucial low level of adult stock. Such systems often work well where the LIFESPAN of a species is relatively short. For example, in some squid and shrimp fisheries, because the lifespan of the harvested species is so short, the sole practicable management objective is to ensure that sufficient adults survive to breed. In other stocks the REGULATION of escapement can be an easy way of managing the fishery without caring about the actual level of effort or catch. For example, in some fisheries the adult stock is monitored and once it reaches a level close to that which is deemed unsafe, fishing is stopped. *See also* FISHERIES, CONSERVATION AND MANAGEMENT; HARVEST METHOD; SUSTAINABLE YIELD. [J.B.]

regulated-percentage harvesting This is an uncommon term, but basically it refers to the simple idea of having a constant level of MORTALITY inflicted on the harvested population. It is closely related to CONSTANT-EFFORT HARVESTING, and is also an implicit part of the analysis that is involved in density-dependence models where a fixed fishing mortality is applied. Theoretically it is an attractive mechanism as long as the percentage harvest taken is below the maximum percentage that the stock can bear. *See also* FISHERIES, CONSERVATION AND MANAGEMENT; HARVEST METHOD; SUSTAINABLE YIELD. [J.B.]

regulating factors MORTALITY (or FECUNDITY-reducing) factors which act in a density-dependent manner, such that the percentage dying increases with POPULATION DENSITY (or the proportion of realized fecundity declines with increasing population density), and this DENSITY DEPENDENCE is responsible, in interaction with density-independent factors, for determining the equilibrium POPULATION SIZE. All regulating factors are density dependent, but not all density-dependent factors are regulating.

Populations lacking any regulating factors would either decline exponentially to extinction or would increase without bounds (strictly speaking, they would increase until they were eventually, but inevitably, subject to regulating factors like COMPETITION for space or shortage of food). Other processes which are potentially regulating include density dependence in the age of first reproduction (females breed first at older ages in dense populations than in sparse) or in the population sex ratio (males make up a greater fraction of the population at high density than at low). *See also* POPULATION REGULATION. [M.J.C.]

regulation Regulation means control according to defined rules or principles, and in ecology regulation usually refers to POPULATION REGULATION. [R.H.S.]

The Regulation and Control of the Deliberate Release of Genetically Modified Organisms *See* TRANSGENIC ORGANISMS.

regulator An organism that can maintain some aspect of its physiology (e.g. body temperature) constant despite different and changing properties of the external environment. *See also* CONFORMER. [V.F.]

regulatory agencies Organizations empowered by governments to implement and enforce environmental legislation. The US Environmental Protection Agency (USEPA) was assembled from components of various federal agencies in 1970 in pursuance of an Executive Order signed by President Nixon. The Environment Agency for England and Wales (sometimes referred to as 'Agency' or EA), and the Scottish Environmental Protection Agency (SEPA) were formed by an Act of Parliament in 1995, again by combining existing bodies. In the case of the former the principal bodies subsumed were the National Rivers Authority (NRA) and Her Majesty's Inspectorate of Pollution (HMIP), while for the latter it was mainly elements of the River Purification Authorities and Her Majesty's Industrial Pollution Inspectorate (HMIPI). *See also* ENVIRONMENTAL (PROTECTION) AGENCIES. [P.C.]

regulatory instruments Methods used by REGULATORY AGENCIES to protect the environment. They can include COMMAND AND CONTROL and economic/market instruments (*see* MARKET INSTRUMENTS). [P.C.]

regulatory organizations/authorities *See* ENVIRONMENTAL (PROTECTION) AGENCIES; REGULATORY AGENCIES.

reinforcement The evolution of increased sexual isolation among incipient species due to selection against hybrids. This process is called secondary reinforcement when it takes place upon secondary contact between populations which diverged in allopatry. If matings between such differentiated populations result in less fit offspring, selection will favour an increase in homogametic matings. As a result SPECIATION will be accelerated and completed by selection in sympatry with evolution of absolute barriers to GENE FLOW.

Speciation by reinforcement is a questionable, but attractive theory that lacks unequivocal examples. HYBRID ZONES, which satisfy conditions for reinforcement — strong selection against hybrids and initial divergence in characters increasing ASSORTATIVE MATING—fail to provide clear evidence of it. The main factors limiting efficacy of reinforcing selection seem to include: recombination between genes lowering FITNESS and genes for preferential mating; gene flow into the zone; or STABILIZING SELECTION on MATE RECOGNITION systems. Theoretical models are similarly restrictive; they require substantial divergence before establishment of sympatry and low hybrid fitness. Perhaps the best evidence for evolutionary significance of reinforcement is greater prezygotic isolation in sympatric than in allopatric species of *Drosophila*, with no such difference for postzygotic isolation. *See also* SYMPATRIC SPECIATION. [J.M.SZ.]

REFERENCE

Butlin, R.K. (1995) Reinforcement: an idea evolving. *Trends in Ecology and Evolution* **10**, 432–434.

reintroduction *See* INTRODUCTION AND REINTRODUCTION.

relative abundance *See* SAMPLING METHODOLOGY/DEVICES.

relative growth rate (RGR) A parameter measuring the instantaneous rate of change in mass. It has the dimension of time^{-1}:

$$\mathrm{RGR} = \frac{1}{W}\frac{\mathrm{d}W}{\mathrm{d}t}$$

where W is shoot dry mass and $\mathrm{d}W/\mathrm{d}t$ is the instantaneous rate of change in shoot dry mass. RGR is typically estimated from the integral by computing the difference in log shoot mass per unit time, from two estimates of mass taken at time 0 and time t:

$$\mathrm{RGR} = \frac{\ln(W_t) - \ln(W_0)}{t} = \frac{\ln\left(\frac{W_t}{W_0}\right)}{t}$$

The longer the time period used for the assessment of RGR, the lower the maximum value of RGR that can be obtained. RGR is the preferred measure for assessing and comparing rates of GROWTH (i.e. increases in mass through time) by statistical analysis (e.g. as the RESPONSE VARIABLE for regression or ANALYSIS OF VARIANCE). [M.J.C.]

relative humidity The WATER content of the air relative to the water content of water-saturated air at the same TEMPERATURE, expressed as a percentage. *See also* HUMIDITY. [J.G.]

relative molecular mass The mass of a molecule relative to the mass of $^1/_{12}$ of ^{12}C. The relative molecular mass is obtained by summing the ATOMIC MASSES of all the atoms that comprise a molecule, for example the relative molecular mass of H_2O is $(1 \times 2) + 16 = 18$. [J.G.]

Relative population estimates

Relative population estimates are used to give an index of POPULATION SIZE. Such indices allow comparisons between sites in space or allow a population to be monitored through time. Unlike ABSOLUTE POPULATION ESTIMATES, a relative population estimate is one where population size is measured in arbitrary units and no attempt is made to determine the actual population size. Relative estimates are usually derived from trapping data, using particular designs of traps for set lengths of time, or from fixed effort SAMPLING methods, such as TRANSECT walking. While the methods are not designed to give absolute densities, it is often possible to use the results as a basis for DENSITY estimates. The distinction between absolute and relative methods for determining population size is not always clear.

The main advantage of relative population estimates over absolute ones is usually reduced effort requirement. Relative methods often produce large collections of data from a small effort, whereas absolute methods often require a large effort to gain single population estimates. For example, using a moth trap every night for mark–release–recapture would require an enormous daily effort, whereas counting and identification requires less effort and may be postponed by storing the catch. In this way very long-term trapping efforts may be sustained at several sites, giving continuous population MONITORING.

Converting a relative estimate into an absolute estimate is fraught with difficulty. This is because the number of individuals caught in a trap or observed from a hide will be dependent on a variety of factors in addition to the actual population size. For example, changes in environment with the season or weather may alter the activity of animals, effectiveness of the trap or the visibility of the target species, thus changing the estimate. Therefore conversions of relative estimates should be attempted with caution. The most successful method is to calibrate a relative estimation technique by simultaneously using an absolute estimation technique on the same population.

There are a variety of methods for obtaining relative population estimates.

Trapping

A huge variety of trap designs have been developed. All have the same purpose: to attract and concentrate individuals of the species or groups in question. The type of trap used will be determined by the type of populations under study. Traps may be based on an attractant, such as a light, sound or pheromones, or baited with food or potential mates. If the population under study is fairly abundant then traps can be simply based on interception, such as sticky traps for flying insects or pitfall traps for beetles.

The relative population estimate can be simply the number of individuals captured by a trap per unit time. However, such simple indices may be more informative if they are corrected for trap efficiency. For example, the effectiveness of sticky traps will change with wind speed. If the number of individuals intercepted could be corrected for this variable it would make the index a better reflection of the true POPULATION DENSITY.

Visual observation

This is the method most frequently used for large-scale bird studies. Observers are given a fixed time in which they record everything that they see. This sort of approach will obviously be affected by the local HABITAT, skill and alertness of the observer, and the weather conditions but may allow huge areas to be surveyed simultaneously. If sampling is repeated, a detailed picture of the population in space and time may be constructed.

Indirect observation

This is a very similar method to visual observation but involves recording sounds, tracks or defecation per unit of search effort rather than direct sightings of the target species.

Group size estimates

In many species that form social groups the size of those groups will change with the population density. Therefore the number of individuals in a group can be used as an index of population size.

Presence/absence

The simplest of scoring methods is to either record a species present or absent from each of a series of traps or observations. The proportion of SAMPLE units in which the species is recorded may be used to provide an index of ABUNDANCE. *See also* SAMPLING METHODOLOGY/DEVICES. [C.D.]

REFERENCES

Caughley, G. (1977) *Analysis of Vertebrate Populations.* John Wiley & Sons, Chichester.

Southwood, T.R.E. (1978) *Ecological Methods.* Chapman & Hall, London.

---

relaxation If a COMMUNITY at equilibrium is perturbed through a reduction in its area, the SPECIES–AREA RELATIONSHIP predicts species extinctions as the community returns to a new equilibrium for its smaller area. This process is termed relaxation and has been documented for the avifaunas of HABITAT fragments and land-bridge islands that became separated from mainland areas during the Pleistocene.

The time taken for the fauna to relax to 1/e (or 36.8%) of the supersaturated value is the 'relaxation time'. Typical relaxation times for island BIOTAS vary with area, isolation and faunas under consideration, and range from tens to thousands of years. *See also* ISLAND BIOGEOGRAPHY. [N.M. & S.G.C.]

relaxation time *See* RELAXATION.

relict A surviving fragment of a plant or animal POPULATION or ASSEMBLAGE that was formerly more diversified (a phylogenetic, evolutionary or taxonomic relict) or more widely distributed (a geographical relict). The term has been applied to populations, species, genera, families, whole floras or faunas, and even communities, and is often qualified by an epithet denoting supposed time of greater diversification or abundance (e.g. Tertiary relict, glacial relict). Characteristics and processes linked (sometimes erroneously) to relicts include poor DISPERSAL ability, primitive appearance, changing environmental conditions, and extinction and isolation. [J.H.T.]

REFERENCE

Holmquist, C. (1962) The relict concept — is it merely a zoogeographical conception? *Oikos* **13**, 262–292.

remediation The process of putting right (i.e. remedying) the effects of POLLUTION, or attempts to do so. *See also* HISTORY OF ECOLOGICAL RESTORATION; RECLAMATION. [P.C.]

remote sensing The gathering/recording of information about a population, ecosystem, area, phenomenon, etc., using any device that is not in physical contact with the subject — for example, by radio, visually by camera, or by satellite imaging. *See also* SAMPLING METHODOLOGY/DEVICES. [P.C.]

renewable energy Energy from sources that are not depleted by the process, for example solar, wind and waves. Cf. non-renewable sources: FOSSIL FUELS. *See also* RENEWABLE RESOURCE. [P.C.]

renewable resource A RESOURCE that is produced by natural ecosystems at rates that can potentially balance rates of removal. Examples of renewable resources include lumber, fish and game. *Cf.* NON-RENEWABLE RESOURCES. *See also* RENEWABLE ENERGY. [P.C.]

REPA *See* RESOURCE AND ENVIRONMENTAL PROFITS ANALYSIS.

repair processes *See* DEFENCE MECHANISMS.

repeatability The closeness of agreement between successive observations employing the same method, test material and laboratory. For example, closeness of results from repeated ECOTOXICITY tests of a single substance. *Cf.* REPRODUCIBILITY. *See also* PREDICTIVE ECOTOXICITY TESTS. [P.C.]

repetitive DNA Small non-coding sequences of DNA in the genome of EUKARYOTES that are expressed many times and whose function is uncertain or not entirely understood. They can be distinguished from tandemly repeated protein-encoding genes (e.g. those encoding ribosomal RNA), and act to separate eukaryotic genes from each other. An example of a repetitive sequence is the Alu family, which is a class of approximately 300 base pairs present in about one million copies and representing about 6% of the total DNA in humans. Bacteria contain very little repetitive DNA (*c.* 0.3%). In eukaryotes, single-copy DNA accounts for about 40–70% of the DNA; 5% is considered to be highly repetitive DNA (i.e. $>5\times 10^4$ copies per genome); and about 20% is considered to be middle repetitive DNA (i.e. $<10^4$ copies per genome). Most genes for PROTEINS belong to the single-copy class; transfer RNA genes and ribosomal RNA genes belong to the middle repetitive class; whereas spacer DNA sequences tend to be highly repetitive. *See also* SELFISH DNA. [V.F.]

replacement diagrams *See* DE WIT REPLACEMENT SERIES.

replication *See* EXPERIMENTAL DESIGN.

reproducibility The closeness of agreement between observations from different laboratories carrying out the same method with the same test material. For example, the closeness of results from different laboratories using the same ECOTOXICITY test (identical protocols) on the same test substance. Reproducibility is assessed by RING TESTS. *Cf.* REPEATABILITY. *See also* PREDICTIVE ECOTOXICITY TESTS. [P.C.]

reproduction (animal) and conservation Animal reproduction is a profligate process. It has to be because it depends wholly on the chemical replication of DNA and subsequent reactions determined by the sequences of submolecules (bases) that make up the DNA. DNA sequences, called genes, replicate, like everything else, in accord with the second law of thermodynamics. This guarantees the process will include entropy, which in turn guarantees a significant number of 'mistakes' will occur in the replicatory reactions. The mistakes, called MUTATIONS, are mostly deleterious. A few, however, are beneficial, especially if the adaptive needs of the animal are changing. Selection for the beneficial mutations is EVOLUTION. Evolution only works because many more offspring are produced than are needed to maintain a species. The surplus are naturally culled, providing for the needs of other species, such as predators. All this works well unless the environment changes more rapidly than the mutation/selection process can produce ADAPTATION, or a species POPULATION is reduced to a number so low that simple bad luck leads to extinction.

A major worry of conservationists working with small populations is INBREEDING DEPRESSION. This bad luck results when too few individuals with deleterious mutations are culled, and therefore breed back into a population of close relatives. Soon, all individuals may be defective. Inbreeding is not a problem, even in very small natural populations, as long as individuals are sufficiently prolific so that loss of defective offspring does not jeopardize the species survival (Remmert 1994). It is a problem in captive or managed populations where defective individuals are artificially maintained and reproduce.

Animals that produce very large numbers of offspring but invest little or no effort in PARENTAL CARE are described as *r*-selected (e.g. tapeworms, cod). Species that produce few offspring but invest considerable care in rearing them are called *K*-selected (e.g. elephants, humans) (*see* *r*- AND *K*-SELECTION). Because more offspring are produced than are needed, and because natural CULLING has been so effectively reduced, reproduction in one *K*-selected species — our own — is the root cause of all conservation problems. *See also* CONSERVATION BIOLOGY. [J.D.LZ.]

REFERENCE

Remmert, H. (ed.) (1994) *Minimum Animal Populations*. Springer-Verlag, Berlin.

reproductive allocation The proportion of energy and materials available to an organism that is allocated to reproduction as opposed to GROWTH or maintenance. Because both reproduction and growth influence the FITNESS of an organism, life-history models can be used to calculate the optimal partitioning of energy and materials into growth and reproduction after costs for metabolism have been subtracted. *See also* REPRODUCTIVE EFFORT. [W.W.W.]

reproductive character displacement *See* SYMPATRIC SPECIATION.

reproductive effort The amount of energy that an organism channels into reproduction. For some organisms this may simply be the amount of energy that is devoted to the production of GAMETES. In many other organisms there are auxiliary patterns associated with reproduction that may consume more energy than that used in gamete production. For example, many plants produce flowers to attract pollinators, and the males of many animals expend a significant amount of energy in attracting females or fighting other males. PARENTAL CARE may also be an important fraction of reproductive effort, particularly in endothermic vertebrates. *See also* REPRODUCTIVE EFFORT MODEL. [D.A.R.]

reproductive effort model Reproductive effort is a major component of an organism's life history (*see also* LIFE-HISTORY EVOLUTION). In modelling the optimal effort it is necessary to consider the age schedule of effort and the allocation of effort per episode. Within this framework the major issues are: (i) SEMELPARITY vs. ITEROPARITY; (ii) GROWTH vs. reproduction; and (iii) survival vs. reproduction. A variety of models have been analysed but they are all based upon the Euler equation and assume that selection maximizes the rate of increase, or if the population is stationary, the expected lifetime FECUNDITY.

A simple model that illustrates the approach to the problem is that analysed by Schaffer (1974). Critical assumptions of this model are: (i) REPRODUCTIVE EFFORT and survival do not change with age; and (ii) fecundity is either constant or grows exponentially with age. Under these assumptions the finite rate of increase, λ (a measure of fitness) is:

$$\lambda = b(E) + s(E)f(E)$$

where $b(E)$ is the fecundity of a 0-year-old, $s(E)$ is the SURVIVAL RATE, $f(E)$ is the fecundity, and all are functions of the reproductive effort E. First, note that for any particular value of λ there is a linear relationship between $b(E)$ and the product $s(E)f(E)$ (upper panel, Fig. R12). Second, $b(E)$ and $s(E)f(E)$ can be plotted as functions of E (left panels, Fig. R12), from which the relationship between $b(E)$ and $s(E)$ can be constructed (right panels, Fig. R12). The OPTIMAL REPRODUCTIVE EFFORT is obtained by superimposing the isoclines on the right-hand graphs: the optimal value is that combination at which an isocline is tangential to the curve.

As can be seen, reproductive effort can be extreme or intermediate. More complex models can be constructed to explore the optimal age distribution of reproductive effort, the optimal allocation between growth and reproduction, etc. *See also* EVOLUTIONARY OPTIMIZATION; OPTIMALITY MODELS; REPRODUCTIVE VALUE; RESIDUAL REPRODUCTIVE VALUE. [D.A.R.]

REFERENCE

Schaffer, W.M. (1974) Selection for optimal life histories: the effects of age structure. *Ecology* **55**, 291–303.

reproductive groups *See* GROUP SOCIAL STRUCTURE.

reproductive isolation The process whereby there is an absence or strong reduction of gene exchange (interbreeding) between two populations (or species) due to various, genetically controlled differences, which are also called ISOLATING MECHANISMS. Two broad categories are distinguished. Pre-mating isolating mechanisms preclude heterotypic matings. Potential mates may not meet since they occupy different habitats in the same area, or have separate breeding or flowering time when syntopic. If they do meet, unique sexual behaviour of each species results in mate discrimination that leads to homotypic matings, or structural differences in sex organs may hinder sperm transfer despite attempted copulation.

Post-mating isolating mechanisms lower the success of HETEROSPECIFIC matings through failure of FERTILIZATION, MORTALITY of zygotes, non-viability of F_1 hybrids, or partial or complete sterility of F_1, F_2 or back-cross hybrids. In some insects, post-mating reproductive isolation is caused by the presence of symbiotic microorganisms in the cytoplasm.

In nature, reproductive isolation between closely related species is often caused by a mixture of several pre-mating and post-mating factors. However, sometimes occasional individuals may hybridize, and the degree of reproductive isolation between pairs of populations may vary geographically. This suggests a gradual build-up of reproductive barriers. Reproductive isolation is thought to be a by-product of local ADAPTATION, SEXUAL SELECTION and drift, rather than a mechanism preventing formation of unfit hybrids. Complete pre-mating isolation between two strains of *Drosophila melanogaster* has been achieved in the laboratory under a multifarious selection regime.

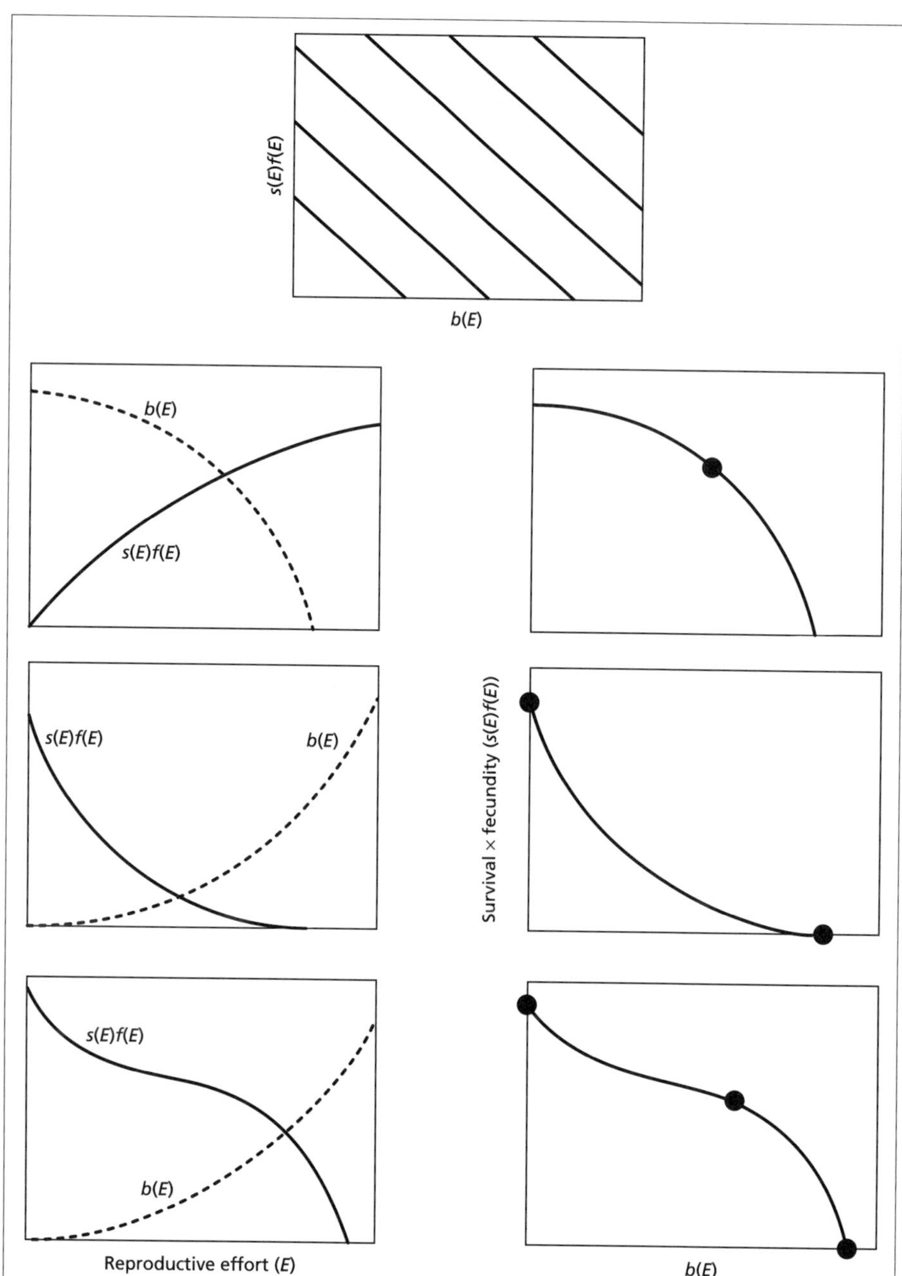

Fig. R12 Predicting the optimal allocation of reproductive effort. The top panel shows the fitness isoclines between the two components *b*(*E*) and *s*(*E*)*f*(*E*). The lower three panels on the left show hypothetical relationships between these two components and reproductive effort *E*. From these can be constructed the curve relating *b*(*E*) and *s*(*E*)*f*(*E*), as shown in the panels on the right. Superimposition of the fitness isoclines shows the possible equilibria, indicated by dots (in each case only one combination maximizes fitness, other equilibria being unstable).

Reproductive isolation is a concept crucial to Ernst Mayr's definition of a biological species (*see* BIOLOGICAL SPECIES CONCEPT; SPECIES CONCEPTS), since together with NICHE specialization it allows independent EVOLUTION of LINEAGES. Some evolutionists equate acquisition of reproductive isolation with SPECIATION. [J.M.SZ.]

reproductive rate *See* INTRINSIC RATE OF INCREASE.

reproductive synchrony The tendency of individuals to carry out some stage of their reproductive cycle at the same time as other members of the POPULATION. Synchronization is achieved by various means, including: environmental cues (e.g. meteorological cycles, temperature, humidity, food availability); internal cues (biological clocks); biological signals exchanged between members of a population; or a combination of these factors.

Reproductive synchrony is widespread among animals and plants. Well-known examples are the periodical cicadas (*Magicicada* spp.), which emerge from the ground to reproduce only every 13 or 17 years; bamboo (Bambuseae), which flowers synchronously at intervals of up to 120 years; or PERENNIAL plants such as oaks (*Quercus* spp.) that show mast seeding. Many marine animals synchronize the release of gametes or larvae with respect to light–dark, tidal or LUNAR CYCLES. For example, coral-reef fishes or crabs spawn eggs or hatch larvae on nocturnal maximum-amplitude high tides near a new or full moon. In temperate climates, flowering of plants can be synchronized by PHOTOPERIOD, whereas some tropical plant species flower after heavy rain showers. Group-living placental mammals often show oestrus synchrony, for example by exchanging olfactorial cues. Communally breeding birds synchronize egg-laying or hatching dates.

The most widely cited explanation for reproductive synchrony is predator-swamping (*see* PREDATOR SATIATION): synchronous breeding leads to a higher survival of the offspring because predators are limited in the number of prey they can eat at any one time. Other explanations involve the facilitation of finding a mate or a pollinator. In colonial birds feeding on unpredictable food patches, information about these patches is distributed most efficiently with synchronized breeding. [W.W.W.]

REFERENCE
Ims, R.A. (1990) The ecology and evolution of reproductive synchrony. *Trends in Ecology and Evolution* **5**, 135–140.

reproductive value The expected lifetime reproduction of a female of a given age (also known as Fisher's reproductive value; symbol, *V*). When juvenile MORTALITY is high, it may be that reproductive value peaks at an intermediate age rather than at birth. Reproductive value is calculated on the basis of age-specific survivorship (l_x) and FECUNDITY (m_x) as follows:

$$V_x = \frac{e^{rx}}{l_x}\int_x^{\infty} e^{-rt} m_t l_t \, dt$$

where r is the INTRINSIC RATE OF INCREASE, l_x is survivorship to age x and m_t is per capita fecundity at age t. Reproductive value is employed in life-history theory as a measure of the importance of individuals of different ages. It is also used in evaluating the costs and benefits of various life-history TRADE-OFFS. *See also* LIFE-HISTORY EVOLUTION; REPRODUCTIVE EFFORT; RESIDUAL REPRODUCTIVE VALUE. [M.J.C.]

REFERENCE
Kozlowski, J. (1993) Measuring fitness in life-history studies. *Trends in Ecology and Evolution* **7**, 155–174.

reprotoxins *See* CARCINOGENS, MUTAGENS, TERATOGENS.

Reptiles, diversity and conservation

History and biology

The chordate class Reptilia goes back in the FOSSIL RECORD some 300 million years to the Late Carboniferous, where it merges into the class Amphibia. The reptiles are the earliest of the amniote vertebrates: i.e. ones possessing a tough, dessication-resistant embryonic membrane, the amnion, thereby obviating the need for an aquatic larval stage. In terms of morphological DIVERSITY, Reptilia is the most remarkable chordate (or vertebrate) group, with forms ranging from long-legged to limbless; turret-eyed to eyeless; syringe-fanged to toothless; hard-shelled to membrane-gliding; PELAGIC to fossorial; egg-laying and live-bearing; weighing less than 1 g to over 1000 kg—and all this in just the living forms. The diversity of the reptilian fossil record is staggering. Considering the enormous popularity of fossil reptiles, it is opportune to point out that meaningful conservation extends beyond living species and should include consideration of the NON-RENEWABLE RESOURCES of fossils and fossil-bearing sites—all too often destroyed or desecrated prior to adequate scientific study.

Diversity

The roughly 5000 living species of reptiles are divided into 4 orders.

Turtles

The turtles (Chelonia), with about 250 living species, resemble the most ancient fossil forms in lacking temporal openings in the skull (our zygomatic arch makes two). However, they are arguably the most specialized vertebrates because their pectoral and pelvic girdles are housed beneath their ribs, within the shell. None have teeth, and all lay eggs on land.

The two major groups are the side-necked turtles (pleurodires), whose necks fold up laterally in the horizontal plane, and the cryptodires, whose necks fold back into their shells in the vertical plane. The leatherback, *Dermochelys coriacea*, is unique in many respects: its neck does not fold at all; its ribs are not fused to its shell; and it is an ENDOTHERM ('warm-blooded').

Crocodilians

The Crocodilia are the last survivors of the ruling reptiles, or archosaurs, a group that included 'dinosaurs'. All belong to the family Crocodylidae, usually divided into gharial (1 species), alligators (2 species), caimans (5 species), and crocodiles (14 or more species, depending upon one's view of several little-known freshwater forms from the East Indies).

Lizards and snakes

The sole surviving representative of the order Rhynchocephalia (meaning 'beak-heads') is the rather iguana-like tuatara (*Sphenodon punctatus*) of New Zealand. All the remaining living reptiles belong to Squamata, the lizards and snakes. Structural diversity in this group is vast. Snakes lack eyelids, ears, forelimbs and—except for rudiments in some primitive forms like boas, pythons and blind snakes—hindlimbs. Some lizards lack some or all of these characters too, so there is no clear distinction. While most squamates lay shelled eggs, many bear live young, and some (e.g. some *Mabuya* lizards) have a very mammal-like pseudoplacental embryo nourishment system. A few (e.g. some *Python* snakes) can be endothermic, at least when incubating eggs. Family and generic level classification of squamates is in chaotic flux, largely driven by 'price list taxonomy' in which proliferation of 'scientific' names is financially rewarded. Reasonable classification schemes and copious ecological, biogeographic and physiological details of Reptilia are provided in Zug (1993).

Conservation

Many conservation problems of reptiles closely parallel those of amphibians. HABITAT destruction leads, followed by human exploitation and environmental POLLUTION. Of the ordinal representatives, only the tuatara seems fairly secure on the few, scattered islands where it survives, but even there it is bedevilled by the anthropogenic pollutant of introduced European mammals such as rats.

Conservation of turtles

Turtles, never very speciose, have come down through the ages virtually unchanged since the Triassic, but today many are imperiled. The meat and eggs of larger species are eagerly sought by people, especially in the tropics, making it near-suicidal for females to nest in many places. Especially affected are the marine turtles of the family Cheloniidae, including the misnamed green turtle (*Chelonia mydas*; derived from green turtle soup—the soup is green, the turtle is brown), and the riverine side-necked turtles (*Podocnemis*) of South America. Smaller species, especially colourful ones, are very popular as pets; entire populations have been wiped out and several species endangered by this deadly expression of human interest. Despite many claims, turtles have never been farmed or propagated successfully on a long-term basis in captivity. Wild stocks continue to supply the markets.

Tremendous effort has gone into the conservation of marine turtles. National and international laws protect them and many of their nesting beaches are patrolled. Tagging, radiotracking, and airplane searches have dramatically increased our knowledge of their movements and migrations. They are even tracked by satellites. There have been successes and failures. Loggerhead sea turtles (*Caretta caretta*) along the Atlantic coast of North America are now abundant, and several other species nest successfully and are increasing.

However, two programmes—nest translocation and head-starting—have proven unwise. Digging up and moving nests entails three major problems: the early embryological development of turtles is easily disrupted if the orientation of an egg is changed with respect to gravity; sex is determined by temperature at a critical time in development, so artificially incubated eggs often hatch all one sex; and finally, hatchling turtles seem to have some sort of genetically encoded 'map' which will enable them to return to their nesting beach as adults, but only if they start from the right place: babies displaced any distance seem unable to find their way back to natal nesting sites. Head-starting, the practice of raising hatchlings to relatively large size before releasing them into the wild, may be the slowest and most expensive way to kill turtles ever devised. Deprived of natural peregrinations of their youth, head-started turtles seem to paddle aimlessly through purposeless lives and die without reproducing.

Oceanic pollution in the form of plastic bags and wrap, styrofoam and tarballs can be deadly to marine turtles, especially the giant leatherback (*Dermochelys coriacea*). These objects seem to be mistaken by the turtles for jellyfish, a major part of their diet. The spinose oesophagus of the turtle, ideal for swallowing jellyfish, precludes regurgitation and the turtle's gut may become clogged with indigestible rubbish. Details on conservation and population status of individual turtle species can be found in Ernst and Barbour (1989).

Conservation of crocodilians

Crocodilians include the largest living reptiles, with gharials and Indo-Pacific crocodiles known to exceed 6 m in length and well over 1000 kg in mass. Most species are persecuted simply because most people perceive such large carnivores as threatening (and several species are indeed 'man-eaters'). At least 15 of the living species are exploited for leather, accounting for 2 million individuals a year by 1990. The products are for the wealthy, with crocodilian leather handbags, for example, regularly retailing for several thousand US dollars. A few species, small ones and those with bony plates in their belly skins, escape hide-hunting pressure, but modern methods of decalcification allow the use of hides today not used a few decades ago. Most countries now have protective laws in place and are signatories to the Convention on the International Trade in Endangered Species (CITES; *see* INTERNATIONAL CONSERVATION CONVENTIONS), but enforcement is difficult and never complete. An emergent major problem with crocodilian conservation is the introduction and farming of exotic species. Warm countries like Taiwan, without native species, can import and raise the animals. Some inevitably escape; if they survive and multiply, one may imagine the effects on native WILDLIFE. Most ominous is the introduction of African Nile crocodiles (*Crocodylus niloticus*) to South Amercia. Should this very large, aggressive, fecund species that evolved for millions of years in sympatry with hominids become established in the wild, NEOTROPICAL ecology will undergo a metamorphosis. There is no animal native to the New World which can match the Nile crocodile's canny ferocity.

Protection and management of crocodiles in Asia and Africa, and of alligators in the USA, has often been spectacularly successful. Some species, like the Siamese crocodile (*Crocodylus siamensis*), would be extinct but for captive propagation. Alligators, not long ago endangered, are now abundant again in the south-eastern USA, forming the basis of a brisk trade not only in hides but meat (however bland and veal-like) for restaurants. Crocodilians are enormously popular tourist attractions. People flock to ogle captives, but this experience cannot compare to the thrill of encountering these magnificent relics in the wild. An excellent reference is Ross (1989).

Conservation of lizards and snakes

The squamate reptiles were largely overlooked by conservationists for decades while birds and mammals held the limelight. However, their problems are legion and similar. The pet trade consumes about 2 million lizards and snakes in the USA alone per year. Reptiles make popular pets because many are colourful and docile, require relatively little food, and can be kept clean with modest effort. The popularity of reptile pets, especially exotic imports, reaches its peak in Europe, especially Germany, probably because there are only a few, rather small or drab native species there. Lizards and snakes are also exploited for leather and, to a lesser extent, meat. While some species have been successfully propagated in captivity, especially for pets, and others can be farmed, like the highly edible iguanas (*Iguana*), most of those consumed are taken from the wild. Depletion of snakes, for example in India for leather, has led to population explosions in their pestilential prey, rats. Rattlesnake (*Crotalus*) 'round-ups' in the USA not only lead to rat proliferation but kill outright large numbers of NON-TARGET SPECIES, including other reptiles, that share burrows with the snakes.

Because reptiles are such excellent over-water dispersers, they have colonized the islands of the warmer parts of the world, nearshore and far. On islands, reptiles have often evolved into remarkable forms, including relative giants. The radiations of iguanid lizards and boid snakes in the West Indies are examples. In turn, these are the sorts of reptiles apt to be most immediately impacted, endangered, extirpated and even exterminated by the human population explosion. People and their exotic, introduced, carnivorous, commensal mammals like cats, dogs, mongooses and even rats, have taken a sad toll. In some areas, like Jamaica and the Virgin Islands, surviving species are making comebacks, in no small measure because of ecotourism. Indeed, lizards like ameivas (*Ameiva*), anoles (*Anolis*) and especially iguanas (*Iguana*) are conspicuous, approachable, attractive, often very colourful and interesting: many of the features that have made birds so popular. They are an increasingly appreciated asset.

Snakes, albeit all too-well loved by pet fanciers, are usually inconspicuous and still feared (sometimes with good reason) by even most ecotourists. Because they lead cryptic lives, snakes provide a distinct set of conservation problems (Dodd 1993). Snakes exemplify, however, the two most pervasive syndromes for all reptilian conservation: our lack of knowledge, especially of their life histories, ecological needs, and details of distribution; and the effects of burgeoning human populations, consuming them and their habitats. *See also* CONSERVATION BIOLOGY. [J.D.LZ.]

REFERENCES

Dodd, C.K. (1993) Strategies for snake conservation. In: *Snakes* (eds R.A. Siegel & J.T. Collins), pp. 363–393. McGraw-Hill, New York.

Ernst, C.H. & Barbour, R.W. (1989) *Turtles of the World*. Smithsonian Institution Press, Washington, D.C.

Ross, C.A. (ed.) (1989) *Crocodiles and Alligators*. Facts on File, New York.

Zug, G.R. (1993) *Herpetology*. Academic Press, San Diego, Calif.

---

repulsion *See* DEFENCE MECHANISMS.

reservoir hosts, parasites A reservoir HOST species is defined in respect to another host species when considering PARASITE transmission. The parasite is ENDEMIC in the reservoir host, but occasional transmission and consequent disease occur in the second host species. Where the second host species is man (*Homo sapiens*), the infection is said to be a zoonosis. Thus, rabies is a MICROPARASITE infection of canids, which are reservoir hosts of rabies for humans and herbivores. *See also* PARASITISM. [G.F.M.]

residence time The period of time for which an entity remains within an ecological unit of interest. For instance, the time for which a food particle or molecule remains within the gut; or for which an organism, nutrient or WATER molecule remains within a stretch of river or within a lake or pond; or the time for which a molecule of a POLLUTANT remains in the ATMOSPHERE. [P.C.]

residual reproductive value (RRV) A concept introduced by Williams (1966) in the context of LIFE-HISTORY EVOLUTION. Williams suggested that REPRODUCTIVE VALUE could be partitioned into a component due to current reproduction and a component due to future reproduction. Consider a population described by an age-classified POPULATION PROJECTION MATRIX, with P_i and F_i the survival probability and fertility of age-class i. The reproductive value of age-class i (v_i) is:

$$v_i = F_i\lambda^{-1} + P_iF_{i+1}\lambda^{-2} + P_iP_{i+1}F_{i+2}\lambda^{-3} + \cdots \quad (1)$$

$$= F_i\lambda^{-1} + P_i\lambda^{-1}v_{i+1} \quad (2)$$

where λ is the POPULATION GROWTH RATE. The first of these terms is current reproduction, the second is residual reproductive value. The powers of λ account for the DILUTION EFFECT of future reproduction in a population whose size is changing. [H.C.]

REFERENCE

Williams, G.C. (1966) Natural selection, the costs of reproduction, and a refinement of Lack's principle. *American Naturalist* **100**, 686–690.

resilience A measure of COMMUNITY or POPULATION stability which has many definitions. The most frequently used is the speed of recovery of a community after a DISTURBANCE from a local equilibrium. This means that a highly resilient community will return very rapidly to local equilibrium. Populations with high growth rates and communities with fewer trophic levels tend to be more resilient. A contrasting definition uses the range of conditions to which a system can be disturbed and still return to local equilibrium as a measure of resilience. Definitions of resilience are often confused with those of RESISTANCE. [C.D.]

resistance

1 The ability of an organism to show insensitivity or reduced sensitivity to a chemical that normally causes adverse effects. Resistance and TOLERANCE have often been used synonymously. However, usage seems to favour retaining the word resistance only when there is evidence of an adaptative change — the chemical having acted as a SELECTION PRESSURE and changed the genetic constitution — whereas tolerance can be used for both this and phenotypic responses, for example increased insensitivity through induction of protective measures. Resistance can also imply a more active response.

2 A measure of COMMUNITY or POPULATION stability that has many different definitions. One definition is that resistance is a measure of the amount of change which can be applied to a system before it is disturbed from its equilibrium. By this definition, systems with high resistance will not move from a local equilibrium when exposed to change. An extension to this definition is to use resistance as a measure of the change of state variables within a system in response to a change in a variable. Resistance and RESILIENCE are often confused.

[P.C. & C.D.]

resource allocation *See* RESOURCE PARTITIONING.

resource and environmental profits analysis (REPA) A model for assessing energy balances and POLLUTANT loads of products along a LIFE CYCLE. *See also* LIFE-CYCLE ASSESSMENT. [P.C.]

resource competition theory A theoretical model for POPULATION

DYNAMICS and COMPETITION in which competition is modelled by explicitly relating growth and multiplication of organisms in relation to the RESOURCES they consume and compete for. Whilst originating in the microbiological literature (e.g. TILMAN'S DIATOM EXPERIMENTS), this line of theory has been used to analyse the determinants of competitive success and community structure in higher plants (*see* TILMAN'S MODEL OF RESOURCE UTILIZATION). There are two basic predictions of resource competition theory: the R^* rule (that the competitor which, at its monoculture equilibrium, reduces the level of a single limiting resource to the lowest level, denoted by R^*, will be competitively dominant), and the RESOURCE-RATIO HYPOTHESIS. [R.P.F.]

REFERENCE

Grover, J.P. (1997) *Resource Competition*. Chapman & Hall, London.

resource defence Resource defence can arise wherever the distribution of RESOURCES in space and/or time is relatively predictable and the benefits of defending all or part of the resource outweigh the costs of defence (the principle of economic defendability; *see* TERRITORY SIZE). Where habitats vary in resource quality, resource defence can lead to a 'despotic' distribution of competitors whereby the best habitats become occupied by the strongest competitors while weak individuals are consigned to the poorer habitats (*cf.* IDEAL FREE DISTRIBUTION). Resource defence is thus a form of INTERFERENCE COMPETITION which reduces the efficiency with which competitors exploit resources. [C.J.B.]

resource-defence polygyny A MATING SYSTEM in which males defend territories containing RESOURCES such as food or nesting sites and thereby gain mating access to females attracted to these resources. Resource-defence POLYANDRY, in which females defend resources, is rarer. Resource defence is a particularly common reproductive strategy in ungulates, and is also widespread in birds, fishes and other animal groups. For example, in Grevy's zebra, males defend territories along the banks of rivers where they can intercept females coming to drink; in rock-dwelling cichlid fishes in Lake Malawi, males defend sheltered spawning sites in caves; and in pheasant-tailed jacanas, females defend mats of floating vegetation where males construct nests. The degree of POLYGYNY (or polyandry) may depend on the variance in quality of male territories in the population, as in the polygyny threshold model. [J.C.D.]

resource-depletion zone The region, immediately adjacent to the surface of plant roots or mycorrhizae, in which nutrient uptake by the plant has locally depleted the supply of a SOIL resource (typically NITROGEN, PHOSPHORUS or POTASSIUM minerals, or WATER). The depletion zone is replenished by diffusion from surrounding, non-depleted zones of the soil, by inputs from above the soil surface, or by local DECOMPOSITION or MINERALIZATION. It is often remarked that plants contest (*see* CONTEST COMPETITION) for LIGHT but scramble (*see* SCRAMBLE COMPETITION) for soil NUTRIENTS, but plants with larger root systems are clearly capable of pre-empting nutrients (especially immobile soil nutrients like phosphorus) from smaller plants and hence introducing ASYMMETRIC COMPETITION below ground (in much the same way as competition for light is asymmetric). [M.J.C.]

resource partitioning

1 The differential use of resource types, by species or other categories of organisms. 'COMPETITION theory' claims that resource partitioning is the major mechanism maintaining COEXISTENCE of potential competitors, referring to the principle of COMPETITIVE EXCLUSION. Resource partitioning presumes the presence of different resource types, i.e. of resource HETEROGENEITY. It is now generally recognized that many kinds of spatial and temporal heterogeneity may also lead to the differential use of RESOURCES (not necessarily of resource types) and thus to coexistence (*see* e.g. AGGREGATION MODEL OF COMPETITION), and that competitive exclusion may also be prevented by predation or disturbances.

2 Also refers to the partitioning of resources from PHOTOSYNTHESIS and FEEDING within organisms between metabolism and PRODUCTION. *See also* OPTIMALITY MODELS; PHYSIOLOGICAL ECOLOGY. [J.G.S.]

resource-ratio hypothesis A model proposed by D. Tilman to explain the broad-scale patterns of PRIMARY SUCCESSION. At the beginning of a primary succession there is plenty of LIGHT but no SOIL (the availability of soil NITROGEN is especially low). At the end of succession there is a CLIMAX forest community in which there is very little light at the soil surface, but a relatively high rate of nitrogen cycling. Tilman's model is based on the assumption that, for any given plant species, there is a TRADE-OFF between adaptations that improve its light-gathering ability (e.g. a taller shoot system) and those that improve nutrient-gathering ability (e.g. a more wide-ranging root system). Tilman sees the process of succession as the stepwise competitive replacement of species that are good competitors for NUTRIENTS (they possess relatively small shoot systems and large root systems) by others which are better competitors for light (they have large shoot systems and relatively small root systems). [M.J.C.]

REFERENCE

Tilman, D. (1988) *Plant Strategies and the Dynamics and Structure of Plant Communities*. Princeton University Press, Princeton, N.J.

resource-utilization curve The curve defining the (relative) extent of usage across a range of RESOURCES. Resource-utilization curves have an important place in the development of theory on COMPETITION, COEXISTENCE, LIMITING SIMILARITY and species packing. [J.G.S.]

resources Materials that are essential for GROWTH and reproduction of living organisms. For plants these include LIGHT, WATER, OXYGEN, NITROGEN, PHOSPHORUS, POTASSIUM, CARBON, CALCIUM, MAGNESIUM, SULPHUR, COBALT, COPPER, MOLYBDENUM, MANGANESE, ZINC, pollinators and other mutualists. For animals they include CARBOHYDRATES, fats, PROTEINS, water, VITAMINS, shelter and breeding sites.

Limiting resources are those whose supply rate currently constrains population growth. The dependence of the rate of population growth on resource availability is the key to understanding resource-based COMPETITION. So long as the resource supply is sufficient that $dN/dt>0$ (*see* INVASION CRITERION), then a population will persist. For two species in competition, there will come a point where the resource supply rate is too low to support the population growth of one species (the inferior competitor) yet is sufficient to support the other species (the superior competitor). The THRESHOLD resource supply rate at which $dN/dt = 0$ is called R^* (R star); the species with the lower R^* outcompetes the species with the higher R^* and competitively displaces it at equilibrium. *See also* RESOURCE PARTITIONING. [M.J.C.]

respiration Respiration is the oxidation of food by organisms to release energy in the form of adenosine triphosphate (ATP; *see also* AEROBIC RESPIRATION). Most organisms require OXYGEN, but some bacteria use either NITRATE or sulphate as the oxidant. In all higher organisms respiration occurs in the mitochondria. The main biochemical pathways involved in respiration are glycolysis, the oxidative pentose phosphate pathway and the tricarboxylic acid pathway (also termed the TCA or Krebs cycle). This latter produces a flow of electrons in mitochondrial membranes which is coupled to production of ATP, termed oxidative phosphorylation. In plants, an additional pathway can be involved termed CYANIDE-resistant respiration or alternate oxidase pathway, which consumes CARBOHYDRATE and oxygen, without production of as much ATP as normal oxidative phosphorylation. Green plant cells in the light can also release CARBON DIOXIDE (CO_2) through a process termed photorespiration, but this is related to the PHOTOSYNTHESIS process, and not to true respiration, which in plant biology is often termed 'dark' respiration.

In very small animals, gases are transferred between the surface and the mitochondria by diffusion alone. In larger animals a circulation transfers gases between the respiratory surface and the tissues, often augmented by a respiratory pigment (e.g. haemoglobin). All respiratory surfaces have large surface areas and small distances for diffusion between the medium and the blood to facilitate gas transfer, but the overall design varies with the environment (i.e. gills in water, lungs in air). Insects have a tracheal system, a series of air-filled, thin-walled tubes that extend throughout the body, taking advantage of the rapid diffusion of gases in air and avoiding the necessity of transporting gases in the blood.

Oxygen is the ultimate electron acceptor. Therefore respiration is dependent on carbohydrate and oxygen supply, and on TEMPERATURE, with rates increasing typically by 50–150% for a 10 degree rise in temperature.

Respiration is also equated with heat loss in the ENERGY BUDGET studies. *See also* ECOLOGICAL ENERGETICS; PHYSIOLOGICAL ECOLOGY. [D.J.R. & J.I.L.M.]

respiratory quotient (RQ) The ratio of the volume of CARBON DIOXIDE (CO_2) expired to the volume of OXYGEN inspired over the same time. Its value reflects the substrate being used in respiratory metabolism (i.e. 1 for CARBOHYDRATES, 0.7 for fats and 0.8 for PROTEINS) and hence is used in calculating the heat loss per oxygen inspired (oxyjoule equivalent), often used to compute heat losses in ENERGY BUDGETS. [P.C.]

response to selection *See* SELECTION RESPONSE.

response variable The *y* variable in statistical analysis; the response (measurement or count) for which we aim to understand the relative importance of various EXPLANATORY VARIABLES in causing variation. Formerly known as the dependent variable. [M.J.C.]

restoration ecology *See* ECOLOGICAL RESTORATION.

restriction map A technique for physically locating sites at which restriction enzymes cut cloned DNA. Restriction enzymes are site-specific nuclease enzymes that recognize short sequences of unmethylated DNA and cleave the double strands at or close to specific sites. The DNA fragments resulting after restriction enzyme digestion can be separated by gel ELECTROPHORESIS since the length of the fragment is inversely related to the distance it moves in an electric field. The order of fragments in the chromosome is determined by analysis of partial digestion products, by successive cleavage with multiple restriction enzymes, by hybridization of individual fragments to reference fragments, by pulse labelling of synthetic DNA, by end labelling of DNA prior to cleavage, or by other mapping techniques. *See also* MOLECULAR ECOLOGY; RESTRICTION SITE POLYMORPHISMS. [V.F.]

restriction site polymorphisms (restriction fragment length polymorphisms, RFLPs) Allelic variants, or polymorphisms, that can be identified (without the need to identify the gene product) because single base changes in the recognition sequence of restriction enzymes can alter the pattern of cuts made in DNA. This gives rise to a detectable variation in DNA fragment length that is inherited in a Mendelian codominant fashion. The variation in the length of DNA fragments generated by the presence or absence of restriction sites is detectable by ELECTROPHORESIS and is used to distinguish different GENOTYPES. In practice, RFLPs are detected by extracting and digesting DNA from different individuals with restriction enzymes, followed by analysis by Southern blotting. Variation in the genomes of different individuals can be detected with probes that detect clusters of repeated base sequences in DNA. RFLPs have been useful in the mapping of loci involved in genetic disorders. [V.F.]

REFERENCE

Knox, B., Ladiges, P. & Evans, B. (eds) (1994) *Biology*. McGraw-Hill, Sydney.

retrogressive succession This term was introduced into ecology in 1964 by the Danish palynologist/ecologist Johs Iversen, who used it to describe the process by which woodland ecosystems could be translated into heathland as a consequence of human activity, particularly burning and grazing. It is a directional development of an ecosystem in which many of the processes normally associated with conventional SUCCESSION are set in reverse. The final outcome may prove persistent and stable while the factors that instigated its development remain operative.

In the case of the regression of woodland into heathland, for example, a number of processes can be seen to have taken place.

1 BIOMASS is reduced, hence LEAF AREA INDEX is also lowered.

2 The gross and net PRIMARY PRODUCTIVITY of the system is reduced.

3 Structural complexity (vegetation architecture) is reduced, leading to simpler profiles and less microclimatic diversity.

4 Overall conditions may, therefore, move from relatively mesic to more extreme and physically stressful.

5 Opportunities for plant and animal life are thus reduced, leading to a lower SPECIES RICHNESS.

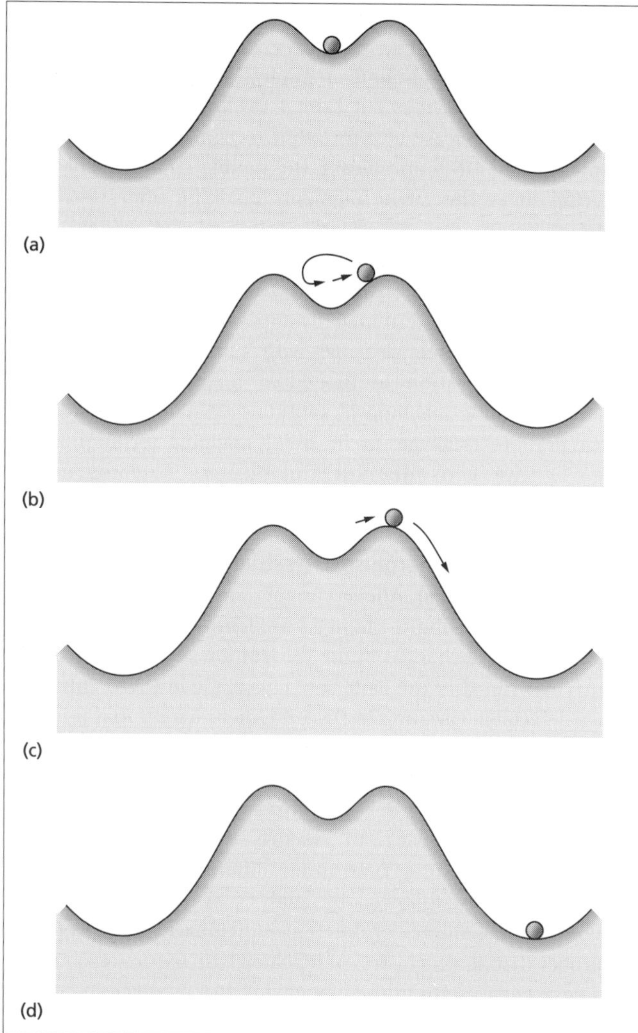

Fig. R13 Retrogressive succession. (a) An ecosystem (represented by the ball) is in a relatively stable set of conditions. (b) Minor perturbation will displace the system to a certain degree, but removal of the disturbing influence will permit the system to return to its original state. (c) More extreme perturbation takes the ecosystem beyond a stability threshold and more permanent changes commence. (d) A new stable condition is achieved at a lower overall energy state than the initial system. The ecosystem has undergone retrogressive succession.

6 The total nutrient capital of the biotic component of the final ecosystem is less than that of the initial system.

One can conceive the system in terms of the billiard-ball model of Lewontin (Fig. R13). A ball resting in a hollow may have a degree of stability, conferred upon it by the surrounding slopes. It possesses inertia (reflected by the steepness of the surrounding slopes), requiring considerable force to set it in motion. It also has the property of RESILIENCE, since it will return to its original position if mildly perturbed. If subjected to more excessive DISTURBANCE, however, there is the possibility that the ball will be displaced beyond the rim of its depression, over a certain THRESHOLD, and will descend until it finds a degree of stability at a new level of potential energy. Such is the model of retrogressive succession.

The concept is applicable to the development of certain types of BLANKET MIRE where woodland under conditions of high precipitation has been subjected to clearance, burning and grazing pressures that have prevented the regeneration of trees and permitted the establishment of *Sphagnum*-dominated ecosystems that have proved self-perpetuating. It can also be applied to some areas of DESERTIFICATION where the local extinction of woody species has rendered the recovery of damaged, more productive ecosystems impossible without reintroduction and rehabilitation. [P.D.M.]

REFERENCES

Iversen, J. (1964) Retrogressive vegetational succession in the post-glacial. *Journal of Ecology* **52** (Jubilee Symposium Supplement), 59–70.

Moore, P.D. (1973) The influence of prehistoric cultures upon the initiation and spread of blanket bog in upland Wales. *Nature* **241**, 350–353.

retrospective analysis (monitoring) It is often necessary to consider if human activity is having an adverse EFFECT on ecological systems. This might be a concern at particular sites that are exposed to regular EMISSIONS of EFFLUENTS or following accidental releases; or it might apply to assessing the general quality of systems subject to possible impacts from a wide range of activities. These are all retrospective analyses: i.e analyses performed after the human activity has occurred.

In general we are rarely able to predict, from first principles, what sorts of ecological ASSEMBLAGES should occur in particular habitats if undisturbed, so retrospective analysis is often based upon comparisons between putatively undisturbed and putatively disturbed systems; i.e. systems separated in space (e.g. upstream vs. downstream comparisons) or time (before vs. after comparisons). Hence, it is usually not straightforward to separate VARIABILITY due to a specific cause (POLLUTANT) from natural variability. How, given that ecosystems are so variable through space and time, can we be sure that the control site would have exhibited ecological characteristics similar to those of the putatively disturbed site had there been no DISTURBANCE? Even carefully designed and replicated SAMPLING programmes cannot completely remove these uncertainties. Therefore, observations and MONITORING should be associated with experimental analysis. For example, does SEDIMENT from the contaminated site have appropriate ecotoxicological properties? Can these properties be removed by stripping out the contaminant? And so on.

Other techniques plant out organisms as *IN SITU* BIOASSAYS. This method has the virtue that initial conditions can be made constant in the control and exposed systems; but it can be open to criticisms of artificiality. Often, especially in general SURVEYS concerned with assessing quality, detailed comparisons are not made; instead, effects are judged in terms of the presence or absence of INDICATOR SPECIES or species complexes that, on the basis of past experience (sometimes captured formally in empirical models), are generally expected to be present in clean or polluted conditions. This is the BIOINDICATOR (biotic index) approach, which itself suffers from a number of problems. *Cf.* PREDICTIVE ECOTOXICITY TESTS. *See also* BIOMONITOR; BIOTIC INDICES. [P.C.]

return time *See* CHARACTERISTIC RETURN TIME.

reuse A form of RECYCLING of WASTE that involves using the object or material again, for the function originally intended. Returned bottles, reused as bottles, are good examples. *See also* WASTE DISPOSAL. [P.C.]

reworking, of sediment The horizontal and vertical movement of SEDIMENT particles as a result of the action of water CURRENTS and of the burrowing and FEEDING activities of sediment-dwelling organisms. Reworking of sediments can complicate stratigraphic analyses because deeper sediment layers may not correspond to older layers. *See also* ECOSYSTEM ENGINEERS. [V.F.]

Reynolds number (*Re*) A dimensionless quantity named after the 19th century scientist, Sir Osborne Reynolds. Fluid mechanics uses the Reynolds number to define whether the flow of fluid is laminar or turbulent. Theoretically, the Reynolds number is defined as the ratio between inertial forces and viscous forces (inertial forces are those associated with the mean motion of the fluid; viscous forces are those associated with frictional shear stress). Mathematically, the Reynolds number is defined as:

$$Re = du/\nu$$

where d is a characteristic length scale, u is the velocity of the fluid and ν is the kinematic VISCOSITY of the fluid. Most atmospheric flows tend to be turbulent because the magnitude of ν is of the order of $10^{-5}\,m^2\,s^{-1}$ and the magnitude of u is of the order of $0.1–10\,m\,s^{-1}$. [D.D.B.]

Reynolds stress The TURBULENT FLOW of a fluid can be decomposed into mean and fluctuating components, for statistical analysis. The Reynolds stress is the COVARIANCE between vertical and horizontal velocity FLUCTUATIONS of a fluid flow. This stress represents the vertical and downward-directed flux of momentum for atmospheric flows over land surfaces. This flux of momentum constitutes a stress because it imposes a deforming force on to the surface.

Vertical gradients of WIND velocity and scalar (i.e. TEMPERATURE, HUMIDITY and POLLUTANT) quantities are proportional to the square root of the magnitude of Reynolds stress (a quantity known as friction velocity). Hence, Reynolds stress is an important variable for modelling atmospheric flows and rates of mass and energy transfer.

The Reynolds stress, like the REYNOLDS NUMBER, is named after the 19th century British scientist, Sir Osborne Reynolds. [D.D.B.]

RGR *See* RELATIVE GROWTH RATE.

rhenium (Re) A metallic element (ATOMIC NUMBER = 75; relative ATOMIC MASS = 186.207; melting point = 3180°C; boiling point = 5627°C), and one of the rarest of the naturally occurring elements (abundance = $0.0007\,mg\,kg^{-1}$ of the Earth's CRUST). Being chemically similar to MOLYBDENUM it is in the sulphide ores of this METAL that it is found in highest concentrations (up to 0.2%), and from which it is obtained by refining of the flue dusts resulting from the roasting of the ore. The vast majority of rhenium is used in the form of PLATINUM/rhenium CATALYSTS for the production of low-LEAD, high-octane petrol. [P.J.B.]

rheophilous Describing a wetland habitat, vegetation or an individual species that demands a flow of WATER to supply its mineral nutrient requirements; literally 'flow-loving'. [P.D.M.]

rheotrophic Describing a wetland ecosystem that receives at least part of its nutrient input through a flow of groundwater; literally 'fed by the flow'. *See also* MIRE. [P.D.M.]

rhithron *See* RIVERS AND STREAMS, TYPES OF.

rhizobenthos *See* BENTHIC HABITAT CLASSIFICATION; WITHIN-STREAM/RIVER HABITAT CLASSIFICATION.

Rhizosphere

Characteristics

The rhizosphere is the narrow zone (*c*. 1–2 mm) of SOIL immediately surrounding the actively growing root, in which the plant has a direct influence on the soil MICROFLORA (and to some extent also, the MICROFAUNA). This influence is exerted through nutrient supply from the plant, via processes including: (i) exudation or passive leakage from living cells; (ii) active secretion; and (iii) release of lysates during autolysis. These NUTRIENTS are: CARBOHYDRATES, including sugars and complex polysaccharides/mucopolysaccharides forming the so-called 'mucigel' (which covers and 'lubricates' the root as it grows through the soil); organic ACIDS; AMINO ACIDS; PROTEINS; as well as VITAMINS and other complex growth factors.

As the root grows forwards by apical extension there is detachment of dead and dying outer cells ('sloughing off'), including damaged root-cap cells and ephemeral root-hair cells. At the same time there are localized changes in pH, and CARBON DIOXIDE (CO_2)/OXYGEN (O_2) balance, and the release of inhibitors affecting specific microbes. The overall result is a marked change in the qualitative and quantitative nature of microbial populations within the rhizosphere compared with the general surrounding soil. These changes become gradually less marked at the edge of this zone, whose limits are not sharply defined.

Microflora

Among the more important members of the rhizosphere microflora are fungi (including mycorrhizal fungi) traversing the region, and bacteria (particularly *Pseudomonas* spp. in certain soils), which may be closely associated with, and embedded in, the mucigel. In the case of fungi, it is interesting to note that soils in general are often inhibitory to the germination of propagules such as spores; this phenomemon, termed mycostasis, or fungistasis, has been variously attributed to the presence in soil of antibiotics, AMMONIA, CO_2 and ethylene from previous microbial activities, but this is now thought to be caused mainly by depletion of nutrients, which are restored within the rhizosphere, thus making germination and active growth possible. Thus, there are particular saprotrophic (saprobic) fungi in the rhizosphere that are dependent to some degree on nutrients/growth factors that are only present in non-limiting amounts in the rhizosphere. Some reports have indicated that bacterial RESPIRATION rates are higher than those of fungi in the rhizosphere, but this situation can be reversed if ectomycorrhizal fungi are present.

The presence of rhizosphere microorganisms is known to increase the rate of loss of assimilated CARBON by plants via the root surfaces, the rate of loss in wheat roots being calculated at up to 0.15% of the total carbon assimilated per week.

At the innermost edge of the rhizosphere, i.e. at the root surface, another ecological microniche is identified, that of the rhizoplane, which is again characterized by a distinctive microflora.

Of the fungi which have been re-isolated from root surfaces some grow readily on simple media but they are often sterile, making identification very difficult; the majority of these are probably vegetative hyphal forms of members of the Basidiomycotina.

Plant–microflora interactions

Potentially pathogenic fungi are frequently re-isolated from the rhizosphere, including species of *Alternaria, Helminthosporium, Fusarium* and *Pythium*. In some cases, it has been shown that infection by such organisms may be inhibited by other, antagonistic organisms living within the rhizosphere; i.e. HOST RESISTANCE here depends upon a beneficial 'rhizosphere effect'.

Strains of certain bacteria belonging to the fluorescent pseudomonads (*Pseudomonas*) colonize roots of plants such as sugar beet, radish and potato, and induce high (more than double) YIELDS in these CROPS. Such bacteria, which have been termed plant growth-promoting rhizobacteria, are known to inhibit the growth of other members of the rhizosphere microflora, including pathogenic microorganisms. One mechanism involves the production of an extracellular siderophore (IRON-CHELATING AGENT) by the bacterium, thus making iron available to the plant, but denying it to the potential PATHOGEN; for example, *P. fluorescens-putida* produces the siderophore 'pseudobactin', which inhibits the growth of the bacterium *Erwinia carotovora*, normally capable of causing 'soft rot' of potato.

Root EXUDATES are also known to stimulate the dormant propagules of important soil-borne root-infecting fungi. Thus, the sclerotia of *Sclerotinia sclerotiorum*, the 'white rot' fungus which is capable of attacking onion roots, remain dormant in soil until 'triggered' to germinate by specific thiol-containing compounds produced in the root exudate of this, and closely-related, hosts. Similarly, the motile zoospores of parasitic fungi such as *Phytophthora* and *Pythium* spp. often demonstrate positive chemotaxis when tested in root exudates, swimming by means of their flagella in the direction of the strongest concentration gradient and towards the root surface which is releasing the active component(s) in the exudate. Ethanol, which is present in exudates of roots undergoing ANAEROBIC METABOLISM in waterlogged soils because of lack of oxygen, is known to attract such zoospores. Similarly, it is known to promote infective structures, such as the rhizomorphs of *Armillaria mellea*, the well-known 'honey fungus', which destructively attacks forest trees, shrubs and other plants.

Finally, it is likely that the metabolic activities of the rhizosphere microflora associated with particular plant species have an important role in contributing to allelopathic interactions between different plants, thus making an important contribution to determining PLANT ASSOCIATIONS. [J.B.H.]

REFERENCE

Lynch, J.M. (1982) The rhizosphere. In: *Experimental Microbial Ecology* (eds R.G. Burns & S.H. Slater), pp. 1–23. Blackwell Scientific Publications, Oxford.

---

rhizosphere microorganisms *See* RHIZOSPHERE.

rhodium (Rh) A very rare metallic element (ATOMIC NUMBER = 45; relative ATOMIC MASS = 102.906; melting point = 1966°C; boiling point = 3727°C), having an abundance of 0.0001 mg kg^{-1} in the Earth's CRUST. It is found in association with the other PLATINUM metals in the nickel-copper sulphide ores of South Africa and Canada, which contain about 0.1% of the METAL. By far the largest use of rhodium is in CATALYTIC CONVERTERS for vehicle exhausts, where it acts in conjunction with platinum to reduce CARBON MONOXIDE (CO) and NO_x EMISSIONS. Compounds of rhodium also find specialist applications as CATALYSTS for hydrogenations and carbonylations of organic compounds. [P.J.B.]

rhyniophytes *See* PTERIDOPHYTES, GLOBAL DIVERSITY AND CONSERVATION.

richness The number of species present at a particular place and time. No assumptions are made about the relative ABUNDANCES (or EQUITABILITY) of these species. If it is not possible to enumerate all species in a COMMUNITY then it becomes necessary to distinguish between numerical species richness, which is defined as the number of species per specified number of individuals or BIOMASS, and species DENSITY, which is the number of species per specified collection area. Colwell and Coddington (1995) provide an extremely useful guide to estimating SPECIES RICHNESS from samples. [A.E.M.]

REFERENCE

Colwell, R.K. & Coddington, J.A. (1995) Estimating terrestrial biodiversity through extrapolation. In: *Biodiversity* (ed. D.L. Hawksworth). Chapman & Hall, London.

Richter scale A measure of seismic activity devised by C. Richter in 1935 at the California Institute of Technology. It measures how much energy is released by ground movement from the centre of a seismic shock. It is logarithmic, so a 6.0 earthquake is 10 times more powerful than a 5.0. The highest ever recording is 8.9. [P.C.]

Ricker curve An important graphical technique for the analysis of single-species POPULATION DYNAMICS, named after the Canadian fisheries biologist W.E. Ricker, who first used it. The *y*-axis shows the POPULATION DENSITY in the next generation $(t+1)$; the *x*-axis shows the population density in the present generation (t). A straight line of slope 1 passing through the origin is the replacement line, where the population in the next generation is exactly the same as the population in the present generation (i.e. $N_{t+1} = N_t$). The equilibrium population density must lie somewhere on this line. The precise location of the equilibrium depends on the intersection of the Ricker curve and the replacement line. When the Ricker curve lies above the replacement line, the population in the next generation will be higher than the population in this gen-

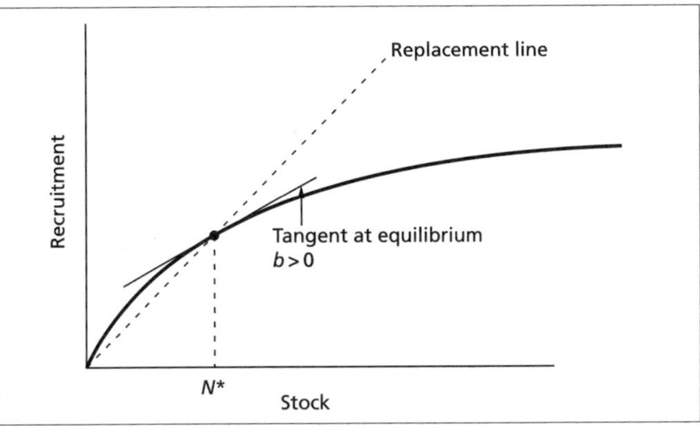

Fig. R14 Ricker curve, also known as the stock–recruitment curve, showing the number of animals entering a fishery as a function of the size of the parental stock that produced them. The straight dashed line is the replacement curve in which $N_{t+1} = N_t$. The recruitment curve lies above the replacement line when stock density is low, and so the population increases (i.e. $N_{t+1} > N_t$). At high stock densities, competition means that the recruitment line falls below the replacement line so population declines (i.e. $N_{t+1} < N_t$). Equilibrium population density, N^*, occurs where the two lines intersect. The stability of the equilibrium and the potential for cyclic or chaotic dynamics is determined by the slope, b, of the tangent to the curve at the equilibrium point. In this case, the slope b is positive, so the equilibrium is a stable point, approached exponentially. Damped oscillations occur when $0 > b > -1$, and chaos is possible when $b < -1$.

eration; i.e. numbers will increase. When the Ricker curve lies below the replacement line, the population will decline. The stability of the equilibrium depends upon the slope of the Ricker curve at the point of intersection (Fig. R14). *See also* SELF-LIMITATION. [M.J.C.]

REFERENCE

Ricker, W.E. (1958) Maximum sustained yields from fluctuating environments and mixed stocks. *Journal of the Fisheries Research Board of Canada* **15**, 991–1006.

riffles, in streams *See* WITHIN-STREAM/RIVER HABITAT CLASSIFICATION.

ring species A geographical DISTRIBUTION of populations in the form of circle overlapping at some point, in which adjacent populations interbreed except at a point where they overlap and are reproductively isolated. Where interbreeding can take place the populations appear to form a single polytypic species consisting of a series of SUBSPECIES. Where overlap and REPRODUCTIVE ISOLATION occur, however, the overlapping populations behave as good species in that they do not interbreed. The best example is a CIRCUMPOLAR group of seagulls (Laridae). The lesser black-backed gull (*Larus fuscus*) and the herring gull (*L. argentatus*) behave as good species in western Europe, where they have clearly distinct plumage and behaviour patterns. Proceeding eastwards across northern Europe and Asia the lesser black-backed gull is replaced by other subspecies, which are in contact across the Bering Strait with the herring gull, extending from Alaska to eastern North America and thence across the Atlantic to Europe. The importance of ring species to evolutionary theory was emphasized by Ernst Mayr, who argued that they show SPECIATION to proceed by accumulation of differences controlled by micromutations, which may build up to the point of reproductive isolation. *See also* SPECIES CONCEPTS. [L.M.C.]

ring test A means of assessing between-laboratory variation in a TOXICITY/ECOTOXICITY test. Single substances are tested according to standard GUIDELINES/protocols, and the results from different laboratories are compared and contrasted. Such procedures are important in the standardization of tests, to ensure that different laboratories achieve the same results for the same test compounds. They are also essential for scientific and legal credibility. *See also* PREDICTIVE ECOTOXICITY TESTS; REPRODUCIBILITY. [P.C.]

Rio Summit The name commonly given to the UN Conference on Environment and Development (UNCED), held in 1992 in Rio de Janeiro, Brazil. *See also* SUSTAINABLE DEVELOPMENT. [P.C.]

rip current A concentrated rapid current moving offshore from a BEACH fronting a longshore current. WATER in rip CURRENTS may reach speeds of 1 m s^{-1}. [V.F.]

riparian Pertaining to the bank(s) of a natural watercourse. [P.C.]

riparian rights The rights of landowners with respect to water on or bordering their land; for example, the right to prevent upstream water from being diverted or misused. [P.C.]

risk assessment In an environmental context, this involves predicting the extent to which the potential of a substance or process to cause HARM is realized under normal and/or abnormal and/or emergency/ACCIDENT conditions. For hazardous chemicals, therefore, this involves a combination of HAZARD assessment (potential to cause harm) with likely FATE of chemicals in the environment and hence EXPOSURE of ecological systems. For industrial processes, risk assessment involves a combination of assessing the potential to cause problems (as a result of ENVIRONMENTAL IMPACT and/or likelihood of litigation) with an assessment of the likelihood of those problems being manifested, which depends upon how the system is operated and managed. Ideally, this should lead to calculation of an explicit probability of a particular magnitude of EFFECT and frequency of occurrence. But often this is not possible in environmental risk assessment due to:

- a lack of understanding of all the complex interactions;
- imprecise definition of the targets one is trying to protect.

So, risk indicators are often used instead. *See also* RISK QUOTIENT. [P.C.]

REFERENCE

Suter, G.W. (1993) *Ecological Risk Assessment*. Lewis Publishers, Boca Raton, Fla.

risk aversion A typical form of preference over PROBABILITY DISTRIBUTIONS of ecological benefits or costs. The term is used commonly in foraging theory. If a forager prefers (i) increased MEAN food intake per unit effort and (ii) decreased food-intake VARIANCE per unit effort, then the animal is risk-averse over probability distributions of food intake. A risk-taker prefers increased levels of both mean food intake and intake variance. *See also* BET-HEDGING. [T.C.]

risk characterization A step in the process of RISK ASSESSMENT.

Thus in European Community (EC) legislation concerning existing and new substances, risk characterization involves comparing PREDICTED ENVIRONMENTAL CONCENTRATIONS (PEC) with PREDICTED NO EFFECT CONCENTRATIONS (PNEC)—a so-called RISK QUOTIENT. [P.C.]

risk communication The exchange of information on likely impairment of human health and/or environment, between risk assessors, risk managers, the public, media, interest group, etc. [P.C.]

risk management In an environmental context, the formulation and enactment of instruments that are intended to control or reduce risks to human health and the well-being of ecological systems from processes and substances. The instruments can involve regulatory (COMMAND AND CONTROL) action; economic incentives, either by direct means (e.g. applying taxes or levies) or by indirect means (by providing appropriate information on performance to customers and stakeholders); and voluntary agreements. [P.C.]

risk minimization models *See* OPTIMAL FORAGING THEORY.

risk perception The awareness or impression of a risk to health or the environment adopted by the public and/or media and/or pressure groups. It may or may not relate to scientific RISK ASSESSMENT, but it is important in influencing policy-makers and regulators and hence is often a driving force in RISK MANAGEMENT. [P.C.]

risk phrases Standard phrases used as labels of substances to identify potential health and environmental hazards (so something of a misnomer; *cf.* RISK ASSESSMENT). Risk phrases are required, for example, by European Community (EC) Directive 67/548/EEC. This includes such phrases as: 'very toxic to aquatic organisms'; 'toxic to aquatic organisms'; 'harmful to aquatic organisms'; and 'may cause long term adverse effects in aquatic environment'. The choice of label depends upon ECOTOXICITY criteria. [P.C.]

risk quotient (RQ) Used in assessment of NEW CHEMICALS and EXISTING CHEMICALS as an indicator of risk (*see* RISK ASSESSMENT). Involves comparing environmental concentrations (PREDICTED ENVIRONMENTAL CONCENTRATIONS, PECs) with threshold concentrations below which no effect is expected (PREDICTED NO EFFECT CONCENTRATIONS, PNECs). So when PEC/PNEC is less than 1, the likelihood of adverse effects is low; when it is greater than 1, an adverse effect is possible and RISK MANAGEMENT is invited. [P.C]

ritualization *See* DISPLAYS.

river continuum concept RIVERS flow, and there are bound to be changes in ecological structure and function from source (springs) to finish (ultimately the SEA). Because of the likelihood of shading from surrounding vegetation, heterotrophic processes often dominate in the upper reaches, whereas primary PRODUCTION becomes more important in the middle reaches. Because of SEDIMENT loading causing shading, heterotrophic processes are likely to become dominant again in the lower reaches. Hence there is a shift in the major sources of energy inputs along the length of a river. Moreover, the energetics of the ecosystems in any one stretch are influenced by the energetics of upstream stretches—for example by import of production and washed-out DETRITUS—and in turn influence the energetics of downstream stretches (by export). Hence there is a longitudinal CONTINUUM of ECOSYSTEM processes and COMMUNITY structure — including functional feeding types (*see* FEEDING TYPES, CLASSIFICATION).

That such a longitudinal SUCCESSION of processes and hence community compositions exists in rivers seems almost beyond doubt. But the river continuum concept (RCC) goes further than this to suggest that there will be general and predictable shifts in these properties; some of the predictions are summarized in Fig. R15.

However, such longitudinal patterns can easily be disturbed by POLLUTION, land management and also, potentially, by local discontinuities in natural features such as geology and sediments. Moreover, flowing-water systems are temporally very dynamic, a consequence of VARIABILITY in physical conditions, especially flow rates. This, in combination with variability in other physical features, is likely to lead to patchiness in susceptibility to scouring by spate, which could result in unpredictable patchiness in the distribution of ecological features compared with the smooth transitions predicted by the RCC. [P.C.]

REFERENCES

Townsend, C.R. (1989) The patch dynamics concept of stream community ecology. *Journal of the North American Benthological Society* **8**, 36–50. [For a critique.]

Vannote, R.L., Minshall G.W., Cummins, K.W., Sedell, J.R. & Cushing, C.E. (1980) The river continuum concept. *Canadian Journal of Fisheries and Aquatic Sciences* **37**, 130–137. [For a description and justification of RCC.]

river invertebrate prediction and classification system (RIVPACS) A model for assessing riverwater quality. The model was constructed empirically from an initial survey of 'clean sites'. It is based on kick samples of benthic invertebrates obtained in a specified way over a specific time. Multivariate discriminant analysis sought CORRELATION between a minimum number of physicochemical variables and taxa present. Then for a particular site the model predicts expected taxa from observations on the same physicochemical variables, and this prediction can be used as a yardstick for the taxa that are actually collected from the site. The extent of divergence between expected and observed is supposed to give an index of river quality (Fig. R16). This kind of approach avoids PSEUDOREPLICATION but is based on correlation analysis, and so until the mechanistic basis of the correlation is defined a divergence from the expected cannot be ascribed with certainty to any particular cause. *See also* BIOMONITOR; BIOTIC INDICES; RETROSPECTIVE ANALYSIS (MONITORING). [P.C.]

REFERENCE

Calow, P., Armitage, P., Boon, P. *et al.* (1990) *River Water Quality.* BES Ecological Issue 1. Field Studies Council, Shrewsbury.

river purification authorities *See* ENVIRONMENTAL (PROTECTION) AGENCIES.

river sampling *See* SAMPLING METHODOLOGY/DEVICES.

rivers Rivers are the natural channels which carry WATER from the land surface to the oceans. Saltwater forms 97% of all the water on

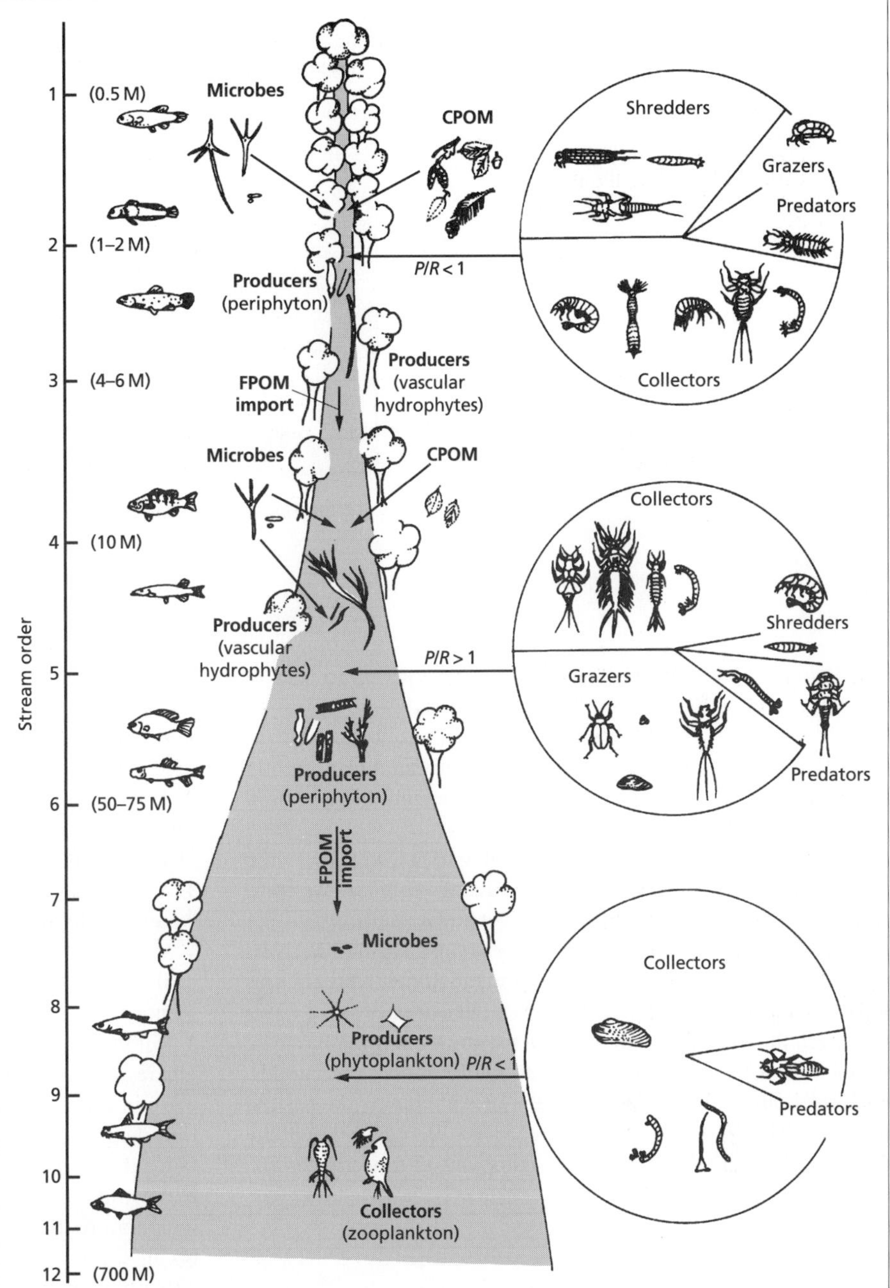

Fig. R15 A generalized model of the shifts in the relative abundances of invertebrate functional groups along a river tributary system from headwaters to mouth as predicted by the river continuum concept (RCC; e.g. Vannote *et al.* (1980)). The river system is shown as a single stem of increasing stream order and width. The headwaters (orders 1–3) are depicted as dominated by riparian shading and litter inputs resulting in a heterotrophic $P/R < 1$. The invertebrates are dominated by shredders which utilize riparian litter as their food resource once it has been appropriately conditioned by aquatic microorganisms (especially aquatic hyphomycete fungi) and collectors that feed on fine particulate organic matter (FPOM). The mid-reaches (orders 4–6) are less dependent upon direct riparian litter input and with increased width and reduced canopy shading they are autotrophic with $P/R > 1$. The shredders are reduced and the scrapers are relatively more important as attached microalgae become more abundant. The larger rivers are dominated by FPOM (and therefore collectors), and the increased transport load of this material, together with increased depth, results in reduced light penetration, and the system is again characterized by a $P/R < 1$. CPOM, coarse particulate organic matter. (After Cummins, K.W. (1975) The ecology of running waters; theory and practice. In: *Proceedings of the Sandusky River Basin Symposium, International Joint Commission, Great Lakes Pollution* (eds D.B. Baker, W.B. Jackson & B.L. Prater), pp. 277–293. Environmental Protection Agency, Washington.)

the planet. Of the 3% that is FRESHWATER, more than 2% occurs as ICE, and of the remaining 1% two-thirds is in groundwater. The remainder comprises lake, river and atmospheric waters, of which rivers contain the smallest proportion — they account for only 0.0001% of total planetary water.

Because all components of the hydrological cycle (*see* WATER (HYDROLOGICAL) CYCLE) on the planet are linked, rivers are necessarily highly dynamic, moving water from land to OCEAN stores. In the course of this movement, rivers have the power to erode the landscape and to carry debris. Rivers carry over 90% of the debris worn from the continents and are responsible for characteristic landforms such as valleys, deltas, flood plains, meanders and braiding. Rivers are also responsible for flooding. Plate 13b (facing p. 398) shows the extent of flood inundation compared to the normal river (Plate 13a). Much flood debris will be deposited on the flooded area thus developing the characteristic flood plain. Velocities, TURBULENCE and load are characteristic of different points in the course from the highlands to the SEA, and these factors influence associated plant and animal communities. Rivers are more intensively used today than in the past, and provide water for communities and industry, form routes for transportation, provide a means of effluent transport, have a large, but not unlimited, self-cleaning capacity and are a focus for recreation.

River channels form patterns in the landscape which reflect the

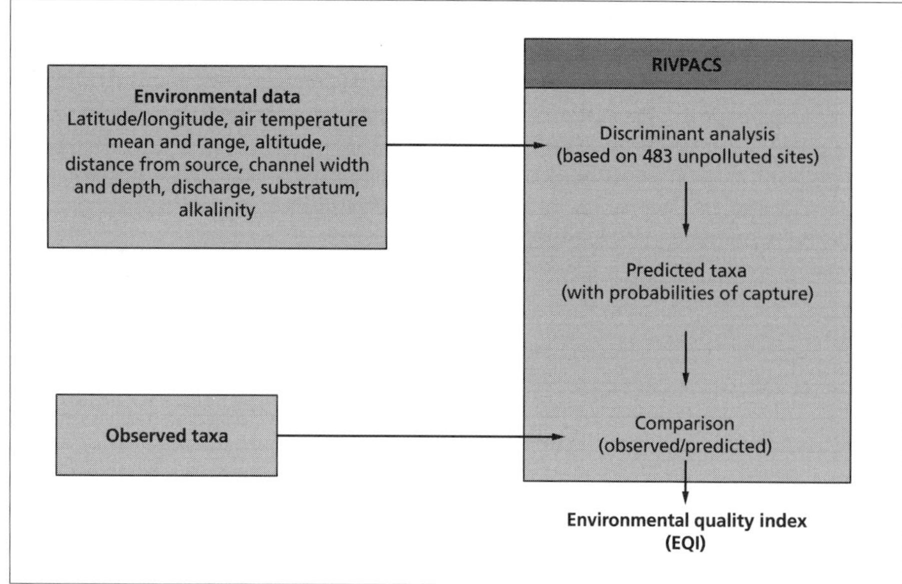

Fig. R16 River invertebrate prediction and classification scheme (RIVPACS). A schematic representation. (After Calow, P., Armitage, P., Boon, P. *et al.* (1990) *River Water Quality.* BES Ecological Issue 1. Field Studies Council, Shrewsbury.)

underlying geology. Rivers normally flow on the surface but in landscapes where there are soluble rocks there can be underground flows. Rivers vary in breadth from less than the pace of a human to many dozens of kilometres. Their influence can extend far out into the ocean, and the SEDIMENT plume can sometimes be seen from space.

The area drained by a river and its tributaries is known as the CATCHMENT (UK) or WATERSHED (US). River catchments are commonly used as the basis for integrated river and land planning—the Tennessee Valley Authority was an early example. The flow in rivers depends upon the CLIMATE of the catchment. Some rivers are seasonal, having periods of high and low flow. In extreme cases there is no flow in the river at certain times. To use the river, water has to be stored at times of high flow for release during low-flow periods. Such rivers are said to be regulated. Because it removes the natural flow variability of the river, regulated rivers are environmentally different and excessive river REGULATION causes concern. *See also* LOTIC; RIVERS AND STREAMS, TYPES OF. [A.M.]

Rivers and streams, types of

The two primary sources of WATER in RIVERS are surface RUN-OFF and groundwater. Rivers may find their original beds in depressions previously developed by an agency other than the river itself (e.g. TECTONIC, volcanic, landslide, glacial, WIND, organic accumulation, meteorites or animals). However, many rivers excavate their own channels, through either physical or chemical erosion, determined by the nature of the bedrock, composition and volume of water, the CLIMATE and the gradient.

A number of classification schemes have been developed to distinguish the different sections along the course of a river, with changes in the BIOTA classified with reference to the physical changes of the river.

A river arising in a mountainous region can be divided into three sections.

1 An upper (or mountain) course with a V-shaped valley in which the water flows fast enough to carry stones.

2 A middle (or foothill) course where the valley is broader in section with stable sides and less prone to erosion, and the water velocity is still fast enough to carry sand and mud in suspension.

3 A lower (or plain) course with a broad, shallow valley and a water velocity slow enough to deposit sediments.

An alternative classification system, originally designed for European rivers (Illies & Botosaneanu 1963), divides the river into two sections.

1 Rhithron — the region extending from the source to the point where the mean monthly water temperature rises to 20°C, the OXYGEN concentration is always high, the water velocity is fast and turbulent, and the bed is composed of rocks, stones or gravel with a fauna that is cold stenothermic and contains no PLANKTON.

2 Potamon—the region of the river where the mean monthly water temperature rises to over 20°C, occasional oxygen deficiencies occur, the water velocity is slow and the substratum consists of either mud or sand with a fauna that is either eurythermic or warm stenothermic and does contain plankton.

In areas outside western Europe, the mean monthly water temperature cut-off point between rhithron and potamon must be adjusted according to the geographic and climatic differences.

A stream or river system consists of a pattern of tributaries joining one another and coalescing to form the main river. Tributaries can be arranged into a hierarchy. First-order streams have no tributaries, but when two first-order streams meet they form a second-order stream. When two second-order streams meet they form a third-order stream and so on. Comparing streams in different orders, there are 3–4 times as many streams, each of less than half the length and draining a little more than one-fifth the drainage area in each successive order.

Vannote *et al.* (1980) suggested that there was a gradual change from source to mouth (first-order to highest order) in which fluvial, geomorphic processes, physical structure and the hydraulic cycle changes are related to patterns of biotic COMMUNITY structure and function and organic matter. This concept is known as the RIVER CONTINUUM CONCEPT (RCC). The RCC focuses on longitudinal resource gradients from the headwaters to the mouth of river systems. There are, however, important interactions between the channel and flood-plain systems, and between the channel and groundwater AQUIFERS, which should also be integrated with a temporal dimension.

Rivers can be permanent, where the water always flows and the WATER TABLE is higher (effluent streams) than the stream level; or intermittent, where the water flows only seasonally (influent streams). Both permanent and intermittent rivers may be interrupted with alternate flow on or below the surface.

Geographic areas from which rivers reach the SEA are termed exorheic, in contrast to endorheic areas, where rivers arise but never reach the sea, and arheic areas, from which no rivers arise. Rivers in endorheic areas tend to have high SALINITY levels, particularly in the lower sections of their courses. *See also* LOTIC. [R.W.D.]

REFERENCES

Illies, J. & Botosaneanu, L. (1963) Problèmes et méthodes de la classification et de la zonation écologiques des eau cairantes, considerées acceptait du point de vie faunistique. *Mitt. Int. Verein. Theor. Angew. Limnol.* **12**, 1–57.

Vannote, R.L., Minshall, G.W., Cummins, K.W., Sedell, J.R. & Cushing, C.E. (1980) The river continuum concept. *Canadian Journal of Fisheries and Aquatic Sciences* **37**, 130–137.

RIVPACS *See* RIVER INVERTEBRATE PREDICTION AND CLASSIFICATION SYSTEM.

road run-off Aqueous drainage from roadways following RAIN or snow/ice-melt. Such RUN-OFF is often loaded with organic and inorganic CONTAMINANTS, derived from motor vehicles, from the road materials themselves, or from materials used to grit the road, particularly for de-icing. *See also* MOTOR VEHICLE POLLUTION. [P.C.]

robust dynamics Dynamics that are relatively insensitive to parameter variation or perturbations. *See also* FRAGILE DYNAMICS; POPULATION DYNAMICS; ROBUSTNESS OF MODELS. [S.N.W.]

robustness of models MODELS are described as robust if the conclusions drawn from them are not very sensitive to the assumptions and PARAMETERS that they employ. This is important in ECOLOGY because of the difficulty of obtaining precise estimates of parameters and the necessary simplification required to make models of real systems tractable. Robustness is an attractive property in that it makes prediction possible using very uncertain information, but at the same time it makes falsification of a model difficult. [S.N.W.]

rocky shores Usually refers to the marine environment, but occasionally can refer to LAKES when there are rocky outcrops on the shoreline. In general, marine rocky shores are the most densely populated and most biologically diverse of all intertidal shores. They are characterized by prominent VERTICAL ZONATION of the dominant species. Horizontal ZONATION also occurs, with the degree of wave exposure having a major influence on the types of species present.

Vertical zonation in temperate and BOREAL rocky shores generally shows a pattern with a black lichen zone at the highest level, followed by a periwinkle zone (these together are sometimes referred to as the supralittoral fringe), a barnacle zone (sometimes called the mid-littoral zone) and a lower zone dominated by various species (e.g. *Mytilus edulis*), depending upon locality (sometimes called the infralittoral fringe). The width of the various zones is controlled by the slope of the shore, the tidal range and the degree of wave exposure. Whereas the upper limit of species zones is generally considered to be controlled by physiological tolerance to ABIOTIC FACTORS (e.g. DESICCATION and temperature extremes), the lower limit is believed to be controlled mainly by interactions with predators or competitors.

A number of studies have demonstrated the important role of top predators (or so-called 'keystone species') in structuring rocky shore communities, and GRAZING by herbivores has also been shown to be of considerable importance in setting limits of algal species distribution. Limiting RESOURCES for which COMPETITION occurs include space, light and food. The relative importance of biological and PHYSICAL FACTORS varies over relatively short distances, both vertically and horizontally. Grazing is one of the main structuring agents of the rocky intertidal on the Atlantic shores of Europe and the Pacific shores of North America. On European shores, the more exposed coastlines tend to be dominated by mussels, barnacles and limpets, whereas on more sheltered shores having fewer grazers, the community is dominated by dense stands of fucoid brown algae (see Plate 14, facing p. 398). Grazing is less important on the Atlantic coast of North America. Grazing declines in importance as one moves polewards in the Atlantic, whereas the importance of grazing increases towards the Equator. Selective SETTLEMENT by larvae of rocky shore organisms may also have an important influence on the resulting community structure. Larval RECRUITMENT is spatially and temporally variable, and this can have an influence on other key processes, such as competition and predation.

In addition to the major zonation patterns, small-scale variations, or patchiness, is a common feature of rocky shores. Local patchiness in organism distribution occurs in response to either physical factors (e.g. local variation in the rock surface, slope or wave action) or biological factors (e.g. territorial or aggressive behaviour). Successional patterns on rocky shores appear to result from a combination of inhibition of newly colonizing organisms by those already present, as well as FACILITATION (i.e. certain species alter the environment to make it more suitable for COLONIZATION by other species). At any time, different areas on the shore may be at a different successional stage, thus contributing to the patchiness. [V.F.]

roentgen The earliest unit used to measure EXPOSURE to RADIATION, named after W. Roentgen, the discoverer of X-RAYS. It preceded the discovery of RADIOACTIVITY, and is applicable only to X-rays and γ-RAYS. Hence it is now superseded by the SI UNIT of exposure, coulomb per kilogram. One roentgen produces 1 esu of electric charge in 1 cm^3 of air at normal temperature and pressure. [R.S.S.]

root exudates *See* RHIZOSPHERE.

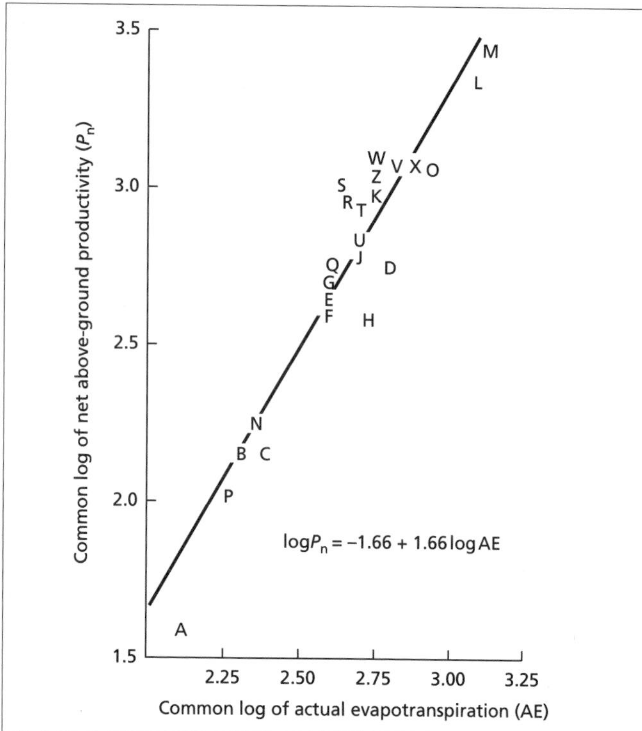

Fig. R17 Rosenzweig's rule. This empirical rule says that log net primary productivity increases linearly with increasing log actual evapotranspiration across different ecosystems. Also, there is often a good correlation between plant productivity and rainfall across years within a given ecosystem. Relationship between net primary productivities of plant communities and the evapotranspiration indices of the sites. Communities are: A, creosote-bush desert; B, arctic tundra; C, alpine tundra; D, tall-grass prairie; E to J, mountain shrublands; K, beech/maple forest; L and M, tropical forests; N, desert sand dunes; O, oak forest; P, annual grassland; Q, fir forest; R and S, spruce/fir forests; T and U, beech forests; V, hemlock forest; W and X, lowland deciduous forest; Z, hemlock forest. (After Rosenzweig, M.L. (1968) Net primary production of terrestrial communities: prediction from climatological data. *American Naturalist* **102**, 67–74. University of Chicago Press.)

Rosenzweig's rule An empirical rule, stating that the logarithm of net primary productivity is linearly related to the logarithm of actual evapotranspiration (Fig. R17). The relationship between PRODUCTIVITY and PRECIPITATION is better described by actual evapotranspiration than by rainfall or potential evapotranspiration because it takes simultaneous account of TEMPERATURE and RELATIVE HUMIDITY. The rule suggests that there is an allometric relationship at the ECOSYSTEM level between net primary productivity and the rate at which water is transpired by plants per unit area per year. *See also* ALLOMETRY. [M.J.C.]

REFERENCE

Rosenzweig, M.L. (1968) Net primary production of terrestrial communities: prediction from climatological data. *American Naturalist* **102**, 67–74.

Rothamsted A research station set up in 1844 by John Bennett Lawes and Joseph Henry Gilbert in Harpenden, near St Albans, in south-east England. Their original research involved agricultural chemistry; they were especially interested in the relationship between MINERAL NUTRIENTS (FERTILIZERS such as NITROGEN, PHOSPHORUS and POTASSIUM) and CROP yield, and they set up several classic long-term experiments, some of which are still running today (*see* BROADBALK EXPERIMENT; PARK GRASS EXPERIMENT). Other early work concerned the chemical composition of farm animals and their products. Rothamsted was a major centre in the development of modern statistical methods (notably the work of R.A. Fisher on the ANALYSIS OF VARIANCE) and statistical computation (the languages Genstat and Glim were developed there). Current work covers crop physiology, soils and biotechnology. [M.J.C.]

REFERENCE

Johnston, A.E. (1994) The Rothamsted classical experiments. In: *Long-term Experiments in Agricultural and Ecological Sciences* (eds R.A. Leigh & A.E. Johnston), pp. 9–37. CAB International, Wallingford.

Rothamsted Insect Survey Ecology plays on a wide stage with processes spanning many decades, large slices of geography or, frequently, both. There are few studies, however, that can provide data on such large temporal and spatial scales since they require the commitment of correspondingly large amounts of time and money. The Rothamsted Insect Survey (RIS) is one of these few exceptions.

The survey was not, however, originally planned as a large-scale ecological study but grew gradually out of light-trap SURVEYS of macrolepidoptera at ROTHAMSTED in the 1930s and late 1940s. In 1960 this survey was restarted and, by recruiting amateur entomologist volunteers, developed by 1968 into a network covering the UK.

As the light-trap network grew a second strand was added to the study by the erection of a '12-metre'-high suction trap at Rothamsted in 1964. This trap was designed to catch the more mobile wind-borne insects, particularly aphids, with the idea of developing an early warning scheme for these insects. A network of the 12-metre traps was therefore set up throughout Britain.

The RIS now comprises a network of 95 light-traps and 15 12-metre suction traps covering much of the UK (Woiwod & Harrington 1994). The survey has, however, used light-traps at 370, and suction traps at 26, sites in total. Together they have yielded over 20 million individuals of more than 1000 species in the target taxa, macrolepidoptera or aphids, and other taxa from the traps remain to be sorted and identified. The RIS material is the largest collection for which the date and place of each individual's trapping is known. It therefore represents an unparalleled ecological resource.

In addition to the practical use of the RIS as an early warning system it has also provided data for the investigation of insect DISTRIBUTION patterns (Taylor 1986), the mapping of populations on a national scale (Woiwod 1986), and studies of the effects of CLIMATE and CLIMATE CHANGE on insect PHENOLOGY (Harrington *et al.* 1995). The RIS database also provided over 5700 POPULATION DENSITY series of longer than 10 years for the most intensive search yet for direct DENSITY DEPENDENCE in insect POPULATION DYNAMICS (Woiwod & Hanski 1992).

The data accumulated by the RIS provide a temporal perspective that transcends the original reasons for setting up the RIS. These data are proving especially valuable in analyses of the effects of GLOBAL WARMING and environmental change, analyses that were not envisaged by the Survey's originators but which emphasize the vital importance of studies carried out on the wide scale of time and space which is the ecological stage. [A.J.D.]

REFERENCES

Harrington, R., Bale, J.S. & Tatchell, G.M. (1995) Aphids in a changing climate. In: *Insects in a Changing Environment; 17th Royal Entomological Symposium 1993* (eds N. Stork & R. Harrington), pp. 125–155, Academic Press, London.

Taylor, L.R. (1986) Synoptic dynamics, migration and the Rothamsted Insect Survey. *Journal of Animal Ecology* **55**, 1–38.

Woiwod, I.P. (1986) Computer mapping in the Rothamsted Insect Survey. In: *Proceedings of the Seminar on Agrometeorology and Pest Forecasting, Fulmar Grange, UK, 1985*, pp. 89–98. CTA/TDRI, Wageningen.

Woiwod, I.P. & Hanski, I. (1992) Patterns of density dependence in moths and aphids. *Journal of Animal Ecology* **61**, 619–629.

Woiwod, I.P. & Harrington, R. (1994) Flying in the face of change: The Rothamsted Insect Survey. In: *Long-term Experiments in Agricultural and Ecological Sciences* (eds R.A. Leigh & A.E. Johnston), pp. 321–342. CAB International, Wallingford.

roughness A measure of the ability of a surface to generate TURBULENCE in a fluid flowing over it. The degree of turbulence will depend on the ROUGHNESS LENGTH of the surface. *See also* REYNOLDS NUMBER. [J.H.C.G.]

roughness length (z_0) A measure of the aerodynamic ROUGHNESS of a surface. Roughness length is defined by the LOGARITHMIC WIND PROFILE equation, such that for conditions of neutral atmospheric thermal stability the windspeed equals zero at height z_0. The magnitude of the roughness length depends on the size and the spacing of the elements making up the surface. For VEGETATION, the roughness length is normally about one-tenth of the average vegetation height. [J.H.C.G.]

round robin test The same as RING TEST. [P.C.]

Royal Commission on Environmental Pollution (RCEP) A UK standing body constituted on 20 February 1970 'to advise on matters both national and international, concerning the POLLUTION of the environment; on the adequacy of research in this field; and on the future possibilities of danger to the environment'. Appointments to the Commission are by Royal Warrant, on the advice of the Prime Minister. The Commission is therefore an independent body, not constrained by government departmental boundaries. The Commission has no specific or restricted task. It is authorized to inquire into any matter on which it thinks advice is needed, and also to inquire into any issues within its terms of reference that are referred to it by any Secretaries of State or Ministers. It has written reports on such matters as toxic pollution, INCINERATION and traffic problems. It has been enormously influential in the development of UK government policies. [P.C.]

RPD *See* REDOX POTENTIAL DISCONTINUITY.

RQ *See* RESPIRATORY QUOTIENT; RISK QUOTIENT.

RRV *See* RESIDUAL REPRODUCTIVE VALUE.

rubber Naturally occurring rubber is a linear POLYMER of ISOPRENE units linked at CARBON atoms 1 and 4. Natural rubber in its raw state is a tacky substance with poor resistance to abrasion or oxidation. Its properties can be improved by vulcanization, a process which introduces a degree of cross-linking into the structure, enhancing resistance to degradation and improving tensile strength and elasticity. The polymer is unsaturated and so two isomeric forms are possible. Natural rubber is the *cis* form whereas the *trans* form can be produced synthetically and has different properties. Various synthetic rubbers are produced by polymerizing ALKENES; for example, nitrile rubber is obtained from butadiene and acrylonitrile. [M.C.G.]

rubidium (Rb) A metallic element (ATOMIC NUMBER = 37; relative ATOMIC MASS = 85.468; melting point = 39°C; boiling point = 688°C) with a natural abundance of 78 mg kg^{-1}. It is extracted as a by-product in the manufacture of LITHIUM. Rubidium has only a few industrial applications, limited to the production of photo-cathodes and other light-sensitive components. Rubidium oxides have been used as CATALYSTS in the synthesis of METHANOL. Rubidium salts have a very low TOXICITY. [C.R.P.]

***Rubus fruticosus–Holcus lanatus* scrub community** *See* SCRUB.

ruderal A plant characteristic of open, disturbed conditions, usually resulting from human activity. Such plants usually allocate a high proportion of their PRODUCTION to the generation of many, normally small, well-dispersed seeds. They usually have a short LIFE CYCLE and are able to expand their populations rapidly following environmental DISTURBANCE. Often demanding in their nutrient and LIGHT requirements, they are sensitive to COMPETITION from more robust species.

Ruderals comprise one of the three plant-strategy classes proposed by J.P. Grime, the other two being competitors and stress-tolerators. [P.D.M.]

rule making The USEPA has powers to interpret and implement and, to some extent, extend legislation by developing formal statements. These are subject to public scrutiny and challenge in the courts. *See also* ENVIRONMENTAL LAW. [P.C.]

rules of thumb *See* BEHAVIOURAL ECOLOGY; MARGINAL VALUE THEOREM.

ruminant A herbivorous animal that has a complex digestive system in which a mutualistic relationship is maintained with anaerobic CELLULOSE-digesting bacteria. These bacteria are found in the greatly enlarged first chamber of the stomach, called the rumen. Several species of protozoans are also present which, in combination with the bacteria, form a diverse multispecies assemblage. In order for food to pass from the rumen into the reticulum (second stomach) it must have a sufficiently small particle size. Often this is not the case when it is first ingested. In consequence the bolus is passed back into the mouth cavity for further mastication. This process is known as rumination and is the defining feature of these animals, all of which 'chew the cud'. After leaving the rumen and passing into the reticulum, the food enters a third stomach, the omasum. Water is absorbed in the second and third stomachs; after this the food enters the fourth 'true' stomach, the abomasum. This stomach contains peptic glands which enable further breakdown of the ingested food. [R.C.]

run-off The fraction of the PRECIPITATION falling on a land area that leaves it as surface or subsurface WATER flow. The term is most appropriately used to describe the amount of water yielded by a drainage basin in a given time period, expressed either as a volume or as an equivalent water depth over the area concerned. Loose usage in the past has led many to believe that the term 'run-off' conveys meaningful ideas about the source of CATCHMENT outflow, or the rapidity with which it is produced. It does not unless prefixed by an appropriate adjective (e.g. overland, groundwater, storm, direct, delayed, baseflow). *See also* ROAD RUN-OFF. [D.C.L.]

Russia and the former Soviet Union, conservation in

The Soviet legacy

With the disintegration of the Soviet Union, Russia has inherited a costly environmental legacy. Economic and environmental mismanagement has resulted in biological wastelands in specific locales (e.g. smelter pollution in Norilsk). In other places, dramatic alterations of systems and processes are evident (e.g. draining of the Aral Sea). Policies in the agricultural and forestry sectors led to depletion of stock, soils and whole ecosystems. Ironically, the extent of exploitation indicated a complete disregard for the official Communist legislation endorsing the principle of sustainable management.

Despite such practices, Russia is unparalleled in BIODIVERSITY riches: 40% of the Northern hemisphere's forests; the deepest and oldest lake (Baikal); the longest continuous coastline in the world; and a considerable portion of the over 165 000 higher plant and faunal species identified as occurring in the former Soviet Union. Of particular interest, however, are Russia's great expanses of intact ecosystems. These relatively untouched contiguous tracts of land, primarily in Siberia, have resisted exploitation due to their distance from industrial centres. Forestry, to date, has affected only one-third of Siberian forests.

Forestry in the former Soviet Union was plagued by inefficient use of harvested timber and a reliance on technology that caused overharvesting. Water-based transportation continues to be the most common mode, and the resultant losses are compensated for through increased HARVESTING. This and a minimal REFORESTATION programme (25% of annual cut) adds to the toll on biodiversity.

Paradoxically, the former Soviet Union developed a peerless system of protected areas. In 1991 the Soviet Union counted 121 strict nature reserves, the *zapovednik*. As of September 1994, Russia counted 89 federal level *zapovedniks*. The total area of these nature reserves is approximately 29 Mha. (1.4% of Russia's land area). These sites account for 40% of all the world's strictly protected areas (IUCN class I). Within *zapovedniks* the norm was unprecedented levels of scientific research with emphasis on CLASSIFICATION and identification. Prevention of visiting and poaching was the principal means of protection. Current economic conditions are causing a breakup of the protection regime and the demise of the research. Data collected over decades are at risk. In general, the integrity of Russia's system of protected areas is eroding. Scientists are not being recruited to replace the established scientists that are leaving. The reserves are pursuing economic ventures to augment their budgets. Ecotourism, limited resource exploitation and entry fees are under consideration for implementation across the system.

National Parks, Special Purpose Reserves (*zakaznik*) and Natural Monuments are other forms of federal-level protected area designations instituted under the Soviet Union. These designations are still in use in Russia and other former Soviet Union countries. Regional-level designations by these same names are also common. In all, 4.1% of the former Soviet Union was under protected designation. Russia, as of the end of 1994, designated approximately 6% of its area as protected.

Russian conservation agencies

In Russia the responsibility for managing the federal nature reserves resides with the Ministry of Environmental Protection and Natural Resources. This Ministry also reviews and sanctions the establishment of other forms of protected areas. The management responsibility of such areas may lie elsewhere. The Federal Forest Service, for example, is the dominant National Park manager while the Department of Wildlife Management and Hunting has jurisdiction over most *zakazniks*.

Shared or rather uncoordinated jurisdiction is another legacy of the former Soviet Union. Some 70 Russian government agencies share responsibility for environmental concerns. Further, parallel government structures reporting to the President add to the fragmentation and to conservation difficulties.

Governing environmental legislation includes carry-overs from the Soviet era and legislation enacted since the creation of the Russian Federation. Contradictions in the legislation are one problem. The major problem, however, is the lack of implementation. Legislation in Russia is often characterized by idealized expressions composed by academics with little chance of attainment. Policy debates as a precursor to law definition are uncommon and regulations for implementing the laws are equally unusual. The chief piece of legislation, passed in 1992, is the Environmental Protection Act. This Act provides for the establishment of protected areas, for environmental assessments, and for COMPLIANCE with environmental standards and regulations. A series of Presidential Decrees have attempted to clarify Sections of this Act, by issuing regulations or simply restating, with more force, responsibilities.

Growth of environmentalism

Since the breakup of the Soviet Union, public involvement in environmental issues has never been more active. Environmental groups have been organizing, some as branches of international organizations. The activities of these new groups vary from publishing newsletters (Biodiversity Conservation Center) to organizing public protests (Greenpeace Russia). Others, such as the Amur Basin Project (Socio-Economic Union), focus on activities involving local communities in restoring endangered species' habitats, and on improving farming and FORESTRY PRACTICES, while involving the local administration. Much now also depends on the level of enthusiasm among local scientists and other individuals. Such scientists, for example, are now the principal agents for maintaining lists of ENDANGERED SPECIES and MONITORING their status. Provision for maintaining such lists, in the form of a 'Red Book' through the work of a Commission, are in the Environmental Protection Act. However, such work has not been a priority of the Ministry of Environmental Protection and Natural Resources.

International interests

World interest in Russian biodiversity is manifest in the GLOBAL ENVIRONMENTAL FACILITY (GEF) funding. Support, through the GEF, is secure for projects which address assistance to protected areas, the integration of biodiversity considerations into economic policies, and a pilot project in integrated resource management centred on Lake Baikal. These projects are part of an Environmental Framework for investment devised through the assistance of the World Bank. Numerous international agencies are actively providing support to the conservation of RESOURCES. However, Russia's lack of modern infrastructure, federal legal instruments in particular covering contractual law and, most of all, investment funds contravene conservation initiatives.

The Russian Far East, for example, because it is accessible to the Pacific rim countries and is largely unexploited, is of particular interest to many. Earlier attempts to exploit resources (Bikin Valley forestry) were halted partly because of an international reaction generated by environmental activists and others. Now, millions of dollars are being spent by US and Canadian agencies, among others, on projects demonstrating SUSTAINABLE DEVELOPMENT principles. Forestry pilot projects, for example, focus on environmental and sustainable exploitation regimes that also protect biological DIVERSITY. However, there is considerable scepticism towards these developments among several Russian environmental groups. The concern is that nefarious and now-opposed North American forestry practices will be imported into Russia. *See also* CONSERVATION BIOLOGY. [N.L.]

Russian conservation agencies *See* RUSSIA AND THE FORMER SOVIET UNION, CONSERVATION IN.

ruthenium (Ru) A metallic element (ATOMIC NUMBER = 44; relative ATOMIC MASS = 101.07; melting point = 2282°C; boiling point = 4050°C), with a crustal abundance of 10^{-4} mg kg^{-1}. It generally occurs as the native metal and has seven natural stable isotopes. The FISSION products ^{103}Ru ($t_{1/2}$ = 39.254 days) and ^{106}Ru ($t_{1/2}$ = 1.06 years) are released to the environment from nuclear facilities, and uptake by edible seaweed (e.g. *Porphyra*) can provide a pathway resulting in low-level human RADIATION exposure. Ruthenium occurs in the +3 oxidation state and has a low solubility in natural waters. [A.B.M.]

Rylands versus Fletcher A ruling in British private or TORT law that has implications for environmental protection. The ruling arose out of a case reported in 1865 and involved the following:

> . . . that the person for his own purposes brings on to his land and collects and keeps there anything likely to do mischief if it escapes, must keep it in at his peril, and if he does not do so, is *prima facie* answerable for all the damage which is the natural consequence of its escape.

This therefore involves an important legal principle: that liability for POLLUTION is strict, i.e. it applies even if there was no intent to do HARM. In principle the implications for environmental protection that follow from this are wide ranging; in practice, however, a number of restrictions placed on the interpretation of the principle in the courts has constrained its effects. One recent development arose out of a decision of the House of Lords in the case of Cambridge Water Co. vs. Eastern Counties Leather in favour of the latter party (ECL), overturning an earlier decision of the Court of Appeal. The facts were that ECL had operated a tannery in such a way as to allow the escape of significant quantities of organochlorine solvent into the groundwater through spillages. This ceased in 1976 but damage to Cambridge Water Company's (CWC's) borehole was discovered only in the 1980s. The House of Lords reasoned that damages in private nuisance or under Rylands vs. Fletcher depend upon foreseeability and that this, despite a previous ruling to the contrary by the Court of Appeal, was not a reasonable interpretation here.

The significance of the ruling is that the highest court in the UK appears to put a barrier in the way of development of strict (i.e. no fault) liability for environmental damage. Nevertheless, strict civil liability for environmental damage embodied in the principle is being considered for application throughout the European Union and forms the basis of Superfund legislation in the USA. Indeed, the decision from the House of Lords on the ECL case actually stated that if strict liability is to be developed within this context it is more appropriate for Parliament to carry out this change than the Courts. *See also* ENVIRONMENTAL LAW. [P.C.]

S

***S/G* selection** *See* SELECTION PRESSURES, CLASSIFICATION.

SACs (special areas of conservation) *See* NATURA 2000.

saeter *See* ALPINE ZONE.

safety margins *See* PREDICTIVE ECOTOXICITY TESTS.

safety (safe) level In an environmental context this implies a level of CONTAMINATION that has zero EFFECT. But logically it is impossible to prove a negative—some effect on a hitherto untested physiological system or individual species, process or ecosystem might arise. So this concept is scientifically not very useful. It is better to specify probabilities of effects, with the term 'safe level' used to indicate a low probability of effect. *See also* PRECAUTIONARY PRINCIPLE; RISK ASSESSMENT. [P.C.]

salic horizons *See* SOIL CLASSIFICATION.

saline lake *See* LAKES; LAKES, TYPES OF.

salinity Salinity is the extent of dissolved salts in waters or soils. Salinization is the process whereby soils accumulate salts over time. Seawater contains about 33 $g l^{-1}$ of dissolved salts, made up as follows (all units $mg l^{-1}$): SODIUM, 10 600; MAGNESIUM, 1300; CALCIUM, 400; POTASSIUM, 380; CHLORINE, 19 000; BROMINE, 65; SULPHUR, 900; CARBON, 28, and many other ELEMENTS at trace levels. Rainwater contains 10–50 $mg l^{-1}$ of dissolved salts.

Salinity of waters and soils is often assessed by measuring the electrical conductivity using a conductivity cell. SOIL is mixed with distilled water to make a paste for these measurements. The conductivity is measured in millisiemens per centimetre ($mS cm^{-1}$). The total dissolved solids (TDS, $mg l^{-1}$) in soil is closely related to the electrical conductivity (EC, $mS cm^{-1}$) by the following empirical relationship, provided the soil is neither very ACID nor very alkaline:

$$TDS = EC \times 640$$

Soil is considered to be saline when the electrical conductivity exceeds 4 $mS cm^{-1}$.

The sensitivity of CROPS to salinity is much studied. Sensitive crops (including apricots, carrots, onions and strawberries) begin to be affected at EC values as low as 1 $mS cm^{-1}$ and their yield may be reduced by 25% for every 1 $mS cm^{-1}$ thereafter. Tolerant crops (e.g. barley, cotton, sugar beet and wheat) begin to be affected at about 7 $mS cm^{-1}$ and thereafter show a reduction of about 5% in yield for every 1 $mS cm^{-1}$.

Salinization of soils may occur by natural or human-induced causes (primary and secondary salinization, respectively). Primary salinization may be the result of natural WEATHERING of the rock materials, or accumulation of salts derived from RAIN, or it may be caused by historical submergence of the soil under seawater. Secondary salinization is the result of salt stored in the profile being mobilized by the extra water from IRRIGATION, which raises the WATER TABLE so that EVAPORATION near the surface leaves salts behind. This is a major problem, affecting about 50 Mha of irrigated land in the world (*see* IRRIGATION).

Associated with soil salinization is alkalization, in which the CLAY fraction becomes sodium-saturated, causing crumb structure to collapse, restricting water penetration and aeration. Soils in which more than 15% of the cation exchange sites are occupied by sodium are termed sodic. [J.G.]

REFERENCES

Ghassemi, F., Jakeman, A.J. & Nix, H.A. (1995) *Salinisation of Land and Water Resources*. CAB International, Wallingford.

McBride, M.B. (1994) *Environmental Chemistry of Soils*. Oxford University Press, Oxford.

salinization An increase in the salt concentration of the soils (or waters) of a habitat, usually resulting from human activity. The term is most frequently used of soils in dry regions which have been artificially irrigated and have become salty as a result of evaporative forces bringing the salts in solution to the SOIL surface. Sodium chloride (NaCl), calcium carbonate ($CaCO_3$), calcium sulphate ($CaSO_4$) and MAGNESIUM salts are often involved in the process.

The formation of gypsum (calcium sulphate) layers in soils leads to concretions and poor drainage, and a high salt concentration results generally in problems for plant life (including crops) as they need to take up water against a steep concentration gradient. In cases of severe salinization, only halophytic plants are eventually able to survive. [P.D.M.]

***Salix lapponum–Luzula sylvatica* scrub community** *See* SCRUB.

salt desert *See* BRACKISH HABITATS.

salt flat A flat, saline area in a DESERT created by the EVAPORATION of a former lake (playa). Often the surface is crusty and firm as a result of salt crystallization. [P.D.M.]

salt marsh A term used to describe the habitat, vegetation and fauna associated with the accretion of fine sediments (mud to fine sand) between high neap- and high spring-tide levels. Bare mudflats are a typical feature below high neap-tide level, but occasionally some algae and flowering plants of the genus *Zostera* (eelgrass) may occur here. Salt marshes are globally confined to temperate to tundra latitudes, being replaced in the subtropics and tropics by MANGROVE swamps (mangals). They form in sheltered conditions on shallow shores where wave action is gentle, typically in estuaries, sometimes on the landward side of shingle spits or BARRIER ISLANDS (see Plate 15, facing p. 398), more rarely on open coasts.

Salt marshes typically show distinct ZONATION of plant communities, associated with changing levels within the MARSH and consequent varying periods of immersion/emersion by saltwater. A zonation of associated fauna, although less obvious, also occurs, involving mainly species of MEIOFAUNA (nematodes, turbellarians, oligochaetes, copepods and halicarids), molluscs, arthropods (insects, arachnids and crustaceans), and some birds and mammals. GRAZING by domestic stock at low tide is sometimes an important factor. Usually the zonation pattern relates to stages in a PRIMARY SUCCESSION (halosere), in which fine sediments carried into the ESTUARY are trapped by the rooted plants thus raising the SEDIMENT level and allowing the SUCCESSION to progress.

Rates of accretion of sediment tend to be greatest at low- and mid-marsh levels, and in general range from one to a few centimetres per year. A salt marsh is also subject to erosion at times of more vigorous wave action or current flow, and mini-cliffs, creeks and pans are manifestations of this.

The plants are halophytes (salt-tolerant plants) and may exhibit succulence. Those of the lower levels of the salt marsh tend to be OBLIGATE halophytes. Both plants and animals exhibit a variety of mechanisms to cope with extreme and wide-ranging levels of several environmental factors, in particular SALINITY, inundation and substrate texture. [B.W.F.]

salt pan A shallow, isolated depression formed in a SALT MARSH and typically devoid of vascular plants. It may form from natural slight depressions of the salt-marsh surface, or from the damming of small tributary creeks. In high-level pans, SALINITY levels tend to be high through EVAPORATION at low tide, but rainwater may accumulate during extended low-tide periods and reduce salinity levels. Such extreme and variable salinity levels may exclude plants, but highly anaerobic conditions formed below the hard-pan surface may also be involved. If pans become part of the drainage system of the salt marsh by linkage to creeks, then recolonization by plants can take place. [B.W.F.]

SAM *See* STANDARDIZED AQUATIC MICROCOSMS.

samarium (Sm) A LANTHANIDE METAL (ATOMIC NUMBER = 62; relative ATOMIC MASS = 150.4; melting point = 1077°C; boiling point = 1791°C) with an average abundance of 6.0 mg kg^{-1} in the Earth's CRUST. The alloy $SmCo_5$ is used in magnets, while the oxide is used in optical GLASS. [J.G.]

sample A statistical term that refers to a subset or selection of all the observations that could be made on a particular item. The summary measurements of samples (STATISTICS) are therefore estimates of population PARAMETERS. They are represented by Roman letters, for example the MEAN ($\bar{x}$) and the VARIANCE (s^2). *See also* POPULATION. [B.S.]

Sampling

In almost all ecological studies it will be impossible to account for every individual in a POPULATION. Therefore it is necessary to examine a subgroup of the total population and extrapolate from this to the whole population. The process by which the subgroup of the population is selected is sampling.

If a population is evenly distributed through a HABITAT then a single small SAMPLE would be enough to gain a good estimate of POPULATION SIZE. However, this is rarely the case and most populations have distributions that are either random or clumped. In such populations a single sample is unlikely to produce a good estimate of population size.

There are a wide variety of sampling strategies in use. It is important to choose a strategy that is appropriate to the population being investigated. There are several steps in the development of a sampling strategy.

Choice of sample unit

A sampling unit may either be defined arbitrarily, such as a QUADRAT or pit-fall trap, or be defined naturally, such as a leaf or individual. Usually, naturally defined sample units will be obvious, but the choice of an arbitrary unit size may be important. If the unit is a quadrat, then there will obviously be a trade-off in effort between the number of sample units that can be observed and their size. The size of sample unit will usually remain constant but may sometimes be variable, especially if the characteristics of a population DISTRIBUTION are being investigated. Methods using variable sample unit size to investigate distribution pattern are called quadrat-variance methods.

Number of sample units

This is nearly always determined by the amount of labour available. However, it is possible to calculate the number of sample units required to produce an accurate estimate of the population size. In general, more sample units will be preferred: as the number of sample units increases the ACCURACY of the population estimate improves.

Positioning of sample units

An unbiased estimate of a population is only possible if the sample units are representative of the total population. The easiest way of achieving this is for each sample unit to contain a RANDOM SAMPLE of the population. If quadrats are being used, then the position of a quadrat within the area under investigation should be chosen at random using random numbers to generate two coordinates that are then used to position a corner of the quadrat. The method of choosing a position using random numbers is to be preferred over the quasi-random techniques, such as throwing a quadrat, which are certain to introduce some observer bias.

Random sample units are ideal in a perfectly homogeneous habitat, but in reality they are likely to produce a biased sample with an estimated variance greater than for the true population. A simple method used to minimize this problem is to take a stratified random sample: the total area is divided into equal plots and an even number of sample units are taken at random from each plot. It is possible to divide the total area into plots of different sizes if there are known to be different habitats in the total area. In this case the number of sample units from each plot should be proportional to the size of the plot.

It might be tempting to conduct systematic sampling with sample units placed at regular intervals across a study area. There is a statistical problem with this strategy as most STATISTICS require that a sample is taken at random from a population. However, some field ecologists suggest that estimates derived from systematic sampling are, on average, better than those from random sampling.

Timing of sampling

Most populations will be affected by season, time of day and local weather conditions. It is very important that timing is taken into account either by sampling strategy or by later analysis. *See also* DISPERSION; EXPERIMENTAL DESIGN; SAMPLING METHODOLOGY/DEVICES. [C.D.]

REFERENCE

Ludwig, J.A. & Reynolds, J.F. (1988) *Statistical Ecology: A Primer on Methods and Computing*. John Wiley & Sons, Chichester.

Sampling methodology/devices

It is rarely possible to count all the species in a COMMUNITY or record all the individuals in a POPULATION. Ecologists try to obtain representative samples that give indicators of these measures, for example in terms of numbers of species per unit space/volume, or numbers/BIOMASS of individuals per unit space/volume. The statistical basis of the design and analysis of these programmes is dealt with elsewhere (*see* EXPERIMENTAL DESIGN; STATISTICS). This entry summarizes some of the methodology, techniques and devices that are used.

One important distinction is that between relative estimates of ecological quantities (relative ABUNDANCE of species or life stages within species) and absolute estimates of ecological quantities (expressed relative to area, volume, habitat). It is therefore possible to identify broad classes of techniques that have potential applicability with respect to the two kinds of estimates. In general, those appropriate for relative estimates will not be applicable for absolute ones. On the other hand, those which are useful for absolute measures can generally be used to provide relative estimates, though the effort required would generally be prohibitive. Table S1 summarizes this classification. What follows is a description of some of the techniques used within each category. For more information see Southwood (1966).

Table S1 Information that can be obtained from different sampling methods.

Sampling method	Relative abundance	Absolute abundance
Effort-based	Y	N
Time-based	Y	N
Traps	Y	S
Space-limited	Y	Y
Volume-limited	Y	Y
Habitat-limited	S	Y
Artificial substrates	S	Y
Mark–release–recapture	N	Y

Y, yes; N, no; S, some situations.

Effort-based techniques

Examples of effort-based techniques would be ones involving observations of numbers of, say, birds, insects or species of flora while travelling a standard path, or taking a number of paces through an area. Kick sampling of stream fauna is another example. A net is positioned downstream of the observer, who then makes a number of standard kicks. The organisms disturbed are caught in the net. Standard beats from vegetation is a similar technique.

Time-based sampling

This is another form of effort-based sampling in which the effort is prescribed in terms of a standard time allowed for making collections or observations.

Traps

These include: air flight traps (e.g. suction traps, light traps (Fig. S1) or sticky surfaces); pit-fall traps for land-dwelling organisms, and drift nets (Fig. S2) for aquatic animals in rivers and streams. Bait can also be an effective trap. The planting of 'virgin' surfaces or volumes may also be considered trapping but this can be used to provide quantitative estimates (see below).

Applicability

Effort-based sampling and time-based sampling are not generally applicable for absolute estimates because the return per unit effort or time will vary with cover and habitat. There are also likely to be appreciable variations between samplers. Similarly, the return per unit trap will depend upon the habitat and the types and condition of organisms being trapped.

Space-limited sampling

This is applied to two-dimensional habitats such as the seashore, or the ground. Quadrats are the basis of most techniques here, and the results are expressed per unit area. Variants of this that use water current to capture animals dislodged from a defined area of the bed

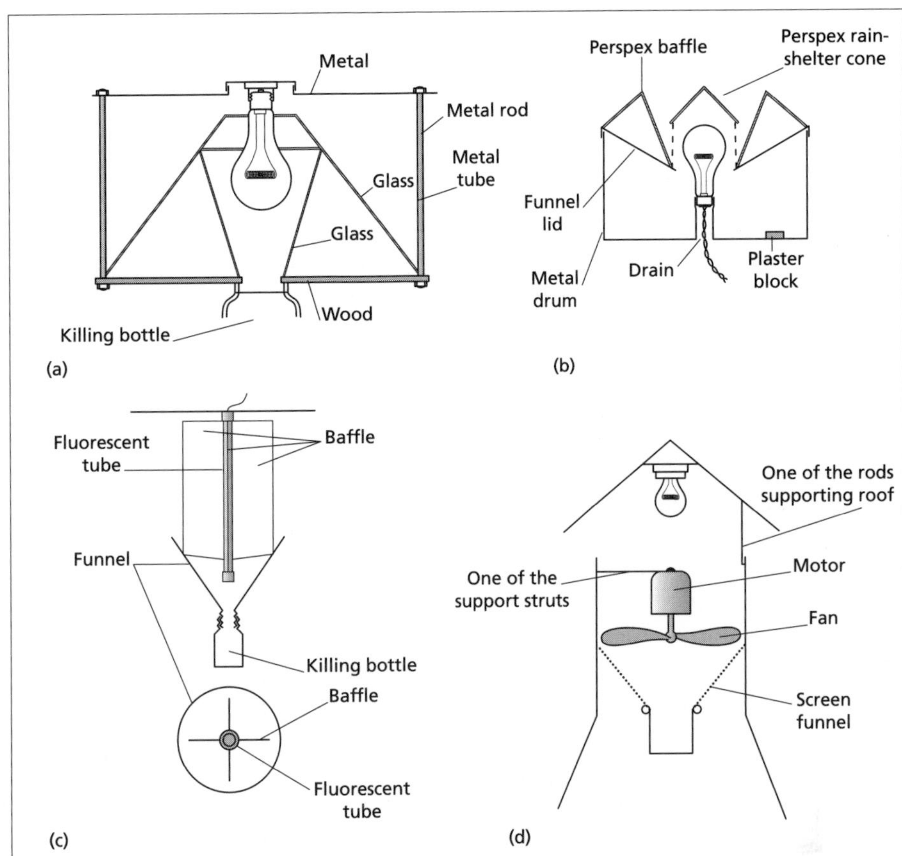

Fig. S1 Light traps: (a) Rothamsted trap; (b) Robinson-type trap; (c) Pennsylvanian trap (sectional view and plan); (d) New Jersey trap. (Figures S1–S6 after Southwood, T.R.E. (1966) *Ecological Methods*. Methuen, London.)

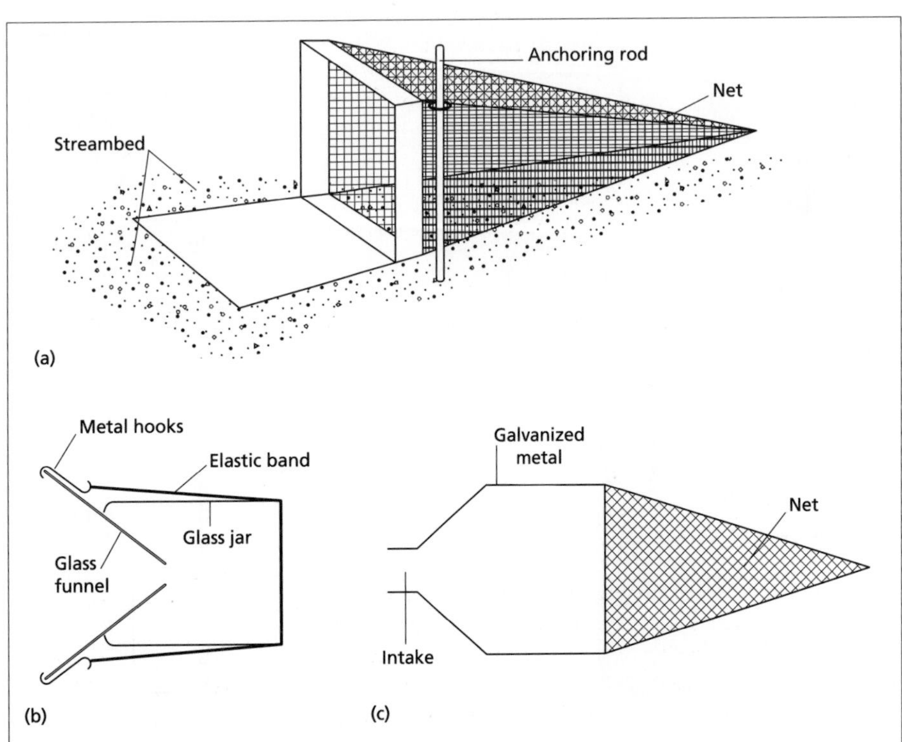

Fig. S2 Aquatic interception traps: (a) for stream drift animals; (b) Pieczynski-type trap for water mites; (c) Cushing-type trap for stream drift animals.

in a river or stream are the Surber samplers (Fig. S3) and Hess samplers (Fig. S4). Dredge sampling can also be used to scoop up organisms from a defined area of the bed in rivers, lakes, estuaries and oceans.

Volume-limited sampling

This is used to take samples from a three-dimensional habitat such as the water column, soil and aquatic sediments. Densities are expressed per unit volume. Techniques here involve devices that enclose volumes of water or solid material, such as corers (Fig. S5) and grabs (Fig. S6). Suction traps can also be used to remove fauna from a given volume of air.

Secondary sampling techniques are often required to remove organisms from the volumes of substrate collected, especially for soils and sediments. Here, methods might involve drying (of soils) to drive organisms out, or chemical treatment (of soils and sediments) to float organisms out.

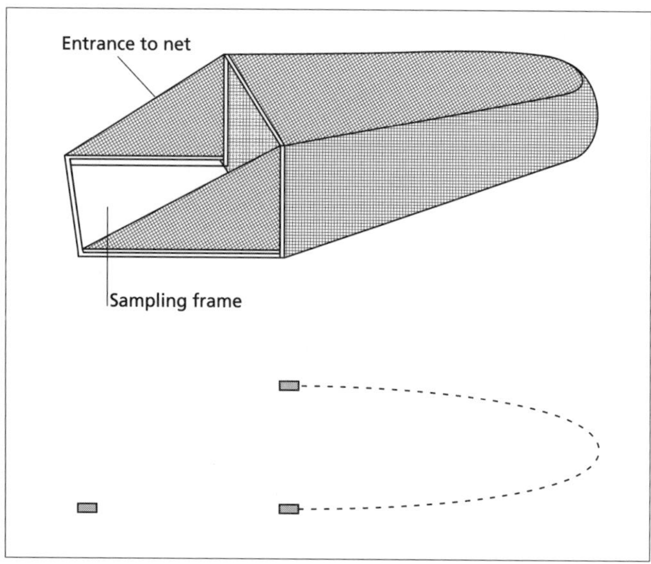

Fig. S3 Surber sampler; sketch and section.

Habitat units

In a sense both area and volume estimates are habitat-based. However, natural habitat units themselves can be used as sampling units; for example, submerged rocks on a rocky shore, a twig or leaf from a tree, an individual host organism if the animals being sampled are parasites. One problem with this kind of sampling is that the natural units may not be entirely equivalent so that presumed replicates may not be the same, hence sampling may not be strictly quantitative.

Artificial substrates as samples

Artificial habitats can be planted and used as a basis of estimating community structure or POPULATION SIZES after a period of colonization. For example, microscope slides have been used to SAMPLE the PERIPHYTON, while trays of stones or sediment may be employed to sample benthic systems. Plastic plants of various degrees of realism have been used to sample flora and fauna in terrestrial and aquatic habitats.

There are a number of problems in using artificial habitats for quantitative work. These relate to what, if any, influence the artificial matrix has on what categories of organisms colonize the matrix, and to what extent, and the optimum time that should be allowed for colonization.

Mark–release–recapture methods

See CAPTURE–RECAPTURE TECHNIQUES; LINCOLN INDEX; MARKING METHODS. [P.C.]

REFERENCE

Southwood, T.R.E. (1966) *Ecological Methods*. Methuen, London.

---

sand dune A term used to describe a mound or hill of sand formed where wind-blown sand accretes (aeolian deposition), typically around surface obstacles, often vegetation (see Plate 16, facing p. 398). Sand DUNES frequently form in coastal situations where specific reference can be made to embryo, fore, mobile and fixed dunes, which are components of the sand-dune SUCCESSION (*see* PSAMMOSERE).

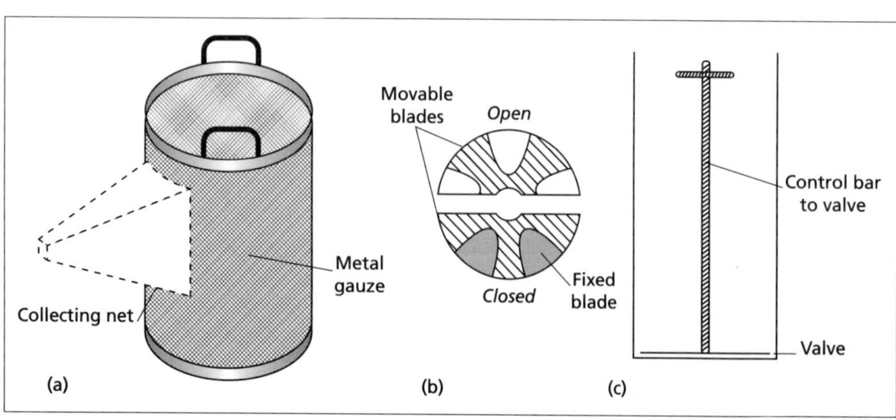

Fig. S4 (a) Hess circular sampler. (b, c) Inner cylinder of the Wilding sampler (simplified).

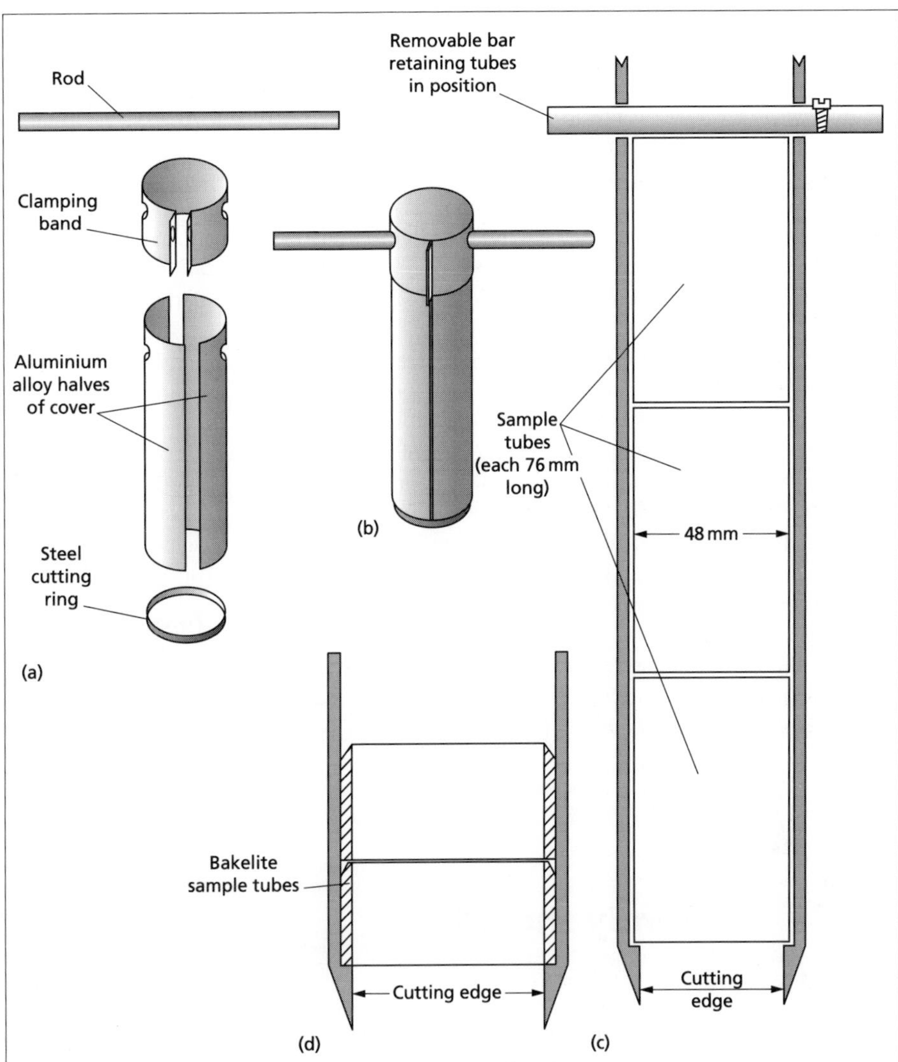

Fig. S5 Soil corers. (a, b) The O'Connor split corer: (a) showing components; (b) assembled. (c) Soil corer with sample tubes. (d) Soil corer for a canister extractor.

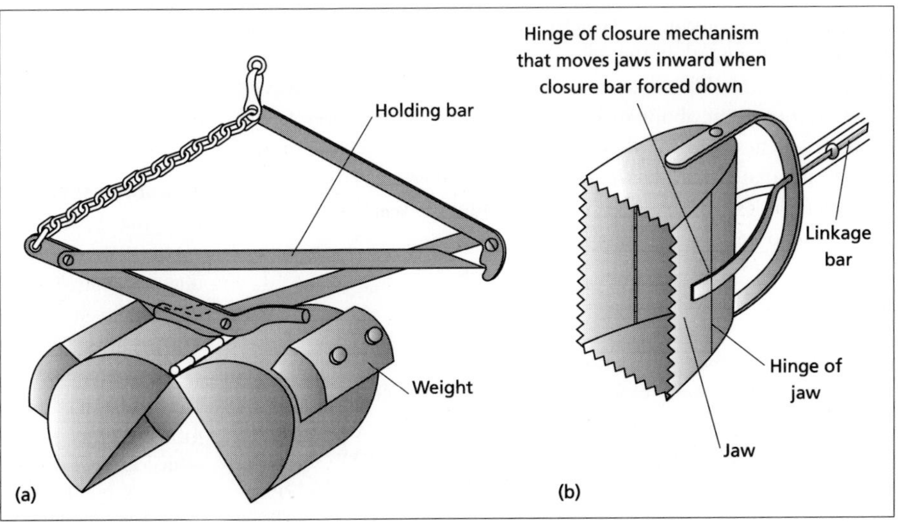

Fig. S6 (a) Petersen grab. (b) Allan grab.

Mobile dunes can be viewed as the CLIMAX of such sand accretion and can reach heights of 10 m or more. Their development is essentially vertical and depends on the ability of certain grasses, like *Ammophila arenaria* (in Europe) and *A. breviligulata* (in North America, but this is perhaps really a subspecies of *A. arenaria*), to grow up with the accreting sand and stabilize it. Other species are involved in the earlier embryo- and fore-dune stages of development, and yet other plants may serve to bind surface sand particles together to help in the stabilization process — for example the moss *Tortula ruraliformis*. Ultimately such dunes can become fixed dunes.

Medano is a term used to describe non-vegetated dunes formed sometimes in coastal but more commonly in inland situations ('DESERT dunes'), where wind is the sole agency involved. These tend be highly mobile, and tend to develop by horizontal movement and may reach heights of 20 m or more. Parabolic dunes (U-dunes) are intermediate in mobility between mobile dunes and medanos, and typically develop from the former, where the ends remain 'fixed' by vegetation but the centre has been blown back (blown out) by the wind. They tend to be very long and rather low, not exceeding 5 m in height.

Finally, dunes may be formed around obstacles other than vegetation, including human artefacts, although the process of aeolian deposition is still essentially natural. [B.W.F.]

Sandoz *See* ECOLOGICAL/ENVIRONMENTAL DISASTERS.

Sandy shores

Physical features

Sandy shores can be categorized along a continuum from dissipative to reflective. Dissipative BEACHES experience strong wave action, have fine particles and a gentle slope, are very eroded, and most of the wave energy is dissipated in a broad surf zone. Reflective beaches tend to experience little wave action, have coarse SEDIMENT and a steep slope, are depositional, and wave energy is reflected off the beach face.

The form and community structure of sandy shores is controlled primarily by wave action, the slope of the shore and the particle size. Particle size and slope appear to be more important than wave action in controlling abundance and diversity on sandy shores. Very exposed beaches, with gentle slopes and fine grains, often are more densely populated than less-exposed beaches having coarse grains and steep slopes. Particle size strongly influences water retention, and fine sand holds more water than coarse sand.

Biota

Most organisms occupying sandy shores live buried in the sediment, where they are to some extent protected against DESICCATION as well as tidal variations in TEMPERATURE and SALINITY. Conversely, OXYGEN can become limiting in sandy sediment and must be replenished by the overlying water. Sandy shores are much more topographically uniform than ROCKY SHORES (see Plates 14 & 17, facing p. 398). In contrast to rocky shores, an important factor influencing the distribution of organisms on sandy shores is substrate movement. Most of the adaptations exhibited by sandy-shore organisms involve adaptations to deal with an unstable substrate. Some organisms bury deeply or very quickly, some are shaped to facilitate burrowing. Many organisms migrate with the tide to maximize feeding opportunities or minimize predation pressure.

An obvious feature of sandy shores is the absence of sessile animals and of large plants. Ephemeral macroscopic algae may dominate during certain seasons on protected sandy shores, whereas on more exposed shores the dominant primary producers are benthic microalgae (e.g. diatoms). Sandy-shore animals consist mainly of polychaete worms, bivalve molluscs and crustaceans.

Whereas biomass tends to be higher on more exposed shores, species diversity tends to be higher on more protected shores. The relative lack of primary PRODUCTION on sandy shores means that most of the fauna feed on material suspended in the water column, on deposited organic debris or on other animals.

Zonation

As on rocky shores, VERTICAL ZONATION of species occurs on sandy shores. However, it is neither as clearly defined nor as well understood in the latter. Although data are limited, sandy shores may be divided into three zones: an upper, or supralittoral fringe, occupied by amphipod crustaceans (temperate shores) or ghost crabs (*Ocypode*) (tropical shores); a broad mid-littoral zone consisting of various species (e.g. sand crabs (family Hippidae)); a lower, or infralittoral fringe, which is the most diverse and is inhabited by a variety of bivalves, gastropods, polychaetes and crustaceans.

Community structure

As particle size decreases suspension-feeders become less abundant, whereas deposit-feeders become dominant. On protected sandy shores predation, and the DISTURBANCE accompanying it, appears to be important in controlling community structure. Compared to rocky shores, much less is understood about the factors controlling community patterns on open sandy shores. From what little is known, it appears that both COMPETITION and predation decrease in importance with increasing wave exposure, and are replaced by physical influences (i.e. wave action, particle size, beach slope) as the main forces determining community structure. [V.F.]

---

saprobien system An early biotic index of WATER QUALITY that recognized four stages in the oxidation of organic matter in FRESHWATER systems: polysaprobic; α-mesosaprobic; β-mesosaprobic; and oligosaprobic. Each stage was identified by the presence/absence of INDICATOR SPECIES. The system was elaborated in the 1950s to take into account the relative ABUNDANCE of organisms in a SAMPLE. The saprobien system has been used widely in continental Europe but has not been taken very seriously in the UK or North America. [P.C.]

REFERENCE

Sládecek, V. (1979) Continental system for the assessment of river water quality. In: *Biological Indicators of Water Quality* (eds A. James & L. Evison), pp. 3.1–3.32. John Wiley & Sons, Chichester.

saprotroph (saprobe) An organism (formerly also known as saprophyte) that derives all its essential energy/NUTRIENTS from dead organic matter and therefore commonly causes the decay/BIODETERIORATION of the latter. In the case of the fungi there is a spectrum of saprotrophs, which includes the following.

1 Independent types that require limited amounts of minerals and a simple CARBON source only; for example, mucoraceous moulds living in SOIL, utilizing simple sugars and capable of fast growth on freshly colonized substrates, such as the dungball of a HERBIVORE.

2 CELLULOSE-decomposing types, which grow more slowly; for example, 'brown rot' fungi (e.g. *Piptoporus betulinus*) on tree stumps or fallen branches; also many Ascomycotina and Deuteromycotina fungi in leaf litter (50% of the leaf litter is carbon) on FOREST floors.

3 Lignin-decomposing types; for example, 'white rot' fungi (e.g. *Coriolus versicolor*), which grow rather slowly but are responsible for the final stages of wood decay and RECYCLING of nutrients.

4 Dependent saprotrophs, found in much narrower ranges of habitats because of their requirement for particular essential compounds, or ecological niches that preclude much COMPETITION, such as the RHIZOSPHERE, rhizoplane and phylloplane of living plants. They usually require exogenous AMINO ACIDS, VITAMINS and complex carbon sources.

The fundamental role of saprotrophs is participation in the CARBON CYCLE. Their activities ultimately release the carbon 'locked up' in dead organic matter in the form of respiratory CARBON DIOXIDE (CO_2). Nitrogen and minerals are recycled in an analogous way. *See also* DECOMPOSITION; NECROMASS. [J.B.H.]

SAR *See* STRUCTURE ACTIVITY RELATIONSHIP; SYNTHETIC APERTURE RADAR.

Sargasso Sea A convergent GYRE in the North Atlantic which, because of its ANTICYCLONIC circulation, tends to deepen the THERMOCLINE and prevent NUTRIENTS from entering the surface water from below. As a result, PRODUCTIVITY is very low. SALINITY is high (36.6‰) because EVAPORATION exceeds PRECIPITATION. The Sargasso Sea is named for the brown, bushy alga, *Sargassum*, which occurs in abundance. [V.F.]

Satellite remote sensing

Satellites carry a range of remote-sensing devices for the production of images of the Earth and its ATMOSPHERE. Here, emphasis is on the five groups of Earth resources satellites that have proved to be of most value in the environmental sciences.

Landsat series

Following the success of 'manned' Earth resources satellites the US National Aeronautics and Space Administration (NASA) and the Department of the Interior developed an experimental Earth resources satellite series to evaluate the utility of images collected from an 'unmanned' satellite. The first satellite in the series was called ERTS, the Earth Resources Technology Satellite, a name that it held until January 1975 when it was renamed Landsat. The main advantages offered by the imagery collected from this satellite were: ready availability, low cost, repetitive multispectral coverage and minimal image distortion. Landsat 1 had a high and fast Sun-synchronous orbit at an altitude of 900 km with a repeat time of 18 days. Landsat 1 lasted for almost 6 years until January 1978 and for part of its life shared the heavens with Landsat 2, which was launched in 1975. Landsats 3, 4 and 5 were launched in 1978, 1982 and 1984. For Landsats 4 and 5 the satellite body was changed to increase stability and payload capability, and the orbit altitude was lowered to 705 km, thus giving a faster repeat cycle of 16 days.

The Landsat satellites all carry, or have carried, two sensors: a multispectral scanning system (MSS) and either a thematic mapper (TM) or return beam vidicom (RBV) television cameras. The MSS records four images of a scene, each covering a ground area of 185 × 185 km at a nominal spatial resolution of 79 m in green, red and near-infrared wavebands. Landsat MSS data are used for classifying land cover, estimating characteristics of the Earth's surface and for MONITORING environmental change.

The thematic mapper (TM) carried by Landsats 4 and 5 records image areas of 185 × 185 km in seven wavebands with a spatial resolution of around 30 m in six of them. The TM is an important sensor in the environmental sciences and is used extensively throughout the environmental sciences. An enhanced thematic mapper is due for launch in 1998.

Return beam vidicom (RBV) television cameras were carried on Landsats 1, 2 and 3. By the time of Landsat 3 they had nominal ground resolution of around 30 m and an image area of 98 × 98 km.

Second generation of Earth resources satellites

Like the Landsat satellites, these carry optical sensors but in a linear array. The satellites so far include the French satellite Système Probatoire d'Observation de la Terre (SPOT). SPOT 1, 2 and 3 were launched in 1986, 1990 and 1993 respectively, and SPOT 4 is due for launch in 1997. The satellites are operated by the Centre National d'Etudes Spatials and have a near-polar, Sun-synchronous, 832 km high orbit, which provide repeat ground coverage every 26 days. They carry two high-resolution visible (HRV) sensors that record an area of 60 × 60 km with a spatial resolution of 10 m in one panchromatic waveband or 20 m in green, red and near-infrared (and in the future middle-infrared) wavebands. These images can be recorded obliquely to give a stereoscopic view of the terrain, avoid patchy cloud and decrease the revisit time.

Japan has launched two Earth resources satellites. The first, in 1990, was the Marine Observation Satellite (MOS-1) and the second, in 1992, was the Japanese Earth Resources Satellite (Fuyo). MOS-1 has an orbit and altitude similar to the Landsat series of satellites and carries three sensors, the most important of which is a linear array multispectral scanner named the multispectral electronic self-scanning RADIOMETER (MESSR). This senses in four wavebands from green to near-infrared, has a spatial resolution of 50 m and an image area of 200 × 200 km. Fuyo carries a stereoscopic linear array sensor and also a RADAR (SYNTHETIC APERTURE RADAR, SAR).

The Indian remote sensing (IRS) satellites carry linear imaging self-scanning (LISS) multispectral scanners with design specifications similar to those of the sensors on board Landsat.

Satellites carrying thermal sensors

Landsats 3, 4 and 5 and the Heat Capacity Mapping Mission (HCMM) carried thermal sensors with a coarse spatial resolution. The HCMM satellite was launched in April 1978 and lasted until September 1980. Its orbit was near-circular at an altitude of 620 km. The satellite contained a scanning radiometer which recorded in a visible and near infrared waveband and a thermal infrared waveband. The data from the HCMM sensors were intended for geological mapping but they have also been used for microclimatology, pollution monitoring and hydrology.

Satellites carrying microwave sensors

The first Earth resources satellite to carry a radar (SAR) was Seasat. This was an experimental satellite designed by NASA to establish the utility of microwave sensors for REMOTE SENSING of the oceans. Images of the land were also obtained giving environmental scientists a synoptic view of the Earth in microwave wavelengths. Seasat had a circular orbit at an altitude of 800 km, sensing the Earth's surface from 72°N to 72°S, orbiting the Earth 14 times a day and passing over the same area every 152 days. The satellite carried two sensors: the one of most interest to environmental scientists was the radar (SAR) type, which produced images of 100 km-wide swaths with a nominal spatial resolution of 25 m. These data have been used for many applications such as the monitoring of sea state and the mapping of vegetation, sea ice and urban form.

The EUROPEAN REMOTE SENSING (ERS-1) SATELLITE carries a radar (SAR) and was launched in 1991 (and ERS-2 in 1995). The Japanese Earth Resources Satellite (Fuyo-1), launched in 1992, also carries a radar (SAR), although this records at longer wavelengths than the radar on board ERS-1.

Satellites carrying suites of environmental sensors

A series of major Earth resources satellites have been developed. These 'polar platforms' will carry suites of sensors that will enable the simultaneous estimation of fundamental environmental processes on land, in the oceans and in the atmosphere. The first two of these satellites are to be launched in 1998/9. The US Earth Observing System (EOS) satellite, AM-1, will carry four sensors, and the European Space Agency (ESA) satellite, Envisat, will carry up to 17 sensors. [P.J.C.]

REFERENCES

CEOS (1995) *Coordination for the Next Decade.* Committee on Earth Observation Satellites. European Space Agency, Paris.

Cracknell, A.P. & Hayes, C.W.B. (1991) *Introduction to Remote Sensing.* Taylor & Francis, London.

Curran, P.J. (1985) *Principles of Remote Sensing.* Longman Scientific and Technical, Harlow.

Foody, G.M. & Curran, P.J. (eds) (1994) *Environmental Remote Sensing from Regional to Global Scales.* John Wiley & Sons, Chichester.

Lillesand, T.M. & Kiefer, R.W. (1994) *Remote Sensing and Image Interpretation,* 3rd edn. John Wiley & Sons, New York.

---

satellite species Those species within a COMMUNITY which are likely to be subject to local extinctions and to periodically recolonize systems. They are likely to be both locally and regionally rare and to have weak interactions with other species in the system. These characteristics make such species likely to be the focus of conservation efforts. *Cf.* CORE SPECIES. [C.D.]

REFERENCE

Hanski, I. (1982) Dynamics of regional distribution: the core and satellite species hypothesis. *Oikos* **38**, 210–221.

satiation The point at which an organism no longer experiences HUNGER and stops FEEDING even when further food resources are available. Like hunger, satiation is a wholly abstract motivational response to underlying physiological processes and one that is only generally attributed to more complex organisms. *See also* FOOD SELECTION; OPTIMAL FORAGING THEORY. [R.C.]

saturation In COMMUNITY ECOLOGY, the point reached by a local (*see* SCALE) COMMUNITY when no more species can invade from the regional species pool and establish to increase the SPECIES RICHNESS. In this situation, all RESOURCES are fully exploited and biotic interactions (particularly COMPETITION) are intense. This is not to say that other species in the regional species pool will be unable to invade and establish in a saturated community, although it is probably unlikely. If a species were to be successful it would have to be at the expense of a resident such that the total species number in the community would remain unchanged (*see* NON-INTERACTIVE COMMUNITY). Saturation, therefore, refers to the degree to which the community is able to admit new colonists from the regional species pool and the extent to which resources (niches) are exploited (*see* NON-EQUILIBRIUM COMMUNITIES). Theoretical studies suggest that the level of saturation may be strongly influenced by the sequence of COLONIZATION (*see* HISTORICAL EFFECTS (COMMUNITY ASSEMBLY)). [S.J.H.]

REFERENCE

Cornell, H.V. & Lawton, J.H. (1992) Species interactions, local and regional processes, and limits to the richness of ecological communities: a theoretical perspective. *Journal of Animal Ecology* **61**, 1–12.

savannah Tropical vegetation, dominated by grasses, but with various amounts of intermixed tall bushes and/or trees (see Plate 18, facing p. 398). [P.C.]

SBS *See* SICK BUILDING SYNDROME.

scale In ECOLOGY the concept of scale is largely concerned with the levels or sizes at which particular ecological entities or processes are considered. Human investigators tend to view phenomena at scales that are relevant to our own perceptions, and we have trouble fully comprehending, for example, the contribution that long time-scale processes such as SUCCESSION and EVOLUTION make to things we observe today. Ecologists have also demonstrated a similar failure to appreciate what goes on at very small space-scales. For example, the ability of plankton species to coexist in an apparently homogeneous environment would not have been considered a paradox if biological oceanographers had focused earlier on fine-scale turbulence and other processes relevant to the lives of individual plankters.

One distinction that is often made is between local, regional and biogeographic scales. Precise definitions for these distinctions are difficult to give and they often need to be interpreted within the context of the problem being discussed. Broadly speaking, however, it is at local scale that ecological processes such as predation, COMPETITION, DISTURBANCE and PARASITISM take place, and it is the biogeographic scale where processes such as widespread extinction, long-distance DISPERSAL and SPECIATION occur. In between these two scales lies the regional scale at which interactions between (and patterns in) the DISTRIBUTION of local communities are interpreted.

Deciding the appropriate scales at which to view particular problems, and devising ways to conduct experiments and make observations at appropriate time- and space-scales are key issues for ecologists. It seems probable, for example, that the point on the continuum between equilibrium and non-equilibrium that a COMMUNITY occupies (*see* NON-EQUILIBRIUM COMMUNITIES) will depend on the spatial and temporal scales at which it is being viewed. Indeed, vociferous disagreements about a number of issues have arisen because conflicting results have been obtained from studies conducted at different scales. A key goal for ecologists is to find appropriate conceptual frameworks for quantifying and understanding the interactions between species and between processes which operate on different scales. It is in this latter area that hierarchical views of ecological organization seem to be making some progress. *See also* FRACTAL ANALYSIS; FRACTAL DIMENSION; HIERARCHY THEORY. [S.J.H.]

REFERENCE

Giller, P.S. & Gee, J.H.R. (1987) The analysis of community organisation: the influence of equilibrium, scale and terminology. In: *Organization of Communities Past and Present* (eds P.S. Giller & J.H.R. Gee), pp. 519–542. Blackwell Scientific Publications, Oxford.

scaling *See* ALLOMETRY.

scandium (Sc) A metallic element (ATOMIC NUMBER = 21; relative ATOMIC MASS = 44.956; melting point = 1539°C; boiling point = 2832°C). Although it was 1960 before the first 500 g of 99% pure scandium was produced, this is not a reflection of its natural abundance (25 mg kg^{-1} of the Earth's CRUST), but rather of its very wide but thin distribution, the only scandium-rich mineral being the rare thortveitite, $Sc_2Si_2O_7$, found in Norway. However, considerable amounts of scandium are now available from the extraction of URANIUM, whose ores contain about 0.02% Sc_2O_3. Scandium has no significant commercial uses, nor is it of biological significance. [P.J.B.]

scavenger organisms Heterotrophic organisms that specialize in FEEDING from the dead bodies of other animals but which do not undergo their whole larval development in the carcass of one animal, as do the specialist carrion-feeding invertebrates such as sexton beetles (*Silphidae*) or blowflies such as *Calliphera*. The most obvious examples of scavenger organisms are vultures (*Gyps bengalis*), which feed on the dead bodies of other large animals but move from one prey item to the next every few days. As with all compartments in any classification of trophic relationships many taxa do not fit neatly within this one compartment. Hyenas (*Crocuta crocuta*) are often thought of as scavenger organisms and they do readily feed on the remains of animals killed by other species such as lions. However, recent research has shown that hyenas are themselves often primary carnivores, actively predating particularly young and/or old and weak herbivores, such as wildebeest (*Connochaetes gnu*) or other antelopes. *See also* FEEDING TYPES, CLASSIFICATION. [M.H.]

scheduled process A process that is listed and subject to environmental legislation. In the UK such processes used to come under the provision of the Alkali, etc. Work Regulation Act 1906; this was superseded by the ENVIRONMENTAL PROTECTION ACT 1990. *See also* PRESCRIBED PROCESSES. [P.C.]

sclerophyllous vegetation SCRUB or WOODLAND vegetation in which many species have EVERGREEN, leathery leaves. Sclerophyllous literally means 'hard leaves' (Fig. S7), and the evolution of this leaf type is related to problems of WATER conservation. Such vegetation is, therefore, usually found in parts of the world with summer DROUGHT, such as Australia, North Africa and California (*see* MEDITERRANEAN SCRUB/FOREST). [P.D.M.]

scope for growth (SFG) The difference between an organism's energy intake and output; i.e. that energy, equivalent to the PRODUCTION term in the ENERGY BUDGET, that is potentially available for somatic production (i.e. GROWTH proper) and reproduction. Because both of these processes potentially influence population RECRUITMENT, this physiological organismic measure is sometimes used as a surrogate for possible population effects. The term derives from Fry (1971). [P.C.]

REFERENCE

Fry, F.E.J. (1971) The effect of environmental factors on the physiology of fish. In: *Physiology of Fishes* (eds W.S. Hoar & D.J. Randall), Vol. 1, pp. 1–98. Academic Press, New York.

scoping A preliminary, low-cost environmental assessment of a business (*see* ECOAUDIT), or product (*see* LIFE-CYCLE ASSESSMENT) or impact of development (*see* ENVIRONMENTAL IMPACT ASSESSMENT). [P.C.]

Fig. S7 An ancient olive (*Olea europaea*) grove at Stari Bar on the Albanian–Yugoslav frontier. The olive is a typical example of a Mediterranean tree with sclerophyllous foliage. (Courtesy of R.M.M. Crawford.)

Scottish Environmental Protection Agency (SEPA) *See* ENVIRONMENTAL (PROTECTION) AGENCIES.

SCP (supercooling point) *See* FREEZE AVOIDANCE.

scramble competition A term first defined by A.J. Nicholson in 1954 to describe resource use, frequently used now to describe both the behavioural process and the ecological outcome of INTRASPECIFIC COMPETITION. Nicholson defined scramble as 'the kind of COMPETITION exhibited by a crowd of boys striving to secure broadcast sweets', where 'success is commonly incomplete, so that some, and at times all, of the requisite secured by the competing animals takes no part in sustaining the population, being dissipated by individuals which obtain insufficient for survival' (Nicholson 1954). Scramble competition involves passive exploitation (*see* EXPLOITATION COMPETITION) rather than direct INTERFERENCE (*see* INTERFERENCE COMPETITION). In the extreme case, pure scramble competition can lead to extinction of a population because no competing individual obtains sufficient resource to allow survival. The ecological outcome of scramble is OVERCOMPENSATION and POPULATION SIZE will tend to fluctuate around an UNSTABLE EQUILIBRIUM in a cyclic or even chaotic fashion. In plant population literature, scramble competition is more commonly referred to as SYMMETRIC COMPETITION. [R.H.S.]

REFERENCE

Nicholson, A.J. (1954) An outline of the dynamics of animal populations. *Australian Journal of Zoology* **2**, 9–65.

scrap Material (usually metals) discarded from manufacturing or products that have reached the end of their useful life. They may be suitable for reprocessing or recycling. [P.C.]

scraper *See* FEEDING TYPES, CLASSIFICATION; FUNCTIONAL GROUP.

Scrub

Characteristics and occurrence

Communities dominated by shrubs or bushes which are intermediate between FOREST and GRASSLAND, often forming an intermediate zone between them. Many shrubs have, or may have, multiple stems, a tendency enhanced by coppicing. Others have a single trunk when left uncut; the distinction between shrubs and small trees is somewhat arbitrary.

Scrub frequently develops where cultivated land is abandoned or forest is extending into grassland; in such instances it is often seral, containing young trees and developing into forest unless checked by fire, felling or heavy GRAZING. The relaxation of rabbit grazing pressure which resulted from the outbreak of myxomatosis in Britain led to wide-scale extension of scrub communities, particularly on SAND-DUNE systems. Conversely the very high deer population levels found in many parts of the Northern hemisphere markedly limit the rate at which seral scrub develops into WOODLAND or forest.

Scrub also results from the climatic limitation of woody VEGETATION on the dry inner edges of forests in continental regions, or between subalpine forest and alpine grassland. The death of mature elms (*Ulmus*) due to outbreaks of Dutch elm disease caused by *Ophiostoma ulmi* led to the development of seral scrub in many English woodlands. A notable example occurred in a woodland south-east of the Ercall, near Wellington, Shropshire, UK, where a valley side became covered with brambles (*Rubus fruticosus* agg.) growing over dead elm logs. Similar situations developed in North America when stands of *Castanea dentata* (American chestnut) were ravaged by Asiatic blight fungus (*Endothia parasitica*).

Scrub communities in Britain

Hawthorn (*Crataegus monogyna*) is the most widespread dominant species in British scrub, flourishing on many SOIL types and occurring in the SUCCESSION to oakwood as well as to beechwood. Dogwood (*Cornus sanguinea*), privet (*Ligustrum vulgare*), blackthorn (*Prunus spinosa*), buckthorn (*Rhamnus catharticus*) and wayfaring tree (*Viburnum lantana*) are very constant in chalk scrub, which has far more species than many other British scrub types. On the South Downs, brambles (*Rubus fruticosus* agg.) and wild roses (*Rosa*) are frequent PIONEER scrub species; *Juniperus communis* seral scrub is also characteristic. Spindle (*Euonymus europaeus*) and elder (*Sambucus nigra*) are common; field maple (*Acer campestre*) and hazel (*Corylus avellana*) flourish particularly on deeper soils towards the bottom of slopes. Traveller's joy (*Clematis vitalba*) is a common and often luxuriant woody climber which often sprawls over regenerating trees, thus helping to perpetuate scrub.

W21 *Crataegus monogyna–Hedera helix* scrub

The National Vegetation Classification (NVC; Rodwell 1991), accommodates British scrub within the six plant communities W20–W25, of which W21 *Crataegus monogyna–Hedera helix* scrub, the typical SUBCLIMAX woody community of circumneutral to base-rich soils throughout the British lowlands, is the most important. *Crataegus monogyna*, *Rubus fruticosus* agg. and *Hedera helix* are constant species and the vegetation is invariably dominated by various mixtures of smaller trees and shrubs, undershrubs, woody climbers and sprawlers. CANOPY height rarely exceeds 5 m and may be little more than 1 m. Although *Prunus spinosa* is less common than hawthorn, it does better than the latter on heavier, moist soils and is more resistant to saline spray. W21 has four subcommunities of which the *Brachypodium sylvaticum* and *Viburnum lantana* subcommunities normally occur on free-draining calcareous soils, whereas the *Mercurialis perennis* subcommunity is on more heavy-textured base-rich soils, especially those which are badly drained. The *Hedera helix–Urtica dioica* subcommunity consists of species-poor scrubs with a dense canopy. The field layer is often very sparse, but ivy may form an extensive carpet.

W20 *Salix lapponum–Luzula sylvatica* scrub

W20 *Salix lapponum–Luzula sylvatica* scrub, which occurs in higher-altitude stands in Scotland, possesses a low, bushy willow canopy in which *Salix lapponum* is by far the commonest species. Other constant species are *Deschampsia cespitosa*, *D. flexuosa*, *Luzula sylvatica*,

Vaccinium myrtillus and the mosses *Dicranum scoparium*, *Hylocomium splendens* and *Rhytidiadelphus loreus*.

W22 *Prunus spinosa–Rubus fruticosus* scrub

Prunus spinosa is the sole woody constant in W22 *Prunus spinosa–Rubus fruticosus* scrub, in which it is almost always overwhelmingly dominant, instead of being frequent and locally abundant as in W21 scrub. *Ulex europaeus* (gorse) occurs occasionally in W22 scrub on soils that tend to be base-poor, while *Corylus avellana* and *Ligustrum vulgare* (privet) may be found where the soil is base-rich. *Prunus* scrub is most characteristic of mesotrophic MULL soils of moderate base status in lowland Britain. The *Hedera helix–Silene dioica* subcommunity is characteristic of drier and more acid soils, whereas the *Viola riviniana–Veronica chamaedrys* subcommunity occurs on more base-rich and somewhat heavier soils. The *Dactylis glomerata* subcommunity is typical of fairly moist mull soils (Rodwell 1991).

W23 *Ulex europaeus–Rubus fruticosus* scrub

The three constants of W23 *Ulex europaeus–Rubus fruticosus* scrub, which is widespread on marginal land in the lowlands and upland fringes, are gorse (*Ulex europaeus*), bramble (*Rubus fruticosus* agg.) and *Agrostis capillaris*. The woody cover of W23 scrub, which is usually between 1 and 2 m high, varies greatly in form. Gorse is normally the dominant species; its bushes are trimmed to rounded hummocks when the young growth is regularly grazed, but become tall and leggy when this does not occur. In some stands broom (*Cytisus scoparius*) is the sole dominant; W23 communities with *Cytisus scoparius* are important in causing a build-up of HUMUS on shingle ridges of the cuspate foreland of Dungeness, Kent, UK. *Ulex–Rubus* scrub occurs on free-draining, moderately to strongly acid brown soils. The *Anthoxanthum odoratum* and *Rumex acetosella* subcommunities are somewhat similar, but the latter is characterized by species common on disturbed or lighter soils, such as *Rumex acetosella*, *Hypochaeris radicata*, *Senecio jacobaea*, *Crepis capillaris*, *Jasione montana* and *Aira praecox*.

The gorse and bramble cover of the *Teucrium scorodonia* subcommunity is usually fairly dense, while the extent and SPECIES RICHNESS of the grassy element are less than in the other two subcommunities. Individual gorse and broom plants have a lifespan of approximately 10 years, and the imposition of grazing or burning regimes may prevent successional development of *Ulex–Rubus* scrub to woodland. It is, however, frequently invaded by trees, including oak (*Quercus*), birch (*Betula*) and Scots pine (*Pinus sylvestris*), which may develop to maturity unless limited climatically as in many cliff communities.

W24 *Rubus fruticosus–Holcus lanatus* scrub

The *Cirsium arvense–Cirsium vulgare* and *Arrhenatherum elatius–Heracleum sphondylium* subcommunities of W24 *Rubus fruticosus–Holcus lanatus* underscrub are typical of abandoned and neglected ground in the British lowlands. Although commonly in a stable form around wood margins or along hedgerows, W24 can represent an early stage in succession to SECONDARY WOODLAND; as with W25, underscrub, shrub and tree cover is low.

W25 *Pteridium aquilinum–Rubus fruticosus* scrub

W25 *Pteridium aquilinum–Rubus fruticosus* underscrub is commonly closely associated with taller woody species of shrubs (including blackthorn, elder and hawthorn) and trees (ash (*Fraxinus excelsior*), sycamore (*Acer pseudoplatanus*), oak and beech). Bracken (*Pteridium aquilinum*) is generally the more abundant of the two constant species and may form a virtually complete canopy, whereas in other places a thick tangle of arching bramble shoots is prominent. The *Hyacinthoides non-scripta* subcommunity is more species-rich than the *Teucrium scorodonia* subcommunity and has a marked vernal aspect; it has the appearance of an oak–bracken–bramble woodland that no longer possesses a cover of trees and shrubs. [J.R.P.]

REFERENCES

Rodwell, J.S. (ed.) (1991) *British Plant Communities*. Vol. 1. *Woodland and Scrub*. Cambridge University Press, Cambridge.

Tansley, A.G. (1939) *The British Islands and their Vegetation*. Cambridge University Press, Cambridge.

---

sea A term often used synonymously with OCEAN or in reference to a subdivision of an ocean, or a very large enclosed body of (usually salt) water. Sea may also describe WAVES generated or sustained by winds within their fetch or a portion of the ocean where waves are being generated by wind. *Cf.* FRESHWATER. [V.F.]

sea-floor spreading The horizontal movement of oceanic CRUST; the process by which LITHOSPHERE is generated at DIVERGENCE zones, such as midocean ridges, and adjacent lithospheric plates are moved apart as new material forms. Convective UPWELLING of magma occurs along the OCEAN ridges and moves the ocean floor away from the ridge axes at rates of from 1 to 10 cm year^{-1}. Crust is destroyed at convergence zones in the deep ocean trenches. The theory of sea-floor spreading was first proposed by H.H. Hess in 1962 and radically influenced the understanding of geological processes. [V.F.]

sea-level changes Alterations in the level of oceanic waters with respect to the neighbouring landmasses. Such changes are caused by the accumulation (causing levels to fall) and melting (causing levels to rise) of glaciers. Sea level is believed to have risen by about 100 m since the melting of the Pleistocene glaciers, approximately 10 000 years ago. It has been estimated that the world mean sea level has been rising at about 1.2–1.5 mm year^{-1} during the first half of the 20th century.

Apart from glacial effects, the volume of OCEAN basins and therefore sea level is controlled by variations in four additional parameters:

1 volume of mid-ocean ridges;
2 volume of SEDIMENT delivered to ocean basins;
3 the extent of oceanic vs. continental CRUST;
4 the volume of SEAMOUNTS.

Storm TIDES (caused by onshore winds and low air pressure) can vary sea level by a few metres, whereas annual tides (related to changes in seasonal heat content, SALINITY, local air pressure, total amount of water in the oceans, and monsoon winds) can alter sea level by 20–30 cm. Sediment compaction and sediment collapse may cause local subsidence of the coast and local rises in apparent sea level. [V.F.]

seamount A submerged, isolated, volcanic mountain that rises from the sea floor and may reach a height of 900 m or more. [V.F.]

searching It is assumed that foragers choose searching modes that maximize the net rate of energy gain. Two aspects have been modelled, the trajectory of the search path and the energetics of searching. Models predict that, when exploiting sources that are evenly spread about the habitat and are depletable within a foraging bout, foragers will maximize the net rate of energy gain by avoiding recently visited sources. This is achieved by 'directional' search, wherein the angle of departure from each point visited has a mean of zero, causing the forager to move straight ahead rather than recrossing its path. Where food occurs in patches separated by expanses of unproductive habitat, the net rate of energy gain will be maximized by directional movement between patches, switching to non-directional movement characterized by non-zero angles of departure causing the forager to meander within the patch. Here, OPTIMAL FORAGING THEORY (OFT) predicts behaviour well known to population ecologists as 'area-restricted searching', for which positive klinotaxis (increased turning rate) and negative orthokinesis (decreased velocity) are established ethological mechanisms.

Energetics models have assumed that increased SEARCHING EFFICIENCY is achieved at greater metabolic cost. Based on the shape of this function and on the relationship between searching efficiency and prey DENSITY, models predict when the forager should switch between energetically costlier and cheaper searching modes. Faster, energetically more expensive searching modes, by raising encounter rates with prey, may be superior to slower modes when conditions are productive. Energetically cheaper modes may be optimal either when food is too scarce to repay the cost of expensive modes or when external forces, such as prey movement, wind or currents, raise encounters with food to a high level even when the forager moves slowly or remains stationary as an ambush, or 'sit-and-wait' PREDATOR. *See also* SEARCHING TIME. [R.N.H.]

searching efficiency A parameter (symbolized, sometimes, by a) that describes the rate at which an individual PREDATOR or PARASITOID encounters prey or HOSTS. It is sometimes called the attack rate, when every encounter leads to an attack. *See also* OPTIMAL FORAGING THEORY; SEARCHING. [H.C.J.G.]

searching time In behavioural ecological models of foraging, and to a lesser degree in POPULATION DYNAMICS, a parameter (usually symbolized as T_S) that describes the time a PREDATOR or PARASITOID spends SEARCHING for prey or HOSTS. The total time available for a forager can be divided into searching time and HANDLING TIME. *See also* OPTIMAL FORAGING THEORY. [H.C.J.G.]

season Any of the four equal periods, marked by the equinoxes and solstices, into which the year is divided. Because the Earth rotates and its axis is inclined in relation to the Sun, and because it revolves around the Sun in an elliptical orbit, the distribution of solar energy over the Earth's surface varies and results in changes of season. At the spring and autumn equinoxes the Equator receives maximum SOLAR RADIATION, whereas at the winter (22 December) and summer (22 June) solstices, the Sun is vertical over the Tropic of Cancer (Northern hemisphere) or the Tropic of Capricorn (Southern hemisphere).

At the summer solstice the days are at their longest and temperatures are high in the north, whereas in the south the reverse is true. When winter prevails, the Sun's rays have to pass through more of the Earth's ATMOSPHERE, as they are oblique, and solar radiation is consequently less. The pattern of seasonal change is repeated annually. Local conditions, particularly TOPOGRAPHY, proximity to oceans, and altitude, will modify seasonal variations, but ultimately the position of the Sun controls the seasons. [S.R.J.W.]

seasonality The nature and degree of climatic (or other) changes associated with changes in SEASON. Generally TEMPERATURE which, with WATER, is the most important climatic factor, decreases from the tropics to the poles. Equatorial areas show little variation through the year, whereas as one moves polewards seasonality becomes more pronounced. Tropical areas have summer rainy seasons, while SUBTROPICAL zones show seasonal temperature variations. Mediterranean areas have winter CYCLONIC rain and hot, dry summers. Temperate zones have hot summers and cold winters with year-round rain. BOREAL regions have cool, wet summers and long, cold winters, and the ARCTIC zones have low PRECIPITATION all year round, a short summer with 24-hour days and a long, cold, dark winter. [S.R.J.W.]

second

1 The SI UNIT of time-interval (symbol: s), equal to $^1/_{60}$ of a minute. It is defined in terms of the resonance vibration of the CAESIUM-133 atom, the interval occupied by 9 192 631 770 cycles.

2 A unit of angular measure (symbol: ″) equal to $^1/_{60}$ of a minute of arc.

[J.G.]

secondary consumer *See* CONSUMER ORGANISMS.

secondary effects of pollution Effects of POLLUTION caused not by a direct impact on survival, development, reproduction, etc., of organisms but via an ecological interaction; for example, the reduction or extinction of a food source; the weakening of a competitor; or the reduction of predation pressure by selective impairment of predators. *Cf.* PRIMARY EFFECTS OF POLLUTION. [P.C.]

secondary forest Forest (either temperate or tropical) which has regenerated following human clearance. The term may also be used in a more restricted sense implying that the land upon which the secondary forest develops has been used for alternative land use, such as pasture or arable. *Cf.* PRIMARY FOREST. [P.D.M.]

secondary host For a MACROPARASITE, a host which is parasitized but which does not actively transport the PARASITE to the definitive HOST. Asexual multiplication may occur within the secondary host, whereas the host within which sexual reproduction occurs is usually defined as the definitive or primary host. Other hosts may be vector species (*see* VECTOR, PARASITES). The TRANSMISSION RATE

OF PARASITES depends on the density of secondary hosts. *See also* RESERVOIR HOSTS, PARASITES. [G.F.M.]

secondary productivity Productivity is the rate at which an individual population, or other ecological unit belonging to the same TROPHIC LEVEL, accumulates BIOMASS or energy by the production of new somatic and/or reproductive tissues. Productivity of AUTOTROPHS is PRIMARY PRODUCTIVITY, and productivity of HETEROTROPHS is referred to as secondary productivity, whether the heterotrophs are members of the grazing or decomposer food webs. Rates of secondary productivity are of fundamental interest because they determine the number of higher trophic levels that can be supported above them and hence the length of food chains and structure of the COMMUNITY. They are also of applied significance when evaluating the extent to which animal populations can be harvested for human use.

Approximate estimates of secondary productivity for whole biomes have been made by starting with more reliable measurements of primary productivity in each system, estimating the exploitation EFFICIENCY of herbivores in those systems — i.e. the proportion of NET PRIMARY PRODUCTION eaten by herbivores—and their ASSIMILATION and production efficiencies to derive the estimates shown in Table S2 for net secondary production. Much interest has been generated by the comparison of productivity of aquatic and terrestrial ecosystems. While the primary productivity of oceans, which is strongly nutrient-limited, is one-third that of terrestrial systems, these estimates suggest that the secondary productivity is nearly as high as in terrestrial systems. This is because of differences in exploitation efficiency of ZOOPLANKTON feeding on PHYTOPLANKTON compared with terrestrial animals feeding on plants having a high proportion of indigestible structural tissues. There are many assumptions in these gross global calculations but they are of very practical economic importance in calculating potential yields of fish further up the aquatic food chains.

At the ECOSYSTEM level much effort has been directed into trying to determine which factors are most important in limiting absolute levels of secondary productivity. A consensus has emerged that for the primary consumers, whether they be herbivores or decomposers, productivity is less often limited by the availability of energy from the primary producers than by the constraints of nutrient limitation. This thesis has been elegantly elaborated by White (1993), who repeatedly tests and finds support for the hypothesis that NITROGEN is a fundamentally limiting nutrient in a very wide range of different systems. This is entirely consistent with the results of the numerous studies which have illustrated increases in secondary productivity following application of nitrogen fertilizer to primary producers, in both natural as well as agricultural ecosystems. *See also* ECOLOGICAL ENERGETICS; ENERGY FLOW. [M.H.]

REFERENCES

Colinvaux, P. (1986) *Ecology*. John Wiley & Sons, New York.

White, T.C.R. (1993) *The Inadequate Environment: Nitrogen and the Abundance of Animals*. Springer-Verlag, Berlin.

Whittaker, R.H. & Likens, G.E. (1973) Carbon in the biota. In: *Carbon and the Biosphere* (eds G.M. Woodwell & E.V. Pecan), pp. 281–302. Washington, D.C.

Table S2 Annual secondary production of principal regions of the Earth. These estimates are very approximate, being based on estimates of animal consumption of plants to compute herbivore consumption, and of assimilation efficiencies to compute gross secondary production. Some of the estimates are little more than intelligent guesses, probably being too high. (From Colinvaux, P. (1986); *Ecology*. John Wiley & Sons, New York. Copyright © 1986. Reprinted by permission of John Wiley & Sons, Inc.)

Ecosystem type	Total net primary productivity (10^9 t C year^{-1})	Animal consumption (%)	Herbivore consumption (10^6 t C year^{-1})	Net secondary productivity (10^6 t C year^{-1})
Tropical rain forest	15.3	7	1100	110
Tropical seasonal forest	5.1	6	300	30
Temperate evergreen forest	2.9	4	120	12
Temperate deciduous forest	3.8	5	190	19
Boreal forest	4.3	4	170	17
Woodland and shrubland	2.2	5	110	11
Savannah	4.7	15	700	105
Temperate grassland	2.0	10	200	30
Tundra and alpine	0.5	3	15	1.5
Desert scrub	0.6	3	18	2.7
Rock, ice and sand	0.04	2	0.1	0.01
Agricultural land	4.1	1	40	4
Swamp and marsh	2.2	8	175	18
Lake and stream	0.6	20	120	12
Total land	48.3	7	3258	372
Open ocean	18.9	40	7600	1140
Upwelling zones	0.1	35	35	5
Continental shelf	4.3	30	1300	195
Algal bed and reef	0.5	15	75	11
Estuaries	1.1	15	165	25
Total oceans	24.9	37	9175	1376
Total for biospheres	73.2	17	12433	1748

secondary sex ratio *See* SEX RATIOS.

secondary sex traits Also known as SECONDARY SEXUAL CHARACTERISTICS.

secondary sexual characteristics TRAITS which have evolved in response to differences among individuals in mating success due to COMPETITION over mates or mate choice (*see* SEXUAL SELECTION). Darwin gave the following examples for males in his original formulation of sexual selection: 'weapons of offence or means of defence against rivals, his gaudy colouring and various ornaments, his power of song, and other such characters'. Thus, secondary sexual characteristics help their bearers when competing for and courting mates but are not strictly necessary for reproduction, as would be the case for testes or ovaries. They are thought to reduce their bearers' survival. [J.D.R.]

secondary succession This is SUCCESSION occurring where there has already been a plant cover, and there is some kind of developed SOIL. It is contrasted with PRIMARY SUCCESSION. Secondary successions may be:

1 natural, for example after fire, flood or wind-blow in forests or woodlands, or after severe DROUGHT, or following mound-building or scratching by animals in grassland; or

2 human-induced, for example on land that is no longer cultivated, grazed, mown or burnt, or in forests that have been clear-felled and abandoned.

Regeneration of late-successional vegetation may or may not be successional in nature. Human-induced secondary successions commonly contrast with natural INTERNAL SUCCESSIONS in involving unnatural mixtures of species recruited from various natural habitats; for example, forest edges against cliffs, rivers and mires or swamps, and from the rocks, alluvia, mires and swamps themselves (Marks 1983). They also sometimes involve species introduced from other continents, as in case of some of the prominent annuals and biennials in the oldfield successions in North America. [P.J.G.]

REFERENCE

Marks, P.L. (1983) On the origin of the field plants of the northeastern United States. *American Naturalist* **122**, 210–228.

secondary woodland Woodland occupying a site which has at some time been artificially cleared of WOODLAND, but which has reverted to woodland by natural SUCCESSION or by planting. Clearance involves the removal of tree cover and use of the site for PASTURE, meadow, arable, building, etc., by which alternative uses a non-woodland VEGETATION has been established. [G.F.P.]

sedentary Describing an organism that remains fixed to a specific location in space throughout its lifespan. [R.C.]

sediment A general term for unconsolidated deposits of either minerogenic or organic origin. The minerogenic deposits are divisible into three groups: (i) sorted, or stratified; (ii) unsorted, or non-stratified mechanical sediments (clastic sediments); and (iii) chemical (non-clastic) sediments.

Till is unsorted glacigenic sediment produced by the eroding, transporting and depositing action of glaciers. It is an unsorted mixture of angular mineral particles of widely different sizes. The distinction between stratified and non-stratified sediments is not always sharp. However, the terms connote a fundamental difference of origin, as stratified minerogenic deposits are formed by the action of either WATER or WIND.

Glaciofluvial sediments are ones that have been formed by meltwater streams either inside or outside a glacier or subglacially. Coarser sediments like sand are not transported over longer distances, in contrast to fine fractions, SILT and CLAY, which become deposited further away, deep in the proglacial water bodies. In the cold, fresh water of proglacial lakes rhythmic deposition leads to a formation of annual varves, which have been widely used in chronological studies. Sea salt causes a FLOCCULATION of suspended sediment, resulting in its settling as homogeneous masses. Regularly varved clays are called diatactic; more or less homogeneous clays are termed symmictic.

The action of wind leads to the formation of fine-grained aeolian sediments, DUNES formed of wind-blown sand and loess of dominantly silt-sized particles.

Chemically (or biochemically) precipitated materials (e.g. calcium carbonate, silica, ferrous compounds, salts) form sediments of greatly varying size.

Biogenic sediments, of mainly organic origin, are divided into two main groups. AUTOCHTHONOUS sediments are formed mainly *in situ*, of material originating from the mother ecosystem. ALLOCHTHONOUS sediments consist partly of transported material. The most important of the autochthonous sediments is PEAT, formed from the remains of local MIRE vegetation preserved usually in anaerobic conditions. Acid MOR consists of plant residues formed on the forest floor. Allochthonous sediments include gyttja, formed in lakes and consisting of often microscopic remains of plants and animals together with various amounts of minerogenic particles. Dy consists almost entirely of HUMUS particles deposited in DYSTROPHIC lakes.

When sediments become consolidated or lithified during the course of time, sedimentary rocks are formed, such as sandstone, shale and limestone. *See also* SEDIMENT TOXICITY; SEDIMENTATION; SOIL. [Y.V.]

sediment toxicity Many TOXICANTS, both inorganic and organic, will adsorb on to dead organic particles, and hence accumulate within aquatic SEDIMENTS and SOILS. Sediment so contaminated may be toxic to the organisms dependent upon it. There is a complex chemical relationship between toxicants adsorbed on to sediment particles and the WATER between particles (so-called pore water). ECOTOXICITY tests designed to study sediment TOXICITY can involve samples of whole sediment and/or samples of extracted pore water (e.g. by centrifugation, elutriation). [P.C.]

sedimentary cycle Any global cycle of an element in which the ATMOSPHERE plays an unimportant role, but in which the geological processes of marine SEDIMENTATION, uplift, erosion, soil WEATHERING, and LEACHING dominate. Elements such as CALCIUM and POTASSIUM have this type of cycle (whereas CARBON, OXYGEN, NITROGEN and SULPHUR, for instance, have important gaseous phases in their global cycling).

At the ecosystem level, the elements that have this type of cycling are supplied mainly by soil weathering or groundwater input. There may be some atmospheric input from DUST or from dissolved or suspended materials from precipitation, but there is no opportunity for the 'fixation' of these elements from the atmosphere, as is the case for carbon and nitrogen. As a consequence, the supply of these elements to an ecosystem is limited by the chemistry of the geological materials underlying the ecosystem or found within its CATCHMENT.

Since several elements with sedimentary cycles are required in some quantity for the growth of plants (e.g. potassium, PHOSPHORUS, calcium), it may be necessary to supplement soil supplies of these elements where plant growth needs to be stimulated, as in horticulture and agriculture. These elements are, therefore, often used in fertilization regimes. [P.D.M.]

sedimentation The tendency of particles in a fluid to fall and settle out under the influence of gravity. A spherical particle soon reaches a constant terminal velocity v_s found by equating the DRAG force (STOKES' LAW) to the gravitational force minus the velocity:

$$v_s = \frac{d^2 g(\rho_0 - \rho)}{18 \upsilon \rho}$$

where d is the diameter of the particle, g is the acceleration due to gravity, ρ_0 is the density of the particle, ρ is the density of the fluid and υ is the kinematic VISCOSITY of the fluid. [J.G.]

seed bank Seed banks are repositories of dormant seeds found either *in situ* in nature, for example SOIL SEED BANKS, or maintained artificially *ex situ* as a conservation tool or for economic uses such as crop breeding. Seed banks of both kinds result from the ability of the seeds of many species to survive long periods of DORMANCY, from several years to many decades.

Seeds of most species fit into either of two more or less discrete categories. Orthodox seeds can survive DESICCATION and frozen storage. Recalcitrant seeds are desiccation intolerant, and thus cannot survive being dried to the point where they can survive freezing. For storage in an *ex situ* seed bank, seeds of orthodox taxa are typically dried to equilibrium at 15°C and 15% RELATIVE HUMIDITY (RH) (or at higher RH; Vertucci & Roos 1993), hermetically sealed in small containers and stored frozen at –20°C. CRYOPRESERVATION refers to storage below –130°C, and some seed banks store seed either immersed in liquid NITROGEN (–196°C), or in the vapour above it (–150°C). [E.O.G. & L.R.M.]

REFERENCES

Cromarty, A.S., Ellis, R.H. & Roberts, E.H. (1982) *The Design of Seed Storage Facilities for Genetic Conservation*. Handbooks for Genebanks: No. 1. International Board for Plant Genetic Resources (IBPGR), Rome.

Leck, M.L., Parker, V.T. & Simpson, R.L. (eds) (1989) *Ecology of Soil Seed Banks*. Academic Press, New York.

Vertucci, C.W., Roos, E.E. & Crane, J. (1994) Theoretical basis of protocols for seed storage. III. Optimum moisture contents for pea seeds stored at different temperatures. *Annals of Botany* **74**, 531–540.

seed dormancy A state in which the metabolism of a seed is reduced to a minimum, allowing survival under adverse conditions. Three types of DORMANCY are generally recognized: enforced, innate and induced.

Enforced dormancy occurs when a seed is deprived of its normal requirements for germination, for example by the lack of sufficient WATER or OXYGEN, or the absence of a suitable TEMPERATURE or LIGHT regime. Innate dormancy is seen in seeds which are incapable of germination when freshly dispersed, even in apparently favourable conditions. Innate dormancy may be due to a number of causes, including embryo immaturity, the presence of a hard seed coat, or the action of germination-inhibiting compounds in the seed. It often wears off with time.

Induced dormancy occurs when the seed acquires special requirements for germination. For example, seeds of many grassland species can initially germinate in the dark, but acquire a light-requirement for germination if they remain ungerminated below ground for a time. Some seeds go through annual cycles of induced dormancy in the SOIL, brought about as a response to FLUCTUATIONS in the soil temperature. The temperature range over which the seeds will germinate expands and contracts through the year, resulting in germination being restricted to narrow seasonal time bands.

The type of dormancy exhibited by a seed is largely under genetic control, but the environmental conditions experienced by the parent plant during seed maturation modify the degree of dormancy shown. In general, warm dry conditions during seed development result in seeds being less dormant. [M.F.]

REFERENCE

Fenner, M. (ed.) (1992) *Seeds: The Ecology of Regeneration in Plant Communities*. CAB International, Wallingford.

seed-limited recruitment Describing RECRUITMENT to a plant population when the input of extra seeds (e.g. by experimental sowing, or by a reduction in the numbers of granivorous animals) leads to an increase in the size of the reproductive plant population. In many ecosystems, adding extra seed has no impact on plant abundance because recruitment is microsite-limited (*see* DISTURBANCE) or herbivore-limited (Fig. S8). All current theoretical models of plant POPULATION DYNAMICS are based on the assumption that recruitment is seed-limited when plant POPULATION DENSITY is low. In the field, however, herbivores may eat any additional seeds that are added experimentally, and a threshold density of extra seed might be required for SATIATION of the seed-feeding animals. [M.J.C.]

REFERENCE

Crawley, M.J. (1992) Seed predators and plant population dynamics. In: *Seeds: The Ecology of Regeneration in Plant Communities* (ed. M. Fenner), pp. 157–191. CAB International, Wallingford.

seek-it-out strategy *See* FEEDING TYPES, CLASSIFICATION.

seek-out feeders *See* FEEDING TYPES, CLASSIFICATION.

seepage lakes *See* LAKES, TYPES OF.

segregation distortion

1 Broadly, the recovery of progeny in significantly non-Mendelian proportions, usually ascribed to the case where at least one parent is heterozygous at the locus concerned.

2 More narrowly (and more precisely), the recovery in a genetic cross of a significant excess of progeny bearing one particular

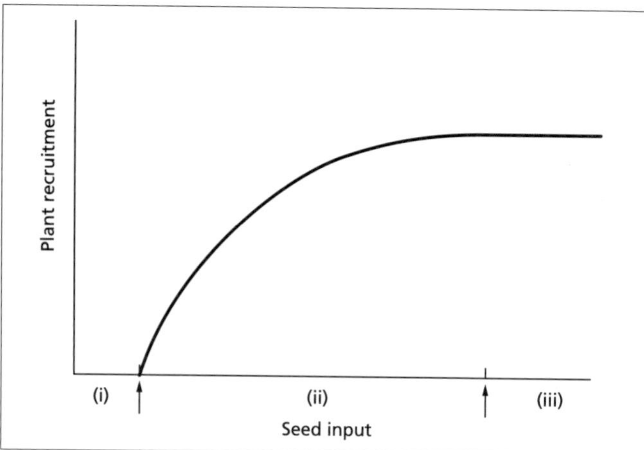

Fig. S8 Seed-limited plant recruitment. The relationship between seed input and the resulting plant recruitment (the number of established plants with a reasonable prospect of reaching sexual maturity) is divided into three zones: (i) herbivore-limited recruitment, where increases in seed input do not lead to plant recruitment because all seeds are eaten by granivores (post-dispersal seed predators); (ii) seed-limited recruitment, where increased seed input does lead to increased plant recruitment; (iii) microsite-limitation, where increased seed input does not lead to increased recruitment, and where substantial granivory can be accommodated without any reduction in plant recruitment.

ALLELE as a result of that allele's prezygotic/gametic effects. The definition is often further restricted to the case where the allele gaining over-representation reduces individual FITNESS (*see* MEIOTIC DRIVE). If deleterious then the allele concerned is deemed to belong to a class of SELFISH GENES (*see* SELFISH DNA). Selfish distortion is generally considered a rare phenomenon but is described through a wide taxonomic range (e.g. ascomycete fungi, insects, mammals, angiosperms).
See also SEGREGATIONAL LOAD. [L.D.H.]

REFERENCE

Lyttle, T.W. (1991) Segregation distorters. *Annual Review of Genetics* **25**, 511–557.

segregational load The relative loss in FITNESS resulting from the segregation of favourable combinations of ALLELES. Also, the genetic disability sustained by a population due to genes segregating from advantageous heterozygotes to unfit homozygotes. *See also* QUANTITATIVE GENETICS; SEGREGATION DISTORTION. [V.F.]

seiche *See* LAKES.

seizers and swallowers *See* FEEDING TYPES, CLASSIFICATION.

seizers, masticators and swallowers *See* FEEDING TYPES, CLASSIFICATION.

selection Usually shorthand for NATURAL SELECTION, though can also be used for ARTIFICIAL SELECTION. [P.C.]

selection arena The inevitable result of the overproduction of zygotes in COMPETITION for restricted space in the confines of a brood chamber. Selection arenas lead to disproportionate success of the fastest growing and most fit offspring, and a corresponding increase in the mean FITNESS of the brood. Hence selection arenas can be selectively advantageous to the maternal parent, especially if zygotes are very cheap to produce. The greater success of the more robust offspring in a selection arena can result from either SELECTIVE ABORTION of slow-growing zygotes by the maternal parent, or greater provisioning of fast-growing zygotes. [C.M.L.]

REFERENCE

Stearns, S.C. (1987) The selection-arena hypothesis. In: *The Evolution of Sex and its Consequences* (ed. S.C. Stearns), pp. 337–349. Birkhäuser, Basel.

selection coefficient In POPULATION GENETICS, the selection coefficient against a GENOTYPE A_iA_j is defined as:

$$s_{ij} = 1 - w_{ij}$$

where w_{ij} equals the relative FITNESS of that genotype ($0 \le w_{ij} \le 1$). *See also* QUANTITATIVE GENETICS. [G.D.J.]

selection differential In QUANTITATIVE GENETICS, phenotypic selection on a QUANTITATIVE TRAIT within a generation can be expressed by the selection differential *S*. The selection differential is defined as (the mean value of the quantitative trait in individuals after selection) minus (the mean value of the quantitative trait in the original population before selection) (Fig. S9). For example, Alatalo *et al.* (1990) measured a total of 8887 young flycatchers. The subsample of those birds which survived for one year had a tarsus length 0.12 mm longer than the group as a whole. The selection differential for tarsus length over this episode of selection is thus +0.12 mm. The selection differential quantifies the effect of selection on the mean value of a quantitative trait within a generation.

The difference between the mean value of the quantitative trait in the population after selection and the mean value ($\bar{x}$) of the quantitative trait before selection has a clear mathematical expression representing the intuitive meaning of selection. Let x stand for any value of the quantitative trait in an individual of the population. Let

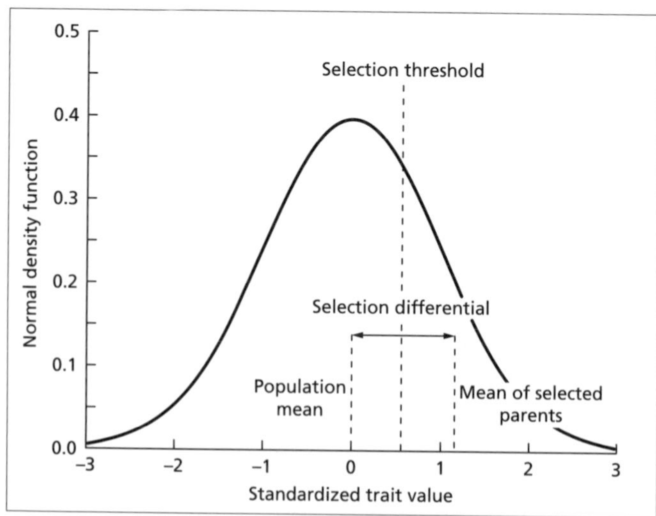

Fig. S9 The selection differential is the difference between the mean of the selected parents and the population mean.

p_x be the frequency of trait value x and W_x be the FITNESS of an individual with trait value x. Then, the selection differential is, by its definition, equal to:

$$S = \sum p_x x \frac{W_x}{\overline{W}} - \sum p_x x$$

$$\overline{W}S = \sum p_x x W_x - \bar{x}\overline{W} = \mathrm{cov}_P(x, W)$$

$$S = \mathrm{cov}_P(x, w)$$

The selection differential equals the phenotypic COVARIANCE between the value x of the quantitative trait and relative fitness $w_x = W_x / \overline{W}$. That is, if high trait values lead to high fitness, the covariance between PHENOTYPE and fitness is positive, and so is the selection differential: selection leads to higher trait values. If high trait values lead to low fitness, the covariance between phenotype and fitness is negative, and so is the selection differential: selection leads to lower trait values. This is part of a general mathematical expression for Darwin's THEORY OF NATURAL SELECTION, one that also involves the ADDITIVE GENETIC VARIANCE and the SELECTION RESPONSE. *See also* SELECTION GRADIENT. [G.D.J. & T.D.P.]

REFERENCE

Alatalo, R.V., Gustafsson, L. & Lundberg, A. (1990) Phenotypic selection on heritable size traits — environmental variance and genetic response. *American Naturalist* **135**, 464–471.

selection experiments and life-history evolution *See* TRADE-OFF.

selection gradient The selection gradient is the slope of the FITNESS PROFILE, the curve giving FITNESS as a function of a QUANTITATIVE TRAIT, either at any particular trait value or over a predetermined range of trait values. The slope of the fitness profile indicates the direction of selection on the phenotypic mean of a quantitative trait. A negative selection gradient indicates a decreasing fitness profile, and therefore selection for lower trait values. A positive selection gradient indicates an increasing fitness profile, and therefore selection for higher trait values. The selection gradient, β, for a trait can therefore be found as the derivative of fitness to phenotypic trait value, i.e. $\beta = \mathrm{d}w/\mathrm{d}x$, or as the linear regression of fitness values on trait values, i.e. $\beta = \mathrm{cov}(x,w)/\mathrm{var}(x)$. The phenotypic selection gradient, β, and the SELECTION DIFFERENTIAL, S, are therefore related as:

$$\beta = \frac{\mathrm{cov}_P(x,w)}{V_P(x)} = \frac{S}{V_P(x)}$$

In the classical model of selection on one locus, with two alleles A_1 and A_2 at frequencies p and $q = 1 - p$, the change in ALLELE FREQUENCY, p, after one generation selection is equal to:

$$\Delta p = \frac{pq}{\overline{w}} \frac{\mathrm{d}\overline{w}}{\mathrm{d}p}$$

which can be written as:

$$\Delta p = \frac{pq}{\overline{w}} \frac{\mathrm{d}\overline{w}}{\mathrm{d}\bar{x}} \times \frac{\mathrm{d}\bar{x}}{\mathrm{d}p}$$

$$\Delta p = \frac{pq}{\overline{w}} \times 2\alpha \times \frac{\mathrm{d}\overline{w}}{\mathrm{d}\bar{x}}$$

(where α is the AVERAGE EFFECT OF A GENE SUBSTITUTION), involving the selection gradient of mean fitness towards mean trait value rather than the slope of the fitness profile. If selection works simultaneously on a number of quantitative traits, the selection gradient vector, $\nabla\overline{w}$, contains all partial derivatives of mean fitness towards the mean trait values. *See also* QUANTITATIVE GENETICS; SELECTION RESPONSE. [G.D.J.]

REFERENCES

Lande, R. (1979) Quantitative genetic analysis of multivariate evolution, applied to brain : body size allometry. *Evolution* **33**, 402–416.

Lande, R. & Arnold, S.J. (1983) The measurement of selection on correlated characters. *Evolution* **37**, 1210–1226.

selection intensity In QUANTITATIVE GENETICS, the relative strength of selection on several QUANTITATIVE TRAITS can be compared by using the selection intensity, i. For this purpose the SELECTION DIFFERENTIAL, S (the difference between the mean trait value of selected parents and the mean trait value of the population), is divided by the standard deviation (σ) of the trait in the population:

$$i = S/\sigma$$

The selection intensity is easily computed if the trait value in the population is normally distributed and selection is by truncation, i.e. all individuals above (or below) a certain threshold value are selected to be parents in the next generation. A direct relation then exists between the intensity of selection and the proportion of the population that is selected. For a given selection threshold, a proportion (p) of the population falls above the selection threshold; the height of the NORMAL DISTRIBUTION density curve at the threshold equals z. Then, $i = z/p$. Given tables of the normal distribution, the selection intensity can be established from the proportion of the population that is selected. [G.D.J.]

REFERENCE

Falconer, D.S. (1989) *An Introduction to Quantitative Genetics*. Longman, London.

selection limits In lines that are artificially selected for a high or low value of a QUANTITATIVE TRAIT, a continuing SELECTION RESPONSE can be expected over a number of generations, after which a further selection response might prove impossible. The larger the number of gene loci that contribute to the selection response, the larger the response after many generations will be in relation to the original ADDITIVE GENETIC VARIANCE in the population. With few loci, the selection limits, which are represented by the most extreme genotypes over all loci, would be well within the range of variation found in the base population. With larger numbers of loci, the extreme genotypes are rare in the base population, and the selection limits are further removed, in units, from the original mean trait value.

The theoretical maximum response over many generations is:

$$R_{\mathrm{max}} = 2N_e R_{\mathrm{generation}}$$

i.e. the theoretical maximum response equals the product of the selection response in the first generation of selection in the base population times twice the EFFECTIVE POPULATION SIZE, N_e. Selection can therefore proceed further in a very large population. *See also* QUANTITATIVE GENETICS. [G.D.J.]

REFERENCE

Falconer, D.S. (1989) *An Introduction to Quantitative Genetics*. Longman, London.

selection pressure A loose term indicating that selection in a particular direction exists or that selection is of any particular strength. Selection pressure might indicate the proportion of the population that is selected to be parents of the next generation, or the SELECTION COEFFICIENT, SELECTION DIFFERENTIAL, SELECTION GRADIENT or SELECTION INTENSITY. *See also* ECOLOGICAL GENETICS; FITNESS; QUANTITATIVE GENETICS; SELECTION PRESSURES, CLASSIFICATION. [G.D.J.]

selection pressures, classification This has most famously been done according to POPULATION DENSITY in terms of *r*- AND K-SELECTION. However, density effects are made up of a number of components, so SELECTION PRESSURES can be more effectively classified according to their effects on FITNESS components; i.e. with respect to survival (*S*), FECUNDITY (*n*) and time between birth and breeding events (*t*) and their age-specific subcomponents. Hence, in principle, the effects of SELECTION and their classification ought to be represented in *n*-dimensional space. In practice, however, it has been more tractable to erect 2- (and sometimes 3-) dimensional classifications that either represent distillations of the *n*-dimensional ones or focus on what are thought to be particularly important dimensions.

A series of such classifications is summarized in Fig. S10 together with a very simplified representation of some of the predicted traits associated with different selection pressures, from high to low STRESS. Given that these are all abstractions from a more complex adaptive space, it is not surprising that there should only be rough equivalence between them. Moreover, some of the components emphasize the characteristics of the environment. For example, FAVOURABLENESS and durational stability — elements of what Southwood (1977) called the habitat templet, rather than the impact on the organisms themselves. Yet the implication must be that the habitat templet has its impact by influencing *S*, *n* and *t*. *See also* LIFE-HISTORY EVOLUTION; OPTIMALITY MODELS; SIZE–NUMBER TRADE-OFF; TRADE-OFF. [P.C.]

REFERENCES

Greenslade, P.J.M. (1983) Adversity selection and the habitat templet. *American Naturalist* **122**, 352–365.

Grime, P. (1979) *Plant Strategies and Vegetation Processes*. John Wiley & Sons, Chichester.

Hildrew, A.G. & Townsend, C.R. (1987) Organization in freshwater benthic communities. In: *Organization of Communities Past and Present* (eds J.H.R. Gee & P.S. Giller), pp. 347–377. Blackwell Scientific Publications, Oxford.

Sibly, R. & Calow, P. (1985) Classification of habitats by selection pressures; a synthesis of life-cycle and r/K theory. In: *Behavioural Ecology* (eds R.M. Sibly & R.H. Smith), pp. 75–90. Blackwell Scientific Publications, Oxford.

Sibly, R.M. & Calow, P. (1989) A life cycle theory of responses to stress. *Biological Journal of the Linnean Society* **37**, 101–116.

Southwood, T.R.E. (1977) Habitat, the templet for ecological strategies. *Journal of Animal Ecology* **46**, 337–365.

selection response If a QUANTITATIVE TRAIT in a population is selected, the observed selection response *R* equals the difference between the mean trait value in the offspring population, $\bar{x}_o$, and the mean trait value in the parental population before selection, $\bar{x}_p$. Hence, $R = \bar{x}_o - \bar{x}_p$.

The regression line between the (mean) PHENOTYPE of the offspring of a pair of parents and the mean phenotype of both parents indicates in how far differences in trait values between parents are found in the offspring. The slope of this regression line equals the NARROW-SENSE HERITABILITY h^2. The predicted selection response (*R*) equals the product of the heritability (h^2) and the SELECTION DIFFERENTIAL (*S*), as can be seen from Fig. S11: i.e. $R = h^2S$. Since $h^2 = V_A/V_P$ and $S = \mathrm{cov}_P(x, W)/\overline{W}$ (where V_A is the ADDITIVE GENETIC VARIANCE, V_P is the PHENOTYPIC VARIANCE, W_x is the fitness of an individual with trait value *x* and $\overline{W}$ is the mean fitness of the population):

$$R = \frac{V_A}{V_P}\frac{\mathrm{cov}_P(x, W)}{\overline{W}} = \frac{1}{\overline{W}}V_A\beta$$

showing that the predicted selection response equals the additive genetic variance (V_A) in the quantitative trait times the SELECTION

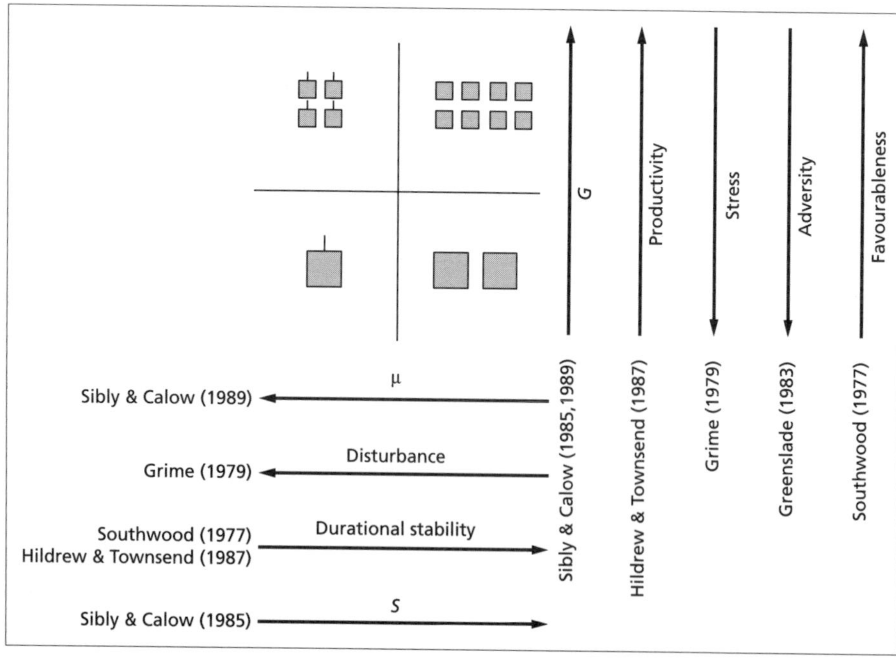

Fig. S10 A very simplified representation of different classifications of selection pressures and their rough equivalence. Arrows representing each variable move from 'low' to 'high'. The blocks represent the kinds of response in reproductive effort (total area of boxes) and propagule size (area of individual boxes) that the classifications predict. *S*, survival probability; μ, mortality rate; *G*, individual growth. The 'spines' emerging from the boxes to the left are intended to represent investment in defence. (After Calow, P. (1989) Proximate and ultimate responses to stress in biological systems. *Biological Journal of the Linnean Society* **37**, 173–181.)

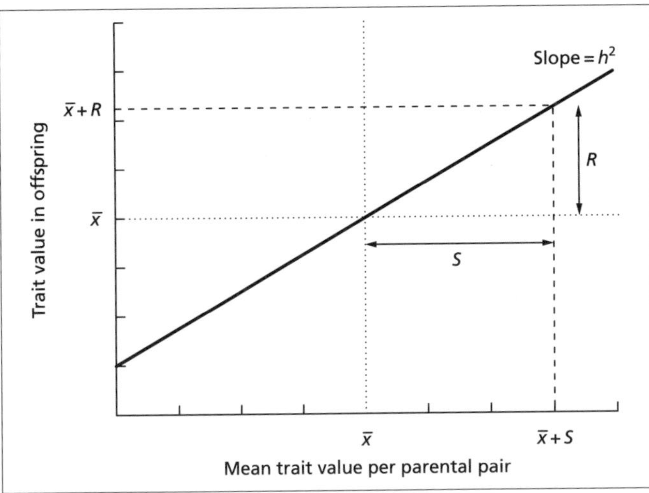

Fig. S11 The predicted selection response in a quantitative trait follows from the regression between parents' and offspring values for the trait and the mean trait value of the selected parents.

GRADIENT (β; the regression between PHENOTYPIC VALUE and FITNESS value, or, the slope of the FITNESS PROFILE). This is a mathematical formulation of Darwin's THEORY OF NATURAL SELECTION: the response to selection (R), depends upon the existence of phenotypic variance in the population (V_P), on the existence of a COVARIANCE between phenotype and fitness ($\mathrm{cov}_P(x,W)$) and on the presence of heritable variation (V_A).

In multilocus genetics, the change in trait mean after one generation of selection is equal to:

$$\Delta\bar{x} = \sum_{\text{loci}} \Delta p_i \frac{\partial \bar{x}}{\partial p_i}$$

where p_i is the allele frequency of allele A_1 at locus i.

By using the definition of the AVERAGE EFFECT OF A GENE SUBSTITUTION (α), the general expression for ALLELE FREQUENCY change (Δp), the multilocus expression for the additive genetic variance (V_A), and the chain rule of differentiation, the selection response becomes:

$$\Delta\bar{x} = \frac{1}{\bar{W}} V_A \beta_m$$

involving the selection gradient β_m of mean fitness towards mean trait value rather than the slope of the fitness profile. Moreover, since the general expression for allele frequency change, Δp, involves the average effect of a gene substitution with regard to fitness, the selection response in a quantitative trait equals the additive genetic covariance between the GENOTYPIC VALUE of the trait and genotypic fitness:

$$R = \Delta\bar{x} = \frac{1}{\bar{W}} \sum_{\text{loci}} 2p_i q_i \alpha_{i,x} \alpha_{i,W} = \frac{1}{\bar{W}} \mathrm{cov}_A(x,W)$$

This relation between the selection response and the additive genetic covariance between phenotype and fitness is known as the Secondary Theorem of NATURAL SELECTION. *See also* HERITABILITY; QUANTITATIVE GENETICS. [G.D.J.]

selective abortion The judicious CULLING of defective embryos in order to promote the development of other embryos, especially in plants. Such a strategy may be of particular adaptive value when self-fertilization or biparental inbreeding is possible and leads to a reduction in expected FITNESS. Inbred or otherwise abnormal seeds may be aborted by the plant, which allows for greater provisioning of the remaining seeds. *See also* CANNIBALISM; INFANTICIDE; PARENTAL CARE; PARENTAL INVESTMENT. [C.M.L.]

selective feeding The ingestion of food items in a proportion different from that available in the surrounding environment. It is usually considered as an adaptation to increase energy intake. *See also* PARTICLE-SIZE SELECTIVITY. [V.F.]

REFERENCE

Barnes, R.S.K., Calow, P. & Olive, P.J.W. (1993) *The Invertebrates: A New Synthesis*, 2nd edn. Blackwell Scientific Publications, Oxford.

selenium (Se) A heavy metalloid element (ATOMIC NUMBER = 34, relative ATOMIC MASS = 78.96; melting point = 217°C; boiling point = 684.9°C) that is an ultra-trace constituent (0.05 mg kg^{-1}) of the Earth's CRUST. Found associated with metal sulphide ores, through isomorphous substitution of SULPHUR, and frequently enriched in shales and FOSSIL FUELS, selenium is used in GLASS manufacture, electronics, pigments and FUNGICIDES.

Selenium is released to the ATMOSPHERE from volcanoes, biogenic sources in the oceans and coal burning, and may be present in the environment in a variety of chemical forms and oxidation states. In soils, it can be present as selenite (Se(IV)O_3^{2-}), which predominates in neutral and ACID soils, and the more readily available and mobile selenate (Se(VI)O_4^{2-}), which is the major form in aerated alkaline soils.

Animals feeding on plants in soils of high selenium concentration and availability can suffer from selenosis (e.g. the 'blind staggers'), and deaths and deformities have occurred in wildlife in contaminated waters. As selenium is an ESSENTIAL ELEMENT, however, insufficient selenium can lead to deficiency conditions (e.g. 'white muscle disease'). The most remarkable example of selenium deficiency in humans occurred in a province of China, where it was manifest as a form of endemic cardiomyopathy known as Keshan disease. Since the 1970s, this has been eradicated by dietary supplementation with selenium. Intimately associated with the enzyme glutathione peroxidase, selenium helps to protect tissue from oxidative damage by peroxides and hydroxyl radicals. Selenium, which can itself be methylated in the environment, also helps to protect against methylmercury (*see* HEAVY METALS). [J.F.]

self-incompatible *See* POLLINATION.

Selfish DNA

A gene or particular stretch of DNA that can increase in frequency in a population, despite not necessarily being advantageous, by virtue of its ability to gain over-representation (i.e. transmission at a rate greater than the Mendelian rate), rather than through causing an increase in individual FITNESS. An absence of significant levels of horizontal transmission is an implicit requirement, and the over-representation is gained typically through manipulation of the host genetic machinery (e.g. by using the machinery to make copies of itself and having these inserted elsewhere into the genome, by inhibiting gametes that do not have a copy of the same selfish DNA and a variety of other techniques).

Analogous terms include selfish gene, ultra-selfish gene, selfish genetic element, self-promoting element, genetic renegade, genomic parasites and parasitic (genetic) element/DNA. These terms are sometimes used interchangeably but sometimes in contrast to each other. In its original meaning (Dawkins 1976) the term 'selfish gene' was intended to apply to any gene/TRAIT that spread in a population under the force of selection on that gene in favour of the trait that it coded for. It was not intended that the selfish gene must be deleterious and spread because of over-representation (this would just have been one subcategory). The usage has since been changed to that described above; i.e. one in which the ability to gain over-representation is a necessary defining feature. In the *Oxford English Dictionary* this latter meaning is given but incorrectly ascribed to Dawkins. The term 'ultra-selfish gene' has been employed to contrast with the original (Dawkins) definition of selfish gene, but is all but equivalent to the later definition (i.e. a genomic component gaining over-representation).

If a distinction exists between SELFISH GENES (later usage, alias ultra-selfish genes) and selfish genetic elements it is that some use 'selfish gene' to include only those genes that are components of the genome (this latter term, however, tends to go undefined and has numerous grey areas). The term 'selfish genetic element' may then either be reserved for extra-genomic obligately vertically transmitted heritable elements (such as vertically transmitted bacteria found in numerous insects) or as a term to encompass both the genomic and non-genomic parasitic elements.

Selfish DNA is sometimes considered as a subclass of selfish elements, which gains over-representation through their ability to ensure that they are copied and the copies are inserted elsewhere within the genome (alias mobile DNA). This class includes a variety of classes of intron (e.g. type I introns such as the ω^+ intron of yeast mitochondrial genomes), TRANSPOSABLE ELEMENTS (a short stretch of DNA with the capability to make copies of itself), genes undergoing biased GENE CONVERSION and 'homing' elements.

Over-representation may be achieved by means other than over-replication. MEIOTIC DRIVE genes (*see also* SEGREGATION DISTORTION) interfere with competitor sister gametes. B chromosomes are unnecessary extra chromosomes that often undergo meiotic drive. Cytoplasmic sex-ratio distorters are uniparentally transmitted elements that manipulate the host sex ratio in favour of the sex that transmits the cytoplasmic genes, typically females. This may be done in a variety of ways. Some convert males into females, some make females reproduce asexually, others kill their host if that host is male.

Selfish elements are similar to conventional parasites in that they tend to be harmful. Unlike conventional parasites, horizontal transmission between individuals is by implicit understanding not a common route of transmission (note that mobile DNA with regular horizontal transmission would be classified as a virus or viroid). However, it is often hard to tell whether selfish elements are actually deleterious. A thorough distinction can none the less be made between elements that could spread were they deleterious and those that could not. In most instances this potential is adequate to define a gene as a selfish gene.

In certain contexts, however, the deleterious nature of the selfish element is of critical importance. The spread in a population of a selfish element that is deleterious creates the context for the invasion of a suppressor acting within the same individual. This difference of interest between the selfish element and the rest of the genome is indicative of a conflict of interests between the two parties.

CONFLICTS instigated by selfish elements, and the build-up of selfish DNA within genomes, are considered to be potentially important forces in the EVOLUTION of genomes and GENETIC SYSTEMS. Their full importance has yet to be resolved. [L.D.H.]

REFERENCES

Dawkins, R. (1976) *The Selfish Gene*. Oxford University Press, Oxford.

Doolittle, W.F. & Sapienza, C. (1980) Selfish genes, the phenotype paradigm and genome evolution. *Nature* **284**, 601–603.

Hurst, L.D. (1993) The incidences, mechanisms and evolution of cytoplasmic sex ratio distorters in animals. *Biological Reviews of the Cambridge Philisophical Society* **68**, 121–193.

Jones, R.N. & Rees, H. (1982) *B Chromosomes*. Academic Press, New York.

Orgel, L.E. & Crick, F.H.C. (1980) Selfish DNA: the ultimate parasite. *Nature* **284**, 604–607.

Shapiro, J.A. (ed.) *Mobile Genetic Elements*. Academic Press, New York.

selfish genes The selfish gene concept was popularized by R. Dawkins in his book *The Selfish Gene* (1976), in which he developed the argument that an organism is a gene's way of making more genes. *See also* SELFISH DNA. [V.F.]

self-limitation DENSITY DEPENDENCE in single-species models of POPULATION DYNAMICS. The key process is intraspecific density dependence. This generally takes the form of COMPETITION for limiting RESOURCES. The dynamics of the population depend upon the nature of this competition; whether it is SCRAMBLE COMPETITION or CONTEST COMPETITION. Population dynamics are likely to be relatively stable under contest competition, while scramble competition tends to produce less stable dynamics (POPULATION DENSITY could exhibit cycles or chaotic dynamics (*see* CHAOS) if the overcompensating density dependence is sufficiently extreme). The stability of the equilibrium in a single-species model depends upon the slope of the RICKER CURVE at the point where it intersects the self-replacement line (the 45° line of $N_{t+1}=N_t$). If the slope is positive, then the equilibrium is exponentially stable; if it is between 0 and –1, the equilibrium is approached by damped OSCILLATIONS; and if the slope is steeper than –1, then the system will exhibit cycles or chaos (with extremely steep slopes implying extreme overcompensating density dependence). [M.J.C.]

self-thinning A form of density-dependent plant MORTALITY that occurs during the development of an even-aged MONOCULTURE, as the individuals grow from germination to reproductive maturity. The relatively small plants die because the relatively large plants grow even bigger and shade them out. The term is sometimes used to describe any density-dependent mortality in plant populations. In many studies of DENSITY DEPENDENCE in field populations of plants, most of the density dependence takes the form of reductions in mean plant size rather than increased death rate. *See also* LAW OF CONSTANT FINAL YIELD; $-^3/_2$ POWER LAW. [M.J.C.]

Semelparity

Also called monocarpy (in plants) or monotely (in invertebrates), semelparity refers to a life cycle characterized by a single reproductive attempt, followed by death. Death ensues because the extreme REPRODUCTIVE EFFORT of semelparous species involves extensive drainage of RESOURCES from somatic tissues, which results in their irreversible damage. All offspring of a given individual are usually produced synchronously; if not, no new resources for reproduction are collected between production of successive clutches. Nevertheless, a population of semelparous species can have a prolonged breeding season if reproduction of different individuals is not synchronized (Fig. S12).

According to this definition, most ANNUAL and BIENNIAL PLANTS are semelparous (some authors consider all annual and biennial species semelparous, irrespective of the length of their reproductive period). Long-lived monocarpic species include many rosette plants of deserts and the tropical subalpine zone — for example, some species of *Agave*, *Yucca* and *Lobelia* — and many bamboo (Bambuseae) species; this last group is known for its synchronous reproduction.

Semelparous species occur in most major animal taxa; in coleoid cephalopods, nereid polychaetes and mayflies (Ephemeroptera) all species are semelparous. Among vertebrates, semelparity is largely restricted to agnathans and fishes, for example lampreys (*Petromyzon*), Pacific salmon (*Onchorhynchus*) and Atlantic eel (*Anguilla*). Males of the dasyurid marsupial mouse *Antechinus* are the only semelparous mammals known.

Early discussion on the advantages and disadvantages of semelparity suggested that evolution towards semelparity would be favoured by low juvenile and high adult MORTALITY, i.e. when the chance of survival to breed again was low due to extrinsic factors (e.g. harsh winter or predation). Other authors pointed out that temporal variation in survival and FECUNDITY would be more important than the averages, and observed that fecundity and juvenile survival are usually more affected by environmental fluctuations than adult survival. Even if fecundity and juvenile survival were high on average but fluctuated unpredictably from year to year, ITEROPARITY would reduce variance of FITNESS, and thus would be favoured as a risk-spreading (BET-HEDGING) strategy. This view is supported by the ubiquity of seed banks, which constitute an alternative risk-spreading strategy, in annual and biennial plants.

Another proposed factor is the relationship between the amount of energy invested in reproduction and the NUMBER OF OFFSPRING produced. If the energetic cost per offspring decreases with increasing offspring number, semelparity is more likely to be favoured. This factor may account for semelparity in long-lived species with high fixed COSTS OF REPRODUCTION, for example ANADROMOUS fishes (lampreys, Pacific salmon) that have to undertake long MIGRATION to reproduce, or desert rosette plants (agaves, yuccas) that produce an enormous flowering stalk to attract scarce pollinators.

Coupled with a short GENERATION TIME, semelparity is also favoured in OPPORTUNISTIC SPECIES that use sparsely distributed but locally abundant resources. Such precocious semelparity is often associated with paedogenesis (reproduction in a larval stage), as in the mushroom midge (*Mycophila speyeri*) and the timber-feeding beetle (*Micromalthus debilis*), in which larvae produce parthenogenetic offspring that hatch within the parent's body and devour the parent from within. *See also* DISPOSABLE SOMA THEORY; EVOLUTIONARY OPTIMIZATION; LIFE-HISTORY EVOLUTION; OPTIMALITY MODELS; TRADE-OFF. [T.J.K.]

REFERENCES

Roff, D.A. (1992) *The Evolution of Life Histories*. Chapman & Hall, London.

Stearns, S.C. (1992) *The Evolution of Life Histories*. Oxford University Press, Oxford.

Young, T.P. (1990) Evolution of semelparity in Mount Kenya lobelias. *Evolutionary Ecology* **4**, 157–172.

---

semi-natural vegetation Genuinely virgin, untouched communities are uncommon, but there is much semi-natural vegetation in which there has been minimal DISTURBANCE, thus allowing the continued existence of natural communities in a relatively unchanged form. The term is also frequently employed for communities, such as traditional hay meadows, which owe many of their characteristic properties to the way in which they have been regulated by humans, often very consistently and over long periods. Such PLAGIOCLIMAX communities are likely to undergo marked changes if traditional management is abandoned. The biota of many British woodlands exploited by traditional methods is believed to be similar to that of the wildwood which existed 5000 years ago. Many traditionally managed GRASSLAND and HEATHLAND communities can also be regarded as semi-natural, although their percentage species composition may have been changed by particular GRAZING or burning patterns (*see* FIRE, ECOLOGY AND MANAGEMENT). [J.R.P.]

semiochemicals *See* INFOCHEMICALS.

semi-social species *See* SOCIALITY, TYPES OF.

senescence

1 A synonym for AGEING. Senescence has advantages as a term for the ageing process, in that its etymology (from the Latin root *senex* or *senis*) indicates deterioration, unlike the term ageing, which can also refer to development or maturation. *See also* DISPOSABLE SOMA THEORY.

2 In botany, the deterioration of deciduous leaves at the end of the growing season, or of flowers at the end of the flowering season. The term is also used to refer to the deterioration of these structures when they undergo intermittent TURNOVER.

[M.R.R.]

sensitivities *See* SENSITIVITY ANALYSIS.

sensitivity analysis

1 In general this could refer to any form of analysis that quantifies the change of one variable in response to changes in another.

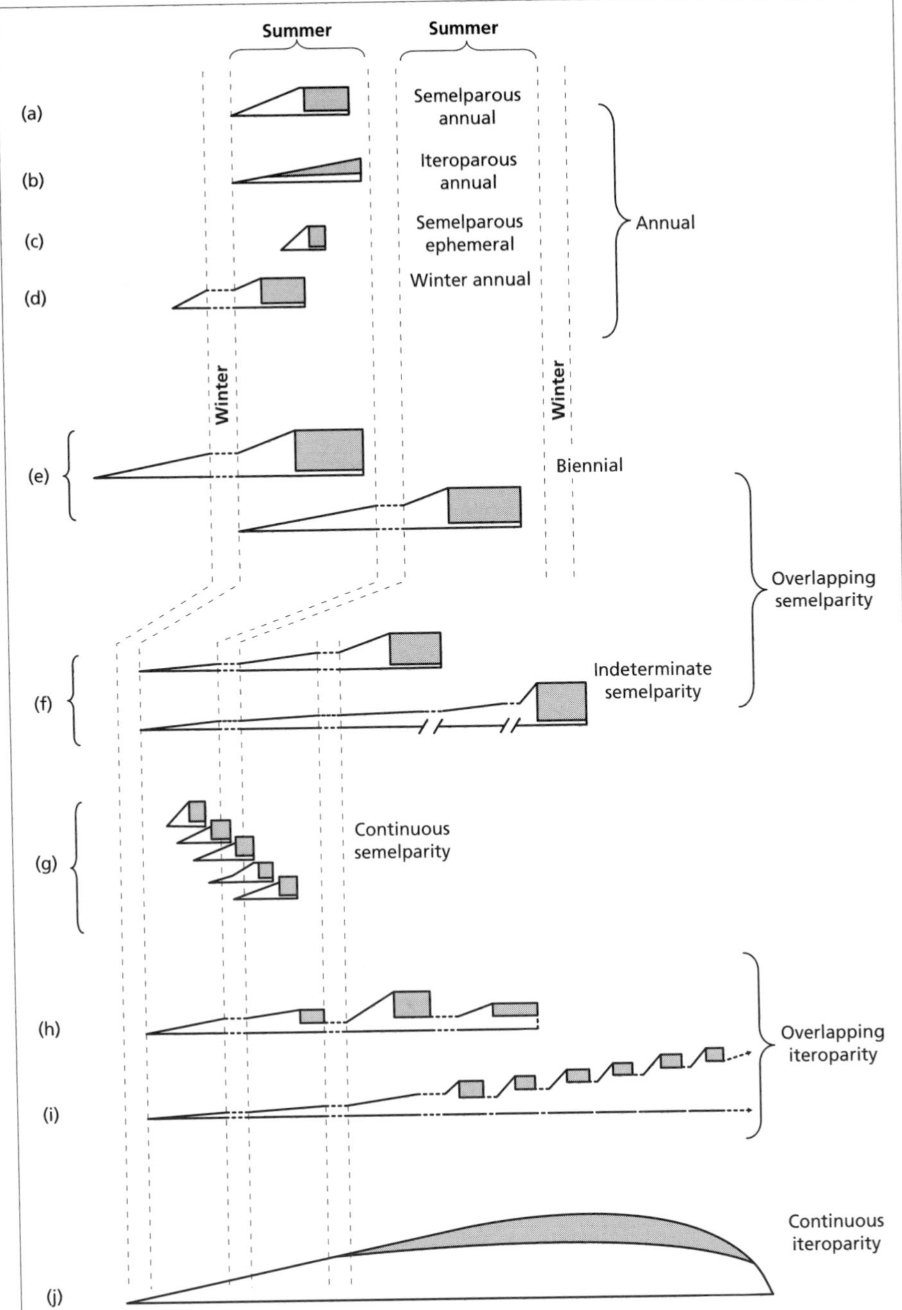

Fig. S12 A range of life cycles. In each case, the length of a bar represents the length of an individual's life. The height of a bar represents individual size (although on a logarithmic scale, so that the exponential growth of an individual early in life appears linear). The shaded region indicates the proportion of available resources allocated to reproduction in illustrative rather than exact terms, i.e. semelparous species, when they reproduce, allocate a large proportion of resources to reproduction and die soon thereafter; iteroparous species tend to allocate less to current reproduction but may survive for further reproductive bouts. Most life cycles should be self-explanatory. In overlapping semelparity, species may have a strict 2-year ((e) biennial) or 3-year, etc. cycle, or they may have a cycle of indeterminate length (f). In overlapping iteroparity, species may be relatively short-lived (h) or long-lived (i), and, if the latter, may continue to grow, apparently indefinitely, as shown here. (After Begon, M., Harper, J.L. & Townsend, C.R. (1996) *Ecology: Individuals, Populations and Communities*, 3rd edn. Blackwell Science, Oxford.)

2 In DEMOGRAPHY, sensitivity analysis usually refers to the sensitivity of POPULATION GROWTH RATE to changes in the vital rates.

Let **A** be a POPULATION PROJECTION MATRIX, with dominant EIGENVALUE λ and corresponding right and left eigenvectors **w** and **v**. The sensitivity of λ to changes in the (i,j) entry of **A** is:

$$\frac{\partial \lambda}{\partial a_{ij}} = \frac{v_i w_j}{\{\mathbf{w},\mathbf{v}\}}$$

where $\{\mathbf{w},\mathbf{v}\}$ denotes the scalar product of the two vectors.

The sensitivity of λ to variables that are not entries of the projection matrix but which affect those entries can be calculated as:

$$\frac{\partial \lambda}{\partial x} = \sum_i \sum_j \frac{\partial \lambda}{\partial a_{ij}} \frac{\partial a_{ij}}{\partial x}$$

The sensitivities can be used to evaluate the effects of UNCERTAINTY in estimates of **A**, to predict the impact of potential management strategies, and as estimates of the SELECTION GRADIENT on a_{ij} in life-history theory. Sensitivities of λ often vary by orders of magnitude from one part of the life cycle to another. In age-classified models, sensitivities to changes in age-specific fertility and MORTALITY decline with age as long as $\lambda \geq 1$.

The elasticity, or proportional sensitivity, measures the proportional change in λ caused by a proportional change in a_{ij}; it is given by:

$$e_{ij} = \frac{a_{ij}}{\lambda}\frac{\partial\lambda}{\partial a_{ij}}$$

The elasticities of λ sum to 1, and can be interpreted as the proportional contribution of each of the matrix entries to λ. Elasticities can also be calculated for other variables that affect the a_{ij}, as:

$$\frac{x}{\lambda}\frac{\partial\lambda}{\partial x}$$

These elasticities do not, in general, sum to 1. The application of elasticities to CONSERVATION BIOLOGY, and the elucidation of intra- and interspecific patterns of elasticities are areas of intense current research activity. *See also* POPULATION DYNAMICS. [H.C.]

REFERENCE

Caswell, H. (1989) *Matrix Population Models: Construction, Analysis and Interpretation.* Sinauer Associates, Sunderland, Mass.

sentinel species Species that are used as indicators of environmental conditions. Such species are usually immobile, and include attached algae, lichens, angiosperms and sedentary/sessile animals such as mussels and enclosed (caged) animals. Various attributes can be used as indicators: for example, uptake of a particular substance; biochemical, cellular and physiological responses; population responses. *See also* BIOMARKERS; BIOMONITOR; RETROSPECTIVE ANALYSIS (MONITORING). [P.C.]

SEPA (Scottish Environmental Protection Agency) *See* ENVIRONMENTAL (PROTECTION) AGENCIES.

sequential comparison index *See* DIVERSITY INDICES.

sequential hermaphroditism (sex change) The reversal of sexual modes during the life of an individual organism. Protandrous HERMAPHRODITES change from male to female mode, and protogynous hermaphrodites from female to male. Sex change is genetically programmed, but its timing may be sensitive to demographic factors, such as population SEX RATIO and survivorship. Sex change may be permanent, occurring once in the LIFE CYCLE, or reversible, occurring in each of several reproductive seasons.

Permanent sex change usually occurs at a threshold body size in animals that continue to grow as adults (INDETERMINATE GROWTH) and may reflect differential size-specific reproductive potential in male and female modes (*see* SIZE-ADVANTAGE HYPOTHESIS). Reversible sex change may be a mechanism for avoiding self-fertilization. *See also* SEX ALLOCATION THEORY. [R.N.H.]

REFERENCE

Charnov, E.L. (1982) *The Theory of Sex Allocation.* Princeton University Press, Princeton, N.J.

sere A unit of SUCCESSION in the terminology of Clements (1916; *see* CHARACTERS IN ECOLOGY); i.e. a kind of succession characteristic of a particular substratum and macroclimate. The term 'successional sere' is sometimes used, but is a tautology. Clements named various types of sere, but the only name to persist commonly in the literature has been the HYDROSERE, the succession from open water via FEN to FOREST or BOG. The term seral stage refers to a structurally or floristically distinct stage in a succession. [P.J.G.]

REFERENCE

Clements, F.E. (1916) *Plant Succession.* Carnegie Institution of Washington, Publication 242.

serotiny *See* CHAPARRAL SHRUBLANDS.

Serozems *See* SOIL CLASSIFICATION.

serpentine soil Serpentine (serpentinite) is an ultrabasic rock dominated by the magnesium silicate minerals, antigorite and chrysotile. These rocks usually contain large amounts of NICKEL, which can be toxic to both plants and animals. The soils vary from deep, highly weathered soils of the tropics to shallow soils occurring mainly in the cool-temperate areas. The latter have a dark topsoil with neutral to weakly ACID pH values, and the exchangeable cations are dominated by magnesium. These properties give rise to distinctive plant communities that are similar to those found on seashores. In humid tropics the soils carry a poor SCRUB forest, due largely to nickel TOXICITY. [E.A.F.]

Sesquisols *See* SOIL CLASSIFICATION.

sessile Describing an organism that is predominantly SEDENTARY but which is also characterized by a mobile DISPERSAL phase, as in many plants and marine organisms such as coral polyps. [R.C.]

seston Particulate matter suspended in seawater. [V.F.]

set-aside Land taken out of AGRICULTURE, usually by financial inducement (e.g. subsidies) to prevent overproduction. Such land may be allowed to return to a natural state. *See also* FARMING PRACTICES, SUSTAINABLE. [P.C.]

settlement A stage in the development of planktonic LARVAE during which they leave the water column to reach a suitable substratum. In order to achieve successful settlement larvae must first develop successfully, maintain their position nearshore, and select an appropriate substratum. Many larvae can temporarily delay settling if an appropriate substratum is not available.

Cues that elicit settling and subsequent METAMORPHOSIS include various physical characteristics of the substratum, the presence of adults of the same species or other predictably co-occurring species, or the presence of various BIOFILMS. Gregarious settling refers to the settlement of larvae that have been attracted to members of their own species. It usually requires direct contact between larvae and adults. The advantage of such behaviour is that a planktonic larva is able to find a MICROHABITAT that is suitable for supporting its species. However, disadvantages include the risk of being eaten by CONSPECIFIC adults. [V.F.]

Seveso *See* ECOLOGICAL/ENVIRONMENTAL DISASTERS.

Seveso Directive A legal instrument of the European Union (Directive 85/201/EEC as amended by 87/216/EEC and 88/610/EEC; substantially amended in Directive 96/82/EC) on major ACCIDENTS attributable to certain industrial activities. Its common name is after the major accident at Seveso, Italy, in 1976, which prompted the legislation. It requires manufacturers using certain dangerous substances above specified THRESHOLD quantities to limit any possible adverse consequences for humans and the environment and to report major accidents. Specifically, manufacturers have to produce: a safety report and on-site emergency plans; a COMPETENT AUTHORITY must produce an official emergency plan; the public must be informed of safety measures and of the correct behaviour in the event of accident. [P.C.]

sewage A dilute mixture of domestic WASTE, industrial (trade) waste, infiltration from the subsoil, and in a greater or lesser extent, RUN-OFF of surface water (Haworth 1984). Its composition is extremely complex and varies considerably depending on the amount of water used per head of population, the degree of infiltration of subsoil water and the nature and proportion of trade EFFLUENTS present. [A.D.]

REFERENCE

Haworth, W.D. (ed.) (1984) *Sewerage and Sewage Treatment, International Practice*. John Wiley & Sons, Chichester.

sewage fungus A visible, plumose slimy growth of heterotrophic organisms found in organically polluted waters. The organisms are usually bacteria and/or protozoans but sometimes microfungi are present. [P.C.]

sewage pollution Pollution from human excretion and defecation. Also, POLLUTION from SEWAGE TREATMENT WORKS, which may be of domestic and/or industrial origin. [P.C.]

sewage sludge Sludge resulting from treatment of raw SEWAGE. *See also* SEWAGE TREATMENT WORKS. [P.C.]

Sewage treatment works

A place that treats WASTE water (containing SEWAGE) from domestic and industrial sources. It is essentially concerned with the reduction of ORGANIC LOADING that would arise if raw sewage were allowed directly into watercourses. The main objective of sewage treatment is to reduce the strength (concentration of pollutants) of sewage to a sufficient degree to be safely discharged to natural waters without causing a nuisance or offence. For practical purposes, the impurities can be divided into three groups: (i) substances suspended in the liquid; (ii) substances dissolved in the liquid; and (iii) extremely fine, colloidal substances. There are several stages in the treatment process.

Primary treatment

This involves physical screening to remove large SUSPENDED SOLIDS, and SEDIMENTATION to allow as much suspended organic solids as possible to settle.

Secondary treatment

This involves biological oxidation, either in percolating filters or in an ACTIVATED-SLUDGE PROCESS. This is usually followed by sedimentation, and most treatment stops at this point.

Tertiary treatment

This involves microstraining with special filters, and possibly reverse OSMOSIS. It may also employ other systems designed to facilitate physical settlement.

However, the type of treatment which is going to be used for sewage purification will depend on the quality objectives for the receiving water and the amount of dilution which is available for that water. [P.C. & A.D.]

REFERENCE

Haworth, W.D. (ed.) (1984) *Sewerage and Sewage Treatment, International Practice*. John Wiley & Sons, Chichester.

---

sex allocation theory The theory that deals with the apportionment of RESOURCES by sexually reproducing parents to male and female offspring or to offspring derived from male and female GAMETES. Sex allocation includes the direct products that contribute to the next generation (i.e. gametes and offspring) as well as expenditure on accessory activities that do not contribute directly to the next generation but which may be substantial (e.g. attraction organs).

The assumption underlying sex allocation theory is that selection determines the allocation of the various structures in proportions such that the FITNESS of individuals cannot be exceeded and is evolutionarily stable. Most sex allocation studies have considered the SEX RATIOS of dioecious populations, but the theory can also be applied to other types of breeding systems. *See also* FISHER'S SEX-RATIO THEORY. [V.F.]

REFERENCE

Michod, R.E. & Levin, B.R. (eds) (1988) *The Evolution of Sex—An Examination of Current Ideas*. Sinauer Associates, Sunderland, Mass.

sex determination *See* HAPLO-DIPLOIDY.

sex, evolution of *See* EVOLUTION OF SEX.

sex ratios Primary sex ratio refers to the relative numbers of males and females at conception. Secondary sex ratio refers to the relative numbers of males and females at sexual maturity. Both terms are applied to organisms, usually animals, with separate sexes (i.e. nonhermaphrodites).

Fisher's theorem states that in a population with an unequal secondary sex ratio, the rarer sex will, in proportion to its rarity, achieve a greater per capita genetic contribution to future generations. Genes promoting that sex will be favoured by NATURAL SELECTION in a frequency-dependent manner. If males and females cost the same amount of energy to produce, FREQUENCY-DEPENDENT SELECTION will drive the secondary sex ratio towards unity. Unequal production costs will interact with frequency-dependent selection, causing the equilibrial sex ratio to be biased in favour of the cheaper sex. If one sex is characteristically less likely to survive to maturity than the other, there may be a compensatory bias in the primary sex ratio, leading to equality of the secondary sex ratio. For example, human male fetuses are less likely to survive than female ones. Correspondingly, sex ratio is biased in favour of males early in ONTOGENY, falling towards unity at sexual maturity.

Ecological factors may influence secondary sex ratio. When local scarcity of food limits body growth, it may be more advantageous to produce males if, at a relatively small body size, they have greater reproductive potential than females. For example, the secondary sex ratio of the endoparasitic nematode, *Romanomeris culicivorax*, is biased towards males when the mosquito host is heavily infected. Here, sex is determined environmentally, malnourished individuals tending to develop into males. Female hymenopterans may adjust sex ratio among their offspring by controlling the proportion of eggs fertilized by stored sperm. Unfertilized eggs are male (*see* PARTHENOGENESIS) and, in certain parasitic wasps, FERTILIZATION is withheld in order to oviposit male offspring in relatively small hosts.

When animals breed in small, self-contained groups, sibling COMPETITION for food (local resource competition) or mates (local mate competition, LMC) may bias sex ratio among a female's offspring in favour of the sex exerting the lesser competitive pressure. In the case of LMC, sex ratio usually favours daughters, since promiscuous sons, being able to fertilize more than one female, compete strongly for mates. *See also* FISHER'S SEX-RATIO THEORY. [R.N.H.]

REFERENCE

Charnov, E.L. (1982) *The Theory of Sex Allocation*. Princeton University Press, Princeton, N.J.

Sex role reversal

The reversal of characteristically male and female secondary sexual traits, in which females are larger, more elaborately armed or ornamented, and more aggressive than males. It occurs in a small percentage of bird, mammal, fish and insect species. Examples include many sea horses and pipe fish, spotted hyenas, and most species of jacanas. In every known case of sex role reversal, it has been found that males expend substantial time and energy caring for the young, and the phenomenon of role reversal has thus been important to the development of theories linking PARENTAL INVESTMENT and SEXUAL SELECTION.

According to the theory developed by Bateman (1948), Trivers (1972), and Clutton-Brock and Vincent (1991), PARENTAL CARE for the young and other forms of parental expenditure such as the production of eggs together determine the maximum rate of reproduction possible for each sex of a species, termed the 'potential reproductive rate'. This rate then determines the ratio of males to females seeking a mating in a population at a given time (the operational sex ratio, OSR), with the slower sex generally in short supply. The OSR determines which of the two sexes will compete among themselves for access to the scarcer sex. Thus, for example, in most mammals, females care for the young and so reproduce much more slowly than males. Because of this difference, more males than females tend to be seeking mates in a population at a given time, so males compete for access to females. Because male–male COMPETITION is a key determinant of FITNESS, males have evolved larger bodies, armaments such as horns, antlers or canines, and higher levels of aggressiveness.

Role-reversed species provide the critical test of this theory. In particular, where females produce costly eggs but males care for the young, the potential rates of reproduction of the two sexes are often similar. Clutton-Brock and Vincent (1991) demonstrated that in these cases males tend to be larger and more heavily armed and aggressive where they can reproduce a bit faster than females, whereas females tend to be larger and more armed or aggressive where they reproduce a bit more quickly. In species such as Majorcan midwife toads and bush crickets, where the relative rate of reproduction is influenced by temperature because of changes in egg maturation rate, the relative aggressiveness of the two sexes may change through the season as the operational sex ratio shifts.

The link between sex differences in ornamentation (e.g. peacocks' trains) and parental care has been less thoroughly investigated, but appears to be similar. The cost to each of the two sexes of choosing a mating partner rather than mating indiscriminantly depends on the time it takes to replace a rejected mate, which is strongly influenced by the operational sex ratio. Thus, where the OSR is male biased, females lose little by rejecting some potential partners whereas the cost to males of rejecting some partners is a proportional reduction in fitness. Thus females tend to be choosier than males, and males are more likely to evolve DISPLAYS and ornaments associated with attracting mates. On the other hand, where the OSR is female biased, it is males that can afford to be choosy and we expect to see greater elaboration of sexual ornaments in females. *See also* SEXUAL DIMORPHISM. [J.C.D.]

REFERENCES

Bateman, A.J. (1948) Intra-sexual selection in Drosophila. *Heredity* **2**, 349–368.

Clutton-Brock, T.H. & Vincent, A.C.J. (1991) Sexual selection and potential reproductive rates of males and females. *Nature* **351**, 58–60.

Trivers, R.L. (1972) Parental investment and sexual selection. In: *Sexual Selection and the Descent of Man* (ed. B. Campbell), pp. 136–179. Aldine Press, Chicago, Ill.

Sexual conflict

Sexual conflict is said to exist if a TRAIT that increases the reproductive success of an individual of one sex can only do so at the expense of the reproductive success of an individual of the other sex (for discussion see Trivers 1985). This is best illustrated by example.

Conflict and parental care

Consider a species with PARENTAL CARE of the progeny. Consider also that the progeny will have a relatively high FITNESS if both mother and father stay behind to help raise the brood. Consider then also a 'fly-by-night' male that leaves the mother holding the infants. This male may now go out and mate with more females. As the progeny of such males have only one parent providing them with RESOURCES, on average the fitness of any given progeny of this male is lower than that of competing males that stay and feed the brood. However, the 'fly-by-night' male will have a higher number of progeny than competing males. If this extra number of progeny more than compensates for the decreased fitness of any given progeny, then the male's trait should spread in the population. This need not maximize the reproductive success of the females that are mated by such males as these females will probably have fewer total progeny (they may well have fewer resources to give to future broods), and the fitness of the progeny that they raise will be lower than that of females that do not mate with fly-by-night males. So, in sum, what is good for a male (to be a fly-by-night male) is not good for a female. Conversely, what is good for the mother (not to mate with a fly-by-night male) is not good for the fly-by-night male. There is a difference, and hence a conflict, of interests.

Conflict versus competition

It is important to distinguish COMPETITION and conflict (*see* CONFLICTS): the new male trait (the fly-by-night trait) is in competition with alternative traits that males may hold (e.g. stay and look after the progeny), but they are not in conflict. If, however, as a consequence of competition (direct or indirect) the reproductive success of the female is reduced, then the male and female are in conflict. Note, the female strategies are not alternatives to the male strategies, hence they are not in competition.

The ultracompetitive female

The ultracompetitive individual need not be the male (as above). Consider, for example, a species in which individuals pair up. Then consider a female that has paired up with a weak (low-ranking) male. It is in the female's interest to find a new mate. Hence there is a conflict between male and female over whether the pairing should be broken.

Pre- and postzygotic conflict

Sexual conflict as illustrated above can either be pre- or postzygotic. If prezygotic, the conflict is over who to mate with, how many eggs to produce, whose sperm should be allowed to fertilize the eggs, etc. If post-reproduction, the conflict is typically over resource allocation to give to progeny. Resource allocation here is used in its broadest sense (i.e. INFANTICIDE may be understood as a form of resource allocation strategy). In general, both parents want the progeny to be fed but neither parent wants to do the feeding, so there exists a conflict between the parents over the relative amount of provisioning.

Multiple paternity

In species in which there is multiple PATERNITY of the offspring of any given mother these types of conflict are particularly intense. Consider, for example, the problem of a father who has sired some but not all the progeny of a given female. It is typically the case that the mother's best interests are to ensure that all her progeny are equally provisioned (or more or less so). In contrast, the father has no interest in the survival of progeny that are not his own (assuming they are not the offspring of his brothers or other relatives). So a male's best interests are both not to feed (or to kill) those brood members that are not his, and to put all his feeding effort into his own progeny (assuming that he is involved in parental care). The killing of progeny may also force the female into redirecting resources into reproduction with the now-resident male.

The fact that in humans the murder of children is done much more often by stepfathers than natural fathers has been interpreted in this light. The same is true of numerous instances of infanticide (e.g. in lions, *Panthera leo*). In many birds (e.g. dunnocks, *Prunella modularis*), however, it seems that males feed all members of the brood, regardless of paternity. Significantly, however, the amount of provisioning is dependent upon the time exposure the male had with the female at around the time of copulation (this is thought to be an indication of the probability that any given offspring was fathered by that male). The fact that the birds do not recognize KIN may itself be an ADAPTATION. If a progeny has a chance of being killed if not related to a given father, then it is in the interests of all parties that the offspring does not reveal its identity (Davies 1992).

Genomic imprinting

The conflicts so far discussed have all involved the actions of males on females, of females on males, or by parents on the progeny. However, even if a male disperses after reproduction a male–female conflict can still go on. Just this sort of conflict is thought to be the explanation for GENOMIC IMPRINTING. Classical genetics supposes that the action of a gene is not dependent upon whether it was inherited from the mother or the father. Imprinted genes contradict this supposition. In mammals and angiosperms some genes switch on only if inherited from the mother. Others switch on only if inherited from the father. Why should this be so?

Fetal provisioning

If a mother has more than one mate, paternally derived genes in any given fetus will not necessarily be related to genes in fellow brood members or subsequent offspring of that same mother. In contrast, a maternally derived gene in any given fetus has a constant high probability (0.5) that its sibs will contain a clonal copy of it. This tripartite asymmetry in relatedness between (i) paternally derived fetal genes, (ii) maternally derived fetal genes and (iii) genes in the mother (and hence in the fetus' sibs), creates a three-way conflict of interest over how much nutrition the fetus should demand from the mother. The mother, as before, would 'prefer' to divide resources more-or-less equitably and so maximize her net fitness, not the fitness of any given progeny. In contrast, if the paternally derived genes in any given fetus are not going to be present in the sibs, then any decrement in maternal fitness resulting in reduced offspring production is, at the very least, irrelevant to these paternally derived genes. Paternally derived genes in a given fetus might thus prefer an allocation in excess of that preferred by the mother. The optimal amount of resources that the maternally derived genes in the fetus should require will be intermediate between the optima for the paternally derived genes (a large amount) and the amount the mother should be prepared to provide (a smaller amount).

It is this difference in optima between the maternally and paternally derived genes in a fetus that is proposed to underlie genomic imprinting. In short, the difference between the father's and the mother's interests in the provisioning of offspring can be exercised by a tussle between the mother's genes and the father's genes in the fetus.

Mammalian examples

Imprinting of murine insulin-like growth factor 2 (IGF-2) and the IGF-2 receptor (alias the mannose-6-phosphate receptor) is the now classic example. Fetally expressed IGF-2 is one of the factors that promote the acquisition of resources from the mother. Early in mammalian embryogenesis, as expected, there is expression of paternally inherited genes that promote the production of IGF-2, while maternally inherited genes that inhibit these growth effects are expressed (Moore & Haig 1991; but see Hurst & McVean 1997). The same logic has been employed to suggest why the mammalian Y chromosome codes for a variety of growth factors (Hurst 1994). The Y chromosome is always paternally derived and hence in a species with promiscuous females will have different optimal allocation of resources than maternally derived genes.

For a system such as imprinting to operate:

1 paternal provisioning of offspring must be sex biased;
2 resource allocation per offspring must be flexible;
3 offspring must be able to manipulate the paternal resource allocation.

Imprinting of the above variety should thus be expected in those groups with intimate maternal/offspring connections. Eutherian mammals are one such group (for which all the best imprinted genes have been described). Most exchange of resources is done across the placenta and it is thus significant that imprinting affects many genes with placental expression. Genomic imprinting is hence not expected in embryonic birds, for example, as MANIPULATION of the mother by the progeny is not possible.

In general, and as with other systems of conflict, if one asks which sexual conflicts are made actual, rather than simply existing as a potential conflict, the answer comes down to details of biology. Most particularly the biology of the interactions between the partners and the ability of one party to manipulate the other. [L.D.H.]

REFERENCES

Davies, N.B. (1992) *Dunnock Behaviour and Social Evolution*. Oxford University Press, Oxford.

Hurst, L.D. (1994) Embryonic growth and the evolution of the mammalian Y chromosome. I. The Y as an attractor for selfish growth factors. *Heredity* **73**, 223–232.

Hurst, L.D. & McVean, G.T. (1997) Growth effects of uniparental disomies and the conflict theory of genomic imprinting. *Trends in Genetics* **13**, 436–443.

Moore, T. & Haig, D. (1991) Genomic imprinting in mammalian development: a parental tug-of-war. *Trends in Genetics* **7**, 45–49.

Trivers, R.L. (1985) *Social Evolution*. Benjamin/Cummings, Menlo Park, Calif.

---

sexual dimorphism Males and females of sexual species differ in the organs required for reproducing themselves which are known as primary sexual characters. They also differ in many sexual characteristics that are not obviously necessary for reproduction *per se*; these are called secondary sexual characters. For example, male deer have antlers, while the females usually do not. If a female has mating access only to a male without antlers, she could just as effectively produce offspring with him as she could with a male bearing antlers. In fact, antlers are used in COMPETITION for mates.

During a given breeding season, a female deer needs to be inseminated only once to produce her offspring, whereas a male's reproductive success is directly proportional to the number of sexually receptive females he mates with. Hence, male deer have evolved weapons that help them to win battles with other males over mating access to females. In many species in which females are found in groups during the mating season, males have evolved large SIZE or weaponry. In other such species, it seems that the females show active rather than passive choice, so that males have evolved elaborate display organs, such as the visually striking tail of the peacock.

Not all such apparent secondary sex differences have evolved to help in competition for mates. For example, in many insect species it seems that a female's reproductive success is often constrained by the number of EGGS she can produce, and that large-bodied females can produce more eggs, so that the body size of females is greater than that of males. Furthermore, if a male and female live together as a monogamous couple in a territory, then they may be able better to exploit limiting food RESOURCES if they differ in morphology; it has been suggested that differences in beak size and shape between the sexes in some species of woodpecker have evolved for this reason, with one sex being able to rip bark from trees and the other exploiting the insects which lie in the crevices beneath. *See also* SEX ROLE REVERSAL. [P.H.H.]

sexual displays *See* COURTSHIP; DISPLAYS.

Sexual reproductive cycles

Also called ALTERNATION OF GENERATIONS, sexual reproductive cycles refer to the extent of the vegetative phase between the two steps of eukaryotic sex, fusion and segregation. These two steps of sexual reproduction are called respectively: syngamy, which doubles the amount of DNA in a cell by GAMETE fusion; and meiosis, which halves it. Three basic LIFE CYCLES are encountered among eukaryotes: diploid, haploid and haplo-diploid (Fig. S13a). They differ by the relative position of syngamy and meiosis and consequently by the relative duration of the haploid and diploid phases.

1 When syngamy directly follows meiosis, individuals are diploids and the life cycle is called a diploid (or diplontic) life cycle.
2 When meiosis directly follows syngamy, individuals are haploids and the life cycle is called a haploid (or haplontic) life cycle.
3 When meiosis and syngamy are separated in time and space, both haploid and diploid individuals occur within the species and the life cycle is called a haplo-diploid (or haplo-diplontic) life cycle.

In addition to the difference listed above, these life cycles differ also by the way gametes are produced: by meiosis in the diploid life cycle and by mitosis in the two other cycles. Moreover, the first two cycles are monophasic while the haplo-diploid is diphasic. Thus, the characteristics of the life cycles shape the mating system of a species and consequently its evolutionary potential. However, this definition of three life cycles is ideal. As soon as one looks at real organisms

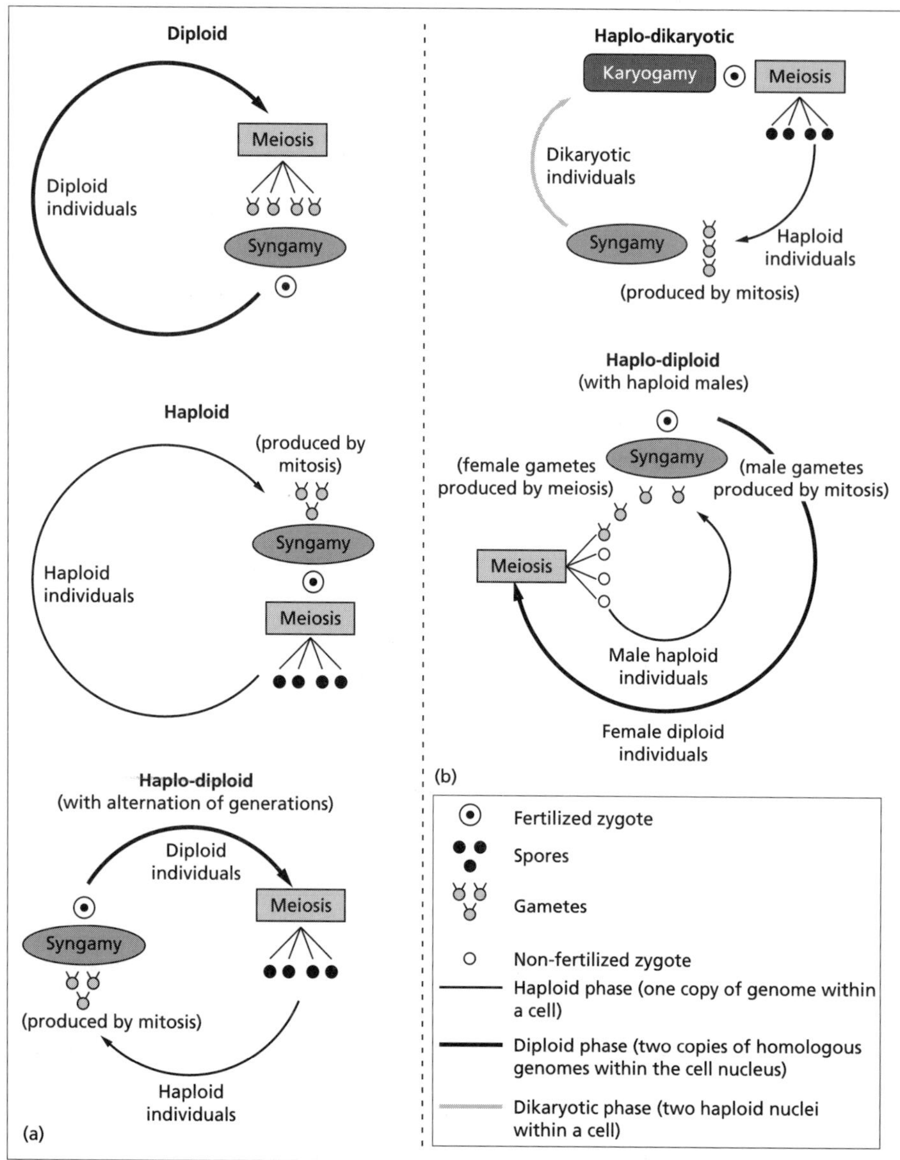

Fig. S13 The principal life cycles. (a) The three basic life cycles encountered among eukaryotes: diploid, haploid and haplo-diploid with alternation of generations. (b) Two examples of particular life cycles found in more restrictive taxonomic groups: haplo-dikaryotic life cycle and haplo-diploid life cycle with haploid males (arrhenotoky). Syngamy refers to gamete fusion and karyogamy to nucleus fusion. In most cases (except in haplo-dikaryotic life cycles), syngamy is immediately followed by karyogamy: as a result, a fertilized zygote is produced. In the haplo-diploid life cycle with haploid males a non-fertilized zygote is produced. This zygote is haploid, in contrast to the fertilized zygote which is diploid.

and their diversity of organization, understanding of the 'length' of a phase becomes rather subjective. Numerous criteria can be used: the number of mitoses; the size and biomass of the individual; the fact that an individual is free-living or growing in one phase, showing a nutritional and energetic autonomy or dependence from the other phase. The problem is easier to solve for unicellular life cycles because the unique criterion of extent of the vegetative phase is the number of mitoses occurring during this phase. For example, in *Chlamydomonas reinhardtii* the life cycle is haploid because mitoses occur during the haploid phase. Mitoses may exceptionally occur during the diploid phase but produce cells incapable of dividing meiotically, which is a dead end for the sexual life cycle. The distribution of these three basic life cycles among taxa shows that ploidy is positively correlated with size and structural complexity. This trend is found at a large scale (among phyla). The most complex organisms are diploid (higher plants and multicellular animals), while many simple organisms are haplo-diploid or haploid (many unicellular protists). It is even more clear in plants, where a decrease of the haploid part of the life cycle is observed from Bryophyta to Pteridophyta and finally to higher plants. However, some taxa like fungi and algae show a great variety of life cycles that are often very complex and in which clear EVOLUTIONARY TRENDS are difficult to ascertain. The alternation of generations was described as a general feature within these last two groups for the first time in 1851. It was only in 1894 that the occurrence of an underlying alternation of nuclear ploidy was discerned. Botanists of this period noted the correlation between habitat and life cycle: most terrestrial plants have a well-developed sporophyte (diploid generation), while in aquatic environments the gametophytic (haploid) generation is often dominant. We still do not know precisely what determines evolution towards one of the three basic life cycles. Because DIPLOIDY is often considered a 'biological success', many arguments have been advanced to explain the evolution of a prolonged diploid phase. The question of the maintenance of the other kinds of life cycle is still open. Concerning the origin of the alternation of haploid and diploid phases, we are not able to determine if it is the cause, the consequence or an associated phenomenon with eukaryotic sex.

Many more complex cycles that do not fit these three basic schemes exist in eukaryotes. It is not possible to mention all of them, but two other life cycles are given in Fig. S13b as examples. Haplo-dikaryotic life cycles are found among higher fungi (Basidiomycota and Ascomycota). Haplo-diploid life cycles with haploid males occur in one out of four arthropod species and in some rotifers. There is a large contrast among taxa concerning their variability of life cycles. Some phyla are quite homogeneous (like the kingdom Animalia); others show a wide variability (like the kingdom Protoctista). In this last kingdom, and red algae in particular, cycle variability can be found at low taxonomic levels, within genera or species. Moreover, variation for life cycles among populations of the same species and even within populations has been observed (e.g. diploid and haplo-diploid life cycles coexist in the red alga *Gracilaria gracilis*). In such species showing variation life cycles can still evolve depending on selective pressures, while in taxa where a life cycle is fixed its evolution depends on phylogenetic CONSTRAINTS. *See also* ARRHENOTOKY. [M.V.]

Sexual selection

Charles Darwin coined the term sexual selection to explain the EVOLUTION of characters, usually possessed by one sex, that appear to be irrelevant or even disadvantageous to the survival prospects of the bearer. Darwin's view was that such characters—for instance, the long, gaudy tails of male birds of paradise or the large antlers of red deer stags—evolved to enhance the mating success of their possessors, the resulting reproductive benefit being sufficient to offset the survival cost of the character.

Sexual selection is thus sometimes regarded as a fundamentally different mechanism of evolution from classical NATURAL SELECTION. However, since natural selection favours strategies of survival only to the extent that they translate into reproductive success, the contrast is merely in the life-history component favoured rather than the mechanism of selection itself.

Competition among males

Sexual selection arises for two reasons:

1 because anisogamous sexual reproduction results in one sex, usually females, investing more heavily in each GAMETE than the other, usually males;

2 because natural selection tends to favour a 1 : 1 ratio of investment in the two sexes within populations, usually leading to COMPETITION among males for access to females.

As a result, sexual selection favours TRAITS that enhance the ability of males either to compete with each other for fertilizations (INTRASEXUAL SELECTION) or to attract the attention of females (intersexual or epigamic selection). In fact the distinction between intra- and INTERSEXUAL SELECTION is redundant since all sexual selection is intrasexual as it occurs among members of the same sex. Since females generally provide a limited resource for which males compete, variance in mating success tends to be greater among males than among females. Where the relative investment of the sexes at the gamete stage is reversed, for instance by males investing expensive protective or nutritional material in their ejaculates, intra- and intersexual selection work in the same way on female traits. Males (or females) may be subject to both forms of sexual selection at the same time.

Intrasexual selection may result in the evolution of characters, such as large size and weaponry, that enhance superiority in aggressive disputes. Alternatively, it may produce traits similar to those evolving under intersexual selection: long, gaudy tails, for instance, may act as status badges to settle disputes between males rather than as decorations to attract females. Intrasexual selection can also continue beyond mating. Where females mate with more than one male, the sperm of rival males may end up competing for fertilizations. SPERM COMPETITION appears to have been responsible for the evolution of a wide range of behavioural, morphological and physiological adaptations among males, from MATE-GUARDING and multiple copulation to vaginal plugs, penis scoops and sperm POLYMORPHISM.

Evolutionary consequences

The evolutionary consequences of intersexual selection are more controversial, partly because of debates about female choice as a cause of non-random mating and partly because it is difficult to test different functional explanations for the evolution of exaggerated secondary sexual characters.

Since each egg is likely to represent a greater proportion of a female's lifetime gametic output than a sperm for a male, females are generally expected to be choosier about their partners. A persuasion/resistance ARMS RACE is thus likely to evolve between the sexes. Female preferences should reflect qualities of the male that are likely to compensate for his initial lack of investment in the zygote. These may be based on evidence of actual or potential paternal investment in offspring (e.g. resources in a male's territory, the quality of nuptial gifts, tolerating a long period of COURTSHIP), or on traits suggesting physical quality that could be passed on to offspring.

'Handicap' hypothesis

Physical quality may be indicated directly by, say, body size, possession of a territory, or age. Alternatively, it may be advertised by so-called 'handicaps', traits that encumber a male in some way and thus demonstrate his ability to survive and compete despite them. To demonstrate quality, handicaps should reflect the limiting SELECTION PRESSURE(S) acting in the male's environment. Thus calcium-squandering antlers might indicate an ability to compete for limited NUTRIENTS, while conspicuous coloration implies survival against the odds in the face of predation. Handicaps may therefore reveal attributes of males not otherwise discernible to the female. In particular, bright colours and intricate patterns might help betray infection with parasites thereby advertising the male's resistance to disease. Furthermore, coevolutionary cycles of host–parasite associations could provide a mechanism for maintaining heritable FITNESS variation in such traits in the face of consistent female choice for bright and gaudy males.

A problem with the handicap hypothesis, however, is that at least two simpler processes could have the same outcome in terms of female preferences and male traits. One, attributable originally to the mathematical geneticist R.A. Fisher, shows that a completely arbitrary preference among females for some male character can result in a runaway elaboration of the character through covariation between the ALLELES coding for the preference and the trait (*see* SEXY SONS HYPOTHESIS). The runaway process continues until the mating advantage of further elaboration is offset by its survival costs. A more mundane explanation for females apparently preferring showier males is that such males are simply more noticeable. Here, non-random mating by females is not due to a preference at all but to passive attraction to whatever stands out from the crowd. Distinguishing between these various selection pressures remains difficult. [C.J.B.]

REFERENCE

Trends in Ecology and Evolution (1996) Vol. 11, No. 2. [Special issue on the ecology and evolution of sexual behaviour.]

---

sexy sons hypothesis The 'sexy sons' hypothesis is a model of INTERSEXUAL SELECTION (*see* SEXUAL SELECTION), attributable to R.A. Fisher, in which female choice drives the EVOLUTION of exaggerated attractive TRAITS in males. What distinguishes it from other explanations for exaggerated traits, such as the 'handicap' hypothesis, is that 'sexy sons' traits may be entirely arbitrary and not relate to any advantageous physical quality of the male.

In Fisher's theory, females initially prefer a male trait, say a longer than average tail (it is important that the preference is relative rather than absolute), because it confers some advantage on the male, say greater agility in flight. However, the initial preference can just as easily arise because the trait is easier to detect or happens to tune in to an arbitrary sensory predisposition among females. The critical point is that the female's offspring, daughters as well as sons, inherit both the ALLELE for the preferred male trait and the allele for their mother's preference, though the trait is expressed in a sex-limited way only in sons and the preference only in daughters.

This combination of COVARIANCE between trait and preference, and the selective advantage of producing attractive (sexy) sons, sets in train a powerful positive FEEDBACK effect leading to a runaway elaboration of the trait. Thus preference for a longer than average tail leads, over successive generations, to an increase in tail length within the population that rapidly outstrips any initial functional advantage of a longer tail. Indeed, the tail may become so long that it is a positive disadvantage for survival and therefore evolves purely through the power of female choice. Eventually, however, the attractiveness advantage is offset by the escalating survival cost and the process comes to a halt. A number of factors, such as the relative strengths of sexual vs. NATURAL SELECTION and the cost of choosing females will determine exactly where the opposing SELECTION PRESSURES cancel out. [C.J.B.]

SFG *See* SCOPE FOR GROWTH.

Shannon diversity index *See* DIVERSITY INDICES.

Shannon–Wiener index *See* DIVERSITY INDICES.

shear stress *See* CRITICAL SHEAR STRESS.

shelf-slope break A line demarcating a change from the gentle CONTINENTAL SHELF to the much steeper DEPTH GRADIENT of the CONTINENTAL SLOPE. [V.F.]

Shelford, V.E. *See* CHARACTERS IN ECOLOGY.

Shelford's law of tolerance An axiom, formulated by V.E. Shelford in 1913 (*see* CHARACTERS IN ECOLOGY), stating that the presence and success of an organism in a place depends upon the completeness of a complex of conditions. Alternatively, the absence or failure of an organism can be controlled by deficiency or excess with respect to any one of several factors. It reflects the fact that organisms differ in their ranges of tolerance for environmental factors. *See also* EURY-; LIEBIG'S LAW OF THE MINIMUM; STENO-. [P.C.]

shelterwood system The practice in FORESTRY of HARVESTING maturing trees on two or more occasions to allow natural regeneration of seedlings from the remaining standing trees. The shelterwood system allows seedlings to become established before the CANOPY is completely removed so providing shelter as well as suppressing competing weeds. In poorly tended FORESTS, inferior trees may be eliminated so that seedling regeneration comes from the better parents. In ideal conditions the last of the mature trees are harvested when the seedling/sapling population has been well established over several years. Where natural regeneration is patchy, seedlings from external sources may be introduced.

Shelterwood systems have been used extensively in beech (*Fagus sylvatica*) forests and in Norway spruce (*Picea abies*) plantations, particularly in southern Germany, Switzerland, Austria and the former Czechoslovakia. With the availability of government grants in the UK to support natural regeneration of woodlands, the shelterwood system may become more widespread in Britain. [G.A.F.H.].

Sherwood number A non-dimensional quantity, commonly used in environmental engineering, that expresses the rate of mass transfer to or from an object. [J.G.]

REFERENCE

Monteith, J.L. & Unsworth, M.H. (1990) *Principles of Environmental Physics*. Arnold, London.

shieling *See* ALPINE ZONE.

shifting balance theory *See* DRIFT, EVOLUTIONARY.

shifting cultivation The practice of bringing into AGRICULTURE previously uncultivated land for several seasons followed by abandonment as part of a human nomadic culture. It is increasingly recognized that the traditional shifting cultivation practised by indigenous peoples, particularly in the tropics, represents a sustainable form of agriculture that is well adapted to natural and semi-natural ecosystems, most notably in RAIN FORESTS. Far from proving an inefficient or wasteful form of agriculture, as was once thought, shifting cultivation is now regarded by many observers as the least harmful to the environment and to the diversity of WILDLIFE. At the level of government and international financial institutions, however, management of forests in a sustainable way through shifting cultivation has yet to be recognized as an environmentally beneficial form of economic activity.

The reasons probably lie in the experiences from 3000 years of European farming practices. It has to be recognized that shifting agriculture as practised in western Europe did lead directly to the non-sustainable exploitation and ultimate destruction of the extensive forests of temperate western Europe, an event widespread in recent decades in North America. For this reason the shifting cultivation of the Amazonian Indians has been looked upon, for much of this century, as inefficient and of irreparable harm to the FOREST fauna and flora. This harm, however, is as nothing compared to the effects brought about by horizon-to-horizon destruction of the same TROPICAL RAIN FORESTS by logging and conversion to grassland for cattle raising, in the past 15–20 years. The switch from sustainable shifting cultivation to contemporary forms of land exploitation has also led to displacement of millions of indigenous peoples from the formerly forested areas and their replacement by communities supporting a considerably smaller human population. *See also* FARMING PRACTICES, SUSTAINABLE. [G.A.F.H.]

Shingle beach

A BEACH made up wholly or in large measure of pebbles (SEDIMENT size 2–200 mm). The formation of such a beach is dependent on three sets of factors.

Source of shingle

Firstly, there must be a source of shingle, typically offshore and laid down in the past, although existing coastal shingle is often eroded and reworked to another part of the coast by a process called longshore drift. In either situation WAVES, WIND and TIDES act together to determine the direction of movement of the shingle and its deposition on the shoreline.

Ridge formation

Secondly, constructive waves throw the shingle up to form a small FORESHORE ridge (or a series of such ridges during the spring to neap tide sequence). Usually these small ridges are destroyed by storm-force waves, but such waves may on occasion produce a larger shingle ridge which becomes permanent.

Nature and aspect of coastline

Thirdly, the shape of the coastline, together with the strength and direction(s) of waves, determines the type of shingle beach formed. Fringing beaches, spits and BARS are the most common and simplest in structure, comprising usually one or more, rarely a few, ridges which tend to smooth the coast outline. Fringing beaches abut directly against the land, whilst spits and bars enclose part of the SEA, in the latter case as a landlocked lagoon which may be brackish rather than saline. APPOSITION BEACHES are much less common and are larger structures which comprise series of sometimes many parallel ridges, formed under conditions of restricted wave approach. Extreme development of this kind can lead to the formation of a cuspate foreland such as the example at Dungeness, on the south coast of England, comprising some five apposition beaches and about 500 ridges in total. Finally, BARRIER ISLANDS comprise large masses of shingle deposited offshore and more-or-less separated from the mainland. Typically, such structures comprise a complex of shingle, sand and SILT.

Colonization

Relatively few plants are able to colonize pure shingle, and then only if it is completely stable; examples inlcude *Crambe maritima*, *Lathyrus japonicus* and *Mertensia maritima*. Such species have to contend with the extreme porosity and very low water retention of the pebbles and need to rely on the mulching effect of shingle reducing evaporative losses. However, where matrix is present with the shingle, many more species can colonize, depending on the nature of the matrix, which can vary from sand to silt to organic in nature. Sand and silt could derive from adjacent SAND DUNES and SALT MARSH, respectively, whilst organic material could be STRANDLINE debris. [B.W.F.]

---

shoots *See* MODULE.

short-term tests TOXICITY/ECOTOXICITY tests carried out over periods of time that are short relative to the LIFESPAN of the organism. Thus a day may be a short span for toxicologists but not for microbes. *Cf.* CHRONIC TESTS; LONG-TERM TESTS. *See also* ACUTE TESTS. [P.C.]

shortwave reflectance Another term for ALBEDO. *See also* RADIATION. [J.G.]

shredders *See* FEEDING TYPES, CLASSIFICATION.

SI units (Système International d'Unités) A system of units for use in science and engineering, consisting of seven base quantities, two supplementary units, and many derived units. The base units are given in Table S3.

Table S3 The seven SI base quantities, with their corresponding units and symbols.

Physical quantity	Name of unit	Symbol
Length	metre	m
Mass	kilogram	kg
Time	second	s
Electric current	ampere	A
Thermodynamic temperature	kelvin	K
Luminous intensity	candela	cd
Amount of substance	mole	mol

Table S4 Some SI derived quantities and their units.

Derived quantity	Unit name	Unit symbol	Base units
Area	square metre		m^2
Volume	cubic metre		m^3
Density	kilogram per cubic metre		$kg\ m^{-3}$
Frequency	hertz	Hz	s^{-1}
Concentration	mole per cubic metre		$mol\ m^{-3}$
Velocity	metre per second		$m\ s^{-1}$
Acceleration	metre per second per second		$m\ s^{-2}$
Force	newton	N	$m\ kg\ s^{-2}$
Pressure, stress	pascal	Pa	$N\ m^{-2}$
Viscosity (dynamic)	pascal second		Pa s
Viscosity (kinematic)	squared metre per second		$m^2\ s^{-1}$
Energy, work, heat	joule	J	N m
Power	watt	W	$J\ s^{-1}$
Electric charge	coulomb	C	A s
Potential difference	volt	V	$W\ A^{-1}$
Electrical resistance	ohm	Ω	$V\ A^{-1}$
Electrical conductance	siemens	S	Ω^{-1}
Capacitance	farad	F	$C\ V^{-1}$
Radioactivity	becquerel	Bq	s^{-1}
Irradiance, heat flux density	watt per square metre		$W\ m^{-2}$
Heat capacity, entropy	joule per kelvin		$J\ K^{-1}$
Thermal conductivity	watt per metre kelvin		$W\ m^{-1}\ K^{-1}$
Molar heat capacity	joule per mole kelvin		$J\ mol^{-1}\ K^{-1}$

The supplementary units are (i) the radian (symbol: rad; the plane angle between two radii of a circle which cut off on the circumference an arc equal in length to the radius); and (ii) the steradian (symbol: sr; the solid angle which, having its vertex at the centre of a sphere, cuts off an area of the surface of the sphere equal to that of a square with sides equal to the radius of the sphere).

There are many derived units. Some of the more useful are given in Table S4.

When quantities are very small or large in relation to the base unit, they may be indicated by prefixes, as given in Table S5. For example, $1\ Mg = 10^6\ g = 1\ 000\ 000\ g$. Prefixes that are not multiples of 3 (h, da, d, c) are usually not necessary and should be avoided.

Table S5 Prefixes used with SI units.

Factor	Name	Symbol
10^{18}	exa-	E
10^{15}	peta-	P
10^{12}	tera-	T
10^{9}	giga-	G
10^{6}	mega-	M
10^{3}	kilo-	k
10^{2}	hecto-	h
10	deca-	da
10^{-1}	deci-	d
10^{-2}	centi-	c
10^{-3}	milli-	m
10^{-6}	micro-	µ
10^{-9}	nano-	n
10^{-12}	pico-	p
10^{-15}	femto-	f
10^{-18}	atto-	a

Current practice is that SI units are in use in all journals, but that exceptions are sometimes allowed. Chemists and soil scientists persist with eq (equivalents); Gt, or billion tonnes, is often preferred to the SI unit petagram Pg; and the litre is commonly used instead of dm^3. In forestry, the hectare (ha) is still more widely understood than multiples of m^2. Other non-SI units that are commonly allowed are minute (min), hour (h), day (d) and degree (°). *See* UNIT for other non-SI units. [J.G.]

REFERENCES

Baron, D.N. (ed.) (1994) *Units, Symbols, and Abbreviations*. Royal Society of Medicine Press, London.
Drazil, J.V. (1983) *Quantities and Units of Measurement*. Mansell, London.
Incoll, L.D., Long, S.P. & Ashmore, M.R. (1977) SI units in publications in plant science. *Current Advances in Plant Science* **9**, 331–343.
Jerrard, H.G. & McNeill, D.B. (1986) *A Dictionary of Scientific Units*. Chapman & Hall, London.
Kaye, G.W.C. & Laby, T.H. (1986) *Tables of Physical and Chemical Constants*, 15th edn. Longman/Wiley, New York.
Royal Society of London (1975) *Quantities, Units and Symbols*, 2nd edn. Royal Society, London.

sib Short for SIBLING. [P.C.]

sib-competition hypothesis Sib-competition is a mechanism that helps to explain why sexual reproduction is maintained by NATURAL SELECTION, even though the requirement to produce male offspring makes sex less efficient than ASEXUAL REPRODUCTION. The hypothesis is concerned with the ecological role of sex, i.e. with the efficient exploitation of a diverse environment. It asserts that in a densely populated environment, where competition is severe, it pays to diversify.

DIVERSITY, in a heterogeneous environment, increases the number of individuals the environment can support by increasing the number of niches that can be exploited. The conditions that promote this advantage of diversity are:

1 the offspring of a single female compete with each other for survival in patches in which only a few organisms can survive and reproduce;

2 a GENOTYPE that is well adapted to one set of circumstances is less well adapted to others;

3 an environment that varies locally.

The latter two conditions are often combined as genotype by environment interactions. These conditions lead to FREQUENCY- and DENSITY-DEPENDENT SELECTION. As a consequence, a diverse range of heterogeneous organisms will outperform a mixture of CLONES, and sexual females can outcompete asexual ones. *See also* EVOLUTION OF SEX. [J.C.K.]

REFERENCES

Bell, G. (1982) *The Masterpiece of Nature. The Evolution and Genetics of Sexuality.* University of California Press, Berkeley, Calif.

Stearns, S.C. (1987) *The Evolution of Sex and its Consequences.* Birkhäuser, Basel.

sib selection A form of KIN SELECTION where a TRAIT is favoured because it confers an advantage on siblings. For example, the first MUTATION of warning coloration in a population of butterflies would probably be eliminated by a naive predator if it arose in only a single individual. But if it arose in the GERM LINE of the parent and was transmitted to the whole brood, the mutant ALLELE carried by all siblings might have a higher chance of survival because the predator that kills a single broodling might, on the basis of that experience, avoid the rest. So the allele has high FITNESS not necessarily by enhancing the survival of its bearer, but indirectly by enhancing the survival of the bearer's SIBLING kin that carry copies of the allele. [P.C.]

siblicide *See* CANNIBALISM.

sibling Any of an individual's brothers or sisters having both parents in common (full-sibling) or only one parent in common (half-sibling). [P.C.]

sibling species Morphologically similar or identical but reproductively isolated species. Their occurrence bears on the debate about mechanisms of SPECIATION. In some cases of sympatric distribution of sibling species, REPRODUCTIVE ISOLATION was acquired in allopatry and evolved more rapidly than did morphological change, thus SYMPATRIC SPECIATION need not be involved. [P.O.]

sick building syndrome (SBS) A condition associated with buildings in which a significant number of occupants (usually $>20\%$) report illness perceived as being due to the internal working environment. The symptoms may include irritation to sensitive surfaces, drying of mucous membranes and skin, headaches, coughs, nausea and faintness. SBS has been attributed to lack of ventilation, building materials and even mites. [P.C.]

sidereal time *See* TIME.

sievert (Sv) The SI UNIT of dose equivalent, used to measure the biological effect of IONIZING RADIATION. The dose equivalent (in sieverts) is obtained by multiplying the dose (in grays) by a quality factor (Table S6), which accounts for the different linear energy transfer (LET) values of different types of RADIATION. The effective dose equivalent is used to evaluate the EXPOSURE of individual organs or groups of organs to ionizing radiation and is determined by multiplying the dose equivalent (in sieverts) by a weighting factor for the organ concerned (Table S7). For irradiation of more than one organ, the total effective dose equivalent is calculated as the sum of the values for the individual organs. The committed dose equivalent is the dose equivalent summed over a 50-year period, taken to be a working lifetime. The collective dose equivalent is used to estimate the collective dose to a population, and is obtained by multiplying the average individual dose equivalent by the number in the population. The former SI unit of dose equivalent was the rem (10^{-2} Sv). *See also* GRAY; RADIOACTIVITY; RADIONUCLIDE. [A.B.M.]

REFERENCE

Kiefer, J. (1990) *Biological Radiation Effects.* Springer-Verlag, Berlin.

sigmoid growth A form of limited population growth occurring where there is COMPETITION for RESOURCES. An increasing population starting at very low POPULATION DENSITY will initially show EXPONENTIAL GROWTH with an average rate of increase per individual *r*. Competition for resources will eventually cause the POPULATION GROWTH RATE to decline as population density increases. When the environment cannot support a higher population density, the size of the population levels off to a plateau, known as the CARRYING CAPACITY *K* (*r* and *K* are parameters in the LOGISTIC MODEL). A graph of POPULATION SIZE plotted against time will have a charac-

Table S6 Quality factors for different types of ionizing radiation.

Type of radiation	Quality factor
X-rays	1
γ-radiation	1
β-radiation	1
Protons	1
Neutrons	20
α-particles	20

Table S7 Weighting factors for calculating effective dose equivalent for different organs.

Organ	Weighting factor
Gonads	0.25
Breast	0.15
Red bone marrow	0.12
Lung	0.12
Thyroid	0.03
Bone (excluding marrow)	0.03
Whole body	1.0

teristic sigmoid shape, providing there is either COMPENSATING DENSITY DEPENDENCE or UNDERCOMPENSATION. Sigmoid growth forms have been used to describe growth of organisms as well. *See also* VON BERTALANFFY EQUATION. [R.H.S.]

signalling *See* SOCIAL STATUS.

silicates Minerals composed of silica tetrahedra, SiO_4, linked in different two- and three-dimensional structures. Most minerals in the soil clay fraction, and primary minerals such as the micas and the serpentine minerals, belong to the group known as the phyllosilicates, with sheet structures that weather relatively easily, releasing plant NUTRIENTS. Other silicate minerals have three-dimensional structures — the tectosilicates. Of these, quartz is very resistant to WEATHERING, but the feldspars (e.g. orthoclase, $KAlSi_3O_8$) weather easily.

Another family of minerals, the inososilicates, have chainlike structures. Examples are the pyroxenes, with single-chain silica tetrahedra, and the amphiboles, the most common of which is hornblende, with double-chain tetrahedra. Layer silicate clay minerals such as smectites have a large capacity to fix ions, such as those of radiocaesium and HEAVY METALS. In wet conditions the mineral layers become separated by a WATER layer, into which the METAL ions can diffuse; when the clay dries out again, the metal ions become trapped between the mineral layers, held by strong electrostatic forces. [K.A.S.]

silicon (Si) A metalloid element (ATOMIC NUMBER = 14; relative ATOMIC MASS = 28.086; melting point = 1410°C; boiling point = 2355°C), which makes up 25.7% of the Earth's CRUST, second in abundance only to OXYGEN. It is found, combined with oxygen, as silica (SiO_2) or SILICATE minerals, which are major components of rocks and soils. Silica is only a minor constituent of plants, occurring mainly as a structural component, but forms the skeleton of aquatic diatoms. Pure silicon, in amorphous or crystalline form, is synthesized industrially for the manufacture of semi-conductor devices and integrated circuits. [J.N.C.]

silt A wet mixture of particles (size range 4–60 μm) found in aquatic environments and sewers, and intermediate between CLAY and mud. [P.C.]

siltation The deposition of fine (4–60 μm diameter) particles of organic matter and CLAY (SILT) in water bodies. In streams, reductions in flow or increases in SEDIMENT load through WATERSHED disturbances (erosion, construction, etc.) can cause silt to fill in the interstices in gravel and rock bottoms, reducing habitat for invertebrates and smothering the buried embryos of salmon and other fishes. [P.B.M.]

silver (Ag) A soft metallic element (ATOMIC NUMBER = 47; relative ATOMIC MASS = 107.868; melting point = 961.93°C; boiling point = 2212°C) with an average abundance of 0.07 $mg\,kg^{-1}$ in the Earth's CRUST. Silver is very widely distributed in the human environment due to its use principally in photographic and electrical applications. Although silver has been suggested to be a water POLLUTANT due to the lack of RECYCLING of mined silver, the current market price for the METAL makes such recycling economic, with approximately 30% of all silver refined being derived from WASTE sources. The TOXICITY of silver for humans is low, but for certain aquatic organisms acute toxicity may occur at concentrations as low as 0.25 $\mu g\,l^{-1}$. Silver has been shown to have antagonistic effects for SELENIUM, COPPER and vitamin E in animals. [P.J.B.]

silviculture The scientific principles and techniques of controlling, protecting and restoring the regeneration, composition and growth of natural FOREST vegetation and its PLANTATION analogues. It is often also referred to as FORESTRY PRACTICES. Silviculture as a discipline is an applied ecological science that rests ultimately upon the foundations of both the natural and social sciences. The application of silviculture should be guided by the forest values determined by the landowner and society. Forest values demanded by people come in the form of a variety of desired products (e.g. timber, fibres, medicinals, spices, foods) and services (e.g. water, microclimate amelioration, wildlife habitat, aesthetics and recreation, genetic diversity). As a comparison, silviculture is to forestry what AQUACULTURE is to fisheries, or agronomy is to AGRICULTURE. *See also* ARBORICULTURE. [P.M.S.A. & G.P.B.]

REFERENCE

Smith, D.M., Larson, B.C., Keity, M.J. & Ashton, P.M.S. (1997) *The Practice of Silviculture: Applied Forest Ecology.* John Wiley & Sons, New York.

similarity coefficient A numerical measure, calculated from multivariate attribute data, of the similarity of two objects. Several well-known similarity coefficients increase from 0, when there is no resemblance, to 1, when there is identity.

In ecological contexts, the objects are often samples or quadrats and the attributes are species abundances. Let x_{ip}, x_{jp} be the abundances of species p in objects i, j. A popular similarity measure for such objects is Czekanowski's coefficient, defined as:

$$s_{ij} = \frac{2\sum_p \min(x_{ip}, x_{jp})}{\sum_p (x_{ip} + x_{jp})}$$

which is equal to 0 when there are no species in common, and to 1 when all species are present in equal abundance.

The nomenclature of similarity coefficients is complex. Czekanowski's coefficient is sometimes called the Steinhaus coefficient and (multiplied by a factor of 100) is familiar to many plant ecologists as the Bray–Curtis per cent similarity. When applied to presence–absence data it is called Sørensen's coefficient. Details of many other coefficients and their nomenclature can be found in standard textbooks (Greig-Smith 1983; Legendre & Legendre 1983; Digby & Kempton 1987).

Given a set of objects, the array of similarities, s_{ij}, calculated for all pairs of objects i, j, can be used to classify or ordinate the objects. For the purposes of ORDINATION, values of a similarity coefficient are usually converted to a distance measure (distance coefficient) by subtraction from d_{max} the maximum possible:

$$d_{ij} = d_{max} - s_{ij}$$

or by means of the formula:

$$d_{ij} = \sqrt{s_{ii} + s_{jj} - 2s_{ij}}$$

derived for the method of principal coordinates analysis by Gower (1966). [M.O.H.]

REFERENCES

Digby, P.G.N. & Kempton, R.A. (1987) *Multivariate Analysis of Ecological Communities*. Chapman & Hall, London.

Gower, J.C. (1966) Some distance properties of latent root and vector methods used in multivariate analysis. *Biometrika* **53**, 325–338.

Greig-Smith, P. (1983) *Quantitative Plant Ecology*, 3rd edn. Blackwell Scientific Publications, Oxford.

Legendre, L. & Legendre, P. (1983) *Numerical Ecology*. Elsevier, Amsterdam.

Simpson–Yule index *See* DIVERSITY INDICES.

Simpson's diversity index *See* DIVERSITY INDICES.

simultaneous hermaphroditism The simultaneous expression of male and female reproductive functions in an individual organism. This involves the simultaneous presence of pollen-producing anthers and receptive stigmas in individual flowers of flowering plants, or of sperm-releasing testes and mature OVA in animals. Self-fertilization may prevail but in many cases is prevented by self-sterility mechanisms. Because all mature individuals can reciprocate sexually, simultaneous HERMAPHRODITISM is advantageous where opportunities for mating are reduced, either by immobility, as in higher plants and SESSILE animals such as barnacles, or by low POPULATION DENSITY. Once having evolved in copulating animals, simultaneous hermaphroditism may be unlikely to revert to the ancestral state of separate sexes because of the stabilizing influence of GAMETE trading. LIFE-HISTORY TRAITS usually make one or other sexual mode more advantageous in terms of potential genetic contribution to future generations. During mating, therefore, individuals should try to 'cheat' by performing only the preferred sexual role. Gamete trading probably has evolved to prevent such CHEATING. It involves bouts of limited performance by each individual in the non-preferred sexual role, so forcing the partner to reciprocate. Egg trading occurs, for example, in the hamlet fish *Hypoplectrus nigricans*, and sperm trading in the sea slug *Navanax inermis*. [R.N.H.]

REFERENCE

Leonard, J.L. & Lukowiak, K. (1991) Sex and the simultaneous hermaphrodite: testing models of male–female conflict in a sea slug, *Navanax inermis* (Opisthobranchia). *Animal Behaviour* **41**, 255–266.

single large or several small (SLOSS) *See* ISLAND BIOGEOGRAPHIC THEORY IN CONSERVATION; ISLAND BIOGEOGRAPHY.

sinking rates The rate (i.e. distance per unit time) at which particles descend vertically in a column of water. The rates at which PHYTOPLANKTON, ZOOPLANKTON and particulate DETRITUS sink influences the transfer of NUTRIENTS and vertical distribution of organic material in the water column. Many aquatic organisms have adaptations (e.g. spines, lipid globules) which reduce their sinking rate. Sinking rates of living phytoplankton are in the range 0–30 m day^{-1}, whereas copepods sink at rates of about 36–720 m day^{-1}. The relative contribution of potential food sources to the DEEP SEA depends in part on their rate of sinking. *See also* STOKES' LAW. [V.F.]

sister group/species Any species or higher monophyletic TAXON that is hypothesized to be the closest genealogical relative of a given species or monophyletic taxon exclusive of the ancestral species of both taxa is known as the sister species or sister group of that species or monophyletic taxon. Thus, when we say that two taxa are sister groups we mean they are hypothesized to share an ancestral species not shared by any other species or taxa. In phylogenetic systematic analysis, the taxon that is most closely related genealogically to the study taxon is its sister group. If known, the sister group of the ingroup is the taxon of first choice for OUTGROUP comparisons, but successful outgroup comparisons do not require such knowledge. The ancestor of the ingroup cannot be its sister because it is a member of the ingroup. *See also* CLADISTICS; PHYLOGENETIC SYSTEMATICS. [D.R.B.]

Sisyphean model *See* EVOLUTION OF SEX.

sit-and-wait feeders *See* FEEDING TYPES, CLASSIFICATION.

sit-and-wait strategy *See* FEEDING TYPES, CLASSIFICATION.

site of special scientific interest (SSSI) Any site designated by nature conservation bodies in Britain as a representative example of British habitats. Each site is seen as 'an integral part of a national series' established with the aim of 'maintaining the present diversity of wild animals and plants in Great Britain'. The selection is on scientific grounds rather than to enhance amenity or provide recreation. Powers for establishing these sites were created by the Wildlife and Countryside Act 1981. Owners and occupiers of SSSIs must seek permission to carry out any potentially damaging operation. [P.C.]

site pre-emption *See* FOUNDER-CONTROLLED COMMUNITIES.

situational sensitivity The sensitivity of FITNESS to a change in some TRAIT gives the SELECTION GRADIENT on that trait. These selection gradients depend on the rest of the life history; S.C. Stearns has referred to this as 'situational sensitivity'. In one sense, all of life-history theory can be viewed as a study of situational sensitivity, since it seeks to understand the conditions under which different combinations of LIFE-HISTORY TRAITS are favoured by selection.

Within formal demographic theory, situational sensitivity has been used to investigate the dependence of selection gradients on the rate of increase of the population. The evolution of SENESCENCE, for example, depends on the decline with age of the selection gradient on survival. All other things being equal, this decline is more severe the more rapidly the population is increasing; if the population is decreasing rapidly enough the selection gradients may actually increase with age.

For populations whose dynamics are described by a matrix population model (*see* POPULATION PROJECTION MATRIX), fitness can be measured by the POPULATION GROWTH RATE, λ, and the selection gradient on the entry a_{ij} of the population projection matrix $\mathbf{A}$ by the sensitivity $\partial\lambda/\partial a_{ij}$. Situational sensitivity can be investigated in this framework using the second derivatives of λ; for example:

$$\frac{\partial^2\lambda}{\partial a_{kl}\partial a_{ij}}$$

which gives the sensitivity of the selection gradient on a_{ij} to a change in a_{kl}. [H.C.]

REFERENCES

Caswell, H. (1982) Life history theory and the equilibrium status of populations. *American Naturalist* **120**, 317–339.

Caswell, H. (1996) Second derivatives of population growth rate: calculation and applications. *Ecology* **77**, 870–879.

Ebert, T.A. (1985) Sensitivity of fitness to macroparameter changes: an analysis of survivorship and individual growth in sea urchin life histories. *Oecologia* **65**, 461–467.

Stearns, S.C. (1992) *The Evolution of Life Histories*. Oxford University Press, Oxford.

size When applied to an organism, means its dimensions. These are often expressed as length, but can be expressed in mass, volume or energy. *See also* GROWTH; SIZE DISTRIBUTIONS. [P.C.]

size-advantage hypothesis The hypothesis that sex change is selectively advantageous because the relationship between reproductive potential and body SIZE is different for male and female functions. Since sperm can be produced in large numbers at relatively low cost, reproductive potential as a male may at first increase rapidly with body size but then decelerate. Because EGGS are energetically relatively expensive to produce and to brood, reproductive potential as a female may remain close to zero at first, but then increase rapidly beyond a THRESHOLD body size. In such a case, the size-advantage hypothesis predicts a sex change from male to female (PROTANDRY) as the individual grows larger.

If males require a breeding territory to attract females and if the ability to defend this territory is strongly enhanced by larger body size, then PROTOGYNY is predicted, whereby the individual changes from female to male on reaching a threshold body size. Protandry occurs, for example, in certain shrimps, prawns (Pandalidae) and limpets (Patellidae), and protogyny is found in territorial wrasses (Labridae) and parrot fishes (Scaridae). The size at sex change may be sensitive to demographic structure and MORTALITY regimes, both through physiological-behavioural processes and through NATURAL SELECTION.

Populations of sequential hermaphrodites may support a small proportion of males that do not change sex. In protandrous populations, non-hermaphroditic individuals that grow into large males may gain advantage through SPERM COMPETITION. In protogynous populations, small, non-hermaphroditic, 'initial-phase' males may achieve reproductive success by 'sneaking' matings within the breeding territories of larger males. [R.N.H.]

REFERENCE

Charnov, E.L. (1982) *The Theory of Sex Allocation*. Princeton University Press, Princeton, N.J.

size at maturity Because SIZE at maturity is frequently related to the age at maturity and reproductive success, it is a critical TRAIT in the evolution of life histories. For females, FECUNDITY typically increases with size and hence selection acting on this aspect alone will favour an increased size at maturity. Similarly, in many species, male success is correlated with size, larger males being more successful at attracting females (or in the case of plants, pollinators) or being able to dominate smaller males. Consequently, selection will also favour larger size at maturity in males. However, the attainment of a larger size requires either an increased GROWTH rate or a longer development time. In the former case, MORTALITY may be increased because of the necessarily higher foraging rate, while in the latter case the probability of surviving to maturity and INTRINSIC RATE OF INCREASE are decreased by virtue of the increased juvenile period. Because of these conflicting effects on FITNESS there will be an OPTIMAL SIZE AT MATURITY. *See also* AGE AT MATURITY IN PLANTS; LIFE-HISTORY EVOLUTION. [D.A.R.]

size distributions A function giving the relative abundance of individuals of different sizes within a POPULATION. The dynamics of the size DISTRIBUTION of a population are determined by the rates of individual GROWTH, MORTALITY and reproduction. They can be described by the continuous McKendrick–von Foerster equation or a discrete size-classified matrix population model (*see* POPULATION PROJECTION MATRIX).

Let x denote size, t denote time, and $n(x,t)$ denote the size density function at time t. The McKendrick–von Foerster equation is:

$$\frac{\partial n(x,t)}{\partial t} + \frac{\partial g(x,t)n(x,t)}{\partial x} = -\mu(x,t)n(x,t) \tag{1}$$

$$\int_{x_0}^{\infty} m(x,t)n(x,t)\mathrm{d}x = g(x_0,t)n(x_0,t) \tag{2}$$

where $g(x,t)$ is the growth rate, $\mu(x,t)$ the MORTALITY RATE, $m(x,t)$ the rate of reproduction, and x_0 the size at birth. Equation 1 describes changes in the size distribution due to growth and mortality; eqn 2 is a boundary condition describing reproduction.

A typical size-classified population projection matrix, shown here for three size classes, is:

$$\mathbf{A} = \begin{pmatrix} P_1 + F_1 & F_2 & F_3 \\ G_1 & P_2 & 0 \\ 0 & G_2 & P_3 \end{pmatrix} \tag{3}$$

where P_i is the probability of surviving and remaining in size class i, G_i is the probability of surviving and growing from size class i to $i+1$, and F_i is the fertility of the ith size class. This form of the model assumes that individuals move no more than one size class.

As is true with AGE STRUCTURES, in a constant environment the size distribution eventually converges to a stable form. The stable size distribution $\hat{n}(x)$ for the McKendrick–von Foerster model in a stationary population satisfies:

$$\frac{\mathrm{d}}{\mathrm{d}x} g(x)\hat{n}(x) = -\mu(x)\hat{n}(x) \tag{4}$$

while for the matrix population model of eqn 3 the stable size distribution is proportional to:

$$\begin{pmatrix} 1 \\ \dfrac{G_1}{\lambda - P_2} \\ \dfrac{G_1 G_2}{(\lambda - P_2)(\lambda - P_3)} \end{pmatrix} \tag{5}$$

where λ is the rate of population growth.

Unlike stable age distributions, which are of necessity decreasing functions of age, stable size distributions can exhibit peaks caused by individuals 'piling up' in certain size classes.

If the rate of population growth (r or λ) and the growth rate function ($g(x)$ or G_i) are known, it is possible to estimate the size-specific mortality rate from the stable size distribution. Similar calculations can be done for the size-classified matrix model in some cases. However, unless growth rate does not vary with size, mortality cannot be estimated directly from the slope of the size distribution, as it can from an age distribution.

Perturbations caused by pulses of RECRUITMENT or FLUCTUATIONS in mortality can cause the size distribution to differ from the stable distribution. Distinguishing such peaks from peaks in the stable size distribution can be difficult. *See also* DEMOGRAPHY; LIFE TABLE. [H.C.]

REFERENCES

Caswell, H. (1989) *Matrix Population Models*. Sinauer Associates, Sunderland, Mass.

Van Sickle, J. (1977a) Analysis of a distributed-parameter population model based on physiological age. *Journal of Theoretical Biology* **64**, 571–586.

Van Sickle, J. (1977b) Mortality rates from size distributions: the application of a conservation law. *Oecologia* **27**, 311–318.

size-efficiency hypothesis A hypothesis proposed by J.L. Brooks and S.I. Dobson to explain the shift in DOMINANCE amongst ZOOPLANKTON species in LAKES with and without fish. Generally, temperate lakes with fish are dominated by small zooplankton species, but in fishless lakes, large species are dominant. The size-efficiency hypothesis assumes that zooplankton compete for small algal cells and that the larger species are more efficient herbivores.

The hypothesis leads to three predictions.

1 Under low predation pressure (no/few fish), large zooplankton species (large cladocerans, calanoid copepods) competitively eliminate smaller species.

2 Under high predation pressure, large zooplankton are eliminated through size-selective feeding by fish; small species (rotifers, small cladocerans and copepods) dominate.

3 Under moderate predation intensity, large zooplankton populations are sufficiently reduced so as to allow COEXISTENCE of small species and enhanced DIVERSITY.

The competitive relationship between large and small zooplankton (prediction 1) is complicated by spatial (MICROHABITAT differences) and temporal (variation in food supply) HETEROGENEITY, which may enhance coexistence of inferior species. Predictions 2 and 3 are supported by data from both laboratory and field manipulation experiments. Historical studies of lake sediments can also reflect the history of fish populations and predation pressure (Fig. S14). Vulnerability to predation tends to be well correlated with zooplankton size and may, in turn, influence the ability of grazing zooplankton to influence algal dynamics in lakes. *See also* SIZE-SELECTIVE PREDATION; TROPHIC CASCADE HYPOTHESIS. [P.S.G.]

REFERENCES

Brooks, J.L. & Dobson, S.I. (1965) Predation, body size and composition of plankton. *Science* **150**, 28–35.

Neill, W.E. (1975) Experimental studies of microcrustacean competition, community composition and efficiency of resource utilisation. *Ecology* **56**, 809–826.

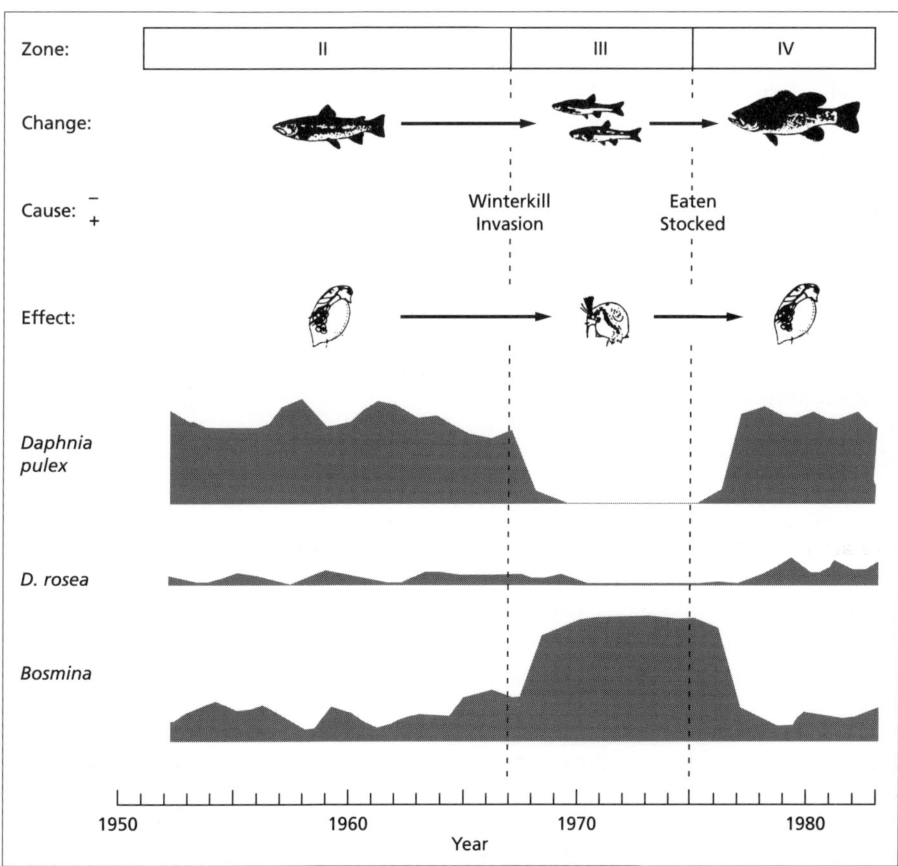

Fig. S14 Response of zooplankton to fish manipulations at Paul Lake (Wisconsin, USA) as recorded by remains in the sediments. The shift from piscivores to planktivores (zone III) led to an increase in small-bodied *Bosmina* after increased predation on the larger and dominant *Daphnia* by minnows. (After Anderson, N.J. & Battarbee, R.W. (1994) Aquatic community persistence and variability: a palaeolimnological perspective. In: *Aquatic Ecology: Scale, Pattern and Process* (eds P.S. Giller, A.G. Hildrew & D.G. Raffaelli), pp. 233–259. Blackwell Scientific Publications, Oxford.)

size–number trade-off The TRADE-OFF between the SIZE and NUMBER OF OFFSPRING. If a parent's REPRODUCTIVE EFFORT is fixed, then if it produces more offspring, each will necessarily be provisioned with fewer RESOURCES. This gives rise to the trade-off. It is generally supposed that it is the existence of this trade-off that prevents the evolution of larger eggs/propagules. Evidence for the existence of the trade-off is not extensive, but is generally supportive. *See also* EVOLUTIONARY OPTIMIZATION; LIFE-HISTORY EVOLUTION; OPTIMALITY MODELS. [R.M.S.]

REFERENCE

Roff, D.A. (1992) *The Evolution of Life Histories*. Chapman & Hall, New York.

size-selective predation PREDATORS generally select prey of certain size classes from amongst a range of sizes. This size-selective predation can operate within or between prey species. Often selective predators consume the largest size classes of the available, encountered prey in preference to smaller ones (e.g. fish, salamanders and many invertebrate predators). Such size selection can be based on purely physical factors, especially where a visual stimulus is the major determinant of prey vulnerability. For planktivorous fish, search-volume size and shape change with prey size, so larger ZOOPLANKTON prey are located visually at greater distances than smaller ones. Alternatively, size selection may be related to optimal foraging, where all else being equal, prey sizes are selected to maximize net energy gain by the predator. The range of sizes utilized may change with relative ABUNDANCE of different size classes and overall levels of resource in the habitat (*see* OPTIMAL FORAGING THEORY).

There is a large literature showing how size-selective predation can dramatically influence COMMUNITY composition. The best examples are drawn from fish–zooplankton systems, where selection of the largest zooplankton species can in turn have dramatic effects on the overall FOOD WEB. *See also* SIZE-EFFICIENCY HYPOTHESIS; TROPHIC CASCADE HYPOTHESIS. [P.S.G.]

skewness A statistic that measures one type of departure from a NORMAL DISTRIBUTION. It is another name for asymmetry. In skewness, one tail of the distribution is drawn out more than the other. Distributions are skewed to the right (e.g. Fig. M6b) or the left depending upon whether the right or left tail is drawn out. In ecological data, distributions skewed to the right are very common. [B.S.]

SLA *See* SPECIFIC LEAF AREA.

slack (dune slack) A low-lying, flat area situated amongst SAND DUNES. It is typically moister, more sheltered and with a higher nutrient status than the adjacent dunes. Slacks may be formed in actively prograding (accumulating) sand-dune systems (primary slacks) or as a result of erosive processes (secondary slacks) but subtle variants of both types exist. They are often subject to flooding in winter, and sometimes have permanent freshwater pools. More rarely, conditions may be brackish or saline where inundation by high TIDES occurs. The vegetation, which is often species rich, reflects mainly variations in SOIL MOISTURE (wet and dry slacks), soil pH (relating especially to calcium carbonate from the shell content of the sand) and slack age (successional processes). Wet slacks contain species typical of either poor FEN where the pH is low, or rich fen/MARSH where the pH is high. Dry slacks of high pH have species in common with chalk or limestone grassland. [B.W.F.]

slash and burn A traditional agricultural system employed by certain nomadic groups. An area is cleared by burning, cultivated for short periods and then abandoned as crop yields fall. Also known as SHIFTING CULTIVATION, it is best suited to low POPULATION DENSITIES. [P.C.]

SLOSS (single large or several small) *See* ISLAND BIOGEOGRAPHIC THEORY IN CONSERVATION; ISLAND BIOGEOGRAPHY.

smog Fog containing air pollutants (e.g. SMOKE, sulphur dioxide (SO_2), unburnt HYDROCARBONS and NITROGEN OXIDES (NO_x)), usually trapped near the ground by temperature inversion. *Cf.* PHOTOCHEMICAL SMOG. [P.C.]

smoke A suspension of fine particles in a gas, usually derived from incomplete combustion or chemical reaction. The particles are usually $<2\,\mu m$ in diameter and are largely of CARBON. Smoke may also contain silica, fluoride, ALUMINIUM, LEAD, ACIDS, BASES and organic compounds such as PHENOLS. [P.C.]

snakes *See* REPTILES, DIVERSITY AND CONSERVATION.

social facilitation The enhancement of a behaviour by the simultaneous performance of the behaviour by a number of CONSPECIFICS or heterospecifics. Social facilitation is a common phenomenon in a number of different contexts. Examples include sexual, comfort and antiparasite behaviour, and foraging. Social facilitation is particularly prevalent among highly social animals that usually have very synchronized DIURNAL and annual cycles.

The FUNCTIONAL EXPLANATION of social facilitation is that the simultaneous performance of a particular behaviour by more than a single individual generates synchronization and thereby provides greater FITNESS benefits for the individual — for instance during communal foraging — than would have been achieved in the absence of social facilitation. Social facilitation during foraging is often achieved by the use of visual signals or specific recruitment calls that can be recorded at a large distance.

The use of social facilitation is context dependent. For example, house sparrows (*Passer domesticus*) only give recruitment calls when a recently detected food source is readily divisible, whereas no calls are given when the food source is small and indivisible. Social facilitation may render foragers less susceptible to predation, because more individuals are scanning for predators in a larger flock, and the latency time to detection of a predator is shorter. Moreover, social facilitation may directly increase the efficiency of foraging because the prey may be more easily captured, when disturbed by a number of different predator individuals. *See also* BEHAVIOURAL ECOLOGY; SOCIOBIOLOGY. [A.P.M.]

social groups *See* SOCIAL ORGANIZATION.

Social insects

Insects that form social groups, usually in the narrow sense of eusociality (*see* SOCIALITY, TYPES OF). They share the characteristics of cooperative brood care, overlap of generations (offspring stay together with parents) and the presence of a reproductive CASTE (only some members of the society reproduce). The insect order Hymenoptera contains most of the approximately 15 000–20 000 species of social insects: the ants (Formicoidea), bees (Apoidea) and wasps (Vespoidea). The order of TERMITES (Isoptera) forms the other large group. In addition, several species belonging to various other taxa, such as sphecid wasps, aphids, and thrips, have been found to be eusocial. Although not insects, social spiders are often discussed alongside the social insects.

Colony structure and reproduction

The habits and lifestyles of social insects are widely different. An idealized hymenopteran society (the COLONY) consists of a single queen (monogynous colony) or several coexisting queens (polygynous) and a large number of workers. Numbers vary from a few dozen to many millions — e.g. colonies of the leafcutter ant *Atta colombica* contain an average of 1–3 million workers. In a generalized hymenopteran social insect (see Plate 19, facing p. 398), the queen starts a nest on her own (independent, or haplometrotic foundation) or sometimes in cooperation with others (pleometrotic). Founding may be in isolation where the queen lives off her stores (claustral; in most independently founding ants), or queens may start to collect food, or be dependent on a host species (as in several formicine ants). In honeybees (*Apis*), sweat bees (Meliponinae), army ants (Ecitoninae) and some other species, a new colony is founded by fission of the old colony, whereby a queen leaves with a retinue of workers. After this initial period, the first new members of the colony emerge and the colony starts to grow in numbers.

Queens normally do all of the egg-laying, but except for a few highly advanced species, workers are fertile and lay their own eggs (which in hymenopterans must develop into sons). During the period of colony growth (the ergonomic stage) eggs are female and develop into more workers. Reproduction takes place when both male and female eggs are laid and develop into male (drones) and female sexuals (daughter queens). Sexuals leave the colony, mate and start a colony of their own. In some cases, mating is within the nest (e.g. in parasitic ants), or daughter queens are recruited back to the maternal nest (in some formicines).

After reproduction in annual species, the mother colony dies and only the fertilized young queens hibernate to the next season to start another cycle (e.g. in most bees and wasps). In perennial species (e.g. in all known ants, termites, some bees and wasps), the mother colony diapauses and resumes growth the next season. Perennial colonies and their queens can be extremely long-lived; for instance, in seed-harvester ants (*Pogonomyrmex*), queen and colony lifespans of several decades have been recorded. Workers, in contrast, live for a few weeks to a few years. In hymenopteran social insects, males are very short-lived and do not normally engage in colony labour.

In termite colonies both males and females are present throughout the entire colony cycle. Colonies are headed by a primary queen and a primary king (the reproductive primary pair), which live together in the royal cell. During their life, queens are repeatedly fertilized by the king. Termite colonies usually produce replacement queens and kings (the secondary and tertiary pairs), which may also be egg-laying at the same time. Thus, termite colonies can be extremely long-lived. Otherwise, the basic colony cycle — foundation, growth and reproduction — is quite similar to that of hymenopterans. In contrast, workers consist of both sexes (males and females).

Caste systems

Workers of social insects can be differentiated into different morphological castes; for example, minors that do the brood care, and majors (or soldiers) that defend the nest. The most complex morphological caste systems are found in leafcutter ants and termites. Morphological caste systems are based on allometric relationships among different body parts (e.g. soldiers have disproportionately large mandibles). With few exceptions, morphological differences are not based on genotypic variation but on differential feeding regimes during larval development or on PHEROMONE cues (as in termites).

In hymenopterans, development is holometabolic with a sequence of larval stages and a final adult form. Consequently, in hymenopterans the larval stages are helpless and must be cared for, while in termites the later nymphal ('larval') stages engage in the work of the colony. But larvae may provide essential services to the colony: for example, silk produced by the larvae is used by weaver ants (*Oecophylla*) to knit leaves together for nest chambers. An age-based DIVISION OF LABOUR (polyethism) is almost universally found among the social insects, where young workers attend the tasks within the nest and older ones are guarding or foraging.

Lifestyles

Social insects occupy almost any ecological niche from the tropics to the high Arctic. Their feeding habits range from processing plant materials (e.g. fungus-growing ants; all termites are consumers of CELLULOSE), to harvesting seeds (many ants), flower nectar and pollen (the bees), predatory habits (ants and wasps), kleptoparasitism (nectar-robbing bees) and social PARASITISM (i.e. exploiting brood care by the host for own offspring, this is known from all major taxa). The ecological success of social insects is often attributed to their robust colony organization and mass actions which provide resilience against environmental fluctuations.

Evolution of eusociality

The EVOLUTION of eusociality can be traced to essentially two different routes. In the parasocial sequence, solitary species are followed by communally nesting ones. Sociality then emerges with quasi-social (with cooperative brood care), semi-social (cooperative brood care and reproductives), and eusocial species. In the subsocial sequence, solitary and primitively subsocial species are followed by intermediate subsocial ones, where first generations overlap and subsequently cooperative brood care emerges, followed by the eusocial stage. While in wasps and bees all of these stages are represented by extant species, all known living ants and termites are eusocial.

Current ideas on the evolution of eusociality emphasize characteristics that favour cooperative brood care and discourage DISPERSAL of offspring. For example, in Hymenoptera sex is determined essentially by the haplo-diploid mechanism, where males are haploid and females diploid, so that sisters are unusually closely related. According to the concept of inclusive FITNESS such relationships make it genetically more profitable to raise a sister than an own daughter. This may explain why eusociality has arisen several times independently in this order, and why ants, bees and wasps are matrifilial societies — ones in which queen(s) and daughters predominate and males are only transient members. However, several other ecological factors have been discussed. Most importantly, in partially bivoltine LIFE CYCLES (as in halictid bees) different generations may overlap and thus favour biased SEX RATIOS, which in turn favours helping a mother. High risks or a lack of suitable nest sites in the habitat also reduce dispersal. In termites (which are diploid), the need to pass on to offspring the endosymbiotic flagellates essential for cellulose digestion is considered to be important. But cycles of inbreeding may also raise genetical relationships to high levels.

Cohesion and conflict

In contrast to classical views, social insect societies are not harmonious entities. The members actively compete against each other over reproduction. In hymenopteran social insects, for example, workers and queens are in conflict over the sex ratio of reproductives. Evidence shows that workers are controlling this ratio and may bias it towards females to the disadvantage of the queen(s). Seemingly obedient societies (as in the honeybees) may be generated by processes such as worker 'policing' whereby workers remove each other's eggs, for each gains most if only the queen reproduces. Hence, the modern picture is one of a society that is conflict-ridden due to different costs and benefits accruing to its members, and whose functioning and reproduction is carefully balanced by the diverging interests through a number of subtle processes. *See also* SOCIOBIOLOGY. [P.S.H.]

REFERENCES

Hölldobler, B. & Wilson, E.O. (1990) *The Ants*. Springer-Verlag, Berlin.

Michener, C.D. (1974) *The Social Behavior of the Bees*. Harvard University Press, Cambridge, Mass.

Ross, K.G. & Matthews, R.W.E. (1990) *The Social Biology of Wasps*. Cornell University Press, Ithaca, N.Y.

Wilson, E.O. (1971) *The Insect Societies*. Harvard University Press, Cambridge, Mass.

Social organization

Social organization of a POPULATION is regulated by reproductive competition among individuals and by the costs and benefits of social interactions. Such interactions lead often to TERRITORIALITY and aggression towards conspecific, competing individuals, but many species also live either permanently or temporarily in social associations. Spatial clumping of individuals can result partly from attraction of individuals to clumped resources and partly from mutual attraction of individuals to each other.

Types of social organization

Social organization can be usefully divided into three main types on the spatial basis: AGGREGATIONS, social groups and colonies.

Aggregations

Aggregations refer to temporary groupings of individuals that are attracted or driven (e.g. by wind or currents) to the same place. The social organization in such aggregations is normally very simple, lacking complex interactions among individuals.

Social groups

Social groups are based on benefits resulting from cooperation. Such groups can be either permanent or temporary, and they vary widely in size and organization. The social structure can be simple as, for example, in fish schools, where the benefits of group living largely come from avoidance of predation. More complex interactions between individuals characterize advanced social groups found, for example, in many mammals (primate troops, lion prides, ungulate herds).

Colonies

Colonies are spatially structured social groups where individuals have fixed places (e.g. breeding bird colonies) or which are centred around a fixed location (e.g. breeding colonies of sea elephants). Invertebrate colonies have often been compared to superorganisms. The colonies of slime moulds, slime bacteria, sponges, jellyfish, corals, moss animals and some related forms have individuals attached to each other in such a way that they become morphologically and functionally specialized—as the cells in a metazoan body. ASEXUAL REPRODUCTION — budding off new individuals from old — keeps the members of such invertebrate colonies genetically identical.

Colonies of SOCIAL INSECTS consist of genetically different individuals that are attached to the same nest. Collaboration of individuals, based largely on the DIVISION OF LABOUR and good COMMUNICATION, leads the COLONY to behave in some respects as a single organism.

Evolution of social groups

The theoretical framework used to explain the EVOLUTION of social groups and colonies is that of KIN SELECTION. The evolutionary success of an individual is measured by its inclusive FITNESS, i.e. the efficiency with which an individual transmits copies of its genes to future generations. An individual can, to a certain extent, forego its own reproduction if it can instead assure genetic benefits by helping a relative.

While the groups are characterized by mutual interests, co-operation and coordination of activities, they simultaneously exhibit competition between genetically different individuals. This competition leads often to social dominance, and individuals have varying social roles.

Dominance

Dominance is a somewhat controversial concept, as the dominance relationships can vary depending on the context. Dominance can be linear (a pecking order), despotic (one dominating the others), or it can include non-linear loops and subgroups. The establishment of a dominance hierarchy reduces the daily CONFLICTS within the group and thus increases the overall efficiency. It is easy to understand that the top dominant individual (α-individual) benefits from this kind of social organization by getting all the benefits from cooperative interactions and by converting them into its own reproduction.

The benefits to a subordinate (as well as to the dominant) individual are dual because its reproductive success has two components. One component consists of the individual's direct reproduction, and the other component includes the inclusive fitness effects, i.e. the effects that the individual has on the reproduction of other group members. These effects are weighted by the genetic relatedness between any two individuals. It becomes easier for an individual to accept a subordinate status if it can help a highly related dominant individual. Genetic studies of social groups have shown that they commonly consist of related individuals. Social dominance within the groups is expected to lead to reproductive skew — unequal reproductive contribution among the group members. The degree of skew depends on the ecological CONSTRAINTS, group size, kinship and fighting ability of individuals.

Social insects

Colonies of social insects are characterized by two important attributes: the number of reproductive females (queens) in a colony, and the number of nests connected to form a single colony. Many social insect colonies are monogynous (i.e. they have a single queen), but polygyny (coexistence of several to many queens) is widespread in ants and swarm-founding wasps, and occurs also in bees and TERMITES. In addition to queens, worker individuals in hymenopteran social insects can commonly lay unfertilized EGGS that develop into males. Some ponerine ants have secondarily lost the morphological queen CASTE, and mated workers do all reproduction.

The ecological success of social insects is based on the benefits of cooperation and coordination of tasks, but the societies face simultaneously intense conflicts between genetically different individuals over reproduction, SEX RATIOS and the allocation of resources in colony growth vs. reproduction. Both cooperation and conflict extend from single nests to colonial networks, which are typical in many ants that produce new nests by budding of old ones. This type of social organization has made some ants ecologically extremely successful, as exemplified by the introduced fire ant (*Solenopsis invicta*) in North America and by mound-building wood ants (*Formica* spp.) in boreal forests. [P.P.]

REFERENCES

Wilson, E.O. (1975) *Sociobiology: The New Synthesis*. Harvard University Press, Cambridge, Mass.

Wittenberger, J.F. (1981) *Animal and Social Behaviour*. Duxbury Press, Boston, Mass.

---

social status Animals living in social groups suffer a certain level of COMPETITION and develop DOMINANCE HIERARCHIES to limit the aggressive interactions between them. An additional way to make life run more smoothly in such groups is by clearly signalling the social status, so that potential contests can be avoided by exchange of information about the social status of the opponent before any interaction. Such social signals are likely to be less rigid than sexual signals, as the social position of individuals is more flexible. Nevertheless, as with all signals, it may pay individuals to signal a status higher than their actual status in order to (possibly) gain greater access to RESOURCES than merited by their true status. Therefore, signals can only be useful if they are honest, and to be honest they need to be costly.

Signalling of social status is most commonly made by using assessment signals that are reliable: i.e. signals that accurately reflect size or athletic ability, and that cannot be faked. An individual's physical power increases with age to a maximum and then decreases. In many animal species, the social status of an individual varies with its physical power, which is then a reliable signal of its social status.

Pretending to be strong is possible in some cases, and we then observe interacting animals using a series of escalating signals, all of which contribute to give a reliable signal of the social status of the actor. In red deer (*Cervus elaphus*), after roaring to each other, males engage in a parallel walk, and only if the contestants remain equally matched at this stage will a serious fight ensue, where they interlock antlers and push against each other. In chimpanzees (*Pan troglodytes*) and mountain gorillas (*Gorilla beringuei*), bluff charges are often used to settle contests, but these bluff DISPLAYS mostly involve vocalizations, rapid running and persist over long periods of time—all aspects being reliable indicators of the social status of the individual. *See also* CASTE; SOCIAL ORGANIZATION. [C.B.]

Sociality, types of

Sociality is the combined properties and processes of social existence. All animal species with two sexes are social, but some are more social than others: many live in societies but some do not. Social behaviour is behaviour directed differentially at members of the same species, thus including sexual behaviour. A society is usually held to be a group of cooperating individuals of the same species, with the interactions between society members involving reciprocal COMMUNICATION and going beyond sexual behaviour.

In discussing sociality, attention must be given to the evolutionary mechanisms maintaining it, because in evolutionary biology many terms regarded as indicating motivation in non-scientific discourse are defined in terms of effects. For example, biological altruism is self-harming behaviour benefiting others, whereas in ordinary speech altruism is usually defined by intention, not by effects. Analogues to sociality occur between species, with the treatment of individuals of other species similarly to CONSPECIFICS termed social MIMICRY. Examples of social mimicry include mixed flocks of birds and mammals (often involving group protection against predators) and the use of the trails laid by one ant species by another. Some such relationships are exploitative: for example, female *Photuris* fireflies use their flashes to lure *Photinus* males and then eat them.

Mutualism and altruism

The effects favouring social behaviour can be divided into MUTUALISM and altruism. In mutualism both parties benefit, but in altruism one party increases the personal FITNESS of the other at a cost to its own.

The benefits of mutualism may not be immediately reciprocal: in social compensation (also termed reciprocal altruism) selection favours aid from one party to another if the eventual reciprocation of this aid is likely. This condition is met best if interactants encounter each other many times during their lives. For example, it is believed that well-fed vampire bats (*Desmodus rotundus*) feed hungry individuals of the same AGGREGATION because these will feed them in turn in future.

Altruism only occurs under KIN SELECTION, namely selection mediated by interactions between relatives. Selection is now understood to maximize not an individual's reproductive success in isolation, but rather its inclusive fitness. Inclusive fitness includes not only the passing-on of an individual's genes through its own offspring but also the effects of its actions on the passing-on of genes identical by descent to its own through relatives. Hence self-sacrifice can be selected for with sufficient improvement in the reproductive success of relatives. In calculating inclusive fitness, the effects of a focal individual's actions on the reproductive success of relatives have to be weighted by the relatedness of these individuals to the focal individual and by their expected REPRODUCTIVE VALUE relative to that of the focal individual.

Levels of social organization

Species can be classified in terms of the degree of reproductive self-sacrifice involved in their societies (Table S8).

Solitary species

Solitary species do not care for their young other than by producing them. Examples include: corals, which expel their gametes into the water column; trigonalid wasps, which lay eggs on leaves to be ingested by host caterpillars; and cuckoos, which leave their young to be raised by others.

Subsocial species

Subsocial species show PARENTAL CARE, occurring as single individuals or mated pairs; these include most birds and mammals, as well as many insects (e.g. burying beetles). Parental care is altruistic in terms of its consequences, although generally 'altruism' is restricted to the benefiting of non-descendants. Consider a beetle faced with two eggs of which it can only rear one: its own, or its mother's. Assuming that its SIBLING is a full sibling, there is essentially no difference whether it cares for one egg or the other in terms of the genes to be passed on, 50% of which in either case are identical by descent to genes in our beetle. Care expended on one egg lengthens the time before another can be produced, thus imposing a cost. That parental care is not generally seen as altruistic probably arises because biologists belong to a species in which parental care has become obligatory (for offspring production) and because most study similar such species (birds and mammals).

Communal and quasi-social species

Communal species resemble subsocial species in having parental care, but form groups of parents, such as for mutual protection against predators, or for concerted hunting. Examples include various bees and wasps that share burrow systems, and in mammals the prides of lions and coteries of prairie dogs.

In quasi-social species, exemplified by many wasp and bee species, and by mammals such as Cape hunting dogs (*Lycaon pictus*), individuals will, on occasion, tend each other's offspring.

Semi-social species

Semi-social species exhibit reproductive DIVISION OF LABOUR; i.e. only some individuals reproduce while others assist in care of the offspring. In mammals (e.g. wolves) and birds (e.g. white-fronted bee-eaters) the assisting individuals are termed 'helpers at the nest' (or den), but in insects (e.g. the halictid bee *Lasioglossum versatum*) they are called workers.

Eusociality

'Eusociality' is restricted to societies with not only reproductive division of labour but also the presence of two or more generations (as when the workers are the offspring of the reproductives). 'Primitive eusociality' denotes species without morphological determination of the roles, and 'advanced eusociality' describes the situation in which reproductives are morphologically differentiated from workers before adult life. Ants, honeybees and hornets are typically advanced eusocial species, as are colonial cnidarians. Naked mole rats (*Heterocephalus glaber*) and some halictid bees (e.g. *Lasioglossum zephyrum*) are examples of primitively eusocial species.

The terms 'primitively eusocial' and 'advanced eusocial' denote functional states and are not necessarily phylogenetic stages. For example, many ant species have lost differentiated queens with reproduction being carried out by morphological workers; these ants are now 'primitively eusocial' but had 'advanced eusocial' ancestors. For this reason, it has been suggested that the terms 'behavioural' and 'morphological' eusociality be used instead (although certainly all advanced eusocial species evolved from primitively eusocial ones). Another source of ambiguity stems from whether the terms in Table S8 refer to species or to specific colonies. In some wasps, for example, a COLONY begins as an association of females with one becoming dominant and producing several worker generations; such a colony has occasionally been described as passing through semi-social and eusocial stages.

Table S8 Reproductive characteristics of different levels of social organization. (From Crozier, R.H. & Pamilo, P. (1996) *Evolution of Social Insect Colonies. Sex Allocation and Kin Selection*. Oxford University Press, Oxford; by permission of Oxford University Press. Data from Michener, C.D. (1974) *The Social Behavior of the Bees*. Harvard University Press, Cambridge, Mass.; Starr, C.K. (1984) Sperm competition, kinship, and sociality in the aculeate Hymenoptera. In: *Sperm Competition and the Evolution of Animal Mating Systems* (ed. R.L. Smith), pp. 427–464. Academic Press, Orlando, Fla.; Wilson, E.O. (1971) *The Insect Societies*. Harvard University Press, Cambridge, Mass.)

Level of social organization	Continued care of young	Cooperative brood care	Reproductive division of labour	Colonies with at least two adult generations	Reproductives morphologically differentiated
Solitary	–	–	–	No colonies	–
Subsocial	+	–	–	–	–
Parasocial					
Communal	+	–	–	–	–
Quasi-social	+	+	–	±	–
Semi-social	+	+	+	–	–
Eusocial					
Primitive	+	+	+	+	–
Advanced	+	+	+	+	+

Kin selection

Although mutualistic interactions affect the form of societies generally, altruism and ultimately eusociality can arise only under kin selection. This restriction follows because in the absence of KIN SELECTION any ALLELES favouring altruism will be lost from the population. Because kin selection is mediated by the costs and benefits to the interactants of altruistic behaviours, the basic biology of the organisms, and their ecological milieu, are also important determinants. Possible contributing factors include the following.

1 Male-haploidy (HAPLO-DIPLOIDY), which gives rise to high levels of relatedness between full sisters (e.g. in Hymenoptera), favours altruism in family groups with high proportions of daughters. Sex allocation is therefore an important factor in the EVOLUTION of eusociality.

2 Split sex-ratios (i.e. family groups in the same population having two or more characteristic SEX RATIOS), may arise in various ways: for example, if some hymenopteran females are unmated they may still produce offspring but these will generally all be males, a circumstance selecting for mated females producing female-biased progenies; in bivoltine species if the males survive from one generation to the next this selects for female-biased progenies in the second generation; differences in the relative costs of producing males and females for colonies of different sizes; and differences in the relatedness structure of colonies (as when some females mate once and others twice).

3 Environmental pressures (e.g. predation) may foster group living; if the groups comprise relatives, differences in reproductive capacity between individuals may lead to selection favouring reproductive and worker roles.

4 Parental MANIPULATION may occur following the fixation of alleles favouring altruistic behaviour for individuals of reduced reproductive capacity: mothers are selected to produce some smaller individuals which then under kin selection stay to assist instead of founding their own colonies.

5 Demographic insurance may arise because an individual assisting in colony reproduction is certain of passing on some genes through its efforts whereas independent colony founding, while having a higher potential payoff, also carries the risk of losing all descendants in the event of the individual's death.

Colony or organism?

The evolution of multicellularity can be seen as an extreme result of eusociality. Colonial protists (e.g. *Volvox*) are generally the descendants of a single cell. Sometimes these are differentiated into reproductive and non-reproductive cells. Further differentiation and integration would cross a perceptual boundary, leading to description of the groups as organisms, not colonies. The question of individuality applies also to cnidarian colonies, which are genetically identical individuals stemming from one founder and which can be regarded as social groups or as individuals with modular parts. Colonies of ants, bees, wasps and naked mole rats are not individuals because the colony members differ genetically. *See also* SOCIOBIOLOGY. [R.H.C.]

REFERENCES

Crozier, R.H. & Pamilo, P. (1996) *Evolution of Social Insect Colonies. Sex Allocation and Kin Selection*. Oxford University Press, Oxford.

Krebs, J.R. & Davies, N.B. (1993) *An Introduction to Behavioural Ecology*, 3rd edn. Blackwell Scientific Publications, Oxford.

Wilson, E.O. (1975) *Sociobiology: The New Synthesis*. Harvard University Press, Cambridge, Mass.

sociobiology

1 An approach to animal social behaviour which is essentially mechanistic and which regards such behaviour as subject to the same selective pressures of EVOLUTION as anatomical features and physiological processes. Social behaviour, therefore, can be explained purely in evolutionary and adaptive terms. The expression is particularly associated with the work and ideas of E.O. Wilson.

2 A logical extension is, of course, that human social behaviour can be treated in the same way. Not all would agree, especially with regard to certain aspects. The debate of NATURE VERSUS NURTURE has been at times heated here.

[P.D.M.]

sodium (Na) A soft silvery-white metallic element (ATOMIC NUMBER = 11; relative ATOMIC MASS = 22.99; melting point = 97.5°C;

Table S9 Well-known sodium compounds and their uses.

Compound	Formula	Use
Sodium azide	NaN_3	Used in explosives
Sodium benzoate	C_6H_5COONa	A food preservative and antiseptic
Sodium bicarbonate	$NaHCO_3$	Used as baking powder and antacid
Sodium carbonate (washing soda)	$Na_2CO_3.10H_2O$	Used as a cleansing agent and in the manufacture of glass, paper and soap
Sodium chlorate	$NaClO_3$	Used in explosives and still sometimes used as a weedkiller
Sodium chloride	$NaCl$	Common salt
Sodium cyanide	$NaCN$	Used in electroplating
Sodium fluoride	NaF	Used to fluoridize drinking water and toothpaste, also as an insecticide
Sodium hydrogen glutamate (monosodium glutamate, MSG)	$HOOC(CH_2)_2CH(NH_2)COONa$	Used to intensify flavour in foods
Sodium hydroxide (caustic soda)	$NaOH$	Used in the manufacture of soap and in the chemical industry, also as a powerful cleaning agent
Sodium nitrate (Chile saltpetre)	$NaNO_3$	Used in fertilizers and as the oxidizing agent in some explosives
Sodium peroxide	Na_2O_2	Reacts with water to give sodium hydroxide and oxygen. Used as a bleaching agent and as an oxidizing agent
Sodium dihydrogen orthophosphate	NaH_2PO_4	Used in dyeing and electroplating
Disodium hydrogen orthophosphate	Na_2HPO_4	Used in dyes, fertilizers and baking powder
Trisodium orthophosphate	$Na_3PO_4.12H_2O$	Used in detergents, paper-making and as a water-softener
Sodium silicate ('water-glass')	Na_2SiO_3	Formerly used to preserve eggs, and used in paper and cement, and as a fireproofing agent in the textile industry
Sodium sulphate (Glauber's salt)	Na_2SO_4	Used in making soap, detergents and dyes
Disodium tetraborate (borax)	$Na_2B_4O_7.10H_2O$	Best known for its use in the borax bead test in the chemical laboratory. Also employed as an antiseptic and a flux, and used in the glass and ceramic industry
Sodium thiosulphate	$Na_2S_2O_3.5H_2O$	Used as fixer or 'hypo' in photography

boiling point=882.9°C) with an abundance of *c.* 2% in the Earth's CRUST. Sodium is highly reactive and must be stored in oil to exclude air and WATER. It tarnishes in air and reacts violently with water to form HYDROGEN (H_2) and sodium hydroxide (NaOH). As a pure METAL it is used in the manufacture of some organic compounds, while the vapour is used in sodium vapour lamps, formerly the main type of street lighting and now used in the glasshouse industry as an energy-efficient illumination. 'Soda' is a very general term for sodium compounds, especially washing soda (sodium carbonate; see below); soda-lime is a mixture of sodium hydroxide and calcium hydroxide ($Ca(OH)_2$), and is used as a drying agent; soda-water is CARBON DIOXIDE (CO_2) held in water under pressure, such that effervescence occurs when the pressure is released. Sodium forms many well-known compounds, these and their uses are shown in Table S9. [J.G.]

soft-bottom habitats *See* SUBLITTORAL ZONE.

Soft selection

The term 'soft selection' has considerably changed in meaning since it was coined by Wallace (1968). Four stages in its evolution can be distinguished.

The term 'soft' selection originated at the end of the 1960s in the context of a discussion concerning whether all newly discovered enzyme polymorphisms could possibly be maintained by selection due to heterozygote advantage, as this would seem to impose a GENETIC LOAD no population could ever sustain.

Wallace (1970) realized FITNESS had to be placed into an ecological context. He distinguished, therefore, two forms of selection, 'hard' selection and 'soft' selection. Soft selection involves a ranking of individuals in the population or a comparison between all individuals present. Of all individuals in the population, a given percentage or a given number will survive or mate, independent of the genotypic composition of the population. An individual's fitness and the fitness of its GENOTYPE therefore depend upon the number and composition of the population, and on the rule that determines how many and in what order the organisms in the population will survive or mate. Soft selection involves no genetic load.

His example of typical 'soft' selection involves COMPETITION between varying numbers of several genotypes for a fixed number of places to survive or pupate. Soft selection thus involves a relative criterion for selection. Soft selection has no direct influence on numbers, but its strength follows from the processes determining numbers.

The next stage in evolution of the terminology is marked by Wallace's (1975) reconsideration of his definition of 'hard' and 'soft' selection in the light of the models of DENSITY- and FREQUENCY-DEPENDENT SELECTION that had appeared between 1970 and 1975. He argued that his definition of 'soft' selection involves selection that is both density- and frequency-dependent. This follows from the earlier definition, namely the presence of a relative criterion of selection for each individual dependent of the number and composition of the population.

Wallace (1970) had given as his example of soft selection a case where a fixed number of positions were to be filled. This led Christiansen (1975) to connect the 'soft' selection of Wallace (1970) with the model of selection in a subdivided environment as described by

Levene (1953), thereby taking the terminology a stage further. In the one-locus model of Levene (1953), genotypic fitness depends upon the subenvironment. Each subenvironment contributes a fixed fraction to the total number of adults; an interpretation could be that in each subenvironment a fixed number of positions was present. ALLELE FREQUENCY change in the total population is computed from the weighted average allele frequency change per environment:

$$\Delta p = \sum_k f_k \Delta p_k$$

where f_k is the fraction of the population from environment k ($k=1, \ldots, n$). Since the change in the allele frequency p of allele A_1 at the locus concerned involves for each environment k the mean fitness $\overline{w}_k$ in that environment, the weighted average allele frequency Δp becomes:

$$\Delta p = \sum_k f_k \frac{pq}{2} \frac{\mathrm{d} \ln \overline{w}_k}{\mathrm{d}p} = \frac{pq}{2} \frac{\mathrm{d} \sum_k f_k \ln \overline{w}_k}{\mathrm{d}p}$$

(where $q=1-p$) it follows that the overall mean fitness in the LEVENE MODEL is given by:

$$\overline{\overline{w}} = \prod_k \overline{w}_k{}^{f_k}$$

Following Christiansen (1975), this multiple-niche polymorphism model of Levene (1953) became known as the 'soft' selection model.

In the most recent stage of the terminology's development, Via and Lande (1985) interpreted Christiansen's (1975) 'hard' and 'soft' selection and the multiple-niche polymorphism models of Dempster (1955) and Levene (1953) as differences in number regulation. In soft selection, they supposed the populations in each environment to be regulated independently, so that each contributes a constant fraction of the total population. However, the difference in number regulation between 'hard' and 'soft' selection is not further worked out and is secondary to the difference in mean fitness and adaptive topography in Via and Lande's development of a model for the evolution of PHENOTYPIC PLASTICITY.

Mean fitness in Via and Lande (1985) depends upon PHENOTYPIC VALUE rather than upon genotypic identity, but the adaptive topography of their 'soft' selection model is the same as the adaptive topography in the one-locus multiple-niche polymorphism Levene model. With soft selection, mean fitness over all environments is given by the GEOMETRIC MEAN over the per environment mean fitnesses:

$$\overline{\overline{w}} = \prod_k \overline{w}_k{}^{f_k}$$

Most usage of 'hard' and 'soft' selection refers at the moment (1998) to multiple-niche models. The difference between 'hard' and 'soft' selection has become a difference in formal modelling: the 'hard' selection model starts with averaging genotypic fitness over environments and leads to ARITHMETIC MEAN fitness; the 'soft' selection model starts with averaging allele frequency change over environments and leads to geometric mean fitness.

Ecological considerations show it to be inappropriate to compute allele frequency change over the total environment by averaging allele frequency change per environment, even if in each environment selection would be 'soft' in the meaning of Wallace (1970, 1975). With spatial variation, geometric mean fitness looks like an artefact. With spatial variation, mean fitness over all environments is given by the arithmetic mean over the per environment mean fitnesses, whatever the type of selection within each environment. This implies that the 'soft' selection multiple-niche model lacks a biological basis. *Cf.* HARD SELECTION. *See also* DEMPSTER MODEL. [G.D.J.]

REFERENCES

Christiansen, F.B. (1975) Hard and soft selection in a subdivided population. *American Naturalist* **109**, 11–16.

Dempster, E.R. (1955) Maintenance of genetic heterogeneity. *Cold Spring Harbor Symposia on Quantitative Biology* **20**, 25–32.

Levene, H. (1953) Genetic equilibrium when more than one ecological niche is available. *American Naturalist* **87**, 331–333.

Via, S. & Lande, R. (1985) Genotype–environment interaction and the evolution of phenotypic plasticity. *Evolution* **39**, 505–522.

Wallace, B. (1968) *Topics in Population Genetics*. Norton, New York.

Wallace, B. (1970) *Genetic Load: Its Biological and Conceptual Aspects*. Prentice-Hall, Englewood Cliffs, N.J.

Wallace, B. (1975) Hard and soft selection revisited. *Evolution* **29**, 465–473.

soft substrata Substrata consisting of sedimentary mineral grains and small organic particles derived from dead and decaying plants and animals. Most of the sea floor, and thus most of the Earth's surface, is covered by marine soft SEDIMENT. Soft substrata are characterized by grain size, sorting and mineral composition. [V.F.]

SOI *See* SOUTHERN OSCILLATION INDEX.

soil The term 'soil' means different things to different people. To the soil scientist, soil is the profile that is exposed when a pit is dug into the surface of the Earth (*see* SOIL PROFILE). To the agriculturist and horticulturist, soil is a medium for plant growth, and when carefully tended will produce good crops. To many others, soil is dirt; hence the term soiled. In reality, without the soil–plant system there would be no animal life on land — humans cannot exist without soil, since most food and clothing comes directly or indirectly from the land.

This simple truth seems to be appreciated by only a minority of humans, such as the native Australians, who consider the soil to be sacred. It would be a great achievement if this attitude could be instilled in humans generally, then we would be spared the appalling soil erosion and POLLUTION that is progressively destroying vast areas of soil. For example, in Belorussia some of the world's best soils have now been polluted for ever with RADIOACTIVITY, following the disaster of Chernobyl.

Thus the definition of soil should be that it is the lifeblood of human existence. In addition to producing food, soil acts as a filter for soluble substances in RAIN or falling directly on to the surface of the Earth. If these substances are harmful, and occur in large amounts, they will percolate through the soil to underground WATER, and may contaminate the domestic water supply. Thus soil, which seems innocuous to many, plays a vital role in the life of both plants and animals, including humans. *See also* PEDOSPHERE. [E.A.F.]

Soil classification

The CLASSIFICATION of soils is one of the most problematical areas of soil science, and despite years of effort there is still no sign of an agreed international system. On the other hand, there are many national systems in use at present, including those of Australia, Canada, England and Wales, France, Germany, Russia, Scotland, South Africa and Zimbabwe.

Zonal systems (Russia)

The early attempts to classify soils were based on soil texture, but V.V. Dokuchaev and his school in Russia discovered that soils have horizons, arranged in many different sequences. Further, the horizons and their sequence vary with parent material, CLIMATE, TOPOGRAPHY, organisms and time—climate apparently playing the dominant role. This led to the development of the zonal system of soil classification, based on climate, in which there are three main categories.

1 Zonal soils: these soils have well-developed profiles, and reflect the main climatic features of the area; they include PODZOLS and CHERNOZEMS.

2 Intrazonal soils: these soils also have well-developed profiles, but display the dominance of a local factor, such as parent material; they include Cambisols and Gleysols.

3 Azonal soils: these are the poorly developed soils, including Regosols and Fluvisols.

This system revolutionized the approach to soil classification, and was used almost exclusively until about 1960. It is still the major system in Russia, where it has gone through a number of stages of development. In the most recent, there are ten levels of categorization:

1 class;
2 subclass;
3 ranges;
4 types;
5 subtypes;
6 genera;
7 species;
8 varieties;
9 categories;
10 phases.

At present, the type is the highest level that is used, since there is as yet no final agreement about the definitions of the higher levels. The types include Arctic soils, Tundra soils, Cryogenic taiga soils, Podzolic soils, Bog-Podzolic soils, Grey forest soils, Brown forest soils, Chernozems, Meadow Chernozemic soils, Chestnut soils, Meadow steppe chestnut, Solonetz, Brown semi-desertic soils, Grey-brown desertic soils, Jeltozem, Cinnamonic soils, Grey-Cinnamonic soils and Serozems.

However, systems based on climate do not work very well in many areas, particularly in the tropics, where the soils are very old and have been subjected to many cycles of soil formation. They also fail in most oceanic areas, such as western Europe. As a result, a number of different systems have been developed, in different countries, to relate specifically to local conditions. Among these systems are those of W.L. Kubiëna and K.H. Northcote.

Kubiëna's system (western Europe)

Kubiëna developed his system to accommodate the oceanic conditions of western Europe and created three main classes.

1 Terrestrial soils, which include Podzols and Chernozems.

2 Semi-terrestrial soils, which include Gleysols and peat.

3 Subaqueous soils, which include soils at the bottoms of ponds and lakes.

Unlike the Russian system, which is based on climate, Kubiëna uses soil characteristics as the basis of his classification. This system has proved to be very useful, and has formed the basis of the current system in Germany. Many aspects of Kubiëna's system are also incorporated in the system by B.W. Avery, for England and Wales.

Northcote's system (Australia)

Northcote developed a system for Australia and has three main classes based on the vertical texture pattern.

1 Uniform, or U soils: the texture remains uniform throughout the depth of the soil.

2 Gradational, or G soils: the CLAY content increases gradually with depth.

3 Duplex, or D soil: there is an abrupt increase in clay in the middle of the soil.

Thereafter the classes are subdivided using letters and numbers to produce codes. For example, a soil with the code Uc 2.20 has a coarse texture, a conspicuously bleached horizon and a dark-coloured middle horizon. This is a Podzol in some other classifications. Northcote's system has proved very difficult to apply and will probably be superseded in the near future.

Soil Taxonomy (USA)

Soil Taxonomy (USDA 1975) is a serious attempt by the United States Department of Agriculture to develop an international system. It is being used in many places, but has failed to replace many of the national systems. Significantly, many new national systems have been produced since its introduction. This system originally contained 10 orders but subsequently two more have been added. The complete list is now: Alfisols, Andisols, Aridisols, Entisols, Gelisols, Histosols, Inceptisols, Mollisols, Oxisols, Spodosols, Ultisols and Vertisols. These orders are subdivided into suborders, great groups, subgroups, families and series.

The names of the orders are coined words, with the common ending '-sol', such as Spodosol. A formative element is abstracted from the name, and used as the ending for all the suborders, great groups and subgroups. For example, in the case of the Spodosols the formative element is -od-. Each suborder name consists of two syllables. The first indicates the property of the class, and the second is the formative element of the order. Thus the Orthods are the common Spodosols (from Greek *orthos,* true + formative element -od). The names of the great groups are produced by adding one or more prefixes to the name of the suborder. Thus the Cryorthods are the cold Orthods (from Greek *Kruos,* cold). The subgroup names consist of the name of the great group preceded by one or more adjectives, as in Typic Cryorthods—the typical Cryorthods.

Diagnostic horizons

A major innovation in *Soil Taxonomy* was the introduction of the concept of diagnostic horizons. Diagnostic horizons have quantitatively defined properties, and are used to categorize soil units. There are 26 diagnostic horizons, very briefly described as follows.

Albic: pale-coloured, leached horizon.

Argic: occurs beneath the plough layer, and has >15% coatings of HUMUS and clay.

Argillic: middle horizon with an accumulation of clay.

Anthropic: upper horizon produced by human activity.

Calcic: middle or lower horizon with an accumulation of calcium carbonate.

Cambic: middle horizon showing some degree of alteration.

Duripan: lower horizon cemented by silica.

Fragipan: hard, loamy, lower horizon with high bulk density and slakes in water.

Glossic: a horizon containing pale-coloured tongues.

Gypsic: middle or lower horizon with an accumulation of gypsum.

Histic: an accumulation of wet organic matter at the surface.

Kandic: a middle, strongly weathered, horizon with more clay than the horizon above.

Melanic: thick, black, humose upper horizon containing large amounts of allophane.

Mollic: thick, dark-coloured, surface horizon composed mainly of faunal faecal material and with high BASE saturation.

Natric: columnar or prismatic middle horizon with an accumulation of clay and exchangeable SODIUM.

Ochric: surface horizon with a low content of organic matter.

Oxic: middle and lower, highly weathered horizons with moderate to high clay content, low cation-exchange capacity and low base saturation.

Petrocalcic: hard accumulation of calcium carbonate.

Petrogypsic: hard accumulation of gypsum.

Placic: thin IRON pan.

Plaggen: upper horizon formed by the continuous addition of material at the surface.

Salic: accumulation of soluble salts.

Sombric: middle horizon, containing an accumulation of translocated humus, but does not have associated ALUMINIUM, as in spodic horizons.

Spodic: middle horizon containing an accumulation of translocated humus and/or iron and/or aluminium.

Sulphuric: pH<3.5 and jarosite mottles.

Umbric: thick, dark-coloured surface horizon composed mainly of faunal faecal material, and with low base saturation.

There are also a number of diagnostic properties including: abrupt textural change; albic material; aquic conditions; coefficient of linear extensibility; durinodes; interfingering of albic materials; linear extensibility; lithic contact; *n* value; paralithic contact; PERMAFROST; petroferric contact; plinthie; sequum and bisequum; slickensides; soft powdery LIME; SOIL MOISTURE regimes; soil temperature regimes; spodic material; sulphidic material; and weatherable minerals.

The diagnostic horizons are use to ascribe soils to different orders. For example, soils with spodic horizons are Spodosols. However, it should be pointed out that some Spodosols do not have spodic horizons; likewise many Oxisols do not have oxic horizons. Thus, this system is not as precise as it appears at first; in fact there are numerous major deficiencies, hence its lack of general acceptance.

FAO/UNESCO system

The second major system is the 'Legend' for the *Soil Map of the World* produced by FAO/UNESCO. Strictly speaking, this is a legend and not a classification but it is treated as a classification. As a result of the need to update the Legend has sprung the *World Reference Base for Soil Resources*. In the Legend, an attempt has been made to use traditional names such as Podzols and Chernozems, but there are also a few newly coined names. The full list of soil names is as follows: Acrisols, Alisols, Andosols, Anthrosols, Arenosols, Calcisols, Cambisols, Chernozems, Cryosols, Ferralsols, Fluvisols, Gleysols, Glossisols, Greyzems, Gypsisols, Histosols, Kastanozems, Leptosols, Lixisols, Luvisols, Nitosols, Phaeozems, Planosols, Plinthisols, Podzols, Regosols, Sesquisols, Solonchaks, Solonetz, Stagnosols, Umbrisols and Vertisols. Each of these classes has up to 10 units, which are named by adding adjectival prefixes to the main class names, for instance Plinthic Acrisols. Like *Soil Taxonomy*, the concept of diagnostic horizons is used, but there are some differences in detail. For example, the oxic horizon of *Soil Taxonomy* has become the ferralic horizon.

Further refinements

The deficiencies in these systems of soil classification led E.A. FitzPatrick to set out the principles that should be used in soil classification. He introduced the concept of reference horizons, which has been accepted, and used in the new *Référentiel Pédologique* produced in France. He also pointed out that soils are not like animals, and cannot be classified in a hierarchical manner. In addition, he suggested a new set of horizon names and symbols, which could be used to produce soil formulas, to give the horizons present in the soil and their sequential arrangement—as was originally suggested by D.G. Vilensky and A.N. Sokolovsky. As an alternative to the term reference horizon, Mazaheri *et al.* (1995) have suggested the use of the concept of centroids. Thus, at present, there are many different systems and approaches to soil classification. *See also* PEDOGENESIS; PEDOLOGY. [E.A.F.]

REFERENCES

Avery, B.W. (1973) Soil classification in the soil survey of England and Wales. *Journal of Soil Science*, **24**, 324–338.

Dokuchaev, V.V. (1883) *Russian Chernozem*. St Petersburg.

FAO/UNESCO (1990) *Soil Map of the World* (revised legend). Food and Agricultural Organization/United Nations Education, Scientific and Cultural Organization, Rome.

FitzPatrick, E.A. (1980) *Soils, Their Formation, Classification and Distribution*. Longman, Harlow.

FitzPatrick, E.A. (1986) A comparison of Soil Taxonomy with FitzPatrick's system of soil designation and classification. *Ciencia del Suelos* **4**, 193–207.

Kubiëna, W.L. (1953) *The Soils of Europe*. Murby, London.

Mazaheri, S.A., Koppi, A.J. & McBratney, A.B. (1995) A fuzzy allocation scheme for Australian Great Groups classification system. *European Journal of Soil Science* **46**, 601–612.

Mückenhausen, E. (1962) The soil classification system of the Federal Republic of Germany. *Transactions of the International Soil Conference in New Zealand*.

Northcote, K.H. (1971) *A Factual Key for the Recognition of Australian Soils.* CSIRO, Australia Soils Division Report, 3rd edn.
Shisov, L.L. (1990) Genetic classification of soils in the USSR. In: *Soil Classification*, pp. 77–93. Dokuchaev Institute, Moscow.
Sibirtsev, N.M. (1985) Genetic classification of soils. *Zap. Novo.-Aleksander Agricultural Institute* **9**, 1–23.
Sokolovsky, A.N. (1930) The nomenclature of the genetic horizons of the soil. *Proceedings of the Second International Congress of Soil Science* **5**, 153–154.
USDA (1975) *Soil Taxonomy Agricultural Handbook No. 436*. US Government Printing Office, Washington, D.C.
Vilensky, D.G. (1927) Concerning the principles of a genetic soils' classification. Contribution to the study of the soils of Ukrania. *First International Congress of Soil Science* **6**, 129–151.

soil ecology *See* ECOLOGY.

Soil moisture

Two aspects of SOIL moisture control its movement and availability to plants and microorganisms: soil moisture content (or water content) and the matric potential of soil water.

Water content

The water content of a soil is the amount of water it contains on a mass or volume basis. For the past 50 years, gravimetric soil water content, *m*, has conventionally been defined as the mass of water per unit mass of oven-dry (105–110°C) soil. Because this definition is on an oven-dry soil basis, a moist peaty soil that can hold more than its own mass of water can have a water content greater than 1 kg kg^{-1}. The definition also represents the standard method for measuring gravimetric water content, in which a representative soil sample is weighed, oven-dried to constant mass (usually overnight or for 24 h), and reweighed. The loss in mass divided by the oven-dry mass is the water content.

Volumetric water content, θ, is defined as the volume of water per unit volume of bulk soil, where the soil volume is measured at the time of measurement. Field measurement requires that undisturbed soil samples be taken or that some other method is used to determine the sample volume. Consequently θ is more time consuming to measure than *m*, which can be measured on a disturbed sample. However, the volumetric water content of the soil is a more useful measure and can be used in soil-water budget calculations, in which the total amount of water entering and leaving a unit cross-sectional area of soil surface can be related to the total sum of changes in volumetric water content of the SOIL PROFILE. The dry bulk density, γ_d, which is the mass of oven-dry soil per unit bulk volume (measured at the time of sampling), can be used to convert from gravimetric to volumetric water content using:

$$\theta = m\left(\gamma_d/\gamma_w\right)$$

where γ_w is the density of water.

Availability of water

The WATER POTENTIAL is a measure of the availability of water to plant roots and microorganisms. It is the sum of the effect of the soil matrix in retaining water (the matric potential) and the effect of dissolved salts (osmotic potential). In all but saline soils the dominant factor is the matric potential, and the effect of salts is ignored. In an unsaturated soil, a suction is required to remove water from the soil and the matric potential (which is always negative or zero) represents the PRESSURE required to do this. Water held at a potential of <-1.5 MPa (called the permanent wilting point) is considered to be too tightly held to be available to plant roots, although this value can only be used as a rough guide. The relationship between the water content and matric potential of any given soil is called its water release characteristic. Because water is more easily released from larger than from smaller pores, the release characteristic gives an indication of the pore size distribution of the soil.

Soil-water movement

Water moves in the soil as a result of differences in pressure (hydraulic potential) between different parts of the soil. In saturated soil the rate of flow is proportional to the saturated hydraulic conductivity of the soil, which can vary from $>10^{-4}$ m s^{-1} for sandy soils to $<10^{-7}$ m s^{-1} for poorly structured clayey soils. Differences in matric potential between different parts of the soil provide the driving force for the unsaturated flow of soil water once any differences in elevation have been allowed for. However, unlike saturated flow, the unsaturated hydraulic conductivity is a variable which decreases over many orders of magnitude with decreasing water content as water has to flow through increasingly more tortuous pathways and through smaller and smaller pores. Below a certain water content the rate at which water is conducted through the soil by mass flow becomes very slow and is exceeded by diffusion of water vapour, which is assisted by TEMPERATURE gradients near to the soil surface.

The theory of soil-water movement and availability is described by Marshall *et al.* (1996), and methods of measurement of soil-water content and other soil-water properties are given by Smith and Mullins (1991) and Klute (1986). [C.E.M.]

REFERENCES
Klute, A. (1986) *Methods of Soil Analysis*, Part 1. American Society of Agronomy, Madison, Wis.
Marshall, T.J., Holmes, J.W. & Rose, C.W. (1996) *Soil Physics*. Cambridge University Press, Cambridge.
Smith, K.A. & Mullins, C.E. (1991) *Soil Analysis, Physical Methods*. Marcel Dekker, New York.

Soil profile

A SOIL profile is a vertical face that is exposed when a pit is dug into the surface of the Earth. It is generally about 1 m wide, and extends down to the relatively unaltered material, which may occur at a depth that varies from < 50 cm to > 5 m. This vertical face usually displays a unique type of banding; the bands are known as horizons, of which there are a very large number. The specific nature of a soil profile is determined by the influence of the parent material, CLIMATE, organisms, TOPOGRAPHY and time—the five factors of soil formation (Fig. S15). These factors vary enormously from place to place, and thus the range of soil profiles is almost infinite (*see* CHERNOZEM; PODZOL).

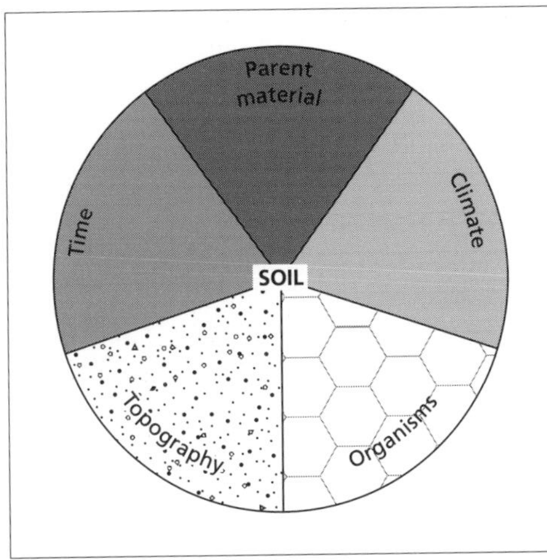

Fig. S15 The five factors of soil formation that interact to form soil.

The processes induced by the factors of soil formation lead to the formation of distinctive horizons, termed diagnostic horizons, or reference horizons (*see* SOIL CLASSIFICATION). Further, the processes leading to the formation of horizons are continuous, so that the horizons in the profile are constantly changing, forming continuum horizons (intergrades) from one to another. The changes may be progressive, and along a fairly fixed path in time. However, there is often a change in the nature of one factor, especially climate, which deflects the path of soil formation in a new direction, causing a new set of properties to be formed, and superimposed upon the previously formed set. This produces compound horizons, which are very common, because of frequent and dramatic climatic changes, particularly during the Pleistocene period.

Over the years, letter and number symbols have been used to designate the various horizons, although diagnostic and reference horizons are given names. The details of the letter and number symbols vary from country to country, but generally the system produced by FAO/UNESCO is used and given below.

Capital letters are used to designate master horizons. Lower-case letters are used to qualify the master horizons together with Arabic numbers to subdivide the horizons. Arabic numbers are also used as prefixes to indicate lithological discontinuities.

Master horizons

H Accumulations of organic matter at the surface and saturated with WATER for long periods of the year.

O Accumulations of organic matter at the surface and not saturated with water for long periods of the year.

A An intimate mixture of organic and mineral material at or near to the surface.

E Pale-coloured medium- to coarse-textured; containing mainly quartz and resistant minerals, as a result of the LEACHING of CLAY, IRON and ALUMINIUM.

B Has one more of the following features:

(a) an illuvial accumulation of clay minerals, iron, aluminium and HUMUS, alone or in combination.

(b) a residual concentration of sesquioxides relative to the source material.

(c) an alteration of the material to the extent that silicate clay is formed, oxides are liberated, or both, or granular, blocky or prismatic structures are formed.

B horizons are very variable and need to be qualified by a suffix; for example, Bt for an accumulation of clay.

C Unconsolidated material similar to the material from which the solum is formed. Strictly this is not a horizon.

R Continuous hard rock. Like the C, this is not a horizon.

Transition horizons

These are horizons in which the properties of two master horizons merge, and are designated by combining the two symbols for the master horizons, for example AE, BC. Horizons that contain mixtures of master horizons are designated by two capital letters separated by a solidus, for example A/E, E/B.

Letter suffixes

These can be used singly or combined, and include the following.

b Buried or bisequal soil horizon (e.g. Btb).

c Accumulation of concretions (e.g. Bck, Ccs).

f Frozen.

g Mottling due to differences in oxidation and reduction (e.g. Bg).

h Accumulation of organic matter in mineral horizons in undisturbed soils.

k Accumulation of calcium carbonate.

m Strongly cemented, consolidated, indurated (e.g. Cmk).

n Accumulation of SODIUM (e.g. Btn).

p Disturbed by cultivation (e.g. Ap).

q Accumulation of silica (e.g. Cmq).

r Strongly reduced (e.g. Cr).

s Illuvial accumulation of sesquioxides (e.g. Bs).

ss Presence of slickensides (e.g. Bss).

t Illuvial accumulation of clay (e.g. Bt).

u Other unspecified properties that have to be subdivided vertically (e.g. Au1, Au2).

v Presence of plinthite (e.g. Bv).

w Altered *in situ* as indicated by clay content, colour and structure.

x Occurrence of a fragipan (e.g. Cx).

y Accumulation of gypsum (e.g. Cy).

z Accumulation of salts more soluble than gypsum (e.g. Az).

Numerical suffixes

Arabic numbers are used as suffixes to subdivide horizons; for example, Bt1, Bt2, Bt3.

Numerical prefixes

Arabic numbers are used as prefixes to distinguish lithological discontinuities; for example, A, B, C, 2C, 3C.

The *World Reference Base for Soil Resources* (*see* SOIL CLASSIFICATION) has wisely abandoned the use of letters for horizons following the suggestions of Sokolovsky (1930), Vilensky (1927) and FitzPatrick (1988).

The soil continuum

A soil profile is simply a two-dimensional section through the soil. However, in reality, soil extends laterally in all directions from the profile, forming a three-dimensional continuum. The lateral extent is usually accompanied by changes in the nature of the profile, giving a continuum of change. Since these changes are gradual, it is usually impossible to produce boundaries in soils. This creates many difficulties in preparing maps, and in soil classification. *See also* PEDOLOGY. [E.A.F.]

REFERENCES

FAO/UNESCO (1990) *Soil Map of the World*, Vol. 1 (revised legend). Food and Agricultural Organization/United Nations Education, Scientific and Cultural Organization, Rome.

FitzPatrick, E.A. (1980) *Soils, Their Formation, Classification and Distribution*. Longman, Harlow.

FitzPatrick, E.A. (1988) *Soil Horizon Designation and Classification*. International Soil Reference and Information Centre, Wageningen.

Sokolovsky, A.N. (1930) The nomenclature of the genetic horizons of the soil. *Proceedings of the Second International Congress of Soil Science* **5**, 153–154.

Vilensky, D.G. (1927) Concerning the principles of a genetic soils' classification. Contribution to the study of the soils of Ukrania. *First International Congress of Soil Science* **6**, 129–151.

Soil seed banks

A soil SEED BANK is the reservoir of viable but dormant seeds which occurs in soils. In most plant communities there exists a hidden reserve of seeds in the SOIL, which remains dormant unless a DISTURBANCE creates suitable conditions for their germination. This seed bank contains seeds showing a range of potential LONGEVITY, from the EPHEMERAL (whose viability may last only a few weeks) to the highly persistent (which may germinate even after many decades). The presence of dormant seeds in the soil explains the sudden appearance of certain species in vast numbers when a field is ploughed after a long period as PASTURE.

Species and age composition

The species composition of the seed bank is normally quite different from that of the current above-ground VEGETATION. In general, persistent seed banks are characteristic of communities subjected to unpredictable disturbance, such as cultivation, fire, unreliable rainfall or fluctuations in water level. In WOODLAND and grassland, they often have a large bias towards species characteristic of early successional stages, which rely upon seeds to persist from one gap-forming disturbance to another.

Seed banks are important in the plant COMMUNITY because they play a key role in the POPULATION DYNAMICS of certain species. The COHORT of seedlings which appears at any one time may be derived from seeds of a range of ages. The older seeds may differ genetically from the recent ones. Cross-fertilization between plants derived from seeds of different ages may act as a brake on GENETIC DRIFT over time by maintaining the same mix of genes in the GENE POOL.

Management

It is often necessary to manipulate seed banks for practical purposes. Farmers attempt to reduce input of WEED seeds into the seed banks of arable fields by various cultural techniques which prevent weed growth. The existing seed bank can be reduced by inducing germination and then eliminating the seedlings by HERBICIDE treatment or by ploughing in. Seed banks are also manipulated in the restoration of degraded land using imported soil. For conservation purposes, vegetation can be managed by creating the appropriate patterns of disturbance to promote the ESTABLISHMENT of desirable species from the seed bank.

Measurement

The accurate measurement of seed banks presents many technical difficulties. Usually the quantities of viable seeds are expressed as numbers per square metre, regardless of depth. Most seeds occur in the top 5–10 cm, with a rapid decline in numbers with depth. However, ploughing can take a substantial proportion to greater depths and upset the natural pattern of vertical distribution.

The horizontal distribution of seeds in the soil is also very patchy. This creates problems when SAMPLING seed banks. The huge variation between samples taken within a few centimetres of each other necessitates the collection of large numbers of samples. Calculations show that most studies in the past have taken too few samples to be sure of arriving at an accurate estimate of seed numbers. Nevertheless, a general picture of the relative numbers in different plant communities has emerged. The following numbers are broadly representative of those found (values are seeds m^{-2}):

- subarctic and alpine forests, 10–100;
- SALT MARSHES, 100–1000;
- tropical forests, 100–1000;
- temperate forests, 1000–10 000;
- GRASSLANDS and moorlands, 5000–20 000;
- arable fields, 20 000–40 000.

Seed characteristics

A characteristic feature of species that have persistent seed banks is the possession of small, smooth seeds. Selection may have favoured seeds which can infiltrate easily into soil, as persistence is only possible if the seeds are buried. Clearly a small, round, smooth seed is more likely to trickle into soil crevices than a larger, elongated seed encumbered with hairs or hooks. A neat study by Thompson (1987) illustrates this in relation to a range of grass species. These characteristics may be of less importance in the case of arable weeds, which are artificially incorporated into the soil by ploughing.

Seed-bank dynamics

Although studies have now been made in most major plant communities to determine numbers of seeds in the soil at any one time, little is known about the dynamics of seed-bank populations: the rates of seed input and output, the longevity of each species under field conditions, the causes of MORTALITY, and the movement of seeds in soils, vertically and horizontally. With more information of this kind, it should be possible in the future to model seed-bank dynamics fairly readily and so predict seed-bank behaviour with sufficient accuracy to enable us to manipulate it effectively for vegetation management. [M.F.]

REFERENCES

Fenner, M. (1995) Ecology of seed banks. In: *Seed Development and Germination* (eds M. Negbi, J. Kigel & G. Galili), pp. 507–528. Marcel Dekker, New York.

Leck, M.L., Parker, V.T. & Simpson, R.L. (eds.) (1989) *Ecology of Soil Seed Banks*. Academic Press, London.

Thompson, K. (1987) Seeds and seed banks. In: *Frontiers of Comparative Plant Ecology* (eds I.H. Rorison, J.P. Grime, R. Hunt, G.A.F. Hendry & D.H. Lewis). *New Phytologist* **106** (Suppl.), 23–34.

soil–vegetation–atmosphere transfer scheme (SVATS) The path of WATER from SOIL through the plant to the ATMOSPHERE can be regarded as a hydrodynamic flow through the various plant tissues, and this paradigm is the basis for the usual soil–vegetation–atmosphere transfer scheme (Whitehead & Jarvis 1981).

The detailed pattern of water movement in the plant tissues, especially the root and leaf, remains an important subject for investigation. The extent to which movement in these tissues is symplastic (i.e. through the cytoplasm) or apoplastic (through the cell walls) is much debated in the literature, but will depend on which pathway offers the least resistance. However, the general pathway is well established.

Water moves vertically upwards in the plant because the total WATER POTENTIAL in the leaves becomes negative when the leaf loses water by TRANSPIRATION, whilst water in moist soil has a water potential close to zero. Water moves from the soil to the plant through root hairs and/or MYCORRHIZAE. Vertical movement in the stems is via the xylem vessels to the leaves, where finally both apoplastic and symplastic movement occurs to the sites of EVAPORATION in the substomatal cavities.

The rate at which water moves cannot be predicted from the difference in water potential, unless the hydraulic resistance is known. This is a measure of the frictional resistance imposed by the water-conducting pathway, and depends on the numbers and diameters of water-conducting cells in the leaves and stems, and also on the root structure. The concept of water moving from the soil to the atmosphere, down a GRADIENT of water potential by passing through a series of resistances was first proposed in 1948 by van den Honert and is still widely used, not only in plant ECOPHYSIOLOGY but also in related fields such as MICROMETEOROLOGY.

In modelling the flow of water, and attempting to evaluate the water potentials in the system from the FLUXES and hydraulic CONDUCTANCES, it should be realized that the root system is a major resistance to flow. The resistance is not easily measured from the physical dimensions of the pathway, because of the geometric complexity and the uncertainty in the pathway, but it can be measured with a suitable pressure probe.

The pipe model, first proposed by Leonardo da Vinci, is a very attractive representation of the plant as a hydraulic system. He observed that 'all the branches of a tree at every stage of its height, when put together, are equal in thickness to the trunk below them'. This places important constraints on the form and carbon economy of plants, and has some biomechanical implications. [J.G.]

REFERENCES

Whitehead, D. & Jarvis, P.G. (1981) Coniferous forests and plantations. In: *Water Deficits and Plant Growth*, Vol. VI (ed. T.T. Kozlowski), pp. 49–152. Academic Press, New York.

Whitehead, D., Jarvis, P.G. & Waring, R.H. (1984) Stomatal conductance, transpiration, and resistance to water uptake in a *Pinus sylvestris* spacing experiment. *Canadian Journal of Forest Research* **14**, 692–700.

Williams, M., Rastetter, E.B., Fernandes, D.N. *et al.* (1996) Modelling the soil–plant–atmosphere continuum in a *Quercus–Acer* stand at Harvard forest: the regulation of stomatal conductance by light, nitrogen and soil/plant hydraulic properties. *Plant, Cell and Environment* **19**, 911–927.

solar altitude (α) The term solar altitude or solar angle describes the elevation of the Sun in the sky (celestial sphere) relative to an observer, as the angle between the plane of the observer's celestial horizon and a line from the observer to centre of the Sun. It is symbolized by α. The solar zenith angle, θ_z, the complement of solar altitude (i.e. $90-\alpha°$), is defined as the angle between the local zenith and the line joining the observer and centre of the Sun. The zenith is defined as the point on the celestial sphere directly above the observer.

Solar altitude can be calculated as a function of latitude (ϕ, north positive), hour angle (ω) and SOLAR DECLINATION (δ), where δ is a function of DAY NUMBER (d_n) (Fig. S16):

$$\sin\alpha = \cos\theta_z = \sin\delta \sin\phi + \cos\delta \cos\phi \cos\omega$$

The hour angle (ω) is calculated with 24 h equalling 360° or 2π radians, morning positive and solar noon zero. The hour angle changes by 15° or $\pi/12$ rad h^{-1} such that $\omega=15(12-h)$ degrees, or $\omega=\pi/12(12-h)$ radians, where h is the true solar TIME or local apparent time.

Solar altitude and SOLAR AZIMUTH angles together describe the location of the Sun in the celestial sphere and are used in calculations of the energy balance of plants, animals and human artefacts. [P.R.V.G.]

REFERENCE

Iqbal, M. (1983) *An Introduction to Solar Radiation*. Academic Press, Toronto.

solar angle *See* SOLAR ALTITUDE.

solar azimuth (ψ) Solar azimuth describes the position of the Sun in the sky (celestial sphere) relative to the observer's location, in terms of its angle east or west of a line running north–south on the celestial horizon. Solar azimuth angles are defined with south zero and east positive, and vary between ±180° or $\pm\pi$ rad. Solar azimuth can be calculated for any combination of latitude (ϕ), SOLAR DECLINATION (δ) and SOLAR ALTITUDE (α) using the following equation, where solar declination is a function of DAY NUMBER and solar altitude is a function of true solar TIME:

$$\psi = \pm\cos^{-1}\left[\frac{\sin\alpha \sin\phi - \sin\delta}{\cos\alpha \cos\phi}\right]$$

The two solutions to this equation result from the symmetrical distribution of solar altitude around true solar noon, with positive values of azimuth (easterly) occurring before noon. [P.R.V.G.]

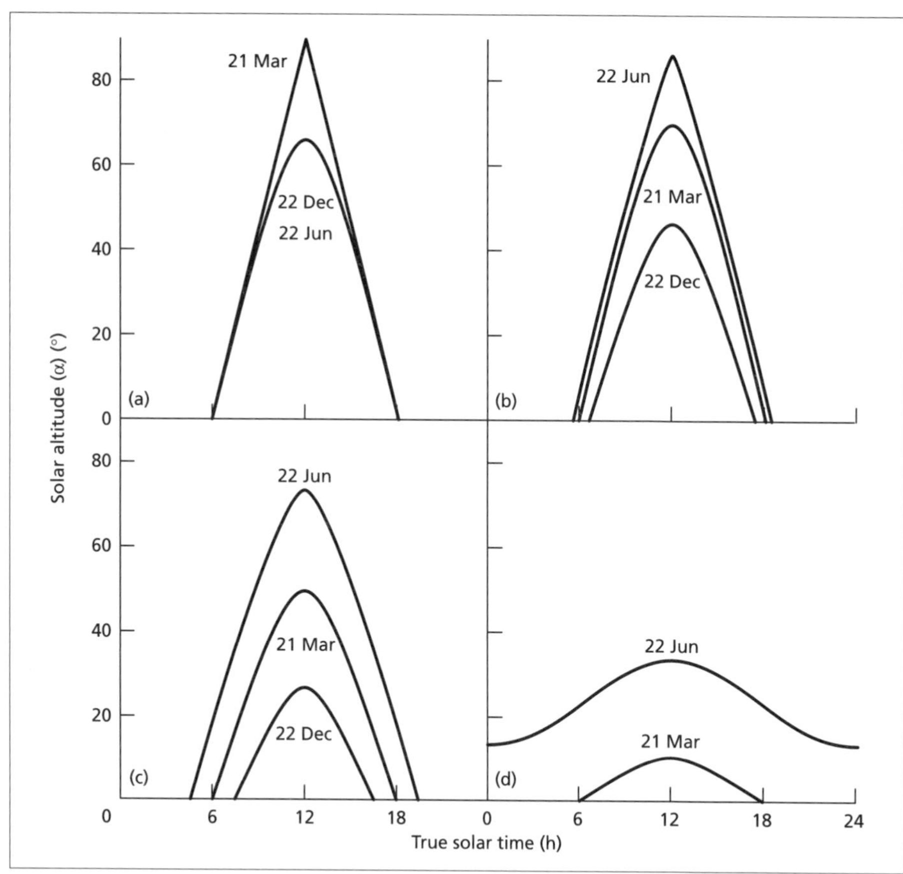

Fig. S16 Solar altitude (α, or solar angle) as a function of latitude (ϕ), true solar time and day number (d_n). (a) ϕ = 0°; (b) ϕ = 20°; (c) ϕ = 40°; (d) ϕ = 80°. The dates shown represent the winter solstice, 22 December, d_n = 356; the spring equinox, 21 March, d_n = 80; and the summer solstice, 22 June, d_n = 173.

REFERENCE
Iqbal, M. (1983) *An Introduction to Solar Radiation.* Academic Press, Toronto.

solar declination (δ) Solar declination in the equatorial system of solar geometry is the angle between a plane defined by the Earth's equator and a line joining the centres of the Earth and the Sun. Solar declination varies annually by ±23.5° resulting from the geometry between Earth's axis of rotation and its orbit around the Sun. Values for solar declination can be obtained from tables (Astronomic Almanac), from diagrams (Fig. S17) or by calculation (equations 1 and 2).

The value of solar declination (δ) measured in degrees can be computed with high accuracy using equation 1:

$$\delta = \frac{180}{\pi}\Big(0.006918 - 0.399912\cos\Gamma + 0.070257\sin\Gamma - 0.006758\cos 2\Gamma + 0.000907\sin 2\Gamma - 0.002697\cos 3\Gamma + 0.00148\sin 3\Gamma\Big) \quad (1)$$

where Γ is the day angle calculated from DAY NUMBER (d_n) as $\Gamma = 360(d_n - 1)/365$ degrees or $\Gamma = 2\pi(d_n - 1)/365$ radians.

An alternative equation, more suited to manual calculation, gives adequate results for many applications:

$$\delta = 23.45\sin\left[\frac{360}{365}(d_n + 284)\right] \quad (2)$$

[P.R.V.G.]

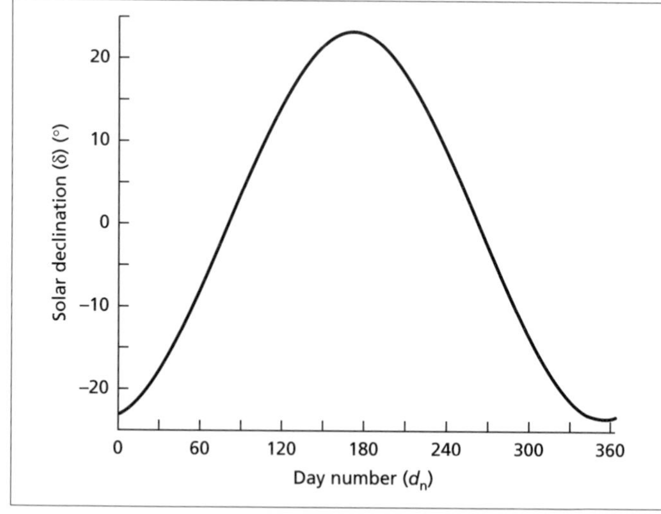

Fig. S17 Solar declination (δ) as a function of day number (d_n), as calculated from equation 1 (see text).

REFERENCE
Iqbal, M. (1983) *An Introduction to Solar Radiation.* Academic Press, Toronto.

solar power Energy from sunlight harnessed to generate electricity and/or to provide heating for water and space. *See also* RENEWABLE RESOURCE. [P.C.]

Solar radiation

Characteristics

The term 'solar radiation' is restricted to electromagnetic RADIATION emanating from the Sun, although the Sun also radiates charged particles (usually referred to as the solar wind) and neutrinos. Most of the energy of this radiation is in the wavelength range 280 nm–4 mm, but the shorter wavelength components (γ-radiation, of wavelength <0.01 nm; X-RAYS, of wavelength 0.01–50 nm; and ultraviolet-C radiation, of wavelength 50–280 nm — these wavelength limits are not universally agreed conventions) are sufficiently intense at the top of the Earth's ATMOSPHERE that they would kill unprotected organisms. Fortunately, all these shorter wavelength components are completely absorbed by the atmosphere. The Sun also radiates radiowaves of a wavelength exceeding 4 μm.

The energy of the radiation in space at the average Earth distance from the Sun amounts to 1.39 kW m^{-2} perpendicular to the radiation. This value, referred to as the 'solar constant', is only constant in the first approximation. It varies slightly (by about 0.1%) with the sunspot cycle and solar activity, and has gradually increased during the evolution of the solar system. It is estimated to have been about 30% lower at the time when life originated on Earth.

Absorption by Earth's atmosphere

When solar radiation reaches the Earth's atmosphere, some of it is absorbed; to what extent depends strongly on wavelength. All components are more or less scattered, partly back into space, the shortwave components undergoing the most intense scattering. CLOUDS reflect a large fraction of the radiation that penetrates down to the 'WEATHER layer', the TROPOSPHERE. The charged particles of the solar wind are deflected by the Earth's magnetic field, which is thus also important for protecting life on Earth.

Of great importance for the CLIMATE is the selective absorption of certain wavelength components by the so-called GREENHOUSE GASES of the atmosphere, the most important being WATER vapour, CARBON DIOXIDE (CO_2), OZONE (O_3) and the artificial CFCs. A large part of the incoming radiation is in the range 400–700 nm (visible LIGHT, also referred to as photosynthetically active radiation), and is only very weakly absorbed by the atmosphere. The ground, on the other hand, being of lower temperature than the Sun, radiates back radiation of longer wavelength, which to a large part is absorbed by the atmosphere. The absorbed radiation heats the atmosphere which, when it in turn emits heat radiation, directs part of it back to Earth. In this way the Earth retains much more of the energy from the solar radiation than would be the case for a planet without atmosphere. When people worry about the 'greenhouse effect' it is, in fact, the enhanced greenhouse effect due to increasing concentrations of greenhouse gases that is the crux of the problem (*see* GLOBAL WARMING).

Ozone is special among greenhouse gases because most of it occurs high up in the STRATOSPHERE, and also because it is a very efficient absorber of ULTRAVIOLET RADIATION. By absorbing radiation it causes the stratosphere to maintain a high temperature and be warmer at higher elevation, in turn causing it to be, as the name implies, stratified.

Table S10 Fraunhofer lines in the daylight spectrum (absorption in solar atmosphere except for oxygen).

Fraunhofer's designation	Element	Wavelength (nm)	Spectral colour
A	Oxygen (terrestrial)	759.4	Far-red
B	Oxygen (terrestrial)	686.7	Red
C	Hydrogen	656.3	Red
D_1	Sodium	589.6	Yellow
D_2	Sodium	589.0	Yellow
E	Iron	527.0	Green
F	Hydrogen	486.1	Blue
G	Iron and calcium	430.8	Violet
H	Calcium	396.8	Ultraviolet-A

Spectral composition

The spectral composition of the radiation emitted by the Sun approximates to that predicted by Planck's radiation law (*see* RADIATION LAWS) for a 'black body' having a temperature of about 5800°C. The spectral composition is modified both by the higher, cooler atmosphere of the Sun and by the terrestrial atmosphere through which it has to pass (*see* Fig. L8). The main effect of the Sun's atmosphere is to remove certain narrow wavelength bands, the Fraunhofer lines (Table S10). Thus, for instance, red light of wavelength 656.3 nm, used up in the solar atmosphere for exciting HYDROGEN atoms to the lowest excited state, is practically absent from daylight.

This fact can be exploited in REMOTE SENSING: at this wavelength, fluorescence from photosynthetic organisms (plants and algae) can be recorded without interference from sunlight. The solar radiation reflected from the Earth is also used in many ways for remote sensing of the BIOSPHERE and our environment. *See also* OPEN SYSTEM; PHOTOSYNTHESIS; PRIMARY PRODUCTIVITY. [L.O.B.]

solar spectrum *See* RADIATION.

solar zenith angle *See* SOLAR ALTITUDE.

solarimeter An instrument for measuring the total energy RADIATION from the Sun, integrated over all wavelengths. The sensor is usually a thermopile consisting of several thermocouples. The SOLAR RADIATION heats every second junction in the thermopile, while the other junctions are shielded and thermally connected to a heat buffer of massive metal.

There are two main types of solarimeters: (i) pyrheliometers, which track the Sun over the sky and measure only the direct beam from the Sun, and (ii) pyranometers, which measure the total radiation falling on a horizontal surface, i.e. the sum of the radiation from the Sun and from the sky. The term 'solarimeter' is not often used nowadays. *See also* RADIOMETER. [L.O.B.]

solitary species *See* SOCIALITY, TYPES OF.

Solonchaks *See* SOIL CLASSIFICATION.

Solonetz *See* SOIL CLASSIFICATION.

solstice *See* SEASON.

solution lakes *See* LAKES, TYPES OF.

soma Early in the ONTOGENY of some multicellular organisms, cell lines giving rise to GAMETE-producing cells, tissues or organs (the GERM LINE) become sequestered from those constituting the rest of the organism (the soma). The soma comprises the part of the organism that is affected directly by environmental influences during its life. MUTATIONS in somatic cells are not inherited by descendants. This relative independence of the germ line from direct environmental influences, dubbed Weismann's barrier, renders EVOLUTION non-Lamarckian. *Cf.* GERM LINE. [D.R.B.]

somatic mutation MUTATIONS may occur either in GERM LINE (the cells that will result in eggs and/or sperm) or in cells that will not become germ line (SOMA). Mutations in the latter are somatic mutations. Although somatic mutations may affect the viability and/or fertility of an individual, they do not affect heritable components of FITNESS as somatic alterations, by definition, will not be transmitted from parent to offspring. Numerous diseases, including various cancers, are the consequence of somatic mutations. [L.D.H.]

somatic polymorphism Although it is the case that every cell within an individual is derived from the same zygote, it need not be the case that all cells are perfectly genetically identical. The differences between the cells that are not GERM LINE (i.e. somatic cells) constitute a POLYMORPHISM within the SOMA of an individual (cf. polymorphism within a population).

Somatic polymorphism has two causes. First, postzygotic MUTATION (i.e. SOMATIC MUTATION). Second, there may be a polymorphism of cytoplasmic factors (e.g. mitochondrial genomes) in the zygote that (as is the nature of cytoplasmic genomes) undergo random segregation into the cell lineages derived from the zygote. Hence, different cells may end up with different cytoplasmic genomes. [L.D.H.]

sombric horizons *See* SOIL CLASSIFICATION.

Some world records for life-history traits

Clutch size

The record for smallest CLUTCH SIZE, i.e. one, is shared by many organisms. The single offspring of a blue whale is as large at birth as an adult African elephant and grows to half its adult mass in the first six months of life, feeding exclusively on fat-rich milk provided by a mother who is herself not able to feed, for the period from birth to weaning is spent in tropical waters poor in food and predators.

Although large, whale offspring are fairly small relative to the SIZE of the mother. Bats have the largest offspring for their body size among the mammals, the record being held by *Pipestrellus pipestrellus*. It normally carries twins whose combined mass at birth is 50% of the mother after parturition—and the mother must fly and feed while pregnant. The kiwi, at 2 kg the mass of a domestic chicken, lays the heaviest egg for its size of any bird. Its 350–400 g egg is more than five times heavier than the largest EGGS from domestic chickens selected for large egg size. The reproductive investment of the kiwi is, however, dwarfed by that of the caecilian *Dermophis mexicanus*, which can give birth to a clutch that weighs as much as 65% of the post-parturition mass of the female.

A clutch size of one is not surprising if the expected NUMBER OF OFFSPRING per lifetime is large. But the small clutches of dung beetles—four or five—are astonishing, for dung beetles are thought to reproduce only once. If so, their survival to maturity is as high as any arthropod's—about 0.5. At the other end of the scale, many organisms have large numbers of very small offspring. The seeds of orchids number in the hundreds of millions to billions. Large bivalves, and some large fishes (e.g. sturgeon, large catfishes, cod) release tens to hundreds of millions of eggs in each SPAWNING. Many trees that live several hundred years produce well over 100 000 seeds per year, and endoparasites, such as tapeworms, produce thousands of eggs daily as adults and can survive for decades in long-lived hosts. The species with the world record for worst juvenile survival is probably an orchid whose seed has a chance of about 10^{-9} of surviving to reproduce. Endoparasites also have very low probabilities of surviving transmissions from host to host.

Lifespan

The longest-lived invertebrates are sea anemones, lobsters and bivalves. The shortest-lived are rotifers, insects and some crustaceans. Parasites of long-lived vertebrates tend also to be long-lived. Species have maximum possible lifespans ranging across four orders of magnitude (days to centuries). The longest-lived molluscs are marine bivalves: *Arctica islandica* can live 220 years, and *Crenomytilus grayanus* can live 150 years. The record for the maximum LIFESPAN in captivity for non-human mammals is held by the African elephant at 57 years, with the domestic horse and the spiny echidna tied for second at 50 years. [S.C.S.]

---

sonde (radiosonde) A free-flying balloon-borne package of sensors used to measure near-vertical profiles of PRESSURE, TEMPERATURE, RELATIVE HUMIDITY and WIND through the TROPOSPHERE and into the low STRATOSPHERE, for WEATHER observation and forecasting. As the sonde rises (at about 4 m s^{-1}), data are transmitted by radio to the launching station, where they are reduced and coded for onward transmission through the global meteorological telecommunication network. The wind profile is obtained from the sonde's successive plan positions, measured by radiolocation or tracking RADAR. After natural bursting of the balloon about 20 km above sea level, descent is controlled by a paper drogue.

In World Weather Watch (*see* METEOROLOGY), several thousand land stations release fully instrumented sondes at 00.00 and 12.00 Z (Universal or Greenwich Mean Time), and sondes for wind profiles only at 06.00 and 18.00 Z. Full details are available from the relevant national weather service. Automatic sonde systems are being developed for installation on very large merchant ships to cover the huge gaps in the network over the oceans. [J.F.R.M.]

soot Finely divided CARBON particles, capable of adhering together, that are left in flues when FOSSIL FUELS are incompletely burnt. [P.C.]

sorting, of sediment The RANGE of scatter of particle sizes about the MEDIAN grain size of a SEDIMENT. Well-sorted sediment shows little variation in grain size. In general, deposits formed by wave action are better sorted than are river deposits, but not as well sorted as wind-blown sands. Glaciers form very heterogeneous unsorted deposits. The degree of sorting in a sediment influences its porosity (ratio of pore volume to total volume). [V.F.]

southern oscillation index (SOI) An index of atmospheric pressure variation in the Southern hemisphere defined as 'the standardized monthly Tahiti mean atmospheric pressure anomaly at sea level, minus the standardised monthly Darwin mean atmospheric pressure anomaly at sea level'. It is derived from atmospheric pressure measurements as an indication of ocean surface temperatures. 'Anomalies' are the amounts by which the mean pressure for a month differs from the long-term average. 'Standardization' is achieved by dividing each monthly value by the standard deviation for that month. *See also* EL NIÑO; ENSO. [J.M.]

REFERENCE

Wright, P.B. (1985) The Southern Oscillation: An ocean–atmospheric feedback system? *American Meteorological Society Bulletin* **66**, 398–412.

space-limited samples *See* SAMPLING METHODOLOGY/DEVICES.

spatial distribution, aggregated *See* AGGREGATED DISTRIBUTION.

spatial distribution, random *See* POISSON DISTRIBUTION.

spatial distributions, clumped *See* AGGREGATED DISTRIBUTION.

spatial distributions, overdispersed *See* REGULAR DISTRIBUTION.

spatial distributions, regular *See* REGULAR DISTRIBUTION.

spawning The deposition or production of EGGS or young, usually in large numbers. Spawning is generally seasonally variable and, in temperate organisms, may be controlled by temperature, light or food availability. Simultaneous release of spawn and eggs by males and females often occurs in aquatic invertebrates with external FERTILIZATION, and facilitates the union of GAMETES. *See also* LIFE-HISTORY EVOLUTION. [V.F.]

Spearman's rank correlation coefficient (r_s) A NON-PARAMETRIC STATISTIC that summarizes the degree of CORRELATION between two sets of observations (x and y). The correlation coefficient (r_s) varies between +1 (perfect positive correlation), through 0 (no correlation), to −1 (perfect negative correlation). The two variables need not follow a NORMAL DISTRIBUTION; it is only necessary that they can be ranked. The data below examine the possibility that the diversity of invertebrates (number of species) is related to pH, in streams.

pH (x)	Diversity (y)	Rank x	Rank y	Difference in rank (d)	d^2
5.0	18	1.5	1	0.5	0.25
5.0	23	1.5	3	−1.5	2.25
5.5	25	3	4	−1	1
5.8	22	4	2	2	4
6.0	33	5	6	−1	1
6.1	34	6	7	−1	1
6.4	40	7.5	10	−2.5	6.25
6.4	36	7.5	8	−0.5	0.25
6.5	32	9	5	4	16
6.9	39	10	9	1	1
					$\Sigma d^2 = 33$

Note that the sum of the d column should equal zero if the ranks have been computed correctly. The Spearman rank correlation coefficient is now calculated as:

$$r_s = 1 - \left[\frac{6\sum d^2}{(n^3 - n)}\right]$$

where n is the number of paired observations and 6 is a constant. For the above data this is equal to:

$$r_s = 1 - \left[\frac{6 \times 33}{(10^3 - 10)}\right]$$

$r_s = 0.8$

This value of the test statistic is compared with the DISTRIBUTION of r_s in the appropriate STATISTICAL TABLES for $n = 10$ (DEGREES OF FREEDOM are not used in this test). For $n = 10$ the 1% value for r_s is 0.794 and therefore our computed value of r_s indicates that there is a very significant positive correlation between pH and invertebrate diversity in these particular streams. [B.S.]

special areas of conservation (SACs) *See* NATURA 2000.

specialist A species, an individual or (in BEHAVIOURAL ECOLOGY) a STRATEGY that uses a relatively small proportion (in extreme cases only one) of the available resource types. *Cf.* GENERALIST. [J.G.S.]

speciation The formation of new species. Neo-Darwinists have based their explanation of this on microevolutionary processes which involve some kind of isolation of POPULATIONS which then diverge under the influence of NATURAL SELECTION to an extent that they become incapable of interbreeding. Classically, isolation is supposed to occur as a result of physical barriers and leads to ALLOPATRIC SPECIATION. Alternatively, in principle, GENE FLOW can be suppressed in populations which exist in the same area (*see also* PARAPATRIC SPECIATION; PERIPATRIC SPECIATION; SYMPATRIC SPECIATION). Whereas few biologists would doubt the importance of the allopatric process, there has been much debate on the importance of other processes. [P.C.]

speciation, allopatric *See* ALLOPATRIC SPECIATION.

speciation, paripatric *See* PARIPATRIC SPECIATION.

speciation, peripatric *See* PERIPATRIC SPECIATION.

speciation, sympatric *See* SYMPATRIC SPECIATION.

species *See* TAXONOMIC CATEGORIES.

species-abundance models In no COMMUNITY are all species equally abundant. Usually a few species are very common and many species are rare. Data on species ABUNDANCE can be depicted as species rank/abundance plots, showing the number of species (*y*-axis) falling into different abundance classes or ranks (*x*-axis) (Fig. S18). In addition to these graphs, species abundance data can frequently be described by one or more of a family of distributions.

Fisher's log series model gives a linear relationship between number of species and log number of individuals. Communities with a small number of abundant species and a large proportion of rare ones are predicted by the model. Light-trap collections and ground flora of conifer plantations produce a reasonable fit.

The LOG NORMAL DISTRIBUTION is derived when the number of species is distributed randomly about some mean value on $\log_2$ scale of abundance ('octaves'; 1, 2, 4, 8, 16, etc.). The log normal successfully models collections of large numbers of taxonomically related species, where a number of independent factors are important to the community, and is the most commonly found DISTRIBUTION in ecological studies.

MacArthur's broken stick (NICHE boundary) model (*see also* BROKEN STICK MODEL) predicts a linear relationship between relative abundance and log species rank (from most to least abundant). The model assumes one factor (e.g. COMPETITION) controls relative abundance distributions. Fits to the model are found amongst small homogeneous groups of long-lived relatively large species such as birds and predatory gastropods.

Whittaker's niche pre-emption or GEOMETRIC SERIES model also assumes that the community is dominated by one factor and predicts a steep linear relationship between log species abundance and species rank. The model has been successfully applied to species-poor, early successional vascular plant ASSEMBLAGES or communities in harsh environments.

The four models follow a progression from the geometric series where a few species dominate the community, through log series and log normal, where species of intermediate ABUNDANCES become more common, to the broken stick, where species abundances are more equal. The models each have a characteristic shape on a standardized rank/abundance plot (Fig. S18). Changes from log normal to geometric distributions have been used as indicators of organic enrichment and pollution, especially in coastal marine benthic communities. *See also* DIVERSITY INDICES; RANK–ABUNDANCE MODELS. [B.S. & P.S.G.]

REFERENCES

Magurran, A. (1988) *Ecological Diversity and its Measurement.* Princeton University Press, Princeton, N.J.

May, R.M. (1975) Patterns of species abundance and diversity. In: *Ecology and Evolution of Communities.* (eds M.L. Cody & J.M. Diamond), pp. 81–120. Harvard University Press, Cambridge, Mass.

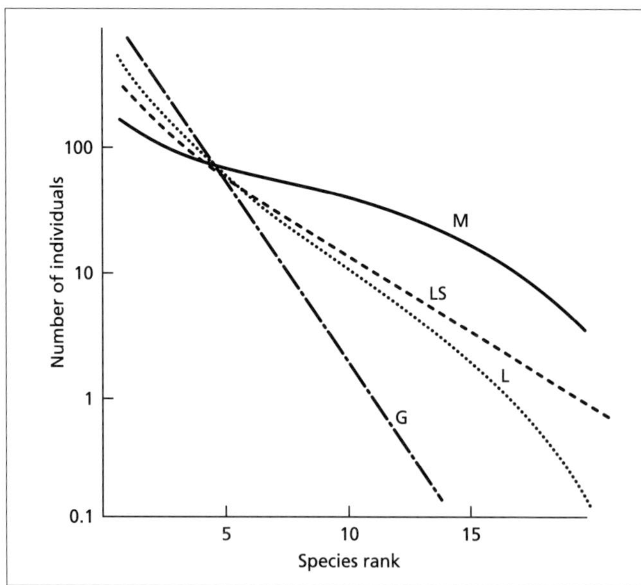

Fig. S18 Species abundance distributions as predicted by the four common models each have a characteristic shape when plotted on a standardized graph of rank vs. abundance. M, MacArthur's broken stick; LS, log series; L, log normal; G, geometric series. (After Giller, P.S. (1984) *Community Structure and the Niche*. Chapman & Hall, London.)

species–area curve *See* SPECIES–AREA RELATIONSHIP.

species–area relationship More species are encountered as progressively larger areas are sampled. This empirical observation is known as the species–area curve and holds whether the samples are islands of an archipelago or portions of a mainland BIOTA. The positive relationship between area and SPECIES RICHNESS is an almost ubiquitous phenomenon. Species–area relationships have been described for oceanic and continental islands, discrete mainland areas such as mountain tops and HABITAT fragments, and for species dependent upon living hosts such as insects on plants. Mathematical descriptions vary from the power log–log plot to semi-log (i.e. exponential) and even linear plots. Theoretically, the power plot (where $S=cA^z$; S=species richness, A=area and c and z are constants) with z=0.25 has held most interest because this can be predicted from the canonical log-normal species abundance curve. Typical z values average around 0.15–0.30.

There are three mechanistic models. Coleman's random placement model acts as a NULL HYPOTHESIS and states that large islands have more individuals and therefore more species by chance alone. The two other mechanisms are the effects of area *per se*, epitomized in MacArthur and Wilson's (1967) equilibrium model, and the effects of habitat heterogeneity. Sampling effects, in which larger areas receive greater sampling effort, may also produce a species–area relationship. The processes generating species–area relationships vary with spatial scale, so that as spatial scale increases, biogeographic and historic explanations may predominate. The species–area relationship has been used in conservation for determining the best design of nature reserves, sometimes paraphrased as the SLOSS debate ('single large or several small' reserves). *See also* ISLAND BIOGEOGRAPHY; ISLAND BIOGEOGRAPHIC THEORY IN CONSERVATION. [A.E.M., N.M. & S.G.C.]

REFERENCES

MacArthur, R.H. & Wilson, E.O. (1967) *The Theory of Island Biogeography.* Princeton University Press, Princeton, N.J.

Rosenzweig, M.L. (1995) *Species Diversity in Space and Time.* Cambridge University Press, Cambridge.

Species concepts

The concepts, or ideas, that underlie what biologists mean by the term 'species'. EVOLUTION produces DIVERSITY at many levels, from genes, races, species, to genera and higher taxa. There are many problems in classifying this diversity, especially with regard to the level in the evolutionary hierarchy in which to place a particular TAXON. The species has often been held in special regard as the only 'real' taxon. However, there are an extraordinary number of different concepts, ideas and definitions of organic species; only a few can be summarized here.

Historical concepts

The ancient philosophical terms 'genus' and 'species' were used in logic to classify objects and ideas as well as living organisms. The species was a collection of objects that had a common underlying 'essence'; in fact the Greek term εἶδος was used for both species and essence. The genus was a group of such species, with its own, broader essence. For biological organisms, this 'essentialism' fitted well with the biblical story of creation, and Linnaeus adopted this essentialist view of CLASSIFICATION in his ambitious project to catalogue every organism known at the time (18th century).

Binomial nomenclature

Linnaeus pioneered the use of binomial nomenclature, employing Latin scientific names which serve as a shorthand for descriptions of species. As an example, the scientific name of the yellow-shafted flicker (a woodpecker) is *Colaptes auratus*, where *Colaptes* is the generic and *auratus* the specific epithet. *See also* BINOMIAL CLASSIFICATION; LINNAEAN CLASSIFICATION.

Morphological species concept

The rise of evolutionary ideas meant that the old creationist essentialism was no longer tenable. Charles Darwin was interested in showing that species evolved, but to do so he obviously had to develop a species concept which depended neither on creation nor on evolution. If they evolved, varieties and species could be continuous in space and time. Thus Darwin wrote: 'hereafter, we shall be compelled to acknowledge that the only distinction between species and well-marked varieties is, that the latter are known, or believed, to be connected at the present day by intermediate gradations, whereas species were formerly thus connected.' To Darwin, the origin of species became the origin of the morphological gaps between populations. This species concept has been called the morphological species concept, although it emphasizes the clustering of members of the same species in morphological space, rather than the fact that morphological characters were used in its implementation.

The subspecies concept

As the geographic representation of well-organized museum collections increased, a major revolution in this Darwinian species concept began to take place (1890–1920). It was shown that some apparently good species of birds and butterflies blend together in areas of overlap, or HYBRID ZONES. Related forms that hybridize and replace each other geographically became downgraded to subspecies and were referred to by a novel extension of the Linnaean system, which now could consist of a trinomial: genus–species–subspecies. For example, the red-shafted flicker (formerly *Colaptes cafer*) is now usually referred to as *Colaptes auratus cafer*, or the red-shafted race of the common flicker. This species also includes the yellow-shafted flicker, *C. auratus auratus*. Species with more than one subspecies became known as polytypic species (see Plate 20, facing p. 398). At the same time, other, more trivial forms or varieties within local populations became excluded from the formal Linnaean TAXONOMY.

Biological species concept

In the 1930s and 1940s, scientists became dissatisfied with the Linnaean/Darwinian character-based definitions of species; they felt that species designations should reflect a real underlying biological phenomenon rather than remaining merely as categories for taxonomic convenience, and they wished to formalize this reality by specifying their idea of the important biological process. Drawing on ideas from G.L. Buffon and other early biologists, E.B. Poulton, Th. Dobzhansky and E. Mayr proposed what is now known as the BIOLOGICAL SPECIES CONCEPT, in which species are thought of as populations which do not interbreed, and are therefore reproductively isolated from other species. These ideas were developed along with (though do not necessarily require) the idea that species were important units of evolution, and that ISOLATING MECHANISMS were protective devices to maintain the genetic integrity of the species.

Alternative concepts

The biological species concept seems to have been largely accepted by zoologists for about 30 years. On the other hand, botanists never fully accepted the idea because plants often had high rates of hybridization, local VARIABILITY and environmentally induced plasticity. In recent years, however, any semblance of agreement about species concepts, even among zoologists, has been shattered. An explosion of new ideas has occurred. One movement, spearheaded by P.R. Ehrlich and P.H. Raven has claimed that populations, rather than species, are the important and real biological units of evolution. Others claim that biological processes do underlie species, but each has supported a different type of process as the important reality. Examples include L. Van Valen's ECOLOGICAL SPECIES CONCEPT, in which species are defined by their ecological niches, and H.E.H. Paterson's RECOGNITION CONCEPT OF SPECIES, in which species are defined by sexual signalling or SPECIFIC-MATE RECOGNITION systems within species (*see also* ISOLATING MECHANISMS). The cohesion concept of species was proposed by A. Templeton to combine REPRODUCTIVE ISOLATION, ecological selection and reproductive compatibility within a single cohesion species concept. The major advantage of this idea was that both hybridizing and asexual species—which could not be classified easily under the biological species concept—could be included.

Evolutionary and phylogenetic species concepts

A completely different approach to species concepts has been to include the idea of evolutionary history as opposed to merely the maintenance of current species. The evolutionary species concept, in which a species is a LINEAGE evolving separately from others, was proposed by G.G. Simpson to allow fossils to be classified as species as well as living organisms. This idea has been formalized recently in various types of phylogenetic species concept, in which the individuals that belong to a species contain all the descendants of a single population of ancestors, i.e. they are monophyletic (*see* MONOPHYLY). This group of ideas was developed by J. Cracraft and others specifically in response to an increase in the use of CLADISTICS in classification. In cladistics, only APOMORPHIES (uniquely derived TRAITS) are used to unite groups; reproductive compatibility and free hybridization supposedly cannot be used in species definitions because they are primitive, or plesiomorphic traits. Unfortunately, hybridization may also allow genes to pass from one taxon to another, and so different genes within groups of organisms may in fact have different phylogenies (phylogenies of single genes are called genealogies).

To get around this problem that a species may have no single, 'true' phylogeny, D.L. Baum and K.L. Shaw have suggested a variant phylogenetic species concept based on the consensus of many estimated genealogies of different genes; this is called the genealogical species concept. Finally, A. Templeton has recently added phylogenetic and genealogical considerations, as well as ecology and reproductive isolation, to his cohesion concept of species.

The problem with species concepts

Most recent species concepts (Ehrlich and Raven's population concept is an exception) attempt to identify the underlying biological 'reality' of species, and are therefore, to some extent, modern examples of essentialist thought. From Mayr onwards, there has been a deliberate attempt to exclude any consideration of usefulness of the term species from the discussions about species concepts. This may be a mistake. The debate can be resolved if we explicitly avoid using evolutionary history or the biological means by which the integrity of a species is maintained, and instead strive for a definition of species which is useful in taxonomy, evolutionary studies and conservation.

Genotypic cluster species

Given that we now have abundant genetic data, we can instead use a genetic version of Darwin's morphological cluster concept, called by J. Mallet the genotypic cluster definition. Genotypic clusters can be identified by the presence of gaps between groups of multilocus genotypes within a local area (Fig. S19 & Plate 20, facing p. 398), in the same way that Darwin's morphological cluster species are identified by morphological gaps; indeed, morphology is often a good clue to GENOTYPE. The genotypic cluster definition reverts back to the taxonomic practice inherent in the polytypic species (in which races are judged CONSPECIFIC by means of abundant intermediates in hybrid zones). Genotypic cluster species are indeed very similar to the practical taxonomic application of the biological species concept. It is, of course, likely that genotypic cluster species will be maintained as distinct entities from one another because of reproductive or ecological traits, and that they will achieve evolutionary, phylogenetic and genealogical separation through time. But genotypic clusters may violate one or more of these biological or evolutionary principles and yet remain distinct. The genotypic cluster definition is therefore related to the cohesion concept in that it allows for multiple means of cluster maintenance and evolution but, instead of arbitrating between every possible process of cohesion, it merely examines the genetic results of the combination of processes.

Species as genotypic clusters are more useful in taxonomy, in conservation, and in the investigation of SPECIATION (the evolution of genotypic gaps that do not dissolve in sympatry) than species based on idealized evolutionary or biological concepts, because biological and evolutionary processes are not assumed before the work starts.

The need for consensus

The profusion of species concepts in the biological literature is currently a major hindrance to the study of biological diversity, and its

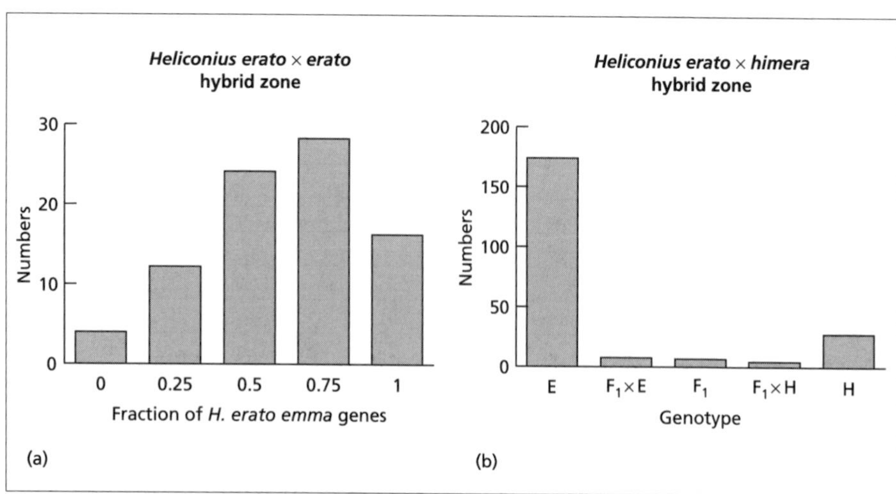

Fig. S19 The frequency of genotypes in two hybrid zones involving *Heliconius erato*. (a) The numbers of individuals with different proportions of colour pattern genes from *H. erato emma* in the centre of a hybrid zone between *H. e. favorinus* and *H. e. emma* near Pongo de Cainarache, San Martín, Peru. (b) The numbers of individuals from one site in a hybrid zone between *H. erato* (E) and *H. himera* (H) plotted against the genotypic class to which they belong; pure, first generation hybrids (F_1) or back-crosses ($F_1 \times E$, $F_1 \times H$). In (a), the forms are considered members of the same species because we can distinguish only one peak or cluster in the genotype distribution. In (b), two genotypic clusters are distinguishable, which are considered separate species.

use in conservation and evolutionary investigations. And yet, biologists no longer argue about 'cell concepts', or 'gene concepts', presumably because the concepts of cells as units of tissue, and DNA as the genetic material, are now broadly understood. We continue to disagree about species concepts because we do not yet understand species very well; at the same time, we need to make decisions because species themselves are becoming extinct at an accelerating rate. These arguments may exist because we are asking the wrong questions (perhaps species do not really exist?), but more probably because we still have much to learn about species. In any case, it is to be hoped that a generally applicable and useful idea of species will soon put an end to the current chaos. Darwin's concept, if updated, may still be the best and most useful solution. *See also* MACROEVOLUTION; MICROEVOLUTION. [J.L.B.M.]

REFERENCES

Berlocher, S. & Howard, D. (eds) (1997) *Endless Forms: Species and Speciation.* Oxford University Press, New York.

Mallet, J. (1995) A species definition for the Modern Synthesis. *Trends in Ecology and Evolution* **10**, 294–299.

Mayr, E. (1982) *The Growth of Biological Thought. Diversity, Evolution, and Inheritance.* Belknap, Cambridge, Mass.

species–distance relationship For an island or habitat patch of given size, equilibrium SPECIES RICHNESS is negatively related to the distance from source mainland populations. The form of the species–distance relationship depends in part on the DISPERSAL ability of species and the effectiveness of the barrier isolating island and mainland areas. In MacArthur and Wilson's (1967) equilibrium theory, distant islands have a lower IMMIGRATION rate, causing the slope of their IMMIGRATION CURVES to be lower, predicting a lower equilibrium species richness. Species–distance relationships only relate to species whose dispersal abilities vary over the spatial scale in question. *See also* ISLAND BIOGEOGRAPHY; SPECIES–AREA RELATIONSHIP. [N.M. & S.G.C.]

REFERENCE

MacArthur, R.H. & Wilson, E.O. (1967) *The Theory of Island Biogeography.* Princeton University Press, Princeton, N.J.

species interactions Interactions between species can be classified and defined in various ways. A very useful method of classification is to use the 'effect' that an individual of one species has upon an individual of another species and vice versa. We ask the question: in the presence of species A (+) does species B (i) increase its numbers (+); (ii) not change its numbers (0); or (iii) decrease its numbers (–) relative to when species A is absent (–)? The same question is asked of species A in the presence of species B. The answers can be conveniently summarized in a table.

		Effect of A on B		
		+	0	–
Effect of B on A	+	++	+0	+–
	0	0+	00	0–
	–	–+	–0	––

Because of the symmetry there are only six types of interaction. These are:

00, NEUTRALISM;
+0, COMMENSALISM;
+–, predator–prey, parasite–host, herbivore–plant interactions;
0–, AMENSALISM;
––, COMPETITION;
++, symbiosis.

See also MUTUALISM; PARASITE–HOST INTERACTIONS; PREDATOR–PREY INTERACTIONS. [B.S.]

species-level diversity, global Despite widespread interest in biological DIVERSITY, estimates of how many species inhabit the Earth are surprisingly vague. Larger organisms are relatively well studied and it is known, for instance, that there are about 9000 species of birds, 20 000 species of fish and 225 000 species of plants (from mosses to angiosperms). Other groups, such as marine MACROFAUNA, fungi and particularly the insects, have only begun to be catalogued. It is, however, evident that insects are by far the most speciose taxon on Earth. At present about 1 million species of insects have been recorded. The total number of 'known' species (across all taxa) is of the order of 1.8 million. Yet, estimates based on the number of beetles in the canopies of certain tropical trees hint at much higher totals, and it has been suggested that there could be as many as 30 million tropical arthropod species. This extrapolation assumes that there are approximately 160 species of canopy beetle per tropical tree species, that beetles represent about 40% of arthropod species, that for every two insect species in the canopy there is at least one existing elsewhere on the tree, and that there are about 50 000 different species of tropical tree. Each step in the logic is fraught by uncertainty, and slight shifts in, for instance, the estimate of the proportion of the fauna specialized on a given tree species can cause the overall total to rise (to above 100 million) or fall (to below 10 million) dramatically. The most conservative estimate puts the global total of invertebrates (and hence the overall species catalogue) at 3 million.

Suggestions that other poorly studied groups, such as marine molluscs, crustaceans, polychaete worms and other benthic macrofauna may contribute another 10 million species worldwide (as opposed to the 200 000 already recorded) are open to debate. It is, however, notable that marine systems contain only about 15% of recorded species but have over 2.5 times as many phyla as terrestrial systems so may actually be much richer depositories of diversity than commonly assumed. The diversity of the microbial world is harder to quantify. Much of it is uncharted, and objective measures of microbial diversity have yet to be produced. [A.E.M.]

REFERENCES

May, R.M. (1990) How many species? *Philosophical Transactions of the Royal Society of London, Series B* **330**, 293–304.

May, R.M. (1994) Conceptual aspects of the quantification of the extent of biological diversity. *Philosophical Transactions of the Royal Society of London, Series B* **345**, 13–20.

Rosenzweig, M.L. (1995) *Species Diversity in Space and Time.* Cambridge University Press, Cambridge.

species recognition This unfortunate term occurs frequently in the literature. The objections to it are several. Firstly, 'species' is a concept peculiar to *Homo sapiens.* Secondly, it is deceptive and misleading because it ostensibly identifies a process which in fact does not occur. To prevent problems, 'recognition' must be defined for use in behaviour. For present purposes we can define 'recognition' as a specific response by one mating partner to a signal from the other. The signal can be auditory, optical, chemical or tactile. The definition of 'recognition' needs modifying if it is to be used, for example, for predators, or habitat or food. The situation that is covered by 'species recognition' is in fact 'SPECIFIC-MATE RECOGNITION': in biparental eukaryotes, mates recognize mates through signals which form part of their FERTILIZATION system, the system which ensures that fertilization occurs as part of sex. Calling such 'mate recognition' 'species recognition' is to misdirect attention away from an evolved process and hence to obscure the causal trail. There can be no justification for obfuscating terms. *See also* ISOLATING MECHANISMS; MATE RECOGNITION; SPECIES CONCEPTS. [H.E.H.P.]

REFERENCE

Paterson, H.E.H. (1993) *Evolution and the Recognition Concept of Species*. Johns Hopkins University Press, Baltimore, Md.

species richness The total number of species present in a COMMUNITY. It is the simplest measure of the taxonomic variety of a community, but is not on its own an adequate measure of species DIVERSITY because it does not take account of the relative ABUNDANCES of different species. For example, if two communities were both comprised of 10 species and 100 individuals, and in one there was 1 individual of each of 9 species and 99 individuals of the 10th species, but in the other community 10 individuals of each of 10 species, then both communities would have the same species richness (10). The second one would, however, be a much more diverse community. For this reason most indices of species diversity have both species richness and EQUITABILITY components. *See also* DIVERSITY INDICES. [M.H.]

species-richness indices *See* DIVERSITY INDICES.

species selection A form of GROUP SELECTION in which sets of species with different characteristics increase (by SPECIATION) or decrease (by extinction) at different rates, because of a difference in their characteristics. This process is not considered INDIVIDUAL SELECTION because it involves the capacity to evolve, which is a population/species characteristic and not a property of individuals. The replacement of some species by others can cause long-term EVOLUTIONARY TRENDS and can influence the species composition of communities in which the evolved properties of species determine whether they can stably coexist. *Cf.* NATURAL SELECTION. *See also* MACROEVOLUTION. [V.F.]

Species Survival Commission (SSC) A body formed in 1949 to provide leadership to species conservation efforts internationally. It is the largest and most active of the six volunteer Commissions of IUCN. At the end of 1997, the SSC network encompassed 7000 volunteer member scientists, field researchers, government officials and conservation leaders from 169 countries. SSC members provide technical and scientific counsel for BIODIVERSITY conservation projects throughout the world and serve as resources to governments, international conventions and conservation organizations.

The SSC's goal is to conserve biological DIVERSITY by developing and executing programmes to save, restore and manage wisely species and their habitats. Toward this goal, the SSC:

- develops and promotes conservation Action Plans, which prioritize steps necessary to ensure the survival of selected species and habitats;
- provides technical information about biological diversity to international treaties, such as the Convention on International Trade in Endangered Species (CITES);
- formulates policy and procedural recommendations on a variety of issues such as sustainable use of wild species, reintroduction of plants and animals to their native habitats, and assessment of the risk of extinction for individual species.

The SSC works primarily through its 109 Specialist Groups, most of which represent particular plant or animal groups that are threatened with extinction, or are of importance to human welfare. A few groups deal with cross-cutting species conservation issues, such as veterinary medicine, CAPTIVE BREEDING, reintroduction of species, invasive species and sustainable use of WILDLIFE. A special task force is investigating the global decline in amphibian populations. *See also* INTERNATIONAL CONSERVATION CONVENTIONS; UNITED NATIONS PROGRAMMES AND BODIES. [G.B.R.]

specific humidity The mass of WATER per mass of moist air. *See also* HUMIDITY. [J.G.]

specific leaf area (SLA) The area of leaf (measured on one surface only) per unit leaf dry mass. It tends to be negatively correlated with UNIT LEAF RATE (ULR). Thin leaves (high SLA) and thick leaves (low SLA) tend to occur in characteristically different LIGHT and WATER environments. [M.J.C.]

specific-mate recognition Specific-mate recognition is the process enabling a sexual organism to recognize another organism of opposite gender as an appropriate mate. Important in this process are the characters of the specific-mate recognition system (SMRS), which is the semiotic part of the FERTILIZATION system, involving a signal–response reaction chain between the partners.

This chain is shortest in sessile organisms where it is restricted to sperm–ovum, or pollen–stigma interactions. In motile animals it includes so-called COURTSHIP behaviour, but does not cover the interactions between, say, an angiosperm and its insect pollen-vector.

Specific-mate recognition is largely genetically determined although learning may be involved. However, even when this is the case, learning of specific signals (e.g. birdsong) is intrinsically facilitated (Marler 1987). Recognition in this sense amounts to response by one partner to a specific signal from the other.

Specific-mate recognition is a key process underlying the RECOGNITION CONCEPT OF SPECIES (*see also* SPECIES CONCEPTS), because the fertilization system fortuitously forms a 'field of gene recombination', a key characteristic of genetical species, as a by-product of ASSORTATIVE MATING. *See also* ISOLATING MECHANISMS; MATE RECOGNITION; SPECIES RECOGNITION. [H.E.H.P.]

REFERENCES

Marler, P. (1987) A sensitive period for song acquisition in the song sparrow, *Melospiza melodia:* a case of age-limited learning. *Ethology* **76**, 89–100.

Paterson, H.E.H. (1993) *Evolution and the Recognition Concept of Species*. Johns Hopkins University Press, Baltimore, Md.

spectral reflectance The reflectance of RADIATION by a surface at a range of wavelengths. The reflectance of any natural surface varies with wavelength. Leaves reflect only about 10% of incident radiation in the ultraviolet and visible wavelengths, rising to over 50% in the near infrared, and declining to almost zero in the thermal wavebands. The precise pattern of reflectance plotted against wavelength is found to differ somewhat between species, and with growing conditions. For example, leaves that have been irradiated by bright sunlight develop a more reflective leaf surface by producing more epicuticular wax. The spectral reflectance of leaves is utilized in SATELLITE REMOTE SENSING, where the reflected SOLAR RADIATION in several wavebands may be used to interpret the nature of the land surface cover. [J.G.]

REFERENCE
Gates, D.M. (1980) *Biophysical Ecology*. Springer-Verlag, New York.

spectroradiometer An instrument used to measure the spectral distribution of electromagnetic RADIATION (usually in $W m^{-2} nm^{-1} sr^{-1}$): variations, ultraviolet; visible/PAR (photosynthetically active radiation); solar infrared; thermal infrared. The total energy flux density is the integral of the spectral energy distribution. Precision spectroradiometers use single sensor(s), with motor-driven gratings to scan the spectrum, as in laboratory spectrometers. Portable spectroradiometers may employ stationary gratings and photodiode detector arrays. Prototype imaging spectroradiometers (spectrographic imagers) using similar principles are already flying in aircraft and are planned for remote-sensing satellites within a few years. [J.A.C.]

sperm *See* SPERMATAZOA.

Sperm competition

The COMPETITION between the SPERMATOZOA from two or more males to fertilize the EGGS of a single female during one reproductive cycle. Sperm competition is a form of male–male competition that occurs between insemination and FERTILIZATION, and is a part of INTRASEXUAL SELECTION. G.A. Parker was the first clearly to conceptualize sperm competition as a subset of SEXUAL SELECTION, and also its implications for insects, in 1970. Since then sperm competition has been demonstrated in a variety of insects, arachnids, crustaceans, gastropods, fishes, amphibians, reptiles, birds and mammals. With few exceptions, sperm competition occurs in animals with internal fertilization.

Sperm competition creates opposing selective forces. On the one hand, selection favours males that succeed in fertilizing the eggs of already-mated females. On the other hand, it also favours mated males that avoid having their sperm displaced by other males.

Multiple mating

Promiscuous mating (multiple mating with different partners) by females is a prerequisite for sperm competition. Males should in principle mate as many times as possible, in order to maximize their reproductive success. The benefits of multiple mating are less obvious for females. Multiple mating may be beneficial to females when the sperm received from one male may not be sufficient to fertilize all the eggs produced, when the viability of sperm stored decreases, or when the quality of the sperm received is low. Females can also receive nutritional benefits from multiple matings by receiving nuptial gifts or nutrients with the spermatophore or the seminal fluid. Multiple mating with different males may also lead to multiple PATERNITY and thus increase the genetic variability within a brood. The benefits of multiple mating might outweigh any costs of copulation, which can include the energy and time of sexual behaviour, an increased risk of predation and the risks of disease and parasite transmission.

Sperm storage

Females of many invertebrate, amphibian and reptile species possess a sperm storage organ, the spermatheca. The morphology of the spermatheca and the ducts serving it may determine whether it is the first or last sperm entering which fertilizes the eggs. In birds, sperm can be stored in sperm storage tubules at the utero-vaginal junction. The storage of viable sperms varies: typical durations are 1–2 days in mammals and 8–20 days in birds, with maximum values of 2 years in salamanders, 3 years in land snails, 4–7 years in snakes, and 10–20 years in queens of SOCIAL INSECTS.

Prolonged sperm storage is a way in which copulation and fertilization can be uncoupled in time. It allows females to fertilize and deposit their eggs under conditions suitable for oviposition, which must not be identical to those of mating. Sperm storage also offers an opportunity for female control of paternity.

Sperm precedence

Sperm precedence is the differential sperm usage from consecutive matings (mating-order effect). It is typically measured as the proportion of eggs fertilized by the second of two mates (the P_2 value). P_2 values range from 0 to 1: a P_2 of 0.5 indicates equal fertilization success for both mates. If P_2 is significantly smaller than 0.5 then there is first-mate sperm precedence, and if P_2 is significantly greater than 0.5 there is second-mate sperm precedence.

Sperm precedence patterns in double-mated females can be studied using the following techniques.

1 Males with different visible genetic markers that are distinctly expressed in the offspring.

2 The STERILE MALE TECHNIQUE (mainly used in insects): one of the two males is exposed to a high dose of X-RAYS that results in a complete hatching failure of eggs fertilized by sperm of this male.

3 Genetic markers assessed by enzyme ELECTROPHORESIS or molecular techniques (random amplified polymorphic DNA (RAPD), microsatellites).

The mechanism by which sperm precedence occurs is only known for some species. Some degree of precedence by the sperm from the latest insemination may occur because, at any point of time, sperm from earlier inseminations have been in the female's reproductive tract longer and have suffered a higher MORTALITY than sperm from the latest insemination. Last-male sperm precedence can simply be a passive consequence of the mortality of sperm in the female's reproductive tract.

Sperm number is important for fertilization success in sperm competition. If sperm from two males are mixed randomly in the spermatheca, their fertilization success is proportional to the numbers of sperm (raffle principle). In this case, sperm precedence depends on sperm number.

Sperm removal and stratification are two mechanisms which can lead to last-male precedence. Sperm removal can be either direct, i.e. males scoop sperm from previous inseminations out of the female's reproductive tract using specially modified parts of their intromittent organs (e.g. in damselflies), or indirectly, when the incoming sperm from the last male flushes out sperm already present (e.g. in the dungfly, *Scatophaga stercoraria*). Sperm stratification occurs when sperm from the most recent copulation overlies that introduced earlier into the sperm storage site so that 'last in–first out' operates (e.g. in the giant waterbug, *Abedus herbertii*). In birds, sperm from the last of a series of inseminations generally have precedence.

Morphological adaptations

The number of sperm that a male can produce is correlated with testis size. Males with relatively large testes (i) may deliver larger ejaculates and a greater number of sperm per ejaculate, and (ii) may be able to mate more frequently and thus provide a greater total number of sperm to several females. In bird and mammal species, in which females mate promiscuously, males have relatively larger testes and/or epididymides (the organs in which the male stores his own sperm) than in monogamous species. The penis itself might have evolved in response to sperm competition, males able to insert sperm further into a female tract and therefore closer to the OVA being favoured. Penis morphology may also be related to sperm displacement and/or plug removal. The enlarged glans penis of many mammals suggests that the penis may have evolved either to remove copulatory plugs, or simply to remove the ejaculate of competing males. These ideas are still in debate.

Behavioural adaptations

Different behavioural patterns and mechanisms have evolved to increase the paternity of a particular male. MATE-GUARDING, i.e. the close following of females by their mates during the female's period of fertility, is an efficient way for males to increase their certainty of paternity. In species with first-male precedence, males will be selected to seek out mates as early as possible in the females' reproductive lives, and may guard them until they become inseminable. Conversely, in species with last-male precedence, males are expected to guard females after copulation has occurred.

Frequent copulation is another form of paternity guard in species in which ecological factors prevent mate-guarding by close following. Frequent copulations can be observed in birds of prey and many colonial breeding birds; for example, goshawk (*Accipiter gentilis*) pairs copulate on average 500–600 times per clutch. The production of copulatory plugs is another way to increase paternity. In many worms, insect, reptile and mammal species copulating males secrete a plug, which partially or fully blocks the female reproductive tract for some time. In general, plugs are not absolutely effective in preventing further matings.

It has been suggested that some sperm (the kamikaze sperm) may be specifically designed not to fertilize eggs but to destroy or block the passage of sperm from other males. The evidence to support the kamikaze hypothesis is controversial at present.

Copulation can be very prolonged in many insect species (ranging from a few minutes to one day). In this way the male himself acts as a mechanical plug. In many species, agents in the male seminal fluid suppress the female's receptivity to further matings. Postcopulatory mate guarding is another way to reduce the probability of sperm competition and hence to assure paternity.

Optimal sperm number is strongly dependent on the intensity of the sperm competition because of the 'raffle principle' where, fundamentally, sperm numbers determine fertilization success. Individual males may adjust their sperm expenditure in response to cues revealing average sperm competition risk, or to risk incurred in a given mating. Adaptive variation in sperm expenditure has been found in butterflies, fishes and primates. Individual adjustment of ejaculate size according to the female's mating status has been recorded in insects, salamanders, rats and humans: males deliver more sperm when sperm from another male is already in the female tract.

In many species, sperm competition may offer females an opportunity to control the fertilization of their eggs. There is evidence for this relatively novel idea of female control of sperm competition. For example, female dunnocks (hedge sparrows, *Prunella modularis*) may extrude a droplet of sperm before copulation. In this way the bird discards sperm from a previous copulation and is willing to accept sperm from another male. [B.B.]

REFERENCES

Birkhead, T.R. & Møller, A.P. (1992) *Sperm Competition in Birds*. Academic Press, London.

Birkhead, T.R. & Møller, A.P. (eds) (1998) *Sperm Competition and Sexual Selection*. Academic Press, London.

Smith, R.L. (1984) *Sperm Competition and the Evolution of Animal Mating Systems*. Academic Press, London.

---

sperm displacement The displacement, by sperm from a later-mating male, of sperm already present in a female's reproductive system from the site of FERTILIZATION or from the best location in the sperm storage organ(s). This is one of the principal mechanisms of SPERM COMPETITION, in which sperm from different males compete within a female to fertilize her EGGS. The COMPETITION takes place because there is almost always a delay between insemination and fertilization. In arthropods, sperm displacement can mean that sperm are pushed further away from the opening of a spermatheca, the sperm storage organ, and so are less likely to be released to fertilize the eggs.

In many species, males may attempt either to physically remove stored sperm or to displace them by flushing the female's reproductive system with their seminal fluids before transferring their own sperm. Females may also influence the rate of sperm displacement, but little is known at present of the mechanisms involved. In many species most of the time spent copulating is used to influence sperm displacement; actual sperm transfer is usually very quick. [P.I.W.]

sperm precedence *See* SPERM COMPETITION.

sperm storage *See* SPERM COMPETITION.

spermatophore A package of sperm, usually wrapped in some protective sheath. Spermatophores occur in several groups of organisms that have internal FERTILIZATION with indirect sperm transfer, such as leeches, cephalopods, onychophorans, myriapods and salamanders. They are also common in insects, although indirect sperm transfer is relatively rare in this group. Spermatophores of some insects contain large quantities of protein that are consumed or absorbed by the female, which is a form of male PARENTAL INVESTMENT. [T.J.K.]

spermatozoa Male GAMETES, also sperm. Adapted for fertilizing female gametes. Hence, often capable of locomotion and penetration of female gamete. *See also* ANISOGAMY, EVOLUTION OF; FERTILIZATION. [P.C.]

spirals of matter BIOTA remove chemical ELEMENTS from the environment but return them to the environment at a later time. This leads to nutrient cycling (*see* BIOGEOCHEMICAL CYCLE). In a flowing-water system (river/stream) a nutrient released at one point will be taken up by biota downstream (Fig. S20). So matter here is said to spiral rather than cycle. [P.C.]

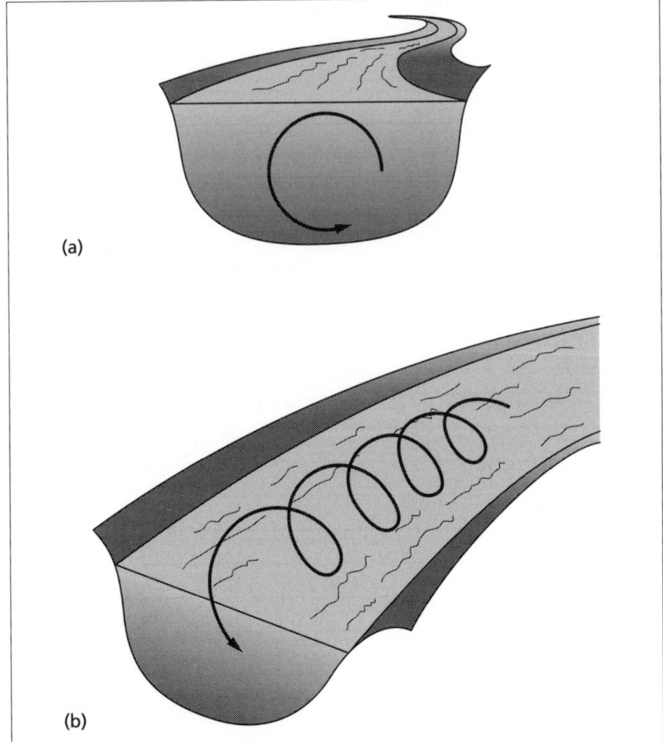

Fig. S20 (a) The nutrient cycle as viewed from a perspective that does not 'see' downstream transport. (b) The nutrient cycle, in conjunction with downstream transport, describes a spiral. (After Newbold, J.D. (1992) Cycles and spirals of nutrients. In: *The Rivers Handbook*, Vol. 1 (eds P. Calow & G.E. Petts), pp. 379–408. Blackwell Scientific Publications, Oxford.)

spit A shingle or sand structure which extends from a point on the solid coastline and remains unattached at its growing point. Shingle spits may comprise several parallel ridges. They typically form across an ESTUARY or indentation of the coastline and invariably possess recurved hooks at the tip and often along their length. Shingle spits may sometimes be covered by SAND DUNES and frequently enclose SALT MARSH on the landward side. [B.W.F.]

spodic horizons *See* SOIL CLASSIFICATION.

sporophyte The spore-producing form of the plant LIFE CYCLE, alternating with the GAMETE-producing GAMETOPHYTE. Sporophytes of homosporous taxa produce one type of spore (which initiates a cosexual gametophyte), those of heterosporous or anisosporous taxa produce microspores and megaspores which, respectively, initiate male or female gametophytes. The sporophyte is initiated sexually as a zygote or asexually by PARTHENOGENESIS, apogamy or adventitious embryony. It is free-living and dominant in the life cycle of most land plants, but remains attached to the gametophyte in bryophytes and in the fern *Anogramma leptophylla*. In algae it may be dominant or co-dominant, or the gametophyte may be dominant. [M.M.]

sport An atypical form of an individual or (in plants) part of an individual, due to SOMATIC MUTATION. The term is somewhat outdated but is still used in horticulture. [P.C. & V.F.]

spray zone (splash zone) The zone above high tide that is kept moist by breaking WAVES. On open coasts, where waves are bigger, wave splash acts to expand the effective INTERTIDAL ZONE and allows organisms to live higher on the shore. [V.F.]

spread Spread, which in everyday usage connotes expansion, diffusion, dissemination, etc., is the preferred term to describe the advance and ESTABLISHMENT of species in a new area (Bennett 1985; Birks 1989). It is recommended by Bennett (1985) in preference to the widely used term 'MIGRATION', which should be reserved for movements that have a specific goal and are achieved in their entirety by each individual rather than being dependent on several generations (e.g. seasonal or sporadic migrations of animals).

The spread of species is complex and involves several processes including DISPERSAL, establishment and finally expansion (Holdgate 1986). A long-term perspective, such as is provided by a well-dated FOSSIL RECORD, is invaluable in understanding present-day patterns of species DISTRIBUTION, and in analysing the role played by factors such as natural barriers (e.g. mountain ranges and open water), availability of suitable habitat, COMPETITION from existing biota, CLIMATE CHANGE, and natural and human-induced DISTURBANCE. [M.O'C.]

REFERENCES

Bennett, K.D. (1985) The spread of *Fagus grandifolia* across eastern North America during the last 18 000 years. *Journal of Biogeography* **12**, 147–164.

Birks, H.J.B. (1989) Holocene isochrone maps and patterns of tree-spreading in the British Isles. *Journal of Biogeography* **16**, 503–540.

Holdgate, M.W. (1986) Summary and conclusions: characteristics and consequences of biological invasions. *Philosophical Transactions of the Royal Society of London, Series B* **314**, 733–742.

spread of species *See* SPREAD.

spring mire A PEAT-forming ecosystem developed over a spring head. Spring mires are RHEOTROPHIC, despite the fact that they have elevated surfaces and convex profiles, because groundwater from below is forced to their surface by pressure from the underlying spring. If the water source is calcium-rich, tufa may form in layers within the peat. [P.D.M.]

square root transformation *See* TRANSFORMATION OF DATA.

SSSI *See* SITE OF SPECIAL SCIENTIFIC INTEREST.

stability analysis In theoretical models of POPULATION DYNAMICS, equilibrium POPULATION DENSITY (N^*) occurs under those conditions where $dN/dt=0$ for all the interacting species. The stability analysis of a given model can be carried out graphically (*see* PHASE PLANE ANALYSIS), analytically (by algebra) or numerically (by computer simulation).

For stable point equilibria, small perturbations lead directly back to equilibrium, while for unstable point equilibria, small perturbations cause the system to move away from equilibrium to a higher or lower density, depending upon the direction of the perturbation (this may lead to local extinction, transition to a higher or lower alternative STABLE EQUILIBRIUM, or unbounded exponential increase, depending upon the precise structure of the model). Some model systems (e.g. the Lotka–Volterra predator–prey model, *see* LOTKA–VOLTERRA MODEL) exhibit neutral stability; this is a pathological behaviour in which perturbations to the system are neither damped nor amplified, simply remembered. More complex forms of stability analysis deal with LIMIT CYCLES and CHAOS. [M.J.C.]

REFERENCES

May, R.M. (1977) *Stability and Complexity in Model Ecosystems*. Princeton University Press, Princeton, N.J.

Turchin, P. & Taylor, A.D. (1992) Complex dynamics in ecological time series. *Ecology* **73**, 289–305.

stability, communities The tendency of a COMMUNITY to return to its original state after a DISTURBANCE or to resist such disturbance. Stability may be measured in terms of the number and ABUNDANCE of its component species or by other community properties or processes. The concept of stability includes the property of RESILIENCE (the speed with which a community returns to its former state after it has been displaced from that state by a perturbation) and RESISTANCE (the ability of a community to avoid displacement in the first place).

Communities may be described as locally stable (able to recover from small perturbations) or globally stable (able to recover from large perturbations). If a community is stable only within a narrow range of environmental or biological conditions it is said to be DYNAMICALLY FRAGILE; if it is stable over a broad range of conditions it is described as DYNAMICALLY ROBUST.

The species composition of a community will be stable whenever the replacement probabilities of the individual species remain constant through time. The CLIMAX community is conceived as the steady-state successional END-POINT of community development which has the property of being highly stable. The stability of a community is proportional to the degree of evolutionary adjustment among its component populations. [V.F.]

stability, population According to the LOGISTIC MODEL of population growth, POPULATION SIZE is stable (i.e. is at equilibrium; POPULATION GROWTH RATE $=0$) at a size corresponding to the CARRYING CAPACITY of the environment (usually symbolized by K). A population is stable if the ABUNDANCE of organisms in it is constant over time. In discrete time logistic models a population exhibits stable behaviour when the net reproductive rate, R, is such that $1<R<3$.

The extent to which populations fluctuate varies widely among species. Density-dependent processes (i.e. processes that cause population size to increase when below a certain level and to decrease when above a certain level) can tend to regulate population size and increase stability. Examples of density-dependent regulating mechanisms include predation, COMPETITION and PARASITISM. *See also* POPULATION REGULATION. [V.F.]

stability–time hypothesis Deep-sea habitats have more or less constant conditions (in temperature, salinity, pressure, pH, etc.) that have existed over a long period of time. The fact that they also possess a high diversity of bivalve and polychaete species in such an apparently spatially homogeneous area, compared to the shallow, more variable, COASTAL marine environments, led Sanders to propose the stability–time hypothesis. This hypothesis suggests that environments that are stable over long periods of time should promote high diversity because extinctions would be rare as population FLUCTUATIONS would be low and evolutionary specialization would be encouraged, leading to a high degree of RESOURCE PARTITIONING, providing that sufficient time elapses without major change (i.e. biologically accommodated communities, Fig. S21). This hypothesis has also been applied to latitudinal DIVERSITY GRADIENTS (tropics more stable and species-rich than temperate areas) and the high diversity of ancient lakes in Russia and Africa (stable over long periods of time).

However, low-diversity communities do occur in stable (but environmentally harsh) environments, and high-diversity ones in some unstable, unpredictable environments subject to intermediate degrees of DISTURBANCE (*see* INTERMEDIATE DISTURBANCE HYPOTHESIS). Also, there may be other explanations for Sanders' original data, including patchiness and HETEROGENEITY on small spatial scales in the DEEP SEA. In general, the stability–time hypothesis is unlikely to explain diversity patterns on shorter time-scales encom-

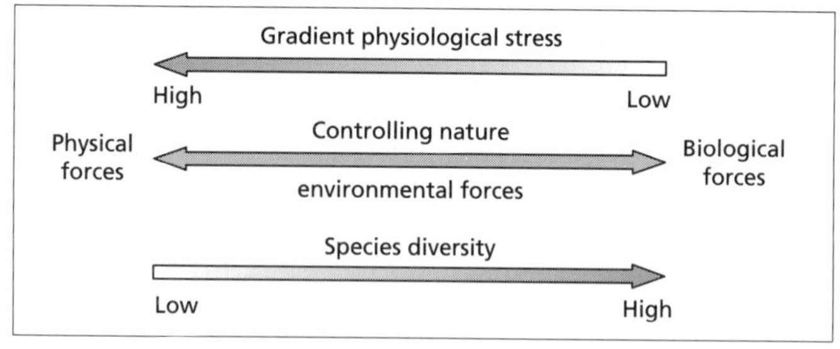

Fig. S21 A schematic representation of Sanders' stability–time hypothesis, that suggests that the underlying cause of high diversity is the persistence, over long periods of time, of stable environmental conditions (with low physiological stress) that allow communities to become biologically accommodated, with smaller, non-overlapping niches. (After Giller, P.S. (1984) *Community Structure and the Niche*. Chapman & Hall, London.)

passing ecological interactions but is consistent with the accumulation of species diversity over evolutionary time-scales. *See also* STABILITY, COMMUNITIES. [P.S.G.]

REFERENCE

Sanders, H.L. (1969) Benthic marine diversity and the time-stability hypothesis. In: *Diversity and Stability in Ecological Systems* (eds G. Woodwell & H. Smith), pp. 71–81. Brookhaven Symposium on Biology, 22.

stabilizing selection A form of NATURAL (or ARTIFICIAL) SELECTION of a character (e.g. body mass) that selects against extreme values of the character, favouring instead some intermediate value or values, as shown in Fig. S22 (also see Fig. D4). *Cf.* DIRECTIONAL SELECTION; DISRUPTIVE SELECTION. [R.M.S.]

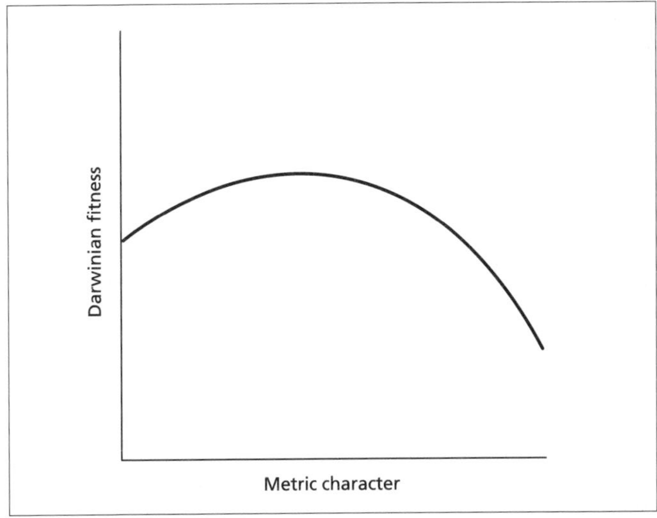

Fig. S22 Stabilizing selection.

stable age distribution The AGE STRUCTURE that is gradually approached by a population in which birth and death rates are constant. A.J. Lotka in 1922 showed that the stable age distribution is unaffected by initial age structure. A population that has achieved stable age distribution will increase at a rate *r*, where *r* is defined by the EULER–LOTKA EQUATION. The stable age distribution can be calculated as the dominant EIGENVECTOR of the Leslie matrix (*see* POPULATION PROJECTION MATRIX). The stable age distribution is very important in population management, where the harvest derived from an exploited population (e.g. fisheries) or the damage caused by a pest population depends on the numbers in particular AGE CLASSES. *See also* DEMOGRAPHY; LIFE-HISTORY EVOLUTION; LIFE TABLE; POPULATION DYNAMICS. [R.H.S.]

stable equilibrium An equilibrium towards which a system returns after perturbation. If the perturbation is small, then return to equilibrium demonstrates the LOCAL STABILITY of the equilibrium. If the perturbation can be of any size, then the equilibrium is globally stable. *See also* STABILITY, COMMUNITIES; STABILITY, POPULATION. [S.N.W.]

stable limit cycle A regular fluctuation in ABUNDANCE, the path of which is returned to after slight displacements from the path. It is a type of behaviour displayed by non-linear systems which involves an oscillation of stable amplitude, so that if the system starts at a point outside the LIMIT CYCLE, its amplitude of oscillation declines, and if inside its amplitude increases, until the stable cycle is achieved. *See also* POPULATION DYNAMICS; POPULATION REGULATION; STABILITY, POPULATION. [V.F.]

stage distribution A DISTRIBUTION showing the relative ABUNDANCE of individuals in different developmental stages. In species that pass through distinct developmental stages, the duration of which depends on environmental factors (e.g. temperature), knowledge of the developmental stage often provides more information about survival and reproduction than does knowledge of chronological age. The dynamics of such a population can be described by a stage-classified matrix population model (*see* POPULATION PROJECTION MATRIX). In a constant environment with no density-dependence, the stage distribution will almost always converge to a stable form. If the population projection matrix **A** has a dominant EIGENVALUE λ, then the stable stage distribution is proportional to the corresponding right EIGENVECTOR **w**, which satisfies:

$$\mathbf{A}\mathbf{w} = \lambda\mathbf{w}$$

[H.C.]

Stagnosols *See* SOIL CLASSIFICATION.

staminate *See* POLLINATION.

stand A general term for an area of reasonably homogeneous VEGETATION which is to be subject to scrutiny, usually for the purpose of phytosociological description. [P.D.M.]

stand cycle The name given by A.S. Watt to the cyclical dynamics exhibited by certain long-lived, patch-forming PERENNIAL plants (e.g. *Calluna vulgaris, Pteridium aquilinum*) where a given patch of ground passes through various stages (pioneer, building, mature and degenerate phases) in a cyclical progression. Seedling RECRUITMENT occurs only during the pioneer stage. The HABITAT as a whole comprises a mosaic of patches in different phases of the stand cycle. The size of individual elements of the mosaic may be determined by tree-falls, fire, herbivory, soil pathogens or other spatially patchy disturbances. [M.J.C.]

REFERENCE

Watt, A.S. (1947) Pattern and process in the plant community. *Journal of Ecology* **35**, 1–22.

standard

1 A requirement in legislation that if breached or contravened is likely to invite prosecution. *See also* ENVIRONMENTAL QUALITY OBJECTIVE; ENVIRONMENTAL QUALITY STANDARD.

2 An exact value, physical entity or abstract definition established by an authority and/or custom and practice as a reference, yardstick, model or rule in measurement procedures, in establishing practices or procedures, or in evaluating and auditing activity. [P.C.]

standard deviation A measure of variation. In a SAMPLE it is represented by the Roman letter *s*, and is an estimate of the POPULATION standard deviation, represented by the Greek letter sigma (σ). It is the square root of the VARIANCE. *See also* STATISTICS. [B.S.]

standard error (of the mean) If we take a SAMPLE of observations from a POPULATION, the sample MEAN ($\bar{x}$) will be an estimate of the population mean (μ). However, it is unlikely that $\bar{x}=\mu$ because just by chance the sample we have taken may contain a non-representative collection of observations. If we repeatedly take a large number of samples, of size n, from a population, the sample means will themselves form a DISTRIBUTION. This SAMPLING distribution of means will be normally distributed, with its mean (the grand mean of means) = μ. This result is called the CENTRAL LIMIT THEOREM. The variation in this sampling distribution of means can be measured by a STANDARD DEVIATION which, to distinguish it from the standard deviation of a sampling distribution of single observations, is called the standard error (SE) of the mean.

The SE will be influenced by two things: (i) the size of the sample (n)—smaller samples will result in more variation in the sampling distribution of means (larger SE); and (ii) the variation in the population (σ)—larger σ will give larger SE. In fact, the precise relationship, worked out by statisticians, is:

$$SE = \frac{\sigma}{\sqrt{n}}$$

In practice σ will be unknown, so we estimate the SE using the formula:

$$SE = \frac{s}{\sqrt{n}}$$

In addition to the sample mean, other sample statistics have SEs but their calculation is frequently much more complex. *See also* REGRESSION ANALYSIS, CURVILINEAR; REGRESSION ANALYSIS, LINEAR; STATISTICS. [B.S.]

standard overcast sky A specification of the brightness of a densely overcast sky, in which the sky is brightest at the zenith and falls to one-third of this value at the horizon. Real overcast skies conform to the standard overcast sky when spot measurements made over a period of an hour are averaged. The standard overcast sky is used by environmental engineers and architects in the design of natural lighting, and by researchers studying the interception of light by plant canopies. [J.G.]

REFERENCE

Grace, J. (1971) The directional distribution of light in natural and controlled environment conditions. *Journal of Applied Ecology* **8**, 155–164.

standard time *See* TIME.

standardized aquatic microcosms (SAM) Multispecies ECOTOXICITY test systems with standardized initial conditions, monitoring and test procedures. *See also* COSMS; MICROCOSM. [P.C.]

REFERENCE

Taub, F.B., Kinding, A.C. & Conquest, L.L. (1986) Preliminary results on interlaboratory testing of standardised aquatic microcosms. In: *Community Toxicity Testing, STP 920* (ed. J. Cairns), pp. 93–120. American Society for Testing and Materials, Philadelphia, Pa.

standardized normal deviate (*d* or *z*) (unit standard deviate; z-score) All normal distributions are unique because of their particular MEAN (μ) and STANDARD DEVIATION (σ). However, they can be standardized by expressing the value of each individual observation as a standardized normal deviate (d) or z-score, using the formula:

$$z(d) = \frac{(x - \mu)}{\sigma}$$

where x is an individual observation, μ is the POPULATION mean and σ is the population standard deviation. In practice, with a SAMPLE of observations, we do not know the value of μ and σ. However, if the sample size (n) is greater than about 30 observations we can use $\bar{x}$ and s^2. In effect, z is the number of standard deviations that an observation is away from the mean. For example, in a sample of observations in which the mean was 10 and the standard deviation was 3, an observation of 13 (one standard deviation above the mean) would have a z-score of +1. Tables of $z(d)$ relate values of standardized normal deviates to PROBABILITY (*see* NORMAL DISTRIBUTION). Usually this gives the probability (P) of getting an observation greater than a particular value of $\pm z$ ($\pm d$) (but *see also* STATISTICAL TABLES), for example:

z(*d*)	1.64	1.96	2.33	2.58	3.29	3.89
P	0.1	0.05	0.02	0.01	0.001	0.0001

See also PROBIT ANALYSIS. [B.S.]

standards organizations There are a large number of standards organizations and their work covers a wide range of subjects. This work is relevant in environmental protection and CHEMICAL CONTROL in that it ensures standardization in methods and methodology of measurement and ecotoxicological tests, and hence facilitates the mutual acceptance of information and classifications between interested parties on an international, national and local scale.

The major players can be organized in a geographical hierarchy, from international — involving the OECD and the International Standards Organization (ISO)—to regional—involving the Comité Européen de Normalisation (CEN) in Europe and the American Society for Testing and Materials (ASTM) in the USA—and, finally, to a national level. Some of the national standards organizations are listed in Box S1. [P.C.]

Box S1 Standards institutions

AFNOR (France): Association Français de normalisation
BSI (UK): British Standards Institute
DIN (Germany): Deutsches Institut für Normung e. V.
ELOT (Greece): Hellenic Organisation for Standardisation
IBN (Belgium): Institut belge de normalisation (Belgisch Instituut voor Normalisatic)
DS (Denmark): Dansk Standardiseringrad
NNI (The Netherlands): Nederlands Normalisatie Instituut
IIRS (Ireland): Institute for Industrial Research and Standards
Luxembourg: Inspection du travail et des mines
UNI (Italy): Ente nazionale italiano di unificazione
CEI (Italy): Comitato elettrotecnico italiano

standing crop The mass of VEGETATION in a given area at one particular time. The term is often used of above-ground vegetation only, but it can include root material also. Although most often applied to plant material, the term may also include animal BIOMASS. [P.D.M.]

startle effect Some animals when disturbed by a potential PREDATOR will expose a bright colour pattern in an attempt to startle a predator. The predator may be frightened away or miss the opportunity to attack. The exposure of the colour pattern is frequently associated with a ritualized display. *See also* DEFENCE MECHANISMS; MIMICRY. [P.M.B.]

stasis The term evolutionary stasis is used to indicate cases in which there has been little or no apparent evolutionary change in a species for a long time. There are no agreed criteria for how long these periods of time must be or how little actual evolutionary change must occur in order for a particular species to be described as 'static'.

Two explanations are routinely invoked to explain stasis. The first is STABILIZING SELECTION: the environments in which the species occur have remained relatively constant over long periods of time, and hence the species living in those environments have remained virtually the same during that time. The second is canalizing selection: the developmental programmes of some species may be so intolerant of variation that no viable options have been produced during long periods of time. The primary evidence we have of evolutionary stasis is so-called 'LIVING FOSSILS', species living today that appear virtually identical to ancient fossil relatives. Examples include horseshoe crabs (*Limulus*, *Tachypleus* and *Carcinoscorpius*), gingkos (*Gingko biloba*) and the coelacanth (*Latimeria chalumnae*). *See also* MACROEVOLUTION; PUNCTUATED EQUILIBRIUM. [D.R.B.]

state of environment reports Reports providing data on physical, chemical and biological indicators of environmental quality. There is, in fact, a whole hierarchy of reports carrying such information, ranging from a global perspective (UNEP reports) to a Europe-wide assessment (European Union reports) to national levels (e.g. UK Department of the Environment reports) right down to regional and local levels. See Box S2.

Two points continue to emerge from these reports. First, there are few indications of grounds for congratulation or complacency—the environment, particularly at the global level, continues to deteriorate. Second, however, in many respects the data on which these conclusions are based are not ones in which too much confidence can always be placed. Too often they are not extensive (either geographically or temporally) or reliable enough. Most agree that more effort should be devoted to the problems of collection and interpretation of environmental BASELINE DATA. There is particular need for reliable, long-term measurements of key environmental indicators at key sites.

The EUROPEAN ENVIRONMENT AGENCY, set up under Regulation 1210/90 of the European Community (EC) and now based in Copenhagen, is likely to play an increasingly important part in preparing reports on the state of the environment in the European Union (EU). Its remit is to provide the EU and its member states with 'objective, reliable and comparable information at European level' as a base for environmental protection measures, to assess the results of such measures and 'to ensure that the public is properly informed about the state of the environment'. [P.C.]

Box S2 State of the environment reports: some examples

World

- *Environmental Data Report 1991/92*, 3rd edn. United Nations Environment Programme/Basil Blackwell, Oxford.
- *The World Environment 1972–1992, Two Decades of Challenge.* (1992) (Eds M.K. Tolba & O.A. El-Kholy), 884 pp. Chapman & Hall, London (for United Nations Environment Programme).

OECD

- *The State of the Environment 1991.* OECD Publication Service, Paris.

Europe

- *The State of the Environment in the European Community.* (1992) (Accompanying the Fifth Action Pro-Programme: *A New Strategy for the Environment and Sustainable Development.*)
- *Forest Condition in Europe.* (1992) Executive Report, International Cooperation Programme on Assessment and Monitoring of Air Pollution Effects on Forests. UNECE and CEC.
- *Europe's Environment.* The Dobris Assessment. (1995) European Environment Agency, Copenhagen.

UK national

- *The UK Environment.* (1992) HMSO, London.

UK regional

- *The Scottish Environment Statistics*, No. 3. (1991) Government Statistics Service, Edinburgh.
- *Environment Digest for Wales*, No. 6. (1991) Government Statistical Services, Cardiff.

state space The set of all possible configurations of a system. A state of a system is its value with regard to one or more state variables, for example the velocity of an electron, the ion concentration in a cell, the age and size of an organism, the density of a population, etc. The state variables represent the dimensions of the state space. A point in state space describes the state of the system by its coordinates. For example, if the state variables are the age and size of an animal, then the state space is the positive quadrant in the two-dimensional space whose axes are time and mass. *See also* MULTIPLE STABLE STATES; SYSTEMS ECOLOGY. [M.D.]

static life table A type of LIFE TABLE (also known as a vertical, or time-specific life table) constructed using data collected at a single time. The AGE STRUCTURE of a sample of individuals from a population is determined and is assumed to be the same as that which would have been found had a single COHORT been followed. This approach is frequently adopted for populations without DISCRETE

GENERATIONS but the assumption is not true if MORTALITY rates or NATALITY rates (birth rates) are increasing or decreasing. [A.J.D.]

stationary age distribution *See* STABLE AGE DISTRIBUTION.

statistical power Statisticians frequently refer to the power of a statistical test. It is the PROBABILITY of rejecting the NULL HYPOTHESIS (H_0) when it is incorrect and the ALTERNATIVE HYPOTHESIS (H_1) is correct. That is, it is the probability of reaching the correct conclusion. Three things affect the power of a test.

1 The difference between the null hypothesis and the alternative hypothesis — it would be easier to distinguish between $H_0=0$ and $H_1=100$, than between $H_0=0$ and $H_1=0.01$.

2 The SAMPLE size—increasing the sample size increases the power of a test.

3 The test used — different statistical methods, testing approximately the same hypothesis, may differ greatly in their power. In general, parametric tests (*see* PARAMETRIC STATISTICS) have greater power than non-parametric tests (*see* NON-PARAMETRIC STATISTICS), when the assumptions of the parametric test are met.

See also ACCURACY/PRECISION. [B.S.]

statistical tables Tables that show the critical values of test STATISTICS for varying levels of PROBABILITY. Such test statistics include Fisher's *F*, *z*, *t*, χ^2, Mann–Whitney *U*, and Wilcoxon's *T*. The probability level in the table is related to the area, in the tail of the PROBABILITY DISTRIBUTION, outside a particular value of the test statistic (*see* STANDARDIZED NORMAL DEVIATE; NORMAL DISTRIBUTION). For some test statistics (e.g. χ^2) this area is only in one tail of the DISTRIBUTION (ONE-TAILED TEST). In other test statistics (e.g. *F* and *t*) it can be in one tail or both tails (TWO-TAILED TEST). [B.S.]

REFERENCES

Conover, W.J. (1971) *Practical Nonparametric Statistics*, 2nd edn. John Wiley & Sons, New York.

Fisher, R.A. & Yates, F. (1963) *Statistical Tables for Biological, Agricultural and Medical Research.* Oliver & Boyd, Edinburgh.

Fowler, J. & Cohen, L. (1992) *Practical Statistics for Field Biologists.* John Wiley & Sons, Chichester.

Rohlf, F.J. & Sokal, R.R. (1995) *Statistical Tables*, 3rd edn. W.H. Freeman, New York.

statistics The summary measurements or characteristics of a SAMPLE. These are represented by Roman letters; for example, the MEAN ($\bar{x}$) and the VARIANCE (s^2). *See also* PARAMETERS. [B.S.]

status, social *See* SOCIAL STATUS.

statutory water quality objectives (SWQO) Legally defined standards—in terms of concentrations of substances—that must not be exceeded as a result of effluent DISCHARGE outside a MIXING ZONE. Such objectives are usually set to avoid POLLUTION outside the mixing zone. *See also* ENVIRONMENTAL QUALITY OBJECTIVE; ENVIRONMENTAL QUALITY STANDARD. [P.C.]

the steady-state economy A human economy characterized by constant population, capital stocks and rate of material/energy throughput. *See also* ENVIRONMENTAL (ECOLOGICAL) ECONOMICS; SUSTAINABLE DEVELOPMENT. [P.C.]

Stefan–Boltzmann constant *See* RADIATION.

Stefan–Boltzmann law *See* RADIATION LAWS.

steno- A prefix denoting narrow, particularly a narrow range of tolerance for some environmental factor; for example, stenobathic for PRESSURE, stenothermal for TEMPERATURE, stenohaline for SALINITY. It is from the Greek *stenos* narrow. *Cf.* EURY-. [P.C.]

stenobathic *See* STENO-.

stenohaline *See* STENO-.

stenothermal *See* STENO-.

stenotopic Describing organisms that are only able to tolerate a narrow range of environmental conditions and hence have very restricted distributions. An example from amongst the British butterflies is the large blue, *Maculinea arion*, which recently became extinct as a native species although it is still widespread in central Europe. This species has very precise requirements as a larva. For the first three instars it feeds on the flowers of wild thyme (*Thymus drucei*), which grows in well-grazed swards. It is then carried by worker ants of just one species, *Myrmica sabuleti*, to the nest where it feeds on young ant larvae, hibernates, feeds further in the spring and pupates. *Myrmica sabuleti* is susceptible to competition from other species of ant, particularly *Myrmica scabrinodis*, and is itself a stenotopic species with very precise thermal requirements which restrict its DISTRIBUTION to heavily grazed grasslands with short swards (Thomas 1980). Hence, it can easily be lost from an area if the grazing regime alters. Thus the large blue has paid the ultimate price of overspecialization, or of being too stenotopic, i.e. extinction, at least from the localities it once occupied at the north-west edge of its range in Great Britain. *Cf.* EURYTOPIC. [M.H.]

REFERENCE

Thomas, J.A. (1980) Why did the large blue become extinct in Britain? *Oryx* **15**, 243–247.

steppe Natural GRASSLANDS of Eurasia, extending in a broad zone from Ukraine in the west to Manchuria in the east. Typical steppe is dominated by tussock-forming bunch grasses, particularly *Stipa* spp. In northern semi-humid zones, forb-rich meadow steppe forms mosaics with steppe woodlands. On the fringes of DESERT regions true steppe is replaced by semi-desert steppe containing xerophilous dwarf shrubs, particularly sagebrush (*Artemisia* spp.). In desert steppe, dwarf shrubs become dominant.

Meadow and typical steppe are homologous to the tall-grass and short-grass prairies of the Great Plains of North America. Sagebrush steppe of North America has many similarities to semi-desert and desert steppe. [J.J.H.]

REFERENCE

Lavrenko, E.M. & Karamysheva, Z.V. (1993) Steppes of the former Soviet Union and Mongolia. In: *Ecosystems of the World 8B: Natural Grasslands Eastern Hemisphere and Resume* (ed. R.T. Coupland), pp. 3–60. Elsevier, Amsterdam.

steradian *See* SI UNITS.

sterile male technique (sterile insect technique, SIT) A method for controlling insect pests by releasing sterilized males in large numbers. Females mate with sterilized males and the EGGS they produce are infertile. It is most effective when ratios of sterile to wild males of 10 : 1 or more can be achieved. The sterile male technique is most often used on recent introductions of exotic pests or in conjunction with other forms of PEST CONTROL such as spraying. Some success has been achieved, notably against the screw-worm, *Cochcliomyia hominovorax*, and the Mediterranean fruit fly, *Ceratitis capitata*. *See also* INTEGRATED PEST MANAGEMENT; PEST CONTROL. [D.A.B.]

REFERENCE

Knipling, E.F. (1955) Possibilities of insect control or eradication through the use of sexually sterile males. *Journal of Economic Entomology* **48**, 459.

Stevenson screen A ventilated, wooden box used to screen thermometers from PRECIPITATION and solar and terrestrial RADIATION, and hence standardize the measurement of shaded air TEMPERATURE near the surface. Developed from early models by Thomas Stevenson, it has been in worldwide use for over a century, and every aspect of construction and exposure is specified by the WORLD METEOROLOGICAL ORGANIZATION. It houses thermometer bulbs 1.25 m (in the UK) above the ground surface, and opens towards the pole to minimize disturbance when reading in daylight. Double roof, walls and floor of white-painted wood provide screening, and extensive louvres maintain adequate ventilation on all but the stillest days. In calculating HUMIDITY from screened wet- and dry-bulb thermometers special screen versions of hygrometric tables are used to allow for poor ventilation. [J.F.R.M.]

sticky traps *See* SAMPLING METHODOLOGY/DEVICES.

stigma *See* POLLINATION.

stochastic dynamic foraging models *See* OPTIMAL FORAGING THEORY.

stochastic dynamic programming *See* OPTIMAL CONTROL THEORY.

stochastic models Stochastic refers to patterns resulting from random factors. In stochastic models there is a random element such that for a given input to the model the outcome is not uniquely determined but takes a range of possible values. Thus, stochastic models predict the outcome of a process as the result of random effects and express events in terms of probabilities. [V.F.]

Stokes' law The relationship between particle size and sinking rate which states that the terminal velocity for a falling sphere increases as the square of the radius. This relationship holds for spheres up to 60 μm, i.e. spheres within the Stokes' range. For PLANKTON, this upper size limit can be extended to 500 μm. For particles outside the Stokes' range, terminal velocity increases as the square root of the radius. This law has been used to understand the dynamics of particles (i.e. organisms, SEDIMENT particles, FAECAL PELLETS) suspended in water. [V.F.]

stol *See* ALPINE ZONE.

storage This is the retention of materials and energy by organisms rather than using these directly in GROWTH, reproduction or metabolism. It can occur outside the body, as in food caches, or within the body either throughout the tissues or in specialized depots, such as the hump of the camel. Storage may be a response to changes in the surrounding environment, for example in anticipation of food shortages associated with drought or winter conditions, and/or to changes associated with the condition of the organism, for example in anticipation of reproduction, migration or hibernation.

There are differences not only in the amount of energy that is stored but in the way that it is stored. Commonly used storage materials are polysaccharides (e.g. starches and glycogen) and lipid, although PROTEINS are sometimes used. LIPIDS are more efficient energy stores than polysaccharides, in that they package more energy per gram—*c.* 40 cf. *c.* 20 J g^{-1}. However, slightly more energy is lost in the formation of lipids from fatty acids and in the transfer of energy to adenosine triphospate (ATP) than in the synthesis and utilization of polysaccharides. Glycogen is more readily available for ANAEROBIC METABOLISM. Proteins are surprisingly poor energy stores: despite having an energy density equivalent to that of polysaccharides (20–25 J g^{-1}), they incur considerably greater costs in transformation. [P.C.]

REFERENCE

Pond, C.M. (1981) Storage. In: *Physiological Ecology: An Evolutionary Approach to Resource Use* (eds C.R. Townsend & P. Calow), pp. 190–219. Blackwell Scientific Publications, Oxford.

storey (stratum) Any of the layers in the vertical PROFILE of a FOREST or WOODLAND. *See also* STRATIFICATION. [G.F.P.]

stormwater discharge Discharge from a device on a combined or partially separate sewerage system, for relieving system of flows in excess of a selected rate in the event of a storm. The excess is discharged, possibly after removal of gross solids, to a convenient receiving water and can be highly polluting. *See also* SEWAGE POLLUTION; SEWAGE TREATMENT WORKS. [P.C.]

stotting A peculiar gait of African antelopes in which the legs are kept almost straight and the individual appears to bounce high off the ground from all four feet simultaneously. Stotting, as well as pronking and leaping, usually occurs in the presence of a PREDATOR, and has puzzled naturalists because it seems to be a particularly inefficient means of fleeing from attack. Recent theoretical work on honest signalling and empirical work on the anti-predator behaviour of gazelles has suggested that antelopes may use stotting to signal to predators their athletic ability, and predators may choose individual prey out of a herd on the basis of stotting performance. *See also* DEFENCE MECHANISMS. [J.C.D.]

strain *See* STRESS.

strandline (driftline) The line of debris left at a high-tide mark, comprising natural (mostly dead seaweed) and materials of human origin. Several such lines may exist during periods of change from spring to neap TIDES, but only the uppermost will have some degree of permanence. This high strandline is regularly colonized by transient halo-nitrophilous plant (strandline) communities, which tend to be distinctive for different habitats. The strandline occurs at the inland (upper) limit of SALT-MARSH communities and at the seaward margin of SAND-DUNE (embryo dune) and shingle communities. COLONIZING SPECIES are typically ANNUALS or sometimes clonal perennials, in which disseminules (seeds and vegetative fragments) are often dispersed by seawater. In the tropics, where DISTURBANCE is less frequent, small woody species may colonize. The term can also be applied to freshwater lake strandlines. [B.W.F.]

strategy A word often used to describe complex ADAPTATIONS—ones involving a large number of interacting and coevolving traits. It is thus applied to life-cycle and behavioural traits (*see also* EVOLUTIONARILY STABLE STRATEGY; TRAIT). Not all are comfortable with the term because it implies forethought and therefore is tainted with elements of ANTHROPOMORPHISM and TELEOLOGY. But it is now widely used and invariably neither anthropomorphism nor teleology are intended. It is sometimes contrasted with tactic—strategy being a broader, longer-term trait, but the distinction is not very sharp. For example, SEMELPARITY and ITEROPARITY might be referred to as strategies, but the reproductive investments that go with them as TACTICS. *Cf.* TACTICS. *See also* ADAPTATION; LIFE-HISTORY EVOLUTION; LIFE-HISTORY TRAITS. [P.C.]

REFERENCE

Alexander, R.McN. (1982) *Optima for Animals*. Arnold, London.

stratification The tendency of FOREST and WOODLAND stands to develop more or less distinct layers in their vertical structure. Forests form many different kinds of stratification, and in many cases the strata are neither vertically discrete nor spatially continuous. Stratification also changes through the growth cycle, becoming more complex as stands grow older. It is commonly possible to recognize CANOPY, sub-canopy, underwood and shrub strata above herbaceous ground vegetation and bryophyte layers. *See also* LAKE STRATIFICATION; LAKES. [G.F.P.]

Stratigraphy

Stratigraphy is the study of the development and succession of stratified horizons in rocks. Its study involves lithostratigraphy, BIOSTRATIGRAPHY and CHRONOSTRATIGRAPHY.

Lithostratigraphy

Lithostratigraphy is the CLASSIFICATION of rock strata on the basis of their physical and chemical natures. Its study permits the establishment of a lithosequence in which one type of rock is described as succeeding another in a vertical column.

Biostratigraphy

Biostratigraphy involves the analysis and description of the fossil sequence in rock strata. The strata may be divided into a series of biostratigraphic zones on the basis of certain fossil types or on whole ASSEMBLAGES of fossils.

Chronostratigraphy

Chronostratigraphy is the study of the time sequence of rock strata. Undisturbed rock sequences will have their oldest sediments at their base, but simple vertical chronostratigraphy can be disturbed by such processes as TECTONIC movements, so this relationship cannot automatically be assumed. Chronostratigraphic relationships can be affirmed only by the use of dating techniques, which may be relative (as in the use of FOSSIL ASSEMBLAGES) or absolute, using physicochemical properties of certain materials (*see* DATING, RADIOCARBON).

Stratigraphic analyses

The outcome of stratigraphic analysis is that separate classifications of the rock strata are produced. It is possible that the lithostratigraphic, biostratigraphic and chronostratigraphic units resulting from the analysis coincide, but this need not be the case. The lithostratigraphy, for example in the case of rock inversion, may have a complex relationship to chronostratigraphy. Also, since the lithostratigraphic sequence is largely determined by the local conditions of SEDIMENTATION when the body of rock was formed, its units may not coincide with those derived from biostratigraphic studies especially if the latter include assemblages of fossils such as pollen and spores that may have travelled considerable distances from the sources. BIOSTRATIGRAPHIC UNITS based on widely dispersed fossils will reflect regional BIOTA that may not be affected by changes in local conditions of sedimentation.

Applications

Stratigraphic studies play an important part in PALAEOECOLOGY since changes in lithostratigraphy and biostratigraphy provide information about past sets of conditions. Studies of recent lake sediments and PEAT deposits have been particularly valuable in research concerning successional sequences of vegetation, since the stratigraphic sequence of strata within these ecosystems retains a FOSSIL RECORD of the history. The strata may also contain evidence of environmental change, such as alteration of climate—which may result in changes in the lithology of lakes as input of water changes in its SEDIMENT load, or changes in the degree of DECOMPOSITION in peat strata resulting from different degrees of wetness in the peat surface layers (*see* MIRE)—or human modification of the landscape (resulting in soil-erosion episodes or the deposition of airborne charcoal particles).

Very recent changes in the environment can be documented either using chemical and physical aspects of lithology (such as HEAVY METAL content in the upper layers of lake sediments) or using biostratigraphic indicators (such as the use of diatoms as indicators of contemporaneous pH conditions in the biostratigraphy of lakes). Stratigraphy is thus an important element in the historical MONITORING of environmental conditions. [P.D.M.]

---

stratosphere The atmospheric layer bounded below by the tropopause (10–15 km above sea level), and above by the stratopause (about 50 km above sea level), and containing 10–30% of the mass of the Earth's ATMOSPHERE. Although deformed by WEATHER systems in the underlying TROPOSPHERE, it plays little direct role in them, and contains almost no cloud except for rare mother-of-pearl CLOUDS and diffuse polar ICE clouds. Much commercial air traffic uses the low stratosphere, and the protective OZONE maximum is maintained in the middle stratosphere by complex photochemistry and gentle air motion. *See also* STRATOSPHERIC CHEMISTRY. [J.F.R.M.]

Stratospheric chemistry

The term refers principally to the chemical processes that determine the concentration of OZONE in the STRATOSPHERE. It also refers to the reactions that oxidize and decompose trace substances released into the ATMOSPHERE from natural and anthropogenic sources. If these are stable enough to escape oxidation in the TROPOSPHERE, they find their way across the tropopause into the stratosphere and are oxidized there (*see* TROPOSPHERIC CHEMISTRY).

The stratosphere

Here we are concerned with the lower part of the stratosphere, from a height of about 15 km at the tropopause to the stratopause, about 50 km above the Earth's surface. In the lower stratosphere the TEMPERATURE increases with height from *c.* −50°C at the tropopause to *c.* 0°C at the stratopause. From a meteorological viewpoint, the region is very stable with little vertical mixing. The PRESSURE falls from *c.* 0.2 atm at the tropopause to *c.* 10^{-3} atm at the stratopause.

Ozone in the stratosphere

The lower stratosphere is relatively rich in ozone with a maximum volume ratio of *c.* 8000 ppb at a height of *c.* 40 km. The total column density of ozone is *c.* 300 Dobson units, which corresponds to a layer of ozone which would be 3 mm thick if it was at 1 atm pressure and 0°C.

The principal concern in stratospheric chemistry is with the reactions that have previously served to maintain the average concentration of ozone at a roughly constant level. The great interest is why there is presently a general decrease of ozone in the stratosphere of about 1% per year, as well as a startling reduction over Antarctica in the spring.

The basic mechanism for ozone formation and removal was proposed by Sydney Chapman some 60 years ago. OXYGEN is photolysed by short-wavelength ultraviolet LIGHT to form oxygen atoms that further react to give ozone:

$$O_2 + h\nu(\lambda < 200\ \text{nm}) \rightarrow O + O \qquad (1)$$

$$O + O_2 \rightarrow O_3 \qquad (2)$$

The ozone is removed by two reactions: photolytic breakdown or reaction with an oxygen atom:

$$O_3 + h\nu(\lambda < 310\ \text{nm}) \rightarrow O_2 + O \qquad (3)$$

$$O + O_3 \rightarrow O_2 + O_2 \qquad (4)$$

Reactions 1 to 4 are in a dynamic steady state in which the ozone concentration depends on the intensity of the SOLAR RADIATION over the two wavelength ranges, and the concentration of oxygen.

When measurements of stratospheric ozone concentrations were made, it was found that the Chapman scheme overestimated ozone concentrations by a factor of about five. It became clear that there are further reactions that remove ozone. The concentrations of other trace substances in the stratosphere are too low (< *c.* 1 ppb) to remove ozone in single-step reactions, so the additional processes have to be catalytic chain reactions.

The NITROGEN OXIDES, nitric oxide (NO) and nitrogen dioxide (NO_2), provide an example. These are formed from nitrous oxide (N_2O) that is produced at the Earth's surface and diffuses slowly up through the troposphere and across the tropopause into the stratosphere. NO reacts with ozone forming NO_2 which, in turn, can react with atomic oxygen to re-form NO.

$$NO + O_3 \rightarrow NO_2 + O_2 \qquad (5)$$

$$NO_2 + O \rightarrow NO + O_2 \qquad (6)$$

$$O + O_3 \rightarrow O_2 + O_2 \qquad \text{Net reaction}$$

The net result is the removal of one O atom and one ozone molecule to form two oxygen molecules, a result equivalent to reaction 4. The concentrations of O and O_3 are always closely related, and the two species are often referred to together as 'odd' oxygen in contrast to 'even' or stable oxygen (O_2). Notice that there is no consumption of the nitrogen oxides in this catalytic cycle of reactions. Such a cycle may allow a single NO_2 molecule to remove up to 100 000 molecules of odd oxygen before the NO_2 is converted to HNO_3 or some other product.

There are two further important examples of catalytic pairs of reactions that remove odd oxygen. The first is the OH/HO_2 cycle:

$$OH + O_3 \rightarrow HO_2 + O_2 \qquad (7)$$

$$HO_2 + O \rightarrow OH + O_2 \qquad (8)$$

$$O + O_3 \rightarrow O_2 + O_2 \qquad \text{Net reaction}$$

and the second is the Cl/ClO cycle:

$$Cl + O_3 \rightarrow ClO + O_2 \qquad (9)$$

$$ClO + O \rightarrow Cl + O_2 \qquad (10)$$

$$O + O_3 \rightarrow O_2 + O_2 \qquad \text{Net reaction}$$

Such reaction sequences do not take place in isolation. In a given air parcel all species are present and the catalytic compounds undergo reactions, not only with odd oxygen but with each other; for example, one important cross-reaction forms the compound chlorine nitrate:

$$ClO + NO_2 \rightarrow ClONO_2 \qquad (11)$$

Such a reaction, which links the NO/NO_2 and Cl/ClO cycles together, serves to reduce the catalytic efficiency of both cycles.

However, $ClONO_2$ can re-dissociate and thus can act as a 'reservoir compound' for the two catalytic chain carriers, ClO and NO_2. Reservoir compounds normally reduce the concentrations of the chain carriers and slow down the rate at which ozone is removed by a particular cycle.

When all the reactions are taken into account, the predicted concentrations of ozone agree well with those found experimentally.

The reduction in stratospheric ozone: the effect of CFCs and stratospheric aircraft

The current decrease in the concentration of stratospheric ozone is attributed to the increased concentration of active CHLORINE (Cl and ClO) released into the stratosphere by the photolysis of CFCs. The physical properties of the CFCs, combined with their chemical inertness, make them ideal for use in refrigerators and air conditioners, and also as drivers for making polymer foams. However, they are also inert to reaction with the hydroxyl radical in the troposphere and thus are not oxidized, as most pollutants are, but simply accumulate in the air. They gradually make their way up through the tropopause into the stratosphere where they are photolysed with short-wavelength ultraviolet light, producing chlorine atoms. With trichlorofluoromethane, CFC-11, for example:

$$CFCl_3 + h\nu(\lambda < 200\text{nm}) \rightarrow CFCl_2 + Cl \qquad (12)$$

The remaining chlorine atoms in the $CFCl_2$ radical are also released into the stratosphere as the radical undergoes further reactions. The atomic chlorine produced increases the destruction of ozone through the Cl/ClO cycle (reactions 9 and 10). The ultimate fate of the chlorine atom is reaction with METHANE to form hydrogen chloride, HCl:

$$Cl + CH_4 \rightarrow CH_3 + HCl \qquad (13)$$

The HCl gradually diffuses down through the tropopause and is washed out of the troposphere. While still in the stratosphere, HCl acts as a reservoir compound for chlorine atoms.

The concentration of stratospheric ozone is likely to continue to decrease. Despite the reductions in the manufacture of CFCs, most of the CFCs produced to date are still in the troposphere. It will take decades for them to diffuse into the stratosphere where they will ultimately be decomposed, releasing chlorine in the process and enhancing the destruction of ozone. Only when the atmospheric concentration of CFCs decreases can one expect the ozone concentration in the stratosphere to recover.

The hydrochlorofluorocarbons (HCFCs), such as CF_3CH_2F (HCFC-134a), now being introduced as substitutes for the CFCs, have hydrogen atoms in the molecule that open them to attack by OH and degradation in the troposphere. They should not, therefore, affect the level of ozone in the stratosphere so severely, but they and their products may pose problems as GREENHOUSE GASES in the atmosphere.

The concern about removal of ozone in the stratosphere originated from plans in the late 1960s for large fleets of supersonic aircraft that would normally fly in the stratosphere. These aircraft would have emitted large quantities of nitrogen oxides and water vapour at a height where their stratospheric concentrations are very low. The two ozone removal cycles involving NO/NO_2 and OH/HO_2 would have been further activated and large amounts of ozone would have been destroyed. The aircraft numbers at that time were overestimated and no problem arose. Recently the idea has been revived and once again there is concern that damage to the OZONE LAYER will result.

The Antarctic ozone hole: complications with surface chemistry

The discovery of the large decrease in the stratospheric ozone concentration over Antarctica, which occurs as the light returns in the spring, was surprising in many ways. It demonstrated particularly that the models of stratospheric chemistry, which are presented in a much simplified form above, were incomplete; there is no way that the reactions so far described could account for the sudden decrease observed in the Antarctic spring.

The probable explanation of the phenomenon involves two particular meteorological factors. First, a large CIRCUMPOLAR vortex of westerly winds forms over Antarctica in the winter and largely isolates the stratosphere over the pole from that over the rest of the Earth. The isolation allows time for the air to be processed and prevents the ingress of warmer ozone-containing air from the north. Second, the extremely low stratospheric temperatures allow polar stratospheric CLOUDS (PSCs) to form during winter. Normally there are no clouds in the stratosphere because water vapour concentrations are too low. PSCs are thought to be formed from crystals of hydrated nitric acid, which does condense at the very low prevailing temperatures.

It is conjectured that during the winter, reservoir compounds such as $ClONO_2$ and HCl can be absorbed on the surface of the crystals in the PSCs. There they react together and form molecular chlorine:

$$ClONO_2 + HCl \rightarrow Cl_2 + HNO_3 \qquad (14)$$

In spring, with the return of the light and a little warmth, molecular chlorine is released into the stratosphere. It is photolysed to chlorine atoms and these react with ozone to form ClO radicals:

$$Cl_2 + h\nu \rightarrow Cl + Cl \qquad (15)$$

$$Cl + O_3 \rightarrow ClO + O_2 \qquad (9)$$

Reaction 9 is not in itself sufficient to set the Cl/ClO cycle in operation since only a few oxygen atoms are present yet in the weak sunlight. However, the ClO radicals can react to form a dimer $(ClO)_2$, only stable at low temperatures, which is then photolysed to Cl atoms. These, in turn, remove ozone and re-form ClO:

$$ClO + ClO \rightarrow (ClO)_2 \qquad (16)$$

$$(ClO)_2 + h\nu \rightarrow 2Cl + O_2 \qquad (17)$$

$$2(Cl + O_3 \rightarrow ClO + O_2) \qquad (18)$$

$$O + O_3 \rightarrow O_2 + O_2 \qquad \text{Net reaction}$$

Here, then, is a catalytic cycle to destroy ozone, which requires low temperatures to form the dimer and is driven by visible light. The reason it is so effective in reducing the ozone concentrations over Antarctica is to be found in reaction 14, which releases chlorine from two reservoir compounds. The formation of nitric acid also removes NO_2 from the stratosphere, preventing the re-formation of the reservoir compound $ClONO_2$, which is normally an inhibitor for the Cl/ClO cycle.

As the season progresses, the temperatures rise, the formation of the dimer is inhibited and the vortex breaks up allowing an influx of air richer in ozone; the concentration of stratospheric ozone is gradually restored.

The sharp decrease in ozone concentrations each spring over Antarctica can be expected to intensify with the years as the CFCs, presently in the troposphere, diffuse into the stratosphere. There is now some evidence that a similar decrease may occur over the ARCTIC in the northern spring.

The discovery of the OZONE HOLE demonstrated to the scientific community that, despite the advances in understanding and the sophistication of experiments and models, the atmosphere still has the capacity to spring surprises. To those responsible for environmental policy, it was a warning that the uncontrolled release of pollutants into the atmosphere, combined with the inherent complexity of the chemistry, can lead to unforeseen effects that are extremely difficult to reverse. *See also* CLIMATE CHANGE. [P.B.]

REFERENCE

Wayne, R.P. (1991) *Chemistry of Atmospheres*. Clarendon Press, Oxford. [See Chapter 4 for an excellent account of stratospheric chemistry.]

---

streams *See* RIVERS AND STREAMS, TYPES OF.

stress The term 'stress' implies an adverse EFFECT, but its precise definition has been somewhat elusive, largely because it is both level- and subject-dependent. It is level-dependent because an adverse response to some environmental variables at, say, a molecular or cellular level may not become manifest at an organismic/population level due to HOMEOSTASIS. It is subject-dependent because there may be GENETIC VARIANCE for stress responses between individuals in a population, and certainly between species. The application of stress to populations, as a SELECTION PRESSURE, may lead to the evolution of TOLERANCE, for example the evolution of tolerance to pesticides and pollutants. It is then debatable if a tolerant GENOTYPE or species can be described as being stressed. Even so, they are adapted to resist stress and it then becomes interesting to consider the extent to which common suites of TRAITS evolve under the same kinds of stress.

Some would restrict stress—by analogy to its definition in physics — to the environmental variables that lead to the response in biological systems — the response being referred to as 'strain'. But this is not general, and in biology the word 'stress' is used both for the cause of a response and the response itself.

Similarly, some ecologists would restrict the definition of stress to environmental factors that impair production processes — growth and reproduction — with the term 'disturbance' applied to factors that impair survival. But again this is not general. Stress is usually employed to describe impairment in general.

The concept of stress has been criticized as being unmeasurable, and as drawing an unreal distinction between favourable and unfavourable environments (the argument being that all environments are unfavourable, because the struggle for existence goes on everywhere). However, multivariate analyses of plant species' traits often produce rankings of species that reflect ADAPTATIONS to soil-nutrient availability (species appear to be ranked along a 'stress axis' from high to low nutrient availability). *See also* COSTS OF RESISTANCE; FAVOURABILITY; GRIME'S TRIANGLE. [P.C.]

REFERENCE

Calow, P. & Berry, R.J. (1989) *Evolution, Ecology and Environmental Stress*. Biological Journal of the Linnean Society, Vol. 37, Nos 1 & 2. Academic Press, London.

stress proteins *See* HEAT-SHOCK PROTEIN.

strontium (Sr) A metallic element (ATOMIC NUMBER=38; relative ATOMIC MASS=87.62; melting point=768°C; boiling point=1381°C), with an average crustal abundance of 384 mg kg^{-1}. It has four naturally occurring stable isotopes, namely ^{84}Sr (0.56% abundance), ^{86}Sr (9.86% abundance), ^{87}Sr (7.02% abundance) and ^{89}Sr (82.56% abundance), and occurs in minerals such as celestite ($SrSO_4$) and strontianite ($SrCO_3$). Strontium has an oxidation state of +2 and is readily soluble in natural waters. Atmospheric testing of nuclear weapons during the 1950s and 1960s led to low-level global CONTAMINATION with the FISSION product ^{90}Sr ($t_{1/2}$=28.5 years: decay product ^{90}Y; $t_{1/2}$=2.671 days) and other radionuclides, with maximum input in 1963–1964. ^{90}Sr and ^{137}Cs account for a large fraction of the initial β/γ activity of irradiated nuclear fuel and fuel reprocessing WASTE, and environmental contamination with ^{90}Sr has resulted from RADIOACTIVE WASTE disposal and accidental releases from nuclear facilities.

^{90}Sr from both atmospheric fallout and nuclear facilities has provided a useful tracer in oceanographic studies. ^{90}Sr is of radiological importance since it has similar chemical properties to CALCIUM and is incorporated in bone. Transfer from soil to grass to milk represents a significant pathway into the human FOOD CHAIN for ^{90}Sr. In geochronological applications, the Rb–Sr dating method is based upon characterizing the decay of ^{87}Rb ($t_{1/2}=4.80\times10^{10}$ years) to ^{87}Sr by measurement of $^{87}Sr/^{86}Sr$ and $^{87}Rb/^{86}Sr$ atom ratios by mass spectrometry. *See also* IONIZING RADIATION; SIEVERT; WASTE DISPOSAL. [A.B.M.]

structure–activity relationship (SAR) The relationship between the structural properties of chemicals and their toxicological/ecotoxicological effects. Determination of this relationship provides some basis for predicting the likely effects of chemicals for which toxicological/ecotoxicological observations are lacking. When these relationships are made quantitative, usually by regression techniques, they are referred to as QUANTITATIVE STRUCTURE–ACTIVITY RELATIONSHIPS (QSARs). This technique applies almost exclusively to organic molecules and is most successful, as far as predicting effects is concerned, for non-polar compounds that have non-specific narcotic effects.

The technique has its roots in the development of drugs such as anaesthetics and antiseptics. Pioneering work was carried out by Hansch and colleagues (e.g. Hansch & Fujita 1964) (for a good recent review see Donkin 1994). The general form of a QSAR equation is:

$C=\alpha+\beta_1\pi_1+\beta_2\pi_2$, etc.

Where C=concentration to produce a particular biological response; α and β terms are constants; and π terms are chemical descriptors that might be obtained either by calculation or observation.

For biological effects QSARs the most important molecular properties are those that relate to solubility in aqueous and lipid components of organisms since they determine transport to and interaction with active sites. If the mode of action is specific, particular steric and electronic factors can be important.

With more than 100 000 artificial chemicals in use and several thousand additional ones being produced annually SAR/QSAR techniques provide the possibility of saving time and effort on ecotoxicological tests. *See also* ECOTOXICOLOGY; EFFECT; FATE; PREDICTIVE ECOTOXICITY TESTS. [P.C.]

REFERENCES

Donkin, P. (1994) Quantitative structure–activity relationships. In: *Handbook of Ecotoxicology*, Vol. 2 (ed. P. Calow), pp. 321–347. Blackwell Scientific Publications, Oxford.

Hansch, C. & Fujita, T. (1964) ρ-σ-π Analysis. A method for the correlation of biological activity and chemical structure. *Journal of the American Chemical Society* **86**, 1616–1626.

struggle for existence *See* NATURAL SELECTION.

subacute/subchronic tests Referring to TOXICITY/ECOTOXICITY tests that are applied at concentrations of chemicals intermediate between the concentrations that are usually applied in acute and CHRONIC TESTS (hence the term subacute), over periods of time that are intermediate between those usually employed for chronic and ACUTE TESTS (hence the term subchronic). Sometimes a subchronic test is distinguished from a subacute test on the basis of its lasting for a much longer time. But all these terms are somewhat imprecise and demand careful usage. *See also* PREDICTIVE ECOTOXICITY TESTS. [P.C.]

subclimax In the scheme of Clements (1936; *see* PROCLIMAX), a type of VEGETATION thought to precede the CLIMAX in SUCCESSION. It involved a great mixture of examples, including forest types maintained naturally by fire (such as those of *Pinus* spp. with cones opening to release seeds only after fire) and vegetation of deeply waterlogged sites (such as reed beds of *Phragmites*, and SWAMP forests of *Larix laricina* or *Taxodium*), all supposed to be able to succeed to the climatic climax if the 'disturbance' of fire, waterlogging, etc. were removed. The concept has been abandoned following the general acceptance of the ideas embodied in the POLYCLIMAX viewpoint. [P.J.G.]

REFERENCE

Clements, F.E. (1936) Nature and structure of the climax. *Journal of Ecology* **24**, 252–284.

subdominant A plant that occupies a significant proportion of the BIOMASS of an ecosystem, but is subsidiary to an even more abundant or influential dominant species (*see* DOMINANCE). [P.D.M.]

subduction The process by which the Earth's outer shell is consumed by descending into the mantle. A subduction zone is an inclined plane descending away from a TRENCH, separating a sinking oceanic plate from an overriding plate. It is a region of high seismic activity. *See also* PLATE TECTONICS. [V.F.]

sublethal response An abnormal response to a chemical at concentrations that do not kill the organism(s) under consideration. The response may be molecular, cellular, physiological, behavioural, developmental, reproductive or a growth response. *See also* PREDICTIVE ECOTOXICITY TESTS. [P.C.]

Sublittoral zone

The depth zone on the shore below the level of mean low tide (or alternatively from the depth of spring low tide; *see* LITTORAL ZONE) and extending to a depth of 200 m. Also known as the subtidal zone. The vast majority of the sublittoral zone of the OCEAN consists of soft SEDIMENT. Rocky sublittoral zones, kelp forests and CORAL REEFS represent other types of sublittoral habitat, but all occur in restricted areas at relatively shallow depths.

Soft substrates

Sublittoral soft-bottom habitats are dominated by suspension-feeders and deposit-feeders, and COMPETITION for food is an important controlling process. Differences among species in their vertical distribution in the sediment may act to reduce competition for space. The distribution of organisms living in soft-sediment subtidal areas is temporally and spatially patchy. The patchiness is due to non-random SETTLEMENT of LARVAE, DISTURBANCE due to water motion or predation, and spatial variability in the time since the last disturbance.

In shallow soft-bottom areas, various seagrasses may dominate. Such seagrass beds cover extensive areas in CONTINENTAL SHELF waters and contribute significantly to the total PRODUCTION of nearshore waters. Although the seagrass is for the most part not directly grazed by herbivores, the DETRITUS derived from decaying seagrass is an important source of organic matter, some of which may be exported a substantial distance from the seagrass bed and hence support SECONDARY PRODUCTIVITY in deeper subtidal communities. Seagrass beds are not only an important source of PRIMARY PRODUCTIVITY in shallow subtidal areas, they also act to stabilize and trap sediment and act as important nursery grounds for a variety of marine species.

Hard substrates

Sublittoral hard substrates are usually densely covered with low-growing plants and SESSILE or SEDENTARY animals, and competition for space is much more intense than in soft-bottom systems. In contrast, food limitation is not as apparent as in soft-sediment areas. Most of the animals occupying subtidal hard substrates are suspension-feeders and are either colonial species able to spread asexually or species having gregarious larval settlement and exhibiting a clumped distribution. The relatively high diversity of hard-substrate subtidal zones is maintained by a combination of biological and physical factors that include storm disturbance, competition for space, GRAZING, RECRUITMENT and MUTUALISM.

Kelps are large brown algae that obtain their NUTRIENTS from seawater (rather than the soil). Kelp forests may dominate hard-substrate subtidal communities in cold-temperate regions of the ocean. Kelp forests are extremely productive, provide a more complex environment with a large number of potential habitats and, compared to other hard substrates, support a greater variety of species.

Coral reefs

Many tropical subtidal communities are dominated by coral reefs, which are composed of deposits of calcium carbonate produced mostly by hermatypic (or reef-building) corals, but also by calcareous algae and a few other organisms. Reefs are restricted to warm, shallow, clear waters, at least partly because the corals rely on the photosynthetic products of their symbiotic plant cells (called ZOOXANTHELLAE). Coral reefs are probably the most complex and most biologically diverse areas in the ocean and provide a habitat for representatives of nearly all phyla and classes. [V.F.]

---

submarine canyon A deep TRENCH cutting through the CONTINENTAL SHELF or slope. Such features are believed to form as a result of erosion by submarine CURRENTS coursing down the CONTINENTAL SLOPES. [V.F.]

submergence marsh The lower part of a SALT MARSH which is subject to (often extended) daily periods of submersion (immersion) at all TIDES, and where the key environmental factors are associated with periods of submersion (immersion). The plants and animals of this level may be considered as mostly species of marine origin which are adapted to surviving limited periods of emersion. *Cf.* EMERGENCE MARSH. [B.W.F.]

suborganismic tests TOXICITY/ECOTOXICITY tests that involve observations at a level below that of the whole organism; i.e. on physiology, anatomy and functioning of organ systems and tissues; cellular structure and processing; or molecular composition and reactions. The extent to which changes at these levels translate into effects at population, community and ecosystem levels is problematic. *See also* PREDICTIVE ECOTOXICITY TESTS. [P.C.]

subsocial species *See* SOCIALITY, TYPES OF.

subspecies The lowest category recognized in TAXONOMY, lying below species. Subspecies can interbreed, species cannot. Given this definition, subspecies must have distinct geographical ranges. Different forms living together within the same area must therefore be a POLYMORPHISM if they interbreed or separate species if they do not. A species composed of numerous subspecies is said to be polytypic. The terms subspecies and race are used synonymously. [L.M.C.]

subspecies concept *See* SPECIES CONCEPTS.

substitutable resource A RESOURCE which is not essential and which can be replaced by a resource of another kind (Fig. S23). Many of the prey species consumed by a GENERALIST predator are substitutable resources. Most resources required by plants are essential, not substitutable; nitrogen cannot be substituted by phosphorus, nor phosphorus by potassium. [M.J.C.]

substitutional load A type of GENETIC LOAD which arises from the loss of FITNESS associated with the complete replacement of one ALLELE by another within a POPULATION or species. Also, the cost to a population in genetic deaths of replacing an allele by another in the course of evolutionary change. Substitutional load is sometimes called 'evolutionary load' or 'the cost of NATURAL SELECTION'. [V.F.]

substitutive experiments *See* DE WIT REPLACEMENT SERIES.

substrate/substratum
1 The material acted upon in a biochemical reaction.
2 The material to which organisms attach and with which they

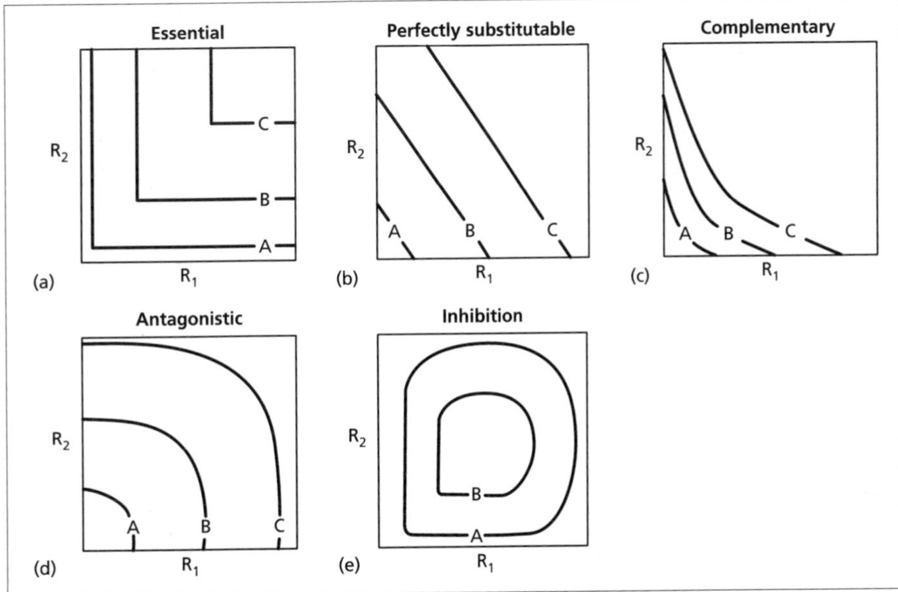

Fig. S23 Classification of resources. Resource-dependent growth isoclines for two resources, R_1 and R_2, showing isoclines separating regions at which a population could persist in the habitat (at higher resource supply rates) from those (below the isocline) where the population could not persist. A has the lowest supply rate, B is intermediate and C has the highest resource supply rate: (a) both R_1 and R_2 are essential resources; (b) the two resources are perfectly substitutable; (c) the resources are complementary; (d) the resources are antagonistic; (e) there is inhibition between R_1 and R_2. (For details see Tilman, D. (1982) *Resource Competition and Community Structure*. Princeton University Press, Princeton, N.J.)

interact to a greater or lesser extent. Sometimes the substrate is used simply as a physical anchor (e.g. rocks for barnacles), whereas in other cases the substrate provides a nutrient source (e.g. SOIL for plants).
[P.C.]

subtidal *See* SUBLITTORAL ZONE.

subtropical The subtropical zones lie mainly polewards of 30°N and 30°S between the tropics and temperate zones. They are characterized by descending air, which is warmed as it descends and so becomes very dry. Daytime temperatures are very high as a result of intense INSOLATION, but at night temperatures fall to near zero because clear skies allow high outgoing RADIATION. Rainfall is low. The subtropical zones are mainly hot deserts and the vegetation consists largely of xerophytes—PERENNIAL drought-resistant shrubs and opportunist EPHEMERALS. [S.R.J.W.]

Succession

Directional change in VEGETATION (or in communities of animals or microorganisms). Succession contrasts with FLUCTUATION, which occurs where change is non-directional. Formerly, some ecologists used 'succession' for any kind of vegetational change, but that usage is now rare. The explicit concept of succession was first enunciated and illustrated by Cowles (1899) on the basis of studies on SAND DUNES near Chicago. An elaborate theory of succession was provided by F.E. Clements but disputed by others (*see* CLIMAX; INDIVIDUALISTIC HYPOTHESIS; ORGANISMIC HYPOTHESIS).

Types of succession

In theory, two kinds of succession can occur: (i) all species are present from the start but different sets of species reach maximum development in successive periods — the 'initial floristic composition' model; and (ii) species invade sequentially — the 'relay floristics' model (Egler 1954).

Successions are usually divided into primary and secondary: PRIMARY SUCCESSIONS are those that occur on substrata without a previous vegetational COVER and without organic matter in the substratum; SECONDARY SUCCESSIONS are those which occur where there has been a vegetational cover and there is a developed SOIL. In general, the relay floristics model applies more often to primary succession, and the initial floristic composition model to secondary successions, although it is almost certain that the relay floristics model applies to some extent to all successions. Certainly the applicability of the initial floristic composition model has been exaggerated by some ecologists; even in the case of succession on abandoned fields, for which it was introduced, only a small minority of the late-successional trees, shrubs and herbs invade very early on.

Another basic categorization of successions has been into ALLOGENIC and AUTOGENIC, respectively brought about by factors originating outside the system, and by the actions of the plants on the habitat. The terms are not often used now, and are better applied to factors than to successions; most successions involve both autogenic and allogenic effects.

Evidence for successions is of many kinds. The most secure evidence comes from historical records; for example, these range from over thousands of years in PEAT (*see* VARIABLE PATHWAYS) to tens to hundreds of years in photographs, maps and written records. Tentative evidence may come from a single survey in which observations are made on which species are growing under others and/or are younger, and from putting together observations on a number of sites to construct an apparent succession. A well-recognized danger is induction of temporal succession where only spatial ZONATION occurs, i.e. the vegetation in each zone is persistent.

Succession may be progressive or retrogressive, respectively involving an increase with time in stature (and usually in species density) or a decrease in both those properties. The widespread occurrence of RETROGRESSIVE SUCCESSION was not generally appreciated before the 1960s.

Characteristics

Various generalizations about trends in succession, for example in PRODUCTIVITY, productivity/biomass ratio (*see* *P*:*B* RATIO), FOOD CHAINS, species DIVERSITY, NICHE breadth, cycles of MINERAL NUTRIENTS, and incidence of symbioses, were made around 1970, but it is now generally accepted that successions are too various in nature for simple generalizations to hold (Drury & Nisbet 1973; Pickett *et al.* 1987). Nevertheless, certain generalizations about early- and late-successional kinds of plants or animals are useful (*see* CLIMAX). Such generalizations are often confused with those about *r*- AND *K*-SELECTION (Grubb 1987).

In general, studies on successions of animals have lagged far behind those on plants (Majer 1989). For microorganisms, successions on the small scale (e.g. on an individual leaf) are better understood than successions at the level of the COMMUNITY (Grubb 1987).

Mechanisms

Much attention has been paid to the mechanisms of succession, often in the hope of generating a general theory, but the latter has proved elusive because successions are so various. Connell and Slatyer (1977; *see* FACILITATION) suggested three models of succession: tolerance, facilitation and inhibition. The last-named phenomenon is not a mechanism bringing succession about, but gives a name to cases where succession is arrested (*see* INHIBITION OF SUCCESSION).

According to the tolerance model, succession occurs simply as a result of the differences between species in life-history characteristics: LONGEVITY, mature height, GROWTH rate, tolerance of shade and dispersibility. In theory, there is no specific effect of any species 'on the habitat' that is needed to drive succession. It is doubtful whether any succession occurs wholly according to the tolerance model. Oldfield successions, which occur where cultivation ceases, are sometimes used to illustrate this type. ANNUALS are succeeded as the most obvious plants by biennials and then perennials; among the perennials herbs give way to shrubs and then trees, and among trees light-demanding species give way to shade-tolerant ones. Under the trees shade-tolerant herbs slowly invade. In general, the

RELATIVE GROWTH RATE of juveniles declines from light-demanding herb to shade-tolerant tree. The dominant annuals and biennials are dispersed through time rather than space, building up large populations of seeds in the SOIL SEED BANK; shade-tolerant trees and herbs appear to be dispersed very slowly in space, and typically have negligible soil seed banks.

It is a moot point whether the casting of shade by the taller plants in such a succession is usefully distinguished from the 'facilitation' effects which occur where plants stabilize the subtratum (as on young DUNES or SALT MARSHES), raise the substratum above the water level (as in young fens), fix NITROGEN (as in many primary successions on well-drained land) or reduce soil pH (*see* FACILITATION). In any case, a clearer example of facilitation in oldfield successions is attraction of fruit-dispersing birds to the first trees and shrubs; inhibition of establishment of tree species in such successions by resident rodent populations may also be important (Myster 1993).

The END-POINT of succession is often uncertain, and depends on the time-scale of interest (*see* CLIMAX). [P.J.G.]

REFERENCES

Connell, J.H. & Slatyer, R.O. (1977) Mechanisms of succession in natural communities and their role in community stability and organization. *American Naturalist* **111**, 1119–1144.

Cowles, H.C. (1899) The ecological relations of the vegetation on the sand dunes of Lake Michigan. *Botanical Gazette* **27**, 95–117, 167–202, 281–308, 361–391.

Drury, W.H. & Nisbet, I.C.T. (1973) Succession. *Journal of the Arnold Arboretum* **54**, 331–368.

Egler, F.E. (1954) Vegetation science concepts. I. Initial floristic composition, a factor in old-field vegetation development. *Vegetatio* **4**, 412–417.

Grubb. P.J. (1987) Some generalizing ideas about colonization and succession in green plants and fungi. In: *Colonization, Succession and Stability* (eds A.J. Gray, M.J. Crawley & P.J. Edwards), pp. 81–102. Blackwell Scientific Publications, Oxford.

Majer, J.D. (ed.) (1989) *The Role of Fauna in Reclaimed Lands*. Cambridge University Press, Cambridge.

Myster, R.M. (1993) Tree invasion and establishment in old fields at Hutcheson Memorial Forest. *Botanical Review* **59**, 251–272.

Pickett, S.T.A., Collins, S.L. & Armesto, J.J. (1987) Models, mechanisms and pathways of succession. *Botanical Review* **53**, 335–371.

successional gradient In relatively few cases the stages in a SUCCESSION occur in series along gradients of environmental conditions, for example at the ends of receding glaciers, and on large flood plains where RIVERS are meandering in such a way as to destroy the VEGETATION on one side of each bend, while depositing SEDIMENT and making it possible for succession to begin on the other side of the bend. [P.J.G.]

sucking *See* FEEDING TYPES, CLASSIFICATION.

suction traps *See* SAMPLING METHODOLOGY/DEVICES.

sulphur (S) A non-metallic element (ATOMIC NUMBER = 16, relative ATOMIC MASS = 32.06; melting point = 112.8°C; boiling point = 444.6°C) that is present at an abundance of 260 mg kg^{-1} in the Earth's CRUST and 900 mg kg^{-1} in the oceans, and constitutes 2–3% of the Earth as a whole. It is an ESSENTIAL ELEMENT for life, being a constituent of important AMINO ACIDS, PROTEINS and enzymes. Although found in many METAL sulphide ores, sulphur is most readily obtained from elemental sulphur deposits, hydrogen sulphide (H_2S) in sour natural gas, organic sulphur from PETROLEUM and pyrite (FeS_2). Its complicated natural BIOGEOCHEMICAL CYCLE (*see* SULPHUR CYCLE) can release sulphur to the ATMOSPHERE in the form of sulphur dioxide (SO_2) or H_2S from volcanoes, H_2S from terrestrial and salt-marsh environments, and dimethyl sulphide ($(CH_3)_2S$) from the oceans.

The most serious anthropogenic release is of SO_2 from the combustion of FOSSIL FUELS, especially high-sulphur coal, resulting in sulphuric acid (H_2SO_4) PRECIPITATION, with detrimental consequences for freshwater lakes, buildings and trees (despite sulphur's role as an essential nutrient for vegetation). The mining of metal sulphide ores and coal can also lead, on oxidation, to unwanted generation of sulphuric acid (H_2SO_4) and LEACHING of metals, which can then enter aquatic systems. [J.F.]

sulphur cycle The movement of SULPHUR (S) between the LITHOSPHERE (the dominant reservoir) and the ATMOSPHERE, HYDROSPHERE and BIOSPHERE, and the associated transformations between different chemical forms (see Fig. A5). Natural release of S into the atmosphere, as sulphur dioxide (SO_2), occurs by volcanic action and by BIOMASS burning (about 30 Tg S year^{-1}). Fossil-fuel burning and smelting of METAL sulphide ores are major anthropogenic sources of SO_2 (80 Tg S year^{-1}).

The biological reduction of sulphate in seawater by PHYTOPLANKTON results in EMISSIONS of dimethylsulphide (DMS; $(CH_3)_2S$). Other biogenic gases released mainly by anaerobic processes are hydrogen sulphide (H_2S), carbon disulphide (CS_2), carbonyl sulphide (OCS) and dimethyldisulphide (CH_3SSCH_3). The total biogenic emission is estimated at about 60 Tg S year^{-1}, two-thirds of which is DMS.

Atmospheric oxidation of these gases creates additional SO_2. Further oxidation by OH and OZONE produces sulphuric acid and sulphates. The latter play an important part in global change: as AEROSOL particles they act as cloud condensation nuclei (CCN) and reduce the absorption of solar energy significantly (*see* SULPHUR OXIDES). Sulphuric acid and sulphate deposition, particularly over polluted industrial regions, is a major component of acid rain, causing ACIDIFICATION of soils and surface waters and mobilizing ALUMINIUM, with adverse consequences for vegetation and fish populations (*see* ACID RAIN AND MIST; pH).

Sulphur in soil organic matter and plant litter is released as sulphate by MINERALIZATION, and is either taken up by plant roots and microorganisms or leached out in drainage water. Microbial reduction of sulphate to sulphides in anaerobic environments produces the insoluble pyrite, FeS_2. Subsequent exposure of pyrite to the atmosphere, for example following land drainage or dumping of mine spoil, leads to reoxidation to sulphate and consequential acidity problems. *See also* BIOGEOCHEMICAL CYCLE. [K.A.S.]

REFERENCE

Charlson, R.J., Anderson, T.L. & McDuff, R.E. (1992) The sulfur cycle. In: *Global Biogeochemical Cycles* (eds S.S. Butcher, R.J. Charleson, G.H. Orians & G.V. Wolfe), pp. 285–300. Academic Press, London.

sulphur cycle over the sea *See* AEROSOLS.

sulphur oxides SULPHUR has two oxides, sulphur dioxide (SO_2) and sulphur trioxide (SO_3). Anthropogenic EMISSIONS of sulphur (S) to the ATMOSPHERE are almost entirely in the form of SO_2, from the burning of FOSSIL FUELS and ore smelting. BIOMASS burning and volcanoes are important natural sources. Oxidation of biogenic S gases by OH radicals adds to the SO_2 burden (*see* SULPHUR CYCLE). Atmospheric oxidation of SO_2 by OH and OZONE produces SO_3, which reacts with cloud WATER to produce sulphuric acid and sulphates, for example ammonium sulphate and ammonium hydrogen sulphate by reaction with AMMONIA in polluted atmospheres. These sulphates play an important role in global CLIMATE CHANGE, forming AEROSOLS that act as cloud condensation nuclei. [K.A.S.]

sulphuric horizons *See* SOIL CLASSIFICATION.

sum of products An analogous term to SUM OF SQUARES. When observations consist of paired measurements (*x* and *y*), the 'sum of products':

$$= \sum(\bar{x}-x)(\bar{y}-y)$$

Like 'sum of squares' (*see* VARIANCE) there is an equivalent, but more convenient formula:

$$\text{Sum of products} = \sum xy - \left(\sum x\right)\left(\sum y\right)/n$$

where *x* and *y* are each paired observations in the SAMPLE, *n* is the number of paired observations in the sample, and the Greek symbol Σ (capital sigma) is the mathematical notation for 'sum of'. When the 'sum of products' is divided by its DEGREES OF FREEDOM we obtain the COVARIANCE. [B.S.]

sum of squares The squared deviations of observations about a MEAN. For example, in the numerator of the formula for the VARIANCE, the 'sum of squares' $=\Sigma(\bar{x}-x)^2$. It has the important property that it can be partitioned or summed. *See also* ANALYSIS OF VARIANCE; REGRESSION ANALYSIS, CURVILINEAR; REGRESSION ANALYSIS, LINEAR. [B.S.]

sunflecks Patches of direct sunlight within a vegetational CANOPY, especially as seen on the floor of a forest when rays from the solar disc filter through the gaps in the canopy of leaves. Sunflecks vary in size, but are typically less than a metre, and often just a few centimetres in diameter. Photon flux varies across the sunfleck, often in the range 100–1000 μmol photons $m^{-2} s^{-1}$. Sunflecks move in arcs across the forest floor as the Sun moves across the sky. They are important in the carbon economy of tree seedlings: in some cases most of the total PHOTOSYNTHESIS is accomplished during brief periods of illumination by sunflecks. [J.G.]

REFERENCE

Chazdon, R.L. (1988) Sunflecks and their importance to forest understorey plants. *Advances in Ecological Research* **18**, 1–63.

sunspot cycles Sunspots are local disturbances on the solar surface, seen as dark spots when an image of the Sun is projected on to a screen. The intensity of sunspots varies on an 11-year cycle causing small variations in the energy flux to the Earth, and associated changes in the magnetosphere and ionosphere. Some people believe that these cycles influence the growth of plants and the behaviour of animals, but the effect is very small if it exists at all. *See also* MICROTINE CYCLES. [J.G.]

sunspots *See* RADIATION.

supercooling (undercooling) The phenomenon by which WATER and aqueous solutions remain unfrozen below their melting point. The melting point is the TEMPERATURE at which the last ICE crystal disappears when a frozen sample is warmed. The supercooling point (SCP) is the temperature at which spontaneous freezing occurs in a supercooled liquid. Supercooling and supercooling point are biological terms synonymous, respectively, with undercooling and temperature of crystallization (or freezing temperature) — terms more commonly used in the physical sciences.

Supercooling is the mechanism by which freeze-avoiding (freeze-intolerant) organisms prevent the formation of ice in their body tissues and fluids which would otherwise be lethal. Freeze-avoiding insects maximize their supercooling ability in winter by evacuating potential nucleators (food particles) from the digestive system and synthesizing antifreeze agents, notably polyhydroxy ALCOHOLS (polyols) and PROTEINS. Mid-winter supercooling points in insects are usually in the range of –20 to –40°C, the lowest values occurring in eggs (Fig. S24). The supercooling point has been widely used as an indicator of the lower lethal limit of freeze-avoiding organisms but can be misleading as an index of cold hardiness if pre-freeze

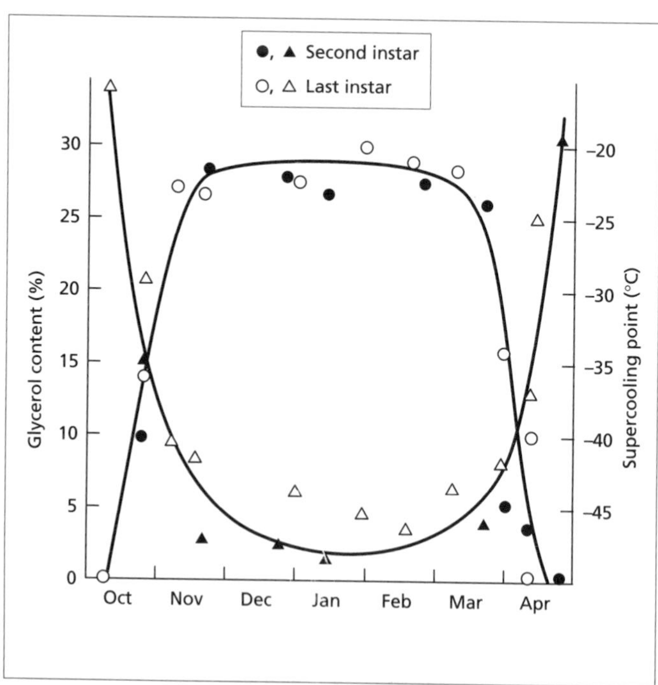

Fig. S24 Seasonal changes in glycerol content (circles; percentage of fresh weight) and supercooling point (triangles) in larvae of *Retinia* (*Petrova*) *resinella*. (After Hansen, T. (1973) Variations in glycerol content in relation to cold-hardiness in the larvae of *Petrova resinella* L. (Lepidoptera: Tortricidae). *Eesti NSV Teaduste Akadeemia Tometised Biologia* **22**, 105–111.)

MORTALITY is the dominant lethal process. *See also* FREEZE AVOIDANCE; OVERWINTERING. [J.S.B.]

REFERENCES

Franks, F. (1985) *Biophysics and Biochemistry at Low Temperatures*. Cambridge University Press, Cambridge.

Leather, S.R., Walters, K.F.A. & Bale, J.S. (1993) *Ecology of Insect Overwintering*. Cambridge University Press, Cambridge.

supercooling point (SCP) *See* FREEZE AVOIDANCE; SUPERCOOLING (UNDERCOOLING).

superfund legislation UNITED STATES LEGISLATION concerned with polluted sites. The legislation is formally designated as the Comprehensive Environmental Response Compensation and Liability Act (CERCLA, or Superfund) 1980, and the Superfund Amendments and Reauthorisation Act 1986 (SARA).

CERCLA was enacted in 1980 to address actual or threatened releases from sites containing hazardous substances. It arose as a direct result of the emergency that occurred in August 1978 when chemicals that had been buried for a long time were discovered seeping out of the ground and into homes in the Love Canal area of Niagara Falls (New York State). Remedial work at the site and relocation of some 200 families cost the State in excess of $30 million; there was no federal programme to provide assistance. Owners and operators of facilities where hazardous substances and wastes are stored, treated or disposed of were required to notify, by 9 June 1981, USEPA of quantities and types of hazardous substances and wastes, and any known suspected or likely releases of these to the environment. USEPA used this information to formulate the original National Priority List (NPL), an inventory of contaminated sites requiring CLEAN-UP. USEPA revises this List from time to time. Currently the NPL has more than 1000 sites requiring immediate remedial attention.

CERCLA gives the federal government power to deal with hazardous substances either by removal or remedial action. It established a substantial billion-dollar trust fund. This is intended to pay for action only on an interim basis or in instances where no party can be held responsible. The primary aim of CERCLA is to require parties connected with a contaminated site, or with the hazardous substances it contains, to pay for the cost of clean-up. Cost estimates for clean-up of the thousand or so waste sites is in the billion dollar range. It is anticipated that responsible parties rather than the superfund will bear much of this expense.

CERCLA imposes strict liability upon any party that owned the site from which the actual or threatened release is taking place; which generated, transported or disposed of the substances at the site; or which operated the site. Liability is joint and several among all such parties. It is no defence that the party's actions may have been lawful—indeed, allowed by permits. Responsible parties are not only responsible for remedial and removal costs but also for damage to natural resources. The President may also issue administrative orders to responsible parties obliging them to take protective measures. SARA added an 'innocent landowner' defence to deal with problems faced by lending institutions and others.

In addition to the amendments to CERCLA, SARA contains a self-contained law that addresses emergency planning and community right-to-know. Amongst other things this requires companies to:

- work with community groups to define how various parties should respond to emergencies involving chemicals;
- notify local groups if an extremely hazardous chemical is released into the environment;
- produce an annual inventory of toxic chemicals that each facility routinely releases into the environment, including data on the average volume of the chemical on hand and the quantity and method of EMISSIONS.

This last provision has reputedly had a significant effect on POLLUTION AVOIDANCE. *See also* TOXIC RELEASE INVENTORY. [P.C.]

supergene A group of tightly linked gene loci on a chromosome which have some form of coordinated function. For an example of the evolution and functioning of a supergene *see* MIMICRY. [P.M.B.]

superior competitor *See* COMPETITORS, GOOD AND BAD.

superorganism A collection of single individuals that together possess the functional organization characteristic of organisms, i.e. sharing a reproduction and heredity system. Social groups of individuals of the same species can operate in this way; ECOSYSTEMS cannot. *See also* MODULAR ORGANISM. [P.C.]

superparasitism The deposition of one or a clutch of EGGS by a PARASITOID on a HOST that has already been parasitized by a member of the same species (termed self-superparasitism if the first egg or clutch is laid by the same individual). Superparasitism was once thought to be due to the inability of the parasitoid to detect previous PARASITISM (lack of host discrimination), but now it is widely considered to be in many circumstances an adaptive behaviour. *See also* HYPERPARASITOID; MULTIPARASITISM. [H.C.J.G.]

supertramp species A term describing a particular pattern of DISPERSAL and COLONIZATION that may be found in certain species. Such species are characterized by high rates of dispersal across newly created habitats—which may result from processes such as terrestrial or marine island formation. Typically, such species are 'fugitives' which only persist during the early stages of colonization or on small 'islands' where more extensive communities cannot become established. In a study of bird populations in the Bismarck Islands (Diamond 1975), it was found that supertramp species had characteristic INCIDENCE FUNCTIONS. These were measured by examining the proportion of islands occupied by a given species across a range of different island areas. Supertramp species' curves show a rapid decline in incidence as island area increases and more dominant communities become established. *See also* ASSEMBLY RULES. [R.C.]

REFERENCE

Diamond, J.M. (1975) Assembly of species communities. In: *Ecology and Evolution of Communities* (eds M.L. Cody & J.M. Diamond). Belknap, Cambridge, Mass.

supplementary units *See* SI UNITS.

supply point In a system with two potentially limiting RESOURCES, the supply point is the point in phase space defined by the maximal amounts of two resources in a given habitat. The rate of supply of a resource is assumed to be proportional to the difference between the supply (S) and the availability (R) of the resource: i.e. $dR/dt = a(S-R)$. This defines a resource supply vector which always points directly to the supply point, and which is used to determine the

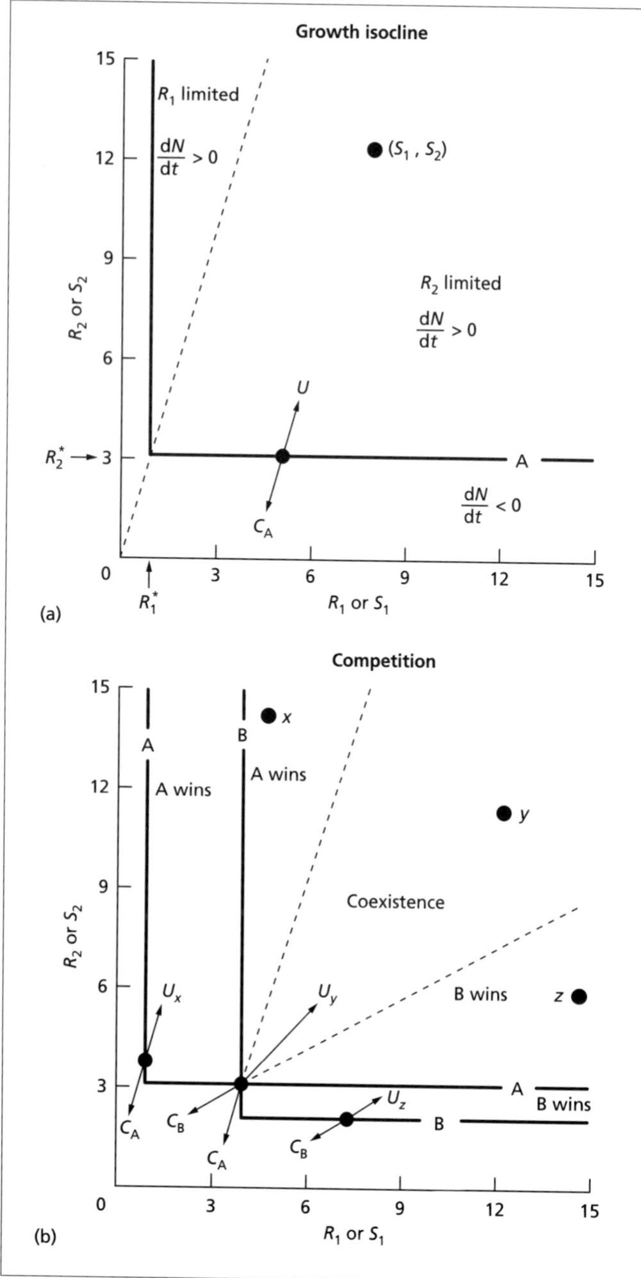

Fig. S25 (a) The solid curve with the right-angle corner is the resource-dependent growth isocline of species A. At all points on the isocline, the reproductive rate of species A equals its mortality rate. Above the isocline it increases in abundance, and below the isocline it declines. Note that the position of the isocline is determined by the values of R_1* and R_2*. The dashed line from the origin through the corner of the isocline shows the amounts of R_1 and R_2 at which the species is equally limited by both resources. The dot on the isocline is the one-species equilibrium point, where the reproductive rate equals the mortality rate, and the resource consumption rate (C_A) equals the resource supply rate (U). (b) The isoclines for species A and B, and their consumption vectors (C_A and C_B), define the types of habitats in which species A is dominant (e.g. supply point *x*), both species coexist in stable equilibrium (supply point *y*), and species B is dominant (supply point *z*). (After Crawley, M.J. (ed.) (1986) *Plant Ecology*. Blackwell Scientific Publications, Oxford; modified from Tilman, D. (1982) *Resource Competition and Community Structure*. Princeton University Press, Princeton, N.J.)

point on the ZERO GROWTH ISOCLINE at which equilibrium occurs. In a two-resource system, this will be at the point where a species is equally limited by both resources (Fig. S25). [M.J.C.]

supply-side ecology An investigative approach to examining factors that influence local POPULATION DYNAMICS and COMMUNITY composition and structure, which stresses the importance of the supply of colonists or RECRUITMENT (often involving physical transport processes) from a regional pool. The term was coined by Lewin (1986) and has largely been developed in relation to marine intertidal communities. These communities are composed largely of SESSILE adults, the competitively dominant species of which have PELAGIC larvae for at least part of their life cycle (e.g. barnacles). In many intertidal communities, SPECIES RICHNESS and community processes appear to be largely determined by the supply of recruits and SETTLEMENT out of the water column on to the hard substrate. Recruitment is, in turn, influenced by offshore processes such as predation on larvae, larval growth rates and physical transport processes such as CURRENTS.

Supply-side ecology can also potentially be applied to PARASITE communities, where availability of parasites could play a major role in determining parasite communities in the HOST species or individual. Supply-side ecology has also been considered in the context of island communities and has a logical counterpart in terrestrial ecology, where mechanisms of PROPAGULE introduction and community fragmentation become important. *See also* DISPERSAL; ISLAND BIOGEOGRAPHY; METAPOPULATIONS. [P.S.G.]

REFERENCES

Lewin, R. (1986) Supply-side ecology. *Science* **234**, 25–27.

Roughgarden, J., Gaines, S. & Pacala, S. (1987) Supply-side ecology—the role of physical transport processes. In: *Organization of Communities Past and Present* (eds J. Gee & P. Giller), pp. 459–486. Blackwell Scientific Publications, Oxford.

supralittoral zone *See* WITHIN-LAKE HABITAT CLASSIFICATION.

surber samplers *See* SAMPLING METHODOLOGY/DEVICES.

surface run-off *See* RUN-OFF.

surface tension The tension that exists in the surface of a liquid as a result of the tendency of molecules to cling to each other. As a result of surface tension liquids behave as if they have a skin. Surface-tension phenomena enable small insects to walk on water, they cause liquids to exist as droplets rather than thin films, and they cause water to rise inside capillaries. Surface tension in water is reduced by the addition of DETERGENTS, also known as wetting agents because they enable water to exist as films instead of droplets on surfaces. [J.G.]

surfactants Substances, also known as surface-active agents or wetting agents, that are the major ACTIVE INGREDIENTS of DETERGENTS. They reduce the SURFACE TENSION of liquids. Surfactants can be anionic, cationic or non-ionic, and they include both soaps and synthetic materials from PETROLEUM as well as natural oils (e.g. from coconut). Large quantities are disposed of down drains, but are often effectively degraded in SEWAGE treatment. Some are ecotoxic. [P.C.]

REFERENCE

Hennes-Morgan E.C. & De Oude, N.T. (1994) Detergents. In: *Handbook of Ecotoxicology*, Vol. 2 (ed. P. Calow), pp. 130–154, Blackwell Scientific Publications, Oxford.

surplus power The energy remaining after standard and active metabolism have been subtracted from the total ENERGY BUDGET. This fraction of the energy budget can be allocated to increased growth and/or reproduction. *See also* PHYSIOLOGICAL ECOLOGY; RESOURCE PARTITIONING; SCOPE FOR GROWTH. [D.A.R.]

surveillance Repeated, standardized measurement of ecological variables at a site or sites to investigate possible trends in time. *Cf.* MONITORING; RETROSPECTIVE ANALYSIS (MONITORING); SURVEYS. [P.C.]

surveys Chemical and/or biological SAMPLING programmes to establish a pattern of spatial variation at one point in time. *Cf.* MONITORING; RETROSPECTIVE ANALYSIS (MONITORING); SURVEILLANCE. *See also* BIOLOGICAL MONITORING. [P.C.]

survival of the fittest *See* NATURAL SELECTION.

survival rate The proportion of the organisms at one age or stage remaining alive at the beginning of the next. It is equivalent to the average probability of an individual staying alive from one age to the next. *See also* DEMOGRAPHY; LIFE TABLE; POPULATION DYNAMICS. [A.J.D.]

survivorship curve A graph of the numbers of an organism against time or age; particularly as produced by plotting a_x, l_x or $\log l_x$ from a LIFE TABLE against x (Fig. S26).Use of log abundances produces curves with the same gradient for populations where the MORTALITY RATES are the same. Curves derived from COHORT LIFE TABLES or from STATIC LIFE TABLES for populations with constant mortality and NATALITY rates will always have negative slope, and the shape of the curve will describe the distribution of mortality with age, mirroring plots of mortality (e.g. q_x or k_x) against age.

Real survivorship curves are rarely smooth but four hypothetical standard curves can be identified (Fig. S26). The type I standard curve represents a population where mortality is concentrated among older individuals and mortality rates increase with age. Type II indicates that a constant number dies in each unit time, and type III that mortality rate is constant, the same proportion dying in each time unit. Type IV represents a population where mortality is most severe among young individuals and mortality rates decrease with time. The type III curve has a constant gradient if $\log l_x$ is plotted on the vertical axis. *See also* a_x; k_x; l_x; q_x. [A.J.D.]

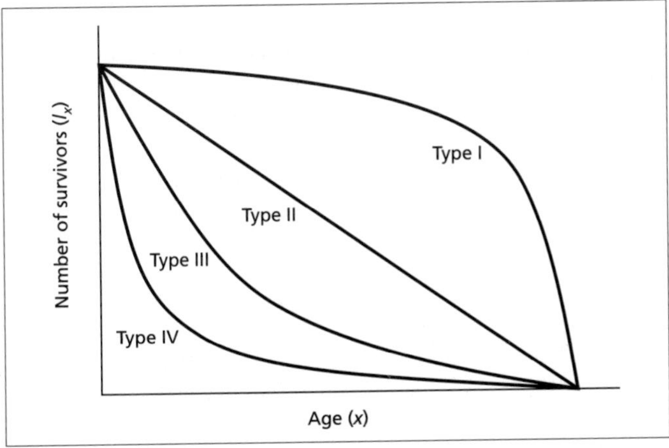

Fig. S26 The four standard survivorship curves of survivorship (l_x) against age (x).

suspended solids Non-living organic particles occurring in the water column of flowing waters, LAKES or the marine environment, and maintained in suspension by physical forces. Unless filtered out by filter-feeding organisms, or decomposed by attached microbes, such particles ultimately sink to form sediments. Suspended solids are defined experimentally as the DRY MASS of solid obtained by filtering a known volume of water through a filter with specifed pore size, and are measured in milligrams per litre. *See also* PARTICULATE ORGANIC MATTER. [P.C.]

suspension-feeder *See* FEEDING TYPES, CLASSIFICATION.

suspension feeding A means of obtaining food by intercepting particles from the surrounding water. Mechanisms of particle capture include sieving, direct interception and gravitational deposition. Suspension-feeders may collect particles on mucus sheets that they have secreted, on mucus-coated cilia, on beating setae or on long tentacles. Particle selectivity usually is for particle size and not particle type, and a wide range of food sources may be utilized including PHYTOPLANKTON, ZOOPLANKTON, bacteria and non-living detrital particles. *See also* FEEDING TYPES, CLASSIFICATION. [V.F.]

sustainability *See* SUSTAINABLE DEVELOPMENT.

sustainable agriculture Methods of AGRICULTURE that do not involve HARM to the farm as an ECOSYSTEM and/or to surrounding ecosystems. *See also* FARMING PRACTICES, SUSTAINABLE. [P.C.]

sustainable city A city designed to last; i.e. one in which infrastructure and social organization are such as to create lasting opportunities for its inhabitants in terms of employment and recreation while minimizing the effects on the environment. [P.C.]

sustainable demand Demand made by customers on business—and hence ultimately by business on environmental RESOURCES for raw materials and the reception of WASTE—that is compatible with the principles of SUSTAINABLE DEVELOPMENT. [P.C.]

Sustainable development

Definitions

According to *Our Common Future* (1987), the report that emerged from the World Commission on Environment and Development set up by the UN in 1987, sustainable development is 'Development that meets the needs of the present generation without compromising the ability of future generations to meet their own needs'. It is neither an obvious logical necessity, nor a principle that is tried and tested. Indeed, it provides much room for interpretation, which is why many pages have been written on it and almost as many clarifying definitions have been offered in place of that given above. It is in reality a statement of faith that we can live in some kind of balance with our environment.

Premises

The sustainable development idea makes a number of assertions:

1 that development is about meeting human needs, implying not only wealth but also welfare — the emphasis is on *our* needs, the approach is frankly ANTHROPOCENTRIC;

2 that we are talking about the needs of everyone in all parts of the world (intragenerational equity) and through time (intergenerational equity);

3 that our response to satisfying these needs can be increased — development implies improvement/progress, certainly not regression — without necessarily impairing our surroundings to such an extent that development will ultimately cease or be reversed;

4 'without compromising' nevertheless recognizes that there will be limitations on how this can be achieved.

Implications and interpretations

So how is this to be understood? Presume that the servicing of needs depends upon wealth. Not all are happy with this materialistic orientation; but a major result of the sustainable development debate is not only an embracing of the link between needs and traditional economic activities, but the fact that social and ecological elements must be factored in as well. So overall wealth (C_T) is made up of the traditional economic capital (C_M), social capital (C_S) in the form of education, technology, medicine, etc., and natural capital (C_N) in the form of environmental and ecological entities. The sustainable development idea goes further and recognizes a delicate and complex interaction between these components (Fig. S27). Thus, increasing C_M uses up RESOURCES and pollutes thereby having a negative impact on C_N; but by providing resources it enhances C_S and so might indirectly benefit C_N by providing people with the 'comfort' they need to worry about environment. It might also have beneficial effects on C_N by providing cleaner technology and methods of REMEDIATION.

As already noted, there are those who object to this economics representation of the problem — as if a sustainable life can be achieved in an economic vacuum. Yet life and living are fundamentally about using and optimizing the use of limited resources of time, effort and raw materials so there can be no escape from economics, whatever the currency. There are, nevertheless, various ways that the balance between the elements of the economy can be understood:

1 C_T can be maintained or enhanced only if C_N is protected — no negative effects allowed;

2 C_N can be subdivided into renewable (C_{NR}) and non-renewable (C_{NNR}) natural capital — OXYGEN, CARBON DIOXIDE and OZONE are renewable components of the environment but species are not — and C_T can only be maintained or increased if C_{NNR} is protected;

3 increases in C_T will inevitably involve negative effects on C_N, both the renewable and non-renewable components, but that the system can be optimized in some way to minimize the effects on non-renewables and to maximize the EFFICIENCY of use of renewables, which will lead to a maintained or increasing C_T;

4 C_T can only be maintained at the expense of negative effects on C_N — but this should not be a concern because increasing levels of C_M and C_S will enable COMPENSATION for further loss of C_N. For example, loss of a non-renewable energy source could be compensated by a development in technology that replaces it. (It would be hard to apply similar reasoning to species extinction, but some might argue that such losses are compensated by the benefits achieved in terms of social welfare.)

These approaches, which need not be as sharply distinct as they are presented, move from what might be described as the 'deep green' and idealistic, to the 'light green' and pragmatic, and finally to the 'non-green'.

Ever since *Our Common Future*, there has never been the view that development entails no change in our surroundings, and some have even admitted that 'Every ecosystem everywhere cannot be preserved intact'. Just how much, if any, change is acceptable for our long-term well-being as a species is open to question and debate. *See also* ENVIRONMENTAL (ECOLOGICAL) ECONOMICS. [P.C.]

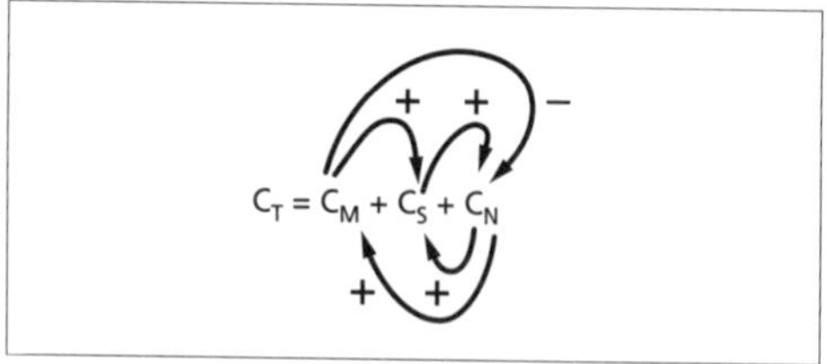

Fig. S27 Total capital (C_T) is made up of man-made capital (C_M), social capital (C_S) and natural capital (C_N). There are interactions (shown by arrows) between them.

REFERENCE

World Commission on Environment and Development (1987) *Our Common Future*. Oxford University Press, Oxford.

---

sustainable yield Removing RESOURCES from the environment (HARVESTING) at a rate which allows balanced replacement by natural processes. The return per unit harvest effort will be optimized at or near this level. However, despite much effort at improving population models, ecologists are still not usually in a position to precisely define sustainable YIELD. So very often the PRECAUTIONARY PRINCIPLE is employed, for example in defining fishing quotas. [P.C.]

SVATS *See* SOIL–VEGETATION–ATMOSPHERE TRANSFER SCHEME.

swamp The definition of this type of wetland varies in different parts of the world, but one important characteristic is that the SEDIMENT surface normally remains below the WATER TABLE throughout the year. Swamp forms a hydroseral stage in SUCCES-

SION following open water with submerged or floating aquatics, and is dominated by EMERGENT PLANT species. In Europe the term is used of ecosystems in which the dominant plants are herbaceous (as in reed swamps of *Phragmites*, sedge swamps of *Carex* or *Cladium*, and papyrus swamps of *Cyperus*), whereas in North America the term is reserved for forested WETLANDS dominated by such tree species as *Taxodium*. Herbaceous swamps in America are termed 'MARSH'.

Swamps are RHEOTROPHIC mires and their sediments are usually rich in clastic material derived from fluviatile erosion of their CATCHMENTS. Their pH and mineral status vary with the chemistry of the catchments and with the rate of flow of water through them. Estuarine swamps may be subjected to BRACKISH WATER input. Their abundant supply of water and nutrient elements means that swamps usually have a high net primary productivity, often exceeding $5\,kg\,m^{-2}\,year^{-1}$.

Swamps are found throughout the world where suitable topographic sites exist but lower-latitude sites may be subjected to occasional drought conditions, as in the marismas of southern Spain. [P.D.M.]

swamping *See* PREDATOR SATIATION.

sweepstake dispersal route The most hazardous and unpredictable of the kinds of pathway that may be available for the MIGRATION of an organism from one habitat to another. Where the migration route is habitable for the organism throughout its length, the term CORRIDOR is used. In this case, the same type of BIOTA can occupy sites during the course of their migration along that corridor, and so use it as an unimpeded DISPERSAL route. If the pathway of migration is of more varied habitat, and is in parts unfavourable for the biota in question, then that pathway is referred to as a filter, inhibiting passage of certain less-versatile organisms (Cox & Moore 1993).

A sweepstake route is the least favourable of potential migration pathways in that it can only be traversed very infrequently and with great difficulty, or only under very rarely occurring circumstances. It is most frequently used in the context of a pathway leading to the COLONIZATION by organisms of an isolated island, where a strong element of chance is involved in any potential colonizer actually arriving from some remote source and establishing itself. A biota which has successfully colonized an isolated island or group of islands by means of sweepstake dispersal will typically have quite diverse elements in it. These elements are likely to be characterized more by their adaptation to long-distance dispersal than to the habitat where they become established. [W.G.C.]

REFERENCE

Cox, C.B. & Moore, P.D. (1993) *Biogeography: An Ecological and Evolutionary Approach*, 5th edn. Blackwell Scientific Publications, Oxford.

swidden agriculture *See* SHIFTING CULTIVATION.

switching The ability of a PREDATOR or PARASITOID to change from attacking one species of prey or HOST to another species as the density of the latter increases or the former declines. Because switching leads to density-dependent MORTALITY, it is a means by which a POLYPHAGOUS natural enemy can contribute to the regulation of the population densities of its prey or host. Switching can give rise to type 3 FUNCTIONAL RESPONSES. *See also* FOOD SELECTION. [H.C.J.G.]

SWQO *See* STATUTORY WATER QUALITY OBJECTIVES.

symbiont An organism that lives in close association with another organism. The relationship is most frequently mutually beneficial (mutualistic) as each 'selfish' individual derives a direct benefit from the association. For example, anaerobic protozoan flagellates in the digestive tracts of TERMITES (isopterans) reduce CELLULOSE to acetic acid, which the termite then absorbs. In return, the protozoans gain a protected environment and constant food supply from which they obtain their own nutritional requirements. *See also* MUTUALISM; SPECIES INTERACTIONS. [R.C.]

symbiosis *See* MUTUALISM.

symmetric competition A term that is generally synonymous with SCRAMBLE COMPETITION, but used most commonly to describe this form of competition in plant populations. Under this mode of competition, no individuals achieve competitive dominance. Instead RESOURCES are divided more evenly amongst competing individuals. The ecological significance of this is that if there is a threshold for survival or reproduction then unstable POPULATION DYNAMICS, such as POPULATION CYCLES or CHAOS may result. [R.P.F.]

Sympatric speciation

A mode of non-allopatric speciation which occurs in the absence of geographic isolation when sister species (*see* SISTER GROUP/SPECIES) evolve within the DISPERSAL range of the offspring of a single DEME. During the course of sympatric speciation the probability of mating between two individuals depends on their genotypes alone. Sympatric divergence is promoted when genetic variation allows a portion of the original species to shift to a new HABITAT or HOST. By exploiting different habitats, COMPETITION for RESOURCES between diverging populations is reduced or avoided. The genetic differences that arise between sister populations as they adapt to different habitats or hosts provide sufficient intrinsic barriers to GENE FLOW to allow each population to establish independent evolutionary lineages (Bush 1994).

Interspecific hybridization

Sympatric speciation in plants is well documented experimentally and from examples in nature. Most commonly, sympatric speciation results from interspecific hybridization between closely related plant species occupying different habitats (soil types, etc.). Hybridization may result in the instantaneous development of a third diploid species or produce a new ALLOPOLYPLOID (or amphiploid) species which has three or more chromosome sets or genomes. The new species consistently occupy intermediate or unexploited habitats. Bread wheat (*Triticum aestivum*), for example, is a tetraploid that has evolved from a series of at least four separate natural interspecific hybridization events. Recent estimates suggest that approximately 47% of angiosperm species and 95% of ferns and fern allies are polyploids, most of which are allopolyploids (Grant 1985). The rapid evolution of new animal taxa by interspecific hybridization is less common, and with the exception of some fishes and amphibians, usually results in asexually reproducing diploids and polyploids (White 1978).

Race formation

The sympatric origin of new, sexually reproducing animal species by mechanisms other than interspecific hybridization is considered rare. However, recent studies of natural populations as well as experimental and theoretical research indicates that sympatric speciation may be more widespread and common in some animal groups than previously thought. Certain kinds of parasitic and PARASITOID insects, mites, nematodes and other animals which use their host, habitat or essential resource as a rendezvous site for COURTSHIP and mating are predisposed to establish distinct races and speciate in the absence of geographic isolation. This may account for the remarkable diversity of insects — estimated to comprise between 10 and 30 million species—of which the majority are plant or animal parasites or parasitoids (Bush 1994).

Several examples of sympatric habitat or host race formation and speciation in PHYTOPHAGOUS insects and other arthropods are now well documented. In about 1860, the hawthorn fly (*Rhagoletis pomonella*) (Fig. S28), a native of eastern North America whose larvae feed within the fruit of several *Crataegus* species, established a population on introduced Old World apples (*Malus pumula*) growing in the Hudson River Valley of New York. The apple population rapidly evolved into a genetically distinct sympatric host race or incipient species. Sympatric populations associated with apple and hawthorn now maintain distinct differences from one another in host preference, summer emergence time, host-associated fitness trade-offs and ALLELE FREQUENCIES in protein-encoding genes even in the face of potential gene flow. Sympatric race formation and speciation in other animals such as molluscs, fish and birds is also reported (Bush 1994).

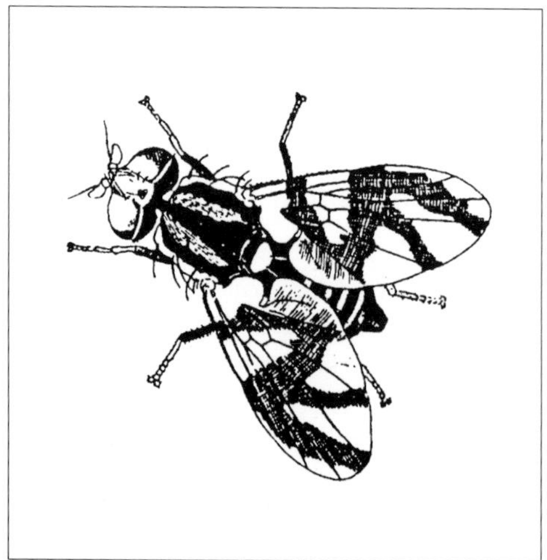

Fig. S28 Female *Rhagoletis pomonella* tephritid fruit fly.

Models

Directed selection

Further evidence for sympatric speciation comes from theoretical treatments and computer simulations which indicate that directed selection can promote ADAPTATION to a habitat in animals that restrict their mating to a preferred habitat or host. As adaptation and habitat fidelity increase, the frequency of positive ASSORTATIVE MATING is pleiotropically strengthened among individuals sharing similar habitat preferences. The result is an incremental reduction and eventual elimination of gene flow, sometimes augmented by SEXUAL SELECTION, between subpopulations adapting to different habitats.

Disruptive selection

The process of directed selection contrasts to models of sympatric speciation based on the evolution of REPRODUCTIVE ISOLATION by DISRUPTIVE SELECTION against the production of inferior hybrids. Sympatric speciation in this case requires that pre-mating barriers to gene flow evolve to strengthen reproductive isolation, a process called reproductive CHARACTER DISPLACEMENT (REINFORCEMENT), rather than as an outcome of adaptation to a specific habitat. Sympatric speciation by reinforcement alone requires unrealistically high levels of NATURAL SELECTION and assortative mating that are unlikely to occur in nature. Sympatric divergence and speciation is likely to occur only when some form of habitat-associated assortative mating and directed selection is involved (Johnson *et al.* 1996).

Overview

Sympatric speciation actually represents one end of a continuum in the degree and pattern of habitat-imposed isolation that may occur between populations as they adapt to different habitats or hosts. Although all individuals potentially have an equal opportunity to meet and mate under ideal sympatric conditions, such conditions seldom exist in nature. The actual level of potential gene flow between habitats or hosts will be determined by their pattern of distribution and level of temporal overlap. The other end of the habitat-imposed isolation continuum exists when sister species evolve while adapting to habitats or hosts restricted to different but adjacent ecological communities or BIOMES. In this situation gene flow between subpopulations is restricted to the transition zone, or ECOTONE, where the two habitats meet or overlap along an extended border *Cf.* ALLOPATRIC SPECIATION. *See also* ISOLATING MECHANISMS; PARAPATRIC SPECIATION; SPECIES CONCEPTS. [G.L.B.]

REFERENCES

Bush, G.L. (1994) Sympatric speciation in animals: new wine in old bottles. *Trends in Ecology and Evolution* **9**, 285–288.

Grant, V. (1985) *The Evolutionary Process; A Critical Review of Evolutionary Theory.* Columbia University Press, New York.

Johnson, P., Hoppensteadt, F., Smith, J. & Bush, G. (1996) Conditions for sympatric speciation: A diploid model incorporating habitat fidelity and non-habitat assortative mating. *Evolutionary Ecology* **10**, 187–205.

White, M.J.D. (1978) *Modes of Speciation.* W.H. Freeman, San Francisco, Calif.

symplesiomorphy Plesiomorphic TRAITS shared by two or more species are symplesiomorphies (*see* PLESIOMORPHY). Although similarity between the species is due to common ancestry in such cases, the shared characters are not necessarily indicative of degree of close phylogenetic relationship. For example, all 'fishes' exhibit two pairs of fins, which are believed to have been inherited from a common ancestor, and yet some fishes, such as lungfishes, are more closely related to tetrapods than they are to other groups of 'fishes'. Those evolutionary taxonomists who would support the recognition of a TAXON including all 'fishes' as a significant evolutionary grade would use a series of symplesiomorphies to diagnose the taxon. *See also* CLADISTICS; PHENETIC SPECIES CONCEPT; PHYLOGENETIC SYSTEMATICS. [D.R.B.]

synapomorphy Apomorphic TRAITS shared by two or more species are synapomorphies (*see* APOMORPH). These traits are the fundamental indicators of natural evolutionary groupings because each one is a homologous trait that originated in the common ancestor of a monophyletic group. For example, xylem and phloem are postulated to be synapomorphic traits linking mosses, gymnosperms and angiosperms in a monophyletic group. *Cf.* AUTAPOMORPH. *See also* CLADISTICS; PHENETIC SPECIES CONCEPT; PHYLOGENETIC SYSTEMATICS. [D.R.B.]

synecology *See* ECOLOGY.

synergism The phenomenon in which chemicals, when mixed, have a greater joint ecotoxicological EFFECT than expected from their effects in isolation. *Cf.* ADDITIVITY; ANTAGONISM. *See also* TOXICITY OF MIXTURES. [P.C.]

synomone *See* INFOCHEMICALS.

synthetic aperture radar (SAR) A specific type of imaging RADAR in which the effective length of the antenna has been enlarged synthetically in order to increase the spatial resolution of the imagery. SARs are carried by aircraft and several satellites, notably ERS (European Remote Sensing Satellite), Fuyo, Almaz and the Space Shuttle. [P.J.C.]

Systematic methods

Definitions

The fundamental job of systematic biology is to organize (systematize) BIODIVERSITY according to some consistent set of principles. Some biologists consider systematics to be the study and CLASSIFICATION of relationships among populations, species or higher taxa. Others, such as Ernst Mayr and Gareth Nelson, equate systematics with comparative biology. Although all comparative biological data are potential sources of information for systematics, not all comparative biologists are practising systematists. For example, biologists studying life-history strategies in several species of poeciliid fishes would generally consider themselves to be evolutionary biologists or ecologists, not systematists.

Evolutionary taxonomy

Three major schools of systematic biology have flourished over the past 50 years. The first of these, 'evolutionary taxonomy', traces its origins to 'the new systematics', a perspective that emerged from the synthetic theory of EVOLUTION in the late 1930s through the 1940s. Evolutionary taxonomists did not produce any methods for inferring phylogenetic relationships. They relied, as had generations of systematists before them, upon the weight of expert opinion in their attempts to distinguish homologies (*see* HOMOLOGY) from homoplasies (*see* HOMOPLASY) in the absence of a phylogenetic hypothesis. Evolutionary taxonomists thus redirected the focus of systematic biology from attempting to document evolutionary patterns towards testing theories of evolutionary mechanisms.

Advocates of this approach proposed that systematic biologists engage in activities other than phylogenetic reconstruction, such as:

1 the analysis of inbreeding and interbreeding relationships in order to determine the nature and limits of species;

2 elucidation of ISOLATING MECHANISMS, which would provide insights into the mechanisms of SPECIATION;

3 increasingly detailed analysis of intraspecific variation and integration of that information into studies of the effects of local selection on local variation.

For evolutionary taxonomists then, systematic activities leading to classifications and activities leading to understanding of evolutionary mechanisms are separate enterprises.

Numerical taxonomy (phenetics)

The second school of systematic biology, 'numerical taxonomy', or 'phenetics', emerged in the 1950s. This movement consolidated the development of statistical methods for analysing univariate and multivariate variation in biological systems with evolutionary taxonomic activities dealing with the documentation and explanation of population variation. Pheneticists differed from evolutionary taxonomists by attempting to integrate such information directly into classifications, represented in the form of 'phenograms'.

NUMERICAL TAXONOMY was founded on the philosophy of logical positivism. Its practitioners therefore asserted that it was impossible to infer evolutionary relationships empirically, so systematists should abandon the effort to reconstruct phylogenies. Pheneticists advocated classifications based on overall similarity rather than phylogenetic relationships, and produced criteria by which information about species could be clustered to produce such classifications.

Phylogenetic systematics

The third, and most recent, school of systematics is 'phylogenetic systematics', which originated in Germany in the 1950s but emerged into the mainstream of systematic biology in the early 1970s following the 1966 publication of Willi Hennig's *Phylogenetic Systematics* (*see* PHYLOGENETIC SYSTEMATICS). This school of thought incorporates elements of both evolutionary TAXONOMY and phenetics. Advocates of phylogenetic systematics believe, like evolutionary taxonomists, that systematists should be engaged in efforts to reconstruct phylogenies and that classifications should reflect phylogenetic information. Like pheneticists, they think that classifications should be derived empirically and should be based on quantitative analysis of as many characters as possible.

Phylogenetic systematists differ from evolutionary taxonomists in two important regards. First, they assert that PHYLOGENY consists only of monophyletic groups rather than a mixture of monophyletic and paraphyletic groups (grades) intuited by the systematist working on the group, and that classifications should reflect that fact. Second, they believe that any understanding of evolutionary mechanisms requires a priori understanding of phylogenetic patterns, and thus a robust and empirically defensible phylogenetic hypothesis is the starting point for all other relevant systematic efforts (such as the search for isolating mechanisms).

Phylogenetic systematists also differ from pheneticists, again in two important regards. First, they contend that phylogeny can be inferred empirically. Second, they believe that SYNAPOMORPHIES are the only appropriate criterion for grouping species and taxa.

Phylogeneticists and cladists

Since 1985, phylogenetic systematics has itself divided into two schools of thought, the 'phylogeneticists' and the 'cladists' or 'transformed cladists'. The phylogeneticists believe that the results of phylogenetic systematic analyses, often called CLADOGRAMS by both groups, should be interpreted as phylogenetic trees and should be explained in terms of evolutionary phenomena. The cladists believe that cladograms should be viewed as pieces of general patterns of natural biological order, without reference to any particular evolutionary phenomena that might have produced the species comprising them. *See also* CLADISTICS; EVOLUTIONARY TAXONOMIC CLASSIFICATION; NUMERICAL TAXONOMY. [D.R.B.]

---

systems constraints *See* CONSTRAINTS.

systems ecology The application of systems theory to ecological processes and entities; i.e. an approach based on the premise that these need to be studied and represented as integrated wholes. The study should be 'top down', because study of parts alone (i.e. REDUCTIONISM) misses the all-important interactive components. Such an approach has involved the application of control theory, information theory, black-box analysis, etc. *See also* HOLISM; ORGANISMIC HYPOTHESIS. [P.C.]

T

t test

This is the small SAMPLE (either n_1 or $n_2 < 30$) version of the $z(d)$ TEST. Basically, the formula for the test is the same, except that we treat the result as t rather than $z(d)$. We also make a combined estimate of the VARIANCE. That is:

$$d = \frac{\bar{x}_1 - \bar{x}_2}{\sqrt{\left(\frac{s_1^2}{n_1} + \frac{s_2^2}{n_2}\right)}} \text{ becomes}$$

$$t = \frac{\bar{x}_1 - \bar{x}_2}{s_c\sqrt{\left(\frac{1}{n_1} + \frac{1}{n_2}\right)}}$$

where $\bar{x}_1$ and $\bar{x}_2$ are the sample means, s_c is the combined estimate of the sample STANDARD DEVIATION and n_1 and n_2 are sample sizes. The combined estimate of the variance is calculated using the formula:

$$s_c^2 = \frac{\left[\sum x_1^2 - \left(\sum x_1\right)^2 / n_1\right] + \left[\sum x_2^2 - \left(\sum x_2\right)^2 / n_2\right]}{n_1 + n_2 - 2}$$

where the terms in square brackets are the SUM OF SQUARES for each sample, and n_1 and n_2 are the sample sizes. Sometimes the above two formulae are shown combined to give:

$$t = \frac{\left(\bar{x}_1 - \bar{x}_2\right)}{\sqrt{\left[\frac{(n_1 - 1)s_1^2 + (n_2 - 1)s_2^2}{(n_1 + n_2 - 2)}\right]\left(\frac{n_1 + n_2}{n_1 n_2}\right)}}$$

The DISTRIBUTION of the test statistic we call t was discovered by W.S. Gossett (1876–1937), a statistician working for the Guinness brewery in Dublin, who published his findings under the *nom de plume* of 'Student'. Consequently this test is often called Student's t test. Gossett found that when sample size was small ($n < 30$) the relationship between $z(d)$ and the area under the curve (PROBABILITY) changed (*see* STANDARDIZED NORMAL DEVIATE). The smaller the value of n, the more the relationship changed (in tables of t, n is replaced by DEGREES OF FREEDOM (v)). Instead of 5% of the distribution being outside $z(d) = \pm 1.96$, with $v = 25$, 5% was now outside $z(d) = \pm 2.06$ and with $v = 5$, 5% was outside $z(d) = \pm 2.57$. To avoid confusion, he replaced $z(d)$ with the new symbol t. Therefore, if the above test formula gives, for example, $t \geq 2.06$ with $v = 25$ (degrees of freedom in the test $= n_1 + n_2 - 2$), the NULL HYPOTHESIS is rejected and the difference between the two means is said to be significant at the 5% level. The t test above is only used if the two sample variances (s_1^2 and s_2^2) are approximately equal, and this equality is determined using the F TEST.

If the two sample variances are very different, and therefore you cannot make a combined estimate of the variance, the formula for the $z(d)$ test is used but the degrees of freedom (taken to the nearest whole number) are calculated as:

$$\frac{1}{v} = \frac{u^2}{n_1 - 1} + \frac{(1 - u)^2}{n_2 - 1} \text{ with}$$

$$u = \frac{s_1^2 / n_1}{\left(s_1^2 / n_1\right) + \left(s_2^2 / n_2\right)}$$

Paired-sample t test (or t test for matched pairs)

A specific form of the t test is used when you have MATCHED OBSERVATIONS. Suppose that you wanted to know if drinking one cup of coffee would increase the heartbeat rate of male students. You could test this by taking two groups of students, giving each of the individuals in one group a cup of coffee, and then measuring the heartbeat rate of both groups. An alternative procedure would be to take one group of students, measure their heartbeat rates, give them a cup of coffee, and measure their heartbeat rate again. In this latter case we have what are called matched pairs of data. With this type of data it is possible to perform a more sensitive comparison using the t test below.

For each pair of observations we record the difference (d), including the sign (+ or −). We then calculate the MEAN and variance of this set of differences using the standard formulae:

$$\bar{d} = \frac{\sum d}{n} \text{ and } s^2 = \frac{\sum d^2 - \left(\sum d\right)^2 / n}{n - 1}$$

where d is the difference between pairs, and n is the number of pairs. The test statistic t is then calculated using the formula:

$$t = \frac{\bar{d}}{s / \sqrt{n}}$$

where $\bar{d}$ is the mean difference between pairs, s is the standard deviation of the difference and n is the number of pairs. Degrees of freedom are $n - 1$.

The t distribution is introduced here as a deviation from the NORMAL DISTRIBUTION. However, in many ways it would be more correct to view the t distribution as the general case, with the normal distribution as a special case of Student's distribution with ∞ (in practice >30) degrees of freedom (v). *See also* CENTRAL LIMIT THEOREM. [B.S.]

---

tactics '[A] set of coadapted TRAITS, designed by NATURAL SELECTION, to solve particular ecological problems.' (Stearns 1976, p. 4.) Widely used in behavioural and evolutionary ecology, usually rather more loosely than the above definition suggests, but sometimes to remind readers that an adaptive hypothesis requires considerable work to support or refute. *Cf.* STRATEGY. *See also* ADAPTATION; LIFE-HISTORY EVOLUTION. [D.A.B.]

REFERENCE

Stearns, S.C. (1976) Life history tactics: A review of the ideas. *Quarterly Review of Biology* **51**, 3–47.

taiga A Russian term denoting a marshy, Siberian WOODLAND and often used to denote the circumpolar BOREAL FOREST. However, it is also used to describe the ECOTONE zone between the boreal forest and the Arctic TUNDRA. This transition zone is marked by more open woodland sometimes confined to areas along river courses where the PERMAFROST zone is deeper in the SOIL and permits root development to a depth sufficient to support tree growth. Such open arctic-fringe forest is also described as lichen woodland. [R.M.M.C.]

tailings Fine-grained WASTE from mining, and processing of ore, usually following washing. Can be polluting. [P.C.]

tangled bank model *See* EVOLUTION OF SEX.

Tansley, A.G. *See* CHARACTERS IN ECOLOGY.

tantalum (Ta) Tantalum (ATOMIC NUMBER = 73; relative ATOMIC MASS = 180.948; melting point = 2996°C; boiling point = 5425°C) is present at a level of about 1 mg kg^{-1} in the Earth's CRUST and is invariably associated with its lighter congener NIOBIUM from which it is difficult to separate. The TOXICITY of the METAL is generally low and it is not considered a health HAZARD. This, along with the physical properties of relative lightness, hardness, ductility, malleability and extreme corrosion resistance, has led to the widespread use of tantalum in the form of wire, plates and mesh in surgery. Due to its high atomic number, tantalum is sufficiently opaque to X-RAYS for its use as an imaging agent in diagnostic radiology. [P.J.B.]

target load *See* CRITICAL LOADS.

target standards *See* ENVIRONMENTAL LAW.

taxon (pl. taxa) The organisms or species that fill a systematic category. The Linnaean taxonomic units are species, genera, families, orders, classes and phyla. Many lineages also include finer divisions such as suborder and superfamily. Taxonomic units above the species level are arbitrary, depending on the attitude of the systematist and the vagaries of EVOLUTION in a LINEAGE. *See also* BINOMIAL CLASSIFICATION; LINNAEAN CLASSIFICATION. [D.A.B.]

taxon-range zone *See* BIOSTRATIGRAPHIC UNIT.

taxonomic categories Biologists recognize a hierarchy of taxonomic categories that expresses the relatedness between organisms. Groups of organisms showing a few major features in common can be subdivided into groups showing these plus other features, etc. Linnaeus thought that this represented a scale of order imprinted on nature by God and specified a series of categories arranged hierarchically to define it, viz:

Kingdom
  Class
    Order
      Genus
        Species

Moving down the series, the categories contain more and more characters in common. Through the years many additional categories have come to be used, but the system is still based on the Linnean hierarchy, although this is now viewed as an outcome of evolutionary processes. Most modern biologists recognize the following categories:

Kingdom
  Phylum (usually Division in botany)
    Class
      Order
        Family
          Genus
            Species

Each category can be subdivided further. Thus at the level of order we can have: superorder, order, suborder, infraorder. *See also* BINOMIAL CLASSIFICATION; LINNAEAN CLASSIFICATION; TAXONOMY. [P.C.]

Taxonomy

Taxonomy is the theory and practice of describing the DIVERSITY of organisms and of ordering this diversity into a system of words, called CLASSIFICATIONS, that convey information concerning kinds of relationships among organisms. In addition, taxonomy encompasses the theory and history of classification. Practitioners of taxonomy are called taxonomists and: (i) make inventories of the world's BIODIVERSITY; (ii) describe and name the species representing that biodiversity; (iii) classify those species in a system of nomenclature; (iv) revise those classifications according to new information and new methods of classification; and (v) build and maintain databases for storing and retrieving all known information about those species. Most taxonomists work in museums, but they may also work in universities and government and private agencies that require knowledge of biological diversity. Taxonomic information is used in all areas of biological endeavours because all biological studies begin with the names of the species being studied in whole or in part.

Virtually all taxonomists subscribe to a common system of nomenclature. Named for the great Swedish naturalist Carolus Linnaeus, the Linnaean system is a hierarchical system of progressively more inclusive groups. The most fundamental units of the Linnaean system are species. Each species receives a binomen (two names—often erroneously called a binominal, or two numbers) from the taxonomist who first recognizes its existence. The binomen consists of a name specifically referring to the species (the specific epithet) preceded by a name placing the species within the next higher category, the genus (the generic epithet). Groups of genera are placed in families, groups of families in orders, groups of orders in classes, groups of classes in phyla, and groups of phyla in kingdoms (Fig. T1). For example, humans are designated as the species *sapiens*, which is a member of the genus *Homo* (hence *Homo sapiens*), which is a member of the family Hominidae, which is a member of the order Primates (primates), which is a member of the class Mammalia (mammals), which is a member of the phylum Chordata (chordates) (and subphylum Vertebrata (vertebrates)), which in turn is a member of the kingdom Metazoa (animals).

Taxonomists have developed a complex set of rules for keeping track of the names of the increasing number of known species. These rules, embodied in several sets of rules of nomenclature (one for animals, one for plants and one for microbes) are designed to resolve conflicts when two or more taxa accidentally receive the same name or when the same species has accidentally received more than one name.

Although the Linnaean system of nomenclature pre-dates evolutionary theory, it has persisted because the hierarchical structure of the Linnaean system conforms well with our ideas about the hierarchical nature of PHYLOGENY. None the less, it is important to remember that biological theory does not include processes acting at any level above the species. Thus, while there is a process called SPECIATION for the production of new species from old species, there is no process of geniation or familiation for the production of genera or families. Even if a supraspecific, or 'higher', TAXON is a monophyletic group (*see* MONOPHYLY), and thus contains all the descendants of a particular ancestor, it is still a construct of a particular taxonomist and not the outcome of a particular evolutionary event. There are thus no guidelines for recognizing the boundaries of a genus, or a family within any particular group of animals, for example, much less across groups such as plants and animals. The frog family Leptodactylidae, for example, contains a number of genera, including one, *Eleutherodactylus*, that encompasses more than 400 nominal (named) species. The family Hominidae, by contrast, includes three genera (*Homo*, *Pan* and *Gorilla*) encompassing a mere five living species. The phylum Acanthocephala (thorny-headed worms) comprises fewer than 2000 species, whereas the phylum Arthropoda, which includes crustaceans and insects, comprises well over a million species.

Today, some believe that we should abandon the Linnaean system in favour of a numerical system more amenable to computerization. They argue that the large number of new species being described each year has created a problem for anyone trying to keep track of biological diversity, and that the manner in which new species can be validly named specified in the current codes of nomenclature do not permit the use of electronic media. Other taxonomists argue that only through the elimination of the Linnaean system of categories can we eliminate the possibility of attaching undue significance to the taxonomic category to which a species is assigned. *See also* SYSTEMATIC METHODS; TAXONOMIC CATEGORIES. [D.R.B.]

Taylor's power law This index of DISPERSION makes use of the fact that in natural populations the VARIANCE is not independent of the MEAN. In most biological populations variance increases with increasing mean. In order to calculate Taylor's power law a series of sets of samples are required. The sets of samples may be of different sizes or from different areas. What is required is a range of SAMPLE means to investigate the relationship between mean and variance. Taylor's power law states that the relationship between mean and variance can be described as follows:

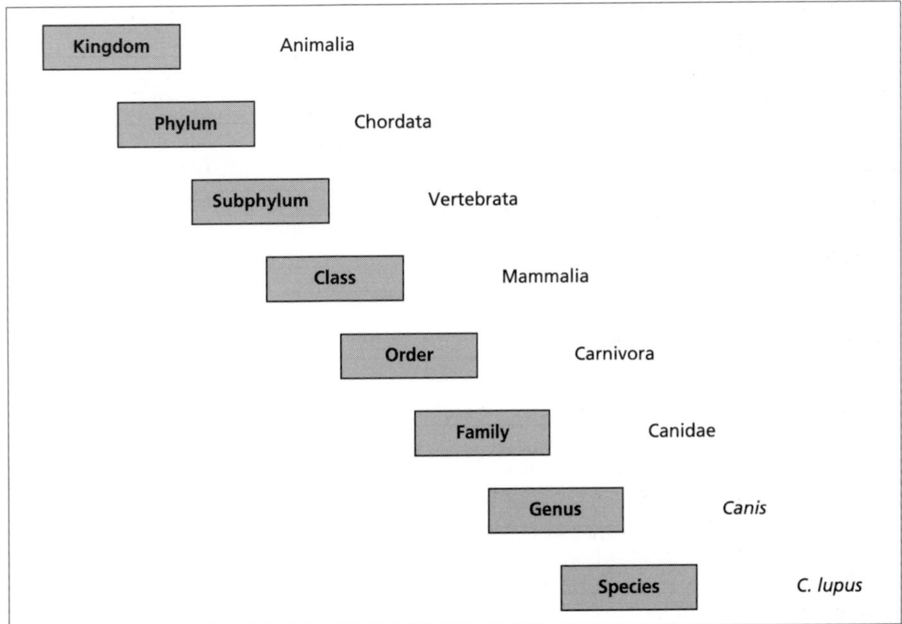

Fig. T1 Each species in a biological classification is a member of a group at each of a succession of more inclusive hierarchical levels. The figure gives a fairly complete classification of the grey wolf *Canis lupus*. This way of classifying living things was invented by the 18th century Swedish biologist who wrote under the Latinized name Carolus Linnaeus. (After Ridley, M. (1996) *Evolution*, 2nd edn. Blackwell Science, Cambridge, Mass.)

$s^2 = am^b$

where a and b are constants, s^2 is the variance and m is the mean.

It follows that there is a linear relationship between log(mean) and log(variance):

$\log s^2 = \log a + b \log m$

The constant a is largely a SAMPLING factor and is usually ignored. The slope of the relationship (b) can be used as an index of dispersion. If populations follow a random (Poisson) distribution, the variance will be equal to the mean at all densities and therefore the slope of the line (b) will be 1. An aggregated or clumped population will have a slope greater than 1, while in an ordered population the slope will be less than 1.

Taylor's power law holds true for most populations and therefore provides an excellent dispersion index. One drawback is that multiple sets of samples are required which requires greater effort. Another possible drawback is that the relationship between log(mean) and log(variance) is not always exactly linear. *See also* INDICES OF DISPERSION. [C.D.]

REFERENCE

Taylor, L.R. (1961) Aggregation, variance and the mean. *Nature* **189**, 732–735.

TBT *See* TRIBUTYLTIN (TBT) POLLUTION.

technetium (Tc) Technetium (ATOMIC NUMBER = 43; relative ATOMIC MASS = 99; melting point = 2200°C; boiling point = 4567°C) occurs naturally only as minute traces of ^{99}Tc ($t_{1/2} = 2.13 \times 10^5$ years) from spontaneous FISSION of ^{238}U. ^{99}Tc occurs more extensively in the environment from weapons-testing fallout and releases from nuclear facilities. ^{99}Tc is also of importance with respect to high-level RADIOACTIVE WASTE disposal. Technetium has oxidation states from +2 to +7 and exhibits soluble behaviour in natural waters. ^{99m}Tc ($t_{1/2} = 6.0$ h) labelled radiopharmaceuticals are used in nuclear medicine for tissue scanning. *See also* WASTE DISPOSAL. [A.B.M.]

technology-based control Legal control of quality of EMISSIONS/EFFLUENTS that refers not (just) to chemical characteristics but also requires the use of specific techniques/technology or makes reference to the use of best available techniques. *See also* BEST AVAILABLE TECHNIQUES NOT ENTAILING EXCESSIVE COSTS; OPTIMUM POLLUTION. [P.C.]

tectonic Applies to the dynamics of the Earth's CRUST and their consequences: warping, folding and faulting. *See also* PLATE TECTONICS. [P.C.]

tectonic lakes *See* LAKES, TYPES OF.

TEF *See* TOXICITY EQUIVALENCY FACTORS.

telemetry The automatic transmission of data from remote sites. Data may be sent by landline, often the public telephone network, or transmitted by radio. Over short distances radio transmission can be by direct link, but transmission via satellite allows the data collection platform (DCP) and the receiving station to be separated by large distances. For example, data collected in South America can be received in Europe. Telemetry is most useful when data are required in real time, for example for flood warning, or when the site is in a remote location and access is difficult. *See also* REMOTE SENSING. [J.H.C.G.]

teleology This comes from the Aristotelian Fourth Cause and implies goal directedness, design in nature, even the future in some way influencing processes of the present. Organisms appear teleological with general 'goals' being provided by NATURAL SELECTION in terms of FITNESS. And there is a sense in which future goals are programmed into the genome of organisms by natural selection, with control systems that direct the processes and behaviour towards these ends. However, scientists do not accept conscious design in nature except when it emanates from the consciousness of humans. *See also* HOLISM. [P.C.]

tellurium (Te) Tellurium (ATOMIC NUMBER = 52; relative ATOMIC MASS = 127.6; melting point = 449.5°C; boiling point = 989.8°C) is widely distributed in the Earth's CRUST, average crustal abundance = 0.005 mg kg^{-1}. It is often associated with COPPER ores from which it is extracted as a by-product after electrolytic refining of copper. Tellurium and its compounds are used as alloys, CATALYSTS, vulcanization agents for RUBBER, solar cells, and thermoelectric cells for the conversion of low-grade heat to electricity. Major sources of tellurium in AMBIENT air originate from coal combustion and electrolytic refining of copper. Hydrogen telluride (H_2Te) and tellurium hexafluoride (TeF_6) are two compounds that are especially toxic by inhalation. A characteristic symptom of EXPOSURE is the odour of garlic on the breath. [C.R.P.]

telolecithal eggs *See* EGGS.

Temperate deciduous forest

These are FORESTS whose trees reduce water loss during the cold winter season by dropping their leaves. Such ecosystems are often termed 'summer-green' or 'cold-deciduous' forests. They are usually strongly stratified (Fig. T2) with distinct tree, shrub, herb/dwarf shrub and cryptogamic layers, the latter being rich in mosses and liverworts. Lichens tend to be less common in the ground cover than in BOREAL FORESTS. Broad-leaved and coniferous trees of the famous lowland forest of Bialowieza, on the borders of Poland and Belarus, reach extraordinary maximum heights. Wild pig (*Sus scrofa*) and European bison (*Bison bonasus*) abound in this largely PRIMARY FOREST, and the complexity of its bird communities is related to the marked STRATIFICATION of its VEGETATION.

Temperate deciduous forests (see Plate 21, facing p. 398) form a broad zone in North America and Eurasia, where the absence of the forest CANOPY during the 'bare half-year' encourages the growth of shade-evading herbs, exemplified in western Europe by *Anemone nemorosa, Hyacinthoides non-scripta* and *Ranunculus ficaria.* Many winter-green shade-tolerating species, including *Lamiastrum galeobdolon* and *Oxalis acetosella*, are also present. These forests are famous for their 'aspect societies' whose species flower in a constantly changing SUCCESSION during the spring and early summer. Expan-

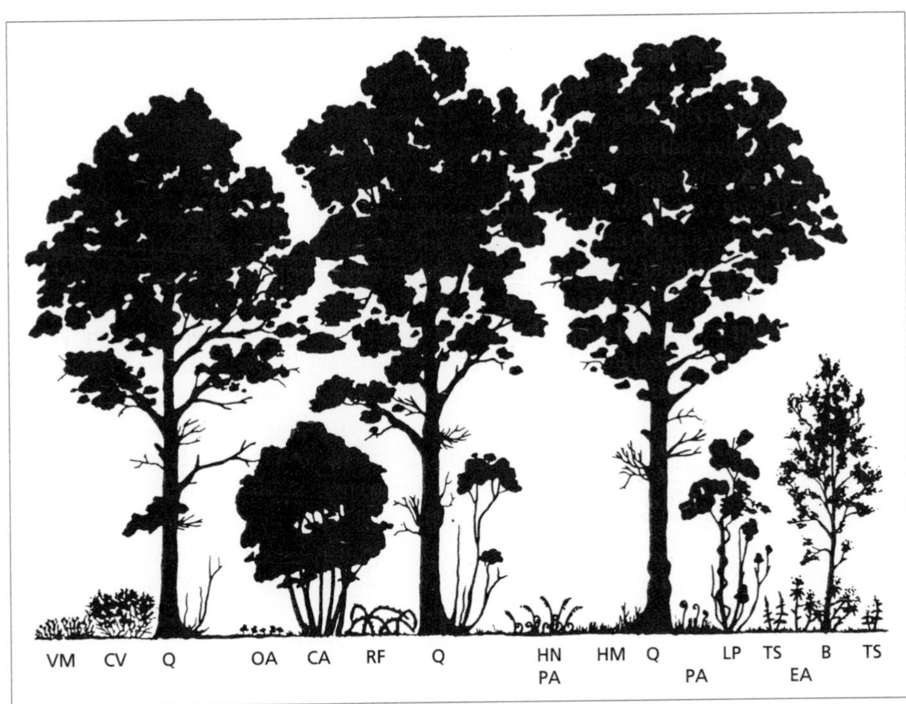

Fig. T2 Stratification of shoot systems in the Wyre Forest, Shropshire, UK. The oak trunks have the basal curve characteristic of individuals which have regrown from coppice stools, while the mosses and liverworts of the bryophyte layer are too small to depict. The area with the most acid soil is on the left, while near the centre is an aspect society in which bluebell, creeping soft-grass and bracken follow each other in a seasonal sequence. VM, bilberry (*Vaccinium myrtillus*); CV, heather (*Calluna vulgaris*); Q, oak (*Quercus petraea* or *Q. robur*); OA, wood sorrel (*Oxalis acetosella*); CA, hazel (*Corylis avellana*); RF, bramble (*Rubus fruticosa* agg.); HN, bluebell (*Hyacinthoides non-scripta*); HM, creeping soft-grass (*Holcus mollis*); PA, bracken (*Pteridium aquilinum*); LP, honeysuckle (*Lonicera periclymenum*) growing over weak oak coppice regeneration; TS, wood sage (*Teucrium scorodonia*); EA, wood spurge (*Euphorbia amygdaloides*); B, silver birch (*Betula pendula*). (After Packham, J.R., Harding, D.J.L., Hilton, G.M. & Stuttard, R.A. (1992) *Functional Ecology of Woodlands and Forests*. Chapman & Hall, London; drawn by P.R. Hobson.)

sion of the forest canopy at the beginning of the 'leafy half-year' is rapid, and the annual patterns of life of many animals are closely geared to this event. Many herbivorous insects, particularly the caterpillars of species such as the green oak roller (*Tortrix viridana*) and winter moth (*Operophtera brumata*), feed on tree leaves and are themselves the prey of the small birds which abound in these forests.

The present DISTRIBUTION of species within the northern temperate deciduous forests can only be understood in terms of geological changes which began during Early Tertiary times, which commenced 70 million years ago (Packham *et al.* 1992). At that time the breakup of the ancient supercontinent, Pangaea, was at a comparatively early stage and many tree genera were dispersed right across the Northern hemisphere. The southward movement of the temperate Tertiary forest was initiated by a drop in temperature in the Late Tertiary and culminated in the Pleistocene Ice Age. Separate western and eastern forests developed in North America because the upthrust of the Rockies created the treeless rain shadow of the Great Plains. In Eurasia the western and eastern forests were separated by a central area whose low rainfall resulted from its wide separation from the sea. These four major forest regions developed differently; in North America the western forests became dominated by conifers while angiosperm trees remained abundant in the east. Because of their east–west alignment, the Alps and the Pyrenees prevented the easy MIGRATION of trees and other plants to and from REFUGES lying to the south of them during successive glaciations in the Pleistocene Ice Age. These events have greatly impoverished many temperate deciduous forests. In the Early Tertiary, for example, the natural distribution of the tulip tree (*Liriodendron*), now restricted to south-eastern USA (*L. tulipfera*) and eastern China (*L. chinese*), included stands in Iceland and Europe. Today, the former magnificence of the rich Arcto-Tertiary forests can be appreciated only in the mixed mesophytic stands of the southern Appalachians or of eastern Asia, areas far richer in trees and UNDERSTOREY species than any other temperate forests.

In Britain, 16 WOODLAND communities (W1–12 and W14–17) are dominated by deciduous trees in the National Vegetation Classification (Rodwell 1991). Successive glaciations have greatly reduced the number of tree species; the most important are *Acer campestre*, *A. pseudoplatanus*, *Alnus glutinosa*, *Betula pendula*, *B. pubescens*, *Carpinus betulus*, *Castanea sativa*, *Corylus avellana*, *Fagus sylvatica*, *Fraxinus excelsior*, *Ilex aquifolium*, *Quercus petraea*, *Q. robur*, *Salix* spp., *Sorbus aucuparia*, *Tilia cordata* and *Ulmus glabra*. The depauperate nature of these British communities, particularly with regard to trees but also in respect to smaller plants and to animals, is emphasized when they are compared with the temperate deciduous forests of the rest of the world (Rohrig & Ulrich 1991). Almost all temperate deciduous forests are no more than semi-natural, and in many of them prey–predator relationships have been disturbed by eliminating or reducing populations of the larger predators. In consequence, successional processes and trees-species composition in the majority of temperate deciduous forests are strongly influenced by excessive deer GRAZING. [J.R.P.]

REFERENCES

Packham, J.R., Harding, D.J.L., Hilton, G.M., & Stuttard, R.A. (1992) *Functional Ecology of Woodlands and Forests*. Chapman & Hall, London.

Rodwell, J.S. (ed.) (1991) *British Plant Communities*. Vol. 1. *Woodland and Scrub*. Cambridge University Press, Cambridge.

Rohrig, R. & Ulrich, B. (eds) (1991) *Temperate Deciduous Forests*, Elsevier, London.

temperate grassland Grass- (Poaceae) or sedge- (Cyperaceae) dominated vegetation found in temperate climatic zones. In the Northern hemisphere such GRASSLANDS occur over large continuous belts of North America and Eurasia. In the Southern hemisphere they are found more discontinuously, principally in southern Africa, South America and Australasia. A defining characteristic is the abundance of cool-season C_4 grasses of the subfamilies Arundionoidae and Pooideae in temperate grasslands. These are replaced in tropical and subtropical grasslands by warm-season C_3 grasses of the subfamilies Chloridoideae and Panicuoideae.

In seasonally arid temperate areas, tree and shrub growth is inhibited by CLIMATE, and grassland is the natural vegetation. The main areas in which these climatic CLIMAX grasslands occur are the prairies of North America, the steppes of central and eastern Eurasia, the high VELD of South Africa and the PAMPAS of South America. More locally natural temperate grasslands occur on the coasts and in mountain areas, where wind exposure and not aridity limits tree growth. Grassland is also found naturally on soils which are shallow, saline, enriched by HEAVY METALS or have other features which prevent tree and shrub growth.

The geographical limits within which temperate grassland is naturally developed are unclear, as natural herbivores and burning by aboriginal peoples are thought to have often extended the grassland beyond its natural climatic limits into the FOREST zone. Due to conversion to AGRICULTURE, all but a very small part of the world's natural temperate grassland has been destroyed and most of that which remains has been modified by GRAZING domestic livestock, burning, introduction of ALIEN species and destruction or modification of natural HERBIVORE communities.

Particularly in western Eurasia, human activity has caused an expansion of grasslands into the natural forest zone due to the clearance of WOODLAND and grazing with livestock. This DEFORESTATION process started in prehistoric times and has resulted in a diverse range of semi-natural grassland types strongly differentiated to reflect differences in substrate, climate, BIOGEOGRAPHY and land management practices. Today, a major part of the grassland of the temperate zones of all parts of the world are highly artificial and dominated by a small number of productive forage species such as ryegrass (*Lolium* spp.) and white clover (*Trifolium repens*) (see Plate 22, facing p. 398). These artificial grasslands are a result of AGROCHEMICAL use and reseeding with commercial varieties of a limited number of forage grasses and legumes. Such artificial grasslands account for a major part of the feed provided to livestock in temperate regions and are of extremely high economic importance. [J.J.H.]

REFERENCE

Coupland, R.T. (1993) *Ecosystems of the World 8B: Natural Grasslands Eastern Hemisphere and Resume.* Elsevier, Amsterdam.

temperate rain forest Temperate forests vary considerably with climate, and those occurring in exceptionally wet conditions are sometimes referred to as temperate rain forests. Areas that have been so termed include the north-west Pacific coast of North America, the windward slopes of the Andes in southern Chile, the southern part of the South Island of New Zealand, and parts of eastern China and Japan.

In Japan and in the Southern hemisphere, the trees are mainly broad-leaved evergreens, while in the Pacific North-West the evergreens are mainly conifers, particularly Douglas fir (*Pseudotsuga menziesii*) and Sitka spruce (*Picea sitchensis*). All of these areas receive abundant rainfall, some sites in Chile having an annual rainfall in excess of 8500 mm spread over 300 rain-days a year. Figures around 2500 mm are more typical in the Pacific North-West and in the South Island of New Zealand. The humid atmosphere of these forests may lead to the development of heavy EPIPHYTE loads, particularly of mosses and lichens, on the canopy trees. *Cf.* TROPICAL RAIN FOREST. [P.D.M.]

temperature Temperature is a property of a body or region of space which can be sensed subjectively as relative hotness. It is a measure of the average kinetic energy status of the molecules or atoms making up the system. If there is a temperature difference between two bodies then heat will flow between them. Temperature can only be measured by observing the behaviour of another substance which varies predictably with temperature, for example the expansion of MERCURY or the electromotive force produced by a thermocouple. Temperature scales are based either on arbitrary points of observable reference such as the freezing point or boiling point of WATER (Celsius scale) or on an absolute scale in which the so-called thermodynamic temperature is a function of the energy possessed by matter (Kelvin scale). *See also* HEAT TRANSFER; THERMOMETRY. [J.B.M.]

temporal caste *See* CASTE.

temporal gradient *See* DIVERSITY GRADIENT.

tephra Materials ejected from a volcano, that fall from the air. [P.C.]

TEQM *See* TOTAL ENVIRONMENTAL QUALITY MANAGEMENT.

terbium (Tb) Terbium (ATOMIC NUMBER = 65; relative ATOMIC MASS = 158.9; melting point = 1356°C; boiling point = 3123°C) is a LANTHANIDE METAL with an average crustal abundance of 0.9 mg kg^{-1}. It is used in phosphors and in laser materials. [J.G.]

Termites

Eusocial and polymorphic species of the insect order Isoptera, living in large family groups composed of reproductive (winged) forms together with numerous apterous, sterile soldiers and workers (see Plate 23, facing p. 398). Inaccurately termed 'white ants', termites should in fact be regarded as a derived form of cockroach which have acquired an OBLIGATE dependence on mutualistic intestinal bacteria and, in some higher forms, externally cultivated basidiomycete fungi. Some termites contain populations of flagellate protozoans in the hindgut; although some of these flagellates have the ability to degrade CELLULOSE and other polysaccharides, their status is uncertain. Termites are the only SOCIAL INSECTS with a true soldier CASTE having no role other than COLONY defence. Primitive forms (lower termites) have a flexible worker caste which may develop into other castes as required, or become reproductives.

Approximately 2600 species have been described in 280 genera and 6 families. Undescribed species exist, especially subterranean forms from tropical forests, but the total number of species is unlikely to exceed 3000. Endemism is high, possibly because the DISPERSAL of reproductives is inefficient and most species are therefore typical of their own biogeographical region and do not occur elsewhere. A relatively small number of primitive forms, loosely termed 'drywood termites', 'dampwood termites' and 'subterranean termites' are able to colonize and/or consume timber used for buildings, furniture, transmission poles and fencing. Such PEST species are distributed by human activities and may attain worldwide distribution within tropical and subtropical regions. Other species are agricultural or silvicultural pests, especially where non-indigenous crops are introduced or where land is cleared of natural plant detritus prior to planting. Termite DIVERSITY is strongly influenced by latitude, with the numbers of naturally occurring taxa declining steadily north and south of the Equator. Some ground-nesting pest species (especially *Reticulitermes* spp.) are active beyond these limits and have been reported in cities as far north as Toronto, Paris and Hamburg. Termites are often reported as a serious pest throughout the tropics and subtropics and in many areas of Australia, Canada, China, Japan, Korea, New Zealand and the USA, as well as parts of Europe and South America. The equivalent of more than US$2 billion is spent every year on termite control throughout the world.

The LIFE CYCLE of termites usually follows a seasonal pattern. Winged reproductives of both sexes are produced in large numbers from mature nests, often during or just before rains (during the summer in more temperate climates). They make short, often rather feeble, dispersal flights (a few species are strong fliers, e.g. *Hodotermes mossambicus*) and pair on the ground after the wings are shed (dealation). Pairs of termites then select new nesting sites by burrowing into the ground or by searching for suitable wooden substrata and effecting an entrance through cracks and fissures, or by excavating an entrance hole. Once established in a suitable site, mating takes place and a first batch of eggs is produced by the female within a few days. Larvae are fed from nutrient-rich salivary secretions produced by both parents and undergo a number of moults until they achieve the adult stage as a sterile worker or soldier. Slow colony growth may continue for many years, accompanied by a gradual increase in the number of individuals, enlargement of the space occupied and many building activities. Constructions made by termites include: (i) numerous galleries tunnelled through wood and SOIL; (ii) underground chambers containing the queen, king, larval nurseries and (in the Macrotermitinae) fungus gardens; (iii) runways attached to the sides of trees, decaying wood and buildings; (iv) soil sheeting covering the surface of the ground; and (v) arboreal and epigeal mound-nests with a complex, highly organized internal structure. Materials utilized by termites include surface soil, sub-soil, compacted faeces, and carton, a lightweight organic-rich mixture of partially digested cellulose, saliva and soil. Epigeal mounds can attain a height of more than 3 m (e.g. *Macrotermes bellicosus*), but the largest termite constructions are probably the lens-shaped subterranean nests of *Odontotermes* spp. in East Africa, which are sometimes more than 30 m in diameter. The number of individuals in a colony varies with species, age and COMPETITION from other termites, but can reach several millions in some forms (e.g. some Macrotermitinae). A large number of species are not primary mound builders, but either exist in diffuse subterranean gallery networks (e.g. some Apicotermitinae) or become established in pockets within the mounds of other species as secondary occupants. Large colonies may include a number of supplementary reproductives, producing eggs to augment or replace the founding queen. Inquilines from other animal groups (e.g. beetles, ants, collembolans, cockroaches, mites and vertebrates) are commonly found in termite nests.

All termites are detritivorous, consuming (in different species) a wide range of freshly dead or decaying plant material including dry grass, leaf litter, sound wood, decaying wood, dung and HUMUS. Lichen and living roots may also be eaten. Through their activities as decomposers and constructors, termites have crucial roles in conditioning soils throughout the tropics and subtropics. It has recently been shown that they can also fix NITROGEN on a significant scale, through symbiotic bacteria, and may thus be the primary providers of this nutrient in many soils and food chains.

Termite abundance varies from less than 50 individuals m^{-2} in arid savannahs to more than 7000 m^{-2} in some African forests. In the tropics as a whole they are estimated to constitute 10% of all animal BIOMASS. The major factors other than latitude influencing termite abundance, biomass and diversity are rainfall (there are more termites in wetter climates, but there is also an interaction with historical factors), altitude (termites are restricted at higher levels), soil organic matter (termites are more abundant in forests than in savannahs) and DISTURBANCE (AGRICULTURE reduces most species but may promote others to pest status).

It is highly probable that termites function as KEYSTONE SPECIES in ecosystems (creating NICHE opportunities for a large number of other organisms) and as ECOSYSTEM ENGINEERS (mediating major biogeochemical processes). They also produce METHANE and some other GREENHOUSE GASES, and may therefore impact CLIMATE CHANGE. Estimates of annual global methane effluxes from termites vary from *c.* 2 Tg to *c.* 300 Tg, although most are less than 100 Tg. An output of 50 Tg $year^{-1}$ would represent about 10% of the worldwide total from all sources. However, most of this methane may be oxidized by soil bacteria before it reaches the atmosphere.

All grass-eating and xylophagous termites so far investigated have been shown to produce their own CELLULASE enzyme complex, independently of their symbionts, which is competent against natural forms of cellulose and sufficient to provide the quantity of substrates required to match metabolic needs. At the same time termites are the only eukaryotes in which there is no measurable activity of pyruvate dehydrogenase, a key enzyme of intermediary metabolism. The mutualistic relationship with intestinal symbionts compensates for this basic biochemical lesion and may provide other benefits, such as a supplementation of dietary nitrogen and partial digestion of LIGNIN or other polyphenolic materials. Macrotermitinae use cultivated specialist fungi (*Termitomyces* spp.) to compost foraged plant litter within the nest, reducing the C:N RATIO and degrading lignin before the material is consumed. These termites have a large direct impact on organic DECOMPOSITION in savannahs in Africa and Asia, but are absent from the faunas of South America and Australia.

Termites are an important indicator group for biogeographical and ecological analysis and MONITORING. This is attributable to their taxonomic tractability, ecological diversification, relatively

SEDENTARY habit, the fact that individuals are present throughout the year, their functional importance in ecosystems and their measurable response to disturbance. *See also* INSECTS, DIVERSITY AND CONSERVATION. [D.E.B. & P.E.]

REFERENCES

Collins N.M. (1989) Termites. In: *Tropical Rain Forest Ecosystems* (eds H. Lieth & M.J.A. Werger), pp. 455–471. Elsevier, Amsterdam.

Edwards R. & Mill A.E. (1986) *Termites in Buildings; Their Biology and Control.* Rentokil, East Grinstead.

terrigenous Referring to OCEAN sediments composed of debris from land erosion. Asia, having the four largest rivers in the world, supplies 25% of the total SEDIMENT discharge from continents to the SEA. *See also* EROSION AND TOPSOIL LOSS. [V.F.]

Territoriality

Territoriality can be said to arise whenever individuals or groups occupying a suitable habitat (*see* HABITAT SELECTION) space themselves out more than would be expected by chance, spacing being active in some way rather than, say, a passive product of underlying HABITAT STRUCTURE or non-random predation. Such an open-ended definition is robust to the easy exceptions plaguing narrower definitions like 'defended area' or 'exclusive area', and encompasses the full scope of scale and mechanism underlying spacing patterns. Separation may range from a few millimetres between neighbouring limpets on a rock to several kilometres between packs of hunting dogs or herds of antelope. Spacing between neighbours may be maintained in several ways, for instance by aggression, as in many birds and fish, or by vocal or olfactory signals as in some primates and ungulates. The secretion of TOXINS into the soil by some plant species can also be viewed as a means of territorial spacing. Territories may reflect the exclusive use of a given area, or the same area may be shared by different individuals or groups using it at different times. Moreover, they can be a permanent or temporary feature of an individual's life history, often existing on a seasonal basis and being restricted to a proportion of competitive individuals (or groups) within a population.

As might be expected, there is no single adaptive function to territoriality. Instead, it reflects a wide range of SELECTION PRESSURES, from COMPETITION for local RESOURCES such as food and mates to avoiding predators. Such selection pressures can result in interspecific territoriality as well as territorial defence among CONSPECIFICS. Where territories are based on resources, they are most likely to occur where resources are predictable and dispersed; unpredictable and clumped (patchy) distributions tending to favour social AGGREGATION. However, this distinction is not absolute, territoriality can occur even within social aggregations, for example among males in leks or kleptoparasites in flocks of birds.

Behavioural ecologists (*see* BEHAVIOURAL ECOLOGY) have attempted to predict the occurrence and size of territories (*see* TERRITORY SIZE) using cost–benefit analysis. Thus, feeding territories are seen as a TRADE-OFF between greater access to food and the time, energy and risk costs of establishing and defending an appropriate site. Similar trade-offs occur in the context of mating, nesting sites and so on. Territories need not be based on resources in the sense of encompassing them within their boundaries. In LEK aggregations, for example, a territory may be nothing more than a small patch of bare ground. Nevertheless, the position of the territory within the lek can be crucial in determining mating success and therefore can become the focus of intense competition among rival males. While territories within leks may not contain resources, leks often form at or near resources frequented by females (so-called 'hotspots'), therefore lek aggregations themselves can sometimes be considered a consequence of resource distribution.

Predation pressure may explain some forms of territorial spacing. For instance, breeding territories in some birds may reduce vulnerability to nest predators. Many predators concentrate their search effort in the vicinity of a recent find (area-restricted SEARCHING) and therefore, by ensuring that nests are a long way from those of neighbours, a territory owner might reduce the chances of their discovery.

Not surprisingly, territoriality can have consequences for population and COMMUNITY structure, although effects at these levels cannot in any sense be regarded as a function of territorial behaviour (as was argued by group selectionists in the early 1960s). Any apparent REGULATION of POPULATION SIZE and DENSITY attributable to territoriality is simply an emergent property of individual decision-making. [C.J.B.]

REFERENCE

Davies, N.B. (1978) Ecological questions about territorial behaviour. In: *Behavioural Ecology: An Evolutionary Approach* (eds. J.R. Krebs & N.B. Davies), 1st edn, pp. 317–350. Blackwell Scientific Publications, Oxford.

territory size TERRITORIALITY is assumed to reflect adaptive strategies of spacing between organisms. If maintaining a territory is adaptive, we would expect the size of territories to reflect the costs and benefits of their defence, all other things (such as physical and environmental CONSTRAINTS) being equal. This is the principle of 'economic defendability', which was first proposed by J.L. Brown in 1964. As territory size increases, so do the potential benefits of defence (e.g. access to more food or mates). However, the costs of defence are also likely to rise because bigger territories are more demanding to police. At the lower end of the scale, the small amount of extra benefit gained is unlikely to offset the cost of defence and so territoriality is not a viable proposition. Similarly, if a territory becomes very large, RESOURCES effectively become superabundant relative to the defender's needs, and so, again, do not make the additional cost of their defence worthwhile. The maximum return per unit cost is therefore likely to lie between the two extremes and thus dictate the optimum size for the territory (although this depends critically on the shape of the cost and benefit curves). The predictions of economic defendability have been supported in a number of comparative and experimental studies in which variation in territory size has been related to resource availability, intruder pressure and other factors affecting the net benefit of defence. In social species, the constraints of economic defendability on single defenders may be overcome by increasing resident group size and thus defender power. Under these conditions, very much larger territories may become economically defendable. [C.J.B.]

REFERENCES

Brown, J.L. (1964) The evolution of diversity in avian territorial systems. *Wilson Bulletin* **76**, 160–169.

Kruuk, H. & Macdonald, D.W. (1985) Group territories of carnivores: empires and enclaves. In: *Behavioural Ecology: Ecological Consequences of Adaptive Behaviour* (eds R.M. Sibly & R.H. Smith), pp. 521–536. Blackwell Scientific Publications, Oxford.

tertiary consumer A tertiary consumer is one which occupies the fourth TROPHIC LEVEL in a COMMUNITY. In the grazing FOOD WEB, herbivores are primary consumers, primary carnivores and secondary consumers, and the animals which eat the primary carnivores are the tertiary consumers. For example, in an oak woodland where the newly emerged leaves are being grazed by a wide variety of caterpillars (lepidopteran larvae), these may be predated by insectivorous birds such as the willow warbler (*Phylloscopus trochilus*) or blackcap (*Sylvia atricapilla*). These primary carnivores are themselves predated by sparrowhawks (*Accipter nisus*) which thus become tertiary consumers in the oak–caterpillar–warbler–hawk FOOD CHAIN. Later in the year a little owl (*Athene noctua*) may catch a common shrew (*Sorex araneus*) on the woodland floor that had been eating the pupae of the caterpillars which had evaded the warblers' attacks. The owl would also be a tertiary consumer, but its next meal might be a vole (*Clethrionomus glareolus*) that had been feeding on plant shoots and seeds, so in this case it would also be a secondary consumer in the grazing food chain. To complicate the picture further, it might also eat a lesser stag beetle (*Dorcus parallelopipedus*) which lives and feeds in rotten wood so it would also be a secondary consumer in the decomposer food chain. If, however, later that night, it captured a large ground beetle, such as *Carabus violaceus*, which had recently eaten a prostigmatid mite that fed on collembola which ate decaying leaf litter, it would again be a tertiary consumer but this time in the decomposer food chain.

This example emphasizes that terms in a TROPHIC CLASSIFICATION are meant to describe the 'functional position' of animals in a community and that any one species can occupy a number of functional positions at the same time. *See also* CONSUMER ORGANISMS; FEEDING TYPES, CLASSIFICATION. [M.H.]

test guideline/protocol Specific, and often very detailed, instructions on how TOXICITY and ECOTOXICITY testing should be executed and updated. Required by GOOD LABORATORY PRACTICE (GLP). Standards organizations, especially the OECD, play a key role in ensuring international standardization of tests so that there is mutual acceptance of data generated from them. This facilitates international control of POLLUTION and also helps to reduce animal testing. [P.C.]

thallium (Tl) A HEAVY METAL (ATOMIC NUMBER = 81; relative ATOMIC MASS = 204.37; melting point = 303.5°C; boiling point = 1457°C) that is an ultratrace constituent (0.5 mg kg^{-1}) of the Earth's CRUST, mostly dispersed as an isomorphous replacement for POTASSIUM. It is found associated with the ores of other ELEMENTS such as lead sulphide and is mainly produced from the flue dusts of smelting. Thallium is used on a small scale industrially, for example in alloys.

Formerly employed, in the form of thallous sulphate (Tl_2SO_4), as a rodenticide, thallium is now mainly released to the environment from the combustion of coal, and in dusts and EFFLUENTS from metal smelters and processing plants, but it cannot be considered a general environmental HAZARD. Thallium as Tl^+ is the stable form in soils and can replace K^+ in CLAYS. It can be accumulated by rape, lettuce and cabbage. The effects on plant metabolism include reductions in PHOTOSYNTHESIS and TRANSPIRATION.

Thallium is one of the most toxic of metals to humans, having severe effects upon the nervous system and gastrointestinal tract. Thallotoxicosis can be difficult to diagnose as many initial symptoms are not unique to thallium, although eventually the resultant alopecia and sensitivity of the skin to touch are taken as characteristic signs. The toxic effects of Tl^+ arise partly from its ability to enter cells instead of K^+, which has a similar ionic radius. The ability to cause hair loss has led to the use of thallium salts in some depilatory creams. [J.F.]

thelytoky The parthenogenetic production of females by dioecious animal taxa. Thelytokous reproduction may be apomictic or automictic, amictic or mictic. *See also* ASEXUAL REPRODUCTION; PARTHENOGENESIS. [M.M.]

theory of natural selection NATURAL SELECTION can be defined as a process occurring under three conditions. If a population has:

1 variation among individuals in some trait;

2 a consistent relationship between that trait and FITNESS, i.e. consistent differences among individuals in fitness;

3 a consistent relationship between parents and offspring for the trait, that is, differences in the trait are inherited;

then:

1 the trait frequency distribution will differ among age classes;

2 the trait distribution will differ between parents and offspring if the population is not in equilibrium.

This description of the theory of natural selection is due to Endler (1986) and is an updated version of Darwin's theory.

The three conditions can be formalized as follows:

1 the PHENOTYPIC VARIANCE in the trait exists, $V_P(x) > 0$;

2 a COVARIANCE between PHENOTYPE x and fitness w exists, $\mathrm{cov}_P(x,w) \neq 0$ or $\mathrm{cov}_G(x,w) \neq 0$;

3 the parent–offspring covariance > 0, i.e. the ADDITIVE GENETIC VARIANCE in the trait $V_A(x) > 0$.

Mathematical formulations for the theory of natural selection are given by:

$$R = \left[V_A(x)\mathrm{cov}_P(x,w)\right] / \left[\bar{w}\,V_P(x)\right]$$

$$R = h^2 S$$

$$R = V_A(x)\beta / \bar{w}$$

where R is the SELECTION RESPONSE, S is the SELECTION DIFFERENTIAL, h^2 is the (NARROW-SENSE) HERITABILITY and β is the SELECTION GRADIENT. [G.D.J.]

REFERENCE

Endler, J.A. (1986) *Natural Selection in the Wild*. Princeton University Press, Princeton, N.J.

thermal imaging A technique providing a visual representation of the incident thermal RADIATION field. Grey tone images can show high temperatures as either bright or dark, analogous to positive and negative images in black and white photography. Alternatively, false colour images correspond to intensity (TEMPERATURE) slices in the grey tone image. Ideally, thermal images represent the spatial distribution of surface temperature (inferred from energy fluxes) in the scene. Qualitative imaging is useful for activities such as SURVEILLANCE, but accurate estimation of temperatures, required for analysis of HEAT TRANSFER, is difficult due to two problems. First, detectors must operate in atmospheric windows (3–5 μm or 8–13 μm) to minimize absorption by GREENHOUSE GASES, and therefore measure only part of the radiation spectrum emitted. Second, the emitted radiation is usually both less than that from a 'black body' (i.e. emissivity $\varepsilon < 1$) and 'diluted' by the proportion of AMBIENT radiation (E_a) reflected from the surface. Since the reflection coefficient is the complement of the emissivity, the incident flux (E_i) is:

$$E_i = \varepsilon E_b + (1 - \varepsilon) E_a$$

where E_b is the thermal radiation emitted by a black body ($\varepsilon = 1$) with the same surface temperature T_s (K). Hence:

$$E_b = \sigma T_s^4$$

where σ ($= 56.7 \times 10^{-9}$ W m^{-2} K^{-4}) is Stefan's constant. Most surfaces of biological origin approximate to black bodies. However, the emissivities of smooth surfaces such as the cuticles of leaves and coat-less animals depend on the angle of view, approaching zero at glancing angles. As a consequence, thermal images are inaccurate at oblique-view angles. *See also* HEAT BALANCE. [J.A.C.]

thermal pollution Heat release into the environment as a by-product of human activity and having a harmful EFFECT on ecosystems. For example, direct thermal POLLUTION of waters may occur at power-stations where more than 50% of the heat content of the fuel may end up as WASTE heat that is removed by the cooling waters and is ultimately released into RIVERS. This can raise the temperature of waters to an extent that favours exotic species. It also reduces DISSOLVED OXYGEN causing metabolic STRESS. [P.C.]

thermal stratification An important feature of water bodies that results when differences in TEMPERATURE occur between the surface water layers and deeper layers. As a result of the temperature difference, upper and lower layers differ in density and are resistant to vertical mixing. Such STRATIFICATION is typical in temperate LAKES and COASTAL marine waters during the summer and limits the availability of NUTRIENTS (which would otherwise be upwelled from depth) to primary producers. *See also* LAKE STRATIFICATION. [V.F.]

thermocline That portion of the water column where TEMPERATURE changes most rapidly with each unit change in depth. Because temperature has an important influence on the density of water, establishment of a thermocline stabilizes the water column and acts to limit the vertical mixing of water. The depth of the thermocline, particularly in shallow, temperate waters, varies seasonally; in some areas the thermocline may disappear entirely during winter months. *See also* LAKE STRATIFICATION; STRATIFICATION. [V.F.]

Thermometry

Thermometry is the measurement of TEMPERATURE. This is achieved by using one of several types of sensor. In this short article we mention only those commonly used in ENVIRONMENTAL SCIENCES.

Mercury-in-glass thermometers are still widespread, being relatively cheap and readable to 0.1°C or 0.5°C, depending on their specification, and are available as maximum–minimum thermometers. The best thermometers are supplied with a CALIBRATION certificate, and usually have scratch marks on the glass at 0°C and 100°C from when they were calibrated using an ice slurry and boiling water. They may be designed for total immersion or for non-immersion. In the latter case, most of the sensitivity is concentrated at the bulb itself, and the glass stem is narrower.

Thermocouples are the most widely used temperature sensor in experimental work, as they are easy to construct, very cheap, small, and produce an electrical signal which can be recorded on a data logger. They consist of wires of different metals, twisted or welded together to make a thermojunction, and depend on the thermoelectric effect (the tendency of electrons to migrate from one METAL to another, at a rate that is a non-linear function of temperature). A thermocouple circuit consists of two thermojunctions, one kept at a reference junction and the other is used to sense temperature. A thermocouple circuit produces very small signals which require a sensitive microvoltmeter to measure (a copper–constantan thermocouple pair produces about 40 μV per degree Celsius of temperature difference; for accurate work it is essential to calibrate thermocouples, as the composition of the metals may vary somewhat from batch to batch). Most modern data loggers used with thermocouples have (i) a very sensitive and stable microvoltmeter and (ii) an electronic reference junction built into the logger, so that the user need use only one thermojunction. In practice, thermocouples used in this way can resolve a temperature difference as small as 0.05°C, and are therefore suitable for PSYCHROMETERS in the measurement of HUMIDITY.

Platinum resistance thermometers (PRTs) are coiled lengths of PLATINUM wire. The sensor exploits the effect of temperature on electric resistance. Platinum is chemically non-reactive and so the sensors are durable and stable over a long time. Instruments supplied as laboratory standards claim an ACCURACY of 0.002°C, and in the field an inexpensive PRT is likely to have an accuracy of 0.01°C.

The thermistor, like the PRT, is a resistance thermometer, but is made of a bead of semi-conductor material. Thermistors have been widely used in environmental science and form the basis for most hand-held electronic thermometers. Integrated circuit sensors are also available, consisting of a semi-conductor diode with trimmed resistors, set up during manufacture so that the signal in millivolts is conveniently equal to degrees Celsius.

Infrared thermometers are now widely available and are excellent for measuring the temperature of natural surfaces. They calculate the temperature from the measured thermal RADIATION emitted by a surface and are usually accurate to within 1°C. The viewing angle may be quite narrow (2°), enabling, for example, the temperature of a leaf to be found very conveniently.

The use of temperature sensors in the natural environment

To measure the air temperature it is necessary to shield the sensor from radiation yet allow the air to circulate over the sensitive part of the instrument. In critical applications, such as the Assmann psychrometer, this is achieved by having a double radiation shield made of concentric brass tubes, the outer one of which is chromium plated to reflect SOLAR RADIATION, and aspirating with air at about 4 m s^{-1} using a fan. In standard meteorological stations, the thermometers are shielded by the familiar STEVENSON SCREEN, a white painted box with louvred sides and a double ceiling. Small variations in the design of the screen are known to produce variations in readings, especially of the daily maximum temperature, that can be as high as 2°C. It is doubtful whether air temperature measured with an unaspirated thermometer can ever be measured with an absolute precision of better than 1°C. The need for radiation shielding in micrometeorological work has been minimized in some cases by the use of thermocouples made from extremely fine wires (0.05 mm diameter). These are so well coupled to the air by CONVECTION that they do not warm up appreciably even in full solar irradiance.

To measure soil temperature, it is best to use thermocouples or thermistors, encased in epoxy resin and further protected by a steel sheath. Meteorological stations still use mercury-in-glass thermometers hung in iron tubes in the soil at 0.5 m or 1 m, and special shallow-soil thermometers with stems bent at right angles.

To measure water temperatures, thermocouples protected by lacquer may be the best approach.

For long-term measurements, cables should be well protected from damage by animals. *See also* HEAT BALANCE; THERMAL IMAGING.[J.G.]

REFERENCES

Marshall, D. & Woodward, F.I. (1985) *Instruments for Environmental Physiology.* Cambridge University Press, Cambridge.

McConnell, D. (1988) Making a simple thermometer screen. *Weather* **43**, 198–203.

---

therophyte A plant life form that survives unfavourable conditions (usually DROUGHT) as a seed. It characteristically has a short LIFE CYCLE (ANNUAL or EPHEMERAL) and is a life form typical of arid environments where soil DISTURBANCE and an infrequent supply of moisture are the norm. It is a life form that has been particularly favoured by human soil disturbance throughout the world and is now typical of arable habitats (*see* RUDERAL). *See also* PHYSIOGNOMY; RAUNKIAER'S LIFE FORMS. [P.D.M.]

thixotrophic The fluid nature of sands and muds as a result of the application of pressure. The burrowing activities of benthic infauna increase the water content of SEDIMENT, such that surface sediment layers in actively bioturbated zones may be 60–80% water. In such thixotrophic sediments, the effort required to move through the sediment decreases with increased burrowing velocity. *See also* ECOSYSTEM ENGINEERS. [V.F.]

thorium (Th) Thorium (ATOMIC NUMBER = 90; relative ATOMIC MASS = 232.0381; melting point = 1750°C; boiling point = 4850°C) has a crustal abundance of 8.1 mg kg^{-1}. ISOTOPES exist from ^{214}Th to ^{236}Th, six of which are members of the natural radioactive decay series. Primordial ^{232}Th ($t_{1/2} = 1.41 \times 10^{10}$ years) is dominant in terms of mass and heads one of the decay series. Thorium, existing only in the 4^+ oxidation state, is particle reactive and highly insoluble in natural waters. Thorium isotopes are extremely useful tracers of natural processes via natural decay series disequilibrium. *See also* RADIONUCLIDE. [A.B.M.]

REFERENCES

Ivanovich, M. & Harmon, R.S. (eds) (1992) *Uranium-series Disequilibrium.* Clarendon Press, Oxford.

Katz, J.J., Seborg, G.T. & Morss, L.R. (eds) (1986). *The Chemistry of the Actinide Elements.* Chapman & Hall, London.

thorn scrub Sparse shrub vegetation characteristic of semi-arid areas which is usually deciduous and richly provided with thorns, thus reducing its palatability to large herbivores. Leaf production is associated with the arrival of rain and the period for which the leaves are retained on the plant is variable, depending on the continuation of a water supply. Thorn SCRUB often forms a belt around the DESERTS, except where high grazing pressure reduces its biomass significantly. It merges into SAVANNAH vegetation as grasses become more abundant.

Studies of thorn scrub vegetation in several parts of the world (e.g. Arizona and Iran) have shown that the main bushes can be evenly (rather than randomly) spaced as a result of intraspecific root COMPETITION for water and possibly the production of germination-inhibiting chemicals by some species. [P.D.M.]

Thornthwaite evaporation In 1948 the American climatologist C.W. Thornthwaite devised a method of calculating potential EVAPORATION from temperature, and this method is sometimes still used. However, it is superceded by methods that are based more firmly on physical principles, such as that devised by H.L. Penman in the UK and developed further by J.L. Monteith. *See also* PENMAN–MONTEITH EQUATION. [J.G.]

REFERENCES

Monteith, J.L. (1965) Evaporation and environment. *Symposia of the Society for Experimental Biology* **19**, 205–234.

Penman, H.L. (1948). Natural evaporation from open water, bare soil and grass. *Proceedings of the Royal Society of London, Series A* **193**, 120–145.

Thornthwaite, C.W. (1948) An approach towards a rational classification of climates. *Geographical Reviews* **38**, 55–94.

threshold In environmental parlance, the divide between when an EFFECT does and does not take place. For example, in ECOTOXICOLOGY, the concentration above which an adverse effect is observed and below which it is not. Cf. a graded increase from zero level

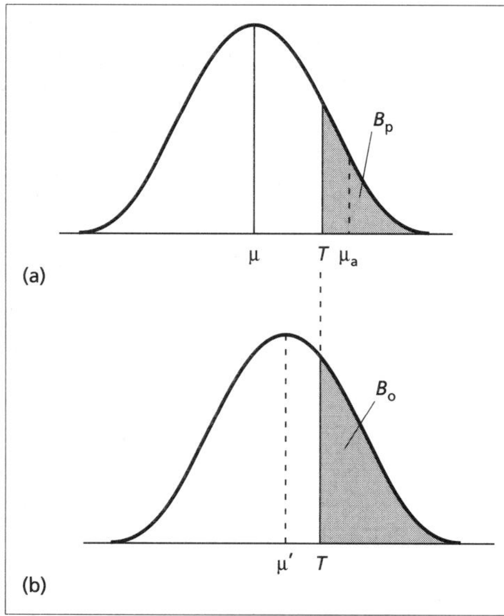

Fig. T3 (a) Distribution of liability assumed for a threshold trait in a hypothetical population. The shaded area denotes individuals having liability above a critical threshold (T) and consequently affected with the trait. B_p is the frequency of affected individuals in the entire population, μ is the mean liability of individuals in the entire population, and μ_a the mean liability of affected individuals. (b) Distribution of liability among offspring who have one parent affected with the trait. μ′ denotes the mean liability among offspring, and B_o is the proportion of affected offspring. (After Hartl, D.L. & Clark, A.G. (1989) *Principles of Population Genetics*. Sinauer Associates Sunderland, Mass.)

(e.g. concentration) upwards. *See also* ACTION LEVEL; PREDICTED NO EFFECT CONCENTRATION. [P.C.]

threshold selection Quantitative characters that come in two discrete classes are THRESHOLD characters. An example is the number of digits on the hindpaws of guinea pigs: some guinea pigs have three toes, others have four toes. Another example is diabetes in humans, with the phenotypes 'affected' and 'not-affected'. Threshold characters are regarded as polygenic TRAITS, with an underlying NORMAL DISTRIBUTION of 'liability' derived from genetic and environmental influences but with expression of the trait only above a certain threshold value of 'liability' (Fig. T3a). Under the assumption of a normal distribution of liability with MEAN 0 and VARIANCE 1, the proportion affected with the trait can be used to find the mean value, μ_a, of the liability in the affected individuals in the population (*see* SELECTION INTENSITY). The proportion affected in offspring of an affected and an unaffected parent can be used to find the mean value μ′ of the liability in the offspring (Fig. T3b). The HERITABILITY of liability is given by $h^2 = 2\mu/\mu_a$. Selection changes the position of mean liability with regard to the fixed threshold. A greater (or smaller) proportion of the population becomes affected. *See also* QUANTITATIVE GENETICS. [G.D.J.]

REFERENCES

Falconer, D.S. (1989) *An Introduction to Quantitative Genetics*. Longman, London.

Hartl, D.L. & Clark, A.G. (1989) *Principles of Population Genetics*. Sinauer Associates, Sunderland, Mass.

thulium (Tm) Thulium (ATOMIC NUMBER = 69; relative ATOMIC MASS = 168.9; melting point = 1545°C; boiling point = 1947°C) is a LANTHANIDE METAL with an average crustal abundance of 0.48 mg kg^{-1}. After NEUTRON irradiation, it emits X-RAYS and so has been used in portable X-ray devices. [J.G.]

tidal energy Energy from the ebb and flow of TIDES. Can use barrages with conventional turbines positioned so that the incoming and outgoing tides generate power, and floating devices in which wave-driven paddles and floats generate the power. *See also* RENEWABLE RESOURCE. [P.C.]

tidal flat Tidal flats are the gently sloping or effectively level expanses of marine SEDIMENT that may be exposed by low TIDES (see Plate 24, facing p. 398). They form the feeding grounds of shorebirds and wildfowl, and of many fish, including juvenile flatfish. The prey species are mostly buried, or in burrows, within the sediment, and include polychaete worms, crustaceans and bivalve molluscs particularly, although surface-dwelling gastropod molluscs are also often abundant. Generally, the cleaner the sand the poorer the BIOMASS and PRODUCTIVITY. Vertebrate predators take mainly adult prey species, whilst the juveniles are consumed by EPIBENTHIC crustaceans and small fish, and by carnivorous worms living within the flat. Primary PRODUCTION from INTERSTITIAL or surface mat-forming algae is often high. [R.S.K.B.]

tides The periodic movement of WATER resulting from gravitational attraction between the Earth, Sun and Moon, the forces of which act unequally on different parts of the Earth. Spring tides occur near the times of the new and full Moon, when the Sun and Moon lie on a straight line passing through the Earth and their influences are additive. Tidal range is greatest during spring tides. Neap tides occur near the first and last quarters of the Moon, when the Sun and Moon lie at right angles to each other and have directly opposing influences. Tidal range is least during neap tides. A DIURNAL tide has one high water and one low water each tidal day. A semi-diurnal tide is a tide with two high waters and two low waters each tidal day, with small inequalities between successive highs and successive lows. A mixed tide is a tide having two high waters and two low waters per tidal day with a marked diurnal inequality. This tide type may also show alternating periods of diurnal and semi-diurnal components. Variability in tides is caused by the configuration of the ocean's bottom TOPOGRAPHY and by seasonal differences in CURRENTS. Tidal currents are caused by the alternating horizontal movement of water associated with the rise and fall of the tide. They are important agents of erosion and significantly affect size, sorting and distribution of SEDIMENT over most of the sea floor. [V.F.]

TIE *See* TOXICITY IDENTIFICATION EVALUATION.

tiered tests TOXICITY/ECOTOXICITY tests that involve increasing levels of detail according either to the results from lower-level tests (e.g. that signal high toxicity/ecotoxicity) or in terms of increasing production and/or marketing and/or importation levels. [P.C.]

tillage systems *See* AGRICULTURE AND SOIL CONSERVATION.

tiller A shoot formed at or near ground level by grasses and sedges. *See also* MODULE. [P.C.]

Tilman's diatom experiments Laboratory tests, using chemostats with limiting SILICATE or PHOSPHORUS, of the theory of single limiting RESOURCES. The prediction is that the species with the lowest R^* (see Fig. S25) will competitively displace a species with a higher R^* in experimental systems with a single limiting resource. *See also* COMPETITION. [M.J.C.]

REFERENCE

Tilman, D. (1982) *Resource Competition and Community Structure.* Princeton University Press, Princeton, N.J.

Tilman's model of resource utilization An optimal foraging model (*see* OPTIMAL FORAGING THEORY) for plant GROWTH which predicts that a plant should consume two essential RESOURCES in the proportion in which it is equally limited by both of them. This proportion defines the consumption vector for the two resources and has important implications for species COEXISTENCE. The model applies in principle to any number of resources. [M.J.C.]

timber line The border between the outmost wood stands and the treeless vegetation of the polar regions, mountains and arid areas. Questions concerning the timber line have been studied especially in subarctic and subalpine regions. Almost everywhere, the level of the timber line has been greatly affected by human activities. A distinction is thus made between the 'potential timber line' which would exist without human interference and the 'actual timber line' of today. The location of the potential timber line is mainly dependant upon various climatic factors: the effective TEMPERATURE sum of the vegetation period, critical temperatures, PRECIPITATION and WINDS.

The altitudinal timber line varies greatly in different latitudes. The highest situated timber line is found on SUBTROPICAL mountains, for example in Pamir at 3800–3900 m above sea level. On tropical mountains, the timber line is lower, for example on the Venezuelan Andes at *c.* 3100–3250 m. It has been suggested that the underlying causes are the SOIL temperatures which, in the tropics, follow the yearly mean temperature becoming at about +8° to +9°C too low for the roots of the trees, whereas in the subtropics the warm summers make it possible for trees to reach higher altitudes. In the temperate regions, for example in the European Alps, the timber line lies at 1900–2300 m. At approximately the 70° northern latitude and on the 60° southern latitude, the timber line descends to the sea level. In the north, the polar timber line is usually lower in oceanic areas and higher in continental areas.

Somewhat roughly, the polar timber line corresponds with the +10° ISOTHERM of the warmest month or with the 600°C effective temperature sum during the growing period (which means the sum of excess of temperature above +5°C during the period). The northern polar timber line, especially, is quite gradual. One can differentiate: (i) the 'economic timber line' indicating the northern limit of FORESTRY (which is approximately the same as the regular limit of generative propagation of the trees); (ii) the 'timber line proper', meaning the northern extension of continuous FOREST (approximately the same as the limit of vegetative propagation of the trees); (iii) the TREE LINE drawn on the basis of isolated bigger or smaller forest stands; and finally (iv) the species limit showing the northernmost occurrence of stunted specimens or seedlings of the trees in question. Between the timber line proper and the tree limit is an intermediate zone of varying breadth known as the forest TUNDRA. Similar zonality also characterizes the alpine timber line (Figs T4–T6).

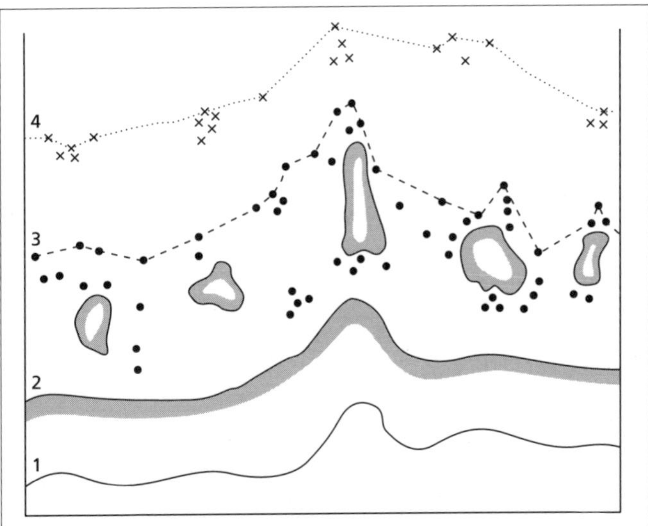

Fig. T4 Diagrammatic sketch of the northern timber and tree line. 1, The 'generative' forest limit; 2, the vegetative forest limit; 3, the tree limit; 4, scattered seedlings or remnants of earlier forest. (After Hustich, I. (1948) The Scotch pine in northernmost Finland and its dependence on the climate in the last decades. *Acta Botanica Fennica* **42**, 1–75.)

There are relatively few tree species forming the northern timber line towards tundra. Birch (*Betula* spp.) is the timber-line tree in Fennoscandia, in Kamtchatka and on the eastern side of Bering Strait. In Alaska, alder (*Alnus* spp.) can occasionally form the timber-line woods. In the European part of Russia, spruce (*Picea abies* and *P. obovata*) forms the forest line, in Siberia spruce is replaced by larch, Siberian larch (*Larix sibirica*) in the west and Dahurian larch (*L. dahurica*) in the east. In North America the main timber-line trees are black spruce (*Picea mariana*) and white spruce (*P. glauca*). [Y.V.]

time Time is a fundamental physical quantity describing the duration of events as a dimension, or the sequence of a number of events. The SI UNIT for time is the SECOND (s) which is defined relative to the frequency of transition between two levels of the ground state of a ^{133}Cs atom. Most other common units of time are defined with reference to the second.

Systems describing time were originally based on the DIURNAL cycle of the Earth's rotation around it's own axis and the seasonal cycle resulting from the orbit of the Earth around the Sun.

True solar time or local apparent time is based upon the apparent motion of the Sun across the sky. Solar noon is defined for any locality as the time when the Sun reaches its highest point or solar angle. This point is known as the local meridian. The period between two meridians is defined as 1 day. The length of the solar

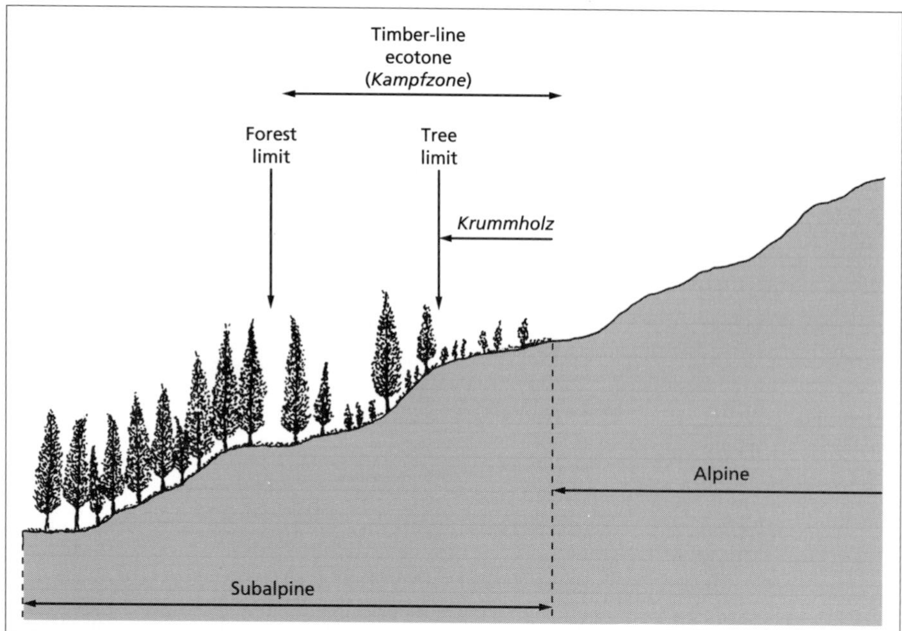

Fig. T5 Diagram showing the relative positions of trees as described by the terms forest limit (timber line), tree limit, timber-line ecotone and krummholz. (After Crawford, R.M.M. (1988) *Studies in Plant Survival*. Blackwell Scientific Publications, Oxford.)

Fig. T6 A timber-line stand of *Pinus cembra* at an altitude of 2300 m near Zermatt, Switzerland. (Courtesy of R.M.M. Crawford.)

day varies by up to 16 minutes during the year because of the position of the Earth in orbit around the Sun. The variation in length of the solar day is described by the equation of time often published as part of ephemeris tables (e.g. Nautical and Astronomical Almanacs). In many biological applications, true solar time must be calculated from local standard time making allowances for longitude, the equation of time and daylight saving. The mean solar day length averaged over a year is used by convention for most purposes. Mean solar time is defined relative to the movement of a hypothetical sun travelling across the sky at the same rate throughout the year. Sidereal time is defined using fixed stars as a reference, with a 'sidereal day' being the period required for the Earth to rotate once relative to a fixed star. The length of a sidereal day is approximately 4 minutes shorter than the mean solar day. Ephemeris time is based upon the annual revolution of the Earth around the Sun, removing the difference associated with the variation in the speed of rotation of the Earth about its axis.

Standard time was developed from mean solar time to avoid complications for travel and communications associated with communities using their own local solar time. The Earth was divided into 24 standard time zones, each separated by 15° longitude. True solar time differs from standard time by 4 minutes for each degree of longitude east or west of the meridian for the time zone. The reference standard time zone is located at the 0° meridian of longitude that passes through the Royal Greenwich Observatory in England. The local mean solar time at the Greenwich meridian is referred to as Greenwich Mean Time (GMT) or Universal Time (UT). Standard time zones are described by their distance east or west of Greenwich but are sometimes modified for political or geographical reasons. Daylight saving is a system where clocks are set 1 or 2 hours forward during the summer months, resulting in an additional period of daylight in the evening.

The GEOLOGICAL TIME-SCALE is a record of the Earth's history defined relative to major geological episodes. Divisions of geological time are based on changes in the FOSSIL RECORD in geological strata. [P.R.V.G.]

REFERENCES

Great Britain Nautical Almanac Office & United States Nautical Almanac Office (1974) *Explanatory Supplement to the Astronomical Ephemeris and the Nautical Almanac*. 3rd impression with amendments. Her Majesty's Stationery Office, London.

Iqbal, M. (1983) *An Introduction to Solar Radiation*, Academic Press, Toronto.

time-based sampling *See* SAMPLING METHODOLOGY/DEVICES.

time-delay *See* TIME-LAGS.

time-domain reflectometry A technique originally used to detect the properties of electrical transmission lines. Time-domain reflectometry (TDR) has recently been used to measure WATER content and electrical conductivity of soils. A step voltage is applied to steel pins embedded in the SOIL. The timing and attenuation of the reflected signal as displayed on an oscilloscope relate to the dielectric (K) and conductivity of the soil, respectively. Water has a high dielectric ($K = 80$) compared to other soil components, air ($K = 1$) and soil grains ($K = 3$–5), thus allowing empirical CALIBRATIONS of volumetric water content using the measured dielectric constant. [J.I.]

REFERENCES

Dasberg, S. & Dalton, F.N. (1985) Time domain reflectometry field measurements of soil water content and electrical conductivity. *Soil Science Society of America. Journal* **49**, 293–297.

Ledieu, J., De Ridder, P., De Clerck, P. & Dautrebande, S. (1986) A method of measuring soil moisture by time-domain reflectometry. *Journal of Hydrology* (Amsterdam) **88**, 319–328.

time-lags The effect of many processes is not immediate, and these are said to be 'time-lagged'. For example, in *Drosophila* populations, egg-laying only produces new adults some time after it occurs: after a time-lag. A model of such a system might look like this:

$$\frac{dN}{dt} = \beta N(t - \tau) - \delta N(t)$$

where β is the birth rate, δ is the death rate, $N(t)$ is the adult population at time t and τ is the time-lag from egg-laying to adulthood.

Time-lags can alter dynamics considerably and are usually destabilizing. In many cases time-lags occur not as a single fixed delay, but as a distributed delay. *See also* POPULATION DYNAMICS. [S.N.W.]

time-series analysis The set of statistical techniques used to suggest likely ecological mechanisms that might have generated a particular run of POPULATION DENSITY data (the time series, Fig. T7). Generally, the time series is de-trended prior to analysis. The principal techniques are graphical.

1 The autocorrelation function (ACF) plots the CORRELATION between successive values in the time series as a function of the lag (the number of years between the POPULATION SIZES being correlated). Cyclic populations exhibit significant and cyclic ACFs.

2 The partial autocorrelation function (PACF) plots the correlation between populations at different lags after factoring out the correlation between populations at intermediate lags (analogous to the partial correlation coefficient). This plot is particularly important because it shows the order of DENSITY DEPENDENCE (1st order density dependence shows up as a significant negative PAC coefficient at lag 1, 2nd order as a significant PAC at lag 2, and so on).

3 The log population change, $\ln(N_{t+1}/N_t)$, is plotted as a function of the log density in year t, $\ln(N_t)$; this shows the shape of the density dependence (e.g. linear, threshold, inverse S-shaped, etc.).

Most ecological time series are 5–40 years long, and this is generally too short to test for chaotic dynamics. For longer time series (many hundreds of points) it may be possible to reconstruct the shape of the chaotic ATTRACTOR (if one exists). [M.J.C.]

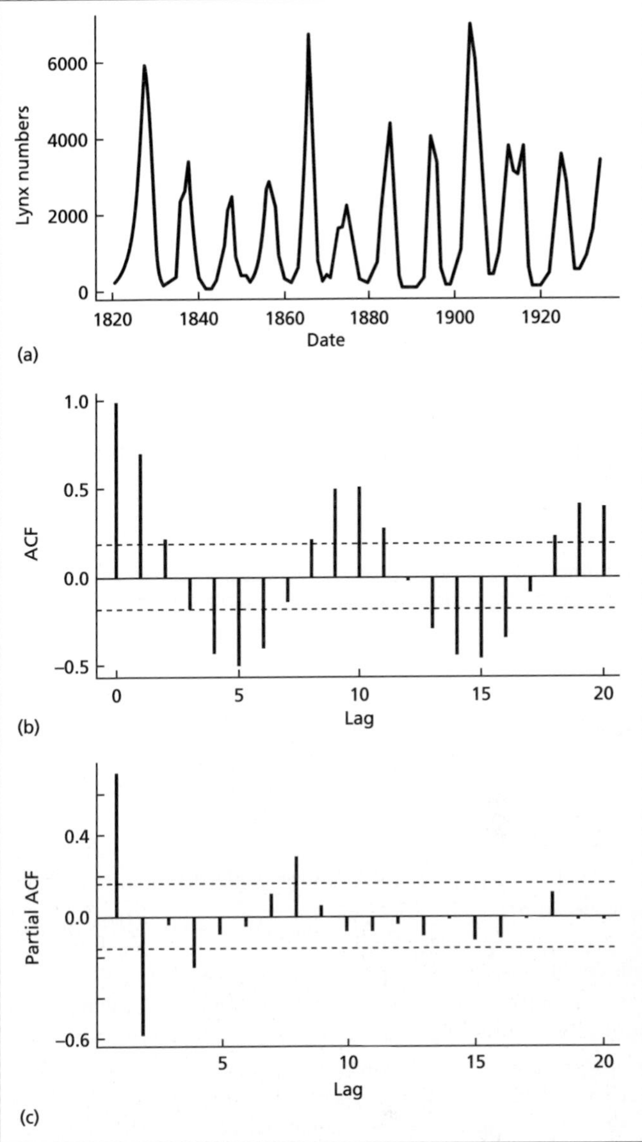

Fig. T7 Time-series analysis. (a) Time series of population size for Canadian lynx (*Lynx canadensis*) over the period 1820–1930 showing highs and lows, and suggesting regular, cyclic dynamics. (b) The autocorrelation function (ACF) for the lynx data, showing the correlation between population sizes at a range of lags. The correlation between points at lag 0 is always 1 but the plot shows that there is a significant (dashed horizontal line) negative correlation between population size at time 0 and the population size in 5 years' time, and a significant positive correlation between the population at time 0 and the size in 10 years' time, suggesting regular, 10-year cycles in abundance. (c) The partial autocorrelation plot is useful for determining the kind of density dependence underlying the pattern shown in the autocorrelation function (ACF) (i.e. both the sign and the lag of the density dependence). (b) In this case, it is the negative density dependence at lag 2 (caused, presumably, by overexploitation of the main prey species, snowshoe hare (*Lepus americanus*)) which is responsible for producing the 10-year cyclic behaviour. No convincing ecological explanations have been offered for the (marginally) significant terms at lags 4 (negative) and 8 (positive) and they may well be spurious.

REFERENCES

Diggle, P.J. (1990) *Time Series: A Biostatistical Introduction.* Clarendon Press, Oxford.

Royama, T. (1992) *Analytic Population Dynamics.* Chapman & Hall, London.

tin (Sn) A soft, HEAVY METAL (ATOMIC NUMBER = 50; relative ATOMIC MASS = 118.69; melting point = 231.9°C; boiling point = 2270°C) that is a trace constituent (2.1 mg kg^{-1}) of the Earth's CRUST. Found mainly as cassiterite (SnO_2), tin has many uses, including coatings, solders, alloys and various chemical compounds such as FUNGICIDES, BIOCIDES, stabilizers and fire retardants. Although tin in inorganic form is largely non-toxic, some of its organic compounds can be highly toxic. Trialkyltin compounds possess strong biocidal activity and the tributyltin compound bistributyltin oxide ($(C_4H_9)_3Sn–O–Sn(C_4H_9)_3$) is used as an ANTI-FOULING agent on the underside of ships. Unfortunately, this has resulted in serious toxic effects upon shellfish in coastal waters and harbours. [J.F.]

titanium (Ti) Titanium (ATOMIC NUMBER = 22; relative ATOMIC MASS = 47.88; melting point = 1660°C; boiling point = 3287°C) is the ninth most abundant element, being present in the Earth's CRUST to the extent of 0.6%. The high whiteness and refractive index of the dioxide (TiO_2, rutile) results in its extensive use as a whitener and opacifier in paints and PLASTICS. The high strength to mass ratio of the METAL, allied with its extreme corrosion resistance, lead to a variety of specialist engineering applications. Titanium is essentially non-toxic, the main environmental problems being associated with the disposal of WASTE products arising from the manufacture of TiO_2 pigments. There is no evidence that titanium has any biological function. [P.J.B.]

tolerance The reduced response to a TOXICANT, BIOCIDE or POLLUTANT as a result of repeated EXPOSURE. This may occur within individuals as a result of physiological ACCLIMATION/acclimatization—for example as a result of induction of protective, metabolic systems such as melallothioneins or MIXED-FUNCTION OXIDASES (MFOs). It may also occur by selection of more tolerant genotypes. *See also* ADAPTATION TO POLLUTION; COSTS OF RESISTANCE; RESISTANCE. [P.C.]

tombolo *See* LAKES, TYPES OF.

top-down controls Important influence on COMMUNITY structure and possibly function from upper trophic levels. Thus removal of a predator releases pressure on prey which has knock-on effects for its competitors and the species populations that form its food. *See also* HOLISM; SYSTEMS ECOLOGY. [P.C.]

top species The species at the ends of FOOD CHAINS, i.e. the pinnacles of trophic pyramids. Also known as top carnivores or top PREDATORS—they are invariably meat-eaters. *See also* FEEDING TYPES, CLASSIFICATION. [P.C.]

topogenous Used of a MIRE that owes its development to the concentration of water in a region by drainage from a CATCHMENT. By definition, therefore, such a mire must be flow-fed, or RHEOTROPHIC, for example BASIN fens and some parts of VALLEY MIRES. [P.D.M.]

topography The surface features of the Earth, including the relief, the terrain, the VEGETATION, the SOILS and all the features within the landscape. It is not synonymous with relief alone. [P.J.C.]

topology The branch of mathematics concerned with properties of objects that are not distorted by stretching, etc. [M.J.C.]

tort Literally, a crooked act. A breach of civil law relating to rights. It has implications for environmental protection from remedies it affords under nuisance, negligence, trespass, RYLANDS VERSUS FLETCHER, and certain RIPARIAN RIGHTS. *See also* ENVIRONMENTAL LAW. [P.C.]

tort law *See* ENVIRONMENTAL LAW.

total environmental quality management (TEQM) One way of representing business activities is in terms of an input–output system that transforms raw materials and energy into products and services (Fig. T8). However, this is only a partial representation. An alternative is to consider a business as a way of transforming information from customers into actions that meet customer requirements (Fig. T9). The major contributions of this total quality management (TQM) approach have been to recognize the primacy of the customer (unless a business is responding positively to appropriate signals from customers, it will cease to survive), to recognize that customer information inputs impinge on all elements of management, and to suggest, therefore, that these inputs should be central to business affairs.

TQM also translates this relationship between external customer and business to internal affairs with internal customer–supplier philosophy between divisions, department, teams, etc. But, the information flows between external customer and business must still remain of prime importance, because it is on the extent to which responses to these are successful that bottom-line profitability will depend.

Environmental consideration can similarly be placed at the heart of business affairs, i.e. with information flowing from the sources of environmental pressure to all elements of business practice as indicated above (Fig. T10). This is because of the following.

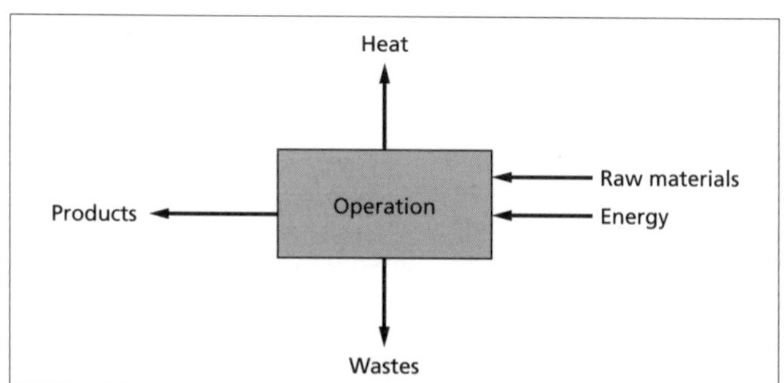

Fig. T8 Factory as input–output system.

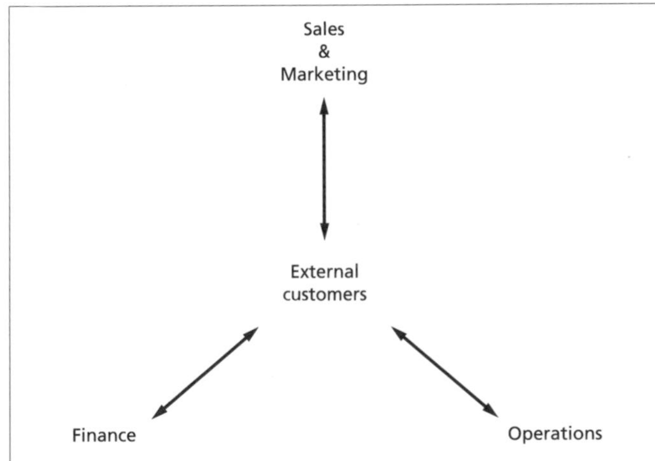

Fig. T9 Total quality management (TQM) philosophy with customers at the centre of things.

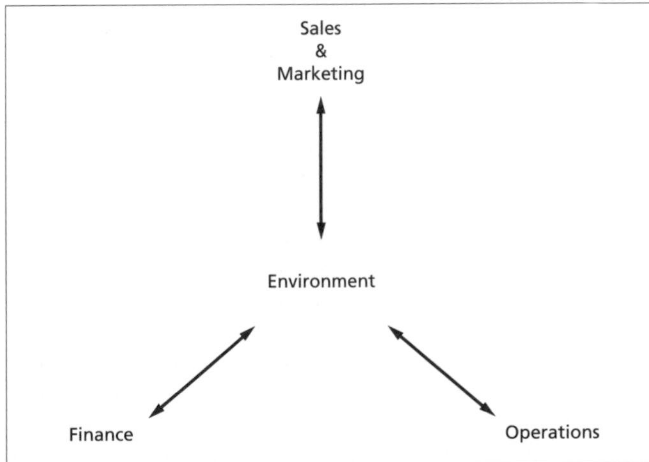

Fig. T10 Total environmental quality management (TEQM) philosophy with the environment at the centre of things.

1 Customers are taking environmental considerations more and more into account in exercising consumer choice. Moveover, REGULATORY AGENCIES are making increasing use of this to effect environmental protection by requiring more and more information on environmental performance to be made public—for example by labels, in inventories, registers, public audit statements, and so on. Therefore, environmental thinking will play an increasingly prominent role in the information flows from customer to business. Not only do traditional customers need to be taken into account, but the views of other stakeholders, such as investors and insurers, will also be of importance. Again, because of the pervasiveness of environmental issues in all managment sectors, these flows of information, from customer and stakeholders to company, should be taken into account by all elements of management.

2 As with all company legislation, environmental legislation has to be taken seriously, for failing to comply can bring bad publicity (negative information flows in **1**), penalties and possible threats of shut-down. However, the influence of environmental legislation is different from most other kinds of legislation, again because of its pervasiveness. Just as environmental issues impinge on all elements of business practice, so environmental legislation is also likely to impinge on all elements of business practice.

TQM puts the external customer at the centre of business practice and recognizes the primacy of information flows from customer to management schemes. TEQM similarly puts the environment at the centre of business practice and recognizes the primacy of information flows from sources of environmental pressure to all elements of business practice. *See also* ENVIRONMENTAL MANAGEMENT SYSTEMS. [P.C.]

toxic bloom One possible outcome of EUTROPHICATION is the appearance of toxic algae, i.e. those algae, most usually blue-greens, capable of excreting substances that are toxic to wildlife and/or domestic animals and/or humans. These algae can dominate algal communities and become visible as coloration—a BLOOM—within the water. This can occur in both inland and COASTAL waters. A well-known example in recent years has been the blooms of dinoflagellates in the coastal areas of Skagerrak and Kattegat, close to the NORTH SEA, that caused considerable harm to the AQUACULTURE industry, and may, by weakening their immune system, have lead to an increased prevalence of disease and death in the seal population in that area. *See also* ALGAL BLOOM. [P.C.]

toxic release inventory (TRI) This is required by the US Superfund Amendments and Reauthorisation Act (SARA). Companies report their stocks of dangerous chemicals, of specified types and exceeding certain specified quantities, to the environmental protection agency (EPA). The EPA is bound to keep all these data, the so-called toxic release inventory (TRI), easily accessible to the public and media as data-based information. It is claimed that the compilation of the TRI has in itself caused cleaner practices in industry. *See also* CHEMICAL RELEASE INVENTORY; LISTS. [P.C.]

Toxic Substances Control Act (TSCA) *See* UNITED STATES LEGISLATION.

toxic units Investigations on the TOXICITY/ECOTOXICITY of mixtures have usually been made with SHORT-TERM TESTS in which CONCENTRATION–RESPONSE curves are obtained from individual chemicals and then compared with the curve for the mixture. LC_{50}s are used to measure the definitive responses because they represent equilibrium conditions. Say the LC_{50} for chemical X is $10\ mg\ l^{-1}$ and for chemical Y is $2\ mg\ l^{-1}$ and that the LC_{50} for the mixture is obtained when the concentration of $X = 5\ mg\ l^{-1}$ and $Y = 1\ mg\ l^{-1}$, then the combined action is additive (*see* TOXICITY OF MIXTURES). If the LC_{50} for a given chemical and test system is given a value of 1 toxic unit (TU) as in the example, the mixtures contained 0.5 TU of X and 0.5 TU of Y, i.e. the combined toxicity was 1 TU. *See also* TOXICITY EQUIVALENCY FACTOR. [P.C.]

toxicant Artificially produced substances that have poisonous effects on humans or wildlife. They may be produced as intended poisons, i.e. BIOCIDES, or they may be produced for other reasons but have toxic/ecotoxic effects. *Cf.* TOXINS. *See also* PARACELSUS' PRINCIPLE. [P.C.]

toxicity The extent of adverse effects on living organisms caused by the substance under consideration. Restricted to effects on individuals, for example impaired survival as judged by LC_{50}/LD_{50}. Should be contrasted with ECOTOXICITY—concerned with ecological systems not just individuals—but often used as shorthand for it. [P.C.]

toxicity equivalency factor (TEF) Factor used to estimate the TOXICITY of a complex mixture of toxicants, most commonly for a mixture of chlorinated dibenzo-*p*-dioxins, furans and biphenyls. In the latter the equivalency factor is based on relative toxicity to 2,3,7,8-tetrachlorodibenzeno-*p*-dioxin (TEF = 1). *See also* ECOTOXICTY TESTS. [P.C.]

toxicity identification evaluation (TIE) Systematic and sequential ecotoxicological analysis of complex chemical mixtures arising as EFFLUENTS from industrial processes with a view to defining the TOXICITY (ECOTOXICITY) of the components. They involve a combination of sophisticated chemistry and ecotoxicity. They can be used in the context of reducing toxicity of effluents, i.e. in toxicity reduction evaluations (TREs). [P.C.]

REFERENCE

Doi, J. (1994) Complex mixtures. In: *Handbook of Ecotoxicology*, Vol. 2 (ed. P. Calow), pp. 289–310. Blackwell Scientific Publications, Oxford.

toxicity of mixtures Refers to the combined EFFECT of two or more chemicals on test systems. Chemicals might combine in a number of ways with respect to toxic effects: (i) additively, when the total effect is equal to the sum of known individual effects; (ii) synergistically when the combined effect is greater than the additive effects; and (iii) antagonistically, when the combined effect is less than the additive effects. [P.C.]

toxicology The science dealing with the effects of poisons (TOXICANTS) on individual organisms (often with humans in mind) (*cf.* ECOTOXICOLOGY). It covers both the routes of EXPOSURE and the mechanisms of the effects and provides a scientific basis for managing risks from toxicants. [P.C.]

toxins Natural poisons produced by organisms, for example plant poisons, snake and spider venoms, and antibiotics. They are usually produced as intended poisons. *Cf.* TOXICANT. [P.C.]

toxkits 'Off-the-shelf' TOXICITY/ECOTOXICITY tests, which usually involve organisms that are derived from resting cysts (e.g. *Artemia*, the brine shrimp, and rotifers) so that no elaborate culturing system is needed. The kits usually provide simplified means of exposing the organisms and recording effects. *See also* PREDICTIVE ECOTOXICITY TESTS. [P.C.]

REFERENCE

Persoone, G. & Janssen, C.R. (1993) Freshwater invertebrate toxicity tests. In: *Handbook of Ecotoxicology*, Vol. 1 (ed. P. Calow), pp. 51–65. Blackwell Scientific Publications, Oxford.

trace element Eight ELEMENTS—OXYGEN, SILICON, ALUMINIUM, IRON, CALCIUM, SODIUM, POTASSIUM and MAGNESIUM—make up nearly 99% of the Earth's CRUST (Benjamin & Honeyman 1992). The remaining elements of the PERIODIC TABLE constitute the trace elements. Occasionally, a trace element may be found in considerable quantities, for example in an ore deposit or mine spoil, but generally their concentrations are low, from $< 1\ mg\,kg^{-1}$ to a few hundred $mg\,kg^{-1}$. The main reservoirs of these elements are the LITHOSPHERE and the oceans. Mobilization occurs by WEATHERING of rocks and volcanic activity, and by anthropogenic activity such as combustion of FOSSIL FUELS and refining of ores. Several trace elements are essential for the growth of living organisms (*see* MICRONUTRIENTS). Others (commonly described as 'HEAVY METALS', e.g. CADMIUM, MERCURY, LEAD) have no known biological function but are potentially toxic, even at very low concentrations. One common way in which potentially toxic elements enter the food chain is their addition to soils in SEWAGE SLUDGE. Maximum permissible concentrations of them have been established for soils in Britain, depending on TOXICITY and BIOAVAILABILITY. These concentrations are $3\ mg\,kg^{-1}$ for cadmium, $1\ mg\,kg^{-1}$ for mercury and $300\ mg\,kg^{-1}$ for lead. *See also* NUTRIENTS. [K.A.S.]

REFERENCE

Benjamin, M.M. & Honeyman, B.D. (1992) Trace metals. In: *Global Biogeochemical Cycles* (eds. S.S. Butcher, R.J. Charleson, G.H. Orians & G.V. Wolfe), pp. 317–352. Academic Press, London.

trace gases A trace gas is one which is found in very low quantities ('trace quantities') in the ATMOSPHERE. There are hundreds of trace gases. They include: (i) the naturally occurring ones, which are present in unvarying concentration, such as NEON, HELIUM, KRYPTON and XENON; (ii) those that are the result of NATURAL processes such as DECOMPOSITION and which may vary over the Earth's surface and be affected by human activities, such as CARBON DIOXIDE (CO_2), METHANE (CH_4) and AMMONIA (NH_3); (iii) those which are given off by vegetation (VOLATILE ORGANIC COMPOUNDS, VOCs); and (iv) those which are entirely synthetic such as CFCS. Some trace gases are GREENHOUSE GASES and have increased in the atmosphere as a result of human activities (Fig. T11). [J.G.]

Trade-off

The exchange of one advantageous character for another. In the evolutionary process there is always NATURAL SELECTION of organisms that have higher growth rates or fecundities, or lower mortality rates. Trade-offs occur when improvement of one of these life-history characters can only be achieved at the expense of another/others. For instance, increasing fecundity may entail a mortality cost, so that there is a COST OF REPRODUCTION.

Many trade-offs have their basis in the laws of physics and chemistry (*see also* PHYSIOLOGICAL TRADE-OFF). Some arise because time, space or energy allocated to one function is not available to others. Thus RESOURCES allocated to defence, lowering mortality rate, may not be available for somatic growth. Defence may be achieved by physical structures such as spines or shells, or by vigilance which is achieved at the expense of feeding. All such defences absorb resources that could otherwise allow faster somatic growth.

The importance of trade-offs is that they represent the limits of what is genetically possible for individuals of a given species. Trade-

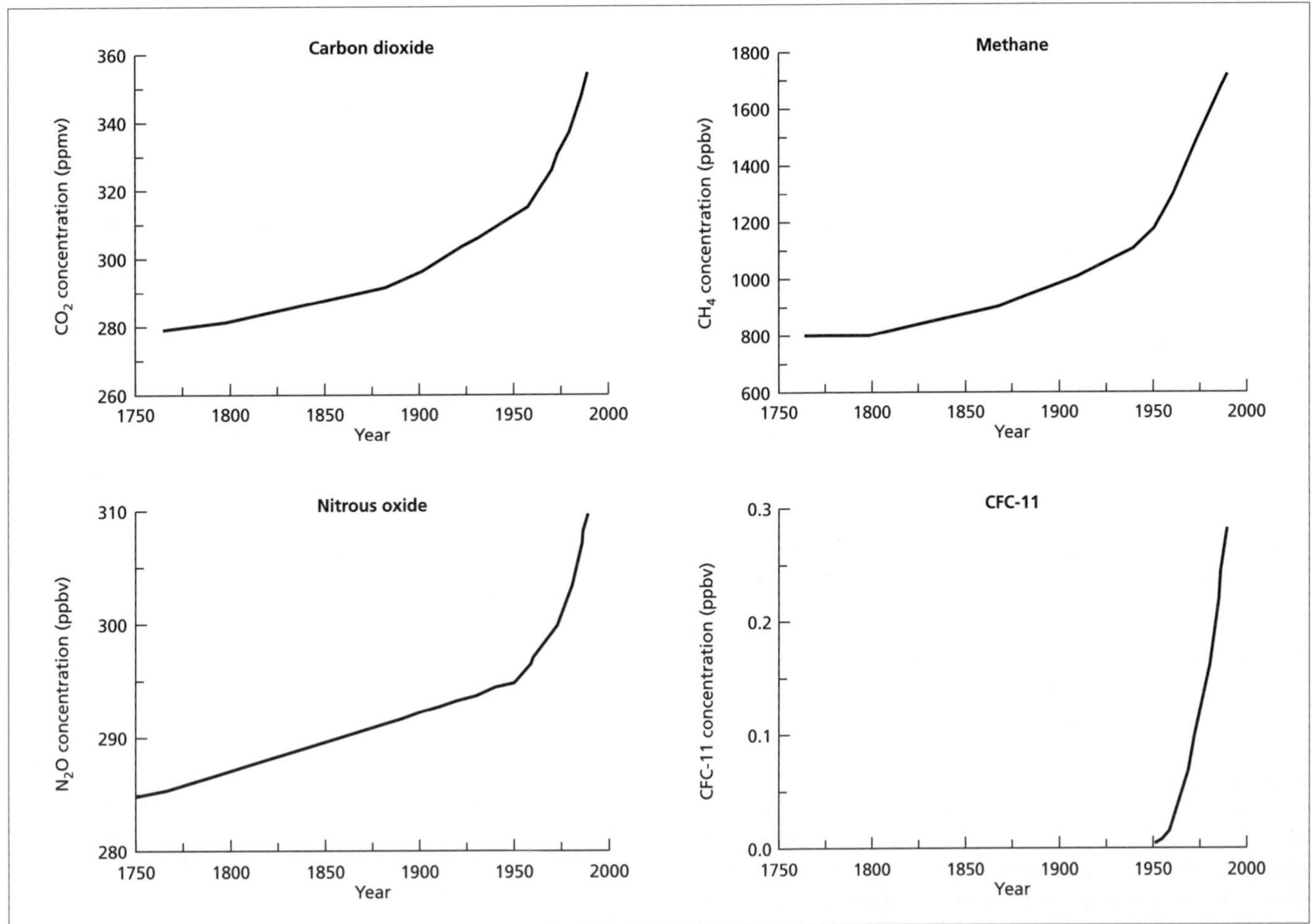

Fig. T11 Concentrations of carbon dioxide and methane, after remaining relatively constant up to the 18th century, have risen sharply since then due to human activities. Concentrations of nitrous oxide have increased since the mid-18th century, especially in the last few decades. CFCs were not present in the atmosphere before the 1930s. (After IPCC (1990) *Climate Change: The IPCC Scientific Assessment*. Cambridge University Press, Cambridge.)

offs may thus have an important role in the evolutionary process. Figure T12 shows what happens in a constant environment. During the evolutionary process some alleles are selected and others are eliminated. The alleles that remain at the end of the evolutionary process are by definition those with the highest Darwinian FITNESS. These alleles occur in a thin band on or beside the trade-off curve, around the optimal STRATEGY (Fig. T12b). These alleles affect more than one life-history character and so are pleiotropic.

Trade-offs may occur between any life-history characters. The number and identity of life-history characters varies between species, but many life histories can be satisfactorily described by development period, fecundity, juvenile and adult mortality rates, and the interval between breeding attempts. Between these five life-history characters there are 10 possible pairwise trade-offs, as follows. Increasing fecundity may have a mortality cost, or may mean a delay before breeding can occur again. Growth rate and mortality rate may be traded off because resources allocated to defence (lowering mortality rate) are not available for production, so development period and interbreeding interval increase and fecundity is reduced. If more offspring are produced, then the survivorship of each may be lower if offspring rely on parents for food and this is in short supply. Adult size and egg size often affect life-history characters, and give rise to trade-offs. For instance, larger adults are often more fecund, but it takes longer to grow to be bigger, and so fecundity is traded off against development period. The same trade-off can occur through variation in egg size. Thus if eggs are smaller, more can be produced, but each has further to grow to reach adult size, so development period is increased. Another possible consequence of smaller eggs is an increased juvenile mortality rate.

Trade-offs may also occur between related individuals because copies of the same gene may occur in both individuals. An obvious example is PARENTAL CARE. Parents may take risks (increasing adult mortality rate) to protect their offspring (reducing juvenile mortality rate). Alternatively, parents may feed juveniles (speeding development) instead of feeding themselves, so that next breeding may

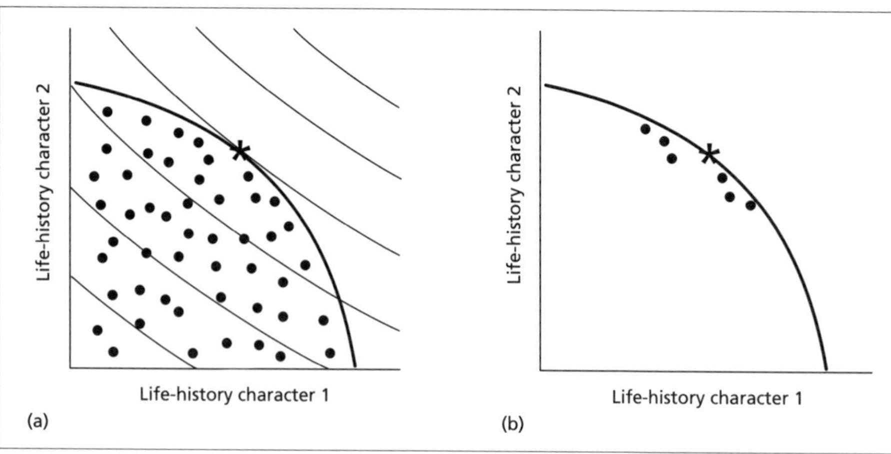

Fig. T12 Trade-off. (a) The operation of natural selection in a constant environment. Environmental effects are assumed small, so that phenotypic characteristics represent genetic effects. Dots represent genetic options. The thick curve represents the trade-off curve. Light lines represent fitness contours, fitness increases towards the top right. (b) Shows which alleles are left after some generations of selection. The asterisk represents the optimal strategy in the study environment.

be delayed, and fecundity may then be reduced. When trade-offs occur between individuals in this way, the calculation of fitness becomes complicated, because account has to be taken of the chance that the gene is present in the helped individual. For instance, genes affecting parental care may or may not occur in the offspring. This uncertainty affects whether or not the genes increase in the population, according to Hamilton's rule. The evolution of altruism between related individuals may also be analysed in this way.

An interesting group of trade-offs affect sex determination. Sometimes characters are advantageous in one sex but disadvantageous in the other. In species in which offspring sex is parentally controlled, daughters can, in principle, be traded for sons. In hermaphrodites, more sperm may mean fewer eggs.

Delineating the shapes of trade-off curves in particular environments has not proved easy. Five methods are used.

1 Between-individual phenotypic correlations might reflect genetic correlations and so provide evidence of trade-offs, as in Fig. T12b. However, individual life histories are greatly affected by the environments individuals grew up in, and the effects of developmental environment generally swamp genetic effects. Phenotypic correlations are therefore considered to give unreliable evidence about trade-offs.

2 Genetic correlations can be measured using the methods of QUANTITATIVE GENETICS, but these generally involve large-scale breeding experiments requiring that the organisms be individually housed. Not only are the experiments difficult to perform, but they can also be difficult to interpret, since the study environment generally differs from the environment in which the animals evolved. The remaining three methods are more useful.

3 Transplant experiments consist of transplanting individuals from a number of habitats to a common environment in which their life histories are measured. Provided that no maternal or other non-genetic effects are carried over from the previous environment, this gives direct evidence about the genetic options set.

4 Experimental manipulation changes individuals' life histories directly, for instance by altering their breeding opportunities, or implanting hormones. Evidence is needed, however, that changes effected by the experimenter could also be produced genetically.

5 Selection experiments alter life histories genetically over several generations by selecting for change in one life-history character. These last two methods both manipulate one life-history character and observe the correlated responses of others. If other life-history characters change, this is evidence that they were involved in a trade-off with the manipulated character.

In the final analysis, differences between species are largely due to differences in the trade-offs that constrain them, together with differences in the SELECTION PRESSURES that act upon the trade-offs. *See also* ANTAGONISTIC PLEIOTROPY HYPOTHESIS; EVOLUTIONARY OPTIMIZATION; LIFE-HISTORY EVOLUTION; OPTIMALITY MODELS. [R.M.S.]

REFERENCES

Partridge, L. & Sibly, R.M. (1991) Constraints in the evolution of life histories. *Philosophical Transactions of the Royal Society of London, Series B* **332**, 3–13.

Sibly, R.M. & Antonovics, J. (1992) Life-history evolution. In: *Genes in Ecology* (eds. R.J. Berry, T.J. Crawford & G.M. Hewitt), pp. 87–122. Blackwell Scientific Publications, Oxford.

---

trade winds The WIND system, occupying most of the tropics, which blows from the SUBTROPICAL highs toward the equatorial lows. Trade winds are from the north-east in the Northern hemisphere and from the south-east in the Southern hemisphere. Trade winds produce a strong equatorial drift of westward-moving water. [V.F.]

tragedy of the commons This is the title of a classic, though much disputed, paper by Garret Hardin (1968). The argument is as follows:

> There is an area of common land on which herders graze cattle. Each herder will seek to graze as many cattle as possible and eventually the carrying capacity is reached. This should result in stability. However, since the positive consequences of adding an additional head go to the herder owning the animal, whereas negative consequences are felt by all herders, the result is a net gain to the individual who increases herd size. This 'inherent logic of the commons remorselessly generates tragedy'

Taylor (1976) developed this metaphor as an example of the Prisoner's dilemma. Consider two users of a water supply who also pollute that supply. The formal analysis shows that it is in the interests of each user to pollute, because the returns from so doing exceed those from not so doing, whatever the other does.

If one accepts this argument, use of commons, whether a local river or the global atmosphere, has to be subject to some kind of control which means that they cannot be treated as freely available commons. Some would argue that this is the ultimate tragedy of the common. *See also* GAME THEORY; SUSTAINABLE DEVELOPMENT. [P.C.]

REFERENCES

Hardin, G. (1968) The tragedy of the commons: The population problem has no technical solution; it requires a fundamental extension in morality. *Science* **162**, 1243–1248.

Taylor, M. (1976) *Anarchy and Cooperation*. Cambridge University Press, Cambridge.

trait Character, usually presumed to be under genetic control and hence open to SELECTION. It can be 'simple' (e.g. eye colour), part of a complex (e.g. life-history feature) or a complex (e.g. semelparous life history). *Cf.* STRATEGY; TACTICS. [P.C.]

trait group A group of features or characters, possessed by an individual, a species or other taxonomic grouping. TRAIT is a term used for a particular PHENOTYPE in human genetics (e.g. sickle-cell trait—the phenotype of the haemoglobin-A/haemoglobin-S heterozygote). [L.M.C.]

trajectory The path in a phase plane graph (*see* PHASE PLANE ANALYSIS) traced by a series of VECTORS (MATHEMATICAL) over time. The trajectory shows the joint densities of the two species over time and the speed of change. [B.S.]

tramp species Animal (usually bird) species found on most tropical and subtropical oceanic islands. These animals often live in habitats created directly or indirectly by human settlement. *See also* ASSEMBLY RULES; DISTRIBUTION. [M.J.C.]

transboundary effects Environmental effects across state boundaries, for example due to active transport (road, rail or air) of substances, or passive transport in the atmosphere, rivers or sea. *See also* ACID RAIN AND MIST; EXPORT OF POLLUTION; PRIOR INFORMED CONSENT. [P.C.]

transect A line along which the DISTRIBUTION of the VEGETATION or fauna of an area is investigated, all plants or animals touching an extended tape being recorded. In a belt transect, a strip, usually 1 m or 0.5 m wide, is sampled between two parallel lines, essentially as contiguous quadrats. Transects are frequently used in zoned vegetation, being positioned across the zone and aiming to describe the maximum variation of vegetation over the shortest distance in the minimum time. Transects may cross ECOTONES, or distinct environmental gradients of, for example, salinity, soil characteristics or TOPOGRAPHY. Belt transect histograms can relate the distribution of species to HABITAT factors. *See also* SAMPLING METHODOLOGY/DEVICES. [A.J.W.]

transfer efficiency *See* ECOLOGICAL EFFICIENCY; LINDEMAN EFFICIENCY.

transformation of data There are three reasons why the original observations recorded by an ecologist need to be transformed or converted into more suitable units. These are:

1 for visual convenience in a graph;

2 to convert a curved relationship into a more convenient straight line; and

3 because the original data is unsuitable for PARAMETRIC STATISTICS.

Population numbers often fluctuate between tens of individuals and tens of thousands of individuals. Plotting such variation on a graph is difficult because a linear scale that allows you to plot 10 000 cannot distinguish between sizes of 20 and 30. The answer is to plot the logarithm of numbers or to plot the original numbers on a logarithmic scale. An example is shown in Fig. T13.

Exponential population growth can be described by the equation $N_t = N_0 e^{rt}$, where N_0 and N_t are the numbers at time 0 and time t respectively, e is the base of natural logarithms (2.71828), t is time and r is the INTRINSIC RATE OF INCREASE. When numbers (N_t) are plotted against time (t) you obtain the familiar exponential curve. However, if we take the natural logarithm of each side of the above equation we obtain $\ln N_t = \ln N_0 + rt$, and if we plot the natural logarithm of numbers ($\ln N_t$) against time (t) we obtain a straight line, with the slope equal to r. Deciding if that data fits a straight line (and therefore in this case shows EXPONENTIAL GROWTH) is easier than deciding if it fits a curve (Fig. T14).

Parametric tests require QUANTITATIVE INFORMATION/

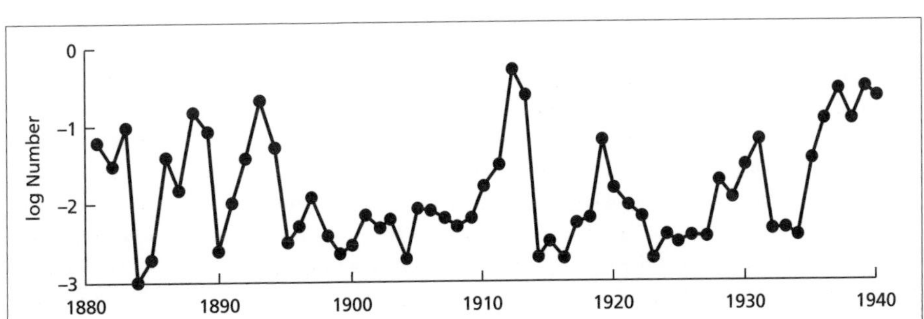

Fig. T13 Numbers of the pine beauty moth (*Panolis flamea*) recorded from a German forest (numbers = larvae/pupae per m² of forest floor).

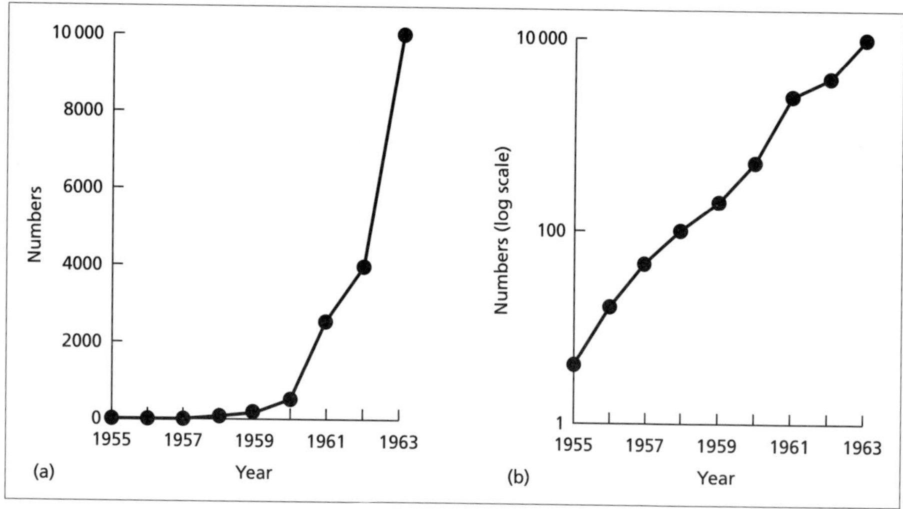

Fig. T14 The growth of the British population of the collared turtle dove (*Streptopelia decaocto*) after its invasion in 1955: (a) untransformed data; (b) numbers on a log scale (= log transformed).

observations to follow a NORMAL DISTRIBUTION and to have similar VARIABILITY (homogeneous variances). In other words, observations must have a DISTRIBUTION that is not too skewed and the VARIANCE must not change with the MEAN. Unfortunately, ecological observations frequently have an asymmetrical distribution and the variance often increases with larger means. In some cases we can correct both these problems simultaneously by transforming the original data to other units and carrying out the parametric tests on these transformed observations. There are three common transformations.

1 Logarithmic: each observation (x) is transformed to logarithms (usually base 10). Therefore the transformed data = $\log(x)$. Data skewed to the right is made more symmetrical and the variance is less dependent upon the mean. If there are zero observations then replace x with $\log(x+1)$.

2 Square root: each observation is transformed to its square root. Therefore the transformed data = $\sqrt{(x)}$. This is frequently used when the observations are whole number counts, for example the number of insects on a leaf. This reduces the CORRELATION between the variance and the mean. If there are zero counts, then replace x with $\sqrt{(x+0.5)}$.

3 Arcsine or angular: used when observations are proportions. In this case the data cannot be normally distributed because it is confined by 0 and 1. The transformed data = $\sin^{-1}[\sqrt{(x)}]$.

See also PROBIT ANALYSIS. [B.S.]

transformation series Transformation series are collections of homologous characters: two homologous characters produce a binary transformation series while three or more homologous characters create a multicharacter or multistate transformation series. Phylogeneticists recognize four categories of transformation series.

1 Ordered transformation series: transformation series in which there is a hypothesis of the particular pathway a character travelled during its evolutionary modification(s); without further information we cannot tell which direction the character moved along the pathway.

2 Polarized transformation series: transformation series in which the relative apomorphic and plesiomorphic status of characters has been determined so that we have a hypothesis of which character state represents the ancestral condition and which represents the derived condition.

3 Unordered transformation series: transformation series in which there are several possible routes open to explain the character changes. Information about the evolutionary direction of character change is provided by polarization.

4 Unpolarized transformation series: transformation series in which the direction of character evolution remains unspecified.

[D.R.B.]

Transgenic organisms

Transgenic organisms are organisms which have been modified by the insertion of 'foreign' nucleic acid to allow the expression of a new TRAIT or PHENOTYPE which would not normally be observed for that organism. A more formal definition, paraphrased below, is given in *The Regulation and Control of the Deliberate Release of Genetically Modified Organisms*—DoE/ACRE Guidance Note 1.

> Transgenic organisms are those in which a new combination of genetic material has been introduced so that the combination could not occur naturally and within which the introduced material will be heritable. This may include integration of the genetic material within the genome of the host organism. The usual method would involve the insertion by any method into a virus, bacterial plasmid or other vector system of a nucleic acid molecule which has been produced 'artificially' outside that vector. Transgenic organisms may also be produced using the techniques of micro-injection, macro-injection, micro-encapsulation or other methods which insert the genetic material directly, or even the fusion or hybridisation of two or more cells to form cells which have new combinations of heritable material. The definition includes the possibility of self-cloning, where nucleic acid is removed from a cell or organism and then

re-inserted into the same cell type or into phylogenetically closely related species.

In most countries the use of such genetically modified organisms is controlled by some form of legal framework. The framework tends to differentiate between the contained (or confined) use of such organisms and their release into the environment. In many countries the confined use of modified organisms is governed by GUIDELINES rather than legislation, whilst in most countries their release is controlled by some form of regulatory regime.

Contained use of modified organisms includes their use in the laboratory and, at a much larger scale, in industry. The regulatory system or guidelines generally assume that the organism carries at least the same HAZARD potential as the parent organisms. Almost all laboratory work with modified organisms is performed using 'good microbiological practice' where the risk of HARM to those using the organism is assumed minimal. This is also true in the industrial environment, where modified organisms are generally used under conditions of 'good industrial large-scale practice' (GILSP) defined by a working group of the OECD. Although organisms may be designed with biological containment, in that in the event of incidental or accidental escape into the environment the probability of survival is minimal, contained use does presume that there are physical barriers limiting access of the organism to the general environment. It follows that any risk associated with the use of such organisms is largely confined to those working with the organisms. The hazard associated with the industrial use of transgenic organisms is exactly the same as that for the use of the same organisms in the laboratory. The risk may be greater, largely due to the scale and hence the possible escape volumes, concentration and duration of EXPOSURE, as well as the possibility that the user is less qualified than those normally found in the laboratory environment. The range of organisms used by industry will have been chosen, in so far as it is possible, with a history of safe use of the host organism as a prime condition so that work may proceed within the confines of GILSP. Pharmaceuticals and therapeutic PROTEINS for human and animal medicine in particular are produced using microorganisms, for example growth hormone, factor IX, insulin and interferons. In most developed countries, many biotechnological products and services have already been placed on the market and are widely used, especially those in the pharmaceutical sector. Currently, more than 1700 clinical trials and more than 1000 field tests are in progress.

The vast majority of work involving transgenic organisms remains that done within containment.

Genetically modified microorganisms

All organisms may be modified to form genetically modified organisms. In the case of microorganisms there are many methods available allowing transformation. The DNA may be a fragment of nucleic acid or a viral genome. In many cases transformation can occur under natural conditions, but in some bacteria and eukaryotic microorganisms it can occur only in cells 'permeabilized' by artificial methods. For rhizobia, for example, protoplasts (cell wall free protoplasmic units) are often made by incubation with lytic enzymes in the presence of osmotic stabilizers. In the presence of calcium chloride ($CaCl_2$) or polyethylene glycol it is then possible to introduce plasmids (into which 'new' genes have been inserted).

In general, natural transformation can only occur at particular periods during the growth of the organism. The fragment of DNA would normally be double stranded and larger than a minimum size (the size depends on the host organism). A short description of transformation of microorganisms is given in the *Dictionary of Microbiology and Molecular Biology* (Singleton & Sainsbury 1987).

Genetically modified plants

The insertion of DNA into plants is generally more difficult. One of the principal hurdles that have prevented the routine modification of CROP plants is the difficulty of introducing foreign DNA into a plant cell. The process of DNA introduction into plants was first achieved in tobacco and has been relatively easy to achieve in other solanaceous species (potato, petunia, various *Nicotiana* species). In certain other crop species, especially the cereals, it has been more difficult. In addition to the protoplast method described for rhizobia which could be adapted for plant cells, a plasmid from agrobacteria has been successfully used in transforming dicotyledonous species.

There are two principal species: *Agrobacterium tumefaciens* and *A. rhizogenes* which, in their wild-type form, are pathogens causing crown gall disease and hairy root disease, respectively. Many dicotyledonous species are susceptible to infection by *Agrobacterium* brought about by the incorporation of genes from an independently replicating plasmid within the *Agrobacterium* cell, which then become incorporated into the host plant. The introduced DNA modifies the phytohormonal levels within the cells and either causes a disorganized proliferation of cells and the formation of a gall, or the production of a mass of roots covered with root-hairs (*A. rhizogenes*). The disease-causing genes are carried between specialized T-DNA (transforming DNA) border sequences on the independently replicating circular plasmid DNA molecule. By recombinant DNA methods, it has been possible to remove the disease-causing genes, so the *Agrobacterium* organism is no longer pathogenic. New vector plasmids have also been constructed which enable foreign genes to be inserted between the T-DNA borders. *Agrobacterium* cells carrying the gene(s) of interest are then incubated with cultured cells of the recipient crop plant and transgenic plants regenerated from them. Only a small proportion of the treated plant cells eventually become transformed, so it is usually necessary to incorporate selectable marker genes (usually conferring RESISTANCE to a particular antibiotic) between the T-DNA border sequences. To select the transgenic plants, the corresponding antibiotic is incorporated into a plant regeneration medium on which only transgenic plants are able to grow normally.

The *Agrobacterium* and protoplast methods have often proved inadequate for the transformation of recalcitrant species. Cereals are not normally hosts to *Agrobacterium* and routine and reliable regeneration of plants from cereal protoplasts is difficult and often GENOTYPE dependant. The technique of particle bombardment was developed in an attempt to overcome some of these problems and involves coating metal particles (usually tungsten or gold particles, 1 mm in diameter) with DNA and shooting them into plant cells capable of subsequent plant regeneration. Particles can be propelled by various means, including 0.22 inch blank cartridges, compressed gases or by the instantaneous evaporation of a water droplet caused by electrical discharge. The small metal particles, with their DNA covering, enter the plant cells and become lodged there. In a tiny proportion of the recipient cells, the DNA becomes incorporated into the genome and transgenic plants can be regenerated from them.

All of these methods result in transformation of a small proportion of the host cells. It is usual to include some form of marker, usually an antibiotic resistance gene such as the gene for kanamycin resistance, so as to allow selection of the modified cells. The use of antibiotic resistance markers has been a cause of concern in many countries.

The number of copies of transgenes that become incorporated into the nuclear genome varies between independently transformed plants, and with the method of transformation used. The integration of one T-DNA copy with the *Agrobacterium* method is common, but higher numbers are observed. Multiple T-DNA inserts do occur and occasionally the presence of 20–50 copies is reported. The position of insertion of T-DNA within the nuclear genome is believed to be random.

There is no evidence, so far, that there are preferred sites of insertion. Most genes are incorporated into the nuclear chromosomes, but there are reports of transformation of the plastid component of the cytoplasmic genome. There is now a wide range of plant promoters for giving transgene expression in specific tissues in the plant. It may be desirable to restrict the expression of a particular protein to those parts of the plant where it is required. For example, in the case of a pest that attacks leaves it should be sufficient to express the pesticide capability solely within the leaves.

Genetically modified animals

The manufacture of genetically modified animals is considerably more difficult. The GENERATION TIME and the NUMBER OF OFFSPRING preclude the use of 'hit-and-miss' selection procedures, and a very high yield for transformation is required. In general, transgenic animals are formed using 'microinjection' where DNA is directly injected into the pronucleus of the fertilized egg. The DNA is often deposited into the cytoplasm. The site of integration is random and the number of copies of DNA that integrate may be unpredictable. Embryonic stem cells may also be transformed more readily. These are derived from the early embryo, can be grown in culture and after transformation using methods similar to those for microorganisms returned to a recipient embryo. During the culture phase, the expression and stability of the transgene can readily be tested. Retroviral vectors have also been used for the transformation of animal cells. The RNA viruses form a DNA copy which is integrated into the host genome. These vectors, modified by the removal of genes for viral proteins and retention only of those genes required for transformation and integration, may be used in the presence of helper cell lines which provide the required proteins. There remain some concerns as to the safety of these vectors and they have not been used in 'animal' experiments, but have been used to produce transgenic poultry. A technique which would revolutionize transformation, if it can be made to work reliably, is 'sperm-mediated DNA transformation' which consists of mixing the DNA containing the required gene with sperm followed by *in vitro* fertilization. It has low reliability and there are concerns as to the stability of the transgenes.

Uses of genetically modified organisms

Transgenic organisms may be designed as 'bioreactors' for the production of high-value chemicals and therapeutic drugs. If cultured cells or microorganisms are used, the product is generally purified before 'release into the environment' and a viable live modified organism would then not be released. One of the disadvantages of this approach is that the required gene-product remains within the cells and methods have to be devised for purification. An advantage is that the 'purified' product is unlikely to carry with it other mammalian proteins which could be the cause of undesirable effects.

The potential uses of modified organisms are vast. Microorganisms may, for example, be modified for deliberate release in order to be used for BIOREMEDIATION and MINERALIZATION of pollutants, or to increase the NITROGEN uptake of soil bacteria. A recombinant rabies vaccine has been used to attempt to deal with the problem of rabies in wild animals.

The vast majority of experimental releases that have been permitted throughout the world involve modification of crop plants to introduce a variety of traits. Tolerance to a variety of herbicides has been engineered into a growing range of crop plants and some of these have recently received, or are about to receive, marketing clearance within the western world. The insertion of herbicide TOLERANCE supplies a means of providing selectivity for herbicides that are quickly degraded in the environment. Pest or disease resistance has been engineered into many plants; in particular, the insertion of genes for the manufacture of *Bacillus thuringiensis* (BT) toxin crystals by a plant has been shown to be at least as effective as the use of the bacteria or the toxin as a pesticide. Viral coat protein genes are frequently used to give protection against particular viruses. Resistance to STRESS conditions such as DROUGHT, saline soils, HEAVY METALS, cold and high temperature are also contemplated, although these phenomena are more complex. Modification of the shelf life of particular crops is an important area of research, as is modification of the colour and shape characteristics of many flowers for the horticultural industry. The first transgenic organisms given approval for marketing within the USA and Europe include tomatoes modified for improved shelf life, tobacco and soya bean containing herbicide tolerance, and maize containing both genes for the formation of a BT toxin and a herbicide tolerance. The pressure to decrease dependency on chemical pesticides is expected to drive the growth of biopesticide production and use. The United Nations has issued a statement indicating some of the research aims for the release of transgenic plants including:

1 increase the YIELD of major crops, livestock and AQUACULTURE species;

2 improve the nutritional value of source crops, animals and microorganisms and reduce food losses;

3 increase the use of integrated pest, disease and crop management techniques to eliminate overdependence on AGROCHEMICALS, thereby encouraging environmentally sustainable agricultural practices;

4 evaluate the agricultural potential of marginal lands in comparison with other potential uses and develop, where appropriate, systems allowing for sustainable PRODUCTIVITY increases, including situations of emerging environmental stresses;

5 expand the applications of biotechnology in FORESTRY and increase the efficiency of NITROGEN FIXATION and mineral absorption by the symbiosis of higher plants with microorganisms. [J.K.]

REFERENCES

The Regulation and Control of the Deliberate Release of Genetically Modified Organisms—DoE/ACRE Guidance Note 1 (1993). A guide to Part VI of the Environmental Protection Act 1990 and the Genetically Modified Organism (Deliberate Release) Regulations 1992. UK Department of the Environment, London.

Dale, P.J. & Kinderlerer, J. (1995) Safety in the contained use and environmental release of transgenic crop plants. In: *Genetically Modified Organisms—A Guide to Biosafety* (ed. G. Tzotzos), pp. 36–60. CAB International, Oxford.

Organization for Economic Cooperation and Development (OECD) Group of National Experts on Safety in Biotechnology (1986). [ISBN 92-64-12857-3]

Powell, D. (1995) Safety in the contained use and release of transgenic animals and recombinant proteins. In: *Genetically Modified Organisms — A Guide to Biosafety* (ed. G. Tzotzos), pp. 110–142. CAB International, Oxford.

Singleton, P. & Sainsbury, D. (eds) (1987) *Dictionary of Microbiology and Molecular Biology*, 2nd edn. John Wiley & Sons, Chichester.

United Nations Organizations Statement following UNCED, the Rio Convention on Biological Diversity in support of Chapter 16: Environmentally Sound Management of Biotechnology of Agenda 21.

---

transition horizons *See* SOIL PROFILE.

transition matrix model of succession (Markov model) System of modelling a successional process based on identifying distinct states (e.g. vegetation types) in the COMMUNITY and estimating the probabilities of transition from any one state to another in a given time period. If there are n states, the current state of the community can be represented as a vector giving the proportion of the system in each of the n states, and there will be an $n \times n$ matrix giving probabilities of the transitions between each pair of states. Multiplication of the vector and the matrix yields another vector containing the predicted composition of the community after one time interval. Repetition of the process with each new vector extends the prediction and the system will eventually converge on a stable composition independent of the initial starting composition (a result that can be derived analytically from the transition matrix). This follows from the principal assumption of simple transition matrix models that transition probabilities remain constant through time. Such models have been applied to many different successional systems. The best known are probably models of species composition in woodlands, where states are canopy-occupying tree species and the transition matrix contains the probabilities of replacement of an individual tree of one species by an individual of any other species in the system in a given time interval. Transition matrix models do not require specification of the processes causing the transitions between states, merely that the probabilities of the transitions can be estimated. *See also* MATRIX MODELS; SUCCESSION. [P.H.W.]

transition mire This term is used of RHEOTROPHIC mires in which the flow of water is slow and the mineral nutrient content of the water is poor, leading to the development of an acidophilous vegetation intermediate between poor FEN and OMBROTROPHIC bog. Often such conditions may lead on to the development of true bog vegetation in the course of further PEAT accumulation. [P.D.M.]

transitive competition This term is applied to multispecies interactions when a fixed or stable hierarchy of INTERSPECIFIC COMPETITION is present. It is used to classify the pattern of interaction present between all directly competing species. For example, species A may displace species B which may in turn displace species C, which may *also* be displaced by species A. The outcome is usually predictable and is often associated with less diverse ecosystems. Successional sequences in temperate regions typically exhibit such patterns on a temporal scale where individuals are displaced by progressively more dominant species. *Cf.* INTRANSITIVE COMPETITION. *See also* COMPETITION. [R.C.]

transmission rates of parasites This is the rate at which hosts become parasitized, it is also known as the incidence of infection and for MACROPARASITES it is the rate of growth of the PARASITE population. The transmission rate is maximum when the parasite is first introduced and is given by the basic reproduction number (*see* EPIDEMIC and OUTBREAKS). For a MICROPARASITE to remain endemic in a HOST population, the average number of hosts that each host infects (the effective reproduction number) must be unity; for a macroparasite, each female parasite produces one reproducing daughter during her lifetime. The transmission rate is reduced from the basic reproduction number to the effective reproduction number by density-dependent processes, especially the immune response within hosts, and the density of susceptible hosts. In EPIDEMIC MODELS, the term defining the transmission rate is non-linear, as it depends on both the numbers of susceptible hosts and the PREVALENCE OF INFECTION. *See also* PARASITISM; POPULATION DYNAMICS. [G.F.M.]

transmission threshold For MICROPARASITES the TRANSMISSION RATE OF PARASITES may be dependent on the DENSITY of susceptible individuals (i.e. those able to be infected), and the survival of the PARASITE in the HOST population is dependent on the rate of supply of susceptible hosts (*see* EPIDEMIC and OUTBREAKS). There is a critical density of susceptible hosts (given as the reciprocal of the basic reproduction number) below which either an epidemic cannot occur or the parasite will be eradicated from the population. The object of immunization and vaccination programmes is to keep the density of susceptible hosts below the critical density, and therefore the transmission rate below the threshold.

For MACROPARASITES the transmission threshold can be defined in terms of the probability of a parasite finding a host, and is thus related to the host density. However, the concept is less widely used in this context. *See also* PARASITISM. [G.F.M.]

transpiration Loss of WATER from the plant surface by EVAPORATION through the stomata. The rate depends on the number and size of the stomata and the gradient for diffusion of water vapour from the inside to the outside of the leaf. *See also* PENMAN–MONTEITH EQUATION. [J.G.]

transplant experiments The transfer of individuals, or samples of populations or communities, from one place to another. Usually done to investigate the importance of the physical/geographical location in characteristics being observed. [P.C.]

transplant experiments and life-history evolution *See* TRADE-OFF.

Transposable elements (transposons)

DNA sequences that are capable of changing their location in the genome. A surprising finding of early studies of eukaryotic genome organization is that not all DNA sequences are unique; some are highly repeated (1000–10 000 copies) or moderately repeated (10–100 copies). Furthermore, the locations of some of these repeated sequences can vary from strain to strain; they are therefore transposable. The moderately repeated fraction of the genome, composed of these transposable elements (TEs), can be quite high. In *Drosophila melanogaster*, for example, approximately 10% of the genome is composed of repeated sequences of about 50 different TE families. TEs are ubiquitous and are found in PROKARYOTES as well as EUKARYOTES. TEs typically encode a transposase protein that is required for transposition. In addition, some TEs also encode repressor PROTEINS that regulate their transposition. Transposition can be non-replicative (transposition to a new genomic location, or site, is accompanied by excision from the old site) or replicative (transposition results in two copies of the TE, at the original site and at the new one).

TEs fall into four structural groups:

1 TEs with long direct terminal repeats;
2 TEs with short inverted terminal repeats;
3 TEs with long inverted terminal repeats; and
4 TEs with no terminal repeats.

Within each structural group exist many TE families defined by sequence HOMOLOGY. The TEs with long direct terminal repeats are also called retrotransposons because of their homology to retroviruses. These include the *copia* and *copia*-like elements of *Drosophila*, *Ty* elements of yeast and endogenous retroviruses of mammals.

The first TEs to be studied genetically, by Barbara McClintock, were the *Activator/Dissociation* (*Ac/Ds*) and *Suppressor-mutator* (*Spm*) controlling elements of maize. These are short inverted repeat type families of elements, and are heterogeneous in structure, consisting of complete (full length) elements and of elements derived from the complete elements by deletion of variable numbers of base pairs internal to the inverted repeats (deletion derivatives). In *Drosophila*, the *P* (short inverted repeats), *I* (no terminal repeat) and *FB* (long inverted repeat) families are likewise heterogeneous in length. Only the complete elements are capable of autonomous transposition, but the deletion derivatives can transpose if transposase is provided by a full-length copy.

Insertion of TEs into novel genomic locations can be mutagenic if the TE disrupts or alters the function of a neighbouring gene. About one-half of the spontaneous MUTATIONS in *D. melanogaster* are caused by insertions of TEs. TEs are therefore examples of 'parasitic' or 'selfish' DNA. Currently, the mechanisms that maintain the characteristic mean copy numbers of TEs in populations in the face of active transposition that tends to increase genomic copy number, as judged by the highly variable sites of insertion between individuals within populations, are not completely understood. Possibilities include (i) self-regulation of copy number, where at equilibrium transposition is balanced by excision of elements, and (ii) transpositional increase in copy number is countered by selection against deleterious effects of insertion and/or selection against lethal duplication and deficiency events produced by ectopic recombination between copies of elements inserted at non-homologous chromosomal sites. Measured rates of transposition are approximately 10^{-4} per element per generation, and rates of excision are at least an order of magnitude less than this, and so (i) is not a likely general mechanism. Because the selection coefficients necessary to balance transposition are of the same order as the transposition rate, direct experimental investigation of the nature of the selection forces is very difficult (if not impossible). Therefore, analyses of the distribution of elements along and between chromosomes sampled from natural populations of *Drosophila* have been used to infer selective mechanisms acting on TEs. These analyses do not unequivocally support or reject the alternative hypotheses of selection against recessive deleterious effects of insertions or ectopic recombination, although more evidence favours the latter mechanism.

Occasionally, transposition of TEs can be elevated to very high rates, such as observed for unstable mutations and for the P-M and I-R systems of hybrid dysgenesis in *D. melanogaster*, associated with, respectively, high rates of transposition of *P* and *I* elements. In hybrid dysgenesis germ-line transposition rates of *P* or *I* elements are increased when males from strains containing complete copies of these elements are crossed to females from strains with no euchromatic copies. The consequences of genomic instability in these dysgenic crosses are high mutation rates, both for single genes and quantitative traits, high frequencies of chromosomal aberrations, male recombination (normally absent in *Drosophila*) and temperature-sensitive gonadal sterility. *P* and *I* elements have recently invaded the genome of *D. melanogaster*. The *P* element is absent in closely related species, but present in distantly related ones, implicating horizontal (cross-species) transfer of these elements by an unknown mechanism.

TEs are useful biological tools. *Drosophila P* elements have been developed into very powerful mutagenic agents, enabling rapid identification of new loci affecting any measurable PHENOTYPE and their molecular cloning using the element as a tag. The *P* elements and TEs in other species have also been developed as transformation vectors, enabling transfer of sequences from one strain (or even species) to another. Finally, TEs are useful genetic markers, and have been used to map QUANTITATIVE TRAIT LOCI in *Drosophila*. [T.F.C.M.]

REFERENCE

Berg, D.E. & Howe, M.M. (1989) *Mobile DNA*. American Society for Microbiology, Washington, D.C.

---

trapping food Some sit-and-wait predators (*see* FEEDING TYPES, CLASSIFICATION) use traps. Web-spinning spiders are an obvious example but there are others. Sit-and-wait planarians lay mucous patches which act as sticky traps and ant-lion larvae (Myrmeleontidae) dig pits and wait at the bottom, with their mandibles open, for prey to fall in.

The cost of building these traps is not insignificant. The metabolic rate of the linyphiid spider, *Lepthyphantes zimmermanni*, increases

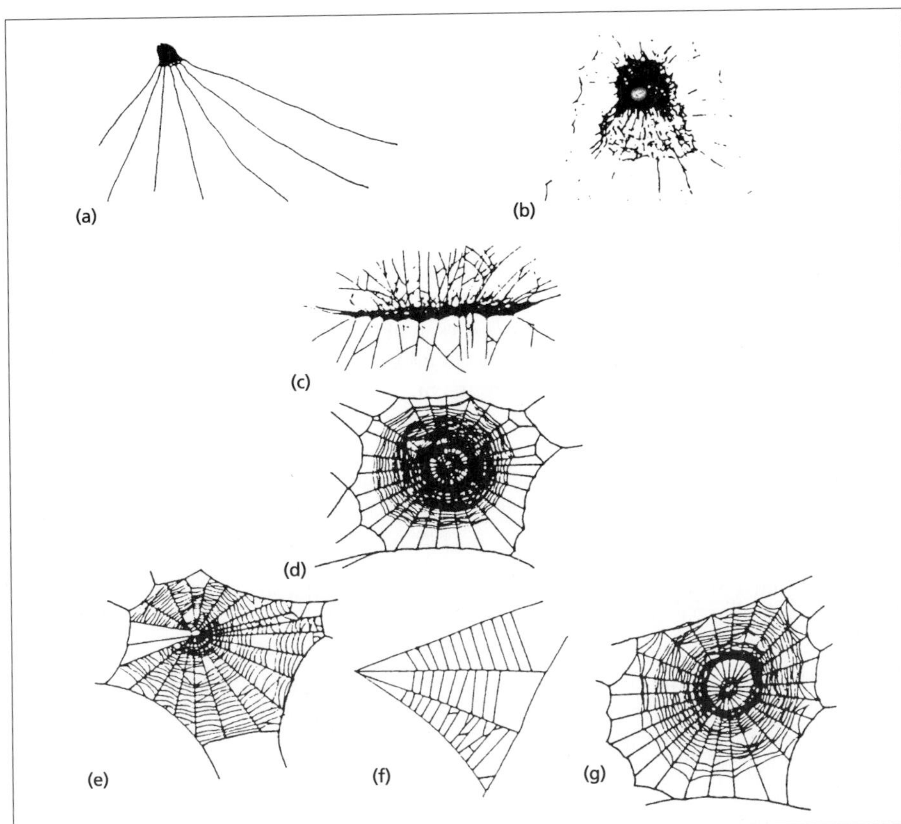

Fig. T15 Spiders' webs. (a) Trip-line; (b) cobweb—prey gets entangled in mesh and spider emerges from tunnel; (c) sheet-web—flying insects strike guy-lines and fall into sheet, spider moves in from beneath or side; (d) orb-web. Several orb-web builders economize by removing segments: (e) one to several omitted; (f) more omitted than produced; (g) central portion omitted. (After Calow, P. (1981) *Invertebrate Biology. A Functional Approach*. Croom Helm, London.)

between two and three times over the resting level when it is building a web and that of the ant-lion, *Morter obscurus*, increases by as much as eight times when it is building a pit. Energy lost as mucus may account for 20% of the energy taken in by a planarian and the PROTEINS lost in silk are likely to be a drain on spiders. Hence it is to be expected that NATURAL SELECTION will have balanced the returns from the traps against these costs and will have favoured economization.

As regards trapping food, most is known about the webs of spiders. In this group, web evolution seems to have been towards the production of more efficient traps with minimum materials (Fig. T15). Primitive spiders rely on 'trip-lines' which are economic on silk but are not very effective traps. The amorphous cobweb is a more effective trap but is much less economical on silk. In contrast, the evolution of two-dimensional nets in the form of the sheet-web of the linyphiids and the orb-webs has achieved both increased trapping efficiency and economy in production. Furthermore, a reorientation from the horizontal to the vertical plane improves the ability to catch flying and jumping prey. There has even been some economy in the manufacture of the orb-web since whole segments of the design are missed out by several species (Fig. T15e–g).

Some orb-web spiders build webs daily but usually eat old ones before starting new ones. Sheet-web builders repair rather than replace their webs and cobwebs may persist for months. Ant-lions often occupy the same pit for the whole of their larval life, enlarging it as they get bigger. Finally, planarians produce a mucus which is resistant to microbial attack and which will retain its stickiness for long periods of time. All of these strategies economize on trap replacement.

Since the costs associated with the construction of traps are significant, natural selection might be expected to favour cheats who use the traps of others rather than their own and this could, at least in principle, lead to the complete evolutionary loss of traps and hence to the downfall of the trappers! In other words, CHEATING is not an EVOLUTIONARILY STABLE STRATEGY (ESS). However, high levels of intra- and interspecific aggression militate against this in ant-lions and spiders and in the latter group there are surprisingly few species of kleptoparasites, i.e. species of spider which specialize in using the webs of others. The factor which ensures that cheating does not do away with traps in planarians is that these animals require a mucous trail for locomotion. The trail lubricates the substratum and provides a point of leverage for cilia which give the motive force for locomotion. Therefore, not even a cheating planarian could do without a mucous trail altogether! [P.C.]

REFERENCE

Calow, P. (1981) *Invertebrate Biology. A Functional Approach*. Croom Helm, London.

traps *See* SAMPLING METHODOLOGY/DEVICES.

tree line The limit of isolated bigger or smaller FOREST stands and single trees towards treeless vegetation in polar regions, on high mountains and in arid continental regions (see Fig. T4). *See also* TIMBER LINE. [Y.V.]

trench Deep and sinuous depressions in the OCEAN floor, usually seaward of a CONTINENTAL MARGIN or a volcanic island arc, where the oceanic crust is destroyed as it is subducted downward to the mantle. Trenches are narrow, relatively steep-sided, and lie at depths greater than 6000 m. Approximately 22 trenches have been identified with depths exceeding 7 km; six of these have depths greater than 10 km (Mariana, Tonga, Kuril-Kamchatka, Japan, Philippine and Kermadec). The deepest part of the ocean lies in the Nero Deep in the Mariana Trench at 11 035 m below sea level. [V.F.]

TRI *See* TOXIC RELEASE INVENTORY.

tributyltin (TBT) pollution Tributyltin is an organometallic substance used in ANTIFOULING paints. It can cause ecological effects at low to sub-nanogram quantities per litre, for example IMPOSEX in gastropods. It is therefore considered to be one of the most toxic kinds of compounds in the aquatic environment. Its use is now strictly controlled in most places. [P.C.]

REFERENCE

Waldock, M.J. (1994) Organometallic compounds in the aquatic environment. In: *Handbook of Ecotoxicology*, Vol. 2 (ed. P. Calow), pp. 106–129. Blackwell Scientific Publications, Oxford.

tritium Tritium (^{3}H, $t_{1/2} = 12.33$ years) is a cosmic-ray-produced radioactive ISOTOPE of HYDROGEN with a natural abundance of about one in 10^{18} hydrogen atoms. Tritium is also produced in nuclear explosions and its environmental inventory was increased by two orders of magnitude by nuclear weapons tests, with peak input in 1963–1964. Tritium is also released to the environment from nuclear reactors and nuclear fuel reprocessing plants. Uses of tritium include tracer studies of water movement in oceanography and HYDROLOGY, production of luminescent materials and labelling of organic compounds for tracer studies. *See also* COSMIC RAYS; RADIOACTIVE FALLOUT; RADIONUCLIDE. [A.B.M.]

troglobitic Cave-dwelling species. [P.C.]

Trophic cascade hypothesis

This hypothesis has developed largely from studies on LAKES and has become one of the most influential and controversial ideas in limnetic ecology. Nutrient input rate, water renewal rate and lake morphology are important PHYSICAL FACTORS that underpin the basic ECOSYSTEM processes in lakes. Strong correlations exist between nutrient loading or concentration and primary PRODUCTION. However, at any given level of nutrient loading, algal BIOMASS or PRODUCTIVITY may differ among lakes by an order of magnitude and this has been attributed to BIOTIC FACTORS under the trophic cascade hypothesis. The hypothesis states that nutrient input level sets the potential productivity of lakes and that deviations from the potential are due to food-web effects. Variations in the intensity of predation by piscivorous fish in lakes are transmitted sequentially (or cascade) from predators to prey through each step of the FOOD WEB such that abundance, composition and productivity of the basal TROPHIC LEVEL (the PHYTOPLANKTON) is constrained by activities of the top level of the food web (Fig. T16). The focus of the hypothesis is thus on the top trophic level rather than at the bottom, and on annual to multiannual time-frames rather than on shorter term seasonal time-scales.

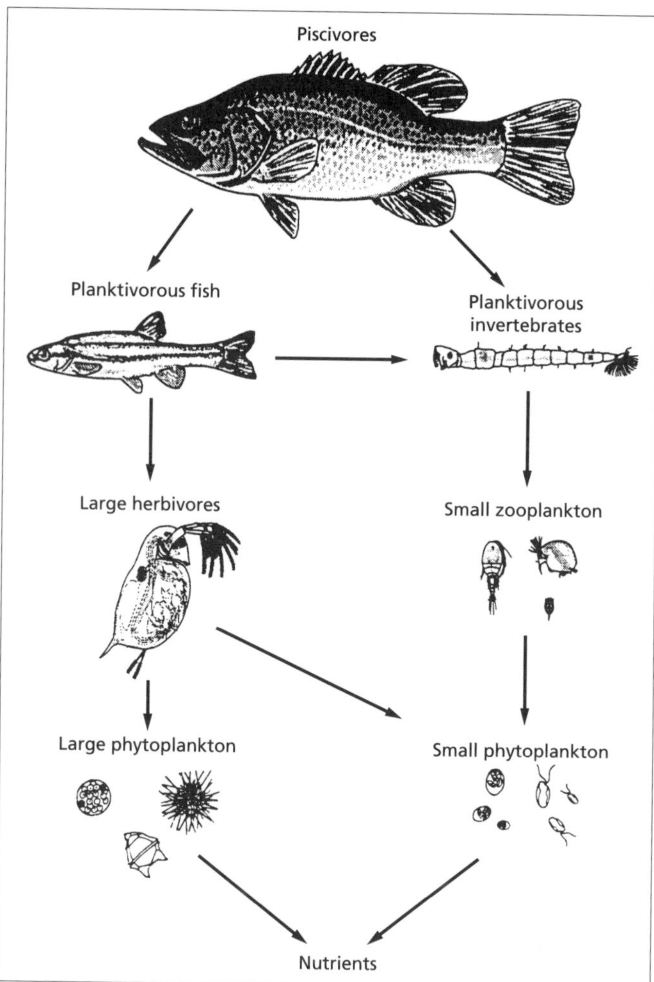

Fig. T16 The basic lake food web showing the major types of interactions of the trophic cascade. Nutrient availability (mainly nitrogen and inorganic phosphorous) is largely governed by inputs from the surrounding catchment and through recycling by animals. (Not to scale.) (After Carpenter, S.R. & Kitchell, J.F. (eds) (1993) *The Trophic Cascade in Lakes*. Cambridge University Press, Cambridge.)

The consumers in lakes are typically selective in the types and sizes of RESOURCES they consume (*see* SIZE-EFFICIENCY HYPOTHESIS; SIZE-SELECTIVE PREDATION). In LENTIC systems, piscivores, when present, can determine the size and composition of the planktivorous fish assemblage beneath them in the food web. In turn, large herbivores are preferentially consumed by planktivorous fish whilst invertebrate predators feed most heavily on smaller ZOOPLANKTON. Thus abundant planktivorous fish can shift zooplankton composition towards DOMINANCE by smaller individuals, but when these fish are absent, predation by planktivorous invertebrates and COMPETITION among herbivores shift the zooplankton community towards larger sized species. These larger individuals, especially large cladocerans like *Daphnia*, have a greater impact on phytoplankton because they consume a broader range of sizes and morphologies of algae, are more efficient grazers and recycle NUTRIENTS at a slower rate than smaller zooplankton HERBIVORE species.

Much of the available nutrients derive from RECYCLING through excretory products of the herbivores. Therefore, according to the trophic cascade hypothesis, algal biomass and primary production would be less in *Daphnia*-dominated lakes (with piscivores, thus lacking planktivorous fish) than in lakes dominated by smaller zooplankton such as *Bosmina*, small calanoid copepods and rotifers (that lack piscivores and possess large populations of planktivorous fish).

Evidence in support of the trophic cascade hypothesis in lakes is largely from systems dominated by a single piscivorous population with a single competitively dominant cohort. Population fluctuations of piscivores (through senescence and replacement of the cohort) can generate strong fluctuations in piscivores and this in turn can cause variations in effects through lower trophic levels over time-scales of perhaps several years (the so-called Wisconsin model of the hypothesis, named after the lake in which such fluctuations were first documented). To test the trophic cascade hypothesis, it is essential to get both processes and time-scales correct; studies over too short or too long time-scales can lose the strong top-down signals that generate the cascade in noise or larger scale (climatic) signals respectively. Thus, most clear-cut experimental tests of the hypothesis usually require strong MANIPULATION of the piscivore numbers and involve lakes that include *Daphnia* as the dominant herbivore. Communities lacking this consumer or its ecological equivalent appear to be less likely to carry a cascade signal from piscivore to phytoplankton.

Although the exact details may vary, the central idea concerning the effect of top predators on communities through trophic cascades has also been documented in other aquatic habitats in addition to ponds and lakes. These include RIVERS and streams (with examples based on predatory bass, grazing minnows or crayfish and algal biomass) and intertidal/subtidal marine communities (with examples based on sea otter–sea urchin–kelp and starfish–sessile invertebrates–benthic algae food webs). The concept has also been successfully applied as a form of BIOMANIPULATION in lakes and reservoirs in order to control nuisance algae.

However, manipulation of top predators in aquatic systems has not always lent support to the trophic cascade hypothesis, and the failure of most terrestrial systems to follow a trophic cascade model is striking. Variation in life-history characteristics and foraging ecologies of piscivorous and planktivorous fish can lead to different consequences of piscivory from lake to lake depending on morphometric and chemical (nutrient) differences. The influence of fish themselves on nutrient dynamics of lakes also complicates the patterns of response. Prey defences and other processes (such as disturbances, especially in intertidal ecosystems) that tend to decouple predator–prey dynamics lessen the cascading effects. Evolutionary variations in prey morphology, physiology or behaviour occur in response to the longer term impact of predation pressure. Shorter term predator-induced adjustments to individual morphology, for example dorsal crests in *Daphnia*, changing body size and shape in other zooplankton and planktivorous or herbivorous fish, or changes in behaviour in the presence of predators, can also reduce the vulnerability of prey on a time-scale of hours, days or weeks. Thus the influence of trophic cascade processes in community organization is still quite a controversial topic. *See also* KEYSTONE SPECIES. [P.S.G.]

REFERENCE

Carpenter, S.R. & Kitchell, J.F. (eds) (1993) *The Trophic Cascade in Lakes*. Cambridge University Press, Cambridge.

trophic classification The majority of FRESHWATER animals are generalists (POLYPHAGOUS) rather than specialists (MONOPHAGOUS or OLIGOPHAGOUS) which has ecological implications regarding COMPETITION and temporal or MICROHABITAT isolation. For example, Cummins (1973) developed a general trophic CLASSIFICATION system for aquatic insects which has become generally used for all aquatic invertebrates. Functional FEEDING groups are analogous to guilds and allows assessment of the dependence of invertebrates on a particular food resource. Using a combination of food categories and feeding mechanisms, the main functional feeding groups are as follows.

1 Shredders (chewers and miners)—feeding on COARSE-PARTICULATE ORGANIC MATTER (CPOM; $>10^3\,\mu m$) such as decomposing allochthonous leaves and the associated bacteria and fungi or living vascular plants (gougers).

2 Collectors—feeding on FINE-PARTICULATE ORGANIC MATTER (FPOM; $<10^3\,\mu m$) and associated bacteria and fungi either suspended in the water (filter- or suspension-feeders) or deposited on the substrate (collector–gatherers).

3 Scrapers (grazers)—herbivores feeding on algae and associated material (PERIPHYTON).

4 Piercers—of living macrophyte cells or filamentous algae.

5 Predators—including those which ingest whole prey or parts (swallowers) and those which pierce the prey and extract cells and tissue fluids (piercers).

6 Parasites—internal and external parasites feeding on living animal tissue.

This classification distinguishes taxa performing different functions within LENTIC and LOTIC ecosystems and reflects CONVERGENT and PARALLEL EVOLUTION. *See also* FEEDING TYPES, CLASSIFICATION. [R.W.D.]

REFERENCE

Cummins, K. (1973). Trophic relations of aquatic insects. *Annual Review of Entomology* **18**, 183–204.

trophic continuum This concept provided an alternative model to the well-established Lindeman TROPHIC LEVEL concept of ECOSYSTEM structure. Often species cannot readily be assigned or confined to a single trophic level and their FEEDING activity spans several levels in a traditionally defined FOOD WEB. Such species have often been described as occupying a trophic CONTINUUM. The trophic continuum model, developed by Cousins (1980), attempted to overcome such problems and is defined by organisms harvesting RESOURCES, where animal size classes, a classification of plant products and DETRITUS provide the resource descriptions of ecosystems. The trophic continuum model is suggested to provide closer links between theories of ENERGY FLOW, species DIVERSITY and ecosystem HETEROGENEITY. Whilst the concept has not met with wholehearted approval in the ecological literature, the importance of body size patterns in food webs has been clearly identified by many authors. *See also* FEEDING TYPES, CLASSIFICATION. [P.S.G.]

REFERENCE
Cousins, S.H. (1980) A trophic continuum derived from plant structure, animal size and a detritus cascade. *Journal of Theoretical Biology* **82**, 607–618.

trophic ecology *See* ECOLOGY.

trophic efficiency *See* EFFICIENCY; ENERGY FLOW.

trophic eggs *See* CANNIBALISM.

trophic group amensalism hypothesis This hypothesis was developed by Rhoads and Young (1970) to explain patterns of low DIVERSITY, high DENSITY and DOMINANCE by certain trophic FEEDING groups in muddy marine benthic habitats. There often appears to be a negative relationship between deposit-feeding invertebrates (especially surface deposit-feeders) and filter-feeding (or suspension-feeding) invertebrates, the former dominating on muddy substrates. The hypothesis assumes that feeding activities of the deposit-feeders resuspends SEDIMENT, thus providing biologically induced instability in the substrate. This inhibits filter-feeders by clogging filtering mechanisms and makes the substrate unsuitable for larval SETTLEMENT and maintenance of filter-feeders, i.e. there is a form of ammensal interaction between the deposit-feeders and the filter-feeders (the latter trophic group being negatively affected by the interaction and the former group unaffected). The DISTURBANCE of the substrate is, in turn, predicted to lead to low-diversity communities dominated by a high ABUNDANCE of deposit-feeders.

The hypothesis was developed for northern and temperate regions but has also been found applicable to some tropical communities. However, data does not always support the trophic group ammensalism hypothesis. A caveat to the original hypothesis was that it should only be applied to areas of CONTINENTAL SHELF where primary PRODUCTION was high and food unlimited. Thus, where such conditions do not prevail, the low-diversity communities dominated by deposit-feeders are not expected to occur. [P.S.G.]

REFERENCE
Rhoads, D. & Young, D. (1970) The influence of deposit feeding benthos on bottom sediment stability and community trophic structure. *Journal of Marine Research* **28**, 150–178.

trophic level A trophic level is a position in a FOOD CHAIN determined by the number of energy transfer steps to that level. Thus, in complex natural communities all organisms which obtain their food from living plants by FEEDING the same number of steps away from the plants along food chains, form a trophic level within that GRAZING food-web COMMUNITY. Similarly for the dead plants and the decomposer FOOD WEB. Green plants form the first trophic level of the primary producers, herbivores form the second trophic level, carnivores feeding on the herbivores the third trophic level, and secondary carnivores which feed on them the fourth trophic level. The concept was fundamental to the functional classification of community structure pioneered by Charles Elton in the 1920s and formed the foundation for the trophic dynamic analysis of communities introduced by Lindeman (1942). Lindeman advocated describing communities not by the number or BIOMASS within each trophic level but by the rate at which energy is transferred through the trophic levels. This led to the development of the whole energy-flow analysis approach to ecosystems ecology.

The greatest practical difficulty in implementing this functional classification is that many species, while they are taxonomically clearly discrete units and the fundamental subjects of NATURAL SELECTION, do not fit neatly into discrete trophic levels. For example, many carnivores feed both on herbivores and predate other carnivores, male horseflies feed on nectar while females seek blood, and a lot of animals, for example slugs, eat both dead and living plant tissues. In any detailed analysis of the trophic relationships in a community it is therefore necessary to know what proportion of the diets of such species come from which level so that in a functional classification of the community, these species can be apportioned between levels appropriately.

The EFFICIENCY with which energy is transferred through a trophic level is always less than 100%, in accordance with the second law of thermodynamics, so there is a finite limit to the number of trophic levels in any ECOSYSTEM. This is rarely more than 4–5 in either the grazing or decomposer food webs.

Animals in successive trophic levels tend to be larger. The structure of an animal puts a limit on the size of food it can ingest and, with some notable exceptions such as the baleen whales with very specialized prey-capture mechanisms, large carnivores cannot live on very small prey items because the metabolic costs of movement involved in catching them outweigh the energy returns for each item captured. *See also* ENERGY FLOW; TROPHIC CONTINUUM. [M.H.]

REFERENCE
Lindeman R.L. (1942) The trophic-dynamic aspect of ecology. *Ecology* **23**, 399–418.

trophic ratios The trophic or PREDATOR : prey ratio is the ratio of the number of species of predators to the number of species of prey within the COMMUNITY. This ratio appears to be relatively constant for a wide variety of communities and independent of SPECIES RICHNESS. Earliest assessments of community food webs broadly defined 'kinds' of predators and prey (TROPHOSPECIES) and found a predator : prey ratio slightly greater than 1. Similarly, the proportion of all species that are 'top' (predators with no predators), 'intermediate' (both predators and prey) and 'basal' (prey with no prey) trophospecies also appears to be relatively constant, irrespective of the total number of species in the FOOD WEB (average proportions 0.285, 0.525 and 0.189, respectively).

When the ratio is examined at the biological species level, lower but still relatively constant predator : prey ratios are found (Fig. T17). Whilst the exact value of the predator : prey ratio varies according to the mode of classification of species (trophospecies versus biological species) and type of community, the apparent constancy within any broadly defined habitat type represents an interesting pattern that has yet to be adequately accounted for. [P.S.G.]

REFERENCE
Pimm, S., Lawton, J. & Cohen, J. (1991) Food web patterns and their consequences. *Nature* **350**, 669–674.

trophic status *See* FEEDING TYPES, CLASSIFICATION.

trophic structure The trophic structure of a COMMUNITY is a way of describing its structure by reference not to its taxonomic species

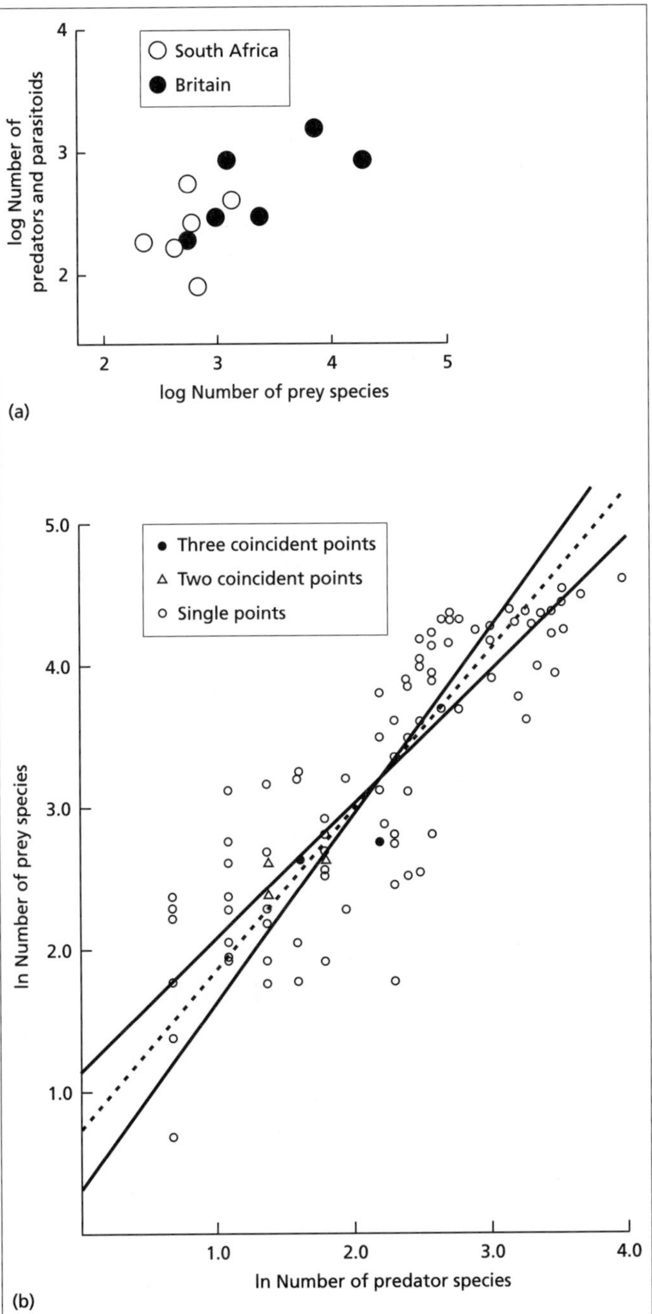

Fig. T17 Examples of predator : prey ratios. (a) On trees in Britain and South Africa, the relationship between number of arthropod predator and parasitoid species and prey species (log scale) has a slope considerably less than 1 (i.e a ratio of less than 1). (After Southwood, T.R.E., Moran, V.C. & Kennedy, L.E.J. (1982) The richness, abundance and biomass of the arthropod communities on trees. *Journal of Animal Ecology* **51**, 635–650.) (b) For freshwater invertebrate communities, the relationship was equally strong, with a predator : prey ratio from 1.2–1.35 depending on species richness, but was robust across a variety of freshwater habitats. Regressions are for both predator or prey species numbers as the independent variable together with the average (dashed line). (After Jeffries, M.J. & Lawton, J.H. (1985) Predator–prey ratios in communities of freshwater invertebrates: the role of enemy free space. *Freshwater Biology* **15**, 105–112.)

composition but by reference to the FEEDING relationships of the organisms within it. It is a functional CLASSIFICATION of the organisms, which groups them according to their positions in the FOOD CHAIN into different trophic levels. Green plants form the first TROPHIC LEVEL which is the base upon which all the others are built. This may either be along the grazing food chain, with herbivores as the second level, primary carnivores as the third and secondary carnivores as the fourth, or from dead plant material along the decomposer FOOD WEB as shown in Fig. T18.

The possible pathways in the GRAZER and decomposer systems are closely parallel except that material is lost from the grazing food web as faeces and urinary waste which passes into the decomposer system. However, faeces and excreta from decomposers simply cycle back within the system and form new substrate for more decomposers until the energy is either entirely metabolized or lost from the system, for example dead bodies of planktonic organisms being lost from the PHOTIC ZONE to the deep ocean floor. Where abiotic conditions limit rates of DECOMPOSITION, organic matter may be accumulated, with some of the dead plant material, as soil organic matter, PEAT or ultimately coal and oil deposits. *See also* FEEDING TYPES, CLASSIFICATION; TROPHIC CONTINUUM. [M.H.]

REFERENCES

Begon, M., Harper, J.L. & Townsend, C.R. (1996) *Ecology: Individuals, Populations and Communities*, 3rd edn. Blackwell Science, Oxford.

Heal, O.W. & MacLean, S.F. (1975) Comparative productivity in ecosystems — secondary productivity. In: *Unifying Concepts in Ecology* (eds W.H. van Dobben & R.H. Lowe-McConnell), pp. 89–108. Junk, The Hague.

trophobiology *See* TROPHIC CLASSIFICATION.

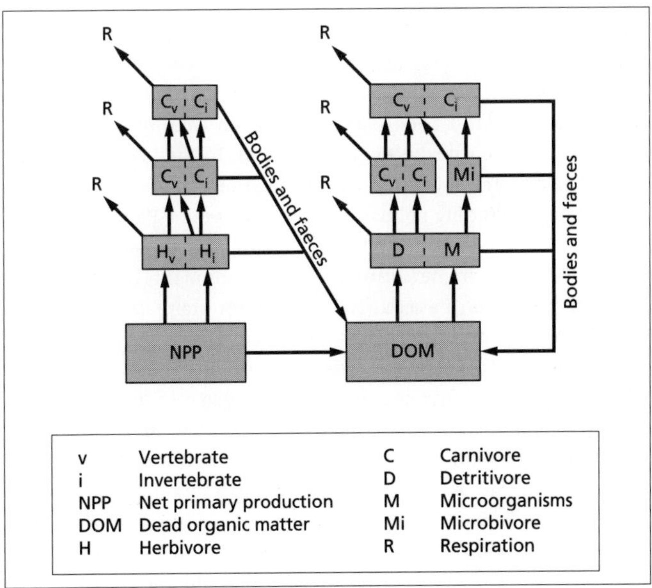

Fig. T18 A generalized model of trophic structure and energy flow for a terrestrial community. (After Heal, O.W. & MacLean, S.F. (1975) Comparative productivity in ecosystems—secondary productivity. In: *Unifying Concepts in Ecology* (eds W.H. van Dobben & R.H. Lowe-McConnell), pp. 89–108. Junk, The Hague.)

trophospecies (trophic species) A concept used to help simplify the description, analysis and modelling of FOOD WEBS. A trophospecies or trophic species is a collection or set of biological species with the same diet and set of predators. Trophospecies can be aggregated into trophic levels. The simplest type of classification has been into 'top' (predators with no predators), 'intermediate' (both predators and prey) and 'basal' (prey with no prey) trophospecies (sometimes including tens to hundreds of biological species) related taxonomically or by resource use within each TROPHIC LEVEL. *Cf.* SPECIES CONCEPTS. *See also* TROPHIC CONTINUUM; TROPHIC RATIOS. [P.S.G.]

tropical lakes *See* LAKE STRATIFICATION.

Tropical montane forest

Above about 1200–1500 m the composition of the tropical FOREST changes. Various kinds of tropical montane forest can be recognized. The most common kind is broad-leaved. Microphyllous or needle-leaved coniferous or ericaceous forest is another kind, and in some places bamboo forest is present. For example, in Malaysia the lowland dipterocarp forest is replaced by a broad-leaved montane forest in which species of oak and laurel are common. At about 1800 m the oak–laurel forest is replaced by a montane ericaceous forest. There is a decline in the height of the CANOPY from about 30–40 m in the dipterocarp forest to 20 m in the oak–laurel forest and 15 m in the ericaceous forest. In the montane forest, the canopy is relatively uniform, with few emergent trees. In many parts of the tropics, where it is particularly wet and light levels are low, there is a CLOUD FOREST or, at higher altitude, an ELFIN FOREST.

With increasing altitude there is a decrease in the frequency of lianes (woody climbers) and trees with buttress trunks, and an increase in vascular epiphytes. On steep slopes, which are common in the varied upland TOPOGRAPHY, the even canopy is broken so that more light reaches to near the ground. In the lower montane forest, the ground vegetation is an impenetrable thicket of spinescent plants including palms, pandans and cycads. Palms or stands of tree-ferns are commonly abundant, found in rock-slides or areas of unstable soil. At higher altitudes, herbaceous plants become more common in the UNDERSTOREY. Tropical mountains are subject to lower temperatures and greater wind than the tropical lowlands. Many tropical mountains regularly experience mist and cloud. Lower temperatures and anaerobic waterlogged soils limit microbial activity so that breakdown of organic matter is low. *Sphagnum* moss grows on a peaty soil. Nutrient levels are low in leached podzolized soils. Sclerophylly, the possession of tough EVERGREEN leaves, is common, but leaves are not notably DROUGHT resistant; the thick cuticle reduces LEACHING and minimizes the attack of epiphyllous fungi. Many leaves have drip-tips which drain the leaf surface, reducing the time for leaching and the potential for growth of epiphyllous plants and fungi.

There is a change in the PHYSIOGNOMY of leaves from mesophyll (medium size) to microphyll (small). Plants are evergreen and sclerophyllous, with tough thick leaves. Plants with pinnate leaves, which are common in the lowland, are absent in the uplands. In an upper montane zone in the ericaceous forest, microphyllous species like conifers, heathers (Ericaceae) and myrtles (Myrtaceae) are commonly present. These produce a slowly decomposing leaf litter. In the subalpine forest of New Guinea and in the Andes, a nanophyllous forest of plants with scale-like leaves has been recognized. Vascular epiphytes are abundant and very diverse in tropical montane forest. They include many ferns and clubmosses, orchids (Orchidaceae), aroids (Araceae), bromeliads (Bromeliaceae) and species from the dicot families Cactaceae, Ericaceae, Gesneriaceae, Melastomataceae and Piperaceae. The bromeliads and epiphytic cacti are common in the neotropics. There is a high degree of endemism in the epiphytes of tropical mountains. Epiphytes show many adaptations to the dry, nutrient-poor niches they occupy. Trash basket and tank plants catch DETRITUS, HUMUS and RAIN. Epiphytes from the outer canopy have xerophytic adaptations.

Seasonally dry mountains have shallow soils. The vegetation varies depending on the length of the DRY SEASON. For example in Panama, where the dry season is less than 3 months, epiphytes are common and there is a thick shrubby understorey below a pine–oak canopy. In areas with a 3–6-month dry season, the pine–oak forest is more open and the there is a herbaceous understorey with grassy glades. Epiphytes are relatively less abundant. [M.I.]

REFERENCES

Mabberley, D.J. (1992) *Tropical Rain Forest Ecology*, 2nd edn. Blackie, Glasgow.
Whitmore, T.C. (1990) *An Introduction to Tropical Rain Forests*. Clarendon Press, Oxford.

Tropical rain forest

The term 'tropical rain forest' was coined by the German plant geographer A.F.W. Schimper in 1898. The lowland equatorial forests are located mainly in South America (about 50% of the world's tropical RAIN FOREST are found in this region), south-east Asia (about 30%) and central Africa (20%). The vegetation is typically dominated by EVERGREEN trees whose buds lack scales or any protection against DROUGHT or cold. Mean annual precipitation usually exceeds 2000 mm and may be greater than 4000 mm. Temperature shows little seasonal fluctuation and usually lies around 26–30°C.

Lowland forest is typified by a multilayered structure with many trees exceeding 30 m in height and some greater than 40 m. The idea that there are distinct layers, or strata, within the forest CANOPY is commonly held; some authors claiming three layers, others more. The concept has been investigated and debated at length and it seems that such a view is an oversimplification of the complexity of forest canopy structure. The detection of distinct layers is often associated with growth phases of certain trees. Indeed, it is helpful to distinguish between LAYERING within the populations of tree species and layering within the community in general, involving many species. The typically closed canopy of lowland forest permits little light penetration, as a result of which undergrowth is sparse and lianes and epiphytes are less common than in some other forms of forest.

TROPICAL MONTANE FOREST occupies the upland equatorial regions and is generally less tall than the lowland vegetation, has a richer ground vegetation as a result of more effective light penetration (see Plate 25, facing p. 398) and has a higher diversity of epiphytes and climbers. Tree-ferns and palms are often present. At high altitudes it is replaced by CLOUD FOREST, which is yet smaller in stature (the canopy is often less than 20 m) and more open in structure, allowing the development of a very rich epiphyte and liane component within the canopy and a denser ground cover of herbs, mosses and ferns.

ALLUVIAL and riverine forests are subject to flooding and disturbance, as a result of which gaps in the canopy are frequent and light penetration is enhanced. Many tree species rely on stilt and buttress roots to provide stability. Herbaceous, epiphyte and liane growth is encouraged in such disturbed situations and, since these were the forests first observed by early explorers, they gave rise to the impenetrable image portrayed of tropical forests. In very wet sites, swamp forest and peat forests are developed (*see* MIRE).

Tropical evergreen rain forests are perhaps the most diverse vegetation types on Earth in terms of the wildlife they support. It is estimated that about half of the world's species of plants and animals are located within the tropical rain forest. Tree species alone can number 40–100 species per hectare and extrapolations concerning invertebrate diversity, based on the analysis of trapping beetles from single trees, run into millions. The uniform climate, the high productivity, the complexity of microhabitats, patterns of disturbance and recovery, geological history and past evolutionary patterns, have all been proposed as factors leading to the high diversity of these forests.

Global destruction of tropical forests is of major concern. Economic gains can be made from the exploitation of the timber RESOURCES of forests and from the reclamation of forest lands for agriculture and pastoralism. During the decade 1980–1990, an average of about 170 000 km^2 of tropical forest was cleared each year, amounting to about 1% of the global area. Some of this, however, has been allowed to recover as SECONDARY FOREST. In certain regions the loss of forest has reduced the remaining stands very significantly; only 12% of Brazil's Atlantic forest remains and only about 40% of Central America's forest is still intact. Most of the countries containing the remaining forest have high rates of population growth and high levels of poverty, which increase pressures for the expansion of agricultural land. The demand for tropical hardwoods from developed nations brings further pressures on remaining forests.

Changing land use in the tropics is likely to continue, and the main concern of conservationists will be the retention of viable sections of forest to permit the survival of as many species as possible. Forest fragmentation is inevitable and studies are being conducted on the impact of such fragmentation on extinction rates. Although some species, particularly top predators and large mammals in general, are placed particularly at risk by the dissection of forests into small, isolated fragments, studies suggest that species survival is generally better than was initially feared in these RELICT fragments. The results certainly justify the policy of maintaining fragment reserves as reservoirs of wildlife. Most experimental studies are still in their early stages, and we need to know the effect of centuries of isolation rather than that of a few years. However, some old fragments do exist and are reasonably well documented, as for example in the Bukit Timah Reserve in Singapore, isolated since 1860. In this 164 ha reserve there remain 50 ha of primary dipterocarp forest supporting over 900 plant species. The BUFFER ZONE around this fragment has undoubtedly contributed to its maintained viability. Debate about the minimum size, the spacing, the selection and the management of such reserves is likely to continue for some time. *Cf.* TEMPERATE RAIN FOREST. *See also* TROPICAL SEASONAL FOREST. [P.D.M.]

REFERENCE

Richards, P.W. (1996) *The Tropical Rain Forest*, 2nd edn. Cambridge University Press, Cambridge.

---

tropical seasonal forest Rainfall is relatively uniformly distributed throughout the year in many equatorial regions, as in south-east Asia generally, and also in much of Brazil. But other areas have an uneven rainfall pattern, such as southern Nigeria, coastal Zaire and parts of Central America. The incidence of distinct dry and wet seasons, particularly as one moves away from the equatorial regions and into the monsoon zone (*see* MONSOON FOREST), leads to new selective pressures on plant and animal species. Seasonal tropical forests may still be dominated by EVERGREEN species of tree, or by species that shed all or a large proportion of their leaves as the DRY SEASON develops. Buds may also be protected by coverings in such species. Climbers and epiphytes are still present, but are less abundant than in the more open of the equatorial rain forests and become less frequent with the increasing severity of the dry season.

The relative advantages of evergreen and deciduous habits in a seasonal climate are finely balanced. Evergreen leaves are generally more cost effective in terms of how much energy goes into their production (they may last several years) and also in the amount of nutrient, particularly nitrogen, needed for their formation. Tropical soils are often poor in the chemical nutrients needed for plant growth and the construction and replacement of leaves represents a heavy demand on soil resources. The economic advantages of the evergreen leaf in terms of energy and nutrients, however, are in part offset by the water loss that will continue from the leaf during the dry season. Deciduous trees reduce water loss and improve their water-use efficiency by shedding leaves, but they are then faced with the competitive disadvantage of a slower rise in productivity when the wet season arrives once more; they are delayed by the necessity to form new leaves. The balance between deciduous and evergreen leaves in tropical seasonal forests is thus determined by local climatic and soil conditions and the competitive balance between the tree species. Some insect species show seasonal migrations between evergreen and deciduous tree species.

These forests tend to have a lower overall SPECIES RICHNESS than their wet evergreen counterparts, and they are often dominated by particular tree species. Several of these species, however, such as teak (*Tectona grandis*), sal (*Shorea robusta*) and various *Terminalia* species, are of considerable economic importance as timber trees. As a result, the seasonal forests are under pressure from loggers. *See also* TROPICAL RAIN FOREST. [P.D.M.]

troposphere The WEATHER-filled part of the ATMOSPHERE, bounded below by the Earth's surface and above by the tropopause

(10–15 km above sea level), and containing 70–90% of its mass (the larger in low latitudes) and almost all of its WATER vapour and cloud. It is continually disturbed by a wide variety of types of weather system, and has typically steep TEMPERATURE lapse rates associated with widespread CONVECTION.

The lowest few hundred metres (the atmospheric BOUNDARY LAYER) is directly affected by the underlying surface through friction, surface shape, heating and cooling. [J.F.R.M.]

Tropospheric chemistry

The chemical processes occurring in the TROPOSPHERE by which trace substances released into the ATMOSPHERE from natural and anthropogenic sources are oxidized. The reactions form stable compounds, such as CARBON DIOXIDE (CO_2) and water, or more soluble compounds, such as peroxides or acids, which are rained out or deposited from the atmosphere. *See also* STRATOSPHERIC CHEMISTRY.

The troposphere

The lower atmosphere, below about 15 km altitude, is termed the troposphere because of its turbulent air motions. A BOUNDARY LAYER of about 1 km thickness is often envisaged as an interface between the Earth's surface and the 'free troposphere'. The troposphere is characterized by a general decrease in TEMPERATURE with height, from *c.* 15°C at the surface to *c.* −50°C at about 15 km. Above this, in the lower STRATOSPHERE, the temperature increases with height, the changeover region being known as the 'tropopause'.

The troposphere contains an enormous range of inorganic and organic trace substances. Many of them, such as METHANE (CH_4), nitrous oxide (N_2O) and dimethyl sulphide ($(CH_3)_2S$), are released by natural processes. Many, such as the NITROGEN OXIDES (NO_x), a variety of VOLATILE ORGANIC COMPOUNDS (VOCs) and sulphur dioxide (SO_2) are released by human activities. Others, such as OZONE (O_3), CARBON MONOXIDE (CO) and nitric acid (HNO_3), are formed in the atmospheric oxidation of the compounds released from both natural and anthropogenic sources.

Initiation of tropospheric oxidation: the hydroxyl radical (OH)

Oxidation of organic compounds occurs through a series of cyclic chain reactions. These are usually initiated by the reactive hydroxyl radical, OH, which is the principal agent of attack on trace substances in the atmosphere. Hydroxyl radicals are formed by the photolysis of ozone at short wavelengths (< 310 nm) in the ultraviolet:

$$O_3 + h\nu(\lambda < 310\,\text{nm}) \rightarrow O_2 + O(^1D) \quad (1)$$

An electronically excited state of the OXYGEN atom, $O(^1D)$, is formed that can then react with methane or water to give one or two hydroxyl radicals:

$$O(^1D) + H_2O \rightarrow OH + OH \quad (2)$$

$$O(^1D) + CH_4 \rightarrow OH + CH_3 \quad (3)$$

Tropospheric ozone itself may either be formed by photochemical oxidation, as indicated below, or it may be brought down by vertical atmospheric mixing, from the stratosphere into the free troposphere or from the free troposphere into the boundary layer.

Tropospheric oxidation of methane

As an example of an oxidation chain reaction, consider the oxidation of CH_4. CH_4 is a major trace constituent, having an average volume ratio of *c.* 1800 ppb in the troposphere. The attack of the hydroxyl radical abstracts a HYDROGEN atom from CH_4 to form water:

$$CH_4 + OH \rightarrow CH_3 + H_2O \quad (4)$$

$$CH_3 + O_2 \rightarrow CH_3O_2 \quad (5)$$

The methyl radical, CH_3, formed at the same time, reacts with oxygen forming a methylperoxy radical, CH_3O_2. In a continental atmosphere during the day, this will probably react with nitric oxide, NO, released by combustion processes, to form a methoxy radical, that in turn reacts with oxygen to form formaldehyde, HCHO:

$$CH_3O_2 + NO \rightarrow CH_3O + NO_2 \quad (6)$$

$$CH_3O + O_2 \rightarrow HCHO + HO_2 \quad (7)$$

Like methane, the formaldehyde molecule itself may be oxidized by OH in a similar sequence of reactions, the end-result being the formation of CO or CO_2 and water.

The hydroperoxy radical, HO_2, from reaction 7 oxidizes NO to NO_2 and regenerates an OH radical:

$$HO_2 + NO \rightarrow OH + NO_2 \quad (8)$$

The reaction sequence 4–8, constitutes a chemical chain reaction with OH as the chain carrier: OH is consumed in reaction 4 and regenerated in reaction 8. The OH radical can thus be regarded as a CATALYST, only a small amount being needed to oxidize much larger amounts of CH_4. In the same reaction chain, two molecules of NO are also oxidized to nitrogen dioxide (NO_2), reactions 6 and 8.

VOCs, such as HYDROCARBONS, are all oxidized by similar mechanisms and so, in sunlight when enough NO_x is present, fuel the synthesis of ozone.

CO, another major trace constituent with an average volume ratio of about 120 ppb in the troposphere, is also oxidized in a chain reaction. Following an initial attack by OH, a hydrogen atom is formed which adds to oxygen to form a hydroperoxy radical, HO_2:

$$CO + OH \rightarrow CO_2 + H \quad (9)$$

$$H + O_2 \rightarrow HO_2 \quad (10)$$

An OH radical is then regenerated by reaction 8.

Ozone formation

In reactions 5 and 7, the peroxy radicals convert a molecule of NO to NO_2. During the daytime, the NO_2 can be photolysed to form an oxygen atom, that can then add to oxygen to form a molecule of ozone, reactions 11 and 12:

$$NO_2 + h\nu(\lambda \approx 400\,\text{nm}) \rightarrow NO + O \quad (11)$$

$$O + O_2 \rightarrow O_3 \quad (12)$$

Reactions 11 and 12 can be reversed by the rapid reaction of NO with ozone:

$$NO + O_3 \rightarrow NO_2 + O_2 \quad (13)$$

The forward and back reactions (11, 12 and 13), with their dependence on the concentration of NO_x ($[NO_x] = [NO] + [NO_2]$) and the LIGHT intensity, determine the local concentration of ozone. On a sunny summer afternoon the ozone volume ratio may be between 60 and 80 ppb in the boundary layer. Overall, the oxidation of VOCs and CO in the continental boundary layer generates ozone, provided that enough NO_x is present.

Oxidation and ozone in remote and background atmospheres

In remote locations on Earth, largely free from anthropogenic emissions and with low concentrations of NO_x (< 1 ppb), trace substances are still oxidized. However, little ozone is formed photochemically. Instead, the peroxy and other radicals may react together, rather than with NO, and form organic peroxides, hydrogen peroxide (H_2O_2) and other compounds, that may be rained out or deposited directly into the BIOSPHERE.

In the free troposphere in the Northern hemisphere, above the boundary layer, the volume ratio of ozone varies between about 30 ppb in winter to 60 ppb in summer. The concentration levels appear to have more than doubled in the last hundred years, with most of the increase being attributed to human activities. The additional ozone is formed photochemically, as shown above, either in the free troposphere itself or in the boundary layer, from which it is transferred upwards by vertical mixing. As in remote areas, the NO_x concentrations in the free troposphere are generally low, so that H_2O_2 and other compounds are formed from additional reactions of the peroxy and hydroperoxy radicals.

Photochemical air pollution: summer smog

In the polluted boundary layer, the concentrations of the primary pollutants, VOCs and NO_x emitted by traffic and industrial sources, are much higher than in the free troposphere above. In the presence of sunlight, high concentrations of ozone and other photo-oxidants are produced, with ozone volume ratios reaching 400 ppb in bad cases. The resulting condition is known as a 'photochemical smog'. It was first characterized in Los Angeles in the 1950s, and is prevalent in cities that have a warm sunny climate. The pollutants are trapped within the boundary layer under a meteorological temperature inversion (temperature increasing with height). The high POLLUTANT concentrations facilitate rapid photochemistry and the inversion prevents the polluted air being diluted by mixing with cleaner air from above. The growth in population and traffic throughout the world means that most subtropical and tropical cities suffer from chronic smog conditions. Summer smog episodes are also regional scale phenomena in the populated temperate parts of the world during sunny periods accompanied by stagnant high-pressure conditions.

A frequent constituent of smog is the powerful lachrymator PAN, peroxyacetylnitrate. It is formed from acetaldehyde, CH_3CHO, a product of the photo-oxidation of ethane (C_2H_6) and other hydrocarbons:

$$CH_3CHO + OH \rightarrow CH_3CO + H_2O \quad (14)$$

$$CH_3CO + O_2 \rightarrow CH_3COO_2 \quad (15)$$

$$CH_3COO_2 + NO_2 \rightarrow CH_3COO_2NO_2\ (\text{PAN}) \quad (16)$$

As PAN can dissociate again to form NO_2, it can act as a transport agent. It is formed in areas of high POLLUTION and can release NO_2 in remote areas far from the pollution source.

Controlling air pollution: problems with complex chemistry

While the reduction of VOC and NO_x EMISSIONS is clearly the way to reduce photochemical smog formation, the complexities of the chemistry make cost-effective measures difficult to devise. For example, in a city in summer, there may be very high VOC and NO_x concentrations, which themselves are undesirable or harmful. Oxidation takes place rapidly by the mechanisms shown above, forming PAN and other products. Very little ozone may in fact be measured in the city itself because it is 'titrated out' as soon as it is formed by the high concentrations of NO, reaction 13. However, downwind of the city, in the suburbs and surrounding countryside, the measured ozone concentrations are greater; there has been more time during transport from the city for the photochemical reactions to occur, and NO_x may be removed faster from the air than the VOCs. Reaction 13 is then less effective and the ozone concentration increases. Analogous effects are likely to be observed when measures to lower NO_x emissions within the city are taken. The concentration of ozone may remain the same or even increase with small reductions of NO_x, and it may require very large reductions to achieve any decrease in ozone at all.

Much current policy-related work is devoted to determining whether the conditions within given localities are 'NO_x limited' or 'VOC limited' and whether, within an area, policy should be directed towards control of NO_x, VOCs or both, to achieve a cost-effective reduction in pollution.

Acid rain: the formation of sulphates and nitrates

Another example of complex chemical transformations during transport through the air is provided by the oxidation of SO_2, emitted from the combustion of FOSSIL FUELS. Sulphuric acid, H_2SO_4, is formed which acidifies rain and snow. The oxidation process involves not only chemical reactions in the gaseous phase, but also reactions within the droplets of cloud water as the air passes through CLOUDS.

The recent reductions in SO_2 emissions in Europe have brought about a decrease in ACIDIFICATION, but the result is not as great as was perhaps hoped. The increasing NO_x emissions from traffic have increased the rain-out of nitric acid, a product from photo-oxidation.

The complexity of tropospheric chemistry: atmospheric modelling

The reactions mentioned are only a few of the very many that occur in the atmosphere. A good chemical model scheme describing the chemistry of the troposphere contains perhaps 50 or so chemical compounds and radicals, and perhaps 200 reactions. The variation of SOLAR RADIATION with time and wavelength must also be included. It is still a simplified scheme and it constitutes one of a number of modules in an atmospheric model.

Such an atmospheric model is intended to simulate, computationally, the production, distribution and transport of pollutants within a geographical region under particular meteorological conditions. As well as the chemical module, there must be a meteorological driver: a module to describe the dynamic motions of the atmosphere, such as ADVECTION, vertical mixing and deposition; modules to include the physics and chemistry of clouds and AEROSOLS; and also an emission module, containing estimates of anthropogenic and natural emissions for a given region.

To be reliable, a large simulation model requires a good scientific understanding of the many parts of this interdisciplinary field. As these models are needed to support modern policy for air-pollution ABATEMENT, it should be no surprise that research is being vigorously pursued in all the areas related to tropospheric chemistry. *See also* AIR POLLUTION; ATMOSPHERIC POLLUTION; CFCS; CLIMATE CHANGE; OZONE HOLE; OZONE LAYER. [P.B.]

REFERENCE

Wayne, R.P. (1991) *Chemistry of Atmospheres*, Chapter 5. Clarendon Press, Oxford. [For an excellent account of tropospheric chemistry.]

---

truncation selection *See* SELECTION INTENSITY.

TSCA (Toxic Substances Control Act) *See* UNITED STATES LEGISLATION.

tsunami Also called a seismic sea wave. A very large sea wave, with a long period (several minutes to several hours), caused by a submarine earthquake, volcanic eruption, etc. In the open OCEAN, tsunamis may travel for thousands of kilometers at velocities of 700–900 km h^{-1}. When they reach the coast, their height may reach 30–70 m and they can therefore cause great damage in coastal areas. [V.F.]

tundra In origin from Finnish or Lappish *tunturi* meaning a treeless hill which has been absorbed via Russian to describe the dwarf shrub, herb and moss vegetation that exists in polar regions too cold to support tree growth (see Plate 26, facing p. 398). The treeless nature is due in part to the presence of PERMAFROST sufficiently near the soil surface which restricts the depth of soil available for tree root development and necessary to support full-grown trees. Exceptions to the exclusion of trees from permafrost zones are the black spruce of North America (*Picea mariana*) and the dahurian larch of north-eastern Siberia (*Larix dahurica*). Both these species are able to extend shallow root systems over a soil ICE horizon and thus support substantial trees in permafrost areas. [R.M.M.C.]

Tundra soils *See* SOIL CLASSIFICATION.

tungsten (W) Tungsten (ATOMIC NUMBER = 74; relative ATOMIC MASS = 183.85; melting point = 3410°C; boiling point = 5660°C) is not a common element (approximately 1 mg kg^{-1} of the Earth's CRUST). However, concentrated ores in the form of scheelite, $CaWO_4$, and wolframite $(Fe,Mn)WO_4$, principally found in China, are available. The METAL is mainly used for the production of specialist heat-resistant alloys and in the form of the carbide, WC, which is extremely hard and wear resistant and so ideal as a tool tip. The most important use of the pure metal is as filaments in electric light-bulbs. About 25% of tungsten used is recycled. Tungsten and its compounds are not considered an important health risk. [P.J.B.]

turbidity Reduced transparency of either ATMOSPHERE (caused by absorption and scattering of RADIATION by solid or liquid particles in suspension) or liquid (caused by scattering of LIGHT by suspended particles). [P.C.]

turbulence The chaotic and seemingly random motion of fluid parcels is known as turbulence. Turbulence has mechanical and convective origins. Shear forces cause mechanical turbulence, while buoyant instabilities (due to the intermingling of fluid parcels with different densities) cause convective turbulence. A distinct property of turbulence in the natural environment is a wide spectrum of motion scales (eddies) associated with the fluid flow. The largest scales of turbulence are produced by forces driving the mean fluid flow. Dynamically instable, these large eddies break down into progressively smaller and smaller scales, via an inertial cascade. This cascading breakdown of EDDY size continues until the eddies are so small that energy is consumed by working against viscous forces that convert kinetic energy into heat. [D.D.B.]

turbulent flow Fluid flow that exhibits chaotic and unstable behaviour is turbulent. Specific properties of turbulent flow include: (i) the motions are unpredictable; (ii) the flows are three-dimensional and possess rotational characteristics; (iii) the characteristic scales of the flow are numerous and break down into smaller and smaller scales; and (iv) the mixing rates are several orders of magnitude greater than rates due to molecular diffusion. [D.D.B.]

REFERENCE

Frisch, U. & Orzag, S.A. (1990) Turbulence: challenges for theory and experiment. *Physics Today*, **Jan.**, 24–32.

turbulent transfer Turbulent transfer is the physical transfer of mass, momentum or energy from one location to another via the chaotic and turbulent motion of fluid parcels. Turbulent transfer is physically distinct from molecular transfer because the rates of turbulent transfer are several orders of magnitude greater than the rate of molecular diffusion. The rate of turbulent transfer is generally proportional to the product of a turbulent diffusivity and the scalar mixing ratio gradient. However, countergradient transfer can occur when the length scales of the turbulent motions exceed the length scales associated with the curvature of the mixing ratio gradients. The rate of turbulent transfer for a scalar has units of mass (or energy) per unit area per unit time. [D.D.B.]

turnover Turnover in ecology broadly refers to the flux of energy and matter (i.e. species, individuals, biomass or nutrients) through an ECOSYSTEM. Species turnover is widely used to describe the replacement of one species by another in an ASSEMBLAGE. The phenomenon of species turnover may be considered in both space and time. Spatially, the replacement of species along ecological and geographical gradients is sometimes termed beta diversity (*see* DIVERSITY, ALPHA, BETA AND GAMMA). Temporal patterns of turnover are predicted by MacArthur and Wilson's (1967) equilibrium model of ISLAND BIOGEOGRAPHY. Examples of assemblages with both high and low levels of species turnover have been documented. *See also* BIOGEOCHEMICAL CYCLE; ENERGY FLOW; LAKE STRATIFICATION; MACARTHUR–WILSON MODEL OF ISLAND BIOGEOGRAPHY. [N.M. & S.G.C.]

REFERENCE

MacArthur, R.H. & Wilson E.O. (1967) *The Theory of Island Biogeography*. Princeton University Press, Princeton, N.J.

turtles *See* REPTILES, DIVERSITY AND CONSERVATION.

two-tailed test In the NORMAL DISTRIBUTION 95% of observations are contained within $\mu \pm 1.96\sigma$ (*see* STANDARDIZED NORMAL DEVIATE) and the remaining 5% of observations are equally divided between the two tails of the distribution, 2.5% in each tail. When we carry out a *z(d)* TEST to examine the difference between two means, and we have no a priori reason to expect a value of the test statistic above or below $z=0$ (either MEAN could be larger), we use the 5% level of significance that corresponds to $z=1.96$. This is a two-tailed test and corresponds to a NULL HYPOTHESIS of H_0: $\bar{x}_1 = \bar{x}_2$ and an ALTERNATIVE HYPOTHESIS of H_1: $\bar{x}_1 \neq \bar{x}_2$. Any statistical test that uses the PROBABILITY in both tails of a distribution is therefore a two-tailed test. The alternative is to use the value of the test statistic that cuts off 5% in just one tail. For the normal distribution this would correspond to $z=1.65$. Using this value as the 5% significance point would imply a ONE-TAILED TEST. [B.S.]

type 1 error The rejection of a NULL HYPOTHESIS when it is actually true. For example, if we use a *t* test to examine the difference between the means of two samples (null hypothesis that they come from populations with the same MEAN: $\mu_1 = \mu_2$) and conclude that they are significantly different when they are not (i.e. conclude that $\mu_1 \neq \mu_2$ when in fact $\mu_1 = \mu_2$). The PROBABILITY of making such a type 1 error, with a *t* test, would be $\alpha = 0.05$ if we accept the 5% level of significance. [B.S.]

type 2 error The acceptance of a NULL HYPOTHESIS when it is actually false. For example, if we use a *t* test to examine the difference between the means of two samples (null hypothesis that they come from populations with the same MEAN: $\mu_1 = \mu_2$) and conclude that they are not significantly different when they are (i.e. conclude that $\mu_1 = \mu_2$ when in fact $\mu_1 \neq \mu_2$).

Scientists are often more worried about type 1 than type 2 errors; making false claims about treatments. However, a situation where there might be a concern about type 2 errors is in accepting no difference between polluted and non-polluted sites when in fact they exist. [B.S. & P.C.]

type 1 functional response *See* FUNCTIONAL RESPONSE.

type 2 functional response *See* FUNCTIONAL RESPONSE.

type 3 functional response *See* FUNCTIONAL RESPONSE.

type 1 predator *See* FEEDING TYPES, CLASSIFICATION.

type 2 predator *See* FEEDING TYPES, CLASSIFICATION.

type I survivorship curve *See* SURVIVORSHIP CURVE.

type II survivorship curve *See* SURVIVORSHIP CURVE.

type III survivorship curve *See* SURVIVORSHIP CURVE.

type IV survivorship curve *See* SURVIVORSHIP CURVE.

typology The taxonomic procedure, going back to Plato and Aristotle, and used in the 17th century by John Ray, in which a set of morphological features is identified which characterize a particular coherent group as a species. Individuals falling outside the group would then belong to another species. This is now superseded because it cannot deal properly with biological variation within and among species. *See also* CLADISTICS; PHENETIC SPECIES CONCEPT; PHYLOGENETIC SYSTEMATICS. [L.M.C.]

U

U soils *See* SOIL CLASSIFICATION.

***Ulex europaeus–Rubus fruticosus* scrub community** *See* SCRUB.

ULR *See* UNIT LEAF RATE.

Ultisols *See* SOIL CLASSIFICATION.

ultracompetitive female *See* SEXUAL CONFLICT.

Ultraviolet radiation

Electromagnetic RADIATION in the wavelength interval between X-RAYS and visible LIGHT, i.e. from *c.* 100 nm to 400 nm (these limits are not universally agreed upon; some scientists set the lower limit at 5 nm or even lower, or the upper limit at 380 nm). Ultraviolet (UV) radiation was discovered in 1801 by the German, Johann Wilhelm Ritter, who found that the radiation outside the violet end of the visible solar spectrum could decompose silver chloride (AgCl).

Together with other kinds of radiation, UV radiation is generated by the Sun and other stars, and can also be produced by artificial sources (see below). UV radiation is subdivided in different ways depending on the discipline. A convenient subdivision for most purposes is: (i) vacuum UV, wavelength less than 180 nm; (ii) UV-C, 180–280 nm; (iii) UV-B, 280–320 nm; and (iv) UV-A, 320–400 nm. Thus defined, the biological effects of UV-A are similar to those of visible light, and this radiation is only very weakly absorbed by the ATMOSPHERE. UV-B is partly absorbed by OZONE (but hardly by any other constituents) in the atmosphere, and its intensity at ground level is strongly dependent on the ozone content. UV-C does not penetrate at all to ground level. Vacuum UV is so strongly absorbed by air and WATER that evacuated equipment is needed for experimentation with it.

Artificial ultraviolet sources

UV radiation can be generated by many types of lamps. Long-wavelength UV is generated in ordinary incandescent lamps, but only in those of the quartz-iodine type does a significant amount of UV-B penetrate the envelope. Shorter wavelengths are emitted by various types of gas-discharge lamps, of which mercury lamps are the most important. Such lamps, with a low pressure of mercury vapour, generate mainly radiation of wavelength 253.7 nm. If the envelope of the lamp is of quartz, this UV-C radiation is emitted from the lamp, and such lamps are used mainly as germicidal lamps. Similar lamps with an envelope that stops the shortwave radiation and has a fluorescent coating on the inside, emit in various wavebands depending on the nature of the fluorescent coating. Ordinary fluorescent lamps used for room lighting are of this type, but lamps emitting in the UV-B band (lamps for solaria and various technical and medical purposes) or the UV-A band ('black light') are very similar in construction. Lamps containing DEUTERIUM are used to produce UV-B radiation over a wide range of wavelengths.

UV radiation is also inadvertently produced by a number of technological processes, such as welding.

Molecular effects

Because UV radiation is efficiently absorbed by most organic substances and the photons are energetic enough to break covalent bonds, it causes a number of photochemical reactions in living cells exposed to the radiation. Among the most important are changes in DNA, the carrier of genetic information; these alterations prevent the proper functioning of the DNA. The most common changes are two kinds of dimerization of pyrimidine bases, but these are efficiently repaired by two enzymes, called photolyases. Photolyases are active only under illumination, and this is one reason why visible light in many cases has an ameliorating effect on damage caused by UV radiation.

Radiation damage to DNA which is more slowly repaired by other enzyme systems are strand breaks, pyrimidine hydration, and bonding to protein. Protein, membrane lipids and other essential substances in the cells can also be altered by UV radiation, and these molecular effects in turn cause the various effects detailed below.

Medical effects

UV radiation has a number of beneficial as well as deleterious effects on the human body. Best known among the former is the production of vitamin D, which requires UV-B radiation. UV-B also ameliorates skin tuberculosis and a number of other diseases. Excessive amounts produce erythema and increase the risk of skin cancer. Although the immediate effect of UV-B is more dramatic than that of UV-A, the risk from the latter is not negligible. It penetrates deeper and affects more cells than does UV-B. UV-B has a negative effect on the immune system, and in contrast to erythema

Table U1 Relative spectral effectiveness factors.

Wavelength (nm)	Relative spectral effectiveness factor (IRPA/INIRC 1991)
180	0.012
200	0.030
220	0.120
240	0.300
260	0.520
280	0.880
300	0.460
310	0.015
320	0.0010
340	0.00028
360	0.00013
380	0.000064
400	0.000030

IRPA, International Radiation Protection Association; INIRC, International Non-Ionizing Radiation Committee.

and skin cancer that affect mostly light (Caucasian-type) skin, all human races are subject to immune suppression by UV-B.

UV radiation can also damage the eyes. As in the skin, UV radiation of longer wavelength penetrates deeper. UV-C radiation damages only the cornea and the conjunctiva. A painful inflammation develops within a few hours of EXPOSURE. The main effect of UV-B is in the lens, where it causes cataract, which may result in blindness. Even small amounts of UV-B probably accelerate the AGEING and yellowing of the lens. UV-A penetrates to the vitreous body. In infants, significant amounts of both UV-A and UV-B may penetrate to the retina, which calls for special caution. Also in the eye, cancer may develop as a result of exposure to UV radiation.

Health protection GUIDELINES have been issued by the International Radiation Protection Association (IRPA) and the International Non-Ionizing Radiation Committee (INIRC). The basic exposure limit for both general public and occupational exposure to UV radiation incident within 8 hours on the skin or eye is 30 effective $J\,m^{-2}$. This effective exposure is calculated by weighting the spectral exposure values (in $J\,m^{-2}$) for each wavelength by a dimensionless HAZARD (relative spectral effectiveness) function and integrating over all wavelengths. The spectral effectiveness factors for a few wavelengths are given in Table U1; a more complete table is given in the Special Issue of *Ambio: Environmental Effects of Ozone Depletion*.

Other biological effects

More important than medical effects may be UV radiation effects on other biological systems, such as PHYTOPLANKTON and terrestrial plant communities. The stratospheric ozone depletion mentioned below has caused concern about the consequences that increases in such effects may have on a global scale.

Ozone depletion and ultraviolet radiation

Much attention has been devoted to UV-B radiation in recent years because its intensity at ground level and in the sea and other waters increases with depletion of the atmospheric ozone. There has been a trend of decreasing stratospheric ozone over the past couple of decades (except near the Equator), almost certainly due to anthropogenic POLLUTION of the atmosphere by CFCS and NITROGEN OXIDES (NO_x). Although this has resulted in international regulations to protect the atmosphere, return to normal conditions will take many decades even if agreements are adhered to. Until now, the effects on UV-B of the depletion of stratospheric ozone have been partly compensated for by other pollutions causing increases of tropospheric (low-level) ozone and haze, and it has not been firmly established that present-day UV-B levels are unnaturally high, except under the ANTARCTIC OZONE HOLE. [L.O.B.]

REFERENCES

Kessler, E. & Björn, L.O. (eds) (1995) *Environmental Effects of Ozone Depletion*. Special Issue. *Ambio* **24**(3), 137–196.

WHO (1994) *Environmental Health Criteria 160: Ultraviolet Radiation*. World Health Organization, Geneva.

Young, A.R., Björn, L.O., Moan, J. & Nultsch, W. (eds) (1993) *Environmental UV Photobiology*. Plenum Press, New York.

umbrella species Species whose occupancy area (HOME RANGE) is large enough and whose habitat requirements are wide enough that if they are adequately protected, other species will automatically be protected as well. *See also* CONSERVATION BIOLOGY. [P.C.]

umbric horizons *See* SOIL CLASSIFICATION.

Umbrisols *See* SOIL CLASSIFICATION.

UNCED (United Nations Conference on Environment and Development) The United Nations Conference on Environment and Development was held in Rio de Janeiro in 1992 and is commonly known as the Earth Summit. It was a follow-up to the World Commission on Environment and Development that first promoted the concept of SUSTAINABLE DEVELOPMENT. The main products of the Earth Summit were as follows.

AGENDA 21—a comprehensive programme of actions needed throughout the world to achieve a more sustainable pattern of development for the 21st century.

Climate Change Convention — an agreement between countries about how to protect the diversity of species and habitats in the world.

Statement of Principles — for the management, conservation and sustainable development of all the world's forests.

International machinery has also been established to follow-up these agreements.

- A new UN Commission on Sustainable Development (CSD) was set up to monitor progress on Agenda 21.
- Each of the conventions will have a conference of parties to monitor progress against commitments and review the need for further advice.
- The CSD will monitor implementation of the forest principles.

See also BRUNDTLAND REPORT; UNITED NATIONS PROGRAMMES AND BODIES. [P.C.]

uncertainty Lack of understanding or control of situations due to:

(i) ignorance; (ii) poor EXPERIMENTAL DESIGN and/or sloppy observation; and/or (iii) stochasticity in the process being studied. [P.C.]

undercompensation Occurs when density-dependent MORTALITY is weak, such that restoration of a population equilibrium following disturbance takes more than one generation. Density-dependent mortality is detected by plotting $k = \log_{10}$(initial density/final density) against $\log_{10}$(initial density); undercompensation occurs if the slope $b < 1$. *See also* POPULATION REGULATION. [R.H.S.]

underdispersed *See* DISPERSION.

underdispersion *See* DISTRIBUTION.

understorey The shrubs and young trees found beneath the trees in woods and forests. Species composition of the understorey may be markedly different from that of the dominant trees; hazel (*Corylus avellana*) is a conspicuous component of the understorey of mixed oak FOREST in Europe. TROPICAL SEASONAL FOREST has an even greater difference, with tall deciduous dominants and an EVERGREEN understorey. In traditionally managed woods part of the understorey consists of young tree suckers, COPPICE poles and, less commonly, POLLARD poles, which are known as the underwood. [J.R.P.]

undifferentiated disjunction *See* DISJUNCT POPULATIONS.

UNEP (United Nations Environment Programme) *See* UNITED NATIONS PROGRAMMES AND BODIES.

unequal crossing-over Unequal crossing-over was first described in *Drosophila* at the *Bar* locus and is a cross-over (i.e. the breakage and reunion of parts of chromosomes during meiosis, that leads to a recombination of ALLELES on homologous chromosomes) between two improperly aligned homologues (e.g. within a tandem repeat), producing one homologue with three copies of a region and the other with one copy of that region. Crossing-over frequently occurs during the first stage of meiosis (i.e. during that stage at which the independent assortment of alleles takes place). Unequal crossing-over may lead to the development of pseudogenes and may lead to changes in the number of repeating units in multigene families among genotypes within species. [V.F.]

uniformitarianism The principle in geology that past events can be explained in terms of phenomena observable today. The history of the Earth is considered to have involved the processes of erosion, deposition, vulcanism, etc., which can be seen at present and not some different, or unique, 'catastrophic' events which, if they alone were responsible, would make interpretation of past EVOLUTION impossible. This position was formulated by James Hutton and propagated in the 19th century by the influential Charles Lyell, as the alternative to the 'catastrophism' of Georges Cuvier and others. The opposition of these two views was important in the history of evolutionary theory, but now both types of process are acknowledged to have played important roles. [L.M.C.]

unimodal gradients *See* DIVERSITY GRADIENT.

Table U2 Common non-SI units.

Quantity and SI unit	Unit name, and symbol where appropriate	Conversion factor
Length (m)	ångström, Ä	0.1 nm
	micron, μ	1 μm
	yard, yd	0.914 m
	foot, ft	30.48 cm
	inch, in	2.54 cm
	statute mile	1.609 km
	nautical mile	1.852 km
	fathom	1.828 m
	parsec, pc	30 857 Tm
Area (m^2)	hectare, ha	10 000 m^2
	acre, ha	404 m^2
	square yard	0.836 m^2
	square foot	9.290 dm^2
	square inch	6.451 cm^2
Volume (m^3)	litre, l	1 dm^3
	cubic yard	0.764 m^3
	cubic foot	0.028 m^3
	cubic inch	16.387 cm^3
	gallon UK, gal	4.546 dm^3
	gallon US	3.785 dm^3
	pint, pt	0.568 dm^3
	barrel, bbl	0.159 m^3
Mass (kg)	tonne, t	1 Mg
	pound (avdp), lb	0.453 kg
	ton (2240 pounds)	1.016 Mg
	ounce (avdp), oz	28.349 g
	ounce (troy)	31.10 g
Density (kg m^{-3})	pound per cubic inch	27.68 Mg m^{-3}
Time (s)	minute, min	60 s
	hour, h	3600 s
	day, d	86 400 s
Velocity (m s^{-1})	foot per second	0.305 m s^{-1}
	mile per hour	1.609 km h^{-1}
	knot	1.852 km h^{-1}
Force (N)	dyne, dyn	10 μN
Pressure, stress (Pa)	bar	0.1 MPa
	lbf in^{-2}	6.894 kPa
	atmospheres, atm	0.101 MPa
	torr	0.133 kPa
	mmHg	0.133 kPa
	inHg	3.386 kPa
Work, energy, heat (J)	erg	0.1 μJ
	electronvolt, eV	0.160 aJ
	calorie, cal	4.186 J
	British thermal units, Btu	1.055 kJ
	therm = 10^5 Btu	0.105 GJ
Temperature (K)	degree Celsius, °C	1 K
	degree Fahrenheit, °F	5/9 K
Radioactivity	curie, Ci	37 ns^{-1}
	röntgen, R	0.258 mC kg^{-1}
	rad, rad	10 mJ kg^{-1}

unit Quantity or dimension adopted as a standard for measurement. The scientific community has adopted SI UNITS (Système International d'Unités) as the standard. Common non-SI units remaining in use, and likely to persist for decades, are listed in Table U2. [J.G.]

unit leaf rate (ULR) A measure of the net rate of CARBON fixation by plants, expressed as the amount of dry matter (e.g. grams) fixed per unit time (e.g. per day) per unit leaf area (e.g. per square metre). Units are g m^{-2} $time^{-1}$. [M.J.C.]

unitary organism A unitary organism has a fixed morphological form which persists throughout it's adult life. Any changes in form are almost always associated with transitions between juvenile and adult stages as is the case in many insect orders. The pattern of development is fixed and is almost totally independent from the surrounding environment. For instance, the morphological form of mammals is fixed at an early embryonic stage and the basic plan is then retained until death. *Cf.* MODULAR ORGANISM. [R.C.]

United Nations Conference on Environment and Development *See* UNCED.

United Nations Convention on Biological Diversity *See* INTERNATIONAL CONSERVATION CONVENTIONS.

United Nations Environment Programme (UNEP) *See* UNITED NATIONS PROGRAMMES AND BODIES.

United Nations programmes and bodies

The United Nations (UN) was established in 1945 to maintain international peace and security and to achieve international cooperation in solving international problems of an economic, social, cultural or humanitarian character. It has spawned a number of programmes and bodies.

Food and Agriculture Organization (FAO)

Established in 1945, the main purpose of the FAO is to raise levels of nutrition and standards of living of all peoples and to secure improvements in the efficiency of production and distribution of all food and agricultural products.

International Labour Organization (ILO)

This was initially established in 1919 under the Peace Treaty of Versailles alongside the League of Nations, and became a specialized body of the UN in 1946. Its main purpose is to promote social progress *inter alia* by adequate protection for the life and health of workers in all occupations; it is therefore largely concerned with the working environment.

International Maritime Organization (IMO)

Established in 1978 as the *International Maritime Consultation Organization* (IMCO) it was renamed as IMO in 1982. It aims to facilitate cooperation among governments on technical matters affecting international shipping, and including special responsibilities for safety of life at sea and for the protection of the marine environment through prevention of POLLUTION at sea.

United Nations Environment Programme (UNEP)

This is a programme, not an agency. Its main aim is to continuously review the world environmental situation, to develop and recommend policies on the protection and improvement of the environment, to promote international cooperation at all levels in the field of the environment, and to catalyse environmental work in and outside the UN system.

World Health Organization (WHO)

This was established in 1946 to promote the attainment by all peoples of the highest possible level of health.

Other United Nations organizations

Another UN organization of relevance is the UN Educational, Scientific and Cultural Organization (UNESCO), established in 1945 and in which the Intergovernmental Oceanographic Commission (IOC) and the Man and the Biosphere Programme (MAB) pay some attention to pollution by chemicals. The UN Industrial Development Organization (UNIDO), established in 1966, encourages and assists developing countries to promote and accelerate their industrializations. Its provisions include hazardous WASTE management, or waste reutilization. Attention is also paid to the safety implications of pesticides. The World Bank, established in 1944, gives loans for large-scale development programmes, but these should take account of ENVIRONMENTAL IMPACT and the potential for chemical pollution.

See also UNCED. [P.C.]

United States legislation There are five main areas of US, federal legislation covering:

1 controls on substances (Toxic Substances Control Act, TSCA, and the Federal Insecticide, Fungicide and Rodenticide Act, FIFRA);

2 controls on environmental quality (the Clean Air Act, CAA, and the Clean Water Act, CWA);

3 controls on current intentional dumping of wastes (the Resource Conservation and Recovery Act, RCRA);

4 powers to address previously CONTAMINATED LAND (Comprehensive Environmental Response Compensation and Liability Act, CERCLA or Superfund, and the Superfund Amendments and Reauthorization Act, SARA);

5 controls on the movements of hazardous materials (Hazardous Materials Transportation Act, HMTA).

Both TSCA and FIFRA require labelling of dangerous substances. RISK MANAGEMENT is done on the basis of RISK ASSESSMENT procedures but (in both TSCA and FIFRA) tempered with a need to balance environmental risks from the product against benefits derived from it in economic and social terms.

Both CWA and CAA define POLLUTION in terms of alterations to the ambient quality—chemical, physical and biological. Both recognize the need for the control of especially dangerous substances and both identify 'priority lists'. Regulation is largely through emission controls (that in CAA may be tradeable) based on ENVIRONMENTAL QUALITY STANDARDS and available technology—with some account of cost of implementation being taken into account. But the Pollution Prevention Act is now putting more emphasis on reduction at source.

RCRA has set up a cradle-to-grave tracking system for WASTE. CERCLA/SARA, in addressing earlier CONTAMINATION, impose strict liability that is joint and several. The Superfund provisions also introduces provisions to guard against major ACCIDENTS and to keep the public informed of potential dangers.

Finally, HMTA makes provision for controlling substances in transit, both nationally and internationally, by requiring appropriate documentation.

Both RCRA and CERCLA are concerned with protection of the soil and in this sense also complement the habitat orientated provisions of CWA and CAA. There is also a series of acts that seek to protect nature (Wildlife and Wilderness Acts). These cover ENDANGERED SPECIES, marine mammals and fish and WILDLIFE. *See also* ENVIRONMENTAL LAW; ENVIRONMENTAL (PROTECTION) AGENCIES. [P.C.]

units of selection Any biological entity that replicates itself and retains its identity from generation to generation is a unit of selection. A unit of selection is a selectable entity. Genetic units of selection include individual genes, chromosomes, genotypes, populations and species. In clonal, or asexually reproducing species, the entire GENOTYPE is reproduced in the offspring, and the individual genotype can be considered to be a unit of selection. In sexually reproducing species, individual organisms consist of unique collections of genes that are not transmitted as a unit to the next generation. However, the single genes upon which selection acts do not replicate independently, but rather depend on the reproduction of the entire organism in which they are carried. Thus the FITNESS of the individual genes is determined by the whole of the genotype and the fitness of an individual gene may, in fact, be heavily dependent on its interactions with other genes carried by the organism. A meaningful measure is a gene's (or allele's) average fitness, which is the average fitness of all of the individual's carrying it (each with their own unique genotype). Populations and species may also act as units of selection in that some will originate or go extinct more rapidly than others, resulting in a shift in the mean PHENOTYPE of the entire group. *See also* DARWINISM; NATURAL SELECTION. [V.F.]

universal time (UT) *See* TIME.

unmatched observations Observations made on different experimental or observed units. For example, if we measured daily egg production (per female) in a group of fruit flies under one set of conditions, and measured daily egg production on another group of females under another set of conditions, the two sets of data on egg production would be unmatched, because the two measurements had been made on different females. *Cf.* MATCHED OBSERVATIONS. [B.S.]

unordered transformation series *See* TRANSFORMATION SERIES.

unpolarized transformation series *See* TRANSFORMATION SERIES.

unstable equilibrium An equilibrium is unstable if a perturbation away from equilibrium tends to increase with time, rather than diminishing. It is sometimes possible that a small perturbation will diminish whilst a large one will increase, in which case the equilibrium is locally stable but globally unstable. Systems with unstable equilibria often show cycles or CHAOS, but may also become extinct or grow without limit. *See also* POPULATION REGULATION. [S.N.W.]

upper infralittoral zone *See* WITHIN-LAKE HABITAT CLASSIFICATION.

upstream–downstream comparison Comparison of the BIOTA above and below an effluent stream emptying into a stream or river. Upstream is used as a comparator for deviations from normality downstream, i.e. it is presumed that the downstream system would be like the upstream in the absence of STRESS —but this need not be so. *Cf.* BACI. *See also* MONITORING; RETROSPECTIVE ASSESSMENT. [P.C.]

upwelling Process by which deep, cold, nutrient-laden water is brought to the surface, usually by diverging equatorial CURRENTS or COASTAL currents that pull water away from the coast. Upwelling is especially conspicuous on the eastern sides of ocean BASINS, and such coastal upwelling regions are the most productive areas in the ocean. The nutrient-rich waters generally promote PHYTOPLANKTON growth which in turn supports rich ZOOPLANKTON and fish populations. Short trophic webs in upwelling areas mean more efficient energy transfers relative to the open ocean where trophic webs are longer. Upwelling areas represent about 0.1% of the ocean surface area but 50% of the the world's fishery. [V.F.]

uranium (U) Uranium (ATOMIC NUMBER = 92; relative ATOMIC MASS = 238.03; melting point = 1132.3°C; boiling point = 3818°C) has a crustal abundance of 2.3 mg kg^{-1}. Identified isotopes exist from ^{226}U to ^{242}U, three of which occur naturally, namely ^{234}U (abundance = 0.005 atom % (the percentage of the total number of atoms present (all isotopes) that are one particular isotope); $t_{1/2} = 2.45 \times 10^5$ years), ^{235}U (abundance = 0.720 atom %; $t_{1/2} = 7.04 \times 10^8$ years) and ^{238}U (abundance = 99.275 atom %; $t_{1/2} = 4.47 \times 10^9$ years). ^{238}U and ^{235}U are the parents of two natural radioactive decay series. Uranium exhibits oxidation states from +3 to +6, but only the +4 and +6 states occur in the environment. U^{4+} is highly insoluble in natural waters, forming minerals such as uraninite (UO_2), whereas U^{6+} is relatively soluble, existing in solution as anionic complexes of the uranyl ion (UO_2^{2+}) such as $[UO_2(CO_3)_2]^{2-}$. The most important uranium mineral is uraninite (or pitchblende) which can be either primary or secondary. Large secondary uraninite 'roll front deposits', formed by cyclical oxidation and reduction at redox boundaries in permeable rocks, are exploited by opencast mining. ^{235}U is fissile and is used for nuclear weapons and nuclear reactor fuel. ^{238}U and ^{234}U are not fissile, and so isotopic enrichment in ^{235}U is required for the use of uranium as nuclear fuel. Natural nuclear reactors operated in uranium deposits at Oklo, Gabon about 1.8×10^9 years ago when the $^{235}U : ^{238}U$ ratio was about 5 times greater than at present. Uranium isotopes are useful tracers of environmental processes via natural decay series equilibrium. *See also* FISSION; RADIONUCLIDE. [A.B.M.]

REFERENCES

Ivanovich, M. & Harmon, R.S. (eds) (1992) *Uranium-Series Disequilibrium*. Clarendon Press, Oxford.

Katz, J.J., Seaborg, G.T. & Morss, L.R. (eds) (1986) *The Chemistry of the Actinide Elements*. Chapman & Hall, London.

urban reclamation *See* RECLAMATION.

USEPA (United States Environmental Protection Agency) *See* ENVIRONMENTAL (PROTECTION) AGENCIES.

uses of genetically modified organisms *See* TRANSGENIC ORGANISMS.

UT (universal time) *See* TIME.

utility functions Utility is a concept originating in 18th-century economics. It has proved a particulary difficult concept to pin down empirically. Attempts to transfer the utility concept to evolutionary biology are misplaced; the standard evolutionary concept of FITNESS should always be used instead. *See also* BEHAVIOURAL ECOLOGY; EVOLUTIONARY OPTIMIZATION. [C.W.C.]

UV-A *See* ULTRAVIOLET RADIATION.

UV-B *See* ULTRAVIOLET RADIATION.

UV-C *See* ULTRAVIOLET RADIATION.

V

vacant niches Niches representing RESOURCES and/or conditions that are not utilized. Vacant niches are taken as evidence against COMPETITION theory, because this theory predicts that the pressure exerted by INTERSPECIFIC COMPETITION selects for NICHE shifts into unused parts of niche space. A persuasive case for the existence of vacant niches can be made when particular resources are not exploited in one area, while they are exploited in other areas, which demonstrates that the niche could be occupied. Convincing examples of vacant niches often concern PHYTOPHAGOUS insects, which is in agreement with the generally weak support for competition theory by this group. [J.G.S.]

vagility Ability to disperse. [P.C.]

Valdez principles Principles developed following the *Exxon Valdez* disaster by CEREs—the Coalition of Environmentally Responsible Economies (Box V1). Businesses are invited to sign publicly to show commitment to environmental protection. [P.C.]

validation A term applied to BIOASSAYS and ECOTOXICITY tests. Implies checking their ability to yield 'true' assessments, but the pursuit of truth is elusive in science. Instead, it generally involves checking the sensitivity of one measure of an environmental signal against another. *See also* RING TEST. [P.C.]

valley bog *See* MIRE; VALLEY MIRES.

valley mires Valley mires, as their name implies, commence their development as TOPOGENOUS wetlands in which the peat-forming vegetation depends upon water draining from the surrounding CATCHMENT. Such mires are often complex in their HYDROLOGY, their vegetation and their TOPOGRAPHY. The nature of the vegetation is dependent upon the chemistry of the drainage water and the rate at which this water flows through the system. Even mineral-poor water can support relatively demanding FEN vegetation if the flow rate is adequate for a constant renewal of supply. However, where the flow rate is slow, acidophilous vegetation may develop and local growth of plants such as *Sphagnum* can elevate the surface of the MIRE until ground water plays relatively little part in the supply of nutrients and the system becomes locally OMBROTROPHIC.

Therefore, in sites with a rich nutrient supply in the ground water, valley mires usually take the form of valley fens. But in regions of poor mineral supply, such as on sandstones and granites, valley mires are particularly complex in their structure. Often a central stream carries the main flow of water, and its immediate environment is well supplied with minerals because of the faster flow; it therefore supports fen vegetation. Lateral gentle slopes and flood-plain surfaces on either side of the central stream have poorer ground-water flow and become more ACID, bearing poor-fen or even BOG vegetation.

Many European valley mires that have been examined stratigraphically show evidence of human impact in their initial sediments. Burning of former vegetation releases fine charcoal particles that erode into valleys and create poor drainage conditions by blocking the soil interstices. The waterlogging that results then initiates the development of valley mire ecosystems. [P.D.M.]

vanadium (V) Vanadium (ATOMIC NUMBER = 23; relative ATOMIC MASS = 50.942; melting point = 1890°C; boiling point = 3380°C) has a concentration in the Earth's CRUST comparable to ZINC and NICKEL (100 mg kg^{-1}). The METAL is invariably obtained as a by-product from the production of other materials since concentrated deposits are rare. Examples are carnotite, potassium uranyl vanadate [$K(UO_2)(VO_4)1.5H_2O$] and vanadinite [$PbCl_2.3Pb_3(VO_4)_2$]. Vanadium is also found in certain crude oils, in particular those from Venezuela and Canada, from which it may be recovered. The metal has been shown to be nutritionally essential in animals, although no deficiency symptoms have been identified in humans. Its biological role remains largely unclear. [P.J.B.]

variability Changes in ecological properties and processes through space and time. Also refers to changes in results from experiments carried out simultaneously (replicability), or at different times in the same laboratory (REPEATABILITY), or at different laboratories (REPRODUCIBILITY). It can be measured as a COEFFICIENT OF VARIATION. *See also* POPULATION DYNAMICS. [P.C.]

variable pathways For many successions it has been established that the pathways from the initial invaders to the last species can vary greatly. Part of the variation results from chance (which species happen to reach the given site) and part results from variation in the local conditions — those that matter depending on the scale of the plant, and being different for (say) lichens and tree seedlings. Unequivocally established evidence for variable pathways comes from the study of macrofossils in PEAT built up in HYDROSERES, summarized for Britain by Walker (1971)—see Fig. V1. Many short-term observational studies have also provided evidence. [P.J.G.]

REFERENCE

Walker, D. (1971) Direction and rate in some British post-glacial hydroseres.

Box V1 The Valdez principles

1 *Protection of the biosphere.* We will minimize and strive to eliminate the release of any pollutant that may cause environmental damage to the air, water, or Earth or its inhabitants. We will safeguard habitats in rivers, lakes, wetlands, coastal zones and oceans and will minimize contributing to the greenhouse effect, depletion of the ozone layer, acid rain or smog.

2 *Sustainable use of natural resources.* We will make sustainable use of renewable natural resources such as water, soils and forests. We will conserve non-renewable natural resources through efficient use and careful planning. We will protect wildlife habitat, open spaces and wilderness while preserving biodiversity.

3 *Reduction of disposal of waste.* We will minimize the creation of waste, especially hazardous waste, and wherever possible recycle materials. We will dispose of all wastes through safe and responsible methods.

4 *Wise use of energy.* We will make every effort to use environmentally safe and sustainable energy sources to meet our needs. We will maximize the energy efficiency of products we produce or sell.

5 *Risk reduction.* We will minimize the environmental, health and safety risks to our employees and the communities in which we operate by employing safe technologies and operating procedures and by being constantly prepared for emergencies.

6 *Marketing of safe products and services.* We will sell products or services that minimize adverse environmental impacts and that are safe as consumers commonly use them. We will inform consumers of the environmental impacts of our products and services.

7 *Damage compensation.* We will take responsibility for any harm we cause to the environment by making every effort to fully restore the environment and to compensate those persons who are adversely affected.

8 *Disclosure.* We will disclose to our employees and to the public incidents relating to our operations that cause environmental harm or pose health or safety hazards. We will disclose potential environmental, health or safety hazards posed by our operations, and we will not take any action against employees who report any condition that creates a danger to the environment or poses a health and safety hazard.

9 *Environmental directors and managers.* At least one member of the board of directors will be a person qualified to represent environmental interests. We will commit management resources to implement these principles, including the funding of an office of vice president for environmental affairs or equivalent executive position, reporting directly to the chief executive officer (CEO), to monitor and report upon our implementation efforts.

10 *Assessment and annual audit.* We will conduct and make public an annual self-evaluation of our progress in implementing these principles and in complying with all applicable laws and regulations throughout our worldwide operations. We will work toward the timely creation of independent environmental audit procedures which we will complete annually and make available to the public.

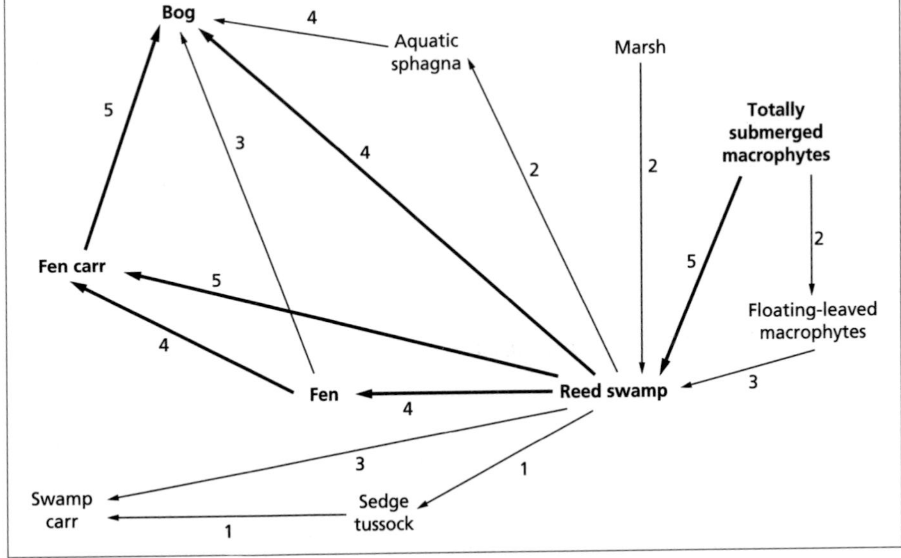

Fig. V1 Hydroseral steps for which there is fossil evidence. The figures indicate the numbers of times that evidence has been found that particular transitions took place. The bold arrows indicate the major pathways. (After Walker D. (1971) Direction and rate in some British post-glacial hydroseres. In: *Studies in the Vegetational History of the British Isles* (eds D. Walker & R.G. West), pp. 117–139. Cambridge University Press, Cambridge.)

In: *Studies in the Vegetational History of the British Isles* (eds D. Walker & R.G. West), pp. 117–139. Cambridge University Press, Cambridge.

variance A measure of variation. In a SAMPLE it is represented by the Roman letter s^2, and is an estimate of the POPULATION variance represented by the Greek letter sigma-squared (σ^2). It is the square of the STANDARD DEVIATION. The formula for calculating the sample variance is:

$$s^2 = \frac{\sum(\bar{x} - x)^2}{n - 1}$$

where $\bar{x}$ is the sample MEAN, x is each observation in the sample, n is the number of observations in the sample, and the Greek symbol Σ (capital sigma) is the mathematical notation for 'sum of'. In the following sample of seven observations or measurements: 2, 3, 3, 5, 9, 15, 16, the $\bar{x} = 6.28$ and therefore:

$$s^2 = \left(4.28^2 + 3.28^2 + 3.28^2 + 1.28^2 + 2.72^2 + 8.72^2 + 9.72^2\right)/(n - 1)$$

$$s^2 = 219.39/6 = 36.56$$

The two components in the above formula for the variance are the numerator, $\Sigma(\bar{x} - x)^2$, called the SUM OF SQUARES and the denominator, $n - 1$, called the DEGREES OF FREEDOM.

The sum of squares is related to the amount of variation in a sample. If the observations (x) were tightly clustered around $\bar{x}$, then each value of $(\bar{x} - x)^2$ would be small and the sum of squares would be small. If the observations (x) were some distance from $\bar{x}$, then each value of $(\bar{x} - x)^2$ would be large and the sum of squares would be large. However, this would also be true if a simple 'sum of deviations' ($\Sigma|\bar{x} - x|$) were used as the numerator. This sum of deviations is not used because the sum of squares has an additional and important property—it can be partitioned or summed (*see* REGRESSION ANALYSIS, CURVILINEAR; REGRESSION ANALYSIS, LINEAR and ANALYSIS OF VARIANCE).

It may appear that the correct value for the denominator should be n (as for the mean), rather than $n - 1$. This is not so. Suppose that we had only one observation ($x = 9$) in our sample, how much information would we have about the population mean and variance? Our best estimate of the mean would be 9 (from one piece of information), but we would have no information about variation. With two observations ($x = 9$ and $x = 11$) we have two (n) pieces of information about the likely value of the mean (our best estimate would be $(9 + 11)/2 = 10$), and one ($n - 1$) piece of information about variation (9–11). Therefore with variance the correct denominator is $n - 1$. There is also another reason for using $n - 1$. With small values of n, the sample variance is biased. It consistently underestimates the real variance of the population ($s^2 < \sigma^2$). This is because in small samples extreme values of x are frequently missing. Statisticians have found that dividing by $n - 1$ corrects this bias.

In practice, the formula for sum of squares given above is rarely used. An equivalent (but less transparent) formula is:

$$\text{Sum of squares} = \sum x^2 - \frac{\left(\sum x\right)^2}{n}$$

where again x is each observation in the sample, n is the number of observations in the sample, and Σ is the mathematical notation for sum of. Notice that Σx^2 (sum of the squares of x) is not the same quantity as $(\Sigma x)^2$ (square of the sum of x). For the sample of seven observation used above: 2, 3, 3, 5, 9, 15, 16, the calculation would be:

$$\sum x^2 = 4 + 9 + 9 + 25 + 81 + 225 + 256 = 609$$

$$\left(\sum x\right)^2 = \left(2 + 3 + 3 + 5 + 9 + 15 + 16\right)^2 = 53^2 = 2809$$

$$\text{Sum of squares} = 609 - \left(2809/7\right) = 207.72$$

$$\text{Variance} = 207.72/6 = 34.62$$

Notice that this value is slightly different from that worked out previously. The latter formula is more accurate (as well as easier to use) because less rounding errors are involved in the computation. [B.S.]

variance : mean ratio *See* INDICES OF DISPERSION.

variance minimization A hypothetical response to random variation in ecological benefits or costs; VARIANCE minimization assumes that reduced VARIABILITY should always be favoured. If each element of a set of available behavioural options has the same average benefit (or cost), variance minimization becomes equivalent to RISK AVERSION. *See also* BET-HEDGING. [T.C.]

vector, mathematical A quantity having both size and direction as opposed to a scalar quantity that has size only. Ordinary numbers (which have only size) are therefore scalars, but a force (which has a size and is applied in a particular direction) is a vector. In ecology they occur in phase plane graphs/diagrams, in which the dynamics of two-species systems (COMPETITION or predator–prey) are depicted. Each point on the graph represents a joint DENSITY of the two species. To each point in this graph we can attach an arrow, or vector, whose length and direction will indicate the dynamics of the system (*see* LOTKA–VOLTERRA MODEL). [B.S.]

vector, parasites Parasites must transmit between definitive hosts (*see* SECONDARY HOST), and if a second species acts to convey the PARASITE between hosts, it is known as a vector species. Parasites may have more or less specificity to different vectors. The vector species may suffer reduced FITNESS when infected by the parasite (but need not). The transmission rate of infection of vector-borne parasites depends on the DENSITY of vectors. *See also* INDIRECT TRANSMISSION, PARASITES; PARASITISM. [G.F.M.]

vegetation The sum total of plants in the broadest sense (tracheophytes, bryophytes and algae) at a site, by convention including certain symbiotic fungi (mutualistic mycorrhizae and mycorrhizae parasitized by orchids and a few other specialized plants) but not the fungi parasitic or saprophytic on the plants.

The vegetation of an area is distinguished from the flora (species list) and requires minimally a definition of the relative ABUNDANCES of the various species. Ideally, the distributions of the species in respect of HABITAT factors will be given. [P.J.G.]

vegetative reproduction Reproduction by somatic division. In plants this widespread method of multiplication may involve, for example, rhizomes, stolons (runners), tubers, bulbs, corms, suckers, gemmae or turions (swollen detached buds of aquatics). Many of these structures also serve as perennating organs. Spread by this method is limited, the formation of CLONES resulting in the local monopoly of space, sometimes huge areas being covered, for example hillsides by bracken (*Pteridium aquilinum*) and reed beds by *Phragmites australis*. Ramets (individuals of a clone) may remain attached for a long time or separate by decay or fragmentation. Widely used artificial means of vegetative reproduction (or propagation) of plants include budding, LAYERING, simple division and rooting of cuttings. Micropropagation (mass rapid clonal propagation), usually from shoot tips, is frequently employed to increase stocks (notably of orchids) and to produce virus-free plants.

The term is sometimes applied to animals, synonymously with ASEXUAL REPRODUCTION. Through continued vegetative reproduction, the LONGEVITY of a clone may exceed that of its members by several orders of magnitude. Most plants and animals reproducing vegetatively are also capable of sexual reproduction. Unless clones are dioecious (either purely male or female), they may be capable of sexual reproduction via self-fertilization. Self-sterility mechanisms, however, may require access to non-clone-mates. [R.N.H. & A.J.W.]

vehicle emissions *See* MOTOR VEHICLE POLLUTION.

veld The temperate GRASSLANDS of the eastern area of the southern part of Africa. Scattered trees and bushes may be present, particularly at lower altitudes. [P.D.M.]

vernal pools Seasonally wet, shallow pools, commonly found in the Central Valley and surrounding lower foothill regions of the coastal and Sierra Nevada ranges in California. These are underlain by an impervious claypan, rocks and silica-cemented horizons (duripan), and support somewhat unique (specialized) biota. A rather unusual combination of SOIL factors, TOPOGRAPHY, HYDROLOGY and the Mediterranean CLIMATE (dry summer, distinct evapotranspiration pattern (Fig. V2), large year-to-year variability) accounts for the unique vegetation and its current status as a highly threatened habitat. There are a large variety of pools varying in depth (10–60 cm), diameter (3–50 m), shape, interpool connectivity, and in FLORISTICS (see Plate 27, facing p. 398). Vernal LAKES, MARSHES and other intermittently wet habitats, some alkaline, share flora with the vernal pools (Mason 1969), but differ in the latter having more showy, spring-flowering ANNUALS.

Another very significant feature lies in the use of island models for analysing species DIVERSITY patterns and the genetic structure as well as SPECIATION mechanisms. Several speciose genera have undergone evolutionary diversification involving the origin of annual, autogamous or polyploid taxa only recently (Jain & Moyle 1984). Fauna includes copepods and ostracods, free-swimming fairy shrimp, and many insects (diving beetles, mayflies, hemipterans and numerous oligolectic pollinating bees and flies). POLLINATION biology shows a critical role of plant–pollinator COEVOLUTION. The role of birds and mammals is known to be little more than foraging, or wading out of the surrounding GRASSLANDS, but more work is needed. Likewise, many algae are found and perhaps play a critical role in physical–chemical changes, but very little is known of them. Vernal pools, even in the DROUGHT years, tend to resist invasion by many non-native taxa and, in fact, show dramatic resilience due to a large SEED BANK of many natives.

Over 120 plant genera occur in the vernal pool habitat of California floristic PROVINCE, of these 16 (e.g. *Downingia, Lilaea, Limnanthes, Orcuttia* and *Pogogyne*) are restricted to this habitat. Nearly 40 genera (e.g. *Isoetes, Lasthenia, Myosurus* and *Veronica*) have at least one species endemic in vernal pools. For the small land area (less than 1%), vernal pools have statewide a much higher proportion of endemics; it is commonly argued (but without much experimental work so far) that the amphibious (rapid cycles of wet and dry) conditions combined with the insular (fragmented, sparse) pattern yield strong selective forces in favour of locally adapted, specialized taxa with annual and highly plastic LIFE-HISTORY TRAITS. Dispersal patterns, DORMANCY, pollination system and the role of ploidy require further research. Physiological work on several aquatic or semi-aquatic genera (e.g. *Isoetes* and *Orcuttia*) has shown a very highly adaptive array of photosynthetic modes, including the CAM and PEP (phosphoenolpyruvate) carboxylase-mediated C_4 mechanisms.

Californian vernal pools have been classified into the Northern and the Southern groups; the former includes Hardpan, Claypan, Basalt Flow and Sandstone Tanks as four types which vary, aside from the soil substrate and impervious layer types, in distribution along the north-east to south-western axis. One can further recog-

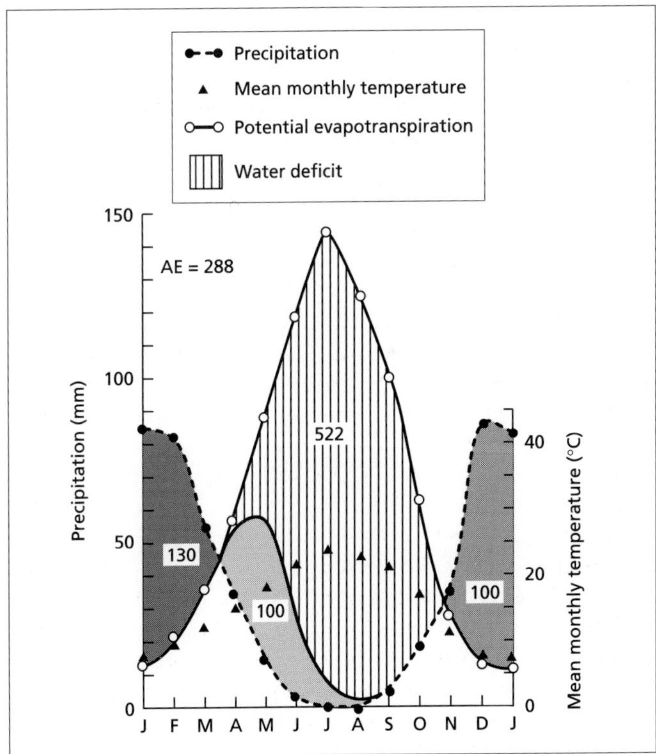

Fig. V2 Mediterranean climate has a characteristic pattern of evapotranspiration/water availability. An example for Davis city is shown here. Actual evapotranspiration (AE) provides a useful measure of water availability for plants, given monthly temperature and rainfall data. Values higher than the level of AE = 288 at Davis represent severe water deficit in dry summer months.

nize hydrologic types (viz. swale, basin, and small and steep versus shallow and wide), each of which has a different set of INDICATOR SPECIES. Volcanic mudflow pools are generally small, rocky, well-drained, with CLAY-loam soils atop an impervious mudflow bedrock. One study found in them 59 plant genera, 76% natives (15% perennials, 38% wetland generalists, 13% obligate pool species) but none endemic to this class of pools.

Using *Limnanthes* as an example, there is an interesting pattern of species distributions in various types of pools. It is remarkable that in only five out of 200 sites, two *Limnanthes* taxa were found living together, whereas on a scale of 10–20 km in a region, several taxa are found among neighbouring sites.

Within-pool diversity shows an interesting phenological sequence; different species occur in concentric circles, and flower sequentially during the spring months (Fig. V3). This relates to moisture requirements, measured in terms of pool inundation periods, and pollinator visits. Vernal pools and their threatened biota have indeed drawn a great deal of attention, to which the discovery of the agricultural potential in using *Limnanthes* seed for unique industrial oils has added an economic dimension. Domestication of *Limnanthes* as an annual winter oilseed has progressed steadily, albeit much breeding and seed-processing work remains (Fig. V4).

Are Californian vernal pools unique? The answer is both yes and no. The habitat has analogues such as granitic domes and vernally moist sandy flats in south-west Australia and granite flat rocks in Georgia and Texas. Pools in Chile even show examples of *Lasthenia* and *Pilularia* disjuncts. But the indigenous flora, and its pollinators, PHENOLOGY, etc. are unique for each area. Zedler (1987) succinctly reviewed many ecological and evolutionary features of vernal pools, with an emphasis also on the ecosystem-level questions. Numerous projects for the habitat mapping and conservation are in progress or planned by various groups in charge of the wildlife, wetland and ENDANGERED SPECIES interests as well as numerous local public groups.

A recent symposium (Witham, in press) reported significant progress in the mapping and classification of vernal pool ecosystems, with an emphasis on devising ways to monitor changes in the critically listed animal and plant species. ECOLOGY and BIOGEOGRAPHY of both microinvertebrates (insects and crustaceans) and macroinvertebrates (frogs, toads and salamanders) still need more attention, as do the phylogenetic and POPULATION ECOLOGY models for the evolution of the fascinatingly varied amphibian life forms. The plant genus *Limnanthes* provides numerous ecogenetic opportunities for measuring selection, GENE FLOW and breeding system parameters. A comprehensive survey of conservation needs and actions was reported (Keeler-Wolf *et al.* 1995) from which one can see the glass as half-full (several protected areas and inter-agency plans are in place) or as half-empty (listed and candidate taxa are continuing to lose ground and the mitigation measures including artificially created pools seem largely unsuccessful). Vernal pools in California have deservedly enjoyed wide media attention and increased academic involvement. [S.K.J.]

REFERENCES

Ikeda, D.H. & Schlisling, R.A. (1990) *Vernal Pool Plants: Their Habitat and Biology.* California State University, Chicago, Calif.

Jain, S. & Moyle, P. (eds) (1984) *Vernal Pools and Intermittent Streams.* Institute of Ecology, Publication No. 28, University of California, Davis, Calif.

Keeler-Wolf, T., Elam, D.R. & Flint, S.A. (1995) *California Vernal Pool Assessment: A Preliminary Report.* California Department of Fish and Game, Sacramento, Calif.

Mason, H.L. (1969) *A Flora of the Marshes.* University of California Press, Berkeley, Calif.

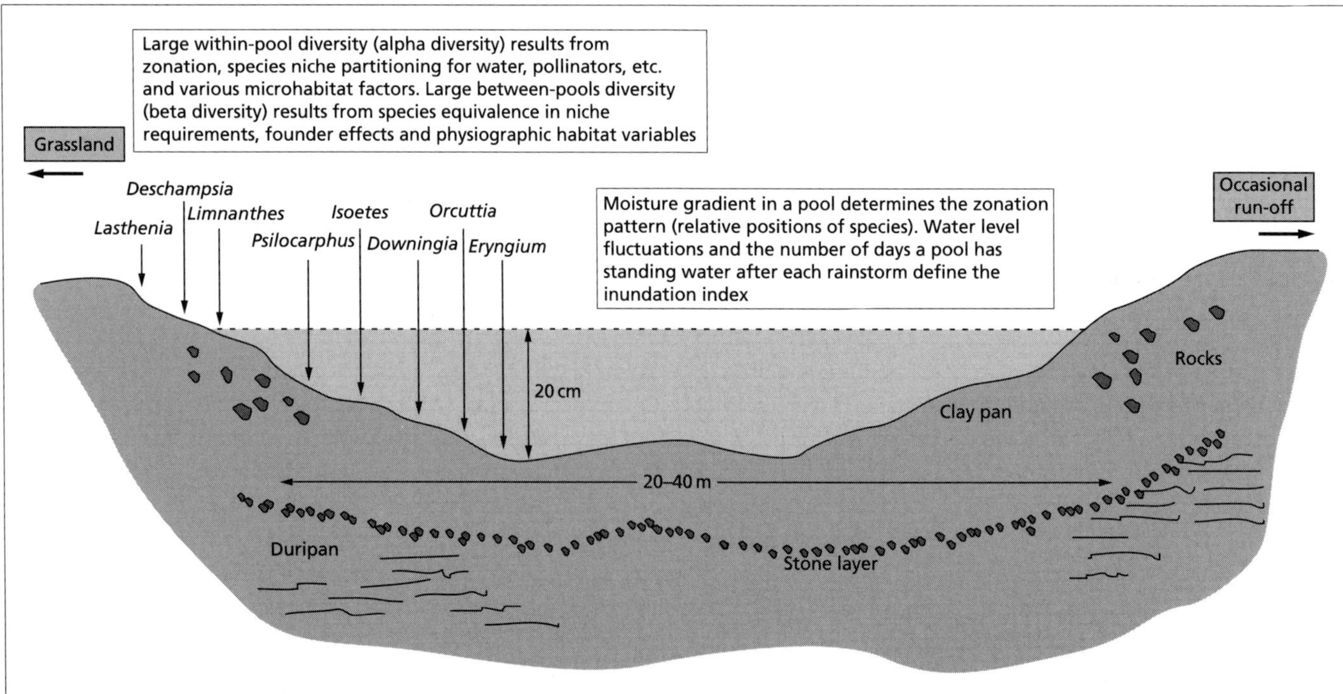

Fig. V3 Diagram of a vernal pool showing some important features.

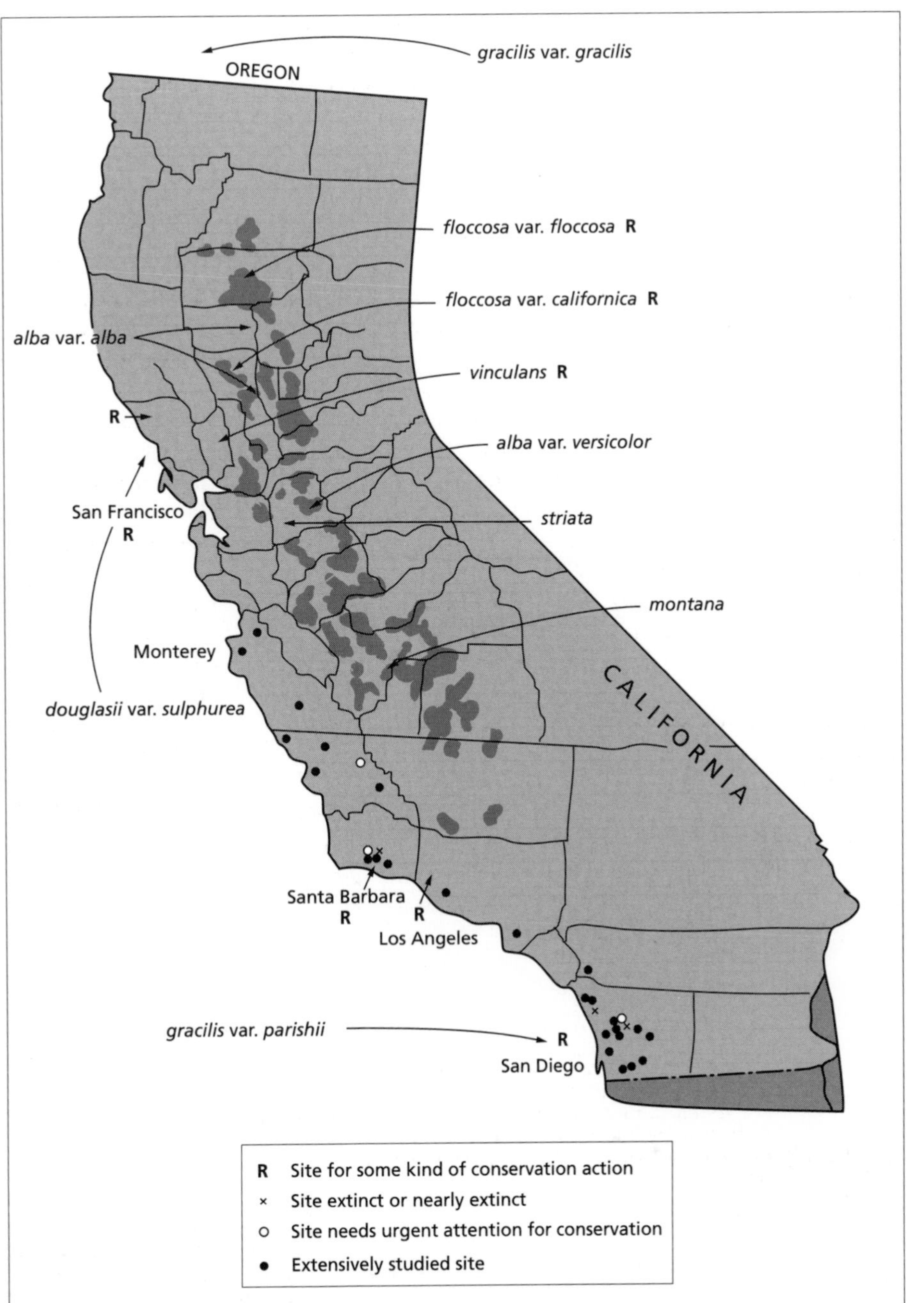

Fig. V4 Distribution of *Limnanthes* shows varying species range and north–south transect separating species distributions.

Witham, C.W. (in press) *Ecology, Conservation and Management of Vernal Pool Ecosystems*. California Native Plant Society, Sacramento, Calif.

Zedler, P.H. (1987) *The Ecology of Southern California Vernal Pools: A Community Profile*. US Fish and Wildlife Service Biological Report No. 85.

vertical migration Movements of PELAGIC organisms on a daily or seasonal cycle that may be associated with FEEDING or breeding. The typical DIEL pattern is movement upward during the night and movement downward during the day. The distance travelled is species dependent, but even small copepods may travel several hundred metres twice in a day. Various explanations have been proposed to explain this behaviour, including escape from visual predators, conservation of energy by time spent in deeper colder water, and exposure to new feeding areas. [V.F.]

vertical zonation The open OCEAN can be divided vertically with respect to the degree of light penetration (*see* EUPHOTIC ZONE; DISPHOTIC ZONE and APHOTIC ZONE) or with respect to depth (i.e. EPIPELAGIC zone = 0–*c*. 200 m; MESOPELAGIC zone = *c*. 200–1000 m; BATHYPELAGIC zone = *c*. 1000–4000 m; ABYSSAL-pelagic zone = *c*. 4000–6000 m; and HADAL-pelagic zone = *c*. 6000–10 000 m). The benthic realm of the ocean can likewise be subdivided into vertical zones (e.g. CONTINENTAL SHELF, slope, rise, abyssal plain, TRENCH). Vertical zonation also refers to the non-random distribution of species with respect to depth in the water column, in the sediment column or in the INTERTIDAL ZONE. For example, in the open ocean, characteristic groups of species are associated with each of the vertical zones listed above. Sediment-dwelling MACROFAUNA can be classified as surface-dwellers, shallow burrowers or deep burrowers, and the major groups of sedimentary bacteria are distributed in repeatable vertical sequences with respect to the efficiency of their metabolic pathways (e.g. aerobic bacteria live above sulphate reducers). In the intertidal zone, dominant organisms occur in distinct bands whose respective upper and lower limits coincide with specific horizontal levels relative to a tidal datum, such as mean low water. The relative position of the bands is controlled by a combination of physiological limits (e.g. tolerance to desiccation) and biological interactions (e.g. competition, predation). *See also* LAKE STRATIFICATION. [V.F.]

Vertisols *See* SOIL CLASSIFICATION.

vesicular–arbuscular (VA) mycorrhizae *See* ENDOTROPHIC MYCORRHIZAE.

vicariance biogeography A type of BIOGEOGRAPHY which is concerned with discovering the commonality of observed distributional patterns of organisms shown by evolutionarily unrelated taxa, which suggests a common and simultaneous process. Hypotheses are based on the assumption that BIOTA are split up (vicariated) by the emergence of barriers. Such barriers could be oceanic straits opened up as a result of CONTINENTAL DRIFT separating originally continuous land areas. They could also be climatic with, for instance, declining temperatures such as during the Pleistocene Ice Age restricting organic distributions to lower latitudes and altitudes, leaving isolated patches of territory as REFUGIA. Vicariance biogeography arose shortly after the general acceptance of continental drift, as reaction against conventional DISPERSAL biogeography which follows from the premise that from a CENTRE OF ORIGIN, species undergo jump dispersal across pre-existing barriers. A later development, known as cladistic biogeography, attempts to apply the cladistic method of TAXONOMY to geographic entities as well. [A.H.]

REFERENCE

Nelson, G. & Rosen, D.E. (eds) (1981) *Vicariance Biogeography: A Critique*. Columbia University Press, New York.

vicarious distribution This is a type of distribution produced after BIOTA are split up (vicariated) by the emergence of barriers through, for example, CONTINENTAL DRIFT, which separate parts of a once continuous biota and alter the proximity of populations or organisms irrespective of their own movements. [A.H.]

vigilance Prey can avoid predators by increasing the amount of time and energy they spend trying to sense nearby predators. Once a PREDATOR is detected, suitable measures can be taken, such as flight, defence or enhanced vigilance. There may well be a TRADE-OFF between predation risk and foraging success mediated by the degree of vigilance. Vigilance is best understood in vertebrates, particularly birds. *See also* PREDATOR–PREY INTERACTIONS. [D.A.B.]

REFERENCE

Lima, S.L. & Dill, L.M. (1990) Behavioral decisions made under the risk of predation: a review and prospectus. *Canadian Journal of Zoology* **68**, 619–640.

viroids *See* MICROBIAL ECOLOGY.

virulence That aspect of PARASITES which measures the harm done to the HOST, although usually only used for MICROPARASITES. Usually meaning the case MORTALITY RATE of a parasite (i.e. the risk that an individual host will die when infected), although, generally, the harm may be considered as a combination of MORBIDITY (disease) and mortality (sometimes referred to as pathogenicity of the parasite).

It is also used to describe the effect of the parasite on the host population level, in which case it also includes the aspects of the parasite's ability to infect hosts, such that a parasite with a high basic reproduction number (*see* TRANSMISSION RATES OF PARASITES) may be more virulent. [G.F.M.]

viruses *See* MICROBIAL ECOLOGY.

viscosity The property of a fluid to resist motion within itself, arising from the intermolecular forces between layers as the fluid flows over a surface. Viscosity is important in problems relating to flow through tubes, soils and other porous media. Newton's law of viscosity states that the shearing force τ between layers of a fluid is proportional to the velocity gradient du/dz. The coefficient of proportionality is the dynamic viscosity η:

$$\tau = \eta(du/dz)$$

where u is velocity and z is distance from the wall. τ has units of $N\,m^{-2}$ and η is in $N\,s\,m^{-2}$ or $kg\,m^{-1}\,s^{-1}$. Fluidity is the reciprocal of the dynamic viscosity. A related measure, the kinematic viscosity is the dynamic viscosity divided by the density of the fluid, and has units of $m^2\,s^{-1}$.

Viscosity of liquids decreases with TEMPERATURE. In the case of gases, viscosity is independent of PRESSURE and increases with temperature. Typical values are given in the Table V1. [J.G.]

viscous forces *See* REYNOLDS NUMBER.

vitalism philosophy *See* HOLISM.

vitamins Organic compounds that are essential for heterotrophs only in minute quantities, but which they cannot synthesize themselves. Vitamins, or derivatives from them, normally have a coenzyme function. [P.C.]

viviparity The production of liveborn young. This form of reproduction is generally coupled with sexual reproduction and internal FERTILIZATION, although some derived species have secondarily evolved forms of ASEXUAL REPRODUCTION, such as PARTHENOGENESIS. Viviparity is distinguished from OVOVIVIPARITY by the continued transfer of nutrients from the mothers to the developing young.

Table V1 Typical viscosities of liquids.

Fluid	η (mN s m^{-2}) at 20°C	η at stated temperature
Water	1.00	0.405 at 70°C
Ethanol	1.20	0.50 at 70°C
n-Pentane	0.23	0.16 at 70°C
Benzene	0.65	0.35 at 70°C
Olive oil	84	12.4 at 70°C
Air	18×10^{-3}	22×10^{-3} at 100°C
Hydrogen	25×10^{-3}	10.4×10^{-3} at 100°C
Carbon dioxide	14.7×10^{-3}	18.6×10^{-3} at 100°C

The transfer of nutrients from the mother to the developing young is accomplished via the placenta in placental mammals, but by a diversity of other structures in other organisms. For example, in some live-bearing fishes it is likely to be accomplished by the secretion of nutrients by the mother into the ovarian fluids, then their absorption by the young via tissues associated with the digestive tract (Goodeidae) or the pericardial membrane (*Heterandris formosa*: Poeciliidae). In many cases, viviparity is inferred from the young gaining mass during development, but the mechanism of nutrient transfer is unknown. Workers with some animal groups, such as reptiles, have argued that one should not distinguish between ovoviviparity and viviparity because there is a continuum of variation, from mothers providing no nutrients to the young after fertilization to most maternal provisioning being provided after fertilization, with the border between the two forms of reproduction being difficult to define. *See also* LIFE-HISTORY EVOLUTION. [D.N.R.]

VOC *See* VOLATILE ORGANIC COMPOUND.

volatile organic compound (VOC) Volatile organic compounds evaporate readily and have adverse effects on human health and the environment. Examples of VOCs include ethylene, propylene, BENZENE, styrene and acetone. *See also* HYDROCARBONS. [P.C.]

volcanic gases A volcanic eruption reduces the pressure on gases which are enclosed in rock cavities and these are often released explosively. WATER vapour (H_2O), CARBON DIOXIDE (CO_2) and sulphur dioxide (SO_2) are the most common. SULPHUR is often also present as the element and is quickly oxidized on exposure to the ATMOSPHERE at temperatures commonly observed in eruptions (700–1200°C).

All these gases are present in the PLUME after an eruption, but large quantities of fine DUST are also commonly emitted and are carried up into the STRATOSPHERE by the buoyancy of the plume (Mt. Pinatubo was a spectacular example of dust at high temperature, and Sakura-jima in Kyushu is a continuous source of these components, while Stromboli emits primarily SO_2 and is therefore called a solfatara). [R.S.S.]

volcanic lakes *See* LAKES, TYPES OF.

Volterra, V. *See* CHARACTERS IN ECOLOGY.

volume-limited sampling *See* SAMPLING METHODOLOGY/DEVICES.

von Baer's law Inspired by the romantic ideas of the early 19th-century German *Naturphilosophie*, the Russian comparative embryologist, Karl Ernst von Baer, put forward the generalization that individual development proceeds from the general features of the archetype, through higher grades of differentiation until the particular features of the species are established. For example, the vertebrate fetus does not pass through the permanent forms of the articulata and mollusca (as would be the case in Haeckel's law of RECAPITULATION). Instead, the embryo of a vertebrate, say a human fetus, is already a vertebrate at the earliest stage of its ONTOGENY. The archetypes were discrete metaphysical entities to this author, who never accepted the contemporary evolutionary ideas; analogous ideas were previously defined by the French comparative anatomist G. Cuvier. *See also* HETEROCHRONY. [P.A.]

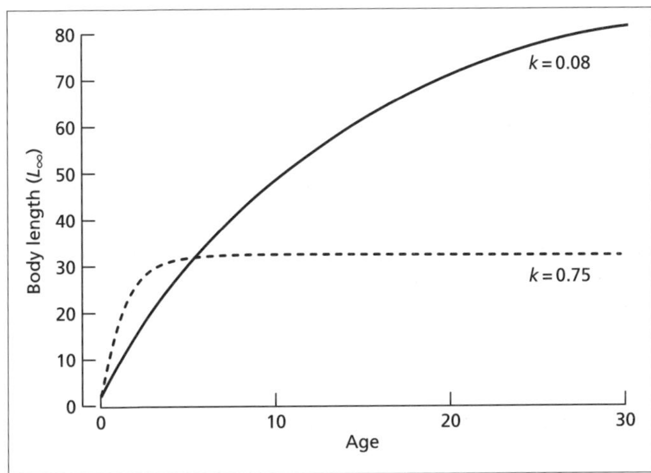

Fig. V5 Plot of two growth curves illustrating the von Bertalanffy equation. The curves have the same initial size, but one has a low value of k and a high value of asymptotic size, L_∞, the other a high value of k and a low value of L_∞.

REFERENCE

Richards, R.J. (1992) *The Meaning of Evolution: the Morphological Construction and Ideological Reconstruction of Darwin's Theory*. University of Chicago Press, Chicago, Ill.

von Bertalanffy equation A simple equation used to describe somatic growth. If W represents body mass and t is age:

$$dW/dt = aW^m - bW$$

where a, b and m are constants. The first term on the right-hand side represents the rate of anabolism—the synthesis of tissues—and the second term represents the rate of their breakdown. The equation is usually used in an integrated form for the case $m = 2/3$, when:

$$L(t) = L_\infty\left(1 - ce^{-kt}\right)$$

where $L(t)$ represents body length at age t, and L_∞, c and k are constants. L_∞ represents asymptotic body length, c depends on initial body length, and k is sometimes called the 'growth coefficient'. k is low if the organism takes a long time to approach its final size, and high if it approaches its final size rapidly (Fig. V5). *See also* ECOLOGICAL ENERGETICS. [R.M.S.]

Vorsorgeprinzip One of the several principles of German environmental policy that roughly translates into the PRECAUTIONARY PRINCIPLE. It was first enunicated by the German Federal Government in 1976, and became more widely used in the rest of Europe in the 1980s. It raises the need to act against risks that are not yet understood or have not yet been proved, and can also include the need to act in the absence of risk, as in maintaining an environmental resource undisturbed. The other principles of action include the principle of proportionality of administrative actions and the prohibition of excessive actions. Action taken under *Vorsorgeprinzip* thus requires a balancing of risks, and costs and benefits. *Cf.* RISK ASSESSMENT. *See also* ENVIRONMENTAL (PROTECTION) POLICIES. [P.C.]

REFERENCE

O'Riordan, T. & Cameron J. (1994) *Interpreting the Precautionary Principle*. Cameron May, London.

Vrba's effect hypothesis From work on fossil antelopes, E.S. Vrba noted the trend that specialists had higher rates of SPECIATION than generalists. She explained this as an effect of the physiologies of the organisms themselves. Narrowly specialized organisms make up species that tend to fragment easily — leading to more and more specialization and hence speciation. [P.C. & V.F.]

REFERENCE

Vrba, E.S. (1980) Evolution, species and fossils: How does life evolve? *South African Journal of Science* **76**, 61–84.

W

WAF *See* WATER ACCOMMODATED FRACTION.

Wallace theories *See* BIODIVERSITY GRADIENTS.

warning coloration *See* APOSEMATIC (WARNING) COLORATION.

waste That for which there is no further use. But this is very much in the eye of the beholder. Thus metal that is SCRAP to one group may form the raw material of another; discarded food materials can form the useful basis of compost; all organic wastes can be used as a source of heat energy. As far as the law is concerned, waste is usually identified relative to the views of the producer rather than user. Wastes can be classified by source (household, industrial, clinical), form (liquid, solid, slurries, etc.) and legal definitions (controlled, special, etc.). *See also* WASTE DISPOSAL. [P.C.]

waste disposal Options for dealing with WASTE. In principle this should only be considered once recovery, REUSE or RECYCLING options are exhausted. This can involve LANDFILL, treatment (biological, chemical and physical), fixation (encapsulation in a variety of matrices), deep underground disposal, DUMPING AT SEA, or INCINERATION. Waste disposal is usually subject under the law to a DUTY OF CARE whereby all those in the chain leading from source, through collection and transport, to ultimate sink are required to adhere to the regulations, thus ensuring the safe handling of waste at all times and enabling the route and disposal of the waste to be tracked. [P.C.]

waste minimization Attempts to reduce losses of WASTE by effective management, increased EFFICIENCY and attending to FUGITIVE EMISSIONS. So-called win–win management: environment gains through reduced impact; business gains by increased efficiency and hence ultimately profits. [P.C.]

waste regulation authorities *See* ENVIRONMENTAL (PROTECTION) AGENCIES.

water Water (H_2O) differs from similar small molecules (e.g. AMMONIA (NH_3), METHANE (CH_4), nitrogen dioxide (NO_2)) in being a liquid as opposed to a gas at 'normal' temperatures because the H_2O molecule is strongly polarized, making the molecules cling together through weak hydrogen bonding (Fig. W1). The energy required to separate the molecules so that they can escape by EVAPORATION is the heat of vaporization, λ. Its value for water is one of the highest known ($\lambda = 2.48$ MJ kg^{-1} at 20°C), and this means that evaporation has a powerful cooling effect.

The hydrogen bonding between the OXYGEN and HYDROGEN of adjacent molecules of water accounts for the high tensile strength of water in capillaries enabling water to exist as a continuum in stems of vascular plants, and to be drawn through the roots, stem and leaves when evaporation at the leaf surface occurs.

Water is an excellent solvent for mineral salts. Ions such as Na^+ and Cl^- are attracted to each other. When placed in water, their attraction is reduced. This capacity of a solvent to influence the attraction between ions is expressed as the dielectric constant. The value for water is 80 whilst that of non-polar organic solvents is less than 2. Thus mineral ions needed for plant growth, and originating from the SOIL, accompany the TRANSPIRATION stream to all parts of the plant, and become available for metabolic uptake in the growing regions of all organs. The solutes in the transpiration stream are not only the mineral elements. Most biochemically important substances in plants are charged, and readily soluble in water. Water molecules become associated with polar groups of organic biochemical constituents, including PROTEINS, polysaccharides and other macromolecules as water of hydration.

Water is a better electrical conductor than other non-metallic liquids as a result of dissociation of H_2O into H^+ and OH^-. At 25°C there are 10^{-7} mol dm^{-3} of each species.

The non-compressible nature of water makes it useful as a hydroskeleton; leaves of many land plants owe their rigidity to the water pressures inside them, which are often in the range 0.3–1 MPa (3–10 atmospheres). This PRESSURE is vital during the cell-expansion phase of growth.

The rate at which oxygen and CARBON DIOXIDE (CO_2) diffuse in water is very low indeed, some 10 000 times slower than these molecules diffuse in air, although both gases are sparingly soluble in water (Table W1). Water is more viscous than most other solvents as a result of the attraction between its molecules which tends to prevent layers of the liquid sliding over each other. Viscosity declines markedly with temperature (Table W2). Thus the tensions during transpiration, when water is drawn through the fine capillaries that constitute the CELLULOSE cell walls and xylem tissues of the plant, are considerable. Water participates directly and fundamentally in biochemical processes. In PHOTOSYNTHESIS the water molecule is split, the H being used to reduce CO_2 to CARBOHYDRATES $(CH_2O)_n$, whilst O is evolved as oxygen gas. This is the origin of the oxygen of the Earth's ATMOSPHERE. *See also* WATER (HYDROLOGICAL) CYCLE. [J.G.]

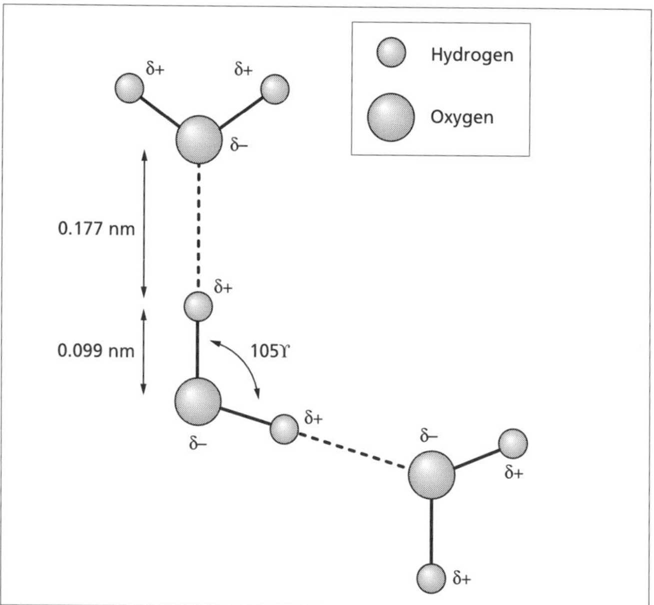

Fig. W1 Adjacent molecules of H_2O clinging together as a result of electrostatic attraction between the positively charged hydrogen and the negatively charged oxygen. Many molecules are associated in this way as 'clusters' in liquid water and they form crystals in ice. (After Jones, H.G. (1992) *Plants and Microclimate*. Cambridge University Press, Cambridge.)

water accommodated fraction (WAF) That fraction of a complex mixture of chemicals (e.g. from oils) that is present in the aqueous phase following a period of mixing. Sometimes referred to as a water soluble fraction (WSF), but WAF is preferred as it is often the case that stable microemulsions of finely dispersed droplets can exist, that do not represent true solutions. WAF is used in ecotoxicological tests of oils. As WAFs form naturally after OIL SPILLS, it is claimed they cause the most important ecological problems. [P.C.]

REFERENCE

Girling, A.E., Makarian R.K. & Bennett, D. (1992) Aquatic toxicity of oil products—some recommendations. *Chemosphere* **24**, 1469–1472.

Water Acts Series of Acts (1989 and 1991) of the UK Parliament establishing controls over the quality of waters from the tap and in the environment, largely for England and Wales. Amongst other things, these privatized the water industry and established the National Rivers Authority (NRA) as the regulatory body for England and Wales. They have now been supplemented by the Environment Act 1995. *See also* ENVIRONMENTAL LAW. [P.C.]

water (hydrological) cycle It was the English astronomer Edmund Halley who, in 1691, first wrote about the natural cycle of EVAPORATION, condensation and rainfall. On the Earth's surface, 97% of the WATER is in the oceans, and most of the remainder is in the ice caps. Only a tiny fraction of the total is in the atmosphere at any time (Table W3). Each day about 1200 km³ of water evaporates from the oceans and 190 km³ is lost from the terrestrial surface by evapotranspiration. Around 1000 km³ falls on the SEA as RAIN and other PRECIPITATION, and 300 km³ falls over the land. A flow of some 110 km³ is returned to the sea via RIVERS, groundwater and meltwater (Fig. W2).

Property		0°	10°	20°	30°	40°
S	NH_3	1130	870	680	530	400
	CO_2	1.7	1.2	0.85	0.65	0.52
	H_2S	4.5	3.3	2.5	2.0	1.6
	N_2	0.023	0.018	0.015	0.013	0.011
	O_2	0.047	0.037	0.030	0.026	0.022
	SO_2	80	57	39	27	19
D	CO_2	0.00158	0.00169	0.0018	0.00190	0.0020
	O_2	0.00176	0.00188	0.0020	0.0021	0.0022
	Heat	0.127	0.135	0.144	0.153	0.162

Table W1 Solubilities and diffusivities in water in the range 0–40°C. *S*, solubility of gas (m³ gas per m³ of water); *D*, diffusivity of gas and heat in water (mm² s⁻¹).

Property	Units	0°C	10°C	20°C	30°C	40°C
Density, ρ	kg m^{-3}	1000*	999.7	998.2	995.6	992.2
Saturation vapour pressure, *e*	kPa	0.61	1.23	2.34	4.24	7.38
Latent heat of vaporization, λ	MJ kg^{-1}	2.50	2.48	2.45	2.43	2.41
Specific heat capacity, c_p	J g^{-1} K^{-1}	4.22	4.19	4.18	4.18	4.18
Viscosity, η	mPa s	1.79	1.30	1.00	0.80	0.65
Surface tension	mN m^{-1}	75.6	74.2	72.7	71.2	69.6
Thermal conductivity	mW K^{-1} m^{-1}	561	580	598	615	630

* Water reaches its maximum density at 3.98°C.

Table W2 Some temperature-sensitive properties of liquid water.

Table W3 Global water reservoirs and fluxes. (After Ward, R.C. & Robinson, M. (1990) *Principles of Hydrology*. McGraw-Hill, Maidenhead.)

	Values (km^3 × 10^3)	Percentage of reservoir
Global reservoir		
Ocean	1 350 000	97.403
Atmosphere	13	0.001
Land	35 978	2.596
Terrestrial reservoir		
Rivers	2	0.006
Freshwater lakes	100	0.278
Inland seas	105	0.292
Soil water	70	0.195
Groundwater	8200	22.792
Ice caps/glaciers	27 500	76.436
Biota	1	0.003
Annual flux		
Evaporation		
Ocean	425.0	
Land	71.0	
Total	496.0	
Precipitation		
Ocean	385.0	
Land	111.0	
Total	496.0	
Run-off to oceans		
Rivers	27.0	
Groundwater	12.0	
Glacial meltwater	2.5	
Total	41.5	

In practice there are numerous short circuits to this cycle. Water, having fallen on to the Earth's surface, may travel back into the atmosphere through any of a number of different pathways. On land, some of the precipitation will be intercepted by the vegetation and evaporated directly back into the atmosphere. Unless the SOIL is saturated, when overland flow may be generated, the precipitation reaching the ground enters the soil, and is either taken up by plants and evaporated as TRANSPIRATION, or drains through the soil to reach rivers and then the oceans, either directly or via groundwater.

Average residence times for the different phases of the hydrological cycle vary from a few days, for the atmospheric phase, to sometimes more than hundreds of years for groundwater. The size of the global water reservoirs and fluxes is given in Table W3. [J.H.C.G. & J.G.]

water potential WATER potential is a measure of the free energy per MOLE of water in the system under consideration, compared with the free energy per mole of pure, free water at the same temperature and pressure. It is expressed in J mol^{-1}, or energy per unit mass of water, J kg^{-1}, or energy per unit volume, J m^{-3}; these latter are also the units of pressure, Pa. Water containing solutes, or contained in a wettable structure where there is SURFACE TENSION or adhesion, or held above a reference height against gravity, will have a negative water potential, compared to pure, free water at the reference height. It is formally defined as the amount of work required to transport, isothermally and reversibly, an infinitesimal quantity of water from a pool of pure water, at a specified elevation and pressure to the system under consideration. Water potential is used in estimating the movement of water in porous systems, such as SOIL, or plant stems, and in characterizing the water status of biological tissues. [J.I.L.M.]

water quality The wholesomeness of WATER to drink, its appearance, its odour, the abundance of fish and other living things within it have all formed part of a subjective assessment of what is meant by 'water quality'. In this sense the concept is a management test rather than having an objective meaning. Neither is 'water quality' synonymous with water chemistry for this can vary, even in uncontaminated systems.

It is impossible to say what water quality should be in any particular circumstances without setting some target that is related to use. Such targets are implicitly biological: safe for drinking; suitable for fish; supporting wildlife in general; suitable for recreation, etc. Water chemistry is important in enabling these biological targets to be met; because chemical measures have proved useful in SURVEILLANCE since they are relatively easy to monitor. However, if water chemistry had no biological impact it would have little intrinsic importance in management. Hence, biological criteria are also important. *See also* RETROSPECTIVE ANALYSIS. [P.C.]

REFERENCE

Calow P., Armitage, P., Boon, P. *et al.* (1990) *River Water Quality*. BES Ecological Issues No. 1. Field Studies Council, Shrewsbury.

water quality objective/standard (WQO/WQS) In UK law, WQO is the target for WATER QUALITY in a stretch of river, for example to supply fisheries, ABSTRACTION for drinking, etc. WQS is a quantitative concentration below which the defined environmental/ecological objectives are likely to be achieved. Elsewhere there is a tendency to use WQO and WQS interchangeably to refer to standards. *See also* ENVIRONMENTAL QUALITY OBJECTIVE; ENVIRONMENTAL QUALITY STANDARD. [P.C.]

water soluble fraction (WSF) *See* WATER ACCOMMODATED FRACTION.

water table The WATER table is the upper envelope of the subsurface zone which is saturated with water, it separates groundwater from the SOIL MOISTURE in the unsaturated zone above. Water tables are dynamic, having depths which are likely to have a seasonal variation in response to the higher rate of groundwater recharge during the winter, when rainfall may be higher and EVAPORATION is less. A perched water table may result from a volume of saturated soil being separated from the groundwater below by a layer of relatively impermeable soil. Although vegetation cannot generally access water that is held below the water table, when the water table is close to the surface, plants may be able to extract water from the capillary zone which is immediately above the water table and which is replenished from below. [J.H.C.G.]

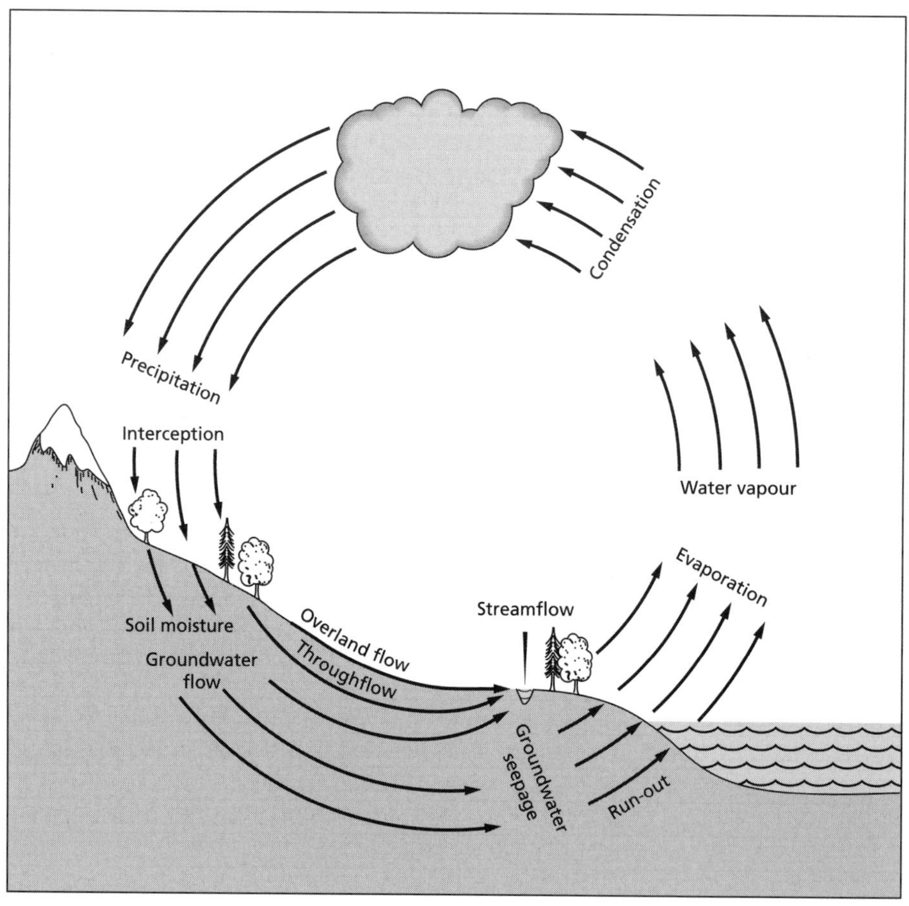

Fig. W2 Schematic diagram of the hydrological cycle. (After Ward, R.C. & Robinson, M. (1990) *Principles of Hydrology*. McGraw-Hill, Maidenhead.)

REFERENCE

Ward, R.C. & Robinson, M. (1990) *Principles of Hydrology*. McGraw-Hill, Maidenhead.

watershed Drainage BASIN. *See also* CATCHMENT. [P.C.]

watt (W) The SI UNIT of power, equal to 1 JOULE of energy per SECOND ($J\,s^{-1}$). Its relation to obsolete units are: 1 horsepower = 0.745 kW; 1 cheval-vapeur = 0.735 kW. [J.G.]

wave power Using energy in the WAVES on seawater to generate electricity. This is a renewable power source. To obtain the equivalent amount of energy as from 1 kg of coal would need 1 tonne of seawater to fall through 3 km—so capital costs would be high. [P.C.]

waves A disturbance of the equilibrium of fluid continuously propagated from point to point without a corresponding advance of the particles in the same direction, by which motion, heat, LIGHT, sound, electricity, etc., are transmitted. Each wave consists of a crest (the highest point of the wave) and a trough (the lowest point of the wave). The vertical distance between a crest and the succeeding trough is the wave height, and the horizontal distance between successive crests or successive troughs is the wavelength. The time that it takes for one wavelength to pass a fixed point is the wave period, and the number of waves that pass a fixed point per unit time is the wave frequency. Longer waves travel faster than short ones. Wave steepness is measured as the ratio of wave height to wavelength; as steepness increases, waves become unstable and break. In deep WATER (i.e. water depth greater than one fourth the wavelength), water particles move in circular orbits but retain the same general position. When deep-water waves enter shallow water, the bottom interferes with the orbital motion of the water particles (the waves 'feel bottom'), causing the wave to slow, steepen, and eventually begin to break. As the wave orbits touch the bottom, the waves change direction through refraction (moving slower in shallow water and faster in deeper water). In the OCEAN, waves are caused by the attraction of the Sun and Moon (i.e. TIDES), by impact (e.g. volcanic explosions and earthquakes cause tsunamis) and WIND (the most common disturbing force acting on the ocean surface). Wave speed and size depend on wind speed, wind duration, fetch (the area of the open ocean over which the wind blows with constant speed and direction, also called the generating area), and the state of the SEA at the time the wind began to blow. The term 'sea' refers to waves generated or sustained by winds within their fetch, whereas the term 'swell' refers to waves that have travelled out of their generating area. Waves dominate BEACH processes primarily through the process of SEDIMENT transport. Water piled up on the shore by incoming waves returns to sea in rip CURRENTS (narrow channels of rapidly flowing water that may pose a hazard to swimmers). Storm surges (sometimes called tidal waves) result in the piling up of water against the shore by wind- and storm-related atmospheric PRESSURE differences. Seiches, or standing waves, are the regular tilting of the water surface that occurs in BASINS as wave energy is reflected from the basin margins. [V.F.]

weather Weather represents the state of the ATMOSPHERE at a particular instant of time and includes a description of its significant features such as anticyclones, fronts and depressions and their progression in time. Its professional and scientific investigation is carried out under the discipline of synoptic METEOROLOGY, the products of which are the familiar charts of present or forecast weather in newspapers or television. Phenomena which we can recognize as part of a description of weather range from small space- and time-scale features such as tornadoes and thunderstorms which last for minutes to hours, through larger phenomena such as fronts which have a space scale of hundreds of kilometres and a lifetime of several days. On a global scale, the jet streams control the movement of depressions and anticyclones and the zonal flow may persist in a specific pattern for several weeks on end, thus resulting in persistent weather patterns for extended periods. *See also* AIR MASS; ANTICYCLONIC; CLIMATE; FRONT. [J.B.M.]

weathering Destruction or partial destruction of rock by thermal, chemical, and mechanical processes. Involves both mechanical disintegration (loss of coherence but maintenance of composition) and chemical DECOMPOSITION (rock decay accompanied by marked changes in chemical and mineralogical composition). Mechanical weathering is caused by freeze–thaw cycles. Growing plant roots, lichens and mosses, burrowing by animals, and microbial activity aid both physical and chemical weathering. Weathering also occurs when CARBON DIOXIDE (CO_2) dissolves in WATER to form carbonic acid (H_2CO_3) which dissolves minerals. [V.F.]

weed (i) A plant where I don't want it; (ii) a plant whose virtues have yet to be discovered; (iii) a plant in the wrong place; (iv) a plant which is the target of HERBICIDE application; (v) a plant for whose removal or destruction someone is willing to pay money or expend effort.

There is remarkably little consensus about the weedy status of many plant species; one man's weed is another man's herbarium specimen. This conflict of interest is seen most clearly in RISK ASSESSMENT studies prior to the import and release of potential weed biocontrol agents (pathogens or insect herbivores). In Australia, *Echium plantagineum* (Boraginaceae) is considered by sheep farmers to be a serious PASTURE weed (they call it Patterson's Curse), but the same plant is a valued source of pollen and nectar to bee keepers (they call it Salvation Jane). Since all plant species have the ability to increase when rare (INVASION CRITERION), attempts to find general traits of weedy and non-weedy species have typically not stood up to critical scrutiny. [M.J.C.]

Wegener, A. *See* CHARACTERS IN ECOLOGY; CONTINENTAL DRIFT.

Weismannian organisms Organisms in which GERM LINE and SOMA become distinguished at an early point in development and are committed to their roles. While the cells in the germ line are specialized to contribute to the next generation, the cells belonging to the soma differentiate into specialized cells, for example nerve cells, and are not heritable. Therefore, any changes to the genes within somatic cells (SOMATIC MUTATIONS) are not transmissible to the next generation. This is often known as Weismann's barrier or Weismann's principle. Only organisms with a clear separation of germ line and soma can be thought of as individuals in the Weismannian sense, as units on which selection operates. Most animals are Weismannian organisms, exceptions include sponges, coelenterates, some annelids and some hemichordates. No plants or fungi are Weismannian organisms. [J.C.K.]

Weismann's barrier *See* GERM LINE; INHERITANCE OF ACQUIRED CHARACTERS; WEISMANNIAN ORGANISMS.

Weismann's principle *See* GERM LINE; INHERITANCE OF ACQUIRED CHARACTERS; WEISMANNIAN ORGANISMS.

Wentworth scale A geometric grading scheme for expressing sedimentary grain diameters. According to the phi (ϕ) scale, grain diameter is expressed as its negative logarithm to base 2, thus permitting percentage differences between small particles to be conveniently compared to percentage differences between particles of very large size. [V.F.]

wetlands A broad term for a wide range of habitats characterized by their abundant WATER supply. The Ramsar Convention (an international meeting concerned with wetland conservation, held in Iran in 1971) defined wetlands as:

> Areas of marsh, fen, peatland or water, whether natural or artificial, permanent or temporary, with water that is static or flowing, fresh, brackish or salt, including areas of marine water the depth of which at low tide does not exceed six metres.

More details of the range of habitats included, all of which are under some measure of threat from human land use and exploitation, are given under the individual terms. [P.D.M.]

wetlands restoration The National Research Council (NRC 1992, p. 2) developed the following definition:

> Restoration means returning an ecosystem to a close approximation of its condition prior to disturbance. Accomplishing restoration means ensuring that ecosystem structure and function are recreated or repaired, and that natural dynamic ecosystem processes are operating effectively again.

The NRC scientists went on to say that restoration differs from creation (construction of a type of ecosystem that is new to the site) and from RECLAMATION (modifying a natural ecosystem to a human use), as these terms refer to alterations of the landscape for some new use.

There are, however, disagreements about the landscape scale within which a system should be considered restored (brought back) versus created (built anew). For example, if the location of a 10-hectare coastal wetland that was filled a century ago is uncertain and a restoration project (excavation of fill) is planned within a 40-hectare site that is thought to include the wetland, the project should not automatically be considered a restoration programme, because it may miss the target. It may place the new wetland on natural upland, rather than fill; it may lack the appropriate HYDROLOGY and substrate qualities; and it may not be able to support the plants and animals that once occupied the wetland site. If this occurs, it should be called a wetland creation project. Yet, at the landscape scale, the project may be considered a restoration effort, especially if the wetland's structure and functioning are thought to mimic the system that was lost nearby. In many cases it is difficult to determine where the lost system was located or what it was like before disturbance. Hence, not all projects are easily classified as restoration or creation efforts.

A second area of debate concerns the time-target for restoration. Ecologists recognize that nature is not static but dynamic, responding to local, as well as global, and short-term, as well as long-term, changes in the environment. Thus, returning a site to some previous condition requires making a choice among various predisturbance states. Very often, records are inadequate and the range of possible targets is unclear. Restorationists must therefore select a predisturbance time-period and predict what structural and functional attributes were in existence. Hence, the specific objectives of restoration are subject to debate.

As an alternative, objectives can be set by selecting one or more nearby undisturbed WETLANDS as reference sites. Where historical information about the lost or damaged habitat is lacking, the restoration project can set out to mimic conditions in an undisturbed site that is considered both desirable and feasible within the landscape setting. Assessment of success is then a matter of comparing the project site with the reference wetland(s).

For most wetland restoration projects, it may never be known whether the functions have been restored to the predisturbance condition. Hence, it seems prudent not to adopt too stringent a definition for 'restoration'. Rather, project proponents and evaluators should draw upon the best science (and scientific advice) available to identify restoration goals and objectives and to determine whether they have been met. The problem with defining restoration too precisely is that one must make assumptions that are usually untestable about the pre-existing habitat. A general interpretation of the term is more realistic and useful. *See also* CONSERVATION BIOLOGY; WETLANDS (TIDAL AND COASTAL), CONSERVATION. [J.B.Z.]

REFERENCE

National Research Council (NRC) Committee on Restoration of Aquatic Ecosystems (1992) *Restoration of Aquatic Ecosystems: Science, Technology, and Public Policy*. National Academy Press, Washington, D.C.

wetlands (tidal and coastal), conservation WETLANDS are ecotones (edge transition zones) between dry land and WATER, with substrates (soils) permanently or seasonally saturated with fresh, saline or marine water. They vary from shallow, transitory ponds or pools to more permanent marshes, BOGS or swamps. The biotic communities in wetlands depend on the attending inundation of water and drying over different time periods (DIEL, SEASON or years) and reflect the permanence, predictability and SEASONALITY of water. A dominant component of wetland communities is emergent aquatic macrophytes (rhizobenthos). Freshwater wetlands are strongly influenced by the hydrological cycle (*see* WATER (HYDROLOGICAL) CYCLE), and tidal and COASTAL WETLANDS are primarily influenced by TIDES and seawater levels.

Wetlands are dynamic, transitory ecosystems dependent on natural perturbations in water availability. With dense vegetation and shallow basins, wetlands are valuable in forming traps for sediments, NUTRIENTS and METALS and in addition provide downstream buffering from both peak- and low-water discharges by storing water or slowly releasing water, respectively. Along coasts, lakeshores and riverbeds, wetlands stabilize the shorelines protecting them from erosion from tides, WAVES or storms.

Until recently, wetlands have been considered undesirable or less valuable than other ecosystems and may have been drained and filled. However, realization of the importance of wetlands, their roles in improving WATER QUALITY, stabilizing shorelines, and the intrinsic ecological diversity of wetland communities and their utilization by resident and migratory birds and amphibians, has led to both their conservation and restoration. *See also* CONSERVATION BIOLOGY; WETLANDS RESTORATION. [R.W.D.]

whaling Although whales have been hunted by mankind since prehistoric times, the term whaling usually refers to the modern commercial catching of whales using tethered, barbed harpoons, which sometimes carry an explosive tip. The technique was developed in Norway in the 19th century, and spread around the world in the 20th century. By the 1970s, most of the stocks of large whales around the world were depleted to small fractions of their original abundance. Blue whales (*Balaenoptera musculus*) in the Antarctic were reduced from an original quarter of a million animals to only a few hundred by the time they were declared a protected species in 1965. The INTERNATIONAL WHALING COMMISSION (IWC), the intergovernment regulatory body for whaling established in 1946, has introduced progressively stricter restrictions on whaling since the 1960s, culminating in a general ban on commercial whaling from 1986 onwards. The IWC has recently been developing scientific approaches to the management of whaling which would eventually enable whale populations to be harvested on a sustainable basis. Although previous approaches to sustainable management of whaling have proven unworkable, the Revised Management Procedure, approved by the IWC in 1994, is considered by most experts to be sound. It will not be implemented until other outstanding issues, such as enforcement of regulations, are resolved. As an additional safety measure, the IWC in 1994 designated most of the Southern Ocean south of 40°S as a sanctuary for whales in which whaling is not permitted. [J.G.C.]

REFERENCE

Cooke, J.G. (1994) The management of whaling. *Aquatic Mammals* **20**(3), 129–135.

White, Gilbert *See* CHARACTERS IN ECOLOGY.

WHO (World Health Organization) *See* UNITED NATIONS PROGRAMMES AND BODIES.

Wien's displacement law *See* RADIATION; RADIATION LAWS.

Wien's radiation law *See* RADIATION LAWS.

wildlife In principle, wildlife refers to all the indigenous flora and

Table W4 The Beaufort scale.

Beaufort number	Descriptor	Speed (m s-1)	Specification
0	Calm	0–0.2	Smoke rises vertically
1	Light air	0.3–1.5	Smoke drifts, but wind-vanes unresponsive
2	Light breeze	1.6–3.3	Wind felt on face, wind-vanes respond, leaves rustle
3	Gentle breeze	3.4–5.4	Wind extends light flag, leaves and small twigs in motion
4	Moderate breeze	5.5–7.9	Raises dust and loose paper, small branches in motion
5	Fresh breeze	8.0–10.7	Crested wavelets form on inland waters, small trees sway
6	Strong breeze	10.8–13.8	Large branches sway, umbrellas used with difficulty, whistling of overhead telegraph lines heard
7	Near gale	13.9–17.1	Whole trees in motion, difficulty walking against the wind
8	Gale	17.2–20.7	Twigs break from trees
9	Strong gale	20.8–24.4	Slight structural damage to roofs
10	Storm	24.5–28.4	Trees uprooted, structural damage widespread
11	Violent storm	28.5–32.6	Widespread damage
12	Hurricane	32.7 and over	

fauna in nature. In practice, its use often intends only animal life. [P.C.]

wildlife management Actions aimed at conserving indigenous WILDLIFE. [P.C.]

win–win management *See* WASTE MINIMIZATION.

wind The ATMOSPHERE in motion. Wind is important for several reasons: (i) because of the catastrophic damage it does to buildings and other structures; (ii) because of its indirect and direct effects on plants, both at macroscopic and microscopic scales; (iii) because of its effect on heat loss from buildings, people and crops; and (iv) as a medium for DISPERSAL of pollens, spores, seeds and pollutants. Wind is strong enough in many parts of the world to have potential as an energy supply. Wind speed is measured with an ANEMOMETER, and direction is measured with a wind-vane. The wind speed is found to increase with height above the ground in a logarithmic manner, depending on the aerodynamic ROUGHNESS of the land cover. At meteorological stations the wind speed is measured well above the ground (10 m). The Beaufort scale (Table W4) describes the wind in a way that relates to everyday experience. [J.G.]

REFERENCES

Coutts, M.C. & Grace, J. (1995) *Trees and Wind*. Cambridge University Press, Cambridge.

Grace, J. (1977) *Plant Response to Wind*. Academic Press, London.

wind-break A barrier or line of trees (shelterbelt) designed to reduce the speed of the WIND to benefit crops or livestock. The effectiveness of such a barrier depends on its height, porosity and structure. A well-designed wind-break should deflect the air upwards and create shelter over about $8h$ where h is the height of the wind-break. A bluff wind-break, such as a brick wall, will not achieve the best results because of violent eddying of the air in the lee of the barrier. Earlier claims that shelter was effective over $20h$ are not well founded, for although the horizontal wind speed is reduced to some extent up to $20h$ from the wind-break, there occurs a downward movement of air at about $12h$, which causes damage to crops.

There is a very large body of evidence to suggest that wind-breaks increase the yield of crops (Grace 1977; Brandle *et al.* 1988), sometimes substantially. Probably the literature distorts the reality, as small increases in the yield of crops, or even deteriorations in yield are unlikely to be published.

It is clear that a reduction in wind speed will influence the energy balance of the crop. In a cold climate the effect is likely to be an increase in the surface temperature, thus bringing forward the onset of growth in the spring and generally speeding up development. In a hot climate the effect may be counterproductive, as the leaves may overheat. Another negative influence of shelter is that insects are more numerous in sheltered localities, and insect pests may do much damage.

One of the most important advantages of shelter from the wind may be the reduction in the abrasion between leaves and fruits. Such abrasion increases TRANSPIRATION and reduces growth, and also may reduce the commercial quality of the crop by making bruises and abrasion wounds on the surface of the fruit. Moreover, plants exposed to mechanical stimulation develop a more fibrous tissue composition, and this is likely to reduce their digestibility to livestock.

Another advantage, particularly in regions where SOIL erosion is likely to occur, is the stabilization of the agricultural landscape.

Traditional landscapes have often contained hedgerows, which do provide limited shelter especially valuable for livestock, but do not exert a noticable effect on the growth of crops. In northern France, the bocage landscape consists of small fields surrounded by large hedgerows, and this arrangement may indeed influence the MICROCLIMATE enough to increase plant growth (however, the reduction in solar irradiance to the crop will have a negative effect at close quarters).

In arid regions of the world the use of agroforestry is an important development. Often, herbaceous crops are grown in between rows of multipurpose trees, often nitrogen-fixing leguminous species. Part of the SYNERGISM between the trees and the herbaceous crops is likely to be the shelter effect. In colder countries, agroforestry has not yet been widely developed, except on an experimental scale, but it seems there may be benefits of systems of hedgerows of fruit-bearing trees interplanted with cereals, especially in flat and windy regions. [J.G.]

REFERENCES
Brandle, J.R., Hintz, D.L. & Sturrock, J.W. (1988) *Windbreak Technology*. Elsevier, Amsterdam.
Caborn, M. (1965) *Shelterbelts and Windbreaks*. Faber, London.
Grace, J. (1977) *Plant Response to Wind*. Academic Press, London.

wind-chill The extent to which a human feels cold depends not only on the air TEMPERATURE but also on the WIND speed. Heat loss from the body surface involves convective transfer of heat, which is a function of wind speed as well as air temperature. It has proved useful to have a scale of wind-chill that encompasses wind and temperature, and enables people engaged in outdoor pursuits to be advised about the chilling effect of the WEATHER. For example, as the table of wind-chill equivalent temperatures (Table W5) shows, human subjects will feel as cold at 0°C with a wind of 15 m s^{-1} as they do at −10°C in calm air. Similar considerations will affect the period of time they can be expected to survive when caught on mountains in bad winter weather. [J.G.]

Table W5 Table of wind-chill equivalent temperatures.

Actual temperature (°C)	Wind speed (m s^{-1}) Calm	5	10	15
0	−1	−2	−7	−11
−5	−4	−9	−13	−16
−10	−9	−13	−19	−25
−15	−13	−19	−26	−33
−20	−18	−26	−34	−42
−30	−28	−37	−50	
−40	−37	−50		

wind-formed lakes *See* LAKES, TYPES OF.

wind pollination *See* POLLINATION.

wind power The use of WIND veins to harness energy from the wind in driving electrical turbines. *See also* RENEWABLE RESOURCE. [P.C.]

wind tunnel A construction for making a controlled flow of air over an experimental object. A WIND tunnel may be 'open' in which the air passes straight through and returns via the room in which the tunnel is placed; or it may be 'closed' in which the tunnel returns on itself, like a doughnut. The latter has the advantage of being unaffected by disturbances in the room, and is the more usual arrangement. Some typical designs are shown in Fig. W3. In the environmental sciences wind tunnels have often been used for model-testing of buildings, forests and for studies of deposition, as well as for calibrating ANEMOMETERS. [J.G.]

REFERENCES
Coutts, M.C. & Grace, J. (1995) *Trees and Wind*. Cambridge University Press, Cambridge.
Pankhurst, R.C. & Hobler, D.W. (1952) *Wind Tunnel Technique*. Pitman, London.

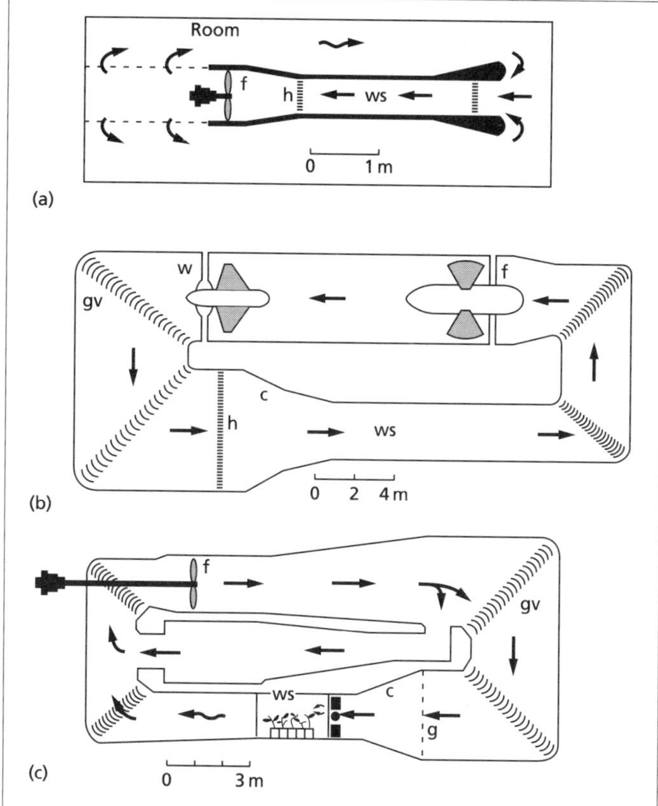

Fig. W3 Wind-tunnel configurations. (a) Open circuit type in which return flow is via the room; (b) closed circuit type producing low turbulence; (c) closed circuit type with by-pass for temperature and humidity control and cross-pieces to generate extra turbulence (Edinburgh University, Department of Forestry and Natural Resources). The diagrams illustrate the following features: c, contraction; f, fan; g, gauze screen; gv, guide vanes; h, honeycombs; w, freely rotating windmill to suppress pulsations; ws, working section. (After Grace, J. (1977) *Plant Response to Wind*. Academic Press, London.)

within-lake habitat classification In LAKES the substrate from the shore down to the deepest parts of the water exhibit a series of zones.

1 Epilittoral—entirely above the water, uninfluenced even by spray.
2 Supralittoral—entirely above the water but affected by the spray.
3 Littoral—from the edge of the water to the point of light penetration where PRIMARY PRODUCTIVITY equals RESPIRATION (COMPENSATION POINT). The littoral is subdivided into:
(a) eulittoral—between the highest and lowest seasonal levels;
(b) upper infralittoral — with emergent rooted vegetation (rhizobenthos);
(c) middle infralittoral—with floating leaves on the rooted vegetation;
(d) lower infralittoral—with submerged rooted vegetation.

Typically, the rhizobenthos forms concentric rings within the LITTORAL ZONE with one group replacing the next as the water depth changes. In the littoral zone the primary producers are of two

types: rhizobenthic and haptobenthic, made up of aquatic macrophytic spermatophytes and phytobenthic algae.
4 The region in the deeper waters which has no rooted vegetation is the profundal where respiration exceeds PRODUCTION, while the transition zone between the littoral and profundal zones is the littoriprofundal.
5 The open water zone is termed the limnetic (or PELAGIC) zone and is equivalent to the EUPHOTIC ZONE, where respiration equals PHOTOSYNTHESIS and the compensation point is usually approximated as 1% of full sunlight.

The profundal community thus depends on the littoral and limnetic zones for its energy sources since respiration exceeds production. While the biotic diversity is not high, the profundal community is of importance in trapping NUTRIENTS and chemical cycling.

Physically, the air–water interface consists of a monomolecular layer of PROTEINS or lipoproteins derived from decomposing plant material which is an important source of food, as well as support, and for entrapping of organisms subsequently used for food. The biological community at the air–water interface is the pleuston which is divided into the macropleuston, which has an upper dry surface and a lower wet surface, and the micropleuston which is associated with either the undersurface (hypomicropleuston) when it is totally wet, or the upper surface (epimicropleuston) when it is totally dry.

Organisms which permanently inhabit the air–water interface are termed eupleuston compared to organisms which only temporarily inhabit this habitat which are termed pseudopleuston. [R.W.D.]

within-stream/river habitat classification There are two major habitat types:
1 in riffles (or rapids) the water is shallow with water velocities high enough to keep the bottom clear of SILT and mud;
2 in pools the water is deeper with velocities low enough that silt and other loose materials can settle on the bottom.

Thus, with a soft bottom, pools are unfavourable to the haptobenthos but favourable to both the herpobenthos and rhizobenthos. Two faunal types correspond to the riffle and pool habitats. However, within these broad categories the type of substrate (i.e. sand, pebbles, CLAY, bedrock, organic content) is very important in determining the nature and composition of the community. The stability of the substrate as judged by its retention in the system and its tendency to shift position is also of great importance in determining community distribution and composition.

As streams and RIVERS change along their lengths in respect to properties such as temperature, water velocity, DISCHARGE and TURBIDITY, they also display longitudinal biotic ZONATION, reflected in the subdivision of rivers into rhithron and potamon.

The hyporheos consists of the interstices between the particles composing the substrate forming a middle zone between the surface waters of the stream and the groundwater (AQUIFER) below.

The hyporheic fauna can be divided into: (i) occasional—consisting of benthic species present on a temporary basis; and (ii) permanent—consisting of species found only in this habitat. It has been suggested that the hyporheos acts as a refuge to the occasional members, by acting as a safety zone during periods of STRESS, such as DROUGHT or very high discharge, and from which they can return to the BENTHOS when recolonization is possible. The hyporheos itself is recolonized mainly from upstream sources, followed by downstream sources and the deeper hyporheos.

While early studies on the hyporheos concentrated on its importance as a potential refuge for invertebrates, more recent studies have shown the importance of nutrient flows between the hyporheos and the water column.

There is also a pleuston community associated with the water–air interface of streams but this is never as well developed as in LAKES. *See also* RIVER CONTINUUM CONCEPT. [R.W.D.]

WMO *See* WORLD METEOROLOGICAL ORGANIZATION.

wood-pasture A semi-cultural type of vegetation with a mosaic of alternating patches of open meadows and small copses of deciduous trees and shrubs. Wood-pastures have been used both as PASTURES for domestic animals, mainly cattle and horses, and as areas for collecting of winter-FODDER such as leaves of trees and grass.

Man-made wood-pastures have been especially typical of Europe and parts of Asia. Many areas of SAVANNAH in tropical and subtropical regions resemble wood-pastures physiognomically and are used in roughly the same way.

The bloom of wood-pastures lasted in Eurasia from the Neolithic to the 19th century. Before the invention of iron as tool material, leaves and twigs from trees were gathered for winter fodder and thus thin woods, which were also suited as pasturelands, were created. The gathering of leaves led to various ways of handling of the trees, such as coppicing and pollarding. Cattle and horses were usually allowed to graze on wood-pasture, only lightly during the spring and again after hay-making had been finished. The existence of wood-pastures was completely dependent upon cultural activities. With the beginning of modern AGRICULTURE these kinds of pasture were abandoned and have, but for a few exceptions, disappeared as a consequence of the invasion of trees.

The main region of wood-pastures in Europe was from the Mediterranean to the hemiboreal FOREST zone. Even in the BOREAL zone, however, similar wood-pastures could be found in an impoverished form. There, mainly former swidden fields were used as wood-pastures and the trees in question were birch (*Betula* spp.) and grey alder (*Alnus incana*).

Wood-pastures with their peculiar physiognomical structure and floristic RICHNESS formed a favourite subject for landscape painters of the Romanticism and can be recalled in most of the famous art galleries. [Y.V.]

woodland Tree-covered land and associated habitats in a matrix of trees. Now usually associated in Britain with small woods consisting mainly of native trees, and with traditional or other non-intensive forms of management in which timber-growing is only one of many objects of management (*cf.* FOREST). In other English-speaking countries, woodland labels open-canopied forest in Australia, but is uncertain in its connotations in North America. [G.F.P.]

World Bank *See* GATT AND ENVIRONMENT.

World Conservation Union *See* IUCN.

World Conservation Union's Commission on National Parks and Protected Areas *See* EXTRACTIVE RESERVES.

World Health Organization (WHO) *See* UNITED NATIONS PROGRAMMES AND BODIES.

World Meteorological Organization (WMO) The WMO is a specialized agency of the UN located in Geneva. Its programmes include World Weather Watch, World Climate Programme, Global Climate Observation System (and its GCOS Data System), and it also has an Education and Training Programme. It has a page on the World Wide Web which documents many of its programmes: http:/www.wmo.ch/. *See also* METEOROLOGY. [J.G.]

World Weather Watch (WWW) *See* METEOROLOGY.

WQO/WQS *See* WATER QUALITY OBJECTIVE/STANDARD.

Wright, Sewall *See* CHARACTERS IN ECOLOGY.

WSF (water soluble fraction) *See* WATER ACCOMMODATED FRACTION.

WWW (World Weather Watch) *See* METEOROLOGY.

X

X-rays (X-radiation) Electromagnetic radiation in the wavelength range between γ-radiation and ULTRAVIOLET RADIATION, i.e. from approximately 0.001 nm to approximately 50 nm (some scientists set the upper limit as low as 5 nm), discovered by Wilhelm Conrad Roentgen in the early 1890s. X-radiation penetrates even materials which are opaque to LIGHT, and a major application is in imaging the internal structures of the human body for medical purposes. Another application is in discovering fractures or other imperfections in materials used in technology. A disadvantage in the medical applications is that X-rays may cause MUTATION and cancer. Therefore, great care is exercised nowadays to prevent unnecessary EXPOSURE, and substitutes such as time-resolved imaging with visible light are being developed. X-rays are also used in many scientific investigations, such as X-ray scattering for determination of molecular structures, and X-ray microscopy. X-rays are generated whenever rapidly moving, charged particles such as electrons are decelerated (or accelerated, or deflected, as in a synchrotron), and in some kinds of radioactive decay of unstable atoms. X-rays are emitted by the Sun, but are completely intercepted by the Earth's ATMOSPHERE before reaching ground level. [L.O.B.]

xenobiotic A foreign substance in a biological system, i.e. a substance not naturally produced and not normally a component of a specified biological system. Usually applied to manufactured chemicals. *See also* ECOTOXICOLOGY; TOXICANT; TOXINS. [P.C.]

xenon (Xe) Gaseous element (ATOMIC NUMBER = 54; relative ATOMIC MASS = 131.30; melting point = −111.8°C; boiling point = −108°C) of the noble gas group. Xenon is a minor constituent of the Earth's ATMOSPHERE, retained since formation of the solar system. The average volume mixing ratio is 8.7×10^{-8} (87 ppb by volume). The element is physically and chemically inert and has no environmental significance, although RADIOACTIVITY in reactors can produce a short-lived RADIONUCLIDE, ^{135}Xe, which 'poisons' NUCLEAR FISSION processes by NEUTRON capture. Xenon is obtained as a by-product in the liquefaction of air and is used in electric discharge signs. [M.R.H.]

xeric Dry or lacking in moisture (*cf.* HYDRIC meaning wet). The adjective xero- is used in a variety of conjunctions, such as xeromorphic denoting plants that are adapted morphologically to prevent WATER loss by EVAPORATION. Xeromorphic adaptations are not confined to plants of dry habitats. Many wetland and wet-heath species (e.g. *Carex, Erica, Juncus*) have xeromorphic features such as hairy or in-rolled leaf margins which reduce TRANSPIRATION. This reduction in transpiration will also reduce water uptake by roots and lower the intake of potentially toxic ions (e.g. Fe^{2+}, Mn^{2+}) and may therefore be of survival value even in the absence of any water shortage (Jones & Etherington 1970). [R.M.M.C.]

REFERENCE

Jones, H.E. & Etherington, J.R. (1970) Comparative studies of plant growth and distribution in relation to waterlogging. I. The survival of *Erica cinerea* L. and *E. tetralix* L. and its apparent relationship to iron and manganese uptake in waterlogged soil. *Journal of Ecology* **58,** 487–496.

xerophile A plant of warm, dry habitats, tolerant of DROUGHT and thriving in dry areas, including deserts and semi-deserts. *See also* XERIC. [R.M.M.C.]

xerophyte A plant with specific adaptations for surviving prolonged periods of DROUGHT. Survival adaptations can be divided into opposing strategies, i.e. poikilohydric, homoiohydric and succulent xerophytes (Walter & Stadelman 1974).

- Poikilohydric xerophytes are plants in which the tissues come into equilibrium with the RELATIVE HUMIDITY of the atmosphere. Plants reacting to drought in this manner lose WATER rapidly during the DRY SEASON and survive the period of drought in a state of intense DESICCATION, for example certain moss and lichen species. There are also numerous accounts of spectacular revivals of certain poikilohydric species of higher plants after years of well-documented desiccation on herbarium sheets. Higher plants showing this capacity in the wild are sometimes referred to as resurrection plants, for example species of *Ramonda* and *Haberlea,* and the most celebrated of resurrection plants which grows to the west of the Victoria Falls *Myrothamnus flabellifolia* which can survive 2 years without rainfall.
- Homoiohydric xerophytes are plants which possess the capacity to reduce water loss and avoid tissue desiccation and drought injury by morphological and physiological adaptations, which not only reduce water loss but also increase water-use efficiency (the ratio of water molecules lost to carbon dioxide molecules gained during periods of photosynthetic activity) (Fig. X1).
- Succulent xerophytes are plants that withstand drought conditions by STORAGE of water. Succulent plants resemble poikilohydric xerophytes in that they are able to survive prolonged periods of drought without any external water supply. The succulent species differ, however, in that they can remain metabolically active during the period of drought whereas the poikilohydric xerophytes are in a state of DORMANCY (see Crawford 1989).

See also XERIC. [R.M.M.C.]

Fig. X1 Stone or umbrella pine (*Pinus pinea*), a typical example of a homoiohydric xerophyte. (Courtesy of R.M.M. Crawford.)

REFERENCES

Crawford, R.M.M. (1989) *Studies in Plant Survival.* Blackwell Scientific Publications, Oxford.

Walter, H. & Stadelman, E. (1974) A new approach to the water relations of desert plants. In: *Desert Biology,* Vol. 2 (ed. G.W. Brown). Academic Press, New York.

xerosere A general term used to describe any SUCCESSION on dry land surface, for example rock (lithosere), sand (PSAMMOSERE) or shingle. [B.W.F.]

Y

Yates' correction for continuity In the CHI-SQUARED TEST the statistic evaluated by the usual formula (a discrete quantity called by some authors X^2) is only approximately equal to the continuous PROBABILITY DISTRIBUTION given in the tables (χ^2). Usually this does not matter, but when DEGREES OF FREEDOM $(v) = 1$, it does. The solution is to use Yates' correction for continuity. This involves subtracting 0.5 from the numerator in the chi-squared formula, before squaring. The formula, including Yates' correction is therefore:

$$\chi^2 = \sum \frac{\left(|O - E| - 0.5\right)^2}{E}$$

where $|O - E|$ indicates that it is the absolute value of the difference from which 0.5 should be subtracted. [B.S.]

yaylak *See* ALPINE ZONE.

yield Return, usually expressed in mass or energy, of natural resources (e.g. fish, lumber) from an area or volume of the environment (*see also* SUSTAINABLE YIELD). Can also refer to a CROP. [P.C.]

yield per recruit model *See* FISHERIES, CONSERVATION AND MANAGEMENT.

Yoda's rule *See* $-^3/_2$ POWER LAW.

Young's modulus (modulus of elasticity) Defined as the ratio of normal stress (σ) to normal strain (ε):

$$E = \sigma / \varepsilon$$

Since stress has units of $N\,m^{-2}$ and strain is non-dimensional, Young's modulus has units of $N\,m^{-2}$. It provides a measure of the stiffness of a material and the product EI, where I is the second moment of area, is a measure of the flexural stiffness of a structure. In plant stems, values of Young's modulus range from 0.1 $GN\,m^{-2}$ for mature lianas which rely on other plants for support, to 2–8 $GN\,m^{-2}$ for self-supporting plants. The value of Young's modulus for the bones of a wide range of animals is close to 20 $GN\,m^{-2}$. [B.A.G.]

REFERENCE

Nicklas, K.J. (1992) *Plant Biomechanics: An Engineering Approach to Plant Form and Function*. The University of Chicago Press, Chicago, Ill.

ytterbium (Yb) Ytterbium (ATOMIC NUMBER = 70; relative ATOMIC MASS = 173.0; melting point = 819°C; boiling point = 1194°C) is a LANTHANIDE METAL with an average crustal abundance of 3.3 $mg\,kg^{-1}$. It is sometimes used in synthetic gemstones. [J.G.]

yttrium (Y) The most important sources of yttrium (ATOMIC NUMBER = 39; relative ATOMIC MASS = 88.906; melting point = 1523°C; boiling point = 3337°C) are monazite sands (3% Y) and bastnaesite (0.2% Y), the same as those for the lanthanides. Yttrium has an average crustal abundance of 30 $mg\,kg^{-1}$. Yttrium has some importance in the electronics industry where it is used as the basis of the phosphors which are used to produce the red colour on television screens. In future, the use of yttrium as a component of superconducting materials may increase its importance. For example, the phase $YBa_2Cu_3O_7$ has a critical temperature of 92 K, attainable by cooling with liquid NITROGEN. Yttrium has not been identified with any biological function, nor is it considered to be toxic. [P.J.B.]

Z

z(d) test A procedure that most commonly tests whether the difference observed between two SAMPLE means is simply due to chance or is real (NULL HYPOTHESIS, H_0: $\bar{x}_1 = \bar{x}_2$, ALTERNATIVE HYPOTHESIS, H_1: $\bar{x}_1 \neq \bar{x}_2$) for both sample sizes ≥ 30 (*see* ONE-TAILED TEST and TWO-TAILED TEST). The formula for the test is:

$$z(d) = \frac{\bar{x}_1 - \bar{x}_2}{\sqrt{\left(\frac{s_1^2}{n_1} + \frac{s_2^2}{n_2}\right)}}$$

where $\bar{x}_1$ and $\bar{x}_2$ are the sample means, $s_1{}^2$ and $s_2{}^2$ are the sample variances and n_1 and n_2 are the sample sizes. In effect, the result of the test is a STANDARDIZED NORMAL DEVIATE, because we are dividing an observed difference $\bar{x}_1 - \bar{x}_2$ (which is an estimate of $\bar{\mu}_1 - \bar{\mu}_2$) by its STANDARD ERROR. That is, we are dividing by the STANDARD DEVIATION of the sampling DISTRIBUTION of the quantity $\bar{x}_1 - \bar{x}_2$. Since this sampling distribution is normally distributed we can use this standardized normal deviate or *z*-score to test the null hypothesis (H_0: $\bar{x}_1 = \bar{x}_2$). A $z(d) \geq 1.96$ is only compatible with H_0 5% of the time. Therefore if $z(d) \geq 1.96$ we reject H_0 and accept H_1. Such a difference between two means is said to be a significant difference, or significant at the 5% level, or that the PROBABILITY of obtaining this value of $z(d)$, by chance, is $P < 0.05$. If $z(d) \geq 2.58$ we reject H_0 and accept H_1. However, such a difference between two means is now said to be a very significant difference, or significant at the 1% level, or that the probability of obtaining this value of $z(d)$, by chance, is $P < 0.01$. [B.S.]

zebra mussel (*Dreissena polymorpha*) *See* BIOFOULING; EXOTIC AND INVASIVE SPECIES.

zeitgebers *See* CIRCADIAN RHYTHMS; ENDOGENOUS RHYTHM.

zero emissions Discharges containing no CONTAMINATION. This is unrealistic on at least two grounds: (i) a technological society could not persist on this basis; and (ii) what is 'zero' depends upon the ability to measure it, 'detection limit' is a more apt descriptor. [P.C.]

zero growth isocline In PHASE PLANE ANALYSIS there are combinations of factors (favourable conditions) that lead to population growth, and other combinations (unfavourable conditions) that lead to decline (see Fig. P15). The line across the phase plane which separates these two regions is called the zero growth isocline (ZGI). The ZGI is the set of points on a phase plane map where $dN/dt = 0$. Equilibria are identified by the intersection of two species' ZGIs. The stability of any such equilibria is assessed by the slopes of the ZGIs at the point of intersection (*see* STABILITY ANALYSIS). [M.J.C.]

zero-sum constraint *See* RED QUEEN'S HYPOTHESIS.

zeroth law *See* HEAT TRANSFER.

zinc (Zn) A HEAVY METAL (ATOMIC NUMBER = 30; relative ATOMIC MASS = 65.39; melting point = 420°C; boiling point = 907°C) that is a trace constituent (76 mg kg^{-1}) of the Earth's CRUST. Zinc is found in ores such as sphalerite (ZnS) and calamine ($ZnCO_3$), and is used in galvanizing IRON and mild steel to prevent corrosion, and in alloys, BATTERIES and ointments.

Zinc is released to the ATMOSPHERE via the burning of coal and other FOSSIL FUELS and the smelting of non-ferrous metals. Leaching of zinc from mining wastes and the application of SEWAGE SLUDGE, which is often highly enriched in zinc relative to soils, can result in the CONTAMINATION of land with consequent harmful effects upon plants. Zinc, however, is an ESSENTIAL ELEMENT for plants, animals and humans. It is involved in producing the plant hormone responsible for stem elongation and leaf expansion, and deficiency causes serious problems for crop growth, especially in tropical regions. In humans it is a constituent of many enzymes and is essential for proper growth and the healing of cuts and wounds. Zinc deficiency in the diet can result in growth failure and sexual infantilism as has occurred in Egypt and Iran where a major component of the diet is unleavened bread prepared from high-extraction wheat flour. Yeast, oysters and meat are important sources of zinc. The biological availability of zinc in food, and cereals in particular, can be adversely influenced by the presence of complexing agents such as phytate and oxalate. [J.F.]

zirconium (Zr) Zirconium (ATOMIC NUMBER = 40; relative ATOMIC MASS = 91.224; melting point = 1852°C; boiling point = 4377°C) is a relatively common element with a natural abundance of 162 mg kg^{-1} in the Earth's CRUST. The main zirconium minerals are zircon ($ZrSiO_4$) and baddeleyite (ZrO_2) which are found in the USA, Canada and Brazil. The major use of zirconium is as an alloy with 1.5% TIN which is used as cladding for uranium dioxide (UO_2) fuel rods in water-cooled reactors. Its excellent mechanical properties and corrosion resistance, allied with its very low absorption of thermal neutrons, make it ideal for this purpose. Zirconium has not been identified with any biological function, nor is it considered to be toxic. [P.J.B.]

zonal soil systems *See* SOIL CLASSIFICATION.

zonal soils *See* SOIL CLASSIFICATION.

zonation A type of pattern in which zones or bands of communities (coenocline) or ecosystems (ecocline) form a GRADIENT across the landscape. The term can apply equally to plants and animals, but is often better exhibited by the vegetation. It can also be applied to species, and even ecotypes within species, and is always associated with abiotic environmental gradients which can be very local (e.g. tidal inundation on shores) or regional (e.g. regional latitudinal climatic variation). Zonation is sometimes associated with SUCCESSION, for example in lakes and ponds (HYDROSERE), sand dunes (PSAMMOSERE) and salt marshes (halosere), where substrate and soil build up and salinity gradients are important. On other occasions it is not, for example in latitudinal (regional) and altitudinal (mountain) climatic zonations and in topographically related zonations reflecting gradients in water-table depth, soil nutrient status and pH, for example the transition from dry heath through wet heath to valley mire. Human interference with the landscape may produce very sharp zonation patterns (e.g. roadside verges). Some of the boundaries associated with zonation patterns may be determined by BIOTIC FACTORS (e.g. competition) rather than by abiotic ones (e.g. the lower limit of *Chthamalus* barnacles on rocky shores). Zonation patterns have been used to focus attention on the controversy between the CLASSIFICATION (discrete ecosystems) and CONTINUUM (merging ecosystems and ecotones) approaches to GRADIENT ANALYSIS. Zonation is sometimes used to describe the pattern of horizons in soils, layers in peat deposits or strata in a forest, where the term STRATIFICATION is perhaps more appropriate. *See also* VERTICAL ZONATION. [B.W.F.]

zoo Abbreviation for zoological garden or park. These have evolved from menageries, as displays of exotic animals, to centres of education, conservation and some research. They are referred to by some as 'modern arks'. [P.C.]

REFERENCE

Page, J. (1990) *Zoo: the Modern Ark.* Facts on File, New York.

zoobenthos *See* BENTHIC HABITAT CLASSIFICATION.

zoocoenosis A term used most frequently in the east European literature to denote the animal component of a BIOCOENOSIS. It is thus equivalent to an animal COMMUNITY within an ECOSYSTEM. [M.H.]

zoogeography The scientific study of the geographic DISTRIBUTION of animals. Its principles are no different from those of BIOGEOGRAPHY as a whole, but whereas PHYTOGEOGRAPHY is almost entirely concerned with terrestrial organisms, zoogeography is as much concerned with marine life as with life on the land. The great pioneers of continental zoogeography were P.L. Sclater and A.R. Wallace who, in the mid-19th century, produced classifications of faunal regions based on birds, and terrestrial vertebrates and molluscs, respectively. The study of marine zoogeography is largely concerned with invertebrates, with the distribution of planktonic larvae as well as benthic and nektonic adults needing to be taken into account. [A.H.]

REFERENCES

Briggs, J.C. (1974) *Marine Zoogeography.* McGraw-Hill, New York.

Darlington, P.J (1957) *Zoogeography: the Geographic Distribution of Animals.* John Wiley & Sons, New York.

Ekman, S. (1953) *Zoogeography of the Sea.* Sidgwick & Jackson, London.

zoological parks *See* ZOO.

zooplankton The animal component of the PLANKTON that live free-swimming or suspended in the water column of marine, freshwater LENTIC and, to a lesser extent, LOTIC systems (where they are called potamoplankton). Many are filter-feeding herbivores and form an important link in nutrient cycling by transferring energy from AUTOTROPHS to predators. In freshwaters, they are dominated by three major taxa, Rotifera, Cladocera and Copepoda, whereas marine zooplankton includes representatives of virtually every group of animals, either as adults or developing stages. *Cf.* PHYTOPLANKTON. [J.L.]

zooxanthellae Single-celled algae that live in a symbiotic association with various cnidarians, including corals. [P.C.]